INNOVATIVE MINE DESIGN FOR THE 21ST CENTURY

PROCEEDINGS OF THE INTERNATIONAL CONGRESS ON MINE DESIGN
KINGSTON / ONTARIO / CANADA / 23-26 AUGUST 1993

Innovative Mine Design for the 21st Century

Edited by
W. F. BAWDEN & J. F. ARCHIBALD
Queen's University, Kingston, Ontario, Canada

A.A. BALKEMA / ROTTERDAM / BROOKFIELD / 1993

Publication of these Technical Proceedings was made possible through a donation from INCO LIMITED.

Cover photo: Automated In-The-Hole drill, being tested at INCO's North Mine in Sudbury. The drill is capable of automatically drilling hole to a preselected depth and of adding and removing drill rods automatically.

The texts of the various papers in this volume were set individually by typists under the supervision of each of the authors concerned.

Published by
A.A. Balkema, P.O. Box 1675, 3000 BR Rotterdam, Netherlands
A.A. Balkema Publishers, Old Post Road, Brookfield, VT 05036, USA

ISBN 90 5410 325 6

Printed in the Netherlands

Innovative Mine Design for the 21st Century, Bawden & Archibald (eds) © 1993 Balkema, Rotterdam, ISBN 90 5410 325 6

Table of contents

9 *Explosives technology and blast monitoring*

10 *Mine management into the 21st century*

Innovative Mine Design for the 21st Century, Bawden & Archibald (eds) © 1993 Balkema, Rotterdam, ISBN 90 5410 325 6

Foreword

Queen's University at Kingston received its Royal Charter, from Queen Victoria, in 1841 and since then has developed into a full degree granting institution with a current enrollment of 13,500 undergraduates and 2,100 graduate students.

The Department of Mining Engineering, established under Provincial Charter, in 1893 as the School of Mining and Agriculture at Kingston, now offers degree programmes in Mining, Mineral Extraction, and Mine Mechanical Engineering at the B.Sc., M.Sc., and Ph.D. levels. Over one half of all Mining Engineering degrees awarded each year in Canada come from Queen's University. During the last 100 years, Queen's Mining Engineering has developed deep roots and traditions in Canada and throughout the world.

The department is one of the few in the world in recent years to have justified a new mining building on the basis of its undergraduate enrollment. In addition to this new facility it operates its own Explosives Test Site outside Kingston, maintains a strong interest in strengthening its industrial associations, and offers a diverse and expanding continuing education program to the mineral industry.

As part of its 100th Anniversary celebrations in 1993, the Department of Mining Engineering is proud to host the International Congress on Mine Design (ICMD). During the past decade the Canadian Mining industry has begun to experience nothing less than the beginnings of an industrial revolution. International market pressures insure that this revolution will continue and probably accelerate through the end of this, and into, the 21st century. The objective of the ICMD is to establish a forum in which state-of-the-art technology, innovative mining practices and anticipated evolution in mining trends can be presented and debated. The international scope of the Congress allows for the introduction of changes in technology which will shape the global future of mining operations.

The technical proceedings consist of over 90 publications covering Slope Stability and Rock Mechanics; Underground Support and Backfill; Grade Estimation and Mine Feasibility; Open Pit Planning and Design; Underground Planning and Design; Machinery, Materials Handling, Automation and Mine Maintenance; Computer Applications and Numerical Modelling; Mine Environment; Explosives Technology and Blast Monitoring; Mine Management in the 21st Century; Research and Development.

The ICMD offers delegates the opportunity to exchange ideas with other mining personnel from around the world. In addition to the technical program, the ICMD includes a trade show where mining suppliers and vendors meet and discuss issues with prospective clients. A very pleasant and comprehensive social program encourages intermingling and casual discussions among delegates.

W.F. Bawden, Congress Co-Chair
P.N. Calder, Congress Co-Chair

1 Slope stability and rock mechanics

Innovative Mine Design for the 21st Century, Bawden & Archibald (eds) © 1993 Balkema, Rotterdam, ISBN 90 5410 325 6

Pit slope designs for the 21st century

P.F. Stacey
Golder Associates Ltd, Burnaby, B.C., Canada

ABSTRACT: This paper considers some of the trends in slope design technology that are emerging as we approach the 21st Century. While the ability to design stable slopes has improved greatly over the past 20 years, trends towards steep short-term slopes for small high-grade orebodies, and very high slopes for mega-pits centered on low grade deposits are starting to indicate limitations in available technology.

Examples of some of the successes and failures of recent slope designs are used to provide a review of the applicability of available technology, the role of artificial support, and the impact of changes in operating procedures on the stability of pit slopes. Areas requiring further investigation such as risk assessment are discussed, in terms of both emerging methodology and practical limitations.

1 INTRODUCTION

As the mining industry progresses into the 21st. Century, the open pit slope design engineer can expect to be faced with an increasing number of challenges. It is likely that the industry will still focus upon small high grade deposits requiring steep slopes of short life span, and the larger, deep open pits whose mining life will stretch into decades. However, many of the deposits will be in less developed countries, and in areas where access will be difficult, and available data upon which to design the slopes may be limited in extent; hence the risks to investment may be greater. At the same time, there will be continued pressure to make mines more cost effective. This will mean that slopes will have to be as steep as possible to minimize stripping, and operators will be striving to use the most efficient mining methods with the minimum of labour.

This paper attempts to review the tools currently available to address these challenges, and to chart the course for future research and development in the area of pit slope design.

By far the largest proportion of the world's minerals is still recovered from open pit mines. In this context Canada is no exception, since over 75% of our metals and coal production comes from open pit mines. These are largely centered at either side of the country, coal and metals in the west, and asbestos and iron in the east. All have different geotechnical problems, but all share the common problems of declining real product prices, and intense competition, both of which result in the need to reduce costs. At the same time, Canadian mining companies are investing heavily overseas in both moderately developed and undeveloped countries. In this, Canada is little different from most of the other developed countries with a tradition of mining.

In the early 1970's, a review of hard rock slopes around the world by height and slope angle put the majority of slopes less than 500 ft in height, with slope angles for the higher slopes typically in the 45° range or less. At that time, slope design technology was still developing, and most of the pits had been designed on the "45 degrees seems reasonable" philosophy. This does not mean that there were no large pit slopes, but the anticipated 700 m high slope at Bougainville were a rare exception.

Now, even small operations are planning slopes in excess of 250 m, at angles as steep as 60°, and there are a significant number of pits in the range in excess of 300 m deep in Canada, and many more throughout the world.

There are also plans for many more mega-pits, with depths in excess of 500 m, including some in Canada. The west wall of the Valley Pit at Logan Lake (Photograph 1) will eventually reach a height of 700 m or more, and the Barrick Goldstrike Pit in Nevada is planned to be in the order of 700 m deep, to name but two.

2. SLOPE DESIGN THEORY AND PRACTICE

2.1 Current Situation

In reviewing the available technology, it is notable that it is almost 15 years since the two bibles of rock slope design aimed at open pits, namely "Rock Slope Engineering" (Hoek and Bray, 1981) and the CANMET Pit Slope Manual CANMET, 1977) were published. These books provided a framework of the technology for open pit slope design throughout the world that has changed little, since in recent years rock mechanics research has tended to focus on the stability of underground openings.

It is only now that many of the pit slopes that were designed using the technology developed in the late 1960's and early 1970's are reaching maturity. It is therefore appropriate to review how these slopes have performed, what we have learnt by applying the technology, and what the future holds.

It is probably fair to say that there have been only limited efforts to use the experience gained in the past two decades to modify the theory significantly.

Rather, we have been using our experience to push the available theories to the limit in the actual operations. In this type of environment, poor operating practice can make, or literally break a slope.

The manuals produced in the 1970's compiled the available theory in terms of structurally controlled failure, and circular (mass) failure. The analytical techniques for each of these failure modes make some pretty broad assumptions, particularly where the failure mode is other than classic. Applications are highly dependant upon a sound understanding of the geology, and here scale comes into play. The theories are generally based upon small rigid blocks in the case of structurally controlled failures, and on relatively homogeneous soil-like slopes in the case of circular failures.

Some of the models, e.g. that for toppling failures, are very simplistic compared with real life, and water pressures are generally assumed to be uniformly distributed. When it comes to rock slopes that are in excess of 200 m in height the validity of these assumptions begin to come into question.

As geology becomes more complex, available theory also becomes less directly applicable. This is the case with the non-catastrophic flexure toppling of high rock slopes with well defined in-dipping structures, such as has been recorded, for example, in several copper porphyry pits, including Lornex and Brenda.

Similarly, where the rock is competent, but there are no structures that would directly result in failure, the potential for some form of rock mass failure must be considered. In this case, the strengths used in design analyses are usually established on an empirical basis using criteria such as those developed by Hoek and Brown, and the failure mode is assumed to be circular. However, there are few if any bench marks against which to calibrate these assumptions.

These comments are not intended to be negative. In fact, it is probably safe to say that the number of unexpected major pit slope failures seems to have decreased significantly in the past two decades. Further, it is now generally possible to

Photograph 1 West wall of the Valley Pit at the Highland Valley operations. This wall will eventually be in excess of 700 m high.

design slopes much closer to optimum than was previously the case, provided that the geology and groundwater conditions are well understood.

These two caveats are critical, since, as slopes become higher and the geology more complex, there is an increased potential to encounter unexpected adverse structures, such as was the case in the failure of the Brenda Mines west wall (Stacey et. al, 1991), and in the recent footwall failures in coal mines in the Kootenay coalfields. Further, there are some indications that other failure modes may affect the higher slopes. A case in point is the failure of a wall in an iron ore mine in Brazil in 1991, see Photograph 2.

The significant features of the Brazilian case were that the instability of the 240 m high slope developed within one month, and that only approximately 15 cm of movement was experienced before the failure, which was itself catastrophic, with the debris moving 500 m across the pit floor to stop against the opposite wall. In retrospect, the control of the failure was obvious, in that there was a very weak sheared bedding plane approximately 20 m behind the slope, dipping parallel to the wall, but the only logical explanation of the failure mode is a compressive strength failure of the material at the toe.

In fact, in both the Brenda and Brazilian cases movement prior to failure was very limited, thereby negating the assumption that failures of large rock slopes are likely to be preceded by significant displacement, and suggesting that stress played a major role.

In extremely competent rocks the slope design tends to be controlled by non-geotechnical factors. A case in point is the Nickel Plate Mine in southern British Columbia, where slopes with inter-ramp angles in excess of 60 degrees have been successfully mined over heights in excess of 200 m in extremely strong rock, see Photograph 3. In fact, the rock is so strong that the steep slopes have remained stable even though they are effectively undercut by mining through earlier underground workings that involved open stopes up to 100 m wide , with only limited pillar support.

The limiting factors in this type of design are not geotechnical, but rather the ability to position the drill against the face for multi-benched slopes, the operating bench height, which is controlled by the reach of the excavating equipment, and the regulated berm minimum width and vertical separation between catch berms.

Photograph 2 Failure of a 240 m high wall in an iron ore mine in Brazil. The failure developed very rapidly, and debris extended 500 m across the pit floor to the wall on the left.

Under these circumstances very carefully regulated operating procedures, particularly controlled blasting and face scaling are required, since there is no flexibility in the design. Any inability to attain the design configuration can only result in stepping in, with the resultant loss of ore at depth.

In higher slopes in very competent rock the more precise monitoring systems which have recently become available are starting to record deformation which cannot be directly explained in terms of evidence of failure. For example, deformations of up to 20 cm in 53° slopes approximately 200 m. high have been recorded in Palabora, with none of the usual evidence of the onset of failure (Martin 1992).

2.2 Artificial Support

In recent years there have been some notable successes in the application of artificial support to pit slopes; there have also been some spectacular failures. The other major developments in this field have been in Australia, with some applications in Canada. For example, in Australia there have been several attempts the steepen major slopes at Mt. Newman and Telford by artificial support means (Rosengren, 1986), while in Canada Smoky River Coal has supported several slopes (Fawcett et al., 1987).

The general experience has been considerable success for slopes of less than 100 m in height, particularly if they are of short life, and for stabilization of bench faces. The success rate has not been as high in the larger slopes for several reasons, including:

- the support is too short to prevent failure on more deep seated geological structure as the slope dimensions (height) increases;
- insufficient understanding of the geological structure and the geotechnical properties of the failure planes;
- lack of understanding of the failure mechanism of the installed reinforcement; and
- severe corrosion of the support (cables or bolts) in acid environments.

There appears to be a practical limit of about 30 m to the length of support cables that can cost effectively be installed on a large scale. This would clearly have some bearing on the depth to which the support can be effective. Until this can be overcome, and the mechanics of artificial

Photograph 3 Slopes at the Nickel Plate Mine have been mine to heights in excess of 200 m at overall angles in excess of 60°.

support is better understood, it is likely that the application of artificial support in slope design will be restricted to its present level.

2.3 Future Trends
Analytical Techniques

One of the main requirements at this time in pit slope design is to re-assess the application of the current technology as it related to large open pit slopes. This in effect will require a further study of similar scope to those that produced "Rock Slope Engineering" from Imperial College, and the CANMET Pit Slope Manual. The main aim of such a study would be to evaluate the wealth of experience that has been assembled over the past 15 years, and to assess the requirements for future technology development.

Future directions may well lie in the area of numerical analysis. In this regard, some research is already in progress directed towards modelling and analysis of progressive mechanism of failure, as an extension of the finite element work on toppling, (Martin 1990, and Pritchard et al. 1990). However, to date it has only been possible to model events in retrospect, and it is likely that numerical modelling will only become a more effective design tool when the ability to characterize the ground conditions is improved.

At this time there is no means of knowing how much deformation can be tolerated in a given high rock slope. In consequence, the design approach must still be to minimize the deformation and damage to the slopes until the impact of stress is better understood.

2.4 Risk Assessment

A tool that is starting to prove extremely useful in communication between the slope designers and the mine planners is risk assessment. This probablistic approach is being developed to address the recognition that there are significant uncertainties in the factors that affect slope design. The uncertainty, which can arise from natural variability and/or lack of data, can be expressed probablistically through either empirical or numerical analysis, see Figure 1.

In this way, the time is approaching when it should be possible to provide a realistic quantitative assessment of the risk of failure of specific slopes, and the probable extent of failure. Any such assessment will, however, largely be dependant upon the certainty that can be placed on the data. In particular, this will require a firm understanding of the geology of the wall rocks, i.e. the distribution of rock types and their structure, as experience suggests that this is one of the largest ares of uncertainty in many slope design projects.

2.5 Slope Monitoring

Slope monitoring methods have continued to develop, and in the past few years there have been some interesting improvements in both the equipment and the data handling technology.

The most recent total station electronic distance measuring (EDM) equipment is now sufficiently accurate that it can monitor very small deformations over the width of even the largest open pits. There are also automated systems available that will provide continuous real-time reports back to the mine control centre through a telemetric system. These systems are relatively expensive, and complex, and it is unlikely that they will be cost effective for basic monthly surveillance monitoring of even large pits. However, where mining is occurring under a moving slope and frequent monitoring is required to ensure safety, e.g. at Brenda Mines, they have clearly been shown to have an application.

Eventually, it is likely that the automated

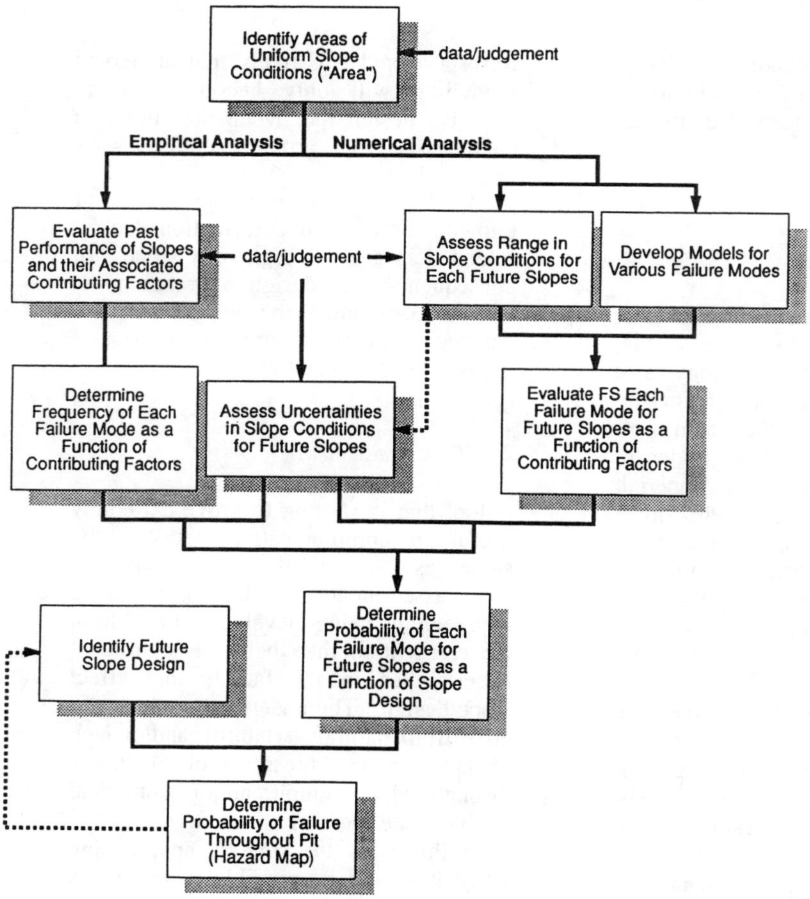

Figure 1 Risk assessment methods for pit slopes

monitoring systems could be coupled directly into the mine dispatch system. An example of this trend is the wireline extensometer system that has been developed by Modular Mining Systems Inc for waste dump monitoring, see Photograph 4.

As the accuracy of global positioning systems (GPS) further improves it is possible that they will also find a role in slope monitoring, although it is unlikely that this type of equipment will replace the much more dispensable survey prism in critical situations.

3 OPERATING FACTORS

As part of the general drive for cost efficiency there have been some significant changes in operating technology. This has included both the use of more powerful blasting agents and the use of larger equipment.

3.1 Blasting

Introduction of higher energy explosives to increase fragmentation and reduce costs has brought other problems related to slope stability. The high detonation velocities associated with the heavy ANFOs and the high density emulsions requires modification of the blast designs and/or initiation sequence if gas fracturing back into the pit slopes along major structures is to be avoided. There have been several instances where significant damage has occurred before this problem was recognized, see Photograph 5.

The more general application of controlled blasting around phase and final walls is essential where optimum slopes are required, as in the case of Nickel Plate. Further, the continuing development of more accurate downhole blasting delays have gone a long way to reducing gas pressure related blast damage to bench faces, and further improvement can be expected.

3.2 Excavation Technology

The increased size of mining equipment can be both an advantage and a disadvantage. Larger shovels permit higher bench faces, which may not be stable where they are excavated in weak rocks and overburden.

On the other hand, the increased power of the larger bulldozers makes it appropriate to consider excavation of bench faces by ripping and dozing in overburden and weak rock, see Photograph 6. This typically results in faces which are less prone to degradation, since blast damage is avoided. It also permits faces to be formed to controlled angles, which are particularly appropriate in the residual soils and highly weathered rock in tropical environments, and in the formation of footwall slopes.

3.3 Slope Depressurization

Horizontal drain holes have become a

Photograph 4 Modular Mining Systems wireline extensometer that can transmit real-time movement data to the mine dispatch system.

Photograph 5 Gas fractures developed along major structures behind a production blast of high density emulsion. These extended into the ultimate wall to the right.

Photograph 6 Dozer cut bench faces in weathered schistose rock.

standard operating tool for slope stabilization in operating mines. Recent improvements in drilling equipment and technology mean that 300 m long holes can generally be achieved in even the worst ground conditions. However, it is interesting to note that the design of drain layouts is almost universally empirical. In other words, more and longer holes are drilled until the flow is reduced and/or the pressures drop.

With improved technology in directional drilling, targeted drain holes and wells of up to 2,000 to 3,000 m in length are not outside the realms of possibility. Thus directed drain holes and well under-drains for a pit cannot be far away.

4 SUMMARY

In summary, there appears to be a

requirement for a further period of intensive research to support pit slope design requirements for the 21st. century. This research should focus on the recognition that the simple gravity block models which currently dominate most of the slope design technology may not be directly applicable to either the mega-slopes that are currently being planned, or to the very steep short term slopes in the smaller pits.

Fundamental elements in this research must be improvements in the methods of characterizing the rock mass, and in understanding the stress-strain relationships that develop in high slopes with complex structure and variable material properties. This will obviously be a long term project, but will be critical to the design of slopes for the 21st. century.

In the short term the most profit can probably be gained by trying to maximize bench-scale stability, and by integrating slope design alternatives with mine design decisions, possibly through risk assessment of the choice of alternatives.

On the operational front, the implications of new mining technology to slope stability must be reviewed before a new product or technique is accepted; the case of gas fracturing from high velocity explosives is a classic example. But new technology will definitely improve our ability to both monitor and optimize the design of slopes.

REFERENCES

CANMET, Mining Research Laboratories, 1977. *Pit Slope Manual*, CANMET Report 77-13.

Fawcett et al., 1987, Footwall Anchoring at Smoky River Coal Limited: 89th Annual General Meeting of the CIM, Toronto.

Hoek, E. and Bray, J.W., 1981, *Rock Slope Engineering*, The Institution of Mining and Metallurgy, London.

Hoek, E. and Brown, E.T., 1988, The Hoek-Brown Failure Criterion - a 1988 Update. 15th Canadian Rock Mechanics Symposium, University of Toronto, p. 1-38.

Martin, D.C., 1990, Deformation of Open Pit Mine Slopes by Deep Seated Toppling: 13th Annual CIM District Six Meeting, Kamloops, British Columbia, September, 1990.

Martin, D.C., 1992, Personal communication.

Pritchard, M.A., Savigny, K.W., and Evans, S.G., 1990: Toppling and deep-seated landslides in natural slopes: Proceedings of the International Conference on Mechanics of Jointed and Faulted Rock, Vienna, p. 937-943.

Rosengren, K.J., 1986: Wall Reinforcement in Open Pit Mining: The Aus IMM/IE Newman Combined Group Large Open Pit Mining Conference, p. 257-265.

Stacey, P.F., Leung, J., and Smith, K.M., 1991: Brenda Pit West Wall - Anatomy of a Failure: 93rd. Annual General Meeting of the Canadian Institute of Mining and Metallurgy, Vancouver, 1991.

Innovative Mine Design for the 21st Century, Bawden & Archibald (eds) © 1993 Balkema, Rotterdam, ISBN 90 5410 325 6

Assessment of the stability of jointed rock slopes

J.P.Ashby
Ashby Consulting Services, Auckland, New Zealand

ABSTRACT: Geological structure plays a dominant role in determining the stability of most rock slopes. The analysis of geological structure by means of stereographic projection is an established technique. Even with the advent of powerful computer graphics, stereographic projection still ranks among the most effective means of presenting and visualising complex combinations of geological structures. The paper reviews simple methods of stereographic projection for analysis of geological structure. The Search Net technique is presented. The Search Net is superimposed on stereographic projections of poles and used to determine likely modes of failure and worst stability conditions (i.e. the lowest Factor of Safety) for jointed rock slopes.

1 INTRODUCTION

Rock slopes, except those which exhibit low intact strength or particularly intense fracturing, tend to fail along discrete discontinuities such as bedding planes, joints and faults.

Most rock masses contain a number of sets of discontinuities. Each set can be composed of planes of varying orientation, frequency, extent and surface properties. The orientation and characteristics of these surfaces are usually studied by means of stereographic projection. Stereographic projection is a powerful aid in the visualisation of complicated 3-D structure.

Analysis of rock slope stability is further complicated by the range of possible failure mechanisms that include sliding along discontinuities, rotation of blocks (e.g. toppling) and failure through intact material. Failure will often involve a combination of these mechanisms. The orientation of major joints plays a major part in determining the probable mode of failure. Basic modes of joint controlled failure are illustrated in Figure 1.

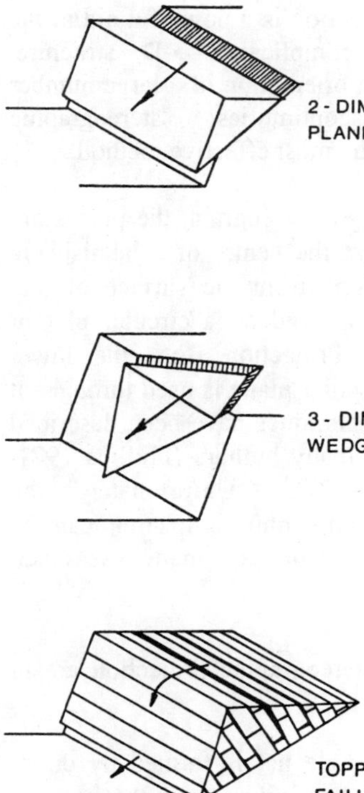

Fig. 1 Common modes of failure

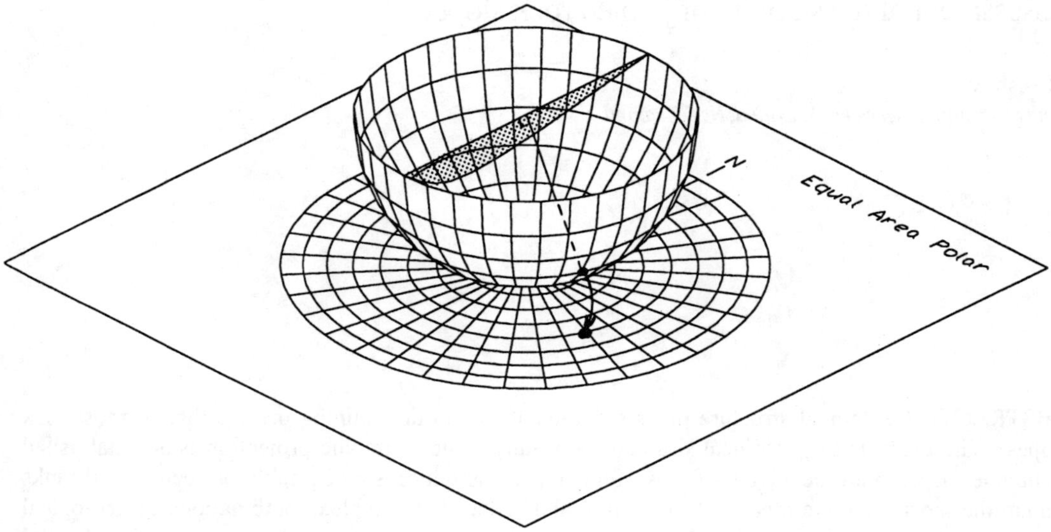

Fig. 2 Lower hemisphere equal area projection of plane dipping 50° azimuth 290°

2 STEREOGRAPHIC PROJECTION

Stereographic projection is a powerful aid in the visualisation of complicated 3-D structure. When studying the orientation of a large number of planar discontinuities, stereographic projection is still the most effective method.

When generating a stereogram, the planes are considered to lie at the centre of a hemisphere and their projections from the surface of that hemisphere are displayed on a circular plot or net (Figure 2). Projection from the lower hemisphere to the polar plane is used throughout this paper. The techniques have been described in more detail by many authors (Phillips 1971; Hoek & Bray 1977). Unfortunately, the techniques of stereographic projection can be rather confusing even to many seasoned practitioners.

Two types of stereographic projection are in common use:

1) The equal angle net is formed by direct projection from the hemisphere such that angular relationships are honoured.

2) The Schmidt or equal area net is generated such that area relationships on the hemisphere are approximated on the projection.

The equal area net is used for plotting poles (or normals) to planes. Each plane, with a given dip and dip azimuth, is represented by a point (as shown in Figure 3).

The traditional method of plotting stereograms manually involved a confusing and cumbersome process of rotating a piece of tracing paper overlaid on a stereonet. A faster method involves direct plotting on a blank form as illustrated in Figure 3. The dip direction (or azimuth) is read off a scale around the circumference of the net and the dip, from a radical scale about the centre. With the advent of PC and readily available and relatively inexpensive software, stereograms can be plotted even more rapidly and to a camera ready quality (Figure 4).

Statistical study of the orientation of discontinuities is best carried out on an equal area net. The density distribution of poles is contoured to define sets or populations of discontinuities. Computer routines are readily available that smooth the contours. A Gaussian normal smoothing routine was used in preparing the contours in Figure 5.

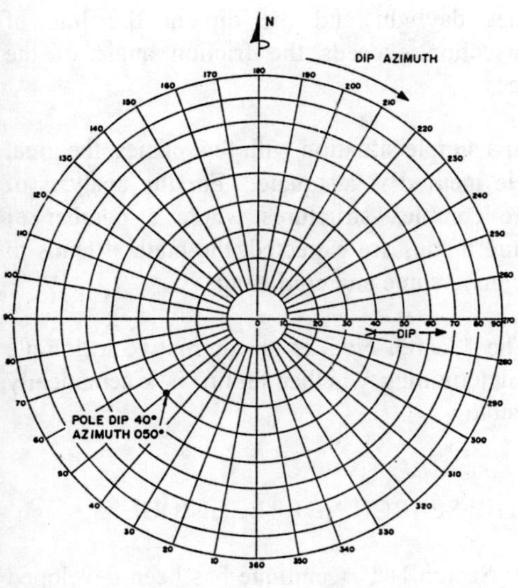

Fig. 3 Stereographic pole plotting form
(polar, equal area, lower hemisphere)

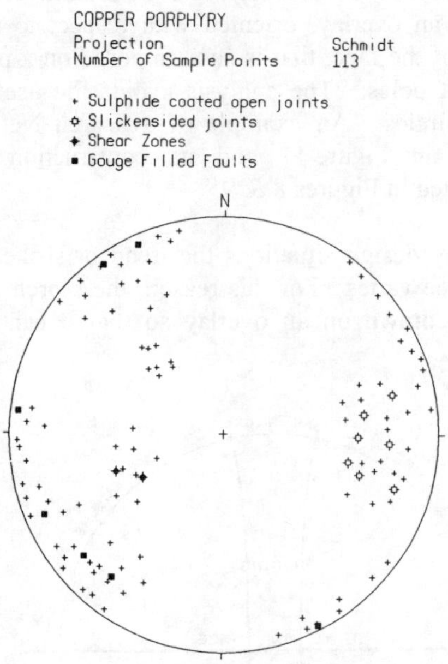

Fig. 4 Computer plotted stereogram of poles

In absence of a computer, manual contouring using a counting circle illustrated in Figure 6 is fast, simple and informative.

The counting circle has an area that is one

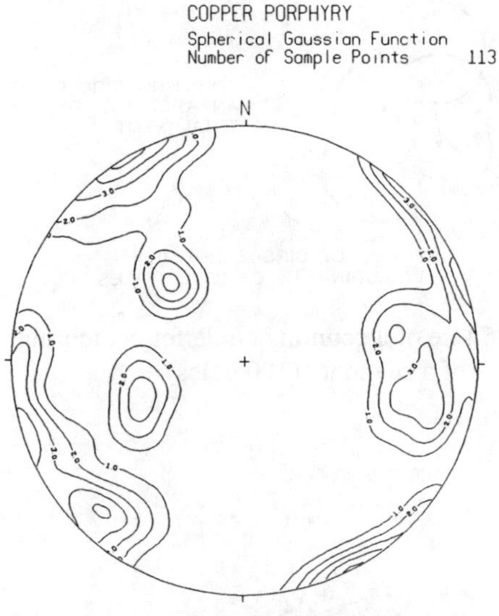

Fig. 5 Computer generated contours
(Schmidt plot 113 poles)

percent of the total area of the projection. The one percent density contour (i.e. one percent of total poles lying in one percent of the area of the net) is drawn by moving the circle around the plot to include one percent of the total poles represented. Higher value contours are then added. By this means a typical contour plot can be completed in a few minutes.

Planes can also be represented by their great circle traces. This is best carried out on an equal angle projection although the error in using an equal area net is not significant, for most practical purposes. Use of great circles enables angular relationships between planes to be measured fairly accurately. Unfortunately, great circle construction tends to be confusing especially when more than a few planes are represented.

Figure 7 shows the great circles of the 113 poles in Figure 4 and is included to illustrate just how confusing great circle plots can be.

Because of the confusion that can result from use of great circles, equal area pole plots are used in the technique presented in this paper.

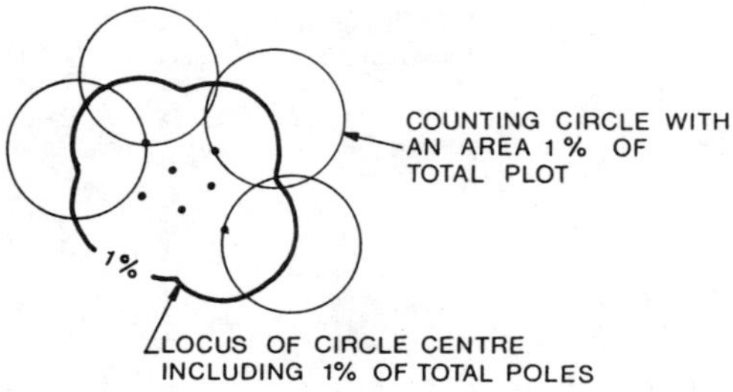

Fig. 6 Use of a counting circle for contouring of a stereonet (100 poles)

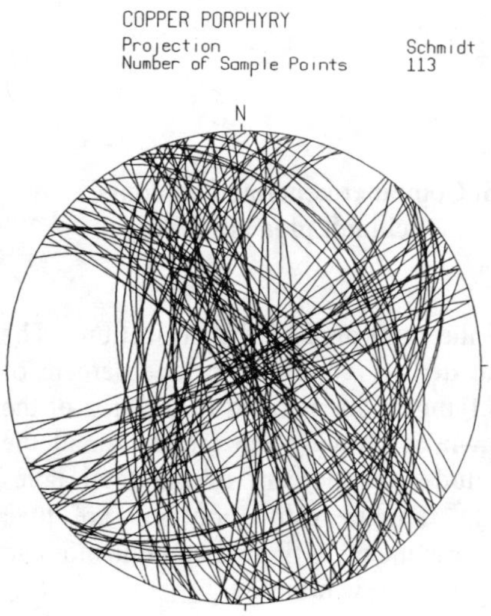

Fig. 7 Great circles of poles from Figure 4. (Schmidt plot 113 poles)

3 GREAT CIRCLE STABILITY ANALYSIS

Great circle construction has been used almost exclusively for analysis of the stability of jointed rock slopes (John 1969; Hoek & Bray 1977). By this method, the angular relationships between the slope face and major discontinuities are determined, enabling likely (kinematically possible) failure combinations to be ascertained. Simple tests are applied to determine whether the discontinuities will undercut or "daylight" the slope. Lines of intersection of planes combining to produce 3-dimensional wedges can be measured and compared to the friction angle using the concept of the friction cone. As a first approximation, failure will occur when the planes daylight and the dip of the line of intersection exceeds the friction angle of the planes.

In a simple situation with few planes, the great circle method is adequate. For the analysis of more complex structures where a number of failure modes are expected, the method tends to be cumbersome and confusing.

The "Search Net" is an alternative technique for determining whether failure is kinematically possible.

4 THE SEARCH NET TECHNIQUE

The "Search Net" technique has been developed over a number of years by the author. The method enables the user to determine the likely mechanism(s) of failure and discontinuities most likely to lead to instability. The method involves use of an overlay, oriented with respect to the strike of the face, that is superimposed on a plot of joint poles. The analysis avoids the use of great circles. An example of a Search Net is shown in Figure 11 and its construction is illustrated in Figures 8 & 9.

In many design situations the trend or strike of the slope varies. For this reason, the search net is best drawn on an overlay so that it can be

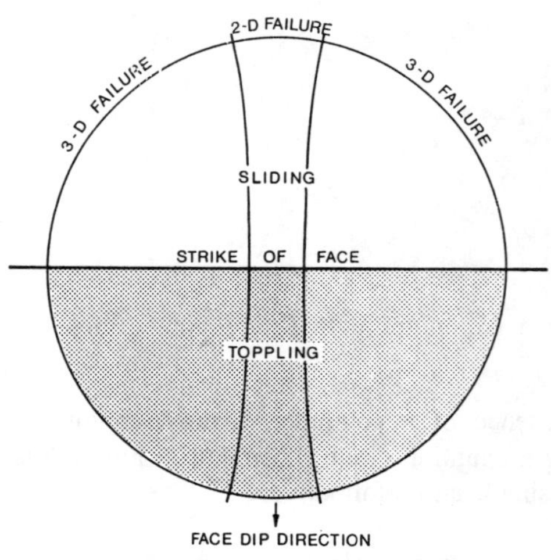

Fig. 8 Search Net, basic construction

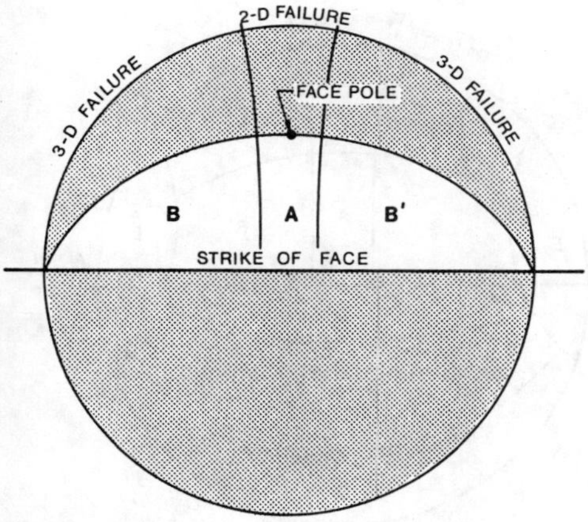

Fig. 9 Search Net, symmetrical "daylighting" planes

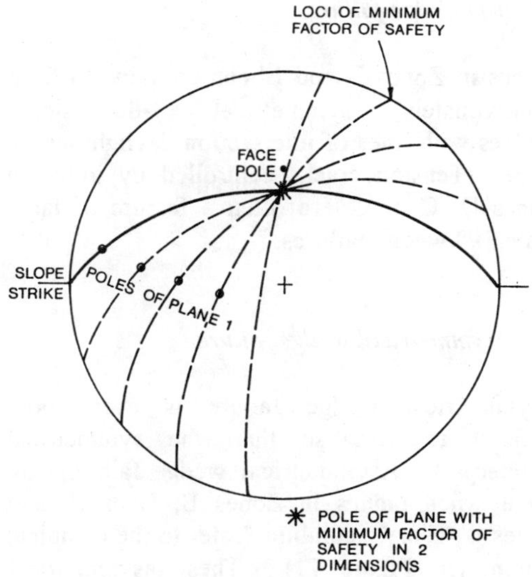

Fig. 10 Minimum Factor of Safety for asymmetrical wedge combinations

rotated to parallel the trend (or strike) of the slope. In Figs. 8, 9 & 11 the slope trend is drawn from left to right. The Search Net is divided into a number of zones which are labelled in Figure 11. Joints that have poles lying within a particular Zone are susceptible to failure in a particular way as follows:

4.1 *Discontinuities dipping <u>into</u> or <u>out of</u> the slope*

The Search Net is divided into two main semi-circular zones oriented about the slope trend. Any discontinuity with a pole falling within the unshaded portion of Figure 8 dips out of the slope and may cause failure by sliding. The shaded area includes discontinuities that form blocks that may topple if undercut or released by displacement at the toe of the slope.

4.2 *2-D and 3-D failure*

The search net is divided into 2-dimensional and 3-dimensional zones as illustrated in Figure 8. The 2-D zone is approximated by a band (formed by two small circles) lying arbitrarily within 10 degrees to the normal to the strike of the face. Any discontinuities with poles lying within the 2-D Zone tend to dominate the stability of the slope. This is because 2-D planes strike approximately parallel to the slope trend and are relatively unrestrained by end-effects.

4.3 *Daylighting discontinuities*

An important division is made on the Search Net by plotting the pole of the slope face. With an equal area equatorial net, a great circle is drawn through the face pole to intersect the slope trend (as shown in Figure 9)

4.4 *2-D plane failure*

Planes with poles lying in the unshaded segment not only dip out of the slope but also undercut of "daylight" the slope. Planes in Zone "A" will fail by 2-dimensional plane failure when the disturbing forces exceed the strength of the planes. 2-dimensional failure is facilitated by the presence of bounding joints in Zone E and tension cracks formed from joints in Zones D, C or C'.

4.5 *3-D wedge failure*

Planes in Zones B and B' can combine to form approximately symmetrical 3-dimensional wedges with lines of intersection daylighting the slope. Tension cracks controlled by joints in Zones D, C or C' are often a feature of large scale 3-D wedge failures.

4.5 *Asymmetrical wedge failure*

Asymmetrical wedge failure is even more difficult to visualise than the symmetrical counterpart. Asymmetrical wedge failure may occur when planes in Zones B, E or C, and Zones B', E or C' combine (refer to the complete search net, Figure 11). These asymmetrical combinations will only "daylight" leading to possible failure of the slope when certain geometrical conditions are met. A simple test for these conditions involves the rotation of the great circle and the unshaded zone in Figure 9 about the face pole as shown in Figure 10.

4.6 *Toppling failure*

It is self evident that toppling as a sole mechanism can only occur when steeply dipping discontinuities in Zone F are undercut. However, a far more common and significant type of failure may take place when steeply dipping discontinuities in Zone F occur in combination with shallow daylighting planes in Zone A, leading to toppling-sliding (Ashby 1971; De Freitas & Watters 1973).

Modelling work (Ashby 1971) suggested that the most critical dip angle for toppling failure probably occurs when the poles of discontinuities in Zone F fall on the great circle through the pole dipping at $(90°-\frac{\phi}{2})$, where ϕ is some function of the friction angles of the sliding and toppling discontinuities. Because of the complexities of toppling-sliding failure, the critical angle should be checked in each situation.

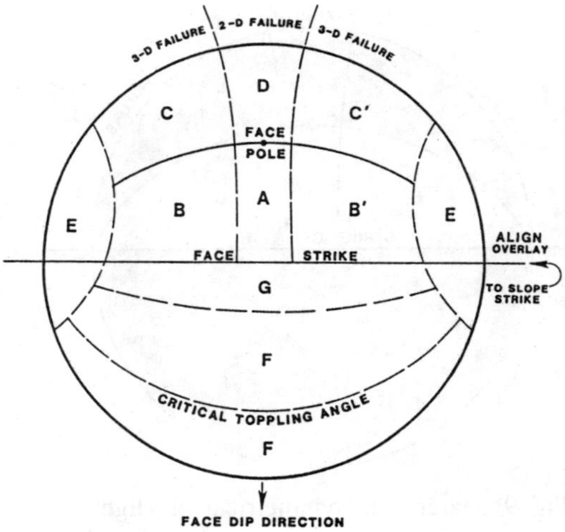

Fig. 11 The complete Search Net showing mode of failure "Zones" (slope angle 45°)

4.7 *Other joint orientations*

Joints lying in Zone G are sometimes involved at the toe of large scale sliding and toppling-sliding failures. These situations, in which movement occurs up-dip near the toe failure, are surprisingly common.

Bounding joints in Zones E and tension cracks in Zones D, C and C' are important planes of weakness that facilitate failure. In strong rock, where the probable failure mechanism is 2-dimensional, the presence or absence of these weaknesses may determine whether failure will occur or not.

A practical trial using the Search Net will clearly illustrate the ease of use and power of the technique in determining combinations of joints that may fail by a number of failure mechanisms.

5 FLOATING PLANE STABILITY ANALYSIS

Having determined the modes of failure that are kinematically possible, slope design requires comprehensive analysis of the various joint combinations. Various methods are available for stability analysis using either graphical or computer solutions (Hoek & Bray 1977). Traditionally, limiting equilibrium analysis has been used. The 2-dimensional analysis of a slope mass sliding on a daylighting discontinuity is straightforward. Geometrically more complex analyses of 3-dimensional wedges can be handled, but are best run on a computer. Unfortunately, toppling-sliding is still difficult to handle rigorously (Goodman & Bray 1976).

Even with rapid computational methods, design of jointed rock slopes can still be complicated by the number of variables entering into the analysis. For each mechanism, the variables include the failure mode, the orientation, frequency, extent and strength of discontinuities, the strength of the intact material, water pressure and external loads. The effect of varying individual design parameters can be presented by sensitivity analysis. However, the effect of varying the orientation of discontinuities is more difficult to display and visualise. The crudest method of representing the results is a printout in the form of a two dimensional array of Factor of Safety for varying combinations of dip and dip azimuth (See Figure 12). The 3-dimensional wedge analysis used in preparing the Figure was based upon the "Comprehensive Wedge" solution (Hoek & Bray 1977). A Fortran program was written in which one plane can be fixed while the other is allowed to "float" in ten degree increments of dip and dip azimuth (Carney 1983).

Presentation of the results using stereographic projection is recommended as an alternative to the rectangular plot of Figure 12. To simplify presentation and to determine the most likely mode of failure (i.e. the minimum Factor of Safety), Factor of Safety for combinations of joints can be contoured on a stereoplot (Figure 13).

The Factor of Safety contour values in Figure 13 refer to the 3-D combination of the floating and the fixed planes. The geometry of the slope and other variables including shear strengths were

AZ	DIP 10	20	30	40	50	60	70	80	90
110.0	26.64	15.59	12.01	10.30	9.41	12.12	-9.99	-9.99	-9.99
120.0	13.46	7.84	6.00	5.17	5.37	-9.99	-9.99	-9.99	-9.99
130.0	9.11	5.25	3.99	3.56	8.25	-9.99	-9.99	-9.99	-9.99
140.0	6.97	3.96	3.00	2.90	-9.99	-9.99	-9.99	-9.99	-9.99
150.0	5.73	3.19	2.41	2.65	-9.99	-9.99	-9.99	-9.99	-9.99
160.0	4.95	2.69	2.02	2.52	-9.99	-9.99	-9.99	-9.99	-9.99
170.0	4.43	2.35	1.74	2.28	-9.99	-9.99	-9.99	-9.99	-9.99
180.0	-9.99	-9.99	-9.99	-9.99	-9.99	-9.99	-9.99	-9.99	-9.99
190.0	4.03	2.16	1.66	1.97	-9.99	-9.99	-9.99	-9.99	-9.99
200.0	4.22	2.32	1.79	1.92	-9.99	-9.99	-9.99	-9.99	-9.99
210.0	4.55	2.54	1.97	1.94	4.78	-9.99	-9.99	-9.99	-9.99
220.0	5.05	2.87	2.22	2.05	2.66	-9.99	-9.99	-9.99	-9.99
230.0	5.83	3.35	2.58	2.30	2.41	4.73	-9.99	-9.99	-9.99
240.0	7.06	4.09	3.15	2.75	2.64	2.97	10.63	-9.99	-9.99
250.0	9.18	5.36	4.13	3.56	3.29	3.25	3.70	20.16	-9.99
260.0	13.51	7.91	6.10	5.25	4.78	4.54	4.49	4.90	20.55
270.0	26.68	15.63	12.05	10.33	9.38	8.82	8.51	8.40	8.65
280.0	*******	*******	*******	*******	*******	*******	*******	*******	*******
290.0	-9.99	-9.99	-9.99	-9.99	-9.99	-9.99	-9.99	-9.99	-9.99
300.0	-9.99	-9.99	-9.99	-9.99	-9.99	-9.99	-9.99	-9.99	-9.99
310.0	-9.99	-9.99	-9.99	-9.99	-9.99	-9.99	-9.99	-9.99	-9.99
320.0	-9.99	-9.99	-9.99	-9.99	-9.99	-9.99	-9.99	-9.99	-9.99
330.0	-9.99	-9.99	-9.99	-9.99	-9.99	-9.99	-9.99	-9.99	-9.99
340.0	-9.99	-9.99	-9.99	-9.99	-9.99	-9.99	-9.99	-9.99	-9.99
350.0	-9.99	-9.99	-9.99	-9.99	-9.99	-9.99	-9.99	-9.99	-9.99
360.0	-9.99	-9.99	-9.99	-9.99	-9.99	-9.99	-9.99	-9.99	-9.99
10.0	-9.99	-9.99	-9.99	-9.99	-9.99	-9.99	-9.99	-9.99	-9.99
20.0	-9.99	-9.99	-9.99	-9.99	-9.99	-9.99	-9.99	-9.99	-9.99
30.0	-9.99	-9.99	-9.99	-9.99	-9.99	-9.99	-9.99	-9.99	-9.99
40.0	-9.99	-9.99	-9.99	-9.99	-9.99	-9.99	-9.99	-9.99	-9.99
50.0	-9.99	-9.99	-9.99	-9.99	-9.99	-9.99	-9.99	-9.99	-9.99
60.0	-9.99	-9.99	-9.99	-9.99	-9.99	-9.99	-9.99	-9.99	-9.99
70.0	-9.99	-9.99	-9.99	-9.99	-9.99	-9.99	-9.99	-9.99	-9.99
80.0	-9.99	-9.99	-9.99	-9.99	-9.99	-9.99	-9.99	-9.99	20.55
90.0	-9.99	-9.99	-9.99	-9.99	-9.99	-9.99	-9.99	23.47	8.65

-9.99 - FAILURE OF COMBINATION NOT KINEMATICALLY POSSIBLE
******* - FACTOR OF SAFETY OVER 1000.00
0.00 - LARGE BUOYANT FORCE CAUSES WEDGE TO "FLOAT"

Fig. 12 Factors of Safety for the combination of a floating plane and a fixed plane (fixed plane dip 70°/100°, Slope dip 45°/180°)

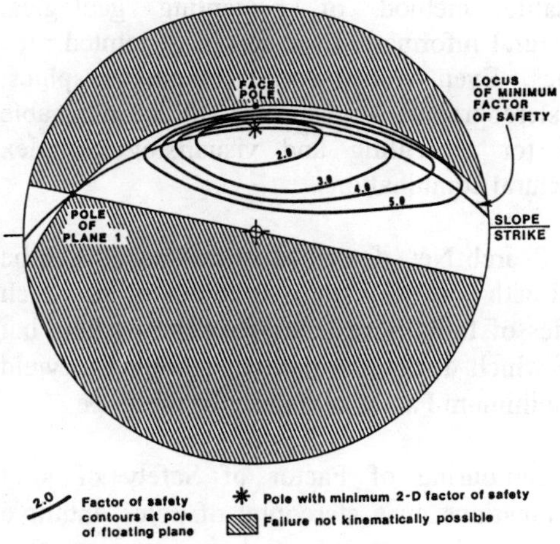

Fig. 13 Stereogram showing Factor of Safety contours for 3-D wedge combinations

fixed. The Factor of Safety contours (Figure 13) are displayed at the location of the floating pole. The contours are somewhat elliptical in shape and lie within the unshaded segment in which failure is kinematically possible. The minimum Factor of safety occurs as the floating plane approaches the 2-D configuration, with a dip that yields the minimum Factor of Safety in a 2-D analysis.

The locus of minimum Factor of Safety for the various combinations of the fixed and floating planes is the great circle passing through the poles of the fixed plane and the plane with the minimum Factor of Safety for 2-D analysis (Figure 10). The segment, containing poles of planes for which failure is kinematically possible, is bounded by:

1) the great circle through the fixed pole and the face pole, and

2) a straight line through the fixed pole and the centre of the stereoplot.

7 CONCLUSIONS

Plotting of poles of discontinuities on an equal area stereographic projection is the most versatile method of presenting geological structural information for design of jointed rock slopes. Even with powerful computer graphics, the stereographic projection is still an invaluable tool for presenting and visualising complex structural combinations.

The Search Net, described in this paper, can be used with pole plots not only to determine which modes of failure are kinematically possible, but also which discontinuities can combine to yield the minimum Factor of Safety for the slope.

By contouring of Factor of Safety of joint combinations on a stereoplot of poles, intuitive patterns and construction techniques emerge that can further simplify the design process.

REFERENCES

Ashby, J.P. 1971. Sliding and Toppling Mode of Failure in Models and Jointed Rock Slopes. *M.Sc. Thesis,* London University, Imperial Coll.

Carney, B.C. 1983 Personal communication.

De Freitas, M. H. & Watters, R. J. 1973. Some Field Examples of Toppling Failure. *Geotechnique (23),* 4, pp.495-514.

Goodman, R.E. & Bray, J.W. 1976. Toppling of Rock Slopes. *Spec. Conf. on Rock Eng. for Found. & Slopes,* Boulder, ASCE, (2), pp.201-234.

Hoek, E. & Bray, J.W. 1977. *Rock Slope Engineering. 2nd Ed.* 402 pp. IMM, London.

John, K.W. 1969. Graphical Stability Analysis of Slopes in Jointed Rock. *J. of Soil Mech. Found. Div.*, ASCE, (94) SM2, pp.497-526.

Phillips, F.C. 1971. *The Use of Stereographic Projections in Structural Geology. 3rd Ed.*, 90 pp., Edward Arnold, London.

Innovative Mine Design for the 21st Century, Bawden & Archibald (eds) © 1993 Balkema, Rotterdam, ISBN 90 5410 325 6

The use of relative probability of failure in the design of open pit mine slopes

Sandro S. Sandroni
Geoprojetos Engineering Ltd, Rio de Janeiro & COPPE/Federal University of Rio de Janeiro, Brazil

ABSTRACT: This paper summaries a simplified engineering approach for estimating the relative probability of failure of slopes in a given mining area. Calculations are based on the procedures described by Christian *et al.* (1992). The basis of this approach is the comprehension of existing local experience on failed and stable slopes. As a result, a design criterion based on the maximum acceptable relative probability of failure is established. As an additional benefit, knowledge is gained on the relative importance of each factor contributing to the total variability of the safety factor, which aids in the engineering decisions.

1.0 INTRODUCTION

Adequate studies of slope stability are essential for the design of open pit mines. In the conventional approach, geotechnical engineers define a minimum value of factor of safety (e.g., FS = 1.2 or 1.3) to be imposed to all slopes in the mine. Deterministic stability calculations are carried out making use of one of the many available methods of limit equilibrium. No consideration is usually given to the probability of failure associated to each individual slope in different regions of the open pit.

Slopes with the same deterministic value of FS may however correspond to significant differences in the probability of failure. These differences arise from proper consideration of the uncertainties inherently associated to loads and materials in geotechnical problems.

This paper summaries the approach followed for estimating the relative probability of failure P[R] of the slopes in a given mining area. Calculations are based in the procedure described in detail by Christian *et al.* (1992). The basis of the approach is the comprehension of existing local experience on failed and stable slopes. An interesting additional benefit is the knowledge that is gained on the contribution of the safety factor.

An example of application to the design of open pit iron ore mines in the state of Minas Gerais, Brazil, is presented and briefly discussed.

2.0 COMPUTATION OF THE PROBABILITY OF FAILURE

The first step for estimating the probability of failure (P[R]) of a slope consists in computing the reliability index (β), which is given by the following expression:

$$\beta = (E[FS] - 1.0) / \sigma[FS] \qquad (1)$$

In this expression, E[FS] is the mean value of the factor of safety (FS) and σ[FS] indicates the standard deviation of FS.

The reliability index β expresses the uncertainty in the stability of the slope. It indicates safety by the number of standard deviation that separates failure from the mean value of FS.

The value of E[FS] corresponds to the best estimate of the factor of safety. It is obtained by a conventional (deterministic) limit equilibrium method of slices, such as Bishop simplified or Spencer. A critical failure surface is obtained considering the mean values of the variables involved in the analysis. In most cases, these variables are the geotechnical parameters: cohesion (c), friction angle (ϕ), unit weight (γ) and pore pressure (u). In some cases, geometrical parameters (such as slope height and slope inclination) may also be considered as variables.

The standard deviation σ[FS] corresponds, by definition, to the square root of the variance V[FS], which is obtained by the following expression (Harr, 1977):

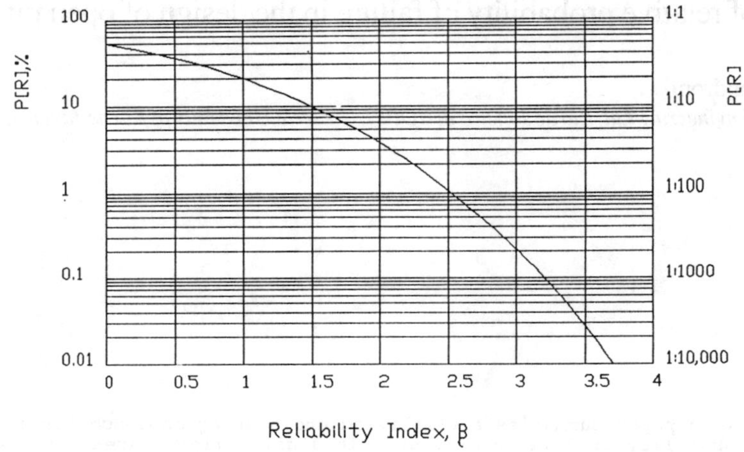

Figure 1: Relation between the reliability index, β, and the probability of failure, P[R]

$$V[FS] = \Sigma^N_1 \; (\delta FS_i \; / \; \delta X_i)^2 \; . \; V[X_i] \qquad (2)$$

In expression (2), X_i represents each of the N parameters considered as variables, δFS_i is the change in FS resulting from a small change (δX_i) in one variable and $V[X_i]$ is the variance of X_i.

The N values of $\delta FS_i / \delta X_i$ are obtained by simply running N times the stability method for the critical surface. In each run, only one of the variables is changed by a small amount.

The values of $V[X_i]$ are computed by simple statistical procedures (Neter and Wasserman, 1974; Christian *et al.*, 1992). An adequate experimental program is needed for estimating the variance of each geotechnical parameter.

The probability of failure P[R] can be related to the reliability index β, provided that the shape of the distribution of FS is known. **Figure 1** presents the relation for the case of a normal (Gaussian) distribution of FS. This relation is practically the same for other common distributions of FS, provided that β is not higher than about 2.5.

Figure 2 shows the relation between FS and P[R] for several values of standard deviation of the safety factor, σ[FS]. It is noted that a given value of P[R] may correspond to a wide range of FS values, as a function of the prevailing value of σ[FS]. In other words, a slope with a low mean factor of safety (e.g., E[FS] = 1.10) may correspond to a smaller probability of failure (e.g., P[R] = 1.3% or 1:80) than other slope with a higher value of E[FS] (e.g., E[FS] = 1.30 and P[R] = 13% or 1:8). This is because the first slope with low safety factor, has a lower degree of uncertainty or variability associated to the parameters considered as random variables in the analysis.

3.0 EXAMPLE OF ANALYSIS OF A SINGLE SLOPE

The procedure for computing the probability of failure of a slope with a height of 200 m and an inclination of 34 degrees, is shown in **Figure 3** (Sandroni and Sayao, 1992).

The slope consists predominantly of residual (saprolitic) soil from a quartz-schist. The piezometric conditions, as indicated by field instrumentation, may be approximated by 2 linear segments. At the top of the slope, the water table is located at an average depth of 80 m below soil surface.

Conventional slope stability computations, considering circular potential failure surfaces, indicated that E[FS] = 1.34. In this example, the top water table was also considered as a variable, together with the soil parameters. The variances of c and ϕ (soil strength parameters) were obtained from the results of 50 direct shear tests on undisturbed specimens of the saprolitic soil. The variance of the water table was based on evaluation of long term piezometric data from an adjacent area (Nunes *et al*, 1985).

Computation of the variance of the safety factor, V[FS], is detailed in **Table 1**. It is seen that V[FS] = 0.0259, corresponding to a standard deviation σ[]FS] = 0.161. From expression (1), the reliability index β = 2.12 is obtained. From **Figure 1**, a relative probability of failure P[R] = 1.8% or 1:60 may be estimated.

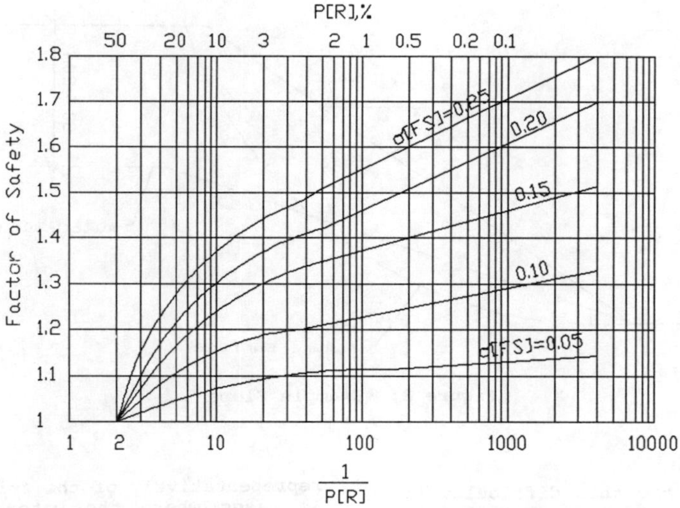

Figure 2: Relation between the probability of failure, P[R], the standard deviation of the safety factor, σ[FS], and the safety factor

Table 1: Example of analysis of a slope
E[FS] = 1.341

X_i	δX_i	δFS_i	$\delta FS_i/\delta X_i$	$V[X_i]$	$(\delta FS_i/\delta X_i)^2.V[X_i]$	
γ_{nat}=28.3	+2.83	-0.004	-0.001	1.96	3.9×10^{-6}	(0.02)
γ_{sat}=29.0	+2.90	+0.022	0.0078	1.96	0.0001	(0.38)
c=25.0	+2.50	+0.004	0.0016	590.0	0.0015	(5.8)
tanφ=0.781	+0.113	+0.188	1.6682	0.0072	0.020	(77.2)
Piezometry	+10m	-0.033	-0.0033	400.0	0.0043	(16.6)
				Total = V[FS] =	0.0259	(100.0)

Table 2: Results from existing Slopes

Slope	H(m)	α(deg)	Material	E[FS]	β	P[R]
failed	30	30	Resid. Soil	1.13	0.67	1:4
	240	42	Iron Ore	1.09	0.90	1:5
	103	29	R.S. & I.O.	0.94	<0.0	>1:2
Stable	234	23	Iron Ore	2.16	4.83	<1:1000
	300	30	Schist	2.30	2.70	1:300
	200	34	Resid. Soil	1.34	2.12	1:60

4.0 ESTABLISHING THE DESIGN CRITERION FOR SLOPE STABILITY

The use of probability of failure in the design of slopes requires the establishment of a maximum (or acceptable) value of P[R] (i.e., a minimum value of β). There exists little experience on $P[R]_{max}$ (or β_{min}) values. Limited information has been summarized by Whitman (1984) for typical applications of risk analyses, suggesting that values of annual probability of failure in the order of 1:10 to 1:50 may be accepted for mine slopes. The establishment of a universally acceptable design value of P[R] faces the same difficulties as those encountered with FS.

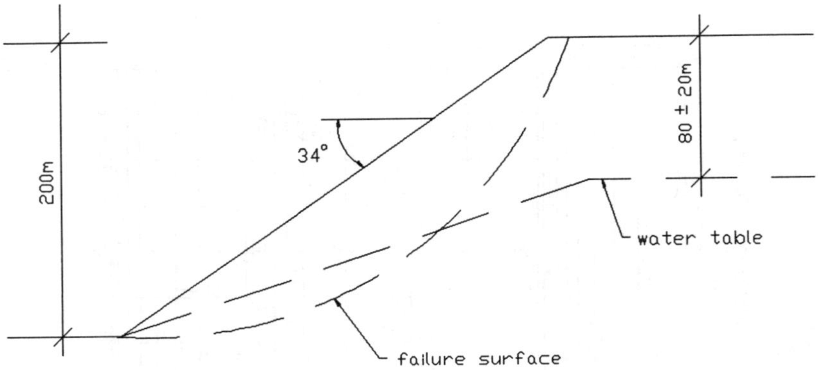

Figure 3: Example Slope

One way to overcome this difficulty is to establish a relative probability criterion by calibrating the procedure in the light of observed local behaviour. This can be done by analyzing a number of stable and failed slopes of a given mining area. **Table 2** summarizes the results obtained in six selected slopes from open pit iron mines in Minas Gerais State. Of these, 3 have failed and 3 are stable. It is noted that failed slopes correspond to β lower than 1 (i.e., $P[R] > 1:7$), while stable conditions correspond to β higher than 2 ($P[R] < 1:50$). In this particular case, a value $P[R]_{max} = 1:50$ (i.e., $\beta = 2$) has been selected as the design criterion.

It should be noted that the relative probability of failure established above refers only to those factors considered as variable in the analysis and that the formulae are approximate. Moreover, averaging of spatial variability has not been considered. The $P[R]_{max}$ criterion must therefore be used only as a nominal indication of stability relative to a given set of studies such as the one described.

5.0 EVALUATING THE IMPORTANCE OF PARAMETERS

The procedure allows the designer to evaluate the contribution of each variable to the stability of the slope under consideration. For the example slope given in section 3, **Table 1** indicates that the friction angle ϕ is the most important factor (among those considered as random variables). Uncertainty in ϕ-value contributes to about 77% of the computed variance of FS. On the other hand, expected variations of the phreatic surface are of lesser importance (about 17%) and variability of unit weights and cohesion are insignificant. It seems, therefore, that particular attention should be given in this case to the quality and representatively of the friction angle. In cases where the water table is a dominant factor, attention should be focused on drainage of the slope.

6.0 CONCLUSIONS

A relative probabilistic design criterion for slope stability can be established from observed behaviour of stable and failed slopes.

For the design of open pit iron mines in the State of Minas Gerais, Brazil, an acceptable maximum value of $P[R] = 1:50$ has been suggested.

7.0 REFERENCES

Christian, J.T., Ladd, C.C. & Baecher, G.B. (1992). "Reliability and Probability in Stability Analysis". Stability and Performance of Slopes and Embankments - II, ASCE, Vol.2, pg.1071. Berkeley, USA.

Harr, M.E. (1977). "Mechanics of Particulate Media: A Probabilistic Approach". Ed. McGraw-Hill. New York, USA.

Neter, J. & Wasserman, W. (1974). "Applied Linear Statistical Models". Ed. Richard D. Irwin Inc. Homewood, USA.

Nunes, A.J.C., Sandroni, S.S., Santana, F.C & Brito Filho, H.G. (1985). "Geomechanical Evaluation for Design of Slopes at Caue Open PIt Iron Mine, Itabira, Brazil". Int. Symp. on the Role of Rock Mechanics in Excavation for Mining and Civil Works, ISRM, Zacatecas, Mexico.

Sandroni, S.S. & Sayao, A.S.F.J. (1992). "Avaliação Estatistica do Coeficiente de Segurança de Taludes". 1a. Conf. Brasileira Sobre Estabilidade de Encostas. ABMS, Vol.2, pg. 523. Rio de Janeiro, Braizil.

Innovative Mine Design for the 21st Century, Bawden & Archibald (eds) © 1993 Balkema, Rotterdam, ISBN 90 5410 325 6

Slope management strategies for successful mining

James P. Savely
Cyprus Miami Mining Corporation, Ariz., USA

ABSTRACT: Slope management is anticipation and control of slope behavior. The objective of slope management is to mine in a safe manner and achieve the mine plan. Mining practices need to be modified in unstable areas and mine plans, schedules, and costs should reflect the effects of necessary strategies. An understanding of the causes of instability and the expected slope behavior is essential to successful mining. Failure geometry, strength properties, and monitoring data are important guides in predicting slope behavior. These guides are discussed in more detail regarding the indicators of stability or instability, and whether mining can proceed safely. In addition, methods of establishing movement rate cutoffs and specific responses that need to be in place should movement rates exceed cutoffs are suggested. Effective slope management requires geologic knowledge and contingency planning. This paper describes some of the technical basis and strategies that have been successful in mining with pit slope movement.

1 INTRODUCTION

Slope management is anticipation and control of slope behavior. It requires a balance of theoretical knowledge and field experience. The objective of slope management is to mine in a safe manner and achieve the mine plan. When losses are incurred from slope movement the problem is often due to the lack of geotechnical evaluation of the mine plan. With proper investigation many slope problems can be anticipated and included as part of the planning process to develop a realistic mine plan. The mine planner has the knowledge of economics and balancing tons and grade with stripping, but he rarely has the geotechnical expertise to determine achievable slope angles.

Geology must be mapped and interpreted ahead of planning and mining. Dewatering devices must be in place in time to provide increased stability and not mined out at a crucial time. Blasting must be controlled on final walls and in sensitive areas to prevent excessive backbreak into the pit walls which can create or aggravate slope instability. Monitoring the slope performance is required for guiding the mining in unstable areas, determining failure geometries, checking design assumptions, predicting slope behavior, ensuring safety, and providing information for redesign when necessary. Periodic field inspection of the slopes in active mining areas is mandated by law and is a part of the monitoring procedure.

Slopes can be designed on stability considerations alone, but stable pit walls do not necessarily mean that the slope has been successfully engineered. A slope that shows no failure may be a too conservative design with maximum profit not realized. Similarly, a slope that appears unstable does not necessarily mean that the slope design is too steep and that safety is being compromised for profit. Economic optimum slopes often have significant instability and there are several examples where mining has been successful and safe while the slope was showing movement (Kennedy and Niermeyer 1970, Brawner, Stacey and Stark 1975, Blackwell and Calder 1982, Savely and Kastner 1982). However, failure must be controlled or contained, and movement must be anticipated to

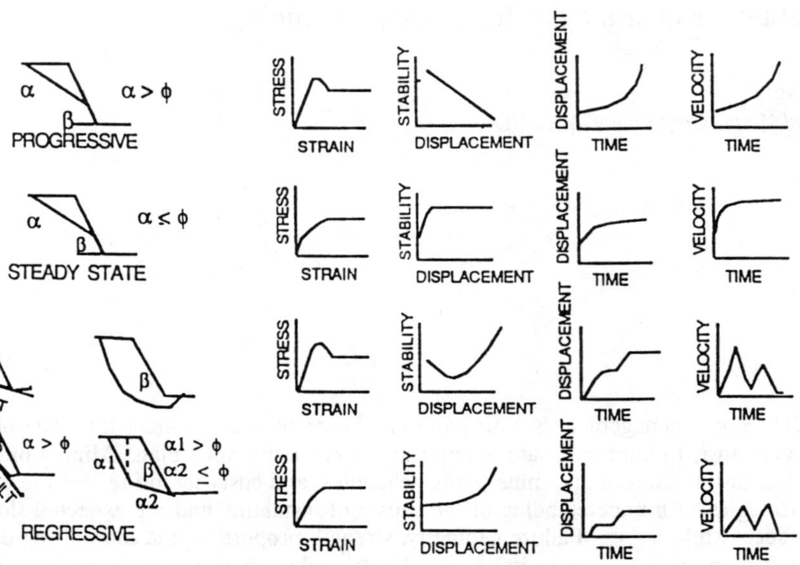

Figure 1. Progressive-regressive failure conditions (from Call 1990).

provide safety and to minimize economic impact. Responses to slope movement should be thought out and included as part of the mining plan rather than becoming crises that balloon costs, reduce benefits, and perhaps threaten the mine life.

2 PREDICTING SLOPE BEHAVIOR

Movement vectors, displacement, and velocity of slope movement are determined from observations of prism targets on the pit slope, wireline extensometers, and occasionally more sophisticated monitoring devices. Velocity is a better measure of slope performance than displacement. Consider stability as the ratio of driving stress to resisting stress. When this ratio is greater than one, driving stress exceeds resisting stress and movement occurs. Technically failure has occurred, but yielding rather than collapse often takes place. At high stress ratios more and faster yielding occurs and velocity will increase. Higher stress ratios indicate higher velocities and the stress ratios are determined by the failure geometry and the changes in the stresses as the slope displaces. Stability is decreased by increasing the driving stress, decreasing the resisting stress, or changing both driving and resisting stress. Therefore, predictions of displacement and velocity are related to the failure geometry and the stresses applied due to mining and water.

Displacement and lost access may be only a mining problem and may not be a failure in terms of the mine plan or safety. But, if toes move faster than the mining can advance along a level, the planned ore can be covered by the slide material, and the mine plan will not be achieved. Economic failure has occurred. The objective is to prevent the slope from reaching a velocity that is unrecoverable and prevents mining, and to give sufficient warning so that men and equipment can be moved from an area near a slope that is approaching an unacceptable velocity.

2.1 Definition of failure stages

Broadbent and Zavodni (1982), defined regressive and progressive failure stages and related these stages to failure geometry. Call (1990) extended their original concept into a useful description of failure characteristics and expected behavior (Figure 1). A slope in the progressive stage tends to become less stable with time. The slope will accelerate and displace to total collapse unless active control measures are taken such as slope support, unloading, or slope flattening. Movement can

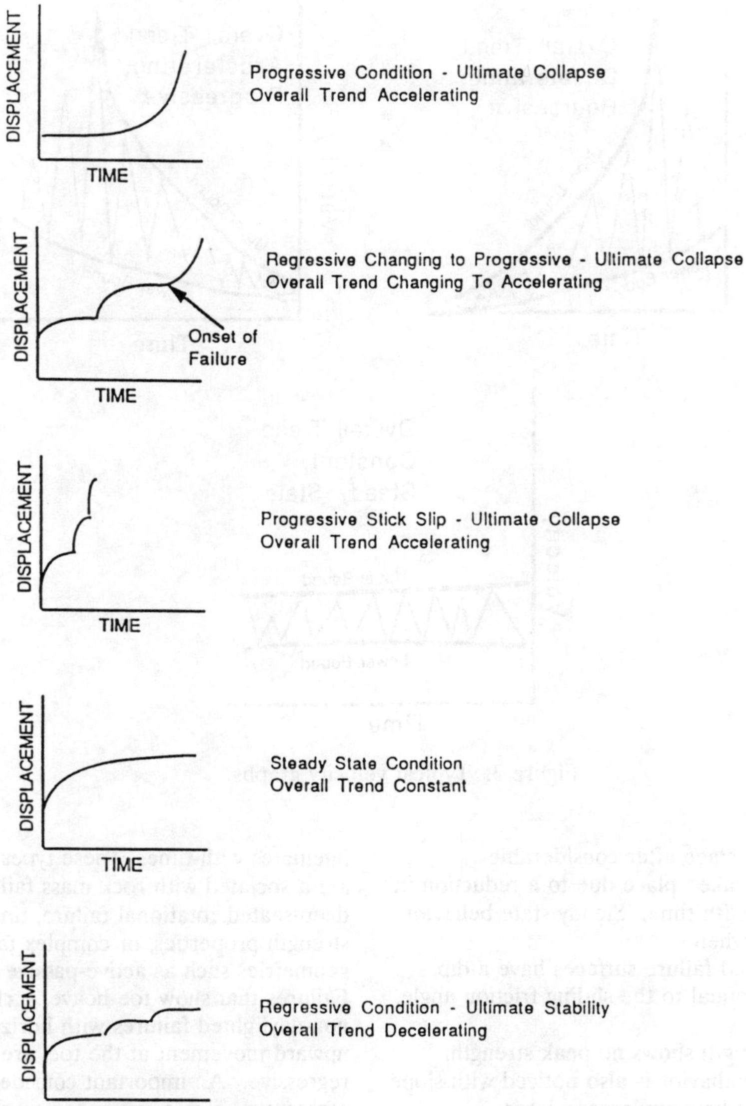

Figure 2. Typical cumulative displacement graphs (from Broadbent & Zavodni 1982).

be sudden. The characteristics of these failures are:

1. the failure surface has a dip direction out of the slope and a dip that is flatter than the slope angle but steeper than the sliding friction angle,
2. the total driving force does not decrease with displacement, and
3. the shear strength drops with displacement (peak shear strength greater than residual strength).

These type of failures are typical of daylighted structures such as plane shear and wedges formed from faults and other major structure.

A slope that continues to creep at constant velocity with a gradual increase in displacement with time is considered to be in a steady state. A steady state slope behavior can change into

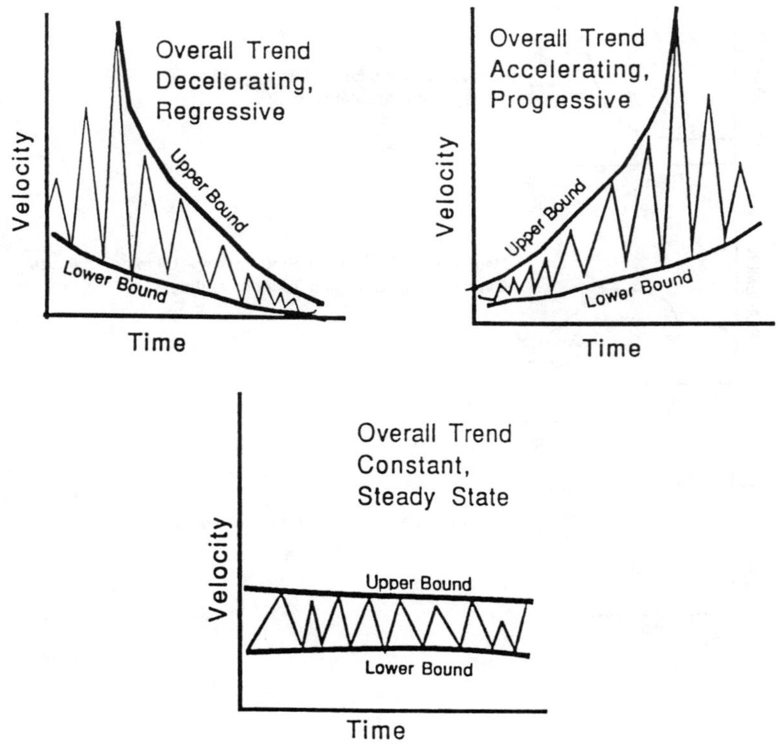

Figure 3. Typical velocity graphs.

the progressive stage after considerable movement has taken place due to a reduction in shear strength with time. Steady state behavior usually occurs when:

1. well defined failure surfaces have a dip approximately equal to the sliding friction angle, and
2. shear strength shows no peak strength.

Steady state behavior is also noticed with slope instabilities that have undergone large displacements over time where failure surfaces have become well defined and the surface worn down to a nearly complete residual strength.

Characteristics of regressive stage failures are:

1. the ratio of driving stress to resisting stress decreases with displacement, and
2. the slope will tend to become more stable with time and show decelerating or stick-slip behavior.

The cumulative displacement graph appears cyclical and the cycles will generally correspond with changes in external forces such as mining activity, blasting, and rainfall. The slope will tend to restabilize if the external forces are removed. The movement rate typically fluctuates with time. These types of movement are associated with rock mass failure, deep-seated rotational failure, time dependent strength properties, or complex failure geometries such as active-passive blocks. Failures that show toe heave, such as non-daylighted failures with horizontal or upward movement at the toe, are typically regressive. An important consideration is that regressive failures can become progressive when:

1. mining daylights a previously non-daylighted failure or new failure surface,
2. the rock mass at the toe breaks up and a shear surface or shear zone is developed,
3. water pressures increase, or
4. when mining is continued or increased without letting the slope regain equilibrium, and the slope accelerates beyond recovery.

Figure 2 shows examples of displacement graphs for the three stages. Figure 3 shows typical velocity graphs. Because of fluctuations in the velocity calculations, it is important to look at a long time history and the upper and lower bounds to determine trends.

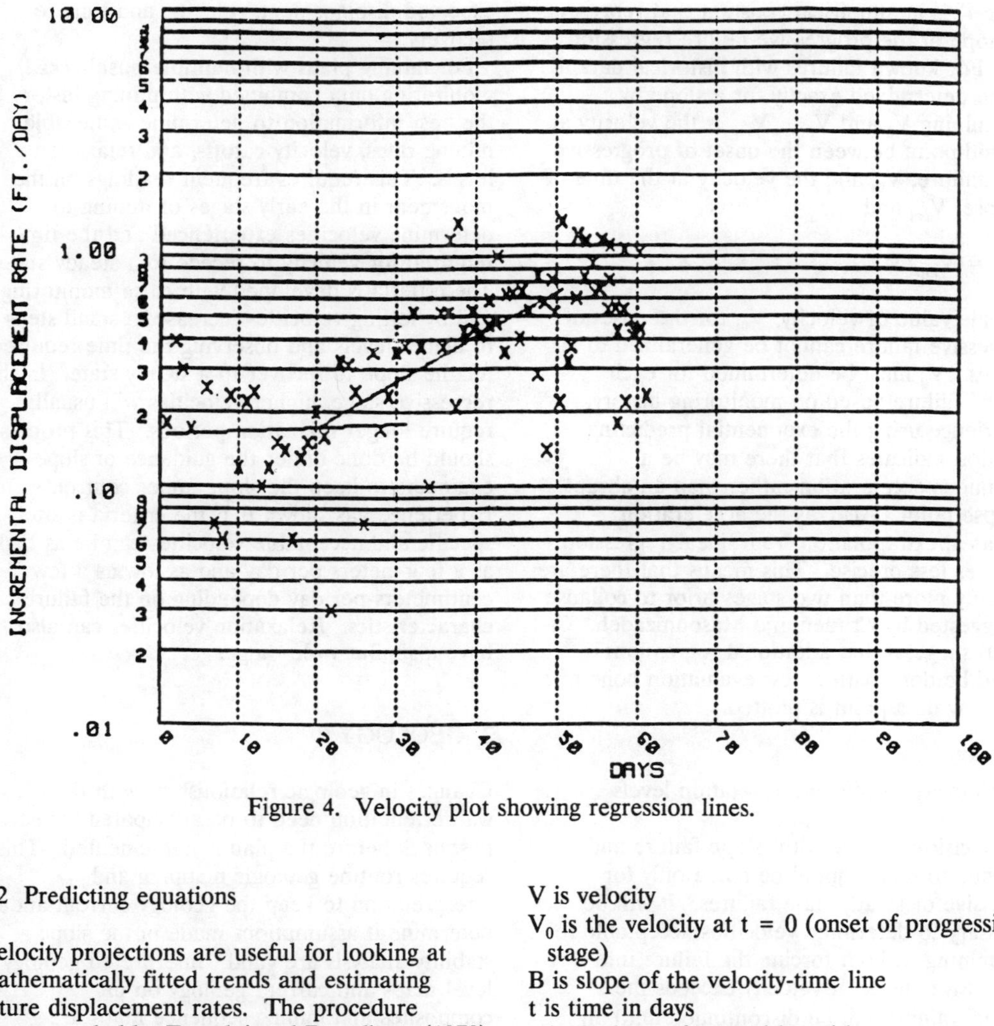

Figure 4. Velocity plot showing regression lines.

2.2 Predicting equations

Velocity projections are useful for looking at mathematically fitted trends and estimating future displacement rates. The procedure recommended by Zavodni and Broadbent (1978) is to plot the log of velocity (incremental velocity or incremental cumulative velocity) versus time. The displacement record appears to be exponential so generally the data plots on a straight line with a change in slope at the onset of progressive stage failure. For example, in Figure 4 the onset of progressive stage failure for the data is interpreted to be at 21 days with a velocity, V_0, of 5.5 cm/day (0.18 feet/day).

Assuming t = 0 at the onset of progressive stage failure, the equation for the progressive stage line is:

1. $V = V_0e^{Bt}$

where:

V is velocity
V_0 is the velocity at t = 0 (onset of progressive stage)
B is slope of the velocity-time line
t is time in days
e is the base of natural logarithm

The plot of the data is examined for a change in slope that is expected to occur at the onset of progressive stage failure. A regression fit is made to the first trend of data points. A second regression fit is made on the second trend of data points. The point where the two regression lines intersect is where t = 0, which determines V_0. Zavodni and Broadbent (1978), developed an empirical relationship to predict the collapse velocity, V_{col}. They plotted data on numerous failed slopes from different mines as log of velocity versus days prior to collapse and estimated that the collapse velocity could be predicted by the equation:

2. $V_{col} = K^2V_0$

where K is an empirical constant that represents the slope of the progressive failure regression line. For known failures with historical data, K can be determined exactly for a slope by determining V_0 and V_{mp}. V_{mp} is the velocity at the mid-point between the onset of progressive stage failure, V_0, and the velocity at the time of collapse, V_{col} and

3. $K = V_{mp}/V_0$

A single value of velocity, V_0, for the onset of progressive failure cannot be generalized to all failures. V_0 must be determined for each specific failure based on monitoring history. Experience using the exponential predicting equation indicates that there may be a continuous acceleration rather than an abrupt collapse point. Also, as the acceleration increases, extrapolation from linear regression becomes less precise. This means that there are probably more than two stages prior to collapse as suggested by Cruden and Masoumzadeh. Call (1990) suggests that additional regressions should be done with a new evaluation done after each new data point is plotted.

2.3 Velocity cutoffs and relaxation levels

The decision to "live with" slope failure and continue to mine should be made only for regressive or steady state failures. It will be necessary to determine velocities acceptable for safe mining without forcing the failure into a progressive stage. If velocity exceeds the criteria, mining will be discontinued until an acceptable relaxation velocity is achieved. The criteria that is developed determines the economics of the mining.

Stability of many slopes can be controlled if only small displacement is allowed before mining is stopped and the slope is allowed to relax. If mining continues until the slope is at a high velocity, it will usually take a longer time for the slope to return to equilibrium. If the slope is pushed too far or if hydrologic conditions change, it may accelerate to the progressive stage and collapse. More frequent pullouts with short mining periods and short time delays may be preferable to long mining periods with fewer pullouts and extended time delays. The geometry and strength characteristics of the failure are indicators of expected displacement, velocity, and time to recovery.

For mining plans with multiple pushbacks, monitoring data combined with mining history is the best information to determine achievable mining rates, velocity cutoffs, and relaxation levels. This requires frequent readings on the movement in the early stages of mining to determine velocities experienced and the time required for velocity to recede to a steady state. The criteria is developed with close monitoring and by letting velocities increase in small steps to higher levels and observing the time required for the slope to recover to a steady state. In the regressive stage, higher velocities will usually require longer relaxation periods. This process should be done under the guidance of slope engineers to keep the slope under control. Experience has shown that the criteria is site specific and acceptable velocities can be as high as a few meters per day and as low as a few centimeters per day depending on the failure characteristics. Relaxation velocities can also have a similar wide range.

3 GEOLOGY

Changes in geologic relationships with the pit wall orientation need to be anticipated for each pushback before the plan is implemented. This requires routine geologic mapping and interpretation to keep the geology current and to determine if assumptions made in the slope stability analysis are valid. Interpreted geology level maps and current geology on pit composites and mining sequence maps are essential tools for evaluating and managing pit slope instability. These maps tell you where you are and where you are going in terms of the slope design.

I have found that by keeping a current composite geology map it is possible to explain stability concerns to operations personnel. In one instance we were able to modify a blasting pattern to reduce backbreak along a major fault. In another instance it was possible to prevent failure of a haul road near the bottom of the pit because the geology projected from the mapping indicated that a failure geometry would be undercut and fail half of the road. Mining concentrated in other areas of the bench and the haul road was intentionally undercut after most of the material had been mined. In another situation it was possible to locate a pump well where it would be more effective for dewatering because the groundwater was compartmentalized between faults that had been mapped and located on the pit map.

4 CONTINGENCY PLANNING

If slope movement is expected, then contingency planning is necessary to minimize the impact of slope failure and to provide a realistic plan. Contingencies that have proven successful are discussed below.

4.1 Provide dual or multiple access to the ore faces and pushbacks

Although this may increase stripping requirements, if access to the ore is lost or production is disrupted the consequences can be costly. With only one access, the slope design must be more conservative to ensure production and this also results in increased stripping and risk to the mine plan. For slopes that have a history of movement even a conservative slope angle may not guarantee continued production, thus making multiple access a better choice.

4.2 Stockpile sufficient ore to maintain production during periods of no mining

The tonnage requirements depend on whether other contingencies are available and the anticipated length of time required for the slope to return to an acceptable velocity for mining.

4.3 Establish additional ore faces before instability occurs

The number of additional ore faces required depends on the extent of stability problems and the orebody characteristics. The objective is to create as many alternative ore sources as possible to provide flexibility.

4.4 Plan wide pushbacks

In large mines wide pushbacks provide the flexibility to simply mine around the failure area and come back later to mine final limits when the movement rate has decreased. The stripping requirements are increased, but this is usually a better option than the delays, constrained mining, and deferred ore that result from failures that affect more confined working areas.

4.5 Plan a stepout

Depending on the orebody configuration it is often possible to plan a stepout for slopes that are expected to show slope movement. The exact level where the stepout is needed may not be known and difficult to determine, but a best guess or an arbitrary level can be taken to account for the most likely event. During mining, if movement requires the stepout, then the plan can still be met. If the stepout becomes unnecessary, then additional ore might be gained. This is a strategy that provides realistic mine plans in historically unstable areas.

4.6 Use dynamic sequencing to add lag times

Most sequencing is done to balance ore tons and grade with stripping to even out capital flow and keep it under control. Most planners do not formally include geotechnical aspects in the sequencing (Couzens 1990). Dynamic sequencing can be done in areas of expected instability where planned periods of slow or no mining are included to provide flexibility and a realistic mining rate. It is similar to lag time in project scheduling where early start/finish and late start/finish can be defined. For some large scale failures it is possible to adjust the sequencing to minimize slope heights and to minimize forces that contribute to instability.

4.7 Provide for failure costs in scheduling and budgeting

This might include estimates of extra tonnage for slide cleanup, additional haulage costs, deferred ore and stripping, production delays, haul road repairs, and stepouts.

4.8 Design to prevent noses in the plan geometry

It is well known that noses are generally less stable than straight or concave pit walls. Most

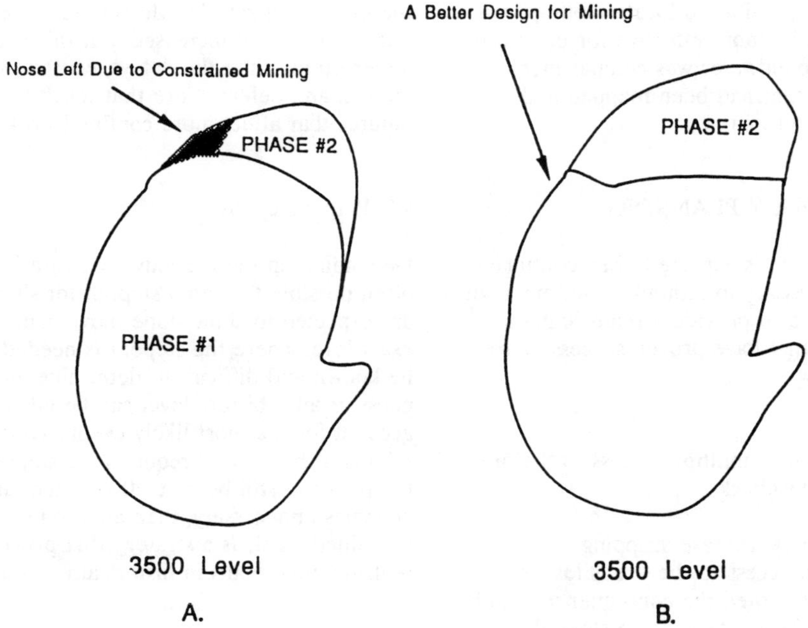

Figure 5. An unplanned nose due to constrained mining.

planners try to minimize noses in their plans, but in mine operations noses can develop because of unachievable geometry. This can happen where different phases of mining join together along a pit wall (Figure 5). An unplanned nose that is left is just as unstable as a planned nose and a needless slope failure can result. Frequently operations cannot mine to the precise corners required by the plan. The planning engineer can anticipate these operations problems and provide for them in the design (Clark 1990).

5 OPERATIONAL CONSIDERATIONS

There are a few specific activities in mine operations that need to be done to improve slope stability and to provide for better mining. Operational considerations include the following.

5.1 Locating sumps in areas where seepage will not affect stability

Usually sumps located in the bottom of a pit away from the toe of the slope are acceptable. In deep mines the water will have to be staged to higher sumps. These sumps should be lined or otherwise sealed to prevent seepage into the pit wall. The number of sumps should be kept to a minimum. Sump placement and dewatering schemes should be planned throughout the life of the mine using the mine sequence maps.

5.2 Eliminating or controlling flows of water that enter into slopes and operating areas from natural sources, leach dumps, tailings dams, processing facilities, and pipelines

Storm runoff should be diverted around unstable areas and the pit perimeter in general. To prevent water from entering tension cracks, the crest and upper slopes of failures should be covered and contoured to provide for fast runoff. Small steady flows into tension cracks or on the slope have a worse effect than occasional large flows from storm runoff. In a typical fractured rock mass, assuming 5 percent of the volume can be considered open space between fractures, a volume of dry slope 150 meter (500 feet) high, 150 meter (500 feet) long, and 60 meter (200 feet) deep (approximately 4 million tons) could hold 72 million liters (19 million gallons) of water. Without drainage and a recharge of

75 liters/minute (20 gpm), all of the available open space will be filled with water in less than two years. Therefore, if there is 75 liters/minute (20 gpm) more going into the slope than is being drained from the slope, the slope will be completely saturated in less than two years and stability will be reduced. It is more likely that failure will occur before the slope is completely saturated.

5.3 Reviewing and implementing controlled blast designs in sensitive areas

Blast damage is generally considered to be excessive backbreak on benches and along major structure. A general statement is that charge per delay, powder factor or energy, and subgrade drilling should all be decreased as the final pit wall or the sensitive slope is approached (Savely 1989).

5.4 Comparing pit geology with mine advance to anticipate slope problems and guide the mining

This is applied geology and it requires that mapping be current and the geology interpreted in time to be useful. The information can be communicated to operations in weekly production meetings.

6 RESPONSES TO SLOPE MOVEMENT

Stewart and Kennedy (1971), discuss actions that can be taken in response to slope instability. Call (1990) has developed these ideas further and suggests eight responses to slope movement.

6.1 Leave the unstable area alone

This option can be taken when instability is in an abandoned or inactive area.

6.2 Continue mining without changing the mine plan

If the velocity is low and predictable, and the area must be mined, living with the displacement while continuing to mine may be the best alternative. This course of action requires a good monitoring system, an understanding of the failure mechanics, and historical information on the slope behavior.

6.3 Unload the slide through additional stripping

Even though unloading has been a common response, in general it has been unsuccessful. In fact, there are situations involving high water pressure where unloading actually decreases stability. The failure geometry and the failure mechanics must be understood to ensure that unloading will stabilize the failure.

6.4 Leave a stepout

Stepouts have been used successfully in several mines. The choice between stepout and cleanup is determined by the trade-off between the value of lost ore and the cost of cleanup.

6.5 Do a partial cleanup

Partial cleanup may be the best choice where a slide blocks a haul road or fails onto a working area. Only that material necessary to get back into operation needs to be cleaned up.

6.6 Mine out the failure

Where the failure is on a specific structure and there is competent rock behind the structure, mining out the failure may be the optimum choice. Structure and failure geometries are often repeating so good geologic interpretation is essential to determine if the new slope will be stable.

6.7 Support the unstable ground with cable bolts

Mechanical support may be the most cost-effective option when a crusher, conveyor, or haul road must be protected.

6.8 Dewater the unstable area

Where high water pressure exists, dewatering is an effective method of stabilization that may be used in conjunction with other options. There are few cases where it is cost effective to live with the effects of high water pressures. An extensive dewatering program that will stabilize slopes can be implemented and maintained for about the same yearly cost as 15 to 20 shifts of stripping. The potential savings in stripping by achieving steeper slopes and reducing failure removal costs because of dewatering can be 1000 shifts or more over the life of the mine. A good discussion of practical dewatering is given in Rantapaa (1990).

7 CONCLUSIONS

Good monitoring information and knowledge of the failure characteristics allow the slope engineer to define the stage of a slope failure as being progressive, steady state, or regressive. For steady state and regressive type movement it is possible to implement a slope management program to "live with" the movement and to continue mining safely. If slope movement is expected, it is necessary to develop contingency plans to minimize the impact of slope movement and to provide a realistic plan. Responses to slope movement should be part of mine planning and alternative responses evaluated with a cost-benefit study. When slope movement occurs the appropriately planned alternative can be implemented instead of a costly, unplanned crisis response that may not deal effectively with the slope problem. Mine operations are also responsible in successful slope management. They should control water and revise blasting practices in sensitive areas.

REFERENCES

Blackwell, G.H. & Calder, P.N. 1982. Practical Aspects of Wall Stability at Brenda Mines Ltd., Peachland, B.C.: Proc. Third Int'l. Conf. on Stability in Surface Mining. Ch. 25, C.O. Brawner, ed., AIME, New York, 573-608.

Brawner, C.O., Stacey, P.F., & Stark, R. 1975. A Successful Application of Mining with Pitwall Movement, paper presented to Western Annual Meeting, Canadian Institute of Mining, Edmonton, Alberta, October 26-29, 1975.

Broadbent, C.D. & Zavodni, Z.M. 1982. Influence of Rock Structure on Stability, Proc. Third Int'l. Conf. on Stability in Surface Mining. Ch. 2, C.O. Brawner, ed., AIME, New York, 7-18.

Call, R.D. 1990. Personal Communication.

Call, R.D. 1982. Monitoring Pit Slope Behavior, Proc. Third Int'l. Conf. on Stability in Surface Mining, Ch. 9, C.O. Brawner, ed., AIME, New York, 229-248.

Cruden, D.M. & Masoumzadeh, S. 1987 Accelerating Creep of the Slopes of a Coal Mine, Rock Mechanics and Rock Engineering, v. 20, 123-135.

Clark, J.W. 1990 Personal Communication.

Couzens, T. 1990 Personal Cornmunication.

Kennedy, B.A. & Niermeyer, K.E. 1970. Slope Monitoring Systems Used in the Prediction of a Major Slope Failure at the Chuquicamata Mine, Chile: Planning Open Pit Mines. Proc. Symp. on Planning Open Pit Mines with Special Reference to Slope Stability. A.A. Balkema, Amsterdam, 215-225.

Rantapaa, M.D. 1990. Mine Dewatering Procedures at Cyprus Miami Mining Corporation, paper presented at Goldtech 4, Sept. 10-12, 1990, Reno, NV (also available in 1990 AIME preprint) 12 pp.

Savely, J.P. 1989. Designing a Final Wall Blast to Improve Stability: Trans. AIME, v. 284, 1869-1877 (also available in 1986 AIME preprint).

Savely, J.P. & Kastner, V.L. 1982. Slope Instability at Inspiration's Mines: Proc. Third Int'l. Conf. on Stability in Surface Mining. Ch. 26, C.O. Brawner, ed., AIME, New York, 609-634.

Stewart, R.M. & Kennedy, B.A. 1971. The Role of Slope Stability in the Econornics, Design and Operation of Open Pit Mines: Proc. First Int'l Conf. on Stability in Open Pit Mining. Ch. 2, C.O. Brawner, ed., AIME, New York, 5-21.

Zavodni, Z.M. & Broadbent, C.D. 1978. Slope Failure Kinematics: Proc. 19th U.S. Symposium on Rock Mechanics. May 1-3. 86-94.

Innovative Mine Design for the 21st Century, Bawden & Archibald (eds) © 1993 Balkema, Rotterdam, ISBN 90 5410 325 6

Slope stability and rock mechanics analyses Chuquicamata mine, Codelco, Chile

Manual Rapiman
Geomechanical Department, Department General Mina, Division Chuquicamata, Chile

ABSTRACT: The Chuquicamata copper mine currently operates at a total production rate of 640,000 tons/day, of which 154,000 tons/day is ore. These production rates are in agreement with the long range program and production will remain at these levels for the near future, with production decreasing in the far future. With the knowledge of the ore reserves to date, the 20 year long range program can be updated annually. To reach the projections defined in the long range program, it is not only necessary to improve the quality of the reserve information at depth, but predict the geomechanic conditions of the pit geometry over the long range life of the mine. The purpose of the geomechanic group at the mine site is to collect geomechanic data, analyze the current pit conditions and to project with relative certainty the stability of the pit as it approaches a depth of over 1,000 metres. The objective of this paper is to outline the geomechanic performance in the Chuquicamata mine. This will include monitoring systems, data collection, support to mine operations and mine design, numerical model results and the geomechanic research program.

1.0 INTRODUCTION

The current depth of the Chuquicamata open pit is 645 metres and by the year 2012 a projected depth of 788 metres should be achieved. The final pit depth is projected to exceed 1,000 metres.

The production rate is achieved with the use of 18 drill rigs, 21 Shovels with bucket capacities ranging from 18 to 34 cubic yards, 90 operating trucks per shift with truck size varying between 190 and 240 tons each and various other service equipment. The equipment in the mine is distributed to several pushbacks in different sectors of the mine.

An open benching method is utilized, where the benches are maintained, to allow access along the benches to various mining sectors where pushbacks are being developed. This method provides access to the orebody at different areas, allowing ore blending which maintains an average head grade going into the concentrator.

In order to maintain extraction in the different mining sectors, narrow pushbacks are used to minimize the time to reach the orebody. This mining method presents several limitations. For example, backbreak is a serious problem since damage to the upper benches will decrease the stability of the necessary haulage routes along the upper benches. The narrow pushbacks also present problems with the shovel truck operations, since space allows room for only two trucks at each shovel.

The final bench heights range from 24 to 26 metres which creates areas of high risk since the shovel buckets are not capable of reaching the crest of the upper bench, leaving zones with high probability of slides.

All the problems which exist in the mine are being studied with operational problems studied on a short term and mine design problems studied on a medium to long term basis. The future of the mine is dependent on the results of these studies. The geomechanical group use the knowledge of the actual rock conditions and the behaviour of different slopes through data from monitoring systems to recommend changes in the mine design, the haulage road grade, work angle and the overall pit angle.

2.0 SLOPE STABILITY

The Chuquicamata mine is bisected by a regional subvertical fault, referred to as "Fault West". The fault defines two structural domains, the West and the East Sectors which exhibit completely different behaviour with respect to slope stability.

2.1 East Sector

The rock in the East Sector is reasonably competent compared to the West Sector, with reasonable stress conditions to date. The rock exhibits an elastic constitutive behaviour with minor deformations occuring.

Over the long term, the ratio of the stress at the toe of the slopes is expected to increase with depth. This aspect is important to the geomechanics group since the east wall will lodge the infrastructure for the ore crushing station. Therefore, it is necessary to maintain a stable upper slope and underground tunnels for this project to proceed.

2.2 West Sector

This area of the mine has been a main concern of the geomechanics group for the last 12 years. A serious stability problem developed in 1980 and has produced many problems for the normal development of this wall to date.

The "Fault West" is located close to the toe of the west wall and has associated with it a zone with poor rock quality exhibiting plastic behaviour. As a result of the stress generated in the wall, large deformations occur within the plastic zone producing a general movement of the wall, with tension cracks developing in the top of the wall and benches. **Figure 1** illustrates this behaviour.

It is important to the geomechanics group to predict the future behaviour of this slope and to define the slope angle for the final pit. Currently, the failure mechanism of the west sector is understood and reproduction of the slope behaviour is possible using the FLAC model. The model has been used to analyze the sensitivity of the slope stability with the changes in the rock properties and the location of the underground water level.

The general movement or creeping of the west wall generates displacements in the surface in a cyclical behaviour, mainly in the areas of the pushbacks. When a pushback is located in the plastic zone, the level of displacement increases, at times exceeding 10 cm/day.

Knowledge of the rock properties and the water table conditions are necessary to predict the future behaviour of the west sector and are monitored through drill hole programs and a hydrologic study.

3.0 MONITORING SYSTEMS

The monitoring system at the Chuquicamata mine has been developed to suit all monitoring requirements, which hay be related to local behaviours or at other times to study the global behaviour of slopes. The monitoring system is also site specific. In areas with small crack development, it is necessary to take measurements with high precision (millimetres) whereas movements of open cracks with large displacements must be measured with reasonable precision (centimetres).

A total station system, used normally for survey control is also used to measure surface displacements. During 1992, a Geodometer System model 140 SMS, see **Figure 2 and 3**, has been used to automatically monitor the displacements in the mine. Over 100 prisms, distributed mainly in the west side, are used as point controls. The equipment has been successful in predicting 5 events in the past year, minimizing loss or damage to equipment or personnel.

The information provided by the Geodometer system has also been used to make short term operational decisions. A wedge displacement in the northwest sector of the mine, refer to **Figure 4**, was detected using the total station system. This early detection provided adequate warning so that the haulage road remained open for two days, and the equipment was moved from below the wedge.

An inclinometer SINCO Digitilt RPP has been used to detect deformation at depth. Refer to **Figure 5** for typical output. Other instrumentation, such as the triaxial box, shown in **Figure 6**, have been designed and constructed by the people involved in the geomechanics program at Chuquicamata.

4.0 DATA COLLECTION

The rock properties necessary for the monitoring program were determined experimentally in a laboratory setting. These laboratory procedures provided typical rock parameters such as cohesion and friction angles, Young's, Poisson's and Bulk's Moduli for each of the rock types. The laboratory tests included triaxial, tension, compression and shear tests measuring deformations. Microlynx software, a graphics data base, was used to interpret the raw data and the analyzed data was then used in the geomechanical models.

The determination of the properties of the rock at depth is imperative since this rock will be the future walls of the pit. The results of the simulations of the slopes and the accuracy in the numerical geomechanics models for the slope stability analysis will depend on the real knowledge of the rock parameters.

Table 1 shows the rock properties obtained from samples recovered from drill holes, diamond drill holes and other geomechanic drill holes. Most of the drilling was completed within the

Table 1
SUMMARY OF ROCK PROPERTIES OF CHUQUICAMATA MINE

	Dens. g/cm³	Cohes. kg/cm²	Ang. Fric. (°)	Comp. kg/cm²	Ten. kg/cm²	Young kg/cm²	Pois.
Granod. Fortuna							
fresh	2.56	134	36	655	19	361000	0.32
alterat	2.43	63	33	328	--		--
Fault West	2.49	51	34	546	37	310000	0.34
Porfid Chuqui							
fresh	2.69	98	36	613	--	--	--
Porfid East							
fresh	2.56	182	36	887	--	--	--
alterat	2.44	72	31	--	--	--	--
Granod. Elena							
fresh	2.59	158	36	811	58	394000	0.33
alterat	2.55	--	--	377	57	315000	0.33

Table 2
REGIONAL IN-SITU STRESS

	Magnitude	Bearing	Inclination
Sigma 1	1444 psi	246.8°	118.8°
Sigma 2	1136 psi	183.8°	41.3°
Sigma 3	815 psi	141.7°	123.1°

ore zone and the ore is therefore well represented. The waste zones, which will eventually form the pit slopes is not well represented. Therefore, an extensive drill hole program must be initiated to determine the rock quality and properties of these waste zones.

In-situ stresses in the mine, generated both by the excavation itself and the tectonic effects, must be determined. These measurements were made in a small underground excavation outside the influence of the Chuquicamata mine. **Table 2** outlines the main orientation and magnitude of the regional stress in the vicinity of the mine.

A program is currently being used to determine the local in-situ stress that exists in the toe of the slopes. This is being used to define the actual stress conditions and to study the change of these stresses as the pit is increased in depth. In turn this information will be used as input to the numerical models for slope stability.

Other geomechanical information is collected from cell mapping, using the methodology recommended by Mr. Richard Call, as will as information from the general geostructural mapping used by the geology department. All this information allows analysis from a geostructural point of view, such as the probabilistic analysis of wedges. In **Figures 7 and 8**, a general representation of the structural domains of the West and East sides can be seen.

5.0 SUPPORT TO MINE OPERATIONS AND MINE DESIGN

The support of the geomechanical group is the result of several years of work. Together with input from the mine operations and mine planning people, the

Table 3: APPLICATIONS OF THE FLAC MODEL

Side	Year	Slope Angle	Rock Properties	Water Conditions	Observations
west	1990	25°PZ 27°OA	C=5 PSI/Phi=20° C=37 PSI/Phi=28°	Wet	Converge
west	1990	Idem	Idem	Dry	Converge
west	1990	Idem	C=5 PSI/Phi=20° C=37 PSI/Phi=25°	Wet	Diverge
west	1993	24°PZ 25°OA	C=5 PSI/Phi=20° C=37 PSI/Phi=28°	Wet	Converge
west	1993	29°PZ 31°IR	Idem	Wet	Diverge Plastic Zone
west	1993	19°PZ 36°IR	Idem	Wet	Non-Convergent Non-Divergent
west	1997	25°PZ 27°OA	Idem	Wet	Non-Convergent
west	1997	25°PZ 27°OA	Idem	Dry	Convergent

Note: PZ = plastic zone
OA = overall angle
IR = interramp angle over plastic zone

Table 4: APPLICATIONS OF THE FLAC MODEL

Side	Slope Angle	Rock Properties	Water Conditions	Observations
west	27°PZ 37°IR	C=0.4 Phi=20° C=2.6 Phi=28°	Wet	Converge
west	27°PZ 37°OA	Idem	Wet	Diverge
west	27°PZ 37°IR	C=0.4 Phi=20° C=4.5 Phi=30°	Wet	Converge
west	24°PZ 25°OA	Idem	Wet	Diverge

Note: PZ = plastic zone
OA = overall angle
IR = interramp angle over plastic zone

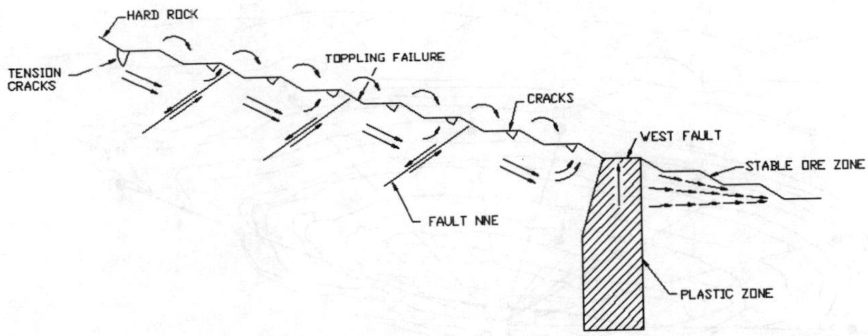

Figure 1: West Side Behaviour

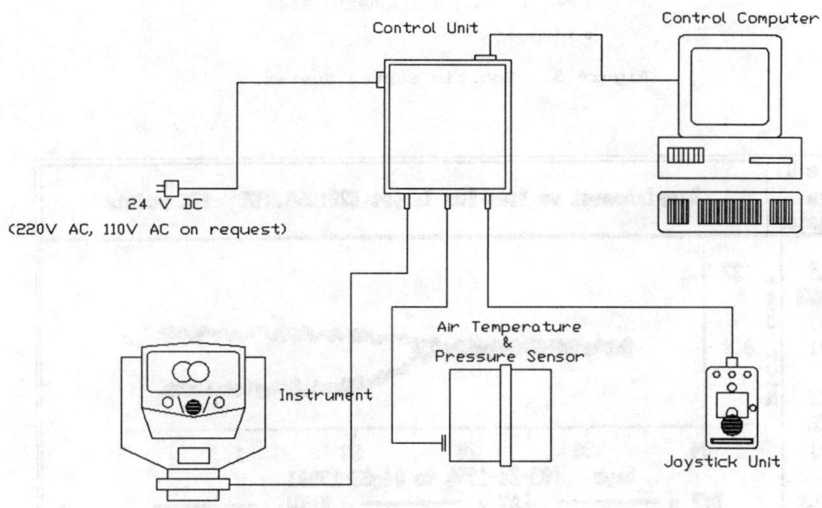

Figure 2: Schematic of Robotic Survey System

best method of determining the needs of mine is decided to optimize the performance of the geomechanics group.

In the mine operations, the detection of risks by wedges and unstable blocks is daily work for the geomechanics group. This is imperative when 21 shovels, 90 trucks and 18 drilling rigs are always working and when the benches are 26 metres in height.

It was deemed necessary to create a special reporting procedure for all geomechanical recommendations. The report would include the name of the addressee, since it is important to define the person responsible for the recommendation and to ensure that the action prescribed will be followed through. It is important that in this process that all conversations between different groups is reported.

The information from the monitoring slope system plays an important role in predicting the unstable rock conditions and to observe the behaviour of the wedges and zones of high risk. For this reason the robotic survey system is used often and starting in March of this year, extensometers (from Modular Mining system) will be transmitting displacement information via radio waves to a computer situated in the geomechanic office.

The main purpose of the geomechanics group with respect to mine design is in defining the slope angle to plan the different pushbacks around the mine.

It is necessary now, to define the slope angles for an open pit design. There are many different angle that are referred to, overall angle, interramp angle, face bench angle and work angle.

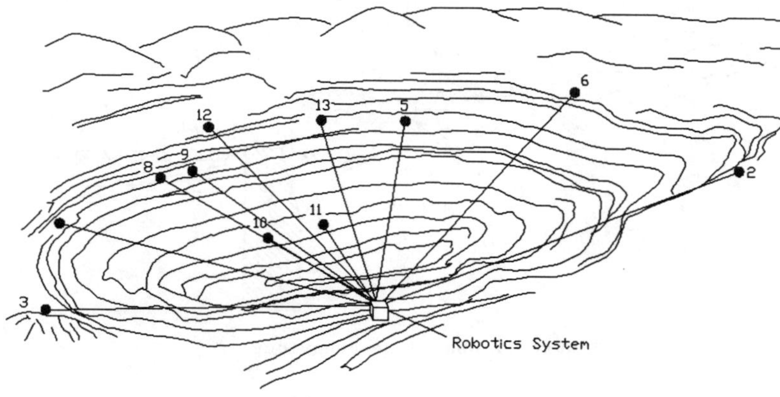

Figure 3: Robotic Survey System

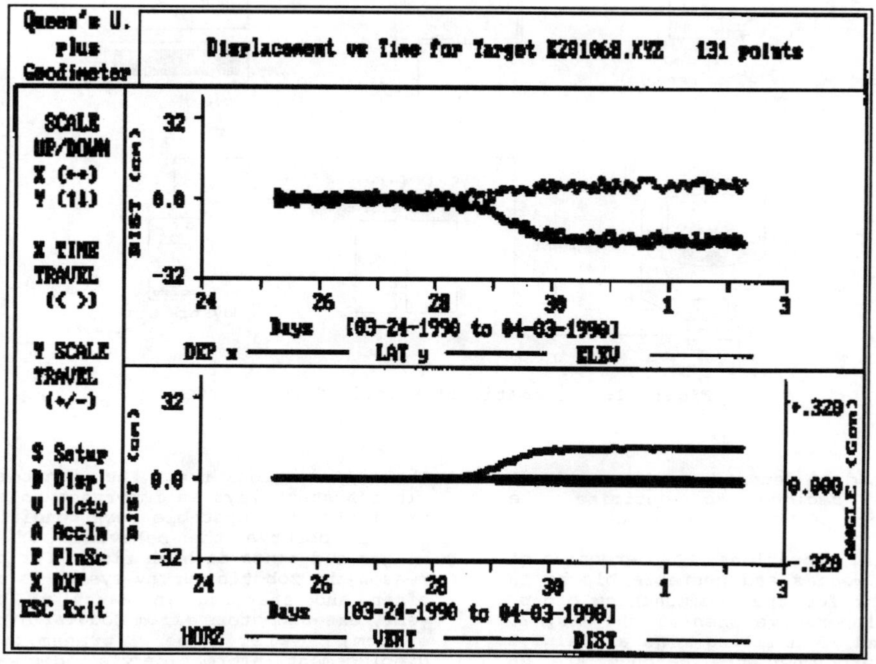

Figure 4: Geodimeter Model 140 SMS Output

It is also necessary to define the final pit angle, which may be represented as an overall or interramp angle.

When an overall angle is recommended, the angle is dependent on the slope stability conditions which in turn depends upon the rock mass quality and to a lesser extent the geostructural conditions. In the West wall, it is possible to define a movement in the slope and therefore the recommendation here is an overall angle of 37°, measured from the west fault to the crest of the slope.

An interramp angle is recommended when the stability depends more on the structural conditions and this angle is defined to be between berms located in

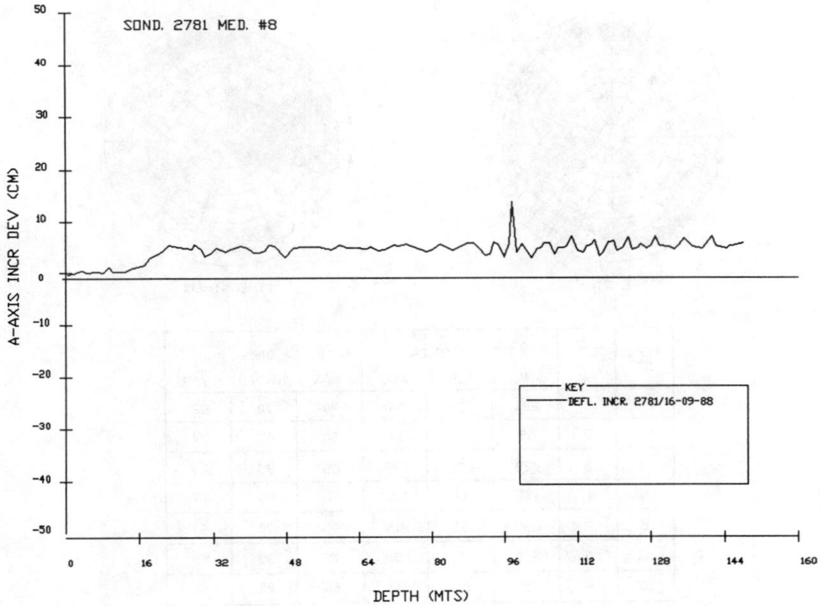

Figure 5: SINCO Digitilt RPP Output

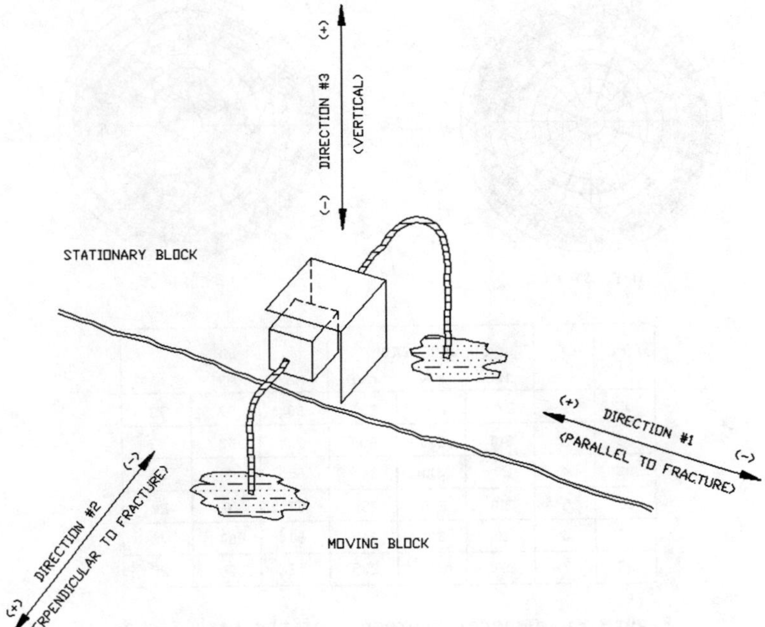

Figure 6: Triaxial Box

the slope. The berms are reached with the ramp locations, with wider benches defined for haul roads or special catch benches. The height of the slope with an interramp angle is dependent on the geomechanic constraint to minimize wedge generation. In the Chuquicamata mine, the interramp angles are defined as follows: the east and southeast walls an angle of 44°, the north wall 40° and the northeast wall 46°, all of which are considered to be the final pit angles.

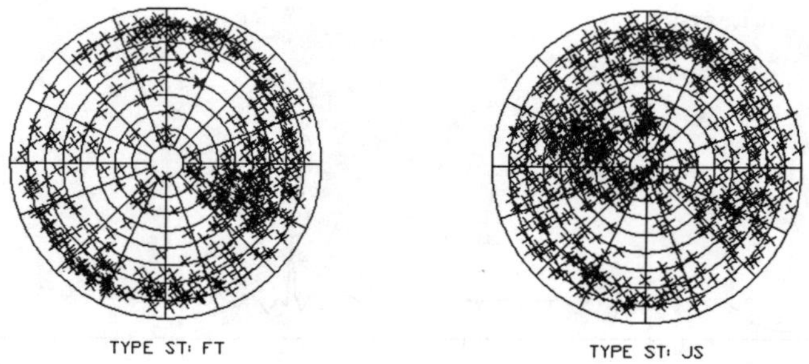

TYPE ST	SET	DIP DIRECTION			DIP		
		MAX	MIN	AVG	MAX	MIN	AVG
FT	1	220	155	200	90	70	82
FT	2	65	40	51	90	60	74
FT	3	320	275	298	65	25	43
JS	1	195	160	181	90	60	77
JS	2	235	195	215	90	70	80
JS	3	320	275	298	75	35	59
JS	4	135	80	113	60	25	41

Figure 7: General Stereonet of West Sector

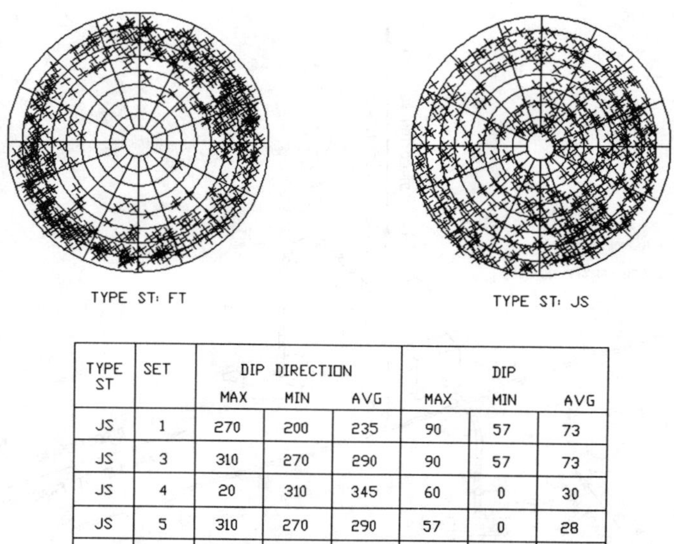

TYPE ST	SET	DIP DIRECTION			DIP		
		MAX	MIN	AVG	MAX	MIN	AVG
JS	1	270	200	235	90	57	73
JS	3	310	270	290	90	57	73
JS	4	20	310	345	60	0	30
JS	5	310	270	290	57	0	28
JS	6	20	310	345	90	60	75
JS	7	270	200	235	57	0	28

Figure 8: General Stereonet of the East Sector

The work angle is related to the actual angle when several pushbacks are being worked in the same wall, generating additional berms due to the width of each pushback. The work angle is measured like the overall angle, but with large berms, it is possible to use the concept of interramp angle between the berms. This is a good way to increase the work angle or to increase the width of the pushbacks thereby reducing costs or improving the mine operations.

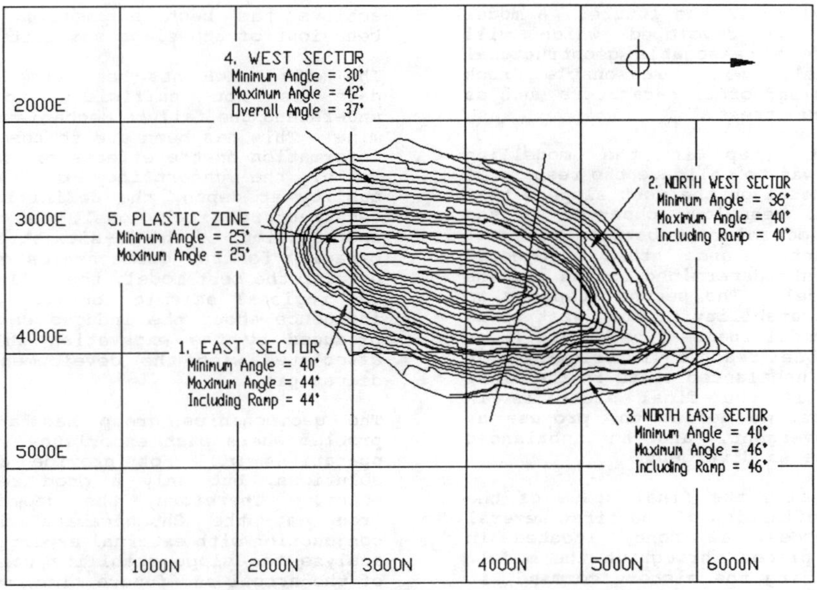

Figure 9: Final Pit Angles

The face bench angle is very important since this angle indicates the backbreak effects from the blasting operations. This angle may be used as a geomechanical parameter since a high angle represents in general a competent rock and a low angle represents a soft rock. This angle is important to the operational conditions, defined by the need to maintain many haul roads due to the narrow pushbacks.

In summary, the purpose of the geomechanic group with respect to mine design is to define the slope angles and backbreak for the different sectors. **Figure 9** illustrates the actual recommendations of the final pit angles for the mine. The pit angles are the first data used by the mine planning people to prepare the long range program for the next 20 years. After the long range program has been set, the pushback geometries are reviewed by the geomechanics group to determine if there is any incompatibility between them from a geomechanical point of view.

6.0 NUMERICAL MODEL RESULTS

The results obtained from the FLAC model represent the behaviour of the main walls in the mine. The west wall was analyzed using FLAC with different combinations of rock properties and water levels. The east wall was analyzed first with a finite element model and subsequently with the FLAC model, with reasonable coincidence between the two methodologies.

Before 1987, numerical modelling was common using finite element versions of bi- and tri-dimensional modelling of both the east and west walls, but there was an insufficient geostructural data base to rely upon. Therefore the actual failure mechanisms in the mine especially in the west wall was not clear, and numerical modelling of the slope stability ceased.

With the initiation of the drill hole program, geostructural work, the cell mapping and testing of recovered samples, it was possible to define rock qualities of the rock at surface and at depth. Therefore in 1990, new numerical models were generated mainly using the FLAC model. This new analysis, with the collaboration of Call & Nicholas Inc., has provided behaviour information in both the east and west wall with different alternatives.

Although both sides have been modelled, the west side is of most concern. If the west slope is not able to achieve a final pit angle of 37°, it will have great consequence to the future of the open pit. A decrease of 1° in the final pit angle of the mine would result in the loss of 100 million tons, confirming the importance of the geomechanic studies of the Chuquicamata pit.

With the actual models available, and with the present technology, the geomechanics group is only starting to understand the actual behaviour of the slopes in the mine and to a lesser extent understanding the projection of this behaviour in the future. A model must now be developed which will incorporate a reasonable geostructural rock model, with reasonable rock properties and other parameters such as the in-situ stress.

The first step in the modelling procedure was to validate the results of the Flac model on the west slope with the actual behaviour observed in the slope. Good relation between the real displacements and the projected displacements determined in the process of the model. The second step was to define the stability of the west slope from the model in its final state (i.e. does the model reach equilibrium or does it show an unbalanced state in its final state). If the final state is in equilibrium, we define the process as being convergent, and an unbalanced final state as divergent.

When analysing the final state of the model, a definition of the first several points (nodes) is done, located in different places throughout the model. When utilizing the history command, it is possible to follow the level of displacements of these points, for each of the steps processed. If the displacements decrease towards zero then the model reaches equilibrium, if the displacements increase for steps over 50,000 then the model is said to be in an unbalanced state.

Table 3 outlines the results of numerical modelling of the west side, for the years 1990, 1993 and 1997. The profile for the year 1990, was analyzed in dry and wet conditions, with different rock properties. This indicates that the model is highly sensitive to small differences in the rock properties and to the underground water table conditions. For the years 1993 and 1997, the profiles were processed for different slope angles and for dry and wet conditions, with similar results as before.

Table 4 shows the results from other geometries for the same west slope profile, with different slope angle and rock properties.

The actual results with the Flac model in Chuquicamata must be reviewed since the model dealt only with static conditions, and the creep behaviour of the plastic material with the blasting affects is a dynamic situation.

7.0 CONCLUSION

The Chuquicamata mine is of great importance to Chile since it is a large mine with a high level of reserves not yet mined. Therefore it is of great importance to evaluate the feasibility of the mine for the future. It is not only necessary to improve operations for an increase in productivity but also to achieve the best estimation of the behaviour of the slope stability.

The work which has been completed to date is not sufficient to fully understand the failure mechanisms in the mine. This has been due to the limited information on the effects of the water table, the uncertainty of the rock quality at depth, the definition of a geostructural rock model, the effects and behaviour of the in-situ stress, the capacity for the geomechanics group to develop the best model, the influence of the regional seismic activity and the influence about the induced seismicity produced by the excavation which are associated with the development of a deeper pit.

The geomechanics group has a unique problem where past experience at other operations will not provide adequate solutions, but only a good reference point. Therefore, the geomechanics group at the Chuquicamata mine in conjunction with external expertise must analyze the slope stability conditions of the actual and future mine, with the best tools available and to achieve a great understanding of rock mechanics.

The geomechanic group of Chuquicamata mine have begun a strong geomechanics program during the past year with several specialists to investigate the important aspects of the mine. The life of Chuquicamata will depend on others aside from the geomechanics group to identify the best mining alternatives to mine this orebody safely and productively on a long range basis.

Innovative Mine Design for the 21st Century, Bawden & Archibald (eds) © 1993 Balkema, Rotterdam, ISBN 90 5410 325 6

Geotechnical analyses for open pit mining in areas of large-scale slope instability

Richard D.Call, Thomas M.Ryan & Ross C.Barkley
Call & Nicholas, Inc., Tucson, Ariz., USA

ABSTRACT: Varying degrees of overall slope instability have developed in the copper porphyries at several mines in sectors with high pit slopes. This instability can occur at relatively flat slope angles (25 to 35 degrees); reducing the slope angle to improve overall slope stability would require a very large stripping program. Slope monitoring and operational experience have demonstrated that displacements of high pit slopes tend to exhibit a regressive character. An economic alternative to flattening the overall slope angle is to modify mine plans and operational procedures so that mining can continue despite these displacements. Mining an active slope requires reliable rock mechanics models to predict the response of the slope to mining. This paper discusses the geological characteristics, deformation behavior, and current stability analysis procedures for modeling of several large-scale open pit slope failures. Operational responses to slope displacement are examined in the context of case history data, and the concept of displacement-tolerant mine planning is presented.

1 INTRODUCTION

Since the 1980's ongoing development of open pit mines at several porphyry copper deposits has resulted in the construction of overall pit slopes in excess of 400 meters in height; long-range mine plans are currently being developed for slopes up to 900 meters in height. Many of these slopes, such as those at the Chuquicamata Mine, are now in excess of 600 meters, and are among the largest rock structures ever to be engineered by man. Because of the size of many of the operations, there is often an economic incentive to design working pit slopes near the optimum overall slope angle; however, experience has shown that a significant amount of slope instability can be expected at these economically optimum slope angles. Since most large open pit operations possess procedural flexibility, considerable slope displacement can be tolerated, provided that the slope "failures" can be adequately modeled, and predictions made as to their size and behavior.

The means of predicting the expected number and tonnage of multibench structurally-controlled failures (wedge, step-path, plane shear and step-wedge) are well-developed, and probabilistic models can be constructed for use in predicting the number, size, and displacement resulting from these types of rock slope failures (CANMET). However, for overall slope instability due to rock mass yield in weaker rock zones, models of rock mass strength, developed stress fields, and rock mass displacement are still in the developmental stages. Where weak zones are present in the pit slopes, overall slope instability and displacement can develop, which will impact mine operational procedures and mine planning. If reliable and predictive rock mechanics models of strength, stress, and displacement can be constructed for the overall pit slopes, then mine plans capable of tolerating overall slope instability can be formulated, which will improve operational safety, increase production efficiency, improve mine economics and extend the mine life.

The geotechnical analysis of large-scale overall slope instability involves these primary areas of study:

1. slope monitoring and kinematic displacement modeling,
2. geological, geotechnical and hydrological analysis, and
3. stress, displacement and stability modeling.

2 SLOPE DISPLACEMENT ANALYSIS

2.1 Regressive versus progressive slope failure

Most post-failure models are primarily based on empirical relationships derived from slope monitoring data (Cruden and Mazoumzadeh 1987, Voight, Orkan and Young 1989). In the past, examination of displacement data has led to the development of two general displacement models: a progressive failure model, for which slope displacement will continue to accelerate to a point of collapse (or greatly accelerated movement), and a regressive failure model, for which the slope will decelerate and stabilize. Zavodni and Broadbent (1982) defined regressive and progressive failure stages for several large-scale open pit slope failures, and related these stages to failure geometry. This work has been expanded by Savely and Call into a useful description of failure characteristics and expected slope behavior (Figure 1).

A slope failure which is in a progressive stage tends to become less stable with time, and has the potential for sudden, large movements. A review by Ryan and Call (1992) of slope failures which have proceeded to collapse indicates that a wide range of pre-collapse velocities exist for progressive slope failures. Savely (1982) and Kennedy (1970) discuss operational and monitoring procedures for minimizing the impact of progressive slope failure on mining.

Regressive slope failures can occur when rock mass yielding is the primary cause of displacement. In the U.S., case histories of regressive slope failures in open pit mining have been documented in the literature since the early 1950's (Bisbee, Butte). Many rock mass movements in high open pit slopes tend to decelerate with time provided that the ratio of driving to resisting forces decreases with displacement. Such large-scale slope instability is most often due to a complex yield condition in which shear and tension failures of both joints and intact rock lead to failure and plastic flow of the rock mass. Surface and subsurface monitoring data from several slope failures show that displacements are extremely variable in direction and in magnitude. Inevitably, this leads to the necessity of a sophisticated monitoring program for large slope areas.

For the remainder of this paper we will discuss the behavior and analysis of regressive, overall pit slope failures which tend to stabilize with displacement and time. Geotechnical analysis and experience with several regressive slope failures demonstrates that mining can continue in areas of large-scale displacement provided that the failure is regressive in character, the failure mechanism is well-defined, monitoring procedures are established and enforced, and an effective slope management program is in place.

2.2 Modeling regressive slope failures

Slope monitoring data consistently indicates that most regressive slope failures occur in response to mining activity. A typical response is shown in Figure 2. As mining proceeds along a level within a zone that is in active yield, confining stresses are reduced, excess strain energy is induced, and displacement of the rock face is produced. These slope displacements can be either elastic or plastic depending upon the state of stress and strength of the near field rock mass. Monitoring data often demonstrates that slope displacement initially develops in lower strength, highly fractured rocks near the toe of an active mine slope. Toe displacement directions are typically at very low angles (5 to 20 degrees from horizontal). Displacements higher in the slope do not develop instantaneously with displacement at the toe, and can have highly variable directions. Since the slope is not acting as a rigid block, displacements will tend to develop in a time-dependent manner, resulting in differing rates of movement from point to point within the slope (Figure 3). This time dependency can also be demonstrated in regard to surface and subsurface displacement.

3 GEOLOGICAL ANALYSIS

Several geologic characteristics common to the documented large-scale regressive slope failures include:

1. low rock mass strength in the toe,
2. clay alteration of fractures, faults and wall rock,
3. a ubiquitous joint set that dips into the pit,
4. high angle faults or continuous joints that form back and side releases for the slope movement, and
5. saturated toe and compartmentalized ground water conditions.

3.1 Low strength rock mass

Unfortunately, thick zones of low quality rock are associated with the processes of ore emplacement. Regional scale tectonics, contact metamorphism and the release of volatiles from the magmatic melt all contribute to the degradation of the rock within and surrounding porphyry deposits.

These zones of low rock mass strength generally exceed 100 meters in thickness and have an RMR value of less than 40 and an RQD of less than 30. Rock mass characteristics range from low RQD zones with clay-filled fractures and moderate wall rock alteration, to pervasively clay-altered zones in which discrete structure has been obliterated by mechanical and hydrothermal actions.

3.2 Clay alteration

Varying degrees of clay alteration are present in low rock mass strength zones, as well as in transition zones that border it. In the low rock mass strength zone, clay alteration varies from a pervasive fracture filling with weak to moderate wall rock alteration, to complete alteration of the rock to clay by mechanical and hydrothermal processes.

Clay alteration extends beyond the low rock mass strength zone, but is usually confined to larger structures. In some cases, a transition occurs between the weak toe rock and fresh rock. Transition zones are characterized by clay-filled fractures and faults; however, the wall rock alteration is diminished and the RQD improves.

Clay adversely affects the stability of slopes for two reasons. First, the shear strength of clay is less than the rock on rock shear strength of the fractures. Clay alteration of intact rock and clay infilling of fractures reduces rock mass strength. Second, clay impedes the natural or induced drainage of the slope, resulting in higher pore pressures within the rock mass.

3.3 Ubiquitous joint set

Ubiquitous joint sets which dip into the pit at 35 to 45 degrees are a common characteristic of these regressive failures. In general, such structures do not daylight in the overall slope, but provide a plane of weakness within the slope.

This ubiquitous plane of weakness reduces rock mass strength in that orientation, and principal stresses tend to realign parallel to the ubiquitous structures with mining (Amadei 1987). This stress realignment enhances slope movement along the ubiquitous joints, and may actually extend them.

3.4 High angle back and side release fault structures

High angle faults that strike parallel to subparallel to the trend of the pit wall form a back release for downward slope movement. These structures diminish the rigidity of the rock mass, form discrete kinematic blocks, and compartmentalize ground water. In addition, the maximum principal stress aligns parallel to these structures on the upslope blocks due to the large contrast in rigidity between the fault structures and the surrounding rock. Clay-filled fault structures or continuous joint systems that strike from perpendicular to oblique to the trend of the pit wall provide side releases for the slope movement. These structures increase the degree of freedom for movement, diminish the rigidity of the slope and compartmentalize ground water parallel to the slope.

3.5 Ground water conditions

Ground water causes a destabilizing effect in all of these large-scale slope failures. In a free-draining slope, the hydraulic gradients are generally low, as evidenced by a gentle drawdown cone and a dry slope. However, clay-filled faults and fractures greatly reduce the permeability of the rock mass comprising the slope. High pressure gradients can occur as evidenced by seepage faces observed at significant levels above the pit bottom. In these cases, the drawdown cone is very steep, resulting in ground water being close to the pit face. This pressure creates a significant buoyant force on potential failure surfaces, which reduces the ability of the slope to resist shear. Driving forces increase further if the water pressure acts along a vertical back release.

Slope movements are quite sensitive to changes in water pressures. The onset of slope failure dilates the rock mass, creating additional storage. This added storage results in a lowering of the water level and a temporary increase in stability. In very dry regions, the lack of recharge may stabilize the slope or cause diminished slope movements for several pushback cycles. However, in areas of high recharge, the increase in stability is short-lived and the slopes should continue to be monitored for increased pore pressure.

4 FAILURE MODELING AND STABILITY ANALYSIS

The geologic characteristics listed in Section 3 of this paper often combine to cause the large-scale regressive slope failures observed. A suggested sequence of events that explains the slope displacements includes:

1. initiation of movement by active mining in the weak toe rock,
2. propagation of movement upslope due to successive release of kinematic blocks, and
3. deceleration to a more stable state.

4.1 Initiation of movement

Mining through low strength rock in the toe reduces the confining stress and creates an overstressed condition. The excess stress is relieved by movement in the toe of the slope and by stress transfer into rock that can carry additional load. The maximum principal stress is commonly horizontal at the toe, resulting in strong horizontal displacement.

Due to the pervasive clay fillings and wall rock alteration in the weak toe rock, natural or induced drainage is limited. Pore water pressures in this low permeability saturated rock can be high due to stress redistribution. The high pore water pressure reduces the ability of this weak rock mass to resist shear. In some cases, this material becomes amenable to drainage if discrete failure surfaces are developed.

4.2 Propagation of movement upslope

Displacements propagate upslope as confinement is reduced in successive upslope kinematic blocks. Upslope movements occur parallel to the non-daylighted joint sets and along high angle release faults.

A commonly observed characteristic of these failures is a differential shear along the high angle faults that form the back releases of identifiable kinematic blocks. Along a particular fault structure, the upslope block drops relative to the downslope block (Figure 4). This differential movement results from a significant vertical stress gradient across the faults. Due to the low strength and modulus of the fault structure, the maximum principal stresses on the upslope block align with the fault. This, coupled with the relief of lateral confinement due to downslope displacements, can result in a stress condition analogous to active earth pressure. The additive effects of the deep shear often result in the observed differential displacement along the release faults.

Clay-filled fault structures provide side releases for the upslope kinematic blocks. The shapes of the kinematic blocks range from rectangular to trapezoidal, depending on the obliquity of the side release faults. Trapezoidal blocks formed by non-daylighted wedges can be responsible for high stress concentrations in the toe. These stress concentrations may contribute to plastic failure of fair to good quality rock in the toe of the slope. Two-dimensional stability analyses do not adequately model this three-dimensional stress concentration.

The high angle release faults that form the kinematic blocks also tend to compartmentalize ground water. Ground water lowers the resistance of the rock mass to shear, resulting in a deeper zone of shearing along ubiquitous joint sets. In addition, water pressure acting on the back planes increases the driving forces.

Lowering pore water pressure in the upslope kinematic blocks greatly reduces overall slope displacements by reducing the driving forces and increasing the shear resistance. Additionally, stabilizing the upslope blocks imparts less stress on the weak toe material. Installation of dewatering systems usually pays dividends because slightly steeper slopes can be achieved and delays in mining operations caused by a moving slope can be reduced.

4.3 Reduction of slope movement

Slope displacements decrease as a result of:

1. halting or lowering the production rate in the weak toe,
2. lowering of the pore water pressure due to dilation of the displacing rock mass, and/or
3. displacing to a more stable geometry.

4.4 Stability modeling

Simple rigid block models are not appropriate for analyzing rock slope stability when the primary mechanism of failure is plastic (toe) to pseudo-plastic (shear along ubiquitous joints) yielding of the rock mass. Limiting equilibrium methods have limited application because they cannot predict strain or the extent of slope deformation. For these reasons, the combination of discrete element and continuum numerical methods are finding widespread application for the overall stability analysis of these high open pit slopes. The sophistication and efficiency of these models has improved rapidly. In the last decade, several software packages have been on the market that are valuable tools for analyzing slope behavior.

When using these numerical methods, it is critical to understand the mechanisms producing the slope failure, the strength characteristics of the rock mass and the pit hydrology. Once the model is defined, it is refined to correspond to the historical slope displacements and measured in-situ stresses.

Parameters that affect the model response include in-situ stress, rock mass strength, elastic properties, structure and hydrology. Because displacements are often not elastic, superposition is not valid, and a careful simulation of historical mining is required for stability analysis. A history match between the model and the slope can often be obtained through several stress paths. Therefore, a large geomechanical database must be developed to define the parameters used in the model. This database should include the following:

1. rock types observed from surface and subsurface geologic mapping,
2. block sizes:
   a. cell mapping,
   b. RQD logging of core, and
   c. surface classification mapping of weak rock masses,
3. intact shear strength:
   a. uniaxial compression tests,
   b. Brazilian disk tension tests,
   c. triaxial compression tests, and
   d. point load tests,
4. fracture shear strength:
   a. small-scale direct shear tests of core, and
   b. large-scale direct shear tests of rock blocks,
5. orientation of geologic structure:
   a. rock fabric mapping,
   b. oriented core, and
   c. geologic mapping of major structures,
6. hydrological characteristics:
   a. geologic mapping and drilling to define character of water-bearing rock,
   b. water level measurement in exploration and geomechanical drill holes,
   c. reporting wet blast holes,
   d. reporting seeps observed on the slope,
   e. slug testing of exploration and geomechanical holes to provide a rough estimate of transmissivity, and
   f. pump tests with observation holes for definition of transmissivity and storage in more homogeneous aquifers, and
7. in-situ stress measurements.

Rock mass strength and elastic properties, and hydrologic parameters are defined from the data mentioned above. These data are used to zone the rock into discrete domains that possess similar engineering characteristics. Quantitative methods have been developed to classify rock masses according to rock type, block size distribution, intact rock shear strength, fracture shear strength, orientation of the geologic structures, pore water pressure, and in-situ stresses. The RMR (Bieniawski 1993) and Q-system (Barton et al. 1974) rock mass classifications have been used extensively to choose appropriate underground mining methods and to estimate support requirements for underground openings, settlements in rock mass foundations and the rock mass shear strength.

Of particular importance in constructing numerical models of pit slopes is the definition of the rock mass shear strength and the elastic modulus. Several empirical classification schemes have been developed to estimate the rock mass strength and elastic properties. Numerical slope modeling experience has shown that the rock mass strength and elastic properties derived from these classification schemes do not always provide model responses that match the measured slope displacements.

More rigorous methods of determining the rock mass strength characteristics are needed. The use of three-dimensional numerical methods to model the mechanical response of the rock mass to various loading conditions show some promise. To be effective, these models must accurately portray the intact rock properties, the orientation of stress fields, the orientation and spacings of discontinuities, and the strength and elastic properties of the discontinuities and/or discontinuity fillings.

Although numerical methods have proven valuable in defining the failure mechanism of these large-scale slope failures and predicting the response of the slope to future mine plans, they are not adequate for modeling the combined responses of mechanical loading, deformation, and water flow. The large size of these slope models places a constraint on the number of sensitivity analyses for various static water conditions which can be performed in a reasonable period of time. Future versions of the numerical methods must represent the mechanical-hydrologic interaction of large pit models efficiently. Topics that updated numerical methods should address include:

1. consolidation,
2. efficient establishment of hydraulic gradients in complex hydrological conditions, and
3. change in hydraulic gradients resulting from rock mass dilation during failure.

4.5 Displacement rates and overall slope angle

A range of overall slope angles can be excavated in regressive slope failures. Steeper slopes experience greater velocities and larger overall displacements. However, there is a maximum limit to the angle that can be excavated before a progressive accelerating slope failure will occur. Steepening the upslope geometry beyond this critical angle results in driving forces that are greater than the resisting forces in the upslope kinematic blocks. These upslope blocks can drive the toe to an accelerated failure condition. One of the fundamental roles of stability modeling is to determine the critical slope angle and identify whether this accelerated condition is beginning to develop. Figure 5, a plot of model velocity histories for three slope geometries, illustrates the use of modeling in defining the critical slope angle.

5 OPERATIONAL PROCEDURES TO MINIMIZE THE IMPACT OF DISPLACEMENT ON MINING

Because of the procedural flexibility in most large open pit mines, slope displacement does not necessarily constitute "failure" from a mine management standpoint. This relationship between theoretical and operational slope failure in mining has been discussed (Munn 1985). In particular, the real hazard to mine operations is often the potential for greatly accelerated movement occurring near equipment and personnel. If this can be mitigated, a significant amount of slope displacement can often be tolerated with routine mine operations such as dozing and additional shovel shifts for cleanup. Mining in areas of large-scale slope instability generally results in unsteady production. Provided that a mine slope failure is regressive in character, slope displacement can be controlled using specific operational procedures: dewatering, additional stripping, control of the excavation geometry, and control of the excavation rate. The effect of these controlling measures can be assessed and predicted with geotechnical analysis.

5.1 Step-out versus unload

At Twin Buttes during the past five years, it has been demonstrated that periodic, small (less than 20 feet) step-outs into the pit could effectively decrease slope displacement in an area of historical slope failure (Ness 1992). There is often a trade-off between the cost of cleanup or additional stripping and the value of ore lost due to a step-out. There also may be an optimum location for a step-out in a pushback, and this can be defined by the geotechnical engineer with stability analysis.

5.2 Controlling excavation rates

Control of the mining rate is another means of maintaining displacement velocity to minimize its impact on mine operations. This technique is difficult to implement, and requires significant flexibility from mine operations. Once displacement velocity of a slope region reaches a limit defined by the rock mechanics staff (through analysis and experience), mining is discontinued until an acceptable relaxation limit is achieved. Savely (1993) discusses how this approach is optimized by maximizing operational efficiency in relation to the moving slope area. At several properties, short mining periods and shorter delays have been found to be more cost-effective than long mining periods and longer delays. This can also be demonstrated mathematically with displacement modeling (Figure 6). Since the vast majority of accelerating and decelerating displacement curves for rock slopes approximate either an exponential or a power function, the relationship between the time spent in excavation and the time required for relaxation is not linear. The higher displacement rates associated with longer mining periods require a substantially greater proportion of time for deceleration. Stability analysis and slope monitoring data can be used to assess optimum extraction rates.

5.3 Pushback width

Optimum pushback widths can be defined from a geotechnical perspective as well as a mining perspective. Practical mining experience suggests that narrow pushbacks within failed slopes are difficult to maintain. This can also be demonstrated numerically with detailed stress and energy analyses of rock slopes. The role of the geotechnical engineer should be to determine whether there is an advantage in changing the pushback geometry due to the state of existing slope instability, or the location of a major geologic structure.

Stability analysis of overall slope failures using continuum models demonstrates that stress concentration occurs at the toe of slopes as they are excavated; this has been validated by in-situ stress measurements at several mines. In the case of the excavation occurring within a plastic zone, a change in the stress and energy state occurs, which results in slope displacement. The stress path for excavation is a combination of lateral extension and axial compression. Strain energy for underground excavations has been analyzed in detail by others (Salamon, Farmer), and similar methods can be applied to pit slopes which are at yield. In general, smaller excavations result in less energy changes and displacement than larger excavations. However, because of the stress concentrations that develop in the toe of a slope, strain energy potential within the slope is not uniform. This leads to a nonlinear relationship between pushback width and strain energy (or maximum shear stress) for varying sizes of incremental mining cuts. When narrow pushbacks are mined, excavation takes place within the zone of stress concentration and the resulting displacements are typically a high percentage of the overall pushback width. This renders narrow pushbacks difficult to maintain, and results in either excessive time spent in additional slide cleanup, or frequent unplanned step-outs into the pit. If step-outs must be taken too often, it may not be possible to complete the pushback, and ore production may be lost. Additionally, if strain softening of the rock mass occurs with displacement, there is considerable geotechnical incentive to "mine out" the existing failed rock as much as possible in order to take advantage of the higher strength associated with a less disturbed rock mass.

For maximum production efficiency, a pushback must be wide enough for double spotting trucks and three lanes of traffic. For a normal rectangular cut with the digging face at right angles to the pit wall, the inside corner is relatively inefficient because a wider swing or single spotting is required. The outer edge is similarly inefficient, in addition to having a less than complete digging face. Thus a narrow pushback, with the majority of the digging time in the inside corner and the outer edge, will produce fewer tons per shovel shift (Figure 7).

Where there is overall slope displacement, the effective pushback width will be reduced by bench sluffing. For example, if a 15 meter high bench dug at 63 degrees sluffs to a 36 degree angle-of-repose, 6.5 meters of the pushback width will be lost. If sluffing of the working bench and the bench below occurs in the same section, the effective width of the pushback would be reduced by 13 meters. Thus, in this case, pushbacks should be designed at least 13 meters wider for a displacing slope than for a stable slope.

In particular, the following sequence is to be avoided:

1. A production schedule is planned with a minimum width pushback and full shovel and truck efficiency.
2. Slope displacement reduces the effective width of the pushback and the tons per shovel shift decreases.
3. At the lower production rate, ore will not be uncovered in time so the pushback width is decreased in an attempt to uncover ore.
4. The reduction in pushback width results in one-way traffic, more single side loading and delays from tension crack offsets. This further reduces the tons per shovel shift.
5. Steps 3 and 4 are repeated until the pushback is not minable.

Another factor that must be considered in production scheduling is the horizontal displacement. If the slope experiences horizontal displacement on some level between one pushback and the next, additional material will have to be mined to achieve the design toe.

5.4 Slope dewatering

Dewatering has been demonstrated in several cases as an effective control for displacing slopes (Argall and Brawner 1979). Cost-benefit analysis indicates that dewatering programs are one of the most cost-effective mechanisms available for improving slope stability for both stable and unstable ground. In many low permeability rock masses, reduction of pore pressure prior to failure is difficult with conventional methods, but is readily achievable after failure. Dewatering below the pit bottom is often required to achieve an acceptable pore pressure condition for the overall slope, and an analysis of aquitards and compartmentalization of ground water must be completed by the geotechnical engineer to ensure that the proper level of effort is focussed on dewatering.

6 CONCLUSIONS

Improvements in continuum and discrete element models have enhanced the capability of engineers to design high overall pit slopes which often experience significant slope instability. These models can be used to predict the development of rock mass movements which previously could only be evaluated empirically. As the sophistication of the stability models improves, more data is required to extend the model; particularly high quality data is necessary for its calibration. Better estimates of the state of stress and pore pressure are needed in addition to improvements in rock mass strength prediction techniques. In particular, stability models should continually be developed which are capable of modeling dynamic processes.

Provided that enough data can be gathered to acquire a firm understanding of the failure mechanism, stability models can be developed which will enable engineers to provide operations personnel with reliable predictions of the impact of mine excavation on the stability of the overall pit slopes. If pit slope analysis can be developed in a spirit of cooperation between the geology, engineering and planning staffs, geotechnical analysis can be run in concert with mine planning to create an optimum slope design.

Experience with several large open pit slope failures indicates that mining can be successful in areas of large-scale slope stability if unsteady production can be accepted in the unstable pit sectors. If such fluctuating production can be tolerated, specific operational procedures can be used to minimize the impact of slope displacement on mine operations. Geotechnical analysis can be used to identify the procedures required to allow mine operations to work within or near areas of large-scale slope instability. Such analyses can greatly improve the possibility of successful mining in these areas.

REFERENCES

Amadei, B., Savage, W.Z. & Swolfs, H.S. 1987. Gravitational Stresses in Anisotropic Rock Masses: Int. J. Rock Mech. Min. Sci. & Geomech. Abstr. 24, 5-14.

Argall, G.O. & Brawner, C.O. 1979. Mine Drainage: Proc. 1st International Mine Drainage Symposium, Denver, Colorado.

Bieniawski, Z.T. 1973. Engineering Classification of Jointed Rock Masses: Trans. S. Afr. Inst. Civ. Eng. 15, 335-344.

Barton, N. Lien, R. & Lunde, J. 1974. Engineering Classification of Rock Masses for the Design of Tunnel Support: Rock Mech. 6, 183-236.

Call, R.D. & Savely, J.P. 1991. Open Pit Rock Mechanics: SME Mine Engineering Handbook, AIME, New York, 860-882.

Cruden, D.M. & Mazoumzadeh, S. 1987. Accelerating Creep of the Slopes of a Coal Mine: Rock Mechanics and Rock Engineering, No. 20, 123-135.

Farmer, I.W. 1986. Energy Based Rock Characterization: Application of Rock Characterization Techniques in Mine Design, M. Karmis, AIME, New York, 17-23.

Kennedy, B.A. & Niermeyer, K.E. 1970. Slope Monitoring Systems Used in the Prediction of a Major Slope Failure at the Chuquicamata Mine, Chile: Planning Open Pit Mines, Proc. Symp. on Planning Open Pit Mines with Special Reference to Slope Stability, A.A. Balkema, Amsterdam, 215-225.

Munn, F.J. 1985. Coping with Wall Failures: Canadian Mining and Metallurgical Bulletin, No. 78-884, 59-62.

Ness, M.E. 1993. Personal Communication.

Ryan, T.M. & Call, R.D. 1992. Applications of Rock Mass Monitoring for Stability Assessment of Pit Slope Failure: 33rd U.S. Symposium on Rock Mechanics, Santa Fe, New Mexico.

Salamon, M.D.G. 1984. Energy Considerations in Rock Mechanics: Fundamental Results: J. S. Afr. Inst. Min. Metall., vol. 84, no. 8, 233-246.

Savely, J.P. 1993. Slope Management Strategies for Successful Mining: Proc. International Congress on Mine Design, Ontario, Canada.

Savely, J.P. & Kastner, V.L. 1982. Slope Instability at Inspiration's Mines: Proc. Third Int'l. Conf. on Stability in Surface Mining, Ch. 26, C.O. Brawner, ed., AIME, New York, 609-634.

Zavodni, Z.M. & Broadbent, C.D., 1978. Slope Failure Kinematics: Proc. 19th U.S. Symposium on Rock Mechanics, May 1-3, 86-94.

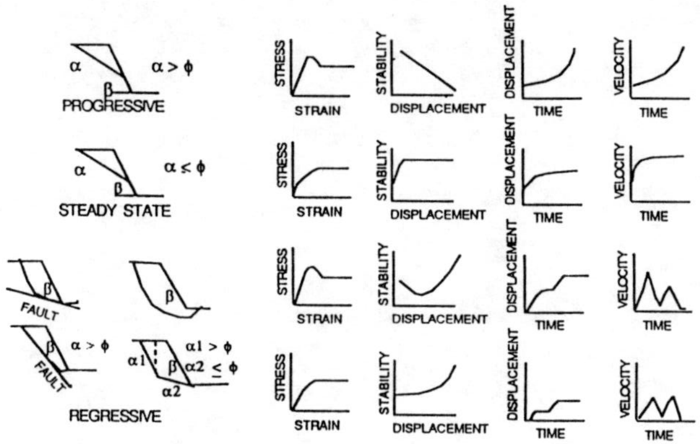

Figure 1. Failure conditions in relation to the stability model.

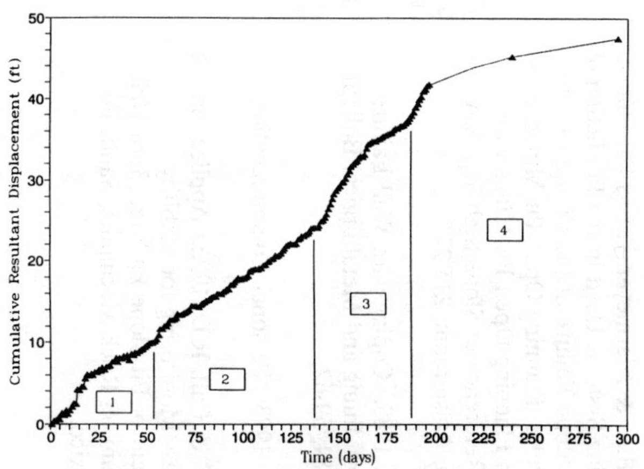

Figure 2. Cumulative resultant displacement of a surface station showing four regressive displacement cycles.

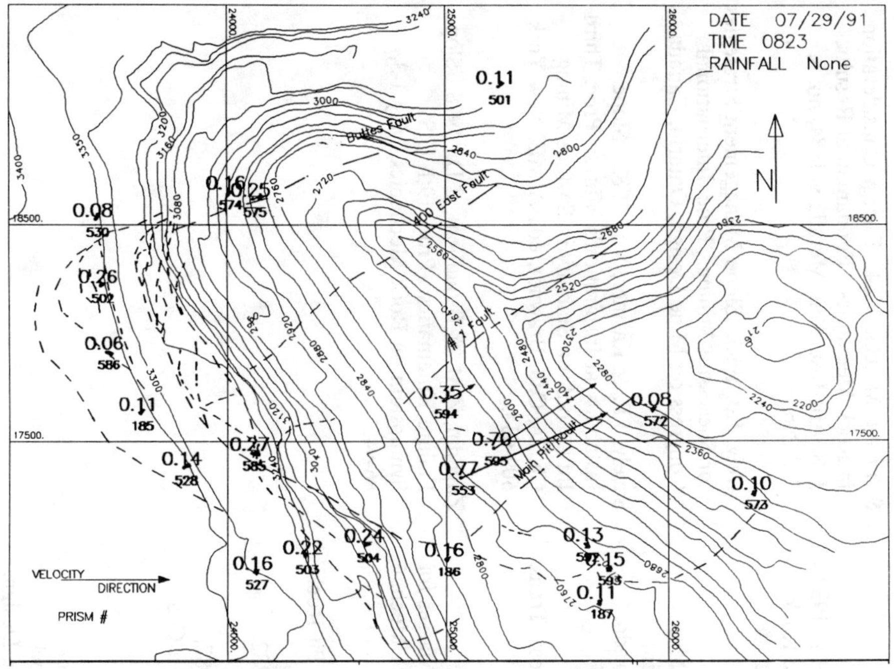

Figure 3. Variable slope velocity of a large-scale slope failure.
Note high velocity directly above the active mine bench at the 2680 level.

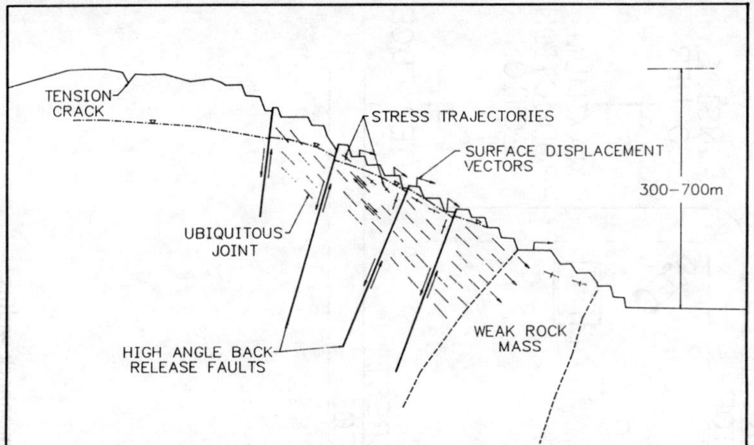

Figure 4. Failure mechanism and geologic conditions for large-scale regressive slope displacement.

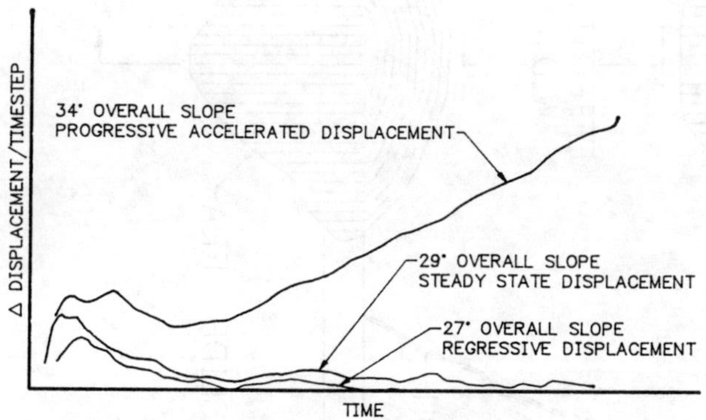

Figure 5. Velocity histories for three slope geometries.

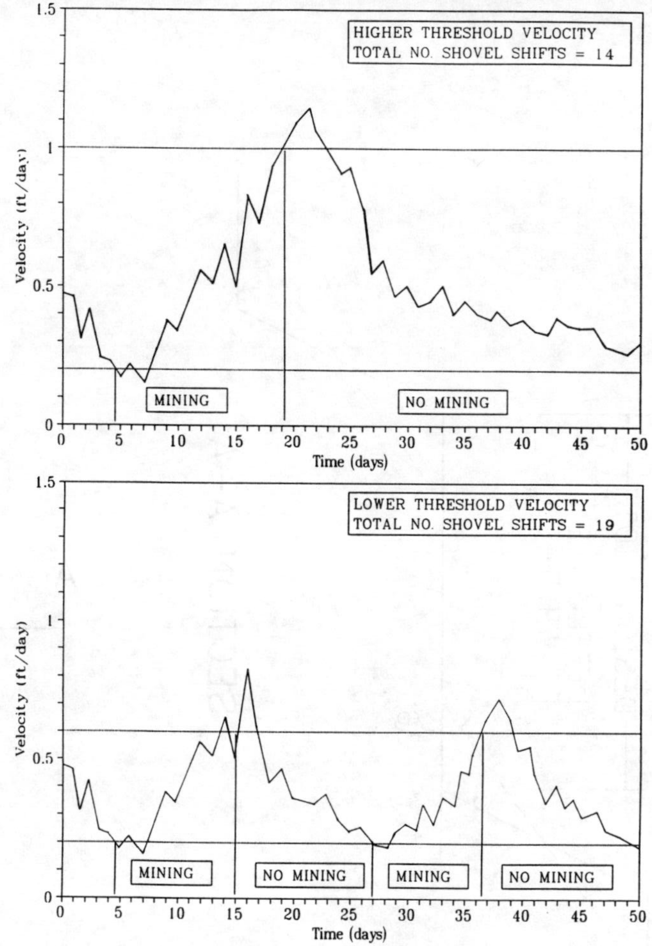

Figure 6. Controlling slope displacement velocity by selective mining in an unstable area.

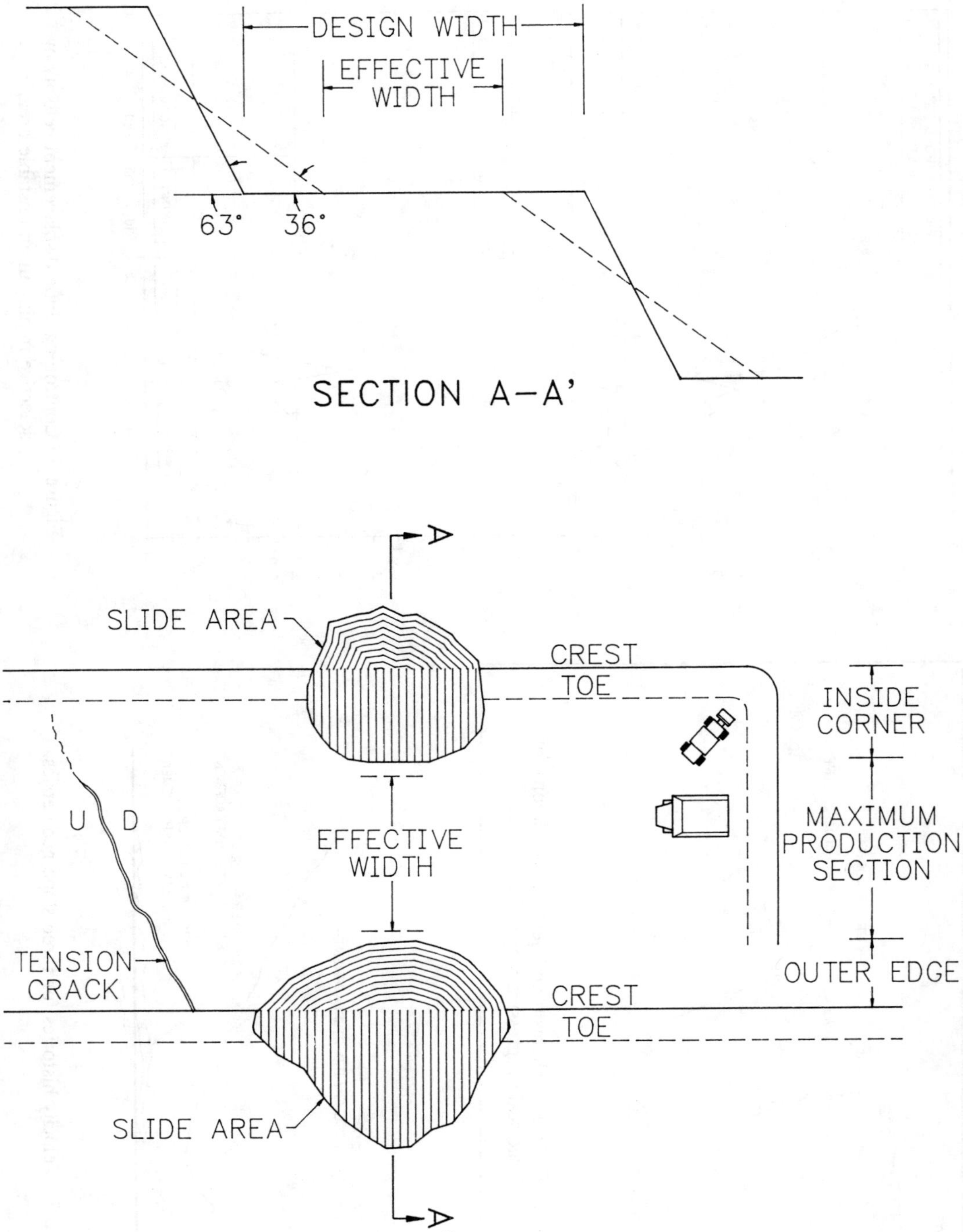

Figure 7. Operational considerations for pushback widths.

Innovative Mine Design for the 21st Century, Bawden & Archibald (eds) © 1993 Balkema, Rotterdam, ISBN 90 5410 325 6

Mining below sea level: Design and performance of the south slope expansion at Island Copper

James I. Mathis & A. MacG. Robertson
Steffen, Robertson and Kirsten Inc., Vancouver, B.C., Canada

R. B. Robertson
BHP Minerals Canada Ltd, Port Hardy, B.C., Canada

ABSTRACT: The pushback of the southern wall of the Island Copper open pit extends the pit wall into Rupert Inlet with the natural ground surface being about 100 ft below sea level. Previous mining established a waste dump in the inlet which reaches an elevation of about 20 ft above sea level. The pit will eventually reach a depth of 1300 ft below sea level. To prevent an inrush of seawater through the waste dump and underlying sediments, a seepage control barrier wall was constructed through these permeable materials and keyed into the underlying tills and sediments. Design issues included seepage, piping, and pore pressures in the slopes, deformations of the slopes and waste dumps due to dewatering, deformation along a major fault which passes through both the barrier wall and the slope, elastic deformation of the barrier and pit wall due to unloading, and the effect of geologic structure and weak rock zones on pit slope stability. Analyses were conducted to predict expected conditions (i.e., groundwater, seepage, deformation). Where possible, testing was done to evaluate specific material properties required for design. Where properties could not be accurately determined, engineering assumptions were made. State-of-the-art, at that time, numerical analyses were conducted to predict tectonic deformation of the fault due to excavation. The compiled data was then used to assist in the development of engineering design and performance criteria for the barrier wall. This paper briefly reviews the design and construction phase of the project and provides a performance review of the wall approximately midway through its design life.

1 INTRODUCTION

Faced with dwindling ore reserves, Island Copper Mine, located on Vancouver Island (Figure 1), British Columbia, Canada, and operated by BHP Minerals Canada Ltd, is currently expanding its existing open pit with a pushback of the present south wall. The pushback will result in the planned recovery of approximately 83 million tons of mill feed, extending the mine life to 1996.

The open pit is located immediately north of Rupert Inlet (Figure 2). The south wall of the open pit was 60 to 100 ft north of the original shoreline. Mine production, which commenced in the 1970's, has resulted in the placement of a mine waste dump extending about 3000 ft south into the inlet. As a result of pushback, the south wall of the pit has advanced about 200 ft south of the shoreline into Rupert Inlet, encroaching into the waste dump and underlying soils. At its new location, the pit wall consists of sections of up to 50 ft of mine waste and 40 to 50 ft of beach and fluvial horizons. With the crest of the pit 8 ft above mean sea level (1008 ft by local datum), and the planned depth of the pit to reach 1300 ft below sea level, the pit is located almost entirely below sea level.

A critical element of the south slope pushback is a barrier wall located in the waste dump and overburden soils to prevent excessive seepage from the sea through to the advancing slope. The location of this wall relative to the Island Copper open pit and Rupert Inlet is shown schematically in plan in Figure 2 and in section in Figure 3. This barrier wall ranges in depth from 6 to 105 ft, or approximately 10% of the depth of the pit. However, as it prevents flooding at the pit rim, it is essential that the barrier wall remains effective until the pit has reached

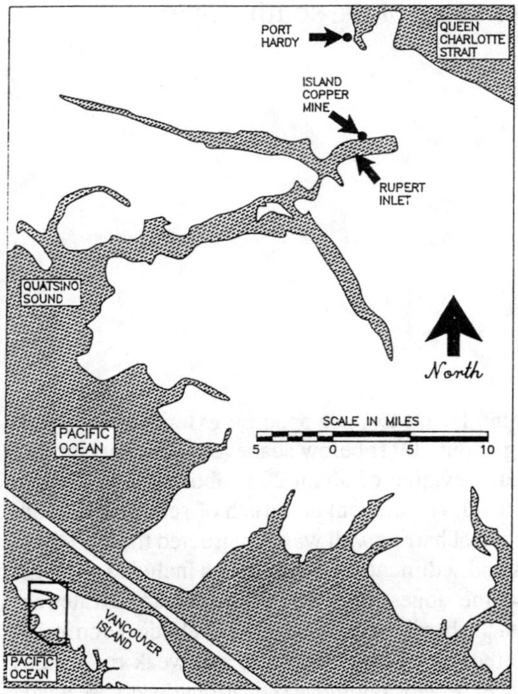

Figure 1. Location of Island Copper mine

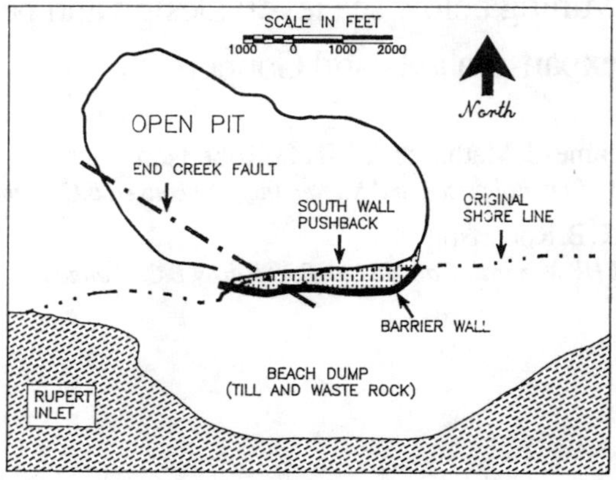

Figure 2. Schematic plan map, south wall

its design depth. This is a requirement not only from a safety perspective, but an economic one as well, as the recovery of the ore located at the base of the pit drives the economics of the pit expansion.

Concerns regarding the pit slope and barrier wall can be divided into three categories:

- seepage through, and under, the wall, giving rise to high slope pore pressures, as well as piping through the barrier wall and soil slopes
- deformation of the pit wall as a result of dewatering and mining excavation with the potential of cracking the barrier wall
- stability of the soil and rock slopes

Each of these will be discussed in subsequent sections in terms of predicted performance and performance to date. For greater detail on design and construction considerations see (Findlay, et. al. 1990; Robertson and Robertson 1991).

2 HISTORY

The barrier wall was installed between April 25 and September 30, 1989. The pushback of the south wall began in December 1989. Overburden excavation was accomplished in two cuts. The pushback initial cut lowered the elevation of the overburden soils about 48 ft. The second cut lowered the elevation an additional 40 ft to bedrock. The end of October 1990 marked the completion of the soils stripping and commencement of rock excavation for the south wall expansion. As of April, 1993, excavation had reached the 320 elevation, or about 600 ft below sea level.

3 INSTRUMENTATION

The components of the dewatering and instrumentation program relevant to barrier wall performance, as well as conditions in the south wall, are as follows:

- dewatering wells equipped with flowmeters
- multiple level piezometer strings located on either side of the barrier wall
- inclinometers/extensometers located within the barrier wall
- levelling pins on, and adjacent to, the barrier wall
- survey prisms on the barrier wall, soil, and rock slopes

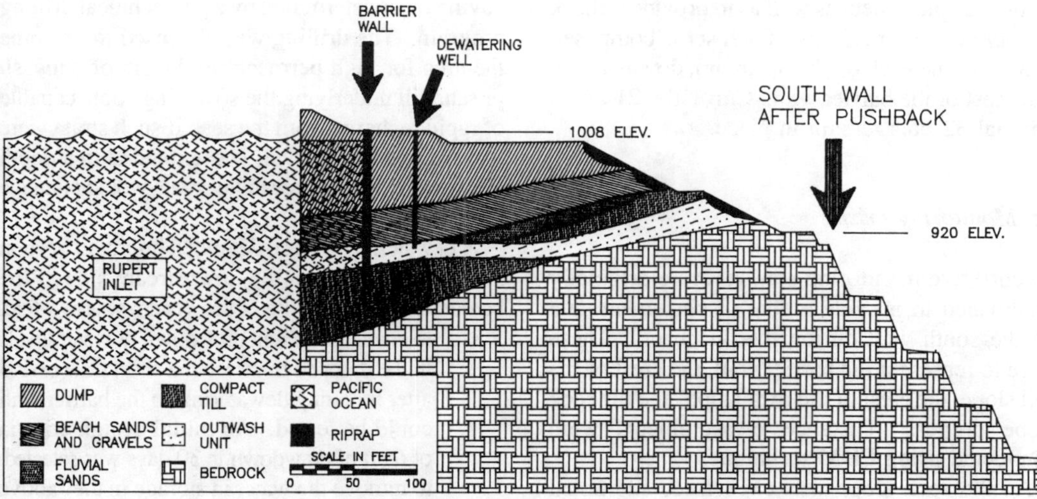

Figure 3. Schematic section showing pit and barrier wall

3.1 *Dewatering facilities*

To enable the satisfactory excavation of the soil slope following the construction of the barrier wall, a dewatering system was installed to drain the formations on the downstream (north) side of the barrier. This was achieved by the installation of dewatering wells, and the digging of sumps, from the surface of the 1008 bench.

Seven dewatering wells were drilled in the initial phase of the project, followed by two additional wells in September 1990. In January 1993, two additional wells were drilled to provide reserve capacity if pumps in adjacent wells stop functioning.

The vertical wells were to be replaced by a series of horizontal drains. Due to economic and geotechnical considerations, this has not taken place. All dewatering remains from the vertical well system.

3.2 *Piezometers*

Fifteen multiple level piezometer strings, each consisting of a standpipe and up to five vibrating wire piezometers, were installed. These piezometers supplement ten standpipe piezometers and one multiple level piezometer installed in 1989. A standpipe piezometer was also inserted inside each of the dewatering wells. In addition, four standpipe piezometers were installed adjacent to potentially sub-standard panels in the barrier wall.

As a result of the pushback, six piezometer nests have been destroyed. The remaining piezometers are operative.

3.3 *Inclinometers/extensometers*

Twelve inclinometer/extensometer installations were completed in the barrier wall. Each was installed in a hole which exceeded the depth of the wall by approximately 5 ft. While most of the installations terminated in till, four of the these, in the vicinity of the End Creek fault, terminated in bedrock.

The objective of the installations was to monitor vertical and horizontal displacement within the barrier wall for comparison with predicted values. Approximate locations of the installations were determined before construction, with final location adjusted based on monitoring considerations of potentially suspect panels within the wall.

3.4 *Levelling pins*

Twelve sets of levelling pins, coincident with each inclinometer installation, were installed to monitor vertical displacement as well as to provide a check on inclinometer readings. Each set is composed of 3 pins, one each on the upstream, downstream, and crest of the barrier wall. Currently, 21 of the original 32 pins remain in place.

3.5 *Monitoring prisms*

Twenty five monitoring stations were originally established to provide global displacement data on the south wall during the pushback. These were to represent the barrier wall, the crest of the soil slope, the crest of the rock slope, and the rock slope proper. Distance measurements were conducted using EDM techniques.

As the rock slope was excavated, additional prisms were installed both on the rock slope in areas exhibiting movement, as well as on, and adjacent to, the barrier wall.

4 SEEPAGE, PIPING, AND PORE PRESSURES

The three items discussed within this section, seepage, piping, and seepage related pore pressure increases, are all intimately related to the amount of water passing through, and under, the barrier wall. However, as their effects on stability, and the wall itself, vary, they will be handled as separate topics.

4.1 *Seepage*

It was accepted during design that seepage could occur both through and under the barrier wall. Seepage through the wall was considered an engineering controlled parameter. Specifications were included in the construction contract which dictated the maximum amount of allowable seepage through any 100 ft section of wall. A "V" notch connector was used at the junction of panels to reduce joint seepage. In addition, continuous inspection by a specialist team during construction, followed by review of construction details, insured that design specified quality was attained.

Under wall seepage was to be controlled by socketing the barrier wall at least 8 ft into till of which 3 ft was to be clay. The depth to, and engineering parameters of, the socketing strata having been determined by a geotechnical drilling program. The drilling was also used to examine the area for high permeability layers of outwash or sand till underlying the socketing strata capable of rapid recharge from the sea. If such strata were located, the barrier wall was extended to a depth such that no high permeability strata would continue under the wall.

A dewatering system, described briefly in Section 3.1, was installed on the downstream side of the barrier wall. The purpose of this system was two-fold.

- to enable sufficient drawdown of water levels in the dumped material in a short time period after initiating dewatering so the barrier wall could be tested. An initial design criteria of 65 ft of drawdown in 30 days was selected.
- to remove the water in storage in the "above till" profile in the shortest time possible. Excavation could not commence until satisfactory slope stability conditions had been achieved. This timing had a direct impact on mine economics.

With these criteria, together with a hydraulic conductivity of $5x10^{-2}$ cm/sec and a drainable porosity factor of 0.25, the required pumping factor was determined to be approximately 4,000 US gallons per minute (gpm) from sumps dug 20 ft into the static water table, operated for up to 10 days and 2,400 US gpm from dewatering wells installed primarily in the dump rock.

Combined dewatering rates from the initial dewatering during October and November 1989 totalled approximately 1,700 US gpm. Thereafter, rates dropped significantly, reaching approximately 200 US gpm in August 1990. The initial sharp reduction in rates reflects the decrease in transmissivity of the profile as water levels were lowered. This led to greater drawdowns in pumping wells and necessitated the reduction of discharge rates.

The rock dump unit was successfully dewatered by the wells, as was a thicker zone of beach materials located near the center of the wall. The beach unit elsewhere, and the underlying fluvial units, have not been dewatered adjacent to the barrier wall.

The 200 US gpm rate achieved in August of 1990 was satisfactory, as the expected flow due to seepage through and under the wall was expected to reach a design maximum of 800 US gpm. However, during the winter of 1990/91 flows began to increase, which was partially a function of precipitation infiltration. As of December 1991, the total well

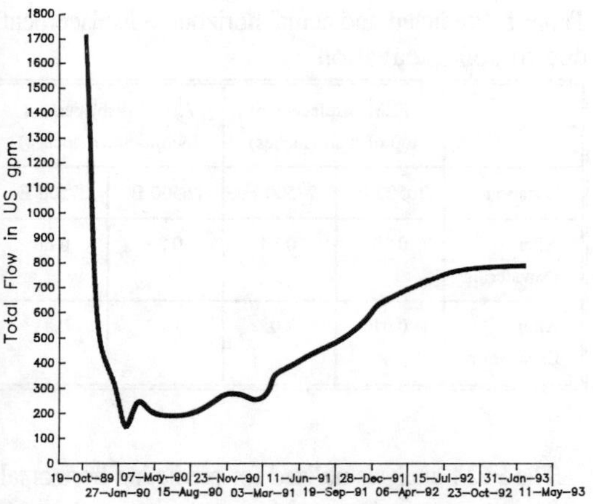

Figure 4. Flow from dewatering wells

pumping rate was approximately 600 US gpm (Figure 4). By late 1992, the pumping rate was approaching the capacity of the operating pumps.

Due to this increase, two additional wells were drilled in January of 1993. These were to act as reserve wells if either flows increased or a pump was incapacitated. The new wells bottomed in the till overlying the outwash strata. Only small flow levels were noticed in these wells.

One well, that which maintains the highest flow rate, penetrates the outwash unit. If this well is down for maintenance, flows increase substantially in the adjacent well.

The conclusion that may be drawn, although unsubstantiated at this point, is that the increase in flows is attributable to three possible factors:

- combined effect of water released from storage and an increase in gradient as the area is dewatered.
- increase in flow through the outwash unit
- a flaw, or perforation, in the barrier wall

As of this date, the pump rate is near the upper design rate of 800 US gpm. As the piezometric head remains below the toe of the soil/rock interface, the stability of the soil slopes does not appear to be compromised. Continuing monitoring and investigations will help in delineating the source of increased flows. At present, they are not critical and seem to be levelling off.

4.2 *Piping*

Concern existed that increased gradients caused by the pushback, if found within undetected zones of high permeability beneath the barrier wall, could cause piping. This, in turn, could lead to slope instabilities and water control problems. As minor piping had occurred in some of the outwash layers in the till during previous excavation, this was a significant concern.

This issue, for design purposes, was addressed with a two-fold process. First, as for seepage control, dewatering wells were installed downstream of the barrier wall. Second, if any high flow layers were discovered after excavation had been initiated, these zones would be grouted. Piezometers were installed both upstream and downstream of the barrier wall to monitor pressures in potentially piping strata. In addition, riprap was installed over filter fabric, covering the soil slopes as they were excavated. This was intended to be a failsafe measure to prevent piping if seepage occurred.

Other than minor piping that occurred after the barrier wall was installed, during the initial dewatering/excavation stage, no piping has occurred. As long as the piezometric surface on the downstream side of the barrier wall remains below the rock head, piping is not expected to be a problem.

5 WALL DEFORMATION

Pit wall deformation was considered to have two components for barrier wall design. The first was deformation due to dewatering the overburden soils. The second was related to response of the rock mass to the excavation of the material found within the south wall expansion. Barrier wall design criteria were established based on the expected magnitude of these events.

5.1 *Dewatering deformation (soils)*

Deformation was to occur on the downstream side of the barrier wall as a result of the increase in effective stresses due to dewatering. This increase in stress, estimated at approximately 65 ft of water, would result in deformation of the barrier wall. The magnitude of the induced shear, compressive, and bending stresses are a function of the magnitude of the deformation and the elastic properties of the wall. Laboratory tests were conducted to obtain the elastic properties of the barrier wall. However, obtaining the properties of the more variable waste rock and till material was more difficult. These values were obtained from a compilation of a large scale field test (pad loading) with in-situ pressure meter testing results.

Consolidation of the mine dump and underlying sediments, would contribute to this deformation as would slope movements associated with lateral deformation induced by soil and rock excavation. The performance of the barrier wall in sustaining these deformations, without cracking, was to be monitored.

A two dimensional finite element analysis (FEM) was conducted by Woodward Clyde Consultants indicated that about 2 inches of lateral movement would occur as a result of soil excavation and dewatering. Approximately one third of the movement would be due to dewatering, the remaining two thirds being a function of the slope's response to excavation. According to the model, the lateral displacement would begin at depth and reach a maximum at the top of the wall. Analyses of the maximum compressive and tensile stresses in the barrier wall indicated that these would not be exceed design parameters.

A comparison of predicted and measured displacement is given as Table 1. The variations from the predicted values can be attributed to the composition of the dump materials and sediments at the observation locations. At 26500 East (local coordinates), beach and fluvial deposits are thicker than dump materials, showing an increase in measured lateral displacement. Conversely, at 27500 East, where the reverse is true, the measured displacements are less than predicted. Measured displacements for most of the inclinometers are well within tolerance limits.

Some inclinometers in the vicinity of the End Creek fault have experienced displacements in excess of 1 inch. The source of this movement is likely related to stress release within bedrock as the rock slope excavation is conducted. The position and orientation of the fault strongly influences the orientation and magnitude of these movements.

Table 1. Predicted and actual horizontal displacement due to soil excavation

	FEM displacement top of wall (inches)		Actual displacement Inclinometer (inches)	
Location	26500 E.	27500 E.	26500 E.	27500 E.
After Dewatering	0.20	0.84	0.39	0.60
After Excavation	0.61	2.05	1.62	0.81

The FEM analysis utilized for predicting horizontal deformation was also used for predicting settlement. Approximately 0.5 inches of settlement was predicted. Of this, approximately 90% was to be attributable to dewatering and the remainder to excavation reaction. Results of selected locations are shown in Table 2.

Table 2. Predicted and actual vertical displacements due to soil excavation

	FEM settlement top of wall (inches)		Actual settlement Extensometer (inches)	
Location	26500 E.	27500 E.	26500 E.	27500 E.
After Dewatering	-0.18	-0.45	0.20	0.35
After Excavation	-0.22	-0.46	0.24	0.47

Measured settlements, although scattered, are less than 1 inch and within tolerances for the design. The display of extension, in Table 2, instead of settlement may be attributable to the bentonite grout used in installing the extensometers. Precision levelling, using levelling pins, confirms that minor amounts of settlement are occurring. However, the settlement magnitude is within survey error, making it difficult to precisely define.

5.2 *Rock slope deformation*

Adjustment of the rock mass due to mining activities results in deformation of the pit slope as well as the immediate vicinity. This deformation is a function of the elastic properties of the rock mass, the shape

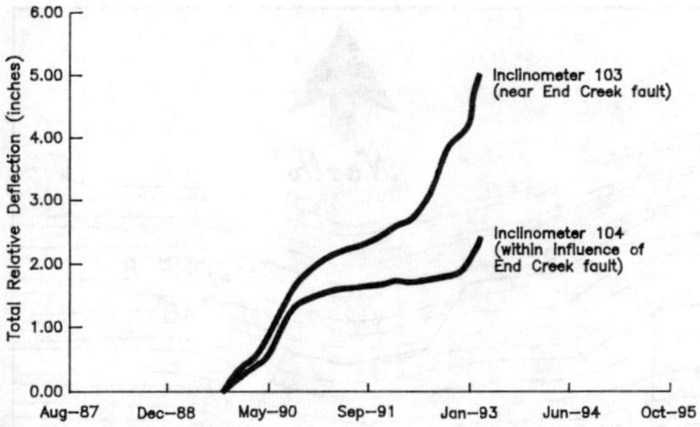

Figure 5. Relative wall deflection as indicated by inclinometers

of the excavation, and the in-situ stresses. Due to the highly fractured nature of the rock in the vicinity of the Island Copper mine, it was deemed unlikely that conventional stress measurement techniques would be successful in accurately assessing the regional stress field. Therefore, the assumption was made that stress magnitudes and orientations were similar to those from the Westmin mine, located on Vancouver Island, to the south of Island Copper mine. Two cases were analyzed, one without the assumed tectonic stresses and one with the horizontal stresses equal to twice the vertical stress. A short description of the envisioned problem follows.

The rock on the west side (footwall) of the End Creek fault is relatively confined, and can only move towards the pit as an elastic response to excavation. Material properties controlling movement of this confined block are those of the stiffer, unfaulted, rock mass.

The east side (hanging wall) of the End Creek fault is far less confined. It can be visualized as a thinning wedge on the pit side of the fault. Movement in this block is controlled somewhat by the elastic response of the rock mass, but more so by the material properties of the End Creek fault. This constraint contrast, together with a weak discontinuity bound (the fault), results in a stress redistribution causing an "unlocking" of the fault. Two factors will tend to mask this motion. The first is that the fault is not a discrete boundary but is, instead, a zone approximately 100 ft wide. The second is the rock/soil stiffness contrast will tend to obscure some of the relative movement.

Three dimensional effects of the pit on the local stress field were analyzed using a three dimensional boundary element program. Due to the size of the pit, the mesh was, by necessity, somewhat coarse. However, as its purpose was to approximate the general stress field for more detailed analysis, this was sufficient.

These results were then combined with a two dimensional finite element program to yield estimates of slippage along the End Creek fault as well as stress effects within the pit wall. Results indicated that slippage, with displacement in the range of 2 - 4 inches near the pit crest, was to be expected along the fault. Displacement midway down the side of the pit was calculated to be on the order of 6 inches.

Relative displacement, as measured by inclinometers in the barrier wall, is in the range of 5 inches as of March 1993, as can be seen from Figure 5. Global movement is unknown at this point due to a repositioning of a survey base station. Indications are that total movement is in the range of 7 - 9 inches. This is the subject of continuing investigation.

Additional evidence of movement is visual. A cement guide wall was placed as an aid in constructing the barrier wall. This guide wall has separated from the barrier wall, on the upstream side of the barrier wall, for a distance of about 100 ft. The width of this crack is on the order of 3 inches. While not being critical, and the subject of monitoring, this is evidence of the deformation taking place within the pit slope along the End Creek fault.

Increases in induced stresses were also indicated by the numerical analyses of the pit wall. However, these will only be verified at greater depths than those the south wall has reached at this point.

6 PIT WALL STABILITY

The pit wall can be divided into two portions: soil and rock. These are discussed separately.

6.1 *Soil slope stability*

Stability of the soil slopes was determined using traditional method-of-slices, equilibrium techniques, assuming pore pressures due to retained water or barrier wall seepage. Static factors of safety of 1.5 were used for design, with a pseudostatic factor

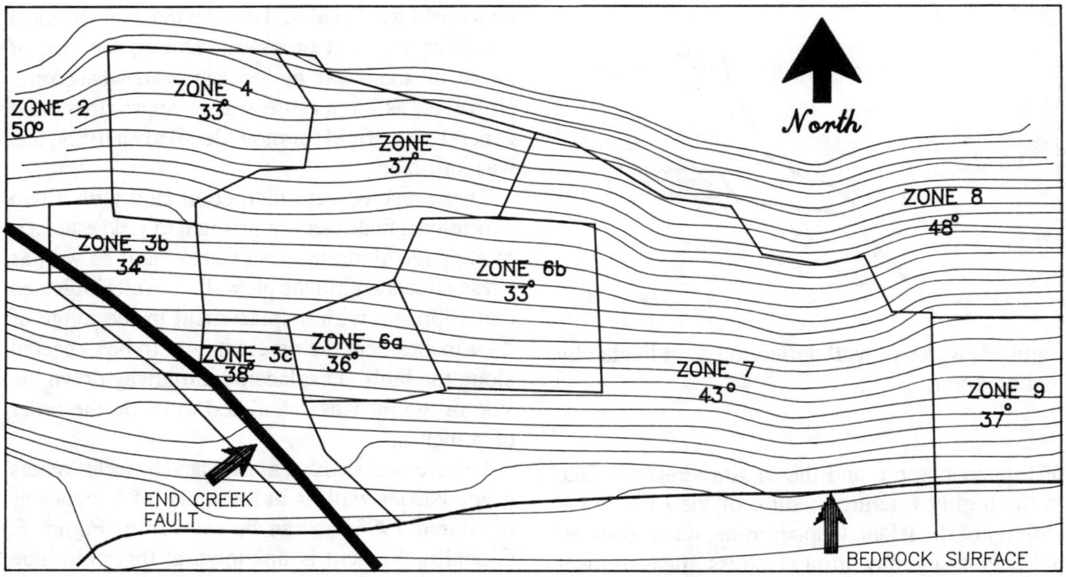

Figure 6. Rock slope design zones and angles

of safety of 1.3 for a design earthquake of 0.21g acceleration.

Soil slopes have behaved satisfactorily to date with no failures occurring.

6.2 *Rock slope stability*

The stability of the pit slopes cut in rock was evaluated in light of stability of discontinuity controlled blocks, weak rock zones, and a combination of both.

Discontinuity characteristics were obtained from bench face mapping, as well as oriented core drilling programs. The data collected was then analyzed to identify zones of structural similarity. Within these zones, analyses were conducted to determine the stability of structurally defined blocks which daylighted. A safety factor of 1.2 was required for potential, structurally defined, failures of substantial size.

Back analyses of numerous failures at Island Copper mine has led to the estimation of weak rock mass strengths associated with a Slope Rock Mass Rating (SRMR). This rock mass rating system has been developed at the mine (Olsen 1987; Robertson, Olsen and Pierce 1987; Robertson and Robertson 1988). SRMR was utilized in this study to determine which areas of the slope could potentially be affected by weak rock failures. Once these were outlined, appropriate slope angles were estimated using limit equilibrium analyses with the SRMR derived strengths. A safety factor of 1.15 was used as the design criteria. This was determined by the incremental nature of the south wall pushback in relation to the large amount of data from the existing south wall.

The design angles, as published in Robertson and Robertson (1991), have been substantially followed to this point. However, one area below the present haul road (zones 6a and 6b in Figure 6) did exhibit failure. This appears to caused by a relatively continuous fault containing weak gouge allowing oblique plane shear failure in the wall. This is combined with probable intact rock failure as a side release. Slopes have been flattened to approximately 30 degrees in this area from the original 33 to 36 degrees. Mining is continuing in conjunction with slope displacement monitoring.

7 SUMMARY

The total volume of water pumped has increased from a low of 200 US gpm to approximately 800 US gpm. Some of this increase is a release of water from storage as bench water levels have fallen. However, this volume is small compared with the total increase. As such, it can be concluded that a real increase in seepage has occurred. Whether this seepage is through the wall or through underlying outwash units is being studied. The pumped volume remains below design volumes and appears to be levelling off.

Water levels within the dump materials remain stable.

The wall has deformed toward the pit in a manner similar to that predicted. The relative proportions of lateral displacements attributable to dewatering is typically greater than expected and conversely, that which is attributable to excavation, less than expected. True dump and soil composition, compared to modelling assumptions, has had an impact on measured displacements when compared to calculated.

Settlements are small, only slightly larger than the accuracy of the monitoring systems, and within tolerances.

Lateral deformation in the vicinity of the End Creek fault is increasing due to rock excavation in the pushback. Excavation of material constraining the hanging wall appears to be allowing "unlocking" and readjustment of the rock in the pit slope. This deformation appears to exceed present calculations but remains within design tolerances. However, as more material will be removed from the hanging wall area, re-analysis of total deformation within the fault zone is being conducted.

8 ACKNOWLEDGEMENTS

The assistance of Cam Scott (SRK) as well as Gordon Clarke and Kevin O'Kane (BHP Minerals Canada) is gratefully acknowledged. Their comments and critique regarding this paper were invaluable.

9 REFERENCES

Findlay, R.B., Robertson, R.B., O'Kane, K.P. 1990. The Island Copper south wall pushback project. International symposium on mine planning and equipment selection in surface mining. University of Calgary, November 7-9

Olsen, R.S. 1987. South wall expansion base case pit slope design. Internal report, Island Copper Mine. Port Hardy, British Columbia.

Robertson, A. MacG. & Robertson, R.B. 1990. Geotechnical considerations in the design of the Island Copper Mine south wall pushback. International Symposium on Mine Planning and Equipment Selection in Surface Mining, University of Calgary, November 7-9.

Robertson, A. MacG. 1988. Estimating weak rock strength. AIME Annual General Meeting, Tucson, January.

Robertson, A. MacG. 1988. Slope control in intensely weathered and fractured rock masses. American Mining Congress, Session Papers, Volume I, Chicago, April 24-28.

Robertson, A. MacG., Olsen, R. and Pierce, G. 1987. Assessment of weak altered rock masses at the Island Copper Mine. Proceedings from SME Annual Meeting, Denver, February.

Innovative Mine Design for the 21st Century, Bawden & Archibald (eds) © 1993 Balkema, Rotterdam, ISBN 90 5410 325 6

High mine rock piles in mountain terrain: Current design trends and experience

David Campbell & Alistair Kent
Golder Associates, Vancouver, B.C., Canada

ABSTRACT: This paper describes and discusses current design approaches for mine rock piles up to 500m high located in steep mountainous terrain at sites in North and South America and Southeast Asia. Innovative consequence-based risk management methodology is being developed in response to environmental criticism of rock pile failures. Potentially, the survival of the coal mining industry in B.C. depends upon a rational acceptance of the possibility of failures, and their consequences.

1 INTRODUCTION

1.1 Brief History

Over the past 15 years or so the coal mining industry in British Columbia (B.C.) has undergone major expansion, resulting in the construction of some of the largest and highest mine waste disposal projects in the world. Waste rock dumps up to 400m high have been advanced onto slopes steeper than 20 degrees. Figure 1 shows the locations of the major coal mines in B.C.

Elsewhere, for example in South America and Southeast Asia, mine rock dumps over 500m high have been constructed or are being planned. Apart from very high dumps, disposal of very weak or highly weathered mine overburden presents challenges with respect to physical stability and environmental control. This paper discusses practical experience, and attendant geotechnical issues, affecting mine overburden waste disposal and management.

1.2 Previous and Ongoing Studies

An increase in the frequency of waste dump failures occurred at the coal mines in B.C.. as production rates increased. This elevated rate of failure also generally corresponded to increases in dump height, and rate of disposal(defined as volume per day per unit crest length). The mining industry, regulatory agencies, consultants, and research groups have responded at various levels to the actual and perceived problems caused by past failures and ongoing dumping practice. Most importantly, there has been an atmosphere of cooperation and exchange of experience between all parties with the objective of learning from past events and evaluating their real effects. A compilation and review of available data for dump performance and failures is presented by Golder (1987). Subsequently, in 1991 and 1992 draft design, operation, and monitoring guidelines were prepared and discussed in open workshops around the Province of British Columbia (Piteau (1991), Klohn (1991), and HBT (1992)). Kent (1992) discussed a number of technical issues which remain unresolved. However, mine waste rock disposal has evolved to the point where safety can be assured and the risks associated with mine waste management can be assessed qualitatively. Quantitative risk analysis of all major dumps is not yet attainable due

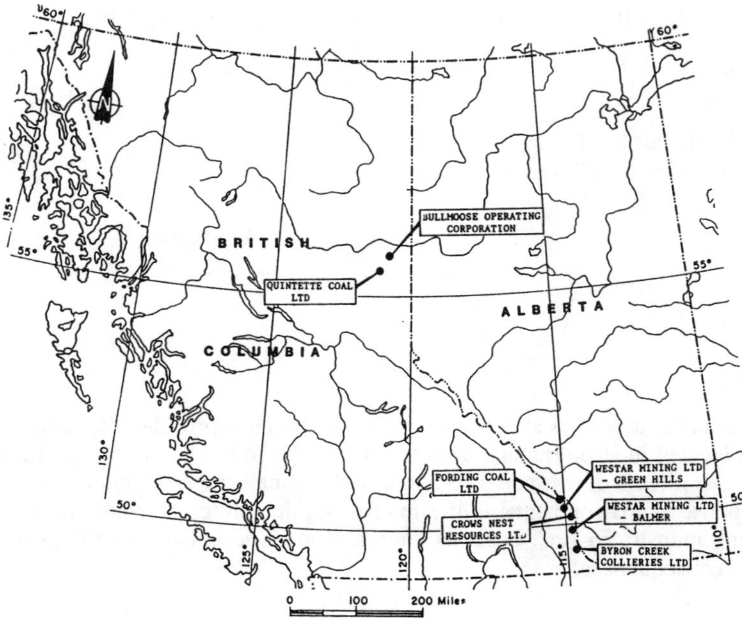

Figure 1 Locations of Coal Mines with waste dumps on steep terrain in British Columbia.

to some unresolved technical issues. Dawson et al (1993) present analyses of postulated failure modes that are as yet unproven.

The following sections describe in more detail the conditions at various dumping sites, their observed performance, and selected failure events.

2 CONDITIONS AND CONSTRAINTS

2.1 Topography

The major valleys in the eastern coal mining areas of B.C. are typically U-shaped as the result of glaciation. Topographic profiles commonly become steeper with increasing elevation.

Golder (1987) indicated that a factor common to many dump failures was a steep foundation slope, often exceeding 25 degrees. Failures that occurred where foundation slopes were flatter could usually be attributed to other adverse conditions such as pore water pressures in the foundation. A dump supported on a slope of more than 25 degrees could normally be expected to have a factor of safety of less than 1.2, corresponding to drained conditions. Relatively small values of pore water pressure acting along, or at shallow depth below the dump-foundation contact could be responsible for de-stabilising such dumps. It should be noted that for reasons of economy mining must often proceed from the top downwards, necessitating high level waste rock disposal during the early stages of mining.

When waste rock is consigned to a steep-sided V-shaped gully the three-dimensional wedging action enhances the stability and can increase the factor of safety by up to about 15 per cent. Thus, a dump situated on 20 to 25 degree slopes may have adequate stability if the layout takes advantage of the wedging effect of gullies.

Clearly, dump stability is enhanced if the dump layout and direction of advance is optimised with respect to the available topography within the available dumping areas. A useful planning task is the development of isopachs of topographic slope. During dumping operations the supervising engineers should have a clear idea of the slope of the current area that is providing support to the toe region of the dump. This appreciation may influence dump management in terms of loading rate or of direction of advance.

2.2 Foundation Conditions

Typically, the foundation areas for the dumps are mantled by colluvium consisting of a broadly graded mix of angular rock fragments and silty sand and gravel and by glacial till. The colluvial soils are usually in a loose to compact state. The degree of saturation of the colluvium may vary widely depending on time of year. Saturation is likely after the dump is in place. The glacial tills typically are a silty sand or sandy silt with a trace of clay. The upper 1 to 2m of the till often is in a compact to dense state, and appears to have been softened by processes such as frost action and by root penetration. At depth the till is hard and very competent. Figure 2 shows the results of shear strength tests on these soils.

2.3 Mode of Disposal

The majority of the coal in B.C. is mined by truck and shovel. Large trucks are used to haul the waste rock to the dump. Commonly the waste rock is dumped over the crest. Loads may be dumped short and dozed onto the face if the rates of deformation in the region of the dump crest are higher than a predetermined limitrender the outer portion of the platform untrafficable.

It is well known that segregation of particle sizes occurs as the waste rock moves down the face of the dumps. The largest and most durable fraction rolls to the dump toe. This zone of coarse segregated rock becomes covered as the dump face advances and results in beneficial basal drainage. Internally, the dump is multi-layered with sequences of alternately coarser and finer material aligned parallel to the dump face. Evidence of this stratigraphy was presented by Campbell (1986). Figure 3 shows the results of a scale model of segregation.

If practicable, dumps are constructed as a series of wrap-arounds which ultimately form a relatively flat overall face amenable to effective and economical reclamation.

2.4 Mine Waste Materials

The strength, durability, and size of waste material varies considerably from site to site, and on individual site geology and mining practices. Broad generalisations as to the expected quality of material can be developed, and the design engineer should attempt to characterise the likely range of material strengths, durabilities and sizes anticipated, and their trends to vary as mining proceeds. Planning of critical phases of dump development can minimise situations where anticipated poorer quality waste might exacerbate other adverse conditions such as steep topography, or might disrupt the formation of a coarse rock drain in the base of a drainage. General pit geotechnical assessments will yield information on rock type, strength, and durability. Observation of actual behaviour of each principal rock type when exposed to the elements should also aid the engineer's judgement. More difficult to assess, particularly for a new project in a new area, is the expected range of particle sizes. However, careful review of the proposed blasting design and comparison with other mines where similar bedrock strata are being excavated will often yield adequate information.

Table 1 lists typical ranges of material properties for the sedimentary rocks in the eastern coal fields of B.C.

As indicated on Figure 3, it is to be expected that fines from one layer would not be able to pass through lower layers. In the field we have not observed any evidence of significant internal erosion within mine dumps.

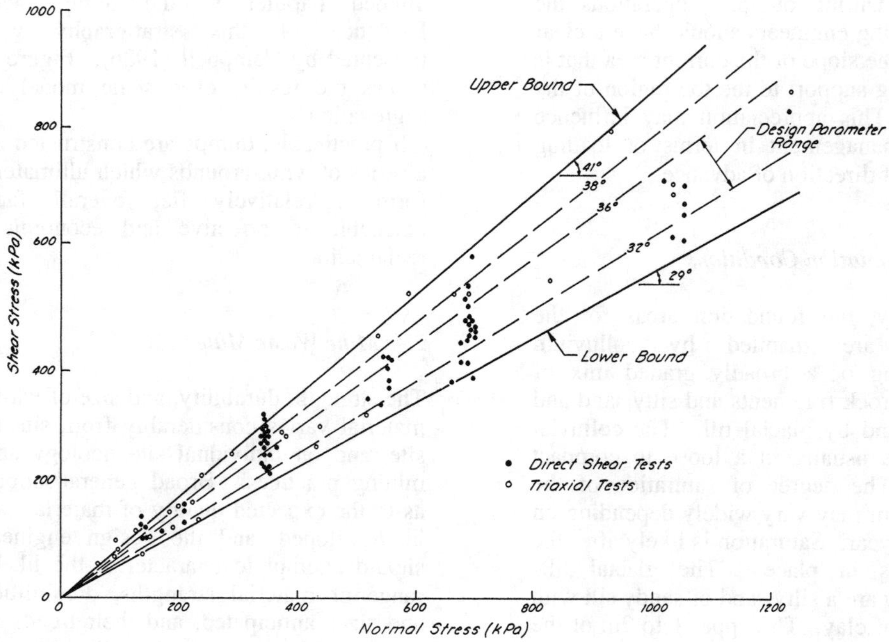

Figure 2 Summary of shear strength test results for foundation soils in B.C.

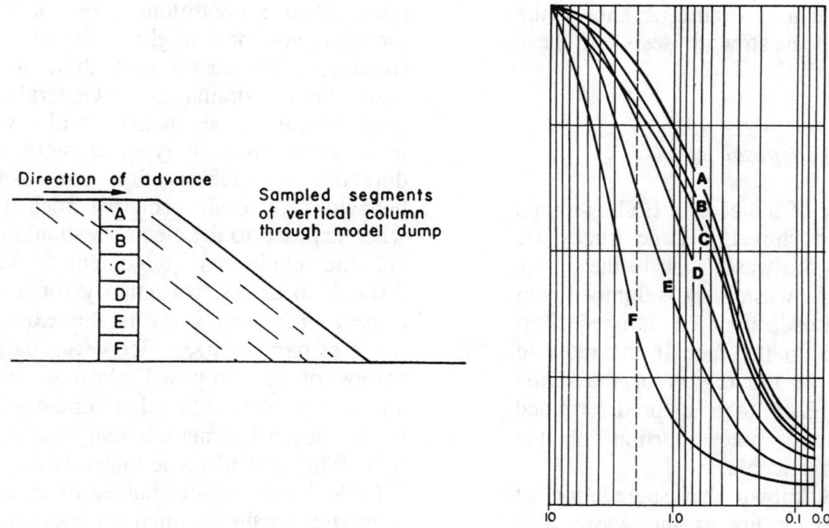

Figure 3 Scale model of waste rock segregation.

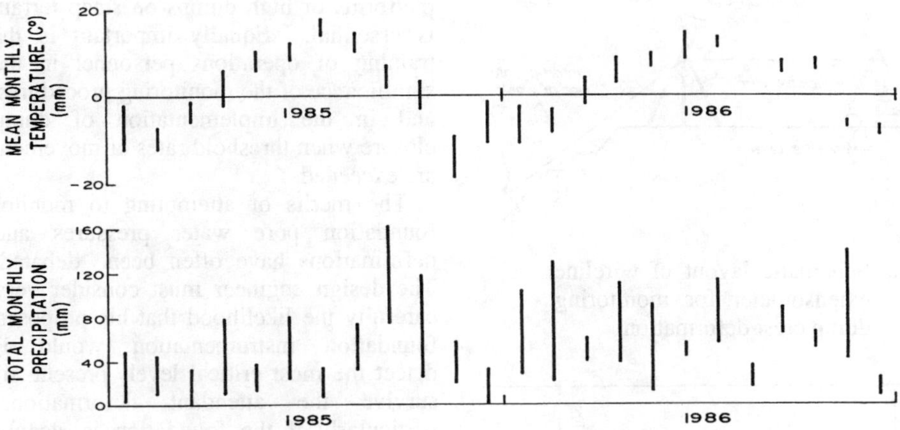

Figure 4 Typical meteorological data for mountain coal mines in B.C.

TABLE 1 ROCK TYPE AND STRENGTH

Mine Property Code	Estimated Proportion Sand Stone	Silt Stone	Mudstone/ Shale	Weighted Average of Unconfined Compressive Strength (MPa)
A	Relatively weak materials			50 to 80
B	Relatively strong			100 to 150
C	40	40	20	74 to 129
D	Relatively weak			50 to 80
E	55	25	20	82 to 137
F	50	30	20	79 to 134
G	20	40	40	58 to 110
			Range	50 to 150

Note: Ranges and averages are approximate, based on available data. Wide variations are to be expected at individual mine sites.

2.5 *Effects of Climate*

The coal mines in eastern B.C. operate under broadly similar climatic conditions, as shown on Figure 4.

The presence of a waste rock dump within a watershed tends to reduce peak flows, and to increase base flows.

In our opinion, heavy rainfall may exacerbate pre-existing conditions which are near to failure. Such conditions may include steep foundation slopes or excess pore pressures generated by strain within foundation soils. The available meteorological records indicate that the incidence of failures increased during spring thaw when recharge and discharge could be expected to peak.

Snowfall normally is not a major problem for active mine dumps. It is conceivable that shallow burial of thick layers of snow could lead to instability but we are aware of few such instances.

In one case an abandoned dump on steep terrain failed, apparently as the result of thawing of old snow and ice. Simple but effective crest monitoring provided ample warning for the protection of men and equipment.

2.6 *Operational Monitoring*

The use of simple wireline extensometers,

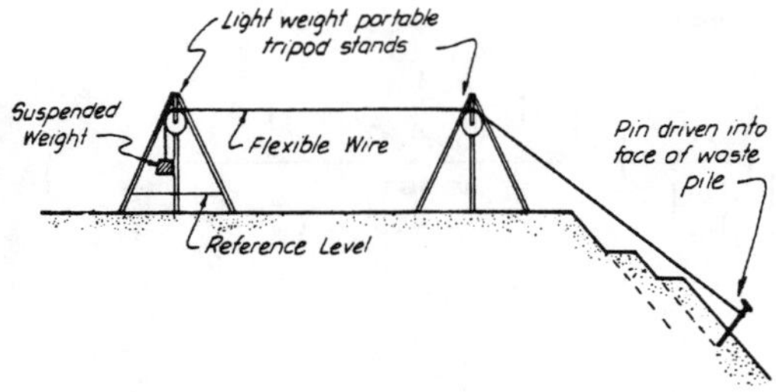

Figure 5a Schematic layout of wireline extensometer for monitoring dump crest deformation.

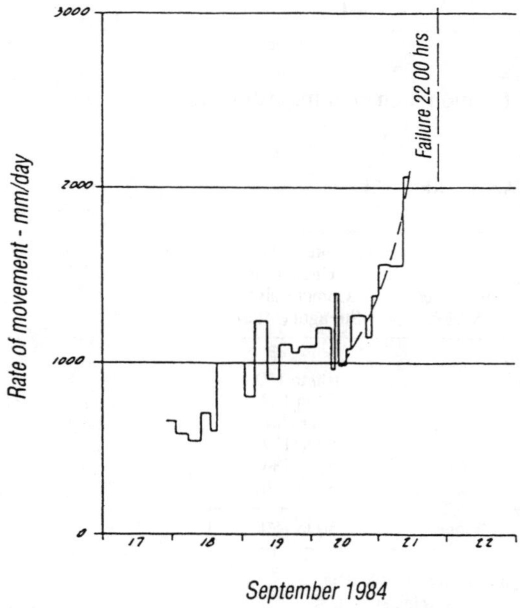

Figure 5b Typical graph of monitoring data preceding a failure event.

as illustrated in Figure 5, have been proven over some 20 years to provide ample warning of impending failures. Intermittent evaluations of other methods of monitoring by engineers and mine operators have not yet found any other methods as practical or effective. A promising modification is the automation of the wireline extensometers wich are linked to radio transmission systems. However, experience shows that continued visual inspection of the platforms of high dumps on steep terrain is essential. Equally important is the training of operations personnel in the significance of the monitoring procedures, and in the implementation of dump closure when threshold rates of movement are exceeded.

The merits of attempting to monitor foundation pore water pressures and deformations have often been debated. The design engineer must consider very carefully the likelihood that his proposed foundation instrumentation would i) detect the most critical levels present, ii) survive the attendant deformations, particularly if the foundation is steeply inclined and iii) permit reliable control of dumping so as to preclude failure. In our opinion, the current state of practice, both analytically and operationally, does not satisfy some or all of these requirements adequately. Field programs which include such monitoring should be regarded as experimental, and not as proven techniques for controlling stability and preventing failures. In critical situations where failure would have unacceptable consequences the design must pro-actively minimise or eliminate those factors that are likely to result in instability. In such cases a high level of effort in the installation and the extent of foundation instrumentation may be justified.

3 OBSERVED PERFORMANCE

3.1 Typical Behaviour and Failure Characteristics

Detailed geometric measurements of the deformations or of the post failure geometries of mine dumps are not commonly available. Eye-witness accounts of dump failures are few. Thus precise definition of failure surfaces is not available. However, field observations over the past 25 years of dump surfaces preceding failure events, and during normal operations have provided meaningful indications of the internal geomechanics.

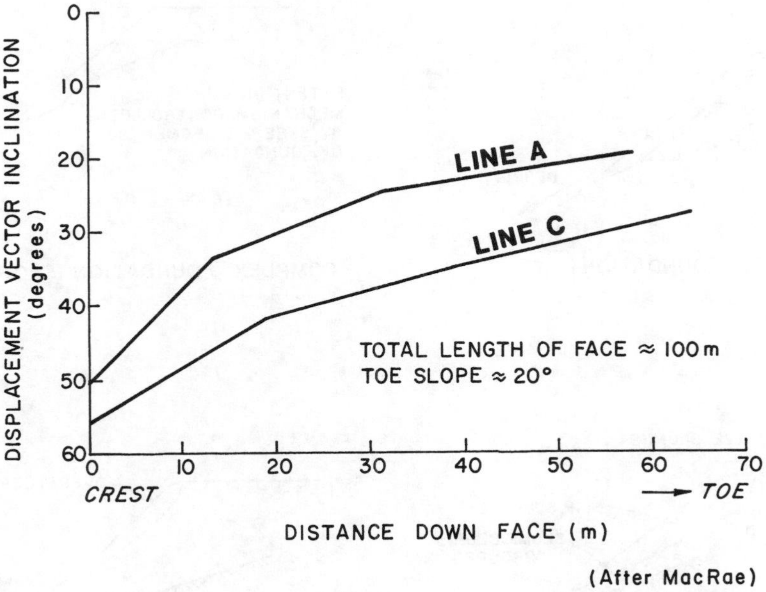

Figure 6 Results of surveys of displacement on a dump face.

Waste rock tends to be in a relatively loose state when end-dumped. Deformation occurs internally and on the surface due to both compressive and shear strain. Surveys of dump surface movement have measured vector directions inclined at between 50 and 60 degrees below horizontal in the vicinity of the dump crests. The magnitude of displacement decreases with increasing distance behind the dump crest, diminishing to negligible values within a setback of about 30 per cent of the dump face height.

Patterns of horizontal striations or ridges are commonly observed over the upper part of dump faces. These patterns are interpreted to be the result of the strain within the upper portion of the dump. associated with Active Rankine stress conditions. These conditions result from both internal compression and shear deformation within the lower portion of the dump.

Figure 6 shows the results of survey measurements on the face of a dump about 100m high, supported on a foundation slope inclined at about 20 degrees. The vectors near the crest were inclined at 50 to 60 degrees, agreeing with model testing, stability analyses, and other observations. The vectors show a trend of flattening with decreasing elevation, probably corresponding to deformation along the dump-foundation contact zone.

Figure 7 shows typical modes of failure inferred from the available records.

A commonly inferred mode is termed the double wedge mechanism, as illustrated in Figure 8. This mechanism is analogous to the development of the active Rankine state in the backfill of a retaining wall. In this analogy the toe region of the dump is the retaining structure, which is capable of deforming sufficiently that the active condition occurs.

Oversteepening of the upper part of dump faces is quite common due to the presence of relatively fine-grained material in the region of the dump platform. Sliver failures of the oversteepened upper face typically

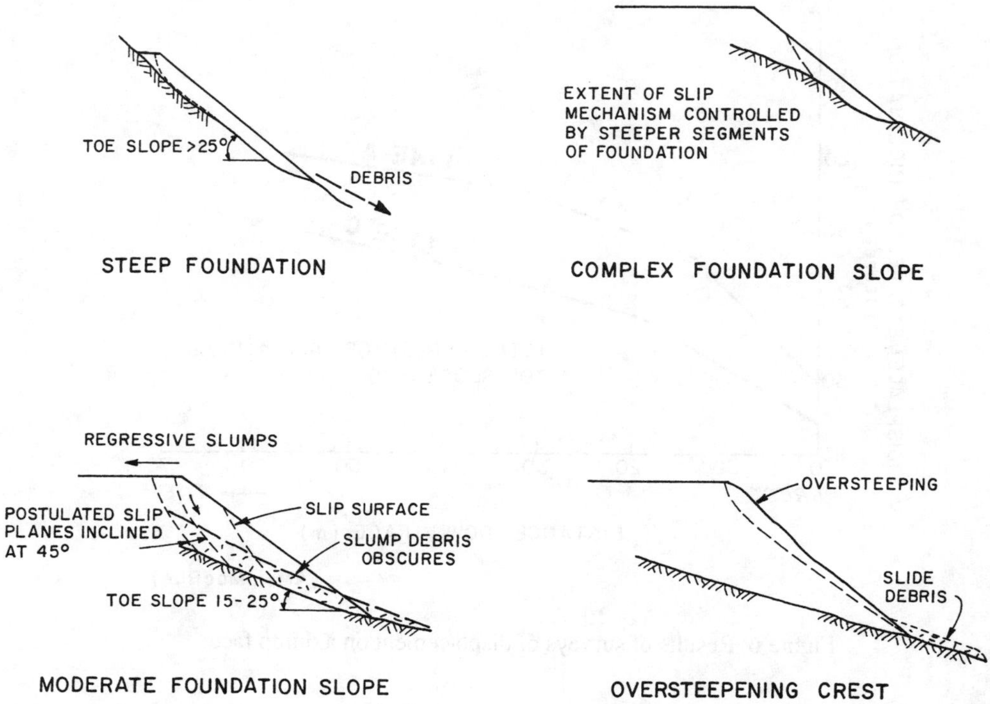

Figure 7 Typical modes of dump failure.

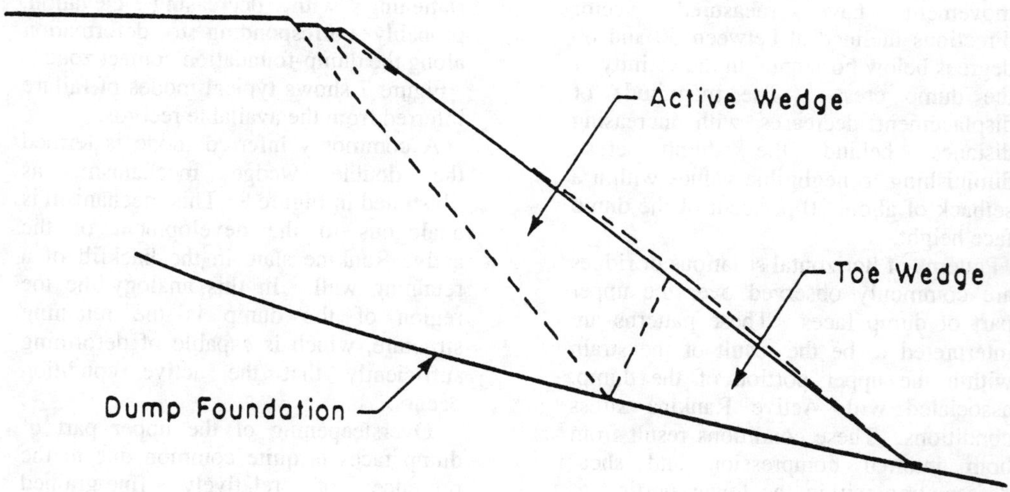

Figure 8 The double wedge failure mechanism.

involves a relatively small volume of material, and the failure runout is minor.

Bulging of the dump face is often detected, with deviations of up to about 5 degrees steeper than the typical angle of repose of 37 to 38 deg. The bulging may progress to a failure of moderate size.

Rapid loading of localised foundation areas mantled by lightly consolidated fine grained soils may cause toe spreading. Highly mobile slides have occurred where larger areas of surficial saturated organic soil have been loaded rapidly by sliding debris.

Deep-seated sliding surfaces have not occurred often because most mountain side-slopes are underlain by overconsolidated soils or bedrock at relatively shallow depths. Usually, potential dump sites underlain by lightly consolidated soils have been avoided.

The following sections briefly highlight several waste dump failures which have occurred at coal mines in British Columbia and several mines overseas.

Golder (1987), (1992) and (1993) provide a database of up to 50 mine dump failures in B.C. over the past 25 years. examined The nature of the failures, the runout of the sliding debris, and the environmental consequences were examined in these references.

For simplicity of reference the following examples have been assigned numerical labels. It should be noted that few, if any, experienced geotechnical engineers have been eye witnesses to these events. Also, comprehensive investigations, such as might be undertaken following civil failures of comparable size, have not been undertaken. Thus, caution is advisable when interpreting accounts of probable failure mechanisms.

Example Dump 1

Dump 1 was developed initially within a V-shaped gully incised in the relatively steep upper and middle portion of the sideslopes of a U-shaped valley. The toe region of the dump at the time of failure, which occurred during the fall, was supported on slopes inclined at 20 to 25 degrees along the axis of the gully. The height of the dump was approximately 420m. Various investigators of the failure have postulated the following possible causes of the slide:

- static liquefaction of layer(s) of relatively fine-grained waste material
- extensive fracturing at particle contacts, resulting in mass commination and rounding of the rock particles, loss of strength, and mass movement
- progressively accumulating pore water pressures within saturated foundation soils at the base of the dump as the result of shear-induced strains

The failure was preceded by a period of elevated rates of movement, detected at the crest by wireline extensometers. Thus, ample warning of the impending failure was provided and the area was safely closed off. The magnitude of the runout of the failure debris was relatively modest, given the dump height, compared to many other events. This outcome may have been due to the relatively coarse and free-draining alluvial gravels present beneath the runout path. Environmentalists claimed that the failure may have resulted in the loss of some 300 fish. However, adverse impacts on water quality in the river were relatively modest because river flows were low and due to the filtering action of the slide debris. Prompt construction of new sedimentation ponds downstream prior to the ensuing spring thaw resulted in acceptable levels of suspended solids being discharged.

Subsequently, the dump has re-activated and has been in operation for over three years. The dump remains stable at present.

Example Dump 2

A dump approximately 20m high triggered a landslide over an area of approximately 700m by 1000m approximately. The cause of the slide was attributed to the widespread transmission of pore water pressure via a thin layer of sand within lacustrine silt and clay. The consequences of the slide appear to have been relatively modest. Although such a failure consequence could never be considered acceptable, the occurrence and lateral extent of the thin layer of sand was very difficult to detect. As noted earlier, typical mountain dumps are not located on such terrain.

Example Dump 3

Dump 3 was constructed in the early 1970's on steep terrain near the top of a mountain range . Initial failure of the dump deposited waste material into a creek. Subsequently, it appears that saturation of the waste material resulted in a highly mobile debris flow which travelled a distance of approximately 3.2km. The relatively long runout was exacerbated by the tightly confined nature of the valley. During the ensuing 20 years the slide debris has been partly reclaimed. Complete equilibrium with respect to erosion and slumping of the creek walls has not occurred yet. Sedimentation ponds located near the junction of the creek and the main river have continued to provide adequate protection.

Example Dump 4

Dump 4 was constructed on steep terrain in such a way that the dump toe remained supported on the valley floor, i.e. the dump crest was advanced in a direction parallel to the topographic contours. The toe of the dump, with a face height of about 260m, advanced onto a deposit of organic soil. A rapid failure occurred, involving approximately 1.5 million cubic metres of waste material The distal portion of the slide debris was unusually mobile coming to rest on a slope of less than 5 degrees. This event illustrated the effect of saturated fine grained soils in the potential runout path of major dump slides.

Some investigators have speculated that inclusion of layers of fines caused localised development of pore water pressure and static liquefaction. However, the overall quality of the mine waste rock was relatively good. The more plausible cause of failure, and its relatively high mobility, was the presence of saturated soils in the foundation and along the runout path.

Example Dump 5

This slide occurred during the early years of the mine development when highly weathered material had to be dumped on relatively steep terrain.

The probable cause of the failure of dump 5 was the combination of moderately steep terrain and poor quality waste material. Part of the waste may have become saturated, precipitating the failure. The slide debris was highly mobile and travelled approximately 2km. It is significant that in the period to date, following the slide, water quality has been adequate, and construction of additional settling ponds has not been required. In fact, the slide debris itself appears to trap and filter dirty water entering the catchment.

Example Dump 6

Initial development of dump 6 successfully advanced across steep terrain, taking advantage of three-dimensional wedging action within the confines of a prominent gully. At the time of failure the toe region was supported on a slope of less than 20 degrees. A possible cause of failure was strain-generated pore water pressures within the upper 1 to 2m of the glacial till foundation, disturbed by frost action. Also, it was possible that the dump toe had advanced over frozen soil. The surficial frozen soil would have impeded the dissipation of pore water pressures due to new loading or to seasonal increases in regional groundwater pressures.

The failure occurred relatively slowly, with estimated velocities of 1 to 2 m/s. Downslope movement of the failure debris was arrested by an embankment that had been constructed in advance, using coarse reject material from the washplant.

Example Dump 7

Initially, the dump was constructed on very steep terrain with the anticipation that failures were likely to occur. Sedimentation ponds were constructed, in advance, downstream of the dump area. A considerable number of failures occurred, always preceded by periods of high rates of deformation at the dump crest. At times large steep scarps formed, necessitating closure of the portion of the dump crest affected. When fully developed the dump extended over a length of about 1.5 km of crest. On several occasions, major slumps occurred in which relatively wide segments of the dump crest slid intact. Most recently, approximately 30 million cubic metres of waste material was involved in such a slide. The probable cause of the failure is strain-generated pore water pressure within the foundation beneath the older slide debris. The velocity of the failure of these massive slides was relatively slow compared to more mobile slides such as examples 1,3,4 and 5.

Example 8 Dump

Relatively fine-grained material was dumped on steep mountain slopes during winter months. Subsequently, another fill was constructed on the slopes immediately below the dump. About 13 months after the dump was completed, failure occurred. resulting in one fatality. The dump was not being monitored. Surveys subsequent to the event showed that up to 8 m of crest settlement had occurred during the interval between cessation of dumping operations and the failure event.

As geotechnical engineers we would argue that the construction and operation of a road beneath a dump supported on steep slopes, without frequent observation and monitoring, is unsafe practice.

The cause of the failure has been debated by investigators. Possible causes include static liquefaction of the slide waste material, and accumulating pore water pressures within foundation colluvial soils subjected to shear stresses which approached the available drained shear strength.

Example Dump 9

Dump 9 was constructed on steep terrain in a region that is subject to high tropical rainfall. The dump face was approximately 200m high and subject to very high losses by erosion and shallow slumping. A failure occurred within adjacent bedrock strata as the result of downcutting of a creek bed. The loss of structural integrity of the bedrock members was the result of reduction in the thickness of the members due to the increased rate of erosion that resulted from the erosive power of bedload continually being eroded from the dump face.

Example Dump 10

Dump 10 was over 500m high and was located on steep mountain slopes ranging from 20 to 50 degrees. Dumping operations continued under conditions of accelerating movement at the crest until operations were suspended by a snowstorm. The dump failed during the snow storm, burying part of an inactive pit below. The cause of failure was considered to be result of accumulating pore water pressures within a layer of fine grained talus that mantled part of the dump-foundation surface of contact. Although the cessation of dumping prior to failure was fortuitous, the dump had been, and subsequently has, continued to be operated successfully. Measurement of displacements in the region of the operating crest, combined with visual inspection, provide sufficient warning of potential failures.

3.2 Monitoring Methods

As discussed in the foregoing examples, simple monitoring and sensible inspection provides the means for safe operations. Implicitly, or explicitly, mine operators and regulators in some cases have accepted the worst case scenarios of operationally safe failures. We consider that attempts to apply more sophisticated monitoring of parameters including foundation pore water pressures and foundation strains probably would not be practicable or cost-effective as an operational procedure. The practicality of achieving reliable measurements is questionable, although the information potentially available through such monitoring is very attractive to the geotechnical analyst.

3.3 Automated Monitoring

At least one operation has decided to automate wireline extensometers in order to reduce the cost of reading and maintenance. Nevertheless, continued vigilance by operations is very important. The data provided by monitoring should be used promptly to support the judgement of operations staff.

A number of mine operations have conducted programs aimed at relating operational factors, such as dumping rate, to dump performance. Other factors such as material quality and terrain have been evaluated. By tracking these types of information, mine engineers have developed site specific assessments of acceptable operating constraints. We encourage mine operators to maintain records such as shown on Figure 9.

4 DESIGN ISSUES

This section discusses some of the technical design issues which we consider do have significant impacts on the performance of major mine waste dumps. Given the operating conditions and constraints described earlier, and the analytical problems discussed below, it is clear that the practical designer must exercise considerable judgement, based on the observational approach.

4.1 Methods of Analysis

Two main types of analysis may be considered for the theoretical prediction of the performance of a proposed dumping scheme.

4.1.1 Limiting Equilibrium Analyses

The principles of limiting equilibrium are commonly used to identify potential sliding surfaces which result in the lowest calculated factors of safety. Realistic analyses rely upon the engineer to propose hypothetical trial sliding surfaces on the basis of his intuition and experience. Conventionally the stability of trial sliding surfaces is assessed in terms of the overall ratio of resisting forces to driving forces acting on the mechanism as a whole.

In reality the stress-strain characteristics of the loose waste rock and the dump foundation soils will likely be different. The observed deformations of dump faces suggest that the behaviour of a pile of loose waste rock on a sloping foundation is analogous to a retaining wall with granular backfill. Wall deflection by about one tenth of one per cent of the wall height is sufficient to fully mobilise the available strength of the backfill. The toe region of a waste dump can be analysed as a structure which retains the upper portion of the dump. Thus, a stability analysis could assume that most, if not all, of the strength of the waste rock is fully mobilised for the portion of the trial failure surface within the dump. An appropriate definition of the factor of safety would be the ratio of resisting to driving forces acting on the toe region of the dump.

4.1.2 Stress-strain Analyses

The practical application of finite element codes to the analysis of mine dump designs is not commonplace. We support continued research into the application of suitable programs. The results of such research may provide further insight into observed patterns of deformation. However, it is not clear that stress-strain analyses will result in designs that are better than those resulting from the use of limiting equilibrium methods of analyses. Practical application must be linked to field observation programs, which should include assessments of rock quality. Although modern computer programs and hardware are now relatively cheap to use, the tasks of mesh generation and parameter selection continue to require considerable experience and engineering judgement.

4.2 Parameter Selection

In general, the assumption of the angle of internal friction for the mine waste rock equal to the observed angle of repose is considered to be conservative. In certain situations, including some high dumps,

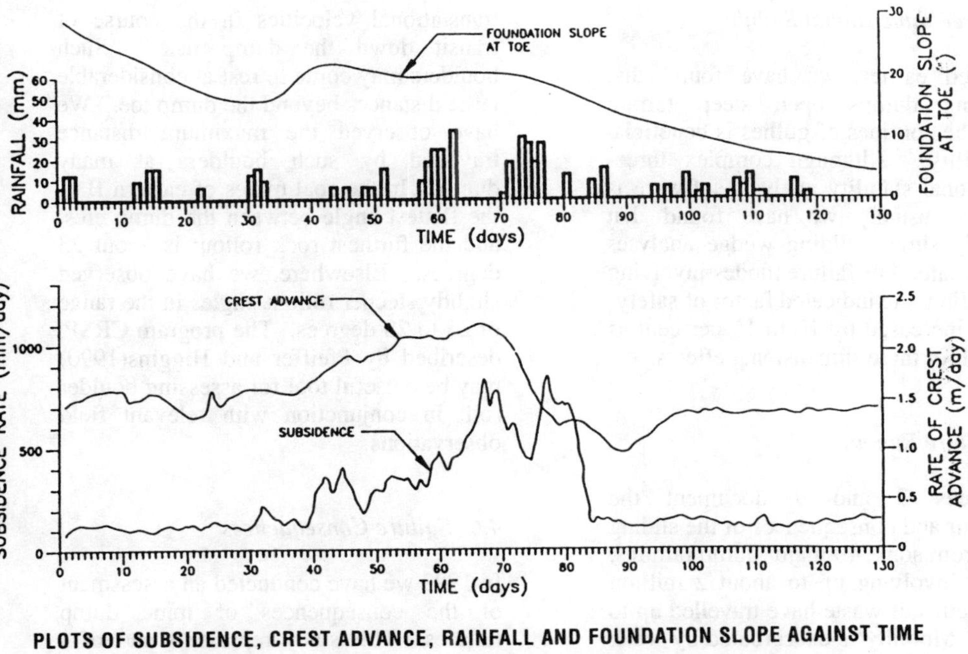

Figure 9 A typical mine dump monitoring record.

more precise characterisation of the non-linear strength of the mine rock may be worthwhile. We have developed a slope stability program assuming a double-wedge mechanism which uses a power function to describe maximum available shear strength in terms of applied normal stress. The program searches for the most critical double wedge and calculates the frictional strength that must be mobilised on the base of the toe region for limiting equilibrium. Selection of the power function can be based on the data assembled by Leps (1970). Alternatively, the adaptation of strength relationships for jointed rock masses to rockfill by Barton and Kjaernsli (1981) is rational and useful. In addition, we have assembled a comprehensive database of test data for a wide variety of rockfill types.

Overall, the engineer must temper theoretical assumptions with the experience garnered through past observations. Of particular importance, are the assumptions regarding pore water pressures. As noted earlier, pore pressure monitoring data is virtually impracticable to obtain and thus virtually non-existent. Stability is believed to be affected to a significant degree by excess pore pressure generated in the foundation. If strains are occurring in a saturated fine grained foundation, the rate of generation exceeds the rate of dissipation, and pore water pressures may rise to critical levels, resulting in instability. The rate of pore water pressure generation probably is also a function of the foundation slope which in turn affects the rate of foundation strain. Ongoing foundation strain will continue to generate pore water pressure after cessation of dumping. The current state-of-practice in such situations is to recognise the possibility and likelihood of pore water pressure generation and to assess the associated risks of failure. In the absence of reliable pore water pressure prediction, pro-active measures to ensure effective drainage, selective removal of foundation material, or control of mine rock quality and placement rate may be considered.

4.3 *Three-dimensional Stability*

As noted earlier, we have found that advancing dumps open steep terrain within the confines of gullies is beneficial to stability. Although complex three-dimensional stability analysis software is available, usually we have found that relatively simple sliding wedge analyses are adequate. For failure modes involving base sliding, the indicated factor of safety may be increased by 10 to 15 per cent as the result of three-dimensional effects.

4.4 *Failure Runout*

References 8 and 9 document the behaviour and consequences of the sliding debris from some 40 major dump failures. Failures involving up to about 2 million cubic metres of waste have travelled up to 2.5 km. Mobility is exacerbated by weak or saturated soil within the runout path, saturated fine grained waste material, steep runout paths, and by tightly confined runout paths. Mobility also may be affected by dynamic effects within the sliding mass. Precise back-analyses of past events is complicated by the foregoing factors. However, selective comparison with relevant cases is useful.

Figure 10 shows a plot of the cotangent of the runout angle with height of the dump face. The runout angle is the angle subtended between the dump crest, prior to failure, and the distal toe of the slide debris. The dashed lines are possible trends for predicting runout on the basis expected conditions.

In addition, the principle of energy conservation can be applied as a general framework for assessing potential runout paths and distances, and the mitigative measures that may be appropriate. Comparison with the runout of major natural mountain rockslides may also be helpful.

4.5 *Boulder Rollout*

Occasionally, large competent particles of waste rock may attain high rotational and translational velocities in the course of transit down the dump face. Such boulders may come to rest at considerable large distances beyond the dump toe. We have observed the maximum distance travelled by such boulders at many dumps. In the coal mines of eastern B.C. the flattest angle between the dump crest and the furthest rock rollout is about 23 degrees. Elsewhere, we have observed slightly steeper rollout angles in the range of 25 to 28 degrees. The program CRSP, described by Pfeiffer and Higgins(1990) may be a useful tool for assessing boulder roll, in conjunction with relevant field observations.

4.6 *Failure Consequences*

In 1993 we have conducted an assessment of the consequences of mine dump failures. This study represents part of a risk-based approach to dump design.

Generally, the impacts of past major dump slides have been fairly well controlled and managed. The purpose of objectively documenting past events is to provide a basis for qualitative project risk assessments. Often mine waste slide debris quickly stabilises and long-term erosion of fines is relatively minor and easily controlled by sedimentation facilities. Fortunately, acid rock drainage is not a common occurrence at the coal mines in British Columbia.

4.7 *Rock Drains*

Campbell(1986) describes the results of monitoring the performance of a rock drain beneath a 50m high waste dump. The results indicate that flow-through capacity(cubic metres per sec. per square metre of gross wetted cross section) decreases with increasing thickness of waste rock above the rock drain. Other instrumented studies of high dumps are in progress. In our opinion, long-term degradation of mine rock in the basal zones of mine dumps is not a major concern. The interior of mine dumps is a relatively stable environment and usually

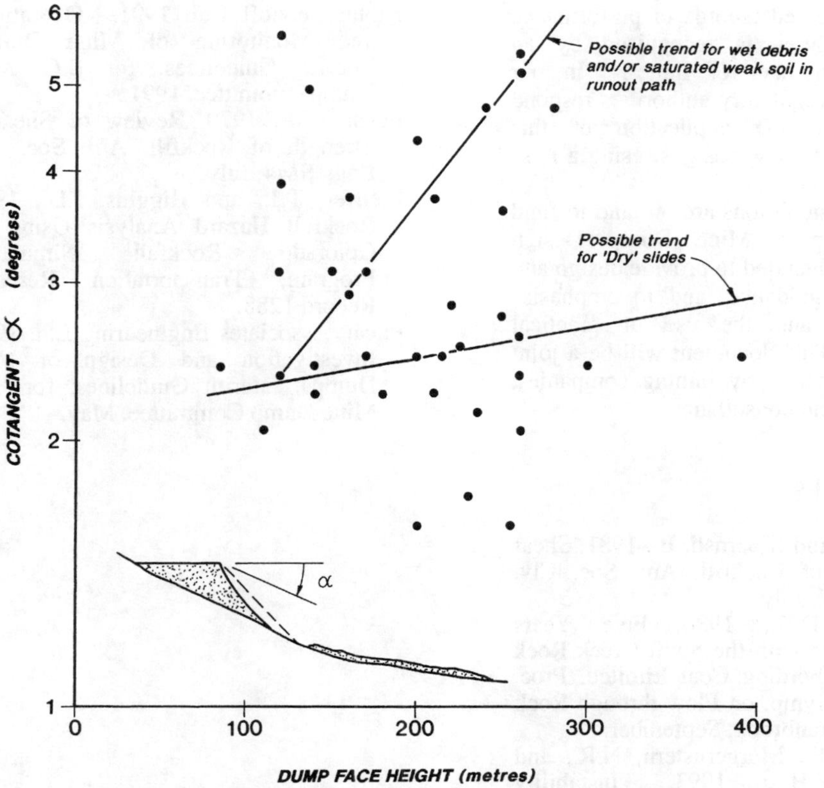

Figure 10 Cotangent of the runout angles and dump face heights for failure events at coal mines in B.C.

the particles which reach the toe are competent and durable. Common sense dictates that major long-term rock drains will not be satisfactory if rock quality is poor. The design of rock drains must be integrated with reclamation plans. Downslope dozing to flatten dump faces may not be compatible with rock drains formed by the natural segregation of coarse competent rock particles, and their accumulation in the region of the dump toe. Construction of ultimate dump surfaces by a series of wrap-around dump platforms is preferable. Over 25 years of observing large mine dumps we have not detected evidence of deep-seated distress among any inactive or reclaimed mine dumps. Nevertheless mine operators should remain vigilant, particularly where infrastructure lies within the runout paths of potential failures.

5 FUTURE TRENDS

Mine dump failures over the past 25 to 30 years have involved about 5 per cent of an estimated 2 billion cubic metres of mine rock that has been produced at the coal mines of B.C. The results of an objective analysis of the consequences of these failures generally indicates that this record is acceptable. However, it is clear that mine regulatory authorities demand detailed explicit assessments of proposed dumping schemes. Unfair bias against the mining industry may result in severe economic penalties if maximum advantage of past experience is not taken. Thus, continued research is essential, with particular emphasis on field observation. Collaborative efforts between industry, consultants, regulators, and university groups is encouraging. The collection of well documented records of performance and of failures is recognised by an increasing number of mines. In our experience, regulatory authorities respond favorably to the application of this information to new designs, using a risk-based approach.

In B.C. negotiations are in hand to fund and develop a Mine Dump Design Handbook, intended to provide design and operational guidance, and to emphasise observation and the use of practical examples. This document will be a joint effort undertaken by mining companies, regulators, and consultants.

REFERENCES

Barton, N., and Kjaernsli, B., 1981. Shear Strength of Rockfill, Am. Soc. Civ. Eng., GT7, July.

Campbell, D.B., 1986, Five Years Experience with the Swift Creek Rock Drain at Fording Coal Limited. Proc. Internat. Symp. on Flow-through Rock Drains, Cranbrook, September.

Dawson, R.F., Morgernstern, N.R., and Gu, W.H., 1993, Instability Mechanisms Initiating Flow Failures in Mountainous Mine Waste Dumps, Phase I Report, Department of Civil Engineering, University of Alberta, November.

Golder Associates, 1987, Regional Study of Coal Mine Waste Dumps in British Columbia, Project Number 862-1231, June.

Golder Associates, 1992, Runout Characteristics of Debris from Dump Failures in Mountainous Terrain, Stage 1, Interim Report, for B.C. Mine Dump Committee, March.

Golder Associates, 1993, Consequence Study, to B.C. Ministry of Energy Mines and Petroleum Resources, March.

HBT Agra Ltd, Methods of Monitoring of Mine Waste Dumps located in Mountainous Terrain, Interim Report, for B.C. Mine Dump Committee, 1992.

Kent, A.H., 1992, Coal Mine waste Dumps in British Columbia, Stability Issues and Recent Developments, Mine Waste Management ews, Vol. 2, No. 3, July.

Klohn Leonoff Ltd, 1991. Operational and Monitoring of Mine Dumps, Interim Guidelines, for B.C. Mine Dump committee, 1991.

Leps, T.M., 1970. Review of Shearing Strength of Rockfill. Am. Soc. Civ. Eng., SM4, July.

Pfeiffer, T.J., and Higgins, J.D., 1990. Rockfall Hazard Analysis Using the Colorado Rockfall Simulation Program, Transportation Research Record 1288.

Piteau Associates Engineering Ltd, 1991, Investigation and Design of Mine Dumps, Interim Guidelines, for B.C. Mine Dump Committee, May.

Innovative Mine Design for the 21st Century, Bawden & Archibald (eds) © 1993 Balkema, Rotterdam, ISBN 90 5410 325 6

The evolution of tailings management practices at Syncrude Canada Ltd

B.R.List & J.M.Buckles
Syncrude Canada Ltd, Fort McMurray, Alb., Canada

ABSTRACT: The Syncrude Canada Ltd. surface mine located near Fort McMurray in northeastern Alberta will produce approximately 128 million tonnes of oil sand feed in 1993 generating approximately 280,000 tonnes of tailings solids daily. All process water used in bitumen recovery and tailings transport is confined within containment facilities. The Mildred Lake Settling Basin, commissioned in 1978, was designed as an all encompassing storage facility, accommodating coarse tailings, fine tails and process water. Fine tails are allowed to settle and consolidate over time producing denser Mature Fine Tailings. The clarified upper water zone is recycled to the plant for re-use in the bitumen recovery process. The Southwest Sand Storage Facility, commissioned in 1991, provides additional coarse tailings storage capacity while the fine tails and process water are returned to the settling basin for consolidation. Commencing in 1995, Mature Fine Tails from the settling basin will be transferred to mined out portions of the base mine. This complex fluid and solids cycle will provide a minimum risk to the environment while optimizing the storage capacity of each of the three tailings facilities. The fluid and solids balance are examined to show how tailings management has evolved since start-up and the new directions to be undertaken in the future. The tailings management philosophy will provide for continued tailings storage into the next century while meeting a long term commitment to the environment.

1 INTRODUCTION

The Syncrude Canada Ltd. surface mining operation currently produces approximately 128 million tonnes of oil sand feed annually which when processed yields approximately 68 million barrels of Synthetic Crude Oil. Bitumen is liberated and recovered from the sand using a froth floatation aqueous slurry process. After the bitumen is recovered, approximately 280,000 tonnes of tailings solids are generated daily. Tailings slurry, comprised of process water, sand, fine clay particles, and trace amounts of unrecovered hydrocarbons, is pumped to the storage facilities via five 600 mm diameter pipelines. Tailings sand is used to hydraulically construct containment dykes and supporting beaches of the storage facilities, while the process water released through the sand structures is retained within the tailings system and re-introduced into the Extraction process. The practice of water containment is in keeping with a corporate policy of not releasing process water to the environment. The existing tailings storage facilities are shown in Figure 1.

The commissioning of the Extraction Plant in May 1978 marked the start-up of tailings deposition and storage at Syncrude. During the first 13 years of operation, all tailings were stored in the Mildred Lake Settling Basin (MLSB). This hydraulically constructed basin is currently comprised of approximately $600x10^6$ m^3 of solids and contains approximately $280x10^6$ m^3 of water with fine tails in suspension. The Extraction Plant currently imports clarified water from the upper 4 m of the settling basin for use in its day to day operation. The fine tails, primarily clay particles, accumulating in the basin are allowed

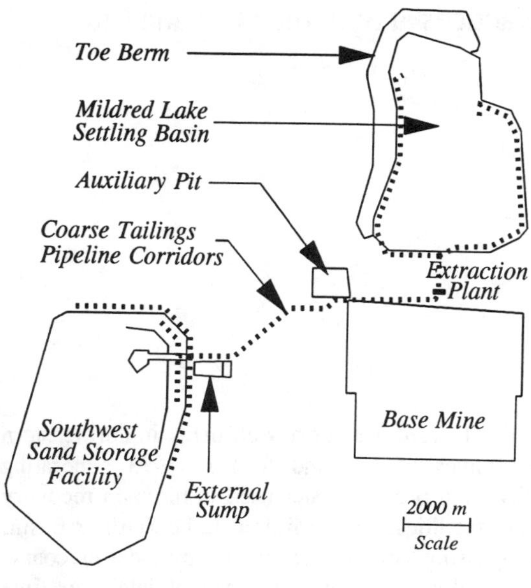

Fig. 1 Syncrude site

to settle and consolidate over time producing a denser Mature Fine Tailings (MFT) zone at the base of the settling basin. The life of this tailings structure is finite, both for the storage of coarse tailings and for the storage of fine tails for consolidation.

Completion of the MLSB containment structure generated the need to store coarse tailings elsewhere which prompted the development of the Southwest Sand Storage (SWSS) Facility which commenced operation in 1991. The fine tails and transport water collected in an internal or external sump are transported back to the settling basin where the fine tails settle out. Designed as a non-fluid retaining structure, stability issues are minimized and the ability to suitably reclaim the structure after completion is enhanced and assured in the SWSS Facility.

Future tailings storage at Syncrude will see the development of a facility in a portion of the mined out pit. In-pit facilities will be developed primarily for the storage of MFT transferred from the MLSB. This transfer is necessitated by the limited fluid storage capacity of the MLSB. Overburden dykes and hydraulically placed sand dykes will be constructed in the mined out pit to provide the required below ground containment. The deliberate placement of fluids and solids in prescribed locations will provide a minimum risk to the environment while optimizing the storage capacity of each of the three individual tailings storage sites.

The establishment of the different tailings storage facilities is in keeping with the long term vision and obligation of creating a stable, self-sustaining landscape. Two possible contrasting types of final landscapes are currently being examined; a wet or dry landscape. Under a wet landscape, the Mature Fine Tails will be capped with a layer of natural water. For a dry landscape environment, the fine tails would be mixed with sand or clay overburden creating a solid base suitable for the growth of vegetation upon reclamation. It is likely that the final landscape will combine these two possible options. Syncrude's continued commitment to the environment will continue to play a major role in the future development and management of tailings facilities.

2 TAILINGS DEPOSITION TECHNIQUES

The design of Syncrude's tailings facilities must meet storage capacity and stability requirements in an optimum and efficient manner. Tailings deposition parameters including production rates, deposition method and the nature of the deposited material are paramount in the design of these facilities. Syncrude has developed an extensive database of the physical properties such as gradation, density, beach profiles developed in above and below water level deposition environments and solids runoff representing tailings deposition through various construction methods. The various characteristics form the basis for the design parameters currently used for ongoing facility development.

Three sand deposition methods; cell construction, contained beaching and overboarding have been used to construct the tailings facilities. Cell construction is used to raise the perimeter dyke surrounding a water retention structure such as the settling basin. Contained beaching is employed to construct the perimeter dyke of the non-fluid retaining structure in the SWSS Facility while overboarding is utilized to construct the beaches upstream of the perimeter containment.

Cell construction is used to construct the

perimeter MLSB containment dyke. These cells, established parallel to the dyke centre line, are typically 50 m wide and up to 700 m long. As tailings are hydraulically introduced, the coarser materials settle out in the cell and are spread and compacted by bulldozers while the finer fractions flow through a spillbox installed at the end of the cell into the settling basin. Solids capture in the cell is controlled by maintaining an optimum water level behind the spillbox weir boards thereby minimizing the accumulation of lower density fine tails. The controlled placement of solids in the cell and accompanying compaction by dozers yields a density of approximately 1655 kg/m^3.

Contained beaching is a modification of the cell construction techniques used in areas such as the SWSS Facility which is not a fluid retaining structure. The cells developed by the contained beaching techniques are generally much larger, comprised of higher lifts and where dozer compaction is minimal in comparison to cell construction.

The majority of tailings are deposited through overboarding; either through a direct single discharge or through spigotting. Overboarding in most often done through a single, 600 mm, discharge point while spigotting is done through numerous smaller discharge points building a steeper beach profile closer to the point of discharge. The beaching techniques employed yield two modes of deposition; Beach Above Water (BAW) and Beach Below Water (BBW). Coarse tailings are the first to settle out from the point of discharge on the way to the water body (settling basin or sump) forming a BAW with a typical density of 1577 kg/m^3. The finer materials eventually settle in the water body forming a BBW with a typical density of 1510 kg/m^3.

3 FINE TAILS BALANCE

The volume of fine tails generated through the extraction process is affected by the fines input rate and density which is directly related to the fines content of the oil sand feed and the amount of fines captured during the deposition process. It is estimated that approximately 50% of the fines are captured in the voids of the coarse sand tails. Both the consolidation rate and the ultimate consolidation density determine the long term fine tails storage requirement. Immature fine tails have a density in the order of 15% solids by weight, whereas MFT achieves a density of approximately 35% after 10 years and is projected to consolidate to a maximum of approximately 40% under the expected loading and drainage conditions.

A simplified water balance for the Syncrude site can be determined by the total raw water imported to the plant less the water removed through plant systems evaporation, storage in the sand voids and storage in the fine tails. This balance assumes that any water introduced to the site, if not removed by one of the above three processes, will eventually require storage in the settling basin as part of the freewater layer as no water discharge from the site is permitted. The settling basin profile, as illustrated in Figure 2, includes a freewater layer, a consolidating fine tails zone and a MFT zone. To maximize the basin volume for fine tails consolidation, the freewater layer must be minimized to a level which allows continuing clarification. In order to sustain the water clarification and fine tails consolidation processes in the settling basin, MFT must be removed to equalize inflow and outflow to a constant basin fluid elevation. The volume of new fine tails entering the settling basin will be balanced by MFT removal. The MFT will be pumped from the settling basin to the mined out pit once adequate initial containment dykes have been established in the mined out pit.

4 TAILINGS STORAGE STRUCTURES

At start-up, Syncrude established the Mildred Lake Settling Basin to contain all the extraction process tailings including coarse sand, clays and process water. The MLSB design has been modified several times to optimize the storage capacity and in response to performance monitoring. As the maximum sand storage capacity was reached while still meeting the future fine tails storage requirements the SWSS Facility was established. This new facility was designed to accommodate the majority of the sand generated from the extraction process well

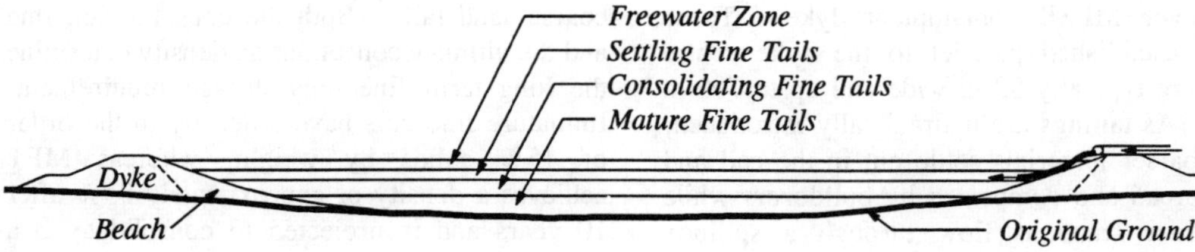

Fig. 2 Schematic Settling Basin cross section

into the next century while the thin fine tails are returned to the MLSB for consolidation. As the fluid storage capacity of the settling basin is filled, a new facility in the mined out pit will be developed. The prime objective of this facility will be to store MFT transferred from the settling basin thus also extending the MLSB life into the next century.

4.1 Mildred Lake Settling Basin

The Mildred Lake Settling Basin was initially sized to accommodate sand storage from the mine production over a period of 25 years accommodating approximately 800x10^6 m^3 of sand. The initial structure was to cover approximately 27 km^2 and be built to a final design elevation of 347 m above sea level; approximately 37 m above the natural ground elevation. Site selection for the basin was based on proximity to the plant site, physical constraints such as roadways, topography and on the basis of the underlying oil sand reserves and foundation conditions.

The current settling basin, shown in Figure 3, has undergone several major design changes in order to optimize overall tailings management. The structure utilization is currently in a transition from an all encompassing storage facility to one in which it will serve primarily as a settling basin for fine tails and a holding area for the clarification of process water to be recycled back to the plant.

Continuous optimization and the need to meet a variety of mining requirements necessitated four major revisions to the initial design. They included a realignment of the dyke centre line, a change in the dyke construction method, an increase in the ultimate dyke crest elevation and a subsequent increase in footprint size through

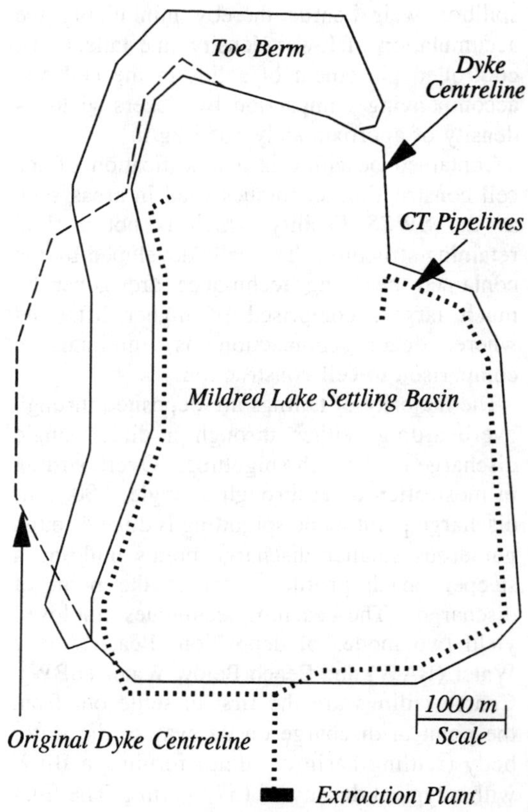

Fig. 3 Mildred Lake Settling Basin

the development of toe berms.

In 1981, the west tailings dyke was realigned to minimize the sterilization of mineable oil sands while maintaining sufficient storage capacity to meet long term production demands. A reduction in surface area to 22 km^2 was made viable by a reduction in the long term water storage requirements and a better understanding of volumetric parameters.

Sand deposition methods were changed in 1982 to a centre line approach in which the perimeter dyke was built independent of the beach. The change to this method required a large step-out of the compacted shell onto the beach to allow for future dyke development. The step-out was achieved by utilizing cell construction over beached sand. This philosophy change increased overall dyke stability as a greater proportion of the total sand was placed in a compacted shell and allowed for fluid storage optimization.

Driven by an increase in plant capacity and economics, the ultimate dyke crest elevation was increased to 352 m in 1984. This elevation was chosen as the best alternative balancing economics with a basin geometry that maximized fluid storage while addressing all environmental issues. The higher settling basin required an extensive geotechnical assessment which resulted in slope flattening in several areas and the completion of internal drainage filters. Subsequent lateral displacement in the underlying foundation materials was addressed through the construction of toe berms adjacent to part of the structure perimeter. This assured ongoing structure stability.

Initial MLSB planning recognized that the observational approach with extensive instrumentation and continuous monitoring was the most practical method to ensure an optimum design and assured structure stability. Syncrude has implemented an extensive monitoring and testing program to support continuous design reassessments providing a successful and ongoing use of this approach.

Geotechnical design considerations for the MLSB were required in three fundamental areas: overall dyke stability, seepage control and liquefaction potential. The underlying clay shale foundation material highlighted potential instabilities underlying much of the MLSB. Various slope angle geometries and rates of construction have been used to assure foundation stability and maintain structure integrity. Considerable experience has been gained over the past 14 years in terms of implementing design changes to respond to foundation deformations. This experience along with diligent monitoring and ongoing risk assessments assures the reliability of the settling basin is maintained.

Seepage control within the dyke has been a major environmental design consideration. Internal strip drains are used to control the phreatic surface within the dyke as well as any potential for internal erosion or piping. The seepage water exiting the settling basin is collected in a sump through a perimeter ditch and then pumped back to the interior of the basin.

The potential for large-scale liquefaction failures is considered to be minimal based on the low seismic risk for the area and the absence of large-scale foundation displacements. The rapid rate of beach construction required in the early years was, however, judged to increase the likelihood of localized upstream beach liquefaction failures. The potential for additional triggering mechanisms, largely related to strains induced by foundation deformations and/or additional loading due to dyke construction, prompted the use of limited explosive densification. To a lesser extent, explosive densification has also been used to densify other areas within the compacted shell.

4.2 Southwest Sand Storage Facility

The approaching completion of the existing settling basin dyke and beach structures, and the need for alternative sand storage site, prompted the development of the SWSS Facility. This new facility was designed to provide sufficient sand storage capacity for approximately 60% of the coarse tailings produced over the next 30 years. The tailings structure, shown in Figure 4, will ultimately extend to cover an area of 30 km^2 and attain a height of 40 m providing storage for approximately $1000 x 10^6$ m^3 of sand. The SWSS Facility site exhibits relatively gentle undulating relief sloping to the northeast. The design of this sand structure developed through hydraulic deposition in a topographically unfavourable site considered detailed hydrological environmental aspects as well as water and solids balance assessment under various construction methods. Water containing fine tails collected in the SWSS Facility is pumped to the settling basin for consolidation.

The unfavourable ground topography and the desire to minimize the initial pumping and piping facilities dictated that initial sand

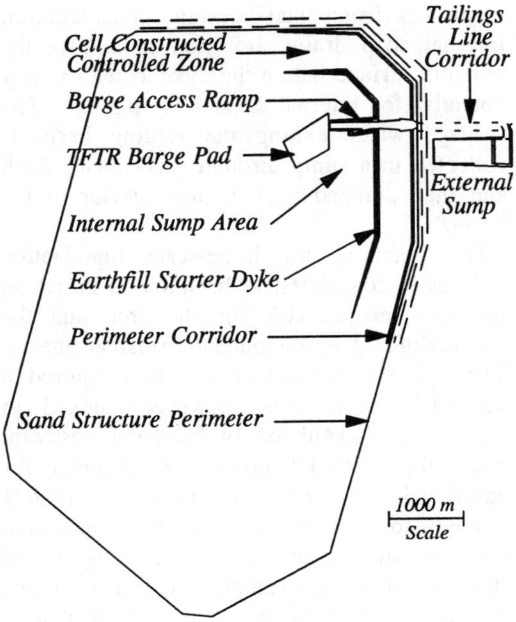

Fig. 4 Southwest Sand Storage Facility

deposition be concentrated in a perimeter-controlled zone parallel to the downstream toe. This zone, constructed using contained beaching techniques, was incorporated into the sand structure design to accommodate the initial rapid dyke construction rate. The controlled zone also serves as a means of assuring stability of the sand structure in the area of the Internal Sump. A sump, external to the deposit footprint, collects the tailings process water during controlled zone construction. The External Sump was designed to collect solids, process water and to accommodate adverse operating conditions and storm events during downstream construction. Upon the transition to upstream construction, the Thin Fine Tails Return (TFTR) Barge will be relocated from the External Sump to the Internal Sump which will be the collection point for the process water. The External Sump will then be used to collect seepage water and runoff from the sand structure. The water collected in the External Sump will be pumped to the settling basin thus preventing the release of process water from the Syncrude site.

The first phase of development after the completion of the controlled zone in 1993 is marked by development of a constant elevation structure. This phase of development is expected to be completed in 1995. The perimeter dyke of the structure will be raised in advance of upstream beach development during the summer months allowing for direct discharge to the interior during the winter. This construction method has been found to be most effective and avoids cell construction or contained beaching under adverse winter conditions.

Construction beyond 1995 will follow along a constant height above original ground philosophy in which the sand structure perimeter is raised in increments until it attains a height of 40 m. The Internal Sump will remain in the same area relative to the sand structure but will progress upwards with the structure development. The fine tails volume will be minimized throughout sand deposition by maintaining thin fine tails transfer to the settling basin.

The geotechnical design of the SWSS Facility sand structure recognized the same three key design factors: foundation and slope stability, liquefaction susceptibility and seepage control. An observational monitoring approach is being taken with this sand storage facility, similar to that taken with the existing settling basin throughout its construction and supported by an appropriate level of geotechnical instrumentation. This will allow actual field performance to be measured and compared to the initial design assumptions. Geotechnical analyses established an initial 20H:1V slope requirement for the 40 m high structure. The governing factor in this shallow slope design is the presence of an underlying Clearwater clay shale foundation material with its low soil shear strength and relatively high pore water pressure response to loading. Through diligent performance monitoring, and confirmation of foundation soil parameters, a greater height and/or slope angle may be attainable, thereby increasing the total sand storage capacity.

The design and construction of the SWSS Facility will be continuously reviewed and optimized from a production, economic and environmental standpoint. Development alternatives could result from production increases, changes in sand storage philosophies, new technology or in response to geotechnical performance.

4.3 In-pit Tailings Storage Facility

The fluid storage capacity of the MLSB will be maximized by 1996 at present production rates. In order to continue to utilize the settling basin for fine tails consolidation beyond 1996 an alternate storage location is required. The rate at which MFT are transferred from the settling basin to an alternate facitly is approximately equal to the inflow of fine tails; primarily pumped from the SWSS Facility. Syncrude will develop a below ground elevation MFT storage facility in the mined-out pit to meet the tailings management requirements. Mature Fine Tails will be pumped from the lower portion of the settling basin to the in-pit facility, balancing inflow and outflow at the MLSB. It is expected that approximately $500x10^6m^3$ of MFT will be contained within the in-pit facility by the year 2010.

Placement of MFT in the mined-out pit has necessitated careful planning of the in-pit overburden berms and waste dumps which will provide the primary containment in the initial years of development. Construction of the initial containment facilities are currently in progress with completion expected in 1994. This initial storage area will support MFT storage requirements for approximately five years.

Subsequent in-pit storage capacity at the east lease boundary will be developed by construction of hydraulically-placed sand dykes. This will be accomplished through the transfer of a coarse tailings line expected in 1997. Thin fine tails from this construction will be retained in-pit to delay or eliminate the need for a return system and to minimize the size of the MFT transfer system. Eventually release water from the consolidating fine tails will be returned to a water-recycle system.

Extensive studies have been and are being carried out to assure that there is no detrimental environmental impact of placing MFT within the in-pit facility. All seepage water from the in-pit facility will be collected and confined within the facitly limits, as is presently the case at the other tailings facilities.

SUMMARY

Syncrude's strong commitment to protecting the environment while minimizing the cost of producing 68 million barrels of Synthetic Crude Oil annually requires ongoing assessment of tailings management practices. Continuous optimization of the Mildred Lake Settling Basin has provided sufficient storage capacity for all tailings for the first thirteen years of production. The need for additional sand storage capacity resulted in the development of the Southwest Sand Storage Facility. This is a unique facility in which the thin fine tailings are not stored within the confines of the structure but are returned to the settling basin for consolidation. Mature fine tailings, recovered from the settling basin, will be pumped into a new facility being developed in-pit within the next two years. This provides ongoing capacity in the settling basin for water storage and fine tailings consolidation well into the 21st century. The transfer cycle of sand, fine tails, MFT and water expected upon the development of the in-pit facility is shown in Figure 5.

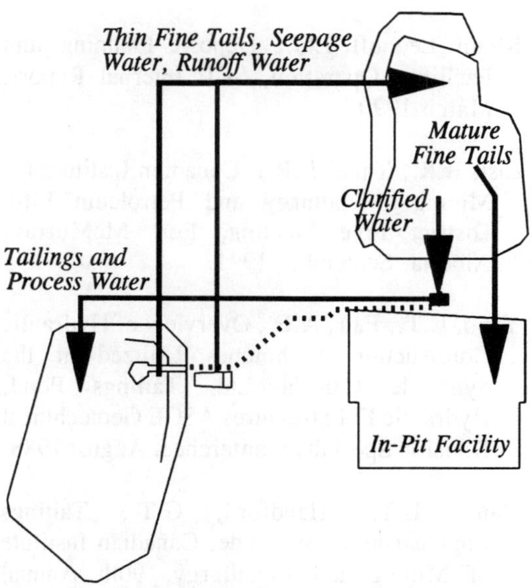

Fig. 5 Tailings material cycle schematic

Reclamation of all tailings facilities is ongoing through development and upon completion. This reclamation may take the form of capping the tailings sand with organic materials able to sustain vegetation and wildlife. Syncrude is currently investigating several techniques for reducing the ultimate fine tails storage requirements, most of which revolve around the idea of capturing fine tails in the void spaces of coarse tails and overburden. Further evolution of Syncrude's tailings management philosophy will occur as this technology becomes available. The current tailings management philosophy and the commitment to the environment will meet the long term vision of creating a stable, self sustaining landscape upon the completion of oil sand mining at Syncrude.

ACKNOWLEDGEMENTS

The authors wish to thank the efforts of the many people who have contributed to this paper from Syncrude Canada Ltd. The continued support from Syncrude in allowing the tailings deposition advancements made within the organization to be shared is also greatly appreciated.

REFERENCES

Klohn Leonoff Ltd., Deposit Planning and Facilities Operation, SCL Internal Report, March 1990.

List, B.R., Yano, L.R., Canadian Institute for Mining, Metallurgy and Petroleum Fifth District Five Meeting, Fort McMurray, Alberta, September 1991.

Yano, L.T., Fair, A.E., Overview of Hydraulic Construction Techniques Utilized at the Syncrude Canada Ltd. Tailings Pond, Hydraulic Fill Structures ASCE Geotechnical Division Specialty Conference, August 1988.

Yano, L.T., Handford, G.T., Tailings Engineering at Syncrude, Canadian Institute of Mining and Metallurgy, 90th Annual General Meeting, Edmonton, Canada, May 1988.

2 Underground support and backfill

Innovative Mine Design for the 21st Century, Bawden & Archibald (eds) © 1993 Balkema, Rotterdam, ISBN 90 5410 325 6

Physical modelling of fracture and flow in mine backfills

R.J. Mitchell
Queen's University, Kingston, Ont., Canada

ABSTRACT: Classified tailings sandfills have been used worldwide for several decades as ground support in large underground mine openings. A major problem with traditional hydraulic sandfilling is that the economics of hydraulic delivery are negated by higher cementation cost due to excess water. Excess delivery water can also cause drainage and potential sandflow problems. During the past decade, considerable research effort has been expended to improve the economics of sandfilling by developing a better understanding of the fill stresses and the cementation requirements for fill stability. This paper briefly reviews the results of physical model studies directed in this area of interest and presents recent contributions to centrifuge stability modelling.

Physical modelling of drainage and resaturation of sandfills is introduced by means of theoretical analyses which demonstrate that accelerated modelling can reproduce real in-stope conditions, including the volumetric water content distribution in a drained fill. The results of backfill resaturation models are presented to demonstrate the application of physical modelling to water flow predictions and to sandflow prevention in uncemented sandfills. Significant water pressures are found to exist in resaturated backfills and precautions should be taken, in practice, to prevent recharge of uncemented sandfills. The application of physical modelling to the transport of contaminants from containment facilities is also discussed and demonstrated in this paper.

INTRODUCTION

Engineers involved in the design of mine backfills and mine tailings impoundments are faced with a wide range of stability problems and a great variety of drainage and contaminant seepage problems. Some of these problems are quite amenable to the traditional geotechnical design approach of laboratory testing to obtain material properties combined with mathematical analysis of the effects of boundary and applied loading conditions. Some are not. In particular, backfill stability problems involving soil arching and backfill resaturation problems are not easily addressed by numerical modelling. A viable alternative is physical modelling.

Physical modelling is carried out on a faithful reproduction of the real (prototype) situation at a model scale, N. Because the driving force in prototype earth stability problems is the planetary gravitational field, a geotechnical centrifuge is necessary to provide simulation of gravity induced stresses. It is not necessary to define or measure any material properties for centrifuge modelling, only to measure the performance of models at different scales. In this respect, centrifuge models may be thought of, simply, as small prototypes.

This paper outlines the use of centrifuge modelling in mine backfill design using examples from previous modelling work and also presents original data from water recharge and contaminant transport modelling.

CENTRIFUGE MODELLING

Centrifuge physical modelling has been carried out for almost a century, beginning with the use, prior to 1900, of small bench centrifuges to model soil drainage. In the early part of this century, larger scale centrifuge modelling of earth pressure and stability problems were initiated in the USA and the former USSR. During the past two decades, interest in centrifuge physical modelling has increased in response to the increasing complexity of the earth stability and contaminant transport problems facing geotechnical engineers. There are about 50 geotechnical centrifuges operating in more than a dozen countries. Mitchell (1991) discusses the current use of modelling centrifuges in engineering design. The following example will serve to introduce the basic principles of centrifuge modelling.

Imagine a prototype tailings dam, 33 m in height and 300 m in base width, somewhere on the surface of planet earth. The downstream slope of this dam is subjected to gravitational acceleration which produces a mass stress, σ, in the structure due to the material density, ρ (i.e. $\sigma = F/A = z \rho g$, where g is the gravitational constant on earth and z is the depth of material mass). Gravity driven flow from the tailings pond also produces lateral seepage forces ($p_s = i_x \times g \rho_w$) in the downstream slope where i_x is the hydraulic gradient, ρ_w is the density of water and x is the distance of flow in the horizontal direction. These forces may cause a slope failure. Imagine a smaller but similar tailings dam, 110 mm in height and 1 m in base width, being rotated at ω radians/s in a geotechnical centrifuge as shown on Figure 1.

All of the particles in this dam are being subjected to an acceleration $a = r\omega^2$. Suppose the radius, r, of rotation is 3 m (so that $z << r$, $x < r$) and the centrifuge speed is 299 rpm, then $a = 2943$ m/s^2 = 300 gravities. Because the dimensions (x , z) have been reduced by 300 and the "simulated

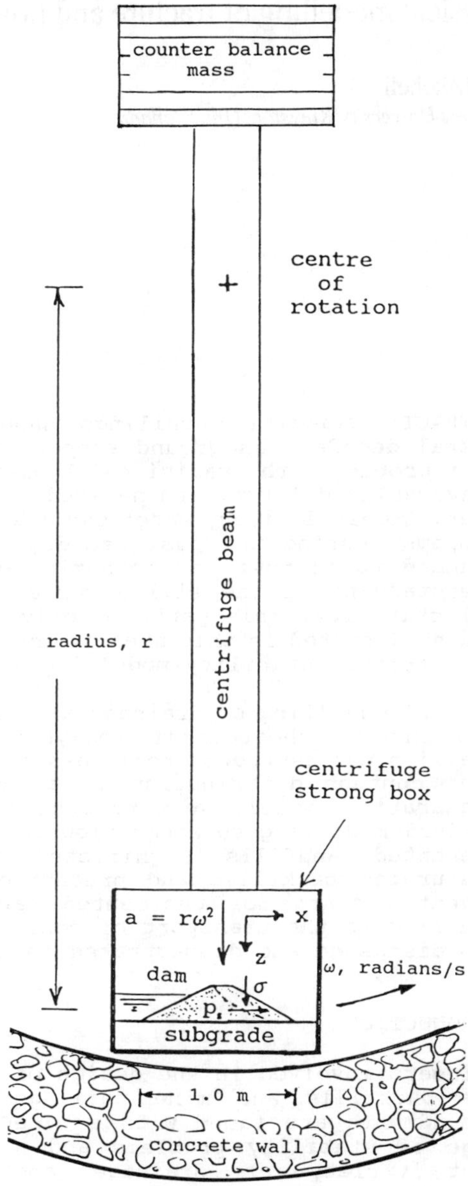

Figure 1

gravity" is now 300 g, σ and p_s are identical at similar points in the two tailings dams (note that the hydraulic gradient, being defined as the height of water over the distance of flow is similar in both). Moreover, the shear strength of the earth material, being given by the Mohr-Coulomb criterion $\tau = c' + \sigma'_N \tan \phi'$, is similar at similar points in both dams (since σ'_N is identical and c' and ϕ' are material constants. Indeed, both c' and ϕ' may be stress dependent in many earth materials and such dependency will be correctly modelled in the centrifuge.

If the small tailings dam (model) fails when subjected to a = 300 g in the centrifuge, the large tailings dam (prototype) will fail if built to 33 m height on the surface of the earth. The maximum safe height of the prototype can be evaluated by building the model and gradually increasing the acceleration until distress is noted. Because the material properties may be unknown (not measured), centrifuge model evaluation is entirely independent and is without assumption regarding material behaviour. Furthermore, if the centrifuge model is built on the prototype subgrade material, the expected bearing capacity performance and the expected settlements (monitored at the linear scale of 1 mm = 0.3 m) could be evaluated. In addition, the designer does not need to speculate on the critical failure mechanism (circular arc, base wedge, etc.) since the model will show how a failure develops.

Suppose the ground surface is partly saturated under the prototype tailings dam. Seepage from the pond may affect the stability of the foundation and contaminants from the pond may eventually reach the ground water. These possibilities will not likely materialize for many years and, at present, are very difficult to evaluate numericallly. Yet, environmental protection legislation requires that these possibilities be addressed. Again, centrifuge modelling can effectively and efficiently provide the necessary data. Because centrifuge accelerations are modelled to produce identical stresses, stress gradients are increased by the scale factor, N (i.e. linear scale = acceleration scale). Thus, all fluid flow velocities are increased by N. Because the flow distance is N times smaller in the model than the prototype, the arrival time of flow processes is increased by N^2. Thus, the time for advective solute transport between similar points will be reduced by a factor of 300^2 in the model tailings dam (i.e. 300^2 times faster than in the real dam). Thus, one day of centrifuge modelling would provide 250 years of performance evaluation with regard to the potential effects of seepage into the ground surface.

The applicability of accelerated modelling, in view of the high seepage velocities, and the modelling of dispersive transport phenomenon were concerns during the early application of centrifuge modelling to flow and transport phenomenon. Appendix A summarizes the results of both theoretical dimensional similitude calculations and experimental evaluations which indicate that centrifuge modelling does accurately reproduce fluid flow and solute transport in the very fine to medium sands encountered as tailings slimes or as classified mine tailings.

MINE BACKFILL STABILITY MODELS

Early physical modelling of backfill stability was carried out under the planetary gravity field (1-g models) using partly-cured cemented fill models to represent fully cured prototype backfills (see, for examples, Mitchell et al., 1982; Smith et al., 1983). Mitchell (1986) reports the first centrifuge model tests on mine backfill stability. Subsequent to this, the 3 m radius centrifuge at Queen's Civil Engineering Department has produced model data on crown pillar subsidence (Wizniak and Mitchell, 1987); stability of high narrow fills (Mitchell, 1988a); stability of layered backfills (Mitchell, 1988b); stability of fills having a variety of wall slopes and closure stresses (Mitchell, 1988c); stability of reinforced mine backfills (Mitchell, 1989); effect of impulse simulated blast loadings on an exposed fill face (Nnadi and Mitchell, 1991) and sill mat stability (Mitchell, 1991). Figure 2 shows some results of the work on backfill stability which has clearly demonstrated that cemented strengths far less than normally used in backfill design can provide adequate fill stability under normal conditions.

Figure 3 shows a typical model after fill failure. Visual observations are one of the major advantages of centrifuge modelling;

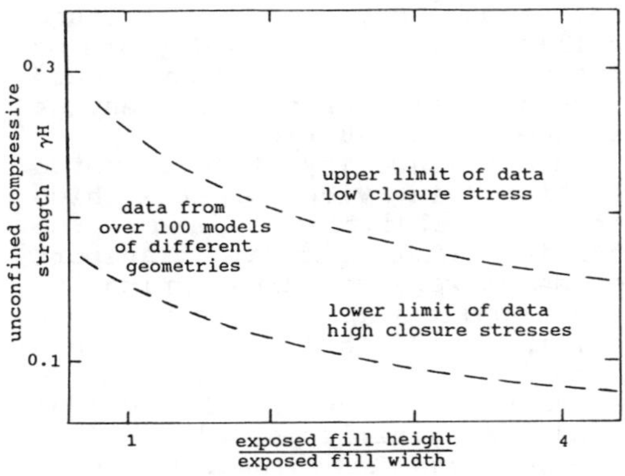

Figure 2. Strength requirements for backfill exposures from centrifuge model testing

Figure 3. Failure of an exposed face of cemented fill; at a g level of 250 gravities, this represents a 70m high backfill

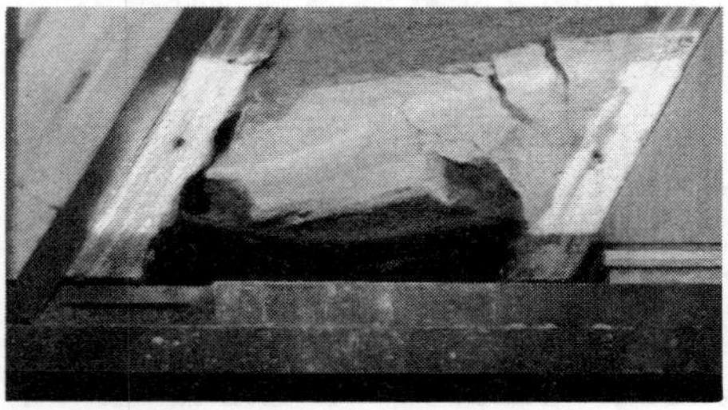

Figure 4. Complex failure in model reinforced sill mat with rough hanging-wall

Table 1. Stresses at the base of models of exposed backfills

Model height to width ratio H_m/L_m	HW-FW slope (degrees)	σ_v measured / γH
1.6	90	0.65
3.2	90	0.45
3.2	75	0.35

the modeller sees, as well as monitors, performance characteristics. Figure 4 shows a unique composite failure in a model cemented sill mat.

In many of the centrifuge model studies noted above, boundary stresses and displacements were monitored in-flight (i.e. while the centrifuge speed was held constant). Table 1 lists measured vertical stress at the base of several models, expressed as a fraction of the maximum possible stress, $\sigma = z\rho g = \gamma H$.

From this data, it is clear that arching and/or boundary shear stress significantly reduces the overall vertical stresses in the backfill. Such stress reduction has also been noted in field (prototype) measurements (see, for examples, Askew et al., 1978; Barrett and Chester, 1980). The boundary stress transfers (stress reorientations) associated with arching are, of course, the reason that stope dimensions (geometry) play such an important role in determining cement strength requirements, as shown by the data on Figure 2.

Figure 5 shows measured subsidence, given as $N\delta$ where N is the model scale and δ is the measured subsidence in the model, and demonstrates the effect of fill depth on subsidence.

DRAINAGE, RESATURATION AND SOLUTE TRANSPORT MODELLING

Mitchell and Roettger (1984) present results from backfill pouring tests carried out under the planetary gravitational field (1-g laboratory column tests) which indicate that

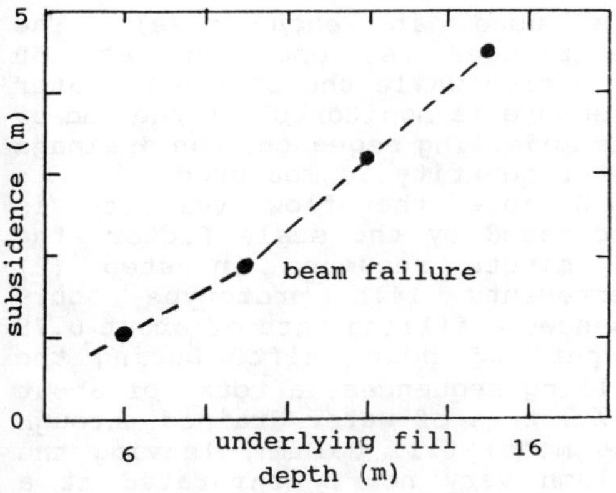

Figure 5. Effect of underlying depth of fill on subsidence of a 2m deep, 14m span crown pillar having a loading of 0.5 MPa

bulkhead effective pressures and water pressures can be modelled, for a saturated sand backfill, without centrifuge acceleration. They used a 1.2 m column height and a filling rate of about 0.2 m/hr to confirm experimentally that good base drainage was the key to maintaining low water pressures in the backfill. Mitchell (1992) reports the results of centrifuge model pours carried out at a scale of N = 15 and indicates that the fill pressures measured in centrifuge models correlate well with fill pressures measured in the 1-g column tests reported earlier.

Partly saturated internal drainage in fine sands cannot be modelled in 1-g experiments (see Appendix A). Centrifuge modelling of the drainage of saturated sands at N = 50 and N = 100, reported by Cooke and Mitchell (1991) confirms the theoretical prediction (Appendix A) that prototype partly drained water content curves can be created at a model scale. Figure 6 shows centrifuge drainage data for a sand having D_{10} = 0.02 mm and a uniformity coefficient of 6, typical of a classified tailings. Conducting centrifuge modelling at two or more scales is termed "modelling of models" and is used to show that observed phenomenon are not scale dependent. Any phenomenon occuring identically at a variety of gravity scales should also occur under the planetary gravity field.

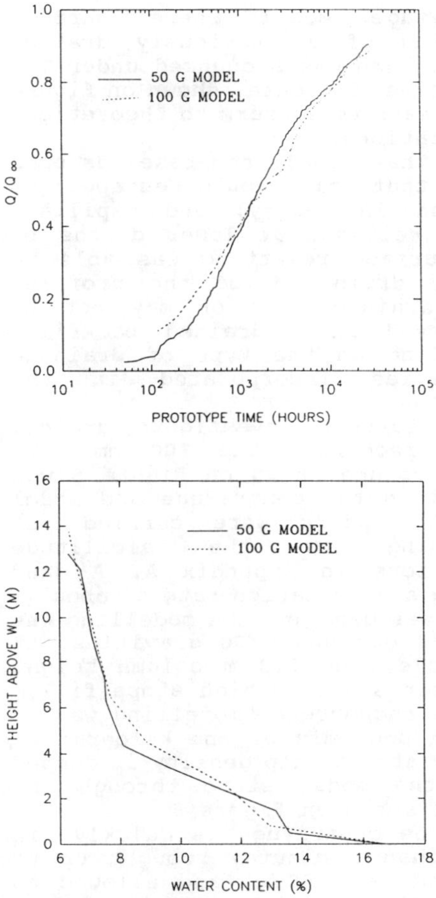

Figure 6. Drainage and residual moisture contents in 50g and 100g models

No experimental research has been found on the recharge of previously drained backfills, despite the fact that flooding of previously drained backfill surfaces can (and does) occur in mine backfill practice. Infiltration has been considered, theoretically, for shallow profiles by hydrologists (see, for example, Whister and Bouwer, 1970) and there is a general concencus that infiltration will occur very rapidly for shallow depths (due to the high suction gradients existing in fine sands) and that the resaturated percolation rate will generally be about half of the saturated percolation rate.

Mitchell and Cooke (1991) reported the result of one

centrifuge model where surface flooding of a previously drained soil column was sequenced under 100 gravities. The data, shown on Figure 7, appears to conform to theoretical expectations.

In that model, the base was open such that air could escape (as assumed in theory) and capillary water released at lower depths by the surface rewetting was able to freely drain out of the profile. Base drainage may or may not be preserved in a drained backfill, depending on the type of drainage facilities incorporated in the stope.

To further investigate drained stope recharge, the 100 mm I.D. model column shown on Figure 8 was placed in the centrifuge and model backfill pours were carried out according to the similitude conditions in Appendix A. A sand having a percolation rate of about 1 cm/hr was used and the modelling was carried out under 50 gravities. At 50 scale, the 0.3 m column height represents a 15 m high stope fill.

The sequence of modelling was:

(1) A pour mix of one kilogram of slurry at 60% pulp density is dumped into the model stope through the funnel shown on Figure 8.

(2) The centrifuge is quickly (in less than a minute) brought up to 133 rpm (a = 50 g) and allowed to run at this speed for 15 minutes, at which time the excess delivery water has drained into the backfill.

(3) Steps (1) and (2) are repeated about a dozen times in order to fill the column. Pore water pressures at the base are monitored during each 15 minute centrifuge run. During all of the above steps the drainage outlet control valve is open.

(4) Drainage water quantity is measured at the end of this 3 to 3.5 hour filling sequence.

(5) The filled column (model stope) is subjected to continuous internal drainage for six prototype months (105 minutes at 50 gravities) with the drainage valve fully opened. Drainage water quantity is measured at the end of this sequence.

(6) The drainage outflow control valve is set in position (four positions were used: fully opened, two-thirds open, one-third open and fully closed) and the surface of the fill is flooded with water (500 cc was used in each case). The centrifuge is operated at 50 gravities while the base pore water pressure is monitored. At the end of the modelling sequence, the drainage water quantity is measured.

Because the flow velocity is increased by the scale factor, the 15 minute sequence in step (2) represents 14.5 prototype hours (hence, a filling rate of about 0.75 m per 7.5 hour shift. During the filling sequences, a total of about 1.7 litres of water drained through the model fill column, leaving the column very nearly saturated at a water content of about 25%. During the 6 month (prototype) drainage

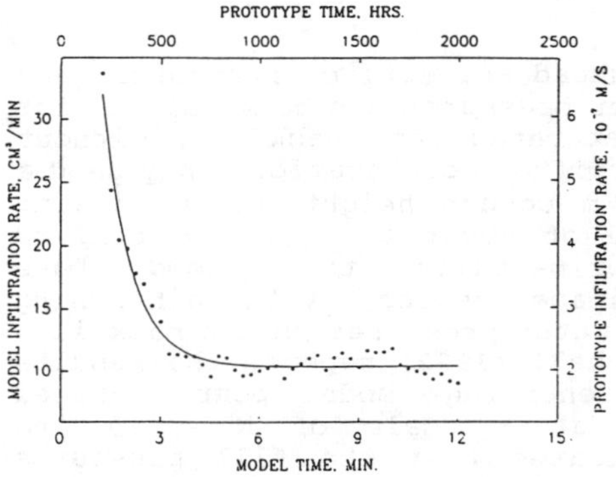

Figure 7. Infiltration into fine sand at 100g

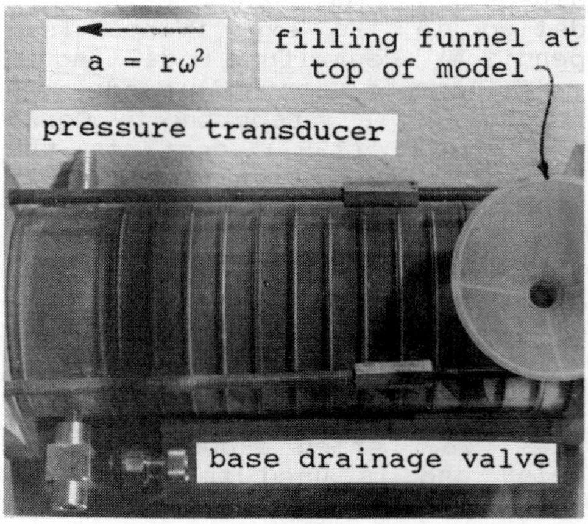

Figure 8. Model backfill pouring and drainage column

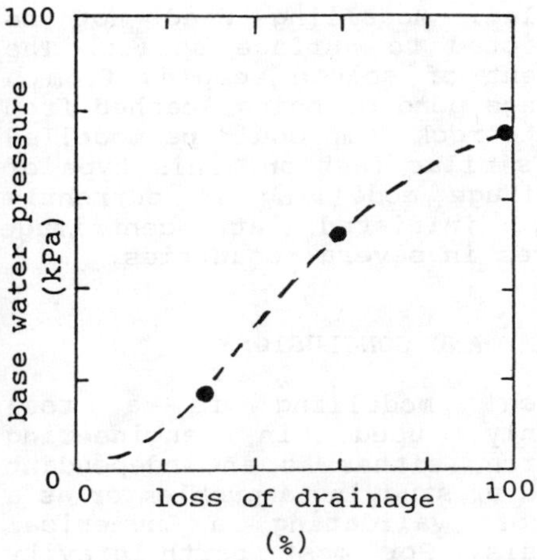

Figure 9. Maximum measured water pressures

Figure 10. Dipping segregation in model pour

period, another 1.6 litres of water drained out leaving the average water content in the column at about 10%. Water balance measurements indicate that the distribution of this water follows the form of the curves on Figure 6.

During the filling sequence, the highest pore water pressure measured at the base of the model stope was 17 kPa. This indicates that the drainage system (which was a 2 cm diameter porous stone inset into the wall of the column diagonally across from the transducer and connected to the valve and drainage line) was, as designed, satisfactory. The porous stone models a 1 m diameter porous opening and is equivalent to a 3 m by 3 m bulkhead that is about 30% porous (a typical free draining timber bulkhead constructed at the brow of a drift). The equivalent prototype stope area being drained by this bulkhead is 50^2 (78.5 cc) = 19.6 m^2.

Figure 9 shows the maximum measured pore water pressures during resaturation of the model backfill. With the drainage fully opened, there appeared to be no pore water pressure generated by the infiltration of 500 cc of water. With the drainage valve completely closed, the pore water pressure at the base gradually increased attaining a peak of about 74 kPa (over 10 psi) after about 20 minutes. This peak corresponds to a water head of about 7.5 m above the bulkhead. At a porosity of 0.4, 500 cc of water would saturate a dry model fill to a height of 0.16 m (8 m prototype). It appears that the extreme condition of no drainage could result in a hydrostatically saturated condition being established in a prototype stope subjected to surface recharge.

The intermediate points on Figure 9 appear to indicate a shallow S-shaped relation between the degree of drainage and the maximum base pressure on recharge with a given quantity of water. If the drainage is reduced by more than 30%, the maximum water pressure exceeds that measured during initial filling. It is quite possible that large reductions in effective drainage could occur due to silting of filters or mousetrap drains. Decant pipes may be quite efficient at draining an initial pour (acting essentially as surface drains) but may be useless at picking up recharge water. Figure 10 shows a section of a model stope after pouring and the dipping segregation is visible through the plastic column wall. Recharge water may flow primarily along the coarser grained layers and very little of this flow would be intercepted by decant pipes.

The recharge modelling outlined above clearly shows the potential danger of allowing drained fills to be subjected to massive recharge flows. Unless the principal stope drainage is located at the stope

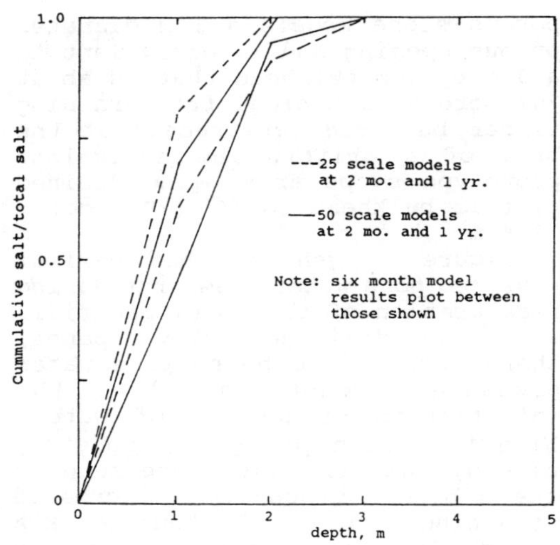

Figure 11. Sodium chloride tracer distributions from 25g and 50g models

base and is preserved at (or close to) 100% efficiency, large pore water pressures may be generated by the recharge water.

Such large pressures could result in bulkhead failure and sandflows from the stope. In most cases, the potential for massive recharge can be eliminated, either by capping drained fills or ensuring that superincumbent pour drainage water is conducted directly to sumps. Alternatively, cemented base plugs in all stopes may be designed to support hydrostatic water pressures.

The solute transport modelling work summarized by Cooke and Mitchell (1991) shows clearly that centrifuge modelling at scales of 50 and 100 is able to reproduce solute transport in partly saturated fine sand profiles. Figure 11 shows results from salt tracer models in which a 2000 litre prototype surface spill of a 10% NaCl solution on a 5.5 m deep partly saturated sand was modelled at scales of 50 g and 25 g.

The data on Figure 11 was obtained by dissection of models following the timed modelling sequences. The dissected portions are subjected to tests designed to determine the solute content of the sand. In this way, the distribution of any solute can be determined for the conditions modelled. Modelling need not be restricted to surface spills. The movement of solute seeping from a tailings pond or being leached from a waste rock dump could be modelled in a similar fashion. This type of centrifuge modelling is currently being initiated at centrifuge centres in several countries.

SUMMARY AND CONCLUSIONS

Physical modelling is a tool commonly used in engineering research, either as an independant method of studying a problem or as a way of validating a numerical analysis. For most earth gravity structures problems, modelling must be carried out in a centrifuge in order to simulate the effects of gravity on fracture and flow phenomenon. The principles of centrifuge modelling, as applied to mine design problems are outlined in this paper. Modelling results are included to show both the breadth of the work that has been carried out and the applicability, to mine backfill design problems, of centrifuge modelling. Recent modelling of water recharge in previously drained backfills is presented and the results of this work clearly demonstrate the potential for large pore water pressures to be manifested during such recharge.

ACKNOWLEDGEMENTS

The support of the Natural Sciences and Engineering Research Council of Canada (NSERC) and the Mining Industry during the construction, in 1984, and the first five years of operation of the geotechnical centrifuge at Queen's University is greatfully acknowledged.

APPENDIX A: FLUID FLOW AND SOLUTE TRANSPORT MODELLING

The flow of fluids through fine to medium sized sands is sufficiently slow to be treated as laminar and is generally found to follow the empirical Darcy flow model, $v_z = k\ i_z$ where k is the coefficient of permeability and i_z is the Darcy dimensionless hydraulically gradient (head loss divided by flow distance). Fluid flow should, then, be modelled correctly in 1-g laboratory experiments: and it is, providing k is a characteristic property of the sand material (it is known to depend on the vicosity of the permeating fluid and the porosity of the sand). When a sand becomes partially saturated, however, k becomes a non-linear function of the volumetric water content and the advective flux is increased locally by suction gradients (i.e. there is no unique hydraulic gradient). Fluid flow in partly saturated (water wetted) fine sands cannot be adequately studied in 1-g laboratory models.

It is interesting to consider the scaling laws governing fluid flow and solute transport through sands in a geotechnical centrifuge. The Darcy proportionality constant, k, combines the effects of fluid properties, sand properties and gravity. The gravitational component may be separated out by using the intrinsic permeability, $K = \mu k/pg$ where μ is the fluid viscosity (Cargill and Ko, 1983). The absolute intrinsic permeability, $K = K/\mu$, is then a function of the sand matrix only. Thus, although the Darcy hydraulic gradient is dimensionless, hence the same in model and prototype, the pressure gradient increases in a centrifuge model and the seepage velocities are related as $v_m = N v_p$, where the subscripts m and p refer to model and prototype, respectively. Combining the increased seepage velocity $v_s = v/n$ (where n is sand porosity) with the reduced model flow distance (scaled to N), the time required for a molecule of water to flow between two similar points in model and prototype are related as, $t_p = N^2 t_m$. A dimensional analysis of the advective-dispersive transport equation (Arulanandan et al., 1988) yields ten dimensionless quantities which govern the process. Two important quantities, the Reynolds Number and the Peclet Number (dv/D, where d is a characteristic microscopic length, v is the seepage velocity and D is the diffusion coefficient) are different in model and prototype because of the difference in seepage velocity. It has been demonstrated by Bear (1979) and Goforth et al. (1991) that similarity between model and prototype is achieved whenever these two numbers are both less than about 0.4. Thus, it is only necessary to select appropriate modelling scale factors; for solute flow and transport in partly saturated fine to medium sands, these conditions are satisfied for all scale factors not greatly in excess of 100. For Peclet Numbers less than 0.4, molecular diffusion dominates dispersive transport and is modelled correctly (see Cooke and Mitchell, 1991).

Of major practical benefit for the modelling of flow and transport phenomenon in partly saturated sands is the fact that the capillary effects number, ρ_w g L l/T, where l is a characteristic micro dimension (average pore size), L is a characteristic macro dimension (stope size) and T is the surface tension, will be the same in model and prototype (since gL = constant). Then it is expected that characteristic parameters such as volumetric water content and suction will be similar in model and prototype. In this respect, centrifuge modelling is necessary for studying flow and solute transport problems in partly saturated sands at a model scale.

The movement of fluids through porous media is a complex phenomenon, particularly when multiphase flow of fluids with differing wetting characteristics is considered. Numerical modellers are, however, beginning to develop solutions and it is important to be able to calibrate and/or validate these solutions. Perhaps the single most important contribution of centrifuge modelling in this area will be to provide data sets, for both simple and complex modelling conditions, in support of the development of numerical solutions to contaminant transport problems.

REFERENCES

Arulanandan, K., P.Y. Thompson, B.L. Kutter, N.J. Neegoda, K.K. Muraleetharan, and C. Yogachandran. 1988. centrifuge modelling of transport processes for pollutants in soils. Journal

of Geotechnical Engineering, ASCE, 114(2), 185-205.
Askew, J.E., McCarthy, P.L., and Fitzgerald, D.T. 1978. backfill research for pillar extraction at ZC/HBHC; in mining with backfill. CIM Spec. Vol. 19, May 1978.
Barrett, J.R. and Chester, G. 1980. post pillar cut and fill mining: comparison of theory and practice. in applications & rock mechanics to cut and fill mining. Lulea, Sweden, June 1980, pp. 272-283.
Bear, J. 1972. dynamics of fluids in porous media. American Elsevier Inc., New York, N.Y.
Cargill, K.W., and H.Y. Ko 1983. centrifugal modelling of transient water flow. ASCE Geotechnical Journal, 109(4), 536- 555.
Cooke, A.B. and Mitchell, R.J. 1991. physical modelling of a dissolved contaminant in an unsaturated sand. Canadian Geotech. J. 28:6:829-833.
Goforth, G.F., F.C. Townsend, and D. Bloomquist 1991. saturated and unsaturated fluid flow in a centrifuge. in: H.-Y. Ko and F.G. McLean (eds.) Centrifuge 91, 497-502.
Mitchell, R.J. 1991. centrifuge modelling as a consulting tool. Can Geotech J. 28:1:162-167.
Mitchell, R.J. 1986. centrifuge model tests on backfill stability. Canadian Geotech. J. 23:3:341-345.
Mitchell, R.J. 1988a. stability of high narrow backfills. CIM Bulletin, 82:921:70-74.
Mitchell, R.J. 1988b. centrifuge model studies on layered backfilll systems. Proc. Centrifuge '88 - International Conference on Geotechnical Centrifuge Modelling. Paris, April, 1988, pp. 225-232.
Mitchell, R.J. 1988c. the effects of stope geometry on the stability of mine backfill. Proc. 41st Canadian Geotechnical Conference, Kitchener, Ontario, October 1988, pp. 8-15.
Mitchell, R.J. 1989. stability of classified tailings backfills containing reinforcements. Proc. 4th International Symposium on Mining Backfill, Montreal, October 1989, A.A. Balkema, pp. 237-246.
Mitchell, R.J. 1991. sill mat evaluation using centrifuge models. Mining Science and Technology 13:301-313.
Mitchell, R.J. 1992. centrifuge model studies of fill pressures on temporary bulkheads. CIM Bulletin 85:960:48-54.
Mitchell, R.J. and Cooke, B. 1991. physical model studies of the hydraulic conductivity of unsaturated porous media. Proceedings CSCE Annual Conf., Vancouver, May 1991, Vol. 1, pp. 116-125.
Mitchell, R.J., Olsen, R.S., and Smith, J.D. 1982. model studies on cemented tailings used in mine backfill. Canadian Geotech. J. 19:1:14-28.
Mitchell, R.J. and Roettger, J. 1984. bulkhead pressure measurements in model fill pours. CIM Bulletin, 77:868:50-54.
Nnadi, G. and Mitchell, R.J. 1991. blast loadings on cemented backfills. CIM Bulletin 84:960:54-62.
Smith, J.D., de Jonge, C.L. and Mitchell, R.J. 1983. large scale model tests to determine backfill strength requirements for pillar recovery. Proc. Int. Conf. on Mining with Backfill, Lulea, Sweden, June 1983, 11 pp.
Whisler, F.D. and H. Bouwer. 1970. comparison of methods for calculating vertical drainage and infiltration for soils. Journal of Hydrology, 10, 1-19.
Wizniak, L. and Mitchell, R.J. 1987. crown pillar subsidence from centrifuge models. Proceedings International Symposium on Prediction and Performance in Geotechnical Engineering, Calgary, June, 1987, Balkema Press, 452 pp.

Innovative Mine Design for the 21st Century, Bawden & Archibald (eds) © 1993 Balkema, Rotterdam, ISBN 90 5410 325 6

On the use of rockfills in overhand cut-and-fill mining

Graham Swan
Falconbridge Limited, Sudbury Operations, Ont., Canada

Behrouz Arjang
CANMET, E.M.R., Elliot Lake, Ont., Canada

David G.F. Hedley
Elliot Lake, Ont., Canada

ABSTRACT: As a result of recent and ongoing work carried out by Lac Minerals Ltd. (Macassa Mine) and Falconbridge Limited (Strathcona Deep Copper Mine) on the use of cemented rockfill in narrow-vein cut-and-fill mining, tentative guidelines for achieving stiffnesses beneficial to the alleviation of rockbursts are being established. This paper summarizes the findings of large-scale laboratory tests, in the literature, as well as ones recently completed at CANMET's Elliot Lake Mining Research Laboratory, on a variety of confined cemented and uncemented rockfills. Results are presented which appear to demonstrate the desirability of the void-filling rather than the cementitious properties of a binder agent. This leads to a discussion of a practical, stiff, rockfill design for use in overhand cut-and-fill mining.

Aside from the matter of fill design, the strategic positioning and underground placement technique is critical to the technical and economic success of rockfill as a backfilling material in cut-and-fill mining. The paper describes these important aspects in the light of experience from the Strathcona Deep Copper Mine.

1 INTRODUCTION

Since the late sixties a significant body of laboratory triaxial test data for rockfill of varying mineralogy, particle size and shape has accumulated, driven by the civil construction industry. Among other things the results have clearly demonstrated a disproportionate increase in strength and stiffness with increasing confining pressures of up to 5 MPa, eg. Al-Hussaini (1983).

Recently, with mine operator interest in broken rock as a viable fill material generally increasing, together with strong evidence for fill stiffness contributing to the alleviation of rockbursts, applications to cut-and-fill mining have been successfully implemented by, among others, Falconbridge Limited at its Strathcona Deep Copper ore body. Early in the engineering design phase the requirement for cement addition to the rockfill, in order to achieve certain in situ mechanical properties, was in question.

In overhand cut-and-fill mining the placed fill is essentially buried and subject, over the course of mining, to fully confined loading conditions (with the exception of sill pillar areas where recovery is planned). Because of this the use of uniaxial compressive test data is inappropriate for design purposes. Unfortunately, only very limited triaxial test data is available on cemented rockfill (CRF) (Gonano et al 1978) and no definitive assessment of the test data from civil work with rockfill (RF) has been made for the mining industry. Such an assessment is made in Sections 2 and 3 of this paper.

Under confined conditions the virtue of adding cement to rockfill must be questioned, particularly where large deformation is anticipated. In order to maximize the fill's long-term stiffness properties, void-filling rather than cohesive strength is paramount. The addition of cement to rockfill can actually be considered on its void-filling merits (Thomas et al 1979) though this is very likely an expensive option. Other options have recently been examined using a one-dimensional laboratory compaction test conducted by CANMET on rockfill mixed with combinations of fine tailings material and cement. The results from this work are summarized in Section 4.

Finally, the paper considers the various innovative aspects to placement technique as practised at Falconbridge's Strathcona Mine. This is a narrow-vein mining operation where conventional bulk truck hauling and dumping of the fill material is impractical.

2 PUBLISHED ROCKFILL DATA

In the civil engineering literature an extensive body of test work has been reported on the triaxial compression of rockfill. The bulk of this work is due to Marachi et al (1972), Marsal (1973), Charles & Watts (1980) and Al-Hussaini (1983).

2.1 *Particle size characteristics*

The wide range of particle size distributions represented in the literature is shown in Figure 1, with maximum particle sizes from 6.3 to 153 mm. In general, the size distributions cover a range of well to poorly graded materials with Uniformity Coefficients from 3.3 to 32.8 respectively. With the exception of Marsal's Pinzanala sand/gravel, the particle shapes may all be described as angular.

Where "stiff" fill qualities are concerned, the packing density or porosity is of some significance. Under fully confined loading conditions, the compressibility of the material will depend strongly upon the "dense" or "loose" packing state. Such states are not only a result of the style of placement but they also depend upon the particle characteristics of the material being used.

As an illustration of this, consider the test results for the crushed Basalt of Al-Hussaini (1983). For the two relative densities of 97% and 74% (representing two placement techniques) and four different particle size distributions (see Figure 1), porosity has been plotted as a function of Uniformity Coefficient in Figure 2. In each case the minimum particle size is 0.6 mm. To effect any further reduction in porosity more fines will need to be added to the size distributions or, in other words, the Uniformity Coefficient must be increased. The Oroville data point (Marachi et al, 1972), also plotted in Figure 2, serves to illustrate this.

It was noted earlier that the addition of cement to RF, while providing some

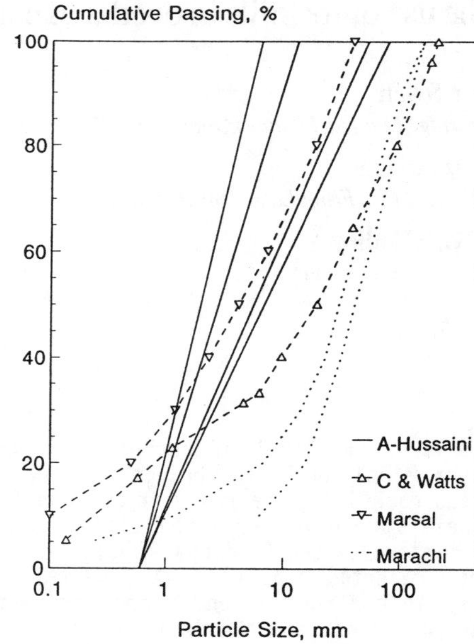

Figure 1 Rockfill particle size distributions taken from published data (see references).

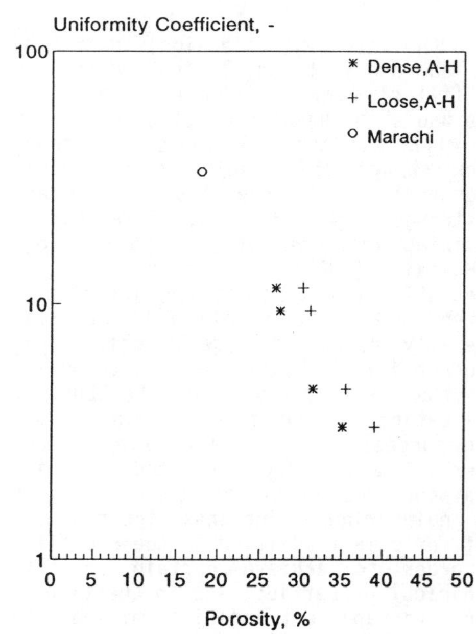

Figure 2 Data showing correlation between porosity and uniformity coefficient.

Table 1 Rockfill summarized data

SOURCE:	MARACHI, 1972	AL-HUSSAINI, 1983				MARACHI 1972	AL-HUSSAINI, 1983				MARSAL, C & WATTS 1973	1980
Description:	SANDSTONE	Dense-Pked BASALT				SANDSTONE	Loose-Pked BASALT				SANDSTONE	
u_{max} (mm)	152.7	76.2	50.8	12.7	6.3	152.7	76.2	50.8	12.7	6.3	200.0	38.0
C_u (-)	32.8	11.6	9.4	4.7	3.3	8.5	11.6	9.4	4.7	3.3	100.0	72.0
Passing -3/8 (%)	24.1	57.8	62.9	91.8	-	12.8	57.8	62.9	91.8	-	40.1	65.8
K-1.0 (GPa)	0.17	0.13	0.10	0.09	0.08	0.06	0.08	0.08	0.05	0.05	-	-
K-2.5 (GPa)	0.19	0.14	0.12	0.11	0.09	0.07	0.09	0.09	0.06	0.06	0.12	-
E-1.0 (GPa)	0.63	0.26	0.23	0.11	0.11	0.16	0.12	0.11	0.07	0.05	0.28	0.38
E-2.5 (GPa)	0.72	0.52	0.41	0.19	0.22	0.17	0.21	0.15	0.13	0.09	0.28	-
σ_c-1.0 (MPa)	5.50	6.00	-	5.15	4.92	4.95	5.54	-	4.72	4.37	5.24	3.83
σ_c-2.5 (MPa)	12.00	11.63	-	10.83	10.25	10.72	11.27	-	10.23	9.63	10.88	7.57
u_{50} (mm)	29.2	6.81	5.41	2.73	2.01	39.8	6.81	5.41	2.73	2.01	19.00	4.10
m_b (-)	1.93	0.794	-	1.28	1.06	1.4	1.09	-	1.41	1.39	1.74	0.33
a (-)	0.812	0.658	-	0.76	0.74	0.8	0.72	-	0.80	0.82	0.81	0.64
Porosity, n (%)	18.0	27.0	27.5	31.5	35.1	30.1	30.3	31.2	35.5	39.0	25.4	20.0
Fric,i-1.0 (dg)	43.3	46.7	45.6	42.3	41.1	40.4	44.3	43.2	39.9	38.6	-	-
Fric,i-2.5 (dg)	40.0	41.2	40.2	37.8	36.7	37.1	39.6	38.7	36.6	35.9	-	-

NOTES:

a). Uniformity Coefficient, $C_u = u_{60}/u_{10}$
b). K-x.x, Bulk Modulus under isotropic loading, confining pressure x.x MPa
c). E-x.x, Initial Tangent Modulus under deviatoric loading, confining pressure x.x MPa
d). σ_c-x.x, Compressive Strength at confining pressure x.x MPa
e). m_b & a, Modified Hoek & Brown failure parameters
f). Fric,i-x.x, instantaneous Friction Angle, confining pressure x.x MPa

cohesive strength may also be thought of as increasing the Uniformity Coefficient or reducing the porosity. For example, an 8% cement addition to Oroville RF will increase the Uniformity Coefficient from 32 to a theoretical high of 276, ignoring chemical effects.

2.2 *Compressibility data*

All data in the literature relating to rockfill compressibility has been derived from large-scale laboratory triaxial testing equipment, with specimen diameters in excess of four times the largest particle size. Unfortunately, for the original purposes of dam design, the data only exists for confining pressures of up to 3 MPa. Theoretical calculations for Strathcona Deep Copper have shown that, for the duration of mining, fill stresses may exceed 30 MPa (Swan et al 1989).

Using the published data, both tangent deformation modulus under deviatoric loading (E) and bulk modulus under isotropic loading (K) were calculated, where possible, for the two confining pressures of 1.0 and 2.5 MPa, see Table 1. In the case of the deformation modulus the data is plotted as a function of porosity in Figure 3. For comparative purposes both CRF and RF data is represented.

2.3 *Strength data*

Recently a comprehensive strength analysis of broken rock and rockfill triaxial data was made by Doruk (1991) using a modified Hoek-Brown failure criterion. The criterion, given by

$$\sigma_1 = \sigma_3 + (m_b\ \sigma_3/\sigma_c)^a\ \sigma_c$$

was used to predict strength at confining pressures of 1.0 and 2.5 MPa of the RF reference data of this study. Table 1 shows the modified Hoek and Brown parameters m_b, a and compressive strength at the two selected confining pressures. Friction angles at the two pressures are also provided in Table 1.

3.0 CEMENTED ROCKFILL DATABASE

Currently the only source of triaxial CRF data is from the work of Gonano et al (1978). As with RF, confining pressure is limited, with a maximum value of 2 MPa. From the test work they performed, five data sets have been examined in detail: the A4, C1, C2, C3 and C4 sample numbers.

Particle size distributions for these materials are shown in Figure 4. Cement contents for each of these sample numbers was 1.4%, 2.2%, 1.9%, 2.3% and 2.0% by weight respectively.

Unfortunately, due to the limited scope of the test work, the data set provided in Table 2 is incomplete compared to that

Table 2 Cemented rockfill summarized data

SOURCE	FALCONBRIDGE			MOUNT ISA MINES				
Description	KIDD	DEEP CU	CRAIG	A4/1	C1	C2	C3	C4
u_{max} (mm)	152.4	152.4	152.4	75.0	75.0	75.0	85.0	75.0
C_u (-)	19.95	33.11	26.10	196.96	213.82	190.70	135.45	213.82
Passing -3/8 (%)	32.50	33.00	18.80	28.10	47.70	45.70	47.70	47.70
K-1.0 (GPa)	-	-	-	-	-	-	-	-
K-2.5 (GPa)	-	-	-	-	-	-	-	-
E-1.0 (GPa)	-	-	-	-	0.40	0.43	1.27	0.84
E-2.5 (GPa)	-	-	-	0.48	1.15	0.63	-	2.13
Cement (% wt)	5.00	5.00	7.00	1.38	2.25	1.88	2.25	2.23
σ_c-0.0 (MPa)	6.40	5.40	6.34	1.26	1.40	1.26	1.40	0.69
σ_c-1.0 (MPa)	-	-	-	3.42	7.74	8.46	8.00	5.38
σ_c-2.5 (MPa)	-	-	-	5.97	-	-	-	-
u_{50} (mm)	20.80	23.40	42.50	22.54	10.64	11.67	10.40	10.40
m_b (-)	-	-	-	-	-	-	-	-
a (-)	-	-	-	-	-	-	-	-
Porosity, n (%)	26.5	31.5	-	21.9	16.0	16.7	16.7	21.9
Fric,i-1.0 (dg)	-	-	-	34.00	-	48.80	-	-
Fric,i-2.5 (dg)	-	-	-	-	-	34.10	-	-

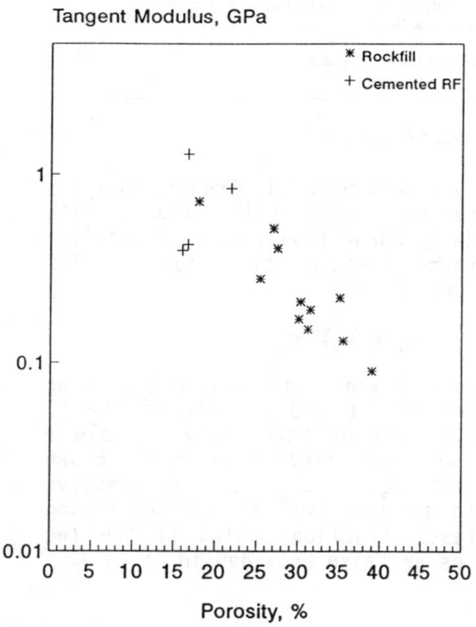

Figure 3 Tangent deformation modulus as a function of porosity, 2.5 MPa confining pressure.

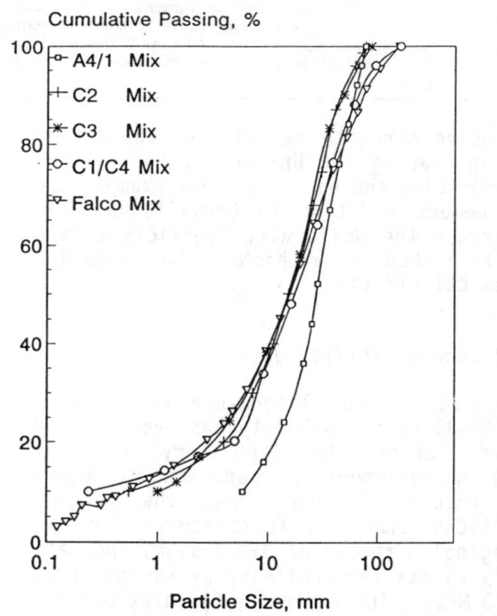

Figure 4 Particle size distributions for Mount Isa rockfills, Gonano et al (1978) and for Falconbridge Deep Copper rockfill, Arjang (1992).

for rockfills. The deformation modulus for these cemented materials is included in Figure 3 with the rockfill data.

4 LABORATORY TESTS

In an attempt to obtain deformation and strength data for RFs and CRFs for a larger range of confining pressures, Falconbridge Limited and CANMET recently performed a series of simple uniaxial compaction tests in rigid steel cylinders (Beaudry 1991) and (Arjang 1992). A further objective of these tests was the verification that void filling and not cement content controls the mechanical properties of confined rockfills for all practical purposes. Some results from the tests described by Arjang (1992) are now presented.

4.1 *Confined compression tests*

A 14 mm thick steel cylinder, length 102 cm, internal diameter 48 cm, was used for specimen confinement and testing. End drainage and low friction walls were provided for in its design, as well as axial and radial deformation measurements of the specimen. Before specimen preparation two samples of rockfill were analyzed for particle size using ASTM procedures. A total of 10 specimens was then prepared in various proportions of rockfill, unclassified tailings and cement, see Table 3. A standard compaction technique was used during specimen placement inside the steel cylinder and placed volumes and weights were noted.

Table 3 Specimen descriptions

SPECIMEN ID	MATERIAL, % by weight R.F.	U.T.	CEMENT	PORO-SITY %	REMARKS
RF-01	100	0	0	25.7	Wet R.F.
RF-02	100	0	0	25.5	Wet R.F.
RFT-03	90	10	0	20.3	R.F. &
RFT-04	90	10	0	19.4	un-class.
RFT-05	80	20	0	16.5	tails @
RFT-06	80	20	0	14.3	50% p.d.
CRFT-07	90	8	2	20.4	As above,
CRFT-08	90	5	5	24.7	cemented
CRF-09	98	0	2	25.4	As above,
CRF-10	95	0	5	21.9	no tails

Each specimen was loaded in compression at a constant rate of 2.5 kN/s to a maximum load of 9 MN (about 50 MPa axial stress). Incremental axial loads, axial deformation and radial deformation were observed for each specimen. Upon completion of each test the specimens were dried and weighed to obtain initial porosity.

4.2 *Results*

Particle size distributions of the rockfill show a poorly-graded material with Uniformity Coefficient in the range of 28 to 45. Axial stress-strain curves for all ten specimens are plotted in Figure 5.

Next, assuming specimen compaction occurs only as a result of reducing porosity, theoretical values of instantaneous porosity were calculated and plotted as a function of fill deformation modulus, Figure 6.

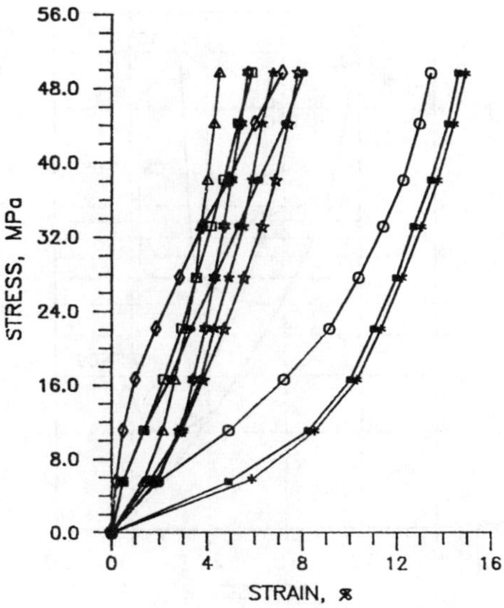

Figure 5 Stress-Strain curves from uniaxial compaction tests.
Legend, top right to left: RF-01; RF-02; CRF-09; CRFT-08; RFT-05; CRF-10; RFT-03; CRFT-07; RFT-04; RFT-06.

During axial compaction the radial confining pressure was estimated using

$$P = Eh\Delta/2\pi r^2$$

where **E** is the elastic modulus for steel, **h** and **r** are respectively the thickness and radius of the steel cylinder and **Δ** is the measured radial expansion.

For all ten tests fill deformation modulus is plotted as a function of axial stress in Figure 7.

Finally, Table 4 gives a summary of certain characteristics of the tested rockfills under confinement, calculated at 4% axial strain.

4.3 *Discussion*

From an examination of the stress-strain curves, Figure 5, a general distinction between the cemented and uncemented fills can be readily seen. Initially, the cemented fills have a higher stiffness which then decreases with increasing stress, indicating a breakdown in the cement bond. The specimen CRF-09, with only 2% cement, has a similar stress-strain curve to those of the pure rockfill specimens RF-01 and RF-02. This probably indicates that 2% cement is

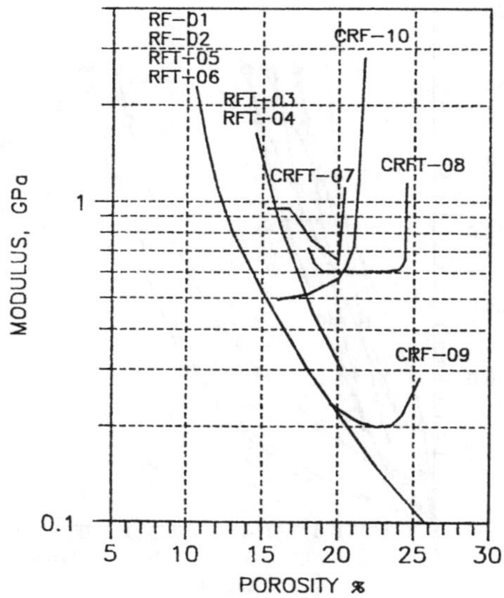

Figure 6 Fill deformation modulus as a function of instantaneous fill porosity.

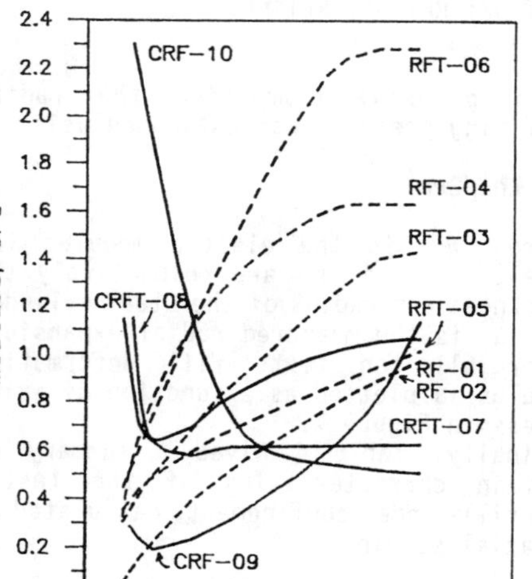

Figure 7 Fill deformation modulus as a function of axial stress.

insufficient to completely coat the rock fragments.

Again, referring to Figure 5, the group of rockfills with tailings addition only shows superior stiffness with axial strains in excess of about 1%. According to Figure 7 the same is true for confining

Table 4 Characteristics of tested rockfills at 4% strain.

SPECIMEN ID	STRESS (MPa) AXIAL	STRESS (MPa) RADIAL	DEFORMATION MODULUS (GPa)	ABSORBED ENERGY (MJ/m^3)
RF-01	4.2	0	0.10	0.02
RF-02	4.2	0	0.10	0.02
RFT-03	20.4	2.0	0.80	0.32
RFT-04	23.4	1.2	1.35	0.34
RFT-05	18.8	1.6	0.64	0.33
RFT-06	40.6	5.4	2.28	0.57
CRFT-07	31.4	3.4	0.96	0.63
CRFT-08	26.6	3.4	0.62	0.59
CRF-09	9.5	0.3	0.20	0.21
CRF-10	34.8	3.8	0.54	0.91

pressures in excess of about 20 MPa, with the exception of RFT-05.

A comparison of the porosity-stiffness relationships for published data (Figure 3) and for the present tests (Figure 6) shows general agreement. In both cases the initial or compacted porosity necessary to achieve a deformation modulus of 1 GPa lies in the range of 12 to 15%.

5 FILL PLACEMENT UNDERGROUND

Aside from questions concerning the achievable properties of rockfills, important issues arise with respect to when, where, what and how to place this material in cut-and-fill mining. In order to fully realize any benefits over conventional cemented hydraulic fill, strategic locations must first be found in the overall mine design. This done, equipment selection and detailed scheduling are crucial to achieving the engineering design.

As a start to understand the role of "stiff" rockfill and where to place it, calculations have been made using simple elastic models, (Hedley 1984), (Swan et al 1989), (Brummer 1991) and (Hedley 1992). These have all shown the viability of utilizing a stiff (1 to 3 GPa) fill material in association with the recovery of a central sill pillar in cut-and-fill mining. The greatest benefits diminish with increasing fill placement position from the sill pillar and with mining widths in excess of 5 m.

Conceptually, stiff fill placed during the initial cuts, proceeding from the sill elevation, eventually acts as a regional load-bearing pillar when reserves are depleted and the sill pillar is mined.

The ability of such a pillar to absorb seismic energy is considered important in alleviating rockbursts (Swan, 1990) and is the subject of continuing research both in Canada and South Africa (Gurtanca et al, 1989). Less easily quantified are the benefits of stiff rockfill on local ground stability. However, experience gained over the three operating years of Falconbridge's Deep Copper Mine suggests that stope back and wall conditions deteriorate at a markedly slower rate.

Methods of placing cemented or uncemented rockfill in the underground environment vary depending upon the mining method and available equipment. In narrow vein mining where a stiff fill makes the greatest sense, conventional bulk truck hauling and dumping of the cemented fill material is impractical. With the object of achieving a reasonable degree of compaction and tightness to the back and walls of the excavation, the ideal fill handling system must be compact and versatile.

A trackless example of such a system is that developed at the Deep Copper Mine in Sudbury (Fuchs and McDonald, 1992). Proceeding at the sill elevation, the method initially required the availability of underground grout tanks to supply 1 and 2 cubic yard scooptrams loaded with development waste. During transportation to the fill site some agitation and percolation of the grout occurred through the rockfill. (In this particular case the in-transit mixing of cement slurry with rockfill was found to provide significantly stiffer in situ properties than a flood and percolation method (Hopkins 1988)).

The critical task of in-stope compaction and tight fill requirement was achieved in part by the scoop operator's placement and bucket handling technique. After placing several buckets or more, a "filldozer" was used to ram the rockfill tight to the sides and back. This equipment features a 4 m long boom with a swivel-action 1.4 m wide blade mounted onto a converted 2 cubic yard scooptram.

6 CONCLUSIONS

Strong technical and economic arguments now exist to suggest the limited but strategic use of stiff rockfill in narrow-vein cut-and-fill mining. With confinement, a desirable stiffness of 1 to 3 GPa can be achieved by rockfills on the basis of the reported laboratory tests. Of critical importance is the aggregate design which must provide for an initial porosity of 12 to 15%. In the case of Falconbridge's Deep Copper rockfill, this has been achieved in the laboratory by the addition of approximately 20% (by weight) unclassified tailings. Where closure strains due to mining are expected to exceed 1%, the use of cement to achieve stiffness is questionable. Experience with the stope placement of these fill materials has emphasized the importance of specialized equipment such as "filldozers" to minimize voids.

ACKNOWLEDGEMENTS

The authors are pleased to acknowledge the strong support of this work from Strathcona Mine's Superintendent, John Vary, and the technical assistance of Mike Beaudry, Mine Technical Services, Falconbridge Limited.

REFERENCES

Al-Hussaini, M. 1983. Effect of particle size and strain conditions on the strength of crushed basalt. *Can. Geotech.* 20:706-717.

Arjang, B. 1992. Large-scale testing on compressibility characteristics of stiff backfill. *CANMET/MRL Division Report*, MRL:92-044 (TR).

Beaudry, M.S. 1991. Small-scale testing of rockfills. *Internal Report*, Falconbridge Limited, November.

Brummer, R.K. 1991. Algebraic modelling of energy balances in backfill/rock interaction. *Golder Ass. Report*, Canadian Rockburst Research Program.

Charles, J.A. & Watts, K.S. 1980. The influence of confining pressure on the shear strength of compacted rockfill. *Geotechnique*, 30, No. 4: 353-367.

Doruk, P. 1991. Analysis of the laboratory strength data using the original and modified Hoek-Brown failure criteria. *MSc Thesis*, University of Toronto, p. 123.

Fuchs, J. & McDonald, J. 1992. Mechanized narrow vein cut-and-fill mining at Falconbridge's deep copper project. *Proc. AIME-SME AGM*, Phoenix, Ariz., February.

Gonano, L.P., Kirkby, R.W. & Dight, P.M. 1978. Triaxial testing of cemented rockfill. *Tech. Report 72*, CSIRO, Australia.

Gurtanca, R.G., Jager, A.J., Adams, D.J. & Gonlag, M.D. 1989. The in situ behaviour of backfill materials and the

surrounding rockmass in South African gold mines. *Proc. 4th Int. Symp. Mining with Backfill*, Montreal.
Hedley, D.G.F. 1984. Utilization of backfill support in longitudinal cut-and-fill mining. *Division Report MRL 84-76(TR)*, CANMET, EMR, Canada.
Hedley, D.G.F. & Quesnal, W.J.F. 1991. Stiff Backfill Project at Lac's Macassa Mine. *MRD Technical Progress Report*, May 15th.
Hedley, D.G.F. 1992. Rockburst handbook for Ontario hardrock mines. *CANMET Special Report SP92-1E*, MRL, Canada.
Hopkins, P.H. 1988. Strathcona deep copper zone consolidated rockfill options - background data. *Internal Report*, Falconbridge Limited.
Marachi, N.D., Chan, C.K. & Seed, H.B. 1972. Evaluation of properties of rockfill materials. *J. Soil Mechs. Fdns Div., ASCE*, Vol. 1:109-200.
Marsal, R.J. 1973. Mechanical properties of rockfill. *Embankment Dam Engineering, Casag rande Volume:* 109-200, J. Wiley & Sons, N.Y.
Swan, G., Steed, C., Espley, S.J., O'Hearn, B. & Allan, G.R. 1989. Strathcona deep copper zone: Geomechanics investigation. *Falconbridge Internal Report*, p. 43.
Swan, G. 1990. Research into rockburst alleviation. *Mintech '90:*42-44, Ed. T.L. Carr.
Thomas, E.G., Nantel, J.H. & Notley, K.R. 1979. Fill technology in underground metalliferous mines. Kingston: *Int. Academic Serv. Ltd.*

Innovative Mine Design for the 21st Century, Bawden & Archibald (eds) © 1993 Balkema, Rotterdam, ISBN 90 5410 325 6

Paste fill and high density slurry backfill

D. Landriault & W. Lidkea
Mines Research, Inco Limited, Ontario Division, Copper Cliff, Ont., Canada

ABSTRACT: The mining industry has recently discovered the many operational and economical advantages of placing backfill underground without any excess run-off water. In the past the only backfill type that could provide this advantage was rock fill. Unfortunately this dry fill required a substantial capital investment to develop the initial raise/haulage system throughout a given mine and had higher operating costs than typical slurry fill systems. Consequently, rock fill was generally used at mines that did not have access to a suitable slurry fill material.

With the development of paste fill at a few mines in the world, the industry slowly realized that there was a potential to obtain many of the advantages of rock fill with less initial capital investment and lower operating costs. This has sparked the recent resurgence of paste fill research in the Canadian mining industry.

While this paper hails this progressive move, it also questions why similar research effort has not been spent on determining what can be done with the existing slurry fill plants to achieve the same end. The maximum pouring density of slurry fill plants for a given mine geometry has been overlooked as a potential solution to the excess backfill water problem.

This paper compares the rheology and material properties of paste fill and higher density slurry fills. The advantages and disadvantages of both are highlighted, along with the limitations for use of both.

1 INTRODUCTION

The Ontario Division of Inco Limited has been conducting research into the development of higher pulp density hydraulic fills that could be transported underground through existing borehole/pipeline slurry fill distribution systems for over a decade. The incentives behind this work were improved backfill cement hydration, reduced water placement and handling underground, shorter backfilling cycle time and superior backfill quality control in the stopes.

This research work has proceeded along two parallel routes. The first and most publicized was the development of a concrete-like ***paste backfill*** that could be transported underground via gravity at low flow velocities without the fear of particle settling and plugging of the pipelines. This high fines content fill type can retain water within its granular matrix to remain slightly beyond its saturation point for extended periods of time. This property allows the material to remain stationary in the underground delivery system for hours and then be remobilized by simply increasing the hydraulic head within the system. It also results in the paste producing no run-off water once placed in the stope.

The second line of research was to determine the maximum pulp density under which the presently used mill tailings and alluvial sand slurry backfills can be transported underground at each given mine. These fill materials were originally designed with a low fine material content to aid the drainage of the carrier slurry water through the granular matrix once the fill was placed in the mined out stope. This fill type, which was transported at pulp densities of 68 to 72 wt% solids, had a well defined critical flow velocity of 1.6 m/sec (5 ft/sec) below which the coarse particles settled out of suspension eventually plugging the pipe line.

It has been generally believed throughout the mining industry that the slurry pulp density should not be increased because it would produce much higher resistance to flow in the distribution system. This would result in in-line pressures that would exceed the operational limits of the pipelines in the delivery system or decrease the flow velocity below the critical limit causing the lines to plug.

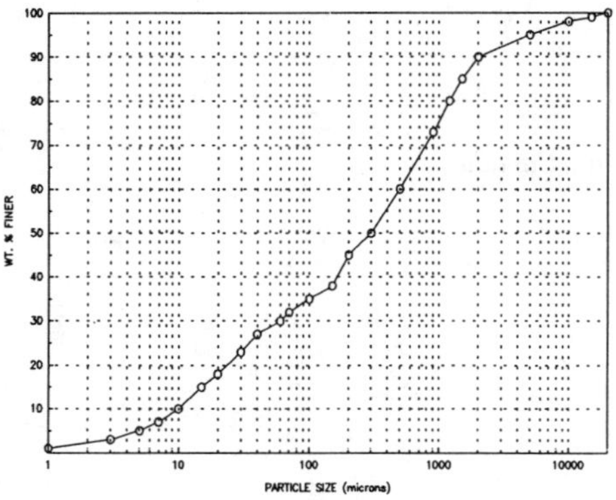

FIGURE; 1
THE PARTICLE SIZE DISTRIBUTION OF 20 MPa PUMPABLE CONCRETE

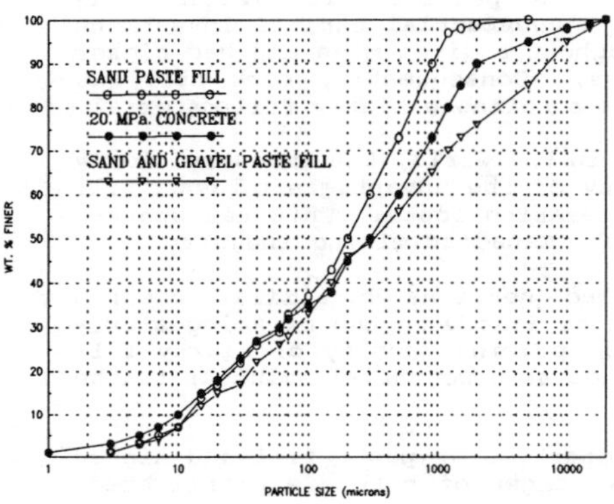

FIGURE; 2
THE PARTICLE SIZE DISTRIBUTION OF ALLUVIAL PASTE FILLS TESTED

Consequently, slurry backfill systems have been operated within these limits for over four decades without understanding the potential for ***high density slurry backfill*** transport through the existing delivery systems.

The major material difference between a paste fill and a high density slurry fill is the fine material content (-20 microns) which is what gives the paste the ability to retain water for long periods of time. High density slurry fills which have a lower fine material content will not retain water within their matrix unless kept in motion. Consequently, high density slurry fill can be gravity transported at solids contents similar to many paste fill materials but must be kept in motion, even if at low flow velocities, to prevent water separation in the pipe and line plugging. Once placed in the stope water will separate from the fill material to produce run-off water but at greatly reduced quantities compared to typical slurry placed fills.

2 PASTE BACKFILL

Paste fill research in Inco's Ontario Division has been focused on two different types of fill materials. There are mill tailings that are presently used as regular slurry fill by the mines adjacent to a mill tailings pipeline distribution system and alluvial sand which is used by the remaining mines as slurry fill. Because of the different processing problems associated with these two different paste fill types the research work was separated into two streams.

2.1 Alluvial Paste Backfill

The approach that was initially taken with this phase of the project was to utilize the technology and rheology developed by the concrete industry to mix and pump ready mix concrete. The size distribution of the alluvial paste fill was initially tailored to that of a pumpable 20 MPa (3000 psi) concrete (Figure 1). This high a uniaxial compressive strength is obviously not required or affordable in underground backfill. The Portland cement that made up the fine size fraction of the pumpable concrete was replaced by the finer fraction of the mill tailings. This material is normally removed in the cyclone overflow during the preparation of the tailings for use as slurry backfill.

The appropriately sized pit run alluvial sand was mixed with a thickened slurry of tailings fines and 3% slag cement by weight at Inco's paste fill test site located at the Garson Mine. Here the rheological properties of this paste fill mix were determined through surface and underground flow loop testing.

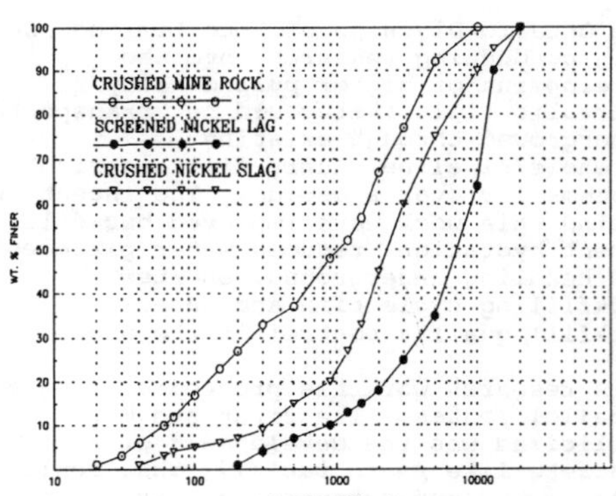

FIGURE; 3
AGGRIGATE WASTE MATERIALS FOR USE AS ALLUVIAL PASTE FILL

The test results demonstrated that alluvial paste fill could be transported from surface via gravity through a borehole/pipeline distribution system at sufficient rates to make it a viable alternative to slurry fill in the Sudbury area mines. When mixed with sufficient water to produce a 180 m (7") slump (North American standard 12" or 30 cm cone slump test) the alluvial paste could be placed through a 150 m diameter delivery system at rates of 200 dry tonnes/hr with a vertical head to horizontal transfer distance ratio of three to one. At ratios of one to one the delivery rates would drop to 100 tonnes/hr. With the orebody geometry of most of the Ontario Division mines, these placement rates were more than adequate to support the existing mining rates.

The paste fill was found to have a higher resistance to flow than the existing slurry fills, as would be expected, but the increased line pressures were manageable within existing pipeline delivery system operating pressure ranges. The above placement rates could be obtained without exceeding operating pressures of 6.9 MPa (1000 psi). The paste was allowed, during many of the tests, to remain idle in the boreholes and pipelines for over three hours without any problems occurring upon resumption of flow. With proper use of cement retardants this could be extended to some unknown length of time.

Purging the distribution system of paste was accomplished by the traditional method of flushing the lines with water as is done with typical slurry fill.

Extensive testing was done with pit run sands of various sizedistributions to determine how sensitive the flow properties of the paste were to material sizing. The results showed that the paste

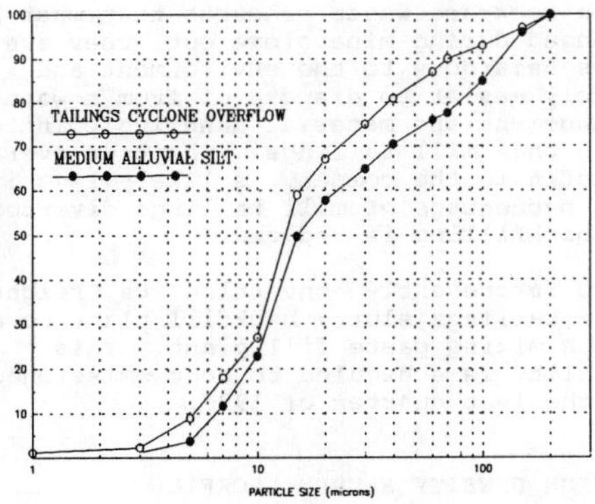

FIGURE; 4
TAILINGS FINES AND ALLUVIAL SILT FOR USE IN ALLUVIAL PASTE FILL

was very insensitive to sizing. Total sand and various ratios of sand and gravels could be used as paste fill provided the basic water holding fine material carrier was present to produce the paste consistency. The sand and gravel content could be varied with little effect on the flow properties of the paste provided that 15 wt% or more of the fill material was finer than 20 microns (Figure 2). There was a limit to the size of the coarse gravel fraction that could be added to the paste which was governed by a pipe diameter-to-largest particle size ratio of 6:1.

This work was extended to determine if some or all of the alluvial sand and gravel used in the paste could be replaced with mine waste such as crushed and/or screened smelter nickel dump slag and mine rock (Figure 3). The results showed that, within the size distribution ranges shown in Figure 2, partial and total substitution of the sand and gravels with these waste produces was possible.

In all the above work the fine fraction of the paste fill was made up of the fine mill tailings that was removed in the cyclone overflow during the preparation of slurry tailings fill. Tests were conducted to determine if this material could be replaced by a naturally occurring alluvial fine silt of a similar size distribution (Figure 4). Flow loop testing demonstrated that the fine silt was as good or better a carrier material than the fine tailings.

Production scale testing at the Coleman Mine saw paste fill transported through the existing hydraulic fill system down to the 3370 level (1030 meters below surface), 915 meters horizontally across the level to the bottom of the orebody, 61 meters vertically up a borehole to the top sill and 65 meters up a 10% grade to the stope. Even though much of the delivery system was made up of 98 m diameter boreholes and 127 m diameter pipe, fill rates of 70 tonnes/hr were obtained. The paste fill was purged from the delivery system after the test by simply adding water from surface to displace the paste from the system.

The existing slurry fill plants at the Coleman and Garson Mines are presently being renovated to allow them to supply these mines with paste fill on a production basis.

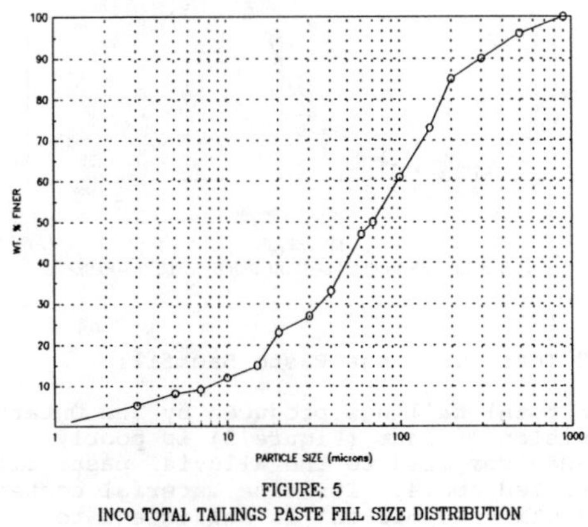

FIGURE; 5
INCO TOTAL TAILINGS PASTE FILL SIZE DISTRIBUTION

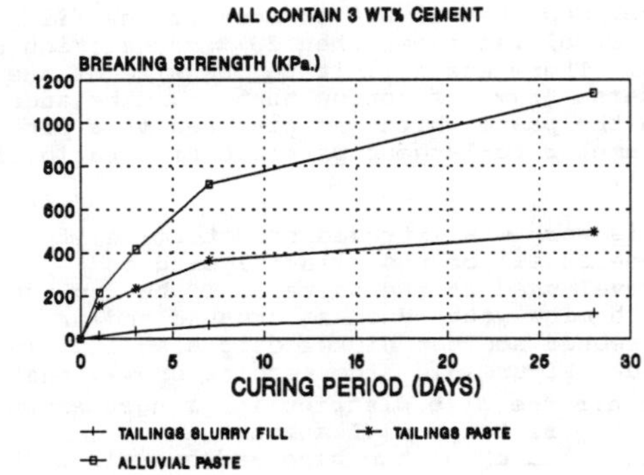

FIGURE: 6
ALLUVIAL VS. TOTAL TAILINGS PASTE FILL

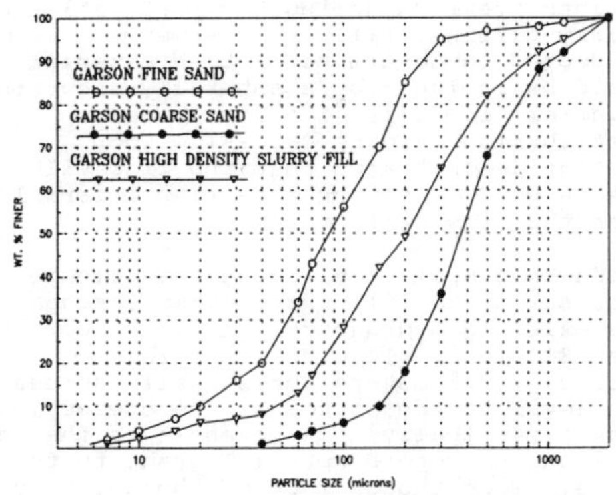

FIGURE; 7
GARSON HIGH DENSITY SLURRY BACKFILL

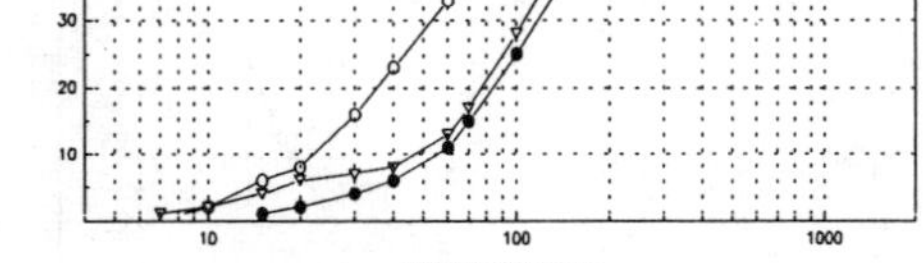

FIGURE; 8
CREAN HILL AND COLEMAN HIGH DENSITY SLURRY BACKFILL

2.2 Total Tailings Paste Backfill

The total tailings produced by the Ontario Division of Inco (Figure 5) is poorly graded compared to the alluvial paste fill reported above. Its fine material content (23 wt% - 20 micron) is sufficient to allow it to be transported as a paste. The flow loop tests conducted with this material demonstrated that its resistance to flow at a 180 m slump consistency was higher than that of alluvial pastes but was sufficient to produce fill placement rates adequate to maintain the present mining tonnages.

The main difference between the total tailings and the alluvial paste fills was the water content necessary to produce the 180 m slump consistency desired for higher flow rates. This consistency was achieved at solids contents of 85 to 88 wt% in the alluvial pastes and 79 wt% in the total tailings due to its higher porosity produced by a poorer gradation.

The negative result of this material property is that there is less benefit gained from the reduction in cement consumption when going from tailings slurry fill to tailings paste fill (Figure 6) as compared to the alluvial pastes. The poorly graded tailings paste also produces a much lower in situ fill stiffness. Testing was conducted in the flow loop to determine how much mine rock and nickel dump slag (Figure 3) could be added to the total tailings paste to decrease its porosity without negatively affecting the flow properties. It was discovered that up to 20 wt% solids of the coarse screened nickel slag and 40 wt% of the crushed mine rock could be added to the tailings before the resistance to flow increased substantially.

While there are material property benefits to be gained from the blending of these waste products they have to be weighed against the cost of diverting more mill tailings to the ponds rather than underground. Even though the slag and mine rock are waste products that must be managed during mine close out, they are less hazardous to the environment and likely easier to dispose of from a waste management and material handling point of view than mill tailings. The true overall benefit to the company is likely less if the percentage of mill tailings diverted to backfilling is reduced.

Inco is presently converting its Creighton Mine tailings slurry backfill plant to a batch mixing paste fill plant. This facility is scheduled to be commissioned in the last quarter of 1993.

3 HIGH DENSITY SLURRY BACKFILL

The research work into high density slurry fill was separated, as paste fill was, between alluvial and tailings materials. The earliest work started with attempts to pump these materials at high pulp densities through a laboratory flow loop using a centrifugal pump. Maximum solids contents in the 76 to 78 wt% range were obtained with those pumps which were abandoned for the larger dual piston positive displacement pump used in concrete pipeline transport. Flow loop tests with the piston pumps proved unsuccessful with the higher density slurries. They were thought to be impractical for underground borehole/pipeline transport until further gravity flow test work at the Garson Mine paste fill test site showed this not to be true.

3.1 Alluvial High Density Slurry Backfill

Flow loop testing was initiated at the Garson Mine site to develop a low water content fill that could be transported via pipeline short horizontal distances to backfill the 6.1 orebody (Whipple, Patterson, 1991). The economics associated with this small orebody dictated that material adjacent to the mine site be used to produce this fill. This eliminated the potential to produce paste fill due to the absence of any -20 micron material.

The surrounding area has an abundance of alluvial sand and gravel but is limited in sources of alluvial silt sized materials. The only quantity of fine material discovered close to the mine site was a fine sand which, because of its limited quantities, was blended at a 50/50 ratio with a medium sand (Figure 7) in an attempt to produce a fill that would be transportable to the 6.1 orebody at high densities.

The laboratory testing of this blended alluvial fill showed that when mixed at solids contents of 80 wt% the material demonstrated colloidal suspension of the fine particles and resembled a high slump paste if kept in motion. At 84 wt% solids the fill had a 180 m slump consistency but, unlike the paste fills, tended to bleed water quickly when left idle. When mixed in a beaker for several minutes water would be formed on the surface and the fill would no longer have the paste consistency attributed to colloidal suspension.

The surface flow loop tests using the positive displacement pump indicated that the blended sand could not be transported any significant distances at pulp densities above 80 wt% solids. Below this solids content the alluvial fill could be transported as a slurry but only at flow velocities above the critical limit. Different results were obtained, however, when the fill was tested in the gravity flow delivery system to the 6.1 orebody.

It was discovered that the blended fill was transportable at 80 wt% solids content and above when sufficient hydraulic head was created in the vertical section (55 m) to overcome the flow resistance in the 185 m horizontal section. The vertical hydraulic head was varied to decrease the flow rate to determine if this material had a critical flow velocity. It was

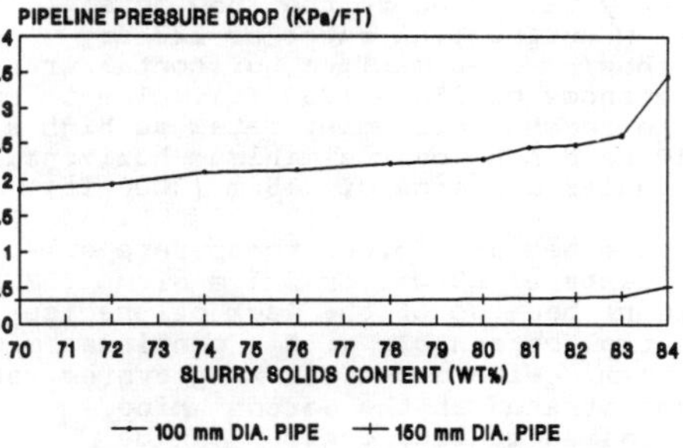

FIGURE: 9
HIGH DENSITY SLURRY FILL RESISTANCE TO PIPELINE FLOW

discovered that the blended fill could be moved at very low flow velocities without particle settling occuring in the horizontal pipe. The flow could be stopped in the line for short periods (5 min.) and then resumed with only slight initial increases in the operating pressures. Long periods of idleness in the line resulted in plugging.

This material property resulted in the fill producing run-off water once placed in the stope. This effect does not occur with paste fills. The quantities of water were minimal compared to typical slurry fills placed at 68 to 73 wt% solids.

The vertical hydraulic head for the 6.1 orebody was insufficient to produce the placement rates desired for the required production so a piston pump was used on surface to add energy to the system and to increase the placement rates. The combination of the pump and vertical hydraulic head proved to be sufficient to overcome the material's adversity to piston pump transport demonstrated in the surface flow loop tests. Approximately 55,000 tonnes of a batch mixed high density slurry fill with a 230 m slump were placed at rates up to 120 dry tonnes/hr in the 6.1 orebody over several years of operation without any significant delivery problems or pipeline wear.

The use of high density slurry fills has been extended to the Crean Hill and Coleman Mine. The Crean Hill Mine uses a coarse silt blended with a medium sand to produce the fill material shown in Figure 8 while the Coleman Mine uses a blend similar to the Garson material. Both mines have converted to high density slurry fill by simply introducing these blended sands to their slurry fill plants and re-programming the continual mixing systems to operate at the higher pulp densities.

The Crean Hill Mine places high density

slurry fill down to the 3650 level (1100 m below surface) at rates of 220 dry tonnes/hr over maximum horizontal transfer distances of 230 m (750 ft). The Coleman Mine reaches placement rates as high as 310 tonnes/hr over a minimum horizontal transfer distance of 915 m (3000 ft).

Both mines are forced to operate at solids contents of 80 wt% or 305 m slump (12 " slump) because of the lack of precise control obtainable with a continual mixing system. With a batch mixing system, as demonstrated at the Garson Mine, increased quality control can be maintained and higher placement densities obtained. With the increased quality control greater savings could be obtained from larger reductions in cement consumption and mine de-watering.

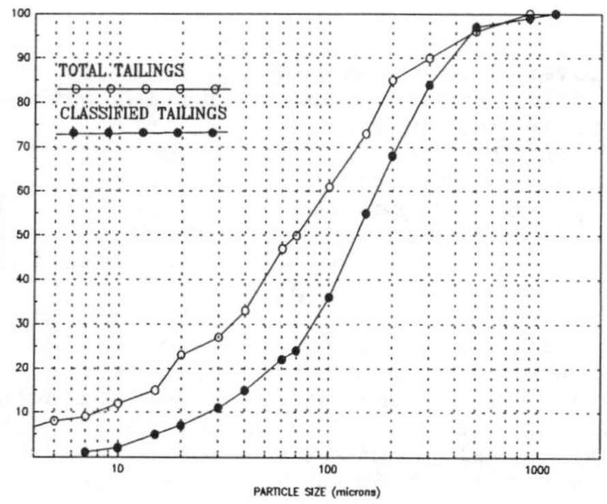

FIGURE; 10
CLASSIFIED TAILINGS HIGH DENSITY SLURRY BACKFILL

The disadvantage of higher placement densities would be a reduction in the high placement rates stated above. As demonstrated in Figure 9 the resistance to flow increases minimally as the alluvial fill goes from a slurry to a high slump high density slurry and substantially as the slump approaches 180 m at 84 wt% solids. A batch mixing system would allow the mines to produce a quality product at any given solids content to meet the demands of the given underground backfilling situation.

The paste fill systems being installed at the Coleman and Garson Mines will be designed to also accommodate the mixing and continual underground flow of alluvial high density slurry fills

3.2 Tailings High Density Slurry Backfill

While conducting flow loop test work with the total tailings paste fill, attempts were made to pump cyclone underflow classified tailings around the surface loop with negative results (Figure 10). It was assumed that classified tailings could not be transported underground at high densities as a result of this work. This was believed until further testing was done in the existing Creighton Mine backfill delivery system with the cyclone underflow material.

Early test work to develop the compaction storage tank de-watering system for total tailings paste fill at the Creighton Mine demonstrated the plant's ability to de-water large quantities of coarser classified tailings to a maximum 79 wt% solids (180 m slump). This system was used to determine the maximum pouring density that classified tailings could be placed underground through the mine's backfill delivery system by slowly increasing the solids content in the continual mixing system as backfill was poured underground.

Underground test pours were closely monitored with pressure transducers on the 1500 and 6000 level. The results demonstrated that the classified tailings behaved like a high density slurry exhibiting colloidal suspension when kept in motion. The classified material was placed throughout the mine at a solids content (78 wt%) and consistency (250 m slump) approaching that of the total tailings paste fill and at a similar resistance to flow. Placement densities equivalent to that obtained with total tailing paste fill can be obtained by converting from a continual mixing system to a batch mixing.

The main difference between the two fill types is the ability to stop flow, leaving the material ideal in the line and not having any problems in resuming flow, and the amount of run-off water once the fill is placed in the stope. Classified tailings, like the alluvial high density slurry fills, must be kept in motion during delivery to eliminate the potential for the fill to bleed water and plug the line.

This same property resulted in run-off water from the fill once it was placed in an underground stope. The amount of water run-off has been estimated at 15 wt% of what is presently experienced with the existing tailings slurry fill placed at 73 wt% solids.

The batch mixing system being installed in the Creighton backfill plant to produce tailing paste fill will also be capable of handling high density classified tailings slurry.

4 ALLUVIAL PASTE FILL VERSUS ALLUVIAL HIGH DENSITY SLURRY FILL

The use of either a paste or high density slurry backfill in a mine can result in major economic and operating benefits over conventional slurry backfills. There are advantages and disadvantages to each and the decision to use one over the other or whether to use both will generally be dependent on the potential fill materials available to the mine site, the fill placement rates required by the mine, the basic mine geometry and the capital investment required to produce the different fill types.

4.1 Backfill Rheology

The major difference in the fill rheology between a paste and high density slurry fill is the length of time they can be left idle in a pipeline without being able to resume flow. The high fines material content of the paste will allow idle periods of many hours while high density slurries in general should not be allowed to come to a stop in the pipeline. The operating benefits of this paste fill property are obvious but how much advantage to the overall mining cycle can be gained from it is mine specific.

The placement rates attainable with high slump, high density slurry fills (220 to 310 tonnes/hr) are the result of being able to maintain a slurry fill flow velocity (2.4 to 3.45 m/sec in a 150 m pipe) with a higher solids content. As indicated in Figure 9, this can only be realized at reasonable operating pressures if the fill consistency is maintained at a 300 m (10") slump or higher.

This is not a problem with high density slurries because their lack of coarse aggregate allows them to be transported at these high slumps even at low flow velocities without segregation of the fill in the line. This is possible due to the colloidal suspension of the fine particles in the fill mix. If a sufficient quantity of fine material (medium to coarse silt) is not present in the high density slurry fill or too great a percentage of coarse material is present, then colloidal suspension in the slurry will not occur increasing the resistance to flow dramatically.

The paste fills that contain coarse aggregate or gravel tend to segregate at high slumps and low velocities resulting in line plugging. To prevent this segregation the paste must be transported above a given velocity which somewhat reduces the advantage of paste fill usage. Pastes without the coarse fraction, like the total tailings described above, could be used like the high density slurries and mixed at high slumps to allow greater tonnage rates to be placed.

The pipeline wear rate associated with high density slurry transport at high velocities has not been defined in the existing operating systems at the Crean Hill or Coleman Mines due to insufficient operating time. The wear, however, is expected to be less than that associated with conventional slurry transport at these flow velocities due to the reduction of particle impact with the pipe walls. The dense packing of the particles and colloidal suspension in the higher density slurries reduce particle movement and wall impact during flow.

The pipe wear associated with paste fill transport has been nil in all production scale tests conducted by Inco with the different paste fill materials. The natural physical properties of plug flow, where the fine particles migrate to the pipe walls to create an annulus of low friction fine material to aid flow, will result in minimal pipe wear compared to the high particle impact with pipe walls associated with existing slurry fill transport.

4.2 Material Properties

The material properties of both fill types are partially governed by their particle size distributions which greatly define the material's underground placement porosity. The high density slurry fills are restricted to a given coarse particle size fraction to maintain colloidal suspension. Consequently they range between a maximum placement density (150 m slump) of 78 to 82 wt% solids depending on their particle size distribution.

The maximum placement density of paste fills ranges between 75 to 88 wt% solids depending on the percentage of -20 micron carrier material present and the width of the particle size distribution. What defines the solids content of both types of fill at a 150 m slump consistency is primarily the volume of voids or porosity of the saturated fill matrix. The porosity of the high density slurry fills tested at Inco range from 38% to 32% while the paste fills vary from 35% to 25%.

The material properties of the different fill products are what define both the economic and operational advantages of each. The maximum reduction in cement consumption and reduced mine water removal will be gained by a lowest porosity fill material similar to the paste fills shown in Figure 2. This is demonstrated by the compressive strength test results shown in Figure 6. Twice the strength gain is generated with the same cement content by the well graded paste fills as compared to the poorly graded total tailings paste and classified tailings high density slurry fill. This superior strength is primarily the result of the lower water/cement ratio obtainable in this lower porosity paste fill matrix.

The lower porosity paste fills will also be least prone to produce run-off water when placed underground. The higher porosity pastes may produce some minor water loss during long term placement but the high density slurry fills will readily bleed water due to their lack of -20 micron material. The maximum water produced by any of the high density fills will still only be approximately 15 wt% of that presently running off the existing slurry fills.

Backfills with lower porosity also produce superior ground support properties underground. The lower the fill porosity the stiffer the fill material which will minimize stope closure once the fill is placed. The well graded paste fills will produce material stiffness properties approaching that of rockfills which have demonstrated superior ground support properties in many mining situations.

4.3 Process Systems

The cost of producing, mixing and transporting a paste or high density slurry fill underground is material dependent and site specific. When alluvial materials are used for either fill type the process problems become those of excavation and material transport, especially in cold climates, and blending.

Alluvial high density slurry material which uses a coarse silt or fine sand blended with a coarse sand can generally overcome the cold weather material handling problems with the finer materials by creating sufficient stockpiles of the blended product in the warm months to get them through the winters. The blended product has proven to be manageable in cold weather (-30 deg C) whereas the fine silt and sand by itself is not.

The process system required to produce the high density alluvial slurry could be that of an existing continual mixing slurry plant or the off-the-shelf concrete technology used to control feed and batch mix in a standard ready mix plant.

The fine alluvial material required to produce an all alluvial paste fill is generally a fine to medium silt which poses much more difficult material handling problems. In both cold and warm weather this material is extremely cohesive when excavated. Consequently, it must be dried to create a fine powder or slurried to 60 to 75 wt% solids to allow it to be added to the coarser sands and gravels to produce paste fill at the rate required for typical mine placement tonnages.

The cost of drying such materials is usually prohibitive. The process technology and equipment required with this fine material are available from the clay mining industry.

If the fine fraction of the alluvial paste is similarly in size distribution to mine tailings from a disposable area the process problem is similar. However, if the fine tailings are recovered from the mill or down stream plant then they are usually in slurry form and the process problem becomes one of de-watering. A standard rake thickener is generally adequate to de-water the tailings to the appropriate solids content to allow it to be mixed with the alluvial materials to produce a paste fill.

When a total tailings is used as a paste fill the de-watering process becomes more difficult and de-watering is traditionally done in two stages using a rake thickener and filters to produce a cake. An alternative one stage system is being researched and developed at Inco which combines material storage and compaction de-watering in a single system called a compaction storage tank. This approach to de-watering tailings in order to create paste fill, along with other methods being researched at Inco, will be the subject of a subsequent paper.

The same processing problems exist for the creation of high density tailing slurry as they do with tailings paste fill except less emphasis has to be placed on capturing as great a percentage of the -20 micron size fraction. The research and development associated with the de-watering of tailings to produce paste fill is also being extended to produce high density tailings slurry fill at Inco.

5 CONCLUSIONS

The development of paste and high density slurry fills will give much of the mining industry more opportunities to optimize mining methods while decreasing operating cost. The most cost effective fill type for a given mine, however, is still site specific. It depends on the available fill materials, the mine geometry and capital available to put the necessary process systems in place to produce a specific fill type.

The ideal situation for any mine is to have a backfill system designed to de-water, mix and deliver slurry, high density slurry and paste fill. The design of such a plant is the present focus of Inco's Ontario Division backfill research efforts.

REFERENCES

Whipple, R. & Patterson R. 1991. High Density Fill at Garson Mine. CIM 10th Underground Operators Conference.

Innovative Mine Design for the 21st Century, Bawden & Archibald (eds) © 1993 Balkema, Rotterdam, ISBN 90 5410 325 6

ARAPREE: A cuttable cable-bolt support system

F.P. Hassani & U.H. Khan
Department of Mining and Metallurgical Engineering, McGill University, Montreal, Que., Canada

ABSTRACT: ARAPREE is an ARAmid PREstressing Element made of Twaron continuous aramid fibers pre-impregnated in epoxy resins. Arapree tendons are manufactured in both round and strip shapes with sanded and helical surface profiles. Arapree tendons have been used as tension elements in the prestress concrete industry. The modulus of elasticity of Arapree is 125 GPa (steel = 200 GPa) and the ultimate tensile strength based on net cross-section of fiber contents is approximately 2800 MPa (prestressing steel = 1700 MPa). Pull-out tests have been conducted on fully grouted Arapree tendons in the conventional steel cable constraints. Additionally, modified configurations of Arapree cable such as double parallel cable and zig-zag cable were also tested. The results show that the pull-out capacity and ductility of this support system are in close proximity to existing steel cable supports. The advantages of Arapree over steel cable are discussed.

1 BACKGROUND

Interior rockbolt reinforcements were first applied to the mining industry in a slate quarry in North Wales in 1872 and were introduced to North America at about the turn of the century. It took over fifty years before bolting was fully accepted as one of the primary underground support systems. In the past twenty years, rockbolting technology has evolved from point anchor mechanical bolt to full contact (fully grouted) bolts as well as friction rockbolts such as split sets and Swellex. Such rockbolts have their own particular mechanical behaviour with associated advantages and disadvantages. It is important to bear in mind that interior support systems which can be employed as passive or active support generally rely on their support element to have been fabricated from steel.

Steel rockbolt as well as cable bolt support systems have contributed greatly to the design and support of both underground and surface rock excavations. However, with demand of industry for more efficient support systems with variable mechanical behaviour as well as a need for cuttable support systems for new, automated mine designs, alternative material for the support element is required. The new material needs to overcome certain constraints and disadvantages attached to present rockbolt or cable bolt systems.

As part of an ongoing research program into support systems for the mining industry by the Rock Mechanics Research Group at McGill University, an extensive research and survey was carried out to evaluate alternative support elements for both hard and soft rock excavations. During the preliminary stage of this work, it was recognized that composite materials could be the likely candidates to replace steel reinforcing elements. Over 350 manufacturing companies around the world were contacted and their products were evaluated (Hassani & Khan, 1991). On the basis of this work, a number of composite materials were identified and further laboratory investigation was conducted to assess their physical and mechanical properties prior to the field investigation.

This paper, for the first time, introduces a special composite material called "Arapree" to the mining industry which has the potential to replace steel rockbolts as well as cable bolts.

The chemical constituents and manufacturing process is briefly introduced and detail of physical and mechanical properties are given. Further research and development for application of such composite materials in the mining industry is also discussed.

2 INTRODUCTION

Twaron is an aramid fiber produced by Enka BV and N.V. Noordekiijke, a joint venture of the Netherlands and Germany. The impregnation of Twaron fibers into epoxy resin resultes in Arapree (ARAmid PREstressing Element) tendons. The manufacturing process is braiding, in which, long fibers are pulled from different spindles and combined to a pre-selected and pre-designed pattern and then passed through an orifice of selected dimensions. The orifice or opening determines the shape of the final product, for example, circular or rectangular. Then in order to obtain various properties, besides dimensional stability, the final product is impregnated into epoxy. In the final stage, the desired surface profile pattern is improvised.

The reasons behind the selection of epoxy resin as the bonding matrix for Twaron fibers are as follows (Schurhoff and Gerritse, 1986):

- Epoxy resins show adhesion to a wide variety of fibers.
- Epoxy resins offer a large range of properties, due to the wide variety of possible formulations of resins and curing agents.
- The curing reaction does not release volatile agents or water, therefore very little shrinkage occurs.
- Resistance to many chemicals, for instance, alkalis.
- Dimensional stability is high.
- Thermally stability is limited and lower than that of aramid fibers.

The properties of epoxy used in the manufacturing of Arapree tendons are shown in Table 1.

Table 1: Properties of epoxy (Schurhoff and Gerritse, 1986)

Thermal stability	150°C
Chemical stability	pH 13
Shear strength	> 30 MPa
Elongation at breaking point	>4 %
Viscosity at RT	< 2500 mPas

The combination of Twaron fibers and epoxy resin result in a composite product which has:

- High strength and high stiffness
- Low creep and relaxation
- High stability in many environments, such as exposure to sea water, oil, calcium hydroxide, chlorides
- Moderate thermal stability
- Low density
- Non-conductivity
- High tensile fatigue resistance

There are other fibers also available in market as shown in Figure 1. Among these fibers, the polymer fiber should be excluded because of its low modulus of elasticity. Polymers have high creep and temperature sensitivity, as well. Carbon fibers (boron fiber, as well) are very insensitive to aggressive environments and are too brittle and do not give adequate warning prior to failure. They are expensive and their applications are limited to aerospace, sports and tools industries. Therefore, the glass fibers and aramid fibers are the likely candidates to be employed as support elements.

In general, glass fibers do not have a great resistance against aggressive environments, however, the recent developments of special techniques, such as special cover treatment of the glass fiber-reinforced tendons offers design resistance to severe environments (i.e. Weldgrip; (Hassani and Khan, 1991)). The aramid fibers, however, have all the properties comparable to steel strands.

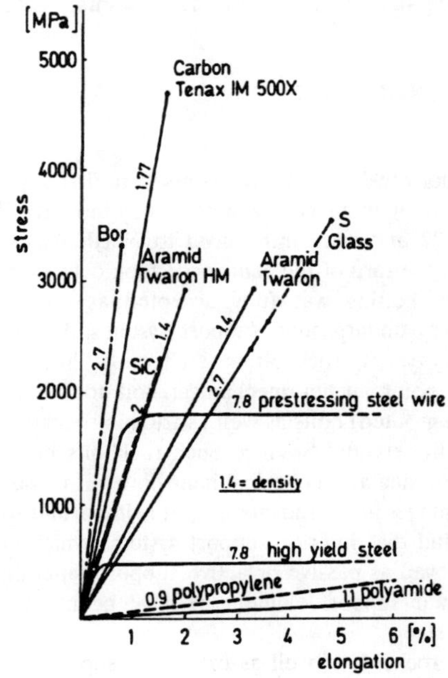

Figure 1: Stress-strain diagram of various fibers (Gerritse et al., 1987)

3 PHYSICAL PROPERTIES OF ARAPREE

The following are the physical properties of Arapree HM:

Geometry

Fibers: -Aramid, type Twaron, endless
- Cross-section of filaments: round, diameter 12×10^{-3} mm.
- shape : rectangular and round, endless.
- type: round and strip like tendon, containing multiples of 100,000 Twaron filaments in epoxy resin, orthogonally anistropic.
- effective aramid cross-section: 11.1 mm^2 per 100,000 filaments.
- composite cross section: n x [1.5 x 20]mm^2 = n x 30 mm^2
- surface pattern: sanded and regular pattern of small nobs.
- Gravimetric composition 40 - 50 % Twaron HM; 60 - 50 % resin
- Volumetric composition 35 - 45 % Twaron HM; 60 - 55 % resin.

4 MECHANICAL PROPERTIES OF ARAPREE

Twaron fibers, based on stiffness, are manufactured in two grades: Twaron, with a modulus of elasticity in the range of 70 kN/mm^2 and Twaron High Modulus (HM) with an E-modulus of approximately 130 kN/mm^2. For the reminder of the paper data refers to Arapree High Modulus material. The mean values of mechanical properties of Arapree HM at 20^oC are given in Table 2. The stress-strain behaviour of Arapree under the name of Aramid Twaron is shown in Figure 1. The cross-sectional area, breaking force and cost of commercially available Arapree tendons are given in Table 3.

Table 2: Strength and Stiffness Properties of Arapree

	E (GPa)	Failure Criterion
Short Term	125.0	f_{st} = 2800 MPa e_{st} = 2.24 %
Long Term	125.0	f_{lt} = 2400 MPa (2800 X 0.85) e_{lt} = 1.92 %
Density	1250 kg/m^3	
Poisson Ratio	0.38	

5 CREEP AND RELAXATION

Creep and relaxation are interrelated material properties of a structural element and both are functions of stress, strain and time. Creep can be described as change in strain as a function of time at constant stress. Relaxation is due to change in stress as a function of time at a constant length. Figure 2 describes the concept and possible relation between creep and relaxation at a certain level (Gerritse, 1987). Since steel is supposed to be replaced by aramid fibers, therefore, steel is used as a standard to compare various properties.

Three possible relationships between creep and relaxation are given by Gerritse et al. (1987). If

Table 3: Different types and shapes of Arapree tendons and associated cost

ARAPREE	
mean axial tensile strength, f_u	2800 MPa
modulus of elasticity, E	125 GPa
mean failure strain, M	2.3 %
ARAPREE	f 100000 round sanded shape
dimension	d = 3.9 mm
Fibre cross-section	A_f = 22.2 mm^2
mean ultimate force	F_u = 33.3 kN (3.33 Tonnes)
Cost	C$ 1.37 per metre
ARAPREE	f 200000 round sanded shape
dimension	d = 7.9 mm
Fibre cross-section	A_f = 22.2 mm^2
mean ultimate force	F_u = 66.6 kN (6.66 Tonnes)
Cost	C$ 2.60 per metre
ARAPREE	f 100000 rectangular shape
dimension	b/d = 1.5/20 mm
Fibre cross-section	A_f = 11.1 mm^2
mean ultimate force	F_u = 33.3 kN (3.33 Tonnes)
Cost	C$ 1.37 per metre
ARAPREE	f 200000 rectangular shape
dimension	b/d = 2.6/20 mm
Fibre cross-section	A_f = 22.2 mm^2
mean ultimate force	F_u = 66.6 kN (6.66 Tonnes)
Cost	C$ 2.60 per metre
ARAPREE	f 400000 rectangular shape
dimension	b/d = 5.0/20 mm
Fibre cross-section	A_f = 44.4 mm^2
mean ultimate force	F_u = 133.2 kN (13.32 Tonnes)
Cost	C$ 5.40 per metre

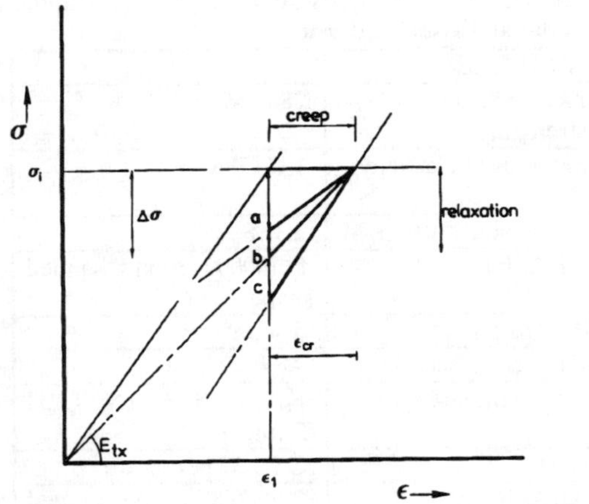

Figure 2: Possible relations between creep and relaxation (Gerritse, 1987)

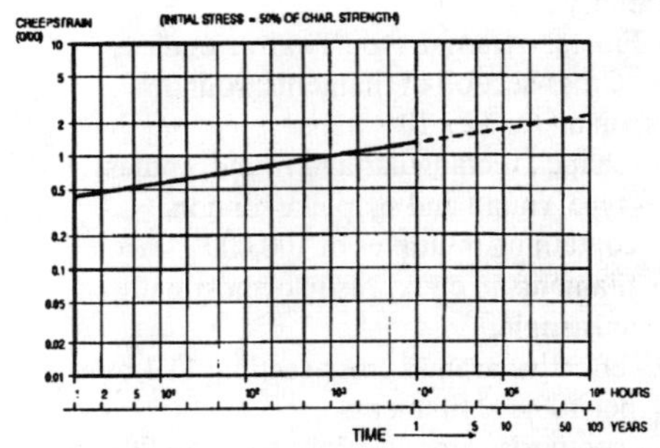

Figure 4: Creep behaviour of Arapree in air (Gerritse and Werner, 1988)

relaxation is simulated by a creep test whereby, after a certain time interval, the measured creep is compensated by a decrease in load, then the obtained relaxation will be according to (a, b, c) in Figure 3. Figure 3(a) reveals the behaviour of steel. This demonstrates that the initial stress level gives a more than proportional increase in creep strain ($\varepsilon_{cr2}/\varepsilon_{cr1} > \sigma_2/\sigma_1$). Relaxation data illustrates a comparable proportionally, from ε_{cr1} to ε_{cr2}. Conclusively, it can be said that, contrary to prestressing steel, an increase in initial stress produces constant or even decreasing relaxation percentages for aramids for the same time interval, $\Delta\sigma_2/\Delta\sigma_1 \leq \sigma_2/\sigma_1$; see Figure 3 (b & c).

Figure 4 shows experimental results presented to evaluate the creep of Arapree. The relationship is approximately linear between applied stress and the creep strain. The values given are determined at an initial stress level of about 50 % of the characteristic strength. Moreover, the test results conducted on relaxation behaviour show that polymeric materials are more or less independent of the applied stress level. Figure 5 shows relaxation of Arapree in an alkaline solution.

6 ANCHORAGE BOND IN CONCRETE

The tests were conducted in the Akzo laboratory to find the bonding strength between the strip shape Arapree and concrete. This figure shows that an embedment length of 6 cm is enough to obtain the breaking strength of strip Arapree. See Figure 6.

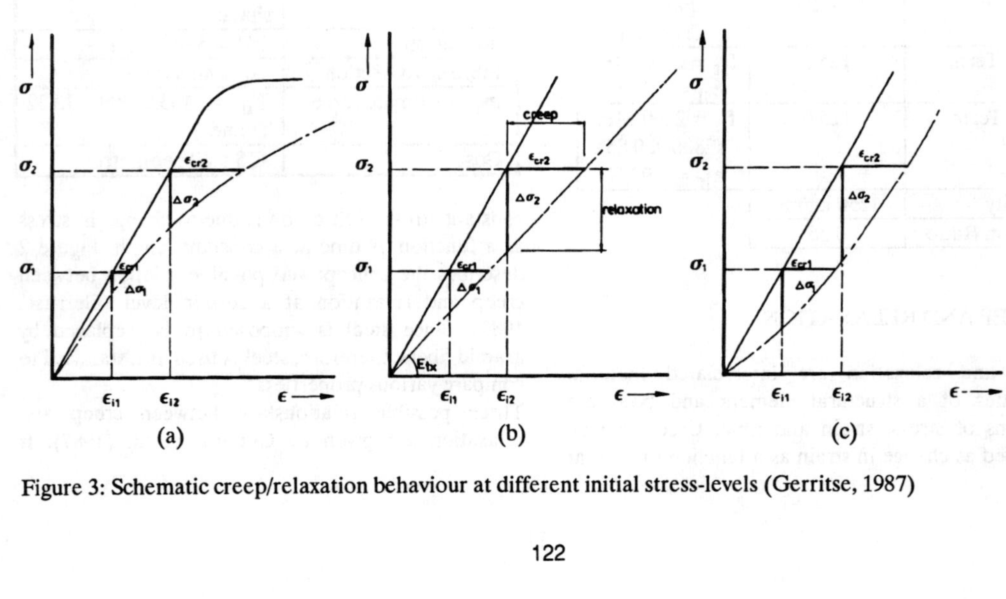

Figure 3: Schematic creep/relaxation behaviour at different initial stress-levels (Gerritse, 1987)

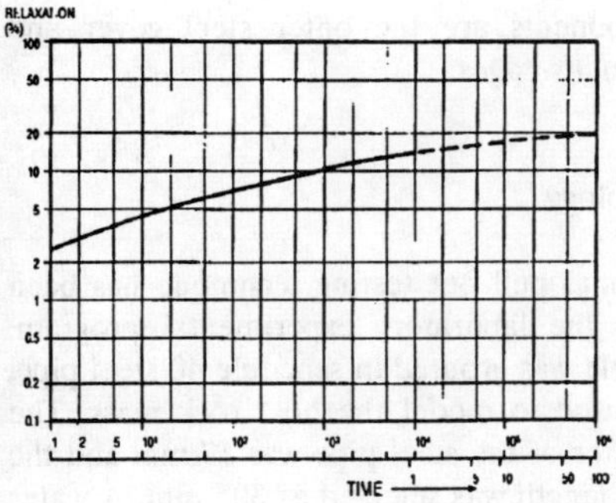

Figure 5: Relaxation of Arapree in alkaline solution (Gerritse and Werner, 1988)

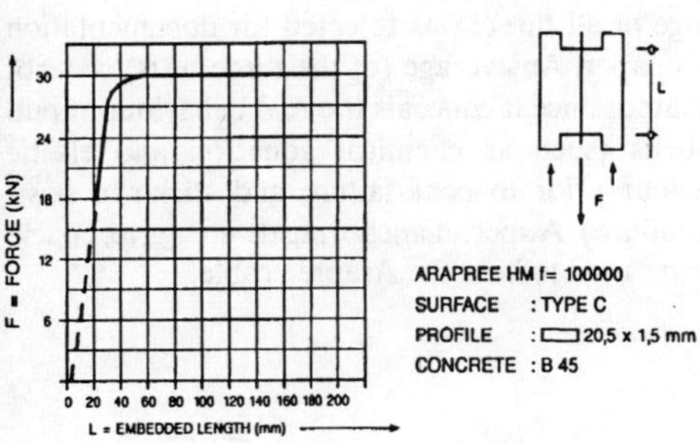

Figure 6: Critical embedment length of Arapree (Gerritse and Werner, 1988)

However, the testing done in mining constraints is presented in the forthcoming sections.

Another series of experiments on concrete elements were conducted in the Akzo laboratory to investigate the behaviour of Arapree in actual practice under flexural loading (Gerritse et al. 1986). Concrete planks were centrically and eccentrically prestressed by Arapree tendons. The cross-section area of the Arapree tendon was 0.25 x 20 mm. Figure 7 shows the loading pattern on planks and the deflection in the planks due to this applied loading. In spite of the very small lever arm of the vertically placed tendons, the tests proved that the tendons were ultimately stressed over the whole section at failure. The test results also showed an ideal pattern of finely divided cracks in bending failure. The slow propagation of such a pattern of cracks demonstrated that adequate warning can be anticipated prior to the failure of the structure supported by Arapree cables.

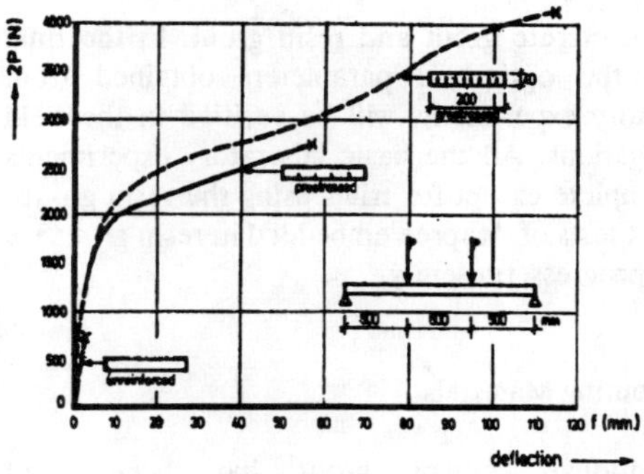

Figure 7: The applied loading on concrete plank and the resulting deformational behaviour (Gerritse et al., 1986)

7 MINING APPLICATIONS

The general mining applications of Arapree tendons could be numerous. The basic potential use of Arapree can be as tensile bearing elements, both as rockbolts and cable bolts. Since Arapree is flexible enough to be coiled in a diameter of approximately 1-meter, therefore, it can be used as flexible cable where adequate head room is not available. Arapree cables can also be used as lacing; to hold loose rock pieces failed due to rock-skin-failure. The failure of pillars due to uniaxial compressive loading can be prevented or slowed-down with the help of Arapree tendons. This can be achieved by winding the Arapree around the pillar. Similarly, continuous Arapree cables can be connected between one drilled hole and another, or U-shaped configurations are also possible. This technique will replace accessories such as nuts, plates, lacing, etc. Arapree cables have the potential to pre-support cut-and-fill stopes.

The major field of Arapree use, which can be foreseen now, is its facility to be used as "yieldable" support. This idea, however, needs much design improvisation at the manufacturing process and further research in mining constraints.

8 EXPERIMENTAL PROGRAM

To evaluate the performance of Arapree in mining environments, a detailed investigation program was initiated. In the first phase, a laboratory investigation was conducted in the conventional steel cable constraints for comparative purposes. The grouting materials were selected to be conventional grout (cement type 10 + water), high strength cement grout, shotcrete grout and resin grout. In the final phase, the optimized parameters obtained from laboratory experiments will be verified in the field investigations. All the basic laboratory experiments are complete except for trials using the resin grouts. Pull out tests of Arapree embedded in resin grout are under progress, presently.

8.1 Grouting Materials

Conventional cement grout has been used successfully for some time as bonding agent for cable bolts and rockbolts. But, there are two problems attributed to the cement grout, these being:

a) Conventional cement grout takes atleast 7 days to cure. The effectiveness of a support to control the excessive deformation of a rock mass could be lost due to this delayed time.
b) The stiffness of the grout is not sufficiently high to transfer the load effectively in the weak rock environments.

Considering these two problems, high strength cement (ordinary cement + water + silica fume + additives) was included and tested in the experimental program. Results showed that high strength grout gains more strength in two days than conventional grout in 7 days (Khan, 1993). The ductility, however, is still maintained; the results in the forthcoming sections show that a displacement of 100 mm was achieved. A shotcrete grout is a combination of coarse aggregates (passing sieve size 9.50 mm (US 3/8) and retaining sieve size 2.36 mm (US 8)) and cement. The presence of coarse aggregates increases the radial stiffness of grout, hence, the load transfer mechanism is believed to be improved (Hassani and Rajai, 1990).

8.2 Anchoring Device.

The ever-existing problem of all the non-metallic tendons is the incompatible holding or anchoring devices. Conventional anchoring devices (grips, chucks, etc.), made of steel, crush the non-metallic testing samples. However, unlike glass-fiber reinforced tendons, Arapree does not get crushed due to line load (the anchoring load). It is also observed that a strip shape is easy to hold without undergoing slipping during testing when compared to the round shape. This is one of the reasons for producing Arapree in the strip shape (Gerritse and Werner 1988). An anchoring device which is presently used in the experiments is shown in Figure 8. Its components are the outer steel cover and polymide inner wedges.

8.3 Methodology

A conventional pull out testing technique has been followed in the laboratory experimental program. Arapree cable was grouted in schedule 40 steel pipe, which was used to model the host rock mass. The inner diameter of the steel pipe was 57 mm and the embedment length was selected as 305 mm. A water cement ratio of 0.4 for cement grout was kept constant for all the samples and all types of grout were tested after 28 days of moist curing. A minimum of three samples were prepared for each category of Arapree cable. One test result, which was close to the average of all three, was selected for documentation in this paper. An average (of the three tests) was not calculated, since it conceals the real behaviour of pull out tests (such as chemical adhesion and elastic behaviour prior to peak-failure and stick-slip post-peak failure). A special anchoring device (grip, chuck, etc.) was used to hold the Arapree cable.

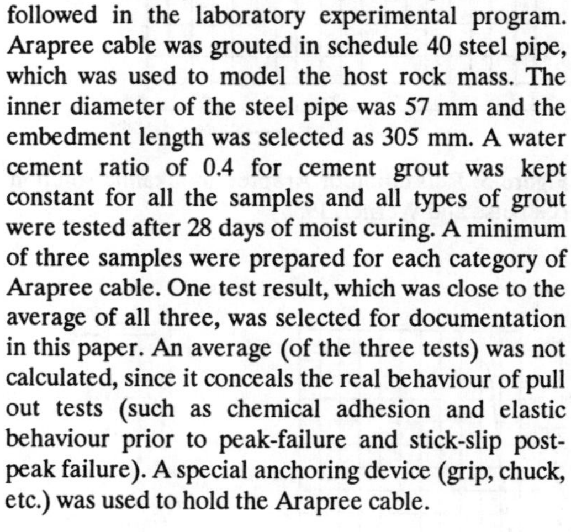

Figure 8: The anchoring device for strip shape Arapree (Gerritse and Werner 1988)

An R. D. P. Howden servo-controlled, stiff testing machine with a capacity of 2500 kN in compression and 1250 kN in tension was used in the experiments. The tests were conducted under constant displacement control mode. The magnitude of the displacement increments was selected to be small enough (i.e. 1 mm/min) to reflect the non-linearity of the response for bond stress between cable and surrounding matrix. The number of increments were sufficiently large to provide approximately 100 mm displacement for each test. Different configurations of Arapree cable, such as single Arapree cable, zig-zag Arapree cable and double parallel Arapree cable, were investigated (see Figure 9). The details concerning these cables are given in the corresponding sections.

9 SINGLE ARAPREE CABLE

In the first phase, all the available types of Arapree tendon, of both round and rectangular shape were investigated in the mining constraints. The pull out tests of all types of Arapree tendons are graphically illustrated in Figure 10 and their test parameters are given in Tables 4 and 5. From this figure, the following conclusions can be drawn:

1. Arapree tendons whose load bearing capacity is less than 10 tonnes are not suitable.
2. The round, sanded shape Arapree failed due to the delamination of its outer skin. It has been noticed that the inner fibers of each tendon remained unaffected i.e. no failure of any fiber was seen. This type of surface profile is not suitable in the mining constraints because neither does it give critical bond strength nor does it provide adequate failure warning.
3. For the rectangular f400000 Arapree, the observed pull out failure was due to the failure of the Arapree-grout interface. The post-failure behaviour showed that the residual strength is approximately 30 kN.

Based on these results, only rectangular f400000 Arapree was selected for future testing. Tests of single Arapree were conducted in conventional grout, high strength cement grout and in shotcrete grout. For the same constraints, conventional steel was also tested in order to compare the results. The results of steel cable tests were shown in the same figures. Figure 11 shows pull out response of cables in conventional grout and Figure 12 shows similar response in high strength grout.

The following conclusions can be drawn from these results:

1. Like steel cable, the first behaviour of Arapree tendons is elastic along with grout elastic dilation i.e. perfect bond deformation.
2. After initial debonding, the residual strength of cable largely depends on the frictional resistance between cable and grout.
3. The performance of the cable support system increases with high strength cement and shotcrete grout.
4. From the visual inspection of pulled out sections of the Arapree cable, no failure of any fiber or outer skin was observed.
5. The post-peak resistance of Arapree is less than that of a steel cable. This capacity stays within a range of 10 to 40 kN. It is not clear why this type of response is obtained.

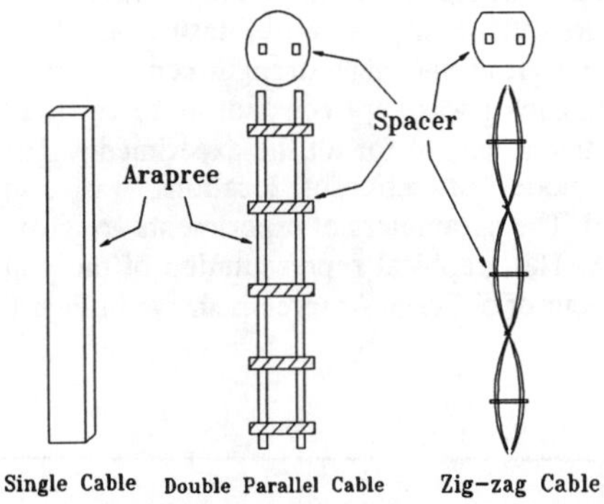

Figure 9: Various configurations of Arapree cable

Table 4: Test parameters of different types of single Arapree

Sample Type	Arapree Type	Arapree Shape	Arapree Dimension (mm)
ANC-SL	f400000	Strip	20 x 5
ANC-SM	f200000	Strip	20 x 2.6
ANC-SS	f100000	Strip	20 x 1.5
ANC-RL	f200000	Round	7.9
ANC-RS	f100000	Round	3.9

Table 5: Test parameters of Arapree and steel cable

Sample Type	Cable Type	Grout Type
NCS-C	Steel	Shotcrete
NCS-0	Steel	Conventional
NCA-C	Arapree	Shotcrete
NCA-0	Arapree	Conventional
HSS-0	Steel	High Strength
HSS-C	Steel	Shotcrete-H
HSA-0	Arapree	High Strength
HSA-C	Arapree	Shotcrete-H

10 ZIG-ZAG ARAPREE CABLE

From the literature (Goris, 1990), it is found that configurations like that of birdcage conventional steel cables, improve the resistance capacity of cable when pulled out from the grout. The reason being that, in addition to mechanical and frictional bonding between cable and grout, the compressive strength of grout (the grout which gets confined in the cage of steel cable) also directly influences the pull out resistance. Two strip Arapree cables were constructed in the zig-zag formation in the laboratory (see Figure 9). Samples were tested in both conventional grout and high strength cement grout. The node spacing was kept constant at 15 cm (half the embedment length) for all the experiments. The effect of nodes at different locations was also considered. The parameters of experiments are given in Table 6. The graphical representation of the pull out behaviour of birdcage Arapree is shown in Figure 13.

Table 6: Test parameters of Arapree birdcage cable

Sample Type	Grout Type	Top Node
BAR-CA	Conventional	Ant-node
BAR-CN	Conventional	Node
BAR-HA	High Strength	Anti-node
BAR-HN	High Strength	Node

The following conclusions can be drawn:

1. The analogy of steel birdcage cable in grout is not identical to that of the Arapree zig-zag configuration. The reason is that Arapree cable slips through the node and grouted confined area after mechanical debonding. It does not squeeze concrete in between strip openings and, as a result, the performance does not improve. The other possible reason for this type of behaviour is that the stiffness per unit length of Arapree is relatively higher than that of the grout,

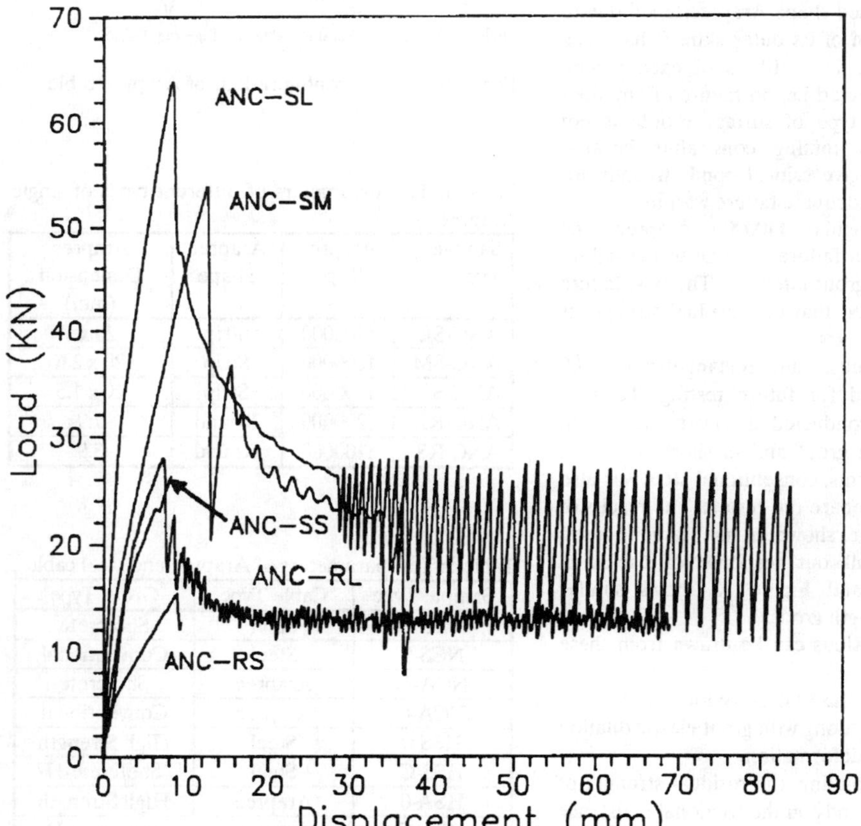

Figure 10: Pull-out behavior of various types of Arapree in conventional grout

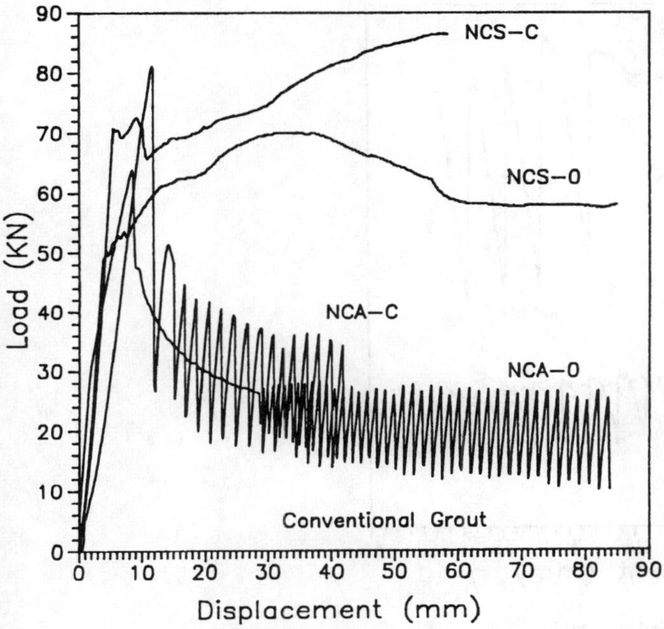

Figure 11: Pull out behaviour of Arapree and steel in conventional and shotcrete grout

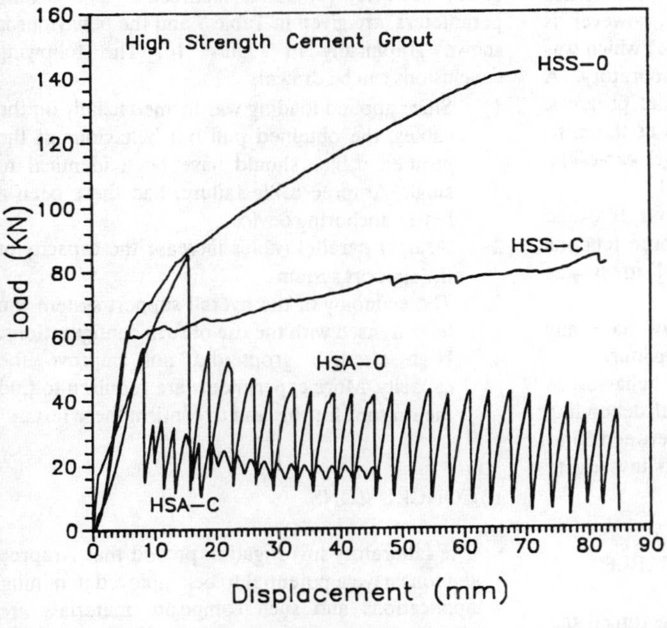

Figure 12: Pull out behavior of single Arapree and steel cables embedded in high strength and shotcrete grout

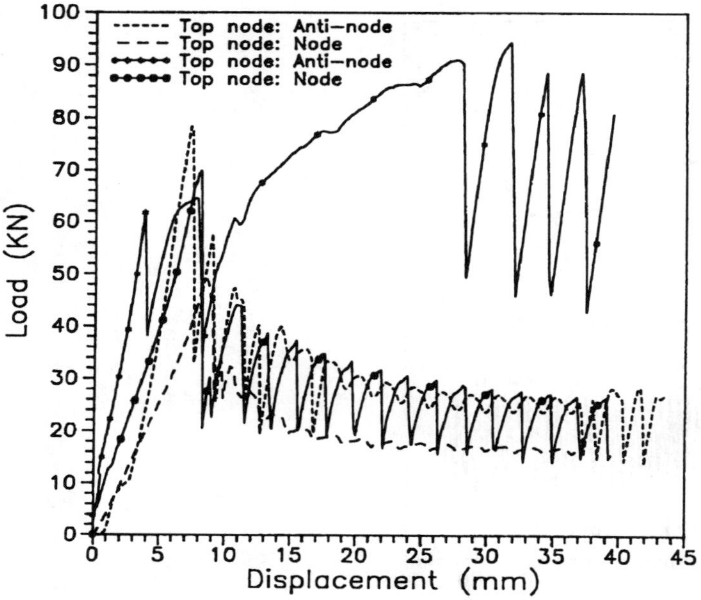

Figure 13: Pull out behavior of Arapree birdcage cable

therefore, the off-axis loading on Arapree breaks the grout wedges before complete debonding. This weakens the system and the Arapree zig-zag cable fails prematurely. This failure, however, is attributed to a loose knot (anti-node) which was manufactured manually in the laboratory. A development of a staple-knot is under progress, which will not let the Arapree slip at the anti-node and hence, a steel birdcage cable-like response can be expected.

2. The visual inspection of pulled out birdcage Arapree cables show that the Arapree tendons remain unaffected (i.e. the applied stress was below its strength).
3. The position of the node does not have any significant effect on the pull out behaviour.
4. Few tests, however, showed unusual behavior, as shown in Figure 15. The mechanical debonding started at about 60 kN and stayed between 50 to 90 kN. More testing is required to investigate this behaviour.

11 DOUBLE PARALLEL ARAPREE CABLE

The pre mature failure of birdcage cable forced the manufacture a twin, parallel Arapree cables. An aluminum button was inserted between strips to make a node. No special grip or anchoring device was available to hold this type of configuration, therefore, the ordinary anchoring device (chuck) was used to pull out the Arapree cable. This chuck failed at about 120 kN of load. The grouted portion of cable and the grout, however, remained unaffected. The testing parameters are given in Table 7 and the behaviour is shown graphically in Figure 14. The following conclusions can be drawn:

1. Since applied loading was formed axially on the cables, the obtained pull out behaviour of the grouted cables should have been identical to single Arapree cable failure, had there been a better anchoring device.
2. Straight parallel cables increase the capacity of the support system.
3. The economy of the overall support system can be increased with the use of such configurations.
4. High strength grout did not improve the capacity. More experiments are required to find the reasons for this and to confirm these tests.

12 CONCLUSIONS

- The laboratory investigation proved that Arapree tendons have a potential to be employed in mining applications and such composite materials are expected to be used as future ground supports.
- In certain applications, where the use of steel tendons have had drawbacks, the Arapree tendons are now a viable alternative.
- The pull out experimental results showed that

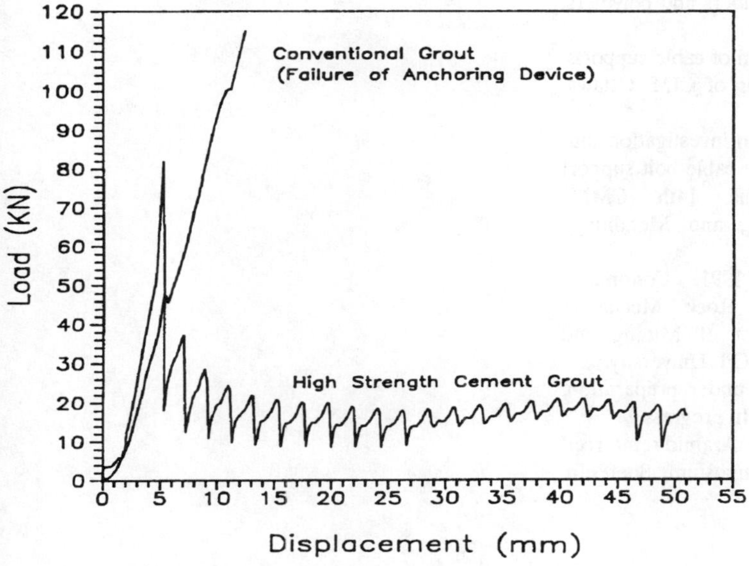

Figure 14: Pull out behaviour of double parallel Arapree cable

Table 7: Test parameters of parallel Arapree cable

Sample Type	Grout Type	Failure Mode
PAR-C	Conventional	Failure of anchoring device
PAR-H	High Strength	Failure of cable-grout interface

Arapree, as rockbolts, has sufficient support capacity to be employed for reinforcing loose rocks near excavated openings.

- The post-elastic limit of Arapree cables is less than that of steel cable but more than that of steel rockbolts. The residual capacity of fully grouted Arapree cables is around 50 kN. This means that the rock mass supported by Arapree cables should give warning prior to failure.
- In mining applications, cable bolts and rockbolts are designed considering their elastic limit under tensile and shear loading. Arapree as a cable bolt and/or rockbolt behaves similar to a steel cable bolt in terms of elastic limits. Therefore, Arapree tendons may be designed to support rock masses in place of steel cables.
- The capacity of fully grouted Arapree cables can be improved by combining two Arapree cables.
- The mechanical properties of Arapree can be altered according to anticipated mechanical behaviour of rocks and, hence, a compatible support system can be "manufactured".
- The Arapree cable is cuttable. This property encourages and contributes to the use of continuous mining methods in both hard and soft rock environments.
- Arapree is a non-corrosive material, thus, it can be used in highly humid and hostile environments
- Arapree is anti-static, so that it can be used near high tension power cables.

13 FUTURE RESEARCH

- Testing of Arapree embedded in commercially available resin grouts.
- The shear capacity of fully grouted Arapree in hard and soft rock constraints should be investigated.
- Mechanical anchoring devices for the Arapree rockbolts should be developed for quick support.

REFERENCES

Gerritse, A., Schurhoff, H.J., 1986. Prestressing with Aramid tendons, FIB 10th congress, New Delhi, India.

Gerritse, A., Schurhoff, H.J., Maatjes, E., 1987. Prestressed concrete structures with Arapree; relaxation. IABSE symposium, Paris, France.

Gerritse, A., Werner, J., ARAPREE, the prestressing element composed of resin bonded Twaron fibers.

Report available from Akzo/Fibers and polymers, Wuppertal, Germany.
Goris, J.M., Laboratory evaluation of cable supports. 92nd Annual General Meeting of CIM, Ottawa, paper no. 162, 170.
Hassani, F.P., Rajai, H., 1990. An investigation into the optimization of a shotcrete cable bolt support system, Proceeding of the 14th CMMI Conference, Inst. of Mining and Metallurgy, Edinburgh, pp.119 - 129.
Hassani, F.P., Khan, U.H., 1991. Continuous Hardrock Support System, Rock Mechanics Research Group, Department of Mining and Metallurgical Engineering, McGill University.
Khan, U.H., 1993. Ph.D. Thesis under preparation, McGill University, Montreal. (In progress)
Schurhoff, H.J., Gerritse, A., 1986. Aramid reinforced concrete, The third rilem symposium, Sheffield, England.

Innovative Mine Design for the 21st Century, Bawden & Archibald (eds) © 1993 Balkema, Rotterdam, ISBN 90 5410 325 6

Optimisation of pillar dimensions in multiple heading layouts in coal mines

Ripu Daman Lama
Kembla Coal and Coke, Wollongong, N.S.W., Australia

Arvind Misra
Sidhgiri Bagh, Varanasi (U.P.), India

ABSTRACT: Mathematical modelling based upon time and motion studies of mining operations in the development of headings using continuous miners in underground coal mines showed that the length of pillars has considerable effect on panel developments rates. Though production decreases with increase in pillar length, panel advance rates increase much more. Panel development rates can be increased to 150 % by appropriate selection of pillar lengths without any change in technology or equipment. Optimum length of pillars depends upon the continuous miner technology used and the number of headings in the layout. It is higher for wider pillars and for panels with larger number of headings. Optimum lengths could vary between 70m and 200m for deep mines requiring high support densities and long support setting times.

1. INTRODUCTION

The retreat system of longwall mining requires preparation of the longwall block before the start of extraction. Total development requirements depend upon longwall block size (Lama & Misra, 1991), dimensions of the pillars (length and width), and the number of headings. To meet longwall panel requirements, required daily development rates are dependent upon longwall face retreat rates. Longwall face production has seen a continuous increase in Australian mines (Lama & Cram, 1988) but development rates achieved have not matched face production increases in most mines. This has required an increase in the number of development unit shifts and increases in manning levels affecting overall productivity. In the last few years the coal mining industry has been experimenting with many different systems (Barrett, 1991) and their variations such as twin panels, cut and flit system, continuous haulage (Richmond & Buender, 1988; Torre, 1990), concurrent cutting and bolting (Underhill, 1992; Halbmayer, 1992), mobile boot end for cutting down belt extension time (Eveleigh, 1992), automation of roof bolting (Nicholls, 1991), although the dominant system still uses one continuous miner (CM) with two shuttle cars for transportation of coal from the back of the CM to the boot end. Use of more than one CM in a panel can increase development rates by 63%-68% (Lama & Misra, 1991) but requires higher capital and manning levels and as such does not add to productivity and lower costs except where negotiations between unions and management are successful. Changes in roof support systems and procedures and the use of especially designed CMs with a capacity to cut and bolt simultaneously can increase development rates 2 - 3 times (Lama, 1992).

This paper discusses the effect of pillar dimensions on the development rate of a conventional system using one CM and 2 electric shuttle cars.

2. LAYOUT AND WORKING OF DEVELOPMENT PANELS

A general layout for development of 2,3,4 and 6 heading panels as practised in Australian mines is given in Fig. 1. In a 2 and 3 heading development system the belt is placed in the side heading (return), with the other headings as intakes. In 4 and 6 heading development panels, the location of the belt may vary, but it is customary to place the belt in one of the intakes with the flanking headings as returns. The sequence of cutting varies from mine to mine. The purpose is to create a good

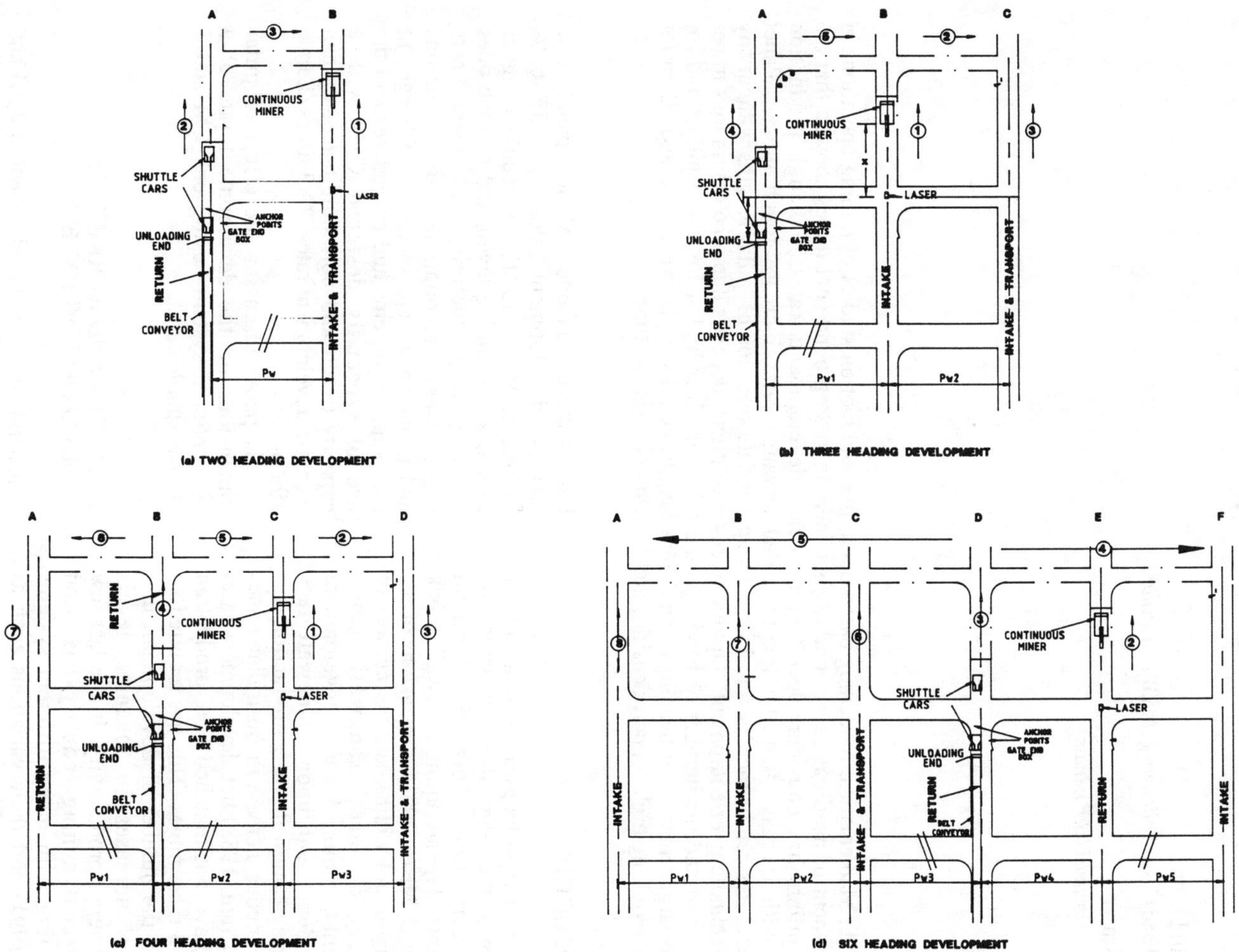

Fig. 1 General Layout of Panels

geotechnical and safe ventilation environment while maintaining less shunting of equipment and thus maximising development rates. The sequence of cutting is so organised that the numbers of turnaways are minimised.

The operation of development of a panel is cyclic. The operation consists of a cutting cycle followed by a support setting cycle. The cutting cycle includes the extraction and transportation of coal from the face to the boot end of the belt. The support setting cycle includes the installation of roof and rib support. This is followed by extension of services such as ventilation duct, water and compressed air lines, electrical cables, etc. The cycles are repeated till the sequence is driven to the designed length. The CM is then flitted to the position of start of the next sequence. Completion of all the required sequences advances the panel by one pillar length. The belt is then extended and the sequential formation of the next pillar is repeated.

Certain headings of a panel require some overdrive . The belt heading usually requires an overdrive distance of 20m to accommodate the CM and the shuttle car. Other headings may need shorter over drive distances except when storage trolleys are stored. However, for the purpose of the analyses, an overdrive distance of 20m is assumed for all headings.

3. MATHEMATICAL MODEL

A mathematical model was built to analyse the individual components of the cyclic operation of the development cycle. Time and motion studies were conducted for the various elements of the cyclic operation and used in the model for calculation of production and advance rates. In a panel, support setting time can be assumed to be constant if no changes are made in support density, technology of support installation and number of machines used. The cutting cycle time however changes as the face advances further and the wheeling distance increases. The increment in wheeling distance after the completion of a cutting cycle depends upon the face advance in the cycle. Within a cutting cycle, the time required for wheeling can be assumed constant unless the face advance per cycle is very high. Total time required to complete a sequence can thus be calculated by using an arithmetic progression. Time required for the completion of a cutting cycle is determined by the amount of coal to be cut which depends upon the width of the heading and the thickness of the seam, advance per cycle , number of trips made by each shuttle car, their filling time determined by the cutting rate of the continuous miner and the efficiency of the CM driver, travelling speed of the shuttle car, distance to be travelled, discharge time, shunting time (this is different for each shuttle car), number of turns and time lost at each turn. Extraction of coal at the turns is slowed down because the CM cuts coal with only a part of the cutting head. All these factors have been taken into consideration in the mathematical model.

Total cycle time is the summation of the cutting time and support setting time. Time required for completion of all the sequences gives the theoretical time. Pillar completion time includes this theoretical time, belt extension time and the flitting time for the CM to flit back to the start of the first sequence. In reality system availability is never 100%. System availability in this analysis is taken to be 80 %. Besides, there is a lunch break during the shift, and at the end of the shift the area must be stone dusted. This reduces the total available time in the shift. Effective shift time, taking into consideration the travel time, is thus reduced. The following parameters have been taken in the mathematical model (Table 1).

4 OPTIMISATION OF PILLAR DIMENSIONS

The effect of pillar dimensions on the development rate and production for 2,3,4 and 6 heading panels has been analysed. Studies were conducted both with and without taking into consideration of the time spent on the extension of belt services. This option was considered as different mines have very different belt extension organisation systems and highly varied belt extension time (from 4 hours/100 m to 3 shifts /100 m) The results of modelling are discussed below.

4.1 Standard system

The effect of pillar dimensions on production rate for a 2 heading development system is shown in Fig. 2. The larger the width of the pillar, the lower is the production rate, though the difference is small, under 9 % for change in pillar width from 12m to 100m when belt extension time is not included. With belt

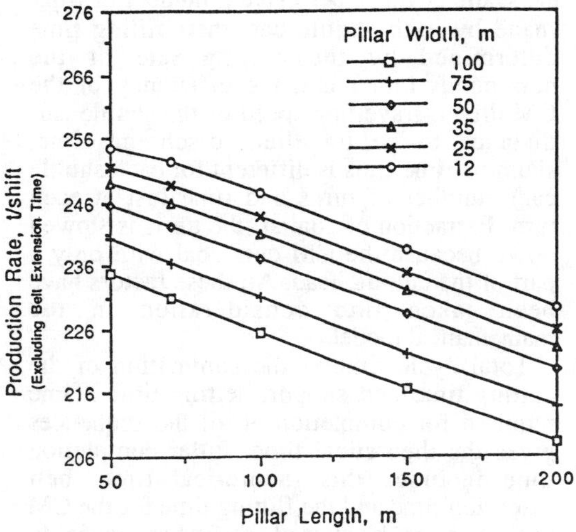

Fig. 2a Effect of pillar dimension on production rate in a standard system for two heading development excluding belt extension time

Table 1. Input values for different parameters

Parameter	Value
Advance per cycle	2.25m
Number of straps per cycle	2
Cutting/loading rate	5.22 t/min
Shuttle car speed	1.54 m/sec
Time lost at every intersection	5 sec
Support installation time	30 min
Roadway width	5.5 m
Seam height	2.75 m
Shuttle car capacity	10 tonnes
Distance of cut-through from boot end	20 m
Radius of curvature at the turn away	7 m
Total time in a shift	360 min
Meal time	30 min
Stone dusting time	15 min

extension time of 3 shifts /100m, the effect is even lower. With an increase in pillar length, the production drops almost linearly. A change in pillar length from 50 to 200 meters results in decrease in production of about 10% in the case where belt extension time is excluded and a maximum of 5% where belt extension time is

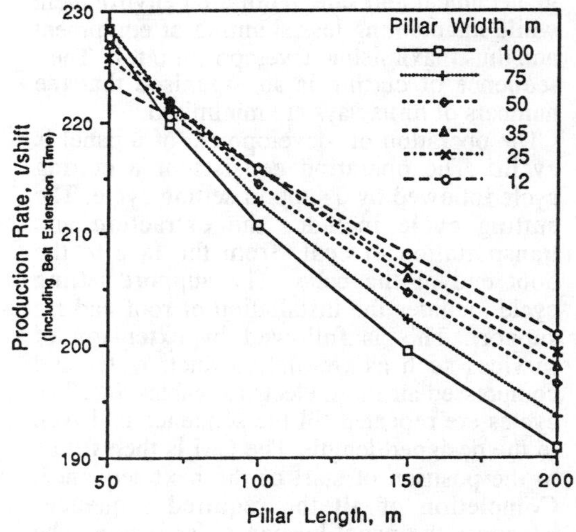

Fig. 2b Effect of pillar dimension on production rate in a standard system for two heading development including belt extension time

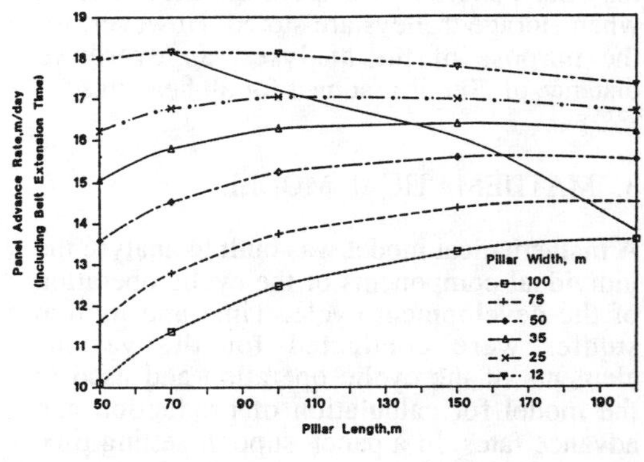

Fig. 3 Effect of pillar dimension on panel advance rate in a standard system for two heading development

included. The curves with belt extension time tend to cross at about a 70m pillar length which indicates that for the conditions modelled, the optimum length of the pillar from a production point of view should be about 70m.

The effect of pillar dimensions on panel advance rate is given in Fig. 3. Panel advance rate increases with a decrease in pillar width due to the effect of lower development requirements of crosscuts, but with an increase in pillar length the wheeling distance increases and therefore cutting cycle time

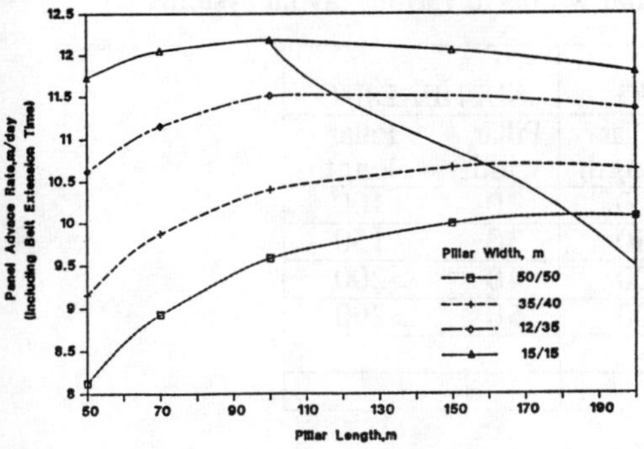

Fig. 4 Effect of pillar dimension on panel advance rate in a standard system for three heading development

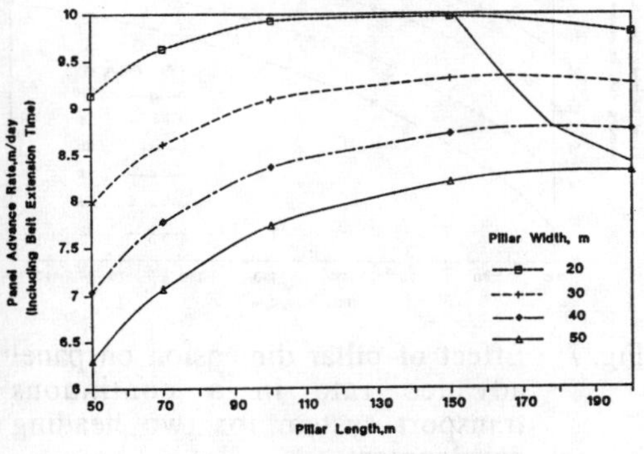

Fig. 5 Effect of pillar dimension on panel advance rate in a standard system for four heading development

increases resulting in decrease in face advance rates and hence panel advance rates. As such for each pillar width, there is an optimum pillar length at which the panel advance rate is maximum.

The shape of the curves for 3, 4 and 6 heading development as a function of pillar dimensions with and without belt extension time is very similar without the cross over point as seen in Fig. 2b. The width of the pillars has a very small effect on production levels (3% for 3 heading, 3.7% for 4 heading and 6% for a 6 heading layout. There is a decrease in production with an increase in pillar length and pillar width but panel advance rate increases with an increase in the pillar length and decreases in pillar width (Fig. 4,5 & 6). The effects of both decrease in pillar width and increase in pillar length is quite substantial. Depending upon the pillar width,

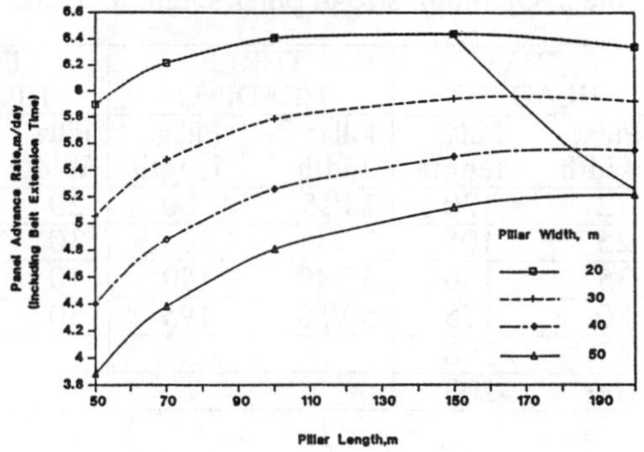

Fig. 6 Effect of pillar dimension on panel advance rate in a standard system for six heading development

optimum values can vary over a large range. Optimum values for different pillar widths for 2, 3, 4 & 6 heading panels are given in Table 2. These results show that optimum length increases with an increase in the width of the pillar. The optimum length is higher for a greater number of headings. Because of wheeling distance and ventilation considerations, the optimum length is usually limited to 100 m.

4.2 Effect of decreasing cutting cycle time

Cutting cycle time can be reduced by: (a) increasing the cutting rate of the CM, (b) cutting down the waiting time of the CM for the arrival of the shuttle car by increasing shuttle car speed, (c) introducing a surge car or (d) using continuous haulage.

An increase in shuttle car speed beyond a certain limit is not possible from a safety point of view. Shuttle car speeds are fixed but are controlled by road surface conditions. This effect has not been considered.

The effect of an increase in cutting rate of the CM is to cut down the cutting time but this does not influence the support setting time, flitting time or belt extension time. The waiting time for the CM when the pillar dimensions are large also remains unaffected. In the standard system discussed earlier, the cutting cycle time varies but the filling time is only 9.12 min per cutting cycle. If the cutting rate of the CM is increased by a factor of 2, it will result in a saving of only 4.56 min. In the total cycle time of cutting + bolting of 42 - 58 min, the cutting cycle time varies between 12-28 min depending upon the pillar dimensions.

Table 2. Optimum size of pillar length for different pillar widths in various layout systems

TWO HEADING		THREE HEADING		FOUR HEADING		SIX HEADING	
Pillar width	Pillar length	Pillar width	Pillar length	Pillar Width	Pillar length	Pillar width	Pillar length
12	70	15/15	80	20	100	20	100
25	105	12/35	105	30	150	30	150
35	130	35/40	150	40	>200	40	>200
50	175	50/50	195	50	>200	50	>200
75	>200						
100	>200						

With the cycle advance of 2.2 m requiring only 5 shuttle cars to be filled per cycle, the reduction of 4.56 min represents a saving of 7.8 - 10.8% in the total cycle time. The effect of this is to move the optimum values of the pillar lengths only slightly towards the lower side.

The introduction of a surge car behind the continuous miner allows it to cut and to store coal on the surge car while waiting for the shuttle car to arrive. Thus filling time of the shuttle car is reduced. This is more effective when the distance between the CM and the shunt position is large and the CM has enough time to cut and to fill a shuttle car. Here the shuttle car loading time is equal to the shuttle car discharge time. When the distances are smaller, shuttle cars arrive before the surge car has been filled. The advantage of introducing a surge car is therefore low for smaller wheeling distances.

The use of a surge car for the CM cutting rate of 5.22 t/min, improves the panel development rate over the standard system by 7-13 % for the 2,3,4 and 6 heading development systems depending upon the pillar dimensions . In the case of a CM with double the cutting rate and with a surge car, the improvement varies between 11-14% over the standard system. The effect of a surge car with a high capacity cutting CM is only 1-2% as for longer wheeling distances, even a lower capacity CM has enough time to cut and fill the surge car.

With a continuous transport system, waiting time is completely eliminated. Improvements in panel development rates can be from 8 - 31% depending upon the pillar dimensions and number of headings in the layout. Optimum pillar lengths increase greatly for the continuous transport system when compared with the standard system . For example, for a 2 heading development system, for a 50m

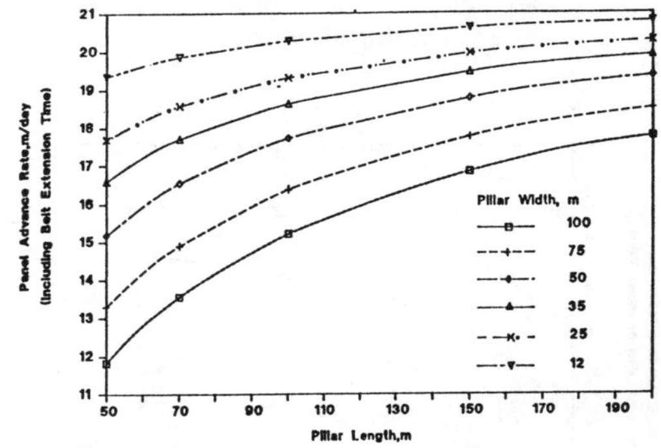

Fig. 7 Effect of pillar dimension on panel advance rate in a continuous transport system for two heading development

wide pillar optimum pillar length is about 160m but in a continuous transport system optimum pillar length exceeds 200m (Fig. 7).

4.3 Twin panel system

In a twin panel system two faces are worked simultaneously. While one face is being cut the other is being supported. This results in overlap of the cutting and supporting cycle. Depending upon the cycle time for the cutting and supporting cycles, one or the other determines the face advance rates and hence the production and panel advance rate. Using the input parameters in Table 1, parametric studies were conducted by varying the dimensions of the pillar to establish the effect of pillar dimensions on panel advance rates. For the range of pillar widths (12 - 100m) when compared with the standard system, a twin system improves panel advance rates by 63 -68%, 71-82%, 80-85% and 86-89% for 3,

Table 3. Optimum pillar length for panel advance rate for different layouts in a twin system

Two Heading		Three Heading		Four Heading		Six Heading	
Pillar width, m	Optimum pillar length, m	Pillar width, m	Optimum pillar length, m	Pillar width, m	Optimum pillar length, m	Pillar width, m	Optimum pillar length, m
12	100	15/15	70-100	20	130-150	20	130-150
25	125-150	15/35	100-130	30	150-200	30	150-200
35	150	35/40	150-200	40	>200	40	>200
50	200	50/50	170-200	50	>200	50	>200
75	>200						
100	>200						

4 and 6 heading development panels respectively. There is an optimum pillar dimension for each layout. Table 3 gives optimum pillar lengths for best panel advance rates for different layouts.

In a twin panel system if the cutting rate of the CM is doubled, development rates improve by about 6 - 9% for the various layouts but the optimum values of the pillar lengths change marginally. With the introduction of a surge car behind the CM, improvements over the standard system range from about 7-12% for the different layouts. In both these cases the level of improvement is higher, the higher the number of headings in the panel.

5. CONCLUSIONS

Results of mathematical modelling indicate that higher production rates are possible by appropriate selection of the pillar dimensions in the layout of multiple headings. Improvement in production and hence face development rates could be 10-20% in panels with different layouts without any change in technology. Pillar dimensions have the maximum effect. Optimum selection of pillar dimensions (width and length)has an enormous effect on panel development rates in particular. For the same width of the pillar, panel development rates can be increased by almost 50 % by optimising the pillar length. For narrower pillars, the length of the pillar is less important than for wider pillars. The optimum length for the narrower pillars is smaller than for wider pillars. Reducing the cutting cycle time by doubling the cutting and loading rate of the CM can increase panel advance rates by about 9%, introduction of a surge car by about 13%, continuous transport by 8-31% and the twin system with a surge car by 89%. It is essential that time and motion studies be conducted to determine the input parameters and analyse the effect of these parameters on the system to design and select optimum pillar lengths.

REFERENCES

Barrett, M.1991. Review of status of new generation full face roadway development machines. Proc. Continuous miner systems: The future, ACIRL, May 31-June 1, Sydney.

Barrett, M., J. Mckendry. 1989. Development rates: Various options. Proc. Seminar Longwall mining, ACIRL, ,Nov. 17-18, Sydney

Eveleigh, R. 1992. Meco mobile boot end. Pro. seminar Underground Proc. Underground mining technology integration, ACIRL, May 21-22, Sydney.

Halbmayer, C. 1992. Simultaneous cutting and bolting: A final solution to longwall roadway development. Proc. Underground mining technology integration. ACIRL, May 21-22, Sydney.

Lama, R. D ,K. Cram. Developments in longwall mining technology in Australia and future trends. Proc. 14th World mining congress, Beijing, vol. 1, pp 3-26.

Lama, R. D., A. Misra, 1991. Development rates using conventional continuous mining technology in deep underground coal mines in Australia. Proc. Int. Conf. on reliability, production and control in coal mines, Wollongong, Sept. 2-6, pp 148-154.

McNamara, P. 1991. Operation of a Joy FCT in two heading gate road development at Southern colliery. Proc. seminar Continuous miner systems-The future. ACIRL, May 31 - June 1, Sydney.
Nicholls, B. 1991. The current status of roof bolting rib reinforcement Proc. Seminar Continuous miner systems-The future. ACIRL, May 31 - June 1, Sydney.
Richmond, A.., W. Buender. Klockner-Becorit continuous haulage system, initial trials. 2nd Conf. on mining machinery, Brisbane, May 9-11.
Torre, D. C. 1990. Overview of continuous haulage, Mining Engineering, pp 264- 267.
Underhill, K. Anderson KB 2-The way ahead. Proc. Underground mining technology integration. ACIRL, May 21-22, Sydney.

Innovative Mine Design for the 21st Century, Bawden & Archibald (eds) © 1993 Balkema, Rotterdam, ISBN 90 5410 325 6

Behavior of a steel-liner-and-bolts system under very high thermal and mechanical loading – The CONVEX Liner Add-On to DIAMOND FORTUNE

F.E. Heuze & R.P. Swift
Lawrence Livermore National Laboratory, Livermore, Calif., USA
Leslie R. Hill & William H. Barrett
Sandia National Laboratories, Albuquerque, N.Mex., USA

ABSTRACT

This work involved the response of a liner-bolt system installed on the wall of the DIAMOND FORTUNE cavity, a 22-m diameter nearly semi-spherical chamber in tuff, at the Nevada Test Site. DIAMOND FORTUNE is a low-yield nuclear test of the Defense Nuclear Agency which was performed in April, 1992. A 1.4-m square, 2.5-cm thick steel plate was anchored by 9-m long bolts: four 2.5-cm diameter bolts at the corners and a 5-cm diameter bolt at the center. The bolt ends daylighted in a tunnel surrounding the cavity, and were tensioned from there. The system was equipped with 20 data channels for strain, acceleration, contact pressure, and temperature. We relate the thermal analyses and the 3-dimensional dynamic analyses performed for this project, and we present the test results which indicated the excellent response of this system to the high dynamic loads and temperatures.

RESUME

Ce project concerne la réponse d'un blindage d'acier ancré par boulons pré-contraints ā la paroi d'une cavité hémisphérique de 22-m de diamètre, sous l'action du tir nucléaire à faible puissance, DIAMOND FORTUNE. Ce test a été effectué dans un tuff volcanique au Centre d'Essai du Nevada, en Avril 1992. Le blindage de 1.4-m sur 1.4-m, épais de 2.5-cm était ancré par 5 barres d'acier longues de 9-m; 4 aux coins, avec un diamètre de 2.5-cm et une au centre, avec un diamètre de 5-cm. Les barres se terminaient dans une gallerie adjacente, d'ou elles pouvaient être tensionées. Le système était doté de 20 canaux d'acquisition de données pour déformations, accélérations, pressions de contact, et température. On présente les analyses thermiques et les analyses dynamiques en 3-dimensions, et on les compare aux résultats de ce test qui révèlent l'excellent comportement du système sous les fortes charges thermiques et mécaniques.

INTRODUCTION AND OVERVIEW

In the early 1980's, the idea was put forth to provide a reusable underground cavity in which a series of low-yield nuclear explosions could be performed for the purpose of research in weapons physics and non-weapon high-energy physics. The concept, now designated as CONVEX (<u>CO</u>ntained <u>N</u>uclear <u>V</u>essel <u>EX</u>periment), is a joint effort of three DOE laboratories (LANL, LLNL, SNLA), and the Defense Nuclear Agency (DNA). Early studies by Heuze (1983) on HEDEF (<u>H</u>igh-<u>E</u>nergy <u>D</u>ensity <u>E</u>xperimental <u>F</u>acility), a precursor of CONVEX, concluded that a steel-lined rock cavity would offer containment advantages over a free-standing steel vessel. In addition, the steel liner would be pre-stressed against the rock by bolts that daylight in tunnels adjacent to the cavity, as proposed by Heuze and Thorpe (1983). This concept was tested recently, as an add-on to the low-yield nuclear test of the DNA, DIAMOND FORTUNE, executed in an 11-m radius hemispherical underground chamber in tuff, in April 1992. Our experiment was called the CONVEX Liner Add-on.

This paper presents a summary of the data obtained on the CONVEX Liner Add-On, describes numerical simulations of the experiment, and gives comparisons of the measured data and results of calculations. The data analyzed include strains on the tensioned bolts and on the steel liner plate, stresses in the

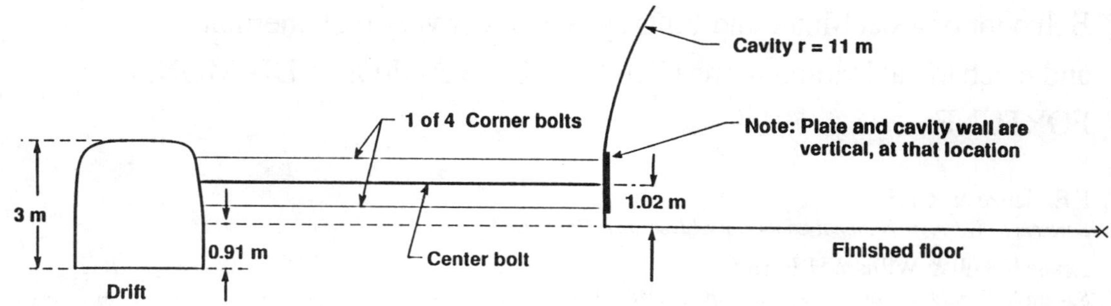

Figure 1: Geometric lay-out of CONVEX Liner Add-On; elevation.

grout/rock backing the liner, temperature on the back of the plate, and accelerations on the plate and in the rock pillar. The quality of the data obtained is very good, and the response features captured by the different data sets are quite similar and reveal the complex nature of the flow field inside the cavity. Calculations were performed using three-dimensional, static and dynamic, finite-element codes, developed at LLNL. NIKE3D, described by Maker et al. (1991), was used to simulate the pre-test bolt tensioning condition. DYNA3D, described by Whirley and Hallquist (1991), provided the dynamic response of the bolt/liner/rock system. Pre-test calculations were performed to guide instrument choice, location and recording ranges, to set the tensioning of the bolts, and generally to produce confidence that the liner system would survive the environment and remain operational. Post-test calculations were performed with the best estimate of the actual cavity pressure history at the liner plate, obtained from measurements and flow field calculations. Experimental measurements showed that the bolts remained in a tensioned state throughout the duration of the test. Hence, the bolt tensioning used (about 330 Mpa) was sufficient to offset the compressive air blast loading. They also showed that the liner plate maintained firm contact with its grout-rock backing, thus fulfilling the main operational objective of the test.

LINER EXPERIMENT

Configuration and Instrumentation

The liner experimental configuration consisted of a steel plate anchored vertically to the cavity wall by four corner bolts and a centerbolt, with the bolts daylighting in an adjacent "run-around" drift (Figure 1). The cavity rock consisted of volcanic tuff located in the P-tunnel complex at the Nevada Test Site. The rock quality was quite variable, as indicated by RQD and in-situ modulus NX jack tests in the pillar between the cavity and the drift (Figure 2). The plate is 1.4 m square, 2.54 cm thick, and made of A36 steel. The bolts, 2.54 cm and 5.08 cm in diameter at the corners and center, respectively, are about 9 m long and made of 1045 steel. A thin layer of rock-matching grout, about 5 cm thick, was used between the plate and rock to provide a smooth bearing surface for the plate. The bolts were tensioned to about 330 Mpa, less than 50% of their yield strength. This initial pre-tension load was selected on the basis of preliminary calculations indicating that the bolts would remain in tension during the dynamic loading. The corner bolts were grouted in their boreholes after tensioning. The centerbolt was encased in a 10 cm diameter schedule 40 pipe casing. The annulus between the casing and the bolt was filled with *Vistanex* (an inert non-toxic hydrocarbon polymer made by EXXON Chemical Company) for containment purpose. The casing was cemented in the pillar.

The set of diagnostics, shown in Table 1, was adapted to a 20 data channel allotment to provide the mechanical and thermal response of the liner/bolt/rock system. The desired measurements included: horizontal and vertical strains, acceleration, and temperature on the back of the plate; axial strains and bolt loads for the centerbolt and corner bolts; contact pressure between the plate and the rock; cavity wall pressure; and acceleration inside the 9 m rock pillar separating the cavity and run-around drift.

The cavity side and drift side of the system ready for testing are shown in Figures 3 and 4.

Summary of Test Results

As indicated in Figures 5 and 6, the salient feature of the test was the fact that the

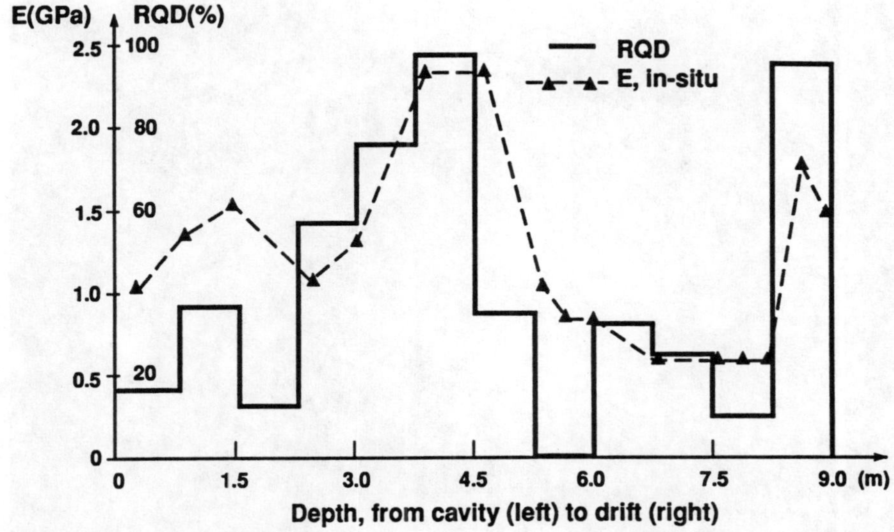

Figure 2: RQD and In-Situ Modulus, of Tuff in Cavity Wall. Note the good correlation between E and RQD.

Table 1. Instrumentation for the CONVEX Liner Add-On.

		Strain	Stress	Accel.	Pressure	Temperature
Plate:	4 Strain Gauges	√				
	1 Accelerometer			√		
	1 Thermocouple					√
Plate/Rock:	2 Toadstools				√	
Center bolt:	4 Strain Gauges	√				
	1 Load Cell		√			
Corner bolt:	2 Strain Gauges	√				
	1 Load Cell		√			
Rock Pillar:	3 Fluid-Coupled Plates				√	
	1 Accelerometer			√		

instruments showed the whole system retaining strong tension and contact with the rock throughout the dynamic phase. Figure 5 is the time history of 4 longitudinal strain gauges on the center bolt within a 30-cm section. The first aspect to note is the excellent reproducibility of the strain records. The second is the fact that the bolt quickly regained up to 80 percent of its original tension, after the first few milliseconds. Such behavior was also true of the only strain-gauged corner bolt (not shown), which retained over 85 percent, of its original tension. Figure 6 shows the time history of the two "toadstool" gauges measuring the stress in the grout perpendicular to and behind the plate. The two instruments again gave consistent results, and the remaining gauge at 80 ms shows a strong contact pressure. Another remarkable feature of the test data is the consistency of various stress and strain records in showing the history of cavity pressure oscillations (Figure 7).

CALCULATIONS

Thermal Calculations

Regarding the thermal environment, the strain-gauges could operate up to a maximum of 175 C.

Figure 3: CONVEX liner plate, cavity side.

Figure 4: Bolts, load cells, and junction box, drift side.

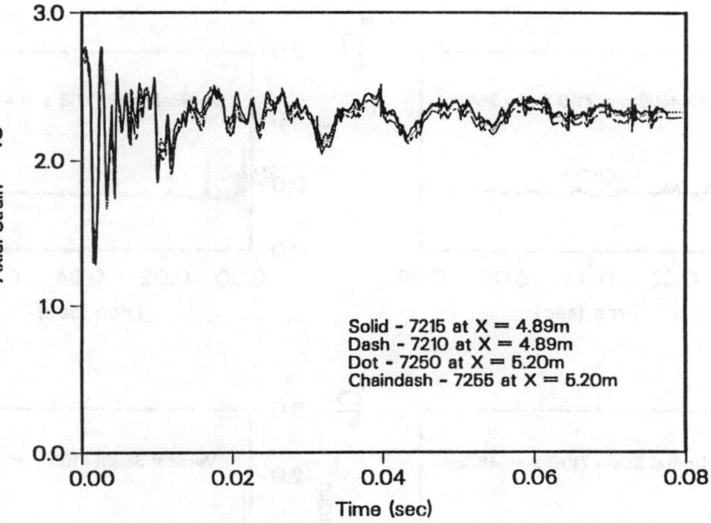

Figure 5: Measured responses of axial strain gauges on the centerbolt.

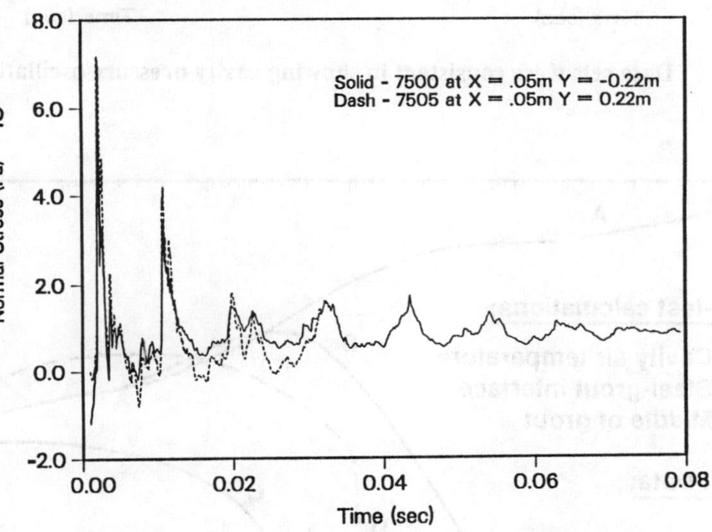

Figure 6: Measured responses of pressure gauges in grout between the liner plate and the rock.

The TOPAZ2D finite element code (Shapiro, 1986) was used to estimate the temperatures in the plate, grout, and rock. For simplicity, the grout and rock were given constant thermal temperatures, although temperature-dependent properties of the grout are available (Moss et al, 1982; Zimmerman, 1982). The rock, grout, and steel thermal properties used in the calculations are summarized in Table 2. Figure 8 shows the input temperature in the cavity, i.e. on the front side of the liner plate (A), and the expected histories on the back of the plate (B), and in the middle of the grout (C). Clearly the temperature gradients are very high. The actual temperature record (D) was well within the range of expectations, even with a simple model. More than 10 seconds elapsed before a temperature of 175 C was reached on the back of the plate. There was no need for temperature correction of gauge readings in the dynamic phase.

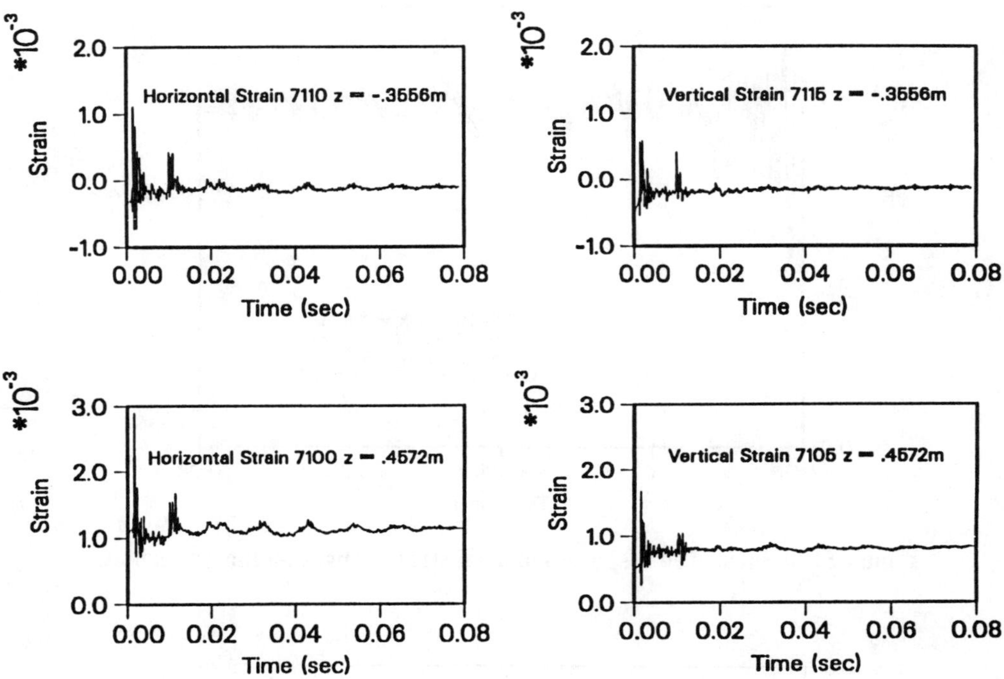

Figure 7: Data sets were consistent in showing cavity pressure oscillations.

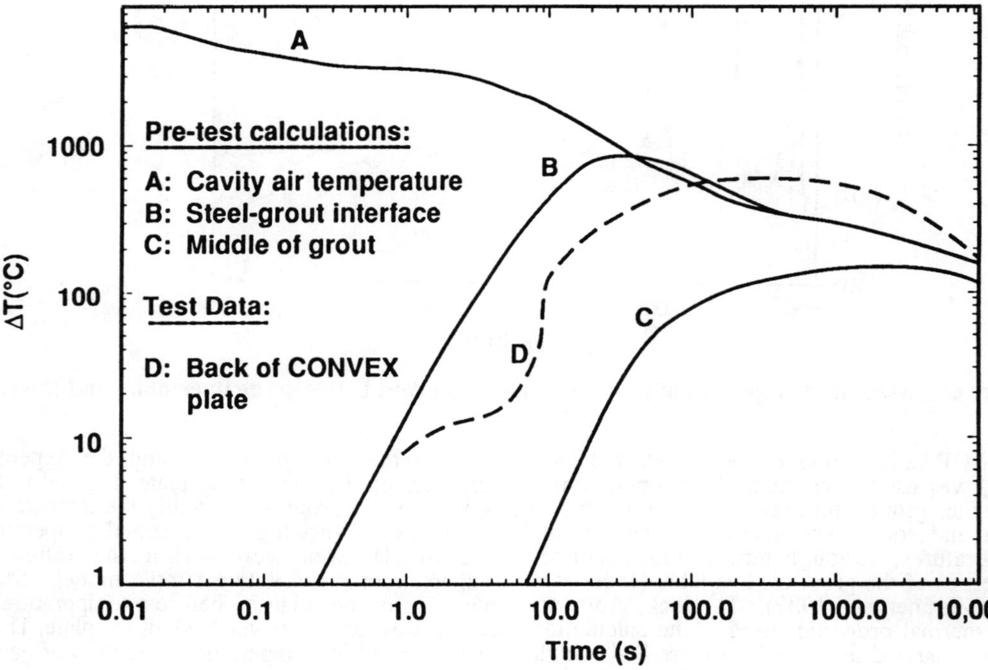

Figure 8: Measured and calculated temperature histories.

Table 2: Properties for Thermal Calculations

	Density (g/cm3)	Heat Capacity (cal/g.°c)	Thermal Conductivy (cal/s.cm.°c)	Melting Temperature (°c)	Latent Heat of Fusion (cal/g)
Steel	7.85	0.116	0.121	n/a	n/a
Rock, or Grout	1.75	0.250	3.58 E-3	1000	80

The 3-D Mechanical Model

The approach to numerically simulate the CONVEX plate/bolt/rock system consisted of defining the equilibrium state due to bolt tensioning, using NIKE3D, and mapping the equilibrium conditions into DYNA3D as initial conditions for the dynamic response calculations. Views of the three-dimensional model are given in Figures 9 and 10, showing the upper-right quadrant of the system that extends back from the cavity face into the rock. The quadrant geometry reflects the symmetry of the experiment about the axis of the centerbolt. The model extends axially outward, 9 m, to the run-around alcove where the wall buffer plates and holding nuts for the centerbolt and cornerbolt are secured. A total of 22,706 solid-brick elements were used in the static model along with 19 slide surfaces. The dynamic model is identical to the NIKE3D model, plus 2139 additional elements and 4 slide surfaces extending the rear of the model into the grout-enclosed run-around region and beyond into the rock to a range of 50 m. Slide surfaces include the steel bolt-steel plate, steel plate-buffer plate, steel plate-grout, steel bolt-grout, and rock-grout interfaces. Both the centerbolt and the cornerbolts were tied to the nuts and plates at the cavity and rear surfaces. The boundary conditions on the top and right outer rock surfaces allowed no vertical and no horizontal motion, respectively.

Estimates for the rock material properties required for numerical modeling were obtained from in-situ borehole-jack tests in one of the cable holes of the pillar, along with wave speed and strength measurements on core samples from the pillar. Wave speed and strength measurements also were obtained on core samples of the HLNCC grout used behind the plate.

The values retained for mechanical analysis were:

Rock:	modulus:	1.70 GPa
	Poisson's ratio:	0.45
Grout:	modulus:	4.0 GPa
	Poisson's ratio:	0.35
A36 Steel:	modulus:	200 GPa
	Poisson's ratio:	0.30
	density:	7.85 g/cm^3
	yield stress:	250 MPa
1045 Steel:	modulus:	200 GPa
	Poisson's ratio:	0.30
	density:	7.85 g/cm^3
	yield stress:	760 MPa

Equilibrium Condition

The equilibrium outward displacements calculated with NIKE3D for the rear surface of the plate adjacent to the grout, and caused by pre-tensioning the centerbolt and cornerbolt, are shown in Figure 11. The pretensioning was done by using initial thermal strains. The maximum displacement is about 0.45 mm in the elements next to the centerbolt. The stress distribution for the centerbolt and plate indicated maximum compressive and tensile axial values of about -300 Mpa and +430 Mpa, respectively, in the region where the nut fastens the bolt to the plate. These stress levels are well below the elastic yield stress, and the entire system is linear. Just behind the plate, at a distance of about 8.5 cm along the length of the bolt, the stress becomes uniform with an axial tension of 325 Mpa (Figure 12) and remains at this level until in the immediate vicinity of the run-around drift wall. This configuration is mapped into DYNA3D as the initial state for the dynamic response calculation. The pseudo-thermal model approach is believed to simulate the stress and strain states associated with bolt tensioning quite realistically. The same conditions would be very difficult to obtain using mechanical force or pressure prescriptions for the bolts, plate and rock exterior surfaces.

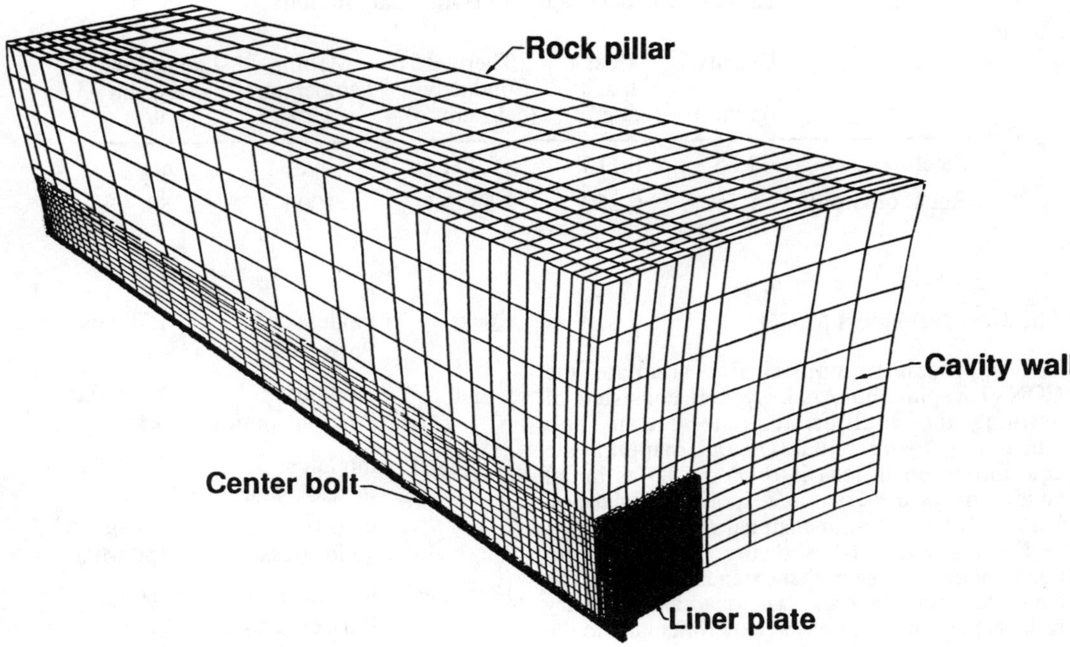

Figure 9: Complete model for 3-D mechanical analysis. Note the double curvature of the cavity wall.

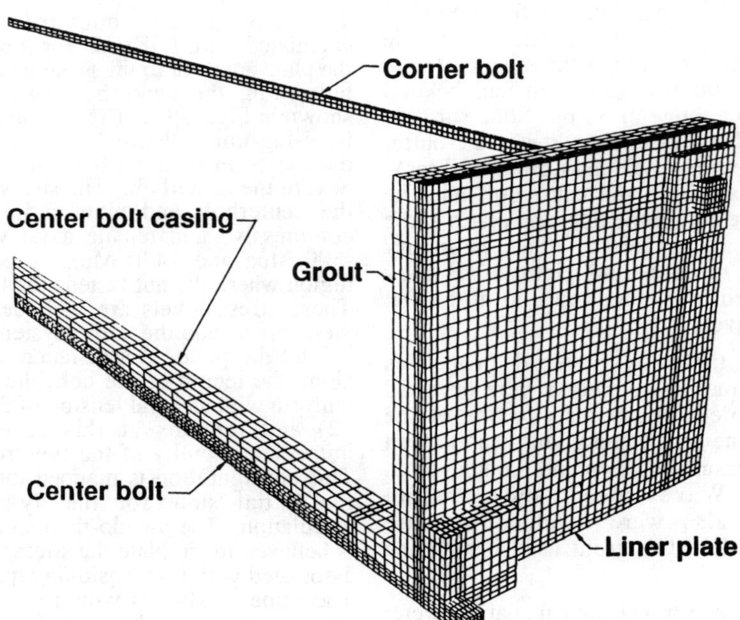

Figure 10: Details of plate, bolts and grout model.

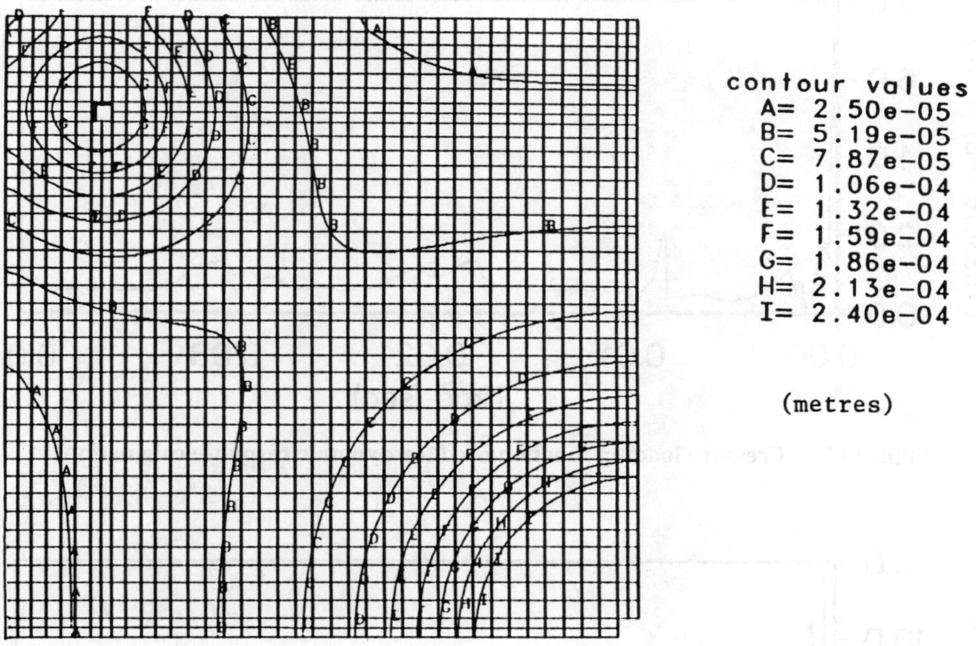

Figure 11: Calculated equilibrium displacement on back of plate, caused by 325 Mpa tension in bolts.

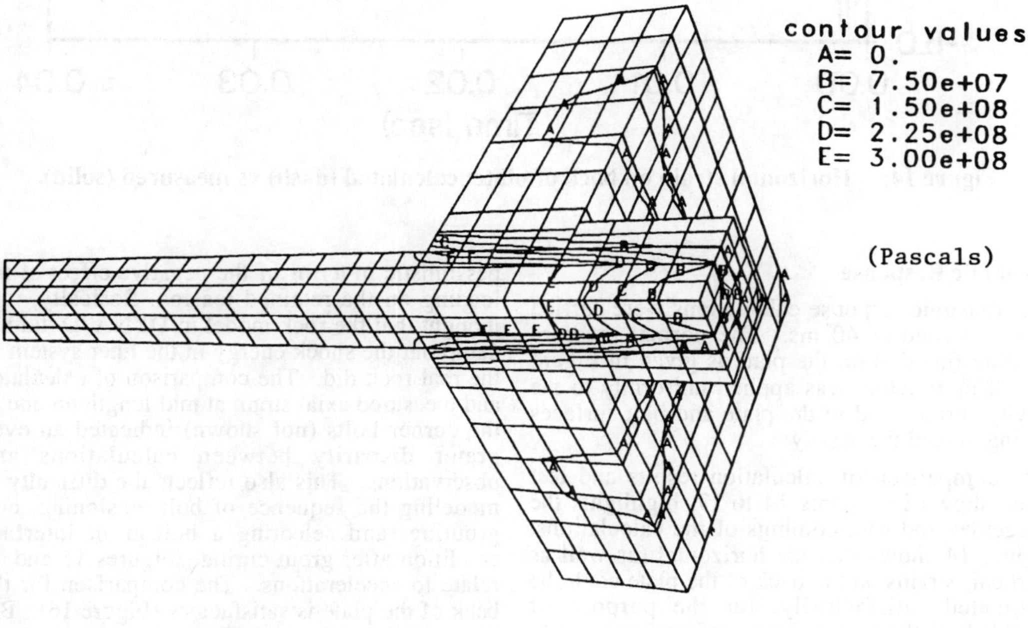

Figure 12: Axial stress distribution in center bolt and plate, at pretensioning.

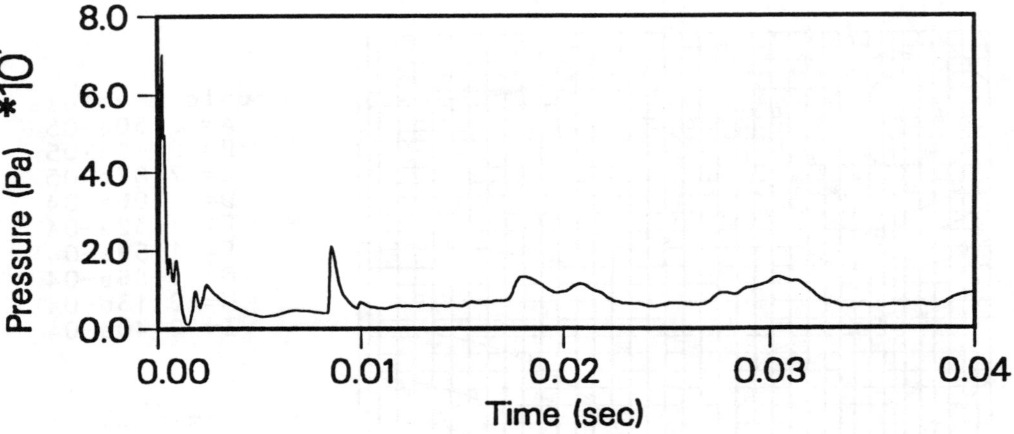

Figure 13: Pressure loading function used for dynamic response calculations.

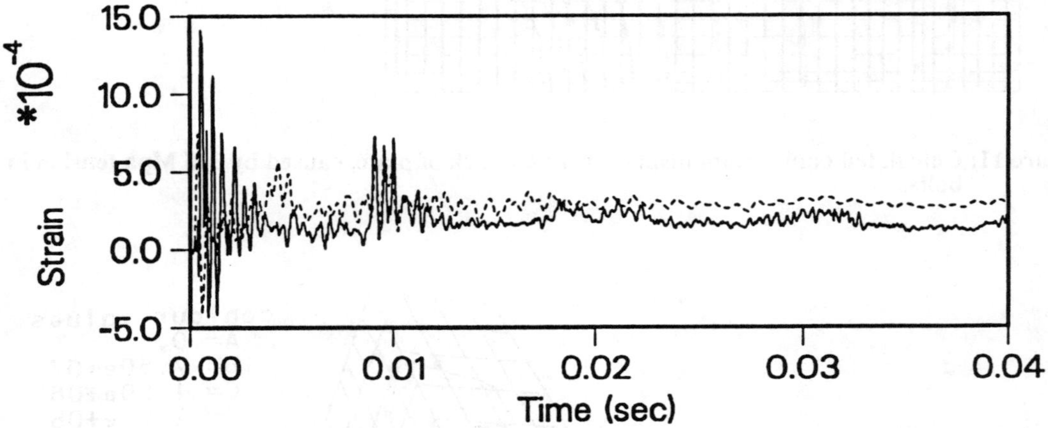

Figure 14: Horizontal strain on back of plate: calculated (dash) vs measured (solid).

Dynamic Response

The dynamic response calculations were carried out to a time of 40 ms. The normal pressure loading function on the plate is given in Figure 13. This function was applied uniformly to the cavity surface and to the plate and bolt surfaces facing toward the cavity.

The comparison of calculation results and test data, shown in Figures 14 to 17, highlights the successes and shortcomings of the calculations. Figure 14 shows that the horizontal (as well as vertical) strains on the back of the plate could be estimated satisfactorily, for the purpose of establishing the type and range of strain gauges to be used. Figure 15 shows that the prediction of center bolt axial-strain history at mid-length was pessimistic in terms of the negative effect of the loading on the retained tension. Basically, it is thought that the rock model in DYNA3D did not damp out the shock energy in the liner system as the real rock did. The comparison of calculated and measured axial strain at mid-length on one of the corner bolts (not shown) indicated an even grater disparity between calculations and observations. This also reflects the difficulty in modeling the sequence of bolt tensioning, bolt grouting, and selecting a bolt-grout interface condition after grout curing. Figures 16 and 17 relate to accelerations. The comparison for the back of the plate is satisfactory (Figure 16). But Figure 17, again, clearly indicates a lack of material damping in the calculations.

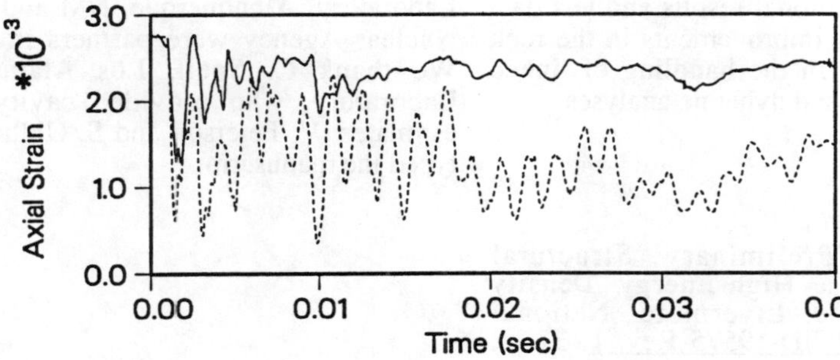

Figure 15: Axial strain on centerbolt, mid length, calculated (dash) vs measured (solid).

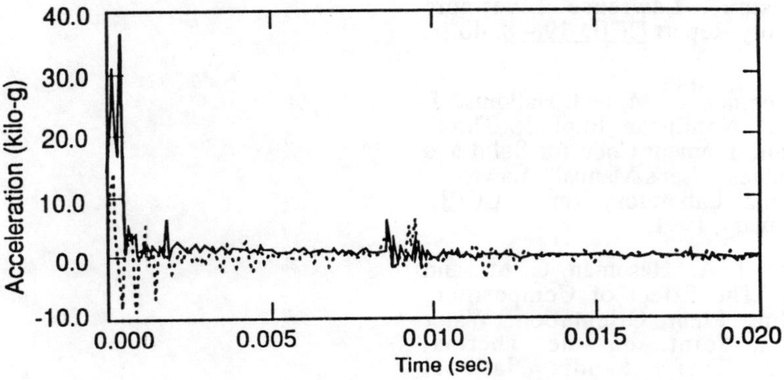

Figure 16: Acceleration on back of plate: calculated (dash) vs. measured (solid).

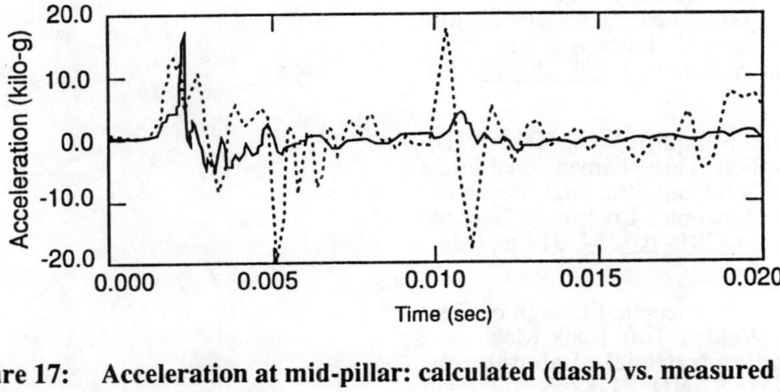

Figure 17: Acceleration at mid-pillar: calculated (dash) vs. measured (solid).

Summary and Conclusions

This combined experimental and modeling project was very successful in proving the validity of the concept of a prestressed steel liner and bolt system, for the reinforcement of rock cavities subjected to extreme thermal and dynamic mechanical loading. The instrumentation performed extremely well and provided the essential diagnostics to determine the residual tensions and liner-rock contact pressure. The two-dimensional thermal modeling was very adequate in predicting the plate temperature. The extensive 3-D mechanical modeling highlighted both the power and some of the shortcoming of these models. The comparison of calculational results and test data suggests the need for improvements in the rock material models and in the handling of linked static (pretensioning) and dynamic analyses.

REFERENCES

Heuze, F. E., "Preliminary Structural Investigations for a High-Energy Density Facility", Lawrence Livermore National Laboratory Report UCID-19575 Rev. 1, 37 p., (June, 1983).

Heuze, F. E., and Thorpe, R. K., "Geological Engineering Considerations for a High-Energy Facility, and a Cost Estimate for the Related Conceptual Design", Lawrence Livermore National Laboratory Report UCID-19848, 40 p., (July, 1983).

Maker, B. N.,. Ferencz, R. M, and. Hallquist, J. O, "NIKE3D - A Nonlinear, Implicit, Three-Dimensional Finite Element Code for Solid and Structural Mechanics- User's Manual", Lawrence Livermore National Laboratory Report UCRL-MA-105268, (January, 1991).

Moss, M., Koski, J. A., Haseman, G. M., and Torney, T. V., "The Effect of Composition, Porosity, Bedding Plane Orientation, Water Content, and a Joint, on the Thermal Conductivity of Tuff", Sandia National Laboratories, Albuquerque, NM, SAND 82-1164, 28 p., (November, 1982).

Shapiro, A. B., "TOPAZ2D: A Two-Dimensional Finite Element Code for Heat Transfer Analysis, Electrostatic, and Magnetostatic Problems", Lawrence Livermore National Laboratory Report, UCID-20824, 110 p., (July 1986).

Whirley, R. G., and Hallquist, J. O., "DYNA3D - A Nonlinear, Explicit, Three-Dimensional Finite Element Code for Solid and Structural Mechanics - User's Manual", Lawrence Livermore National Laboratory Report, UCRL-107254, 314 p., (May, 1991).

Zimmerman, R. M., "Conceptual Design of Field Experiments for Welded Tuff Rock Mechanics Program", Sandia National Laboratories, Albuquerque, NM, SAND 81-1768, (October 1982).

ACKNOWLEDGMENTS

This work was supported in part by the Department of Energy under contract W-7405-Eng-48 with the Lawrence Livermore National Laboratory. Sandia National Laboratory, Albuquerque, NM and the Defense Nuclear Agency were partners in this project. We thank C. Snell, Los Alamos National Laboratory, who provided cavity flow field estimates. V. Peterson and S. Uhlhorn skillfully typed the manuscript.

Innovative Mine Design for the 21st Century, Bawden & Archibald (eds) © 1993 Balkema, Rotterdam, ISBN 90 5410 325 6

Towards the practical application of ground reaction curves

Trevor G.Carter & José L.Carvalho
Golder Associates Ltd, Toronto, Ont., Canada

Graham Swan
Falconbridge Ltd, Sudbury Operations, Ont., Canada

ABSTRACT : The trend towards an increasing reliance on yielding support in highly stressed ground and the need to find a more cost-effective use of support systems in general has focused attention on achieving a better understanding of the interaction of support elements with the surrounding rock mass. One approach for achieving this is by means of ground reaction curves developed with the more practical user in mind. The reaction curve itself establishes the known dependence of supported and unsupported rock mass strength upon deformation. However, for a given excavation geometry, such curves are also rock quality dependent.

This paper presents an attempt at using standard rock mass characterization techniques in combination with the ground reaction curve concept to better identify appropriate ground support in general. Examples are given which suggest that opportunities exist to optimize support costs in a variety of ground conditions.

1. INTRODUCTION

For a number of very good reasons the practical design of support for underground mine excavations continues to favour empirical methods. An explanation for this has to do with the difficulty of characterising rock in a deterministic manner over the lifetime of an excavation.

By now it is generally well-accepted that few support systems act independently of the rock mass in which they are installed. Furthermore, for the purposes of mining economy, support systems are required to create an excavation with adequate stability, rather than to prevent rock failure. This means that theoretical design procedures must be able to quantify the fully-coupled support/rock mass interaction over time, inclusive of rock failure effects. In an active mine the "time" effect often creates drastic changes to the local rock quality in the form of progressive failure. It is these effects which are difficult to quantify both in general terms, and theoretically.

The difficulty with empirical methods is that they are only as good as the limited experience upon which they are based, which often results in misunderstandings. In this paper a concerted effort is made to build on the concept of the ground reaction curve as a basis for support design. Application of current approaches as formulated by Ladanyi (1974), Daeman (1977), Brown et al (1983), Gyenge (1987), Detournay and St. John (1988) and others is often quantitatively limited in the mining context because of the assumptions used in describing the excavation - materially, geometrically and temporarily. This tends to be a serious drawback given the common requirement to re-support non-circular excavations in ground which has been disturbed by adjacent mining over time scales of tens of years.

Because of these many uncertainties, over-design is common and support costs typically escalate with late-stage mining. It is therefore, at the very stage when a decision is required on what is the most cost-effective form of support that the means of achieving a satisfactory design are the least reliable.

With advances in computational ability using coupled finite element/boundary element codes (Brady and Wassyng, 1981), modelling of complex ground reaction behaviour with different rock quality and support methods becomes possible. Attempts have been made in this paper through numerical analysis and review of actual case experience to calibrate and understand ground reaction curves with emphasis on mature mine excavations.

Traditionally this interaction type of analysis has been based on ground reaction curves developed from modelling rock support as an equivalent internal pressure applied to the excavation boundary. This approach can only correctly model support which is external to the rock mass and which reacts to the absolute displacement of the excavation boundary. In addition, being an externally applied load, it is independent of, and does not react to rock movement. Support in the form of installed elements in the rock mass (e.g., bolts, dowels, cables, etc.) can therefore not be correctly modelled by an assumed equivalent internal pressure.

To overcome this limitation and to more accurately reflect the reaction of support elements to differential movements in the rock mass and not to the absolute displacement of the boundaries they are installed from, extensive use has been made of the program PHASES© (Hoek, 1992). Use of this hybrid finite element/boundary element program, which has plastic and reinforcement capabilities, has allowed detailed insight to be gained of support behaviour within the framework of rock quality defined by means of standard rock mass classifications. The validity of results has then been verified through back-analysis of a documented support/rock mass failure for which rock conditions have been described in terms of standard rock classification schemes as well as where the effect of installed support has been considered as an equivalent improvement in rock quality.

2. ROCK MASS QUALITY ASSESSMENT

Good geological appreciation of underground conditions is basic to achieving an adequate assessment of rock quality in order to undertake any numerical analysis. For design, the geological data-base has to be as extensive and accurate as feasible. Even the most sophisticated analysis can become a meaningless exercise if the geological information on which it is based is inadequate or inaccurate.

As classification methods themselves provide only a measure of the global quality of the rock mass, the fabric of the rock mass also has to be quantified. This is important not only for design but also so that changes that may be brought about by subsequent mining can be recognized and designs modified accordingly. Use of computer programs such as DIPS© (Hoek, 1992) provide one of the best approaches to tackling the basic statistical summarization and presentation of the masses of discontinuity data that are often collected from underground mapping in a mining environment.

A contoured stereonet provides the most useful analytical tool for synthesising structural geological data so that primary joint trends can be picked for further analysis purposes. This standard approach however can, on occasion, be misleading as often in actual underground situations, some joints (and not always the commonest set) may be of more importance than others. Further, frequently not all joints are represented by the mean, neither are all joint combinations represented by intersections of the mean planes.

Often considerations of persistence and surface roughness play a more important part in controlling stability than joint attitude. Assessing the influence of continuous, adversely oriented smooth joints in an otherwise competent rock mass is often a situation where classifications fall short. Even though joint information is usually implicitly included in the data compiled for developing the classification indices, often the influence of structure is not adequately evaluated by use of global classification indices alone. Therefore, as a pre-cursor to the analytical approaches explored in this paper some discussion is provided on the limitations of using the common rock mass classifications solely as a means for design.

2.1 Q and RMR Methods

It is correct that classifications should rely heavily on data from field mapping, core logging and observation. However care is required to obtain the correct values especially where ground conditions are problematic. The selection of input parameters for the two most common systems can be fraught with problems. For example, even when defined in the format originally specified by Barton, 1974 for the Q system or by Bieniawski, 1973, 1976, 1989 for the RMR system, the actual classification approaches themselves exhibit some inherent differences simply due to the way that each system defines different facets of the rock mass.

The Q system tends to examine the rock mass primarily from the viewpoint of joint parameters, defining rock quality on a logarithmic scale from 0.001 for extremely poor to 1000 for extremely good rock with parameters defined from:

$$Q = \frac{RQD}{J_n} \cdot \frac{J_r}{J_a} \cdot \frac{J_w}{SRF}$$

Where:

RQD = rock quality designation (Deere, 1963)
J_n = joint set number
J_r = joint roughness number
J_a = joint alteration number
J_w = joint water reduction factor
SRF = stress reduction factor

In the Q system, each of the basic terms have some physical significance. The term RQD/J_n represents the average block size of the mass; J_r/J_a the peak shear strength of controlling joints and the J_w/SRF term the relative importance of water pressure and the effects of high stress, squeezing, swelling, or loosening of the rock mass.

The Geomechanics Rock Mass Rating (RMR), by contrast, includes some evaluation of rock material strength and covers a linear scale from 0 to 100 corresponding to a range of rock quality from very poor to very good. This classification also differs from the Q system in that there is no specific attempt to define physical significance to individual parameters other than by crude weighting. The various parameters considered are:

- Uniaxial compressive strength of intact rock.
- RQD (rock quality designation).
- Spacing of discontinuities.
- Condition of discontinuities.
- Groundwater conditions.
- Orientation of discontinuities.

2.2 Applicability and Limitation

Comparison of the input details for these two principal classifications shows that the weightings of the parameters in the RMR classification are slightly different than those applied within the Q system. The RMR approach places more emphasis on intact strength and RQD measurement than on jointing patterns. While this makes RMR assessments for rock material strength related failure straightforward, and determination from borehole core somewhat easier than with the Q approach, it does tend to limit the sensitivity of the RMR method when attempting to classify subtle, but important joint controlled effects.

These inherent differences, rather than being a limitation to classification of rock masses, in fact provide a significant advantage in that when both systems are applied and the deduced RMR and Q values are compared some confirmation of the reliability of the field observation data is obtained. Not only can inherent rock mass variability be assessed, but uncertainty in parameter definition can also be better appreciated. Used together, therefore, in conjunction with other specific information on discontinuity orientation, persistence and surface conditions, such classifications provide a good basis for examining rock support interaction behaviour for both low and high stress regimes.

In a mining environment assessing rock mass conditions around excavations is not always straightforward as conditions often change with time due to adjacent mining-induced stress changes. Estimating controlling rock mass classification parameters, such as RQD is also not trivial in such situations. Again, some redundancy in parameter definition is appropriate in order to limit uncertainty or error in the assessed values. Comparison of RQD determinations from core with values based on fracture frequency measurements from field observations of scanlines is often important in the mining environment as core is not always available for areas of the rock mass for which a classification is required. In such cases average RQD values can be estimated from discontinuity mapping using average fracture frequency data, based on the exponential decay relationship proposed by Priest and Hudson (1976) correlated to other published trends (including Deere, 1964), ie.

$$RQD = 100e^{-0.15\lambda}\ (0.15\lambda + 1)$$

where: λ is the mean fracture frequency (fractures/metre)

Values computed from this approach based on mapping underground can be checked using the method proposed by Palmström (1982); ie.,

$$RQD = 115 - 3.3\,J_v$$

where J_v is a measure of the number of joints crossing a cubic metre block of rock.

3. FAILURE CRITERION AND ANALYSIS PARAMETERS

3.1 Hoek-Brown Criterion

The well known Hoek-Brown failure criterion in terms of principal stresses at failure in two dimensions is expressed as follows:

$$\sigma'_1 = \sigma_3' + \sqrt{m\sigma_c\sigma_3' + s\sigma_c^2}$$

where:

σ'_1 and σ'_3 are major and minor effective principal field stresses, and

σ_c is the uniaxial compressive strength of the intact rock.

In this relationship the m and s empirical parameters describe the rock material fabric and rock mass fabric respectively; where m; for intact rock ranges from values of the order of 5 for rocks with highly ordered generally weakly interlocked crystalline microfabrics to 25 or more for rocks with polymineralic highly interlocked, inherently strong crystal microstructures; and s ranges from 1 for intact rock to 0 for a completely disaggregated rock mass.

For the numerical analyses involving plasticity discussed in this paper, the three dimensional form of the Hoek-Brown failure criterion has been used with a non-associative flow rule (tension positive).

At yield,

$$F = \frac{m\sigma_c I_1}{3} + 4J_2\cos^2\theta + m\sigma_c\sqrt{J_2}\left[\cos\theta - \frac{\sin\theta}{\sqrt{3}}\right] - s\sigma_c^2 = 0$$

...and...

$$Q = \frac{\alpha\sigma_c I_1}{3} + 3J_2 + \frac{\sqrt{3}\,\alpha\sigma_c\sqrt{J_2}}{2} = 0$$

where

F = failure criterion
Q = strain potential
I_1 = first spherical stress invariant
J_2 = second deviatoric stress invariant
Θ = lode angle, and
α = rock mass dilation parameter
and m, s and σ_c are as defined previously

Evaluation of the plastic behaviour of the rock mass at yield requires understanding of the dilation parameter, α, included in the above expression. Some estimate can be made from the Barton-Bandis model of dilation behaviour for single natural joints knowing the dilation angle:

$$d_n = JRC\ \log\ (JCS/\sigma_n)$$

where:

σ_n = normal stress across the joint
JRC = Joint roughness coefficient, and
JCS = $JW_a \cdot \sigma_c$ = Joint wall strength, with
JW_a = Joint wall alteration number

Knowing the dilation angle the one dimensional dilation ratio for a single joint set can be established from:

$$\frac{\Delta\varepsilon_{normal}}{\Delta\varepsilon_{shear}} = \lambda\tan(d_n) = \lambda\tan\left[JRC\ \log\left[\frac{JW_a\sigma_c}{\sigma_n}\right]\right]$$

where λ = fracture frequency (ie. 1/joint spacing)

Assuming for modelling purposes, that three orthogonal sets of joints with the same surface characteristics exist within the rock mass, then the 3-dimensional dilation ratio can be approximated by:

$$\frac{\Delta\varepsilon^p}{\Delta\Gamma^p} = 3\lambda\tan\left[JRC\ \log\left[\frac{JW_a\sigma_c}{P}\right]\right]$$

where:

$\Delta\varepsilon^p$ is the volumetric plastic strain

$\Delta\Gamma^p$ is the deviatoric plastic strain

This expression in turn can be equated to the dilation ratio for the Hoek-Brown criterion, ie.:

$$\frac{\Delta\varepsilon^p}{\Delta\Gamma^p} = \frac{\alpha}{\frac{3\sqrt{J_2}}{\sigma_c} + \frac{\sqrt{3}\,\alpha}{4}}$$

This then allows calculation of the dilation parameter, α, in terms of the dilation angle on each of the three assumed orthogonal joint sets, ie.

$$\alpha = \frac{9\tan(d_n) \cdot \frac{\sqrt{J_2}}{\sigma_c}}{\frac{1}{\lambda} - \frac{3\sqrt{3}}{4}\tan(d_n)}$$

where d_n = dilation angle

Realizing that when plastic flow occurs, the state of stress must lie on the yield surface, the normalized deviatoric component in the above expression can then be defined in terms of the Hoek-Brown failure criteria, ie;

$$\frac{\sqrt{J_2}}{\sigma_c} = M = \frac{1}{2}\sqrt{\left(\frac{m}{4}\right)^2 + \frac{mP}{\sigma_c} - s} - \frac{m}{8}$$

where M is the normalized deviatoric stress.

Substituting, an expression is developed relating the Hoek-Brown dilation parameter, α, to the Barton-Bandis joint roughness parameter, JRC, ie.;

$$\alpha = \frac{9M\tan\left[JRC\log\left(\frac{JW_a\sigma_c}{P}\right)\right]}{\frac{1}{\lambda} - \frac{3\sqrt{3}}{4}\tan\left[JRC\log\left(\frac{JW_a\sigma_c}{P}\right)\right]}$$

This expression now allows plastic three dimensional dilation of the rock mass to be numerically evaluated at any appropriate stress level knowing the surface roughness characteristics of the controlling joints.

3.2 Parameter Assessment

In order to obtain a valid model of the behaviour of a rock mass in a mining environment it is essential that representative data be gained regarding shear and compressive strengths, deformation properties, and Poisson's ratio for the rock mass. Shear and normal stiffness data for the controlling discontinuities within the rock mass are also required for support assessment modelling.

It is often extremely difficult to define such parameters precisely. However, it has now become viable to generically estimate many of the required parameters based on various empirical and semi-empirical correlations (Carter et. al., 1990). Definition of controlling input parameters for the modelling undertaken for this paper has therefore been carried out by reference to the classification data in Tables 1 and 2 which are based on various published correlation relationships between RMR values or Q indices and controlling geotechnical parameters, (Bieniawski, 1973, Barton et al., 1974 and others).

3.3 Deformation Parameters

For deformation properties specifically, rock mass parameters can be derived explicitly from the following relationships:

$$\text{Modulus of deformation, } E_{rm} = 10^{\left(\frac{RMR-10}{40}\right)} \text{ GPa}$$

$$\text{Shear modulus, } G_{rm} = \frac{E_{rm}}{2(1+\nu_{rm})}$$

$$\text{Bulk modulus, } K_{rm} = \frac{E_{rm}}{3(1-2\nu_{rm})}$$

where ν_{rm} is the Poisson's Ratio for the assumed transversely isotropic jointed rock mass, ie.,

$\nu_{rm} = \nu_i(E_{rm}/E_i)$,

...and...

ν_i is the Poisson's Ratio for the intact rock material.

Table 1 Descriptive Assessment of Rock Quality Indices and Joint Spacings

Rock Mass Quality	CSIR RMR	NGI Q	RQD %	Joint Spacing (mm)	Characteristics Using Terzaghi, (1946) Terminology	Typical Rock Condition
Exceptionally Good Quality	100	500	100	≥ 3000	Massive, intact "sound" rock	Virtually non-jointed high strength rock mass free from significant rock defects.
Very Good Quality	85	100	100	625	Massive to moderately jointed	Tightly interlocking, jointed, undisturbed rock with unweathered joints spaced at ± 3 metres. Block sizes range from several metres to tens of metres in dimension. RQD usually above 85%. Joint surfaces usually rough.
Good Quality	65	10	75	350	Moderately blocky and seamy	Fresh to slightly weathered rock, often with stained discontinuities, spaced at 1-3 metres. Controlling block size in the 1.0 - 5.0 metre range. RQD's typically 75-85%.
Fair Quality	44	1	50	160	Moderately block and seamy	Several sets of weathered or low strength discontinuities, spaced at 0.3 - 1 m. Block sizes typically about 0.5 m. RQD's typically 40-60%.
Poor Quality	23	0.1	35	80	Very block and seamy	Numerous sets of weathered or low strength discontinuities spaced at 30 mm to 500 mm. Some with gouge filling. Rock material often showing penetrative staining and alteration. RQD's typically 25-40%.
Very Poor Quality Exceptionally Poor Quality	3 0	0.01 0.001	≤ 20	≤ 25	Very blocky and seamy to completely crushed	Closely jointed, or foliated, with controlling block size in the mm. to cm. range. RQD's typically 0-25%. Sugar cube jointing or paper thin foliation. In near surface environments more competent parent rock materials may exhibit extensive weathering and closely spaced, heavily weathered discontinuities.

Table 2 Representative Deformation and Strength Parameters for Various Rock Mass Qualities

Rock Conditions	Hoek-Brown Parameters		RMR	Q	Poisson's Ratio	Rock Mass Moduli			Ubiquitous Joint Stiffnesses (Goodman)	
						Deformation	Bulk	Shear		
	m/m_i	s			v_{rm}	E_{rm} (GPa)	K_{rm} (GPa)	G_{rm} (GPa)	J_{kn} (GPa/m)	J_{ks} (GPa/m)
Very Poor Quality	0.03	0.00002	3	0.01	0.04	0.7	0.2	0.3	30.0	15.0
Poor Quality	0.06	0.0002	23	0.1	0.10	2.1	0.9	1.0	42.0	21.0
Fair Quality	0.14	0.002	44	1	0.16	7.5	3.7	3.2	115.0	56.0
Good Quality	0.29	0.02	65	10	0.22	23.7	14.0	9.7	400.0	200.0
Very Good Quality	0.59	0.2	85	100	0.19	75.0	40.5	31.5	430.0	220.0

3.4 Stiffness Parameters

Appropriate normal and shear stiffness parameters for a wide range of rock mass conditions are harder to estimate as comprehensive data on joint stiffnesses are not readily available. However, some information has been published, for example by Bandis et al.(1983), Barton (1976 and 1981), Swan (1983 and 1985) and by Kulhawy (1975). With this information as a guide, some estimate of shear and normal stiffnesses can also be determined explicitly, if information is available on the spacing and character of the joints and on the elastic modulus of the intact rock and rock mass. The governing relationships for a ubiquitous joint approach are given by Goodman (1989); as follows:

$$\text{Joint Normal Stiffness: } J_{kn} = \frac{E_{rm} \cdot E_i}{S(E_i - E_{rm})}$$

$$\text{Joint Shear Stiffness: } J_{ks} = \frac{G_{rm} \cdot G_i}{S(G_i - G_{rm})}$$

where:

E_i, G_i and E_{rm}, G_{rm} = elastic and shear moduli for the intact rock and rock mass respectively, ...and...

S = controlling joint spacing

Alternatively individual joint stiffnesses can be expressed in terms of the Barton-Bandis empirical constitutive model which relates stiffnesses to joint wall strength and roughness.

With this approach values of normal stiffness (K_n) at different levels of normal stress can be calculated using the following equation:

$$K_n = K_{ni}\left[1 - \frac{\sigma_n}{V_m K_{ni} + \sigma_n}\right]^{-2}$$

where V_m, = maximum irrecoverable joint closure. This is given by:

$$V_m = A + B(JRC_o) + C\left[\frac{JCS_o}{a_j}\right]^D$$

The initial stiffness can be derived from:

$$K_{ni} = 0.02\left[\frac{JCS_o}{a_j}\right] + 1.75\ JRC_o - 7.15$$

and where JCS_o is in (MPa)

a_j is the initial aperture in (mm) for each cycle of loading for four incremental stages

V_m is in (mm), and

the constants A, B, C and D, corresponding to the four cycles of loading through to the assumed "undisturbed" *in situ* condition are as tabulated below:

Constants for Closure Calculations				
	First Cycle i = 1	Second Cycle i = 2	Third Cycle i = 3	Fourth Cycle i = 4
A_i	-0.296	-0.1005	-0.1031	-0.1031
B_i	-0.0056	-0.0073	-0.0074	-0.0074
C_i	2.241	1.0082	1.135	1.135
D_i	-0.245	-0.230	-0.251	-0.251

With joint properties known shear stiffnesses can then be derived from:

$$K_s = \frac{1}{\delta} \cdot \sigma_n' \tan\left[JRC_L \log\left(JCS_L/\sigma_n'\right) + \phi_r'\right]$$

where JRC_L is the scaled roughness for the length of sheared joint and the joint displacement δ is given by...

$$\delta = \frac{L}{500} \cdot \left[\frac{JRC_L}{L}\right]^{0.33}$$

It will be noted that the shear stiffness is not only stress dependent it is significantly scale dependent, whereas normal stiffness, although stress dependent is effectively independent of scale. Thus, in addition to knowing joint surface conditions, some idea must be had of controlling joint spacing and persistence. Estimates can however be readily made based on published correlations (Barton, 1989, Swan 1974).

4. JOINT ROCK INTERACTION BEHAVIOUR

4.1 Stress State

For late stage mining applications, it is important to be able to evaluate the effects of stress changes on support and on jointing as these factors control stability. Commonly this involves determining changes of stress that occur in the crown zone above a drift. This can be readily accomplished for any required geometry by means of PHASES© or by boundary element or other numerical modelling approaches. Alternatively, changes in stress state surrounding the immediate vicinity of a mined drift can be simplistically examined analytically by means of the Kirsch solutions for a circular opening.

By either means it then becomes possible to determine the normal and shear stresses at a point on a hypothetical joint that may be located centrally within the crown area of the drift. This then allows comparison to be made to the available rock mass strength, which can be defined in terms of the Hoek-Brown failure criteria. This available strength can be readily determined for various rock qualities for undisturbed rock conditions based on published correlations to RMR and Q, ie.;

$$m/m_i = \exp\left[(RMR - 100)/28\right]$$

$$s = \exp\left[(RMR - 100)/9\right]$$

where $RMR = 9\,\mathrm{Log}_e Q + 44$ or $Q = \exp\left[\dfrac{RMR - 44}{9}\right]$

and m_i is the Hoek-Brown frictional parameter for the *intact* rock (Hoek and Brown, 1988).

This approach allows development of a conceptual and analytical model to evaluate the effects of joint roughness and stress shadowing on the stability of drifts adjacent to larger mined openings. As an example, consider the excavation shown on Figure 1. When initially developed, this excavation was stable in relatively highly stressed ground. With development of large stopes 60 m away the drift became shadowed and initial stresses decreased as a result of excavation geometry effects. If this stress decrease is expressed as a ratio of the initial stress σ_I at the crown of the drift to the final stress after excavation of the adjacent opening σ_F, and these stresses are resolved into their normal and shear components, the influence of the normal stress drop on available shear strengths within the rock mass can be explored in terms of joint roughness using the Barton-Bandis JRC concept and the previously discussed failure criteria.

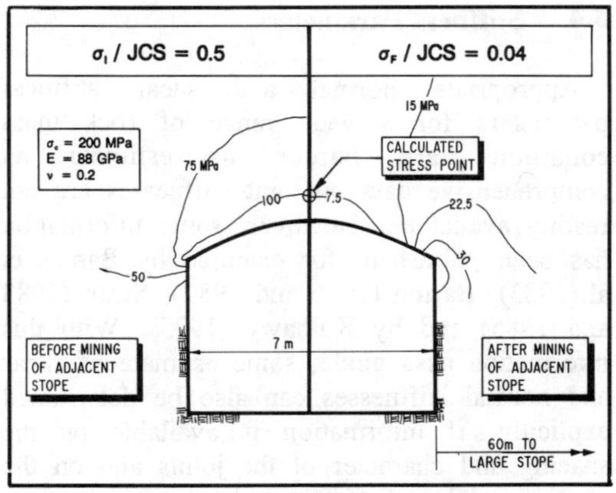

Figure 1 Stress Change Induced by Adjacent Mining of Large Stope 60 m Away from Drift Location

4.2 Mobilized Strength

To examine the relationship between mobilized equivalent roughness on the joints at failure, it is appropriate to take the non-linear Mohr-Coulomb failure envelope expressed in terms of the Hoek-Brown frictional parameter, m, ie.

$$\tau = (\mathrm{Cot}\,\phi' - \mathrm{Cos}\,\phi')\,\frac{m\sigma_c}{8}$$

and calculate the instantaneous peak friction angle ϕ' at the appropriate normal stress, existing across the controlling joints within the crown of the drift, ie.,

$$\phi'_{peak} = \mathrm{Arctan}\,\frac{1}{\sqrt{4h\mathrm{Cos}^2\theta - 1}}$$

$$\text{where;}\quad h = 1 + \frac{16(m\sigma_n' + s\sigma_c)}{3m^2\,\sigma_c}$$

$$\text{and,}\quad \theta = \left[1/3\,\left[90° + \mathrm{Arctan}\left[\frac{1}{\sqrt{h^3 - 1}}\right]\right]\right]$$

This then allows the instantaneous cohesive strength c' to also be derived, ie.,

$$c' = \tau - \sigma'_n \,.\, \tan\phi'$$

If it is then assumed that these peak c', ϕ' parameters govern failure of any interlock on incipient controlling joints within the rock mass it becomes possible to relate Hoek-Brown peak strengths to the strength of the controlling joints

Assume then, that for any given rock mass, the residual friction angle for shear on the controlling joints within the mass can be related to J_a. The value of ϕ_r therefore can be derived from inspection of the descriptive tabulation for the Q system parameters (Barton, 1974) or can be estimated from the relationship:

$$\phi_r = \left[\frac{J_a}{6000}\right]^{-0.42}$$

Alternatively a relationship between residual friction angle, ϕ_r, and discontinuity conditions can be characterized using shear box test data or can be inferred from the empirical relations based on Schmidt hammer and tilt test data (Bandis, 1981). Taken together with the second term of the Q parameter classification ie. J_r/J_a, which provides an estimate of the peak shear strength (friction angle) for the controlling joints within the rock mass, a relationship can be developed between the Hoek-Brown derived ϕ_{peak} values and J_r and thence using some correlations published by Barton, 1984, JRC (the NGI-Q joint roughness parameters); ie.,

$$J_r = J_a \tan \phi_{peak}$$

...and...

$$JRC = 5J_r \left[1 - \left[\frac{12}{14.5 + J_r^8}\right]\right]$$

This then allows the dilation component d_n and hence α for the controlling joints to be evaluated knowing joint wall strength based on:

$$JCS = JW_a \cdot UCS$$

where JW_a is a weathering index, or joint wall alteration parameter that can be expressed in terms of J_a as follows:

$$JW_a = 1.05 - \tanh\,[(J_a - 0.9)/2]$$

This weathering index, JW_a, is introduced by reference to standard weathering description and classification work, simply as a means to derate σ_c to match natural weathering behaviour of joints which suggests that as weathering increases (and JW_a decreases), the joint wall strength decreases asymptotically to some lower bound value appropriate to the characteristics of the joint infill material (Table 3).

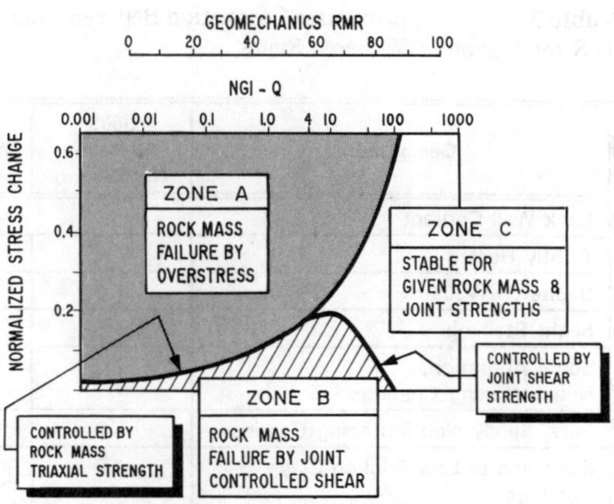

Figure 2 Conceptual Behaviour of Rock Mass and Joint Controlled Failure due to Stress Change Effects

With the connections made between the instantaneous Hoek-Brown ϕ_{peak} value and Barton-Bandis equivalent JRC's for any given rock quality, a complete matrix of controlling roughness values can be determined for variations in rock mass quality defined by the complete range of the Q or RMR classification systems. This then allows a family of curves to be developed for any given stress drop from in situ controlling stresses to a destressed situation. The conditions for which the rock mass fails in triaxial compression can thus be defined, as also can the zone in which shear occurs on ubiquitous jointing, (Figure 2).

4.3 Required Strength

If it can be assumed that gravitational release of material within the crown of an excavation can be related to the controlling rock mass quality, either by means of support pressure correlations (as inferred from by modelling, or as indicated by Barton, 1977 or Unal, 1974 or using by relationships such as proposed by Rose, 1981 based on Terzaghi's classic Rock Load Concept), it becomes possible to compare the frictional resistance inherent within the rock mass (as would be represented by the instantaneous peak friction angle deduced from Hoek-Brown relationship) to the frictional resistance and interlock required to maintain stability against gravitational ground falls for that given rock mass.

The required frictional resistance or shear interlock of the joints or the rock mass can thus be expressed in terms of an equivalent joint

Table 3 Approximate Correlation Between Joint Alteration Numer J_a and Joint Wall Rock Parameters JW_a and JCS for Various Weathered States

Description	Joint Alteration Number, J_a	Weathering Grade	Joint Wall Alteration Parameter JW_a	Typical Wall Rock Strength JCS
Rock Wall Contact				
Tightly Healed	0.75	Ia	1.1	>50 MPa or rock strength
Unaltered Walls	1.0	Ia	1.0	>25 MPa or rock strength
Slight Staining	1.5	Ib	0.8	
Some Alteration, Non-Softening Coatings	2.0	Ib	0.55	25 - 50 MPa
Silty, Sandy Non-Softening Coatings	3.0	Ib	0.27	15 - 30 MPa
Softening or Low Friction Clay Coatings	4.0	II	0.14	5 - 15 MPa
Contact for 10 cm Shear				
Sandy, Disintegrated Zone	4.0	III	0.14	5 - 15 MPa
Strongly Overconsolidated Non-Softening Clay <5mm thick	6.0	III-IV	0.08	0.25 - 10 MPa
Medium Consolidated Clay <5mm thick	8.0	IV-V	0.05	0.25 - 5 MPa
No Contact				
Gouge Zones of Sandy/Silty or Crushed Rock Material	8.0	III-IV	0.10	0.25 - 5.0 MPa
Gouge Zones with Clay Matrix	10.0	IV-V	<0.05	0.1 - 5.0 MPa

Notes:

1. Values of ϕ_r, the residual friction angle, JCS, and JW_a are intended as approximate guides related to wall rock surface conditions, assuming mineralogical properties of the alteration products, if present, modified from values presented by Barton et al., 1974.
2. Weathering grade data modified from ISRM 1978, QJEG, 1972, 1977, BS 5930, 1990, Brown, 1981 and Martin & Hencher, 1989.
3. Approximate empirical relationships between major properties: - $JCS = JW_a \cdot \sigma_c$; $JW_a = 1.05 - \tanh[(J_a - 0.9)/2]$; $J_a = 6000\ \phi_r^{-2.4}$

roughness needed for stability, in an analogous manner as has been done for deriving mobilized roughness. If the support pressure expression of Unal (1974) is used, ie.,

$$p = \frac{100 - RMR}{100}\gamma B$$

and the rock mass is analysed to examine the required frictional resistance on the controlling joints within the crown zone, under the defined stress state a relationship can be developed between JRC, the maximum required clamping stress and the controlling rock mass quality.

4.4 Empirical Stress Change - Rock Quality Classification Chart

If now the results of the evaluation for $JRC_{required}$ and $JRC_{mobilized}$ with respect to the Hoek-Brown relation at yield for a given m_i (intact rock material) are plotted with respect to rock quality, Q or RMR, three failure zones can be defined (Figure 2). Zone A on the left side of the diagram, in which the rock mass strength of the jointed, fractured rock mass is insufficient in triaxial compression to prevent failure under the previously defined in situ stresses; a central portion on the chart, Zone B, where mobilized rock mass strength is insufficient to prevent failure by shear on discontinuities without significant roughness existing on the discontinuities characterizing the rock mass, and Zone C where, for the known joint roughness, the inherent rock mass strength is well in excess of the demand placed upon it.

In concept therefore, (assuming that the rock is characterized by an idealized set of discontinuities) the restraint provided by the rock mass in Zones B and C is mainly provided by shear interlock on the discontinuities as well as by the intrinsic strength of the rock material between the discontinuities. A rock mass with a low rock

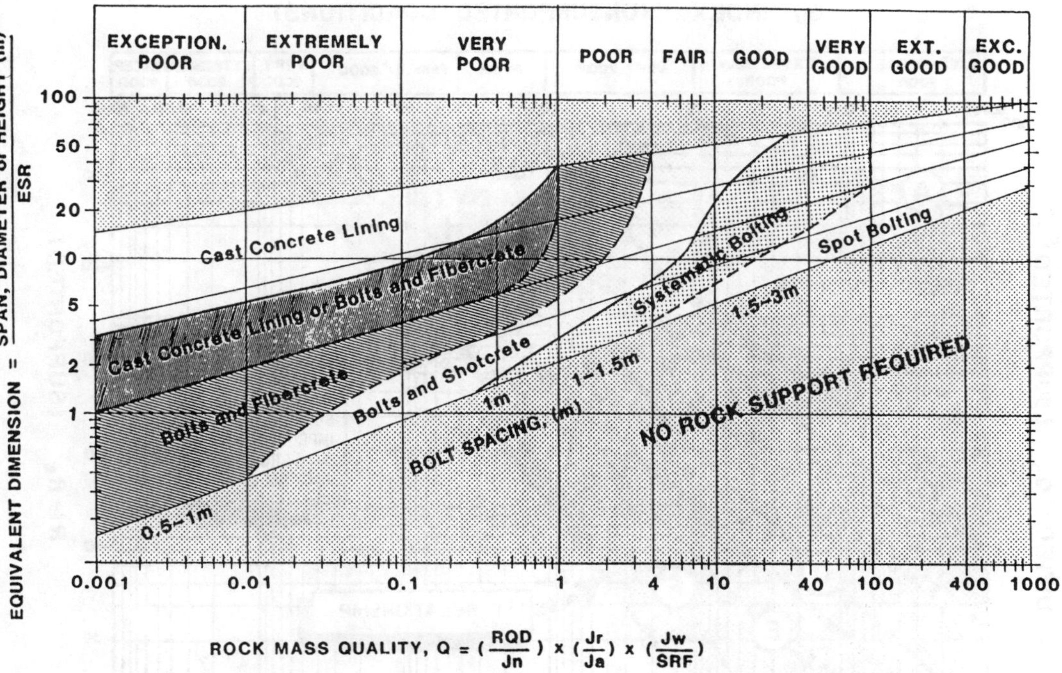

Figure 3 Recommended Ground Support for Q System

quality can be considered to only have available a certain level of mobilizable frictional resistance to gravity failure. This frictional resistance, when expressed in terms of the Mohr-Coulomb instantaneous friction parameters for the non-linear Hoek-Brown envelope, can also be equated to equivalent roughness on controlling discontinuities in terms of the Barton-Bandis model. By this means, charts of the form shown in Figure 2 can be readily prepared for any given m_i values, noting that the residual strength for any given rock mass is rock material dependant and therefore consistent across the rock quality scale.

5. EMPIRICAL SUPPORT GUIDELINES

The influence on opening stability of the effect of support (in the form of bolts, cables, shotcrete, etc.), is well documented in a variety of empirical guidelines included in the mining and civil engineering rock mass classifications. The mining schemes - the Mathews method - Golder Associates, 1981, Potvin et al., 1989 and the MRMR system, (Laubscher, 1977) provide information on cable support and capability, while the basic civil engineering schemes - (the Q system, Barton, 1974, and the RMR scheme, Bieniawski, 1973, 1976, 1989) provide a range of information on support types ranging from bolts to concrete arches (Figure 3). Although much of the support doctrine has been formulated empirically it is amenable to being addressed more quantitatively and thus guidelines can be rationally extended to provide calibration for modelling purposes. This can be most readily achieved if the influence of support is considered not as a component part of the engineered opening, but rather as an effective improvement in rock mass quality (Carter, 1992).

The following example illustrates this concept: Suppose, based on Barton's 1989 guide for tunnelling support (Figure 3), it is required that a 2.5 m wide opening be excavated and supported in a very poor quality rock mass (Q = 0.1, RMR = 23), then bolts and fibrecrete are recommended for support. As this same dimension span is suggested from Barton's data to be stable without any support in a rock mass with a Q rating of about 2, the application of the recommended support can be considered to effect an improvement in rock quality from Q = 0.1 to

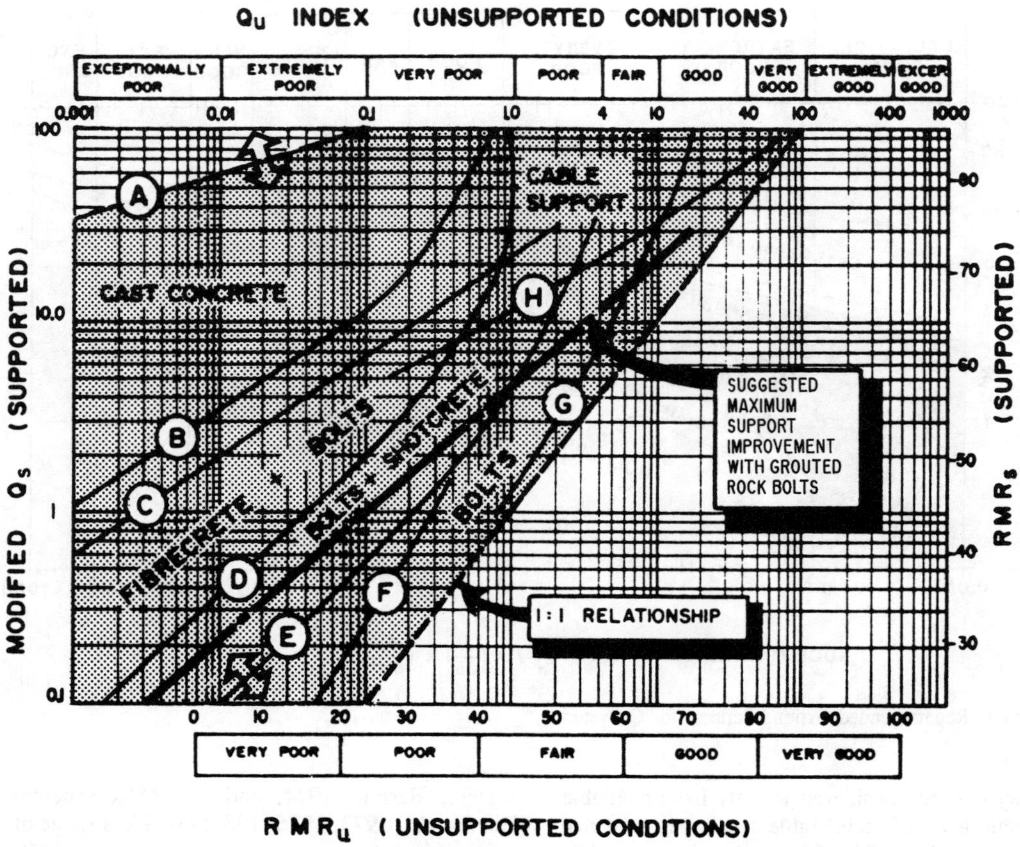

Curve	Support Method	Typical Support Pressure (T/m^2) (Hoek, 1980)	Derived Relationship	Source of Data
A	Maximum Support Limit		$Q_s = 150Q_u^{0.2}$	Barton, 1974
B	Concrete 0.3-0.5m Thick	100-600	$Q_s = 40Q_u^{0.5}$	Barton, 1974,1989
	Heavy Steel Sets	50-250	-	Hoek & Brown,1980
C	Cable Bolting @ 1 per m² density	30-50	$Q_s = 20Q_u^{0.5}$	Bawden,1988 Potvin,1989
D	Fibre/Shotcrete + Bolts	20-60	$Q_s = 8Q_u^{0.65}$	Barton, 1974
E	Onset for Mining Support	-	$Q_s = 2.5Q_u^{0.67}$	Laubscher, 1977
F	Patterned Resin Bolting @ 1.0m. Spacing	30	$Q_s = 5Q_u^{1.25}$	Barton, 1974,1989
G	Patterned Mechanical Bolts @ 1.5m. spacing	10	$Q_s = 1.5Q_u^{1.25}$	Barton, 1974
H	Mining Limit for Cable Support	-	$Q_s = 12.5Q_u^{0.45}$	Potvin, 1989

Figure 4 Equivalent Improved Rock Mass Quality Relationships for Various Support Methods

about 2. If this latter value is termed Q_s for the improved, supported condition, and this same approach is taken to examine all Barton's support curves, these can be replotted as illustrated on Figure 4. This same methodology can then also be applied to the support lines proposed for use with other classifications (eg., Laubscher, 1977, Bawden, 1988, and Potvin, 1989) as has been done on the Figure.

In this form, this chart can be used to provide some guidance for modelling deformation behaviour of supported ground. As support is generally needed more at the poor rock quality end of the spectrum, regression fits have been computed for the linear sections of the tails of each of the various curves shown on Figure 4. These derived expressions are tabulated below the Figure.

6. EVALUATION OF BEHAVIOUR OF BOLTED GROUND

6.1 Model Arrangement

In order to consider a wide range of in situ stresses without the need to generate a large number of runs, analyses were carried out using the program PHASES©. A single excavation shape, namely, a 5 m square drift with a curved roof was analysed. The type of support considered was end anchored patterned bolts as are commonly used in many of the mines with the stiffness of the support system ranging from 5 MN/m to 40 MN/m per m² of roof area corresponding with typical ranges used in the field. The ranges were modelled both by changing the stiffness of the elements themselves and by changing the spacing between them.

6.2 Input Conditions

- **Stresses**

To achieve a relatively wide range of rock mass strength to stress conditions for a range of in situ stresses, the governing principal stresses q^o and σ_3^o were split into their spherical (confining stress) and deviatoric (driving stress) components. These stresses were then normalized with respect to the rock mass uniaxial unconfined compressive strength ($\sqrt{s}\sigma_c$) ie.:

$$P_o^* = \frac{\sigma_1^o + \sigma_3^o}{2\sqrt{s}\ \sigma_c} \ and\ S_o^* = \frac{\sigma_1^o - \sigma_3^o}{2\sqrt{s}\ \sigma_c}$$

For the analyses, a range of in situ stresses with normalized spherical components (P_o^*) ranging from 1 to 2.5 times the rock mass strength were analysed in order to provide a reasonable assessment of typical field stress conditions. Further, four levels of maximum allowable in situ normalized deviatoric stress S^*_{oall} were evaluated to match typical obliquity conditions in virgin stress states; with this deviatoric stress being evaluated with respect to the Hoek-Brown failure criterion constants at yield, ie.,

$$S_{o_{all}}^* = \sqrt{\frac{m^2}{64s} + \frac{mP_o^*}{4\sqrt{s}} + \frac{1}{4}} - \frac{m}{8\sqrt{s}}$$

It should be appreciated that these allowable stresses are the computed maximum stresses on the failure envelope, representing the yield stresses at which the rock mass would fail under in situ conditions (prior to any excavation). An opening under this stress state would therefore fully mobilize any installed support.

- **Rock Mass Quality**

Conditions were then analysed for each in situ stress condition for three rock qualities (RMR = 23, Q = 0.1; RMR = 44, Q = 1; and RMR = 65, Q = 10) for each in situ stress condition (ie. 16 combinations of P_o^* and S_o^*)

- **Support Cases**

Five individual support conditions were examined representing typical conditions experienced in many mining development headings:

- Unsupported
- (B1) End anchored 1" diameter, 10 ft. long. (mechanical shell) bolts each with a stiffness k_b=35MN/m @ 1.45 m centre to centre (c/c), (ie. 17.5 MN/m/m² of roof area),
- (B2) End anchored bolts each with a stiffness k_b=35N/m @ 1.0 m c/c (ie. 35 MN/m/m² of roof area),
- (B3) End anchored bolts each with a stiffness k_b=35MN/m @ .71 m c/c (ie. 70 MN/m/m² of roof area), and
- (B4) End anchored bolts each with a stiffness k_b=35MN/m @ 0.5 m c/c (ie. 140 MN/m/m² of roof area).

For these analyses, rather than considering the force provided by each bolt ($T = k_b \cdot \Delta u$), which depends on the differential displacement between anchor and collar, the ground reaction curves computed from the analyses have been expressed in terms of the average support load provided by the pattern. It should be noted that this support load acts as a restraint on dilation of the rock mass between the end anchored shell location and the face plates, thus inducing confinement, which results in strength gain.

6.3 Support Interaction Results

The results of the analyses show that these types of bolts exert little control on the absolute displacement of the roof although they can significantly reduce the extent of the failed (plastic) zone. This behaviour was also found for comparative runs with fully column bonded bolts and cables. This is not unexpected since a large percentage of rock mass displacement observed on the roof/walls of the analysed excavation reflects elastic expansion of the rock mass for which differential elastic displacements (anchor to collar) are not sufficient to mobilize large forces in the bolts.

It is therefore required that plastic deformation (due to failure of the rock mass) occur to mobilize support. The magnitude of this plastic displacement is largely controlled by the dilation characteristics of the rock mass. Typically, the middle quality range of rock mass exhibit the largest dilation ratios. This is because the poorest quality rock masses tend to crush rather than dilate and the very good quality rock masses with smooth joints, also do not produce much dilation. If the dilation parameter, α for the Hoek-Brown criterion is estimated by means of appropriate JRC values for various roughnesses and rock qualities, it can be found that α ranges up to values of 4 or 5 for fair to good quality rock, going to zero for poor rock. (It also can approximately be related to m/4).

Further, it should be appreciated that absolute dilation will be highly stress dependent, decreasing with an increase in $\sqrt{J_2}$ (driving stress). This is the correct trend since higher values of $\sqrt{J_2}$ and hence lower values of α can only occur with increases in I_1 (confinement).

The typical ground reaction curves for three rock qualities and the range of stresses are shown in Figure 5 for point A (at the midpoint of the drift roof).

Because the bolts only experience differential displacement, the curves shown on Figure 5 are plotted in terms of the absolute displacement of point A. The conventional reaction curves for a circular opening under isotropic stress conditions are also shown for the poor rock case for comparison.

These curves are all plotted for the stress conditions at which significant plastic deformation occurs, ie. highly stressed ground relative to the rock strength. Actual confining, P_o, stress values therefore vary from a comparative low level for the poor rock case to a very high value for a good rock condition.

The influence of stress can be seen on Figure 6 where the variation in support reaction is plotted for the various bolting densities for each of the rock mass quality conditions analysed. The behaviour of the good rock at this single stress level is essentially elastic while at the poor rock quality end of the scale, significant plastic deformation occurs, fully mobilizing the strength of the bolt system.

6.4 Equivalent Rock Mass Quality Method

As a comparison to the models where actual bolts were installed, some comparative runs were undertaken to evaluate the accuracy of the Q supported, or improved rock quality assessments, based on the chart in Figure 4.

Two support cases were investigated for poor rock (RMR=23, Q=0.1) assuming a normalized in situ stress condition of $P_o^* = 2.5$, $S_o^* = 1.35$. For this rock mass quality, with $s = 0.0002$, actual field stresses would be approximately $\sigma_H = 0.05\sigma_c$ and $\sigma_V = 0.02\sigma_c$. The first support case was analysed with end anchored, patterned bolts @ 0.5 m c/c with a nominal stiffness of $k_b = 35$ NM/m (ie. 140 MN/m/m^2 of roof area). The second case analysed assumed the same bolt pattern but with the addition of a 75mm thick layer of shotcrete. The results of analyses of the support systems were then compared to the results achieved by substituting a zone of equivalent improved rock mass according to the chart in Figure 4; where the following guidelines were used:

- F line - for bolts only - providing equivalent supported rock mass quality improvement from $RMR_u = 23$ to $RMR_s = 33$.

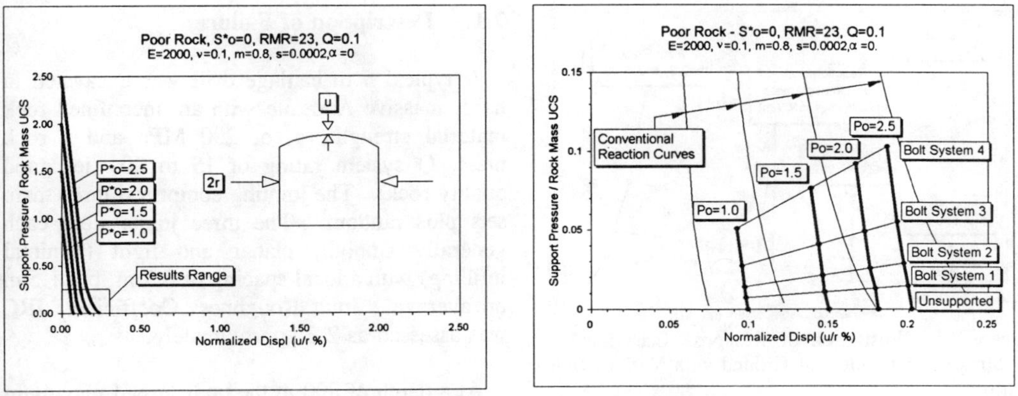

a) **Ground Reaction Curves for a Bolted Excavation in Poor Rock Conditions**

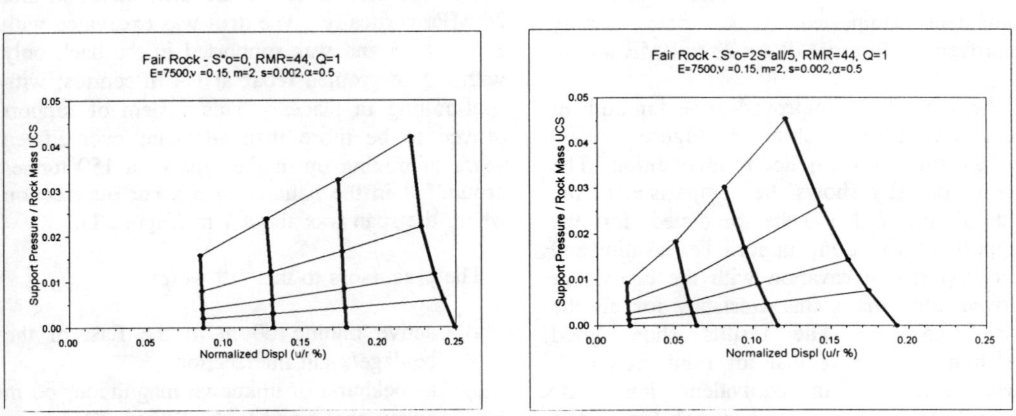

b) **Ground Reaction Curves for a Bolted Excavation in Fair Rock Conditions**

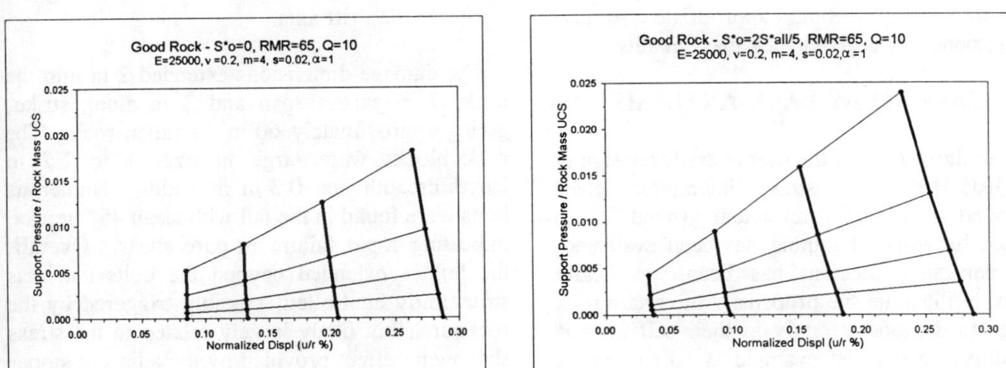

c) **Ground Reaction Curves for a Bolted Excavation in Good Rock Conditions**

Figure 5 Illustration of the Influence of Bolt Density on Ground Reaction Behaviour for Mining Excavations within Rock of Different Qualities

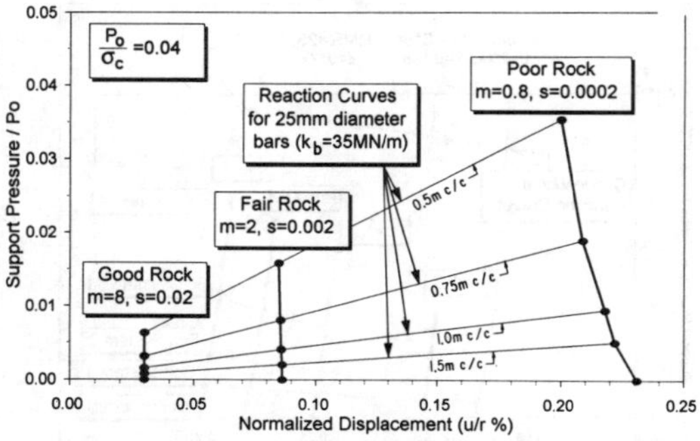

Figure 6 Illustration of Response Behaviour at Low Stresses of Reinforced Ground with Various Bolt Densities

- D line for bolts and shotcrete - providing equivalent supported rock mass quality improvement from $RMR_u = 23$ to $RMR_s = 50$.

The extent of the improved zone for both of these support cases is shown in Figure 7 along with the results for the bolted roof condition. The figure graphically shows the comparison of the extent of the failed zone predicted for the unsupported excavation, supported excavation and the unsupported excavation with the equivalent improved rock mass characteristics for all the analysed cases. The results show good correlation, suggesting that for reinforced rock masses, assessing an equivalent improved, supported rock quality rating based on Figure 4 is an appropriate and quick method both for modelling the influence of installed support as well as for selecting appropriate support requirements for a given rock mass quality.

7. CASE STUDY BACK-ANALYSIS

As an illustration of the usefulness for design of the JRC/H-B rock stress interaction chart illustrated on Figure 2 an actual ground failure induced by adjacent mining has been evaluated. This particular incident has been specifically chosen to illustrate the propensity of certain rock masses to degrade strongly in their self-support capability. Since the example is taken from a mining context, the degrading capability, strong or weak, is not simply a time-dependent effect but more importantly a reduced confinement effect. An understanding of these effects is critical to substantiation of recent claims that rock excavations always tend to stabilize over time and become less dependent on the bearing capacity of support systems, Helfrich (1989).

7.1 Description of Failure

A typical 5 m haulage drift was excavated in hard, massive Andesite with an unconfined rock material strength, σ_c of 200 MPa and a rock mass, Q system rating of 15 to 20, (ie. good quality rock). The jointing comprised three main sets plus random. The three joints were each generally smooth, planar and tight (minimal infilling) with a local spacing of 0.5 m, but 1.5 m on average. Joint Roughness Coefficient, JRC was assessed as 2-3 approximately.

At a depth of 800 m the undisturbed maximum principal stress was estimated at 56 MPa, horizontal and normal to the drift direction and 24 MPa vertically. The drift was excavated with a 1 m arch and was supported in the back only with 1.5 m grouted rebar at 1.4 m centres, with spot-bolting in places. This system of support proved to be more than adequate over fifteen years of mining up to the time of a 150 tonnes groundfall in the haulage at a x-cut intersection where the span was about 7 m (Figure 1).

The precursors to this fall were:

a) active mining 60 m to the East of the haulage/x-cut intersection;
b) a rockburst of unknown magnitude, 60 m away, above and to the West; and
c) some joints opening, principally of the vertical set which strikes E-W orthogonal to the drift axis.

The damage dimensions extended 2 m into the back, 7 m across span and 7 m along strike, giving approximately 60 m^3 of fallen rock. The rock blocks were large in size, 1 to 1.2 m length/breadth and 0.3 m in width. Numerous bolts were found in the fall with clean 45° breaks, indicating rapid failure in pure shear. Overall, the failure extended beyond the bolts and was structurally controlled, possibly triggered by the rockburst, but likely largely related to the stress shadowing effect provided by the adjacent stope.

Because of the character of this particular rockmass, prior to the groundfall, the existing support system was considered generally adequate. The blocky nature of the debris had not been readily apparent from an inspection of the intact excavation prior to failure, presumably because of the tight structure. It was also evident

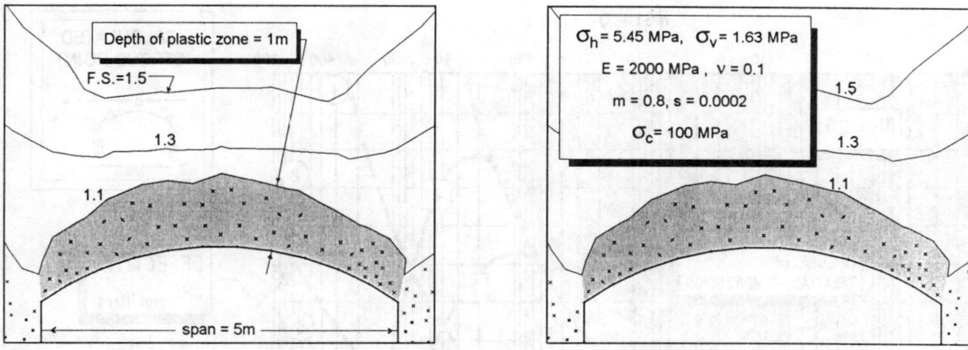

a) Initial Rock Mass Conditions, Q_u = 0.1, RMR_u = 23

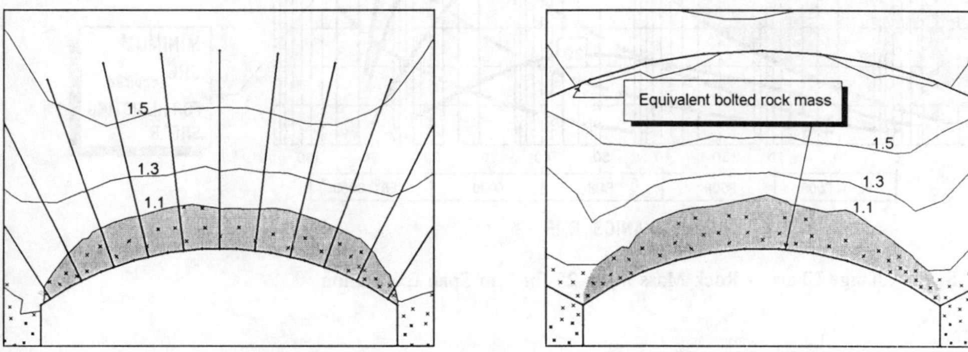

b) Bolted Rock Mass, and Equivalent Supported Rock Mass, Q_s = 0.3, RMR_s = 33

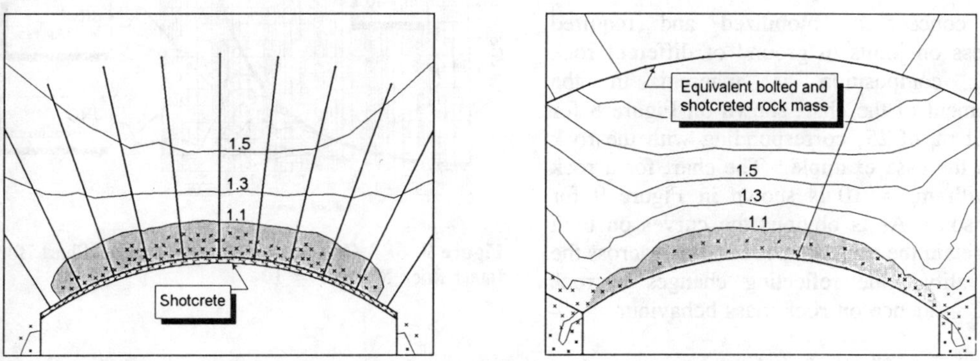

c) Supported Rock Mass (Bolts & Shotcrete), and Equivalent Supported Rock Mass, Q_s = 2.0, RMR_s = 50

Figure 7 Extent of Failure Zones for Supported and Unsupported Conditions as Predicted from PHASES Analyses

that the smooth, planar character of the joints was a controlling factor in the size of the failure, which was considered large compared to experience elsewhere under comparable mining conditions. In fact, as a result of this failure, and evaluation of rock quality in more detail it was concluded that the rock quality for almost the entire drift has needed to be downgraded to a Q of 0.5. Such a conclusion has serious economic consequences with respect to rehabilitation costs and must also be justified in view of the contradictory statements made elsewhere, Helfrich (1989).

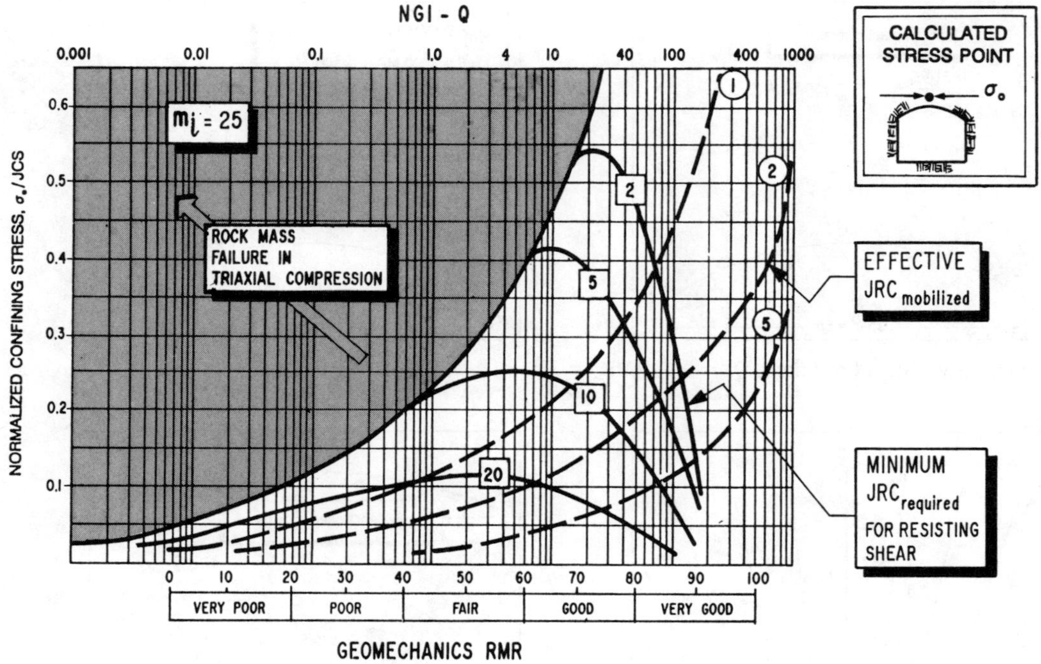

Figure 8 Stress Change Chart of Rock Mass m_i = 25 for 7 m Span Excavation

7.2 Evaluation of Stability Using Stress Change Chart

The concept of mobilized and required roughness on joints in ground of different rock material composition is inherent in the development of the chart shown on Figure 8 for an intact m_i of 25, corresponding with the rock mass in the case example. The chart for a rock mass with m_i = 10 is shown in Figure 9 for comparison. As is obvious the curves on both graphs retain the same shapes but move across the rock quality scale reflecting changes in rock material influence on rock mass behaviour.

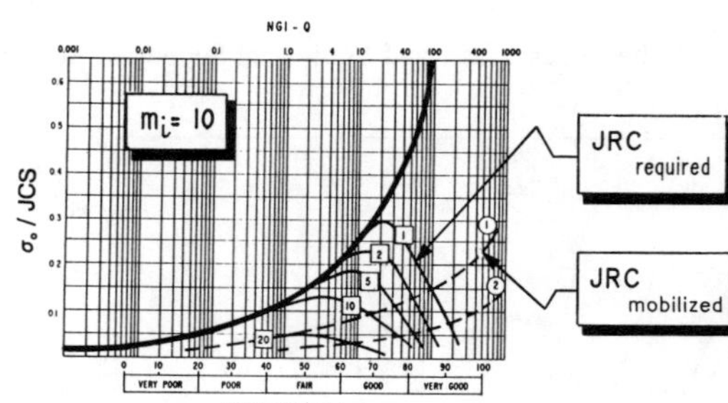

Figure 9 Comparison Stress-Change Chart for Intact Rock Mass m_i = 10

Referring to Figure 8 and the annotated chart on Figure 10 for an intact rock m_i of 25, first the initial Q value for the rock mass is plotted as a vertical line through point A on Figure 10 corresponding with the cross point for the initial normalized stress state (ie. at σ_o/JCS = 0.5). For this case, it will be noted that for the rock mass at the stress point considered, ie. just into the crown above the drift (Figure 1) the rock is fairly close to the yield line (both as indicated by the position on the chart on Figure 10 and by the results of the PHASES© analysis). With installed support the Q_u improves to a new Q_s location at point B (ref. Figures 4 and 10) and, incidentally as it takes the apparent rock quality past the $JRC_{required}$ = 2 line, it stabilizes the excavation.

If the drift now undergoes a marked stress drop to σ_o/JCS =0.04, corresponding to the influence of the excavation of the adjacent stope 60 m away, as illustrated on the right hand part of Figure 1, then the apparent stability condition drops from the plotted rock quality/stress point B in Figure 10 first to point C, down the line BC

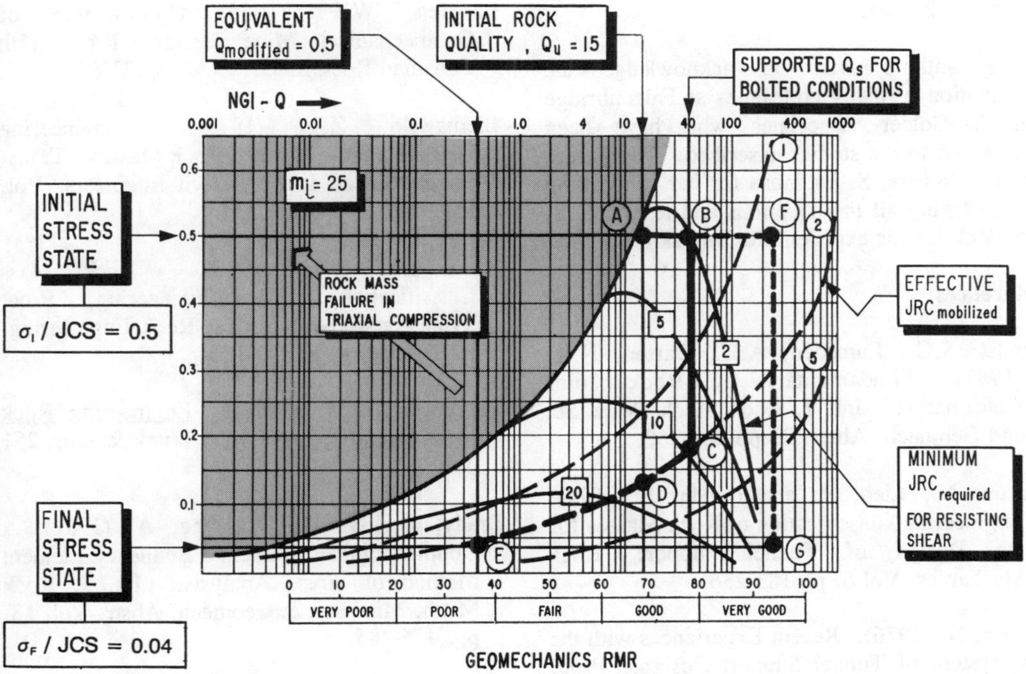

Figure 10 Illustration of Use of Stress-Change Chart for Design of Support for Ground Likely to be Subjected to Significant De-Stressing

corresponding to the vertical line for improved quality Q_s until striking the $JRC_{mobilized} = 2$ curve. The rock at this stress level fully mobilizes the joint strength for the smooth JRC = 2 joints, and as such, stability decreases with further stress drop, as inferred by travelling along the JRC = 2 curve past point D to point E, corresponding to the final stress point. From Figure 3 it can be seen that at this rock quality (Q = 0.5) for stability, bolts and fibrecrete are required as a minimum. If Figure 4 is used to estimate the effective improvement in quality for this derated rock mass a supported Q_s of 10 would be indicated (line D on Figure 4). This brings the effective rock mass quality back close to the original. If the same apparent Q_s improvement were to have been applied by installing fibre-crete and bolts in the original drift at the original rock quality state, point B would move to point F reaching an improved Q_s of about 250 (RMR = 95). For this apparent improved rock quality, derating due to a marked stress-drop would occur entirely to the right hand side of the $JRC_{required}$ = 2 line down to point G. Thus no failure would be predicted. This therefore points the way to being able to pre-design the level of support needed to prevent ground falls associated with significant stress relaxation effects.

It should be appreciated that in reality, as is reflected by this analysis, any increase in joint roughness helps dramatically in maintaining the stability of the rock mass. If the joints were even a little rougher than were actually experienced, say with a JRC of 5, then with the original Q of 15 (RMR = 68 approximately) no significant increase in support would be required due to the effects of the stress drop, as derating from σ_o/JCS of 0.5 to σ_o/JCS of 0.04 would occur almost completely above the curve for $JRC_{mobilized}$. This again provides an indication of the usefulness of this type of chart for design, allowing the required equivalent derating of the rock mass to be computed. It also shows the extreme sensitivity of rock masses with smooth joints to creating problems as a result of de-stressing effects during mining.

Acknowledgements

The authors wish to acknowledge the contribution of many colleagues at Falconbridge and at Golder Associates who have been contributed to the studies described. Thanks are also due to Mrs. S. Espinoza for her painstaking care in typing all the equations and to Mr. R.B. Craddock for his excellent drafting skills.

References

Bandis, S.C., Lumsden, A.C., Barton, N.R. (1983). Fundamentals of Rock Joint Deformation. Int. J. Rock Mech. Min. Sci and Geomech. Abstr. 20,pp.249-268.

Barton, N., Lien, R. and Lunde, J. (1974). Engineering Classification of Rock Masses for the Design of Tunnel Support, Rock Mechanics. Vol.6, pp.183-236.

Barton, N. (1976). Recent Experiences with the Q-system of Tunnel Support Design. Proc. Symp. on Exploration for Rock Engineering, Johannesburg, pp.107-117.

Barton, N.R. (1976). The Shear Strength of Rock and Rock Joints. Intnl. J. Rock Mech. Min. Sci., Vol. 13, No. 10, pp. 1-24.

Barton, N.R. and Choubey, V. (1977). The Shear Strength of Rock Joints in Theory and Practice. Rock Mechanics, Vol. 10, pp.1-54.

Barton, N.R. (1981). Estimation of In Situ Joint Properties. Nasliden Mine. Proceedings of the Conference on Application of Rock Mechanics to Cut and Fill Mining, Institute of Mining and Metallurgy, London, England, pp. 186-192.

Barton, N.R. and Bandis, S. (1982). Effects of Block Size on the Shear Behaviour of Jointed Rock. Issues in Rock Mechanics, 23rd Symposium on Rock Mechanics, pp. 739-759.

Barton, N. (1989). Cavern Design for Hong Kong Rocks. Proc. IMM. Symposium on Rock Caverns, Hong Kong. Ed. A.W.Malone and P.G.D. Whiteside, pp 179-202

Bawden, W.F. (1988) Optimization of Geomechanical Mine Design. Proc. 15th Canadian Rock Mech. Conf. pp.73-87

Bieniawski, Z.T. (1973). Engineering Classification of Jointed Rock Masses. Trans. South African Inst. of Civil Engineers, Vol. 15, No. 12, pp. 335-344.

Bieniawski, Z.T. (1976). Rock Mass Classification in Rock Engineering. Proc. Symp. on Exploration for Rock Engineering, Johannesburg, pp. 97-106.

Bieniawski, Z.T. (1989). Engineering Rock Mass Classifications. John Wiley & Sons, 251 pp.

Brady, B.H.G. and Wassyng, A. (1981), A Coupled Finited Element - Boundary Element Method of Stress Analysis. Int. J. Rock Mech. Min. Sci. & Geomech. Abstr. Vol. 18, pp. 475-485.

Brown, E.T., Bray, J.W., Landanyi and Hoek, E. (1983). Characteristic Line Calculations for Rock Tunnels. J. Geotech. Engr., ASCE, 109, pp. 15-39.

Carter, T.G., Yuen, C.M.K., Wong, J.S.K. and Barrett, S.V.L. (1990) Evaluation of Stress Conditions within Surface Crown Pillars. Proc. Conf. on Stresses in Underground Structures, Ottawa, pp.163-174

Carter, T.G. (1992). A New Approach to Surface Crown Pillar Design. Proc. 16th Canadian Rock Mechanics Symposium, Sudbury, pp. 75-83.

Daeman, J.J.K. (1977). Problems in Tunnel Support Mechanics. Underground Space, 1, pp. 163-172.

Deere, D.U. (1963). Technical Description of Rock Cores for Engineering Purposes. Rock Mech. and Eng. Geology, 1, 1822.

Detournay, E. and St. John, C.M. 1988. Design charts for a deep circular tunnel under non-uniform loading. Rock Mech. and Rock Engr., 21, pp. 119-137.

Golder Associates (1981). Report 802-1571 to Canmet on "Prediction of Stable Excavation Spans for Mining at Depths below 1000 meters in Hard Rock". Canmet Contract No. 17SQ.23440-0-9020.

Golder Associates (1990). Report 881-1739 to Canmet on "Crown Pillar Stability Back-Analysis". Report #23440-8-9074/01-SQ, Canada Centre for Mineral and Energy Technology, pp. 90.

Goodman, R.E. (1989). Introduction to Rock Mechanics John Wiley & Sons, 2nd.Ed.

Gyenge, M. 1987. Design aspects of bolting systems. Underground Support Systems, CIM Special Vol. 35, pp. 7-12.

Helfrich, H. and Finkel, M. (1989). The durability of rock bolts - investigation at the sandstone mine at Kvarntorp. BeFo Report 246:1/89, Swedish Rock Engineering Research Foundation, Stockholm, (in Swedish).

Hoek, E. and Brown E.T. (1980). Underground Excavations in Rock. IMM, London U.K. 527pp.

Hoek, E. (1983). Strength of Jointed Rock Masses. 23rd Rankine Lecture, Geotechnique, Vol. 33 No. 3, pp. 187-223.

Hoek, E. & Brown E.T. (1988). The Hoek-Brown failure Criterion - a 1988 update. Proc. 15th Canadian Rock Mech. Symp.,Toronto. pp. 31-38.

Hoek, E. (1992). Support Design for Hard Rock Mining - A Progress Report. Int. Conf. on Rock Support in Mining and Underground Construction, Sudbury, pp. 3-15.

Kulhawy, F.H. (1978). Geomechanical Model for Rock Foundation Settlement. J. Geotech. Eng. A.S.C.E. Vol.104. pp 211-227.

Ladanyi, B. 1974. Use of the Long-Term Strength Concept in the Determination of Ground Pressure on Tunnel Linings. Proc. 3rd Congr., Intnl. Soc. Rock Mech., Denver, 2B, pp. 1150-1156.

Laubscher, D.H. and Taylor, H.W. (1976). The Importance of Geomechanics Classification of Jointed Rock Masses in Mining Operations. Proceedings of the Symposium on Exploration for Rock Engineering, Johannesburg, pp.119-128.

Laubscher, D.H. (1977). Geomechanics Classification of Jointed Rock Masses - Mining Applications. Trans. Instn. Min. Metal. (Sect. A) Vol. 86. pp. A1 - A8

Laubscher, D.H. (1984) Design Aspects and Effectiveness of Support Systems in Different Mining Conditions. Trans. Instn. Min. Metal. (Sect. A) Vol 93 pp. A70-A81.

Laubscher, D.H. (1990) A Geomechanics Classification System for the Rating of Rock Mass in Mine Design. J.S.Afr.Inst. Min. Metall. Vol 90 #10 pp. 257-273.

Potvin, Y., Hudyma, M.R. and Miller, H.D.S. (1989) Design Guidelines For Open Stope Support., CIM Bulletin, Vol 82, #926, pp. 53-62.

Swan, G. (1983). Determination of Stiffness and Others Joint Properties from Roughness Measurements. Rock Mechanics and Rock Engineering, 16, pp. 19-38.

Swan, G. (1985). Methods of Roughness Analysis for Predicting Rock Joint Behaviour. Proc. of the Int. Symp. on Fundamentals of Rock Joints, Björkliden, pp. 153-161.

Zonggi, S. (1985). Asperity Models for Closure and Shear. Proc. of the Int. Symp. on Fundamentals of Rock Joints., Björkliden, pp. 173-183.

3 Grade estimation and mine feasibility

Innovative Mine Design for the 21st Century, Bawden & Archibald (eds) © 1993 Balkema, Rotterdam, ISBN 90 5410 325 6

Ore reserve definitions and classification systems

B. Haystead
Canadian Imperial Bank of Commerce, Corporate Finance Division, Toronto, Ont., Canada

ABSTRACT: The need to review ore reserve definition and classification in Canada has been highlighted by a number of authors citing the recent history of mine failures and pointing to reserve reporting and estimation problems as factors. Changes to the Australian Code and the recently proposed changes to the U.S. system further emphasized the requirement for establishing clearer standards for the reporting and classification of mineral resources and ore reserves in Canada.

Following a motion of CIM council dated May 1, 1991, a committee reporting to the Mineral Economics Society of the CIM was set up to develop standard definition of ore reserves for the Canadian Industry. During the committee's review, numerous submissions from industry professionals containing descriptions of classification systems and proposals for public and internal standards were received. Existing systems in use in Australia and those proposed for the U.S. and Britain were also reviewed.

This paper summarizes the various classification systems and reporting standards considered in the committee's deliberations and briefly outlines the general conclusions.

1 INTRODUCTION

A number of authors have written about the need for change in the way mineral resources and ore reserves are defined and reported within Canada. Dr. Owen Owens in his paper, "ORE RESERVES: WHAT NEEDS FIXING?" presented at the 1991 CIM Annual General Meeting discussed in detail many problems with reserve definitions and estimation practices. Dr. Owens argued that more complete definition of classification categories referencing more specific criteria would improve public reporting of ore reserves. Dr. Owens further commented that "Any looseness of definitions in securities laws can allow calculations to be made which ignore important details and yet the determinations may fall within regulatory requirements." David Harquail of Beutel Goodman and Company, in April 1992, listed some 40 North American gold mining failures, all of which were "vetted by company managements, their consultants, bankers, security exchanges, underwriters and their analysts. All of these industry experts made recommendations that in the final analysis were wrong". Although Mr. Harquail did not classify the 40 projects based on the root cause of the failure, we can assume that faulty reserve estimation and or reporting can be cited in many cases.

2 CIM SPECIAL COMMITTEE ON RESERVE DEFINITIONS

In response to the observed problems, a resolution was passed at a CIM executive meeting on May 1, 1991 authorizing the Mineral Economics Society to form a Special Committee of the CIM to develop new standard reserve definitions for use in Canada. An Operating Committee consisting of six people representing the Geology Society, the Mineral Economics Society and the Metal Mining Division of the CIM was set up to act as the administration and executive of the Special Committee. An Advisory Committee consisting of 39 industry professionals was formed. This group has the responsibility of reviewing and commenting on material prepared by the Operating Committee. To ensure maximum involvement by the mining community at large, submissions were requested through advertisements in the fall 1991 issues of the CIM Bulletin. As a result of the submissions, Advisory Committee comments and other communications, a great deal of information has been complied on mineral resource and reserve classification, terminology and definition. In its deliberations, the Operating Committee has reviewed this material, current usage, and, as well, the recent standards proposed by the Institution of Mining and Metallurgy (IMM) for Britain, the recent standards proposed by Working Party

#79 of the Society of Mining, Metallurgy, and Exploration (SME) for public reporting in the United States, the Resource/Reserve Classification for minerals adopted by the U.S. Bureau of Mines and the U.S. Geological Survey (USGS Circular 831), and the recently adopted AUSIMM Code for public reporting in Australia.

3 CURRENT USAGE

National Policy No. 2A, (the Guide for Engineers, Geologists and Prospectors submitting reports on mining properties to Canadian Provincial Securities Administrators), keys on the word "Ore" in the classification of mineral deposits. Ore is defined as follows:

"Ore" means a natural aggregate of one or more minerals which, at a specified time and place, may be mined and sold at a profit, or from which some part may be profitably separated.

The terminology "Possible", "Probable" and "Proven" is used to describe degrees of geologic and technical assurance. For example, "Probable Ore" is defined in the following manner:

"Probable Ore" or "Indicated Ore" means that material for which tonnage and grade are computed partly from specific measurements, samples or production data, and partly from projection for a reasonable distance on geological evidence, and for which the sites available for inspection, measurement and sampling are too widely or otherwise inappropriately spaced to outline the material completely or to established its grade throughout.

The policy states that where the word "ore" may not properly be used, such terms "mineralization", "mineralized bodies" or "concentrations" should be used.

The Association of Professional Engineers of Ontario (APEO), in Guidelines for Professional Engineers Reporting on Mineral, Oil and Gas Properties, keys on the word "Reserve" in the classification of mineral deposits. For example, "Probable Reserves" or "Indicated Reserves" are defined in the following manner:

"Probable Reserves" or "Indicated Reserves" are those materials for which tonnage and grades are computed partly from specific measurements, samples, or production data, and partly from projections for a reasonable distance on geological evidence. The sites available for inspection, measurement and sampling are too widely or otherwise inappropriately spaced to outline the material completely or to establish its grade throughout. It should be stated whether the tonnage and grade of "Probable" or "Indicated" reserves are in situ or extractable. Dilution factors and cut-off grades, if used, should be clearly explained and the vertical and horizontal projections from intersections or sample points should be given.

The Guideline cautions that the term "ore" should be used with "discretion and prudence, generally only with feasibility studies or in conjunction with discussions of reserves of operating mines".

The APEO Guidelines define the term ore as follows:

"Ore" is a natural aggregate of one or more minerals which under current conditions, may be produced and sold at a profit.

It should be noted that while the classification, terminology and definitions recommended by the APEO and stated in the National Policy 2A are very similar, definition of the term "ore" is different. The APEO Guideline stipulates that to be "Ore" the mineralization must be profitable "under current conditions". National Policy 2A defines ore as that which is profitable "at a specified time and place". This represents a major philosophical difference as the commodity price assumption is a major variable in the determination of what is "ore". In the APEO guidelines, it is also unclear as to how the term "ore" relates to the classification system. For example, is the term "ore" added to "Probable Reserves" (i.e., Probable Ore Reserves) to denote economic as well as technical feasibility?

4 THE SUBMISSIONS AND ADVISORY COMMITTEE COMMENTS

It is clear, from the submissions and the Advisory Committee comments that reserve practitioners and authors of public reports find the current standards somewhat problematic. Two of the major classification issues raised were as follows:

1. The current standards do not accommodate reporting of information as a logical progression from exploration information through to material that is expected to be mined. In fulfilling the requirement of "full and complete disclosure", there is a need to report information as it becomes available. At present, there is no accepted standard for reporting of information collected at the various stages of deposit delineation.

2. Confusion exists in the minds of investors and the general public over what a company reports as "Resources" or "Mineral Inventory" and what a company actually expects to mine.

The legitimate need to report information as it becomes available has resulted in the development of a number of "ad hoc" classification systems used with various terminology to describe the status of mineral deposits in Canada. Figure 1 is an example of a classification system along with terminology commonly seen in company and public reports in the past.

Geological Reserves	Ore Reserves
Possible	Possible
Probable	Probable
Proven	Proven

Fig.1

The submissions reviewed by the committee contained everything from discussions of reserve issues to proposals for specific classification systems and terminology. Most contributors proposed either a two or three axis classification system. Figure 2 shows an example of a two axis system. In the 3 axis system, technical and economic assurance and split to form a 3rd axis.

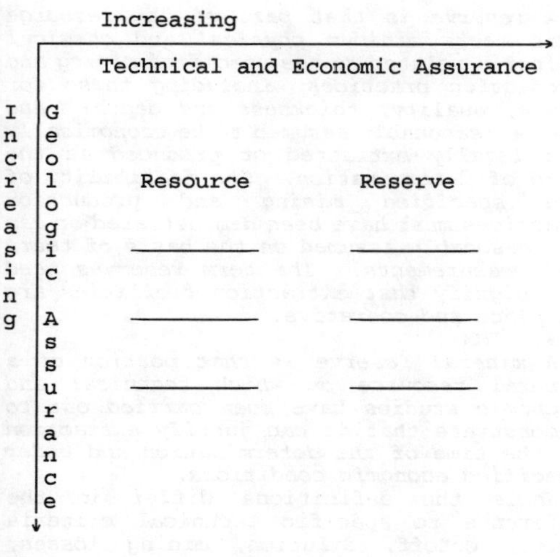

Fig.2 2 Axis System

In the submissions, the words "Resource" and "Reserve" were the most common terms used to describe degrees of technical and economic assurance.

There were a number of other issues raised in the submissions:

1. The question of "who is competent to do reserve work". A number of contributors commented on the need for specific experience in the type of deposit being reviewed, and the fact that "Ore Reserve" estimates required the application of a number of professional disciplines.
2. The fact that in Canada there is no National body with the mandate of monitoring compliance to reporting standards set.
3. While beyond the scope of the committee's work a number of contributors saw the need to address setting of standards for a whole range of calculation issues from sampling density to interpretation and estimation methodology. One group suggested that the classification scheme must be accompanied by a set of guidelines relating to various aspects of estimation procedures.

5 THE POTENTIAL FOR AN INTERNATIONAL STANDARD

Mr. Jack F. McOuat in his presentation "Establishing Mining Operations at Home and Abroad: Requirements for Global Reserve Standards and Practices" delivered at the Annual Meeting of the Prospectors and Developers Association in March of 1993, discussed the need for international standards for the reporting of reserve information. He argued that the "multinational aspect" of Canadian companies with deposits in a number of different countries and listings on "Canadian, U.S., European, Australian, Hong Kong and Japanese stock exchanges", created a definite need for the establishment of international standards for the classification and reporting of mineral resources.

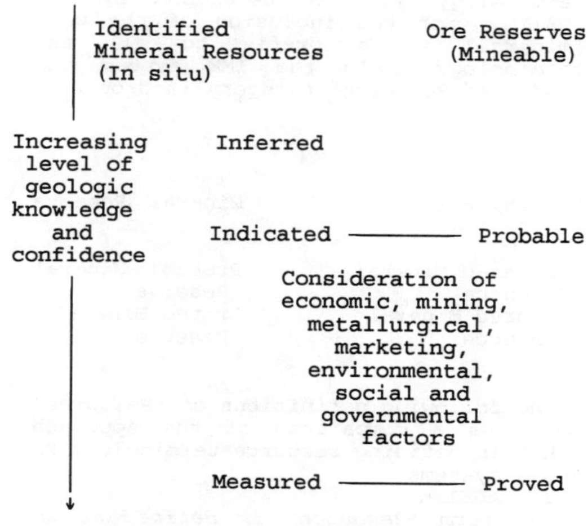

Fig.3 The Australian System

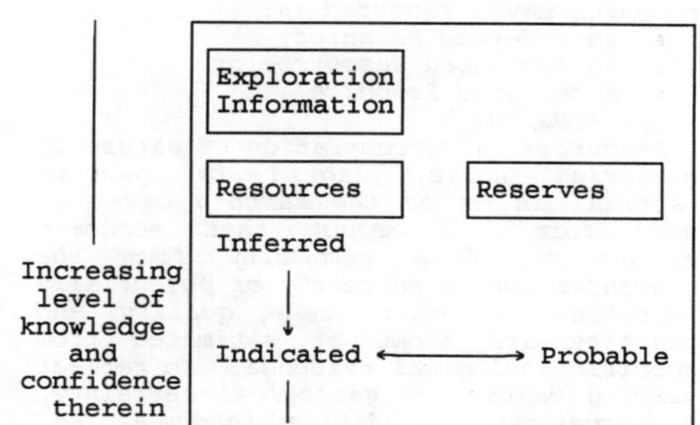

Fig.4 Working Party #79

The committee believes that the classification systems, terminology and to a certain extent definitions in use in Australia and proposed for use in the U.S. and Britain are sufficiently similar as to "set the stage" for an international standard. Figure 3 shows the classification system recently adopted for use in Australia. Figure 4 shows the classification system recommended by Working Party #79 of the AIME/SME.

With the exception of the added "Identified Mineral" before Resources and "Ore" before Reserves, the two systems are identical in form and terminology. The SME system expands on the Australian system by providing a category for "Exploration Information". The IMM proposes essentially the same system and terminology as that developed by the AUSIMM, with the inclusion of the word "mineral" as a prefix to all the terminology. In the IMM system the "Inferred Resource" category is dropped:

IMM

Mineral Resources	Mineral Reserve
Indicated Mineral Resource	Probable Mineral Reserve
Measured Mineral Resource	Proved Mineral Reserve

The following definitions of "Resource" serve as a comparison of the approach taken in defining resource terminology in the 3 systems.

1. AUSIMM

The term "Resource" is defined as an identified in situ mineral occurrence which excludes "Pre-Resource" mineralisation, from which valuable or useful minerals may be recovered. A resource maybe reported as:

- *an inferred Resource;*
- *an indicated Resource; or*
- *a Measured Resource.*

2. AIME/SME

Resource: a concentration of naturally occurring solid, liquid or gaseous material in or on the Earth's crust in such form and amount that economic extraction of a commodity from the concentration is currently or potentially feasible. Location, grade, quality, and quantity are known or estimated from specific geological evidence. To reflect varying degrees of geological certainty, resources can be subdivided into measured, indicated, and inferred.

3. IMM

A mineral resource is a tonnage or volume of rock or mineralization or other material of intrinsic economic interest whose grades, limits and other appropriate characteristics are known with a specified degree of knowledge.

The following definitions of "Reserve" serve as a comparison of the approach taken in defining reserve terminology in the 3 systems.

1. AUSIMM

The term "Ore Reserve" means that part of a Measured of Indicated Resource, which could be mined including dilution and from which valuable or useful minerals could be recovered economically under conditions realistically assumed at the time of reporting. Ore Reserves should be reported as:

- *Probable Ore Reserves; or*
- *Proved Ore Reserves.*

2. AIME/SME

A reserve is that part of the resource that meets minimum physical and chemical criteria related to the specified mining and production practices, including those for grade, quality, thickness and depth: and can be reasonably assumed to be economically and legally extracted or produced at the time of determination. The feasibility of the specified mining and production practices must have been demonstrated or can be reasonably assumed on the basis of tests and measurements. The term reserves need not signify that extraction facilities are in place and operative.

3. IMM

A mineral reserve is that portion of a mineral resource on which technical and economic studies have been carried out to demonstrate that it can justify extraction at the time of the determination and under specified economic conditions.

While the definitions differ in the reference to specific technical criteria (i.e., cutoff, dilution, mining losses, etc.) the following common points are noted:

1. In all cases, "Resources" can be estimated by a geologist, based on geological data.
2. In all cases, to be classified as a "Resource" the mineralization described must be of potential economic interest.
3. In all cases, the transfer from "Resources" to "Reserves" is based on the determination of technical and economic feasibility.

6 CONCLUSIONS

It is apparent from the submissions and the comments of the Advisory Committee that there is almost unanimous support for the development of a new Canadian Classification system and terminology having basis in the Australian system and those proposed for Britain and the U.S.

The Special Committee is moving in this direction. In the November 1992 draft of the report to the CIM that was circulated to the Advisory Committee for comment, the following classification system and terminology were proposed:

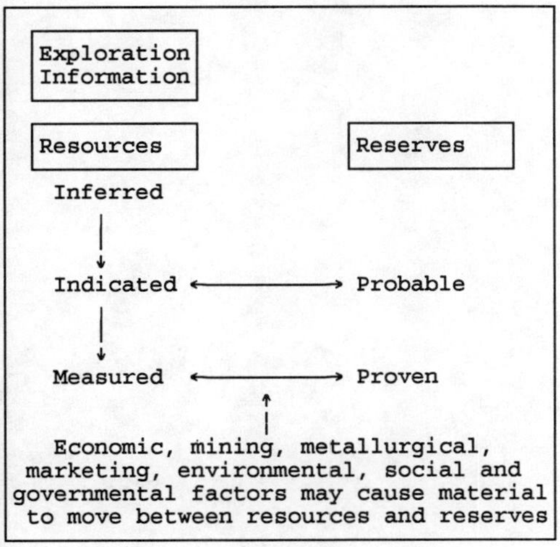

The Committee is currently revising the definitions proposed in the report based on the comments received from the Advisory Committee.

ACKNOWLEDGEMENT

The committee would like to acknowledge the effort of those who took the time to prepare submissions and the work of the Advisory Committee in reviewing the draft report.

REFERENCES

AUSIMM, 1989. Australasian code for reporting of identified mineral resources and ore reserves. Report of the Joint Committee of the Australasian Institute of Mining and Metallurgy and Australian Mining Industry Council.

AUSIMM, 1990. Guidelines to the Australasian Code for reporting of identified mineral resources and ore reserves.

HARQUAIL, D.E. 1992. Survival at the top of a gold company in the 90s. Text of presentation to the February 1992 Annual Meeting of the AIME.

IMM, 1991. Definition of terms for reporting of asset.

MCOUAT, J.F. 1993. Establishing mining operations at home and abroad: requirements for global reserve standards and practices. Text of presentation to the 61st Annual Meeting of the Prospectors and Developers Association.

OWENS, O.E. 1991. Ore reserves: what needs fixing? Text of presentation to the 93rd AGM of the CIMM.

SME, 1991. A Guide for reporting exploration information, resources, and reserves. Report of Working Party #79, SME Planning Committee.

Innovative Mine Design for the 21st Century, Bawden & Archibald (eds) © 1993 Balkema, Rotterdam, ISBN 90 5410 325 6

Defining the resource and the reserve

William E. Roscoe
Roscoe Postle Associates Inc., Toronto, Ont., Canada

ABSTRACT: In mine feasibility studies, the mineable ore reserve is often one of the most sensitive parameters, in particular the estimated head grade. Defining the mineable reserve is a two-stage process. First, the in situ resource must be established, based on data collected in the field, usually from drill holes. Second, the mineable portion of the in situ resource is estimated by imposing mining and economic parameters.

In order to make a realistic ore reserve estimate, reliable data, a reasonable geological interpretation, practical mining considerations, and economic criteria are needed. In many cases, drill hole information alone may not be adequate, and bulk sampling is required for reliable head grade estimation, metallurgical testwork, and to establish deposit geometry and continuity.

INTRODUCTION

Ore reserves are among the key assets of a mining company. Reliable estimation of reserves is critical, both to the economics of an ongoing mining operation and as input to feasibility studies on a potential operation. Reserves are often one of the most sensitive variables in economic analyses of potential operations, particularly the estimated head grade.

Reserve estimation is viewed as a two-stage process. First, the in situ resource is an estimate of the quantity and quality of in place mineralization based on data collected in the field, usually drill holes. Second, the mineable reserve is an estimate of the part of the in situ resource that can be exploited economically. The first stage involves mainly geological input. The second involves mining, processing, environmental, economic and other related parameters.

It should be noted that resources and reserves are always estimated, and subject to possible error. The reliability of the reserve estimate increases with the level of knowledge of the deposit and the amount of data. Thus, a reserve may move from the probable (indicated) category to the proven (measured) catagory as more knowledge and data accumulate.

This paper describes the parameters and key points relative to defining the in situ resource and mineable reserve, illustrated with examples from metallic mineral deposits. Reserve estimation techniques are not covered.

DEFINING THE IN SITU RESOURCE

An in situ resource estimate is based chiefly on sampling, assays, and geological interpretation (Table 1). In the following sections, key points of sampling and assaying are outlined, and a great deal of emphasis is placed on establishing the geological framework. The most important part of an in situ resource estimate is preparation of the information prior to the actual estimation process (Wrigglesworth, 1991).

Table 1. In situ resources - key points.

SAMPLING
- Representative
- Adequate

ASSAYS
- Sample preparation
- Appropriate, reliable method
- Consistent, reproducible

GEOLOGICAL FRAMEWORK
- Data presentation
- Interpretation of sections
- Geological controls
- Ore deposit model
- Section to section correlation
- Longitudinal section - 2D
- Block model - 3D

RESOURCE ESTIMATION

Sampling

Key points for sampling are representativeness and adequacy. Sample sites must be chosen and samples must be taken in such a way that a sampling bias is not introduced. This applies to all scales of sampling, from initial drilling through to large scale bulk sampling. For example, a bulk sample is commonly taken from the best part of a deposit, and may not be representative of the deposit as a whole for the purpose of grade estimation and metallurgical testwork.

Drilling is the most common method of sampling a deposit and collecting information about the geological setting, shape and dimensions of the deposit. Other types of sampling, such as underground and surface channel or chip sampling, are also very useful and can provide critical information on continuity and geometry of mineralization, as well as material for metallurgical test work. Grab sampling or "character" sampling is not representative and should never be used in reserve estimation.

The advantage of drilling is that holes can be planned to test and sample a mineralized body at regular intervals, and can be drilled normal to the long dimensions of the deposit. Drilling offers more flexibility and much lower costs than underground openings or surface excavations. Some disadvantages to drilling are the relatively small sample size, and the possibility that some types of drilling, such as reverse circulation and percussion, can introduce sampling bias and provide less geological information than diamond core drilling. In addition, drilling provides only indirect evidence of continuity of mineralization, although in many cases this is sufficient.

At what point does one know when sampling of a mineralized body is adequate? This is a difficult question and depends on a number of considerations. The first is adequacy for what purpose? In this paper the purpose of sampling is for reserve estimation related to feasibility studies. This means that reserves should be in the probable (indicated) and proven (measured) categories. For other purposes, such as initial delineation of a deposit, the amount and spacing of drilling may be adequate for possible (inferred) resources, but not for feasibility work.

Some considerations for adequacy of sampling are:

- o Drill hole spacing can be different for different deposit types and commodities, and must be assessed for each deposit. For example, a vein gold deposit may require 15 metre drill hole spacing, whereas holes spaced at several hundreds of metres may be adequate for a tabular potash or coal deposit.
- o Drilling must be adequate to determine that the mineralization forms a discete mineralized body, or deposit, at the scale of mining contemplated. Otherwise, it is not possible to estimate a resource or reserve. Figure 1 is a drill section showing gold intersections that are difficult to correlate with any confidence, even where holes are close together. Drilling is therefore not adequate in this example to establish a discrete deposit.
- o Drilling must be adequate to establish the geological

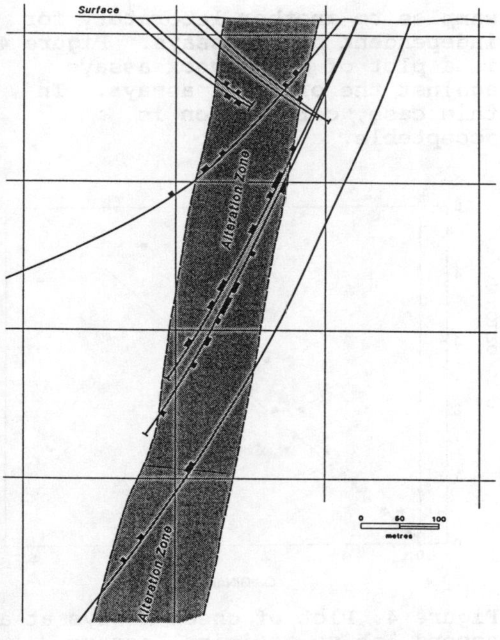

Figure 1. Difficult to correlate gold intersections.

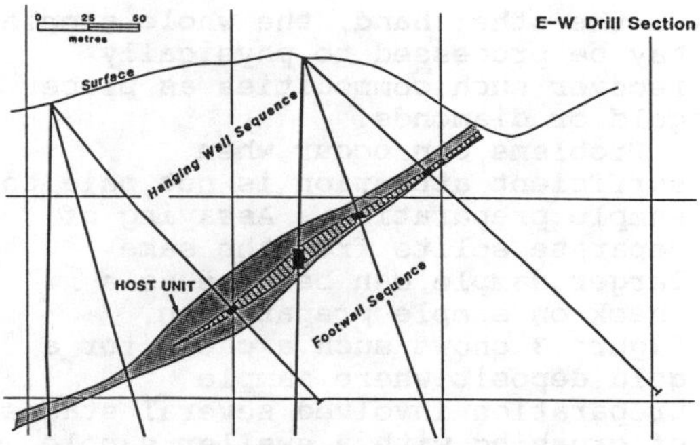

Figure 2. Readily correlatable gold intersections.

continuity, geometry and dimensions of the deposit. These are required to estimate the volume of the deposit. In contrast to the example in Figure 1, the drill section in Figure 2 shows that intersections can be correlated and that a discrete gold deposit is present. From this section, the drill hole spacing appears adequate to estimate the volume of the deposit.

- o Sampling must be adequate to establish density of the in situ mineralization in order to reliably convert volume to tonnage.
- o Sampling must be adequate to assess grade variability within the deposit in order to produce a reliable grade estimate. This is often an area subject to professional judgement, and requires expertise and experience in the particular deposit type and commodity.
- o Sampling must be adequate to develop reliable data on mining, metallurgical, and economic parameters in order to estimate mineable reserves. These include geotechnical data for underground mining methods or pit slopes, metallurgical test work data for process design and estimated recoveries, assays for metals and other materials for possible environmental impact, and other information related to cost estimates.

Assays

Assays should be one of the most reliable components of a reserve data base, since the magnitude of possible error is relatively small. Steps should be taken to ensure the reliability of assay data. Key points (Table 1) are sample preparation, appropriate method, and reliability of results.

Sample preparation is very important for metals which occur in relatively small concentrations, particularly where the metal occurs as coarse grains, or "nuggets", irregularly distributed through the deposit. Examples are precious metals deposits, such as vein gold deposits, placer deposits, and diamond deposits.

In general, assays are done on a small sample taken from the larger sample supplied to the assay laboratory. Sample preparation often involves crushing and pulverizing the larger sample to homogenize it so that the smaller sample split is representative of the larger sample. More than one stage of sample preparation and splitting may be done to reduce a very large sample to assay size.

On the other hand, the whole sample may be processed to physically recover such commodities as placer gold or diamonds.

Problems can occur when sufficient attention is not paid to sample preparation. Assaying of separate splits from the same larger sample can be done as a check on sample preparation. Figure 3 shows such a check for a gold deposit where sample preparation involved several stages of crushing with a smaller sample split off at each stage and the final sample sent for assay. In Figure 3, assays of the second split are plotted against assays of the first split of the same original sample. The lack of correlation between assays of the two splits points to a sample preparation problem. Other studies indicated that the assays were reliable.

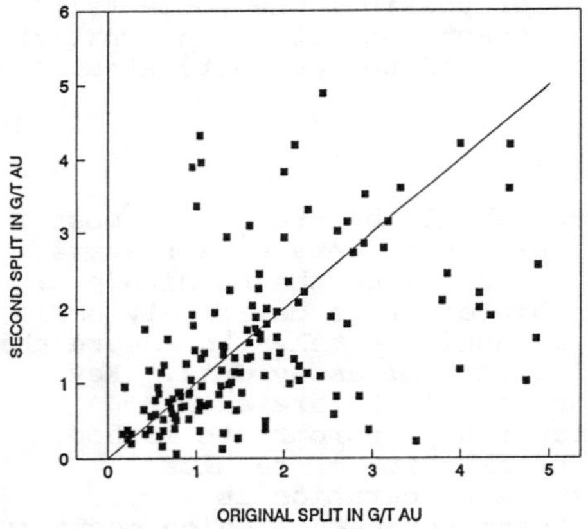

Figure 3. Assay of second split vs. first split from drill hole sample.

Assay methods must be appropriate for the types of sample and for the level of assay values expected. For example, geochemical analyses are designed to detect low concentrations and are not appropriate to use as assays.

Reliable commercial laboratories and mine site laboratories have internal procedures for assuring quality and consistency of assays. These include assaying check samples at regular intervals and the use of standard samples. It is common practice, in addition to these checks, to submit suites of samples to another laboratory for independent check assays. Figure 4 is a plot of such check assays against the original assays. In this case, correlation is accepteble.

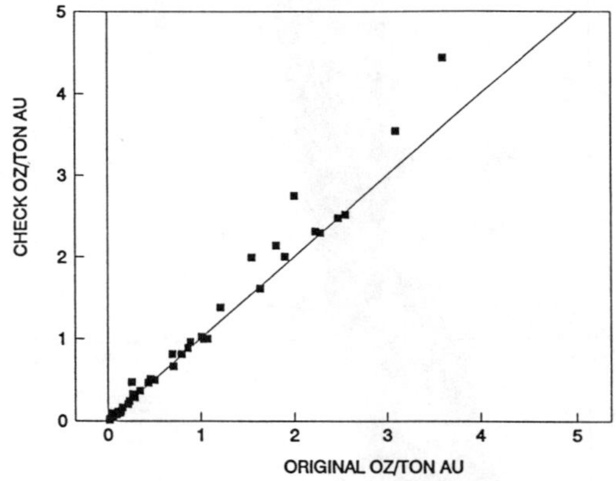

Figure 4. Plot of check assays at a second lab vs. original assays.

In precious metals deposits, erratic high grade assay values can have a major influence on the grade, and hence on the assessment of economic viability. If they are excluded, the grade of the deposit may be underestimated; if they are included, they may have the opposite effect. A realistic estimate of the grade should allow all legitimate values to be incorporated without giving them undue influence on the average grade.

For gold reserve estimates, high grade values are often cut back to some prescribed value, such as one ounce per short ton. Cutting high assays in effect reduces the influence of high grade values, but it is difficult to determine what level to cut high values to without operating results. Other ways to reduce the influence of high grade values include reducing the area or volume they affect, and use of the median instead of the mean for the highest class of value.

Geological Framework

Even if sampling and assaying are correct and unbiased, the estimated tonnage and grade of the in situ resource can vary considerably depending on the geological framework within which the estimate is made (Roscoe, 1991). The geological framework is established by fitting the mineralization into a reasonable geological model so that continuity, geometry and dimensions can be determined. It is necessary to establish whether one is dealing with a discrete mineral deposit or with an uncorrelatable "collection of assays". A mineral deposit has physical limits and is constrained by the geological framework.

A reasonable geological framework can be a stratigraphic unit, a host rock type, an alteration zone, a mineral association, or a stuctural feature such as a fault or fold axis. Typical ore deposit model types include vein gold deposit, volcanogenic massive base metal sulphide deposit, porphyry copper deposit, magmatic nickel-copper deposit, sedimentary iron deposit, evaporite gypsum deposit, etc.

The establishment of the geological framework of a deposit is also a matter of scale. For example, in a large bulk mining situation, the geological framework may be at a much larger scale and more generalized than in a selective underground mining situation. The minimum grade for inclusion (cut-off grade) will also be lower in a lower cost, bulk mining situation. Thus, when interpreting the geolgical framework, it is necessary to consider general levels of minimum grade and width requirements for the scale and type of mining being contemplated. For example, mineralization grading in the order of 2 grams per tonne gold may be of interest in an open pit situation but may be irrelevant in an underground mine.

Of paramount importance in establishing the geological framework of a mineral deposit is logical and appropriate treatment and presentation of the data. This is usually done by plotting sets of sections and plans showing the factual information, such as drill hole traces with geology and assays. The orientation of the sections should be chosen both to maximize the information available and to cross-cut the deposit.

It can be convenient and efficient to plot data using computer software designed for this purpose. Computer aided plotting requires careful checking to make sure the input data are error free, since errors may not be obvious. It has the advantages of speed and the ability to plot data in any orientation.

The geologist interprets the factual information on sections in order to establish hole to hole and section to section correlation of mineralization. In addition to using assay data, this is done using as a guide the interpreted geological contacts, structures, alteration envelopes, mineralogy and other relevant information. It can be misleading to interpret the shape and orientation of a mineralized body using only the assay data, without knowledge of the geological setting and mineralization controls. This can become an exercise in "connecting the dots" and can create the illusion that a discrete deposit exists when the actuality is an unrelated array of assays.

The initial interpretation must be tested by transferring it to another set of sections or plans in a different orientation. This is necessary to confirm section to section correlation and continuity. Modifications can be made until the interpretation fits both sets of section and plans. Computer plotting and data management capability can speed up this cross checking process. The key to confidence in the interpretation is predictability. The way that the outline of the mineralization is drawn on the sections is important since it impacts on the estimated tonnage. The outlines therefore should be in keeping with the geologist's best judgement of reality.

Figure 5 shows two very different interpretations of the same drill section. The lower, more recent interpretation benefits from two additional drill holes and from observations made in a test pit. It can be seen that the different interpretations have implications for tonnage estimation. For the more recent interpretation, estimated tonnage for the section is 27% lower and the grade 15%

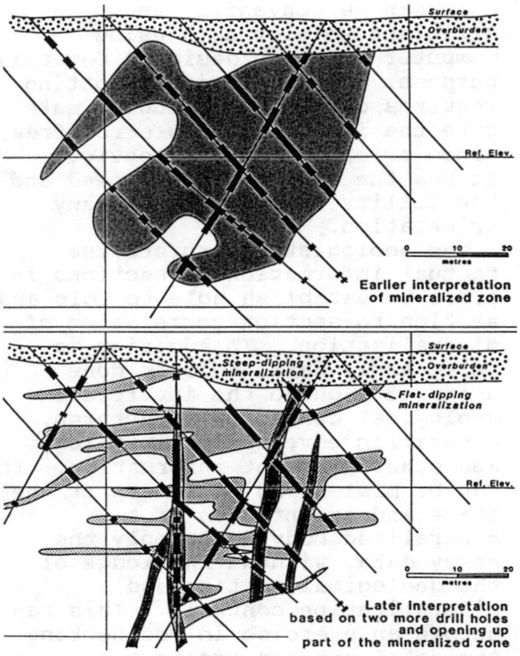

Figure 5. Two very different interpretations of the same drill section.

higher.

Once the geological framework is established, a longitudinal section can be constructed, for tabular or planar deposits that can be viewed in two dimensions. The longitudinal section is the primary tool for subsequent estimation of reserves. A separate longitudinal section must be constructed for each individual mineralized deposit or lens.

The purpose of the longitudinal section (or plan) is to depict drill hole intersections and other data as they occur within the plane of the mineral deposit. Since the longitudinal view is commonly used to do the actual estimate of tonnage and grade, it is critical that only correlatable information be plotted on a single section to fairly represent distribution of mineralization within the deposit. It is equally important that all of the data points, including low grade intersections, be plotted on the longitudinal section. Structures, such as fault traces or fold hinge lines, can also be plotted on longitudinal sections, as possible controls or constraints for the reserve estimation.

Some deposits are not amenable to two dimensional treatment, and connot be depicted in a longitudinal view. For example, the deposit depicted in Figure 5 cannot be resolved into meaningful longitudinal sections, and some other approach must be taken to estimate reserves, such as cross sections or three dimensional block modelling.

For computer-based reserve estimates, such as geostatistics and inverse distance interpolation, it is just as important to establish the geological framework as it is for manual estimation methods. This is commonly done by constructing a digital block model, whereby each block is assigned a rock or alteration code based on the geological interpretation. This way different geological units can be treated separately for computerized grade estimation methods which utilize interpolation and extrapolation of assay data. This is particularly important where mineralization is restricted to particular geological units and where grade distribution is different in different units. The geological framework is used to constrain in situ resource blocks to specified geological units, thus preventing extrapolation of grades to units interpreted to be unmineralized.

DEFINING THE MINEABLE RESERVE

A mineable reserve estimate is based on the in situ resource overlain with mining, processing, environmental, economic and other related factors (Table 2). In general, a mineable reserve is that portion of the in situ resource that can be extracted and processed at a profit. The list of factors in Table 2 can be thought of as hurdles that must be overcome to get to a mineable ore reserve, and are commonly addressed in feasibility studies.

In some cases, the database and information may prove inadequate to make a reliable reserve estimate. More drilling may be required, or possibly bulk sampling to establish deposit geometry and continuity,

Table 2. Mineable ore reserves - key points.

IN SITU RESOURCE
AMENABILITY TO MINING
SCALE AND METHOD OF MINING
MINING DILUTION AND RECOVERY
AMENABILITY TO PROCESSING
METALLURGICAL RECOVERY
PRODUCT MARKETABILITY, PRICE
ENVIRONMENTAL CONSTRAINTS
TAXATION CLIMATE
LEGAL ASPECTS, PERMITTING
POLITICAL RISK
PROJECT ECONOMICS

estimate head grades, and obtain material for metallurgical test work.

Amenability to Mining

The in situ reserve must be physically mineable in terms of deposit geometry and continuity. This consideration is also a matter of scale and method of mining contemplated, as noted previously. In an underground mining situation, deposit geometry and continuity must be such that stopes can be visualized and at least a conceptual mining plan developed. Minimum width and cut-off grades can be applied, and mining dilution and recovery factors estimated.

For open pit situations, deposit geometry and dimensions must be considered in relation to the selective mining units, which are determined by bench height, equipment type and size, and mining selectivity. The block size in the deposit in situ block model should be more or less matched to the selective mining units. The in situ block model resource is the basis for computerized open pit design and optimization. Mining, recovery, and economic criteria are applied to the block model to produce a pit design and to determine which blocks are economically mineable.

Cut-off Grade

Cut-off grade is an important concept in determining project economics. For underground mines, cut-off grade is combined with minimum mining width, which is the smallest opening from which mineralized material can be practically extracted. Cut-off grade is the minimum grade that can be profitably extracted at the marginal or breakeven cost, and is a function of the cost of mining and processing, metallurgical recovery, and metal price. Different cut-off grades and minimum widths may apply to different mining methods and areas of a mine. Applying the cut-off grade over the minimum width in an underground mine is commonly used to separate ore from waste.

The concept of cut-off grade can be more complicated in open pit mining. For example, in bulk mineable gold deposits, one cut-off grade may be used in pit design and optimization, whereas other cut-off grade levels may be used during mining to separate high grade mill ore from low grade mill ore or from heap leach material.

Reconciliation Studies

One way of testing the validity of reserve estimates at producing mines is to compare them to actual production figures. This is an extremely useful exercise that can provide information on whether or not estimates are biased high or low as well as information on precision of the reserve estimates. For underground mines, reserve estimates for individual stopes can be compared with actual production figures. For open pits, reserve estimates can be compared with actual production on a bench by bench basis, or over some suitable time frame, such as monthly. One problem is that it may be difficult to determine the actual production figures for stopes or benches, since mine records may not be set up to capture the appropriate information.

Figure 6 is a reserve reconciliation plot for an underground gold mine. It shows estimated reserve grade vs. actual mined reserve grade for 41 stopes. There is a large scatter of data points, indicating poor estimation precision. Sixteen of the 41 points fall below the 1:1 reference line,

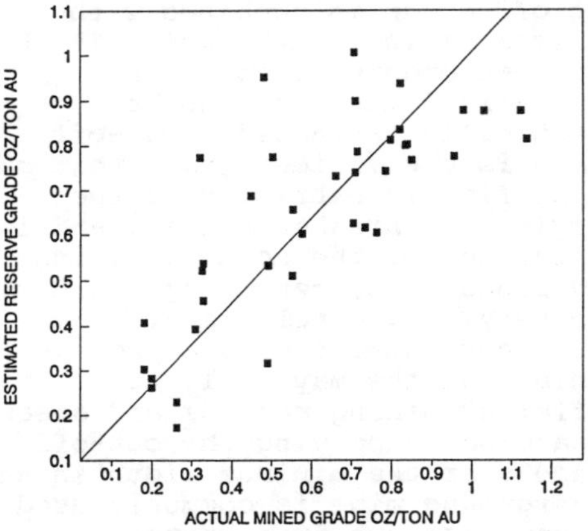

Figure 6. Reconciliation plot of estimated reserve grade vs. actual mined grade for 41 stopes in an underground gold mine.

indicating that their grade was underestimated. The other 25 stopes had their grade overestimated. On average for the 41 stopes in Figure 6, the grade was overestimated by 7%. This study indicates that, although the overall accuracy is reasonable at 7% error, the estimation precision is poor for individual stopes.

CONCLUSIONS

A reliable estimate of the in situ resource depends on adequate number and spacing of drill holes; representative and unbiased sampling and assays; reasonable and coherent geological interpretation of the deposit geometry and continuity; and appropriate resource estimation methods.

Adequacy of drill holes and spacing depends on type of mineralization; deposit geometry, continuity and complexity; grade variability and sensitivity to high assays; variation in mineralogy and metallurgy; reliability of sampling and assays; and mining method and scale contemplated.

Estimation of the mineable ore reserve from the in situ reserve must consider deposit geometry and continuity relative to the mining method and scale contemplated. This determines the minimum practical mining width and unit (stope or block). Cut-off grade, the minimum grade that can be profitably extracted at the marginal or breakeven cost, is a function of cost of mining and processing, metallurgical recovery, and metal price.

REFERENCES

Roscoe, William E. 1991. Geological Framework for Reserves. Preceedings of a seminar offered by the Prospectors and Developers Association of Canada, March 23, 1991, Toronto, Canada: 55-64

Wrigglesworth, Lorne A. 1991. Basic Parameters for Ore Reserve Calculations. Preceedings of a seminar offered by the Prospectors and Developers Association of Canada, March 23, 1991, Toronto, Canada: 65-75

Innovative Mine Design for the 21st Century, Bawden & Archibald (eds) © 1993 Balkema, Rotterdam, ISBN 90 5410 325 6

Ore reserve fundamentals

Owen E.Owens
Cominco Ltd, Vancouver, B.C., Canada

ABSTRACT: The estimation of tonnage and grade of a mineralized body entails determination of the character of the mineralization; assembling sufficient data to represent the mineralization on plan and sections; and determination of mineralization outlines and continuity, all in advance of reserve or resource calculation and classification. Ore outline and calculation considerations differ for ore bodies to be mined underground from those suitable for open pit mining.

INTRODUCTION

The estimation of the tonnage and grade of a mass of rock on the basis of drill hole intersections or bedrock sampling is, by the limited and widely spaced nature of the hard information normally available, an imprecise process. The first step is to make a quantified estimate and the second step is to find terminology to qualify the conclusion so the degree of confidence in the estimate is demonstrated. The process of arriving at the size and grade of an ore body whether it is (a) a newly-discovered body of mineralization, (b) a potential ore reserve, (c) an ore body proposed for mining, or (d) the ore reserves in a mine, requires fundamental geological, and geometrical determinations, as well as the normal physical and chemical analyses, and calculations.

There are three principal components of ore reserves: grade, dimension and continuity of mineralization. The latter requires accurate evaluation of what should logically be projected to what, to outline a <u>mineable</u> unit. Selection of the ore boundary is a key determination and must be made on the basis of geological and/or correlatable rock geometry. It should be something the miner can follow in mining. Assays alone are poor guides to determine a mineable limit particularly in an underground mine. In gold especially, but in all deposits, it is the continuity of mineral grade values which is usually the most difficult question. The calculation should reflect and come very close to the true distribution of the values present in the rock. This is where the role of clear geological understanding is so important since the metal distribution is determined by geological processes and controls.

Basic considerations in ore calculation include (1) geological distribution of metal values; (2) sampling and assay procedures; (3) treatment of unusually high values; (4) reasonableness and correlatability of the ore boundary in relation to the overall ore character and its continuity; (5) similarity in ore correlation from section to section or level to level; (6) any disturbance structures (discontinuities or structures which interrupt the continuity); (7) mineablity of the outlined ore shape; (8) intentional dilution (smoothing); (9) provision for unintentional dilution due to ground stability conditions and geological unknowns; (10) densities of each ore type and waste; (11) metallurgy; (12) marketable product; and (13) deleterious substances.

The open pit miner can select low grade portions of the deposit on a blast by blast basis to be shipped to waste, and has more control over the grade of rock shipped to the concentrator. The underground miner rarely has this opportunity. Open pit cutoff grades are readily treated by in-pit selections but this approach is nearly impossible to apply underground to achieve calculated tonnes and grade in production, except on a large mineable unit basis. While geostatistical methods alone may give satisfactory results when estimating the reserves of a large disseminated orebody, to be mined by open pit methods, usually they look to geostatistics to provide all the answers. The difficulty in applying such new ore calculation methods developed primarily for open pit porphyry copper deposits to underground deposits is that the geometry is quite different and the mining approach requires different constraints.

The key issues of ore reserves is that they are intended as a clear as possible statement of metal in place in a mineable ore body. Precise, widely understood definitions help convey understanding of the figures, but accuracy requires careful geological input before the calculation is made and a definition is applied. The ore reserves or preferably *Reserves*, of a mining company are usually its most valuable asset. The officers of a company have an obligation to advise the public on the reserves, and these reserves are a key component in its financial report. They should be determined by a person with adequate experience in the deposit type involved, normally a geologist. At the same time, ore reserves are an unusual asset that they are often imperfectly known and must be estimated.

There are three principal requirements in ore reserve determination:-

(a) Need for geological work and know how to outline a sufficiently continuous ore body suitable for mining.
(b) Estimation of quantity and grade of the mineable body.
(c) Presentation of the information to interested parties through the use of an ore classification system which is precise and unambiguous.

The question is what geological information is needed to outline the ore. Is there sufficient information or is more drilling or other work needed? If there is sufficient information, does the geologist have an adequate understanding of the ore type and general mining considerations to outline the mineable zone and then calculate its size and grade? Another question is whether the calculation is being made for a deposit being considered for immediate production. If so, ending with an imprecise calculation, and an imprecise ore category, probably won't be sufficient. If there is apprehension in this regard, the need for more drilling or other ore testing work is evident.

In the author's experience, there are 5 key elements to be considered in arriving at adequate estimations of ore reserves or resources. These can be grouped under Character of Mineralization, Representation on Plan and Section, Continuity of Mineralization, Ore Calculation and Classification.

Character of Mineralization

1. What is the nature of the mineralization, or the character of the ore?
2. What walls of the zone make geological and mineable sense?
3. What is its appearance and how will it be identified in mining? Can it be identified visually or only by assays. What is the host rock structure?
4. Will the ore be mined underground or in open pit? In an open pit situation, blast hole sampling and trucking internal low grade to waste is possible but this is rarely possible underground so the emphasis in underground mining must be on a sufficiently continuous body to be readily mineable?
5. How is the miner likely to identify ore and determine the walls of the stopes?

Representation on Plan and Section

1. Drill hole, surface, and any underground information should be assembled on plans and cross-sections at appropriate scales.
2. The information should include principal geological features, potential ore mineralization units, and assay data.
3. The structure or continuity of the geological features and potential ore mineralization should be interpreted on section and plan.
4. Is there sufficient geological and mineralization data and understanding to make meaningful interpretation of continuity of units from drill hole to drill hole (or other points of information) or is more information required? The information spacing should be sufficient to demonstrate geological and particularly mineralization continuity on plan and sections through the potential ore deposit.

Continuity of Mineralization

1. What type of continuity, from drill hole to drill hole, makes geological and geometrical sense on the basis of ore character and mineable regularity?
2. Should the mineralization be sub-divided into zones of different ore character and continuity? For instance, disseminated and stringer mineralization is normally separated from massive mineralization.

Calculation

1. The calculation of tonnage is made on the basis of the outlined ore body, thickness, extent, grade and specific gravity.
2. The calculation should be of a body with a visualized and demonstrated outline on plan and section in a mineable configuration.
3. Great care and discrimination is required if assay walls alone are used

to determine ore limits, especially for underground mines. In the latter case, the ore outline should be smoothed to a reasonable mineable shape and any waste added by smoothing added into the calculation.

4. Once the potential orebody of mineralization is outlined on plan and cross-sections there is a clearer understanding of the deposit and particularly the distribution of metal values within the deposit.

5. Even for the calculation of potential open pit mines reserves, the author recommends: (a) outlining the potential ore on plan and cross-sections; (b) visually inspecting assay grade distribution, looking for any particular irregularities; (c) averaging the grade of individual holes; (d) averaging the grade for each section; (e) calculating the tonnage for each section half way to adjacent sections; and then (f) summing the section tonnages and averaging grades. Assuming sufficient drill data is available, final ore reserves will probably be calculated by block modelling and geostatistical methods which can make allowances to exclude internal low grade as may be possible by eventual in-pit selection. However, the outlining of ore on plan and sections serves the useful purposes of: (a) visual presentation of the deposit and any important irregularities therein; (b) a check on the geostatistical calculation; and often (c) help in determining the ore and open pit limits which may be used in geostatistical block models.

6. For underground mines it is the author's experience that calculating ore reserves on the basis of outlines on plan and sections is critical to a meaningful result. The dimensions and grade of the orebody calculation must be realizable in the mining process. This requires outlining a mineable shape on section which is correlatable as a reasonably smooth outline from section to section. The miner will attempt to have a sharp limit to his stopes and pillars but they will have to be reasonably regular. All material within this outline must be included in the grade and tonnage calculation. The average grade for each drill hole can be calculated and the tonnage determined for the volume of rock half way to adjacent holes calculated (polygonal method), or all holes on a section multiplied by half the distance to adjacent sections. Computers and geostatistical methods may be used but are only effective if the methods can deal with the ore outline as referred above. Geostatistical methods which only deal with assays and their geometric location rarely deal with mineable outlines. They normally search for ore on the basis of grade ignoring important geometric continuity detail, ore wall irregularity and necessary internal dilution smoothing. The calculation may also reach outside a mineable outline and include low grade values in arriving at average grades.

7. Appropriate allowances must be made for unintentional dilution at least in the case of reserves as contrasted to resources. Even at the resource stage, it is appropriate to think about recovery and dilution and, if a reasonable mining scheme can be visualized at this stage, they should be included in the calculation. The advantage is that, the first released numbers on an orebody are the ones commonly remembered. Dilution will vary on the basis of the ore regularity, wall rock competency and mining method.

Classification

1. An ore definition category should be assigned to the reserves to convey: (1) the amount of data available; (2) the confidence appropriate on the basis of the consistency and correlation of the data; and (3) whether metallurgical, mining and economic evaluation data is available to demonstrate a reserve which can be mined at a profit, or whether economic consideration is limited to very preliminary review, and value is considered more in relation to similar occurrences elsewhere.

2. The author recommends that the term "ore reserves" or preferably *Reserves* should be restricted to figures supported by metallurgical, mining and economic evaluation showing the potential for a profitable operation. Other calculations would be referred to as *Resources*. The terms *Resources* or *Reserves* should be qualified by terms like *Inferred, Indicated, Probable* or *Possible*. The terms *Measured* or *Proven* can imply a degree of confidence only present where they refer to material: (a) opened up underground in stopes or openings on at least 2 sides; or (b) in the case of large open pit deposits, where very close-spaced surface drilling is available. Such closely-spaced information may only be necessary where distribution of ore values is erratic. It is suggested that *Measured* or *Proven* reserves not be grouped with other reserves. The author believes most *Reserves* have a degree of definition or confidence more properly designated as *Indicated*.

3. In the case of *Reserves*, the author believes the geologist is obligated to use his best efforts, including more drilling if necessary, to arrive at a close estimate of tonnes and grade. In such a circumstance, figures in which their confidence is only plus or minus some percentage (even as low a figure as 10% for near marginal deposits) are rarely meaningful to those planning a new mining operation.

Innovative Mine Design for the 21st Century, Bawden & Archibald (eds) © 1993 Balkema, Rotterdam, ISBN 90 5410 325 6

Use of geostatistics: Current situation and future perspectives

N.Champigny
Coopers & Lybrand, Toronto, Ont., Canada

ABSTRACT: Since the treatise of G. Matheron in the early 1960's, the discipline of geostatistics has developed a wide range of methods and approaches designed to assist in solving three fundamental mining-related problems: (1) estimation of global in-situ and global recoverable reserves normally performed as part of a mining project's feasibility analysis, (2) estimation of local in-situ and local recoverable reserves which is usually carried out as part of pre-production planning and detailed mine design, and (3) prediction of the grade and other ore characteristics of individual mining units as part of the grade and production control process.

Grade and production control is the stage at which geostatistics has enjoyed the most success, particularly for open pit mines. Although the underlying principles for the application of geostatistics at the grade control stage are identical to those used for global and local estimation, there is still significant resistance within the mining community to systematically applying geostatistics for the purposes of pre-production estimation.

Based on recent surveys conducted amongst senior international mining organizations this paper attempts to provide a snapshot of the current application of geostatistics and discusses how and why the use of geostatistics is expected to grow. According to the survey a much higher acceptance of geostatistics can be achieved if: (1) geostatistics is not applied in isolation but rather makes maximum use of available geological information as well as sampling information, (2) close collaboration is established between geologists, mining engineers and geostaticians, (3) computer-assisted design tools to model orebodies that possess a complex geometry, for example many of the Canadian underground gold mines, are integrated with existing geostatistical methods, (4) misconceptions about the usefulness of geostatistics are dispelled through the education of industry decision makers, and (5) well packaged geostatistical software for education is available.

1 INTRODUCTION

In the year of 1778 Benjamin Franklin wrote "man is a tool-making animal" and less than a century later Thomas Carlyle, a Scottish historian and essayist stated that "man is a tool-using animal". These two brilliant individuals encapsulated in a few words, what man, over time, has been occupied with, and become better at tool-making and tool-using. This paper described some of the tools, geostatistical tools, that are designed and applied to improve efficiency of mining operations. Many companies are using geostatistics, but some are still apprehensive because of some common misconceptions. Specifically, this paper:

1. Provides a brief historical review of the evolution of geostatistics;
2. Discusses how reserve/grade estimation problems are being addressed using geostatistical methods.
3. Compares some of the misconceptions about geostatistics in its current usage by mining companies for reserve estimation, mine planning and grade control.
4. Summarizes the future trends in the application of geostatistics.

2 HISTORY

Based on an examination of the historical review papers prepared by Matheron and Kleingeld (1987), Verly (1988) and Dowd (1992), the evolution of geostatistics may be divided in three phases described below:

2.1 Linear geostatistics 1945-1965

It is generally recognized that the beginnings of geostatistics are attributed to the work on South African gold mining projects by H.S. Sichel and later by D.G. Krige. In 1962 the concept of random function which underlies the theory of geostatistics was introduced in the forestry field by B. Matern. G. Matheron first used the name geostatistics in his treatise published in 1962 and his 1965 doctoral thesis is the foundation of the theory of linear geostatistics including the estimation method of ordinary kriging. Ordinary kriging will be referred simply as kriging for the remainder of this paper.

2.2 Non-linear geostatistics 1966-1974

Over this period and under the guidance of G. Matheron the theoretical basis of non-linear geostatistics was laid out which were developed for the estimation of recoverable reserves. In addition A.G. Journel performed a significant amount of work on the simulation of deposits. Other important contributors include M. David, Ch. J. Huijbregts, P. Delfiner, P. Chauvet and J.P. Chiles.

2.3 More widespread application and additional developments 1974-1993

More widespread use of geostatistics in the mining industry started in the late 1970's probably triggered by the availability of low-cost, high performance personal computers and relatively user-friendly geostatistical software. The practice of geostatistics involved mainly variograms and kriging, for estimating global in-situ reserves and performing grade control as indicated by the proliferation of papers published and illustrated by the selective bibliography of Dowd (1992). This period saw (1) the introduction and application of advanced geostatistical methods including disjunctive kriging and indicator kriging and (2) the enhancement of simulation approaches. Geostatistics also gained acceptance in other fields of application such as petroleum, geotechnical engineering, hydrology, geochemistry, environmental sciences, agriculture and fisheries.

3 ESTIMATION PROBLEMS

Sample grades provide an incomplete picture of a deposit. Without this picture, it is not possible to perform modelling for the purposes of reserve estimation, mine planning and grade control. A grade interpolation tool allows the prediction of what is happening between sampled locations. It cannot be over-emphasized that interpolation must be bounded by recognizable geological frontiers. To be accurate and efficient a grade interpolation tool has to take into account:

1. The areas of influence of neighbouring samples.
2. The "clustering" of samples in high grade areas.
3. That the distribution of blockgrades, which are being estimated, is different from the distribution of sample grades.

Geostatistics has now become a grade interpolation tool widely used for estimating reserves, and performing mine planning and grade control. In the mining community there are many supporters of geostatistics, but there are also those who have important reservations. Two important reservations are not directed towards geostatistics, but rather, towards geostatisticians:

1. Geostatisticians are often accused of using large doses of mathematical jargon barely understandable to the average geologist or engineer.
2. Geostatisticians are blamed for leaving out geological information when building a reserve model.

These reservations are addressed below by describing the benefits and drawbacks of geostatistics for four types of estimation problems usually encountered during the life of a mining project:

1. The estimation of global in-situ (i.e. geological) reserves.
2. The estimation of global recoverable (i.e. mineable) reserves.
3. The estimation of local in-situ and recoverable reserves.
4. The estimation of local recoverable reserves as part of grade control.

It is to be noted that the term reserve is used here in a broad sense and may include in some cases, the term resource as reflected in current and proposed definitions.

3.1 Global in-situ reserves

Estimation of global in-situ reserves is the first step of the project's feasibility study and the end result is often controlled by the amount and accuracy of the grade information and the quality of the geological interpretation more than by the estimation method used (polygonal, inverse distance, or geostatistical).

The objective of this estimation is to obtain a reasonable approximation of the grade-tonnage curve within a deposit confined by recognizable geological boundaries or mineralized envelope, and from which representative samples have been obtained. With an intelligent use of geology and reliable grade information, polygonal, inverse distance and kriging often yield estimates that are not vastly different.

Clustering is a higher concentration of samples over a certain area of the deposit which is frequently the highest grade area. When this occurs, the polygonal estimation approach tends to depart significantly from other methods. This clustering effect often results in an overestimation of the average grade. It is the experience of many reserve estimation practitioners that when large differences are observed between the reserve figures provided by geostatistical and conventional methods, it is due to the lack of geological input to the geostatistical estimate, and not the mathematical dissimilitudes of the two approaches. Ignoring geology led to scepticism of the applicability of geostatistics in its early days but with the increased collaboration between geologists and geostatisticians, the application of geologically-controlled geostatistics is now the rule rather than the exception.

At the global in-situ reserve estimation stage, production information is not available and therefore it is impossible to compare the effectiveness of any estimation method with "reality". Even with the provision of production data such comparison would be misleading: an in-situ estimate leaves out external and internal dilution and other mining constraints, all of which directly impact on tonnage and grade of the ore produced.

Without the "real" in-situ reserve to compare with, the benefits of using geostatistics are difficult to measure. There are however indirect benefits:

1. Variograms will help to determine the reasonableness of geologically inferred directions of preferential grade continuity, and of geologically inferred mineralization domains.
2. Variogram parameters are useful for determining appropriate ranges of influence and search distances assumed by other methods.
3. Geostatistical results will provide valuable insights on the parameters used downstream, such as for the estimation

of recoverable (i.e. mineable) reserves and for grade control.

Drawbacks of geostatistics at this stage are:

1. Significant up-front effort is necessary to implement the method (i.e. variogram analysis, increased computation time).

2. Danger of using geostatistics in isolation (i.e. geological input is vital).

3.2 Global recoverable reserves

Estimate of global recoverable reserves is the determination of the amount of mineable material that can be recovered from the in-situ reserves using given:

1. Cutoff grades.
2. Selective mining unit size.
3. Sampling configuration at the time of ore/waste selection.
4. Technical constraints specific to the mining method envisioned.

Global recoverable reserves are the basis for the feasibility analysis of the project. The ability to mine selectively, such as for many open pit operations, is often essential to the project's viability. At the in-situ reserve estimation stage, the grade-tonnage curve generated is for blocks or mining units of size roughly equal to the borehole spacing, whereas the size of the selective mining unit is normally much smaller. Therefore, the grade-tonnage curve of the large blocks must be corrected to obtain the grade-tonnage curve of the selective mining units. This can only be done using geostatistics because conventional methods are of little use.

After this first adjustment to the grade-tonnage curve, a second adjustment is performed to account for imperfect selectivity when a decision must be made whether material is ore or waste. Imperfect selectivity accounts for the fact that the "true" grade of individual selective mining units is unknown, but instead is estimated using an ultimate sampling configuration (i.e. closely spaced samples in and around each unit). An example of this ultimate sampling configuration is blastholes in an open pit. Imperfect selectivity produces misclassification of ore and waste blocks. This results in a downgrading of potential selectivity, commonly known as internal dilution. Again, geostatistics provides an appropriate correction of the grade-tonnage curve by quantifying the estimation error based on the proposed ultimate sampling configuration. Again, conventional methods do not provide for this adjustment.

Technical constraints subsequently applied are practical mining considerations which bring the ideal theoretical selectivity to practical and attainable levels. These constraints include external dilution from wall rocks, haulage roads in open pit operations, and access, rock mechanics, ventilation and scheduling considerations in underground mining. Grade-tonnage curve adjustments to take into account these technical constraints are normally based on experience and are fairly crude. In some instances they can be quantified using sophisticated, but not geostatistical, simulation models. In all cases, the presence of technical constraints makes the decision to select ore versus waste more difficult and less reliable.

Usually, several cutoff grades are selected to form the basis for economic analysis and optimization.

The drawbacks of using geostatistics for the estimation of global recoverable reserves at this stage are outweighed by the benefits. Nevertheless two potential drawbacks, or more appropriately cautions, must be issued:

1. It is common to jump the step of the adjustment of the grade-tonnage curve to the smaller selective mining unit size, and krige each small unit using widely spaced data. Although kriging will provide an estimate for each unit, these estimates are over-smoothed and show a false impression of a relatively uniform grade distribution. In other words resolution using kriging cannot be pushed to that level.

2. There are many geostatistical approaches to calculating global recoverable reserves, and choosing the most appropriate one may be difficult. Each approach is based on underlying assumptions and the relevance of these assumptions must be evaluated for each reserve study by the reserve estimation team which should include an experienced geostatistician.

3.3 Local in-situ and recoverable reserves

Estimation of local in-situ and recoverable reserves is performed as part of pre-production planning, detailed mine design and scheduling. At this stage more accurate estimates of the grade of individual mining blocks are required. Global reserve figures are no longer useful. This is where conventional estimation methods tend to provide less accurate estimates.

The polygonal estimation method for example, applies the grade value of each individual sample to its surrounding area of influence, and allocates this value to all individual mining blocks lying within this area. The distribution of the blocks will be similar to the distribution of the drillholes. Using a single grade value, the polygonal method gives a very poor estimate of local grade. It ignores the fact that individual mining blocks will not have the same value as the mining blocks containing the original sample point (usually a drill core sample).

Inverse distance uses several neighbouring samples to perform local estimation, a definite improvement over polygonal, but it still does not take into account the size of the mining blocks being estimated.

Kriging provides improved local in-situ estimates. However, in order for the estimates to be meaningful there should be at least one sample in each mining block being estimated. Kriging should not be used to interpolate values into small blocks lying between sampled locations, as cautioned above for the estimation of global recoverable reserves. Better local estimates are obtained by:

1. Kriging blocks of size similar to the wide-spaced drilling grid.

2. Estimating local grade-tonnage curves that define the distribution of selective mining units within the larger blocks.

Advanced geostatistical techniques such as indicator kriging, disjunctive kriging and conditional simulation are available to estimate local recoverable reserves.

At this stage of a mining project, geostatistics clearly outperforms other methods by improving significantly the accuracy of local estimates. As in the case of the estimation of global recoverable reserves, it is important to caution against kriging directly the individual selective mining units. This short cut is tempting but it is also a meaningless exercise. Instead, large blocks are first kriged and then advanced techniques are applied to provide an estimate of local recoverable reserves.

There are three drawbacks to this approach:

1. The relative complexity of the computation.
2. The necessity of being guided by geostatistical experts;
3. The end result is an estimate of the proportion (i.e. tonnage) of the selective mining units, and of the average grade of these units, above cutoff. The grade of each unit is not provided. With only widely spaced data available at this stage, this is as much as one can hope for!

3.4 Grade control

Grade control or the local estimation of recoverable reserves during production is the stage at which geostatistics has enjoyed the most success, particularly for open pit mines. Detailed sample grade information is available during production to perform the last estimation before a final decision is taken as to whether the block or unit should be mined as ore, or treated as waste. In the presence of closely spaced sampling data and with the necessary superposition of geological information, kriging performs well, and surpasses conventional methods. Reconciliation of the kriging results with production results is normally done without too much difficulty. Based on published studies and the experience of the author, the differences very rarely exceed 10% and for some mines, are less than 1%.

The underlying principles for the application of geostatistics at the grade control stage are identical to those used for earlier stages of a mining project. Since geostatistics performs well for grade control it is not unreasonable to assume that it would perform equally well for reserve estimation and mine planning.

Misuse of the method, such as ignoring geological input and the kriging of small mining blocks is the main reason in some of the scepticism for the use of geostatistics at the grade control stage.

4 Does geostatistics work?

Among the management of some mining companies, a common perception is that the application of computer-aided tools, including geostatistics, has as many drawbacks as benefits. This belief is based on a number of misconceptions as demonstrated below.

A common view is summarized by the statement from an unnamed mining analyst:

> "Part of the problem was the rocket scientists with their computer models. The models were developed for the porphyry copper deposits and do not apply when it comes to gold".

This misconception is often further generalized to the statement that "geostatistical reserve models work for large open pit mines but not for underground situations" in spite of the fact that early successful applications were in South African gold mines. Other views expressed are:

1. Reserves estimated using geostatistics ignore the geologist's knowledge base, it is strictly a number crunching exercise.
2. The use of geostatistical reserve estimation hinders collaboration between the geologist and mining engineer.

If in fact, these views are correct (which the author does not subscribe to) than the geostatistical estimations will be fruitless at best and very misleading at worst. In recent years geostatistical practitioners have been stressing the utmost involvement of geologists and mining engineers. It should also be noted that estimates made by other methods without the involvement of an experienced geologist and collaberation with engineering would be equally suspect.

In the gold mining industry, a survey was carried out in 1989 of 19 of the world's major gold producers (Champigny and Armstrong 1993). As shown in Figure 1, geostatistics remains the most common tool during production and is also used at the feasibility and exploration stages.

When asked about agreement between reserve estimates and production, the surveyed companies stated that good agreement was obtained often a difference of less than 10% (Figure 2).

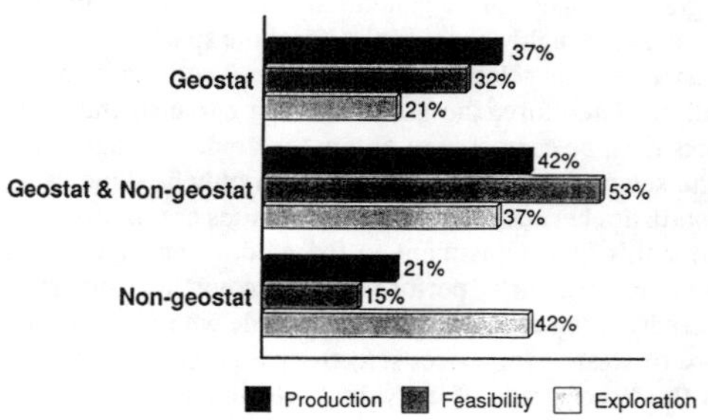

Figure 1

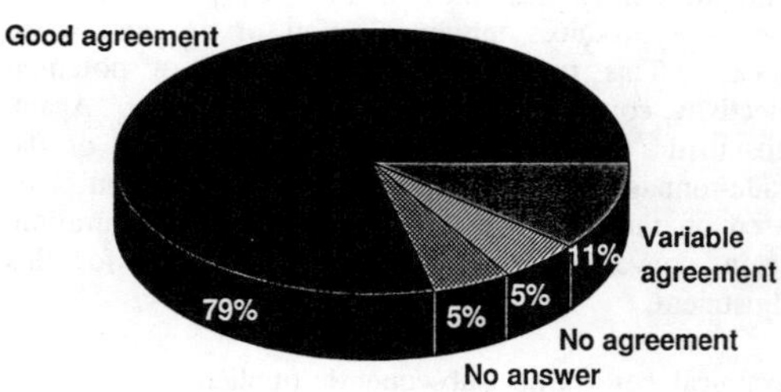

Figure 2

Based on (1) a series of discussions by the author during the first half of 1993 with major mining organizations operating in Canada, the U.S.A. and Australia, (2) and numerous geostatistical assessments of mining projects worldwide carried out by the author, the current use of geostatistical tools can be summarized schematically as follows:

	OPEN PIT	UNDERGROUND
EXPLORATION & FEASIBILITY	Frequently	Occasionally
PRODUCTION	Routinely	Rarely

5 CONCLUSIONS

The mining industry recognizes that grade estimation tools are essential for global in-situ and recoverable reserves estimation, mine planning and grade control. Geostatistics is a powerful grade interpolation tool but its correct application depends largely on the constant use of geological information. Geostatistical tools allow analysis of more effective what if situations on orebodies and on variations in metal prices and changes of technology.

In spite of the greater acceptance of geostatistics, there is still some hesitancy in its application. An apparent obstacle is that some mining industry stakeholders believe that it is too mathematically complex and thus will not use it. However, many other disciplines have equally complex mathematics (e.g. rock mechanics) but face no objections for their application.

In the opinion of the author, geostatistics lacks popularity for underground situations, not because it is not applicable, but rather because powerful computer-aided geometric tools need to accompany geostatistics. Until the mid-1980's appropriate geometric tools were not available and this has slowed down its implementation for underground mines.

The overview presented of geostatistics current practices shows that:

1. Geostatistical tools are being successfully used by many mining companies worldwide.
2. Their applications are expected to grow.
3. Some misconceptions about their usefulness are being rapidly dispelled.

When geostatistics was in its infancy, the geostatistician often worked in isolation. Now the geostatistician tends to work in close collaboration with geologists and mining engineers. This insures incorporation of geological information and of mining parameters, all of which are essential to the development of "realistic" reserve models.

While geostatistical training often consists on having a set of tools increasing attention must be and is paid to integrate geological assumptions and geostatistical concepts. Such integration is probably the most significant challenge for the future evolution of geostatistics.

ACKNOWLEDGEMENTS

The author is grateful to the management of mining operations that have provided their views and experiences on the user of geostatistical methods. Without their help it would be difficult to suggest to other mining project operators to also become users of geostatistics.

The kind assistance of Dr. P.H. Grimley and C. Bilokrely of Coopers & Lybrand is gratefully appreciated.

REFERENCES

Champigny, N. & M. Armstrong 1993. Geostatistics for the estimation of gold deposits: a review and survey of current practice up to 1989. *Mineralium Deposita* - in press. (4 pages & figures).

Dowd, P.A. 1992. A review of recent developments in geostatistics. *Computers & Geosciences*, v. 17, No. 10: 1481-1500. Great Britain: Pergamon.

Matheron, G. & W.J. Kleingeld 1987. The evolution of geostatistics. *Proc. 20th APCOM 87*, v. 3 Geostatistics: 9-12.

Verly, G. 1988. Pour une meilleure diffusion de la géostatistique dans l'industrie minière. *Proc. CAMI*: 173-179. Rotterdam: Balkema.

Innovative Mine Design for the 21st Century, Bawden & Archibald (eds) © 1993 Balkema, Rotterdam, ISBN 90 5410 325 6

Modelling and representation of spatial data using artificial neural networks

E.Clarici, D.B.Owen & S.Durucan
Department of Mineral Resources Engineering, Royal School of Mines, Imperial College, London, UK

ABSTRACT: Preliminary experiments were carried out to investigate the potential of neural networks as a tool for analysing spatial data in the minerals industry. A data set of 60 records, each with fields giving position and contaminants were used in experiments. 48 of these records were used by the neural network to analyse underlying patterns and then were used by the neural network to predict contaminants of each of the other 12 records given only by their position. The degree of scatter of predictions were compared with those obtained using kriging. The neural network was then used to predict values for a grid for the area under investigation.

1 INTRODUCTION

This paper describes preliminary experiments carried out to examine the potential of neural networks in the analysis of spatial data for solving problems in the minerals industry.

Spatial data which is used for geological modelling and resource estimation is often non-ideal in one respect or another. Due to many factors, e.g. the data's complex nature, the presence of clusters, the subjectivity of interpretation, limitations in access, time, money or equipment etc., earth scientists are usually presented with incomplete, imprecise and high noise content data sets. Although these characteristics have made spatial data resistant to many traditional methods, these are the very properties amenable to artificial neural network solutions.

Artificial neural networks, a specialist field of Artificial Intelligence (AI), are mathematical tools which seem very promising for a wide range of modelling problems. Neural models attempt to achieve good performance via dense interconnection of many nonlinear computational elements, operating in parallel and arranged in patterns related to biological neural networks (Lippmann, 1987). Most research in neural networks has focused on developing classifiers and tools for pattern recognition. As the result of an enormous volume of research over the past few years, neural networks' applications have become as diverse as image processing, robotics, signal processing, optics, medical diagnosis, manufacturing systems, business and marketing, speech and image recognition, modelling chemical process systems, all in all an attempt to set new standards of performance by looking at old problems in a new way.

The data set used in the experiments consisted of 60 records, each with fields of Northings, Eastings, parts per million (ppm) of Arsenic, ppm of Lead, and ppm of Cadmium. 48 of these records were used to train a commercial network simulator. After training, the network was then used to predict the Arsenic, Lead and Cadmium content, given only the Northings and Eastings, of the remaining 12 records.

The same 48 records used for training the network were then kriged and from these results the assays contents of the twelve records were again predicted.

The degree of scatter of the predicted values about the true values was summarised by calculating the moments of inertia for both network and kriging. The measurements of performance of the neural network were, in these preliminary and limited experiments, superior.

It should be emphasised that in using real data sets, the key to the interpolation of the degree of pollution or the grades of mineral deposit rests on an understanding of processes involved in the original pollution or the genesis of the mineralization. The indications are that neural networks have considerable potential as a tool to help the engineer towards this understanding.

2 ARTIFICIAL NEURAL NETWORKS

Artificial neural networks consist of a large number of simple, but densely interconnected, processing units, called neurons. They attempt in a greatly simplified way, to mimic the behaviour of their biological counterparts. Of all the interesting characteristics of artificial neural networks, none captures the imagination like their ability to learn. Furthermore this is one of the main characteristics that makes them so different from traditional computing. Instead of performing sequences of commands, neural networks try to learn patterns from examples. They learn during training.

The objective of training is that the application of a set of inputs produces the desired (or at least consistent) set of outputs. Each such input (or output) set is referred to as a vector. Training is accomplished by sequentially applying input vectors, while adjusting network weights according to a predetermined procedure. During training, the network weights gradually converge to values such that each input vector produces the desired output vector. There are two groups of training algorithms, i.e. *supervised* and *unsupervised*. This paper will not go deeply into the neural network theory. Interested readers are refereed to the list of references.

The multilayer perceptron, which learns via the backpropagation algorithm, has been chosen as the most suitable type of neural network for the problem of spatial data modelling.

2.1 *Multilayer perceptron with backpropagation*

Multilayer perceptrons are artificial neural networks with one or more layers of neurons between the input and output layer. These additional layers contain hidden neurons that are not directly connected to both the input and output layer.

Backpropagation is a systematic method for training multilayer artificial neural networks.

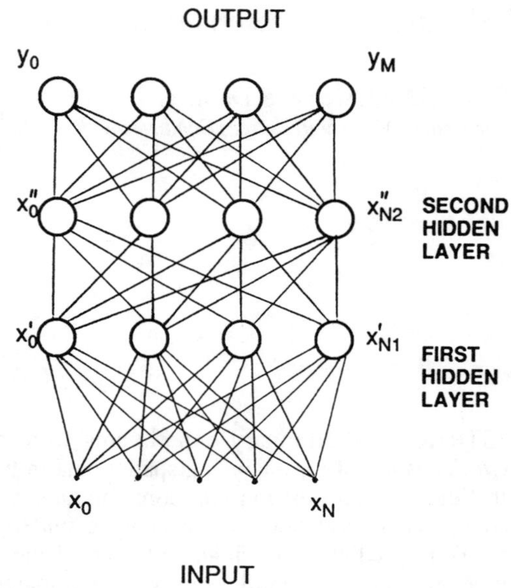

FIGURE 1 Three layer perceptron (after Lippmann, 1987)

Figure 1 shows a multilayer perceptron suitable for training with backpropagation.

Output from first hidden layer is calculated as:

$$x'_j = f\left(\sum w_{ij} x_i - \theta_j\right) \qquad 0 \le j \le N_1 \qquad (1)$$

Output from second hidden layer is:

$$x''_k = f\left(\sum w'_{jk} x'_j - \theta'_k\right) \qquad 0 \le k \le N_2 \qquad (2)$$

Neural network output is:

$$y_l = f\left(\sum w''_{kl} x''_k - \theta''_l\right) \qquad 0 \le l \le M \qquad (3)$$

In the formulas above x'_j and x''_k are the outputs of neurons in the first and second hidden layer, θ'_k and θ''_l are thresholds in those neurons. w_{ij} is connection strength from the input to the first hidden layer, and w_{ij} and w''_{ij} are the connection strengths between first and second hidden layers and between second hidden and output layer respectively. Figure 2 shows the neuron with activation function used for backpropagation.

Table 1 The backpropagation training algorithm (after Lippmann, 1987)

The backpropagation training algorithm is an iterative gradient algorithm designed to minimise the mean square error between the actual output of a multilayer feed-forward perceptron and the desired output. It requires continuous differentiable nonlinearities. The following assumes a sigmoid logistic nonlinearity is used as the function $f(\alpha)$

$$f(\alpha) = \frac{1}{1+e^{-(\alpha-\theta)}}$$

Step 1. Initialise Weights

Set all weights to small random values.

Step 2. Present Input and Desired Outputs

Present a continuous valued input vector $x_1,...,x_N$ and specify the desired outputs $d_1,...d_N$. The input could be new on each trial or samples from a training set could be presented iteratively until weights stabilise.

Step 3. Calculate Actual Outputs

Use the sigmoid nonlinearity from above and formulas in Figure 1 to calculate outputs $y_1,...,y_N$.

Step 4. Adapt Weights

Use a recursive algorithm starting at the output nodes and working back to the first hidden layer. Adjust weights by

$$w_{ij}(t+1) = w_{ij}(t) + \eta\delta_j x_i^{'}$$

In this equation $w_{ij}(t)$ is the weight from a hidden neuron i or from an input to neuron j at time t, $x_i^{'}$ is either the output of neuron i or the input, η is training rate coefficient (0.01-1.0), and δ_j is an error term for neuron j. If neuron j is an output neuron, then

$$\delta_j = y_j(1-y_j)(d_j-y_j)$$

where d_j is the desired output of node j and y_j is the actual output. If neuron j is an internal hidden neuron, then

$$\delta_j = x_j^{'}(1-x_j^{'})\sum_k \delta_k w_{jk}$$

where k is over all neurons in the layers above neuron j.

Step 5. Repeat by going to Step 2

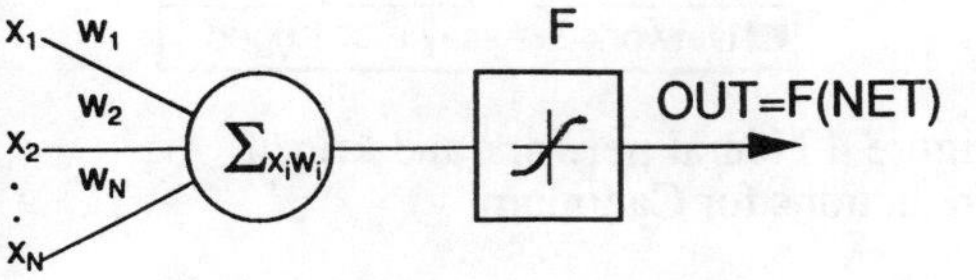

Figure 2 Neuron with activation function

$$\sum_{i=1}^{N} x_i w_i = x_1 w_1 + x_2 w_2 + . . . + x_N w_N \qquad (4)$$

$$OUT = F\sum_{i=1}^{N} x_i w_i .... \qquad ....(5)$$

The capabilities of multilayer networks stem from the nonlinearities used within neurons, produced by this activation function.

The backpropagation algorithm described in Table 1 is a generalisation of the least mean square algorithm. LMS minimises the mean square error between the desired output of a perceptron network and the actual output. It is also called the Widrow-Hoff algorithm (Maren et al, 1990). An essential component of the algorithm is the iterative method that propagates error terms required to adapt weights back from neurons in the output layer to neurons in lower layers.

3 CASE STUDY

The data set consisted of 60 records of soil samples, each with the fields Easting, Northing, ppm Arsenic, ppm Cadmium and ppm Lead. This data set is in the public domain and had been provided to illustrate the operation of geostatistical software, i.e. GEO-EAS. The neural network simulator used in the experiments was 'BrainMaker'.

The experiment used 48 of the records to train the neural network to recognise the underlying pattern in contamination. After training, the neural network was used to predict the values of the contaminants of the 12 records whose assays values were known to us but not known to network.

Table 2 Moments of inertia of the predicted values about the true values

	Arsenic	Cadmium	Lead
Network 12 readings	0.0381	4.369	70.798
Network 9 readings	0.0383	5.401	83.397
Kriging 9 readings	0.0885	5.276	112.42

The neural network architecture, a three-layer perceptron with backpropagation, consisted of an input layer of two neurons (Eastings and Northings), 2 hidden layers of 15 and 20 neurons respectively and an output layer with 3 neurons, one for each pollutant.

The same principle of 48 records to predict the remaining 12 was used with kriging. The kriging method was not able to provide any prediction for three of the 12 positions due to the methodology of kriging.

The results of the predictions for both the neural network and the kriging predictions are illustrated in the Figures 3, 4 and 5. Only the 9 values which were estimated by kriging are plotted in order to facilitate comparisons of predictions by kriging and networks on the scattergraph.

Table 2 shows the values for the moments of inertia of the predicted values about the true values (i.e., the 45° line).

It can be seen that the degree of scatter as measured by the moment of inertia is the least for the 12 network predictions.

After the validation of the network performance, it was then used to predict pollutant concentrations for a grid 20 x 20 feet, for the area Easting 260-500 feet, and Northing 120-320 feet. The results are shown in Figures 6, 7, 8, 9, 10, 11.

4 CONCLUSIONS

Values predicted using neural networks show less scatter, as measured by the moment of inertia about the true values, than those predicted using kriging.

Neural networks are able to generate a fuller output for a larger range of positional data than the kriging method used.

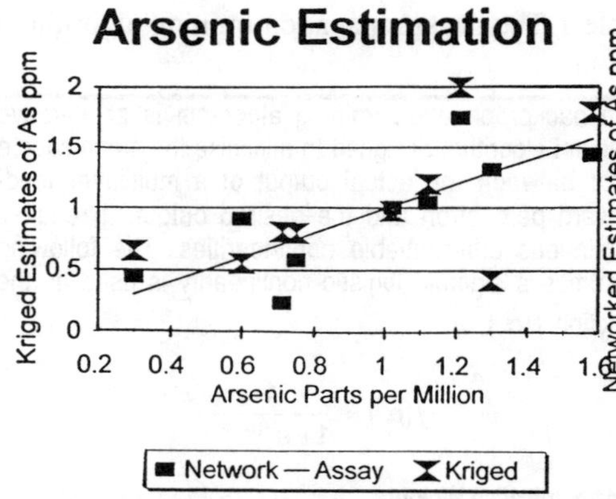

Figure 3 Neural network and kriged predictions for Arsenic

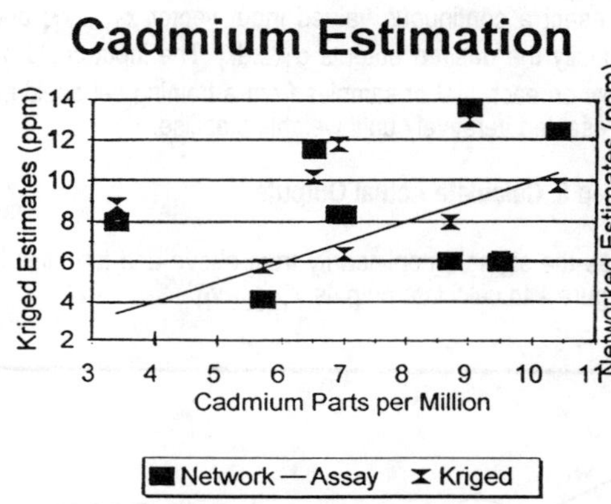

Figure 4 Neural network and kriged predictions for Cadmium

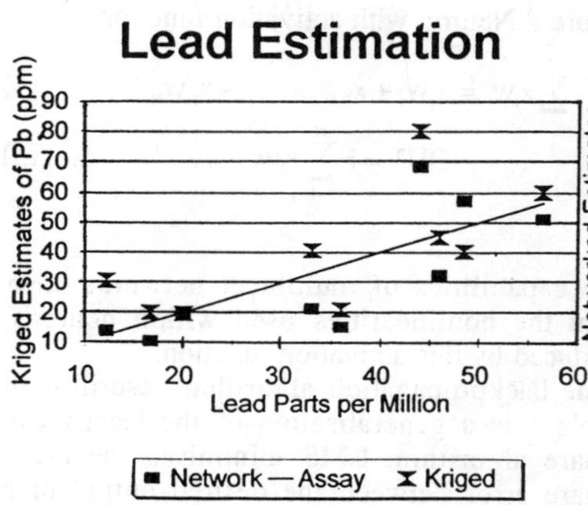

Figure 5 Neural network and kriged predictions for Lead

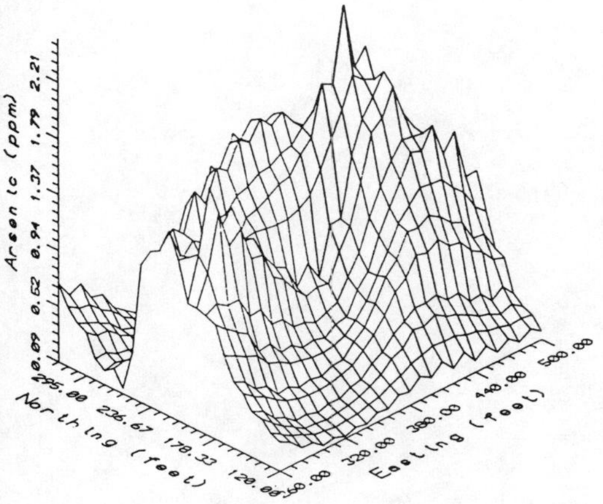

Figure 6 Network predictions for Arsenic

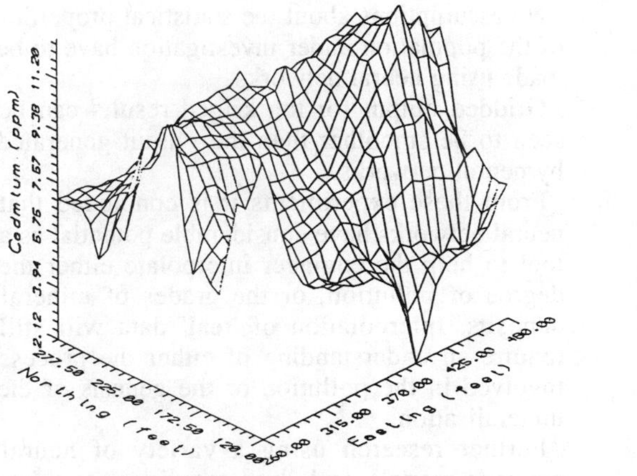

Figure 9 Kriged estimates for Cadmium

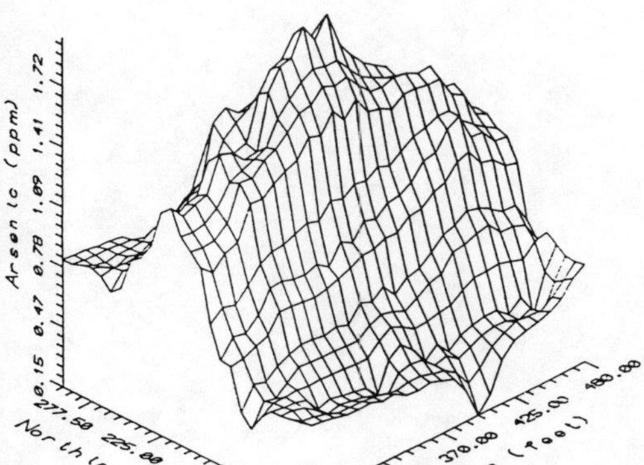

Figure 7 Kriged estimates for Arsenic

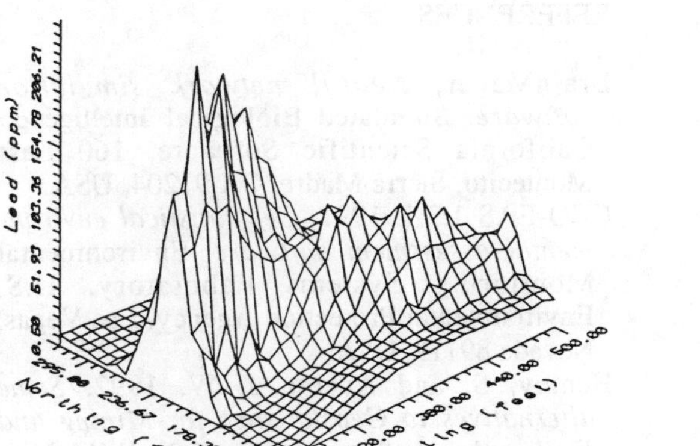

Figure 10 Network predictions for Lead

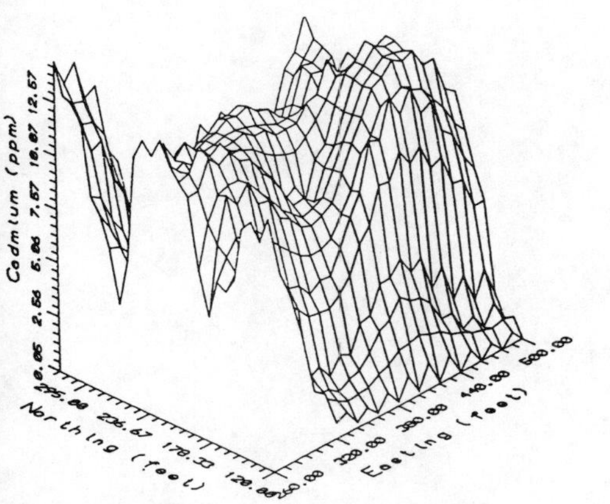

Figure 8 Network predictions for Cadmium

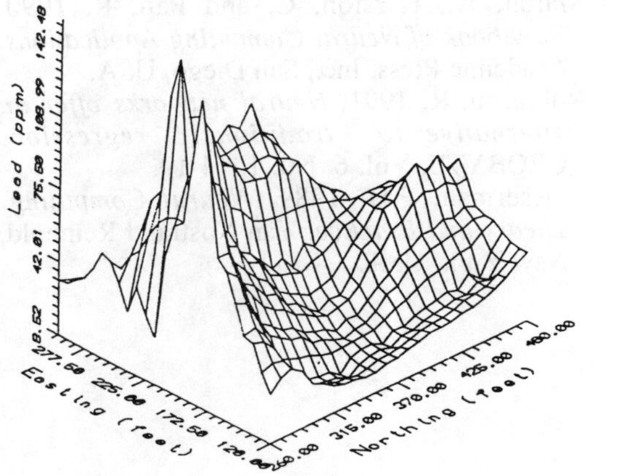

Figure 11 Kriged estimates for Lead

No assumptions about the statistical properties of the population under investigation have to be made using neural networks.

Gridded output for teh kriged results can be seen to be smoother than the output generated by neural networks.

From these experiments it is concluded that neural networks have considerable potential as a tool to help the engineer interpolate either the degree of pollution, or the grades of mineral deposits. Interpolation of 'real' data will still require an understanding of either the process involved in the pollution or the genesis of the mineralisation.

Further research using a variety of neural network models and data sets is now being planned.

REFERENCES

BrainMaker, *Neural network simulation software*. Simulated Biological Intelligence, California Scientific Software, 160 East Montecito, Sierra Madre, CA 91204, USA.

GEO-EAS 1.2.1. 1991. *Geostatistical environmental assessment software*. Environmental Monitoring Systems Laboratory. U.S. Environmental Protection Agency, Las Vegas, Nevada 89119, USA.

Henley, S. and Aucott, J. W. 1992. *Some alternatives to Geostatistics for Mining and Exploration*. In Transactions of the IMM, Vol. 101, No A1-74, A36-A40.

Lippmann, R. P. 1987. *An Introduction to computing with neural nets*. IEEE ASSP Magazine, Vol. 4, No 2., 49-67.

Maren, A., Harston, C. and Pap, R. 1990. *Handbook of Neural Computing Applications*. Academic Press, Inc., San Diego, USA.

Robinson, R. 1991. *Neural networks offer an alternative to traditional regression*. GEOBYTE, Vol. 6, NO. 1, 14-19.

Wasserman, P.D. 1989. *Neural Computing. Theory and Practice*. Van Nostrand Reinhold, New York, USA.

Innovative Mine Design for the 21st Century, Bawden & Archibald (eds) © 1993 Balkema, Rotterdam, ISBN 90 5410 325 6

Quantification and analysis of mineral venture management alternatives using derivative asset valuation

Jerry M.Whiting & Samuel Frimpong
Mining, Metallurgical and Petroleum Engineering Department, University of Alberta, Edmonton, Alb., Canada
David G.Laughton
Finance and Management Science Department, University of Alberta, Edmonton, Alb., Canada

ABSTRACT: The parameters of a gold property are defined. Two questions related to mine venture development alternatives are asked: (1) What is the effect on mine present value of various options to do more studies -- drilling, evaluation, testing, refinement of values, documentation -- all of which could provide venture stakeholders with data useful in making better decisions. (2) Options exist to delay development by doing studies, waiting, or "dithering," before choosing to proceed with development, or even abandoning the project. What is the present value of such options? The answers to these questions are quantified and analyzed by modern financial methods, particularly derivative asset valuation. The answers comprise a series of graphs, and their derivations are outlined as a basis for applying this method to similar questions.

1 INTRODUCTION

Many decisions related to the development of mineral ventures are made by analyzing values obtained from discounted cash flow analyses. Typically, such analyses calculate the net present value of the venture's projected cash flow using a chosen discount rate, or they define that discount rate at which net present value is zero. The latter value is referred to as the discounted cash flow return on investment, or the venture's internal rate of return. The insight and understanding provided by completing DCF analyses are important in understanding the financial nature and sensitivity of a mineral venture; nevertheless, the DCF methods are not suited to help answer many questions of vital interest to the investors and management of new ventures.

However, additional tools for analysis and evaluation are available to those who must make numerous decisions to manage often large investments in mining ventures. The tools might be referred to as modern financial analysis methods, or as modern financial engineering methods, because best use of them is made by a person, or a team, who are informed about both finance and engineering. The insights provided by modern financial analysis have had profound effects on the development of many parts of the financial market, such as those involving options and collateralized securities. They give traders the ability to calculate values for complex assets, and provide methods for designing hedging positions for firms and other economic agents as they seek to extract value from their financial structures. The question which can be posed is: "Can these insights, ideas and techniques form the basis of a set of financial analysis methods that will aid in the management of non-financial assets such as potential mining ventures, and, if so, how?"

This paper is the result of collaboration between a finance/management professor, a mining professor with many years of experience related to new mining ventures, and one of their mutual interests, a Ph.D. student in mining engineering. Without the patient guidance of the finance professor, the mining engineers would have gotten lost in a swamp trying to develop methods to answer questions they believe are very typical and important to venture managers. However, without the latters' interests and insights, perhaps the finance professor would not have devoted the very considerable time required to create the software needed to generate answers to their questions.

The purpose of this paper is to demonstrate the application and results of applying modern financial methods -- specifically derivative asset valuation (DAV) -- to determining the present value of a few common venture management alternatives. Typically, such options include whether or not to commit to a venture feasibility study of a certain cost and duration, given a number of factors which need further definition or resolution; whether to wait to begin venture development or to undertake development immediately; whether to begin environmental studies of certain cost and duration to define existing conditions and to estimate the ultimate effects of future operations; and whether to undertake additional exploration to define ore reserves more precisely.

With modern financial engineering methods, the sensitivity of such determinations to changes in key venture parameters (such as product price, feasibility study cost and duration, different possibilities for capital and operating costs, expected reserve tonnage and grade, and the amount of environmental variability) can be readily examined through "what if" computer runs which are limited only by the analyst's imagination. Because of space limitations, the writers examine only two questions in this paper, but their treatment and the results are illustrative of using DAV as a tool for quantifying mine venture management options.

2 "REAL-WORLD" PROPERTY DESCRIPTION

For the purposes of this paper, it would be helpful to imagine a surface-mineable gold deposit. Enough exploration core drilling has been carried out at this property to define the three-dimensional location and configuration of a large, near-surface volume of gold-bearing rock.

Sufficient cores have been recovered and assayed to estimate that the average grade of the deposit is close to 0.075 Troy oz of gold per metric tonne throughout about 30 million contiguous tonnes of rock. The present cumulative judgement of venture staff familiar with the deposit is that this estimate of total metal content (about 2.25M oz) has a "low" probability of either increasing or decreasing by more than four million tonnes of "effective ore", if assay data from additional coreholes were to be included in a thorough geostatistical analysis of the deposit, and the results -- plus other new data -- proved some present assumptions to be incorrect.

(Note: "Effective ore" is that quantity (in tonnes) which, when multiplied by net cash flow (output price less "effective average cost"), yields the venture's (project's, mine's) present value.)

Venture staff have completed preliminary estimates of mining and mineral processing costs. At this stage, the total "effective average cost" (defined below) per tonne of ore is estimated to be $20 per tonne, including all capital and expense (operating) charges. Preliminary plans specify mining by conventional open pit methods. Gold recovery is envisioned by gravity concentration of metallic gold particles, plus flotation of gold-bearing sulfide concentrates followed by roasting, leaching, precipitation, and subsequent reduction of the precipitates to metallic gold.

Project engineers estimate, on the basis of information collected to date, experience and judgement, that effective average cost has a "low" probability of increasing more than $4 per tonne due to further project definition and assessment through various studies, bench-scale testing, bulk sampling and even prototype-plant testing.

The model examined in this paper defines the total value of the venture (project, mine) as a linear function of the output price at the time development is initiated, as is illustrated by the dashed line (the "now-or-never" option) in Figure 1, Basic Mine Valuation. Note that output (gold) price in this paper is stated in terms of revenue dollars per tonne of ore. For example, it takes about 13.33 tonnes of rock to produce an oz of gold; therefore, at a price of $266.67 (Cdn), one tonne of rock contains $20 of gold. Figure 1 shows the mine present value is zero at this price. As the price of gold doubles to $533.33 per oz ($40 per tonne of ore), mine present value approaches $600 million. (The curved solid line, representing the "timing option" will be discussed below, under the heading Mine Development Model.)

The slope of the dashed line is a measure of the the effective amount of ore; the price intercept on the X axis denotes the effective average cost. (For example, the slope of the dashed line increases or decreases in relation to factors which increase or decrease the tonnage of rock which can be mined and processed for a positive value [i.e., ore]. The cost intercept does not change, unless there is a change in gold mining and processing cost. Note: The effective amount of ore, and the effective average cost, would be the actual amount of ore and the actual average cost, if the output of the mine could be produced and sold instantaneously at the time of development.)

(More generally, the effective amount of ore increases with the actual amount of ore. It also increases with earlier production. The effective average cost increases with cost levels. If the convenience yield, which is the rate of return in the revenue annuity, is greater than the risk-free rate of return, which is the rate of return in the cost annuity, then the effective average cost increases, if production is pushed back in time. Otherwise, it decreases. Finally, an increase in the amount of ore decreases the actual average cost at that time, if it increases production at a given time, and if there are increasing returns to scale. It also stretches out

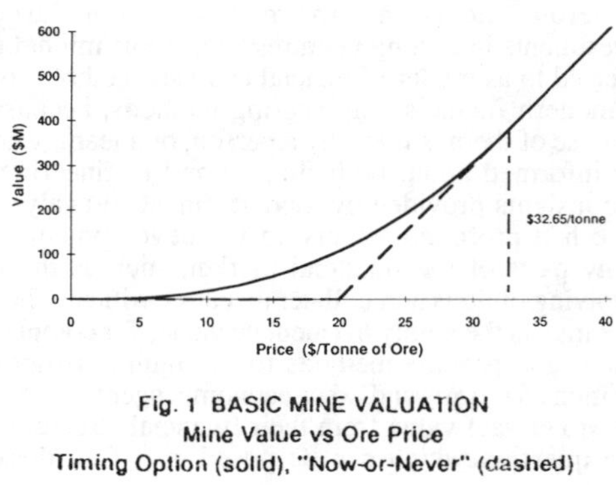

Fig. 1 BASIC MINE VALUATION
Mine Value vs Ore Price
Timing Option (solid), "Now-or-Never" (dashed)

production, if there are capacity constraints. If the convenience yield is less that the risk-free rate, an increase in the amount of ore unambiguously decreases the effective average cost under these circumstances. However, if the convenience yield is greater than the risk-free rate, the scale effects must be large enough, compared to the effects of a longer production life, to make the effective average cost decrease with an increase in the actual amount of ore.)

3 ADDITIONAL ASSUMPTIONS

The following additional assumptions are used to demonstrate the principles of applying the DAV method, and to illustrate the application of its output to the venture outlined above:

1. A variety of concurrent and sequential technical and economic feasibility studies could be undertaken beyond this moment of the project's evolution to better define and document current estimates of: (a) the grade, volume and configuration of the ore deposit, (b) optimum mining and processing methods, and (c) mine present value. Such studies are described below:

1A. Additional drilling at carefully-selected portions of the deposit, along with limited feasibility studies based mainly on experience, file information and judgement. Cost -- one million dollars ($1M).

1B. Twice as much of the same type of exploratory, conceptual, limited-detail feasibility work could be done. Cost -- two million dollars ($2M).

1C. (1) Significantly greater exploration of the deposit could be accomplished for $5M. This could result in greater certainty about the presently-estimated quantity of effective ore, and/or it might add a significant volume of ore-grade reserves (i.e., 6M, 10M tonnes) through discovery of presently unknown deposits. (Depending upon the geologic setting, the latter possibility could range from nil to good; however, it is not considered further in this paper.)

1C. (2) The sum of $5M could also be used to fund further research, fact gathering, documentation, geostatistical analysis, refinement of designs and values, computer simulations and the like, based primarily on the intellectual efforts of various professionals, as well hardware-intensive studies such as bench-scale and pilot-plant-scale tests of metallurgical processes. In fact, the latter type of studies could be essential in providing a highly-reliable estimate of the percentage of the deposit's metal content which would be recoverable by any selected metallurgical process design. These kind of studies could change the effective amount of ore and/or effective average cost.

1D. For $10M, it is assumed that virtually all questions could be answered to the degree useful, and to the satisfaction of any present or potential stakeholder, and that the operating mine and process facilities would match final feasibility report projections of cost, recovery, capacity and profitability.

2. This paper considers that feasibility studies could take zero time. (That is, the information could be instantly available at a cost. For example, the venture might be able to buy a complete "package" of information from earlier investigators.) The authors also consider more usual cases where completion of various geological, mining, and processing studies etc., as described above, would take one year or two years.

3. Mainly for comparison purposes, consideration is given to the possibility that studies also might have zero (incremental) cost, (or at least negligible cost). Practical examples of this might be related to some residual obligation of a previous property owner, or to transfer of costs to another non-venture account.

4. The authors also consider the effect on mine value of options to "dither," that is, either to delay making studies, to delay starting development, or to abandon the venture. However, once a certain magnitude of study is started, the assumption is that it cannot be stopped with either a cost or time savings.

4 MANAGEMENT QUESTIONS AT THIS TIME

The management of this hypothetical gold venture has several questions at this moment which are related to projections of future cash flow. Some of these questions will be answered, in part, by carrying out traditional discounted cash flow analyses using a selected array of the values derived from the various studies outlined above. In general, two types of analyses will be done: (A) Determination of the project's internal rate of return (IRR), and/or (B) determination of the project's net present value using a selected discount rate. The results will be compared to other investment alternatives.

However, preceding these final DCF analyses, much time-consuming and expensive work could be authorized, based largely on experience, judgement and perceived needs. The main questions which need to be answered at this time, are those which impact near-term decisions and activities:

Question 1. What is the quantitative effect on mine present value of various options to do more studies -- drilling, evaluation, testing, refinement of values, documentation -- all of which could provide venture stakeholders with data useful in the process of making better decisions.

Question 2. Options exist to delay development by either doing studies, or by simply waiting, or "dithering," before choosing to proceed with development, or even abandoning the project. How can one quantify and analyze the value (positive or negative) of such options ?

5 THE MODEL

5.1 Output price and valuation model

A key proposition to use of derivative asset valuation is that valuation may be carried out, to a good approximation, as if the financial market were efficient and free of transaction barriers. In such a market, different assets with the same cash-flow consequences have the same price. Moreover, it is possible to replicate the cash-flows from a complex asset, such as a right to develop a specific gold deposit, by trading portfolios of more basic and usually simpler assets, such as gold forward contracts. Because such replication is possible, the value of a complex asset can be determined from the valuation of the basic assets. Finally, all asset prices are determined by the overall risk preferences of investors. Therefore, the basic assets that provide the information about risk discounting are those directly associated with key future macroeconomic variables, such as the forward gold contracts in the example just given, which are related to corresponding future prices of gold.

The insights provided by modern financial analysis have had profound effects on the development of many parts of the financial market, such as those involving options and collateralized securities. They give traders the ability to calculate values for complex assets, and provide methods for designing hedging positions for firms, and other economic agents, as they seek to extract value from their financial structures.

Brennen and Schwartz (1985) were the first to look at a stylized model for an operating mine. At any time, their mine can be closed, at a cost, if open; at a different cost, if closed -- or, it can be abandoned. There is no uncertainty or time dependence in the production rate at the mine when open, or the cost of operating the mine, open or closed. The amount of ore left to be mined is also known at any point. The only uncontrollable uncertainty in the cash-flows is due to the output price.

Their model model of this uncertain price, and of the valuation of claims to the price, is also fairly simple. It is based on an initial price and a pattern of possible price changes at subsequent times. The price changes in their model, while uncertain, are generally (but not always) small over small time periods, with a known and constant proportional amount of uncertainty (or volatility). The current term structure of the forward output prices exhibits a constant exponential decline with respect to the forward term. Finally, the term structure of the forward prices that will occur in any possible situation in the future is taken to have the same decline; only the spot price is uncertain.

They showed how, under these circumstances, ignoring the options to open, close and abandon a mine can leave the mine undervalued, and how ordinary project valuation based on discounted cash-flow methods can give misleading comparisons among mines. While they offered some insight in what the new theory might contribute to the practice of project valuation, they did not provide a blueprint for an operational tool that an organization might use in the evaluation of real projects.

Results from the Brennan and Schwartz type price model modified for this paper are shown in Figure 2. In this specific example, the risk-free rate is 3% per year, the rate of decline in the term structure of forward prices is 2% per year, and the price volatility is 20% in annual terms. The figure shows the 80% confidence interval for the term structure of prices (solid lines), if the term structure of price medians is flat, and scaled to be 1 (the central dotted line). It also shows the term structure of current values for claims to prices in the future. This term structure has a decline of 5% per year, which is a combination of the rate of decline in the forward prices (2% per year) and the risk-free rate (3% per year). Note carefully that results reported in the remainder of this paper would differ with changes in input parameters.

Considerably more space would be needed to explain the reasoning and derivation of algorithms used to model output (gold) price. This is not done here, because conveying an understanding of the price model is secondary to illustrating other aspects of this method of analysis, and the type of results obtained from it. Those who are particularly interested in this subject, and want to examine the more of the reasoning and algorithms behind the results shown in this paper, should contact Dr. D. G. Laughton, Faculty of Business, or Dr. J. M. Whiting, Mining, Metallurgical and Petroleum Engineering Department, University of Alberta, for a more complete background of Figure 2 (and many

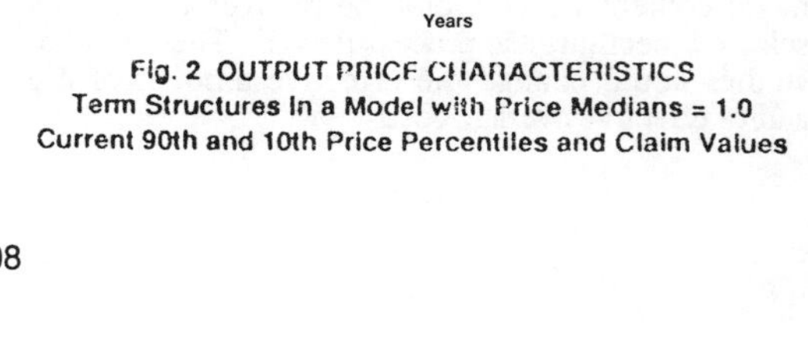

Fig. 2 OUTPUT PRICE CHARACTERISTICS
Term Structures In a Model with Price Medians = 1.0
Current 90th and 10th Price Percentiles and Claim Values

other facets of this paper), as included in Laughton, Frimpong and Whiting paper 4-92 (1992).

5.2 Mine development model

The mine development model used to define mine (venture, project) present value at the time of development makes this value a function of the contemporaneous price of gold. Costs and other factors are considered to be independent of the economic state. That is, regardless of what is happening to the economy and the price of gold, it is not reflected in costs, reserves, production rate, process alternative, or the like. Therefore, net cash flow (revenue minus cost) changes only in direct relation to changes in gold price.

Uncertainty in costs, reserves, production rate, process alternatives, etc., are not correlated with uncertainty in price, or any macroeconomic risk factor. The value of future revenue is the product of the expected rate of production and the value of the claim to the output price at that future time, and it is proportional to gold price at the time of development. The value of the cost is projected cost cash flow discounted at the risk-free rate of return. The is independent of price at the time of development.

Mine value at the time of development is the sum of the following cash flows: Positive -- The price, at the time of development, times the integral of incremental quantities of production, discounted at the convenience yield. Negative -- The cost of producing these incremental quantities, discounted at the risk-free rate. The resulting value is linear in the price at the time of development. (See the dashed line in Figure 1.)

In this paper, the effective mine value parameters are specified directly, rather than through examination of the determinants. Value parameters are variations around an amount of 30M tonnes, and a cost of $20 per tonne. While these two basic parameters are derived from the example outlined earlier, they merely set the basic price and value scales to be considered in the analysis, and could be changed, under appropriate scaling of the results, without loss of generality.

In the present example, development should be undertaken at any price above the effective average cost, if there were no option to wait (called the "now-or-never" situation). However, if there is an option to wait (the "timing-option" situation), the value from development must be greater than the value of this option before development should take place. The price at which this occurs is proportional to the effective average cost. For the price and valuation model used in this paper, the proportionality factor is 1.632. (An explanation for this phenomenon is too long to be included here. However, for a given risk-free rate, this ratio increases with increasing volatility and decreasing yield. It is one, if there is no uncertainty, and, if there is uncertainty, it is infinite, if the convenience yield is not positive. What this means, in the basic mine model, is that development should occur at prices above $32.65 per tonne (to the closest $0.05 per tonne) . The value of the basic mine is graphed in Figure 1 as a function of output price expressed in $/tonne of ore. (Merton 1973, Laughton 1992).

The dashed line shows the "now-or-never" development value, while the solid line shows the value if there is a "timing option" available for management to delay, indefinitely, the development of the mine. Notice the now-or-never value and the timing-option value are the same above a price of $32.65 per tonne, where the option to delay development would not be taken. However, an option to delay development significantly increases mine value below this value, especially in the output price range of, say, $15/tonne to $25/tonne range. The maximum effect is at $20/tonne, where the additional present value is about $110M.

5.3 Feasibility study model

Feasibility studies are defined by cost and duration. If the study duration is greater than zero, consideration is given to whether or not the study can be abandoned. An abandonment option may be valuable, if there is a price where immediate development, without the information provided by the study, would be more valuable than waiting.

All analyses in this paper consider a set of alternatives where the feasibility study has been done (the "known-state" situations), cannot be done (the "no-study" situation), and can be done instantaneously at zero cost (the "free-study"situation). The practicality of such situations was commented on earlier. These last two situations are the opposite extreme cases of no information and free information. Next, a set of studies, each of which takes zero, one or two years, and has a cost of zero (if the study duration is not zero), $1M, $2M, $5M and $10M. The value gained over the no-study

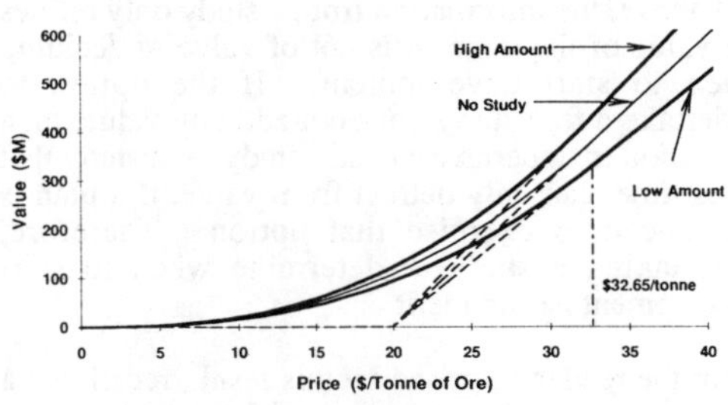

Fig. 3 4 M TONNE UNCERTAINTY IN EFFECTIVE AMOUNT OF ORE
Mine Value vs Ore Price
Known States and No Study
Timing Option (solid), "Now-or-Never" (dashed)

situation, and the value lost, compared to the free-study situation are quantified. For studies which have both a cost and a positive duration, the value lost, compared to the no-cost study of the same duration, is also quantified.

The benefits of each study are used to describe them. For example:

5.3.1 Better timing: Reducing uncertainty in the effective amount of ore

Figure 3 shows the effect on mine present value of a study which reduces uncertainty about the effective amount of ore. Note that this study does not affect the expectation of this parameter (30M±4M tonnes), nor does it reveal any new information about the effective average cost ($20/tonne).

Three pairs of results are shown in Figure 3. The dashed line in each pair of curves defines mine value in a given physical information state, if the mine must be developed immediately, or not at all (the "now-or-never " value). The solid line shows the value, if there is an option to begin development at a time of the developer's choosing. The outer pairs of lines represent mine values in the two possible known physical states that can occur, if the feasibility study is done. If the timing of development is optional, the mine will be developed in either state, as it would be for the basic potential mine outlined in the previous section, if and when the price is above $32.65/tonne (see above). (As in Figure 1, the timing-option and the now-or-never value are the same above this price.)

The central pair of lines shows the value of the deposit if no study is possible. The developer must develop a mine without knowing which physical state will be realized, the high-amount or the low-amount. Notice that the effective average cost, by construction, is $20.00/tonne, just as it is for the two possible known physical states. Therefore, the critical price for development is again $32.65/tonne.

Returning to the fact that the timing-option and now-or-never values are the same at $32.65/tonne and above, the information from a study only refines the value of the mine; it is not of value in deciding when to start development. If the option to undertake a free study does not add any value, then the option to undertake a costly study, or a study that takes time, can only detract from value, if a plan is introduced to exercise that option. Therefore, undertaking a study to determine when to start development has no merit.

For the reader surprised by this result, recall that a revised, reliable determination of the actual amount of ore (as opposed to the effective amount) may result in a revised effective average cost, as well as the effective amount of ore (i.e., if there are fixed costs of development or production). What is shown here is that research about the effective amount of ore alone is worthless; but this does not apply to research about the actual amount of ore that may also affect knowledge of the effective average cost.

5.3.2 Better timing: Reducing uncertainty in the effective average cost

In the following set of examples, a feasibility study would resolve an equiprobable uncertainty of ±$4/tonne about an expected cost of $20/tonne, but would accomplish nothing else.

The scope of the problem is shown in Figure 4, where the value of the venture in various situations is graphed against the output price. The dashed lines again show the value in now-or-never situations, while the solid lines show the values for timing options. The outer pairs of curves are again the values in the two after-study known physical states, with a low effective average cost of $16/tonne, and a high effective average cost of $24/tonne, respectively. The central dashed line, and the lower of the two central solid lines are the values of the no-study situation. The other central solid line shows the values for the timing option with a free study.

If waiting is allowed, the mine would be developed in the low-cost state at any price above $26.10/tonne, in the high-cost state at any price above $39.20/tonne, and in the no-study situation at $32.65/tonne. (Note that with these parameters, the effective average cost for the no-study situation is the same for the basic mine.) Therefore, without a study, development might be undertaken at lower prices than would be desirable (within the price range between $32.65/tonne and $39.20/tonne), should cost prove to be $24/tonne; or development might be delayed at higher prices than would be desirable (between $26.10/tonne and $32.65/tonne), should cost prove to be $16/tonne. Therefore, the option to undertake a study to better understand the physical state may have value. Indeed, the free-study value is greater than the no-study value at prices below $39.20/tonne.

The spread between the no-study and free-study timing option curves shows the maximum

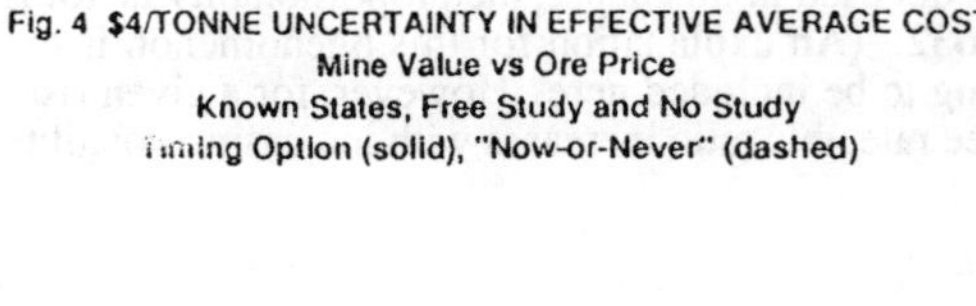

Fig. 4 $4/TONNE UNCERTAINTY IN EFFECTIVE AVERAGE COST
Mine Value vs Ore Price
Known States, Free Study and No Study
Timing Option (solid), "Now-or-Never" (dashed)

incremental value that can be gained from the option to do a study, or, equivalently, the maximum value that is lost, because a study involves a cost or delay.

Notice that this difference goes to zero at zero price. This is because all values are zero at zero price, where there is no probability that the price will change from zero to allow the development of the mine. In other graphs, however, the difference is positive at low positive prices where the developer should wait to act. In such cases, this incremental value is positive, because the study option enhances the value of the payoff for which the developer is waiting.

The incremental study-option value is zero at prices above $39.20/tonne, which is the critical price for immediate development in the high-cost physical state. Immediate development is optimal at those prices, regardless of the study results, which makes the option to undertake the study worthless.

Options to undertake studies that are costly, or take time, are examined below, beginning with the costly studies which take no time.

5.3.2.1 Instantaneous studies

Figure 4A shows the incremental positive value of the instantaneous study options (costing $0, $1M, $2M, $5M or $10M) over the no-study value. The top curve is for the study that has no cost (the free study). It is the difference between the free-study and no-study lines (the two central timing-option curves) in Figure 4.

The value is positive from a price of zero to the critical price of the high-cost mine, $39.20/tonne, given the free-study situation. In the high-price region, where the study value is zero, development is undertaken immediately without doing a feasibility study to define whether effective average cost is either $16 or $24/tonne, or something in between. The critical price decreases with the cost of the study, because the trade-off between the study option and immediate development shifts toward development as the study cost increases.

The curve for the free study shows that most value which can be added by having a study option, is just over $20M, and it occurs at a price of just under $28.50/tonne. There would be no value in doing a feasibility study to define effective average cost between $16 and $24/tonne, if it cost more than $20M. Finally, the price where each study option has its maximum value is the same for all studies in this series.

5.3.2.2 Study time as a cost

The time that the study takes is also a costly feature of the study. This is shown in Figure 4B for studies

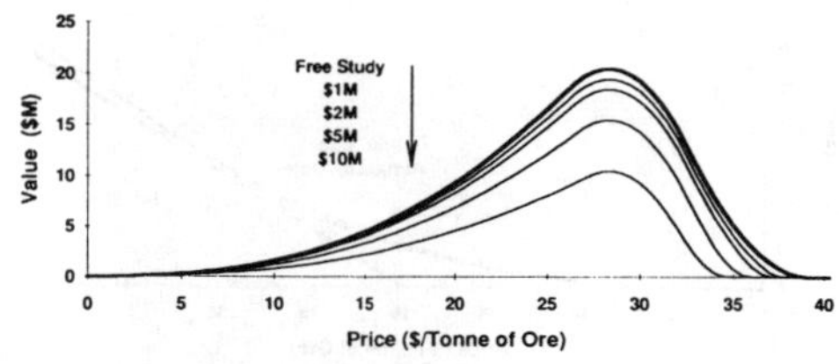

Fig. 4A $4/TONNE UNCERTAINTY IN EFFECTIVE AVERAGE COST
Study Option Value vs Ore Price
Free Study and
Study Time = 0 yrs Study Cost = $1, 2, 5, 10M

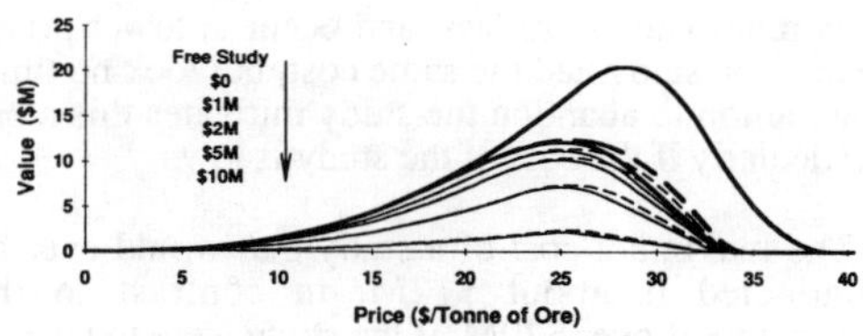

Fig. 4B $4/TONNE UNCERTAINTY IN EFFECTIVE AVERAGE COST
Study Option Value vs Ore Price
Free Study and
Study Time = 1 yrs Study Cost = $0, 1, 2, 5, 10M
With (dashed), Without (solid) Study Abandonment Option

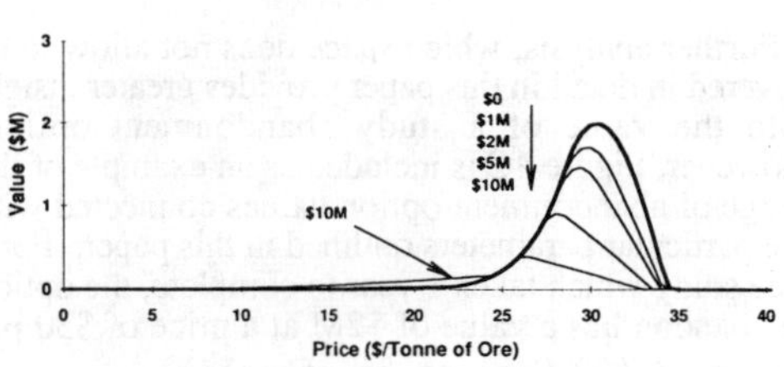

Fig. 4C $4/TONNE UNCERTAINTY IN EFFECTIVE AVERAGE COST
Study Abandonment Option Value vs Ore Price
Study Time = 1 yrs Study Cost = $0, 1, 2, 5, 10M

that take one year to complete. Figure 4B should be compared to Figure 4A.

Figure 4B shows the incremental value of the free (no cost, zero duration) study option over the no-study situation. It also shows the incremental value of the study options where the study takes a year, and costs $0, $1M, $2M, $5M, or $10M. The dashed lines show the values if the study may be abandoned, and the solid lines if it may not. The effects of the abandonment option are discussed in further detail below.

The free-study value is the top curve, and is the same as in Figure 4A. The value of each of the one-year studies is less, and decreases with the study cost. Comparison with Figure 4A shows that the development region begins at lower prices, and the

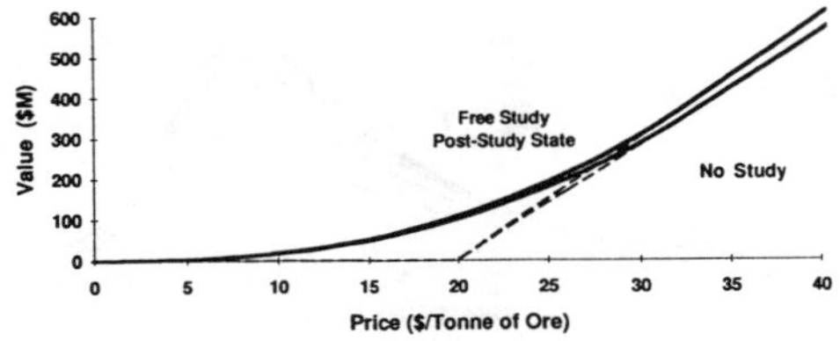

Fig. 5 2M TONNE LOSS OF EFFECTIVE ORE
Mine Value vs Ore Price
Post-Study State, Free Study and No Study
Timing Option (solid), "Now-or-Never" (dashed)

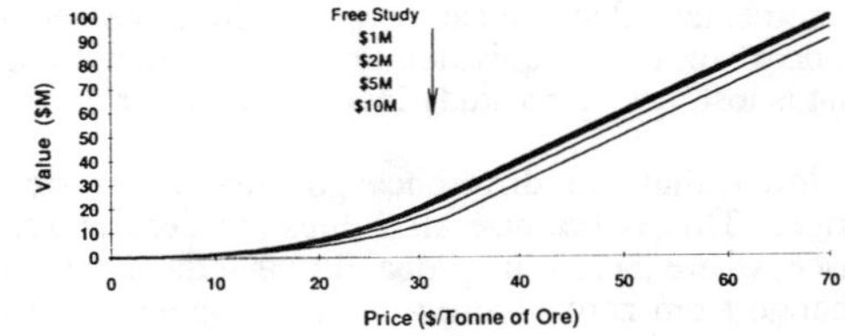

Fig. 5A 2M TONNE LOSS OF EFFECTIVE ORE
Study Option Value vs Ore Price
Free Study and
Study Time = 0 yrs Study Cost = $1, 2, 5, 10M

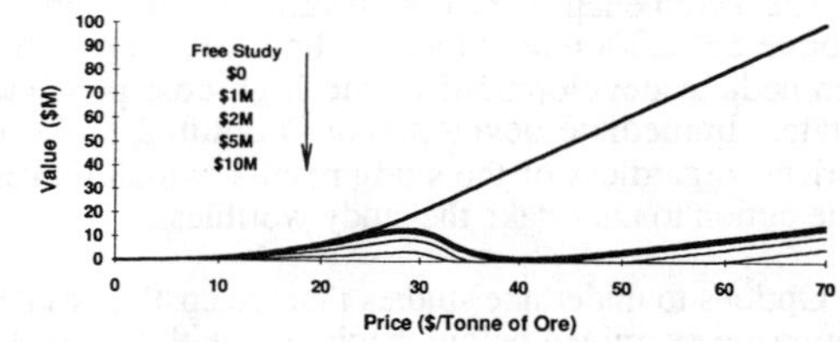

Fig. 5B 2M TONNE LOSS OF EFFECTIVE ORE
Study Option Value vs Ore Price
Free Study and
Study Time = 1 yrs Study Cost = $0, 1, 2, 5, 10M

maximum values are less, and occur at lower prices than if the study had the same cost, but took no time. The option to abandon the study mitigates this a bit, particularly if the cost of the study is low.

The maximum cost of a study that would ever be conducted is about $12M, in contrast to the maximum of over $20M, if the study were to take no time. All of these facts reflect the cost of the study duration, and the shift, with increased study time, away from doing a study toward either waiting or immediate development. Finally, the price where the maximum value of the study option occurs decreases with the cost, from over $26/tonne to about $25/tonne.

Further analysis, which space does not allow to be covered in detail in this paper provides greater insight into the value of a study abandonment option. However, Figure 4C is included as an example of the range of abandonment option values connected with the particular parameters outlined in this paper. For a free study which takes 1 year to complete, the option to abandon has a value of $2M at a price of $30 per tonne ($ 400 Cdn per oz of gold).

5.3.3 More value at development: More effective ore

In the following set of examples, we examine the effects of study options, with the same costs and durations as those examined above in situations where the study does not resolve any uncertainty about the cash flow, but does allow the developer to avoid losing 2M tonnes of effective ore from the 30M tonnes that is otherwise available. An example could be a study which proves optimum mine and/or mill gold recovery techniques.

Figure 5 shows the value of the venture with no study option, and with a free study available. (Because there is no uncertainty resolved in the cash-flows, and because venture management is always, at worst, indifferent about undertaking a free study, the free-study values are the same as those that would be obtained after any feasibility study.)

5.3.3.1 Studies with No Delay

Figure 5A shows the incremental value (over the no-study situation) of the instantaneous studies. The value of the study option increases with the output price, and deceases with the cost of the study. The value increase continues into the high price regime. This is due to the increasing value of the ore that is not lost if the study is undertaken. The present value of the free study option at a gold price of $266.67 (Cdn) per oz ($20 per tonne ore) is about $8M. At a $40 per tonne price, the value increases to about $40M.

5.3.3.2 Study Time as a Cost

The pattern of optimal behavior is again more complex, if the studies take time. This is shown in Figure 5B for studies that take 1 year to complete. (Compare to Figure 4B.)

Figure 5B shows the incremental value of options to undertake the studies that take 1 year, and of the free-study option (no cost, no study time), over the no-study situation. The free-study value is again the top curve, and is the same as in Figure 5A. The abandonment option has no value in these situations, and the values with and without this option cannot be distinguished. Once the cost of a costly study is sunk, there is no incentive to abandon the study, if, in the presence of the option to undertake the equivalent no-cost study, development before undertaking the study would not be contemplated at any price.

A 1-year study option increases in value with price at high prices, but at slower rate with respect to price than the instantaneous study of the same cost. The year delay in production caused by the study is equivalent to an effective loss of ore, through the extra discontinuing of the revenue to be received at the time of the sale of the ore. This loss brings the effective amount of ore with the study down to 28.5M tonnes at high prices at the commencement of the study. Therefore, the net gain at high prices from doing the feasibility study decreases from 2M tonnes of effective ore to 0.5M tonnes. This is the slope of the high price part of the study value graphs in Figure 5B.

The price intercept of the asymptote to the high-price part of the value function for each of the study options is also different for the 1-year studies. At high enough prices, it is best to do a study as soon as possible, if one is allowed, and to develop the mine immediately, if no study is allowed, or immediately after a study, if one is allowed. Therefore, the asymptote of the value function at high prices is the value of the study option in the absence of any waiting or abandonment options.

As an example, consider the no-cost studies. For the instantaneous study, the price intercept without waiting is the effective average cost. The value of the mining opportunity is zero there, whether a study is possible or not. However, if the study causes a delay in the development payoff, this delay has different effects on the value of the costs compared with the value of the revenues. The costs are discounted less than the revenues under our price and valuation model, and therefore decrease in value less with the delay of 1 year due to the study. For a 1-year study, this shifts the effective average cost up from \$20/tonne, the level with no delay, to \$20.40/tonne. Thus, we have a situation, where, with no waiting or abandonment allowed, the mine without a study has a value function with a lower slope than the mine with the study, 28M tonnes as opposed to 28.5M tonnes, and also a lower price intercept \$20/tonne as opposed to \$20.40/tonne. The two value functions intersect at a price of \$41.48/tonne. Above this price, if waiting and abandonment are not allowed, the study option has higher value, while below it, immediate development dominates. This critical price is shifted up with the cost of the study, as the trade-off between doing the study and immediate development shifts, with the extra study costs, toward development. Note that this is the behavior of the intercepts of the asymptotes to the high-price part of the study value functions in Figure 5B.

The value of the 1-year study options has a complex structure at low to medium prices, if the option to wait before acting is allowed. For each type of study, it increases with price at low prices. An increase in the value of the payoff for an investment timing option (in this case, stemming from the option to do the study) always increases the incremental value of the option when it is far enough "out of the money". Because at zero price, the incremental value is zero, this means there must be region at low prices where the incremental value is increasing in the price. However, as the price enters the region in which the developer would want to develop the mine if no study were available, the effect of the delay caused by the study starts to decrease the incremental value of the study. This continues until the price reaches the high price regime, where, as is shown above, the option to undertake the study increases in value when compared to immediate development.

Note that low cost studies are still preferred to immediate development at prices below the critical prices, at and above \$41.48/tonne, found for situations that do not allow waiting or abandonment. This is because undertaking a study allows a subsequent development timing option after the study is completed, which might be used to delay development in the event that price does not increase enough while the study occurs. However, if the cost of the study is high enough, there is a range of prices where the advantage of this subsequent timing option is nullified. The incremental value of the study option is zero at these prices, and immediate development is optimal.

5.3.3.3 Longer study times

If the study time is long enough, the effective loss of ore at high prices due to the production delay increases enough to overwhelm the benefits of the feasibility study. The option to undertake 2-year studies, in contrast to the 1-year studies, has no value at high prices.

However, some 2-year studies may have some value at lower prices, because at prices in a region below the critical price for developing the no-study mine (\$32.65/tonne), the time taken to do the study

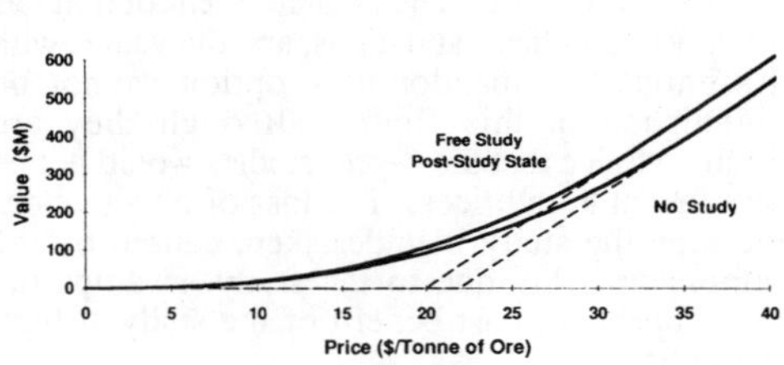

Fig. 6 \$2/TONNE INCREASE IN EFFECTIVE AVERAGE COST
Mine Value vs Ore Price
Post-Study State, Free Study and No Study
Timing Option (solid), "Now-or-Never" (dashed)

would otherwise be spent, at least to some extent, in waiting for development to occur until the price reaches this level. Note that the maximum value of the no-cost 2-year study is less than $10M, which means that the $10M 2-year study option does not have any incremental value at any price.

5.3.4 More value at development: Lower effective average cost

This set of examples, examines the effects of study options that allow the developer to avoid an increase in effective average cost of $2/tonne above the base level of $20/tonne.

Figure 6 shows the value of the venture with no study and with a free study available. It is similar to the situation where the study permits the developer to avoid an effective loss of ore, except that the incremental value of the free-study option is constant at high prices.

5.3.4.1 Studies with no delay

Figure 6A shows the incremental value (over the no-study situation) of the instantaneous studies. The value of the study option increases with the output price up to the price region where the study is undertaken and the cost saving is realized. Unlike the situations where the study results in an improvement in the amount of ore, the study value does not increase further at higher prices. Nevertheless, as in the last set of instantaneous study examples, immediate development is not optimal at any price.

5.3.4.2 Study Time as a Cost

The pattern of optimal behavior is more complex again, if the studies take time. This is shown in Figure 6B for studies that take a year to complete.

Figure 6B shows the incremental value over the no-study situation of the free study (which has no cost and takes no time), and of the 1-year studies. The free-study value is again the top curve and is the same as in Figure 6A. The abandonment option has minimal value in these situations, and the values with and without the abandonment option cannot be distinguished in this figure, although they are different. Notice that the 1-year studies would not be undertaken at high prices. The loss of effective ore at the time the study is undertaken, caused by the development delay due to the study, swamps the price-independent cost benefit of the study at high enough prices.

5.3.4.3 Longer Study Times

If the study time is made longer, the effective loss of

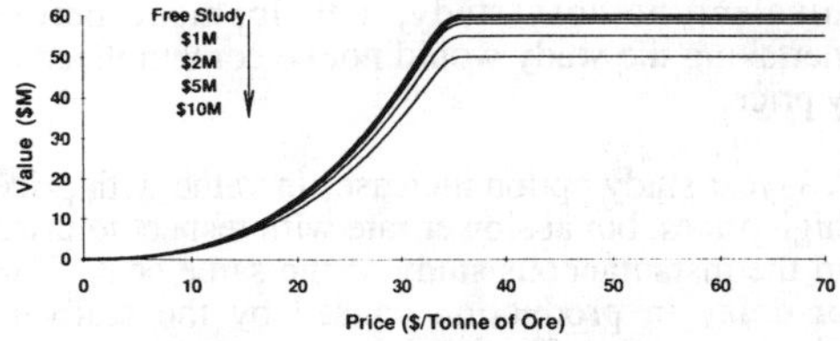

Fig. 6A $2/TONNE INCREASE IN EFFECTIVE AVERAGE COST
Study Option Value vs Ore Price
Free Study and
Study Time = 0 yrs Study Cost = $1, 2, 5, 10M

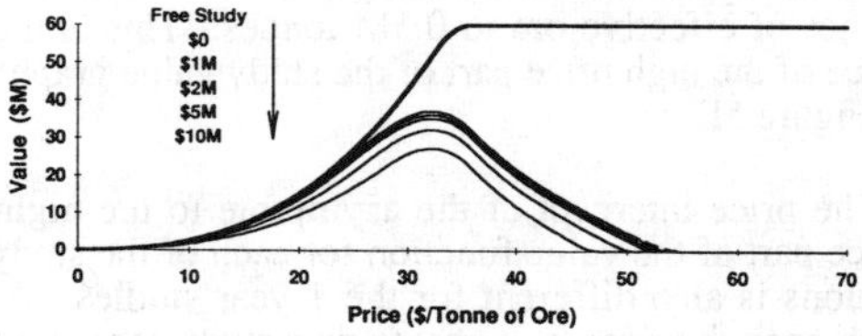

Fig. 6B $2/TONNE INCREASE IN EFFECTIVE AVERAGE COST
Study Option Value vs Ore Price
Free Study and
Study Time = 1 yrs Study Cost = $0, 1, 2, 5, 10M
With (dashed), Without (solid) Study Abandonment Option

ore at the initiation of the study increases and overwhelms more readily at high prices the benefits of the feasibility study. The maximum value of the no-cost 2-year study is less than $10M, which means that the $10M 2-year study option has no incremental value over the no-study situation.

6. CONCLUSIONS

6.1 General

Modern asset pricing theory can be used to analyze the basic structure of mining feasibility study management and value. The potential benefits of such studies can be classified into a two-by-two matrix. One dimension considers the aspect of the future development decision that is affected by the study: better development timing, which results from more refined information about the development value (considered as a function of the contemporaneous output price), and a better mine design, which results in a higher development value function. The other dimension considers the parameter of the development value function that is affected by the study: the effective amount of ore, which measures the degree of dependence of the development value on the contemporaneous output price, and the effective average cost, which is price where the development value is zero.

Each type of benefit has been examined separately, and shown how they give different patterns of study values and management, if these are considered as functions of the output price, study duration and study costs.

The first step in analysis of a given type of benefit is to graph the incremental value of the free-study option over the no-study situation. If this is anywhere positive, one can then analyze other study options by study duration, and then, for each duration, by study cost.

6.2 Better Timing: Reducing Uncertainty in the Effective Amount of Ore

A reduction in uncertainty of the effective amount of ore alone would not influence the development policy, and the option to achieve such a reduction has no value under any circumstances.

6.3 Better Timing: Reducing Uncertainty in the Effective Average Cost

A reduction in uncertainty of the effective average cost can have value. If there is an upper limit on the possible effective average cost, the value of a free study is zero above a critical price, which marks the lower boundary of the price region where immediate development would be undertaken, even if the results of the study were known. This critical price, the study option value at prices below it, and the location of the price region where the study is actually begun, all decrease with duration for the no-cost studies. Probably, the critical price is positive for any finite duration.

The maximal value of the no-cost study, for a given duration, sets the upper limit on the cost of studies, with the same duration, that have any positive value. For studies of a given duration, the critical price for immediate development, the study option value below that pice, and the size of the study region decrease with increasing cost to zero at the upper-limit cost just mentioned.

Finally, the authors conjecture that any no-cost study has positive value in a price region of positive size for any positive amount of uncertainty to be resolved. Moreover, at levels of uncertainty where a given study is valuable at some prices, the extent of that price region, and the value there, increases with the amount of uncertainty to be resolved.

6.4 More Value at Development: More Effective Ore

If a study prevents an effective loss of ore, then, if it is free, its value increases indefinitely with the price. This is also true of all the costly instantaneous studies. However, there is a critical study duration where the value of the amount of ore found by the study is lost at high prices through the delay induced in selling any of the ore. The no-cost studies with a duration above this critical duration have a critical price above which immediate development is more valuable than undertaking the study. Likely, this critical price decreases with duration. The analysis of costly studies with these high durations is the same as that for studies where the benefit is a reduction in cost uncertainty.

Studies, with positive durations below the critical duration, have a more complex structure. For low enough study cost, there is a region of increasing value at low prices, which is followed, at intermediate prices, by decreasing value before a high price regime is reached, where value increases again. If the cost is high enough, there is a part of the intermediate price region where immediate development is optimal, and the study option has no value.

The value of the free study increases with the effective amount of ore that would be lost, if the study were not done. The critical duration also increases with the amount of the potential loss. For any given study duration and cost, the extent of the range where immediate development would be optimal, if one exists, would decrease as the loss increases.

6.5 More Value at Development: Lower Effective Average Cost

If a study allows the developer to avoid an increase in the effective average cost of the mine, then, if it is free, it has an increasing incremental value up to the post-study critical price for development. Above this price, the value is constant at the total value of the costs avoided.

If the studies are costly (with a cost less than the avoided post-study cost increase) and instantaneous, they have a similar pattern of value. The level of the high-price value plateau is at the difference between avoided post-study cost increase and the study cost. It is reached at a price that increases with the study cost. If the study cost is greater than the avoided post-study cost increase, then the study has zero incremental value, and is equivalent to having no study option at all.

The no-cost studies of positive duration have zero value at high enough prices, because the avoided post-study cost increase is dominated at high enough prices by the study delay costs. The lower boundary of the high-price regime, where the study value is zero, decreases with the duration of the study, as does the value of the study below that price. Likely, for any finite duration, the boundary is positive.

The addition of a direct cost further decreases this price boundary, and the value below it, until a duration-dependent upper limit is reached where the study value is uniformly zero. The actual study region also decreases in extent with the cost of the study up to this upper limit. The upper limit decreases with study duration, but likely is positive if the duration is finite.

Finally, the incremental value of a study option, if it has any value, increases with the increase in effective average cost that would occur if the study were not done. If the study is not free, the region where the study would be undertaken also increases in extent, and if the study has positive duration, the critical price for the onset of immediate development increases.

6.6 Further Work

While the qualitative structure of the comparative statics with respect to the level of study benefits and costs seems fairly clear, a quantitative exploration is in order. Moreover, any real example of a study option would present a combination of benefits. With the structure of the basic classes of benefits understood, combinations should be examined. Lastly, we have not examined in this paper are the comparative statics of study value and management with respect to the price and valuation model, which is parameterized by the amount of price volatility, the convenience yield of the output, and by the amount of time discounting. Although the general structure of the effects of these parameters on an investment timing option overall is known (Merton 1973, Laughton 1992), the detailed incremental effects, when these parameters are considered in interaction with the mine and study parameters, has yet to be explored.

A more basic exploration of the technology choice dimension of the development decision, and how it is influenced by a study option, is also in order. The feasibility study model itself should be expanded, so that there is choice amongst possible studies, as well as more internal structure within a given choice. The interaction of the study process, the initial development decision, and subsequent flexibility in mine operations is an area that is worthy of exploration. There are also tax and regulation issues that may be addressed by aggregating from project-level analyses such as this. These can include not only the effect of different royalty structures on venture management, but also the effect of environmental regulation, which can require background studies that can take long, uncertain periods of time to complete before development is permitted to begin.

Finally, mining feasibility studies are directly analogous to commercial research activities. The applicability of this general approach to research planning and evaluation should be addressed for other classes of commercial knowledge-creation.

REFERENCES

Brennan, M. J. and E. S. Schwartz 1985. Evaluating natural resource investments. Journal of Business. 58: 135-157.

Merton, R. C. 1973. A rational theory of option pricing. Bell Journal of Economics and Management Science. 4: 141-183.

Laughton, D. G. 1992. Dithering. University of Alberta Institute for Financial Research working paper 3-92.

Laughton, D. G., S. Frimpong and J. M. Whiting 1992. Mining feasibility studies: Development timing and value versus study duration and cost. University of Alberta Institute for Financial Research working paper 4-92.

Authors' mailing address: 606 CME Building
University of Alberta
Edmonton, AB T6G 2G6

Innovative Mine Design for the 21st Century, Bawden & Archibald (eds) © 1993 Balkema, Rotterdam, ISBN 90 5410 325 6

Geostatistical estimation of grade and dilution, Snip Gold Mine, Northern British Columbia

A.J. Sinclair & Z.A. Radlowski
The University of British Columbia, Vancouver, B.C., Canada

R.M. Srivastava
FSS International, Vancouver, B.C., Canada

A. Samis
Cominco Limited, Vancouver, B.C., Canada

ABSTRACT

Snip deposit, an operating gold mine in the Iskut River area of northwestern British Columbia, is in a metamorphosed sequence of sedimentary rocks (feldspathic greywacke with lesser siltstone) correlated tentatively with the Lower Jurassic Hazelton Group. The principal ore deposit, the Twin Zone, is a gold-bearing, shear vein system characterized in places by a calcite-quartz-pyrite assemblage (calcite or quartz can dominate), a biotite rich assemblage and elsewhere by massive pyrite-pyrrhotite. Twin zone is a roughly tabular zone striking about 120° and dipping moderately but variably (30° to 70°SW). Thicknesses vary from less than 1m to about 10m (mine average is 4.2m). Published reserves are about 940,000 tonnes grading 28.5 g Au/t.

Critical to reserve estimation is the so-called BSU dyke, roughly tabular in form, which has intruded subparallel to and in places within the Twin Zone, in many places dividing the ore zone into two parts separated by waste (hence the name Twin zone). Our geostatistical study included the following steps: (i) a 2-dimensional approach to mineral inventory (block) estimation within the shear vein, (ii) development of an indicator-based procedure for quantifying internal dilution resulting from the BSU dyke, and (iii) estimation of dilution resulting from Main Zone (Twin Zone plus internal BSU) widths less than the minimum mining width.

Available gold assays from 116 underground drill holes and 2 surface drill holes (levels 180m to 300m) are the data base for our study which entailed data analysis, variogram modelling, crossvalidation and comparative estimates by kriging, polygons and inverse squared distance. A multicomponent, practical geostatistical reserve estimation procedure has been developed; limited comparative estimates indicate that the method is highly satisfactory for mineral inventory studies.

1 INTRODUCTION

Snip is a small, complex gold vein deposit in the Iskut River area, northwestern British Columbia, Canada. The property, presently producing, is owned jointly by Cominco Ltd. and Corona Corporation (now Homestake MInerals Ltd.), both of Vancouver, B.C.; Cominco is the present operator. Production began in January 1991 and by the end of February had attained 8000 ounces (272,000 grams) of gold. Published reserves are about 1 million tonnes grading approximately 30 g Au/mt with minor amounts of silver and copper.

Snip Mine is near the British Columbia-Alaska border, about 110 km northwest of Stewart, B.C. The ore deposit, known as the Twin Zone, is south of Iskut River on the northern slope of Johnny Mountain, and crops out between elevations 600m to 650m. The deposit was the subject of a feasibility study during the 1980's that resulted in the present mining operation at an initial ore production of 100 tons daily which quickly rose to 300 tpd, with an average gold grade of 28.5 g/t. The accompanying silver and copper (15 g/t and 0.06% respectively) are not considered in reserve estimation at the minesite due to their low grades and discontinuous nature.

This study is designed to compare ore reserve estimates using several techniques with emphasis on geostatistical methodology. Our aim is to identify an optimal procedure of reserve estimation that could be adapted onsite, as well as to produce guidelines useful in the evaluation of other precious metal deposits with similar characteristics.

2 GEOLOGY OF SNIP DEPOSIT

The area is underlain by a sequence of stratified rocks which correlates with the lower Jurassic Unuk River Formation of the Hazelton Group. This sedimentary and volcanic package consists of feldspathic greywackes, argillites, andesitic flows and tuffs, intruded by feldspar porphyry stocks and dykes. The meta-sedimentary rocks, mainly feldspathic greywackes interbedded with fine siltstones at the Snip property, are poorly bedded to massive, with a dip of about 33°N.

The metasedimentary unit is intruded by a porphyry body and "lamprophyre" dykes in the southeastern corner of the property. Biotite lamprophyre dykes dip about 85° west and clearly cross-cut the gold bearing structure. Recent work by Rhys and Godwin (1992) indicate that these dykes were emplaced late in the mineralization and structural history of the Twin vein.

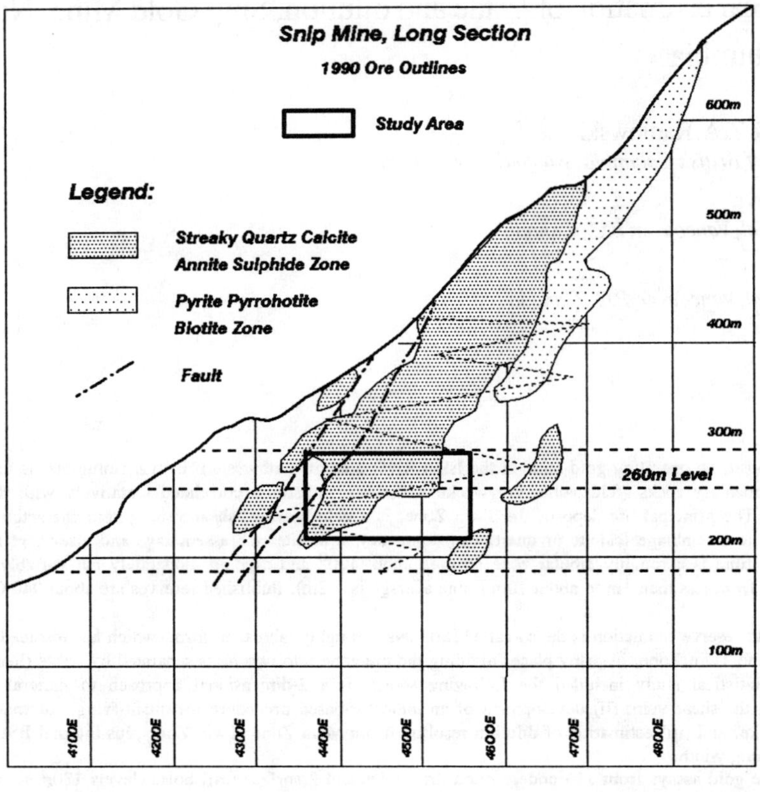

Fig. 1: Longitudinal section of Snip mine showing position of study area relative to the dominant ore types of the Twin zone.

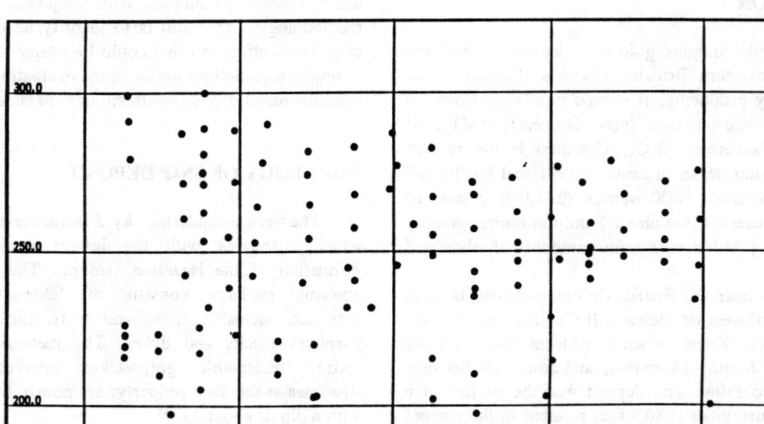

Fig. 2: Longitudinal section of study area, Twin zone, showing pierce points of underground drill holes (AX core) used in this study.

The area is cut by several systems of faults with dominantly NW and NE strikes. The gold bearing Twin Zone strikes north-northwesterly, dipping about 50° toward the south. Twin Zone is a shear vein system which cuts discordantly through a massively bedded feldspathic greywacke-siltstone sequence. Gold mineralization is mesothermal in character and of several mineralogical types, particularly calcite-quartz-($\pm$pyrite) and massive pyrite-pyrrhotite. The quartz-carbonate mineralization is very continuous and constitutes the bulk of the ore reserves. It occurs in centimetre to metre scale alternating layers of massive, streaky calcite with disseminated to massive pyrite, and so-called "crackled quartz", with thin chlorite streaks and local fault gouge development. Thick parts (>3 metres) locally contain layers of weakly to highly altered but low grade greywacke. Pyrite-pyrrhotite mineralization forms the eastern margin of the deposit (Fig. 1). This type of mineralization is much less continuous than the quartz-carbonate type, merges into quartz-carbonate on the west, and pinches out into biotite altered greywacke toward the east.

Twin zone strikes 120° with moderate but variable dips to the south. The steepest part of the zone is between 180 and 300 metre levels (average dip 60°); the dip changes progressively to about 30° at higher levels. Twin zone has been pierced by drilling for a strike length of about 1000 metres and a vertical range of 500 metres (levels 150 to 650).

Contacts of the mineralized zone are fairly easily defined visually; gold grade drops dramatically on the edge of the zone to about 100 ppm in the immediate footwall and hangingwall. The biotite spotted unit (dyke) is spatially associated with the Twin Zone. The dyke roughly parallels the Twin zone partly in the adjoining hangingwall, but in places, dividing the Twin zone into two tabular sheets (hence, the name Twin zone).

3 DATABASE

Our study is based on exploration drillhole data for a part of Snip deposit centred approximately on the 260 level (Fig. 1), bounded by sections 4350E and 4550E (easting in mine coordinates rotated 30° clockwise from true north). The data set for 116 underground and two surface drillholes includes coordinates of drill hole collars, hole orientation data, down-hole survey data, geochemical and assay results and lithological information.

Drill holes are oriented roughly north-south; 116 of them are within 10° of local grid north. Plunges vary from -83° (directed downward from the collar) to $+85^{o}$ (directed upward from the collar) with no particular plunge predominating. Exploration drill core is entirely AX and recovery through mineralized sections invariably approaches 100 percent.

True thicknesses of Twin zone intersections were integrated into our data base. In addition to the drill holes used by mine personnel for reserve estimation, we have included pierce points of seven drill intersections of narrow stringer zones that are too thin to develop as ore but which are important for our evaluation of physical continuity of the Twin zone. Of the 118 drill holes for which data were provided only 105 completely pierce the study zone (Fig. 2).

Sampling of drill core was done by traditional manual core splitting methods with individual samples representing half-core lengths of 0.2m to 3.3m. Fire assays were obtained on a final subsample of one assay tonne. A total of 2272 core samples were analyzed for one to three elements: gold, silver, and copper. Initially, as a cost saving technique, all samples were analyzed geochemically and only those samples with Au contents greater than 1000 ppb were fire assayed. Eventually, all samples from the mineralized zone were sent directly for gold fire assay.

3.1 Data evaluation

Sample assay and location information includes analytical data for one to three elements (Au, Ag and Cu) with gold by fire assays and/or geochemical analysis, and silver and copper by geochemical analysis only. Our data include 307 samples from Twin Zone itself (i.e. vein material) for which the distribution of gold grades is highly positively skewed. These Au assay data have a systematic relation between local squared mean and local variance, that is, a proportional effect (Radlowski et al, 1993). Other sample parameters such as lengths and orientations of samples are not uniform but do not contribute significantly to the variance of gold grades.

3.2 Internal structure of the mineralized zone

Grade profiles show abrupt discontinuities at shear vein margins and sharp changes of grade between vein material and internal, weakly mineralized rock (greywacke and BSU). For estimation purposes the Twin zone was defined as vein material plus internal low grade greywack lenses (horses). The BSU occurs either internal to, adjacent to, or some distance into the hangingwall from the Twin zone, as shown schematically in Fig. 3. Where present internally the dyke rock is potential dilution at the mining stage. We choose to estimate thickness and gold grade of the vein material as one component of our study, and to assess independently the contribution of the BSU to dilution.

A significant positive correlation exists between gold vs silver ($r = 0.606$). Thus, silver grade can be estimated easily and quickly by a simple linear relationship with gold:

$$Ag(g/t) = 0.284Au(g/t) + 8.45$$

Thicknesses of Twin zone range from 0m to 7m, with a global mean equal to 2.3m. The average grades for Twin zone drill intersections were calculated as weighted averages of from 1 to 8 assays; together with thicknesses these data form the data base for block estimates presented here.

4 VARIOGRAMS

We have defined the mineral inventory estimation as (1) 2-dimensional block estimates of average grade, thickness and metal content of the Twin zone; (2) estimation of internal dilution by the BSU; and (3) estimation of a component of external dilution that arises due parts of the Twin zone having thicknesses less than the minimum mining thickness of 1.5m. The 2-dimensional approach requires that we consider thickness and metal accumulations for entire drill intersections of the Twin zone. Relative variograms are used

Table 1: Summary of mean errors and dispersions of errors for crossvalidation of Au accumulation and thickness at pierce points of study area (see Fig. 2) by three estimation techniques (point kriging, polygons and inverse squared distance.

Method	Kriging		Polygons		ISD	
Parameter \ Variable	Err(AuAccum)	Err(Thickness)	Err(AuAccum)	Err(Thickness)	Err(AuAccum)	Err(Thickness)
N	100	100	101	101	101	101
Mean	2.25	-0.022	-7.49	-0.160	-2.62	0.043
StDev	87.40	1.3065	105.19	1.61	84.11	1.3178

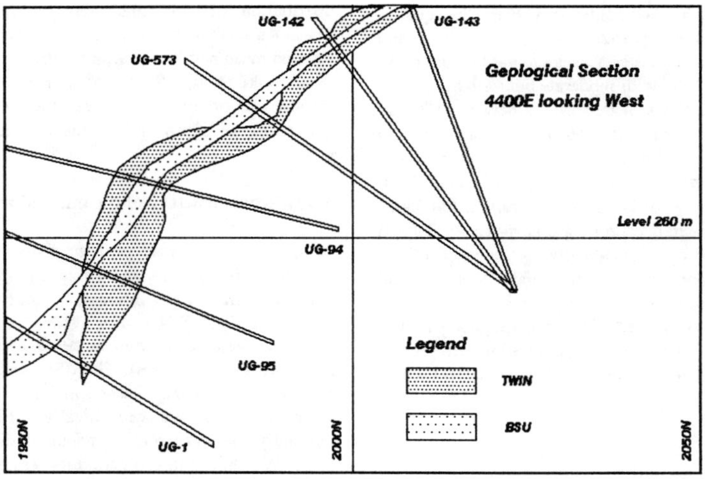

Fig. 3: 4400E Section, Snip mine, looking west. Labelled paired lines are drill holes (e.g. UG-94). Note variable position of BSU dyke to Twin zone (mineralization).

because they accomodate the proportional effect shown to exist in the data.

Directional variograms for thickness of the Twin zone show an isotropic continuity represented by a spherical model:

$$\gamma(h) = 0.067 + 0.348\ Sph_{28}(h)$$

Directional variograms for Au accumulation indicate a zonal anisotropy with a range of 38 m along the strike and 25 m down dip: the general variogram model for Au accumulation is

$$\gamma(h) = 0.40 + 0.87\ Sph_{25}(h_t) + 0.35\ Sph_{38}(h_s)$$

where Sph_{25} is the spherical model for a range of 25m, h_t is the true distance between two points and h_s is the strike component of distance between two points.

4.1 Cross validation

The anisotropic variogram model for Au accumulation with a minor axis/major axis ratio of 0.66 provides a guideline for the shape of a search ellipse with which to isolate data for individual block estimates. Cross validation with a variety of search ellipse sizes (all with an axial ratio of .66) show that minimum error of estimation for Au accumulation is obtained for a 33m x 20m search ellipse. Search ellipse size is not critical for thickness estimation.

Point kriging results (based on an optimal search ellipse) are compared with two other commonly used estimation procedures, polygonal (nearest neighbour) and inverse squared distance or ISD (based on a 33m search radius). Control points (i.e. known pierce points, cf Fig. 3) were estimated using the 3 techniques; means and dispersions of errors are summarized in Table 1. All three estimation methods appear unbiased andproduce roughly comparable results although polygonal estimates have the largest average error and the largest dispersion of errors. Kriging and ISD produce comparable results although kriging is marginally better. The ISD method, however, does not compensate for smoothing inherent in block estimation, as does kriging.

5 ESTIMATION OF ORE RESERVES

Blocks were estimated by 2-dimensional ordinary kriging, for each of two block sizes, 15m x 15m and 25m x 15m. The total area of longitudinal section covered by 25m x 15m blocks is about 5% larger than that for 15m x 15m blocks; a comparable discrepancy is expected in related volume and tonnage estimates of reserves.

Three sets of information are considered separately (i) total thickness and grade of Twin zone, and (ii) thickness of BSU (internally in Twin zone) which is assumed to be of zero grade for purposes of determining dilution, and (iii) thickness greater than the minimum mining width necessitate external dilution (zero grade) to bring the thickness to the minimum mining width. Undiluted reserves and two components of dilution were estimated separately, and results were combined to produce "internally diluted" and externally diluted estimates.

5.1 Undiluted resources

Block kriging was performed separately for gold accumulation and thickness. Gold grade averages were assigned to each block by dividing the kriged gold accumulation by the kriged thickness. These results can be used to estimate average metal accumulation and average thickness (hence, tons and average grade) for any combination of blocks e.g. those that approximate the geometry of a particular stope. Similarly, blocks with an average thickness less than the minimum mining width are readily identified and can be considered in mine planning as pillars, dilution, etc., as the case may be.

Results are summarized in Table 2. Tonnages calculated for the 25m x 15m block model are about 3% higher at similar average grade and thickness, relative to the 15m x 15m block model.

5.2 Internally diluted resources

A relatively few, thin horses of weakly mineralized greywacke have been included in our Twin zone intersections; hence, this small source of dilution is included in what we present as "undiluted" reserves. In contrast, the BSU dyke, which contributes substantially to internal dilution, is a well defined, continuous structure whose spatial distibution and relation to the Twin zone can be analyzed quantitatively. All BSU intersections are coded as one of three geometric situations: (1) BSU within Twin zone, (2) BSU adjacent to Twin zone, and (3) BSU distant from Twin zone (Fig. 4). Indicator kriging was used to produce a contour map of the probability that condition (1) above, prevailed, i.e. the probability that the BSU occurs within the Twin zone. The indicator variogram model is

$$\gamma(h) = 0.001 + 0.160\ Sph_{30}(h)$$

Overlaying a block model on such a contour map allows those blocks with internal BSU to be identified.

The amount of internal dilution by BSU depends on its average thickness where it occurs within the Twin zone. BSU thickness was kriged for both block models considered (Radlowski et al, 1993) using the following variogram model for BSU thickness:

$$\gamma(h) = 0.34 + 0.77\ Sph_{38}(h)$$

Diluted block grades are weighted grades over the combined Twin Zone-BSU thickness. Results for 15m x 15m blocks are included in Table 3. The total internally diluted reserves are 203,456 tonnes at an average gold grade of 23.16 g/t and an average thickness of 3.17m. Internally diluted reserves based on a cut-off grade of 13 g/t are 134,254 tonnes averaging 30.3 g/t gold and 3.08m thickness. The internal dilution is thus 17.5% of in situ resources, and 16.5% of reserves above the stipulated cut-off grade.

In the case of internally diluted reserves the larger block model produces estimates that are about 10% lower than the results reported in Table 3 for 15m x 15m blocks. Detailed results are given by Radlowski et al, 1993.

5.3 Externally diluted reserves

To account for some of the external dilution during mining we have attempted to estimate that component of external dilution that results because some parts of the Twin structure are less than the minimum mining width. That is, we have diluted such narrow blocks (i.e. internally diluted blocks thinner than the minimum mining width) by sufficient zero grade material to bring the average thickness up to the minimum mining width. This procedure is certainly a low estimate of the eventual external dilution but it is the component that we can quantify easily.

Results of these calculations for 15m x 15m blocks are compiled in Table 3. Total resources, diluted to the minimum mining width, are 204,274 tonnes at an average gold grade of 23.0 g/t and an average thickness equal to 3.19m. Reserves above the 13 g/t cut-off grade are 134,951 tonnes that average 30.0 g/t gold and 3.10m thick. Estimated total dilution is thus 18.0% of total resources, and 17.0% of reserves above the specified cut-off grade.

6 COMPARISON OF RESERVE ESTIMATION METHODS

The results of block kriging calculations are compared with estimates by polygonal and inverse squared distance (ISD) methods (Table 3). Kriged total tonnages are 5-8% higher than results by the other methods and the kriged average gold grade is 2 to 4% lower.

The diluted reserves estimated by kriging and ISD are very similar; diluted reserves calculated by polygons are more than 20% less than those by the other two methods.

Assays of underground samples taken during production from 2647 stope were provided as a detailed test case. The data are distributed in nine horizontal rows, each about 2m apart vertically, with samples taken approximately 2 metres apart along each row. Gold grade and thickness were determined at each measurement point. A comparison of

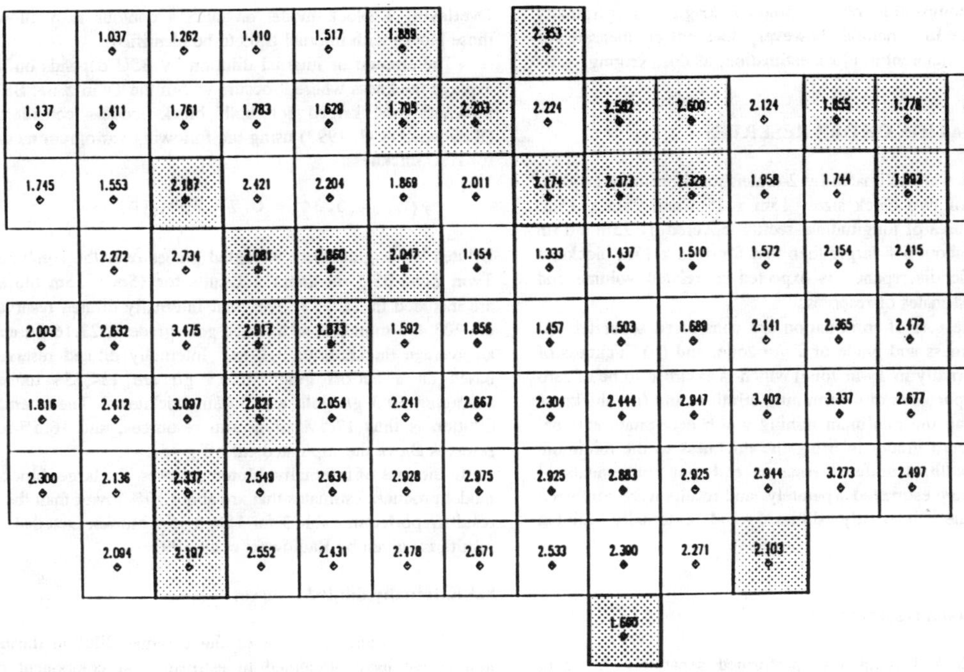

Fig. 4: Example of 15m x 15m block model (vertical, longitudinal projection). Central figure in each cell is kriged average thickness of BSU dyke for the cell. Shaded cells are those for which indicator kriging estimated a probability >0.5 that the BSU dyke occurs internally within the Twin zone.

Table 2: Resource/reserve estimates for our study area, Snip mine. Results are listed for two block models, in each case for zero cutoff grade and for a cutoff grade of 13g Au/mt.

Block Size	Cut-off g/t Au	Undiluted Reserves tonnes	Averages Gold Grade g/t	Averages Thickness metres
15m x 15m	none	173,120.9	26.88	2.64
	13	115,227.7	34.44	2.47
25m x 15m	none	184,221.4	26.23	2.69
	13	119,878.2	33.95	2.46

stope estimates by several methods was done for the 45m x 20m rectangular block that coincides almost exactly with the stope outline provided. Results, summarized in Table 4, show that kriging produces a close estimate of reality for stope 2647, whereas, reserves calculated by polygons and inverse square distance methods are about 40% above reality in this single test case. Such a large discrepancy among results of various estimation procedures probably stems from (i) smoothing, which is taken into account best by kriging, and (ii) the effects of anisotropy which did not enter the polygonal and ISD methodologies.

Table 3: Summary of mineral inventory, study area, Snip mine, for 15m x 15m block model. Results are listed for three estimation methods, in each case for zero cutoff grade and for a cutoff grade of 13g Au/mt.

Reserves	Estimation Method	tonnes	Averages Gold Grade g/t	Thickness metres
Cut-off = none				
Undiluted	Kriging	173.120.9	26.88	2.64
	Polygons	163,138.3	26.87	2.49
	ISD	158,562.7	29.74	2.42
Internally Diluted 1/	Kriging	203,455.8	23.16	3.17
	Polygons	193,473.2	23.51	3.02
	ISD	188,897.6	25.89	2.95
Externally Diluted 2/	Kriging	204,273.5	22.96	3.19
	Polygons	200,738.9	21.90	3.15
	ISD	189,808.8	25.63	2.97
Cut-off = 13 g/t				
Undiluted	Kriging	115,227.7	34.44	2.47
	Polygons	96,394.5	44.01	2.61
	ISD	123,617.9	35.19	2.36
Internally Diluted 1/	Kriging	134,254.3	30.33	3.08
	Polygons	102,408.3	41.68	3.12
	ISD	130,080.3	32.26	2.76
Externally Diluted 2/	Kriging	134,951.1	30.05	3.10
	Polygons	105,713.3	38.71	3.23
	ISD	130,863.9	31.92	2.78

1/ with BSU

2/ up to 1.5m

Table 4: Comparison of underground sampling "reality" (based on close space sampling described in text, for part of stope 2647, Snip mine, with estimates by 3 methods using local exploration data.

Estimation Method	tonnes	at Averages Gold Grade g/t	Thickness metres
Underground	9,402.3	30.20	3.37
Kriging	10,657.8	27.85	3.82
Polygons	13,866.3	52.35	4.97
Inverse Square Distance	14,731.2	36.05	5.28

7 CONCLUSIONS

(1) Twin Zone is a single vein system in places divided into two tabular zones by the BSU dyke; locally unmineralized greywacke inclusions occur within the vein. Twin vein is interpreted by mine staff as a physically continuous structure where this study is centered.
(2) Thickness was modeled by an isotropic spherical relative variogram (range of 28 m). The variogram for gold accumulation revealed an anisotropy with a range of 38m along the strike and 25m down the dip; continuity was modelled with a relative, spherical variogram.
(3) Crossvalidation showed that a data search ellipse with a major axis of 33m in an easterly direction and a minor axis of 20m down dip is optimal for both thickness and Au accumulation estimation by point kriging. Point kriging, polygonal (nearest neighbour) and inverse squared distance (ISD) estimates show that all three methods are relatively unbiased but the polygonal method has the greatest dispersion of errors and kriging is marginally better than ISD.
(4) The undiluted reserves were estimated by 2-D ordinary kriging of gold accumulation and thickness for two block sizes, 15m x 15m and 25m x 15m, using relative variogram models; average grade for each block was determined from block estimates for gold accumulation and thickness.
(5) Dilution by the BSU dyke where the dyke occurs within the mineralization zone (so-called "internal dilution") was estimated by a two step procedure. Average BSU thickness for each block was estimated by ordinary kriging. To determine which ore blocks are to be diluted, indicator block kriging was performed using an indicator for presence or absence of the BSU internally within the mineral-ization zone. Blocks for which the probability was greater than 50% for BSU occurrence within the mineralized zone were diluted by the estimated amount of BSU taken at zero grade.
(6) Next, we estimated the dilution to a minimum mining width equal to 1.5m (so-called "external dilution").
(7) Ordinary kriging estimates, in general, are comparable with ISD and are better than polygonal results.
(8) Underground data from stope 2647 were used to estimate ore reality which was compared with estimates by ISD, polygonal and geostatistical methods. Under-ground data are closely in agreement with kriging estimates; very large discrepancies exist for the other estimation procedures.
(9) The geostatistical approach developed here warrants consid-eration for implementation at the mine site.

ACKNOWLEDGEMENTS

This project was first suggested by Dr. O. Owens, Vice-President Exploration, Cominco Limited. We have benefited substantially by the close cooperation of the mine staff. Funding for the project was provided by the Science Council of British Columbia. This is paper #029 of the Mineral Deposit Research Unit, Dept. of Geological Sciences, The University of British Columbia.

REFERENCES

Radlowski, Z. A., R. M. Srivastava, A. J. Sinclair and A. Samis, 1993, Mineral inventory studies of pre-cious metal deposits in British Columbia: Snip mine; Mineral De-posit Research Unit, Tech. Rept. No. 5, Dept. of Geological Sci-ences, The Univ. of Brit. Columbia.

Rhys, D. A., and C. I. Godwin, 1992, Preliminary structural interpretation of the Snip mine: in Geological Fieldwork 1991, B. C. Ministry of Energy, Mines and Petroleum Resources, Victoria, Paper 1992-1, pp. 549-554.

Innovative Mine Design for the 21st Century, Bawden & Archibald (eds) © 1993 Balkema, Rotterdam, ISBN 90 5410 325 6

Solving open pit grade control problems

Mark Springett
Boulder, Colo., USA

ABSTRACT: The first step in solving open pit grade control problems is to identify the problems and to attempt to quantify the impact of each. Problems in grade control arise from several sources. These include the disparate data bases used to generate the models used for long term mine planning and the models used to carry out ore - waste definition. Sample acquisition and preparation is often of lesser quality than that used for exploration purposes despite the fact that the results will be used to make a unique and irrevocable decision. Data processing techniques often do not pay sufficient heed to the quality (or lack of quality) of the data being used for defining and valuing the ore and waste. Operational factors in the pit include surveying practices, destination signalling and haul truck driver error. The significance of these issues are often under-rated by management in terms of the economic impact on the operation.

1. INTRODUCTION

Grade control is the practice of classifying material into categories based on various operational criteria. One of the most prevalent criteria is the eponymous "grade". Other characteristics, for example, might be metallurgical nature. Most of the comments in this paper are directed toward grade but can be applied to other chacteristics. Grade control presents one area where improved performance can make for very definite economic benefit. Unfortunately this is often difficult to demonstrate in a convincing manner partly due to the fact that typically this "quality control" practice is a responsibility shared by more than one department and while the benefit of improved practices and therefore more efficient ore-waste selection is shared by all as represented by improved profitability, the cost of improved practices is born primarily by the Mining Department.

1.1 Metal accounting versus decision

There is a very fundamental difference between the two ways that blast hole samples are used. Firstly they are used to determine the metal content of the material delivered to process, stockpile and waste. Secondly - and vitally - they are used to determine the character and destination of the selected material. Importantly, Francois-Bongarcon (1988) makes the point that "...No efficient grade control can be implemented when people believe that its primary objective is to reconcile the various estimates and reserve accounting figures reported at the various stages of the production process.". The point is, I hope, obvious. Reconciliations can be relatively easily achieved by such artifacts as the historic South African "Mine Call Factor"; however when they simply serve to balance the books and disguise real problems that, when solved, can increase profitability, such factors are counter-productive. The approach that should be followed is to identify the causes behind the descrepancies and to use that information to resolve the problems in a cost effective manner.

2 THE TECHNICAL PROBLEMS

The key technical problems that deserve better treatment than is currently typical are the method of acquisition of the sample, the number and locations of the samples, the preparation and handling of the sample and its assaying. Once assayed, the manner in which the technical data are integrated to provide an effective database and then used to provide the outlines of panels, dig lines and their associated grade estimates. The methods used for calculating the grade control boundaries and the estimates of their contained attributes can have a significant impact on the way that discrimination is made, and what grade is assigned. The classic paper by Raymond (1979) deserves more attention despite its vintage. He clearly demonstrates the impact of the fundamental problems through his resampling tests and his comparisons between Kriged and conventional data processing methods. Unfortunately, the conventional polygonal approach to blast hole data processing remains popular as "industry standard", although increasing use is being made of indicator kriging.

2.1 The sampling problems.

Springett (1989) demonstrated that even an unbiased sampling error distribution introduces a clear and quantifiable bias when the samples are used to truncate a second distribution. Any reduction in the sampling error distribution variance will help reduce this bias and hence improve selection. Improvements can be made by acquiring a better and more representative sample. It should be noted that the typical blast hole sample represents a twenty foot run from an open hole whereas the typical exploration sample is recovered from a reverse circulation or diamond core drill with four individual five foot samples representing the twenty foot increment. The corresponding fourfold increase in samples and their higher quality is often clearly apparent when the nugget variance of the downhole variograms are compared with the variograms of blasthole data. Similarly the very large number of blasthole samples that are delivered to the mine assay laboratory can result in inadequate sample preparation. The assay laboratory usually is under the administration of the plant department and plant head samples have been seen to receive far more rigorous preparation than the blasthole samples. There is commonly a failure to recognize that plant head samples do not get used to separate material into different categories and that overall in that situation an unbiased sampling error will result in unbiased estimates. Similarly it is common practice for blasthole samples to atomic adsorbtion assays performed on 10 or 15 gram aliquots whereas plant or exploration samples are assayed by fire assay on one or even two assay ton aliquots.

There are clearly improvements that can be made in sample aquisition, sample preparation and assaying practices for blasthole data that will result in more efficient grade control and improved profitability. Each mining operation has different practices and different styles of mineralization. The areas that merit attention and improvement need to be identified by inspection of the entire grade control and sampling process.

2.2 Sample spacing and location.

The blasthole is regarded primarily as a receptical for explosives for breaking ground. The blasthole spacing that is used is traditionally determined by identifying the ideal compromise between minimizing the number of holes that are drilled and attaining the right amount of fragmentation and minimizing pit wall damage. Barnes (1989) introduces an interesting and simple model for examining the appropriate economic compromise for blasthole spacing taking into account fragmentation requirements, geostatistical data requirements and the cost/benefit consequences. The model has the potential to be expanded into a more comprehensive economic tool for exploring all the areas where improvement in practices at a quantifiable cost can be weighed against the economic benefit.

2.3 Data processing.

Many mines continue to use a polygonal approach to grade control and selection, essentially constructing panels of different grade materials by visually and manually contouring the raw blasthole assay values. There is a psychologically overwhelming tendency to believe the actual assay value printed on the map without appreciation that it is an estimate subject to substantial sources of error as discussed above. Raymond (op. cit.) made an early and powerful arguement for the benefits of ordinary kriging as well as the importance of proper sampling proceedures based on his experience at Similkameen, British Columbia. Acceptance of the benefits of geostatistical methods of grade control has been painfully slow since then. Renewed interest was generated by a small group of practicing geostatisticians from Amax, City Resources, Gold Fields and Homestake that made a detailed examination of current practices in September 1987.

Indicator Kriging for grade control has become increasingly used within the gold mining industry during the late eighties. Indicator Kriging overcame some of the disadvantages of Ordinary Kriging such as inappropriate treatment of outlier values and the results lend themselves to more rigorous interpretation (Guardiano et al. 1990) which in turn has the potential to improve decision making through the use of loss functions.

Conditional simulation is a tool that has the potential to provide the most effective solution to the grade control data processing challenge. Srivastava et al. (1992) discuss some of the ways that conditional simulation can be applied. This important paper also addresses other components of the overall problem. Conditional simulation further has promise to solve some of the more complex grade control problems that will become of greater significance in the nineties as deeper refractory ores with complicated metallurgical requirements for blending become of greater economic significance in, for example, northern Nevada.

3 OPERATIONAL CONSIDERATIONS

Operational considerations include the correct identification and flagging of the dig lines in the pit from the information provided by the grade control engineer. Colored flagging is often used for marking the dig lines. Care should be taken that appropriate colors are used. Colors that may contrast very effectively in daylight may be more confusing in artificial light at night time. As part of equipment operator job selection a check should be made for color blindness.

Equipment operator training should include exercises on the economic importance of grade control and its vital role in making an efficient operation. The importance of the communications from loader operator to truck driver in determining truck destination is self-evident and operators should be encouraged to query each other if in doubt.

Proper blasting practices can help minimize dilution and make it easier to identify the correct dig line after blasting. Davis et al. (1989) describes techniques for examining the effect of blasting practices and the magnitude of blast movement.

4 CONCLUSIONS

Grade control is fraught with many potential problems. By a comprehensive study of the entire process it is possible to identify the individual components of these problems and to assess its contribution to the overall problem. Once that has been achieved the solutions to the individual problems can be assessed in terms of their cost of implementation and the benefit to be realized if enacted in terms of improved profitability. The proper case can then be recommended to management based on a firm economic footing rather than a purely qualitative basis.

REFERENCES

Barnes, R.J. 1989. A geostatistical approach to Operational sample design for ore/waste selection in surface gold mining. 21st APCOM Proceedings. Alfred Weiss (ed) pp.262-274

Davis, B.M., J. Trimble & D. McClure. 1989. Grade control and ore selection practices at the Colloseum gold mine. Preprint 89-47. SME Littleton.

Francois-Bongarcon, D. 1988. Personal Communication.

Guardiano, F.B. & H.P. Knudsen 1990. Indicator kriging for grade control - is it better? SME Goldtech 4

Raymond, G. 1979. Ore estimation problems in an erratically mineralized orebody. CIM Bulletin, Vol. 72, No 806 pp. 90-98

Springett, M.W., 1989. Biasing unbiased data - mine versus mill. 21st APCOM Proceedings. Alfred Weiss (ed) pp.286-296

Srivastava, R.M., D.R. Hartzell & B.M. Davis 1992. Enhanced metal recovery through improved grade control. 23rd APCOM Proceedings. Y.C. Kim (ed) pp.243-249

Innovative Mine Design for the 21st Century, Bawden & Archibald (eds) © 1993 Balkema, Rotterdam, ISBN 90 5410 325 6

Calculation of geological reserves contained in the Louvicourt copper-zinc-gold deposit

H.R. Stockford
Aur Resources Inc., Toronto, Ont., Canada

ABSTRACT: The Louvicourt copper-zinc-gold deposit, discovered by Aur Resources Inc. in 1989, is widely acknowledged to be one of the most important exploration successes in the mining industry in Canada in recent years. This volcanogenic massive sulphide (VMS) deposit was discovered as a result of strategic land acquisition, a sound geological model and careful compilation of past exploration work, followed by systematic deep drilling spaced at intervals selected on the basis of economic criteria (minimum size and grade) and with the knowledge that borehole geophysics (PEM) and lithogeochemical data would enlarge the effective target area. Various cut-off grades were utilized to manually calculate geological reserves based on geological, assay and specific gravity data derived from deep surface drillholes spaced at roughly 60 metre intervals. At a 2% Cu equivalent cut-off grade, the undiluted geological reserves are 27.9 million tonnes at an average grade of 4.3% Cu, 2.1% Zn, 27.4 g/tonne Ag and 1.06 g/tonne Au, between depths of 400 and 850 metres below surface. The deposit is currently being developed for production at a rate of 1.4 million tonnes per year. The Louvicourt project is operated by Aur under the terms of a joint venture agreement between Aur, Teck Corporation and Novicourt Inc. The current ownership is Aur 55% and Novicourt 45%, with Teck Corporation earning a 25% interest in the project from Aur. This paper will discuss the method used in calculating the geological (in situ) reserves.

1 INTRODUCTION

The Louvicourt volcanogenic massive sulphide (VMS) deposit is located within the Abitibi Greenstone Belt, 25 km east of the City of Val d'Or, Quebec. The deposit consists of six stacked lenses of semi-massive to massive pyritic sulphides containing variable quantities of chalcopyrite and sphalerite, the principal copper and zinc sulphides, and by-product levels of gold and silver largely in the form of electrum. Disseminated and stringer-style chalcopyrite mineralization occur in significant quantities between the lenses. The deposit extends over a strike length of 500 metres and is deep, extending from about 400 metres to 850 metres below surface and below which only limited exploration has been carried out to date.

The deposit was delineated by diamond drilling from surface following discovery in mid-1989 until early 1991, by which time some 60 drillholes and wedge cuts had been completed in and around the deposit at spacings of some 60 to 90 metres between drillhole intercepts.

A detailed geological interpretation was made by Aur's exploration staff on cross-sections and plans, initially spaced at 200 ft. (61 m) intervals. Vertical longitudinal projections were then prepared, oriented at 060° 45', for each mineralized zone and sub-zone including zones A1, A2, A3, A4, A5, A5N, A5S and A6. The polygon method was then used to manually calculate the undiluted geological (in situ) reserves using a variety of cut-off grades ranging from 1% Cu-equivalent to 8% Cu-equivalent. The resulting reserve figures were published in March, 1991. By November 1991 new sections and plans had been prepared with some revisions to the detailed geological interpretation. New longitudinal projections were then prepared, oriented at 089° 16', paralleling the azimuth of the NAD 83 provincial grid and more closely paralleling the E-W strike direction of the deposit. Geological reserve (in situ) figures are given for both scenarios in Tables 1 and 2.

A 17 year production plan was prepared as part of a feasibility study for a 4000 tonne per day operation completed in June 1992 by Fluor Daniel Wright. The production schedule was based on utilizing a declining 4%-2% Cu-equivalent cut-off grade and projected stope tonnages were arrived at by superimposing transverse and longitudinal blasthole stope outlines on the in situ reserve polygon boundaries and adding between 10% and 30% dilution factors. The resulting production forecast assumes that 24 million tonnes grading 3.89% Cu, 1.95% Zn 0.985 oz/tonne Ag and 0.039 oz/tonne Au will be mined (mostly from zones A4, A5 and the A5N and A5S sub zones). The mine is currently being developed for mill startup in July 1994 and commercial production to commence in October 1994. A detailed definition drilling program designed to upgrade the in situ reserves to the proven and probable categories, and to provide the basis for a detailed mine plan and hence, to establish mineable reserves, will be carried out during the period June 1993 to October 1994 using a dedicated hanging wall exploration drift as a drilling platform. Geostatistical methods will be used in addition to manual (cross-sectional) methods to calculate the mineable reserves.

2 REGIONAL GEOLOGY

The Louvicourt deposit is enclosed within Archean-age volcanic rocks of the Abitibi subprovince of the Pre-Cambrian Shield of Canada which is host to a number of important "Noranda-type" volcanogenic massive sulphide (VMS) deposits. In the Val d'Or area the volcanic sequence is comprised of thick tholeiitic mafic and ultramafic flows of the Lower Malartic Group overlain by calc-alkalic intermediate to felsic rocks of the Upper Malartic Group. The felsic sequence which occurs at the base of the Upper Malartic Group, attains a thickness of up to 2 km, occurs over an east-west strike length of more than 30 km, and is the host to numerous copper-zinc occurrences including four former base metal deposits which have collectively produced over 30 million tons of ore. The Louvicourt deposit was discovered by systematic, deep drilling, based upon a sound geological model, of the favourable mineralized stratigraphy, for approximately 1 km to the east of the past-producing Louvem mine.

The Upper Malartic Group stratigraphy is intruded by two principal synvolcanic bodies, the Bourlamaque Batholith, to the west of the Louvicourt deposit, and the smaller, Bevcon pluton to the east. A number of significant gold deposits are associated with these intrusions.

The dominant east-west fault structure in the area, the Cadillac Break, lies 5 km to the south of the Louvicourt deposit. Many of the known deposits in the Abitibi Greenstone Belt are spatially related to this major structure.

3 LOCAL GEOLOGY AND MINERALIZATION

The Louvicourt deposit consists of at least six stacked lenses of semi-massive to massive pyritic sulphides with accompanying disseminated and stringer-style mineralization. It is principally a copper deposit with lesser zinc, gold and silver. The main ore minerals are medium to coarse grained chalcopyrite and sphalerite which exhibit excellent metallurgical characteristics.

The sulphide zones are generally conformable and occur within an east-west striking, steeply north dipping, overturned homoclinal sequence of felsic volcaniclastic sediments (VSED) underlain to the north by quartz-veined andesitic flows with intercalated felsic flow breccia and fragmentals and overlain to the south by andesitic (cap-rock) flows followed by more felsic volcaniclastic and intermediate flow units.

The mineralized zones exhibit a steep to moderate easterly plunge and are separated by at least three metres of barren rock except where the sub zones of zone A5 coalesce.

The host VSED unit displays varying degrees of sub-conformable hydrothermal alteration, dominated by chloritization and, to a lesser extent, sericitization.

Overall the mineralization of economic importance extends over a strike length of over 500 metres, a similar dip length (as known to date) and may exceed 100 metres in thickness on the eastern extremities. The deposit exhibits metal zonation typical of many other VMS deposits, with a copper-rich base and zinc/precious metals-rich top. The individual zones typically consist of a core of semi-massive to massive pyrite-chalcopyrite, with attendant stringer and disseminated chalcopyrite in the stratigraphic footwall and with sphalerite bands in the cherty hanging wall rocks. Some remobilization of chalcopyrite has occurred in the stringer zones producing very rich sub-zones within the deposit (e.g. A5S); the geometry and distribution of this style of mineralization is poorly defined given the relatively wide-spaced drilling and more detailed information is eagerly awaited.

Zones A4 and A5, including sub-zones A5N & A5S, contain over three quarters of the ore at the 2% Cu equivalent cut-off grade. Early production development will be concentrated in these zones, but care will be taken not to destroy access to other zones in the structural hanging wall.

At the 2% Cu cut-off, zone A4 occurs intermittently over a strike length of 600 metres and averages 14.5 metres in horizontal thickness, accounting for 27% of the in situ reserves. Zone A5 occurs over a strike length of 600 metres and averages 24 metres in horizontal thickness, reaching a maximum of 56 metres and contributing 34% of the in situ reserves. Zones A5N and A5S extend over strike lengths of 300 metres and 500 metres respectively, averaging 5 metres and 18 metres in horizontal width, accounting for 1.6% and 11.5% of the in situ reserves.

Two gently north-dipping fracture or fault zones of unknown displacement cross the upper part of the mine sequence. These structures may influence the shape of the upper extremities of the main ore zones. A large northeasterly lineament with an esker aligned along it passes to the immediate east of the deposit. The lineament may be the manifestation of a paleostructure which has influenced the geometry of the eastern end of the ore zones. Although no direct evidence of isoclinal folding has been observed in the drill core the stratigraphy does exhibit a certain symmetry to the north and south, about an east-west axis; hence the possibility of folding cannot be ruled out and this question may be answered in the next phase of drilling and underground development.

4 RESERVE DATABASE

Aur used information from 60 drillhole and wedge cut intercepts to calculate the geological (in situ) reserves. Assay data was collected from over 11,000 samples. The samples were produced by cutting the BQ/NQ sized core into two halves using a diamond saw. One half was sent for assay at a commercial lab (Chimitec Ltée) in Quebec City. Sample lengths were selected on the basis of the visual characters of the mineralization, but did not exceed 5 feet. Check assays were performed on approximately 200 samples at two other commercial labs. Samples were systematically analyzed by standard AA methods for percent copper and zinc and by fire assay/AA methods for gold and silver. Where visible galena was observed (a local occurrence only) core samples were also assayed for lead. Other analyses performed for purposes other than reserve calculation included systematic wholerock analysis for major oxides and trace elements. In addition to the check assaying mentioned above, head grade assays performed during metallurgical testwork on 17 composite samples of concentrates provided added confidence to supplement core assay data. Chimitec's assays compared well with the other checks performed and results were well within the range of error normally expected from sub-sampling.

Specific gravity measurements were carried out manually by Aur on every assayed core interval within the boundaries of the in situ reserve blocks. Specific gravity checks were also performed on the 17 composite metallurgical samples from the deposit, using a picnometer, at the Centre de Recherches Minérales in Quebec City. Those measurements made on -200 mesh pulps averaged 4% less than the specific gravities of Aur but measurements made on -10 mesh pulps compared very well with the manual measurements.

Rock Quality Designations (RQD's) were recorded systematically as part of the core logging procedure. These measurements as well as preliminary rock strength test results indicate that the ore and enclosing rocks are of medium strength and that ground conditions can be controlled using standard ground support systems. The large scale blasthole mining methods proposed in the current mine plan envisage using a stiff backfill to fill stoped areas.

Downhole surveying of drillholes was treated seriously since considerable deviation was experienced in all drilling programs that have been completed to date on the Louvicourt property. Even navigational drilling methods failed to keep holes straight although this method was used to successfully obtain wedge cuts or to re-orient drillholes in some areas. The drillholes which ranged up to 1,350 metres in length were each surveyed using a variety of methods to help resolve discrepancies. Magnetic methods (Sperry Sun Single and Multishot) and light beam methods (Light-Log) were combined with gyroscopic surveys and a seismic triangulation method using two blasts per hole and 24 surface geophones to arrive at final plots. In an audit conducted by Roscoe Postle Associates Inc (RPA). it was stated that "in RPA's opinion Aur has addressed the drillhole survey problem in a careful and professional manner and has made maximum use of available data to determine drill hole locations as accurately as possible". The accuracy of plots will ultimately be determined as a result of drilling closely spaced holes in the upcoming underground definition drilling program. All surface holes which intersected the Louvicourt deposit have now been cemented so that surface groundwater will not interfere with underground workings. The cement employed to fill drillholes was colour coded using various dyes so that holes encountered in the underground workings can be identified and the accuracy of downhole surveying methods determined.

Detailed geological interpretations were completed on 1:1200 scale cross-sections and level plans spaced at 200 ft (61 m) intervals and then the reserve blocks were transferred to individual longitudinal sections for each zone.

5 RESERVE ESTIMATE METHODOLOGY

Aur Resources' exploration staff calculated geological (in situ) reserves for the six mineralized zones and sub-zones (A1, A2, A3, A4, A5N, A5, A5S and A6) as well as the old Louvem No. 6 Zone in the upper levels, using the polygon method, at six different cut-off grades.

Lithological units, alteration and mineralized zones were plotted and interpreted on 1:1200 scale horizontal plans and cross-sections spaced at 200 ft (61 m) intervals oriented at 150° 45' (Louvem Mine grid). Data up to 30 metres above and below levels were projected up or down to the levels at the average dip angle of 70° and a careful geological interpretation was carried out. Mineralized zones were correlated between the plans and sections. A subsequent series of 1:500 scale metric plans and sections were prepared using the NAD 83 Quebec grid oriented at 179° 16' which direction is more normal to the strike of the lithologies than the earlier series.

Vertical longitudinal sections were prepared normal to the cross-sections for each mineralized zone firstly on the 60° 45' azimuth and 1:1200 scale of the Louvem Mine grid and subsequently on the 89° 16' azimuth NAD 83 grid and at 1:1000 scale. On the initial series of longitudinal sections the limits of each zone were outlined using a 1% Cu equivalent cut-off grade and utilizing grade-width contours to define the plunge trends. The NAD 83 Grid longitudinal sections employed 2% Cu equivalent cut-off outlines.

Polygons were drawn by computer. The surface area of each polygon was calculated by computer and verified manually and found to be within 2% of the computer calculations.

Continuity between ore intersections for each zone was assumed on the basis of reasonable and coherent geological interpretation and the maximum area of influence for a drill hole was limited to 300 feet for the initial (Louvem Mine grid) calculations or to 100 metres for the NAD 83 grid calculations. Horizontal widths of intersections were calculated perpendicular to the longitudinal sections and a minimum horizontal width of 8 ft (2.43 m) was used. The specific gravity for each individual core assay interval was used to determine the specific gravity of each reserve polygon or block. Intersections were averaged by weighting assays of Cu, Zn, Ag and Au by specific gravity and core length and determined for each zone by taking into account interpreted geological continuity and copper equivalent cut-off grades of 2%, 3%, 4%, 6% and 8% for both sets of calculations as well as 1% for the initial (Louvem Mine grid) calculations. Copper equivalent grade was calculated using Cu grade plus 0.7 times the Zn grade but excluding gold and silver values. Where intersections exceeding the cut-off grades were separated by more than 20 feet or 6 metres, separate zones were estimated (e.g. A5, A5N).

In the case of the Louvem Mine grid calculations, the same polygon areas on longitudinal sections that were constructed using the 1% Cu equivalent cut-off were used to estimated reserves at each of the 6 cut-off grades and in the case of the NAD 83 grid calculations the 2% Cu equivalent cut-off outlines were used.

Horizontal widths of the polygons were recalculated for each cut-off grade and polygonal blocks with average grades less than each Cu equivalent cut-off grade were excluded in the in situ reserve estimate for each relevant cut-off.

The volume of each polygonal block was calculated from the area on the longitudinal section multiplied by the horizontal width for each specified cut-off grade. The tonnage of each block was then calculated using the weighted average specific gravity for the drill hole intersection within the polygon. The average grade used for each polygon was the weighted average of Cu, Zn, Ag & Au for the drillhole intersections at that cut-off and finally the total for each zone was calculated by weighting all of the polygon grades above each cut-off by the relevant polygon tonnage.

The results of the Louvem Mine grid and NAD 83 grid reserve calculations at the various cut-off grades are given below (see Tables 1 and 2). No allowance has been made for dilution and mining losses in these calculations.

Table 1. Undiluted geological (in situ) reserves - Louvem mine grid (060° 45')

Cut-off %	Total Reserves Metric Tonnes (000)	Specific Gravity	Cu (%)	Zn (%)	Ag (g/t)	Au (g/t)
1	39,209	3.470	3.30	1.78	21.7	0.83
2	27,961	3.615	4.32	2.11	27.6	1.07
3	20,385	3.721	5.36	2.52	33.9	1.32
4	16,448	3.769	6.15	2.69	38.3	1.51
6	9,449	3.943	8.32	2.78	49.8	1.90
8	5,236	3.961	10.46	3.64	60.7	2.66

Table 2. Undiluted geological (in situ) reserves - NAD 83 grid (089° 16')

2	26,474	3.630	4.22	2.15	28.1	1.10
3	19,046	3.740	5.27	2.59	35.2	1.38
4	15,373	3.793	6.04	2.79	40.0	1.58
6	8,684	3.973	8.11	2.97	53.4	2.01
8	4,889	3.965	10.18	3.80	64.4	2.90

Roscoe Postle Associates Inc. (RPA) independently estimated the in situ reserves of zone A5 (including A5N & A5S) using the contouring method based on Aur's longitudinal sections oriented along the Louvem Mine grid (060° 45') at cut-off grades of 2% and 3%. The in situ reserves by RPA are given in short tons and correspond reasonably well with Aur's figures for the same zone(s) given the different methods used and the relatively wide drillhole spacing at this stage. The comparisons are shown in Table 3.

The following is an excerpt from RPA's report and describes the method used for their estimate:

"The RPA reserve estimate excludes parts of subzone A-5N, A-5S and Hole 309-112, which are included in the

Table 3. Comparisons of in situ reserves - A5 zone (RPA vs Aur)

Cut-off	Tons	Cu	Zn	Ag	Au
	RPA (Contouring method)				
%	(000)	(%)	(%)	(oz/t)	(oz/t)
2	10,326	4.45	1.81	0.77	0.04
3	7,610	5.53	2.46	0.86	0.04
	Aur (Polygon method)				
2	10,331	5.10	2.03	0.97	0.04
3	7,833	5.89	2.42	1.14	0.05

Aur estimate. To compare the two estimates, these have been added to the RPA estimate of zone A-5"...

"The RPA reserves were estimated using the contour method as described below. Intersection averages were plotted on longitudinal sections depicting horizontal width (T), horizontal width divided by tonnage factor (T/TF) and grade x thickness divided by tonnage factor (GT/TF) for each metal. The density assigned to each mineralized intersection was based on data supplied by Aur. Assays were averaged by RPA. Aur's horizontal widths and coordinates on longitudinal section were checked, accepted and used by RPA. A minimum horizontal width of 8 ft. was used.

The T, T/TF and GT/TF values for each metal were contoured manually on the longitudinal sections for each cut-off grade. The contours were digitized and the areas between contours were calculated by using AutoCAD software. Areas between each set of contours were multiplied by the average T, T/TG or FT/TF value for each contour interval and summed.

Tonnage was calculated from the total T/TF x area measured on the longitudinal section. Grade was calculated by dividing the total GT/TF x area for each metal by the total tonnage."

RPA classified the undiluted geological reserves estimated (on the Louvem Mine grid basis) by Aur according to categories defined in National Policy 2-A adopted by the Ontario Securities Commission and in RPA's opinion:

"the Louvicourt reserves can be classified as in situ reserves almost all of which are in the probable category and the rest in the possible category, at the 2% and 3% Cu cut-off grades. The possible reserves consist mostly of material inferred from isolated drillhole intersections".

The reserve estimate by Aur is considered to be a "Geological or In Situ Reserve" rather than a "Mineable Reserve". For the purpose of mine planning and feasibility studies a separate production life of mine forecast was made by superimposing mining outlines upon the geological reserve outlines and adding between 10% and 30% dilution, at a declining 4%-2% Cu equivalent cut-off grade. While not yet considered to be a "Mineable Reserve", this estimate totals 24 million tonnes averaging 3.89% Cu, 1.95% Zn, 0.985 oz. Ag/tonne and 0.039 oz. Au/tonne.

A program of detailed underground definition drilling will be carried out during the pre-production period to provide drill intercepts at a maximum spacing of 30 m throughout the main parts of the deposit. Geostatistical reserve calculation methods will be employed in addition to manual methods to provide updated mining reserve estimates to acceptable confidence limits. Cut-off grades will be based on projected smelter netback values and operating costs.

6 CONCLUSIONS

Geological (in situ) reserves calculated using the polygon method represent a reasonable initial estimate for use as a basis for planning underground work at Louvicourt. More detailed drillhole and development information is required to upgrade the reserves into the proven and probable ore reserve categories and new reserve calculations will be carried out after completion of the planned underground definition drilling program as well as a number of development headings driven in ore.

REFERENCES

Boisvert, G., Mannard, G.N., Britt, C.N., Bubar, D.S., Rougerie, Y. December 1991. Geology of the Louvicourt deposit and results of the 1990-1991 phase IV and V exploration programs. Unpubl. report by Aur Resources Inc.

Britt, C.N. and Boisvert, G. March 1991. Louvicourt deposit and Louvem zone 6 ore reserve calculations (Louvem grid). Unpubl. report by Aur Resources Inc.

Britt, C.N. and Boisvert G. November 1991. Louvicourt deposit and Louvem zone 6 NAD 83 polygonal ore reserve calculations. Unpubl. Report by Aur Resources Inc.

Delisle G. 1991. Centre de Recherches Minérales, Québec. Metallurgical characteristics of different zones from the Louvicourt deposit. Unpubl.

Fluor Daniel Wright. June 1992. Louvicourt project feasibility study for 4000 tonnes per day. Unpubl.

Roscoe Postle Associates Inc. 1991. Reserve Audit of the Louvicourt Cu-Zn Deposit of Aur Resources Inc. and La Societe Minière Louvem inc. Unpubl.

4 Open pit planning and design

Innovative Mine Design for the 21st Century, Bawden & Archibald (eds) © 1993 Balkema, Rotterdam, ISBN 90 5410 325 6

Modelling, planning and mining a geologically complex coal deposit

D.Clarke & K.Sharman
Quintette Operating Corporation, Tumbler Ridge, B.C., Canada

ABSTRACT: Quintette Operating Corporation (managed and 33% owned by Teck Corporation) operates a large open pit coal mine located in northeastern British Columbia, Canada. The mine is comprised of 3 main pits (Mesa, Wolverine and Shikano) which employ truck/shovel mining techniques to produce 4.35 million tonnes of metallurgical coal annually. The geological structure of the area is extremely complex. A high degree of folding and imbricate thrust faulting within the main syncline and anticline features has resulted in coal seams with dips ranging from nearly flat to overturned. Mineable seam thicknesses vary from 0.6 m to 10 m. The complexity of the deposit has made interpreting, planning and mining a challenge and has necessitated the adaptation of the most up to date 3-dimensional modelling and planning technology available to Quintette's unique operation. The application of these engineering tools facilitates the diverse mining methods incorporated at Quintette.

1 INTRODUCTION

Quintette Operating Corporation operates a large open pit coal mine located in northeastern British Columbia, Canada (see Figure 1).

Quintette's complex deposit and significant production requirements (300,000 tonnes/day total production) have always demanded thorough geological interpretation, fast and iterative planning, and creative approaches to mining. Historically, these objectives were met by utilizing a VAX mainframe based cross-sectional geology model and planning system. This approach, although initially adequate, proved to be an ineffective engineering tool as it was inflexible, labour intensive, slow and difficult to update.

In 1989, following an extensive software search, test reserves based on one of Quintette's more geologically complex subpits were successfully modelled by Lynx Geosystems Inc. of Vancouver, B.C. The Lynx system, which is based on up to date 3-D modelling technology, was subsequently purchased in 1991. The system was installed on 6 Hewlett Packard 9000 series work stations and networked with several PC's for AutoCad interfacing.

The focus of this paper is to outline this modern modelling system and its integration with existing mine planning software, enabling Quintette geologists and planning engineers to effectively model, plan and ultimately mine this geologically complex coal deposit.

2 GEOLOGY

Coal seams mined at Quintette are contained in the Lower Cretaceous Gates Formation, a sequence of nearshore, delta plain, and alluvial sediments deposited along the eastern margin of the developing Rocky Mountains. Mineable seams range in thickness from 0.6 m to 10 m with interburden thicknesses of 1.5 m to 30 m.

The regional structure of the area is typical of

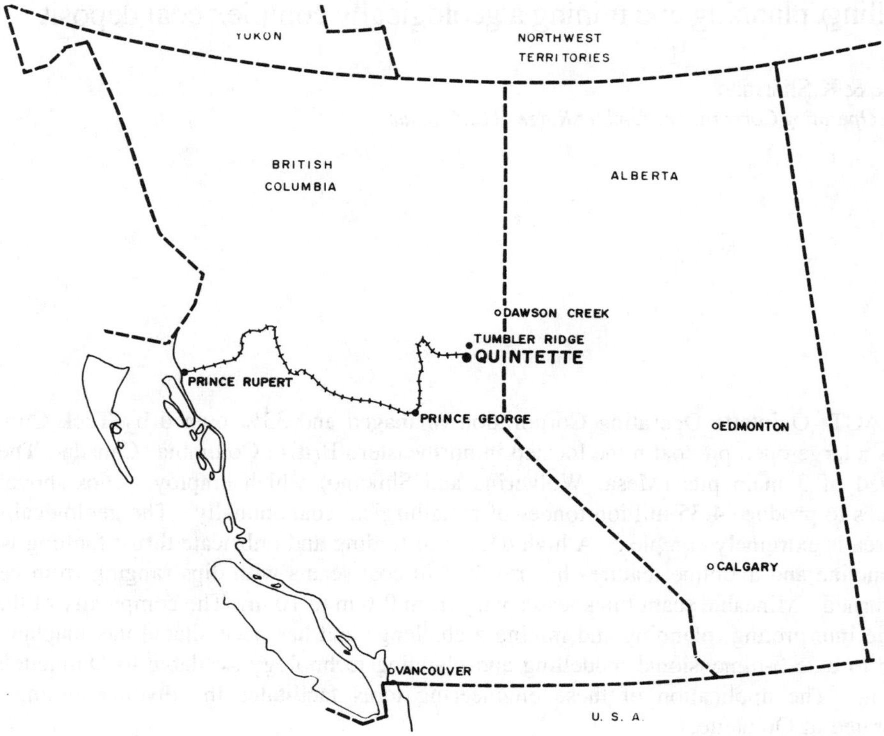

Fig. 1 Location map

foreland fold and thrust belts with open to tight folds, local overturning, and imbricate thrust faults. A wide range of structural complexity is shown by the three active pits of the mine area.

The Wolverine pits are developed on both limbs of a broad syncline (see Figure 2). The structure of the Wolverine South pit is the simplest of any area at Quintette. The average seam dip is 28 degrees. Faults are rare and are of less than 5 m displacement.

The Wolverine North pit, located on the north limb of the syncline, is more complex structurally. Dips are steeper (45-80 degrees) and thrust faults repeat parts of the coal section. Where these faults cut through bedding they cause shearing, small scale folding, and local overturning of bedding.

The structure of the Shikano pit is more complex than Wolverine (see Figure 3). It consists of a main anticline-syncline pair as well as a smaller fold structure. Faulting occurred both before and after folding and faults with 2 m to 40 m displacement are found. Structural thickening of coal seams has occurred in fold hinges.

The Mesa pit is extremely complex structurally (see Figure 4). A series of anticlines and synclines are cut by major faults with up to 100 m displacement. Folding has resulted in dips ranging from nearly flat to overturned. The seams are also cut by numerous closely spaced minor thrust faults.

3 MODELLING

With such complex geology, methods of interpretation and presentation must satisfy several key requirements:

1. The method must be able to handle the large number of data points needed to unravel the structure.
2. Complex geology must be able to be illustrated well in 3-D.
3. Updates as new information is available

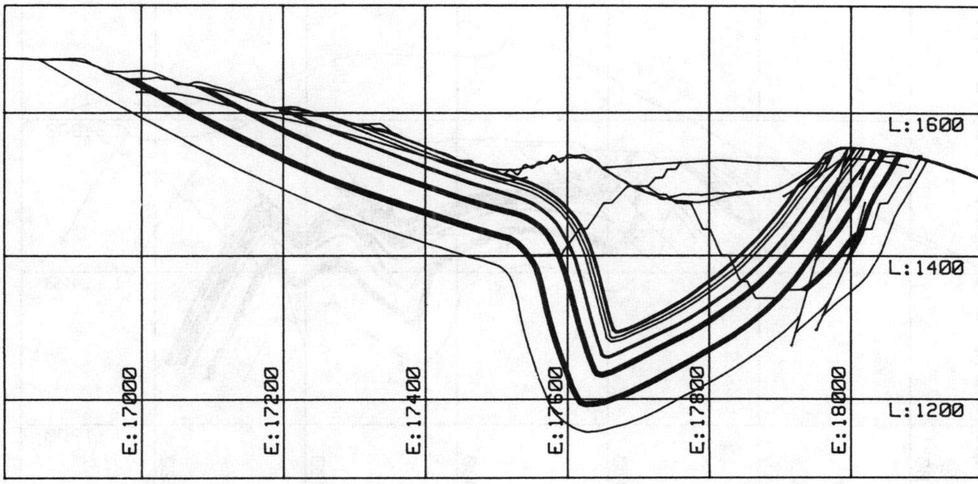

Fig. 2 Typical Wolverine pit cross section

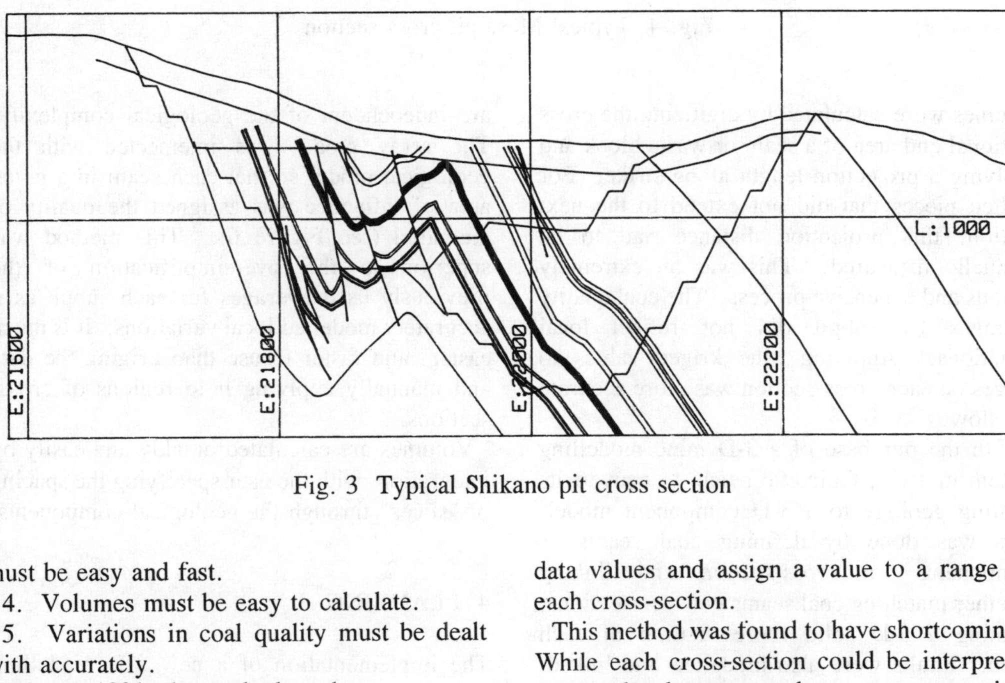

Fig. 3 Typical Shikano pit cross section

must be easy and fast.

4. Volumes must be easy to calculate.

5. Variations in coal quality must be dealt with accurately.

Prior to 1991, the method used was:

1. Enter exploration data (drillholes, mapping points etc.) into MINEPAK software on a VAX mainframe.

2. Plot paper cross-sections and interpret by hand using structure contours where necessary to gain continuity from section to section.

3. Group quality data together by subpit and use arithmetic averages for each seam, or krige data values and assign a value to a range on each cross-section.

This method was found to have shortcomings. While each cross-section could be interpreted accurately, the structure between cross-sections was not as well interpreted. Structure contours were found to be very time-consuming and, in complex faulted areas, almost impossible to decipher. When new information from drilling and mapping was available, new paper sections had to be plotted - another time consuming task. Updates were not performed as often as they should have been. Coal and waste

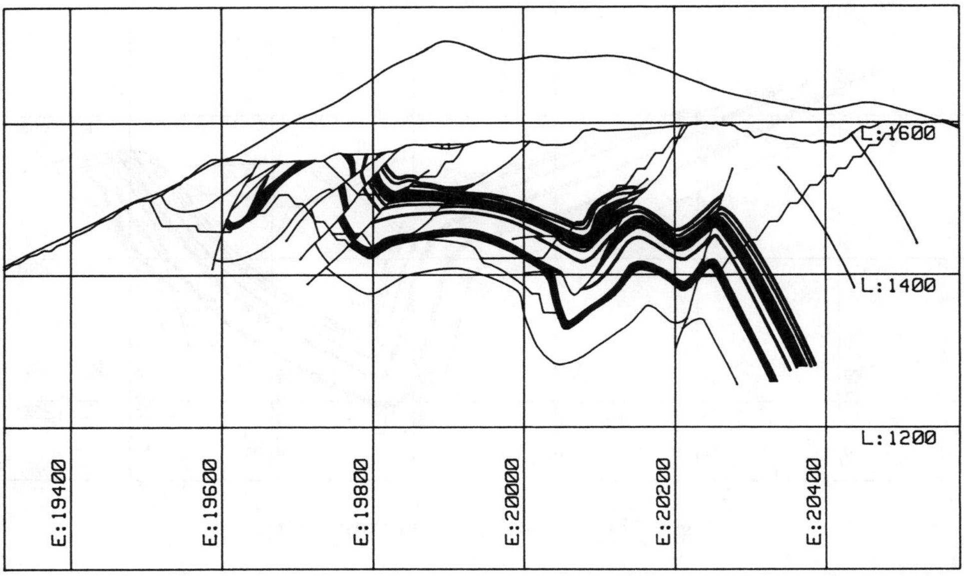

Fig. 4 Typical Mesa pit cross section

volumes were calculated by digitizing the cross sectional end area of a seam or waste block and applying a projection length along strike. For faulted pieces that did not extend to the next section, this projection distance had to be manually measured. This was an extremely tedious and expensive process. The coal quality averages by subpit did not reflect local variations. Applying the kriged values to ranges on each cross-section was more accurate but slow.

With the purchase of a 3-D mine modelling system in 1991, Quintette began to convert its existing geology to a 3-D component model. This was done by defining coal seams as components on cross-section and linking together matching coal seams between sections, forming a 3-D solid representation of each seam. Faults were also modelled and linked. The resultant model can be sliced in any orientation to generate cross-sections, longitudinal sections, and plans. Figure 5 illustrates the 3-D linking system.

Quality data was treated by inverse distance weighting data points for each seam to produce values that were assigned to each element in a 3_D array. The size of the areas of influence in this array can be varied by the geologist and are independent of the geological complexity. The array model was intersected with the geological model so that each seam in a given area of influence was assigned the quality of that area (see Figure 6). This method was superior to the oversimplification of the previously used averages for each subpit as it accurately modelled local variations. It is much easier and faster to use than kriging the data and manually applying it to regions of cross-sections.

Volumes are calculated quickly and easily by integration, with the user specifying the spacing of "slices" through the geological components.

4 PLANNING

The implementation of a new 3-D modelling system was a positive advancement toward the more thorough understanding and interpretation of Quintette's complex geology. The speed and flexibility of the modelling process was also a significant improvement. The challenge now facing mine planning engineers was how to exploit this information and manipulate it into meaningful production plans.

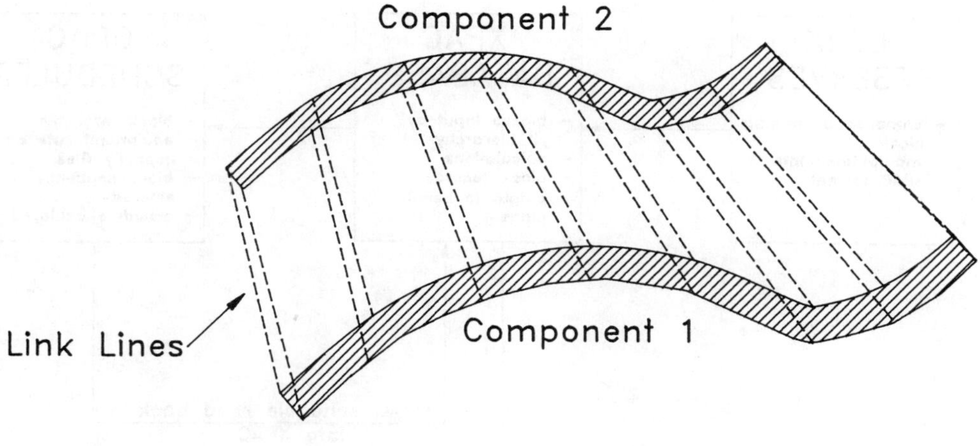

Fig. 5 Linking of two components

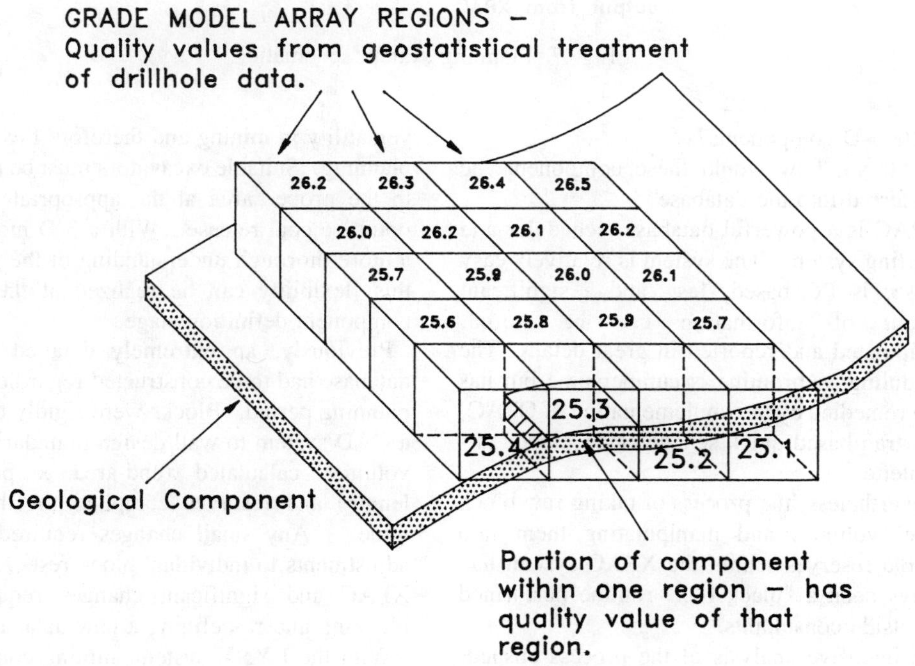

Fig. 6 Grade model schematic

4.1 Converting to 3-D

Because of Quintette's difficult geology and significant production rates, extensive planning is a necessity. Equipment locations and priorities must be set and followed for coal release to be met. Efficient and iterative planning is therefore required.

Planning engineers were generally satisfied that the existing XPAC reserve database and scheduling system met these requirements. However, to maintain this system, the planners were faced with two main problems:

1. Could the XPAC system be utilized to

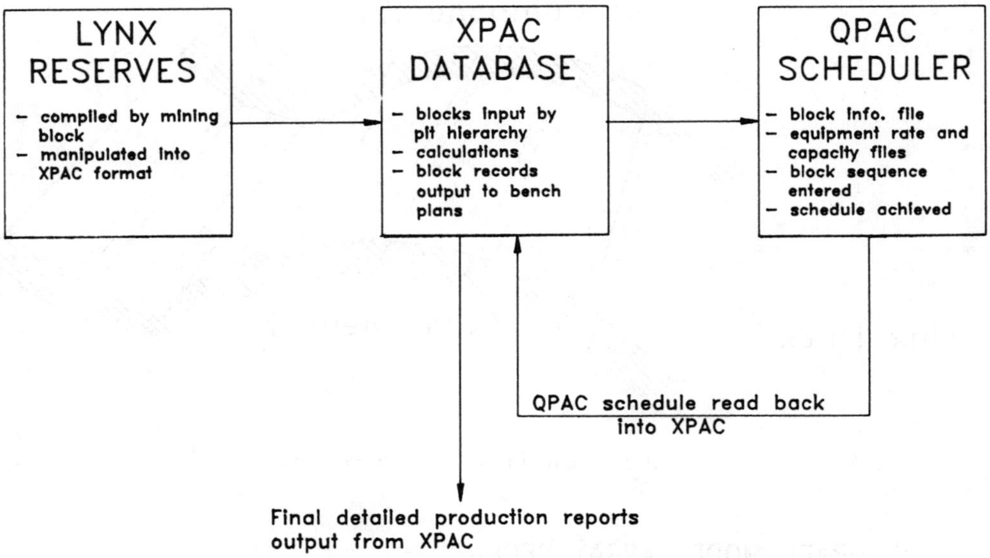

Fig. 7 Planning system schematic

handle 3-D components?

2. If so, how would these components be transferred into the database?

XPAC is a powerful database, scheduler and reporting system. The system is relatively easy to use, is PC based, fast, and a significant amount of information can be stored, manipulated and reported in great detail. The scheduling subroutine is cumbersome but has been remedied by the implementation of QPAC, a Fortran based quick scheduler developed at Quintette.

Nevertheless, the process of taking raw block model volumes and manipulating them into useable reserves within the XPAC system had always been a "black box" routine performed by outside consultants.

An intensive analysis of the process ensued. The many calculations performed and overall structure of the reserve database were simplified. Conversion programs were developed by Lynx Geosystems Inc. to assist in the transfer of data. The final system is illustrated in Figure 7.

4.2 Planning in 3-D

Quintette's geological complexity demands versatility in mining and therefore flexibility in planning. Suitable excavators must be allocated to the proper area at the appropriate time to optimize coal release. With a 3-D model, and a more thorough understanding of the geology, this flexibility can be realized at the mining component definition stage.

Previously, an extremely detailed reserves database had to be constructed regardless of the planning period. Blocks were rigidly traced on a CAD system to wall design boundaries, their volumes calculated (end-area x projection length) and reserves compiled for the entire mine. Any small changes required manual adjustments to individual block reserves within XPAC and significant changes required re-blocking and re-defining a new database.

With the LYNX system, mining components (blocks) are created in only as much detail as is indicated by the planning period. By intersecting these components with the appropriate geological component model, reserves by mining component are established. Any significant changes to geology or wall design are handled by simple component modifications and a re-definition of the reserve database. The process is rapid and far less labour intensive. Moreover, any modification can be completed on-site with Quintette geologists and engineers.

The mining component size and shape is entirely flexible and is dictated by planning period, geological complexity and anticipated mining method. For example, for long range planning, mining components representing entire benches are created.

For shorter range plans, more detailed information is required. In areas of consistent and relatively straight forward geology, separate coal and waste components are created since coal and waste in these areas would typically be mined with different excavators at different rates (see Figure 8).

In extremely geologically complex areas, the coal and waste are usually mined concurrently with a selective excavator such as a backhoe. In these areas the coal and waste are simply blocked together in single, somewhat smaller and regular shaped blocks (see Figure 9).

Once created in plan, mining components are then checked by slicing through both the geology and mining models in section and, if necessary, longitudinally (or at any desired viewplane).

The blocking flexibility and practical link to probable mining method along with continually working in a 3-D environment is a significant advantage. A great deal of thought and discussion among planning engineers often accompanies the creation of mining components in certain areas. This analysis results in a very thorough understanding of an area's geology and sets much of the groundwork required for practical scheduling.

5 MINING

"Plan the mine, mine the plan!" has always been a popular adage used at Quintette. As previously mentioned, realistic and well-defined plans based on an accurate geology model are necessary for Quintette to function effectively. Using the 3-D blocking system, the probable mining method to be used in an area is conceived months in advance. A wide range of mining equipment is used to implement the plan (see Table 1).

Mining methods employed at Quintette vary with seam thickness, dip, and physical mining constraints. The mineable seams vary from 0.6 m to 10 m thick resulting in a relatively high average strip ratio of 10:1 (bank cubic m waste:clean coal tonnes). Because of this combination of high strip ratio and a wide

Table 1 Equipment list

Equipment Type	Description	Capacity	Number
SHOVELS	P&H 2800 Cable Shovel	26 m3	4
	P&H 2300 Cable Shovel	14 m3	3
	P&H 2100 Cable Shovel	11 m3	2
	P&H 1900 Cable Shovel	8 m3	2
BACKHOES	Demag 241 Backhoe	14 m3	4
	Hitachi EX1100 Backhoe	6 m3	2
	Cat 245 Backhoe	3 m3	2
LOADERS	Letourneau L-1000 Loader	12 m3	8
DRILLS	Gardner Denver 70	31 cm dia.	1
	Gardner Denver 120	31/38 cm dia.	6
	Ingersoll Rand DMH	31 cm dia.	4
HAUL TRUCKS	Wabco 170/190	154 tonnes	68
SUPPORT	Cat D-9N Dozer		7
	Cat D-10N Dozer		14
	Cat D-11N Dozer		2
	Cat 824C RTD		2
	Cat 834 RTD		5
	Cat 16-G Grader		7

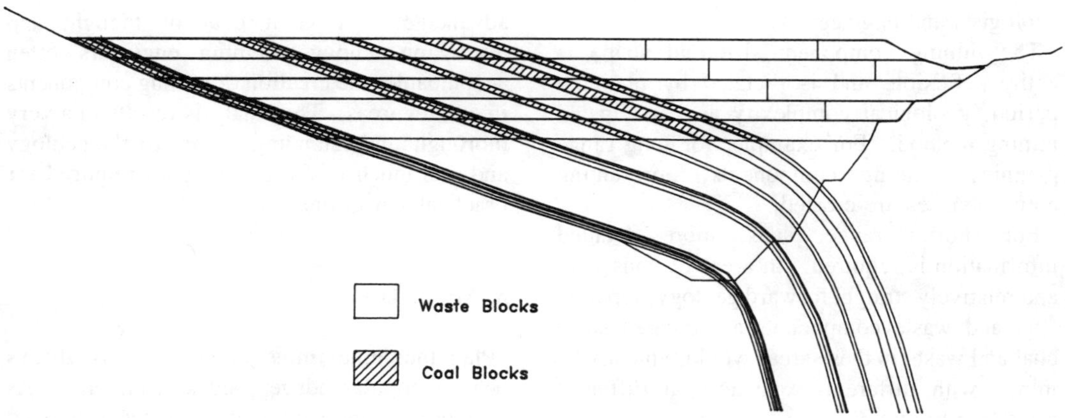

Fig. 8 Separate coal and waste blocks

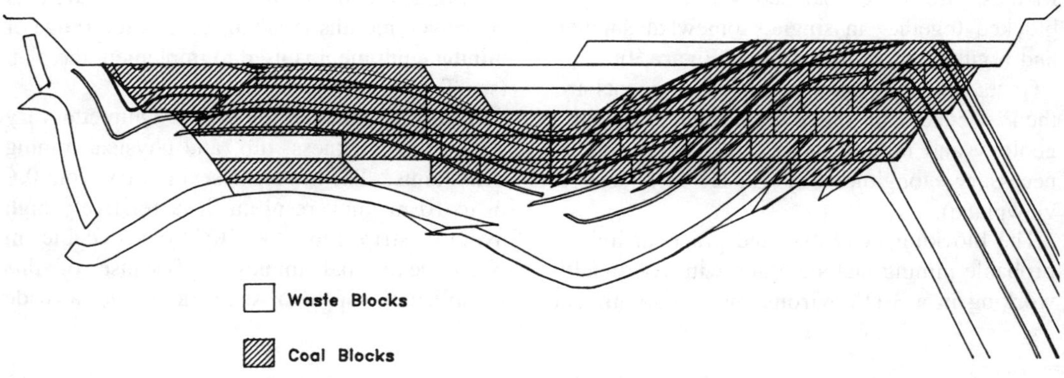

Fig. 9 Coal and waste blocked together

variation in seam thicknesses, production equipment capacities must range from massive to quite small. The result is a very large number and range of machines located in a relatively tight space.

6 BLASTING

In areas of simple structure and moderate dips such as Wolverine and parts of Shikano, the waste above the seam is shot, then mined; the coal is recovered; the waste to the next seam is shot and so on. In more complex areas, both the coal and waste are drilled and shot together with a 2 m explosives standoff below the coal and a 0.5 m standoff above. Crushed rock (or plant refuse at Shikano) is used as stemming through the coal itself.

Blast pattern sizes are established as short range mining components during monthly planning sessions. These sizes depend on current blasted inventory levels, planned excavator type, and coal seam complexity. Pattern volumes are calculated using LYNX and form the basis for weekly coal forecasts. The weekly plans are also used to direct all drilling and blasting operations with final pattern limits cut or extended as required.

7 EXCAVATION

In general, mining at Quintette can be broadly

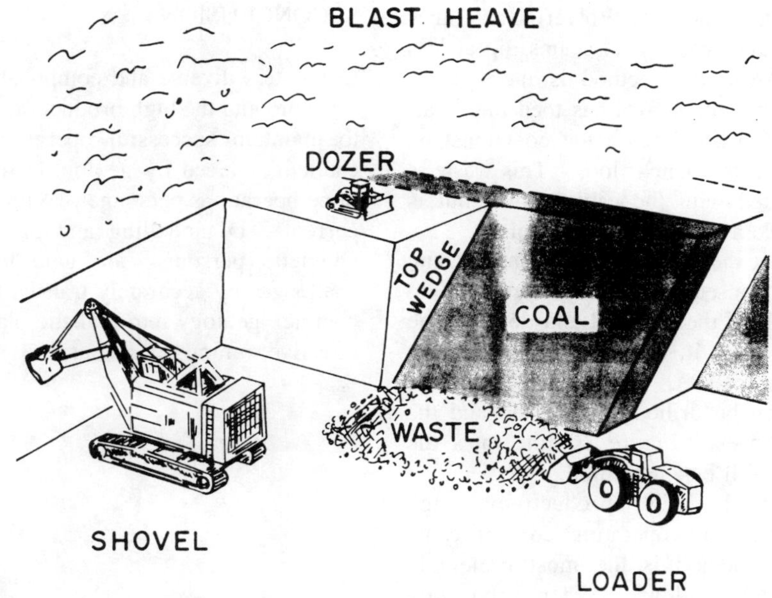

Fig. 10 Cable shovel/full bench method

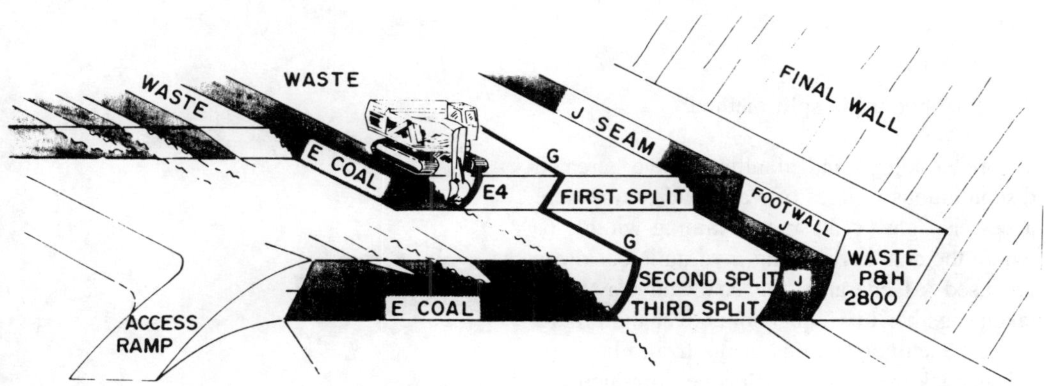

Fig. 11 Backhoe triple split method

categorized into two main equipment type/methods:

1. Cable shovel/full bench method
2. Backhoe/triple split method

7.1 Cable shovel/full bench method

This method is generally practised in Wolverine and most areas of Shikano as well as areas of Mesa that have simple geological structure. A cable shovel, the size of which is dictated by waste block size (stripping ratio) and physical mining constraints (minimum mining width), advances along strike and toward the coal seam until it toes into coal taking full bench height waste.

In areas such as parts of Shikano and Mesa where seams dip steeply (between 65 and 90 degrees), the coal is taken with the shovel itself. Where the seams dip more moderately (45-65 degrees), some dozer assistance in the way of side-cutting is sometimes necessary to "peel" the coal away from the footwall.

In other areas, especially Wolverine and other parts of Shikano where coal seams dip at less than 45 degrees, the method is more dozer intensive. Once the shovel has toed into coal, the wedge of waste left on the coal must be dozed down to the bench floor. This waste is often excavated with the cable shovel but is sometimes taken with loaders as well.
Similarly, once the waste wedge is removed and the coal is exposed, dozers are then used to push the coal off the footwall and down to the bench floor where it, too, is taken with either the shovel or loaders. The waste to the next seam can then be drilled and blasted and the process continues. Figure 10 illustrates the cable shovel/full bench method.

Because of the high productivity, high availability and low operating cost of cable shovels, this method is the most preferable where conditions allow. When production engineers and operating personnel analyze areas to be mined or at the LYNX blocking process, all efforts are focused on trying to adapt this method.

7.2 Backhoe/triple split method

Where folding and faulting have created discontinuous wedges of waste and coal, where access is tight (<30 metres mining width), or where the coal is flat lying, hydraulic backhoes are used. Generally, the coal and waste are taken together in 3 equal splits over the full 12 metre bench height (plus heave from blasting).

The backhoe is very selective, producing a high coal recovery, but it is also slow and expensive. The challenge often lies in identifying backhoe areas and exposing these early in the advancement of the bench. Otherwise benches can become "stacked" with drilling and the subsequent advancement of underlying benches being delayed. Smaller backhoes, with bucket capacities of 6 m3 or less, are used in areas so geologically complex that even the larger backhoes have difficulties in recovering the faulted pieces of coal. Figure 11 illustrates the backhoe/triple split method.

8 CONCLUSION

Quintette's diverse and complicated geological structure and the high production rates required to maintain successful operations create the challenges faced by geologists, engineers and mine operations personnel. With the use of the current 3-D modelling and planning system, Quintette personnel are able to meet these challenges by accurately translating the mine's complex geology into realistic plans and effective mining operations.

Innovative Mine Design for the 21st Century, Bawden & Archibald (eds) © 1993 Balkema, Rotterdam, ISBN 90 5410 325 6

Alternatives to shovel/truck overburden pre-stripping: Three case studies

Donald L. Ewart, Jr, Richard R. Marston & Stephen M. Pries
Marston and Marston, Inc., St. Louis, Mo., USA

ABSTRACT: The purpose of this paper is to identify and examine alternative methods (other than shovel/truck stripping) of effecting the pre-stripping of consolidated overburden, in conjunction with dragline stripping operations, at surface coal mines. The analysis will be case study oriented with practical mining considerations and comparatives economics being evaluated. Alternatives that will be examined include blast casting, dozer push, and multiple pass dragline sequences.

In surface coal mining it is sometimes necessary to remove a portion of the overburden material in advance of the primary stripping unit(s) to facilitate the logical progression of mining and/or accommodate physical machine limitations. Pre-stripping with shovel/truck fleets is common industry practice (with the term "shovel" being used to refer to electric mining shovels, hydraulic excavators, and/or front-end loaders). However, in some instances a more efficient and/or cost effective alternative to shovel/truck pre-stripping can be employed as detailed in the individual case studies.

1 INTRODUCTION

Dragline stripping is a principal surface coal mining method currently employed in the United States. In its various configurations (i.e. area stripping, mountain-top removal, and, to a lesser extent, contour stripping), dragline mining has been applied in all the major coal producing regions of this country.

In simple applications, draglines are employed exclusively to remove all consolidated overburden overlying the mineable coal seam(s). In mines with relatively constant overburden depth and fixed coal production, a dragline(s) may provide sufficient stripping capacity for the mine life. However, as overburden depths are seldom uniform and coal production rates are subject to change, it often becomes necessary to supplement dragline stripping capacity to maintain or increase the mine's production capabilities. This requirement to augment dragline stripping operations can arise from the need to

1) maintain and/or achieve the desired rate of coal release given fixed dragline capacity as deeper cover is encountered and/or the target rate of coal production is increased;
2) eliminate coal removal bottlenecks caused by stripping constraints; and
3) compensate for physical machine limitations as design overburden depths are exceeded.

Historically, dragline(s) capacity is augmented by pre-stripping operations, which are designed to remove sufficient overburden in advance of the dragline to: facilitate the desired rate of coal release; permit the logical progression of unit mining operations; and/or effect practical cut configurations.

It has long been an accepted industry practice to accomplish pre-stripping requirements with shovel/truck fleets. The term "shovel" includes electric mining shovels, hydraulic excavators, and front-end loaders. Due to their mobility and flexibility, shovel/truck fleets can be readily deployed to redress dragline stripping shortfalls without the need for extensive design and planning.

However, the acquisition of a shovel/truck fleet generally represents a sizeable capital outlay which may not be warranted by the level of pre-stripping required. Shovel/truck stripping operations are also fairly labor intensive and exhibit relatively higher equipment operating costs than incurred with other forms of stripping (e.g. dragline or dozer).

Fortunately, alternatives to shovel/truck pre-stripping do exist and in certain instances these alternative methods can prove to be more efficient and cost effective than shovel/truck pre-stripping. For example, pre-stripping required to overcome dragline stripping shortfalls can be effected through the use of dozer benching and/or a properly implemented blast casting program. Physical dragline operating limitations can be overcome using a multiple pass stripping configuration.

The application of these alternative pre-stripping methods will be examined in this text on a case study basis. The case studies presented are based on actual operating practices at currently operating coal mines. The data and diagrams presented in this paper represent MARSTON's original work effort generated either for various engineering studies performed recently for the respective operations, or developed especially for this paper. In several instances, current operating practices, at the mines used as examples, have evolved from recommendations submitted by MARSTON. As some of the data presented herein could be construed as proprietary information, the study mining operations have been designated as to generalized producing region rather than by name.

The three case studies to be examined; denoted as Case 1, Case 2, and Case 3 for ease of identification; consist of the following:

Case 1 - Southern Powder River Basin coal mine that utilizes dozer push stripping to remove overburden in excess of the dragline stripping capacity. Dragline capacity is provided by a single machine equipped with a 39 cubic yard bucket. Only one thick, flat lying coal seam is recovered in any given pit.

Case 2 - Single coal seam, single dragline, northern Powder River Basin area mine that is required to increase production capacity in excess of the current dragline stripping capacity. The operation employs a single dragline with a 75 cubic yard bucket. The coal seam is generally flat lying and averages 23 feet in thickness.

Case 3 - Multiple coal seam, multi-dragline, multiple pit northwestern U.S. area mine which employes multiple-pass dragline stripping in conjunction with shovel pre-stripping in overburden depths exceeding the dragline stripping capacity to achieve production requirements. Three draglines equipped with 58 cubic yard buckets operate in three separate pits in which three to six coal seams ranging in thickness from 3 to 35 feet are recovered.

2 CASE 1

The Case 1 operation is designed to release coal from two seams with each seam mined separately. Because the interburden between the two seams is too thick to allow mining of both seams in a given area, a series of roughly parallel pits, following the respective seam sub-croplines, has been developed.

The Case 1 mine is a captive operation which supplies coal to a mine-mouth power plant. Initially the mine was designed to release approximately 4.0 million tons per year (MTPY) of coal with mining progressing in from the respective seam sub-croplines to develop approximately 9 miles of highwall paralleling the strike of the seams. An electric walking dragline equipped with a 39 cubic yards (rated capacity) bucket was selected as the primary stripping machine. This machine has an effective annual production capacity of 9.75 million cubic yards. Due to the relatively shallow overburden depths encountered during initial development, the dragline was originally capable of effecting all stripping requirements.

The mine has now been in operation for over twenty years and mining has progressed into significantly higher stripping ratio areas than were encountered initially. The mine's annual coal order now stands at approximately 2.8 MTPY due to reduced plant requirements. Currently, virgin stripping ratios ranging from 2.5 to 6.0 are encountered with an average ratio of 5.2 reported for 1991. The effective stripping ratio (which encompasses rehandle requirements) was roughly 6.3 for 1991. Note that stripping ratios are expressed in terms of bank cubic yards (bcy) of stripping per ton of coal delivered.

In 1991 the mine was required to strip a total of about 17.3 million cubic yards of material to effect desired coal release. Because the dragline did not have sufficient capacity the operation relied on a combination of dragline and dozer stripping, with some supplemental shovel/truck stripping, to facilitate required material movement.

2.1 *Comparison of alternative pre-stripping techniques*

As deeper cover was encountered and dragline capacity first began to be exceeded, the mine initially attempted to compensate for dragline stripping shortfalls through the use of shovel/truck pre-stripping. However, the mine's management soon concluded that dozer pre-stripping represented a more cost effective alternative to shovel/truck methods.

As previously mentioned, virgin stripping ratios encountered at the operation currently range from 2.5 to 6.0, which correspond to stripping depths of roughly 65 to 160 feet. The mine has developed different stripping configurations for different depths of cover.

The choice of method is also dependent on projected annual stripping requirements with dragline bench heights adjusted to facilitate desired material movement while insuring optimal dragline utilization. That is, to avoid exceeding annual dragline capacity, dragline bench heights for a given cover depth must sometimes be reduced below the design maximum limit (thereby allocating more material to the dozer fleet). Generally, dragline benches are established at the highest practical height which will sustain the desired level of coal production. As pits are generally not continuous, the use of different methods in different areas does not constitute a problem.

In shallow cover (i.e. stripping depths less than or equal to 85 feet) a combination of single spoil-side pass (from a cross-pit bench) and extended bench configurations are employed, as depicted on Figures 1 and 2 respectively, to effect desired coal release. With the cross-pit bench method (Figure 1), approximately 56% of the total virgin cut material is pre-stripped by the dozer fleet. The extended bench method (as shown on Figure 2) is more dragline stripping intensive with only roughly 36% of the cut pre-stripped with dozers. The shovel/truck pre-stripping alternative is also illustrated on Figures 1 and 2.

A comparative economic analysis of the alternative pre-stripping methods (i.e. cross-pit bench and extended bench dozer pre-stripping and shovel/truck pre-stripping) suitable for shallow cover applications is presented in Table 1. All Case 1 economic comparisons are based on estimated per unit stripping volume mining costs; which encompass labor, supply and power, and depreciation costs; for stripping and drilling and blasting functions only. Unit mining costs; for drilling and blasting, dozer pre-stripping, shovel/truck pre-stripping, and dragline stripping; represent average estimated costs over a twenty-five year period and reflect a mixture of estimated consumables consumption rates and actual unit costs and depreciation allowances. Note that the Case 1 unit costs may not accurately reflect conditions at other operations due to dissimilar machine applications, intensity of equipment usage, and depreciation treatment. For comparative purposes, unit stripping costs are assumed to be constant for all scenarios. However, drilling and blasting unit costs are variable between dozer and shovel/truck pre-stripping alternatives due to the different powder factors employed (as per historical mine data). Specifically dozer/dragline waste is shot at a powder factor of 0.3 pounds of explosives per bcy ("lbs./bcy") while shovel/truck material is shot at a 0.65 lbs./bcy powder factor.

Table 1. Comparative mining costs — cut basis: shallow cover

Item/ Description	Dozer Pre-Stripping: Cross-Pit Bench	Dozer Pre-Stripping: Extended Bench	Shovel/ Truck Pre-Stripping
Comparative mining costs (per ft of highwall);			
Drilling & Blasting (DL/DZ) @ $0.117/bcy	$49.40	$49.40	$37.38
Drilling & Blasting (S/T) @ $0.164/bcy	N/A	N/A	$16.84
Dragline Stripping @ $0.224/bcy	$58.98	$82.01	$71.57
Dozer Stripping @ $0.181/bcy	$46.64	$28.45	N/A
Shovel/Truck Stripping @ $0.754/bcy	N/A	N/A	$77.44
TOTAL MINING COST (Per Lineal ft of Highwall)	$155.02	$159.86	$203.23
Cost per Virgin bcy	$0.367	$0.379	$0.481

As can be seen from Table 1, total unit stripping volume mining costs (per lineal foot of highwall) with dozer pre-stripping in shallow cover are roughly $0.10 to $0.11 per bcy lower than the shovel/truck pre-stripping alternative. On an annualized basis, assuming all production at the depicted (Figures 1 and 2) shallow cover stripping ratio, the use of dozer rather than shovel/truck pre-stripping could effect annual cost savings of $0.92 to $1.02 million (depending on whether extended bench or cross-pit configurations are employed). On a cost per delivered coal ton basis, the above variances equate to savings of $0.33 to $0.37 per ton.

The various dozer and shovel/truck pre-stripping configurations applicable at stripping depths between 85 feet and 110 feet (medium cover) are illustrated on Figures 3 and 4. As was the case in shallower overburden, both cross-pit bench (Figure 3) and extended bench (Figure 4) dozer pre-stripping methods can be employed in medium cover. Note that unlike the shallow cover configuration, a spoil-side pass is required for the medium cover extended bench sequence.

The comparative mining economics of the various pre-stripping alternatives for medium depth cover are outlined in Table 2.

Table 2. Comparative mining costs – cut basis: medium cover.

Item / Description	Dozer Pre-Stripping Cross-Pit Bench	Dozer Pre-Stripping Extended Bench	Shovel/ Truck Pre-Stripping
COMPARATIVE MINING COSTS (Per ft Of Highwall);			
Drilling & Blasting (DL/DZ) @ $0.117/bcy	$64.34	$64.34	$37.38
Drilling & Blasting (S/T) @ $0.164/bcy	N/A	N/A	$37.79
Dragline Stripping @ $0.224/bcy	$87.81	$136.08	$71.57
Dozer Stripping @ $0.181/bcy	$60.76	$18.75	N/A
Shovel/Truck Stripping @ $0.754/bcy	N/A	N/A	$173.72
TOTAL MINING COST (Per Lineal Foot Of Highwall)	$212.91	$219.17	$320.46
Cost Per Virgin bcy	$0.387	$0.399	$0.583

Table 3. Comparative mining costs — cut basis: deep cover

Item / Description	Dozer Pre-Stripping Cross-Pit Bench	Dozer Pre-Stripping Extended Bench	Shovel/ Truck Pre-Stripping
COMPARATIVE MINING COSTS (Per ft Of Highwall);			
Drilling & Blasting (DL/DZ) @ $0.117/bcy	$85.01	N/A	$37.38
Drilling & Blasting (S/T) @ $0.164/bcy	N/A	N/A	$66.76
Dragline Stripping @ $0.224/bcy	$177.30	N/A	$71.57
Dozer Stripping @ $0.181/bcy	$82.92	N/A	$0.00
Shovel/Truck Stripping @ $0.754/bcy	N/A	N/A	$306.95
TOTAL MINING COST (Per Lineal Foot Of Highwall)	$345.22	N/A	$482.67
Cost Per Virgin bcy	$0.475	N/A	$0.664

Per lineal foot of highwall, total unit mining costs with dozer pre-stripping are approximately $0.18 to $0.20 per bcy lower than the costs incurred with shovel/truck pre-stripping (at medium depths of cover). Assuming full year production at the medium cover stripping ratio, the use of dozer pre-stripping (rather than shovel/truck) could result in annual cost savings of $2.16 (extended bench) to $2.29 (cross-pit) million; or $0.77 to $0.82 per delivered ton of coal.

In deep cover (i.e., greater than 110 feet), extended bench configurations are generally not practical given annual coal production requirements. Due to the greater dragline stripping requirements associated with extended bench sequences, as compared to the cross-pit bench alternative, extended bench dragline stripping is incapable of effecting desired coal release in deeper overburden. Consequently, only one dozer pre-stripping configuration (cross-pit bench), as depicted on Figure 5, for deep cover has been included in MARSTON's analysis. The alternative shovel/truck pre-stripping configuration in deep overburden is also illustrated on Figure 5.

In comparison, unit mining costs with dozer pre-stripping in deep cover are $0.19 per bcy lower than total mining costs with shovel/truck pre-stripping (refer to Table 3). Note that the unit mining cost savings attributable to dozer pre-stripping is not as great as was experienced in medium cover due to the significantly higher rehandle requirements incurred in deep cover. On an annualized basis, the variance in unit mining costs could theoretically translate into a savings of $2.92 million per year (however, an additional dragline would be required to support this level of stripping). Correspondingly, per ton coal costs would be $1.04 per delivered ton less with the dozer pre-stripping option.

Thus dozer pre-stripping represents a lower cost alternative to shovel/truck pre-stripping, at all encountered overburden depths, for the Case 1 operation.

3 CASE 2

The Case 2 mine is a relatively low stripping ratio, single seam, northern Powder River Basin operation. A single 75 cubic yard dragline is used to uncover a coal seam that averages about 23 feet in thickness. The coal seam is flat to gently rolling, and stripping ratios vary from less than 2 bcy per ton to over 10 bcy per ton. The currently mineable reserves are generally accepted as coal reserves with a stripping ratio of less than 6.0 bcy per ton.

Extensive near surface sub-crops of mineable coal are present within the property which allows low ratio stripping to begin by opening boxcuts along these sub-crops. A pit design which parallels the sub-crop permits the production of low stripping ratio reserves early in the mine life.

Although initially designed to produce 5.6 MTPY of coal, the mine has been required to increase production levels in excess of 6.5 MTPY and ongoing coal orders of 6.0 MTPY and greater are anticipated in the future. In concert with the greater coal production requirements, increasingly deeper cover will be encountered as mining progresses in from the sub-cropline; which will ultimately result in existing dragline capacity being surpassed.

Alternatives available to the mine in light of the higher coal orders and increasing stripping ratios are

1) to supplement dragline stripping capacity with shovel/truck pre-stripping;
2) to implement a blast casting program that will increase the production of the current equipment fleet; or,
3) to reduce the coal output as the stripping capacity of the current equipment is exceeded.

As may be expected, the mine operator does not accept the failure to meet increased coal orders as a viable alternative and an analysis of the remaining options is thus required.

3.1 Study parameters

The mine is required to produce 6.0 MTPY of coal using the most cost effective means possible. The 75 cubic yard dragline presently employed has an average annual total capacity of 18,750,000 cubic yards. The dragline is the only stripping unit used to uncover the coal.

The dragline operates primarily in a simple side cast mode and rehandle rates are expected to be eight percent of the virgin material. This rehandle estimate is designed to account for overburden that must be handled twice, or more, by the dragline to provide for spoil-side access ramps, cross-pit bridges, dragline ramps, and spoil placement at the ends of each pit. The net effect is that the dragline has the capacity to strip 17,361,000 bcy of virgin overburden annually. This equates to a maximum stripping ratio of 2.89 bcy per ton of coal produced. Assuming an average coal thickness of 23 feet and 95 percent recovery, the maximum practical average dragline stripping depth at this ratio is 68.6 feet.

Where the average overburden depth exceeds 68 feet, additional stripping capacity is required to ensure that the 6.0 MTPY production target is reached. The use of average values for depth and coal thickness is consistent with a mining sequence designed to progress uniformly from low to high stripping ratio areas which, necessarily, produces the most economical reserve available at any time. Local variations in depth and thickness which occur, are sufficiently minor that they do not preclude the use of average values for design purposes.

3.2Stripping Alternatives

3.2.1Shovel/Truck Pre-stripping

The addition of a shovel/truck fleet to meet changing overburden stripping requirements provides a flexible and simply engineered method of assuring that production targets are achieved. The dragline can be readily positioned at a fixed height above the coal seam to release the required production. In addition, as the mine matures additional production units can be added to the stripping fleet.

Figure 6 provides an indication of the average dragline bench height required to uncover a particular annual coal quantity. At the higher coal production rates contemplated, the dragline will be positioned at less than 70 feet above the seam. This will insure that the dragline continually operates in its most productive simple side cast mode. In this example, the dragline bench will be established approximately 68 feet above the seam while the shovel/truck fleet pre-strips above this limit in order to produce 6.0 MTPY.

Figure 7 illustrates the planned operation at an overburden depth of 90 feet. In this case the pre-stripping fleet must remove an average of 22 feet of material in advance of the dragline. At the 6.0 MTPY coal production level the shovel/truck fleet is required to remove a minimum of 5.4 million bcy annually.

The required unit mining operations are as follows:

1) Topsoil is removed to a depth of two feet in advance of mining. This topsoil is stockpiled for later placement or placed directly on graded spoil.
2) A diesel powered drill is to be used to drill 8-3/4 inch holes for blasting of the shovel waste. The shovel waste is to be blasted using an ANFO/emulsion mix at a 0.40 pound per bcy powder factor.

3) A 28 cubic yard capacity shovel is planned for loading 150 ton payload haul trucks. The shovel has been sized to accommodate increasing pre-stripping requirements. The loaded trucks will drive around the pit ends or cross a spoil bridge to dump the waste between the dragline spoil peaks. Placement of the spoil will be directed to facilitate reclamation to the required post reclamation topography.

4) The dragline overburden is drilled using an electric powered rotary drill to drill 12-1/4 inch diameter holes. Blasting is effected with an ANFO/emulsion mix at a powder factor of 0.35 to 0.40 pounds per bcy of overburden.

5) Dragline stripping in a single side cast mode is used to uncover the coal. The maximum overburden depth accommodated with this method is approximately 80 feet with a resulting coal release of 5.1 MTPY. The dragline bench will be lowered to allow higher production rates to be achieved.

6) The coal is cleaned, to upgrade the coal quality, drilled and blasted. ANFO at a 0.25 pounds per ton powder factor is employed.

7) The coal is loaded with a 25 cubic yard capacity mining shovel into 160 ton capacity bottom dump coal haulers for transportation to the tipple.

3.2.2 Dragline and blast cast stripping

As in the shovel pre-stripping method, the analysis of dragline and blast cast stripping derives conclusions from an analysis of average overburden depths and coal thicknesses. Again, this is only valid for an operation in which the pit design generally corresponds to a low to high ratio sequence which does not have large variations in stripping depth along the length of the cut.

With the implementation of blast casting the stripping capacity of the mine operation is increased with little or no additional capital investment. Plans for the blasting program anticipate that blast casting, combined with dozer bench preparation, will place 25 percent of the virgin bank material directly into the previously mined cut (i.e. rehandling of this material by the dragline will not be required). Effectively, the stripping capacity is increased by one-third over the dragline capacity alone by substituting explosive energy for mechanical energy.

From a review of Figure 6 it is seen that an overburden depth of up to 91.4 feet can be accommodated with the blast cast and dragline system at the anticipated average coal thickness and production rate. A range diagram is provided as Figure 7 to illustrate the removal of approximately 90 feet of overburden utilizing blast casting. At this depth an annual estimated production rate of approximately 6,100,000 tons can be expected.

The primary unit operations required to implement the blast casting program are as follows:

1) Topsoil is removed and replaced as in the previous method.

2) Benches are surveyed and prepared for drilling with 12-1/4 inch diameter holes. Holes are drilled at an angle of 20 degrees from the vertical to provide a consistent burden over the length of the hole. A powder factor of approximately 1.0 pounds of ANFO/emulsion blend per bcy is estimated to produce an effective direct cast of 25 percent of the bank material.

3) A Caterpillar D11 dozer is used on the cast material to develop a level working bench for the dragline.

4) The dragline operates from a position above the edge of the coal to be uncovered and digs and places the spoil in its final position. A dozer is used to assist the dragline in cleaning the spoil from against the new highwall. The dumping height of the dragline combined with the relatively low overburden depths allows the dragline to effect spoil placement from the blast cast pad without requiring spoil-side stripping.

5) Coal drilling, blasting and loading are completed as in the previous method.

3.3 Economic analysis

The estimated operating costs associated with each mining method to produce 6.0 MTPY of coal are tabulated in Table 4. In the case of this analysis it should be noted that only operating and ownership costs associated with overburden stripping have been included. The costs of topsoil removal, coal blasting, loading and hauling and other costs common to each method have been excluded. A brief discussion of the costs included follows:

Table 4. Comparative mining costs per lineal foot of cut.

Item/Description	Volume	Estimated Cost		
		Drilling & Blasting Cost	Stripping Cost	Total Cost
SHOVEL AND DRAGLINE				
Shovel Truck Stripping	123.0 bcy/ft	$23.60	$75.84	$ 99.44
Dragline Stripping (Virgin)	381.0	37.49	92.62	130.10
Dragline Stripping (Rehandle)	30.5		7.41	7.41
SUBTOTAL	504.0			236.95
Virgin Stripping Ratio	3.83 bcy/ton			
Effective Stripping Ration	4.06 bcy/ton			
TOTAL PER BANK CUBIC YARD				0.470
TOTAL PER TON				1.799
DRAGLINE AND BLAST CASTING				
Blast Casting	504.0	96.37		96.37
Dragline Stripping (Virgin)	378.0		91.89	91.89
Dragline Stripping (Rehandle)	30.2		7.35	7.35
SUBTOTAL	504.0			195.60
Virgin Stripping Ratio	3.83 bcy/ton			
Effective Stripping Ratio	4.06 bcy/ton			
TOTAL PER BANK CUBIC YARD				0.388
TOTAL PER TON				1.485
COST SAVINGS				
Dollars per bank cubic yard				0.082
Dollars per ton				0.314
Dollars per year @ 6.0 MTPY Production Rate				$1,883,786 .00

3.3.1 Drilling and blasting

Three distinct drilling and blasting costs have been estimated to forecast the costs of blasting shovel/truck, dragline only, and blast cast overburden. All labor, materials, ownership, and supply costs are included in each category as are the estimated maintenance labor costs.

3.3.2 Shovel/truck costs

The cost of operating a 28 cubic yard mining shovel and four 150 ton capacity haul trucks is included in this analysis. The average one way haul distance is 4,100 feet. The cost of operating a Caterpillar D9N dozer is included to provide for bench preparation, cleanup around the shovel and dump construction.

3.3.3 Dragline costs

The operating cost of the 75 cubic yard capacity dragline includes all labor, maintenance, supplies and depreciation associated with the dragline's operation. The operating cost of a Caterpillar D10N dozer has been included to prepare the dragline bench and to assist the dragline in cleaning the highwall.

The costs provided in Table 4 provide an estimate of the annual savings that can be realized through the implementation of a blast casting program and the deferral of shovel/truck stripping. In the example chosen, the unit costs of stripping 90 feet of overburden are $0.467 per bcy for the shovel/truck alternative and $0.301 per bcy for the blast casting alternative. The estimated savings are $1,807,000 per year.

The analysis provided is specific to an overburden depth of 90 feet which approaches the maximum depth at which the blast casting alternative can provide 6.0 MTPY. Where overburden depth is less than 90 feet the savings that accrue to blast casting are reduced and eventually eliminated as the need for shovel/truck pre-stripping is eliminated. The applicability of the blast casting method to this mine is apparent from an analysis of the mineable coal reserve. Above the single dragline cutoff of 68 feet of overburden (at 6.0 MTPY), to an arbitrary reserve stripping ratio limit of 6.0 bcy per ton, the mineable reserve contains an estimated 130 million tons. Included in the 130 million tons are 44 million tons that can be produced at 6.0 MTPY using the dragline and blast casting only. The remaining 86 million tons requires shovel/truck pre-stripping to supplement the dragline/blast casting program. It is during the production of the final 86 million tons during 14 years of production that the full benefits of blast casting are realized as 5.8 million bcy per year are effectively cast versus moving this volume with the shovel/truck fleet.

Figure 6 provides a measure of the shovel/truck pre-stripping capacity required to supplement dragline and dragline/blast cast stripping to achieve the 6.0 MTPY production rate. For example, at an average overburden depth of 110 feet, 10.5 million bcy per year are required to support the dragline only operation while only 4.7 million bcy are needed to support the dragline/blast cast operation.

In summary, the planning and implementation of a blast casting program can reduce or eliminate the need for shovel/truck stripping as a supplement to dragline stripping. Substantial cost savings can be realized as blasting is used to replace shovel/truck stripping requirements.

4 CASE 3

Coal production at the Case 3 operation is primarily from six seams which are encountered in each of the three dragline pits and a small shovel/truck pit. Three to four additional coal seams, which exist at higher elevations in advance of the current highwalls, are also recovered when encountered during normal pre-stripping operations. The development of a fourth dragline pit is anticipated due to the depletion of economically mineable reserves in one of the existing dragline pits. A comparison of the alternative stripping methods applicable for mining of the new pit will be examined to demonstrate the value of a multiple pass, high rehandle dragline operation relative to a shovel/truck and dragline operation.

Due to the variable dip and orientation of the coal seams, dragline cuts have been aligned so as to minimize pit floor slopes to enhance spoil stability and limit coal haulage grades along the pit bottom. The apparent dip of the coal seams within the current and proposed pits ranges from approximately 5.0 to 9.5 percent. A generalized stratagraphic column showing the spatial relationship and thickness of the mineable seams and interburden thicknesses within the proposed pit is depicted on Figure 8. Note that recovery of the six seams present will result in an overall pit depth of roughly 300 feet. These conditions are similar to those which exist in the three currently operating dragline pits.

Due to the relatively thick overburden and interburden encountered at the mine, a combination of dragline and mobile equipment (primarily shovel/truck) stripping is required to effect desired coal release of approximately 6.7 MTPY (4.3 MTPY of clean coal delivered to the power plant). Three electric walking draglines,equipped with 58 cubic yards rated capacity buckets, are being utilized as the primary stripping machines. These machines have effective stripping capacities ranging from 12.4 to 12.8 million cubic yards annually per dragline. The virgin stripping ratios encountered within the three dragline pits in 1991 were reported to range from 7.0 to 9.3 bcy per clean ton (bcy/ton) with an average ratio of 7.8 for the year. Inclusion of the rehandle volumes incurred results in an effective stripping ratio of 12.0 to 18.0 bcy/ton.

In order to effect desired coal release in 1991, approximately 33.3 million cubic yards of waste were stripped at the mine by three draglines, three electric mining shovels and a scraper fleet. The rehandling of material required during stripping activities necessitated the removal of an effective waste volume totaling roughly 51.6 million cubic yards, of which 71.5 percent was moved by the draglines.

Following closure of one of the dragline pits at the mine, a new production source will be required. A multi-pass dragline stripping sequence has been developed to provide this production requirement based on the anticipated geologic and mining conditions within the proposed pit.

A comparison of the proposed multi-pass dragline stripping method, and a method which incorporates increased shovel/truck stripping is outlined below. The analysis is predicated on providing coal production at the most cost effective stripping means available.

4.1 Comparison of alternate stripping techniques

Recall that, stratagraphically, the three to four uppermost coal seams mined are uncovered when they exist by mobile stripping equipment during pre-stripping activities in advance of the dragline operations. An average cover depth of approximately 160 feet exists between the lowermost coal seam stripped by mobile equipment and the upper seam uncovered as a result of dragline stripping (the dragline bench height being dictated by coal release requirements and the anticipated stripping conditions). Currently, the draglines remove from 140 to 180 feet of material from above the thickest coal seam (Seam 1). The proposed stripping operation as presented in Figure 9 has similarly been developed with the dragline stripping approximately 140 feet of overburden material above Seam 1 in 3 highwall passes. Due to highwall stability concerns, the dragline excavates the material overlying the upper coal seam at a highwall angle of 45 degrees. A 55 feet wide berm is incorporated into the highwall design at each bench elevation to allow for dragline tail clearance (necessary for keycut excavation) and to further enhance highwall stability. An additional highwall pass is then initiated from an extended bench at an elevation roughly equal to the Seam 1 bottom. The Seam 2/Seam 3 interburden is removed and a portion of the previously spoiled waste is rehandled from this position in effecting the release of Seams 3 and 4.

Following the release of Seam 3 and 4, the dragline crosses the pit and begins a series of spoil-side passes to create sufficient spoil room required in effecting the release of the lower two coal seams. The thinner interburdens are removed through the use of mobile stripping equipment (i.e. shovel/truck or scraper fleets). At this stripping depth a dragline rehandle factor of 139.6 percent is incurred. Dragline rehandle is defined within this study as spoil (waste) material which must be handled more than once before final deposition in the spoil pile by the dragline. Percent dragline rehandle is computed by means of the following expression:

Equation 1

Percent Dragline Rehandle = (DLRH/DLVS) x 100

Where: DLRH = dragline rehandle volume (bcy)

DLVS = dragline virgin stripping volume (bcy)

Note that pre-stripping, topsoil and parting volumes removed by mobile equipment are excluded from the dragline rehandle calculation.

An alternative to the multiple highwall pass dragline stripping method is depicted in Figure 10. In this method, the overburden is pre-stripped to a manageable dragline bench height above the uppermost coal seam from which two highwall passes are executed with the dragline. The initial dragline bench elevation, in the case depicted, is established at 50 feet above the top of Seam 1. Approximately 64 percent (90 feet) of the waste material overlying Seam 1 is allocated to the shovel/truck stripping fleet. The shovel/truck pre-stripping benches are typically developed one to one and one-half cut widths in advance of the pit due to the excessive overall pit depth and highwall stability concerns. The material excavated by the shovel must be transported from the pre-stripping benches, via the waste haul road located along the highwall directly above the dragline bench, to the dump located within the dragline spoils (roughly two to three spoil piles behind the active pit). Truck cycle times ranging from 12 to 15 minutes are typically incurred resulting in relatively high shovel/truck stripping costs.

Although the amount of virgin material handled directly by the dragline is reduced by approximately 34 percent, several spoil-side passes are required, as in the previous method, to reduce the overall spoil angle and to effect the release of the coal seams located at lower elevations within the pit. The dragline rehandle incurred with this method is reduced to 79.4 percent, a reduction of 60.2 percent of the dragline virgin volume.

A comparative economic analysis of the alternative methods, developed on a per foot of pit basis, is summarized in Table 5. The cross-sectional volumes calculated from the two stripping method range diagrams provided in Figures 9 and 10 were utilized to determine the average cost per lineal foot of pit. The unit mining costs utilized in the analysis (i.e. drilling and blasting, shovel/truck pre-stripping and dragline stripping) include direct labor and supply and power charges. The costs are based on the 1991 Mine Plan developed by MARSTON and historic data supplied by the mine. The following unit costs, exclusive of depreciation charges, were used:

Dragline stripping	\$0.208 per bcy
Shovel/truck stripping	\$0.900 per bcy
Drilling and blasting	\$0.136 per bcy

For comparative purposes, a summary of waste volumes and coal tonnages projected for each alternative are presented in Table 5. A virgin stripping ratio of 8.80 bcy per clean ton of coal has been calculated with effective stripping ratios varying from 19.23 bcy per ton to 12.34 bcy per ton, respectively, for Method 1 (dragline removes 140 feet of material overlying Seam 1) and the reduced dragline bench height stripping method (Method 2). The considerable variance in effective ratio between the two methods is attributable to the notable difference in dragline rehandle requirements (139.6 percent versus 79.4 percent). Estimated average stripping costs are \$0.694 and \$0.768 per bcy of waste for Methods 1 and 2, respectively. Similarly, average coal costs of \$6.105

Table 5. Comparative production summaries and mining costs per lineal foot of cut

Item/Description	Method 1 Multi-Pass Dragline Stripping – Dragline Bench at 140 Feet	Method 2 Shovel/Truck Pre-Stripping – Dragline Bench at 50 Feet
PRODUCTION SUMMARY		
Drilling and Blasting (bcy)	1759	1759
Virgin Dragline Stripping (bcy)	1495	893
Dragline Rehandle (bcy)	2087	709
Shovel/Truck Stripping (bcy)	263	866
Total Virgin Stripping (bcy)	1759	1759
Total Effective Stripping (bcy)	3846	2468
Clean Coal Production (tons)	200	200
Virgin Stripping Ratio (bcy/clean ton)	8.80	8.80
Effective Stripping Ratio (bcy/clean ton)	19.23	12.34
COMPARATIVE MINING COSTS		
Drilling and Blasting ($)	$ 239.22	$ 239.22
Dragline Stripping	745.16	333.13
Shovel/Truck Stripping	236.70	779.40
TOTAL STRIPPING COST	$1,221.08	$1,351.75
Cost per Virgin Bank Cubic Yard	0.694	0.768
Coast per Clean Coal Ton	6.105	6.759

(Method 1) and $6.759 (Method 2) per clean ton have been estimated. Further analysis of these results reveals that a cost benefit of $0.074 per bcy, or $0.654 per clean ton of coal, can be realized through the implementation of stripping Method 1 at a dragline rehandle rate of 139.6 percent. In this analysis, annual savings of approximately $465,000 in stripping costs for the recovery of 715,000 clean coal tons would be realized at the average virgin stripping ratio of 8.80 bcy per ton.

Thus, multi-pass dragline stripping, even in high rehandle applications, can be a viable economic alternative to conventional shovel/truck pre-stripping.

DOZER PRE-STRIPPING
(CROSS-PIT BENCH)

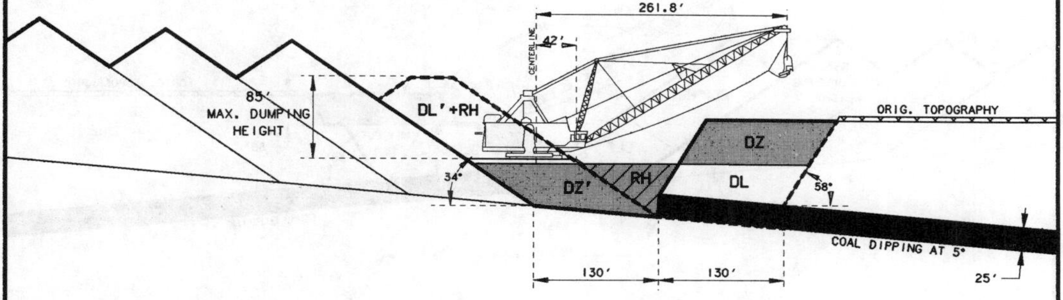

DESCRIPTION	DESIGNATION	VOLUMES		% OF
		B.C.Y.	L.C.Y.	TOTAL
TOTAL CUT VOLUME		422.2	548.9	100.00%
TOTAL VIRGIN DOZER PUSH	DZ	234.9	305.4	55.64%
TOTAL VIRGIN DRAGLINE STRIPPING		187.3	243.5	44.36%
DRAGLINE SPOIL SIDE PASS	DL	187.3	243.5	44.36%
TOTAL REHANDLE		98.8	128.4	23.40%
DRAGLINE REHANDLE	RH	76.0	98.8	18.00%
DOZER REHANDLE		22.8	29.6	5.40%

DOZER PRE-STRIPPING IN CONJUNCTION WITH DRAGLINE STRIPPING

SHOVEL/TRUCK PRE-STRIPPING

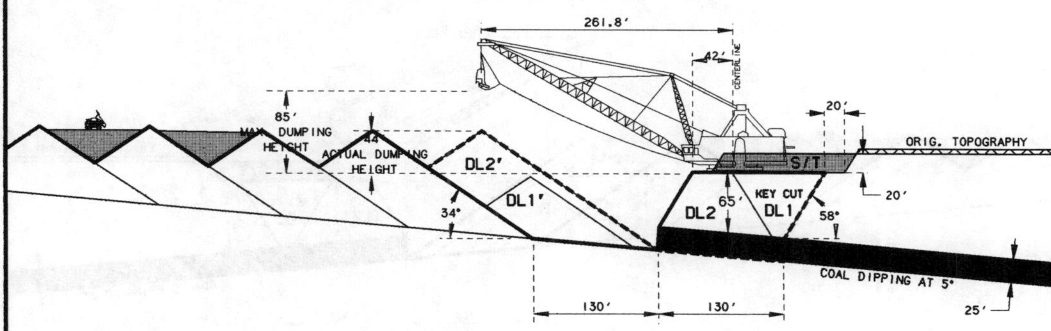

DESCRIPTION	DESIGNATION	VOLUMES		% OF
		B.C.Y.	L.C.Y.	TOTAL
TOTAL CUT VOLUME		422.2	548.9	100.00%
TOTAL VIRGIN S/T PRE-STRIPPING	S/T	102.7	133.5	24.32%
TOTAL VIRGIN DRAGLINE STRIPPING		319.5	415.4	75.68%
DRAGLINE KEY CUT	DL1	137.9	179.3	32.66%
DRAGLINE HIGHWALL SIDE PASS	DL2	181.6	236.1	43.02%

SHOVEL/TRUCK PRE-STRIPPING IN CONJUNCTION WITH DRAGLINE STRIPPING

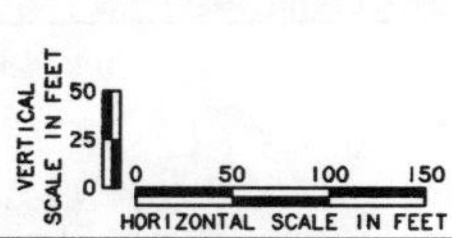

LEGEND

- DOZER OR SHOVEL/TRUCK PRE-STRIPPING
- DL1 DRAGLINE
- COAL
- REHANDLE
- PREVIOUS PROFILE
- TOPSOIL
- FINAL PROFILE

NOTE: STRIPPING DEPTH = OVERBURDEN LESS TOPSOIL DEPTH

Marston & Marston, Inc.
St. Louis, Missouri

COMPARISON OF ALTERNATIVE PRE-STRIPPING METHODS – DOZER VS. SHOVEL/TRUCK AT 85 FT. STRIPPING DEPTH

FIGURE 1

MAY 1992

DOZER PRE-STRIPPING
(EXTENDED BENCH)

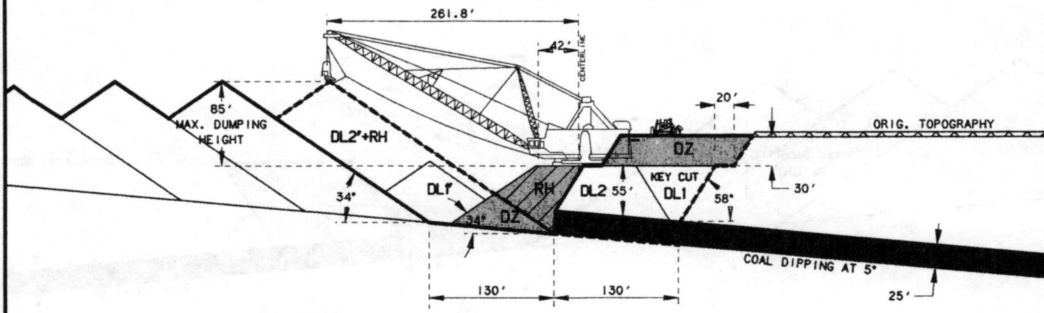

DESCRIPTION	DESIGNATION	VOLUMES B.C.Y.	VOLUMES L.C.Y.	% OF TOTAL
TOTAL CUT VOLUME		422.2	548.9	100.00%
TOTAL VIRGIN DOZER PUSH	DZ	150.9	196.2	35.74%
TOTAL VIRGIN DRAGLINE STRIPPING		271.3	352.7	64.26%
DRAGLINE KEY CUT	DL1	99.5	129.4	23.57%
DRAGLINE HIGHWALL SIDE PASS	DL2	171.8	223.3	40.69%
TOTAL REHANDLE		101.1	131.5	23.95%
DRAGLINE REHANDLE	RH	94.8	123.2	22.45%
DOZER REHANDLE		6.3	8.2	1.50%

DOZER PRE-STRIPPING IN CONJUNCTION WITH DRAGLINE STRIPPING

SHOVEL/TRUCK PRE-STRIPPING

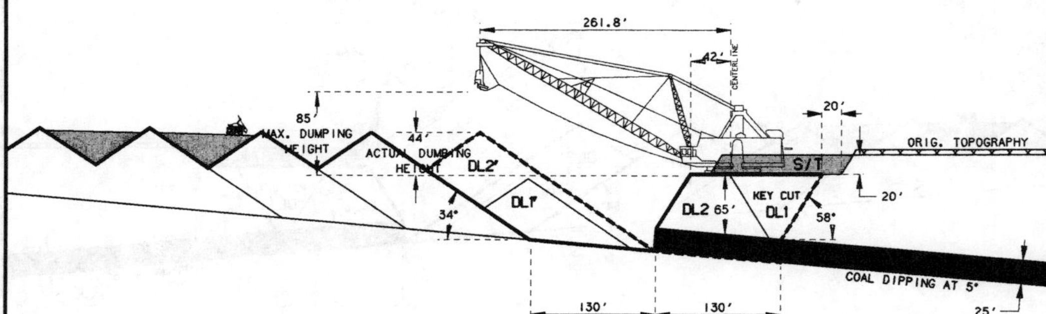

DESCRIPTION	DESIGNATION	VOLUMES B.C.Y.	VOLUMES L.C.Y.	% OF TOTAL
TOTAL CUT VOLUME		422.2	548.9	100.00%
TOTAL VIRGIN S/T PRE-STRIPPING	S/T	102.7	133.5	24.32%
TOTAL VIRGIN DRAGLINE STRIPPING		319.5	415.4	75.68%
DRAGLINE KEY CUT	DL1	137.9	179.3	32.66%
DRAGLINE HIGHWALL SIDE PASS	DL2	181.6	236.1	43.02%

SHOVEL/TRUCK PRE-STRIPPING IN CONJUNCTION WITH DRAGLINE STRIPPING

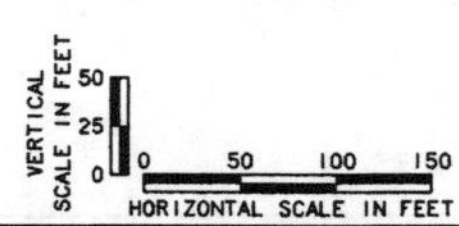

LEGEND

- DOZER OR SHOVEL/TRUCK PRE-STRIPPING
- DL1 DRAGLINE
- COAL
- REHANDLE
- PREVIOUS PROFILE
- TOPSOIL
- FINAL PROFILE

NOTE: STRIPPING DEPTH = OVERBURDEN LESS TOPSOIL DEPTH

Marston & Marston, Inc.
St. Louis, Missouri

COMPARISON OF ALTERNATIVE PRE-STRIPPING METHODS - DOZER VS. SHOVEL/TRUCK AT 85 FT. STRIPPING DEPTH

FIGURE 2

MAY 1992

DOZER PRE-STRIPPING
(CROSS-PIT BENCH)

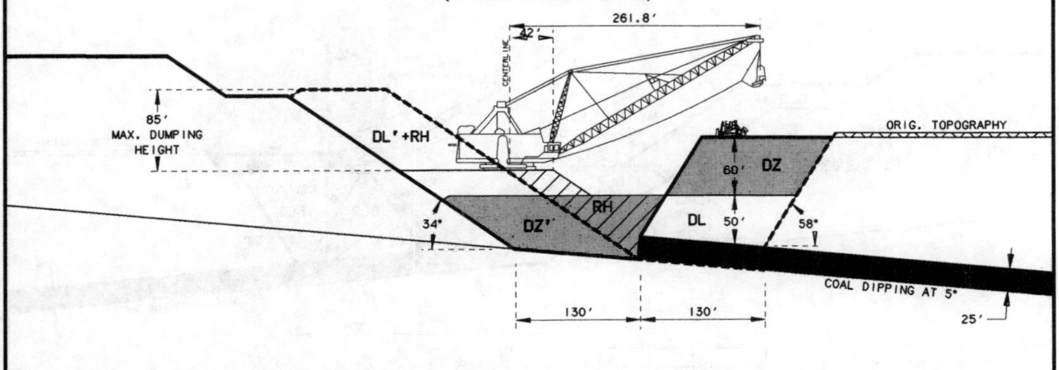

DESCRIPTION	DESIGNATION	VOLUMES		% OF
		B.C.Y.	L.C.Y.	TOTAL
TOTAL CUT VOLUME		549.9	714.9	100.00%
TOTAL VIRGIN DOZER PUSH	DZ	306.0	397.8	55.65%
TOTAL VIRGIN DRAGLINE STRIPPING		243.9	317.1	44.35%
DRAGLINE SPOIL SIDE PASS	DL	243.9	317.1	44.35%
TOTAL REHANDLE		177.8	231.1	32.33%
DRAGLINE REHANDLE	RH	148.1	192.5	26.93%
DOZER REHANDLE		29.7	38.6	5.40%

DOZER PRE-STRIPPING IN CONJUNCTION WITH DRAGLINE STRIPPING

SHOVEL/TRUCK PRE-STRIPPING

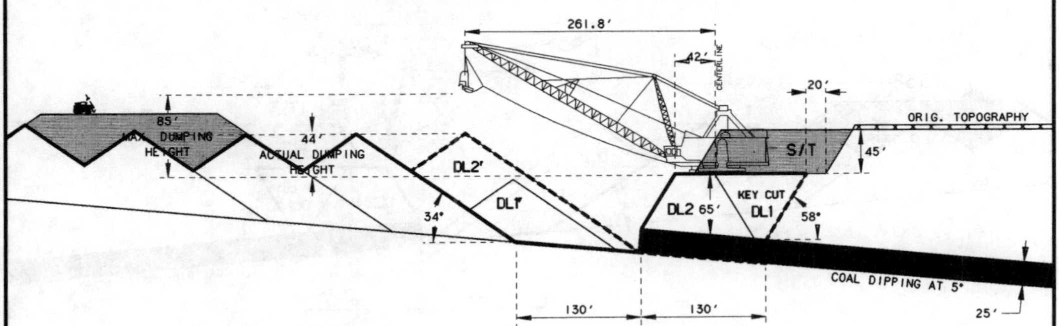

DESCRIPTION	DESIGNATION	VOLUMES		% OF
		B.C.Y.	L.C.Y.	TOTAL
TOTAL CUT VOLUME		549.9	714.9	100.00%
TOTAL VIRGIN S/T PRE-STRIPPING	S/T	230.4	299.5	41.90%
TOTAL VIRGIN DRAGLINE STRIPPING		319.5	415.4	58.10%
DRAGLINE KEY CUT	DL1	137.9	179.3	25.08%
DRAGLINE HIGHWALL SIDE PASS	DL2	181.6	236.1	33.02%

SHOVEL/TRUCK PRE-STRIPPING IN CONJUNCTION WITH DRAGLINE STRIPPING

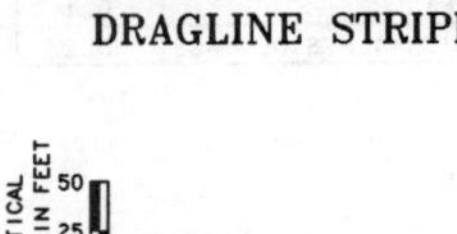

LEGEND

- DOZER OR SHOVEL/TRUCK PRE-STRIPPING
- DL1 DRAGLINE
- COAL
- REHANDLE
- ——— PREVIOUS PROFILE
- TOPSOIL
- - - - - - FINAL PROFILE

NOTE: STRIPPING DEPTH = OVERBURDEN LESS TOPSOIL DEPTH

Marston & Marston, Inc.
St. Louis, Missouri

COMPARISON OF ALTERNATIVE PRE-STRIPPING METHODS - DOZER VS. SHOVEL/TRUCK

AT 110 FT. STRIPPING DEPTH

FIGURE 3

APRIL 1992

DOZER PRE-STRIPPING

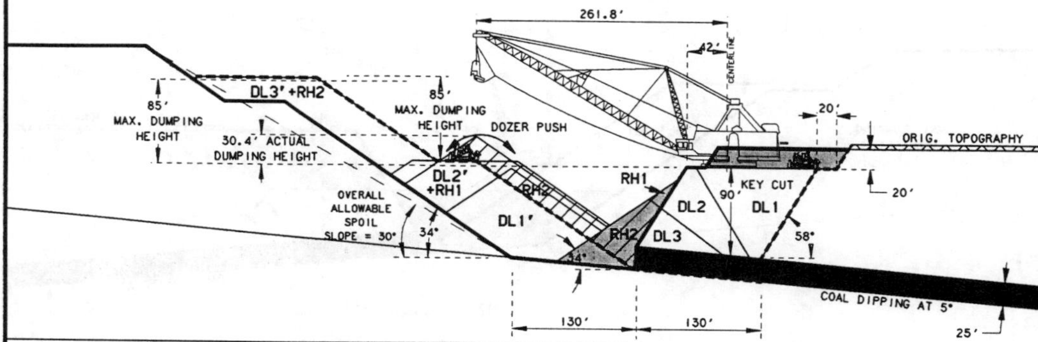

DESCRIPTION	DESIGNATION	VOLUMES B.C.Y.	VOLUMES L.C.Y.	% OF TOTAL
TOTAL CUT VOLUME				
TOTAL VIRGIN DOZER PUSH	DZ	548.7	713.3	100.00%
TOTAL VIRGIN DRAGLINE STRIPPING		103.4	134.4	18.84%
DRAGLINE KEY CUT	DL1	445.3	578.9	81.16%
DRAGLINE HIGHWALL SIDE PASS	DL2	239.5	311.4	43.65%
DRAGLINE SPOIL SIDE PASS	DL3	135.6	176.3	24.71%
TOTAL REHANDLE		70.2	91.2	12.80%
DRAGLINE REHANDLE		161.2	209.6	29.38%
DOZER REHANDLE	RH1+RH2	161.2	209.6	29.38%

DOZER PRE-STRIPPING IN CONJUNCTION WITH DRAGLINE STRIPPING

SHOVEL/TRUCK PRE-STRIPPING

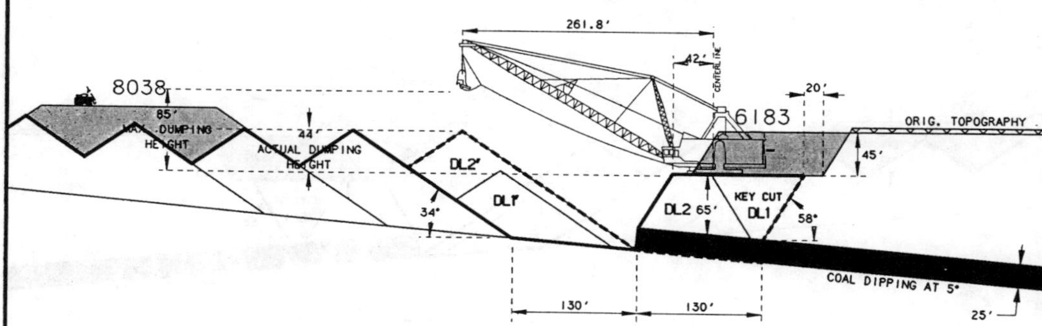

DESCRIPTION	DESIGNATION	VOLUMES B.C.Y.	VOLUMES L.C.Y.	% OF TOTAL
TOTAL CUT VOLUME		545.6	709.3	100.00%
TOTAL VIRGIN S/T PRE-STRIPPING	S/T	229.0	297.7	41.97%
TOTAL VIRGIN DRAGLINE STRIPPING		316.6	411.6	58.03%
DRAGLINE KEY CUT	DL1	136.7	177.7	43.18%
DRAGLINE HIGHWALL SIDE PASS	DL2	179.9	233.9	56.82%

SHOVEL/TRUCK PRE-STRIPPING IN CONJUNCTION WITH DRAGLINE STRIPPING

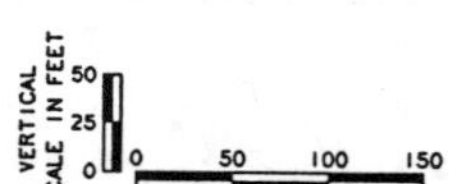

LEGEND

- DOZER OR SHOVEL/TRUCK PRE-STRIPPING
- DL1 DRAGLINE
- COAL
- REHANDLE
- PREVIOUS PROFILE
- TOPSOIL
- FINAL PROFILE

NOTE: STRIPPING DEPTH = OVERBURDEN LESS TOPSOIL DEPTH

Marston & Marston, Inc.
St. Louis, Missouri

COMPARISON OF ALTERNATIVE PRE-STRIPPING METHODS - DOZER VS. SHOVEL/TRUCK AT 110 FT. STRIPPING DEPTH

FIGURE 4

APRIL 1992

DOZER PRE-STRIPPING

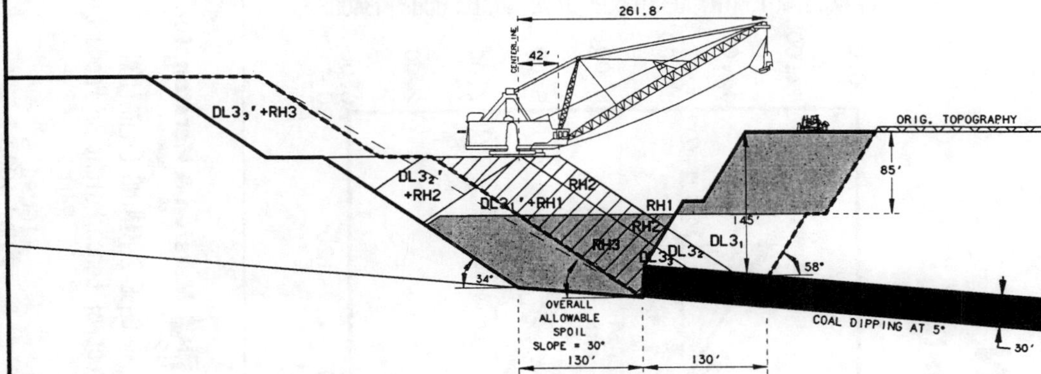

DESCRIPTION	DESIGNATION	VOLUMES B.C.Y.	VOLUMES L.C.Y.	% OF TOTAL
TOTAL CUT VOLUME		738.7	960.3	100.00
TOTAL VIRGIN DOZER PUSH	DZ	443.8	576.9	60.08
TOTAL VIRGIN DRAGLINE STRIPPING		294.9	383.4	39.92
DRAGLINE SPOIL SIDE PASS	DL3	294.9	383.4	39.92
TOTAL REHANDLE		511.6	665.1	69.26
DRAGLINE REHANDLE		465.4	605.0	63.00
DOZER REHANDLE		46.2	60.1	6.26

DOZER PRE-STRIPPING IN CONJUNCTION WITH DRAGLINE STRIPPING

SHOVEL/TRUCK PRE-STRIPPING

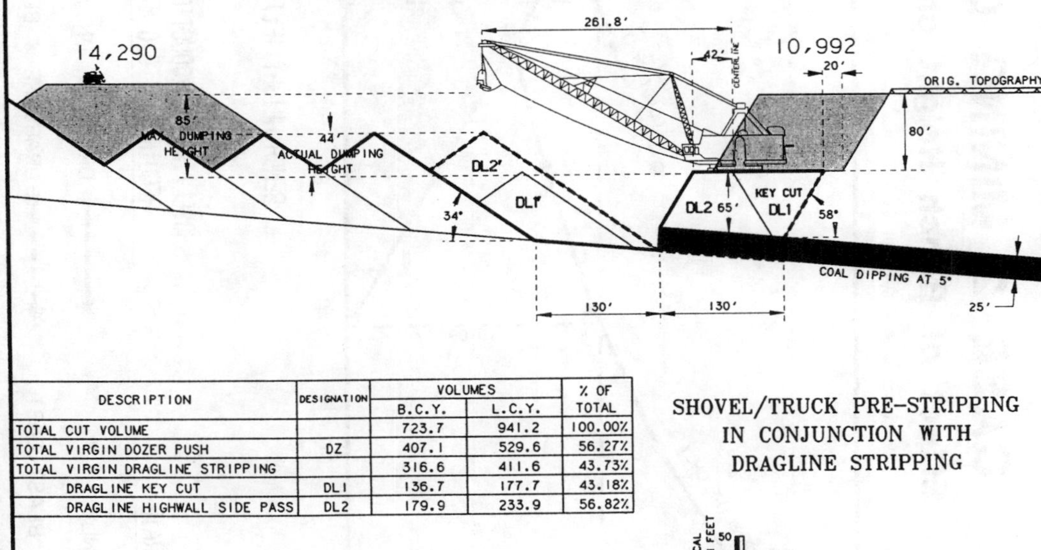

DESCRIPTION	DESIGNATION	VOLUMES B.C.Y.	VOLUMES L.C.Y.	% OF TOTAL
TOTAL CUT VOLUME		723.7	941.2	100.00%
TOTAL VIRGIN DOZER PUSH	DZ	407.1	529.6	56.27%
TOTAL VIRGIN DRAGLINE STRIPPING		316.6	411.6	43.73%
DRAGLINE KEY CUT	DL1	136.7	177.7	43.18%
DRAGLINE HIGHWALL SIDE PASS	DL2	179.9	233.9	56.82%

SHOVEL/TRUCK PRE-STRIPPING IN CONJUNCTION WITH DRAGLINE STRIPPING

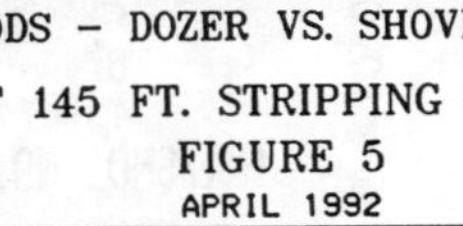

LEGEND

DOZER OR SHOVEL/TRUCK PRE-STRIPPING

DL1 DRAGLINE

COAL

REHANDLE

PREVIOUS PROFILE

TOPSOIL

FINAL PROFILE

NOTE: STRIPPING DEPTH = OVERBURDEN LESS TOPSOIL DEPTH

Marston & Marston, Inc.
St. Louis, Missouri

COMPARISON OF ALTERNATIVE PRE-STRIPPING METHODS - DOZER VS. SHOVEL/TRUCK AT 145 FT. STRIPPING DEPTH

FIGURE 5

APRIL 1992

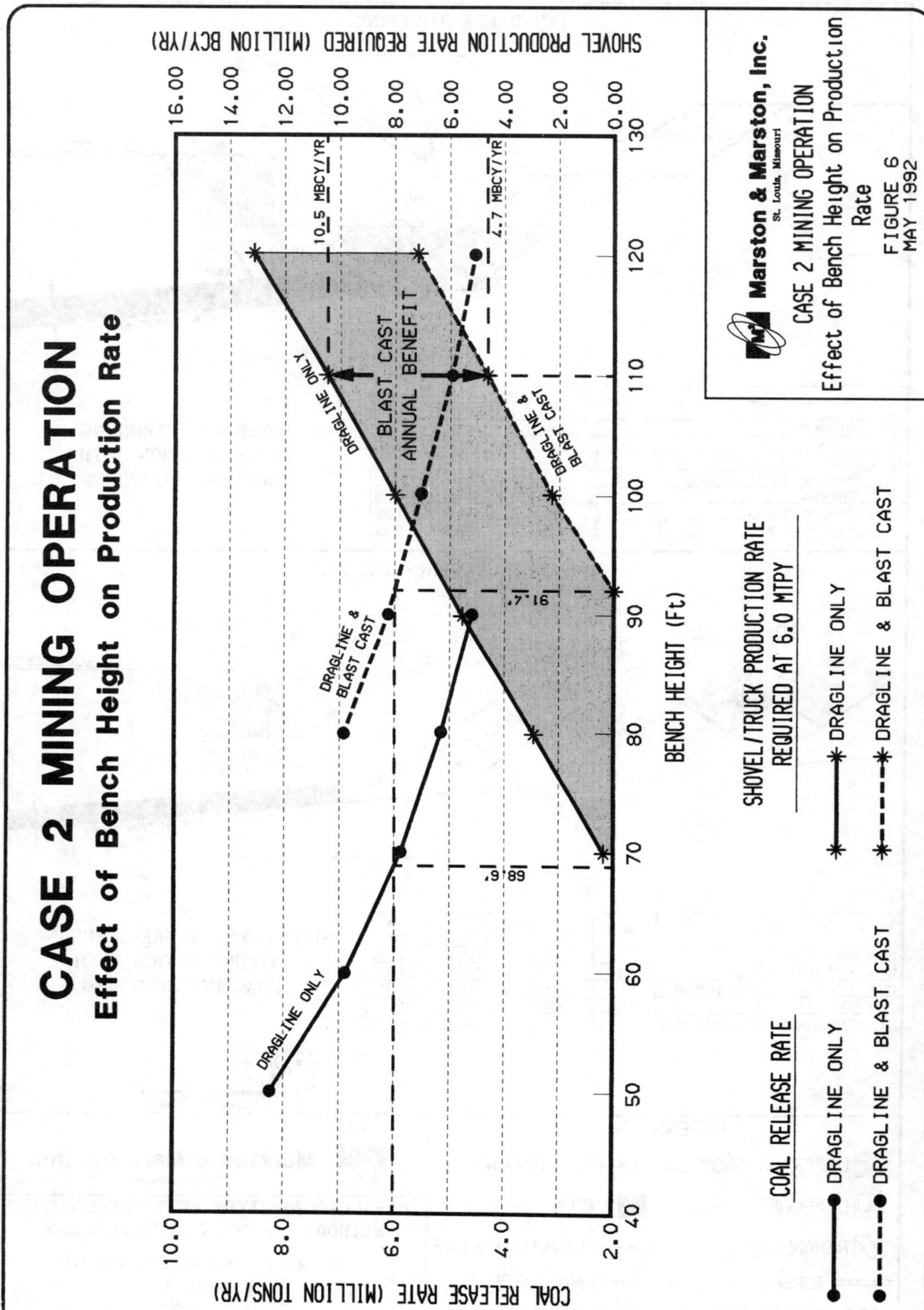
CASE 2 MINING OPERATION
Effect of Bench Height on Production Rate
COAL RELEASE RATE (MILLION TONS/YR)
SHOVEL PRODUCTION RATE REQUIRED (MILLION BCY/YR)
BENCH HEIGHT (Ft)
10.0
8.0
6.0
4.0
2.0
16.00
14.00
12.00
10.00
8.00
6.00
4.00
2.00
0.00
40
50
60
70
80
90
100
110
120
130
DRAGLINE ONLY
DRAGLINE & BLAST CAST
BLAST CAST ANNUAL BENEFIT
10.5 MBCY/YR
4.7 MBCY/YR
68.6'
91.4'
COAL RELEASE RATE
DRAGLINE ONLY
DRAGLINE & BLAST CAST
SHOVEL/TRUCK PRODUCTION RATE REQUIRED AT 6.0 MTPY
DRAGLINE ONLY
DRAGLINE & BLAST CAST
Marston & Marston, Inc.
St. Louis, Missouri
CASE 2 MINING OPERATION
Effect of Bench Height on Production Rate
FIGURE 6
MAY 1992

SHOVEL/TRUCK AND DRAGLINE STRIPPING – 150 FT. PIT WIDTH
OVERBURDEN DEPTH – 90 FT.

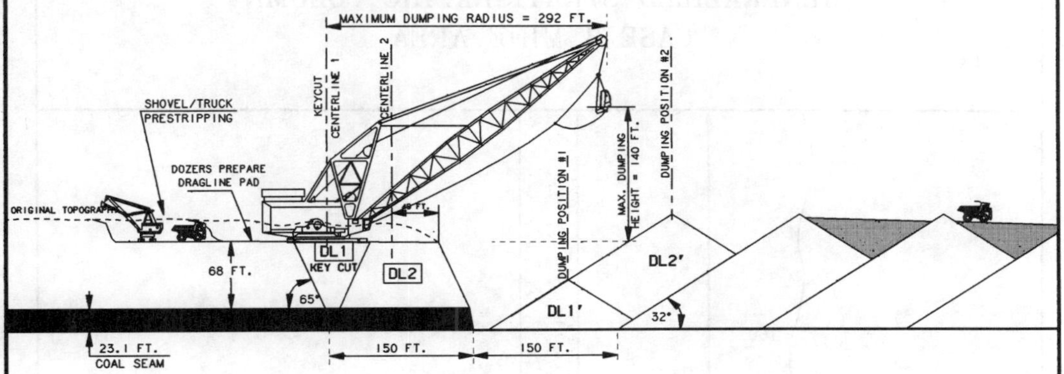

DESCRIPTION	DESIGNATION	VOLUMES		% OF TOTAL
		B.C.Y.	L.C.Y.	
TOTAL CUT VOLUME		504.0	645.1	100.00%
TOTAL VIRGIN S/T PRE-STRIPPING		123.0	157.4	24.40%
TOTAL VIRGIN DRAGLINE STRIPPING		381.0	818.9	75.60%
DRAGLINE KEYCUT	DL1	120.0	153.6	23.80%
DRAGLINE HIGHWALL SIDE PASS	DL2	261.0	334.1	51.80%

BLAST CAST STRIPPING – 150 FT. PIT WIDTH
OVERBURDEN DEPTH – 90 FT.

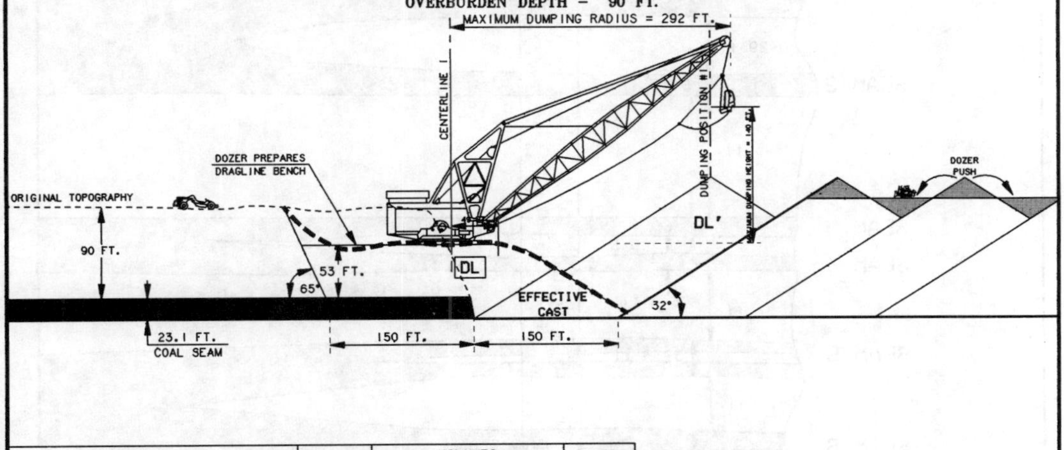

DESCRIPTION	DESIGNATION	VOLUMES		% OF TOTAL
		B.C.Y.	L.C.Y.	
TOTAL CUT VOLUME		504.0	645.1	100.00%
TOTAL DRAGLINE STRIPPING	DL	377.5	483.2	74.90%
EFFECTIVE CAST		126.5	161.9	25.10%

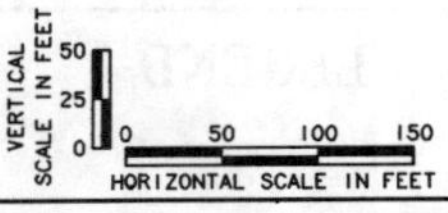

LEGEND

- DOZER OR SHOVEL/TRUCK ROUGH GRADING SPOIL
- BLAST CAST PROFILE
- DL1 DRAGLINE CUT DESIGNATION
- COAL
- ORIGINAL TOPOGRAPHY

Marston & Marston, Inc.
St. Louis, Missouri

COMPARISON OF ALTERNATIVE
STRIPPING METHODS
OVERBURDEN DEPTH – 90 FT.
FIGURE 7

MAY 1992

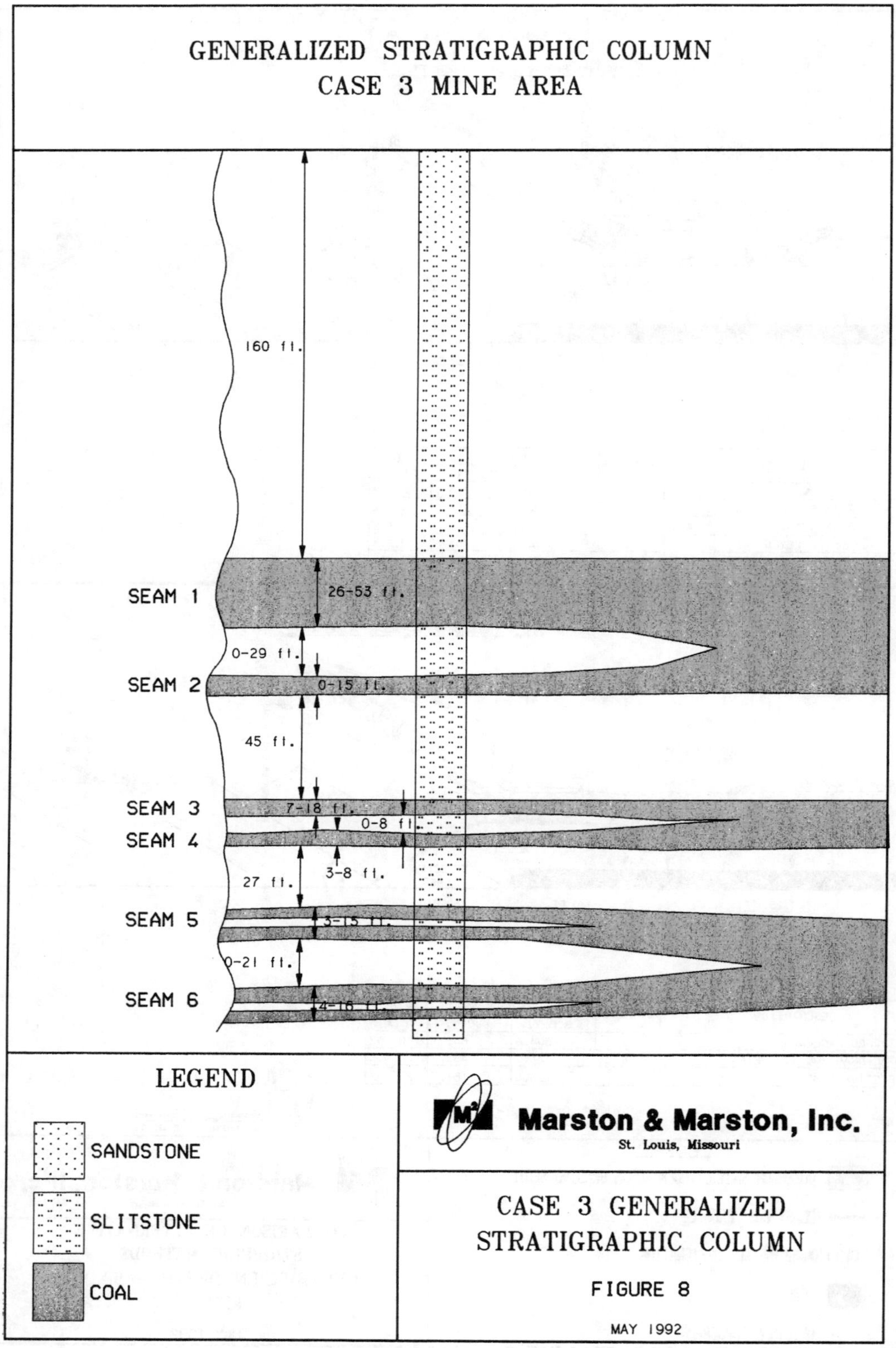
GENERALIZED STRATIGRAPHIC COLUMN
CASE 3 MINE AREA
160 ft.
SEAM 1
26-53 ft.
0-29 ft.
SEAM 2
0-15 ft.
45 ft.
SEAM 3
7-18 ft.
0-8 ft.
SEAM 4
3-8 ft.
27 ft.
SEAM 5
3-15 ft.
0-21 ft.
SEAM 6
4-16 ft.
LEGEND
SANDSTONE
SLITSTONE
COAL
Marston & Marston, Inc.
St. Louis, Missouri
CASE 3 GENERALIZED
STRATIGRAPHIC COLUMN
FIGURE 8
MAY 1992

	DESIGNATION	VOLUMES		% OF TOTAL
		BCY	LCY	
TOTAL VIRGIN CUT VOLUME		1758.9	2172.2	100.00%
TOTAL VIRGIN S/T STRIPPING	S/T1 + S/T2 + S/T3	263.5	324.6	14.98%
TOTAL VIRGIN DRAGLINE STRIPPING	C1+C2+C3+C4+C5+C6+C7	1495.4	1846.8	85.02%
TOTAL DRAGLINE REHANDLE	R1+R2+R3+R4+R5+R6	2087.1	2577.6	118.66%

$$\% \text{ DRAGLINE REHANDLE} = \frac{2087.1 \text{ bcy}}{1495.4 \text{ lcy}} \times 100 = 139.6\%$$

MULTI-PASS DRAGLINE STRIPPING METHOD
DRAGLINE BENCH HEIGHT AT 140 ft.

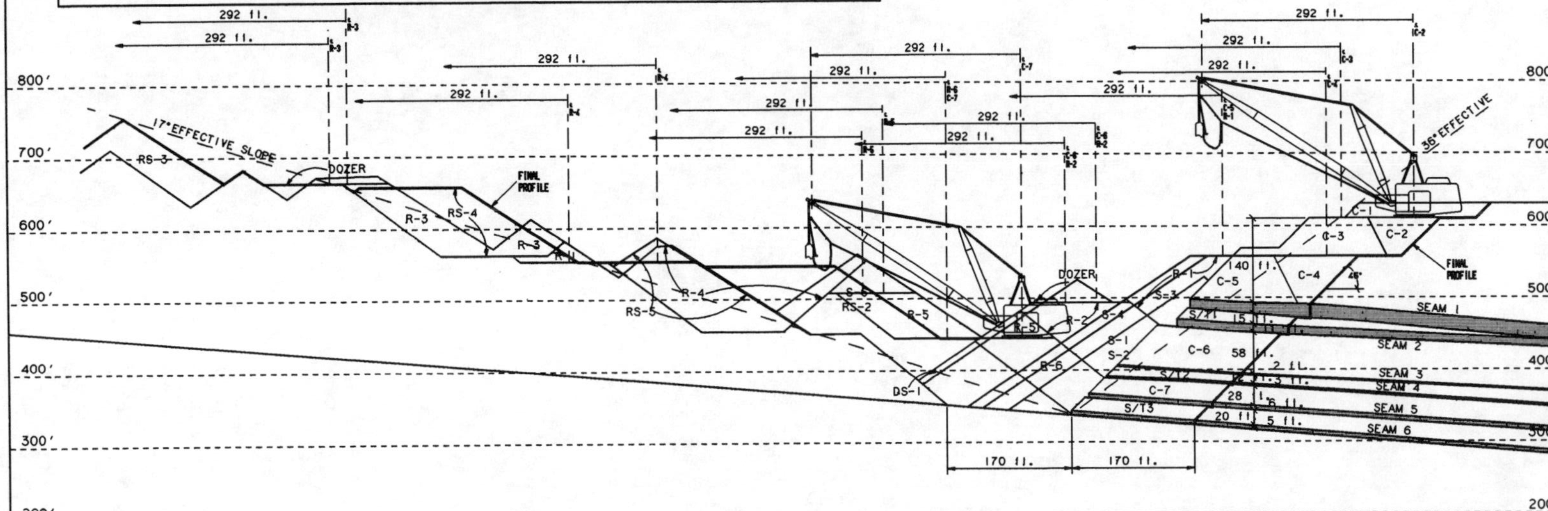

Marston & Marston, Inc.
St. Louis, Missouri

MULTI-PASS DRAGLINE STRIPPING METHOD
AT 140 ft. STRIPPING DEPTH

FIGURE 9

MAY 1992

0 50 100 150 200
SCALE IN FEET

Innovative Mine Design for the 21st Century, Bawden & Archibald (eds) © 1993 Balkema, Rotterdam, ISBN 90 5410 325 6

Fording River Operations' Eagle Mountain Project

D.A. Kennedy & R.A. Thibault
Fording Coal Limited, Fording River Operations, Elkford, B.C., Canada

ABSTRACT: Fording Coal Limited's Fording River Operations has been producing coal for export since 1972 and is today the largest metallurgical coal producer in Canada. In the early 1980's, Fording introduced major revisions to its mine plan and undertook an ambitious program to develop the Eagle Mountain Project. Floating cone analysis was applied to the large Eagle 3-D block model with a total reserve of 357 million metric tonnes clean coal (MTCC) to derive a relatively insensitive design comprising 181 million MTCC in three distinct volatile products. Objectives of the mine plan and schedule include production rates of six million MTCC per year, consolidated access and infrastructure, large efficient work areas, and flat or downhill waste hauls. All objectives are consistent with the application of economies of scale. A steadily decreasing strip ratio ensures future viability in the face of global oversupply and weak prices.

1 INTRODUCTION

Fording Coal Limited was incorporated in 1968 to develop the extensive land holdings of CP Rail in western Canada. Originally, there was a 60% and 40% partnership between CP Limited and Cominco, respectively. Fording Coal Limited is today a wholly owned subsidiary of CP.

The Fording River Operations, located 26 km north of Elkford, in southeastern British Columbia (Figure 1), commenced production in 1972 with a 15 year 3.0 million metric tonne clean coal (MTCC) per year metallurgical contract with the Japanese Steel Mills. FRO is currently the single largest metallurgical coal mine in Canada with production capacity exceeding 6 million MTCC per annum.

Corporately, Fording Coal Limited has expanded to include a contract thermal coal operation (Whitewood, Alberta), a joint venture thermal coal operation/powerplant (Genesee, Alberta) and a contract overburden stripping operation at the Fort McMurray Alberta oil sands. In December 1992, the Greenhills Operations, a 3 million MTCC metallurgical producer adjoining the Fording River Operations to the south, was acquired. Coal and gas leases throughout western Canada are controlled from the corporate office located in Calgary, Alberta. A subsidiary, NYCO Minerals Inc., mines industrial minerals wollastonite and tripoli from its U.S. base near Willsboro, New York.

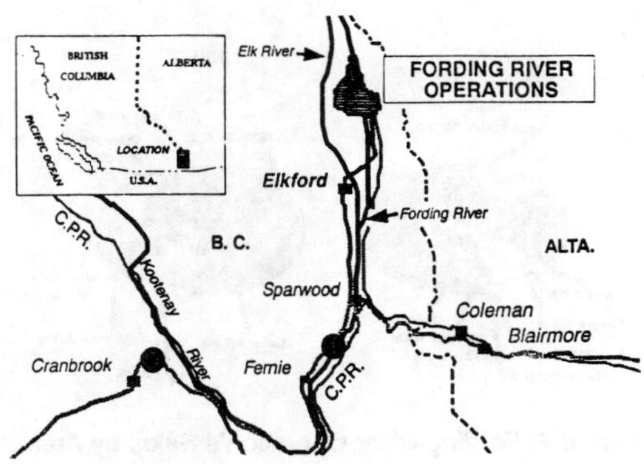

Figure 1. Map

1.1 Geology

The general geology of the Fording River Operations consists of two north-south trending synclines in the non marine Kootenay Group, with one on each side of the Fording River (Figure 2). These synclines are separated by the Ericson fault, a normal fault with a vertical displacement of 300 metres up to the east. The Fording River Valley is a result of the erosion of the intervening anticline along the locus of the Ericson fault.

The majority of rock in the mining areas is

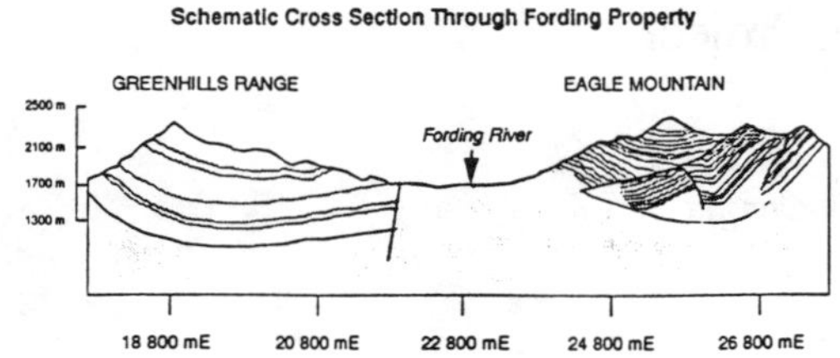

Figure 2. Schematic Cross Section Through Fording Property

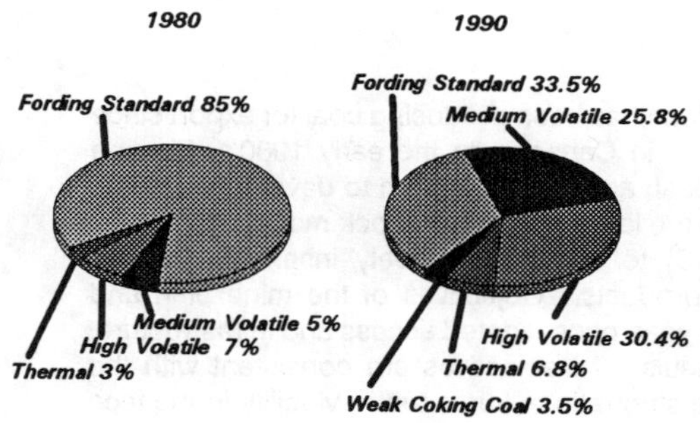

Figure 3. Fording River Operation's Sales by Product

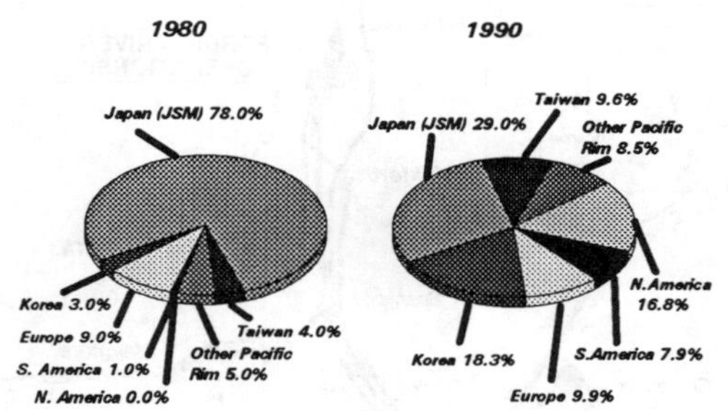

Figure 4. Fording River Operation's Sales by Area

comprised of a 500 m thick sub unit of the Kootenay Group known as the Mist Mountain Formation. Rock within this formation range from mudstones with uniaxial compressive strengths of 25 MPa to coarse grained sandstones with compressive strengths as high as 200 MPa. Joint sets are approximately orthogonal to bedding with spacing ranging from 10 cm to several metres. Bedding thickness is extremely variable.

Over eighteen mineable seams varying from 0.6 m to 10 m thickness are mined on the property. Coal seams exist at dip angles from 0° to 45° requiring a range of mining techniques. The coals range in rank from medium to high volatile bituminous (19 - 32% volatiles:adb), representing the largest range of coals from a single property in Canada.

Total property reserves consist of 470 million MTCC in the proven, probable and possible categories. Approximately 60% of the 360 million total reserve within Eagle Mountain are classified as proven and probable.

1.2 Markets

Fording River Operations commenced production in 1972 with a 3.0 million tonne per year contract for Fording Standard product (a blend of the lowest volatile coals), exclusively with the Japanese Steel Mills (i.e.: single product, single customer). Although the Fording Standard product still comprised 85% of sales in 1980, product diversification was significant by 1990, with the Fording Standard product comprising about one third of sales. Medium and high volatile coal from the diverse geological base, as well as a thermal and weak coking coal were marketed thus ensuring total sales were maximized (Figure 3).

The customer base was diversified in the same time frame. The Japanese steel mills still accounted for 78% of sales in 1980, but were reduced to only 29% of sales by 1990. Since 1980, significant markets have been developed in Korea, Taiwan, South America, Europe and the United States (Figure 4).

1.3 Mining

A Marion 8400 walking dragline with a 46 m^3 bucket has been used, since 1972, to mine primarily Fording Standard in subcrop pits. This machine is the largest single piece of mining equipment operating in a mountaineous environment in North America. A truck/shovel fleet has historically provided prestripping capacity for the dragline in the relatively complex mountainous terrain. The flexibility of the truck/shovel fleet has also provided additional production capacity. More recently, the flexibility of truck/shovel mining has supported the multi staged development of Eagle Mountain to enable the concurrent release of all three volatile products. The combined dragline and truck/shovel fleets have typically moved 50 million bank cubic metres of waste (BCMW) annually.

With the majority of mining activity now concentrated on Eagle Mountain, the fleets have changed significantly as economies of scale have been applied over the last decade. The truck/shovel fleet has recently been responsible for about 80% of the coal and waste production at Fording River Operations. Figure 5 lists a comparison between the fleets used in 1983 and that of 1993.

The Marion 8400 walking dragline removes both blasted rock and coal from mining cuts where coal seams are relatively flat lying. Mining cuts can be as deep as 45 metres. The dragline mining sequences are very complex, requiring detailed planning to control rehandle and spoil room.

The truck/shovel operation mines blasted waste from 15m high benches. The shovels toe and clean hanging wall waste, exposing coal that is on the footwall. The coal seam is then pushed off the footwall by dozers and handled by loaders into coal haulers. Parting material or footwall is then removed to expose more coal. Waste rock is hauled to designated waste dumps.

2 MINE PLANNING

The long range mine plan has evolved over a period of time, changing in response to geology, market conditions, mining methods and corporate goals. Originally, the Marion 8400 dragline was utilized to contour mine on the east flank of the Greenhills Range with truck/shovel activity envisioned as a series of perimeter pits around the base of Eagle Mountain. The approach provided relatively low strip ratios initially and a blend of predominantly Fording Standard coal, consistent with initial contracts.

By the late 1970's however, conditions had changed. Pushbacks of the dragline cuts implied higher strip ratios, deeper cuts, adverse water conditions, and increased truck/shovel prestripping with increasing waste hauls. An increased demand for higher volatile coal and the potential to expand production led to a review of Eagle Mountain. Initially, the planned approach was to continue with the perimeter pits and underground mine the mountain core. In 1978, an underground adit was initiated in No. 15 Seam, the uppermost major high volatile seam on Eagle Mountain. The adit provided markets with an indication of high volatile blend potential and Fording River Operations with a review of underground mining technology. By early 1981, the underground methodology was abandoned in favour of open pit mining due to poor resource recovery, lack of expertise with the technology, and risk. The open pit method was deemed to provide the lowest possible unit cost mining due to economies of scale while providing flexibility for blending and production rates.

2.1 The Eagle Mountain Project

The decision to mine Eagle Mountain by open pit methods was made in a relatively short period of time. Although drill density was limited, overall seam structure was reasonably well understood and seam continuity was expected to be good. Preliminary plans were developed and in 1982-83 access, power and initial prestripping was undertaken. Apart from

1983	1993	Equipment Type	
		Shovels	
2	--	BE 280B	11.5m^3 bucket
3	1	P&H 2100	11.5m^3 bucket
3	4	P&H 2800	23.0m^3 bucket
--	1	Marion 301M	44.0m^3 bucket
		Draglines	
1	--	Lima 2400	3.8m^3 bucket
1	1	Marion 8400	45.8m^3 bucket
		Haulage Trucks	
10	--	Unit Rig M36-Waste	154 tonne
7	--	Wabco 120B-Waste	108 tonne
31	23	Wabco 170D-Waste	154 tonne
--	12	Haulpak830E-Waste	218 tonne
14	--	Wabco 120B - Coal	108 tonne
--	10	Wabco 170C - Coal	154 tonne
		Front End Loaders	
5	--	Letourneau 800	11.5m^3
--	4	Caterpillar 992C	15.0m^3
--	1	Caterpillar 994	35.0m^3
		Drills	
3	--	BE-45R	25cm diameter
3	5	BE-60R/61R	31cm diameter
1	--	Drillteck D80K	31cm diameter

Figure 5. Heavy Equipment Listing (For 1983 and 1993)

the physical difficulties attributed to mountainous terrain and climate conditions, the biggest challenge was to increase overall property productivities to counter the high strip ratios inherent in the Eagle development. The difficulties were exacerbated by the economic recession and falling coal prices.

The venture was successful. The first high volatile coal was hauled from Eagle Mountain in August 1984 following more than 2 years of prestripping activity. Property clean coal strip ratios peaked by 1985 at 11.5:1 (BCMW/MTCC) as increased emphasis was placed on the Eagle development, and have decreased steadily since that time.

2.2 The Eagle Mountain Long Range Mine Plan

The longevity of the Eagle Mountain project was recognized from the onset, although geological confidence was lacking at depth. Extensive exploration programs returned a total of 40,000m between 1980 and 1983 substantially increasing the geological database.

In 1983, a mine planning software package was purchased and utilized in conjunction with the existing in-house geological ore reserve evaluation system (drillhole database). Initially a mainframe version, the planning system has been converted to a PC based version in recent years. The current 3-

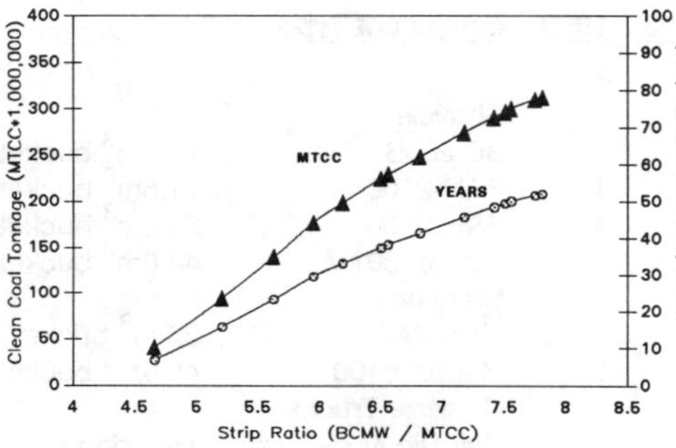

Figure 6. Eagle Mountain Clean Coal Tonnage/Years versus Strip Ratio

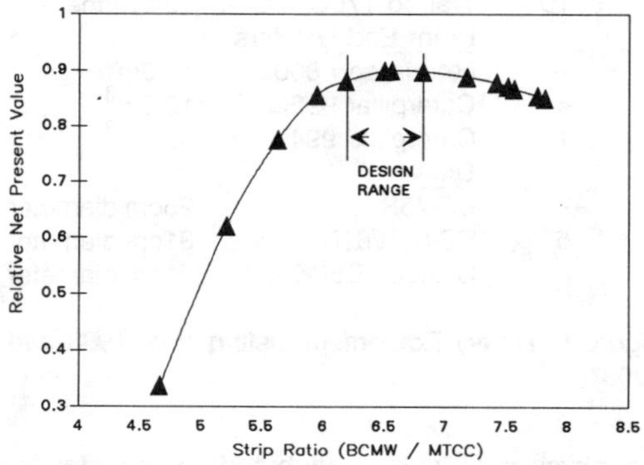

Figure 7. Eagle Mountain Net Present Value versus Strip Ratio

D block model is comprised of approximately 662,000 active blocks sized at 25m x 25m x 15m vertical (corresponding to bench height), representing a model 3,250m N x 4,000m E x 780m vertical.

Conceptually, the long range plan has not changed significantly since inception in the early 1980's, a tribute to the geological integrity, consistency of objectives, and foresight of early planners. An ongoing exploration program has added detail to the geological model and although no major differences have been noted, adjustments have been made particularly to areas east of the main Eagle syncline where the two major thrust faults have been refined. Mining parameters within the model have been adjusted to reflect actual experience on Eagle. Currently, the long range block model utilizes a 0.9m minimum mineable seam thickness, 0.7m minimum removable parting, 0.3m mining loss and 0.3m dilution. Clearly, these represent reconciled averages given the range of seam and interburden qualities and dips.

The objectives of planning have been consistent with the goals of economies of scale, high productivity, and longevity:

1. Relatively detailed, long range mine plan for market and workforce security and capital payback.
2. Systematic release of coal products with a degree of flexibility.
3. Large consolidated mining areas with common access and infrastructure.
4. Level or downhill waste hauls.
5. Steadily declining strip ratio.

The initial manual long range mine plan generated in 1983 yielded 216 million MTCC at a clean coal strip ratio of 7.1:1 (BCMW/MTCC). The plan identified several logistical requirements including averaging the barren waste above 15 seam and scheduling coal production to correspond to existing and anticipated market demands. The solution, still valid today, was a phased approach to the mining sequence:

1. Accelerate the entire south flank of Eagle Mountain where the coal was higher due to the north trending phase.
2. Mine the peak area overlying the syncline in a series of north-south slices.
3. Develop the Brownie area to the east concurrently.

The long range plan is the subject of ongoing reviews as the geology model is updated and design details are refined. The ultimate pit limits have been adjusted as recently as 1990 following a major model rebuild. In this case, a sensitivity analysis applying a floating cone algorithm was utilized to determine a minimum/maximum positions for the final highwall design. Input parameters included dollars available for mining (net value), waste mining cost, coal mining cost, exchange rate, recovery factor (S.G. x plant yield), current topography and pit slope angle. It was assumed that mining costs would increase proportionally to the coal selling price, negating the effect of inflation. A condensed block model derived from the detailed geological model was utilized to generate equivalent value surfaces by varying coal selling price or net value by a minimum/maximum 20%.

The results of the runs were used to generate reserves and further economic analysis. Figure 6 represents clean coal tonnage versus strip ratio. The resulting years of mining at a nominal production rate of six million MTCC per year is also shown. The relative net present value of these various cases is demonstrated in Figure 7.

Economic pit designs with clean coal strip ratios ranging from 6:2 to 6.8:1 met the initial objectives and provided maximum net present value (NPV). The surfaces provided the basis for more detailed manual designs. The manual designs necessarily included consideration for physical smoothing of pit limits, geotechnical highwall design, access development, and waste volume/spoil balancing. The resulting design contained 180 million MTCC at

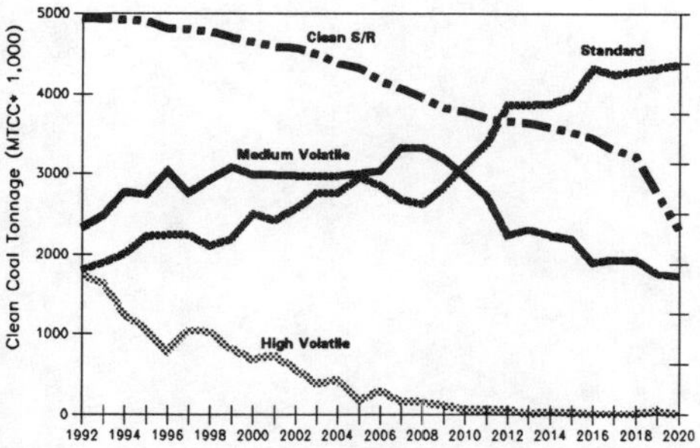

Figure 8. Fording River Operation's Long Range Schedule at 6.2 Million MTCC/Year

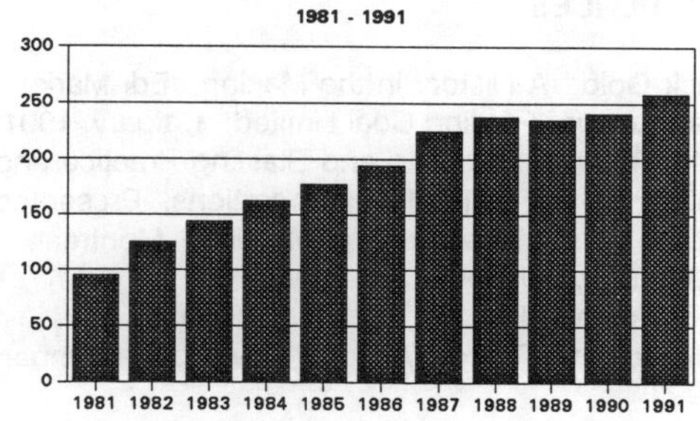

Figure 9. Total Material (BCM) Per Manshift

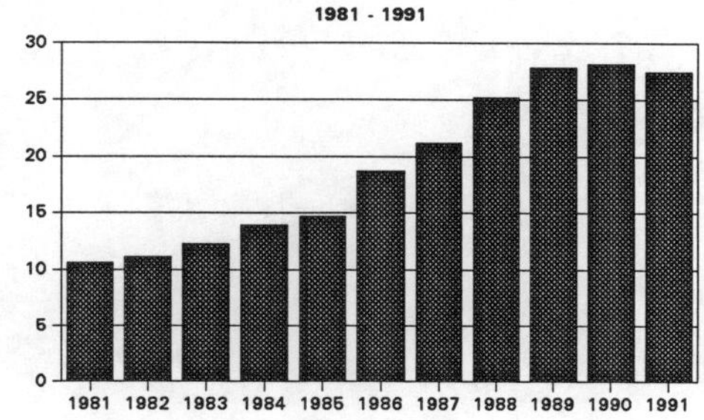

Figure 10. Clean Coal (MTCC) Per Manshift

a clean coal strip ration of 6.3:1.

Minor adjustments have been introduced to the long range plan in recent years. In 1992, reserves including dragline mining until 1997, currently scheduled for production are 181 million MTCC at an average clean coal strip ratio of 6.5:1. This provides 29 years of mining at a nominal scheduled rate of 6.2 million MTCC per year. The substantial remaining property reserves have not been utilized in the existing schedule. Figure 8 demonstrates the steadily declining property strip ratio and the scheduled release of products.

Waste disposal has provided a challenge in Eagle Mountain planning considering the large volume and steep terrain. Large valley fills have been designed to accommodate the 1.1 billion BCMW mined over approximately thirty years. Two primary spoil complexes are utilized: Brownie Spoil to the east and South Spoil to the south. Both spoils are designed to provide flow-through rock drains for the existing creeks with mean annual flows of up to 8 m^3/sec.

Haulroad and spoil designs are integrated with decreasing pit elevations to ensure optimized hauls. The significant scale of mining of the Eagle Mountain project necessitates longer hauls over time as the consolidated mining areas are developed and spoil platforms expand.

3 RESULTS

The benefits of a relatively detailed long term mine plan for Eagle Mountain has allowed for the introduction of new technologies and economies of scale by minimizing risk and providing capital payback. New equipment, mining methods, work standards and a consolidated mine plan have contributed to increased productivity and profitability in an increasingly competitive global market. Figures 9 and 10 document the total material and clean coal productivity per eight hour manshift respectively. Further productivity increases are anticipated with the recent introduction of the next generation of large mining equipment; the Marion 301-M shovel and Haulpak 830E trucks.

Ongoing reviews assure the applicability and currency of the mine plan. Although generating higher strip ratios initially, the steadily declining strip ratio within the plan ensure the continued viability of operation. Total clean coal production exceeded six million MTCC in 1990 and 1991 for the first time ever. Although 1992 was affected by a lengthy labour dispute, projected production for 1993 is again more than 6 million MTCC. Although manpower has decreased exclusively through attrition over the last decade, work force stability and security has resulted from the long term direction.

4 CONCLUSION

The Eagle Mountain long range plan meets the original planning objectives and provides for a reliable, long term supply of metallurgical coking coal. The plan has enhanced Fording River Operations reputation as a competitive, world class metallurgical coal producer.

REFERENCES

Black Gold: A History in the Making. Ed. Mark Ladouceur. Fording Coal Limited: Calgary, 1991.

Gold, Robin D.: Drilling and Blasting Practice and Initiatives at Fording River Operations. Presented at the CIM Annual General Meeting: Montreal, 1992 (unpublished).

Medsystem: User Manual Vol. III Release 10. Mintec Inc.: Tuscon, Arizona, Revised September, 1992.

Innovative Mine Design for the 21st Century, Bawden & Archibald (eds) © 1993 Balkema, Rotterdam, ISBN 90 5410 325 6

Improvement of blasting productivity at Carol Lake

Barbara J. Mossop
Iron Ore Company of Canada, Labrador City, Nfld, Canada
John Pinto
ICI Explosives Ltd, Labrador City, Nfld, Canada

ABSTRACT: To survive in today's economic climate, mining and other industries bear the challenge to reduce costs and increase productivity while maintaining or improving quality. Drilling and blasting is the basic operation upon which all subsequent mining operations are crucially dependant. Continual testing and research is being done at the Carol Project by the operations, engineering and ICI technical personnel. This paper outlines recent innovations and objectives for quality and continuous improvements in the explosives manufacturing and the drilling and blasting operation at the Carol Lake Project.

1 INTRODUCTION

1.1 *I.O.C.'s Carol Lake Project*

The Iron Ore Company of Canada operates an iron ore mine, concentrator and pelletizing plant in Labrador City, Newfoundland, 322 km north of Sept Iles, Quebec. Open pit mining began in 1962 and has been developed to handle an annual production capacity of 40,000,000 tonnes of crude ore.

Geologically the Carol Lake Project lies in the south-west extremity of the Labrador Trough, within the area of the Grenville province of deformation (see Figure 1). The highly metamorphosed sedimentary and volcanic rocks of the Labrador Trough have a high iron content which, in localized areas constitute mineable iron-ore deposits. The ores lie in a stratigraphic sequence from hard, brittle fine grained quartz-magnetite and quartz-magnetite specularite to a softer, more elastic, coarser grained quartz-specularite. The deposits are overturned and tightly folded. The stratigraphy strikes approximately north-south and dips at 40 - 50 degrees to the east. The average grade is 39 percent iron. Country rock and waste inclusions are predominately quartzite and quartz-carbonate.

Currently the ore is mined from two main pits and two smaller pits in 13.7 meter benches.

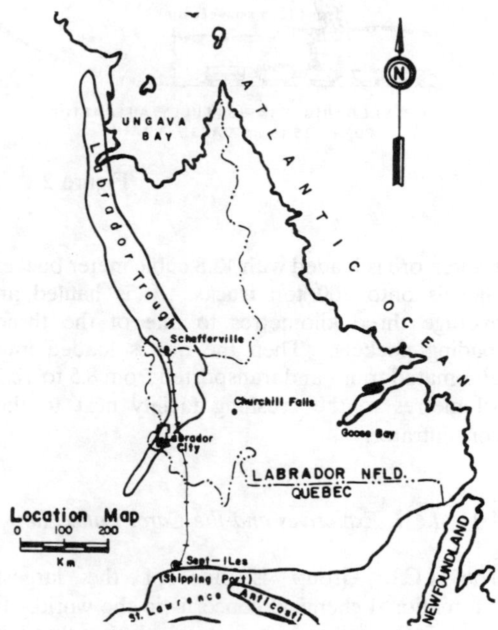

Figure 1 Location Map

Drilling is done with single pass rotary drills using 15 and 16 inch diameter bits. Explosives are manufactured on site and delivered to the blastholes in bulk by ICI Explosives Limited. The

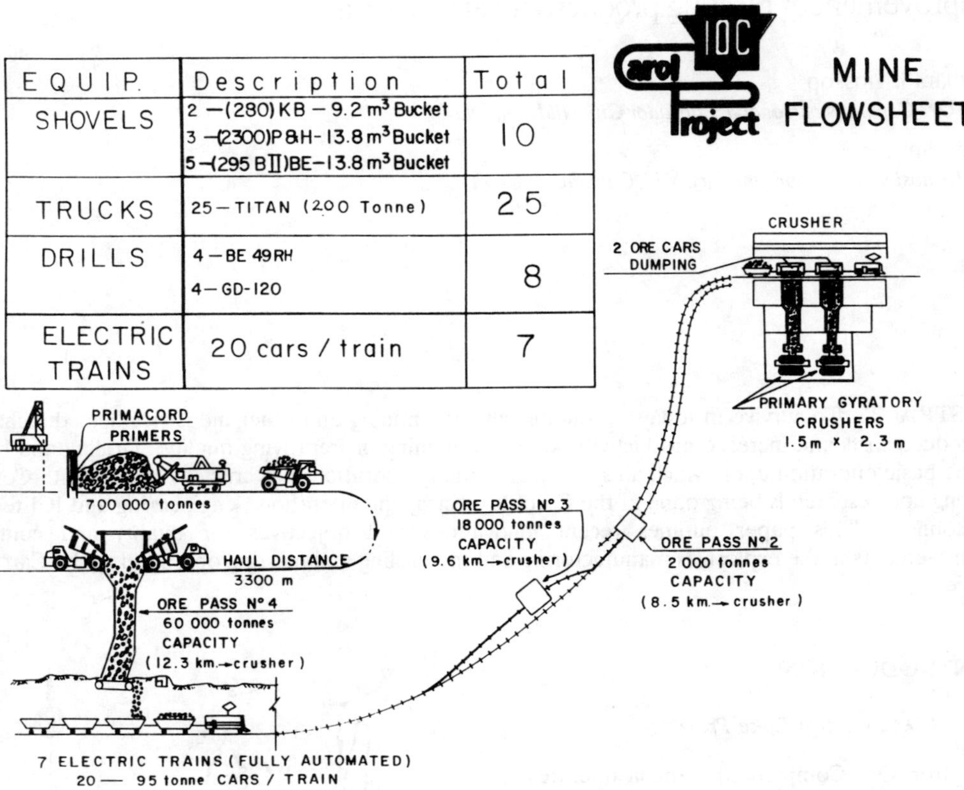

EQUIP.	Description	Total
SHOVELS	2 –(280) KB – 9.2 m³ Bucket 3 –(2300) P&H – 13.8 m³ Bucket 5 –(295 B II) BE – 13.8 m³ Bucket	10
TRUCKS	25 – TITAN (200 Tonne)	25
DRILLS	4 – BE 49RH 4 – GD-120	8
ELECTRIC TRAINS	20 cars / train	7

Figure 2 Mine Flowsheet

broken ore is loaded with 13.8 cubic meter bucket shovels onto 200 ton trucks. It is hauled an average three kilometres to one of the three loading pockets. Then the ore is loaded into automated trains and transported from 8.5 to 12.3 kilometres to the crushing facility next to the concentrator.

1.2 *I.C.I. Explosives and the Carol Lake Plant*

The I.C.I. Group is one of the largest international chemical concerns in the world. It is a multifaceted organization, which aims through research and development to provide a unique and high tech service to its customers.

The ICI Explosives Plant is located at the Carol Lake Project, approximately 6 km from the present open pit mining sites. The plant has been in operation since the mid sixties, then operated by Canadian Industries Limited (CIL Inc.). In 1990, ICI obtained full ownership of CIL Inc. The plant has been modified over the years to accommodate the changing requirements of explosives required for iron ore production blasting at the Carol Lake Project.

2 MINE PLANNING

The mine planning engineers are responsible for both the long and short range planning. A computerized mine planning system designed by Queen's University is being used. It allows for the engineering staff to create and review alternate mining plans in a timely manner.

Mining occurs in four pits: Humphrey Main, Humphrey South, Humphrey North and Lorraine. The active drilled and broken reserves are kept between 4 and 6 million tonnes for ore and between 400 and 750 thousand cubic meters for waste. The blast size and location is based on production and ore blending requirements. The geologist uses basic diamond drill information, face samples and experience from the upper benches to determine the individual blast grades and structural features.

The blasting engineer designs the blast on plan. Consideration into the pattern and bit size is made dependant on the material being blasted and the type of drill to be used. The pattern size ranges from 7.6m x 7.6m to 9.1m x 9.1m. The standard pattern for ore is 8.0m x 8.0m.

Holes are drilled to an average depth of 15.2 meters (13.7 meter bench height plus 1.5 meters of subgrade). The average blast consists of 300 holes and breaks over 600,000 tonnes of material.

3 PRODUCTION DRILLING

Eight drills, four GD-120 and four BE-49RH units drill an average 690 meters per day with either 15 or 16 inch diameter bits. Pulldown plus rotation of the tri-cone bits must provide sufficient force to overcome the compressive strength of the rock. The drills normally operate at a pulldown of 80,000 pounds and a rotation of 80 rpm. Compressed air is passed through the stem for cooling and evacuating the cuttings.

The drill crew consists of 18 drill operators working on four and five crew shifts following a 28 day schedule. Drill operators are required to fill out a drill report each shift to record the meters drilled and drilling time.

The BE 49RH drills are equipped with footage indicators and the drill operators are encouraged to drill to the designed depth to reduce over and under drilling.

3.1 *Recent Changes in Production Drilling*

3.1.1 *Bit Monitoring Program*

In an effort to improve bit performance the current bits were analyzed as to their effectiveness. A simple database stores the information gathered on each bit. The two top performing brands of bits on the project were isolated and given the contract to supply bits for 1992. Performance increased from 733 meters per bit in 1991 to 840 meters per bit in 1992. Likewise the penetration rates increased from 10.1 meters per hour to 10.3 meters per hour.

3.1.2 *Selected Use of 16 Inch Diameter Bits*

In 1992, 16 inch diameter bits were tested in a difficult drilling area of the mine. The average bit life in this area increased from 522 meters per bit with the 15 inch diameter bits to 820 meters per bit with the 16 inch diameter bits. The penetration rates also increased from 11.6 meters per hour to 12.5 meters per hour.

The larger bits were tried in other areas of the mine and had comparable performances to their 15 inch counterparts. In areas where the drilling patterns can be expanded to maintain the powder factor without compromising fragmentation, 16 inch diameter bits are now being used.

3.1.3 *Program Drilling Control*

The BE 49RH drills are equipped with Program Drilling Control (PDC) packages. This package partially automates the drilling process. The pulldown pressure and rpm are set and the hole depths are entered. With PDC, the amount of over and under drilling is reduced. To protect the drill when there is excessive vibration, the PDC will shut it down. A conscientious operator can outdrill the PDC, however the PDC may provide the opportunity for one drill operator to operate multiple drills in an area.

3.1.4 *Electronic Drill Monitoring*

Instrumentation is installed and is currently being debugged on the BE 49RH drills. This instrumentation will automatically monitor the pulldown pressure, bit air pressure, rotation, rotary torque, meters drilled and the drilling and standby times. This information will be recorded at the central dispatch. Accurate penetration rates and drilled meters per bit will be recoverable using this system. The drilling techniques of the best operators will be identifiable and passed on. Overall drilling productivity is expected to increase.

3.2 *Future Trends in Production Drilling*

IOC is actively involved in joint government-industry research projects. One of the current projects that IOC is interested in is with the Mining Industry Council of Canada. It is a study to determine the suitability of using dedicated microwave radio systems for positioning drills. A drill positioning system using microwave radio or satellite based signals (Global Positioning Systems, GPS) would reduce or eliminate the need for blasthole surveying. The benefits of an automated drill positioning system are believed to be:

- reduces surveying and resurveying requirements
- eliminates problems of sighting stakes
- provides consistent accuracy
- provides summary reports showing actual hole locations, date and time drilled, hole depth etc.
- reduces risk of drilling off pattern, thus reduces extra fill in holes and pattern sizes may be increased

The accuracy of current off the shelf microwave radio positioning systems is + or - six inches when in line of site of the microwave beacons. The beacons can broadcast up to 10 km.

Ultimately there would be an automated link between the planning computer models and the drill operator.

4 BULK EXPLOSIVES SUPPLY

Since 1987, bulk emulsion explosives have been used exclusively on a production basis at the Carol Lake Project. Presently the bulk explosives employed at this site is exclusively Magnafrac B-9000,(TM) a non-aluminized bulk emulsion. It has an extremely efficient detonation reaction, high velocity of detonation (VOD) and excellent water resistance. These features make it very suitable for the hard rock, wet borehole blasting predominantly encountered at this mine.

Before 1987 and for a period of about 20 years, Hydromex (TM), a high density, TNT based slurry explosive, was developed and used at this site. However, the escalating cost of raw materials had greatly reduced the cost effectiveness of the product. Magnafrac B-9000 is a cost and energy effective explosive, and proved to be an ideal replacement for Hydromex slurries. Initially aluminized emulsions were used to emulate the energy of aluminized Hydromex. In 1988, by a joint effort between IOC and ICI Explosives, actual field data input processed by ICI's Sabrex blast simulation model proved that aluminum could be eliminated while maintaining the same blasting parameters. This significantly reduced explosives cost without compromising performance.[1]

The Magnafrac B-9000 employed at Carol Lake is a water and oil emulsion. It has an extremely efficient detonation reaction due to the intimate relationship between the fuel and oxidizer, both of which are liquids. Magnafrac B-9000 is a booster sensitive, chemically gassed bubble sensitized emulsion.

MAGNAFRAC B-9000 SPECIFICATIONS:

Average in Hole Density (g/cm^3)	1.25
Velocity of Detonation (m/sec)	5500
Relative Weight Strength (rws)	81
Relative Bulk Strength (rbs)	121
Water Resistance	Excellent

Magnafrac B-9000 is highly suitable for blasting in massive hard rock type geology. Its sleeptime, even in wet conditions, is over three months.

4.1 *Recent Changes in the Explosives Supply*

4.1.1 *Explosives Manufacturing:*

In keeping with the philosophy of customizing the explosives production operation to meet customer requirements, a completely new and revolutionary bulk explosives manufacturing facility was built by ICI Explosives in early 1992. The new state of the art system manufactures emulsion at the rate of 1200 kg/min. The employment of strict computer programmed controls on raw materials flowrates, ensures that the emulsion obtained from ICI's special Elk mixer technology is of the highest quality. Since long sleeptimes of up to three months in the borehole is a requirement to permit the loading of large blasts at Carol Lake, emulsion quality cannot be under estimated.

Also of importance, especially from a cost point of view, is that the new technology permits a one man operation of the manufacturing facility. The production of 26 million kilograms of bulk emulsion was realized in 1992. If required the new manufacturing facility could produce twice this volume, necessitating only the increase in the delivery vehicles employed.

4.1.2 *Explosives Delivery*

A new explosives delivery vehicle was built in

1992. It has a stationery 13 tonne reservoir, and was designed for the Carol Lake Operation. Prior to this, the explosives were mixed on the delivery truck using its modified 11 tonne cement mixer type bowl. This bowl also served as the reservoir for the delivery vehicle. This is an excellent system having proved its value worldwide for over a decade, and is still the logical choice of many mining operations.

Up to 150 tonnes of explosives per day are required by the Carol Lake Mine. Two new explosives pump truck vehicles have replaced the four old style trucks. This makes the bulk explosives part of the operation, more efficient and economical, since it is less labour intensive.

The delivery vehicles designed for Carol Lake, must also be reliable for the extreme cold weather (-40ºC) winter application. The explosives pump trucks are designed to perform under these difficult conditions.

4.1.3 *Safety*

In the quest for continuous improvement ICI have not neglected the safety aspects. Safety devices to control excessive temperatures and pressure hazards have been installed where necessary throughout the manufacturing and delivery systems. Strict maintenance standards for equipment, including an ICI pump maintenance and control program, are employed.

4.1.4 *Business Review and Development*

Joint IOC and ICI business review and development meetings are held on a quarterly basis. These meetings have four main objectives:

- Provide a forum for the exchange of ideas between IOC and ICI blasting Personnel.
- Establish priorities for the projects undertaken.
- Review the results obtained.
- Look for opportunities where improvements can be made.

These meetings are the core of the quality and continuous improvement program endorsed by IOC and ICI.

4.2 *Future Trends in Bulk Explosives*

Recent geological trends show some soft rock or sandy areas. In these section a 30% ANFO doped emulsion may provide more heave and improve diggability. Tests using Magnafrac BR-9075 will be performed in the near future.

MAGNAFRAC BR-9075 SPECIFICATIONS:

Average in Hole Density (g/cm^3)	1.33
Velocity of Detonation (m/sec)	5500
Relative Weight Strength (rws)	88
Relative Bulk Strength (rbs)	131
Water Resistance	Excellent

Magnafrac BR-9075 is particularly suitable for medium, soft and friable wet rock blasting. It can be loaded at higher densities than MAGNAFRAC B-9000, which together with it's higher RWS allows some pattern expansion possibilities. The IOC sleeptime requirements of three months should still be possible with BR-9075. This will be proved in-situ at the Carol Lake Mine.

5 PRODUCTION BLASTING

At the current production level the annual explosives consumption exceeds 20,000 tonnes. The average powder factor is 1.72 kg/cu.m. The drilling-loading-blasting cycle on an average 600,000 tonne blast is approximately four weeks.

The loading of explosives is accomplished by 4 blasters, 4 blasting helpers and a stemming loader operator. Two cordline detonators with one pound primers are placed in each hole, one at grade elevation and the other at mid column. The blasters tape all drill holes to determine the hole depth and to advise the explosives truck driver the amount of product required. Once loading of the holes is complete the blasters tape the collars to ensure that sufficient product has been loaded. The normal delay time between drilling and loading a hole is one day. The least delay possible is essential for maintaining uniform daily explosives costs and for ensuring that the blast is ready for blasting soon after drilling is complete. In the winter the short delays are required especially for the wet holes which will have a tendency to freeze at the collar.

The blast tie-in design is based on the strike of the bedding and joint planes and the position of the most favourable burden. The blasts are initiated using the "V1" cut. In 1992, the blasters tied in and shot an average of 2 production or presplit blasts per week.

5.1 *Recent Changes Made in Production Blasting*

5.1.1 *Increased Blast Size*

The average blast size has grown from 300,000 tonnes to over 600,000 tonnes in the past three years. The ability to have large blasts came when the explosives were given a guaranteed sleep time of 60 days. The benefits of the larger blasts are fewer equipment moves, fewer interruptions to production and fewer boulders originating from the free face (since the amount of free face relative to the blast size becomes smaller). The disadvantages of going to the larger blasts are that the powder factor is higher for the same blast hole pattern and there is a potential for more traffic on the blast site and for holes to buried in the snow or otherwise damaged.

5.1.2 *Detonation Systems Testing*

The standard in-hole and surface delays which have been in use since the early 1980's is 280 milliseconds and 75 milliseconds respectively. Field observations and VOD monitoring have indicated that there is interhole interference during the blast in some areas of the operation. In order to reduce this interference and to reduce the scatter time of the delays, shorter delays (100 ms in the hole and 25 ms between rows) are being tested.

The shock tube detonation systems have better accuracy than the cordline detonators and were tested at this operation. They were found to be unsuitable for the harsh conditions of extreme abrasion and cold and for the larger blast sizes.

6 BLAST EVALUATION

6.1 *Fragmentation Analyses*

Presently, extremely time consuming less than precise grid photogrammetric fragmentation analysis methods are being employed at Carol Lake. In the near future the introduction of a computer based imaging technology for fragmentation analysis, will be developed for Carol Lake. This technique will provide a basis for prompt blast evaluation, and generate more precise fragmentation data for use in the ICI's Sabrex computer blast modelling program.

The advent of computerized controls and monitors on the shovels will generate practical means for evaluating fragmentation. Diggability indexes based on power draw may be created. Bucket factors and cycle times may also serve as valuable measurements. Computers retrieving, storing and analysing the shovel operating parameters will become a useful tool in fragmentation analyses.

6.2 *Velocity of Detonation Recorder (VODR)*

Recently VODR equipment was obtained for the in-situ measurements of explosive performance. The equipment was developed by Blasting Analysis International. Although only limited experience with the VODR has been obtained to-date at Carol Lake, the equipment has remarkable potential. It has the capacity of recording not only multiple holes for the velocity of detonation in the explosive column, but also through the stemming material. The VODR results can also show discontinuities in the explosive due to contamination, loss of confinement or desensitization due to dynamic pressure from previous borehole detonating. Intensive VODR studies in difficult geological zones are planned for this year (1993).

6.3 *Computer Based Three Dimensional Image Technology*

ICI is in the process of developing computer based 3-D imaging techniques. The three dimensional image will enable a more precise study of the face velocity and fly rock than the present two dimensional film permits.

7 FINAL WALL PROGRAM

Presplitting and buffer blasting at the final excavation limits is essential for safety and highwall stability. I.O.C. began a final wall control program in 1975. With controlled blasting techniques, the hangingwall of the pit has been redesigned from 48° to 60° thus creating major savings in stripping costs and ore recovery.

The ultimate design of the Humphrey Main and Humphrey South pits exceed 350 meters in depth. The overall pit slopes on the hangingwall side of these pits are designed to 60 degrees and on the footwall side they are designed to be between 37 and 41 degrees. The basic design is three 13.7 meter stacked benches followed by a 12.5 meter berm. To maintain the steep slopes on the hangingwall side of the pits, preshearing and control blasting techniques are used. Short staggered buffer holes are occasionally used on the footwall side to remove potentially unstable wedges.

Large diameter holes are used for preshear wall control using a modified air-deck technique. To maximize drill productivity, the final wall hole size is dictated by the size of the production blast holes. Fifteen or sixteen inch diameter holes are drilled at a 4.0 meter spacing on the final pit limit to the bench grade. The holes are loaded with 200 kg bulk emulsion explosives. The use of air bags and stemming were deemed to be unwarranted as the open pits are over ten kilometres from the town and the plug is mainly to suppress noise and airblast related complaints.

The advantages of using the large diameter holes over the classic small diameter hole preshearing are as follows:

- simpler technique
- uses production size holes and equipment
- less labour intensive
- less expensive
- eliminates scheduling problems with drilling contractors

The buffer row is responsible for ensuring that the toe of the final wall is pulled during the blast without creating excessive damage to the final wall. The current practice is for either 15 or 16 inch diameter holes (dependent on the hole size used on the main blast pattern) to be drilled 4.0 meters away from the final wall. They are drilled on an 8.0 meter spacing to 1.5 meters below grade. The distance of the buffer row from the production blast is 6.0 meters. The holes are loaded the same as the production holes, usually to a 6.0 or 7.0 meter collar.

7.1 *Current Final Wall Test Work*

Most of the holes drilled at the Carol Lake mining operation are filled with water. This water is believed to confine the charge, thus increasing the energy at the base of the hole and decreasing the energy throughout the upper portion of the column. Dewatering the holes is not a practical solution at this operation since the holes fill as quickly as they are emptied. Polyethylene bags and sewer pipe are being tested for the purpose of providing an even distribution of explosives in the final wall holes.

A minimum 12.7 cm pipe or bag diameter is required in order to maintain the borehole pressure and the appropriate explosives load. The bulk emulsion explosives has a critical diameter of 10.2 cm, which is well within the design. The load is 200 kg with 12 meters loaded.

Bags are simple for loading into the holes. A pocket is included at the bottom of each bag so that rocks can be inserted to weigh it down for ease of loading it into the hole. The strength of the bags are the problem, the bags must be able to withstand a 200 kg load at 68°C, the loading temperature of the explosives.

The sewer pipe is strong enough to handle the load. It has built in connections with seals to aid in the assembly and loading of the hole. The base of the pipe will be sealed with stemming material once the pipe has been loaded into the hole.

8 SUMMARY

New technology, careful monitoring and controls have been combined to improve the productivity of the drilling and blasting operations at Carol Lake.

The BE 49 RH drills have programable drilling control packages and computerized monitoring packages which link to the central mine dispatch. Monitoring bit lives and penetration rates have resulted in significant improvements. In the future blast plans will be designed on computer and automatically fed to either microwave radio or satellite positioning systems for positioning the drills.

The installation of the new ICI emulsion explosives mixing plant has resulted in a threefold increase in the rate of production. This has allowed ICI to reduce the number of delivery vehicles from four 11 tonne vehicles to two 13 tonne vehicles. When 10 delivery truck loads in a shift are required, the demands are met by employing only one plant operator and two delivery truck operators. Thus ICI is capable of supplying IOC economically with quality bulk explosive products for their present and future blasting requirements.

The blasting improvements show up mainly in the effects to the rest of the mining operation. Shovel and truck productivities improve and there is less bridging in the loading pockets and primary crushers. With larger blasts, the overall mining productivity has increased due to fewer equipment moves and production delays due to blasting. Blasts continue to be monitored, evaluated and modified using various evaluation techniques such as borehole velocity of detonation recording equipment, high speed filming and fragmentation analyses. Final wall controlled drilling and blasting has become part of the production process using the simpler airdeck techniques.

REFERENCE

Mason, R., Lemieux, M., & Tremblay D. 1989 Iron Ore Company of Canada (IOCC) Blast Optimisation with Emulsion Explosives. Proceedings of the Society of Explosives Engineers 14th Conference of Explosives and Blasting Technologies.

Innovative Mine Design for the 21st Century, Bawden & Archibald (eds) © 1993 Balkema, Rotterdam, ISBN 90 5410 325 6

Considerations in the selective placement of stripmine overburden

J. Lyall Workman
Calder & Workman, Inc., Washburn, N.Dak., USA

ABSTRACT: When stripping to uncover coal, or other mineral commodities, large quantities of waste strata are excavated and dumped in a previously mined portion of the pit. During this process the material is mixed and frequently overturned in stratigraphic sequence. A variety of problems may result from near surface boulders to acidity, alkalinity, sodicity and alteration of groundwater flow patterns. Situations exist whereby maintaining the dumped strata in a similar sequence to that found in the undisturbed ground would be beneficial. Similarly, eliminating boulders or placing them in a more suitable location would assist the reclamation effort. Mining methods are discussed which can provide such selective mining results. These are methods that are in use, or are adaptations of such systems. Methods suitable for use in various geological and topographical settings are provided.

1 INTRODUCTION

While regulations may vary, land reclamation is an important goal whenever stripmining is performed. This is especially true in North America where area mining to recover coal has been carefully scrutinized and regulated for more than twenty years. Further, reasonable expectations are for additional examination and regulation in this decade.

A variety of problems may exist with the waste rock removed from above a seam deposit (Workman, 1981). These include acidity, alkalinity, sodicity, trace metals, altered groundwater flow patterns, compaction and boulder production. Different areas of the United States and Canada experience these problems to different degrees.

Typical mining methods may exaberate one or more of these problems by mixing and overturning the spoil material. Interburdens are often deposited above the initial overburden, as is frequently the case in multiple seam dragline operations. In general, mining methods and equipment are selected for reasons other than how the waste may be deposited in the dumps. However, integrating overburden removal and subsequent reclamation and environmental requirements, through the appropriate selection of equipment and mining techniques, can be beneficial when specific sequences of waste deposition are desired to avoid subsequent environmental problems.

Where undesirable material exists that will adversely affect reclamation efforts two broad classes of remedial action are available.

1. Mitigate the problem after mining by using various chemical and/or physical amendment techniques.

2. Resolve the problem during mining by selective placement of the deleterious strata.

It is the author's view that selective placement must generally be considered on a macro scale. Some prior studies have examined the placement of certain strata in specific locations within the spoils (Dolhopf, et al, 1977). Material was deposited in thin layers with equipment such as scrapers. This approach can be useful in handling small isolated lenses of material. If applied to a full scale mining operation the approach would be very costly. However, much can be done with the large scale selective handling of waste, at more reasonable costs.

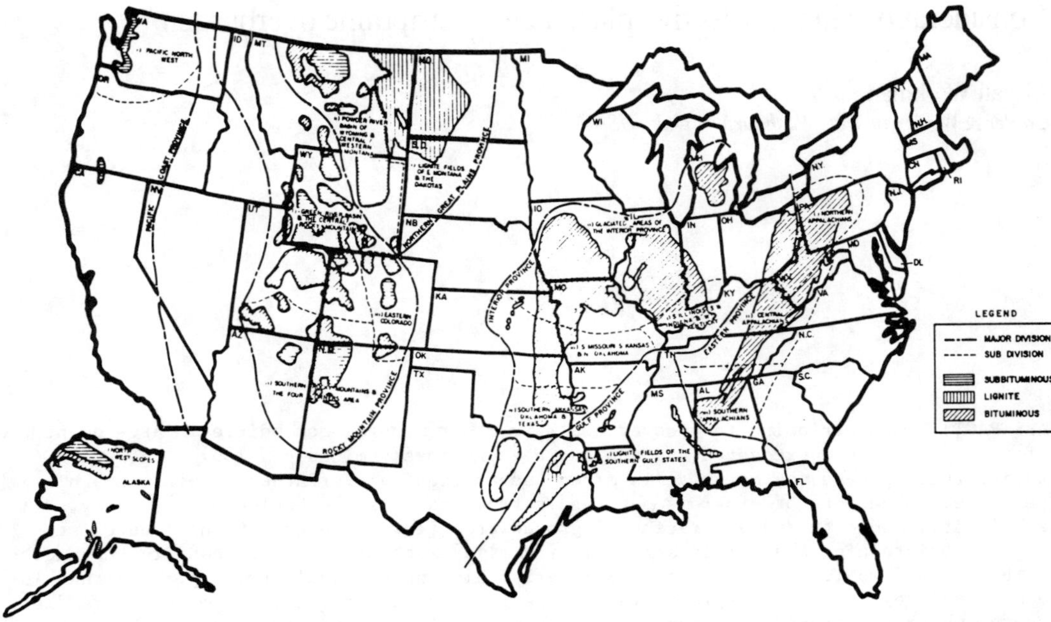

Fig.1 Classification of U.S. coal fields by geographical region according to overburden characteristics

The cost associated with selective placement is an important consideration.

If a method is prohibitively expensive it is not useful today no matter how technically superior the technique may be. Where methods may be used that are nearly the same as those already in use the costs may be comparable. When mining methods must be substantially changed costs may rise significantly, but this is not necessarily true. It is important to consider the comparative costs of remediating a given reclamation problem. It may be that selective handling raises mining costs but reduces reclamation cost to an extent that offsets the increased capital and operating costs.

It is true in the U.S., with extensive coal fields throughout the nation, that the different regions have different reclamation problems and different topographic and geologic settings. This is indicated in figure 1 which shows the coal mining regions of the United States. Methods of selective handling must be matched to the condition in the different regions if a successful result is to be achieved.

2 THE PROBLEM

Study of the major coal mining regions of the U.S. shows that there are three principal physical environmental concerns involved (George, et al, 1986). These are:

1. The increased erodibility of materials. This is particularly true of unconsolidated strata or those that break down easily upon exposure to the elements.

2. The uncovering of overburden layers which have associated problems such as acidity, sodicity and metal toxicity. Also the exposure of boulders at the top of the spoil which adversely effect regrading and redistributing of topsoil, especially in areas designated as prime farmland.

3. The altering of the hydrological regimes that existed prior to mining.

Within any given region, as shown in figure 1, the degree of impact is dependent upon the interplay of several factors. These are topography, geology, overburden, hydrology, climatic conditions and stripping techniques all of which contribute to the uniqueness of a given site. The only controllable factor in this list is the stripping technique. Therefore, one would expect that suitable modification of mining methods to include selective handling should lead to reduced severity of environmental impact.

The various regions differ considerably in terms of the factors mentioned above. Therefore no one method can be expected to be generally applicable. Mining techniques suitable for selective placement must be assessed within the context of the given region and site. Figure 2 shows various settings in which seam deposits may occur.

3 SELECTIVE HANDLING MINING METHODS

3.1 General discussion

Existing methods are not generally selected and designed for the purpose of selectively handling waste rock. If such placement occurs it is usually a side effect of the method, which has been chosen for geometric, production and cost reasons. However, some techniques currently in use can selectively handle materials if operated appropriately.

For selective placement a prime characteristic is that the mining system be flexible. Material can then be transported to various locations in the spoil designated by the reclamation and mine planning staff.

The most flexible system is an excavator and trucks. The excavator may be a cable or hydraulic shovel or a front-end loader. Scraper and dozer systems may also be quite flexible but are not as generally applicable. The flexibility of excavator-truck systems results from the possibility of multiple loading and dumping locations.

Dragline systems are not very flexible in general. Waste material is usually mixed or overturned by these methods. Fig. 3 is an example for a multiple seam operation where the interburden is placed above the overburden spoil. Using draglines alone this sequence cannot be avoided, unless a machine with much greater operating radius than currently available is manufactured.

In a detailed study including technical, operational and economic analysis of 21 existing mining methods it was determined that five existing mining methods show promise (Workman, 1981). These are:

1. Dragline - Shovel/truck area mining
2. Dragline - Bucket wheel excavator area mining
3. Shovel/truck open pit mining
4. Mountaintop mining
5. Haulback contour mining

At least one of these methods is applicable to each of the different surface mine settings and regions in the United States. Table 1 summarizes the settings, regions and applicable methods from among the five listed above.

Draglines operated alone are not flexible systems and do not have good potential for selective placement in most cases. An exception would be when a dragline is used to take an advanced bench of strata such as glacial till and place it back on top of the spoils to aid plant growth and the burial of boulders.

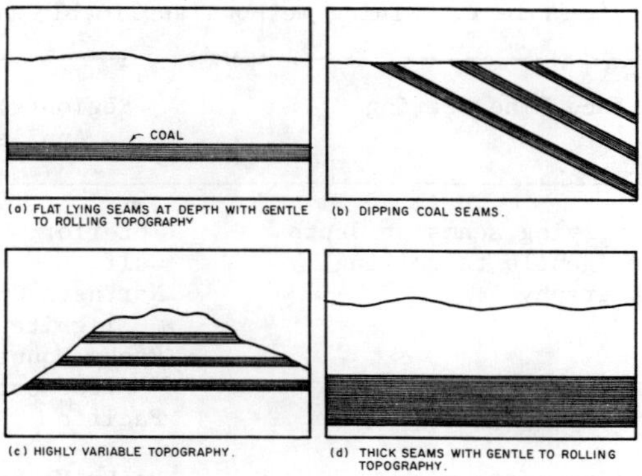

Fig.2 Various topographic and geological settings found in strip coal mining

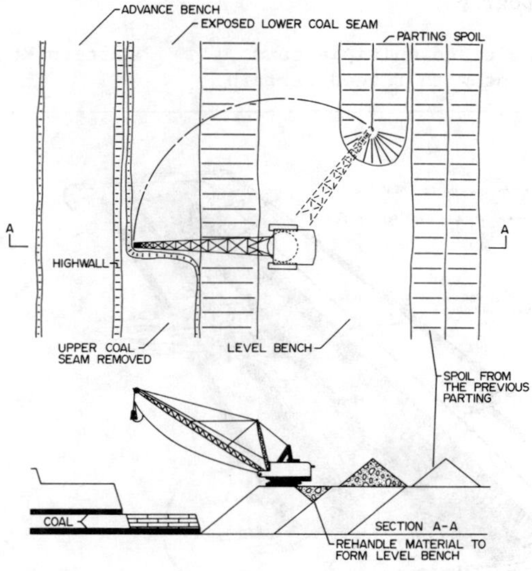

Fig.3 Example of overturned strata in an operation using a dragline to mine two coal seams

Table 1. Mining methods applicable to each surface mine setting and region

Surface Mine Setting	Region(s) Setting Applies To	Applicable Mining Methods
Flat lying seams at depth with gentle to rolling topography	Interior Gulf Northern Great Plains-- lignite subregion Rocky Mountain-- four corners subregion Pacific	Dragline - shovel/truck Dragline - BWE
Dipping seams with rolling to rugged topography	Rocky Mountain Pacific	Shovel/truck Shovel/truck - Dragline
Thick flat lying seams with gentle to rolling topography	Northern Great Plains-- Powder River Subregion	Shovel/truck Shovel/truck - Dragline
Single and multiple seam mining with rugged terrain	Eastern Region	Haulback Contour Mountaintop Mining

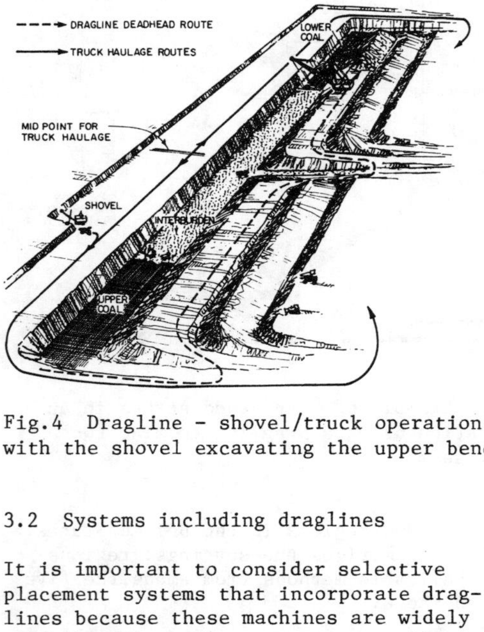

Fig.4 Dragline - shovel/truck operation with the shovel excavating the upper bench

3.2 Systems including draglines

It is important to consider selective placement systems that incorporate draglines because these machines are widely used throughout the country, as table 1 readily indicates. Therefore, combination systems that utilize a dragline as well as other equipment are proposed.

3.2.1 Dragline - shovel/truck system

This method attempts to take a rather rigid mining method (dragline mining) and introduces additional flexibility. The shovel/truck operation allows a portion of the waste to be transported to a designated area of the dump not defined by pit and machine geometry.

In the first case the shovels and trucks are used to mine one or more benches of waste at the top of the overburden. The remaining material is mined by dragline and deposited in the bottom of the pit. The trucked spoil is placed above the dragline dump in a sequence similar to that found in the original bank. With suitable strata in the upper benches the approach can aid the burial of boulders and toxics and provide a better material, directly beneath the topsoil, for plant growth. If used in a multiple seam context, with the dragline mining some overburden and the interburdens, the interburdens will be located above the dragline overburden, but beneath the shovel/truck waste, thus some overturning does still occur. Figure 4 illustrates this situation.

A second case is where a pit containing more than one seam has toxic interburdens. In this instance the shovel/truck operation could be used to excavate the interburden and keep the material in the bottom of the pit. The waste would be trucked to a location behind the advancing coal face of the lowest seam, in the current cut, with dragline waste (which would be a mixed product) deposited above. Thus the toxic materials could be kept away from plant growth zones or groundwater flows that would be adversely affected. Fig. 5 is a sequence diagram for the method. Similar diagrams have also been developed for the other methods.

The dragline - shovel/truck method has the following advantages:

1. Flexibility in material movement is introduced.

2. A detrimental layer anywhere in the bank could be handled, at least in theory.

3. A boulder-free surface may be more readily achieved.

4. Spoil regrading requirements are reduced if truck spoils are the upper layer.

5. It is a combination of two existing, well understood methods.

3.2.2 Dragline - bucket wheel excavator mining

This system may be used in some area mining operations to selectively place a portion of the strata. As in the previous system flexibility is introduced but to a lesser degree. This is true because the BWE can only dig weak material, it cannot work independently of the dragline and it will usually be restricted to the removal of an upper layer of overburden.

In this case the discussion is limited to BWE systems that are integrated units where one machine performs all the functions of dig, transport and dump. Units with discharge radius up to 210 meters have been produced. BWE units with around pit or cross pit conveying can also be considered however.

If more than one seam is mined, with the dragline excavating a portion of the overburden and the interburden, some material will be overturned. It may be possible to consider a BWE mining the interburden, but this is less feasible than an excavator and trucks. In any event the upper overburden will be deposited at the top of the spoil where it will bury boulders and aid plant growth. Figure 6 is an illustration of this method.

This method scores well as a selective handling technique for the following reasons.

1. It adds flexibility beyond what can be attained by dragline alone.

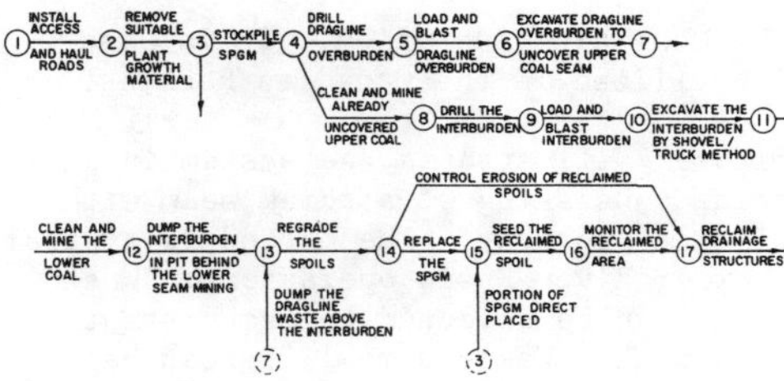

Fig.5 Sequence of events for dragline - shovel/truck mining with the shovel excavating the interburden

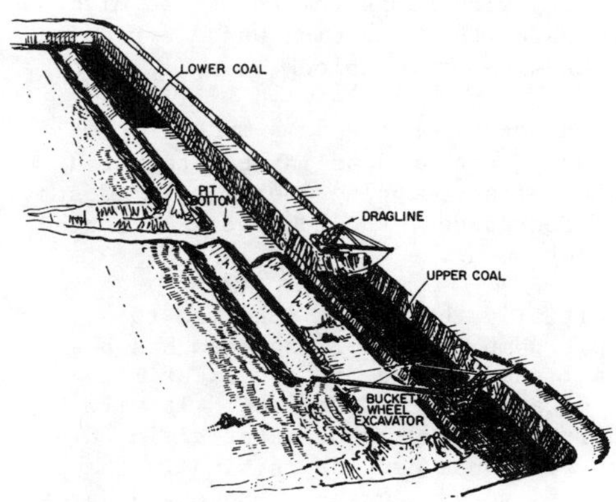

Fig.6 Dragline - Bucket-wheel excavator mining on two seams

2. Upper overburden can be replaced in a location similar to that from which it was extracted.

3. Boulders are buried as are toxics.

4. Spoil regrading requirements are reduced.

5. The method is already used, albeit not for selective placement necessarily, and therefore well understood.

3.3 Methods using excavators and haul trucks

3.3.1 Excavator and truck open pit mining

Excavator and truck mining is probably the most flexible method for handling waste materials above a seam deposit. The excavator may be a cable or hydraulic shovel, or a front-end loader. Seams may be flat or dipping. The method is often used in the Powder River Basin and Green River Basin areas of the Northern Great Plains region and in the Central

Rocky Mountains subregion. It is also well utilized in the Canadian Rockies.

Figure 7 illustrates the system in a setting consisting of a thick seam with rolling topography. The method is typical of Powder River Basin operations. Waste benches can be selected to match strata in the bank. Waste dump lifts can be designed so that the overburden can be dumped in a sequence related to its occurence in the undisturbed ground.

Haulage access is shown on each bench. However, with ramps some of these might be eliminated thus reducing waste stripping to the side of the block.

With dozers on the dump much of the regrading can be done in an integrated manner with stripping operations. Costs will be reduced. Boulders can be buried as they occur.

Initially the waste is hauled to an ex-pit dump until working room has been established in the pit and backfilling of the excavation can commence. It will be difficult to segregate material in this dump for subsequent deposition in the pit in the desired sequence. Separate dumps may be needed. When dumping is advancing in the pit, material will be displaced horizontally relative to its position in the bank but will be similarly located in the vertical sense.

Dipping seam operations also allow for selective placement. The process is not as straightforward and is likely to be less exact because the strata would need to be placed in the dump on a slope equal to the material angle of repose to approximate the dipping planes in the ground. Folding and faulting will also contribute to the complexity of selective placement.

More waste may need to be stored ex-pit during initial mining since the pit will

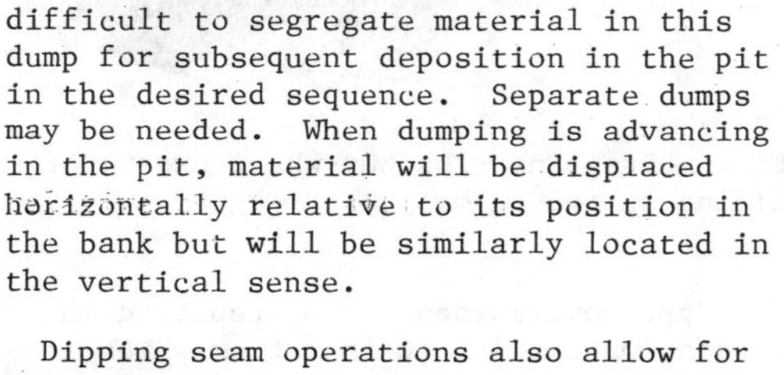

Fig.7 Shovel/truck terrace pit mining of a thick seam

usually be mined to greater depth on multiple benches than in the flat seam case. Backfilling cannot commence until the hanging wall has been fully developed and working room established at the pit bottom. However, the problem is no greater than would be the case in the absence of selective placement except that initial waste storage may be more complex.

The excavator and truck method should be considered, in the appropriate settings, for the following reasons:

1. It is the most flexible of existing methods.

2. Various strata can be catered for by adjusting, within reasons, the shovel bench heights.

3. The system is widely applicable and capable of mining the full spectrum of strata.

4. It is a well understood system that is already in use in areas where the method is most applicable.

5. Waste dump stability can be easily maintained which is important to avoid alteration of the sequence through dump failures.

6. Toxics and boulders are readily dealt with and post-mining regrading is minimized.

3.3.2 Haulback contour mining

In contour mining overburden is removed into a steeply sloping hillside. The coal and overlying strata are generally flat lying. Ratios of overburden cubic meters to tonnes of coal increase rapidly to a point of breakeven economics. Pit widths therefore, seldom exceed 50 meters.

In haulback contour operations the overburden is excavated, usually with a frontend loader, and loaded into trucks. The waste is hauled back along the pit to a point behind mining and augering operations, and is redeposited in the mined out cut. Thus, the hillside is restored.

Therefore, haulback mining is an excavator/truck method but it differs from the method previously discussed to a significant extent. It is applicable in the Eastern Region, but not in others.

Usually the overburden is mined in one

lift and the backfill is advanced in one lift as well. This does not necessarily provide a selective placement result. However, some operations have selectively buried acid bearing materials in the past while employing this method.

To selectively place materials one would employ benches in the overburden. The material would be hauled back to similar elevations in the dump. Two or three benches could be utilized. The overall distance within which mining is completed would be longer than for the typical case. Figure 8 is a sequence diagram for this method.

The haulback contour mining method was selected for the reasons below:

1. It is the only contour mining system widely applicable under current regulations.

2. It is well suited to mining conditions in the eastern region.

3. The system was developed to meet reclamation concerns.

4. It is a system with the flexibility to selectively handle waste.

5. The method is well understood in the region where it applies.

3.3.3 Mountaintop removal

This method is primarily found in the Eastern Region. However, it is applicable wherever one finds rugged topography and multiple, relatively flat lying seams. The technique may be operated as one that uses only excavators and trucks. Alternatively the method can be utilized as a combination excavator/truck-dragline method. Mountaintop mining involves the removal of a mountain or hilltop to extract a series of included seams which outcrop around the hillside. Unlike contour mining, which only advances a short distance in the hill, mountaintop mining allows for complete extraction of the seams. As mining advances the mountaintop is restored using the waste rock, which is deposited in a series of lifts in the waste dump.

Figure 9 is an illustration of mountaintop mining utilizing shovels and trucks exclusively. A series of benches are mined with the coal seams being extracted when encountered. The mountaintop is restored in a series of lifts using the overburden. A valley fill has been constructed to store the excess material resulting from the swell that occurs in the strata during mining.

The primary requirement, if this material is to be used in a selective way, is

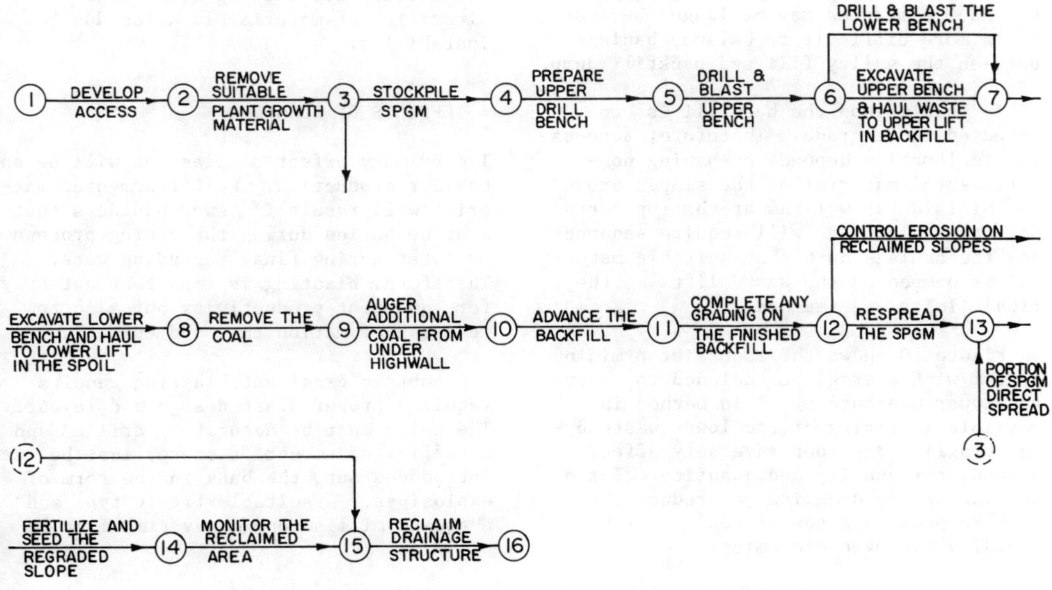

Fig.8 Sequence of events for a contour haulback operation utilizing two benches in the waste above a single coal seam

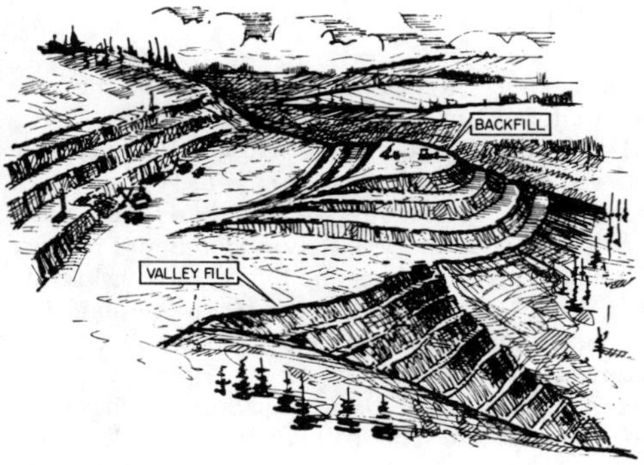

Fig.9 Mountaintop mining using excavators and a haul truck fleet

to plan the waste haulage in order that the mined strata are successfully placed in the chosen location in the dump. To do so also requires careful planning of the advance of the overburden benches. Detailed short-term planning is required to successfully place the material while insuring the correct sequencing of mining advance and balancing haulage distances.

The valley fill is a quite environmentally sensitive part of mountaintop mining. Therefore material placement within the fill should be carefully planned. It may be advised to exclude any particularly toxic waste from the valley fill to avoid groundwater related problems. With proper planning this is possible to accomplish, but haul distances may be longer because it is more difficult to balance haulage between the valley fill and backfill dump.

In this method the backfill is constructed above grade. Therefore, successful reclamation depends on having non-detrimental material on the slopes around the hillside as well as at the top surface. Selective placement will require sequencing the haulage such that suitable material is dumped on the waste lifts as the final limits are reached.

Figure 10 shows the mountaintop mining system with a dragline included to remove the lower overburden. This method is possible if mixing of the lower waste by the dragline does not adversely affect groundwater quality and quantity. Introduction of the dragline may reduce the cost to produce a ton of coal, thereby yielding improved economics.

In the example shown the dragline mines on a 25 meter bench. The shovel benches are 12 meters in height. The dragline moves about one-third of the waste rock. The dragline bench is fixed by the interburden thickness between the second and third seams.

Prior to dumping trucked spoil above the dragline dump the peaks are rough graded to maintain the sequence of the strata. The upper spoil is dumped in lifts to further restore the strata to near the pre-mining location. Truck haul distances may be three quarters of a mile one way. The distance could be reduced by the use of bridge ramps across the pit to allow the trucks across pit access as well as access around the end. This approach will be successful in longer pits where there will be less interference.

There are several reasons why this method was chosen as a good selective placement technique:

1. It provides the flexibility needed for selective placement of the waste.

2. It is a major mining method in the Eastern Region.

3. Mining and reclamation activities are partially integrated.

4. The method is well known.

5. Boulder problems are reduced and the stepped dump contributes to spoil stability thereby eliminating subsequent alteration of material location due to instability.

4 EFFECTS OF BLASTING

The primary effect of blasting will be on boulder production. Well fragmented material will result in fewer boulders that must be buried during the mining process or later during final regrading work. Therefore, blasting is important not only for equipment productivity but also to reduce reclamation time and cost.

To obtain excellent blasting results requires proper blast design and layout. The holes must be accurately drilled and a sufficient amount of energy must be introduced into the bank in the form of explosives. A suitable tie-in type and adequate millisecond delay time must be provided.

Methods are available which allow fragmentation to be predicted for blasts of reasonable design. One method which is

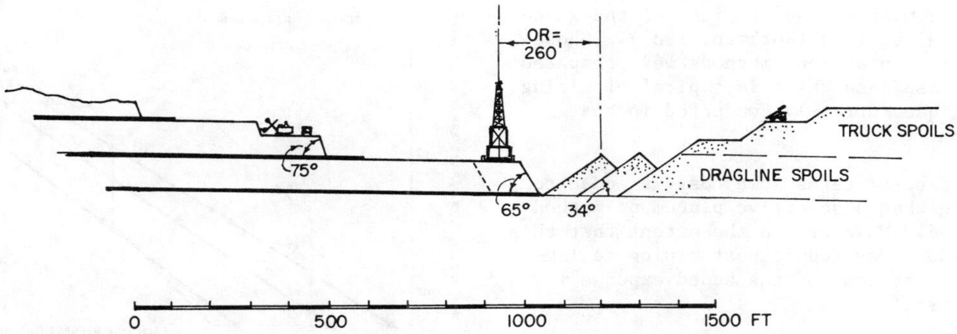

Fig.10 Mountaintop mining combining the excavator/truck operation with a dragline lower bench

often used is the Kuz-Ram model (Cunningham, 1987). Using this technique an estimate of the amount of oversize produced by a blast can be obtained. Different patterns and explosives can be analyzed to determine what combination will minimize boulder production.

Cast blasting may also be helpful in some cases where draglines are used. In this method high energy is used in the bank to throw material over into the previously mined pit. As the material displaces in a relatively horizontal fashion into the prior pit, some waste will be located in a similar location in the spoil to the in-situ sequence.

It is of interest to note that a form of cast blasting has been used in quarries to selectively separate shot rock from waste material in a single bench (Stevenson and Stuessy, 1991). In this application the lower portion of the bench consisted of rock suitable for quarrying, while the upper portion of the bench consisted of weak, weathered rock unsuited for use.

In the cases described the good rock is loaded to a high powder factor but no explosive is loaded in the waste rock. Upon detonation using suitable delays the good rock moves out from under the altered strata which falls vertically into the void generated, thereby separating the materials.

This application would appear to have some potential in dragline operations where competent rock directly above the coal is overlain by unconsolidated material such as glacial till, provided the contact plane is sufficiently well defined.

5 MINE PLANNING

A key to the successful utilization of selective placement at reasonable cost is careful planning of the stripping operations. The greater the flexibility of the operation the more this is true. Scheduling the day by day mining and dumping sequence is essential for effective operation.

Computer aided mine planning therefore becomes an important tool for successful operation. These programs will allow the sequencing of material movement to be examined in detail within a reasonable time frame. Several scenarios may therefore be studied and the best approach selected.

Long-term planning which studies material movement for a year, quarter or month is important to the effort. However, equally important will be the ability to plan shovel locations and haulage routes on a day-to-day basis. Therefore, programs that include a short-term day-to-day planning feature will be particularly powerful when studying selective placement needs.

The current status of mine planning software leads one to conclude that selective placement is becoming more feasible when it is now possible to plan and control the process quite exactly. It is considered that this will be very helpful in controling costs when selective placement is utilized.

6 ECONOMIC CONSIDERATIONS

The economics of the various systems selected has been extensively studied using example mining cases typical of the given regional setting (Workman, 1981). Chosen selective placement methods were compared to a base case which is typical of mining technique currently conducted in that region.

In general terms some cost is added to mining when a selective placement method is used. However, to the extent that this procedure may reduce post mining reclamation cost some of the added expense may be offset.

When performing project planning studies it must be considered that some variation will occur between the project estimates and actual results. Therefore, variations in cost and production of ±15 percent were considered for each method. For given rates of return calculations were performed to determine what combinations of variations between competing methods would lead to a net present value (NPV) of zero.

This information could be plotted as in figure 11. If the resulting line lies within the envelope of expected variation in cash flow parameters then one cannot clearly distinguish economic advantage. If it lies outside the sensitivity envelope one or the other method is clearly favored.

In the example shown one line lies at the bottom of the envelope. Therefore, the two dragline method is more attractive economically when the rate of return is 20 percent, which might be a typical desired return for mining projects. However, at higher rates of return economic advantage between methods cannot be clearly established at this stage of analysis. As the discount rate increases the annual operating costs have less impact on the result while differences in capital costs become increasingly important. In this case this causes the economics to trend toward the dragline - shovel/truck combination method which requires less capital.

The economic studies performed did tend to indicate more costly operation when selective handling was introduced. However, the results were frequently within a zone where possible variations in production and cost numbers for the case studies meant that clear economic advantage could not be determined between methods. Therefore, it should not automatically be assumed that selective placement techniques will be more costly. In some cases there is no effect result-

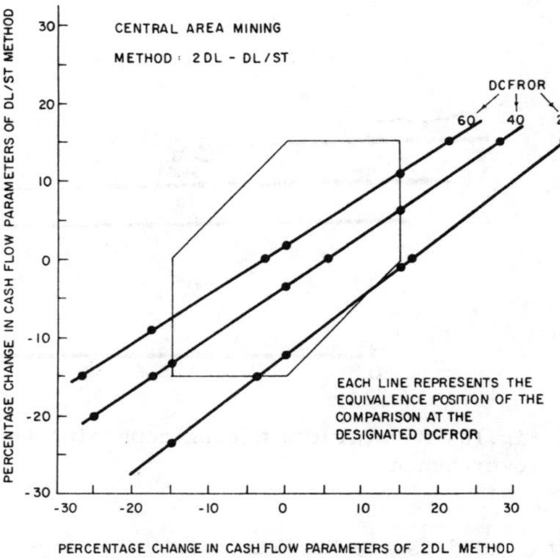

Fig.11 The influence of discount rate selection on the comparison of the two dragline method versus the dragline - shovel/truck method for the interior region case study

ing from the introduction of selective handling, in others the effect is unclear because it relates to risk and in others it may make a specific project unviable.

7 CONCLUSIONS

Some selective placement of waste material has been done in the past. This has involved the burying of boulders, the burial of toxics and selective placement of glacial till material back on top of the spoils when advance benching with draglines. However, the practice has not been widespread.

If selective placement is to be successful it must be performed on a large scale using mining equipment designed to move large volumes of rock at reasonable cost. In most instances, if attempted on a "micro" scale the procedure will make otherwise viable properties non-economic.

Perhaps the single most important factor is that the system used be flexible in waste handling. Methods that employ an excavator and truck haulage are among the most flexible. Combination methods that involve a dragline and other equipment can also be flexible enough to successfully segregate waste.

It has been found that there are mining

systems and equipment already in use that will achieve selective handling goals. Therefore, the procedure can be instituted in many cases while using well understood equipment and methods.

Fragmentation is an important issue. Well fragmented material means less boulders to be dealt with, an especially important issue where land has been designated prime farmland. Blasting techniques can have other important benefits as well such as partial selective placement through blast casting techniques.

It is believed that accurate and complete mine planning is an important requirement for successfully achieving selective placement goals. The requirement is basically one of material management and placement control. The advent of sophisticated mine planning software is a key development for selective placement planning and sequencing. Programs that provide both long-term and short-term day-to-day planning will be most suitable.

From an economic standpoint selective placement will often add cost to the operation. However, done on the proposed macro scale the increase may be small. In fact the economic analysis of typical regional case studies has indicated that when potential variation in the input data is accounted for it is hard to distinguish economic advantage in many cases.

Perhaps most difficult to assess is the effect that selective overburden placement will have on post-mining reclamation work and costs. However, burying boulders, keeping toxics away from plant growth zones, producing less rugged spoil profiles and avoiding placing toxics in zones where one expects underground flows to reestablish is bound to have a beneficial effect on the reclamation effort. Cost savings in post-mining reclamation can be expected to offset some increased mining cost that might accrue to the use of a selective placement method.

ACKNOWLEDGEMENTS

The author wishes to thank the United States Bureau of Mines for financial support of the study upon which this paper is based. Also to Marston and Marston, Inc. to whom the study contract was awarded and who provided project management and much useful information.

REFERENCES

Cunningham, C. 1987. Fragmentation estimates and the Kuz-Ram model-four years on. Proceedings of the Second International Symposium on Rock Fragmentation by Blasting, Keystone, Colorado: 475-487.

Dolhopf, D.J., Hall, W.D., Schafer, W.M., Depuit, E.J. and Hoddler, R.L., 1977. Selective placement of coal stripmine overburden in Montana. USBM Contract H-262032, Interim Report I - Database.

George, H., Meech, J. and Workman, J.L., 1986. Towards reducing the physical environmental impact of North American surface coal mines; a review of potential selective overburden handling techniques. Mining Science and Technology, 3: 81-94.

Stevenson, R.H. and Stuessy, V., 1991. Selective mining using explosives. Proceedings of the Seventeenth Conference on Explosives and Blasting Technique, Las Vegas, Nevada: 109-117.

Workman, J.L. 1981. Selective placement of stripmine overburden in the United States. Final report, USBM Contract J0199040, Marston & Marston, Inc.

Innovative Mine Design for the 21st Century, Bawden & Archibald (eds) © 1993 Balkema, Rotterdam, ISBN 90 5410 325 6

Application of Wenco International's Production Monitoring and Control System for grade control and productivity at Wabush Mines

S.G.Burns
Wenco International Mining Systems Ltd, Vancouver, B.C., Canada

W.H.Bell & T.G.Zwirz
Wabush Mines, Labrador, Nfld, Canada

ABSTRACT: Wabush Mines in Labrador Newfoundland and Wenco International Mining Systems Ltd. of Vancouver B.C. collaborated to solve several unique operating challenges at Scully Mines through the application of a computerized mine monitoring and dispatch system. Wabush Mines is an iron ore producer located in Western Labrador, Canada. The ore body mineralogy is characterized by localized highly variable quantities of manganese as well as significant variations in silica and weight recovery. Quality control parameters for mill feed grades are however, very restrictive. These elements combined with low operating tonnage and limited blending locations, lead to a complex balance between ore grading and mine production. A decision was reached to proceed with the implementation of a computerized mine monitoring and dispatch system customized to deal with the grading and production problem. The Wenco system was designed to incorporate detailed planning and scheduling capabilities to simultaneously address the opposing objectives of grade control and mine production. The system utilizes a detailed shift planning module coupled with an automated non-linear dispatching algorithm to dynamically meet the desired objective of maximizing equipment productivity while meeting tight quality control constraints. The computerized mine monitoring and dispatch system has generated a number of positive results. A number of primary and secondary benefits that have been achieved are described.

MINE CONFIGURATION AND OPERATION

Wabush Mine's Scully Mine is located in the province of Newfoundland, Canada, about 1,000 kilometers north west of the city of St. Johns. The Scully mine is operated by Wabush Mines; a consortium comprised of two Canadian participants, Stelco and Dofasco, three American participants, Acme Steel Co., Inland Steel Co. and Cleveland Cliffs Inc., and one international partner, Finsider International of Italy. The project is managed by Cliffs Mining Company, a subsidiary of Cleveland Cliffs Inc. Since 1964 it has been producing 6 million long tons of pellets annually, and mines 26 million tons of total material. In recent years the operation has been downsized to 4.5 million long tons of pellets.

The ore body is part of the Labrador Trough geosyncline. The ore types are best described as quartz specularite hematite schists. The ore body is heavily folded along three main axes and is bisected by a reverse fault at 60 degrees of dip.

Conventional open pit equipment is used. The loading fleet consists of six 8 cubic yard shovels, one 14 cubic yard shovel, and one 10 cubic yard loader. Haulage trucks are a mixed fleet of 150 long ton units. The relatively high ratio of loading equipment to hauling equipment is dictated by the complex ore grade control practice used at the mine and caused by the wide variations in ore quality. This results in the concurrent operation of a number of pits each with several working faces.

THE CHALLENGE

Mill feed grades are dictated by the market specifications for iron ore products and the milling constraints required to achieve the specifications. Currently, five separate products are being produced; a 1% and 2% manganese flux pellet, a 1% and 2 % manganese acid pellet, and a 2% manganese concentrate.

Each of these products requires specific mill feed grades, that must be delivered at an established feed rate.

A number of conditions combine to present a difficult operational problem in meeting the mill requirements:

1. The ore body is highly variable with respect to manganese, silica, and weight recovery.
2. Operating tonnage is low.
3. Available digging locations are limited.
4. Product grade specifications are strict, due to owner quality requirements.

Table 1 contains the principle assay parameters at Scully mine in comparison with a typical mill feed specification for a 2% manganese plan.

It is apparent that any attempts at controlled blending would require a form of feedback within a time frame that enables any operating change to effectively impact on the near term results. A manual tracking system would be quite capable of reporting on the net results of material blended from multiple sources. However, the decision making for actions necessary to correct blends "drifting" away from the targets are highly complex and do not lend themselves well to manual systems.

Table 1. In pit mining assay parameters at the Scully Mine, 2% manganese plan.

Quality	Min.	Max.	Targ.
Crude Fe	30.0	45.0	35.8
Crude Mn	0.02	20.0	1.95
Crude Magnetics	0.15	40.0	2.50
Wt. Recovery	25.0	60.0	37.0
Crude Silica	0.40	7.00	2.65

Experienced mine operations and engineering staff have been successful in dealing with the idiosyncrasies of the mine in maintaining a blend. The breakdown in a manual system comes from the frequency with which blending decisions must be made and monitored. Ideally, an analysis and decision should take place at every change in;

1. the number of trucks hauling;
2. the number of shovels running;
3. conditions in the digging faces.

The combination of changes in all grade impacting conditions can lead to hundreds of occurrences requiring grade adjusting actions. In addition to the identified equipment changes, the grading engineer is faced with highly variable materials within a given face, requiring a regeneration of a "blend plan" with each frequent shovel move. Figure 1 illustrates a typical mining face.

While attempting to balance all of the grade impacting events, mine operation's

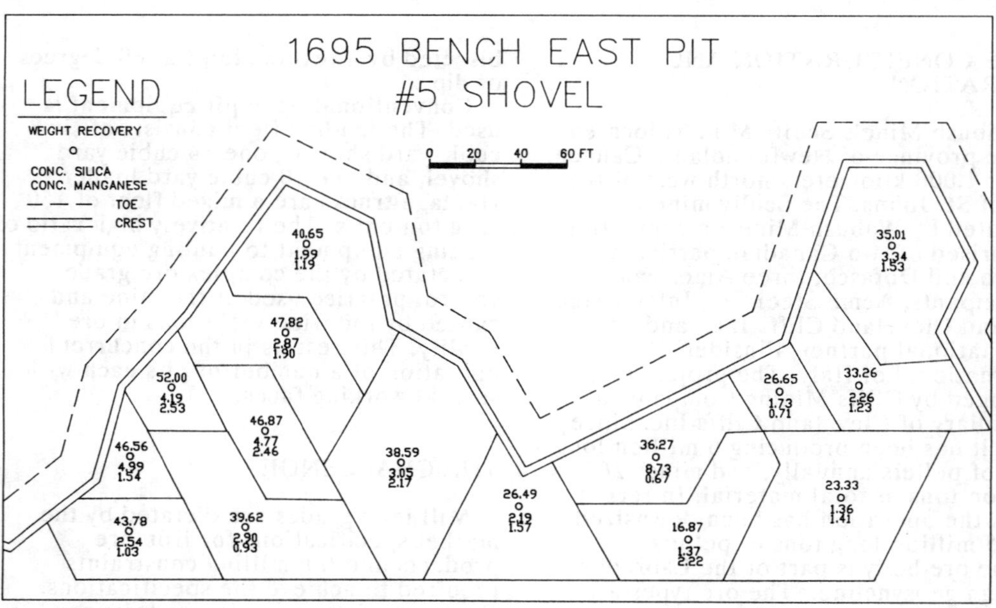

Figure 1. Typical mining face.

personnel must also still consider the material movement plans, and effective utilization of mine equipment. Clearly, this requires decisions that must simultaneously address a number of the competing objectives.

This led to an ideal opportunity to apply an information based management tool (as used at other Cliffs operating mines) to improve the quality and consistency of ore blending and equipment utilization.

A computerized mine monitoring and dispatch system could accomplish this by:

1. Collecting real time location and status information of all mine operation's equipment through a data radio network.
2. Defining a set of operating objectives and specifications, for mill feed rates, grade ranges, and waste removal.
3. Integrating all the information into a decision making dispatch algorithm.
4. Informing the production equipment about the required dispatch assignments.

The effect of continually and optimally adjusting truck assignments for both ore and waste materials is to provide an even distribution in the utilization of equipment. This produces the required grade control while maximizing ore and waste production.

THE SOLUTION

System Requirements

Seven specific requirements for a computerized mine monitoring and dispatch system were identified.

1. As a primary requirement the system must dispatch trucks to the crusher to meet a specified dumping rate (crusher capacity) and a specified material blend of up to ten quality constraints.
2. As a secondary requirement, the system should dispatch trucks to the waste shovels to maximize waste production using any trucks not currently required to meet the ore production plan. When desired, it must be possible to set a specific production rate for any shovel, that is, to prioritize shovel production or force production from a given face or area.
3. Operation of the dispatching system must not require added personnel, and the allocation of new duties to existing personnel must be minimal. To accomplish this and allow the pit foremen to spend the same amount of time in the field as before, the system must provide a mobile control centre. Essentially, the foremen must have the capability of monitoring and controlling the system from a vehicle in the pit as well as from a base control centre.
4. The system must be reliable and relatively simple to operate and maintain. The system is to be maintained by mine engineering staff and run by mine operation's staff who have little or no previous experience with computers.
5. The reporting function of the dispatching system must be fully integrated with the mine's existing computerized data processing and reporting system. This includes the transfer of data with a remote main frame computer.
6. Data accuracy should be validated at the earliest possible point in the process. Essentially this means that the data coming from the trucks should not depend solely upon a manual source such as the equipment operator. Where possible, this data should be verified or augmented by data from an automated process such as roadside radio beacons at critical points in the truck routes and by dump switches on the trucks.
7. Due to the critical and variable nature of the ore grading problem, the system should provide the facility for the grading engineer to generate alternate production dispatch strategies in advance to meet a variety of ore grading and production criteria. These plans would then be available to the mine foremen for instant use in the dispatch algorithm as dictated by changing circumstances.

System Configuration

To satisfy the specified system requirements, Wenco International Mining Systems Limited were chosen, to customize and install their Production Monitoring and Control System 3000 (PMCS 3000).

PMCS 3000 manages pit production requirements, controls blending, and accumulates critical real time operation's information. The system utilizes distributed processing on a PC based Local Area Network (Walshe and Cimolini 1989) a radio data network including a repeater station, and mobile field hardware, and a Mobile Supervisor's Terminal.

The PMCS computer system is a Local Area Network with a dedicated file server and multiple workstations. Wenco uses its own proprietary Inter Process Communication (IPC) protocol specifically designed for the truck dispatch system to enable communications between the various real-time processes. Data from the truck dispatch system is also transferred to an external data processing system that performs management reporting functions.

The radio network utilizes two different radio modulation schemes sharing the same radio channel. A Hexadecimal Sequential Coding (HSC) protocol handles communication with the MDT's in the trucks and shovels and a Minimum Shift Keying (MSK) scheme is used to serve the Mobile Supervisor's Terminal (MST). The radio channel available for data use at Wabush is a single frequency in the VHF band. A 150 watt radio repeater located in the crusher building ensures full radio coverage for the mine property.

The field hardware consists of Mobile Data Terminals (MDT) on board all production equipment and roadside beacons strategically located to provide critical equipment location information. MDTs both provide and receive information from the central computer network. The information contains details on the unit's status, location, and materials.

The Mobile Supervisor's Terminal (MST) is located in a pickup truck used by the mine foremen. The MST is a rugged industrial type microcomputer mounted in a single case with a specialized radio modem and a data radio. The radio modem allows it to communicate with the computer network. The foremen monitor and control the truck dispatch system from the field by using the MST in essentially the same manner as using one of the office PMCS computers.

System Software

Four primary computer programs or modules make up Wenco's truck dispatch system.

1. Monitor
2. Dispatch
3. Planning and Scheduling
4. Information Systems

Monitor

The core of the system is a series of programs called Monitor. This program interacts directly with the trucks and shovels through the radio modem. It maintains the real-time data bases, coordinates the Inter Process Communications, and manages the radio data communications between the field equipment and the processes on the network. The additional versions of Monitor employed are called Foreman and Observer. These programs co-ordinate the foremen and mine controller interaction with the system as well as providing all users a large number of selectable mine information displays.

Dispatch

The dispatch program is responsible for maintaining the planned crusher feed rates and specified grade ranges. Dispatches for haul trucks are calculated at key points in the operating cycle and then communicated to the individual units. Since this is a dynamic process, dispatch responds to the moving in pit conditions, resulting from changes in travel and loading times, and changes when equipment becomes available or is withdrawn from service.

The algorithm matches measured equipment event times with planned times that are established by the planning module to meet the mining objectives set for the current shift. The Dispatch program calculation for each truck assignment considers not only the present situation but also projects up to nine future events for the affected equipment, to ensure that this move will result in an optimal solution.

Planning

The planning module is used to generate operating plans for the dispatch program. Its algorithm uses a combination of linear, non-linear, and deterministic programming to calculate optimal plans. In addition to preparing operating plans for the use of the Dispatch module, the grade control engineers use this module to simulate different production scenarios using a variety of equipment combinations or changing grade and production constraints.

A mine description program is used to maintain the equipment lists and describe the road network.

Information Systems

PMCS 3000 contains a number of information systems. A real time graphic display, provides graphed details on current shift or past shift production, grade, and maintenance performance. The Operation's Information System (OIS) allows detailed equipment, material, and time reporting for shift periods. The Management Information System (MIS) summarizes the mine data on various time bases and generates a series of customized reports and graphs. This data is also used in building a comprehensive query driven database reporting system to meet specialized information needs.

OUTCOMES

Over the course of the past three years improvements have been realized in:

1. Productivity
2. Information Utilization
3. Equipment Utilization

It is always difficult to evaluate pre and post computerized mine management and dispatching system (dispatch system) results for at least two major reasons.

Firstly, the basis of measurement is changed from an operator dependent batch recording system to a computerized real time collection system.

Secondly, it is virtually impossible to create a controlled environment in which the dispatch system is the only variable being altered for a given period of time.

However, tracking the trends in a number of key performance indicators can provide a basis for measurable improvement resulting from the dispatch system. In addition the changes in mine operating practices and mine engineering decisions resulting from the utilization of the information systems can offer further proof on the benefits realized from a computerized mine monitoring and dispatch system.

Productivity Gains

One of the anticipated benefits from a dispatch system is an improvement in the allocation of the available hauling resources. This can often be quantified by evaluating the total truck fleet ton mile per hour, before and after the installation of truck dispatching.

To properly represent the results, however, the average haul distances must first be considered. Increases in average hauling distances themselves will cause an increase in TMPH results, since the portion of time a truck spends in a hauling status increases relative to the constant times associated with loading and dumping.

Figure 2 displays the long ton mile per hour (LTMPH) results for the two years (January 1988 to November 1989) prior to the installation of truck dispatching, and for the two and half years (July 1990 to December 1992) following truck dispatch. The best fit lines through the two data sets reveal an increase in LTMPH of 4.2 % at a typical median haul distance of 1.75 miles.

An additional truck utilization measurement developed to evaluate the dispatch system performance is the ratio of the truck hauling time to empty time. Figure 3 illustrates the increase of truck hauling time over truck empty time from July 1990 to December 1992. In 1992 the average hauling time to empty time ratio was 0.890, which is a 35% increase over the 1990 six months hauling time to empt time starting average 0.659.

Gains in the hauling time to empty time performance can be attributed to improved dispatching decisions as both the system and users proceeded along a learning curve. The dispatching logic was modified to address specific site requirements, and operations and engineering personnel increased the utilization of mining set ups that were conducive to dynamic dispatching.

Grade Control

The second major area that a benefit from computerized truck dispatching is expected is in grade control. Grade control at Scully Mine is focused on providing the mill with a consistent feed of known assay values. Providing this consistent feed then results in higher recovery, reduced energy consumption, and reduced mill set up costs to accommodate grade variance in the feed.

For most mines any improvement in grade control usually comes at some cost to productivity from the mining operation. Conversely, the ability to increase productivity while maintaining the existing level of grade variability can also be viewed as an improvement in grade control.

The mine is responsible for the manganese content in all of the iron ore products. Figures 4 and 5 illustrate the annual train manganese grades, and

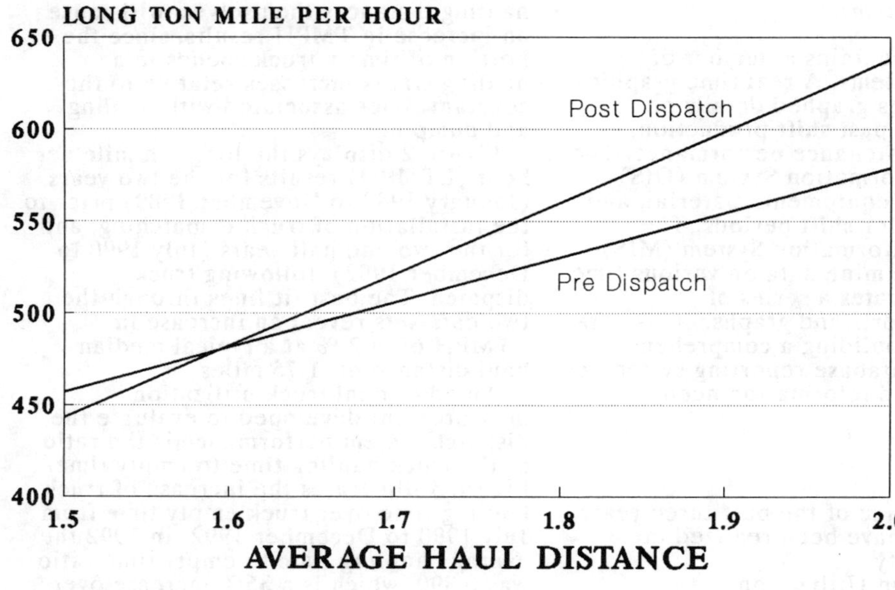

Figure 2. Long ton mile per hour at equivalent haul distances.

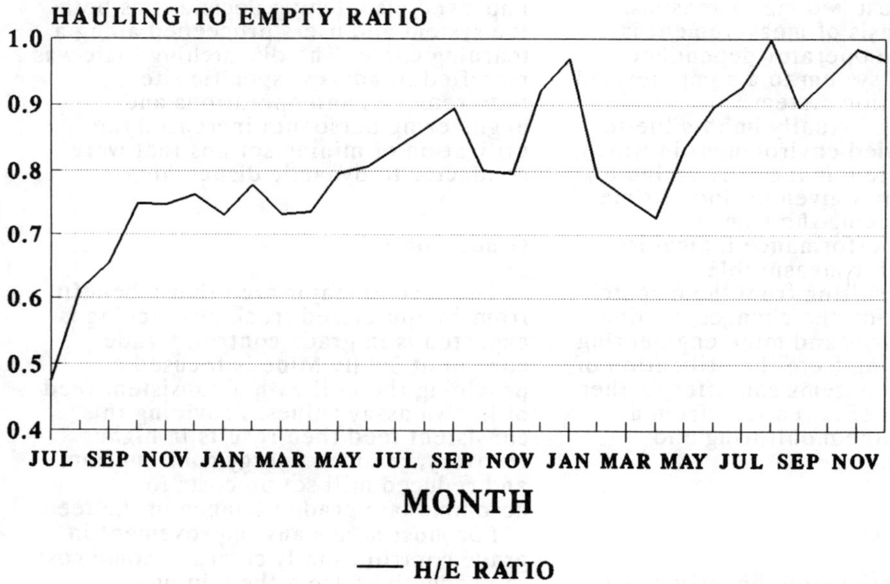

Figure 3. Hauling time to empty time ratio, July 1990 to December 1992.

percentage of trains within the grade specification (compliance) for 1988 to 1992.

Ideally, the evaluation of the dispatch system grade control would be based on mill feed grades only. However, the most reliable and most available grade data is based on the final concentrate grades for trains. These results incorporate all other grade influencing variables inherent throughout the milling process, allowing only an intuitive, and not a quantifiable

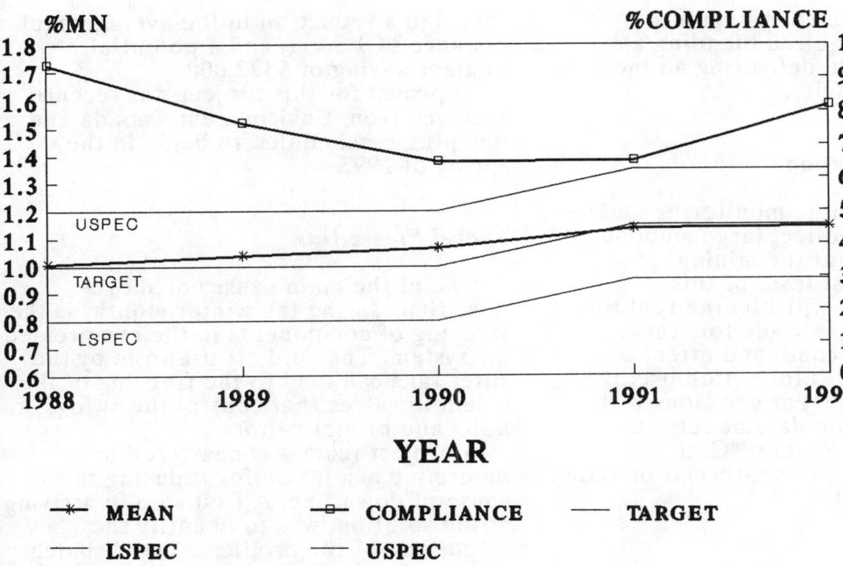

Figure 4. Train Mn grades 1.00% plan, 1988 to 1992.

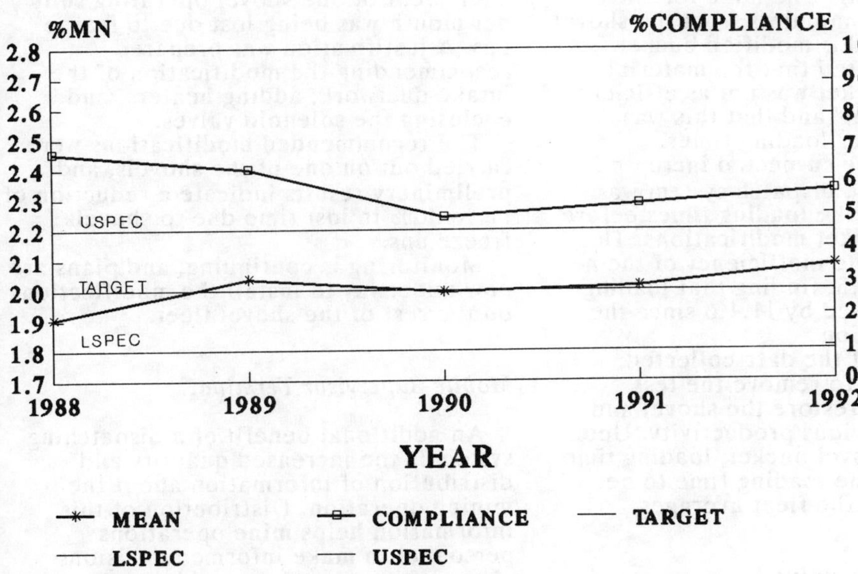

Figure 5. Train Mn grades 2.00% plan, 1988 to 1992.

evaluation of the grade control performance.

Given the preceding restrictions, it is difficult to establish if the installation of the dispatch system has led to any improvement in the manganese grade average or percentage of trains in compliance with the specification.

However, has been no significant decline in the grade control for the period following the dispatch system installation, despite a reduction in the number of loading units being utilized for blending. Therefore, in conjunction with the measurable improvement in LTMPH (as previously described), we can infer that the dispatch system has maintained the required blending and grade control while delivering an increase in mining productivity.

Information Utilization

Computerized mine monitoring and dispatch systems collect large amounts of real time data about the mining operation. The first level of this information use is typically the real time automated decisions made for truck dispatches. The second, and often as significant level of information use, is in supporting management decisions with factual, quantifiable data about the mining operation. Several "Case Examples" utilizing this information data base are described.

Shovel Loading Time Investigation

A new style of teeth and adaptors had been installed on one of the mine shovel buckets for testing. The mine foremen received many complaints from the shovel operators using this modified bucket. Operators indicated that the material flow into the bucket was not as efficient as the "old bucket" and that this was resulting in longer loading times.

To evaluate the suspected increase in loading times, the dispatch system was used to evaluate the loading times before and after the bucket modifications. The data supported the inefficiency of the new adaptors and teeth, finding that loading times had increased by 11.4% since the bucket modification.

On the basis of the data collected, action was taken to remove the test adaptors, and to restore the shovel and bucket to its previous productivity. Upon re-testing the shovel bucket, loading time data confirmed the loading time to be back in line with the fleet averages.

Quartzite Lake Dumping

Quartzite lake is a small lake located adjacent to the ore body and surrounded by overburden dumps. Changes in the mining conditions offered the possibility of filling in Quartzite lake to reduce the average overburden haul distance, reducing overburden haulage costs.

Calculations for savings were made based on 1993 budgeted production levels and the dispatch system's historical dumping data. The resulting calculations revealed a reduction in the average haul distance of 1.5 km, and a potential haulage saving of $322,000.

Approval for this project was recently received from Environment Canada and dumping is scheduled to begin in the spring of 1993.

Shovel Freeze Ups

One of the main causes of shovel downtime during the winter months is the freezing of components in the compressed air system. The cold air drawn in by the filter fans can lead to the freezing of the solenoid valves that control the swing, hoist and propel motors.

A project team was chartered to determine a solution for reducing this cause of downtime. A first step in arriving at the solution, was to quantify the magnitude of the problem. The dispatch system was configured to collect specific records of shovel downtime due to freezing .

The results indicated that the equivalent of one shovel operating shift per month was being lost due to freeze ups. A justification was prepared recommending the modification of the intake ductwork, adding heaters, and enclosing the solenoid valves.

The recommended modifications were carried out on one of the shovels, and preliminary results indicate a reduction of up to 30% in lost time due to shovels freeze ups.

Monitoring is continuing, and plans are now underway to install the modifications on the rest of the shovel fleet.

Mobile Supervisor Terminal

An additional benefit of a dispatching system is the increased quantity and distribution of information about the mining operation. Distribution of this information helps mine operations personnel to make informed decisions about actions to take to address changes in the operating conditions.

One of the best indicators of the "value" of this real time operating information, is that the mine now operates with a single mine foreman per shift. The mine foreman is connected to the dispatch system through a Mobile Supervisor's Terminal (MST) placed in the mine foreman's truck.

The MST allows the foreman to observe and update the location and status of the mining equipment. In addition he can observe and override truck assignments if he sees a condition requiring immediate action, or wants to dump material in a specific location.

This functionality has eliminated the need for the mine foremen to be in the immediate area of the dispatch network in order to update or affect changes in the dispatching system.

CONCLUSIONS

Clearly, from the quantifiable productivity gains, and the sample usage of the collected system data, benefits have been derived from the installation of the computerized mine monitoring and dispatch system in at least two major areas.

1. Real time haulage optimization.
2. Information utilization.

The magnitude and area of benefit will often reach beyond the initial conception and justification of haulage optimization and grade control. Potentially, the greatest savings to be realized, will lie in the utilization of the detailed information that the system gathers.

Obtaining the benefit from both, the real time dynamic dispatching, and the collected information requires a structured approach to solving the mine's operating problems. For the effective utilization of dynamic dispatching, mine management must support and encourage mining practices and setups conducive to dynamic dispatching. In the utilization of the collected information, mine management needs to seek out supporting factual data and incorporate this into their mine operating decision making.

It is only through this continued effort and commitment that the full spectrum of benefits will be derived from a computerized mine monitoring and dispatch system.

REFERENCES

Walshe, P.M., & P.F.Cimolini 1989. Truck dispatch - a practical and progressive method. *Society of Mining Engineers Annual Meeting:*Preprint number 89-46.

Innovative Mine Design for the 21st Century, Bawden & Archibald (eds) © 1993 Balkema, Rotterdam, ISBN 90 5410 325 6

Computer-aided achievement of mine planning and production goals at Barrick Goldstrike Mines Inc.

C.J. Bostwick & T.L. Buchanan
Barrick Goldstrike Mines Inc., Elko, Nev., USA

ABSTRACT: Over the past five years American Barrick Resources has implemented an extensive suite of systems designed to assist in achieving an aggressive 415,000 tpd mining program at the Goldstrike operation near Elko, Nevada. Goldstrike employs a mix of software to undertake reserve modeling, long-range planning, equipment evaluation, short-range planning, ore control and daily production optimization and accounting. The key to integrating these functions is utilizing the best off the shelf software currently available and developing custom applications to Goldstrike's specifications when software does not exist to meet the requirements.

1 INTRODUCTION

One of the most commonly cited advantages of computer-assisted mine planning is that the "engineer is rapidly able to assess multiple variations and strategies". This point has become obvious to most professionals in the mining industry; the current challenge is to minimize the time required to optimize a plan so that the engineer can proceed to the next cost saving/optimization project. Competent personnel with an intimate knowledge of the project and state-of-the-art software tools in hand help to reduce the number of iterations necessary to arrive at an optimal solution.

2 BETZE DEVELOPMENT PLAN

American Barrick Resources Corp., acquired the Goldstrike Mine, located on the Carlin Trend, 50 miles north-west of Elko, Nevada, in December 1986 (Figure 1). Goldstrike at the time was a 40,000 ounce per annum heap leach operation exploiting multiple small pits. A deep drilling program, started in 1987, has since revealed an extensive sulfide orebody, which at the beginning of 1993 was estimated to contain a total of approximately 21 million ounces of economically mineable gold on Barrick property. A detailed development study was undertaken and used as the basis for financing the $500,000,000 capital cost of the project by various means, most notably a 1,050,000 ounce gold loan.

What followed was a controlled expansion of mine operations from 87,000 tpd in 1988 to 415,000 tpd in 1993. In addition, a phased construction/expansion of milling facilities resulted in a 15,000 tpd sulfide circuit being completed in March 1993. Tonnage and ounce projections outlined in the four-year expansion phase (1989-1992) of the Development Plan have been consistently exceeded (Figure 2). Planned mine production and ore deliveries have been achieved or surpassed through the use of large scale mining equipment (Figure 3) and aggressive operating and engineering

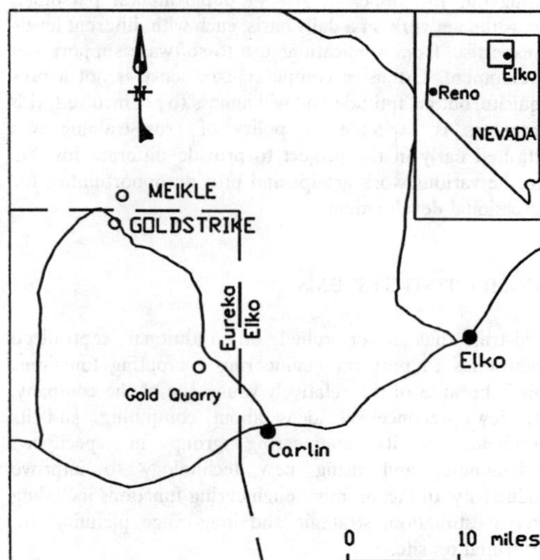

Figure 1. Goldstrike Location

procedures. As a result, Goldstrike has evolved into one of the largest hard rock metal mines in the world and in 1992 was the single largest producer of gold in North America.

3 ENGINEERING ORGANIZATION

The primary function of mine engineering is to directly support all facets of the mine's operations. Mine operations, engineering and maintenance make up the Mine Division, the largest of four divisions at Goldstrike. The division employs over 750 staff and hourly personnel. The Engineering Department is organized around work groups, with five

Year	Betze Development Plan		Actual Production		Year End
	Total Tons	Rec. Oz.	Total Tons	Rec. Oz.	Reserves Oz.
1987	----	----	16,472,498	40,144	10,760,668
1988	----	----	31,256,306	119,418	16,109,900
1989	62,474,000	185,000	70,234,802	207,264	18,373,100
1990	91,583,000	315,888	127,067,574	352,880	18,378,000
1991	100,317,000	407,000	127,759,262	546,146	18,741,000
1992	100,368,000	944,000	140,480,258	1,108,219	20,620,000
1993	99,390,000	987,000	151,000,000 *	1,300,000 *	

Figure 2. Betze Development Phase - Yearly Production * forecasted

senior engineers reporting to a Chief Mine Engineer (Figure 4). Each individual within a group has specific responsibilities and each group performs express functions within the department.

Although definite roles are played, the four main groups, long-range planning, short-range planning, ore control and geologic modeling have constant interaction due to the natural flow of information between them. A significant amount of this information flow comes from data interchange through an extensive computer network. Thus computers play an indispensable role within the department and, in general, throughout the minesite. All 44 departmental personnel access the network on a daily basis, each with different levels of expertise, from applications use to software support and development. Extensive computer experience is not a pre-requisite, but an aptitude and willingness to get involved with computers is expected. A policy of cross-training was instituted early in the project to provide different insights into the various work groups and provide opportunities for professional development.

4 COMPUTING SYSTEMS

Goldstrike has never relied on traditional centralized mainframes to perform engineering computing functions. This is because of the relatively young age of the company, with few preconceived ideas about computing, and its confidence in its engineering group in specifying, implementing and using new technology to improve productivity. In fact all mine engineering functions including reserve estimation, strategic and long-range planning are performed on site.

In 1988 the core computing equipment consisted of a MicroVax II, with 12 terminals and two personal computers. The VAX served as the principal platform for ore control, survey, cone and reserve runs; and was remotely linked to the metallurgical lab for assay tracking. It was evident that this original hardware and software was inadequate in power and functionality for the planned size, complexity and aggressive development of the Goldstrike operation. Goldstrike thus had the advantage of starting from scratch in implementing a technical working environment when major changes were underway in engineering computing systems.

In August 1989, two Sun Microsystems SPARCStations arrived on-site to service the Truck DISPATCH system. The power, functionality and utility of the SPARCS was immediately realized and an additional machine was purchased for software development. The original SPARCS on site each possessed over 12 times the processing power of the MicroVAX and featured a multitasking/multiuser environment, superb networking abilities and high-end graphics capabilities.

The first major application to be ported to the SPARC, other than DISPATCH, was the Q'Pit mine planning software. Other major applications were added when SPARC/UNIX software ports became available. As of June 1993, Goldstrike's Engineering Department had 16 SPARCStations running over an ethernet based network. All peripherals including printers, plotters, and storage devices, are linked to all CPU's on the network. In addition, the department's 12 personal computers have full network access. Goldstrike's Engineering Department, over the last four years, has implemented a distributed computing environment.

By putting the power of the computer in the hands of the users, while retaining responsibility for software maintenance, training and hardware in the hands of individuals in the same working environment, users become more productive in their work and proactive towards a computer-assisted engineering environment. It may seem that the hidden agenda is to put a workstation on every engineer's desk. There are many that believe this is how it should be, but the objective at Goldstrike is to have the necessary equipment available to get the job at hand done in a timely, consistent and aggressive manner. Engineers should not schedule their time around the availability of workstations.

The success of mine engineering's implementation of a networked distributed computing environment is reflected in the adoption of such a system for the entire minesite's administrative computing requirements. A network of over 200 SUN servers and workstations, X-terminals, and P.C.'s came fully online in 1993 to replace an overloaded mainframe. This new network is devoted to accounting, purchasing, payroll, warehousing and mill and mine maintenance functions.

1988	1993
3 992C loaders	3 23 yd hydraulic shovels
2 13 yd hydraulic shovels	4 42 yd electric shovels
1 16 yd hydraulic shovels	18 100 ton trucks
21 85 ton trucks	46 190 ton trucks
12 100 ton trucks	

Figure 3. Primary Mining Equipment

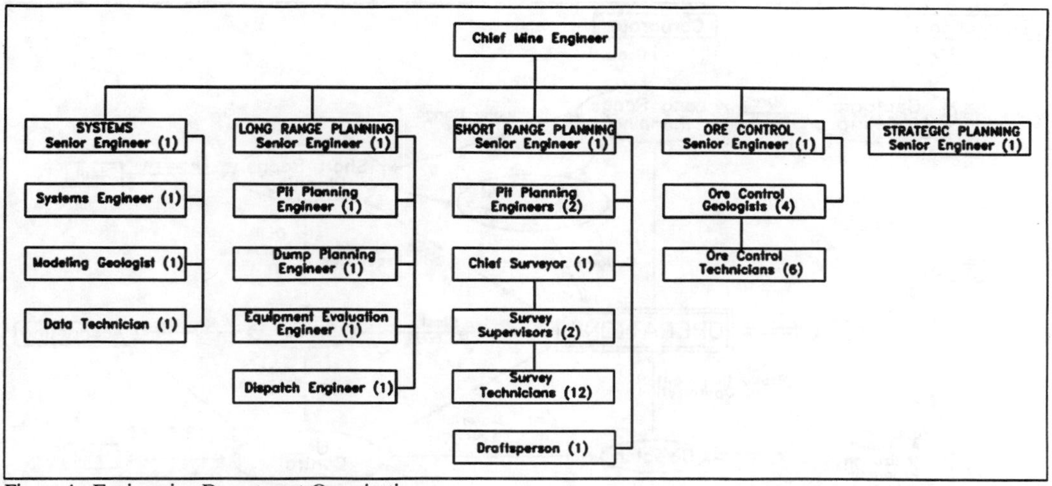

Figure 4. Engineering Department Organization

5 SOFTWARE APPLICATIONS

Unfortunately for the mining industry, software functionality has not kept pace with advances in computing power. As desirable as it may seem, there is currently no one system available within the mining software industry that can meet all of Goldstrike's requirements in each of the prime areas of our operation. Each of the over one-dozen major software packages has an area of specialty and uniqueness on which the system is marketed but each also, unfortunately, has areas of considerable weakness that negates it's consideration as a "complete" system. Areas of weakness can be found in the lack of functionality or in the inability to cater to the size of an operation such as Goldstrike. Through necessity, Goldstrike has opted to utilize separate software packages in each of the main areas of work, namely orebody modeling, pit planning and scheduling, ore control, surveying and technical drawing. Thus the strengths of each package are utilized while the weaknesses are not encountered.

Some of the obvious disadvantages of operating multiple software packages are training and supporting users in multiple systems, the inter-transferability of data of different formats and the cost of acquisition. If the cost of acquisition can be justified by the benefits brought by each system, a commitment to the required user training and support can be made by various internal and external sources. This leaves data transferability as an area of concern. Extensive, and largely user transparent, transfer routines have been developed at Goldstrike to ensure that both drawing and numerical data can be readily accessed by all applications. Figure 5 depicts some of the data interactions both within mine engineering and between various departments. The following is a discussion of some of the major software applications:

5.1 Mine Modeling

Before any deposit can be planned and mined, the best possible representation of the orebody must be produced. The orebody modeling modules of MEDSystem are used exclusively in Goldstrike's modeling efforts. Two other systems had been used at various times in the past, but have been discontinued due to a lack of functionality with respect to Goldstrike's requirements. Problems discovered with other software, not necessarily unique to Goldstrike and largely ignored by many vendors, are:

1) the complexity of the modeling methodology
2) the size of the modeled area
3) the amount of input data.

Some of the requirements with respect to complexity include:

- 3-dimensional kriging
- kriging boundaries and zones
- solids modeling of geologic zones
- multiple kriging parameters
- interactive variogram modeling
- blasthole integration
- flexible input/output routines

Goldstrike's database has been developed over a number of years and has grown to a considerable size. As a result, data storage and manipulation requirements must be able to handle:

- over 5000 exploration holes (increasing by 200 per year) with over 100,000 assay intervals and up to 8 assays per interval
- over 1,000,000 blastholes (increasing by 10,000 per month)
- 3,000,000 mining blocks with 20 parameters and 10 descriptive attributes per block

The primary modeling effort is separated into three areas:

5.1.1 Reserve Modeling

The gold reserve model is prepared on an annual basis and is used primarily for long-range pit planning and securities regulatory review of reportable reserves. As such it receives considerable attention and is used as the basis for all other modeling efforts. Work on the reserve model is undertaken throughout the year with the drill hole database being updated as new logs are received. Models are generated for internal use in planning and forecasting with an official

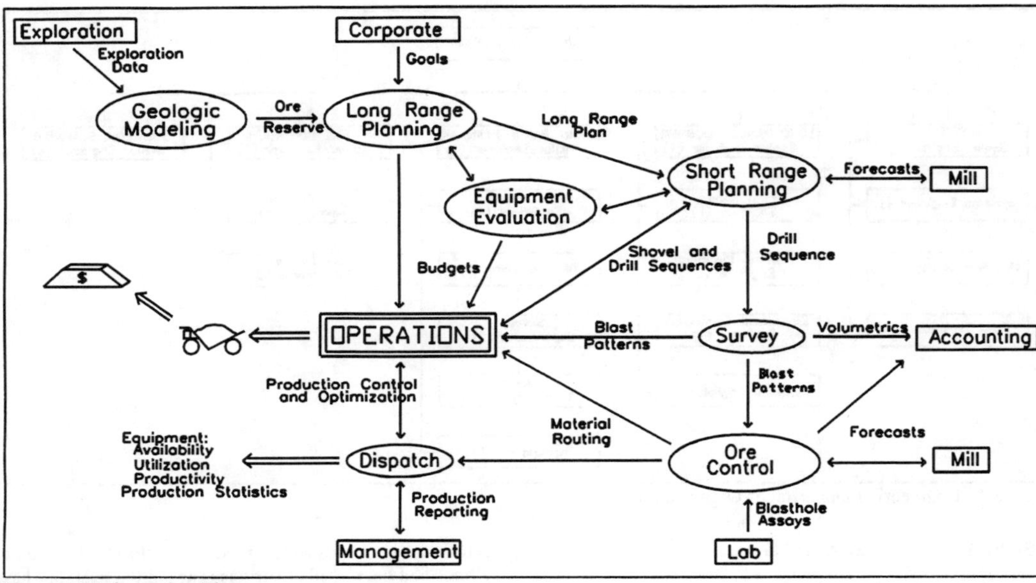

Figure 5. Information Flow

reserve model generated in December of each year.

The reserve model currently uses standard 3-dimensional kriging within seven distinct zones of unique variography. Six values are estimated and three are calculated for each 50x50x20 foot block. These, along with statistical kriging parameters (ie. variance, distance to nearest composite etc.), are used to estimate the contained gold, recovery, process type (ie. leach, bioleach or autoclaving) and reserve category. The reserve is audited on an annual basis and the auditor's recommendations are incorporated into subsequent efforts.

Current drilling and modeling work is concentrating on delineation of waste/ore contacts in the footwall, as some final walls will soon be established, and infill drilling to tighten the drill spacing and define internal high-grade areas. Goldstrike has historically over-estimated the tons and under-estimated the grade due to smearing of values. As a result, extensive work is being undertaken in new modeling techniques to make the grade/tonnage curve more realistically reflect the actual blasthole values.

5.1.2 Short-Range Modeling

Since the orebody is structurally controlled and there are well defined, although not necessarily apparent in the exploration data, hard cutoffs between ore and waste, further definition of the ore zones is required for short-range pit planning. Three-month mining forecasts are given to the modeling/ore control geologists on a monthly basis and areas for additional work are outlined.

The exploration and blasthole databases are merged for the targeted areas. Structural control, determined from field mapping and interpretation, is entered. Blastholes from adjacent previously mined benches or cuts, if available, are used in the estimation. The variography is determined and 25x25x20 foot blocks are kriged accordingly. In addition, ore control frequently targets areas of high-grade ore and areas of potentially poor metallurgical properties for partial pre-production drilling with 80-foot blastholes. This provides additional information on vertical or steeply dipping structures and small high-grade pods that may be as deep as three benches below the current mining horizon.

Short-range modeling has been proven to decrease the variance in block estimation of mining cuts and has increased the accuracy of weekly and monthly ore forecasts.

5.1.3 Waste Modeling

Acid mine drainage from waste dumps has recently been the focus of much attention in the North American Mining industry. Although the nature of the materials being mined at Goldstrike have little or no potentially acid-producing capability, Goldstrike has developed a comprehensive model of the acid-generating properties of waste within the planned mining limits. Using static base data from sulfur and carbonate assays, the acid-neutralizing and acid producing-properties of all waste blocks have been estimated. This model, in conjunction with blastholes and kinetic test data for calibrating static data, will be used to route waste to specific locations within the waste dump to mitigate the potential for acid mine drainage. It is believed that no other operation has undertaken such an effort to identify potential problems before mining actually occurs. Typical environmental regulations require only that quarterly samples be submitted, and then, only after dumping has taken place.

5.2 Long-Range Planning

Early in the project it was realized that a flexible system would be required to meet the day-to-day pit planning and evaluation requirements at Goldstrike. An informal software search did not yield a suitable product nor any suppliers

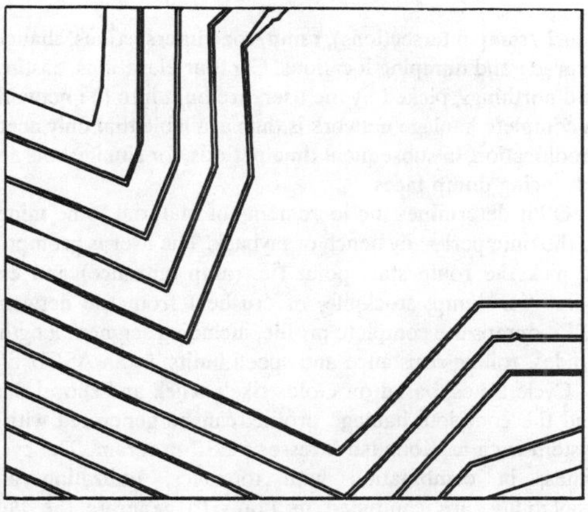

Figure 6a. Original Surface

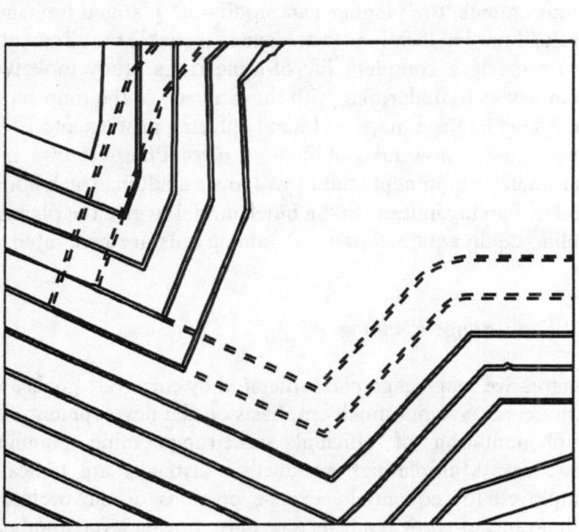

Figure 6b. Projection of Cut

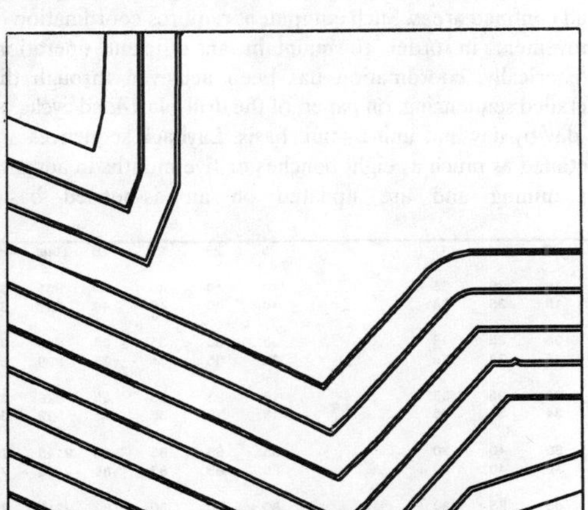

Figure 6c. Surface after Applying Cut

Figure 6. Projecting and Applying Cuts

willing to make the modifications required to meet our standards. Attention came to the Q'Pit system, initially developed as a commercially funded university research project, that had the potential to meet Goldstrike's requirements. The system was conditionally acquired in January 1990 with the understanding that the developer would meet a defined functional specification. The specifications were met and additional functionality has continuously been added.

Q'Pit uses a standard floating-cone algorithm, with a user-defined revenue function, to generate economic shells. The power of Q'Pit, however, can be found in phase development, ramp generation and sequence derivation. These are facilitated through a powerful graphical interface that provides the user with the drawing, display and computational tools needed to quickly and efficiently develop and evaluate a mine plan.

Goldstrike produces "phases" by floating cones at various gold prices. The phases are then smoothed and ramps added to produce minable shapes. The phases dictate the economic development of the pit and are used as limits in layback development. The current pit shell has six laybacks, or expansions, to be developed through the life of the mine.

Cuts, laybacks or pushbacks, which are either input or selected by the planner, are projected up or down a specified number of benches. The prevailing slope angle, bench face angle, and catch-bench width are maintained during the projection. The projected cuts can be modified at this point or "applied" against the original surface (Figure 6). When applied, the projections cut the original surface leaving a new surface of the desired geometry.

When applying a cut, a limit reference can be used to restrict the bench-by-bench extent of the cut. This would be used to cut benches back to a pre-designed layback, final wall or defined surface (ie. property line, power line etc.). Any line on any pit, limit reference or boundary displayed on the screen can be edited and independently saved.

Ramps are automatically generated by specifying a top elevation, bottom elevation, start point and ramp direction.

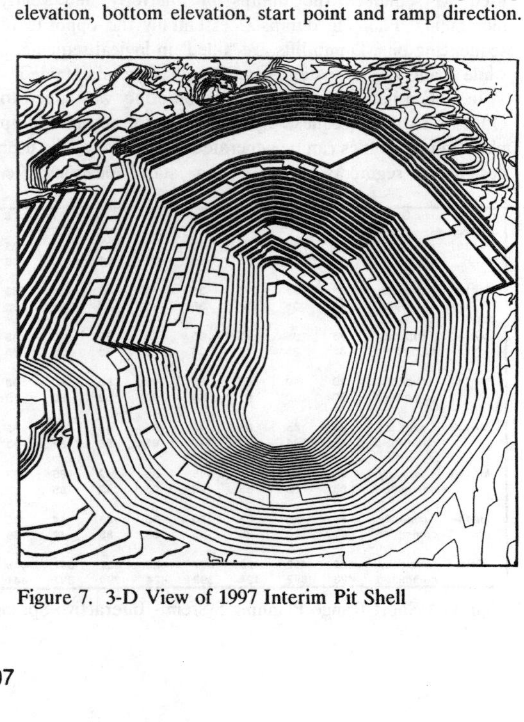

Figure 7. 3-D View of 1997 Interim Pit Shell

Ramp parameters such as width, percent inside and outside of limit, grade and slope angle are maintained while the cuts and fills are generated.

Numerous utilities are available to the engineer to act as aids during the planning process. Any geologic model can be displayed in plan or section. Blocks can be color coded by material type, grade range, metallurgical properties, with any number of kriged assay values displayed within the block. Surfaces such as rock-type contacts, faults, or boundaries such as property or claim groups can be displayed. Online tonnage summaries, in a user-defined format, can be displayed by cut, bench and/or time period. Detailed hardcopy reports can also be generated. Scaled hardcopy plots can be generated of any graphics image through screen dumps or formal plotting routines. Pits can be displayed or plotted in 3-dimensional views (Figure 7) and can be exported in formats suitable for AutoCAD, MEDS and Surpac.

One of the most useful time-saving utilities has been found to be the UNDO facility. This enables the planner to revert any modified line, projection, or cut, to it's previous position. This can be accomplished in either stepwise fashion or to any selected line or set of lines.

5.3 Equipment Evaluation

An integral part of the pit planning process is the determination of equipment requirements. Goldstrike runs a truck dependent operation and thus utilizes all available truck hours. Given a constant quantity of equipment, with varying material centers of gravity between time increments, achievable production can vary from month to month and year to year.

As pit sequencing is driven by the need to expose and produce ore, flexibility in the haulage cycle is attained by varying the length of the dump portion of the haul. Q'Pit is used to sequence the dumps in quarterly and yearly increments. Planning dumps is essentially the opposite of sequencing pits. Dump lifts are "filled" in logical sequences, while roadways, ramps and catch benches are maintained.

In 1993 a haulage profile generation utility was added to Q'Pit at Barrick's request. By combining the pit and dump plans, route profiles can be generated interactively. The user designates segments by selecting start locations (ie. bench/ramp intersections), ramp/road intersections, changes in grade and dumping locations. Contour elevations, eastings and northings, picked by the user, are built into the network. A complete haulage network is then available that only needs modification, in subsequent time periods, for sinking cuts and advancing dump faces.

Q'Pit determines the increments of material to be mined in the time period by bench or layback. The user is prompted to pick the route start point (ie. ramp entrance) and end point (ie. dump, stockpile, or crusher) from the network. Q'Pit dumps the complete profile, including segment lengths, grades, rolling resistance and speed limits, to an ASCII file.

Cycle times, based on Goldstrike's truck and shovel fleet and the complete haulage profile, can be generated with a system such as Komatsu-Dresser's ESP program. The cycle times, in combination with tonnages, utilization and availability, are combined in Lotus to generate the fleet requirements for that time period. After reviewing the fleet requirements, the planner can modify the planned tonnages or the dumping locations to achieve reasonable requirements.

Formerly a complete life-of-mine truck study took two man-weeks to undertake, with the majority of the time spent analyzing haulage maps and hand entering profiles into ESP. Such studies now take about two days. Phase II of semi-automated equipment studies will involve adding truck hours and drillability indexes to the block model to give the planner online equipment estimates as mining cuts are generated.

5.4 Short-Range Planning

Aggressive long-range plans, dictated by corporate goals and strategies, have put much emphasis on the development and implementation of efficient short-range mine planning practices. Multiple-high production laybacks are typically employed to sequentially expose portions of the orebody. High rates of vertical advance require moving large numbers of drilling and loading equipment through relatively small and confined areas. Such equipment requires coordination of movement in order to maintain an efficient operation. Historically, coordination has been achieved through the detailed sequencing, on paper, of the drill-blast-load cycle, on a day-by-day and unit-by-unit basis. Layback sequences are detailed as much as eight benches or five months in advance of mining and are updated on an as-needed basis.

SHOVEL #	DATE	1	2	3	4	5	6	7	8	9	10	11	12	....	25	26	27	28	Total	Avg.
MA131	capacity	18	35	35	35	35	35	35	35	35	18	35	35	....	40	40	40	40	901	32
	committed	18	34	35	35	36	35	35	36	35	18	35	34	....	40	40	40	40	901	32
MA132	capacity	35	35	18	35	35	35	35	35	35	35	35	18	....	35	35	35	35	929	33
	committed	35	35	18	35	35	36	35	35	35	35	34	18	....	35	35	35	35	929	33
MA135	capacity	35	35	35	35	18	35	35	35	35	35	35	35	....	18	35	35	35	929	33
	committed	35	35	35	36	18	35	35	35	35	34	35	35	....	18	35	35	35	929	33
MA141	capacity	80	40	80	80	80	80	80	80	80	80	40	80	....	85	85	85	85	2145	77
	committed	80	40	81	80	80	80	80	79	80	79	40	80	....	85	85	85	85	2145	77
MA142	capacity	85	85	85	42	85	85	85	85	85	85	85	80	....	80	80	80	80	2212	79
	committed	85	85	85	42	85	86	85	85	85	85	85	79	....	80	80	80	80	2212	79
MA143	capacity	85	85	85	85	85	42	85	85	85	85	85	85	....	85	42	85	85	2251	80
	committed	85	85	85	85	85	42	85	85	85	85	85	85	....	85	42	85	85	2251	80
MA144	capacity	85	85	85	85	85	85	42	85	85	85	85	85	....	85	85	85	42	2251	80
	committed	85	85	85	85	85	85	42	86	85	85	85	85	....	85	85	85	42	2251	80
Total	capacity	423	400	423	397	423	397	397	440	440	423	400	418	....	428	402	445	402	11618	415
	committed	423	399	424	398	424	399	397	441	440	421	399	416	....	428	402	445	402	11618	415

Figure 8. Short-Range Planning System - Interactive Spreadsheet

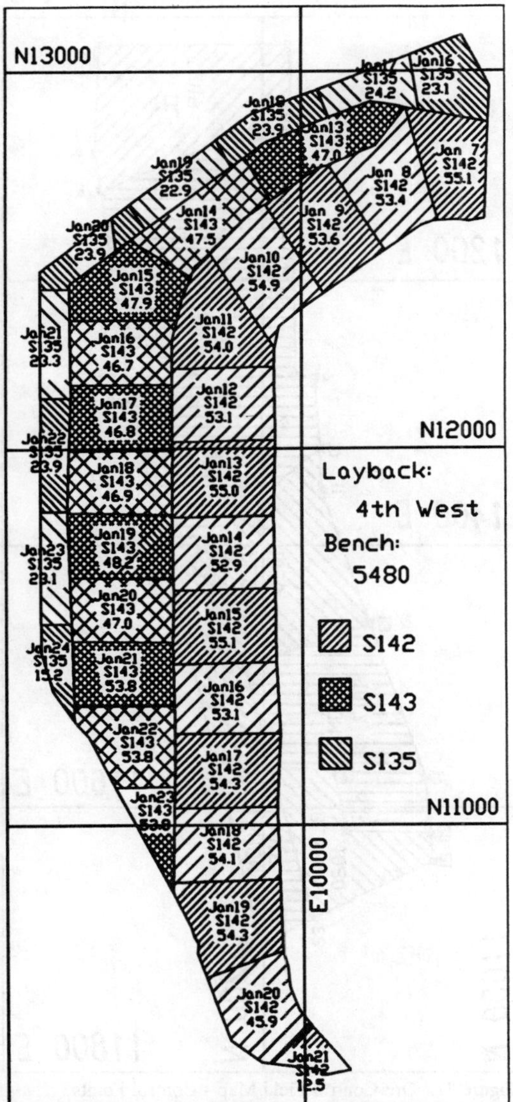

Figure 9. Screen Dump of Shovel Sequence

Communication of the sequences among the engineering, operations and maintenance staffs ensures that the plan is achievable and all potential problems are discussed and resolved.

The emphasis placed on short-range planning and the time-consuming efforts required to develop and update plans made this function an obvious candidate for some form of computerization. An extensive software search indicated that the proven long-range planning system employed at Goldstrike, Q'Pit, should be used as the basis for development of the short-range system. The conceptual design provided for mechanization of the repetitive tasks which the short-range planner must perform to develop and evaluate a shovel sequence. A system was developed that allows the engineer to begin with an actual surveyed pit configuration and interactively develop daily, weekly and monthly shovel sequences based on the long-range mine plan and production constraints. The developed sequences can be evaluated against the reserve model or the short-range ore control model to produce accurate ore forecasts.

The short-range system is initialized by identifying the shovels to be sequenced, the length of the sequence period, and the default daily productive capacity for each of the loading units defined. Q'Pit then produces an interactive spreadsheet (Figure 8) which gives the productive capacity for each shovel for each day of the sequence period (up to 31 days displayed). This spreadsheet can be edited to reflect planned maintenance on the loading units; the productive capacity for each unit can be reset on any day to any desired figure.

Surveyed pit configurations are imported into the system directly from the survey software or through digitizing. Long-range plans are utilized to produce "initial cuts", which represent the volumes of material remaining on the benches to be mined in the current long-range plan. The user identifies the bench on which to begin the sequence. The "initial cut" for the bench is interactively divided into "strips" representing the major cuts to be mined - typically these are crest, production and trim cuts. Tools within the software allow these strips to be sized appropriately. A direction of advance, or guide line, is interactively specified for a strip and a shovel selected along with a start date. The strip is then automatically split into increments representing one day's mining for the specific shovel. The splitting into increments can be accomplished for the entire strip at once, for a specified number of days, for one day at a time or for a specific gross tonnage. The spreadsheet then displays the "committed" mining quantities for the shovel which is being

BEGINNING STOCKPILE INVENTORY:	Week of 10/1						
		R1	R2	R3	R4	R5	Total
	TONS (x1000)	250	221	165	227	198	1061
	Au (oz/t)	0.189	0.293	0.430	0.226	0.456	0.306
	CO3 (%)	1.211	2.320	2.380	8.237	7.580	4.316
	S2 (%)	2.100	0.889	0.543	1.122	1.349	1.256
MINED BY SHOVEL:							
MA131	TONS (x1000)	30	5	0	27	36	98
	Au (oz/t)	0.177	0.266	0.000	0.277	0.452	0.310
	CO3 (%)	0.590	3.410	0.000	7.100	9.200	5.690
	S2 (%)	0.890	0.300	0.000	1.320	1.200	1.092
MA142	TONS (x1000)	12	0	0	0	5	17
	Au (oz/t)	0.173	0.000	0.000	0.000	0.670	0.319
	CO3 (%)	4.300	0.000	0.000	0.000	9.600	5.859
	S2 (%)	0.910	0.000	0.000	0.000	2.120	1.266
MA142	TONS (x1000)	0	15	37	24	0	76
	Au (oz/t)	0.000	0.251	0.455	0.320	0.000	0.372
	CO3 (%)	0.000	3.560	5.230	9.200	0.000	6.154
	S2 (%)	0.000	0.890	1.200	2.120	0.000	1.429
TOTAL MINED:							
	TONS (x1000)	42	20	37	51	41	191
	Au (oz/t)	0.176	0.255	0.455	0.297	0.479	0.336
	CO3 (%)	1.650	3.523	5.230	8.088	9.249	5.890
	S2 (%)	0.896	0.743	1.200	1.696	1.312	1.242
PROCESSED:							
	TONS (x1000)	25	25	25	15	15	105
	Au (oz/t)	0.178	0.289	0.476	0.257	0.489	0.331
	CO3 (%)	2.100	1.930	5.200	9.120	8.230	4.676
	S2 (%)	0.890	0.910	0.860	0.900	1.190	0.932
END STOCKPILE INVENTORY:							
	TONS (x1000)	267	216	177	263	224	1147
	Au (oz/t)	0.221	0.357	0.563	0.267	0.523	0.369
	CO3 (%)	1.590	2.923	4.046	9.198	8.944	5.401
	S2 (%)	2.191	1.084	0.879	1.349	1.512	1.454

Figure 10. Ore Mining/Stockpile Inventory Spreadsheet

sequenced. The next strip or portion of a strip can then be sequenced in the same manner utilizing any of the shovel-days not yet "committed". A shovel can mine two or more increments on the same day as long as the productive capacity of the shovel on that day is not exceeded. Figure 9 is a completed shovel sequence map as seen on the graphics screen. Each days mining increment is shown with the date, shovel number, and tonnage indicated. Shading, or color, is unique for each shovel. The spreadsheet, which now indicates planned shovel production per day, can be output as well.

A fully or partially completed sequence can be evaluated against a geologic model to indicate the resulting ore/waste production. User-designed reports are employed to produce output that can be imported to a spreadsheet. Figure 10 is an example of this output.

If planning is interrupted, the sequence, as developed, can be stored and retrieved at a later date to continue with the planning process. The geologic model can be displayed at any time during the sequencing process. Additionally, any other data resident on the network pertinent to the design can be displayed: existing topography or planned mining lines on any bench, power lines, dewatering pipelines, property boundaries, etc.

A Beta release version is currently being tested at the minesite with full implementation planned for early 1993. One of the main implementation goals is to decrease the time spent in developing and modifying layback sequences and enable the planning engineers to spend more time in the field doing quality-control work. Enhancements being considered include interactive design of blast patterns; as the design and sequencing of blast rounds is an integral part of the short-range mine planning process at Goldstrike.

5.5 Ore Control

The primary function of the ore control group is to estimate the grade of blasted rounds and assign the material, in minable units, to the destination dictated by the current routing logic. Ore control receives electronic copies of the blastholes from survey, and the assay data from the lab. Blasthole locations and assays are automatically merged and combined with blastholes from adjacent, previously mined patterns, in preparation for kriging. Kriging, performed with MEDSystem, utilizes dig lines from short-range planning, 3-d kriging, zonal variography and structural control.

After kriging, each 10x10x20 foot block is color coded according to broad material categories (ie waste, leach, sulfide) and numbered according to the detailed routing scheme. Since it is not feasible to individually stake and route ten-foot blocks, ore control uses the MEDS Interactive Graphics Planner (IGP) to group blocks into minable units. Ore control interactively draws polygons on the screen around blocks of material with similar grade and metallurgical properties. Loading equipment size, direction of mining and minimization of grade dilution are considered when delineating the size and shapes of ore and waste polygons. An online spreadsheet is presented to the user depicting the weighted recovered ounces for each possible routing, highlighting the optimal material routing. After ore control is satisfied that the most optimal polygon layout has been achieved, a map with control points and routing

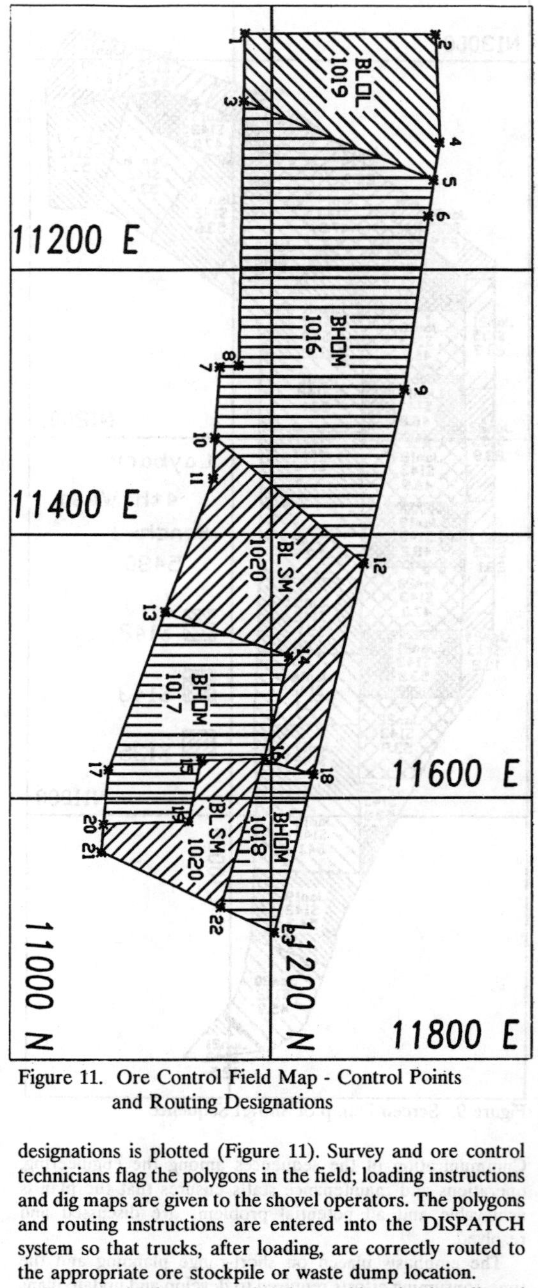

Figure 11. Ore Control Field Map - Control Points and Routing Designations

designations is plotted (Figure 11). Survey and ore control technicians flag the polygons in the field; loading instructions and dig maps are given to the shovel operators. The polygons and routing instructions are entered into the DISPATCH system so that trucks, after loading, are correctly routed to the appropriate ore stockpile or waste dump location.

Computerized ore control and record keeping has allowed more time to be concentrated on optimal ore routing, field work, quality control and project work.

5.6 Truck Dispatch

The benefits of computerized truck allocation systems, particularly Modular Mining Systems' DISPATCH system, are well documented in technical literature. The DISPATCH system was commissioned at Goldstrike in October, 1989, and has resulted in a large productivity increase within the mining operation. The system was installed for the obvious productivity reasons and to provide a means of operational control during the rapid expansion phase.

All loading, hauling, and drilling equipment is equipped with DISPATCH hardware and all other mine mobile equipment is tracked for availability within the system's database. As such, DISPATCH is used as the sole method for day-to-day production, ore routing and availability reporting.

Goldstrike has added numerous extensions to the system in an effort to maximize the return on the capital invested in the mine fleet. All drills are equipped with DISPATCH interfaces so that engineering and operations can track drilling advance and performance by round and area. Drillability of rock units is used by short-range planning in sequencing of patterns and drills, and by operations in budgeting for and selecting drill bits. Drill rounds are cross-referenced to shovel dig rates to determine optimum spacing and explosive requirements.

Work is ongoing in implementing the vital signs subsystem. When commissioned, the subsystem will monitor equipment performance and alert maintenance to problem conditions. Interfaces developed between DISPATCH and third party systems, such as Detroit Diesel's DDEC system, are being evaluated. Performance to date on vital signs and other monitoring systems has been very encouraging.

Pit production by shovel, bench and layback is tracked extensively by short-range planning to provide the most current and up-to-date planning parameters. The mine maintenance department tracks availability and down-time events by equipment ID, class and type, as well as by down reason, time and duration. Equipment histories can thus be generated and potential problem maintenance areas identified. Similarly, haulage fleet tires are tracked in a separate DISPATCH database, with tire hours, distance, and ton-miles per hour directly cross referenced to the equipment's production and performance.

6 GENERAL COMMENTS

Goldstrike's management and staff have taken a positive approach to computerization and view it as a primary means of enabling engineers to concentrate on other potentially cost-saving work. The key is not to force the application to mimic the methodology previously employed, but rather to allow for different approaches, while achieving the same or better results. Computerization often compels the user to follow preset actions which forces consistency and allows standards to be maintained. At the same time, the software must be sufficiently flexible to allow the engineer to react to non-standard conditions.

Goldstrike's openness to new ideas, and the size and complexity of the operation, has enabled it to become a Beta test site and excellent proving ground for new products and updated systems. The mine planning group has taken on a partnership approach with respect to design and needs assessment with the developers of the Q'Pit planning software. This has allowed our needs to be quickly addressed while increasing the marketability of the product. Goldstrike works closely with Modular Mining Systems in the implementation of new or modified software tools. Such upgrades are generally useful to Modular's user base and implemented at no charge. A software vendor's willingness to work hand in hand to meet Goldstrike's specifications is a requirement and the failure to do so has often resulted in Goldstrike not considering purchase of or further use of a supplier's product.

The interface between the planning groups requires short-range planners review long-term plans to see if the plans are achievable (ie. mining widths, rate of vertical advance) and long-range planners review short-term sequences and forecasts to see if targets are being met. Face positions and equipment performance by area are monitored for future plan and sequence modifications. It is evident to the engineers on-site that the ability to maintain the high level of productivity achieved in the planning process is directly related to the availability of flexible and functional planning software.

7 CONCLUSIONS

Changing environmental regulations, fluctuating metal prices and high market expectations are placing new demands on today's mining companies. Successful companies will be those that are adaptable, flexible in their approach to business decisions, and able to react rapidly to new situations. American Barrick Resources' success is evident in the aggressive implementation of the Betze Development Plan and Goldstrike's position as North America's single largest gold producer.

The mining community has been criticized, both internally and externally, for being resistant to changing attitudes and technological advances. Goldstrike's engineering group has been able to overcome much of this resistance with it's ability to react quickly to events and it's ability to implement solutions to problems before the problems become an issue. Much of this achievement can be attributed to the group's structure and leadership, site and corporate management's support and commitment and the availability of powerful and flexible tools.

Automated truck allocation and long-range mine planning software, through DISPATCH and Q'Pit, were the first to benefit from Goldstrike's strategy for computer system development. These systems have seen much support and development, both internally and from the respective suppliers, and have undoubtedly contributed to the success Goldstrike has seen in meeting and surpassing production goals.

A networked suite of software systems has enabled Goldstrike's engineering group to attain the high productivity levels and enhanced flexibility required of today's modern mining operations. When possible, the individual system strengths of off-the-shelf software packages have been implemented. When functional systems are not available they are custom developed to Goldstrike's specifications.

Innovative Mine Design for the 21st Century, Bawden & Archibald (eds) © 1993 Balkema, Rotterdam, ISBN 90 5410 325 6

Decisions in open pits and their relation to present value

E. Koniaris
Q'Pit Inc., Kingston, Ont., Canada

ABSTRACT: A descriptive approach is followed in the examination of some of the decisions routinely taken in long term open pit mine planning. A condition for the maximisation of present value is presented, based on the selection cutoff grades through time. This condition is subsequently used to explain some of the widely accepted practices in open pit mines and states that immediate mining costs are not relevant in setting a cutoff grade based present value maximisation policy. The established and required mine capacity are discussed in relation to present value. Examples are used to demonstrate that the gain in present value and the incentive that a mining company has in modifying its current operating environment and the established mine capacity, are measurable and, sometimes, small.

1 INTRODUCTION

The objective of the open pit mine design process is to define the best strategy for the depletion of an economic ore reserve. Strategy or policy decisions of a mining company are usually taken on the basis of the long term plan. Of the many objectives sought, much emphasis has been placed to date on present value as the figure of merit that should be maximized. This investigation examines whether some of the decisions and rules from common mining practice in open pit long term planning are in agreement with the objective of present value maximisation. It establishes the mine capacity as the primary quantity of interest and examines in particular present value as the incentive the mine operator has in modifying the current operating environment.

The reasons strategy or policy decisions of mining companies are examined on the basis of the long term plan may be attributed to the following:

In the long term, all costs can be treated as variable. The elimination of fixed costs simplifies the problem and the financial situation of producers may be described by the long term plan, or in other words, the determination of the sequence of volume removal from an orebody.

The short term response of a company may be quite constrained due to the logistical and contractual obligations. Alternatively it may depend on intangible factors, which are expected to average out in the long term.

The primary reason though for dealing with the long term is the fundamental requirement for analyses of company behaviour with a profit generating criterion. The convenient assumption is that a company must expect to generate some profit in the long term, in order to have an incentive to commence and/or sustain any activity. Cost reduction may be an alternate overriding criterion in the short term, which does not necessarily guarantee making a profit.

1.1 Related work

Of the reported investigations of the long term plan, one may distinguish two approaches, the cutoff grade based analyses and the solution derivation methods. The basic premise of cutoff grade based analyses is that the order of removal of the material (i.e. the long term plan) is known or may be neglected. As may be observed in many decision models, the underlying assumptions and the formulation is quite simple. Solution derivation methods on the other hand, concentrate on indicating the long term plan which would realize the optimum value of an objective. Models are in general more elaborate with this approach, usually based on the block model concept for the description of solutions. There appears to exist no widely accepted methodology in posing the long term planning problem in open pits, in antithesis to the universal acceptance for graph theory and maximum flow formulations for the solution to the pit limit problem as a net revenue maximisation problem.

The cutoff grade based applications originate from the seminal work of K. F. Lane(1963) who reported that, for the maximisation of present value, cutoff grades should decline with time. The model after Lane, shown in Figure 1, was used in work reported my many researchers, with notable the work by Taylor(1972), Jordi and Currin(1979), Shaap(1984), and for underground orebodies, Yi and Strugul(1987). This model is used as the basis of this investigation as well.

The relationships used to describe this model as presented by Lane are:

The concentrator level cutoff is given by

$$g_c = r\,p\,/\,c_p \qquad (1)$$

The mine level cutoff is given by

$$g_m = r\,p\,/\,(c_m + c_p) \qquad (2)$$

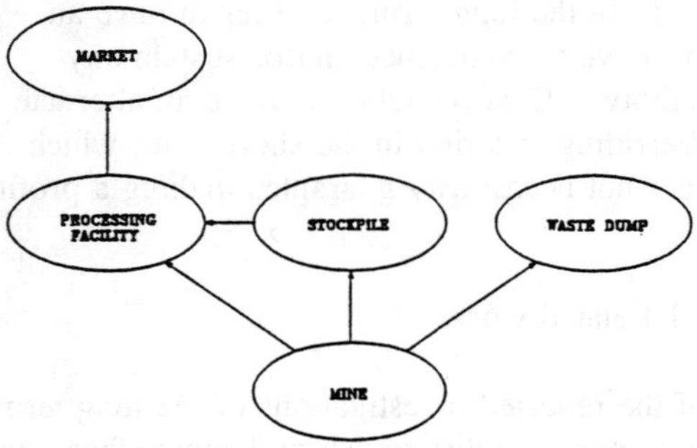

Figure 1. The model of a mining operation

With

c_m : mining cost

c_p : the processing cost

r : recovery

p : price

In order to have a sustainable economic activity, the mine must operate on average above the mine level cutoff grade at some point in time in the life of the property. The concentrator level cutoff grade is the grade at which costs are minimised for the last unit volume of material removed from the mine, or as usually stated, is the cutoff for the last time period of an operation.

In addition to the preset order of removal of the material, the assumptions in this formulation include:

1. The information contained in an orebody model is certain and complete.
2. Mine capacity may vary at zero cost.
3. No additional costs are incurred in identifying and re-routing materials of different grades.
4. The mineral or metal recoveries in the region of interest may be described with linear relationships.
5. The only reasonable estimate for the price of the mine commodity is the selling price at the time of the decision. This permits neglecting storage of "higher grades" for later processing and the related arbitrage questions.
6. There is only one tax rate.
7. Also mine closing costs, effect of economies of scale, blending and different methods of processing can be neglected.

1.2 The incentive to highgrade

The incentive to highgrade is examined in the model presented in Figure 2. The two volumes of material V_0 and V_k are indicated in the long term plan in periods 0 and k respectively. The mineralisation values are NV_0 and NV_k with $NV_k > NV_0$. We also define the mining cost, c_m, the rehandling cost, c_{rh}, and the discount factor for period k, δ^k, with, $c_m, c_{rh}, \delta > 0$. For notational simplicity the processing cost is subtracted from the mineralisation value. It should be noted that the order of mining in this model is fixed. In order to change the sequence of processing, i.e. realize NV_k in the current time period, volume V_0 must be mined first. The mine operator has an incentive to mine and stockpile V_0 for processing in time period n, when the following relationship holds:

$$(NV_k-c_m)\ (1-\delta^k) > NV_0 - (NV_0-c_{rh})\delta^n \qquad (3)$$

with $k=n>>1$, $d^k = d^n \rightarrow 0$ and

$$NV_k - NV_0 > c_m \qquad (4)$$

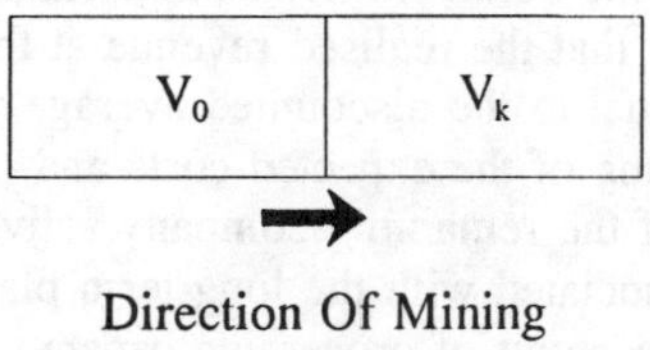

Figure 2. Order of mining (Fixed)

NV_k will in general be the mineralisation at the average grade of the material which is routed through the processing facility. Then (4) states that the incentive to highgrade will exist as long as the difference between the value at the cutoff grade and the average value of the ore or processed material (marginal revenue) differ by more than the mining cost (marginal cost). Expression (4) describes the single period profit maximisation operating point. With long mine life, this will hold with or without provision for a stockpile in the model.

If one examines a property in which all material above the mill cutoff is processed, then the average value of the deposit and the average value of the processed material is the same. As mentioned earlier, for a sustainable and always profitable mining activity, the average grade must be greater than the mill-mine cutoff described by expression 2. With the notation in this section, the latter is expressed with

$$NV_k - c_m > NV_0 = 0 \qquad (4)$$

Then expression (4) states that for a profitable operation and no matter how close to the end of life, NV_0 will always exist and be a positive mineralisation value.

Without provision for a stockpile and $k=n=1/12$, $\delta = .0153$,

$$NV_0 < (NV_{1/12} - c_m)/\ 66.34$$

This states that, without provision for a stockpile, the cutoff mineralisation value will be close to zero.

With $\delta = 1/(1+d)$, $d = .2$ and $t=1$, expression (3) gives that the mineralisation at the cutoff is

$$NV_0 < NV_1 - c_m + 5\ c_{rh} \qquad (5)$$

This states that rehandling is feasible in a sustainable operation one year before the end of mining operations when c_{rh} is sufficiently small. In a model with a stockpile and assuming that the rehandling charge is negligible, expression (5) denotes that highgrading will be attractive in a profitable operation, even in the last year of operation. In this case, the stockpile only serves as the means to perfectly order the sequence in which material is processed. Stockpiling gold ores is a typical example in this case, with the rehandling cost being considerably lower than the processing cost. The convergence point again is the single period profit maximisation point.

2 THE MAXIMISATION CONDITION

A maximisation condition may be derived based on the model above and the discussion presented by Lane, if the bound on the mine capacity is removed. The simplest way to look at the cutoff grade is to assume that enough material of that grade exists to feed the processing facility at capacity for one time period. This is descriptive when the time period is made quite small i.e. the time required to process one truckload of material. Assume then that the cutoff grade exists and the mineralisation value at that grade is NV_0. By the definition of cutoff grade, the mine operator should be indifferent in terms of present value by the following two actions:

1. Mine and process the material in the current time period. The mining and stripping costs are incurred. NV_0 is realised by processing the material and the remaining costs and revenues do not change.

2. Mine and discard (or stockpile) the material and the associated stripping in the current time period. This means that the mining and stripping costs are incurred, even though the material is not processed. NV_0 is "lost" but now there exists a unit of unused processing capacity. This permits moving ahead by one time period all subsequent activities to the end of the mine life. The statement of this model avoids mentioning that any additional mining is taking place in the current time period. This is included in the statement that the mining process is accelerated by one time period.

Let:

PV_1, PV_2 : the present values to be realised if courses 1 and 2 respectively are chosen

PV_k : the present value from activities k to tf

NV_0 : the mineralisation value at the cutoff grade at the current time period.

Q : the processing capacity (fixed).

c_m: the mining cost.

sr : the stripping ratio.

δ : the discount factor, with $\delta = e^{-it.}$

i : the interest rate.

t : a unit time, which in this example is the time required to process 1 unit volume of material and

tf: the last time period in which mining takes place

The present value for the two actions is:

$$PV_1 = Q\ NV_0 - Q\ c_m\ (1 + sr(NV_0))\delta\ PV_k \qquad (6)$$

$$PV_2 = -\ Q\ c_m\ (1 + sr(NV_0)) + PV_k \qquad (7)$$

The indifference condition at the cutoff suggests

$PV_1 = PV_2$ and

$$NV_0 = (1 - \delta)\ PV_k\ /\ Q \qquad (8)$$

Expression (8) is the present value maximisation condition with unbounded mine capacity. Compared with the similar condition presented by Lane, this formulation does not include the mining costs and therefore it will give lower cutoff grade mineralisation values.

The physical meaning of expression (8) is derived as follows. If a mining company processes material at the cutoff value only and without any mining costs, then the yearly revenue is given by

$YP = Q\ NV_0$

The present value of an infinite series of equal yearly payments, YP, at a discount factor δ, is also given by expression (8). Then the maximisation condition may be expressed as follows:

With fixed processing capacity and unbounded mine capacity which may be established and decommissioned at zero cost, at every point in time, the mine operator who maximises present value, sets the cutoff mineralisation value is such a way that the realised revenue at that grade is equal to the discounted average value of the remaining of the expected costs and revenues of the remaining company activities that are associated with the long term plan, expressed per unit of processing capacity. The discounted average value is to be considered as the expected value of a mineralisation which, with infinite time and no mining costs would give a present value equal to the present value the company expects from the remaining of its activities.

This condition does not require that the calculated present value be the result of an optimising policy. It requires though that the present value be known. It also implies that the pit limit be fixed, as the calculation of present value requires that the sequence of mining of all the remaining material be known at each point in time. In addition, it requires the association of activities with a parcel or block of material. As the long term plan describes the sequence of material volume removal, activities and their associated costs and realised revenues are moved forwards or backwards depending on the operating cutoff grade. The commencement of a new phase is a typical example. The time mining commences in a new phase is affected by the cutoff values during preceding phases. For the development of a satellite orebody, the time the related expenditures are incurred and the realised present value from its depletion are determined by the cutoff during the depletion of another orebody. This case as well as mining equipment replacement (capital) costs should also be included for the calculation of present value of the remaining activities. It is interesting to note the increased present value a mining company calculates after every capital expenditure. Condition (8) indicates that the cutoff grade mineralisation value must decrease just before and increase just after the time of a capital expenditure. The result may be compared to the formulation on the "learning by doing" problem discussed by Tirole (19).

Condition (8) holds and provides a unique cutoff grade for constant recovery or tailings grade as was stated in the model assumptions. When the recovery is a function of grade, the same mineralisation value may exist for two or more grades. This problem includes a blending decision for determining how much material is mined in each time period and is not covered under the assumptions of this model. Also, since the operation in production, the calculation of present value does not include fixed in time capital expenditures (constants).

The discussion above describes, explains and verifies the validity of the common process of generating cash flow tables for different long term planning strategies which include, in addition to the cutoff grades, equipment purchase schedules, production rates etc.

3 ON LASKY'S RELATIONSHIPS

Shaap (1984) suggested as the operating point for a mining firm the single period profit maximisation point. For an orebody which follows Lasky's relationships (in other words when the underlying probability density function is exponential), the single period profit maximisation point is expressed as:

$$T_{mx} = Q\, K\, r\, p/((1 + sr)c_m) \qquad (10)$$

T_{mx}: The ore output with infinite mine life
Q : The processing capacity
K : Lasky's constant, the difference between the cutoff and average grades and
sr : the "external" stripping ratio
The exponential distribution requires that

$$K = g_0\text{-}g_{c0} = g\text{-}g_c \text{ or } (g_0\text{-}g_{c0})/(g\text{-}g_c) = 1$$

with g_c, g_{c0} cutoff grades and g, g_0 the respective average grades of the material above each cutoff. Taylor (1972) reports that the distribution is common in copper deposits with the ratio of the difference between cutoff and average grades varying from .75 to 1.3.

The ore output for the last time period is equal to the processing capacity, with the cutoff grade being the mill level cutoff. With large reserves, the single period profit maximisation is the convergence point, as it is the only capacity that leaves present value unaffected. All other options reduce the calculated present value.

An orebody following Lasky's relationships provides a convenient medium for examining and testing the following observations, made on the basis of expressions (8) and (10). These observations will hold with infinite reserves and as the convergence points for properties with long mine life, under the assumption that present value does in fact govern decisions of mining companies in establishing the long term plan.

The maximisation condition requires that two companies that mine orebodies with similar grade distributions and similar processing and mine technologies have comparable mine operating budgets. If the stripping ratio is the only difference in the two properties, the two companies will have the same mine capacity. If one examines this argument for a single mine operator with an orebody with reserves sufficiently large and with the stripping ratio increasing with time, then the mine capacity should remain constant with time. A typical example may be made for a vertical "seam", mined in phases with increasing stripping ratios for say 40 years and δ=0.8.

If technology reduces mining costs at one site, the mine operator has an incentive to increase the mine capacity, as is intuitively reasonable. The incentive to maximize present value induces the more efficient mine operator to mine more material and operate at a higher cutoff grade than the less efficient operator. The constant factor is the mine operating budget which, at convergence, should equal the revenue at the expected value of the deposit (mineralisation).

A similar comparison for the effects of technology change on the mill leads to a conclusion which is not intuitively reasonable. If one assumes that a change in processing technology reduces the "tailings" grade of the deposit only while processing costs remain unaffected, the cutoff and average grades will decrease in absolute terms. The average mineralisation value however remains constant. This again requires that, at convergence, the mine capacity remain the same. Consequently, the cutoff grade in absolute terms will also remain the same.

It can be inferred that with a uniform distribution or with any distribution and a cutoff grade at a region of the grade probability density function with a positive slope, a decrease in the mill cutoff always decreases the average grade. However, since the cutoff is also lower, the expected money value of the mineralisation increases. This in turn provides the incentive to increase the mine capacity. The only remaining question then is the behaviour of the model when the cutoff is at a region with negative slope in the grade probability density function. The exponential distribution provides the limit at which the expected value of the mineralisation and the mine capacity at convergence are the same. In this case, expression (10) requires that irrespective of the tailings grade and the stripping ratio, the mine capacity at two sites mining ore of the same distribution be proportional to the established processing capacity.

The recovery must be treated the same as price, if it is the proper measure of technological innovation at the mill. If the recovery increases by 10% (i.e. 80% to 88%), the mine capacity at convergence must increase by the same ratio. The observation that higher grade materials should be processed in higher prices is also due to Taylor (1972). Development of an open pit in phases represents an implementation of this concept, with discounting increasing the incentive for early presentation of higher grades or highgrading. Accessing the higher grade materials first is always the preferred option under the assumptions of this investigation, as there exists only one decision price i.e. today's market price for each commodity. Higher prices require that close to the end of mine life, the cutoff grades decrease in absolute terms. The same applies to the total mine capacity. However, with mine life sufficiently large, the mine capacity must increase.

3.1 An example for discussion

All cutoff grade based decision models are based on the assumption that cutoff grades will vary. Also, that the cutoff grade for the last time period in the model is the mill level cutoff. This implies that the mine life cannot be fixed. In a decision model for maximizing present value, fixing the mine life is not "reasonable". Assume that the recovery increases by 10% in a property. In an unconstrained model, the mine capacity must increase by 10% at convergence and decrease close to the end of mine life, as the mill cutoff decreases.

Let us examine a constrained model close to the end of the mining activity and without provision for a stockpile. Cutoff grades and mine capacity are planned to be decreasing. For small profitability changes and possibly for sales obligations, the existing pit limit and mine life are fixed. The latter assumption fixes the model completely. If the cutoff grade and by extension the revenue realised and the mine capacity are fixed for the last time period, then an increase in recovery (or price) increases the calculated present value by the same factor. At the same time, the value at the cutoff grade is increased by the same factor and condition (8) requires that nothing change. If this operation was planned to operate with an optimal set of cutoff grades in an unconstrained model, then this set is still optimal in the constrained model. The incentive the mine operator has in modifying the mine capacity is nil.

This example is not intended to question the validity of assumptions or constraints. It merely points out that an operator who maximises present value may be constrained to keep the same operating environment even when the profitability of the operation changes. It should also be noted that the future expected revenues increase in this example. Intuitively, the observation will not hold if the price decreases and the operation for the last time period(s) must process material bellow the concentrator level cutoff.

3.2 When is the stockpile ore

In the discussion on the physical meaning of the

cutoff mineralisation value, the present value a company calculates through time is the primary quantity of interest. Present value drops when a high stripping cost phase commences, capital is expended or prices drop. Some of the material in the stockpile becomes ore in these cases. This problem may be reasonably neglected close to the end of mine life of the operation. If the average grade of the ore mined is larger than the grade of the stockpile and there is no significant amount of stripping required, varying the mine capacity is likely to be avoided for its logistical implications.

The example bellow on Lasky's relationships verifies this argument with long mine life (at the convergence points of the mine capacity).

Let p_1 be the first decision price. The value at the cutoff grade $V(g_{c1})$, is equal to the revenue per unit of processing capacity R_1 or

$$R_1 = V(g_{c1}) = Krp_1 \ln (Krp_1/(1+sr)c_m) \quad (11)$$

The value of the highest grade of the stockpile is also $V(g_{c1})$. At the second decision price p_2, with $p_2 < p_1$

$$R_2 = V(g_{c2}) = Krp_2(\ln (Krp_2/(1+sr)c_m) \quad (12)$$

At price p_2, the mineralisation value at g_{c1} is

$$R_{21} = Krp_2 \ln(krp_1(1+sr)c_m) \quad (13)$$

$$R_{21} > R_2 \quad (14)$$

This states that some of the stockpiled material must be processed, which causes a double logistical problem. The mine capacity ought to decrease for the new convergence point. Also, the excess capacity for stripping is not required for the amount of material to be rehandled. While reclaiming the stockpile may not be evident from reactions of mining companies to prices, present value maximisation explains the use of the stockpile and lowering the cutoff grade when a large stripping ratio phase starts, as a means of controlling the mine capacity.

4 EMPIRICAL WORK

The policy of contracting the mining activity is more common lately, especially in smaller gold properties in the United States. This may be cited as evidence for the incentive to use mine capacity as the control variable and also avoid the capital and logistical implications of varying the mine capacity. Empirical work was undertaken to examine present value in absolute and relative terms with the grade distribution of a copper-molybdenum orebody with present value and the mine capacity both addressed as objectives. The multi criteria approach recognizes that, at some point, it is not possible to improve on one objective without forcing a deterioration on the value of one or more other objectives(Stadler, 1988).

The orebody is developed in 4 phases, with the higher grade material mined mostly with the first two phases. Mining is assumed to proceed within the phases on a bench by bench basis, without provision for selective mining within each bench. The mill capacity is 6.0Mt/annum. The processing life is 14.2 years when a provision for the stockpile exists. A computer program implementing condition (8) to solve the problem with processing capacity fixed, unconstrained mine capacity and without a provision for a stockpile. The solution for the required yearly mine capacity at a discount rate of 15% is shown in Figure 3. The mine capacity varies from 6.02Mt to 14.2Mt. Figure 4 shows the yearly cutoff grades, with notable the increase of the cutoff grade after stripping for phase 3 is completed after year 8.

The next question is what would be the effect of placing an upper bound on the mine capacity as well as providing for a stockpile in the model. This problem was solved using a linear programming formulation but without provision for "stripping in advance". When the mine capacity constraint is active, the cutoff is the best available grade. Figure 5 shows the portion of maximum present value achieved if mine capacity is bounded. At the decision price, the maximum possible present value is \$56.6M. This model is also evaluated for a higher copper price, with maximum possible present value of \$169.9M. It is interesting to note that more than 99% of maximum possible present value is achieved with mine capacity of 9.5Mt/annum for both prices. While the incentive to increase the mine capacity further definitely exists, the total cost of implementing

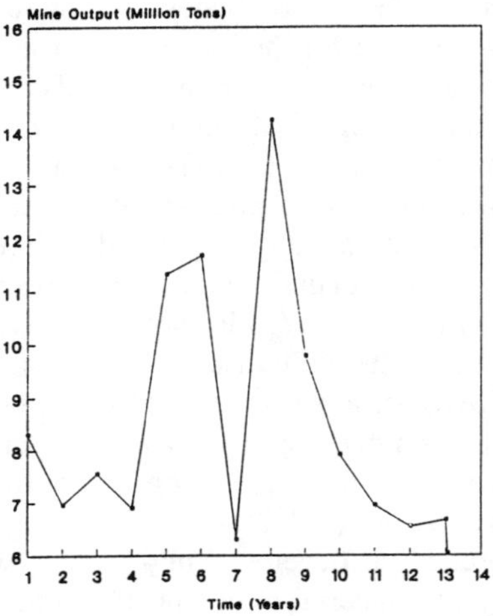

Figure 3. Mine Capacity Profile

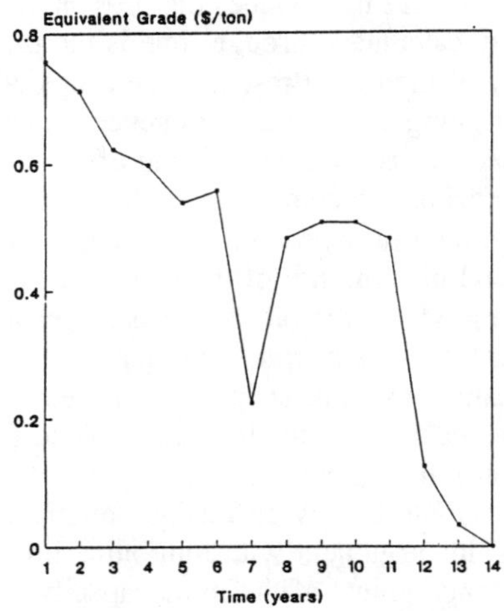

Figure 4. Cutoff Grade Profile

such a policy should be quite low. For both prices and presumably for every price in between, the possible gain in present value cannot be more than $1.0M, which would make the addition of one more truck to the fleet unattractive. The main reason for the limited variation of present value is the natural ordering of the presentation of higher grades with the proper development of the laybacks, which creates a rather stable operating environment at the mine level, irrespective of the commodity price.

5 CONSEQUENCES

The following observations may be made on the effects of factors on present value and its direct relationship with the cutoff mineralisation value.

The cutoff mineralisation value depends on the grade distribution, commodity price and costs. Mine capacity must vary, irrespective of the immediate stripping ratio, to achieve the mineralisation value indicated by condition (8). This is a consequence of the fact that revenue generation only depends on the rate in which material is processed. This is the case for the long term model presented above, under a set of assumptions which may or may not be descriptive. Any stripping requirements, preproduction or at commencement of a new phase must be made in an instant. This requires that no bound exist on the mine capacity.

The statements for declining cutoff of cutoff grades must be qualified with a provision of continuing profitability of the operation in terms of present value. This means that as one moves from the end of mine life to the present, present value must increase. While this is true for a series of equal positive annual payments, it does not, in general, hold. Commencement of development of a new phase in the mine is a good case in point and an intuitively reasonable example. If the amount of waste that is to be removed in this phase is significant, then the perceived present value at commencement of the phase stripping is lower than when the stripping is completed. The cutoff mineralisation value must decline before the phase commences. This is also the case any time a capital expenditure tied with the long term plan is incurred, for example when capital is expended.

Processing material bellow the concentrator level cutoff is indicated by the present value maximisation condition when the mine operator expects a negative present value from the remaining activities, the mine closing costs providing a typical example.

While the processing capacity affects the "present value of the remaining activities", it appears reasonable to suggest that the size of the reserve and the processing and mining capacities will have some predictable relationship. With large reserves, doubling the processing capacity will likely be linked with doubling of the mine capacity. The present value of the remaining activities will be the same (similar) per unit of processing capacity. By extension, cutoff grades will also be similar on a year by year basis.

There is no direct relationship between cutoff mineralisation values and discount rates. Cutoff grade mineralisation values will reflect the calculated present value. With equal or declining profitability of the operation, a higher discount rate will force the model to converge faster. However, the effect of the discount rate when present value drops with expending of capital or the beginning of a new phase is case specific. In an operation which is profitable at all times, a smaller discount rate is more likely to leave the calculated present value monotonically decreasing as one moves from the present to the end of life. In this case, a cutoff grade based present value maximisation policy will indicate declining cutoff grades.

With large reserves and a constant grade distribution, mine capacity converges to the single period profit maximisation point, with or without a stockpile. The provision for a stockpile should make the model converge faster.

With an exponential density function and reserves sufficiently large, the ratio of mine and processing capacities depends only on the average value of the distribution and the operating factors of recovery, the mining costs and varies linearly with price.

Undiscounted profit maximisation explains the use of prevailing mathematical formulations in the determination of the pit limit. The pit limit is established at a "zero profit line", which does not necessarily contradict a present value maximisation objective. The use of present value alone in establishing a solution to the long term planning problem appears insufficient, as present value varies predictably and, with long mine life, only marginally. The additional criteria that may be examined in explaining long term decisions and policies of mining companies include:

Minimising costs, with present value or other economic criteria as additional and likely secondary objectives. It is demonstrated in our

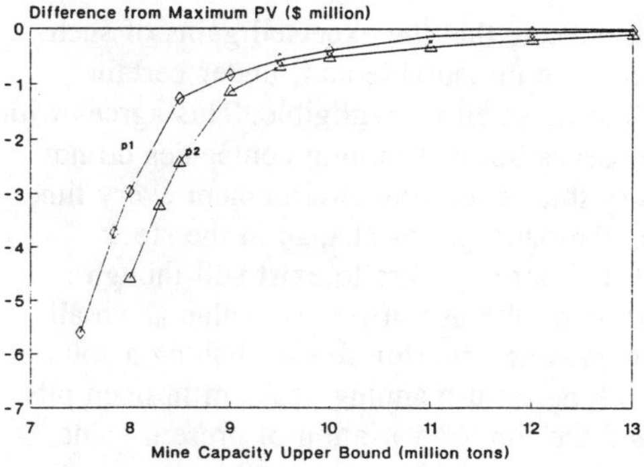

Figure 5. PV vs bounded mine capacity

empirical work that present value gains at some point appear to be small enough to warrant comparison with capital and operating costs of establishing and decommissioning mine capacity as well as the logistical costs in distinguishing between different grades.

It is most likely that the social function of a company and the objectives of survival and growth will indicate sustaining mining activities for as long as possible. With the exception of properties where stockpiling is a viable option, present value maximisation conflicts with this objective.

Company growth is only one of the managerial objectives. The interests of upper and middle management may form an applicable criterion in long term planning.

Finally, the effect of mining technology and company experience should be examined in as much as they affect the mine operating costs.

6 CONCLUSION

This investigation established that in a cutoff grade based present value maximisation policy, the cutoff mineralisation value is a function of the present value a mine operator calculates at each point in the life of a property. As such, it does not necessarily decline continuously with time. In addition, some conditions and constraints were identified under which present value considerations indicate keeping the same operating environment for a mine-mill model in open pit mines. The analytical and empirical work indicate that the expected gains of such policies are measurable and, under certain conditions, small or negligible. This agrees with the observation that mining companies do not modify their operating environment every time the commodity prices change in the stock market. There appears to exist still though a question on the use of present value as an all encompassing criterion in establishing a solution to the long term planning problem in open pits. Should the limited variation of present value prove the case for blending problems as well, it appears that at least one alternate or additional criterion for the analysis of the long term decisions of operating mining companies is required, possibly the reduction of operating costs. It is submitted that a widely acceptable solution to the long term planning problem in open pits will be difficult, until a satisfactory explanation of current practices is first established.

REFERENCES

1. Carlisle, D. 1954. The Economics of a Fund Resource with Particular Reference to Mining, *American Economic Review*, September , pp. 575-616.
2. Dagdelen, K. 1985. Optimum Multi-period Open Pit Mine Production Scheduling, *PhD Thesis, Colorado School of Mines*, Golden, Colorado.
3. Dagdelen,K. and Johnson, T.B. 1986. Optimum Open Pit Mine Production Scheduling by Lagrangian parametrirization, *Proceedings of the 19th APCOM Symposium*, The Pennsylvania State University, pp. 127-142.
4. Davis R.E., Williams C.E. 1973. Optimisation Procedures for Open Pit Mine Scheduling, *Proceedings of the 11th APCOM Symposium*, Tuscon, Arizona, pp. 127-142.
5. Elbrond, J., Lizotte, Y and Plasse, C. 1982 Use of an Interactive Dynamic Program System as an Aid to Mine Valuation, *Proceedings of the 17th APCOM Symposium*, Littleton, Colorado, pp 463-474.
6. Fytas, K., Calder, P.N., 1986. A computerized Model for Open Pit Long and Short Term Production Scheduling, *Proceedings of the 19th APCOM Symposium*, The Pennsylvania State University, pp. 109-119.
7. Johnson, T.B., 1968. Optimum Open Pit Production Scheduling, PhD Thesis, University of California, Berkeley.
8. Jordi, K.C. and Currin, D.C., 1979. Goal Programming for Strategic Planning, *Proceedings of the 16th APCOM Symposium*, Tuscon, Arizona, pp. 296-302.
9. Koniaris, E. 1991. A Positive Approach to Long Term Planning in Open Pit Mines,*PhD Thesis, Queen's University, Kingston*.
10. Lane, K.F. 1963. Choosing the Optimum Cutoff Grade, *Quarterly of the Colorado School of Mines*, 59 (ii), pp. 811-829.
11. Lasky, S.G., 1950, How Tonnage and Grade Relationships Help Predict Ore Reserves, *Enginnering and Mining Journal*, pp 81-85.
12. Lerchs, H. and Grossman I.F. 1965. Optimum Pit Design for Open Pit Mines, *CIM Bulletin*, vol 58, pp 47-54.
13. Lizzote, Y. and Elbrond, J., 1982 Choice of Mine Mill Capacities and Production Schedules Using Open Ended Dynamic Programming, *CIM Bulletin*, vol 75, pp 154-163.
14. Milner, T.E. 1977. Long Range Mine Planning at Gibraltar Mines Limited, *CIM Bulletin*, Oct , pp ...
15. Rudenno, V. 1979. Determination of Optimum Cutoff Grades, *Proceedings of the 16th APCOM Symposium*, Tuscon, Arizona, pp. 261-268.
16. Shaap, W.1984. Objective Funtion and Methodology in Cutoff Grade Theory, *Proceedings of the 18th APCOM Symposium*, London, England, pp. 575-581.
17. Stadler, W., Editor, 1988. *Multicriteria Optimization in Engineering and the Sciences*, Plenum Press, New York.
18. Taylor, H.K.,1972 General Background Theory of Cutoff Grades, *Institute of Mining and Metallurgy, Transactions*, Section A, v 81, July , pp A160-A179.
19. Tirole, J., 1989 The Theory of Industrial Orginisation, The MIT press, Cambridge, Massachusetts.
20. Yi, R. and Strugul, J.R, 1987. Analysis of Cutoff Grades Using Optimum Control Theory, *Proceedings of the 20th APCOM Symposium*, Johannesbourg, Vol 3, pp 263-269.

Innovative Mine Design for the 21st Century, Bawden & Archibald (eds) © 1993 Balkema, Rotterdam, ISBN 90 5410 325 6

Integrated short range production planning at the Mont Wright operation

Pierre Lestage & Laura Mottola
Ad Opt Conseillers en Informatique et Logistique Inc., Montréal, Que., Canada
René Scherrer
Compagnie Minière Québec Cartier, Mont Wright, Que., Canada
François Soumis
École Polytechnique de Montréal, Que., Canada

ABSTRACT: The short range production planning (SRPP) problem considered in this paper is that of a large open pit truck and shovel iron ore operation. Current computerized methods utilized to perform SRPP display several limitations. In particular, shovel displacements cannot be evaluated by systems based on shovelling capacities. The SRPP process consists in fulfilling the extraction requirements defined by the long range plan, while respecting the operating constraints as well as improving grade control and equipment utilization. The SRPP objective is to suggest an optimal mining sequence given an initial pit configuration. In order to realistically attain this goal, a new software package was developed by Ad Opt in conjunction with the Quebec Cartier Mining Company. Based on dynamic programming techniques, it provides a daily solution over a three month period comprising a precise drilling and blasting sequence, shovel displacements and utilization in time, as well as the type and quantity of material to be mined. The SRPP software package was installed at the Mont Wright operation and is presently being utilized by the mine planning team to obtain suggested planning strategies as well as to evaluate the impact of what-if scenarios. Automated interfacing with the long range planning and truck dispatching systems in place is currently underway.

RÉSUMÉ: Cet article traite du problème de la planification court terme de la production dans le contexte d'une exploitation de minerai de fer à ciel ouvert utilisant pelles et camions. Les systèmes informatisés effectuant cette planification comportent plusieurs lacunes. En particulier, les déplacements de pelles ne peuvent être considérés par ceux utilisant une approche basée sur la capacité de pelletage. Le procédé consiste à réaliser le plan trimestriel découlant de la planification long terme tout en respectant les contraintes liées aux opérations, en facilitant le contrôle de la qualité et en maximisant l'utilisation des équipements. L'objectif vise à proposer un plan d'extraction optimal étant donné l'état initial de la fosse. Ad Opt, en collaboration avec le personnel de la Compagnie Minière Québec Cartier, a développé un nouveau logiciel permettant d'atteindre cet objectif. A l'aide de la programmation dynamique, le système fournit une solution journalière sur un horizon de trois mois, comportant une suite précise de forages et dynamitages, déplacements et utilisation des pelles, ainsi que la quantité et le type de matériel à extraire. Le logiciel, installé à la mine Mont Wright, est présentement utilisé par le personnel en charge de la planification afin de définir la stratégie d'extraction et d'évaluer l'impact de divers scénarios. L'intégration avec les système de planification long terme et d'assignation de camions en place est en cours.

This project has been subsidized by the National Research Council of Canada.

1 INTRODUCTION

Conventionally, mine production planning is broadly categorized as long, medium, and short range, depending on the time interval studied as well as on the nature of the planning problem addressed. Typically, long range planning (LRP) involves estimating ore reserves and determining the final pit limit and mining strategy based on the long term investment. Medium range planning (MRP) comprises the design of haul roads as well as considering equipment investment and determining the mining sequence of accessible reserves over one year. Finally, short range planning (SRP) involves scheduling drilling, blasting, and shovel displacements as well as defining shovel priority and utilization on a daily basis over a period of three months. Recent developments in truck dispatching systems brought about the introduction of a new category of mine production planning which can be viewed as very short range planning (VSRP), since the ore grade fed to the mill is controlled on a shift basis.

The short range production planning process offers an opportunity to bridge the gap between the estimated medium-long term plans and the detailed production schedule generated by the truck dispatching system. The integration of the two processes is achieved by including specific operating conditions within the resolution of a three month plan in order to ensure a more accurate fulfilment of the one year plan.

2 SHORT RANGE PRODUCTION PLANNING (SRPP)

The main objectives of short range production planning reside in:

- fulfilling the three month plan;
- ensuring minimum daily ore production;
- determining detailed daily production;
- maintaining the grade within established limits;
- ensuring sufficient ore exposure;
- utilizing equipment efficiently;
- determining shovel displacements;
- respecting conventional mining practices.

Methods currently utilized to perform SRPP display certain limitations:

(i) manual approach: the mining engineer determines the production strategy on a weekly basis; the short-sighted nature of this approach can lead to substantial deviations to the three month plan which often jeopardize its fulfilment;

(ii) simulation tools and CAD techniques: the trial and error nature of this method requires time consuming user interaction; the principle of the optimal follower does not necessarily lead to the optimal solution;

(iii) mathematical programming techniques include:

(a) linear programming: the problem of shovel displacements cannot be addressed since no variables are linked to the equipment; blast hierarchical constraints cannot be considered satisfactorily; multiple period planning is difficult and often leads to unrealistic solutions;

(b) dynamic programming: the states do not contain information regarding the equipment; daily production is based on overall equipment capacity and does not consider specific equipment operating conditions such as maintenances and production rates.

Current methods, restricted to shovelling capacities, fail to associate a particular equipment to the material extracted. Consequently, shovel displacements cannot be evaluated, nor can different shovelling rates be utilized and scheduled maintenances respected.

A new system, developed by Ad Opt in conjunction with QCM personnel, was designed to assist the planner in his/her task by suggesting an optimal mining sequence for a given initial pit configuration. It provides a daily solution over a three month period comprising a precise drilling and blasting sequence, shovel displacement and utilization in time, as well as the type and quantity of material to be mined. The software can easily be run on a weekly basis with little user interaction and updated data provided by other systems if available.

3 A NEW COMPUTERIZED APPROACH TO SRPP

The SRPP software package was designed to integrate the production planning process. This was achieved by establishing the blasting sequence, the shovel displacements and priorities, the amount of material to be extracted daily, as well as the working faces. The ultimate goal of the program consists in: fulfilling the three month production plan while minimizing the shovel displacements and the cumulative deviation from the required stripping ratio; ensuring the minimum production requirements of ore; and providing access to new working faces to facilitate grade control. Constraints related to particular aspects of the operation, such as equipment safety clearance when blasting, are also taken into account.

The short range planning problem is solved by first constructing the block model and then performing dynamic programming. The block model is built in three phases, namely data input, validation, and storage, while the resolution is performed by defining the problem, solving it per period, and storing the solution.

4 THE BLOCK MODEL

4.1 Blocks

The model is based on blocks 14 m high, 10 m wide, and 10 m deep. Within a bench, a block is positioned with respect to the (X,Y,Z) coordinates of its geometric centre. The set of blocks defines the fundamental characteristics of the material to be mined within the three month period. To each block is associated a number of key parameters, such as grade, recovery, etc., the values of which can be obtained by interfacing with LRP systems in place. The information pertaining to the blocks can be subsequently revised upon analysis of the drill cuttings and other drilling data.

4.2 Blasts

For practical purposes the blocks are grouped in blasts. The final shape of a blast is determined manually by the mining engineer since the criteria governing the shape are hardly quantifiable and often subjected to the field experience of the engineer. Every blast in a given bench is defined by a closed polygon representing its outline. All the blocks inside the polygon are assigned to the associated blast.

4.3 Entrance Segments

A shovel can start digging a blast only from sides of the polygon which are designated as entrance segments predefined by the user. Blocks that are intersected by entrance segments are identified as entrance blocks as well as immediately neighbouring blocks.

5 DATA STRUCTURE AND VALIDATION

This phase performs the construction of the data structure utilized to solve the planning problem. The validation ensures that each and every block is associated with one and only working face and one and only blast. SRPP determines the set of blocks erroneously defined and informs the user of the possible faces and blasts with whom these blocks may be associated. The data structure is corrected upon the decision made by the user. The extraction sequence for the blocks in each face is then constructed. The blocks are then chained in a list starting with an entrance block, or the nearest block to an entrance block.

Finally, the priority between blasts and working faces within a blast are determined.

5.1 Block Assignment

The assignment of a block to a particular blast is performed by tracing the blast outline on a one metre grid bitmap and sweeping the lines and columns to determine whether the block area is at least 51% within the outline. The set of blocks within a bench is distributed among blasts. Unassigned blocks must be assigned by the user.

5.2 Generating Working Faces

The generation of working faces within a given blast can be performed manually using the polygons or automatically by means of an algorithm. In the first case, each block is assigned to a working face depending on whether it belongs to a predefined working face polygon. On the other hand, the automated process creates clusters that minimize the deviation in grade among the blocks of a group. Each working face must have a designated block from which extraction may be started. If the working face comprises an entrance block, then it is designated as a starting block, otherwise the nearest block to an entrance block is designated. The set of blocks assigned to a working face are chained by pointers from the starting block on.

5.3 Prioritizing Blasts

The SRPP software determines the blast hierarchy, i.e. a list of predecessors and successors of a given blast and for all the blasts in the problem definition. A predecessor of a blast can be categorized as vertical or horizontal.

5.3.1 Vertical Predecessor

The software determines the blocks within the intersection of the area covered by the current bench and that covered by the overlaying bench. For each one of these blocks in the intersection, the software determined to which blasts the immediate overlaying block and its eight neighbours belong to, and adds these blasts to the list of the predecessors of the blast in the current bench.

5.3.2 Horizontal Predecessor

If all the entrance blocks of a given blast have no neighbouring blocks assigned to another blast, then the given blast has no horizontal predecessors, otherwise that blast is a horizontal predecessor of the given blast.

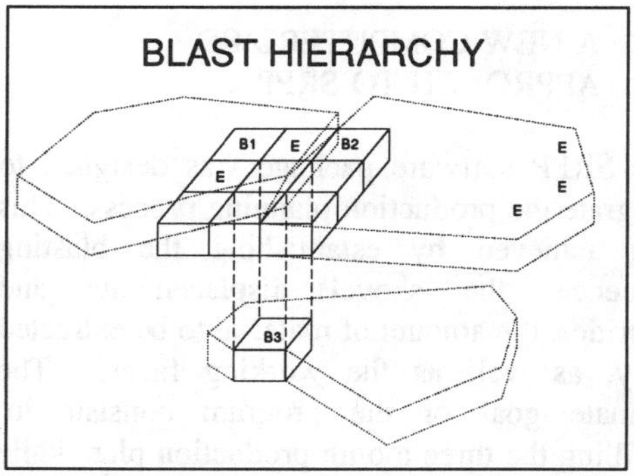

6 RESOLUTION

For each blast the software computes the drilling period, i.e. start and end dates, the blasting date, and the date at which the blast is cleared. For the overall pit, it computes the total shovel displacements, the production ratio, the total amount of material extracted, as well as the penalties cumulated throughout the three month resolution period. For each day and blast the user is provided with the status, the volume of ore and waste extracted, the average ore grade, the remaining volume to be extracted, the coordinates of the blocks removed, the identification and status of the shovel operating in the blast.

Dynamic programming is the technique used to solve this SRPP problem. A state describes a given pit configuration, comprising the set of shovel and blast statuses and positions for the pit. Given an initial state, this method consists in defining and evaluating all the possible states at each one of the 90 periods, and, upon completion of the three month resolution, in conserving the least penalized state. The path leading to this state represents the solution sought. The set of possible states for a given period is deduced from the set of states for the previous period and from the set of rules governing the changes in shovel and blast statuses. It is important to note that this method offers a great flexibility for adding new rules to the system.

6.1 Blast Status

The status of a blast is one of the following:

- unaccessible: blast cannot be reached by the drill;
- accessible: blast can be drilled;
- drilled: blast is ready to be blasted;
- blasted: blast is ready to be extracted;
- clear: no material is left in the blast.

6.2 Shovel Status

The status of a shovel is one of the following:

- operating: shovel is active;
- idle: shovel is inactive;
- down: shovel is down for scheduled maintenance;
- mechanical maintenance: shovel has reached the maximum number of operating hours since the last mechanical maintenance;
- electrical maintenance: shovel has reached the maximum number of operating hours since the last electrical maintenance.

6.3 Problem Definition

At this stage the user defines the initial pit configuration and other parameters necessary for the resolution of the SRPP problem, including the planning objectives and the resources available.

6.3.1 Blast Initialization

For each blast, the user is asked to define:

- its current status;
- the inventory, i.e. the amount of ore and waste remaining;
- the drilling rate, expressed in terms of a ratio to the average rate.

A given blast may contain a temporary or even a permanent ramp. In the first case, part of the material is considered as inactive inventory for as long as the ramp is operational. In the second case, access is restricted to half of the total blast volume.

6.3.2 Shovel Initialization

For each shovel, the user is asked to define:

- its current position and status;
- the rate of advance in ore and waste;
- the time intervals allocated to scheduled maintenance.

6.3.3 Network

A network is defined by connecting the dumping points to the blasts and to the pit entrance. A segment is defined by two points in the coordinate system of the mine. The shortest path between two given points in the network is calculated only once and stored in a matrix. Each blast is connected to the network through a predefined junction point. When displacing a shovel between two blasts, it is first moved in a straight line to the junction point of the starting blast, and then through the network to the junction point of the destination blast.

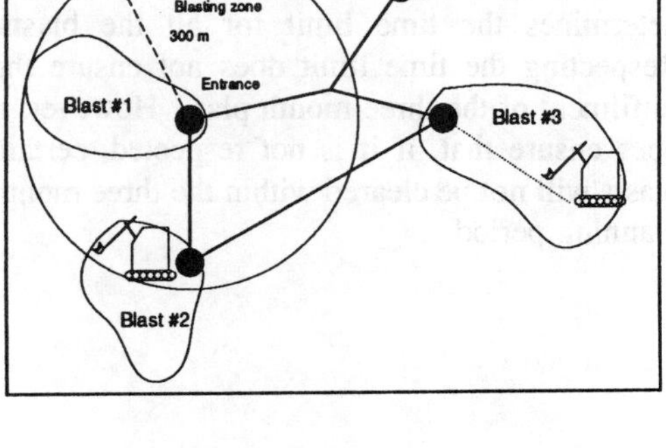

6.3.4 Objectives

The user is asked to provide the following information:

- production ratio over 90 days;
- targeted shovel utilization [%];
- a geological parameter to be controlled;
- targeted grade for each geological parameter per period;
- required number of working faces in ore per period;
- targeted number of operating shovels per period;
- number of blasts allowed per period;
- number of electric and mechanical maintenances allowed per period.

6.4 Time Limit of a Blast

The time limit of a blast is defined as the latest possible date at which the blast should be blasted if the three month plan is to be respected. The extraction period represents the time required to clear the blast and is calculated based on the average advance rate and availability of the shovel, the total volume of the blast, the time required to drill the blast, and the number of predecessors of the blast.

The time limit is determined by going through the list of successors of a given blast. If a given blast has no successors, the latest time at which it can be blasted is the length of the extraction period before the end of the 90 day period. If a given blast precedes others, then the latest time at which it can be blasted is the length of the extraction period before the earliest time limit of its successors. This iterative procedure determines the time limit for all the blasts. Respecting the time limit does not ensure the fulfilment of the three-month plan. However, it does ensure that, if it is not respected, certain blasts will not be cleared within the three month planning period.

6.5 Rules for Generating New States

6.5.1 Blast Rules

The following rules govern the generation of new states from a given parent state. The status of a blast can undergo the following changes:

- unaccessible → accessible: all the vertical predecessors must be cleared and the horizontal predecessors must be blasted or cleared;

- accessible → drilled: the time period since the blast became accessible must be larger than the drilling time;

- drilled → blasted: the blasting must be permitted at this particular period and the total remaining volume in the set of predecessors must not exceed 10% of the volume to be blasted;

- blasted → clear: the remaining volume is smaller than one quarter of the daily shovel production rate.

Note that the status changes from unaccessible to accessible to drilled, and from blasted to clear are compulsory if the conditions described above are satisfied.

6.5.2 Shovel Rules

The displacements and changes in statuses of the shovels are performed in the following manner. First, the blasts which are ready to be extracted and the shovels which can operate are identified. Then a state is generated for each blast-shovel combination that minimizes the displacements for each group of blasts where the shovels can operate.

The following rules are applied:

- a shovel can only move towards a blast which is ready to be extracted;
- a shovel is considered to be down for the entire duration of its scheduled maintenances;

- a shovel is allowed to continue operating in a given blast if and only if there remains enough material to be extracted;
- if a blast is scheduled to occur at a given period, then the displacements outside the safety zone must be taken into account.

6.5.3 Shovel Utilization

Given a set of shovels and their associated blasts defined in the current state, the system determines the amount of material to be extracted from each working face as they become exposed. The amount of material extracted is determined in order to ensure that the blast is cleared before the time limit while targeting the average utilization specified by the user. Electrical, mechanical, and scheduled maintenances must be respected when required and permitted. The minimum targeted ore production must be achieved while respecting the overall production ratio in the pit. When possible, the working faces are selected in order to control the average grade fed to the crusher.

6.6 Evaluation of a State

The objectives described earlier are measurable quantities. The evaluation of a state is based on the deviation from these objectives at each period. Thus, the value of a given state is determined using proper weighing according to the following:

- the cumulative displacements of all the shovels;
- whether the time limit was exceeded;
- the number of working faces in ore;
- the overall average ore to waste ratio for the pit;
- the variation in grade;
- the amount of ore extracted per period;
- the number of operating shovels.

7 INSTALLATION AT MONT WRIGHT

At Mont Wright, a large open pit truck and shovel iron ore operation, short range production planning was a manual process undertaken by the mining engineer, whereas long and very short range planning were performed by the Q-Pit and Charlemagne systems respectively.

Q-Pit, developed at Queen's University and in use at Mont Wright for several years, comprises a block model and interactive graphics to perform optimal pit design and ore reserves calculation. At this level, the calculation of ore and waste productions is solely based on average equipment capacities and speeds.

Charlemagne, developed by Ad Opt, is a truck dispatching system comprising a module responsible for planning equipment utilization and providing grade control on a shift basis. The main objective of this type of module is to maximize production while respecting quality constraints by selecting optimal truck itineraries between shovels and dumping points and by computing shovel production rates.

The Quebec Cartier Mining Company needed to bridge the gap between the LRP and VSRP tools already in use at Mont Wright. To this effect, a new software was required to aid the planner in his/her task by suggesting an optimal mining sequence for a given initial pit configuration and over three months.

In order to minimize the workload required to define the production planning problem, the block coordinates and associated geological parameters are obtained from Q-Pit, while the daily volumes extracted in each face are available from Charlemagne. New blast outlines are taken from an AutoCad representation which is updated weekly. A separate AutoLisp module has been implemented to facilitate the retrieval of this information by interfacing Autocad with SRPP and generating compatible data files regarding the pit configuration to be analyzed. Cuttings analysis is readily available in Lotus form while hole position can be determined using drill pattern specifications and hole numbering, thus permitting to update the parameters associated with the blocks corresponding to these locations.

Project specification and problem definition were initiated in 1992 in collaboration with QCM mining personnel on site. The first prototype was installed at Mont Wright in late January 1993. The software, written in C, runs on a RISC 6000 workstation operating under UNIX. The initial data acquisition process is performed in approximately 3 hours. A solution is obtained within 15 minutes and the results are exported to Autocad for graphical representation. Automated interfacing with the Q-Pit and Charlemagne systems is currently underway. At Mont Wright, the SRPP tool is utilized to suggest planning strategies as well as to evaluate the impact of what-if scenarios.

8 CONCLUDING REMARKS

The short range production planning problem involves several levels of decision making procedures which, when handled manually, can lead to substantial deviations from the long term plan. Currently available computerized planning tools, restricted to shovelling capacities, are unable to evaluate shovel displacements, utilize different shovelling rates, or respect scheduled maintenances, because no particular equipment is associated with the material extracted.

The dynamic programming procedure applied to a state comprising specific shovel and blast configurations inevitably leads to a more realistic solution closer to the actual mining operation. In contrast with the block model approach, the identification of a particular shovel operating in a particular blast allows the integration of specific operating constraints, such as shovel maintenances, security clearing when blasting, and shovel displacements.

The SRPP software was designed to suggest an optimal mining sequence to the mine planning team at Mont Wright. This computerized tool provides a daily solution over a three month period comprising a precise drilling and blasting schedule, shovel displacement and utilization, as well as the type and quantity of material to be mined. In addition, it enables to prioritize certain areas of the pit by evaluating what-if scenarios and anticipating bottlenecks due to a tight physical hierarchy or equipment availability.

9 REFERENCES

Baron, J. 1987. Évolution de Charlemagne II après une année d'utilisation. Compagnie Minière Québec Cartier.

Chanda, E.K.C., Wilke, F.L. 1992. An EDP-Model of open pit short term production scheduling optimisation for stratiform orebodies. 23rd APCOM. Application of Computers and Operations Research in the Mineral Industry.

Mann, C., Wilke, F.L. 1992. Open pit short term mine planning for grade control - a combination of CAD-techniques and linear programming. 23rd APCOM. Application of Computers and Operations Research in the Mineral Industry.

Soumis, F., Ethier, J., McInnis, D. 1986. Optimisation algorithm for truck dispatching in an open pit mine. Proceedings AFCET, Paris, p.181-184.

Innovative Mine Design for the 21st Century, Bawden & Archibald (eds) © 1993 Balkema, Rotterdam, ISBN 90 5410 325 6

Large scale hill top mining with concern for the environment

A.G. Paithankar
Visvesvaraya Regional College of Engineering, Nagpur, India

ABSTRACT: The mine is producing 22.6 million tonnes of iron ore per year. It is situated at an altitude of about 1200m above M.S.L. The general topography of the area is hilly. The iron ore deposit is of sedimentary origin. The mine area is subjected to heavy rainfall. Large quantity of solid waste in the form of soil, ore and waste flow down the hill sloped and pollute the tributaries and ultimately the main river, specially in the rainy season. The mine management have therefore constructed two pollution control dams. Additionally a 1.2 km long diversion channel is constructed from upstrem point of one of the tributaries to divert water to the main river. Also, a earth fill dam is constructed against tailing flows. Mansoon stockpiles are covered. Stations have been fixed along the main river and tributaries for collections of water samples. To minimise air pollution wet drilling is adopted. Mine roads are sprinkled with water in the dry season. Dust extraction systems are provided in each crusher house. Conveyor belt is covered. A massive reforestation programme is underway.

1 INTRODUCTION

The Kudremukh Iron Ore Mine, the largest in India,is located on top of a mountain range, which runs almost parallel to west coast. The mountain range extends in length to about 1200 km. The topmost bench of the mine is located at an attitude of 1248 m. above sea level.

The mine is designed to produce 22.6 million tonnes of crude ore per year with a daily production of 82,000 tonnes. The annual rainfall varies from 6000 to 7000 mm., mostly received as a concentrated dose during monsoon period extending from June to September. During this time the mountain is also lashed with high speed wind. The highest recorded velocity is 125 km/hour.

The mountain top location of the mine coupled with its exposure to high speed wind and heavy concentrated rain has given rise to peculiar challenging environmental problems which the company has very successfully solved. The mine is now perceived as one of the most environmentally conscious mines in the world.

2 GEOLOGY

The iron ore deposit is of sedimentary origin and is of pre-combrian age. It occurs as banded magnetite quartzite with an average thickness of iron formation extending up to 100 m. The ore body has a dip varying from 35° to 80° towards the east occasional sills of chlorite schists occur within the formation amphibolite schists form the basement rock. The thickness of the overburden is almost nil. Surface weathering and oxidation have partially altered the original magnetile to hemalite in the top 30 to 50m of the formation. This has produced a soft and friable ore termed "Soft Weathered Ore" which quickly becomes airborne when fragmented. The weathered zone is underlain by a transitional hard weathered zone" about 30m thick, below which lies the primary are zone of another 30m thickness. The present scheme of mining is designed to exploit only the weathered ore which has an average iron content of 38%. This is beneficiated to produce concentrate of 67% iron.

Kudremukh deposits are known to be one of the largest in the world. In one deposit alone, where the mining is currently carried out, reserves of 700 million tonnes of weathered ore and 450 million tonnes of primary BMQ have been proved. Besides, there are other deposits in the vicinity with probable reserves of 4000 million tonnes.

3 PRE-MINING SCENARIO

3.1 Water

The Bhadra River passes right through the middle of the leasehold area. A number of tributaries originate within or adjacent to the leasehold and join the Bhadra River. The Bhadra River and all other tributories are perenial.

There is an irrigation dam, 55 km down stream of Bhadra River which is used for irrigation as well as for generating hydroelectric power.

3.2 Soil

The soil is red in colour , turning to lateritic on exposure, a characteristic of the Western mountain range. In valleys and planes the soil is rich in humus and support agriculture. The thickness of soil varies from 1 to 2m in valleys, hardly half to one metre on hill slopes and still less on hill tops. Because of the high speed wind and the torrential rains the soft soil on hill tops is washed away and only bald hill tops appear, devoid of trees or shrubs. Trees and vegetation grow only in valleys and in the plains.

3.3 Wild life

As the leasehold and the neighbouring areas are devoid of thick vegetation, it was observed that there is hardly any wild life worth special mention. However, in the nearest Bhagwati forest (12km) bisons,sambers, bears and deers are living and occasionally panthers are located. Birds are also few but reptiles are abundant.

3.4 Human settlement

There were two small villages adjacent to the leasehold areas and total of 73 families lived in these two villages. Agriculture was their main occupation. The low lying plain land along the valleys was used for agriculture. There were no schools in the villages and the nearest school was at a distance of 50 kms. The villagers were never exposed to education nor to the modern amenities of civilised life.

3.5 Transport and communication

The transport and communication system was very poor. The nearest highway junction was at a distance of 68 kms, the nearest railway station at a distance of 150 kms and nearest post office at a distance of 25 kms.

4 MINING, PROCESSING, SHIPLOADING AND PELLETISATION

4.1 Mining

Kudremukh, the largest iron ore mine in India, is designed to produce 22.6 million tonnes of crude ore per year with a daily production of ROM of 82,000 tonnes. The mine is planned for a three shift operation.

4.2 Process

The mined ore is transported from benches to two gyratory crushers capable of crushing the ore to minus 7" size.The crushed ore is screened into coarse and fine fractions before feeding to autogenous mills. The ore ground to minus 20 mesh size is subjected to primary and secondary magnetic separation to recover the magnetic concentrate, and spiral separation to recover non- magnetic concentrate. Both the concentrates are re-ground in ball mills. The reground concentrate is pumped through a 67 km long slurry pipeline to Mangalore where the slurry is filtered to produce filter cake with about 8 percent moisture.The entire operation, using the most sophisticated equipment from crushing to shiploading, is fully automatic and computer controlled. Computerised mine planning techniques are being used to maintain strict quality control.

4.3 Shiploading

Iron ore concentrate stored in two stockpiles having a total capacity of

400,000 tonnes is reclaimed by two reclaimers, and loaded into ships through a shiploader with a rated capacity of 6,000 tonnes an hour. A loading rate of 35,000 to 40,000 tonnes a day is guaranteed for Panamax type vessels.

A captive iron ore berth has been provided to handle ships of 65000 DWT with a loaded draft of 12.5 meters. With the provision of an exclusive berth, waiting time for ships is eliminated.

4.4 Pelletisation

A three million tonnes per year Pellet Plant based on technical knowhow of M/s Lurgi is under operation at Mangalore Pellet Plant.

5 ENVIRONMENTAL PROBLEMS OWING TO HILL TOP MINING AND SOLUTIONS

5.1 Water

As mentioned earlier the mine at high altitude is subjected to heavy rains, accompanied by high speed winds during monsoon season. Large quantities of solid waste in the form of soil, iron ore and waste rock specially from the soft weathered zone of the mining area flow down the hill slopes at high velocity.

In order to minimise the pollution of Bhadra river, a consultancy organisation was engaged by the company to find out ways for achieving the obove objective. Following suggestions were made by the experts

1. Cutting two contour trenches, each 10 m wide and 2m deep, all along the mining boundry and planting trees in between the trenches.

2. Developing a 200m wide green belt by afforestation along the Bhadra river on either bank.

The management however apprehended that contour trenching and afforestation may not be adequate and hence undertook following measures in lieu of the above.

a. Construction of two mini dams of rock fill type in two major valleys. One of these dams is against the catchment basin of crusher No.2 and the other against a tributory of Bhadra river. These dams catch the flowing materials from the ore body, allowing only the clear water to join the Bhadra river.

b. Construction of a 1.2 km long diversion channel from a suitable upstream point of the main tributory to river Bhadra, before it reaches the active mining zone.

This has resulted the in diversion of a major quantity of clear water to the Bhadra river. The mineral concerntration plant at Kudremukh requires approximately 20 litres/t of ore for processing. Since most of the water is recycled ,the effluent quantity is small. The makeup water is drawn from the main tailing dam called as Lakya dam by gravity. This year in July-August 1992, the furious monsoon downpour assisted by the steep slopes of the mountain ranges, caused damage to the spillway over Lakaya reservoir endangering the life and property of the people on the down stream side. Almost all the concrete slabs on the downstream sides of the Lakaya spillway have been damaged. The retaining wall of the spillway and the dyke however was in sound condition. Protection work by placement of sandbags on the upsteream side of the spillway diverted the water flow. A crane was placed on the dyke and large size tyres were placed in the valley created by the spillway flow so that further erosion does not take place. As an additional precaution people living in the low lying areas were evacuated.

The Lakya dam was built in 1978 at a level of 855 m above mean sea level. It was however found to be inadequate to accomodate the additional iron ore tailings and hence it was decided to increase the height of the dam by 35 m. It will require about one year to increase the height to the desired level of 890 m.

The heavy rains cause erosion of the ore body during the monsoon season. The solids get arrested at the rockfill dam and are reclaimed periodically and fed into the beneficiation plant together with ROM for processing.

Permanent stations have been fixed at strategic locations along Bhadra river and its tributaries for collections of water samples at fixed intervals of time. These samples are tested at the company's laboratory and also in the govt. laboratory.

5.2 Air

Because of the high altitude and high wind velocity the dust produced due to mining activity is likely to get dispersed over long distances. The company has adopted the following measures to successfully combat the menance of dust pollution.

A unique transportation system for ore which is a combination of truck,covered belt conveyor and slurry pipe line. Truck transport is restricted to a short distance from benches to crusher.The crushed ore is transported from crusher by covered belt conveyor upto beneficiation plant which protects the ore from wind and rain.

Water sprinkling over mine roads is carried out every hour by three high pressure sprinklers each of 28000 litres capacity. Water sprinkling is done only during dry months of December to April when about 66000 litres of water is consumed per hour for the purpose.

Dust extractors are provided in each crusher hall, conveyor transport point and the monsoon stock pile.

From the beneficiation plant ore is successfully transported as slurry through pipe lines by gravity, over a distance of 67 km, upto Mangalore port.

The notable merits of hydraulic transportation are:

a. Since the mine is on the top of the mountain, gravity assists the flow of slurry through pipeline upto Mangalore port, thus resulting in great saving in the transportation cost.

b. The pipeline transport doesnot disturb the environment and ecology.

5.3 Land

The area receives heavy rains, 4000 to 9700 mm per annum spread over four months from June to September. A considerable amount of soil and the soft weathered ore is washed off down the mountain slopes at high velocity every year in the monsoons seasons.

Blasting operations which result in fragmentation and loosening of the rock have further increased the rate and quantity of the washoff.

In order to minimise the damage of cultivable land in the plains, due to silting of the river and consequent flooding of the land, one major dam and two mini dams have been built across the tributories and one of the valleys. The management is also monitoring the rate of siltation of the Bhadra Reservoir.

Reforestion over the overburden dump is in progress. Out of 480 ha of mineralised area 2.4 ha has been disturbed by mining activity so far. Mine roadways occupy another 10 ha. Concentrates, crushing plants, warehouse for monsoon stockpile etc is spread over 71 ha. Lakya dam catchment covers an area of 2000 ha. The two minidams and diversion channel cover a catchment area of about 445 ha. Thus a total of 2750 ha of land is occupied by mining and other related activities. The management has planned for reforestation in mined out areas.

5.4 Socio-economic impact solutions

There were two villages within the leasehold area before commencement of mining activities. The main occupation of villagers was agriculture, but their land was acquired by the mining company. Major portion of the land acquired was utilised for constructing the township, which is now considered as one of the most modern mining town. The township has well constructed houses equipped with electricity and tap water supply, tarred roads, hospitals, educational institutes, post offices , all weather roads connecting it to Mangalore port and and a helipad.

Most of the eligible residents of these villages have employment in the mining company. They have also been adequately compensated for the loss of their land.

It is, therefore, observed that the mine has produced an extremely favourable socio- economic impact on an area which was othrewise backward and devoid of any civic amenities.

6 CONCLUSIONS

The high altitude location of the mine, steep mountain slopes, high concentrated rain fall and high velocity wind during the monsoon season have given rise to very special problems of environmental pollution such as soil wash off, siltation of the river and transportation problems. However, all these problems have been successfully combated by the unique measures adopted by the company such as the construction of dams, covered conveyors and hydraulic transportation systems. In fact the high altitude location of the mine has proved to be a boon in disguise for the hydraulic transport of ore. The mine has also created an extremely favourable socio-economic impact.

Innovative Mine Design for the 21st Century, Bawden & Archibald (eds) © 1993 Balkema, Rotterdam, ISBN 90 5410 325 6

Application of stochastic and probabilistic techniques at Malanjkhand Copper Project

A.R. Leventhal, C.S. Barker & L.P. de Ambrosis
LongMac Associates Pty Limited, Crows Nest, N.S.W., Australia

Abstract:
The Malanjkhand Copper Mine is located in central India in the state of Madhya Pradesh and is India's largest base-metal open-cut mine. The existing open pit is currently about 100 metres deep and produces annually about 11 million tonnes of waste rock and 2 million tonnes of chalcopyrite ore from a wide quartz reef in a granite environment.

As part of a study for the optimum exploitation of the deposit, geotechnical and hydrogeological studies were undertaken to determine open pit batter slopes for a possible open-cut to 200 metres depth and conditions for underground mining to 600 metres. The investigation extended to depths of 700 metres. The statistical methods employed to characterise the rock mass and the probabilistic techniques used to predict open-cut batter stability and underground openings performance are discussed, together with the statistical methods used in the prediction of groundwater flow into the open-cut and underground mine workings.

The preferred underground mining method requires promotion of hanging wall instability for economic mining of the resource. The assessment of the onset of hanging wall failure in a 'slot and mass-blast' mining method involving 0.75 x 10^6t blasts also is discussed.

1. INTRODUCTION

Malanjkhand Copper Project (MCP) is currently the largest open cut base metal mine in India. The mine is operated by Hindustand Copper Limited (HCL), a Government of India enterprise, and is located 90km north-east of Balaghat in the state of Madhya Pradesh. The open cut mine presently is about 100 metres deep with a strike length of 1.5km and annually produces about 11 x 10^6 tonnes of waste rock and 2 x 10^6 tonnes of chalcopyrite ore from a quartz reef, varying in width though typically 60 to 100m wide, in a granite environment.

Bishimetal Exploration Co., Ltd. (BEC) of Japan was awarded a contract for Technical and Consulting Services for Optimum Exploitation of Malanjkhand Copper Project, subsequent to a global tender called by HCL. A part of the study involved the geotechnical and hydrogeological investigations for a feasibility report on expansion of the MCP mine. BEC awarded responsibility for these investigations to LongMac Associates Pty Limited (LM) of Australia. This paper explains some of the techniques adopted for the investigations and analyses which were conducted.

2. PROJECT BACKGROUND

The mine is located on a hillock arising about 50m above the regionally flat relief (at about 575MRL). The geological reserves have previously been estimated at about 60 million tonnes, averaging a grade of 1.2 percent Cu at a 0.45 percent cut-off. Oxidised copper ore is present in the upper zones which are weathered to depths of 40 to 80 metres. Pre-stripping commenced in 1979.

The current mine plan allows for continuation of the open pit to depths in excess of 200 metres (376MRL) over a pit length of about 1,200 metres. In part, the project undertaken by BEC was to determine the optimum layout for both the open-pit and underground mines and at which depth mining operations would change to underground methods.

3. TECHNICAL INVESTIGATIONS

3.1 Philosophy

The aim of the investigation was to provide representative engineering parameters for the open pit and underground mine design. Accordingly, the major thrust of the fieldwork programme was to gain sufficient and statistically relevant data for stability analyses, rock mass quality determination and hydrogeology modelling.

3.2 Fieldwork

The fieldwork was conducted between May 1990 and March 1991. In broad terms, this involved: drilling

FIGURE 1: This view of the open pit hanging wall (left hand side of photo) is from the rim of the main haul road across the pit's northern end wall. Geotechnical Domain DGT4 contains the hanging wall Aplite intrusion and the orebody is contained within DGT2, DGT5, DGT8 and DGT11 in this photo. The orebody dips steeply to the left (east). (February 1991)

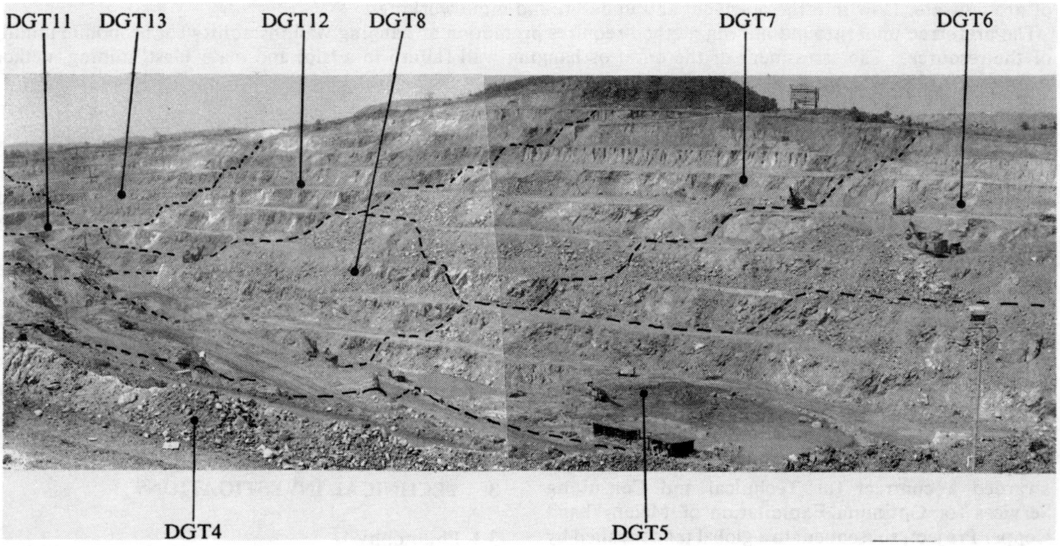

FIGURE 2: View of the open pit footwall from the northern end of the hanging wall. The orebody is contained within Geotechnical Domains DGT5, DGT8 and DGT11 in this photo. The base of the pit is at about RL 496 and the sump at about RL 484 whilst the crest of the pit is typically about RL 590. (February 1991)

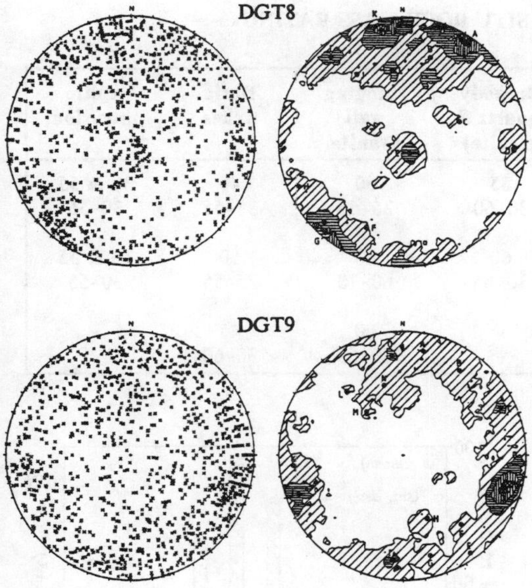

Figure 3: Stereonets for domains from orebody (DGT8) and hanging wall (DGT9).

and logging of 6.7km of oriented core; logging 6.1km of unoriented exploration core; piezometer installations to 600m depth; packer testing; recovery tests in standpipes down to 200m depth; engineering geological appraisal and 2.1km of scanline mapping (over 7600 data points) within the existing pit; and measurement of in-situ stresses by the hydraulic fracturing method.

Subsequent laboratory testing of drill core included, inter alia, multi-stage triaxial testing and Hoek shear box testing of natural and 'artificial' saw-cut joints. These and other rock mechanics tests were conducted both in India and Australia.

The details of this work are discussed elsewhere (Leventhal et al, 1992).

4. GEOTECHNICAL ENVIRONMENT

4.1 Geological Setting

It is not the place of this paper to provide detailed geological assessment of the setting or occurrence of ore at MCP as this is well covered elsewhere (MECL, 1984).

However, by way of simplistic description, the orebody is closely associated with an arcuate quartz reef, about 60 metres wide, broadly running north-south over a strike length of about 2km and steeply dipping (at about 70 degrees) towards the east. The host rocks are predominantly granites which form the hanging and footwall to the orebody. Chalcopyrite mineralisation is associated mainly with the quartz reef, though some mineralisation of the granites occurs. Several episodes of dyke intrusion have been identified within the pit by BEC. These sub-vertical dykes transect both the reef and host granite, are of variable thickness and are probably associated with contemporaneous faulting. A late stage barren aplite intrusion occurs approximately at mid-length on the hanging wall side and apparently may be the source of mineralisation.

4.2 Engineering Geology

The current open pit was appraised by detailed observation and visual assessment by which boundaries between visually homogeneous zones were identified. Frequently, the boundaries between domains consisted of persistent sheared dykes which transect the orebody and somewhat typically these dykes were narrow. Other boundaries were related to lithology changes.

Thirteen geotechnical design domains were identified and an indication of their extent within the setting of the current pit can be gained from Figures 1 and 2. Each of the geotechnical domains in the pit is statistically different in regard to joint set orientations and the orientation of respective joint sets are variably rotated across domain boundaries.

Scanline mapping was performed in preference to area mapping as a result of site constraints and safety considerations. The consequent bias against short persistence defects is acknowledged. A bias against sub-horizontal discontinuities (due to quasi-horizontal scanline mapping) was also expected. This notwithstanding, a significant presence of near horizontal defects was collected particularly within the orebody domains - see Figure 3.

Joint sets within each domain were identified not only on the basis of stereonet contouring but also joint set characteristics. Structural orientation information was manipulated on lower hemisphere equal area projections together with histograms of joint spacing, persistence, termination and wave length of contoured joint sets. Up to 9 joint sets were identified on the medium scale with a wide spread of orientation data (as seen on Figure 3) for each domain. This figure also demonstrates the virtually ubiquitous nature of the joint distribution within the rock mass on a large scale basis.

The extrapolation of the geotechnical domains with depth is perceived with less precision than domain identification from mapping of the pit.

4.3 Rock Mass Quality

In broad terms, the rock mass at MCP can be described as "fair". This broad description applies for both the hanging wall and footwall granites and the orebody itself. In general terms, the intrusive dykes within the rock mass and the aplite intrusion (principally in the hanging wall) are of significantly lesser quality than the other rocks at MCP due mainly to the frequency of defects. Naturally, significant variations do occur within the rock mass and statistical methods have been adopted for the distribution of rock quality.

The rock quality was assessed independently using 3

TABLE 1 GENERALISED IN-SITU ROCK MASS RATING

Mining region		Footwall granite	Orebody (quartz & granite)	Hanging wall granite	Basic dykes	Aplite intrusion
Open pit	typical value	65	55	60	40	30 & 55
	general range	55-75	30-60	35-65	35-55	30-55
Underground	typical value	60-65	60	60	50	35 & 55
	general range	55-70	30-65	60-70	25-55	30-55
Far-field	typical value	60-65	-	60	45	-
	general range	45-75	-	30-65	30-60	-

techniques - Rock Mass Rating (RMR) (after Bieniawski, 1984), Norwegian Geotechnical Institute Q-system (after Barton et al, 1974) and the method of Laubscher (1977). Good correlations, in agreement with published data for RMR & Q and with a 1:1 relationship for RMR & Laubscher's method (as might be expected), were observed. The RMR method was subsequently adopted.

In-situ values of RMR were computed from the database created from the drill core logs. Adjustments to the rating system include the use of $I_S(50)$ in place of UCS and application of the work of Priest & Hudson (1976) to relate the imprecise RQD to natural fracture spacing.

An example of the distribution of RMR within a geotechnical borehole is presented in Figure 4. The distribution of in-situ RMR for the deposit, which resulted from a general appraisal of the drill core, is shown in Table 1.

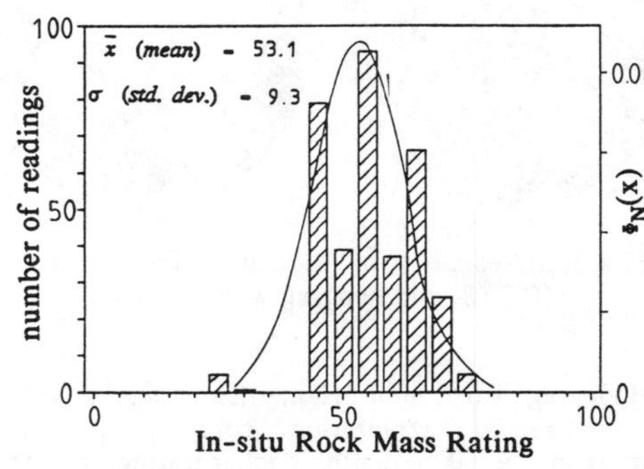

Figure 4: Distribution of rock mass rating in BH LM 12 in the footwall granite of the deposit.

4.4 Pervasive Stress Field

From the results of the testing employing the hydraulic fracturing technique, the direction of major principal stress was inferred roughly parallel to the orebody strike. The major horizontal principal stress and the average horizontal stress were inferred greater than vertical stress, the principal horizontal stress being about twice vertical stress at 600m. In a world context, the magnitudes of the pervasive stresses agree well with reported trends (see Brady & Brown, 1985, Brown & Windsor, 1990, and Arjang, 1991).

5. HYDROGEOLOGY

MCP exists in a monsoonal climate and is located at a topographic high on the boundaries of several sub-catchments of the Banjar and Son Rivers. Neither a high head recharge nor a recharge catchment exist up gradient of MCP.

The hydrogeological regime at MCP was assessed on the basis of: monitored standpipes and vibrating wire piezometers installed during the fieldwork; downhole packer testing during drilling; recovery tests performed in standpipes; records of pumping from the pit; and an assessment of the regional groundwater by survey of village wells in proximity to the orebody. The aim of the study was to predict seepage inflows out of, and pore pressure distributions within, the rock slopes.

The packer test interval was typically 30m to provide mass permeability of the rock units. It was concluded that the permeability inferred from this testing (within the restrictions of the test method) provided the best estimate of the overall permeability of the rock mass and these results should not be biased by shadowing of joint sets due to the relative orientation of the borehole and the joint sets themselves. No significant difference in the broad range of back-figured permeabilities between packer testing and recovery tests was detected, though some of the lowest permeability results were recorded for the recovery tests.

Broad rankings of hydraulic conductivity on the basis of rock units proved most amenable and a summary of inferred permeability values is presented on Figure 5. Ranking of the permeability test results by RMR of the test section or position in the pit proved inconclusive at MCP.

Assessment of the piezometer monitoring in and around the pit showed a piezometric surface little

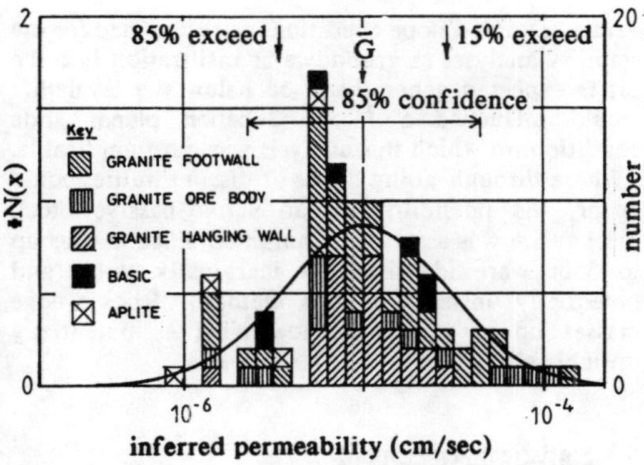

Figure 5: Distribution of field permeability results.

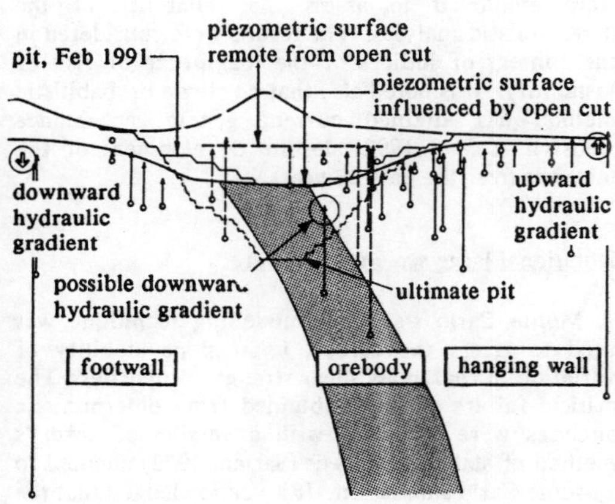

Figure 6: Hydrogeological type-section.

influenced by mining operations a short distance along strike from the pit. Various downward and upward hydraulic gradients were identified which suggest a general movement of groundwater from due south towards due north, perhaps modified by the orebody itself and by lineaments and faulting in the area. Remembering its arcuate shape, this means a groundwater trend which crosses the orebody. Further, in general terms, the piezometric pressures were suppressed on the footwall side more so than on the hanging wall side, as seen in Figure 6. Differences of up to 30 metres in pressure between boreholes on the footwall side fit the model of depressed water table associated with the presence of the pit and a downward drainage gradient in proximity of the quartz reef.

Seepage inflow into the pit and the phreatic surface were established for critical mining stages (ultimate open pit and two levels of underground mining) using this data together with a computer program package (Baalam & Booker, 1990) which performs a two-dimensional finite element seepage analysis.

The assessment of seepage inflow to the pit employed a statistical technique (Rosenblueth, 1975 & 1981) based upon point estimates (at fixed offsets of one standard deviation either side of the mean value) for the various permeability data. This produced an estimate of the mean and standard deviation for the total population. Results from this analysis are presented in Figure 7 for the model of the existing pit. From available historical pit pumping records, a correction factor of about 5 was necessary on mean value estimates of seepage.

A relatively simple statistical method combined with a reasonably small number of mathematical models, provided 'base case' estimates to which future field measurements can be compared.

6. OPEN PIT SLOPES

'Ultimate' pit profiles previously developed by RTZ Consultants Limited (1978) were used as the start point for the overall pit batter analyses. Recognition and acknowledgment of those profiles is made here.

6.1 Mining Constraints

Typical final individual bench geometry may consist of a 12m high bench with 6m wide berms and 70° slope angle for the individual benches. Alternatively, an option employing double benches 24m high, again with a 6m wide berm, may be considered. When allowance is made for the inclusion of safety berms and haul roads, these two geometries result in overall batter slopes of between 50 and 65 degrees.

6.2 Kinematic and Deterministic Analyses

Kinematic and deterministic analyses were conducted for rock slopes within each of the geotechnical domains, using relatively traditional techniques.

Rotational large scale failures

The geotechnical logging and scanline mapping demonstrated the random and heavily jointed nature of the rock mass at MCP (see again Figure 3). When considered in large scale, ubiquitous jointing is a reasonable assumption. Consequently, the rock mass was regarded as sensibly isotropic and homogeneous and it follows that philosophically compatible homogeneous rock mass strength criteria could be adopted. The use of the Hoek-Brown empirical rock mass strength criterion (Hoek & Brown, 1988) provides a suitable model to describe the jointed rock mass strength for this situation. Hoek & Bray (1977) report that rock mass strengths estimated by this method are reasonable when used in slope stability studies.

The assessment of large scale, deep-seated or mass failure of the ultimate pit involved analysis of geotechnical type-sections for each domain, which thereby provided a spatial variation of rock properties.

Use of limit equilibrium stability analyses involving the Spencer-Wright method of analysis permitted both circular and non-circular failure surfaces to be analysed. The analyses directly employed Hoek-Brown failure criteria generated from frequency histogram distributions of RMR and $I_s(50)$ for each domain. By way of example, the adoption of representative shear strength parameters for the type-section on the footwall (see Figure 8) produced a factor of safety (FOS) = 1.71 for overall mass failure with 65° slopes between haul roads.

Planar and wedge failures

On the single or double bench scale, planar block or wedge failures were assessed for each geotechnical domain using kinematic analysis of the stereographic projections and histogram data produced from the scanline mapping of the current pit. Statistical mean orientations of the various joint sets or clusters identified as 'problem' sets were used for the stability which employed limit equilibrium (deterministic) methods.

Joint shear strength was determined from an empirical method (Barton et al, 1974) to manipulate the laboratory determined joint shear data. Scanline mapping data provided typical joint set roughness

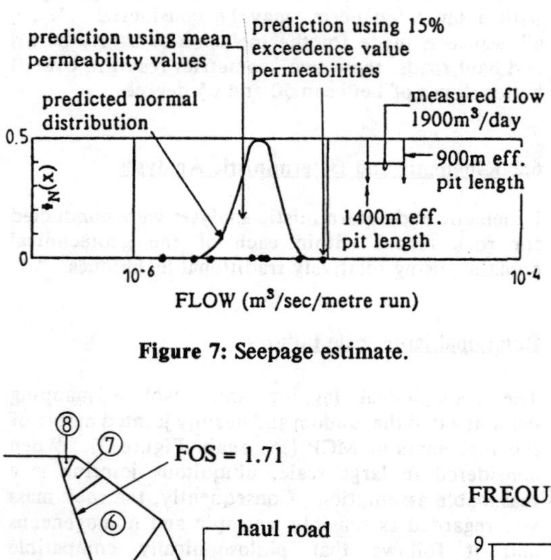

Figure 7: Seepage estimate.

values. Drained slope conditions were assumed for the stability analyses as groundwater infiltration into the pit is expected generally to be below the level that would influence a 1- or 2-bench planar slide condition, to which this analysis was most applicable.

Where through-going dykes or discontinuities could occur, the possibility of an active-passive block mechanism was assessed. Combined block masses up to 350t were identified as marginally stable and potentially unstable by this method. Rock wedge masses up to 125t were identified as potentially unstable also.

6.3 Statistical Assessment

To cater for the inherent uncertainties within the description of material properties, stochastic methods were employed to assess the reliability of the deterministic analyses. The results were considered in the context of indices of the magnitude of risk of instability. It is noted also that no single probabilistic method has attained current general acceptance (Mostyn & Soo, 1992) despite development of the methods over the past 20 years.

Rotational large scale mass failure

A Monte Carlo statistical modelling technique was used to assess the effects upon slope stability of variation in rock mass shear strength properties. The critical failure surfaces obtained from deterministic analyses were modelled with a version of Sarma's method of stability analysis (Sarma, 1973) adapted to a Monte Carlo simulation. It is acknowledged that the critical deterministic failure surface may not coincide with a truly random generation. However, within the context of a sensitivity study, this limitation is believed of minor consequence to a practical solution.

For the assessment of shear strength within each realisation, instantaneous cohesion and friction values (Priest & Brown, 1983) were generated for the rock mass, given $I_s(50)$, RMR and the empirical strength

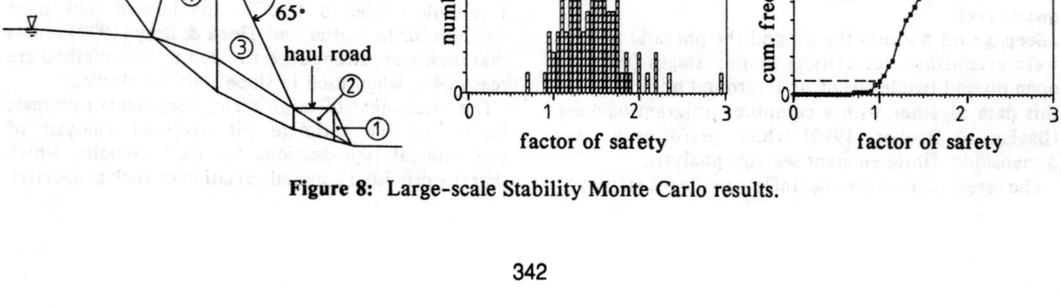

Figure 8: Large-scale Stability Monte Carlo results.

parameter for intact rock (m_i). Each realisation was thus comparable in formulation to the deterministic analysis.

$I_s(50)$ and RMR are not independent variables. To overcome this, a partial rating (RMR*) was established which has the rating for $I_s(50)$ removed. Following independent selection of $I_s(50)$ & RMR* (wherein auto correlation is typically low), RMR was reconstituted. On a large scale basis, m_i for the various granites and orebody rocks was mathematically modified for the proportion of basic dyke intrusions throughout the rock mass. Potential large scale failure mechanisms typically are not adversely influenced by the attitude of the basic dykes. Accordingly, this was considered a mathematically convenient method of incorporating the influence upon overall rock mass shear strength of the variable presence of the lower shear strength basics. This estimate of shear strength conservatively ignored auto correlation (see Mostyn & Soo, 1992).

Typically, a truncated normal distribution best represented both the observed distributions of rock intact strength and quality for the various rock types throughout the pit domains. Measurements of the percentage of basic dyke within the rock mass from borehole observations showed that a modified form of normal distribution achieved the best fit to the data. An example of the results of the analysis for the same portion of footwall analysed above (FOS = 1.71) is presented on Figure 8. Here, the probability of failure is about 7% which in terms of a 'yes/no' index test with auto correlation conservatism is accepted as a very low probability.

Planar sliding

Statistical assessment of the stability of planar slides as they might influence bench stability were conducted using methods (after Coates, 1977, and McMahon, 1974) which are somewhat standard techniques and which will not be discussed further here. The probability of failure for 24m high double benches was about half that of a 12m high bench, when considered in terms of restriction to bench access.

Wedge sliding

A statistical assessment of stability was conducted upon wedge slides which previously had been identified from deterministic modelling, again as may influence bench access. In this instance, the probability was assessed for a joint or discontinuity within a joint set or cluster to have an orientation other than the mean orientation value for that particular joint set. By fitting a statistical function to the observed distribution of joint orientations, a Monte Carlo simulation could be utilised to assess FOS in respect to sliding.

Selection of the fitted distribution primarily depends on the shape of the joint cluster on the stereonet. Typically a two-dimensional normal distribution was assumed in the directions of mean dip of the joint set and dip angle. This assumes dip and dip-direction are mutually independent which may well not be the case. Some curtailing of the joint set extent on the stereonet also was necessary due to joint set interaction. Theoretically, this may influence the probability outcome.

Combinations of pairs of joints were input into a wedge limit equilibrium model employing Barton's empirical joint strength criteria as previously discussed.

In addition to two distributions of joint persistence (for the joint planes forming the rock wedge), a distribution of joint set spacings is required to predict probability of wedge failure. A Monte Carlo simulation was used to produce a cumulative frequency distribution plot for each model (Figure 9).

For an assumed single bench geometry of a 12m high 70° sloping bench with a relatively narrow 4m berm width, loss of any bench width due to a wedge failure impinges upon access. These analyses demonstrated probabilities of failure ≤1% for joint controlled wedges in granite domains. The situation of a through-going dyke occurring in combination with a joint set significantly increased probabilities to values that suggest regular failures could be expected.

7. UNDERGROUND MINING

7.1 Mining Method

The mining method preferred for MCP consists of 'slot and mass blast'.

In general terms, this mining method involves the creation of vertical blocks in the orebody from which broken ore is drawn out from its base following mass blasting. The method relies upon stability of initial slots (created by open stoping and which may be up to 200 metres high) within the block from which the first recovery of ore is achieved and which provide the void into which the subsequent mass blast can be fired. More importantly, the method relies on the progressive caving of the hanging wall so as to not create unsafe working conditions within the mine, particularly in regard to air blast from sudden massive failures. Consequently, a major tenet of this mining method is that the hanging wall must cave and follow downwards the withdrawal of ore from the mass blasted blocks.

The mining method adopts mass blasting to provide broken ore in a size that can be conveniently handled by bulk material methods. Though the size of the mass blasted blocks could be up to 0.75×10^6t, the ability of the ore itself to cave is not of importance to the mining method.

The major geotechnical controls upon the mining method include:

- The 'quality' of the rock mass - in this instance, in particular, the quality of the hanging wall rocks in regard to caving.
- The stress regime within the rock mass as it may influence caving, particularly in regard to arching due to compression.
- The orientation of the mining openings in regard

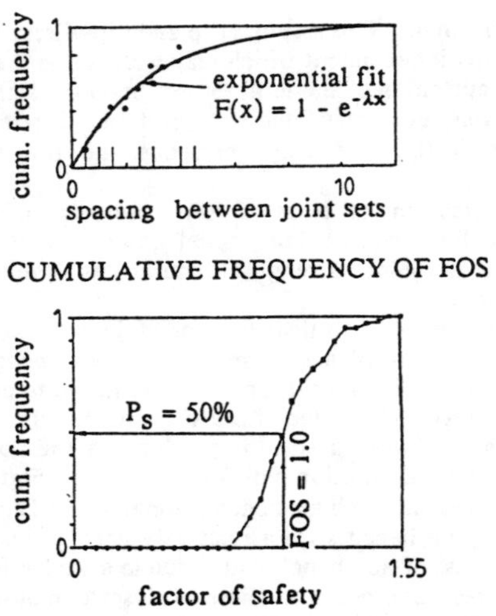

Figure 9: Results from Monte Carlo analysis of wedge model.

to the pervasive stress field.

As may intuitively be expected, virtually any rock mass can be made to cave provided it is undercut by a sufficiently large area. The challenge was to predict this undercut area. Support requirements for permanent underground openings and for production areas are not discussed here.

7.2 Mathematical Modelling

The broad thrust of the analysis has been through use of the boundary element technique. The approach has been to create a mathematical model incorporating the pervasive stress fields, which was then adjusted for the creation of the 'ultimate' open pit and for underground mining. Within the constraints of the plane strain mathematical technique, out-of-plane stresses were incorporated.

The accuracy of the boundary element analysis is somewhat a function of scale. In the context of the modelling of the underground openings (in particular the mass blast blocks and the behaviour of the hanging wall), a homogeneous assumption is considered reasonable and the broad assumptions within the analysis are considered appropriate.

The computer package adopted is discussed elsewhere (Carter & Alehossein, 1991). In this package, a continuum representation of the jointed rock mass is adopted for the analysis of boundary value problems in two-dimensional elastostatic space. The strength parameters were specified for the continuum using the failure criteria suggested by Brady & Brown (1985).

7.3 Hanging Wall Caving

The philosophy of analysis of the hanging wall caving was to determine the likelihood of caving in the cross-section plane strain situation and to then determine sufficient longitudinal length to permit this caving to occur. In this context, caving was modelled as that part of the rock mass that had a factor of safety in regard to shear strength that was ≤1.0. A constant value of elastic modulus, E = 20GPa was adopted for the analyses as little modification to the induced stresses was expected as a result of moduli contrasts between domains within the mathematical model.

Sensitivity studies were adopted as the preferred method of analysis due to program limitations which permitted the use only of homogeneous material failure criteria throughout the model. Initially, material properties for the hanging wall granite were selected on the basis of 'average' and 'mean + one standard deviation' values. This latter value was to permit conservative modelling of the potential for hanging wall caving, in that the stronger rock mass is a conservative estimate in the prediction of the failure of the rock mass and hence the onset caving.

Underground mining was modelled for two depth increments below the ultimate pit, each of 200 metres, which reflect the preferred mining system. The results of analyses for mean values of material properties are presented in Figure 10. Therein, it can be observed that failure of the hanging wall is predicted and therefore the onset of caving is predicted. The same, however, cannot be said for the higher strength (mean + one standard deviation) hanging wall granites as full caving to the base of the underground mining was not predicted for the first level of mining.

As a consequence of this, Monte Carlo simulations were undertaken to permit variation in material properties to be incorporated in the analyses. The prediction of material properties not only took into account the variation of rock mass quality and strength in the determination of RMR, but also variation in the percentage of basic rocks as they might influence shear strength of the rock mass.

The results of these simulations were that an 88% probability for hanging wall caving to occur was demonstrated for randomly generated material properties for the first level of underground mining. This is interpreted to imply that the 'mean + one standard deviation' has a probability of occurrence of less than 12%. A 96% probability was demonstrated for the second level of underground mining. It of course is noted that the modelling was of homogeneous elastic continua and as such did not take into account the presence of joints which no doubt will increase the probability of caving.

An indication of the extent of caving and subsequent subsidence of the surface can be gained by the shape of the failure surface in Figure 10. The extent of caving away from the orebody on the hanging wall side is somewhat more extensive than might intuitively be expected. This may be a function of the mathematical modelling or equally well may be an influence on the high horizontal stresses that run normal to the plane strain model and which have been included in the assessment of factor of safety.

Having demonstrated the viability of caving for plane

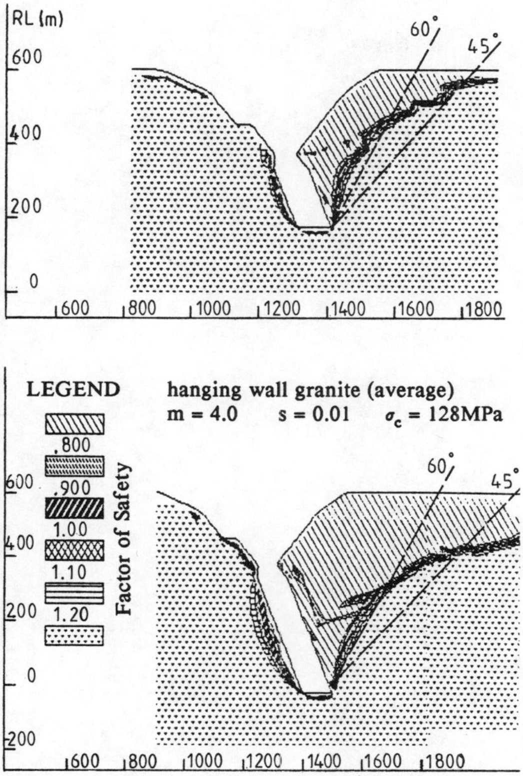

Figure 10: Example of output from boundary element analysis of cross sections through first and second level of underground mining (upper and lower respectively). The areas filled with shading above the number '1.00' in the legend have failed or yielded. Significant straining of the rock mass and movement into the mining void is predicted for these zones of the hanging wall.

strain conditions, the next step in assessing sustainable caving was to determine the size of production openings within the plane normal to the first analysis. This was undertaken by modelling multiples of the standard mass blast block (36 metres in dip direction by 48 metres along the strike). The analysis was conducted at various depths below the current surface level. Depths of 200m 300 and 400 metres were adopted as representative of the first level of underground mining. Following the results of the analyses discussed above, 'average' and 'mean + one standard deviation' failure criteria were adopted for sensitivity studies.

It was demonstrated by these analyses that significant failure on the hanging wall occurred readily at depths of 300 and 400 metres, though was incomplete for the depth of 200 metres. This was presumed directly attributable to the high horizontal stresses that concentrated beneath the base of the 'ultimate' pit. It is not until a significant opening was achieved, as represented by 6x2 and 4x3 blocks - hydraulic radii (area/perimeter) greater than 30m - that significant sustainable failure within the hanging wall was predicted by the modelling. This compared favourably with empirical methods (such as Laubscher, 1977) and reported performance of mines with sustainable caving.

8. CONCLUSION

The use of statistical methods in open pit stability, underground caving and hydrogeological predictions have proven valuable in assessment of the reliability of deterministic analyses and as indices of the magnitude of risk.

9. ACKNOWLEDGMENT

The authors wish to thank the management of BEC & HCL and in particular Messrs H. Fuchimoto, H. Fujita, M. Kusakabe and Shri M. Narasimhan of Bishimetal Exploration Co. Pty Ltd and Shri Y. Subramanyam (Chief Geologist at MCP) as well as our many other acquaintances in HCL for the opportunity to work with them on a daily basis for the project at MCP.

From LongMac Associates, the authors especially wish to acknowledge the work of Mr Greg Kotze whose field appraisal and interpretative work delineated the geotechnical domains at MCP.

REFERENCES

Arjang B. (1991), Pre-Mining Stresses at Some Hard Rock Mines in the Canadian Shield, *CIM Bulletin*, Vol. 84, No. 945.

Balaam N.P. & Booker J.R. (1990), *Finite Element Analysis of Steady State Seepage*, Centre for Geotechnical Research, School of Civil and Mining Engineering, University of Sydney, Australia.

Barton N., Lien R., Lunde J. (1974), Engineering classification of rock masses for the design of tunnel support, *Rock Mechanics*, Vol. 6.

Bieniawski Z.T. (1984), *Rock mechanics design in mining and tunnelling*, A.A. Balkema Publishers, Netherlands.

Brady B.H.G. & Brown E.T. (1985), *Rock Mechanics for Underground Mining*, George Allen & Unwin (Publishers) Ltd.

Brown E.T. and Windsor C.R. (1990), Near Surface In Situ Stresses in Australia and Their Influence on Underground Construction, *VII Australian Tunnelling Conf.*, Sydney.

Carter J.P. & Alehossein H. (1991), *Program BEAN - Two-Dimensional Boundary Element Analysis of Anisotropic Media*, Centre for Geotechnical Research, University of Sydney, Users' Manual for Program BEAN, Version 1.3.

Coates D.F. (1977), *Pit Slope Manual, Chapter 5 - Design*, Canada Centre for Mineral and Energy Technology, CANMET Report 77-5.

Hoek E. and Bray J.W. (1977), *Rock Slope Engineering*, Inst. Min. Metal., London.

Hoek E. and Brown E.T. (1988), The Hoek-Brown failure criterion - a 1988 update, *15th Canadian Rock Mech. Symp.*, Toronto.

Laubscher D.H. (1977), Geomechanics classification of jointed rock masses - mining applications, *Trans. Inst. Mining & Metallurgy*, Vol. 86.

Leventhal A.R., Barker C.S. & de Ambrosis L.P. (1992), Malanjkhand Copper Project - Overview of the Geotechnical Investigation for Optimum Exploitation, *ISRM Regional Symposium on Rock Slopes*, New Delhi.

McMahon B.K. (1974), Design of Rock Slopes Against Sliding on Pre-existing Fractures, *Intl. Soc. Rock Mech., 3rd Int. Cong.*, Denver.

Mineral Exploration Corporation Limited (1984), *Exploration Report (July 1980 - February 1984), Malanjkhand Copper Ore Deposit.*

Mostyn G.R. and Soo S. (1992), The Effect of Auto-correlation on the Probability of Failure of Slopes, *Proc. 6th ANZ Conf. on Geomechanics*, Christchurch.

Priest S.D. and Brown E.T. (1983), Probabilistic Stability Analysis of Variable Rock Slopes, *Trans. Inst. Min. Metal, London*, Vol. 92.

Priest S.D. and Hudson J.A. (1976), Discontinuity spacings in rock, *Int. J. Rock Mech. Min-Sci & Geomech Abstr.*, Vol. 13.

Rosenblueth E. (1981), Two-Point Estimates in Probabilities, *Appl. Math. Modelling*, Vol. 5.

Rosenblueth E. (1975), Point Estimates for Probability Moments, *Proc. Nat. Acad. Sci. USA*, Vol. 72, No. 10.

RTZ Consultants Limited (1978), *Geotechnical Review, Malanjkhand Project*, report prepared by Golder Associates in full collaboration with RTZC for Hindustan Copper Limited.

Sarma S.K. (1973), Stability Analysis of Embankments and Slopes, *Géotechnique* 23, No. 3.

Innovative Mine Design for the 21st Century, Bawden & Archibald (eds) © 1993 Balkema, Rotterdam, ISBN 90 5410 325 6

Planning shovel truck fleet requirements for open pit mining operations

P.N.Calder
Department of Mining Engineering, Queen's University, Kingston, Ont., Canada

ABSTRACT: The analysis of shovel truck fleet requirements is a significant aspect of open pit mine planning. This paper describes the use of simulation techniques to conduct specific studies, then goes on to evaluate the use of a simplified modelling procedure which can provide mine planning engineers with quick reliable answers concerning truck requirement and fleet operating costs under a wide variety of changing conditions.

1 BASIC HAULAGE NETWORKS

Figure 1 illustrates a simple haulage network for an open pit mine. Beginning from the crusher, a road leads out and down into the pit. Normally two way traffic is involved and in addition to haulage trucks, many types of service vehicles also use these roads. Due to their size, haulage trucks are not permitted to pass one another. Often trucks with different performance characteristics are present in the fleet and the speed of the slower units affects the overall fleet.

1.1 Truck dispatching

As trucks proceed down the haulage ramp they encounter branch roads leading to the current working benches. These roads also branch out to the various shovel locations on that bench. The decision as to which branch to follow is controlled in a variety of ways. The simplest is fixed truck assignment in which the driver is instructed at the beginning of the shift to load at a particular shovel. Other methods including a dispatcher contacting the driver by radio as the truck approaches a road junction and the use of a computer controlled electronic display. Automated computer controlled dispatch systems are in widespread use today at large scale operations.

1.2 Queuing

Upon arriving at a shovel the truck will join the line up "queue" if other trucks are waiting to be loaded. Methods such as double back up are often used to minimize the spotting times at the shovel between loads. Queuing also occurs at crushers and dumps. On completion of loading, the truck will proceed back along the haulage network to the assigned dumping destination.

1.3 Factors controlling truck velocity

Truck velocity at any point is determined by a number of key factors. The performance characteristics of the engine and the braking system are the important truck parameters. The mine will establish speed limits in many situations to ensure safe operating conditions. Travelling downgrade loaded and road intersections are examples of areas where truck velocity needs to be restricted. The grade(inclination) of the haulage road, the rolling resistance of the road surface and the weather conditions, including visibility, are all important. The speed at which different drivers chose to proceed under a variety of conditions is a major factor. Drivers will often not accelerate the truck utilizing the maximum available rimpull just as most prefer not to drive an automobile that way. Comfortable rates of maximum acceleration need to be defined before we can use raw rimpull data to calculate realistic truck speeds for a variety of conditions. For this reason, truck velocities based strictly on truck rimpull data often give optimistic results.

1.4 Truck cycle times and "match" factor

Truck cycle time is the average time lapse for the passage of the truck around its circuit. The match factor represents the ideal number of trucks that should be assigned to a shovel. This is equal to the total cycle time divided by the mean loading time.The cycle time for each trip is affected by the queues that are encountered at the loading and dumping points and by interference from encountering slower vehicles which can not be passed and general traffic congestion.

Loading times at the shovel are often highly variable due to changing digging conditions and the need to reposition the shovel, etc.The ore dump point, usually a crusher or loading pocket, is another source of queuing. Crushers , for

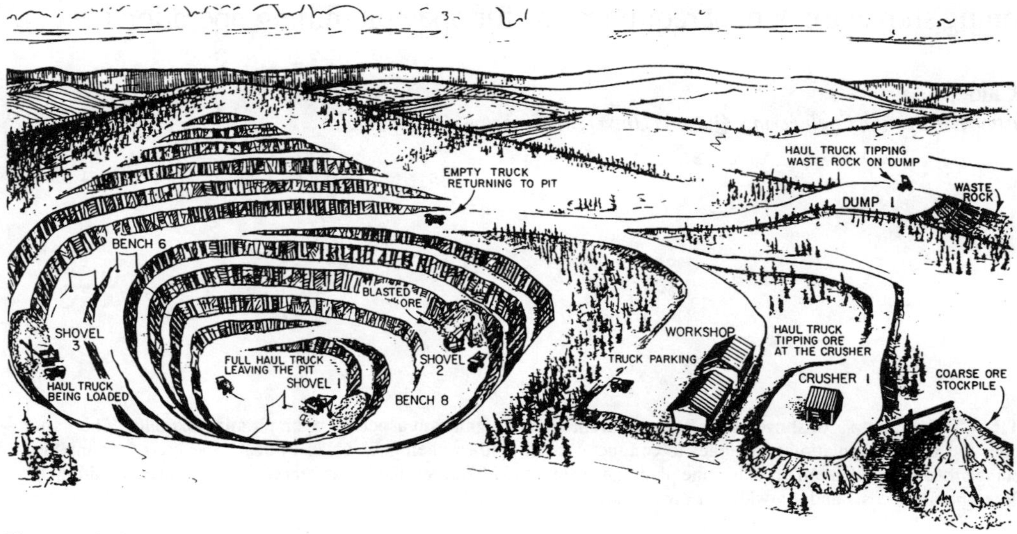

Figure 1. A simple open pit haulage network

example, are frequently plugged by large pieces of oversize rock and dumping is halted until it is cleared. Since all trucks on the ore haul usually utilize a single crusher this is more disruptive than the breakdown of a single shovel.The net result is that haulage cycle times exhibit a great deal of scatter.One consequence of this is that it is not possible to accurately predict shift production simply from a knowledge of truck cycle and loading times.

1.5 Operating delays

Procedures for truck re-fuelling, shift change and lunch breaks effect overall efficiency, as do scheduled maintenance, unplanned equipment breakdowns and the availability of spare equipment. If shift change and lunch breaks occur at a common time and location, trucks are grouped together and system efficiency decreases. Some difficulties of this nature are unavoidable.

1.6 Estimating fleet production and costs

Given the dynamics of the situation, a mine planning engineer needs a way to accurately forecast the requirements for trucks and shovels and estimate the operating costs. As mining progresses at a rapid rate in modern large scale operations, these parameters will change dramatically from period to period. This paper examines methods of reliable analysis for these systems and presents a method for estimating fleet requirements which can be integrated into the computerized systems currently being used by the planning engineer to define mining areas by time period.

2 SIMULATION MODELS

The modelling of open pit haulage networks using computer simulation has been in wide spread use for many years (Calder & Waring 1965), (Calder & Bauer 1973), (Chick,1980) and (Fytas & Calder 1984). These models have been developed in a variety of ways including time study data, calculation based on manufacturers performance curves and real time data generated by computerized truck dispatch systems. Descriptions of these methodologies is well documented in the current literature and will be dealt with here very briefly.

2.1 Time studies

In this method the independent times of the various movements and operations are recorded. For example, the time it takes a certain type of haulage unit to traverse a haul road segment when it is not interfered with by slower trucks in front directly measured by an observer in the field. Interference between slow and fast trucks, queuing, etc. is accounted for in the simulation procedure. Travel times for each truck type, loaded and empty, are required for each road segment. Similarly, loading and dumping times are required for each truck type for the various shovels and

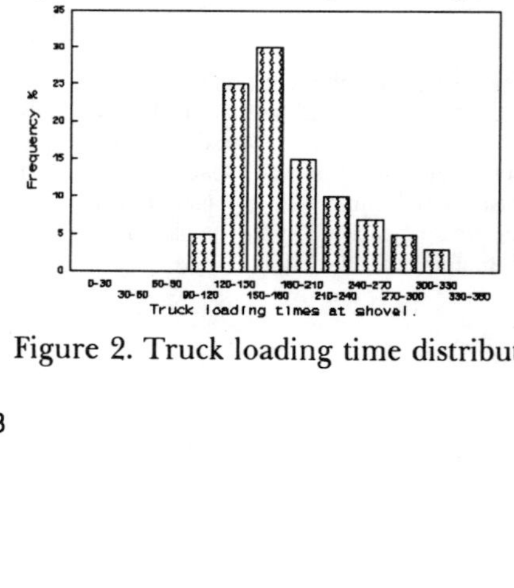

Figure 2. Truck loading time distribution

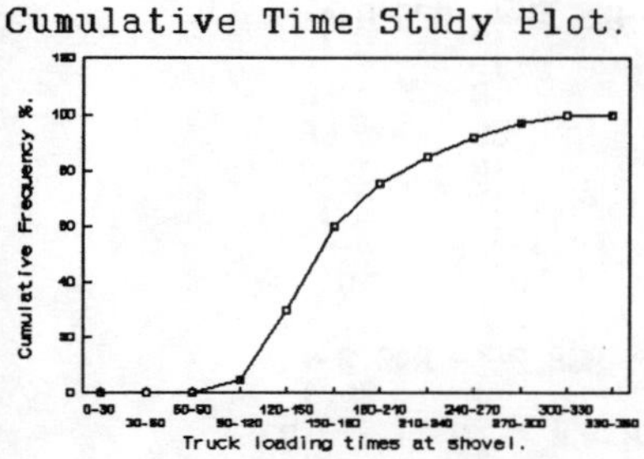

Figure 3. Cumulative distribution

dump points. This data is often represented in the form of a histogram, Figure 3 is an example of a typical time study histogram of truck loading times at a shovel. A time increment is selected and the percentage of the total number of readings falling in each increment is plotted. These distributions are generally log-normal in shape.

The next step is to plot a cumulative time distribution as illustrated in Figure 3. The value for each time increment is the sum of the percentage times within that increment and all preceding increments. During the simulation process trucks are cycled through the haulage network following a series of rules regarding shovel assignment, etc. When a truck enters a road segment it is randomly assigned a travel time based on the time study data. This technique is known as Monte Carlo Simulation because of the random way the data is selected. By generating a random number between 0 and 100, the selected time is determined by reading horizontally along the cumulative frequency distribution to the curve, then down to the time axis. An example is given in Figure 3. While it is possible to complete a simulation manually, it would be very time consuming to do so and should not be contemplated in today's computer age. I point this out to emphasize that the simulation procedure itself is a very simple brute force technique. The role of the computer is to be an efficient number cruncher, nothing theoretical is involved in the simulation process, we are simply moving trucks through a network according to rules we have established and the observed performance characteristics of the units involved. A good computer simulation program will perform this quickly and keep track of the required study outputs. Time study based simulation has several major disadvantages relating to the conditions and configuration of the haulage road network. These studies are useful when selecting equipment for a new mine. The configuration of haulage road networks change frequently, maintaining current data is time consuming and impractical if the data is collected manually. Estimating travel times through a calculation procedure, while still retaining the ability to select times randomly from a realistic frequency distribution is preferable. This is discussed in the following section.

2.2 Performance curve calculations

The velocity of a truck moving along a haulage road segment can be calculated from a knowledge of the rimpull force which the truck can deliver as a function of speed. Rimpull force is the force acting parallel to the road surface resulting from the tractive power of the engine during acceleration or the retarding power of the brakes during deceleration. The basic equations of motion (Newtons law) can be used with these charts to calculate truck movements. Loading and dumping times can be estimated or observed, a good approach is to assign a mean and a statistical variation which is determined randomly. When used in combination with practical speed limit rules and comfortable driver acceleration limits, and including some randomness in loading and dumping times as discussed, this method provides a very good modelling system which can be readily implemented to handle most situations.

2.3 Modelling with real time information

Modern computer dispatch systems keep continuous track of vehicle movements and create a real time computer database of haulage fleet movements. This could provide a powerful method of updating the model based on current shovel locations, road conditions, etc. This system gets its information from the times trucks pass electronic beacons along the haul road and from other communications with the driver and shift foreman. It would be necessary for this kind of system to identify all types of delays, for example if a truck were delayed on the haulage road due to an accident, this would have to be recognized in all situations and the basic cycle time observations adjusted. Cycle times without interference are required by simulation models. Based on my observations, these systems have not evolved to a state where real time information can be easily used to forecast future truck requirements.

2.4 Fleet analysis using simulation

Determining the required number of trucks and shovels to meet the production objectives is an important part of any mine plan. This is required when opening a new operation and when planning for future time periods. In both of these situations real time data is not available, performance curve calculations modified by common sense rules and incorporating an element of randomness in loading and dumping times provides the best method. Simulation studies can be used to evaluate proposed additions to the fleet such as incorporating a computerized truck dispatch system or adding new trucks to the fleet which have different performance characteristics. Examples of such studies will be presented in the following sections.

3 THE PIT-S SIMULATION SYSTEM

PIT_S is a one second time increment Monte Carlo simulation system for open pit mines. It is written in the "C" programming language utilizing dynamic memory allocation. Structures are used to represent entities including trucks, loaders, crushers, dumps and roads. All entities are queues. Linked lists are used to control queues at loaders, crushers and dumps, and on the roads.Truck rimpull data is used to calculate motion parameters. A number of user specified speed limit rules are also applied in controlling truck velocities under various conditions. A schematic diagram,

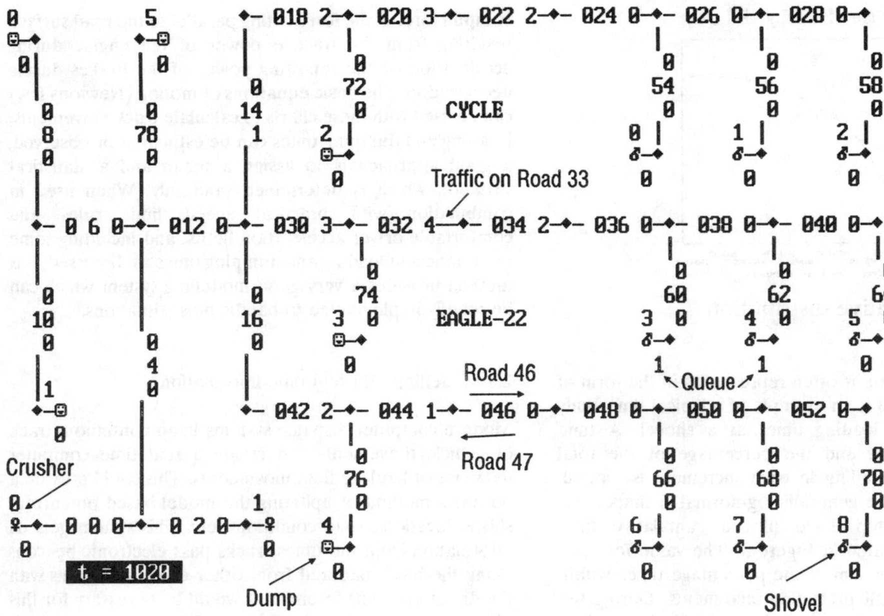

Figure 4. A general road network diagram.

utilizing text mode is produced on the screen. The diagram illustrates the haulage network, truck movements are indicated on boxes within the diagram. These boxes illustrate the number of empty and loaded trucks for each road, loader, crusher and dump and are updated whenever a truck moves from one entity to another. Figure 4 is an example of this type of road network diagram.

3.1 Control system with files

The geometry of the haulage network and most of the parameters controlling truck movements, such as road rolling resistance, grade, length, speed limits, loading and dumping times are stored in a series of files which can be edited by the user. The program writes two files for review, recording all truck movements, and produces a summary file on the screen.The program now simulates one eight hour shift (28,800 sec.). When running the simulation program an option is provided to close any of the loaders for that run. This, in combination with the ability to change road lengths etc. by editing files, allows for a variety of situations to be handled without creating a new road network each time.The program applies several rules concerning truck velocities internally. For example, a truck exiting a road connected to a loader, crusher or dump must have an exit velocity of 0. The truck will begin to decelerate some distance before the end of the road, so that rate of deceleration will not exceed a predefined limit. If a maximum road exit velocity is specified a truck will begin to decelerate at an appropriate distance so that it will attain this exit velocity at the end of the road without decelerating at a rate greater than that specified. A faster truck upon overtaking another truck on a haulage road will be required to maintain a distance 50 feet behind the truck in front and not allowed to pass.

3.2 Simulation study output

Table 1 is a summary of the road-truck output which is produced on the screen at the end of the run. It indicates, for each truck, the cumulative waiting time, the percentage time travelling up-down-empty and up-down-full, percentage time loading, percentage time dumping, percentage time for other activities, the operating cost per hour, the number of loads for each truck and the cost per load. A line is printed

Table 1 Summary shift report for shovels, crushers and trucks

PIT_S LOADER REPORT

SHIFT	UNIT	LOADS	GRADE	WAIT	DELAY
1	L- 0	115	39.1	711	1720
1	L- 1	108	44.0	484	3600
1	L- 4	75	25.0	9521	1200
1	L- 5	82	39.1	6828	3080
1	K- 0	240	37.6	9581	960
1	K- 1	132	37.8	14521	2920

PIT_S TRUCK REPORT

TRUCK	WAIT	E-UP	E-DN	F-UP	F-DN	LOAD	DUMP	OTHER	$/HOUR	LOADS	$/LOAD
0	6	6	18	30	10	14	4	12	123	21	46
1	4	6	17	30	11	15	4	13	124	23	43
2	2	6	18	32	10	15	4	13	127	22	46
3	11	5	17	29	9	13	3	13	118	19	49
4	10	6	15	27	10	15	4	13	117	22	42
5	9	6	16	28	10	14	4	13	119	21	45
6	15	6	15	25	9	13	4	13	112	20	44
7	4	6	17	31	9	13	4	16	123	20	49
8	5	6	16	29	10	15	4	15	121	22	44
9	16	5	14	25	9	13	3	15	110	19	46
10	5	6	17	30	10	15	4	13	123	23	42
11	19	5	14	24	9	13	3	13	109	19	45
12	5	6	16	29	10	15	4	15	121	22	44
13	5	6	16	29	10	15	4	15	121	22	44
14	8	6	16	28	10	14	4	14	119	21	45
15	4	6	17	30	10	14	4	15	123	21	46
16	14	6	14	23	10	15	4	14	110	22	40
17	4	6	17	30	10	14	4	15	123	21	46

380 loads were produced at a combined shovel truck cost of $ 60/load .

Note: L-1 ... shovel #1, K-0 ... crusher #0
E-UP, E-DN ... % time travelling empty up grade, empty down grade
F-UP, F-DN ... % time travelling full up grade, full down grade
WAIT, LOAD, DUMP, OTHER .. % time spent waiting, loading etc.

Table 2 Data used in determining shift operating costs

Truck Travelling	Fuel Consumption litres/hr	Operation cost/hr Including Fuel $
Empty-up	115	110
Empty-down	70	84
Full-up	255	214
Full-down	125	130
Loading	20	98
Waiting	20	42
Dumping	20	87

Shovel operating cost .... $180/hr.
Fuel Cost .... $0.35/litre.

at the bottom indicating the total number of loads produced and the combined shovel-truck cost per load.The loader truck report, Table 1, provides an estimate of operating cost in dollars per hour and dollars per load for trucks, and a combined shovel truck cost for the shift. The cost assumptions on which these calculations are based are given in Table 2.

3.3 Operating delays

At the start of the shift there is a 600 second delay for truck inspection following which trucks leave the crusher at 10 second intervals. The loaders, crushers and dumps cease operating for a 20 minute lunch break from 13,800 seconds into the shift until 15,000 seconds. Trucks would continue to the nearest facility and be held up there while the loader, etc., was closed.Later arriving trucks would be at the rear of the queue.The loaders, crushers and dumps are closed at 27,600 seconds, 20 minutes before the end of the shift. Trucks continue to operate until held up by a closed facility. One breakdown occurs during the shift for each loader and crusher. The time and duration of these breakdowns is affected by the random number selected at the beginning of the run and are reported on the screen. Trucks do not break down, it is assumed that replacements are available. Truck time spent waiting at a broken down facility is included as "other" in the truck summary report, as are shift start-up, lunch and shift-end delays.

3.4 Truck dispatching

The program has a facility for automatic and fixed truck dispatching.In the automatic mode the time is computed for the truck to travel over each of the alternative routes and return loaded to the decision point. The effect of traffic on the forward roads , roads immediately behind, and at the loaders is included, as are loaders experiencing a maintenance delay. The truck is then dispatched along the alternative route offering the minimum cycle time. In the fixed mode the truck always loads at the assigned shovel.

4 CASE STUDIES WITH PIT-S

The use of the program will now be demonstrated in three situations: comparing automated and fixed truck dispatching, studying the effect of mixing trucks with different performance characteristics on the same network and in estimating network match factors. In all cases a typical 200 ton electric wheel truck was used.

4.1 Automated vs fixed truck dispatching

For this study, a pit approximately 200 m in depth with four shovels and two crusher bays was used. Two of the shovels were located 100 m below the others. The crusher was located approximately 700 m from the pit entry. Average loading and dumping times were 200 and 60 seconds respectively. The fleet size was varied from 4 to 36 trucks as indicated on Figure 5. Both automatic dispatching and fixed assignment of trucks were tested. The results are presented in Figure 5 as $/load and total loads vs the number of trucks in the fleet. This study indicates that when truck assignments are correctly matched to shovel capacity, productivity gains for the automated system amount to approximately 2% and the operating costs for the two systems are relatively even. During situations where the shovel is under trucked there is a significant advantage for the automatic dispatching system. It should be noted that these automated systems do provide very useful statistical reporting on key aspects of fleet operation. Also, because the automatic dispatch systems all pit supervisors to closely scrutinize the actions of truck drivers, the actual percentage improvement in an operating may be higher than indicated here.

4.2 Mixed truck fleet

The same network was used as in the previous study. The

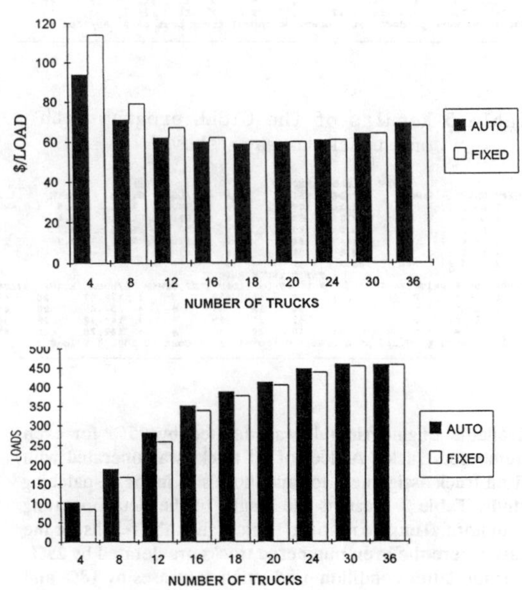

Figure 5. Automated vs fixed truck dispatching

Table 3 Study output for 20 fixed assignment trucks with normal performance

LOADER REPORT

SHIFT	UNIT	LOADS	GRADE	WAIT	DELAY
1	L- 0	115	39.1	646	1720
1	L- 1	107	39.1	690	3600
1	L- 4	93	33.3	5810	1200
1	L- 5	89	44.4	4978	3080
1	K- 0	249	38.8	10082	960
1	K- 1	144	39.1	13620	2920

TRUCK REPORT

TRUCK	WAIT	E-UP	E-DN	F-UP	F-DN	LOAD	DUMP	OTHER	$/HOUR	LOADS	$/LOAD
0	7	7	15	24	11	16	5	15	115	24	38
1	5	7	15	26	11	16	5	15	118	24	39
2	2	5	19	36	9	13	3	13	131	19	55
3	10	4	17	32	8	11	3	15	121	17	56
4	12	6	14	22	11	15	4	16	109	23	37
5	16	6	12	21	10	15	4	16	106	22	38
6	6	5	18	34	8	12	3	14	126	18	56
7	4	5	19	36	9	13	3	11	131	19	55
8	13	6	13	22	11	15	4	16	109	23	37
9	18	6	12	20	9	14	4	17	103	21	39
10	6	5	18	34	8	12	3	14	126	18	56
11	13	4	17	32	8	11	3	12	121	17	56
12	7	7	15	24	11	16	5	15	115	24	38
13	19	6	12	20	9	14	4	16	103	21	39
14	5	5	19	34	8	12	3	14	126	18	56
15	11	4	17	32	8	11	3	14	121	17	56
16	14	6	13	22	11	15	4	15	109	23	37
17	12	6	13	23	10	15	4	17	110	23	38
18	8	5	18	32	8	11	3	15	122	17	57
19	15	4	17	30	7	11	3	13	117	16	58

404 loads were produced at a combined shovel truck cost of $ 60/load .

Table 4 Study output for 20 fixed assignment trucks, half de-rated by 75%

LOADER REPORT

SHIFT	UNIT	LOADS	GRADE	WAIT	DELAY
1	L- 0	97	39.1	4532	1720
1	L- 1	95	39.1	3434	3600
1	L- 4	69	33.3	10724	1200
1	L- 5	69	44.4	8963	3080
1	K- 0	189	38.7	12743	960
1	K- 1	129	39.4	17490	2920

TRUCK REPORT

TRUCK	WAIT	E-UP	E-DN	F-UP	F-DN	LOAD	DUMP	OTHER	$/HOUR	LOADS	$/LOAD
0	7	5	12	37	9	13	4	13	131	20	52
1	15	5	11	29	8	13	3	16	115	19	48
2	3	4	14	50	7	9	2	11	148	14	84
3	4	4	15	45	7	10	3	12	141	15	75
4	3	6	12	37	9	13	4	16	131	20	52
5	13	5	11	28	9	13	4	17	115	20	46
6	4	4	14	50	7	9	2	10	148	14	84
7	12	4	14	40	6	9	2	13	130	13	80
8	1	6	13	39	9	13	4	15	135	20	54
9	6	6	12	34	9	14	4	15	127	21	48
10	1	4	14	50	7	9	2	13	148	14	84
11	15	3	13	38	6	9	2	14	126	13	77
12	9	5	11	35	9	13	3	15	126	19	53
13	13	5	11	29	8	13	3	18	115	19	48
14	6	4	14	47	6	9	2	12	142	13	87
15	10	4	14	43	6	9	2	12	135	14	77
16	12	5	11	33	8	12	3	16	121	18	53
17	18	5	10	27	8	12	3	17	111	18	49
18	5	4	14	47	6	9	2	13	142	13	87
19	13	4	14	37	6	9	2	15	125	13	76

330 loads were produced at a combined shovel truck cost of $ 80/load .

Table 5 Results of the CYCLE program with one truck to each shovel

PIT_S LOADER REPORT

SHIFT	UNIT	LOADS	GRADE	WAIT	DELAY	$/LOAD	CYCLE	MATCH
1	L- 1	26	11.1	82	0	98.11	1115	5.6
1	L- 2	25	22.2	82	0	101.72	1151	5.8
1	L- 3	22	33.3	85	0	118.04	1340	6.7
1	L- 4	22	44.4	84	0	117.42	1342	6.7
1	K- 0	46	32.3	89	0			
1	K- 1	46	21.2	89	0			

PIT_S TRUCK REPORT

TRUCK	UNIT	WAIT	E-UP	E-DN	F-UP	F-DN	LOAD	DUMP	OTHER	$/HOUR	LOADS	$/LOAD
0	L- 1	0	7	19	36	11	18	5	4	138.67	26	42.
1	L- 2	0	7	20	35	12	17	5	4	137.69	25	44.
2	L- 3	0	6	21	41	10	15	4	3	144.42	22	52.
3	L- 4	0	6	21	40	10	15	4	4	142.70	22	51.

95 loads were produced at a combined shovel truck cost of 108.15 $/load .

available engine rimpull was derated by 25% for even numbered trucks. A fleet of 20 trucks was operated with fixed truck assignment to four shovels as in the dispatching study. Table 3 presents the results of the fleet operating with normal engine rimpull, Table 4 gives the results for the case where the even numbered trucks are derated by 25%. For the latter condition production decreases by 18% and operating costs per load are increased by 33% . The faster trucks are held up and forced to spend a significantly higher

Table 6 CYCLE results with the match numbers from Table 5 rounded up

PIT_S LOADER REPORT

SHIFT	UNIT	LOADS	GRADE	WAIT	DELAY	$/LOAD	CYCLE	MATCH
1	L- 1	142	11.1	0	0	53.48	1207	6.0
1	L- 2	141	22.2	1	0	54.85	1214	6.1
1	L- 3	141	33.3	2	0	62.31	1426	7.1
1	L- 4	140	44.4	1	0	64.76	1433	7.2
1	K- 0	274	33.2	38	0			
1	K- 1	276	22.1	40	0			

PIT_S TRUCK REPORT

TRUCK	UNIT	WAIT	E-UP	E-DN	F-UP	F-DN	LOAD	DUMP	OTHER	$/HOUR	LOADS	$/LOAD
0	L- 1	6	6	17	33	10	16	5	7	129.99	24	43.
1	L- 2	4	6	18	34	12	16	5	5	133.89	24	44.
2	L- 3	4	5	20	38	9	14	4	6	136.72	21	52.
3	L- 4	5	5	20	38	9	13	4	6	136.16	20	54.
4	L- 1	7	6	17	33	10	16	5	6	129.99	24	43.
5	L- 2	5	6	18	33	11	16	5	6	131.29	24	43.
6	L- 3	5	5	20	38	9	13	4	6	136.16	20	54.
7	L- 4	5	5	19	38	9	13	4	7	135.74	20	54.
8	L- 1	8	6	17	33	10	16	5	5	129.99	24	43.
9	L- 2	6	6	18	32	11	16	5	6	129.57	24	43.
10	L- 3	6	5	20	38	9	13	4	5	136.16	20	54.
11	L- 4	7	5	19	38	9	13	4	5	135.74	20	54.
12	L- 1	10	6	17	32	10	16	5	4	128.27	24	42.
13	L- 2	7	6	18	32	11	15	4	7	128.56	23	44.
14	L- 3	7	5	19	38	9	13	4	5	135.74	20	54.
15	L- 4	8	5	19	38	9	13	4	4	135.74	20	54.
16	L- 1	10	6	17	32	10	15	4	6	127.26	23	44.
17	L- 2	7	6	18	32	11	15	4	7	128.56	23	44.
18	L- 3	8	5	19	38	9	13	4	4	135.74	20	54.
19	L- 4	8	5	19	37	9	13	4	5	134.02	20	53.
20	L- 1	10	6	17	32	10	15	4	6	127.26	23	44.
21	L- 2	8	6	18	32	11	15	4	6	128.56	23	44.
22	L- 3	9	5	19	37	9	13	4	4	134.02	20	53.
23	L- 4	10	5	19	36	9	13	4	4	132.30	20	52.
24	L- 3	9	5	19	36	9	13	4	5	132.30	20	52.
25	L- 4	11	5	19	36	9	13	4	3	132.30	20	52.

564 loads were produced at a combined shovel truck cost of 59.05 $/load .

Table 7 CYCLE results with the match numbers from Table 5 rounded down

PIT_S LOADER REPORT

SHIFT	UNIT	LOADS	GRADE	WAIT	DELAY	$/LOAD	CYCLE	MATCH
1	L- 1	127	11.1	11	0	54.02	1125	5.6
1	L- 2	123	22.2	14	0	55.43	1162	5.8
1	L- 3	126	33.3	12	0	64.93	1363	6.8
1	L- 4	126	44.4	12	0	64.93	1371	6.9
1	K- 0	242	33.4	46	0			
1	K- 1	246	22.1	44	0			

PIT_S TRUCK REPORT

TRUCK	UNIT	WAIT	E-UP	E-DN	F-UP	F-DN	LOAD	DUMP	OTHER	$/HOUR	LOADS	$/LOAD
0	L- 1	0	7	19	36	11	18	5	4	138.67	26	42.
1	L- 2	0	6	19	35	12	17	5	6	136.59	25	43.
2	L- 3	0	6	21	39	10	14	4	6	140.42	21	53.
3	L- 4	1	6	21	39	10	14	4	5	140.42	21	53.
4	L- 1	1	7	19	35	11	18	5	4	136.95	26	42.
5	L- 2	2	6	19	35	12	17	5	4	136.59	25	43.
6	L- 3	1	6	20	39	10	14	4	6	140.00	21	53.
7	L- 4	1	6	20	39	10	14	4	6	140.00	21	53.
8	L- 1	2	7	19	35	11	17	5	4	136.39	25	43.
9	L- 2	3	6	19	34	12	17	5	4	134.87	25	43.
10	L- 3	3	5	20	39	10	14	4	5	139.32	21	53.
11	L- 4	2	5	20	39	10	14	4	6	139.32	21	53.
12	L- 1	4	7	18	35	11	17	5	3	135.97	25	43.
13	L- 2	4	6	19	34	12	16	5	4	134.31	24	44.
14	L- 3	4	5	20	39	9	14	4	5	138.44	21	52.
15	L- 4	4	5	20	39	9	14	4	5	138.44	21	52.
16	L- 1	4	6	18	35	11	17	5	4	135.29	25	43.
17	L- 2	4	6	19	34	12	16	5	4	134.31	24	44.
18	L- 3	5	5	20	39	9	14	4	4	138.44	21	52.
19	L- 4	5	5	20	38	9	14	4	5	136.72	21	52.
20	L- 3	6	5	20	38	9	14	4	4	136.72	21	52.
21	L- 4	6	5	20	38	9	14	4	4	136.72	21	52.

502 loads were produced at a combined shovel truck cost of 59.69 $/load .

percentage of their time travelling up grade loaded. In addition to losing production, the operating costs per hour of these trucks is also increased. A similar series of tests using automatic dispatching did not produce significantly better results. It is obvious that trucks with reduced speed performance will greatly reduce the efficiency of faster units in mixed fleet situations.

4.3 Match factor studies

The match factor represents the ideal number of trucks that should be assigned to a shovel. This is equal to the truck cycle time, when it is not interfered with by slower vehicles, breakdowns, queue delays etc., divided by the mean loading time. To determine match factors all of the delays were eliminated from the PIT_S program, including shift start up, lunch break, equipment breakdowns etc. Then by running a single truck from each source to each destination match factors can be calculated. This version of the PIT_S program, with the delays removed, is referred to as CYCLE. Table 5 gives the results of a run of the cycle program with conditions similar to the previous studies. As expected loader waiting time is very high as only truck is assigned to each shovel, truck waiting time is zero and combined costs are high due to under trucking. Truck cycle time, in seconds, the match factor and a number of productivity and cost factors for the shovels and trucks are reported. Other information such as the net revenue per hour can readily be calculated. The next step is to assign the number of trucks indicated by the match factor to each shovel and run the shift again. The purpose of this is to determine if the match factors alone can determine the correct number of trucks to assign to each shovel or whether vehicle congestion will introduce significant errors. In order to assign the trucks the match number must, of course, be rounded up or down. For the first test the match numbers are rounded up. The results are given in Table 6. A second test is run with the match numbers rounded down, the results are given in Table 7. The best results in this case occur with the match factors rounded up. Loader waiting is minimal and the combined shovel truck cost is lower. Notice in Table 5 that the match numbers, if rounded off to a single digit, would round up in this example. This suggests that rounding off to a single digit is the best practise. The match numbers reported in Tables 6 and 7 have changed from Table 5 due to circuit congestion. When more than one truck is assigned to a shovel, these values are useful in comparing with the single truck numbers but are not true match numbers. Taking the number of loads produced by each loader with only one truck assigned, from Table 5, and multiplying by the number of trucks assigned when rounding up or down , as in the studies reported in Tables 6 and 7, in each case the production has decreased by roughly 8%. It can be concluded that a haulage circuit, operating with close to optimal truck capacity, will achieve a production of approximately 92% of that estimated using a single truck to determine match factors and multiplying the single truck production by the match factor. Since match factors can be generated rapidly this relationship may provide a quick yet accurate method of estimating truck requirements.

Table 8 PIT_S results with the match numbers from Table 5 rounded up

PIT_S LOADER REPORT

SHIFT	UNIT	LOADS	GRADE	WAIT	DELAY
1	L- 2	121	11.1	389	1480
1	L- 3	110	22.2	407	3320
1	L- 4	112	33.3	2042	1200
1	L- 5	101	44.4	2369	3080
1	K- 0	266	26.8	7694	960
1	K- 1	166	27.3	15215	2920

PIT_S TRUCK REPORT

TRUCK	WAIT	E-UP	E-DN	F-UP	F-DN	LOAD	DUMP	OTHER	$/HOUR	LOADS	$/LOAD
0	16	5	14	25	8	12	3	17	109	18	48
1	22	4	13	24	8	11	3	15	106	17	49
2	10	5	17	31	8	11	3	15	120	17	56
3	10	5	17	30	8	11	3	16	118	17	55
4	16	5	14	26	8	13	3	15	111	19	46
5	19	5	14	24	8	11	3	16	107	17	50
6	8	4	17	31	8	11	3	18	119	17	56
7	7	5	18	31	8	12	3	16	121	18	53
8	13	5	15	27	8	13	3	16	113	19	47
9	22	4	13	22	8	11	3	17	102	16	51
10	7	5	17	32	8	12	3	16	122	18	54
11	14	4	17	30	7	11	3	14	117	16	58
12	21	5	13	24	7	11	3	16	105	17	49
13	23	4	13	22	8	11	3	16	102	16	51
14	8	5	18	31	8	11	3	16	120	17	56
15	16	4	15	29	7	11	3	15	114	16	57
16	19	5	13	25	8	12	3	15	108	18	48
17	23	4	13	22	8	11	3	16	102	16	51
18	10	4	17	31	8	11	3	16	119	17	56
19	8	5	17	32	8	12	3	15	122	18	54
20	16	5	14	25	8	12	3	17	109	18	48
21	26	4	13	22	8	11	3	13	102	16	51
22	13	4	16	29	7	11	3	17	114	16	57
23	10	4	16	31	8	11	3	17	119	17	56
24	12	4	16	30	7	11	3	17	116	16	58
25	12	4	16	31	8	11	3	15	119	17	56

444 loads were produced at a combined shovel truck cost of $ 65/load .

Table 9 PIT_S results with the match numbers from Table 5 rounded down

PIT_S LOADER REPORT

SHIFT	UNIT	LOADS	GRADE	WAIT	DELAY
1	L- 2	115	11.1	1272	1480
1	L- 3	104	22.2	1461	3320
1	L- 4	93	33.3	5744	1200
1	L- 5	94	44.4	3618	3080
1	K- 0	245	27.0	10373	960
1	K- 1	150	26.2	13296	2920

PIT_S TRUCK REPORT

TRUCK	WAIT	E-UP	E-DN	F-UP	F-DN	LOAD	DUMP	OTHER	$/HOUR	LOADS	$/LOA
0	8	6	16	30	9	14	4	13	121	21	46
1	16	5	15	25	9	12	3	15	110	18	48
2	6	5	18	33	8	12	3	15	124	18	55
3	10	5	18	32	8	11	3	13	122	17	57
4	13	5	15	28	9	13	4	13	116	20	46
5	17	5	15	25	9	12	3	14	110	18	48
6	4	5	19	33	9	13	3	14	126	19	53
7	12	4	16	31	8	11	3	15	119	17	56
8	9	6	16	28	9	13	4	15	117	20	46
9	17	5	14	25	9	12	3	15	110	18	48
10	8	5	17	32	8	12	3	15	122	18	54
11	5	5	18	35	9	13	3	12	129	19	54
12	9	6	16	28	9	13	4	15	117	20	46
13	18	5	14	25	9	12	3	14	110	18	48
14	20	4	15	28	7	11	3	12	112	16	56
15	6	5	18	33	9	13	3	13	126	19	53
16	12	6	16	28	9	13	4	12	117	20	46
17	19	5	14	24	8	11	3	16	107	17	50
18	5	5	18	34	9	13	3	13	127	19	53
19	5	5	18	33	8	12	3	16	124	18	55
20	6	5	18	32	8	12	3	16	123	18	54
21	4	5	18	33	8	12	3	17	124	18	55

406 loads were produced at a combined shovel truck cost of $ 65/load .

5 COMPARING SHIFT SIMULATION AND CYCLE STUDY ESTIMATES OF FLEET PRODUCTION

The PIT_S program was used to simulate a shift of production, including all of the delays due to shift start up, breakdowns etc. The network conditions, ie number of shovels and trucks, road lengths etc. were identical to the previous study using the CYCLE program. The results, presented in Tables 8 and 9 are directly comparable to the results presented in Tables 6 and 7 respectively. The average total delay per shift for the PIT_S program is approximately 5300 seconds or 18% of total shift time. These delays include shift change, the average shovel breakdown duration per shift and half of the average truck cycle (in the cycle program trucks begin the shift at their assigned shovel). Examining this data it can be seen that a one shift 26 truck simulation with PIT_S (Table 8) produces 444 loads as compared to an identical study with CYCLE (Table 6) which produced 564 loads. This is a reduction of 21%. A one shift 22 truck simulation with PIT_S (Table 9) produces 406 loads as compared to an identical study with CYCLE (Table 7) which produces 502 loads, a reduction of 19%. In both of these studies a minimum reduction of 18% would be expected due to the delays included in PIT_S. Some additional reduction would be expected due to the grouping together of trucks at lunch time which occurs with PIT_S and not with CYCLE. It can be concluded that the CYCLE program can be used to predict the output of the PIT_S program by reducing the shift time by the average total shift delay in PIT_S and by an additional small amount (1.5%) to account for inefficiencies due to lunch break etc.

6 CONCLUSIONS

The following conclusions can be drawn from the full shift simulation and truck cycle time studies that have been conducted:

1. Simulation provides a practical modelling technique for the analysis of shovel truck productivity and operating costs.

2. When comparing automated truck dispatching with fixed truck assignment, the automated system generated a productivity gain of only 2% when close to the optimum number of trucks were operated. During situations where the network was under trucked the gains for the automated system were somewhat higher.

3. Mixing trucks with different speed characteristics within the same fleet can seriously reduce fleet efficiency and increase operating costs.

4. Match factors based on average truck cycle and loading times, generated by a single truck assigned to each shovel, can be used to predict the production for the full fleet when the matched number of trucks is assigned to each shovel. This is done by decreasing the projected production based on the single truck by 8% to allow for fleet congestion.

5. The production results of a full shift simulation, including shift start up and end procedures, lunch break and shovel breakdowns, etc., can be accurately predicted based on average cycle times and match factors. This is done by decreasing the total shift time by an amount equal to the average total shift delay with an additional small adjustment for inefficiencies due to shift change, etc.

6. For any given network it would be possible to determine factors to project full shift production from cycle times. These would not change significantly for normal expansions of the haulage network. This will allow the planning engineer to determine truck requirements from a single truck cycle study which could be embedded into the computerized mine planning the engineer uses to determine volumes and sequences. Accurate estimates of productivity rates, truck requirements and costs could obtained in a few seconds.

REFERENCES

Calder, P.N. & Waring, R.H 1965. The Carol Mining Simulator, APCOM symposium, University of Arizona.

Bauer,A. & Calder, P.N. 1973. Planning Open Pit Mining Operations Using Simulation, 10th. APCOM symposium, Johannesburg S.A.

Chick, P. 1980 The Interactive Simulation of Open Pit Haulage, M.Sc. thesis, Queen;s University.

Fytas, K. & Calder, P.N. 1984. An Interactive Computer Simulation Model of Open Pit Haulage Systems, proceedings Society for Computer Simulation Congress, USA.

5 Underground planning and design

Innovative Mine Design for the 21st Century, Bawden & Archibald (eds) © 1993 Balkema, Rotterdam, ISBN 90 5410 325 6

Research into the design and implementation of hardrock mine automation

G.R. Baiden
Inco Limited, Mines Research, Ontario Division, Sudbury, Ont., Canada

M.J. Scoble
Canadian Centre for Automation and Robotics in Mining, McGill University, Montreal, Que., Canada

S. Flewelling
Fraser-Strathcona Complex/Lockerby Mine, Falconbridge Limited, Ont., Canada

ABSTRACT: It has only been in the last few years that the underground hardrock mining sector has begun to recognize both the feasibility and the necessity of automation. It is recent advances in information technology that have created the potential for hardrock mine automation.

This paper considers what was seen as the priority challenge to the implementation of mine automation, namely: effective infrastructure to support voice, data and video communication for both stationary and mobile locations. It reports on underground studies that have evaluated the design criteria for effective communication infrastructure and the economic potential. In conclusion the paper reviews a strategy for mine automation and a new initiative in Canada for mine automation research.

1. INTRODUCTION

As the beginning of the 21st century approaches it will not be enough to just manage more effectively, our industry must learn how to manage constantly changing technology as we move toward mine automation. The need for mine automation has not been more apparent than today, higher grades and lower labour rates around the world have thrown new challenges to Canadian mining which will effect our industries' very ability to survive.

The mining industry must undergo radical change to the production process if it is to survive. This begs the question "What change". To answer this question we can refer to other similar industries and observe their changes with time for a clue. The most logical industry we should observe is manufacturing. Womack, Jones, and Roos[1] discuss production technologies through history as it relates to the automobile industry. They describe three basic production technologies:

* Craft Production
* Mass Production

and

* Lean Production

Recent work in the manufacturing industry, has added a fourth production technology, "Agile Production". This section will describe the four production technologies and discuss their characteristics.

Craft Production - Before the industrial revolution began the main form of production technology was "Craft Production". This technology depended on the skill of a highly trained workforce. Examples of the kind of skilled individual available were ship-builders or railroad engineers where each unit made was distinctly different but improved as the craftsman gained more skill. This form of production could be characterized by high quality and low production of units.

Mass Production - Henry Ford in the 1930's observed the Craft Production techniques and invented the concept of "Mass Production". This concept combined the standardization of skills with the mass production of standardized parts. Production with this methodology was considerably more productive but, as time was to prove, lacked proper quality measures as production numbers grew in the 1960's and 1970's. This problem resulted in poor quality products when compared to the newer concepts of "lean production".

Lean Production - At this point the Japanese, specifically Toyota came to North America and observed the waste from Mass Production and returned to Japan with ideas that would reshape the very form of production technologies. The concepts are now known as "Lean Production". This can be characterized as the reduction of the mass production batches to more manageable, smaller work sizes that when measured using statistical tools, such as those developed by Demming[2], would allow high quality and high production rates. This kind of production required a highly skilled workforce, capable of analysis in an independent fashion. Agile Production - A great deal of effort is currently focused on a new production technology called "Agile Production"[3]. With the advent of considerable computing power at low cost, the possibility of intelligent manufacturing systems has become a reality. In the future, manufacturing technologies will take the craftsman-like quality and ingenuity and combine it with the repeatability of computerization. This new production technology will allow custom made products tailored at the individualism of the consumer.

2. AUTOMATION

To apply automation we must first define it. The following is an attempt to define automation:

> "The automatic transfer of one unit of a complex industrial assembly to a succession of machines, each of which completes another stage in manufacture".

To automate a mine we should regard our process (delineation, development, production and backfill) as an automated manufacturing process. If we can apply automation technology then the impact will be beneficial, as indicated by the success of German companies applying automation, see figure 1. Such companies have quoted a manufacturing cost

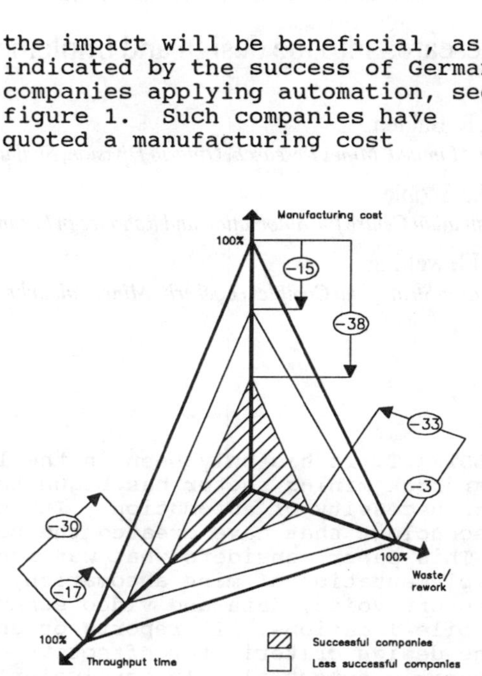

Figure 1 :Manufacturing Automation Benefits, after Rommel (4).

reduction of 38%, a throughput time reduction of 30% and a reduction in waste/rework of 33%. Even the less successful companies have achieved significant gains such as a manufacturing cost reduction of 15%, a throughput time reduction of 17% and a reduction in waste/rework of 3%. Such results are being sought by the mining industry in Canada in its effort to compete in the future.

Another point to consider is that before the manufacturing businesses automated they typically experienced manufacturing costs in which the labour component amounted to 50-60%, i.e., equivalent to those that the mining business is currently trying to deal with. This manufacturing cost labour component is now, with automation, typically 5-10%.

3. MINE AUTOMATION

The results achieved by the manufacturing business should be obtainable if the mining business adopts a systematic logical approach. The necessary elements are seen to comprise:

* Research and Development
* Strategic Alliances
* Workforce Education
* Creative Management Structures and Policies
* Entrepreneurship

The PRECARN Associates Mining Automation Plan[5] Project led by Inco Limited and Falconbridge Limited is a first step towards a long range automation goal. This project combines the talents of industry, academia and government towards a focused goal that will develop a framework for the Automated Mine of the future using Agile Production technologies.

To achieve this goal, strategic alliances will need to be formed creating secondary industry to support the automated mining systems to be developed. As the industry moves toward Agility the current and new mining workforce will require education for these new production technologies. The need for workforce participation and skill is essential to success in this production technology. Many mining companies are starting to change dramatically as increasingly cheaper labour is driving management and unions to solve their long standing differences and work together, simply to remain in business. Another important consideration if automation is to be successful is the creation of progressive management structures and policies. As the labour component of the mining business is reduced then so too does the need for the current layers of management. This transformation can already be seen at many mining organizations. The final critical point is the need for entrepreneurship. This characteristic has been and will continue to be a mainstay in the mining business. New challenges will require new levels of entrepreneurship if mining automation is to be applied and profitability is to be maintained.

4. TELECOMMUNICATION INFRASTRUCTURE

Recent work at Inco Limited has demonstrated the capability of telecommunication infrastructure through the teleoperation of an LHD from surface[6] and the combining of teleoperation of an LHD with guidance[7]. These two results were achieved with a high bandwidth communication infrastructure, figure 2, that combined voice, data and video over a hardwired and radio system. This has the capability to support multiple machine teleoperation similar to figure 3.

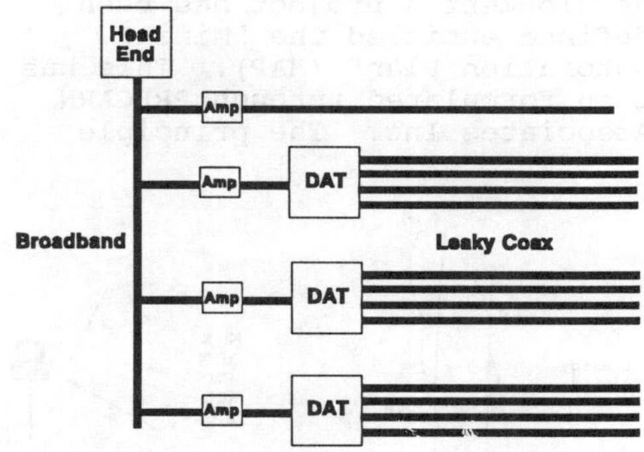

Figure 2 : Mine Telecommunication Network

Figure 3 : Teleoperation of LHD on 3000 Level

5. MINING AUTOMATION STRATEGY

In the mining automation strategy the development of communication infrastructure is seen as a foundation for automated subprocesses. The next step was to develop applications according to a prioritized list based on economic impact and using the measurement subprocesses. The next step was to develop applications according to a prioritized list based on economic impact and using the measurement parameters (mining cost, waste/rework and throughput time) established by the manufacturing industry (figure 1). With the telecommunication infrastructure as a base the potential to move toward a Numerically Controlled (NC) mining environment is feasible. In order to undertake the next phase of automation research and development a project has been defined entitled the "Mining Automation Plan" (MAP). This has been formulated through PRECARN Associates Inc. The principle

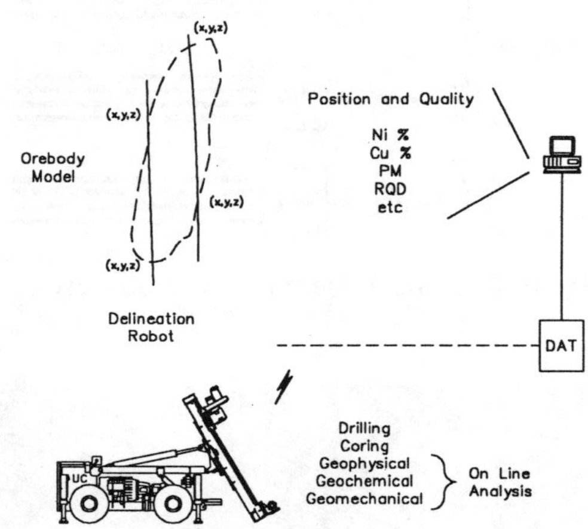

Figure 4 : Delineation Process

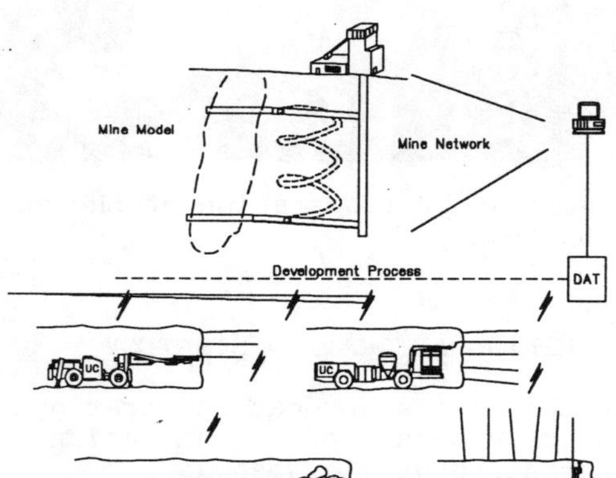

Figure 5 : Development Process

partners in the venture are Inco Limited, Falconbridge Limited, Ainsworth Automation Inc., Canadian Centre for Automation and Robotics in Mining (CCARM) and the Queen's University Centre for Resource Studies (CRS).

The vision of the project is to perform strategic research through three projects (delineation system, environment monitoring and social and economic impacts) that will enable NC mining. NC mining requires the combining of engineering, process control and information systems with mining process equipment (drills, LHDs, ground support, hoisting, pumping and ventilation) to create a "Computer Integrated Manufacturing" (CIM) environment. This is best demonstrated by figures 4,5,6 and 7[8].

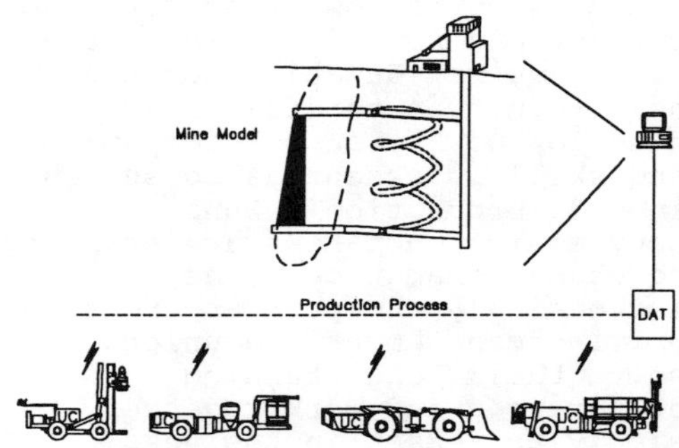

Figure 6 : Production Process

These four figures together represent conceptually how to integrate the mining process using computer modelling and process monitoring and control systems. Figure 4 is a representation of a delineation drill connected to the communication infrastructure that links it to the orebody model so that the data collected can be analyzed as it is available. This same data once analyzed is the knowledge required to develop the access method for the ore after the mine engineering is complete using a CAD (Computer Aided Design) and/or CAE (Computer Aided Engineering). The data generated will be sent via the process monitoring and control system to operate the mining equipment (drill jumbo, explosive loading, material

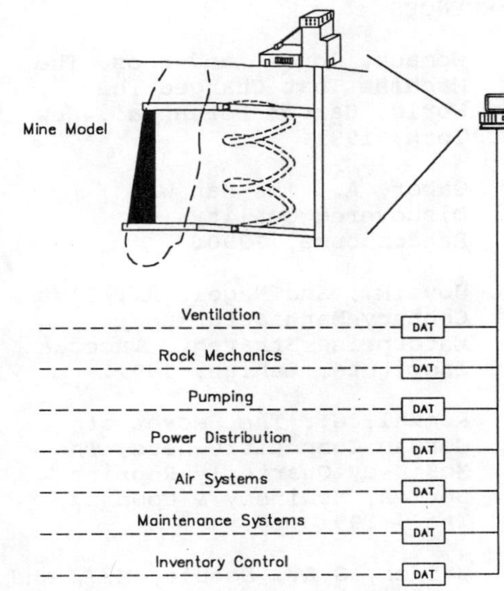

Figure 7 : Mining Support Systems

handling and ground support) that will develop the orebody as shown in figure 5. MAP is based upon the current mining approach, with no account taken of radical change in mining equipment or methods. With the access created, the production engineering process will begin on the orebody model. The data generated by production engineering must be sent via the communication infrastructure to operate the production equipment such as drilling, explosive loading, material handling and ground support as shown in figure 6. Figure 7 illustrates the mechanism that the backfill and other support processes will use for monitoring and control. All these processes can be improved significantly by real-time access to process information.

6. ECONOMIC MODELLING

To demonstrate the viability of NC mining a model of a complete mining operation was developed, based on a hypothetical orebody. The orebody was then modelled using a spreadsheet analysis that tracked over the mine life the revenue and costs by process, workforce allocation, mining rates and maintenance costs. This model was then used with actual operating data from a mine using Vertical Crater Retreat (although any mining method could be modelled). Then the model was analyzed based on improvements in mining cost, quality and throughput time resulting from varied levels of automation.

The mining cost improvement was based on gains seen in productivity in both management and labour. In this analysis energy, engineering and supplies costs were seen to remain relatively constant. The productivity improvements achievable are one person operating three or four machines, for example LHDs[7]. The quality component was measured using an increase in product value of 20% and a reduction in waste/rework of 15%. Throughput time was measured through an increase in utilization of 56.4% (due to teleoperation from surface and operating three-shifts/seven days per week) and a mining rate increase of 50%. The improvements that will result from automation can be seen in figure 8.

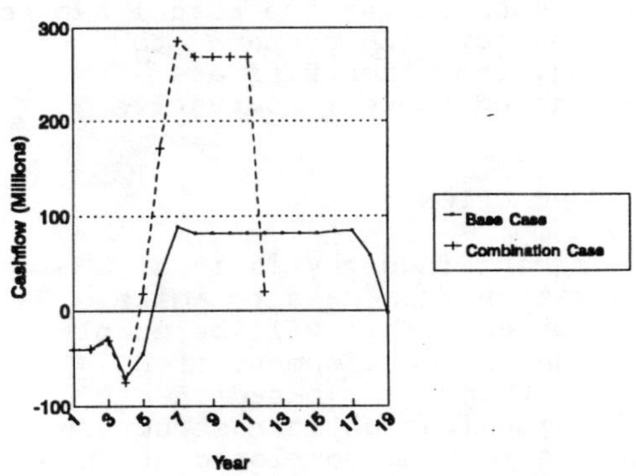

Figure 8 : Comparison of Cashflows.

This shows two cashflows a base-case representing current mining practices and a reasonably achievable case[8] based on the improvements in mining cost, quality and throughput time as discussed above. In this comparison, it can be seen that the cashflow improves significantly and the project life is reduced by seven years. These parameters when modelled yielded the potential

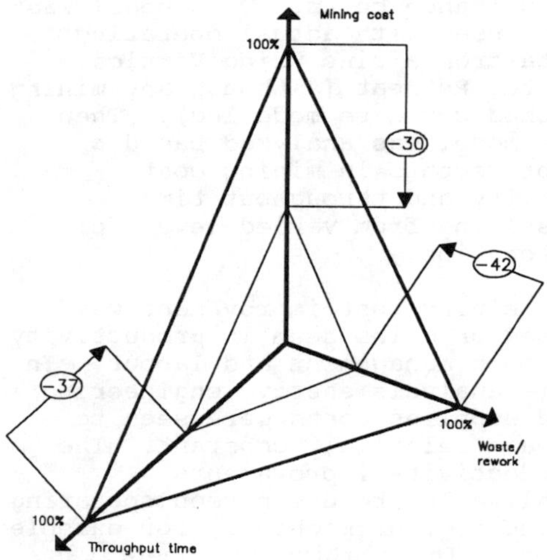

Figure 9 : Mining Automation Potential Benefits based on Economic Modelling

improvement in quality, as illustrated in figure 9. These improvements seem large, but, relate realistically with the benefits achieved the benefits achieved in manufacturing. In fact, considering the actual nature of manufacturing compared to mining, these benefits are considered to be conservative.

7. CONCLUSION

The mining industry is in a transition from mass to agile production. This will be enabled through the development of NC mining techniques based on a telecommunication infrastructure such as the one developed at Inco Limited.

The benefits of automation have a significant impact in the areas of mining cost, quality improvement and throughput time reduction that positively affect mining project cashflow. These benefits when fully applied will assist the mining industry to survive in the future.

REFERENCES

1 Womack, Jones, and Roos, The Machine That Changed The World, Harper Perennial, New York, 1991.

2 Gabor, A., The Man Who Discovered Quality, Randomhouse, 1990.

3 Dove,R., and Nagel, R., 21st Century Manufacturing Enterprise Strategy, Iacocca Institute, Lehigh, 1991.

4 Rommel, G., The Secret of German Competitiveness, The McKinsey Quarterly Reprint Series, McKinsey & Company, Inc., 1991, pp. 38.

5 Baiden, G.R., Scoble, M.J. and Flewelling, S.B., Robotic Systems Development for Mining Automation, Canadian Institute of Mining and Metallurgy, AGM, May, 1992.

6 Baiden, G.R. and Scoble, M.J., Mine-Wide Information System Development at Inco Limited, Canadian Institute of Mining and Metallurgy, AGM, May, 1991.

7 Baiden, G.R., Combining Teleoperation and Vehicle Guidance to Improve Productivity, Canadian Institute of Mining and Metallurgy, AGM, April, 1993.

8 Baiden, G.R., A Study of Underground Mine Automation, McGill University, Ph.D. Thesis, March, 1993.

Innovative Mine Design for the 21st Century, Bawden & Archibald (eds) © 1993 Balkema, Rotterdam, ISBN 90 5410 325 6

Probabilistic risk assessment in mine pillar design

R.J.Pine & I.G.T.Thin
Camborne School of Mines, Cornwall, UK

ABSTRACT: Continuing developments in other areas of risk assessment and quality assurance indicate the need for a more formal approach to risk assessment in mine design. Benefits should include a greater awareness of design limitations and areas of inadequate design information, and a better basis for design decision making by mine management. In this context the design of hard rock pillars using an existing model and of soft rock (coal) pillars using a new model is extended to include probabilistic risk assessment. A variety of simulation and analytical methods are compared and a design strategy developed for relatively simple and more complex pillar models. The strategy is applicable to other design problems such as depths of tension zones and compression failures and hangingwall spans.

1 INTRODUCTION

Recent consulting assignments, a forthcoming Code of Professional Practice for Chartered Engineers in the UK (UK Engineering Council, 1992) and a report by the Royal Society (1992) have emphasised the need to make a systematic approach to risk assessment in engineering design. One of the consulting assignments related to the selection of pillar dimensions and hangingwall spans at South Crofty Mine and this has motivated research into and refinement of design methods using a probabilistic assessment of risk (Pine et al 1992).

The aim of this paper is to review the various numerical methods available for a probabilistic approach, applied specifically to individual pillar design. Examples are given for a hard rock pillar where the basic strength model can be adequately described with a semi-empirical formula and for a soft rock (coal) pillar where the strength behaviour is complex.

Provided that the inputs to the pillar design are from a continuous distribution it is little more difficult to apply a probabilistic rather than a deterministic approach. More judgement is required for the input values, notably their variability, but a good design should address this in any case. In a probabilistic approach the designer or manager accepting the design, are forced to confront the variability of the data in a formal manner. This often raises important issues regarding the adequacy of the input data (is more investigation required ?), the design model and risk acceptance. These may be inadequately considered by deterministic methods with conventional sensitivity analysis.

Although this paper considers pillar strength examples in detail, the methods described are applicable to many aspects of underground mine design. Further examples include the depth of compression or tension zones in critical structural elements, the design of hangingwall spans (Pine et al, op cit) and a probabilistic approach for block release incorporating key block theory (Tyler et al 1991).

2 MEASURES OF SAFETY AND ACCEPTANCE CRITERIA

2.1 Measures of safety

The most common form of safety measurement is the Factor of Safety, F which can be described as the ratio between Capacity (C) and Demand (D). In a deterministic analysis the mean values of C and D are used to give mean F. The failure value for F is 1. In pillar design, capacity and demand are usually pillar mean strength and mean imposed stress respectively.

The Safety Margin (SM) is given by capacity minus demand (C-D) and typically has the dimensions of load or stress in a pillar design. The failure value for SM is 0.

The probability of failure, can be described by the probability of F being less than 1.0 [p (F<1)] or SM being less than zero [p(SM<0)]. A further, useful measure of the fitness of the structure is the Reliability Index, β, the ratio of the mean and standard deviation of SM.

As rule of thumb $p(F<1.0) = p(SM<0) = 10^{-\beta}$

A reliability index of 2 is approximately equivalent to $P(F<1.0) = P(SM<0) = 1\%$.

In the pillar design examples presented in this paper it is usually found that the distribution of the output for capacity, C is log normal but the distribution output for demand, D is normal. This is because C is a relatively complex combination of inputs and D is relatively simple (dominated by variance of one variable). This can be checked by the simulation methods described in section 3. The factor of safety, F = C/D is approximately log normal and safety margin, SM = C-D the same, but the latter can be closely approximated by a normal distribution. The reliability index, β can be calculated on the assumption of normal or log normal distribution for SM. In this paper natural logs are used or implied throughout, but only the normal distribution assumption has been used for calculating β.

2.2 Acceptance criteria

After a risk or probabilistic analysis the acceptability of the risk values must be considered. In Table 1 some acceptance criteria are suggested for consideration, based on values given by Priest and Brown (1983) and modified by Pine (1991). These values are for guidance only. Priest and Brown also gave guidelines regarding designs which fulfil or only partly fulfil these criteria. In any design the economic and non-economic consequences of failure must be balanced against cost of safety. The balance will vary from case to case and the final decision is the responsibility of management.

The first line in each category includes the original criteria and the second line the modified criteria. The two values for p(F<1.5) are for normal and log normal factor of safety distributions respectively. Although these criteria relate to slope design, they have also been used for pillar and hangingwall design (Pine et al, op cit).

3 AVAILABLE METHODS OF PROBABILISTIC ANALYSIS

The following methods of probabilistic analysis are all applicable for design models with continuous input distributions:

- o Monte-Carlo simulation (MCS)
- o Latin Hypercube simulation (LHS)
- o General Point Estimate Method (GPEM)
- o First Order Second Moment (FOSM)

This list is ranked according to the computer power usually needed with MC requiring the

TABLE 1. Acceptance criteria after Priest and Brown (1983), with suggested modifications.

Category & consequences of failure	Examples	Acceptance Criteria Maximum values(%) p(F<1.0)	p(F<1.5)	Minimum values Mean F	ß
1	Non critical	10	20	1.3	
Not serious	benches	10	80/75	1.3	1.4
2	Semi-permanent	1.0	10	1.6	
Mod. serious	slopes	2.0	36/39	1.6	2.3
3	High/permanent	0.3	0.05	2.0	
Very serious	slopes	0.3	8/12	2.0	3.2

most and the FOSM the least. All methods except LHS are described by Harr (1987).

Monte-Carlo simulation is completely general and can allow any input and output distributions. It requires a large number of cycles (at least 5000) to adequately sample the tails of the input distributions. It is accessible as a spreadsheet add-in, @RISK (Palisade Corp. 1988).

Latin hypercube simulation is a refinement of MCS incorporating a form of stratified sampling. It is more efficient, requiring fewer cycles to sample adequately the tails of the input distributions (a factor of 5 to 10 less than MCS, Startzman and Wattenbarger, 1985). It is also available as an option in @RISK.

The General Point Estimate Method (GPEM), developed by Rosenblueth (1981), is applicable to symmetrical input distributions (eg, normal, uniform, symmetrical triangular etc) and combines sample points at ± 1 standard deviations above and below the mean inputs. The total number of sample combinations is 2^n where n is the number of variables. Output is in the form of a mean and variance or standard deviation of all the samples. Probabilities are determined by assuming normal or log normal distributions. GPEM provides a very fast, efficient analysis, and within the input limitations gives comparable results to MCS and LHS. A general program is easy to write and an example was provided by Pine (1991).

First Order Second Moment method (FOSM). This fundamental method is based on of determining the variance of output based on means and variances of inputs. It is based on the Taylor series expansion of the output expressed as a function of the inputs, retaining only the first order , linear terms. It is necessary, strictly, to be able to differentiate the output with respect to each input at the mean output value. However, the method can still be used where the required differentials can be approximated numerically.

The variance of the output is given by

$$\sigma_f^2=\sum_{p=1}^{n}(\sigma_p\frac{\partial f}{\partial p})^2 \qquad ...(1)$$

The numerical form is

$$\sigma_f^2=\sum_{p=1}^{n}(\sigma_p\frac{\Delta f}{\Delta p})^2 \qquad ...(2)$$

where

σ_f = std. deviation of the output

σ_p = std. deviation of the input parameter, p

$\partial f/\partial p$ or $\Delta f/\Delta p$ = slope of f vs parameter p

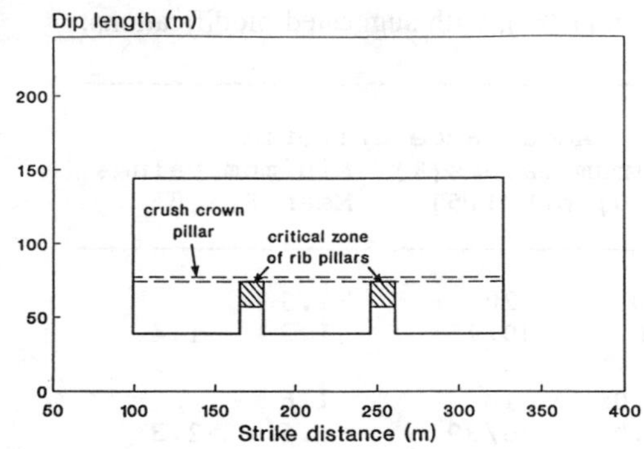

Figure 1. Rib pillar layout at South Crofty mine

all evaluated at the mean values. The mean output is determined from mean input values. In these equations, f may be C, D, F or SM.

Because of the simplicity of equation (2), FOSM is suitable for a probabilistic sensitivity analysis of a complex model. The mean output, f is evaluated first for a combination of mean inputs. Then for each of the inputs in turn the model is re-run with different input values, and the $\Delta f/\Delta p$ slopes evaluated. The magnitudes of the Δp must be sufficient to cause a substantial (detectable) Δf in the model. Using Δp of about 1 standard deviation should be suitable in most cases. The σ_p values are determined by field measurement or by estimation.

Once the mean and variance of the output are known, probabilities can be determined based on an assumed normal or log normal distributions. Examples and comparisons are given later.

Where output (C, D, F or SM) is log normally distributed the mean and variance of the basic output can be used to generate log normal parameters and interpretation.

If L = log f, and is normally distributed with mean μ_L and standard deviation σ_L, then $\mu_L \approx \log(\mu_f)$ and using equation (1), $\sigma_L \approx (dL/df)\,\sigma_f = \sigma_f / (\mu_f)$.

MCS, LHS and GPEM are most suitable where the critical output (eg F, SM or C) can be expressed in a series of closed form equations which can be readily incorporated in a spreadsheet or computer program. The equations may be analytical or empirical or a combination of the two. FOSM is equally applicable to closed form equations or a complex model for which such equations may not be possible. The main limitation with FOSM is that if the coefficients of variation of the inputs are too large it becomes progressively more inaccurate.
This is examined further in an example later.

4. RIB PILLAR DESIGN AT SOUTH CROFTY MINE

The background to this design was described in detail by Pine et al (1992). The selected design layout is shown in Figure 1, where the critical loading develops at the top of the rib pillars. The equations for the pillar capacity (C) and demand (D) were given in Pine et al (op cit) and are summarised as follows:

$$C=\sigma_c \exp\left(\frac{RMR-100}{9}\right)(0.778+0.222We) \quad ...(3)$$

where

$$We=\frac{2W\frac{L}{W}}{(1+\frac{L}{W})} \quad ...(4)$$

and

$$D=\sigma_n(\gamma+\delta W) \quad ...(5)$$

Here, σ_c is the uniaxial compressive strength, W and H are the pillar width and height, σ_n is the field stress normal to the top of the pillar, L is the pillar length and the parameters γ and δ are used to define the variation of induced stress with change in pillar width.

The values of γ and δ were determined by multiple runs of the displacement discontinuity

TABLE 2. Input parameters for hard rock pillar example.

Parameter	μ	Cov(%)	σ
σ_c	190.0 MPa	16.0	30.40 MPa
RMR	80	5.0	4.0
W	22.0 m	7.0	1.54 m
H	3.0 m	10.0	0.30 m
σ_n	37.0 MPa	10.0	3.75 MPa
γ	4.55	10.0	0.46
Constant values:			
δ	$-0.07\ m^{-1}$		
L/W	2.33		

TABLE 3. Output for hard rock pillar.

Method	Param.	μ	σ	CoV(%)
GPEM	C (MPa)	197.4	57.1	28.9
LHS 5k		196.4	53.6	27.3
MCS 25k		196.2	53.3	27.2
FOSM (a)		191.2	55.2	28.9
FOSM (b)		191.2	55.5	29.0
GPEM	D (MPa)	111.4	20.7	18.5
LHS 5k		111.4	20.6	18.5
MCS 25k		111.2	20.6	18.5
FOSM (a)		111.4	20.8	18.7
FOSM (b)		111.4	20.8	18.7
GPEM	SM (MPa)	86.0	61.4	71.4
LHS 5k		85.0	58.3	68.6
MCS 25k		85.0	57.6	67.8
FOSM (a)		79.8	59.6	74.7
FOSM (b)		79.8	59.9	75.1
GPEM	F	1.839	0.652	35.5
LHS 5k		1.832	0.640	34.9
MCS 25k		1.831	0.634	34.6
FOSM (a)		1.716	0.597	34.8
FOSM (b)		1.716	0.605	35.2
GPEM	log F	0.548	0.350	63.8
LHS 5k		0.548	0.340	62.3
MCS 25k		0.549	0.336	61.2
FOSM (a)		0.540	0.348	64.4
FOSM (b)		0.540	0.353	65.3

program, MINTAB. Note that the pillar is vertical with W and L measured in the plane of the seam.

The input values for the design, mean μ, coefficient of variation of (CoV) as a percentage, and standard deviation σ, are included in Table 2.

The inputs were modified from the original design to examine the effect of a higher coefficient of variation and lower mean value for the rock mass rating (RMR). This meant that a greater pillar width was necessary to achieve comparable results.

The capacity and demand equations have been used in probabilistic analyses as follows:

- o MCS and LHS in a spreadsheet using @RISK.
- o GPEM in the program by Pine (1991)
- o FOSM (a) using equation (1) for variance
- o FOSM (b) using equation (2) for variance

The LHS and MCS were run with 5,000 and 25,000 samples of each variable input respectively. Results are given in Tables 3 and 4.

The basic results in Table 3 are reasonably close for the different methods used, but with noticeable differences between FOSM (non-simulation) and the other (simulation) methods. This is a consequence of the output distributions being mainly log normal, such that the mean of the output samples is not the same as the central value (using the mean values for all inputs). Differences between parameters for log F are much smaller than for F. An example output distribution for F from LHS is shown in Figure 2 showing a clear log normal shape. The skewness and kurtosis for log F from the LHS and MCS runs were very close to expected values for normal distributions (0 and 3 respectively).

There are significant differences in the derived results in Table 4 for some parameters.

However the differences for p(log F <0) and p(log F < 0.405) are relatively minor for all methods. The latter is the log normal equivalent for P(F<1.5). In Table 4 all probabilities from the LHS and MCS have been read direct from cumulative distribution curves without any assumption regarding distribution form. As a consequence the results are internally consistent (compare p (F < 1), p(SM < 0) , p(log F < 0 for either LHS or MCS).

5. FLAC MODEL OF COAL PILLAR CAPACITY

In this model the finite difference program FLAC, version 3.01 (Itasca 1991) was used to determine the capacity of a strain-softening coal pillar subject to a progressive displacement- controlled loading. The work was part of the Ph.D programme of the second author and has been partly reported elsewhere (Trueman et al 1992). As a result of the modelling a new relationship has been determined for the variation of pillar capacity with height to width ratio, including squat pillars (width to height ratio greater than 4). The relationship is a modified version of the Bieniawski empirical formula (see Bieniawski 1987) and is given in equation (6).

$$C=1.15\sigma_c\sqrt{s}(0.64+0.36R)(\exp(\frac{W}{H}-R)^N) \qquad ...(6)$$

where if W/H < 8 then R = W/H, else R = 8. B and N are dimensionless constants used to fit the model data and R is the critical width to height ratio for the onset of squat pillar behaviour. Other parameters are as defined earlier. The Hoek and Brown constant, s is given by:

Table 4. Derived results for hard rock pillar.

Method	Derived Values	
	p(F<1.0)	p(SM<0)
GPEM	9.95 %	8.06 %
LHS 5k	5.34	5.32
MCS 25k	5.09	5.07
FOSM (a)	11.56	9.16
FOSM (b)	11.77	9.06
	p(log F<0)	
GPEM	5.87 %	
LHS 5k	5.32	
MCS 25k	5.07	
FOSM (a)	6.38	
FOSM (b)	6.23	
	β	$10^{-\beta}$
GPEM	1.402	3.96 %
LHS 5k	1.479	3.45
MCS 25k	1.481	3.34
FOSM (a)	1.339	4.59
FOSM (b)	1.332	4.66
	p(F<1.5)	p(log F<0.405)
GPEM	30.2 %	34.1 %
LHS 5k	33.5	33.5
MCS 25k	33.7	33.7
FOSM (a)	35.9	34.9
FOSM (b)	36.0	35.1

Note (a) derived from equation
(b) derived from assumed linear sensitivity

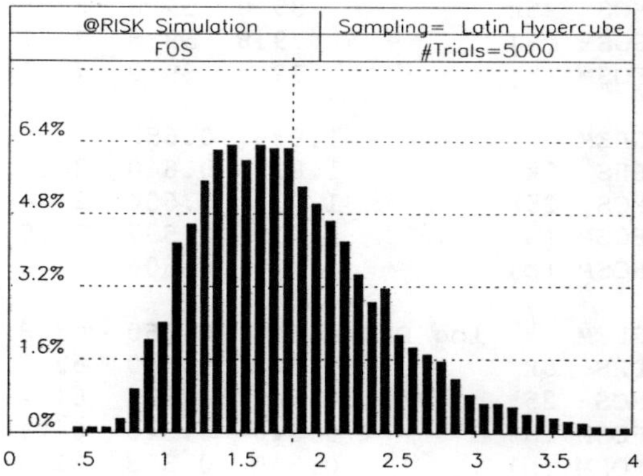

Figure 2. Factor of safety distribution from LHS

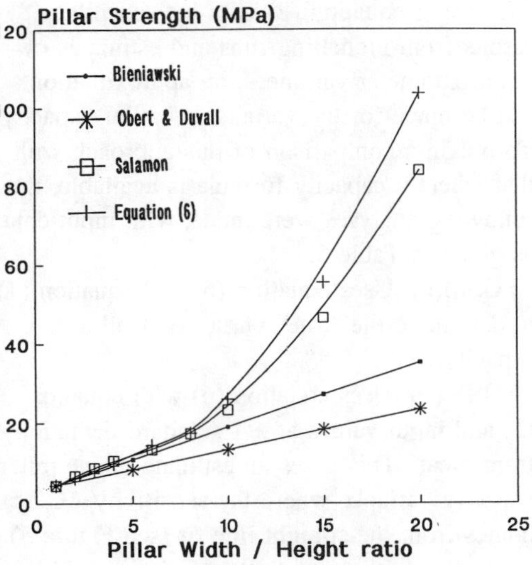

Figure 3. Comparison of pillar capacities vs W/H ratio.

$$s=\exp(\frac{RMR-100}{9}) \qquad ..(7)$$

Equation (6) is compared graphically in Figure 3 with the Salamon (1982) squat pillar formula and the Bieniawski and Obert and Duvall (1967) formulae.

The constant 1.15 is used to correct uniaxial compression strengths from cylinder tests to equivalent values for cubes. The peak strength for the FLAC model was based on a non-linear envelope using Hoek and Brown rock mass constants of 0.013 and 1.99 for s and m respectively. These were derived from rock mass rating values, RMR. Mohr-Coulomb parameters c and ϕ were needed for the strain softening model in FLAC and these were determined by obtaining local values from linear fitting to the Hoek and Brown envelope, after Hoek (1990). These values were confining-stress dependant.

The residual strength envelope used was linear and based on the unconfined compressive strength of the broken rock mass and the brittle ductile transition, assumed to be at a principal stress ratio of 3.4 after Mogi (1966).

The post-peak deformation response was based on a plastic strain at which residual strength was reached, $\varepsilon_r = 0.001\ \sigma_3$, (σ_3 in MPa) controlled by model fitting to existing empirical formulae with W/H less than 4.

A typical mesh, boundary conditions and the removal of some failed elements are shown in Figure 4.

The modelling sequence was as follows:

1. Run an elastic version of the model at a given total vertical load to determine confining stress distribution within the pillar.

2. Determine equivalent c and ϕ values dependent on range of confining stresses.

3. Re-run as strain-softening Mohr-Coulomb model

4. Examine state of elements across mid-height of pillar and ensure vertical stresses are consistent with confining stresses and failure envelopes. Some load adjustments and element removal is necessary as outer elements fail progressively inwards.

5. Determine total pillar load from summation of mid-height element vertical stresses.

6. Apply further vertical load increment and repeat from 3.

Pillar failure occurs at the maximum total load where all mid-height elements have vertical and confining stresses consistent with strain softening strength envelopes. The sequence was repeated for pillars of varying width to height ratios from 1 to 20 from the results of which equation (6) was developed.

In the first attempt at the modelling considerable manual intervention was necessary. A later version of FLAC allowed much greater, but not complete, automatic inspection of and alteration to individual element post-peak properties.

Table 5. Inputs to coal pillar strength example .

Parameter	μ	Cov(%)	σ
σ_c	30.8 MPa	15.0	4.62 MPa
RMR	61	5.0	3.05
W	10.0 m	10.0	1.00 m
Constant values:			
H	2.0 m		
B	0.296		
R	8.0 m		
N	0.744		

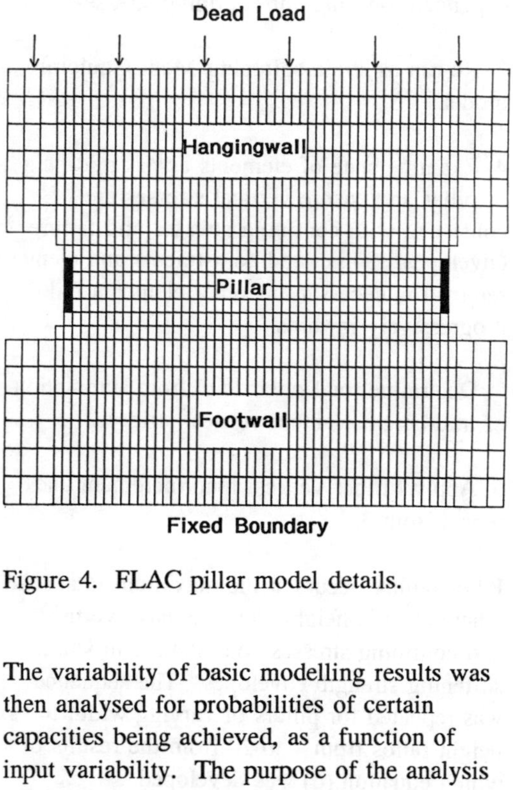

Figure 4. FLAC pillar model details.

The variability of basic modelling results was then analysed for probabilities of certain capacities being achieved, as a function of input variability. The purpose of the analysis was to simulate the situation where a pillar capacity formula had not been developed, but certain limited modelling results were available. (Note the approach is also being considered for other underground design aspects such as depths of tension and overstressed zones).

By using equation (2) with the sensitivity results from modelling runs and estimates of input parameter variances, an approximation can be made for the variance of pillar capacity. To obtain a comparison of this approach with that where a capacity formula is available, the following analyses were made, with input data as given in Table 5.

FOSM(a): Uses equation (6) and equation (1) to determine the "true" variance of pillar capacity.

FOSM(b): Uses equation (6) with equation (2) and input values at ± 1 standard deviation from mean. This gives an estimate of the pillar capacity variance where the sensitivity $\Delta C/\Delta p$ comes from the straight line fit (see Figure 5). For example the sensitivity $\Delta C/\Delta\sigma_c$ would be determined from using σ_c of 25.18 and 34.42 MPa.

FOSM(c): As with FOSM(b) but using the values of C determined from the mean input values and + 1 standard deviation (Fig.5). For example $\Delta C/\Delta\sigma_c$ would be determined from using σ_c of 30.8 and 34.42 MPa.

Results are shown in Table 6. Differences between FOSM (a), (b), and (c) show the level of error expected by taking results for progressively fewer model runs. Ideally many runs are made, a formula developed and FOSM(a) or one of the other methods used. At the other extreme for FOSM(c) only 4 runs are needed (mean and + 1 standard deviation for σ_c, RMR and W).

Again the LHS and MCS can be taken as the "true" results in the sense that no assumptions are made for the output distribution. (The quality of the LHS and MCS results is discussed later). The probabilities, $p(C < 16 \text{ MPa})$ were read directly from their cumulative output distributions. LHS and MCS are in quite close agreement with each other.

FOSM (a) to (c) show a progressive decrease in accuracy, but this is less evident with the log normal interpretation. GPEM results are also closer to LHS and MCS for a log normal distribution of C. The FOSM results are close enough together to confirm that FOSM(c) should give useful results.

Note that if the demand, D on the pillar were 16 MPa, without any variance, the pillar mean F would be about 1.5 (23.8/16). However, with the input variances as assumed in Table 5, there is quite a large probability of "failure", about 10%. The probability would be even higher if there was the expected variance in demand. The consequences of "failure" would depend on the local and global mine and pillar stiffnesses and the effect of load transfer on adjacent pillars.

6. DISCUSSION AND CONCLUSIONS

The essence of probabilistic risk analysis in the current context is shown by the FOSM equations (1) and (2). These provide the variance of the output , f as a combination of input parameter sensitivity (Δf/Δp) and variance. The other methods provide similar results by different means.

The analysis has two obvious critical aspects. The basic model relating inputs to output must be correct and appropriate to the design task. The variances of the inputs must be determined.

From the results presented, it is clearly desirable to develop an equation (model) relating the output to all the inputs, in which case any or all of the methods can be used, depending on software available. The equation may be analytical or empirical. This had already been achieved for the hard rock pillar and a new equation was developed for the coal pillar capacity. It should also be possible to develop equations for other critical partial design aspects (depth of tension zone, depth of compressive failure, critical hangingwall span etc) based on existing models.

If the model is more complex, a limited sensitivity analysis can be undertaken and usable results obtained (eg FOSM (b) and (c) for coal pillar capacity). The input parameter range of ± 1 standard deviation appears

TABLE 6. Results for coal pillar.

Method	μ_c	σ_c	CoV_c (%)	p(C<16) (%)
Normal distribution:				
GPEM	23.90	7.06	29.5	13.2
LHS 5k	23.82	6.81	28.6	10.5
MCS 25k	23.83	6.83	28.7	11.3
FOSM (a)	23.45	6.85	29.2	13.9
FOSM (b)	23.45	6.88	29.3	14.0
FOSM (c)	23.45	7.15	30.5	14.9
Log normal distribution:				
GPEM	3.131	0.294	9.04	11.2
LHS 5k	3.130	0.286	9.14	10.5
MCS 25k	3.130	0.288	9.20	11.3
FOSM(a)	3.155	0.292	9.25	9.6
FOSM(b)	3.155	0.293	9.30	9.6
FOSM(c)	3.155	0.305	9.66	10.5

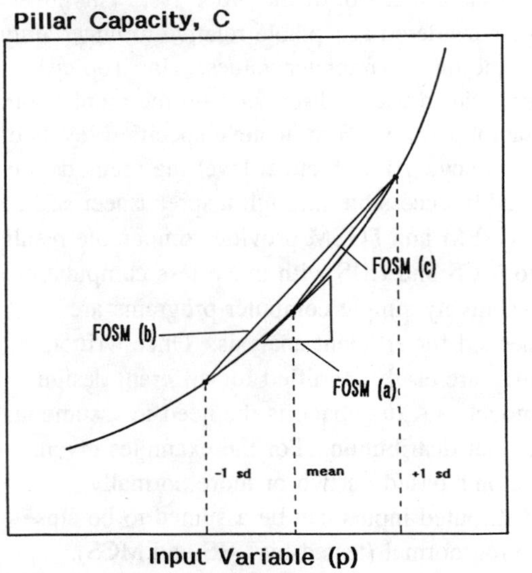

Figure 5. Approximations for ∂f/∂p in FOSM methods.

suitable, provided the output can be accurately determined at each stage. If there is uncertainty in the model output this will affect the accuracy of the Δf/Δp. A wider range than

± 1 s.d. would be useful where the relationship between f and p were linear. Some judgement is needed.

The variances or standard deviations of the inputs should be determined by measurement, case history study or assumption in descending order of quality. A combination may also be appropriate. Assumptions (or "educated guesses") are better, in the authors' opinion, than making no attempt at considering variability.
A more detailed examination of this aspect, was provided by Pine (1992).

There are advantages and disadvantages with using each of the analysis methods. MCS and LHS provide an unbiased estimate of the output distributions and with large numbers of samples can be considered quite accurate. In the current cases sample sizes of 5,000 for LHS and 25,000 for MCS led to differences in key parameters in separate simulations, using the same method, of the order 1%. This must be considered acceptable relative to uncertainty in the input parameter values. Harr (op cit) provides a useful discussion on the number of samples for MCS to achieve specified levels of accuracy. At a practical level the methods are readily accessible through a spreadsheet add-in.

GPEM and FOSM provide comparable results to MCS and LHS with much less computation. Relatively simple computer programs are needed for efficient analysis. Once written, they are easily modified for different design models. A drawback is the need to assume an output distribution. For the examples given, outputs based on two or more normally distributed inputs can be assumed to be close to log normal (tested by LHS and MCS).

A suitable design approach would be to use LHS or MCS, if available to determine the output distribution for a given type of problem, otherwise assume a log normal output for a multiple input. If repeated analyses are needed then GPEM would be more efficient and provides results closest to LHS and MCS. Given the very large numbers of samples needed for the latter, the GPEM results may be as accurate in most practical cases. FOSM is a useful alternative, especially where the $\Delta f/\Delta$ are linear or nearly so. It has a sufficiently simple input that a probabilistic risk analysis can be developed from sensitivity analysis of a complex model.

One aspect not covered in the models presented is the variability of the individual inputs to RMR. This could and perhaps should be incorporated probabilistically in further model refinements. Some judgement is necessary to limit which parameters to vary and which can be considered essentially constant (see inputs in Tables 2 and 5).

The levels of probability which can be accurately assessed needs some consideration. As the probabilities become lower and the tail of the output distribution is approached, there is much less achievable accuracy with all the models. Categories 1 and 2 (Table 1) should be assessable without great difficulty, particularly where $p(F<1.5)$ is used as an indicator (but it should not be used alone). Category 3 has relatively low probability values and it may be necessary to take a conservative (safe) view to be confident of fulfilling the acceptance criteria. Priest and Brown (op cit) provide further recommendations on the use of the criteria in combination.

Public and professional awareness of risk assessment management is increasing and professional bodies have responded through review studies and codes of professional practice. The mining industry has a broad spectrum of risks some of which can be assessed with probabilistic methods as described in this paper. The application of the methods is not difficult and provides management and designers with advantages regarding quality assurance and design decisions. Their use should be further encouraged.

ACKNOWLEDGEMENTS

The authors wish to acknowledge the financial support given by the UK Science and Engineering Research Council and support in kind given by Carnon Consolidated Ltd (owners of South Crofty Mine). Dr Nigel Powell of the Camborne School of Mines has provided assistance with some statistical aspects of the work and review of the paper.

REFERENCES

Bieniawski Z.T. 1987. *Strata control in mineral engineering.* A A Balkema, Rotterdam, 59-90.

Engineering Council 1992. *Engineers and risk issues - Annex to Code of Professional Practice,* draft 5, July 1992. U.K. Engineering Council, London, 51 pp.

Harr M.E. 1987. *Reliability based design in civil engineering.* McGraw-Hill, New York, 290 pp.

Hoek E. 1990. Estimating Mohr-Coulomb friction and cohesion values from the Hoek-Brown failure criterion. *Int. J. Rock Mech. Min. Sci. and Geomech. Abstr.,* Vol 27, No.3, 227-229.

Itasca Consulting Group, Inc. 1991. *FLAC manual (ver 3.01).* Minnesota, 55414, USA. (2 Vols).

Mogi K. 1966. Pressure dependence of rock strength and transition from brittle fracture to ductile flow. *Bull. Earthquake Res. Inst.,* Tokyo Univ., Vol 44, 215-232.

Obert L. and Duvall W.I. 1967. *Rock mechanics and the design of structures in rock.* Wiley, New York, 542-545.

Palisade Corporation (1988) *@RISK* Analysis and Simulation Add-In for Lotus 1-2-3.

Pine R.J. 1991. Risk assessment design applications in mining. *Proc. Conf. Research applications in the mining industry.* Univ. Nottingham, UK, 17 Oct. 1991, 13 pp.

Pine R. J., Lay S., Randall M. M., and Trueman R. 1992. Rock engineering design developments at South Crofty mine. *Trans. IMM, Sect. A,* Vol 101, A13-A22.

Priest S.D. and Brown E T. 1983. Probabilistic stability analysis of variable rock slopes. *Trans. IMM. Sect. A,* Vol 92, A1-A12.

Rosenblueth E. 1981. Two-point estimates in probabilities. *J. Appl. Math. Modelling,* Vol 5, 329-335.

Royal Society 1992. *Risk: Analysis, perception and management.* Report of a Royal Society Study Group, The Royal Society, London, 201 pp.

Salamon M.D.G. 1982. Unpublished work to Wankie Colliery.

Startzman R.A. and Wattenbarger R.A. 1985. An improved computation procedure for risk analysis problems with unusual probability functions. *Proc. SPE Hydrocarbon Economics and Evaluation Symp.,* Dallas.

Trueman R., Thin I.G.T. and Tyler D.B. 1992. Rock mass classification as an aid to estimating the strength of coal pillars. *Proc. Conf. Ground control in mining.* Univ. Wollongong and Aus.IMM, July.

Tyler D.B., Trueman R.T. and Pine R.J. 1991. Rockbolt support design using a probabilistic method of key block analysis. *Proc. 32nd US Symp Rock Mech.,* Norman, Oklahoma.

Innovative Mine Design for the 21st Century, Bawden & Archibald (eds) © 1993 Balkema, Rotterdam, ISBN 90 5410 325 6

Neural networks for optimal design of underground opening support systems

Xia-ting Feng & Yun-mei Lin
Department of Mining Engineering, Northeast University of Technology, Shenyang, People's Republic of China
Li-na Wang
Shenyang Institute of Gold Technology, People's Republic of China

ABSTRACT: Correct identification of an opening support system is an important problem in mine design and production because of the complexity of rock masses and mine production, a great amount of hidden, complex geotechnical features and insufficient knowledge. We have presented in this paper a novel approach. Our approach makes use of a neural-network-based machine learning algorithm and uses the first neural network to identify the class of rock masses; the second neural network to identify the type of rockmass structures; then the third neural network that uses these results and other geotechnical features to determine the type and dimensions of opening support system. Our approach has extracted higher order features and hidden, complex relationships, and produced a system that makes use of all features which may include qualitative descriptions. It offers a method of gaining knowledge from difficult cases and conducts uncertain reasoning mainly based on adaptive, self-learning and associative procedures. Simulating results show that the proposed approach produces good decision results with less than 5% error on average.

1 INTRODUCTION

Efficiency of rock support systems in mine opening and tunnelling has a direct bearing on economy and safety. Therefore, it is imperative that optimal opening support systems be designed.

Many researchers have contributed to the design of opening support systems and have studied different decision mechanisms. With improved computer technology new approaches, such as the finite element method, boundary element method and expert system technologies (Fairhurst, 1985; Feng, 1991), have been introduced to help this decision process.

It is well known that a rock mass is a material whose components are extremely complex. In many cases, rock behavior has not been clearly understood and the data acquired are often incomplete. Insufficient knowledge of the design factors for opening support systems and lack of characterizing information restrain existing approaches from achieving optimal results.

Numerical analysis technology, including finite element and boundary element methods etc., seeks an equivalent mathematical model to represent varied relationships between design factors and the choice of support parameters. These equivalent models are often simplified because the exact relationship among design factors is thus far not clearly understood. On the other hand, these models can not include input variables of qualitative descriptions which are important to support system design. Therefore, numerical solutions sometimes do not satisfy actual requirements. This is why even though a lot of in-situ experiments and numerical analyse have been done, empirical design often needs to be used.

The knowledge-based expert system technique plays an important role in making use of previous experience and knowledge, but knowledge acquisition is known to be problematic to expert system growth and, although machine learning is advancing as a field, the development of robust knowledge acquisition tools is still a chellenging problem.

In these cases a new technology, the neural network, is ideally suitable.

Neural network is an information handling system that simulates the working of the human neural system. It reflects some basic features of the human brain. It is a kind of high nonlinear dynamic system. Neural networks can be used to acquire uncertain knowledge from training samples and extend this knowledge to solve new problems. Neural networks can be applied to process associative inference tasks and adaptive decision tasks.

In this paper we apply a neural-network-based machine learning approach to design optimal support systems for rock openings. Our method uses a neural network to appraise first the categories of rock masses, another neural network to identify categories of rockmass structures and then the third neural network that uses these appraisals and other rock features to design support type and parameters.

Various rock features and geological situations, including hidden ones, are extracted from the training samples and used to train three multilayer feedforward neural networks using a Back-Propagation algorithm. Over 257 sections in rock openings and their support systems have been collected from several different sources and divided into two sets. A set including 230 samples is first used to train the network and then another sample set including 27 cases is used to test the trained networks. The simulation studies show that once the networks are trained they are able to predict support requirements with an average error that is less than 5%.

2 GEOTECHNICAL FEATURES AND EXTRACTION

To determine support type and dimensions, one needs to identify the key features that characterize the rock masses. Our research has indicated that following features have a direct effect on the identification of opening support system:

- mechanical properties of rock such as uniaxial compressive strength (Rc), young's modulus (E), poisson's ratio (υ), angle of internal friction (Φ), cohesive force (C);
- discontinuity/joint features (DF) such as spacing, the set number, strike, dip, infilling, roughness, persistence, aperture, type;
- groundwater inflow;
- lithological characters (IC) (unitary, interbedded, complex);
- average thickness of strata (AVT);
- regional tectonism (RT);
- weathering gradation (WG);
- strike of opening (SO);
- overburden depth of opening center (ODO);
- type of opening (TO);
- opening life (OL);
- span of opening (SOO);
- height of opening section (HO);
- major principal stress (MPS);
- stress concenstration caused by residual pillars (CS);
- vibration acceleration from blasting (VA);
- class of rock masses;
- type of rockmass structure.

Clearly, although some of the geotechnical features, i. e., mechanical properties of rock, joint features etc. can be specified by detailed geological investigations and measurements in the laboratory, other features such as the classification of the rock mass and type of rockmass structure are unknown. Therefore, the unkown features have to be accounted for either directly using appraising techniques or indirectly by treating them as two hidden features.

The features we have described above are by no means independent of each other. The category of rock mass classification is related to uniaxial compressive strength, joint features and groundwater action. The category of rockmass structures is related to joint features, lithological characters, average thickness of strata, regional tectonism, weathering gradation of the rock and so on. Statistical correlation analysis also indicates this dependence between some parameter pairs. For example, the correlation coefficient of uniaxial compressive strength and point load strength of rock is greater than 0. 9

Therefore, it is necessary for support system decision procedures to consider the interrelations between features and to extract hidden or complex relationships that exist between them. This paper presents techniques that can extract these hidden and complex relationships and produces a system that makes use of all features. This approach is based on a back propagation algorithm applied to an appropriate neural network, with the known features presented at its input and its output representing the support type and dimensions.

3 APPROACH OF NEURAL NETWORK RECOGNITION OF SUPPORT SYSTEM

3. 1 *Neural network decision system*

Our approach to extract the hidden, complex relationships and decide support system is based on learning and adaptive identification of neural networks . It makes use of the capability of neural networks to generalize, i. e., once a neural network is trained it is able to decide the type and dimensions of support system that it has seen previously using the information it has learned from the trained data.

Our neural network identification is carried out in two steps. First a neural network trainer

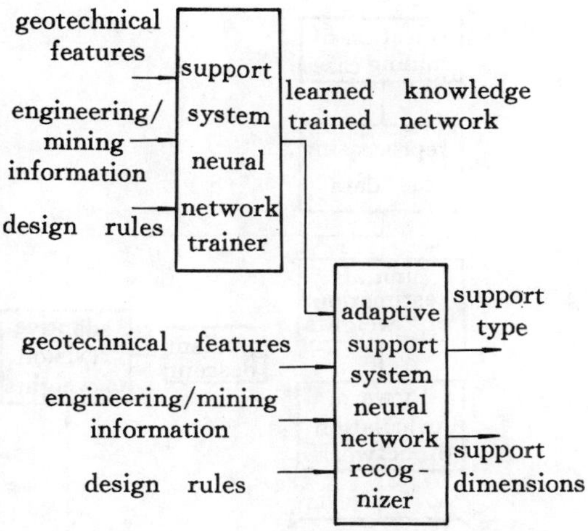

Fig. 1 Learning and adaptive identification of neural networks for opening support system

is constructed to gain knowledge from training samples and to distribute this on all interconnections of the trained network. Next, an associative neural network recognizer performs associative identification of support type and dimensions when similar geotechnical features, mining information and engineering features discussed above are used as input for the trained network. This identification procedure, based on neural network learning, is shown in Fig. 1.

To perform the above decision procedure, we constructed three subsystems as follows

(1) rockmass classification neural network subsystem (RMCNN);

(2) rockmass structure classification neural network subsystem (RSCNN);

(3) support type and dimension decision neural network subsystem (OSDNN).

These three subsystems are used as components of the overall system of opening support design. RMCNN determines the class of rock masses, while RSCNN produces the category of rockmass structure. The output of the RMCNN and RSCNN are fed, together with other features to the OSDNN, to produce the decision of support type and dimensions. A block diagram of the overall system is shown in Fig. 2.

3.2 *Model and algorithm of neural network learning*

A multilayer feedforward neural network, for each subsystem discussed above was constructed. Each multilayer network is composed of many interconnected identical simple processing units called neurons as shown in Fig. 3. These neurons per layer are all sigmoidal. The range of the sigmoidal function is between 0 and 1, with common nonlinear behavior. The sigmoidal function is responsible for the nonlinear mapping of features. An individual neuron sums its weighted inputs and passes the result through a threshold unit. In this multilayer network the neurons are organized into layers with no feedback or lateral connections.

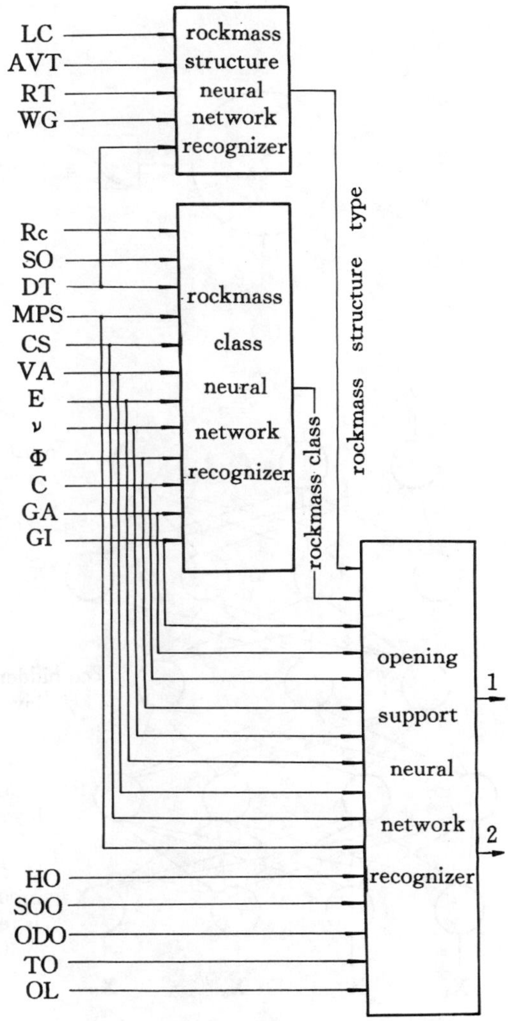

1: support type

2: support dimensions

Fig. 2 A block diagram of the support decision system.

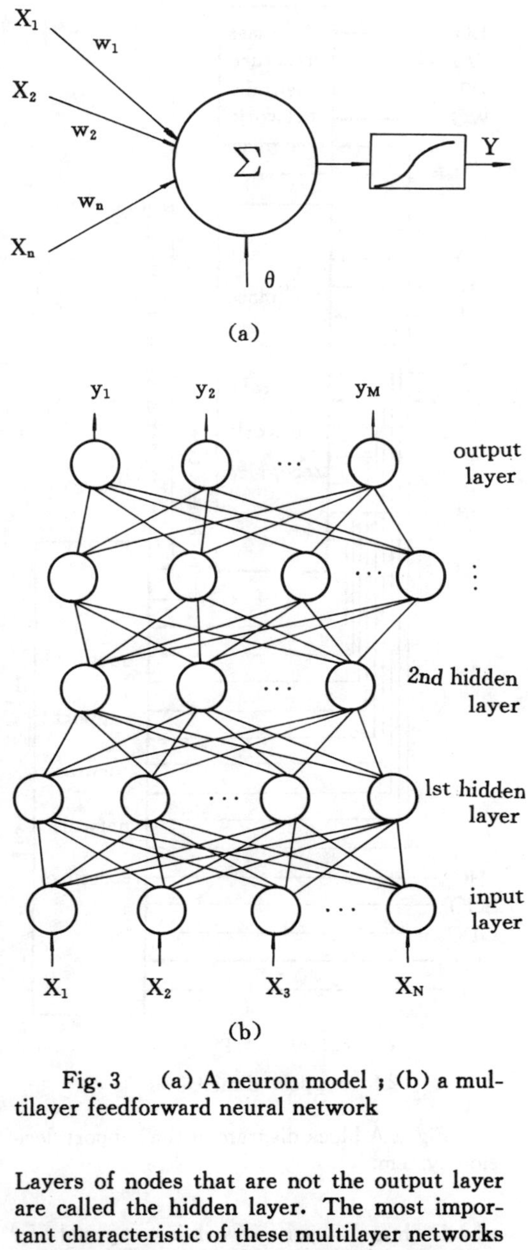

Fig. 3 (a) A neuron model ; (b) a multilayer feedforward neural network

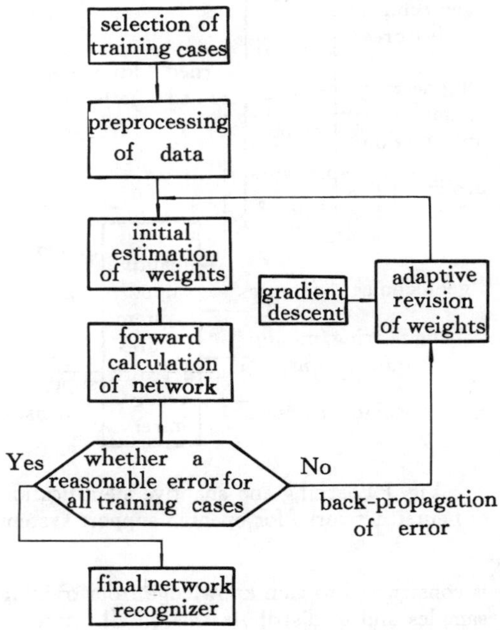

Fig. 4 Flowchart for back-propagation learning algorithm used in the training of the support design system

Layers of nodes that are not the output layer are called the hidden layer. The most important characteristic of these multilayer networks are that they can learn complex mapping through a simple adaptive process that updates the weights during repeated representations of the training samples.

The algorithm used to train these networks is back-propagation learning (Rumelhart, 1986). This choice emphasized the following two historical points: 1) back-propagation has achieved great success and influence as a supervised learning paradigm, and 2) it has been applied almost exclusively to layered, feedforward networks of analog perceptrons (Lacher, 1992).

Back-propagation learning is more precisely described as steepest descent supervised learning using back-propagation of error. This makes an initial estimation of interconnection strengths (weights) of network. Then, repeated revisions will be made based on a gradient descent algorithm until either a total mean square error is minimized or all patterns produce the desired output with a predefined acceptable error margin. The procedure of back-propagation learning is shown in Fig. 4.

4 CASE STUDIES

A large number of case studies have been conducted to evaluate

- network architectures including the number and size of hidden layers for achieving good performance;
- learning parameters (coefficient α, learning ratio η, error precentage margin) for good training and generalization;
- validity of neural network decision for rock opening support.
- role and effects of extracted hidden, complex geotechnical features.

4.1 *Training and testing data*

Our training and testing data were collected from several different sources. These data sources include undergound railways, coal mines, iron mines and gold mines distributed in 7 provinces in China. The case histories were chosen so as not to bias the network twoward any sectional shape (semi-circle arch, semi-elliptic arch, trapezium-shaped arch), opening type (extraction drift, tunnel, headgate, entry etc.) or load type (static, dynamic loads). The actual solutions such as support type and amount for these selected openings, are often known. Altogether over 257 cases (257 sections in 150 openings) have been assembled. 230 of them have been used as training sets of networks and the rest were used as testing sets.

The hidden, complex geotechnical features and mining information were extracted from the training samples and used to train the three multilayer networks using a back-propagation learning algorithm.

In training input data the desired vectors of the network were preprocessed in advance. The desired data of the RMCNN and the RSCNN are designed into Boolean values 0 and 1 as shown in Table 1 and Table 2 respectively. The desired outputs of the OSDNN are arranged into floating-point values between 0 and 1 which are transformed by the constructed afferent rules.

4.2 *Network training*

The three networks were trained separately using above data. To start training, all connection weights were initialized to small pseudorandom numbers. An initial training set of 200 sample vectors was formed for each of the three networks. Then, training was carried out by presenting the 200 vectors successively. Normally, many iterations were needed before the network reached a stable configuration. Now all 200 vectors are correctly identified by our three networks and all output errors reach an acceptable margin. Then, we took the next 30 sample vectors , added them to the initial training set, and repeated the training procedure. This training, carried out on all 230 vectors for each network, as included in the training set. The networks were allowed to train for more iterations, in order to push the network outputs even closer to the desired outputs. This completes the training phase. The connection weights were than ready for support decision.

Table 1 The desired outputs of rockmass classification neural network

Class no.	Description	Desired outputs				
I	Very good rock	1	0	0	0	0
II	Good rock	0	1	0	0	0
III	Fair rock	0	0	1	0	0
IV	Poor rock	0	0	0	1	0
V	Very poor rock	0	0	0	0	1

Table 2 The desired outputs of rockmass structure classification neural network

Category	Desired outputs
Intact structure	1 0 0 0 0 0 0 0
Massive structure	0 1 0 0 0 0 0 0
Bedded structure	0 0 1 0 0 0 0 0
Thin-bedded structure	0 0 0 1 0 0 0 0
Mosaic structure	0 0 0 0 1 0 0 0
Bedded Kataklastic structure	0 0 0 0 0 1 0 0
Kataklastic structure	0 0 0 0 0 0 1 0
Granular structure	0 0 0 0 0 0 0 1

4.3 *Network testing and simulation*

The collected 27 sample sets were first used to test separately the three neural network subsystems to evaluate their performance on their specific tasks. Then, this testing was combined to evaluate the performance of the overall support design system.

For the RMCNN and the RSCNN the true classification was identified using the following formula, according to their true output values respectively.

$$\begin{aligned} &|1 - y_{max}| < \delta_{max} \\ &y_j < \delta_{max} \quad for \ \forall\ j, y_j \neq y_{max} \end{aligned} \qquad (1)$$

Where $y_{max} = \max_i (y_i)$

δ_{max} —an acceptable error.

The support type and dimensions determined by the OSDNN were obtained with an transformation of the true outputs. This transforms inversely the true outputs (numerical vectors) to the original representations as follows

$$o_i = efferent(y_i) \qquad for\ all\ i \qquad (2)$$

The testing results of the RMCNN are shown in Table 3. Only 20 opening test sections are shown to illustrate performance of the system. The classification accuracy of 96. 30%

Table 3 Testing results of rockmass classification neural netwcrk subsystem

No.	Actual						Identified					
	Desired outputs					Class no. *	True outputs					Class no.
#1	1	0	0	0	0	I	0.945	0.000	0.047	0.002	0.001	I
#2	0	0	1	0	0	Ⅲ	0.002	0.030	0.957	0.070	0.051	Ⅲ
#3	0	1	0	0	0	Ⅱ	0.000	0.990	0.009	0.011	0.007	Ⅱ
#4	0	0	1	0	0	Ⅲ	0.000	0.019	0.987	0.019	0.019	Ⅲ
#5	1	0	0	0	0	I	1.000	0.001	0.000	0.000	0.000	I
#6	0	0	0	1	0	Ⅳ	0.006	0.017	0.023	0.963	0.037	Ⅳ
#7	0	0	0	1	0	Ⅳ	0.090	0.001	0.002	0.916	0.053	Ⅳ
#8	0	0	0	1	0	Ⅳ	0.007	0.009	0.073	0.930	0.017	Ⅳ
#9	0	1	0	0	0	Ⅱ	0.003	0.932	0.025	0.038	0.071	Ⅱ
#10	0	1	0	0	0	Ⅱ	0.039	0.949	0.016	0.029	0.008	Ⅱ
#11	0	0	0	0	1	V	0.000	0.001	0.019	0.026	0.971	V
#12	0	0	0	0	1	V	0.009	0.000	0.017	0.019	0.993	V
#13	0	0	0	0	1	V	0.007	0.010	0.023	0.032	0.989	V
#14	0	0	0	1	0	Ⅳ	0.000	0.023	0.032	0.961	0.003	Ⅳ
#15	0	0	1	0	0	Ⅲ	0.007	0.000	0.913	0.091	0.000	Ⅲ
#16	0	1	0	0	0	Ⅱ	0.000	0.901	0.101	0.031	0.011	Ⅱ
#17	0	0	1	0	0	Ⅲ	0.000	0.037	0.907	0.098	0.073	Ⅲ
#18	0	0	1	0	0	Ⅲ	0.006	0.019	0.938	0.042	0.017	Ⅲ
#19	0	0	0	1	0	Ⅳ	0.000	0.007	0.029	0.106	0.901	V
#20	0	0	0	0	1	V	0.009	0.010	0.017	0.005	0.999	V

* On-site appraisals suggested by a group of experts who are familiar with features and behaviours of the investi-gated excavations

Table 4 Testing results of rockmass structure classification neural network subsystem

	Actual									Identified								
	Desired outputs								Category	True outputs								Category
#1	1	0	0	0	0	0	0	0	IS	0.998	0.021	0.000	0.006	0.001	0.001	0.000	0.003	IS
#2	0	0	1	0	0	0	0	0	BS	0.017	0.037	0.991	0.03	0.015	0.001	0.007	0.008	BS
#3	1	0	0	0	0	0	0	0	IS	1.000	0.009	0.000	0.000	0.001	0.000	0.001	0.009	IS
#4	0	0	0	0	1	0	0	0	MOS	0.000	0.003	0.019	0.07	0.975	0.019	0.001	0.019	MOS
#5	1	0	0	0	0	0	0	0	IS	0.975	0.390	0.019	0.009	0.006	0.005	0.001	00.000	IS
#6	0	1	0	0	0	0	0	0	MS	0.001	0.935	0.006	0.001	0.079	0.009	0.010	0.000	MS
#7	0	0	0	1	0	0	0	0	BS	0.000	0.910	0.026	0.993	0.007	0.067	0.030	0.023	BS
#8	0	0	0	0	0	1	0	0	BKS	0.000	0.001	0.000	0.031	0.091	0.971	0.023	0.069	BKS
#9	0	1	0	0	0	0	0	0	MS	0.013	0.979	0.073	0.001	0.031	0.001	0.007	0.017	MS
#10	0	1	0	0	0	0	0	0	MS	0.039	0.965	0.023	0.015	0.079	0.013	0.019	0.023	MS
#11	0	0	0	0	0	1	0	0	BKS	0.000	0.009	0.003	0.013	0.037	0.093	0.931	0.029	KS
#12	0	0	0	0	0	0	1	0	KS	0.000	0.007	0.006	0.013	0.033	0.047	0.953	0.031	KS
#13	0	0	0	0	0	0	0	1	GS	0.000	0.000	0.000	0.007	0.019	0.019	0.013	0.999	GS
#14	0	0	0	0	1	0	0	0	TS	0.000	0.000	0.001	0.003	0.973	0.013	0.023	0.009	TS
#15	0	0	0	1	0	0	0	0	TS	0.001	0.003	0.019	0.989	0.009	0.023	0.021	0.002	TS
#16	0	1	0	0	0	0	0	0	MS	0.003	0.998	0.006	0.007	0.001	0.000	0.000	0.000	MS
#17	1	0	0	0	0	0	0	0	IS	0.921	0.047	0.016	0.013	0.000	0.000	0.001	0.007	IS
#18	0	1	0	0	0	0	0	0	MS	0.019	0.973	0.023	0.007	0.000	0.000	0.000	0.001	MS
#19	0	0	0	0	0	0	1	0	KS	0.000	0.000	0.000	0.000	0.001	0.983	0.039	0.001	BKS
#20	0	0	0	0	0	0	0	1	GS	0.000	0.000	0.000	0.000	0.001	0.001	0.009	0.999	GS

IS=intact structure, MS=massive structure, BS=bedded structure, IS=thin−bedded structure, MOS=mosaic struture, BKS=bedded kataklastic structure, KS=kataklastic structure, GS=granular structure

Table 5 Actual support type and dimensions of testing samples

No.	Sprayed concrete	Rock	bolt				Reinforcing		net		Trestle		
	Thickness (cm)	Type	Diameter (cm)	Length (m)	Spacing a (m)	Spacing b (m)	Type	Diameter (cm)	Spacing a (cm)	Spacing b (cm)	Type	Model	Spacing (m)
#1													
#2		CSB	16	1.8	1.0	1.0							
#3	8.0												
#4	10	CSB	18	2.5	1.5	1.5							
#5	5.0												
#6											STRS	12#I—steel	0.8
#7	15	CSB	18	3.1	0.8	0.8							
#8	10	CSB	18	1.8	0.8	0.8	metal mesh	6	40	40			
#9		CSB	16	1.8	1.0	1.0							
#10		CSB	16	1.8	1.0	1.0							
#11	15	PFB	40	1.8	0.5	0.5	metal mesh	6	20	20			
#12	18	CSB	22	2.0	0.8	0.5	metal mesh	8	20	20			
#13	18	CSB	22	2.0	0.5	0.5	metal mesh	12	20	20			
#14	12	PFB	40	1.8	1	0.8	metal mesh	6	30	30			
#15	6	CSB	16	1.8	1.0	1.0							
#16	6	CSB	16	1.8	1.2	1.2							
#17											SYS	29U	0.6
#18											SYS	25U	0.7
#19	14	CSB	18	1.8	0.8	0.5	metal mesh	8	20	20			
#20	15	CSB	18	1.9	0.8	0.5	metal mesh	10	25	25			

CSB=cement-sand-grouted steel bolt, STRS=steel trapezium-shaped rigid support, PEB=piped frictional bolt, SYS=steel arched yielding support

Table 6 Testing results of the OSDNN using experts' appraised class and structure type of rock masses

No.	Sprayed concrete	Rock	bolt				Reinforcing		net		Trestle		
	Thickness (cm)	Type	Diameter (cm)	Length (m)	Spacing a (m)	Spacing b (m)	Type	Diameter (cm)	Spacing a (cm)	Spacing b (cm)	Type	Model	Spacing (m)
#1													
#2		CSB	16.03	1.81	1.12	1.17							
#3	8.07												
#4	10.19	CSB	17.73	2.47	1.53	1.47							
#5	5.00												
#6											STRS	12#I—steel	0.85
#7	14.70	CSB	18.32	2.85	0.81	0.87							
#8	11.30	CSB	18.17	1.91	0.80	0.79	metal mesh	6.00	29.30	29.50			
#9		CSB	15.99	1.73	1.10	1.00							
#10		CSB	16.10	1.81	1.11	0.97							
#11	15.90	PFB	40.00	1.81	0.51	0.53	metal mesh	7.30	20.10	20.00			
#12	17.30	CSB	21.70	2.09	0.81	0.53	metal mesh	8.30	20.10	20.09			
#13	17.50	CSB	21.70	2.00	0.51	0.50	metal mesh	10.93	21.01	19.81			
#14	12.70	PFB	39.75	1.79	0.95	0.51	metal mesh	6.10	25.09	25.09			
#15	6.90	CSB	16.01	1.80	1.00	1.03							
#16	6.01	CSB	16.00	1.80	1.15	1.01							
#17											SYS	29U	0.60
#18											SYS	25U	0.68
#19	13.10	CSB	16.90	1.80	0.70	0.49	metal mesh	6.01	29.3	29.7			
#20	15.10	CSB	18.01	1.87	0.79	0.51	metal mesh	8.09	20.50	21.09			

Table 7 Testing results of the OSDNN using classes and structure types of rock masses learned by networks

No.	Sprayed concrete Thickness (cm)	Rock bolt Type	Diameter (cm)	Length (m)	Spacing a (m)	Spacing b (m)	Reinforcing net Type	Diameter (cm)	Spacing a (cm)	Spacing b (cm)	Trestle Type	Model	Spacing (m)
#1													
#2		CSB	16.03	1.81	1.12	1.11							
#3	8.07												
#4	10.11	CSB	17.73	2.47	1.53	1.47							
#5	5.00												
#6											STRS	12#I—steel	0.83
#7	14.70	CSB	18.30	2.83	0.81	0.81							
#8	11.30	CSB	18.17	1.90	0.80	0.80	metal mesh	6.00	29.30	29.50			
#9		CSB	16.00	1.73	1.10	1.00							
#10		CSB	16.00	1.81	1.10	1.01							
#11	15.00	PFB	40.00	1.81	0.51	0.50	metal mesh	7.30	20.10	20.00			
#12	17.30	CSB	21.70	2.09	0.81	0.53	metal mesh	8.30	20.00	20.09			
#13	17.50	CSB	21.70	2.00	0.51	0.50	metal mesh	11.89	20.00	19.93			
#14	12.70	PFB	39.75	1.83	0.99	0.49	metal mesh	6.10	25.10	25.10			
#15	6.35	CSB	16.01	1.81	1.01	1.01							
#16	6.07	CSB	16.00	1.80	1.19	1.10							
#17											SYS	29U	0.61
#18											SYS	25U	0.69
#19	14.50	CSB	18.09	1.80	0.79	0.50	metal mesh	8.00	19.9	19.8			
#20	15.10	CSB	18.05	1.87	0.79	0.50	metal mesh	8.09	20.01	20.01			

was reached. The identified categories of rock masses were obtained using above rule according to the true output values .

Testing results of the RSCNN were also obtained with an accuracy of 92. 59% (see Table 4). Testing results of the OSDNN using on-site appraised class of rockmass and rockmass structure given by experts are shown in Table 6. Support types of various openings were identified with a full accuracy of 100%. The highest error on support dimensions is about 50% and the lowest one is 0. Testing results of the combined RMCNN, RSCNN and OSDNN are shown in Table 7, which also shows a comparison between the performance of the network on the expert's appraised class of rockmass and rockmass structure and on that of the identified ones of networks. The testing results show that the performance of overall system rises on average with respect to support dimensions, with the highest error descending from about 50% to 25%.

An interesting discussion is identification of the section #19. The error on diameter of reinforcing net drops from about 25% to 1. 5% and that of spacings of reinforcing net drops from about 50% to 0 by substituting network learning for experts' appraisals. The drop in the identified diameter and spacings is explained by the more reasonable identified class and rockmass structure for this section of opening. This performance of neural network identification shows it can help the recognizer to correct the wrong identifications given by experts.

4.4 *Discussion*

We discuss here why our three network subsystems have gained good performances.

The performance of neural network decision is related to architecture of the networks. Therefore to gain better performance, it needs more suitable network architecture. Because the number of neurons on input layer and on output layer is permanent for a specific task only the number and size of hidden layer are able to be adjusted to gain better decision performance. Many architectures were evaluated in our three network training. It indicates that the performance of networks improves as increase of the number of hidden layers and then remains almost stable. Our network architectures for three subsystems having achieving good performance for our given training data are shown in Table 8.

Learning parameters such as η and α have also influence on convergence ratio and generality of network learning algorithm. Many learning ratio η and coefficient α have been evaluated for our three networks. The suitable parameters η

Table 8 Suitable network architectures and learning parameters

	The number of neurons					Parameters	
	Input layer	1st hidden layer	2nd hidden layer	3rd hidden layer	Output layer	η	α
RMCNN	19	19	10	8	5	0.5	0.943
RSCNN	12	12	10	6	8	0.5	0.949
OSDNN	14	14	11	8	13	0.5	0.931

and α having achieving good convergence ratio and generality for our given training sets are shown in Table 8.

The extraction of higher order features improves performance of network decision. It is the adopted neural network supervised training approach combined with the multilayer architecture who is perfoming such higher order feature extraction. This is verified by comparison of testing results shown in Table 6 and Table 7.

5 CONCLUSIONS

This paper has proposed an efficient, novel approach. Our approach is based on learning and adaptation of multilayer feedforward neural network. Three subsystems such as the RMCNN, RSCNN and OSDNN were developed on SUN 80486 computer in GCLISP, and used in cascade as components of overall system of rock opening support design. The proposed approach can be expected to produce much better results than other reported techniques for the important reasons such as extraction of higher order features, input of many qualitative descriptions, self-learning, adaptation and associative inference of networks and nonlinear dynamic processing. It reduces the limitation of insufficient knowledge and lack of characterizing information that have to occur in the process of rock support design.

In the futher work, we will focus on further testing on a larger test set and apply our approach further in actual design. We will intend to link the develpped neural network system to intelligent CAD system of rock opening design.

REFERENCES

Fairhurst, C. & Lin D. 1985. Fuzzy methodology in tunnel support design, *Proc. of 26th US Symposium on Rock Mechanics*: 269-278. Rotterdam: Balkema.

Feng, Xia-ting. 1991. *Intelligent systems for optimal support design of underground excavations*. Ph. D. thesis. Shenyang: Northeast University of Technology.

Kohihen, T. 1988. *Self-organization and associative memory*. 2nd edition. Berlin: Springer-Verlag.

Lacher, R. C., Hruska, Susan I. & Kuncicky David C. 1992. Back-propagation learning in expert networks. *IEEE Transactions on neural networks*. 3: 62-72.

Lippmann, R. P. 1987. An introduction to computing with neural nets. *IEEE ASSP Magazine*. 4-22.

Rumelhart, D. E. & McClelland, J. L. 1986. *Parallel distributed processing*. Cambridge, MA: MIT Press.

Innovative Mine Design for the 21st Century, Bawden & Archibald (eds) © 1993 Balkema, Rotterdam, ISBN 90 5410 325 6

Development of a computerised mine planning system at Mount Isa Mines

Brian E. Hall
Mount Isa Mines Ltd, Qld, Australia

ABSTRACT: Mount Isa Mines has been developing its own computerised mine planning system for four years, based on Intergraph workstations and the MicroStation CAD package. Initially, a purchased geology and design package and in-house database facilities were used. Subsequent developments have involved the establishment of an ORACLE database for the geology, survey, and mine design functions. Utilities developed enable both regular and occasional users to access the system on both UNIX workstations and DOS PCs through a common user interface for operations ranging from simple viewing of portions of the mine to complex geological interpretation and mine design procedures. Use of industry standard CAD and database software as the core of the system, combined with in-house developments and porting and integration of existing standalone mainframe facilities is providing an integrated system which is able to evolve with computing industry developments, yet be tailored precisely to the needs of a variety of users within the company.

1. INTRODUCTION

Mount Isa Mines Ltd., (MIM) located in north west Queensland, Australia, produces annually some 12Mt of copper and silver/lead/zinc ores from 30 orebodies by underground mining at Isa and Hilton Mines. The technical functions supporting these operations work from a database of 18 000 diamond drill holes from which are generated, together with underground mapping results, geological interpretations of structures and orebodies and geostatistical grade estimates. Design and survey functions deal with 30 000m of development and 600 000m of blasthole drilling each year. The Isa Mine covers an area 4.5 km long, up to 1 km wide, and has workings from near surface to 1500m deep. Several dozen managerial and technical staff deal directly with these mine planning functions, with numerous others having indirect or infrequent contact.

In the mid 1980s, investigations indicated that a change from paper plans to a computer based planning system was becoming economic because of reducing costs and increasing capabilities of graphical computer systems. The various data flows within the overall mine planning environment were identified. (Fig. 1) These were all on paper, such as plans, tabulations and memos, with duplicated but often inconsistent databases established in the various departments. Substantial benefits would be achieved if all the information were in one database accessible to all users, with as many linkages as possible being electronic rather than paper. (Fig. 2)

The capabilities of commercially available mining software were investigated, and it was concluded that no system fully addressed the identified needs of the Isa mining operation. Each available system fell short of the identified ideal, and projected developments did not seem to be moving in the direction of MIM's ultimate target.

It was decided in 1989 to develop an Integrated Mine Planning System (known as IMPS) in-house. This is based on the MicroStation computer aided design (CAD) package on Intergraph UNIX workstations and DOS personal computers. CADMIN, developed by Wright Engineers of Vancouver, Canada, to run as a UNIX MicroStation application, provided some initial geological and design facilities. The first stages of the project would be concerned with integrating the three key areas from which all else flowed, the Geology, Survey, and Mine Planning functions.

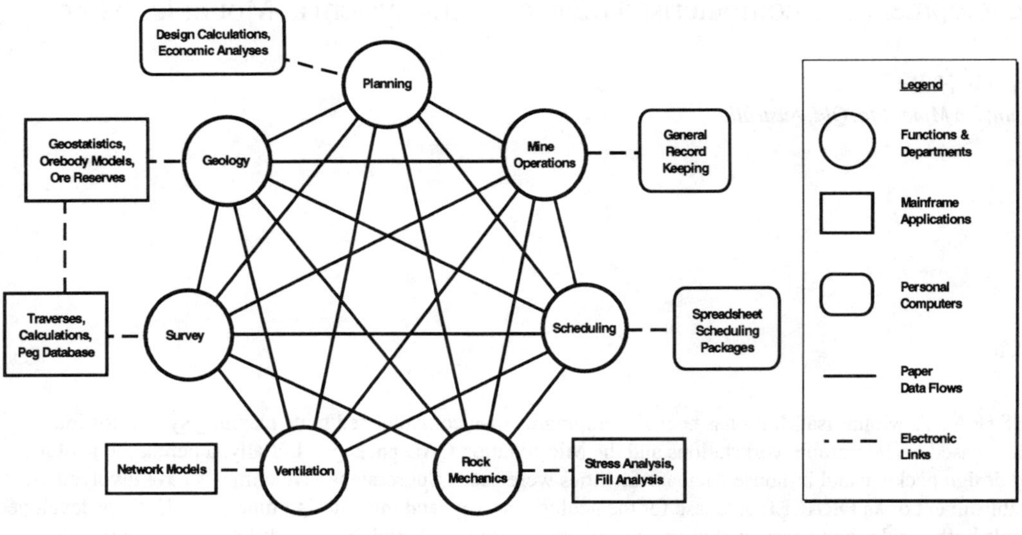

Fig. 1 Data flows and computerisation identified prior to IMPS development

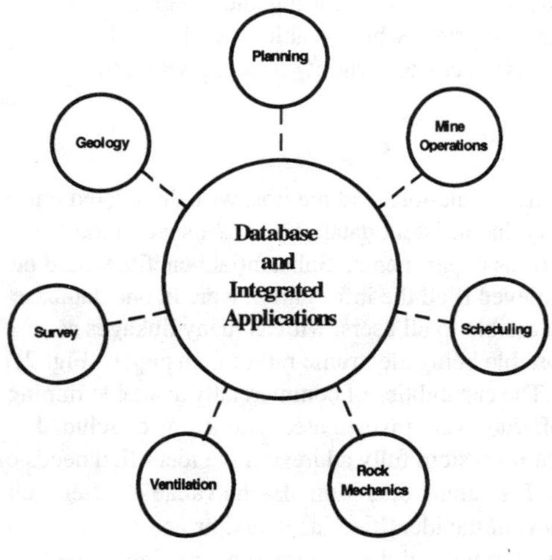

Fig. 2 The ideal data flows

A number of utilities were developed and the system released for use in the Mine Office a year later in 1990. Since that time, additional facilities have been developed, and the initial simple database was replaced with ORACLE database software. Development is continuing to provide the facilities required by users, with regular meetings of senior user representatives setting development priorities, and overall management of the project vested in a steering committee of senior managers of user and support departments.

This paper describes development progress, features of the overall system, and facilities provided for users in each functional area.

2. OVERALL PROJECT DEVELOPMENT

There have been a number of distinct phases in the development of IMPS, as described in the following sections. Additional details of some of the software described appear in later sections devoted to various functions within IMPS.

2.1 *1989-1990 - Initial developments by a small project team*

After the decision to proceed in 1989, a project team consisting of a mining engineer, two geologists and two computer analysts was formed to prove the feasibility of a prototype system. A project area was established with an Intergraph file server, three workstations with single 19" screens, an A0 digitising table, and a 36" Versatec colour electrostatic plotter.

The initial work on small sections of the mine quickly proved the feasibility and desirability, and the study evolved into the first stage implementation. (Bower *et al* 1991) The engineer and geologists developed relevant procedures and, together with their professional colleagues, identified facilities which would be needed in the workplace for a first stage system.

The analysts began to develop these utilities and establish the basis of the underlying system. They were faced with steep learning curves in several new fields. These included the UNIX operating system, the MicroStation CAD package, its programming library routines, known as MicroCSL, and the C programming language.

At this stage, the team concentrated more on initial graphical representations and techniques, even though the early investigations had indicated that the overall project was in reality a major database project. This was necessary to get an early financial return on the capital invested, and it was seen that this would come by having technical staff using the three dimensional (3D) CAD facilities of MicroStation as quickly as possible. This created results which were not only real benefits in planning and design, but also highly visible to all levels of management. It was also necessary to trial various features before committing money and development time to a final system.

CADMIN provided graphical representations of diamond drill holes, a data structure for orebody models, and other facilities. Since MIM already had a substantial investment in mainframe kriging and surveying packages, corresponding facilities in CADMIN were not used, and early development work concentrated on routines to download mainframe information from the diamond drill hole, orebody block model and survey peg databases into IMPS and CADMIN data structures. Utilities were also developed by the analysts to assist with generation of 3D surfaces constructed from interpretations of drilling sections, and to provide sections at any orientation through the 3D models.

The initial work concentrated on proving and establishing the system in limited but representative areas of the mine. Geological developments were concentrated in two areas. The southern 1100 Orebody, a disseminated copper orebody with a structurally controlled basement contact zone at the Isa Mine, and the initial mining block of the stratiform lead/zinc orebodies at the Hilton Mine provided experience in structural and orebody interpretations and modelling, and in ore reserve estimation techniques.

Mine planning and design work was also done in the southern 1100 Orebody, an area with minimal previous mining. This not only enabled the interaction between geological and design work to be developed, but also permitted design work to be more easily done in an area where the strike of the orebody is significantly different from the directions of the mine grid on which diamond drilling cross sections are aligned. This capability alone was sufficient to justify the initial capital investment.

By the end of the first year, basic file structures were in place, and some utility routines had been developed. Kriged block models of portions of the copper area were in place, and surfaces representing the basement contact had been interpreted. Mine designs in a limited area were being done on IMPS, but were transcribed onto the existing paper based system for plan publication. System security and controls necessary for a system with several dozen users when fully operational had been developed.

2.2 *1990-1991 - Initial implementation in the workplace, and utilities developed*

After the first year's project work, the immature system was released into the work place. The number of analysts was increased to four. The original Interserve 305 file server was moved to the Hilton Mine, and replaced at Isa Mine with a more powerful Interserve 6105. Additional workstations were acquired for work to proceed in geology, survey and mine planning at Hilton and Isa. These were Intergraph 6000 series workstations with a variety of configurations: some with integral A1 digitising surfaces, some with dual 19" screens, and others with single 27" screens. Several 25MHz 80386 based dual screen PCs with the PC version of MicroStation were also installed on the network.

In the 3000 Orebody area, a new section of the mine with very little development in place, the decision was made that all design plans would be created on IMPS and issued electronically to users on the network, and via the IMPS plotter for hardcopy output. Existing survey master plans in this part of the mine were digitised, and routines were developed for surveyors to enter their pickups of mine development directly into the system.

Diamond drill holes and associated grade and rock type information were accessible in CADMIN generated files. Kriging procedures were established on the IMPS hardware, and results were dumped into the CADMIN block modelling data structure. The basics of a mine planning system were available , but geological interpretations of structures were still done off line and transferred into the system.

By mid to late 1991, the system was in widespread use throughout the mine, with design work being done almost exclusively in IMPS, though plan approval and issuing in the established parts of the mine were still done through the paper based system.

All survey development pickups were being entered interactively, and survey peg and diamond drill hole data were downloaded regularly from the existing mainframe systems into IMPS. Facilities for the geologists, however, were still quite rudimentary.

2.3 *1991-1992 - Frustrating software upgrades, database development*

In the latter half of 1991, a number of significant events occurred in the system development and implementation.

Intergraph released Version 4 of MicroStation. This has a number of features of great value to the IMPS project, some of which are essential to progress towards the ultimate goals of the development. The UNIX and DOS versions are effectively identical. Version 4 included a new programming language known as MDL, which meant another learning curve for the already busy analysts.

MDL provides a common development package on both DOS and UNIX machines, so that programs developed on one platform can be ported to the other with generally no modifications. Previous attempts to develop identical applications on both platforms had caused many problems and delays, and in some cases these were abandoned. MDL also provides a consistent user interface, enabling in-house developments to have the same look and feel as the basic MicroStation package. Many of the database querying facilities required by the geologists for text information while doing on screen interpretations in an interactive graphics session could not be easily provided without it.

The early releases of version 4 contained some serious bugs (from MIM's point of view), and a lot of time was lost getting previously operating software to run. In DOS, a great deal of difficulty was experienced getting a compatible set of linkers and compilers.

ORACLE software was also acquired, and the task of formal database development began. This too was new software for the analysts, in both DOS and UNIX. The structure of the database is described below. Routines were developed for loading the database with pre-existing survey and diamond drilling text data from the mainframe, adding new text and graphical information in all areas, and retrieval of information by users who will range from regular to occasional clients. These facilities are also described below.

For faster data capture of existing plans, a scanning system was acquired. This consists of a scanner, raster image editing facilities, and raster to vector conversion software generating MicroStation design files. This was used to input 1:2500 scale plans of the whole mine to make available electronic plans as a base for general work in all areas. For more accurate plans, suitable for a design base, a program of scanning, editing and vectorising the 1:500 scale survey master plans is in progress. This is described in more detail below.

During this time, a lot of the work being done by the analysts was not visible to system users, and many of its results will also remain transparent to them. With most of the analysts involved in behind the scenes development, users were frustrated by slow provision of utilities. However, the portion of the mine being actively worked on in IMPS continued to grow. As more information was added to the system, hardcopy plan output done exclusively through IMPS in a large part of the mine, and technical and operating personnel became more familiar with the system. The process of agreeing on a design evolved to become an informal gathering of the relevant parties around one of the 27" screen workstations, looking at the design in 3D from various angles, and modifying it immediately on the screen as changes were agreed upon.

2.4 *1992-1993 - Database implementation, and mainframe systems ported to IMPS*

Recent development work has been almost entirely directed at the ORACLE database and related user access facilities. At the time of writing, the database is being populated with information from sources such as the mainframe for survey and diamond drilling, and the original IMPS databases for the graphical work done since the project began. A number of utilities will become available to users, especially geologists, when the change to ORACLE occurs in mid 1993.

The remainder of the year will be dedicated largely to the provision of a number of geology utilities which are dependent on the database, and the complete transfer of the existing geology and survey functions from the company's mainframe into the IMPS environment. This will eliminate interfaces and use of multiple systems, and also meets corporate objectives to eliminate the central mainframe system and migrate to a distributed processing environment by 1994/95.

3. THE IMPS SYSTEM DESIGN

3.1 *Hardware and networking*

IMPS is a network of UNIX workstations and DOS PCs. At Isa Mine there are 15 Intergraph 2000 and 6000 series Interpro and Interact workstations on the network. Each workstation has its own hard disk, of 350 to 670 Mb capacity, intended to be used mainly for users' working files.

The central data is located on the Interserve 6105 file server, which has 4.4Gb of magnetic disk space. This server also controls the Versatec plotter, and has tape drives for backing up data. Incremental backups are done on a daily basis on the server and all workstations, and weekly full backups are scheduled throughout the week for all machines. For security, weekly backup tapes are stored in a fireproof vault in the Mine Office, and monthly transfers of tapes are made to a vault in the General Office, one kilometre away.

PCs form an integral part of IMPS. The initial design envisaged that the more powerful UNIX workstations would be used with large graphics data files and for procedures requiring substantial floating point processing power. Less demanding operations could be handled by PCs. Eight dual screen 33MHz 80486 based PCs are on the network in the survey, planning and design areas. It is also envisaged that users will access the system for minor queries via PCs used mainly for normal PC tasks such as spreadsheet and word processing work. It is therefore essential that general user facilities and most design facilities be available in DOS.

The network runs on thin ethernet, using TCP/IP protocols and Sun NFS (Network File System). UNIX disks are generally able to be mounted by any other UNIX machine. PCs are able to mount up to eight UNIX drives as logical DOS disks. The physical location of the disks being accessed is transparent to the user, and the system operates virtually identically on PCs and workstations.

3.2 *IMPS objects*

Every graphically described item in the mine is defined as an "IMPS Object". Each object falls naturally into one of three classifications represented by the functional user groupings of Geology, Survey, and Planning. Within each class of object are several types. Geology objects may represent diamond drill holes, faults, or orebodies. Planning objects may be headings, vertical openings, stopes or service objects, and survey objects include pegs as well as the same types as planning. All users can view all objects, but authority to submit a new or modified object to the database is limited to a few senior users or managers in each user group.

3.3 *Database structure*

Each object has a unique name so that it can be readily retrieved from the database when required, and a unique linkage number assigned by the system and used to link graphical and text data. The MicroStation graphic construction is held in a database table as raw binary code taken directly from the MicroStation design file structure. Other text information about the object is held in other tables. The volume of the mine is divided into 25m cubes aligned with the mine survey grid of northings, eastings, and reduced levels (RLs). A spatial index records which cubes any IMPS object impinges on, permitting rapid retrieval from the database of all objects falling within any user defined volume.

The principles of the database structure are fairly simple, but the detailed design and implementation is much more complex. The database design allows for later expansion of the system to encompass all areas shown in Fig. 1, even though only three of these are initially being linked by IMPS.

3.4 *Database access facilities for users*

A number of user facilities are being provided in the initial development of the database.

VIEW is available for all users, and has two forms. ViewVolume requests entry of minimum and maximum northings, eastings and RLs or mine levels. The process retrieves objects impinging on the spatial index cubes partially or wholly within the specified volume. The user can specify which classes of objects - geology, survey and/or planning - are required. ViewObject prompts the user to specify an IMPS object and uses the extremities of that object to define the volume extracted.

MicroStation has the facility to attach to the design file in which a user is working a number of other design files. These "reference files" can be manipulated with various viewing commands within MicroStation, but cannot have their contents altered.

The IMPS VIEW facility places copies of the database objects in design files in the user's working directory, one file for each class of object requested,

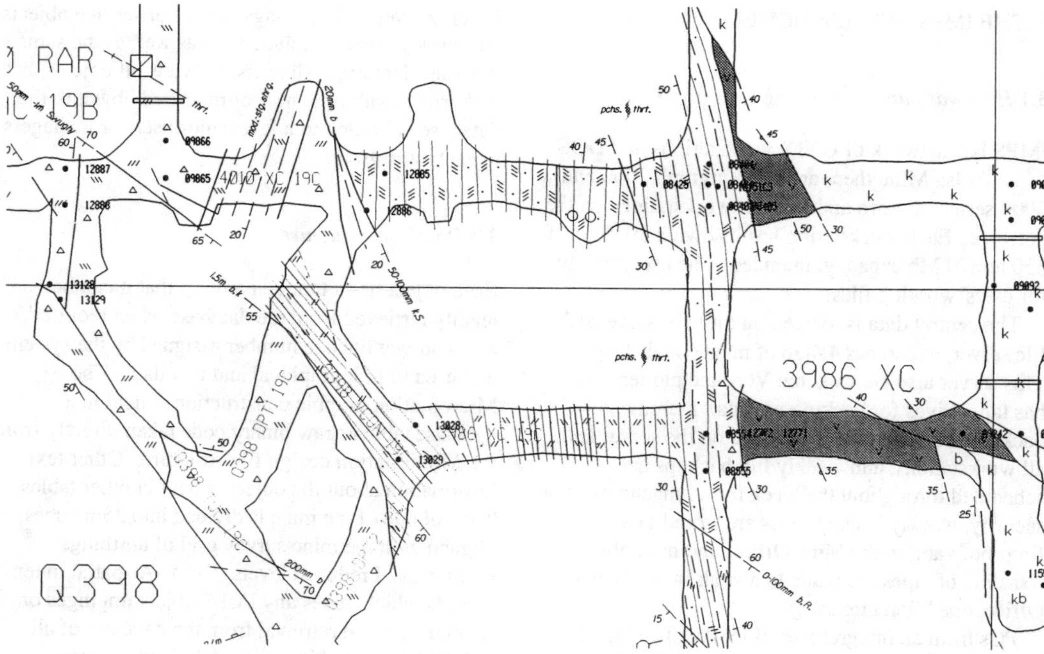

Fig. 3 A typical IMPS geological mapping plan

and attaches these as reference files to an active design file. The user is then free to do whatever he wishes in his own design file using standard MicroStation viewing and design commands. VIEW permits users to work with files containing only those parts of the database which are needed for the job at hand. Minimising the file sizes and locating them on the user's workstation gives minimised screen update times, and only requires access to the database across the network when the reference files are first created or regenerated to get the latest information.

The other facilities are for users who may be in a position to modify the database.

SUBMIT adds the graphic representation of a new or modified object. Survey objects are compared with the planning object of the same name to check that they are in the same general location in 3D space, to avoid gross errors.

BORROW makes a copy of an object in a user's design file for purposes of modification. The original object remains in the database and is not altered until the revised object is SUBMITted. Only one user may BORROW an object, and later attempts by others generate warning messages.

User managers may delegate authority to their team members to update the database. The system administrator establishes controls so that only authorised personnel may complete transactions which update the database, and then only for the class of objects for which they are responsible.

4. GEOLOGY IN IMPS

4.1 *Geological mapping of mine development*

Mapping is essential for detailed interpretations after development has been done. Items typically displayed include the strike and dip of structures and rock types. In a 3D graphical environment, it is necessary to place mapping information at the correct location in 3D space in order to construct accurate 3D interpretations of geological features.

Geologists use the survey plans as a base, and a variety of push button menu functions to place standard mapping symbols. A typical plan of mapping plotted using IMPS is shown in Fig. 3.

4.2 *Diamond drilling*

In the mainframe system, core logging and assay information were written on logging cards and hand keyed into the mainframe database. Downhole survey results were transferred from the survey system, and entries were not generally made until all information was available. The result was a delay, sometimes of months, between the completion of drilling a hole and its appearance in the database.

The greater power of the ORACLE database now makes it easy for the various aspects of diamond drill hole information to be entered as soon as they are available. The path of the hole can be entered as a straight line as soon as the hole is drilled, or a "mock survey" can be used to place a curved hole path according to the average deflections obtained in surveys of holes of similar bearing, dip, and length over the years. These can be replaced with the actual results when the hole has been surveyed.

Geological core logging of rock types, structures and the like has been computerised. Geologists enter logging information into notebook PCs in the core shed. This information, which can also contain visual grade estimates, is transferred to the IMPS database. Assay results are entered into a computerised database in the assay laboratory. Until the IMPS and laboratory networks are linked, this information is transferred on magnetic tape, but will eventually be directly downloaded into IMPS.

Information in the drill hole database can be displayed as colour coded line segments, histograms, or text along the path of the drill hole. Facilities are being developed to display database information in a text window during the graphics session, so that clicking on a graphical element will bring up the corresponding text information.

4.3 *Surface generation*

An important part of a geological system is the ability to create surfaces to represent interpretations of structures and boundaries. CADMIN provided a basic facility to connect parallel planar lines or shapes. Unfortunately, these conditions are rarely seen in practice. Diamond drill holes, though nominally lying in parallel planes, always deflect, so an early development in IMPS was the provision of a utility to generate a surface from lines or shapes only approximately parallel and planar.

Geologists use MicroStation to create interpretations by joining points on diamond drill holes on the same geological section. The surfacing utility is then used to join these to create a 3D surface. Guide lines can be added between the lines to be joined, constraining the process to follow the geologist's interpretation of structures in complex areas. Delaunay triangulation was added, and the routine proved quite useful for overfolded surfaces which could not be handled by an off-the-shelf digital terrain modelling package. The utility has been pressed into service in a number of other situations for which it was not designed, and is being rewritten as a general surface construction facility.

The other major type of surface required is an interpolated surface within the orebody block model, representing, for instance, a 3D grade "contour" or envelope. A utility written for this task can be applied to any numerical value stored in the block model, such as ore grade, kriging variance, or number of bedding plane breaks per metre. A triangulated surface is generated by interpolating between values in the cubic block model, and can be run multiple times to create, for example, ore limits at a series of grades.

A further utility was developed to enable a section to be generated through these surfaces at any orientation. This is particularly useful in parts of the mine where the strike direction of the orebody has swung significantly away from the mine grid, and the standard geological sections do not give a true representation of the orebody cross section.

4.4 *Kriging and block modelling*

MIM has used geostatistics on the mainframe for orebody modelling and ore reserve estimation in the disseminated 1100 copper orebody for many years. The system, and the variograms on which it is based, have proved to be very robust. At first, mainframe kriged estimates were transferred into the CADMIN block model in IMPS.

The kriging routines have since been upgraded and transferred onto the IMPS hardware, and interfaces were established to translate sample grade and position data and graphically created control surfaces into forms suitable for the kriging algorithms. Interfaces were established to load kriged estimates into the block model, which has 5m cube blocks for copper orebodies.

Lead/zinc orebodies are stratiform and (at Isa) stratabound. These had traditionally been manually interpreted on cross sections, and sectional methods had been successfully used for ore reserves calculations. Previous investigations had indicated that geostatistics could be suitable for these orebodies. With the successful transfer of copper geostatistics into IMPS, the practicalities of doing the same with lead were investigated, and the resulting system is now in regular use for the lead/zinc orebodies at both Isa and Hilton.

The stratiform nature of these orebodies requires

greater constraints in the horizontal direction perpendicular to strike, where the block size is only 1m compared with 5m vertically and along strike. More control surfaces are required than in copper, and it has been necessary to develop routines to follow folds in the orebodies. As in copper, interfaces pass the sample data and significantly more graphically generated control surface information to the kriging algorithms, as well as recording the results in the block model.

The geostatistical processes are working well and are producing extremely valuable information, especially in the lead/zinc areas where rapid generation of orebody models at various combined cutoff grades has only been a dream in the past. However, the system, while workable, is decidedly user unfriendly. A goal for the near future is the full integration of the geostatistical processes into the user friendly IMPS graphical environment, so that rekriging of small areas affected by new data becomes a normal part of each geologist's work, rather than infrequent large updates being done by one or two expert users.

4.5 *Tonnage and grade estimation*

All the above facilities are useless if an estimate of tonnage and grade of an orebody or defined volume cannot be calculated. CADMIN provided a facility which was satisfactory for volumes which were large relative to the block size. The IMPS analysts have developed a utility which will handle any practical size, from whole orebodies down to a few metres of development heading. Orebody model blocks which lie fully within the enclosing volume are fully accumulated. Blocks which are intersected by the boundary of the volume can be subdivided into smaller blocks which again are included if wholly within the volume.

This process can be repeated for up to five stages of subdivision at the user's discretion. There is a trade off between accuracy and processing time. As with all similar procedures, it is up to the user to identify when increasing mathematical precision of the calculation process becomes meaningless because of the inherent sampling and estimation errors.

5. SURVEYING IN IMPS

5.1 *Survey traverses and peg locations*

Most of the survey computing functions still reside on the company mainframe. They were written in-house to meet company requirements many years ago. Field book measurements are entered into various input screens, and the results of relevant calculations are displayed. Co-ordinates of pegs and other relevant data are stored in a database on the mainframe. This data has been downloaded into the IMPS database, and graphical IMPS objects have been created to represent each peg in 3D space. There are some 66 000 survey pegs on record.

A significant part of the development effort this year will be taken up with porting all required survey facilities from the mainframe to IMPS. Hilton mine has made the transition to total stations for survey work, and the IMPS developments will allow for downloading total station data, which at present is being handled at Hilton by using SURPAC software, with interfaces to IMPS. The IMPS survey developments will fully integrate all survey user requirements into the IMPS environment, ensuring a common user interface for all personnel at each location. It is likely that Isa Mine will also change to total stations once these developments are in place.

5.2 *Development offsets*

Surveying of advancing development is done in different ways at Isa and Hilton mines.

At Isa, the traditional method of extending a tape along the centre line and measuring to the sidewalls at intervals along the drive is still in use. Field book measurements were transferred to the plans in the office. Routines were written to mimic this process in MicroStation. After specifying the survey pegs being used for control, the surveyor is prompted to enter the distance from the controlling peg and offsets left and right. Graphic representations of the sidewalls are generated on screen as this occurs. Information from the peg database is used to place offsets immediately at the correct elevation and gradient in the design file.

At Hilton, total stations are used to pick up the four corners of the drive profile. These are also downloaded into SURPAC, then translated into IMPS by the interface routines. This method is generally superior to the other, as no assumptions are being made about the back and floor profiles. The lines representing the four corners will also provide better information for creating a 3D solid representation of the drive, a process which has yet to be developed.

5.3 *Stope surveys*

The nature of the stoping methods makes accurate surveys of stope walls an important requirement for future designs. Stope survey techniques have evolved with the development of electronic distance measuring equipment, but still involve an instrument being set up on the edge of the stope, and a surveyor directly operating it. Since IMPS has been in use, information has been fed into the surface generation software described above, and some success has been achieved in generating 3D volumes.

Unfortunately, the most critical stope wall for future designs is often the one which cannot be seen, being the one which the heading being used for the survey penetrates. MIM is watching developments in several proprietary stope survey instruments which can be located inside the stope and pick up all walls, in some cases automatically. Associated software generates a 3D solid representation. This process could be interfaced with, or integrated into, IMPS. Developments are very encouraging.

5.4 *Plan maintenance*

Systems are now in place to capture all ongoing survey data. However, there are sixty years' worth of existing plans on linen, paper, and transparencies, and many of the old areas are being reactivated to investigate recovery of high grade remnants, quite apart from currently active areas. There are still portions of the mine which have not been metricated since the conversion from imperial measurements in the late 1970s. Most, if not all, of this data needs to be electronically captured.

Three main sets of plans have been used. The surveyors' working plans are at a scale of 1:250. Drafting staff then transfer this information to the survey master plans at 1:500 scale, the base for most planning activities, and to 1:2500 scale plans which are used for small scale location plans covering a large area, and as a base for recording ventilation flows and controls, compressed air, water, and electrical reticulation, and emergency facilities.

Initially, digitising of the 1:500 survey master plans was trialed. Results after checking and editing were acceptable, but the time taken was excessive for capturing the whole mine in a useable form. A PC based scanning system was then acquired. A Contex FSS 3012 scanner produces 400 dots per inch greyscale raster files. These are edited, transformed to remove distortions from the scanning process or from inaccuracies in the original plans, and vectorised into MicroStation design file format. All of this process occurs in 2D files.

All the 1:2500 plans were processed to create an electronic plan base of the whole mine. These are suitable for general use in plan view, but are accurate only to within a couple of metres, and therefore unsuitable for detailed design. Each file was converted to a 3D file with the planar development outlines located at the approximate RL of the level, so that a plan view with a view depth covering the standard range of RLs for a given level in the mine would display the appropriate level plan.

Similar processes were then started to capture more accurate and detailed plan information from the 1:500 scale survey master plans. Once scanned, edited, transformed and vectorised in 2D, further vector editing is done using purchased PC software to consolidate many small vector elements in the file into larger composite elements, thereby reducing the size of the file substantially. These digital survey master files are carefully compared with the original plans, and any adjustments required are made to ensure that these plans are as accurate as the nature of the surveying methods and the original plotting permit. While the 1:2500 plans cover areas of 2.5 km x 1.5 km, the 1:500 plans cover blocks 500m x 300m. Once an electronic survey master plan is passed as accurate by a senior surveyor, the area covered by it is deleted from the less accurate file generated from the 1:2500 plans.

These files are also converted to 3D with the development placed at the average RL of the level represented. An IMPS routine is then used to move the planar representation vertically to the correct elevations and gradients in 3D space in the design file using information from the survey peg database.

The graphic representation of the development is thus accurately located in all three dimensions. It can be used as a base for accurate design procedures, as well as providing useable isometric or other non-orthogonal views, albeit of just two sidewall lines. Busy current mining areas are being processed in whole plan blocks. Portions of the plan block are moved to the correct elevations on an as required basis in less busy areas.

IMPS survey plans of actual development may therefore show up to four classes of development information:

1. Planar information from the 1:2500 scale scanned plans.

2. Planar information from the 1:500 scale scanned plans, accurate in plan view.

3. Information from the 1:500 scale scanned plans,

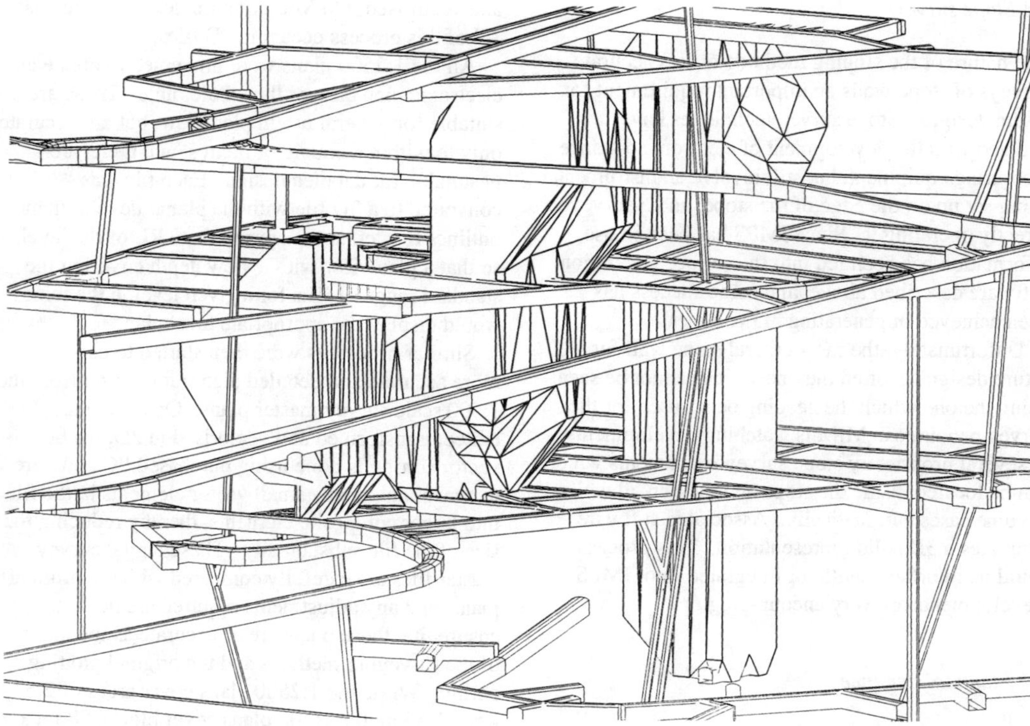

Fig. 4 3D view of stoping and development design

accurate in three dimensions, and

4. Directly entered new survey pickups.

These are distinguished by being shown in different colours on the screen and in colour hardcopy output.

Existing vertical openings are also being loaded into the system. Stopes shown on the scanned and vectorised plans are being joined between levels using the surface generation routines to create 3D volumes. Details of other vertical features such as raises and drill holes have been extracted from the mainframe database, and routines have been written to represent them as 3D graphical objects.

6. MINE PLANNING IN IMPS

Most of the stope and development design in IMPS is done using standard MicroStation CAD functions. Development is represented with a centre line, two sidelines, and a 3D representation generated by projecting a shape along the line of the drive. MicroStation has the facility to place elements on different layers which can be turned on or off as required for viewing purposes. The three components of the development design are placed on different layers to facilitate viewing in 3D and plotting of plan views. Also, different categories of development are placed on different layers to make it easier to view, for instance, the orepasses or main ventilation system without other clutter in the view.

A number of utilities have been requested by users to improve the design process, but these are low priority items, as the standard MicroStation facilities are fully adequate in most situations. All the geology and survey utilities described above are available for planning purposes if relevant. A typical 3D view of development design and associated stopes is shown in Fig. 4. Since the planning engineers have been able to do most of their work using MicroStation alone, the use of IMPS is progressing fastest in this area.

7. OPEN STOPE BLASTHOLE RING DESIGN

In the late 1970s, an attempt was made to computerise the ring design process on the company mainframe. For a variety of reasons, this project failed, not least because of the inadequacies of graphical interface facilities at that time. The project was revived in 1987, this time making use of the graphical facilities then available on PCs, and a system was put into use for the copper mine ring designs.

When work on IMPS was beginning, it became necessary to design rings not catered for by the PC system. It was decided to port it into IMPS and upgrade it to handle the extra requirements. While it would have been possible to do ring designs with MicroStation as a CAD system, it was decided to develop Ring Design as an integrated application so that future links to other parts of IMPS would be facilitated. In this way, the system "knows" that certain lines are holes and others are explosive charges, aiding editing of designs and compilation of statistics. Also, the algorithms automatically place holes and charges rapidly, and satisfactorily in the vast majority of cases. This leaves the designers free to concentrate on the broader principles of the stope blast design, or on the detail of the rare areas which are not handled well by the algorithms.

The system allows the designer to specify a variety of input parameters, such as drill rig type (which is used to place the location of the drilling centre in the correct location in a heading), hole size, and explosive type. The designer places two shapes, one to define the volume for tonnage calculations, the other to define the drilling boundary for the hole placement algorithms. The two are frequently not identical, for example, where holes are planned to stop short of the fill in an adjacent stope, but it is intended that all the ore up to the interface will be broken.

The designer can place holes individually, let the system place holes between other existing holes, or let the system design the ring completely. When the hole placements are satisfactory, the system can then be used to design the charging pattern. A typical lead stope ring is shown in Fig. 5.

Routines have also been written to extract design information for input to a PC blast evaluation program to examine explosive distributions. This is presently used on an occasional project basis, with findings flowing back into the standard blast design parameters. It is envisaged that it will eventually be integrated into IMPS and become a routine design tool, but this is not a high priority.

8. PROJECT CONTROL, COSTS AND BENEFITS

One of the prime reasons for the perceived success of the IMPS project is the high degree of

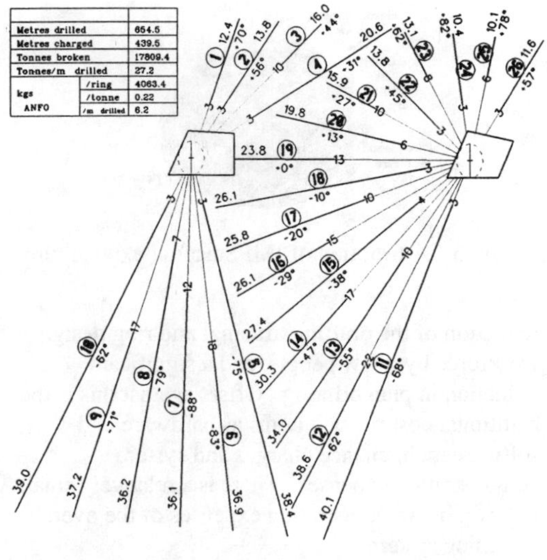

Fig. 5 An IMPS blasthole design

commitment by senior management. A steering committee of senior managers from the various user departments and the Information Services section meets on a regular basis to review progress and decide on major policy issues.

A regular meeting of senior people from the various user sections is convened to reach a consensus on the priorities for development, and to agree on specifications for parts of the system affecting more than one area. This ensures that the analysts continue to build the system in a way which will maximise the value to all users, and avoids conflicting requirements being given to them on an ad hoc basis.

It is significant that the main push for the system has come from the engineers and geologists, rather than the drafting staff. It is seen as a technical design and database system, not just a CAD system. The geologists, engineers, and surveyors are the main users of the system, with the drafting staff filling an important support role.

A major investment of capital has been made, both for purchase of hardware and software, and in the commitment of analysts' time. (Fig. 6) The cost of the analysts is thus a relatively small proportion of the total, and if an off-the-shelf system had been acquired, similar amounts would still have been required for hardware, and probably more for software.

While it is easy to identify the costs of such a system, the benefits are much more difficult to quantify. Some savings are readily seen, such as the

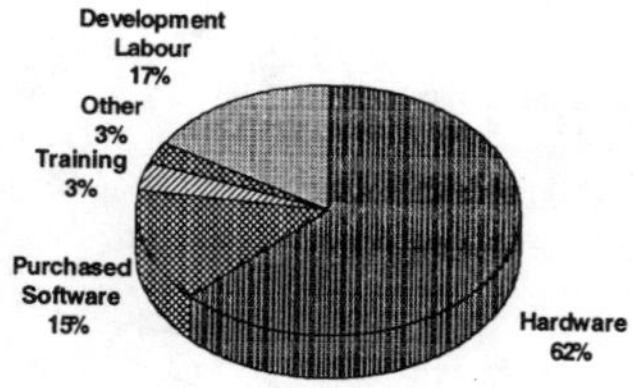

Fig. 6 Distribution of IMPS capital expenditure

reduction of the drafting, tracing, and ring design workforce by a few people, and a significant reduction in plan printing. Offset against this is the additional cost of such items as hardware and software maintenance charges and system administration expenses. There is a relatively small net gain in direct operating expenses of the overall "planning system".

The big gains, however, come in the mine operations through improved designs. These benefits are difficult to determine, and involve assessing the difference between what has been done with IMPS and what might have been done with the manual drafting system. This becomes more difficult still as IMPS procedures become the norm, and the old ways are forgotten. Nevertheless, it is estimated that the capital costs have been recovered five to ten times over. Some of the benefits are immediate "one off" savings, while others are the net present values of ongoing savings from improved designs. IMPS has certainly more than paid for itself already.

9. CONCLUSION

The development process involved in a project like IMPS is not something which can be undertaken by a small mine with a short life, but for an operation with sufficient resources there are significant benefits to be gained. A system based on industry standard CAD and database software can keep pace with the rapid developments in the computer industry, and is not subject to the fluctuating fortunes of the mining industry for significant enhancements.

The development of IMPS is by no means complete, and there are significant features of systems available in the marketplace which have yet to be developed in IMPS. The development has had its problems. There are areas where the users are still justifiably dissatisfied. Communications with users of other systems indicate that similar comments apply to those systems also. Overall, MIM users are convinced that the correct approach has been taken for local circumstances, and IMPS is perceived to be a success.

The key ingredients for this success are:

1. A strong commitment by senior management.

2. The perception that the project is developing a technical design and database system, not "just" a CAD system.

3. Highly competent systems analysts and programmers building a robust system meeting the users' needs, and

4. "Champions" with vision to see what the system can be, pushing it to the limits of what is now possible rather than complaining about what is not yet possible, and producing the results which maintain the strong management commitment.

ACKNOWLEDGMENTS

The permission of Mount Isa Mines to publish this paper is acknowledged. The author expresses his gratitude to those who have helped with illustrations for the paper, and to those who read early drafts and made constructive comments and suggestions. Thanks are due to all the people who have contributed to the success of the project: the management and staff of Intergraph Corporation, especially the Queensland office; the analysts and programmers who have built the system at Mount Isa; all the users at the Isa and Hilton mines, who have supported the project and come up with the ideas; and especially the "champions" who have kept it all going when things looked bleak.

REFERENCE

Bower, J.V., Edwards-Davis, P.C., Hall, B.E. & Horsley, T.P. 1991. Development of an integrated mine planning system at Mount Isa Mines. *Proceedings of the Second Australian Conference on Computer Applications in the Mineral Industry:* 78-85. Sydney: University of Wollongong.

Innovative Mine Design for the 21st Century, Bawden & Archibald (eds) © 1993 Balkema, Rotterdam, ISBN 90 5410 325 6

Double post mining (DPM)

C.M.Gryba
Moneta Porcupine Mines Ltd, Timmins, Ont., Canada

ABSTRACT: Double Post Mining (DPM) is a new undercut and fill mining method. The innovative use of pre-posting and double posting combined with cemented backfill creates a very safe, standardized mining environment suitable for large scale conventional or automated mining equipment. In larger orebodies DPM approaches a continuous mining system as the cyclical interference of each mining function is effectively eliminated. As a result DPM provides the ideal platform for a wide range of productivity improvements, including introduction of automated equipment or even alternate methods of breaking rock. The totally engineered system using concrete posts and floors provides a selective stoping environment independent of ground conditions. The system intrinsically optimizes geological control and facilitates computerized mine planning. As a result the effects of dilution are minimized and ground control problems become less of an issue. The author will describe DPM and compare its economics and productivity to VCR and VBM bulk mining methods using cemented rock fill.

1 BACKGROUND

The Canadian hard rock mining industry has, over the past ten years, spent a great deal of time and effort automating individual pieces of mining equipment. Concentrated R & D programs were targeted to achieve a ten fold increase in mining productivity within a ten year period. The primary aim has been to lower mining costs by improving the productivity of bulk mining methods such as the VCR or VBM methods developed by Inco. In parallel, major blocks of research funds have been directed at trying to eliminate the cyclical nature of drill and blast methods by evaluating other methods of breaking rock. Both areas of study have more or less assumed that current drill-blast mining methods could not be substantially improved.

Transporting of broken ore from the stoping blocks through to surface has been successfully automated using PLC controls. Automation of actual breaking of intact ore has been disappointing. Total efforts to date have resulted in overall mine productivity of 15 to 35 tonnes per manshift for individual mines.

Increasing the stope height of large hole blasthole stopes has the best potential for improving productivity and reducing costs but at the risk of incurring additional dilution. Noranda, for example, estimates the cost of waste rock dilution to be $25.00 per tonne (Whiteway, 1991).

In many instances bulk mining methods can not be implemented due to geometry of individual orebodies, weak ore, and/or incompetent wall rocks.

The author has reviewed all the factors that influence underground mine profitability. For large, competent orebodies, a Raise Bore Mining (RBM) method has been developed whereby all drill-blast functions are performed from a lined raise bore hole. Stope height can be increased to 200 meters or more. Most of the individual stoping functions can be automated with current technology. Studies indicate that 100 to 200 tonnes per manshift productivity for all underground work is possible. To minimize dilution a method of rebarring stope walls has been developed. Primary patents have already been granted by Canada and Australia and the patent process is continuing.

The much more difficult problem is to improve mining productivity from orebodies that are not ideal. What is required is a selective stoping system that is adaptable to a broad range of orebodies and ground conditions. This new method should also provide a framework for ongoing innovation, automation, improvements in safety and overall mine profitability.

2 INNOVATION OVERVIEW

Double Post Mining (DPM) is a new breakthrough in mining method technology. DPM is an undercut and fill mining method whereby a continuous concrete floor slab is pre-posted on a regular grid pattern prior to extracting the next lower lift. A double posting system has been developed which, in conjunction with the flexible floor slab, supports the cemented backfill of the previously mined lifts. See Figure 1.

DPM may be likened to a room and pillar mining operation performed under a concrete roof. Pre-inserted concrete posts provide the initial support and act as temporary pillars. Tight filled mined out rooms become the equivalent of permanent pillars left in a room and pillar operation. The advantage of DPM is that 100% of the ore is extracted in a single pass at room and pillar productivity rates.

If five meter high posts are installed on a 8 meter by 8 meter grid, each "room" between four posts would contain 800 to 1200 tonnes depending on the specific gravity of the ore.

Initiating DPM requires top slicing through the

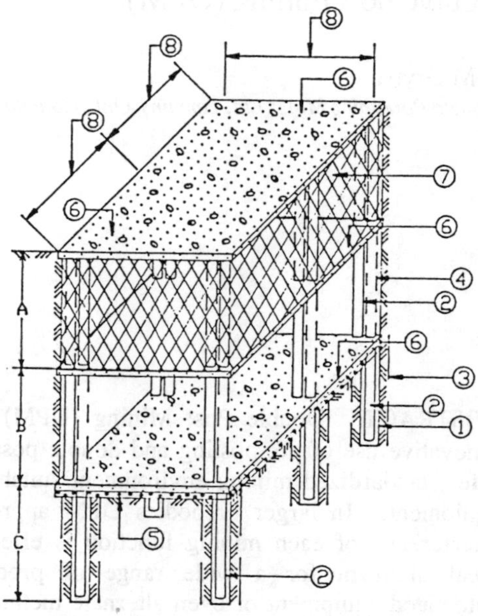

Fig.1 Isometric drawing DPM

orebody on regular intervals to allow drilling of post holes, installation of concrete posts on a grid pattern and setting up the concrete floor slab. Concrete posts are inserted into 4.7 meter deep post holes prior to pouring a concrete floor slab. The remainder of the drift above the floor slab is tight filled. Figure 2 shows DPM top slice development which is basically a standard drift and fill operation.

Post holes are drilled with a suitably weighted 12 inch diameter DTH drill hammer. A 6,000 lb weight provides the proper loading on a .56 meter (22 inch) diameter bit. Ingersoll Rand drill tests indicate a drill rate of seven feet per hour is possible with 175 psi mine air. On completion of the drilling, a .45 meter (18 inch) diameter by 5 meter long concrete post is inserted into the hole. Next, six to eight helper holes, 50 mm (2 inch) diameter x 5 meters deep, are drilled and blasted to prebreak the immediate area around the post (approximately 2 meters in diameter).

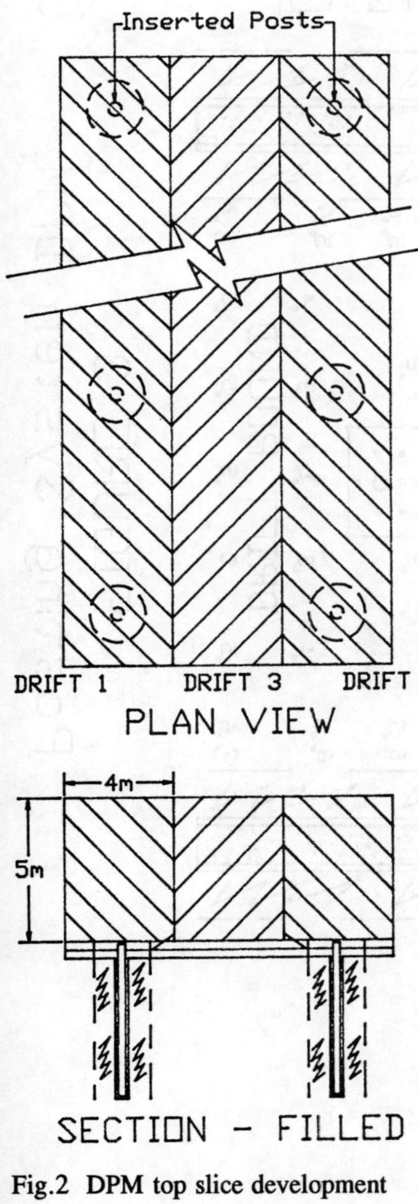

Fig.2 DPM top slice development

The above posting process would work for the first lift, but when one tries to post for the next lower lift, it becomes obvious that it is impossible to drill directly below the previous post. Hence, a double posting procedure has been developed.

Figure 3 illustrates the posting system, a backfilled room, and drift round drilled off under the concrete floor. In general, post 1 is the post inserted from the previous lift. Post hole 2 is drilled with an offset of approximately six inches. Post 2 is then inserted. After pre-breaking around post 2, the preparation for pouring the concrete floor is completed. Post 3 is then stood. Post 3 supports the bottom of post 1 from two lifts above. Pouring the concrete floor then ties together posts 1, 2 and 3. Thus by alternating a double post system, a very safe continuous support system is created.

The concrete floor slab is reinforced with continuous rebar and screen and is cushioned from blast damage below by 200 to 300 mm of broken ore. Mined out rooms are tight filled with cemented rock fill. Tight filling of cemented rock fill is achieved using a portable slinger unit.

Continuous alignment of the double posts above and below the concrete slab eliminates the problem of posts punching through the floors. The concrete floor is designed as a flexible member that ties the posts together and is a safety roof to prevent "cemented rock fill loose" from becoming a problem when mining the next lower lift. Similarly safety is improved if dynamic loading of the fill occurs due to large seismic events.

A section of concrete floor is removed (or not poured) where internal ramps are required to access the next lower lift. Temporary and permanent ramps can both be used to maximize mining flexibility of any particular lift of rooms. See Figure 4 for a ramp layout.

A major benefit of DPM is its flexibility. The engineered posting system means multiple rooms can be left open and multiple faces exposed. Individual drill-blast-muck operations can be

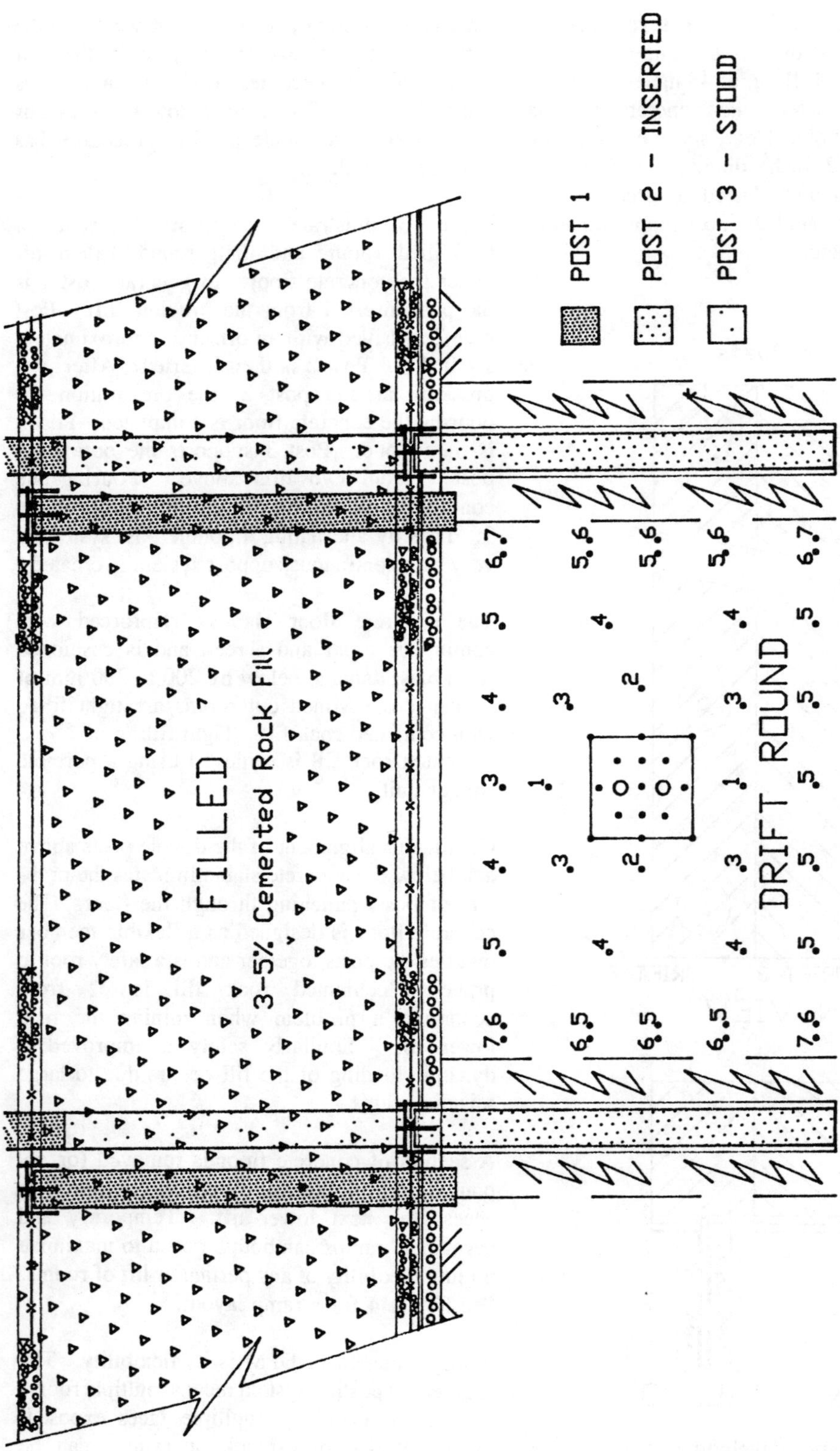

FIGURE 3
Posting System DPM

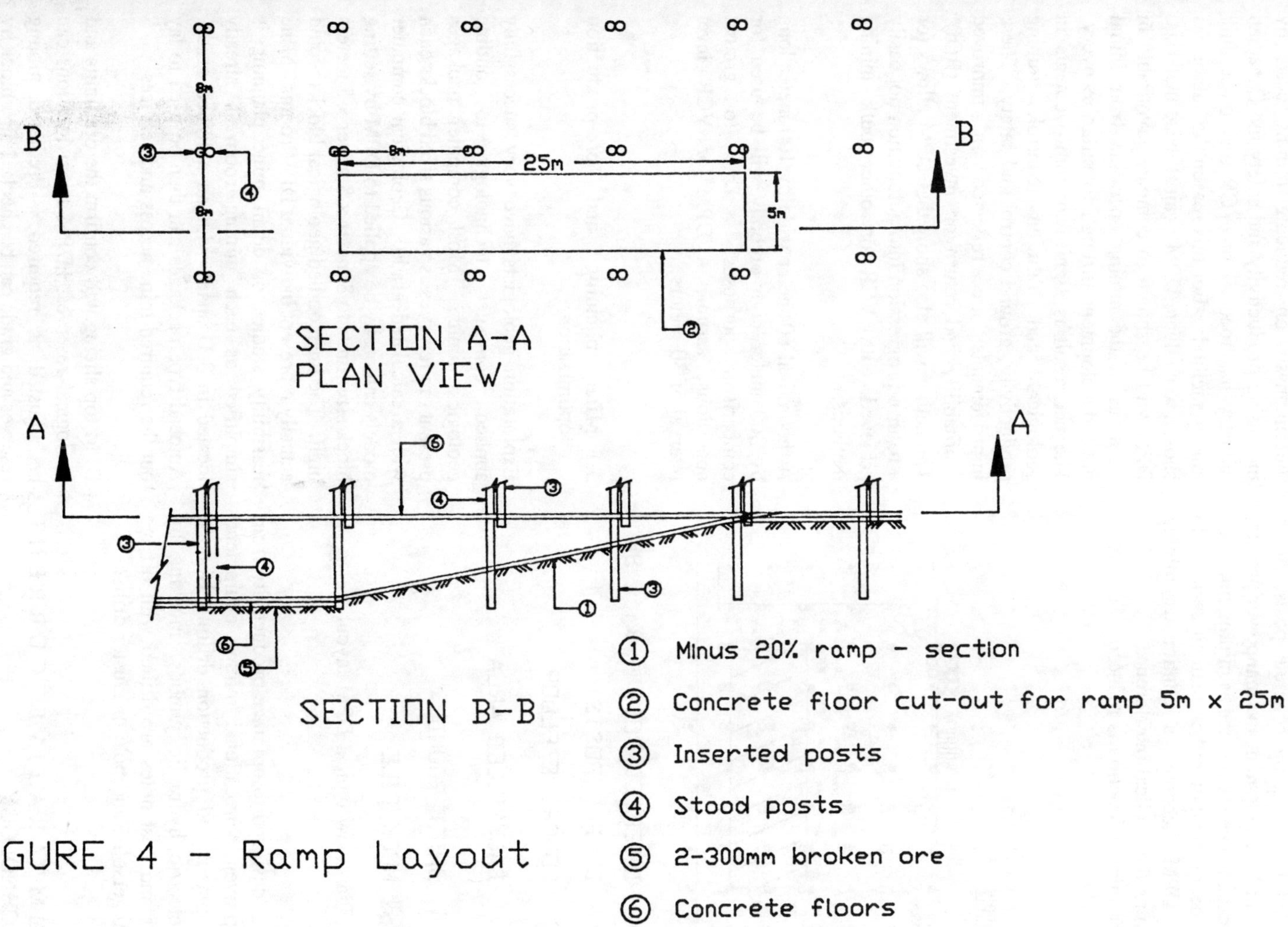

FIGURE 4 - Ramp Layout

separated from placing concrete posts, floors and backfill. Equipment and manpower productivity can then be maximized as mining functions are done in parallel rather than in series. In effect DPM becomes a continuous mining system as the cyclical interference of each mining function is reduced. See Fig 5

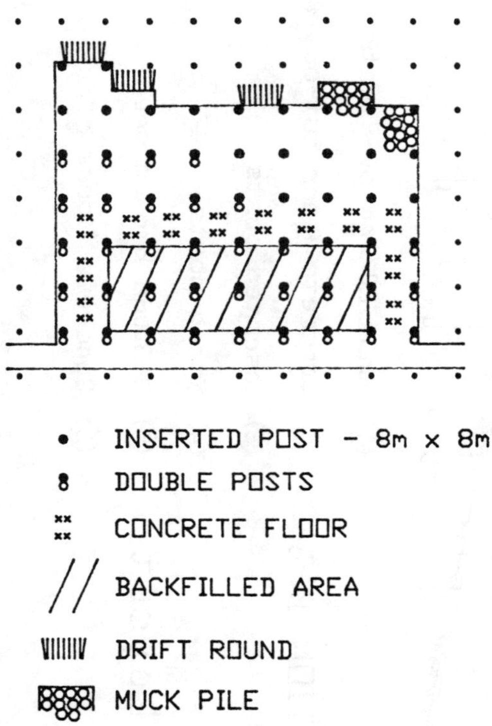

Fig.5 Plan view typical DPM layout

DPM is designed to optimize productivity from well proven conventional mining equipment. Development and production drilling is done with electric hydraulic jumbos, mucking by LHD's and post holes are drilled with off the shelf Ingersoll Rand pole or cluster drills.

3 COMPARATIVE CURRENT TECHNOLOGY

Depending on the individual characteristics of an orebody DPM will be competitive with a wide range of current mining methods. Because of the emphasis on reducing mining costs by increasing productivity and safety the Canadian industry has focused on VCR and other bulk mining methods. For this reason the author has chosen to evaluate DPM against these methods. DPM will be shown to compare favourably in all areas of the mining operation from initial mine development through to pillar recovery. The net result is significant improvements in geological and dilution control, mining productivity, ground control and safety. These improvements are reflected in improved profitability and return on investment (ROI). The author will also show that DPM allows for a natural progression into automation not easily achieved with VCR or other bulk mining methods.

A theoretical 60 meter wide by 100 meter long by 300 meter high orebody will be used for comparison purposes. Due to ground conditions, assume a DPM or VCR level interval of 30 meters.

3.1 Mine planning and pre-production optimization

Exploration geologists have very successfully applied computer technology to outline geological limits. Most computer packages present ore reserves as various sized blocks each with a calculated grade. Geological computer models can easily be applied to DPM by setting block size limits to 8 meter x 8 meter x 5 meters high. The unsmoothed geological blocks would actually represent individual DPM rooms. Mine feasibility studies or detail mine planning is simplified as each mining room is already located in 3-D space with an assigned grade. Autocad type packages can then be used to lay out the required mine access and services.

DPM top slicing will confirm the ore limits and geological reserves. DPM room locations can be adjusted as required. Predicted mining reserves and grade can be checked on a block by block basis once individual blocks have been mined. This degree of geological correlation is impossible with VCR or VBM mining and engineering costs are reduced.

A very necessary step required for blasthole mine planning of any orebody is detail definition drilling prior to laying out the VCR or VBM stopes. Blasthole stopes, 30 meters high by 15 meters wide by 30 meters long would each contain about 50,000 tonnes. Each 50,000 tonne stope has to be carefully located to minimize wall dilution. Stoping has to be sequenced properly to optimize ore production. A comprehensive pre-production drill program can add six months or more to the mine development time. Ongoing detail diamond drilling is a major expense.

Much less definition drilling of the orebody is required with DPM as each room has reserves of about 1,000 tonnes. Each room is mined usually by two drift rounds. The hanging wall or footwall location of the orebody then becomes non-critical. If a particular ore contact moves 5 or 10 meters, one just adds or deletes a series of rooms. Narrow widths, off-shoots etc. can be followed by drift and fill methods at appropriate elevations. Internal waste, faults, shear zones etc. can also be easily handled by DPM whereas they present major problems for blasthole stopes. Initial DPM top slicing is equivalent to an exploration development program except that it is a production cycle function rather than a time delaying exploration program.

Thus compared to blast hole mine planning, DPM requires much less definition drilling and a shorter production lead time. Reduced preproduction time, exploration capital and interest charges are reflected in improved ROI calculations.

3.2 Grade control and dilution

DPM proceeds from the top down in 5 meter lifts. Fill dilution is eliminated. Wall dilution can also be reduced to nearly zero. Face assaying and mapping on a round by round basis can separate ore-waste contacts. Waste rounds (or sub ore grade rounds) can be used for backfill thus even internal dilution can be minimized. There is adequate time for face sampling and mapping as there are virtually unlimited faces available for drilling. Assay turn around time is not as critical as for single heading advancing faces.

DPM requires drilling .56 meter diameter x 4.7 meter deep holes in the floor to insert posts for the next mining lift. By collecting and assaying drill cuttings, the grade of the next elevation of geological blocks (or mining rooms) is checked one to three months in advance of mining. Quarterly grade predictions will be accurate and production planning can be optimized.

Grade control in VCR stopes is limited. Once the block is developed and drilled off, the final stope grade is the grade of the block plus dilution from wall rock or adjacent backfill. Even well cemented rock backfill can add 3-5% dilution. Ongoing grade control in blasthole stopes is limited to assaying muck samples.

Meaningful grade control is much more practical and achievable with DPM than by bulk methods. Dilution has a significant negative impact on mine-mill profitability (Bawdin, 1989).

3.3 Mine development

DPM requires access to the orebody at each top slice elevation. From that point on all development is within the ore limits.

DPM can be likened to excavation of an underground parking garage from the top down except that the spaces between the posts are filled with cemented rock, sand or paste fill. By not filling the main travelways, a permanent ramp and access drift system is left behind "caste" into the fill. The walls of the opening could be cemented rock fill and floors and roofs 200 to 300 mm thick concrete slabs. The floor and ramp surfaces could be further reinforced if they are to be part of a main production truck haulage system. See Figure 4 (Ramp drawing).

The ability to keep production openings in the mined out area is a major cost saving as it minimizes preproduction and ongoing waste development. In areas of poor ground or rock-burst prone ground, a production ramp within the cemented rock fill could result in substantial ongoing maintenance savings as well. Ventilation - mine escapeway openings and other services can also be caste into the backfill as required. DPM allows zone ventilation. There could be refrigeration/power savings in hot or deep mines when compared to the cooling of individual headings or drawpoints.

VCR mining requires a great deal of preproduction development as the tops and bottoms of each stope must be developed. Permanent production drifts and drawpoints for blast hole stopes are excavated in waste most of which cannot be used as backfill because of timing problems. To obtain an efficient stoping and fill sequence, stopes have to be mined on strike in a 1,5,9 sequence. Thus preproduction development is also extensive in a lateral sense.

DPM represents a significant reduction in capital when compared to VCR or blast hole stoping. Most stope development is within the ore limits and only a minimum amount of preproduction development is required. In most cases only one access drift is required to the top of each DPM stoping block with a single ramp connecting main levels for moving equipment. All other waste development is eliminated.

3.4 Stope production rates

Detailed stope planning plus production realities at Kidd Creek Mine indicate that it takes an average of six to seven months to develop, mine and backfill a 50,000 tonne blasthole stope. This results in an average production rate of 350 tonnes per day over the complete stoping cycle.

DPM productivity is a combination of top slicing and mining horizontal lifts of rooms under posted concrete floors. If stope blocks are 30 meters high, there would be one top slice per five sets of rooms. Similarly stope blocks 50 meters high would have one top slice per nine DPM lifts. Top slice productivity is dependent on the ground conditions.

Assuming Kidd Creek ground conditions, detailed scheduling of DPM top slicing indicates an average production rate of 350 tonnes per day.

Fig 5 shows approximately the area top sliced or the number of rooms that have to be mined and backfilled to be equivalent to a blasthole stope. Because DPM rooms are standard , mine estimating is simplified as the details for one room will apply to all other rooms. The time to mine and back fill a standard room is shown in Table 1. DPM rooms in a massive sulphide orebody would each contain about 1200 tonnes.

Analysis indicates that five manshifts are required to mine a DPM room and seven manshifts to complete the backfill cycle (see Table 1). Mining and filling a 1200 tonne room in 12 shifts results in a productivity of 100 tonnes per manshift. If production personnel are 50-60% of the mine payroll then overall mine production can be in the range of 50 to 60 tonnes per manshift.

Detailed time scheduling of the mining sequence shown in Fig 5 indicates that it would take 25 days (50 shifts) to mine 50 rooms. Backfilling proceeds in parallel but lags the breaking cycle by about 20 rooms. It would take an additional 20 days or so to complete those rooms and perhaps five days to ramp down to the next lift. The average production rate would be 50 rooms x 1200 tonnes/room divided by 50 days equals 1200 tonnes/day.

One top slice plus five mining lifts would result in an average production rate of 1050 tonnes per day for an orebody volume equivalent to six blasthole stopes.

DPM has the potential to triple the average production rate of blasthole stopes.

3.5 Ground Control

DPM eliminates most ground control problems by providing an engineered environment. Mining proceeds downward under a concrete roof. Temporary ground support is limited to initial top slice development and perhaps some peripheral wall control in the lower lifts. The

Table 1. DPM room productivity

	EQUIPMENT	PRODUCTION ESTIMATES	MANSHIFTS PER ROOM
MINING CYCLE			
1) Drilling	2 boom electric/ hydraulic Jumbo	2 rounds/shift	2
2) Loading	Powder vehicle	2 rounds/shift	2
3) Mucking	6 1/2 yard scooptram	100 buckets/ shift	1
BACKFILL CYCLE			
4) Drilling post holes	Ingersoll Rand pole drill	3 holes/shift	1
5) Helper holes- drill and blast	Jumbo - extension steel	2 sets/shift	1
6) Inserting and standing posts	Service vehicle with crane	6 sets/shift	1
7) Rebar, screen and plastic	Service vehicle with crane	3 rooms/shift 3 man crew	1
8) Pouring concrete floors	LHD or truck	2 rooms/shift	1
9) Backfilling	LHD or truck plus slinger	2 shifts/room	2
TOTAL			12 manshifts

cost of concrete posts, postholes and concrete floors is offset by significant savings in conventional ground control costs. The uncertainty factors of scheduling ground control is eliminated. The framework for continuous mining is created at no additional cost.

One of the perceived advantages of VCR mining is safety, as for most of the production cycle, the miners remain out of the stope. In reality each VCR stope requires a top slice and undercut both excavated to final stope limits. Stopes are typically 10 to 20 meters wide by 30 to 40 meters long. The back of the stope must remain stable for at least six months while production blasting, ore extraction and backfill functions are completed.

The back must survive the vibration effects of five to ten production blasts. Because of their close proximity to production blasting, the access openings are subject to extensive blast vibrations in addition to vagaries of rock strengths, redistribution of ground stresses, geological structures, and in deep mining, rockbursts. Falls of ground continue.

In comparison, DPM requires only 50% of the conventional development required by VCR stoping because the orebody has only to be top sliced. Individual top slice drifts have to remain stable for only two to three weeks while the initial posts and concrete floors are installed and the drifts tight filled. Thus the conventional ground control requirements of DPM (loose detection, scaling, rock bolting) is much less onerous. Cable bolting can, in most cases, be totally eliminated.

DPM has major advantages in mining bursting

ground. The stope walls and floor can be destressed by proven Inco techniques (Hedley,1992). The combination of stiff cemented fill plus double posts tied together by flexible concrete floor slabs, internal ramps, and de-stressing potentially makes DPM one of the safest mining methods possible in highly stressed ground. In addition numerical modelling of ground stresses is optimized as DPM rooms are already in a block model format.

Basically, if the orebody is competent enough to drift though, DPM can be implemented. Weak walls, faults, multiple joint sets, clay alteration, etc. will not affect the DPM cycle or increase dilution. The same cannot be said for bulk mining methods.

4 SAFETY AND TRAINING

DPM is a very safe mining method. Outside of the first lift, which is mainly conventional mine development, all subsequent mining is done under pre-supported concrete floors.

The intrinsic advantage of DPM is that mining is standardized and a safe mining environment is created independent of ground conditions. Each DPM room is exactly the same. Miners, supervisors, safety and labour inspectors can easily check to see that each job is being done correctly. That is not to say DPM would become boring. Each mining lift or orebody will be different. It will be much like playing chess, the chess board remains the same but each chess game will be different.

An added advantage of DPM is that for new mines, a test or training mine can be set up on surface. A new work force can be trained prior to even having to access to the underground orebody. It is very difficult to model VCR or any other bulk stoping method on surface.

5 SUITABILITY FOR AUTOMATION

Automation of production stopes to date has been limited to automating the individual pieces of equipment required for current bulk mining methods. Total automation of a mining method such as VCR is virtually impossible. For example, repeated loading and blasting of VCR blastholes can create "candling" of drill hole collars. An uneven, floor defeats most types of automated equipment.

DPM may provide the framework to solve many production automation problems.

The drill, blast, muck functions are all totally standardized. The backfill cycle is also mechanized virtually eliminating manual work. Even functions such as mucking out lifters and levelling concrete floors can be mechanized by using small backhoes. Hiab type cranes mounted on service vehicles can be used to handle concrete posts and floor reinforcing steel.

DPM rooms provide a totally engineered environment. Posts are installed on a grid pattern which matches the unsmoothed ore reserve blocks. Every computerized DPM room has the same dimensions, is located in 3-D space and is properly sized to allow efficient operation of automated equipment.

For example, with a 5 meter x 8 meter x 8 meter room mining can proceed in any direction under the concrete floor. Two or three boom automated drill jumbos have an ideal production environment. Every drill round is exactly the same, there is no scaling or loose problems and rock bolting is eliminated. Perhaps just as important, a number of faces are available for drilling.

Similarly, 6 1/2 yard or larger scooptrams can muck remotely with current technology. The flat concrete roof plus concrete posts spaced at regular intervals facilitates the installation of remote control and feedback systems. Compare the ease of DPM remote control mucking versus remote mucking around a 50 tonne chunk of oversize, located one third of the way into a VCR blasthole stope.

One of the most difficult functions to automate is ground control. Loose detection, scaling, rock bolting, cable bolting, etc. are all too difficult to totally automate. DPM either eliminates or greatly reduces the need for ground control and therefore automation of this function is not an issue.

6 SUMMARY

Open pit mining productivity has continuously improved over the past twenty years whereas underground mining costs continue to increase. VCR is the most recent development in major underground mining methods. That technology is now twenty years old. Millions of dollars have been and will continue to be spent on trying to rectify the problems associated with big hole, bulk mining techniques. The results have been disappointing and mining costs continue to increase. Perhaps, the mining and research industry should investigate new approaches to mining rather than focusing on VCR type methods.

Preliminary engineering and scheduling indicates that DPM can at least double the productivity attained in most Canadian blast hole mines. The safety of mining under concrete will certainly match any bulk mining method. The standardization of DPM rooms makes the introduction of automated mining equipment much easier and more practical.

Alternate methods of breaking rock will still require ground control. Breaking of the actual ore within the DPM post system could be done with mechanized road headers, plasma blasting or even radial axial splitters if there is a cost advantage over conventional explosives.

DPM production and backfill costs on a per tonne basis will be very similar to VCR stoping using cemented rock fill. However stoping costs are only one component of mine economics. A savings of one or two dollars per tonne on mining costs can easily be offset by incurring extra dilution. All of the factors that go into ROI calculations, including development and capital costs and interest expense associated preproduction have to be included. When all factors are considered DPM results in a substantially better ROI that VCR type methods. The DPM framework may be the required platform for the next major breakthrough in mine safety and productivity. DPM patents have been applied for in Canada and patenting will be extended to all major mining countries.

REFERENCES

Bawdin, J.W., J. Nantel and D.Sprott June 1989. Practical rock engineering in the optimization of stope dimensions - application and cost effectiveness. CIM Bulletin, volume 82, no.926.

Hedley, D. 1992. Rockburst Handbook for Ontario Hardrock Mines.

Whiteway, P. April 1991. Made-in-Canada Mining Technology. Canadian Mining Journal.

Innovative Mine Design for the 21st Century, Bawden & Archibald (eds) © 1993 Balkema, Rotterdam, ISBN 90 5410 325 6

The nutcase cable bolt

A.J. Hyett & W.F. Bawden
Department of Mining Engineering, Queen's University, Kingston, Ont., Canada
R. Powers & P. Rocque
Hemlo Gold Mines Inc., Golden Giant Mine, Marathon, Ont., Canada

ABSTRACT: Both laboratory and *in situ* cable pull tests have demonstrated a significant improvement in cable bond strength using nutcase cable compared to conventional 7-wire strand. The advantage is most pronounced in poor quality, destressed or failed rock masses, namely under ground conditions for which effective cable bolt support is often critical. Furthermore, the bond strength of nutcase cable is less sensitive to the grout *water:cement* ratio, thus relaxing the installation quality control requirements. In cases where the cable-grout bond strength was sufficient, tensile rupture of the cable occurred after 25-40*mm* of slip at the cable-grout interface, and consequently, the nutcase cable bolt may still be considered as a *yielding* support system especially in poor ground.

Optimal performance requires a maximum nut size of less than 1/2" or 12.7*mm* (across the flats) located periodically at 12-16" (300-400*mm*) intervals along the cable. A cement paste in the range 0.35-0.40 *water:cement* ratio is recommended for minesite installation.

1. INTRODUCTION

Recent laboratory tests conducted in a modified Hoek cell (MacSporran, 1993) indicated that very low radial dilations (in the order of 10-100μm) are responsible for the mobilization of frictional bond strength for standard 7-wire strand making the ultimate capacity highly sensitive to:

(*i*) low quality grout (high water:cement (*w:c*) ratios);
(*ii*) low modulus, poor quality, destressed, or failed rock masses; and,
(*iii*) poor operator installation quality control (*e.g.* empty breather tubes).

One potential solution to these inadequacies is to induce much larger radial dilations, and thus generate higher ultimate capacities, through the implementation of cables with modified geometry.

At Minnova's Winston Lake Division, in the summer of 1991, a modified cable geometry was the object of considerable interest amongst the engineering staff. It comprised a hexagonal nut set over the central of a standard 7-wire strand, and interwound within the six peripheral wires so that each rested on a flat. The consensus of opinion was that this cable, by virtue of its geometry would significantly increase the bond strength at the cable-grout interface, which was the limiting factor for 7-wire strand cable bolts that were presently being used. Even so, skepticism was expressed, primarily concerning the lack of test data for the new modified cable geometry, but also relating to several additional uncertainties; most specifically:

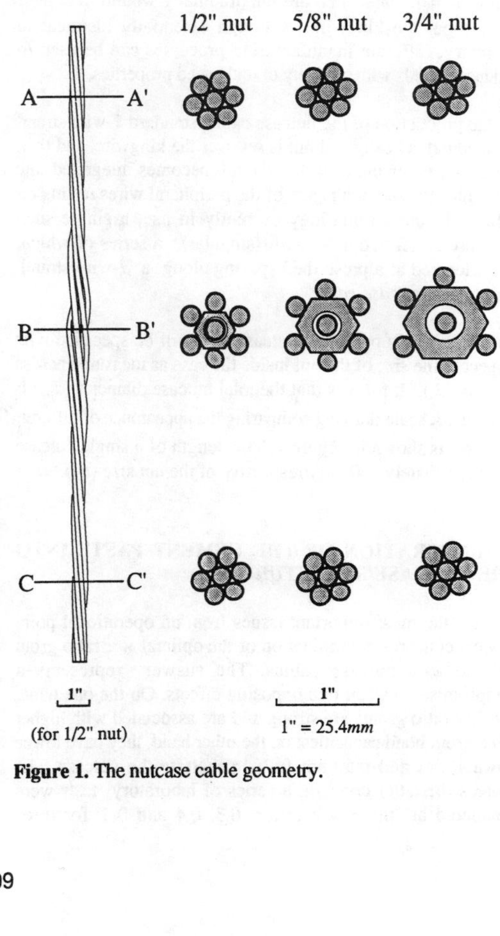

Figure 1. The nutcase cable geometry.

(*i*) what *w:c* ratio should be used in conjunction with the cable and could penetration of the cement paste into the nutcase structure be guaranteed?
(*ii*) would the cable overcome the quality control problems associated with mixing and pumping cement grouts?
(*iii*) what was the smallest nut size that could be used for the nutcase and would this be small enough to enable installation with standard diameter (2.125"or 53.9*mm*) cable bolt holes?
(*iv*) how well would this cable work in poor quality ground, and would it be too stiff for good ground?

Appreciating its potential, and due to the enthusiasm expressed, it was decided to include a limited number of tests as part of an ongoing test programme being conducted by Queen's University at Hemlo Golden Giant Mine. The results, although limited in number, indicated that the new modified cable geometry (by this time being referred to as "*nutcase* " cable) achieved higher strengths than either standard 7-wire strand cable or birdcage cable (see Section 5.4). Subsequent field and laboratory test programmes, which will also be described in this paper, have confirmed this initial potential, so that the nutcase cable is now regarded as close to the optimal cable bolt option in poor quality ground.

2.GENERALITIES

Standard 7-wire strand consists of one straight, central *kingwire* ($d_k = 5.41mm$) and six slightly smaller *peripheral* wires ($d_p = 5.16mm$) that are longitudinally wound in helical form about the kingwire. Although essentially identical in geometry, different manufacturing processes can be used to produce strands with a variety of enhanced properties.

In the production of the nutcase cable, standard 7-wire strand is unwound, a hexagonal nut is set over the kingwire, and then the strand is rewound so that the nut becomes integrated into the cable structure with each of the peripheral wires resting on a flat. In the terminology currently in use, a single such structure is termed a *nutcase* (singular), a series of which, when located at a prescribed spacing along a 7-wire strand, comprise a *nutcase cable*.

Throughout this paper the nutcase size will be specified with respect to the size of the nut inside the case as measured across its flats (d_n). It follows that the total nutcase diameter (d_{nc}) is $d_n + 2d_p$. A scale drawing comparing the appearance of different nut sizes is shown in Figure 1. The length of a single nutcase is approximately 150*mm* irrespective of the nut size (d_n).

3. PENETRATION OF THE CEMENT PASTE INTO THE NUTCASE STRUCTURE.

One of the most important issues from an operational point of view concerns determination of the optimal *w:c* ratio grout for use with nutcase cables. The answer represents a compromise between two opposing effects. On the one hand, low *w:c* ratio grouts are strong and are associated with higher cable-grout bond capacities: on the other hand, they have lower flowabilities and may not fully penetrate the nutcase. To address the latter concern, a series of laboratory tests were conducted at three *w:c* ratios, 0.3, 0.4 and 0.5, for three

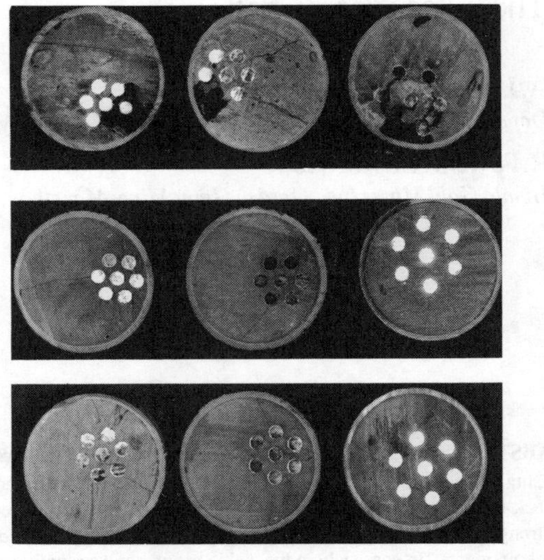

Figure 2. Results of the grout penetration tests. 1/2" (12.7*mm*) to the left; 0.3 *w:c* ratio at the top.

Table 1: Results of the grout penetration tests.

	Nut Size		
w:c ratio	1/2"	5/8"	3/4"
0.3	N	N	N
0.4	Y	Y	Y
0.5	Y	Y	Y

different nutcase sizes, 1/2"(12.7*mm*), 5/8"(15.8*mm*), and 3/4"(19.1*mm*) (*i.e.* nine tests in total) to establish whether different nutcase sizes, 1/2"(12.7*mm*), 5/8"(15.8 *mm*), and 3/4" (19.1*mm*) (*i.e.* nine tests in total) to establish whether different cement pastes would penetrate into the nutcase structure. The samples were grouted, left to cure, sectioned and then photographed (see Figure 2 and Table 1 for results).

For 0.4 and 0.5 *w:c* ratio pastes full penetration was achieved for all three of the nutcase sizes. However, only partial penetration was possible for a 0.3 *w:c* ratio, unless the cable was vigorously agitated (impossible in practice) so inducing liquification and improved flowability. In view of these results, and because of the limited number of tests to be conducted, the 0.3 *w:c* ratio tests were omitted from the pull test programme.

4. LABORATORY PULL TESTS

An outline of the laboratory test programme is shown in Table 2. The experimental setup is shown in Figure 3. The intention was to investigate three key factors:
(*i*) the effect of *w:c* ratio of the grout (see Table 3 for details);

Table 2. The laboratory test programme.

Test #	*w:c*	Confinement	Nut Size	L_e
1 (A-D)	0.5	Sch 80 - Alum	5/8" (15.8*mm*)	300*mm*
2 (A-D)	0.5	Sch 80 - Steel	5/8" (15.8*mm*)	300*mm*
3 (A-D)	0.4	Sch 80 -Alum	1/2" (12.7*mm*)	300*mm*
4 (A-D)	0.4	Sch 80 - Steel	5/8" (15.8*mm*)	300*mm*
5 (A-D)	0.4	Sch 80 - Alum	5/8" (15.8*mm*)	300*mm*
6 (A-D)	0.4	Sch 40 - Alum	5/8" (15.8*mm*)	300*mm*
7 (A-D)	0.4	Sch 80 -Alum	3/4" (19.1*mm*)	300*mm*

(*ii*) the effect of radial confinement (see Table 4 for details); and,
(*iii*) the effect of nut (or nutcase) size.

All laboratory tests were conducted with a 300*mm* embedment length; the nut being located at the midpoint (*Note:* 250*mm* embedment length was used for the conventional 7-wire strand pull tests). Four tests were conducted for each combination of parameters(A-D). The selected standard *w:c* ratio grout was 0.4 (*i.e.* that offering the optimum comprise between strength and flowability); the standard radial confinement was a Sch. 80 aluminum pipe (*i.e.* that for which the performance of nutcase cable relative to standard cable was expected to be most impressive); and the standard nut size was 5/8"(15.9*mm*), with tests also conducted at 1/2" (12.7*mm*) and 3/4"(19.1*mm*).

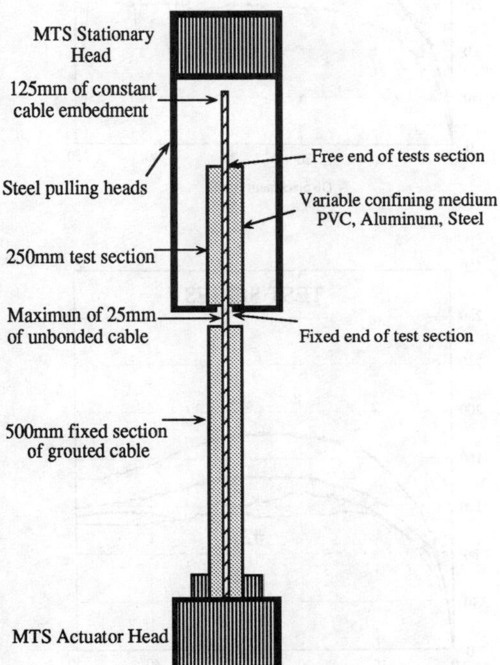

Figure 3. The laboratory pull test setup (after Reichert et al., 1992).

Table 3. Physical properties of the cement paste.

w:c ratio	ρ_d (28 day) (*g/cm3*)	UCS (*MPa*)	*E* (GPa)	ν
0.38 - 0.40	1.95 - 1.97	58 - 63	9.2 -10.1	0.18
0.47- 0.51	1.85 - 1.81	39 - 44	11.8 12.6	0.18

Table 4 . Specifications for the confining pipes.

Pipe Spec.	I.D. (*mm*)	O.D. (*mm*)	Kr (*MPa/mm*)	Equiv Rock Mass Modulus (*GPa*)
Sch. 80 - *Steel*	49.30	60.13	1604.2	49.7
Sch. 80 - *Alum*	49.02	60.45	599.5	17.7
Sch 40 - *Alum*	52.22	60.38	332.5	10.3

4.1 Laboratory Pull Test Results

The full load-displacement plots are presented in Figure 4, with comparable results for a standard 7-wire strand shown dashed. Appropriate test parameters for each test series (1-7) correspond to those listed in Table 2.

4.2 Interpretation of Results - Failure Mechanisms

Following testing, and in view of the inconsistent results obtained, each of the failed samples was sectioned (*i.e.* cut perpendicular to the cable axis) for observation. This revealed that failure had occurred by two different mechanisms.

Failure Mechanism #1 involved straightforward pullout of the nutcase structure.

Failure Mechanism #2 involved slip and twist of the cable through the nutcase structure, so that although the cable had been pulled in excess of 50*mm*, the nut remained in its original position and the cement grout in the vicinity of the nutcase was left completely undamaged.

Figure 5 schematically illustrates the principal features of both. In all cases it was found that one or other of the two failure mechanisms occurred, but never a combination of both.

Based on these observations, the laboratory pull test results results can be summarized as follows:

(*i*) in all cases for which the nut did not slip (failure mechanism # 1) a peak strength close to, or in excess of, 240*kN* (the yield point of the cable) was achieved, and in the majority of cases the cable ruptured. Assuming that these tests are representative of the general behaviour of a nutcase cable, some encouraging points can can be made.

- In Sch 80 aluminum the 1/2" (12.7*mm*) nutcase confinement gave results 250% higher than those predicted for standard cable: likewise for two of the tests (5A and 5C) with the 5/8" (15.9*mm*) nut.

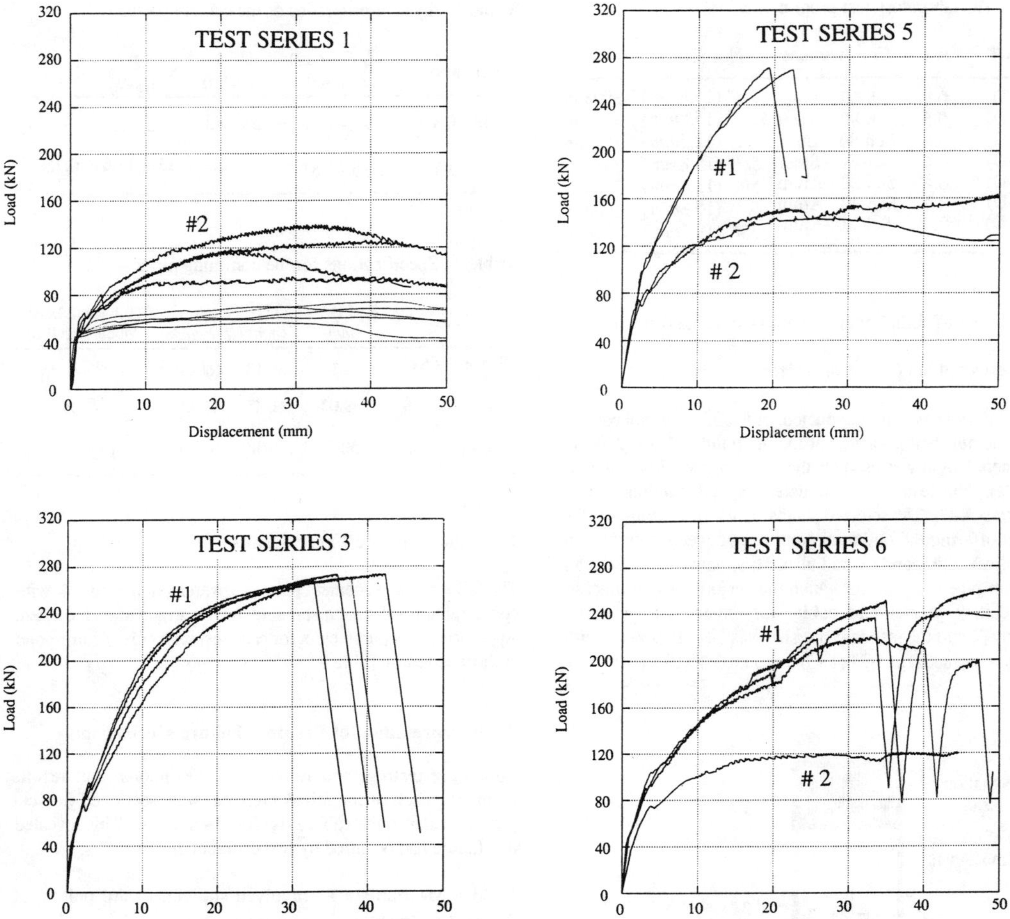

Figure 4. The laboratory cable pull test results. The *test series 2* results were virtually identical to those for *test series 1*. # 1 and #2 refer to the failure mechanism occuring. The test series correspond to those in Table 2.

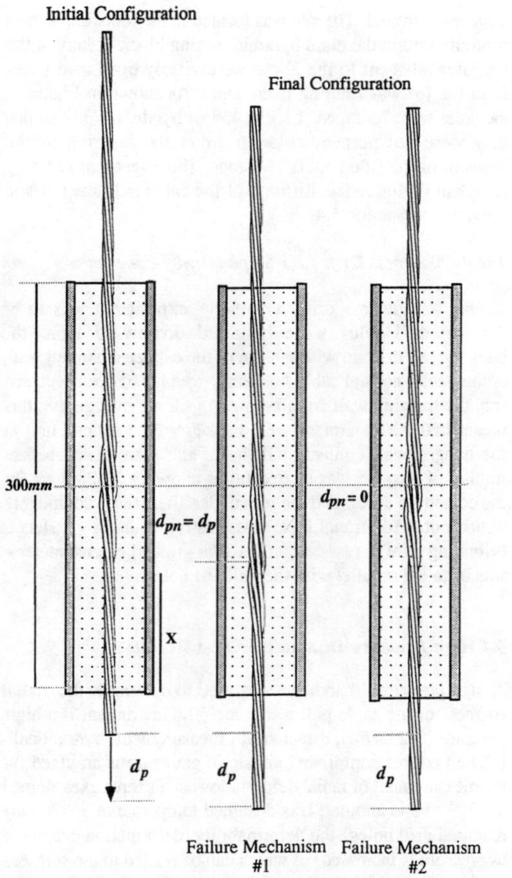

Figure 5. Schematic representation of the two failure mechanisms. X is referenced in the text.

- The smaller nut size resulted in a less stiff response, most particularly at loads above 200*kN*, so while the cable ruptures after 25-30*mm* for the 5/8" (15.9*mm*) nut, it is delayed until 40-50*mm* for the 1/2" (12.7*mm*) nut. This may offer a degree of flexibility in certain rock masses and mining situations where the nutcase cable is suspected of being too stiff.
- In cases of very low confinement, although the cable did not rupture, high capacities were achieved (approx. 240*kN*). Although no standard 7-wire strand results exist for Sch. 40 aluminum such capacities are predicted to be only in the order of 30% those that were achieved for the nutcase (see Hyett *et al.* (1992) for a comparison). These results suggest that the nutcase cable will provide virtually the optimum response in poor quality (low modulus or low strength) rock masses subjected to relatively large (25-75*mm*) deformations.

(*ii*) For tests in which slippage did occur (failure mechanism #2) capacities on average 25% greater than for the equivalent standard cable were obtained. The trends with changing confinement and *w:c* ratio are consistent with those observed (or predicted) for standard cable. In effect, for this failure mechanism, the length of cable below the nutcase structure (X in Figure 5), rather than the nutcase through which the individual wires are slipping with minimal resistance, may predominantly control the capacity.

Although, the mechanics that determine whether failure occurs by mechanism #1 or mechanism #2 are uncertain (and no doubt complex), the related evidence is as follows:

i) the 1/2" (12.7*mm*) tests (*i.e.* the smallest nutcase) did not slip;
(*ii*) the 3/4" (19.1*mm*) nut (*i.e.* the largest nutcase) slipped in all cases;
(*iii*) the 5/8" (15.9*mm*) nut was inconsistent; sometimes slipping, sometimes not.

Consequently, it appears that the failure mechanism which occurs is primarily controlled by nut size, and that, by restricting this to below 5/8" (15.9*mm*), and preferably closer to 1/2" (12.7*mm*), the occurrence of nut slippage will be eliminated.

5. FIELD PULL TESTS - HEMLO GOLDEN GIANT MINE

5.1 Field Pull Test Procedure

Until recently, no coherent set of guidelines or suggested methods existed for the field testing of cable bolts. Previous tests involved grouting a short embedment length of cable (usually 250 - 375 *mm*) some distance down a borehole, and pulling on the cable using a barrel-and-wedge grip. Owing to the deformability of the cable above the test section, this procedure allows rotation at both ends of the embedded section of cable, enabling failure to occur by an unrealistic 'unscrewing' mechanism. In order to prevent this the pull test setup illustrated in Figure 6 was devised. Comparative field test results (see Bawden *et al.*, 1992) indicate that when rotation is prevented cable capacities increase between 200% and 300%. This may explain to a large extent the anomalously low field test results obtained by some previous workers. Therefore, tests were conducted according to the procedure outlined in the Technical Note for the *in situ* testing of cable bolts presented by Bawden *et al.* (1992). The test section was 250*mm* long for the standard 7-wire strand samples and 300*mm* for the nutcase. For the nutcase cables, the nut size was 1/2"(12.7*mm*).

Three test sites were chosen, with each fulfilling a specific experimental objective:
(*i*) the 4775 level: location well away from active mining: tests conducted in ore and hanging wall;
(*ii*) the 4500 2W footwall X-cut: location in destressed ground; and,
(*iii*) the 4490 access drift : location in the hanging wall to the H5 stope which was to be extracted during the test programme (*i.e.* "mineby" experiment).

At each, the holes for the cable pull tests were 2.5"(63*mm*) percussion drilled, the collars of which were reamed with a 6" (150*mm*) flat faced bit. Parallel with these a 3"(76 *mm*) diamond hole was drilled for borehole dilatometer tests and borehole camera survey.

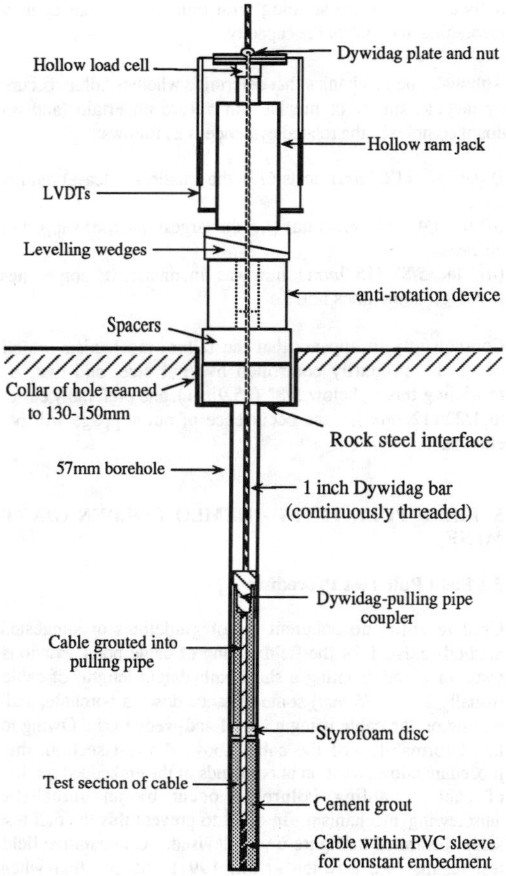

Figure 6. The *in situ* pull test setup (after Bawden *et al.*, 1992).

5.2 Properties of the Type 10 Portland Cement

The specimens were grouted using a Minepro pump. A number of cement control specimens were collected during sample installation. Both the wet density (ρ_{av}=2.06 g/cm^3) and 28 day density (ρ_{av}=2.08 g/cm^3) indicated that the cement paste was in the range 0.32 - 0.35 *w:c* ratio or slightly thicker than recommended for routine installation.

5.3 Test Site Locations and Rock Mass Characteristics

The 4775 Level : Undisturbed Ground Remote from Mining.

A comparison was made between standard 7-wire strand, birdcage and nutcase cable. Pull tests were conducted in both the hanging wall sericite and biotite schists and the feldspathic units that comprise the ore.

The 2W X-Cut : Destressed Ground.

In the 2W X-cut, cable pull tests were to be conducted in destressed ground. The site was located in the footwall, almost centrally within the main pyramid mining block . Many of the fractures adjacent to the X-cut were visibly open, and water from the fill was running from some. As shown in Figure 7, the holes were "dumped" back at 45^0 with a dip of -30^0 so that they were not perpendicular to either the layering or the footwall of the filled 4500 2W stope. The effect that this may have had on the radial stiffness of the cable pull test holes is discussed in Section 5.4.

The 4490 Access Drift : H5 Stope Mineby Experiment.

In the 4490 access drift, a 'mineby' experiment was to be conducted. Cables were installed downward from the hangingwall drift to within 4-5*m* of the expected hangingwall, amongst the normal cable bolt array which had been recessed into the hanging wall from below. As shown in Figure 8, this meant that the specimens were back from the sericite unit at the hangingwall contact. Of these, half were tested before mining of the H5 was started (after allowing 26-28 days for the cement to cure), and the remainder after it was completely mined out. Additional rock mass instrumentation - a Herget (vibrating wire) stress cell, and a 10*m* grouted extensometer - was installed parallel with the pull test holes.

5.4 High Pressure Dilatometer Test Results

A high pressure dilatometer was used to determine the radial stiffness of the cable pull test hole*s*. The instrument is a high pressure (0-25*MPa*), direct strain measurement, hydraulically inflated packer comprising six strain gauge arms arranged for the measurement of radial deformation across three axes aligned at 120^0 to one another. It is designed to operate in 3" (76*mm)* diamond drill holes, the deformability (deformation occurring as pressure is increased) of which can be related to the stiffness of the 63*mm* (2.5 *inch*) cable bolt holes (K_c)through the equation:

$$K_c = \frac{d_d}{d_c} K_d$$

where K_d is the stiffness of the dilatometer hole, d_d is the diameter of the dilatometer hole, (*i.e* 76*mm*), and d_c *is* the diameter of the cable bolt hole (63*mm* at Hemlo).

The results, as determined from the dilatometer hole and as predicted for the cable bolt hole from the equation above, are tabulated in Table 5 . No bladders were blown at any site suggesting that those fractures intersecting the pressurized section remained closed during testing up to 20*MPa*.

At each location, a 3" (76*mm*) diamond drill hole was included in the array of 2.5" (63*mm)* cable pull test holes, and was used for a high pressure dilatometer (HPD) survey to determine the radial deformability of the borehole wall. Recent research (Reichert *et al.*, 1992; Hyett *et al.*,1992) demonstrated that the radial stiffness of the borehole wall is a critical factor controlling cable bolt capacity, and furthermore that its determination enabled a correlation with laboratory results.

Table 5. Borehole wall stiffnesses at Hemlo Golden Giant mine.

Location	d_d (mm)	d_c (mm)	K_r (MPa/mm) Dilatometer	K_r (MPa/mm) Cable bolt	Equivalent Rock Mass Modulus (GPa) $v=0.23$
4490	76	63	1333	1586	62.3
4500 2W	76	63	571	679	26.6
4775 HW	76	63	1461	1753	68.3
4775 Ore	76	63	>1500	>1920	>70.1

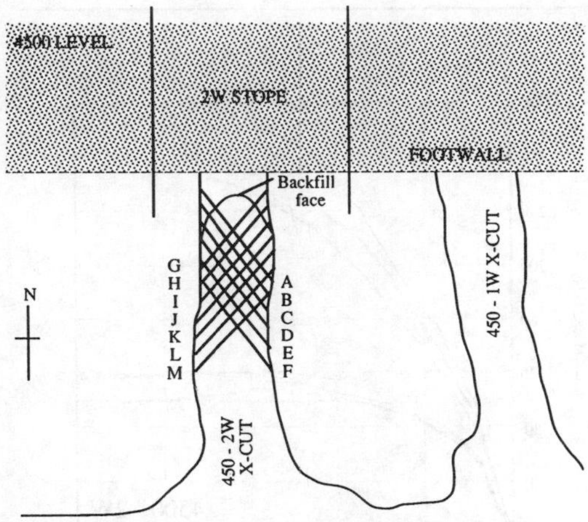

Figure 7. The 4500 - 2W test site at a depth of 825m.

Table 6. A comparison of E calculated by various methods.

Location	E_{Intact}	$E_{Rock\ Mass}$ (Q')	$E_{Dilatometer}$
4775 HW	54.8	22.3 (9.1)	66.6
4775 Ore	64.2	26.8 (13.5)	>68.4

On the 4775 level, the results for both hangingwall and ore agreed well with those obtained from UCS testing of the core (Table 6), suggesting that for undisturbed, undamaged rock, E_{intact} provides a better estimate of the rock mass modulus than a value based on a rock mass classification (*Q'*).

The pre-mineby results for the 4490 HPD test were essentially similar (though perhaps slightly lower) to the results obtained for the HW at 4775. Due to limited accessibility and poor ground conditions, dilatometer testing was impossible after the H5 mineby.

The results in the 2W X-cut were considerably lower than at the undisturbed/stressed sites. Several factors may combine to cause this.

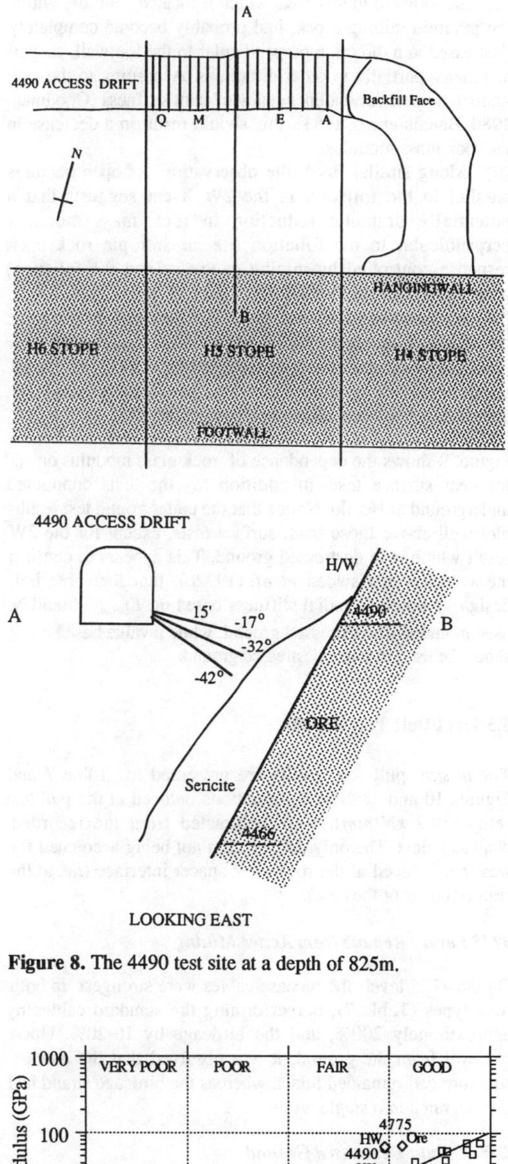

Figure 8. The 4490 test site at a depth of 825m.

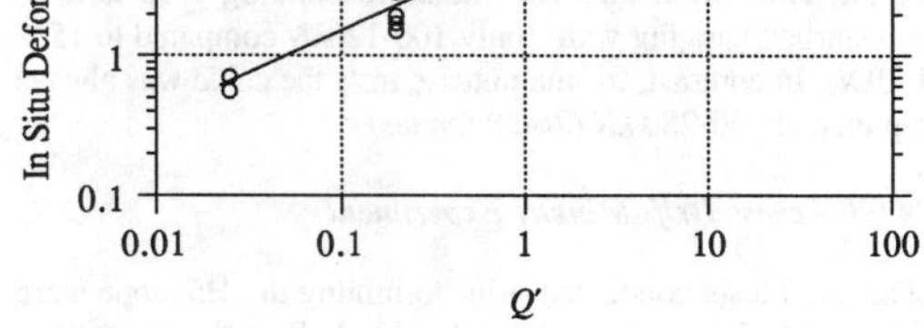

Figure 9. The relation between *in situ* deformation modulus and *Q'*. The Hemlo results are indicated. (see Hyett *et al.*(1992) for further details).

(*i*) The footwall in this area, which is located centrally within the pyramid mining block, had probably become completely destressed in a direction perpendicular to the footwall, as well as at least partially in other directions. According to the well known normal stress dependency of joint stiffness (Goodman, 1980, Bandis *et al.*, 1983), this should result in a decrease in the rock mass modulus.

(*ii*) Along similar lines, the observation of open fractures parallel to the foliation in the 2W X-cut suggests that a potentially dramatic reduction in rock mass modulus perpendicular to the foliation (*i.e.* anisotropic rock mass response controlled by mining geometry and the foliation) occurs as a result of footwall relaxation.

(*iii*) As mentioned above, the 2W pull test holes were orientated obliquely to the layering, and consequently their radial stiffness may have been affected by a more significant component of the modulus reductions described in (*i*) and (*ii*), than if they had been aligned perpendicular.

Figure 9 shows the dependence of rock mass modulus on Q' for near surface tests in addition to the tests conducted underground at Hemlo. Notice that the underground test results plot well above those from surface tests, except for the 2W result which is in destressed ground. This appears to confirm the assertion of Bawden *et al.* (1992b), that for cable bolt design, a borehole radial stiffness based on E_{intact} should be used in undisturbed/stressed ground, while a value based on Q' should be used in failed/destressed ground.

5.5 Field Pull Test Results

The *in situ* pull test results are presented in Table 7 and Figures 10 and 11 .The deformations induced in the pull test setup (39.2 kN/mm), were subtracted from the recorded displacements. The only deformation not being accounted for was that induced at the rock-steel spacer interface (*i.e.* at the reamed collar of the hole).

4775 Level : Remote from Active Mining

On the 4775 level the nutcase cables were strongest in both rock types (Table 7), outperforming the standard cables by approximately 200%, and the birdcage by 10-20%. Upon retrieval from the ground, it was observed that the nutcase structure had remained intact, whereas the birdcage strand had disintegrated into single wires.

2W X-Cut : Destressed Ground

Pull test results from the 2W X-cut are shown in Figure 10. The capacities obtained for the standard 7-wire strand were significantly lower than for the corresponding 4775 tests in undisturbed hanging wall (only 100-120kN compared to 150-160kN). In contrast, for the nutcase tests the cable was always ruptured at 260-280 kN (26-28 *tonnes*).

4490 Access Drift Mineby Experiment

The 4490 tests conducted prior to mining the H5 stope were expected to be representative of cable bolt performance in a stressed, though undamaged, rock mass, or in other words virtually the optimum rock mass conditions encountered at Hemlo with respect to cable bolt performance. The results are

Table 7. Cable pull test results on the 4775 level.

NUTCASE		BIRDCAGE		STANDARD CABLE	
Hole #	Peak Load (*tonnes*)	Hole #	Peak Load (*tonnes*)	Hole #	Peak Load (*tonnes*)
HW1	30.4	HW3	27.7	HW7	15.3
HW2	30.3	HW4	26.8	HW8	14.3
		HW5	25.8	HW9	16.5
AVG	30.35	AVG	27.7	AVG	15.37
ORE1	27.4	ORE 9	24.2	ORE2	19.2
ORE 12	27.8			ORE5	17.4
				ORE6	17.8
				ORE7	17.2
AVG	27.6	AVG	24.2	AVG	17.9

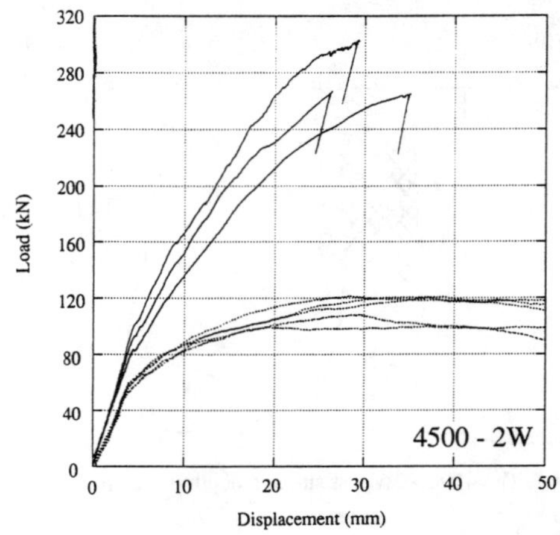

Figure 10. The 4500 - 2W cable pull test results.

presented in Figure 11a. For the standard 7-wire strand, and nutcase cable, the ultimate capacities were in the same range as those previously obtained under similar rock mass conditions in the hanging wall at 4775; 140-160kN for 7-wire stand, 240kN+ for nutcase. Lower capacities were attained for the 7-wire strand pull test result in ring A (hole 1) than elsewhere along the drift. The corresponding nutcase (ring A-hole 2) test result was unaffected. This result may be related to a loss in confinement caused by the 90^0 change in orientation of the 4490 HW drift just beyond ring A (see Figure 8). Linear elastic stress analysis using *Examine* ^{3D}indicated no significant destressing in the vicinity of the pull tests: however localized failure may have occurred in this area.

Following mining of the H5 stope, access to the 4490 hanging wall drift was precarious. The roof had failed and assumed a slabby structure (slabs were 125-200mm thick and controlled by the foliation), and pattern rock bolting appeared to be the only restraint to widespread unravelling.

Apart from the problem of limited access, mining resulted in damage to the already imperfect, reamed collars of the holes, several of which failed during pull testing. These and related

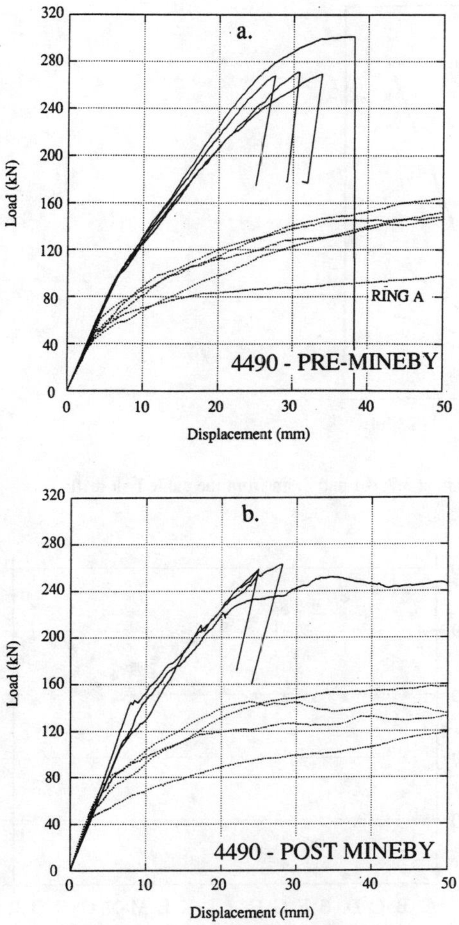

Figure 11. The 4490 cable pull test results (a) pre-mineby (b) post mineby.

problems explain why only 60% of the post mineby pull tests were successful.

From the post-mineby results obtained, no significant differences in capacity could be distinguished between those tests conducted before and after mining of the H5 stope (compare Figures 11a and 11b). Capacities of 120-150*kN* were attained for standard cables, and all tests for nutcase resulted in ruptured cables. In retrospect and as will be discussed in detail below, these results are consistent with both instrumentation data and numerical modelling results, both of which suggest that during mining of H5 no significant destressing occurred in directions radial to the test holes, and hence no reduction in cable capacity should have been expected.

5.6 Interpretation of the H5 Mineby Experiment

The Herget vibrating wire stress cell (self plucking variety) installed in the hanging wall adjacent to Ring G at a depth of 5*m*, and approximately parallel with the cable pull test holes

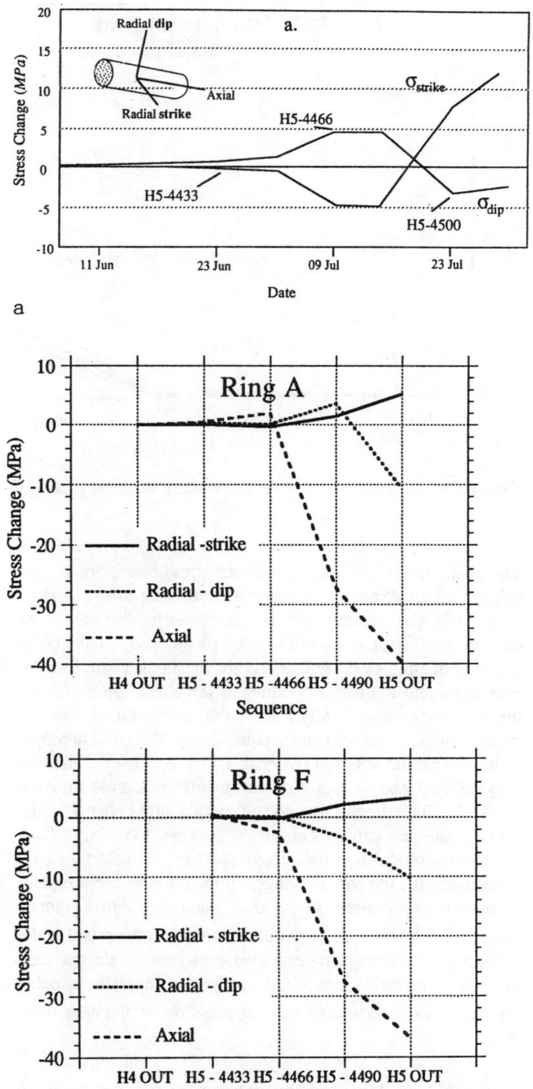

Figure 12. (a) Stress cell results. (b) Stress changes predicted using *Examine 3D*.

measured stress changes in two orthogonal directions radial to the hole (see inset in Figure 12a). The changes measured as mining progressed are presented in Figure 12a. The trends agree well with those predicted from elastic boundary element modelling using *Examine3D* (Figure 12b). Results from the borehole camera survey conducted in the 4490 in the 3" (76*mm*) diamond drill hole (between rings J and K) following mining, indicated that the hangingwall had undergone significant relaxation perpendicular to the hangingwall despite the limited stress changes in the plane parallel to it. These observations are in agreement with the extensometer results shown in Figure 13.

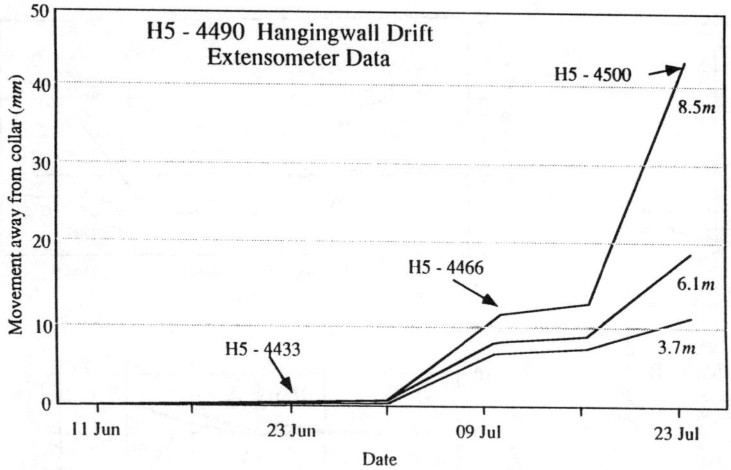

Figure 13 Extensometer results. The hole is adjacent to Ring G, with anchors at 3.7, 6.1 and 8.5m from the cable bolt drift.

The effect that such mining induced stress changes have on cable support depends on the corresponding stress change induced in the cement annulus surrounding the cable. As detailed by Kaiser *et al.* (1992), the stress change induced in the cement annulus depends on both the stress change in the rock mass and the relative values of the elastic properties of the rock and cement . Only if the latter are identical does the stress change in the rock mass equal that in the grout annulus. Using the elastic solution (Duncan-Fama & Pender, 1980) for a cylindrical elastic inclusion in an infinite elastic medium (*i.e.* full 3-D treatment), the corresponding stress change in the cement annulus can be calculated (Figure 14). Note that, because the modulus of the cement paste is lower than the rock mass modulus, the stress changes in the annulus are less than those in the rock mass. Also, σ_{strike} and σ_{dip} almost counter balance each other throughout, so that, conceptually, the reduction in cable capacity expected in response to the decrease in σ_{dip} is probably cancelled out by the increase in cable capacity associated with extra "clamping" due to the increase in σ_{strike} .

Although the 4490 mineby experiment did not result in mining induced stress changes in directions critical to cable bolt pull test performance it does represent an estimate of the performance of the hanging wall cables during an operational mining sequence. As with the other test sites at Hemlo, the results suggest that the nutcase cable offers a highly effective cable support option.

6. Practical Implications

A combination of laboratory and field testing has been used as a preliminary comparison of nutcase and conventional 7-wire strand cable.

In the laboratory it was determined that optimal performance of the nutcase cable requires maximum nut size of less than 1/2" or 12.7*mm* as measured across the flats. Also to guarantee

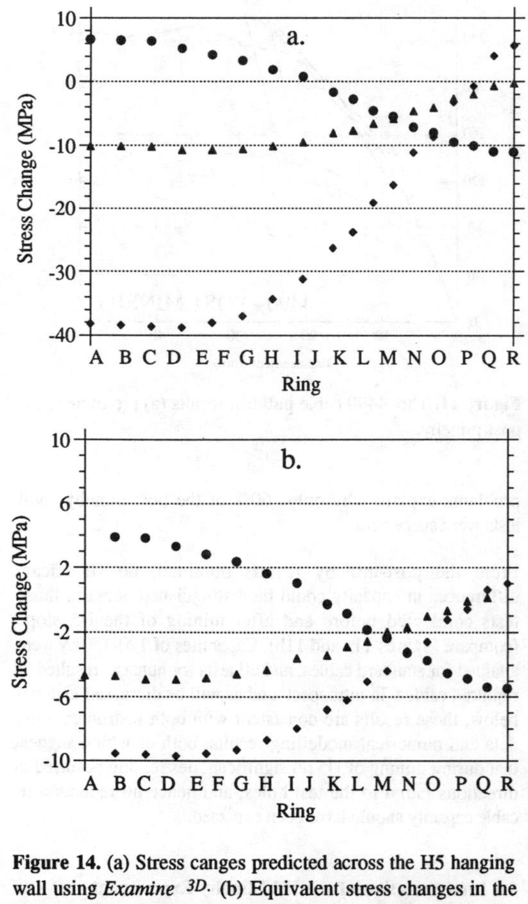

Figure 14. (a) Stress canges predicted across the H5 hanging wall using *Examine* ^{3D}. (b) Equivalent stress changes in the grout annulus: circles - sig-strike; triangles - sig-dip; diamonds-sig-axial

penetration of the nutcase a cement paste in the range 0.35-0.40 *w:c* ratio is recommended for installation.

Based on *in situ* pull test results at the Hemlo Golden Giant mine the following key results can be drawn.
(*i*) conventional 7-wire strand capacities at Hemlo vary from approximately 10 *tonnes* in fully relaxed ground to 15 *tonnes* under ideal ground conditions;
(*ii*) the use of nutcase geometries results in a dramatic improvement in bond capacity (*i.e.* 180% - 250%) under all ground conditions; and,
(*iii*) rupture of the nutcase cable occurred after 25-40*mm* of axial deformation, and consequently, it may still be considered as a *yielding* support system especially in poor ground.

The implications of the above for support design are very significant. Clearly, in moderate to low quality rock masses, nutcase cable will offer an effective alternative to conventional 7-wire strand in all cases of (*i*) stress relaxation, (*ii*) significant mining induced destressing in directions radial to the cable bolt holes, and (*iii*) the support of failed ground.

Although, the significant performance enhancement demonstrated for nutcase cables makes a compelling case for their widespread use, one must keep in mind that such the support system will be stiffer than a 7-wire strand cable support system (see Figures 10 and 11). Conditions may then occur where such support could conceivably become overstressed and rupture. Such conditions would most likely to occur in cases where stress induced failure occurs after the support is placed, and can potentially be predicted using numerical stress analysis so that if deemed necessary appropriate modifications to the design could be introduced. Consequently, at any minesite, the first location in which nutcase cables are installed should be instrumented in order to verify the ground-support interaction characteristics under such conditions.

A second advantage of using modified cable geometries is that it is possible that they may, at least partially replace the need for surface fixtures (*i.e.* plating). In cases of recessed bolting of remote support (*i.e.* hanging wall support placed from a hanging wall drift where plating is impossible), the modified cable geometries are expected to control spalling and unravelling (*i.e.* short embedment length) failures better than conventional strand.

A third advantage of modified cable geometries, is to reduce operator reliance on grout quality control. Use of a 0.4 *w:c* ratio grout is recommended with the nutcase cable, in order to ensure penetration into the case structure, in other words slightly thinner than the 0.35 *w:c* ratio recommended for the installation of standard 7-wire strand. Pumping a grout in the range 0.35-0.4 should be routinely possible using most grout pumping systems.

On the bottom line, the advantages of using nutcase cable may be translated into (*a*) reduced dilution, (*b*) wider cable bolt patterns and hence reduced drilling and installation costs, and/or (*c*) larger stable stopes. Optimization of the cable system as an integral component of the overall stope design will involve a trade-off between these benefits: whichever option is pursued, both the operational and financial dividend associated with an effective cable bolt system may be orders of magnitude greater than the extra cost of the nutcase cable.

ACKNOWLEDGEMENTS

Adam Coulson manufactured the first Queen's nutcase after inspecting a sample at Winston Lake, and has provided valuable advice throughout. Peter Lausch was an invaluable asset in the field, where Donna Cortolezzis, Mike and Dale Toppi provided operational support. Adam Allan at DSI funded the laboratory testing, and provided Dywidag bars for the field test setup.

REFERENCES

Bandis S., Lumsden A.C. & Barton N.R. 1983. Fundamental of rock joint deformation. *Int. J. Rock Mech. Min. Sci.* Vol. 20. 249-268.

Bawden, W.F., Hyett, A.J. and Lausch, P. 1992. An experimental procedure for the in situ testing of cable bolts. *Int. J. Rock Mech. Min. Sci.* Vol. 29, No. 5, 525-533.

Bawden, W.F., Hyett, A.J. & Cortolezzis D.M. 1992b. Towards a methodology for performance assessment in cable bolt design. In: *Rock Support*. Kaiser, P.K. & McCreath, D.R. (eds) 277-284.

Duncan Fama, M.E. and Pender, M.J. 1980. Analysis of Hollow Inclusion Technique for measuring in situ stresses. *Int. J. Rock Mech. Min. Sci.* Vol 17, No. 3. 13-146.

Goodman, R.E. 1990. *Introduction to Rock Mechanics* Wiley, New York .

Hyett, A.J., Bawden, W.F. & Reichert, R.D. 1992. The effect of rock mass confinement on the bond strength of fully grouted cable bolts. *International Journal of Rock Mechanics & Mining Science*. 29, 5, 503-524.

Kaiser, P.K., Yazici, S. & Nose, J. 1992. Effect of Stress Change on the Bond Strength of Fully Grouted Cables. *Int. J. Rock Mech. Min. Sci.* Vol. 29, No. 3. 293-306.

MacSporran, G.R. 1993. An Empirical investigation into the effects of mione induced stress change on standard cable bolt capacity. MSc Thesis. Queen's University, Kingston.

Reichert, R.D., Bawden, W.F. & Hyett, A.J. 1992. Evaluation of Design Bond Strength for Fully Grouted Cable Bolts. *Bulletin of the Canadian Institute of Mining & Metallurgy*. 85, 110-118.

Innovative Mine Design for the 21st Century, Bawden & Archibald (eds) © 1993 Balkema, Rotterdam, ISBN 90 5410 325 6

Planning the Porgera underground mine

F.E. Kaesehagen
MIM Holdings Limited, Brisbane, Qld, Australia

G.G. Northcote
Renison Goldfields Consolidated Limited, Sydney, N.S.W., Australia

ABSTRACT: The Porgera Gold Mine has successfully met the challenging targets set for its construction and development with the result that, in the first two years of operation, gold production exceeded 2.4 million ounces. Planning of the underground mine was based on results of closely spaced diamond drilling from an exploration adit which was also used for initial access and production haulage. Using grade and geotechnical data from this drilling, a stoping method, using small dimension stopes and cemented rock fill, was selected in order to maximise early production from ultra high grade areas without sacrificing overall recovery of the high grade resource. The filling method also provided flexibility in meeting the production schedules. In order to meet the requirement to treble production to 4,500 t/d, haulage of ore by truck has been replaced by underground crushing and a vertical conveyor hoisting system.

INTRODUCTION

The Porgera Gold Mine is located in Enga Province in the highlands of Papua New Guinea approximately 130 km west of Mount Hagen and 600 km northeast of Port Moresby (Figure 1). It is situated in rugged, forest covered, mountainous terrain at an elevation of 2,300 metres. The area has a mild climate due to its elevation, with an annual rainfall of 3,600 millimetres.

The Porgera Feasibility Study, which was based on initial production from underground mining of high grade reserves in order to enhance profitability, was presented by the Porgera Joint Venture (PJV) partners, Placer (P.N.G.) Pty Limited, Highlands Gold Properties Pty Limited and RGC (Papua New Guinea) Pty Limited, to the Papua New Guinea Government in June 1988. Following discussions with the Government and other interested parties, including the Porgera landowners and the Enga Provincial Government, a Mining Development Contract between the Joint Venturers and the Government was signed in May 1989. The Government also exercisedits right to equity participation through Mineral Resources Porgera Pty Limited which obtained a 10% interest in the mine by transfer of equal parts from the Joint Venture Partners.

Development of the underground mine began with the commencement of an exploration adit in December 1985 and during the period covering the preparation of the Feasibility Study and the negotiations with the Government, the PJV continued underground development and activities such as upgrading of 70km of the access road in order to facilitate early commencement of operations. These initiatives allowed the successful development of the underground mine which commenced production at 1,500 t/d in June 1990 and is on target to reach 4,500 t/d in the June Quarter of 1993. Production during the first two years of operations yielded in excess of 2.4 million ounces of gold.

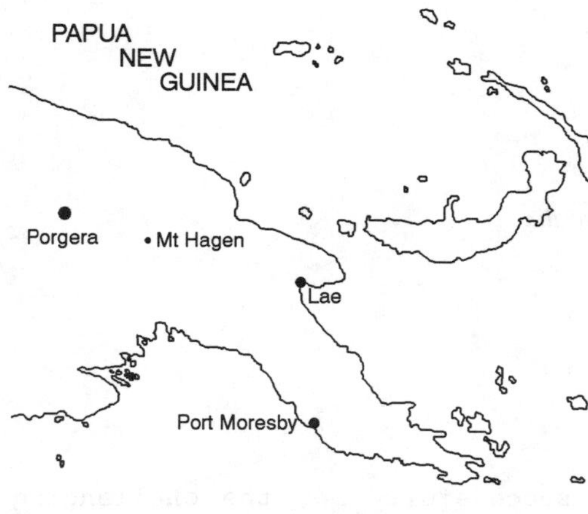

Fig.1 Location map

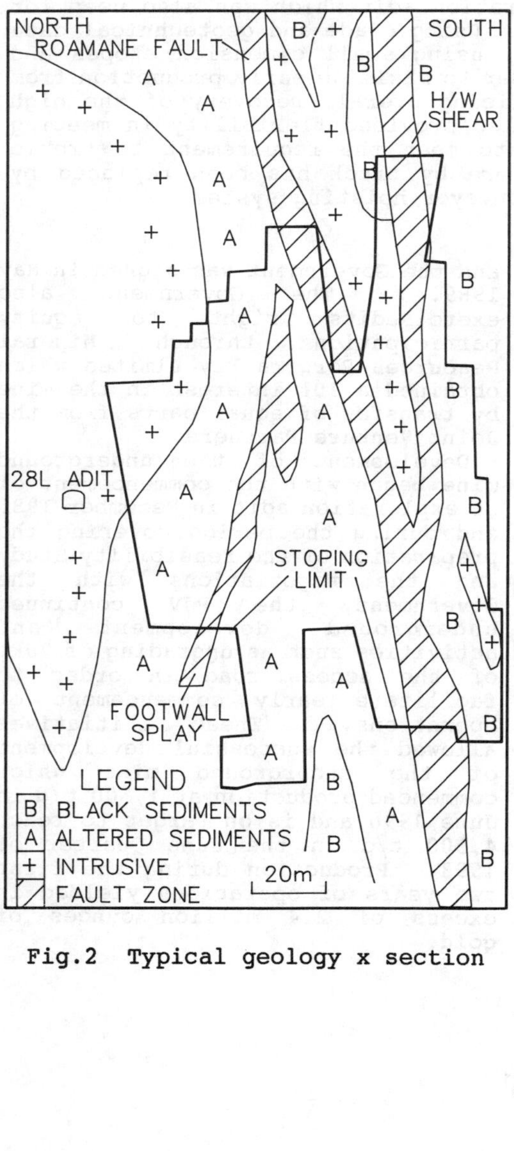

Fig.2 Typical geology x section

GEOLOGY AND MINERALISATION

The Porgera Zone VII high grade gold deposit is an epithermal style orebody hosted within thermally metamorphosed sediments of the Cretaceous Chim Formation and the associated Porgera Diorite Intrusive Complex of Miocene age. The known orebody extends for up to 430m along strike, with a maximum width across strike of 100m, but often no more than 50m. The geology of the Porgera gold deposit is described in detail by Handley and Henry (1990).

The orebody is closely associated with the three dominant structural trends indicated in Figure 2. The underground mine is focussed on the Roamane Fault Zone, which is generally up to 12m thick and strikes in an E - W direction, dipping 75° to the south. The Hanging Wall Shear intersects the Roamane Fault between 28 level and 31 level and is a subparallel feature dipping 85° to the north. The Footwall Splay trends away to the northwest at about 15° to the Roamane Fault, and also dips to the north at 85°.

The Roamane Fault zone forms a loose boundary between Black Sediments in the hanging wall and Altered Sediments in the footwall. These sediments have been thermally metamorphosed and metasomatised to varying degrees, resulting in two distinct lithologies. The Black Sediments are slightly indurated weakly calcareous mudstones and clayey siltstones and within the fault zone they are extremely sheared with common friable calcareous gouge and clay bands. The Altered Sediments which abut the intrusives are more intensely metamorphosed than the Black Sediments, and are consequently more competent. They are also slightly dolomitic with extensive breccia zones, which are associated with the mineralising events occurring throughout the unit.

The intrusives tend to be concentrated towards the footwall of the deposit and the complex is comprised of many individual stocks and dykes. The rocks are competent but they tend to be brittle, and in the vicinity of the orebody, are extensively veined and brecciated.

UNDERGROUND EXPLORATION PROGRAMME

Until late 1984 the Porgera exploration programme had focussed on surface diamond drilling with the objective of proving up an open pit reserve. Following the discovery of the high grade section of Zone VII, which extended to the west some 400 to 600m below Mount Waruwari, this programme was reviewed and a decision was made to undertake an underground exploration programme. It was also evident that in order to make the required financial impact on the project this high grade ore would need to be extracted by underground methods. Consequently, a decision was made to develop an underground access which would serve both the underground exploration programme and also provide access for an initial underground mining operation. This resulted in the development of a 1476m long 4.5m x 4.8m adit at a +2% grade situated in the footwall of the high grade ore zone. Development commenced in December 1985 and was completed in July 1987.

Drill recesses were developed off the adit to allow underground drilling to proceed in parallel with development and drilling commenced in March 1987. This work was done by contractors using three rigs to drill fans of holes from +75° to -80° with depths ranging from 70 to 200m. Two hundred and ninety four holes, comprising 32,000 metres of drilling were completed during 1987 and 1988 to drill the high grade zone, which measured 430m along strike and 220m down dip, on a 15m x 15m pattern.

As well as providing a quality geological model for ore reserve estimation, the drilling information was used to prepare a comprehensive geotechnical model.

All core was photographed and geotechnically logged and the following data recorded:

* rock Strength (pointload basis)
* RQD (Rock quality designation)
* fractures per metre
  shearing
  planarity
  roughness
  coating/infilling
  angle to core axis
  width of structure
* comments (including weathering)

Mean and maximum strength values are shown in Table 1.

The intrusives and altered sediments strength histograms showed a broad spread either side of the mean, while the black shales and fault zone were markedly skewed with the bulk of the samples having less than mean strength.

Figure 3 is an example of the RQD histograms which were plotted for each drill hole and provided an excellent structural picture.

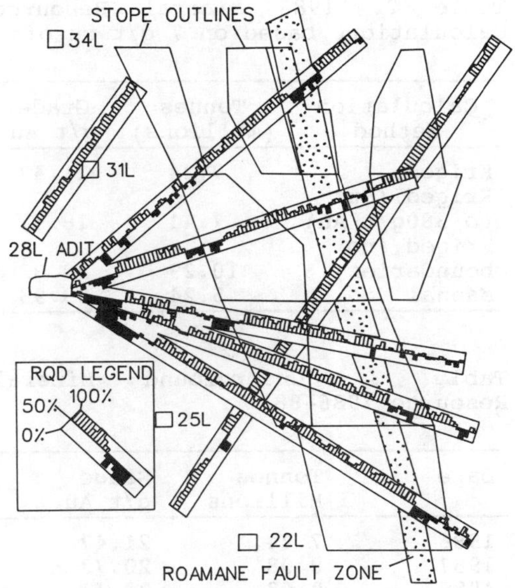

Fig.3 RQD histogram

Table 1. Rock Strength

Rock Type	Mean Strength MPa	Maximum Strength MPa
Intrusives	110	246
Altered Sediments	72	198
Black Sediments	42	198
Fault Zone	34	176

MINERAL RESOURCE

Early mineral resource estimates were calculated using inverse square

Table 2. 1986 Mineral Resource Calculations based on 7 g/t cutoff

Calculation Method	Tonnes (millions)	Grade g/t Au
Kriged	7.24	21.47
Kriged (cut to 380g/t Au)	7.41	18.5
Kriged (no boundaries)	10.29	20.3
Manual	5.24	23.93

Table 3. Underground Mineral Resource 1986-88

Date	Tonnes Millions	Grade g/t Au
1986	7.24	21.47
1987	8.08	20.73
1988	9.63	23.11

Table 4. Underground Mining Reserve

Date	Tonnes Millions	Grade g/t Au
October 1987	3.4	29.2
January 1989	6.64	26.1

techniques. However as the quantity of data increased, geostatistical methods were employed. The first kriged resource estimate, based on sixty surface drill intersections at 30m x 40m spacing, was produced in December 1986. The block dimensions selected were:

* east west direction 7.5m
* north south direction 2.5m
* vertical direction 10m

The 7.5 metre blocks were centred on the planned 15 metre drill sections and the 2.5 metre block size was selected to reflect the steeply dipping, structurally controlled, mineralisation. Geological boundaries were used to control the availability of data to the search ellipsoids. Several options were also run to test the effects of various cutting limits, the elimination of geological boundaries, and the use of probability kriging. A manual sectional resource was also calculated for comparison. (Refer Table 2)

In October 1987 the resource estimate was updated by the inclusion of the results of 57% of the underground drilling programme. This increased the overall tonnes by 12% and decreased the grade by 3%. At this time the first underground mining reserve was calculated from actual stope design outlines and the possibility of producing ore at twice the average mineral resource grade, 60g/t Au, for at least the first two years of mining, was demonstrated. This mining reserve was used in the Feasibility Study presented to the Government in May 1988.

In November 1988 the resource estimate was further revised by including the results of the now completed underground drilling. As a result of this the tonnes increased by 19% and the grade increased by 11%, representing an overall increase in total contained gold of 33% as shown in Table 3.

In January 1989 stope outlines were revised to take into account the new data and a new mining reserve was calculated. A significant increase in both tonnes and grade occurred as shown in Table 4.

PROJECT CONCEPTS

During the period 1985 to 1987 a number of studies were completed which examined various options for developing the Porgera resource. The favoured plan was a nominal 8,000 t/d project, with ore mined from both open pit and underground, milled to produce a sulphide concentrate which would be processed through a pressure oxidation plant and then treated by cyanide leaching and carbon-in-pulp methods to recover the precious metals.

The underground mine was scheduled to produce at the rate of 1,500 t/d for a period of 6 years when known reserves would be exhausted. The open pit would then continue to supply the total required ore production. This plan had a preproduction time of 39 months.

The new ore reserve data plus some new concepts led to a review of the

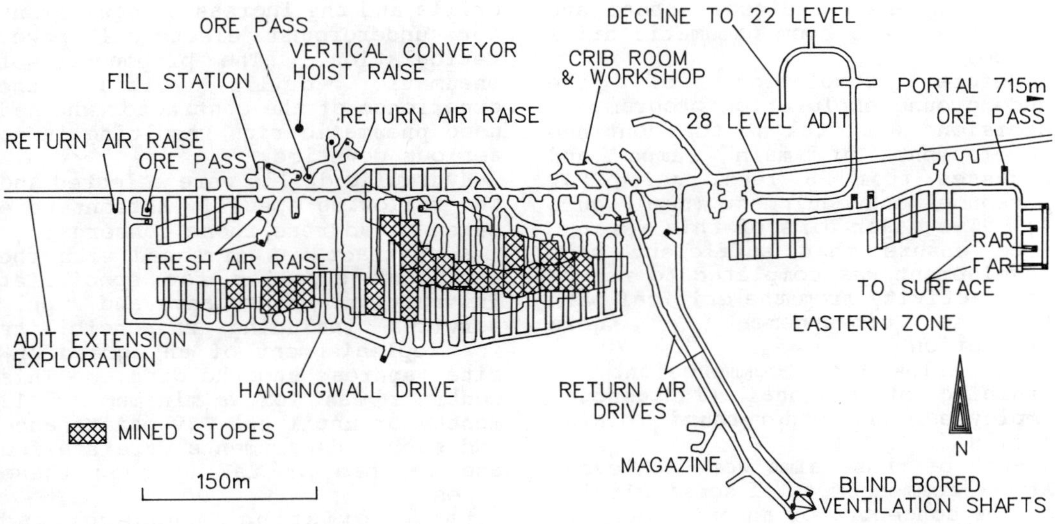

Fig.4 28 level adit

plan during 1987-89. The key factors were:
* the increase in underground reserves.
* early stope planning indicated that a number of 20,000 tonne stopes with grades in the range of 80 to 200 g/t Au were situated within 80m horizontally of the exploration adit
* 95% of gold from the high grade zone could be recovered in the sulphide concentrate
* 65-70% of gold contained in the sulphide concentrate could be extracted by the cyanide leaching and carbon-in-pulp process
* residue from this process could be stored for retreatment in the pressure oxidation plant when commissioned
* technical problems associated with open pit waste disposal and pit slope angles
* unknowns associated with autoclave performance

The outcome of these reviews resulted in a number of changes to the project. Firstly in May 1987 a decision was made to bring forward underground production to reduce the time to first gold production to 15 months. To achieve this a 1,500 t/d concentrator would be constructed to treat the ore by cyanidation of the sulphide concentrate and the tailings stored for retreatment when the pressure oxidation plant was commissioned. Secondly in January 1990 further changes were made including deferring open pit production by 12 months with underground production increased to 3,500 t/d to compensate. Construction of the process plant was also staged to fit the new production schedules. The result of these changes was a significant improvement in project economics as gold production was improved whilst capital expenditure was deferred. This plan also increased focus on the underground mine and its performance was critical to the project success.

MINE DEVELOPMENT

During the time the exploration adit was developed:
* road access to the site was limited to 4WD vehicles with a maximum capacity of 6 tonnes
* telephone communication was restricted to 2 lines with poor reliability
* spares holdings in Papua New Guinea for underground mining and drilling equipment were either non existent or at best remote from site

The mining contractor coped well with these conditions and completed this development with suitable second hand trackless equipment which included a diesel loader and trucks and a 3 boom pneumatic drill jumbo.

Before completion of the underground exploration programme a decision was taken to continue development of main ramps and accesses from 28 level using PJV personnel and equipment (see Figure 4). Two major aims of this were:

* to ensure that sufficient mine development was completed to remove this activity from the critical path to the commencement of gold production
* to allow the commencement of training of national (indigenous) employees in underground mining skills.

Both of these aims were achieved. At the commencement of commissioning of the concentrator an ore stockpile in excess of 100,000 tonnes was available and production from stopes had commenced. Furthermore, the mine was able to match increases in the concentrator throughput of 13% above design capacity in the first three months of operations rising to 25% above design capacity over the next 12 months.

Training of miners, which commenced with "on the job" training of the initial crews under the direct supervision of experienced expatriate miners, has been successfully modified over time to meet the increased requirement of the expanding mine. A more formal training program, conducted by specialist training supervisors and taking from 8-12 months to complete, is now in use.

In selecting equipment for this initial development it was necessary to recognise both the existing poor communications and access to site and the longer term need to use the latest technology and equipment. Loading and hauling equipment, comprised of 3.5 cubic metre capacity diesel LHDs and 25 tonne capacity diesel trucks, was selected as it was suitable for both development and production.

The selection of development drilling equipment was surrounded by a debate on the merits of pneumatic versus hydraulic rock drills. Whilst it was accepted that hydraulic drills were the latest with superior technology, there was concern over both the specialised maintenance requirements for the drills and the increased requirement for underground electrical power reticulation. The proponents of pneumatic drills cited the experience of the contractor who had used pneumatic rigs resulting in no serious downtime.

Hydraulic drills were selected and the following specific actions were taken to address these concerns:

* a contract was arranged with the equipment supplier for specialist instruction, maintenance and supply services including responsibility for replenishment of an agreed on-site spares stockholding. This contract was for a minimum of 12 months or until the PJV maintenance and supply departments were staffed and capable of taking over these roles
* the existing underground electrical reticulation equipment was replaced with equipment capable of being integrated into a system designed for the long term operation of the mine including possible expansion
* emphasis was placed on the early employment of electrical supervisors and tradesmen with underground mine experience.

These initiatives proved to be successful and since that time all additional development, rockbolting and production drilling rigs purchased have been electric/hydraulic.

VENTILATION

The establishment of a good ventilation system early in the mine development was perceived as critical to the achievement of the tight schedule for initial production. Consequently, in October 1988, a decision was made to develop airway connections to the surface early in the mine development program. Completion of these connections, plus the provision of an alternative means of egress, was also a prerequisite to the expansion of the trainee underground crews.

The location of ventilation shafts was complicated by the fact that an open pit operation was planned to commence while underground mining was in progress. This resulted in the ventilation shaft site being selected 280m into the hanging wall of the orebodies. The first crosscut to this site had been established by this date and had encountered weak black sediments for its full length and required steel set support.

Test drilling indicated ground conditions for the proposed shafts would be similar or weaker. Raise drilling was eliminated as an option and tenders were called for the sinking of a 4.2m diameter concrete-lined shaft, partitioned to provide both downcast and upcast airways. These tenders indicated the facility would not be available until August 1989. As timing was critical, the option of blind boring three 2.5m diameter shafts, lined with hydrostatic steel liners, was investigated and a contract was awarded to a shaft drilling contractor in November 1988. Drilling commenced in January 1989 and the first shaft was completed within 4 weeks. The contract, which now included a fourth shaft, was completed by May. Completion of the first shaft in February allowed the expansion of underground training of the local workforce to continue as planned and with the four shafts completed prior to project approval date, a rapid build up of development rates was facilitated. It is estimated that this decision brought forward underground production by 6 months.

The decision to develop a dedicated return air level (26 level) between the first two production levels (see Figure 5) further improved the quality of the ventilation system.

MINING METHOD

The following factors were considered when selecting the mining method:

* the ore grade distribution and the need to maximise production from the ultra high grade zone during the first 2 years
* the poor ground conditions associated with ultra high grade zone
* the need to minimise ore dilution and to maintain the integrity of the ore surrounding this zone for future mining.

A panel fill stoping system was selected as the appropriate method. This involves mining small open stopes 30m high, 15m along strike and 15 to 25m across strike. Figure 6 shows the stope layout.

The key aims of the method are:

* keep the strike dimensions small
* use cable support to maintain hanging wall and back stability
* extract the stope quickly
* fill the empty stope immediately with a high quality consolidated fill.
* extract ore blocks adjacent to filled stopes as soon as fill was cured, that is, after 1 month.

Each stope was accessed by an

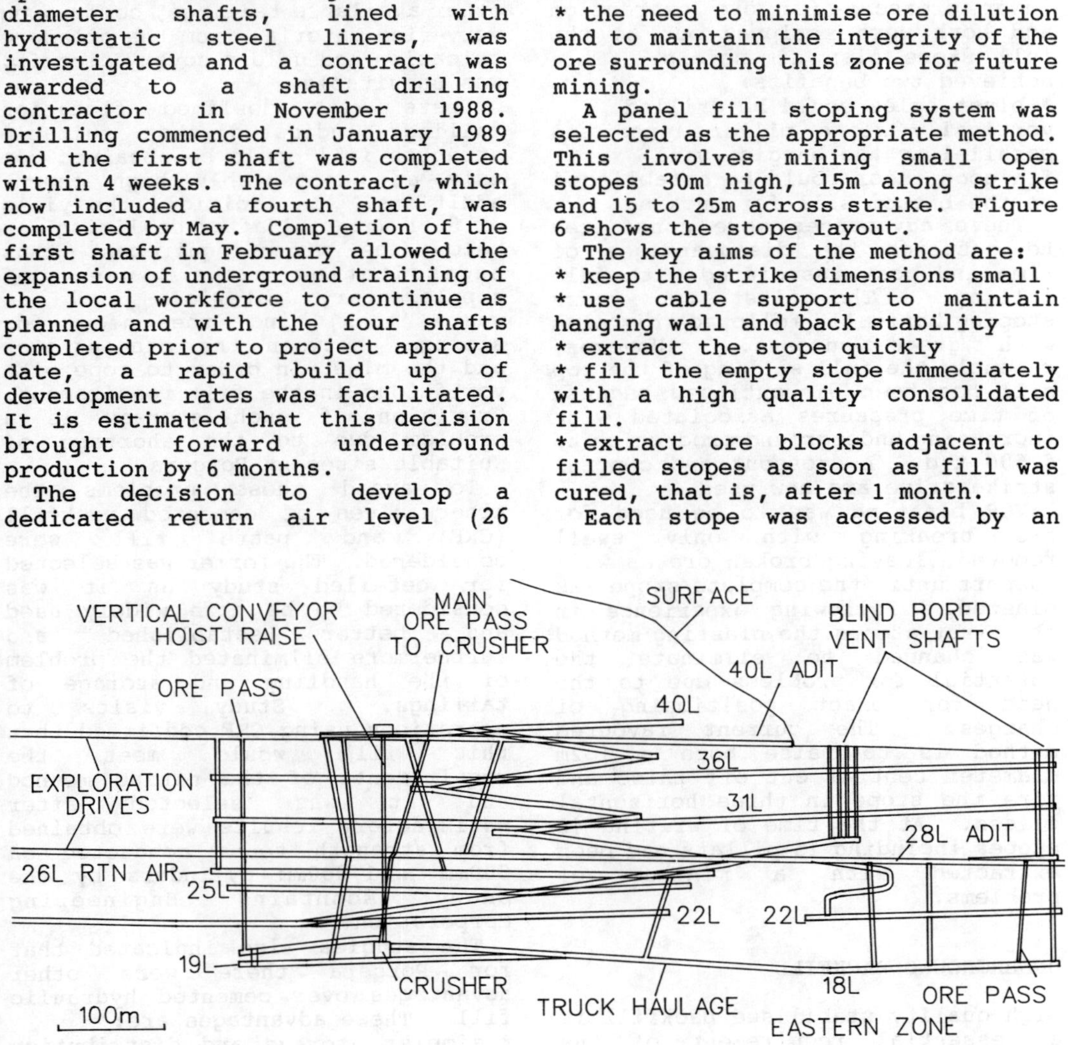

Fig.5 Longitudinal projection

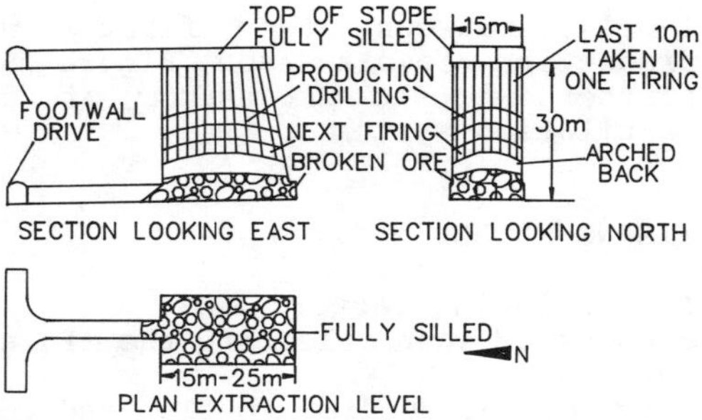

Fig.6 Stope layout and blasting method

upper crosscut for drilling and a lower crosscut for ore extraction and both were stripped out to the full stope plan dimensions. This achieved two benefits:

* blast holes could be drilled vertically along pillar lines and parallel to the hanging wall
* a good floor could be established on the bottom sill for mucking.

These advantages were considered to outweigh the disadvantages of time and cost associated with full silling. The first two years stoping basically followed this plan with great success. However, recently the full silling of the top sill has been discontinued because of time pressures associated with increasing underground production to 4,500 t/d. A crosscut and central strike drive are now used.

VCR blasting was to be used for ore breaking with only swell removed, leaving broken ore as wall support until the complete stope was blasted. Following experience in the first stopes the blasting method was changed to eliminate the potential for problems due to the need for exact positioning of charges. The current favoured method is to raise bore a 1.2m diameter central cut off raise and fire the stope in three horizontal slices. At the time of writing 40 stopes including 15 pillars had been extracted with a minimum of problems.

UNDERGROUND BACKFILL

High quality stabilised backfill is an essential requirement of the selected mining method in order to allow complete extraction of the high grade orebodies. In addition to providing wall support the fill is required to have sufficient inherent strength to:

* minimise the amount of fretting or sloughing from free-standing faces of fill during extraction of the adjacent ore
* provide a hard upper working surface to reduce penetration of blasted ore into the fill and dilution by removal of fill during ore loading.

It was initially planned to use cemented hydraulic fill and testwork was conducted on the flotation tailings which comprise 70% to 80% of the total concentrator feed. The other available tailings, comprising very fine material from the pressure oxidation and gold recovery circuit, was unsuitable.

Tests with deslimed flotation tailings indicated that suitable strength fill could be obtained but only at unacceptably high cement additions. In addition, the yield of fill material after desliming was below the mine requirement thus necessitating the use of supplementary material (such as waste rock) and the need for storage, between the concentrator and the mine, in order to cope with variations in the fill requirement. Provision of such storage is a problem due to the shortage of suitable sites at Porgera.

To avoid these problems the alternatives of cemented rockfill (CRF) and paste fill were considered. The former was selected for detailed study as it was considered to be more widely used and better established and furthermore eliminated the problem of the handling and storage of tailings. Study visits to operations using CRF confirmed that this fill would meet the requirements of the mining method and it was selected after satisfactory results were obtained from strength tests conducted on 500mm x 1,000mm cylinders by the Snowy Mountains Engineering Corporation.

The studies also indicated that for Porgera there were other advantages over cemented hydraulic fill. These advantages are:

* simpler storage and distribution system
* fill bulkheads, drainage and

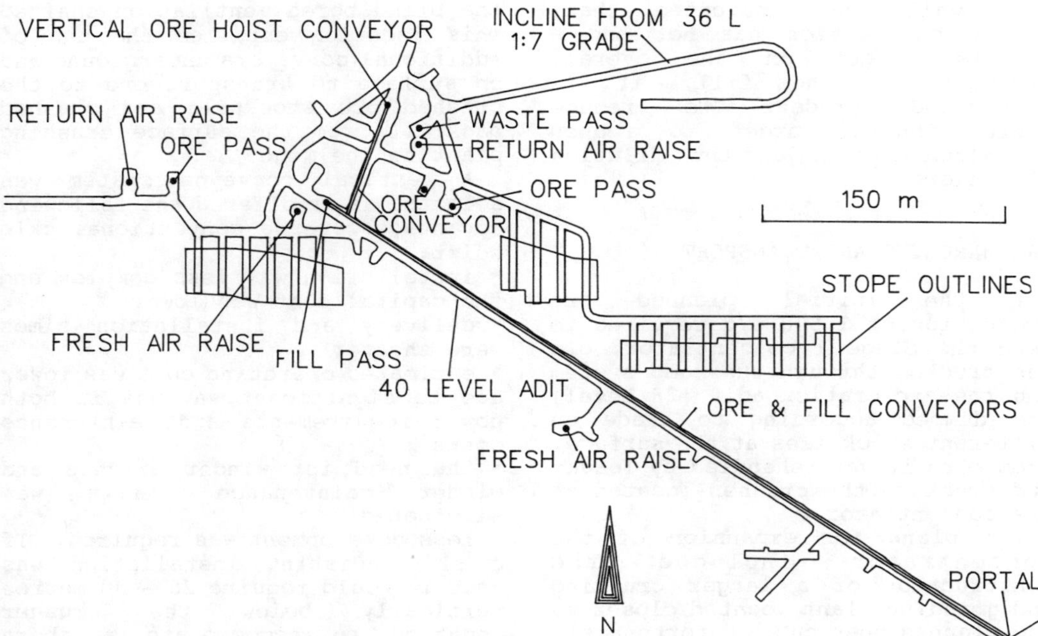

Fig.7 40 level plan

percolation systems required for hydraulic fill are eliminated.

The time saving from the latter is a significant factor in achieving the fast turn-around of small tonnage stopes which is required at Porgera.

CRF aggregate is obtained from intrusive rock and calcareous sediments sourced from underground development and open pit waste. The rock feed is screened to remove the minus 40 mm fraction and then crushed and screened. The screen fractions are blended to produce a well graded minus 100 mm product with a minus 9.5 mm fraction of between 20% to 30% and a dry void ratio of less than 30%. Size analyses are carried out daily to ensure product quality. The water cement slurry comprises type 'C' Portland cement, mixed to a pulp density of 60% solids, and added to the aggregate at rates ranging from 4.5% to 7.0% in the total mass. A measured quantity of cement slurry is sprayed onto the stream of graded aggregate as it is fed from the fill pass chute into the waiting truck and the mixture is then trucked to the stope. Mechanical mixing or agitation of the mix is not required.

A further advantage of using CRF and placing it by truck was that the fill handling system could be introduced in stages to suit the development of the mine. For the first stage of production a temporary arrangement was used in which a section of the slurry mixing plant was placed at 28 level adit and slurry added to the graded aggregate as it was placed in the trucks by loader. This arrangement was used until the 40 level fill conveyor, the fill pass from 40 level to 28 level and the slurry mixing and reticulation system, were installed and commissioned (see Figure 7).

Initially all CRF was end-dumped from the crosscut into the stope void. This method creates a long active rill face which results in segregation with the coarser material reporting to the base of the rill. Wherever possible CRF is now placed through the bored cut-off raise of the stope above reducing the length of the rill face and consequently the amount of segregation. Observation of the fill walls has indicated that, whilst segregation has not had a serious effect on the overall strength of the fill, it is considered prudent to reduce segregation in order to reduce localised fretting and thus dilution of the ore.

ORE HANDLING AND TRANSPORT

For the initial planned ore production of 1,500 t/d required to feed the Stage 1 concentrator, ore was trucked through the ramp system and the exploration adit (28 Level) and dumped according to grade on different stockpiles at the surface. From here it was rehandled by loader and truck to the crusher located at the concentrator.

The plans for expansion of the concentrator included the construction of a larger crushing and grinding plant located closer to the planned open cut. Unfortunately this trebled the surface haul distance for underground ore from the 28 level adit. In order to reduce the haul distance to this location, and to eliminate double handling, an incline haulage ramp was planned from 2,800m RL to the crushing plant to allow ore delivery direct to the crusher. However, following the decision to expand the underground mine production to a minimum of 3,500 t/d and also to cater for the change from cemented hydraulic fill to CRF, it became obvious that truck haulage should be replaced by underground crushing and ore hoisting either by skip hoist or conveyor.

The preferred location of the underground crushing installation and the hoist raise was within the competent intrusive rocks in the footwall. This placed any surface winding or conveyor installation within the open pit limits and necessitated the use of an underground hoist chamber and conveyor adit. The steep topography and predominance of unstable conditions on the surface, which had previously limited the choice of sites for the ventilation shafts, ruled out the possibility of locating the portal of the conveyor adit adjacent to the crushing plant. The only suitable location was the "island" of stable ground used for the blind bored ventilation shafts. This location entailed the use of additional conveyors underground and on surface to transport ore to the crushed ore stockpile and crushed backfill from the surface crushing plant to the mine.

A vertical conveyor system was selected as it offered the following advantages over a conventional skip hoist:

* installation was less complex and the capital cost was lower
* delivery and installation times were shorter
* estimated operating cost was lower due to significant savings in both power requirements and maintenance costs
* the need for winder drivers and winder maintenance skills was eliminated
* less development was required. If a skip hoisting installation was used it would require 25 - 30 metres vertically below the crusher location to accommodate a short pass, the loading station and shaft bottom equipment with a similar amount required above the 40 level conveyor for the sky shaft and head sheaves.

The first revised project schedule required expansion by January 1993. In this instance the crusher station was to be located on 16 level and two 120 metre lift Flexowell vertical conveyors in series were required to convey the ore to 40 level. Shortly after the commencement of production it became apparent that increased underground ore production would be required by September 1992. In order to meet this requirement, the crusher location was changed to 19 level to reduce the time required for access development. This new location allowed the selection of a single lift vertical conveyor installation which further simplified the design. Figure 8 shows the crushing and conveying system flow sheet.

This system was commissioned ahead of schedule in July 1992. Examination of the times achieved during the construction program confirms that the timing advantages of a vertical conveyor were achieved, as indicated by the following key dates:

* crusher excavation complete 13 February
* 40 level hoist room excavation

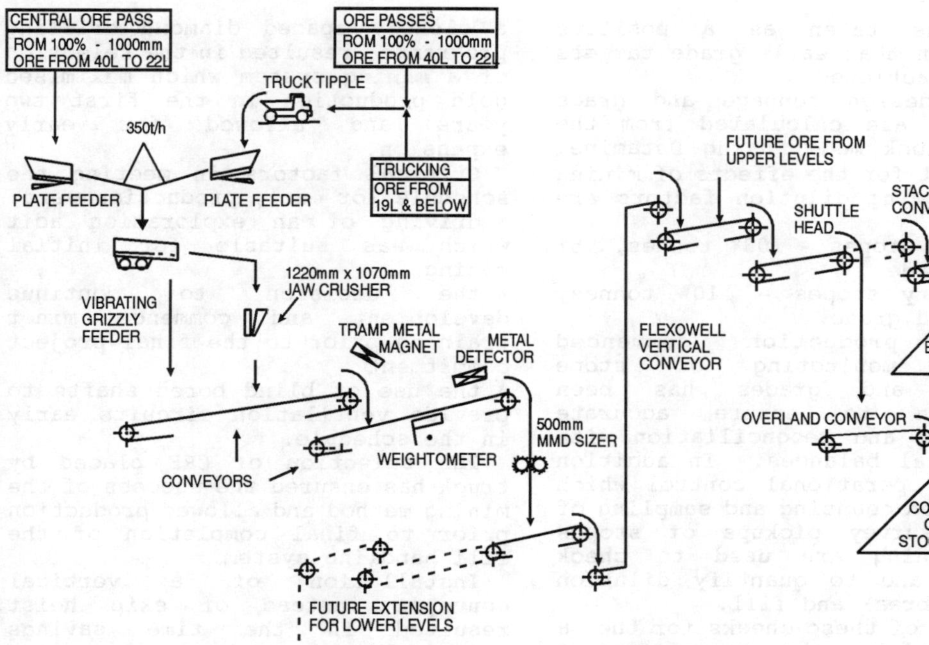

Fig.8 Ore handling flowsheet

complete 10 March
* vertical conveyor belt delivered to site 15 May
* commissioning of crusher station and vertical conveyor commenced 11 June.

Final commissioning of the system was delayed by late delivery of the surface stacker conveyor.

Critical to the achievement of this tight schedule were the detailed planning and engineering discussions held between Placer Dome Technical Services (Construction Manager) and the conveyor supplier on the handling, transport and installation procedures for the belt.

The belt minus sidewalls was supplied in one piece and was packed in two 10 metre long steel shipping containers joined end to end. This dual container which had a total weight of 50 tonnes was transported from the port of Lae to site on special heavy load transportation equipment previously used for large process plant items. It took 7 days to traverse the 654 km to site. At site the two containers were disconnected and a special articulation frame installed. With this frame fitted, the two containers were moved underground along the 40/L conveyor adit to the raise position using a combination of a bulldozer and LHD, a process which took only several hours.

PRODUCTION RECONCILIATION

As the mine was scheduled to produce preferentially from the ultra high grade areas in order to boost early revenue there was a degree of concern about the ability to achieve the grade targets and prior to production this was tested in two ways.

Firstly, closer spaced drilling (on 7.5 m x 7.5 m pattern) was conducted in the area of one very high grade stope. Secondly, geostatistical cross-validation of the estimated grade for the volume enclosing the first year of production was done by removing the data for entire drill holes on a one by one basis and kriging the grade at the location using only nearby drill holes. Both of these checks indicated that there appeared to be no systematic overstatement of grade within the area selected for mining, which was taken as a positive indication that early grade targets would be achieved.

Stope design tonnage and grade estimates are calculated from the kriged block model using Datamine. To account for the effects of mining the following dilution factors are applied:

* primary stopes - 103% tonnes, 95% kriged grade
* secondary stopes - 110% tonnes, 90% kriged grade.

Since production commenced detailed monitoring of stope tonnages and grades has been undertaken to ensure accurate accounting and reconciliation with plant metal balances. In addition to daily operational control which is based on counting and sampling of trucks, survey pickups of stopes after mining are used to check tonnages and to quantify dilution from overbreak and fill.

Results of these checks for the 88 week period from the commencement of operations indicates that cumulative mine tonnes and average diluted kriged grade was 1,212,196 tonnes grading 59.18 g/t Au compared with mill throughput of 1,174,169 tonnes grading 61.86 g/t Au.

On the basis of the kriged design the 31 stopes mined in this period were estimated to contain 925,074 tonnes grading 65.9 g/t Au. The reconciled estimates indicate that these stopes as mined contained 941,997 tonnes grading 65.5 g/t Au. These results also indicate that although some quite large differences have been observed for individual stopes, the errors have cancelled and the earlier conclusion that there was no systematic overstating of the grade within the high grade volume along the Roamane Fault was correct.

Dilution as calculated from stope volume surveys was 4.5% in primary stopes and 8.9% in secondary stopes which compares closely with the values selected.

CONCLUSIONS

The economics of the Porgera project were improved by early production from underground mining of the high grade Zone VII orebody. Detailed planning, which was based on grade and geotechnical data obtained from a closely spaced diamond drilling programme, resulted in the selection of a mining system which maximised gold production in the first two years and allowed for early expansion.

Critical factors in meeting the schedule for early production were:

* driving of an exploration adit which was suitable for initial mining
* the decision to continue development and commence miner training prior to the final project commitment
* the use of blind bored shafts to provide ventilation circuits early in the schedule.

The selection of CRF placed by truck has ensured the success of the mining method and allowed production prior to final completion of the fill handling system.

Installation of a vertical conveyor instead of skip hoist resulted in the time savings necessary to meet the programme for mine expansion.

ACKNOWLEDGEMENTS

The authors wish to acknowledge the work contributed by the PJV management and mining staff and Placer Dome Technical Services in the successful planning and development of this highly successful state of the art underground mine. They also wish to thank the PJV Manager for permission to publish this paper.

REFERENCES

Handley, G.A. & Henry, D.D. 1990. Porgera Gold Deposit. Mineral Deposits of Australia and Papua New Guinea, p.1717-1729. Australasian Institute of Mining and Metallurgy.

Innovative Mine Design for the 21st Century, Bawden & Archibald (eds) © 1993 Balkema, Rotterdam, ISBN 90 5410 325 6

Geotechnical principles governing mine design at the Williams Mine

D. Bronkhorst & J. Rheault
Williams Operating Corporation, Marathon, Ont., Canada
G. M. M. Ley
Golder Associates Ltd, Mississauga, Ont., Canada

ABSTRACT: The Williams Mine is Canada's largest gold producer, with annual production of 2.1 million tonnes of ore from a highly mechanized underground operation. The objective at the mine is to achieve total extraction of the orebody and mine design from the outset was based on sound geotechnical principles. As mining has progressed, basic parameters such as stope dimensions, sublevel interval, and sequencing have been refined to compensate for changing ground conditions. Effective ground control is maintained by using appropriate support systems and backfilling practices, and by controlling blast vibration damage. The results of computer modelling have been calibrated to observed conditions and measured stress levels. Modelling predictions have been combined with accumulated geotechnical data to identify areas where new mining methods will need to be considered. The geotechnical principles employed at the mine have allowed for near total extraction of the orebody to date and lend validity to high expectations of a long life for the mine.

1 BACKGROUND

The Williams Mine is one of three mines operating in the Hemlo gold deposit, located forty kilometres east of Marathon, Ontario, along the north shore of Lake Superior. Mine production is six thousand tonnes per calendar day.

The orebody extends from surface to a depth of 1300 metres, dipping approximately 70 degrees to the north. The average width of the orebody is 25 metres, but varies from 45 metres on the eastern property boundary to about 6 metres at the probable western limit and at depth. Proven and probable reserves currently stand at 32 million tonnes grading 6 grams per tonne.

The orebody on the Williams property has been divided into three mining zones (Figure 1). Original ore production was from the A-zone open pit while the A-zone and B-zone underground workings were being developed. The mining method employed underground is blasthole open stoping with delayed rockfill.

The mine is expected to operate well into the 21st century. Considerable effort has been given, therefore, to attaining the highest possible extraction with the least impact on ground stability.

This paper describes the mine planning process and the evolution of current design parameters with a focus on the role of geotechnical principles.

2 GEOLOGY

The Williams Mine lies on the south side of the east-west striking Heron Bay belt of metamorphosed Archean aged rocks (Muir 1982). The orebody lies along the contact between overlying metasedimentary rocks and underlying felsic volcanic rocks and dips north at 60 to 70 degrees.

The hangingwall rocks are banded, fine grained, pelitic metasedimentary rocks. A weak muscovite schist band, which varies from 1 to 15 metres thick, often separates the

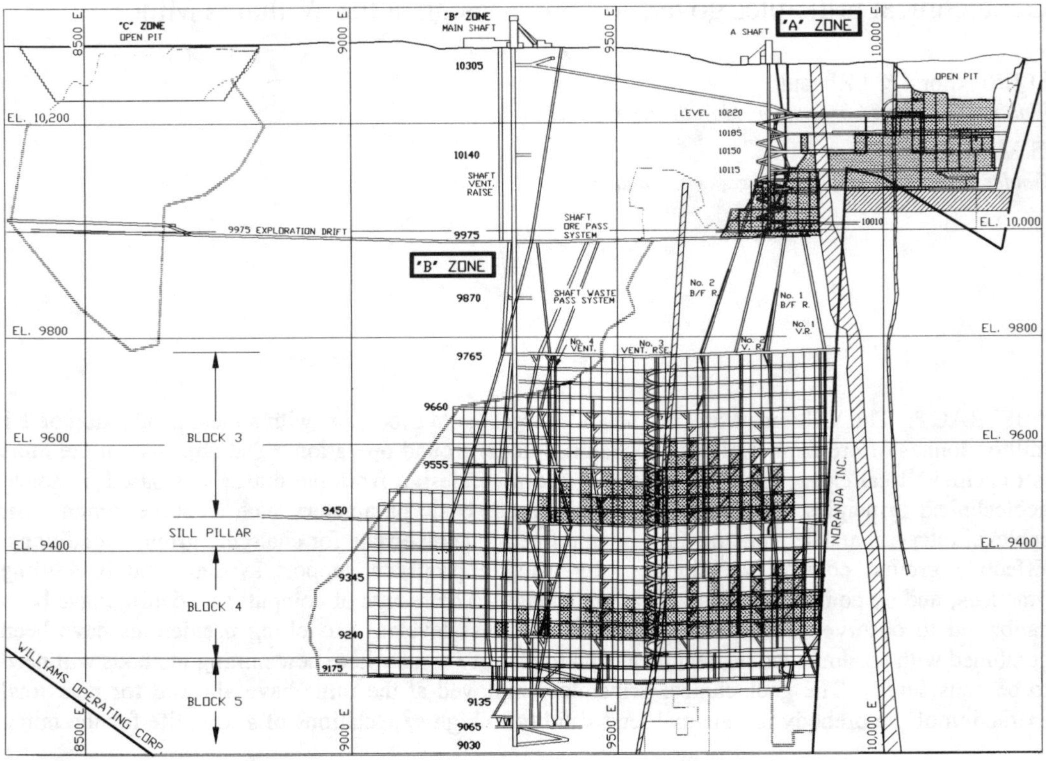

Figure 1. Orebody long section

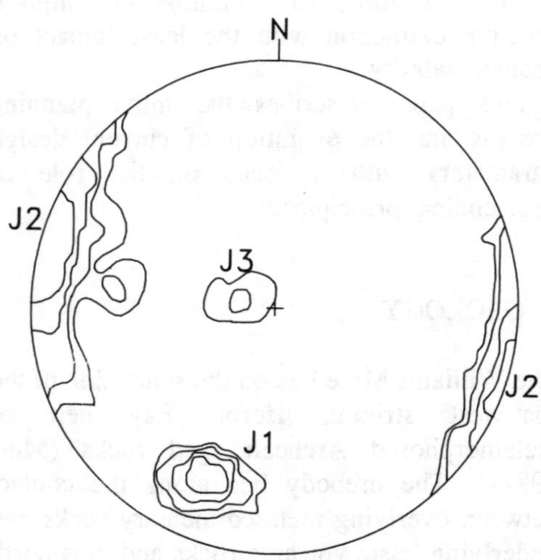

Figure 2. Lower hemisphere equal area polar contour plot of major discontinuities

orebody and the metasediments.

The orebody is composed of a fine grained, feldspar rich groundmass with fracture-controlled pyrite and barite mineralization running parallel to foliation. It is concordantly intruded by late felsic dykes and may also contain muscovite schist bands.

The footwall rock is quartz-eye muscovite schist or felsic quartz-eye porphyry.

3 ROCK MASS PROPERTIES

The initial rock mass classification program was undertaken in 1983. Mapping of discontinuities was performed on outcrops, highway cuts and surface sampling trenches. A lower hemisphere, equal area polar projection of the major discontinuity groups is shown on Figure 2.

The mapping indicated three major discontinuity sets:

Table 1. Rock mass properties

Rock Type	NGI (Q)	CSIR (RMR)	m_u	s_u	UCS (MPa)	Elastic Constants
FW. Quartz-muscovite schist	3.9	56.3	2.52	.0077	84	Representative values for all rock types: E = 35 GPa ν = 0.23
ORE. Baritic feldspathic schist	3.4	55.1	2.82	.0068	100	
HW. Muscovite schist	1.3	46.4	1.47	.0026	85	
HW. Banded sediments	3.5	55.4	3.05	.0070	115	

1. foliation (J1), parallel to the orebody, striking east-west.
2. a subvertical joint set (J2), striking north-south.
3. a weaker, sub-horizontal set (J3), striking east-west.

Mapping of underground openings, as they became available, confirmed the existence of these discontinuities throughout the orebody.

Laboratory testing to determine the intact rock strength was initially carried out on core selected from exploration boreholes. Subsequent testing has been carried out on core obtained at depth during mining.

The major sequences of hangingwall and footwall rocks are of high strength, but the band of sericitic schists containing the orebody is of only medium strength, with a gradational increase in strength towards the centre of the orebody. There are significant variations within the same rock unit, depending on the degree of silicification.

Values for the NGI 'Q' value (Barton 1976) and the CSIR rock mass rating (Bieniawski 1976) were determined from underground mapping and RQD values.

Rock mass strength parameters 'm' and 's' (Hoek and Brown 1980) were also determined.

Representative rock mass properties are summarized in Table 1.

4 IN SITU STRESS

In situ stress measurements were made at depths of 900 metres and 1000 metres, employing the overcoring technique and using diametral displacement (USBM) gauges and hollow inclusion (CSIRO) strain cells.

The combined results from 16 successful tests showed that the vertical stress is slightly greater than the calculated overburden weight and that the ratio of the average horizontal stress to the vertical stress ranges from 0.9 - 1.1 to 1, i.e. that the in situ stress regime at depth in the Williams Mine is approximately isotropic.

5 MINE DESIGN PARAMETERS

The value of the resource dictated a mining plan aimed at total extraction. Numerical modelling, using three dimensional and two dimensional methods, was initially used to examine various mining alternatives.

The conclusions drawn were:

1. Within the top third of the orebody (including all of A-Zone), great flexibility of layout and mining sequence was possible.

2. Induced stresses when mining at depth and at high extraction ratios in the B-Zone would be similar to, and probably exceed, the measured intact rock strength.

3. The sequence of mining should be upwards

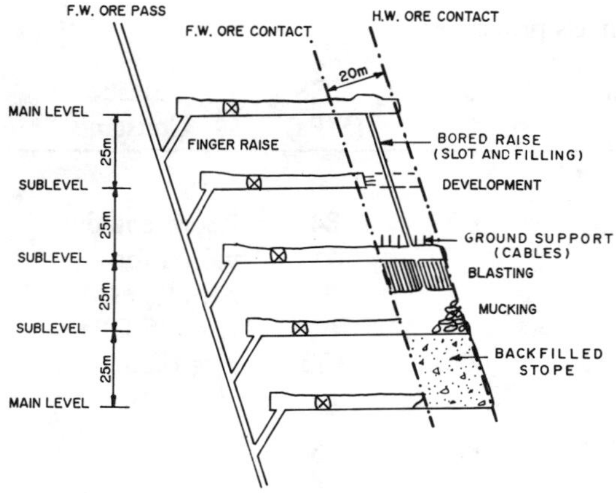

Figure 3. Cross section

from depth, as far as practicable, so that ground in the highest stress environment would be mined when the overall extraction ratio was low and the stabilizing effect of unmined abutments was greatest.

4. Pillars, which would attract high stresses, should be mined as soon as possible after primary mining.

5. The effect of mining close to property boundaries would be modest in terms of overall mining, but a co-operative approach between operators would be needed to optimize recovery, particularly at the boundary.

5.1 *A-Zone mining*

Some 3 million tonnes of ore were recovered from the A-Zone, which has now been completely mined. As expected, mining in the A-Zone provided invaluable information for developing optimum mining practices for the B-Zone at depth, which contains the bulk of the gold reserve and is in a higher stress environment.

Many design parameters were modified to improve recovery and mining efficiency, including sublevel interval, stope strike length, orebody development, drilling, blasting, backfilling and ground support.

5.2 *B-Zone mining*

B-Zone production started in the first quarter of 1987. Since that time, some 9 million tonnes have been recovered. Figures 3 and 4 illustrate the basic mining method for the B-Zone.

Sublevel interval

The sublevel interval was reduced to 25 metres from 35 metres to accommodate a reduction in blast hole size from 152 mm (6") to 115 mm (4 1/2"). This change improved drilling accuracy and increased the stability of the hangingwall by reducing its area and shortening the time required for backfilling.

Figure 5 shows a Mathews analysis (Golder Associates 1981) of three stopes: a double lift (i.e. mined two lifts before backfilling) A-Zone stope, a double lift B-Zone stope and a single lift B-Zone stope. The stability predictions correlated well with the observed behaviour of the three stope hangingwalls. The 3-16 stope (A-Zone) and the double lift 12 stope (B-Zone) experienced significant failure of the hangingwall, while the 10 stope stood well.

Double lift stopes had two problems. Not only was the hydraulic radius of the exposed hangingwall greater, but the stand-up time required was longer. Mucking time and backfilling time were about double those of a single lift stope. Despite the substantially greater availability of broken ore that the double lift strategy provided, it was abandoned in favour of greater stability and lower dilution.

Stope strike length

After serious failure of the hangingwall during an unsuccessful attempt to mine a strike length of 30 metres in the A-Zone, 22 metres was established as the standard stoping length for the A-Zone, and generally proved stable.

Consideration was given to 15, 20 and 25 metre strike lengths for stoping blocks in the B-Zone. A 25 metre strike length was considered too prone to an unacceptable level of dilution considering the weak nature of the hangingwall

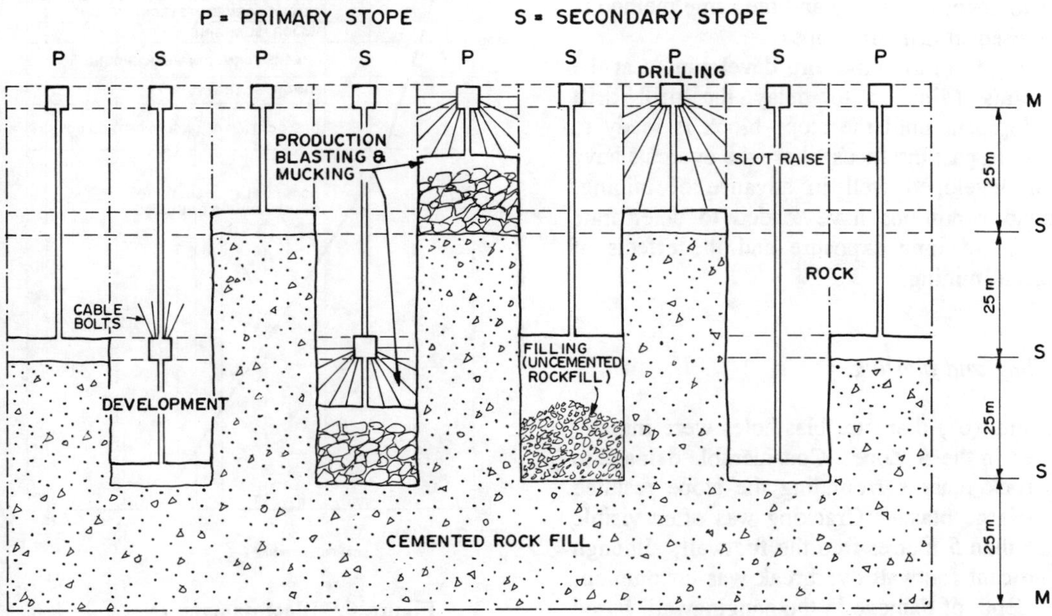

Figure 4. Longitudinal section

schists. A 15 metre strike length would not maintain an adequate pillar between adjacent drawpoints to provide a stable mucking access. Therefore, a 20 metre strike length was adopted as the standard for the B-Zone.

Attempts at longitudinal mining have proved to be inefficient in both the A-Zone and the B-Zone.

It was found that any backs more than about 8 metres wide which were opened along strike more than 30 metres were prone to failure. Longitudinal mining methods may have to be developed, however, for narrow ore at the extremities of the deposit.

Development in ore

Originally, primary stopes were silled to a full 20 metre strike length and from hangingwall to footwall. This allowed for parallel drilling of blastholes, resulting in cost efficient drilling and low powder factors.

Secondary stopes were initially silled at a strike length of 12 metres from hangingwall to footwall. This has been decreased to 8 metres and then to 6 metres in response to more difficult ground conditions. The significant increase in rock mass fracturing in secondary stopes, resulting from greater mining depth and higher overall extraction, can be more easily managed at the narrower crosscut width.

Fan ring drilling is used in the transverse drill drift. While this reduces drilling and blasting efficiencies, it has greatly improved the overall

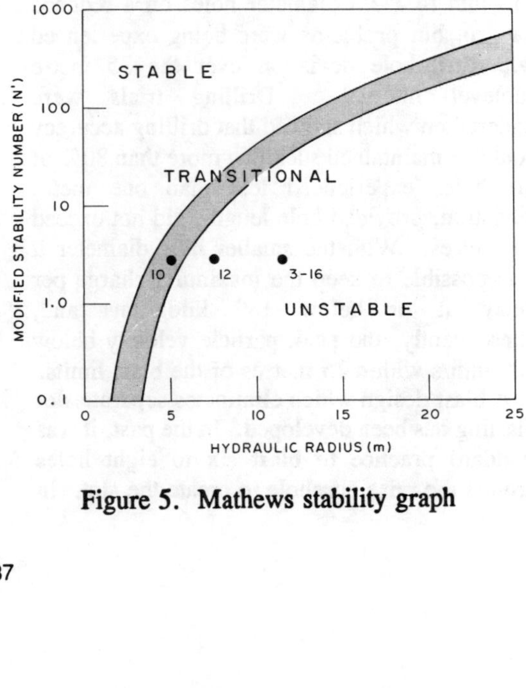

Figure 5. Mathews stability graph

development efficiency and the same method is now used in primary stopes.

The key to effective ore development at the Williams Mine is to delay the drill drift development until the stope block is ready to mine. In past instances where the overcuts have been developed well in advance of mining, ground conditions have tended to deteriorate because of long exposure and the effects of adjacent mining.

Drilling and blasting

152 mm (6") diameter blastholes were initially drilled in the A-Zone. Considerable damage to the rock mass surrounding the stope resulted from large blasts. Cracking was often visible more than 5 metres into the footwall, although significant footwall overbreak was uncommon. The depth of damage in the hangingwall, based on the observed degree of spalling, was usually less but invariably resulted in dilution.

Using vibration measurements from various blasts, and observations of resulting effects on the rock mass, a rock damage criterion was developed, as shown on Figure 6. For this rock environment, the onset of damage to the rock mass occurs at a peak particle velocity of around 500 mm/s. With the large diameter holes, it was not possible to reduce the charge weight per delay sufficiently to achieve this vibration level close to stopes.

By late 1988, drilling had been converted to 115 mm (4 1/2") diameter holes on a reduced pattern, but problems were being experienced with drill hole deviation over the 35 metre sublevel interval. Drilling trials were undertaken which showed that drilling accuracy could be maintained such that more than 80% of the holes experienced less than one metre deviation, provided hole lengths did not exceed 25 metres. With the smaller hole diameter it was possible to keep the maximum charge per delay at or below 180 kilograms and, consequently, the peak particle velocity below 500 mm/s within 15 metres of the blast limits.

A blast design which eliminates separate slot blasting has been developed. In the past, it was standard practice to blast six to eight holes around the raise borehole to create the slot. In

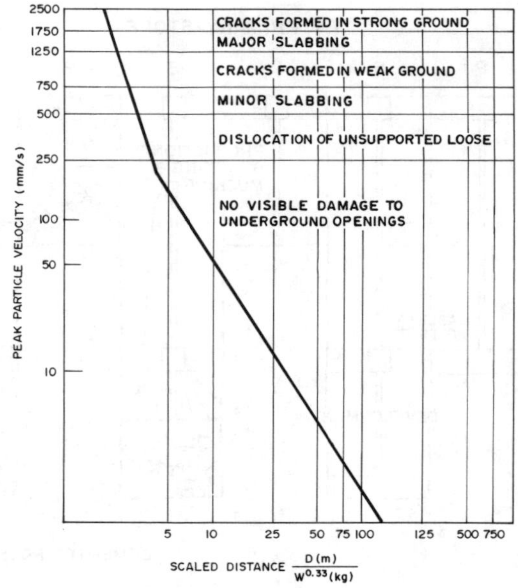

Figure 6. Blast damage curve

secondary stopes, this often caused substantial failure of the undercuts. In the current method, the blast is sequenced radially from the raise borehole, encompassing the entire undercut.

Two production blasts are taken over the 25 metre stope height. The first blast is just large enough to create sufficient void to contain the mass blast of the remainder of the stope block. This no slot/two blast method proved so effective that it has also been implemented in the primary stopes.

Backfilling

One of the keys to successful total extraction is the rapid placement of high quality backfill. In primary stopes, cemented rockfill controls hangingwall convergence and allows for total extraction of the secondary stope.

Graded rockfill with 5% Portland cement/flyash slurry addition is utilized in all primary stopes and in secondary stopes that are directly over a sill pillar.

Much has been done to establish and maintain a high quality backfilling procedure at the mine site. An elaborate quarrying, crushing, screening and storage operation on surface supplies the underground with a product consistent in size distribution (nominally 25% minus 19 mm) and moisture content. A fully automated surface batching plant separately provides the underground with slurry that is invariably within water:cement ratio specifications. Backfill trucks load rockfill and a measured quantity of slurry successively at underground fill stations and thorough percolation occurs during transport.

Backfill is placed with 26 tonne trucks dumping directly into open stopes via a production raise bore hole from the fill distribution level.

The operational flexibility of the trucks, high placement rates and the high quality of cured backfill have proven the effectiveness of this system. Dilution from backfill is below 2 percent.

The combined efforts to improve drilling, blasting, and backfilling have decreased the combined dilution from waste rock and backfill to between 6 and 10 percent.

Ground support

During attempts to mine multiple lifts, hangingwall support was installed in every overcut. It consisted of arrays of 5 inclined 10 metre long cable bolts at 2 metre horizontal spacings. The cables were linked by a rectangular pattern of steel straps and were tensioned to about 2 tonnes.

Since the move away from multilift mining, however, the use of hangingwall cable bolts was abandoned.

Cable bolting is applied routinely in the backs of all stopes using rows of four 7 metre, fully grouted, slightly fanned cables on a nominal 2 metre by 2 metre pattern at the toes. The cables are plated with 150 millimetre square steel plates and tensioned to 4 tonnes.

Frictional stabilizers are utilized, with 100 millimetre welded mesh where necessary, to provide primary support of openings in stope overcuts, while close following cable bolt installations provide deep seated anchorage of the rock mass.

Cable slings and steel fibre reinforced shotcrete have been applied on a trial basis in some long term openings or in particularly poor ground. To date, ground conditions have not required the routine application of such extensive surface support, although it is recognized that this may become necessary at some point.

6 INSTRUMENTATION

Many types of instrumentation are used on the property, but the most commonly employed types are vibrating wire stress cells and ground movement monitors.

Ground movement monitors are used as single point extensometers, anchored at roughly 7 metres depth, to detect near-surface movement in the rock mass. They are generally installed across visible structure in stope overcuts and monitored regularly for rate of movement, which is considered indicative of the localized state of stability.

Vibrating wire stress cells are strategically placed throughout the orebody to measure stress changes induced by mining activities in areas of concern. They have been used in secondary stopes to observe the redistribution of stresses as the adjacent primary stopes were extracted, and in primary stopes to indicate the response to approaching mining. Stress cells have been a valuable tool for calibrating the mine numerical models in order to improve the reliability of modelling predictions.

7 NUMERICAL MODELLING

Numerical modelling of proposed mining sequences has been undertaken periodically using Golder Associates' NFOLD code. The NFOLD program analyzes the three dimensional stress state in pillars and abutments in a tabular ore body as well as the changes in stresses and displacements resulting from mining activities.

Initial modelling assumed simple linear elastic rock mass behaviour, since the objective was to compare alternative mining strategies. The most recent modelling (November 1991) took into

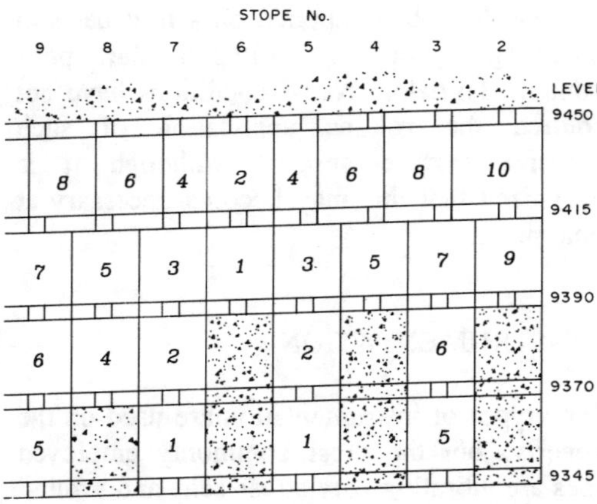

Figure 7. Sill pillar mining sequence

account peak and residual rock mass strengths, the strengths of rockfill and the measured in situ stress tensor.

Variations in observed ground conditions in active stopes were used to calibrate the model, and influenced the peak rock mass strength (60 MPa) selected.

The current five-year mining plan was modelled, taking into account the effects of projected mining on the adjacent property.

Modelling is also routinely carried out on site on microcomputers using the EXAMINE^{2D} and EXAMINE^{TAB} codes. Permanent and infrastructure openings, and any areas of general concern, are modelled to help assess ground support needs.

The long range schedule is also modelled on site and results correlate well with the NFOLD modelling results.

8 FUTURE MINING

When mining began in the B-Zone in 1988, two separate mining blocks were established in order to sustain the desired production rate of 6000 tonnes per day. The lower block (block 4) was started at the deepest level of proven ore. Since that time, exploration has proven the limit of the orebody below that area. A new area (block 5) is scheduled to come into full production in 1994.

Although a great deal of work has been done to refine the mining parameters to optimize the extraction of the orebody, some areas will require a different approach for full recovery of the ore.

One is the 50 metre high sill pillar between block 3 and block 4 (Figure 1). Modelling of the long range schedule indicated that a primary-secondary extraction sequence would not be a viable option in this area and work was done to devise a more appropriate mining sequence for this ore block.

A second area of concern is block 5 (Figure 1). This will be the deepest mining on the property and, with most of block 4 extracted, the most highly stressed.

The average ore thickness in block 5 is approximately 6 metres, but is as much as 45 metres in the block 3 sill pillar.

8.1 *Sill pillar mining*

Continuing the current primary-secondary sequence (Figure 3) in the sill pillar will induce a stress level which probably exceeds the peak strength of the rock mass in both the sill pillar and in parts of block 4. Changing to a pillarless retreat sequence (Figure 7) will reduce stress levels significantly throughout block 4.

A primary stope close to the east boundary will be the lead stope to punch through the sill pillar. To achieve a pillarless configuration, the adjacent secondary stope on the west side will have to pass the primary. The change in sequence involves opening a forty metre strike length at the overcut level. To minimize the potential ground support problems which this could cause, the pass will occur two lifts below the sill pillar, in a more favourable stress regime.

Subsequently, mining would retreat, using the newly established pillarless sequence, to the east and west. Further west, the ore thins considerably and the footwall drill drift method, discussed later in this paper, can be used.

One of the most interesting problems posed by mining the last lift of the sill pillar will be driving the drill crosscut. Optimum ground conditions are anticipated to result from driving the crosscut in the top of the rock, with the back of the crosscut in backfill. The cemented rockfill used in all the stopes over the sill pillar was of higher binder content than that normally used (7% versus 5%).

The fill is expected to be stable, not requiring elaborate support. Mined openings within the rock mass, however, are expected to require extensive support, designed to knit together a failed rock mass.

In the pillarless retreat method, all mining takes place in a highly stressed abutment. Ground conditions, however, are expected to remain more or less constant throughout the mining for all stopes in the sill pillar, allowing consistent application of mining and ground support procedures.

8.2 *Block 5 Mining*

Block 5 (Figure 1) consists of three 25 metre high lifts with an average ore thickness of 6 metres. It is proposed to maintain a primary/secondary sequence throughout the block, starting at the eastern property boundary and retreating to the west.

Modelling indicates that the last lift will be very highly stressed and this, combined with the ore thickness, raises some geotechnical concerns. The pillars (secondary stopes) will be squat (width to height ratio greater than 3) and will, therefore, attract a high level of stress. They may tend to fail suddenly rather than yield gradually. In order to avoid the danger to men and equipment caused by direct exposure to highly stressed and possibly burst-prone ground, the last lift will be mined remotely. Rather than establishing a transverse drill drift in the ore, the stope will be drilled off from a footwall drift as shown in Figure 8.

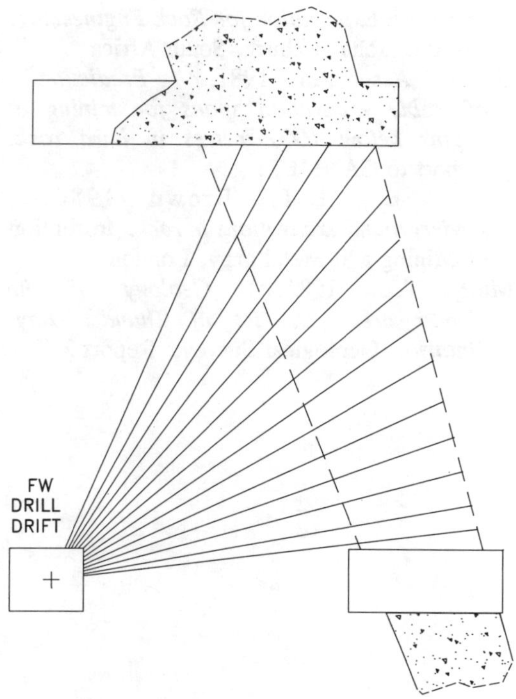

Figure 8. Footwall drill drift

9 CONCLUSIONS

The management of Williams Mine is committed to the safe and economical recovery of this major orebody. Sound geotechnical principles have been applied to mine design and operations from the earliest planning stage. Operating practices have been modified in response to changing ground conditions. As outlined in this paper, further changes will be necessary to meet the difficulties presented by high extraction ratios and deeper mining. Geotechnical input will play an increasingly important role in maintaining safe and cost-effective mining conditions and ensuring a long life for the mine.

10 ACKNOWLEDGEMENT

The authors thank their respective companies for permission to prepare and publish this paper.

11 REFERENCES

Barton, N. 1976. Recent experiences in the Q-system of tunnel support design. *Proc. Symp. on Exploration for Rock Engineering*. 107-117. Johannesburg, South Africa

Bieniawski, Z.T. 1976. Rock mass classification in rock engineering. *Proc.*

Symp. on Exploration for Rock Engineering. 97-106. Johannesburg, South Africa

Golder Associates 1981. *Prediction of stable excavation spans for mining at depths below 1000 metres in hard rock.* Report to CANMET.

Hoek, E., E.T. Brown 1980. *Underground excavations in rock.* Institution of Mining and Metallurgy, London

Muir, T.L. 1982. *Geology of the Hemlo area, District of Thunder Bay, Ontario.* Geological Survey, Report 217.

Innovative Mine Design for the 21st Century, Bawden & Archibald (eds) © 1993 Balkema, Rotterdam, ISBN 90 5410 325 6

Maximum span and stand-time of underground excavations

J.A. Franklin & M. Palassi
Earth Sciences, University of Waterloo, Ont., Canada

ABSTRACT: The stability of mine stopes, drifts and tunnels depends on time and the size of the opening, factors that continue to be ignored by most analytical solutions. Empirical design offers a promising alternative. Some concepts and rules for empirical design are proposed. The Lauffer-Bieniawski type of span-stand-time diagram is potentially a very powerful tool, but the rock quality contours need to be extrapolated and relocated. A family of logarithmic curves with asymptotes to the time and span axes appears to provide a satisfactory fit to the data, as a replacement for straight line RMR contours. More data are needed. The maximum span that can be mined without support is discussed. Interchangeability of RMR, Q and other systems of classification seems now to be well-established, and suggests the possibility of a unified and simplified rock mass classification. Examples are given of progress toward the development of expert systems for underground excavation design and control.

1 INTRODUCTION

Two of the most important questions facing the designer of underground works are the maximum size of opening that will remain stable indefinitely without support in rock of any given quality, and the "stand-time" during which a larger excavation than this will remain stable while support is installed.

Continuum mechanics solutions fail to address these questions let alone answer them. They predict stresses and displacements but say little about stand-time or the size of opening that can be mined. The solutions are usually dimensionless and independent of time. They ignore two fundamental principles well-known to the miner and tunnelling contractor, namely:

- a small excavation is more stable than a large one, and remains stable for longer;
- the size of opening that can be mined depends on the quality of the rock.

This is why large stopes, caverns and tunnels in poor rock are excavated in stages by first mining and stabilizing a small top-heading, then benching or slashing to the full required dimensions. The self-supporting capability of the rock is fully mobilized by selecting a mining method appropriate for the ground. The choice ranges through a broad spectrum from open stoping in good rock, through room and pillar, cut and fill, VCR, shrinkage stoping, and caving methods as ground conditions deteriorate. Each method has its characteristic range of unsupported opening sizes.

Anchoring, bolting and other forms of support are employed only as a last resort, when smaller unreinforced openings or stopes supported by broken ore or backfill are considered uneconomic.

To choose a mining method, design openings and select blasting and bolting patterns we still rely almost entirely on judgement and empiricism. Unfortunately, empirical design methods have long been neglected in the search for more "scientific" analytical alternatives, and as a result, the empirical methods remain of only moderate use to the miner. In the following discussion, we examine some limitations of present empirical methods, and how they might become more useful if placed on a sounder scientific basis.

2 STEPS IN EMPIRICAL DESIGN

Empirical design is the systematic use of knowledge gained from previous works to make decisions on present and future works in similar ground. Good empirical design goes beyond the judgement of a single individual by making available the experience of others who have operated in a much broader range of ground conditions. Empirical design therefore requires three essential components:

- Input on ground quality by a quantitative and reproducible classification system.
- Output for a full range of design parameters such as stand-time, support requirements, and optimum excavating methods.
- Correlations of input with output by compilation of case histories for a full spectrum of ground conditions. Contours are drawn separating "successes" from "failures". The correlations are progressively improved by adding further data.

This can be expressed by a functional relationship as follows:

$$P(D, E, R) = f(G, S)$$

where:

P =	performance parameters
D =	opening dimensions
E =	excavating methods
R =	reinforcement and support methods
f =	functional relationship determined from experience (case histories)
G =	ground quality parameters (ground classification)
S =	desired safety factor based on acceptable risk, skill and experience of personnel, etc.

A basic premise is that *if the "quality" of the ground can be adequately defined, so can its "behaviour". For every type of ground there exists an optimum mining method, and this can be determined.* We can define a number of desirable attributes of an empirical design that will lead to superior predictions: *the selected parameters should be relevant, necessary and sufficient, clearly and unambiguously defined, quantitative, and easy to measure. The functional relationship should be based on an adequate number of relevant data; quantitative, probabilistic, and easy to apply.*

3 SPAN-STAND-TIME DIAGRAM

3.1 Lauffer and Bieniawski charts

The questions "how large an opening can I excavate", and "for how long will this stand without support" were addressed by Lauffer (1958) who developed a rock mass classification in the form of a "span-stand-time diagram". This was employed as a basis for the New Austrian Tunnelling Method (NATM). His classification was qualitative. It derived rock quality from the observed behaviour of a tunnel, and without a tunnel, the rock could not be classified except in a broad and descriptive sense.

Bieniawski reversed this process by classifying the rock according to observable properties, and predicted tunnel performance from the results. He developed the Rock Mass Rating (RMR) system with an accompanying Lauffer-type plot of roof span versus stand-time for various rock mass classes (Bieniawski 1973, 1974). Originally, 49 case histories were investigated. 62 coal mining case histories were added by 1984 and a further 78 tunnelling and mining case histories by 1987 (Bieniawski, 1989).

Figure 1 shows Bieniawski's 1989 diagram which records stand-time, span and RMR for cases of observed roof collapse in tunnels and mines. His observations fall within upper- and lower-bound curves which he defines as "limits of applicability" for the data. The bounding curves do not attempt to predict stable or unstable conditions; they serve only to outline the current limits of the data. With few exceptions, the data fall within a rather limited region of the diagram where stand-times range from 5 hours to 1 year, spans from 3 m to 15 m, and rock qualities from RMR 35 to RMR 70.

As further data are collected and added, the bounding curves will need to be relocated. Many larger stopes and natural caverns exist, even a few larger than 200 m, and many openings in rock of superior quality with RMR 70 - 100. Further data can be collected for unsupported openings in rock of both good or bad quality which can remain stable for many years or just a few minutes depending on the span.

This clarification is needed because the annotation "no support required" towards the lower right of Bieniawski's chart can easily be misunderstood to imply that the lower-bound curve predicts the *ultimate unsupported span* S_∞, defined here as *the largest span that will remain stable for an infinite time without any form of support or rock treatment* (these and subsequent definitions are a bit pedantic but precise terminology may help to eliminate some of the present confusion).

Bieniawski's lower bound curve has nothing to do with S_∞, as can be seen from the stand-times, from one hour to 100 years, that pertain along the curve. Similarly, in spite of the annotation "immediate collapse" in the upper left of Bieniawski's chart, the upper bound curve should not be confused with the *instant collapse span* S_0, defined here as *the largest unsupported span that can be mined without instant collapse.*

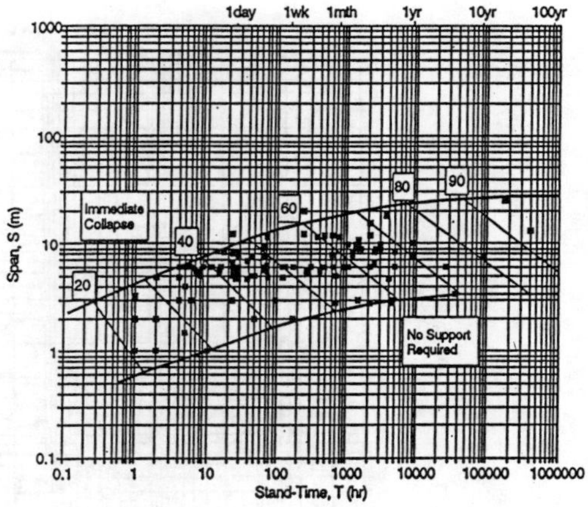

Fig. 1 Bieniawski's 1989 span-stand-time diagram

3.2 Rock quality contours

Values for S_∞ and S_0 correspond to intersections of rock quality contours with vertical lines at stand-times of infinity and zero. These theoretical extreme values are located at infinite distances to the right and left of the design chart and are of no great interest to the mine engineer. Two values of more practical interest for design are the intersections of a given rock quality contour with the right and left edges of a design chart such as in Figure 2. We will call these the *100-year unsupported span* S_{100y} and the *1 hour unsupported span* S_{1h}.

After application of an appropriate safety factor, S_{100y} may be considered an *acceptable maximum span for long-term stability*; this value at 100 years is likely to be very similar to S_{10y} or S_{1000y} in most cases. Only for mine stopes and temporary engineering works with an operating life of a few months would we expect S_p, the

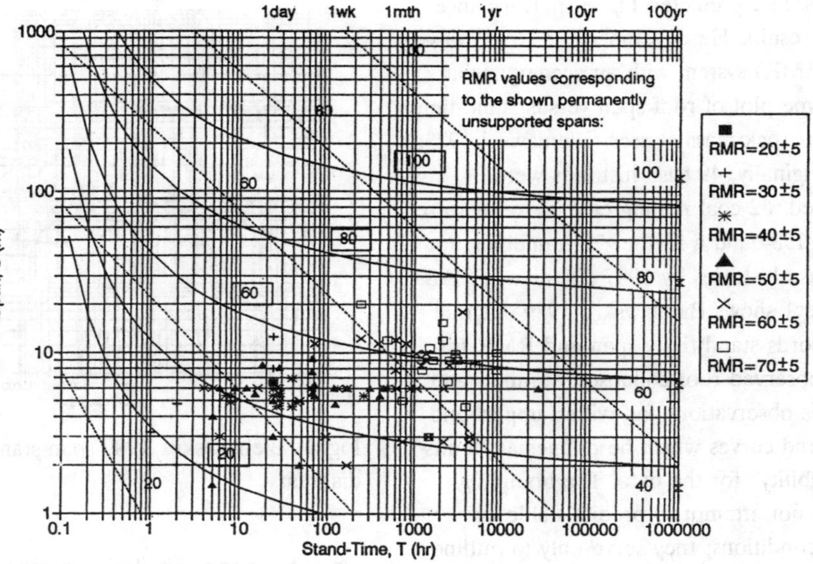

Fig. 2 Modified span-stand-time diagram. Bieniawski's straight-line RMR contours (extrapolated) are compared with the curvilinear RMR contours proposed in the present paper

permissible unsupported design span, to increase significantly. In other words, at the long-time end of the rock quality contour, the stable span is likely to approach a constant value as stand time tends to infinity. Hence we suggest that *the rock quality contours (whether expressed in the RMR, Q or any other rock classification system) are likely to be hyperbolic or logarithmic in shape, becoming asymptotic to the vertical at T = 0, and to a horizontal S_∞ line as the stand-time increases.* Other constraints apply. For example, the contours cannot curve in a U-shape, because this would lead to two or more alternative stand-times for a given size of opening in a given quality of rock.

Properly located rock quality contours are the key to converting a span-stand-time diagram into a practical and useful design tool. Contours are difficult to obtain when only quite a small region of the diagram is populated by data as at present, and when only failures, not successes (stable openings) have been recorded. However, with the help of the above hypothesis regarding asymptotes, we were able to obtain a more realistic set of contours by fitting a logarithmic surface to a three-dimensional data set (S, T, RMR) which included Bienawski's 1989 data plus Barton's (1976) data for long-term stable spans. The latter were converted from Q to RMR using Bieniawski's (1976) correlation. After testing and selecting the form of equation, parameters in the equation were quantified by a non-linear regression analysis using the "SYSTAT" program (SYSTAT Inc., 1990).

$$Log\,S = \frac{-Log\,T}{Log\,T + 1.70} + 0.027\,RMR \qquad (1)$$

where

S = Span of the opening (m),
T = Stand-time (hr), and
RMR = Rock mass rating.

This family of curves, derived from a single expression, is plotted in Figure 2. Clearly the curves and straight lines give different predictions. The curves provide an improved fit even within the region of Bieniawski's data, but particularly in the extended region that includes long-term span information such as Barton's. We therefore propose using curved contours not straight lines in future applications.

To further examine the fit of Bieniawski's RMR lines to his data, we employed the SYSTAT multiple regression to fit a planar surface to the data in three-dimensions. The resulting linear best-fit RMR contours turned out to be steeper than Bieniawski's, with a large discrepancy in some cases. The lines became less steep when Barton's long-term data were added to the set.

The straight line contours are sensitive to the addition or subtraction of just a few data points, particularly if these lie outside the region of Bieniawski's present data. Even using curved contours which are less sensitive, further data clearly are needed in thinly populated regions of the chart. Data for longer stand-times in the range 10 - 100 years are available from records of natural caverns and abandoned mine stopes and drifts. Cases of short and moderate stand-times in poor quality rock can be studied in mines employing caving or longwall methods. The data set should cover stable conditions as well as collapses; the most reliable contouring is obtained by separation of failed and stable cases.

4 MAXIMUM UNSUPPORTED SPANS

4.1 Barton's and Carter's Predictions

The unsupported spans predicted by various researchers can be plotted and compared on a single span-stand-time diagram as shown in Figure 3, making use of the correlations discussed in 4.2 below. Although most researchers make no attempt to predict stand-time, both span and stand-time are essential pieces of information, and without one, the other is insufficiently defined.

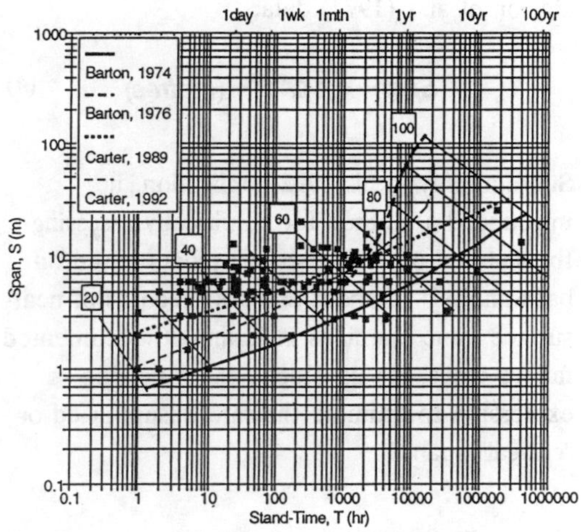

Fig. 3 Predictions of "maximum unsupported span" by Barton and Carter, transposed to a span-stand-time diagram and compared with the Bieniawski data set

Barton et al. (1974) established a plot of the equivalent dimension of the opening D_e (span, diameter, or height divided by ESR) versus rock mass quality Q, where the excavation support ratio ESR is an inverse factor of safety selected according to the use of the opening. An ESR=1 has been employed to plot the Barton (1974) curve: the maximum unsupported span (metres) is expressed by:

$$D_e = 2\,Q^{0.4}$$
$$or\quad Span = 2\,ESR\,Q^{0.4} \qquad (2)$$

Barton (1976) pointed out that with only a few discontinuous joints and Q>500, the maximum unsupported span can exceed 200 m and is limited only by the ratio of rock stress to rock strength. The limiting envelope is more or less linear and is represented in metre units by:

$$Span = 2\,Q^{0.66} \qquad (3)$$

Carter (1989) developed a limiting span relationship based on Bieniawski's (1973) and

Barton et al.'s (1974) data:

$$Span = 4.4\,Q^{0.32} \quad (metres) \qquad (4)$$

Subsequently in his 1992 publication, he introduced a method for empirically assessing the stability of surface crown pillars based on back-analysis of about 200 case records of near-surface mine openings including 30 documented failure cases. He developed the following expression to estimate the safe unsupported or "critical" span S_c:

$$S_c = 3.3\,Q^{0.43} * [Sinh^{0.0016}(Q)] \quad (metres) \qquad (5)$$

4.2 Conversion from Q to RMR

To allow comparison of predictions based on the Norwegian Q system of rock quality classification with Bieniawski's span-stand-time data which uses the RMR system, Q has been converted to RMR by applying the correlation that exists between the two systems. The correlations proposed by Bieniawski (1976), Rutledge and Preston (1978), Abad et al. (1983), and Kaiser and Gale (1985) are respectively:

$$RMR = 9\,Log_e Q + 44 \qquad (6)$$

$$RMR = 13.5\,Log_e Q + 43 \qquad (7)$$

$$RMR = 10.5\,Log_e Q + 41.8 \qquad (8)$$

$$RMR = 8.5\,Log_e Q + 35 \qquad (9)$$

Bieniawski's correlation based on 111 civil engineering case histories has been confirmed by Jethwa et al. (1982) for 12 case histories in India. Rutledge and Preston's correlation based on seven tunnelling projects in New Zealand was found by Brumleve and Maier (1981) to apply also at the Henderson mine in Colorado. Abad's correlation established from measurements in 187 coal mine roadways in Spain was confirmed by Moreno Tallon (1982) for rock conditions in four Spanish tunnelling projects.

The predictions of maximum unsupported span as a function of rock quality, converted to RMR using Bieniawski's correlation, are compared in Figure 3 which shows the predictive equations superimposed on Bieniawski's data and RMR contours. The curves of Barton and Carter are similar to each other and also pass through the central to lower-bound trend of Bieniawski's data. They diverge in the region of superior rock quality, RMR>70, for which Bieniawski gives few data.

Note the broad range of unsupported stand-times corresponding to the "maximum spans" proposed, which can be just a few hours or days in poor quality rock. Bieniawski has recorded many cases of short-term roof collapse in excavations that Barton and Carter would have predicted as stable in the long-term. This suggests that the Barton and Carter equations may predict not the long-term unsupported stable span S_{100y}, but a value closer to S_p, the maximum span that can be excavated while leaving enough time for support installation. Alternatively if we take the Barton and Carter predictions as correct for long-term stability, they should plot along the S_{100y} axis, requiring a change to the RMR contours on Bieniawski's diagram.

5 UNIFIED CLASSIFICATION

The interchangeability of RMR and Q classifications demonstrated by the existence of a substantial degree of correlation between the two suggests that *existing classifications can be unified and simplified with little loss of predictive power*. If so, this will greatly facilitate the further development of empirical design tools, and make them much easier to use.

This close correlation exists in spite of some differences in the rock quality attributes on which the Q and RMR systems are based and how these are combined, suggesting that these and other systems may have much in common. In the early 1970s a survey conducted by the ISRM Commission on Rock Mass Classification found that most classifications rely heavily on just two parameters, joint spacing and compressive strength of the rock, i.e. the size and strength of the "blocks" that comprise the rock mass (Franklin 1970, 1976).

Although a two-parameter size-strength classification is simple to apply, a unified classification should probably include a third parameter namely the shear strength of block boundaries, which is an important independent variable in both the Q and RMR systems. The three attributes together (block size, intact and boundary strength) are likely to account for by far the greatest part of rock quality variation. The addition of further parameters is unlikely to prove worthwhile. Several researchers have suggested that water and stress be omitted, as in the Mathews modification of the Q system. Groundwater and stress are transient properties and difficult to define and to measure, and can be taken into account separately when needed.

A unified rock mass classification is an essential step towards developing improved empirical predictive tools. Correlations between existing systems and between parameters within systems suggest how unification can be achieved. The aim is not to develop yet another classification, but to simplify and combine existing systems that contain the required elements. This will reduce rather than add to the present complexity, and will allow retention and use of the extensive existing databases relating Q, RMR and other parameters to observed rock behaviour.

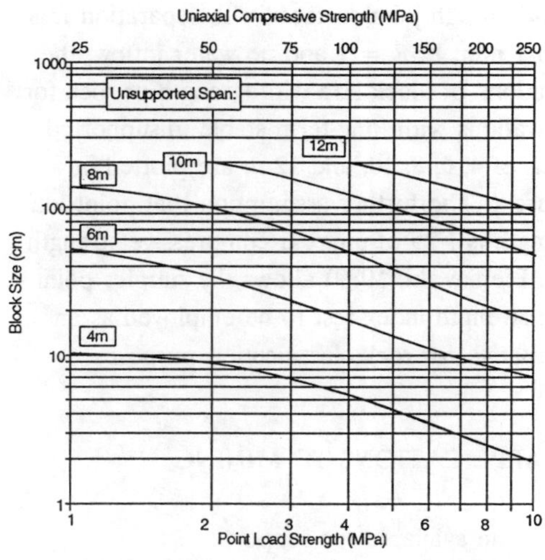

Fig. 4 Contours of maximum unsupported span S_{100y} plotted on a size-strength classification diagram

6 SPANS ON SIZE-STRENGTH AXES

As a further example of conversion between classification systems, Figure 4 shows contours of long-term unsupported span plotted on a Size-Strength classification chart. This has been obtained by combining Equation (2) which gives the Barton et al. (1974) prediction of span for permanently unsupported excavations, with Equation (6) which is Bieniawski's (1976) correlation between RMR and Q systems:

$$Span = 2\,ESR[\exp(\frac{RMR-44}{9})]^{0.4} \qquad (10)$$

According to Priest and Hudson (1976), RQD is correlated to joint spacing (block size) as follows:

$$RQD = 100\,e^{-0.1\lambda}(0.1\lambda+1) \qquad (11)$$

where λ is the average number of joints per metre along a sampling line (e.g. drill core), and $1/\lambda$ is the average joint spacing. Using this approach and taking the common condition of slightly rough joint surfaces with separation less than 1 mm, ESR = 1 and no water inflow, the variations of block size with strength of rock for excavations with long-term stable unsupported spans of 4, 6, 8, 10, and 12 m are plotted in Figure 4. The further assumption that point load strength is 1/25 of uniaxial compressive strength (e.g. Bieniawski, 1989) allows the simpler point load strength index test to be employed as an alternative for rock classification purposes.

7 APPLICATIONS TO MINING

7.1 Time available for support

Span-stand-time predictions have important applications to the planning of support installation. The permissible unsupported design span S_p corresponds to the maximum span that can be excavated while leaving enough time for rock reinforcement. Larger openings take longer to support; a 3 m span advanced 10 m by blasting can for example generally be rockbolted in a single 8-hour shift, whereas a 5 m span might require two shifts or a second crew working back from the face. In drill-and-blast operations, depending on the predictions of unsupported stand-time, the size of opening may need to be reduced, working in top-heading and bench through sections of difficult ground to maintain a stable heading and to ensure that support keeps pace with advance.

If a tunnel is to be bored, stability at the selected diameter combined with the estimated rate of advance will determine the need or otherwise for a shielded machine, the possibility of squeeze bringing a machine to a halt or impairing its efficiency, and the ground pressure on the final liner.

Larger excavations take more time to support not only because of their greater size, but also because they require different types and intensities of treatment, e.g. longer and more closely spaced bolts, thicker and more extensive shotcrete. Planning becomes an iterative procedure where first the support requirements are predicted from rock quality and excavation dimensions using an empirical support prediction tool such as the one proposed by Grimstad and Barton (1988). Time required for installation can then be estimated and compared with the unsupported stand-time obtained from the span-stand-time diagram or formula. If the available time for support is insufficient, stability can be maintained by reducing or staging the size of excavation, limiting the rate of advance, or increasing the numbers of ground control personnel.

7.2 Expert systems

The complex interactive decision-making typical in underground work is an ideal application for an expert system. For those unfamiliar with the rapidly developing field of artificial intelligence, an expert system is an intelligent computer program that uses knowledge and inference procedures to solve problems that are difficult enough to require significant human expertise for their solution (Walker and Miller, 1990). It contains a knowledge base of facts and rules associated with the problem; an inference engine for solving the problem; and a working memory (global database) for keeping track of the problem's status. The inference engine receives data from the user and applies the knowledge base to solve the given problem.

Expert system programs are modular, so that additions and changes can be made to one module without affecting the workings of others. They can contain and assist in the management and application of a variety of empirical and analytical tools. As the knowledge of more and more human experts is incorporated in the program, the level of expertise rises and soon can exceed that of any specific human. The program can pinpoint inconsistencies between one expert and another. It becomes an excellent teaching tool, as well as a practical one to apply.

However, to develop a really useful expert system for underground mining, the classifications and correlations employed as a basis for prediction will need to be improved. Empirical methods also will need to address a much broader range of practical questions, not only those related to span, stand-time and support requirements, but also to choice of excavating methods and prediction of rates of advance, water inflows, pumping and muck handling requirements.

7.3 Classex

An expert system called "Classex" was developed by Butler and Franklin (1990), which incorporates three empirical methods of rock tunnel design based on the Q, RMR, and RQD rock mass classification systems. Classex uses VP-EXPERT software and consists of eighteen separate knowledge bases which allow the user to arrive at best-estimates of Q and RMR, then to apply these to tunnel design. Classex is available as part of the ISRM educational software collection, through the ISRM Secretariat in Lisbon, Portugal.

The main problem with Classex and similar expert systems is that if wrong input is given to the program, the mistake is not easily corrected and the user must start the consultation again from the beginning. Also, Classex was designed to rigorously apply the existing rock mass classifications and predictive correlations with all of their current limitations. This system, therefore, requires modification to improve its use and also to expand its data base and the scope of output.

7.4 Spreadsheets

To overcome the inflexibility of a linear expert system we are experimenting with interlinked

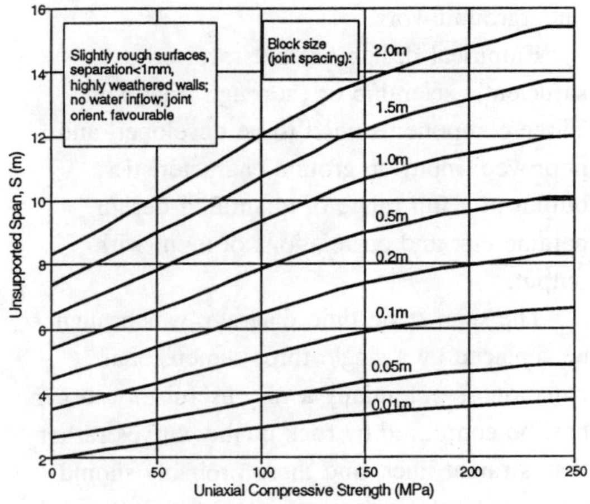

Fig. 5 Example of a sensitivity analysis using interlinked Quattro-Pro spreadsheets

spreadsheets which allow input adjustments while observing on the spreadsheets and in graphs the effects of these modifications. Predictions are rapidly optimized, for example estimation of RMR from rock mass attributes, and the user can then move on to further spreadsheets that apply the results to estimation of stable spans, stand-times, and optimization of excavation and support methods. The program makes use of Quattro-Pro spreadsheets, the RMR system, and correlations between parameters such as outlined in this paper.

The example given in Figure 5 demonstrates use of this spreadsheet program to investigate the sensitivity of unsupported span to variations of intact strength and block size. This graph was prepared in about 5 minutes, whereas Classex takes much longer to produce similar results.

8 CONCLUSIONS

Small excavations are more stable and remain stable for longer: although often overlooked, this basic rule of ground behaviour is fundamental to the choice of mining method and to the dimensioning and sequencing of underground works.

Empirical design has not received a sufficiently scientific or thorough treatment. Three components need to be developed and improved: input on ground characteristics; output of a full range of quantified design parameters; and correlations of input with output.

The span-stand-time diagram, which might be replaced by a single three-dimensional equation, is potentially a very useful tool. It can best be contoured by rock quality curves rather than straight lines, and these probably should satisfy asymptotic requirements at $T = 0$ and $T = \infty$. Systematic statistical methods (multiple regression) can be used to locate these contours on the diagram. One form of contouring equation has been suggested; others may be possible. The curved RMR contours are tentatively located and should be checked and revised after gathering further data particularly in thinly populated regions of the design chart.

The Barton and Carter equations of "maximum span" are similar but appear to give anomalous stand-times when converted to a span-stand-time format. We should reconsider whether they predict S_{100y}, the long-term unsupported stable span, or a value closer to S_p, the maximum span that can be excavated while leaving enough time for support installation.

The interchangeability of RMR and Q classifications demonstrated by the existence of a substantial degree of correlation between the two suggests that these and other classifications can be unified and simplified with little loss of predictive power. Derivation of a unified rock mass classification is an important step in developing improved empirical predictive tools. Empirical methods will need to address a much broader range of practical questions, not only those related to span, stand-time and support requirements, but also to choice of excavating methods, and prediction of rates of advance, water inflows, pumping and muck handling requirements. The complex interactive decision-making that is typical of underground work is an ideal application for an expert system approach.

REFERENCES

Abad, J., Celeda, B., Chacon, E., Gutierrez, V., and Hidalgo, E., 1983. Application of geomechanical classification to predict the convergence of coal mine galleries and to design their supports. Proc. 5th Int. Cong. Rock Mech., ISRM, Melbourne, Vol. 2, pp. E15-E19.

Barton, N., Lien, R., and Lunde, J., 1974, Engineering classification of rock masses for the design of tunnel support. Rock Mech., Vol. 6, pp. 189-236.

Barton, N., 1976, Recent experiences with the Q-system of tunnel support design, Proc. Symp. Exploration for Rock Engineering, Johannesburg, A.A. Balkema, Vol. 1, pp. 107-117.

Betournay, M.C., 1987, A design philosophy for surface crown pillars of hard rock mines. CIM Bul., Vol. 80, No. 903, pp. 45-61.

Bieniawski, Z.T., 1973. Engineering classification of jointed rock masses. Trans. S. African Inst. of Civ. Engrs., Vol. 15, pp. 335-344.

Bieniawski, Z.T., 1974, Geomechanics classification of rock masses and its application in tunnelling, Proc. 3rd Int. Cong. Rock Mech., Denver, ISRM, Vol. IIA, pp. 27-32.

Bieniawski, Z.T., 1976, Rock mass classifications in rock engineering, Proc. Symp. Exploration for Rock Engg., Johannesburg, A.A. Balkema, Vol. 1, pp. 97-106.

Bieniawski, Z.T., 1989, Engineering Rock Mass Classifications, John Wiley and Sons, New York, 251 pp.

Brumleve, C.B. and Maier M.M., 1981. Applied investigation of rock mass response to panel caving, Henderson mine, Colorado, U.S.A. Design and Operation of Sublevel Stoping

Mines, E.D. Stewart ed., A.I.M.E., New York, pp. 223-249.

Butler, A.G. and Franklin, J.A., 1990. Classex: an expert system for rock mass classification. Proc. ISRM Int. Symp. Rock Mech., Mbabane, Swaziland, pp. 73-80.

Carter, T.G., 1989. Design lessons from evaluation of old crown pillar failures. Proc. Int. Conf. on Surface Crown Pillars for Active and Abandoned Metal Mines, Timmins, Ontario, pp. 177-187.

Carter, T.G., 1992. A new approach to surface crown pillar design. Proc. 16th Canad. Rock Mech. Symp., Sudbury, Ontario, pp. 75-83.

Duffaut, P. 1982. Les facteurs qui limitent la taille des cavités. Proc. Int. Symp. Rock Mech. Caverns and Pressure Shafts, Balkema, Rotterdam, Vol. 1, pp. 245-253.

Franklin, J.A., 1970. Observations and tests for engineering description and mapping of rocks. Proc. 2nd Int. Cong. Rock Mech., Belgrade, Vol. 1, pp. 11-16.

Franklin, J.A., 1976. An observation approach to the selection and control of rock tunnel linings. Proc. Conf. Shotcrete for Ground Support, ASCE, Easton, Md., pp. 556-596.

Grimstad, E. and Barton, N., 1988. Design methods of rock support. In: Norwegian Tunnelling Today, Norwegian Soil and Rock Eng. Assoc., Pub. 5, pp. 59-64.

Jethwa, J.L., Dube, A.K., Singh, B., and Mithal, R.S., 1982. Evaluation of methods for tunnel support design in squeezing rock conditions. Proc. 4th Int. Cong., Int. Assoc. Eng. Geol., Delhi, Vol. 5, pp. 125-134.

Kaiser, P.K. and Gale, A., 1985. Evaluation of two rock mass classification systems. Proc. Mexroc '85, Int. Symp. Rock Mech., Mexico D.F., pp. 1-7

Kendrick, R. 1970. Induction caving at the Urad mine. Min. Congr. J., Vol. 56, No. 10, pp. 39-44.

Lauffer, H. 1951. Classification of rocks for tunnel construction. Geologie-und Bauwesen 1951 Heft 1.

Maury, V. 1977. Environmental protection, monitoring and operation at the May sur Orne underground oil storage facility. Proc. Rockstore 77, Int. Symp., Pergamon Press, Oxford, Vol. 1.

Moreno Tallon, E., 1982. comparison and application of the geomechanics classification schemes in tunnel construction, Proc. Tunnelling '82, 3rd Int. Symp., Inst. Min. Metal., London, pp. 241-246.

Priest, S.D. and Hudson, J.A., 1976. Discontinuity spacings in rock. Int. J. Rock Mech. Min. Sci. & Geomech. Abstr., Vol. 13, pp. 135-148.

Rutledge, J.C., and Preston, R.L., 1978. Experience with engineering classifications of rock. Proc. Int. Tunnelling Symp., Tokyo, pp. A3.1 - A3.7.

SYSTAT Inc., 1990. SYSTAT manual. 1800 Sherman Ave., Evanston Il. 60201-3793.

Walker, T.C., and Miller, R.K., 1990, Expert Systems Handbook, The Fairman Press Inc.

Innovative Mine Design for the 21st Century, Bawden & Archibald (eds) © 1993 Balkema, Rotterdam, ISBN 90 5410 325 6

Mine design an empirical approach

R. Pakalnis
University of British Columbia, Vancouver, B.C., Canada

S. Vongpaisal
CANMET, Ottawa, Ont., Canada

ABSTRACT: The design of underground mine structures is a major area of focus in the Mining & Mineral Process Engineering Department at the University of British Columbia. A mine structure, whether it be a mine pillar or an opening, is generally influenced by numerous blocks of intact rock. Individually their properties and behaviour can be readily assessed on a laboratory scale. However, when analyzed on a minewide scale, the interaction of the rock blocks and the rock mass behaviour is difficult if not impossible to predict employing solely a deterministic approach.

This paper describes three design curves developed for purposes of a) estimating dilution for open stope operations (Ruttan Mine, HBMS), b) establishing stable cut and fill spans for wide ore bodies (Detour Lake Mine, Placer Dome Inc.) and determining optimum pillar dimensions for open stopes and cut and fill operations at the Myra Falls Operation (Westmin Mines Ltd.). In all three cases, the studies were sponsored by the individual operations and CANMET in developing design tools that are readily available and usable to the mine operator. The design curves are based upon the factors critical to the opening/pillar design such as geometry, rock mass rating, prevailing structure, the rate of extraction, stress levels, blasting influences, production constraints among others. The curves are individually related to a database for the operation that consists in excess of 200 observations which have been calibrated visually as well as through instrumentation. The design curves are employed by the respective operations for purposes of mine design in terms of establishing optimum open stope span dimensions, pillar geometries, and maximum stable spans for cut and fill operations. The paper focuses on a methodology towards design that is of applied use to the mine operator.

1 INTRODUCTION

A factor unique to the design of mine structures in jointed materials is the difficulty of determining strength or stress values directly from laboratory and numerical analysis. This is due largely to the difficulty of identifying the jointed nature of the material, assigning properties thereto and establishing input parameters for subsequent numerical codes. Estimates are made and subsequently calibrated to mine behaviour, however, they tend to change from stope to stope and consequently cause difficulty in relying on these interpretive models on a production minewide scale. The design methods that are available for assessing the stability of mine structures, whether they be mine openings or pillars, can be categorized as follows:

- Analytical
- Observational
- Empirical

Analytical methods are based on the analyses of stresses and deformations around mine structures. Observational methods rely on the monitoring of ground movement during mining to detect measurable instability. Empirical methods assess the stability of mines by the use of statistical

DILUTION APPROACH DESIGN EQUATIONS

ISOLATED STOPES(61 OBS)

$$DIL(\%) = 5.9 - 0.08(RMR) - .010(ER) + 0.98(HR)$$
$$r = \pm 0.78 \, , \hat{S} = \pm 3.3\%$$

ECHELON STOPES(44 OBS)

$$DIL(\%) = 8.8 - 0.12(RMR) - .018(ER) + 0.8(HR)$$
$$r = \pm 0.85 \, , \hat{S} = \pm 2.3\%$$

RIB STOPES(28 OBS)

$$DIL(\%) = 5.9 - 0.08(RMR) - .01(ER) + 0.98(HR)$$
$$r = \pm 0.78 \, , \hat{S} = \pm 3.3\%$$

where:

DIL(%) - Stope Dilution(%), ie. 10%, DIL(%) = 10
RMR - CSIR Rock Mass Rating(%), ie. 60%, RMR = 60
ER - Exposure Rate as Volume removed(m^3)/mth/stope width(m)
HR - Hydraulic Radius(m) of exposed stope wall

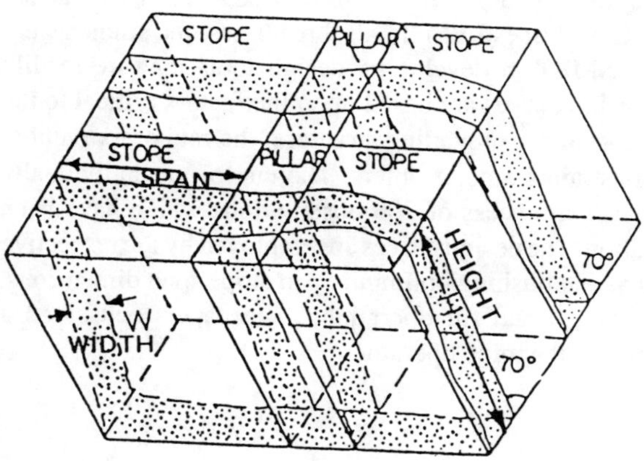

DATABASE

Stope Dilution = 10% ± 6%
Hydraulic Radius = 11m ± 3m
Stope Width = 15m ± 8m
Stope RMR = 56% ± 20%
ER = 180m^2/mth ± 90m^2/mth
Excavation Rate = 2700m^3/mth ± 1300m^3/mth
Span(Strike Length) = 31m ± 13m
Height = 68m ± 20m
Stope Depth = 360m ± 48m below surface
Stope Inclination = 68° ± 9°

Jointing Parallel to Hanging Wall
Hanging Wall in Relaxation

Figure 1: Dilution Approach to Open Stope Design

analysis of underground observations. Analytical methods are primarily employed for comparative design and parametric studies. The observational approach would require a large database and would have to be implemented in the initial stages of mine development in order to achieve some reliable measurement of stability. Empirical derivations have gained acceptance over the last ten years since they have been implemented in mine operations and have proven to be successful amongst practitioners in terms of prediction (Potvin,1988; Hoek,1980; Pakalnis, 1991).

2 OPEN STOPE SPAN ESTIMATION

Approximately 51% of all ore production by underground metal mines in Canada is derived directly from open stoping operations. This method of mining requires that large excavations remain open until the ore is extracted under a minimal acceptable dilution.

The "Dilution Approach" for estimating open stope dimensions was a culmination of a five (5) year joint research effort between the Ruttan Mine of Hudson Bay Mining & Smelting Inc., CANMET and the Department of Mines of Manitoba (Pakalnis, 1989). The overall objective of the project was to determine ground stability guidelines for large open stopes. The study evaluated the factors influencing the stability of large, open stopes for an existing mining operation. The Ruttan mine is a 6000 tpd underground operation located in northern Manitoba, Canada. The Ruttan copper-zinc orebody is a multi-lensed, steeply dipping (70°) en echelon deposit. The ore is massive/semi massive sulphides with the hanging and footwall located within strong acid sediments having an unconfined compressive strength (ucs) of 135MPa and ranging to weaker chlorite talc schists (ucs of 25MPa). Individual lenses have a maximum strike length of 350m with widths varying from 7m to 61m. The mining method is open stoping with delayed backfilling and extends from surface to 860m below surface.

Stope spans (along strike) at Ruttan were designed initially employing classical beam theory. Failures at the mine were found to be controlled primarily by structure, and could not be explained by a deterministic homogeneous model. Figure 1 summarizes the database of information derived from this open stoping operation.

The Ruttan mine has mined by open stoping methods since 1979. This has resulted in a large data base of information which includes forty-three (43) stopes at various stages of extraction, thereby yielding 133 observations of:

- Rock Mass Rating (footwall, hanging wall, ore)
- Stope Dimensions (height, width, length)
- Observed Dilution
- Rate of Excavation
- Stope Configuration (isolated, rib, echelon)
- Mining Sequence/Method

In addition, the observations were supplemented with in-situ measurements, structural mapping, stress and deformation monitoring and historical observations. A test stope was monitored and the results employed for purposes of verification of design guidelines presented in this study. It was concluded through numerical modelling, observation and measurement that the hanging wall and footwall of the individual stopes are in a state of relaxation, thereby enabling structural blocks to be released. This would generally be the case for most stope geometries whose major in-situ stress direction lies perpendicular to the long dimension of the opening. Consequently, the critical parameters were quantified in terms of their effect on dilution by employing multivariate statistical analysis. The relationships derived were entirely confined to the Ruttan operation as they were empirically delineated and quantified.

2.1 Design methodology

The parameters most critical to design were established after an extensive statistical and observational approach showed that for a particular stope that the resultant dilution (tonnes of external slough/ tonnes of ore blasted) was largely a function of:

- the rock mass rating of the hanging wall
- the exposed surface area of the hang ing wall
- the rate of extraction
- the stope configuration (isolated, rib or echelon)

Figure 1 shows the derived relationships which were subsequently modified into design charts shown in Figure 2. The design equations correlated ($r \simeq 0.8$) extremely well with observed dilution as recorded for individual stopes. Figure 2 shows the resultant dilution employing an exposure rate equal to zero. This was employed so as to show the original database superimposed upon the dilution

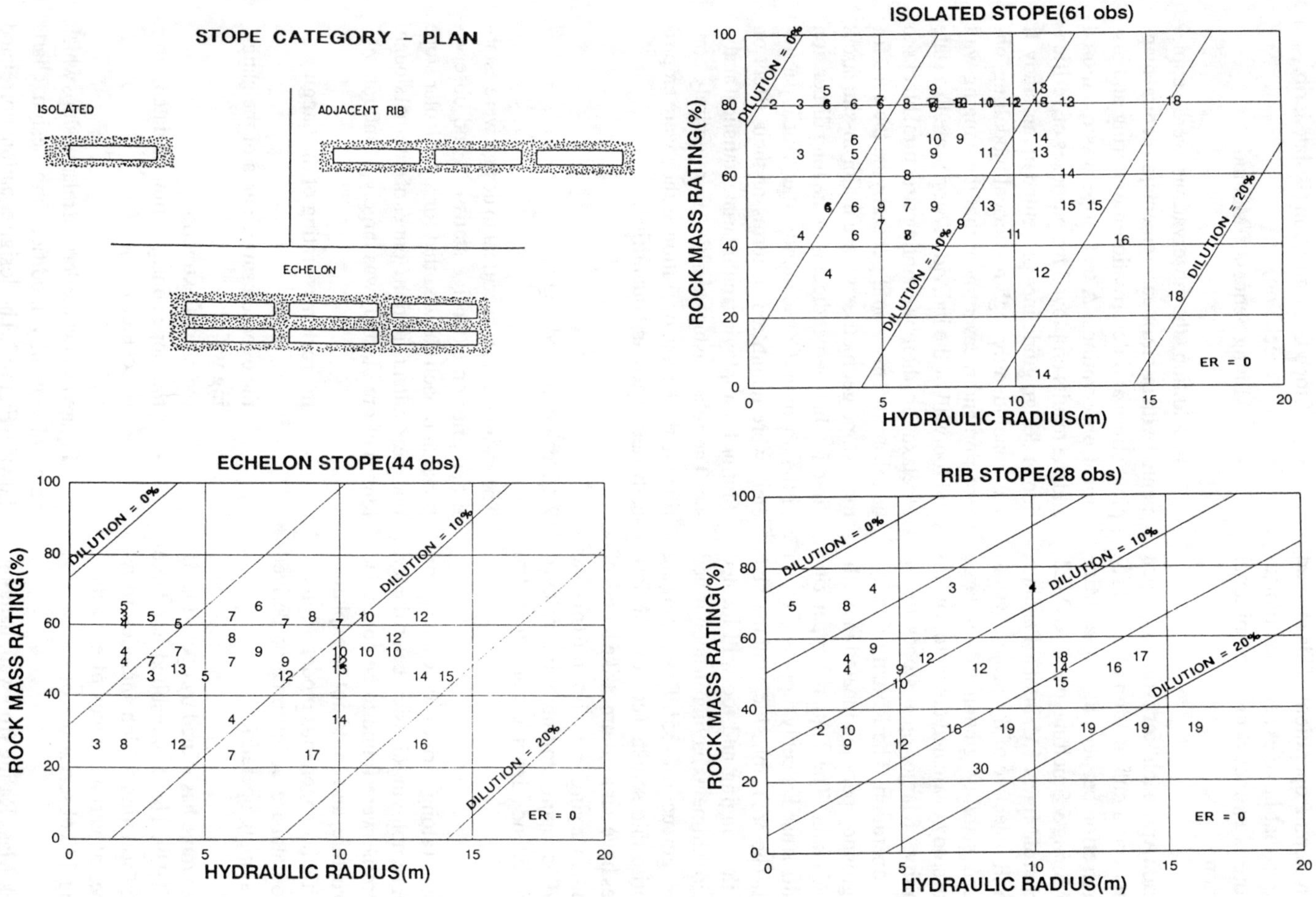

(5) numbers refer to dilution measured

Figure 2: Design Chart - Dilution Approach to Open Stope Design(ER = 0) curves shown in Figure 2. The method consists of determining:

Step 1: Determine rock mass rating of the hanging wall (RMR). This paper assumes that the hanging wall is the source of the majority of the observed dilution, if the footwall is more critical this wall is employed for purposes of input.

Step 2: Determine the hydraulic radius for the hanging wall which is defined as the exposed surface area divided by the perimeter in meters.

Step 3: Determine the exposure rate in terms of the volume (m^3) excavated per month divided by the average stope width (m). This results in an exposure rate (ER) which relates the rate that the hanging wall is exposed (m^2/m).

Step 4: Establish the stope configuration as shown in Figure 2. This factor incorporates the influence of adjacent stopes.

Step 5: Determine the predicted dilution from Figure 2 which assumes an ER = 0. This method has been employed successfully at numerous operations (Pakalnis,1991). The resultant dilution should be reduced by an equivalent ER x coefficient shown in Figure 1 since an increase in mining rate reduces the overall dilution. In addition the above does not include blast induced dilution which at Ruttan has been estimated as 3%.

2.2 Discussion

The above approach can be related to the Modified Stability Graph (Potvin, 1988) by employing a relaxed hanging wall (A=1), parallel structure to hanging wall (B=0.3) and a 70° inclined stope wall (C=5.9). This method must be calibrated, as all empirical methods, to the user's database whether it be site specific and/or inherent to his/her past observations and design approaches. This method enables mine design personnel to estimate dilution levels albeit once calibrated and to relate the results to alternative methods. The dilution levels have been determined for a database whose average stope width is 15m - 8m. The potential use of this approach is governed largely by the adherence to the original data base and subsequent calibration. The reader is referred to Pakalnis (1987) for a more detailed description of this method.

3 CUT & FILL SPAN ESTIMATION

Cut and fill mining among other entry type mining methods accounts for 35% of the tonnage produced from underground metal mines in Canada. Cut and fill mining poses a further requirement as opposed to non-entry methods such as open stoping. It requires that a stable environment exist where uncontrolled rock falls cannot occur due to the presence of mine personnel and machinery. Falls of ground are the single highest cause of fatalities in the mining industry. A stable open stope wall can be considered as being stable and still have a 1% dilution (Pakalnis, 1987). This dilution if present in a cut and fill back cannot be tolerated due to the hazard imposed on mine personnel working within the stope area.

There exists no accepted design methods predicting cut and fill spans in jointed materials. Beam and plate theories, Voussoir block analysis, and numerical models have been employed in the past, however, they are generally restricted by homogeneous, isotropic, and elastic assumptions of the rock mass. Empirical design methods have been developed for the design of non entry stopes as well as for general span design. In the latter case, much of the database consists of civil (Barton, 1974) engineering case histories which require long term stability and higher safety factors than that employed for cut and fill, which is open for a much shorter duration (three months).

This paper summarizes a study that was jointly sponsored by Detour Lake Mine of Placer Dome Inc. and CANMET. The Detour Lake Mine is a 2200 tonne per day gold mine located in northern Ontario, Canada. Production commenced in 1984

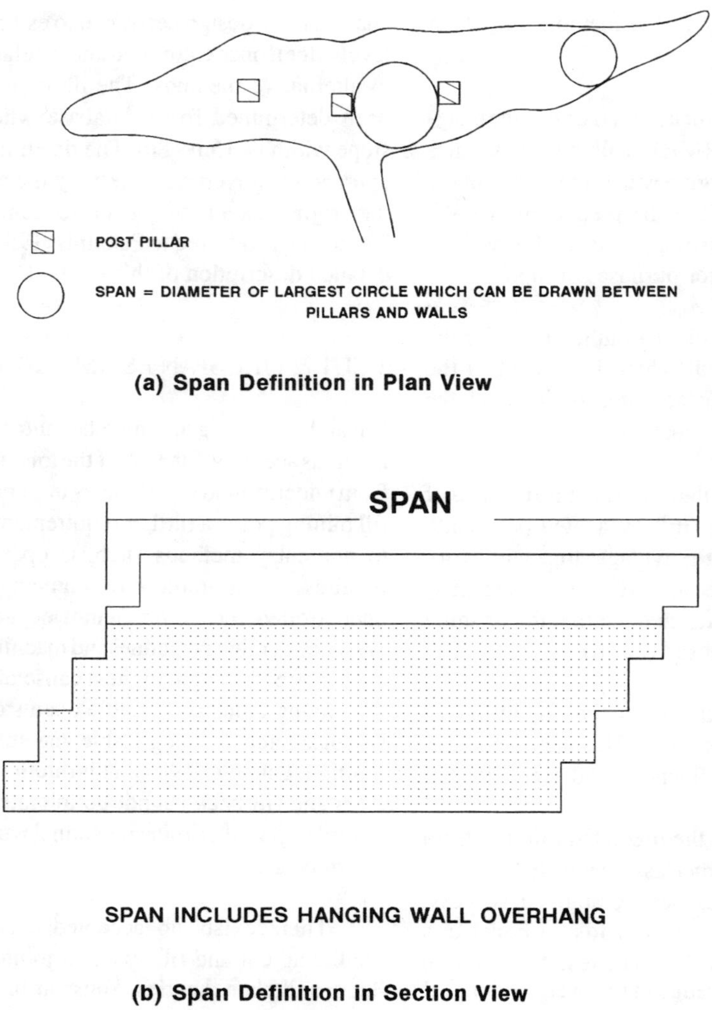

Figure 3: Span Definition - Cut and Fill

from an open pit and continued until 1987 thereafter continuing by underground methods. Mining is carried out using mechanized cut and fill, captive cut and fill and longhole. Mechanized stopes range in spans (footwall to hanging wall) from 3m to over 35m with individual lift heights being 5m. The inclination of the orebody varies from a low of 45° to an average of 60°. The orebody plunges to the west at 45° and has an average strike length of 200m. Mining is from surface to 660m below surface. The ore is comprised of strong mafic material (andesite) having an unconfined compressive strength in excess of 170MPa to weaker Talc schists having a ucs of 40MPa.

The overall objective of this study was to develop stability guidelines for the mining of wide ore bodies employing cut and fill mining methods. Design guidelines were developed at Detour Lake Mine that enabled individual stopes to be designed and extracted under controlled and predicted conditions. This study evaluated the factors that influence the stability of large cut and fill stopes having spans ranging from 3m to over 35m. The Detour Lake Operation has mined by mechanized cut and fill methods since 1987 yielding a large database of information. This includes fifty (50) mining lifts at various stages of extraction, thereby yielding 172 observations of:

- Rock Mass Rating (back)
- Structure (back)
- Stope Dimensions (back)
- Observed Stability Condition (Stable, Potentially Unstable, Caved)
- Stope Configuration
- Mining Sequence
- Stress
- Support
- Monitoring/Observation of Mining Behaviour

The above critical parameters were quantified in terms of the exposed span (back), rock mass rating and the potential for instability. Relationships have been developed whereby for a given rock mass rating one is able to derive a maximum critical span. The span is defined as the maximum circle that can be derived from a given back geometry.

3.1 Design methodology

The Detour Lake Mine database consists of 172 records. These observations were recorded from the commencement of underground mining in 1987 to the present date. Each record consists of the location, RMR-CSIR rock mass rating, stability, span, support and if major structures or flat jointing are present.

3.1.1 Definition of span

The term span refers to the diameter of the largest circle which can be drawn within the boundaries of the excavation viewed in plan, Figure 3. This definition was adopted for two reasons. Firstly, cut and fill stopes can commonly have width (FW to HW) to strike length ratios greater than 0.5 making plane strain assumptions suspect. Secondly, post pillars and the irregular stope boundaries make the calculation of hydraulic radius as employed previously difficult and misleading. The term span includes the overhang material that has not been contained by the fill, Figure 3. The design span refers to spans which have used no support and/or spans which include pattern bolting (1.8m long bolts on 1.2m x 1.2m) for local support. It does not include spans which are supported by more intensive ground support such as cable bolts, post pillars or timber sets. The "local support" is placed to confine potential blocks (loose) which may become open due to nearby blasting and/or stress redistribution caused by subsequent mining activity. This is generally scaled when the face is first exposed, however, in the long term the degree of visual checking on a regular basis is not nearly as intensive as the inspection of the first exposed face. This necessitates the need for support to ensure further opening of potential structures is restrained. Barton (1974) indicates that for rock masses in excess of 65% RMR, unsupported spans in excess of 20m are possible. There does not exist a mine in Canada where man entry methods of these widths are unsupported (no bolts) due to potential of loosening of structures within the immediate back.

3.1.2 Definition of stability

Bieniawski (1976) has shown there to be a relationship between span, rock mass rating, and stand-up time. Larger spans will result in lower stand up times for a given rock mass rating. In this study, most of the data has been obtained from cut and fill stopes where the required stand up time rarely exceeds three months. Therefore a stable excavation is defined as one which has remained stable for at least three months.

The stability of an excavation is classified into three categories:

a) Stable Excavations

- there have been no uncontrolled falls of ground
- if instrumentation is available, no movement of the back has been observed
- there were no extraordinary support measures implemented

b) Potentially Unstable Excavations

- the opening may exhibit faults/shears having orientations that form potential wedges in the back

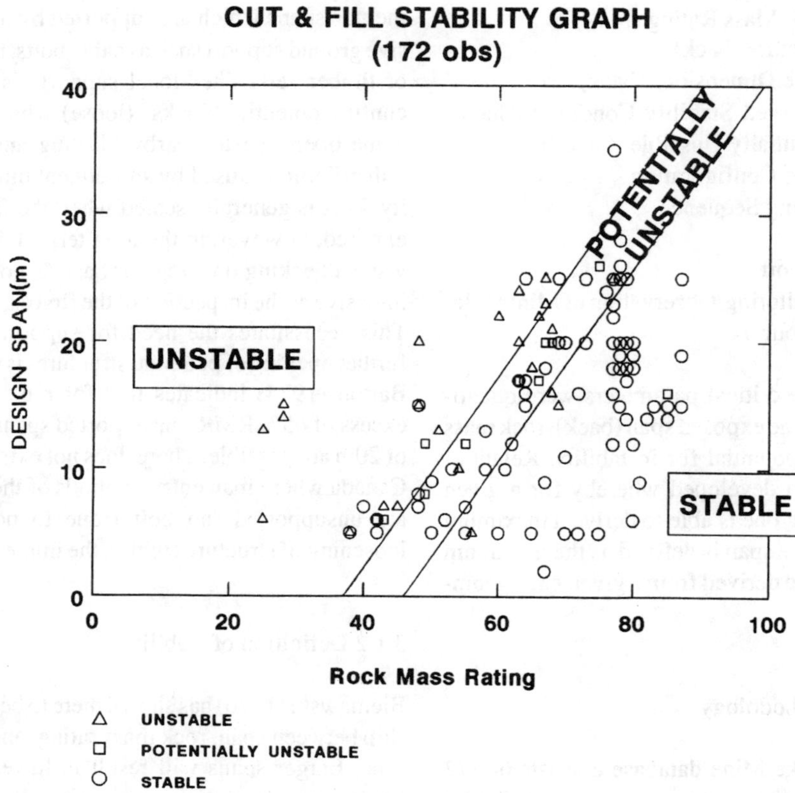

Figure 4: Span Design Curve - Cut and Fill

- extra ground support may have been installed to prevent potential falls of ground
- instrumentation installed in the back has recorded continuing back movement
- there may be increased frequency of noise (ground working)

c) Unstable Excavations

- uncontrolled fall of ground has occurred
- stope caved

Figure 4 shows the above data plotted onto a design curve. A further correction to the rock mass rating is made by reducing the RMR by 10% where unfavourable structural orientation is recorded which may include a continuous open joint or fault within the back dipping between 0° and 60° or where a flat joint/joint set is observed within the back (0° to 30° dip). This value has been suggested by Bieniawski (1976) as a correction factor for adverse structure and has been found to correlate well with the data presented in this study. Figure 4 has been statistically derived employing discriminant analysis in order to define three zones; stable, unstable and potentially unstable.

Design employing Figure 4 at Detour consists of the following:

Step 1: Determine the rock mass rating of the back of the stope to be designed. Determine if adverse structures exist in order to assess a representative RMR for the back. If a major wedge exists within the back, span design is based upon either supporting this wedge, modification to span so as not to cause the wedge to daylight and/or employing a post pillar.

Step 2: Employ Figure 4 to determine the maximum span that can be exposed. If the ore width is greater than the recommended value then a post pillar, sequencing and/or cables should be implemented. If the ore width is approximately equal to the critical span but falls in the potentially unstable area, then monitoring is recommended and possibly support. If the ore width is much less than the critical span then only local support is required in order to confine any potential loose.

3.1.3 Discussion

The empirical span design method shown in Figure 4 does not directly account for stress. The state of stress within the immediate back for stope geometries employed at this operation were recorded/modelled as being low or within a relaxed state. The method is employed where ground falls due to structural release pose the major problem and not burst induced falls of ground. The cut and fill stability graph shown in Figure 4 is a significant improvement over existing methods of predicting stable spans in cut and fill stopes. The limitations of its use can be summarized as follows:

- where discrete wedges have been identified these must be supported prior to employing the design curve shown in Figure 4
- design span refers to backs that have local support
- stable refers to short term stability (at least 3 months)
- graph is considered applicable over the RMR range 40 to 85%
- stress induced failure does not occur within the back
- the back is horizontal

A detailed description of the above method is found within Lang (1991).

4 EMPIRICAL DETERMINATION OF PILLAR STRENGTH

This paper updates a study that is jointly sponsored by the HW mine of Westmin Resources Ltd. and CANMET. The HW mine is a 4000tpd base metal mine located on the west coast of British Columbia, Canada. Production is derived primarily from the HW orebody which consists of massive sulphide deposits (Main Zone, North Zone) located at depths in excess of 600m below surface. The east-west striking lenses range from 2m to 65m in thickness over a plan dimension of 300m x 1200m along strike. Production from the HW mine commenced in 1985 by open stope, cut & fill and room & pillar mining methods. Approximately 15% of the ore within the Main Zone is designated as barrier pillars. These pillars range in size from 15-25m in width, 50-200m in length and 20-50m in vertical height. The objective of this study, which was initiated in 1991, is to develop guidelines for the optimum extraction of mine pillars at the Myra Falls operation. This requires that a database be developed whereby factors relevant to the design of mine pillars be established and related to the applied stress. The approach employed is to relate the Westmin database, which has in excess of fifty (50) pillars of various dimensions, stress levels, rock quality and levels of deterioration, to strength/stress levels. This empirical approach is used to generate pillar design curves.

The HW Main Zone includes ten barrier pillars with widths varying between 15 to 25m. The barrier pillars provide orebody access and as a result have numerous openings developed within. These openings result in the development of smaller dimension pillars (4-5m height) which are comprised of drawpoint and rib pillars. They are referred to as discrete pillars.

The pillar database is comprised of 141 barrier pillar observations which includes sixty-five discrete pillars and have been characterized according to a failure rating shown in Figure 5. In addition, the following were recorded:

- Rock Mass Rating of pillars and all accessible areas

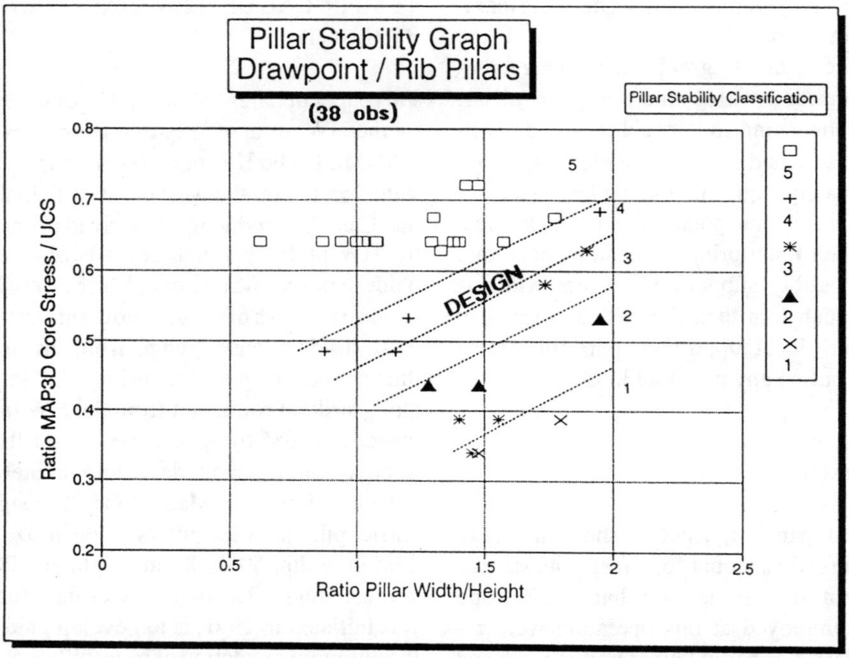

PILLAR CLASSIFICATION

CLASS	DEFINITION
1	No sign of stress induced fractures
2	Some fractures, no spalling of walls
3	Fractures < 1/2 pillar height, some spalling, fractures <5mm open
4	Fractures > 1/2 pillar height, more wall spalling, fractures > 5mm open
5	Disintegration of pillar, fractures > 10mm open, blocks falling out

Figure 5: Pillar Stability Graph

- Intact rock strength properties
- In-situ stress measurements
- Geotechnical mapping of all areas
- Characterization of all pillars in terms of geometry, level of stabillity
- Integrity of pillars established (observation, instrumentation)

4.1 Design methodology

The approach to pillar design is one whereby factors critical to design have been assessed and combined to arrive at a design tool to enable pillar strengths to be established. It is based upon calibration of prediction to pillar behaviour and subsequent modification.

Rock Mass Classification employing the CSIR (Bieniawski, 1976) system was performed throughout the barrier study area at approximately 10m intervals. Generally the rock mass within the pillars ranged from a fair quality rock mass to a good quality rock mass. It must be noted that the RMR represents the current rating within the pillars and not the pre-mining ratings. Generally areas of the mine deemed to be under lower stress conditions were observed to yield higher rock mass ratings.

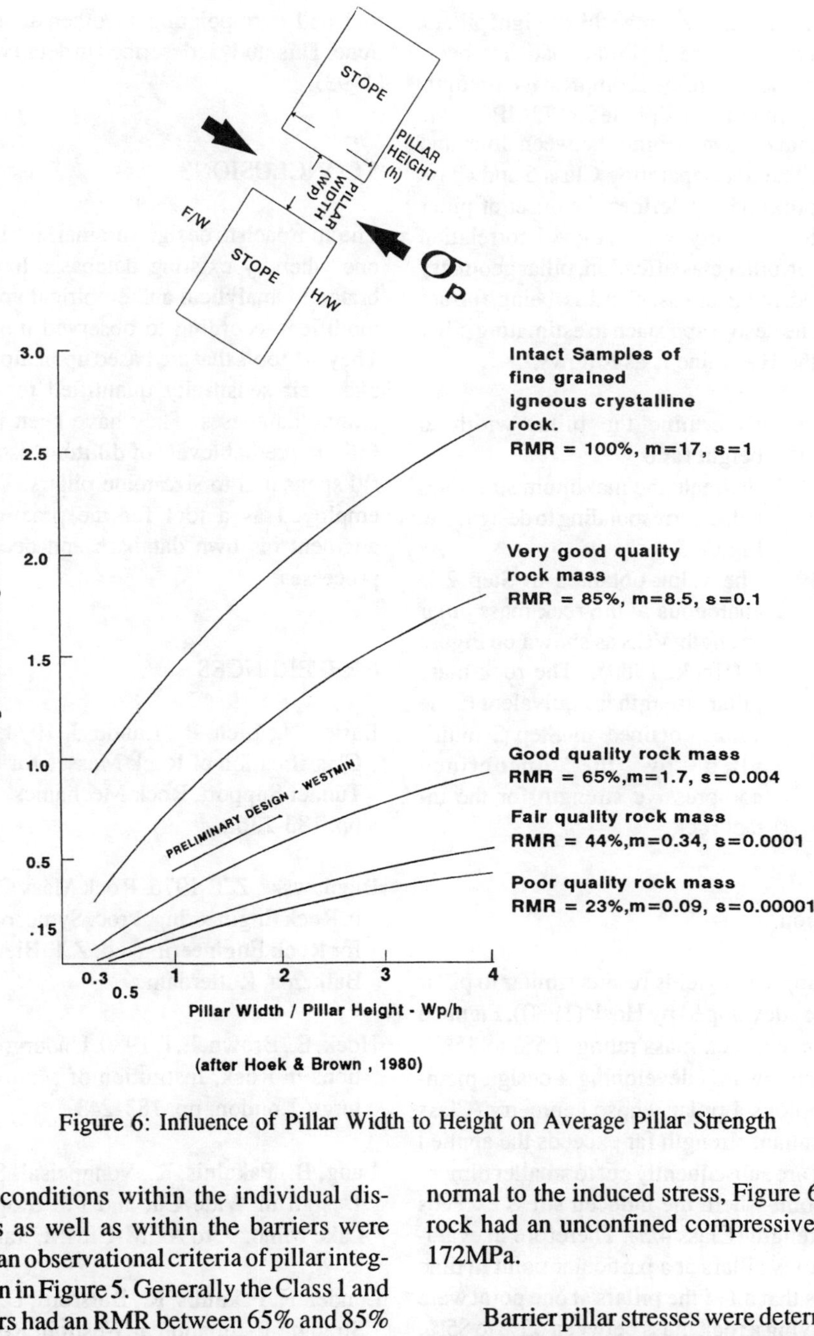

Figure 6: Influence of Pillar Width to Height on Average Pillar Strength

Pillar conditions within the individual discrete pillars as well as within the barriers were based upon an observational criteria of pillar integrity as shown in Figure 5. Generally the Class 1 and class 2 pillars had an RMR between 65% and 85% whereas the class 4 and class 5 pillars were between an RMR of 30 to 55%.

The thirty-eight (38) pillars were comprised of eight rib pillars and thirty drawpoint pillars. Pillar height was recorded in the direction parallel to the induced principal stress and the pillar width normal to the induced stress, Figure 6. The intact rock had an unconfined compressive strength of 172MPa.

Barrier pillar stresses were determined using MAP3D (Wiles, 1991), a three dimensional elastic boundary element numerical modelling program. Stresses within individual pillars located within the larger barrier pillar were determined based upon the induced stress levels within the barrier and related to the extraction ratio in the vicinity.

Preliminary results for the thirty-eight pillars are presented in Figure 5. Pillar load has been normalized to the unconfined compressive strength of the intact massive sulphides (172MPa). All pillars presented have heights between four and five meters. The line separating Class 5 and Class 4 pillars approximately defines the onset of pillar failure (factor of safety = 1). A good correlation exists between pillar classification, pillar geometry and predicted pillar stresses and is being further evaluated. The design approach to estimating pillar strength at the HW mine is as follows:

Step 1: Determine the pillar width to height ratio

Step 2: Estimate the maximum stress/ucs value corresponding to design line, Figure 5

Step 3: The value obtained in Step 2 is analogous to the rock mass pillar strength/VCS as shown on Figure 6 (Hoek, 1980). The rock mass pillar strength is equivalent to the value obtained in Step 2 multiplied by the unconfined compressive strength for the intact rock.

4.2 Discussion

The above approach yields results similar to pillar design curves developed by Hoek (1980), Figure 6 for an equivalent rock mass rating (65% to 85%). The approach towards developing a design methodology employs a pillar whose geometry (Class 1,2) and resultant strength far exceeds the applied loads which are subsequently cut to smaller dimensions to a point where the induced stress exceeds the pillar strength (Class 4/5). Therefore in evaluating numerous pillars at a particular point in time one assumes that all of the pillars at one point were a Class 1 having a rock mass between 65% to 85%. The design line shown in Figure 5 should be considered as relating the pillar geometry to the rock mass strength for a 65 to 85% rock mass rating. The pillar design curve shown in Figure 5 and Figure 6 enables a verification and confidence to be developed in terms of acceptance of a design tool and extrapolating to other areas within the mine. This study is described in detail within Lunder (1993).

5 CONCLUSIONS

The approach to design summarized in this paper is one whereby existing databases have been calibrated to analytical and empirical approaches and modified according to observed mine behaviour. They are tools that are based upon input parameters and their sensitivity quantified to the variances among databases. They have been used successfully to predict levels of dilution, estimate cut and fill spans and to size mine pillars. They are to be employed as a tool for the practitioner and to augment his own database and decision making processes.

6 REFERENCES

Barton N., Lien, R., Lunde, J. 1974. Engineering Classification of Rock Masses for the Design of Tunnel Support, Rock Mechanics, Vol. 6, No. 4, pp. 183-236.

Bieniawski, Z.T. 1976. Rock Mass Classifications in Rock Engineering, Proc. Symp. on Exploration for Rock Engineering, ed. Z.T. Bieniawski, A.A. Balkema, Rotterdam.

Hoek, E., Brown, E.T. 1980. Underground Excavations in Rock, Institution of Mining and Metallurgy, London, pp. 183-243.

Lang, B., Pakalnis, R., Vongpaisal, S. 1991. Span Design in Wide Cut and Fill Stopes at Detour Lake Mine, 93rd AGM-CIMM, Vancouver.

Lunder, P., Pakalnis, R., Botsford, D. 1993. Pillar Strength Estimation at Westmin Resources Ltd. - HW Mine, 95th AGM-CIMM, Calgary.

Pakalnis, R., Miller, H., Vongpaisal, S., Madill, T. 1987. An Empirical Approach to Open Stope Design, ISRM Proceedings, Montreal.

Pakalnis, R., Vongpaisal, S., Madill, T., Lantz, B. 1989. Ground Stability Evaluation with Particular Reference to an En Echelon Lensed Orebody - Design Guidelines, Dept. of Energy, Mines, and Resources, Canada, DSS File Number: 14SQ.23440-4-9147-1, 190p.

Pakalnis, R., Vongpaisal, S., Lang, B., Cook, K. 1993. Ground Stability Guidelines for Cut & Fill Mining of Wide Ore Bodies, Dept. of Energy, Mines, and Resources, Canada, DSS File No. 14SQ.23440-49-9145, pp. 274.

Pakalnis, R., Tenney D., Lang, B. 1991. Numerical Modelling as a Tool in Stope Design, CIMM Bulletin, July, pp. 64-73.

Potvin, Y. 1988. Empirical Open Stope Design in Canada - PhD thesis, University of British Columbia.

Wiles, T. 1991. MAP3D Mining Analysis Program. Mine Modelling Ltd., Sudbury, Canada.

Innovative Mine Design for the 21st Century, Bawden & Archibald (eds) © 1993 Balkema, Rotterdam, ISBN 90 5410 325 6

The excavation effects program at the Waste Isolation Pilot Plant

Christopher T. Francke & Liane J. Terrill
Waste Isolation Division, Westinghouse Electric Corporation, Carlsbad, N.Mex., USA

ABSTRACT: Excavation effects are the observable evidence of damage incurred by the rock mass due to the removal of material during mining. Excavation effects of primary concern to ground control and maintenance engineers are the displacement of the rock and the formation of various types of fractures and separations near the excavation. Well established and relatively simple techniques already exist for measuring rock displacements. Characterization of fracture systems surrounding excavations is a much more difficult problem, especially in creeping materials such as halite for which the extent and orientation of fracturing may change continually with time. Many indirect geophysical methods, such as ground probing radar, are available for this purpose, but they are usually quite costly and require equipment and expertise normally not available at a production mine.

An inexpensive technique developed for characterizing fracture systems surrounding underground excavations at the Waste Isolation Pilot Plant (WIPP) provides a quick and systematic method that does not require special expertise to use successfully. WIPP is being developed near Carlsbad, New Mexico, for the disposal of transuranic nuclear wastes in underground excavations in salt 2150 feet below the surface.

1 INTRODUCTION

Ground control is a basic, fundamental problem that must be considered by every underground operation. The magnitude of ground control problems at a facility directly affects production rates, operational costs, facility design, engineering effort, and operational safety. Underground operations in salt at the Waste Isolation Pilot Plant (WIPP) are especially sensitive to ground control concerns where they affect criteria for operational safety and waste retrievability in an ever changing operational environment. Programs were implemented at WIPP to monitor the response of the salt to excavations. The information from these programs is important to the continued safe operation of the plant, as well as to the mining industry as a whole.

1.1 *Waste Isolation Pilot Plant*

The WIPP was authorized by Congress in 1979 (Public Law 96-164) to provide "...a research and development facility to demonstrate the safe disposal of radioactive wastes resulting from the defense activities and programs of the United States exempted from regulation by the Nuclear Regulatory Commission" (U.S. DOE, 1992). The WIPP is intended to receive, handle, and permanently dispose of transuranic waste. To fulfill this mission, the U.S. Department of Energy (DOE) is constructing a full-scale facility to demonstrate both technical and operational principles of the permanent isolation of transuranic waste. The facility is also designed for in-situ studies and experiments in salt.

1.2 *Location*

The WIPP is located in the United States in southeastern New Mexico about 50 km (30 miles) east of Carlsbad. The surface facilities have been built on the flat to gently rolling hills that are characteristic of the Los Medanos (sand hills) area. The underground facilities are being excavated approximately 655 meters (2,150 feet) beneath the surface. Figure 1 shows a plan view of the underground facilities at the WIPP site.

1.3 *Geology*

The WIPP facility excavations are located in the Permian Salado Formation, a thick sequence of bedded evaporite deposits. The formation is composed predominantly of halite containing

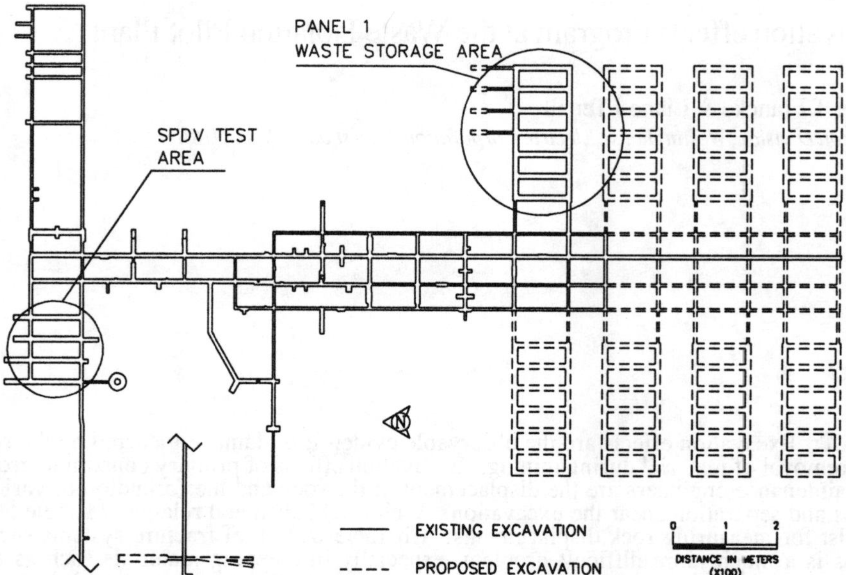

Figure 1. Waste Isolation Pilot Plant (WIPP) Mine Plan

minor amounts of clay, anhydrite, and polyhalite. Several areally persistent layers of anhydrite and polyhalite occur within the Salado, locally denoted as marker beds.

The facility horizon lies within a 12 meter (40 feet) thick unit of halite, argillaceous halite, and polyhalitic halite, intercalated with thinner beds of anhydrite (Figure 2). The anhydrite beds and associated clay seams are the most structurally important units. Marker Bed 139, a persistent bed of anhydrite, lies about two meters (five feet) below the floor in most areas of the facility and is underlain by clay E. Anhydrite "a" occurs about four meters (13 feet) above the roof in most excavations. Anhydrite "b" occurs about two meters (seven feet) above the roof.

Marker Bed 139 and the clay layers can have a significant impact on the geomechanical performance of excavations. The clay layers provide surfaces along which slip can occur, whereas Marker Bed 139 acts as a strong, brittle unit that does not deform plastically with time. Undulations along its top resist shear movement along the interface with the overlying salt.

2 STRUCTURAL EVALUATION OF EXCAVATIONS

In order to determine if the underground excavations are meeting design criteria and safety standards, the openings are regularly inspected and monitored in a variety of ways.

2.1 *Geomechanical Instrumentation System*

Hundreds of geomechanical instruments have been installed in the underground excavations at WIPP. Multiple point borehole extensometers (MPBX) and various convergence measuring instruments are the primary devices, although inclinometers, stress meters, crack meters, time domain reflectometry cables, and rockbolt load cells have also been used. These instruments are read on a regular basis and provide information on the performance of the excavations. Increasing rates of displacement generally indicate the development of instability in the rock surrounding the excavations. MPBX's provide indirect information on displacements of fractures and separations at stratigraphic contacts.

2.2 *Visual Inspections*

The simplest methods of monitoring excavation conditions are visual. The conditions of the roof, floor, and walls of the excavations are regularly assessed at WIPP. Areas of spalling and heave are noted and compared to earlier inspections to determine how quickly conditions are changing. The type and extent of ground control installed in an area is also noted. This qualitative method primarily serves to locate areas of concern which may require more detailed analysis.

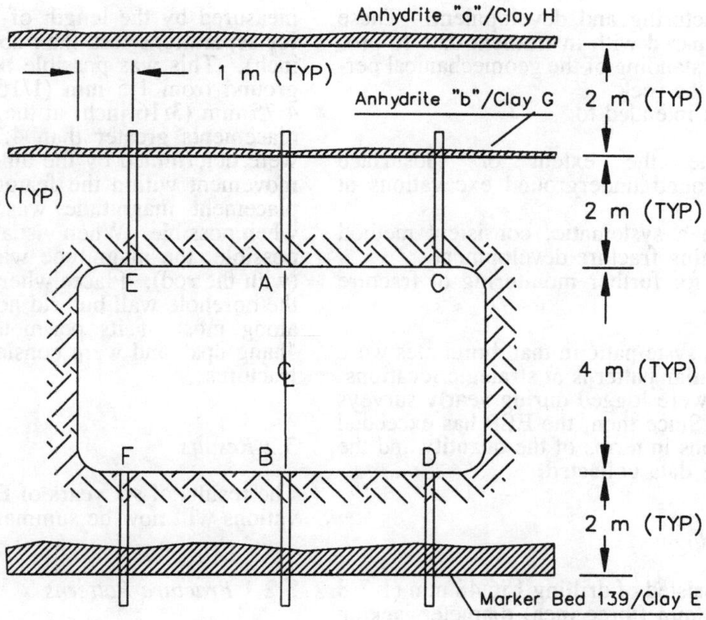

Figure 2. Stratigraphy Around Underground Openings at WIPP

2.3 *Geophysical Methods*

Several geophysical methods have been employed at WIPP to determine structural conditions. Ground probing radar has been successfully used to delineate large fractures and clay seams within a few meters of the excavation surfaces. Radar is the only reliable way to observe the condition of the rock without drilling boreholes. However, this method is very expensive, requiring special equipment and training for personnel and is very labor intensive in terms of both fieldwork and data reduction and analysis. In addition, this method does not reveal small fractures which may be as important structurally as the larger fractures it is capable of discerning.

Resistivity methods have been used at WIPP with limited success. Resistivity is not capable of discerning individual fractures. There is also a good deal of ambiguity in resistivity results, as both the competence of the rock and the amount of water in the rock mass effect the resistivity of the ground. Like radar, resistivity requires an investment in specialized equipment and is labor intensive. At this time, resistivity is of only academic interest and does not give practical results.

2.4 *Borehole Inspections*

Borehole inspections have been the most successful method for determining the condition of the rock immediately surrounding excavations. Fracture logging of open boreholes has been done on an informal basis since the first holes were drilled at the facility horizon in 1983. Concern over the extent and nature of fracturing near excavations increased in 1985 when a well developed fracture system was discovered in the floor of Site and Preliminary Design Validation (SPDV) Test Room 3. This fracturing spanned the room from rib to rib and extended into the thick anhydrite Marker Bed 139 (MB-139). Open spaces across fractures as great as 15 cm (six inches) were observed. This activity heightened interest in fracturing at WIPP and lead to the establishment of the Excavation Effects Program (EEP) in the following year.

3 EXCAVATION EFFECTS PROGRAM

The EEP was initiated shortly after the discovery of the large fracture system in SPDV Test Room 3.

The purpose of the EEP is to study fracture development as a result of underground excavations at the WIPP. As part of the Geotechnical Monitoring Program, the EEP was developed to provide consistent documentation and monitoring of fracture development. The program was begun in 1986 to meet these needs.

The EEP utilizes the inspection of drill holes in the excavations to provide information on the extent of fracturing and development. These data are compared with instrument data to provide an understanding of the geomechanical performance of the rock.

The EEP is intended to:

1. determine the extent of subsurface fracturing around underground excavations at WIPP,
2. establish a systematic, consistent method for documenting fracture development,
3. provide for further monitoring of fracture development.

The EEP is systematic in that boreholes were drilled in similar patterns at strategic locations. These holes were logged during yearly surveys since 1986. Since then, the EEP has exceeded all expectations in terms of the quantity and the quality of the data collected.

3.1 *Methodology*

The EEP consisted of drilling 156 48 mm (1-7/8 inch) and 76 mm (three inch) diameter jackleg drill holes at 30 locations (hereafter referred to as arrays) in the SPDV Test Rooms and along selected drifts and intersections. Holes were drilled a nominal 2.75 meters (nine feet) into the roof and floor at most of the arrays. A typical array layout is shown in Figure 2. The 2.75 meter depth was chosen because it would penetrate the first major clay seam in both the roof and floor in most locations. Economics and limitations of the drilling method were also factors.

3.1.1 *Mapping Procedure*

The arrangement of the boreholes in each array was chosen to intercept locations in the roof and floor which were thought to be more likely to have fracturing. In some instances, pre-existing boreholes were utilized when they were close enough to the desired locations to make drilling a new hole unnecessary. The centerline holes at most arrays were drilled with a 76 mm (three inch) diameter to facilitate inspections with a borehole camera if necessary.

All holes were inspected using an aluminum probe rod with a flattened nail or screw about 1.5 mm (1/16 inch) wide attached normal to one end (Figure 3). Features were identified by scratching the nail along the sides of the borehole while applying moderate pressure. Features were located when the nail caught on the borehole wall at the same depth on all sides of the wall.

For the purposes of this program, a fracture is defined as any discontinuity with a mappable vertical or horizontal component of displacement. Vertical displacement across features was measured by the length of the nail penetration for separations less than about 4.75 mm (3/16 inch). This was possible because the nail was ground from 1.5 mm (1/16 inch) at the tip to 4.75 mm (3/16 inch) at the base. Vertical displacements greater than 4.75 mm (3/16 inch) were determined by the amount of vertical rod movement within the feature. Horizontal displacement magnitude was visually estimated when possible. When visual estimation was not possible, the magnitude was estimated by feel (with the rod). Places where the nail caught on the borehole wall but did not penetrate the wall along most of its perimeter were designated "hang ups" and were considered to be possible fractures.

3.2 *Results*

The results of six years of EEP borehole observations will now be summarized.

3.2.1 *Fracture Patterns*

One of the greatest insights from the EEP data is the fact that regardless of the size or stratigraphic location of an excavation, the fracture pattern in the roof and floor of the excavation is very similar. This allowed us to design a roof-bolt pattern that not only addresses the fracture pattern but is universally applicable throughout the mine, regardless of the drift width.

A schematic depicting the typical fracture pattern found in the roof and floor of excavations at WIPP is shown in Figure 4. Diagonal fractures tend to develop near the rib-roof intersection which, in the extreme cases, terminate at the clay seam about two meters into the back. In the more advanced cases, subhorizontal fractures begin to develop near the centerline of the excavation just below clay G. Subhorizontal spalls also develop within the first few centimeters of the excavation surfaces.

In the floor, the dish-shaped fractures first seen in SPDV Test Room 3 are prevalent throughout the mine.

3.2.2 *Stratigraphic Offsetting*

The EEP borehole observations have shown us that significant horizontal offsets develop at clay seams near excavations. In most cases, the beds nearer to the excavation move towards the centerline of the excavation relative to the deeper beds. The offsetting is usually symmetrical around the centerline of the room, with the greatest offset magnitude near the ribs (Figure 4). The rate of offset of the beds has been determined from the EEP observations to range from about 3 to 12.5 mm/year (1/8 to 1/2 inch/year), about 3 mm/year (1/8 inch) on aver-

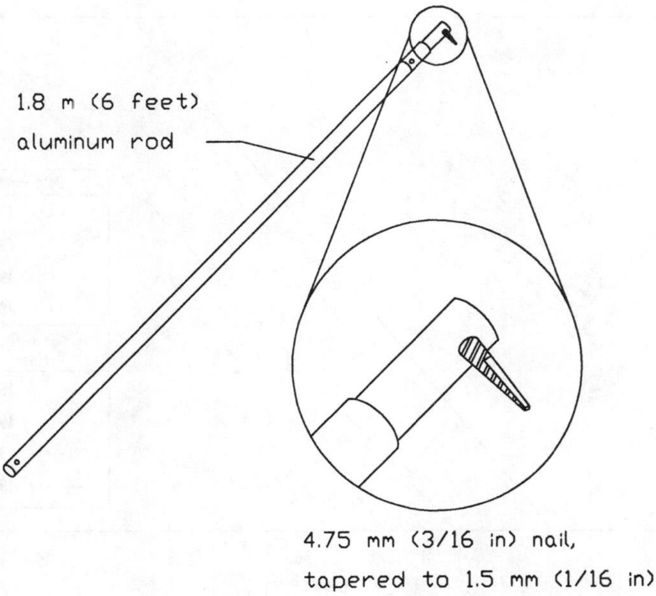

Figure 3. Excavation Effects Program Probe

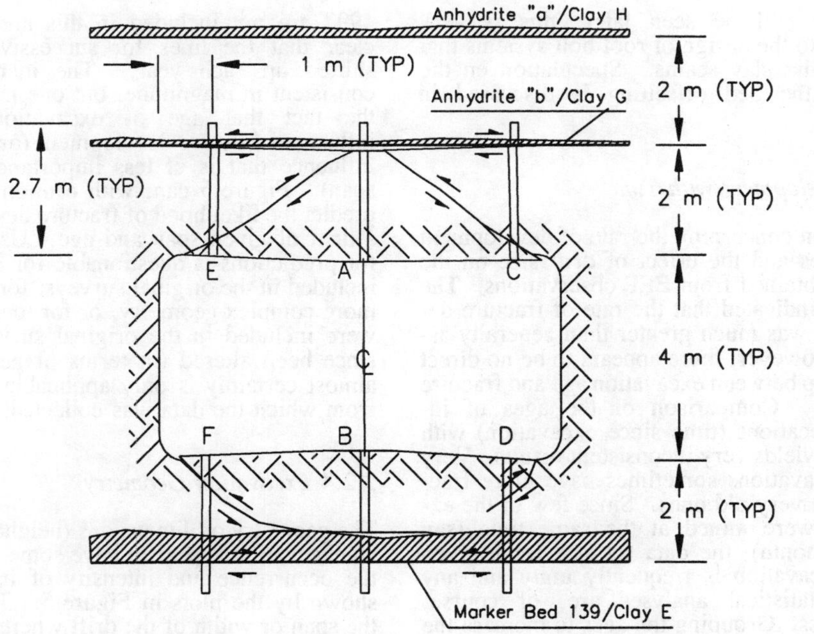

Figure 4. Typical Deformation Patterns Around Underground Openings at WIPP

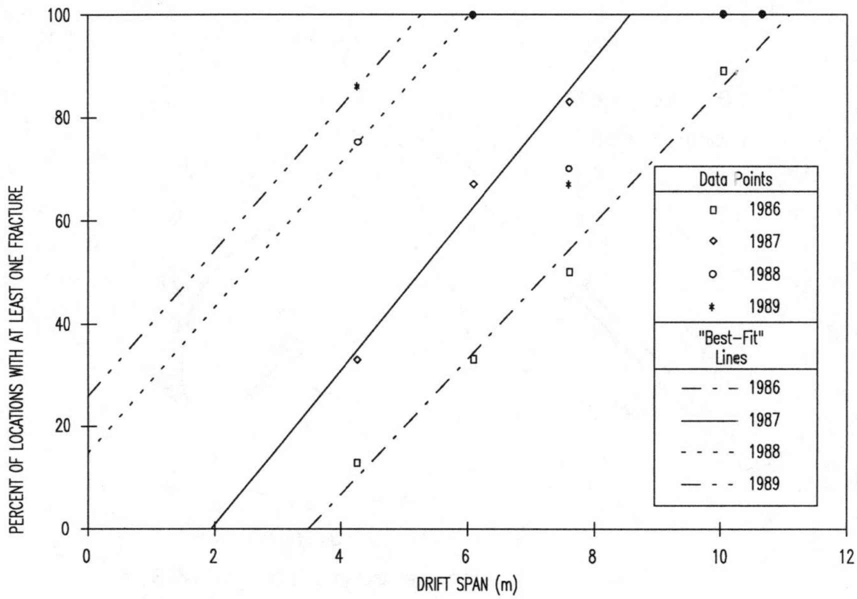

Figure 5. Drift Span Versus Fracturing

age. As will be seen later, this is very important to the design of roof bolt systems that penetrate the clay seams. Speculation on the cause of the bed offsetting is discussed in Section 4.

3.2.3 *Development with Time*

Information concerning the rate of development of fractures and the effect of drift size on the rate was obtained from EEP observations. The EEP data indicated that the rate of fracture development was much greater than generally assumed. However, there appears to be no direct relationship between excavation age and fracture frequency. Comparison of the ages of individual locations (time since excavation) with fracturing yields very inconsistent results. Very young excavations sometimes have more fracturing than very old ones. Since few of the excavations were mined at the same time (say within a month), the data set size for a given age of excavation is frequently unity and any ensuing statistical analyses are, of course, meaningless. Grouping the ages to increase the set size proves very difficult and is difficult to justify. However, when the age of the entire data set is considered rather than that of individual locations, a relationship between age and fracturing becomes evident. Figure 5 contains plots of drift span versus fracturing for each of the first four surveys. After 1989, fracturing was found at nearly all borehole locations or the holes were inaccessible, so the data for 1990 to 1992 are not included in this analysis. It is clear that the lines for successive years are shifted up each year. The increase is not consistent in magnitude, but one must consider the fact that age of excavation indirectly influences fracture development (or has a direct influence that is of less importance than drift span). Figure 5 can, with caution, be used to predict the likelihood of fracture development in a drift of given span and age. Using this plot for predictions is questionable for locations not included in the original surveys, for locations of more complex geometry, or for locations which were included in the original survey but have since been altered (in terms of geometry). It almost certainly is only applicable to the mine from which the data was collected.

3.2.4 *Excavation Geometry*

The geometry or dimensions (height X width) of an excavation appear to have some influence on the occurrence and intensity of fracturing, as shown by the plots in Figure 5. The x-axis is the span or width of the drift where the array is located. The y-axis is the percent of arrays of a certain span for which fracturing was observed in at least one borehole. There appears to be a linear relationship between fracturing and drift span.

The problem with this analysis is that the data set is not well distributed across the range of drift spans. For example, one third of all arrays have 10 meter (33 feet) spans while only one tenth have 6 meter (20 feet) spans. Therefore, one must consider the number of arrays with a particular span when using Figure 5. Despite the statistical questionability, one must conclude that drift span has an influence on fracture development.

When analyzed in the same manner as drift span, there appears to be no correlation between drift height or cross-section with fracture frequency.

3.2.5 *Correlation with Instrument Data*

Convergence data can suggest the location and extent of fractures. Convergence instruments have shown that the deformation across the width of the drift is rarely uniform. In many cases, one side of a drift shows significantly higher convergence rates than the other. Frequently, there is greater fracturing on the side with higher convergence rates.

Multiple point borehole extensometers are perhaps more efficient at detecting fractures. In the SPDV Test Rooms, the development of large separations at clay seams was first detected through extensometer readings and confirmed through EEP borehole inspections.

4 CONCEPTUAL MODEL FOR EXCAVATION EFFECTS

The stratigraphy immediately above the Test Rooms (and the waste storage panels) typically comprises two layers of halite, each about two meters (seven feet) thick, with a thin anhydrite bed and clay seam above each layer (Figure 4). The clay seams have virtually no structural competence, thus the two halite layers can be considered as separately acting beams. The observations of offsetting at clay G from the Excavation Effects Program inspections and the unusually high strains across clay G seen in the MPBX analyses confirm this.

The compression of the pillars imposes horizontal compressive forces to the roof and floor beams above and below the excavation. This force, the vertical loads imparted by the overlying material, and the self-weight of the beams cause the beams to deflect downward, more so towards the center of the beams. The greater the horizontal movement of the ribs, the larger the drift span, and the smaller the flexural rigidity of the beam, the greater the deflection of the beam. Separation between the beams is the result of one beam deflecting more than the other. Because the lower beam has undergone more displacement, it can be assumed that the lower halite layer has been subjected to more strain than the upper layer. It can further be assumed that the lower beam has less flexural rigidity than the upper layer due to the greater structural weakening caused by the greater strain and the lack of confinement. This bending would lead to horizontal displacement near the ends and vertical displacement near the center of the contact between layers (Figure 4). This pattern is frequently observed in the EEP borehole arrays.

Although the beam ends probably displace both vertically and horizontally, it is assumed that the vertical movement of one end relative to the other is negligible since the magnitudes of vertical creep of ribs on opposite sides of an opening are nearly equal.

The lower beam is apparently being carried towards the room along with the ribs as the ribs creep inward. The rates of horizontal displacement at clay G derived from the EEP surveys are equal to 75% or more of the horizontal displacements of the surfaces of the ribs. The lower beam being deflected by creep of the ribs in addition to its own weight and the weight of the overlying beam would account for the increasing separation at the clay seam seen in the bay strains from roof extensometers

The beam deflection/buckling theory accounts for much of the EEP and instrumental observation. However, there are certainly other forces at work which involve changes to the material properties of the halite. Fracturing in halite, rather than creep, is induced by strain rates exceeding a critical value for a given temperature and stress situation. It can be assumed that there exists a zone immediately surrounding all excavations which is practically de-stressed or stress relieved. In this zone, it must also be assumed that strain rates remain relatively high due to the continued movements of the rock farther from the opening where stress levels are appropriate for creep. It is therefore inevitable that strain rates near the surfaces of the excavation will exceed the level required for fracturing to develop. This causes spalling and similar phenomenon seen so often on the surfaces of the drifts.

5 APPLICATION OF RESULTS

The EEP borehole surveys have significantly advanced our understanding of the deformation mechanisms and the development of fractures surrounding the underground excavations at WIPP. Several practical applications of this knowledge have improved operations at WIPP.

5.1 *SPDV Test Room 1 Roof Fall*

The SPDV Test Rooms in the north end of the repository were excavated to monitor the closure, fracture development, and rock behavior in the underground. The information gathered from the test rooms enable design modifications to enhance the life of the rooms in the waste storage panel.

Observations in these rooms, including EEP data, have been made over the past eight years. Low angled shear fractures had, with time, formed in the roof of Test Room 1 and propagated along the length of the room. In addition, a separation occurred at the base of anhydrite "b", approximately seven feet above the roof. This led to the formation of an arch-shaped slab up to two meters thick (6.9 feet) in the center which separated from the overlying salt along the fractures and at anhydrite "b". On February 4, 1991, an arch-shaped section of roof weighing approximately 6.2 MN (700 tons) separated from the overlying salt and fell into the room. The section of rock which fell measured approximately 43 meters (140 feet) in length, ten meters (33 feet) in width, and has a maximum thickness of about two meters (seven feet).

The development of the roof failure was closely monitored. Indications of roof instability were observed in the EEP data approximately five years before the roof fall, and in instrument data from SPDV Test Room 1 approximately three years before the roof fall. An extensive program to map the fractures and upgrade the instrumentation in the Test Rooms was initiated. Analysis of the data resulted in the timely abandonment of the room. Remotely read instrumentation allowed convergence measurements to be made to within ten minutes of the roof fall. The geomechanical data from this room provided a unique study of roof behavior above an excavation in bedded salt from its construction until immediately preceding a large failure. Test Room 1 provides the most detailed example of the performance that can be expected from other rooms having similar geometries.

5.2 *Roof Support Systems*

The roof support systems used at WIPP were designed to address the fracture patterns revealed by the EEP borehole observations. The EEP-influenced designs saved labor and material costs that would have been incurred by constructing more conservative roof support systems which would have been required without the knowledge gained from the EEP.

5.2.1 *General Mine Rockbolt Pattern*

Before the first underground openings were excavated at WIPP, it was thought that rockbolting would only be required for remediation of isolated sections of bad ground. However, the EEP borehole observations made it clear that large scale fracturing of the roof of the excavations was likely throughout the mine. Therefore, a rockbolt pattern was devised to specifically address the fracture patterns observed in the EEP.

The general rockbolt pattern used in most of the mine is divided into three sections (Figure 6). The center third of the drift uses three meter bolts on a 1.2 x 1.2 meter offsetting square pattern. The outer two thirds are on an 1.2 x 1.8 meter pattern. The rockbolt pattern was designed to support a triangular slab with a height of two meters and a ten meter base. The concentration of the support is at the center of

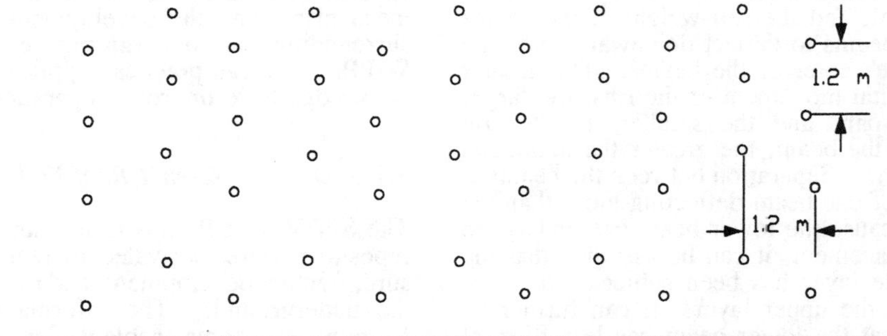

Figure 6. General Mine Rockbolt Layout

the drift cross-section because, according to EEP data, the thickest part of the roof slab is in the middle. The intent of the rockbolt design was not to beam-build but to simply support the dead weight of the slab predicted by the EEP data.

This pattern has been installed in all accessible drifts in the mine and has performed well for short term support. However, as the installations aged, the frequency of failures of rockbolts has increased. Many bolts have failed at the level of clay G. Apparently, the horizontal offsetting at the clay seam imparts an axial load to the bolt, causing tensile failure. To date, this has happened to only a very small percentage of the installed rockbolts.

5.2.2 *Panel 1, Room 1, Rockbolt Pattern*

The environmentally sensitive mission of WIPP places requirements on mine operations that are not normally encountered in mining. The design of the roof support system for Panel 1, Room 1, the first room scheduled for receiving transuranic waste experiments, was subjected to intense scrutiny. It was determined by external organizations that the support system in this room must guarantee stability for the duration of the waste experiments (U.S. DOE, 1991). This would mean the room would be at least twelve years old at the time the experiments were scheduled to terminate. Unfortunately, the roof fall discussed earlier occurred in a similar room less than eight years after its excavation. Therefore, an extremely conservative support design was required.

The design of the roof support system in Room 1 was once again influenced by fracture patterns seen in the EEP. This rockbolt design incorporates information concerning the rate of horizontal offsetting of bedding planes, which had been suspected of causing most rockbolt failures at WIPP. The support system consists of the general mine rockbolt pattern using three meter long mechanical anchor bolts, which was installed before the stringent new specifications were developed. A set of four meter (13 feet) long, 25 mm (one inch) diameter Dywidag steel anchor bolts were added to the support system. Once again, the bolts are distributed across the width of the room according to the weight of the slab predicted by the EEP. The bolts are installed on a square pattern with rows (width) about three meters apart. The bolts are 0.61 meters (two feet) apart in the center of the drift, and range from 0.76 meters (2.5 feet) to 0.91 meters (three feet) as the ribs are approached. These bolts have 0.91 meter (three feet) grout anchors. Two types of wire mesh and a set of steel cables running longitudinally along the room were also added to this system. In order to reduce the risk of bolt failure from horizontal offsetting at the clay seam, the ungrouted section of the bolt holes were oversized to 76 mm (3.0 inches) diameter. This is intended to allow horizontal offsetting to occur without impacting the rockbolts.

6 CONCLUSIONS

The experience with borehole fracture observations and their applications gained from the Excavation Effects Program at the Waste Isolation Pilot Plant has led to the following recommendations and conclusions.

1. A borehole observation program must use arrays of boreholes with consistent layouts in order to provide statistically meaningful results.

2. Because borehole observations are frequently subjective, the best data is obtained when the same person completes each survey.

3. Boreholes destroyed by mining should be replaced immediately to allow continuity of data.

4. The EEP is an inexpensive, fast method for determining the condition of the ground surrounding underground excavations.

5. The EEP provides information necessary to efficiently design a roof support system in bedded rock.

6. The EEP enabled the project to predict the size and shape of a large roof fall, and in combination with instrument analyses, formed the basis for an accurate prediction of the time of the roof fall.

REFERENCES

U.S. Department of Energy (DOE), October 1991, Waste Isolation Pilot Plant Supplementary Roof Support System, Underground Storage Area, Panel 1, Room 1, DOE/WIPP 91-057.

U.S. Department of Energy (DOE), March 1992, Geotechnical Field Data and Analysis Report, DOE/WIPP 92-010.

Work supported by the
U.S. Department of Energy
under DOE Contract No.DE-AC04-86AL31950

Innovative Mine Design for the 21st Century, Bawden & Archibald (eds) © 1993 Balkema, Rotterdam, ISBN 90 5410 325 6

Underground oil mining at the Underground Test Facility

H.G. Stephenson & D.A. Wright
Norwest Mine Services Ltd, Calgary, Alb., Canada

R.W. Luhning & R. Coutts
AOSTRA, Calgary, Alb., Canada

ABSTRACT: The paper describes a concept for recovering bitumen from oil sands using underground tunnels. The concept is being tested at an Underground Test Facility (U.T.F.) near Ft. McMurray, Alberta. Mining techniques used to design, construct and operate such a facility, the hazards encountered and the safety procedures used to mitigate them, are described. The potential for large, underground operations using this very promising technology is evaluated.

1 THE CONCEPT

1.1 *Potential of the Alberta Oil Sands*

Deposits of bitumen in the Alberta oil sands are believed to be 250 billion m^3 (E.R.C.B. 1986). This is one of the world's largest deposits of liquid hydrocarbons

Although two very large surface mines are successfully exploiting the shallow sands only approximately 5% of the reserve is amenable to surface mining. Recognizing this, in 1975, the Alberta Government set up the Alberta Oil Sands Technology and Research Authority (AOSTRA) with the primary responsibility of developing the technology for alternative commercial operations in oil sands.

1.2 *Underground Oil Recovery*

The system employed at the U.T.F. utilizes tunnels, constructed in limestone 13m below the oil sands, to drill and complete pairs of horizontal wells into the reservoir. By injecting steam into the upper well of each pair, bitumen is mobilized and gravitates through the lower well to a pipeline system in the tunnel. The bitumen is then pumped to the surface. This is known as the Shaft and Tunnel Access Concept (SATAC). The process employed is Steam Assisted Gravity Drainage (SAGD). Figure 1 shows how the concept works.

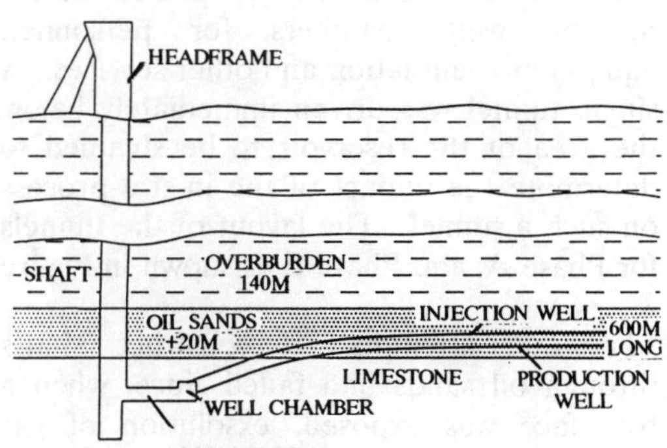

Figure 1. Shaft and tunnel access concept for underground oil recovery (not to scale).

1.3 *Design and Construction of the Underground Test Facility*

Studies carried out by AOSTRA in the 1970's indicated that this concept could be the key to exploiting the vast reserves of oil sand beyond surface mining limits. Consequently, in 1982, AOSTRA decided to construct the U.T.F. for pilot testing the SATAC concept using the SAGD process. Norwest Mine Services Ltd. were assigned the task of costing and designing the Access Stage of the U.T.F. and project managing construction of this stage.

1.4 *Design of The Access Stage*

It was decided that The Access Stage would consist of two, 213m vertical shafts with insets at the 183m level. The shafts would be 3m in diameter and equipped with steel liners capable of withstanding full hydrostatic head, and ground swelling pressures.

No.1 intake shaft would be equipped with a 186kW hoist handling a skip cage combination with the capacity to hoist 100tph of tunnel waste.

No.2 return shaft would be equipped with an emergency access hoist and would accommodate the pipe ranges for steam, bitumen, etc.

Almost 1km of tunnel would provide access to the well chambers for personnel, equipment, ventilation and other services. A single tunnel was driven immediately below the area of the reservoir to be steamed to determine the impact of the in-situ process on such a tunnel. The layout of the tunnels for Phase A, and Phase B, is shown in Figure 2.

Previous attempts to construct shafts through oil sands had failed since, when a free face was exposed, exsolution of gas occurred causing the sand to collapse.

For this reason blind shaft drilling was specified for sinking the two shafts, the pressure of drilling fluid being utilized to support the shaft walls prior to liner installation.

In anticipation of successful results from the pilot stages each element of the U.T.F. was designed to form part of a commercial operation if such a development was implemented.

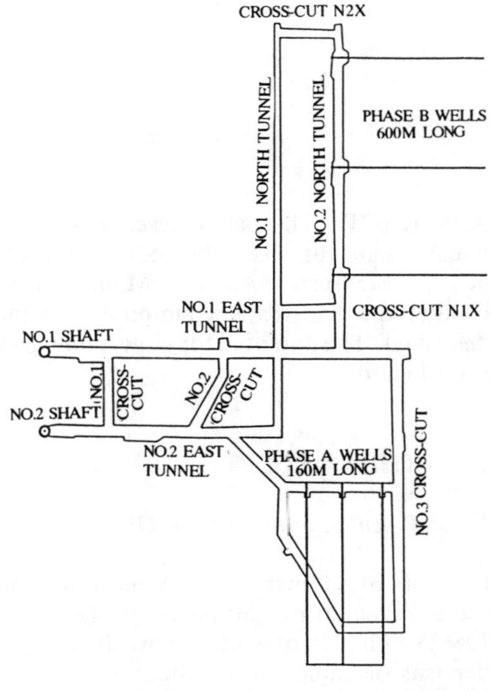

Figure 2. Layout of tunnels and wells for Phases A and B.

1.5 *Construction*

Engineering and costing was completed in January 1984 and in June 1984 AOSTRA decided to go ahead with construction. A conventional deep hole oilfield rig, with a 500t hook load capacity, was employed to drill the shafts using a 250t drill bit, 4m in diameter. Both shafts were drilled successfully, following which the liners were installed and grouted. The time required to drill, line, grout and equip both shafts consecutively was 16 months.

Tunnels and well chambers were constructed by conventional drill and blast techniques.

Construction of the shafts, tunnels and other mining facilities has been described by Stephenson et al 1985 and Patterson et al 1985.

All mining facilities required for Phase A

Figure 3. The head frame over No.1 shaft.

testing were completed by mid 1986. Norwest Mine Services Ltd. was appointed as mining operators, in charge of mine safety and maintenance.

1.6 *Phase A and Phase B Pilot Operations*

Phase A began with the drilling of three horizontal wellpairs 160m long. Adjacent wellpairs were 24m apart and each had a 60m completion zone in oil sand and 5m vertical separation between the injector and producer. Steaming began in December 1987 and continued until March 1990, producing a total of 20,600m^3 (130,000 barrels) of bitumen. The main purpose of Phase A was to test the performance of the SAGD process which has been described by Edmunds 1987.

Numerical simulation of process mechanisms carried out during the

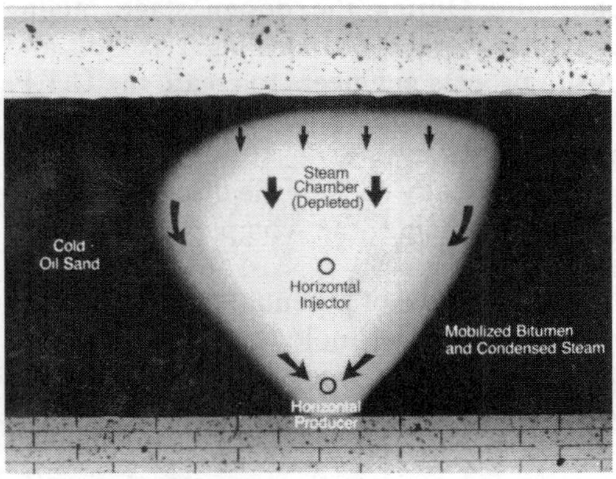

Figure 4. The steam chambers formed by the SAGD process.

implementation of Phase A indicated that recoveries would be almost 50% of bitumen in place. Actual recoveries have exceeded 60% for each of the three well pairs. This recovery was achieved over approximately two years, considerably less than the four year period expected. Steam oil ratios have been extremely low, averaging 2.5 compared to 4 or more for most other in-situ operations.

Details of the performance in Phase A have been described by Haston et al 1989.

Phase B was then conceived to test the process with well geometry suitable for a full scale commercial facility. Three wellpairs spaced 70m apart, with 600m completion zones in the reservoir and 5m vertical separation between injector and producer, were installed. After completing the horizontal wells, steaming of individual wellpairs began in October 1991 and continued until March 1992 when all wells were shut-in to allow for construction of larger steam generation facilities on surface. In December 1992, steaming resumed and presently continues.

2 HAZARD IDENTIFICATION AND SAFETY PROCEDURES

2.1 *A Technological Mix*

The mix of disciplines required by an Underground Oil Recovery operation is unusual. During the design stage mining engineers worked with drilling, reservoir and in-situ process engineers to create the U.T.F. and a means of operating it safely. The main problems were:

- The difficulty each of the disciplines faced in understanding the complexity of other areas; and
- The existence of potential hazards created by the use of technology such as oil well drilling and in-situ processing in and adjacent to underground tunnels.

Considerable effort was expended to identify and mitigate the hazards inherent in such an operation

2.2 *Strata Control During In-Situ Operation*

Strata control in the underground tunnels requires both a rock support system and rock deformation monitoring using geotechnical instrumentation. The design of both systems depends on:

- The rock properties and geology of the limestone;
- The geometry of the tunnels; and
- The geometry of the horizontal wells.

A major concern is the impact of the in-situ steam process applied to the reservoir, involving the application of steam at 1500kPa and 200°C, on the stability of the tunnel immediately below.

The tunnels are located in Devonian limestone of the Waterways formation which consists of alternating units of argillaceous and massive limestone, the strength of these units ranging from 19 to 84MPa.

Limestone with a high argillaceous content is weak, being highly fractured in all directions. The massive limestone is very strong. It has non-persistent, near vertical fractures with an average spacing on the order of 5m.

The tunnels were driven with the roof following a bedding plane in a 5m thick layer of massive limestone. The upper section of this bed provided a natural roof support. The lower 3m section formed the wall of the tunnel and provided a strong foundation for the wellhead.

The tunnels and well chambers range from 4 to 10m in width and 5 to 7m in height. The well chambers are wider and higher than most of the access tunnels.

2.2.1 *Support of Tunnels and Well Chambers*

The tunnels are supported by 1.8m long Swellex rock bolts installed on a 1.2m square grid spacing. At sensitive or oversized areas, such as the well chambers and the ventilation overpass, 2.4m long Swellex rock bolts on 1.0m spacing are used. Because of the wider and thinner layer of competent bedded limestone at these areas, twinned, 15mm diameter, 6m long steel cable anchors were installed and grouted to provide additional support. These cables have a minimum breaking strength of 53 tonnes. All rock bolts and cable anchors are attached to a face plate and, in the high roof areas, the rock is covered with chain link wire mesh to prevent loose material from falling.

To eliminate failure caused by fracture sets in the bedded limestone, steel straps were fastened with mechanical rock bolts on the upper walls of the No.2 North tunnel. Shotcrete was applied to the tunnel walls in high traffic areas to seal exposed argillaceous limestone and reduce weathering. The shotcrete was whitewashed to improve visibility and to facilitate detection of weathering or fracture development.

2.2.2 *Geotechnical Instrumentation and Monitoring Procedures*

The purpose of the geotechnical monitoring system is to detect any unusual movement, temperature or pressure in the roof or sides of the tunnels which could eventually lead to failure of the tunnel perimeter or the well heads.

Table 1 below summarizes the geotechnical instrumentation in both Phases.

Table 1

Instrument	Number Installed In Phase A	Number Installed In Phase B
Monitoring Stations	18	9
Convergence arrays	18	9
Multi-point extensometers	23	11
Single point extensometers	31	3
Thermocouple strings	13	12
Piezometers	0	8

Figure 5 is a cross section of a typical Phase B wellpair showing the convergence pins, extensometers, thermocouple strings and the relationship of these instruments to the geology and the horizontal wellpairs. Piezometers in the roof of the tunnels and well chambers monitor changes in hydraulic pressure in the argillaceous limestone unit above.

During Phase A of the project, the extensometers were read using a micrometer. In Phase B, the new extensometers were constructed with Linear Voltage Displacement Transducers (LVDT's) attached to the extensometer rods at the collar of the hole. These are read manually using an electronic readout unit.

During Phase A, all thermocouples were read manually using an electronic readout unit. During Phase B, they were wired to the Supervisory Control And Data Acquisition (SCADA) system. This system is used to control the steam injection and bitumen production process and has been programmed to continuously scan the underground thermocouples once per hour.

2.2.3 *Conclusions Regarding Strata Control*

The results from the Phase A monitoring program are summarized below.

Wellhead movement:

- Injector wellheads moved 10-16mm inward
- Producer wellheads moved <4mm inward

Extensometers:

- Tunnel wall movement ranged from 0-0.8mm inward.
- Tunnel and well chamber roof movement ranged from 2.1-4.7mm inward except for the roof immediately below the A3 wellpair which lowered 8.0mm.

It is believed that the movement of the tunnel walls is entirely accounted for by thermal expansion. Roof movements were caused by a combination of thermal expansion and hydraulic pressure in argillaceous limestone.

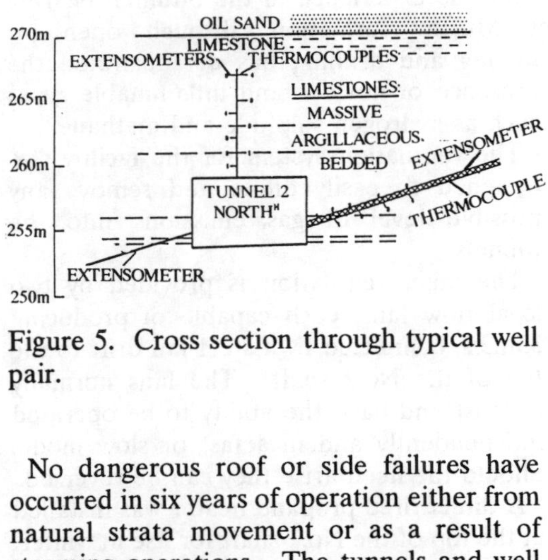

Figure 5. Cross section through typical well pair.

No dangerous roof or side failures have occurred in six years of operation either from natural strata movement or as a result of steaming operations. The tunnels and well chambers, even those immediately below the pressurized wellpairs, provide a very safe working location.

If this experience is repeated during the Phase B operation we can confidently assert that existing strata support systems are more than adequate for use in a commercial operation.

More information regarding the design of the underground instrumentation is given in Madsen, 1992, Collins, 1991 and Laing et al, 1988.

Figure 6. The underground well chamber.

2.3 *Ventilation and Environmental Monitoring*

Previous experience in the bitumen bearing McMurray formation, through open pit mining and drilling, has demonstrated the presence of noxious and inflammable gases such as hydrogen sulphide and methane.

The ventilation system for the facility was designed to easily dilute and remove any possible level of gas emissions into the tunnels.

The main ventilation is provided by two axial flow fans, each capable of producing 35m3/s, connected by a steel fan drift to the top of the No.2 shaft. The fans normally exhaust and have the ability to be operated independently and in a fast or slow mode. Should the need arise they can be reversed.

A direct fired propane heater was installed at the top of the No.1 shaft for use in winter.

A multi-tube bundle system was installed in Phase A to take air samples from various points in the mine and continuously analyze them. In addition an electronic monitoring system was installed to monitor the status of the ventilation plant, atmospheric pressure, air temperature, relative humidity, shaft pump and water level.

During Phase A it became clear that the multi tube bundle air/gas analyzer system could not be relied on to give consistent, accurate results and the electronic monitoring system, whilst operating well, could not easily be expanded.

Consequently a Supervisory Control and Data Acquisition system was designed to continuously monitor all environmental and safety parameters, as well as all process control functions. This system, consisting of personal computers linked to programmable logic controllers and intrinsically safe detectors positioned throughout the tunnel system, is capable of monitoring, every eight seconds, the following parameters:

Methane
Hydrogen Sulphide
Carbon Monoxide
Carbon Dioxide
Oxygen content
Air temperature
Relative Humidity
Ventilating air velocity

Alarm points are typically set on a two stage basis with the initial alarm activating audible and visual devices at a relatively low harmless level, and the secondary alarm initiating an Emergency Shut Down (E.S.D.) situation. The E.S.D. situation effectively shuts in all processing functions to a fail safe status. Manual readings are also taken regularly.

Fortunately no noxious gases have ever been detected in the ventilation air in the facility, although both methane and hydrogen sulphide have been detected in the horizontal wells prior to processing. These gases are present in the produced fluids and are flared and safely vented at the surface process facility.

During processing the high temperature of the insulated steam and product pipelines and valve gear increases the temperature of the ventilation air. This has not caused any problems, but will have to be addressed as the facility grows. In fact the effect of thermal lift in the number two shaft has been sufficient to provide adequate natural ventilation for the facility, thereby allowing us to safely discontinue the operation of the mine air fans.

2.4 *Codes of Practice and Safe Operating Procedures*

In Alberta the only specific legislation in place for the control of underground mines is the Coal Mines Safety Act and Regulations which are inappropriate for underground oil recovery.

After discussions with the Energy Resources Conservation Board and the Department of Occupational Health and Safety, it was agreed that the facility would be exempted from the current Coal Mines Safety Act and Regulations provided that acceptable Codes of Practice, prepared specifically for the facility, were applied.

These Codes of Practice were based on relevant Coal Mine and Oilfield Regulations, on field experience, and on good work practices. The following Codes of Practice are currently in force at the facility:

1. Ventilation
2. Training
3. Precautions Against Inrushes
4. Explosives
5. Emergency Response Organization
6. Underground Haulage and Transport
7. Electricity
8. Fire Precautions
9. Hoisting No.1 and No.2 Shafts
10. Tunnel Design and Support
11. Safe Work Permits
12. Horizontal Well Drilling

Where the Codes of Practice are not specific enough, or more detail is required to perform a particular function in a safe manner, Standard Safe Operating Procedures are used. Some examples are:

- Inspection of braking systems on the No.1 and No.2 hoist and conveyance;

- Starting and synchronizing of the standby or emergency generators; and

- Installation of strata bolts and support systems.

2.5 *Drilling and Completion of Horizontal Wells*

Accurate, safe drilling and completion of wells is critical to the SAGD process. The wells were drilled from the underground well chambers using a specially designed, track mounted, electro-hydraulic self propelled unit capable of drilling in a potentially explosive atmosphere. The unit can drill 222mm diameter horizontal wells to 600m and provides 100 tonnes of thrust. Direction of the hole is surveyed, and may be modified, during drilling operations. A back pressured circulating mud system removes cuttings from the bit and all drilling is performed through a blow-out preventer connected to casing cemented into the wall of the well chamber.

More information regarding the design of the wells may be found in Edmunds et al 1987.

Figure 7. The drilling machine.

2.6 *Safety Record to Date*

Since the beginning of construction of the facility in 1984, up until the time of writing, there has never been a lost time accident associated with the underground facility.

The relatively small number of people involved makes it easier to educate and train personnel. This does not detract from the efforts of the small dedicated team that have developed the standards regarding personnel selection, training, engineering and maintenance procedures for this unusual underground oil recovery operation.

It is also important to recognize the contribution of the Health and Safety Committee whose members monitor and develop standards. The committee actively investigates all concerns, however minor they may seem, and ensures that corrective action is taken to reduce potential loss at any level.

Our philosophy is to be proactive rather than reactive to the general health and safety of all employees, equipment and systems.

3 COMMERCIAL APPLICATIONS

3.1 *Commercial Application In Oil Sands*

The economics of a commercial operation producing 1600m^3/d (10,000 Bbl/day) from the U.T.F. lease were described by Haston et al (1989). Capital costs for this operation were estimated at C$155 million and annual operating costs at C$22 million.

The production cost (inclusive of capital and operating expenditures) was C$68.80 per m^3 ($10.88 per bbl) of bitumen.

Using a price for West Texas Intermediate of US$19.00 per Bbl. and discounting this by 65% gave a value for bitumen in Ft. McMurray of C$12.45.

These economics will be re-examined in the near future to determine the potential for a commercial operation based on the U.T.F. The results are expected to be even more attractive than those quoted above.

A concept for underground oil recovery using SAGD in reserves under shallow overburden just beyond surface mining limits was described by Stephenson 1989. The idea was to take advantage of infrastructure at one of the existing oil sands operations and supply bitumen to the upgraders for conversion to synthetic crude. At these shallow depths vertical shafts could probably be replaced by inclined access tunnels.

Further studies of this concept were carried out by AOSTRA and Syncrude and described by Luhning et al 1991.

The study examined underground oil recovery under the existing tailings pond and in two surface mineable areas. One of these was a "grass roots" area where the economics of underground oil recovery were compared with surface mining. Three production rates ranging from 20,000 to 50,000 Bbls (3,200 to 8,000m^3) of bitumen daily were examined. The study indicated that the SAGD scheme could have a significant cost advantage over surface mining for a grass roots operation and should be seriously considered when planning a new oil sands project. Moreover the SAGD process has obvious environmental advantages over surface mining.

Figure 8. Surface buildings and pipelines at the U.T.F.

3.2 *Application to Heavy and Conventional Oil*

3.2.1 *Heavy Oil*

Recoveries of heavy oil in place when applying in-situ methods through surface wells are usually in the range of 5-25%. Energy costs are usually very high. In certain heavy oil reservoirs much higher recoveries may be possible at favourable oil/steam ratios if the SATAC technology is applied.

3.2.2 *Conventional Oil*

There are conventional oil reservoirs in Canada and the United States which lie within the range for which SATAC may be economically viable, up to 700m deep, and where the oil remaining in place justifies consideration of the SATAC approach.

4 CONCLUSIONS

The construction and operation of the U.T.F. has removed all doubts concerning the feasibility of:

1. Constructing shafts through the oil sands and overburden.
2. Constructing tunnels below the oil sands and maintaining them safely immediately adjacent to a sealed and pressurized reservoir.
3. Drilling and completing commercial length horizontal wells to the level of accuracy required for SAGD.
4. Operating the SAGD process at in-situ recovery rates and steam consumption levels which are likely to prove economic.
5. Safe management of an underground oil recovery operation which involves a wide range of technologies.

Phase B operations will determine whether the performance of commercial length wells is as economically attractive as the short wells used in Phase A. If the answer is positive then AOSTRA, and its nine industrial partners, will have developed a technology which can produce bitumen from a large proportion of the oil sands reserves beyond surface mining limits at costs which make the development of commercial operations viable at oil prices of US$19.00 per bbl and above.

This pilot project alone has gone a long way towards fulfilling the goals set for AOSTRA by the Alberta Government in 1975.

If Phase B is successful bitumen from the SAGD process could be produced and upgraded to synthetic crude at costs, and in quantities, which could make Canada self sufficient in oil using known reserves, and at costs much lower then high risk operations in the Arctic or elsewhere.

5 ACKNOWLEDGEMENT

The authors wish to thank AOSTRA and the partners in the U.T.F. project for permission to present this paper.

REFERENCES

Collins, P.M., 1991. *Design of the Monitoring Program for AOSTRA's Underground Test Facility, Phase B Pilot*, Pet. Soc, of CIM and AOSTRA 91-87.

Edmunds, N.R., *The Use of Gravity Drainage Processes In The U.T.F.*, AOSTRA Conference on Advances in Petroleum Recovery and Upgrading Technology; Edmonton, June 2-3, 1987.

Edmunds, N.R., Wong, A., McCormack, M.E.,
and Suggett, J.C., 1987. *Design of Horizontal Well Completions, AOSTRA Underground Test Facility*, 4th Annual Heavy Oil & Gas Symposium, University of Calgary, February 18.

Energy Resources Conservation Board, *Alberta's Reserves of Crude Oil, Oil Sands, Gas, Natural Gas Liquids and Sulphur*, St. 86-18.

Haston, J.A., Luhning, R.W. and Gittins, S.D.,
AOSTRA Underground Test Facility, Progress and Potential, CIM District 5 Meeting, Calgary, October 1989.

Laing, J.M., Scott, J.D., Stokes, A.W., Suggett,
J.C., and Wood, D.F., 1988. *Geotechnical Instrumentation of the AOSTRA Mine Assisted Underground Steaming Trial*, 4th Intl. Conf. on Heavy Crude and Tar Sands Proc., AOSTRA UNITAR/UNDP Information Centre for Heavy Crude and Tar Sands, Edmonton, August 7-12.

Luhning, R.W., Komery, D.P., Phillips, R.S., McIntosh, J.R., Fair, A.E. *A Cost Comparison Study of Open Pit Mining Vs. In Situ Steam Assisted Gravity Drainage*. CIM District 5 Meeting, Ft. McMurray, 1991.

Madsen, D.A., Hammet, R.D., and Collins, P.M. 1992. *Geotechnical Monitoring of the AOSTRA Underground Test Facility*, 94th CIM Annual General Meeting, Montreal.

Patterson, M.A., Cuthbert, L.G. and Tremble, D., *Blind Boring Through Oil Sands - A Trail Blazer*, CIM A.G.M Vancouver, April 1985.

Stephenson, H.G., *Exploitation of Shallow Oil Sands Beyond Surface Mining Limits*, CIM Fourth District 5 Meeting, Calgary, 1989.

Stephenson, H.G., Owen, H.D., and Turner, L.R., *The AOSTRA Underground Test Facility - An Update*, CIM A.G.M. Vancouver, April, 1985.

Wright, D.A., Phillips, R.S., *Safety and Merging Technologies at the AOSTRA Underground Test Facility*, CIM District 5 Meeting, Ft. McMurray, 1991

6 Machinery, materials handling, and mine maintenance

Innovative Mine Design for the 21st Century, Bawden & Archibald (eds) © 1993 Balkema, Rotterdam, ISBN 90 5410 325 6

Formalizing design of mine shaft conveyances in a new design code

G.J. Krige
University of the Witwatersrand, Johannesburg, South Africa

ABSTRACT: The legal requirement for the design of conveyances in use in the underground gold mines of South Africa is that they should be of "substantial construction". For many years there has been much debate as to exactly what constitutes substantial construction, with different mining companies following different philosophies in design. A need was thus recognized to formalize the design procedures and requirements, and in 1986 the drafting of a new design code for conveyances in mine shafts was initiated. It defines the requirements for design procedures which may be adopted, materials which may be used, the loads to which conveyances are subjected, and the strength and fatigue requirements for different combinations of these loads. The development of this code of practice is discussed and the technical content is summarised.

1 INTRODUCTION

For many years, all conveyances in use in South African mine shafts, have been designed in accordance with the statutory requirement of the Minerals Act (1991), which is that they should be of substantial construction. Over the years however, different Engineers, working with different mining companies, have placed rather different interpretations on this requirement. At the one extreme, small companies with very little operating capital have produced conveyances which have only marginal levels of safety, whereas at the other extreme, unrealistically heavy designs have been produced because unnecessarily severe requirements have been placed on limiting states such as fatigue performance. Some of these discrepancies are discussed in more detail below. Many conveyance designs have also been based on a prototype approach, where a previous design has essentially been copied with proportionally larger or smaller main members, to allow for higher or lower payloads. This has led to a progressively changing safety margin, which may or may not be calculated.

In 1986 a forum, initiated by the South African Institute of Steel Construction and including senior Engineers from the mining companies as well as conveyance suppliers, concluded that a recognised code of practice for the design of conveyances was necessary. The main reasons for reaching this conclusion, and the design philosophy and main technical requirements of this code, designated as SABS0208 Part 3 (1993), are discussed below.

2 THE NEED FOR A DESIGN CODE FOR CONVEYANCES

2.1 Consensus throughout the industry

The past few years have seen the introduction of increasingly tight legislation covering industrial safety and the working environment.

This has made it important that there is broad consensus in the mining industry as to what constitutes normal and acceptable engineering practice in many fields, including that of conveyance design. This consensus is perhaps best achieved and reflected in a code of practice.

2.2 Quality control

The mining industry has introduced strict quality control procedures to many items of equipment, including conveyances, over the past 20 years. Disputes have, however, arisen between suppliers and mines in this regard. For instance, a requirement that design shall comply with accepted practice is likely to cause some misunderstanding where supplier and mine follow differing practices. Quality control is only meaningful if there is a well defined and widely accepted standard against which quality can be measured, and if it is applied to the entire design - manufacture - operation - maintenance spectrum. The best yardstick for quality control in design is a code of practice. The South African quality assurance code, SABS 0157 (1987), following ISO 9000 (1987), carefully specifies the documentation and agreement necessary for the design process. The availability of a design code immediately satisfies most of these requirements. SABS 0157 (1987) also lists requirements for such factors as uses and misuses, safety, reliability, and failure modes which must be considered. These have all been rationally assessed and, where necessary, included within the requirements of SABS 0208 Part 3 (1993).

2.3 Consistent level of reliability

Smith (1992) reported wide discrepancies between different design procedures, which have been recognised as a problem for some time. Some Engineers apply a safety factor against yielding of the material, whilst others apply the factor against ultimate failure. There are thus different limit states considered. In some cases, Engineers apply the safety factor as a load factor, so that the design load is increased but normal stresses are used, as shown in equation 1.

$$\text{Resistance} \geq \text{Load x Safety Factor} \quad . . 1$$

This has sometimes led to unreasonable deflection and fatigue requirements which are based on the design load, rather than on the actual load. In other cases, Engineers apply the safety factor as a material factor, so that the actual load is used for design, but the allowable stresses are reduced, as shown in equation 2.

$$\text{Resistance / Safety Factor} \geq \text{Load} \quad . . 2$$

The safety factors used also vary widely, between 5 to 6 on older conveyances, and 9 to 10 on newer conveyances. Different conveyances may thus have strengths which differ by a factor of as much as 3.5.

The introduction of a design code can eliminate this variability, and ensure that all conveyances are designed to the same reliability.

2.4 Rational basis for assessing new equipment

Some difficulties have arisen in the implementation of new concepts in conveyance design, because current practice is not clearly defined. An example of this is the introduction of circular skips, which has been proposed in several quarters (for example Penning 1988, Thomas 1990). The pressures resulting from filling, hoisting, and emptying the ore have not been well defined, so it has not been easy to determine the strength requirements for circular skips. Another example is fatigue cracking which has been experienced in the region of the door hinges of certain skips constructed in aluminium. Lack of knowledge regarding the stresses at this point have led to this difficulty in applying a new material.

Such difficulties can be eliminated, or at least reduced, if all the loading and operational conditions are understood and rationally and comprehensively reflected in a code of practice. Many years of experience, which might otherwise be lost, can be incorporated into a code of practice, to provide this type of guidance.

2.5 The link between design and operation and maintenance

A properly formulated code should define the operating and maintenance requirements for shaft alignment and conveyance condition, which are implicit in many design assumptions. This link has typically been either assumed, or overlooked, by designers, and not clearly recognised by operations and maintenance personnel. Due benefit can then be given in design if shafts are very well maintained, or on the other hand lower maintenance levels can be accommodated by a more stringent design requirement. This facilitates a more holistic approach to the entire spectrum of shaft engineering, from design through to maintenance.

3 DESIGN PHILOSOPHY

In the development of this new design code, it was important that the basic philosophy to be adopted for design should be agreed by all parties, and then adhered to throughout the code. The overall design philosophy had to comply with the Minerals Act (1991), which requires that conveyances must be of substantial construction. This was interpreted as meaning that conveyances should be capable of carrying not only their design payloads with impact, but also those forces which may arise due to emergency conditions, poor maintenance, or known levels of misuse during normal hoisting operations. It was also recognised that the development and application of new conveyance equipment can significantly influence the nature and magnitude of the forces which new conveyances must withstand, and that the design philosophy must thus make allowance for new developments wherever possible.

3.1 Normal hoisting operations

The essential design philosophy for the normal operating conditions is that conveyances must function reliably over long periods of time. This means that deformation of the conveyance must be limited, fatigue must be eliminated, and there must be an acceptable margin of safety. In order to satisfactorily meet this design philosophy, it is necessary to define all the forces to which the conveyance may be subjected. Four categories of forces are recognised.

The first category of forces is that derived from known loads, such as the self weight and the payload. These are straightforward.

The second category of forces are those generated during loading and unloading of the conveyance. It is recognised that some level of misuse of conveyances is inevitable. It is, however, not possible to specify reasonable design loads to cover any possible misuse. For example, a survey of typical operating practices (SAISC 1987) showed two particular areas of concern. The first was the vertical alignment between a cage deck and the station. Ideally, the cage deck should be aligned exactly at the station, but in practice this is difficult to achieve. One mine accepted a misalignment of up to 700 mm, whereas most mines expected that misalignment did not exceed about 50 mm. The second was where kepping devices were used to hold cages at stations. Quite commonly such devices are used to stop a slowly moving cage at the correct level. One mine would use keps to stop a cage travelling as fast as 3 m/s, whereas most mines would only attach the kep to a cage travelling at less than 0.5 m/s, or once the cage was already stationary. If design loads were to be specified to allow for impact from the extremes quoted above, the result would be very heavy conveyances, and perpetuation of unacceptable operating practices. Accepting that no design code can allow for any conceivable level of misuse, the philosophy adopted in the code is to allow for "normal" operating practices. "Normal" is here used to refer to what is most commonly found to apply at South African mines. Operating practice must then be altered where necessary, to comply with these design assumptions.

The third category of forces are those required to guide the conveyance as it travels up or down the shaft. The major factors in determining the magnitude of these forces is the misalignment of the shaft guides, the shaft geometry, and the conveyance velocity (Thomas 1990). The shaft geometry and conveyance velocity are determined by the Engineer, but the misalignment is dependent on the shaft installation and maintenance. The different mining companies adopt differing practices regarding the installation and maintenance of shaft steelwork. It was decided that the lower forces, resulting from more careful guide installation and better maintenance, and vice versa, should be recognised by the code. The design philosophy thus links the operating forces to the accuracy of installation and maintenance.

The final category of forces is that arising from manoeuvring the conveyance into, or out of, the shaft. Provision is made for this in the layout of the headgear, but it may also be necessary to attach holding or winching ropes to the conveyance. Additionally it is possible for load reversal to occur, if the conveyance is supported at its base. The philosophy adopted here is that this only involves the self weight, so the loads are not large, and it never occurs under operating conditions, so that these loads can be ignored. Certain recommendations regarding minimum side plate thickness and corner angle sizes are however included to ensure a sufficiently small likelihood of damage occurring.

3.2 Emergency conditions

The essential design philosophy under emergency conditions is that injury to people and damage to other shaft equipment should be minimized. This philosophy accepts severe deformation of the conveyance, provided such deformation does not crush occupants of a cage, and that the conveyance does not break up and plunge down the shaft.

3.3 Design procedures

Design of conveyances is typically carried out by Structural Engineers, so the design procedures should be those with which Structural Engineers are familiar. The adoption of limit states design procedures internationally, and their improved ability to predict structural reliability, led to the specification of limit states design procedures for conveyances.

3.4 Incorporation of new equipment

There is an ongoing quest for improved hoisting equipment for South Africa's mines. The possibility of new equipment requiring the specification of new design loads or procedures can thus not be excluded. Design codes have frequently been criticized for restricting the application of new concepts. A deliberate philosophy in developing this code has therefore been to anticipate improvements and make specific provision for them. In some cases, this has been more explicitly done in a commentary to the code (Krige 1993).

4 SUMMARY OF THE TECHNICAL CONTENT

A brief summary of some of the main technical content of the new design code follows.

4.1 Materials specification

The code at present specifies the use of structural steel grades or structural aluminium alloys, with certain more stringent requirements. In the case of structural steel, brittle fracture is a potential problem in the low temperature, high impact conditions pertaining in mine shafts. The typical lowest temperatures experienced in the mining areas of South Africa, are of the order of -10ºC, so that a Charpy V-notch impact resistance of 27J at 0ºC is specified. The structural steels which have this impact resistance are difficult to obtain in the small quantities required for conveyances, so various solutions have been adopted by different mining companies. Conveyances may be built of plate which is available in the impact resistant steel grades, rather than of rolled sections. Alternatively, large quantities of impact resistant steel sections may be purchased and held in stock, but at current high interest rates, this has serious financial implications. Most commonly, steel of ordinary grades is subjected to extra impact testing, to establish a satisfactory impact resistance.

In the case of structural aluminium, technical grade material, which is subjected to extra quality control tests, is required.

4.2 Definition of loads applied

The loads to which conveyances are subjected, are defined in three categories. The first category is those loads which are common to all different types of conveyances. Loads in this category include the self weight of the conveyance, loads induced by underslinging of heavy or large equipment, the loads from tail ropes where they are used, and the loads generated between the guides and the conveyance as it travels in the shaft.

The loads generated between the conveyance and the guides have been shown to be dependent on a number of factors, such as hoisting speed, guide steelwork geometry, and alignment of the guides (Thomas, 1990). The alignment of the guides depends on the state of the shaft, which is determined by the level of maintenance. Thomas (1990) follows Krige (1985) and Loubser and Bull (1962) in classifying shafts into the range of good, average, or poor shafts, according to the extent of guide misalignment. The magnitude of these forces is high for poor shafts, but much lower for good shafts. The Engineer is free to select whatever shaft category is appropriate, and can be justified on the basis of the maintenance schedule. The shaft alignment requirements, and typical guide forces, are shown in table 1.

The second category of loads is that which is applicable to one type of conveyance only. For example, man loads are only applicable to man cages, and rock loads are only applicable to skips. Impact loads caused by the loading of materials and material cars are defined for material cages, based on the actual operating practices. Ore pressure loads are defined following measurements which have been made by Krige and Hofmeyr (1991).

Table 1: Shaft classifications.

	Shaft Classification Good	Average	Poor
Max. local misalignment (mm)	5	10	20
Max. guide gauge error (mm)	3	5	10
Realignment interval (months)	3	6	12
Typical guide forces (kN)			
low	5	10	20
high	10	20	40
No. of buntons between slamming forces	20	10	5
Maximum lateral accel. (m/s^2)	5	10	20

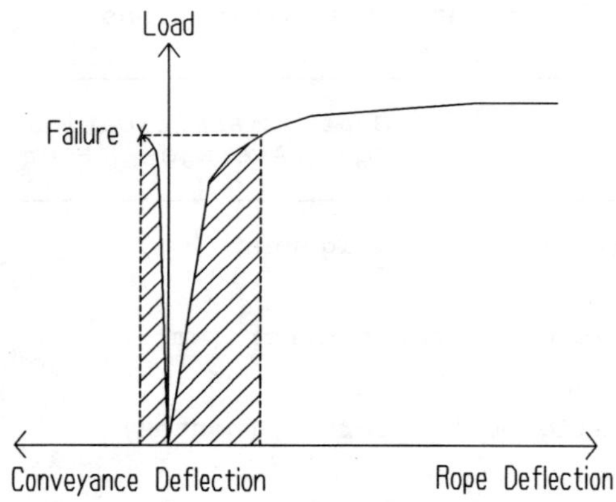

Figure 1a: Energy absorption capacity for conveyance failure.

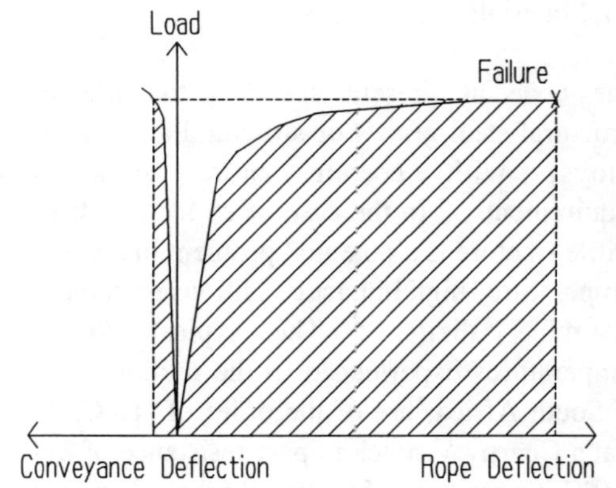

Figure 1b: Energy absorption capacity for rope failure.

The third category of loads is the emergency category. These loads are also applicable to all types of conveyance. Two emergency loading conditions are considered. The first results from an overwind and subsequent rope detachment. The conveyance then falls back onto jack catches, inducing high impact forces on the conveyance. These forces are difficult to calculate, as they depend on the distance fallen, and the stiffness of the catches, neither of which is easy to assess. They have earlier been defined for headgear structures, in SABS0208: Part 1 (1986), as being equal to the rope breaking force, and this definition is retained in the conveyance design code. The second emergency condition results from the conveyance jamming, or more commonly from the rope becoming slack followed by the conveyance falling, in both cases leading to the possibility of the rope breaking. The legally required capacity factor, or ratio of rope strength to total conveyance weight plus payload, is 9 for skips and 10 for cages, and earlier conveyance design has thus traditionally provided a conveyance with an equivalent safety factor of 9 or 10. In the deep gold mine shafts in South Africa, the capacity factor is, however, higher than 10. It could be as high as 14 or 15. A safety factor of 10 for the conveyance design would then lead to the conveyance breaking, rather than the rope. If the rope becomes slack and the conveyance falls, the likelihood of failure depends on the ability of the rope and conveyance to absorb the energy released. The total energy absorption capacity of the rope and conveyance is rather small if the conveyance fails first, as it is very stiff, as shown in figure 1a. If, however, the conveyance is made to be a little stronger than the rope, then the rope fails first with a much greater energy absorption capacity, as shown in figure 1b. This higher energy absorption capacity will reduce the likelihood of the rope breaking. The code thus requires that the main structural members should be capable of resisting a load equal to the rope breaking load multiplied by a load factor of 1.05.

A further innovation regarding the design loads, is the specification of how, and where the emergency loads are applied. The emergency rope break loads may result from slack rope, or from the conveyance becoming stuck in the shaft, so the manner in which these emergency loads are applied is uncertain. The extent to which the entire conveyance must be designed for the emergency loads has also been the subject of much debate. It is thus necessary to define the application of these emergency loads exactly.

4.3 Safety, Load Factors and Load Combinations

In common with all limit states design, the overall safety factor is separated into load factors and material factors. The load factors defined in the code are generally in accordance with those defined in the South African loading code, SABS0160 (1990). These are summarised in table 2.

The code requires the use of the ultimate load factors for strength determination, and the serviceability load factors for fatigue life calculations. The material factor normally applied is 0.9 as defined in most clauses of the structural design code. In special circumstances it may be increased to 1.2 as shown in table 3 for emergency loads, which allows for the effects of strain hardening of the material. This reflects the philosophy that under emergency loads extensive deformation is acceptable, provided the conveyance does not break.

Load combinations are dealt with by defining a matrix of all possible interactions between loads. This is considered appropriate because the possible combinations are fairly limited, but quite complex. The normal limit states procedure of combining gravity load, a main live load, and other loads with a reduction factor, is not permitted, as the interaction of loads is not significantly dependent on their actual statistical distributions.

4.4 Design procedures

The design procedures specified are based on the current South African structural design practice, using limit states design codes. For the design of conveyances in steel, this is specified by SABS0162 : Part 1 (1992). For the design of conveyances in aluminium, this is specified by BS8118 : Part 1 (1992). Both of these codes are intended for use in building structures, so certain extra requirements are imposed. Typical material factors are as shown in table 3.

Table 2: Typical load factors.

Type of Load	Load Factors (main limit states) Ultimate	Service
Self weight	1.2	1.0
Live load which cannot be exceeded	1.3	1.0
Personnel load	1.4	1.0
Nominal live load	1.6	1.0

Table 3: Typical material factors.

Strain hardening possible	Material Factor Normal loads	Emergency loads
No	0.9	0.9
Yes	0.9	1.2

Under emergency conditions, the stresses for steel sections are allowed to exceed the yield stress and enter the strain hardening range of the material, on condition that local buckling does not occur. This is reflected by the higher material factor permitted.

5 APPLICATION OF THE CODE IN DESIGN PRACTICE

5.1 Implications for conveyances

Smith (1992) has recently completed a survey of skip design practice in South African gold mines. This study includes a consideration of

Table 4: Implication of new code for skip design.

(a) Typical self weight changes	Maximum	Minimum
Main Structural Members		
Top Transom	+ 50%	+ 10%
Bridles	+ 20%	None
Bottom Transom	+ 30%	None
Body of Conveyance	+ 3%	- 4%
Overall	+ 7%	None

(b) Typical geometric changes	Maximum	Minimum
Main Structural Members		
Top Transom	30% deeper	None
Bridles	None	None
Bottom Transom	15% deeper	None
Body of Conveyance	None	None
Overall	1% deeper	None

the implications of applying the new code to typical skip design. His findings, and those of other comparative calculations which have been done, are summarised in table 4. These values are indicative values only, as they vary widely because of a strong dependence on the previous design procedure and the materials used.

From table 4 it can be seen that the design implications of the new code for typical conveyance self weight are that the top transom is heavier, the bridle may be heavier particularly on larger conveyances, the bottom transom is usually heavier, and the body of the conveyance is unlikely to change. The overall result is likely to be a small increase in self weight. Similarly, the conveyance is likely to be very slightly deeper when designed using the new code.

Perhaps the more important implications are likely to emerge as the new code gives impetus to various innovations. An example of this is the development of an energy absorption device on the conveyance (Hymers, 1991). This device consists of two steel strips which are attached to the rope and run through a set of rollers. In the event of high impact forces being applied to the conveyance or rope, these strips are deformed through the rollers, absorbing the impact energy and reducing the maximum force to perhaps 40% of the rope break force.

5.2 Adoption of the code by engineers

Most engineers responsible for the design of conveyances have already adopted the provisions of this new code for conveyance design. A few had in fact started using the basic philosophy reflected in the code prior to the development of the code, and have used the draft code as it has emerged. This has proved to be most valuable, as realistic comment has been possible regarding the clarity and ease of application of the proposed code requirements. Present indications are that the new code will soon be used almost universally in South African conveyance design practice. Trends showing increasing application of the code are indicated in figure 2, which gives a rough assessment of the number of new conveyances designed by the major companies, for use in South African gold and platinum mines over the last 8 years and the number which were designed using the new code, or the philosophy which it embodies.

Where there has been reluctance to implement the new code, two main reasons have been given. The first has been uncertainty about the status of old conveyances, where the design does not comply with the new code. Should these historical designs no longer be acceptable, the mines operating them are faced with high replacement costs. The detailing requirements are such that headgear and station

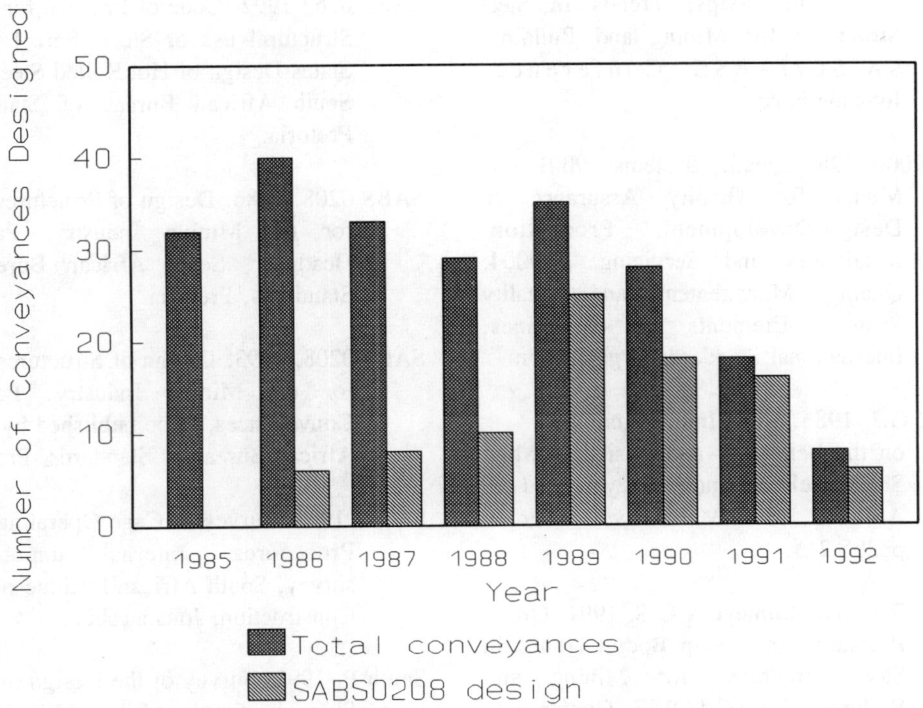

Figure 2: New conveyances designed and use of SABS 0208 (1993).

modifications may be needed as well as new conveyances if the new design is to be implemented. This difficulty still has to be satisfactorily resolved. The second reason is the increase in self weight resulting from the new design philosophy. This is a particularly important consideration in the deep gold mine shafts in South Africa, as ropes are already utilised near their capacity, so that any increase in self weight must be compensated by a reduction in payload, resulting in reduced production capacity. In some new shafts this can be offset by using larger ropes or conveyances of lighter materials, but modification of existing shafts is difficult.

6 CONCLUSION

A new code of practice for the design of conveyances has been implemented in South Africa. This development is justified on the basis of formalising design practice for quality control purposes, and of generally improving conveyance design practice. The technical requirements of the code have been developed around a philosophy of ensuring safety and reliable operation, and include a careful definition of the loads applied and how these are combined on operating conveyances. Although the implementation of the code does generally lead to slightly heavier conveyances, it has been widely accepted as the appropriate design standard in South Africa.

REFERENCES

BS 8118 1992. Code of Practice for the Limit States Design of Aluminium Sections, British Standards Institution, London.

Hymers T. 1991. An Energy Absorbing

Device for Skips, Trends in Steel Structures for Mining and Building, SAISC/IABSE Conference, Johannesburg.

ISO 9000. 1987. Quality Systems. 9001: Model for Quality Assurance in Design/Development, Production, Installation and Servicing. 9004: Quality Management and Quality System Elements - Guidelines, International Standards Organisation.

Krige G.J. 1986. Some Initial Findings on the Behaviour and Design of Mine Shaft Steelwork and Conveyances, J. S. Afr. Inst. Min. Metall., vol 86 no 6, pp205-215.

Krige G.J. and Hofmeyr A.G.S. 1991. Ore Pressures on a Skip Body. Trends in Steel Structures for Mining and Building, SAISC/IABSE Conference, Johannesburg.

Krige G.J. 1993. Commentary on SABS 0208: Part 3 Conveyances, to be published by South African Institute of Steel Construction, Johannesburg.

Loubser R.S. and Bull J.R.M. 1962. Loading Conditions in Guide and Bunton Systems : The Problem and Initial Findings, S. A. Mech. Eng., pp259-285.

Minerals Act 1991. (Act 50 of 1991) and Regulations, Lex Patria Publishers, Johannesburg.

SABS 0157. 1987. Quality Systems. Part I: Model for Quality Assurance in Design/Development, Production, Installation and Servicing. Part IV Quality Management and Quality System Elements - Guidelines, South African Bureau of Standards, Pretoria.

SABS 0162 1992. Code of Practice for the Structural use of Steel. Part 1: Limit States Design of Hot Rolled Steelwork, South African Bureau of Standards, Pretoria.

SABS 0208. 1986. Design of Structures for the Mining Industry. Part 1: Headgear, South African Bureau of Standards, Pretoria.

SABS 0208. 1993. Design of Structures for the Mining Industry. Part 3: Conveyances, to be published by South African Bureau of Standards, Pretoria.

SAISC 1987. Survey of Cage Operating Procedures, Internal unpublished survey, South African Institute of Steel Construction, Johannesburg.

Smith P. 1992. Survey of the Design of Skips in South Africa, MSc Thesis, University of the Witwatersrand, Johannesburg.

Thomas G. 1990. Design Guidelines for the Dynamic Performance of Shaft Steelwork and Conveyances, COMRO, Johannesburg.

Innovative Mine Design for the 21st Century, Bawden & Archibald (eds) © 1993 Balkema, Rotterdam, ISBN 90 5410 325 6

The application of a computerised mine maintenance system to the mechanised mining operations within the deep gold mines of South Africa

R. Morris
TM³, Open Pit Mining and Technical Services Division, Johannesburg Consolidated Investment Company, Limited, (JCI) Johannesburg, South Africa

ABSTRACT: In order to capture the performance of some 846 pieces of mechanised mining equipment, it was recognised that a computerised mine maintenance system was required to meet the demands of scheduling, resource allocation, component failure analysis and costing analysis. The paper describes this process within a deep, mechanised gold mining company in South Africa.

1 INTRODUCTION

Traditionally within the mining industry, the maintenance function has been regarded as a movement out of the promotional main-stream and into a backwater. In fact maintenance activities were often neglected because it was possible to cement over the cracks in maintenance effectiveness by the retention of an excessive spare parts inventory, the use of stand-by or spare equipment or an increase in overtime (Atkinson, Cassapi and Morris 1988).

The trend towards higher technology in mining has resulted in steep rises in capital investment which emphasizes a much greater sensitivity to machine breakdown, and a need for a workable relationship between the mining and engineering departments of a Trackless Mechanised Mining Operation (TM³).

2 THE ROLE OF THE ENGINEERING DEPARTMENT WITHIN THE TM³ OPERATION

2.1 The relationship between mining and engineering

It must be accepted that on a mechanised mine, the maintenance engineers assume the same importance as the production engineers. A degree of cross-training in the disciplines becomes necessary to promote harmony of effort. The respective responsibilities are well defined - the engineers provide the machines and the production managers put them to best use. The fleet is under the care of the maintenance department and is loaned to the production schedule. This division of responsibility ensures that:

a) Maintenance schedules can be fulfilled.
b) The temptation to use TM³ equipment in non-productive work is avoided, and
c) It instills accountability for non-availability of equipment or shortfall in production effort. Investigation within the Gold and Uranium Division of JCI has found that of equal and complementary importance to the sound mining practices and procedures is a high standard of maintenance (Scott-Russell, Morris and Ball, 1988).

2.2 The Relationship of Engineering Cost Within an Operation

Figure 1 depicts the breakdown of costs within a mechanised gold mining operation in South Africa (Morris and Cox, 1992).

Thus it is important to note that engineering cost/ton equates to 25% of the final cost per ton milled. This then is by far the most important area of a mechanised mine which can indeed make an effective contribution to the profit or indeed to the loss of an operation.

The importance of the cost/ton engineering factor can be further explained as follows:

i) Tons/Hour

The related tons/hour is made up of the performance of the units within a TM³ project area and is affected by:

- Operator skills
- Distances to be travelled
- Gradients to be negotiated
- Roadway conditions
- Size of ends
- Size of haulages

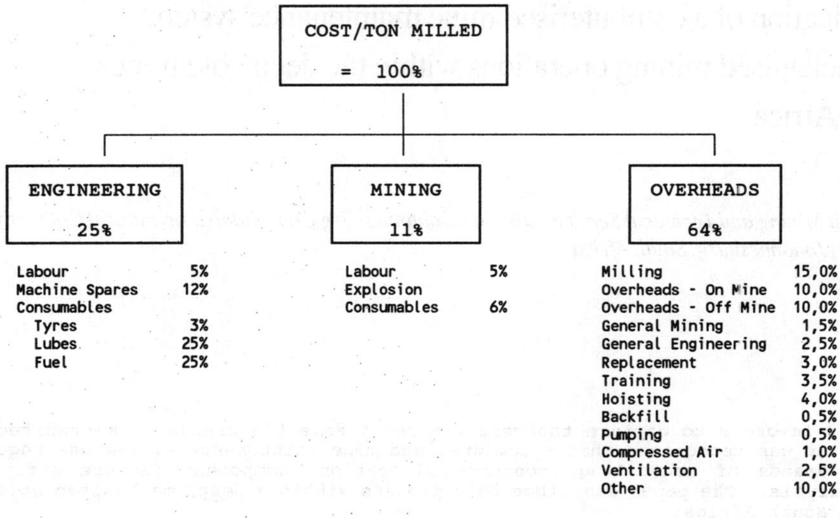

Figure 1.

- Environmental conditions
- Matching of equipment, and
- Selection of equipment.

ii) Cost/hour

The cost/hour aspect is fully in the hands of the engineering personnel and is made up of:

- Labour to maintain the equipment
- Tyre costs
- Fuels and lubricants and complementary importance to the sound mining practices and procedures in the high standard of maintenance (Scott-Russell, Morris and Ball, 1988).
- Spares, and
- Major repairs

Which is affected by:

- Damage
- Operator skills, and
- Environmental conditions

Thus the final engineering costs for fleet management can be shown to be a function of both mining and engineering as follows:

$$\text{Cost/ton}_{\text{(engineering)}} = \frac{\text{Cost/hour}_{\text{(engineering)}}}{\text{Tons/Hour}_{\text{(mining)}}}$$

and as previously depicted in Figure 1, this $\text{Cost/ton}_{\text{(engineering)}}$ is 25% of the Total Cost/Ton Milled.

Thus the key parameter is to produce tonnage for every hour that a machine runs otherwise the cost will be exorbitant. Examples of this is where the machinery is run at expected cost/hour but if the unit does half its tonnage the cost is doubled. The control and containment of this engineering factor is achieved in part by the application of a computerised mine maintenance system (Klintworth and Cox 1988).

3. THE COMPUTERISED MINE MAINTENANCE SYSTEM

Within the Gold & Uranium Division of JCI, a package known as Cybergem, marketed by Control Data (Pty) Limited, was selected by the engineering staff and subsequently tested and installed at each shaft within the Division.

3.1 Objective of a Planned Maintenance System

3.1.1 In the trackless mining environment the objective of the maintenance function is to provide optimal machine availability. Availability of mining machinery has a direct impact on production tonnages and hence gold (or revenue) generation (Purdue 1987).

3.1.2 Machine availability is created by the application of the resources controlled by the maintenance function, i.e. labour, materials and money. These resources are

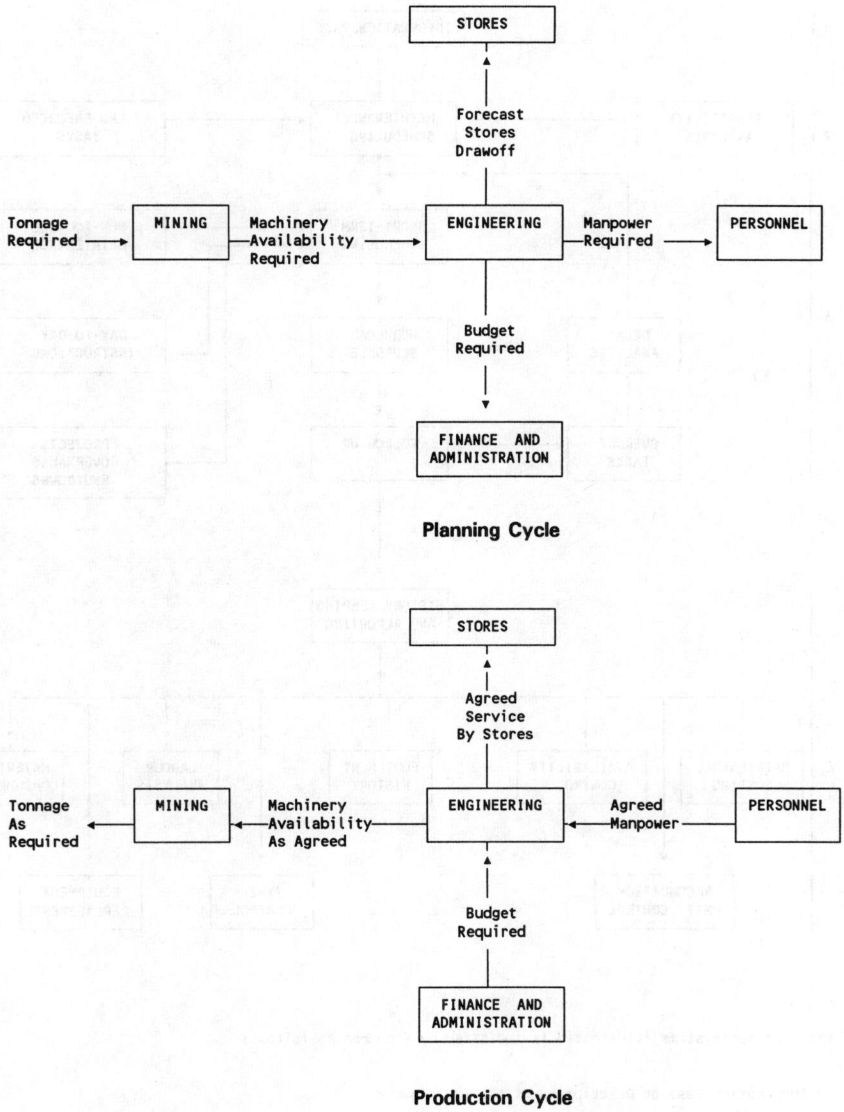

Figure 2.

typically provided by other departments in the organisation, i.e. personnel, stores, finance and administration (refer Figure 2).

3.1.3 The purpose of the Cybergem system is to collect history relating to the application of these resources to the job of machine maintenance so that the maximum cost efficiency can be achieved in future maintenance. The objective is to measure the past to be able to plan more efficiently in the future. This is depicted in Figure 3.

With respect to workshop management of an underground TM^3 prospect, Figure 4 depicts the functional flow diagram of the Cybergem System.

3.2 Functionality Overview of Cybergem on Trackless Mining

3.2.1 Equipment Database

The system records all the TM^3 equip-ment that is down the mine, where it is and how long it has been there.

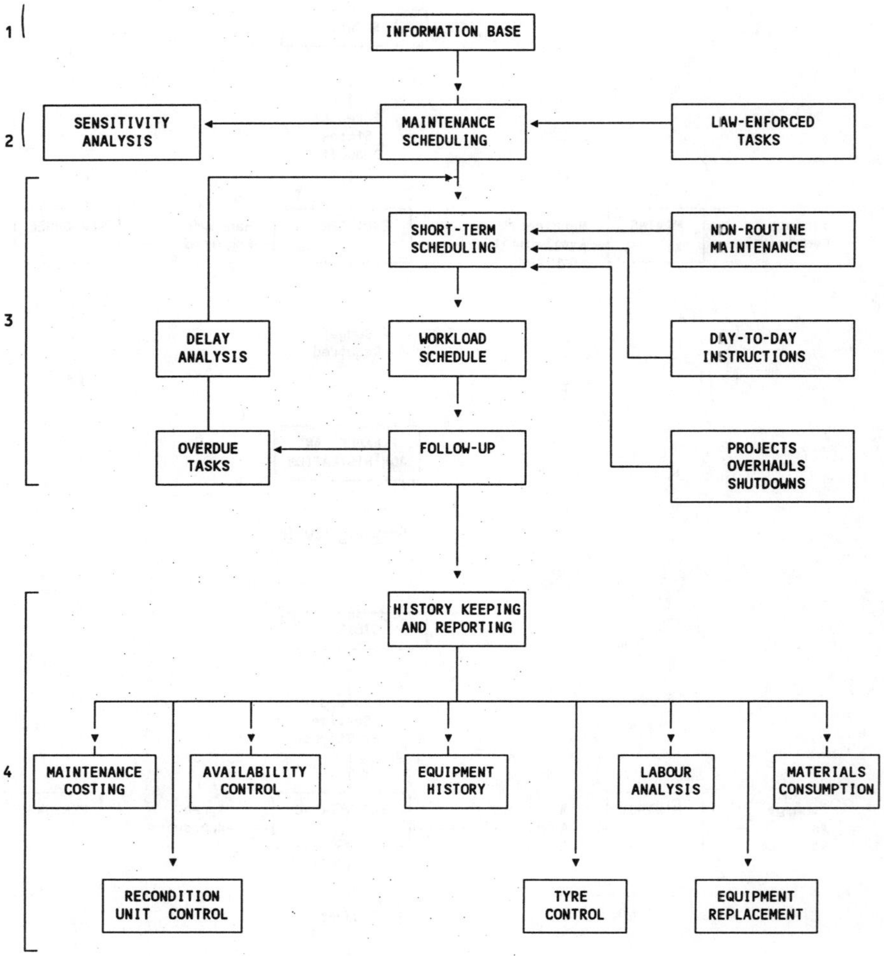

Each of the four sub-systems illustrated is indicated by a number as follows:

1 = Information Base or Description of the Environment

2 = Maintenance Scheduling or Services and Master Tasks

3 = Short-term Scheduling or Work in Progress

4 = History Keeping and Analysis

Figure 3.

3.2.2 Scheduled Maintenance

Effective management of scheduled maintenance is one of the key functions delivered by the Cybergem system. All the set-frequency, scheduled maintenance tasks (services, brake checks, oil sampling, etc) for all the TM³ machines are logged in Cybergem. They are then scheduled and job cards printed as the tasks fall due. The job cards are distributed weekly to the respective foremen and the feedback data captured as the job cards are returned. The system is then used to monitor the outstanding services to keep tight control on the effective scheduled maintenance. As has been proved frequently on the JCI sites good preventive maintenance is the key to low-cost, successful trackless mining. Depending on the size of the site 200 - 400

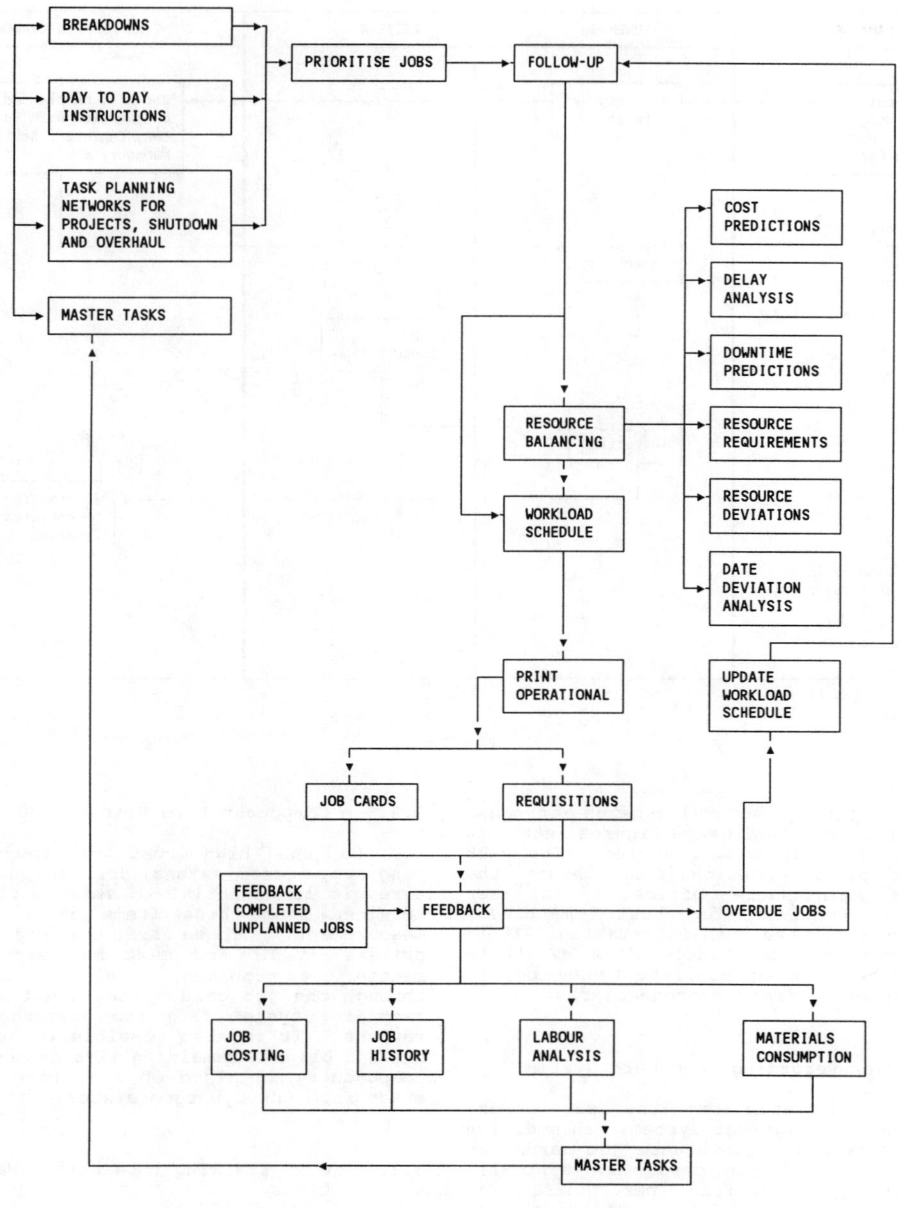

Figure 4.

planned job cards are issued weekly. Keeping control of scheduled maintenance in an underground environment given the number of machines, tasks and geographic separation is a logistical problem. The situation can rapidly turn into a vicious circle of breakdowns, missed services, more breakdowns and endless problems which occurred during the early days of TM³ before the appropriate management systems were developed (Fynn 1993).

3.2.3 Equipment Availability and Utilisation

Further key functionality is provided in terms of recording and processing equipment status changes and running hours from operator's shift reports. From these inputs equipment availability and utilisation figures are processed. On a three-shift operation between 150 - 250 operator shift reports are generated daily on a TM³ site

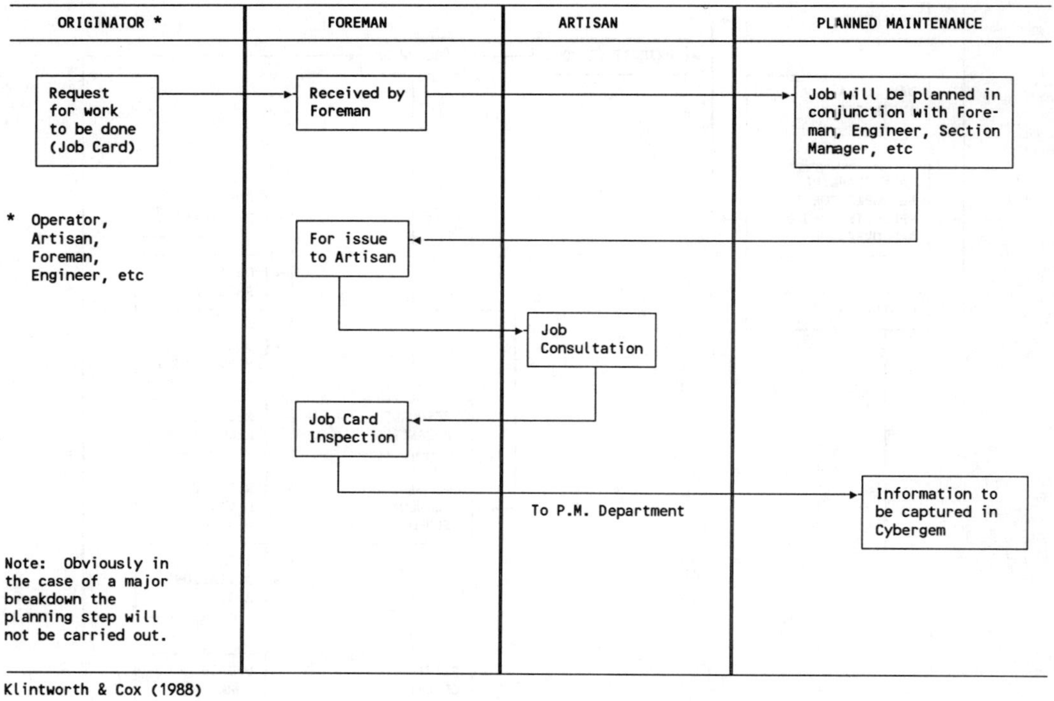

Figure 5.

each containing several status changes, meter readings, fuel usage figures, etc. It is apparent that on a trackless site vast amounts of information flow though the Planned Maintenance Office. To keep effective control of their machines, managers must have this information. It is only practical to handle data in these volumes with the appropriate technology to avoid an army of error-prone clerks.

3.2.4 Task Recording (Job Card System)

The job card system forms the heart of any maintenance management system. In addition to the scheduled maintenance job cards, an additional 2 - 3 job cards are typically generated per artisan per shift for breakdown or ad hoc jobs. Although the Scheduled Maintenance job cards consume more manhours it is these non-routine tasks which utilise the costly spares for replacement and changeouts (Condition-based Maintenance.) These jobs are recorded into Cybergem and drive various other control systems such as component life measurement, tyre life, visibility on manhour usage, etc. Thus in addition to the Operators Shift Reports and Scheduled Main-tenance job cards, 200 - 300 other job cards pass through the Planning Office daily.

The system is depicted as shown in Figure 5.

3.2.5 Component Life Measurement

Due to the high cost of componentry (engines, transmissions, drifters, etc) and tyres it is essential to measure the life obtained from these items as a control measurement. These are the major cost drivers on TM^3 and must be measured to manage out problems. This is achieved through the job card system and the meter recording system from the operator shift reports. It is also possible to tell the age and planned remaining life on the major componentry in place on a machine at any stage from the Cybergem system.

3.2.6 Cost Tracking to Machine, Machine Class

Spares usage data is drawn from the Gold Division Stores system on the mainframe into Cybergem to link these items to specific machines and tasks. These cost figures are linked to the machines in Cybergem along with the machine running hours and tons to provide cost per ton and cost per hour figures per machine and hence machine class. These figures cannot be obtained anywhere else and are critical figures to assess in terms of mine planning and equipment selection. Thus in summary Cybergem links spares used to individual machines along with tons and running hours for purposes of comparative or ratio reporting.

3.2.7 Fuel, Lube Linked to Machine Machine Class

Fuel, lube and hydraulic oil are linked to the individual machine in Cybergem via the job card and operators shift report.

4. CONCLUSION

To assess the benefits of running an effective Maintenance Management System on a trackless section one should consider trying to manage that section **without** the following:

1. Effective planned maintenance.
2. Availability and utilisation figures.
3. Cost per ton/hour figures and trends for each machine, machine class.
4. Lives of major components on failure.
5. Tyre lives.
6. Fuel and oil usage per machine, machine class.
7. Running hours on each machine, major component in situ.
8. Where last month's money was spent.
9. Where last month's manhours were used.

A trackless project producing 40 000 tons per month can be spending in the order of R5 million per year on material and producing ore which will generate some R65 million per annum in revenue. Managing an operation involving these kinds of numbers without effective systems or appropriate management information is not logical. Consider that a 2% increase in production equates to R1,3 million in revenue and a 5% reduction in material costs amounts to R250 000,00. Extrapolation of these figures to a whole shaft makes the case for strong computerised mine maintenance systems **imperative**.

REFERENCES:

Atkinson T, Cassapi V B and Morris R. 1988. The Management of Synergistic Wear in Mineral Exploitation. *'Antiwear 88'*. Royal Society, London, 1988.

Fynn C C J. 1993. Maintenance Management Systems in JCI Gold Division. Internal Report JCI, February 1993.

Klintworth R I and Cox L J. 1988. Maintenance of TM³ Equipment. Association of Mine Managers of South Africa. *Trackless Mining Symposium*. February 1988, Johannesburg, South Africa.

Morris R and Cox L J. 1992. Trackless Mining Statistics within the Gold and Uranium Division. JCI Internal Report.

Purdue J. 1987. The Introduction of the Cybergem Planned Maintenance System to the JCI Gold Division. JCI Selected Technical Papers, Volume 4, 1987. Ed: R Morris, C J Fauconnier and N P du Plessis. Published by JCI Company Limited.

Scott-Russell H, Morris and Ball J C. 1988. The Refinement of TM³ at the Randfontein Estates Gold Mining Company (W) Limited. Association of Mine Managers of South Africa. *Trackless Mining Symposium*. February 1988, Johannesburg, South Africa. (AMMSA Gold Medal Paper - 1988.)

Innovative Mine Design for the 21st Century, Bawden & Archibald (eds) © 1993 Balkema, Rotterdam, ISBN 90 5410 325 6

Hardrock underground mining in the 21st century

J. Pathak & J. E. Udd
Mining Research Laboratories, CANMET, Ottawa, Ont., Canada

ABSTRACT: Because of the 'unstructured' environment of underground hardrock mining, it has been accepted that robotization of many of the activities, in particular production, pose challenges for engineers. Nevertheless, despite difficulties, great strides have been made in these areas since the mid-1980s. This paper describes recent advances in automation and robotization and identifies areas in which more effort is needed. It also gives our view of underground mining in the 21st century.

INTRODUCTION

Since World War II, the quantum leap experienced in the field of electronics caused by the transistor, a device now relatively simple to understand, has changed the whole materialistic life of men. The impact of this development is so great that there is no field left in which electronic devices are not used. The sophistication achieved through this advancement is mind boggling.

Although mining has been a latecomer in implementing advances in electronics, the strides made in their applications to equipment design is rapidly bringing the industry in line with other commercial fields. It is not that mining engineers are not in need of electronics or are very conservative, but mining, by its nature, is not as easily adapted for change as all other industries. Mining has inherent problems which do not allow the ready application of this sophisticated technology. Despite these drawbacks, tomorrow's mines will not be recognizable to to-day's mining engineers.

A mine without a miner underground might seem far-fetched, but tomorrow's mines will come close to that concept. A very high degree of automation with full managerial support assisted with automated decision making using knowledge-based systems will virtually eliminate middle management. Record keeping using paper and pencil will be obsolete. Electronic portable key pads or voice recorders will give direct input to interpreting machines, thereby eliminating manual data entry to computers. The entire mine will be displayed on large electronic screens in order to give underground superintendents a view of the real-time status of the mine. It will also form the communications centre for the whole mine. The same information will be relayed and displayed simultaneously at the head or corporate offices, which could be located on other continents.

Only a few years ago the idea that robotics might have a role to play in mining was not credible to many mining engineers in spite of the common and sometimes indispensable functions that robotics play in manufacturing.

Technological changes accompanying the introduction and use of high-tech will have a profound effect on the way mines are managed, operated and, especially, staffed. The spectrum of educational requirements for mine employees can be expected to change significantly and may prove to be a bottle-neck. The effect of new technology on the input/output parameters of mine operations is given in Figure 1. (1).

THE SHAPE OF THE FUTURE MINE

Based on the present mining technological and mine equipment research activities, it is possible to envisage the shape of the 'future mine'.

If Canada has to lead the world in high-tech applications, in automation, and in robotics in hardrock mining, the direction of the research effort put into the development and application of new technology for mine production process improvement has to be closely looked into. The current total value of the research efforts in Canada is difficult to obtain, and varies from source to source. Canada does not have a good record of R&D, however, as compared with Japan, or some European countries, or even the USA. The figures quoted are that 1.36%, 1.9% and 3% of the GDP is spent on R&D by Canada, the USA, and Japan, respectively. In North America, only 19% of R&D goes to improving production processes, whereas Japan spends almost 73% for this or to apply technology developed by others. Improvements to production processes are what significantly improve competitiveness and are responsible for world-wide market captures for the

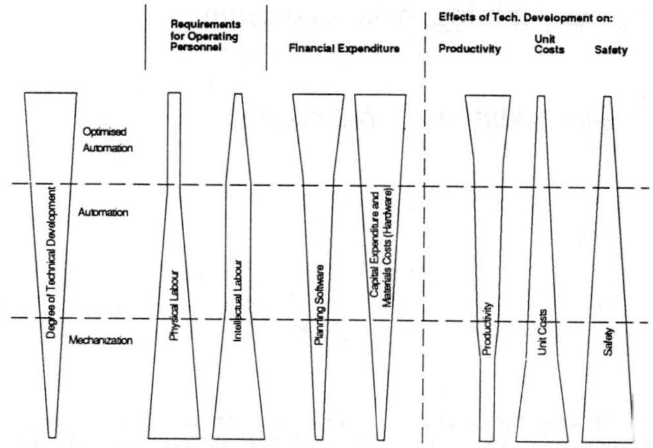

Figure -1 Mine Mechanization/Automation vs Input/Output Parameters.

products involved (2). In Finland, 58% of the R&D funds for mining goes for product development-mostly for mining equipment. It is the latter part of the research efforts for production process improvement which will make a significant change to mining technology so that it may remain competitive with that in developing countries-which have both the availability of cheap labour and richer ore bodies.

RESEARCH AND DEVELOPMENT EXPENDITURES IN MINING

The Canadian mining industry spends, on the average, only 0.7% of its sales on R&D. The leaders in this area are INCO, with 1.1%, and Rio Algom and Noranda with 1.0% and 0.8%, respectively (3). Only INCO and Noranda have large R&D sections for mines research; Rio Algom allocates most of its research to mineral and metal processing and very little to mining. INCO had very little mining research, per se, before 1982. The Noranda Technology Centre (NTC, previously known as the Noranda Research Centre, Point Claire, Quebec) has developed its mining research division since 1985. At present, it is the largest section in the NTC. Both of these centres have significantly changed Canada's position in mining technology compared with that in Sweden and Finland; which have been known for some time as the international leaders. It will be corporate strategy and thinking in favour of automation and robotics in mining that will keep the Canadian mining industry alive and competitive. During the recession of 1990 - 1992, M. Sopko, Chairman of INCO, reiterated that it will be R&D in the area of mine automation and robotics that will continue to enhance productivity and safety and to reduce production costs in mines (4). This statement is contrary to conventional practice where R&D sections are the prime target for down-sizing, and sometimes elimination, during dire times.

The participation of industry, government, mine equipment manufacturers, and mine operators has a significant influence on the direction of, and the targets for R&D for mining. For example, in Sweden and Finland the domestic market consumes only from 10 to 15% of the mining equipment manufactured in those countries, but, nevertheless, their manufacturing industries have taken the lead in new equipment R&D. The mine operators have provided the testing grounds. In Finland, financial support in equal measures is obtained from both manufacturers and mining companies (50%) and Government (50%). The Canadian practice is different. R&D is led by the mining industry with some financial support being obtained from the federal and provincial governments - The contributions are: mining companies 94.6%, federal government 3.9%, provincial government 0.1%, and others 1.4% (3).

The R&D network between the mining industry and the Noranda Technology Centre (NTC) is shown in Figure 2. The INCO network is similar to NTC's. In Finland, possibly because the mining industry and the country are small, there is but one clearing house for research - the Technology Development Centre, TEKES, - which also funds projects. Canada has more clearing houses (+7), and most do not deliver research themselves: They only identify areas of R&D and provide assistance to broker funding arrangements and mechanisms; a function which could possibly be done, by a fewer organizations.

THE MINING PRODUCTION PROGRESS

The purpose of the automation and robotization of underground mining production processes is to enhance productivity, improve safety and reduce costs. Mining, as is any manufacturing process, is composed of many sub-systems that are involved in making the final product; in this case, crushed run-of-mine

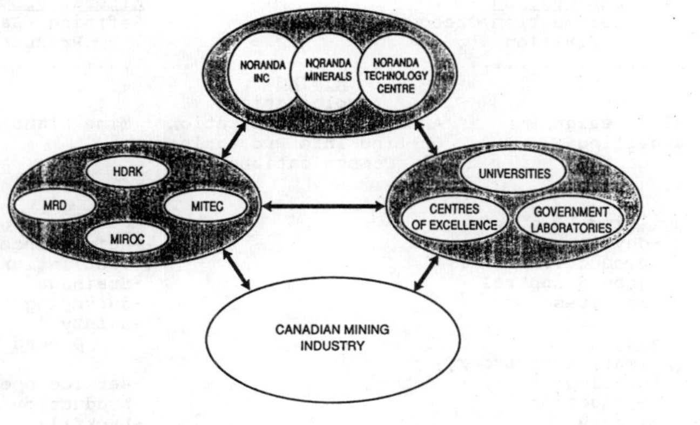

Figure 2 NTC - Mining Industry R&D Network

muck in a surface bin. Figure 3 shows the different systems in an underground mine that comprise the total process of production. Before descriptions of the state of the industry in each area are presented, and a view of the future is given, the current status of each must be described.

The costs of mining average about 60% of the total mine-site costs (including milling, surface and General and Administration and excluding royalties, taxes and capital repayment). Mining is also the most difficult activity in the total mineral production system to automate because of the discrete nature of the operation. Mills,, on the other hand, have reached a stage where these can now be operated and controlled from a control room with practically no operating personnel on the mill floor. Only a few personnel are required for control and for routine maintenance.

In-stope production and ground control are very important to any mine operation, but are difficult to automate. These are the two factors which vary the most significantly from mine to mine. Loading, intermediate and long-distance haulage, and hoisting, are fairly comparable from mine to mine on a per-unit basis.

Further, as mines attain greater depths, costs accelerate. The cost acceleration is due to the mines' centre of activity moving away from the main access point. Haulage ways, service lines, and ventilation circuits become longer. As travel times increase the effective work day becomes shorter. The greater depths also mean greater ground stresses and higher costs for ground control.

As a typical example, between 1986 and 1990, the average depth of the Trout Lake operation, Manitoba, (a cut-&-fill mine), increased by 84% to 350 metres. This resulted in an actual drop in in-stope productivity of 30.8% to 2.3 t/Mh (12 hour-shift basis), and a significant increase in the in-stope production costs; to which the additional ground support requirements contributed greatly. From 1986 to 1990 the rockbolt footage increased by 16.7% to 280,000 feet, with an additional 175,000 feet being required for cable-bolting, which was introduced only in 1988 (5).

The above scenario hints at the future problems that the Canadian industry will face as depth increases. These problems will be the most difficult aspects of the mining production system to automate and robotize, and could defeat the introduction of complete manufacturing technology underground. These activities can, at most, be highly mechanized.

One of the great breakthroughs in mining methods came with the development of the Vertical Crater Retreat (VCR), or (vertical retreat method (VBM)) for bulk mining, in 1975/76 at INCO, in Sudbury. The advantage, besides high production stoping, was safety - miners remain outside the stope. It is possible to achieve more than 60 t/MS productivity for all underground workers by using this method. This could equal the productivity achieved in soft rock mining (Potash) with its present levels of mechanization and automation. The standard stope height in Canadian hardrock mining for any bulk method other than VCR is approximately 66 metres (200 ft). In VCR mining, because of the lack of accuracy of drilling and concerns for hanging wall control, the original stope heights were 33 metres with additional intermediate levels for drilling having been required. Gradually, with experience and the development of more accurate drilling and hanging wall control techniques, the stope heights were increased. The target is probably a Canada-wide average of 100-150 metres. Such a height should result in a productivity of more than 200 t/MS for the entire stoping cycle, and 100 t/MS for underground. Productivities of 1,000t/MS can be achieved for some individual elements in the mining cycle, namely; drilling, loading and mucking (6).

NEEDS FOR AUTOMATION

To be profitable, future mines must be designed for productivities above these

Exploration
Delineation/Orebody Definition

Mineral Processing
Refining, Saleable Product

MINING
(Exploitation)

Mine Design & modelling

Automation/Robotization
On-Line Info Processing
Communications

Mine Planning

Drilling
- development
- production
- ground control
- services

Blasting
(primary & secondary)
- development
- production
- sundry

Loading
- development
- production

Haulage & Transport
- intermediate
- main

Services
- maintenance
- ventilation/heat'g
- drainage
- surveying
- safety

Ground control
- service openings
- production open'gs
- backfill

Hoisting

Crushing

Figure 3 Sub-systems for an u/g production

figures. At present, though, it seems that the productivity curve is flattening out and a breakthrough is needed. If mining engineers closely look into the technology, there is only one avenue left to obtain this - automation of the mine - by considering mining systems rather than by automating individual pieces of mine equipment; which has already been achieved. The ratio of non-production workers to production workers in a mine presently stands at about 50:50. With the application of mine-wide automation and robotization, the production worker component would be reduced drastically. Simultaneously, there is the possibility that the non-production component could increase.

On the other hand, with mine-wide automation, fewer production machines would be required because of increased availability. These could be running entirely independent of operators. Fewer machines, although very sophisticated, would also mean fewer, but more highly skilled, maintenance people; a claim disputed by many mining engineers.

INCO attributed a reduction of man-power and an increase in production of nickel lbs/MS to the implementation of new mining and process technologies, as shown in Figure 4, (7). It must be noted that, for evaluating mining productivity, the lbs/MS of nickel is not the best measure since it is highly influenced by both the grade of the ore and the productivity in the mill and refinery. In INCO's case, although the grade has changed, it has not done so significantly. The improvement in productivity and the reduction in manpower is remarkable, and is due to the improved technology.

Referring to Figure 3, it should also be noted that the delineation or definition of an ore body is not a part of the underground production process but is a prerequisite for designing an efficient mine. Therefore, a few comments here are appropriate.

Our present day method of delineating an orebody depends principally on core samples obtained by diamond drilling and chemical analyses of these. It is a laborious, and often manual, process of interpretation. The subsequent total mine lay out depends upon the shape and the geological and mineralogical parameters of the ore body. Future research will probably be concentrated on powerful geophysical methods which, from surface, and from a few small-diameter boreholes, will delineate the ore body for a subsequent sophisticated mine design program, thus avoiding considerable time-losses between data acquisition, interpretation, and mine design. Presently, a concentrated research effort in this area is lacking. In the future, the quantitative analysis of metal content from a drill core will probably be a real-time process.

The automation of individual equipment for underground hardrock mining has been very successful in Sweden, Finland and in Canada. The best examples are: drill jumbos; DTH drills; cable bolters, which are available but are not in widespread use; out-of-sight remote control of LHDs; and an automated haulage truck (AHT). In addition, NTC's optical guidance system and the Mintronics coded tape-laser (Opti-Trak) guidance system are now undergoing field trials. These have been described elsewhere and, although very sophisticated and highly automated, these are still, in

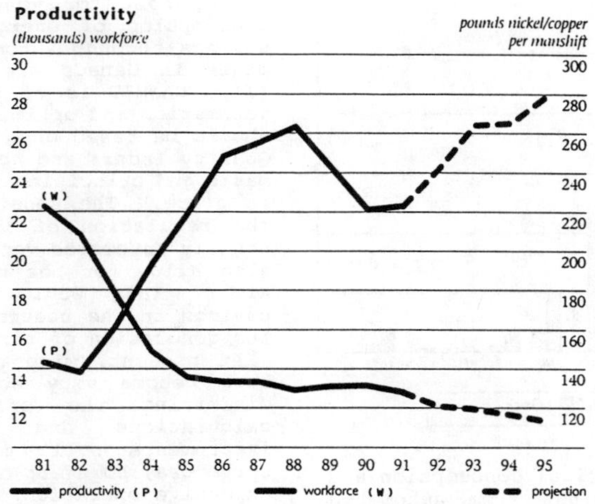

Figure 4 Productivity at INCO 1981-1995 (Actual up to 1991, projected to 1995)

most cases, manned with an operator.

The first stage in robotization would be to remove the operator from the equipment. This has already been done in the case of INCO's automated haulage truck (AHT) - a 70-ton truck in use at the Little Stobie Mine in Sudbury, which can haul 2500 t/d (8). Control of LHD's from the surface has also been tested and demonstrated in Sweden, Finland and at INCO's Research Mines in Sudbury.

The above examples are of individual units, which, while automated, are still (except for the AHT) not autonomous. A totally autonomous unit would drastically improve availability as the machine would attain its maximum utilization. This will be the biggest advantage to be realized from the automation and robotization of equipment.

Next, electrical and electro-hydraulic equipment have obvious advantages over diesels. The trend in underground metal mining is towards all-electric mines. These are easier to control, cleaner, and more efficient. For example, an electric LHD can load from 20 to 30% more, with no exhaust, and, therefore, with less ventilation being needed. Further benefits are that less heating of air is required during the winter while visibilities underground are improved. All of these contribute to safer working conditions. In Canada, mine automation and robotization towards the all-electric mine has already begun. INCO's AHT is a totally electric truck and its Crean Hill Nickel Mine is an all-electric mine. In Finland, the Pyhasalmi Mine, an all electric mine, is targeted for maximum automation under their 5-year, US$ 12 million, "Intelligent Mine" Program (9).

As previously mentioned, the most difficult element of a mining system to automate is the actual production of broken rock, with predictable fragmentation, from stopes. The use of explosives seems to be the ultimate method of breaking hardrock efficiently but it is an intermittent process which requires the evacuation of personnel during blasting and the stoppage of work in order to replenish fresh air back to the workings afterwards. Drilling, explosives, loading and detonation have all have been significantly improved in both quality and techniques. Nevertheless, the blasting component of rock breaking remains a process which can be neither automated nor robotized. Unpredictable fragmentation presents an insurmountable challenge for designing an autonomous muck loader.

If a mine is to be fully automated and if all of the mining systems are to be controlled from a control room, the breaking of rock by explosives will have to be replaced by mechanical cutting. This would result in continuous production at a homogenous fragmentation. The broken ore could be directly loaded onto a continuous haulage system.

Attempts are being made to develop such continuous excavating machines. The compact Underground Borer (CUB) and Robbin's Mobile Miner are examples. Presently, these machines are being used only for mine development. If efficiences and economics matching those obtained through rock excavation by explosives can be achieved, however, production methods (stoping) will have to be redesigned in order to use these machines for achieving continuous production. This would give a tremendous advantage for mine-wide automation and would permit all interdependent sub-systems to be operated autonomously by a single operator from a control room. It would also make ground control much easier and less expensive. The great hardness of rock, the short lengths of drifts, and high ground pressures, however, make the operation of such a machine difficult and uneconomical at present.

In Canada, HDRK Hardrock Mining Research

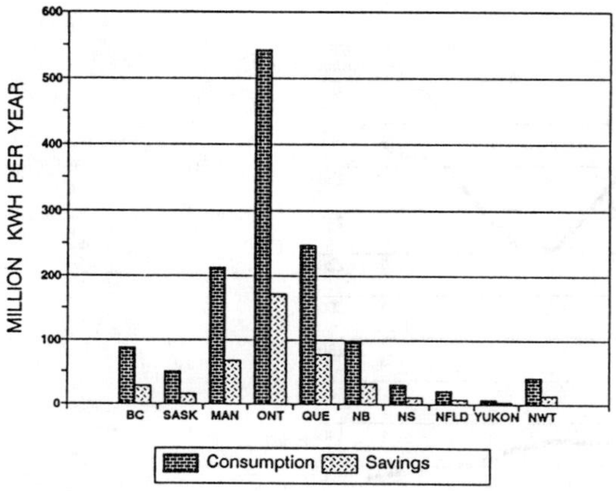

Figure 5 Annual electrical consumption and potential energy savings from automated ventilation control system

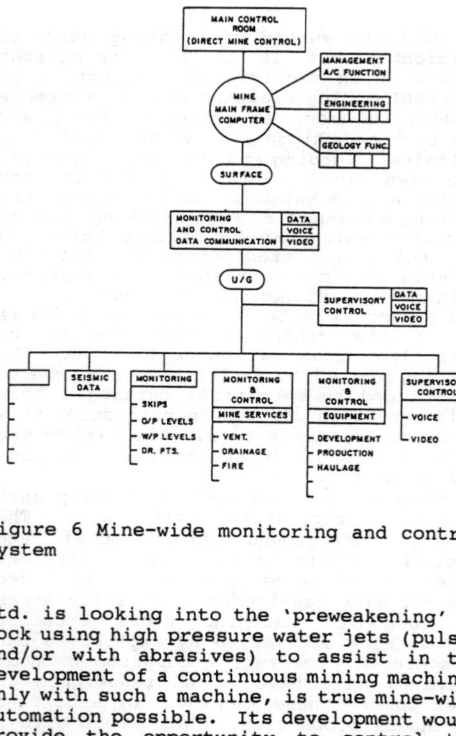

Figure 6 Mine-wide monitoring and control system

Ltd. is looking into the 'preweakening' of rock using high pressure water jets (pulsed and/or with abrasives) to assist in the development of a continuous mining machine. Only with such a machine, is true mine-wide automation possible. Its development would provide the opportunity to control the entire underground mine operation from a control room, whose location could be discretionary.

On the service side of mining operations, the preeminent target for full automation is ventilation. Some estimates indicate that the total electrical energy used for ventilating hardrock mines in Canada could be about 2,800 GWh per year; accounting for over $ 120 million (10). It is estimated that a fully automated ventilation system could result in savings of 31.5% of the energy cost for ventilation. The total consumption of energy and the potential savings through automated ventilation for mines in Canada are shown in Figure 5, (11). CANMET is of the opinion that the automation and optimization of ventilation should be based on air quality (AQI - Air Quality Index) and not on air volume. The basis and principles for such a program are in place. The sensing of pollutants and the regulation of ventilating air in a totally automated ventilation system would also allow for the recirculation of mine air. This would provide significant savings in the heating of mine air and in the generation of air volumes.

At present, mine surveying instruments have become very sophisticated, thereby minimizing the efforts required for calculations and plotting. These instruments still must be deployed by surveyors, however; a labour intensive and expensive operation. It is envisaged that in the future, permanent surveying stations will be established at strategic locations underground from where survey information and data will be sent automatically to engineering office computers for direct calculations and plotting.

The prerequisite for any mine-wide automation of the sub-systems indicated in Figure 3 is mine-wide underground communications. Due to the geometries of mine openings and the electro-magnetic characteristics of the ore and the host rocks, the development of a versatile, simple, reliable, trouble-free, and efficient communications system is making slow progress. The flow of information, for monitoring and controlling multiple and interdependent activities has to be on a real-time, two-way basis. Although voice communication underground is simple and is found commonly in major Canadian mines, high speed data and video signal transmission is not yet available to the extent that it should be. This is fundamental for real-time information flow from sensors to computers and for control signals to the equipment. Line of sight remote control is fully developed and available. The target for automation and robotization is for control of multiple machines and out-of-sight control of many interdependent and individual machines. The schematic for a mine-wide monitoring and control system is shown in Figure 6 and for a communications system in Figure 7, (12).

Thus far, this paper has discussed mining production process improvement by the use of automation or robotization. In a different approach, the US Bureau of mines has developed an 'in situ mining system' for fractured porphyry copper. It is a new low cost technology to exploit small, deep seated, low grade deposits (13). A pilot plant has been set up to study the practicality and economics of the concept (Santa Cruz Project, Asarco Inc.), in a large oxidized porphyry copper deposit at Casa Grande, Arizona. At the site, a dilute acid solution is pumped into a fractured ore zone, where it selectively dissolves copper. The solution is then

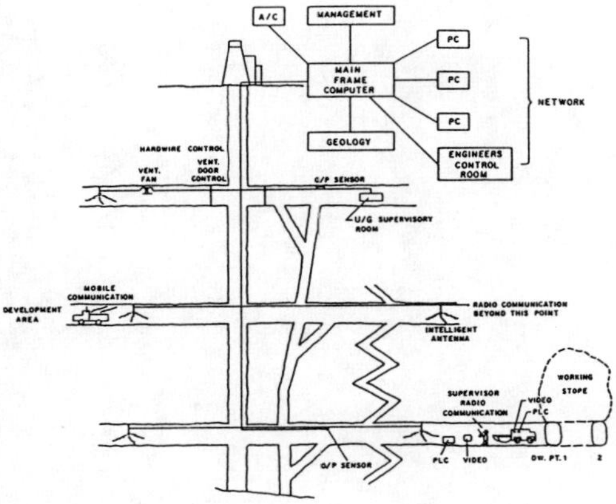

Figure 7 Mine-wide hardwire/wireless communications systems

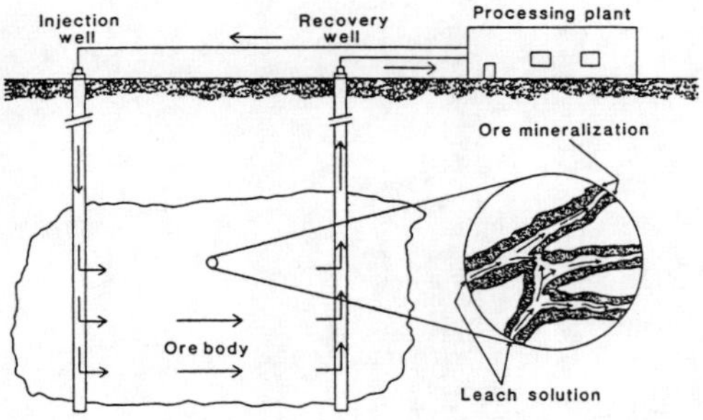

Figure 8. Cross section of an in situ mining system in which ore is leached via a system of vertical wells

collected through recovery wells and sent to a solvent extraction/ electrowinning plant on surface, Figure 8 (14). This low capital cost method is very simple. There is no pit, no underground operation, dumps or tailings and the method has little impact on the surface enviroment.

CONCLUSION

In conclusion, it can be said that the mine of the future will be highly sophisticated, and comparable to any automated manufacturing processes. The control room technology developed for manufacturing will be the state-of-the-art for underground mine production, with simultaneous on screen displays of: the mine in 3-D; vital statistical data; and information for the efficient control of the day-to-day operation of the mine from anywhere in the world. Mining will become an operation which will be run by a team of specialists from the various fields of engineering, science, computer software and hardware, mathematics, artificial intelligence, and knowledge-based systems. It will never be the domain of generalists as it has been in the past. The Canadian mining industry is ready for it.

REFERENCES

1. Pathak, J. 1984. Potential application of advanced technology in underground hardrock mining. 12th World Mining Congress, New Delhi, India, Vol.II, 4.12-1-14.
2. Willis Andrew 1992. Mining Intelligence Financial Times, September 2, 1992.
3. Marilyn Scales, editor, 1990. 1990 Mining Sourcebook. Published by Canadian Mining Journal.
4. Sopko, M. 1992. Quoted from Globe & Mail, October 15, 1992.
5. Cluett, J. Lewis, 1992. Personal communications and Cable bolting practices at Trout Lake - our successes and failures. CIM 93rd Annual General Meeting, Vancouver B.C., April 1991.
6. C. Gryba, 1992. Personal communication.
7. Inco 1991 Annual Report.
8. Baiden, G.R., Chaumont, I. 1992. Automatic haulage truck design and development and mine implementation. 5th Canadian Symposium on Mining Automation, Vancouver, B.C., September 27-29, 1992.
9. 'Intelligent Mine' technology program 1992-1996 by TEKES, Technology Development Centre, Finland.
10. Gangal, M.K. et.al 1992. Mine monitoring and automated Air Quality Control System. 5th Canadian Symposium on Mining Automation, Vancouver, B.C., September 27-29, 1992.
11. Gangal Mahendra, Pathak, Jayant. 1992. A case for conservation of electrical energy in Canadian underground mines. 11th WVU International Electrotechnology in Mines Conf. Morgantown, WV., USA. July 1992.
12. Pathak, J. 1989. Automation of mining equipment - a position paper. Unpublished EMR/CANMET internal document.
13. Staff, Bureau of Mines 1989. In situ leach mining. IC 9216 Proceedings:Bureau of Mines Technology Transfer Seminars, Phoenix, Az, April 4, 1989.
14. Graybeal F.T. & Larson W.C. 1989. In situ copper mining at the Santa Cruz Deposit, Casa Grande Arizona: A cooperative research project between the Santa Cruz joint venture and the United States Bureau of Mines. Presented at American Mining Congress Convention in San Francisco, Sept. 18, 1989.

Innovative Mine Design for the 21st Century, Bawden & Archibald (eds) © 1993 Balkema, Rotterdam, ISBN 90 5410 325 6

Coefficient of wall friction for settling slurry flow

H. Sato
Akita University, Japan

K. Otsuka
Institute for Mining Research and Studies, Japan

Y. Cui
Fuxin Institute of Mining and Technology, People's Republic of China

ABSTRACT: This paper reports experimental study of the static and kinetic coefficient of friction, μ_S and μ_K, using 1-in. transparent or translucent pipes of four different materials; acrylic resin, glass, vinyl chloride, and polyvinylidene fluoride. Four different sizes of solid particle from 0.86mm to 2.18mm and sliding velocities from 1cm/s to 11cm/s were investigated. To determine these coefficients of friction, original apparatus were built. Friction tests were administered to measure angles of an inclined plane on which solid particles start sliding for μ_S and to measure friction forces of the particles moving at constant speeds for μ_K. The results of these are compared with the coefficients of sliding friction between a submerged sliding bed of solids and a pipe wall for saltation flow in a test pipeline constructed of the same materials as those for the friction tests of μ_S and μ_K.

1 INTRODUCTION

In pipeline transport of settling slurry at low velocities solid particles contained in it proceed in discontinuous jumps, slides or rolls. This is referred to as the saltation or sliding bed regime. The Coulombic friction between particles and pipe wall is a very important parameter in predicting pressure gradients in this flow because sliding of solid particles on the pipe wall is a dominant resisting process. However, few investigations have been concerned with the coefficient of sliding friction, so that serious disagreement among the correlations that have been previously reported for pressure gradient and critical deposit velocity cannot reasonably be explained. In other words the pipe material should be considered as an operating parameter during the pipeline design.

There have been essentially two kinds of friction of solid particles moving on the pipe wall, kinetic and sliding friction. As the flow velocity increases the particles are at least partially suspended by fluid lift forces, then kinetic friction may occur at the limited low velocity flow conditions near the critical deposit velocity. Moreover, it is, in practice, difficult to define the threshold from kinetic friction to sliding, and also the coefficient of sliding friction f is seldom distinguished from that of kinetic friction μ_K. Table 1 shows proposed values of μ_K for the solid particles, as compared with coefficients of static friction μ_S. The wide deviations in their empirical results suggest that the coefficient of sliding and/or kinetic friction may be influenced by the method and the friction conditions applied to measure it, but we have had no succinct conclusion.

The present comparative study revealed not only the difference between μ_K and μ_S but the effects of operating parameters such as sliding velocity, particle diameter, and kind of pipe material on the sliding friction.

2 EXPERIMENTAL: FRICTION MEASUREMENTS

Experiments were conducted with four types of solid particles, sand or gravel,

Table 1. Comparison of the reported test results of the Coefficients of kinetic and static friction

Material of Solid Particle	Particle Diameter d, mm	Specific Gravity δ, -	Surface Type	Surface Condition	Kinetic Coefficient μ_k, -	Static Coefficient μ_s, -	Investigator	Observations
Gravel	3.0~3.5	2.56	Steel Plate	In the Air		0.757	Uematsu & Tsuchiya (1951)	
			Blass Plate			0.526		
			Glass Plate			0.426		
			Blass Plate			0.270		Equiv. Rolling Coefficient
Hard Coal			Steel Pipe	In the Water	0.1 ~0.2		Smoldyrev (1963)	
Crushed Coal					0.2 ~0.32			
Gravel & Soft Coal					0.32~0.46			
Moderate Hard Rock					0.46~0.56			
New Crushed Rock					0.56~0.7			
Gravel	1.02~4.60	2.64	Hard Vinyl Pipe	In the Water	0.314		Kuzuhara (1964)	
				In the Water		0.567		
				In the Air		0.570		
Sand	0.53~1.57	2.64	On the Deposit	In the Water	0.4 ~0.8		Shook & Daniel (1965)	
Spherical Lead	2.07	10.87	& Steel Pipe	In the Water	0.4 ~0.8			
Coal	7.62~13.60	1.478	Galvanized Steel Plate	In the Air		0.406	Ayukawa & Ochi (1967)	
	7.62~13.60	1.478	Polycarbonate Plate	In the Air		0.308		
Crushed Rock	3.42~10.01	2.594	Galvanized Steel Plate	In the Air		0.697		
Crushed Rock	3.5, 6.9	2.65	Stainless Steel Pipe	In the Water	0.4~1.7		Noda et al. (1970)	*V_m= 1~2.6 m/s μ_k increases, as V_m decreases.
Coal		1.28	Aluminum &	In the Water	0.26		Terada (1973)	
Gravel		2.65	Steel Drums	In the Water	0.26			
Coal Shale	4.35~4.80	2.13~2.20	Perspex	In the Water	0.40		Televantos et al. (1979)	d= d_{50}
Gravel	3.7	2.59	Perspex	In the Water	0.32			
	4.23	2.58		In the Water	0.32			
Coal			Steel Pipe	In the Water		0.5	Eyler & Lombardo (1980)	μ_s=-0.69(d/D)+0.39 for coal particles in PVC pipe
			PVC Pipe	In the Water		0.25~0.35		
Crushed Rock			Steel Pipe	In the Water		0.58		
			PVC Pipe	In the Water		0.4		
Sand	1.7		Glass Pipe	In the Water	0.65	0.64	Briscoe et al. (1983)	**V= 0.47 m/s
				In the Water	0.22			V= 1.07 m/s
Coal	6.4~50.8		Epoxy-lined pipe	In the Water	0.376	0.416	Hudson (1984)	
			Steel Pipe	In the Water	0.452	0.507		

* V_m: Mean velocity of slurry flow, ** V: Sliding velocity of solid particles

Table 2. Type of Testing Pipe

Pipe Material	Pipe Diameter (cm)
Acrylic Resin	2.69
Glass	2.65
Vinyl Chloride	2.84
Polyvinylidene Fluoride (PVDF)	2.72

of different sizes in transparent or translucent pipes listed in Table 2. The particles were sized in Ro-tap shaker with JIS type sieves. The specific gravities of particles are shown in Table 3.

2.1 Static friction

Figure 1 presents a schematic diagram of the apparatus used in the static friction tests. The sand bed, of mass m, with two acrylic resin cells, of mass m_C, holding it on both sides, was placed in a inclined pipe at an angle θ. The angle was increased at the angular velocity of $9.7\text{x}10^{-3}$ rad/s until

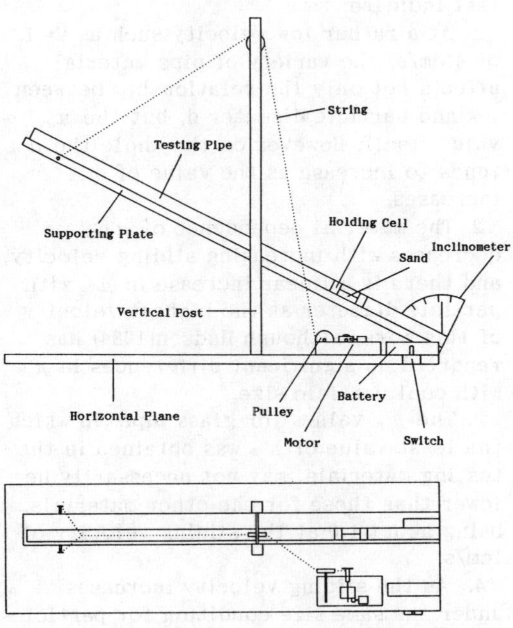

Figure 1. Schematic diagram of apparatus for static coefficient

Table 3. Properties of Solid Particle

No.	Screen Sizes (mm)	Particle Diameter d(mm)	Specific Gravity δ (-)
1	0.71 ~ 1.00	0.86	2.66
2	1.00 ~ 1.41	1.21	2.65
3	1.41 ~ 2.00	1.71	2.65
4	2.00 ~ 2.36	2.18	2.67

the sand moved down the pipe. The ratio of the bed thickness to the pipe diameter was 0.22 to 0.24. At the slip point, where the sand bed starts moving as illustrated in Figure 2, we have an equilibrium equation

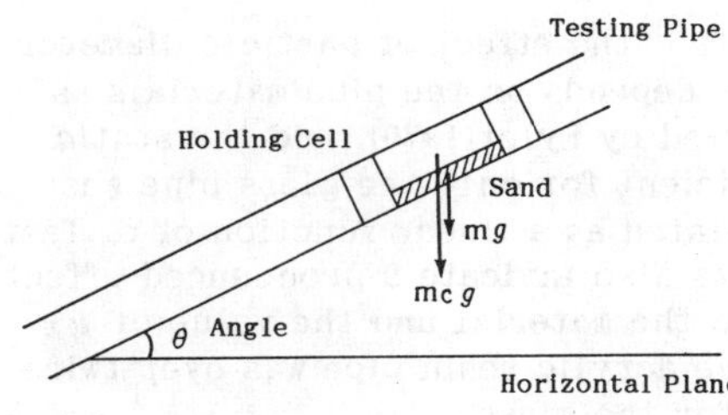

Figure 2. Sketch of instance of moving down for the sand with holding cells in a testing pipe

of the components: gravitational forces along the pipe axis and friction forces.

$$mg \sin\theta + m_C g \sin\theta = \mu_S mg \cos\theta + \mu_{CS} m_C g \cos\theta \quad (1)$$

It gives the following expression for the static coefficient μ_S of sand particles.

$$\mu_S = \frac{m + m_C}{m} \tan\theta - \frac{m_C}{m} \mu_{CS} \quad (2)$$

where μ_{CS} is the static coefficient of holding cells, determined by blank pretests, and each of m and m_C was fixed to 5.0 and 8.0 g in this friction test.

Figure 3 presents the experimental results of the static friction in air for acrylic resin, glass, vinyl chloride, and PVDF(polyvinylidene fluoride) pipe surfaces as a function of particle diameter d; each point plotted being the average of twenty determinations. These values of μ_S have been calculated using Eq.(2). It can be

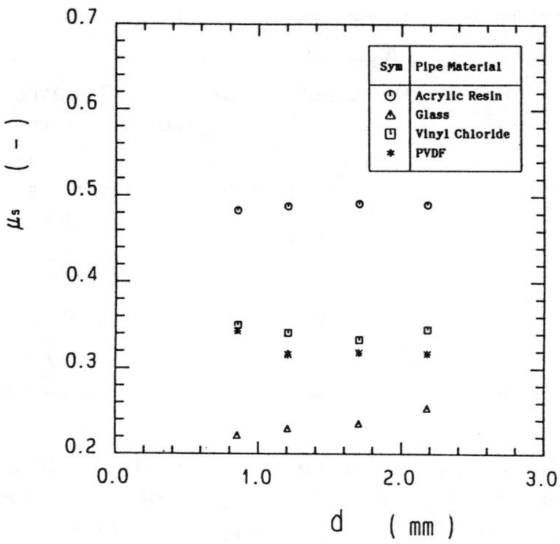

Figure 3. Coefficient of static friction μ_s versus particle diameter d for the various pipe materials

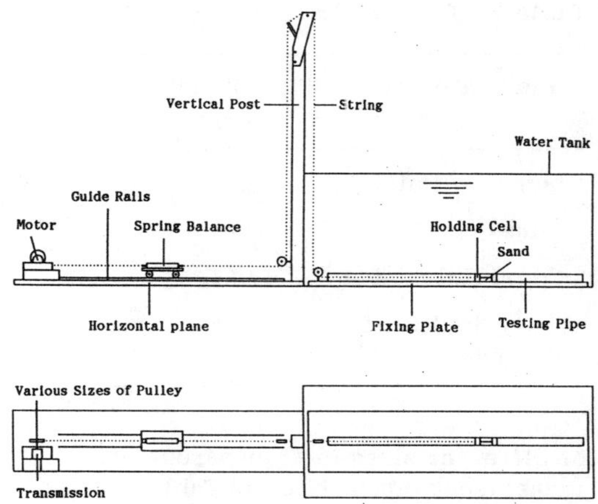

Figure 4. Schematic diagram of apparatus for kinetic coefficient

seen that the effect of particle diameter on μ_s depends on the pipe materials as reported by Eyler(1979), and the static coefficient for only the glass pipe was correlated as a linear function of d. Test results also indicate a pronounced effect due to the material and the value of μ_s for the acrylic resin pipe was over twice that for the glass.

2.2 Kinetic friction

A schematic diagram of the experimental equipment designed for the measurements of kinetic coefficient μ_k is shown in Figure 4. A sand bed was set with two acrylic resin cells in a test pipe of the same experimental conditions as those of static friction. The holding cells were then pulled at four different velocities by an electric motor. These velocities were varied by changing of the gear ratio in the transmission of the motor and the diameters of the driving pulley connected to it: nominal velocities being 1, 4, 7, and 11cm/s. The μ_k values were measured from the tensions of a spring balance attached to the shuttle, which could be moved smoothly along guide rails. Blank testing was also performed in advance with no sand bed to find the effective tension F, i.e., the friction force only for the particles. It was necessary to exercise caution during the experiment to avoid catching particles in the clearance between the cell and the pipe wall. Assuming that normal load N of particles on the wall equals their submerged weight, the ratio μ_k=F/N was calculated.

Figure 5 shows the relationship of particle diameter and sliding velocity to μ_k for the four pipe materials, each point plotted being the average of twenty determinations. The results of this friction test indicate:

1. At a rather low velocity such as V= 1, or 4 cm/s, the variety of pipe material affects not only the relationship between μ_k and particle diameter d, but the μ_k value itself. However on the whole the μ_k tends to increase as the value of d increases.
2. The material dependence of μ_k decreases with increasing sliding velocity, and there is a linear increase in μ_k with particle diameter at the highest velocity of this work, although Hudson(1984) has reported no significant differences in μ_k with coal particle size.
3. The μ_k values for glass pipe ,in which the least value of μ_s was obtained in the testing materials, may not necessarily be lower than those for the other materials, being about 1.0 at the sliding velocity of 1cm/s.
4. As the sliding velocity increases under the same size condition for particles, μ_k approaches the constant value with a decrease for the glass and PVDF pipes,

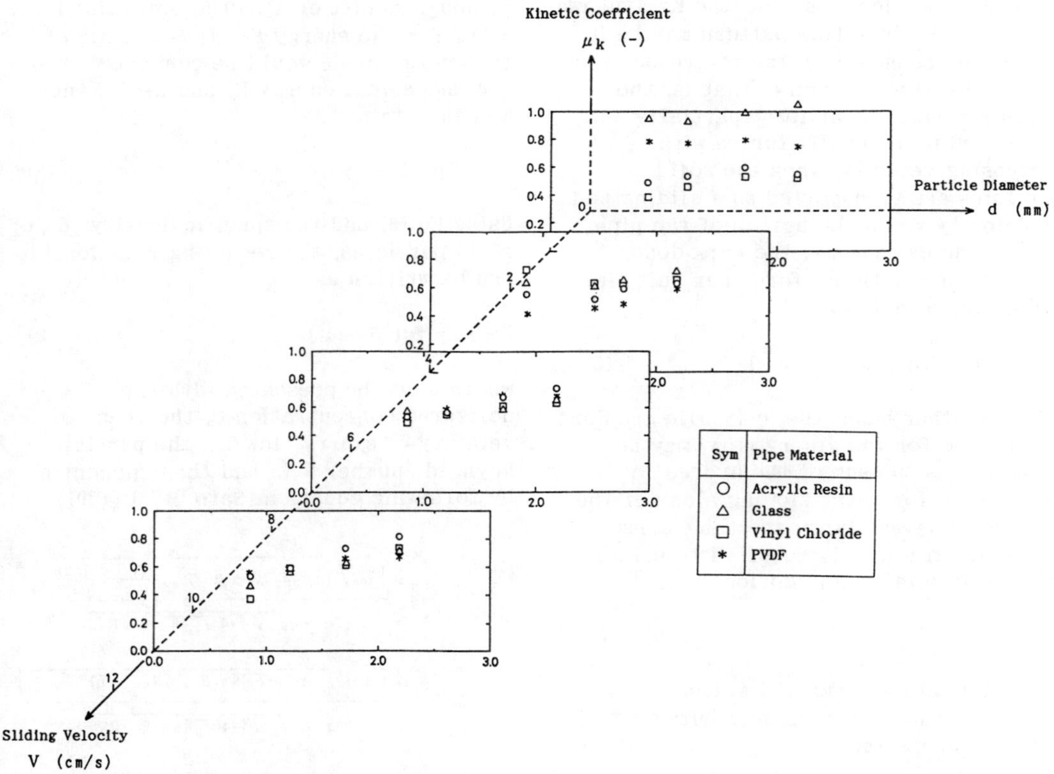

Figure 5. Relationship of particle diameter and sliding velocity to μ_K for various surface materials

while with an increase for the acrylic resin and the vinyl chloride.

3 SLIDING FRICTION IN SLURRY FLOW

3.1 Basic concept of sliding friction

As the velocity of settling slurry flow decreases under the velocity for the heterogeneous flow in a pipeline, the flow pattern will change to saltation, as illustrated in Figure 6. In this case, it can be assumed that solid particles are transported mainly in sliding movement. Hence the effective submerged weight W_{ap} of solid particles contained in a unit length of the pipeline with a cross-sectional area A is given by

$$W_{ap} = qA(\gamma_s - \gamma) \qquad (3)$$

in which: q= in situ concentration, γ_s and γ = specific weights of solid particles and water. Using the coefficient of sliding friction f of solid particles on the pipe wall, the frictional force operated at the surface F_f can be represented by

$$F_f = f \cdot W_{ap} = fqA(\gamma_s - \gamma) \qquad (4)$$

It should be noticed here that the coefficient f does not coincide with the

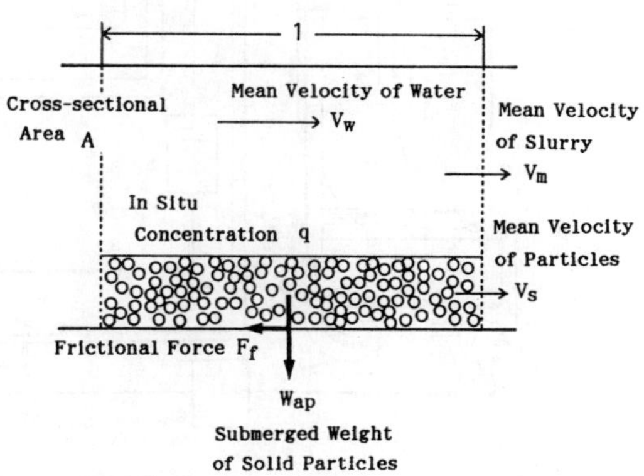

Figure 6. Sketch of saltation flow

kinetic coefficient, as reported by Ayukawa & Ochi(1967), since flow pattern may be in principle dependent on the the velocity in even the saltation regime, that is, the particles could be at least partially suspended by fluid lift forces with increasing velocity. When the solid particles are transported as a sliding bed of velocity V_s on the bottom of the pipe, the mechanical energy, the work done against the frictional force per unit time, can be represented by

$$E = F_f \cdot V_s = fqV_sA(\gamma_s - \gamma) \qquad (5)$$

On the other hand, the hydraulic gradient i, required for the slurry flow, may be estimated as the sum of the hydraulic gradient i_w for water flowing alone at the same velocity as slurry and the excess hydraulic gradient i_s required to maintain the particle in sliding motion, i.e.,

$$i = i_w + i_s \qquad (6)$$

According to the practical motion of particles, the energy E_s attributable to i_s can be represented by

$$E_s = i_s \gamma AV_s \qquad (7)$$

although Newitt et al.(1955) determined $i_s \gamma AV_m$ for the energy E_s. Thus , if all of the energy of E_s would be converted into the mechanical energy E_f and E_f=E_s, then we can obtain

$$i_s = fq(\gamma_s - \gamma)/\gamma \qquad (8)$$

Using Eq.(6) and the specific density δ of solid particles, the resulting relationship can be written as

$$i - i_w = fq(\delta - 1) \qquad (9)$$

where q can be presented with the delivered concentration C, the mean velocity of slurry flow V_m, the particle Reynolds number R_{ep}, and the exponent n by following equation (Sato et al.(1991)).

$$\left[\frac{\bar{q}-C}{(1-\bar{q})}\ \frac{V}{V_t}\right]^2 \frac{\left[\alpha \sqrt{\alpha^2+4\sqrt{48}\alpha\beta/R_{ep}\frac{V_t}{V}\frac{\bar{q}(1-\bar{q})}{\bar{q}-C}}\right]^2}{\{\alpha+\sqrt{\alpha^2+4\sqrt{48}\alpha\beta/R_{ep}}\}^2} + (1-\bar{q})^{2(n-1)} \frac{\left[\alpha+\sqrt{\alpha^2+4\sqrt{48}\alpha\beta/\{R_{ep}(1-\bar{q})^{n-1}\}}\right]^2}{\{\alpha+\sqrt{\alpha^2+4\sqrt{48}\alpha\beta/R_{ep}}\}^2} - 1 = 0 \qquad (10)$$

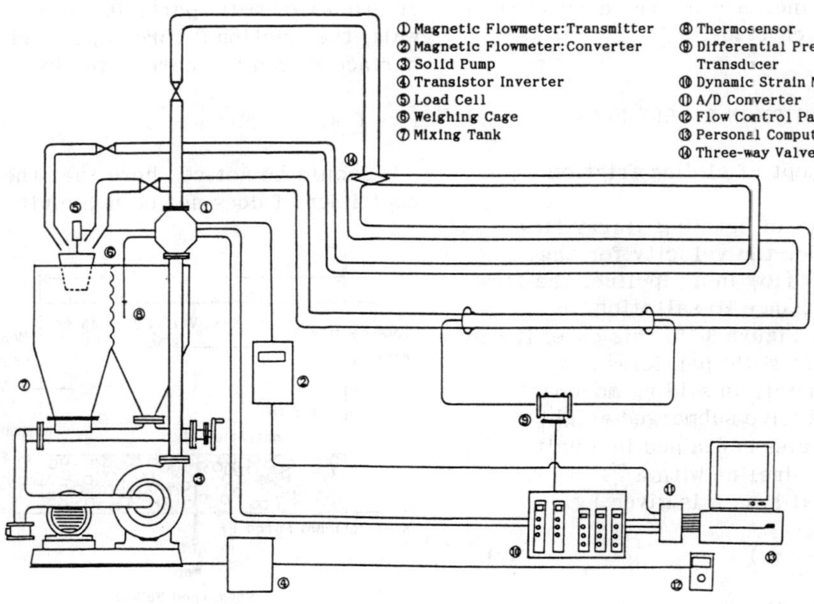

Figure 7. Schematic of the flow circuit

4 EXPERIMENTAL PROCEDURE OF FLOWING TESTS

Figure 7 presents a schematic diagram of the flow circuit. It consists of two different testing pipelines which can be easily changed by a three-way valve. The materials of pipelines and their inside diameters are the same as those listed in Table 2. The flow rate was controlled by changing the revolutions of the motor of solid pump through an inverter and the velocities were measured with a magnetic flowmeter. The pressure taps were mounted at 2m intervals in the return pipe of the circuit and the pressures from them were led to the differential pressure transducer. The test pipes were joined with couplings and seal tapes to produce the full length of the line. Delivered concentration C was obtained both by collecting solid particles with a cage at the pipe discharge and by measuring the corresponding time interval. The weights of the cage were measured by a load cell.

The outputs from these instruments were connected to the personal computer through an A/D converter, and data was analysed and recorded by the computer. Then, to obtain further data, the flow rate was decreased by small intervals until it reached the critical deposit velocity, which was determined by visual observation.

4.1 Sliding friction by Smoldyrev's method

Smoldyrev(1963) has reported that the sliding coefficient was independent of flow velocity V_m and pipe diameter D, and could be determined by the characteristics of solid particles as shown in Table 1. He analysed the experimental data with Eq.(9). Figure 8 shows a typical relationship between $i-i_w$ and $q(\delta-1)$ for the four pipe materials, based on the Smoldyrev's analytical method. The sliding coefficients can be given as the slopes of average lines of the data, so that the values of f determined in this way are nearly dependent on the empirical ranges of flow velocity V_m and in situ concentration q.

Figure 9 shows the sliding coefficient f as a function of the particle diameter d for various materials of pipe. The deviation of f at d=2.18mm in the glass pipe

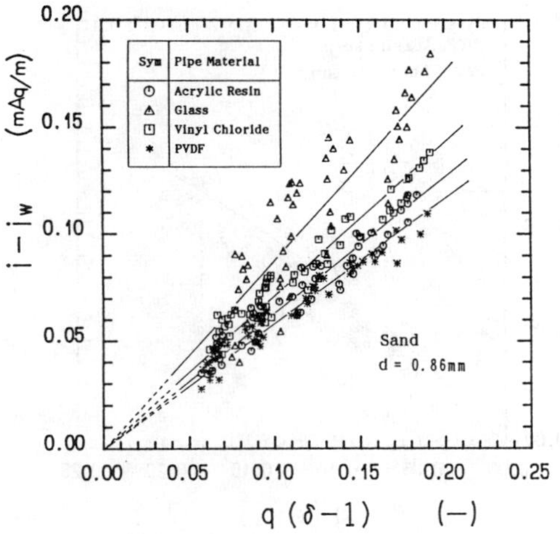

Figure 8. Comprehensive plot of $i-i_w$ versus $q(\delta-1)$ for various pipe materials — runs with sand of the diameter d=0.86mm

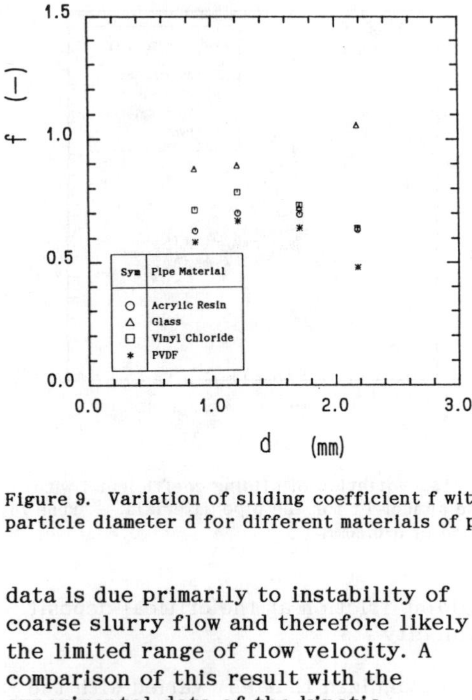

Figure 9. Variation of sliding coefficient f with particle diameter d for different materials of pipe

data is due primarily to instability of coarse slurry flow and therefore likely to the limited range of flow velocity. A comparison of this result with the experimental data of the kinetic coefficient μ_K in Figure 3 reveals that there is not a serious discrepancy between f and μ_K, while the pipe materials have a different effect on f values from that on μ_K values.

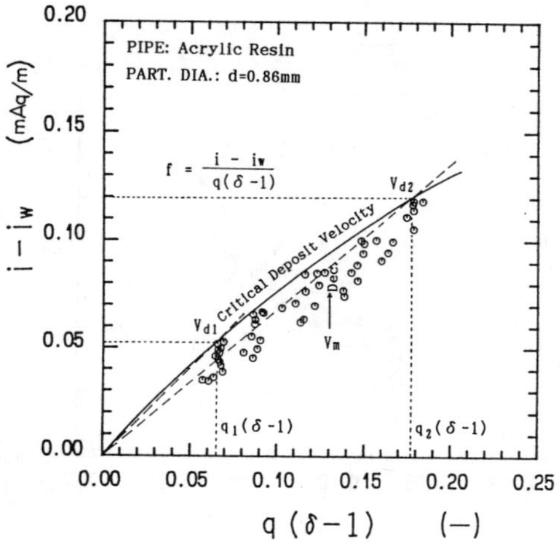

Figure 10. Typical dependency of sliding coefficient on the velocity V_m and the concentration q

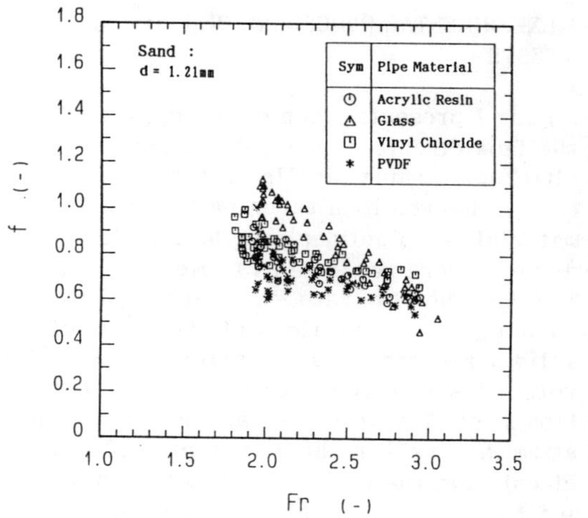

Figure 12. Variation of sliding coefficient f with Froude number Fr for the pipe materials – runs for the sand of d=1.21mm

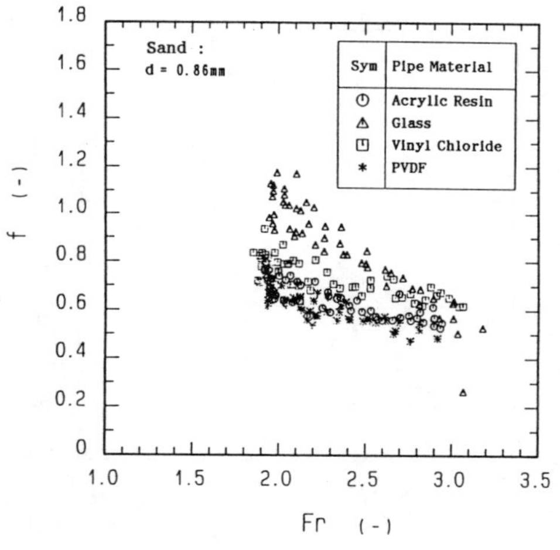

Figure 11. Variation of sliding coefficient f with Froude number Fr for the pipe materials – runs for the sand of d=0.86mm

4.2 Sliding friction at the critical deposit velocity

It was recognized that $i-i_w$ varied with the velocity V_m for a steady concentration q in the flowing tests. Figure 10 is a typical plot of $i-i_w$ versus $q(\delta-1)$ for the acrylic resin pipe, drawn from Figure 8 to make this aspect clear. Based on visual observations, the maximum values of $i-i_w$ corresponded to the critical deposit velocities V_d. Hence, the values of f for the critical points, at which solid particles start to settle out on the bottom of a pipe, can be presented as the slope of the line passing through the two points: the origin and the point on the upper-bound curve enveloping the data. This implies that the sliding friction, calculated by Eq.9, increases with decreasing velocity V_m, as reported by Noda et al.(1970), to reach the maximum value at the critical deposit velocity.

Furthermore, the form of the upper-bound curve indicates that the values of f decrease with increasing the in situ concentration q.

4.3 Froude number Fr dependency

Given that the intensity of flow required to move the solid particles in a pipe is determined by the Froude number, propsed by Ayukawa & Ochi(1967) and Terada(1973),

$$Fr = V_m/\sqrt{gD} \qquad (11)$$

, then it is reasonably estimated that the coefficient of sliding friction f, predicted by Eq.(9), may be as a function of Fr, since the flow patterns of slurry such as heterogeneous flow and saltation flow are changed by the intensity. Figures 11 to 14 show that an increase in Fr has less of an influence on the f value to the upper limit of the flowing tests, Fr ≒ 3.0, than the pipe material. This trend corresponds to that of the experimental results of the kinetic coefficient μ_K, shown in Figure 5. It is interesting that at low velocities the value of f was greater than 0.8 for the gravel of 2.18mm on the glass surface, like the results of μ_K, and at times just greater than 1.0 for the special case of frictional conditions, different from the coefficient of the conventional friction between solid bodies in air.

5 CONCLUSION

1. Definite relationships between the kinetic coefficient μ_K in water and the static coefficient μ_S in air could not be found in the present investigation, and particle dependence of μ_S varied with the materials.
2. At low velocities the kinetic coefficient was significantly affected by the materials, increased with increasing the particle diameter, and had a tendency of approaching the constant values as the velocities increased.
3. The sliding coefficients determined by the Smoldyrev's method were dependent on the experimental regions of th velociy V_m and the concentration q, but they were in serious disagreement with the kinetic coefficient μ_K.
4. The sliding coefficient reached the maximum value at the critical deposit velocity, and decreasesd with increasing velocity and concentration.
5. The dependence of the materials on the sliding coefficient decreased with increasing velocity, similar to the experimental results of μ_K.
6. The sliding coefficient was greater than 1.0 in some frictional conditions. It hence seemed that frictional phenomena of solid particles on the pipe wall was different from that between solid bodies in air.

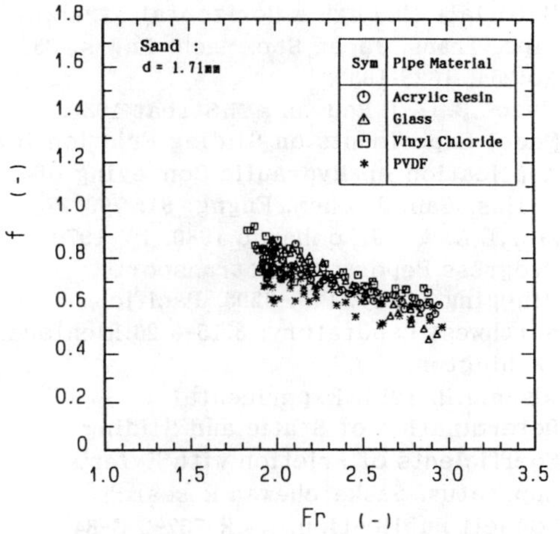

Figure 13. Variation of sliding coefficient f with Froude number Fr for the pipe materials — runs for the sand of d=1.71mm

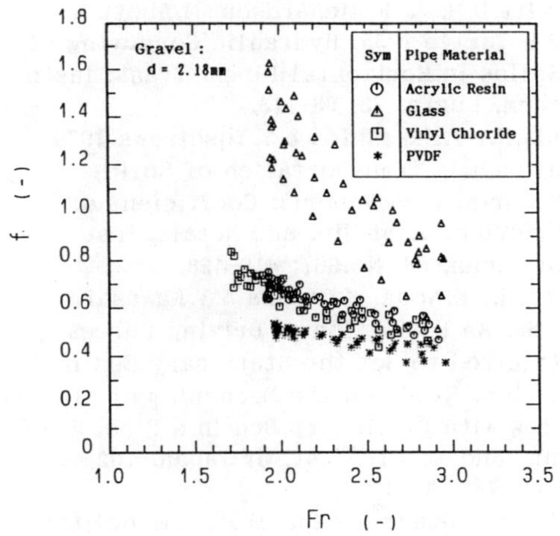

Figure 14. Variation of sliding coefficient f with Froude number Fr for the pipe materials — runs for the gravel of d=2.18mm

ACKNOWLEDGMENTS

This work was supported by A Grant under The Monbusho International Scientific Research Program; Joint Research. This support for the project number 03044024 is gratefully acknowledged. We wish also to thank Mr. William E. Wright for his encouragement and aid in the preparation of this paper.

REFERENCES

Ayukawa,K. & J.Ochi 1967. Pressure Drop in the Hydraulic Conveyance of Solid

Materials through a Horizontal Straight Pipe. Trans. Japan Soc. Mech. Engrs. 33, No.254: 1625-1632.
Briscoe,B.J., H.Radwan & M.Streat 1983. Model Experiments on Sliding Friction for Application in Hydraulic Conveying of Solids. Can. J. chem. Engng. 61: 769-775.
Eyler,L.L. & N.J.Lombardo 1980. FY-1979 Progress Report: Hydrotransport Plugging Study. PNL-3203. Pacific Northwest Laboratory: 3.15-3.26.Richland: Washington,
Hudson,J.H. 1984. Experimental Determination of Static and Sliding Coefficients of Friction with Eyler's Apparatus. Saskatchewan Research Council Publication. No.R-752-2-C-84
Kuzuhara,S. 1964. Falling, Rolling or Sliding of Gravel in Rest Water in a Vertical or Inclined Smooth Pipe. Trans. Japan Soc. Mech. Engrs. 30, No.213: 594-598.
Newitt,D.M., J.F.Richardson, M.Abott & R.B.Turtle 1955. Hydraulic Conveying of Solids in Horizontal Pipes. Trans. Instn. chem. Engrs. 33: 93-113.
Noda,K., T.Kawashima & Y.Yoshizawa 1970. Hydraulic Transportation of Solid Material (3rd Report): Coefficient of Friction. J. of Min. and Metall. Inst. of Japan. 86, No.987: 419-423.
Sato,H., K.Noda, K.Otsuka & T.Kawashima 1986. An Estimation of Driving Forces Required to Set the Stationary Bed in Motion: Study on the Mechanism of Slurry Flow with Stationary Bed in a Pipe. J. of Min. and Metall. Inst. of Japan. 102, No.1176: 73-83.
Sato,H, K.Otuka & Y.Cui 1991. A Model for the Settling Slurry Flow with a Stationary Bed in a Pipe. Proc. 1st ASME-JSME Fluid Engng Conf. FED-Vol.118. Liquid-Solid Flow: 53-63.
Shook,C.A. & S.M.Daniel 1965. Flow of Suspensions of Solids in Pipelines: Part I. Flow With a Stable Stationary Deposit. Can. J. chem. Engng. 43: 56-61.
Smoldyrev,A.E 1963. Pipeline Transport., Trans. Ed. from Russ. to Jap. 181-183. Gijyutsu-shoin: Tokyo
Televantos,Y., C.A.Shook, A.Carlenton & M.Streat 1979. Flow of Slurries of Coarse Particles at High Solids Concentrations. Can. J. chem. Engng. 57:255-262.
Terada,S. 1973. Pipeline Transport: Flow Mechanisms and Pressure Drop Predictions. in Jap. 186-203. Rikoh-tosho: Tokyo
Uematsu,T., K.Tsuchiya & S.Okamura 1951. Packing and Friction of Powder. Trans. Japan Soc. Mech. Engrs. 29, No.56: 72-77.

Innovative Mine Design for the 21st Century, Bawden & Archibald (eds) © 1993 Balkema, Rotterdam, ISBN 90 5410 325 6

Combining teleoperation with vehicle guidance for improving LHD productivity at Inco Limited

G.R.Baiden
Inco Limited, Mines Research, Ontario Division, Sudbury, Ont., Canada

ABSTRACT: This paper discusses the application of Load-Haul-Dump (LHD) machine teleoperation in combination with a guidance system for single operator control of multiple LHDs. A system accomplishing these goals has been tested at Copper Cliff North Mine of Inco Limited. This development uses a high speed, high capacity production network to teleoperate an LHD during loading by supplying 3D video and teleoperation control. Following loading, the operator dispatches the LHD using an automatic guidance system to traverse the drifts to the orepass. Manual dumping occurs at the orepass and then the machine returns to repeat the procedure. This has significant productivity gains and improvements in safety. Productivity is improved since the operator is only involved in the loading and dumping part of the process. This frees the operators to utilize their time more effectively loading or dumping a particular machine while other machines with guidance systems will be continuing to work travelling and/or automatically dumping ore in the orepasses. Safety is improved as the operator is located on a remote control station and only involved in the material handling process remotely.

1. INTRODUCTION

This paper describes the research currently underway at Inco Limited to combine LHD teleoperation with onboard guidance systems. Teleoperation of a piece of equipment is the remote control of a mining machine from an operation point where the machine is not in direct line-of-sight. These systems will provide mining operations with significant productivity gains while reducing cost and improving safety. The work was carried out at Copper Cliff North Mine, Ontario Division, as an extension of the telecommunications network development, FOREMAN[1], shown in figure 1.

LHD teleoperation was the first identified application for the telecommunications network recently developed at Inco limited that would have significant productivity

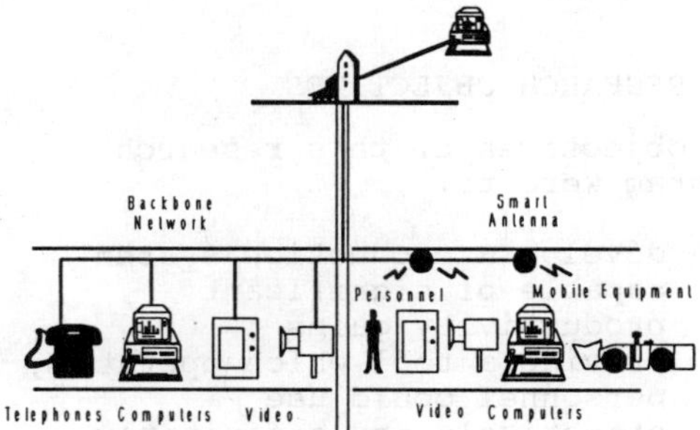

Figure 1 : Underground Communication Network Concept.

gains. Initial experimentation with that network, proved the feasibility of one person operating a machine from surface or anywhere in the mine, figure 2. The purpose of this work was to provide multiple LHD operation. Multiple

Figure 2 : Teleoperation of LHD, 3000 foot level at Copper Cliff North Mine of Inco Limited.

LHD operation would be provided when three facilities were completed. These are 3-D video for depth perception, a teleoperation station for multiple machine control and vehicle guidance to eliminate operator involvement during the tramming portion of the cycle. With these characteristics and the telecommunications network, which would eliminate the mine geographical constraints, one operator could muck with more than one machine.

2. RESEARCH OBJECTIVES

The objectives of this research program were to:

a) develop a production system capable of significant productivity gains
b) provide a tool which operating personnel could use effectively and efficiently
c) provide the ability to use and apply this technology quickly to all Inco operations.

3. MINE EXPERIMENTATION

The technologies discussed here were implemented at Copper Cliff North Mine. This section discusses underground telecommunication network, 3D vision systems, teleoperation control systems, vehicle guidance and machine monitoring.

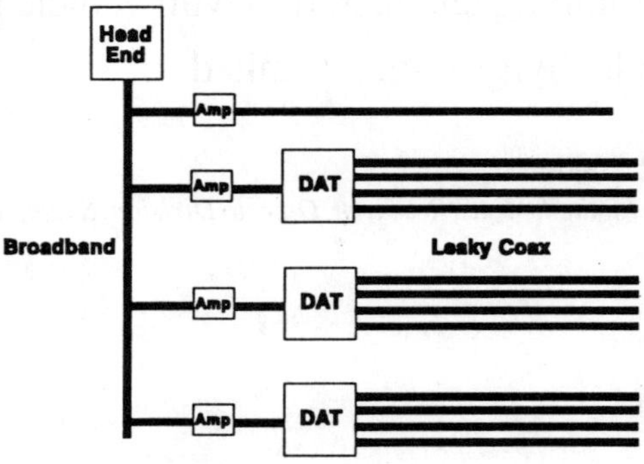

Figure 3 : Mine Telecommunication Network

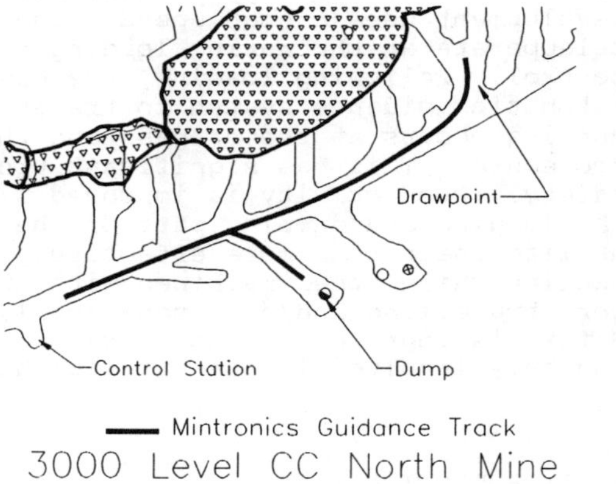

Figure 4 : 3000 Level Test Area

4. COMMUNICATION SYSTEM

Copper Cliff North Mine has the first high bandwidth underground telecommunication network developed by Automated Mining Systems (AMS) to be applied at Inco Limited. A system schematic is shown in figure 3. The headend of the system is located on surface and a broadband trunk cable system interconnects the various levels including 3000 level were the main experiment took place. On 3000 level, the trunk system is split into the level distribution system and a "Distributed Antenna Translator" or DAT. The DAT is located at a central location on the level near the garage. Leaky Coaxial Antenna is then distributed from the DAT around the level as shown in figure

4. Broadband distribution cable is also available on the level although not necessary unless the control station is moved onto the level as in this experiment. The interconnection of the various subsystems was performed using MODEMs and radios.

5. 3-DIMENSIONAL VIDEO

A serious problem in the initial teleoperation experiments, in 1991, was the lack of depth perception for the operator trying to run the machine. He addressed this problem by the installation of a bright coloured tape three inches in width. The tape was installed along the walls of the drift and hung at the loading and dumping points. While this was adequate for the testing it would not be suitable for an actual production unit. Therefore, work was started in the development of 3-D video systems.

The 3-D video systems work has three components onboard the machine, telecommunication network transmission and display at the teleoperation station. 3-D video is obtained by using a pair of cameras set at the correct angles and focal length. These cameras provide a pair of video signals that when interlaced can be displayed on a 3-D, no-glasses video monitor. This display provided the operator with a meaning to simulate the third dimension. This supplied the operator with depth perception.

The telecommunications network provides the system capability to move the pair of video signals over the network accepting the signals from the mobile LHD. The signals are then transferred via the leaky coaxial network through the DAT and onto the broadband system. This allows the connection of the teleoperation control-station directly to the LHD through an RF link.

The final work has been in the development of 3-D display systems to present these pictures to the operator. Two choices were evaluated (displays that use polarized glasses and nonglasses displays). Either of these can be applied, however, the decision was made to work towards nonglasses displays. These displays were deemed to offer the most potential for use in an underground operating environment. The display used was provided by Dimension Technologies Inc.

6. TELEOPERATION CONTROL SYSTEMS

The remote control system used in the original experiment was a typical unit used for LHD control at Inco Limited provided by MOOG Limited. This remote control proved difficult to operate as it was not designed for teleoperation of equipment from a desk both ergonomically and electronically.

The operator of the system during the original experiment had many comments that have been addressed in this work. These concerns were the remote control unit being in an awkward position for operation, the unit sliding around on the desk, the joysticks being too close together and some commands of the system were not intuitive. A teleoperation chair was created for operation of multiple machines and to address the previous comments. The chair is shown in figure 5 & 6.

Figure 5 : Teleoperation Chair

The chair combines the electronics in the MOOG radio with the telecommunication system electronics for a portable teleoperation station. The

Figure 6 : Teleoperation of LHD on 3000 Level

joysticks were installed into the chair on either armrest that was tilted for operator comfort. Foot pedals were added to allow more natural control of the machine (similar to a car) instead of the current switches on top of the joysticks. Finally, the forward-reverse command was switched with the steering control to provide a more intuitive control of the LHD using the video display.

7. VEHICLE GUIDANCE SYSTEM

To provide LHD guidance a system was purchased from Mintronics Limited. The basic system was designed for trucks although it was felt that the system could be modified for use with an LHD. The system was purchased and installed on an LHD at Copper Cliff North Mine.

The system consists of a pair of LASERS mounted on either end of the machine pointed straight up to read a tape on the back. This generated angles to the computer system that calculated a signal that was provided to LHD steering valve used by the remote control unit. To provide and action a further marker is added to the system for change in commands such as change direction, change speed, enter loading area or enter dump area. As well, a microwave detector was added to the front and rear of the machine to detect personnel, wearing a tag on their hardhats, stopping the machine.

8. RESULTS

The three subsystems required to test multiple LHD operation were installed on an LHD on 3000 level at North Mine. The cameras, remote control and guidance system are shown in figure 7. These systems were then integrated on the machine and to the communications network, which was previously installed, for the operation of the system. A total of 300 feet of track was installed on the back for the guidance system as shown in figure 8. Along with the track is a bar code installed for the LHD to get a command.
Successful combined teleoperation and guidance have been achieved for one person operating one LHD at

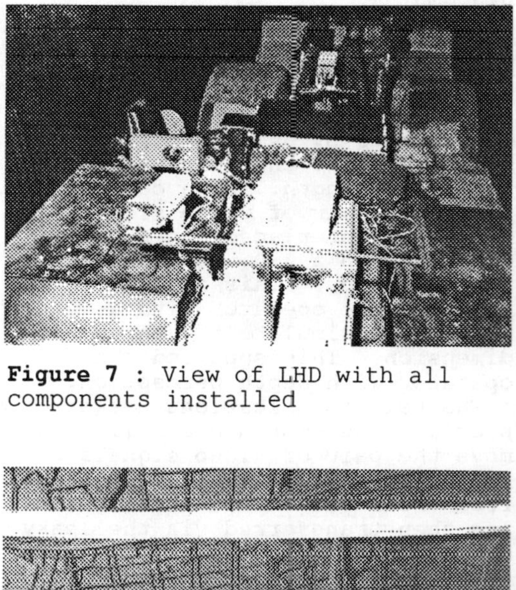

Figure 7 : View of LHD with all components installed

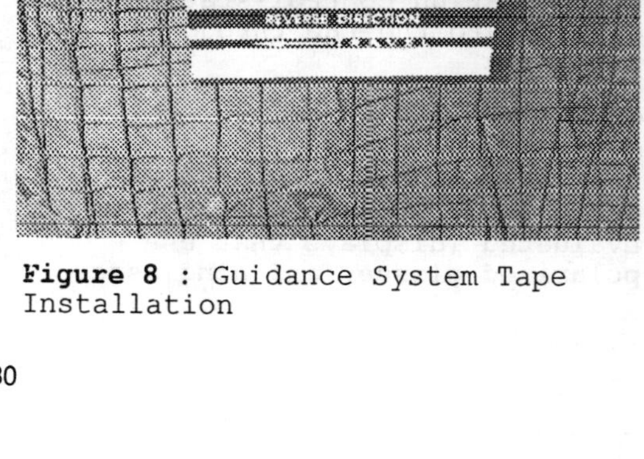

Figure 8 : Guidance System Tape Installation

North Mine as shown in figure 8. The operator used the system from the seat with no noticeable problems. Ergonomics of the system proved to be suitable for multiple LHD teleoperation.

The 3-D video system has not yet been attempted in production although initial testing underground has shown the usefulness of this capability. The teleoperation chair has become an absolute necessity for the operation of an LHD in this fashion. While functional, the guidance system had some limitations for LHD guidance that are:

* too slow - only speed one full throttle has been attained and speed two full throttle is required
* the LASER orientation caused a problem in the steering algorithm that resulted in difficult setup of the track system.

With an increase of speed the system will be suitable for truck guidance although the ability to guide an LHD is questionable with this sensor set.

9. BENEFITS

The benefits of multiple LHD operation are significant in terms of productivity, safety and cost. As well, the same technology offers large potential for difficult operating environments such as deep mining with ground problems and hazardous mining (high grade uranium for example).

With the telecommunications infrastructure and some limited automation productivity improvement is practical. The level of productivity improvement is dependent on loading and travel time. For example, if an LHD has a total cycle time of two minutes and if takes 30 seconds to load it, it is feasible for one operator to run four machines. A graph has been developed to show the potential of tons/personshift and productivity improvements versus LHD/person, figure 9. A reasonable level of automation to expect for LHDs is

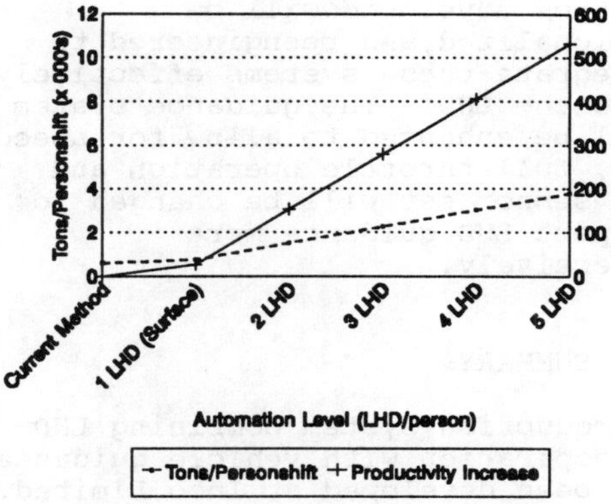

Figure 9 :Productivity Improvement vs Level of Automation

between two or three. Therefore, productivity increases of 200% to 300% can be expected.

As a result the geographical constraints of mining have been eliminated with the telecommunications network. It will be practical to run equipment further away from the actual mining face thus reducing personnel exposure time improving safety. Safety benefits include the ability to remove the man from the machine entirely reducing the potential for injury of the operators backs and ears while operating and other injuries that occur during climbing on and off the machine. This will occur as now the operator located in a chair located in a more ergonomically suitable environment.

Cost reduction will result from increased utilization of the LHD, reduced personnel requirements and more consist operation and maintenance. Increased utilization will occur for two main reasons: Eventually, the LHD can be operated from surface saving approximately 25% of shift time since operator can start sooner and since the operator is not on the LHD the operation time can be extended.

10. FUTURE WORK

Future work will be to outfit another LHD and run two machines simultaneously on the network. The systems developed will be rationalized and reengineered to integrate these systems effectively onto the LHD. The guidance system will be enhanced to allow for speed two, full throttle operation and the sensor set will be changed to support LHD guidance more effectively.

11. SUMMARY

A production system combining LHD teleoperation with vehicle guidance has been developed at Inco Limited. This system is capable of being retrofit onto any LHD and multiple machine operation is possible with the AMS telecommunications network installed. This production system is the first application for the telecommunication network. The system achieves large productivity gains (200% to 300%) and improves safety while reducing cost.

REFERENCES

1 Baiden G.R. and Scoble, M., Mine-Wide Information System Development at Inco Limited, Canadian Institute of Mining and Metallurgy, 1991.

2 Brophy, G. and Euler, D. The Opti-Track System, Mining Automation - 5th Canadian Symposium, September 1992.

Innovative Mine Design for the 21st Century, Bawden & Archibald (eds) © 1993 Balkema, Rotterdam, ISBN 90 5410 325 6

Equipment condition monitoring programs improve production equipment reliability at the Iron Ore Company of Canada

Leslie McHattie
Iron Ore Company of Canada, Labrador City, Nfld, Canada
John H. Burrows
J. H. Burrows Electronics Inc.

ABSTRACT: Vibration monitoring and analysis programs are being developed for a wide variety of equipment including: grinding mill and crusher drives and gearboxes, process fans, process water pumps (vertical), tailings pumps, locomotives, shovels, drills and haulage trucks. These programs have enabled personnel to reduce maintenance costs through early detection of potential component failures and have highlighted the need for strictly controlled equipment installation and maintenance procedures. Periodic analysis of lubricating oil provides information on component wear and alerts maintenance personnel of system contamination. Infrared thermographic scanning has improved maintenance planning capability and effectiveness for both electrical and mechanical components. In the near future, expert systems and/or neural networks will be incorporated into these programs adding structured diagnostic capability to facilitate efficient and consistent analysis of the data collected.

1 INTRODUCTION

The Iron Ore Company of Canada (I.O.C.C.) operates an open-pit mine, with concentrating and pelletizing plants, in Labrador City, Newfoundland. The Mine sends approximately 85,000 tonnes of crude ore to the Concentrator each day for processing. At the mine, electric rotary drills are used to drill 15" blast holes. Once a blast is shot, 18 cubic yard shovels are moved into the area to dig the blasted ore and waste rock for loading into 200T haulage trucks. The trucks move the waste to dump areas located outside of the pit limit, while the ore is transported to one of three loading pockets. These pockets are underground storage bins excavated in the waste rock. A feeder at the base of each pocket loads the ore into 95-tonne capacity, side-dump rail cars on the Automatic Train System. Trains deliver the ore a distance of 11.25 km to two (2) gyratory cone crushers.

After crushing, conveyors transport the rock to a 300,000 T capacity ore storage building. This building provides feed to four (4) grinding mills, two (2) wet and two (2) dry. Once ground, the feed goes to a three-stage spiral process. A secondary gravity separation stage, the Reichert cones, handle a slightly finer feed. The final stage of processing, magnetic separation, maximizes the recovery of magnetite. The concentrate product is shipped to I.O.C.C. dock facilities in Sept-Iles, Quebec or transported by conveyor to the pelletizing plant located adjacent to the Concentrator.

Long-term corporate objectives set by the Iron Ore Company of Canada during 1992 include:

1. reducing the unit cost of production;
2. increasing production equipment availability; and
3. improving equipment reliability.

All new initiatives undertaken within the company must be directed toward meeting these objectives.

Traditionally, maintenance personnel performed component maintenance and replacement work based upon operating hours accumulated. This resulted in:

1. Components being changed out when they had reached their useful "life" (in operating hours) whether they were worn out or not (eg. truck engine life standardized at 12,000 hours during 1980s, by 1992 extended to average of 18,000 hours).

2. Unexpected premature failure of some components before they reached the specified useful "life" (eg. early failure of bearings in gear cases usually resulted in extensive damage to shafts and gears).

This approach contributed to unnecessarily high maintenance costs and equipment downtime. The need for improved control of cost and availability necessitated a change in the traditional maintenance philosophy. Research indicated that various technologies had been developed which could provide data on equipment and component condition on an on-going basis. It was determined that the implementation of these technologies would improve the ability of maintenance personnel to accurately predict component failure and effectively prevent it.

2 FAILURE PREVENTION AND PREDICTION

Preventive Maintenance involves actions taken to prolong the operating life of production equipment and to eliminate premature component failure through:

1.effective equipment inspection and service, and
2.equipment condition monitoring.

These actions reduce the frequency of unexpected breakdowns and, thereby allow more maintenance work to be planned and scheduled. Additional benefits realized by an effective Preventive Maintenance Program include:

1.reduced equipment downtime resulting in increased production,
2.reduced overtime and more effective utilization of manpower, and
3.reduced maintenance costs through early detection of problems.

Regular equipment condition monitoring provides maintenance personnel with the ability to effectively detect and diagnose problems at an early stage of development. The appropriate corrective action can then be planned during a scheduled equipment shutdown. This minimizes both maintenance repair cost and equipment downtime. Equipment inspections should include:

1.Visual inspection of components and structures.
2.Documentation of current operating characteristics obtained from the equipment operators.
3.Collection of data on equipment condition parameters through such techniques as vibration monitoring of rotating equipment, lubrication oil analysis, infrared thermographic scanning of electrical and mechanical components to identify hot spots and ultrasonic testing to evaluate the integrity of major components and structures.

The results of equipment inspections should be closely analyzed by maintenance and inspection personnel. On-going trending of inspection data enables maintenance personnel to forecast when a component will fail. It also provides the opportunity to extend component life by tracking the effect of maintenance or design modifications on the condition trend of the component.

Critical to the success of the Predictive/Preventive Maintenance Program is the assurance of response to maintenance recommendations made by the inspection group. Policies for follow-up must be established prior to the implementation of the program and they must be strictly enforced through accountability.

3 EQUIPMENT CONDITION MONITORING

As discussed in the previous section, effective Preventive Maintenance must include condition-based monitoring of production equipment. A review, conducted during 1992, of existing Preventive Maintenance programs revealed that equipment condition monitoring was being used effectively in some areas. Maintenance departments had implemented those techniques commonly associated with the type of equipment maintained in their area (ie. plants developed vibration monitoring and Mobile developed oil analysis). The Maintenance Engineering Department was directed to investigate the potential of the various tools and technologies available for condition monitoring and to apply these in all areas where they could feasibly improve equipment reliability. I.O.C.C. maintenance departments are responsible for a large variety of production equipment as outlined in Table 1.

Table 1. Iron Ore Company of Canada maintenance departments and production equipment maintained.

Department	Equipment maintained
Mobile	Haulage trucks - Titan 2200 (25)
	Loaders - Cat 988, 992, 930, 936 & L-1100
	Tractors - Cat D10N, Komatsu D375
	Graders - Cat 16G
	Rubber tire dozers - Cat 834B
	Mobile cranes - 50T to 120T capacity
	Other - Rock Jack, Terminator, Fuel Truck, etc.
Field	Shovels - B.E.295BII(5), P&H2300(2), B.E.280KB(2)
	Drills - B.E.49RH (4), GD120 (2)
	Loading pockets (3)
	Calcium plant and mine rockcrusher
A.T.O.	Locomotives (9)
	95T capacity side-dump ore cars (150)
	Track (46km) - A.T.O. & Yard
	Switches (67) - A.T.O. & Yard
	Wayside signal system
	Auxiliary rolling stock - Catenary cars (3), Flat cars (2)
Concentrator	Mills, conveyors, pumps, spirals, crushers, car dumpers, motors, piping, electrical distribution, etc.
Pellet Plant	Regrind mills, process fans, tripper car, conveyors, pumps, motors, electrical distribution etc.

4 PROGRAM DEVELOPMENT

This section outlines the growth of existing condition monitoring programs and their development in new departments since 1992.

4.1 Vibration Monitoring and Analysis

The Concentrator and Pellet Plant have had vibration monitoring and analysis programs in place since the late 1980's. Some restructuring of the program in the Concentrator was required in 1992 to improve its effectiveness. Prior to the changes which took place, the ability to trend and analyze data effectively was limited because data was being collected from too much equipment. There was not sufficient time to perform proper analysis and the database had become unmanageable.

The first step in addressing this problem was to identify the critical, life-line equipment in the area. The program was then modified to ensure that this equipment was given priority for data collection and analysis. The less critical equipment has been removed until additional resources are added to support the program. The vibration crew is now able to effectively handle both the data collection and analysis roles effectively.

The response of maintenance personnel to recommendations made by the vibration crew has also been improved through the establishment of formal procedures for reporting and follow-up. Recommended maintenance action is now communicated through the computerized maintenance management system. The vibration crew puts a work request on the system which outlines the required maintenance work. The maintenance planner then reviews the request and plans the maintenance job for the department. The maintenance system allows the vibration crew to track their work requests and therefore monitor the action taken by maintenance personnel. In cases where further discussion is required, meetings are held between the maintenance and vibration groups to assess the need for corrective action and to plan the work.

These two (2) changes have resulted in improved maintenance performance within the Concentrator area. The maintenance personnel are now taking more advantage of the vibration monitoring and analysis technology available to them.

The Maintenance Engineering Department has since expanded the application of this condition monitoring technique to include equipment in other departments. The plant process water and tailings pumps are now being monitored and the vibration data trended. The analysis of this data led to the following recommendations:

1.Pump impellers must be properly balanced in the shop prior to installation.

2.Some pump foundations must be repaired to reduce equipment vibration which could potentially lead to premature failure.

3.Supplier recommended installation and maintenance procedures and precautions should be obtained and followed precisely by pump maintenance personnel.

During December 1992, the Central Shops balanced pump impellers for the first time. Proper shop balancing, both separately and as an assembly, improved the quality of pump rebuild work and resulted in smoother running machines.

Pump maintenance personnel have developed a plan to make the required pump foundation repairs as units are changed out. This is expected to significantly reduce the previously high vibration readings.

Pump suppliers were contacted to obtain recommended installation and maintenance procedures and operating tolerances. The provision of this information to the tradesmen has resulted in improved maintenance workmanship in both the field and the shop. The vibration technologist provides support as required in this area and performs a vibration check following all installation and repair jobs. The supplier information has also been provided to the I.O.C.C. Functional Training Department for incorporation into the appropriate training manuals. The quality of maintenance performed on these pumps has also been improved through the provision of alignment training to the area tradesmen.

Vibration levels on the locomotives on the Automatic Train system are now being monitored. Problems such as unbalance and misalignment of the drive motor have been successfully identified and corrected before failure of the motor occurred. Currently, data collection can only be performed when the locomotive is in the shop because all data collection points cannot be accessed with the hood in place. However, permanent installation of accelerometers on the locomotives will enable personnel to collect data during equipment service and operation. All nine (9) locomotives will be instrumented by the end of 1994.

Equipment currently being monitored for vibration at I.O.C.C. rotates at constant speed. Therefore, the concepts, procedures and analysis equipment are well established. However, the Maintenance Engineering Department plans to implement a program which will involve monitoring of Mine production shovels. The drives on this equipment are variable speed and therefore, traditional procedures and equipment do not meet our needs. During 1992, the company approved the provision of capital to purchase data collection and analysis equipment and software required to conduct a study aimed at developing an effective method of gathering, trending and analyzing vibration data from variable speed equipment. A digital audio tape recorder collects real time data for a number of cycles of operation. Simultaneously, a tachometer monitors the rotational speed throughout the data collection process. Full cycle data is then analyzed on a real time vibration analyzer. The procedure for effective data collection and analysis should be established and implemented by the end of 1993. This research and development project is a joint endeavour of I.O.C.C. and J.H. Burrows Electronics of Montreal.

Another element of this research project involves the evaluation, extraction and identification of useful patterns and trends in the vibration signals collected with the aim of classifying them into distinct categories which would indicate the mechanical state of the equipment. The aim of this activity is to collect sufficient data to train a neural network to accurately and reliably predict potential component failure and to track wear patterns.

There are two techniques that can be used in the decision-making process for evaluating the condition of a machine and its components. They are:

1. A rule-based expert system
2. A heuristic artificial neural network system

The two approaches process the data in different ways (see figure 1).

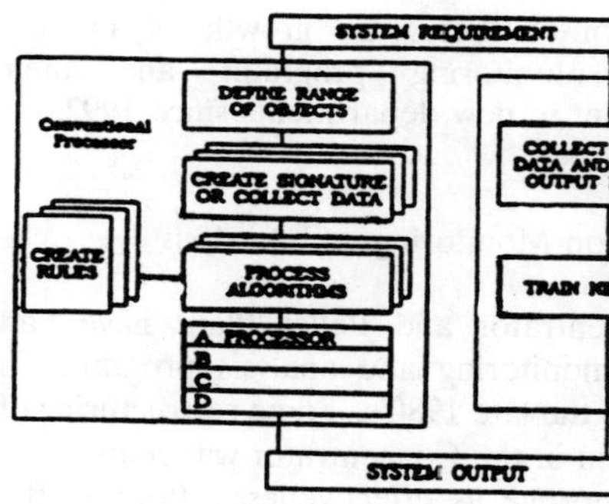

Fig.1 A comparison of processes

The rule based expert system requires an intimate knowledge of the processes and the data in order to generate algorithms that can be used to program a computer to make decisions. The heuristic artificial neural network system does not require an intimate knowledge of the system because artificial neural networks are massively parallel, extremely fast and intrinsically fault tolerant. They can be taught to perform complex tasks without programming (see Figure 2). These systems learn from experience, generalise from examples, and extract essential characteristics from noisy data.

The aim of the Maintenance Engineering Department will be to train the neural network (see Figure 3) to perform the majority of the equipment problem diagnoses in order to ensure efficient and consistent analysis of the data collected.

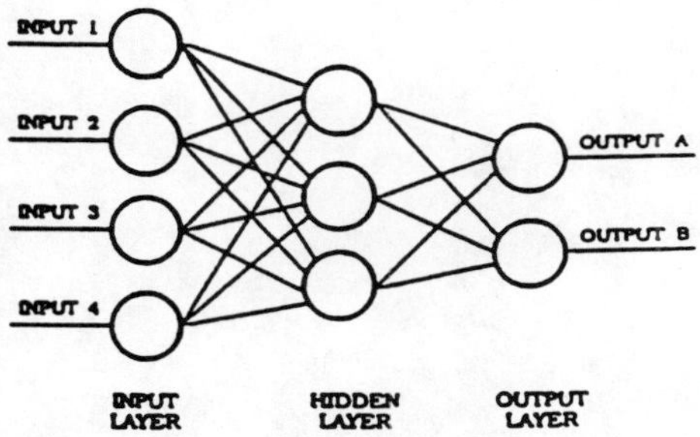

Fig.2 A three level neural network

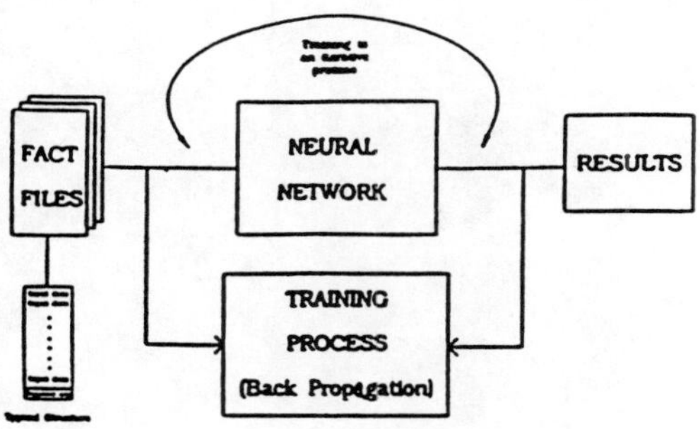

Fig.3 Back propagation training process

4.2 Lubrication Oil Analysis

Engine and hydraulic oil is sampled and analyzed for the 200 tonne haulage trucks and the auxiliary mine equipment on 300 hour and 1500 hour intervals. Oil analysis has been an integral part of the Mobile Maintenance Preventive Maintenance Program since the 1980's. Samples collected from the engine, transmission, bull gears and planetaries are sent to the company laboratory facilities in Sept-Iles, Quebec for analysis. The results provide details on the quantity of metal (eg. Fe, Pb, Cu, etc.) and/or other contaminants (eg. glycol) found in the oil which would signify component problems. The current turnaround time on samples sent to Sept. Iles is approximately three (3) days. This significantly exceeds the requirements of Mobile Maintenance personnel who need results before the equipment leaves the service line (ie. within 8 hours).

In order to improve the oil analysis service to the mine, I.O.C.C. plans to purchase an oil analysis spectrometer to establish a laboratory on-site during 1992. In order to further justify the high capital cost of developing this oil testing facility, other applications for this condition monitoring method were identified. By April of 1994, lubrication oil samples from shovel and drill gear cases, grinding mill gear boxes, locomotive traction motors and numerous other components will be regularly sent to the laboratory for analysis. The results will be trended to monitor the equipment condition on an on-going basis.

4.3 Infrared thermographic scanning

Field Maintenance implemented infrared thermographic scanning of electrical cabinets on production shovels in 1991. This technique enabled electrical maintenance personnel to identify "hot spots" resulting from faulty thyristors, fuses and electrical contacts. This technology has since spread to include numerous other applications throughout I.O.C.C. Some examples include:

1.Monitoring of bearing temperature to detect overheating of bearings.
2.Scanning of pellet furnace refractory to identify areas requiring maintenance of the refractory material.
3.Scanning of mine power lines and the A.T.O. catenary to identify poor connections, damaged insulators, etc.
4.Scanning of dry mill cyclones in the Concentrator to track ceramic liner wear and to proactively plan for overhaul work required.

The equipment purchased by I.O.C.C. is being used successfully in all areas of the project. Further development of the program will occur as more applications are identified.

5 CONCLUSION

High reliability of production equipment is critical in the increasingly competitive iron ore mining and processing industry. The introduction and development of condition monitoring tools and techniques, as an integral part of the overall Preventive Maintenance program, will continue to improve maintenance effectiveness and equipment reliability. Increased control of maintenance activities brought about by the ability to proactively plan jobs will result in lower maintenance costs. The use of neural networks to support condition monitoring will ensure consistency of decision-making and will also enable results from various monitoring techniques to be simultaneously analyzed during equipment condition assessment.

Through the use and development of currently available technology, the maintenance departments within the Iron Ore Co. of Canada continue to move toward their target of maintenance excellence.

REFERENCES

Burrows, J.H. 1992. Strategies, techniques and tools for improving the decision making process of plant maintenance.

Burrows, J.H. 1991. Optimizing the use of periodically collected data for machine condition monitoring. 3rd International Congress on Condition Monitoring and Diagnostic Engineering and Management.

McHattie, L.D. and El-Alfy, S.E. 1992. Maintenance improvements through the introduction of reliability concepts. CIM Bulletin, October 1992.

Innovative Mine Design for the 21st Century, Bawden & Archibald (eds) © 1993 Balkema, Rotterdam, ISBN 90 5410 325 6

Dynamic simulation and computer control of a boom-type tunnelling machine

H.Wu & Laeeque K.Daneshmend
The Canadian Centre for Automation and Robotics in Mining (CCARM) & Department of Mining and Metallurgical Engineering, McGill University, Montreal, Que., Canada

Abstract

This paper investigates the dynamic simulation & control of a vertical turret, boom-type tunnelling machine. Models of the actuators, kinematics, rigid body dynamics, and cutter-rock interaction are formulated. These models are validated through computer simulation. Controllers are then designed for the actuators using the pole-placement technique for discrete-time (sampled-data) control. The controllers are tested in a variety of operational modes, using the computer simulation of the machine model. Results show that good tracking and disturbance rejection is achieved.

1 Introduction

The general goals in tunnelling are natural ore support, precise tunnel shape, high productivity, and low cost. According to the hardness of the rock and the desired roadway (tunnel) shape, different tunnelling and excavation methods can be chosen. For each tunnelling and excavation method, there are several types of machines that can be employed.

Boom-type roadheaders (also known as boom-type tunnelling machines) are generally used in softer rocks, but are increasingly favoured for harder rock applications. They can typically deal with strata of a compressive strength upto 12,000 $1\mathrm{bf}/in^2$ ($82\ \mathrm{kN}/m^2$) [14] [15]. Recent prototypes with horizontal turrets can tackle rocks of twice this strength. They can cut arc, circle, square, trapezoidal D-shape and various profiles of roadway. Generally speaking, although a boom-type tunnelling machine has a lower rate of production than normal drum or rotary-pick-type continuous miners, it has many advantages over the those machines. It can tunnel selectively and cut any shape or size of drivage (excavation) within a fixed envelope. In addition, because it can concentrate the entire power and thrust of the machine at a single location (rather than spreading it across the width of the cutting head), it can cut very much harder rock than that which can be attacked successfully by other types of tunnelling machines.

These machines are classified into three basic types: (*a*) Vertical turret machines, (*b*) Horizontal turret machines and (*c*) Double-Jointed-Arm Mechanisms (DJAM).

[1] This work has been partially supported by the Canadian Federal Centres of Excellence Program, via Project C-2 of the Canadian Institute for Robotics and Intelligent Systems (IRIS), and by an FCAR team grant from the Government of Québec.

The vertical turret machine has a boom with two directions of movement, i.e. vertical elevation and horizontal traversing. The horizontal turret mounting machine boom has two free movements: radius of displacement (i.e.ranging) and turret rotation. The DJAM consists of two links, again giving it two degrees of freedom. The primary arm is attached to the machine axis. The secondary arm pivots on an axis at the end of the primary arm, and it holds a cutting drum at the other end.

This paper focuses on vertical turret boom-type machines, however, the control and automation methodology presented here should be applicable to the other types of road-header as well. The design and operation of this type of machine is first analyzed, and control objectives are then formulated. The methodology of modelling, simulation, and controller design is then presented, followed by the specific models for the actuators, kinematics, and pick-rock interaction, leading to an integrated model which forms the basis for simulation. The controller design is then briefly reviewed, and simulation results presented. Finally conclusions are formulated and directions for future work are suggested.

2 Machine Design & Operation

2.1 Kinematics

The motion of the roadheader consists of four parts. First, the whole machine is moved forward on its tracks, sumping the cutting head into the work face at a controlled rate. Second, the cutting head rotates and cuts the rock. Third, the motion of the vertical turret causes the cutting head to move on the profile parallel to the horizontal axis of the machine. Fourth, the up-down motion of the boom causes the cutting head to move on the profile along the vertical axis. The third and fourth motions can be integrated to achieve various straightline and curved profiles on the work face.

2.2 Structure

A track-mounted roadheader machine is typically electro-hydraulically operated. It consists of the main frame, cutting assembly, twin crawler assembly and various components of the electric and hydraulic subsystems. At the front end of the main frame, there is the cutting trunk assembly which includes the cutting boom, lifting cylinders, arcing cylinders, cylindrical cutting head and cutting head motor. At the rear end of the main frame there is a oil tank, a hydraulic pump, operating control panel and an electric motor.

2.3 Actuators

The actuators serve two different types of functions: the rotary cutting head actuator provides the cutting action; the positioning actuators provide motion along the profile. In addition, there are the crawler, gathering head and conveyor actuators. In this paper, the topics of interest are the control of the cutting head and position, therefore these two aspects will be introduced briefly as follows:

(1) The cutting head actuator: the cutting head motor is designed to supply two types of power, i.e. normal running loads and torsional shock loads. The latter is required if the equipment suddenly slows or jams. The method used to solve this problem is to make the actuator have sufficiently large inertia, which will have a flywheel effect and make the system overcome the unfavourable torsional shock loads. Experience has shown that AC or DC electric motors show good inertia effect, but that hydraulic motors have undesirable inertia behaviour. This is a major reason why the majority of cutting head motors are AC or DC electric. After the type of motor is chosen, the question remains of how to achieve a compromise between having sufficient inertia to ensure the equipment will continue working in a non-homogeneous geological environment and overcome the torsional shock loads, while avoiding oversizing of the motor which increases power consumption and mechanical design requirements [2] [6].

(2) The positioning actuators: some roadheaders use electric AC or DC motors as the positioning actuators, but since the costs of electric actuators are higher than those of hydraulic actuators, most roadheaders use hydraulic motors for positioning drives. The major advantages are a slow speed, high torque characteristic, and automatic hydraulic braking. The negative points are that they are less efficient than electric motors, equipment is more complicated in terms of hydraulic supply and hoses, and maintenance is more frequent.

2.4 Operation

The operational cycle of the machine consists of three steps: sumping, cutting, and advance/conveying. Their physical significance is described below.

2.4.1 Sumping

Sumping is the part of the operational cycle where the roadheader cutting head is inserted into the work face. This step must be completed before the machine head cuts the next stratum. Most tunnelling machines sump their cutting heads by advancing the whole machine, others are equipped with a reciprocating sumping head, which can insert the head directly into the profile without moving the crawlers. The advantage of using a sumping head is that it reduces the sumping damage to the machine from the tramming and spinning of the crawlers.

2.4.2 Cutting

The cutting head is typically driven by an electric AC motor, or several AC motors (depending on the number of booms, usually, the number of AC motors is the same as the number of the booms). The motion of the cutting head is parallel to the work face so that the cutting head has only half its face working for every rotational period. The whole boom, including the cutting head, is positioned by the hydraulic positioning cylinders in the vertical and horizontal planes.

2.4.3 Advance and Conveying

The machine must advance to enable it to commence cutting a new work face while maintaining the desired profile. This action is accomplished by a pair of crawlers, driven by hydraulic motors. The conveying mechanism consists of the gathering head and the conveyor. The gathering head collects the cutting debris into the conveyor, and the conveyor works continuously to transport the rock out of the machine. The conveyor and the gathering head are both hydraulically powered.

3 Control Objectives for Boom Type Roadheader

The aim of mining automation is to reduce the physical effort of the underground workers and reducing the danger to which they are subjected. In addition, like other industries, the mining industry needs to increase its rate of production, increase product quality(i.e.reduce ore dilution) and decrease the cost per unit product. For attaining these aims, mining automation is imperative. A substantial part of this effort lies in tunnelling machine automation.

For tunnelling machine control, the functions to be automated cover four aspects: control of the power in the head; profile shape and accuracy control; productivity rate; reduction in vibration for machine integrity. These four aspects are expanded on below.

3.1 Power in Head

The head power is defined as the work done in excavating unit volume or mass of rock [4]. It is a very useful concept for measuring the cutting efficiency. The goal of the control is to satisfy the production requirements with the least power. The approach taken is to vary the speed of the cutting motor when the force acting on the cutting head is constant, so that the power is controlled. The cutting motors of traditional tunnelling machines were equipped with gearboxes allowing only two speeds, selected by the operator. Recent work [12] has shown that variable cutter head speed is required for optimizing cutting efficiency. This implies a variable speed drive is used, together with appropriate automatic control.

3.2 Profile Shape and Accuracy

The profiles most often desired in mining and civil engineering applications are circles, squares and D shapes. Most tunnelling machines are monitored by an operator to achieve the work profiles. Usually, the profile shapes cut are not accurate, so a large amount of additional work must be done after every cutting cycle is finished. In order to increase the accuracy of the profile and reduce manpower requirements, a kinematic-control strategy can be employed. It consists of controlling the slewing and lifting electrohydraulic actuators to make the boom move on the work face with the desired trajectories.

3.3 Productivity

The productivity rate may also be viewed as a control issue. Maximum productivity and minimum cost are always the ideal aims. The techniques used rely on increasing the speeds of the cutting head, traction and crawler, as well as the quality of the profile.

3.4 Reduction in vibration for machine integrity

The vibration of the whole machine structure arises mainly from the cutting action of the head against the work face. It is transmitted from the head into the various parts of the machine, including the mechanical structure, AC motor, as well as hydraulic positioning actuators. For the mechanical structure, the mechanical designer must consider the requirements for avoiding exessive vibration and resonances. For the rotary head and positioning actuators, control methods can be used to attain the goal of eliminating vibrational effects.

4 Methodology

The approach to investigating the control of system will be based on the fundamentals of dynamic modelling, simulation and control system design. The steps in the methodology are described below, and the structure of the remainder of the paper follows these steps.

4.1 Dynamic Modelling

In the case of the tunnelling machine, complete dynamic modelling should include the actuators dynamics, rock-cutter interaction, cutting forces and normal forces on the cutting head. This is because the driving parts of the tunnelling machine are various actuators which accomplish boom lifting and slewing, as well as the cutting head rotation. The head attacks the work face so that rock and cutter interact. The interaction forces, i.e. the cutting force and normal force, are transmitted to the various parts of the machine. In order to effectively control the machine, a thorough understanding its dynamics is imperative. This paper restricts its analysis to rigid body modelling of the mechanical structure of the machine.

4.2 Simulation

To test dynamic models, which are formulated in the previous phase, software simulations are implemented and time domain behavior investigated in open-loop the conditions. The various aspects of the dynamics are tested and the behavior of the system analyzed. This analysis is the foundation on which control system design is based.

4.3 Controller design and Validation

The controllers for the various parts are designed to attain the desired system dynamics. The controllers include the continue time controllers based on the dynamics and the discrete time controllers based on the dynamics and computer control. The simulations are built to test the controller responses to reference inputs and their disturbance rejection. ACSL(Advanced Continue Simulation Language) [11] is used as the simulation tool. Simulated effects, which include A/D and D/A converters, computational delay, anti-aliasing filtering of sensors, as well as sampling-period, make the simulation approximate the real behaviour of the process under real-time computer control.

5 Modelling

5.1 Actuators

Based upon modelling of electro-hyrdraulic systems [1] [13], the boom-joint motors 1 and 2 can be modelled as:

$$\ddot{\theta}_{1,2} = -\frac{C_{1,2}+H_1}{I_{1,2}}\dot{\theta}_{1,2} + \frac{K_{1,2}H_2}{I_{1,2}}i_{1v,2v} - \frac{1}{I_{1,2}}T_{1d,2d}$$

The AC cutting head motor can be modelled as [9]:

$$\ddot{\theta}_3 = -\frac{C_3+C_5}{I_3}\dot{\theta}_3 + \frac{K_3C_4}{I_3}u - \frac{1}{I}T_{3d}$$

5.2 Kinematics

The kinematic coordinates sets for the roadheader are shown in Figure 1. The terminology and theory of the kinematic formulation is presented at length in a number of papers and books [10] [3].

5.2.1 Forward Kinematics

The forward kinematics problem is: given the joint angle vector (θ_1,θ_2, θ_3), find the position and orientation of the end effector of the vertical turret machine in the work profile coordinate system, which is with respect to a reference coordinate system.

Table 1 describes the link parameters, in Denavit-Hartenberg notation.

Table 1: Link Parameters

i	α_{i-1}	a_{i-1}	d_i	θ_i
1	0	0	L_1	θ_1
2	90°	0	0	θ_2
3	90°	0	L_2	θ_3

Computing the individual transformations for each link which defines frame $\{i\}$ relative to the frame $\{i-1\}$, one can use the formula,

$$ {}^{i-1}_{i}T = \begin{pmatrix} c\theta_i & -s\theta_i & 0 & a_{i-1} \\ s\theta_i c\alpha_{i-1} & c\theta_i c\alpha_{i-1} & -s\alpha_{i-1} & -s\alpha_{i-1} d_i \\ s\theta_i s\alpha_{i-1} & c\theta_i s\alpha_{i-1} & c\alpha_{i-1} & c\alpha_{i-1} d_i \\ 0 & 0 & 0 & 1 \end{pmatrix} $$

which, after much manipulation, yields:

$$X_4 = s_2 L_2 + L_1 - Z_0$$

$$Y_4 = -s_1 c_2 L_2 + Y_0$$

$$Z_4 = c_1 c_2 L_2 - X_0$$

5.2.2 Inverse Kinematics

The inverse kinematics problem of the vertical turret tunnelling machine : is given the desired cartesian position and orientation of the last link of the machine relative to the work profile, how does one calculate a set of joint angles which can achieve the desired position and orientation.

$$ {}^0_3T = {}^0_1T\, {}^1_2T\, {}^2_3T $$

$$ = \begin{pmatrix} -c_1 s_2 c_3 + s_1 s_3 & c_1 s_2 s_3 + s_1 c_3 & c_1 c_2 & c_1 c_2 L_2 \\ -s_1 s_2 c_3 - c_1 s_3 & s_1 s_2 s_3 - c_1 c_3 & s_1 c_2 & s_1 c_2 L_2 \\ c_2 c_3 & -c_2 s_3 & s_2 & s_2 L_2 + L_1 \\ 0 & 0 & 0 & 1 \end{pmatrix} $$

$$ = \begin{pmatrix} r_{11} & r_{12} & r_{13} & P_x \\ r_{21} & r_{22} & r_{23} & P_y \\ r_{31} & r_{32} & r_{33} & P_z \\ 0 & 0 & 0 & 1 \end{pmatrix} $$

which yeilds:

$$\theta_1 = \arctan(\frac{P_y}{P_x}) = \arctan(\frac{Y_0 - Y_4}{\sqrt{L_2^2 - X_4^2 - Y_4^2}})$$

$$\theta_2 = \arcsin(\frac{P_z - L_1}{L_2}) = \arcsin(\frac{Z_0 + X_4 - L_1}{L_2})$$

5.3 Pick-Rock Interaction

Based on the formulation by Hurt, Morris, and Macandrew, [8] [7] normal and cutting forces on each pick are given by:

$$F_c = k_c s d_{max} sin\theta_3$$

$$F_n = k_n s d_{max} sin\theta_3$$

$$T = \sum F_c \cdot r$$

where k_c and k_n are constants(k_c varies according to different rock types; k_n varies according to rake angle α). and d_{max} is the depth of cut. Considering the number of picks on each line and the pick sequence number on the cutting head, we can obtain:

$$T_{3d} = \sum F_c \cdot r = \sum_{i=1}^{m}\sum_{j=1}^{n} k_c s_i d_{max} sin(\theta_3 + \frac{2(j-1)}{j}\pi) r_i$$

$$F_y = \sum_{i=1}^{m}\sum_{j=1}^{n} \frac{k_c s_i d_{max}}{2} sin(\theta_3 + \frac{2(j-1)}{j}\pi) cos(\theta_3 + \frac{2(j-1)}{j}\pi) + \sum_{i=1}^{m}\sum_{j=1}^{n} \frac{k_n s_i d_{max}}{2} sin^2(\theta_3 + \frac{2(j-1)}{j}\pi) cos\phi$$

$$F_x = \sum_{i=1}^{m}\sum_{j=1}^{n} \frac{k_n s_i d_{max}}{2} sin(\theta_3 + \frac{2(j-1)}{j}\pi) cos(\theta_3 + \frac{2(j-1)}{j}\pi) + \sum_{i=1}^{m}\sum_{j=1}^{n} \frac{k_c s_i d_{max}}{2} sin^2(\theta_3 + \frac{2(j-1)}{j}\pi)$$

Based on the kinematics, we can calculate force and torque transmission to joint 1 and 2:

$$T_{1d} = -L_2 c_2 (s_3 F_x + c_3 F_y) + s_2 T_{3d}$$

$$T_{2d} = L_2 (c_3 F_x - s_3 F_y)$$

5.4 Integrated Model of Actuators & Cutter-Rock Interaction

Combining the above models, yields an integrated model of the system:

$$ \frac{d}{dt}\begin{pmatrix} x_1 \\ x_2 \\ x_3 \\ x_4 \\ x_5 \\ x_6 \end{pmatrix} = $$

$$ \begin{pmatrix} 0 & 1 & 0 & 0 & 0 & 0 \\ 0 & -\frac{C_1+H_1}{I_1} & 0 & 0 & 0 & 0 \\ 0 & 0 & 0 & 1 & 0 & 0 \\ 0 & 0 & 0 & -\frac{C_2+H_1}{I_2} & 0 & 0 \\ 0 & 0 & 0 & 0 & 0 & 1 \\ 0 & 0 & 0 & 0 & 0 & -\frac{C_3+C_5}{I_3} \end{pmatrix} \begin{pmatrix} x_1 \\ x_2 \\ x_3 \\ x_4 \\ x_5 \\ x_6 \end{pmatrix} + \begin{pmatrix} 0 & 0 & 0 \\ \frac{K_1 H_2}{I_1} & 0 & 0 \\ 0 & 0 & 0 \\ 0 & \frac{K_2 H_2}{I_2} & 0 \\ 0 & 0 & 0 \\ 0 & 0 & \frac{K_3 C_4}{r} \end{pmatrix} \begin{pmatrix} i_1 \\ i_2 \\ u \end{pmatrix} $$

$$+\begin{pmatrix} 0 & 0 & 0 \\ -\frac{1}{I_1} & 0 & 0 \\ 0 & 0 & 0 \\ 0 & -\frac{1}{I_2} & 0 \\ 0 & 0 & 0 \\ 0 & 0 & -\frac{1}{I_3} \end{pmatrix} \begin{pmatrix} T_{1d} \\ T_{2d} \\ T_{3d} \end{pmatrix}$$

where,

$$T_{1d} = -L_2 c_2 (s_3 F_x + c_3 F_y) + s_2 T_{3d}$$

$$T_{2d} = L_2 (c_3 F_x - s_3 F_y)$$

$$T_{3d} = \sum_{i=1}^{m} \sum_{j=1}^{n} k_c s_i d_{max} sin(\theta_3 + \frac{2(j-1)}{j}\pi) r_i$$

6 Controller Design

Based on the previously stated control objectives, and the machine constraints, three different control modes were investigated:

- Variable velocity along profile with constant cutter speed.
- Constant velocity along profile and variable cutter speed.
- Variable velocity along profile and variable cutter speed.

Discrete-time pole-placement controller design [5] [16] was utilized on all three actuators, resulting in the following controllers, which are illustrated in Figures 2 and 3.

On the cutting head motor a digital compensator:

$$D_1(z) = 0.01767 + \frac{0.007967z}{z-1} = \frac{0.025637 - 0.01767z^{-1}}{1 - z^{-1}}$$

which yields

$$u(k) = D_1(Z)(\omega_{3r}(k) - \omega_3(k))$$

On the hydraulic positioning actuator a digitalcompensator:

$$D_2(z) = \frac{6.7387 - 5.7312z^{-1}}{1 - z^{-1}}$$

which yields

$$i_2(k) = D_2(z)(x_{3r}(k) - x_3(k)) + K_{24}(x_{4r}(k) - x_4(k))$$
$$= 0.5372(\omega_{2r}(k) - \omega_2(k))$$
$$+ (\frac{6.7387 - 5.7312z^{-1}}{1 - z^{-1}})(\theta_{2r}(k) - \theta_2(k))$$

7 Simulation Results

The step responses of the cutting head AC motor under the above defined control are shown in Figures 4 and 5. Disturbance rejection is good, even when faced with a periodic cutting force.

Figures 6 and 7 show the response for one of the positioing actuators under the above defined control. Again, disturbance rejection is good. The validity of the simulation can be verified by the eaxmining the nature of the disturbance torque in Figure 7: it is a modulated sinusoid as is to be expected from the dynamic model.

The ultimate test of the system is tracking a predefined profile. Figure 8 shows the desired trajectory that is to be followed. Figures 9 and 10 show how the actual profile that is simulated varies, depending on velocity along the profile.

8 Conclusions

This paper applied concepts from robot dynamics, kinematics, and control, to the control of a boom-type vertical turret tunnelling machine.

- A physical model of the complete machine was derived, based upon rigid body assumptions. This included the positioning actuators, cutting head motor, and kinematics as well as dynamics.
- A physical model of the cutter-rock interaction was derived, based upon the layout of the cutter picks and the rock properties. This yielded formulas for cutting force and normal force.
- The kinematic model was used in conjunction with the cutter force model to derive a model for the corresponding disturbance torques seen at the positioning actuators during cutting.
- An integrated model of the complete process, i.e. the machine and its interaction with the rock, was constructed based on the previous three models.
- The individual models, as well as the integrated model, were implemented in software, using $ACSL^{tm}$, and verified by simulation.
- Based upon the physical models, pole-placement controllers were designed for single-input single-output (SISO) control of the actuators. Both continuous-time and discrete-time designs were formulated.
- The controller designs were first validated individually via simulations, and then the complete control system was tested by simulation of the cutting of a D-shaped profile.
- Three different operational modes were investigated. They were: constant velocity with variable speed, variable velocity with constant speed, and variable velocity with variable speed.

This paper has shown the relevance of applying manipulator dynamics and kinematics concepts to tunnelling machines. It has also shown that the control of such mining machines also requires a thorough understanding of the cutter-rock interaction. Software simulation has proved to be a very powerful tool for validating the dynamics, kinematics, and cutter-rock models. Linear decoupled control has been shown to be a valid solution to the control of this type of machine.

8.1 Suggestions for Future Work

Several improvements can be made upon the modelling and simulation of the machine and the of cutting process:

- The models could be extended to include non-rigid body effects of the machine structure.
- The current work relies upon physical modelling and scant empirical data from previously published research. Comprehensive empirical modelling, based upon data acquisition during field tests followed by mathematical model building using regression and system identification techniques, is highly desirable. Such empirical models would serve to either refute the physical models, or provide greater insight into, and parameterization of, the physical models.

In the area of controller design, a number of avenues need to be explored:

- The control problem can be posed as a non-decoupled, multivariable, problem. This is a more accurate reflection of the coupled nature of the process, and requires the formulation of an appropriate cost function as the basis of a Linear-Quadratic (LQ) optimal controller design.
- Since the rock properties are changing with respect to time, as is the kinematic disturbance coupling, adaptive control may be a relevant technique for stabilization in a range of different rock types.
- Concepts from manipulator force control, such as impedance control, appearly highly relevant.

Since the design, and success, of a control system is intimately related to the controlled process, several aspects of tunnelling machine design also appear worthy of investigation:

- Variable speed cutting head motors appear highly desirable from the viewpoint of control. The extent of speed variation is limited in practice, due to the actuator technology. Different types of cutting head actuators should be evaluated in this regard.
- Rethinking of structural design, e.g. parallel instead of serial kinematics would greatly enhance machine rigidity and reduce disturbance loading of actuators, while likely reducing weight and size.
- Integration of sensing of rock properties, e.g. using microseismic techniques or ground penetrating radar, would provide additional feedback and enable the application of techniques such as preview control.

References

[1] Anderson,W. *Controlling Electrohydraulic Systems*, Marcel Dekker, INC., 1988.

[2] Cotton,B.B. *Electrics and Hydraulics on Roadheader Machinery*, IMechE, 1985, pp.13-22.

[3] Craig,J.J. *Introduction to Robotics*, second edition, Addision-Wesley Publishing Company,INC. 1989.

[4] Farmer,I.W., Garritty,P. *Prediction of Roadheader Cutting Performance From Fracture Toughness considerations*, Proceedings of Sixth International Congress on Rock Mechanics, Canada, August 1987, pp.621-624.

[5] Franklin G.F., and Powell J.D.,Workman,M.L. *Digital Control of Dynamic Systems*, Addison-Wesley Publishing Company, 1990.

[6] Goodall,R.M. and Whitfield,A.H. *Automation of a Wood-Turning Machine, Hydraulic or Electric?*, IMechE 1985, pp.59-65.

[7] Hurt,K.G., and Morris,C.J. *Computer Designed Cutting Heads Improve Roadheader Performance*, Tunnels and Tunnelling, March, 1985. pp.37-38.

[8] Hurt,K.G.,Morris,C.J.and MacANDREW,K.M. *The design and operation of boom tunnelling machine cutting heads*, Rock Breaking and Mechanical Excavation, 14th Canadian Rock Mechanics Symposium, Vancouver, British Columbia, May 13-14, 1982, pp.54-58.

[9] Kosow,I.L. *Control of Electric Machines*, Englewood Cliffs, N.J.:Prentice-Hall, 1973.

[10] Lee,C.S.G. *Robot Arm Kinematics, Dynamics, and Control*, IEEE Computer 1982, pp.62-80.

[11] Mitchell and Gauthier Associates, concord, Mass. 01742. *Advanced Continuous Simulation Language(Reference Manual)*.

[12] Neil, D., Ozdemir L. *Considerations for Development of Hard Rock Roadherders*, Proceedings of International Symposium on Mine Mechanization and Automation, Volume I, Golden, Colorado, June 10-13,1991.

[13] Pippenger,J.J., Hicks,T.G. *Industrial Hydraulics*, third edition, Gregg Division, McGraw-Hill Book Company, 1979.

[14] Roxborough,F.F. *Cutting Rock With Picks*, The Mining Engineer, June 1973, pp.445-455.

[15] Sadreddini,S.M. *Model and Control of Underground Tunnelling Machine*, University of Sheffield,Dept.of control Engineering,Ph.D thesis,1987.

[16] Savant,C.J., Stefani,R.T. *Design of Feedback Control Systems*, N.Y.:Saunders College Pub., 1989.

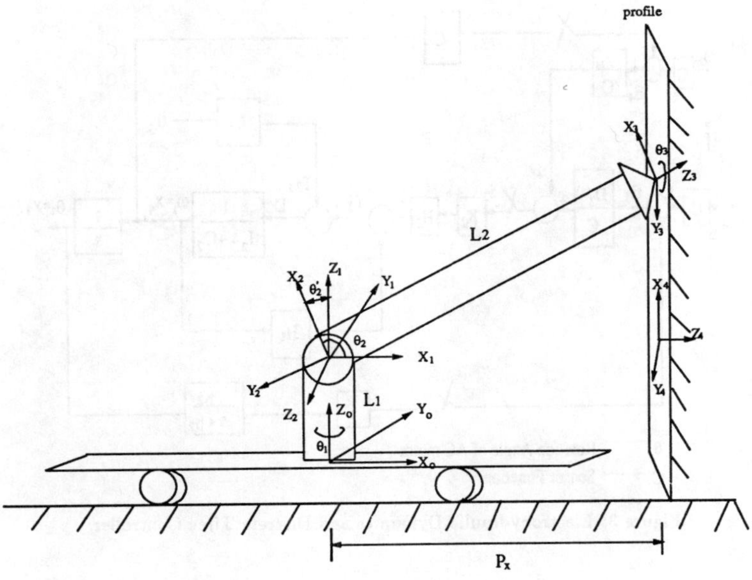

Figure 1: The Coordinate Set Diagram

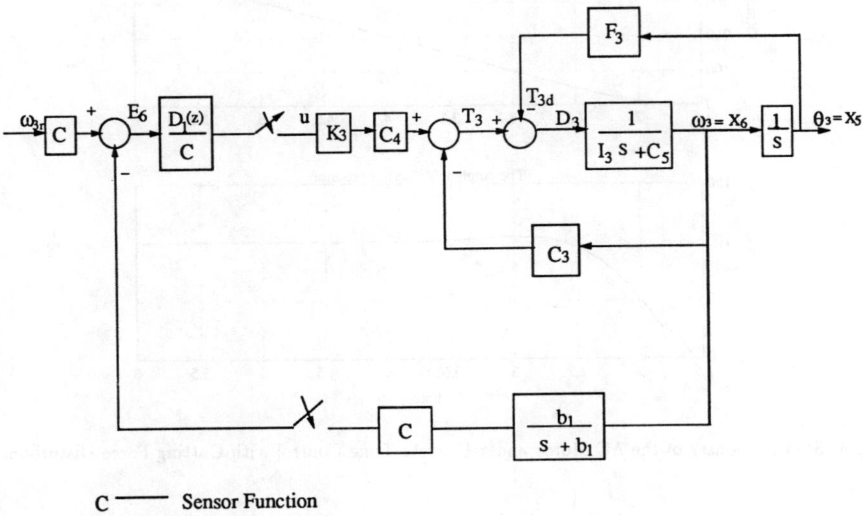

Figure 2: Cutting Head Dynamics and Discrete Time Controller

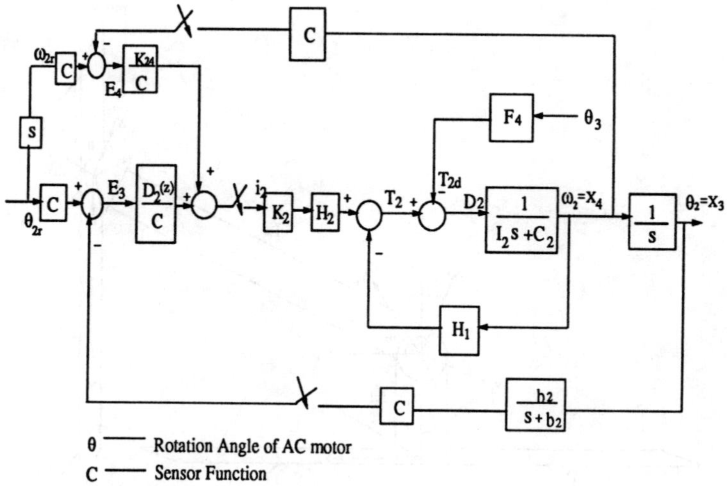

Figure 3: Electrohydraulic Dynamics and Discrete Time Controller

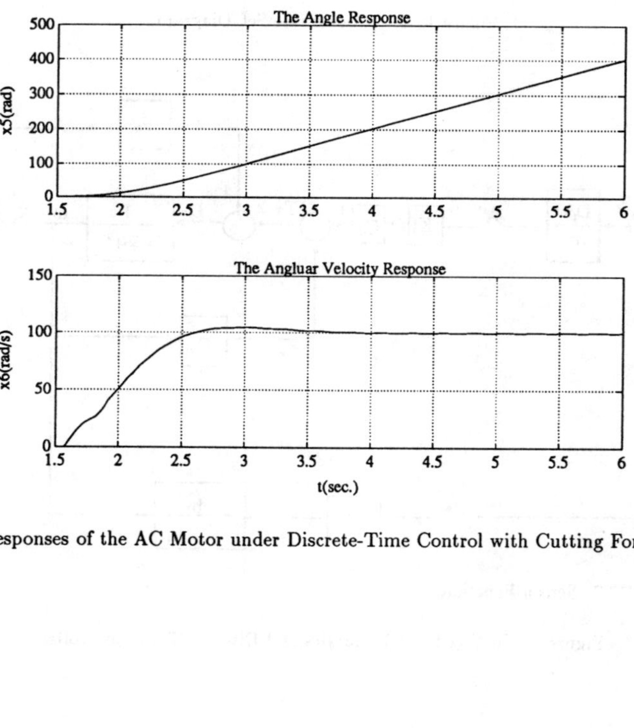

Figure 4: Step-Responses of the AC Motor under Discrete-Time Control with Cutting Force Disturbance

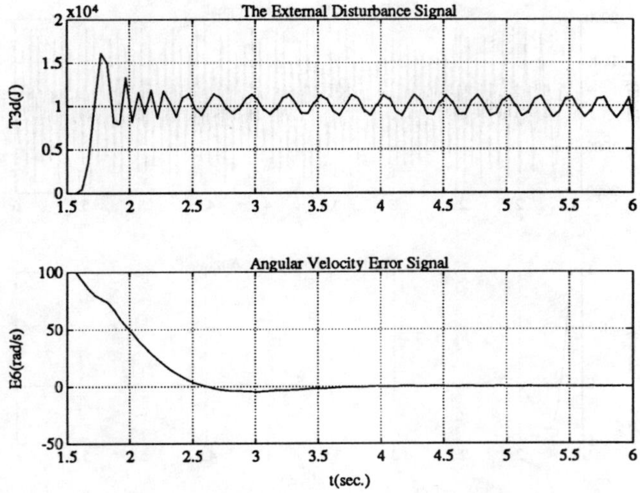

Figure 5: Disturbance and Error Signal of the AC Motor under Discrete-Time Controller

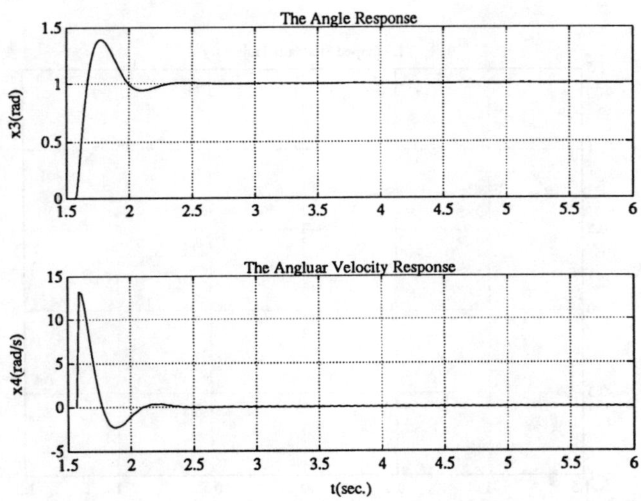

Figure 6: Step-Input Responses for the Electrohydraulic Positioning System under Discrete-Time Control, with cutting Force disturbance

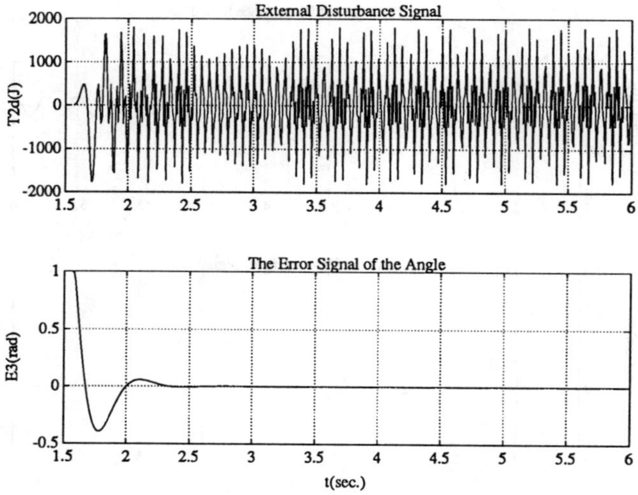

Figure 7: Disturbance Signal and Error Response for the Electrohydraulic Positioning System under Discrete-Time Control, with cutting Force disturbanc

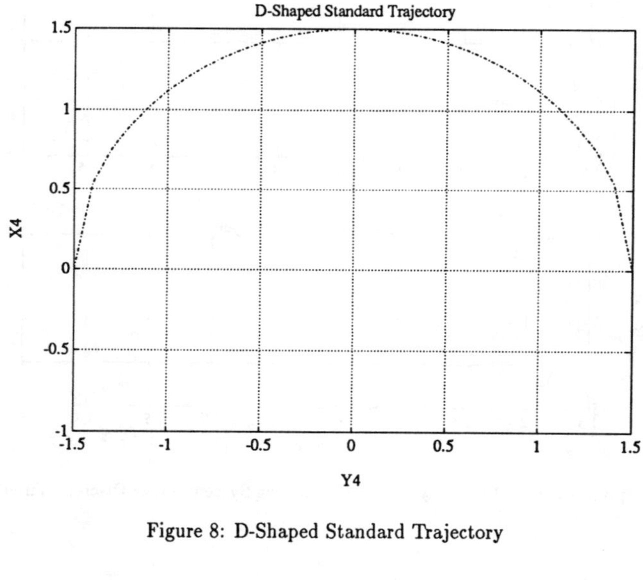

Figure 8: D-Shaped Standard Trajectory

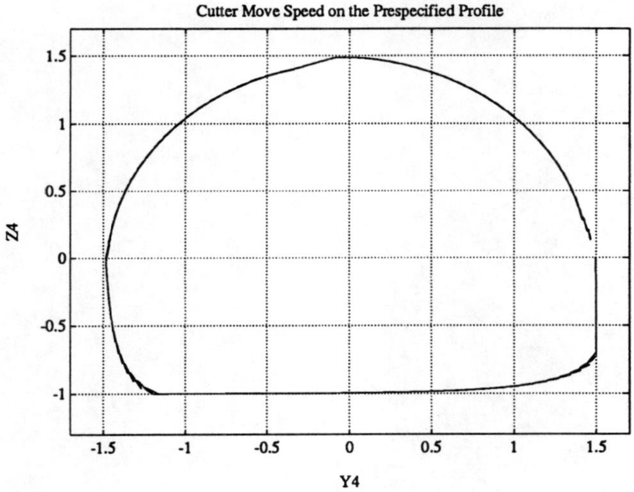

Figure 9: Cutting profile at 0.025 m/s

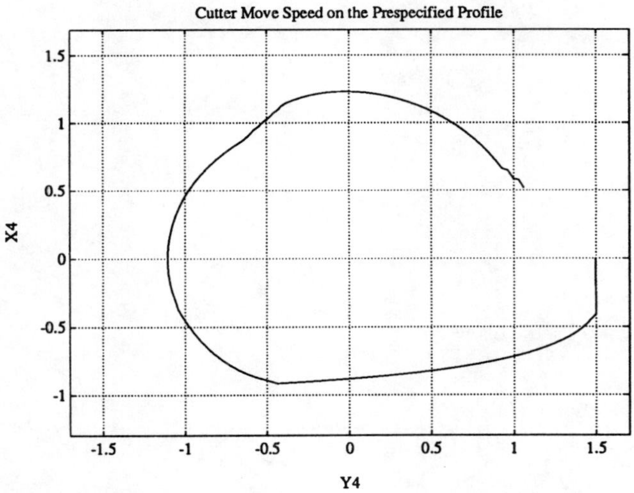

Figure 10: Cutting profile at 0.0357 m/s

Innovative Mine Design for the 21st Century, Bawden & Archibald (eds) © 1993 Balkema, Rotterdam, ISBN 90 5410 325 6

Design of a mining shovel simulator

L. Daneshmend, C. Hendricks, S. Wu & M. Scoble
The Canadian Centre for Automation and Robotics in Mining (CCARM) & Department of Mining and Metallurgical Engineering, McGill University, Montreal, Que., Canada

ABSTRACT: The successful automation of loading equipment is dependant upon an understanding of the relationships which exist between digging dynamics, material characteristics and operational practices. This paper builds upon recent work (Hendricks 1993) on the development of kinematic and dynamic models, and a software simulator, for electric mining shovels. The present focus is an analysis of shovel dynamics based upon the iterative Newton-Euler formulation. Prior research by the authors has shown shovel monitored data (work motion motor responses and digging trajectory) to provide an accurate reflection of muckpile conditions, and hence, the efficiency of fragmentation by blasting. While it is appreciated that both the shovel and its operator will respond differently, according to the geometry and demands of the fragmented material (size, looseness etc.), a definitive understanding of these operational relationships, and how they combine to determine machine productivity, is lacking. The availability of a shovel simulator will greatly enhance the capability to explore the influence on productivity of; different shovel front end geometries, operator practices (trajectories) and various shapes and size distributions of fragmentation. The completeness and calibration of the simulator (i.e. front-end filtering, state and parameter estimation (load weight), as well as diggability and trajectory assessment) will be undertaken through subsequent field performance monitoring of a mining shovel.

INTRODUCTION

The work reported upon in the paper is part of a broader effort aimed at improving the performance and productivity of electric mining shovels as used in the surface mining industry. This initiative will include machine performance monitoring, system modelling, signal processing and trajectory analysis.

This paper reports specifically on recent work towards the development of a dynamic model and simulator of an electric mining shovel. The model specifically considers all of the components which contribute towards hoist motor torque; a shovel parameter understood to be central to the determination of muckpile diggability, shovel productivity and assessment of operating practices. The dynamic model, and prior knowledge of the various dynamic and kinematic parameters which characterize the model, enable us to compute several items of interest:

1. Production; During free motion of the shovel (i.e. during the swing cycle) it will be possible to calculate the load suspended in the dipper, based upon measurement of the changes in the geometry of the digging attachment (i.e. dipper position) and hoist motor torque.

2. Operator Practice; It will be possible to calculate the forces acting on the digging attachment during the dig cycle (in the cartesian coordinate frame of choice), and to

assess the influence of various operating practices (digging trajectories) on machine performance; i.e. under what conditions of trajectory and cut depth, is digging effort minimized while maintaining maximum production (fill factors). In this capacity, the simulator has a direct application in operator training.

3. Diggability; Given adequate insight into the interaction of the shovel with the muckpile, it will be possible to simulate the influence of various muckpile conditions on both operating practice and productivity.

BACKGROUND and MOTIVATION

The mining shovel occupies a critical role in governing the productivity of surface mines. The rate and efficiency with which these pieces of equipment operate in loading haul trucks will largely determine the efficiency of the overall mining operation. Hence, most surface mining operations tend to evaluate their productivity based on the performance of their primary loading equipment. There exists considerable scope for productivity enhancement in surface mine extraction through the optimized use and operation of current excavation equipment.

Between 1987 and 1989 the authors (Hendricks et al., 1990) undertook a series of field experiments which involved the instrumentation and performance monitoring of a P&H 2800XP electric mining shovel, at the Fording River Mine in British Columbia, Canada. In this initial work, shovel performance data from the excavation of three test blasts (each specifically engineered to offer varying digging conditions and comprising in excess of 500000 m^3 of waste rock) was intensively analyzed.

A diggability index (DI) was established through this work as a measure of digging effort, based on analysis of the shovel's hoist motor signals. The DI clearly demonstrated that the quality of the blasting process governed the digging productivity of the shovel. The DI correlated strongly with observed digging conditions during the excavation of the test blasts, as well as with variations in rock fragmentation size distributions, as determined by photographic monitoring and subsequent use of image processing and analysis.

Records of dipper digging trajectories provided a definitive characterization of each shovel operator's loading practice and allowed several critical conclusions to be drawn:

- Variations in operating practice exist even amongst groups of highly skilled shovel operators with more than 15 years of experience.
- Variation in digging trajectory has a pronounced effect on the response of the shovel's work motion motors (especially hoist). Logically, any measure of digging effort based upon the response of these motors will also be compromised by variation in digging trajectory. It is the path followed through the muckpile rather than the actual character of the material being handled (i.e. its looseness or fragmentation size distribution) which exerts the greatest control over shovel performance.
- Having identified that variation in operating practices did exist, the productivity of individual operators was examined further. A productivity factor based on average tons of rock loaded by each operator, per minute of actual digging time during each shift was calculated, and offered some indication of dipper fill factors for each operator. An alarming disparity, of up to 30%, existed amongst the individual operators' characteristic productivity factor.

The issues of shovel operating practice and operator training are as nebulous as that of "diggability", in the sense that no direct quantification has traditionally been applied to either. Although a knowledge of a dipper's trajectory is the most definitive measure of operating practice, they have yet to be analyzed in a any detail. There exist certain subtleties in shovel operation, however, which are apparent only through records of digging trajectory.

While unapparent to shovel training personnel (who rely on simple observation), they obviously have significant influence on machine performance and productivity.

This initial work indicated a clear requirement for an effective model of the shovel digging process to both improve future operator training effectiveness and to serve as a foundation for longer term analysis and development of automated digging strategies.

BASICS OF ELECTRIC MINING SHOVEL DESIGN

Figure 1 illustrates the key components of a generic mining shovel, which will be considered in the following discussions. It is important to appreciate that although two motors; hoist and crowd, combine to facilitate the dig cycle, the crowd motor is not a work producing motor in the sense that it does not contribute significantly to dipper filling. The true effort associated with crowd is restricted to the start of the dig cycle where it is responsible for providing penetration, to a suitable depth, into the muckpile. Hoist motor power is therefore much higher than crowd, i.e. on a large shovel hoist may deliver 1450 hp (1082 kW) versus only 400 hp (298 kW) for crowd.

KINEMATIC ANALYSIS

A kinematic analysis of the mining shovel is first required, in order to formulate a dynamic model of the system; this is now presented. Following sections then describe evolution of the dynamic model and it software simulation.

Kinematics is the science of motion: however, kinematics treats motion without consideration of the forces responsible or its cause (Craig, 1986). As such, kinematic analysis is useful for studying the relationships between position, velocity and acceleration variables in a mechanism. Since a dynamic model of a mechanism requires the use of all three of these relationships in its formulation,

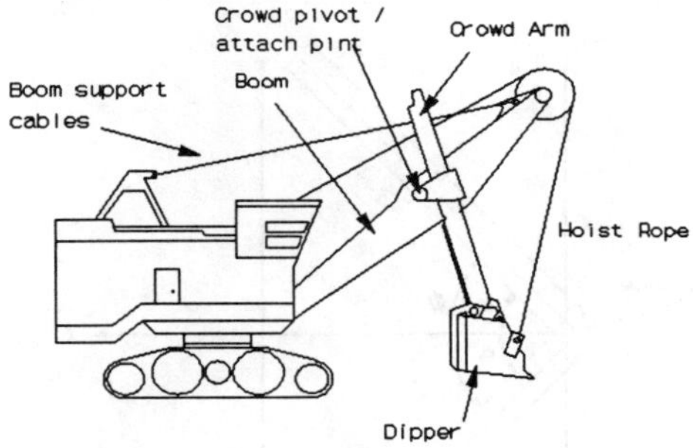

Figure 1 - Electric mining shovel - basic elements.

kinematic analysis is a prerequisite to dynamic analysis.

In the case of an actuated mechanism, such as a shovel, kinematic analysis allows for the development of relationships between motion of the actuators, i.e. front end geometry of the shovel and motion of the dipper.

Thus the kinematic analysis establishes relationships between motion in front end joints and motor angle coordinates, with motion of the end effector in some cartesian coordinate frame. The cartesian coordinate frame chosen for the end effector is termed the "task frame".

The "forward kinematics" of a mechanism, is the equation which yields the end-effector task-frame position and orientation, given the joint or motor angle coordinates. having established the forward kinematics it is possible to derive the relationships between joint and task frame velocities and accelerations.

The following equations establish the forward kinematics of a generic electric mining shovel. The relationships between joint and task frame velocities and accelerations are subsequently determined.

In reviewing the development of the kinematic model, the reader is directed to Figure 2, in which: R is the length of the shovel boom from

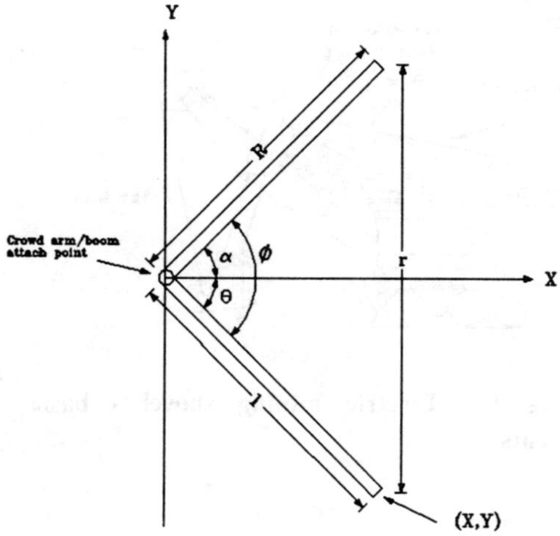

Figure 2 - Elements of kinematic analysis.

the crowd arm attach point to its end (at the point sheaves) - R is a fixed value; r is the length of the hoist rope at any time (t) and l is the amount (length) of crowd arm extension at any time (t).

Using basic geometry:

$$\phi = \alpha + \theta \tag{1}$$

$$\theta = \phi - \alpha \tag{2}$$

$$r = r(\theta_1(t)) = c_1\theta_1(t) \tag{3}$$

$$l = l(\theta_2(t)) = c_2\theta_2(t) \tag{4}$$

Where θ_1 and θ_2 are the hoist and crowd motor angles respectively.

$$x = l\cos(\theta) = l(\theta_2(t))\cos(\theta) \tag{5}$$

$$y = l\sin(\theta) = l(\theta_2(t))\sin(\theta) \tag{6}$$

$$\phi = \arccos\frac{r^2 - R^2 - l^2}{-2Rl} \tag{7}$$

$$\theta = \arccos\frac{R^2 + l^2 - r^2}{2Rl} - \alpha \tag{8}$$

$$x = l(\theta_2(t))\cos(\arccos\frac{R^2 + l^2(\theta_2) - r^2(\theta_1)}{2Rl(\theta_2)} - \alpha) \tag{9}$$

$$y = l(\theta_2(t))\sin(\arccos\frac{R^2 + l^2(\theta_2) - r^2(\theta_1)}{2Rl(\theta_2)} - \alpha) \tag{10}$$

Substituting (3) and (4) into (9) and (10), we get

$$\begin{aligned} x(t) &= c_2\theta_2(t)\cos(\arccos(R^2 + {c_2}^2\theta_2(t)^2 \quad (11) \\ &\quad -{c_1}^2\theta_1(t)^2)(2Rc_2\theta_2(t))^{-1} - \alpha) \end{aligned}$$

$$\begin{aligned} y(t) &= c_2\theta_2(t)\sin(\arccos(R^2 + {c_2}^2\theta_2(t)^2 \quad (12) \\ &\quad -{c_1}^2\theta_1(t)^2)(2Rc_2\theta_2(t))^{-1} - \alpha) \end{aligned}$$

Equations (11) and (12) define the forward kinematics of the shovel, in terms of position. To define the relationships for velocity and acceleration:

$$W = \frac{R^2 + {c_2}^2\theta_2(t)^2 - {c_1}^2\theta_1(t)^2}{2Rc_2\theta_2(t)} \tag{13}$$

$$\dot{W} = \frac{A - B}{C} \tag{14}$$

$$\ddot{W} = \frac{(\dot{A} - \dot{B})C - (A - B)\dot{C}}{C^2} \tag{15}$$

Where,

$$A = 2Rc_2\theta_2(t)(2{c_2}^2\theta_2 t\dot{\theta_2}(t) - 2{c_1}^2\theta_1(t)\dot{\theta_1}(t)) \tag{16}$$

$$B = 2Rc_2\dot{\theta_2}(t)(R^2 + {c_2}^2\theta_2(t)^2 - {c_1}^2\theta_1(t)^2) \tag{17}$$

$$C = 4R^2{c_2}^2\theta_2(t)^2 \tag{18}$$

Hence,

$$\begin{aligned} \dot{x} &= c_2\dot{\theta_2}(t)\cos(\arccos W - \alpha) \\ &- c_2\theta_2(t)\sin(\arccos W - \alpha)\frac{-\dot{W}}{\sqrt{1 - W^2}} \quad (19) \end{aligned}$$

$$\begin{aligned} \dot{y} &= c_2\theta_2(t)\sin(\arccos W - \alpha) \\ &+ c_2\theta_2(t)\cos(\arccos W - \alpha)\frac{-\dot{W}}{\sqrt{1 - W^2}} \quad (20) \end{aligned}$$

$$\begin{aligned} \ddot{x} &= c_2\ddot{\theta_2}(t)\cos(\arccos W - \alpha) \\ &- c_2\dot{\theta_2}(t)\sin(\arccos W - \alpha)\frac{-\dot{W}}{\sqrt{1 - W^2}} \end{aligned}$$

$$
\begin{aligned}
&- \quad c_2\dot{\theta}_2(t)\sin(\arccos W - \alpha)\frac{-\dot{W}}{\sqrt{1-W^2}} \\
&- \quad c_2\theta_2(t)\cos(\arccos W - \alpha)\frac{-\dot{W}}{\sqrt{1-W^2}} \\
&- \quad \{c_2\theta_2(t)\sin(\arccos W - \alpha) \\
&\quad \times \frac{-\ddot{W}\sqrt{1-W^2} - \frac{W\dot{W}^2}{\sqrt{1-W^2}}}{1-W^2}\Bigg\} \qquad (21)
\end{aligned}
$$

$$
\begin{aligned}
\ddot{y} \quad = \quad & c_2\ddot{\theta}_2(t)\sin(\arccos W - \alpha) \\
+ \quad & c_2\dot{\theta}_2(t)\cos(\arccos W - \alpha)\frac{-\dot{W}}{\sqrt{1-W^2}} \\
+ \quad & c_2\dot{\theta}_2(t)\cos(\arccos W - \alpha)\frac{-\dot{W}}{\sqrt{1-W^2}} \\
- \quad & c_2\theta_2(t)\sin(\arccos W - \alpha)\frac{-\dot{W}^2}{1-W^2} \\
+ \quad & \{c_2\theta_2(t)\cos(\arccos W - \alpha) \\
& \times \frac{-\ddot{W}\sqrt{1-W^2} - \frac{W\dot{W}^2}{1-W^2}}{1-W^2}\Bigg\} \qquad (22)
\end{aligned}
$$

The cartesian velocity of the dipper is given by:

$$\vec{V} = \frac{d\vec{r}}{dt} = \frac{dx}{dt}\vec{i} + \frac{dy}{dt}\vec{j} \qquad (23)$$

$$\vec{V} = V_x\vec{i} + V_y\vec{j} = \dot{x}\vec{i} + \dot{y}\vec{j} \qquad (24)$$

$$|\vec{V}| = \sqrt{\dot{x}^2 + \dot{y}^2} \qquad (25)$$

The cartesian acceleration of the dipper is given by:

$$\vec{a} = \frac{d\vec{V}}{dt} = \ddot{x}\vec{i} + \ddot{y}\vec{j} \qquad (26)$$

$$|\vec{a}| = \sqrt{\ddot{x}^2 + \ddot{y}^2} \qquad (27)$$

DYNAMIC ANALYSIS

Using the kinematic relationships established in the preceding section, formulation of the dynamic model can be addressed (Asada, 1986). The dynamic analysis in this paper is based upon the iterative Newton-Euler formulation (Craig 1986), in contrast to the Lagrangian formulation presented in (Hendricks 1993). The iterative formulation lends itself to efficient software implementation, which is of importance for real-time simulation and parameter estimation.

In the present dynamic analysis of our generic mining shovel, the following assumptions have been made:

- All system components, including ropes, are approximated as rigid bodies.
- Backlash and friction torque components are negligible in comparison to load torques.
- The mass distribution in the crowd arm is uniform and known.

The dynamic analysis will make reference to Figure 3, in which:
$\{X_i, Y_i, Z_i\}$ = the coordinate frames of the various links & members, as well as the bucket ($X_5 Y_5 Z_5$) and task ($X_6 Y_6 Z_6$) frames.
L_i = the fixed link lengths and offsets.
d_4 = the variable offset between the crowd arm attach point and the dipper.

The vectors which locate the center of mass for each link are:

$$^1P_{c1} = 0\hat{x}_1, \qquad ^2P_{c2} = \frac{L_2}{2}\hat{x}_2 \qquad (28)$$

$$^3P_{c3} = \frac{L_3}{2}\hat{x}_3, \qquad ^4P_{c4} = \frac{d_4}{2}\hat{z}_4 \qquad (29)$$

Considering that each link only rotates around each fixed axis, we have inertia tensors as follows:

$$^{c1}I_1 = \begin{pmatrix} 0 & 0 & 0 \\ 0 & 0 & 0 \\ 0 & 0 & I_{zz1} \end{pmatrix}, \qquad ^{c2}I_2 = \begin{pmatrix} 0 & 0 & 0 \\ 0 & 0 & 0 \\ 0 & 0 & I_{zz2} \end{pmatrix} \qquad (30)$$

$$^{c3}I_3 = \begin{pmatrix} 0 & 0 & 0 \\ 0 & 0 & 0 \\ 0 & 0 & I_{zz3} \end{pmatrix}, \qquad ^{c4}I_4 = \begin{pmatrix} 0 & 0 & 0 \\ 0 & 0 & 0 \\ 0 & 0 & I_{zz4} \end{pmatrix} \qquad (31)$$

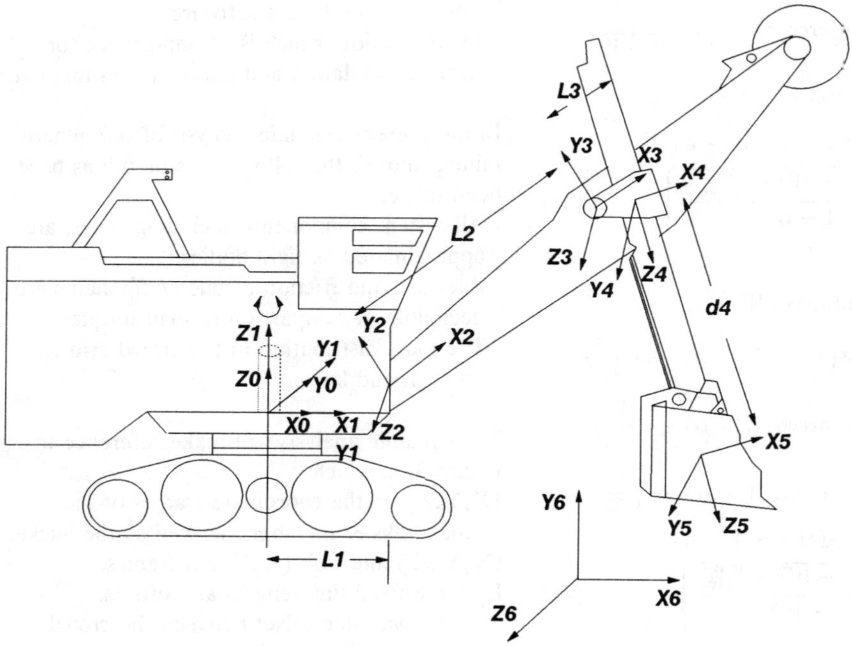

Figure 3 - Elements of shovel dynamic model

The rotations between neighbouring link frames are:

$$ {}^0_1R = \begin{pmatrix} c_1 & -s_1 & 0 \\ s_1 & c_1 & 0 \\ 0 & 0 & 1 \end{pmatrix}, \qquad {}^1_2R = \begin{pmatrix} \frac{\sqrt{2}}{2} & -\frac{\sqrt{2}}{2} & 0 \\ 0 & 0 & -1 \\ \frac{\sqrt{2}}{2} & \frac{\sqrt{2}}{2} & 1 \end{pmatrix} \tag{32} $$

$$ {}^2_3R = \begin{pmatrix} c_3 & -s_3 & 0 \\ s_3 & c_3 & 0 \\ 0 & 0 & 1 \end{pmatrix}, \qquad {}^3_4R = \begin{pmatrix} 1 & 0 & 0 \\ 0 & 0 & -1 \\ 0 & 1 & 0 \end{pmatrix} \tag{33} $$

$$ {}^4_5R = \begin{pmatrix} 1 & 0 & 0 \\ 0 & 1 & 0 \\ 0 & 0 & 1 \end{pmatrix} \tag{34} $$

The gravity effect is calculated as:

$$ {}^0\dot{v}_0 = g\hat{z}_0 \tag{35} $$

The base of the shovel is assumed to be fixed, and hence not rotating, during operation, hence we have:

$$ \omega_0 = 0, \qquad \dot{\omega}_0 = 0 \tag{36} $$

By the analysis at the bucket, we can conclude the effects of the forces and moments are as follows:

$$ f_5 = \begin{pmatrix} f_{51} \\ f_{52} \\ f_{53} \end{pmatrix} = \begin{pmatrix} f_{51} \\ 0 \\ f_{53} \end{pmatrix}, \qquad n_5 = \begin{pmatrix} n_{51} \\ n_{52} \\ n_{53} \end{pmatrix} = \begin{pmatrix} 0 \\ 0 \\ 0 \end{pmatrix} \tag{37} $$

We can now proceed to obtain the iterative Newton-Euler dynamics algorithm for the shovel.

<u>Outward iterations: i: 0 → 3:</u>

$$ {}^{i+1}\omega_{i+1} = {}^{i+1}_iR\,{}^i\omega_i + \dot{\theta}_{i+1}\,{}^{i+1}\hat{Z}_{i+1} \tag{38} $$

$$ \begin{aligned} {}^{i+1}\dot{\omega}_{i+1} &= {}^{i+1}_iR\,{}^i\omega_i \times \dot{\theta}_{i+1}\,{}^{i+1}\hat{Z}_{i+1} \\ &\quad + {}^{i+1}_iR\,{}^i\dot{\omega}_i + \ddot{\theta}_{i+1}\,{}^{i+1}\hat{Z}_{i+1} \end{aligned} \tag{39} $$

$$ \begin{aligned} {}^{i+1}\dot{v}_{i+1} &= {}^{i+1}_iR({}^i\dot{\omega}_i \times {}^iP_{i+1} \\ &\quad + {}^i\omega_i \times ({}^i\omega_i \times {}^iP_{i+1}) + {}^i\dot{v}_i) \end{aligned} \tag{40} $$

$$
\begin{aligned}
{}^{i+1}\dot{v}_{c_{i+1}} &= {}^{i+1}\dot{\omega}_{i+1} \times {}^{i+1}P_{c_{i+1}} \\
&\quad + {}^{i+1}\omega_{i+1} \times ({}^{i+1}\omega_{i+1} \times {}^{i+1}P_{c_{i+1}}) \\
&\quad + {}^{i+1}\dot{v}_{i+1}
\end{aligned} \tag{41}
$$

$$
{}^{i+1}F_{i+1} = m_{i+1}{}^{i+1}\dot{v}_{c_{i+1}} \tag{42}
$$

$$
\begin{aligned}
{}^{i+1}N_{i+1} &= {}^{c_{i+1}}I_{i+1}{}^{i+1}\dot{\omega}_{i+1} \\
&\quad + {}^{i+1}\omega_{i+1} \times {}^{c_{i+1}}I_{i+1} + {}^{i+1}\omega_{i+1}
\end{aligned} \tag{43}
$$

<u>Inward iterations: i: 4 → 1:</u>

$$
{}^{i}f_i = {}^{i+1}_{\;i}R^{i+1}f_{i+1} + {}^{i}F_i \tag{44}
$$

$$
\begin{aligned}
{}^{i}n_i &= {}^{i}N_i + {}^{\;\;i}_{i+1}R^{i+1}n_{i+1} \\
&\quad + {}^{i}P_{c_i} \times {}^{i}F_i + {}^{i}P_{i+1} \times {}^{\;\;i}_{i+1}R^{i+1}f_{i+1}
\end{aligned} \tag{45}
$$

$$
\tau_i = {}^{i}n_i^{T\,i}\hat{Z}_i + b_i\dot{\theta}_i \tag{46}
$$

The complete computation of the dynamics of the shovel based upon the above algorithm is quite lengthy. In order to illustrate the power of the algorithm, while avoiding unnecessary detail and length, only the dynamics for link 3 (crowd ar rotation) will be presented here.

<u>Outward iterations for link 3 (crowd arm rotation):</u>

$$
\begin{aligned}
{}^{3}\omega_3 &= \begin{pmatrix} c_3 & s_3 & 0 \\ -s_3 & c_3 & 0 \\ 0 & 0 & 1 \end{pmatrix} \begin{pmatrix} \frac{\sqrt{2}}{2}\dot{\theta}_1 \\ \frac{\sqrt{2}}{2}\dot{\theta}_1 \\ \dot{\theta}_1 \end{pmatrix} \\
&\quad + \begin{pmatrix} 0 \\ 0 \\ \dot{\theta}_3 \end{pmatrix} \\
&= \begin{pmatrix} (c_3 + s_3)\frac{\sqrt{2}}{2}\dot{\theta}_1 \\ (c_3 - s_3)\frac{\sqrt{2}}{2}\dot{\theta}_1 \\ \dot{\theta}_1 + \dot{\theta}_3 \end{pmatrix}
\end{aligned} \tag{47}
$$

$$
\begin{aligned}
{}^{3}\dot{\omega}_3 &= \begin{pmatrix} c_3 & s_3 & 0 \\ -s_3 & c_3 & 0 \\ 0 & 0 & 1 \end{pmatrix} \begin{pmatrix} \frac{\sqrt{2}}{2}\ddot{\theta}_1 \\ \frac{\sqrt{2}}{2}\ddot{\theta}_1 \\ \ddot{\theta}_1 \end{pmatrix} \\
&\quad + \begin{pmatrix} c_3 & s_3 & 0 \\ -s_3 & c_3 & 0 \\ 0 & 0 & 1 \end{pmatrix} \begin{pmatrix} \frac{\sqrt{2}}{2}\dot{\theta}_1 \\ \frac{\sqrt{2}}{2}\dot{\theta}_1 \\ \dot{\theta}_1 \end{pmatrix} \times \begin{pmatrix} 0 \\ 0 \\ \dot{\theta}_3 \end{pmatrix} \\
&\quad + \begin{pmatrix} 0 \\ 0 \\ \ddot{\theta}_3 \end{pmatrix} \\
&= \begin{pmatrix} \frac{\sqrt{2}}{2}(c_3 + s_3)\ddot{\theta}_1 + \frac{\sqrt{2}}{2}(c_3 - s_3)\dot{\theta}_1\dot{\theta}_3 \\ \frac{\sqrt{2}}{2}(c_3 - s_3)\ddot{\theta}_1 - \frac{\sqrt{2}}{2}(c_3 - s_3)\dot{\theta}_1\dot{\theta}_3 \\ \ddot{\theta}_1 + \ddot{\theta}_3 \end{pmatrix}
\end{aligned} \tag{48}
$$

$$
\begin{aligned}
{}^{3}\dot{v}_3 &= \begin{pmatrix} c_3 & s_3 & 0 \\ -s_3 & c_3 & 0 \\ 0 & 0 & 1 \end{pmatrix} \left(\begin{pmatrix} \frac{\sqrt{2}}{2}\ddot{\theta}_1 \\ \frac{\sqrt{2}}{2}\ddot{\theta}_1 \\ \ddot{\theta}_1 \end{pmatrix} \times \begin{pmatrix} L_2 \\ 0 \\ 0 \end{pmatrix} \right. \\
&\quad + \begin{pmatrix} \frac{\sqrt{2}}{2}\dot{\theta}_1 \\ \frac{\sqrt{2}}{2}\dot{\theta}_1 \\ \dot{\theta}_1 \end{pmatrix} \\
&\quad \times \left(\begin{pmatrix} \frac{\sqrt{2}}{2}\dot{\theta}_1 \\ \frac{\sqrt{2}}{2}\dot{\theta}_1 \\ \dot{\theta}_1 \end{pmatrix} \times \begin{pmatrix} L_2 \\ 0 \\ 0 \end{pmatrix} \right) \\
&\quad \left. + \begin{pmatrix} \frac{\sqrt{2}}{2}(g - L_1\dot{\theta}_1^2) \\ \frac{\sqrt{2}}{2}(g + L_1\dot{\theta}_1^2) \\ g - L_1\ddot{\theta}_1 \end{pmatrix} \right) \\
&= \begin{pmatrix} \dot{v}_{31} \\ \dot{v}_{32} \\ \dot{v}_{33} \end{pmatrix}
\end{aligned} \tag{49}
$$

where

$$
\begin{aligned}
\dot{v}_{31} &= \left(\frac{\sqrt{2}}{2}(g - L_1\dot{\theta}_1^2) - \frac{3}{2}L_2\dot{\theta}_1^2\right)c_3 \\
&\quad + \left(\frac{\sqrt{2}}{2}(L_1\dot{\theta}_1^2 + g)\right. \\
&\quad \left. - \frac{1}{2}L_2\dot{\theta}_1^2 + L_2\ddot{\theta}_1\right)s_3
\end{aligned} \tag{50}
$$

$$
\begin{aligned}
\dot{v}_{32} &= -\left(\frac{\sqrt{2}}{2}(g - L_1\dot{\theta}_1^2) - \frac{3}{2}L_2\dot{\theta}_1\right)s_3 \\
&\quad + \left(\frac{\sqrt{2}}{2}(L_1\dot{\theta}_1^2 + g)\right. \\
&\quad \left. - \frac{1}{2}L_2\dot{\theta}_1^2 + L_2\ddot{\theta}_1\right)c_3
\end{aligned} \tag{51}
$$

$$
\dot{v}_{33} = g - L_1\ddot{\theta}_1 + \frac{\sqrt{2}}{2}L_2\dot{\theta}_1^2 - \frac{\sqrt{2}}{2}L_2\ddot{\theta}_1 \tag{52}
$$

$$
{}^{3}\dot{v}_{c_3} = \begin{pmatrix} \frac{\sqrt{2}}{2}(c_3 + s_3)\ddot{\theta}_1 + \frac{\sqrt{2}}{2}(c_3 - s_3)\dot{\theta}_1\dot{\theta}_3 \\ \frac{\sqrt{2}}{2}(c_3 - s_3)\ddot{\theta}_1 - \frac{\sqrt{2}}{2}(c_3 - s_3)\dot{\theta}_1\dot{\theta}_3 \\ \ddot{\theta}_1 + \ddot{\theta}_3 \end{pmatrix} \tag{53}
$$

$$\times \begin{pmatrix} \frac{L_3}{2} \\ 0 \\ 0 \end{pmatrix} + \begin{pmatrix} (c_3+s_3)\frac{\sqrt{2}}{2}\dot\theta_1 \\ (c_3-s_3)\frac{\sqrt{2}}{2}\dot\theta_1 \\ \dot\theta_1+\dot\theta_3 \end{pmatrix} \times (\begin{pmatrix} (c_3+s_3)\frac{\sqrt{2}}{2}\dot\theta_1 \\ (c_3-s_3)\frac{\sqrt{2}}{2}\dot\theta_1 \\ \dot\theta_1+\dot\theta_3 \end{pmatrix} \times \begin{pmatrix} \frac{L_3}{2} \\ 0 \\ 0 \end{pmatrix}) +^3 \dot v_3 = \begin{pmatrix} v_{31} \\ v_{32} \\ v_{33} \end{pmatrix}$$

Where,

$$v_{31} = -\frac{L_3}{2}(\frac{1}{2}\dot\theta_1^2(c_3-s_3)^2 + (\dot\theta_1+\dot\theta_3)^2) + (\frac{\sqrt{2}}{2}(g-L_1\dot\theta_1^2) - \frac{3}{2}L_2\dot\theta_1^2)c_3 + (\frac{\sqrt{2}}{2}(L_1\dot\theta_1^2+g) - \frac{1}{2}L_2\dot\theta_1^2 + L_2\ddot\theta_1)s_3 \quad (54)$$

$$v_{32} = \frac{L_3}{2}(\ddot\theta_1+\ddot\theta_3+\frac{1}{2}\dot\theta_1^2(c_3^2-s_3^2)) - (\frac{\sqrt{2}}{2}(g-L_1\dot\theta_1^2) - \frac{3}{2}L_2\dot\theta_1)s_3 + (\frac{\sqrt{2}}{2}(L_1\dot\theta_1^2+g) - \frac{1}{2}L_2\dot\theta_1^2 + L_2\ddot\theta_1)c_3 \quad (55)$$

$$v_{33} = -\frac{L_3}{2}(\frac{\sqrt{2}}{2}\ddot\theta_1(c_3-s_3) - \frac{\sqrt{2}}{2}\dot\theta_1\dot\theta_3(c_3-s_3)) + \frac{L_3\sqrt{2}}{4}\dot\theta_1(c_3+s_3)(\dot\theta_1+\dot\theta_3) + g + \frac{\sqrt{2}}{2}L_2\dot\theta_1^2 - (\frac{\sqrt{2}}{2}L_2+L_1)\ddot\theta_1 \quad (56)$$

$$^3F_3 = \begin{pmatrix} F_{31} \\ F_{32} \\ F_{33} \end{pmatrix} \quad (57)$$

Where,

$$F_{31} = m_3 v_{31}, \qquad F_{32} = m_3 v_{32} \quad (58)$$

$$F_{33} = m_3 v_{33} \quad (59)$$

$$^3N_3 = \begin{pmatrix} \frac{\sqrt{2}}{2}I_{zz3}\dot\theta_1(c_3-s_3)(\dot\theta_1+\dot\theta_3) \\ -\frac{\sqrt{2}}{2}I_{zz3}\dot\theta_1(c_3+s_3)(\dot\theta_1+\dot\theta_3) \\ I_{zz3}(\ddot\theta_1+\ddot\theta_3) \end{pmatrix} = \begin{pmatrix} N_{31} \\ N_{32} \\ N_{33} \end{pmatrix} \quad (60)$$

Inward iterations for link 3 (crowd arm rotation):

$$^3f_3 = \begin{pmatrix} 1 & 0 & 0 \\ 0 & 0 & -1 \\ 0 & 1 & 0 \end{pmatrix} \begin{pmatrix} f_{51}+m_4v_{4c1} \\ m_4v_{4c2} \\ f_{53}+m_4v_{4c3} \end{pmatrix} +^3 F_3 = \begin{pmatrix} f_{51}+m_4v_{4c1}+F_{31} \\ -f_{53}-m_4v_{4c3}+F_{32} \\ m_4v_{4c2}+F_{33} \end{pmatrix} = \begin{pmatrix} f_{31} \\ f_{32} \\ f_{33} \end{pmatrix} \quad (61)$$

$$^3n_3 = \begin{pmatrix} N_{31}+n_{41}-\frac{d_4}{2}m_4v_{4c2}-\frac{L_3}{2}F_{32}+L_3f_{53}+L_3m_4v_{4c3} \\ N_{32}-N_{43}+\frac{L_3}{2}F_{31}+L_3f_{51}+L_3m_4v_{4c1} \\ N_{33}+N_{42}+\frac{d_4}{2}m_4v_{4c1}+d_4f_{51} \end{pmatrix} = \begin{pmatrix} n_{31} \\ n_{32} \\ n_{33} \end{pmatrix} \quad (62)$$

Finally, the expression for the joint torque is arrived at:

$$\begin{aligned} \tau_3 &= N_{33}+N_{42}+\frac{d_4}{2}m_4v_{4c1}+d_4f_{51}+b_3\dot\theta_3 \\ &= (I_{zz3}+\frac{d_4}{2}+\frac{d_4}{2}m_4L_2s_3)\ddot\theta_1 \\ &+(I_{zz3}+\frac{d_4}{2})\ddot\theta_3+(\frac{I_{zz4}}{2}(c_3^2-s_3^2)+\frac{d_4^2}{2}m_4s_3c_3 \\ &-\frac{d_4^2}{2}m_4+d_4c_3(\frac{\sqrt{2}}{2}m_4gL_1 \\ &+\frac{\sqrt{2}}{4}m_4L_1^2-\frac{3}{4}m_4L_2) \\ &+d_4s_3(\frac{\sqrt{2}}{4}m_4L_1-\frac{1}{4}m_4L_2))\dot\theta_1^2-d_4^2m_4\dot\theta_1\dot\theta_3 \\ &-\frac{d_4^2}{2}m_4\dot\theta_3^2+b_3\dot\theta_3 \\ &+(\frac{\sqrt{2}}{4}m_4c_3g^2+\frac{\sqrt{2}}{4}m_4s_3g+f_51)d_4 \end{aligned} \quad (63)$$

This expression for torque allows the computation of all the components of the torque which are due to the free motion of the shovel. Hence such an expression can be used to compute what component of measured motor torque exceeds this value, leading to computation of digging forces.

SIMULATOR DESIGN

Based on the nature of the models derived above, the simulator must be capable of providing dynamic simulation of nonlinear, multivariable, time-varying, differential equations. This requirement for nonlinear system simulation rules out the sue of simple analytical or recursive techniques, which are only applicable to linear time-invariant systems. Numerical integration techniques are the only tools which are adequate for this type of simulation.

A number of off-the-shelf software packages with the required simulation capabilities are currently available. SIMULINK^{tm} was selected for the purposes of this project, due to its excellent graphical user-interface, proven numerical software, as well as its integration into the MATLAB^{tm} package. MATLAB^{tm} [Moler, Little, and Bangert, 1987] provides a comprehensive environment for signal processing, system identification, and control system design tools. These other capabilities of MATLAB^{tm} will be used in later phases of this research, e.g. identification of motor dynamics based on input-output responses.

The system block diagram of the simulator, as currently implemented, is shown in Figure 4. The various blocks are currently being developed and validated, e.g. the interaction with the muckpile is still modelled in a rudimentary manner at present.

IDENTIFICATION OF SHOVEL DYNAMICS BASED ON MONITORED FIELD DATA

The formulation of the dynamic model presented above depends on precise knowledge of the various kinematic and dynamic parameters which characterize a particular shovel. For purposes of building a simulator, these parameters can be calculated from first principles, based on manufacturer supplied data such as link lengths and moments of inertia.

However, an analytically derived dynamic simulator fails to address certain aspects of shovel performance which are difficult to quantify in terms of parameters, e.g. friction and losses in the transmission are dependant on the wear on a particular machine, actuator control circuit dynamics are nonlinear, etc.

Hence, an alternative approach to formulating a dynamic model is to utilize empirical data, e.g. from field monitoring, to "identify" the dynamics. Such approaches essentially rely upon using least-squares techniques to fit parametric models to the input-output data sequences (Ljung 1987).

A number of proven system identification techniques exist for identifying the dynamics of manipulators and mechanisms based upon measurements of actuator torques/forces and joint position data. In this paper, the authors apply the real-time nonlinear robot dynamics modelling technique of (Izaguirre 1992) to shovel data, to derive empirical models of actual shovels. Software developed previously by Izaguirre et al. is being utilized for this purpose

The comparison between empirical data, simulations based upon least-squares estimation, and simulation based upon analytical models, will provide significant insight into the validity of the analytical modelling assumptions and the extent of nonlinearity of the shovel dynamics.

CONCLUSIONS & CONTINUING WORK

The complete dynamic model is to be validated by comparison of the simulated performance with actual field test data from an instrumented mining shovel. The assumptions that friction and backlash effects are negligible, may not be justified. However, well proven techniques exist, with which to estimate these effects towards their inclusion within the dynamic model.

The next step in this research program is the

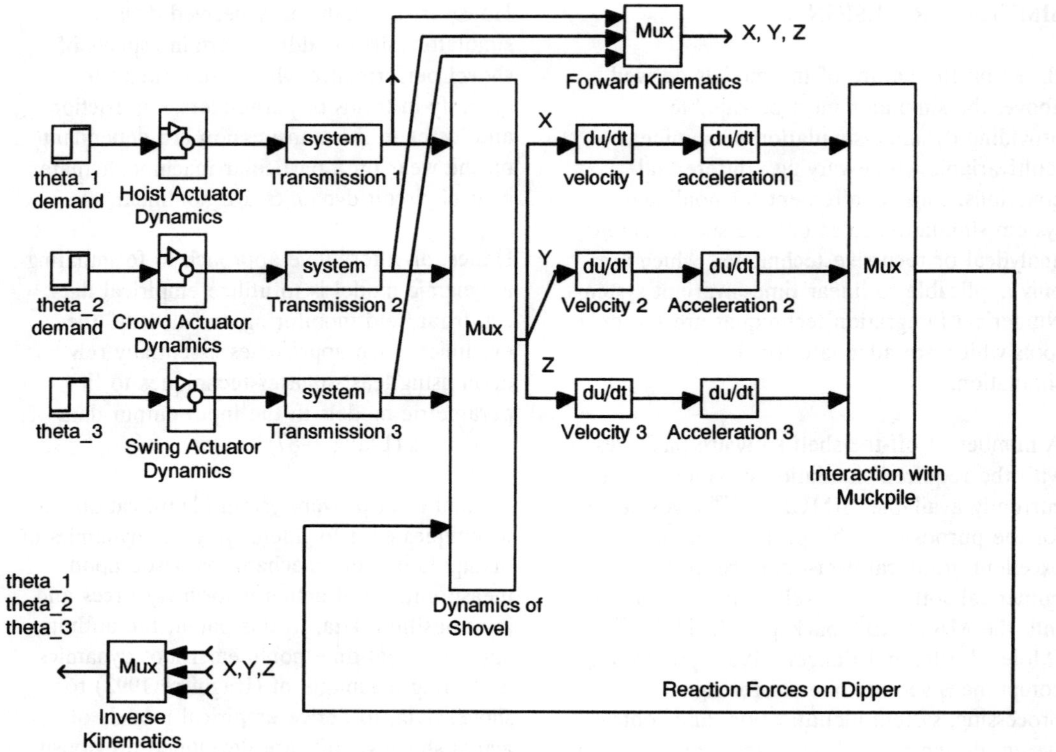

Figure 4 - System block diagram for the shovel simulator

characterizing, modelling and simulation of the interaction between the shovel and the muckpile. This will eventually enable complete simulation of the shovel operation, as well as assessment of muckpile digging conditions based on on-line measurements of hoist motor torque and dipper position.

Recent work relating to the automation of loading equipment has focused on the partial automation of the production cycle; underground LHD (Load Haul Dump) units have been instrumented with computer-based vision systems that permit their automated tramming to and from drawpoints, (e.g. Brophey, 1992); at the other end of the spectrum computer engineering has been applied to the automation of the swing and dump cycles on draglines, (Godfrey et al. 1989). The complete automation of loading equipment, of any form, is dependant upon analysis and understanding of the relationships which exist between digging dynamics, material characteristics and operational practices. These remain the focus of continued work by the authors.

ACKNOWLEDGEMENTS

The work reported on in this paper has been supported by a Strategic Research Grant from the Natural Science and Engineering Research Council of Canada, NSERC. Additional support has been provided under Project C-2 of the IRIS Centre of Excellence grant.

REFERENCES

Asada,H., and Slotine,J.J., 1986. *Robot Analysis And Control*, Wiley-Interscience, 1986.

Brophey, D.G. and Euler, D.W., 1992. The Opti-Track System, a System for Automating Today's LHD's and Trucks. *Proc. 5th Canadian Symposium on Mining Automation*, Vancouver, Sept. 27-29, 1992. pp 291-298.

Craig,J.J., 1986. *Introduction to Robotics: Mechanics and Control*, Publ. Addison-Wesley, 1986.

Godfrey, N and Susanto, A., 1989. Partial Automation of a Dragline Working in Conjunction with a Hopper/Crusher/Conveyor Overburden Removal System. *Proc. Second Large Open Pit Mining Conf.*, Latrobe Valley, Australia, April 3-6, 1989, pp 83-90.

Hendricks C., Peck J. and Scoble M., 1990. Integrated Drill and Shovel Performance Monitoring for Improved Fragmentation and Productivity. *Proc. Third International Symposium on Rock Fragmentation by Blasting*, Brisbane, Australia, August 26-31, 1990, pp 9-20.

Hendricks, C., Daneshmend, L., Wu, S., and Scoble, M., 1993. Design of a simulator for productivity analysis of electric mining shovels, *Proc. Second International Symposium on Mine Mechanization and Automation*, Luleå, Sweden, June 7-10, 1993.

Izaguirre, A.., Hashimoto, M., Paul, R., and Hayward, V., 1992. A New Computational Structure for Real-Time Dynamics, *The International Journal of Robotics Research*, Vol. 11, No. 4, August 1992.

Ljung, L., 1987. *System Identification: Thoery For The User*, Prentice-Hall, 1987

Moler,C., Little,J., and Bangert,S., 1987. *PRO-MATLAB Users Guide for Sun Workstations*, Version 3.2, The MathWorks Inc., Sherborn, MA., 1987.

Innovative Mine Design for the 21st Century, Bawden & Archibald (eds) © 1993 Balkema, Rotterdam, ISBN 90 5410 325 6

A detailed comparison of double boom and single boom hydraulic jumbo drill

R. Morris
TM[3] and Technical Services Division, JCI, Johannesburg, South Africa
J.C. Ball
Technical Support Services, JCI, Johannesburg, South Africa

ABSTRACT: The authors discuss in detail the research carried out over a four year period in an underground mechanised mining operation in order to determine the best drilling combination in a deep gold mining environment in South Africa.

1 INTRODUCTION

Johannesburg Consolidated Investment Company Limited played the dominant role in the introduction of trackless mechanised mining methods (TM^3) into the South African Gold Mining Industry. The two large mines of the Gold Division are Western Areas and Randfontein Estates, milling at the time of introduction some 300 000 tons per month and 700 000 tons per month respectively from conventional sources. At the peak of trackless activity which was achieved in 1986 each mine was producing some sixty per cent (60%) of ore broken from the trackless operations. The build up from initial operations to in excess of 600 000 tons per month combined from trackless sources was extremely rapid especially in view of the new and unfamiliar technology. Conventional tonnage was decreased commensurate with the trackless tonnage build up. Mining operations are currently taking place at depths ranging from approximately 1 000 metres to in excess of 2 500 metres in the metamorphic rocks of the West Witwatersrand field. The introduction of TM^3 in JCI was characterised by the universal installation of double boom hydraulic jumbo drill rigs. The initial fleet of 73 double boom rigs was comprised of Atlas Copco 'Boomer' H127 and Tamrock Minimatic rigs in approximately equal numbers. The domination of double boom rigs in the initial operations derived from a number of factors: they were the most readily available "off the shelf" at the rate required to support the rapid production build up schedule; they offered a greater cross-sectional coverage potential than single booms; and in the event of failure of one or other of the drifters or booms the other unaffected drifter offered some residual drilling capacity to complete the round. The original equipment was supplied against a specification requiring the capacity to drill blast holes of 3,2 - 3,4 metre length and diameter of 45 millimetres and stab holes of 64/76 millimetres.

In the course of operations as the technical staff progressed along the learning curve of the new technology various unsatisfactory technical and economic results motivated the Technical and Gold Divisions to re-examine the relative merits of double and single boom equipment together with the efficiency of the original drifter and feed beam selections.

In consequence several single boom configurations and alternative drifters were introduced to facilitate a comparative evaluation.

2. COMPARATIVE AVAILABILITY DOUBLE BOOM AND SINGLE BOOM HYDRAULIC JUMBO DRILL RIGS

The hoses and hydraulic components of the boom, feed mechanism and drifters constitute those parts of the rig most vulnerable to damage from scaling rocks during the drilling process.

Similarly the booms, ancillary equipment and attachments are the most maintenance intensive parts of the rig. The following analyses are based on the two shift drilling practice common throughout the Gold Division of JCI.

2.1 Attendance Time

This represents the hours during which the rig is manned in the course of a normal month of 24 double shifts. In consequence of the long travelling distance from surface to the underground project sites, often via a secondary shaft system, and the legal constraints on the underground hours worked by any individual, the effective shift length at the work site averages some six and one half hours (6½ hours). Therefore:

6,5 hrs x 24 shifts/month x 2 shifts/day
= 312 hrs attendance time

In the course of this attendance time the rig must be serviced, maintained and repaired in order to achieve the required production call.

This attendance time is identical for both double and single boom configurations.

2.2 Scheduled Maintenance Requirements Double Boom Rig

This requirement derives from the time necessary to carry out the recommended service requirements in accordance with the predetermined frequencies. JCI operates a fixed interval schedule, i.e. independent of production duty:

Shifts per month = 24 days x 2 shifts
= 48 shifts/mth
Monthly schedule - frequency one per month
Service duration 9,25 hrs x (1) = 9,25 hrs
Weekly schedule - frequency three per month
Service duration 4,00 hrs x (3) = 12,00 hrs
Daily schedule - frequency twenty per month
Service duration 1,6 hrs x (20) = 32,00 hrs
Pre-operational check alternate shifts
- frequency twenty four per month
Service duration 0,5 hrs x (24) = 12,00 hrs
Pro-rata provision for 3 monthly, 6 monthly and annual service = 6,25 hrs

Total monthly scheduled maintenance requirement = 71.50 hours

The maintenance schedules are constructed in such a manner that they are cumulative, i.e. the weekly comprises the elements defined by the daily plus additional elements deemed to require a lower frequency than daily. The elements of the weekly are in turn the initial portion of the monthly service procedure.

2.3 Non-Scheduled Maintenance Requirement Double Boom Rig

This represents the average monthly time requirement, in excess of the scheduled service time, required to effect repairs necessitated by breakdown, damage and abuse. Based on JCI history (Cox 1990) non-scheduled time = 48 hours.

2.4 Potential Effective Operational Hours (POH) Double Boom Rig

POH = Attendance time - (Scheduled + non-scheduled maintenance time)

= 312 hrs - (71,50 + 48,00)
= 192,5 hrs per month.

This represents the production hours available during which all travelling, drilling and ancillary activities must be completed.

One hundred per cent (100%) availability
= attendance time - scheduled maintenance time

= 312 hrs - 71,50 hrs

= 240,50 hours

Practical availability

$$= \frac{POH}{240,50} \times \frac{100}{1}$$

$$= \frac{192,50}{240,50} \times \frac{100}{1}$$

$$= 80,04$$

All JCI rig activity is planned around an availability of 80 per cent.

2.5 Scheduled Maintenance Requirements Single Boom Rig

Monthly schedule - frequency one per month
Service duration 6,75 hrs x (1) = 6,75 hrs
Weekly schedule - frequency three per month
Service duration 3,50 hrs x (3) = 10,50 hrs
Daily schedule - frequency twenty per month
Service duration 1,33 hrs x (20) = 26,66 hrs
Pre-operational check alternate shifts
Service duration 0,40 hrs x (24) = 9,60 hrs
Pro-rata provision for 3 monthly, 6 monthly and annual service = 4,50 hrs

Total monthly scheduled maintenance requirement = 58.01 hours

2.6 Non-Scheduled Maintenance Requirements Single Boom Rig

From JCI history (Cox 1990):
Average non-scheduled maintenance requirement = 37.00 hours.

2.7 Potential Effective Operational Hours (POH) Single Boom Rig

POH = Attendance time - (Scheduled + non-scheduled maintenance time)

= 312 hrs - (58,01 + 37,00)

= 216,99 hrs per month.

One hundred per cent (100%) availability
= attendance time - scheduled maintenance time

= 312 hrs - 58,01 hrs

= 253,99 hours

Practical availability

$$= \frac{POH}{253,99} \times \frac{100}{1}$$

$$= \frac{216,99}{253,99} \times \frac{100}{1}$$

$$= 85,00$$

2.8 Relative Availability Double and Single Boom Rigs

The JCI experience has indicated that in consequence of the single boom configuration and the decreased exposure to damage of the vulnerable boom/drifter component the single boom rig exhibits a five per cent (5%) advantage in availability over the double boom rig in the prevailing circumstances. This means that the single boom has an additional potential of some 24,49 hours of productive time, i.e. 12,70 per cent more drilling hours relative to the double boom, and engineering artisan input is reduced by the same potential. This reduction in engineering time is significant in an environment where hydraulic skills are at a premium.

It could be argued that maintenance economics and perhaps maintenance intensity would be reduced by a schedule based on machine duty, i.e. either tonnage duty or alternatively percussion hours. JCI has in recent times given serious consideration to effecting such a change in maintenance policy. Similarly it could be argued that scheduled maintenance should be carried out on the third, non-drilling shift. Several factors mitigate against this practice; the third shift is invariably a short shift due to observance of a general four hour re-entry period; supervision levels and logistical support systems are geared down during the short shift.

Table 1. Cycle time double boom and single boom rig 20 m^2 cross sectional excavation 55 x 3.2 m x 45 mm blast holes 2 x 3.2 m x 76 mm stab holes

		Double Boom Rig (Minutes)	Single Boom Rig (Minutes)
Travel to or between sites		20.00	20.00
Set up and stabilise rig		10.00	10.00
Drill blast holes			
Drifter 'A' penetration (0.95 m/min)		134.40	246.40
Drifter 'B' penetration (1.01 m/min)		128.40	235.40
Drifter 'C' penetration (1.70 m/min)		89.70	164.50
Drill stab holes			
Drifter 'A' penetration (0.60 m/min)		13.00	13.00
Drifter 'B' penetration (0.65 m/min)		12.10	12.10
Drifter 'C' penetration (1.08 m/min)		8.20	8.20
Change drill bits during cycle		5.20	5.20
Stand down machine on completion		10.00	10.00
Total cycle time	Drifter 'A'	192,60	304,60
	Drifter 'B'	185,70	292,70
	Drifter 'C'	143,10	212,70

3. THE IMPACT ON PERFORMANCE OF INSTALLED DRIFTER CAPACITY

It is not the object of this publication to make any comparison of the technical and financial advantages or relative disadvantages of any particular drifter of the spectrum of drifters available commercially from alternative suppliers. Suffice it to say that the JCI experience has demonstrated significant differences between machines of the same class and rating sourced from different suppliers (van Niekerk 1990). Such differences derive from design characteristics and different operating parameters. The discussion in this section is confined to comparison of different classes and power ratings with no reference to the supplier or make.

During the initial operational phase of TM^3 various sites within JCI experimented with alternative blast hole diameters ranging from 42 millimetres to 48 millimetres. Ultimately 45 millimetre diameter blast holes were adopted as the general standard. The cycle times indicated in Table 1 refer to three drifters 'A', 'B' and 'C' which are rated at 7,5 kw, 11 kw and 15 kw respectively and from alternate suppliers. The cycle times are derived from observations over a protracted operational period in alternative project sites at both Western Areas and Randfontein Estates. It is evident from the table that the double boom rig exhibits a reduced cycle time relative to the single boom rig of similar capacity drifter. This advantage is decreased as the drifters mounted become more powerful thus reducing the proportion of the variable (percussion) time element in the total cycle.

It is of particular note that the cycle time of the 15 kw percussive power mounted on the single boom configuration is only some 10,4 per cent longer than the 15 kw percussive power represented by the two 7,5 kw drifters mounted in the double boom mode and is only at a 14,5 per cent disadvantage relative to the double boom mounting 11 kw machines. The cycle time of the two rig configurations mounting the alternative drifter options is represented in Table 2 over a range of cross-sectional dimensions. The variety of cross-sectional dimensions mined varies to accommodate a variety of factors including reef thickness, ground conditions, depth, rock mechanics considerations, etc. The average cross-sectional area is in the region of 24 m^2.

The relative advantage of the double boom rig over the single boom at the various dimensions is plotted in Figure 1 overleaf where the relative advantage at each cross sectional area is defined by:

Relative advantage DB

$$= \frac{\text{Cycle time single boom}}{\text{Cycle time double boom}} \quad \ldots (1)$$

Table 2. Cycle time in minutes double and single boom rigs at various end dimensions

End Dimensions (m^2)	10,5 Mins	16,0 Mins	18,0 Mins	20,0 Mins	21,0 Mins	24,0 Mins	28,0 Mins	30,0 Mins	35,0 Mins	40,0 Mins	45,0 Mins	50,0 Mins
Double Boom												
Drifter A	135	167	181	193	199	218	245	254	286	318	350	402
Drifter B	131	161	174	186	192	209	236	244	275	305	326	387
Drifter C	103	125	134	143	147	159	177	184	205	226	248	290
Single Boom												
Drifter A	195	254	282	305	318	355	409	432	496	560	644	704
Drifter B	187	244	271	293	305	340	393	415	476	537	618	675
Drifter C	143	183	202	213	227	251	288	303	346	390	453	493

Table 3. Blasted tonnage at various end dimensions 3,2 m x 45 mm blast holes

Dimensions (m^2)	10,5	16	18	20	21	24	28	30	35	40	45	50
Tons/Blast	88	134	151	168	176	201	235	252	294	336	377	419
Blast holes (1)	31	44	50	55	58	66	78	83	97	111	125	138
Stab holes (2)	2	2	2	2	2	2	2	2	2	2	2	2
Drill Metres (1)	99,20	140,80	160,00	176,00	185,60	211,20	248,60	265,60	310,40	355,20	400,00	441,60
Drill Metres (2)	6,40	6,40	6,40	6,40	6,40	6,40	4,60	6,40	6,40	6,40	6,40	6,40
Tons per Drill Metre	0,83	0,91	0,91	0,92	0,92	0,92	0,92	0,93	0,93	0,93	0,93	0,94

Table 4. Production profile double boom rig with alternate drifters 45 mm x 3,2 m blast holes

End Dimensions (m^2)	10,5	16,0	18,0	20,0	21,0	24,0	28,0	30,0	35,0	40,0	45,0	50,0
Double Boom Rig												
Drifter 'A'												
Cycles/month	85,6	69,2	63,8	59,8	58,0	53,0	47,1	45,5	40,4	36,3	33,0	28,7
Face Advance (m)	261	211	195	182	177	162	144	139	123	111	101	88
Tons/month	7533	9273	9634	10046	10208	10653	11068	11466	11878	12196	12441	12025
Drifter 'B'												
Cycles/month	88,2	71,7	66,4	62,1	60,2	55,3	48,9	47,3	42,0	37,9	35,4	29,8
Face Advance (m)	269	219	203	189	184	169	147	144	128	116	108	90,9
Tons/month	7762	9608	10026	10432	10595	11115	11492	11919	12348	12734	13345	12528
Drifter 'C'												
Cycles/month	112,1	92,4	86,2	80,8	78,6	72,6	65,3	62,8	56,3	51,1	46,6	39,8
Face Advance (m)	342	282	263	246	240	221	199	192	178	156	142	121
Tons/month	9865	12382	13016	13574	13834	14592	15345	15826	16552	17170	17568	16676

The figure confirms that the 15 kw drifter mounted singly has a smaller relative disadvantage than do the lighter 7,5 kw and 11 kw drifters to their respective double boom counterparts. The general trend of the graphs indicates that the relative advantage of the double boom rigs increases with increasing cross-sectional dimensions of the excavation. The findings confirm the logical conclusion that provided the double boom rig can access the excavation it must have a shorter potential cycle time than a single boom rig mounting one drifter of comparable class.

4. PRODUCTION POTENTIAL OF DOUBLE AND SINGLE BOOM RIGS

In the preceding discussions it has been indicated that in consequence of higher reliability deriving from the reduced vulnerability of sophisticated components and the reduced maintenance requirements the single boom rig has a higher potential effective operating time than does the double boom. The point at which the break-even production potential of the two configurations must therefore occur is the point where this advantage is offset by the decreased cycle time of the double boom rig, i.e.

Table 5. Production profile single boom rig with alternate drifters 3,2 m x 45 mm blast holes

End Dimensions (m²)	10,5	16,0	18,0	20,0	21,0	24,0	28,0	30,0	35,0	40,0	45,0	50,0
Single Boom Rig												
Drifter 'A'												
Cycles/month	66,8	51,3	46,2	42,7	40,9	36,7	31,8	30,1	26,2	23,2	20,2	18,5
Face Advance (m)	204	156	141	130	125	112	97	92	80	71	62	56
Tons/month	5878	6874	6976	7174	7198	7376	7473	7585	7703	7795	7615	7751
Drifter 'B'												
Cycles/month	69,6	53,4	48,0	44,4	42,7	38,3	33,1	31,4	27,4	24,2	21,1	19,3
Face Advance (m)	212	163	146	135	130	117	101	96	84	74	64	59
Tons/month	6124	7156	7248	7459	7515	7698	7778	7912	8056	8131	7954	8087
Drifter 'C'												
Cycles/month	91,0	71,1	64,5	61,1	57,4	51,9	45,2	43,0	37,6	33,4	28,7	26,4
Face Advance (m)	278	217	197	186	175	158	138	131	115	102	88	80
Tons/month	8008	9527	9740	10265	10102	10431	10622	10836	11054	11222	10820	11062

RELATIVE ADVANTAGE DOUBLE BOOM : SINGLE BOOM CYCLE TIMES OVER A RANGE OF EXCAVATION DIMENSIONS

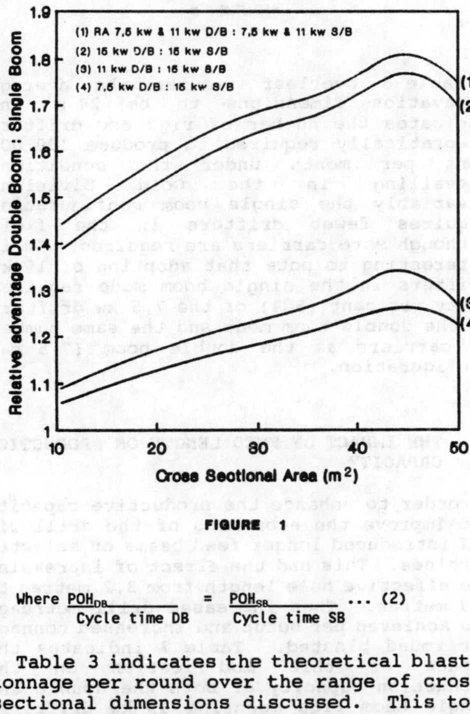

FIGURE 1

$$\text{Where } \frac{POH_{DB}}{\text{Cycle time DB}} = \frac{POH_{SB}}{\text{Cycle time SB}} \quad \ldots (2)$$

Table 3 indicates the theoretical blasted tonnage per round over the range of cross-sectional dimensions discussed. This tonnage assumes no overbreak and a mean socket length of 0,15 metres; in practice overbreak of 5 - 7,5 per cent is experienced and sockets vary from zero to 0,50 metres.

From these data a production profile for each of the double boom and single boom configurations has been developed and is represented in Tables 4 and 5 respectively.

The basis of these production profiles is illustrated below:

$$\text{Cycles/mth} = \frac{\text{Potential effective operating hrs}}{\text{Cycle time per round}} \quad (3)$$

$$\text{Production/month} = \text{Cycles/month} \times \text{tons/blast} \quad (4)$$

The indicated production capacities of the various configurations have been plotted in Figure 2. The plot indicates the superior production capacity of the double boom rig over its single boom counterpart of the same class and furthermore once again demonstrates that this advantage is enhanced with increasing end dimension. In addition the plot illustrates the overwhelming superiority of the 15 kw drifters in both double boom and single boom mode. At the lower end of the dimensions range the single boom rig mounted with the 15 kw drifter outperforms both the double boom fitted with 7,5 kw drifters and that mounting 11 kw drifters. The break-even production point occurs at round 17 square metres excavation dimensions for the 7,5 kw machines and 21 square metres for the double boom mounting 11 kw machines.

The performance of the 11 kw drifters is somewhat puzzling in that it was anticipated that performance from this drifter would be approximately half way between the performances of the 7,5 kw and 15 kw machines. In practice it demonstrates a 4 per cent (4%) advantage over the 7,5 kw drifter as opposed to the anticipated 12 per cent (12%) advantage. The 15 kw machine exhibits a 30 per cent (30%) advantage over the 7,5 kw machine. It is possible that the poor relative performance of the 11 kw machine is derived from its inability to sustain the stresses imposed by the rigours of hard rock drilling. In consequence due to rapid deterioration of physical condition it was

PRODUCTION PROFILE
DOUBLE AND SINGLE BOOM RIGS
3,2m x 45mm BLAST HOLES

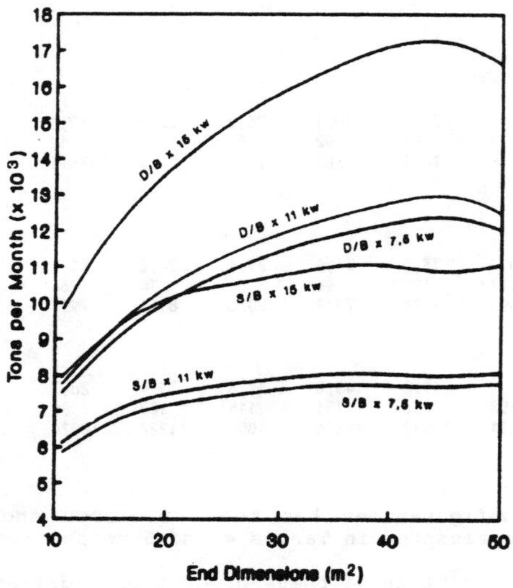

FIGURE 2

COMPARISON OF EQUIVALENT DRIFTER
CAPACITIES MOUNTED IN THE
DOUBLE BOOM OR SINGLE BOOM MODE
(3,2m x 45mm BLAST HOLES)

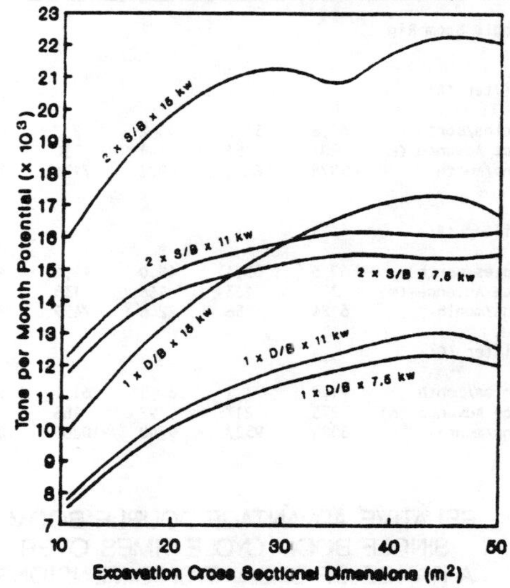

FIGURE 3

Table 6. Double boom and single boom requirements per 100 kiloton at $24m^2$ end dimension

Drifter Rating	7,5 kw (DB)	7,5 kw (SB)	11 kw (DB)	11 kw (SB)	15 kw (DB)	15 kw (SB)
Rigs per 100 kilotons	10	14	9	13	7	10
Drifters	20	14	18	13	14	10
Tons per drifters	5000	7142	5556	7692	7142	10000

unable to deliver its nominal power for sustained periods. This hypothesis is supported, to some extent, by the fact that the overhaul interval for the 7,5 kw machine was some 200 percussion hours, the interval for the 11 kw machine some 400 hours and the 15 kw drifter 1000 - 1100 hours.

5. ALTERNATIVE OPTIONS

The production potential of the double boom rig is demonstrably greater than the single boom rig of the same drifter class over the complete range of dimensions.

Figure 3 indicates the comparison when the same drifter capacity represented by the sum of the paired drifters in the double boom mode is distributed over two single boom configurations. The figure demonstrates that the sum of the production potential of the two single boom units exceeds the production potential of the same installed percussive capacity in the double boom mode.

Table 6 overleaf assumes the average excavation dimensions to be 24 m^2 and indicates the number of rigs and drifters theoretically required to produce 100 000 tons per month under the conditions prevailing in the Gold Division. Invariably the single boom configuration requires fewer drifters in the field although more carriers are required. It is interesting to note that adoption of 15 kw drifters in the single boom mode requires fifty per cent (50%) of the 7,5 kw drifters in the double boom mode and the same number of carriers as the double boom (7,5 kw) configuration.

6. THE IMPACT OF FEED LENGTH ON PRODUCTION CAPACITY

In order to enhance the productive capacity and improve the economics of the drill rig JCI introduced longer feed beams on selected machines. This had the effect of increasing the effective hole length from 3,2 metres to 3,9 metres. Thus increased drill metreage was achieved per setup and increased tonnage per round blasted. Table 7 indicates the impact of this modification on the production capacity of both the double and single boom rigs mounting 15 kw drifters. Similar results are demonstrable for the 7,5 kw and 11 kw configurations.

Averaged across the entire range of end dimensions the increase in cycle time from the 3,2 metre to 3,9 metre hole is some 10,1 per cent in the case of the double boom and some 10,6 per cent for the single boom.

Table 7. Production profile double boom and single boom (15 kw drifters) 3,9 m x 45 mm blast holes

Dimensions (m^2)	10,5	16	18	20	21	24	28	30	35	40	45	50
Tons/Blast	108	165	186	206	217	248	289	309	361	413	464	516
Blast holes (1)	31	44	50	55	58	66	78	83	97	111	125	138
Stab holes	2	2	2	2	2	2	2	2	2	2	2	2
Drill metres (1)	120,9	171,6	195,0	214,5	226,2	257,4	304,2	323,7	378,3	432,9	487,5	538,2
Drill metres (2)	8,2	8,2	8,2	8,2	8,2	8,2	8,2	8,2	8,2	8,2	8,2	8,2
Tons/drill metre	0,84	0,92	0,91	0,92	0,92	0,93	0,93	0,93	0,93	0,94	0,94	0,94
Cycle time S/B 'C'	155	200	221	238	249	277	318	336	385	433	502	547
Cycle time D/B 'C'	110	134	144	151	158	172	193	200	224	249	273	318
Cycles/month S/B	84,0	65,1	58,9	54,7	52,3	47,0	40,9	38,7	33,8	30,1	25,9	23,8
Cycles/month D/B	105	86,2	80,2	76,5	73,1	67,2	59,8	57,8	51,6	46,4	42,3	36,3
Tons/month S/B	9072	10741	10955	11268	11349	11656	11820	11958	12201	12431	12018	12280
Tons/month D/B	11340	14223	14917	15682	15863	16666	17282	17860	18627	19163	19627	18730

Table 8. Engineering maintenance costs double and single boom drill rigs

		Per Percussion Hour (R's)
Engineering Spares Boom and Fittings		35,17
Engineering Spares Carrier and Fitting		44,68
Drifter Spares and Fittings	15 kw	21,28
	11 kw	34,90
	7,5 kw	46,55
Lubricants and Hydraulic Fluid per Boom		3,39
Engine and Transmission		4,54
Dieseline	Double Boom	0,66
	Single Boom	0,34
Engineering Labour	Double Boom	79,28
	Single Boom	70,80

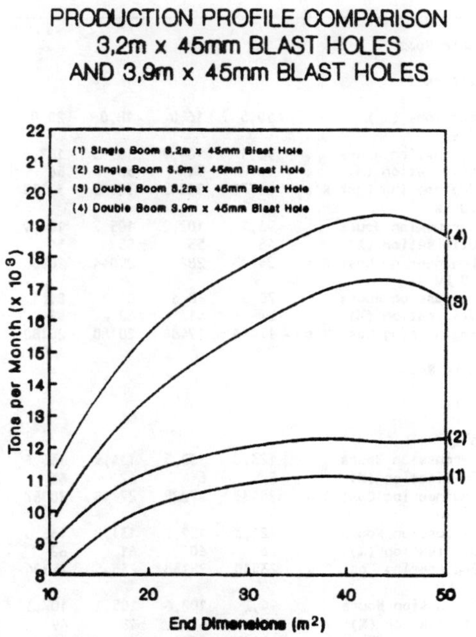

FIGURE 4

The increase in cycle time is compensated by an average 23 per cent increase in the tons pulled per round and the percussion time gained by a reduced number of moves and setups. The net result of the interaction of the variables is an 11,4 per cent increase in productive capacity of the single boom despite a 9,2 per cent reduction in cycles. Similarly productive capacity of the double boom increases by some 13,4 per cent with a 7,0 per cent reduction in cycles drilled.

Figure 4 represents a comparison of production capacity of 3,2 metre and 3,9 metre effective hole length over the range. It is evident that the longer feed beam and consequently effective hole length has the required effect of enhancing productive potential of the rig.

7. DOUBLE BOOM AND SINGLE BOOM DRILL RIG COSTS

This cost comparison of the two configurations is based on actual engineering running costs accumulated during the investigation period and the observed manning during the trials costed at the prevailing wage rates.

The costs compared comprised engineering mining costs, operating labour costs and zero based modular replacement costs for major items. Capital has been depreciated straight line over the life cycle tonnage of the machine. Drill string costs have been left out of account as have tyres both having been regarded as neutral in the debate between double and single boom rigs.

Table 9. Engineering cost double and single boom rigs rands per percussion hours

	7,5 kw Drifters		11 kw Drifters		15 kw Drifters	
	Double Boom	Single Boom	Double Boom	Single Boom	Double Boom	Single Boom
Boom Spares and Fittings	70,34	35,17	70,34	35,17	70,34	35,17
Carrier Spares and Fittings	44,68	44,68	44,68	44,68	44,68	44,68
Drifter Spares and Fittings	93,10	46,55	69,80	34,90	42,56	21,28
Lubricants and Hydraulic	6,78	3,39	6,78	3,39	6,78	3,39
Engine Oil and Transmission	4,54	4,54	4,54	4,54	4,54	4,54
Dieseline	0,66	0,34	0,66	0,34	0,66	0,34
Engineering Labour	79,28	70,80	79,28	70,80	79,28	70,80
Cost per percussion Hr R's	299,38	205,47	276,08	193,82	248,84	182,20

Table 10. Potential percussion hours and monthly engineering costs double and single boom rigs : 3,2 m x 45 mm holes doubleshift operation

Double Boom												
Drifters												
Dimensions (m^2)	10,5	16,0	18,0	20,0	21,0	24,0	28,0	30,0	35,0	40,0	45,0	50,0
7.5 kw												
Percussion Hours	96,1	104,9	107,5	107,4	110,7	113,1	116,3	117,5	120,2	122,3	124,1	119,2
Utilisation (%)	50	54	56	56	58	59	60	61	62	64	64	62
Engineering Cost R's	28770	31405	32183	32153	33141	33860	34758	35177	35985	36614	37153	35686
11.0 kw												
Percussion Hours	93,1	102,2	105,2	104,9	108,1	110,9	113,6	114,9	117,5	120,1	125,2	116,4
Utilisation (%)	48	53	55	54	56	58	59	60	61	62	65	60
Engineering Cost R's	25703	28215	29044	28960	29844	30617	31363	31722	32439	33157	34565	32136
15.0 kw												
Percussion Hours	70,3	78,3	81,1	81,1	83,8	86,6	90,1	90,6	93,6	96,2	98,0	92,4
Utilisation (%)	37	41	42	42	44	45	47	47	49	50	51	48
Engineering Cost R's	17493	19484	20180	20180	20852	21550	22420	22544	23291	23939	24386	22993
Single Boom												
Drifters												
7.5 kw												
Percussion Hours	123,8	132,5	134,9	136,6	137,8	140,1	142,8	143,6	145,6	145,7	144,1	145,4
Utilisation (%)	57	61	62	63	64	65	66	66	67	67	66	67
Engineering Cost R's	25437	27225	27718	28067	28314	28786	29341	29505	29916	29936	29608	29875
11.0 kw												
Percussion Hours	121,3	129,7	131,8	133,6	135,3	137,5	139,8	140,9	143,3	144,4	141,5	142,7
Utilisation (%)	56	60	61	62	62	63	64	65	66.0	67	65	66
Engineering Cost R's	23510	25138	25545	25894	26224	26650	27096	27309	27774	27988	27426	27658
15.0 kw												
Percussion Hours	94,2	102,6	105,2	107,3	108,0	110,7	113,4	114,7	116,8	118,4	114,4	116,0
Utilisation (%)	43	47	48	49	50	51	52	53	54	55	53	53
Engineering Cost R's	17163	18694	19167	19550	19678	20170	20661	20898	21281	21572	20844	21135

7.1 Engineering Costs

The costs associated with the rigs over the review period are detailed in Table 8. The aggregate costs have been reduced to a cost per percussion hour which figure is regarded as the measure of engineering efficiency.

It is argued that engineering department influence over cost per ton and cost per metre is diluted due to the effect of utilisation which is a mining responsibility remote from engineering control. In Table 9 the costs of the individual rig configurations are illustrated and indicate that on an hourly basis the single boom rig is invariably less expensive, in terms of aggregate cost to operate than the double boom of the same class. Similarly in the JCI experience increase in percussive power of the drifter is accompanied by a progressive decrease in operating cost of both configurations (Ball 1991).

Assuming that the various machines are fully utilised at design capacity throughout the production month the cumulative engineering costs are of the order of magnitude indicated in Table 10 overleaf. With regard to both configurations cumulative engineering costs increase progressively with end dimensions reflecting the greater proportion of operating time represented by percussion hours as the larger ends facilitate less travelling and fewer setups.

Table 10 also reflects the utilisation potential where utilisation is indicated as the percentage of potential operating hours represented by percussion hours. Utilisation is reflected as a variable which is dependent *inter-alia* upon penetration rates (percussive power) and drill metres per setup (end dimensions). Regarding 24 square metres as the average end dimensions utilisation of between 47 and 65 per cent could be anticipated in an efficient project depending on the rig configuration. In practice utilisation in the Gold Division ranges between 22 and 44 per cent with individual rigs occasionally recording utilisation as high as 60 per cent. JCI set an average target norm of 40 per cent as representing acceptable levels. In addition to the factors inherent to the drilling system which affect utilisation any external factors which retard face availability result in reduced utilisation, e.g. fall of ground, cleaning and support delays, services failures, etc.

Table 11 illustrates the relationship of the aggregate engineering costs at 24 square metres and regarding the monthly aggregate running cost of the twin boom mounted with 15 kw drifters as a cost index of 100.

The table indicates the impact of inadequately powered drifters on escalation of engineering costs. These costs increase significantly with reduction in percussive power.

7.1.1 Engineering Cost Per Ton

The engineering costs are of value in the determination of engineering efficiency through comparison with predetermined norms. Table 12 below illustrates the comparison of cost per ton produced when the engineering norms are allied with the production potential of the various configurations. The data is represented graphically in Figure 5 which indicates that the single boom rigs mounting 7,5 kw and 11.0 kw

Table 11. Relative cost matrix cumulative engineering costs

	15 kw	11 kw	7,5 kw
Double boom	100	143	157
Single boom	94	124	134

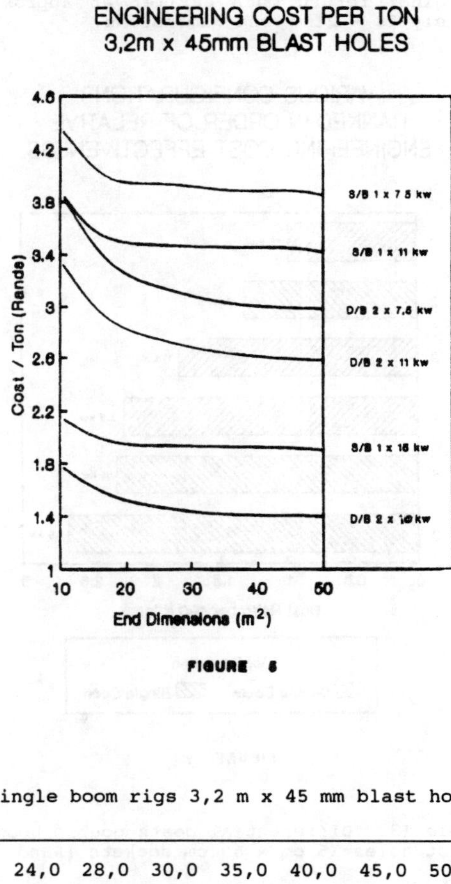

FIGURE 5

Table 12. Engineering cost per ton double and single boom rigs 3,2 m x 45 mm blast holes

Dimensions (m^2)	10,5	16,0	18,0	20,0	21,0	24,0	28,0	30,0	35,0	40,0	45,0	50,0
Double Boom												
Drifters												
7,5 kw R/t	3,82	3,39	3,34	3,20	3,24	3,18	3,14	3,07	3,03	3,00	2,99	2,98
11,0 kw R/ton	3,31	2,94	2,90	2,78	2,82	2,75	2,73	2,66	2,63	2,60	2,59	2,57
15,0 kw R/ton	1,77	1,57	1,55	1,49	1,51	1,48	1,46	1,42	1,41	1,39	1,39	1,38
Single Boom												
Drifters												
7,5 kw R/ton	4,33	3,96	3,97	3,91	3,93	3,91	3,93	3,89	3,88	3,84	3,89	3,89
11,0 kw R/ton	3,84	3,51	3,52	3,47	3,49	3,46	3,48	3,45	3,45	3,44	3,45	3,42
15,0 kw R/ton	2,14	1,96	1,97	1,91	1,95	1,93	1,95	1,93	1,93	1,92	1,93	1,91

drifters are consistently less cost effective than any of the double boom rigs in respect of engineering inputs. The single boom rig mounting a 15 kw drifter is significantly more cost effective than either of these two double booms and is only out-performed by the double boom mounting similar drifters.

It is also noticeable that the double boom rigs manifest a reducing cost per ton between 10,5 m^2 and 35,5 m^2 end dimensions beyond which upper limit the rate per ton exhibits only marginal reduction. The single boom rigs achieve this point of marginal return much earlier at approximately 18 - 21 m^2 end dimensions.

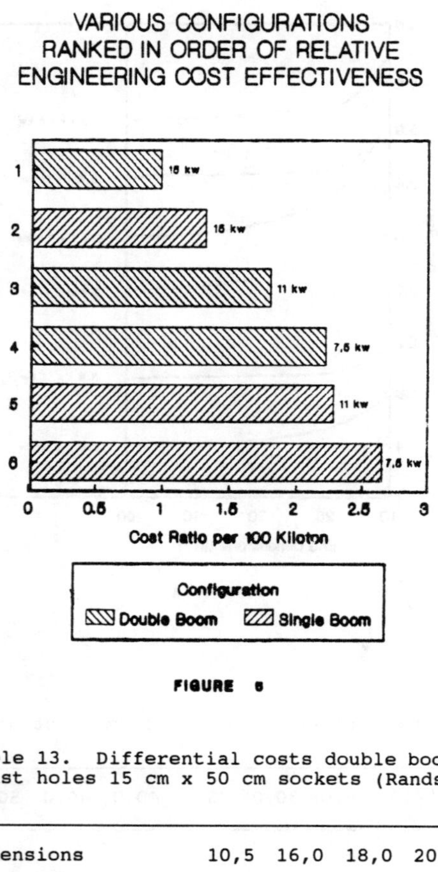

FIGURE 6

7.1.2 Sensitivity of Cost Per Ton to System Inefficiency

Given circumstances where system inefficiencies result in reduced tonnages but demand the same metreage drilled, i.e. percussion hours, the reduction is reflected in an increased engineering unit cost. Table 13 below indicates the impact on cost for the 11 kw drifter configurations where tonnage has been reduced due to an increase in socket length caused by bad breaking.

The unit cost of the single boom is increased by a larger value than that of the double boom to the extent of some R0,09 per ton across the spectrum of end dimensions. Similarly the differential between the double and single boom is expanded to the further disadvantage of the single boom.

Thus in circumstances such as those described the single boom cost is more sensitive to system inefficiency than is the unit cost of the double boom.

A similar increase in disadvantage for the single boom is demonstrable at the other drifter capacities

7.1.3 Relative Fleet Effectiveness

Figure 6 represents the ranking of the various configurations in terms of relative engineering costs per 100 kilotons using the fleet concept outlined in paragraph 7.5 under alternative options.

This indicates that the original fleet composition (applying to 50 per cent of the original fleet) ranks only fourth of the six options discussed. The cost ratio of one (1.0) in the figure represents the most preferred situation and the figure also indicates that one of the single boom options is preferred to the existing double boom (2 x 7,5 kw drifters).

The differential between the optimum configuration and the existing situation is

Table 13. Differential costs double boom : single boom rig 11 kw drifter : 3,2 m x 45 mm blast holes 15 cm x 50 cm sockets (Rands per ton)

Dimensions	10,5	16,0	18,0	20,0	21,0	24,0	28,0	30,0	35,0	40,0	45,0	50,0
Differential Cost/ton (R's)												
Double Boom	0,43	0,37	0,36	0,35	0,36	0,36	0,35	0,35	0,34	0,35	0,33	0,34
Single Boom	0,49	0,48	0,45	0,44	0,45	0,45	0,46	0,45	0,45	0,45	0,44	0,43
Δ D/B : S/B (1)	0,53	0,57	0,62	0,69	0,67	0,71	0,75	0,79	0,82	0,84	0,86	0,85
Δ D/B : S/B (2)	0,59	0,68	0,71	0,78	0,76	0,80	0,86	0,89	0,93	0,94	0,97	0,94

(1) Represents differential between double and single boom cost per ton at 15 cm sockets
(2) Represents differential between double and single boom cost per ton at 50 cm sockets

Table 14. Operating labour costs double boom rigs (Rands per ton)

End Dimensions	10,5	16,0	18,0	20,0	21,0	24,0	28,0	30,0	35,0	40,0	45,0	50,0
Double Boom												
Basic Wage (R's)	5744	5744	5744	5744	5744	5744	5744	5744	5744	5744	5744	5744
Production Tons												
7,5 kw	7533	9273	9634	10046	10208	10653	11068	11466	11878	12196	12441	12025
11,0 kw	7762	9608	10026	10432	10595	11115	11492	11919	12348	12734	13345	12528
15,0 kw	9865	12382	13016	13574	13834	14592	15345	15845	15826	16552	17170	16676
Bonus Rate R's/ton	0,20	0,20	0,20	0,20	0,20	0,20	0,20	0,20	0,20	0,20	0,20	0,20
Cost per ton R's												
7,5 kw	0,86	0,72	0,70	0,67	0,66	0,64	0,62	0,60	0,58	0,57	0,56	0,58
11 kw	0,84	0,70	0,67	0,65	0,64	0,62	0,60	0,58	0,56	0,55	0,53	0,56
15 kw	0,68	0,56	0,54	0,52	0,52	0,49	0,47	0,46	0,45	0,43	0,43	0,44

some R187 7500 per 100 kilotons. Some 323 000 tons per month are sourced from the 7,5 kw rigs throughout the Gold Division which relates to an additional cost of R604 000 per month or R7,248 million per annum. This emphasizes the potential for loss associated with sub-optimal equipment.

7.2 Operating Labour Costs

Operational labour costs comprise a base wage, which is essentially a monthly fixed cost per rig, and a production bonus based on rig output. The predominant manning levels are one operator per boom although there are limited occasions where a single operator has acquired the necessary skills to operate both booms efficiently. In such instances an assistant is provided to assist with rod changes, bit changes, setups and ancillary work. This assistant also constitutes an operator in training. Single boom rigs similarly are allocated an assistant in addition to the operator.

The wage rates applicable are indicated below:

Rig operator R59,83/shift
Rig assistant R24,76/shift

It is proposed to consider only the dominant situation where two operators are allocated per double boom.

Thus the monthly operating labour cost for each configuration is as follows:

Double Boom:
2 operators x 2 shifts x 24 days x R59,83
= R6 743,68 plus production bonus

Single Boom:
1 operator x 2 shifts x 24 days x R59,83
1 assistant x 2 shifts x 24 days x R24,76
= R4 060,32 plus production bonus

Tables 14 and 15 indicate the respective labour costs of operating the double and single booms under the specific conditions described. There is a small advantage in favour of the single boom rigs relative to the equivalent booms at end dimensions less than 28 square metres. Beyond this dimension the unit labour costs are fairly even.

7.3 Modular Overhauls

In addition to engineering running costs modular replacement of major components is carried out consequent to these components achieving pre-determined life cycles. The costs associated with such component replacements are carried as provisions accumulated in terms of cumulative percussion hours (Wynne 1990).

In order to simplify the calculations the costs are here calculated at the average cross sectional dimensions of twenty four square metres (24 m^2).

Component	Component Life Standard	Replacement Cost R's
Engine	10 000 hours	R15 890,00
Transmission	10 000 hours	R 9 868,00
Differential	10 000 hours	R38 308,00
	TOTAL	R64 066,00

This replacement total is based on use of reconditioned components at an average of 60 per cent of new cost.

7.3.1 Major Component Replacement - Drilling Unit

7,5 kw drifter	2000 hrs	R 25 430,00
11 kw drifter	3000 hrs	R 67 800,00
15 kw drifter	4000 hrs	R 84 750,00

Table 15. Operating labour costs single boom rigs (Rands per ton)

End Dimensions	10,5	16,0	18,0	20,0	21,0	24,0	28,0	30,0	35,0	40,0	45,0	50,0
Single Boom												
Basic Wage (R's)	4060	4060	4060	4060	4060	4060	4060	4060	4060	4060	4060	4060
Production Tons												
7,5 kw	5878	6874	6976	7174	7198	7376	7473	7585	7703	7795	7615	7751
11,0 kw	6124	7156	7248	7459	7515	7698	7778	7912	8056	8131	7954	8087
15,0 kw	8008	9527	9740	10265	10102	10431	10622	10836	11054	11222	10820	11062
Bonus Rate R's/ton	0,063	0,063	0,063	0,063	0,063	0,063	0,063	0,063	0,063	0,063	0,063	0,063
Cost per ton R's												
7,5 kw	0,75	0,65	0,64	0,63	0,63	0,61	0,61	0,60	0,59	0,58	0,60	0,59
11 kw	0,73	0,63	0,62	0,61	0,60	0,59	0,58	0,58	0,56	0,56	0,57	0,57
15 kw	0,57	0,49	0,48	0,46	0,46	0,45	0,45	0,44	0,43	0,42	0,44	0,43

Table 16. Modular overhaul costs - Rands per percussion hour

Configuration Percussion Hours pa		Carrier	Drifters	Boom	Pump	Total	Cost/hour
Double Boom 2 x 7,5 kw Drifter 1357	Component Fraction	0,1357	0,6785	0,6785	0,4523		
	Cost R's	8694	78204	170208	14456	271562	200,12
Double Boom 2 x 11 kw Drifter 1331	Component Fraction	0,1331	0,4437	0,6655	0,4437		
	Cost R's	8527	60161	166947	14180	249815	187,69
Double Boom 2 x 15 kw Drifter 1039	Component Fraction	0,1039	0,2598	0,5195	0,3463		
	Cost R's	6656	44028	130322	11069	192075	184,87
Single Boom 1 x 7,5 kw Drifter 1681	Component Fraction	0,1681	0,8405	0,8405	0,5603		
	Cost R's	10770	48438	105424	8954	173586	103,26
Single Boom 1 x 11 kw Drifter 1650	Component Fraction	0,1650	0,5500	0,8250	0,5500		
	Cost R's	10571	37290	103480	8789	160130	97,04
Single Boom 1 x 15 kw Drifter 1328	Component Fraction	0,1328	0,3320	0,6640	0,4427		
	Cost R's	8508	28137	83286	7074	127005	95,64

Boom	2000 hrs	R125 430,00
Hydraulic Pump	3000 hrs	R 15 980,00

Similarly these costs represent reconditioned replacement components at 60 per cent of new cost. In Table 16 overleaf the aggregate costs have been worked back to a cost per hour based on percussion time. The table illustrates the heavy financial bias toward maintenance of the boom and drifters; the carrier is relatively inexpensive to operate.

The modular overhaul costs per ton broken are represented in Table 17 overleaf and again indicate that costs decrease with percussive capacity of the machines and with increasing end dimensions up to a limiting cross-sectional area beyond which unit cost reduction becomes marginal. This limit is in the region of 28 to 35 square metres for the double boom rigs and 18 to 21 square metres for the single boom.

7.4 Capital Costs

Due to the practice of modular overhauls the life cycle of the machines is greatly extended; in fact as has been previously mentioned, this practice essentially constitutes capital replacement on working cost.

For the purposes of this and accommodating

Table 17. Cost of modular overhauls per ton produced

Dimensions (m^2)	10,5	16,0	18,0	20,0	21,0	24,0	28,0	30,0	35,0	40,0	45,0	50,0
Double Boom												
7.5 kw												
Aggregate Cost R's	19232	20993	21513	21493	22153	22634	23274	23514	24054	24475	24834	23854
Cost per Ton R's	2,55	2,26	2,33	2,14	2,17	2,12	2,10	2,05	2,03	2,00	2,00	1,98
11.0 kw												
Aggregate Cost R's	17744	19182	19744	19699	20289	20814	21322	21566	22054	22542	23499	21847
Cost per Ton R's	2,29	2,00	1,97	1,89	1,91	1,87	1,86	1,81	1,79	1,77	1,76	1,74
15.0 kw												
Aggregate Cost R's	12996	14475	14993	14993	15492	16010	16657	16749	17303	17784	18117	17082
Cost per Ton R's	1,32	1,17	1,15	1,10	1,12	1,10	1,09	1,06	1,05	1,04	1,03	1,02
Single Boom												
Drifters												
7.5 kw												
Aggregate Cost R's	12784	13682	13930	14105	14229	14467	14746	14828	15035	15045	14880	15014
Cost per Ton R's	2,17	1,99	2,00	1,97	1,98	1,96	1,97	1,95	1,95	1,93	1,95	1,94
11.0 kw												
Aggregate Cost R's	11771	12586	12790	12965	13130	13343	13566	13673	13906	14013	13731	13848
Cost per Ton R's	1,92	1,76	1,76	1,73	1,71	1,72	1,71	1,73	1,73	1,72	1,73	1,71
15.0 kw												
Aggregate Cost R's	9009	9813	10061	10262	10329	10587	10846	10970	11171	11324	10941	11094
Cost per Ton R's	1,13	1,03	1,03	1,00	1,02	1,01	1,02	1,01	1,01	1,01	1,01	1,00

Table 18. Life cycle tonnage various rigs at 10 000 percussion hours

Rig Configuration	Life Cycle Months	Tons per Months	Life Cycle Tons
Double Boom			
7,5 kw	88,42	10 653	941 938
11 kw	90,17	11 115	1 002 240
15 kw	115,50	14 592	1 685 376
Single Boom			
7,5 kw	71,38	7 376	526 499
11 kw	72,73	7 698	559 876
15 kw	90,33	10 431	942 232

Table 19. Capital cost per life cycle ton various rig configurations

Rig Configuration	Capital Cost Rs x 10^6	Life Cycle Tons	Cost/ton R's
Double Boom			
7,5 kw	1,25	941 938	1,33
11 kw	1,29	1 002 240	1,29
15 kw	1,34	1 685 376	0,80
Single Boom			
7,5 kw	0,75	526 499	1,42
11 kw	0,77	559 876	1,38
15 kw	0,79	942 232	0,84

the overhaul policy a total life cycle of 10 000 percussion hours per rig has been assumed. Table 18 below indicates the potential life cycle tonnage of the respective configurations using a cross-sectional area of 24 square metres.

In Table 18 the life cycle tons have been derived by dividing 10 000 percussion hours by the monthly percussion hours at 24 square metres end dimension.

Eliminating differences in prices from alternative suppliers, i.e. assuming the rigs are standard with the exception of percussive power the relative capital cost per life cycle production unit is illustrated in Table 19.

In both configurations unit capital reduces with increased productive capacity and there is a degree of overlap; the 15 kw single boom machine is more capital efficient than either of the 7,5 kw or 11 kw double boom rigs. Figure 7 represents the impact on unit capital cost of increasing end dimension which again has direct impact on the productive capacity of the various rigs.

7.5 Total Costs

Total costs have been derived by addition of the previously discussed components, i.e. engineering mining costs, operational labour costs, modular overhaul and replacement costs and life cycle capital costs.

These costs have been plotted in Figure 8 which indicates that in the hard rock conditions of the gold mines the rigs fitted with 15 kw drifters are significantly more economic than the rigs with lesser percussive power.

A single boom rig fitted with a 7,5 kw drifter is totally uneconomic while the

RELATIONSHIP BETWEEN UNIT CAPITAL
AND END DIMENSIONS
DOUBLE BOOM RIG (2 x 15 kw DRIFTERS)

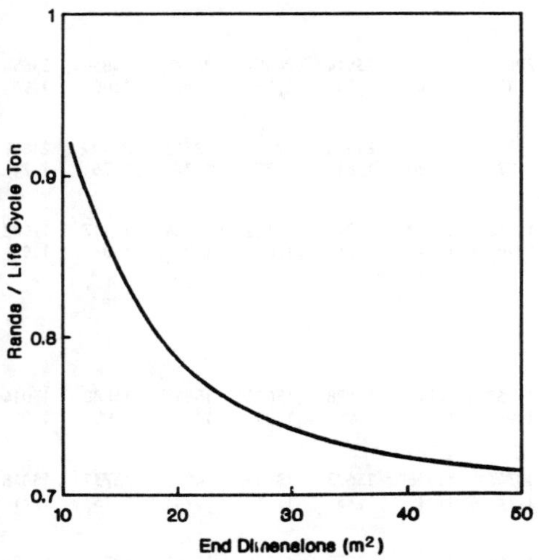

FIGURE 7

TOTAL COSTS
DOUBLE & SINGLE BOOM RIGS
3,2m x 45mm BLAST HOLES

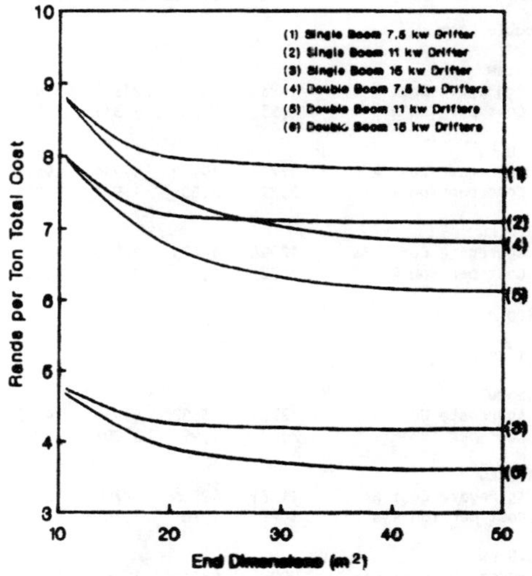

FIGURE 8

single boom with an 11 kw drifter is economically comparable with the double boom mounting two 7,5 kw drifters up to end dimensions of 28 square metres after which the advantage reverts to the double boom. The single boom mounting a 15 kw drifter is only at a disadvantage relative to the double boom mounting similarly powered drifters.

By utilising double booms mounting 7,5 kw drifters JCI has 'lost' some R11,8 million per annum potential saving which could have been effected by utilising single boom rigs mounting 15 kw drifters in preference to the double booms mounting 7,5 kw drifters. The 'loss' relative to the use of a twin boom mounting 15 kw drifters is some R13,3 million per annum. There is no financial justification for using single boom rigs in preference to double booms of the same class over the range of end dimensions examined, i.e. provided the dimensions of the ends permit access to both configurations the double boom, class for class, holds the advantage. There are conditions however in which single boom rigs with higher percussive power offer advantage over double booms mounting lighter drifters.

8. DRILL STRING CONSIDERATIONS

There are a number of advantages associated with the drilling of larger diameter blast holes.

The Gold Division has moved toward the substitution of 45 mm standard diameter bits by 48 mm bits and it is the opinion of the authors that 51 millimetre bits will prove to be the ultimate standard in the prevailing conditions. The 48 mm diameter hole represents the practical upper limit of the 7,5 kw drifter and this larger diameter is more easily accommodated by the greater percussive power of the 11 kw and 15 kw machines. In addition to the benefits of reduced drill metres and explosives economics deriving from the larger diameter hole a heavier more cost effective drill string can be adopted and the larger starting gauge facilitates a more robust bit construction with larger margins for sharpening. Table 20 illustrates the increased drill string performance of the heavier drill rod. With the 32 mm rod the shank end is increased to 38 mm and a male shank substituted for the female shank normal with the 28 mm rod. The data indicates the male shank to be superior to the female. The combined effect of the heavier drill string is a saving of some R0,81 per drill metre or 28,5 per cent. This equates to a potential annual saving of R5,13 million per annum across the Gold Division and once again emphasizes the potential for loss associated with the selection of sub-optimal equipment.

9. CONCLUSION

Double boom drill rigs are potentially more productive and more cost effective to operate than the equivalent single boom drill rigs. Thus provided the end dimensions are such that both machines can access the face the normal choice would tend to favour the double boom configuration. The twenty square metre

Table 20. Drill string performance and costs

Unit/ Configuration	Drill Metres	Component	Dimensions	Consumption	Unit Cost R's	Aggregate Cost R's	Cost/Drill Metre R's	Metres per Component
Single Boom		Drill Bit	Ø 48 mm	445	152	67640	1,36	111,45
1 x 15 kw	49595	Rod	H32 R38 R32	54	475	25650	0,52	918,43
		Coupling	R38	22	78	1716	0,03	2254,31
		Shank	Male R38	15	384	5760	0,12	3306,33
TOTALS	49595			-	-	100766	2,03	
Double Boom		Drill Bit	Ø 45 mm	665	145	96425	2,13	67,95
2 x 11 kw	45188	Rod	H28 R32 R28	85	353	30005	0,66	531,62
		Coupling	R32	32	52	1664	0,04	1412,13
		Shank	Male R32	20	384	7680	0,17	2259,40
TOTALS	45188			-	-	135774	3,00	
Double Boom		Drill Bit	Ø 45 mm	2534	145	367430	2,07	69,98
2 x 7,5 kw	177337	Rod	H28 R32 R28	292	353	103076	0,58	607,32
		Shank	Female	87	384	33408	0,19	2038,36
TOTALS	177377			-	-	503914	2,84	

H32 = 32 mm Hexagonal drill rod R38 = 38 mm Rope thread R32 = 32 mm Rope thread
H28 = 28 mm Hexagonal drill rod R32 = 32 mm Rope thread R28 = 28 mm Rope thread

breakpoint quoted to JCI as a criteria for selection of single or double boom equipment is a purely arbitrary figure having no economic or production justification.

The advantage of double boom rigs is enhanced by increasing end dimensions within the range of dimensions investigated. There are conditions where single boom rigs may out-perform double boom rigs mounting drifters of lesser percussive power; where the percussive power of the double boom is mounted on two separate carriers the productive capacity is enhanced; however there is an economic penalty for the increased production. When making the selection of double or single boom equipment it is therefore totally inadequate to base such selection on some or other arbitrary end dimension. There are a number of interrelated factors which must be considered including end dimensions, reef thickness, required production rate, geological and rock mechanics factors, end availability, layout and economics.

The sensitivity of economic production to the selection of the correct equipment is emphasized by the JCI experience where fifty per cent of the original drilling equipment suite was equipped with inadequate percussion power for the hard rock conditions encountered. The combined effect of underpowered drifters and sub-optimal drill string have cost JCI an estimated R2,5 per ton which equates to an increased cost per kilogram of gold produced of some R600.

In addition to the sub-optimal drifter and drill string JCI initially neglected to optimize feed length which offers a powerful means to optimize drilling costs.

In the hard rocks of the deep-level gold mines of the Witwatersrand the indications are that the minimum drifter required is one rated at a percussive power of 15 kw especially if the operator wishes to exploit the potential benefits of larger diameter blast holes.

Similarly the minimum feed length, ground and rock mechanics considerations permitting, should be some 4,0 metres. The introduction of single boom rigs mounting 15 kw drifters and 4,2 metre feed length by JCI as a substitute for the double boom 7,5 kw drifter machines is a positive step to redress the deficiencies represented by the latter. Double boom rigs mounting 15 kw drifters could be an even more cost effective measure. The 11 kw drifters offer only a marginal benefit relative to the 7,5 kw drills.

References:

BALL, J C 1991. A Comparison of Conventional and Hydraulic Drilling in Deep Level Hard Rock Mining. M.Phil Dissertation, University of Nottingham, England.

COX, L J 1990. Operating Cost Control Parameters TM³ Management. JCI Selected Technical Publications, Vol 7. pp 800-893. Edited Morris R, Fauconnier CJ and du Plessis NP.

VAN NIEKERK, C 1990. Drifter Analysis, Randfontein Estates Gold Mining Company (Witwatersrand) Ltd Internal Report July 1990.

WYNNE, V E 1990. Calculation of Zero Based Total Costs per Operating Hour - Drill Rigs. JCI internal Report, April 1990.

Innovative Mine Design for the 21st Century, Bawden & Archibald (eds) © 1993 Balkema, Rotterdam, ISBN 90 5410 325 6

Remote operation of LHD and pick hammers in rock bursts conditions at El Teniente Mine

Alejandro F. Moyano & Julio C. Vienne
Codelco-Chile, El Teniente Division, Rancagua, Chile

ABSTRACT: The successive rock bursts occured during the starting up of the Sub 6 operational area of El Teniente Mine has given commencement to a series of studies associated to primary ore exploitation. Among these, stand out the implementation of the mining process telecommand system, which starts by the remote control operation of the LHD and pick hammers equipment watching it through a video. This paper shows this project and the progress obtained to date.

1. INTRODUCTION

The El Teniente Mine, one of the four Codelco-Chile's orebodies, is located in Chile's central zone, on the Andes mountain range and approximately 130 Kms. to the Southeast of Santiago city.

El Teniente mine began exploitation in 1906, starting by the highest levels and closer to the surface, constituted basically by secondary ore, soft ore of good grade and easy extraction, progressing the exploitation toward the lower levels as the first ones are exhausted. This situation has obligated to increase the primary ore extraction, which has meant a lower grade, increase of the ore hardness, changes in the mineralogical composition, and the increase of geomechanics phenomenon.

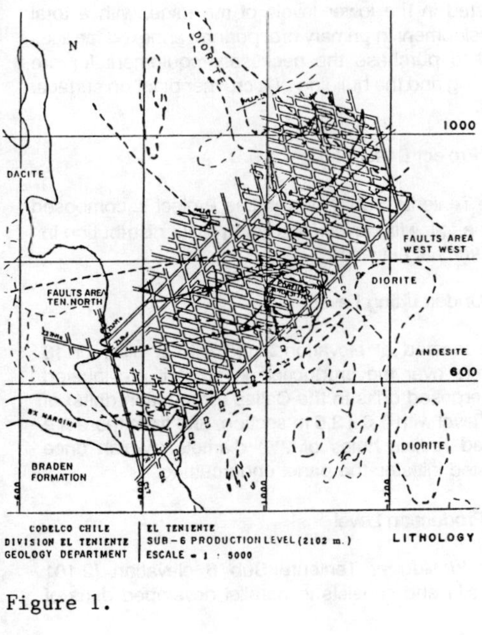

Figure 1.

In order to sustain the El Teniente production capacity appears the North Mine Project, Teniente Sub 6, which due to its big ore reserves and large extension is the largest level put in production at El Teniente Mine in all its history.

However, successive rockbursts have obligated to stop the normal production in Teniente Sub 6 and leave it as an experimental area, in which new support techniques, extraction, LHD equipment, pick hammers, and auxiliary systems will be tested by remote control.

This paper gives a description of the Teniente Sub 6 Project and the analysis of its initiation, emphasizing the factors that obligated its redefinition as experimental area. Following it is detailed the mining equipment telecommand project and its progress to date.

2. GEOLOGY AND TENIENTE SUB 6 RESERVES

Figure N°1 shows the geological plan of the production level of the Sub 6 North Area. It can be appreciated that it is a lithological heterogeneous sector with Andesite and Diorite predominium and a central nucleus constituted by Braden Formation.

The Andesite as Diorite present proper characteristics of the El Teniente primary ore, with high cohesion, waterproof, of big hardness and mechanic resistance.

Besides, the conditions of high rigidity that the primary solid massive rock shows, it prevents the

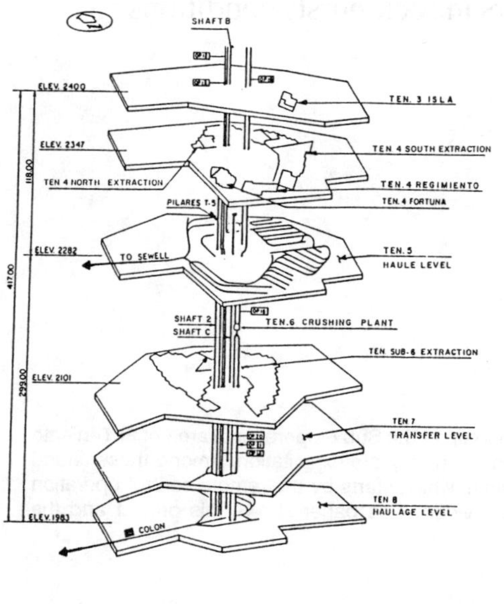

MINE LEVELS SCHEMATIC DRAWING

Figure 2.

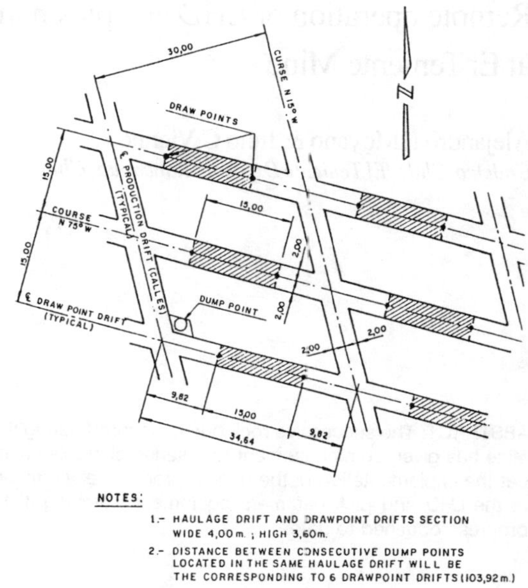

PRODUCCION LEVEL - TYPICAL MODULE

Figure 3.

relaxation of stresses, originating high energy concentrations, which may be gradually liberated through rock falling in the drifts, or violently when rock bursts.

The reserves located over the present haulage level, Teniente 8, are superior to the 2.000 millions of tons, grade over 0,95% The Teniente Sub 6 level is the major present ore reserve, estimating that reaches to 1.200.000 millions of tons and 40 years will be required to be exhausted.

3. PRODUCTION LEVELS

At present El Teniente mine has 6 production areas, which are being exploited by Block Caving or Panel Caving. They are: Teniente 5 Pilares (1994), scrapers extraction; Teniente 3 Isla (2003), manual extraction; Teniente 4 Fortuna (1995); Teniente 4 Regimiento (2003); Teniente 4 South (2009); and Teniente Sub 6, mechanized extraction with LHD equipment. The date between parenthesis indicates the exhausting year of the area. Figure 2 shows the relative location of each one of the previous production sectors.

There are two more projects under preparation, which useful life is: Teniente **5 C**entral from 1994 to 1999, and Quebrada Teniente from 1994 to 2004, with LHD and manual extraction, respectively.

As it can be appreciated, the contribution to production of the areas where LHD equipment is used has been increasing since its implementation in 1982, and in the medium term will be the 100% of El Teniente Mine production.

4. TENIENTE SUB 6 PROJECT

With the purpose to deliver since 1989 the necessary tonnage to satisfy the El Teniente Mine production goals, it was necessary to prepare a new sector located in the lower levels of the mine, with a total development in primary ore, perform annexed facilities and to purchase the necessary equipment for ore hauling and the building of a crusher plant on surface.

4.1 Project Description

The Teniente Sub 6 North Mine Project is composed of the following levels according to its distribution in the vertical elevation.

- **Undercutting Level**

It is located at elevation 2.119 with a pillar of 18 meters over the production level. In it are located superposed drifts to the Calles (production drifts) of this level with 3,6 x 3,6 m section, and from there are drilled radials holes of 2½" diameter, which once blasted initiates the panel undercutting.

- **Production Level**

It is located at Teniente Sub 6 elevation (2.101 m.a.s.l.) and consists in parallel developed drifts of

4.0 x 3.6 m section (LHD calles) and 30 m separated, which are intersected with 60° angles and every 17.32 m by drifts of 4.0 x 3.6 m section (drawbells), which constitute the ore fragmented drawpoint. The distances between Calles and gathering Zanjas (drawbells) defines a mesh with an influence of 259,5 m^2 for every drawpoint (See figure N°3).

- **Ventilation Sub Level**

It is located 20 m below production level, consists in 4,5 x 4,5 m drifts connected to surface by Adit 62, 6,3 x 6,3 m section tunnel. From this drifts fresh air is injected through 2,0 m diameters chimneys to the production level.

The forced air injection is made through the V-62 fan of 4,0 m diameter, with a 1200 HP motor and a volume of 800.000 CFM capacity, located in Adit 62.

- **Ore Transfer Level**

It is located at 2.066 m elevation and 35 m below production level. The ore arrives to this level through an ore pass system in Y form of 3,0 m diameter and 35 m long of each arm, coming from the adjacents Calles in the production level, which converge to a pick chamber which has a stationary electro-hydraulic pickhammer, through which the ore size is reduced to 30" to be passed to a main ore pass of 3,0 m diameter and 83 m long toward the haulage level (See figure N°4).

- **Haulage Level**

It is located at main haulage level of the Mine, Teniente 8, (1.983 m.a.s.l.) and consists of 5,0 x 6,0 m parallel drifts, separated 120 m each other.

At this level the ore is loaded from the ore passes through a pneumatic side loading chute to the 80 tons ore cars, which haul the 30" size ore to the primary crusher located at surface, at a distance of approximately 10 Kms.

4.2 Initial Supporting

The Ventilation, Transfer and Haulage levels were at the beginning supported with 2.80 m, 22 mm diameter, L-type bolt, grouted to complete column, with mesh and 10 cms of shotcrete in all drift section.

The production level, besides the 2.4 m, 22 mm diameter, L-type bolt, with mesh and 10 cms. of shotcrete, incorporated 1" diameter steel cables of 6-10 meters long, distributed in the roof. Furthermore, in the sides of the Calles it was considered confining walls with steel and concrete until 2.5 m high.

The drawpoint, apart of bolts and cables, has 5 steel arches, covering a zone of 3 meters with steel and concrete. On the floor are installed 3 rails to avoid the erosion of the concrete.

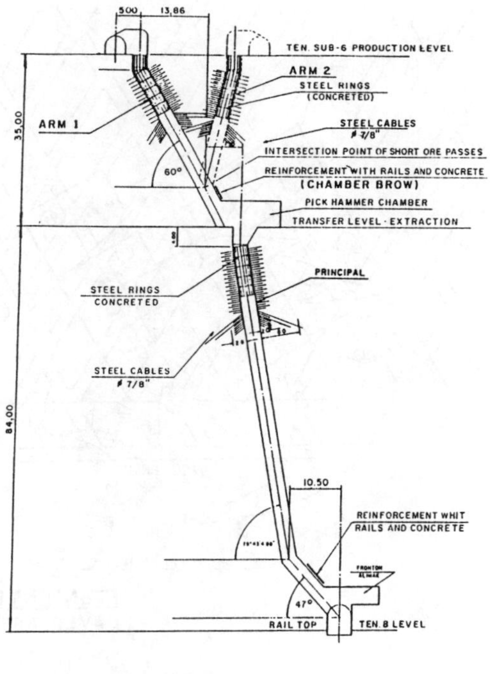

TEN. SUB-6 ORE PASS SUPPORTING SYSTEM

Figure 4.

4.3 Main Facilities

The sectors has a series of installations such as: main electric substations, compressors room, underground service station, daily maintenance shops, radiocommunication system, warehouses, powder magazines, offices, first aid post, loading and unloading equipment stations, laboratories, and a casino able to attend the feeding of 220 persons simultaneously. All these are located at Teniente Sub 6 elevation and in the Braden Formation sector.

The daily maintenance of LHD, jumbos and other automotive equipment is done in the underground maintenance shops of Teniente Sub 6. The rest of the maintenance as the weekly, semi-monthly or major repairs will be made in the shops of La Junta sector, for which the equipments are transported to the surface using ramp trucks.

5. STARTING-UP

The Teniente Sub 6 started its ore extraction on July 18, 1989, alternating gathering zanjas blasting system with undercut blasting method during 1989 and part of January 1990, until reach a total of 10.622 m^2 of undercut area and 29 excavated zanjas, for a total of 58 drawpoints.

The initial production of 1.000 tons/day was gradually

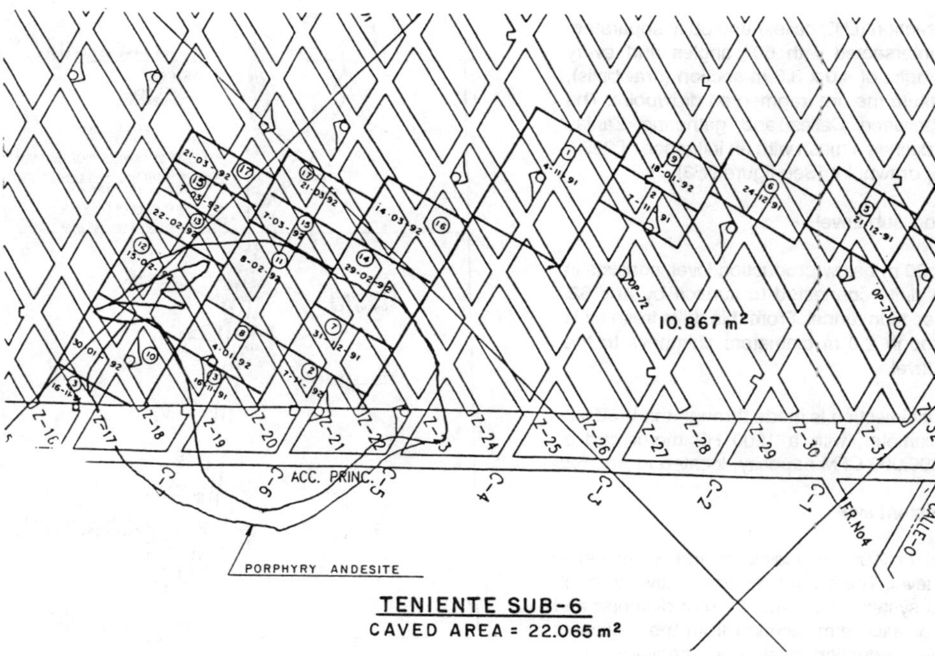

Figure 5.

increased, together with the undercut area, until reach 3.500 tons/day in January 1990.

However, on January 18, 1990, a rock burst obligated to stop the production activity of the sector, compromising the haulage drift stability. This drift was recuperated on April of the same year, re-starting production until reach 3.200 tons/day at the end of June.

On July 2, 1990, a new rock burst made Management decide to stop again the operations activities of the area in order to make a systematic resupporting of all levels, to allow a minimum distance of 100 meters between the activities of preparation and production, and basically to allow a safer place for the personnel that work in development and construction area, establishing a safety zone around the caved area in all levels.

The definitive supporting of all levels, in the sector 100 meters ahead of the undercutting front, was finished at the end of March 1991, therefore Teniente Sub 6 restarted its operation activities beginning April 1, 1991, reaching a production of 5.500 tons/day on May 22, 1991. Furthermore taking in account the normal behavior of the sector in April, it was decided to restart the undercutting process beginning May, incorporating 2 gathering Zanjas on May 10 and 245 m^2 in undercutting on May 20.

However, on May 23, 1991, again this area was shaken by a rockburst causing damage to the resupporting area. This situation made Management decide to stop the operations activities, taking out Teniente Sub 6 level of the production programs, and leaving it as an experimental area to study the designing and operational methods, extending it to future sectors emplaced under primary ore.

Consequently, the Teniente Sub 6 extraction activities were paralyzed in order to define a future work planning strategy and repair the damage structures, specially floor concrete.

Beginning November 1, 1991, it was implemented an undercutting plan to reach a total of 25.000 m^2 in a period of 7 months, 10.000 m^2 located at the West side of the area, according to a work methodology to difference the Undercutting and Extraction activities of the break ore to give the free face to the following blasting, evacuating and isolating the level after each blasting or when the geomechanics instrument detects an increase in the seismic activity of the sector.

The caved area in the West sector will allow to reach the minimum hydraulic radius to cause the undercutting and ore breakage, and in this way to create a crater to the surface.

In the period November 1991 to March 1992 a total of 17 blastings were blasted, incorporating an area of 11.198 m^2 to reach an accumulated area of 22.065 m^2 (See figure N°5).

On March 25, 1992 another rockburst took place which caused severe damage to the main access of the production level, and obligated to stop the operations in order to repair, resupport the sector and define operations methods to minimize or eliminate the presence of personnel in the different levels of the Teniente Sub 6 Project, when it is operating.

The repair and resupporting stage considered the massive use of 10 tons post-stressed steel cables in the roof in all intersections of Calle-Zanja, in a length of 10 meters, steel arches in the sectors with more damage and mesh and lacing support in the production and transfer accesses levels.

The use of post-stressed steel cables in walls and roofs is with the purpose to get a better support, due that it yields but do not cut off when a rockburst take place, and consequently a safer place to work is attained.

The Teniente Sub 6 starting on stage, comprised between July 1989 and March 1992 allowed to detect some problems, being the following the more importants:

- Overexcavation of the horizontal drifts since its development.
- Overexcavation of the ore passes up to compromise the upper levels stability.
- Intense seismic activity and presence of rockbursts which puts in danger the personnel life, equipment and installations.
- Need of a stronger supporting as well as in horizontal developments as in ore passes, which difficults and defers its execution and rise the costs.

These problems are being caused for the following reasons:

- In Teniente Sub 6 there is a big in-situ stress, which is the main reason for the presence of rockbursts.
- From the geological point of view, the rock bursts occurred at Teniente Sub 6 have evidenced that the geological structures, mainly faults, has played an important roll in the ocurrence of this geomechanics phenomenons.

  This because in Teniente Sub 6 there is a structural behavior of the rock mass differente to the upper levels, which was unknown at the beginning of the project and for this reason it was not considered in the design. Today it is consensus that the actual structural geological arrangement is very important, and must be considered in any work to be performed in the sector.

  This structural geological arrangement allows wedge formation, in the horizontal as well as in vertical developments, which lead to the blocks formation that when falling makes drifts and ore passes overexcavation.

- The area geology shows the presence of several intrusives dioritic bodies and Andesitic porphyry immerse in the Andesite, which characterize because its major rigidity, existing also two major structural features: Faults zone West-West and Teniente North faults that intersect the sector in undercut process.

  The geological characteristics determine a high lithological heterogeneity and a marked structural anisotropy, which added to the actual high stresses in situ, is converted in a complex area to be exploited and exposed to rockburst.

6. TENIENTE SUB 6 EXPERIMENTATION PLAN

Regarding the antecedents obtained in the start-on stage in the Teniente Sub 6 area, due to its experimental character, studies were begun to evaluate new support systems, investigate the characteristics of the primary ore subsidence, prosecute the behavior evaluation of the gravitational ore passes, observe and quantify the stresses redistribution concerning the production area, which would provide information to evaluate the possibility to re-start development and construction activities, as well as the definitive reincorporation of Teniente Sub 6 level to the mine production programs.

The proposed mining plan consists to extract ore from the present undercut sector with the purpose to obtain block complete subsidence. This extraction will be performed in a sector of 13.200 m^2, which column altitude reaches 240 m, located West side of the area present caved Calle 5 to Calle 8. In the East sector, Calle 1 to Calle 4, due that it has a column superior to the 800 meters high, is not possible to obtain at short term the global subsidence in a virgin ground (See figure N°6).

The objective to attain the complete subsidence, is based in the experience of others mine sectors, in which once attained clearly improved the geomechanics conditions for the area expansion through "Panel Caving".

However, due to previous antecedents, is necessary to assume that during all the undercutting propagation process in height virgin ground, it will be maintained the stresses field, acting in a structural geology environment and
geotechnical complex propense to rockburst occurrence.

The above mentioned compels to consider extraction alternatives that allow safe work for the area personnel, specially in the production and pick levels,

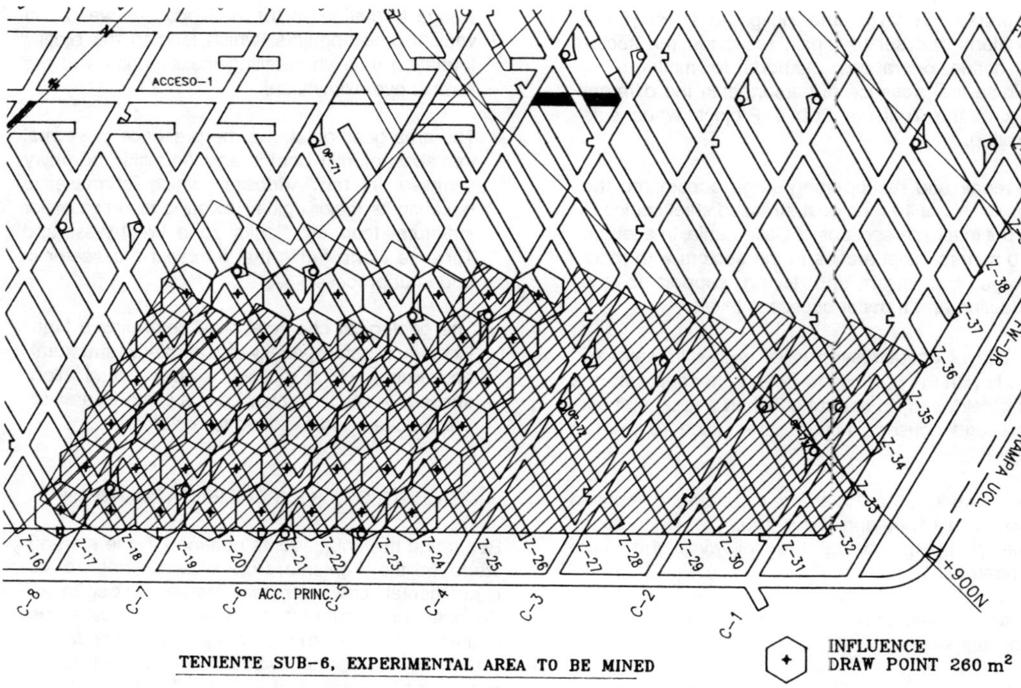

Figure 6.

affected by rockbursts every time the area is being extracted.

Considering this appear the telecommand project of mining process at El Teniente, with the purpose to remote the equipment and installation operation, involved in the production activities which will be performed in Teniente Sub 6 and Teniente 7 and, in a second stage, the loading activities to ore cars at Teniente 8 level.

7. MINING PROCESS TELECOMMAND PROJECT

On performing an analysis of the different activities involved in the production process, we find the following according to the corresponding level:

Production Level:

- Ore extraction with LHD: Load, loaded transport, unload, empty return.
- To open and close valves of the sprinklers to decant the suspension dust.
- To operate auxiliars fans.
- To close and/or open ventilation chimneys.
- Secondary reduction: To break oversize rocks and to loosen hung drawpoints.
- Floor cleaning with miniloaders.
- Drainage pumping.
- Installation maintenance.
- Maintenance and/or retreat of automotive equipment with mechanical problems.

Transfer Level

- Hydraulic hammer operation to reduce ore size.
- Open and close water sprinklers.
- Bar down ore passes over the pick hammer chamber.

Haulage Level

- Ore train goes in empty and returns loaded.
- Loading ore train with pneumatic chute.
- Bar down ore passes over the empty chute.

Due to the diverse equipment that should be operated or actions to be taken, the telecommand project has been divided in 5 subprojects as follows:

Subproject A: "Automatation of LHD Operation".

Subproject B: "Remote Operation of Station Pick Hammer".
Subproject C: "Secondary Reduction Alternatives Studies".
Subproject D: "Telecommand of Equipment and Minors Installations".
Subproject E: "Telecommand of Loading to Ore Trains".

The subproject C considers oversize rock break, and the bar down of drawpoints and ore passes present a high grade of complexity, and it is in a conceptual stage and tests of bulks size reduction such as: Sudden temperature changes, drill and use of a hydraulic expansion system, mobile hydraulic pick hammer, etc.

In this reduction work, as well as the bar down of drawpoints and ore passes, alternatives that allow to perform operation without using explosives are analyzed.

The subproject D considers the remote operation of miniloaders, valves remote action, ventilation regulators, drainage pumps and fans already have in operation the on/off systems, and the miniloader will be operated according to LHD telecommand results.

In the subproject E that will be developed in a second stage, for the remote control handling of ore train, supported by a system that gives the real positioning in real time of the equipment and video, the telecommanded operation of the loading bin will be tested through proportionals valves replacing the pneumatic systems by hydraulics if it is necessary.

The ore train remote operation would be made firstly only in the loading crosscuts sector, approximately 1 kilometer. If the test succeed, it could be extended until surface operation, with an additional distance of 9 Kms.

The A and B subprojects which correspond to the automation of the LHD and remote pickhammer operation, respectively, following are described in its conceptual approach and todate results (February 1993).

7.1 LHD Equipment Operation Automation

7.1.1 Operation Description

El Teniente Mine for primary ore extraction use Block Caving exploitation system with 6 yd^3 LHD haulage, (trademark PUMA), model TEN-8000D, designed and manufactured by Codelco Chile-El Teniente Division.

Considering that from the geological point of view, El Teniente orebody has a triangular form, with a sterile body called Braden Formation, in which never has happened rockbursts. Due to its rock characteristics, regarding stability, the great part of the mine infraestructure (shops, offices, personnel and materials shafts, compressors, etc.) are located in this area. When the LHD and pickhammers equipment operation system are implemented with remote control, the central command zones will be placed in this area at 1 Km distance from the production area, or at 10 Km in the surface.

Due to that the mining design for LHD equipment El Teniente applies, in which the drawpoints are placed in opposite, the LHD must reverse its direction to load in the opposite points, operation that makes between 3 to 4 times in a shift. This, plus the cable handling problems, do not allow fast telecommand operation of the electrical LHD equipment, therefore all the design is being done for diesel equipment.

Once the LHD has the bucket loaded it moves through the calle toward the drawpoint located at the side of it, selecting the nearest to the loading point, and for this reason in normal conditions its haulage distance should not be superior to 60 meters.

The general operations conditions of the equipment in Sub 6 level are the following:

- Height over sea level: 2.101 meters.
- Relative humidity: 70% average
- Temperature: 16°C average

- Access to level by a right drift of 4,0 x 3,6 mts or by a zig-zag drift of 3,6 x 3,6 mts.

- Free section of the calles in the production level: 3,60 mts wide x 2,60 mts high.

- LHD dimensions: 2,47 mts wide x 2,29 mts high.

- Concrete road of the LHD equipment: High strength concrete.

- The usual work system is only one LHD equipment operating in a calle.

- Initial simultaneous operation of maximum 4 LHDs.

- The LHD equipment should be retreated of the production level for fuel filling and/or maintenance, at least once every two shifts.

- It is considered that due to the extraordinary dispersion regarding size and location of the broken ore in the drawpoint, the ore loading is not automatically operated, being feasible to make for the rest of the process, loaded transfer, unload, empty transfer, which will be studied in a more advanced stage of the project.

- In the mean time the equipment operation is not being automated, the operation of every LHD is considered through remote control operated by a laborman.

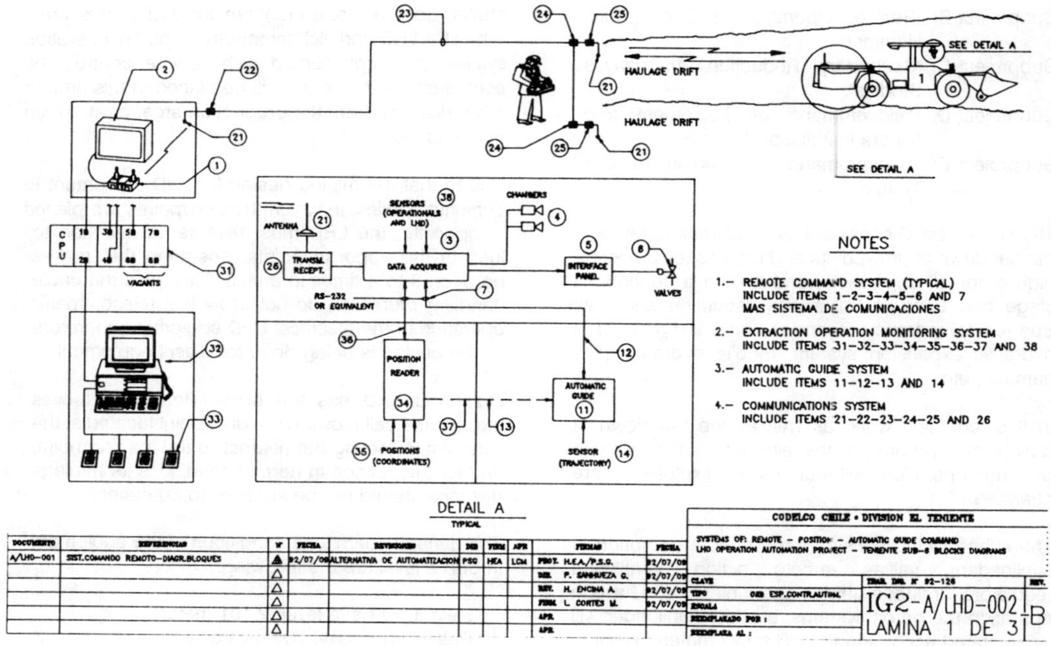

Figure 7.

7.1.2 Advantages and Disadvantages of LHD Remote Telecommand

The main advantages of the LHD remote telecommand system are:

- To allow reinitiate the production activities of the Teniente Sub 6 area.

- More safety for the operator due that is not exposed to accidents, such as collisions, falls to ore passes, rockbursts, etc.

- Better environment conditions because the operator is not exposed to dust, noise, vibrations.

- Better utilization of the working hours because is not necessary the operator moves to and from the work front.

- Better draw control because the LHD operators are concentrated in an operations room, instead of being distributed in a big geographic area, which undoubtedly will make the ore be drawn effectively from where is programmed.

  This can be supported by the positioning system that registers the points, where the LHD equipment effectively loaded the mineral, and the weight system and transported bucket counting, allowing to compute the production registration of every LHD.

The possible disadvantages to telecommand the equipment may be:

- Less schedule productivity of the LHD when down the velocities due to visual loss on performing the activities only through the TV screens.

- To work with a LHD mechanically and electronically more complex, which increase the faults possibilities.

- Use of advanced technology which may result in maintenance problems, due to short experience and limited information.

- Investment in every LHD is over US$50.000 for the control systems, video, transmission, etc.

7.1.3 Plan of Action and Progress to Date

The plan of action designed to implement the automation of LHD consider several steps tending to test the system operation, according to calles and zanjas mining design of el Teniente Mine. The steps and results to February 1993 are as follows:

a) To operate LHD through remote control and operator direct vision.

   Due that in El Teniente Mine equipment commanded by remote control hasn't been used, this test allowed to prove that in spite of the narrow of the drifts, the personnel is able to operate the equipment without problems. However, it is necessary to improve the ventilation, lighting and placement of reflecting guides to support the operation.

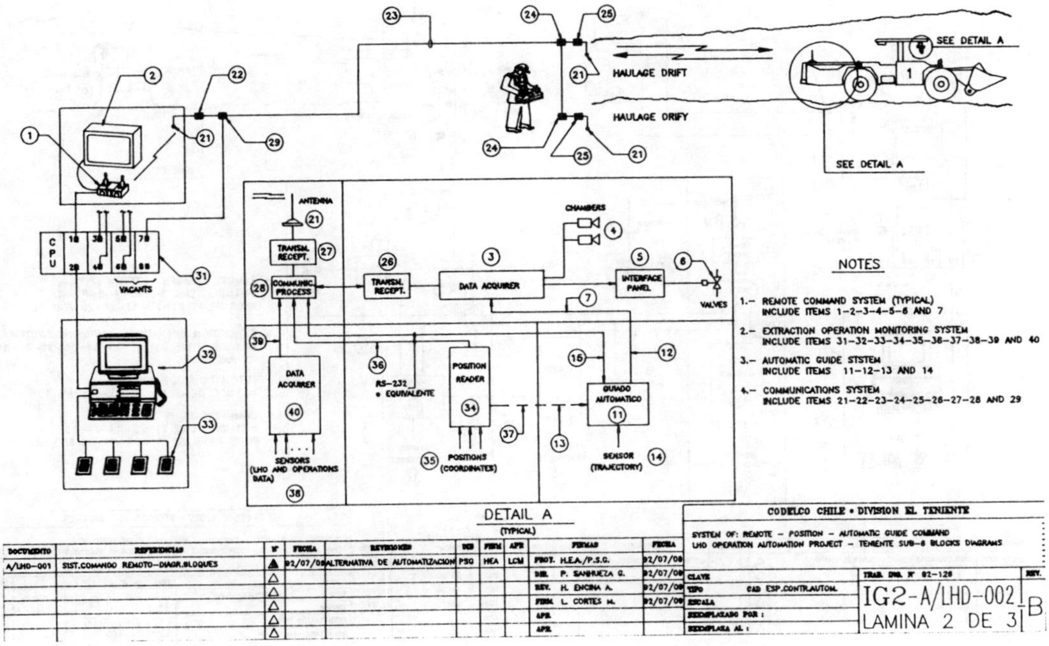

Figure 8.

b) To operate LHD through local remote control and vision through monitor.

Through cameras in the drift it was proved that its use does not constitute a solution for a normal production, due that operator loses vision, productivity goes firmly down because the vision difficulties, and it is required a high quantity of cameras located close to the drawpoint, which causes maintenance problems and damage if secondary blasting is used.

c) Use of wireless cameras mounted on LHD.

This test allowed to prove signals transmission feasibility so video as equipment command in a distance rate that allow direct communication with the LHD, which allows to study the signals transmission systems toward a control room located at 1 Km. distance, operating the LHD in a central way to distance.

d) Remote reading of parameters.

In a simultaneous way to operation and vision in the LHD it is necessary to consider the reading and transmission of equipment operational parameters, such as motor temperature, oil and fuel level, RPM, battery tension, etc., and as production parameters such as ore loaded in every bucket, bucket counter, drawpoint location in operation, etc., for which it has been developed and is being tested an electronic board and a weigher and bucket counter system which will allow to monitor the LHD draw operation, and will be connected to the remote command system in protocol RS-232.

e) Global remote operation.

This means integration of equipment guiding, captivation and processing of the operational parameters and LHD production to attain operation of all LHD in central way and to distance. For this are in study the signals transmission systems and in development the process software and operation and production display for maintenance and ore drawn control.

f) Automatic guide.

This stage will be developed once the remote telecommand is obtained, however it is under study the positioning systems which allow to locate the LHD in the field, in order to control the production parameters and to support the equipment operator in its field location.

7.1.4 Requisites and System Configuration

All the remote operation system should allow, in a first stage, the simultaneous work of 4 LHDs, and to have the capacity to expand to 6 LHD, likewise should transfer information to a personal computer for its processing. Is necessary to consider some safety systems as obstacles detection to persons

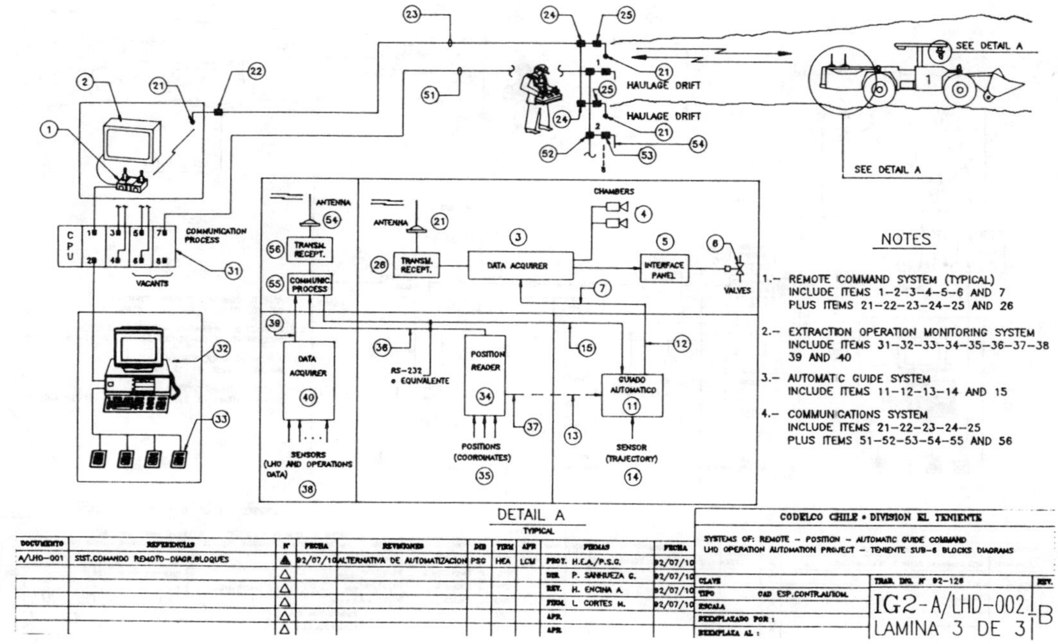

Figure 9.

and the remote activation of fire extinction semiautomatics systems which may occur in the LHD.

Furthermore, all the assembly should be of simple installation and maintenance, to have an adequate protection against the environment dust effects, acid water from dropping and vibrations or impacts for collisions or in the loading stage.

The blocks diagram of remote command system and monitor system of the draw operation is shown in figures N° 7, 8 and 9.

In order to facilitate the operator work and to improve the equipment productivity, it has been considered to use:

- Proportional control systems that allow a more fine movement in the curves.

- Fixed chambers installed in pairs ahead and behind the equipment, one of them of the big angular type and the other provided with zoom, both in colors, and automatic cleaning disposal to avoid the problems produced by the dust and water dropping.

- Use of color TV monitor, in 19" as minimum, in NTSC norm.

- Motor RPM display and alarms activation and display possibility, in case that some of the remainder operations parameters (Temperature, levels, etc.) is in levels out of normal.

- Support systems to vision as reflecting guides in walls and roof, besides changes in the lighting systems as in LHD as in field.

- Capacity, if the operator wishes, to listen the sound caused by LHD operation in field.

- Furthermore, the LHD remote control equipment, apart of the usual movements as to advance, move backward, turnings, bucket movement, etc. should have the capacity to act over the equipment gear box, allowing to work to high, middle or low speed according to operation necessity.

7.1.5 Positioning and Communication Systems

The mining design in which the LHD equipment will work, production drifts with "zanjas" every 17,32 m, in which explosives eventually are used, do not permit the cables use or a wire transmission system, due that the installed equipment on the LHD should have the capacity to transmit and receive command signals, data, video, etc. toward or from an antenna located up to 200 mts distance in the drifts intersection of main acces, place from where a cable with amplifiers arrive to the control room located in the center of the mine. (See Figure N°10).

Considering that in the exploitation system by Block Caving applied by El Teniente, an important point is the regularity in the ore draw. In order to avoid a premature dilution is under development simultaneously to the weight control, a remote

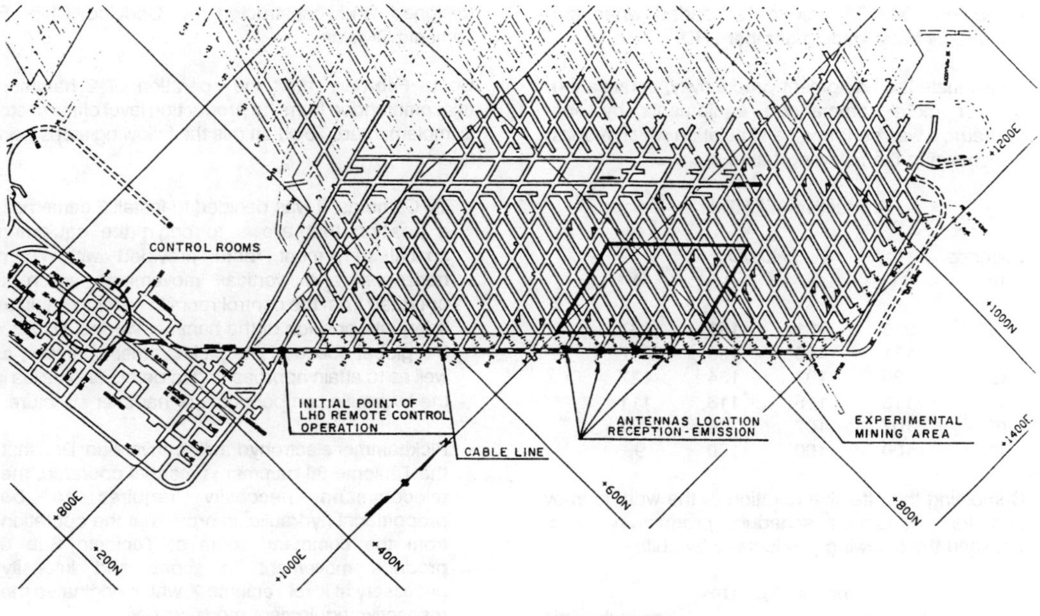

Figure 10.

identification system from the drawpoint, in which the LHD has been loaded. This system means a great quantity of fixed points to be identified, which should be captivated by the LHD through a very little window to avoid identification duplicities when captivating simultaneously the opposite drawpoints. This information is basic for the production data processing made at shift end, in which the production is calculated as well as the corresponding Cu. grade and, furthermore, this positioning will serve as base for an automatic future guided of the LHD equipment.

7.1.6 Productivity of LHD Equipment

In order to determine the impact that the implementation of a telecommanded or semiautomation system will have in the LHD equipment productivity, simulation tests were made of both alternatives and its comparison with the present situation of manual operation. The simulation model was developed in SLAM language, with some routines in FORTRAM, using the SLAMSYSTEM software.

Considering the operational situation of El Teniente mine, for this simulation the following considerations were made:

- The productive time of the LHD operator increase 21%, because the transportation to and from the production area is eliminated.
- In the telecommand situation, all the operations are performed at remote control considering that equipment velocity is diminished up to 80%, in comparison with present situation due to visual problems.
- In the semi-automation situation, only the loading is teleoperated and the velocity diminish to 60% of present, due to drift narrowness.
- Being 104 m the distance between dump points, the simulation is made for intermediate distances rank of drawpoint location.

According to above, the schedule productivities would be:

	Productivity T/LHD-H		
Distance m	Present Situation	Telecommand	Semiautomatic (1 LHD/Op)
15	340	268	245
30	250	195	171
45	216	167	139
60	186	147	116
75	174	134	107
90	171	132	100

In the semiautomatic operation an operator may be in charge of more than one LHD equipment, taking advantage of the free time that remains when the equipment perform some activities of the cycle in automatic form. In this situation, some of the equipments should wait to be loaded when the operator is busy loading another LHD.

To evaluate the equipment productivity, considering one or more commanded equipments by one operator, it was performed a simulation which results are shown in the following table:

Productivity T/LHD-H

Distance	LHDs/Op			
m	1	2	3	4
15	245	231	181	137
30	171	168	159	137
45	139	137	134	127
60	116	115	113	111
75	107	107	106	105
90	100	100	100	98

Combining the effective duration of the work journey and the equipment schedule productivity, it is obtained the following productivity by shift:

Productivity T/LHD-H

Distance m	Present Situation	Telecommand	Semiautomatic (1 LHD/Op)
15	1468	1474	1282
30	1078	1073	0895
45	0924	0920	0727
60	0803	0809	0607
75	0748	0737	0560
90	0737	0726	0523

Of the values obtained, it may conclude that:

- In telecommand mode, the LHD has a minor schedule productivity to the one obtained with the present operation way, with operator on board; however, considering the shift productivity, if the displacement velocity is superior to 80% of present, the telecommand mode is more productive due to major duration of the effective work journey.

- In semiautomatic characteristic, the LHD middle productivity in general diminishes when an operator handles several equipment. This effect is larger while smaller is the haulage distance and it is due that equipments have to wait for loading and the operator is busy loading another LHD.

- In semiautomatic characteristic, the LHD is less productive than in the present operation, even considering a work effective shift longer.

7.2 Station Pickhammer by Remote Control

As has been indicated, the pickhammers are located at the base of the ore passes, 35 meters below the production level and should transfer the ore toward the lower haulage level, reducing previously its size to 30". The equipments installed are of the TTE 86 type, designed and constructed by Codelco Chile El Teniente Division.

Up to February 1993 the operation of 2 hammers telecommanded from the production level office sector is implemented. To attain this the following steps were taken:

- TV Cameras. It was decided to install 2 cameras in a position that allows to origin the equipment spatial movement, both provided with zoom, horizontal and vertical movement, which is operated from the control room in order to attain an adequate position of the hammer wedge, visualize the global space of the pickhammer chamber, as well as to attain approaching to detect anomalies in the hydraulic componentes or hammer structure.

- Pickhammer electrohydraulic conversion: Due that the Teniente 86 hammer is manual operated, the telecommand neccesity requires to be proportional hydraulic, in order that the operation from the command room at Teniente Sub 6 produce movement in sense and intensity necessary at level Teniente 7, which originated the respective equipment modifications.

- Signal transmission to distance, which were divided in those corresponding to TV cameras, vision and movement of same and operation and pickhammer control owing to receive in control room the data of power unit, such as oil level, temperature, cooling operation, etc. and send the orders for pickhammer operation.

  Due that in general the acting final elements are of big accuracy in admission, as analogicals as digitals, the transmitted signals from the control room maintains value when acting over the pickhammer located at 1 Km from it.

The telecommand implementation system of the pickhammers allows:

- More safety and comfort for the operator.

- Possibility to operate simultaneously severall hammer with only one operator.

- To decrease the picking chamber size because the equipment reduce the volume not contemplating the operation cab.

- Improvements in the maintenance procedures as a consequence to measure and evaluate computationally the hammer operations parameters.

8. CONCLUSIONS

The efforts tending to telecommand the productives processes at El Teniente Mine are directed basically to improve the safety conditions for the operations personnel, and specially those that work in Teniente Sub 6 in the mine.

During 1993, the year in which the telecommanded systems will be operating, extraction tests will be carried out in this level, tending to attain the subsidence and change the area geomechanics conditions to re-initiate production and sector expansion.

At this moment the massive use of LHD equipment telecommanded in production will be decided, or if only it will be used in preparation activities and mine development.

The remaining activities will be analyzed also for definitive implementation. However, at the light of good results obtained in the remote operation of pickhammers, their telecommand will be implemented in the whole mine since 1993.

Teniente Sub 6 and the remaining mine lower levels, emplaced in primary ore, represent the major part of present reserves of El Teniente orebody, and if not exploited this will mean in the middle term production gradual reduction to arrive to antieconomic levels in short time, which will depend of the copper market at that moment.

For this reason, the big challenge for the mine engineers of El Teniente Division is to be able to set methods and designs that allow exploit this level in a secure and economic way, therefore all our efforts and knowledge must be toward to solve the Sub 6 problems and exploitation of primary ore in general rockburst conditions to assure the future of El Teniente Division.

9. REFERENCE

1. Vienne, J. Mine North Project Teniente Sub 6. "Reverbero" Magazine N°90, July 1991, National Copper Supervisors Association.

2. The Rockburst Problems at El Teniente Mine. Internal report of Studies and Operationals Methods Dept., Codelco Chile-El Teniente Division 1990.

3. Gjakoni, L."Test on LHD Operated by Remote Control", internal Mine Management report, October 1992.

4. Contreras F., Neves L., Wire Telecommand Implementation for Teniente 86 Pickhammer, X Congress of Automatic Control Chilean Association, October 1992.

5. Espinoza R., LHD Equipment Teleoperation and Semiautomation Analysis. Internal report of Industrial Engineering Dept., Codelco Chile El Teniente Division, September 1992.

6. Espinoza R., Transportation Equipment Simulation at El Teniente Mine. X Congress of Automatic Control Chilean Association, October 1992.

7 Computer applications and numerical modelling

Innovative Mine Design for the 21st Century, Bawden & Archibald (eds) © 1993 Balkema, Rotterdam, ISBN 90 5410 325 6

64-bit computing: New technology that will revolutionize mine planning and mine design

J.Jonkman
Digital Equipment Corporation, Salt Lake City, Utah, USA
K.Dagdelen
Colorado School of Mines, Golden, Colo., USA

ABSTRACT: The recent introduction of 64-bit computing is revolutionizing the application of mine planning and design to a much greater extent than was experienced in the late 1970's with the introduction of 32-bit computing. The ability to address 4 billion times more data and achieve computational speeds far greater than similarly configured and priced 32-bit system will unleash applications previously not imagined and provide answers to problems which just could not be solved.

This paper will introduce the new computing architecture as it applies to mine planning and design and will discuss 21st century applications being made possible by the 64-bit architecture. Technical details will be provided to assist software developers in optimizing the use of the new architecture.

1. APPLICATION OF COMPUTERS IN MINE PLANNING AND MINE DESIGN

Demand for raw materials is increasing as a result of the world's population growth and an increase in the standard of living. By adopting the latest advancements in applicable technologies, the mining industry has been able to respond by finding additional ore reserves and extracting minerals from lower grade deposits. In recent years the emphasis has been on investing in cost-reduction technologies, including computer systems, in order to lower production costs.

Computing technology has been integrated into the mine planning function and the extraction process of every surface and underground mine. Computer systems currently used for mine planning and design can be classified into three categories of origin and use:

1. In-house developed systems, which have been modified over the years.
2. In-house adaptations of generic commercial database or computer aided design software.
3. Off-the-shelf software packages, often integrating many aspects of mine planning and design.

1.1. Overcoming current architectural limitations

A careful observation of the type of computer systems in use reveals that for the most part:

1. The inherent computer techniques are primarily adaptations or computerizations of manual methods.
2. The systems are not used to integrate various facets of the mining and milling operation.

Basically, the mining industry has taken advantage of computers as fast and accurate calculators and as map generators. These applications were all developed within the limitations of previous and current generation computers. An improvement in turnaround time will result if the computer technology introduced in this paper is applied to these existing applications.

However, the application of computers, and in particular the upcoming generation using the 64-bit processors, can also introduce us to a whole spectrum of new approaches to problem solving that was not possible before. Development of new solution algorithms that take advantage of the pure power of these computers and optimize the mining operation as a whole, is going to dominate the next era in computer applications development in the mining industry.

2. ADVANCEMENTS IN COMPUTER TECHNOLOGY AS APPLIED TO MINE PLANNING AND DESIGN

The introduction of 32-bit departmental minicomputers in the late 1970's revolutionized the way computers were used for mine planning and design. Mining engineers no longer were dependent on a central staff of computer experts and a "batch-type" central computer to accomplish their work. The users of these fast, low-cost, departmental computers not only saw their productivity improved dramatically, they also were able to choose from an increasing number of off-the-shelf software applications.

Departmental computing evolved into desktop computing during the 1980's with the introduction of personal computers and workstations. Interactive mine design became feasible by using dedicated processors, which also could render excellent visualization of the results of the mine planning activity.

The introduction of 64-bit computers in the early 1990's will revolutionize mine planning and design in an unprecedented way. As will be described in section 3, the Alpha AXP™ architecture developed by Digital Equipment Corporation not only performs operations between 64-bit registers allowing much faster computers to be built, but also addresses memory through 64-bit addresses, which eliminates many addressability barriers.

With each generation of mine planning software becoming increasingly sophisticated, it would be difficult to imagine a computer that is too fast, that is too inexpensive, or that has too much memory to find application in the mining engineering discipline. The new 64-bit architecture excels in each of these three areas.

2.1. A springboard for 21st century computing

The Alpha AXP architecture will open up new areas of computing in the mining industry and specifically make new applications for mine planning and design affordable and practical for widespread use. Very large programs which used to require expensive mainframes can now run on the mining engineer's desktop. Analyses which were done in seconds can now be done instantaneously. Creation of complex models which used to take over a weekend to complete, will eventually be done over a lunch hour. Answers to what-if scenarios will be provided during a meeting when the answers are needed and while the results are still relevant.

3. GROUND-BREAKING APPLICATIONS IN MINE DESIGN AND MINE PLANNING

After the original introduction of mainframe computing for commercial applications in the mining industry, it was the mining engineering departments that often introduced new computing technologies into their companies. When minicomputers became widely available, they were applied to off-loading the mine planning software from the mainframes. When interactive graphics and powerful workstations became affordable, the mining engineering departments led in the migration of applications to utilize this new technology. Each new introduction of technology led to the development of breakthrough applications. It is anticipated that the introduction of 64-bit computing will again lead to the emergence of applications which are now only imagined. The authors are presenting the following topics to stimulate the development of such applications.

3.1 Open pit design optimization

Most open pit mines in are either designed by the heuristic moving cone algorithm or by the Lerchs-Grossmann algorithm (Lerchs and Grossman, 1965) which determines the optimum ultimate pit boundary based on the break-even analysis. For most pit design problems the number of blocks in a given model is quite large and as a result the implementation of the Lerchs-Grossmann algorithm requires reduction of the size of the block model by re-blocking into larger size blocks if one expects to solve the problem in a reasonable amount of time. Doing this results in an approximate solution since the optimizing algorithm of Lerchs-Grossmann can not be applied directly to the problem at hand due to the memory and speed limitations of the computers currently being used.

With the implementation of faster 64-bit computers and expanded addressability, one can not only focus on break-even optimum pit limits, but also work towards developing the algorithms that will provide definition of ultimate pit limits by incorporating time value of money. More accurate definition of open pit mining limits will eliminate the removal of unnecessary waste material to reach the marginal ore. This in turn increases the economic viability of a given mine, and minimizes the environmental damage by eliminating unnecessary disturbance to the land.

3.2 Open pit scheduling

One can follow many different sequences to extract ore from a given deposit. The general open pit scheduling problem is to find a path through the block model such that when followed will provide required tonnage and grade to the mill, subject to equipment requirements and slope constraints (Dagdelen, 1986). When formulated as a mathematical optimization, this problem becomes a very large scale optimization, including over a million variables and several million constraints.

The existing computer solution to this problem is based on manual techniques and, at most, allows engineers to generate only one or two alternative solutions within a reasonable time period. Therefore, the engineer is left with a great degree of uncertainty regarding the difference between the obtained plan and the optimum plan when the results of such planning are put into operation.

The algorithm that has been developed for solving this problem as true optimization requires the development of hundreds of ultimate pit limit solutions. Each of these solutions currently takes up to four hours to compute on a computer with a rated speed of 20 MIPS. This kind of mathematical optimization will become possible and practical with the application of new and fast computers such as those using the Alpha AXP processor.

The sheer size of the computational effort required for pit optimization has been a barrier to the development of efficient statistical approaches and simulation techniques for true pit optimization. It is foreseen that the elimination of this barrier will encourage software developers to approach this problem with a different mindset leading to much greater certainty of the optimization. Knowing exactly if a given block of material should be mined or not and if it is going to be mined, then knowing when it should be mined, has significant economic benefits in terms of net present value considerations.

3.3 Open pit production optimization

The use of truck dispatching systems that route trucks to different shovels and other destinations in open pit mines is another example of the successful application of computers in the mining industry (Anon, 1990, and Arnold & White, 1988). Such systems incorporate up-to-date computer hardware and software in providing short-term production scheduling solutions at open pit mine sites. (See Figure 1). The solution to the problem is provided in terms of allocating trucks to the shovels so that truck wait time at a given shovel is minimized, subject to travel times and truck availability.

The new generation of computers will make possible the integration of a long-term scheduling algorithm as mentioned above with truck dispatching systems. These new systems could not only allocate trucks according to equipment availabilities, but also according to modeled ore characteristics and blending requirements of the mill. Much faster computers which are now becoming available will also enable the complete integration of these systems with material flow optimization in the mill which would provide further improvements in cost reduction and efficiency of the operation.

The envisioned integration of individual mining systems with one another and solving these larger problems at once for total system optimization will become feasible as software designers start to comprehend the power of 64-bit computing.

3.4 Geographic information systems

Geographic information systems (GIS) have been developed mostly independent of mine planning software to manage and display spatial data to a much greater extent than traditional mapping software could do (Katz, 1989). Many of the features of GIS can be used for mine design. Specific application areas include the

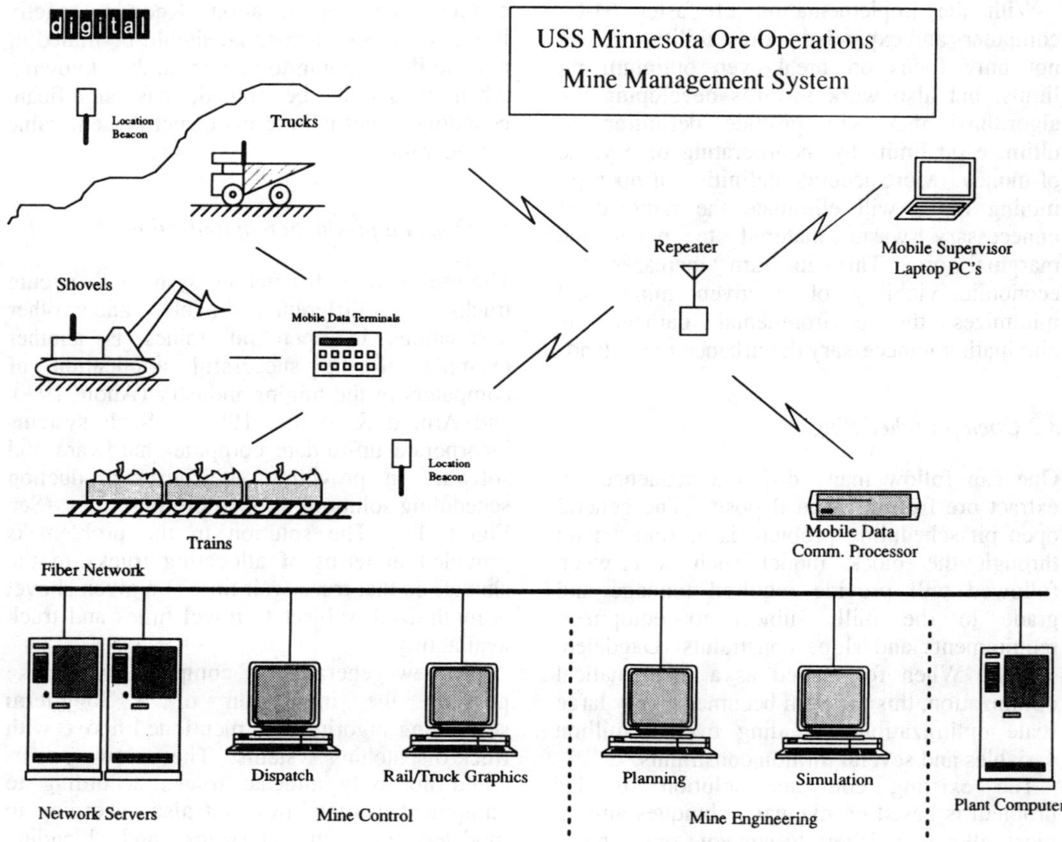

Fig.1 Integration of equipment dispatch system with plant operations and mine planning

planning of facilities, roads, pipelines, and slopes as well as the management of land ownership, water reservoirs, and revegitation projects. For a GIS to function effectively it needs access to large amounts of data, in vector and raster form, and needs fast computers to provide reasonable response times. Since turnkey GIS systems are now either limited in scope or expensive, their use in mine design has not yet become widely accepted.

New computer architectures, such as Alpha AXP, will allow software designers to address these limitations and make full functional software available on fast inexpensive computers. This will make it practical to incorporate GIS functionality as modules in mine planning software.

The computerization of underground mining operations is still in primitive stages. Up until most recently, the speed of most workstations was not sufficient to make 3-dimensional rendering possible. Incorporation of various data bases with the images of three dimensional openings in 3-D GIS systems will become reality in the near future for more sophisticated 3-D geological, mine, and groundwater analysis, as workstations based on the Alpha AXP processor are being applied.

Improved graphical visualization of spatial data in a timely fashion will significantly enhance the functionality of GIS and its usability for mine planning. The quality of mine plans will improve as a result of improved accuracy and underground mines will then benefit from computer technology to the same extent as has been made possible for surface mining.

4. DEVELOPMENTS IN TECHNICAL COMPUTING

The following technical detail is provided to

give software developers a better understanding of the power that is available from this new technology. This will help them in developing 'ground-breaking' applications for mine planning and design which fully benefit from this new architecture. It should also help the users of these new systems understand the technology which enables these applications to be developed.

4.1. Development of 32-bit computer technology

Although some computers which used 32-bit integers and pointers were introduced during the 1960's, they could compute addresses only to 24 bits. This limited the addressability to 16 megabytes (MB). This was acceptable then, as virtual memory was not yet supported and very few computers were sold with more than 1 MB of memory. It would now be considered very limiting if one had to keep file sizes within 16 MB. Database mapping and file management would be more complex. Programming would be more difficult and more expensive.

The first truly 32-bit computer was introduced in 1978 by Digital Equipment Corporation as the VAX™ 11/780. With its ability to compute 32-bit addresses, it could address 4 gigabytes (GB) of memory, virtual or real. When this computer was first introduced, very few people could imagine ever needing that much memory. However an empirical rule-of-thumb has been that memory requirements for technical applications increase 1.5 times per year, increasing the demand for addressability by .6 bits each year. Computers based on this 32-bit architecture quickly became popular for numerous mining engineering applications.

4.2. Development of RISC technology

The application of Reduced Instruction Set Computing (RISC) technology became popular for compute-intensive processing during the 1980's. Because of simpler hardware, improved price/performance was achieved with RISC processors. This resulted in the porting of many mine planning systems to RISC-based platforms.

In this architecture the instruction set was reduced to only the most frequently used single-stage instructions and the language compiler was given greater responsibility to handle the more complex operations.

RISC processors were optimized for the UNIX® operating system. For example, they provided hardware instructions for manipulating a UNIX view of virtual memory management. Even though the concept of "open systems" was becoming popular in association with UNIX and RISC, each vendor's RISC hardware would only run its own proprietary derivative of UNIX.

4.3. Development of the DECchip™ 21064 and the Alpha AXP architecture

In 1988, an engineering task force at Digital Equipment Corporation was assigned to develop a computer architecture for 21st century computing. Code-named Alpha, the project team was able to start with a clean sheet and consider all the lessons learned from previous RISC implementations and overcome the limits of 32-bit architectures. As part of developing an advanced computer architecture, the team designed a new 64-bit processor-on-a-chip, the DECchip 21064.

Ranked by the Guinness Book of Records (Guinness Publishing, 1993) as the fastest computer chip available when the processor worked at a clock speed of 150 MHz, it has since been improved to run at 200 MHz. This translates into a clock cycle time of 5 nanoseconds, or about the time it takes a light beam to travel 5 feet. Because of its design, the processor can actually execute two 32-bit instructions per clock cycle, resulting in program execution speeds of up to 400 million instructions per seconds (MIPS).

Alpha AXP was designed as a true 64-bit architecture. All operations are performed between 64-bit registers and memory is addressed through 64-bit addresses. Other so-called 64-bit computers available up to now do not have 64-bit addressability. The advantage for mining industry applications is that systems based on this architecture can address 4 billion times more data than was possible with 32-bit systems. If memory requirements continue to increase at the current rate then this increased capacity will last well into the 21st century. However, past experience has demonstrated that the creative minds of mine planning and designs software developers will find ways to use up the power and memory space that is given to them and then ask for more. This will lead to applications which have been imagined, but have not been feasible within the limits of 32-bit computing architectures.

4.4. The advantages of an open architecture

While previous RISC designs were optimized for particular UNIX operating systems, the RISC-based Alpha AXP architecture was designed to be independent of any operating system and be a true open architecture. This is achieved through the use of an instruction library, which is analogous to the BIOS firmware of a personal computer. The library consists of subroutines of RISC instructions which are optimized for the DECchip 21064 hardware. The subroutines are activated as needed by instructions which specify that certain operating system primitives be performed. Operating systems which are now running on the Alpha AXP architecture include OpenVMS™, DEC™ OSF/1® (which is Digital's implementation of UNIX as defined by the Open Systems Foundation), and Microsoft's Windows NT™. Library subroutines can be created to optimize any operating system and it is foreseeable that specialized operating systems will be developed for applications such as simulation, modeling, machine learning, etc. The open systems architecture of Alpha AXP mandated a scalable design, specifically to achieve a performance improvement by a factor of 1000 to 400 billion instructions per second within the next 25 years. Initially running at 200 MHz, the chip architecture will allow for speeds up to 2,000 MHz. Initially performing two instructions per clock cycle, the development of smart compilers and special-purpose operating systems will allow the processor to be effective at executing up to ten instructions per clock cycle. Another aspect of the open systems architecture of the Alpha AXP chip is that the chip design is licensed to other companies. For example, Cray Research has announced a new line of massively parallel supercomputers that will include up to 1,000 Alpha AXP processors each for a performance rate as high as 200 gigaflops. Personal computers which include the Alpha AXP processor have also been announced, making this high-performance computing technology widely available.

5. CONCLUSION

Numerous computer applications which could significantly benefit the mine planning and design function have been stretching previously available computers to their 32-bit limits. The Alpha AXP architecture, developed by Digital Equipment Corporation, which includes the first true 64-bit computer available, offers virtually unrestricted freedom to software developers.

This paper lays the foundation for future development by suggesting several application areas which could benefit from this revolutionary technology. It is expected that the mining engineering community will take advantage of the greater speed and addressability of the new computer architecture and develop applications, now only imagined, which will greatly reduce the cost of mining.

REFERENCES

Anon. 1990. Striking gold with open systems. *Workstation News* 1:3, 110-11.

Arnold, M. J. & J. W. White 1988. Advances in computer dispatching. *Proc. MINExpo Intional:* 279-293.

Dagdelen, Kadri 1986. Optimum open pit mine production scheduling by Lagrangian Parametrization. *Proc.19th Symposium on Applications of Computers and Operations Research in the Mineral Industry. SME, Littleton, CO:* 127-141.

Guinness Book of Records 1993. Guinness Publishing Ltd. Enfield, Great Britain.

Katz, Solomon S. 1989. Geographic information systems; more than just another computer mapping program. *Proc. 21st Symposium on Applications of Computers and Operations Research in the Mineral Industry. SME, Littleton, CO:* 597-602.

Lerchs, H. & I. F. Grossman 1965. "Optimum Design of Open Pit Mines". *C.M.I. Bulletin* Vol. 58, No. 633: 47-54.

Innovative Mine Design for the 21st Century, Bawden & Archibald (eds) © 1993 Balkema, Rotterdam, ISBN 90 5410 325 6

Thermomechanical analysis of a waste disposal drift in rock salt

A. Pudewills & E. Korthaus
Kernforschungszentrum Karlsruhe GmbH, Institut für Nukleare Entsorgungstechnik, Germany

ABSTRACT: This paper summarizes the results of the numerical analysis of thermomechanical effects in the near field of a disposal drift within the nuclear waste emplacement panel. The temperature fields in the disposal area, the drift convergence followed by the long-term compaction of the backfill material and the resulting stresses in rock salt were investigated. The results of a parametric study to determine the influence of the pillar width and of the disposal depth on the drift stability are also presented.

The calculations were performed with finite element codes specially developed for solving the repository problems, taking into account the nonlinear, temperature and time dependent behaviour of the rock salt. With regard to the backfill material an isostatic, temperature dependent creep consolidation was adopted.

1 INTRODUCTION

In recent years several repository concepts have been developed for the disposal of both high-level radioactive waste from reprocessing and spent fuel elements in salt formations (Bechthold et al., 1989). According to the three-level drift emplacement concept, large self-shielded packages (POLLUX casks) will be emplaced on the floor of the disposal drifts at three horizons between 870 m and 1170 m. Subsequently, the drifts will be backfilled with dry crushed salt. At the emplacement time, the heat generation of a POLLUX cask will be up to 7.5 kW, depending on the duration of intermediate storage.

The increase in temperature in the disposal drifts will affect the thermomechanical behaviour of the backfill material and of the surrounding host rock. Furthermore, due to the high lithostatic pressure at the lower emplacement horizons (1020 and 1170 m), problems may arise in the stability of the drifts. On the other hand, due to thermally induced convergence of the disposal drift, the crushed salt will compact sufficiently to support the surrounding rock salt and will serve as a seal for the waste containers. In this context, geomechanical modelling is of great importance in order to design and construct a repository which is stable during the operational phase, and to predict the long-term thermomechanical behaviour of the rock salt for the performance assessment of the repository concept.

The purpose of this paper is to document the results of the predictive calculations of thermomechanical phenomena in the near field of a disposal drift within the waste emplacement panel for each of three levels. These specifically include numerical modelling of the temperature fields, the drift closure followed by the consolidation of the backfill material and the resulting stresses in the surrounding salt. The porosity of the backfill material and the pressure increase on the POLLUX casks were estimated with regard to the long-term behaviour of the back-filled disposal drift.

The numerical analyses were conducted in two steps. The first step consisted in the thermal analysis. The FAST-BEST finite-element code (Ploumen,1979), specifically developed for studying drift disposal, was used to calculate the temperature fields. The second step involved the thermomechanical analysis with the MAUS code (Albers,1984) which includes suitable material models for rock salt and crushed salt.

2 CONSTITUTIVE MODELS

2.1 Rock salt

The thermomechanical behaviour of rock salt was approximated in the finite-element analysis by a thermoelastic material model with temperature dependent steady state creep. The transient creep certainly affects the starting period of the excavation. However, this has not been considered in the calculation because sufficient data are not available.

The thermoelastic properties used in the calculations (Langer,1982) are the following:

Modulus of elasticity: 24000 MPa
Poisson's ratio: 0.27
Coefficient of thermal expansion: $4.2\ 10^{-5}$ 1/K

The constitutive relation governing the steady state creep deformation was determined from laboratory tests (Wallner et al. 1979):

$$\dot{\varepsilon}_c = A\, \sigma_{eff}^5 \exp(-Q/RT)$$

where

$\dot{\varepsilon}_c$: effective creep strain rate (1/s)

σ_{eff} : effective stress (MPa)

A : $2.06\ 10^{-6} MPa^{-5} s^{-1}$
T : temperature (K)
Q : 54.21 kJ/mol
R : $8.314\ 10^{-3}$ kJ/mol/K

A strength criterion allowing to determine whether or not the rock salt has failed, was considered in numerical modelling. For compression the strength criterion given by Diekmann et al.(1986) was assumed:

$$\tau_{o,f} = a\, |\, \sigma_m|^b$$

$\tau_{o,f}$: predicted octaedric shear stress upon failure (MPa)

σ_m : mean stress (MPa)

a = 2.329 and b = 0.747 (material constants)

A "factor of safety" was defined as the ratio of the current value of the octaedric shear stress ($\tau_o = \sqrt{2/3}\ \sigma_{eff}$) to the predicted shear stress at failure:

$$FS = \tau_o \,/\, \tau_{o,f}$$

Failure will occur when FS > 1.

For tensile stress the following criterion was used:

$$\sigma_t > 1 \text{ MPa}$$

with σ_t being the uniaxial tensile strength.

2.2 Backfill material

In the MAUS finite-element code a material model to describe the isostatic, temperature-dependent creep consolidation of the backfill material is already available. This formulation was adapted to the recent laboratory tests carried out by Diekmann and Stührenberg (1990) at temperatures up to 200^0C. The rate of creep consolidation of the crushed salt is given by the following preliminary relation:

$$\dot{K} = c_1\, (\eta_0 - K)^6\, (P - c_3 K^{C_4})^2$$

$$c_1(T) = c_1 \exp(c_2\, (\, 1 \,/\, T_{ref} - 1 \,/\, T))$$

$$c_3(T) = c_3(c_5 + c_6(T - T_0) + c_7(T - T_0)^2)$$

where

$\dot{K}$: rate of compaction (1/d)
η_0: initial porosity (31%)
P: compaction pressure (MPa)
T: temperature (K)
$T_0 = 273K$, $T_{ref} = 392K$

$c_1 = 170$ (1/d), $c_2 = 4000K$,
$c_3 = 160$, $c_4 = 1.8$,
$c_5 = 2.6$, $c_6 = -0.021K^{-1}$,
$c_7 = 510^{-5}K^{-2}$
are fitting parameters.

This formulation represents a transient creep compaction rate. The thermal properties of rock salt and crushed salt i. e. the specific heat capacity and the coefficient of thermal conductivity as functions of the temperature are given in Table 1.

3 FINITE ELEMENT MODELS

In the repository concept, the POLLUX casks (diameter 1.5 m, length 5.5 m) with heat- generating waste will be placed one behind the other on the floor of long parallel drifts in rock salt. Upon emplacement, the heat generation of a cask will be up to 7.5 kW. Then the open space in the drift will be backfilled. The disposal drifts, 4.5 m (width) by 3.5 m (heigt), and about 200 m long will be excavated into the emplacement panels at a distance of 17 m from each other. The disposal horizons will be located at a depth of 870, 1020 and 1170 m, respectively.

In preliminary thermal calculations, it was found that for a maximum temperature of 200^0C at the package surface the spacing between two casks will be 5 m, with a 12.5 m salt pillar between the drifts. Furthermore it was assumed that all disposal drifts in the emplacement panel were excavated and filled with POLLUX-casks and crushed salt instantaneously at time zero. Due to the geometric symmetry and the mentioned assumptions, a three-dimensional model of a representative portion of the disposal drift was defined for the thermal analysis. The initial temperature before the heating was taken to be 35^0C. For the two dimensional thermomechanical calculations the temperatures are interpolated from a vertical section passing through the center of the model geometry.

A plane strain finite-element model was chosen for the thermomechanical investigation. The vertical section right through the center of a POLLUX cask represents only a half of the disposal drift and of the pillar between adjacent drifts. In Figure 1 the finite element model and the boundary conditions are presented. Symmetry conditions are assumed on the left and the right border. The upper boundary of the model was loaded with 18, 22 and 25 MPa lithostatic pressures, respectively (corresponding to the depths of the three emplacement horizons).

4 NUMERICAL RESULTS

The numerical simulation starts with the calculation of the isothermal convergence and the stress distribution around the disposal drift prior to the emplacement of heat generating POLLUX casks. After about 0.7 years the backfill material and the temperature development in the drift were taken into account. The thermomechanical analyses were continued over a period of 20 years.

4.1 Analysis of the disposal drift during the operational phase

Firstly, the convergence of the disposal drift during the operational phase (i.e., the isothermal phase from excavation of the drift until loading with POLLUX casks and crushed salt) was investigated. Figure 2 shows the histories of volumetric convergence (closure) of the drift at 870, 1020 and 1170 m depths. The large values of drift closures for the lowest horizon are caused by the high lithostatic pressure and, on the other hand, by the small pillar width between the drifts.

Table 1: Thermal properties (ref. Korthaus at al.,1989)

	T (°C)	0	50	100	150	200	250
Rock salt	λ (W/mK)	6.10	5.02	4.19	3.57	3.11	2.78
	ρc_p(J/m³K)			$1.90 \cdot 10^6$			
Crushed salt	λ (W/mK)	0.55	0.75	0.90	0.99	1.03	1.01
	ρc_p(J/m³K)			$1.37 \cdot 10^6$			
POLLUX-Cask	λ (W/mK)	14.1	14.8	15.6	16.3	17.1	17.9
	ρc_p(J/m³K)			$3.77 \cdot 10^6$			

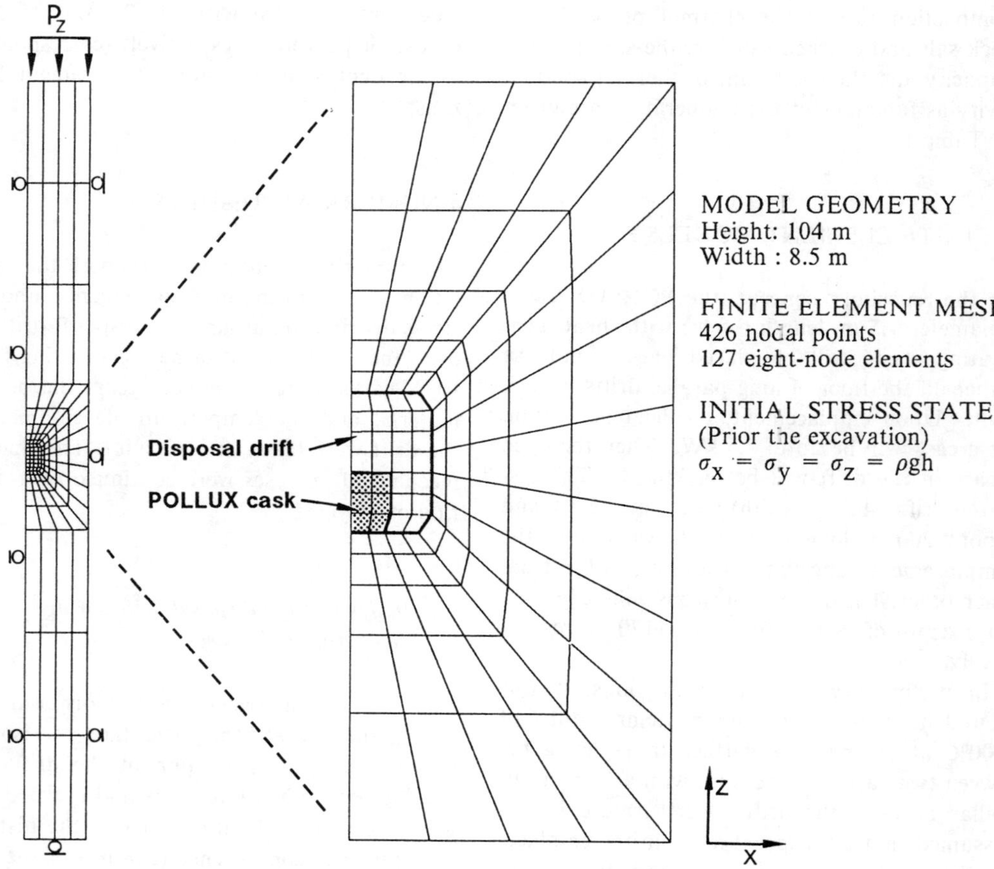

Fig.1 Finite-element model and boundary conditions for thermomechanical calculations

Further model calculations were performed in order to investigate the influence of pillar width on the disposal drift closure. The width of 12.5 m (reference design data) was increased to 20, 40 and 80 m, respectively. The volume closures as a function of the pillar width for two emplacement horizons (870 m and 1170 m) are represented in Figure 3. These results show that for an assumed pillar width of 20 m or above, the closures (especially at depth of 1170m) are clearly lower (by a factor of 0.5) compared to the reference case. After excavation of the disposal drift, the initial lithostatic stress is significantly disturbed. A stress concentration occurs around the drift with the largest magnitude at the lowest disposal horizon. The resulting effective and mean stresses along the pillar midheight at the end of the operational phase are shown in Figure 4. It was also found that the highest compressive stress approached 65% of the assumed strength criterion. Tensile stresses in the drift walls do exist but are of a lower magnitude (σ_t < 1 MPa).

4.2 Long-term behaviour of the disposal drift

After emplacement of the POLLUX casks and backfill of the disposal drift, the volume closure and hence the compaction of the backfill material are mainly determined by the temperature development and the thermomechanical behaviour of the crushed salt. This causes faster creep deformations in the surrounding rock salt and also in the backfill material.

For the reference case with a pillar width of 12.5 m the histories of the calculated temperatures at three caracteristic points are shown in Figure 5. In Figure 6 the thermally induced volume closures are given for the three disposal levels. The decrease of the porosity of backfill material, averaged over the drift cross-section, is shown in Figure 7. The development of the compaction pressure in the backfill material is given in Figure 8. which also shows the influence of the disposal depth (i.e., the rock pressure).

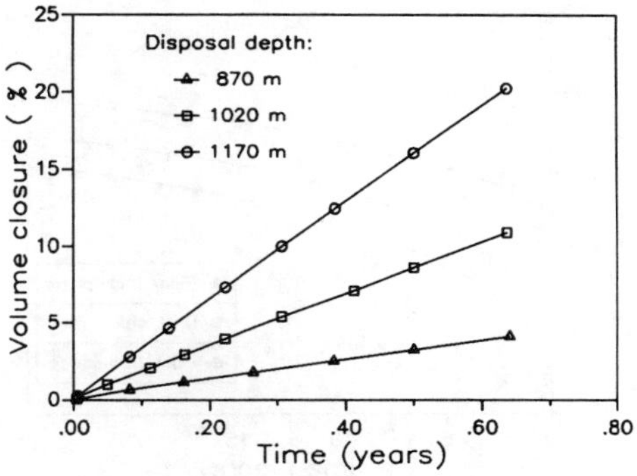

Fig. 2 Volume closure of the disposal drift for three emplacement levels

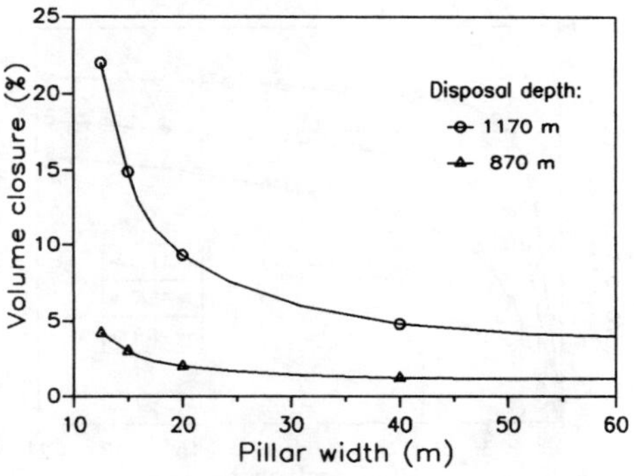

Fig. 3 Volume closure of the disposal drift as a function of the pillar width for two levels

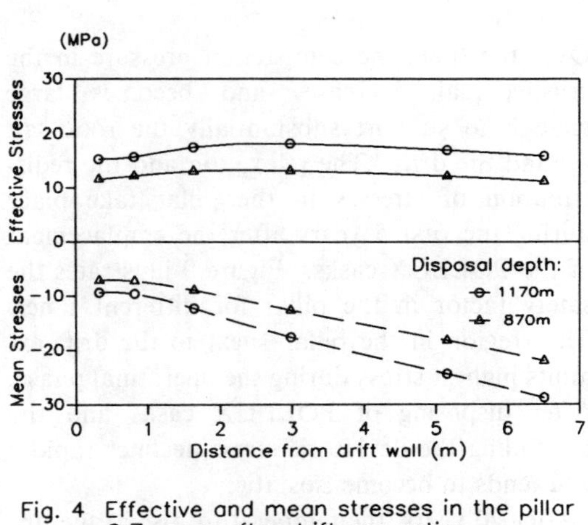

Fig. 4 Effective and mean stresses in the pillar 0.7 years after drift excavation

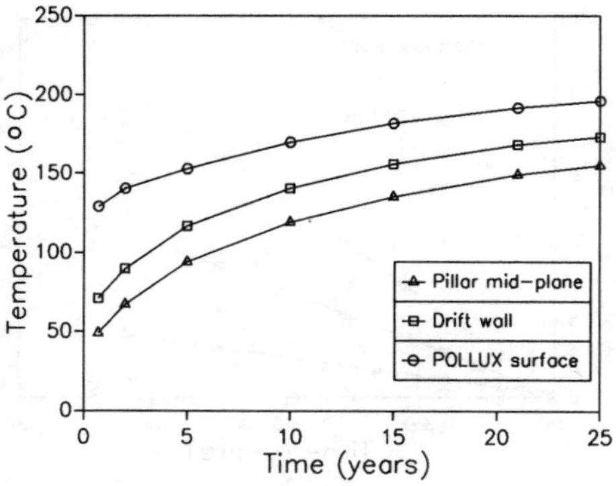

Fig. 5 Temperature histories at different positions

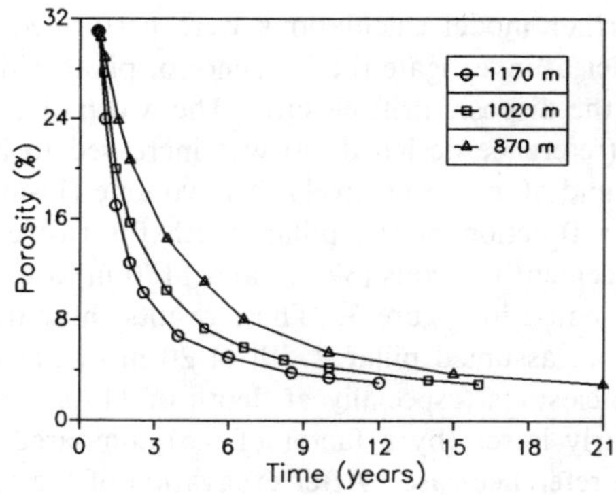

Fig. 7 Development of the porosity in backfill material for the three emplacement horizons

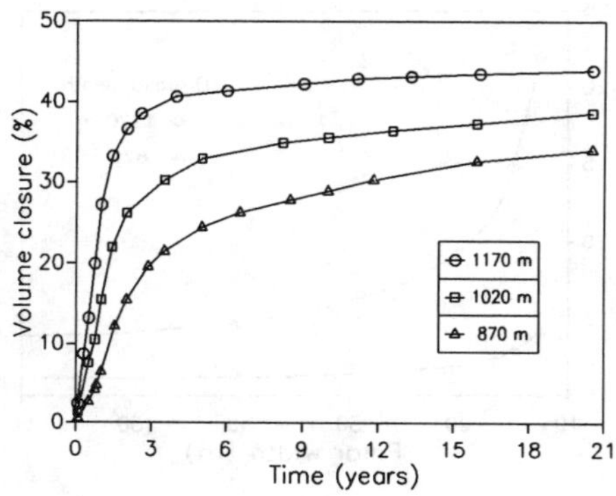

Fig. 6 Volume closure of the backfilled disposal drift for the three emplacement horizons

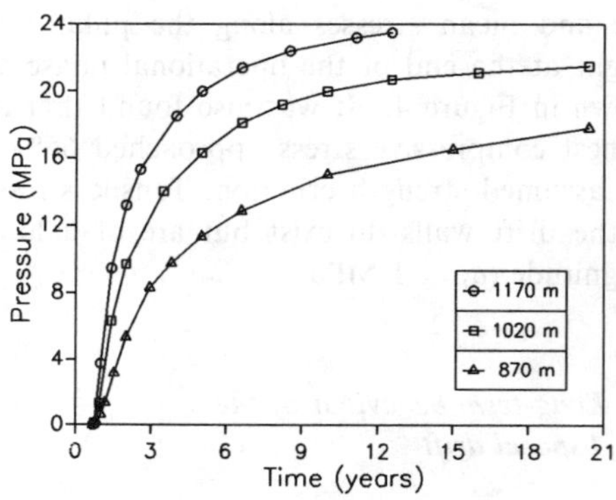

Fig. 8 Development of the compaction pressure in the backfill material

Over the time, the compaction pressure in the crushed salt increases and becomes large enough to support substantially the rock salt around the drift. The relaxation and the redistribution of stresses in the pillar take place during the first 5 years after the emplacement of the POLLUX casks. Figure 9 illustrates the safety factor in the pillar for different times. The region in the pillar next to the drift exhibits highest stress during the operatinal phase. After disposing of POLLUX casks and the backfilling the deviatoric stress declines rapidly and tends to become isostatic.

After 20 years, the temperature rise at the upper boundary of the geometry indicates that the numerical model becomes inadequate for this period of time. For the numerical analyses relating to the long-term periodes, the far field thermomechanical effects have to be considered.

5 CONCLUSIONS

From the results of the numerical analyses the following conclusions can be drawn:

- The convergences of the disposal drifts are crucially dependent on the disposal depth, i.e., on the prevailing lithostatic stress. It appears from the results that the volume closure at a depth of 1170 m during the operational phase is about four times the one of the 870 m level. However, there are no criterion for judgement of the higher convergence rates on the two lowest horizons.
- Furthermore, the closure of the disposal drift is considerably influenced by the width of the pillar between adjacent drifts. For instance, doubling of the specified pillar width of 12.5 m reduces the convergence by 5%.

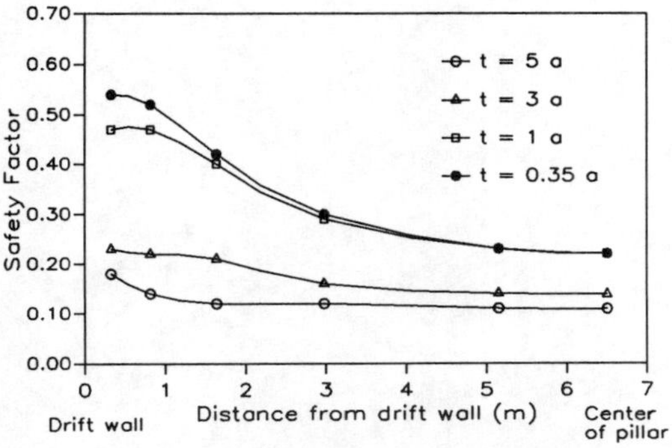

Fig. 9 Development of the safety factor in the pillar for different times (disposal level: 870 m)

- High compressive stresses were calculated in the vicinity of the drift during the mining phase. However, the ratio of the actual compressive stress to the assumed failure criterion remains less than unity, i.e., no cracks are expected.

- The results related to the long-term behaviour of the disposal drift show that already after a few years the porosity in the backfill material has become greatly reduced. By increasing the compaction pressure, the supporting effect of the backfill material is improved, which means that the surrounding rock salt is relieved.

REFERENCES

Albers, G. 1983. MAUS - A Computer Code for Modelling Thermomechanical Stresses in Rocksalt, in Computer Modelling of Stresses in Rock, Proc. Tech. Session, Brussels, EUR9355 EN

Bechthold, W., W. Braun, C. Brückner, K.D. Closs, U. Knapp, R. Papp 1989. Systemanalyse Mischkonzept, Hauptband, KWA-2190 A1, Kernforschungszentrum Karlsruhe GmbH.

Diekmann, N., U.Hunsche, D. Meister 1986. Über das geomechanische Verhalten von Steinsalz bei erhöhten Temperaturen, *Z. dt. geol. Ges.*,Vol.137, p.29-46

Diekmann, N., D. Stührenberg 1991. Untersuchungen zum Kompaktionsverhalten von Salzgrus mit der Oedometerzelle TRE3002, BGR-Laborbericht zum Forschungsvorhaben KWA 58019, Teilproject II, October 1990, Archiv-Nr.:107586

Korthaus E., W. Bechthold, K.D. Closs, A. Pudewills 1989. Thermische und thermomechanische Analysen des Endlager-Nahbereichs und Ermittung des Flächenbedarfs des Endlagers, Systemanalyse Mischkonzept, Tech. Anhang 5, KWA-Nr. 2190 A1, Kernforschungszentrum Karlsruhe GmbH

Langer, M. 1982. Geotechnical Investigation Methods for Rock Salt,*Bulletin of Int. Assoc. of Eng. Geol.*, No.25, p.155-164

Ploumen, P. 1980. Numerische Langzeitrechnung dreidimensionaler Temperaturfelder mit Hilfe eines speziellen Finite-Element-Verfahrens am Beispiel der Endlagerung hochradioaktiver Abfälle im Salzgestein, Diss. at the RWTH-Aachen.

Wallner M., C. Caninenberg, H. Gonther 1979. Ermittlung zeit- und temperaturabhängiger mechanischer Kennwerte von Salzgestein, Proc. 4. Int. Congr. Rock Mech., Montreux, Sept.2-8, 1979, Vol.1,p.313-318, Rotterdam, Balkema

Innovative Mine Design for the 21st Century, Bawden & Archibald (eds) © 1993 Balkema, Rotterdam, ISBN 90 5410 325 6

Application of two and three dimensional numerical models for intact salt rock

L. Rothenburg
University of Waterloo, Ont., Canada

M.A. Frayne
Sifto Canada Inc., Goderich, Ont., Canada

D.Z. Mraz
Mraz Project Consultants Ltd, Saskatoon, Sask., Canada

ABSTRACT: An extensive research program has been conducted to provide a technical basis for the use of halite backfill in underground potash mines in Canada. Constitutive relationships to predict the behavior of intact salt rock and a methodology to determine material parameters were developed using results from in-situ monitoring and laboratory testing. The constitutive relationships have been incorporated in two and three dimensional numerical models. Results of the numerical simulations were compared to in-situ convergence rates, pillar stress and subsidence rates. Excellent agreement was obtained between predicted and actual measurements for both models in their respective ranges of applicability.

1 INTRODUCTION

This paper presents a brief account of an extensive in-situ monitoring, laboratory testing and numerical modeling program carried out over a number of years at the Denison-Potocan Potash Company (DPPC) mining site in Sussex, New Brunswick, Canada (CANMET, 1990). The project was conducted under the auspices of the Canada Centre for Mineral and energy Technology (CANMET). The research was conducted to provide a technical basis for the use of halite backfill in underground potash mines in Canada.

The emphasis of the paper is on the methodology of determining creep parameters of potash from a combination of laboratory testing and in-situ observations on virgin ground openings. These parameters were further used to model numerically an existing large scale mining layout. The results of modelling are presented in the paper and compared with in-situ observations to validate the methodology of determining material parameters and computational techniques. The paper presents a detailed account of this effort and a description of the 2D and 3D computer codes involved.

2 IN-SITU MONITORING PROGRAM

The Clover Hill deposit is confined to the Marchbank Syncline, which is the most southerly major structural feature of the Moncton Sub-basin. The deposit is approximately 10 km in length along the strike and has a maximum plan width of 3.5 km. It has an average thickness of 12.5 meters with an average depth of 880 meters below ground surface. The room and pillar mining method is utilized to extract the ore uses both continuous mining machines as well as drilling and blasting techniques. Figure 1 shows the existing mining layout.

The significance and value of in-situ rock mechanics data to mine design was recognized early with collection and analyses of the data beginning in 1982. The instrumentation ranged from simple inexpensive manual measurement systems to electronic measuring devices and programmable data loggers. Prior to any significant mining, the test room and stope were excavated to provide virgin ground field data (Figure 1). As mining progressed into the production phase, emphasis was placed upon the monitoring of active mining sites. Monitoring stations were constructed in Stopes 11200 and 11240, and in a three room entry system located in Panel 12 (Figure 1).

The information collected from virgin ground provided a tool for monitoring localized and global mine stability.

3 CREEP BEHAVIOR OF POTASH

In general, a comprehensive characterization of salt rock and potash creep requires an extensive laboratory testing in a wide range of stresses and temperatures. Since creep is extremely slow at low shear stress, laboratory testing at stresses and temperatures relevant to mining conditions is time consuming and expensive. Even though behavior of potash at ele-

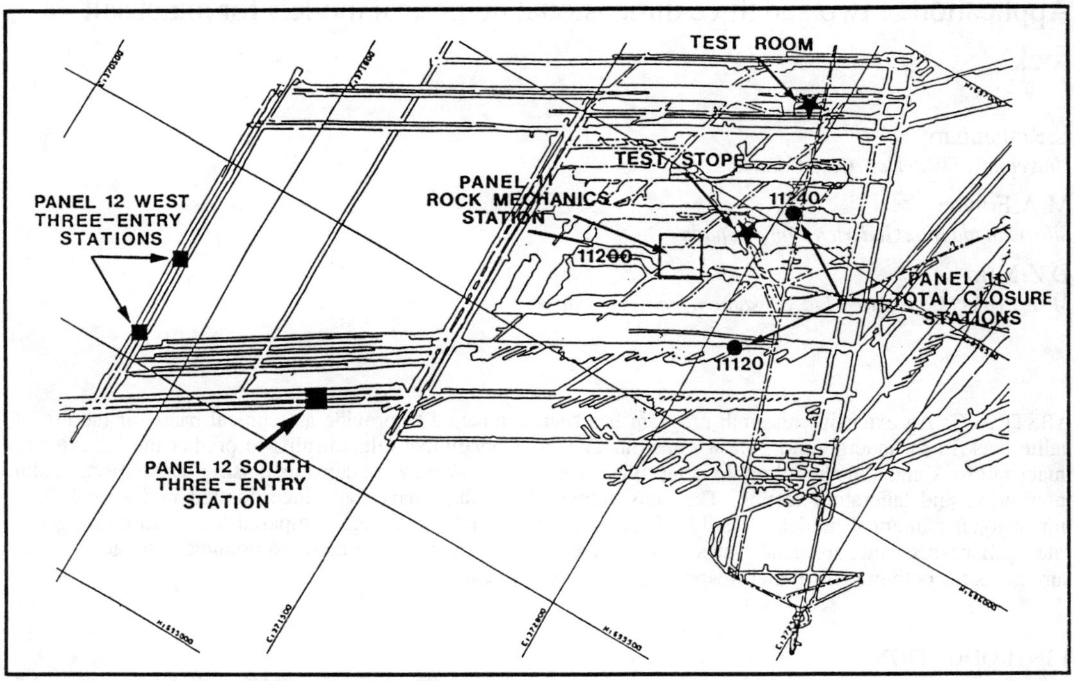

Figure 1: Rock mechanics instrumentation stations

vated temperatures is usually irrelevant in mining situations, determination of material parameters for a comprehensive material model (e.g. Munson and Dawson, 1982) is impossible without tests under elevated temperatures. Very few comprehensive data sets are available for salt rock and there appears to be none for potash. As a result, other methodologies have to be pursued in mining situations.

According to modern theoretical views, creep in polycrystalline materials is primarily due to the presence of lattice defects and discontinuity in lattice orientations across grain boundaries. Impurities in natural materials and small quantities of brine in salt rock are additional imperfections that may enhance or possibly even control creep deformations. In general, the rate of creep depends on the nature of defects, their concentration and mobility under given conditions.

The full complexity of the mechanisms by which materials behave in creep became apparent only in the last decade. What was once thought to be a simple mechanism has now revealed itself as a hierarchy of interacting mechanisms, each dominant over a certain range of stress, strain rate and temperature. A mechanism map in Figure 2 developed for salt rock (Munson, 1979) describes, in a generic sense, currently perceived dominant mechanisms controlling the steady state creep of salt. The observed steady state creep is a sum of contributions from different mechanisms. Because of the exponential nature of creep rates for individual mechanisms as a function

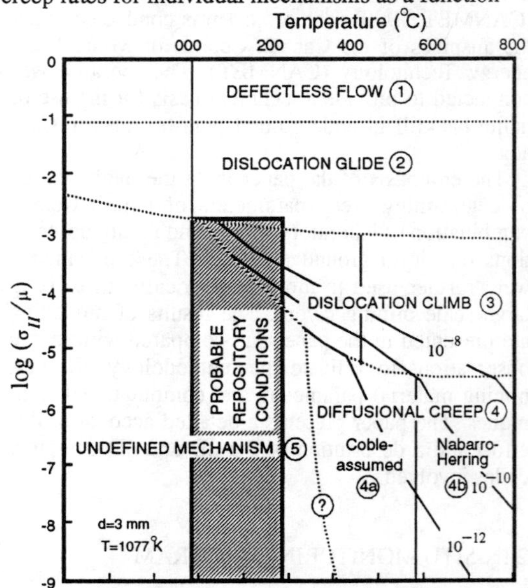

Figure 2: Creep mechanisms (After Munson, 1979)

of shear stress, not all mechanisms are significant contributors into the overall strain rate in a particular range of stresses and temeratures.

3.1 Two mechanism creep law

As Figure 2 illustrates, only two mechanisms are thought to operate at typical mining temperatures and stress conditions. These are dislocation glide and still unidentified mechanism. This leads to the concept of a two-component creep law reflecting a change of a power law exponent at some stress level between the two regions.

Analytically the concept of a two component steady state creep law can be expressed as follows:

$$\dot{\varepsilon}_{ss} = \begin{cases} \dot{\varepsilon}_0 \left(\dfrac{\sigma_{eff}}{\sigma_0}\right)^{N_1} & \text{for } \sigma_{eff} \geq \sigma_0 \\ \dot{\varepsilon}_0 \left(\dfrac{\sigma_{eff}}{\sigma_0}\right)^{N_2} & \text{for } \sigma_{eff} < \sigma_0 \end{cases}$$

where $\dot{\varepsilon}_{ss}$ is the steady state creep rate, σ_{eff} is the 'effective stress', σ_0 is the shear stress at the boundary between mechanisms controlled by exponents N_1 and N_2, $\dot{\varepsilon}_0$ is the strain rate at $\sigma_{eff} = \sigma_0$. The above model implies a sharp change in strain rate between the two mechanisms, which is unlikely to occur in reality. Nevertheless, the advantage of the model is in its simplicity and it represents a step forward from a single mechanism creep law that has been used in practice for some time.

In order to apply the two mechanism creep law in practice it is necessary to develop a realistic procedure for assessing material parameters.

4 DETERMINATION OF CREEP PARAMETERS

The creep law is characterized by four parameters, two power law exponents N_1,N_2, the transition stress σ_0 and the corresponding strain rate $\dot{\varepsilon}_0$. These parameters were established based on a combination of laboratory tests, available information on creep of salt rocks and in-situ observations.

4.1 Power law exponents

The stress exponents N_1 characterizes creep at high shear stress and can be determined from laboratory tests that are reasonably fast. A group of such tests corresponding to $N_1 = 6.9$ is illustrated in Figure 3.

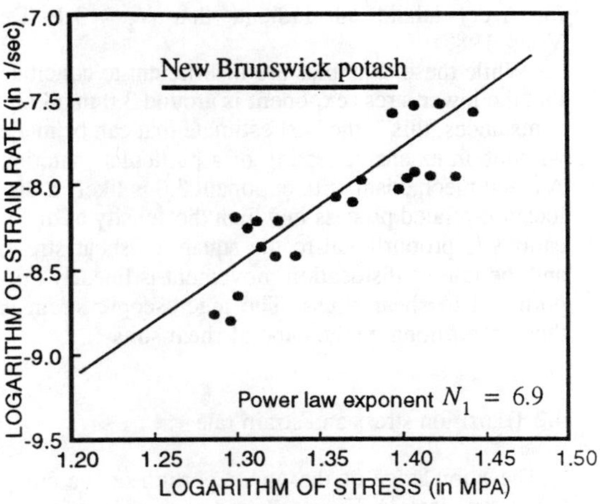

Figure 3: Results of creep tests at high stress range

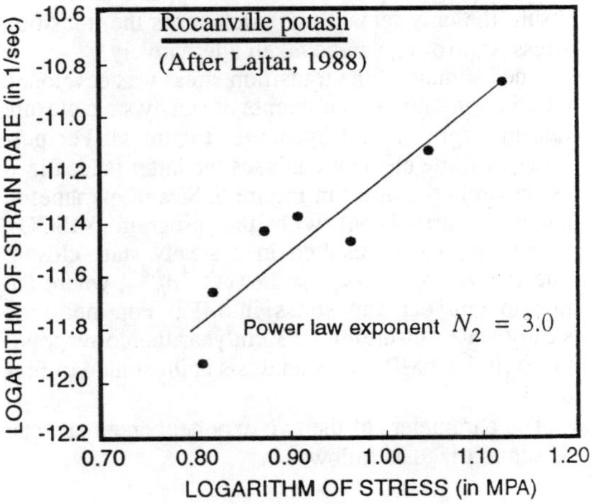

Figure 4: Results of creep tests at low stresses

At low shear stress and room temperature, a single creep test may take in excess of half a year to complete. Considering that at least half a dozen tests are required to determine the stress exponent, such lengthy testing is usually not feasible.

For potash there appears to be only one comprehensive set of uniaxial compression tests at low stress (Lajtai, 1988). The tests carried out on Rocanville potash and linear regression of the data (Figure 4) suggest an exponent $N_2 = 3.0$

More experimental information on creep at low stress is available for salt rock. A set of 34 tests on Palo Duro Salt (combined time in test cells 6.5 years) gives the low exponent of 3.0 when parameters for the steady-state portion of Munson-Dawson model are estimated (Rothenburg, 1988). Similar estimates for Avery Island salt indicate that $N_2 = 3.1$ (De Vries, 1988).

While these examples are insufficient to conclude that the lower stress exponent is around 3.0 in all circumstances, this is the best estimate that can be made without an extensive testing of a particular material. A creep mechanism with exponent 3.0 is likely a dislocation-related process in which the density of dislocations is proportional to the square of shear stress, and the rate of dislocation movement is linearly proportional to shear stress. The macroscopic strain is then proportional to the cube of shear stress.

4.2 Transition stress and strain rate

The knowledge of the upper portion of the creep law from data in Figure 3 and the assumption of $N_2 = 3.0$ for the lower branch leaves only one parameter of the creep law unknown. This is because the continuity of creep rates at $\sigma = \sigma_0$ links σ_0 and $\dot{\varepsilon}_0$ according the power law of the upper branch. As a result, the only remaining parameter is the transition stress, σ_0. (or $\dot{\varepsilon}_0$ can be taken alternatively).

The estimate of the transition stress was developed based on in-situ measurements of steady state closure rate in virgin ground openings (Figure 1). For purposes of finite element analyses the latter is idealized as shown in the insert in Figure 5. Several parametric analyses carried out with the program VISCOT (INTERA, 1983) resulted in a steady state closure rate $\dot{u}_{ss}$ vs. σ_0 as $\dot{u}_{ss} = 1.67x10^{-4}\sigma_0^{3.8}$, where the rate in cm/year and stress in MPa. For measured steady state closure of 1.13 cm/year the corresponding σ_0 is 10.2 MPa. This analysis is illustrated in Figure 5.

The parameters of the two exponent creep law can be summarized as follows:

High streess branch exponent:	$N_1 = 6.9$
Low stress branch exponent	$N_2 = 3.0$
Transition stress	$\sigma_0 = 10.2MPa$
Transition strain rate	$\dot{\varepsilon}_0 = 3.24x10^{-11}\sec^{-1}$

5 ASSESSMENT OF CREEP PARAMETERS

The consitency of the parameter calibration based on the observed closure rate was further assessed by comparing measured salt rock displacement rates around the opening with numerically computed displacement rates. In connection with this comparison it is worth discussing some features of numerical solutions with the adopted two-mechanism creep law.

One important qualitative implication of the above creep law is that effective stress around deep openings should be close to σ_0 as stresses in excess of σ_0 lead to high creep rates that result in shear stress redistribution reducing the effective stress to the level close to σ_0. In turn, this implies that the strain rate in

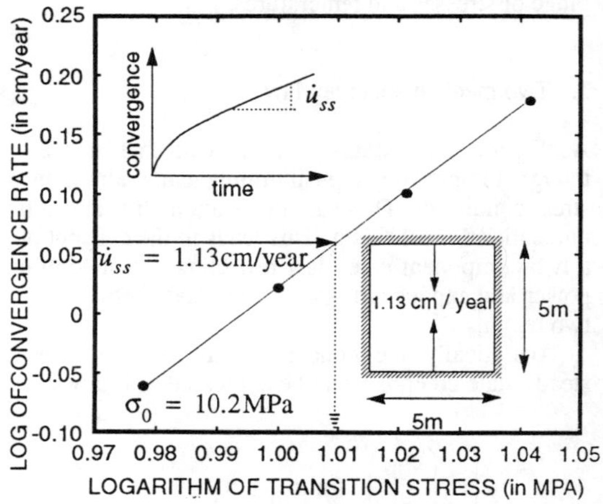

Figure 5: Illustration of the procedure for determining salt creep parameters from in-situ data and FEM

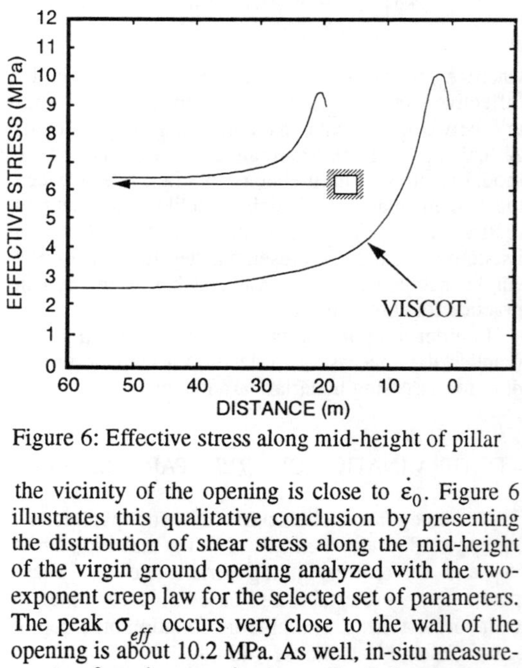

Figure 6: Effective stress along mid-height of pillar

the vicinity of the opening is close to $\dot{\varepsilon}_0$. Figure 6 illustrates this qualitative conclusion by presenting the distribution of shear stress along the mid-height of the virgin ground opening analyzed with the two-exponent creep law for the selected set of parameters. The peak σ_{eff} occurs very close to the wall of the opening is about 10.2 MPa. As well, in-situ measurements of strain rates in the wall of the test room resulted in strain rates of the order of $\dot{\varepsilon}_0$.

Exact correspondence with in-situ measurements cannot be expected since the strain rate measurements with extensometers give only averages of creep rates over a distance between two anchors. However it is possible to compare salt rock displacement rates, measured with extensometers, with computational results. Figures 7 and 8 illustrate this comparison for extensometers located in the roof and wall of the test room. Note that the vertical closure rate of 1.13 cm/year quoted above relates to roof-to-floor measurements while roof and floor displacement rates in Figures 7 and 8 are about half of that values.

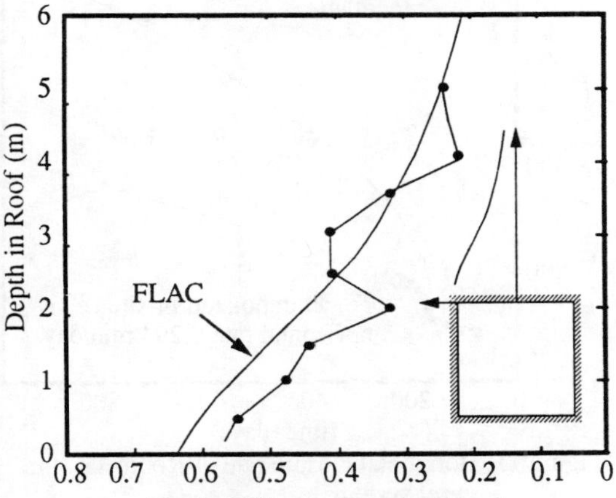

Figure 7: Vertical closure rate of test room

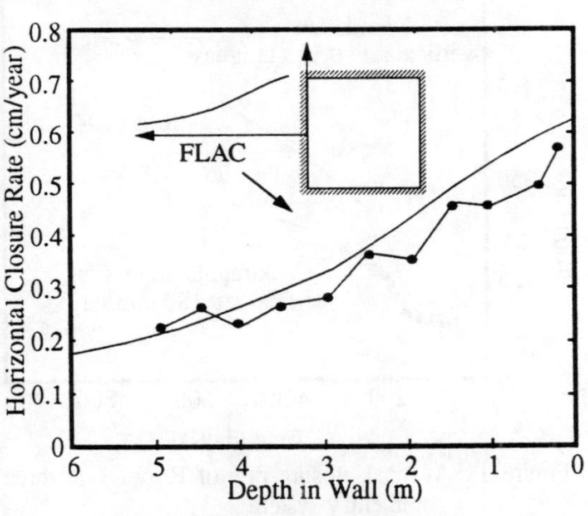

Figure 8: Horizontal closure rate of test room

The numerical data in Figures 6, 7 were obtained with the well known 2D finite difference code FLAC (ITASCA, 1989) which accounts for geometry changes during creep. The parameter calibration was carried out with VISCOT which uses a small strain formulation. Results of both runs are numerically indistinguishable as far as steady state rates are concerned.

All computations carried out during this study neglect a material component of transient creep. The transient stress effects captured in the described analysis are related to stress redistributions that occur from the moment the room is excavated (instantaneously in the above analyses) and until steady state is reached. Since neither the sequence of excavation not the transient creep in the material law are accounted for, the total computed closures of the excavations do not match field measurements (Figure 9). Nevertheless, the steady state rates are represented adequately.

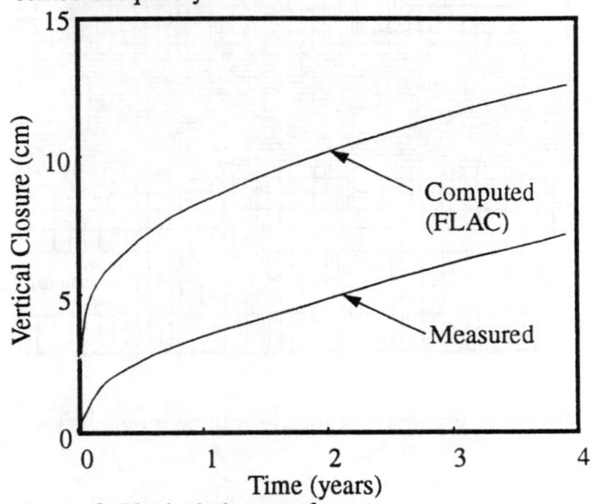

Figure 9: Vertical closure of test room

Overestimation of in-situ closures is, most likely, due to the assumption of instantaneous excavation which initially generates rapid creep. Also, field data usually miss some closure that occurs before extensometers are installed.

5.1 Two dimensional modelling of an isolated three room entry system

Closure monitoring data from various parts of the DPPC mine provide a unique opportunity to compare results of numerical modelling with observations. In most areas of the mine the layout (Figure 1) is too complex for 2D modelling. However, in the south of the panel 12 the geometry of a three room entry system can be adequately represented in plain strain approximation. Another advantage of this geometry is that during the monitoring period these rooms were sufficiently remote from areas of active mining. The geometry of rooms, idealized for FLAC analysis, is shown in Figures 10 and 11 depicting the two stages in excavation of these entries.

The first stage of development and its numerical analysis relates to excavation of rooms 1 and 3 (Figure 10). The FLAC analysis with the previously selected parameters matched in-situ creep rates adequately. However, when the excavation of the middle entry was introduced, numerical results overestimated the observed closure rates by a factor of 1.6

The excavation of the middle entry apparently resulted in some overload of pillars and their actual plastic response (not accounted for in the numerical model) resulted in additional stress redistributions that lead to a reduction in-situ creep rate.

Other areas where closures were monitored

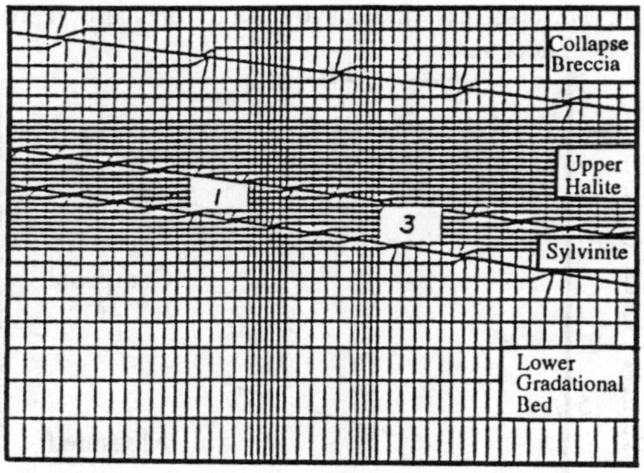

Figure 9: Fragment of mesh for three room entry system. Stage 1

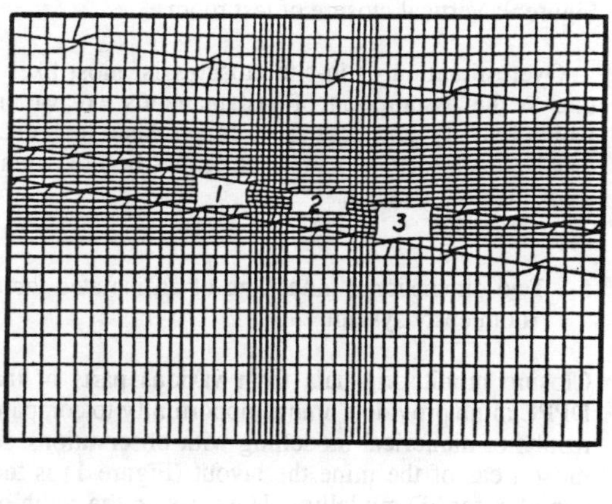

Figure 10: Fragment of mesh for three room entry system. Stage 2

included test stopes in panel 11 (Stopes 11120, 11200 and 11240, Figure 1). These locations are in the areas of very complex geometry and an attempt to carry out a two-dimensional analysis resulted in significant overestimation of convergence rates. Much better results were obtained in the course of three-dimensional modelling described below.

6 THREE DIMENSIONAL MODELLING

The three-dimensional modelling was done using the program VELMINA (Visco-Elastic Mine Analysis). The program was designed for three-dimensional analysis of large scale mining layouts that involved large numbers of rooms and pillars in visco-elastic rock. From a technical perspective, the program is an implementation of the Displacement Discontinuity method for visco-elastic media. Various implementa-

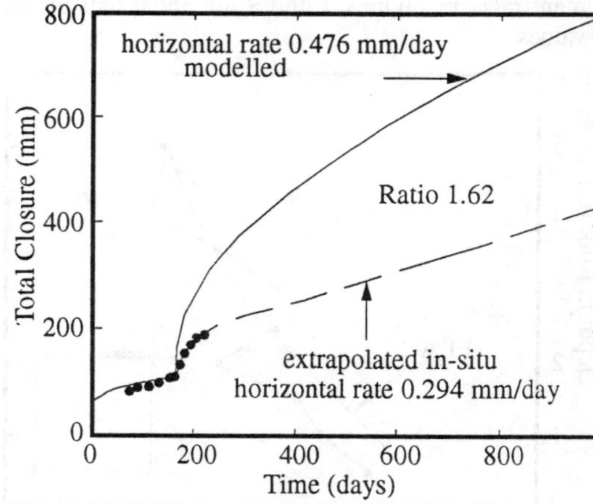

Figure 11: Horizontal closure rate of Room 1 of three entry system

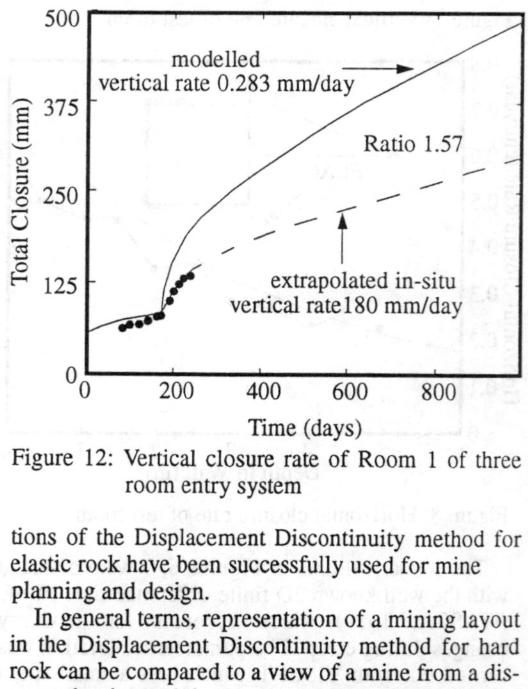

Figure 12: Vertical closure rate of Room 1 of three room entry system

tions of the Displacement Discontinuity method for elastic rock have been successfully used for mine planning and design.

In general terms, representation of a mining layout in the Displacement Discontinuity method for hard rock can be compared to a view of a mine from a distance that is much larger compared to sizes of rooms and pillars in the mine. From this distance the mine openings would appear as a set of discontinuities in the host rock. If one is to trace displacements of rock along a line crossing the mining plane, the impression would be that displacements are discontinuous across the mine. The Displacement Discontinuity Method is a mathematical idealization of that situation.

In all existing implementations of the Displacement Discontinuity method for hard rock mining, pillars are described in terms of equivalent gross properties like gross elastic stiffness, collapse loads, etc. From a computational point of view, a typical pillar in the mine is represented by a set of elements that support the roof and transfer load to the mine floor. Such elements "feel" only stress normal to the roof and deform in response to this stress only. However, deformational and strength properties of a pillar are dependent on confining stresses within the pillar. The latter are not a part of the analysis and a relationship between stress directly involved in computations and confining stress in pillars must be established *a priory*. Ideally, a series of finite element analyses is required for this purpose. When pillars creep, determination of confining stress in pillars is unavoidable since creep rates in pillars are dramatically affected by confining stress.

The current mathematical formulation expressed in the VELMINA code is a rigorous implementation of the Displacement Discontinuity method for a linearly visco-elastic rock outside of the mining seam. In-seam rock properties can be described by an arbitrary non-linear steady state creep law. Computational efficiency is achieved for the "bi-linear" creep law with individual exponents pertaining to each branch. Pillars can also be assigned non-linear elasto-visco-plastic properties (i.e. creep, limiting load and softening). Both transient and steady state conditions can be analyzed. However, determination of transient response requires considerable computing effort. Multi-level mining and sequential excavation can also be analyzed. The capability to model sequential excavation includes placement of backfill.

Non-liner creep properties of host rock outside of mining plane cannot be modelled directly and equivalent linear properties must be assigned. The program has a capability of doing an iterative analysis to determine matching equivalent linear properties for a specified non-linear creep law of the host rock.

The entire mining layout is discretized into a series of rectangular elements that can represent either mined-out area or a part of a pillar. Each element is associated with a distinct set of material properties. For elements representing pillars with visco-elastic properties, a ratio of vertical to effective stress is assigned.

The ratios of vertical to effective stress are assigned based on finite element analysis of pillars. Although it is technically possible to couple finite element discretization of pillars with the displacement discontinuity code and solve for stresses in pillars, this is not feasible in practical applications that involve hundreds of pillars. To cope with this difficulty VELMINA has a set of built-in functions that assign effective to vertical stress ratios based on the distance of the element from the edge of the pillar. These functions were developed based on an extensive finite element study of rectangular cross-section pillars with different width to height ratios. In case of openings of other shapes, a separate finite element study is required.

The automatic assignment of effective to vertical stress ratio and determination of equivalent viscosity for out-of-seam material are the most significant technical elements of the VELMINA code. These are described in Appendix A together with other technical aspects of VELMINA. As well, verification of the code based on comparison of its performance in conditions of plane strain with results of finite element computations is presented.

6.1 Simulation of 1988 mining layout

The status of mining excavations in late 1988 is shown in Figure 1. The first mining panel, Panel 10, has been only partially mined out due to problems associated with geological complexities. Panel 11 was the primary object of modelling. In order to reflect the complexity of the ore body correctly in the computer simulation, the ore thickness information has been used based on the ore thickness isopach map.

Since VELMINA utilizes a uniform mesh of rectangular elements, the size of elements must be chosen carefully in order to properly represent the excavations within the ore zone. The size of the elements selected for simulation was 7.67 m by 7.67 m. This element size represented the best compromise for simulation of various rooms, stopes and pillars throughout the mine layout.

The mesh of the mine geometry consisted of 30,000 elements. Figure 13 illustrates the mesh used in analysis. The mine development openings were excavated to the height of 4 meters.

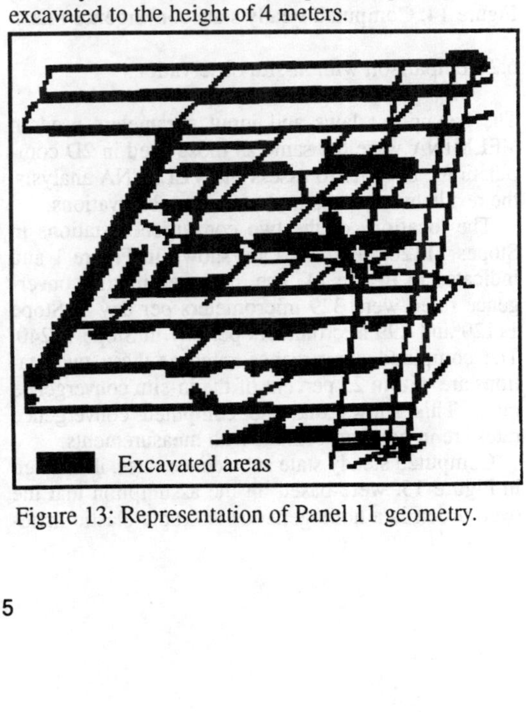

Figure 13: Representation of Panel 11 geometry.

Figure 14 illustrates simulated contours of steady state convergence rates and compares simulation results with in-situ measurements. Figure 16 compares actual and computed steady state surface subsidence rates across the mining layout.

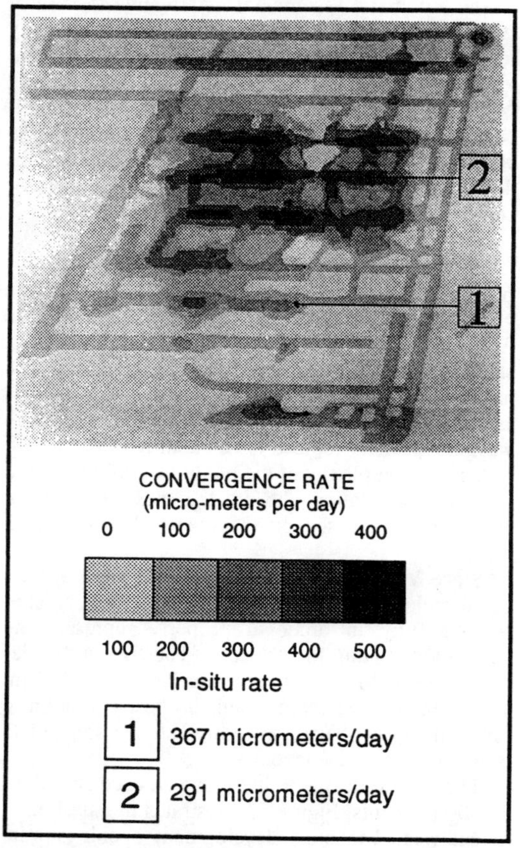

Figure 14: Computed steady state convergence rates.

6.2 Comparison with in-situ observations

The behavioral laws and input parameters used in VELMINA were the same as those used in 2D computations. In order to assess the VELMINA analysis, the results were compared to in-situ observations.

The locations of the two convergence stations in Stopes 11120 and 11240 are shown in Figure 1 and indicated in Figure 14. The measured in-situ convergence rated were 329 micrometers per day in Stope 11120 and 299 micrometers per day in Stope 11240. The computed convergence rated at these two stations are within 25 percent of the in-situ convergence rates. This shows that the computed convergence rates are in good agreement with measurements.

Computed steady state subsidence rates, illustrated in Figure 15, were based on the assumption that the overburden is a homogeneous isotropic elastic material and the ground surface is a flat horizontal plane at the level of shaft collar. Actual stratigraphy of the overburden and the topography of the ground surface are very complex. Nevertheless, the computed subsidence

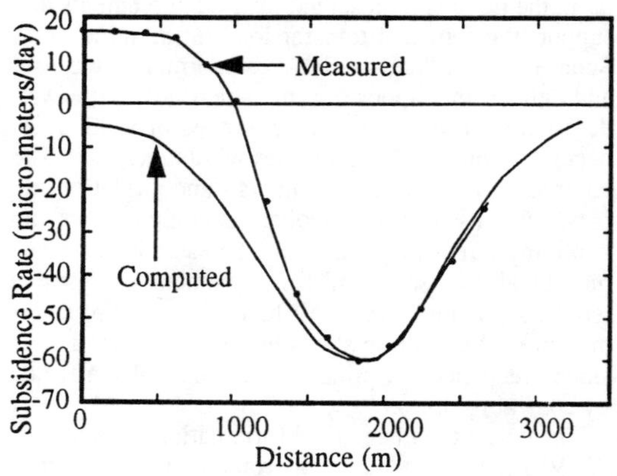

Figure 15: Subsidence rate for 1987-1988, East-West

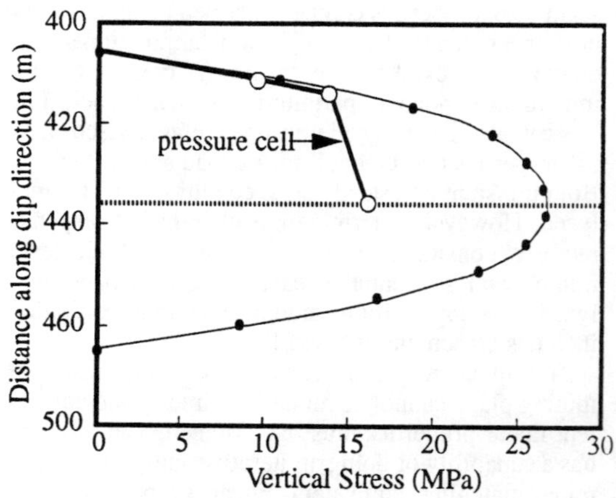

Figure 16: Vertical stress in pillar in Stope 11200

rates are in excellent agreement with actual subsidence rates despite the introduced simplifications.

The north pillar of the rock mechanics station in Stope 11200 have been instrumented with pressure cells. The comparison of readings from three undamaged cells and the computed stresses within the pillar is shown in Figure 16. It would appear that the cells within the high exponent region reflect the stress reasonably well. However, the final reading of the cell located in the center of the pillar, in the low exponent region, indicates that it did not reach full pressure before monitoring stopped.

In terms of absolute numbers, the convergence rates are the most reliable data in verifying the computer simulation. In spite of mining disturbances at the time of measurements, the predicted and measured convergence rates compare well and confirm the validity of the input parameters and material laws used. The fact that the subsidence rates, and to some extent vertical stress measurements, compare well with the predicted values increases the level of confidence in this modelling.

7 CONCLUSION

The methodology described in this paper was effectively used to determine creep parameters of potash from a combination of laboratory testing and in-situ observations of virgin ground openings. These parameters were successfully used to numerically model an existing large scale mining layout. Comparison of actual and computed convergence rates and stresses showed good correlation. Calibrated models represent a useful mine design tool that can be used for assessment of mining layouts.

REFERENCES

CANMET, 1989. Use of backfill in New Brunswick potash mines - Phase II. Report prepared by Denison Potocan Potash Company, Mraz Project Consultants Ltd. and University of Waterloo, for the Canada Centre for Mineral and Energy Technology, Dept. of Energy Mines and Resources, Ottawa, ONT

De Vries, K.L. 1983. Visco-plastic laws for Avery Island salt. Topical Report RSI-0333 (Draft), Re/Spec Inc., prepared for ONWI, Rapid City, SD

INTERA, 1983. A two-dimensional and axisymmetric nonlinear transient thermo-viscoelastic finite element code for modelling time-dependent viscous mechanical behavior of rock mass, Technical Report, INTERA Environmental Consultants Inc., Houston, TX

ITASCA, 1989. User manual for FLAC - Fast Lagrangian Analysis of Continua, Version 2.2, Itasca Consulting Group Inc., Minneapolis, MN

Lajtai, E.Z. and Duncan, E.J., 1988. The mechanism of deformation and fracture in potash rock, Final Report for CRD Project 0038997, University of Manitoba, Winnipeg, MAN

Munson, D.E. and Dawson, P.R., 1982. A transient creep model for salt during stress loading and unloading, SAND82-0962, Sandia National Laboratories, Albuquerque, NM.

Rothenburg, L., 1988. Constitutive behavior of salt rock and thermo-mechanical aspects of waste package performance in bedded salt, Report for Stone & Webster Engineering Corporation under contract with US Department of Energy, Boston, MA

APPENDIX A

A1 TECHNICAL ASPECTS OF VELMINA CODE

The basic formulation in the VELMINA code assumes linearly visco-elastic behavior of host rock outside of the mining horizon. This assumption makes it possible to use the entire mathematical apparatus of the Displacement Discontinuity Method for linearly viscous materials by using the well-known formal analogy between displacement fields in linearly elastic incompressible solids and velocity fields in a Newtonian (linearly viscous) fluid.

The origin of the analogy between behavior of incompressible elastic solids and Newtonian fluids is related to the fact that the strain rate $\dot{\varepsilon}_{ij}$ in a linearly viscous media is related to stress deviator σ'_{ij} as $\dot{\varepsilon}_{ij} = \sigma'_{ij}/2\mu$, where μ is the dynamic viscosity. This relationship implies incompressibility of flow, (i.e. $\dot{\varepsilon}_{kk} = 0$ since $\sigma'_{kk} = 0$) and it is mathematically analogous to generalized Hooke's law for elastic material with Poisson's ratio $\nu = 0.5$ and shear modulus $G = \mu$. In this analogy, the strain rate corresponds to elastic strain. Due to this formal correspondence, a velocity field in a linearly viscous fluid can be found by determining the displacement field in an incompressible solid for the same boundary conditions as in the flow problem. As a result, the displacement discontinuity solution of linear elasticity taken for $\nu = 0.5$ is sufficient to describe the discontinuity in velocities of viscous flow across the mining plane. In a sense, the described method can be referred to as the "velocity discontinuity method".

A1.1 The velocity discontinuity method

Technical details of the VELMINA code can be best outlined for the case when steady state stresses and rates are computed bypassing determination of transient response. Further description follows rather standard logic of the displacement discontinuity method in application to mining.

Let the mining plane be subdivided into M "face elements" representing mine pillars and excavated areas. The steady state creep rate in the host rock is discontinuous in crossing the mining plane from roof to floor of the mine. Let the magnitude of rate discontinuity associated with element i at location r_i be d_i. As a result of this discontinuity, stresses are induced everywhere in the media. In particular, stress at centers of other elements located at r_k and acting normal to the face elements can be determined as $\sigma_k = d_i S(r_i - r_k)$ where S is a function representing the displacement discontinuity solution of linear elasticity for $\nu = 0.5$. The total stress at any location includes the pre-mining stress σ_k^0 and the sum of contributions from all displacement discontinuities.

The rate discontinuities d_i are unknown quantities and must be determined as a part of solving the problem. In excavated areas, stresses induced by rate discontinuities must be negative of the pre-mining stress so that the total stress is zero. In elements representing pillars, the roof-floor discontinuity rate d_k must be the same as the rate of pillar compression $\hbar_k$. The latter is a function of the vertical stress σ_k at location r_k as well as other stress components at the location of the element k. These are dependent on the element position within the pillar, i.e. $d_k = \hbar_k = \hbar(\sigma_k, r_k)$ where $\hbar(\sigma, r)$ is a function representing behavior of pillars.

The set of equations for determining rate discontinuities are as follows:

$$\sum_{i=1}^{M} d_i S(r_i - r_k) + \sigma_k^0 = 0 \quad \text{(Excavated areas)}$$

$$\sigma_k = \sigma_k^0 + \sum_{i=1}^{M} d_i S(r_i - r_k) \quad d_k = \hbar(\sigma_k, r_k) \quad \text{(Pillars)}$$

The above system of M equation with M unknown displacement discontinuities is solved iteratively.

A 1.2 Mathematical model for pillars

In the described displacement discontinuity formulation, the mechanical response of pillars is represented by a mathematical model reflected in the relationship $d = \hbar(\sigma, r)$. By its physical nature, the function $\hbar(\sigma, r)$ describes the steady state compression rate of an element of a pillar that is under normal stress σ and is located at position r in the mine. In actual applications it is assumed that $\hbar(\sigma, r)$ depends only on the location within the pillar itself and is independent on the arrangement of other pillars in the mine. With this assumption $\hbar(\sigma, r)$ can be derived by doing finite element analysis of a single pillar. The method of deriving this function is described below assuming for simplicity a single mechanism power law of the form:

$$\dot{\varepsilon}_{ij} = \varepsilon_0 \frac{3}{2} \frac{\sigma'_{ij}}{\sigma_{eff}} \left(\frac{\sigma_{eff}}{\sigma_0}\right)^N \quad \text{where } \sigma_{eff} = \sqrt{\frac{3}{2}\sigma_{mn}\sigma_{mn}}.$$

In the displacement discontinuity idealization of mining conditions only vertical stress $\sigma = \sigma_{11}$ is involved (when the mining plane is horizontal). The actual response of a pillar expressed in terms of $\hbar(\sigma, r)$ is determined by complex stress distributions that involve all stress components, not only σ. Since in the displacement discontinuity formulation only σ is involved, it is necessary to reconstruct other stress components based on finite element analysis. To make such analysis generic and applicable to any stress level it is assumed that all stress components are locally proportional to vertical stress with which the displacement discontinuity analysis operates. In this case stress ratios from finite element calculations can be treated as characteristics of pillar geometry.

Consider first a long rectangular pillar of height H and width B that can analyzed in plane strain approximation. Let x refer to a coordinate describing the location within the pillar and selected it in such a way that $x = 0$ in the middle of the pillar. The vertical compression rate of the pillar $\hbar(x)$ at the location x can be given by integrating vertical strains as follows:

$$\hbar(x) = \int_0^H \dot{\varepsilon}_{yy}(x, y)\, dy = \int_0^H \frac{3}{2}\varepsilon_0 \frac{\sigma'_{yy}}{\sigma_{eff}} \left(\frac{\sigma_{eff}}{\sigma_0}\right)^N dy$$

The above relationship can be rewritten in the form similar to the creep law by introducing average vertical $\sigma(x)$ and effective stresses $\sigma_{eff}(x)$ along the height of the pillar:

$$\frac{\hbar(x)}{H} = \frac{3}{2}\varepsilon_0 \bar{D} \left(\frac{\bar{\sigma}_{eff}}{\sigma_0}\right)^N, \text{ where } \bar{D} = \frac{1}{H}\int_0^H \frac{\sigma'_{yy}}{\sigma_{eff}} dy$$

For the above two relationships to be identical, the average effective stress must be understood in the following sense:

$$\bar{\sigma}_{eff} = \left[\frac{1}{H\bar{D}}\int_0^H D\sigma_{eff}^N dy\right]^{1/N}, \quad D(x, y) = \frac{\sigma'_{yy}}{\sigma_{eff}}.$$

Although the expression for $\bar{\sigma}_{eff}$ is somewhat unusual, actually computed values are within a few percent from arithmetic averages. However, since creep is very sensitive to the effective stress, the precise definition of averages for determining pillar compression rate is essential.

Figure A1 illustrates average of the ratio of vertical stress deviator to effective stress, D, for several of width to height ratios between 1.25 and 5.625.

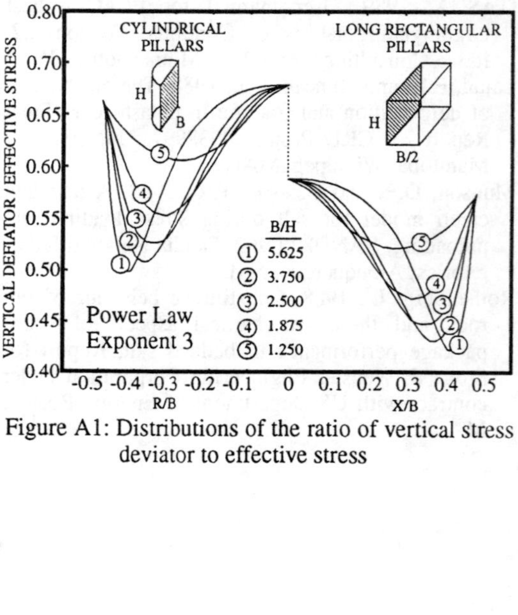

Figure A1: Distributions of the ratio of vertical stress deviator to effective stress

Data obtained from FEM runs were generalized in analytical form to arrive at functions involved in the creep law and others needed for determining stress components not directly involved in the creep law, e.g. ratios of horizontal to vertical stress for infinitely long and circular pillars, $\sigma_{xx}/\sigma = f(x/B, B/H, N)$ illustrated in Figure A2.

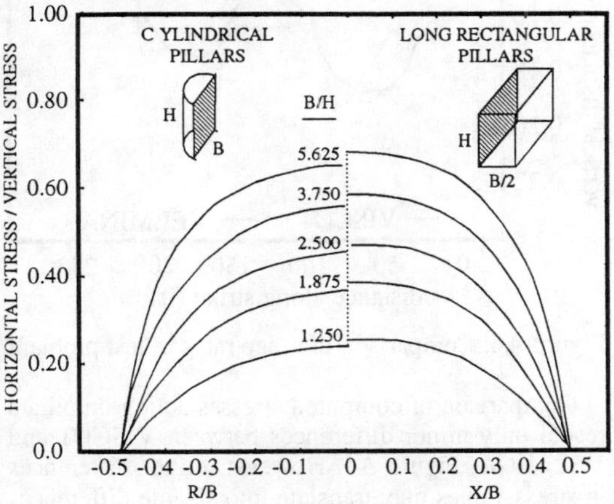

Figure A2: Distribution of horizontal to vertical stress ratio for pillars of different width to height ratio

Based on the study of stress distributions in infinitely long and circular pillars, the procedure implemented in VELMINA for describing confining stresses in rectangular pillars is as follows.

For each element, minimum and maximum distances of the element to excavated areas are determined. Stress ratios for the element are determined by linearly interpolating between values corresponding to circular and infinitely long pillars, i.e.

$$\frac{\sigma'_{yy}}{\sigma_{eff}} = \frac{B}{L}\left(\frac{\sigma'_{yy}}{\sigma_{eff}}\right)_{circular} + \left(1-\frac{B}{L}\right)\left(\frac{\sigma'_{yy}}{\sigma_{eff}}\right)_{infinite},$$

where B is the minimum distance between excavated areas along the line crossing the element and, similarly, L is the maximum distance between excavated faces. As Figure 2 suggests, the difference between these ratios for circular and infinitely long pillars is small, although must be accounted for due to sensitivity of creep rates to stress.

A 1.3 Determination of equivalent linear viscosity

Observations on many finite element computations that involve non-linear creep suggest that both in pillars and in areas adjacent to pillars there is a tendency to equalize shear stress. This effectively minimizes differences in creep rates in different areas of a mine. In VELMINA, the equivalent viscous properties of the roof and floor material are assigned in such a way that vertical strain rates in pillars and the roof are equal on average over the entire layout. If σ_{eff} computed for some element, the associated effective strain rate is $\dot{\varepsilon}_{eff} = \dot{\varepsilon}_0 (\sigma_{eff}/\sigma_0)^N$ (for the single mechanism creep law) and linear viscosity μ is calculated to match the stress rate for the given σ_{eff}. This computation is done for every element and average of viscosities computed for each element is determined. The problem is re-analyzed until μ does not change from iteration to iteration. This procedure is rather laborious and is rarely used. In practice, an initial assessment of average effective stress in pillars is done by computing vertical stress based on tributary areas, then the average effective stress is assessed based on geometry-related vertical to effective stress ratios. Linear viscosity is then assigned to match creep rates at this effective stress. After the first iteration with the assigned viscosity, new σ_{eff} is calculated. In several iterations one can determine a relationship between the matching stress and σ_{eff}, and by graphical interpolation determine the point when the matching effective stress for viscosity calculation is equal to the average effective stress in the mine. This procedure is illustrated in the following test problem.

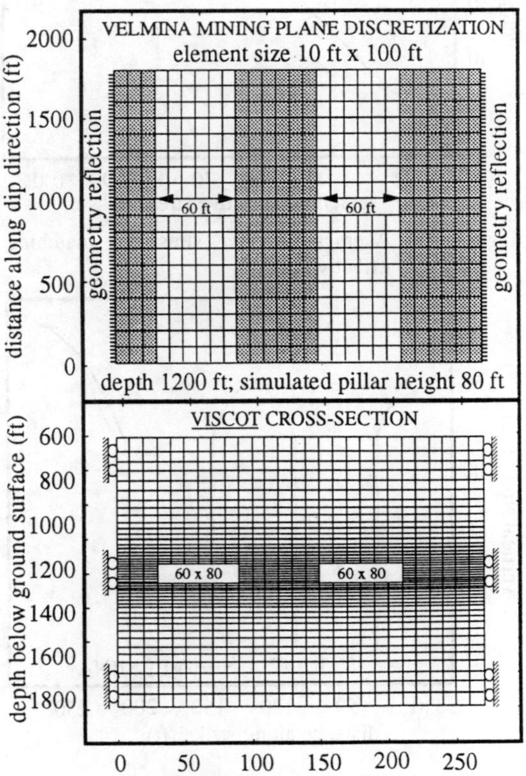

Figure A3: Test problem

A 1.4 Test problem

To test the validity of heuristic assumptions of the VELMNA code, the same plane strain geometry was analyzed using VELMINA and FEM code VISCOT.

Figure A3 illustrates an infinite sequence equivalent to four 60 feet wide and 80 feet high rooms separated by 60 ft wide "yield" pillars and a 120 ft wide "bearing" pillar between each sequence of fourth rooms.

Figure A4 illustrates a relationship between the average effective stress for the pattern versus the matching effective stress assumed for roof and floor viscosity calculations.

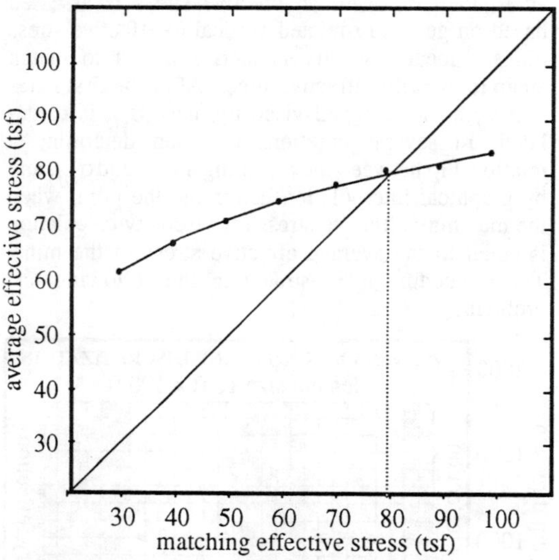

Figure A4: Average effective stress vs. matching effective stress

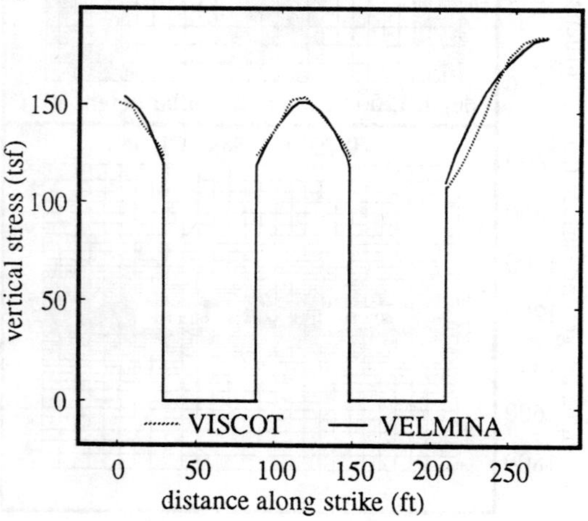

Figure A5: Comparison of vertical stresses on pillars in test problem

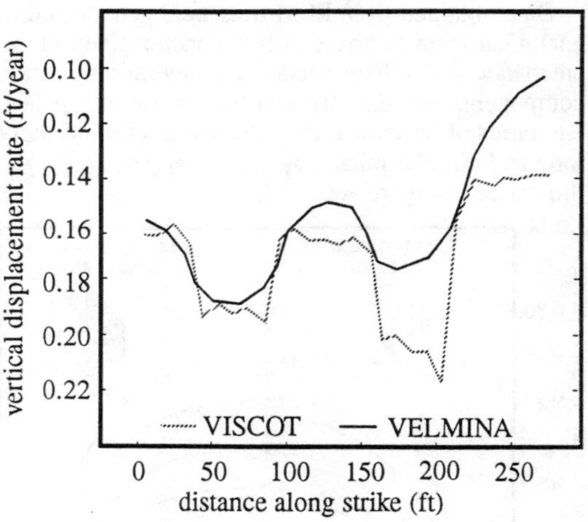

Figure A6: Comparison of creep rates in test problem

Comparison of computed stresses acting on pillars reveal only minor differences between VISCOT and VELMINA, Figure A5. However, minor differences in stress values may translate into sizable differences in steady state creep rates, as Figure A6 indicates.

Innovative Mine Design for the 21st Century, Bawden & Archibald (eds) © 1993 Balkema, Rotterdam, ISBN 90 5410 325 6

Using seismic data to enhance numerical stress analysis procedures: A case study

M.W.Grabinsky & J.H.Curran
Department of Civil Engineering, University of Toronto, Ont., Canada

ABSTRACT: A case study, involving the extraction of a pillar at Strathcona mine, is reviewed to show how seismic data may be of use in developing enhanced stress analysis models for mining. The enhanced models may then be used to generate an improved understanding of rock mass failure mechanisms and strength properties, and hence better mining strategies. The seismic data considered includes event location, event density and seismic velocity. The stress analysis tools reviewed encompass elastic, elasto-plastic and discrete element modelling. The kinds of problems traditionally encountered when attempting to use numerical modelling for mine design are reviewed, and the potential to overcome these problems by using seismic data is illustrated. Consideration is then given to the modelling issues that will likely become most relevant with this new approach to the use of numerical stress analysis in geomechanical mine design.

1 INTRODUCTION

Modern mineral processing facilities must be supplied by mining operations that are capable of maintaining reliable, high production rates. This in turn puts pressure on the mine designer to develop mining strategies that will meet production demands without compromising the safety of crew and equipment. As economic constraints become more critical and mining conditions become more severe, many designers are questioning the suitability of the empirical design methods that have traditionally been used, and are looking to the application of more advanced technologies to assist in developing better mining strategies. Two such technologies are numerical stress analysis and seismic methods.

Seismic methods make measurements essentially throughout an entire volume of rock, and may therefore provide useful complementary data to the data obtained from point-wise measurements made using conventional geomechanical tools (e.g., stress cells and extensometers). This combined data then provides better input to numerical modelling tools, which may be used to predict rock mass response to various alternative mining strategies.

A research project has been carried out over the last two years with the objective of investigating the integrated use of conventional geomechanics technologies, numerical stress analysis and seismic methods for geomechanical mine design purposes. Several companion articles resulting from this research project can be found in the present Proceedings. In this paper, consideration is given specifically to the ways in which significant improvements in the quantity and quality of input data could make numerical modelling exercises more quantitative, and the impact this would have on the specific modelling techniques used in practice. The ultimate significance of these changes would be an improved understanding of general rock mass behaviour, and hence a capacity to execute safer and more productive mining strategies.

The background leading to the design philosophy adopted for the project is first provided, including a description of a conceptual model for linking the seismic behaviour of a rock mass with its states of stress and stiffness. A case study involving the extraction of a pillar at Strathcona mine is then used to illustrate the traditional problem of using numerical modelling in a data-limited type of analysis, and to show how seismic data can complement conventional geomechanical data and provide the essential information required to justify undertaking more sophisticated analyses (e.g., three-dimensional nonlinear analyses). Finally, and perhaps most importantly from a modelling point-of-view, consideration is given to a host of issues that will take on a renewed significance as a result of our expanded ability characterize and monitor whole rock mass volumes.

2 BACKGROUND

Geomechanical design of underground mines includes the selection of mining methods, the dimensioning of stopes and pillars, the layout and sequencing of excavations, specifying rock reinforcement for production and service excavations, locating service excavations relative to the stopes, using destressing techniques to control the build up of strain energy in the rock mass, and so on. In addressing any of these design issues, the designer attempts to find a balance between the competing objectives of minimizing costs and maximizing productivity. For example, increasing stope size may facilitate greater production rates, but also usually necessitates more elaborate and expensive systems of rock reinforcement.

Finding the balance between minimizing costs and maximizing productivity is seldom a precise exercise in mine design, due to the fact that operational considerations (e.g., abrasion or the requirement to withstand blasting in proximity) are often just as significant as, say, the intensity of the local stresses or the strength of the rock mass. Further, the very nature of mining facilitates a design-as-you-go approach. As a result, many practical design decisions must ultimately be made on the basis of a considered *risk assessment*, and it should not necessarily be expected the kinds of procedures used to design more traditional structures (such as buildings and machinery) are reasonable or even desirable for mine design.

Recognizing the role of empiricism in the development of mine design procedures, it must also be appreciated that significant changes at many mining operations (e.g., increased mechanization, mining at greater depths and elevated stresses, increased extraction ratios, and increased international competition) are necessitating the development of more rational analysis and design approaches in order to achieve a tighter balance between costs and productivity. It is proposed that realizing better design methodologies requires the following:

1) generating an improved understanding of (reinforced) rock mass behaviour;
2) developing calibrated tools for characterizing and monitoring whole rock mass volumes; and
3) improving the integration of practical analysis tools in the design process.

Further, it is believed that seismic methods and a proliferation of user-friendly numerical stress analysis programs make it now possible to realize these objectives to a degree that will significantly affect mine design.

In the characteristic manner in which mine design procedures evolve, it should be anticipated that the above three points will be somewhat interdependent. For example, new investigative and monitoring tools based on seismic methods are currently actively being developed; these generate better data, which will spur the development and application modelling techniques; better data and more thorough analysis will result in a better understanding of rock mass behaviour; this improved understanding may lead to enhanced analysis and design procedures, or to the development of further tools for examining different aspects of rock mass behaviour, and so on.

From a modelling point-of-view, integrating the newly-acquired seismic data into the analysis process requires extending the traditional idealizations of rock mass behaviour to include the relevant seismic parameters. The next section presents a preliminary working model of the linkages between a rock mass's mechanical and seismic behaviours.

3 MODELS OF ROCK MASS BEHAVIOUR

3.1 *Theoretical background*

In linking the mechanical and seismic behaviours of a rock mass for mine design purposes, two aspects of the mass's seismic behaviour may require consideration: seismic velocity and acoustic emission. Seismic velocity is the speed of propagation of compressional and shear stress waves through the rock mass. Acoustic emissions are associated with damage incurred by the rock mass during progressive failure.

Theoretical and experimental (laboratory and field) work has confirmed that discontinuities have an influence on the propagation of seismic waves through the rock mass. A useful review of such work can be found in Cook (1992). In general, increasing normal stress across joint surfaces will tend to increase the stiffness of the system and therefore increase the velocities of seismic wave propagation. However, examinations of single joints quickly become theoretically and practically complex as consideration is extended to the nature of reflected, transmitted and converted waves across interfaces comprised of displacement discontinuities and velocity discontinuities possessing elastic as well as viscous coupling (see Pyrak-Nolte (1988) and Pyrak-Nolte and Cook (1987)). Attempting to further consider the effect of *systems* of *nonlinear* joints on the propagation of seismic waves through rock masses, then, poses an even more daunting theoretical and practical challenge.

Similarly, works as early as Kaiser (1953) have investigated the phenomena of acoustic emission resulting from damage of brittle materials. With respect to mining induced seismicity, attempts have been made to study this phenomenon within the theoretical framework developed for the analysis of earthquakes, and a review of some of the practical advances in this approach can be found in Young et al. (1989). It is becoming evident, however, that not all seismic events associated with mining conform to the basic idealizations currently typically used in seismic analyses, and more detailed investigations are being pursued to quantify the significance of the observed discrepancies.

3.2 *Working model*

Given the practical difficulties just noted of developing complete theoretical models of coupled mechanical and seismic behaviours, it was decided that for this project a more empirical approach to analysis and design should be adopted. The working model of coupled seismic and mechanical behaviours is, therefore, basically a collection of concepts and heuristics commonly employed in rock mechanics and seismology. These basic ideas are then applied to practical problems and tests are conducted to assess the validity of the various assumptions made. In this way it is hoped that essential information may be collected which will then facilitate formulating a rational approach to developing more rigorous models of coupled seismic and mechanical rock mass behaviour. The following description of the working model is taken from Bawden et al. (1992).

In many Canadian mines, the orebodies are tabular and dip subvertically, the major principle stresses (σ_1 and σ_2) are subhorizontal and the minor principle stress (σ_3) is subvertical, and open stope mining methods are used. Therefore, the mining configuration shown in Figure 1 will be used as an example here, although the concepts can be generalized to other mining configurations as well. Generally, stresses concentrate around the top and bottom of the stope, and a zone of stress relaxation is formed in the stope's sidewalls. Assuming more-or-less homogeneous and isotropic rock mass properties, if the stresses at the stope boundaries do not exceed rock mass strength then it is expected that seismic velocity will be high (relative to ambient values) in the rock at the top and bottom of the stope and low (relative to ambient values) in the stope's sidewalls, and there will be insignificant induced seismicity. This situation is probably typical of early mine life, where rock mass instabilities are usually inconsequential.

For increasing in situ stress magnitudes, the stress concentrations at the stope boundaries will begin to cause failure: fracturing, spalling and/or joint slip at the top and bottom of the stope, and block unraveling in the sidewalls. Initially this will produce a seismic velocity distribution similar to before (high velocity at the top and bottom of the stope and low velocity in the sidewalls), and seismicity will become significant in the rock at the top and bottom of the stope due to stress induced fracturing.

If failure progresses, appreciable changes may occur in the rock at the top and bottom of the stope and three zones may be identified in these areas:

1) immediately adjacent to the stope boundary lies a zone of failed, dilated rock that has low stiffness and seismic velocity and is aseismic;
2) next is a zone of actively failing rock where there are large gradients of stress, stiffness and seismic velocity, and where seismic activity concentrates; and
3) further still into the rock mass lies an intact, highly stressed zone with high stress and seismic velocity; this zone exhibits lower rates of seismicity, but the events are likely different in character and magnitude than the events in the previous zone.

These zones are shown in Figure 2.

The basic model just described may also be extended to consider the effects of rock mass anisotropies, and the influence of prominent discrete structures such as faults and shear zones. The next section demonstrates application of the model to a practical analysis and design problem.

4 CASE STUDY

4.1 *Introduction*

Strathcona mine is a Falconbridge owned and operated copper nickel mine located in the north ridge of the Sudbury basin. It has been operating continuously since 1968 at mining rates of up to 2.1 million tons per year. The

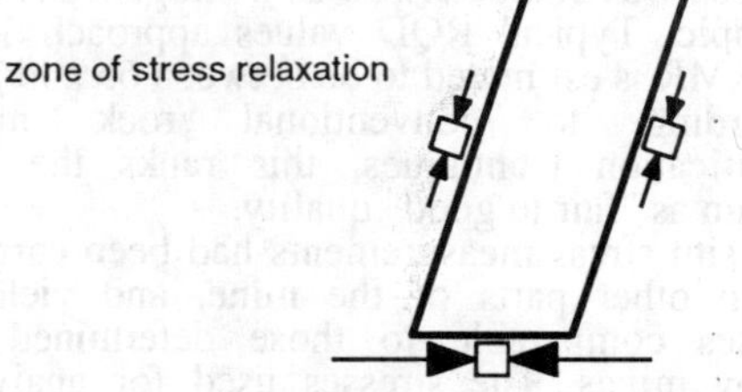

Fig.1 Transverse section of example stope showing trends in stress concentrations at the stope boundaries

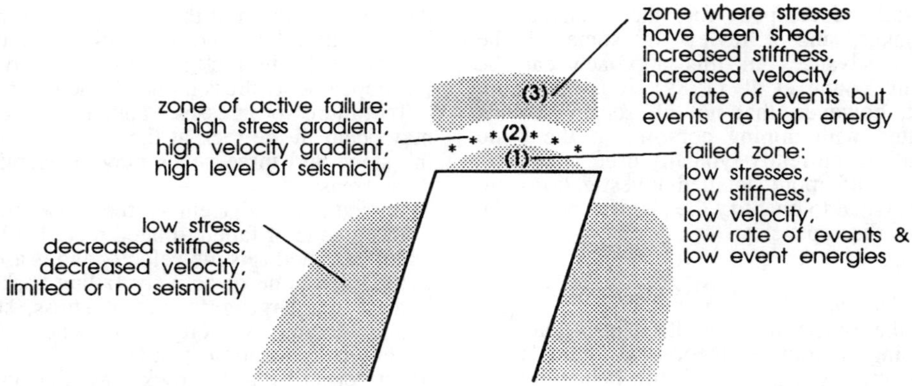

Fig.2 Stress-stiffness-velocity-seismicity relations around an example stope for which the rock mass is progressively failing

original ore reserve was nearly 80% extracted without significant ground problems, but the mine has since been subjected to significant damaging bursts.

The main orebody is located between 570 m and 900 m below surface, with a N-NE strike of about 800 m and a dip of 45-80 degrees S. The width of the mineralized zone varies between 10 m and 100 m, with an average of 50 m. The study area is located in the 25-39-D4 (or D4) stope of the Eastern Regional Pillar (ERP), and is shown in Figure 3.

Predominantly cut-and-fill mining methods have been used in extracting the stopes in proximity of the ERP and in the D4 stope itself.

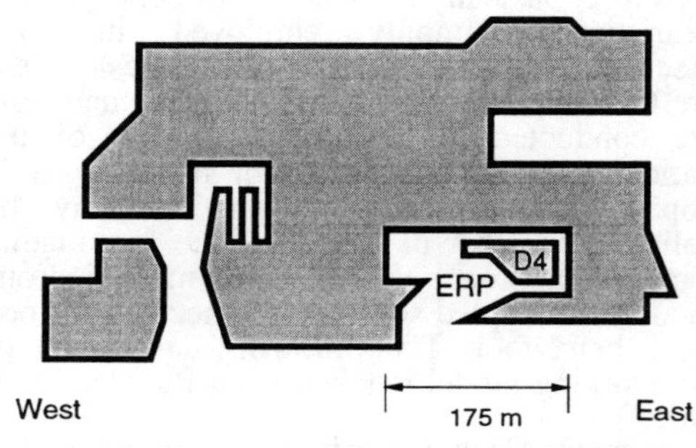

Fig.3 Longitudinal view of Strathcona mine showing the 25-39-D4 stope study area, located within the Eastern Regional Pillar (ERP)

4.2 *Geomechanics studies*

Geomechanical mapping of the D4 area identified two predominant subvertical joint sets striking S and SW. A third subhorizontal joint set was observed intermittently, but was probably under-represented in the mapping studies due to mapping bias (i.e., limited vertical exposure). All joints had similar surface characteristics. As a result, the rock mass in the D4 area was characterized as homogeneous and isotropic. Typical RQD values approach 100, and RMR is estimated to be between 65 and 75. According to conventional rock mass classification techniques, this ranks the D4 domain as "fair to good" quality.

In situ stress measurements had been carried out in other parts of the mine, and yielded stresses comparable to those determined in nearby mines. The stresses used for analysis purposes were (stress in MPa and depth in m):

$\sigma_1 = 0.0440 \times \text{depth}$ strike W-E
$\sigma_2 = 0.0383 \times \text{depth}$ strike N-S
$\sigma_3 = 0.0267 \times \text{depth}$ vertical.

The Hoek-Brown failure criterion was assumed. Intact rock unconfined compressive strength was 110 MPa, and Hoek-Brown m and s parameters were 2.8 and 0.006, respectively.

4.3 *Mining history*

Logistical problems at the site precluded the use of any conventional field instrumentation for this study. The available evidence consists of data from seismic analyses (discussed in the next section) and visual observations of rock mass performance.

The ERP was originally left as a regional stabilizing pillar during mining of surrounding areas. Although some large events ($m_N > 1.0$) were recorded during the mining of these surrounding areas, none resulted in any operational problems. However, it was subsequently noted that a drift located in the upper half of the ERP had undergone significant stress fracturing, indicating that the induced stresses in this area were now potentially severe.

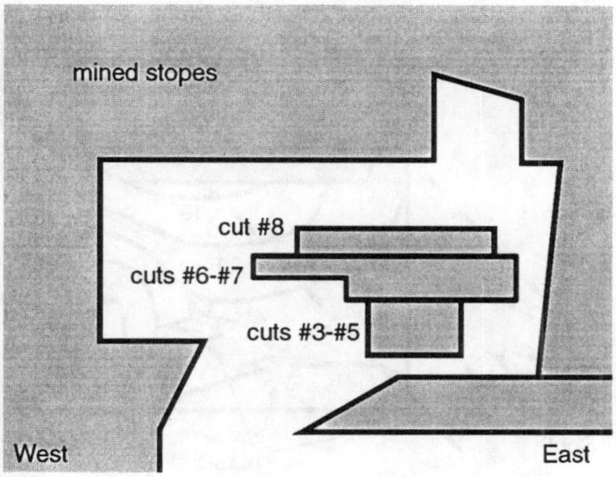

Fig.4 Mining cuts of the D4 stope in the ERP (longitudinal view of the ERP)

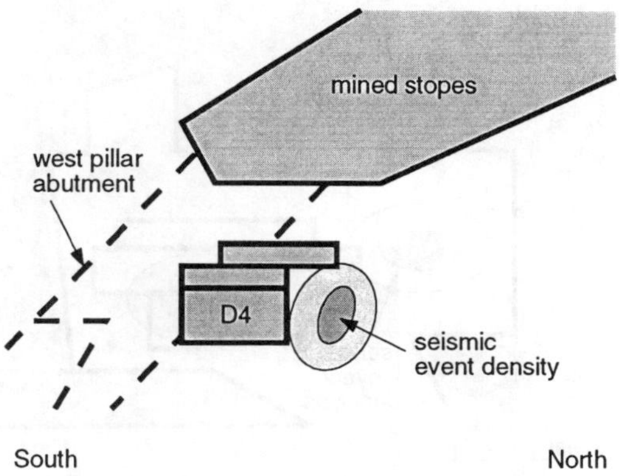

Fig.5 Location of m.s. event concentrations in the ERP (transverse view through D4 stope)

Extraction of the D4 stope was the first attempt at mining directly within the ERP. The configuration of the D4 stope with respect to the ERP is shown in Figure 4. Initial mining (cuts #3-#5) proceeded without operational problems and did not appear to generate any large seismic events. Mining cut #6 coincided with three large events ($m_N > 1.0$), but no significant damage resulted. However, while mining cut #7 three further bursts occurred and resulting damage was observed in service excavations and haulageways. Subsequently, a m_N 2.6 rock burst occurred while mining cut #8, resulting in significant damage to service excavations in areas (*i*) near the D4 stope and (*ii*) in the footwall of the Western portion of the ERP.

The damage caused by these latest events warranted temporary cessation of mining in order to rehabilitate the affected areas and reconsider the present mining strategy. It was then decided that a destress blast should be performed. The destress blast, however, led to even further damage in proximity to the blast location: it was estimated that the amount of energy released as a result of the destress blast was in excess of that available from the blasting charges, and thus it was suspected that the destress attempt had in fact triggered yet another rockburst.

Ultimately, the effect of the destress blast could not be quantified by any conventional means. This fact, combined with the uncertainty of how to mine safely in the burst-prone areas, resulted in the decision to suspend mining in the ERP indefinitely.

4.4 *Seismic studies*

An MP250 system, used to source locate events, was already available at the site and was used throughout the life of the case study. Additionally, a Queen's whole waveform m.s. monitoring system was installed and was used to collect data for mining cuts #6 and beyond. As a result, it was possible to use the Queen's system to better calibrate the MP250 system, and then reanalyze the MP250 data in order to increase confidence in the source locations of events generated prior to the use of the Queen's system.

Regardless of the time period considered, the seismic activity is found to concentrate in the footwall of the ERP just below pillar mid-height, as shown in transverse view in Figure 5. Longitudinally, the concentration appears to segregate into two clusters, one near the D4 stope, and the other just to the West in the ERP (Figure 6). These locations correspond to the areas where stress fracturing was observed in haulageways, as noted in the previous section.

Initial mining (cuts #3-#5) generated only moderate m.s. activity. This activity occurred predominantly in the footwall of the D4 stope, amongst haulage drifts and ore transfer raises (OTR's). Event density increased by a factor of 2 to 3 during extraction of cut #6, and located both near the D4 stope and further to the West in the ERP. The three large events noted during mining of cut #6 (see previous section) did not locate in the ERP, according to the m.s. system. To this point in time, the determined pattern of seismicity is consistent with the lack of observed damage in service excavations located within the pillar.

Seismic data from mining cut #6 was used to generate a three-dimensional passive seismic velocity survey. The velocity data demonstrated a consistent trend along the strike of the pillar: the lowest velocity areas are at the top of the pillar, and the highest velocity areas are in the footwall of the ERP just below pillar mid-height (Figure 7).

During mining of the remaining cuts, event density increased by a factor of 3 to 4 over that

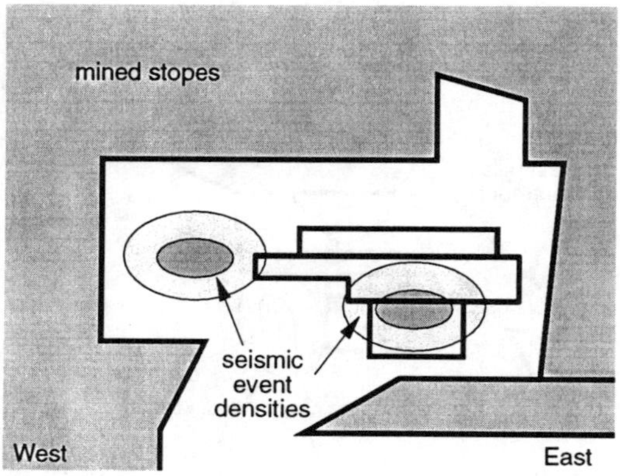

Fig.6 Location of m.s. event concentrations in the ERP (longitudinal view of the ERP)

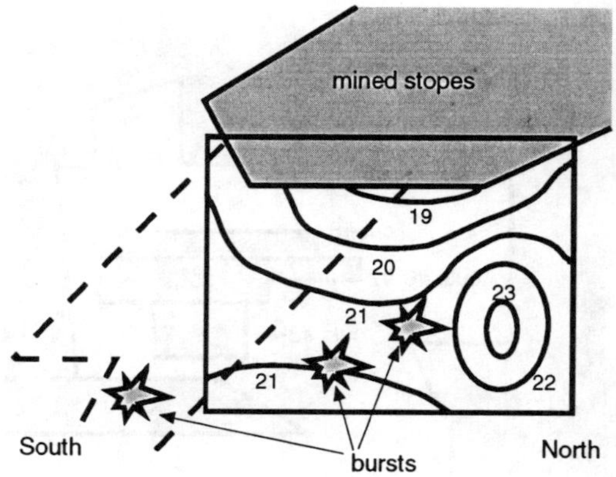

Fig.8 Location of rockbursts during cuts #7-#8, superimposed on seismic velocity from Figure 7

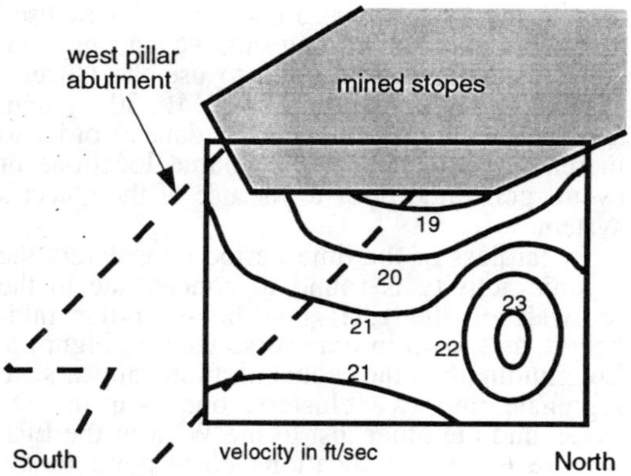

Fig.7 Results of seismic velocity survey after mining cut #6 (transverse view through Western portion of ERP)

observed during cut #6, and occurred uniformly throughout the footwall of the ERP just below pillar mid-height. The large rockbursts that occurred during mining cuts #7-#8 (noted in previous section) were located by the m.s. system in the Western half of the ERP at or below pillar mid-height (Figure 8). The location of the largest burst, at pillar mid-height, is consistent with the area of observed damage to service excavations.

4.5 *Numerical stress analysis*

An elastic analysis was used first to provide an essential idea of the potential distribution of stresses (magnitudes and directions) through and about the ERP, prior to excavation of the D4 stopes. Since geomechanical mapping of the D4 area identified three orthogonal joint sets with similar mechanical properties, a homogeneous and isotropic elastic model was initially sufficient. More advanced models are used later as deemed appropriate on the basis of interpretation of the elastic modelling results. The program *Examine^{3D}*, developed at the University of Toronto, was used to determine the three-dimensional elastic stress distribution. The values of the various input data (in situ stresses, elastic properties and Hoek-Brown strength parameters) were varied between plausible limits to ascertain the sensitivity of the predicted extents of zones of rock failure to the value of a given input parameter. It was found that the amount of predicted failure was most sensitive to the values of the strength parameters. However, changing the values of the strength parameters from worst- to best-case limits did not change the predicted amount of rock mass failure enough to alter the engineering interpretation of the results, described next.

Based on the assumed elastic stress distribution just prior to mining the D4 stope, the predicted zones of rock mass failure were located predominantly at the top of the ERP, essentially uniform along the strike of the pillar (Figure 9). There are three indications that the rock mass in this area had indeed failed, and further was probably behaving in a post-peak manner (with reference to the model presented in section 3.2):

1) the stress fracturing observed in drifts in the upper portion of the pillar, just prior to mining the D4 stope, suggests rock failure;
2) the predicted failed zone was *aseismic* during mining of the D4, suggesting the rock mass had already failed and was now responding in a post-peak manner; and
3) the predicted failed zone is coincident with the low velocity determined after mining cut

#6, suggesting dilated fractures and low stress levels.

In addition, the uniform distribution of seismic velocities along the strike of the ERP provides the clue that failure is pervasive through the top of the pillar, and is not localized around a given drift.

The next logical step would be to use the available data to calibrate the strength parameters and run a plasticity analysis to see what effect rock mass failure has on stress redistribution. To this end, a three-dimensional finite element model was generated (Figure 10) using I-DEAS, a commercial FE program. Details of the mesh generation process can be found in Grabinsky and Curran (1993).

A variety of continuum plasticity models were available for use in this study, including Drucker-Prager and Mohr-Coulomb (ABAQUS, 1989), Hoek-Brown (Shah, 1992) and Elastic-Brittle-Ductile (Ofoegbu and Curran, 1991) models. As a result, nonlinear and strain-softening types of behaviours could be represented. However, increased modelling flexibility is obtained at the price of a greater number of modelling parameters for which representative field values must be found.

In order to expedite the nonlinear modelling, further sensitivity analyses were carried out using elastic model results. The predicted elastic stresses were post-processed to determine the distribution of strength factors based on the various possible failure criteria. In general, it was found that the extent of predicted rock mass failure was most dependent on "cohesion" and relatively insensitive to "friction angle" (where the particular parameters used for cohesion and friction angle are dependent on the material model). It was deduced that the observed dependencies were due to the geometry of the excavated stopes overlying the pillar, which resulted in high gradients of σ_1 and low gradients of σ_3 at the top of the pillar. Hence, the different failure criteria produced similar predicted extents of rock mass failure for equivalent values of model cohesion.

The next issue in plasticity modelling involves describing the post-peak behaviour of the rock mass. Two extreme assumptions are perfectly-plastic response (the load bearing capacity at peak strength is maintained as the rock mass deforms) and brittle response (the load bearing capacity rapidly diminishes as the rock mass deforms). These extremes were modelled by (*i*) using plasticity models with perfectly plastic flow rules, and (*ii*) using further elastic models in which the assumed zone of failed rock mass was removed to simulate complete loss of load bearing capacity.

For the perfectly plastic model, the predicted zone of rock mass failure (after progressive failure had been simulated and stresses redistributed) is shown in Figure 11. The stress

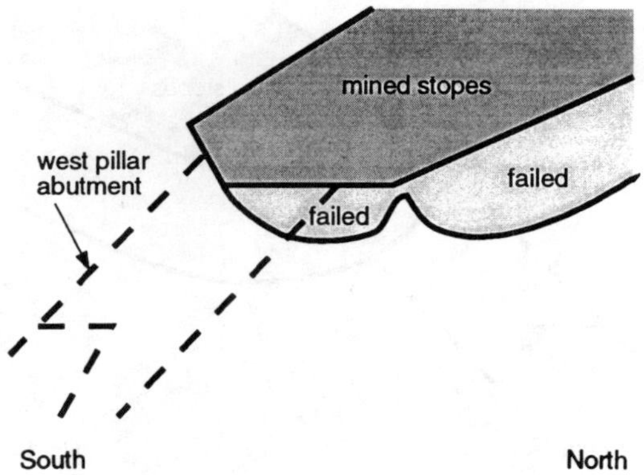

Fig.9 Predicted zone of failure based on elastic modelling results, prior to mining the D4 stope (transverse view through ERP)

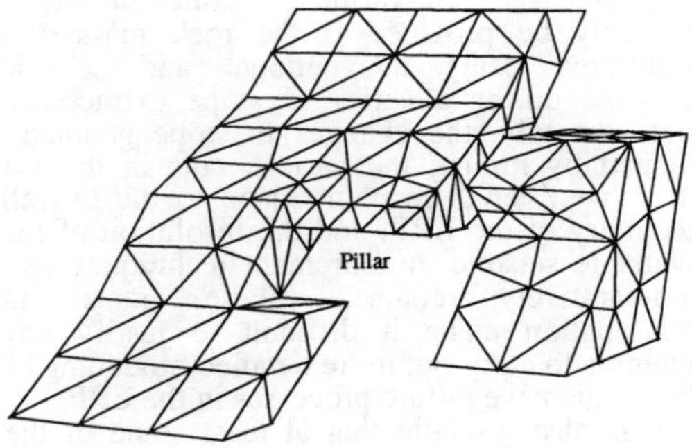

Fig.10 Perspective view of modelled excavation surfaces, from three-dimensional finite element plasticity model

redistribution associated with this plastic straining, however, was not significant in that contour plots of stresses before and after failure were essentially identical. This is attributed to the significant post-peak load bearing capacity of the rock mass.

In removing the predicted zone of failed material to simulate perfectly brittle post-peak response, the shape of the excavation at the top of the pillar changes: the original knife-edge excavation profile becomes more rounded and smooth. This shape change results in a redistribution of stresses that is generally more favourable to rock mass stability. If the process is repeated, however, the excavation will continue through the area occupied by the D4 stopes.

The actual post-peak response of the rock mass, then, lies somewhere between the two extremes modelled here. However, without further information it is not possible to unambiguously determine what the actual

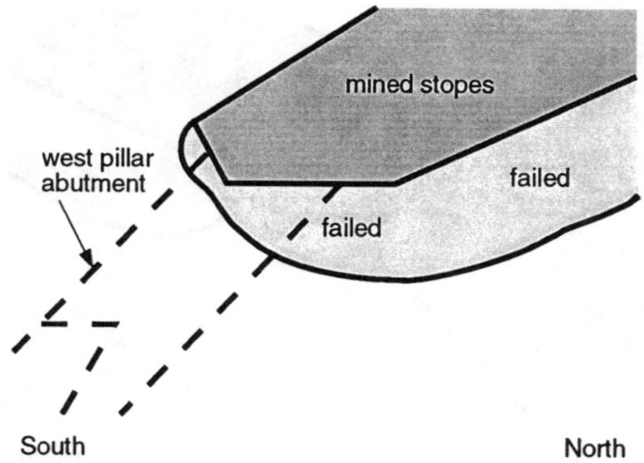

Fig.11 Predicted zone of rock mass failure based on assumed perfectly plastic post-peak rock mass response

response is. Such further information would probably be provided if the rock mass was monitored using conventional and seismic methods during a series of stope extractions. Unfortunately, the changes in stope geometry created by mining the various cuts in the D4 stope were small in comparison to the overall geometry of the ERP, and the resolution of the available seismic measurement techniques was comparatively coarse. These facts in combination make it difficult to justify any attempts to carry out more detailed modelling of the progressive failure processes in the ERP.

It is also possible that at least some of the major rockbursts in the ERP could have been the result of slip on existing discontinuities. The consequences of this mode of failure were investigated using three-dimensional distinct element modelling (Tinucci and Hart, 1990). Apparently, however, joint mapping had not been sufficient to provide quantitative data regarding the orientations and characteristics of major discontinuities, required as basic input for such an analysis. Hence, probable values were hypothesized and appropriate caveats applied to the interpretation of analysis results. Nevertheless, the investigators felt that the potential for slip was considerable enough to warrant further investigation. They also noted that the stress changes associated with these slippages could be small. This might suggest difficulties in using changes in seismic velocity to identify the region in which a slip occurred.

5 LESSONS FROM THE CASE STUDY

One of the most challenging tasks in geomechanical analysis can be determining the mode of failure of the rock mass, and the spatial extent of the failure process. In this case study, the integrated interpretation of seismic velocity, seismic source location, conventional geomechanical and numerical modelling data provided strong evidence that the rock mass at the top of the ERP was responding in a post-peak manner, and that this response was essentially uniform across the top of the pillar. Further, the boundary of seismic activity around the low velocity zone at the top of the ERP provided a useful indication of how deep the failure extended down into the pillar. This interpretation would not have been as conclusive had any of the data been missing.

In this particular case study, however, it was difficult to determine which nonlinear models might be most suitable. Ambiguities also arose in the process of trying to quantify values for many of the modelling parameters. It is believed that an appropriate combination of seismic coverage and sequential removal of stopes could provide the kind of data required to better select and calibrate models of rock mass failure. Ultimately, the process of selecting and calibrating models would become enhanced as experience in applying these design techniques is gained. Interestingly, linear elastic modelling, and post-processing of the elastic analysis results to consider the consequences of various forms of failure, provided the majority of useful analysis information. Generalizing this observation, the mine designer would be well advised to consider the careful use of parametric analyses based on the results of elastic analyses, with the objective of gaining insights into the essential mechanisms linking geometry, stresses and rock mass behaviour, before attempting to undertake any resource intensive nonlinear analyses.

Three zones of seismic/mechanical rock mass behaviour were inferred in the ERP:

1) a low velocity, aseismic zone at the top of the pillar, which coincides with a region where stress fracturing was observed;
2) a high velocity gradient area around the above zone, in which the majority of m.s. activity was observed; and
3) a high velocity zone, essentially void of m.s. activity, which appeared to coincide with good ground.

This generally corresponds to the conceptual model presented in section 3.2. However, the major rockbursts in the ERP were located in and adjacent to zone 2, as well as at locations more remote from zone 2. These more remote bursts might be attributable to particular structural features. Nevertheless, the case study illustrates the difficulties of attempting to quantify when and/or where rockbursts will occur. In particular, the conceptual model would have been very useful in identifying the fringe zone (zone 2, above), and in suggesting that the most likely burst locations lay in or adjacent to this zone. However, additional information would be required to suggest whether a burst was more likely to occur on the hangingwall or footwall side of the ERP, and it predicting when such a burst would occur (except, possibly, relative to stope sequencing) would be tenuous at present. As a result, it is more likely that the designer will find the techniques described in this paper useful if they are applied in the context of *seismic risk management* as opposed to rockburst prediction.

6 CONCLUSIONS

Seismic data is gathered from whole rock mass volumes, and measurements can be made remotely. As such, seismic data provides a powerful complement to conventional geomechanical data. The enhanced datasets also enable the designer to more confidently develop numerical models, and to more quantitatively interpret the analysis results. Consideration of further seismic parameters (e.g., fault plane solutions and source parameters) will facilitate the development of even more comprehensive and quantitative design methods.

Mathematical models have been developed for representing a wide range of behavioural phenomena, and the power of current computers makes it possible to run these models in a practical design environment. However, determining appropriate values for the various model parameters is seldom an unambiguous process, as illustrated in the case study given here. Nevertheless, the enhanced data that is now available (through the use of seismic methods) enables the designer to run parametric studies and determine probable parameter values within much tighter limits than was previously possible. As a result, further experience in applying integrated seismic-geomechanics-modelling design approaches will lead to an improved ability to quantify rock mass behaviour. (Note that such an empirical approach to calibrating design procedures has been used successfully elsewhere in rock engineering, for example, rock mass classification schemes, stope design methods, and rock mass failure criteria.)

Given our current capacity to quantify and predict the linked mechanical-seismic behaviour of rock masses, the designer should adopt a design philosophy based on risk management rather than exact prediction of rock mass behaviour.

ACKNOWLEDGMENTS

This work was carried out under a Collaborative Research Grant jointly sponsored by the Natural Sciences and Engineering Research Council of Canada, the Mining Research Directorate and Atomic Energy Canada Limited. Seismic data was provided by the Engineering Seismology Laboratory, Queen's University; geomechanical data was provided by Mining Engineering, Queen's University.

REFERENCES

ABAQUS User's Manual, Version 4.9, 1989. Hibbitt, Karlsson & Sorensen, Inc., Pawtucket, Rhode Island.

Bawden, W.F., Curran, J.H. and Young, R.P. 1992. Enhancing Geomechanical Mine Design Procedures Through Linkage of Stress Analysis and Seismic Techniques: Methodology and Rationale. Report MD#001 to the Mining Research Directorate, 45 pp.

Cook, N.G.W. 1992. Jaeger Memorial Dedication Lecture - Natural joints in rock: mechanical, hydraulic and seismic behaviour and properties under normal stress. *Int. J. Rock Mech. Min. Sci. & Geomech. Abstr.* 29:198-223.

*Examine^{3D} User's Manual, 1993. Rock Engineering, Department of Civil Engineering, University of Toronto, 199 pp.

Grabinsky, M.W. and Curran, J.H. 1993. Efficient mesh generation procedures for finite element analysis of underground structures. Submitted for publication to *Int. J. Rock Mech. Min. Sci. & Geomech. Abstr.*

Kaiser, J. 1953. Erkenntnisse und folgerungen aus der messung von gerauschen bei zugbeanspruchung. *Archiv fur das Eisenhuttenwesen* 24:43-45.

Ofoegbu, G.I. and Curran J.H. 1991. Yielding and damage of intact rock. *Can. Geotech. J.* 28:503-516.

Pyrak-Nolte, L.J. 1988. Seismic visibility of fractures. Ph.D. Thesis, University of California, Berkeley.

Pyrak-Nolte, L.J. and Cook, N.G.W. 1987. Elastic interface waves along a fracture. *Geophys. Res. Lett.* 14:1107.

Shah, S. 1992. *A study of the behaviour of jointed rock masses.* Ph.D. Thesis, Dept. Civil Eng., Univ. Toronto, 252 pp.

Tinucci, J. and Hart, R.D. 1990. Development and testing of the three-dimensional distinct element method - case study 5: examination of mining-induced seismic events at the Strathcona mine. Itasca Consulting Group report prepared for Falconbridge Ltd.

Young, R.P., Talebi, S., Hutchins, D.A. and Urbancic T.I. 1989. Source mechanism studies of mining-induced microseismic events at Strathcona mine, Sudbury, Canada. *Special Issue of Pure and Appl. Geophys. on Mining Induced Seismicity* 129:455-474.

*Program available from J.H. Curran, Department of Civil Engineering, University of Toronto, Toronto, Ontario, Canada M5S 1A4

Innovative Mine Design for the 21st Century, Bawden & Archibald (eds) © 1993 Balkema, Rotterdam, ISBN 90 5410 325 6

3D modelling of the central pillar in the Pyhäsalmi mine

Juha Antikainen & Ari Simonen
Helsinki University of Technology, Laboratory of Rock Engineering, Finland
Ilpo Mäkinen
Outokumpu Finnmines Oy, Pyhäsalmi Mine, Finland

ABSTRACT: At the Pyhäsalmi mine, a central pillar was left under the open pit between levels + 135 and +210 next to the inclined ramp. Mining of the central pillar may jeopardize both the stability of the main shaft and the concentrator. On the other hand, the pillar contains 1.2 Mt of ore which is over a years production. Modelling of the central pillar excavation involves purely a three dimensional problem geometry which cannot be solved by two dimensional modelling. Therefore a 3D-distinct element method was chosen. The modelling work contained the following phases: an empirical analysis, code testing, planning of the mining alternatives, determining the significant geological features and rock mechanical parameters, rock mechanical monitoring, modelling and evaluation of the results. According to the modelling, the pillar has been affected by high stresses but they have been released and the pillar is now in the post-failure state.

1 INTRODUCTION

Outokumpu Finnmines Oy Pyhäsalmi mine is located in central Finland (Figure 1). The ore was discovered in 1958 by a local farmer while he was digging a well in his back yard and hit his shovel directly on the ore. Outokumpu Oy immediately commenced geological and geophysical exploration. Diamond drilling proved the well being located almost at the centre of the deposit. After a one year intensive prospecting period Outokumpu Oy made the decision to go ahead with the mine development.

A two and half year construction period followed which saw the mine commence operations on 1. of March 1962. Mining was started by open-pit which final depth of 125 - 135 m was reached in 1967. The same year saw the phasing in of underground operations. The total production from the open pit was 6.8 Mt of ore and 5.6 Mt of waste rock.

In 1967, after extension and automation of the processing plant, the production rate was increased from 600.000 tpy to 800.000 tpy.

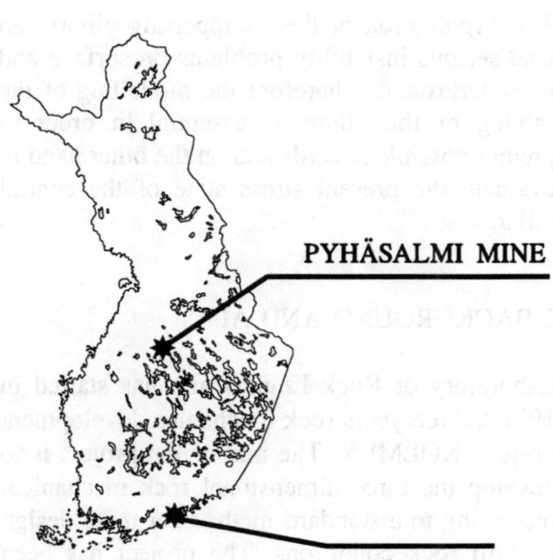

Fig. 1. The location of Pyhäsalmi mine

During this phase the quality of concentrates improved and recoveries increased.

The current annual mining of the Pyhäsalmi Mine is 1,040,000 tonnes of ore. The mill produces annually over 700,000 tonnes of pyrite concentrate, 40,000 tonnes of zinc concentrate, and 35,000 tonnes of copper concentrate. The proven ore reserves will be mined out by 1999.

The ore is located in a volcanogenic schist belt consisting of felsic and mafic metavolcanics. The ore is developed in altered sericite schists and mica and cordierite gneisses (Figure 2). The orebody dips steeply with a length of 200 m - 650 m, a thickness of 5 m - 40 m . The deepest ore intersections are at 900 m below surface. The contact between ore and waste rock is sharp. The ore itself is stiff, but it is enveloped by an altered rock formation as sericite schist and mica rock, which are very weak.

Maximum horizontal stress below 400 m level is two and half times as high as the vertical gravity load. 3 Mt of ore is in crown pillars and in a central pillar, which acts as a barrier pillar to the main shaft and the open pit (Figure 3).

The exploitation of these supporting pillars can lead serious instability problems on surface and in underground. Therefore the modelling of the mining of the pillars is essential in order to predict possible hazards and on the other hand to evaluate the present stress state of the central pillar.

2 BACKGROUND AND AIMS

Laboratory of Rock Engineering has started in 1991 a three years rock mechanics development project, KOLMEX. The aim of the project is to develop the three dimensional rock mechanical modelling to a standard method for mine design in hard rock conditions. The project has been divided to three parts: 1) 3D-properties of intact rock and models with modelling parameters for rock mass without joints, 2) 3D-properties of joints and joint models with modelling parameters for joints, 3) three dimensional modelling development by solving actual mining problems with 3D distinct element code 3DEC. 3D-modelling of the central pillar in the Pyhäsalmi mine is an essential part of KOLMEX project.

The central pillar contains 1.2 Mt of ore which is over a years production of the Pyhäsalmi mine. The geometry of the central pillar is very complicated with the surrounding weakness zones and backfilled stopes. It was clear that 2D-modelling would not give a reliable results. The mine was doing its long term planning including the mining of the central pillar and needed a reliable analysis of effects of mining the central pillar.

The modelling of the central pillar started in the beginning of 1991 and was completed early in 1992. The primary aims were:

- testing the 3DEC code by a simple and known problem in Pyhäsalmi mine
- after the code approving building up a model of the mine in three stages: global, regional and local models
- determining the stress stage of the pillar and the possible instability problems caused by mining and find out the basement for the best mining sequence for central pillar
- final report within the year

The approach was to divide the tasks between the mine and the Laboratory of Rock Engineering :

task	laboratory	mine
empirical analysis		x
testing of the code	x	
mining alternatives of the pillar		x
geological basis for modelling		x
measurements and follow up in the mine		x
modelling of the central pillar	x	x
analysis of the results	x	x
final report	x	

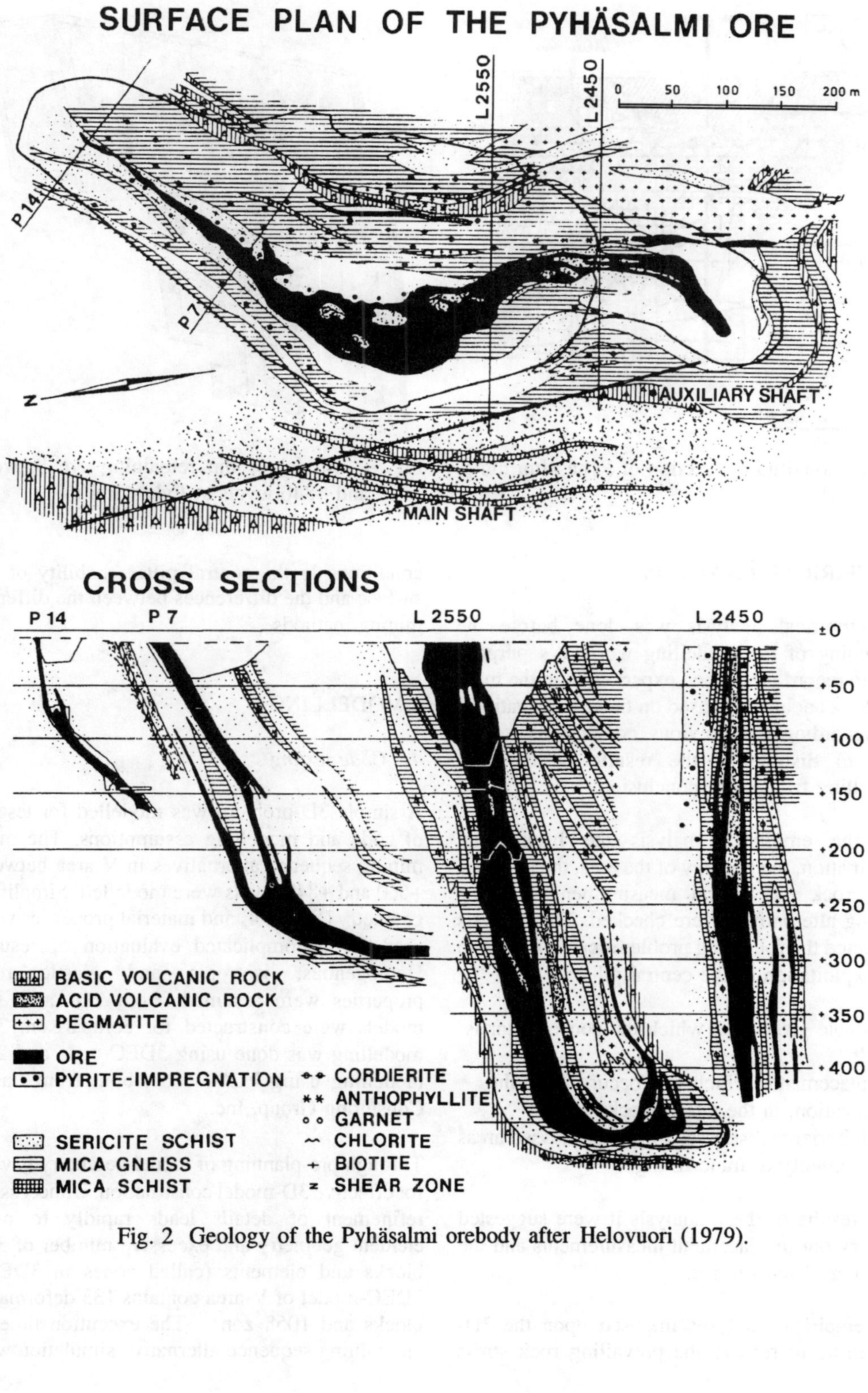

Fig. 2 Geology of the Pyhäsalmi orebody after Helovuori (1979).

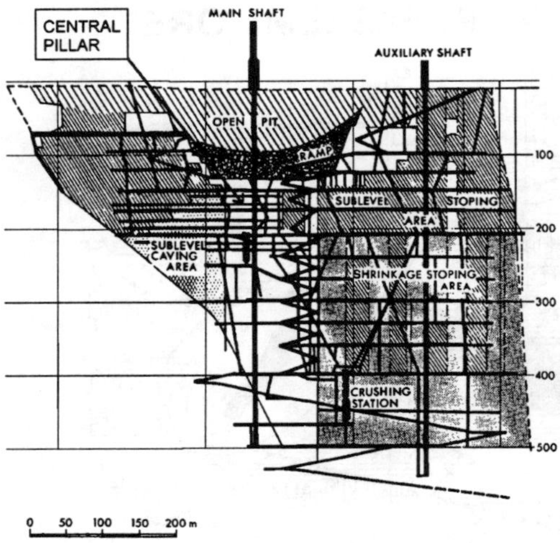

Fig. 3. Longitudinal section of Pyhäsalmi mine.

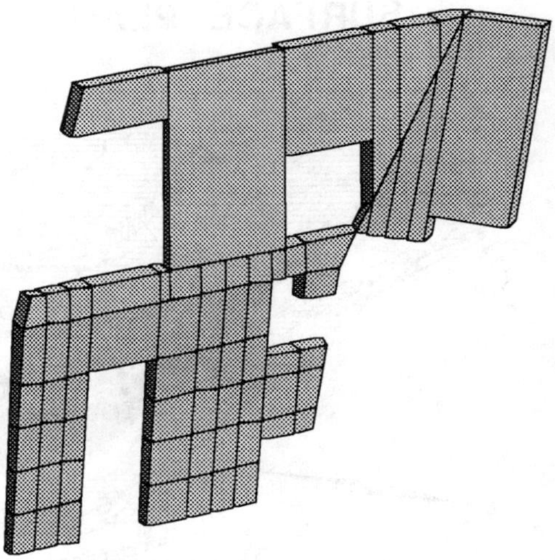

Fig. 4 Model of the remaining ore between +530 and +840 levels at Pyhäsalmi mine.

3 EMPIRICAL ANALYSIS

An empirical analysis was done before the beginning of the modelling work. It's purpose was to record the gained experiences in the mine and the conclusions based on those observations. By recording the previous conclusions it was able to distinguish the results of the 3D-modelling from those conclusions.

For the empirical analysis the background information, the geology of the area, the previous done rock mechanical measurements and the mining alternatives were checked. The analysis predicted the following problems can arise due to the exploitation of the central pillar:

- unstable ground on which the concentrator is built
- displacements, which will occur due to the excavation, in the altered rocks
- high horizontal stresses or/and destressed areas
- poor quality of mine fillings

As a results of these analysis it were suggested to carry out displacement measurements and the levelling of the surface.

The empirical analyses imposed upon the 3D-modelling to resolve the prevailing rock stress conditions in the central pillar, stability of the surface and the differences between the different mining methods.

4 MODELLING

4.1 Code testing

A simple 3D-problem was modelled for testing of code and modelling assumptions. The main mining sequence alternatives in V-area between +660 and +840 levels were modelled. Simplified geometry (Figure 4) and material properties were used for uncomplicated evaluation of results. Homogenous, isotropic and elastic rock properties were assumed. Both 2D- and 3D-models were constructed for comparison. 3D-modelling was done using 3DEC code and 2D-modelling using FLAC-code, both from Itasca Consulting Group, Inc..

Through pre-planning of model geometry is vital for effective 3D-model construction. Unnecessary refinement of details leads rapidly to poor element geometry and excessive number of 3D-blocks and elements (called zones in 3DEC). 3DEC-model of V-area contains 133 deformable blocks and 1058 zones. The execution time of one mining sequence alternative simulation with eight mining steps took about 48 hours using 25 MHz 486-microcomputer with Weitek 4167 co-processor.

The result of 3D-modelling was as expected: the influence of out-of-plane excavations and pillars to stresses and displacements was clearly demonstrated by comparing 2D- and 3D-modelling results.

4.2 Global model

As a first step, a model with very large dimensions and quite simple geometry was built for finding out adequate size and realistic boundary conditions for more detailed model. The selected size of global model was 5 km x 5 km in horizontal section and the height of model was 1.5 km.

The modelling was done using fixed boundaries. Initial stresses were linearly varying by depth. Maximum horizontal stress was east-west direction and 2.5 times and minimum horizontal stress 1.25 times the gravitional vertical stress.

When the equilibrium of calculation was achieved, the ore properties were replaced by properties of unconsolidated backfill. The dimensions of regional model were set according the changes of directions and magnitudes of stresses. By selecting 1.4 x 1.4 x 1 km rectangular shape (Figure 5) for regional model, it was possible apply the boundary stresses as linearly varying values on the boundaries.

4.3 Regional model

Displacements around excavations are usually more uniform and have longer range of influence than stresses have. Therefore the shapes of excavations as well as boundaries between materials must be modelled with higher accuracy if estimate of stress changes is required, compared to a situation, where only displacements are main point of interest. Qualitatively detailed models are required in case of failure along rock discontinuities, because the

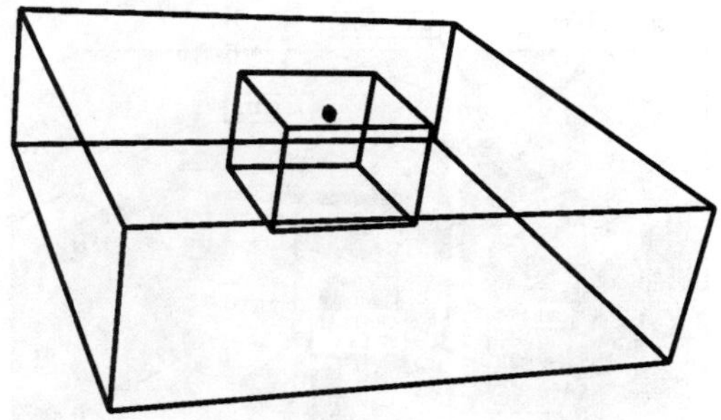

Fig. 5. The shapes of global model (5 x 5 x 1.5 km), regional model and the central pillar.

actual failure mode might not become detected with oversimplified model.

Calculations using the regional model at Pyhäsalmi mine were executed to provide answers to three questions:

1) stress state in central pillar ?

2) displacements at main shaft ?

3) surface subsidence at footwall, especially at concentrator ?

As described above, detailed model is required for realistic modelling in this case. Because of the folded and twisted shape of ore formation, quite extensive work was to be performed for building a model of the entire mine. The boundaries of rock types as well as stopes were constructed from planning drawings. The cross sections produced from final 3DEC-model matched closely with geological maps.

The final regional model (Figure 6) consist of 1241 fully deformable blocks and 29226 zones (elements). The building of the model took about six weeks, due to many cross-checks of geometry of structures as well as reforming of numerically unfavorably shaped blocks.

The material properties were estimated using values from laboratory testing. Scaling down the strength properties of rock was made by method

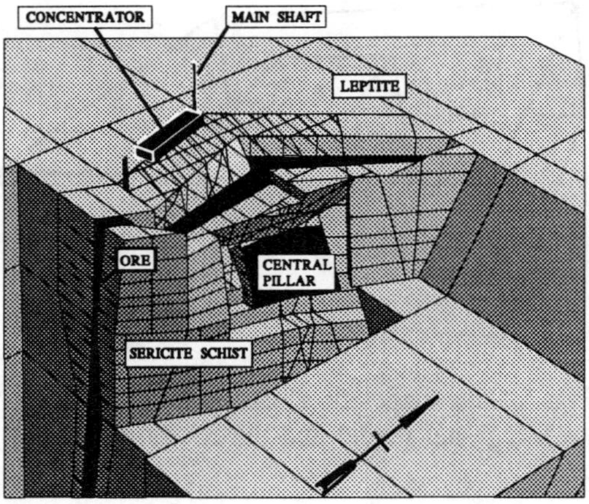

Fig. 6 Central part of the regional model. Blocks have been removed for exposing the central pillar.

suggested by Hoek (1990).

Joint shear stiffness was reduced using chart of Barton & Bandis (1983). Joint normal stiffness was reduced in proportion of joint filling thicknesses between laboratory samples and joints observed in field. Material properties used in modelling are presented in table 1 and joint properties in table 2.

The modelled footwall rock jointing and location of headframes and the concentrator is demonstrated in Figure 6. Although the RAM memory of the microcomputer was 17 Mb, the number of modelled joints had to be restricted. Because the main area of interest is the footwall side rocks near the surface, the actual jointing there was modelled as close to reality as possible.

The mining methods utilized during history of Pyhäsalmi mine have been open pit mining, sublevel caving, sublevel stoping, shrinkage stoping and bench cut-and-fill stoping. The main mining method today is small scale sublevel stoping. Use of the expensive hardening backfill have been avoided whenever possible. Thus most of stopes are filled with classified tailings, rock fill or caved rock. This fill is mostly poorly compacted due to scattered mining sequence. The open pit is filled with rockfill up the level +60.

According the mining strategy of the mine, no pillars will be left besides few small stopes with negative cashflow. So quite large rock displacements are expected during and after mining. In hangingwall there is no vital underground or surface structures. All the connections, shafts, crushers and ramp are situated in footwall. At surface, the headframes, concentrator, railway and concentrate stockpiles are located close to the open pit.

Besides elastic response of rock to excavating and filling of stopes, both plastic strains and block sliding along joints will cause displacements. Both modelling and field observations suggest, that between quite early stage of underground mining and the far future, tens or even hundreds of years, there is no equilibrium state. This brings problems in modelling, because nowadays there is no universal method for predicting deformation speed of full scale hard rock mass. With intensive rock monitoring program, a site-specific model could be constructed.

In modelling described here, time between excavation steps was determined according the maximum unbalanced force. Explicit timestepping was continued until major stress changes had taken place. Therefore the years presented in modelling results should be considered as a reference to mining sequence rather than years of real time.

The mine has continuous levelling program for surface monitoring. These results might be used in future for improving the time-dependency estimate of model. In modelling presented here, no attempts were made to adjust modelled and measured displacements.

Stress state in central pillar

The central pillar has experienced large changes of stress state during history. In model, maximum initial stress before mining was about 15 MPa. As a result of open pit and underground mining, large stresses gathered themselves into the central pillar which caused the yielding of the pillar. Pillar is now in post-failure state, and general stress state is low, average maximum stress in the model is about 5 MPa. The H/W ratio of pillar increases during future mining, and the maximum horizontal stress decreases due to further yielding of the pillar.

The field observations agrees with the modelling result. There have been no evidence of high stresses in central pillar since early 1980's. The originally very stiff (Young's modulus about 100 GPa) ore is now softened. An attempt to install vibrating wire stress cells in pillar failed, because the pretensioning of cells was lost in few months. Laboratory test of new core samples from central pillar demonstrates cohesion loss of intact ore. The uniaxial compressive strength is now about 50 % of pre-mining strength.

The stability of openings in central pillar has been still fair, mainly due the ductile nature of the ore.

Displacements at main shaft

Modelled displacements at main shaft, since beginning of the mining, indicate 35 mm horizontal movement at the bottom of the shaft, with virtually no displacement at surface. Explanation to this somewhat surprising behavior is the geometry of the formation. At surface, the main shaft is situated near a corner of ore formation (Figure 4). At depth, the position of shaft, in middle of almost tabular portion of formation, is less favorable. So larger displacement occurs at the bottom of shaft, in spite of longer distance between the main shaft and stoping area.

Modelled horizontal displacements by end of the mining at 1999 are 100 mm at the bottom of the shaft and 50 mm at surface. Thus relative displacement between the bottom and the collar of main shaft is 50 mm.

Surface subsidence at footwall

The concentrator is situated at footwall along the contact of weak and jointed sericite schist and strong, compact leptite (Figure 6) and the railway and concentrate stockpiles are between concentrator and open pit. The stability of these structures is essential for continuous operation of the mine.

Modelled surface subsidence at footwall by 1991 (Figure 7) varies from 100 mm at the edge of open pit to below 10 mm at leptite under the concentrator. It is significant to note the irregular subsidence of separate blocks. This is more harmful to surface structures than uniform displacement is.

The mining of central pillar began at end of 1991. In modelling, the speed of subsidence increased definitely, and current displacements are predicted double themselves by end of mining at 1999 (Figure 8). The subsidence will continue after the end of mining, and at same speed, if not faster.

The net of levelling points at footwall is nowadays sparse, because most of the points near open pit have been destroyed in during last two

Table 1. Material properties

property	ore	F/W	H/W	leptite	fill
K GPa	22	9	9	9	0.01
G GPa	17	7	7	7	0.006
density kg/m³	4400	2900	2900	2900	1600
cohesion MPa	0.22	0.031	0.037	5.0	0
friction angle	66°	63°	63°	69°	30°
tensile strength kPa	1	0	0	1	0

Table 2. Properties of joints.

property	joints	faults
normal stiffness	20 GPa/m	1 GPa/m
shear stiffness	2 GPa/m	0.1 GPa/m
cohesion	0	0
friction angle	35°	20°

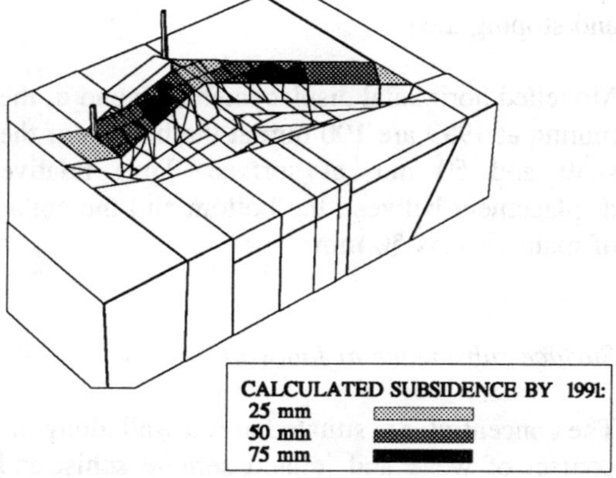

Fig. 7. Modelled subsidence at footwall by 1991.

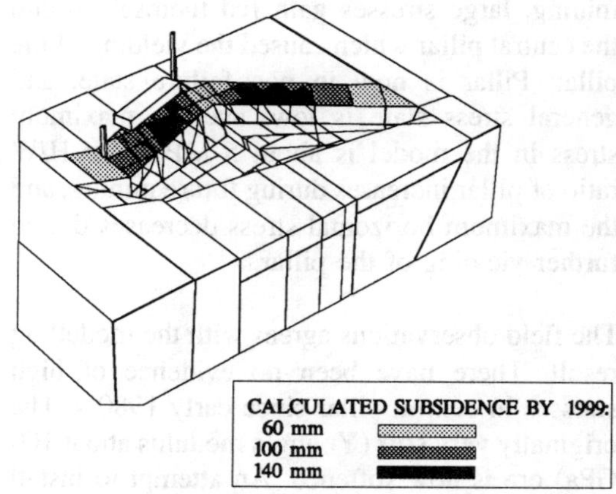

Fig. 8. Modelled subsidence at footwall by 1999.

decades. Three closely spaced points at middle of weak footwall rocks have survived. The mean subsidence of these points and calculated subsidence at same place agrees quite well (Figure 9), especially if we take account the fact that modelling parameters were not changed according the measured displacements.

4.4 Local model

As stated before, the regional model is designed to give realistic response of rock excavation. Because it's complex structure it's less suitable for detailed modelling of central pillar. Also the long computing time, two weeks to simulate the entire mining of Pyhäsalmi mine, clearly requires a more versatile model. The local model of central pillar (Figure 10) was built for future use in optimization of mining sequence and modelling of cable bolting.

At beginning, same initial boundary stresses as in regional model were applied. This resulted to high horizontal stress field in the central pillar and the footwall subsidence was 40 % smaller than obtained from regional model. When the boundary conditions were set according the results of regional model in mining phase in question, the stresses and displacement in both models were close each other. This demonstrated positively, that the stepwise modelling from global model to regional model, and from regional model to local model, is necessary to collect realistic boundary conditions for detailed modelling.

5 DISCUSSION AND CONCLUSIONS

Empirical analysis

Empirical analysis gave a valuable estimation of the potential problems in mining out the central pillar. The analysis was done by the mine planning personnel including technicians and foremen with a long experience in Pyhäsalmi mine. All the questions aroused in the analysis were answered during the project and also a totally new point of views were discovered. It was found very advisable to make an empirical analysis before modelling and it is always necessary to compare modelling results with the actual observations in the mine.

Code testing

The code testing showed that 2D modelling gave a higher maximum stresses than 3D modelling which was expected. The influence of out-of-plane excavations and pillars to stresses and displacements was demonstrated clearly. A good geometry pre-planning is vital for a successful 3D modelling.

Global model

By modelling in two or three steps it was possible to proceed to more detailed models with a linearly varying boundary stresses. Each step provides boundary conditions for next step. In Pyhäsalmi case the size of the regional model was determined to be 1.4 x 1.4 x 1 km (width x length x depth).

Regional model

Three main questions were expected to be answered by regional modelling:

1) stress state in central pillar
2) displacements at main shaft
3) surface subsidence at footwall, especially at the concentrator

The final regional model (Figure 6.) consist of 1241 fully deformable blocks and 29226 zones (elements). The building of the model took about six weeks, due to many cross-checks of geometry of structures as well as reforming of numerically unfavorably shaped blocks. The material properties were estimated using values from laboratory testing which were scaled by method suggested by Hoek (1990). Joint shear stiffness was reduced using chart of Barton & Bandis (1982) and joint normal stiffness was reduced in proportion of joint filling thicknesses between laboratory samples and joints observed in field. Material parameters were determined before modelling and they were not changed during modelling.

All the previous mining sequences were modelled starting from the open pit and ending to the closure of the mine. Because the main area of interest was the footwall surface and main shaft on footwall, the actual jointing there was modelled as close to reality as possible. The computing time for the simulation of the entire mining of the Pyhäsalmi mine was two weeks. The used microcomputer was 486/25 MHz with 17 Mb RAM and Weitek co-processor.

The results of the regional model showed that the central pillar is now in post-failure state, and

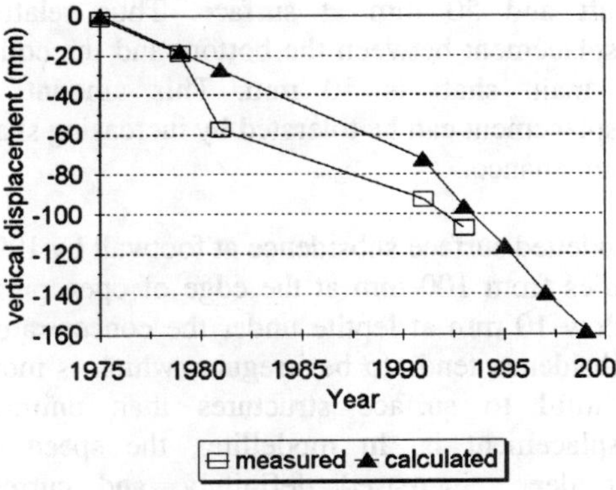

Fig. 9. Measured and calculated subsidence at middle of weak footwall rocks.

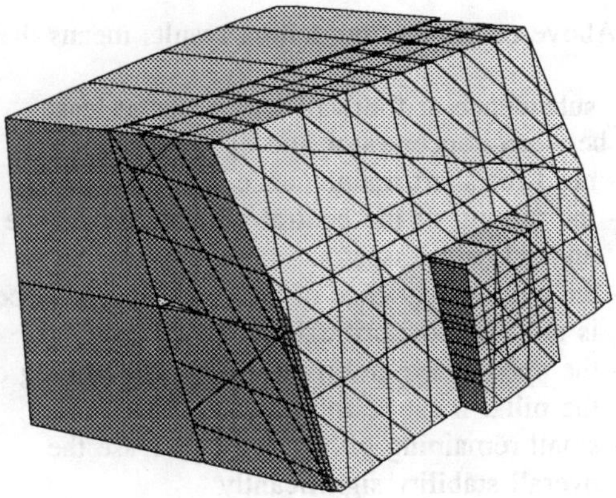

Fig. 10 The local model. The hangingwall and filling materials are removed to expose the central pillar.

general stress state is low, average maximum stress in the model is about 5 MPa. The field observations agrees with the modelling result. There have been no evidence of high stresses in central pillar since early 1980's. The originally very stiff (Young's modulus about 100 GPa) ore is now softened. Laboratory test of new core samples from central pillar demonstrates cohesion loss of intact ore. The uniaxial compressive strength is now about 50 % of pre-mining strength.

Modelled horizontal displacements by end of the mining at 1999 are 100 mm at the bottom of the shaft and 50 mm at surface. Thus relative displacement between the bottom and the collar of main shaft is 50 mm. This amount of displacement can be tolerated by increasing shaft maintenance.

Modelled surface subsidence at footwall by 1991 varies from 100 mm at the edge of open pit to below 10 mm at leptite under the concentrator. Subsidence tends to be irregular which is more harmful to surface structures than uniform displacement is. In modelling, the speed of subsidence increased definitely, and current displacements are predicted double themselves by end of the mining at 1999. The subsidence will continue after the end of mining, and at same speed, if not faster.

Above mentioned modelling results means that:

- subsidence at footwall will occur at least between concentrator and open pit
- need for shaft maintenance will increase
- the mining of the central pillar will be done under low stresses
- the mining sequence of the central pillar stopes is not the most critical factor
- the open space and roof area during mining of the pillar must be minimized
- small remaining pillars don't increase the overall stability significantly

The 3D modelling of the central pillar showed that the three dimensional distinct element modelling can be an effective tool for mine planning. The obtained results don't disagree with the field observations. The mean subsidence of levelling points at footwall on surface and calculated subsidence at same place agrees quite well. Also laboratory tests of the central pillar ore agrees with the calculated result of the post failure stage of the pillar. There is, however, development potential. The most time consuming part of the modelling is one of the most vital parts: building up the model geometry.

Acknowledgements

The authors wish to acknowledge the Ministry of Trade and Industry of Finland and Outokumpu Finnmines Oy for their strong support to make this project possible. The personal information given by prof. Raimo Matikainen and the competent help of the staff of the Pyhäsalmi mine is greatly appreciated.

REFERENCES

Barton, N.R. and Bandis, S., Effects of Block Size on the Shear Behavior of Jointed Rock. 23rd U.S. Symposium on Rock Mechanics. Berkeley, California 1982. pp. 739-760.

Helovuori, O., Geology of the Pyhäsalmi Ore Deposit, Finland. Economic Geology, Vol 74 (1979), pp. 1084-1101.

Hoek, E., Technical Note: Estimating Mohr-Coulomb Friction and Cohesion Values from the Hoek-Brown Failure Criterion. Int. J. Rock Mech. Min. Sci. & Geomech. Abstr. Vol. 27, No. 3. pp. 227-229,1990 Great Britain.

Innovative Mine Design for the 21st Century, Bawden & Archibald (eds) © 1993 Balkema, Rotterdam, ISBN 90 5410 325 6

Model input parameter selection through instrumentation and back analysis of large excavation failure in a deep level South African gold mine

M.P. Raffield
Randfontein Estates Gold Mining Co. Ltd., South Africa

J.V. James & I. Humphreys
South Deep Exploration Co. Ltd., South Africa

A.K. Isaac
University of Wales, College Cardiff, UK

ABSTRACT: The South Deep Exploration Company is undertaking feasability studies to develop a new mine to exploit Witwatersrand conglomerates at depths between 2,4 and 3,3 km.

At this depth, tunnels are severely damaged by stress despite the good rock quality encountered in the Witwatersrand quartzites.

In order to design the correct profiles and support for large permanent excavations on the mine the FLAC numerical modelling program is used. A large excavation, the crusher station of the underground backfill plant, was monitored during excavation and the results of the monitoring programme were compared with the numerical modelling to modify the input parameters which were obtained from laboratory testing.

The input parameters so derived are more suitable for design of further excavations in the same or similar rock mechanics environments.

1 INTRODUCTION

The planned South Deep gold mine will lie to the South of the existing Western Areas Gold Mine about 40 km south west of Johannesburg in South Africa. The orebody to be exploited is a typical Witwatersrand sedimentary placer deposit consisting of gold bearing quartzitic conglomerate bands in quartzitic host rocks forming the Upper Formations of the Witwatersrand Supergroup. Directly overlying the topmost conglomeratic reef, the Ventersdorp Contact Reef (VCR), lie tuffaceous basaltic lavas of the Ventersdorp Supergroup.

To the west these conglomerates become thinner until ultimately, only the VCR is developed as a payable orebody with channel widths ranging from 0 to 2,5 m

The ore body extends over 12 km at depths ranging from 2,4 to 3,3 km.

South Deep will thus be a deep level mine with all the inherent problems of potential seismicity and high field stresses with associated damage to sidewalls of all excavations.

In this environment stress related fracturing occurs immediately after development of any excavation, if a correct profile and adequate support is not installed then unravelling of the sidewall will continue, ultimately damaging the hangingwall as well.

In medium to long term excavations it is thus necessary to design the support and profile to ensure their stability through the life of the mine.

South Deep is currently developing the future shaft reef area on VCR, for mining and backfilling, prior to shaft sinking (Raffield and James, 1993). Access is via the adjacent Western Areas Gold Mine's South Shaft system and a 3,5 km access tunnel and return airway.

The excavations, for the backfill plant and underground waste crushing plant, have been developed from the Main Intake Haulage and are the site of the project described herein.

Profile and support design of large excavations in the SA Mining Industry has previously been carried out using two dimensional, boundary element, elastic computer programs. Design generally followed the principle of eliminating areas of high stress concentration where failure was likely to occur.

This is analogous to designing tunnels to the

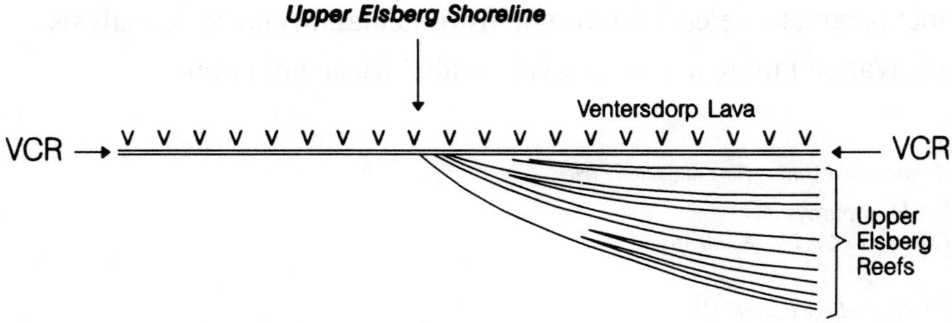

Figure 1 Geological Cross Section of the South Deep Ore Body

profile closest to the shape with which the failure of sidewalls would ultimately stabilize to (Richards and Byorkman, 1978). This can also be expressed as minimizing the shear stresses about a tunnel.

In civil engineering design, finite element analysis and boundary element analysis have both been used to define and then minimize and support the failed, or relaxed zone about an excavation commonly using a Mohr-Coulomb or Hoek and Brown failure criteria (Ho-Yeong Kim et al, 1990). The input parameters for such analysis were often obtained by back analysis of instrumentation results from a test excavation or previous excavations in the same environment (Bowcock et al, 1976; Sharp et al, 1979).

To obtain a more realistic conceptual model of the mechanisms of failure and movement in the rock about large excavations in the strong but brittle quartzitic rocks at the depths described, it was necessary to employ a modelling package which would simulate the non-elastic post-failure behaviour of the rock and the interaction of this rock mass with various forms of tunnel support.

The aim of the project was to compare this modelling with instrumentation results for a newly excavated cavern and thus to derive general input parameters which can be used for support and profile design of future large excavation in similar environments.

2 THE SOUTH DEEP CRUSHER EXCAVATION

It was necessary to identify a location on South Deep where a monitoring program could be set up. The crusher excavation of the underground

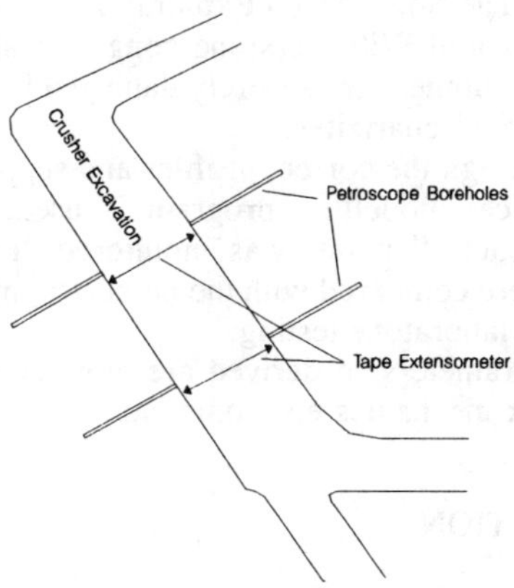

Figure 2 Plan of the Underground Backfill Plant

backfill plant was chosen for the following reasons:-

(i) At the time of initiation of this project the excavation had not begun, it was thus possible to commence a monitoring program to obtain a complete history of the rock behaviour around the excavation.

(ii) It is the largest of the excavations planned in the short term and as such was expected to show the greatest degree of movement and failure.

(iii) Once completed, the excavation would not be in use immediately, this allowed the continuation of the monitoring program until the excavation stabilised.

The excavation will eventually house the waste rock crushing plant, for use in the production of backfill. The excavation is thus classified as permanent and must be designed and supported

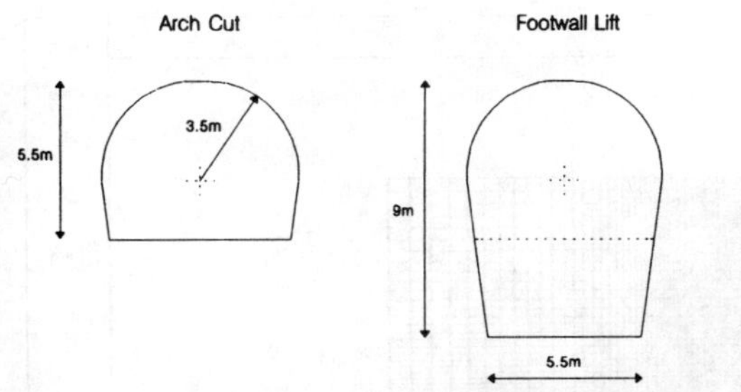

Figure 3 Profile of the Crusher Excavation After the Two Stages of Mining

so as to remain standing for the life of the mine. No mining will be carried out in the proximity of the backfill plant so it is not necessary to consider large changes in the stress regime. The excavation is located in homogeneous footwall quartzites, approximately 100 m below the reef horizon. Figure 2 is a plan view of the underground backfill plant, and Figure 3 shows the dimensions and mining sequence for the excavation which was mined in two cuts.

The first cut was the arch which was approximately 5,5 m high and 7 m wide. The footwall was then lifted, increasing the height to approximately 9 m.

3 NUMERICAL MODELLING OF EXCAVATION FAILURE

3.1 *Generation of the FLAC Model*

As explained previously, it is necessary to use elasto-plastic modelling methods in the near zone of a large excavation in order to obtain accurate failure and deformation results.

FLAC (Fast Langrangian Analysis of Continua) is a two-dimensional, finite difference modelling package used to predict rock behaviour.

In order to construct a FLAC model the following areas must be addressed:-

(i) Grid generation
(ii) Boundary conditions
(iii) Initial conditions
(iv) Choice of constitutive model
(v) Material properties

Figures 4 a) and 4 b) show the FLAC grid used for the modelling of the crusher excavation. The large dimensions of the grid in comparison to the actual excavation size are required in order to limit the effect of the boundary on the stress distribution around the excavation (ITASCA, 1992).

To reduce computer run times the elements further away from the excavation have been enlarged by progressively increasing their size towards the edge of the grid. The grid was set up with half symmetry down the centre of the excavation, for the same reason.

To simulate the in-situ stress conditions it was necessary to apply internal stresses and boundary conditions to the model.

Rock fracturing around excavations on South Deep suggests that a K-ratio (horizontal stress:vertical stress) of 0,5 is present. Vertical stress increases lithostatically with depth, thus at the depth of the excavation modelled, 2650m, it is approximately 70 MPa.

Boundary stresses are applied around the sides of the grid as shown in Figure 5. Frictionless rollers are applied to the centre line of the excavation to provide symmetry conditions.

There are several models available in FLAC which can be used for modelling elasto-plastic behaviour. Each of the models has been developed to represent a specific type of constitutive behaviour associated with geologic materials:-

i) The Mohr-Coulomb plasticity model is used for materials which yield when subjected to shear loading.
ii) The ubiquitous joint model applies to a Mohr-Coulomb material which exhibits a well defined strength anisotropy.
iii) The strain-softening model is based on the Mohr-Coulomb model, but it allows the material to show a degradation of shear strength when loaded beyond the initial failure limit.
iv) The double yield model extends the strain-softening model to simulate irreversible compaction as well as shear yielding.

In choosing the most representative model it was necessary to look at the application of the various models. The basic Mohr-Coulomb model is suitable for loose and cemented granular material such as soil, rock and concrete. The ubiquitous joint model applies to thinly laminated materials exhibiting strength anisotropy, for example, slate. The double yield model is suitable for lightly cemented granular material in which pressure causes a permanent volume decrease, for example backfill. The strain-softening model is applicable where it is

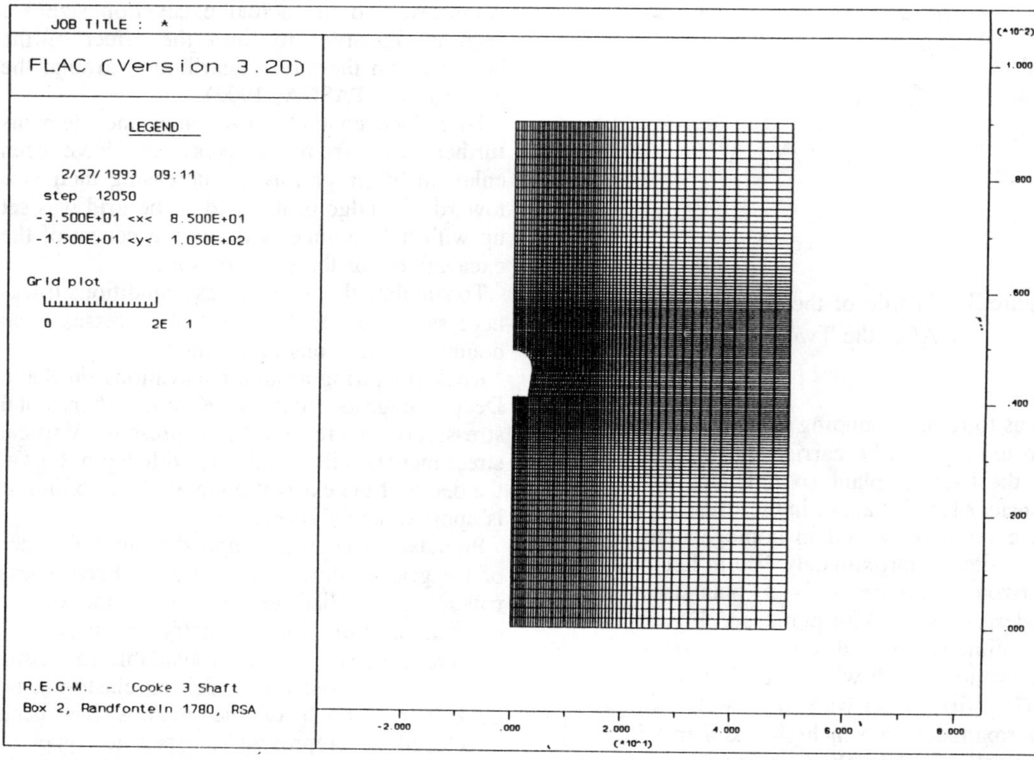

Figure 4 a) Complete FLAC Grid

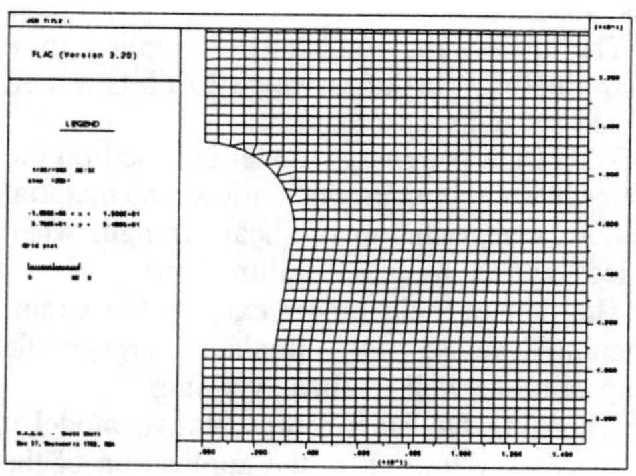

Figure 4 b) Close-up of the FLAC Grid around the Crusher Excavation

necessary to define the non-linear post-failure properties of rock.

The following factors favour the use of the strain-softening model:-

(i) The rock around the crusher excavation is composed of brittle, homogeneous, quartzites, thus preventing the use of the ubiquitous joint or double yield models.

(ii) The rock in the near excavation zone does not keep a constant strength once it has failed, rather it becomes progressively weaker as fracturing increases.

The theory behind the strain-softening model and the material properties used to define it are addressed in the following section.

3.2 *The Strain-Softening Model*

As with all the constitutive models, the initial load/deformation response is elastic and is defined by the elastic properties, shear and bulk moduli. The shear and bulk moduli are directly related to Young's Modulus and Poisson's Ratio, however they are thought to give a more accurate stress/strain relationship (ITASCA, 1992).

Material failure in the strain-softening model is defined by the linear Mohr-Coulomb failure criteria (Figure 6).

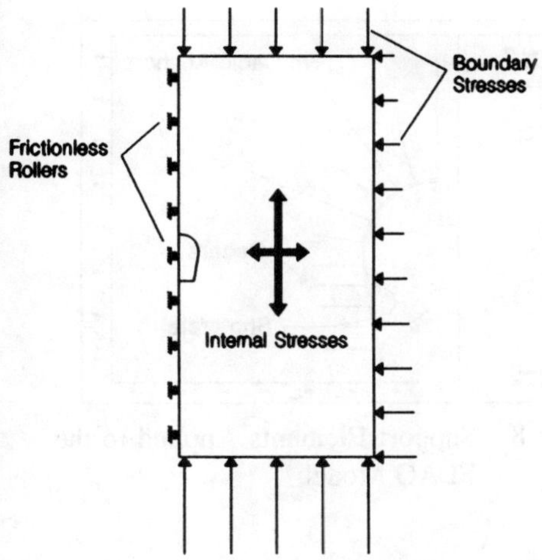

Figure 5 Boundary and Internal Stresses Applied to the FLAC Model

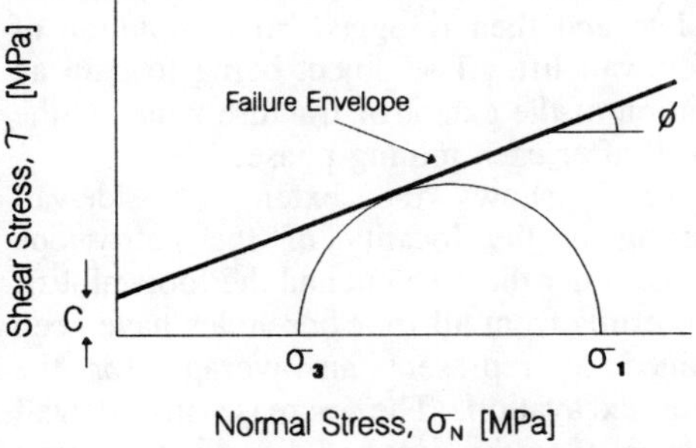

Figure 6 Mohr-Coulomb Failure Criteria

The shear yield function, f_s, is defined as:-

$$f_s = \sigma_1 - \sigma_3 N_\phi + 2C\sqrt{N_\phi}$$

where:
$N_\phi = (1 + \sin\phi)/(1 - \sin\phi)$
ϕ = friction angle
C = cohesion.

Shear yield is detected if $f_s < O$, remembering the σ_1 and σ_3 are negative for compressive stress. The post-failure properties of a rock mass are defined by cohesion, friction angle and dilation angle. The dilation angle defines the shear dilatancy of the rock, that is, the change in volume which occurs with shear distortion of the material. When the strain-softening model detects $f_s < O$ it allows plastic flow to take place

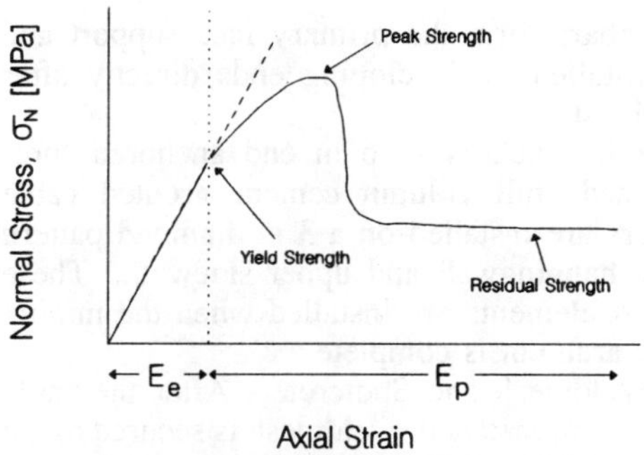

Figure 7 Typical Stress/Strain Curve for a Laboratory Tested Rock

which reduces cohesion, friction and dilation values thus lowering the Mohr-Coulomb yield criteria and restoring $f_s = O$. as the high strain areas of the model undergo failure, so friction, cohesion and dilation are progressively lowered reducing the rock strength in these areas.

Figure 7 shows a typical stress/strain curve for a laboratory tested rock specimen. The elastic strain portion of the curve, E_e, is related to the stress by the elastic parameters, bulk and shear moduli. The plastic strain portion of the curve, E_p, is related to the stress by values of cohesion, friction and dilation defined as functions of total plastic strain.

The physical values for bulk modulus, shear modulus, cohesion, friction and dilation, can be determined from triaxial testing of rock specimens. However, due to scale effects, these laboratory results do not describe the behaviour of the rockmass on a large scale. Discontinuities lead to the rock mass exhibiting a far lower strength than predicted in laboratory tests. Input parameters obtained from the laboratory testing results must be significantly modified to accurately simulate the underground environment.

3.3 *Support Modelling*

FLAC has the capacity for support elements to be included in the modelling process. For the purposes of modelling the crusher excavation, rebars, cable anchors and shotcrete elements have been used to simulate an integrated support system.

(i) Rebars - 2,4 m full column resin grouted rebars are installed on a 1,5 m diamond pattern.

The rebars form the primary face support and are installed in developing ends directly after each blast.

(ii)Cable Anchors - 6 m end anchored, pre-tensioned, full column cement grouted cable anchors are installed on a 3 m diamond pattern in the hangingwall and upper sidewall. These support elements are installed when the mining of the arch cut is complete.

(iii) Weldmesh and Shotcrete - After the cable anchors are installed, weldmesh is secured to the rebars and the whole excavation is sprayed with 50 mm of steel fibre reinforced shotcrete. It is not possible to model the weldmesh as a separate entity in FLAC, however, its effect is to significantly reinforce the shotcrete and this is accounted for in the modelling.

Figure 8 shows the support elements and their position in the FLAC model.

The diamond configuration of the rebar and cable elements leads to two different support patterns when the excavation is viewed in section. Both of these patterns are projected onto the FLAC model so that the effect of all the support elements are accounted for. The material properties of these elements are then reduced to account for the fact that they are not present at every point in the tunnel.

The timing of the support elements being installed in the model must accurately reflect their emplacement in the real situation, for example the rebars are installed only hours after excavation, whereas the cable anchors and shotcrete are installed weeks after excavation. This situation is represented in the model by leaving equilibration steps between installation of the various support types.

4 IN-SITU EXCAVATION FAILURE MONITORING

In order to provide information for the back analysis of the numerical modelling, two forms of monitoring were used: a borehole petroscope qualitatively measures fracture development in the sidewalls; and a tape extensometer is used to quantify closure across the excavation.

Two monitoring stations were set up in the crusher excavation. Figure 2 shows the position of the petroscope boreholes and the displacement monitoring points.

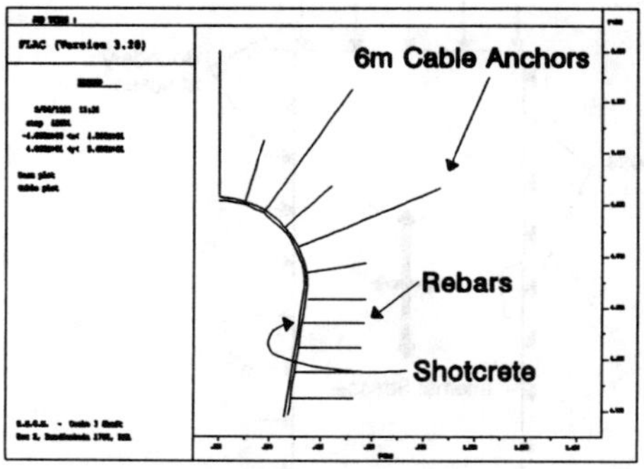

Figure 8 Support Elements Applied to the FLAC Model

4.1 *The Borehole Petroscope*

This instrument is used for optical logging of fracture development in a borehole. The four petroscope holes were logged when the arch was complete and then relogged on completion of the footwall lift. The object being to gain an indication of the extent of fracture zones in the sidewall after each mining phase.

Figure 9 shows the extent of sidewall fracturing in the locality of the petroscope boreholes after the arch cut and the footwall lift.

The results from all four boreholes have been combined to represent an average for the crusher excavation. The increase in sidewall fracturing after the footwall lift was observed and is shown to be significant. The depth of fracturing was 1.8 m after the arch cut, increasing to 4.1 m after the footwall lift.

4.2 *Tape Extensometer*

This instrument was used to measure the distance between rock anchored points at various depths to a high degree of accuracy. In the crusher excavation reference points were anchored on the surface, at 1 m and at 2 m depth. Readings were taken weekly so the movement could be quantified as a function of time. Two measuring stations were used and Figure 10 is a synthesis of the results from those two stations.

The points anchored onto the surface of the excavation showed a far greater degree of movement than the points anchored 2 m into the sidewall. This is due to inelastic buckling and spalling failure of the immediate sidewall.

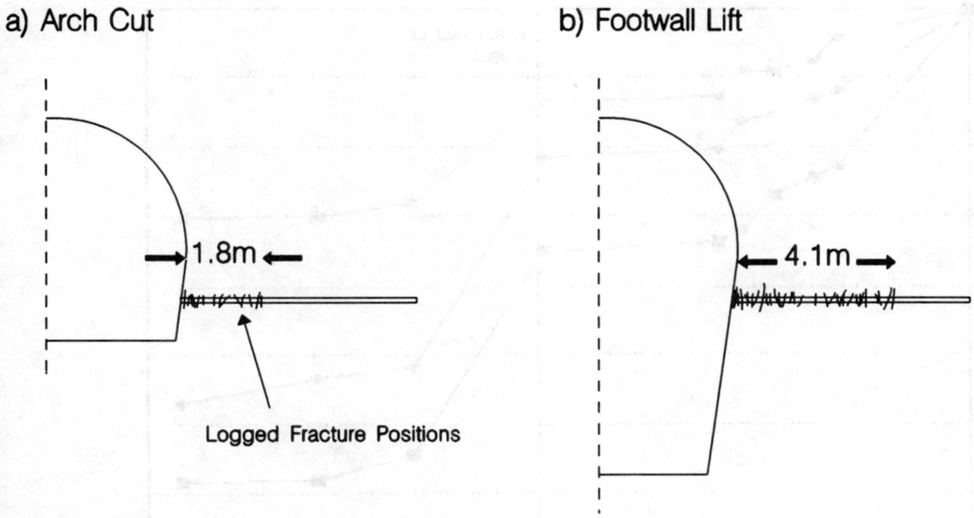

Figure 9 The Extent of Sidewall Fracturing Indicated by the Borehole Petroscope

The results show that before the footwall lift was taken, the rock mass had stopped moving and attained equilibrium. Movement of the sidewalls due to the footwall lift were more extensive than those measured in the arch cut due to the increased height of the excavation. At the end of the monitoring period the whole excavation seemed to be in equilibrium and it was thus considered acceptable to use the results for comparison with the modelling.

5 BACK ANALYSIS MODELLING OF EXCAVATION DESIGN

5.1 *Introduction*

A comparison between the actual failure monitored in the crusher excavation and the failure predicted by the FLAC modelling was made.

FLAC modelling was carried out with input parameters derived from laboratory testing of intact specimens and the necessity for accurate detuning in order to simulate reality was then highlighted.

The comparison between monitored and modelled results was carried out as follows:-

(i) Borehole petroscope results which qualitatively defined the extent of the fracture zone, were compared with the zone of plastic strain on the model. Zones of modelled plastic strain signify that plastic failure has occurred, this can be equated to fracturing in the real situation.

(ii) Extensometer results defined the movement of the rock surrounding the excavation at various depths in the sidewall. Reference points were set up on the FLAC grid (in the same positions as the end anchored rods) to keep histories of horizontal movement with FLAC time steps.

It is important to note that FLAC time steps cannot be directly compared to real time. However, the trend is similar with rapid initial elastic movement, followed by plastic failure and finally equilibrium.

5.2 *Input Parameters Derived from Laboratory Testing of Intact Rock Specimens*

Esterhuizen and Valenta (1988) carried out triaxial testing on all the South Deep rock types, including the footwall quartzites in question. The derivation of the strain-softening parameters from triaxial stress/strain curves is a complex exercise. Ryder et al. (1992) have written two programs which simplify the process. The function of the TRIAX program is to transform the complete laboratory post-failure behaviour of a rock into the terms of FLAC's strain-softening parameters (the evolution of cohesion, friction angle and dilation angle with increasing plastic

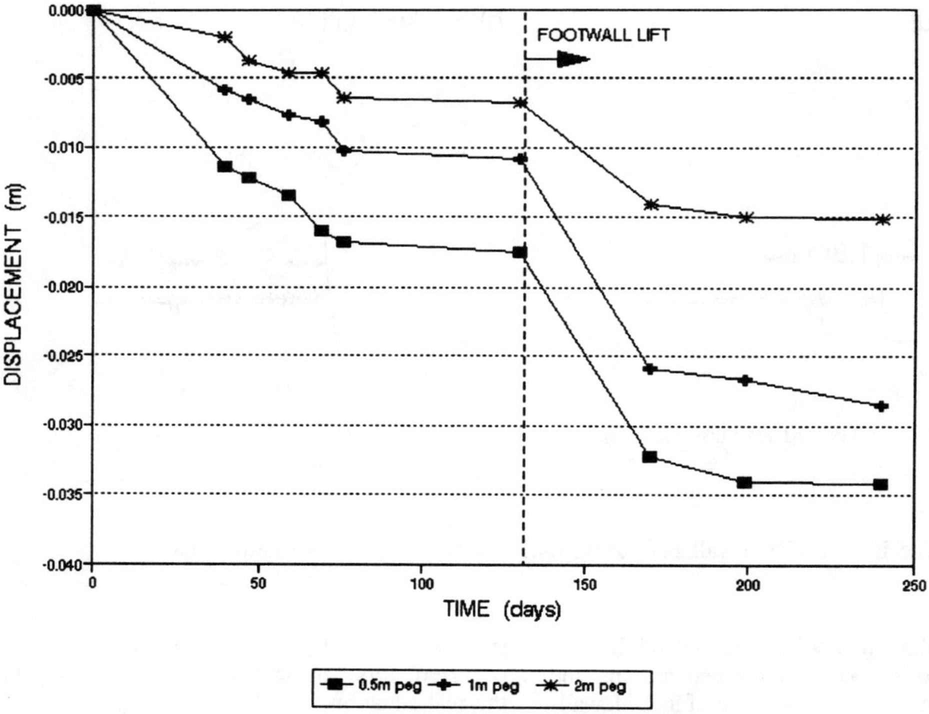

Figure 10 Sidewall Deformation as Measured by the Tape Extensometer

Table 1: Intact Rock Properties of Footwall Quartzites

Young's Modulus, E (GPa)	Poisson's Ratio	Bulk Modulus, K (GPa)	Shear Modulus, G (GPa)	Density, (Kg/m³)
70	0.18	36	30	2700

		Plastic Strain, ϵ_p (strains)					
		0	0.005	0.010	0.015	0.020	0.1
Cohesion, C	(MPa)	32	20	9	3	1	1
Friction Angle, F	(deg.)	53	49	45	43	40	40
Dilation Angle, D	(deg.)	25	18	11	5	0	0

strain). The SS program has functions which are inverse to the TRIAX program, i.e the FLAC input parameters are used to define a stress/strain behaviour. Used in conjunction, the TRIAX program analyses the raw laboratory data and the SS program is used to simplify and refine the data to check that the input parameters into FLAC define realistic behaviour.

Table 1 summarises the intact rock properties as derived from the laboratory testing of the South Deep footwall quartzites. The results of

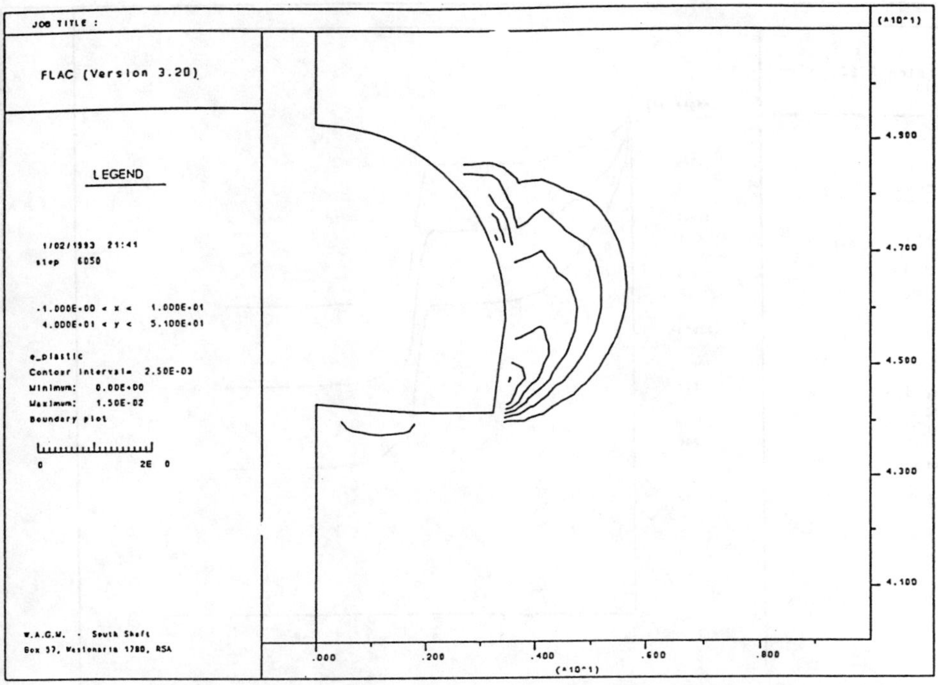

Figure 11 a) Detuned Model Plastic Strain Results for the Arch Cut

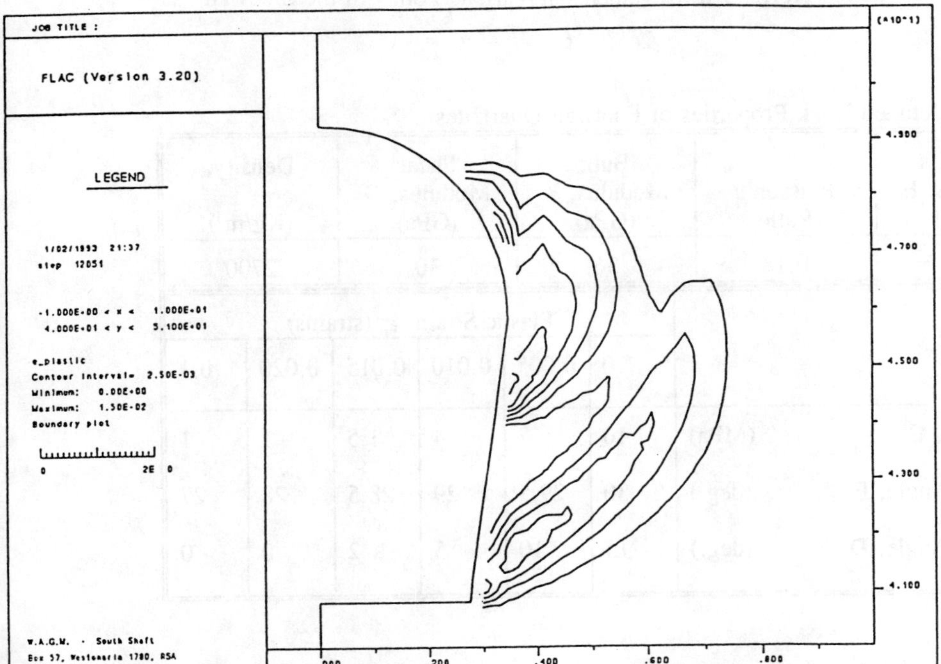

Figure 11 b) Detuned Model Plastic Strain Results for the Complete Excavation

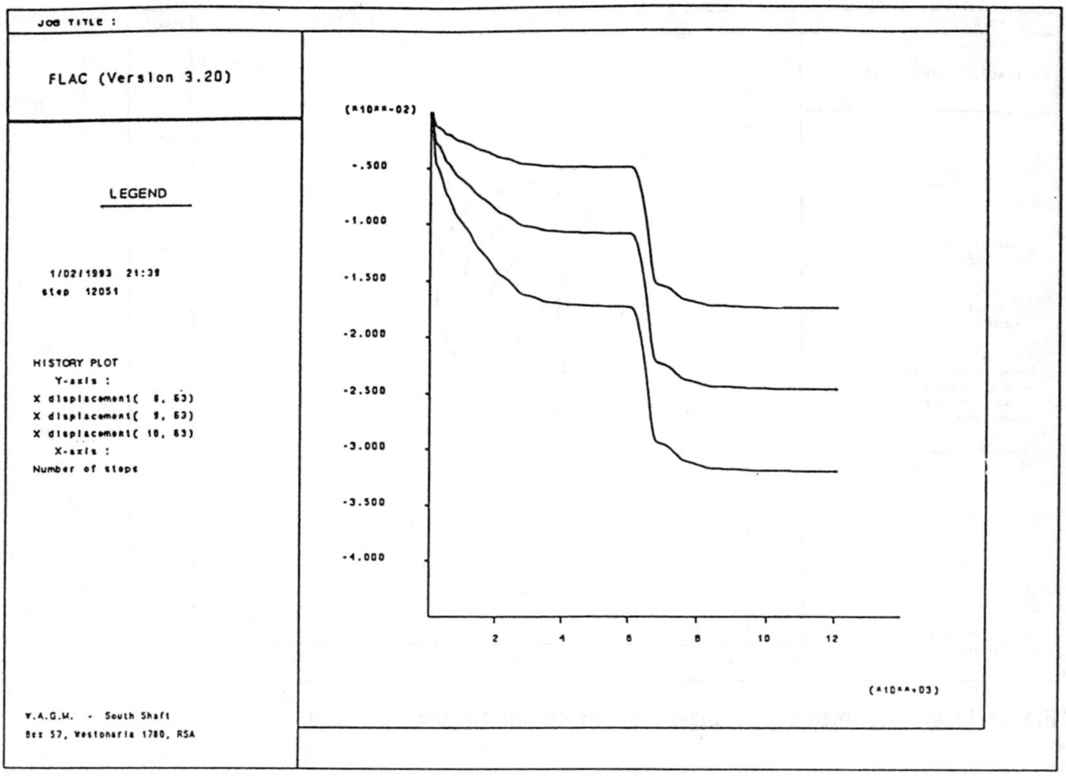

Figure 12 Modelled Horizontal Movements at Various Points in the Sidewall

Table 2: Detuned Rock Properties of Footwall Quartzites

Young's Modulus, E (GPa)	Poisson's Ratio	Bulk Modulus, K (GPa)	Shear Modulus, G (GPa)	Density, (Kg/m^3)
50	0.18	36	30	2700

		Plastic Strain, ϵ_p (strains)					
		0	0.005	0.010	0.015	0.020	0.1
Cohesion, C	(MPa)	10	7	4	1.5	1	1
Friction Angle, F	(deg.)	30	29.5	29	28.5	28	27
Dilation Angle, D	(deg.)	15	10	5	2	0	0

the modelling using these properties predicts a zone of plastic yield around the crusher excavation which extends only 20 cm into the sidewall. Petroscope monitoring of the crusher excavation (Fig 9) shows that the fracture zone extends at least 4 m into the sidewall after the footwall lift. The results of sidewall displacement also show a large discrepancy between modelled and actual movements.

5.3 *Modification of Laboratory Results to Simulate the Observed Rock Behaviour*

It is obvious that the intact rock properties do not accurately simulate the observed behaviour.

It is thus necessary to reduce the strength parameters of the rock to allow greater failure and movement in the model.

Table 2 details the final input parameters which were used to model the crusher excavation. These parameters are derived through a combination of practical experience, underground monitoring, empirical detuning methods and trial-and-error.

Practical experience leads to a qualitative approximation of scale effects and underground monitoring can assist in quantifying these effects. Ryder *et al* (1992) provides empirical methods for detuning the cohesion, friction and dilation values. He also emphasises that in practice numerical experimentation and back analysis are of the utmost importance. Finally, logical trial-and-error is required with all the input parameters to fine tune the model to behave as closely as possible to the real situation.

Figures 11 a) and b) show the modelled results of zones of plastic strain for the detuned input parameters.

The modelled plastic zone extends approximately 2 m into the sidewall after the arch cut (Figure 11 a)) and 4 m after the footwall lift (Figure 11 b)) both of these are a good correlation with the petroscope results shown on Figure 9.

The modelled horizontal displacements at three depths in the sidewall, for the arch cut and footwall lift, are depicted in Figure 12. If these results are compared with the actual movements on Figure 10 again a good correlation is noted.

6 CONCLUSIONS

The support and profiles of large excavations are often designed using empirical rules of thumb or more recently two dimensional elastic analysis. These approaches generally work well but by the nature tend to be conservative and non-specific. More accurate design is possible using the sort of analysis described in this paper, however suitable input parameters for programs such as FLAC cannot be directly obtained from laboratory tests.

Back analysis of monitored excavations can provide suitable input parameters and indicates within specific environments the amount of detuning and modification to laboratory rock testing results required to obtain input parameters. As more results of this type are obtained and published it may be possible to modify inputs suitably by reference to the available literature.

The authors feel that the results obtained within specific environments will alter the amount of detuning or modification to laboratory rock testing results required to obtain input parameters. As more results of this type are obtained and published it may be possible to modify inputs suitably by reference to the available literature. The authors feel that the results obtained from the modelling, monitoring and back analysing exercise on South Deep correlate extremely well and can be used for future design of excavations on the mine. They may also be suitable for design of excavations in nearby mines with similar rock types and stress environments.

Further work of this type should be carried out as opportunities occur on South Deep, and also by other users of FLAC and similar programs, in order to build up a database of input parameters.

ACKNOWLEDGEMENTS

The authors would like to thank the management of the South Deep Exploration Co. and JCI for the permission to publish this paper.

REFERENCES

Richards, R. and Bjorkman, G.S. 1978. Optimum shapes for unlined tunnels and cavities. Engineering Geology, Vol. 12, No. 2.

Ho-Yeong Kim, Kyung-Won Lee, Yoshiaki Mizuta,and Hi-Keun Lee 1990.

A study on determination of the relaxed zone around an excavated tunnel.
International Symposium, ISRM, Swaziland, R Brummer (ed.).

Bowcock J.B., Boyd J.M., Hoek E., and Sharp J.C. 1976.
Drakensberg pumped storage scheme - Rock engineering aspects.
Proceedings, Symposium on Exploration for Rock Engineering, Jo'burg, Vol. 1.

Sharp J.C., Richards L.R. and Byrne R.J. 1977.
Instrumentation considerations for large underground trial openings in civil engineering.
Field Measurements in Rock Mechanics, Proceedings of the International Symposium, Zurich, K Kovari (ed.), AA Balkema.

ITASCA 1992
FLAC, Version 3.2., Itasca Consulting Group, Inc. Minneapolis.

Esterhuizen G.S. and Valenta A.I. 1988.
Triaxial testing results for the South Deep rockmass. JCI Internal Report.

Ryder J.A. 1991.
Conversion of triaxial data to strain softening input format.
COMRO Report.

Raffield M.P. and James J.V. 1993
Computer modelling of shaft movement induced by pre-extraction and backfilling of the shaft reef area on the South Deep mine.
SAIMM Backfill Conference, Jo'burg, August (in press).

Innovative Mine Design for the 21st Century, Bawden & Archibald (eds) © 1993 Balkema, Rotterdam, ISBN 90 5410 325 6

In-situ stress measurements and use in the design of a deep gold mine

P.R.Smallbone & J.V.James
South Deep Exploration Company Ltd, JCI, South Africa

A.K.Isaac
University of Wales, College Cardiff, UK

ABSTRACT: The orebody being investigated by the South Deep Exploration Company is approximately 58 km South-west of Johannesburg and lies on the West Wit's line. At depth stress is a major problem and the determination of the in-situ stress field is an important pre-requisite for any computer simulations. Instrumentation included both the "doorstopper" and later the CSIR triaxial strain cell, the results are included. CSIRO cells were also installed as part of a continuous monitoring programme. The results have been used in computer simulations, to predict the magnitude of displacements, strains and stresses within the shaft barrel during mining. The results show that no damage to either the shaft lining or steelwork will occur if the recommendations are adopted in the near reef zone.

1. INTRODUCTION

The South Deep orebody is to the south of the Western Areas gold mine orebody. At present the mine is accessed for shaft pillar extraction from Western Areas south shaft on 265 level (2650 metres below surface). The access has been developed with twin ends, a main intake and return airway, some 3,5 km long (figure 1). Development of a waste crushing and /backfill complex and access to the Ventersdorp Contact Reef (V.C.R.) at the shaft positions is complete, stoping operations are planned to extract the shaft pillar prior to shaft sinking and equipping (Valenta, 1990).

At this depth all excavations which are developed suffer immediate stress fracturing around them. The changes in stress field can also affect discontinuities within the rockmass.

The aim of this paper is to present the instrumentation programme, the results and initial data from the computer simulation of stress, strain and displacements in the proposed shaft barrel.

2. IN SITU STRESS MEASUREMENT

Field measurements have been carried out at two sites within the shaft reef extraction area (Figure 1). The sites are located in cross-cut 11 and refuge chamber no.4 (Figure 2). These sites were chosen using constraints noted by Pallister (1968):

1. The ground in which the stress tensor is to be measured should be affected as little as possible by mining operations.
2. The sites should be located away from any major geological discontinuities.
3. The surrounding country rock should be as homogenous and massive as possible (continuous core of at least twice the overcoring diameter is necessary)
4. At the site there should be a large enough area to pull rods, and services required for drilling.

The two sites also satisfied the requirements of being close to the future mining for the continuous monitoring programme.

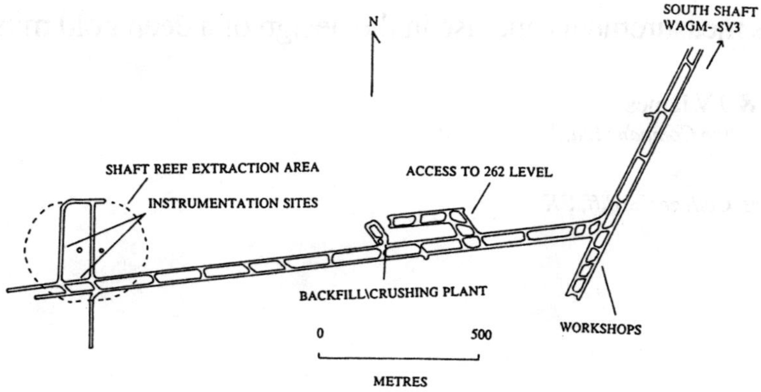

Figure 1. South Deep Project layout.

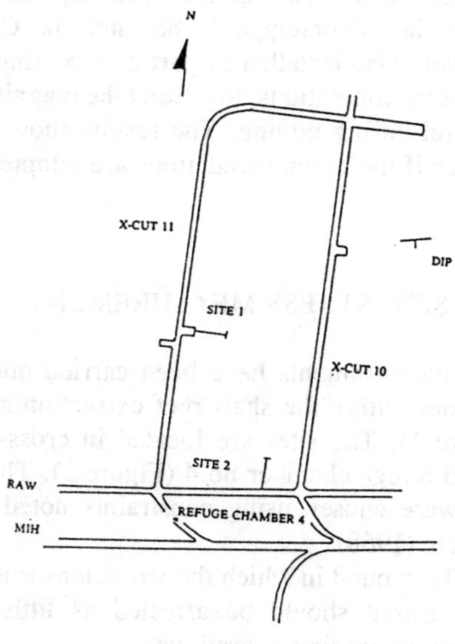

Figure 2. Instrumentation sites.

2.1 *Instrumentation*

The initial programme of instrumentation was based on the use of the "Doorstopper" method developed by Leeman (1965) and Cahnbley (1965,1991) This method of measurement is two dimensional and thus requires three perpendicular holes to be drilled to establish the three dimensional stress field. The method was chosen because of its relative simplicity and drilling requirements when compared to other methods. However, in practice the method suffered from several problems; which are cited below:

1. Excessive vibration of the drill string during overcoring.
2. Difficulty in polishing and smoothing the borehole end, and visually inspecting it.
3. Discing of the core.
4. The inability to see joints and fractures just behind the borehole end.
5. No speed control on the drill.
6. Poor core retrieval technique.

The above in the majority of cases caused the "doorstopper" to become detached or the results meaningless. Due to these problems other methods were considered, and the CSIR triaxial strain cell was adopted. The measurements were taken from the same sites as before. The CSIR strain cell consists of a cylindrical plastic body into which three rosettes are incorporated, these can be expanded and glued to the sidewalls.

The method of installation is briefly described below and shown in figure 3. Initially an NXCU borehole is drilled to a depth of 12 metres (2.5 times the diameter of the excavation) and then ground flat. An EX borehole is then drilled concentrically into the end of the NXCU borehole for a distance of approximately 450 mm. The EX borehole is then cleaned and primed. A thin layer of glue is applied to each of the rosettes and the cell is installed using an installation tool containing the orientation device.

The glue is left to cure for 8 hours or more and then a set of readings is taken. The installation tool is then removed and the EX hole is overcored. The cylinder of core is removed and the installation tool is reconnected. The two are left to equilibrate with the ambient temperature and then a second set of readings is taken.

Core samples are removed from the site as well for modulus determination in a laboratory. These are then used to compute the stress tensor.

3 RESULTS

During both methods of instrumentation problems with core retrieval caused major problems. Core discing during overcoring was particularly prevalent at site 1, with only one cell remaining intact and the others being either partly or completely destroyed. However, overcoring operations in the second hole proved more successful with the length of the overcored specimen being determined by joint spacing.

The totally different behaviour of the core may be related to the orientation of the borehole with respect to lithology. However, drilling technique cannot be wholly ruled out.

The results are taken from the two cells retrieved at site 2 and are presented above (tables 1 and 2) and are shown graphically on a schmidt equal area net (figure 4).The accuracy of the underground measurements is approximately 50 microstrains, with a tested rock moduli of between 55 and 65 GPa this gives a measured accuracy of 3 to 4 MPa. The calculation of the stress tensor is by the method of least squares and for greater statistical accuracy three cells were placed in each hole. Results of the continuous monitoring programme are still being collected using the CSIRO cells.

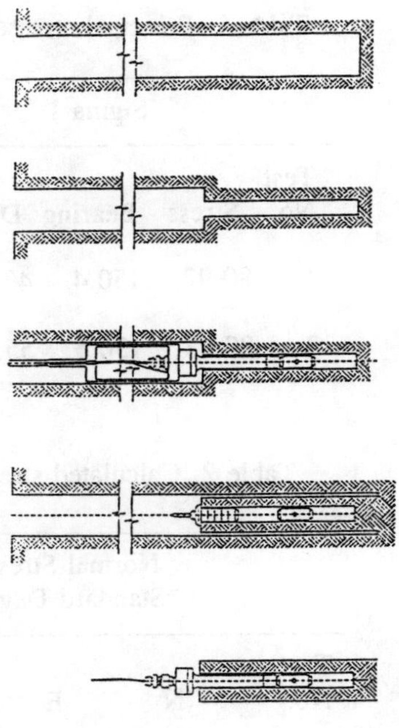

Figure 3. Installation Procedure.

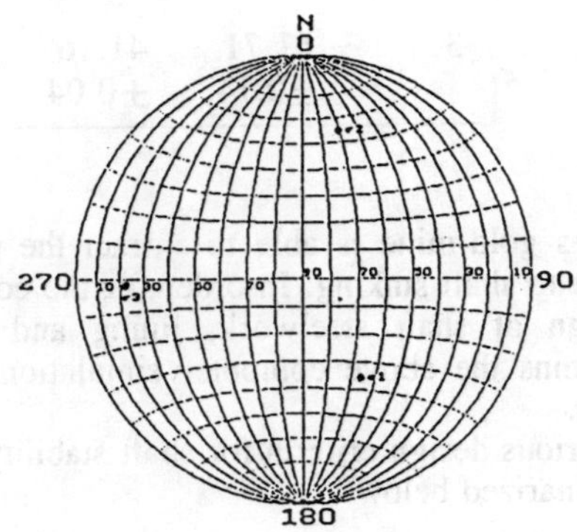

Figure 4. Lower hemisphere equal area net showing poles of principal stresses.

4 NUMERICAL MODELLING

The in situ results form part of the input file for the BESOL\MSCALC suite of programs. The model which is designed to predict stresses, strains and displacements in the shaft barrel is mined in five large blocks (Figure 5) using detuned backfill to mimic multi step mining (Raffield,1991). Windows are placed vertically down the shaft barrel at set intervals. Results are then transferred via a post processor to a spreadsheet for graphical presentation.

4.1 *Discussion*

The extraction of shaft pillars as an alternative to the traditional protection systems is well documented (McKinnon,1989) The South Deep project due to its access from the Western

Table 1. Principal stresses and directions in common coordinate system.

	Sigma 1			Sigma 2			Sigma 3		
Test No.	Stress	Bearing	Dip	Stress	Bearing	Dip	Stress	Bearing	Dip
1	80.92	150.4	40.1	57.39	13.6	30.2	39.05	267.9	22.5
3	80.63	136.5	55.2	60.52	352.3	29.4	33.53	262.4	16.9

Table 2. Calculated stress components in common coordinate system.

	Normal Stresses *Standard Deviation			Shear stresses *Standard Deviation		
Test No.	N	E	V	NE	EV	VN
1.	67.32 ±0.25	45.65 ±0.11	71.20 ±0.10	-6.21 ±0.13	13.45 ±0.06	-12.53 ±0.11
3.	61.71 ±0.14	41.16 ±0.04	71.81 ±0.04	-10.37 ±0.04	13.65 ±0.02	-4.57 ±0.03

Areas gold mine is able to extract the pillar prior to shaft sinking. In order for the correct design of shaft steelwork, lining and pipe columns the above computer simulation was used.

Various design criteria for shaft stability are summarized below :

Shaft lining	0.2 mϵ
Shaft steelwork	0.4 mϵ
Tilt	10 mϵ
Field stress	100 MPa
Stress change	20 MPa

The above are the maximum values which are allowable before serious damage and expensive remedial maintenance is required.

It should be noted that the graphs show the differential values, which is the movement after step one, as the shafts will be sunk after this initial mining.

The results are shown below (graphs 1 to 3). In terms of the design criteria stated, vertical

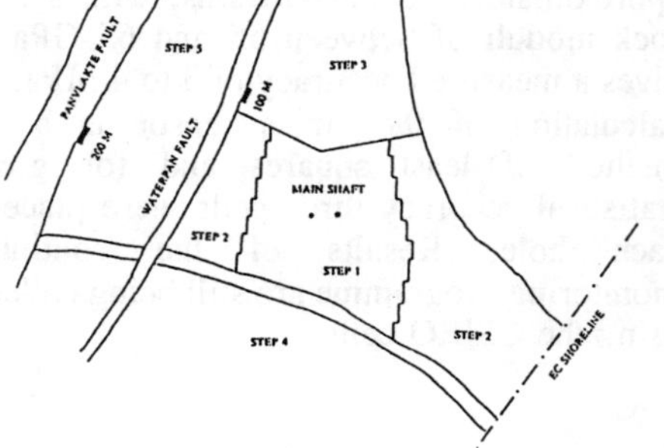

Figure 5. Diagrammatic representation of computer model.

strain is exceeded for the shaft lining approximately 200 metres above and 150 metres below the shaft reef intersection. In this zone different support will be considered incorporating shotcrete, weldmesh and cable anchors. The criteria for shaft steelwork has not been exceeded at any place in the shaft barrel.

However suspended steelwork is considered necessary to accommodate the reef plane closure of approximately 40 cm.

The suspension of the steelwork from 50 metres above and 50 metres below is recommended.

5 CONCLUSIONS

Stress measurement should be carried out by experienced practitioners. All methods however, rely on good drilling technique and practice. Suitable machines where revolution speed can be controlled and a twin tube core barrel should be used. Planning is also essential to minimise delays and additional equipment should always be available. Stress measurement at depth is difficult with the core deteriorating rapidly once overcored.

The in situ stress measurements forms part of an ongoing instrumentation programme on South Deep. A successful set of readings was obtained. The use of the new stress tensor has in comparison with a K ratio of 0.5 has slightly increased the zones over which different support, to the traditional concrete lining, is required.

ACKNOWLEDGEMENTS

We would like to thank the Management of the South Deep Exploration company and JCI for permission to publish. The Authors are also grateful to colleagues on South Deep and to Charles Janssen Schmidt of the CSIR for his work under difficult conditions.

REFERENCES

Anon, 1990. MINDATA field manual of CSIR\CSIRO Triaxial cell. pp.120

Cahnbley, H.K.R. 1965. The measurement of stress in rock. SAIMM 65,7:408-418.

Cahnbley, H.K.R. 1991. Pers. Comm.

CSIR, 1988. In situ stress measurement on the Western Areas gold mine.

CSIR, 1992. In situ stress measurement on the South Deep gold mine.

Leeman, E.R. 1965. The measurement of stress in rock. SAIMM 65, 7:408-418.

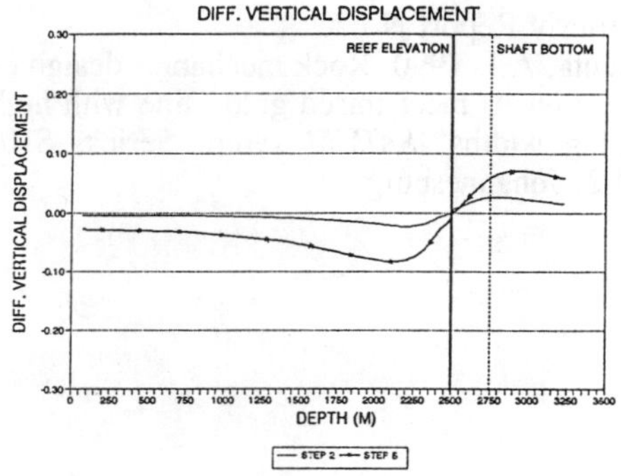

Graph 1. Differential vertical displacement.

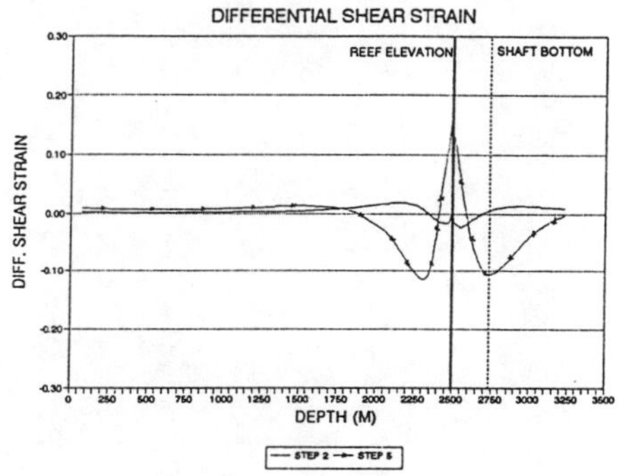

Graph 2. Differential shear strain.

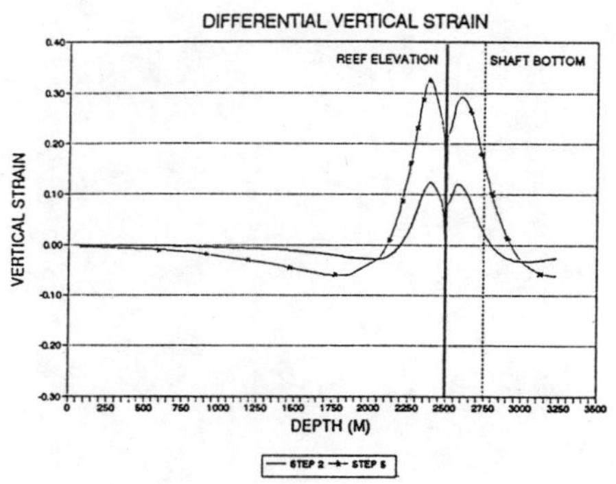

Graph 3. Differential vertical strain.

McKinnon, S.D. 1989. Protection of deep vertical shaft systems. PhD thesis. University of the Witwatersrand.

Pallister, G.F. 1968. The measurement of virgin rock stress. COMRO Report. No.6/59.

Raffield, M.P. 1991. South Deep Project,

Quarterly Report No.2.
Valenta, A.I. 1990. Rock mechanics design of a deep level mechanised gold mine with high stoping widths. *SAIMM* symp. Series, S10, Vol.2. Johannesburg

Innovative Mine Design for the 21st Century, Bawden & Archibald (eds) © 1993 Balkema, Rotterdam, ISBN 90 5410 325 6

Models and drawings in CAD-based mine design

Peter S.Campbell & Bernard F.Smith
Falconbridge Limited, Sudbury Operations, Ont., Canada

ABSTRACT: In mine design, very much more than geometric information must be considered. Information like tonnes and grade of the ore body by zone, stope or blast is very important but is typically not included in a CAD-*drawing* system. In order to be a useful design tool, designers of mines require a complete *model* of the mine consisting of geometric and non-geometric data. Such a model must contain all information that an engineer might wish to consider in the course of designing a mine. This information might consist of diamond drill hole data, geological interpretations, ore reserves, structural geology, ground stresses, costs, schedules as well as geometric data from existing and planned excavations. At best, this non-geometric information is very difficult to link with entities in a *drawing* system.

A CAD-*design* system must strive to build a complete *model* of all data in the design process. To be of practical use, non-geometric information must be as accessible as geometric information is. Falconbridge has developed such a model-based mine design system. The system was built on a CAD platform that uses solid modelling. The solid modeller is object-oriented, uses a hierarchial data structure, has an internal database for storing and retrieving object attributes, and is programmable in 'C'.

This paper will focus on Falconbridge's modelling approach versus a more conventional drawing methodology.

1.0 INTRODUCTION

Does CAD stand for Computer-Aided *Design* or Computer-Aided *Drafting*? Design work is most often associated with engineering while drafting is usually associated with documentation of the design. Computer-aided *drafting* systems have certainly helped many engineers with their design work. Historically, this can largely be attributed to the earlier availability of drafting systems.

By definition, the end product arising from the use of a computer-aided *drafting* system is the paper drawing which is used to communicate with others. Certainly the digital representation of the drawing which exists in the computer can be edited and reproduced at various scales. However, this is a very incomplete model of the design since it only contains geometric information.

Consider the two geological sections illustrated in Figures 1a an 1b. Both have been constructed with a CAD system, yet, one is a drawing while the other is a model. Which is which? Does it matter? What does this have to do with mine design? The answers to these questions will be examined herein.

2.0 MODELS AND DRAWINGS

Upon closer examination of Figures 1a and 1b, there does not appear to be any differences between the two. So how can one determine which is the drawing and which is the model?

Webster's New World Dictionary[1] defines drawings and models:

drawing — 1. the act of one that draws; specif., the art of representing something by lines made on a surface with a pencil, pen, etc. 2. a picture, design, sketch, etc. thus made. ...

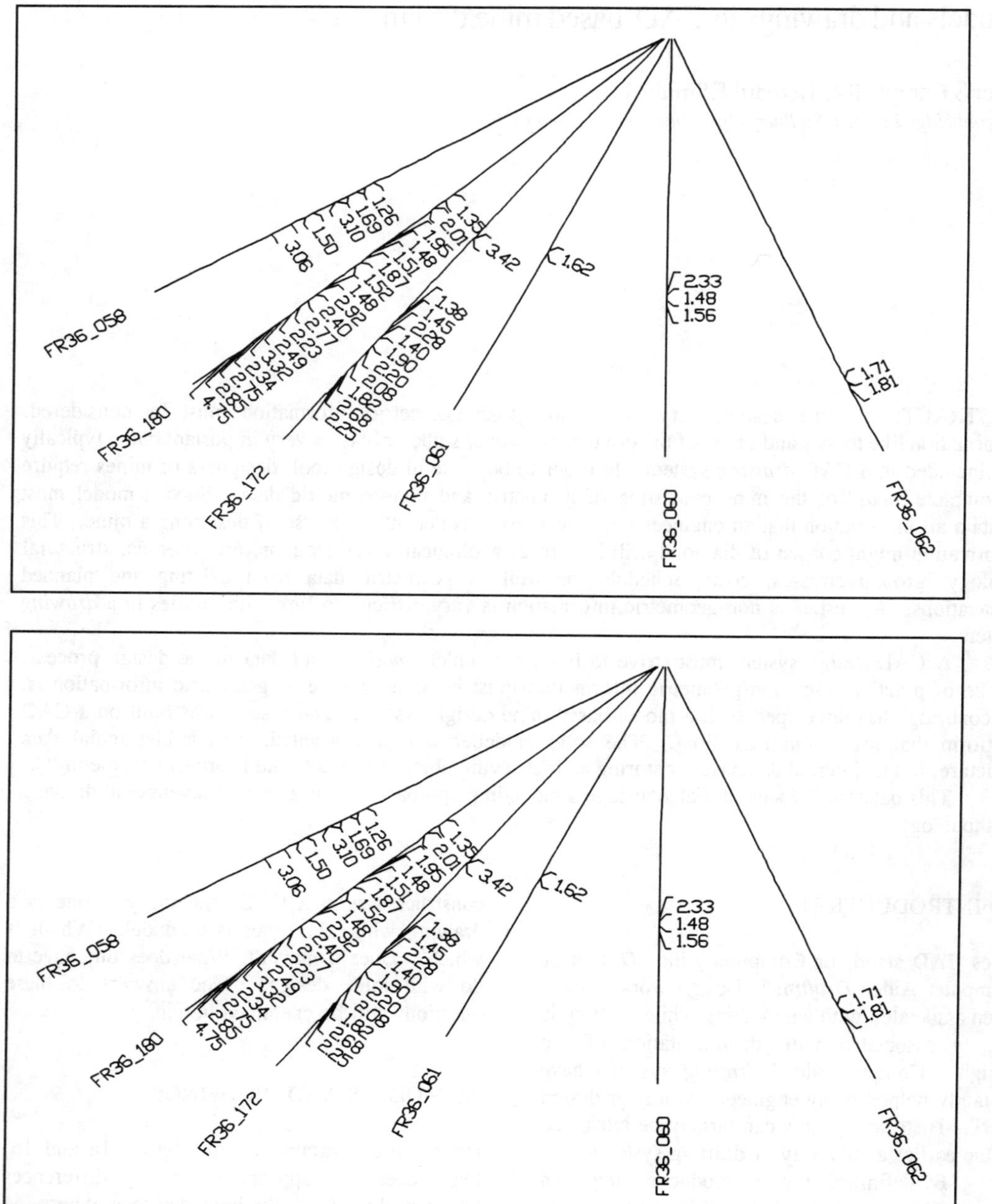

Figure 1 a,b - Diamond drill hole sections

model — 1. a) a small copy or imitation of an existing object, as a ship, building, etc. made to scale b) a preliminary representation of something, serving as the plan from which the final, usually larger, object is to be constructed ...

Perhaps these definitions begin to shed some light on the apparent subtlety of the question: Which is which? To begin to answer this question, consider the following definitions derived from above:

drawing — a schematic diagram of something drawn to scale constructed by putting lines on paper; usually annotated to convey additional information

model a copy of any object usually represented at a different scale; a perfect copy is indistinguishable from the original while practical copies vary only in level of detail

Armed with these new definitions, can any light be shed on the identity of the two diamond drill hole sections in Figure 1a and 1b? From these definitions it may well seem that both figures satisfy the definition of "drawing". So the question can be simplified to: Which of these two figures is the model?

3.0 CONSTRUCTING THE MODEL

In the definition of "model" proposed above, "a copy of any object" serves to indicate something deeper than "lines on paper" used in the definition of "drawing". In a drawing of a diamond drill hole section, lines are adequately used to schematically represent the trace of a diamond drill hole. But this is not a model. What are the important characteristics of a diamond drill hole that are required to make a "perfect copy" of such an object (i.e. to construct a model of a diamond drill hole)?

Obvious characteristics of a diamond drill hole (i.e. ddh) are its geometric attributes. A ddh is collared at a point in 3D space and is drilled at some azimuth and dip to a final depth. In practice, ddh's rarely remain perfectly straight so additional down-hole measurements are taken to measure deviation. Deviation measurements are usually in the form of azimuth and/or dip at a distance from the collar. With this geometric information, an accurate drawing of the ddh can be constructed in any (oblique) view either manually or with a CAD system. However, as any geologist will agree, a ddh contains information far more important than just its cartesian position.

The fundamental purpose of a ddh is to allow us to peer into solid ground. Consequently, the essential information afforded by ddh's are their non-geometric data. Non-geometric ddh data primarily consists of geology expressed as rock types and analytical data expressed as grades. In fact, in the case of ddh's, non-geometric data is far more important than geometric data. If the non-geometric information is of such great importance, then why do geologists plot sections instead of working from logs?

Despite the fact that ddh logs are constructed sequentially (i.e. from collar to toe), analyzing ddh data from more than one ddh requires the data from multiple holes to be presented in relation to each other. The relationship is the spatial distribution of non-geometric ddh data. Clearly there is a need to display the non-geometric data graphically.

In a drawing, non-geometric ddh data are treated as annotation. For example, at a distance of 100m from the collar, a 2m length of ddh might contain 1.5% Ni and 0.9% Cu. In a typical drawing, such as that in Figure 2a, these non-geometric values are "written" next to the ddh at the appropriate location defined by the geometric data. Whether done manually or with a CAD system, the annotation is textual.

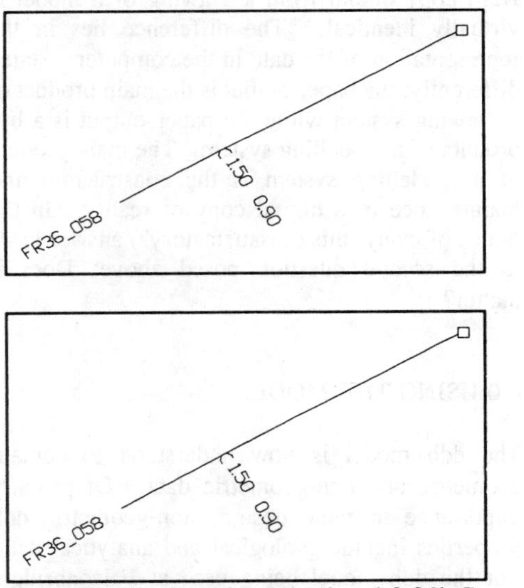

Figure 2 a,b - 2m of 1.5% Ni and 0.9% Cu 30m from the collar.

In "real life", the ddh does not have this text written next to it. We, as users of these data, annotate the ddh as a matter of convenience. In "real life", nickel and copper grades (i.e. non-geometric data) are *properties* of this ddh segment in Figure 2a. In a model, which we have defined to be an indistinguishable copy of the original, non-geometric ddh data must also be treated as a property. Such a model can only be constructed with an appropriate CAD system. In a practical sense, properties of a graphical object are treated as attributes. Figure 2b illustrates a *model* of the same ddh segment as in 2a. As before, the pictorial representations are identical. However, the grade properties of the graphical object have been attached as attributes. As in "real life", these properties cannot be seen but are none-the-less a part of the real object. For convenience, the CAD system has displayed the value of these attributes next to the hole so we may observe them. Removing the text from Figure 2a (i.e. the drawing) results in a hollow representation of reality. Removing the text from Figure 2b (i.e. the model) is inconsequential because the object retains its properties despite the fact that the properties are momentarily "unobservable".

To respond to the first of the questions posed by Figure 1a and 1b -- which of these two figures is the model? -- the answer is indeterminate from the information provided. Hard copy output from a drawing or a model is virtually identical. The difference lies in the representation of the data in the computer. Stated differently, the paper output is the main product of a drawing system while the paper output is a by-product of a modelling system. The main product of a modelling system is the construction and maintenance of a digital copy of reality. In the minds of many, this (unsatisfactory?) answer leads to the second question posed above: Does it matter?

4.0 USING THE MODEL

The ddh model is now understood to contain geometric and non-geometric data. Of primary importance in mine design, non-geometric ddh properties include geological and analytical data. For the ddh model being used at Falconbridge, these properties explicitly include rock type and grades (i.e. Ni, Cu, Co, S, Ag, Au, Pt, and Pd) for every ddh segment. The geometric data is implicitly handled by the CAD system. It has already been shown that the CAD system can be used to display the text corresponding to any or all of these properties. However, this is hardly different from many "clever" drawing systems.

One of the first considerations in mine design, is the determination of "what is ore?". A very common and simple way of estimating an ore reserve is through the use of sectional polygon modelling. This simple example will serve to highlight the power of a modelling approach.

Sectional polygon modelling of an ore reserve involves the placement of polygonal shapes over a ddh section. The limits of the polygons are used to define the limits of the ore as interpretedby the geologist. Figure 3 shows a typical ddh section with polygons on it. Again this step could just as easily have been done manually on paper or with a CAD-drawing system. However, the similarities end when calculating the grade of the polygon.

Whether manually or with a CAD-drawing system, grade information is only present as text. The actual grade values reside elsewhere and there is no link between the polygon and the grades contained therein. Only the user can "read" the text. Polygon grades, if they are to be calculated with a drawing system, must be calculated either manually or externally in the ddh database. In the latter case, there must be some link between the polygon and the assays it contains. This is usually done by the user who must explicitly define the assays he wishes to use in the calculation. With a CAD-modelling system, grade information has been attached as attributes to each ddh segment. Grade information exists as numeric values and this information is present and available for further calculation. Calculating the grade of the polygon is a simple process of instructing the CAD-modelling system to calculate the length-weighted average of the grade attributes for each ddh segment inside the polygon. In this case, the CAD-modelling system "knows" which ddh segments lie inside the polygon (from a simple geometric test), it "knows" the grade attributes of these ddh segments, it "knows" the length of each ddh segment (from a simple geometric calculation), and it can evaluate any mathematical expression with these data (e.g. length-weighted average).

5.0 EXPANDING THE MODEL

Calculating the grade and rock type for any polygon is fast and accurate with a ddh model. However, this is not the entire model. Now there are polygons in the model. Like ddh segments, these polygons have attributes of grade and rock type. Not only has the CAD-modelling system internally calculated these values, but it has attached the grade and rock type as attributes to each polygon. In one very simple and elegant

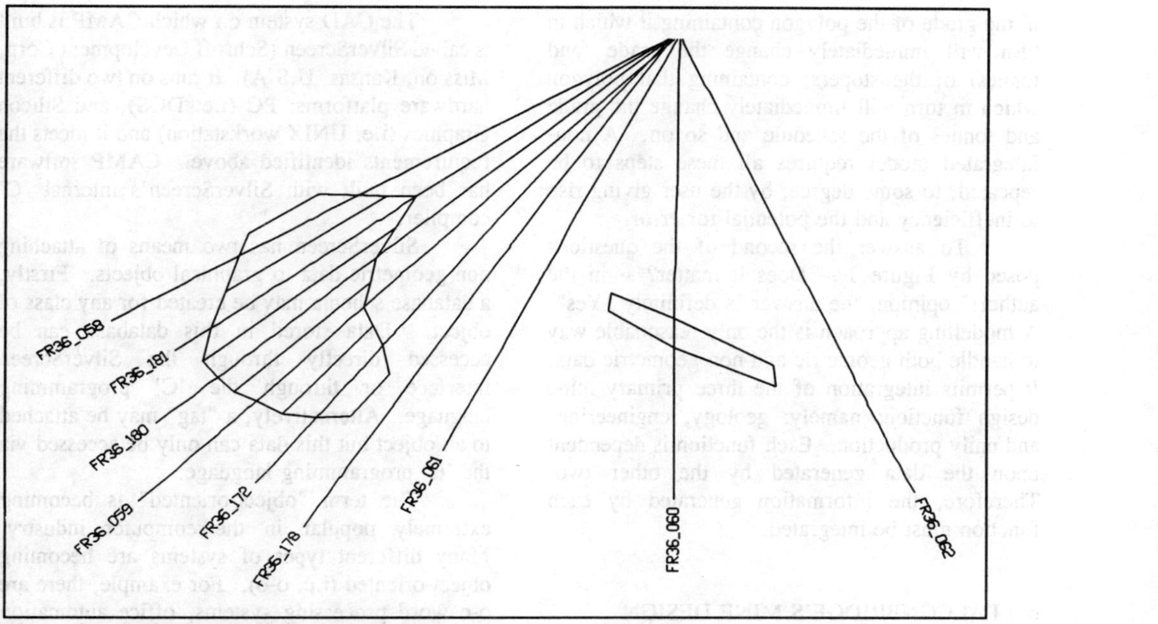

Figure 3 - Geological section with polygons

step, polygon grades have been calculated and the model updated with this new information, thus, an ore reserve model is constructed. From this ore reserve model, stopes can be designed.

Stopes are geometric entities (i.e. 3D shapes) with non-geometric data attached. In this instance, stope attributes come from the ore reserve model. Stopes have much the same non-geometric data as do polygons. The power of a modelling system is in providing the ability to iteratively query the ore reserve model to optimize the stope shape and extraction sequence for grade, tonnage, recovery, cash flow, etc. Once a stope has been optimally designed, the next step in the process is to lay out development.

The CAD-modelling approach permits development to be laid out on the current model of the stope design. Any changes made to the stope design model are immediately available to the development design process. Development itself is but another aspect of the same mine model. And development is yet another example of a graphical object which has non-geometric information attached. Development may have a workplace name, account code, priority, cost per metre, ground support requirements, etc. all of which are non-geometric data which also form part of the model.

Eventually, excavation will begin and there will be a need to model these excavations. Surveying is the usual form of measurement. Stopes, development, and ddh's are surveyed to obtain their position relative to where they were designed to be. Surveying is actually construction of an as-built model and is the primary means of engineering control. Surveys are usually thought of as being primarily geometric in nature. However, they too contain non-geometric data such as workplace name, account code, date, etc. Of course geometric data, such as area and/or volume, are handled by the CAD system. It is essential that the as-built model be a part of the total mine model. For example, a survey "pick-up" of a ddh is far more accurate than down-hole measurement. Relocating a ddh as a result of a survey is simple to achieve in an integrated model. This change is immediately available to the entire model. However, the primary use of surveying is to ensure that development is being built where it is designed to be and that stopes are built where the ore is.

The mine design model presented here is an *integrated* model. At any time, the model reflects our complete understanding of the real-life situation. All aspects of geology, planning, and surveying are handled with one all-inclusive model. There is a downstream "chain reaction" effect when even the smallest change is made to the model. For example, changing an erroneous grade in a ddh will result in an immediate change in the grade of the polygon containing it which in turn will immediately change the grade (and tonnes) of the stope(s) containing that polygon which in turn will immediately change the grade and tonnes of the schedule and so on. A non-integrated model requires all these steps to be repeated, to some degree, by the user giving rise to inefficiency and the potential for error.

To answer the second of the questions posed by Figure 1 -- Does it matter? -- in the authors' opinion, the answer is definitely "Yes". A modelling approach is the only reasonable way to handle both geometric and non-geometric data. It permits integration of the three primary mine design functions namely: geology, engineering, and daily production. Each function is dependent upon the data generated by the other two. Therefore, the information generated by each function must be integrated.

6.0 FALCONBRIDGE'S MINE DESIGN MODEL

Falconbridge has put these modelling concepts of mine design into a system called CAMP which stands for Computer-Aided Mine Planning. In this context, the term "mine planning" embodies the essential elements of mine design; geology, engineering, and daily production.

6.1 CAD Platform

The most critical element of this project was the selection of an appropriate CAD system. The CAD system had to be a modelling system and not a drafting system. An extensive review of CAD systems was undertaken. Everything from low-end PC CAD to high-level workstation systems were investigated. The results of this study indicated that there is a trend in the evolution of CAD systems. All "serious" CAD systems show a definite trend toward:

- solid modelling
- object-oriented data organization
- internal database for non-geometric object attributes
- 'C' programming language

Since this project represents a long-term commitment, it was essential that the CAD platform embody these long-term principles. Furthermore, it is unlikely that such an ambitious project could be completed without these qualities.

The CAD system on which CAMP is built is called SilverScreen (Schroff Development Corp, Mission, Kansas. U.S.A). It runs on two different hardware platforms: PC (i.e. DOS), and Silicon Graphics (i.e. UNIX workstation) and it meets the requirements identified above. CAMP software has been built with SilverScreen's internal 'C' compiler.

SilverScreen has two means of attaching non-geometric data to graphical objects. Firstly, a database schema may be created for any class of object. Data stored in this database can be accessed directly through the SilverScreen interface or through the 'C' programming language. Alternatively, a "tag" may be attached to an object but this data can only be accessed via the 'C' programming language.

The term "object-oriented" is becoming extremely popular in the computer industry. Many different types of systems are becoming object-oriented (i.e. o-o). For example, there are o-o word processing systems, office automation systems, general ledger systems, and, of course, CAD systems. An o-o system is one that encapsulates various types of data into logical units, for example, geometric and non-geometric data. "Larger" objects can be composed of an assembly of several "smaller" objects, thus, a hierarchy is built up. SilverScreen uses an internal hierarchy, similar to a DOS sub-directory structure, to organize objects. This preserves the principle of inheritance where daughter objects inherit any modification made to the parent object. For example, rotating a parent object also rotates its daughter objects.

There are many ways in which different CAD systems mathematically handle geometry. There are 2D CAD systems, wire-frame systems, surface modellers, and solid modellers. Solid modelling is the most universal type of CAD system. Aside from being able to model solids, a solid modeller can also be used to draw in 2D, construct wire-frames, and model surfaces. It is easy to confuse a 3D wire-frame system with a solid-modelling system because each can produce similar-looking drawings despite the differences in the drawing database. A solid-modeller applies a rigorous test to solids to ensure they are valid. There is no need to be as rigorous with a wire-frame system since all wire-frames are valid. With valid solids, a solid modeller can easily perform boolean operations such as union, difference, intersection, and trim. These operations are *impossible* with a wire-frame system.

7.0.0 THE CAMP SYSTEM

The CAMP system is a mine model composed of three main sub-models: geology, engineering, and daily production. These sub-models are further composed of several sub-sub-models.

7.1.0 Geological model

The geological model is composed of four main sub models: the ddh model, geological interpretation, ore reserve model, and underground mapping model.

7.1.1 Diamond drill hole model

The ddh model is a 3D model of geometric and non-geometric ddh data. All data is stored internally and is managed by SilverScreen. The fundamental building block is a single straight line segment representing the smallest interval of geological interest. This is determined by the geologist at the time the core is logged. This object is tagged with the following non-geometric data: %Ni, %Cu, %Co, %S, Ag ppm, Au ppm, Pt ppm, Pd ppm, and rock type. The name of this object is determined by its distance from the collar. All objects belonging to the same ddh are stored together as a super class. This super class is given the name of the ddh. In SilverScreen, the word "object" is reserved for the lowest level entity. An assembly of objects is called a "block". Therefore, ddh "blocks" drilled on the same section are stored together to construct a higher class of blocks. This block is given the name of the section it is drilled from. Finally, all sections drilled at the same azimuth are stored together to construct a still higher class of blocks. This block is given the name of the azimuth.

Note that the only graphical data are the individual ddh segments which are stored as objects. The hierarchical organization in SilverScreen creates the abstract entities "hole" and "section". A typical address for a particular object such as that identified in Figure 2 might be as follows:

\azi90\dip0\e7975\tl13_045\dp300
where:

dp300 is the depth from the collar (i.e. 30m)
tl13_045 is hole name (i.e TL13-045)
e7975 is section name (i.e. 7975 East)
dip0 is the *normal* of the section dip (i.e. vertical)
azi90 is the azimuth of the section (i.e. looking east)

This naming convention could just as easily be appiled to a DOS sub-directory system.

Despite the fact that every single ddh for a particular zone is loaded, one can easily work with a single section or a single hole by simply "isolating" the appropriate block. For example, to work with section 7975 East, the block with the name "\azi90\dip0\e7975" can be isolated. All daughter blocks and objects inherit the parent's visibility.

7.1.2 Geological interpretation

Geological interpretation is the process of trying to make sense of the information provided by the ddh model. Essentially, this is a matter of isolating a section in the ddh model (described in sec. 8.1.1) and using the CAD system to construct lines representing geological contacts or other regions of interest. This is exactly the same process as is usually done manually on paper except the interpretation now becomes part of the overall model. Geological interpretations would be put into a block called "interp" and would be located under the section block: \azi90\dip0\e7975\interp. A typical geological interpretation is illustrated in Figure 4.

7.1.3 Ore reserve model

Within the CAMP system, there are two ways in which an ore reserve model can be created: sectional polygons and block modelling. Sectional polygons have been described in section 4.0. The sectional polygon model fits into the ddh model in much the same way as do geological interpretations. Recall the polygons in Figure 3. Each polygon is an object and polygons belonging to the same geological zone are grouped into a block. In Figure 3, these polygons belong to Zone 4B and each polygon is given an arbitrary, sequential name. The first polygon drawn is called p1 and its full name would be:
\azi90\dip0\e7975\polygon\z4b\p1
where:

polygon is a block containing polygonal reserves
z4b polygonal reserves for Zone 4B
p1 is the polygon object

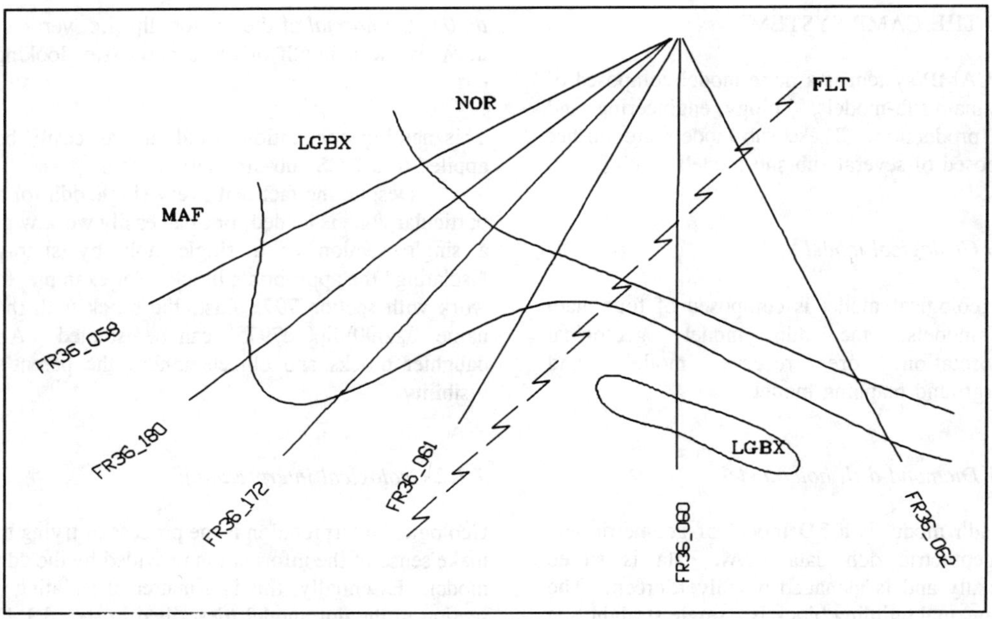

Figure 4 - Geological section with interpretation

The other method of creating an ore reserve model is through the use block modelling. Block modelling is a 3D method. The block model requires the ddh model to estimate grades.

To run the block model, a 3D solid model of the ore zone is created using CAD tools. This is usually done in conjunction with existing data in the overall model such as ddh's, interpretations, or polygons. The solid model is then discretized by slicing it into three orthogonal directions as is shown in Figure 5. As an intermediate step, all ddh objects falling outside this solid are "turned off" so that their grades do not influence the value of the blocks inside the model. Finally, the grade of each block is calculated on the basis of inverse-power-of-distance. The grades are attached as tags to each element of the block model. The validity of the block model can easily be verified against any of the other data in the current model. A more thorough discussion of the geological block model used in CAMP can be found in references 2 and 3.

7.1.4 Underground mapping model

Underground geological mapping is a big part of the cut-and-fill mining method used by Falconbridge. It is the primary means of verifying ore reserves and exercising grade control. Underground back mapping is digitized into the model. This data is in a horizontal plane. However, it is most valuable when displayed on section with the ddh model. A CAMP program has been written to perform that task. Whenever one wishes to use a ddh section, any geological back mapping on that section is automatically "cut" and displayed on that section. This is an important way to improve and modify geological projection into the rock.

7.2.0 Engineering model

The engineering model is composed of four sub-models: stope reserves, stope design, blast design, and development design.

7.2.1 Stope reserve model

The stope reserve model is used to calculate the tonnes and grade for a given stope geometry. It is a simple matter to perform a "what if" analysis by changing the shape or sequence of the stope to determine the effect on tonnes and grade. The methodology of building this model is dependent on the type of ore reserve model.

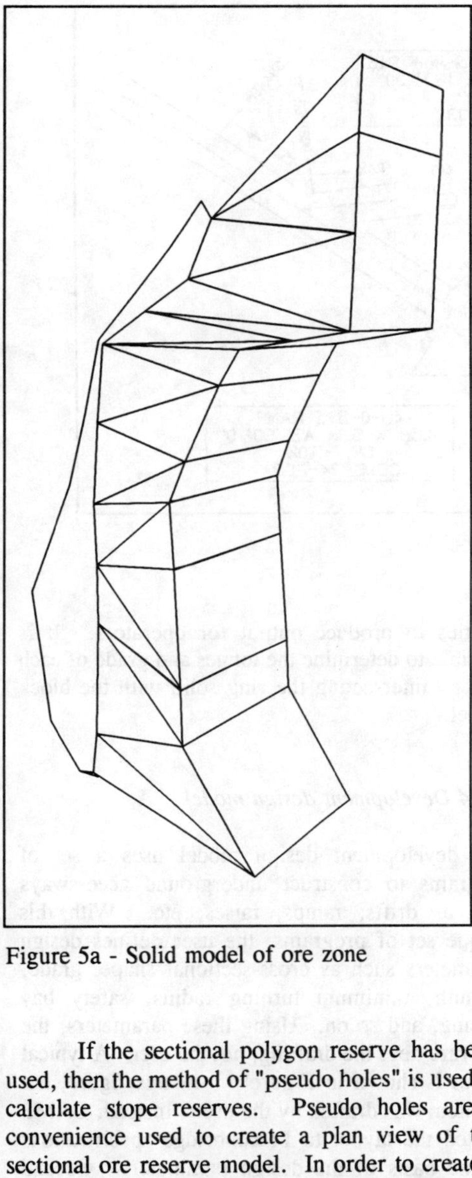

Figure 5a - Solid model of ore zone

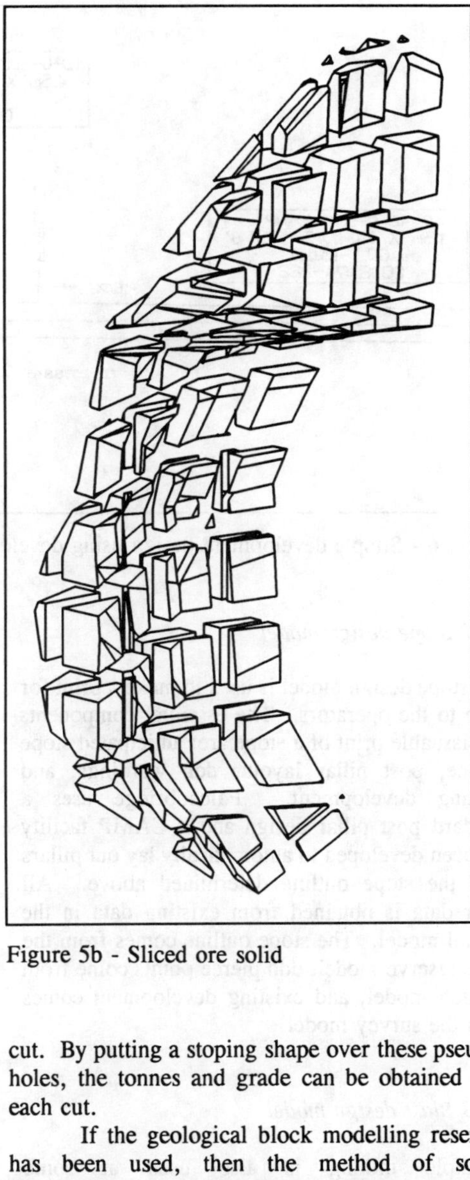

Figure 5b - Sliced ore solid

If the sectional polygon reserve has been used, then the method of "pseudo holes" is used to calculate stope reserves. Pseudo holes are a convenience used to create a plan view of the sectional ore reserve model. In order to create a plan view, a horizontal cutting plane is defined at the desired elevation. The intersection of the horizontal cutting plane with the sectional polygons results in a line segment at the desired elevation. The line segments "pick-up" the attributes of the polygon from which they arise. Since these line segments have the same attributes as ddh's (i.e. %Ni, %Cu,...) they are referred to as "pseudo holes". By grouping all pseudo holes at the same elevation, a plan view of the ore reserve can be obtained. In cut-and-fill mining, these plan views are usually produced for each cut. By putting a stoping shape over these pseudo holes, the tonnes and grade can be obtained for each cut.

If the geological block modelling reserve has been used, then the method of solid intersection is used to create the stope reserve model. Initially, the engineer will examine the block model to locate "profitable" mineralization. This may be done by querying the block model to display ore-blocks exceeding a certain grade, display a grade-tonnage calculation, or by slicing through the model in various directions. Finally, 3D stope shapes are constructed and the grade and tonnes of the stope are calculated by intersecting the stope model with the ore reserve model. Varying the shape of the stope permits a "what if" analysis.

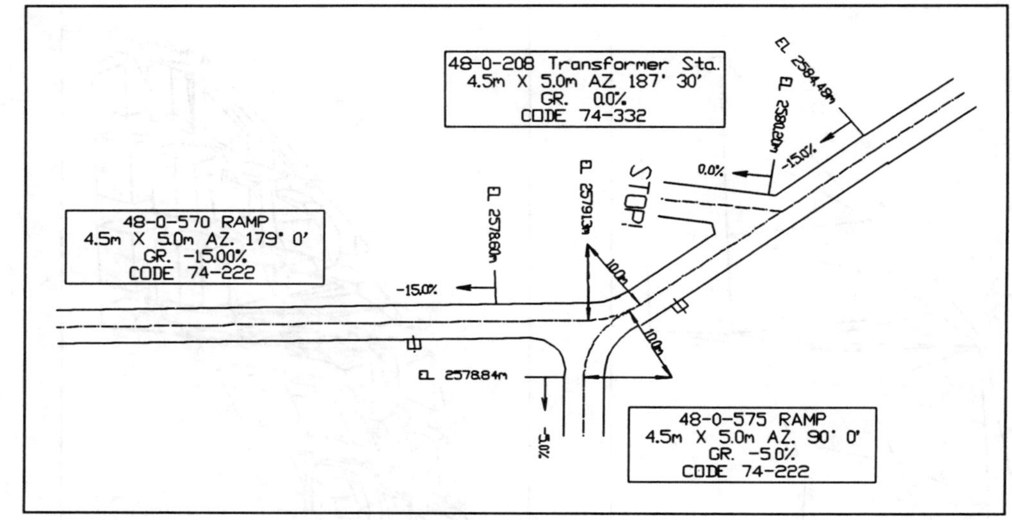

Figure 6 - Simple development design using development tool kit

7.2.2 Stope design model

The stope design model is used to make a print for issue to the operators. The essential components of a issuable print of a stope are: anticipated stope outline, post pillar layout, ddh warnings, and existing development. Falconbridge uses a standard post pillar design and a CAMP facility has been developed to automatically lay out pillars over the stope outline determined above. All other data is obtained from existing data in the overall model. The stope outline comes from the stope reserve model, ddh pierce points come from the ddh model, and existing development comes from the survey model.

7.2.3 Blast design model

Blasthole mining is also used at some Falconbridge operations. Both VCR and conventional fan-drilled holes are used depending on the situation. Firstly, a solid model of the stope is required. One is available provided the geological bock model was used; otherwise, one will need to be created. Upon completion of constructing development into the stope, the stope reserve model and the surveying model can be sliced at the desired ring spacing. CAMP will automatically lay out the desired holes on each ring on the basis of a set of design parameters specified by the user. Each hole is tagged with a cap number, powder type, etc. There are several utilities to produce output for operators. It is possible to determine the tonnes and grade of each ring by intersecting the ring solid with the block model.

7.2.4 Development design model

The development design model uses a set of programs to construct underground accessways such as drifts, ramps, raises, etc. With this unique set of programs, the user defines design parameters such as cross-sectional shape, grade, azimuth, minimum turning radius, safety bay spacing, and so on. Using these parameters, the software does the drafting; not the user. A typical layout is shown in Figure 6. In this figure, not one item was drafted by the user. Instead, CAMP drafted the layout to Falconbridge specifications on the basis of the design parameters. Special effort has been put toward appropriate "hooking up" of development in 3D at junctions. A model of the development is thus built up with additional non-geometric data such as account code for each segment of the design. Additional tools are provided for slashes and optimizing the location of raises.

7.3.0 Daily production model

The daily production model consists of two sub-models: layout model and survey model.

7.3.1 Layout model

The layout model is derived from the development design model. This model is used to produce production layouts for excavation of the design. To construct this model, the design is "cleaned up" to meet internal standards. Clean up is accomplished through the use of CAD commands (e.g. rotate text, move text, etc.) or through the use of CAMP. CAMP programs will place ddh warnings, grids, title blocks, north arrow, and so on into the model so that the output will *appear* as if it had been drafted.

7.3.2 Survey model

The survey model is used to construct an as-built model. Raw survey information is input directly to the CAD system. The CAD system reduces the field notes to world coordinate data and organizes the data into its proper position in the model hierarchy. CAD tools are used to interactively edit or correct the survey and to properly "hook it up" to existing development. After the survey has been verified, CAMP is used to "expand" the 2D survey into a 3D solid. The solid has attributes of workplace, volume, account code, date, etc. Month-end reconciliation is simply a matter of querying the model to tally workplace volumes by account code for a given month.

The solids created in the survey model are also used in a number of other ways. Obviously, this process creates a 3D solid model of the entire mine. This model can be viewed and rendered in any orientation. The solid model also provides a basis from which future designs can be made. The most useful purpose of the solid model is to produce "openings-on-section". For example, when plotting or viewing a geological ddh section, it is desirable, and sometimes required by law, to also display any excavations that lie on that section. Producing these openings-on-section is simply a matter of slicing through the mine solid model at the viewing plane. Openings-on-section are as up to date as the survey model is.

8.0 SUMMARY

The CAMP system was built on a CAD platform and that system is SilverScreen. It has been shown that geometric and non-geometric data are used in all aspects of mine design and the combination of both data types results in a model. The overall mine model is composed of three sub models: geology model, engineering model, and a daily production model. Each of these sub-models are themselves composed of sub-sub-models. These models are mostly created and managed separately but information freely flows between them. The CAMP software provides tools for the creation, management, and querying of the mine model. All design data resides in a single model. This is an integrated system.

REFERENCES

Webster's New World Dictionary, second college edition. D.B. Guralnik Ed. Prentice Hall, New York, 1986.

Campbell, P.S. & B.F. Smith 1989. An Integrated Mine Planning System For Geology, Surveying, And Engineering. 2nd CAMI, Vancouver, B.C.

Campbell, P.S. & B.F. Smith 1992. Mine Planning Using Solid Modelling. 82nd AGM of the CIM, Montreal, P.Q.

Schroff Development Corporation. SilverScreen Reference Documentation V2.0. Mission, Ka.

Innovative Mine Design for the 21st Century, Bawden & Archibald (eds) © 1993 Balkema, Rotterdam, ISBN 90 5410 325 6

Computer modelling at the Sullivan Mine: A practical application in a complex operating environment

R.M.Van Ooteghem & P.K.Rodenstein
Cominco Ltd, Kimberley, B.C., Canada

ABSTRACT: This paper describes the practical application and implementation of a 3-D computer modelling system as a mine planning and engineering tool in a complex mining environment. The Sullivan Mine is one of the oldest and largest underground base metal mines in the world and has evolved into one of the most complex. Currently, ore is being recovered primarily from rib pillars. Pillar recovery requires careful and accurate mine planning to avoid interferences between existing and new development and, at the same time, ensure the safe development and optimum production and profitability of a mining block. Use of the Lynx 3-D computer modelling system has improved the speed, accuracy and flexibility of current mine planning and design techniques. The one-time storage and recall of geological and mine models, the ability to perform quick interactive design and modification, and the capability of 3-D visualization are the key features which can optimize and accelerate the completion of mine plans. The use of Lynx has removed the routine, repetitive and time-consuming activities of planning, and had allowed more productive and efficient time to be spent on the creative and problem-solving aspects of mine design. The application of a computer modelling and design system at the Sullivan Mine has enhanced the skills of our mining technicians and engineers.

1 INTRODUCTION

The Sullivan Mine, located in Kimberley, B.C., is an underground operation which currently produces 7,300 tonnes per day of zinc-lead-silver ore. The mine has been in operation for over 83 years and for the last 30 years, ore has been recovered primarily from rib pillars. Mining methods include single and multistage slot and shell, sublevel caving and open stoping incorporating both tracked and trackless mining equipment. Total ore production from the Sullivan Mine is about 136 million tonnes with 18 million tonnes of ore remaining in reserves.

To date, more than 560 kilometres of headings have been developed including track drifts, subs, raises and ramps. Current annual activities include 9,000 meters of new development and 240,000 meters of longhole drilling to maintain an acceptable level of broken ore to meet annual production requirements. Detailed and accurate mine planning is essential to the pillar recovery process. Mine design features must incorporate or contend with original stope infrastructure development. The yearly activity levels add to the enormity of the mine planning task.

The Lynx 3-D Computer Modelling System was initially investigated by Sullivan Mine personnel in late 1988 for its suitability as a mine planning aid. In 1989, an intensive six month test was conducted on-site to evaluate the full potential and capabilities of the system for the specific application of mine planning in our unique and complex mining environment. The system proved favourable and was subsequently installed. An implementation plan was then formulated to assess hardware and training requirements for optimum utilization of the system.

Hardware and software needs are continually being assessed and enhanced with the growing number of users and the increasing volume of different types of data being stored and manipulated. Currently, Lynx is being used regularly by a number of planning technicians and engineers as an aid to improve the quality and speed of engineering designs.

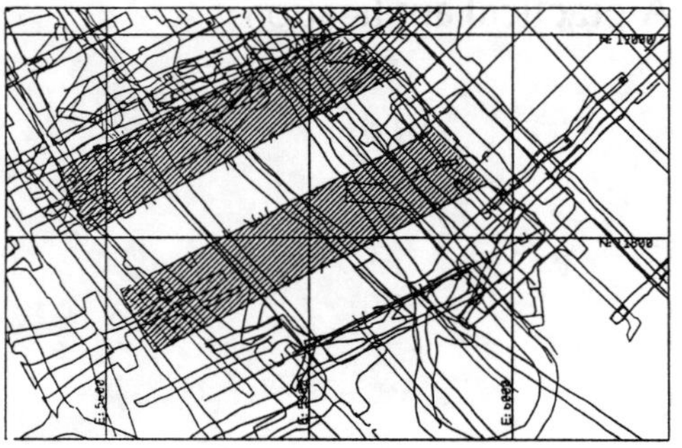

Fig. 1 Typical mine plan of pillars (shaded) and associated infrastructure

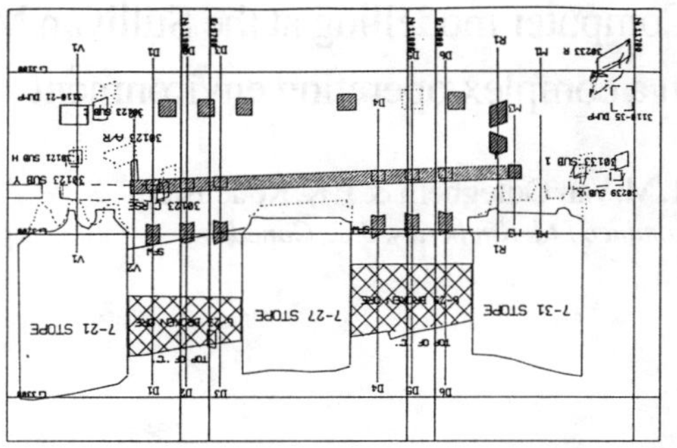

Fig. 2 Cross-section through pillars and old stopes showing integration of new and existing development.

2 WHY 3-D MODELLING?

The highly developed nature of the mine requires precise and accurate mine planning to avoid interferences with existing mine openings and structures. Each pillar is a unique situation and multiple section-cutting, although routine, is the most critical and most time-consuming function a planner performs. Planners may be required to cut as many as 100 sections in a typical pillar to show existing development and stopes for various design options. Ultimately, sections relate to the axis of chosen development. Section-cutting alone constitutes approximately 50% of the time to complete a planning drawing. A planning drawing, comprised of selected sections and a composite plan for a pair of pillars, typically requires three man months to complete. Recognizing the complexity of multi-level plans as shown in Figure 1, a simplified survey working drawing is prepared independently to serve as a check and to facilitate implementation of development. This drawing is updated as development is driven and is used as the basis to generate additional and usually systematic sections and plans for longhole drilling. For a typical pillar, a total of up to 140 manshifts are expended to provide these drawings using conventional means.

Designing the optimum location of production points to minimize dilution and to maximize the recovery of ore being produced is also a critical and time-consuming mine planning function. Tight tolerances to existing development as shown in Figure 2 often require slight changes in production and drill locations necessitating the cutting of additional sections to indicate potential interference and drawdown coverage. It is not uncommon for new development to be placed within two to three meters of existing headings.

A 3-D computer modelling system such as Lynx provides several benefits. The one-time investment of data input and verification allows the subsequent recall of stored data for any purpose and can therefore significantly reduce the time to complete mine plans, survey drawings, longhole drill layouts and any revisions. As a direct consequence, lead time for preparation of such plans is similarly reduced. Once data is verified for accuracy and completeness, plans and sections generated from this data are no longer subject to human error and interpretation.

The time-consuming and routine nature of section-cutting is eliminated with the use of Lynx. Quick section-cutting at any orientation and interactive design capabilities provides added flexibility to develop unlimited design options and virtually eliminates drafting time (see Figure 3). Removal of the conventional task of section-cutting allows quality time to be focused on the design challenges and associated creative problem solving which are fundamental to pillar mining.

Three dimensional maps of complex mine plans are valuable to help engineering and operating personnel identify areas of close tolerance and to better visualize the ideal location of new development, drill sites and production points for optimum ore recovery (see Figure 4).

3 WHY LYNX?

The Lynx 3-D Computer Modelling System is a state-of-the-art Unix-based modular software package developed specifically for mine design and geological applications. Lynx is designed to accurately and precisely represent complex geological and mining structures as true 3-D solids models in various colours and at any orientation in both 2-D and 3-D configurations. The modular structure of the software provides the versatility to select a software package suitable for a specific application. Lynx fully supports and upgrades their software. Enhancements are made by specific user request and by general improvements applicable to the industry at large.

3.1 Lynx testing and evaluation

The Lynx 3-D Computer Modelling System was tested using an IBM-RT Model 130. The objective of the test was to determine the suitability of Lynx as an effective tool to facilitate mine planning. The specific goal was to generate a planning drawing consistent with our current practices and stringent standards. The decision to commit to Lynx was to be based not only on satisfying these criteria but also on user friendliness. Acceptance by users was considered to be a critical issue.

The software modules acquired for the evaluation were:

1. System Management
2. Drill Log Management and Display
3. Map Data Management and Display
4. Drill Log Compositing
5. 3-D Solids Modelling
6. 3-D Block Modelling
7. 3-D Underground Mine Design

Testing was primarily dedicated to the use of the 3-D solids modelling module and the 3-D underground design module. These modules contain the features directly pertinent to mine planning. The full geotechnical capability of Lynx was not a consideration. Before testing began, planning technicians and engineers were solicited for input and asked to rate specific parameters which would make the system a useful aid from their point-of-view. The following list summarizes the survey parameters:

1. One-time input of data into 3-D model.
2. Easy and quick recall and update of data files with new or revised information.
3. Easy use and learning of the system (i.e. planners can quickly feel comfortable using the system).
4. A reasonable drawing quality is achieved (final drafting may be required).
5. Present 3-D planning conventions and practices are maintained and can be applied in 3-D.

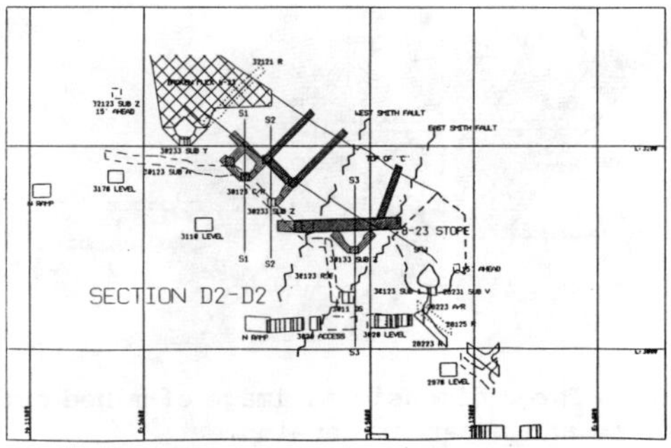

Fig. 3 Typical section through pillar

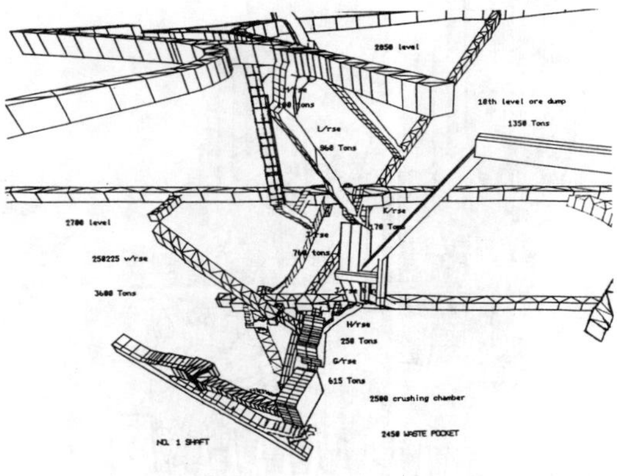

Fig. 4 3-D visualization of a complex ore pass system

6. Easy manipulation of 3-D images.
7. Quick interactive design and editing or revision capabilities.
8. Quick and easy provision for cutting sections at any orientation.
9. Accurate and reliable plots of plan and sections.
10. Credible and reliable support by Lynx personnel.

These parameters became a major part of the evaluation. To quantify the evaluation of the Lynx System, specific test criteria were defined and rated based on bench marks including ease of use, performance, ease of learning and importance to the function of planning. The strengths of the Lynx System were determined to be:

1. Speed, accuracy and versatility in section-cutting.
2. Flexibility in design.
3. Easy creation and manipulation of 3-D displays.
4. Short training time.
5. Easy and quick access, recall and editing of stored data.

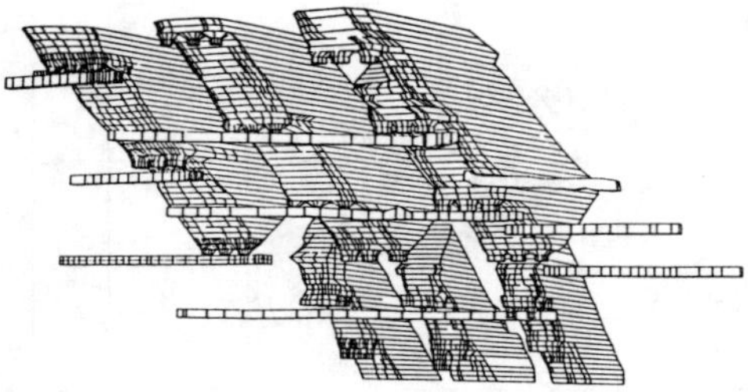

Fig. 5 Three dimensional image of mined out stopes and undercut development

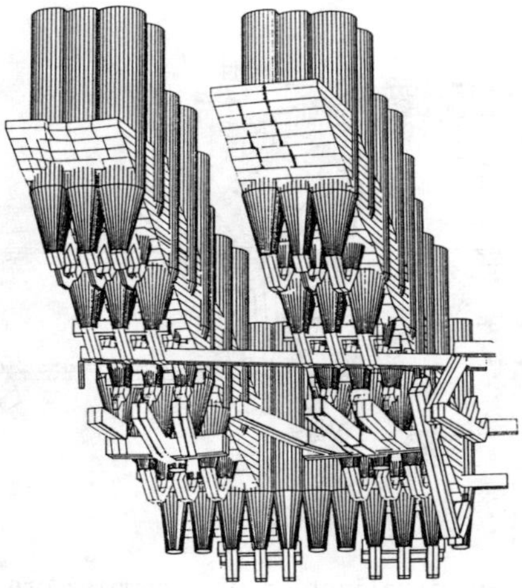

Fig. 6 Three dimensional view showing drawdown coverage

6. Repeatability, reliability and accuracy of plots.
7. Quick and accurate volume calculations.

With the initial software version, testing identified a number of specific weaknesses. These included poor labelling features, poor final drawing preparation and assembly capabilities, cumbersome user documentation and onerous and time-consuming digitizing functions.

These problem areas have been satisfactorily enhanced with upgraded versions of the software. The addition of the Underground Production Control Module has improved the speed and performance of our initial system. This module, developed in part with suggestions from our evaluation, accelerates the digitizing function and allows development or irregular shapes at varying grade or dip to be more accurately modelled in 3-D. This was critical to planning considering the close tolerances normally experienced between new and existing development. Lynx planning drawings can be exported to Autocad for labelling and final assembly of drawings for improved presentation quality.

The IBM-RT Model 130 workstation was not suitable for expansion purposes and was replaced with an IBM RS6000 computer.

4 HOW IS LYNX USED?

An initial and critical decision was made to capture and store existing mine data as required for new design projects. Any attempt to simulate the advanced state of development of an 80-year old major mine was out the question.

The first step is data input to create geological and mine models. At the Sullivan Mine, geological contours are provided for ore footwall and hanging wall horizons and fault structures. Previously mined stope contours are used to define pillar walls. This data is digitized into the system to create the 3-D geological model. In conjunction with this, 3-D mine models are created to include all existing development and infrastructure related to the specific pillar (see Figure 5). Initial construction of geological and mine models requires considerable time as all data must be digitized from existing drawings. However, once this information is entered, data can be recalled for planning of adjacent pillars and for any necessary revisions.

Following data input and checking for accuracy and completeness, development for optimum ore recovery can be designed. The interactive design and section-cutting capabilities allow the planner to work on-screen to develop unlimited design options for selection of the most practical and cost effective plan. Drawdown cones can be superimposed onto the screen in any possible production location as shown in Figure 6 and volumes calculated to determine the potential ore recovery from each draw point. Drill locations can also be designed for the most practical and effective drill pattern to facilitate a successful blast (see Figure 7). Manway accesses, muck raises and ventilation circuits can be designed in safe locations to best suit the location of existing development. The final design can be portrayed in 3-D to determine areas of unacceptably close tolerances which may have otherwise been overlooked or found difficult to visualize in two dimensions.

Unlimited section-cutting capabilities and 3-D visualization coupled with flexible design features allow for fine-tuning of production and infrastructure development to refine planning drawings far in excess of traditional methods. Preferred options can easily be stored for reference and review until the design work is complete. Plan and sections of the chosen design are then plotted and assembled for final presentation in a format consistent with established Sullivan Mine planning practices and conventions.

Survey working drawing and longhole drilling layouts are prepared using the same geological and mine models. As development or drilling is completed, data models can be revised and updated to reflect the "as driven" and "as drilled" activity.

Although the Lynx System was initially intended for pillar planning, its use has been applied effectively in an open stoping area on the fringe of the main ore body where the geology is not as well defined. The system has been used to modify and update existing plans as new geological information is available and as mining progresses (see Figure 8).

Even though Lynx is largely graphics oriented, proper data management is the key to its effective use. Duplicate data entry and accidental data deletion must be avoided at all costs. Clearly the requirement becomes higher as more users are introduced.

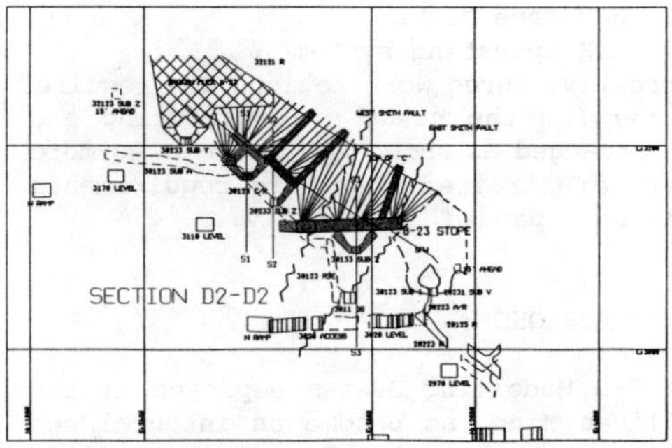

Fig. 7 Typical drill pattern

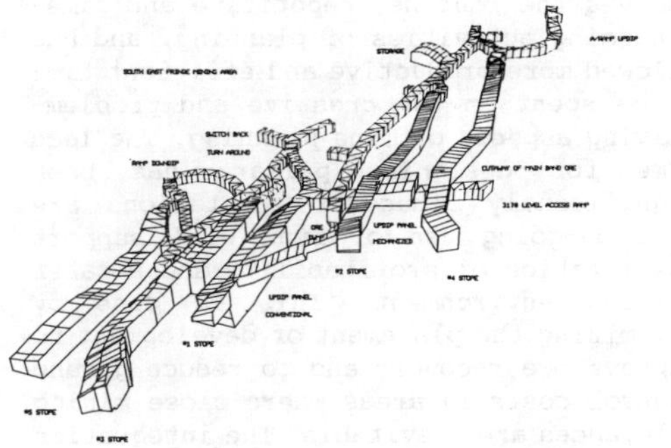

Fig. 8 Plot of the current mining status in an open stoping area

5 TRAINING AND IMPLEMENTATION

The strategy for training was to develop detailed procedures for the use of Lynx by completing an actual mine plan and then to develop an in-house training manual for specific engineering functions (i.e. planning, surveying and drilling). The goal of the program was to initially train two or three senior planning technicians in-house and establish a smooth flow of computer usage from planner to surveyor to longhole drill technician. To date, three planning technicians and five planning engineers have been trained and use Lynx almost exclusively as an aid to generate mine plans. Surveyors and drill technicians will be trained to use the system for their specific application. Full implementation of Lynx for all mine engineering functions will lead to maximum and continuous use of the system.

Training and implementation of the Lynx System has been most challenging. Traditional approaches are not changed easily, particularly with a mature work force. The users are highly skilled mining technicians but they have had very little exposure to computers of any kind, least of all, a 3-D computer modelling system such as Lynx. A philosophical and psychological change in thinking had to take place. Although training has progressed somewhat slower than anticipated and has extended our overall implementation program, acceptance has generally been good. The strengths of the system are well recognized. Only a few pioneers are needed to create interest at large.

6 HARDWARE

The IBM RS6000 workstation was used for training and drawing output and includes the following list of accessories:

1. 2 IBM X stations linked to RS6000
2. 300 MB Hard Disc Storage
3. 1 laser jet printer
4. 1 large back-lighted digitizer (Calcomp)
5. 1 large size 8-pen colour plotter (IBM)

6. 8mm tape drive
7. AIX operating system

Currently, three workstations are required to service the needs of engineering and are arranged as shown in Figure 9. As more users are trained, hardware requirements will be expanded accordingly.

7 CONCLUSIONS

The 3-D Modelling System employed at the Sullivan Mine has become an integral and effective tool in the planning of pillar recovery. Use of Lynx improves the accuracy and flexibility of conventional mine planning and design techniques. It removes the routine, repetitive and time-consuming activities of planning, and has allowed more productive and efficient time to be spent on the creative and problem-solving aspects of mine planning. The lead time for designing pillars has been significantly reduced. Expectations are that on-going use of Lynx will support maximization of profitability and a safer working environment. This is done by optimizing the placement of development to improve ore recovery and to reduce ground control costs in areas where close mining tolerances are inevitable. The integration of a highly specialized computer modelling system into an old and established operation proved to be a formidable challenge which has benefited the mine and has enhanced the skills of our technicians and engineers.

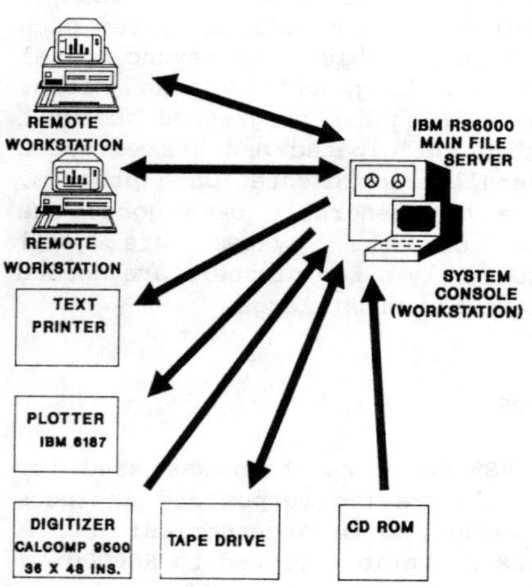

Fig. 9 Arrangement of Lynx computer components

8 ACKNOWLEDGEMENTS

The authors wish to thank the management of Cominco Ltd., Kimberley Operations for their support and commitment in promoting the use of Lynx from its inception in 1989 and for providing the opportunity to present this paper. In particular, the efforts of Dave McMurdo (Mine Manager) for initiating the acquisition of Lynx and for his critical review of this paper are very much appreciated. Special thanks is extended to Gerry Goosen (Management Information Systems Coordinator) for his technical support and to Wally Maluta (Planning Technician) for his assistance in the intensive in-house testing. We wish to express our thanks to Lynx Geosystems for their cooperative support during testing and installation and their ongoing involvement to provide system enhancements and training.

8 Mine environment

Innovative Mine Design for the 21st Century, Bawden & Archibald (eds) © 1993 Balkema, Rotterdam, ISBN 90 5410 325 6

Design of an auxiliary ventilation system for the exploratory developments of the Yucca Mountain Project

Felipe Calizaya & Pierre Mousset-Jones
University of Nevada, Reno, Mackay School of Mines, Nev., USA

ABSTRACT: The North and South ramps and the drifts of the Yucca Mountain Exploratory Studies Facilities (ESF) Project, if approved, will result in approximately 14 km of tunneling requiring multiple fans and rigid ducts to ventilate a number of headings developed simultaneously. The tunneling would start with an intake ramp at the northern end of the proposed repository site, continue with two parallel long drifts on two seperate levels, and complete the excavation with an exhaust ramp at the southern end. It is proposed that this would be developed with tunnel boring machines, mobile miners, and roadheaders in excavations varying from 7.6 m in diameter to 2.7 m by 4.9 m in cross section. Currently, the proposed plan is to have at some point in the project at least four to six headings being developed simultaneously on two levels (Topopah Spring and Calico Hills). This requires the implementation of a multiple fan auxiliary ventilation system.

The study includes an estimation of airflow requirements, percent leakage, selection of ducts, and the determination of fan operating points for a variety of input parameters. The study also includes a procedure to estimate the resistances of leakage paths along the duct, and the impact of selecting the appropriate leakage path resistance on the final design.

1 SALIENT FEATURES OF THE VENTILATION SYSTEM

1.1 Layout of the Ventilation Plan

Figure 1 shows a line diagram of the underground excavations for one of the proposed plans to develop the ESF. It includes one main access way, the North Ramp, and six development headings. During the construction, the North Ramp is to split into two drifts: Topopah Spring (TS) and Calico Hills (CH). From each drift, three headings are developed simultaneously. They all are to be ventilated by means of an auxiliary ventilation system. Table 1 shows a summary of the airway parameters needed for the ventilation design.

1.2 Estimation of Flow Requirements

A review of past experience on tunnel ventilation has shown that respirable dust is the major contaminant to be controlled during the development stage. The problem is present not only when the opening is developed with roadheaders and mobile miners, but also when tunnel boring machines are used. To overcome the problem, the auxiliary ventilation system is designed to maintain continuously a required quantity of fresh air at the workings. Often, this quantity is determined based on a minimum face air velocity and the cross-sectional area of the opening. Based on the O.S.H.A regulations, currently, this velocity is set at 0.15 m/s (30 fpm), (CFR-29, 1992). In this study, based on surveys of flow requirements in existing construction projects and in consultation with manufacturers and users of tunnel borers (Marks, 1991; Jurani, 1991; Moffat, 1992, and USBR, 1987), a face air velocity of 0.51 m/s (100 fpm) was chosen. Based on the above air velocity and the airway cross sections shown in Table 1, for this project, the following flow requirements were determined:

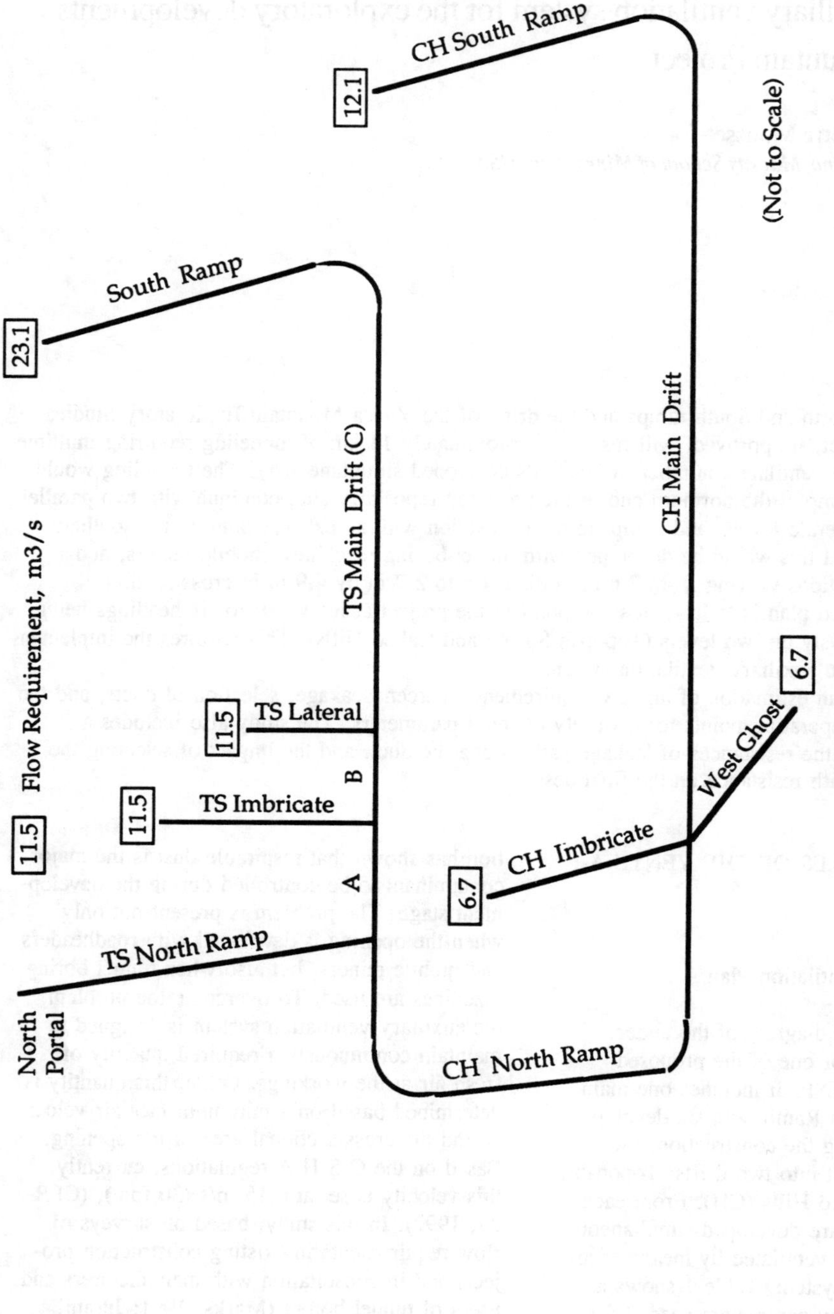

Figure 1. Line Diagram of the Proposed Underground Openings - Yucca Mountain E.S.F.

	Area, m^2	Quantity, m^3/s
Topopah Spring Level:		
- TS Main Drift	45.4	23.1
- TS Imbricate	22.6	11.5
- TS Lateral	22.6	11.5
Calico Hills Level:		
- CH Main Drift	23.8	12.1
- CH Imbricate	13.2	6.7
- CH West Ghost	13.2	6.7

Total Quantity Requirement, Q_o: 71.6 m^3/s (152,000 cfm)

2 STATEMENT OF THE PROBLEM AND SOLUTION

2.1 Network Formulation

Figure 2 shows the ventilation network used in this study, which represents the worst case scenario shortly before through-flow ventilation is established. This network was built to represent all the underground excavations and the ducts used in the ventilation system. Here, it is assumed that the fresh air is directed to the headings through ramps, drifts and cross-cuts, and the contaminated air returned to the surface through an auxiliary duct ventilation system. The network includes 83 nodes and 121 branches. To represent the losses of fresh air through duct joints and fittings, the network includes a set of high resistance air paths known as leakage branches.

In this network, each branch is represented by three parameters: airway resistance, pressure and quantity. At the initial stage, the resistances and a few airflow quantities are the only parameters that can be readily calculated. The other parameters, pressure and quantity are determined analytically by solving the ventilation network.

To simplify the calculations, only two types of resistances are considered: duct resistance, and leakage path resistance. The duct resistance was calculated using Atkinson's equation. This equation is given by:

$$\mathbf{R} = \frac{\mathbf{k\ Per\ L}}{\mathbf{A^3}} \tag{1}$$

where:

R = resistance, Ns^2/m^8
k = friction factor, kg/m^3
Per = cross-sectional perimeter, m
A = cross-sectional area, m^2

Table 1. General Information for Six Development Headings of the Yucca Mountain Project (Jurani, 1991)

Name of Airway	Length (m)	Cross section Width (m)	Height (m)	Remarks
North Ramp	2000	d = 7.6		Main Access
TS Main A	400	d = 7.6		" "
TS Main B	200	d = 7.6		" "
TS Main C	3000	d = 7.6		" "
South Ramp	2400	d = 7.6		1st Heading
TS Imbricate	800	3.7	6.1	2nd "
TS Lateral	400	3.7	6.1	3rd "
CH Ramp	2100	d = 5.5		Main Access
CH Main	3000	d = 5.5		" "
CH S-Ramp	2200	d = 5.5		4th Heading
CH Imbricate	800	2.7	4.9	5th "
CH West-Ghost	400	2.7	4.9	6th "

d = tunnel diameter (m)

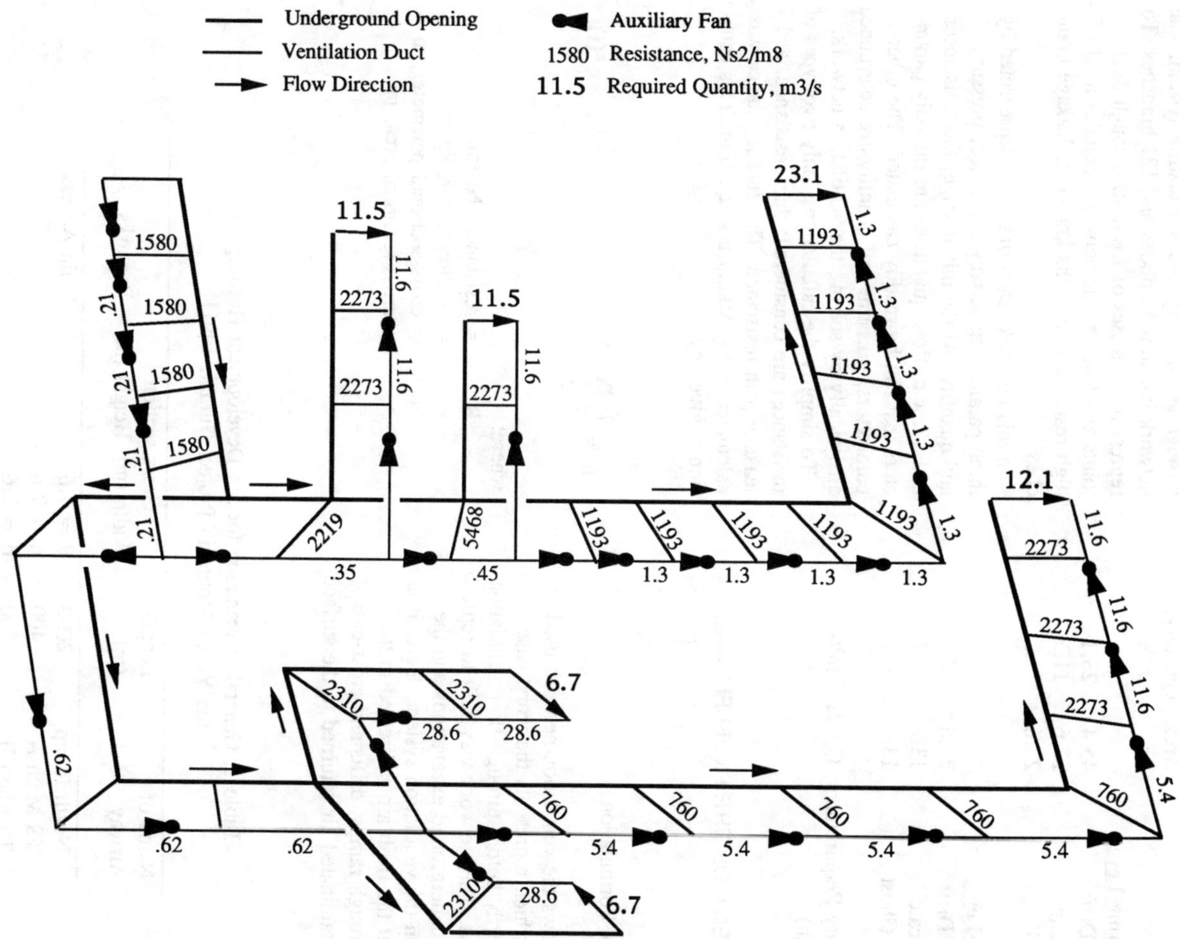

Figure 2. Ventilation Network for the Proposed Underground Openings- Yucca Mountain E.S.F.

In the estimation of resistances, the duct friction factor was set equal to 0.0028 kg/m³ (15 E-10 lb min²/ft⁴). The duct diameter was determined based on the flow requirements and a dust transport velocity of 12.7 m/s (2500 fpm).

The equivalent resistance of a leakage path was calculated by combining the resistances of a set of joints, 100 m apart, arranged in series/parallel combination. In this study, it was assumed that for each set, this resistance is equal to 10000 Ns^2/m^8 per 100 m-long duct.

For two airways with resistances R_1 and R_2, arranged in series, the equivalent resistance R_e is calculated from:

$$R_e = R_1 + R_2 \tag{2}$$

For two parallel airways, this resistance is given by:

$$\frac{1}{\sqrt{R_e}} = \frac{1}{\sqrt{R_1}} + \frac{1}{\sqrt{R_2}} \tag{3}$$

Since the cross-sectional area of an underground opening is considerably larger than that of a duct, in this study, the drift resistances were made equal to zero.

Figure 2 shows the airway resistances used in this study.

2.2 Fan Selection Procedure

The procedure used in this study is that developed for long single development headings (Vutukuri, 1984; Calizaya, 1993). To illustrate the major steps of this procedure, a sample auxiliary ventilation problem is solved (Figure 3). This is a two fan ventilation network. Here, it is assumed that the air contaminants generated at the face are exhausted through an auxiliary ventilation system. It is also assumed that the fans are of equal size (diam = 1.52 m) and installed at a constant interval (600 m) along the ductwork. The duct joints are represented by leakage paths with an equivalent resistance of 10000 Ns^2/m^8 per 100 m-long duct.

By applying equations 1, 2 and 3, for each segment, the following resistances were calculated:

$R_d = 1.34\ Ns^2/m^8$; $R_l = 1193.00\ Ns^2/m^8$

Where

R_d = duct resistance
R_l = leakage path resistance

The leakage path resistance was calculated by combining the resistances of six sets of ducts and joints arranged in a series/parallel combination.

According to this procedure, the fan operating points are determined by solving a set of two mesh-single fan networks. The calculations are started by solving the ventilation network for the innermost segment (Segment 1), then moving to the next segment outbye of the working face, updating the flow requirements, and repeating the calculations. The procedure is completed when all the fan operating points are determined.

Using this procedure, the network of Figure 3 can be solved in two stages as follows:

Stage 1. Start with the ventilation network for the innermost duct segment (Figure 4). Figure 4a shows the first segment of a long auxiliary duct system, see section A-A on Figure 3. This is a two mesh-single fan network. In this network, branches 1, 2, and 3 represent the opening, the working face, and the ventilation duct respectively. Branch 4 represents the leakage path, and branches 5, 6, and 7 are dummy branches used to close the ventilation network. In this system, the required quantity of air at the face is equal to 23.1 m³/s, and the resistances for the duct and the leakage path are 1.34 and 1193 Ns^2/m^8, respectively. The drift resistance is neglected.

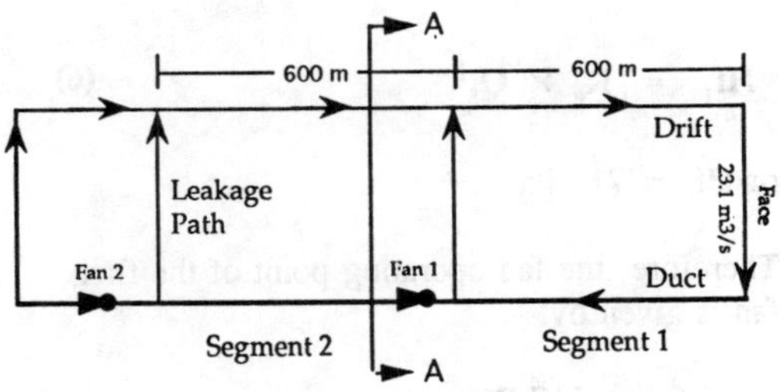

Figure 3. Line Diagram of an Auxilary Ventilation System

Based on the above information, the problem can now be reduced to that of determining the leakage flow quantity, and the fan operating point.

The problem is solved in four steps.

Step 1. Determine the equivalent resistance for the network (R_e).
For the parameters of Figure 4a, equation 3 yields: $R_e = 1.26\ Ns^2/m^8$.

Step 2. Determine the leakage quantity (Q_4). This quantity is calculated from Kirchhoff's second law (energy conservation), which for mesh 1 in Figure 4b is represented by $P_1 + P_2 + P_3 - P_4 = 0$. Since the resistances of branches 1 and 2 are equal to zero, then $P_3 = P_4$ or $R_3\ Q_3^2 = R_4\ Q_4^2$. Then,

$$Q_4 = \sqrt{\frac{R_3}{R_4}} \times Q_3 \qquad (4)$$

Which for the above parameters yields $Q_4 = 0.78\ m^3/s$.

Step 3. Determine the total air quantity (Q_5). This quantity is calculated from Kirchhoff's first law (flow conservation) as follows:

$$Q_5 = Q_3 + Q_4 \qquad (5)$$

or $Q_5 = 23.88$ m3/s.

Step 4. Determine the fan pressure (Hf_1)
This pressure is calculated from Atkinson's equation:

$$Hf_1 = R_e \times Q_5^2 \qquad (6)$$

or $Hf_1 = 717$ Pa.

Therefore, the fan operating point of the first fan is given by:

$Hf_1 = 717$ Pa
$Qf_1 = 23.88\ m^3/s$

Stage 2. The ventilation network used in this stage is similar to that shown in Figure 4 except for the flow rate in the fixed quantity branch which has been increased from 23.10 to 23.88 m^3/s. The difference represents the volume of air lost in the first segment of the ductwork. The network is solved using the same procedure outlined in stage 1. Since the network resistances in this stage are the same as those considered in stage 1, the equivalent resistance will remain unchanged ($1.26\ Ns^2/m^8$). However, the leakage quantity, due to an increased air pressure in the duct, will increase. For the new conditions, the following results are obtained:

- Leakage Quantity = $0.80\ m^3/s$.
- Fan Operating Point (2nd fan):
  $Hf_2 = 766$ Pa
  $Qf_2 = 24.68\ m^3/s$

Figure 5 shows the final air pressure/ quantity distribution for this example. Based on this

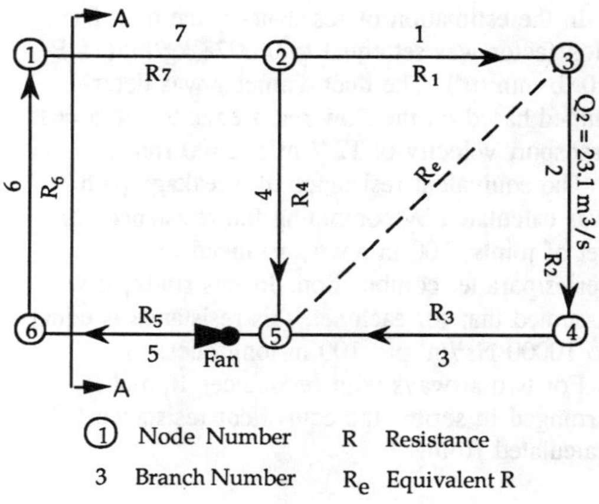

Fig.4a. Resistance Diagram

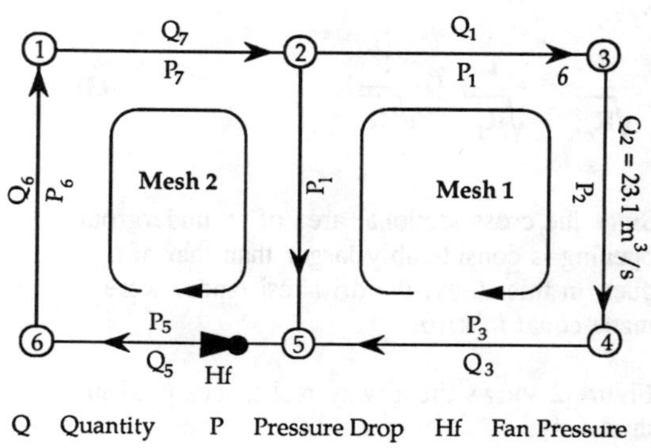

Fig.4b. Presssure /Quantity Diagram

Figure 4. Ventilation Network for Segment 1 of Fig. 3

Figure, the total leakage flow rate is equal to 1.58 m³/s, which represents 6 percent of the total quantity of air supplied by the ventilation system.

In this example, it is assumed that the duct length is equal to 1200 m, and there are two fans along the ductwork. If this length were equal to 5400 m, and nine fans were used to satisfy the flow requirements, the new system would have to supply a total quantity of 31.09 m³/s. Of this, a total volume of 7.99 m³/s (26 percent) would be the lost in the form of leakage flow which is highly dependent on the values of R_4 or the quality and maintenance of the ducts and joints.

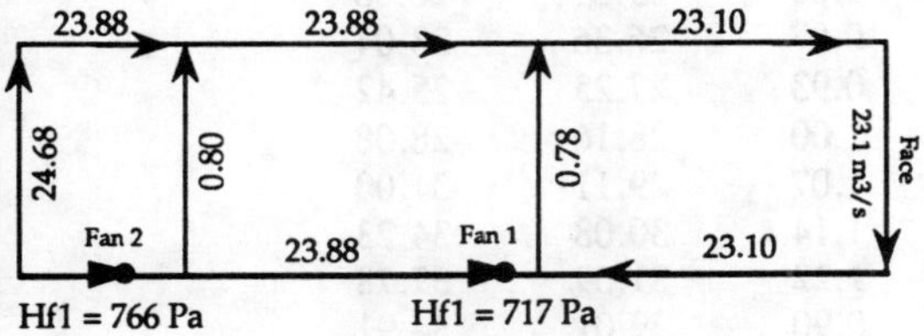

Figure 5. Flow Distribution for the Network of Fig. 3

2.3 Solution to the Problem

The procedure outlined previously is now used to solve the network problem of Figure 2. To this purpose, the network is partitioned into 11 segments of finite length. Here, it is assumed that each segment represents a ventilation system of a single development heading. It is also assumed that the energy required to circulate the air in the system is developed by a set of fans, installed in series, and equally spaced along the ductwork. The duct diameter is determined based on the required quantity of air at the heading, and the minimum air velocity within the duct. When applicable, the flow quantity is calculated by applying Kirchhoff's first law.

In the formulation of the problem, the following additional data were used:

a) Duct Diameters:

To be calculated based on the following transport velocities:

- Maximum transport velocity: 30.5 m/s (6000 fpm)
- Minimum transport velocity: 12.7 m/s (2500 fpm)
- Maximum fan operating pressure: 2490 Pa (10 in. w.g.)

b) Resistance for leakage paths:

To be determined based on a nominal resistance of 10000 Ns²/m⁸ per 100 m long duct.

For the above parameters and the resistances of Figure 2, the following results were found:

Required Quantity, Q_o:	71.6 m³/s (152,000 cfm)
Supplied Quantity:	103.4 m³/s (219,000 cfm)
Leakage Quantity:	31.8 m³/s (67,000 cfm)
Percent leakage:	31 %
Number of fans:	32
Required Air-power:	1646 kW (see Table 2 for details)
Power Cost/yr (@ 5 c/kWh and 70 % fan efficiency):	$ 1.0 million

A commercially available ventilation simulator (VNETPC) was used to check the above results. For the conditions stated previously, the results obtained were practically the same as those shown in Table 2.

The 32 fan locations in Figure 2 were arrived at through a trial and error procedure to ensure that the air pressure difference between the duct and a tunnel does not exceed 2490 Pa (10 in.w-.g.).

3 MAXIMUM AIR VELOCITY IN THE SYSTEM

Considering that the North Ramp will also be used to transport the broken rock produced at the headings, this velocity was calculated as follows:

Maximum Velocity = Total Quantity/Ramp cross-sectional area

For a quantity of 103.4 m³/s and an effective area of 41 m², this velocity is equal to 2.5 m/s (496 fpm). This is the maximum air velocity that would be expected at the portal of the North Ramp. This figure is below the maximum velocity of 3 m/s (600 fpm) recommended for this

Table 2. Proposed Ventilation System for the Yucca Mountain E.S.F.- Summary of Results

Fan No.	Location	Duct Diam. m	Pressure kPa	Quantity m^3/s	Air power kW
1	North Ramp	2.13	2.02	99.92	201.94
2	" "	2.13	2.07	101.06	209.00
3	" "	2.13	2.12	102.22	216.19
4	" "	2.13	2.16	103.39	223.74
5	TS Main A	1.82	1.15	57.73	66.39
6	TS Imbricate	0.92	1.54	12.32	18.96
7	" "	0.92	1.77	13.20	23.33
8	TS Main B	1.52	0.84	43.81	36.97
9	TS Lateral	0.92	1.54	12.32	18.96
10	TS Main C	1.52	0.72	23.88	17.12
11	" "	1.52	0.77	24.68	18.90
12	" "	1.52	0.82	25.51	20.86
13	" "	1.52	0.87	26.36	23.04
14	" "	1.52	0.93	27.25	25.42
15	" "	1.52	1.00	28.16	28.08
16	" "	1.52	1.07	29.11	31.00
17	" "	1.52	1.14	30.08	34.23
18	" "	1.52	1.22	31.09	37.78
19	CH N-Ramp	1.82	0.90	39.01	34.91
20	" "	1.82	0.94	40.02	37.70
21	" "	1.82	0.99	41.06	40.73
22	CH Imbricate	0.76	1.29	7.45	9.58
23	" "	0.76	1.59	8.28	13.14
24	West-Ghost	0.76	1.29	7.45	9.58
25	CH Main	1.22	1.19	16.14	19.24
26	" "	1.22	1.40	17.50	24.50
27	" "	1.22	1.65	18.97	31.23
28	" "	1.22	1.93	20.57	39.78
29	" "	1.22	2.27	22.30	50.68
30	CH S-Ramp	0.92	1.70	12.97	22.10
31	" "	0.92	1.96	13.90	27.19
32	" "	0.92	2.25	14.89	33.46
	Total air-power		1645.73 kW		

Fan diameter is equal to duct diameter

type of opening. However, due to the fact that the North Ramp will be used to both inject fresh air in one direction and extract the muck in another direction (antitropal ventilation), some dust related problems are expected, since the preliminary design for the conveyor belt speed is 2.5 m/s (500 fpm), which results in a net air velocity over the belt of 5 m/s (1000 fpm).

In order to evaluate the effect of the leakage path resistance on the air velocity, the nominal leakage path resistance per 100 m-long duct was decreased from 10000 to 5000 Ns^2/m^8, and the fan selection procedure repeated. For the new conditions, it was found that the leakage quantity was estimated to be as high as 41 percent of the total quantity circulated through the system ($120.78\ m^3/s$). This would increase the air velocity in the North Ramp to 3.0 m/s with a net air velocity over the belt of 5.5 m/s (1100 fpm).

4 CONCLUSIONS

For the conditions stated in this study, the simultaneous development of six headings for the proposed Yucca Mountain ESF would require a total of 32 fans. Of this number, only one would be installed on the surface, the remainder would be located underground in an auxiliary duct ventilation system. To satisfy the flow requirements, these fans would need to be designed and installed at appropriate locations to circulate a flow quantity of 103.4 m^3/s of air through the system without exceeding a differential pressure between duct and tunnel of 2490 Pa (10 in. w.g) anywhere in the duct system.

The air-power requirement in the system would be equal to 1646 kW. For fan efficiencies of about 70 percent, this would represent over 1.0 million dollars in operating cost.

The results shown in Table 2 were found for design conditions in which the nominal leakage path resistance was equal to 10000 Ns^2/m^8 per 100 m-long duct. If this resistance is reduced to 5000 Ns^2/m^8, a figure also reported in practice (Mathias, 1980), the leakage quantity in the system would increase from 31.8 to 49.2 m^3/s, and the air-power from 1646 to 2400 kW.

The ventilation design parameters included in this study are not adjusted for changes in air density (compressibility of air). However, taking into account the gage pressures to be maintained (< 2490 Pa), it is anticipated that this factor would not affect the results substantially.

An important parameter in the study is the leakage path resistance which depends on the design, installation, and maintenance of the ducts and couplings. More field measurements are needed in currently operating long, auxiliary ventilation systems to determine what is a reasonable value to use over the life of the ESF auxiliary ventilation system.

5 ACKNOWLEDGMENT

This paper was prepared with financial support of the U.S. Department of Energy, project number DE-FC08-90NV10891. Their assistance and cooperation are gratefully acknowledged. Special thanks are extended to Mr. R. Waters, Mining Engineer, DOE-Yucca Mountain Project, Las Vegas, Nevada for his assistance throughout this study.

REFERENCES

Calizaya F., & Mousset-Jones P., 1993. A Method of Designing Auxiliary Ventilation Systems for Long, Single Underground Openings. To be published in the Proc. of the 6th U.S. Mine Ventilation Symposium, Salt Lake City, 6 p.

CFR-29, 1992. Code of Federal Regulations, Title 29, US Government Printing Office, Washington D.C., July 1st (updated annually).

Jurani R., 1991. Preliminary Ventilation Analysis, ESF Construction Phase, Raytheon Services Nevada, Las Vegas, 60 p.

Marks J. R., 1991. Ventilating a Long, Deep-Level Exploration Drift. Proceedings of the 5th US Mine Ventilation Symposium, SME, Littleton, CO, pp.376-383.

Mathias P., et al., 1980. A Straight Blowing Auxiliary Ventilation System for A 7500 ft Single Heading Exploration Drive. Proceedings, 2nd Mine Ventilation Congress, SME of AIME, New York, NY, pp. 743-749.

Moffat B., 1993. Telephone communication.

USBR, 1987. US Bureau of Reclamation, Construction Safety Standards, Section 23.3.2.i (4).

Vutukuri S. I., 1984. Design of Auxiliary Ventilation Systems for Long Drivages. Proceedings, 5th Australian Tunnelling Conference, The Institution of Engineers, Sydney, Australia, pp.73-79.

Innovative Mine Design for the 21st Century, Bawden & Archibald (eds) © 1993 Balkema, Rotterdam, ISBN 90 5410 325 6

Remedial mine ventilation planning – Tracer gas definition of leakage routes

S.G. Hardcastle
Mining Research Laboratories, CANMET, Energy, Mines and Resources Canada, Elliot Lake, Ont., Canada

G. Klinowski
Mining Research Laboratories, CANMET, Energy, Mines and Resources Canada, Sydney, N.S., Canada

D. Mchaina
Dickenson Mines Limited, Balmertown, Ont., Canada

ABSTRACT: A common scenario for a mine ventilation engineer is the inheritance of a ventilation system with multiple working and development areas that are at the geographical extremities of the orebody. The mine is traversed by numerous access levels which in turn are intersected by raises. In such a situation controlling leakage becomes one of the engineer's primary concerns as it reduces the available air to dilute contaminants and the speed at which they are removed. Leakage is an expensive loss, as it has consumed fan power and possibly heating.

Although leakage is known to exist in ventilation systems, it can be difficult to locate and quantify. This especially true when it occurs mid-shaft and through caved areas. Therefore, to some extent leakage has been left unchecked as bulkheading every conceivable leakage route is not an economic solution.

CANMET has developed a tracer gas based simple test that has already shown the extent of inter-shaft leakage in a multi-level mine. The first test indicated a 22% loss of fresh air to just one of the mine's exhaust routes within the first 570 m, that was prior to the first production area. The test also indicated the general location of the leakage, and further iterative use of the test will pin-point the exact leakage route and its associated volume. Such tests allow the ventilation engineer to design remedial action selectively correcting the leakage in his system (i.e., through the construction or improvement of bulkheads) in the most cost effective manner. The in-mine element of this test can be executed by mine personnel, once trained, with CANMET performing sample analysis and data interpretation.

This method of interrogating a mine ventilation system has been applied at an Ontario gold mine. Here the first tracer gas tests indicated a 22% loss of intake air via three primary routes to one of the mine's exhaust shafts. This leakage costs the mine $50k p.a. in main fan power, when compared to $4k to construct a bulkhead, the potential cost benefits of isolating and limiting leakage can be quickly realised.

1 INTRODUCTION

Over the past 15 years CANMET's Mining Research Laboratories have been both developing and promoting the use of tracer gases as an analytical tool for interrogating mine ventilation systems. Early applications of tracer gas were primarily designed to provide a qualitative indication of airflow conditions. However, in-house development of gas release systems, sampling methods, tracer gas analysis and data interpretation have greatly improved its potential(Kennedy 1987; Stokes et al. 1989; Kennedy 1989). Tracer gas has been used to determine both quantitative and qualitative measurements of airflow and also as a pollutant surrogate. For example, tracer gas methods have been used in the following assessments

(Hardcastle et al. 1992):

1) Measurement of multiple airflows from a single injection.

2) The determination of shaft flows or other non-accessible airway from a remote location.

3) To establish in situ fan operating points.

4) The evaluation of auxiliary ventilation systems, their resistance and leakage characteristics.

5) The appraisal of auxiliary ventilation efficiency with respect to the removal of pollutants.

6) Tracing dust transportation from specific sources such as crushers in a processing plant.

7) Testing the integrity of escape routes for fire contamination.

Newer applications of the tracer gas technology permit the remedial design of ventilation systems so facilitating the optimum use of air. These include the determination of mine-wide leakage patterns and the blast clearance times. This paper details a simple test developed by CANMET to assess inter-shaft leakage.

A mine ventilation engineer rarely has the opportunity to design an efficient ventilation system from the mine planning stages and then develop it as the mine expands. At worst, the engineer inherits a ventilation system and is confronted with multiple working and development areas that are at the geographical extremities of the orebody. The mine is traversed by numerous access levels directly off the main ventilation shafts which in turn are intersected by raises. Once the levels become non-productive such airflow routes become potential short-circuit or leakage paths. Although widely dispersed working areas is the extreme case, in most instances such a scenario is unavoidable as the development has been dictated by production demands which were in turn driven by market price. In such a situation, the use of tracer gas to identify such ventilation parameters as residence times and to highlight leakage within the ventilation system is very valuable. Leakage is one of the ventilation engineer's primary concerns as it reduces the available air to dilute contaminants generated at the workplace. Leakage is also expensive as it consumes fan power and possibly heating, plus in certain instances it can promote dangerous oxidization processes.

Although the mine ventilation engineer knows that leakage exists, without such tracer gas analyses it can be difficult to locate and quantify. This is especially true when it occurs mid-shaft and through caved areas. Therefore, to some extent, leakage has been left unchecked as bulkheading every conceivable leakage route is not an economic solution.

Using the described tracer gas test to pin-point the exact leakage routes and their associated volume, allows the ventilation engineer to design a remedial action plan that selectively corrects the leakage in the ventilation system (i.e., through the construction or improvement of bulkheads) in the most cost effective manner. In addition, repeated use of tracer gas can be used to monitor the long-term integrity of the corrective measures through periodic repeat tests.

2 TRACER GAS METHODOLOGY

Sulphur Hexafluoride (SF_6) is the preferred tracer gas. It is non-toxic, stable, mixes readily with air. It is normally absent from the mining environment, has low detection limits, and its concentration can be accurately determined. Hence, it poses no risk to the workforce, sample analysis can be performed at leisure with an electron capture gas chromatograph, and only small amounts of the gas are needed to dope the ventilation system.

Over the last 10 years CANMET has developed high accuracy gas chromatographs, calibration systems and gas release equipment necessary for tracer gas based interrogation of ventilation systems. Use of these, in association with various release, sampling and analysis protocols has allowed the diverse applications introduced earlier.

Recently, CANMET developed the following simple test to define leakage in a ventilation system. This investigation can be predominantly performed by mine personnel with CANMET providing sample analysis and result interpretation.

The test entails instantaneously releasing a known volume of tracer gas into the intake of a mine ventilation system. This investigation employed gas sampling bags that had been filled with gas using a 1 L syringe, as the tracer gas source. Then sampling for that gas at numerous locations downstream, both on the intake and exhaust sides of the ventilation system. Air samples are collected at regular time intervals commencing with the gas release in disposable polypropylene 30 cc syringes fitted with airtight caps. Tracer gas concentration profiles are obtained from post-test gas chromatographic analysis of the air samples. Comparison of the gas profiles at the various locations will then highlight any anomalies in the ventilation system such as significant leakage routes or short circuits. The simplicity of this analysis is shown in the following two assessments at a test mine.

3 LEAKAGE INVESTIGATION RESULTS

Two tests were performed to determine the extent of inter-shaft leakage in a multi-level mine. A simplified diagram of the ventilation system is shown inset within Figure 1. The mine predominantly employs series ventilation above the 23 Level. The intake working locations, 16-017 and 22-977, are directly off the main ventilation system. They are supplied air by auxiliary fans which then returns to the main intake flow. For 23 Level and below, there is some splitting of the airflow into a parallel system to serve each major production level, for example, 23 and 28 Levels. However, on each level there is series ventilation of multiple working areas. On the exhaust side of the mine, working areas at 7-134 and 15-2528 are again series ventilated. The sampling station at 21-1058 is also in the exhaust but represents a location where major flows from 23 to 28 Levels combine.

3.1 Test #1 - Fresh Air Simulation

The tracer gas release and sampling locations for the first test are shown within the inset of Figure 1. This test was designed to determine how quickly fresh air from surface reached six locations in the mine that were representative of the general ventilation system. The tracer gas was released into the intake of the surface intake fan. The gas was sampled for at two intake locations: 16-017 the first working area, and 23 Complex the first multiple working area. A sampling station at 28 Alimak was chosen to represent the exhaust side of the multiple workings between 23 and 28 Levels. Two sampling stations were also established at 15-2524 and 7-134, adjacent to working locations in the exhaust part of the ventilation system and the final sampling point was at the exit of the largest of two major surface exhausts.

Air samples were collected every three minutes for a two-hour sampling period commencing at the time of the surface gas release. Figure 1 shows the tracer gas concentration profile at all six locations for the first injection. From these profiles it is immediately apparent that tracer gas reached the monitored surface exhaust and 7-134 prior to arriving at 16-017, the first intake working location. This indicated that fresh air was short-circuiting across the ventilation system to these two return locations. For the remaining sampling stations, the gas arrived in a logical order. Tracer gas was detected at 23 Complex after 16-017 and at 28 Alimak after 23 Complex. The gas arrived at 15-2548 after being detected at 23 Complex but before reaching 28 Alimak, this was due to air reaching 15-2548 quicker from 23 Complex than from the lower routes through the mine such as via 28 Alimak. Theoretically, if it were not for the short-circuits, or leakage, tracer gas would not have arrived in any significant concentrations at 7-134 and subsequently at the surface exhaust until after it had passed 15-2548.

Integrating the area under the concentration profiles at the surface exhaust and 7-134 up to the arrival of time of tracer gas at 16-017 provides an indication of the magnitude of the leakage. Firstly, the initial peak arriving at the monitored exhaust, prior to gas arriving at 16-017, represents 7% of the tracer gas released into the intake fan. Secondly, the area under the partial concentration peak for 7-134, prior to gas arriving at 16-017, represents another 1% of the gas released into the intake. Therefore, in association with the tracer gas, a total of 8% of the test mine's fresh air is also short circuiting to the exhaust ventilation system prior to any fresh air reaching 16-017 from surface.

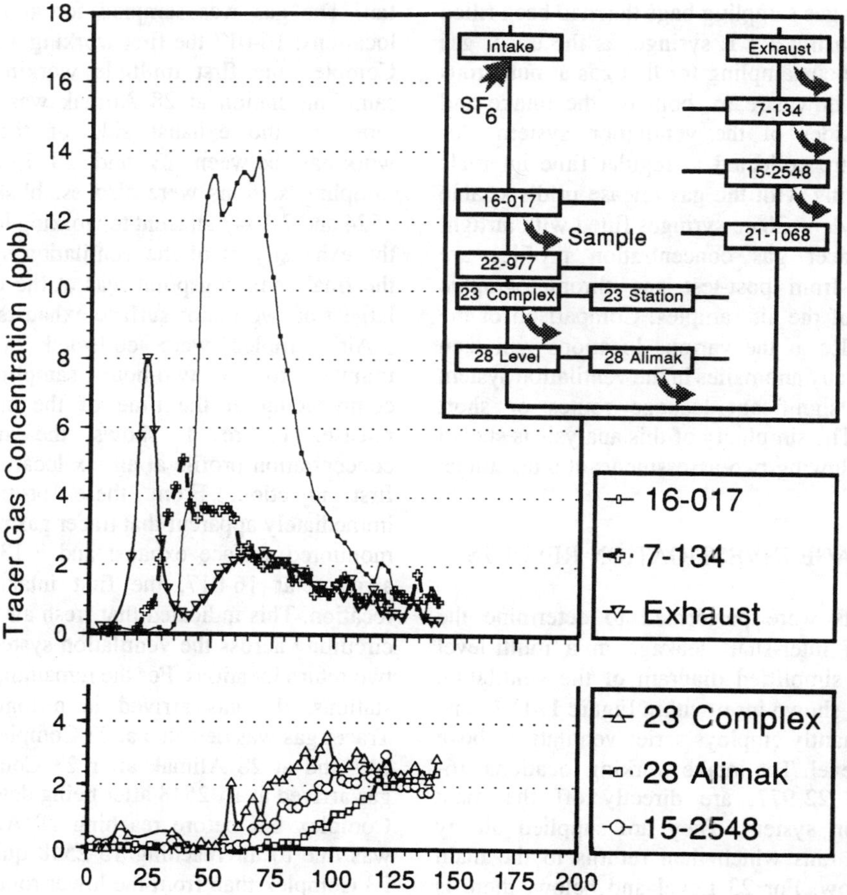

Fig.1 Tracer gas profiles - fresh air release

This test only provides a first order estimate of the leakage short circuiting the ventilation system as it does not take account of air reaching the return side by conventional routes. Nevertheless, the test does indicate that 7% of the mine's air very quickly passes from the intake to one surface exhaust. The loss of this air volume immediately impacts on the available air in the rest of the mine to dilute pollutants.

3.2 Test #2 - First Working Location

To determine the full magnitude of the leakage in the test mine's ventilation system a second test was necessary. In this instance the tracer gas was introduced at the first working/sampling location in the intake system, namely at 16-017, and then sampled for at six locations downstream as shown in Figure 2. In this test the remaining five sampling locations from Test #1 were maintained plus another at 23 Station.

The concentration profiles in Figure 2 show that air passing 16-017 progresses logically through the remainder of the ventilation system. Tracer gas arrives at 23 Complex first, that is the next location downstream on the intake. It then arrives at 23 Station and 28 Alimak at similar times. This indicates that the airflow to 28 Alimak is relatively faster and/or larger when

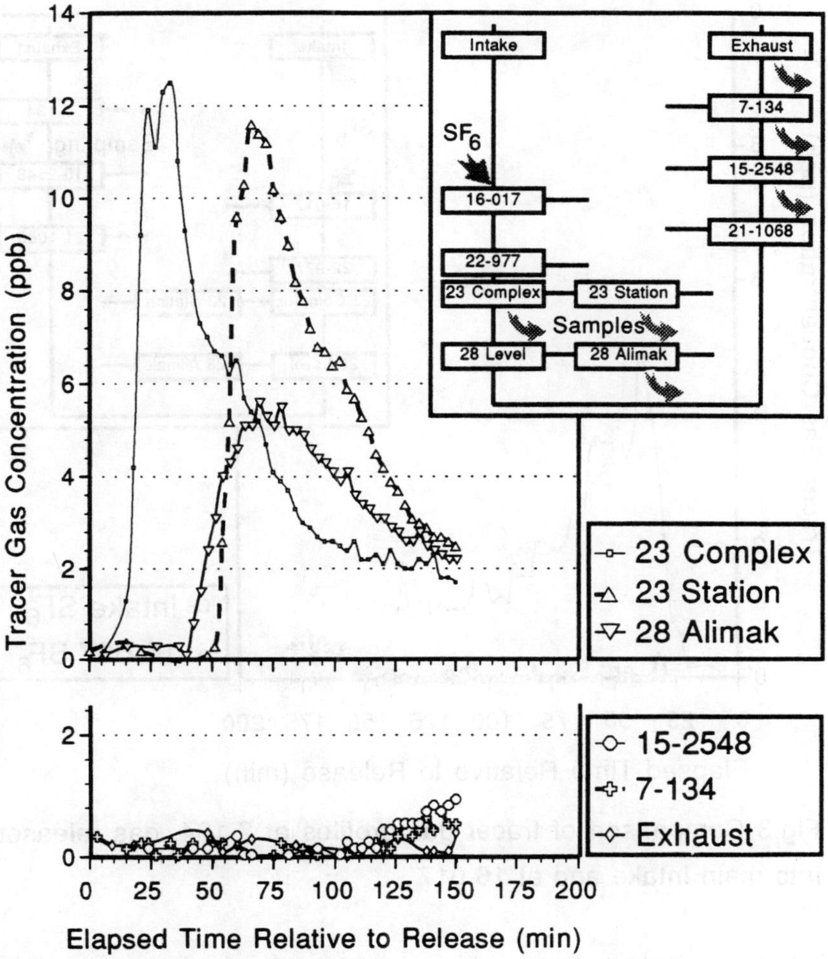

Fig.2 Tracer gas profiles - 16-017 release

compared to that traversing across to 23 Station. The tracer gas then arrives at 15-2548, 7-134 and a surface exhaust after reaching 23 Station and 28 Alimak. However, due to dilution it is not possible to ascertain precisely which location it reaches first.

Comparison of the results from the two injections provides more detail of short circuits in the mine ventilation system. Figure 3 shows the gas profiles at 7-134 from the intake fan and 16-017 tracer gas injections. Note, the latter profile has been retarded by 32 minutes, the time it takes tracer gas to reach 16-017 from the intake fan. These two profiles show that effectively all the gas monitored at 7-134 from the surface intake injection, the upper profile, arrives before any gas would have travelled from surface to 16-017 and then through the rest of the mine by conventional routes. Therefore, the whole peak is a result of gas short circuiting to 7-134 prior to reaching 16-017. Integrating the area under the profile indicates that 7% of the intake air is leaking across the mine to this sampling location.

Figure 4 compares the tracer gas profiles at the monitored surface exhaust from the intake fan and 16-017 gas injections. Here the tracer gas arrived at the exhaust in two peaks prior to any air travelling trough the mine via 16-017. Integration of the area in each peak indicated

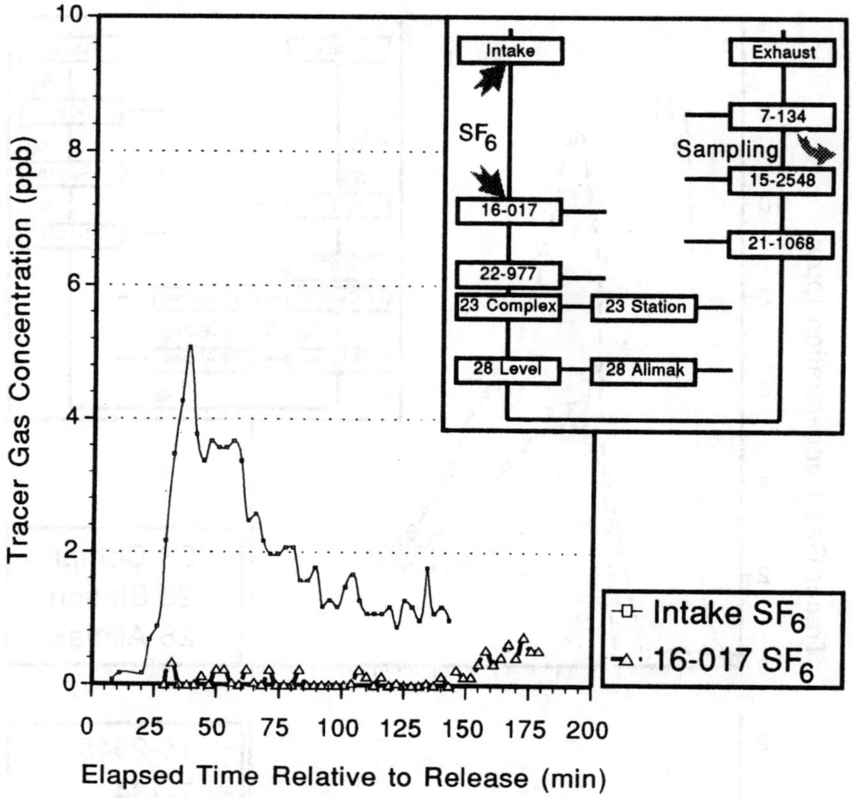

Fig.3 Comparison of tracer gas profiles at 7-134, gas released into main intake and at 16-017

that the first peak contained 8% of the tracer gas released into the intake fan and the second peak contained another 14% of the release. In this instance the analysis of the first peak provided a larger result than under Test#1 because it was not bounded by the arrival time of gas at 16-017. As the second peak occured after the peak arriving at 7-134 in Figure 3, it could in part contain the 7% of tracer gas released that was detected at that lower point in the mine.

In total, Figures 3 and 4 show that 22% of the tracer gas released into the surface intake fan short-circuits. This is either to 7-134 or to the monitored surface exhaust and occurs prior to gas following the desired route to 16-017 and then travelling through to the exhaust.

Figure 5 summarizes the three main short-circuits and also details the associated ventilation cost of each air leakage.

4 FUTURE TRACER GAS INJECTIONS

At this time the exact location of each leakage route and the extent of leakage to the other main surface exhaust are unknown. These may be defined with further injections. In future tests the mine should consider parallel sampling at both major exhausts to provide a more complete picture of existing leakages.

Furthermore, pin-pointing the location of leakage routes in the mine can be achieved with the iterative application of releases between the surface intake and 16-017, the first intake working location. In these future tests, other sampling locations should be considered both in the intake, between surface and 16-017, and in the return above 15-2548.

To facilitate these additional tests, the mine personnel were trained during the initial

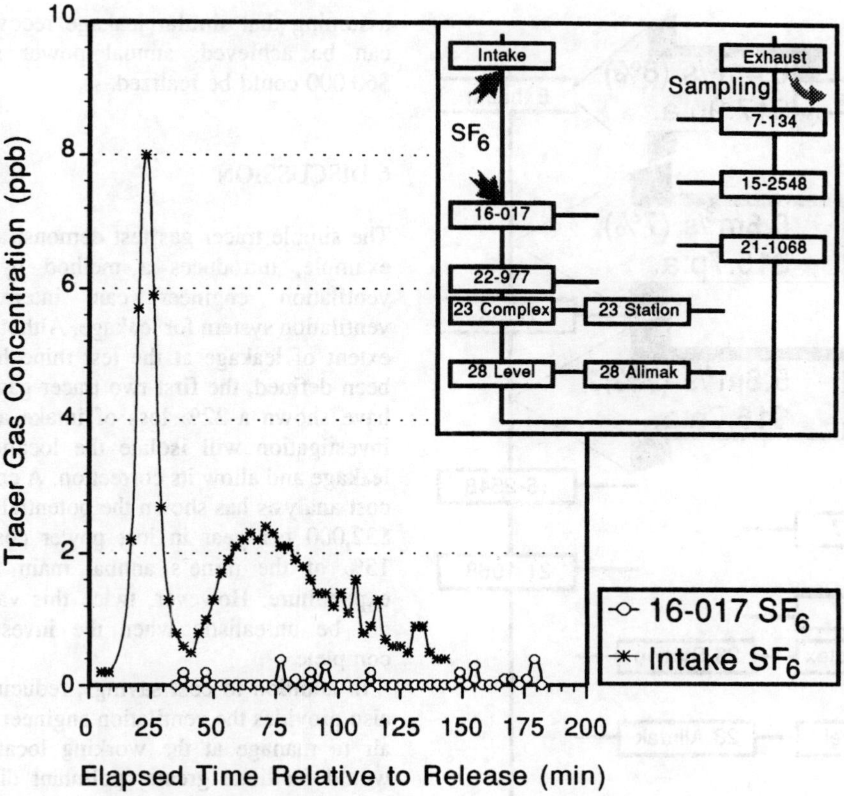

Fig. 4 Comparison of tracer gas profiles at exhaust, gas released into main intake and at 16-017

investigation by CANMET specialists. The mine will then send the air samples back to CANMET for analysis and interpretation. Training the mine personnel to perform its own tests and CANMET supplying the analysis as a laboratory routine service greatly reduces the cost of the procedure.

5 LEAKAGE PREVENTION COST ANALYSIS

The test mine estimates that the annual power cost for their main fans at ≈\$2,800/m³/s air. The following analysis compares this cost against that of constructing 3 m x 3 m permanent bulkheads at ≈\$4,000 (including labour and materials) to limit air leakage.

So far, this investigation has shown that the test mine is not using up to 22% of its intake air to the benefit of its workforce. Of this amount, 8% directly short-circuits from the intake fan through to one surface exhaust. The mines intake air volume is 80 m^3/s, therefore the mine has the potential to recoup up to 17.6 m^3/s. This may be used to both dilute pollutants and accelerate their removal from the mine.

As the first leakage tracer gas peak arriving at the surface exhaust is very narrow it would be reasonable to assume that it is the product of a single leakage path. Therefore, it could potentially be corrected with a single bulkhead at a cost of \$4,000. In this instance, assuming 90% leakage recovery the lost power savings would be \$16,000 per annum. Therefore, on comparing these two costs the pay-back period would be 3 months.

The other leakage peaks, the second at surface

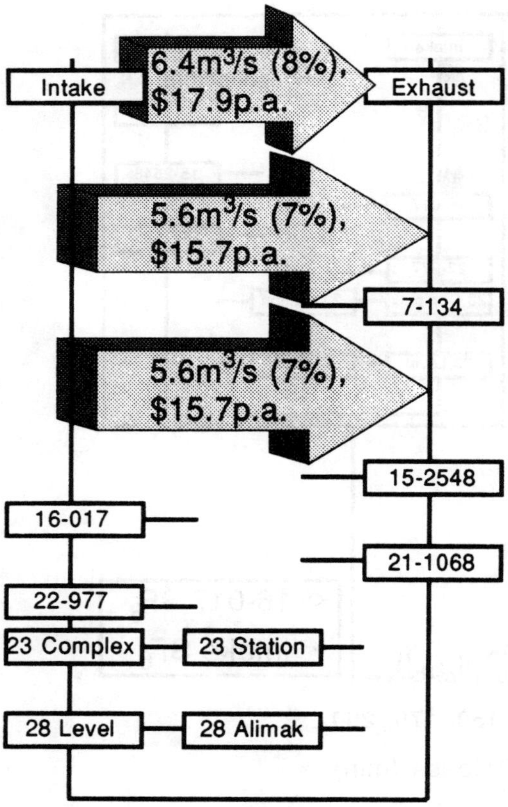

Fig.5 Primary inter-shaft leakage

and the one monitored at 7-134 are not as narrow. Therefore, they are probably the result of more than one leakage route. As such they will require more than one bulkhead for leakage control and the air recovery would probably be less. For example, assuming four bulkheads are required, their cost would be $16,000, and assuming only 50% leakage recovery, the lost power savings would be $16,000 per annum. That would a pay-back period of 1 year.

Combining the two leakage reduction exercises, the cost would be $20,000, and potential savings $32,000 per annum, i.e., a 7.5 month pay-back period.

As mentioned earlier, these initial tests only investigated the leakage to the main surface exhaust which handles ≈55% of the intake air. It would be reasonable to assume that air is also short-circuiting to the other exhaust routes. If the leakages are of the same magnitude, and assuming that similar leakage recovery values can be achieved, annual power savings of $60,000 could be realized.

6 DISCUSSION

The simple tracer gas test demonstrated in this example, introduces a method by which a ventilation engineer can interrogate his ventilation system for leakage. Although the full extent of leakage at the test mine has not yet been defined, the first two tracer gas injections have shown a 22% loss of intake air. Further investigation will isolate the location of this leakage and allow its correction. A conservative cost analysis has shown the potential for saving $32,000 per year in lost power costs, that is 15% of the mine's annual main fan power expenditure. However, twice this value would not be unrealistic when the investigation is complete.

In addition to cost savings, reducing leakage also provides the ventilation engineer with more air to manage at the working locations. This would facilitate greater pollutant dilution and faster clearance times. Such parameters can reduce existing re-entry time limits which can also impact on the productivity of the mine.

CANMET recommends that mines should consider such a leakage interrogation analysis prior to embarking on developing a simulation model. Controlling leakage is good ventilation practise and its presence can be identified with this simple tracer gas test. Although mines should employ models to optimize their mine ventilation planning, developing a ventilation model can be complex and time consuming. Also the initial exercises will similarly show the degree of leakage in the system. In such a situation the tracer gas method is quicker and cheaper than the simulation route.

REFERENCES

Hardcastle, S.G., Klinowski, G. & Grenier, M. 1992. Tracer gas - An essential tool for ventilation and airborne pollutant investigations. Proceedings Safety, Hygiene and Health in Mining, PtII, Harrogate, U.K.,

November 1992, T.I.M.E.

Kennedy, D.J. 1987. An analytical method for determining trace levels of sulphur hexafluoride, bromotrifluoro-methane and dichlorodifluoromethane in mine air samples. Division Report CRL 87-76(TR), CANMET, Energy, Mines and Resources Canada.

Kennedy, D.J. 1989. A fixed point calibration curve generator for gas analysis. Division Report CRL 89-58(TR), CANMET, Energy, Mines and Resources Canada

Stokes, A.W., Kennedy, D.J. & Hardcastle, S.G. 1987. A real time tracer gas analyzer - An investigational tool for mine ventilation studies. Min. Sci. Tech. 5:187-196.

Innovative Mine Design for the 21st Century, Bawden & Archibald (eds) © 1993 Balkema, Rotterdam, ISBN 90 5410 325 6

Unusual work schedule analysis for risk assessment and occupational exposure evaluation

D.M. Mchaina
Dickenson Mines Ltd., Balmertown, Ont., Canada

M.G. Grenier & S.G. Hardcastle
Elliot Lake Laboratory, CANMET, Ont., Canada

ABSTRACT

An overview of a process to be employed to evaluate the impact of a 10-hour shift work schedule on safety and health objectives of an underground mine is provided.

In many industrial operations, unusual work schedules have been introduced to permit more efficient use of human resources, improve productivity, permit adequate time off the job, permit better equipment utilization and maintenance plan, and increase productive time. Whenever unusual work schedules are proposed, uncertainties exist regarding their effect on exposure to various workplace contaminants. It is accepted that existing Time Weighted Average Exposure Limits have to be reduced to provide protection for employees working unusual schedules. Taking into account such aspects as increased hours of exposure and decreased hours of recovery for a 10-hour shift work schedule it is suggested that a reduction factor of 0.70 be applied to all existing limits.

The implementation of extended schedules is currently prohibited under the Ontario Occupational Health and Safety Act and Regulations for Mines and Mining Plants. The reason for the prohibition is based on a concerns that unusual work schedules in underground mines may result in an increased lost time and medical aid accident frequencies.

There are a number of models that can be used to set up operating exposure levels in the event that an unusual work schedule is introduced. The discussion in this paper will be limited to three models namely, Pharmacokinetic, Brief and Scala, and OSHA.

Historical exposure data should be evaluated and compared with the modified limits to determine whether the health of a worker may be affected by exposure to the contaminants. Modified limits may result in significant adverse technical and economic impacts.

An assessment determining the economic impact resulting from bringing exposures into compliance with reduced TWAE Limits is presented. It is recommended that an impact assessment be performed to determine the effect of the modified limits on the overall company safety standards. This should be incorporated in all feasibility studies related to the introduction of unusual work schedule.

INTRODUCTION

In many industrial operations, unusual work schedules have been introduce to permit more efficient use of manpower, improve productivity, permit adequate time off the job, permit better equipment utilization and maintenance plan, and increase productive time. The interest in utilizing extended workday schedules in the mining industry is gaining popularity due to the benefits mentioned above. Operations located in remote areas are already utilizing extended workday shift schedules to minimize travel time to work. In general, the present trend is that extended shifts are preferred to the usual 8-hour day, 40-hour per day, work week.

The implementation of extended schedules is currently prohibited under the Ontario Occupational Health and Safety Act and Regulations for Mines and Mining Plants. The reason for the prohibition is based on a concerns that unusual work schedules in underground mines may result in an increased lost time and medical aid accident frequencies. Operations located in remote areas are exempted due to economical reasons.

Whenever extended work schedules are proposed, uncertainties exist regarding the effects of exposure to workplace contaminants. A 10-hour or 12-hour shift schedule are typical examples of unusual work schedules. The existing Threshold Limit Values presented in documents such as American Conference of Governmental Industrial Hygienists (ACGIH) and the "Ontario Regulation respecting Control of Exposure to Biological or Chemical Agents - made under the Occupational Health and Safety Act" represents "conditions under which it is believed that nearly all workers may be repeatedly exposed day after day without adverse effect". The values refer to "Time-Weighted Average Exposure (TWAE) concentrations for a 7-or 8-hour work-day and 40-hour work week".

The current operating exposure limits have to be modified to provide protection for the exposed workers. The limits should be reduced by some amount to take into account aspects such as increased hours of exposure and decreased hours of recovery. The modified limits together with close medical surveillance should adequately protect the workers.

There are a number of models that can be used to set up operating exposure levels in the event that an unusual work schedule is introduced. This paper will investigate the application of three models namely, Pharmacokinetic, Brief and Scala, and OSHA. However, these models do not take the following parameters into account:

- Variation in Pollutant Biological Half Time;
- Individual Susceptibility;
- Hypersensitivity;
- Pre-existing Medical Conditions;
- Synergistic, Additive and Antagonistic Effects of Individual Pollutant;
- Biotransformation of the Pollutant in Question;
- Absorption and Elimination Kinetics of the Chemical;
- Of-the-Job Exposure Effect; and
- Work-Rest Cycle.

Historical exposure data should be evaluated and compared with the modified limits to determine whether the health of a worker may be affected by exposure to the contaminants. Modified limits may result in significant adverse technical and economic impacts.

MODIFIED EXPOSURE LIMITS

The existing limits must be reduced by some amount to take into account the following aspects:

- Increased Hours of Exposure ("$RF_{(IHE)}$");
- Decreased Hours of Recovery ("$RF_{(DHE)}$").

The outcome is a "Reduction Factor (RF)" which is a product of the two factors - "$RF_{(IHE)}$" and "$RF_{(DHE)}$" calculated for the work-day or work week.

Reduction of TLV

= Effect of Increased Exposure Time x Effect of Decreased Exposure - free Time

There is limited information on the impact of working long hours on the health of employees. In addition, the TLV Committee of the ACGIH recognizes that the normal 40-hour work week basis for TLVs may need to be adjusted for unusual work schedules but at present there is no unilateral system in place to do so. However, in the 1989-90 TLV Preface it is stated that, "Application of TLVs to workers on work schedule markedly different from the conventional 8-hour day, 40-hour week requires particular judgement in order to provide, for such workers, protection equal to that provided to workers on conventional work shifts". The ACGIH then goes on to state - As tentative guidance, field hygienists are referred to the "Brief and Scala model", and the "OSHA model".

The Ontario Ministry of Labour existing legislation does not have exposure limits for unusual work schedules.

Many researchers have proposed mathematical formulae or models for testing occupational exposure limits for use during unusual work schedules. These researchers have developed models which modify exposure limits for varied exposure regimens. Among these models are those of OSHA, Brief and Scala, and the Pharmacokinetic model.

Both the OSHA and the Brief and Scala models are easier to use than some of the more complex models which are based on Pharmacokinetic effects. The Brief and Scala model will recommend the lowest limit and the pharmacokinetic model will recommend the highest.

The models are compared below.

Brief and Scala > OSHA > Pharmacokinetic

MODELS

Brief and Scala Model

TLV Reduction Factor

= $[(8)/(h)] x [(24 - h)/(16)]$

= $[(8)/(\text{Hours Worked})] x [(\text{Hours Off Work})/16)]$ .... (I)

Where h

= hours worked per day

Excursion Factors (EF) for TWA Limits

= $[(EF_8 - 1)RF + 1]$ .... (II)

Where EF_8

= Excursion Factor for 8-hour TWA

EF

= Desired Excursion Factor

RF

= TLV Reduction Factor

TLV Reduction $(RF_{(W)})$

= $[(40)/(H)] x [(168-H)/(128)]$ .... (III)

Where H

= hours exposed per week

OSHA Model

By comparison, the OSHA model (Equation IV) if applied to other than 8-hour shifts, could be recast in its simplest form using the Brief and Scala symbols:

R.F.

= $[(8 \text{ hours})/(h)]$ .... (IV)

Where h

= hours of exposure per day

It is important to recognize that these models do not account for uptake rate.

Table 1 gives the respective reduction factors derived from the three models discussed above. Based on the models discussed above, the Scala and Brief Model is more conservative than the other two. Due to this it would be prudent to derive the new Reduced TLVs using Brief and Scala model. The modified Time-Weighted-Average Exposure Levels are calculated from the limits as detailed in the Ontario Regulation respecting Control of Exposure to Biological or Chemical Agents - made under the Occupational Health and Safety Act. The Reduction Factors for a 10-hour shift work schedule as applied to regulated mine pollutants are calculated using the Brief and Scala method and are given in Table 2.

Table 1. Comparison of Reduction Factors Derived from Various Models

Work Schedule	Ontario M.O.L. Current Practice*	Worst Case Pharmaco-kinetic Model, F_P	Brief & Scala Model, R.F.	OSHA Current Practice	OSHA Reduction Factor for Extended Shifts
Five 8-hr Days per Week	1	1	1	1	1
Four 10-hr Days per Week	N/A**	0.84	0.70	1	0.80
Three 12-hr Days per Week	N/A	0.75	0.50	1	0.67

* M.O.L. = Ministry of Labour

** N/A = Not Available

Table 2. Calculated Reduced TLVs for 10-Hour Shift Work Schedule

PARAMETER	8-Hour TLV*	Reduction Factor	Reduced TLV
Carbon Monoxide	35	0.70	24.50
Carbon Dioxide	5,000	0.70	3,500
Nitrous Fumes (NO_x)	25	0.70	17.50
Nitrogen Dioxide	3	0.70	2.10
Ammonia	25	0.70	17.50
Silica	0.10	0.70	0.070
Arsenic	0.20	0.70	0.140
Lead	0.150	0.70	0.105

* All parameters in parts per million except Silica and Arsenic which are in milligrams per cubic metre.

CASE STUDY

Shown in Table 3 are dust exposure concentration levels from personal sampling for a case study mine.

When the upper range for workplaces given in Table 3 are compared with the reduced TWAE given in Table 4, the following conclusions can be made:

General silica concentration levels are a concern at workplaces R and T. At workplace T, the levels are higher than the calculated TWAE.

Specific arsenic concentration levels at workplace S would need addressing because the highest concentrations level recorded were higher than the reduced limits.

Table 3. Workplace Exposure Data During 1990

Workplace	Silica, mg/m³		Arsenic, mg/m³		Lead, mg/m³	
	Range	Avg.	Range	Avg.	Range	Avg.
P	0.0-0.10	0.0284	0.009 - 0.056	0.0073	-	-
Q	N/A	0.0400	0.0042 - 0.0107	0.0025	-	-
R	N/A	0.0700	0.0007 - 0.0016	0.00117	0.012 - 0.275	0.0965
S	0.01-0.05	0.0200	0.0076 - 0.2850	0.0718	0.0013 - 0.0240	0.0072
T	N/A	0.1100	N/A	0.0146	-	-
U	N/A	0.0000	0.0008 - 0.0045	0.0027	-	-

Table 4. Reduced Limits Impact Assessment Based on Current Limits

Work Place	Silica, mg/m³			Arsenic, mg/m3			Lead, mg/m³		
	A	B	C	A	B	C	A	B	C
P	0.10	0.07	0.028	0.20	0.14	0.0280	0.15	0.105	-
Q	0.10	0.07	0.040	0.20	0.14	0.0025	0.15	0.105	-
R	0.10	0.07	0.070	0.20	0.14	0.0012	0.15	0.105	0.0965
S	0.10	0.07	0.020	0.20	0.14	0.0720	0.15	0.105	0.0072
T	0.10	0.07	0.110	0.20	0.14	0.0146	0.15	0.105	-
U	0.10	0.07	0.000	0.20	0.14	0.0027	0.15	0.105	-

Legend A = Current TWAE
B = Reduced TWAE, Based on Scala's Model
C = Average Workplace Exposure Levels

Lead concentration levels at workplace R should also be reduced as they are near the reduced limits.

PROPOSED ARSENIC EXPOSURE LIMIT IN ONTARIO

Arsenic and its compounds are regulated in Ontario as designated substance in all workplaces except mines and concentrators. There is a proposal to remove the exception for mines and concentrators. If the exception is removed, the mining industry will be required to adopt the proposed exposure limit of 0.01 mg/m³ over normal work schedules (8 hour shifts, 5 day weeks).

To determine the impact of the proposed limit on extended work day, an impact assessment is presented in Table 5. The reduced limit based on the proposed limit is compared with the average exposure level for the workplaces detailed in Table 3.

The data shown in Table 5 indicates that workplaces T, S and P will not meet the proposed limit of 0.01 mg/m³ and the reduced TWAE. All the average concentration levels recorded for these workplaces are above the proposed and the reduced limits.

As these three areas represent the majority of the operations, workforce compliance with either the proposed limit or the modified limit for extended hours will result in significant

Table 5. Reduced Limit Impact Assessment Based on Proposed Limit Impact Assessment

Workplace	Arsenic, mg/m3				
	D	E	F	COMPLIANCE	
				G	H
P	0.01	0.007	0.0280	No	No
Q	0.01	0.007	0.0025	Yes	Yes
R	0.01	0.007	0.0012	Yes	Yes
S	0.01	0.007	0.0720	No	No
T	0.01	0.007	0.0146	No	No
U	0.01	0.007	0.0027	Yes	Yes

Legend D = Current TWAE
E = Reduced TWAE, Based on Scala's Model
F = Average Workplace Exposure Levels
G = Compliance with the proposed TWAE (Column D Compared to Column F)
H = Compliance with the Reduced TWAE (Column E Compared to Column F)

adverse technical and economic impact. It is therefore, important to conduct a feasibility study to investigate the impact of modified limits on the company's health and safety standards and policies.

In considering the technical and economical aspects of achieving compliance with the modified exposure limits, the company should take into account issues such as the operations or processes which give rise to the exposures of concern, the decrease in exposures which would be required to bring exposure into compliance with the modified limit, changes in equipment or operations necessary to achieve and maintain compliance with the modified limits, costs of such changes, and an evaluation of conditions or circumstances which make it technically and economically feasible.

CONCLUSIONS AND COMMENTS

The Brief and Scala model provides more worker protection, it has also received wide acceptance among Industrial Hygienists. Due to this fact this it would be prudent to use the Brief and Scala model for setting TWAE Limits.

Presently, the Ontario Ministry of Labour which presides over this mine does not have any specific guidelines for the extended shift hours, therefore it is difficult to come up with acceptable TWAE Limits for this case study.

An assessment demonstrating the economic impact resulting from bringing exposures into compliance with the reduced TWAE Limits should be incorporated in all feasibility studies related to the introduction of unusual work schedule.

It should be recognized that the general model approach is a first approximation only. It should be pointed out that a close medical surveillance must be maintained for some reasonable period until it is assured that modified limits do indeed protect the workers.

References

1. Brief, R.S., and Scala, R.A. 1975. Occupational Exposure Limits for Novel Work Schedules. American Industrial Hygiene Association Journal,36:467 (June 1975).

2. Mchaina, D.M., Grenier, M., and Hardcastle, S.G., 1992. An Assessment of Arsenic Exposure Levels at a Canadian Gold Mine. New Technology in Mine Health and Safety, Editor: Khair, A.W. Proceedings of the Symposium held at the SME Annual Meeting, Phoenix, Arizona(February, 1992).

3. Ontario Ministry of Labour, "Regulation respecting Arsenic - Made under the Occupational Health and Safety Act". Revised Statutes of Ontario, Chapter 321, Ontario Regulation 176/86, 1980.

4. American Conference for Governmental Industrial Hygienists, ACGIH, "1990 -91 Threshold Limit Values for Chemical Substances and Physical Agents and Biological Exposure Indices"; ACGIH, 6500 Glenway Avenue, BLDG. D-7, Cincinanti, OH 45211-4438.

5. Ontario Ministry of Labour, "Occupational Health and Safety Act and Regulations for Mines and Mining Plants". August, 1988.

6. Ontario Ministry of Labour, "Regulations respecting Control of Exposure to Biological or Chemical Agents - Made under the Occupational Health and Safety Act".; Revised Statutes of Ontario, Chapter 321, 198, Ontario Regulation 654/86 as Amended by Ontario Regulation 77/86 and Ontario Regulation 339/87.

7. Wagner, J.A., 1990. A Comparative Study of 8 and 10-Hour Shift Schedules at an Underground Gold Mine. Presented at Mines Accident Prevention Association of Ontario (MAPAO) 59th Annual Meeting and Technical Sessions. Toronto, Ontario, Canada (May 30 - June 1, 1990).

8. Ontario Ministry of Labour, "Regulations respecting Silica - made under the Occupational Health and Safety Act". Revised Statutes of Ontario, 1980, Chapter 321. Occupational Health and Safety Division, Toronto, Ontario, Canada.

9. Wagner, J.A., 1989. Safety and Health Aspects of Extended Workdays. Presented at Mines Accident Prevention Association of Ontario (MAPAO) 58th Annual Meeting and Technical Sessions. Toronto, Ontario, Canada (May 24-26, 1989).

Innovative Mine Design for the 21st Century, Bawden & Archibald (eds) © 1993 Balkema, Rotterdam, ISBN 90 5410 325 6

The importance of monitoring system in controlling environment in OAL Mine

T.Güyagüler
Mining Engineering Faculty, METU, Ankara, Turkey

H.Karaman
OAL Mine, Beypazarı, Turkey

ABSTRACT : Dynalink micro minos environmental monitoring system provides control and monitoring facilities of the performance of machinary and environmental conditions. The System was first realized as a part of mechanization project of the Middle Anatolian Lignite Mine (OAL). Monitoring System has vital importance if coal seam is liable to spontaneous combustion. Close control of carbon monoxide (CO), which is the indication of seam fire, is only possible by monitoring system. During the application of the system several fires have been detected at early stages and all were extingueshed before. In this article technical description and the problems realized in the application of dynalink micro minos environmental monitoring system will be given brieflybecoming disaster. Also the importance of the system in controlling mine fires will be illustrated as a case study.

1 INTRODUCTION

Underground environmental monitoring system is widely used in coal mines today. The application of monitoring system provides solid advantages in terms of safety, reliability, productivity and economy (Hartman, 1982). As far as the environmental control is concerned, it is accepted that the monitoring system is more advantageous comparing to conventional measuring techniques (Michell, 1990). In this paper after brief description of OAL mine, the application and the advantages of dynalink micro minos environmental monitoring system together with some problems observed in mining practice will be explained. Being very important for OAL mine, special emphasize is given to mine fires. A real fire is given as a case study indicating the importance of monitoring system on fire prevention. Also some general criticisms are brough considering the problems seen during the application of the system.

2 GENERAL DESCRIPTION OF OAL MINE

OAL mine which is the first fully mechanized underground coal mine in Türkiye, lies in the western part of Ankara province.

A total of 500 million tons of coal reserve has been found in the basin. Two coal seams with 1.50 and 1.70 m thicknesses are separated by a rock band having average thickness of 1 m. Coal contains 30 % moisture, 27.5 % ash and 4.65 % sulphur as mean values. The lowest calorific value observed is 2800 kcal/kg. The coal extracted mostly consumed by the thermic station to produce electricity.

Company employs 2438 workers, 112 technical and 251 administrative personnel. The yearly production in 1990 and 1991 were 1.68 and 2.00 millions of coal respectively.

Upper and lower seams are mined simultaneously applying retreating longwall method.

The performance and the functioning of the underground equipment and machinary are followed continously by dynalink micro minos environmental monitoring system.

The signal from the transducers which are fitted on various equipment such as coal shearers, roadheaders, conveyors etc. and transmitted to the central station and them analysed. The line diagram of the transportation and haulage is continously followed on visual display unit.

The system also monitors carbon monoxide, methane content of mine air and airflow velocity from underground measuring station.

The mine is accepted as non-gassy mine where methane emission is very law. But the coal seam is very liable to spontaneous combustion which caused several mine fires in the past. These fires are detected in the early stages by the monitoring system and extinguished before becoming a disaster.

3 APPLICATION OF DYNALINK MICRO MINOS MONITORING SYSTEM IN OAL

This system controls and monitors equipment, machinary and environmental conditions underground to the central station located on the surface.

The system composed of four main parts, namely central station, transmission system, outstations and the transducers. The general view of the system is shown in Figure 1.

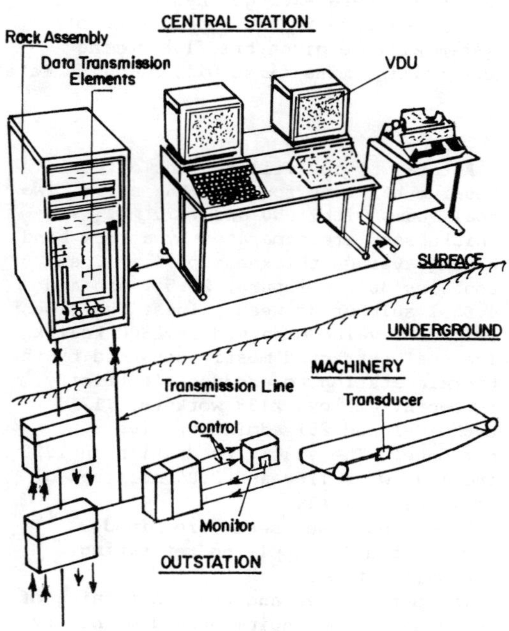

Figure 1. Elements of micro-minos *monitoring system*

The central station controls the operation of the transmission system, collects and analyses the monitored data. Station comprises a microcomputer two visual display units, a keyboard, a graphic and log printers, 50 line comminication system, two remote terminal and stand by power supply.

Dynalink is a continous two way data transmission system between the computer and the outstations which are connected in series. Outstations convert the transducer signals into data transmission signals. They contain a steel sheet enclosure providing accornodation for the intrinsically safe (IS). The number of outstations that can be connected to each transmission ring is 16 each of which has a capacity to monitor 16 digital and 8 analog input and 3 control relay. The system used in OAL has 7 rings capacity. But now only 4 rings with maximum number of 11 outstations are in use. The length of a ring cable is limited to 10 km. Figure 2 shows the ring configuration and explanation on mine plan map (Hekimoğlu, 1989).

As far as the mine environmental control is concerned airflow velocity, methane and carbon monoxide content of the mine air are monitored continously.

Velocity measurements are done by sieger model BA5 monitor which has velocity ranges of 0-2, 0-5, 0-10 and 0-20 m/s.

Methane monitoring is provided by sieger model BM1 monitor which is a flexible modular system for the continous measurements of methane.

Continous monitoring of carbon monoxide is done again by sieger model BC01 monitor which has the measuring ranges of 0-50 ppm and 0-200 ppm.

System contains 26 off outstation, 4 off BA5 sieger air velocity monitor, 4 off BC01 sieger carbon monoxide, 4 off BM1 sieger methane monitor. 23 off outstation, 10 off air velocity monitor, 10 off carbon monoxide monitor 4 off methane monitor are stored as spares.

4 ADVANTAGES AND THE CRITICISMS OF THE MONITORING SYSTEM

In this section comparison of monitoring and conventional systems will be made, the importance and the problems observed in practice during the application of the system will be discussed.

4.1 Comparison with conventional measuring technique

Monitoring system was first installed in OAL mine in 1985. Before that harmfull gases have been measured instantaneously by hand detectors or by colorimetric tubes. Air velocity measurements have also been done by the similar way using

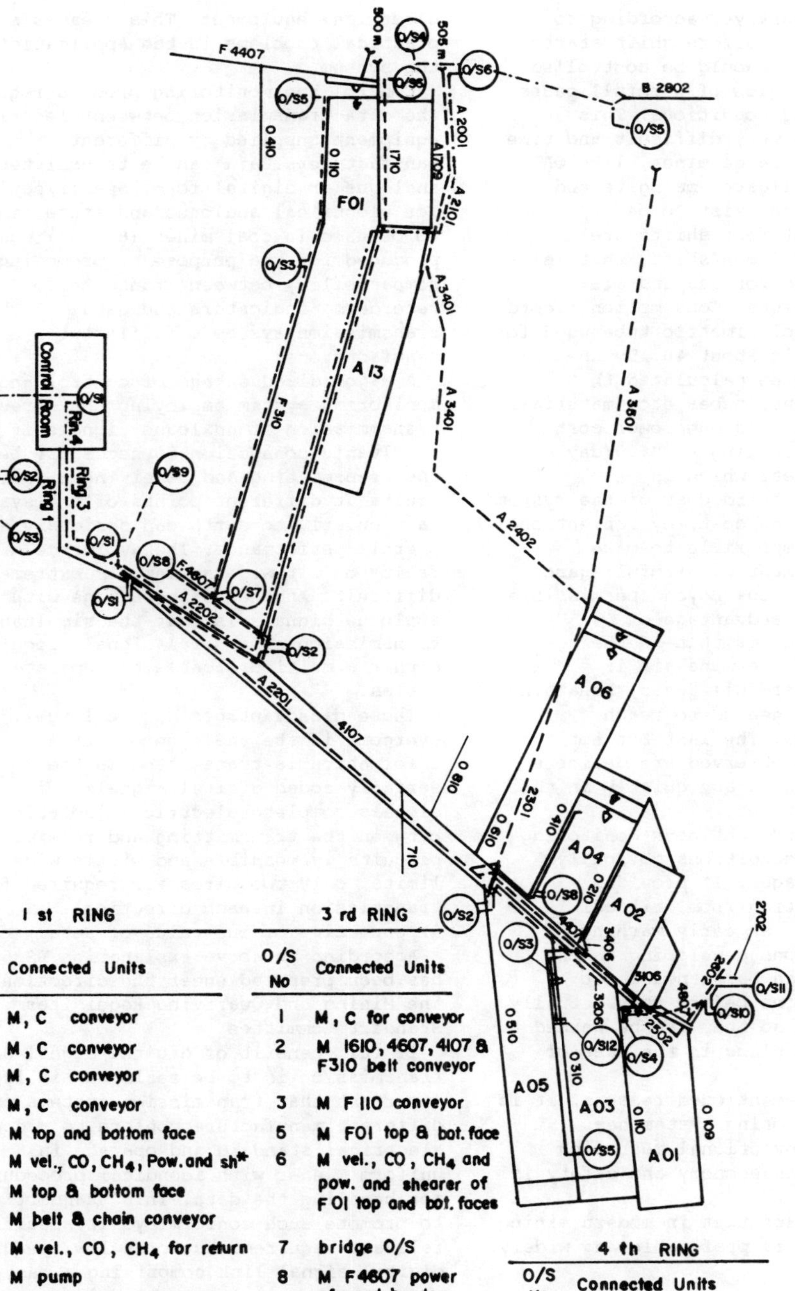

1 st RING

O/S No	Connected Units
1	M, C conveyor
2	M, C conveyor
3	M, C conveyor
4	M, C conveyor
5	M top and bottom face
6	M vel., CO, CH_4, Pow. and sh*
7	M top & bottom face
8	M belt & chain conveyor
10	M vel., CO, CH_4 for return
11	M pump
12	M belt & chain conveyor

3 rd RING

O/S No	Connected Units
1	M, C for conveyor
2	M 1610, 4607, 4107 & F310 belt conveyor
3	M F 110 conveyor
5	M F01 top & bot. face
6	M vel., CO, CH_4 and pow. and shearer of F01 top and bot. faces
7	bridge O/S
8	M F4607 power of road header
9	M power plant

4 th RING

O/S No.	Connected Units
1	M coal screen
2	M conveyor & weigher
3	M conv. & weigher
4	M pump
5	M conveyor

sh*: shearer for top & bottom face

M: Monitor

C: Control

Figure 2. OAL Ring Arrangement and explanations

anemometers. In Türkiye, according to safety regulations, before shift starts each working place should be controlled from the point of view of harmfull gases and unsafe working conditions. This procedure becomes very difficult and time consuming for the large mines, like OAL where 10 km of tailgate, maingate and development heading exist to be controlled. In OAL four shifts are applied. Assuming 3 men/shift, that is 12 men/day are needed for gas and air velocity measurements. Consumption records show that daily colorimetric tube used for gas measurements is about 40. By the company, it has been calculated that including equipment, tubes etc. material cost reaches $ 150 and men power cost reaches to $250 totaling to $400/day or simply $146.000/year which is approximately one third cost of the system itself. Besides high cost, by conventional technique it is impossible to make continous measurement of harmfull gases which become dangerous beyond permissible limits. Another disadvantage of conventional method is that in the emergency cases where mine air is contaminated by harmfull gases, breathing apperatus will be needed to reach measuring stations. The last but not the least it has been observed are delicete and may lose its accuracy quickly in the underground conditions.

On the other hand continous monitoring of environmental conditions obviously offer many advantages. It provides continous information from critical locations and give an early warning of dangerous situations developing in the mine. Personnel can be warned about dangerous situations before they actually enter the mine. Also they can be warned to leave the working place in the case of emergecy.

Due to the aforementioned reasons, it is obvious that monitoring system has superiority on conventional measuring technique as far as economy and safety is concerned.

It is a known fact that in modern mining monitoring system is preferable and widely used in practice.

4.2 General criticisms of the system

Nearly all electrical equipments used in the mine were supplied by a company (Davis of Derby Equipment). But the environmental monitoring system was purchased from another company (Hawker Siddelay Dynamic Engineering Ltd.). For this reason it is not completely compatible to the electrical equipment. This creates some practical problems in the application of the system.

Control and monitoring process require the data transmission between electrical equipment supplied by different manufacturers.Data can be transmitted in analogue or digital form. Specification for electrical analogue and state signals to be used in coal mines (BS5784) was produced for the purpose of promoting compatibility between transducers, recorders, indicators and data transmission system of different manufacturers.

A major disadvantage of control and monitoring system employing the direct transmmission of analogue signals is that a galvanic connection is necessary between the transmitting and receiving circuits, faults at different points of the system, particularly to earth can seriously effect overall performance. The avoidance of such faults on pit wide systems is extremely difficult. Another disadvantage with analogue signals is that the simultaneous transmission of several signals requires either a multicore cable or separate cables.

These disadvantages can be largely overcome if the analogue or state information is transmitted in the form of serially coded digital signals. With such signals complete electrical isolation between the transmitting and receiving circuits is possible and within wide limits, only two wires are required for transmission in each direction irrespective of the value of data.

According to above explanation BS 6556 has been prepared under the direction of the Mining and Quarrying Requisites Standart Committee.

If full benefit of digital signal data transmission is to be realised, it is important that transmission systems of different manufacture conform to a common electrical standard and operate in a uniform manner with idendical procedure for handling the data. This standart aims to promote such conformity. It specifies relevant requirements for a low speed digital signal link comprising a master and or one more slaves which interconnects the intrinsically safe circuits of other apparatus in such away that units of different manufacture can be in interchangeably coupled at the transmission line terminals.

Hawker Siddeley Dynamics Ltd. Environmental Monitoring system does not conform BS 6556. It is also very important to support service and spare parts to costumer after selling the system which can be achieved easily if standard equipment is used.

4.3 Importance of the system and problem observed in practice

It is well known that monitoring system is the most effective tool to detect a developing mine fires. Previous fires will be given as case studies to illustrate the importance of the monitoring system together with the problems involved in the application.

In the fire which occurred recently (2.7.1992), operator observed that the level of CO at the return airway of panel F01 (see Figure 1) has suddenly increased to beyond 50 ppm within few seconds. Operator telephoned to the conveyor maintenance men, working very near to fire zone, to find out the cause of this sudden increase of CO. After a short period of time, it was understood that there was a fire in the drive unit of F0110 belt conveyor. Immediately fire was informed to the manager and related units specifying the type of the fire. Fully equipped rescue team went to fire place as soon as possible, attached to the fire by means of extinguishers and within less than 1 hour the fire was out. The early detection of fire by monitoring system saved the mine and workers from further troubles. During the fire the measurement recorded by the system is shown in Table 1.

Table 1. The measurements recorded during the fire.

Time (hrs)	Carbon Monoxide (ppm)	Methane (%)
16.55	45	1.38
17.17	48	0.07
17.27	46	0.06
17.30	37	0.05
17.32	33	0.05
17.41	25	0.05
17.51	17	0.04
18.07	9	0.07
18.21	6	0.02
18.31	5	0.02
18.47	4	0.02

After the fire was completely extinguished the location effected by the fire was examined by engineers and it was observed that only 6 m. of 4 off cable line, the top belt were burned. Drive motor, the brake unit and gear box were slightly affected. It is obvious that if fire could not be detected on time the hazard would be enormous.

This case has been chosen as an example because of the emission of methane or any other gas which is monitored by the system as methane during the fire. It increased up to 1.38 % which is higher than MAC value of methane (1 %).

As it mentioned before mine is a non-gassy mine with very limited methane emission. Therefore, it is understood easily that methane release is due to the burning of the belt. But if it were a gassy mine, a question will arise how can it be decided that methane is coming from burning of some material or it is due to the sudden gas emission from strata.

Another fire has started in the caved part of the face due to the spontaneous combustion. During this fire relative humidity and temperature of return air were measured. Relative humidity was approximately 100 % and the temperature was higher than 40 ^{o}C. These values are higher than measuring ranged stated by specifications of CO monitor that are 20-90 % and 0-40 ^{o}C respectively.

Another practical difficulty observed is about range selection of CO monitor which has ranges of 0-50 and 0-200 ppm. Range selection has to be done by adjusting it from central station and from the monitor unit found at the underground measuring station. Usually 0-50 ppm range is in use. In this range, measurement error is ± ppm and alarm level is set at 10 ppm. But in the case of fire measuring range should be adjusted to 0-200 ppm. The system should provide the facility of range selection directly from central station.

Sometimes, during the fire, CO concentration exceeds 200 ppm for some period of time. Since this value is out of the measuring range, no measurements could be taken from monitor. In this period measurements were done by using conventional method.

Therefore, it is advisable to add the CO measuring range of 0-400 ppm to follow the trend of CO monitor having formation.

5 CONCLUSION

The investigation done about the application of Dynalink micro minos environmental monitoring system in OAL mine, in spite of its drawbacks, showed that monitoring system is a part of modern mining and it can not be given up.

The problems related to monitoring system and problems observed during the application of system are as follows.

- System does not conform related modern specifications such as British standards (BS 6556).
- Monitoring system is not completely compatible to electrical equipments used in the mine.
- Methane monitor should be designed or modified not to be effected by the gases liberated during the burning of belt conveyor.
- System should have the facility to adjust measuring range of carbon monoxide monitor directly from the central station.
- Increasing the measuring range of carbon monoxide monitor to 400 ppm is suggested.
- Monitors should be capable to take measurements in hot atmosphere with 100 % relative humidity.
- It has been observed that sensors are affected by the dusty atmosphere of the mine. The resistance of the sensors against mine dust should be increased. It should be kept in mind that, no matter either it is in use or kept as spare in the stock, the life of sensors is 2 years. The possibilities to increase the life of sensors should be searched.
- Addition of Carbon dioxide monitor to the system is suggested from the point of view of fire and environmental control.

REFERENCES

Hartman, H., 1984. Mine ventilation and air conditioning New York, John Wiley.
Hekimcioğlu, Z. and others, 1989. Mine monitoring system in OAL mine, Mining engineering magazine of Türkiye, v. 28, No.3, 13-22.
Mitchell, D., 1990. Mine fires, Chicago, Maclean Hunter.

Innovative Mine Design for the 21st Century, Bawden & Archibald (eds) © 1993 Balkema, Rotterdam, ISBN 90 5410 325 6

Analysis of underground ventilation networks utilizing the skyline nodal pressure method

K. Sasaki, Y. Fukuda & H. Miyakoshi
Akita University, Japan

T. Onozuka
Hanaoka Mining Co. Ltd, Japan

ABSTRACT: A calculation system for the analysis of underground ventilation networks by means of a nodal pressure method using the skyline modified Choleski's decomposition scheme, named 'skyline nodal pressure method', has been studied. The generalized linear equations for solving steady-state ventilation airflow have been developed using a connective function between nodes. The skyline decomposition scheme has an advantage on rapid convergence and the sake of computer memory space as compared with other schemes. Five examples of ventilation networks have been calculated using a popular personal computer. A convergence criterion, that the maximum error of the summed airflow-quantity in continuity at each node is less than 0.1 m^3/min, was used for the calculations. Under the conditions that the fan-pressure is given as a constant value and that the acceleration factor is set at 1.4, the number of the iterations until convergence was not more than 13 for each network. The calculation processing time by the skyline nodal method is roughly in proportion to the total number of nodes to the 3/2 power.

1 INTRODUCTION

In recent years, many reports have been released with respect to analytical systems for underground ventilation networks to predict flow-quantity and air conditions related to secured maintenance of mines using small computers (Wallace and Brunner, 1989). The authors have also developed a calculation code *'MIVENA'* for underground metal mines (Sasaki et al., 1990, 1992). Wang et al.(1985) give an excellent review of the ventilation network theory.

Bhamidipati & Procarione (1985) have compared two well-known methods, the mesh flow-rate and the nodal pressure (head) methods. They have suggested that the nodal pressure method is a superior process due to its simple approach for computer implementation which does not require complex procedures such as the selection of meshes used in the mesh flow method.

The nodal pressure method is based on an approximated linear calculation system incorporating nodal pressures as primary variables. The nodal pressures have to be improved by iterative calculations in order to satisfy Kirchhoff's 1st(current) law.

The nodal methods previously proposed are classified into two types. The first is a conventional method used to improve each of the nodal pressures by a simple equation which incorporates neighboring nodal pressures. Inoue et al.(1989) have recently applied this method to mine ventilation analysis. This method, hereinafter referred to as the 'conventional nodal method', has the merit of requiring extremely small memory space for data and computer programs. However, its convergence process becomes unstable and the processing time increases as the number of nodes increases. This means that attention has to be paid to the convergence process.

The second type is an analytical method which uses a direct matrix operation to solve the simultaneous linear equations. This method, hereinafter referred to as the 'matrix nodal method', has an advantage of achieving solutions with a stable convergent process by handling whole nodal pressures. However, it requires comparatively large working memory space. Bhamidipati & Procarione (1985) have analyzed mine ventilation networks using the matrix nodal method. This method uses the modified Choleski's decomposition scheme(hereinafter referred to as a 'MCD') with a band matrix to solve an admittance matrix as well as the pipe network flows(Isaacs & Mills, 1980).

They have formulated equations related to nodal pressures (not correction factors) as unknown variables. Sasaki et al. (1990) have reported a matrix nodal method used to solve mid-size networks (<200 nodes) by applying the MCD scheme. This method has an exceptional memory usage with a popular personal computer (Melosh et al. 1967, Melosh & Bamford 1969).

The purpose of this study is to introduce a nodal pressure method with rapid calculation, stable convergent process and smaller memory space for the analysis of middle or large size of ventilation networks.

In this study, the fundamental equations of the nodal matrix method are based on linear equations which consider the correction factors for nodal pressure as unknown variables. By introducing a function which expresses the connections between nodes, the linear equations are provided in a more general form and the iterative calculation schemes are explicitly shown.

The calculation schemes used to solve the admittance matrix of the linear equations are based on four kinds of schemes: *a)* original MCD, *b)* band matrix MCD, *c)* skyline MCD and *d)* ICCG scheme. The skyline MCD scheme is an improved MCD method in limiting calculation and memory space of the matrix components by using one dimensional memories. The ICCG scheme refers to incomplete Choleski's decomposition and iterative calculations using the conjugate gradient scheme, which requires relatively small memory space for large scale networks. When the skyline scheme was adapted, the calculation speed was accelerated at least 6 times as compared to that of the original MCD scheme for middle size networks (≈140 nodes).

Two illustrative small ventilation networks and three actual ventilation networks of underground metal mines were developed to evaluate the four schemes. This study permitted a comparative analysis concerning the number of iterations, the calculation speed and the amount of required memory space for the four schemes.

2 ANALYTICAL METHOD

2.1 Definition of symbols and fundamental flow equations

The underground ventilation network dealt with in this study is consisted of *a)* airways, *b)* nodes connecting airways and *c)* junctions connecting some airways to the atmosphere. The junctions of *c)* (intake portals and fan inlets) are named 'boundary nodes'. Ducting or diffusers from exhaust fans are omitted in this study because these have negligible resistances as compared to mine airways.

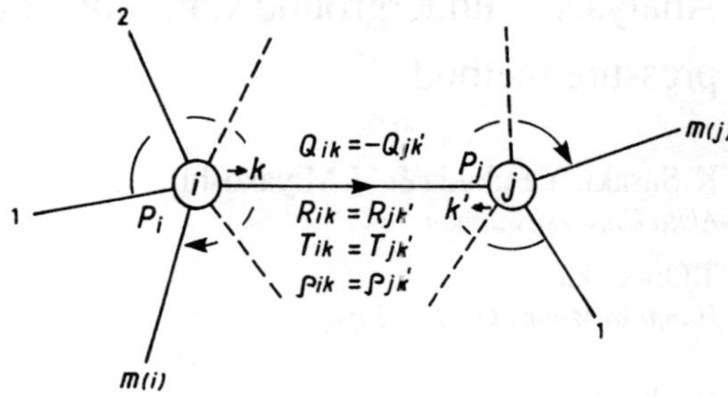

Fig. 1 Schematic diagram indicating connective relations between nodes

Consider a network consisting of a total number of nodes N and let a nodal number be i (=1~N). Figure 1 shows a schematic configuration representing airways connecting two nodes, i and j. The total number of airways connected to node i is represented by $m(i)$. The subscript $k(=1\sim m(i))$ is used to identify a connected airway or node. The nodal numbers connected to node i are defined using with a function denoted by $J(i,k)$. For example, when the nodal number of the kth node connected to node i is j (see Fig.1), it is defined as,

$$j = J(i,k) \tag{1}$$

Thus all connective relations between nodes are represented by $J(i,k)$. This concept is similar to the connectivity matrix in the finite-element method (Reddy 1984). The use of $J(i,k)$ is also more effective to save computer memory space as compared with the incidence matrix used in the mesh flow method. While only $2N\cdot m(i)_{max}$ of memory space is used in the case of $J(i,k)$, N^2 of memory space is required for the incidence matrix (Wang 1982). For N=100 and $m(i)_{max}$ = 5, the ratio $2N\cdot m(i)_{max}/N^2$ becomes 0.1. Furthermore, the authors have developed a relational data management system to assist the development of input data. This system follows the procedure developed by Ng (1989). This management system has convenient functions used to delete or add nodes and airways by reevaluating $J(i,k)$ and $m(i)$.

The nodal pressure at node i, p_i(Pa), is defined in this study as the static pressure induced by fans and natural ventilation not including atmospheric static pressure due to elevation difference and airflow dynamic pressure. This is because the elevation difference does not induce airflow and the dynamic pressure is considerably small as compared to the static pressure. Furthermore let the thermal natural ventilation pressure upon an airway from i to j be $\Delta\Phi_{i-j}$(Pa). Then $\Delta\Phi_{i-j}$ is given by an integral value along the airway as,

$$\Delta\Phi_{i-j} = -\int_{h_i}^{h_j} g\cdot[\ \rho(h,\Theta,P_{a0})-\rho_0(h,\Theta_0,P_{a0})]\,dh \tag{2}$$

where h(m) is the height from the datum level, g(m/s^2) is the gravitational acceleration, ρ(kg/m^3) is the air density at each airway and ρ_0(kg/m^3) is the density of the atmosphere.

The total pressure difference, denoted by ΔH_{i-j} or ΔH_{ik} (Pa), to drive airflow from i to j is given by,

$$\Delta H_{i-j} = \Delta H_{ik} = p_i - p_j + \Delta\Phi_{i-j} \tag{3}$$

If the differences in level and density are given by $\Delta h_{ik}(=h_i-h_j)$ and $\Delta\rho_{ik}$, respectively, then $\Delta\Phi_{ik}(=\Delta\Phi_{i-j})$ can be approximated as,

$$\Delta\Phi_{ik} = g\cdot(\rho_{ik}-\rho_{0ik})(h_i - h_{J(i,k)}) = g\cdot\Delta\rho_{ik}.\Delta h_{ik} \tag{4}$$

where ρ_{ik}(kg/m^3) is the average airflow density upon the airway ($i\rightarrow J(i,k)$) and ρ_{0ik} (kg/m^3) is the atmospheric density at h_{ik} [$=(h_i+h_j)/2$]. The airflow quantity, Q_{ik} (m^3/s), is given by

$$\begin{aligned} Q_{ik} &= \delta_{ik}|\Delta H_{ik}/R_{ik}|^{1/n} \\ &= \delta_{ik}|p_i-p_{J(i,k)}+g\cdot\Delta\rho_{ik}.\Delta h_{ik}|^{1/n} \end{aligned} \tag{5}$$

where R_{ik}(Pa/(m^3/s)^n) is the resistance of the airway, and δ_{ik} is the sign expressing the flow direction as follows:

$$\delta_{ik} = \begin{cases} +1 : \Delta H_{ik} \geq 0 \\ -1 : \Delta H_{ik} < 0 \end{cases} \tag{6}$$

Thus Q_{ik} is positive when the airflow is directed from i to $j=J(i,k)$. The index n in Eq.(5) is usually equal to 2 for fully developed turbulent flow (n=2 was used in this study).

In addition, $\Delta\rho_{ik}$ in Eq.(5) is given by,

$$\Delta\rho_{ik} = 1.293\ \frac{P_{a0}\cdot\exp(-h_{ik}/8620)}{101.3} \times\frac{\Theta_{ik}-\Theta_0}{273+(\Theta_{ik}+\Theta_0)/2} \tag{7}$$

where Θ(°C) is the airflow temperature, Θ_0(°C) is the atmospheric temperature and P_{a0}(kPa) is the absolute pressure at the datum elevation (= 101.3 kPa at h_{ik}=0 m).

2.2 Linear equations for correction factors of nodal pressure

The linearized flow admittance, T_{ik}(m^3/s·Pa), between ΔH_i and Q_{ik} is defined as,

$$T_{ik} = (R_{ik}^{1/n}\cdot|\Delta H_{ik}|^{(n-1)/n})^{-1} \tag{8}$$

Equation(5) can thus be rewritten as,

$$Q_{ik} = T_{ik}\cdot\Delta H_{ik} = T_{ik}\cdot(p_i-p_{J(i,k)}+g\cdot\Delta\rho_{ik}.\Delta h_{ik}) \tag{9}$$

This equation provides the approximated linear relationship between Q_{ik} and ΔH_{ik}. When $|\Delta H_{ik}|$ in Eq.(8) is less than 10^{-10}(Pa), the calculation is treated as $|\Delta H_{ik}|=10^{-10}$(Pa) for the sake of the computer implementation.

According to the Kirchhoff's 1st law, the sum of the flows, Q_{ik}, flowed into node i, provided by Eq.(9), is given by,

$$\begin{aligned} q_i &= \sum_{k=1}^{m(i)} Q_{ik} = (\sum_{k=1}^{m(i)} T_{ik})\cdot p_i \\ &- \sum_{k=1}^{m(i)}(T_{ik}\cdot p_{J(i,k)}) + \sum_{k=1}^{m(i)}(g\cdot\Delta\rho_{ik}.\Delta h_{ik}) \end{aligned} \tag{10}$$

where q_i(m^3/s) is the air quantity supplied into the node i from external air-sources such as a compressed air line which is independent of airflow from intakes. In general, q_i is given as 0 on each node.

The improved nodal pressures of the ($M+1$)th iteration, $\{p_i\}^{M+1}$, are obtained from the Mth nodal pressures $\{p_i\}^M$, and the Mth correction factors $\{\Delta p_i\}^M$(Pa):

$$\{p_i\}^{M+1} = \{p_i\}^M + \{\Delta p_i\}^M \tag{11}$$

By substituting Eq.(11) into $\{p_i\}$ in Eq.(10), the following equation is obtained (Sasaki et al., 1990),

$$[A_{ij}]\{\Delta p_i\}^M = \{B_i\}^M \tag{12}$$

where the admittance matrix $[A_{ij}]$, of order $N\times N$, and the vector $\{B_i\}$, of order N, are expressed as,

$$\left.\begin{aligned} A_{ii} &= \sum_{k=1}^{m(i)} T_{ik} \quad (\ i=1\sim N\ ) \\ A_{ij} &= -\ T_{ik} \quad (\ j=J(i,k)\ ) \\ A_{ij} &= 0 \quad (\ j\neq J(i,k)\ ) \\ B_i &= q_i - \sum_{k=1}^{m(i)} Q_{ik} \end{aligned}\right\} \tag{13}$$

Equation(12) expresses the simultaneous linear equations relating to $\{\Delta p_i\}^M$ as unknown values. If $\{p_i\}^M$ are treated as unknown values as suggested by Bhamidipati et al.(1985), a large amount of double precision variables are needed for the computer implementation (refer to Sasaki et al., 1990).

Assuming that $j=J(i,k)$, defined by Eq.(1), and concurrently that the k'th node connected to node j in opposite side is defined as $i = J(j,k')$ (see Fig. 1), then the relation between A_{ij} and A_{ji} is given as,

$$A_{ij} = A_{iJ(i,k)} = -T_{ik} = -T_{jk'} = A_{jJ(j,k')} = A_{ji} \quad (14)$$

Above Eq.(14) shows that $[A_{ij}]$ is a symmetric matrix (Rao 1987). The vector $\{B_i\}$ in Eqs.(13) is composed of the errors associated with the continuity of flow quantity at each node i after the Mth iterative calculation. Equation (12) takes the structure that $\{p_i\}^{M+1}$ converges, i.e. $\{\Delta p_i\}^M \approx \{0\}$, as the errors of flow quantity at whole nodes decrease.

Assume that the total number of the boundary nodes is ΔN and that the nodal number at one of the boundary nodes is τ. Since ΔN pieces of inflow and exhaust air-quantities, q_τ(m^3/s), at the boundary nodes are unknown, the ranking of Eq.(12) is $N-\Delta N$. Thus $\{\Delta p_i\}^M$ cannot be obtained by directly solving Eq.(12). At these boundary nodes, however, the nodal pressures, $\{P_\tau\}^M$, are given and $\{\Delta p_\tau\}^M$ are set as 0, i.e.

$$\left.\begin{array}{l} A_{\tau\tau} = 1 \\ B_\tau = 0 \\ A_{\tau j} = 0 \quad (j \neq \tau) \end{array}\right\} \quad (15)$$

For example, the pressure at an intake-portal node is given as 0, and the pressure at a fan-inlet node is determined on the fan characteristic curve. The ranking of Eq.(12) increases by one, however, $[A_{ij}]$ in Eq.(12) is no longer a symmetric matrix. In order to convert it to a symmetric matrix, conversions of $[A_{ij}]$, given by,

$$A_{i\tau} = 0 \quad (i \neq \tau) \quad (16)$$

are used, and the result is expressed as follows,

$$\begin{bmatrix} A_{11} & \cdots & 0 & \cdots & A_{1N} \\ \cdot & \cdots & \cdot & \cdots & \cdot \\ \cdot & \cdots & 0 & \cdots & \cdot \\ 0 & \cdots 0 & 1 & 0 \cdots & 0 \\ \cdot & \cdots & 0 & \cdots & \cdot \\ \cdot & \cdots & \cdot & \cdots & \cdot \\ A_{N1} & \cdots & 0 & \cdots & A_{NN} \end{bmatrix} \begin{Bmatrix} \Delta p_1 \\ \cdot \\ \cdot \\ \Delta p_\tau \\ \cdot \\ \cdot \\ \Delta p_N \end{Bmatrix} = \begin{Bmatrix} B_1 \\ \cdot \\ \cdot \\ 0 \\ \cdot \\ \cdot \\ B_N \end{Bmatrix} \quad (17)$$

(column τ marked above the matrix)

If the all pressures of the boundary nodes are settled, the ranking of Eq.(17) should be N.

2.3 Skyline modified Choleski's decomposition scheme

In solving large-scale simultaneous linear equations using a personal computer, attention should be paid to calculation speed and effective use of memory space as com-pared with the conventional nodal method, because the direct matrix operation needs a comparably large memory space for the case of large-scale networks.

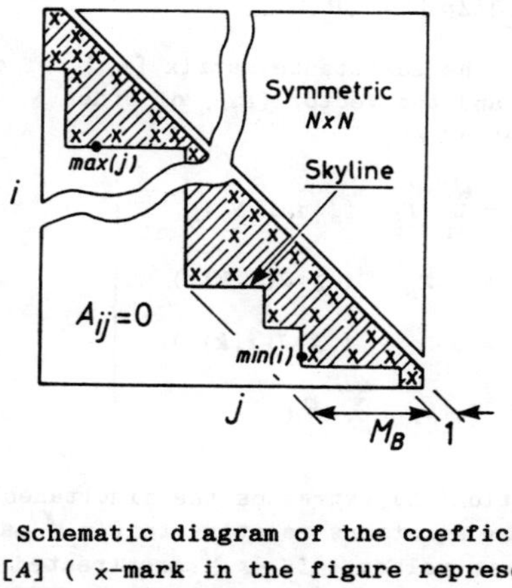

Fig. 2 Schematic diagram of the coefficient matrix [A] (x-mark in the figure representing the matrix elements of $A_{ij} \neq 0$)

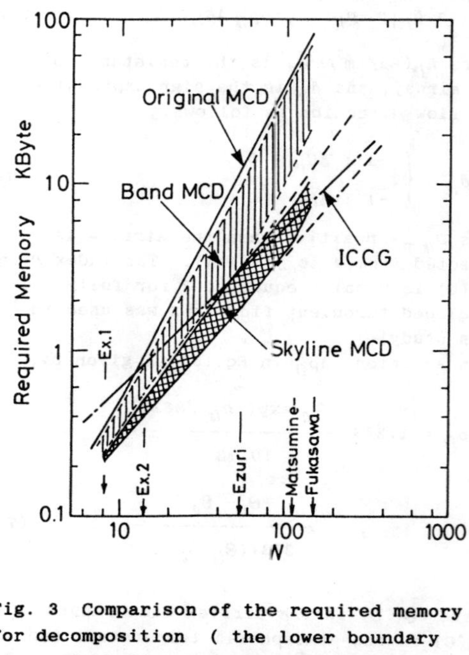

Fig. 3 Comparison of the required memory for decomposition (the lower boundary lines of skyline and band MCD shows the case of the optimal nodal numbers)

In general, since it is relatively easy with the finite-element method to make a half band width be small, only a band matrix with elements that are nonzero in the neighborhood of diagonal elements is considered to save computer memory (Collins, 1973). However in the case of ventilation networks, there is a possibility of an extreme case in which the number difference between connected nodes is almost N (see Table 1). When the continuous arrayed memory is restricted within an extent of memory space, the termination and the numbering of nodes are considered not to exceed the limited width.

Four decomposition schemes of the symmetric admittance matrix were compared in this study. Furthermore the calculations of the decomposition are restricted to the regime containing nonzero elements in $[A_{ij}]$, as illustrated in Fig. 2. The minimum column number of each row, $j_{min}(i)$, and the maximum row number of each column, $i_{max}(j)$, are defined to express the boundary elements of the nonzero regime. For the case of Fukasawa Mine (N=141, Fig. 4(d)), the occupied area ratio of the nonzero regime is approximately 17% in the whole $[A_{ij}]$, and it should be smaller for a case of larger N. Thus, calculation speed is particularly enhanced by restricting numerical calculations to the regime in the under triangular matrix represented by,

$$A_{ij} \; ; \quad i = j \sim i_{max}(j),\; j = j_{min}(i) \sim i \qquad (18)$$

This method is usually called as the skyline method or the wave front method (Melosh & Bamford 1967). The one dimensional arrayed memory for the nonzero regime, expressed by Eq.(18), is used due to the economy of memory space. Figure 3 shows a comparison of the memory quantity required for the matrix operation as function of N in each scheme. The skyline MCD scheme is favorable for N < 500.

2.4 Fan characteristics

When a main fan characteristic exhibiting the relation between the air quantity Q_f (m^3/s) and the fan-inlet static pressure P_f(>0, Pa) at a standard rotational speed n_0(rpm) is approximated with a Γth degree polynomial, obtained by the least squares method, the relation is provided as,

$$P_f(Q_f, n_0) = \sum_{\gamma=0}^{\Gamma} C_\gamma \cdot Q_f^{\gamma} \qquad (19)$$

where C_γ(Pa·s/m^3) is the γth order coefficient. Since a fan can be controlled by changing its rotational speed, P_f on a driving rotational speed n_d(rpm) is given as

$$P_f(Q_f, n_d) = \beta^2 \cdot \sum_{\gamma=0}^{\Gamma} C_\gamma \cdot (Q_f/\beta)^{\gamma} \qquad (20)$$

where, $\beta = n_d/n_0$ (Γ=3 was used in this study).

2.5 Initial values of nodal pressures

The initial values of the whole nodal pressures, defined as $\{p_i\}^0$, were given as a same constant pressure,

$$p_i = -K_0 \cdot P_f(0, n_d) \qquad (i \neq \tau) \qquad (21)$$

where K_0 is an initial value factor and $P_f(0,n_d)$ is the fan inlet pressure at Q_f =0, obtained from Eq.(20). On the other hand, initial values upon boundary nodes are given by,

$$\left.\begin{aligned} p_{\tau 0} &= -P_f(0, n_d) \ \text{at fan-inlet nodes} \\ p_{\tau 0} &= 0 \quad \text{at intake-portal nodes} \end{aligned}\right\} \qquad (22)$$

2.6 Improvements of fan pressure

The flow quantity Q_f and the static pressure P_f at the fan inlet are improved in accordance with the following method.

With the flow quantity Q_f^M calculated from the Mth iterative nodal pressures $\{p_i\}^M$ and fan inlet pressure P_f^M, the equivalent admittance of flow T_f^M between Q_f^M and P_f^M is defined as,

$$T_f^M = P_f^M / Q_f^M \qquad (23)$$

When the improved Q_f is expressed as Q_f^*, the $P_f(Q_f^*, n_d)$ determined from the Eq.(20) is equal to the total pressure loss of the network $T_f^M \cdot Q_f^*$. Thus an algebraic equation concerning Q_f^* is obtained as follows:

$$T_f^M \cdot Q_f^* - \beta^2 \cdot \sum_{\gamma=0}^{\Gamma} C_\gamma \cdot (Q_f^*/\beta)^{\gamma} = 0 \qquad (24)$$

The Q_f^* is determined by Newton-Raphson's method. Finally, the $(M+1)$th improved P_f^{M+1} can be obtained by substituting Q_f^* to Eq.(20). The convergence of P_f synchronizes with the convergence of $\{p_i\}$. The nodal pressure p_τ^{M+1} at the fan-inlet node τ for the $(M+1)$th iteration is given as,

$$p_\tau^{M+1} = -P_f^{M+1} \qquad (25)$$

In the case that a principal fan is

installed in the network, the whole nodal pressures, p_i^M, are corrected by,

$$p_i^{(M+1/2)} = p_i^M \cdot P_f^{M+1} / P_f^M \quad (i \neq \tau) \qquad (26)$$

where $p_i^{(M+1/2)}$ is the adjusted nodal pressure and becomes the approximated value for the sake of the $(M+1)$th iteration. Equation (26) is a simple method to adjust the nodal pressures due to the influence of the fan pressure with an assumption that the flow between an individual nodal pressure and the fan node pressure is approximately constant.

Booster fans installed in series to provide the required air quantity are treated by the conventional method used by Sasaki et al. (1990). Consider the approximate resistance based on pressure loss near the booster fan, predicted from its pressure ΔP_f(Pa) and flow quantity ΔQ_f (m^3/sec), defined as,

$$\Delta R_f = \Delta P_f / (\Delta Q_f)^2 \qquad (27)$$

Then ΔR_f and $\pm \Delta P_f$ are added to the terms of R_{ik} and ΔH_{ik} respectively. The sign of ΔP_f is decided taking into consideration of the flow direction. The airway flow rate with the booster fan whose flow direction is set as $i \to j$, is given by,

$$Q_{ik} = \delta_{ik} \left| \frac{\Delta H_{ik} + \Delta P_f}{R_{ik} + \Delta R_f} \right|^{1/n} \quad ; \; j = J(i,k)$$

$$Q_{jk'} = \delta_{ik} \left| \frac{\Delta H_{jk'} - \Delta P_f}{R_{jk'} + \Delta R_f} \right|^{1/n} \quad ; \; i = J(j,k') \qquad (28)$$

2.7 Convergence acceleration factor

In order to accelerate the convergence, an acceleration factor ω is effective(Inoue et al., 1989). The Mth solution $\{p_i\}^M$ is given using ω as follows,

$$\{p_i\}^M = \{p_i\}^{M-1} + \omega \{\Delta p_i\}^{M-1} \qquad (29)$$

However if ω is too large, the convergent process becomes unstable. Taking into account such a situation, the optimum value of ω was decided by preliminary calculations. The fastest convergence has been obtained with most of the networks dealt with in this study, when ω was set at 1.4 (refer to §3.1).

2.8 Convergence criterion

The calculation processing time until convergence is varied depending on the convergence criterion used. In this study the error quantity in continuity upon individual nodes, denoted as ΔQ_i(m^3/s), was calculated as (Bhamidipati(1985) and Sasaki et al.(1990)),

$$\Delta Q_i = q_i - \sum_{k=1}^{m(i)} Q_{ik} \qquad (30)$$

The maximum absolute value of ΔQ_i denoted by $|\Delta Q_i|_{max}$ could be regarded as a reasonable convergence criterion. The minimum limit of $|\Delta Q_i|_{max}$ used to verify convergence should be set in accordance with the amount of total inflow quantity in each network. However, in this study a fixed convergence criterion

$$60 \times |\Delta Q_i|_{max} < 0.1 \text{ m}^3/\text{min} \qquad (31)$$

was applied to all networks, and the number of iterations until satisfying Eq.(31), denoted by M_c, was obtained. The criterion of Eq.(31) is believed to be a relatively strict one compared with 6 m^3/min used by Bhamidipati et al.(1985).

2.9 Procedures of iterative calculations

The solution can be obtained by the following iterative calculation procedure:

1) Read the network data and fan characteristics, and calculate natural ventilation pressures, $\Delta \Phi_{ik}$.

2) Set the initial pressures, $\{p_i\}^0$, using Eqs.(21) and (22).

3) Find the fan operating point Q_f and P_f by Eqs.(24). (in case of one main fan, adjust p_i using Eq.(26))

4) Calculate T_{ik} and derive $[A]$ and $\{B\}$ using Eqs. (8) & (13).

5) Set the boundary nodal pressures and adjust $[A]$ and $\{B\}$ using Eqs.(15) & (16) as Eq.(17).

6) Compute the solution $\{\Delta p_i\}$ using a decomposition scheme and correct $\{p_i\}$ by Eq.(29), Q_{ik} using Eq.(9) and ΔQ_i using Eq.(30).

7) Iterate calculations in 3)~ 6) until the solution satisfies the convergence criterion in Eq.(31).

8) Save or print out the solutions (p_i and Q_{ik}).

2.10 Examples of ventilation networks

Two illustrative networks and three actual underground ventilation networks, which are trackless copper mines (the Ezuri Mine, the Matsumine Mine and the Fukasawa Mine) run by the Hanaoka Mining Co. LTD., were developed to evaluate the developed method. Fig. 4(a)-(e) represent the ventilation networks

Ex. 1

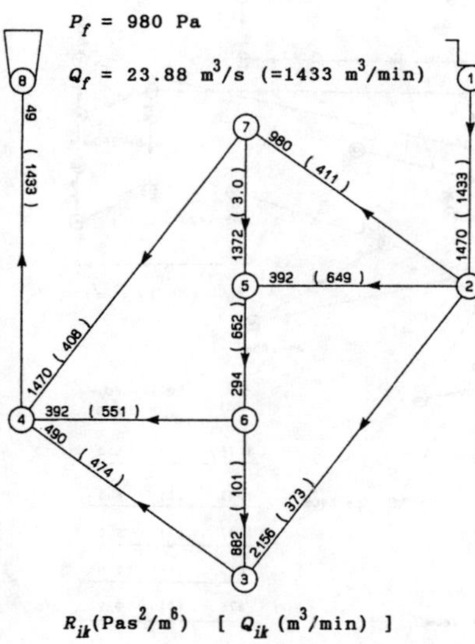

(a)

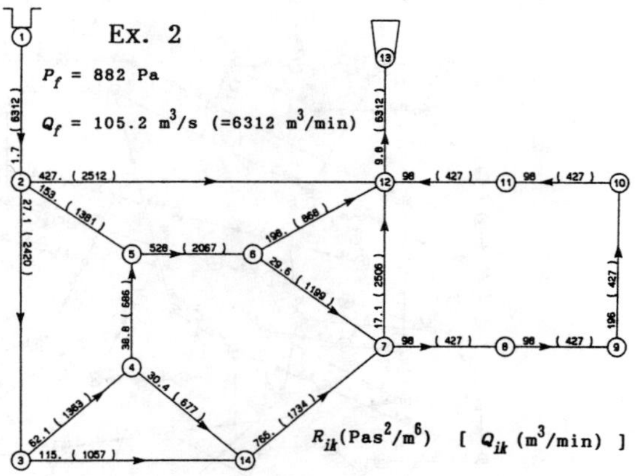

(b)

analyzed. Placement of nodes in the actual networks was determined in consideration of airway junctions and types of the airways. The nodal numbers were manually given. Table 1 gives the total number of nodes N, the total number of airways $L(=\Sigma m(i)/2)$, the half band width M_B (see Fig. 2), the average number difference between connected nodes (denoted by ΔI), the fan characteristics and air quantities for the developed examples. The ΔI is defined as follows:

$$\Delta I = \sum_{i=1}^{N} \{ \sum_{k=1}^{m(i)} |i-J(i,k)| \} \Big/ \sum_{i=1}^{N} m(i) \qquad (32)$$

In the case of the skyline MCD and ICCG schemes, the calculation speed is affected by ΔI. The networks indicated by '*' in Table 1 represent the cases with optimal distribution of nodal numbers to reduce ΔI and the half band width by means of the optimization method (Sasaki et al.,1990).

The characteristic curves of the surface main fans are illustrated in Fig. 5. With respect to the analyses of the three actual networks, the airflow temperatures upon individual airways are incorporated in the calculation as input data and the term $\Delta\Phi_{ik}$ was considered in the numerical calculations. The airflow temperatures(Θ_{ik}) were determined using a forecast analytical system for temperature and humidity developed by authors (see Sasaki et al. 1992).

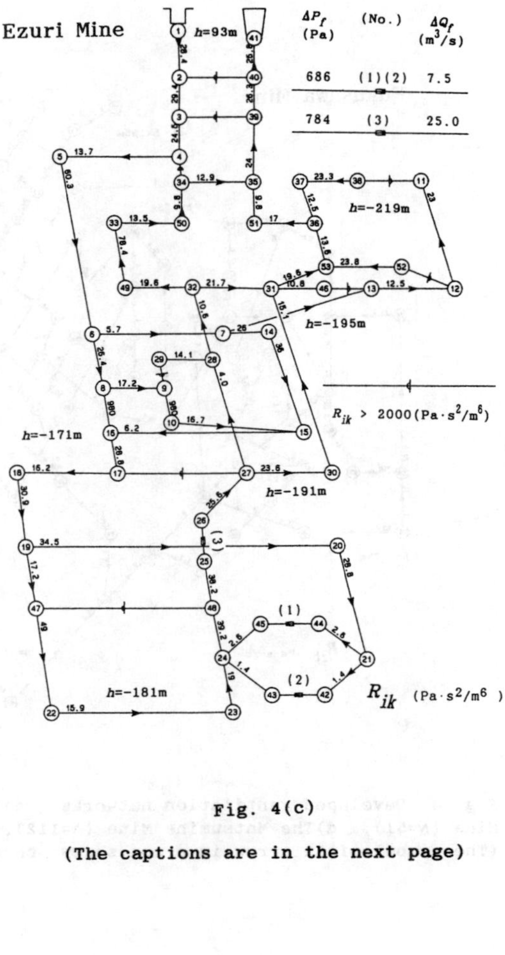

Fig. 4(c)

(The captions are in the next page)

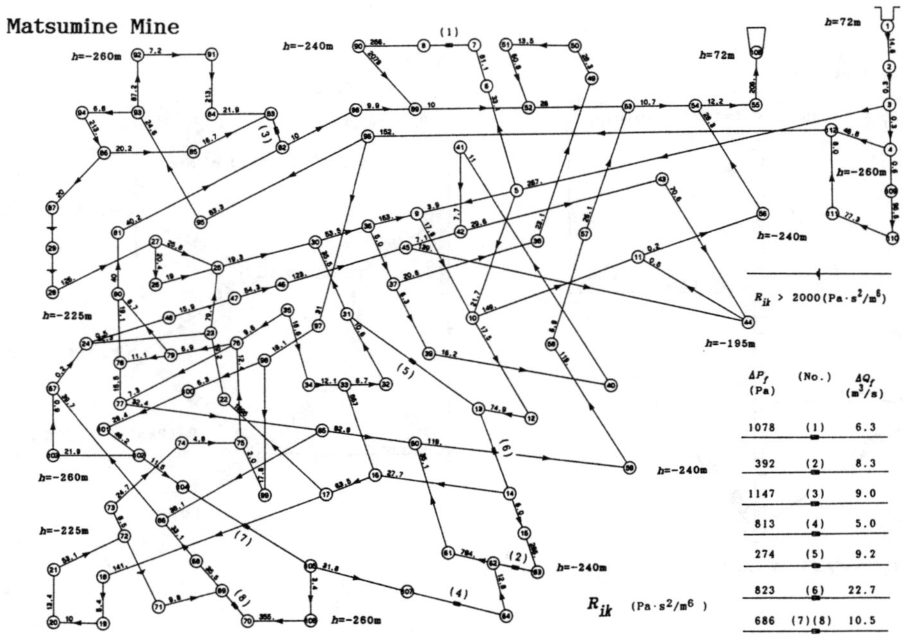

(d)

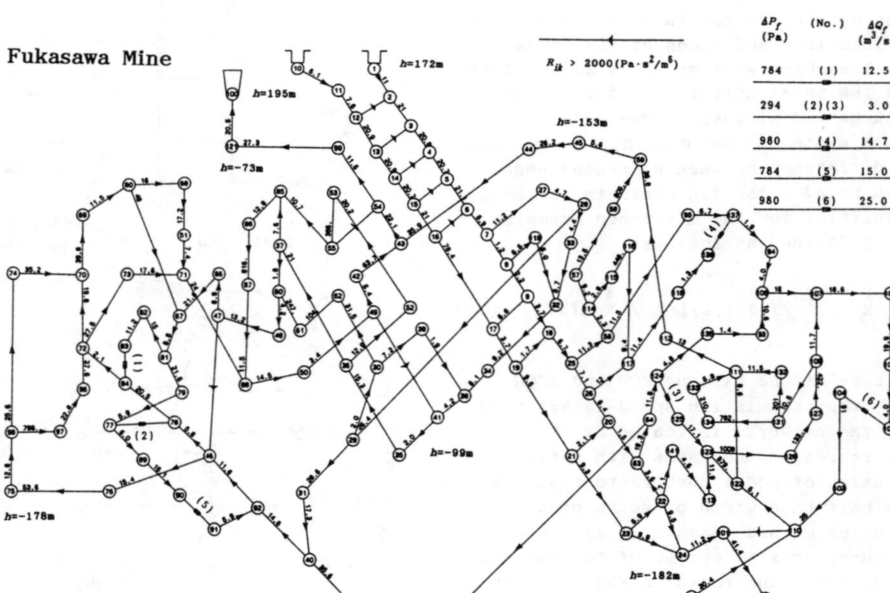

(e)

Fig. 4 Developed ventilation networks, a)Example 1 (N=8), b)Example 2 (N=14), c)The Ezuri Mine (N=51), d)The Matsumine Mine (N=112), e)The Fukasawa Mine(N=141)
(The symbol of an arrow in a box shows booster fans)

Table 1 Characteristics of the ventilation networks

Network	N	L	M_B+1	ΔI	P_f	$60\cdot Q_f$
	–	–	–	–	Pa	m^3/min
Example 1	8	11	6	3.3	980	1433
*			4	1.7		
Example 2	14	19	12	3.4	882	6312
*			7	2.2		
Ezuri Mine	51	67	41	9.6	≈880	≈3900
*			11	2.9		
Matsumine Mine	112	137	109	8.4	≈1580	≈4900
*			26	6.5		
Fukasawa Mine	141	179	120	13.0	≈910	≈5200
*			35	5.2		

(* ; the cases with the optimized nodal numbers)

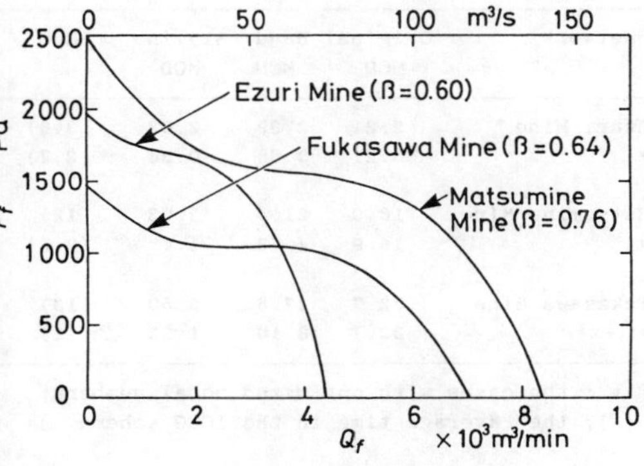

Fig. 5 Characteristics of the main surface fans in the three actual mine networks

3 RESULTS AND DISCUSSIONS

The solutions of Q_{ik} for Ex. 1 and Ex. 2 are showed in Figs. 4(a) and 4(b). The calculated airflow rate agree well with measured values in the actual networks within ±18% of the total intake airflow-quantities. It should be noted that the uncertainty of the measurements was estimated to be more than 20% because the mines were running at the time of measurements.

In this study, calculations were performed using the matrix nodal method and a personal computer. The program was written in the Microsoft BASIC Ver.7.1™.

3.1 Convergence acceleration and initial-value factors

Figure 6 shows the value of $|\Delta Q_i|_{max}$ and the maximum absolute value of Δp_i, $|\Delta p_i|_{max}$, plotted against respective iterations of Mth degree for the Fukasawa Mine. It is noticed that both $|\Delta p_i|_{max}$ and $|\Delta Q_i|_{max}$ are reduced exponentially with M. This means that the calculation system is reasonable and its convergence process is very stable. In addition when the convergence criterion expressed by Eq.(31) is satisfied, $|\Delta p_i|_{max}$, is in the range of 10^{-3}~10^{-2}(Pa).

The effects of ω and K_0 are shown in Figs. 7 and 8 respectively. M_c decreases as ω becomes larger as long as the value of ω remains less than 1.4. It is also noticed that the solutions converge with almost the same M_c for any K_0 given in the range of 0.1~0.9 at ω=1.4. Thus the convergence process is almost not influenced by K_0.

When ω=1.4, K_0=0.3 and the fan pressure is given as a constant value, M_c is no more than 13 for each of the networks analyzed.

3.2 Effect of the fan characteristics on the number of iterations

Table 2 shows the variation of M_c for the following conditions

i) the fan pressure P_f is kept constant

ii) the fan characteristic with the operation point unsettled is given as $P_f(Q_f)$

iii) the individual nodal pressure is adjusted using Eq.(26) simultaneously with the $P_f(Q_f)$ in case of *ii)*.

In case *i)*, the convergence has been completed with $M_c \le 13$. However in case *ii)*,

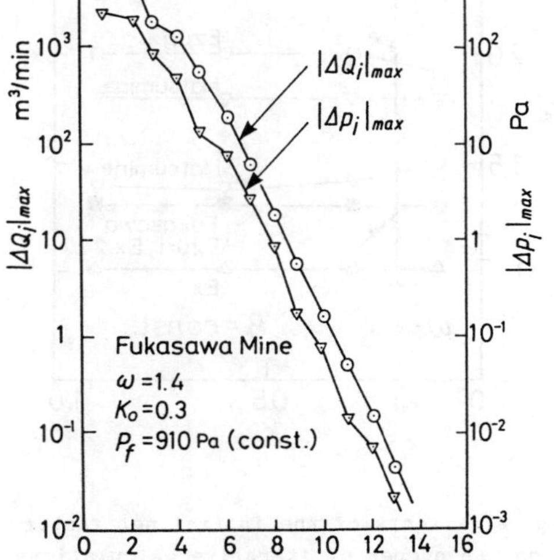

Fig. 6 A convergence process (the Fukasawa Mine)

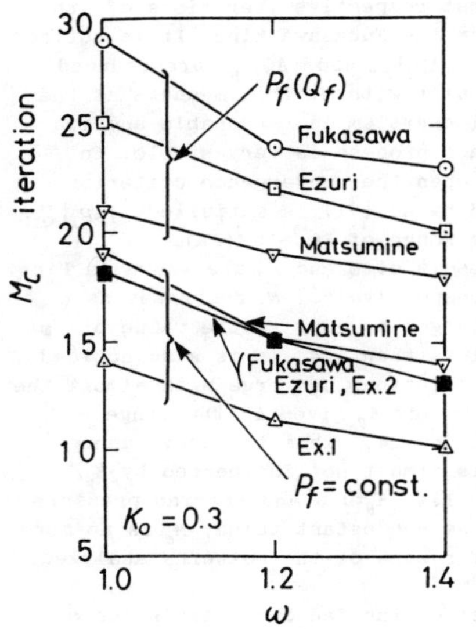

Fig. 7 Effects of the convergence acceleration factor ω on the number of iterative calculations until convergence (M_c)

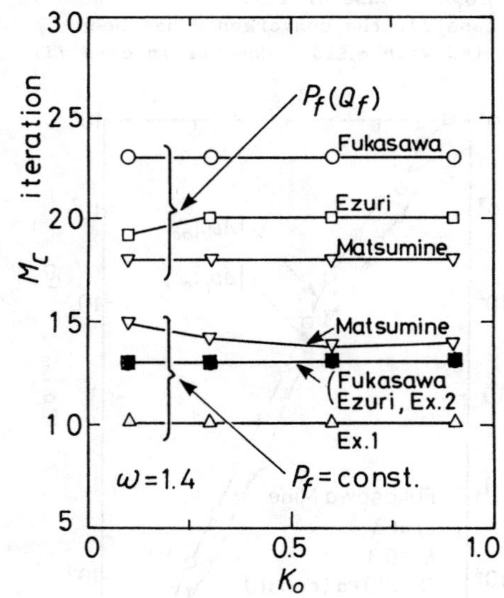

Fig. 8 Effects of the initial set factor K_0 on the number of iterative calculations until convergence (M_c)

Table 2 Effects exerted by the characteristics of fans and Eq.(26) on the number of the iterations M_c (original MCD, band MCD and skyline MCD schemes)

Network	P_f =const.	$P_f(Q_f)$ (Fig. 5) -	$P_f(Q_f)$ (Fig. 5) use Eq.(26)
Example 1	10	-	-
Example 2	13	-	-
Ezuri Mine	13	56	18
Matsumine Mine	13	45	15
Fukasawa Mine	13	40	17

(unit; iteration)

Table 3 Comparisons of computer processing time an iteration, Δt_{CPU}(sec/iteration), among four decomposition schemes

Network	Original MCD	Band MCD	Skyline MCD	ICCG
Ezuri Mine	2.21	2.39	2.03	(3.6)
*	2.21	0.68	0.38	(2.7)
Matsumine Mine	16.9	21.4	2.03	(12)
*	16.9	4.79	1.47	(9.5)
Fukasawa Mine	32.7	27.6	3.50	(13)
*	32.7	8.10	1.83	(11)

[* ; the cases with optimized nodal numbers, ( ); the average time in the ICCG scheme.]

applied to the actual networks, M_c increased to 40~56. This was due to the fact that extra iterations were necessary to make the effect of the adjusted P_f distribute throughout the network nodes. In case *iii)*, M_c is reduced to 17~18, less than the half of ii), by adjusting the effect of P_f upon the individual nodes utilizing Eq.(26). The adjusting method using Eq.(26) is a simplified one, but it is effective in reducing the iterations for a network with a main fan.

3.3 Calculation processing time

The reducing process of $|\Delta Q_i|_{max}$ against M is almost independent from the type of hardware and programming language used. However, in order to distinguish the merits of the actual system, the calculation processing time is practically important.

Table 3 shows a comparison of the calculation processing time required for an iteration Δt_{CPU} (sec) by four decomposition schemes. The results of Δt_{CPU} for cases with optimized nodal numbers are also shown for comparison. The calculation by the skyline MCD is the fastest among the four schemes for every network. When the skyline MCD was used, the solutions could be gained with Δt_{CPU} = 1.8~3.5(sec/iteration) even in the case of Fukasawa Mine(N=141). In the case of the fan characteristics given as $P_f(Q_f)$, as shown in Fig. 5, the whole calculation time until convergence is 31~61(sec), which is obtained by multiplying M_c(=17) by Δt_{CPU}.

The difference of assigning optimizing nodal numbers is expressed as the difference of the calculation time. In the case of Fukasawa mine, the calculation time for optimized nodal numbers is almost 1/2 of that with manually numbering (see Table 3). Accordingly, with a large scale network of almost N=500, it is expected that the overall calculation time becomes smaller by incorporating a routine of automatically reassigning nodal numbers as in the case of finite-element methods.

Figure 9 shows Δt_{CPU} plotted against the total number of nodes N. The difference in Δt_{CPU} between the skyline MCD and other schemes becomes larger as N increases. In

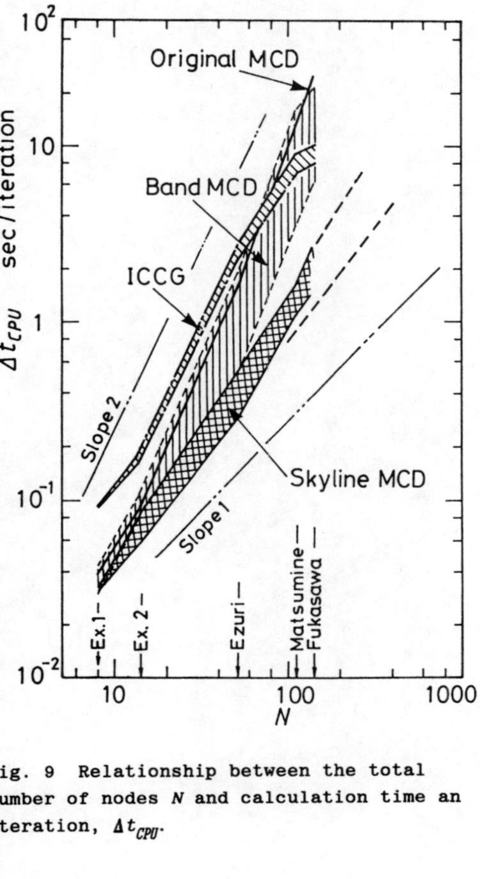

Fig. 9 Relationship between the total number of nodes N and calculation time an iteration, Δt_{CPU}.

the case of the original MCD or the band MCD, Δt_{CPU} is almost multiplied in proportion to N^2. In the case of the skyline MCD, Δt_{CPU} is almost multiplied in proportion to $N^{3/2}$. For the skyline MCD, Δt_{CPU} is believed to be proportional to N as long as ΔI does not change. However, ΔI unavoidably becomes larger as the connections between nodes become more complicated or with the increase of N (see Table 1). Thus, Δt_{CPU} increases with increasing values of N.

4 CONCLUSIONS

The nodal pressure method has been utilized with some success to predict ventilation network flow distribution. The calculation system of the linear equations for the correction of nodal pressures as unknown variables was described. By introducing the connective function between nodes (Eq.(1)), the fundamental equations for the networks were expressed in a more general form, and the calculation schemes were explained. The use of the connective function is effective to save computer memory as compared with the case of incident matrix used in the mesh flow method.

The influence exerted by the different decomposition schemes, the total number of nodes, fan characteristics and optimization of nodal numbers upon the number of iterations until convergence, M_c, has been investigated. Calculations were made with two simple illustrative networks and three actual mine networks as examples of analysis. From the comparative investigations, the skyline modified Choleski's decomposition scheme has advantages on calculation speed and memory space as compared with the other schemes. A condition that the maximum absolute error in continuity of airflow quantity upon nodes is less than 0.1m^3/min was employed as the convergence criterion.

Convergence was achieved with $M_c \le 13$ for the condition of constant fan pressure with a convergence acceleration factor of 1.4. In case of the skyline decomposition scheme, the relationship between the calculation time an iteration Δt_{CPU} and the total number of nodes N is $\Delta t_{CPU} \propto N^{3/2}$ for the examples dealt with in this study.

ACKNOWLEDGMENTS

The present study was supported by a Grant under the Monbusho International Scientific Research Program from the Ministry of Education, Science and Culture of Japan.

The authors would like to express their thanks to Dr. Euler De Souza (Queen's University, Canada) for his helpful discussion.

REFERENCES

Bhamidipati S.S. & Procarione, J.A. 1985. Linear Analysis for the Solution of Flow Distribution Problems in Mine Ventilation Networks. Proc. of 2nd US Mine Ventilation Symp., Reno, Nevada: 646-654.

Collis, R.J. 1973. Bandwidth Reduction by Automatic Renumbering. Int. Journal for Numerical Methods in Engineering 6: 345-356.

Inoue, M. & Uchino, K. 1989. Development of multi-purpose ventilation analysis system "AVENT" and some example of its application. Proc. of MMIJ/IMM Joint Symp., Kyoto : 461-466.

Isaacs, L.T. & Mills, K.G. 1980. Linear theory methods for pipe network analysis. Proc. of ASCE , Journal of the Hydraulics Division 106-HY7: 1191-1201.

Melosh, R.J. & Bamford, R.M. 1969. Efficient solution of load-deflection equations. Proc. of ASCE, Journal of the Structural Division: 661-676.

Melosh, R.J., Lang, T., Schmele, L. & Bamford, R.M. 1967. Computer analysis of large structural systems. AIAA Paper No. 69-955.

Ng, D. 1989. Mine ventilation computer code for personal computers. U.S. Bureau of Mine, PA, Report No. IC-9251.

Rao, B.V. 1987. Finite element analysis of flow networks. Engineering Analysis 4-1: 35-39.

Reddy, J.N. 1984. An introduction to the finite element Method, p.64-193. McGraw-Hill Book Co.

Sasaki, K. Miyakoshi, H., Onozuka, T. & Kinoshita, H. 1990. Nodal head method with matrix operation in analysis of mine ventilation networks. SHIGEN-TO-SOZAI (J. of the Mining and Materials processing Inst. of Japan) 106-10: 587-594. (in Japanese)

Sasaki,K., Miyakoshi, H., Onozuka, T. & Kinoshita, H. 1992. A Prediction system for airflow temperature and humidity in underground ventilation networks. SHIGEN-TO-SOZAI (J. of the Mining and Materials Processing Inst. of Japan) 108-5: 378-388. (in Japanese)

Wallace, G.W. & Brunner, D.J. 1989. The use of microcomputers in mine ventilation planning. Proc. 4th US Mine Ventilation Symp. : 425-433.

Wang, Y.J. 1982. A computer program for mine ventilation network analysis. In Hartman, H.L. (ed.), Mine ventilation and air conditioning 2nd ed., p.483-516. Wiley-Interscience.

Wang, Y.J., Hartman, H.L. & Mutmansky, J.M. 1985. Recent developments in mine ventilation network theory. Proc. of 2nd US Mine Ventilation Symp., Reno, Nevada: 667-675.

Innovative Mine Design for the 21st Century, Bawden & Archibald (eds) © 1993 Balkema, Rotterdam, ISBN 90 5410 325 6

Aspects of environmental protection provided by thickened tailings disposal

S.L.Barbour & G.W.Wilson
Department of Civil Engineering, University of Saskatchewan, Saskatoon, Sask., Canada

R.J.Salvas
Department of Civil Engineering, Ryerson Polytechnic University, Toronto, Ont., Canada

L.St.Arnaud
Noranda Technology Centre, Montreal, Que., Canada

D.Bordin
Kidd Creek Division, Falconbridge Ltd, Timmins, Ont., Canada

ABSTRACT: This paper describes the enhanced environmental protection provided by tailings placed by the thickened tailings disposal technique. Whereas conventional tailings disposal facilities rely on "engineered" control structures such as water retention dams, cutoffs, and liners; in thickened tailings disposal the properties and placement of the tailings themselves are "engineered" in order to minimize effluent seepage rates and prevent acidification of the tailings. Key features of the thickened tailings disposal system are the low hydraulic gradients, low hydraulic conductivity, high capillarity and spatial homogeneity of the tailings. These key features are illustrated using data from the Falconbridge Kidd Creek Division mine in Timmins, Ontario, Canada.

1 INTRODUCTION

As we move towards the 21st century, ever increasing public concern demands that acidic drainage from mining operations be better controlled to protect the environment. Acidic drainage is the largest single environmental problem facing the mining industry today. Conservative estimates of the reclamation costs for Canadian acid generating tailings sites exceed 2 billion dollars (Feasby et al 1991).

In general, past practices are no longer acceptable. Public concern is expressed through ever-tightening environmental regulations. These regulations are becoming so rigid and economically prohibitive, that capital from developed countries is being directed to ventures in countries where the environmental regulations are less restrictive. More recently, there has been considerable effort expended in finding better solutions for problems of acid drainage caused by conventional tailings disposal practices. These include the use of wet barriers (underwater disposal and flooding), as well as dry covers and re-vegetation.

2 BACKGROUND

The conventional surface disposal of tailings requires the formation of surface impoundments that will retain both the tailings and the mill wastewater effluent. Provisions are made for settling of the tailings and clarification of the mill effluent for return to the mill. Traditionally, surface impoundments have been formed by building one of two types of perimeter structures: water retention dams or raised embankments. These perimeter embankments have steep downstream faces with a settling pond on top of the tailings. Water retention dams are based on traditional dam designs. Raised embankments are formed by either downstream, upstream or centerline construction techniques. The raised embankments rely on segregation of the coarse tailings fraction in order to provide construction material for progressively raising the embankment above the original ground surface and the natural groundwater table. Although operations containing such structures may include recycling of all or most of the 'visible' perimeter seepage during the active life of the mine, long-term closeout remains difficult. The principle design requirements for these impoundments have been to provide adequate storage capacity and stability at minimum cost.

In recent years, increasing design requirements have been placed on these impoundments. In addition to stability concerns, the design must prevent or minimize the rate of release of contaminated effluent, it must prevent the acidification of the tailings through oxygen diffusion into the tailings, and it must provide for progressive decommissioning towards a "walk-away" condition.

It is difficult for the typical conventional tailings disposal system (CTD) to meet all of these requirements. Furthermore, the typical CTD basin contains a superimposed sedimentation pond that promotes high seepage gradients during active operations. This is undesirable as it does not provide a self-draining slope to promote runoff after termination of operations. It also allows an oxidizing environment to develop around the perimeter of the pond where the phreatic surface is drawn down as a result of lateral seepage. Conventional spigoting procedures invariably result in the accumulation of coarser, more pervious tailings around the perimeter of the basin. Seepage is therefore high, particularly at the perimeter dams, due to both high gradients and high hydraulic conductivities.

The conventional tailings pond cannot be progressively reclaimed during operations, as it is continually being filled. It is also difficult to leave it in a 'walk-away' condition that will not require effluent treatment long after operations have ceased. Furthermore, it is difficult to blend the high perimeter dam into the topography so that it has a natural appearance after decommissioning.

The design of CTD systems has been limited by one particular constraint: that the nature and character of the tailings stream produced in the milling circuit controls the design of the waste management system. In order to meet stringent environmental design requirements, additional "engineered" features such as liners, cutoffs and covers have been added.

An alternative technique for the disposal of potentially acid-generating tailings is reviewed in this paper. The conceptual basis for design is that the tailings themselves should be "engineered". The environmental protection provided by this technique does not rely on specially constructed barriers or covers but on the nature and character of the tailings stream itself. "Natural" control of effluent releases and acidification is provided by controlling the hydrogeology of the tailings themselves.

The properties of the tailings stream are managed in order to optimize tailings placement and environmental protection. This involves controlling the properties of the tailings stream so that the tailings act as a moisture retention barrier that prevents the ingress of oxygen while minimizing infiltration and effluent release rates.

This paper provides a review of the tailings properties required for optimum performance and illustrates these key features using data obtained for the Kidd Creek mine in Timmins, Ontario.

3 KIDD CREEK - A THICKENED TAILINGS DISPOSAL (TTD) SYSTEM

The thickened tailing disposal (TTD) system was developed by Dr. E. I. Robinsky in the late 1960's for a mine in Ontario, Canada. The method was first described by Shields (1975), then in more detail by Robinsky (1975) and more recently by Salvas (1989). Briefly, the method consists of thickening the tailings to a heavy slurry consistency prior to discharge from an elevated location to permit the development of a sloped deposit. The amount of thickening is controlled so that the deposited tailing forms a gentle, pre-determined slope. Because the tailings are self-supported, the need for surface impoundment is either eliminated or minimized.

Details of the TTD system have been published by Robinsky (1976), (1978), (1982) and (1986). Specific applications have also been published by Robinsky (1981). To date, thickened mine tailings disposal studies have been carried out for more than forty mines in Canada and in other countries (Robinsky 1982, Robinsky 1986, and Wood and McDonald 1986).

Some of the advantages of the thickened tailings disposal system include:

1. Lower initial capital investment,
2. Lower operation costs,
3. Greater placement volume for a given

height of containment dyke,

4. Elimination of the construction and maintenance of high perimeter dams,

5. Elimination of slimes ponds and decant systems, and

6. Low reclamation costs.

The TTD system has been operating at Kidd Creek Mines Ltd. in Timmins, Ontario for approximately twenty years. The mill tailings pass through a basin-type thickener, with flocculent added to achieve the appropriate consistency, and are then discharged through spigots from a ramp running to the center of the tailings area. During the life of the operation, the slopes have been gradually increased from 0.5 percent to 1.5 percent following a pre-selected schedule of thickening to produce the required slopes.

The slope of the deposited tailings is governed primarily by the percent solids. This relation was clearly established by laboratory tests before commencing operations. Figure 1 shows the laboratory curve for the Kidd Creek Mines Ltd. tailings in 1977.

Using this curve, it would have been necessary to thicken the tailings to approximately 60.5% to produce 1.5 percent slopes. However, as the characteristics of the ore and therefore the mill tailings slurry change, it is necessary to redefine the deposition curve. These curves are also shown in Figure 1. The curve shows that the tailings produced in 1990 must be thickened to more than 66 percent to achieve 1.5 percent slopes. At present, the 1.5 percent slopes are being produced by a combination of thickening and the addition of lime. It has been shown by laboratory tests that the slope of the deposited tailings can be increased by the addition of lime to the pre-thickened tailings. The lime also acts to neutralize acidity that may be generated by the tailings and therefore serves a dual purpose.

At present, the deposited tailings have built up a conical ridge approximately 23 m higher than the perimeter dikes. A recently completed off-shoot ramp at right angles to the central ramp will permit a higher storage capacity and will permit progressive reclamation to commence in the next few years at the far end of the new ramp, even as active deposition continues in the area. No perimeter dikes have been built since 1975 and no new dikes for confining tailings

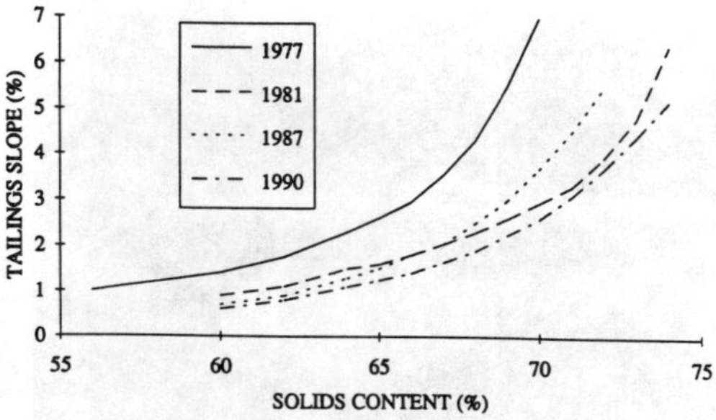

Fig. 1 Slope of deposited thickened tailings versus slurry solids content

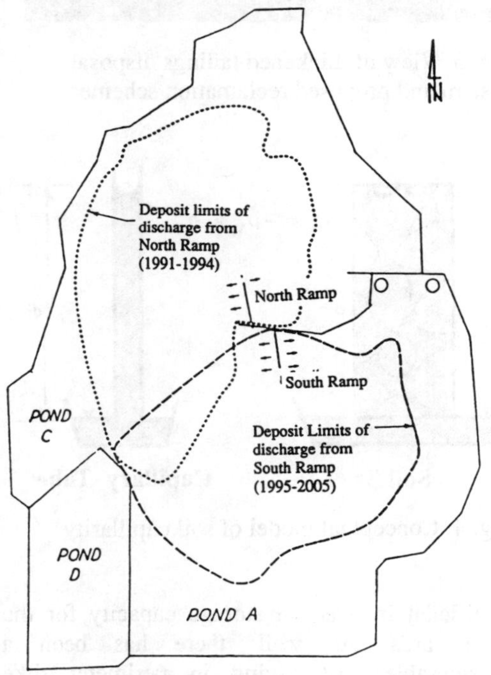

Fig. 2 View of thickened tailings disposal operations

will be required. A lime station has been built at the far end of the off-shoot ramp for this purpose. Figure 2 is an aerial view of the central ramp, the off-shoot ramp with the lime station, and the fan-like shape of the deposited tailings.

It is proposed to build a second off-shoot ramp in the future and to follow the same operational procedure on the south side of the central ramp as shown in Figure 3. The use of the TTD system at this site has resulted in a

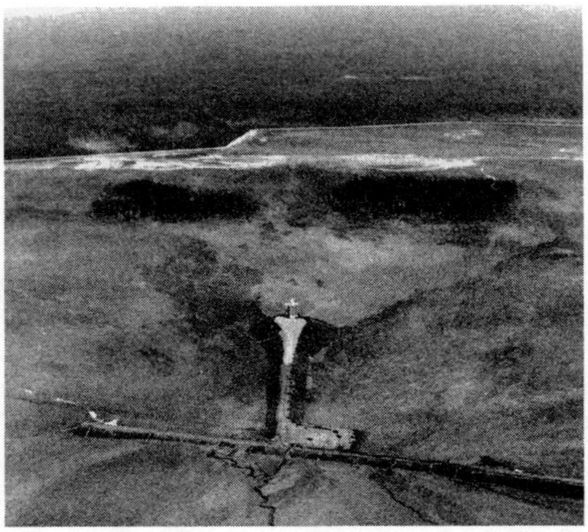

Fig. 3 View of thickened tailings disposal system and proposed reclamation scheme

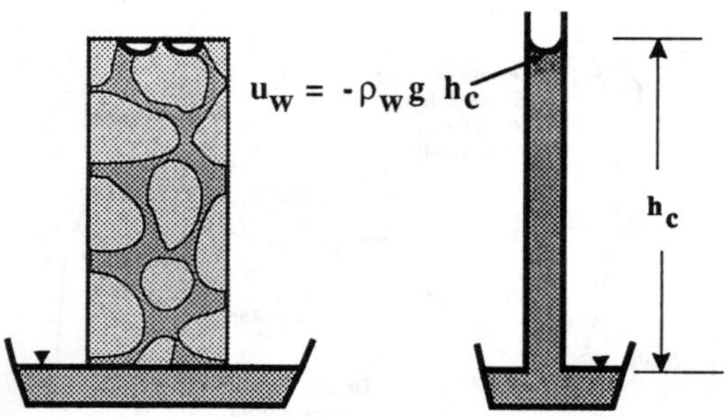

Fig. 4 Conceptual model of soil capillarity

significant increase in storage capacity for the given area. As well, there has been a considerable cost saving in perimeter dike construction.

4 EVALUATION OF TTD PERFORMANCE FOR ENVIRONMENTAL CONTROL

The key question to be addressed is whether or not the TTD scheme provides advantages in meeting the key environmental requirements for potentially acid generating tailings which include:

1. The prevention of acidification by maintaining saturated conditions to surface,
2. The reduction of effluent seepage and release rates, and
3. The ability to support progressive reclamation and decommissioning.

These requirements are in some apparent contradiction in that if we want to maintain saturated conditions on an elevated tailings mass, we may have some difficulty in maintaining low effluent release rates. Similarly, maintaining a saturated tailings mass may cause difficulties in providing the trafficability required for progressive decommissioning.

This apparent contradiction can be overcome if the tailings have the correct hydraulic characteristics. For example, the requirement of a low effluent flux requires that the tailings mass have both low hydraulic gradients and a low hydraulic conductivity. If these are maintained, then the infiltration and generation of effluent from the tailings are minimized. In addition to low hydraulic gradients and hydraulic conductivities, the tailings should have the ability to remain saturated while under negative water pressures. Under these conditions, a shallow water table will be accompanied by a capillary-saturated zone which will maintain saturated conditions at the tailings surface. The development of negative water pressures at surface will also increase the strength of the tailings and help provide the trafficability required for decommissioning operations (Robinsky 1986).

Figure 4 illustrates a conceptual model of soil capillarity. For small diameter capillary tubes, the tubes will remain saturated at significant heights above the water table. Pores of approximately 0.01 to 0.006 mm in diameter would be required for a 3 to 5 meter thick tension saturated zone (Means and Parcher 1963). This pore diameter is similar to that of a typical silt.

The benefits described above could not be developed throughout the tailings mass unless they were homogeneous. Non-homogeneous conditions would allow the development of areas of higher hydraulic conductivity or low capillarity. Hazen's approximation for the hydraulic conductivity of a well sorted soil is as follows:

$$k = (d_{10})^2, \qquad (1)$$

where k is the hydraulic conductivity (cm/sec.)

and d_{10} is the grain-size of the 10% finest fraction (Means and Parcher 1963). This suggests that even a change from a silt to a fine sand (an increase in d_{10} of approximately one order of magnitude) would increase the hydraulic conductivity by two orders of magnitude. This same change in soil type would also reduce the thickness of a tension-saturated zone by one order of magnitude, from meters to centimeters in height.

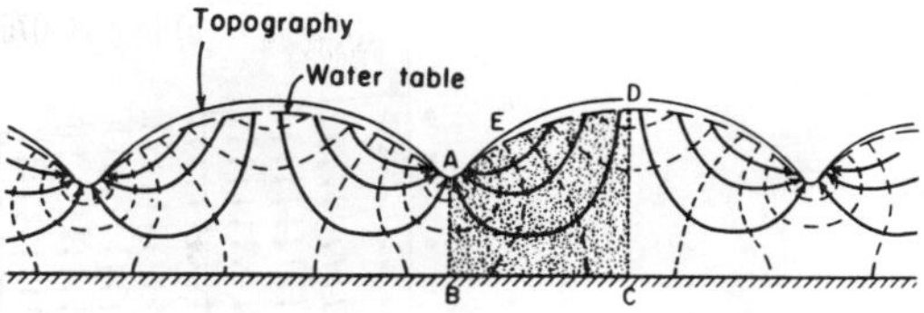

Fig. 5 Topographically controlled groundwater flow systems (after Freeze and Cherry 1979 and Hubbert 1940)

4.1 Natural Analogue

A natural analogue for this preferred behavior for the tailings mass is present in the topographically controlled groundwater flow systems that develop within fine grained deposits. Freeze and Cherry (1979) provide summaries of the work of Hubbert (1940) and others which show that in these types of systems, the distribution of hydraulic heads is controlled by the surface topography. In other words, the water table forms a shallow reflection of the topographic surface (Figure 5).

For a fine textured soil, and with the climatic conditions present in many parts of Canada, only a small fraction of the available climatic flux actually enters the flow system as infiltration. The fine grained nature of the material also promotes the development of saturated conditions near the ground surface.

The key aspects of this type of flow system are as follows:

1. Low hydraulic gradients controlled by gently sloping topography,
2. Low hydraulic conductivity,
3. High capillarity due to the fine grained nature of the sediments, and
4. Spatial homogeneity.

Figure 6 illustrates how critical these features are by showing the effect that changes in hydraulic conductivity, climatic flux, or topographic slope have on the flow system and water table profiles for a topographically controlled groundwater flow system. With a proper combination of hydraulic conductivity, surface flux, and topographic slope, the water table remains near the surface (Figure 6a) even under low infiltration rates or steepened slopes (Figure 6b and 6d). Increases in the hydraulic conductivity or heterogeneity of the deposit will however, produce lower water tables and higher rates of seepage (Figures 6c and 6e).

Over the last 3 years, a series of field and laboratory investigations have been undertaken on the Kidd Creek site which demonstrate that many of the features described previously are present in the Kidd Creek tailings. The long term effectiveness of the Kidd Creek tailings deposit in providing the "natural" flow system as described in Section 4.1 is still under study. A short summary of test results illustrating these features is presented in the remainder of this paper. The data that is reported here is based on lab and field studies conducted by Robinsky (1990), St. Arnaud and Woyshner (1992), and Barbour et al (1992). Sample locations and piezometer installations referred to in the discussion are illustrated on the site plan in Figure 7.

4.2 Topographically Controlled Low Gradients

The maximum lateral gradients within the tailings mass are controlled by the average slope of the tailings. The average slope for the Kidd Creek tailings is approximately 1.5%. Vertical gradients at the site are also small because of the presence of a natural clay barrier underlying the tailings pile that has a low hydraulic conductivity. St. Arnaud and Woyshner (1992) found that the lateral gradients near the center of the tailings cone ranged from 0.1% to 2%, the range of fluctuation being due in part to changes in climatic conditions.

The flow system that develops under these conditions is illustrated by the water table surface through the tailings measured at different times during the year. Figure 8 illustrates a North-South cross-section through the tailings. It is apparent that a shallow water table develops across the tailings, reflecting the tailings surface

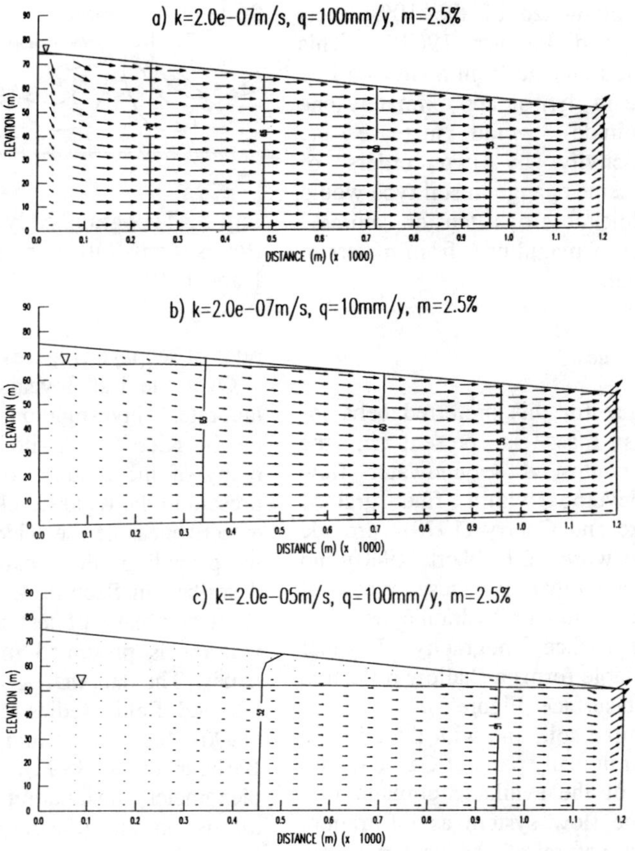

Fig. 6 Effects of changes in hydraulic conductivity (k), climatic flux to the soil surface (q), and topographic slope (m) on groundwater flow systems

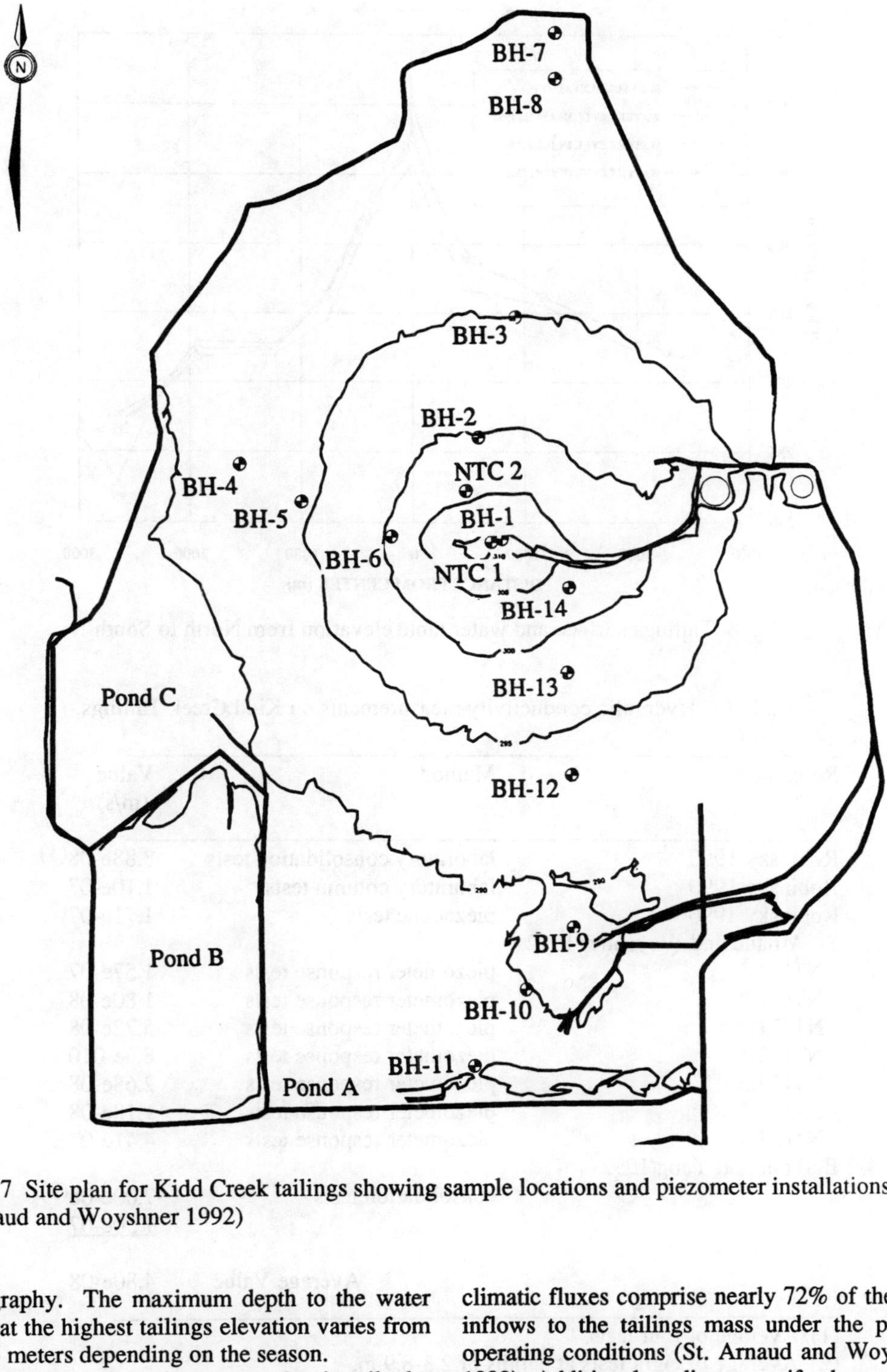

Fig. 7 Site plan for Kidd Creek tailings showing sample locations and piezometer installations (St. Arnaud and Woyshner 1992)

topography. The maximum depth to the water table at the highest tailings elevation varies from 0 to 5 meters depending on the season.

This flow system is supported primarily by climatic precipitation. Preliminary water balance studies on the tailings mass have indicated that climatic fluxes comprise nearly 72% of the total inflows to the tailings mass under the present operating conditions (St. Arnaud and Woyshner 1992). Additional studies to verify these values are ongoing.

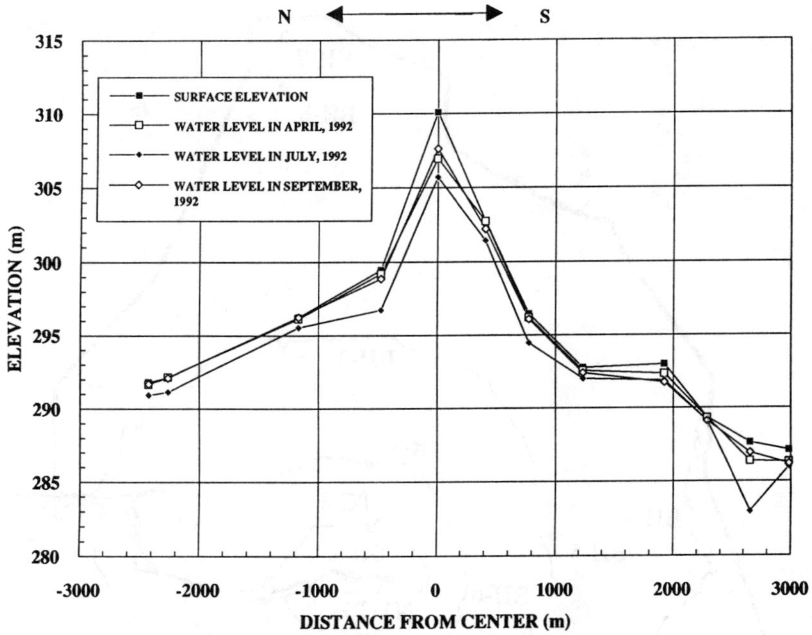

Fig. 8 Tailings surface and water table elevation from North to South

Table 1 - Hydraulic conductivity measurements on Kidd Creek Tailings.

Reference	Method	Value (m/s)
Robinsky 1990	laboratory consolidation tests	3.88e-08(1)
Robinsky 1990	laboratory column tests	1.10e-07
Robinsky 1990	piezocone tests	1.71e-07(2)
St. Arnaud and Woyshner 1992		
- NTC 1.1	piezometer response tests	1.57e-07
- NTC 1.2	piezometer response tests	1.80e-08
- NTC 1.3	piezometer response tests	5.72e-08
- NTC 2.1	piezometer response tests	8.6e-010
- NTC 2.2	piezometer response tests	2.68e-08
- NTC 2.3	piezometer response tests	1.18e-08
- NTC 2.4	piezometer response tests	4.41e-07
Barbour and Yang 1992		
	consolidation tests -	7.00e-08(3)
		1.50e-07
	Average Value	4.80e-08

(1) Average of four tests.
(2) Average of eight tests at depths of 2.3-8.9 m.
(3) Values taken over range of field void ratios of 0.65-0.8.

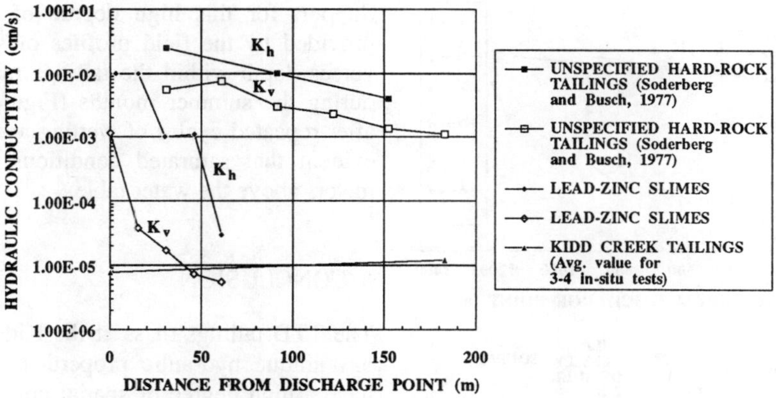

Fig. 9 Variations in hydraulic conductivity with distance from discharge for well segregated tailings beaches (after Vick 1990), with Kidd Creek data superimposed

4.3 Hydraulic Conductivity

A series of both lab and field measurements of the hydraulic conductivity of the tailings have been made over the past 3 years. Table 1 summarizes some of these test results. The average hydraulic conductivity for the tests is approximately 5.0e-08 m/s with a standard deviation of less than 1 order of magnitude (based on an assumed log-normal distribution for hydraulic conductivity). The consistency of these results is quite unusual considering that these tests have been conducted with different techniques and at different times and locations.

Figure 9 illustrates variations in hydraulic conductivity with distance from the spigot point for conventional segregated tailings as presented by Vick (1990). The field measurements of hydraulic conductivity for the Kidd Creek tailings are superimposed on this plot. Based on this comparison, the consistently low values of hydraulic conductivity across the Kidd Creek tailings are quite unusual when compared with conventional deposits.

4.4 Homogeneity

The level of homogeneity reflected in the hydraulic conductivity and water table measurements discussed previously is also evident in the results of direct sampling of the tailings for grain-size and water content. Figure 10 shows a series of grain-size curves for samples of the tailings taken up to 1200 meters from the discharge ramp. The consistency of the gradation over this distance is quite remarkable. Figure 11, taken from Vick (1990), shows typical variations in gradation with distance from the spigot for typical metal mine tailings after conventional deposition. Comparisons of the grain size curves of samples obtained from the Kidd Creek site indicate how little the gradation within the Kidd Creek tailings vary, even over distances greater than 1000 meters from the spigot point.

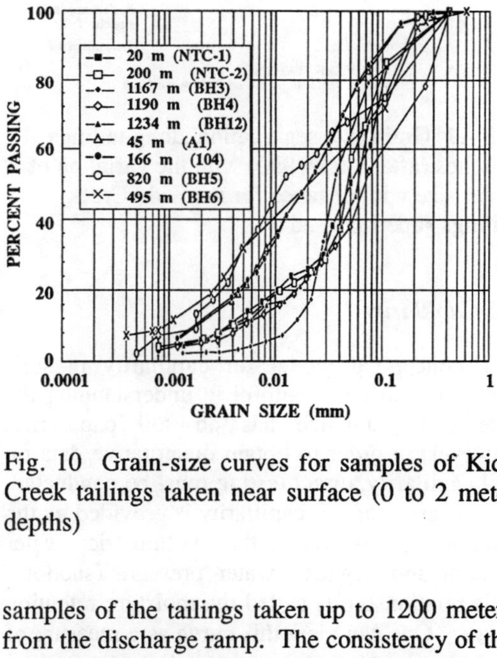

Fig. 10 Grain-size curves for samples of Kidd Creek tailings taken near surface (0 to 2 meter depths)

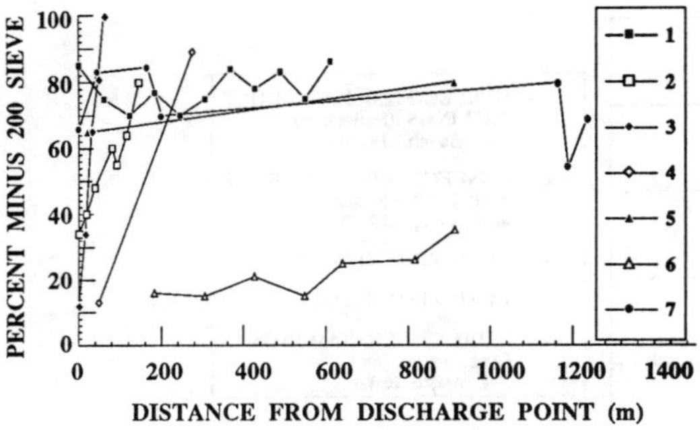

CURVE	TAILINGS TYPE	G	% - 200 MILL DIS-CHARGE	PULP DENSITY OF MILL DIS-CHARGE	SOURCE
1	gold				Blight and Steffen, 1979
2	copper	2.7	45	45	Volpe, 1979
3	lead-zinc slimes	3.4	75	<40	Unpublished
4		2.7	38	30	Soderberg and Busch, 1977
5		2.7	60	50	Soderberg and Busch, 1977
6	copper	3			Unpublished
7	KIDD CREEK TAILINGS (Sampling from 0 to 6 feet below surface)				

Fig. 11 Grain size segregation along tailings beaches (after Vick 1990) with the variation of grain-size with distance for the Kidd Creek tailings superimposed

4.5 Capillarity

The conceptual model for capillarity in soils presented earlier is helpful in understanding the role that grain-size has on soil capillarity; however, in order to obtain quantitative data on soil capillarity, direct testing must be conducted. The measure of soil capillarity is provided by the relationship between the volumetric water content and negative water pressure (suction). This relationship is called the moisture retention curve. One can view this curve as a measure of moisture content within the tailings at various heights above the water table under conditions of hydrostatic equilibrium after drainage to a lower water table.

Figure 12 illustrates typical moisture retention curves for the Kidd Creek tailings. The curves illustrate that these tailings will not become unsaturated by drainage until the negative water pressure in the tailings exceeds approximately 70 kPa. This is equivalent to a capillary saturated zone approximately 7 meters thick. Support for this high degree of capillarity is provided by the field profiles of water content versus depth within the tailings mass measured during the summer months (Figure 13). Even after repeated cycles of wetting and drying, it is evident that saturated conditions exist several meters above the water table.

5 CONCLUSION

The TTD tailings mass at the Kidd Creek mine has unique hydraulic properties. The tailings have a high degree of spatial homogeneity with respect to grain-size gradation and hydraulic conductivity. Furthermore, the tailings have a uniformly low hydraulic conductivity and a high capacity to retain water even under high negative water pressures. These hydraulic properties result in the formation of a shallow water table which mirrors the gently sloping topography of the tailings surface. In general, the tailings tend to remain saturated to surface above the water table. This is a favorable condition with respect to the decommissioning of the tailings mass in that it minimizes the ingress of oxygen. This should reduce the long-term potential for oxidation and the resulting development of acid drainage.

Similar benefits might be realized for other tailings sources when placed using the TTD system. The grain-size, hydraulic conductivity, and moisture retention characteristics for the tailings may vary; however, these tailings could still be placed in gently sloping, homogeneous deposits. The water table configuration, seepage and effluent release rates, and depth of capillary saturation must be predicted for each site based on its local climatic conditions and hydrogeology.

In conclusion, it is recommended that consideration should be given to using the TTD system for new mines in the future in order to reduce the impact on the environment, including a reduction in acid drainage. The system also permits the conversion of an existing conventional disposal basin, thus providing not only additional storage capacity, without raising the perimeter dams, but also gaining substantial environmental advantages.

ACKNOWLEDGMENTS

Support for the studies described in this paper was provided by the Kidd Creek Division of Falkenbridge Mining Ltd. and by the Mine Evaluation Neutral Drainage (MEND) program. This support is gratefully acknowledged.

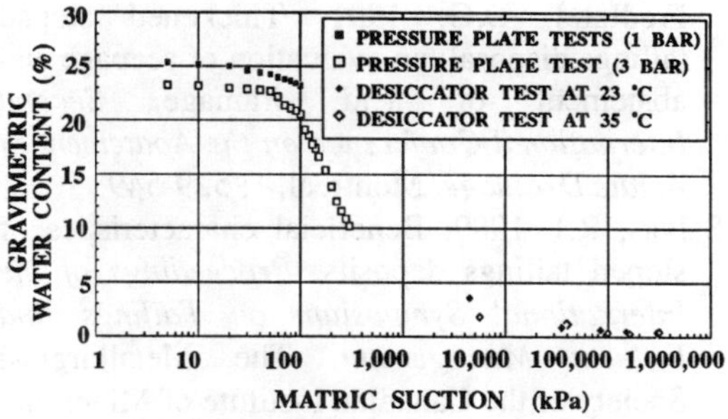

Fig. 12 Moisture retention characteristics (moisture characteristic curve) of Kidd Creek tailings

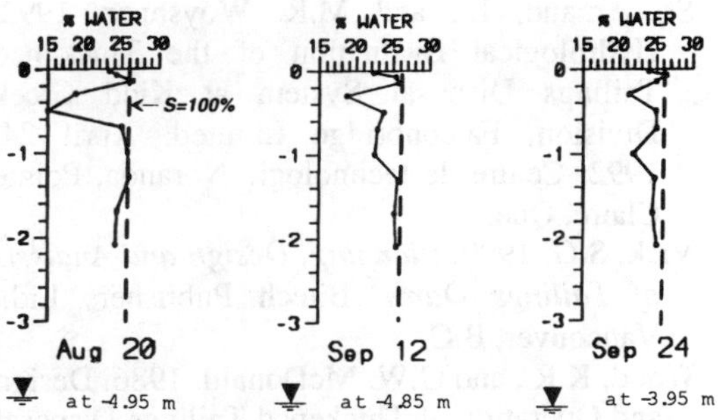

Fig. 13 Variations in tailings water contents (St. Arnaud and Woyshner 1992)

REFERENCES

Barbour, S.L. 1990. Discussion of "Reduction of acid generation in mine tailings through the use of moisture-retaining cover layers as oxygen barriers". *Canadian Geotechnical Journal*, 27(3): 398-401.

Barbour, S.L., Yang, N., and Wilson, G.W. 1992. A Laboratory Characterization of the Properties of the Kidd Creek Thickened Tailings Required for Evaporative Flux Modelling. Geo-Environmental Engineering Group, Department of Civil Engineering, University of Saskatchewan, Sk.

Feasby, G., Tremblay, G., and Blanchette, M. 1991. The mine environmental neutral drainage (MEND) program. *Second International Conference on the Abatement of Acidic Drainage*, vol. 1: 1-26.

Freeze, R.A. and J.A. Cherry. 1979. *Groundwater*. Prentice-Hall Inc., Englewood Cliffs, N.J.

Gardner, W.R. 1958. Some steady-state solutions of the unsaturated moisture flow equation with application to evaporation from a water table. *Soil Science*: 228-232.

Hubbert, M.K. 1940. The theory of groundwater motion. *J. Geol.* 48: 785-944.

Means, R.E. and J.V. Parcher 1963. *Physical Properties of Soils*. Charles E. Merril Books Inc. Columbus, Ohio.

Robinsky, E.I. 1975. Thickened discharge - a new approach to tailings disposal. *CIM Bulletin*, 68(764), December.

Robinsky, E. I. 1976. Advantages of Tailing Disposal by Thickened Discharge. 8th Annual Meeting of Canadian Mineral Processors: 289-310.

Robinsky, E.I. 1978. Tailing disposal by the thickened discharge method for improved economy and environmental control. *2nd International Tailing Symposium*, Denver, CO, May.

Robinsky, E.I. 1981. Uranium Tailing Disposal By The Thickened Tailing Discharge System. *Nuclear Energy Agency, OECD Proceedings*, Fort Collins, Colorado, U.S.A.

Robinsky, E. I. 1982. Thickened Tailing Disposal in Any Topography. *Proceedings, 111th Annual Meeting of AIME*, Metallurgical Society, Dallas, Texas, U.S.A.

Robinsky E.I. 1986. Solving some difficult geotechnical problems with the thickened tailing disposal system. *3rd Conference on Construction and Mining in Difficult Soils*, Sudbury, Ontario.

Robinsky E.I. 1990. A Comprehensive Study for Tailings Disposal and Progressive Reclamation - Kidd Creek Mines. July-August 1990. Prepared for Falconbridge Limited, Kidd Creek Division, Timmins, Ontario by E.I. Robinsky Associates Limited, Toronto, Ontario.

Robinsky, E., Barbour, S.L., Wilson, S.L., and

Fredlund, D.G. 1991. Thickened sloped tailings disposal: an evaluation of seepage and abatement of acid drainage. *Second International Conference on the Abatement of Acidic Drainage*, Montreal, 1: 529-549.

Salvas, R.J. 1989. Beneficial characteristics of sloped tailings deposits. *Proceedings of the International Symposium on Tailings and Effluent Management,* The Metallurgical Society of the Canadian Institute of Mines and Metallurgy, 14, Pergamon Press Inc.: 215-226.

Shields, D. 1975. Innovations in tailings disposal. *Canadian Geotechnical Journal* 12: 320-325.

St. Arnaud, L., and M.R. Woyshner. 1992. Hydrological Evaluation of the Thickened Tailings Disposal System at Kidd Creek Division, Falconbridge Limited. April 24, 1992. Centre de technologie Noranda, Pointe Claire, Que.

Vick, S.G. 1990. *Planning, Design and Analysis of Tailings Dams*. Bitech Publishers Ltd., Vancouver, B.C.

Wood, K.R., and G.W. McDonald. 1986. Design and Operation of Thickened Tailings Disposal System at Les Mines Selbaie. CIM Bulletin, November: 47-51.

Innovative Mine Design for the 21st Century, Bawden & Archibald (eds) © 1993 Balkema, Rotterdam, ISBN 90 5410 325 6

Innovative concept for waste rock management

François Bautin
Operations Cogema Canada Limited, Saskatoon, Sask., Canada

ABSTRACT: In the Cluff Mining operations, the last ore body found, Dominique-Janine Extension (DJX) is located partly under the shore of Cluff Lake, partly under the lake. The project for mining DJX comprises the building of a dam with a cut-off-wall to isolate the pit from the lake. This project could provide 5,000 tonnes of Uranium from a 10-million BCM open pit operation and is scheduled to last 7 years.

In order to include the decommissioning plan into the project design, the management of waste rock has been cautiously studied. The solution retained, with which many experts agree, is to dispose of the waste rock in an underwater stockpile. The advantages are:

- Less oxidation meaning less acid generation;
- Easier decommissioning.

To promote this innovative concept in Saskatchewan, long term impact modelling studies have been carried out.

CLUFF MINING PRESENTATION

Cluff Mining is a partnership managed from its creation in 1979 to 1992 by AMOK whose major shareholder has been COGEMA CANADA. As of January 1, 1993, AMOK and COGEMA CANADA have been amalgamated, so COGEMA CANADA now is the operator of Cluff Mining.

The mining and milling facilities are located in Northern Saskatchewan, near Cluff Lake, approximately 75 km south of Lake Athabasca and 25 km east of Alberta border.

The drainage basin of the complex is governed by interconnected lakes and rivers which discharge into the Douglas River. The Douglas River flows in a northwesterly direction and discharges into Lake Athabasca, Slave River, Great Slave Lake and into the MacKenzie River (see figure 1).

CLUFF MINING HISTORY

Since the discovery of the first orebody in the late 1960's several deposits have subsequently been found and mined (see figure 2).

- the "D" orebody, small but with a relatively high grade of about 3% U, was mined as an open pit from 1980 to 1981.

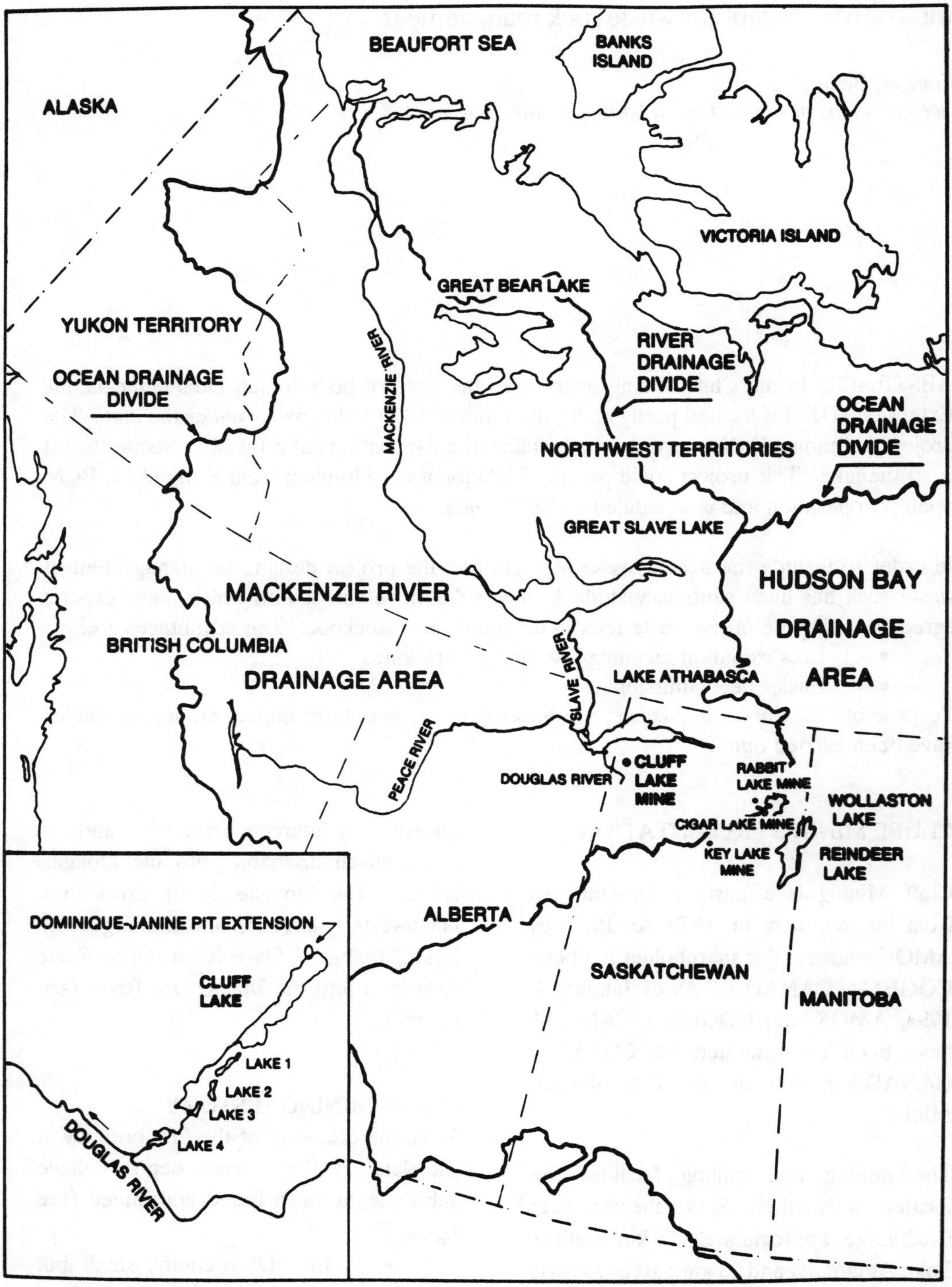

DRAINAGE AREAS OF NORTHERN SASKATCHEWAN

Figure 1

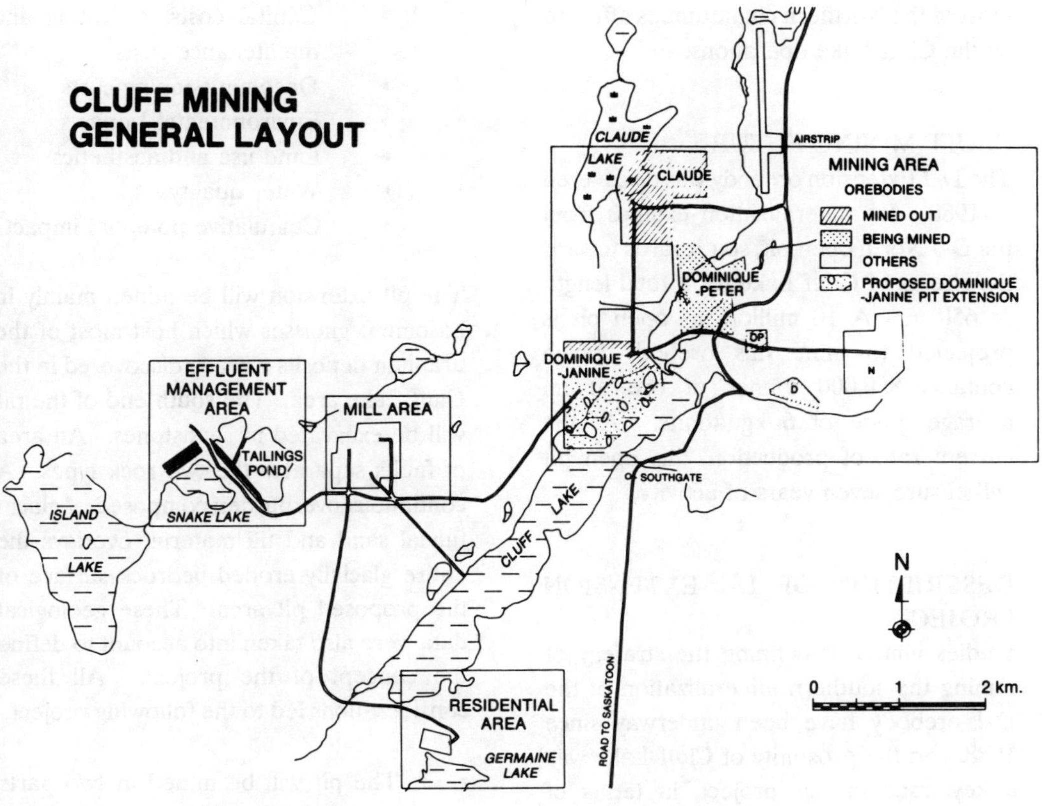

Figure 2

- The "Claude" orebody was mined in a 4 million m³ open pit from 1982 to 1989.
- The Dominique-Janine North orebody was then open pit mined from 1989 to 1991.
- the Dominique-Peter orebody has been mined underground since 1984. The undercut and fill method is used because of the sinuosity of the veins and the poor quality of the rocks.

Prior to the commencement of mining activities at Cluff Lake, the project was reviewed by the Cluff Lake Board of Inquiry established by the Province of Saskatchewan whose Chairman was Mr. Justice E.D. Bayda. The final report issued in 1978 approved the proposed mining and milling operations, subject to certain conditions and recommendations. The Board also recommended that further expansion of uranium mining activities beyond the initial Cluff Lake proposal be allowed. Following developments were approved after review by governmental agencies.

Amok was required by the Bayda Commission report to employ more than 50% Northerner's at the mine site. This was successfully done and allowed AMOK to have the confidence and support of most of the Northern communities affected by the Cluff Lake operations.

CLUFF MINING FUTURE

The D-J Extension orebody was discovered in 1988. Its mineralization extends from the D-J North open pit, southwards toward the shores of Cluff Lake for a total length of 650 m. A 10 million m^3 open pit is projected to mine this orebody which contains 850,000 tonnes of ore at an average grade of 6 kgu/tonne. At the current rate of production, this open pit will ensure seven years of activity.

DESCRIPTION OF D-J EXTENSION PROJECT

Studies aimed at defining the strategy of mining the southern mineralization of the D-J orebody have been underway since 1988, and the proximity of Cluff Lake was a key issue in this project, in terms of potential environmental impact, as the ultimate south wall of the proposed open pit will be situated beyond the present shoreline of the lake. Various scenarios have been evaluated before finalizing the current project and a number of technical studies have been conducted to evaluate the engineering, environmental, and socio-economic issues. The water management aspect is critical as the proposed pit area has to be kept dry. These studies underlie the proposal to mine the southern mineralization of the D-J orebody by extending the existing DJN to the south.

The following criteria were used for screening the various options:

- Technical Operation
- Implementation time and/disruptions
- Capital costs, operating and maintenance costs
- Decommissioning costs
- Environmental Impacts
- Land use and aesthetics
- Water quality
- Cumulative potential impact.

The pit extension will be mined mainly in basement gneisses which host most of the uranium deposits already discovered in the Cluff Lake area. The south end of the pit will be excavated in sandstones. An area of faults separates the two rock types. A continuous overburden composed of glacio fluvial sand and till material overlays the entire glacially eroded bedrock surface of the proposed pit area. These geological data were also taken into account to define the concept of the project. All these considerations led to the following project.

- The pit will be mined in two parts (see figure 3). The first part entails strip mining of waste rock and ore in the northern end of the Extension, where the bedrock elevation (325m) is above the elevation of Cluff Lake (317.5 m). In addition, Part one is distant enough from Cluff Lake to prevent an inflow of lake water into the pit. The second part is in the southern half of the orebody where the bedrock slopes down from above lake level to 30 m below.

- In order to control lake water inflow into the pit, a perimeter dam will be constructed around the southern rim of the pit using the materials extracted during the mining of the first part (see figure 4). The dam will be composed of an overburden

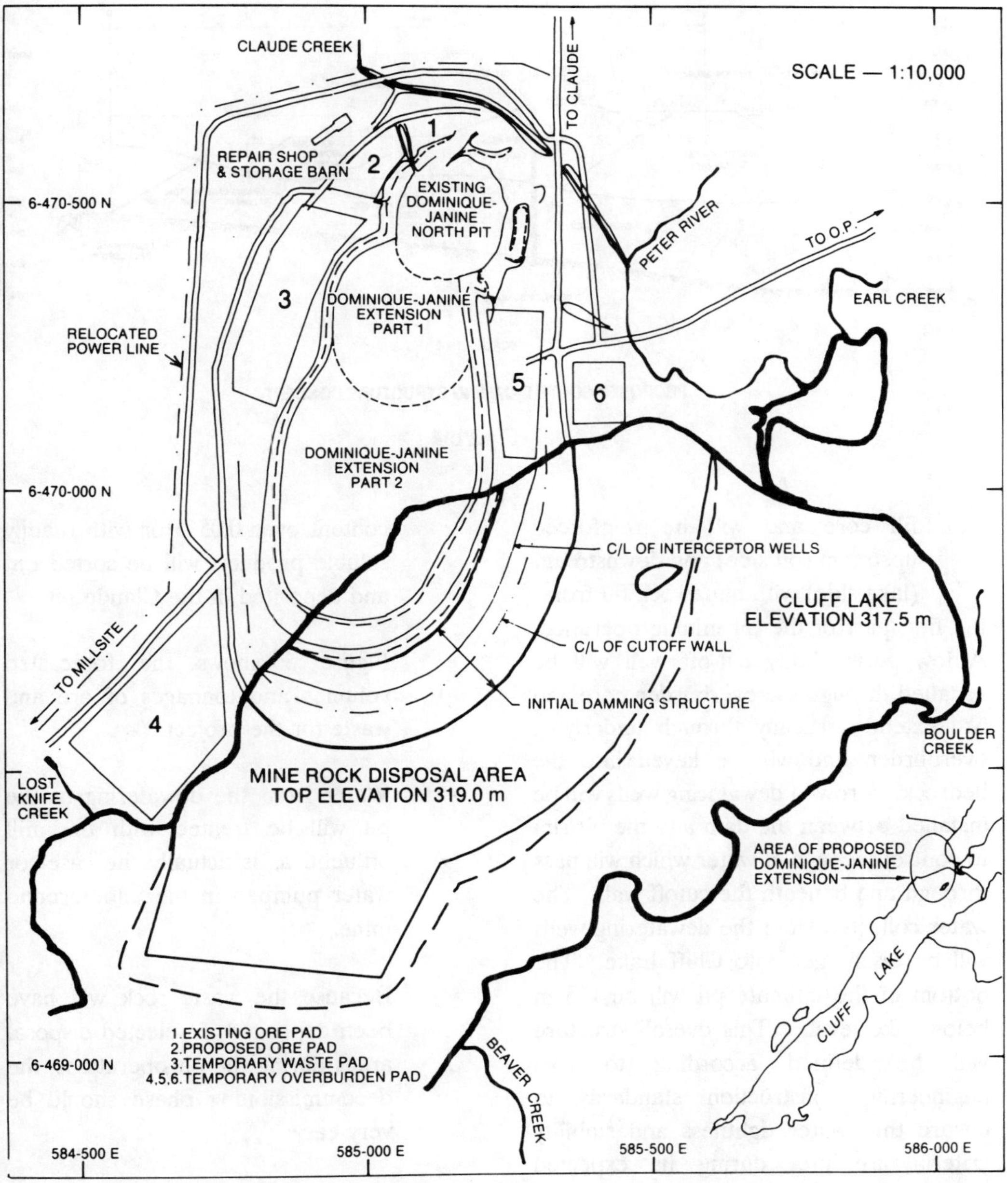

DOMINIQUE-JANINE PIT EXTENSION/GENERAL ARRANGEMENT

Figure 3

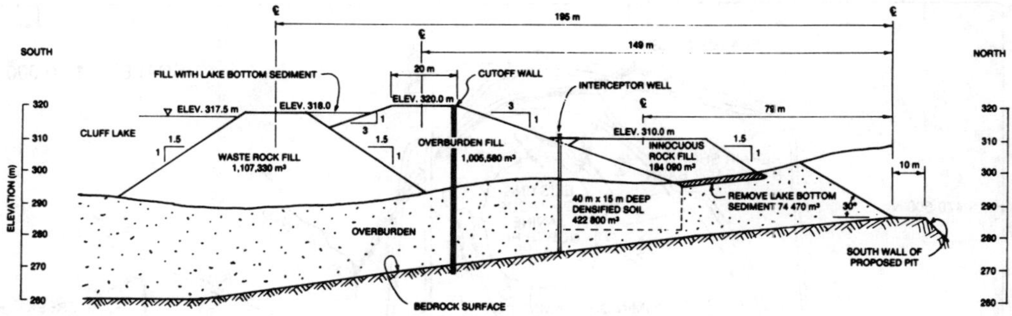

SCALE: 1:1000

PROPOSED SOUTH DAMMING STRUCTURE CONCEPT

Figure 4

fill core and will be reinforced upstream (pit side) and downstream (lake side) with mine rock fill from the first part of the pit mining operation. A low permeability cut-off wall will be installed through the overburden core and will extend vertically through underlying overburden and will be keyed into the bedrock. A row of dewatering wells will be installed between the dam and the pit rim to control seepage of water which will pass through and beneath the cutoff wall. The water collected from the dewatering wells will be discharged into Cluff Lake. The bottom of the ultimate pit will be 125 m below lake level. This overall structure will be defined according to civil engineering construction standards to ensure that water tightness and stability criteria are met during its expected lifetime.

- Additional waste rock mined from the second part of the extension will be placed under water in Cluff Lake downstream from the initial dam, resulting in a waste pile expanding to the south. However, the "special waste", which includes rocks with U content over 0.05% or with readily soluble products will be sorted out and deposited in the Claude pit.

- Figure 5 shows the forecasted volumes and tonnages of ore and waste for the project.

- Water from the dewatering of the pit will be treated with the mill effluent, as is actually the case for water pumped in the underground mine.

- Because the waste rock will have been deposited in selected disposal areas during the pit operations, the decommissioning phase should be very easy:

    - surface fixed equipment and installations will be removed.
    - ground surface areas disturbed by pit operations, including the part of the waste pile which emerges out of Cluff Lake will be made amenable to revegetation.
    - the dewatering of the pit will

Dominique-Janine Extension Pit
Project Volumes and Tonnages

	Overburden	Ore	Waste	TOTAL
Volume in BCM				
Part One (1)	678,000	95,000	2,161,000	2,934,000
Part Two (2)	1,641,000	245,000	5,845,000	7,731,000
Total Pit	2,319,000	340,000	8,006,000	10,665,000
Volume in LCM				
Part One (1)	793,200	123,500	2,809,300	3,726,000
Part Two (2)	1,920,000	318,500	7,598,500	9,837,000
Total Pit	2,713,200	442,000	10,407,800	13,563,000
In Tonnes				
Part One (1)	1,220,400	235,500	5,402,500	6,860,400
Part Two (2)	2,953,800	612,500	14,614,500	18,178,800
Total Pit	4,174,200	850,000	20,015,000	25,039,200

BCM = Bench cubic metres
LCM = Loose cubic metres

Figure 5

be discontinued to allow the pit to flood. The flooding rate of the pit will be accelerated by siphoning water from Cluff Lake. After mine closure and rehabilitation of the pit, Cluff Lake will be preserved as a lake and the reduction in the surface area of Cluff Lake will be compensated for by the surface area of the flooded D-J pit. The retaining structure between Cluff Lake and the pit will be maintained to provide a separation between flooded pit water and Cluff Lake. The pit water quality will be monitored until such a time as the water quality is deemed to be acceptable, then the pit will be linked with the lake by a channel constructed through the dam.

RATIONALE FOR UNDERWATER DISPOSAL OF WASTE ROCK

The project was studied and designed to minimize environmental impact in the long term, by limiting potential acid generation

PREDICTED MEAN ANNUAL CONSTITUENT CONCENTRATIONS IN CLUFF LAKE ([4])

Constituent	Units	Water Quality Objective	Year 1	Year 5	Year 10	Year 100	Year 500
Arsenic	(mg/L)	0.050[1]	0.0002	0.0073	0.0057	0.0010	0.0002
Copper	(mg/L)	0.010[1]	0.0016	0.0037	0.0031	0.0016	0.0013
Lead	(mg/L)	0.010[2]	<0.0001	0.0007	0.0004	0.0001	<0.0001
Molybdenum	(mg/L)	-	<0.0001	0.0302	0.0236	0.0011	0.0002
Nickel	(mg/L)	0.025[1]	0.0017	0.0034	0.0031	0.0013	0.0013
Sulphate	(mg/L)	500[2]	1.07	4.17	3.39	1.25	1.14
Uranium	(mg/L)	0.100[2]	0.0005	0.1010	0.0744	0.0026	0.0009
Zinc	(mg/L)	0.050[1]	0.0087	0.0101	0.0096	0.0081	0.0066
Lead-210	(Bq/L)	-	0.0098	0.0428	0.0299	0.0113	0.0098
Radium-226	(Bq/L)	0.11[3]	0.0091	0.0490	0.0373	0.0140	0.0107
Thorium-230	(Bq/L)	-	0.0082	0.0191	0.0127	0.0078	0.0065

Note:

1 Saskatchewan Surface Water Quality Objective for protection of aquatic life and wildlife.

2 Health and Welfare Canada drinking water quality guideline.

3 Saskatchewan Surface Water Quality General Objective.

4 This modelling was done for a scenario with special waste in the Dominique-Janine North pit.

Figure 6

and leaching of stored contaminants from the waste.

Previous experiences with on-land rock piles have often indicated that acid generation and leaching occur and control measures need to be addressed for future mine rock. The state of the art methodology for control of acid generation is underwater disposal, although there is only a few historical information on successful subaqueous disposal of waste rock in freshwater. However, at the Skorovas Mine in Norway, highly oxidized waste rock was placed in Dausjoen Lake in 1990. Early indications in 1991 showed that copper and zinc leached from the waste rocks had been reduced by 80% and 70% respectively. The results are quite encouraging and could be partly attributed to underwater disposal which prevent air entry by convection since the voids between mine rock are filled with water. Oxygen transfer to the mine rock is also limited due to the low coefficient of diffusivity of the water. That is why the underwater disposal of fresh and unoxidized waste rock was chosen for this project.

MODELLING OF THE IMPACT ON CLUFF LAKE WATER

The waste rock placement will impact the water of the lake in the following way:

- discharge of water of the interceptor wells during the operating period;
- contact and exchange between the lake and the rock after the operating phase;
- construction of the dam and the waste pile.

During the operational phase the estimated seepage picked up by the interceptor well and discharged into Cluff Lake is estimated by a hydrogeological model at 160 m^3/hr. After the operational phase, exchange between the pore water of the waste pile and the lake is taken into account to determine the long term effect. A total of 11 water quality parameters were modelled. The predicted annual mean concentration of the constituents is shown in figure 6.

The predicted contaminant levels reach a peak around years 4 to 6. The results of the predictions indicate that the levels of most of the elements of concern will be acceptable and well below surface water quality objectives for the protection of aquatic life. The overall impact on Cluff Lake during the operational phase is assessed to be moderate while after decommissioning the impact is expected to be minor.

Activities related to construction of the dam and the waste pile will use appropriate deposition techniques that will minimize turbidity and suspended solids impacts. Essentially, mine rock will be dumped near the crest and pushed over the edge by bulldozer to allow the material to slide as a mass down the slopes rather than by end dumping over the crest.

OTHER IMPACTS

There will be a loss of fish habitat as 10% of the Cluff Lake area will become land. Therefore, an habitat compensation agreement will be prepared and agreed upon with both federal and provincial Fisheries and Oceans following the assessment of existing fish habitat. In fact Cluff Lake is typically considered as a northern oligotrophic lake with low productivity and does not support commercial fishery activities. It appears that the new shoreline that will be created with the waste pile will be longer than the initial one and selective rock materials may offer a good fish habitat.

A thorough analysis of potential radiation exposure pathways was carried out to assess the potential radiation doses of hypothetical receptors. The pathways analyses considered both air and water emissions of radionuclides coming from several sources (namely the mill site, the tailings area and the pit itself) and numerous potential exposure pathways. Doses are predicted to be small both with respect to dose limits for the public and natural variations in background dose levels; this is obviously consistent with actual results in the Cluff Lake area where open pit mining and milling operations have been on going for more than ten years.

CONCLUSION

The underwater waste rock disposal concept is aimed at minimizing the effects of unsaturated oxidizing environment which currently exists in some above-ground piles.

Consideration was also given to integrating the decommissioning aspects and waste rock management at the design stage. A "design for closure" label would be appropriate for this project. In so doing, long term potential liabilities and impacts will be alleviated or minimized.

A comprehensive monitoring program will be implemented. It will assess the accuracy of the predictions and we are confident it will demonstrate that the overall impact of the underwater disposal of waste is less significant than that of above ground waste piles.

Cogema is confident that this project will "lead the way" to other projects, and that the disposal of wastepiles underwater may well become an environmental standard for uranium mines in the future.

Innovative Mine Design for the 21st Century, Bawden & Archibald (eds) © 1993 Balkema, Rotterdam, ISBN 90 5410 325 6

Passive treatment and control methods for Acid Mine Drainage

L.H. Michaud
Department of Mineral Engineering, The Pennsylvania State University, University Park, Pa., USA

ABSTRACT: Conventional methods of controlling Acid Mine Drainage (AMD) have primarily been limited to collecting the drainage and treating the flows by adding alkaline material. These methods are labour intensive, expensive and require continual and long term management. They do not provide a permanent solution to the AMD problem.

This paper will review the technology that is presently available and is currently being developed for the control of the formation of AMD and for the passive treatment of the AMD. This includes the use of constructed wetlands and anoxic limestone drains. Several techniques can also be utilized to decrease the volume of water that enters a system, including encapsulation of spoil material with an impermeable material, sealing old mine workings, flooding mines, grouting and injection with an alkaline material. Techniques that have shown promise at inhibiting the formation of AMD will also be reviewed. These include phosphate addition, bactericide treatment, alkaline addition, selective mining and spoil placement, in-situ aeration and the use of connecter wells.

1 THE FORMATION OF ACID MINE DRAINAGE

Through mining activities, both coal and metal deposits, large volumes of sulphide bearing rock are exposed to the atmosphere and to the surrounding surface water and ground water. The result is the oxidation and hydrolysis of the sulphides, the production of sulphuric acid and the formation of Acid Mine Drainage (AMD). Common characteristics of AMD includes high acidity, which a possible pH value below 3, and high sulphate and metal concentrations. Typically iron, aluminum and manganese are in high concentrations, although other metals may be present depending on the mineralogy of the surrounding host rock.

AMD can be formed at both surface mining operations and from underground mining. With future and existing mining operations, the prevention of AMD is possible through proper isolation and containment of the sulphide bearing rock. However, most AMD problems are a result of past mining operations. In the United States, it is estimated that more than 20,000 km of streams and rivers are impacted by AMD, with the majority of this related to abandoned mines (Skousen et al., 1992).

The presence of AMD can be devastating. It affects the surface and subsurface hydrology and extends its impact to the surrounding land. The acid and the high metal concentrations are toxic to the aquatic live, wildlife and the vegetation. Sedimentation of metals precipitates affects the visual aesthetics of the environment and adversely affects the water chemistry and aquatic life. Finally, unless adequately treated, the resulting water quality is unsuitable for local commercial and residential use.

Conventional methods of treating Acid Mine Drainage (AMD) include collecting the water after the AMD is formed and neutralizing the acidity by the addition of alkaline material. Any readily available highly alkaline material such as lime, limestone, soda and sodium hydroxide are common sources of alkalinity although other alkaline chemicals have also been successfully used (Ackman et al., 1991a). Regulations also require the removal of metals in excess of established limits. Upon neutralization these metals are precipitated as oxides or oxyhydroxides. Although this method of treatment is effective, it is expensive, both in labour and chemicals, and requires continual maintenance. It must be continued for an indefinite period (Skousen et al., 1992).

In Pennsylvania, increased restrictions of environmental legislation have prohibited mining if AMD will result. Other states have already instituted similar restriction or will likely follow suit in the near future. As well, because the chemical treatment is not considered a permanent solution to the problem, environmental regulators have instituted a policy of blocking the release of reclamation bond for mine sites where conventional methods of treatment are being used.

AMD is a serious environment problem that is directly impacting the coal mining and metal mining industries. To insure the long term future of these industries and the health of our environment, permanent solutions must be found to the AMD problem.

2 PASSIVE TREATMENT SYSTEMS

The goal of AMD research is to provide long-term, zero or low-maintenance methods of abating AMD. Although some human intervention may be required, these processes have been termed passive treatment systems because they provide solutions that are significantly more self-sustaining when compared to conventional chemical treatment techniques.

The most significant recent developments in passive treatment systems that have proven to be effective at abating AMD include the use of aerobic wetlands, anaerobic or compost wetlands and anoxic limestone drains. Although these methods of treatment are presently being utilized in the field, the technology is still in a rapid state of evolution. Research is currently being carried out by universities, government agencies, private researchers and mining companies to further understand and optimize the technology.

2.1 Constructed Wetlands

It was initially discovered that AMD water quality was improved after passing through natural wetlands (Huntsman et al., 1978)(Wieder and Lang, 1982). Since then, hundreds of wetlands have been constructed for the treatment of AMD from mine sites (Hedin, 1991). During this time, the design and technology used to construct the wetlands has evolved in order to increase the efficiency of the systems. Research is continually being carried out on the chemical and microbiological activities that occur in the wetlands and on the interactions of the various parameters in the system. Other areas that are being investigated include the effectiveness of different vegetation mixes and substrate material and the design and sizing of the wetlands, including the depth of the substrate, water retention times, effects of different water levels and fluctuations in the water level. These parameters all affect the optimization of the mechanisms that occur within the wetlands.

Passing AML through wetlands can have the effect of decreasing the concentration of dissolved metals in the water, decreasing the acidity (and therefore raising the pH) and trapping sediments. The ability of wetlands to remove contaminants from water is a result of the chemical, physical and biological processes that occur in the plant/substrate/water environment.

The factors that take an important role in treating AMD have been summarized by Skousen (1992). The basic mechanisms that have been identified for the retention metals include the accumulation of metals in living plants and in the organic substrate, exchange and organic complexion reactions with the substrate and chemical and microbiological oxidation and reduction reactions that lead to precipitation. These processes are all interdependent.

It is known that the ability of specific plant species to accumulate metals varies. However, because plant tissue is renewed on a yearly basis, metal that is taken up by the plants is eventually accumulated in the substrate material. Certain aspects such as the long term effect of metal accumulation in the plant material and in the substrate are still unknown.

The wetland plants also serve to stimulate microbial processes by providing sites for microbial attachment, by releasing oxygen from their roots and by supplying organic matter (Kleinmann, 1991). The organic matter in the substrate serves to removes metals from solution by adsorption and supports chelation and cation exchange processes (Weider et al., 1986). Over 50% of metal retention is a result of these chemical and physical processes that do not require microorganisms. In effect, the wetland is acting as a physical filter. This filtering capability is however finite and quickly becomes saturated.

The most important reaction for the removal of metals is the oxidation and hydrolysis reactions that result in the formation of insoluble precipitates. These processes account for 40% to 70% of the total iron removal in wetlands. Although there is no theoretical maximum to the iron precipitate formation, it is limited by the physical density of the produced sediment that may clog the system. The long term stability of the metal precipitates that accumulate in the wetlands is unclear. Ferric hydroxides may be reduced to ferrous iron by anaerobic bacteria present in the wetlands, resulting in high ferrous concentrations in the effluent (Skousen et al., 1992).

Another important reaction is bacterial sulphate reduction that occurs in the anaerobic environment in the substrate. This process forms an insoluble precipitate that is contained in the substrate. It also consumes acidity and raises the water pH. Up to 30% of the iron may be reduced and may be combined with sulphides (Skousen et al., 1992).

There are two types of constructed wetlands, aerobic wetlands and anaerobic wetlands. Although most of the chemical and physical processes will occur to some degree in both of these systems, their principal purposes differ.

2.1.1 Aerobic Wetlands

Aerobic systems consist of simple wetlands of cattails or more diverse vegetation growing in a clay soil or mine spoil substrate. Clay or synthetic liners may underlay the substrate in order to isolate the wetland from the ground water system. The substrate depth is usually shallow (<30 cm) with a water depths above the substrate of 30 to 45 cm. These conditions enhance oxygen levels in the substrate and therefore oxidation reactions (Hedin et al., 1992). Intermixing shallow wetlands with deeper areas (>60 cm) increases vegetation and aquatic diversity.

The goal of aerobic wetland systems is to aerate the water and to retain the water for a sufficient length of time for oxidation and precipitation reactions to occur. The main reaction in aerobic wetlands is the oxidation and precipitation of metals as oxides and hydroxides. These wetlands are best suited for the treatment of alkaline water or for the precipitation of metals from acidic water after alkalinity has been added, either chemically of by the water passing it through an anoxic limestone drain.

Aerobic wetlands are often constructed in series with interconnecting simple aeration structures such as rock-lined ditches or 'waterfalls'. Often limestone is used in the construction of these unit to add alkalinity to the system and to raise the pH. One aeration unit can provide enough oxygen to oxidize 50-70 mg/l ferrous iron, which then precipitates in an insoluble ferric form.

2.1.2 Anaerobic Wetlands

Anaerobic wetlands, or compost wetlands, are constructed with a thick (30-45 cm) organic substrate (Hedin et al., 1992). Various materials rich in organic matter have been used as substrate material, including hay bales, manure, wood chips and peat moss (Hedin et al., 1992). The most effective substrate material that has been found to date is spent mushroom compost. It contains high concentrations of organic matter, is an excellent microbial substrate and contains limestone. The compost may be mixed with up to 50% spoil material to decrease the density of the material and to increase the permeability (Hellier, 1992).

The vegetation in an anaerobic wetland is similar to that used in aerobic wetlands, consisting mainly of cattails and other emergent species. These plants serve to stabilize the substrate and provide organic matter for the processes that occur in the system.

Unlike aerobic wetlands where the presence of oxygen is important, the main goal of anaerobic wetlands is to create an environment where the oxygen content is restricted. This environment promotes anaerobic sulphate reduction. To promote this, the flow regime is contained mainly within the substrate. Ideally, the water level will stay at or just below the top of the substrate.

2.2 Anoxic Limestone Drains

Anoxic limestone drains (ALD) consist of high quality limestone that is buried in a lined trench and covered with clay or a synthetic liner to insolate it from the atmosphere. When AMD is diverted through an ALD it picks up alkalinity and increases the pH by dissolution of the limestone. The main restriction in the effectiveness of limestone addition in open systems, during conventional chemical treatment of AMD, is the armouring of the limestone with iron oxides, effectively blocking any reactive surface on the limestone. The anoxic state in the trench prevents the oxidation of the iron and the subsequent precipitation of ferric iron in the trench (Nairn et al., 1991)(Nairn et al., 1992)(Skousen, 1991).

ALDs are not intended to treat AMD on their own. They provide a means of generating required alkalinity without the risk of biological limitations (Brodie et al, 1990). They must be followed by an aerobic system in which metal oxidation, hydrolysis and precipitation reactions will occur. The addition of the alkalinity results in an increased effectiveness of the metal removal activities that occur in the wetlands.

The usefulness of ALD are limited to situations where the dissolved oxygen (DO) content and the ferric iron and Al concentrations are low. Skousen (1992) places these limits at a DO value of less than 2 mg/l and Fe^{3+} and Al^{3+} concentrations at less than 25 mg/l. Hedin and Nairn (1992) have stated that <u>any</u> levels of DO are sufficient to promote oxidation of iron within the drain, potentially causing armouring and plugging. Likewise, the presence of <u>any</u> Fe^{3+} and Al^{3+} in the water will result in the hydrolysis and precipitation of the metals, which will also lead to armouring of the limestone. Although the acceptable limits of DO, Fe^{3+} and Al^{3+} may not be accurately known, their presence in any quantity will probably decrease the effective lifetime of the ALD.

3 ACID MINE DRAINAGE VOLUME CONTROL METHODS

Methods of treating AMD after it's formation were described in the previous section. Alternatively, the problems associated with AMD can be resolved by preventing the formation of the AMD. Ideally, this is the preferred method and this route can be instituted to a great degree at new mining operations. Unfortunately, as was stated previously, the majority of our AMD problems are a result of past mining operations.

When examining any mine site and designing a AMD treatment system, the total site should be viewed as a unit. The formation of AMD requires three components: sulphur bearing material, water and air. By eliminating any one of these components, AMD formation will not occur.

There are several treatment techniques have been identified that can result in significantly decreasing the volume of water that enters the system or that will inhibit the formation of

AMD. Most of these techniques are still in the experimental or field trial stage; however, they do show promise as useful methods of abating the AMD problem.

3.1 In-situ Inhibitors

Several methods of treating broken waste rock have been determined that inhibit the acid producing reaction (Ahmed, 1991). Some of these methods that are presently under investigation are described in the following section.

3.1.1 Phosphate Addition

It has been determined that phosphate can form an extremely insoluble surface coating of ferric phosphate on pyrite, thereby inhibiting the oxidation reaction (Evangelou, 1992)(Rermers et al., 1992)(Chiado et al., 1988). This process has been tested in a laboratory environment using a solution of KH_2PO_4 and H_2O_2, and has been found to be very effective in slowing the rate of oxidation. In principle, phosphate could be added in solution or as a soluble salt to the surface of mine spoil, possibly along with an oxidant, and it would migrate downward into the spoil, coating the pyrite and inhibiting the oxidation.

To date, no field trials have been carried out and the practicality and economics of this method on spoil piles have not been demonstrated. One possible problem with this method is the enhancement of bacterial activity that can be generated by the addition of phosphate, which acts as a nutrient for bacteria (Francis et al., 1989).

3.1.2 Bactericide Treatment

Ferric iron is recognized as the dominant oxidant in the formation of severe AMD. The slow step in the process, under acidic conditions, is the oxidation of ferrous iron (Fe^{2+}) to ferric iron (Fe^{3+}) (Singer et al., 1970). This reaction, which is very slow under acid conditions by purely inorganic pathways, is catalyzed by the presence of <u>Thiobacillus ferroxidans</u> bacteria. Bacterial viability is susceptible to various bactericides, and experiments have shown that AMD generation can be slowed by the addition of bactericides to pyritic spoil (Kleinmann et al., 1981)(Sobek et al., 1990).

A potential limiting factor with this technique is that it would probably be necessary to add the bactericides continually or at least periodically. A single treatment is not permanent, even with the use of slow release formulations.

3.1.3 In-situ Alkaline Addition

Another possible method is the addition of an alkaline substance to the spoil material. This can serve to neutralize acidity as well as inhibit the acid-forming reaction. Several methods of accomplishing this inhibition have been attempted. Considerable evidence indicates that the formation of acid is slow at near-neutral pH. Addition of limestone to the surface of spoil material can contribute alkalinity to water percolating downward through the spoil. However, sufficient alkalinity can rarely be added using this method to affect serious AMD problems to a significant degree.

Alternatively, limestone, lime or rock phosphate can be imported and mixed with the spoil as it is mined. This can be used to maintain near-neutral pH conditions under which the formation of acid is slowed, or where the acid that is produced is neutralized within the spoil. This method, however, is not applicable to previously mined material.

A third possibility is the injection of alkaline solutions or slurries (eg. NaOH) into the spoil. This method may be temporarily effective in slowing the acid forming reaction or in neutralizing the AMD, but the reactants are soluble and gradually leach away, allowing the reactions to resume.

3.1.4 Selective Mining and Spoil Placement

At existing or future mine operations, the pyritic material can be effectively isolated from air and water by separating it from the remainder of the spoil by selective mining methods. This material can then be placed on a pad of impervious material (clay or synthetic liners) and covered with an impervious barrier. However, if large amounts of acid producing waste are present, or if suitable local material is not present to construct impervious barriers, the expense of this option may be prohibitive.

Pyritic material can also be selectively spoiled by layering it with alkaline overburden material found at the site. If sufficient alkalinity is present, the acidity in the spoil will be neutralized.

3.1.5 In-situ Aeration of AMD Within the Spoil

If enough alkalinity is available in the spoil to neutralize a significant part of the acidity, wells could be drilled through which air could be injected below the ground water level. This would lead to precipitation of ferric hydroxide within the spoil and the outflow water would therefore contain little or no iron and acidity. It may be necessary to combined this treatment with the injection of alkalinity, as discussed previously. The method is the subject of a current research proposal (Parizek et al., 1991), but has not yet been attempted in field scale tests.

3.1.6 Connecter Wells

With this method, also in the research stage, water can be rapidly removed from an acid-generating sedimentary layer, commonly a coal seam, by provision of one or more wells which will drain the water into deeper zones before much acid has been generated (Parizek et al., 1984). For this method to be successful, a high permeability lower aquifer is required.

3.2 Reduction of the Volume of AMD by Hydrologic Intervention

At abandoned surface mining operations, AMD results from exposing large volumes of fractured pyritic spoil material to air and to water. Water enters the system in the form of direct rainfall or snowfall into the area, surface runoff and ground water flow. Because of variations in the volumes of any of these components over the natural yearly cycle, the total volume of contaminated water that will require treatment can vary significantly over the year. In areas of high snowfall, spring runoff can result in an increase of more than five times the AMD volume. Because all treatment systems must be designed to accommodate the maximum flow volume, if the volume of AMD can be decreased, the cost to treat the water can also be significantly reduced.

In abandoned underground operations, AMD results from the exposure of old mine workings to air and the alteration of subsurface flow patterns, causing exposed rock to come into contact with the ground water. This contaminated water then surfaces as AMD seeps near the mine workings, flows directly into the surrounding streams or stay in the substrate flow.

For both mining methods, there are several methods of reducing the volume of water entering the disturbed zone that should be considered and incorporated into the design of a total AMD abatement system.

3.2.1 Diversion of Surface Flow

Diversion ditches constructed around the perimeter of the site can be used to collect any surface runoff entering the site from the surrounding area. At large mine sites, the surface topography and limits of the local drainage basin should be examined in order to accurately determine surface flow patterns.

In extreme cases, stream channels may lead directly into spoil material or old underground mines, or the stream flow may infiltrate its bed and gain access to acid-generating spoil or abandoned underground workings. Appropriate modification of the stream bed, including relocation of the stream channels and the construction of impervious stream beds by grouting or lining, should be carried out. Studies have shown that this can result in a considerable reduction in the infiltration volume (Ackman et al., 1991b). It is usually sufficient to direct this flow through settling ponds before discharge off-site.

3.2.2 Grouting and Injection

At abandoned underground mine sites with poor overburden competency, air can enter the mine working through fractures in the overburden. Grouting or injection of alkaline material into fractured bedrock can also be used to divert ground water from acid-generating zones (Ackman, 1990)(Ladwig, 1982). Materials that are presently being investigated for sealing cracks in the overburden include flyash, concrete, flyash/concrete mixtures and plastics. Because flyash is a waste product of power generation, its availability is high and its cost is low.

3.2.3 Capping of the Mine Wastes with an Impermeable Material

Infiltration of surface runoff can be minimized by sealing entire spoil piles with an impermeable layer. As well, this has the effect of minimizing the access of oxygen to the spoil material. Several potential capping materials are presently under investigation, including dry and wet layers. Potential wet capping materials include bog peat, sewage sludge and wood chips (Dave et al., 1991)(Lundgren et al., 1991). Although these materials can be effective, they must be kept saturated in order to form a seal and usually require the installation of an expensive irrigation system.

Dry capping materials that are being studied include flyash, clay, sludge, glacial till with bentonite, cement mixtures, mixtures of tailings and flyash, plastics, asphaltic and bituminous materials, geotextiles and geopolymers (Scheetz, 1992)(Murray et al., 1988)(Gerencher et al., 1991)(Malhotra, 1991). The major thrust of the research is to identify a suitable capping material with long term stability. The cost of capping large areas can be high (up to $35/m^2). However, in areas where a significant volume of the drainage is due to direct melting and infiltration of snow and rainwater, these techniques may result in reducing the maximum flow of AMD by over 50%. This therefore decreases the necessary size of all further treatment facilities.

A capping material must be able to support vegetation without sacrificing it's impermeability or it must be covered with soil for revegetation. The effective establishment of a vegetative cover with a dense root system can, in itself, reduce infiltration and contain the majority of the flow in the upper horizons of the soil. This results in a significant decrease in the percolation of water through the underlying spoil material (Michaud, 1981).

3.2.4 Flooding of the Pyritic Material

Underground mines or surface spoil material can also be effectively isolated from air by flooding the material with water. Surface mine spoils can be inundated by placing them in water filled pits. Underground mines can be sealed and allowed to flood naturally. Acid generation is minimal under these conditions because of the low volume of dissolved oxygen present in water. Diffusion rates in water are also very slow compared to rates in the air present in the unsaturated zone (Dyer, 1984).

Although very effective, this method is only suitable under certain conditions. The overburden and surrounding rock must be competent enough to contain the water, the mine design must be such to allow effective sealing of all mine openings, and the water volume must be adequate to insure 100% flooding of the pyritic material on a year round basis. If the pyrite is exposed for a portion of the year, AMD formation will result.

4 PROPOSED ABATEMENT AND TREATMENT PROGRAM

It is not possible to determine the 'best' technique to mitigate the AMD problems at a site until a detailed site investigation has been carried out. Parameters that must be investigated include an analysis of the local soil and overburden mineralogy, hydrologic flow patterns of both the surface and subsurface flows, location and sampling of all AMD seeps and an analysis of their chemistry and flow rates over an annual cycle, and the location of all previous mining operations. Based on this information, the AMD problem can be delineated and recommendations for the optimum method of AMD mitigation can be formulated.

As the first stage in the development of an overall AMD treatment program, methods of reducing the volume of AMD should be considered. The remaining drainage from the site will include any subsurface flow that infiltrated prior to entering the spoil area and any water that contacts the pyritic material due to the extension of the water table into the mine spoil. With the elimination of infiltration of water and oxygen through the surface of the spoil piles, the quality of the remaining seeps may improve significantly.

The optimum treatment of the remaining AMD may include a combination of passive treatment techniques that were described in the first part of this paper, including aerobic wetlands, anaerobic wetlands and anoxic limestone drains.

The sizing and design of the constructed wetlands is dependent of the AMD properties and on the site characteristics of the mine. This includes the topography and available land area, the water quality, flow volumes and variations in volumes, AMD discharge points and exit points into the surrounding hydrologic system and the available hydraulic potential to drive the wetland treatment processes.

Passive treatment systems can have a significant impact on the water quality. However, this technology may not remove all of the contaminants. Where the effluent water quality does not meet required standards, post treatment unit are necessary as the final stage of treatment. This may includes a final settling pond to insure removal of any sediment. This pond can also provide a site for testing of the final effluent prior to discharge off-site and, when needed, for temporary additional chemical treatment, such as the addition of alkaline material or flocculants.

REFERENCES

Ackman, T.E., 1990. Locating and abating sources of acid mine drainage at reclaimed strip mines. In Proceedings of the 1990 Mining and Reclamation Conference and Exhibition, Vol. II, Charleston, WV, April, p.503.

Ackman, T.E. and Kleinmann, R.L.P., 1991a, 'An In-line System for Treatment of Mine Water', International Mine Waste Management News, Vol 1, No. 3, p.1-4.

Ackman, T.E. and Hustwitt, C., 1991b. Repairing stream channels to reduce water loss into underground mines. Technology News, no. 388, U.S. Bureau of Mines, Pittsburgh Research Center, Pittsburgh, PA, July, pp.2.

Ahmed, S.M., 1991. Electrochemical and surface chemical methods for the prevention of the atmospheric oxidation of sulfide tailings. In Second International Conference on the Abatement of Acid Drainage, Montreal, Canada, September.

Brodie, G.A, Britt, C.R., Taylor, H., Tomaxzewski, T. and Turner, D., 1990, 'Passive Anoxic Limestone Drains to Increase Effectiveness of Wetlands Acid Drainage Treatment Systems', Proceedings of the 12th National Association of Abandoned Mined Land Conference, Breckinridge, CO.

Chiado, E.D., Bowders, J.J., and Sencindiver, J.C., 1988. Phosphatic clay slurries for reducing acid mine drainage from reclaimed mine sites. U.S. Bureau of Mines IC 9183, vol I: Mine Water and Mine Waste, Washington D.C., p.44-50.

Dave, N.K. and Michelutti, R.E., 1991. Field evaluation of wet cover alternatives of high sulfur tailings final report. In Second International Conference on the Abatement of Acid Drainage, Montreal, Canada, September, p.61-81.

Dyer, K.L., 1984. Water, friend or foe in the control of acid mine drainage. In Proceedings of the Fifth Annual West Virginia Surface Mine Drainage Task Force Symposium, Morgantown, WV, March, pp.16.

Evangelou, V.P., 1992. Personal Communication. Department of Agronomy, U. of Kentucky, Lexington, KY.

Francis, A.T., Dodge, C.J., Rose, A.W. and Ramirez, A.J., 1989. Aerobic and anaerobic microbial dissolution of toxic metals from coal waste: Mechanism of action. Environmental Science and Technology, v 23, p.435-441.

Gerencher, E.H., Wong, J.Y., van Dyk, R., Konasewich, D.E., 1991. The use of mine tailings in concrete surface covers to control acid mine drainage in waste rock dumps. In Second International Conference on the Abatement of Acid Drainage, Montreal, Canada, September, p.69-84.

Hedin, R.S. and Nairn, R.W., 1991, 'Contaminant Removal Capabilities of Wetlands Constructed to Treat Coal Mine Drainage', Proceedings of the International Symposium on Constructed Wetlands for Water Quality Improvement, Pensacola, FL, Oct.

Hedin, R.S. and Nairn, R.W., 1992, 'Designing and Sizing Passive Mine Drainage Treatment Systems', Internal report, U,S, Bureau of Mines, Pittsburgh Research Center, Pittsburgh, PA, pp.11.

Hellier, W.W., 1992, DER - Bureau of Mining and Reclamation, personal conversation.

Huntsman, B.E., Solch, J.B. and Porter, M.D., 1978, 'Utilization of a Sphagnum Species Dominated Bog for coal Acid Mine Drainage Abatement', 91st Annual Meeting of the American Geologic Society -Abstracts, Ottawa, Canada.

Kleinmann, R.L.P., Coerar, D.A., Pacelli, D.A., 1981. Biogeochemistry of acid mine drainage and a method to control acid formation. Mining Engineering, p.298-305.

Kleinmann, R.L.P., 1991, 'Biological Treatment of Mine Water - an Overview', Proceedings of Second International Conference on the Abatement of Acidic Drainage, MEND, Montreal Canada, p.27-42.

Ladwig, K.J., 1982. Delineation of zones of acid mine drainage using surface geophysics. In Symposium on Surface Mining Hydrology, Sedimentology and Reclamation, Lexington, KY, December, pp.9.

Lundgren, T. and Lindahl, L., 1991. Remedial action programs for sulphidic waste from ancient mining in central Sweden. In Second International Conference on the Abatement of Acid Drainage, Montreal, Canada, September, p.403-416.

Malhotra, V.M., 1991. Fiber-reinforced high volume fly ash shotcrete for controlling aggressive leachate from exposed rock surfaces and mine Tailings. In Second International Conference on the Abatement of Acid Drainage, Montreal, Canada, September, p.505-513.

Michaud, L.H., 1981, 'A Manual of Reclamation Practices', International Academic Services Ltd., Kingston, Ontario, 1981, pp.345.

Murray, E.W., Goudey, S.P., McCready, R.G.L., and Salley,J., 1988. Laboratory and field testing of a salt-supplemented clay cap as an impermeable seal over pyritic slates. U.S. Bureau of Mines IC 9183, vol I: Mine Water and Mine Waste, Washington D.C., p.52-58.

Nairn, R.W., Hedin,R.S., and Watzlaf,G.R., 1991. A preliminary review of the use of anoxic limestone drains in the passive treatment of acid mine drainage. In Proceedings of the 12th Annual West Virginia Mine Drainage Taskforce Symposium, Morgantown, WV, p.23-38.

Nairn, R.W., Hedin, R.S. and Walzlaf, G.R., 1992, 'Generation of Alkalinity in an Anoxic Limestone Drain', paper presented at the 1992 National Meeting of the American Society for Surface Mining and

Parizek, R.R., Buss, D.R., Yackin, G. and Diaz, J., 1984. Demonstration of the corrector well method of reducing leakage and abating acid mine drainage from an underground coal mine, Tioga County, Pennsylvania. Final Report to PA DER, in cooperation with the U.S. Environmental Protection Agency, pp.148.

Parizek, R.R. and Rose,A.W., 1991. Demonstration of two insitu methods of reducing or treating acid mine drainage within reclaimed surface mines. Proposal Submitted to the Minerals Industry Waste

Treatment and Recovery Center, University of Nevada-Reno, NV, pp.12.
Rermers, G.W. and Franke, D.F., 1992. Effect of additives on pyrite oxidation. U.S. Bureau of Mines, R.I. 9353, pp.13.
Scheetz, B.E., 1992. Personal Communication. Senior Scientist and Professor of Materials, Materials Research Laboratory, Penn State University, University Park, PA.
Skousen, J., 1991, 'Anoxic Limestone Drains for Acid Mine Drainage Treatment', Green Lands, Vol 21, No. 4, p.30-35.
Skousen, J., Sextone, A., Garbutt, K. and Sencindiver, J., 1992, 'Wetlands for treating Acid Mine Drainage', Green Lands, Vol 22 No. 45, Fall, p.31-39.
Singer, P.C. and Stumm, W., 1970. Acid mine drainage: the rate determining step. Science, v 167, p.1121-1125.
Sobek, A.A., Benedetti, D.A., and Rostogi, V., 1990. Successful reclamation using controlled release bactericides: Two case histories. In 1990 Mining and Reclamation Conference and Exhibition, Charleston, WV, April, p.33-41.
Wieder, R.K., and Lang, G.E., 1982, 'Modifications of Acid Mine Drainage in a Freshwater Wetland', Symposium on Wetlands of the Unglaciated Appalachian Region, West Virginia University, Morgantown, WV, p.43-53.
Wieder, R.K., and Lang, G.E., 1986, ' Fe, Al, Mn and S Chemistry of Sphagnum Peat in Four Peatlands with Different Metal and Sulfur Input', Water Air amd Soil Pollution, Vol 29, p.309-320.

9 Explosives technology and blast monitoring

Innovative Mine Design for the 21st Century, Bawden & Archibald (eds) © 1993 Balkema, Rotterdam, ISBN 90 5410 325 6

Continuous velocity of detonation measurements in full scale blast environments

R. Frank Chiappetta
Blasting Analysis International, Inc., Allentown, Pa., USA

ABSTRACT: An explosive's velocity of detonation (VOD), can be used to indicate a number of important characteristics regarding the product's performance under specific field and test conditions. A number of new characteristic and transient VOD curves have been identified in the field, which can be used to evaluate explosive performance, control ground vibration amplitudes and frequencies, select the correct amount and type of stemming for use at the collar and in stem decks, eliminate explosive desensitization, evaluate primer performance, design air deck based blasts, evaluate contaminated explosives and to overcome post blast noxious fumes.

Tests were conducted over a six year period in single and multi-hole blasts using laboratory and full scale blast environments. Explosives tested ranged from pure Emulsion to Anfo and various grades of Emulsion/Anfo blends. Field test parameters were; borehole diameter (1 1/2 - 30 inches), hole depths (10 - 120 feet), primer size (0.5 - 6.4 pounds) and the blast environment varied from soft, jelly-like tar sands to some of the hardest iron ore formations. Most tests were instrumented with an array of blast monitoring instrumentation systems consisting of continuous velocity of detonation recorders, high-speed 16 mm cameras, laser-surveying instrumentation and seismographs which were placed in the near and far fields.

1. INTRODUCTION

With the growing economic pressures faced by mining operators today, blast optimization programs are beginning to demand ever more attention than before. There are no other unit mining operations, other than major equipment and/or plant failures, that can affect the total mining system costs more. Inadequate or malfunctioning blast designs can have grave consequences throughout the entire mining cycle, beginning from the direct blast result itself to the loading, hauling, crushing and processing cycles. It has also been well recognized that additional expenditures and emphasis on the direct drilling and blasting operations can usually result in an overall system cost decrease. This can only occur when the situation is carefully evaluated and a suitable blasting program is implemented according to the site specifics, blasting applications and mine objectives.

Techniques of assessing blast results are quite varied, with the most common and widely used approach still being visual assessment immediately following a blast. Although this technique can qualitatively point out gross differences between blasts, it has never been effective in producing substantial productivity improvements nor in identifying potential problem areas in the blast for further improvements.

Operators are also finding that one or two blast designs to encompass all blasting situations at the mine may not be in the best overall economic interests for the operation. Site-specific and varying custom blast designs are required to accommodate the different blast environments, applications, geologic settings, environmental concerns and safety considerations. With new PC based blast monitoring instrumentation systems and recently introduced new analytical techniques of analysis, the operator can easily eliminate most of the guess work involved in designing efficient blast designs and in evaluating explosive performance in any mining environment.

Although the primary focus of this paper is on continuous velocity of detonation measurements in varying full scale blast environments and their impact on blast results, a discussion on the complete methodology to characterize the in-situ environment, material response under dynamic loading and explosive characteristics is included for completeness.

2. MONITORING PRACTICE

Detailed monitoring on hundreds of single hole and multi-row blast designs were completed in varying blast environments for specific analyses using high-speed photography and/or videography, laser surveying, refraction seismographs, conventional seismographs to measure ground vibration/airblast and continuous VOD instrumentation. The majority of field tests were instrumented with multi-instrumentation systems in order to generate the most meaningful interpretation of the results for use in blast design optimization. Very rarely was only one type of instrumentation system used on full scale tests because it would severely restrict the scope of information required to generate site-specific custom blast designs, (Chiappetta, 1991). For example, those that promote a single piece of equipment such as a seismograph to measure a single disturbance point in all of space, and then attempt to use the result to redesign a blast for improving on fragmentation, selecting optimum delay intervals, advising on explosive selections and recommending general blast design parameters, etc. can be misleading. Likewise, redesigning a blast based strictly on high-speed photography results, and assuming that the explosives, detonators and primers performed perfectly in the production blast can also have grave consequences. Stand alone velocity of detonation data, if not coupled to information obtained from the other instrumentation systems can also be meaningless from a design standpoint, unless obvious explosive malfunctioning or performance is evident. The point here is that unless the data are reviewed collectively from all of the instrumentation systems, true blast design optimization might never be realized, and the resulting design could be flawed.

3. CONTINUOUS VELOCITY OF DETONATION MEASUREMENTS

Until recently, large volumes of data pertaining to velocity of detonation measurements in multiple holes were not possible in the commercial explosive sector, (Chiappetta, 1992 a). Furthermore, data were either restricted to relatively short lengths or sampled with discontinuous systems that missed critical information, or did not have the resolution to record transient VOD's along the entire explosive length, including the important primer region. Knowing how and when explosives function can assist in making intelligent decisions regarding explosive, detonator and primer selections.

Tests were performed with a portable, continuous VOD recorder (VODR-1), which was developed and fielded as a collaborative venture between EG&G Special Projects, Inc. and Blasting Analysis International, Inc. The VODR-1 is a commercial version of the CORRTEX System, which was originally developed by Los Alamos National Laboratory to measure shock wave propagation in nuclear tests to determine nuclear explosive yield. In the measurement of explosive performance, a coaxial cable is placed in an explosive filled borehole with the far end of the cable taped on the primer. Operation is similar to that of RADAR where a pulse of radio waves is sent out and a echo or reflected pulse is returned to give ranging information. The VODR-1 uses standard coaxial cables to carry a fast rise time electrical pulse back and forth every 5 - 200 μs, which travels between 70 - 85% the speed of light. (Refer to Figure 1) Accurate displacement versus time data can be obtained for the entire length of explosive column. Information can be processed into tabular or graphical forms to analyze detonation rates, timing information and disturbance velocities through stemming and compared to expected performance.

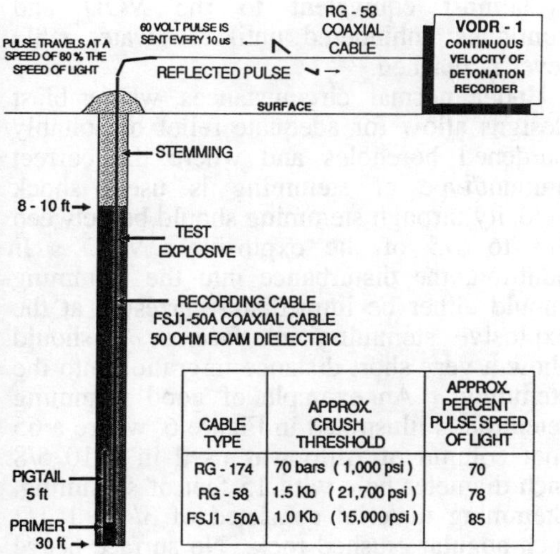

CABLE TYPE	APPROX. CRUSH THRESHOLD	APPROX. PERCENT PULSE SPEED OF LIGHT
RG - 174	70 bars (1,000 psi)	70
RG - 58	1.5 Kb (21,700 psi)	78
FSJ1 - 50A	1.0 Kb (15,000 psi)	85

Figure 1. Typical set-up for single hole measurements

3.1 Single - multiple hole VOD measurements

A single hole example illustrating the detailed, raw and unfiltered data obtained from the VODR-1 system is presented in Figure 2 as a displacement versus time plot. The data were

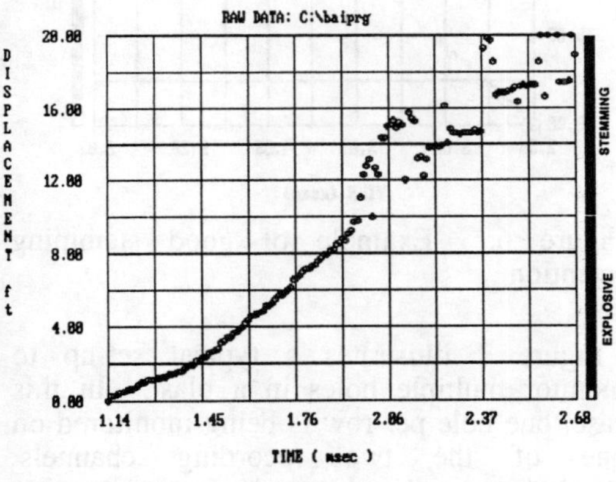

Figure 2. Displacement versus time showing raw unfiltered data.
Sample rate = 10 μs

sampled at 10 μs for a 10 foot continuous column of a gassed Emulsion/Anfo blend in a 20 foot, 10 inch diameter borehole.

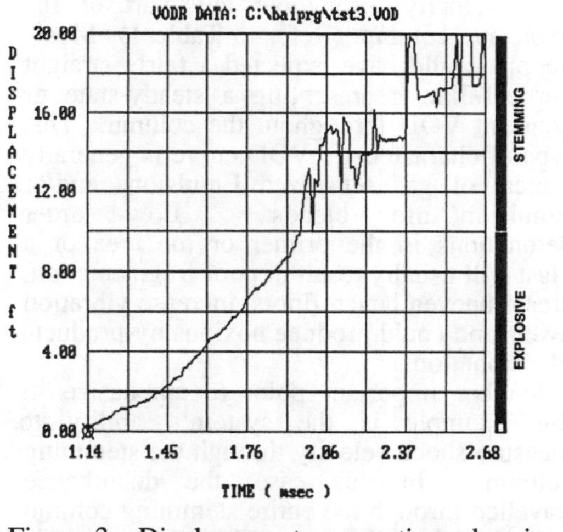

Figure 3. Displacement versus time showing data points connected

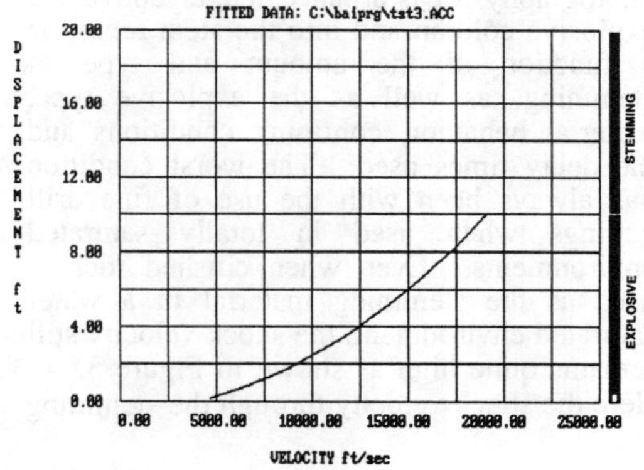

Figure 4. Displacement versus VOD

Numerical and graphical analysis were performed with the VODR-PC1 software that can easily manipulate the raw data to perform curve fitting, zooming, filtering and other graphical analyses. Figure 3 illustrates that same data as a continuous function in the time domain and Figure 4 shows the numerical calculations and curve fitting results as a plot of displacement versus VOD. Note that the VOD in this case started at about 5,000 ft/sec in the primer region and increased uniformly along the explosive column to about 19,000 ft/sec. In this particular environment
and with this borehole diameter and explosive formulation, the VOD did not reach a steady

state velocity throughout any part of the explosive column. (Refer to Table 1) Most people would have expected a fairly straight vertical line, representing a steady-state or constant VOD throughout the column. This type of characteristic VOD curve is generally typical of gas sensitized Emulsions and/or Emulsion/Anfo blends. Low order detonations in the primer or toe area of a blast will usually result in poor fragmentation, create uneven bench floors, increase vibration levels and could produce noxious by-products of detonation.

Another important point to emphasize in this example is the system's ability to measure shock velocity through the stemming column. In this case, the disturbance travelled through the entire stemming column for a length of 10 feet. This observation is usually indicative of violent stemming ejections, cratering and airblast generation as was substantiated with the high-speed photography. Disturbance length above the explosive column and into the stem region is a function of the amount and type of stemming, as well as the explosive type, material behavior, confining conditions and the delay times used. The worst condition has always been with the use of fine drill cuttings when used in totally saturated environments. Even when crushed rock is used as the stemming material in a water saturated environment, the shock velocity still remains quite high as shown in Figure 5. Here the shock velocity through the stemming

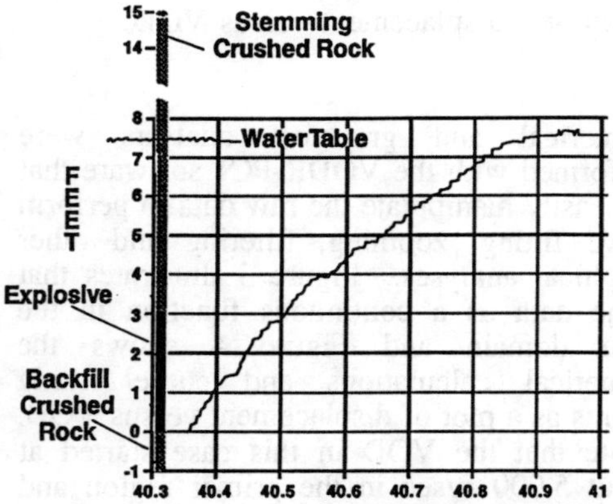

Figure 5. Displacement versus time illustrating VOD and shock velocity through water saturated stemming

is almost equivalent to the VOD and continues unhindered until the water table level is reached.

Under normal circumstances where blast designs allow for adequate relief of suitably burdened boreholes and where the correct amount/type of stemming is used, shock velocity through stemming should be between 1/3 to 1/5 of the explosive's VOD. In addition, the disturbance into the stemming should either be immediately arrested at the explosive stemming interface or it should show a very short distance migration into the stemming. An example of good stemming retention is illustrated in Figure 6, where a 65 foot column of Anfo was used in a 10 5/8 inch diameter hole with 15 feet of stemming. Stemming material consisted of 3/4 - 1 1/2 inch angular crushed rock. No surface heave was experienced on this blast.

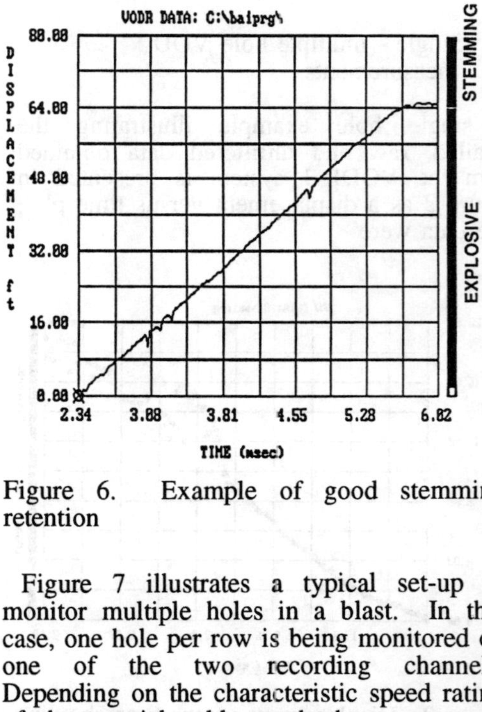

Figure 6. Example of good stemming retention

Figure 7 illustrates a typical set-up to monitor multiple holes in a blast. In this case, one hole per row is being monitored on one of the two recording channels. Depending on the characteristic speed rating of the coaxial cable used, the system can record from 2,500 - 4,825 feet per channel, excluding the minimum take-up cable of 850 feet going to the VODR-1 recorder. Under the one channel set-up, shown in Figure 7, detailed information on each hole can be obtained for the continuous VOD, crush velocity through the stemming, exact

Table 1. Measured and calculated numerical values obtained from the VODR-1 continuous velocity of detonation recorder

Time (msec)	Displacement (ft)	VOD (ft/sec)	Acceleration (ft/sec^2)
1.120	0.00	----	-----
1.140	0.17	4667	17573
1.188	0.42	5515	17573
1.236	0.71	6362	17573
1.285	1.03	7210	17573
1.333	1.40	8058	17573
1.381	1.81	8905	17573
1.429	2.26	9753	17573
1.478	2.75	10601	17573
1.526	3.28	11448	17573
1.574	3.86	12296	17573
1.622	4.47	13144	17573
1.671	5.12	13991	17573
1.719	5.82	14839	17573
1.767	6.56	15687	17573
1.815	7.33	16534	17573
1.864	8.15	17382	17573
1.912	9.01	18229	17573
1.960	9.91	19077	17573

Table 2. Measured VOD,s and firing time of holes shown in Figure 8

Hole no.	Average VOD (ft/sec)	Cumulative firing time (msec)	Delta firing time between holes (msec)
1	11,480	0.00	
2	11,970	34.83	34.83
3	12,150	59.97	25.14
4	11,985	74.72	14.75
5	12,460	102.41	27.69
6	11,275	152.03	49.62
7	11,995	157.12	5.09

Time resolution = ± 10 microseconds

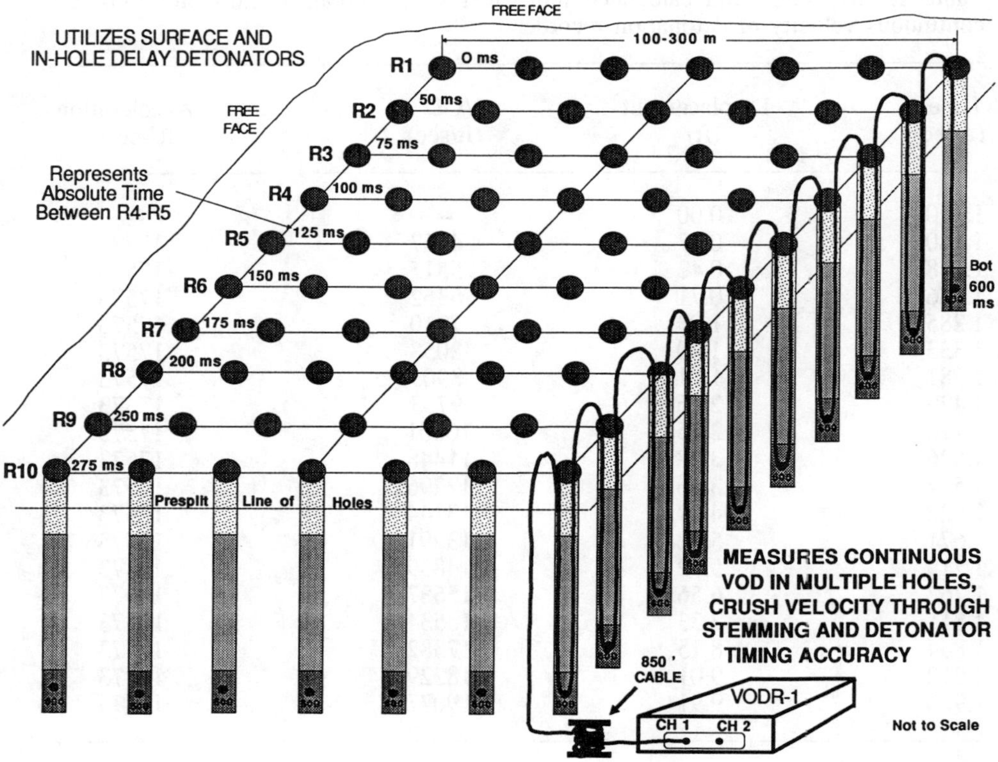

Figure 7. Typical field set-up for multiple hole measurements

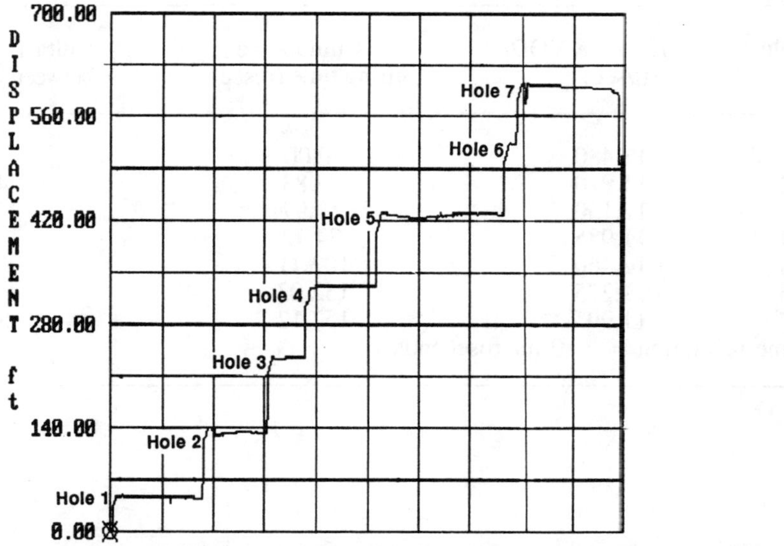

Figure 8. Results of a 7 hole, one row shot

detonator firing time, run up VOD's above the primer and the exact hole and location within any hole where massive ground shifts may have occurred.

Figure 8 illustrates the results of a 7 hole, one row shot. One to three cartridges of pure emulsion cartridges were placed at the hole bottoms to raise the bottom load above the water level. Free poured Anfo was used as the top load in the 4 1/2 inch holes, drilled to depths ranging from about 30 - 40 feet. Stemming varied between 10 - 15 feet, for the primary control of airblast. Timing between holes was expected to be 25 milliseconds, relative to hole Number 1. The y-axis represents the distance from the end of the coaxial cable and the x-axis represents time after shot initiation relative to hole Number 1. Results are listed in Table 2.

Each hole is marked in Figure 8 in the area of the VOD slope line. The stepped vertical increases along the y-axis from one hole to another represent cable distances between the crush disturbance through stemming of one hole to primer initiation in an adjacent hole. It is obvious that the detonators did not fire at their rated times as shown by the irregular time lengths along the x-axis representing adjacent holes. Because the control and reduction of blast induced ground vibrations were critical in this operation and the currently used detonators consistently exhibited unpredictable scatter times, recommendations were made to change initiation systems immediately. Trying to control ground vibrations and reducing complaints with this initiation system, as the mine had been attempting to for years, was concluded as being nothing less than a futile attempt. Detonator timing results obtained for three full scale blasts and five separate detonator series tests on surface showed similar erratic timing, including severe overlaps in 5 out of the 8 tests. It was unfortunate that the mine had attempted just about every blast design change conceivable to alleviate this problem, but never once suspected detonator timing problems. It was also one of the reasons why our full scale seismic simulations did not correlate very well. This case clearly illustrates that a complete blasting analysis with multi-instrumentation systems are required to first identify where the real problems are before trying to implement any correctional measures.

3.2 Types of characteristic VOD curves

A number of characteristic and transient VOD curves have been identified to date and are summarized as shown in Figure 9 (a - j). The most common, generally expected result is illustrated in Figure 9 (a) where the VOD remains constant with no variation throughout the entire explosive column. In Figure 9 (b) the VOD can either increase linearly, beginning in the primer region throughout the column or it can decrease linearly as shown in Figure 9 (c). In Figure 9 (d), the VOD increases at an increasing rate and in Figure 9 (e), the VOD decreases at an increasing rate.

In Figure 9 (f), the VOD starts low in the primer region, increases at an increasing rate to about 30 feet into the explosive column, and then stabilizes at about 15,000 ft/sec. for the remainder of the column between 30 - 47 feet. In Figure 9 (g), the VOD increases throughout the column to about half way up the borehole and then begins to decline towards the top half of the borehole. The opposite effect can also occur as shown in Figure 9 (h) where the VOD decreases in the bottom half of the borehole and then begins to increase in the last half of the borehole. In Figure 9 (i and j), the VOD can be either low or high beginning in the primer region, and alternates continuously between high and low (or vise versa) throughout the column. Various combinations or derivatives of the characteristic VOD curves presented in Figure 9 can also occur.

The results presented in Figure 9 were obtained in varying full scale, blast environments which varied from soft, jelly like tar sands to some of the hardest, most competent iron ore formations and everything in-between. Explosives varied from pure Emulsions to Anfo, including different grades of Emulsion/Anfo blends with sensitized and unsensitized agents. Borehole diameters varied from 4 1/2 - 30 inches, hole depths varied from 10 to 120 feet, primers varied from 1/2 to 6.4 pounds and the initiation systems included detonating cord, shock tube, electric and electronic detonators. Measurements also included single stand alone holes and multi-hole measurements in the same blast to study the effects of VOD performance on adjacent holes.

Similar results, in terms of the characteristic

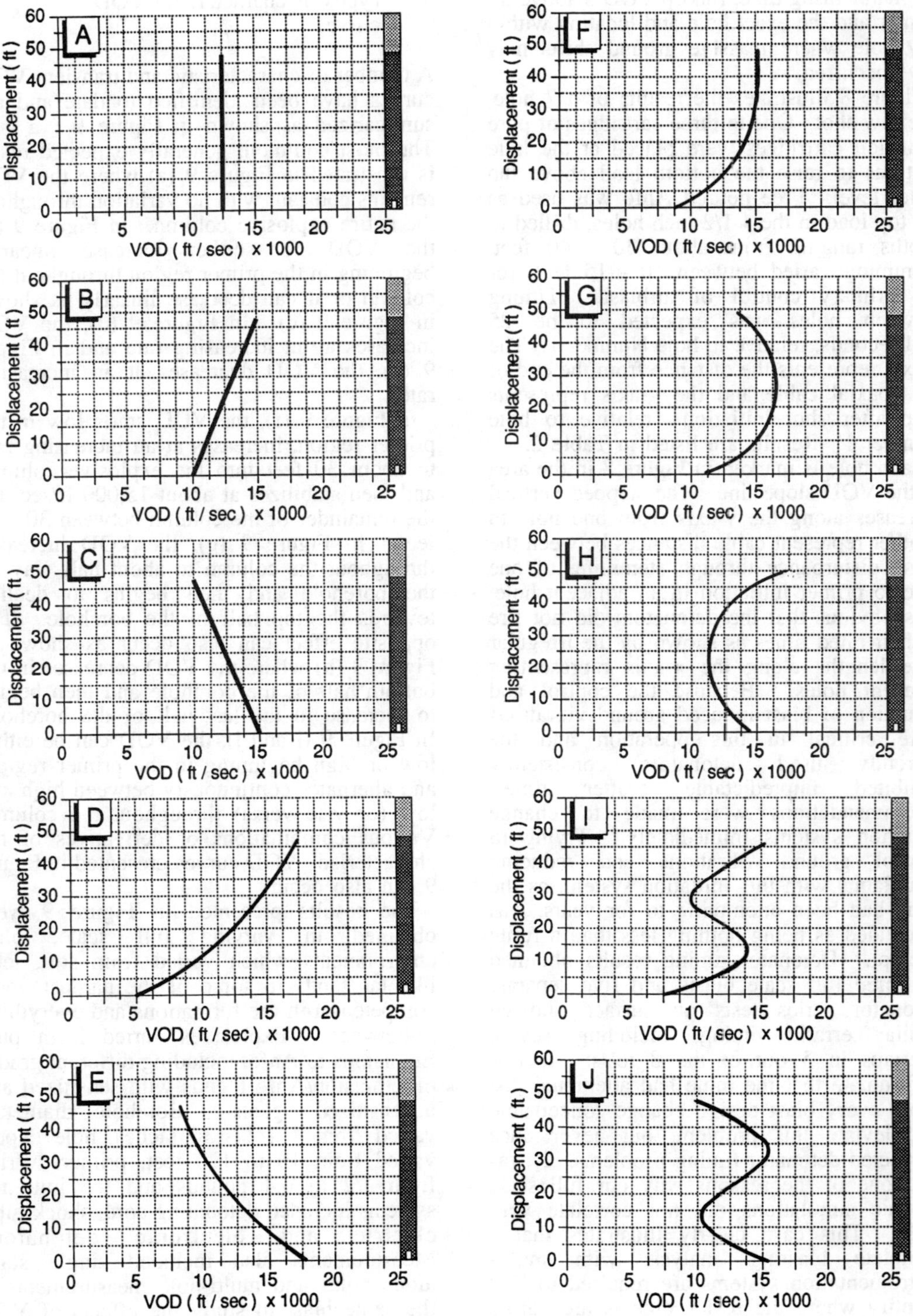

Figure 9. Typical characteristic VOD curves

VOD curves presented in Figure 9, were also documented in well controlled, confined and unconfined tube tests which were fired on surface. Numerical values from laboratory results differed and did not correlate very well with field values when the laboratory test parameters, (i.e. borehole diameter, primer, confining conditions), deviated greatly from the field parameters. For example, the results of laboratory tests in 10 inch tubes did not correlate very well with production size holes between 18 - 30 inch diameters. A good correlation was observed, however, between 7 3/4 to 12 inch production holes with results from the 10 inch tubes. Likewise, laboratory tests using small diameter holes in the range of 1 1/2 - 4 inch did not correlate well with the larger 6 1/2 to 15 inch production holes. This was attributed directly to scaling parameters and the performance characteristics of various explosives when they were subjected to specific field conditions.

Although over 600 tests have been completed to date, it is still not completely clear as to why or how these types of characteristics and transient VOD curves occurred. Detonation performance was found to be dependent on considerably more sensitive variables than was previously believed. In addition to the physical and chemical characteristics of the explosive formulation, the most predominant condition was that of the blast environment itself. Given different blast environments, the same explosive (Anfo, emulsions or emulsion/Anfo blends) can behave with unpredictable performance. General blast design parameters, such as burden, spacing, timing, hole depth, blast geometry and the amount and type of stemming can also affect the detonation characteristics. In-situ borehole confining conditions, loading procedures, dilution effects, water and bridging, etc. were also equally important. In spite of the field complexities identified, there were a number of interesting observations that have broadened our understanding of what actually occurs during detonation and what can be achieved to eliminate and/or alleviate poor performance.

3.3 VOD characteristics of different explosive formulations

Figure 10 illustrates the characteristic VOD transient curves for a number of different Emulsion/Anfo blends. These were performed under very controlled standardized test conditions in 10 inch diameter, 10 foot long thick walled tubes. The priming system was held constant using 2 pound cast primers. Note, that the VOD along the explosive column length is not constant but increases at varying rates beginning at the primer region. Also, the characteristic curves are essentially identical, but shift to the right or left depending on the formulation. When the same formulations were tested in the full scale environment consisting of a very soft formation, completely different characteristic curves were obtained. This emphasizes once again that there is no substitute for full scale, in-situ characterization tests.

Figure 11 compares the results of different Emulsion/Anfo formulations for individual large diameter boreholes. The explosive column for all tests was approximately 13 feet and is drawn to scale on the right hand side of the graph. Both explosive 1 and 5 resulted in partial detonations that failed about one third of the way up the column. Explosive 2 produced a full detonation for the entire length of column. Explosive 4 resulted in a low order detonation for the bottom third

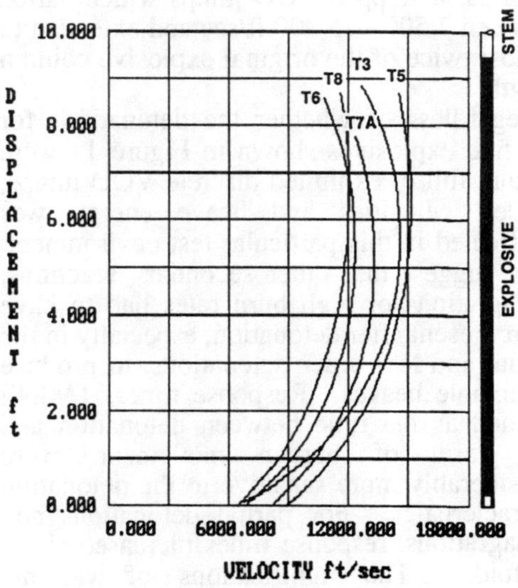

Figure 10. VOD results for a number of different Emulsion/Anfo blends

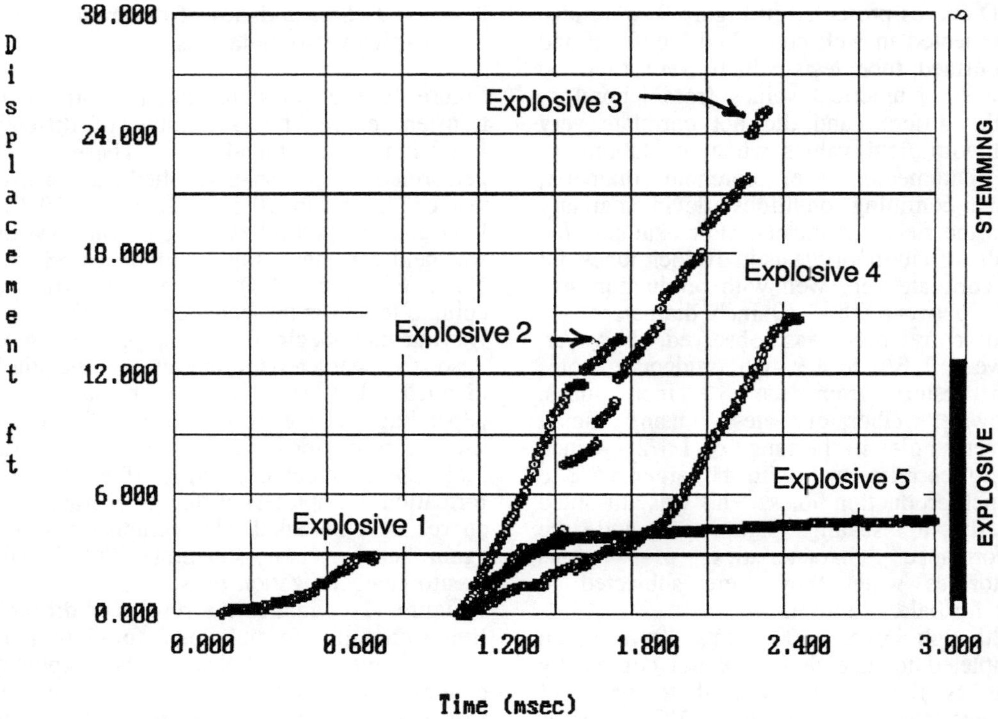

Figure 11. VOD results for a number of different Emulsion/Anfo blends from large diameter production holes

of the explosive column and then suddenly kicked into a full order detonation for the top 2/3 of the explosive column. Explosive 3 resulted in stepped VOD jumps which varied between 7,500 - 15,800 ft/sec and extended to almost twice of the original explosive column length.

Regardless of whether the detonations for the five explosives shown in Figure 11 were partial, full or exhibited discrete VOD jumps, burden velocities and heave energy was unaffected in this particular test environment. This suggest that other secondary reactions, deflagrations or high burn rates had to have been present after detonation, especially in the partial and low order detonations, to produce acceptable heaves. Response times, TMIN's defined as the time between detonation and the onset of burden movement, were considerably more sensitive to the detonation characteristics. For partial detonations and deflagrations, response times increased about 3 fold. The ramifications of varying detonation characteristics in full scale blasts are that the delay times selected between

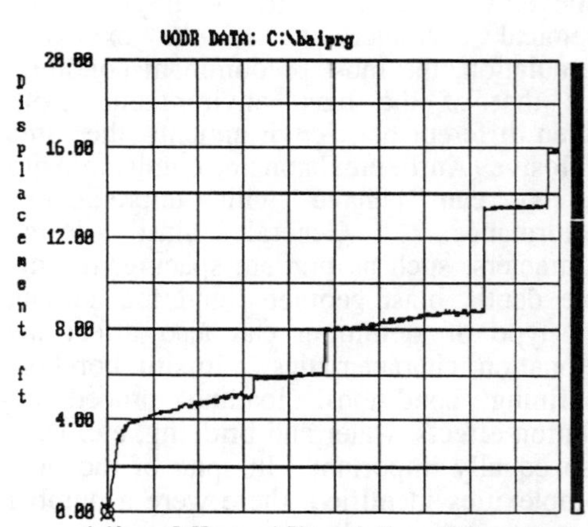

Figure 12. Partial detonation and late time deflagrations

holes and rows of holes can be quickly negated. This can alter the heave appearance, muck pile shape, fragmentation and the overall blast result. Figure 12 illustrates an excellent test result of a partial 3 foot detonation over a 13 foot explosive column and at least 3 deflagration sections following detonations in late time along the column.

3.4 Site characterization using VOD measurements

When the correct coaxial cable is selected for measurements in a particular environment, the crush threshold of the cable can provide excellent information on the integrity, structural and geomechanic condition of the surrounding rock mass. Consider the zoomed-in, detailed VOD results of hole 1 and hole 3 from the 7 hole, one row shot presented in Figure 8. Holes 1 and 3 are shown in Figures 13 and 14, respectively. Both VOD records are raw, unfiltered data. The obvious difference between these two VOD records is that hole 1, (the first hole on the shot), produced a very noisy record, while that of hole 3, (the third hole fired on the shot), produced a very clean record with virtually no noise. In both holes, performance details for the bottom load emulsion and top load Anfo explosives are quite evident and identifiable. Stable detonations and respectable VOD values were also achieved in both holes.

The absence of electronic noise from hole 3 indicates quite clearly that the rock mass bounded by the length of the explosive column was quite intact with very minor structural discontinuities and that no soft layered formations existed. In hole 1, the electronic noise correlates extremely well to those areas which are not intact, and consist of weak and/or soft layered formations. In reference to the y-axis of hole 1, shown in Figure 13, the two main weak areas are bounded between 4.8 - 7.4 feet and between 10.6 - 13.5 feet. Maximum resolution is obtained when the coaxial cable is placed adjacent to the borehole wall for this purpose. In very large diameter holes, (15 - 30 inches), some details were not as evident when the cable was placed directly in the center of the explosive column.

There are many reasons why sections of the rock mass in the bench were less intact than others, and could be due to just naturally occurring, backbreak from previous blasts, ground shifting and pulverizing effects of a primer, etc. Nevertheless, this information is

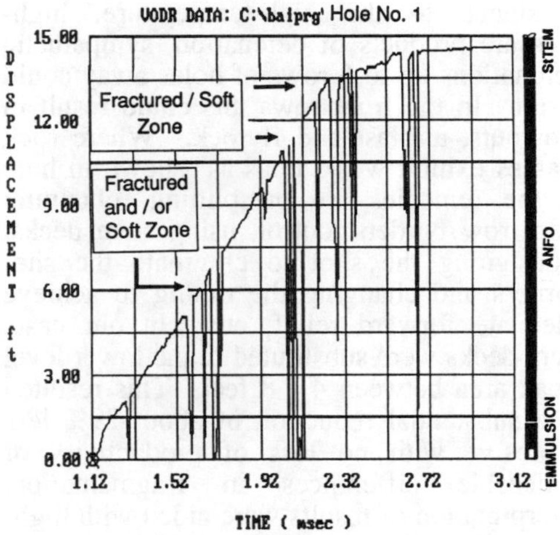

Figure 13. Rock characterization using VOD measurements--weak sections of ground

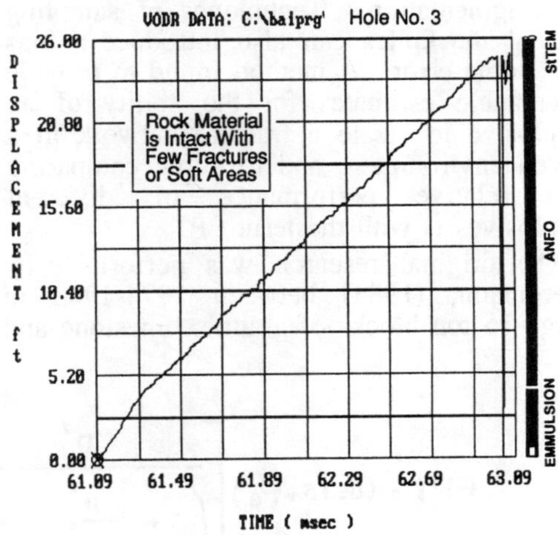

Figure 14. Rock characterization using VOD measurements - Material is intact

invaluable in assessing the degree of confinement, optimum placement of explosive and stem decks, identifying soft and hard band layers, evaluating the consistency of burdens and spacings on adjacent holes during the shot sequence and their effect on explosives performance. In many cases, the confining conditions on an explosive column in the back row corner holes can be so disruptive that product failures or partial detonations are experienced. In soft layered bands of material, offering little or no resistance to the high-temperature, high-pressure products of detonation, sympathetic detonations up to 6 rows of holes away could result. In the front rows this could result in blowouts, airblast and flyrock. Where rock masses exhibit weak areas as shown in hole 1, the remedies are presplitting, obtaining front row burden control, using stem decks, redesigning the shot to eliminate the shot corners and changing the timing to achieve adequate forward relief, etc. In our case, stem decks were substituted in the lower level weak area between 4 - 8 feet. This resulted in a substantial reduction of about 29% less explosive with no loss of productivity or noticeable differences in fragmentation. Interpretation of results were aided with high-speed photography, sonic velocity measurements, seismographs and laser surveying of the bench face.

Other than sieving an entire muckpile, there are no precise methods to estimate the degree of fragmentation. Techniques of sampling partial muckpiles can also introduce a bias and some error. A method found to provide acceptable estimates for the ability of an explosive to create a fracture network in a given environment, and to allow comparing the relative performance of different explosives is with the term EPT.

The original research was performed by Bergmann, (1983) between 1974-1983 in large 15 ton blocks of granite, limestone and sandstone. Large sizes were necessary to minimize end effects and to retain realistic scaling ratios. The EPT (Explosive Performance Term) indicates that fragmentation is not controlled by a single property such as energy but by a combination of explosives energy, detonation velocity, density, degree of decoupling between explosive and borehole wall, explosive volume to borehole volume, sonic velocity of the rock, powder factor and shot geometry.

Equation 1 has been tested and modified over the years by BAI to reflect more realistic inputs from the full scale environment. It is critical that sonic velocity values and continuous VOD's are measured in-situ, in the full scale environment and not obtained from book or general data. Sonic velocity is a function of Young's modulus (a measure of the material's elasticity), Poisson's ratio (a measure of the material's brittleness and density (a measure of mass per unit volume). Sonic velocity is also characteristic of the material's inherent integrity in terms of large, small and hairline cracks, grain size, structural geology, moisture content and porosity. Measured VOD reflects precisely how the product performed in the blast environment under a given blast design, field conditions and loading procedures. By using thermodynamic codes such as the new commercial TIGER code, version 2.0 and manipulating the original explosive

$$\mathrm{EPT} = (0.36+\rho_e)\left[\frac{D^2}{\left(1+\dfrac{D^2}{V_R^2}-\dfrac{D}{V_R}\right)}\right]\frac{1}{R}\left(\frac{E_M}{E_T}\right)\rho_e \qquad \text{Equation 1}$$

Where:

EPT = Explosive performance term
ρ_e = Explosive density (g/cc)
V_R = Sonic velocity in rock (Km/sec)
D = Velocity of detonation (Km/sec)
R = Decoupling ratio (hole vol:expl vol)
E = Calculated maximum expansion work of explosive in (Kcal/g)
where:
E_M = non-ideal value
E_T = theoretical

formulation by either substituting inert material, fixing or freezing parts of the formulation through an iterative process, a realistic value for the maximum expansion work of the explosive can be obtained. The ratio of E_M/E_T in Equation 1 provides a much better estimate for the nonideal behavior of the explosive when compared to theoretical calculations. Calculations of EPT values suggest that the blast environment can, in many cases, be more influential in the final fragmentation outcome than by the selection of various grades of explosives as claimed by some suppliers. Furthermore, the highest energetic and/or fastest reacting explosives may not always achieve the best results in given instances. In some blast environments, a low order detonation and/or deflagration were concluded as being more desireable from an overall performance standpoint.

3.5 Stemming retention and explosive efficiency

When all blast design parameters are held constant except for the explosive type, the disturbance or crush velocity through the stemming column can be used as a reliable heave index for relative explosive performance and in predicting crater results, (Chiappetta, 1993). Figure 15 illustrates the results of four different explosive formulations superimposed on a graph of displacement versus time. Three of the explosive formulations consisted of the same Emulsion/Anfo blend, but with different quantities of Aluminum in quantities of 0, 3 and 9 percent. The fourth explosives consisted of standard Anfo and was used as the control explosive. Since most mine operators work on a volume basis, the explosive length was kept constant at 6.0 feet. Each hole was detonated separately and in total confinement to eliminate any end boundary effects or disturbances from adjacent detonating boreholes.

As expected, the explosive containing 9% Aluminum resulted in the best overall performance, followed by the 3% Al, 0% Al and Anfo. The disturbance length above the top of the explosive column into the stemming correlated very well to the surface measurements in terms of the crater radius, heave volume, material displacement and uplifting velocities. In this series of tests, a total of 24 single holes were detonated for the purposes of obtaining some statistical significance and repeatability. High-speed photography, seismographs and manual measurements of the surface effects were also used to substantiate and correlate results.

It was concluded in this particular environment that the minimum and maximum crater heave heights and crater radii could be predicted for the different explosives based solely on VOD results. Performance in terms of crater height and crush distance into the stemming and radius are presented in Table 3 and Figure 16. Since substantial improvements were evident with the addition of metallic elements to the explosive formulation, they were highly recommended.

Table 3. Crater test results

Explosive	Crater Heave Height (ft)	Crater Radius (ft)
Anfo	5 - 10	25 - 30
E/A @ 0%	10 - 25	30 - 35
E/A @ 3%	20 - 35	35 - 40
E/A @ 6%	30 - 40	40 - 45
E/A @ 9%	40 - 55	45 - 50

3.6 Effect of water on Anfo

Figure 17 illustrates VOD results for two boreholes loaded in the same environment with Anfo. Hole 1 resulted in a fairly stable detonation at about 10,000 ft/sec throughout its explosive column length. Hole 2 resulted in a low order detonation of about 6,000 ft/sec for the first 3 feet above the primer and then immediately went to a full order detonation of about 12,200 ft/sec for the remainder of the column. Both holes were fired within about 15 minutes after loading. The most likely cause of the partial low order detonation in the bottom third of Hole 2 was water contamination and/or absorption by the prill. It is interesting to note that the transition zone between the low order detonation to high order detonation occurred almost instantaneously.

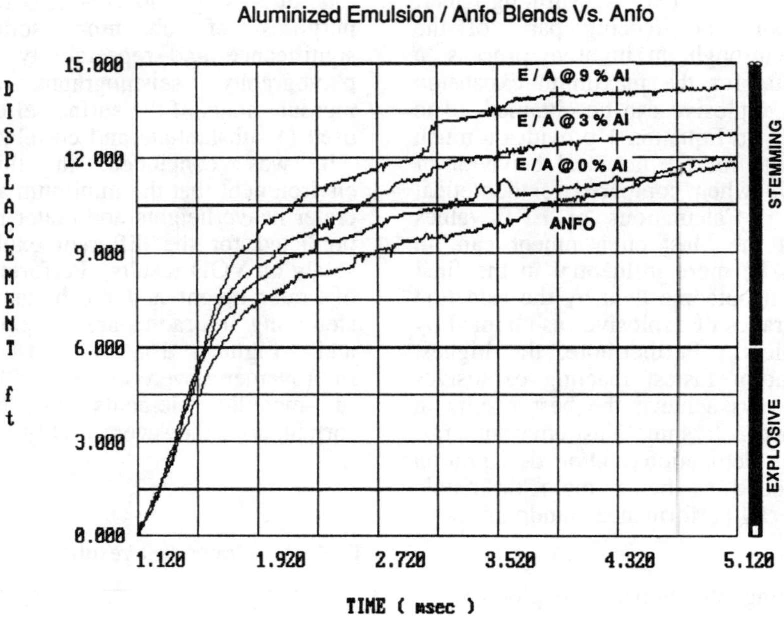

Figure 15. Results of aluminized Emulsion/Anfo blends versus Anfo

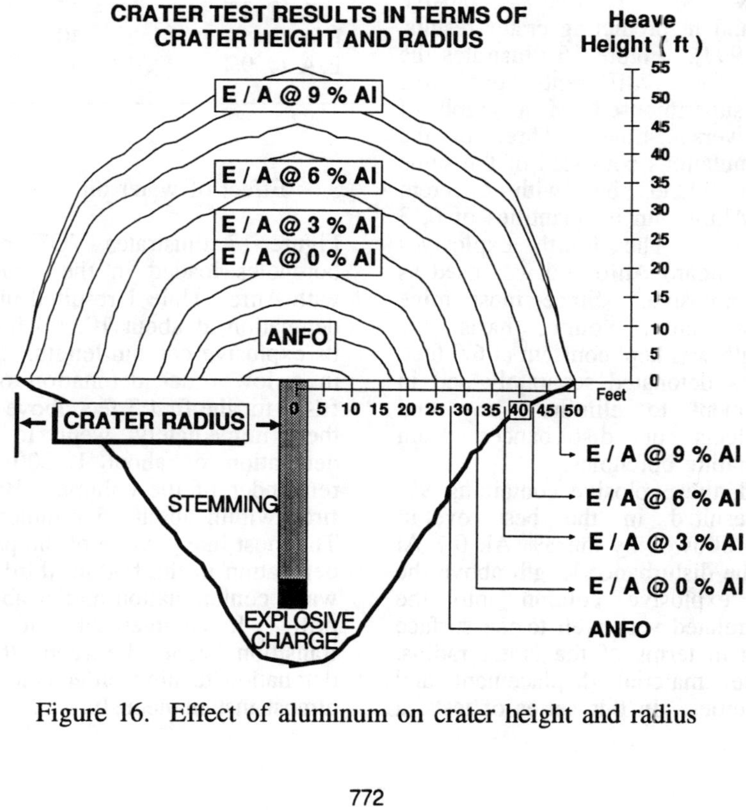

Figure 16. Effect of aluminum on crater height and radius

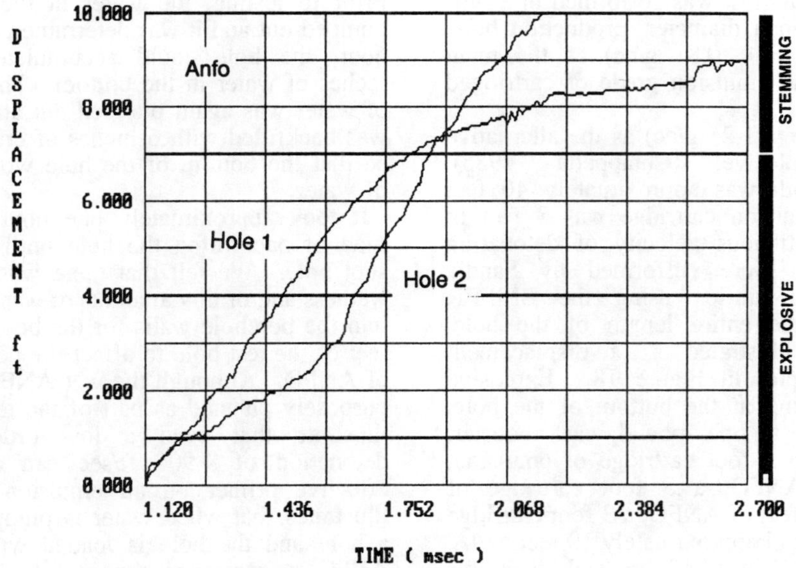

Figure 17. Effect of water on Anfo

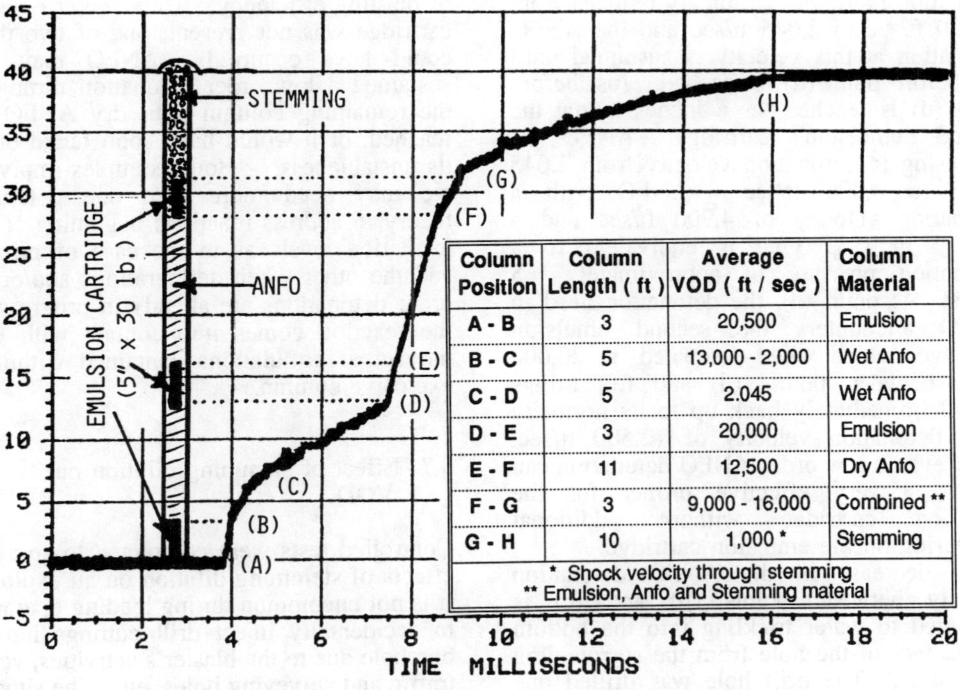

Column Position	Column Length (ft)	Average VOD (ft / sec)	Column Material
A - B	3	20,500	Emulsion
B - C	5	13,000 - 2,000	Wet Anfo
C - D	5	2,045	Wet Anfo
D - E	3	20,000	Emulsion
E - F	11	12,500	Dry Anfo
F - G	3	9,000 - 16,000	Combined **
G - H	10	1,000 *	Stemming

* Shock velocity through stemming
** Emulsion, Anfo and Stemming material

Figure 18. Example of multiple transient VOD's and the effect of water on Anfo

Another VOD test was performed in a full-scale, 6 1/2 inch diameter, production hole, with ANFO (ρ = 0.81 g/cc) as the main charge, and an emulsion grade of cartridged explosive
(5" x 30 lb, ρ = 1.25 g/cc) as the alternative velocity explosive, (Chiappetta, 1985). Borehole depth was approximately 40 feet and each emulsion cartridge was 3 feet in length. Continuous velocity of detonation measurements were performed by Sandia National Laboratories using the SLIFER System for the entire length of the hole. Results are illustrated as a displacement versus time plot in Figure 18. Explosive loading, starting at the bottom of the hole, consisted of a one pound cast booster, followed by a 3 foot cartridge of emulsion, 10 feet of ANFO, a 3 foot cartridge of emulsion, 11 feet of ANFO, a 3 foot cartridge of emulsion, and approximately 10 feet of 1/2 inch - 1 1/8 inch of crushed rock for stemming.

Many important and interesting points are noteworthy in the results. Between points (a) and (b), the velocity of detonation for the 3 foot length of emulsion cartridge is 20,500 ft/sec. Between (b) and (c), the velocity of detonation in ANFO is quickly reduced from 13,000 ft/sec to 2,045 ft/sec and the ANFO detonation at this velocity is sustained until just before point (d) is reached. Just before point (d) is reached, or 6 inches below the second emulsion cartridge, ANFO is increasing in detonation velocity from 2,045 ft/sec to 4,900 ft/sec. ANFO with a detonation velocity of 4,900 ft/sec and a density of 0.81 g/cc is equivalent to a detonation pressure of approximately 4.5 Kbars. At point (d), the detonation head in ANFO encounters the second emulsion cartridge, which when detonated, at 20,000 ft/sec between points (d) and (e), brings ANFO immediately back up to its normally rated detonation velocity of 12,500 ft/sec. Thus, even a low order ANFO detonation can act as a very effective primer for the emulsion cartridge without additional boostering on the emulsion cartridge.

The decrease in the ANFO detonation velocity between points (b) and (c) is attributed to water trickling into the bottom 12 1/2 feet of the hole from the surrounding rock mass. The drill hole was drilled one week in advance of testing and it had accumulated approximately 3 feet of water. Prior to testing, the water in the hole was pumped out and it was determined that in one hour, the hole could accumulate about 2 inches of water in the bottom. The 2 inches of water was again pumped out and the hole was backfilled with 6 inches of crushed rock so that the bottom of the hole would be out of water.

It took approximately one hour from the time of backfilling the hole until firing the shot hole. We felt that there was sufficient wetness and/or tiny amounts of water trickling into the borehole walls for the bottom 12 1/2 feet of the test hole to affect the performance of ANFO. Although the wet ANFO was not purposely planned as part of the test, it does illustrate that even a low order ANFO detonation of 4,900 ft/sec can act as an effective primer on an emulsion. It also illustrates that when water is pumped out of a hole and the hole is loaded with ANFO, ANFO performance can still be drastically affected just by the wetness on the inside walls of the borehole. Inadequate priming at the bottom of this hole would have probably resulted in a failure.

Although ANFO can tolerate up to 10% water saturation level, it does so at the cost of blasting efficiency. If the center emulsion cartridge was not present, one of two things could have occurred. ANFO may have sustained a low order detonation throughout the remaining column until dry ANFO was reached, or it would have soon failed due to its unstableness. Both examples imply that we may need more than one detonation theory to address reactions beginning at zero excitation levels as in the case of priming, and the other when deflagrations and/or low order detonations are already in progress and the reaction comes into contact with other unreacted formulations contained within the explosive column.

3.7 Effect of stemming dilution on VOD

Controlled tests were completed to study the effects of stemming dilution on an explosive. It is not uncommon during loading operations to accidentally inject drill cuttings into the borehole due to the blaster's activities, vehicle traffic and surveying holes, etc. The situation usually worsens in extreme weather conditions and when the blast crew is rushing to meet

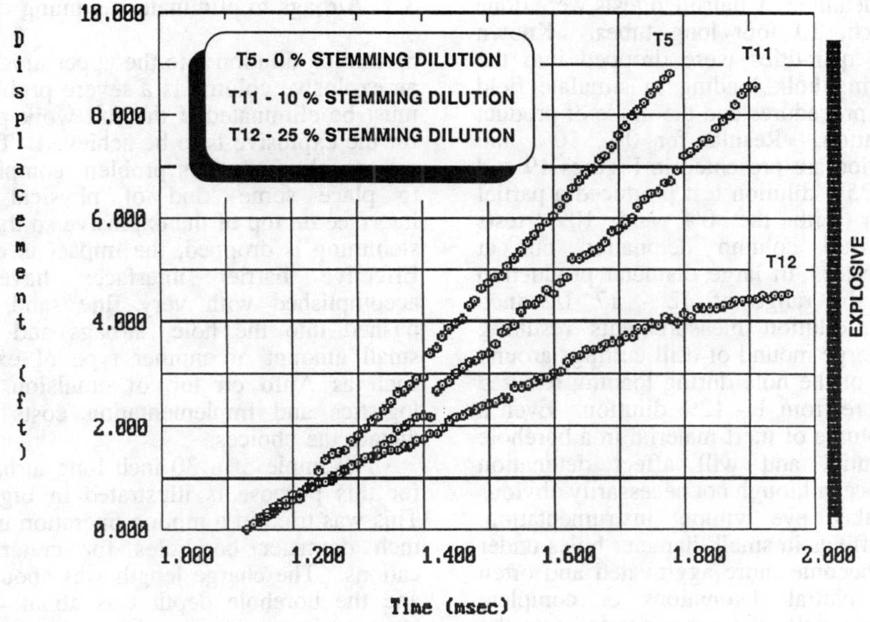

Figure 19. Results of stemming dilution tests as a function of displacement versus time

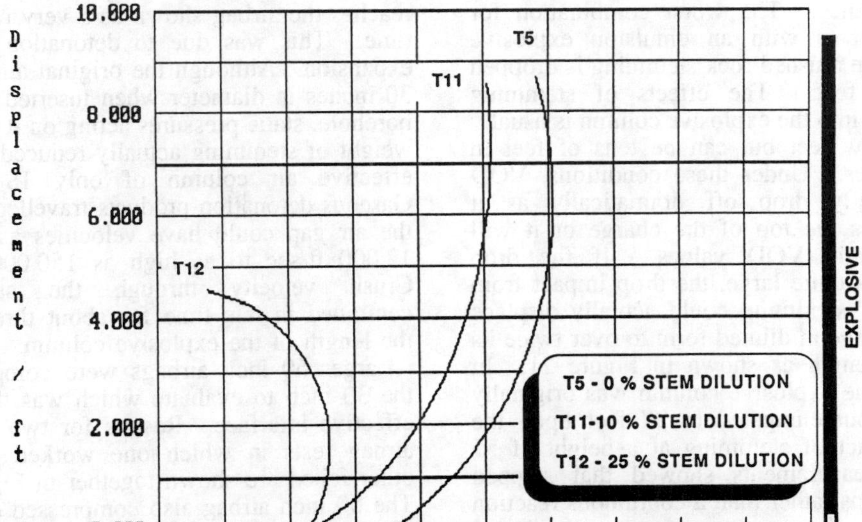

Figure 20. Results of stemming dilution tests as a function of VOD

the blast deadline. Controlled tests were done in 10 inch, 10 foot long tubes. Known stemming quantities were dropped into the tubes during bulk loading to simulate field condition procedures and the mode of product contamination. Results for 0%, 10% and 25% dilution are presented in Figures 19 and 20. The 25% dilution test produced a partial detonation with the 0% and 10% tests reaching full column detonation but at different rates. In large diameter production holes in the range of 12 - 17 1/2 inch diameters, dilution measurements resulting from the large mound of drill cuttings around the collar of the hole during loading resulted in anywhere from 1 - 12% dilution. Even a 5 - 6% volume of inert material in a borehole is substantial and will affect detonation performance, although not necessarily obvious to the naked eye without instrumentation. Dilution effects in small diameter holes under 4 inches become more aggravated and often result in partial detonations or complete failures, especially if large rocks fall into the explosive column.

The worst stemming dilution always occurs within the top part of an explosive column, when stemming is dropped into the loaded hole. The degree of stemming migration into the explosive depends on the explosive viscosity, type and size of stemming and the drop distance. The worst combination for this effect is with an emulsion explosive when large crushed rock stemming is dropped over 15 feet. The effects of stemming migration into the explosive column is usually only a few feet but can be tens of feet in some cases. Under these conditions, VOD will usually drop off dramatically as it approaches the top of the charge or it will show erratic VOD values. If the drop distance is quite large, the drop impact from clumps of stemming could actually displace the explosive in diluted form to over twice its original length as shown in Figure 21. In this test the explosive column was originally 12 feet, but elongated to 27 feet upon the drop impact of stemming at a height of 55 feet. Measurements showed that stepped VOD jumps rather than a continuous reaction occurred along the new redistributed explosive column.

3.8 Airbags to eliminate stemming dilution

Stemming dilution into the upper most part of an explosive column is a severe problem and must be eliminated if the full work potential for the explosive is to be achieved. The only way to eliminate this problem completely is to place some kind of physical barrier interface on top of the explosive so that when stemming is dropped, the impact is arrested. Effective barrier interfaces have been accomplished with very fine sand, plastic pushed into the hole, airbags and even a small amount of another type of explosive such as Anfo on top of emulsion. Field logistics and implementation costs usually dictate the choices.

An example of a 30 inch long airbag used for this purpose is illustrated in Figure 22. This was tried in a mining operation using 30 inch diameter boreholes for crater applications. The charge length was about 4 feet and the borehole depth was about 40 feet. Our testing prior to the use of airbags concluded that under these constraints, stemming dilution into the top part of the explosive averaged about 20%, but could vary from 10 - 100%. This was part of the problem why blast results were not always consistent. Detonation throughout the explosive column is fairly steady, until it reaches the airbag shown as a very rapid rise time. This was due to detonation product expansion. Although the original airbag was 30 inches in diameter when inserted into the borehole, static pressures acting on it with the weight of stemming actually reduced it to an effective air column of only 18 inches. Gaseous detonation products travelled though the air gap could have velocities as low as 12,000 ft/sec to as high as 150,000 ft/sec. Crush velocity through the stemming continued in late time for about three times the length of the explosive column.

Larger 60 inch airbags were compared to the 30 inch to evaluate which was the more effective interface. Results for two 60 inch airbag tests in which one worked and the other failed are shown together in Figure 23. The 60 inch airbag also compressed down to 18 inches, which was equivalent to the compression of the 30 inch airbag. In some cases, the static pressure was so great that the 60 inch airbag actually failed, (Figure 23). It was concluded that the 30 inch airbags were more durable and suitable for the application and they were highly recommended.

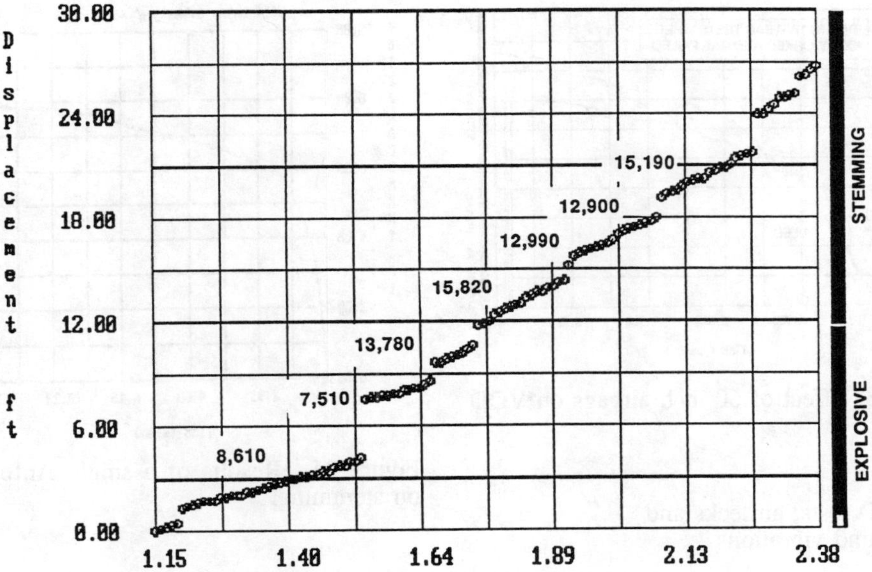

Figure 21. An excellent example of stepped VOD jumps and the effects of dropping large clumps of stemming onto the explosive

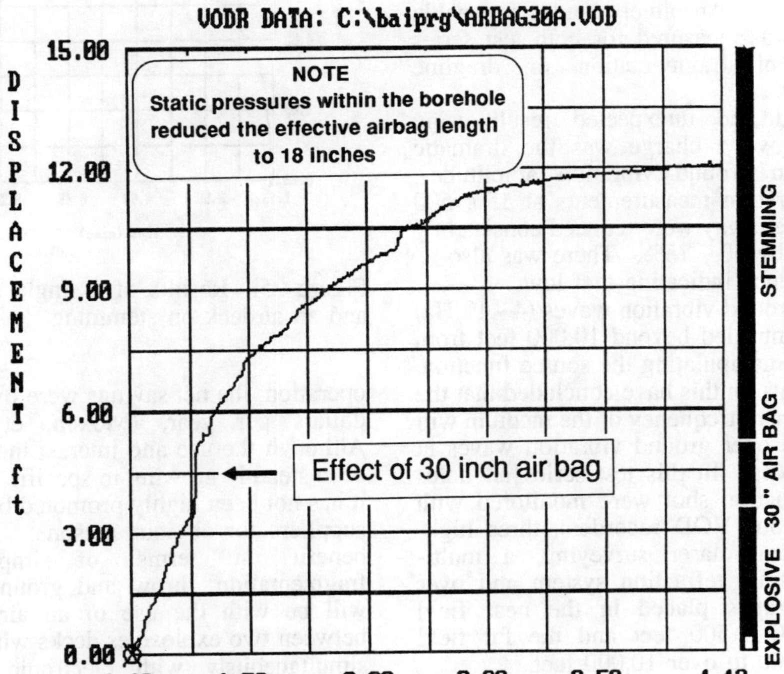

Figure 22. Effect of 30 inch airbags on VOD results

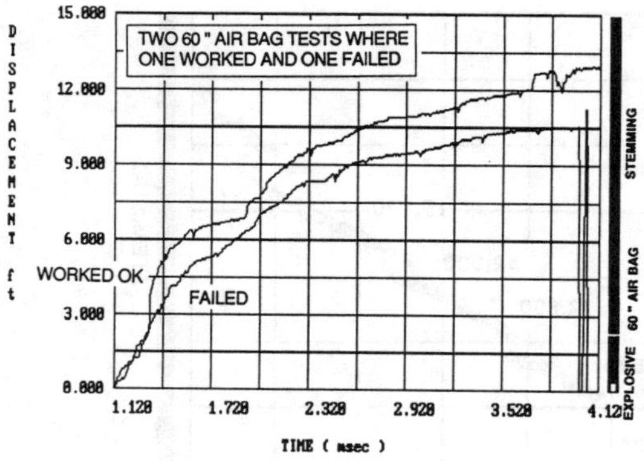

Figure 23. Effect of 60 inch airbags on VOD results

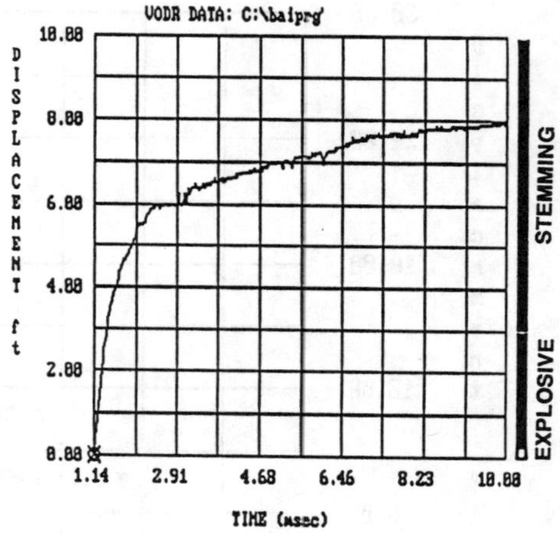

Figure 24. Results of a single Anfo charge on stemming.

3.9 VOD tests, airdecks and ground vibrations

Figures 24 and 25 compare VOD results for a continuous column charge of Anfo and the same charge with a 3 foot airdeck situated directly above the Anfo, respectively. Both charges produced shock disturbances of equal length beyond the top of charge, but Anfo's VOD with the airdeck was approximately 10% less. Absolutely no noticeable difference was measured for both test series in terms of fragmentation or dragline productivity.

An added and unexpected result using airdecks above a charge was the dramatic reduction in ground vibration amplitudes. Ground vibration measurements at 150, 500 and 1000 feet away were reduced consistently on average by 50 - 75%. There was also strong evidence indicating that low frequency ground vibration waves (4 - 12 Hz) could be controlled beyond 10,000 feet from the shot by manipulating the source function. Prior thoughts on this have concluded that the natural resonant frequency of the medium will predominate over ground vibration waves at these distances. In this test series all holes in the production shot were monitored with two continuous VOD recorders, three high-speed cameras, laser surveying, a multi-channel seismic refraction system and over 30 seismographs placed in the near field between 100 - 500 feet and the far field between 1,000 to over 10,000 feet.

Similar tests with airdecks in the USA, Australia and South Africa resulted in 10 - 50% less consumption of explosives. In one operation, the net savings were over 2 million dollars per year, (Moxon, et al, 1991). Although the use and interest in airdecks has been steadily growing in specific applications, it has not been highly promoted by explosives suppliers for obvious reasons. The ultimate benefit in terms of improving on fragmentation, throw and ground vibrations will be with the use of an airdeck placed between two explosives decks which are fired simultaneously with electronic detonators, (Chiappetta, 1992 b).

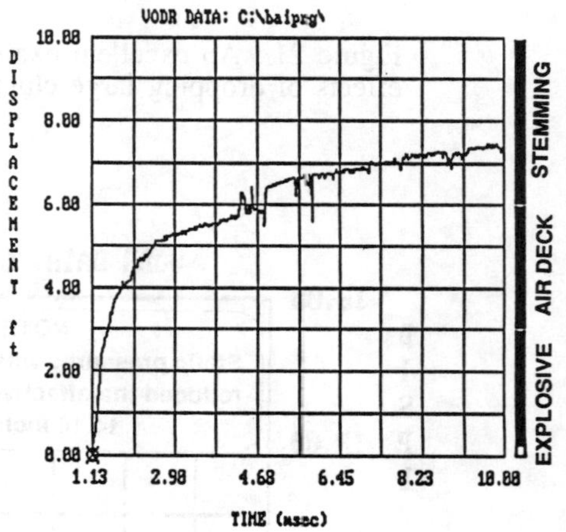

Figure 25. Results of a single Anfo charge and an airdeck on stemming.

The type of explosive and column length can also be used to control ground vibration amplitudes and the predominant frequencies in a given environment. A series of single, full column (21 - 25 feet) and decked charges (8 feet) were tested in a different environment to compare pure emulsions and Anfo, as illustrated in Figures 26 and 27, respectively.

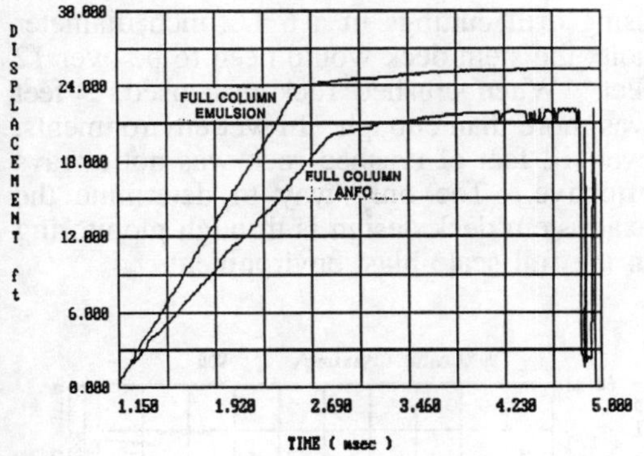

Figure 26. Full column emulsion charge versus a full column Anfo charge

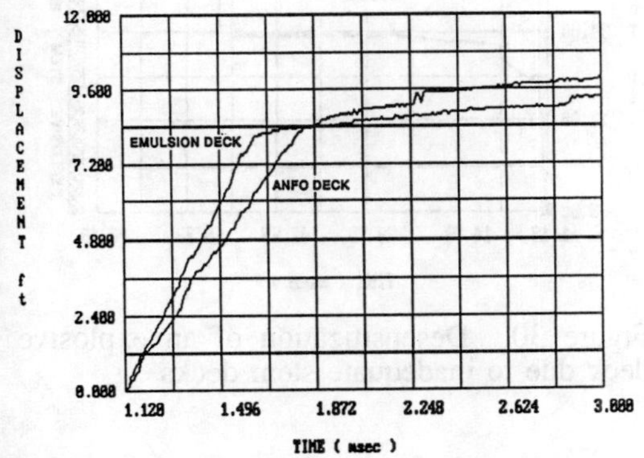

Figure 27. Decked emulsion charge versus a decked Anfo charge

Generally, the shorter the column for a given explosive, the higher the predominant frequencies. For example, an eight foot Anfo deck compared to a 21 foot Anfo column would generate dominant frequencies of approximately 75 - 250% higher in the near field within 200 feet from the shot. Beyond 500 feet, dominant frequencies were almost identical. Likewise, Emulsion explosives (full column or decked) always produced considerably higher dominant frequencies than the equivalent Anfo charge in the near field. In critical, close-in and sensitive blasting conditions next to residential homes, Emulsions or fast reacting explosives and decked charges are preferred over other less energetic explosives and longer column lengths.

The most important design parameters in the control of ground vibration amplitudes and low frequency waves are still controlled by delay intervals between detonations. As such, accurate detonators are critical in assuring a detonation order, with minimal deviation. Detonator firing times could be measured with the continuous VOD recorder by taping a series of detonators at a constant distance (2 - 10 feet) from each other along a length of coaxial cable. Results for a series of fairly accurate nonelectric detonators is shown in Figure 28. The vertical step

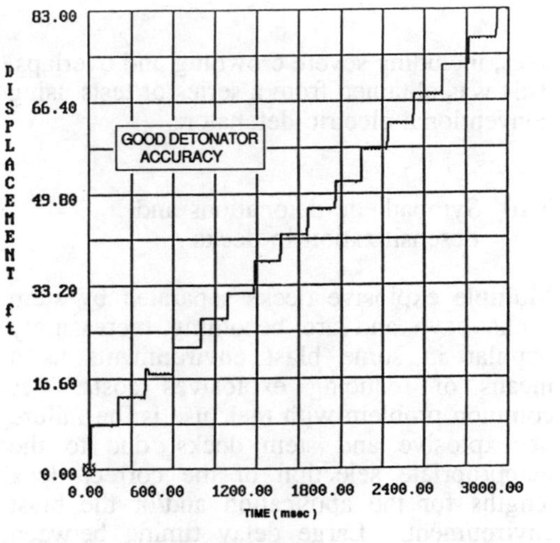

Figure 28. Typical result of a detonator test showing fairly accurate firing time, (non-electric system)

increases along the y-axis represents the placement of each detonator at constant distance along the cable. The horizontal step increases represent the firing times of each detonator. Even at a glance, the uniform step increases along both axes indicates that the nonelectric detonator system tested fired fairly accurately. Figure 29 illustrates what a graph looks like when detonator accuracy has significant timing errors. Inconsistent and nonuniform step increases are evident on both

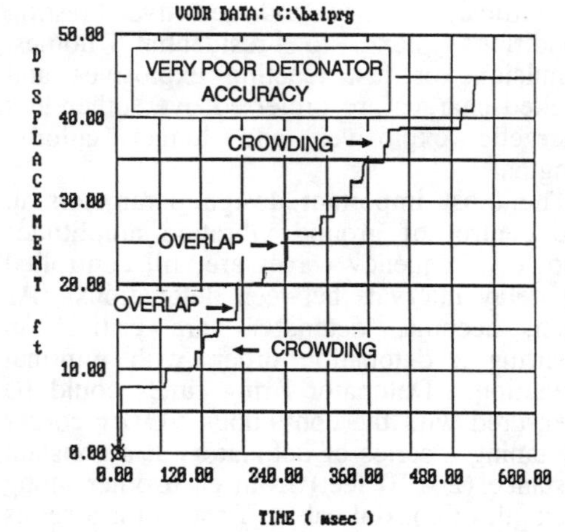

Figure 29. Typical result of a detonator test showing poor result with crowding and overlaps

axes, including severe crowding and overlaps. This was obtained from a series of tests using conventional electric detonators.

3.10 Sympathetic detonations and desensitization in decks

Multiple explosive decks separated by stem decks have and are becoming increasingly popular in some blast environments as a means of reducing explosives costs. A common problem with their use is the failure of explosive and stem decks, due to the inappropriate selection of the correct deck lengths for the application and/or the blast environment. Large delay timing between explosive decks is another problem which is often overlooked in view of the shock velocity generated into the stemming as shown in Figures 12, 15 and 24. An example of a stemming deck failure and desensitization of the explosive deck immediately above it is illustrated in Figure 30. In this case, a 6 foot stem deck comprising of drill cuttings in a 10 5/8 inch diameter hole was ineffective. Under the right conditions, sympathetic detonations could also occur between explosive decks, (Katsabanis, 1991). The effectiveness of a stem deck is very site-specific and depends on the rock material's behavior under dynamic loading, length, type of stemming material, explosive, water conditions and timing between explosive decks. In over 75% of our tests, conventional rules of thumb on the correct length of stemming to use in dry or wet conditions did not apply and many problems were uncovered. For example, in some cases it was determined that for a stem deck to remain effective in a dry environment using drill cuttings in a 6 1/2 inch diameter hole, the stem deck would need to be over 12 feet. When crushed rock was used, 2 feet was more than enough. In wet environments, even 12 feet of crushed rock was not always effective. The only way to determine the exact stem deck design is through monitoring in the full scale blast environment.

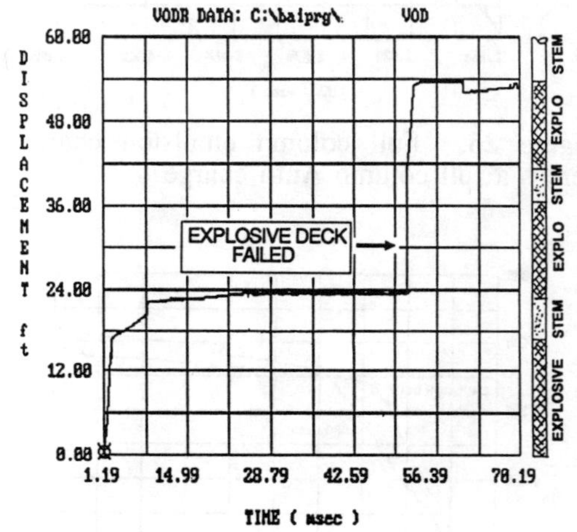

Figure 30. Desensitization of an explosive deck due to inadequate stem decks

4. RUN-UP VOD'S IN PRIMER REGION

A reliable method of determining the start or run-up VOD in the primer region to obtain sufficient details for primer selection and how the primer affects explosive performance along the column is with continuous VOD measurements. Table 4 lists the results of 5 different sizes of cast primers ranging from 0.5 - 5.0 pounds in 10 inch, 10 foot unconfined tube tests, using an Emulsion/Anfo blend. Run-up distances to a relatively stable detonation ranged from 3.0 - 4.0 feet and the product's VOD over the total detonation length varied between 11,190 - 13,030 feet. As expected, the greatest

Table 4. Primer test results for 10 inch tubes and 10 foot columns detonated upright

Primer size (lb)	Explosive column (ft)	Recorded detonation length (ft)	Run up distance to a relatively stable VOD (ft)	Start VOD at primer (ft/sec)	Average VOD over detonation length (ft/sec)	Lowest and highest VOD (ft/sec)
0.5	9.5	7.8	3.0	3,950	11,920	3,950 14,430
1.0	9.5	9.2	3.0	6,810	12,840	6,810 14,610
2.0	9.5	9.0	3.5	8,320	11,190	5,250 17,100
3.0	9.5	9.0	4.0	10,130	12,910	10,130 15,540
5.0	9.5	8.7	3.0	10,950	13,030	10,950 15,050

Note: Primers were placed at bottom of charge

difference in results was the start VOD in the primer region, which was lowest for the 0.5 pound primer @ 3,950 ft/sec and highest for the 5.0 pound primer @ 10,950 ft/sec. The greatest relative improvement was observed in the smallest three primers when going from the 0.5 - 1.0 - 2.0 pound sizes. Primer sizes beyond 5.0 pounds did not show any significant performance increase in terms of the product VOD's. Generally, the higher the start VOD was in the primer region, the more reliable and steady was the detonation throughout the main column charge. In field tests, some results correlated very well, while others did not and indicated that other elements were at play. For reliable and consistent detonation of the main explosive charge, a primer should never be selected solely on its ability to get a product up to its rated VOD, because this usually only applies to ideal conditions. In order to overcome the unknowns in the field and to compensate for rough field and borehole conditions, the next one or two primer sizes should always be selected. A primer may also serve to enhance fragmentation, especially in difficult toe breaking environments. Again, full scale monitoring will eliminate much of the guess work in the selection process.

Figure 31 illustrates the results of two identical primers of different composition on an Emulsion/Anfo blend. Distinct start VOD's are noticeable between primer 1 versus primer 2. Primer 1 immediately gets the main charge of explosive up to its rated VOD, while primer 2 struggles to get the product going until about 200 microseconds after detonation. Although such a time frame is insignificant in terms of overall detonation performance of the main product, the danger is that the start VOD is weak and could result in a failure where rough borehole conditions are encountered, such as water and slimes.

In a different series of tests consisting of over 80 primer tests, it was concluded that the intimate contact of the primer surface area was more important than the primer size. In many cases, a 3/4 pound primer was as effective as a 4.0 pound primer. No significant differences were found between the same weight of primer and variations in the primer shape, regardless of whether the primer was cylindrical, flat, shaped or spherical. Primers surrounded and housed in heavy plastic containers did not perform as well as primers in which their composition was exposed or housed in thin housings.

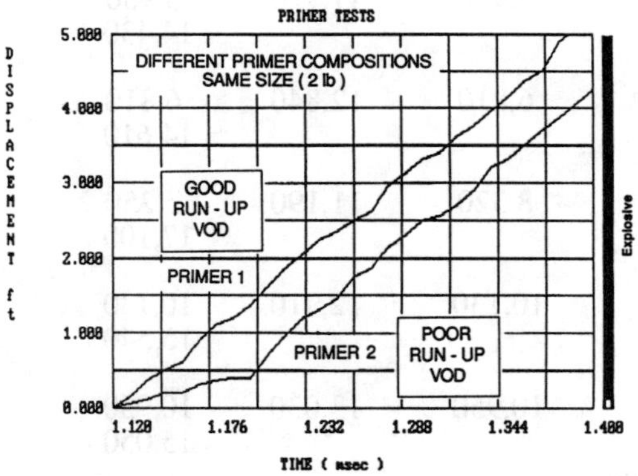

Figure 31. Example showing good and poor run-up VOD's above the primer

5. NOXIOUS FUMES

Undesirable fumes generated from blasting are a growing worldwide concern in terms of safety, performance and environmental agencies. Deep red-orange fumes are nitrogen oxides in the form of NO_2, N_2O, and NO. They are usually detectable by their colors which are deep red, orange, red-brown and orange-brown. They can also be sensed by odor and taste, giving an irritating odor and/or a bitter taste. These are by-products of detonation, when the reaction has not achieved sufficient temperatures in the primary and/or secondary reaction zones. A small concentration of nitrogen oxides of only 0.005% is fatal. The human body can also accumulate nitrogen oxides over time, similar to mercury, and can lead to illness or fatalities. There is no permanent cure for body accumulation and there have been cases where people have accumulated enough to die during sleep. The legal ramifications for explosive suppliers and mine operators could have grave consequences in the event of illness or death to nearby homeowners and/or mine workers alike.

Figure 32 illustrates the results of two tests using the same explosive formulation where one produced post blast oxides of nitrogen and the other only clean, white smoke. It is beyond the scope of this paper to present all of the detailed results consisting of a few hundred tests. These will be presented in future papers and thus only a general overall summary of the things which have been uncovered and clearly identified as contributing to fumes are presented here. Generally, detonations which have not achieved their design equilibrium temperatures will cause noxious fumes. This can result from partial detonations, low order detonations, deflagrations, poor formulations, foreign material injected into the explosives during manufacturing, careless field controls, water, and poor blast designs, etc.

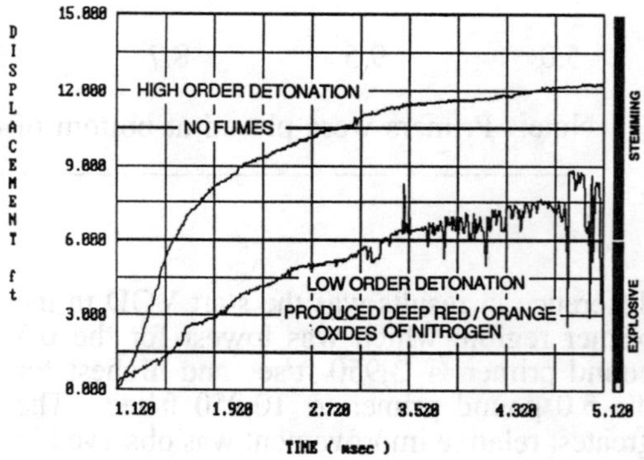

Figure 32. Low order detonation versus a high order detonation

CONCLUSIONS

1. Continuous velocity of detonation measurements can provide a wealth of information for use in blast design optimization. The most meaningful data is always obtained with continuous VOD measurements for the entire time history, including disturbances into and beyond the stemming region. When VOD results are coupled with information obtained from other blast monitoring instrumentation systems in the full scale blast environment, potential problem areas with the product, blast design parameters, field conditions and loading

procedures can easily be identified and remedial actions taken.

2. Continuous VOD measurements in a production environment can be used to:

a) Evaluate an explosives "health", consistency, reliability and repeatability.
b) Select minimum primer size.
c) Evaluate product performance.
d) Design more efficient blast designs.
e) Choose the correct amount and type of stemming.
f) Eliminate explosive desensitization and/or sympathetic detonations.
g) Select an explosive in terms of its abilities to create a fracture network, maximize heave, control ground vibration amplitude and frequency.
h) Evaluate initiation systems and detonator accuracy.
i) Evaluate the effects of a water saturated environment on blast design parameters.
j) Determine the causes of post blast nitrogen oxides, and
k) Determine if massive ground shifts are occurring and the exact location within a shot hole.

3. A number of new, characteristic VOD curves have been identified under varying field conditions. Contrary to popular belief, a steady-state velocity was not required to produce acceptable blast results in specific environments. In many cases, VOD oscillations throughout the explosive column produced better results than steady-state VOD's. In addition, partial detonations and/or deflagrations could also produce good results. This implies that in many blast environments, particularly in soft, highly fractured and/or closely jointed materials, slower reacting explosives are preferred over very fast and overly energetic explosives.

4. VOD results are as dependent and unique to the blast environment as they are to the explosive formulation and composition. There is still no real substitute to evaluate explosive performance and blast designs than in the full scale blast environment, including today's computer modelling capabilities. Although small scale laboratory testing, theoretical predictions and blast simulations have their place, they should be used with caution when extrapolating to a production environment. In some cases, recommendations from these avenues were so far from reality that their predictions in the field were completely reversed!

5. Run-up VOD's above the primer region increase at a diminishing rate from primer sizes ranging from 0.5 - 5.0 pounds. No significant increases were found between 5.0 - 6.4 pound primers under controlled test conditions. In full scale tests, the 3.0 - 5.0 pound primers performed consistently better and the results were considerably more noticeable than primers of a lesser size. In general, the higher the start VOD was in the primer region, the higher and more consistent was the overall VOD throughout a given explosive column.

6. No differences were found in blast results from different shapes of primers, including the recently introduced spherical primers. Primers with thick plastic housings, which were introduced primarily for appearance and safety considerations, did not perform as well as bare primers of equivalent size. For example, a bare 3/4 pound cast primer was found to be as effective as a 4.0 pound primer housed in thick plastic. In full scale tests, this implies that the intimate surface area of a primer on the main charge is quite important.

7. Explosive suppliers are encouraged to eliminate the common, one absolute value regarding the VOD of an explosive, unless the exact test conditions accompanying their data sheets for that particular product are disclosed. VOD for a given product can vary substantially in different blast environments and conditions. It would be better to specify an acceptable VOD range for a product, rather than a single, absolute VOD value. For example, Anfo could be rated as being acceptable from 8,000 - 13,000 ft/sec in 6 1/2 inch holes. The best alternative which would provide a user with the most useful and realistic design criteria, however, would be for the suppliers to represent a product's characteristic VOD as outlined in this paper and it must include the complete time history, including disturbances into the stemming.

8. The commonly accepted rules of thumb pertaining to the amount and type of stemming required at the collar and in stem

decks between two explosive decks did not correlate to our results. Blast design parameters, explosive type, stemming type, stemming amount, confining conditions, delays between holes and rows of holes, blast geometry, detonator accuracy, response times and forward relief, etc., all have a direct bearing on stemming retention. The only way to determine the extent that each contributes to stemming retention is with continuous VOD measurements in the blast itself.

9. Although partial detonations, low order detonations, deflagrations and lesser reactive explosives may in some cases be more desireable, they can affect TMIN or the response time up to three fold of normal. This can negate the initiation delay sequence for its intended purpose and must be compensated for with appropriate delay intervals.

10. The use of airdecks in production shots has shown considerable promise in lowering explosive consumption by 10 - 50%, lowering overall mining costs and reducing ground vibration amplitudes by up to 75%, while still maintaining acceptable productivity.

11. Fairly good correlations were found between the crush/shock disturbance distance into the stemming with differing aluminum grades of Emulsion/Anfo blends to the crater height and crater radius. Thus VOD measurements could, in specific environments and applications, be used to predict explosive efficiency in terms of a crater's height and radius.

ACKNOWLEDGEMENTS

We wish to express our sincere gratitude to the following organizations for their support, technical discussions, permission to publish results and for their encouragement throughout this continuing research effort: AMAX Coal Company; Atlas Powder Company (now ICI Explosives USA); Blastech; Detonic Research; Hawaiian Cement; IRECO, Inc.; Los Alamos National Laboratory; MSI-Mining Services International; Optimum Colliery; Sandia National Laboratories; SMX Explosives: Suncor Inc., Oil Sands Group; Thiokol Corporation and Trans Natal Colliers Ltd.

A special thank you is deserving of Mr. Pat McLaughlin and Mr. Mike McDougall of Suncor Inc. for their excellent support during specific field tests, and Mr. Gerald (Butch) Maki of IRECO, Inc. for sharing VODR-1 records with us.

REFERENCES

Bergmann, O.R. Effect of explosives properties, rock type and delays on fragmentation in large model blasts, First International Symposium on Rock Fragmentation by Blasting, Lulea, Sweden, August, 1983, pp 70-78.

Chiappetta, R. Frank. Alternate velocity loading techniques and detonations in a production environment, presentation Second Pennsylvania Blasting Conference, University Park, PA, USA, November 1985

Chiappetta, R. Frank. Generating site-specific blast designs with state-of-the-art blast monitoring instrumentation and PC based analytical techniques, proceedings 17th Conference on Drilling and Blasting Techniques, International Society of Explosives Engineers, Las Vegas, Nevada, USA, 1991

Chiappetta, R. Frank, Vandenberg B., Pressley, J.R. Portable, Multi-channel and continuous velocity of detonation recorders, proceedings Fourth High-Tech Seminar on State-of-the-Art Blasting, Technology, Instrumentation and Explosives Applications, Nashville, TN, USA, June 1992 (a).

Chiappetta, R. Frank. Precision detonators and their applications in improving fragmentation reducing ground vibrations and increasing reliability - a look into the future, proceedings Fourth High-Tech Seminar on State-of-the-Art, Blasting Technology, Instrumentation and Explosives Applications, Nashville, TN, USA, June 1992 (b).

Chiappetta, R. Frank, Aluminized explosives characterizations and stemming contamination studies, (internal, unpublished report prepared for Suncor, Inc., Oil Sands Group., Fort McMurray, Alberta, Canada, 1993

Katsabanis, P.D. (Takis) Sympathetic detonations - causes and solutions, Third High-Tech Seminar on Blasting Technology, Instrumentation and Explosives Applications, San Diego, CA, USA, June 1991

Moxon, N.T., Armstrong, L.W., Mead, D. Reducing blasting costs using air-decks: the do's and don'ts, proceedings, Third High-Tech Seminar on State-of-the-Art Blasting Technology and Explosives Applications, San Diego, CA, USA, June, 1991.

Innovative Mine Design for the 21st Century, Bawden & Archibald (eds) © 1993 Balkema, Rotterdam, ISBN 90 5410 325 6

Developments in the computer based 3-D analysis system

Stephen H. Chung
ICI Explosives Technical Service Group, North York, Ont., Canada

Andrew DuPlessis
ICI Explosives Group Technical Centre-B, McMasterville, Que., Canada

ABSTRACT: Photographic and surveying systems are being used today as standard tools to document blasting activities and geometrical changes in mining. Some examples include high speed framing cameras, video recorders and laser profilers. Although these systems yield accurate data, most cameras capture only two-dimensional images while laser profilers provide discrete 3-D data. Attemps to extract 3-D information from photographs have been unsatisfactory. Recent advances in electronic imaging and computer programming have made it possible to develop a 3-D computer based analysis system which can capture and measure the geometry of a 3-D image quickly and accurately. This paper describes the technology that creates quality 3-D images of objects of different sizes displayed at various distances. Measured values compare well with actual data. The potential application to mining operations is discussed.

1 INTRODUCTION

The explosives and mining industries are strong supporters of computer blast modeling. Both groups see computer simulated blasting as a valuable predictive tool which can greatly increase productivity by eliminating the need for costly and time-consuming trials. For example, SABREX, the computer blast model developed by ICI Explosives, has helped ICI chemists develop explosives to meet specific requirements for fragmentation or muckpile profile. It has also been used by mine operators around the world to optimize blast results.

However, model predictions are affected by inaccurate input parameters such as burden, spacing and velocities of rock mass movement in blasting. Conventional monitoring techniques are often used in practice to measure a limited number of data points for depicting a blast. For example, a high-speed camera such as LOCAM can only record 2-D images of rock mass movement during a blast. The analyst must then extrapolate 3-D information from this 2-D data by laborious computations. Even so, these 3-D interpretations of 2-D data are not actual measured values. Moreover, data is only obtained for a few target points on the face where markers are hung. Events occurring between or outside these markers cannot be measured because the scale is not known. ICI started in late 1990 to develope a practical 3-D imaging system which not only captures the action of blasting in 3-D, but for the results to be digitized as well. (Chung & Burchell 1992)

2 THE PRINCIPLE AND EARLY FORM OF 3-D IMAGING

Normal human vision involves processing of images received from both left and right eyes. The brain then fuses the two images to produce stereopsis which is one of the most important cues for determining distance and depth perception to give a vivid 3-D sense. This differs greatly from the so-called 3-D

Figure 1. DTI 100M Auto-stereoscopic 3-D Graphics Monitor

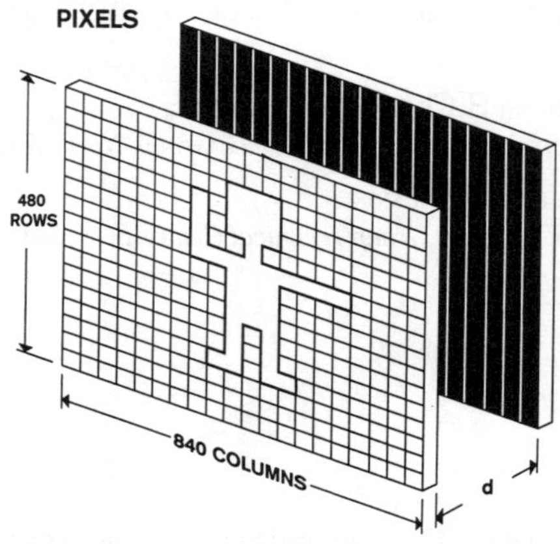

Figure 2. Auto-stereoscopic system

impression presented in a painting which uses the perspective cue, shading and superposition to convey merely an imaginary depth.

The first photographic systems to create 3-D images were developed in the early 1800's. These older systems involved photographing a scene twice, from two perspectives about 64 mm apart. This is approximately the distance between the eyes. The two images were then studied through a stereo viewer so that the left eye could see only the left photo and the right eye the photo on the right. If color-filtering or polarizing apparatus were used to take the two photos, then similar devices were also needed to view the images when they were projected on a flat screen. Specifically, viewers had to wear color-filtering or polarizing glasses which duplicated the scene as photographed to separate the two images for viewing with each eye.

3 CURRENT TECHNIQUES

Newer systems use glasses with "shutters" synchronized with a monitor that alternates the image for each eye. Although this produces a 3-D image, measurement cannot be made from it.

The recently developed Auto-stereoscopic 3-D Graphics Monitor, the DTI 100M, shows two images at the same time. These images correspond to the left and right views and can be fused into one image file by software so one can see a 3-D image from the monitor without wearing special glasses. (Figure 1)

As shown in Figure 2, a thin illumination plate is mounted behind the LCD screen which can show 16 shades of gray. When the back-light passes through the illumination plate, it generates a set of 320 equally spaced, very thin, bright vertical lines. The columns of lines are spaced in parallel so that the left eye sees only the odd columns and the right eye sees only the even columns of lines. When both eyes view the image simultaneously, the brain perceives a three-dimensional image. (Eichenlaub 1991; Caver 1991) The viewing zone, which is the area in which the observer sees stereo images, is quite large. However, there is a defined relationship between distance d and the viewing distance D as shown in Figure 3.

While the monitor is displaying a 3-D image, the cursor can be controlled by a mouse, to move in and out of the screen (the Z-direction), as well as vertically and horizontally (the X- and Y-directions). This unique feature makes it possible to measure the actual position of an object. As the cursor is moved, changes in the X, Y and Z coordinates are displayed on the screen and they can be printed to a data file in ASCII format. Once the scale is defined, measurements can be converted into actual

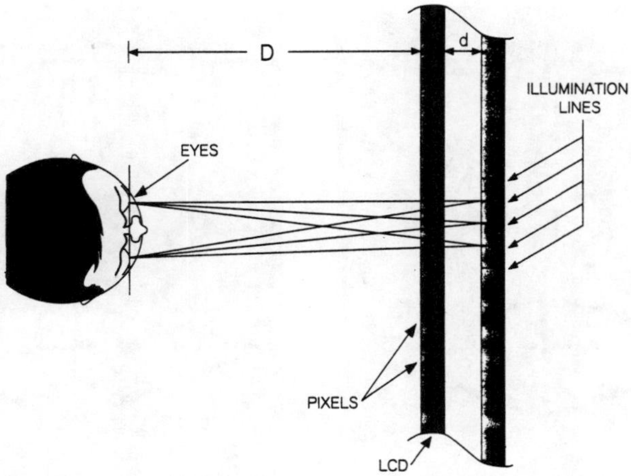

Figure 3. View from above. Auto-stereoscopy makes use of a transmissive LCD generated line by line, and then a thin illumination plate

distances and volumes with the supporting software.

The lightweight, portable monitor is driven by a PC computer equipped with an additional Yamaha DisplayMaster Extended EGA card. This card can be used in parallel without causing interference with the computer's existing monochrome graphic board.

The basic technique for using a single film camera to photograph the left and right eye views as described previously can still be applied. The stereo-pair of photos just needs to be converted into video signal and input to the Yamaha card. The supporting software will then combine the two images into one 3-D image which can be archived in a PC Paintbrush Plus (PCX) format.

The advanced technique involves the use of two video cameras equipped with CCD, Charge-Coupled-Device, and a gen-lock system. The output video signals are mixed by a multiplex and then fed into two CORTEX-I frame grabber boards installed in the PC computer. This setup allows the 3-D live video images to be shown on the DTI 100M monitor, captured and archived.

Figure 4 shows two pairs of stereo images taken with a single camera. The right image of the top pair was taken at a view point translated 64 mm side-ways from the left view point while the bottom pair was taken at two camera view points separated by three feet. Figure 5 shows two photographs taken at a fixed view point of an object being moved and rotated. Although the right image was captured after being moved for 64 mm, a stereo image cannot be made from this pair of photographs because the effect of rotation of the object is not the same as the rotation of the camera. However, a proper rotation of the camera after translation can enhance the depth of view. In order to achieve a good quality stereo pair of images, using either a single camera or two mono cameras, they must be mounted on a level track to ensure both the left and right images are taken on the same platform. There are cameras such as the stereo 35mm Nimslo and the stereoscopic video camcorder Toshiba SK-3D7 which have a fixed twin lens built in to eliminate the alignment problem. But the built-in fixed lens separation limits the 3-D effect when shot at a far field object such as a blast or a muck pile.

4 EVALUATION AND APPLICATIONS

The first quantitative test of the 3-D imaging technique was done with a single 35mm camera.

In the experimental set-up, an X-Y co-ordinate grid was laid out as shown in Figure 6. This was then photographed at 50 and 100 m distances from the grid center (0,0) from a

Figure 4. Top stereo pair: images of a wire cube taken with a single camera. Bottom stereo pair: images of a muck pile taken at two camera view points separated by three feet

position 10 m above grid elevation as shown in Figure 7. This dimension was selected to simulate a practical situation for picturing fragmentation during mucking in a quarry operation.

The grid was photographed at 1 m spacing between camera locations. A 105 mm lens was used to provide acceptable coverage of the grid area in each picture for evaluation. In all cases the X and Y results proved to be repeatable to within 10 cm while the Z values were repeatable to approximately 30 cm. The Z values were found in best agreement with the actual values when camera separations of about 4 m for the 50 m target and 8 m for the 100 m target were used. There were difficulties encountered in positioning the cursor in the Z direction, i.e. in and out of the screen. It appears that better accuracy can be obtained by using a longer focal length lens to zoom into a smaller area. However, a further study has yet to be done.

3-D technology has never been used in blast analysis. However, a successful 3-D shot of a quarry blast using two video cameras as shown in Figure 8 indicates that it can be applied effectively to determine the geometry of a blast, face velocity, and the directions where rocks move to. Its ability to measure the dimensions

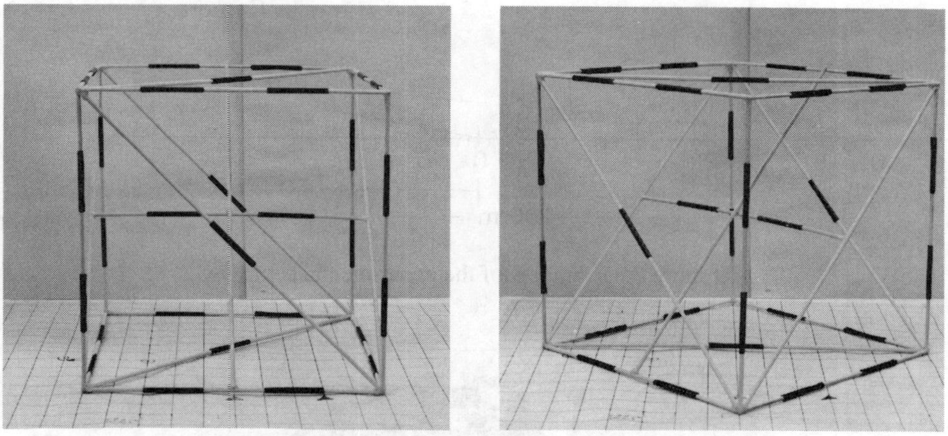

Figure 5. Photographs taken at a fixed view point. The right image shows the result of rotation of the object

and profile of a muck pile requires no further mention. It can also provide 3-D information about fragment size. This has long been a goal of blast analysis. However, when using the 2-D photographic method, fragmentation can be evaluated but not volume because the depth information is not provided. With the current 3-D technology in place, the difficulty in determining muck volume can be overcome.

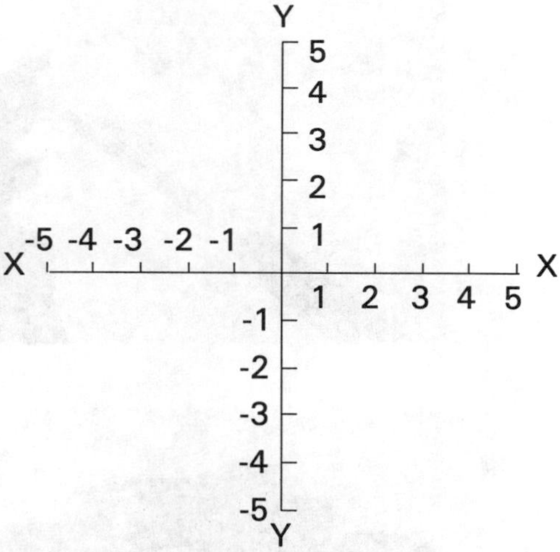

Figure 6. Plan view of a 5 m X 5 m grid

5 DISCUSSIONS

All these techniques require a pair of images taken at a distance D meters from two view points separated by S meters. To get the best view of the 3-D display, the distance to separation ratio, D/S, should be proportional to d/s where d and s are, respectively, the viewing distance from the 3-D monitor and the average separation between the left and right eyes.

In human vision, depth perception decreases with distance. For example, visual depth perception beyond 600 m is very poor. (Hyzer 1989) This human limitation is not eliminated by the monitor. Even though the in-and-out movement of the cursor along the z-axis can determine depth, as the distance increases, resolution starts diminishing and makes it more difficult to locate the desired points for analysis. Also, since the z-axis is relatively short, small movement along the z-axis can represent considerable distance. All these factors make the task of locating a distant point open to time consuming trial and error.

Equations exist to calculate optimum camera separation. They are based on the lens focal length, the depth of the object, the distance to the object and the spatial resolution of the photographic system. For still objects, such as the bench face of a blast or a muck pile, a single camera mounted on a track can be used to capture images from different perspectives. As

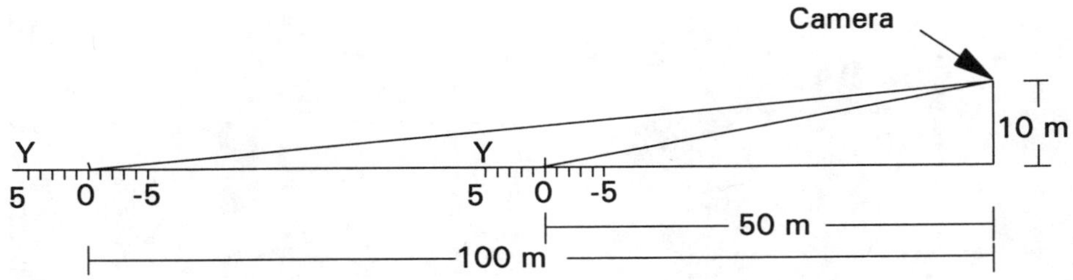

Figure 7. Side view of the experimental setup

Figure 8. Setup of a single 35 mm and two digital video cameras used for stereo imaging

described above, the sliding track ensures that both images are taken on the same platform.

To study rock mass motion in blasting, two synchronized CCD cameras are used to record the left and right images so they can be matched for the same instant in time. During the play back of the 3-D video on the monitor, each frame can be captured and digitized in time sequence. DTI has just come out with a new color monitor which allows live action in 3-D to be viewed, but no supporting software has been developed for the image analysis.

6 CONCLUSIONS

Computer based instrumentation has led to many advances in blast analysis in the last ten years. These innovations have furthered the development of blast modeling. Analysis with 3-D techniques will greatly improve understanding of fragmentation and the formation of the muck pile during a blast. The DTI 100M has proven its value in blast assessment. Further experience in image capture techniques will improve the quality of the 3-D images and thus the accuracy of the measurements.

REFERENCES

Carver, D. E. et al 1991. Development Of A Stereoscopic 3-Dimensional Drawing Application, SPIE Vol 1457, Stereoscopic Display and Applications II.

Chung, S. H. & Burchell, S. 1992. Computer Based 3-D Blasting Analysis System. SEE 18th Annual Symposium on Explosives and Blasting Research. Jan. 22-23, Orlando, Florida, USA.

Eichenlaub, J. B. 1991. Progress In Autostereoscopic Display Technology At DTI SPIE vol 1457, Stereoscopic Displays and Applications II.

Hyzer, W.G. 1989. Scientific 3-D Photography, Photomethods, January.

Innovative Mine Design for the 21st Century, Bawden & Archibald (eds) © 1993 Balkema, Rotterdam, ISBN 90 5410 325 6

Innovations in oilsands blasting

P.D. McLaughlin
Suncor Oil Sands Group, Fort McMurray, Alb., Canada

ABSTRACT: In nearly all mining operations, blasting plays an integral part in the extraction of ore. In addition, blasting is the initial stage in sizing of the ore to prepare it for final processing.
For these reasons, improvements and innovations in drilling and blasting techniques should be carried out as a matter of course and not as a matter of neccessity.
This paper will outline some of the more important steps in arriving at present blasting techniques at Suncor. Some of these steps were taken out of necessity, but the majority were due to proactive measures of the people involved.
Of particular interest is the almost exclusive application of crater blast design and techniques, and its evolution in the Suncor environment.

1 INTRODUCTION

Very early in Suncor's history, it was recognized that blasting would be required to economically dig the oil sands. The bucketwheel excavators (BWE's) were encountering lenses of sand- and siltstones, which induced heavy shock loads on the machines, and were having extreme difficulty digging in frozen faces during winter months.
For these reasons Clifton W. Livingston of Barodynamics, Inc., was asked to do a preliminary study and submit recommendations.
A summary of Livingston's study follows :
1. Large displacements would be required for fracture of rock and hard lenses at depth.
2. Response characteristics of blast craters in oilsands would vary due to oil content.
3. Temperature of the oilsand would affect blasting response.
4. Recommended hole diameters of from 610 to 1065 mm, using crater blast designs.
5. Recommended an explosive with high gas volumes and low rate of energy release - in effect, AN/FO.
6. Predicted some form of water control mechanism would be required.
7. Recommended caisson-type drill rig to achieve the hole diameters and depths needed.
8. Outlined the economic benefits of blasting to the mine, including an analysis of BWE productivity, availability and maintenance costs, and the effect of blasting on each of these.
9. Predicted blast vibration concerns at the process plant.
10. Outlined a test program to determine optimum design criteria for production scale blasting.

Mr. Livingston's recommendations and predictions have proven remarkably accurate, and his initial work set the stage for the next twenty-five years of research and developement in the drilling and blasting program at Suncor.

2 ECONOMICS

The earliest recorded study done on the actual economic impact of blasting showed a high sensitivity of the BWE's to any loss in production, with as little as a 5% loss being sufficient to warrant blasting (Livingston, 1967).

Further work also outlined the delays and expenses that could be expected from not blasting, thereby reaffirming the need to blast as closely as possible to final design (Bauer, et al, 1968).

2.1 Bucketwheel excavators in oilsand

Later, in-house studies concluded the following :
1. In medium-coarse oil sands, productivity increased by 16% and tooth consumption decreased by 69% in blasted ground.
2. In coarse grained oil sands, productivity increased by 12% and tooth consumption decreased by 15% in blasted ground.
3. Benefits would be much more pronounced in winter conditions.
4. Blasting cost $0.10 per ton, saved $0.08 per ton in tooth costs alone.
5. Further savings would be realized through reduced equipment maintenance, repairs and replacement costs.

2.2 Bucketwheel excavators in overburden

Internal studies comparing ripping to blasting for frozen ground in overburden operations for a BWE concluded :

1. Drilling and blasting was approximately 20% less expensive than ripping for frost breakage.
2. There was a 20% increase in digging rates in blasted material
3. Blasting resulted in a 25% decrease in power consumption
4. Expect up to 80% reduction in tooth costs in blasted material
5. Improvements in BWE availability due to blasting was actually higher than predicted
6. Blasting costs on a unit basis were lower than predicted.

2.3 Shovels in overburden

In a later overburden shovel study, it was determined that blasting gave an overall saving of $0.054 per BCM on equipment requirements alone, with reduced maintenance and longer equipment life possibly doubling this figure.

3 PITWALL BLASTING CRITERIA

An early concern was the design of slope angles for pit faces and final walls, and the effect of blasting on their stability. In addition, it was neccesary to determine the significance of these slope angles with respect to mining methods.

3.1 Damage envelope

In 1975 a report on the effect of blasting on stability of oil sand slopes was completed that tied probable damage limits to PPV's and predicted shear stresses as a function of distance and known charge weight (Bauer, 1975). This was further refined in 1977, when it was calculated that stresses that induce failure are limited primarily to within the crater zone and should not induce failure outside this region (Bauer and Crosby, 1977).

3.2 Early stand-off criteria

Recommended safe scaled distances were developed on a very conservative approach, based on distances to underlying berms and adjacent design walls. The scaled distance for maximum crater radii at surface had been essentially doubled and applied through-out the entire length of the hole. In addition, specific critical conditions were identified as qualifiers to these criteria (Hardy and Associates, 1977). These included the presence of specific types of clay lenses, formation dip, pore pressures and previously blast damaged areas.

These basic criteria were maintained until quite recently, when a need for reassesment was foreseen.

3.3 Current practice

The general mine plan for late 1992 and early 1993 involved cutting the final wall on the west end of the middle bench. Final pit wall designs were reviewed at this time with the aim to maximize ore recovery along this wall. Blasting criteria were included in this review, primarily to minimize the amount of unblasted material left for the BWE.

Recommendations included a change to the final slope and elimination of a mid-face berm. New blasting criteria were based on actual rupture zones as observed and measured over the history of the mine and used the scaled distances to the extreme rupture limits as measured at depth, essentially doubling these. The result was a reduction in unblasted material from up to 30 m. from the final pit wall to as close as 5 m.

The wall in question has been entirely excavated to date and has shown no signs of instability due to blasting. In fact, a geotechnical analysis of the oilsand pitwalls concluded that blasting as close as possible to final contours will improve stability by allowing dissolved gas in the bitumin to escape thus gradually relieving in-situ pore pressures. The degree of pore pressure relief would be a function of time between blasting and excavation and, to a lesser extent, to ambient barometric pressure.

4 PATTERNS AND LOADS

During the earliest test program, a full spectrum of single and multiple hole crater shots had been taken, with a variety of explosives in hole diameters from 125 mm to 475 mm (Bauer, et al, 1968). The results were scaled up to the design blasthole diameters and depths for production overburden and mine blasting, using crushed high-density AN/FO as the primary explosive.

Patterns and loads for oilsand blasting were developed based on Livingston's Crater Theory, using 475 mm and 795 mm diameter holes.

Patterns for early overburden work were developed as a function of borehole diameter, following the widely accepted Ash Formulae, using 230 to 305 mm diameter holes in typical bench blasting.

In 1976, porous prill was introduced and adopted as the standard. This resulted in a slight change in calculated scaled depth of burial (SDOB), primarily due to its lower density, and a slight reduction in patterns.

4.1 Transition to heavy-AN/FO, drop 795 mm diameter holes

Porous AN remained the standard until the introduction of heavy-AN/FO in 1984. At this time, the 795 mm holes were dropped and 475 mm adopted as the standard.

Again, a reduction in SDOB resulted, but this time it was due to the actual detonation characeristics of the product, despite its higher density and increased water resistance. Patterns, however, remained constant, with the maximum attainable burden in the 10.7 m. deep holes. Decking was used for any holes deeper than this, at a constant pattern.

4.2 Reintroduction of 795 mm diameter holes

In 1987, three areas of the mine were targetted for the testing of 795 mm holes, using heavy AN/FO, each having a distinct oilsand environment as functions of grade and percent fines/clays. These tests were intended to determine the feasibility of drilling larger diameter holes as well as to determine any advantages to the explosive performance. For this reason, two worst case scenarios were deliberately set up - one in which badly squeezing ground could hamper drilling and one in which serious flowing ground water conditions were expected.

Larger diameter holes also introduced the possibility of using deeper holes while maintaining crater blast geometry, so one test area involved 18 m. deep holes (maximum capacity of drills on site).

Immediate blast results were considered excellent in all the test areas, and rated from equal to superior to previous practice.

Serious problems were encountered on the 18 m holes, but only while drilling the last 1.5 m. From this, it was determined that 16.75 m. was the maximum economic capacity of drills being used.

The other test areas had been done with 14 m. holes, with a variety of patterns and loads, based primarily on powder factors. Near optimization was achieved in one pattern, which on investigation was found to be very close to a geometric scale-up from an optimized 475 mm pattern and load.

Assessment of the work with 795 mm holes from 1987 was completed in 1988. Those areas blasted with 795 mm holes in 1987 were monitored as dug, with visual and subjective appraisal on the middle and bottom benches and in-depth analysis of the deeper blast area on the west end of the top bench.

Digging conditions showed improvement on the lower benches, being considered "as good as the best previous conditions or better".

Top bench monitoring included current draw and voltage records coupled with digging rates for a full spectrum of material, including unblasted, standard blasted (475 mm holes, 14 m. deep with decked charges) and the test area. Digging rates went from 2500 tons per hour (tph) in unblasted through 4500 tph in standard to 7500 tph in the test area with only marginal changes in total power consumption.

4.3 Standardization

Over the years there was a tendency to change loads to compensate for varying blast results in differant areas of the mine without adjusting the patterns.

One area on the east end of the bottom bench in the mine had given perennial digging problems despite many such attempts. Another area on the top bench had consistently given good blast results despite an equal variance in powder factors. Based on this, powder factors were eliminated from all design criteria in mine and overburden blasting in 1988.

Instead, crater characterization studies - careful measurement of visible (surface) crater limits coupled with blast assessments and face conditions - were used to determine optimum designs. By determining common factors in good blasts and using these to adjust poor or acceptable blast patterns/designs, it was possible to standardize all the patterns to good to excellant results. From this, standardized patterns were developed on basis of SDOB and scaled burden (SB). This approach has given consistently good results since its full inception, and has allowed greater freedom in scaling and conceptual design.

Problems digging on the east end of the bottom bench continued, however, so another approach was attempted. This involved increasing the SDOB:SB ratio to allow more forward movement and concentrate the gas chambers at depth, allowing overlap before they began to break through to surface. Delay criteria were also developed to ensure time for the chamber overlaps before the following holes shot.

These delay criteria were then expanded to cover all ore and overburden types in the mine, with the fastest being in the "brittle" hard clays and the slowest being in the highly plastic high-grade ores.

4.4 Open face overburden blasting

An opportunity arose in the fall of 1989 to use a production blast as a test in a deep overburden cut to determine maximum pull at toe - this was to an open, unblasted face, cut out by shovels. Varied burdens were used across the face to determine at what scaled burden actual breakage would occur and the degree of movement.

The results from this study were used to develop casting criteria, design patterns and loads for overburden material.

That later year a particularly high overburden ramp had to be removed. This ramp had frozen to 4 m. in depth from the surface and 3 m. in from the face. The cut ran from 23 m. on the west end to 7.5 m. on the east end, with an effective width on the top of 30 m. and a length of 230 m. Standard practice would have involved taking the ramp out in two lifts of 11 m. each, but this was considered impractical due to the ramp being open on the south side and too narrow for efficient use of the shovel and trucks. It was decided to attempt a casting type of shot, to push the material out onto the floor to the south.

Despite problems with out-of-sequence firing, and product failure in some of the holes, the blast was considered a success, verifying the design criteria used.

By incorporating the data from the previous studies, spreadsheets were developed for pattern generation in overburden and oil sand material for the various cuts, hole diameters and types of blasts used at Suncor. Additional spreadsheets were then developed to computerize blast recording, incorporating all aspects of the drill/blast cycle.

4.5 Open face oilsand blasting

Past practice whenever an unblasted face in oilsand had been exposed was to back off at least one full row from the face and use this as the "free face" when designing and laying out further pattern.

These occurences are not uncommon, and frequently occur during the winter months, when a BWE system has advanced further than anticipated. This results in long unbroken strips of oil sand, frequently frozen, for the BWE's to chew through before they can get into blasted material.

In an attempt to eliminate these unblasted strips, studies were begun to determine the actual radius that could be broken at grade, as a function of charge weight and depth, to an unblasted face. An opportunity arose in a particularly large blast on the east end of the top bench in the mine. The exposed face to the south and off the east end was entirely unblasted, and up to 27 m. in height. In addition, the conveyor system had not been moved up to the toe, leaving almost 100 m. of free space south of the blast. The nearest risk off the east end was a haul road that could be cleared quickly of any fly material.

The entire front row was designed to break just to the face, allowing the material to fall, and the east face was designed to throw out for a distance equal to the face height. Both faces moved as predicted, with no material falling outside the expected range.

The design has been improved and incorporated into all open face blasting carried out in the mine. A table of loads was drawn up, with additional tables to allow the blasters to calculate actual burden at grade based on stand-off from the crest, face angle and hole depth. In this way, the surveyors simply stake the holes as per design and the blasters can field adjust depending on face conditions.

4.6 Scale up to 1250 mm diameter holes

The highlight of the 1991 season was the use of 1250 mm diameter holes in a deep cut on the middle bench. To maintain spherical charge geometry and crater design at the depths being shot, a minimum hole diameter of 1250 mm was required. Since only two of the four available drills could achieve the required hole depth, the pattern and penetration rates had to be double the standard production on a ton per hour basis. The very large patterns developed for the 1250 mm holes made this possible.

During drilling, several of the holes on the front row began caving badly. This was attributed to back-break from the previous blast, and extended to a maximum of 18 m. in depth. Normal practice when in caving ground was to restake the area for smaller diameter holes, but in this case the same drills simply dropped the 1250 mm augers and replaced them with 795 mm. The crew then drilled two holes to depth and loaded each with half the design load for a larger hole.

These smaller holes were drilled at 3 m. out along the row line on either side of the original stake. This distance was based on calculated chamber dimensions of a 2630 kg charge in a 795 mm diameter hole at greater than critical depth. The intent was to have the chambers overlap, allowing the two charges to function as one single point charge. The resulting crater was indistinguishable from a single 1250 mm hole crater, and no discernable difference was seen between the two charge configurations when digging through the area.

The success of the 1991 blast with 1250 mm diameter holes lead to a second, larger shot in 1992. This one was 3 m deeper, and taken in a single blast, to a predominantly open face. The results were even better than in the 1991 shot, primarily due to the fact that it was shot "open". The area was mostly dug out in late 1992, with excellant digging throughout.

4.7 Twinning with smaller diameter holes

The success of the twinning process while drilling the 1250 mm hole pattern lead to a further application in standard production blasting. This has the potential of saving over 10% in operations time and money, and also improving blast results in caving or squeezing ground. These conditions have historically been very difficult to drill resulting in poor to un-acceptable blasts.

Past practice had been to use 475 mm holes for all cuts of 10.7 m. or less and 795 mm holes for cuts greater than 11 m. The main problems had arisen when drilling the 795 mm holes in unstable ground. In severe cases, the area would be restaked for a 475 mm hole pattern, drilling to full depth and decking. The drawbacks with this solution were that decked holes have a tendency to eject and pull poorly at grade in oilsand blasting, and the smaller pattern resulted in approximately 250% higher drilling and priming requirements. In addition, the patterns take longer and are more expensive to tie in.

A scaled down version of the solution used for the 1250 mm holes was tried and proved very successful. The 475 mm augers were replaced with 580 mm, and new patterns and loads developed. These resulted in using 580 mm holes for all cuts of 15 m and less and 795 mm holes for 17 m. to 21 m. cuts. When the ground conditions warranted drilling with smaller holes, the drillers would simply offset 2.6 m. with 580 mm holes on either side of the stake location and "twin" the original hole. An added bonus of using this method was dramatically improved blast results - far more than anticipated. This is believed to be primarily a function of effective charge geometry, as the charge has been "spread out" over a plane normal to the direction of the burden, but maintained within an essentially spherical orientation. Further work with this concept in arrays of three or more holes had been planned for applications in long slopes with excessive toe burdens.

5 EXPLOSIVES

Originally, crushed high density AN/FO had been used as the primary bulk explosive. This was loaded into lined boreholes from bulk explosive trucks equipped with hammer mills, and was in use from 1968 to 1976.

Despite its inherant fuel retention problems, it proved a highly effective product, having the highest scaled depth of burial and resultant largest patterns of all the explosives used at Suncor over the years. Its effective bulk strength for oilsand blasting was due to a combination of detonation characteristics and the fact that it was a "dry" product - there was no formulation water to reduce reaction temperatures and gas expansion.

In 1976 porous prill AN/FO was introduced. This didn't need crushing for sensitization and oil retention properties, so was considered a more reliable and cost effective product. A slight reduction in effective bulk strength resulted, but this was acceptable and the product continued in use until 1984.

5.1 Methane explosion

Perhaps the most dramatic and far-reaching event in Suncor's blasting history was a methane explosion in 1983 that injured two members of the blasting crew. This resulted from a sequence of events, starting with misfire problems and ending with an accident that would lead to major changes in the blasting operations and procedures.

Misfires, a long-term concern at Suncor, had been attributed to problems with the integrity of single wall plastic borehole liners. In an attempt to reduce or eliminate the problem lighter gauge double-wall liners were introduced, and these were very successful.

Unfortunately, the double wall also reduced the ability of any static charge build-up to dissipate to ground. As a result, static build-up in one liner discharged to the borehole wall and ignited methane in the borehole.

Since methane gas is common in coal and oilsands mines, and AN/FO in liners is a standard explosive in both these industries, the provincial government and Suncor initiated an intense study to find a way to eliminate the static discharge risk. The problem was approached from a number of fronts, including :

1. Gas detectors to determine if individual holes contained an explosive mixture of air and methane
2. Compressed air to purge the methane from the holes
3. A salt water spray was used to ensure blasters and equipment could be properly grounded
4. An elaborate system was developed to ground personnel and equipment in the vicinity of the hole.

Overall, safety for the blasting crew was improved but the productivity was notably impaired - 20% to 25% drop.

5.2 Heavy AN/FO

In 1984 a second initiation of methane occured in a mine blasthole despite the precautions implemented since the 1983 accident. In addition, mis-holes were again becoming a problem due to liner damage from the purging air-hose.
Tests were initiated with Heavy-AN/FO, an emulsion and AN/FO blend, without liners. This product was more expensive on a per meter loaded basis than straight AN/FO but its increased bulk strength and improved water resistance were expected to help offset this. The immediate results of conversion to this new product were:

1. Improved safety and streamlining of loading operations.
2. A marked decrease in misfires - from up to 10% of the holes down to approximately 5%.
3. Hole diameters were decreased from 795 to 475 mm based on the theoretical bulk strength of the product, resulting in higher penetration rates.
4. Patterns were reduced, though, resulting in lower tons per drill hour and increased drilling requirements.
5. On balancing the benefits against losses, the overall increase in price per ton blasted was acceptable.

5.3 Misfire concerns, solutions

Despite the best efforts of the explosives supplier and the Suncor blasting crews, there were still some problems with misfires and incomplete detonations. Dr. William Crosby of Mining Resource Engineering, Ltd., was asked to do a site visit to assess these problems and submit recommendations. These were as follows:

1. Suggested 795 mm dia. holes for 10.7 m and deeper blasting, discontinue decking
2. Triple primer weight from 450 gm to 1360 gm
3. Increase viscosity of emulsion to maximum attainable
4. Reduce temperature of emulsion phase to reduce damage to prill
5. Two day drill/blast cycle to reduce effect of water on product
6. Ongoing testing program to ensure quality control of product
7. Suggested trying liners with reduced emulsion content heavy-AN/FO
8. Identified the combination of water and methane as a primary site condition damaging the explosive.

This last point had been recognized for several years at Suncor, but no specific solution to the problem had been presented. It is strongly suspected that this same phenomenon is prevalent in coal operations that have high levels of methane accompanied by water. Damage is caused by water carried up the borehole by percolating methane gas, stripping the emulsion matrix off the prills. The emulsion is then carried on up into the stemming material and the exposed prill is dissolved.

Crosby recommended higher viscosity emulsions to resist this stripping effect. He also noted that slurries or watergels, with their crosslinked, gelled structure, would be much less susceptible to stripping. On this basis, and on the basis of their detonation characteristics, he suggested that slurry blends would be more applicable to Suncor blasting, but also warned that they may be more expensive.

Suncor's bulk explosives supplier was also carrying out extensive product tests during 1987, with a range of pumpable and augered blends from approximately 75% emulsion through to 50/50 emulsion-AN/FO. Gassing and microballon sensitization was tested, as were a variety of fuel combinations, and VOD's were recorded for 475 mm and 795 mm diameter holes with standard product. In addition, a limited amount of pumped slurry was tried in one area of the middle bench.

The final recommendation after all these product tests were completed was to stay with the current product/blend and concentrate on controlling the water-methane problem that is inherant to oil sand blasting.

Misfires continued to be a concern, so an intense study was initiated to find the cause, and therefore hopefully the cure. During this study, many misfires were actually excavated and examined.

In the majority of the misfired holes, the burn mark from the detonating cord downline was clearly visible in the column as was the more extensive burn from the primer, but invariably it was against the wall of the hole. This area was also found to be contaminated with loose material and/or water. In some the deterioration along the wall was too extensive to identify any trace from the cord. It was hypothesized that this zone was therefore desensitized to the point that a healthy detonation could not be initiated.

The mine engineering department then designed and patented a device for positioning and holding the primers in the center of the column. This was produced at a local plastics fabricating shop and introduced into the loading procedures. Since the time of its introduction, recorded misfires at the mine have dropped from 5% to less than 1%. The mine engineering department and the blasting crews believe this is mostly due to the centering device. It should be noted, however, that it is impossible to attribute all the success to the device, since the blasting personnel are now much more aware of the importance of proper primer placement and take greater care of the downlines while loading as well as when tying in.

These centering devices were originally designed for 475 mm and 795 mm diameter holes but they can be readily adapted to the standard range of quarry and open pit mining blast-holes. A commercial application of this technology was therefore initiated by Suncor which led to the device being produced by a specialty blasting accessories firm, and it has been further refined to allow adjustment for a variety of primers and hole sizes.

A point of interest is that the product failure noted previously for a cast-type shot on a frozen ramp was theorized to be due to an extreme temperature differential between the prill and the emulsion as they were blended during loading. The prill was estimated to be at approximately -40°C and the emulsion to be at +50°C when the two came together. When everything had stabilized, they would both have been at +4°C, ambient ground temperature. This constituted several phase changes in the prill, and a quite dramatic and rapid drop in the temperature of the emulsion matrix. This theory supports Crosby's recommendation to reduce the temperature of the emulsion matrix prior to blending with the prills.

5.4 Explosive trials

In 1989 and 1990, an extensive supplier/product trial was undertaken, involving six established bulk explosive suppliers. Products were tested in each of the three main oilsand types and assessed.

Suppliers were asked to submit calculated Relative Bulk Strengths of the test products and the method used to obtain these figures. This was to allow scaling and adjusting of loads to ensure a fair assessment. Some discussion ensued with the various suppliers on whether the scaling laws used by Suncor could be applied - this was primarily due to the variety of methods used by different suppliers to rate explosives. A standard was suggested by Suncor for all the suppliers to use - namely, moles of gas generated per kilogram at STP, and percent by weight water in the liquid phase of the blend. Again a few of the prospective suppliers balked at giving this information and two in particular outright refused. The rest, however, were quite willing to supply this limited information on their products when the need for accurate scaling was explained. For those explosives for which the values were withheld, an educated guess was made, and final comparisons based on this.

It was understood that these two values alone could not be used for scaling across explosive types - ie slurry and emulsion blends - but they were found to be very accurate tools for comparing within explosive families in the Suncor environment.

From these trials, several conclusions were drawn :

1. Higher solids (AN) content blends were more effective, at the risk of water resistance

2. Higher viscosity emulsions were less susceptible to damage from squeezing ground and water-methane stripping

3. Optimum densities for emulsion AN/FO blends were in the range of 1.2 - 1.3 gm/cc

4. Maximum desirable emulsion content was approximately 40% by volume

5. Sensitization of emulsion with gassing or microballons had no discernable effect on fume generation

6. Only one emulsion-AN/FO blend shot completely clean of fumes, all others exhibited noticeable to severe NOx fumes.

7. Emulsions with the highest percentage of AN (vs CN) had the most heave.

8. Total water content had a distinct effect on heave - higher water content reduced heave

9. Slurry-AN/FO blends were clearly better than emulsion-AN/FO blends, provided the slurry-AN/FO blend fit the next criteria below (10.)

10. Minimum desirable solids (AN) content of slurry blends was approximately 40% by volume

11. Auger delivered slurry blends with solids content up to 60% by volume had sufficient water resistance

12. Slurry blends should be fully cross-linked within 2 minutes of loading and preferably sooner.

13. The slurry blends consistently gave lower (indiscernable) fumes.

14. Effective bulk strength of explosives in the Suncor environment is a function of total gas generation, water content of blend and explosive type

5.5 Optimizing energy partition

A major change in 1991 in production blasting was the nature of the explosive used. This was the result of blend modifications to substantially change the energy partition characteristics by shifting the detonation and reaction zone components. As a result, more energy was released in the later stages of the detonation process, maintaining a load on the ground for a longer time, giving larger (at depth) and more uniform craters, with reduced throw. To date, digging conditions have shown a marked improvement over the full depth of exposed faces.

To ensure that this new blend was being properly initiated, and achieving full healthy detonations, an extensive study was undertaken. This included VOD and high speed photographic analysis to accurately measure the effect of different size and shapes of cast primers, both in a variety of hole diameters and in unconfined conditions.

Good correlation was found between the 250 mm diameter unconfined charges and 475 mm diameter production blastholes, but this could not be extended to the 795 mm holes. This was attributed to methods of loading, higher static borehole pressures and effect of longer free-fall on the augered blends in the larger, deeper holes. Stemming contamination of the top zone of the charge was also more pronounced in the larger diameter holes, showing up quite clearly on the VOD traces.

From these studies, it was determined that steady state detonations could be reliably initiated with 450 gm primers, but the run-up distance was too large a percentage of the total length for the spherical charge geometry used at Suncor. For this reason, 900 gm primers were recommended and are being used.

5.6 Stemming contamination

Stemming contamination of the top one to two borehole diameters of charge had long been suspected, but the full extent of this contamination had not been known until the studies done in 1991.
In very long explosive columns, this contaminated region may only comprise a small percentage of the total explosive charge, but in spherical charges, one borehole diameter of contamination equates to a minimum of 15% of the total. For these reasons, the depth of contamination is critical at Suncor.

In an attempt to quantify the damage from contamination, four holes with relatively short columns were loaded and the top of the column protected with an air bag. These air bags rested firmly against the top of the explosive, and their sole purpose was to prevent stemming contamination. The bags were inflated sufficiently to fill the entire cross-section of the hole, and the holes stemmed. Each hole had been traced for VOD and high speed films were taken of the shots.

Visually, there appeared to be a slight differance in the uniformity and height of the crater domes when the airbag was used, but the most definite evidence of improvement was shown in the VOD traces - a clean transition from explosive to stemming at the top of the column.

These gave clear indications, when combined with the previous year's traces in standard holes, that one borehole diameter of charge was being severely damaged or even entirely lost to the detonation process. In short, if the airbags or a similar barrier to stemming contamination was properly used, better blasting results with less explosives would result. Initially, explosives charges would be reduced in length by one borehole radius, and if sufficient improvement is seen, this could be extended to one borehole diameter, depending on design loads for the specific depth of hole.

5.7 Aluminum tests

The last stage of product studies involved using varying percentages of aluminum of 20 and 30 mesh size. Again, VOD and high speed photographic analysis was used, in conjunction with simple physical measurements of apparent craters and extreme rupture limits. When the aluminum content exceeded 3%, the physical measurements became too subjective to use, as the ejecta masked the usual surface features associated with these zones, so the final analysis was dependent on the high-speed photography.

Visual assessment of the shots indicated that at about 5% aluminum, of either grade, crater dimensions maximized. Higher percentages of aluminum appeared to simply eject the material higher, faster and further.

Analysis of the film, however, showed quite clearly that crater dimensions and total displacement continued to increase with higher aluminum content. The mesh size of the aluminum being tested seemed to have little effect, but it is believed that substantilly larger mesh sizes would show further improvement.

An additional concern would have to be addressed if larger mesh sizes or higher aluminum contents were to be considered - after-burn and fires. During the tests, large fireballs were caused by the 9% aluminum blends. These actually ignited escaping methane and the oilsand itself. It is hypothesized that the fireballs were a result of the intense heat generated by the burning aluminum in the reaction zone (gas chamber) igniting vaporized bitumin as the expanding gas reached atmosphere and oxygen. If this is the case, and methane ignition is secondary in the process, then the phenomenon would be universal throughout the oilsand deposit and not restricted to high methane areas. Future benefits of aluminization, then, would have to be tempered by control of this risk.

A further benefit of aluminization was evident in the lack of visible fumes generated from the blasts. Historically, all emulsion-AN/FO blends had given high levels of NOx fumes in the richer bitumin ore, grading to little or no visible fumes in the low bitumin, high fines ore and overburden materials. The aluminum tests had deliberately been carried out in an area that was known to have extremely rich ore, since this is usually the most difficult to obtain good craters in and because it was an area prone to high NOx fume generation. As with crater dimensions, fume control improved with increasing aluminum content, to the point where none was evident at the 9% aluminum level.

Review of the test results is not complete yet, but it is expected that optimization will be at approximately 5% aluminum by weight in the current blends.

6 LIMESTONE

The earliest standard for limestone blasting established a 200 mm diameter x 7.5 m deep x 4.5 m x 4.5 m pattern with a powder factor of 1.12 kg/BCM. This had been developed while still using porous prill AN/FO and liners.

6.1 Pattern expansion

When the mine had changed over to heavy AN/FO, the drill pattern in the limestone pit was maintained, but the hole diameter was changed to 170 mm to maintain the same powder factor. Manouvering on this tight a pattern was difficult, and the explosives consumption was definitely too high for the rock being broken, so it was decided to maximize the pattern, and reduce to a minimum the amount of heavy AN/FO required, using standard AN/FO where-ever possible. An interim pattern of 5 x 5.8 m with 220 mm holes loaded completely with AN/FO was tried, followed by the present 6.5 x 7 m. pattern with 250 mm holes, to a depth of 12 m, again loaded completely with AN/FO except in wet holes, or when exceptionally tight conditions are encountered at grade. In these cases, a toe load of approximately 1.25 m of heavy AN/FO is used, with or without a liner as required. These changes have resulted in doubling the drill productivity and cutting the overall cost of limestone blasting by almost 50%.

6.2 Vibration criteria

Concerns with vibration levels at in-pit dumps and earthfill structures from limestone blasting lead to developing a fairly comprehensive set of spreadsheets for design of patterns, loads and delays. These were based on previously successful design criteria, but needed flexibility in hole diameters, explosive densities, varying column load densities, hole depths, and delay configurations. These were further refined to calculate costs based on varying prices for explosives, accessories and drilling, and are still in use today.

6.3 Relocation and deepening of pit

In the past, all pits had been worked on the east end of the property, but this was changed due to congestion from haul roads, dyke construction and conveyor systems coupled with increased stripping requirements. Past pits had also been established with a 7.5 m. cut, and simply extended as mining progressed. As a result, large surface areas had to be stripped to obtain any volume of limestone. In addition, the top three meters of the pits were usually poor material, grading to good at the bottom. The material was generally not excavated selectively, resulting in an overall fair to poor quality.

Extensive upgrading of existing haul roads and new road construction for the truck and shovel system as well as ongoing requirements led to developing a new limestone pit on the west mine floor. Total limestone requirements for 1993 were estimated at 153,000 BCM, and the pit was designed to yield this exact amount, including the material from the ramp. An 11 m. cut was chosen to maximize drill utilization and reduce stripping requirements as well as to reduce the impact of the top layer of lower quality material. The ramp was designed to be completely outside the main excavation, running at an 8% slope, at 32 m. width by 152 m. long.

The pit was opened with a sinking cut, pulling the entire volume in one lift, with a total of 463 holes loaded with 120,500

Table 1. Percent loss as function of charge length, depth of contamination, in borehole radii

Charge length, in radii	Radii of contamination 1	2	3	4
2	50.00%	100.00%		
4	25.00%	50.00%	75.00%	100.00%
6	16.67%	33.33%	50.00%	66.67%
8	12.50%	25.00%	37.50%	50.00%
10	10.00%	20.00%	30.00%	40.00%
12	8.33%	16.67%	25.00%	33.33%

Table 2. Patterns and loads for various thickness of frost

Frost thick-ness	Hole depth	Hole diameter	Charge	Burden	Spacing
0.75	7.5	581	285	10	11.5
1.00	7.5	581	285	10	11.5
1.25	7.5	581	285	10	11.5
1.50	1.15	140	1.0	1.50	1.75
1.75	1.25	140	2.0	1.90	2.20
2.00	1.50	140	2.5	2.05	2.35
2.25	1.70	140	3.0	2.15	2.50
2.50	1.90	140	3.5	2.30	2.65
2.75	2.10	140	4.0	2.40	2.75
3.00	2.25	140	5.0	2.60	3.00

kgs of explosives. Excavation volume was calculated at 154,200 bank cubic meters, and stockpile volume estimated at 230,000 loose cubic meters. This was at least twice as large as any previous limestone blast at Suncor, and despite the fact that it was a sinking cut, it had the lowest unit cost.

7 FROST BLASTING

In 1983, Steve Chung of CIL conducted high speed photography in shallow crater holes in frozen tarsand to quantify the response of this material at differant scaled depths of burial. This was initiated by CIL to help in pattern designs for winter blasting.

Past attempts at production blasting in frozen oilsand had met with mixed (mostly poor) success, resulting in either very large frozen lumps still in place or very large frozen lumps a long way from their point of origin - neither was acceptable. This particular study did not solve all the frost blasting design problems, but did show quite clearly that large diameter holes drilled to production depths would not be the optimum method. Future work involving the use of small diameter satellite holes in the production patterns showed promise, and this was continued on a sporadic basis until the adoption of a new frost blasting design in 1985 using 200 mm dia holes on 3.5 m x 3.5 m pattern, 2 - 2.5 m deep.

8 RECENT CHANGES

In 1993, Suncor will be changing from its present fleet of four bucketwheels to a truck and shovel operation in the mine. Trucks will be dumping into permanently mounted sizers at the ends of the main feed conveyors where any large material (rocks or frozen lumps) will be broken down for the conveyors. These sizers will be able to handle blocks of up to 1.5 m. on a side if necessary, but not on a continuous basis.

It is anticipated that production blasting will not be necessary for the new shovels, but occasional frost blasting may be.

To prepare for this possibility, past frost blast patterns were reviewed and fine-tuned, and then compared on a cost per ton basis to present production blasting patterns. From this, a table of patterns and loads was developed for frost blasting that is based on the thickness of the frost layer. Any frost less than 1.5 m. thick would be shot with standard 7.5 m. deep auger holes, and any equal to or thicker than 1.5 m. would be drilled with 100 to 125 mm. diameter rotary holes.

Total cost per ton of material to be dug is approximately the same for both approaches, but the choice of which to use is driven by the size of lumps that will be formed.

CONCLUSIONS

A philosophy of continual research and improvement over the last twenty-five years in Suncor blasting programs has resulted in numerous benefits to the production operations.

These benefits include :

1. Improved digging conditions for both BWE and shovel operations
2. Reduced equipment requirements and maintenance
3. Improved blasting control while reducing blasting costs
4. Improved pitwall stability
5. An overall reduction in the cost of extracting ore
6. The tools to design, implement and optimize future blasting applications

It is believed that the information gathered during this time can also benefit others in the blasting industry, whether in mining or construction, and it is hoped that these benefits are wide-ranging.

It is also believed that Suncor's simple philosophy is the key to present and future blasting solutions, and should be incorporated into any mining operation that relies on blasting as an integral part of ore extraction. This philosophy must be a matter of course, since inevitably it will become a matter of necessity.

REFERENCES

The following are all internal unpublished reports:

Adams, R. C. 1978. Overburden blasting tests.

Bauer, A. & W. A. Crosby 1975. The effect of blasting on tarsand slopes.

Bauer, A. & W. A. Crosby 1980. Blast vibration and airblast test report.

Bauer, A. 1982. Blast designs to minimize ultimate wall blasting damage at Suncor.

Chiappetta, R. F. 1991. Explosives and primer characterizations.

Chiappetta, R. F. 1992. Aluminized explosives testing and stemming contamination study.

Crosby, W. A. 1986. Visit to Suncor Inc., March 1986.

Crosby, W. A. 1987. Visit to Suncor, Inc., June 1987.

Hardy Associates 1986. Recommendations for safe blasting: Dyke slopes and pitwall.

Livingston, C. W. 1967. Excavation and blasting requirements for G.C.O.S.

Livingston, C. W. 1968. Procedure and cost estimate, feasibility phase of blasting field tests.

McLaughlin, P. D. 1988. Improvements to blasting at Suncor using thirty inch diameter holes.

McLaughlin, P. D. 1988. Basic design criteria for crater blasting in oilsand.

Pearson, M. 1968. Blasting studies in tarsand.

Innovative Mine Design for the 21st Century, Bawden & Archibald (eds) © 1993 Balkema, Rotterdam, ISBN 90 5410 325 6

Blasting with electricity: Paving the way into 21st century mining

M. Hamelin, L. Asikainen & F. Kitzinger
Noranda Technology Centre, Pointe-Claire, Que., Canada

ABSTRACT : An electrical method for breaking rock is presented. The method, called plasma blasting, exploits the electrohydraulic effect to fracture rock in a clean and well controlled manner. Due to the important advantages of such a process over the use of chemical explosives, plasma blasting will have numerous applications in mining and civil engineering. For example, machines for continuous hard rock mining and autonomous rock breakers for secondary blasting are being considered for underground developments. Other applications such as building roads in remote locations and blasting in populated areas also generate a significant level of interest in the related fields. In this paper, test results obtained with the first prototype self-contained rock breaker are presented. Also, background information on the early developments of the technology is reviewed and estimation of some of the operating parameters of a plasma blasting based continuous mining machine are introduced.

1. PLASMA BLASTING PROCESS

Economic as well as environmental factors have recently prompted the mineral industry to increase the research effort in the field of continuous excavation techniques for hard rock. With the development of safe and practical mining equipment as a final objective, the use of electrical energy to fracture rock has been evaluated in several countries during the last three decades (Kutter 1969, Kennedy 1971). However, until the recent development of the plasma blasting process (Nantel 1990, Kitzinger 1991, 1992), no electrical rock fragmentation method has approached industrial scale. This progress has been made possible in part by the availability of high energy density capacitive storage units.

In the plasma blasting process, stored energy is released in the form of an electrical discharge through a small amount of electrolyte confined in a borehole drilled in the rock. Electrical breakdown occurs between two electrodes, establishing a plasma channel in the first microseconds of the blast. Pressure is developed in and around the channel by the rapid expansion of hot ionized electrolyte vapor. This process is sustained during roughly a tenth of a millisecond by the dissipation of electrical energy in the plasma channel. As the rapidly expanding gas impacts the surrounding material, a steep rise of pressure on the borehole walls generates a shock wave that, in conjunction with cracks propagating from the blast zone, fractures the rock.

With the currently used industrial scale prototype equipment, a 5-8 cm diameter, 50-80 cm deep hole is first drilled in the rock. The electrolyte is then injected in the hole, and a rigid coaxial conductor assembly is introduced. A high voltage (12-20 kV) pulse is applied to the inner rod of the conductor (which represents the anode) to produce the blast. The coaxial pair of electrodes, hereafter referred to as the blasting probe, fulfils two crucial functions: it conducts the electrical current to the blast zone and also acts as a containment barrier that confines the hot plasma during the first microseconds of the discharge. For this reason, the blasting probe needs to be simultaneously a good electrical conductor and an extremely resistant impact absorber.

The energy is stored in a capacitor bank and transmitted to the electrodes through coaxial spark gap switch and cable assembly. Typical discharge time is in the 50-150 μsec range and the delivered power level is routinely in excess of two gigawatts. Systems with various energy levels have been used in laboratory and field testing.

2. EARLY WORK

The underlying mechanisms of the plasma blasting process as well as the potential use of the technology for rock fragmentation has first been studied at low energy levels (5-16 kJ) in the laboratories of the Noranda Technology Centre. These experimentations with desktop-size equipment provided fundamental insight into critical parameters such as the required energy and power levels, the influence of blasting electrodes geometry, the effect of the burden on the fragmentation, etc. All subsequent phases of the project were based on these findings.

A first series of tests was performed with two sets of concrete and granite cube shaped samples. For samples of the same dimension (23 cm side), it was found that a blast of a given energy level produced approximately the same level of fragmentation in the two materials, even if granite has a significantly higher uniaxial compressive strength than concrete (80 MPa for the granite used in the tests and 50 MPa for concrete). The threshold energy (below which the samples were not fragmented) was found to be up to 50% higher for granite than concrete, but once fragmentation was produced, the fragment distribution was comparable. Small fragments tended to be ejected with more kinetic energy when granite was blasted. Since concrete is more homogeneous and easily obtainable than hard rock, it was used for most of the subsequent tests.

The level of fragmentation expressed as the average size of the resulting pieces was measured, and was well correlated to the input energy in the 5-16 kJ range. As expected, when the same energy was used in crater or bench type blasts (i.e., when fewer free surfaces were located in the proximity of the blast point), the resulting fragmentation was considerably less.

Small scale plasma blasting tests in free standing rock samples were also compared with chemical blasts of similar energy level. The chemical explosive used for these tests was PETN. Fragmentation measured in terms of average size of the pieces was equivalent in both cases, the only remarkable difference being that the plasma blasting event produced fewer fine particles than the chemical blast.

Similar studies were repeated during a succession of tests with increasing available energy, and all the facts stated above proved to hold, at least qualitatively, up to a net blast energy of 400-500 kJ.

3. COMPARISON OF PLASMA BLASTING TO CHEMICAL EXPLOSIVES

As mentioned in the preceding section, experimental results tend to indicate that the overall efficiency (i.e., the amount of energy required to induce comparable damage to a rock sample), of the Noranda plasma blasting process is comparable with that of a high explosive. However, one should be careful not to extend this similarity to the physics of the phenomena involved, since they considerably differ in certain aspects.

For example, when referring to the energy of

a plasma blasting discharge, it is important to make a distinction between the stored (nominal) electrical energy and the net spent energy. In effect, part of the energy stored in the capacitor bank is dissipated in the transmission components of the system during the discharge, and thus never reaches the rock. The two plasma blasting units used for most of the tests described in this paper consist of a 300 kilojoules nominal energy system with an electrical efficiency of approximately 60% and a one megajoule nominal energy system with an efficiency of 40-50%.

These efficiency figures represent averages, since for each individual blast, the electrical behavior of the plasma channel will determine the amount of energy effectively spent for the explosive expansion of the hot ionized gases. Moreover, experience showed that the efficiency of the energy transfer is related to the mechanical response of the surrounding rock. In effect, the burden and the mechanical characteristics of the rock influence the pressure conditions in the blast zone, and thus the electrical response of the system.

The efficiency of a plasma blasting unit can be optimized through careful adjustment of the value of the electrical components and proper evaluation of the amount of energy required for a given burden. Plasma blasting technology therefore considerably differs from chemical blasting on this respect.

Compared with chemical explosives, a plasma blasting discharge generates a small quantity of gas. This fact has important consequences both on the fragmentation mechanism and on the environmental aspects of the method. The absence of large amount of gases suggests that the damage induced in the rock results mainly from the shock wave that is generated at the tip of the blasting probe. In the milliseconds following the blast, the fragments are pushed away from the impact point by the momentum transferred from the expanding plasma, and gases appear to play a marginal role in the ejection of fragments. Fortunately, unlike the large chemical explosive blasts, plasma blasting is a relatively low energy process that requires little energy to remove the fragmented rock after the impact. Due to the small burden distance used in an average plasma blasting discharge (when compared with chemical explosives), improper removal of the rock mass due to fragment interaction is unlikely to occur.

In fact, the lack of gaseous by-products represents a significant benefit of the method since it contributes to reduce the problem of ground vibrations and fly rocks. When the blast energy is properly dosed, one can stand at few tens of meters from the blast without special protection. Moreover, the absence of toxic residues or fumes is an obvious advantage for underground operation.

4. YIELD ESTIMATES FOR A PLASMA BLASTING BASED CONTINUOUS MINER

The productivity of a secondary rock breaker using the plasma blasting process is currently assessed with an industrial scale rock breaker prototype. With this equipment, the limiting factor on the repetition rate is the positioning and set-up of the machine near the boulder to be fractured. Consequently, the number of blasts per day of operation will heavily depend on the accessibility and configuration of the site. The following calculation is an attempt to evaluate in the most realistic manner the productivity of a continuous mining machine used for tunneling. More work on both the modeling and experimentation aspects of this problem will be required to refine these estimates, but it is the opinion of the authors that currently available data allow one to establish lower bound for some of the operating parameters of an eventual plasma blasting based continuous miner.

Based on crater type blasts of 200 kJ net energy and on tests in concrete blocks at 400 kJ net energy, the fragmentation produced in a rock face with a 600 kJ net blast energy machine will be estimated in the next

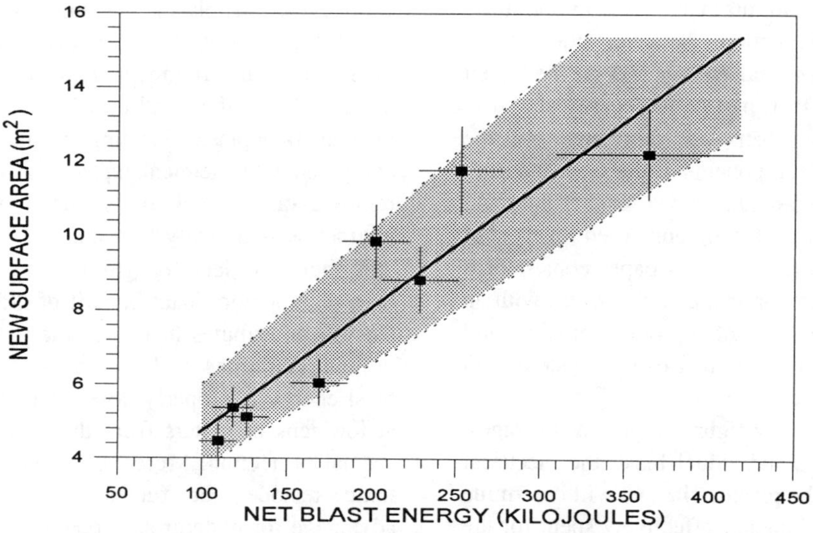

Figure 1 : Fragmentation in terms of new surface area as a function of net blast energy for a series of plasma blasting tests on high strength concrete blocks.

paragraphs. To this date, hard rock as well as concrete free standing samples averaging roughly one cubic meter have been properly fragmented with a net energy in the 150-200 kJ range. In at least one occasion, a large boulder weighing approximately 18 metric tons was split with a net 200 kJ discharge. Crater blasting in a rock face resulted, as expected, in a significantly smaller fragmented volume of roughly 0.2-0.4 m^3 for the same energy.

A series of tests has been performed at higher energy levels on concrete blocks to establish the relationship between input energy and fragmentation for the plasma blasting process (Hamelin 1993). A capacitor bank capable of storing 1 MJ was used to deliver the electric discharge through eight parallel coaxial cables and one hydraulically operated spark gap switch. The tests were performed at three nominal energy levels (300 kJ, 600 kJ and 1 MJ) with three sizes of cubic high strength concrete blocks (0.83 m^3, 1.7 m^3 and 2.58 m^3). For each test, the fragmentation was evaluated in terms of new surface area created in the sample, and the net blast energy was calculated. The relationship between net energy and fragmentation is shown in figure 1. These results indicate that the process is scalable to a net energy of 400-500 kJ in a roughly linear fashion.

For a single crater type blast, experience has shown that a cone shaped volume separated in several fragments is extracted from the rock face. Assuming that a constant fraction of the available energy is spent to create new surfaces, i.e., that the newly created surfaces increase linearly with the energy, the scaling of the extracted volume for an increase by a factor of three of the input energy will depend on the shape of the fragments. One extreme case is represented by a situation where the resulting fragments are numerous and of elongated or quasi-plane shape. Based on the assumption stated above, the fragmented volume then increases proportionally with the blast energy. In the other extreme ideal case, if only one conical fragment (with a top angle of 90°) is produced, it is easy to calculate that the extracted volume increases by a factor of five when the energy is tripled.

A linear increase of the volume with the input energy is considered for the calculations presented here. Of course, the assumption that the newly created surfaces increase linearly with the input energy is an approximation based on the hypothesis that the other forms of energy, such as the kinetic energy of the rock fragments after the blast, remain proportionally constant while the total energy is increased (Whittaker 1992). In the case of the plasma blasting process, it is relatively safe to consider that the fraction of the input energy resulting in residual kinetic energy of the fragments is small, and therefore will not considerably affect the total energy balance.

Using a conservative figure for the volume fragmented with 200 kJ, we can then assume that a single crater blast resulting from a 600 kJ net energy discharge could be realized with a 0.8 m deep hole creating a cone with a top angle of 90°. This represents a volume of 0.56 m^3, or roughly 1.5 metric tons for a typical hard rock. For the present yield estimate, we will consider a simple fracture model in which the fragmented volume is defined by a minimal distance between the impact source and a free face. Stated more accurately, a blast with a given energy will fragment the rock in the directions along which the distance to the free face is smaller than a radius related to the cubic root of the blast energy. This simplified model is currently being validated with field tests and computer modeling of shock wave and fracture phenomena.

Let us now consider a group of holes arranged in a square grid pattern in a rock face. The holes are positioned such that the influence zone of one blast extends to the nearest hole along a diagonal in the square pattern. The holes are thus 80 cm deep and separated by 80 cm. Such a pattern insures that when the holes are blasted, the totality of the rock layer will be fractured. This case represents a very conservative arrangement since over fragmentation will occur along the parallels to the sides of the square in the pattern. It is thus obvious that the hole pattern geometry will have a critical influence on the productivity of the machine. With the pattern described above, 35 holes located on five rows comprising seven holes each will be required to remove roughly one 80 cm layer in a 3.2 m X 4.8 m tunnel. Since the pattern is symmetrical, the results are readily scalable to other dimensions.

The volume extracted by each blast can then be calculated, and by knowing the repetition rate, it is possible to obtain the rate of advance of the tunnel. Since a continuous mining machine evolves in a more predictable environment than a rock breaker, it is expected that the limiting factors on the repetition rate for such equipment will be related to it's capability to charge the capacitor bank, drill a hole and inject the electrolyte in a short time. This should be accomplished in an efficient way by a dual drill-and-blast arm machine as described in section 5.

5. APPLICATIONS AND EQUIPMENT

Despite the fact that the plasma blasting technology is not yet fully developed, many applications are being considered in both mining and civil engineering. Underground, this technology is particularly promising for tunneling and mine development, trenching, secondary breaking and for all type of mining except bulk extraction methods.

Above ground, the most interesting applications include road building in remote locations, pipeline works, rock breaking near existing structures and various civil engineering works where ground vibration and explosives handling cause problems. The availability of low energy mining methods not using chemical explosives in politically unstable regions also represents a major benefit for public and workers' safety.

Machines could be designed for specific applications or for general use. Two main concepts are presently under study. The first one, a general purpose secondary rock breaker, is presently at the industrial scale prototype stage, and the second one, a continuous mining machine for hard rock is at the preliminary design stage.

Figure 2 :Sequence of photographs of a plasma blasting discharge taken during the rock breaker prototype commisionning.

The Noranda plasma blasting rock breaker prototype is an autonomous machine based on a standard diesel powered carrier. A pulsed power module consisting of a 300 kJ capacitor bank switched by a piezo-electrically triggered spark gap, a coaxial cable and a blasting probe, was designed for underground use and integrated in the carrier. Electrical power is generated by an alternator to allow autonomous operation. The capacitor bank is charged with an 8 kJ/sec high frequency switching power supply adapted for this application.

The blasting probe is mounted on a four degrees of freedom drill-and-blast arm that allows suitable positioning for rock breaking as well as free face blasting (see photograph in figure 2). An indexing mechanism was designed to facilitate the insertion of the blasting probe in the borehole after the drill is removed.

The capacitor bank and the other high voltage components were specifically designed to resist severe environmental conditions such as 100% condensing humidity and vibrations produced by riding on uneven terrain.

The overall dimensions of the prototype (shown in figure 3) are as follow: 11.3 m length, 2.16 m height and 2.44 m width. Total weight is 21 000 kg, and the minimum turning radius is four meters. These characteristics allow access to nearly any locations in a typical underground mine.

While a secondary rock breaker based on plasma blasting presents many advantages over the use of conventional explosives, the real benefits of this development are expected from continuous mining machines for hard rock. Beside providing improved ground control and automation potential, significant increases in tunnel advance rate are anticipated from such machines. The expected reduced operating costs of a plasma blasting based continuous miner combined with the associated environmental benefits make the technology highly attractive.

Figure 3 : Prototype rock breaker based on plasma blasting developed by the Noranda Technology Centre, Pointe-Claire, Que.

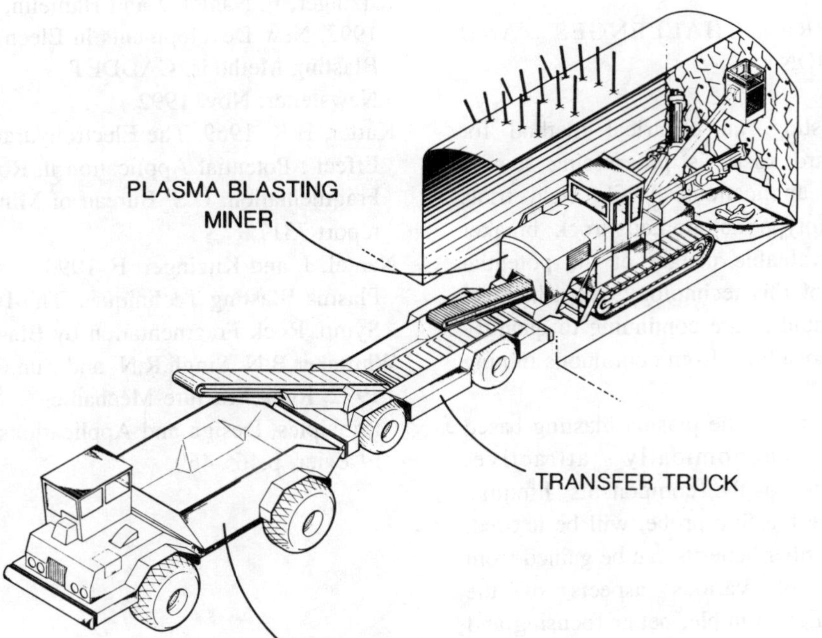

Figure 4 : Concept drawing of a plasma blasting continuous mining system.

When considering continuous mining based on plasma blasting, one must think in terms of a mining system. To be efficient, such a mining system should include all the support services to ensure continuous operation: Ground support, muck removal, power, etc. A concept drawing of a plasma blasting mining system is shown in figure 4. It consists of a crawler mounted mining machine that includes two drill and blast booms linked to one capacitor bank. To optimize the capacitor bank availability, one boom would be used for drilling while the other is blasting. The machine is also equipped with a muck handling system which transfers the rock fragments from the face to a transfer truck located at the back of the system. This transfer vehicle, placed between the plasma blasting unit and the haulage truck, acts as a buffer to insure continuous operation. A rock bolting tool and a man basket boom for scaling and inspection complete the accessories. The number of haulage trucks required will depend on the hauling distance.

6. FUTURE CHALLENGES AND CONCLUSION

Plasma blasting, an electrical method for breaking hard rock, is progressing toward commercial applications. While field trials with a prototype autonomous rock breaker provided invaluable insight of the potential advantages of this technology over chemical explosives, studies are continuing to quantify the benefits resulting from continuous mining of hard rock.

In order to make the plasma blasting based equipment economically attractive, improvements to the components lifetime, especially the blasting probe, will be needed. Moreover, further benefits can be gained from optimization of various aspects of the technology. For example, better focusing and orienting of the shock wave could greatly improve the effective use of electrical energy to fracture rock in a crater type blast.

ACKNOWLEDGMENT

Permission of Noranda Minerals for presentation of this work is gratefully acknowledged.

The authors would also like to thank G. Sauriol and D. Mack for their constructive comments.

REFERENCES

Hamelin, M. and Kitzinger, F. 1993. Hard Rock Fragmentation with Pulsed Power. Ninth Intl. Pulsed Power Conf.

Kennedy, P.A. 1971. An Analysis of Rock Breakage by High Voltage Electrical Discharge. Ph.D Thesis, Queen's University of Belfast.

Kitzinger, F. and Nantel, J. 1991. U.S. Patent 5 106 164, Plasma Blasting Method.

Kitzinger, F. Nantel, J and Hamelin, M. 1992. New Developments in Electrical Blasting Methods. CADDET Newsletter, Nov. 1992.

Kutter, H.K. 1969. The Electrohydraulic Effect : Potential Application in Rock Fragmentation. U.S. Bureau of Mines, report 7317.

Nantel, J. and Kitzinger, F. 1990. Plasma Blasting Techniques. Third Intl. Symp. Rock Fragmentation by Blasting.

Whittaker,B.N. Singh,R,N. and Sun,G. 1992. Rock Fracture Mechanics, Principles, Design and Applications. Elsevier, p.465-469.

Innovative Mine Design for the 21st Century, Bawden & Archibald (eds) © 1993 Balkema, Rotterdam, ISBN 90 5410 325 6

Overpressure generation and control in tunnel blasts

G.W. Kuzyk & D.P. Onagi
AECL Research, Underground Research Laboratory, Pinawa, Man., Canada

Q. Liu
Mine-Laboratoire, CANMET, Val d'Or, Que., Canada

ABSTRACT: Overpressure or air shock waves from underground blasting are recognized sources of destruction in underground excavations. This paper describes the mechanisms for overpressure generation and propagation caused by tunnel blasting, and suggests blast designs and protection methods to reduce the effects of overpressure.

The formation of air shock waves by tunnel blasting is illustrated from an aerodynamics perspective according to the Mach shock waves on the walls of tunnels. The empirical relationships of overpressure attenuation during propagation are derived on the basis of the available literature and were applied to estimate the effects of blasting carried out at AECL's Underground Research Laboratory (URL). The model used to predict shock overpressures takes into consideration factors such as the mass of explosive charge, the travel distance, the cross-section area of the tunnel, the roughness of the tunnel walls, and the effects of intersections and changes in cross section.

The prevention of strong air shock waves is discussed, with an emphasis on the proper use of explosives in blasting. General protection principles are discussed and a design is presented for a timber curtain to protect the instrumentation at the URL's 240 Level during the blasting in Room 214. Work is recommended to improve the understanding of the air shock waves and effective protection designs.

1 INTRODUCTION

One of the safety concerns of using explosives is the destructive power of the air shock waves generated by blasting. When the explosives detonate in blastholes, most of the explosive energy is used to fragment the rock and push the rock to form a muckpile. However, there is always some energy wasted or lost during blasting. Most of the energy lost is either transmitted to the rock mass as seismic energy or is transmitted to the air surrounding the blast. The former creates the problem of ground vibration and the latter a problem of air vibration (shock waves) that might be destructive to underground facilities and/or personnel.

Air shock waves, or overpressures, from blasting have been studied extensively for surface blasting but information on underground blasting is very limited in the available literature. Scaling laws have been used to calculate both overpressure and vibration levels from surface blasting and their application to underground vibrations has found some success. However, it is not appropriate to use scaling laws to calculate overpressure since the propagation of an air shock wave in a tunnel is one-dimensional, not two- or three-dimensional as is the case of ground vibration. As a result, the rate of pressure attenuation with distance is much slower in underground blasts than in open-pit blasts where the shock wave on the surface propagates into air with a spherical (three-dimensional) geometry. Therefore, the scaling laws of overpressure derived from open-pit blasting do not apply to underground tunnelling. In particular, the scaled distance $D/W^{1/3}$ (Olson and Fletcher 1971) should not be used to scale the overpressure in tunnel blasting. In addition, the positive duration of the shock wave is much longer in tunnels than that in the open air. Thus, the shock impulse is greater for the same overpressure.

AECL's Underground Research Laboratory (URL) is a major geotechnical research facility developed as part of the Canadian Nuclear Fuel Waste Management Program to assess the concept of nuclear fuel waste disposal in a vault deep in the stable plutonic rock of the Canadian Shield (Dormuth and Gillespie 1990). The excavation of tunnels at the URL often requires the protection of instrumentation and sensitive equipment from air shock waves generated by blasts. In some cases, experimental equipment on the 240 Level has been installed less than 50 m from a tunnel blast round. Specially designed obstacles were required to attenuate the overpressure before it reached the target. This paper presents a method of calculating the overpressure. The mechanism of shock-wave generation and wave propagation in tunnels is reviewed and details for the design of protective structures are discussed.

2 OVERPRESSURE

An air shock wave is a pressure pulse propagating in air. It can be produced by the detonation of an explosive or by a burst of compressed air. A shock wave is characterized by its sharp increase in overpressure. The velocity of an air shock wave is always faster than the sonic speed of air. The shock-wave velocity and its overpressure have the following relationship (Kinney and Graham 1985):

$$\Delta P = \frac{7}{6} P_0 \left[\frac{D^2}{C_0^2} - 1\right] \qquad (1)$$

where ΔP is the overpressure (kPa),
P_0 is the atmospheric pressure (kPa),
D is the velocity of air shock wave (m/s), and
C_0 is the sonic velocity of atmospheric air (m/s).

The sonic velocity changes slightly with temperature:

$$C_0 = 20.05 \sqrt{T_0} \qquad (2)$$

where T_0 is the temperature of atmospheric air in degrees Kelvin.

The sonic velocity is also affected by the humidity of air. However, the effect is negligible (<0.53%) for temperatures below 25°C (Hiismaki and Kamarainen 1987).

Figure 1 shows a typical pressure-time curve of a shock wave. For a typical cut blast with 5 kg ANFO explosive per delay, the overpressure would be approximately ΔP = 20 kPa at a distance of 40 m from the face, with a positive duration of t_d = 40 ms. This wave form with a sharp pressure rise is usually the first peak re-

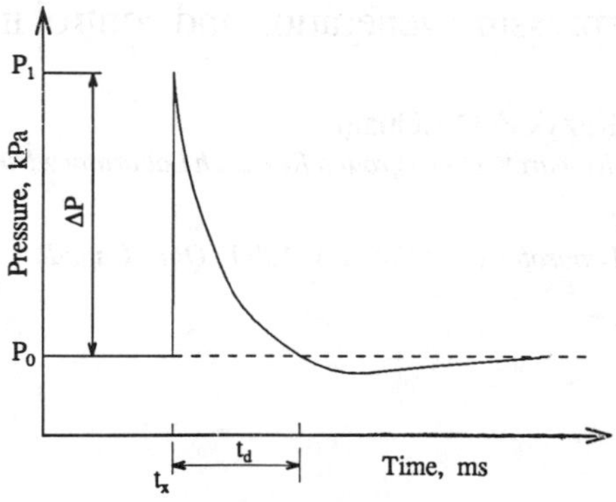

Figure 1. Typical pressure-time curve for an air shock wave

corded by a pressure transducer if multi-delay periods are used in the tunnel blast. Because of the irregularities of the geometric configuration of excavated openings, the shock front may be disturbed somewhat by wave reflections on the uneven surface of rock walls or at intersections. As a result, the shock front may not keep the overpressure rise as sharp as in the case of a surface blast. However, as the shock wave proceeds along the tunnel, the shock front tends to build a sharp rise again, as illustrated in Figure 2. This is a typical case showing the nature of shock-wave formation by compression.

As shown in Figure 2(a), points A, B and C are the wave pressures at different locations on a compressive wave. Point A is at atmospheric pressure, B is a middle overpressure on the wave front and C is the peak overpressure. The wave at point C is travelling in the air compressed by the wave at point B. Because the pressure, density and particle velocity in region B-C are higher than that in region A-B because of higher pressure, the wave at point C travels faster than at point B. Thus, as shown in Figures 2(b) and 2(c), point C tends to catch up with point B. In a similar way, point B tends to catch up with point A. After a certain time, all three points are lined up and propagating with the same velocity, Figure 2(d). The shock front is formed at this moment, and propagates as a normal shock wave. The peak overpressure is also increased somewhat because of pressure cumulation, that is, the compressive duration becomes shorter and the overpressure becomes higher while the impulse remains unchanged significantly. This process is a self-adjusting feature of a shock wave during its propagation in tunnels.

The sharp overpressure from a tunnel

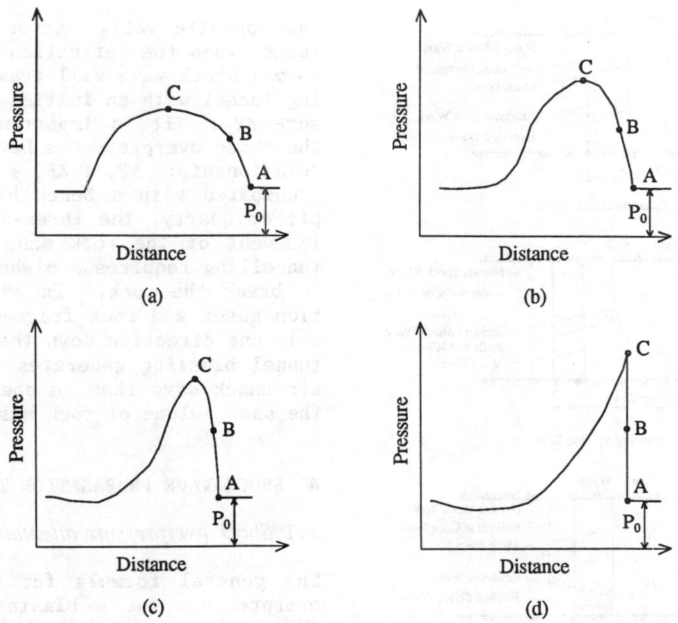

Figure 2. Formation of shock front from compressive-wave propagation

blast with delay detonators is observed only with the first peak overpressure because it is travelling in undisturbed air. Subsequent shock waves from delayed charges cannot maintain the sharp overpressure because they are travelling in the air disturbed by the previous shock. The movement of the shocked air may strengthen or weaken the shock overpressure. As a result, the wave forms have slower pressure rise and slower pressure drop. From the protection point of view, the long duration of the overpressure must be considered because the pressure impulse from the subsequent delays may exceed that of the first pressure pulse even though the peak overpressures might be lower. Thus the measurement of a shock wave should include both the overpressure and the shock impulse in order to identify its damaging power (see Section 5).

3 THE FORMATION OF AIR SHOCK WAVES FROM BLASTING

When a blasthole is located at the geometric centre of a round, such as the hole shown in Figure 3(a), the detonating explosive in the blasthole breaks the rock and pushes the broken rock fragments towards the "free" face and into the tunnel. At the same time, gaseous products from the detonation are ejected into the tunnel, compressing the air at the face to form a shock wave. Initially, the shock wave expands spherically from the face to the excavated walls of the tunnel. An air shock wave reflection will occur when the wave front meets the tunnel wall. The reflection becomes an irregular reflection called the Mach reflection at a certain distance from the face. Another shock front called the Mach stem is produced as a result of the reflection, and travels faster than the original spherical shock wave. Thus, the Mach stem keeps growing with the distance it travels. After a certain distance, when the Mach reflection is completed, a normal plane shock wave is formed and propagates freely in the tunnel. As the shock wave travels, it loses energy by adiabatic heating of the air in the shock front and from friction on the walls of the tunnel. Thus, the shock overpressure decreases with distance travelled.

The same phenomenon occurs with blastholes at other locations in the round. Figure 3(b) shows the shock-wave formation for a blasthole located on one side of the round, and Figure 3(c) shows the shock-wave formation for blastholes initiated simultaneously on both sides of the round. The length of the reflection zone is usually 4 to 8 times that of the tunnel width.

When the shock wave passes an intersection, as shown in Figure 4, it sends a new incident shock wave down the joining tun-

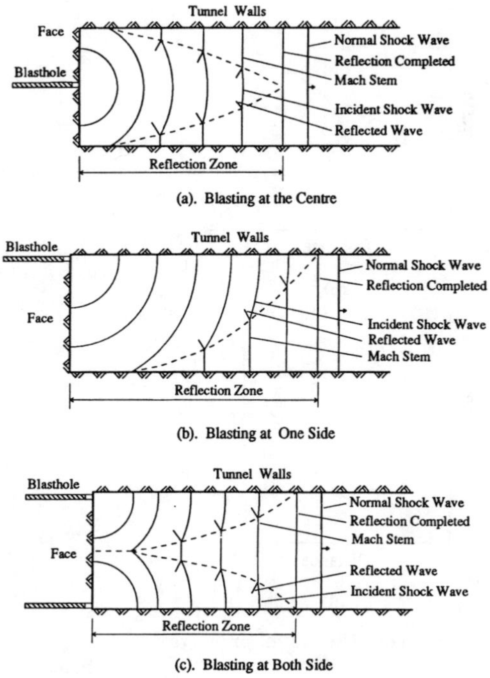

Figure 3. Formation of air shock waves from tunnel blasting

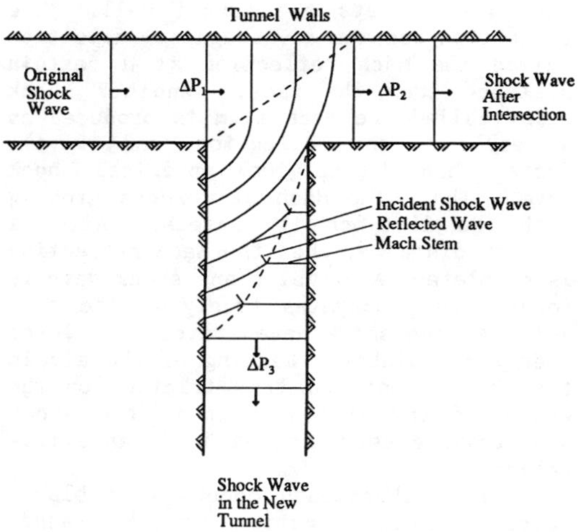

Figure 4. Shock wave at intersection

nel. The shock wave is weakened because some of its energy is required to form the new incident shock wave; thus, the overpressure is reduced from ΔP_1 to ΔP_2 after passing the intersection. The new incident shock wave travelling down the joining tunnel will have a Mach reflection on the opposite wall. After a certain distance, when the reflection is completed, a normal shock wave will travel in the joining tunnel with an initial shock overpressure ΔP_3. It is important to note that the three overpressures have the following relationship: $\Delta P_2 + \Delta P_3 \le \Delta P_1$.

Compared with a bench blast in an open pit or quarry, the three-dimensional confinement of the rock mass in underground tunnelling requires a higher powder factor to break the rock. In addition, detonation gases and rock fragments can move in only one direction down the tunnel. Thus, tunnel blasting generates a much stronger air shock wave than an open-pit blast for the same volume of rock mass broken.

4 SHOCK-WAVE PROPAGATION IN TUNNELS

4.1 *Shock overpressure attenuation*

The general formula for calculating the overpressure for a blasthole loaded with ANFO explosive is shown below. The formula is derived from empirical formuli believed to have been developed by Y.S. Yakovlev in 1961:

$$\Delta P = \left[2900 \frac{m\,W_A}{D\,S} + 733 \sqrt{\frac{m\,W_A}{D\,S}}\right] e^{-nD/d} \qquad (3)$$

where ΔP is the overpressure (kPa),
- m is an effectiveness ratio associated with blast design:
  - Secondary blasting (buried charges), $m = 0.30$
  - Cut holes in tunnel blasting, $m = 0.1 \sim 0.15$
  - Other holes in tunnel blasting, $m = 0.05 \sim 0.1$,
- W_A is the mass (ANFO equivalent) of explosive used per delay (kg),
- S is the cross section area of the tunnel (m^2),
- D is the distance from blasting face (m),
- n is the frictional coefficient of tunnel wall:
  - $n = 0.02$ for smooth walls and floor
  - $n = 0.03 \sim 0.04$ for rough walls and floor, and
- d is the equivalent diameter of the tunnel, which is determined by

$$d = 2\sqrt{\frac{S}{\pi}}\ . \qquad (4)$$

If another type of explosive is used, the equivalent mass relative to ANFO must be

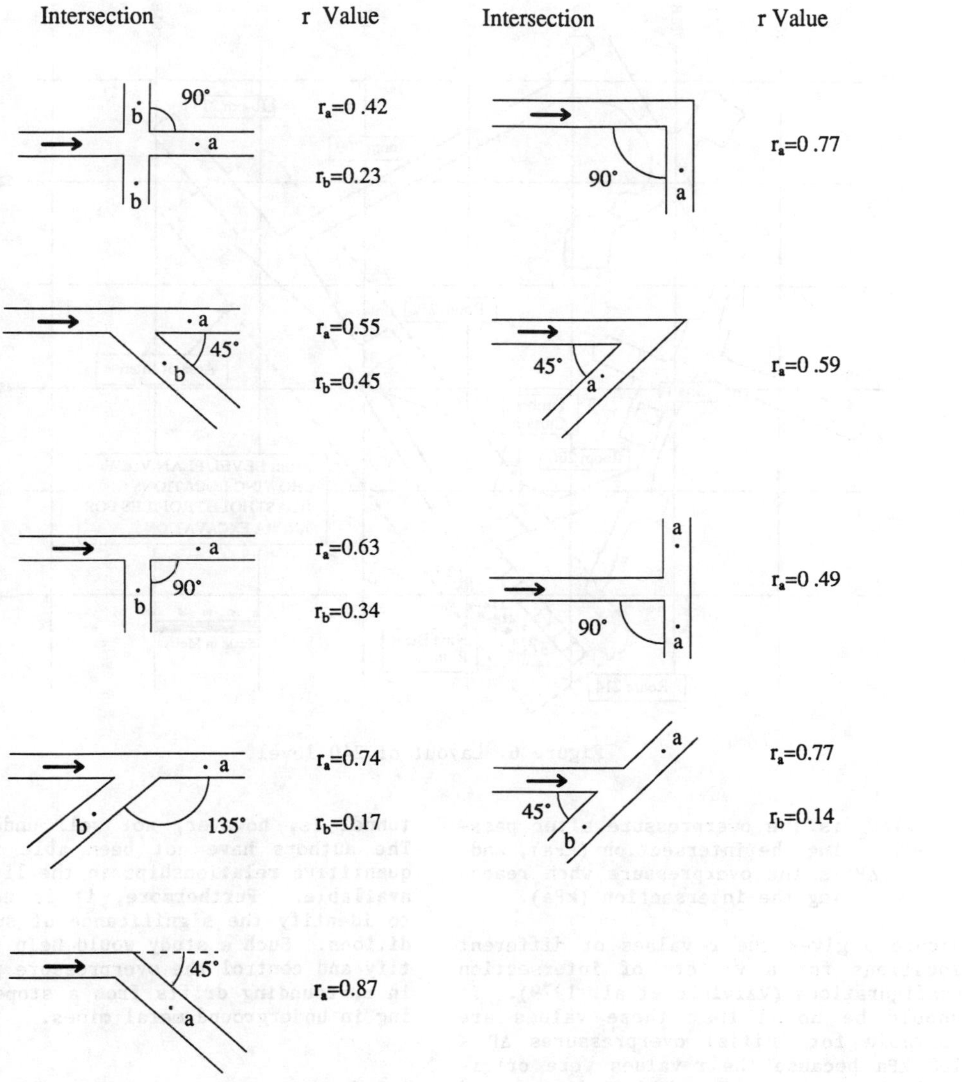

Figure 5. Ratio of overpressure reduction with different types of intersection

calculated regarding its relative weight strength:

$$W_A = \frac{S_w}{100} W_t \quad (5)$$

where S_w is the weight strength of the explosive used,
100 is the weight strength of ANFO, and
W_t is the total mass of explosive used per delay (kg).

The accuracy of estimating the overpressure (ΔP) is strongly affected by the "effectiveness ratio," m, which in some cases needs to be evaluated for a certain type of blast design.

Tunnels frequently intersect each other or change direction. In such cases, the overpressure is always reduced after passing the intersection or change in direction. The ratio of overpressure reduction after an intersection is defined by:

$$r = \frac{\Delta P_x}{\Delta P} \quad (6)$$

where r is the ratio of the overpressure reduction,

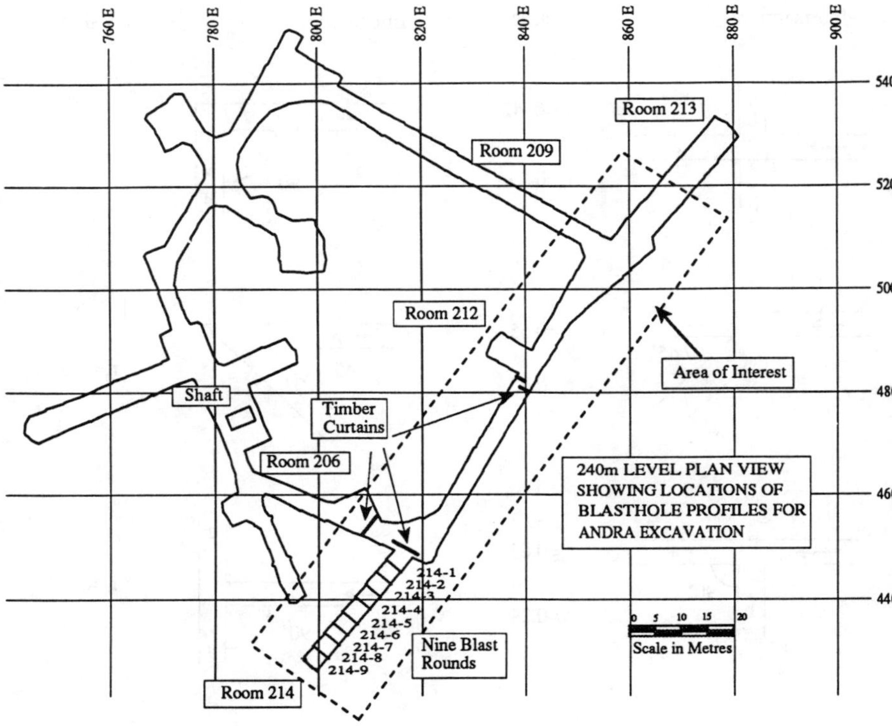

Figure 6. Layout of 240 level

ΔP_x is the overpressure after passing the intersection (kPa), and
ΔP is the overpressure when reaching the intersection (kPa).

Figure 5 gives the r values of different locations for a variety of intersection configurations (Valwinko et al. 1979). It should be noted that these values are suitable for initial overpressures $\Delta P < 100$ kPa because the r values were originally determined under this condition.

If a shock wave encounters a change in the cross-section area of the tunnel, the overpressure is also changed. When a shock wave travels from a small tunnel into a larger one, the reduction of overpressure is related to the cross section area of the two tunnels (Yakovlev 1961):

$$\Delta P_1 = \Delta P_0 \left[\frac{S_0}{S_1}\right]^{0.8} \qquad (7)$$

where ΔP and S represent the overpressure and cross-sectional areas respectively, and subscripts 0 and 1 represent the small and the large tunnel respectively.

The change from a larger tunnel to a small tunnel is, however, not well understood. The authors have not been able identify quantitive relationships in the literature available. Furthermore, it is necessary to identify the significance of such conditions. Such a study would help to quantify and control the overpressure produced in surrounding drifts from a stope blasting in underground metal mines.

4.2 *Case study*

Overpressures were estimated at different distances from the blasting face to illustrate the method described above. A typical blast in Room 214 on the 240 Level of the URL is used to demonstrate the calculation procedure. Figure 6 shows the layout of the 240 Level. The area framed with dashed lines is the area of concern for shock-wave propagation and protection. Upon reviewing the blasting records and overpressure plots of the ANDRA project (Potier et al. 1992), it was found that the blastholes initiated with delay #12 in Blast 214-3 had the highest charge mass of explosives of all the blastholes fired in the nine blasts monitored. Therefore, this shot is used as an example for the

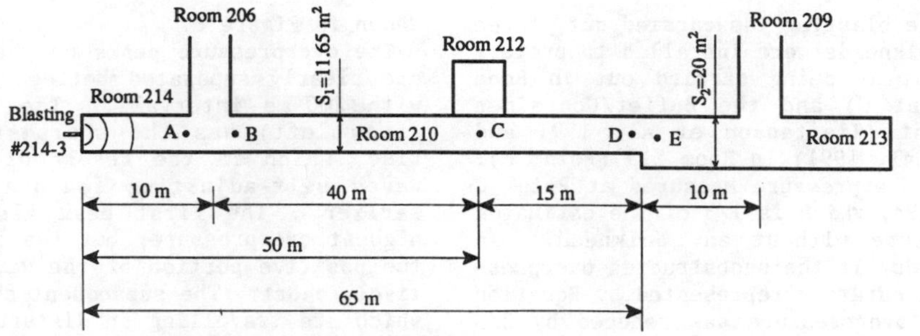

Figure 7. Simplified tunnel configuration used to estimate overpressure

Table 1. Charge mass used and ANFO equivalent mass

Explosive	Mass (kg)	Relative Weight Strength	ANFO Equivalent (kg)
Xactex	4.5	78	3.5
Forcite 75%	10.7	103	11.0
			Total: 14.5 kg

calculation. The blasthole contained 4.5 kg Xactex and 10.7 kg Forcite 75%, which are products of ICI Explosives Canada. Electronic detonators were used to provide timing for the blast.

Figure 7 shows a simplified tunnel configuration for calculating the overpressures at various locations in Room 210. The cross-section area (S_2 = 20 m²) is larger from Point E to Room 213 than the previous tunnel length from Point A to Point D (S_1 = 11.65 m², d = 3.85 m). Protection from overpressure was required for experimental instrumentation at Points C and E.

The charge mass used and the ANFO equivalent mass are shown in Table 1.

Since the blastholes fired on delay #12 were the cushion and production holes, the effectiveness ratio is taken as m = 0.08. The tunnel walls, with about 80% half barrels (Kuzyk et al. 1987), are considered smoothly controlled: n = 0.02.

Equation (3) is used to calculate the overpressure at various distances from the blast. At Point A the overpressure is:

$$\Delta P_A = \left[2900 \frac{0.08 \times 14.5}{10 \times 11.65} + 733\sqrt{\frac{0.08 \times 14.5}{10 \times 11.65}}\right] e^{-0.02(10/3.85)}$$

$$= 97 \text{ kPa} = 14 \text{ psi}.$$

After the shock attenuation through the intersection, the overpressure at Point B becomes:

$$\Delta P_B = 97 \text{ kPa} \times 0.63 = 61 \text{ kPa} = 8.8 \text{ psi}.$$

At Point C:

$$\Delta P_C = 0.63\left[2900 \frac{0.08 \times 14.5}{50 \times 11.65} + 733\sqrt{\frac{0.08 \times 14.5}{50 \times 11.65}}\right] e^{-0.02(50/3.85)}$$

$$= 18.7 \text{ kPa} = 2.7 \text{ psi}.$$

At Point D:

$$\Delta P_D = 0.63\left[2900 \frac{0.08 \times 14.5}{65 \times 11.65} + 733\sqrt{\frac{0.08 \times 14.5}{65 \times 11.65}}\right] e^{-0.02(65/3.85)}$$

$$= 14.9 \text{ kPa} = 2.2 \text{ psi}.$$

The cross section changes between Point D and Point E. In the tunnel with the larger cross-section area, the overpressure is reduced to

$$\Delta P_E = 14.9\left(\frac{11.65}{20}\right)^{0.8} = 9.7 \text{ kPa} = 1.4 \text{ psi}.$$

When the blasting was carried out, three timber bulkheads were installed to protect an experiment being carried out in Room 212 (Point C) and the Buffer/Container Experiment (Kjartanson et al. 1991 and Kuzyk et al. 1991) in Room 213 (Point E). The peak overpressure measured at Point C was 6.2 kPa, which is 1/3 of the estimated overpressure without any bulkhead. In other words, if the unobstructed overpressure is accurately represented by Equation (3), the overpressure was reduced by 2/3 by using the timber bulkheads.

5 PROCESSING OF RECORDINGS

Figure 8(a) shows a typical overpressure recording from Blast 214-5, which was recorded by ICI Explosives Canada. This is only a part of the total record duration, or about one second out of three. The transducer location was at Point C in front of Room 212 on the 240 Level shown in Figure 7. Therefore, the shock waves were attenuated through the two bulkheads shown in Figure 6.

The overpressure peaks on the recording are clearly separated between the delays with 200-ms intervals. The first peak (from left) has the sharpest pressure rise, which is the result of the shock waves self-adjusting feature discussed earlier. The first peak also has the highest overpressure, but the duration of the positive portion of the wave is relatively short. The subsequent shock waves, which are travelling in disturbed air, do not have this sharp pressure rise and the durations of the positive portion are generally longer, as can be seen in Figures 8(b) and 8(c).

As mentioned earlier, the two most important parameters of a shock wave are the overpressure and the impulse of the shock wave. The overpressure can be read directly from the recorded pressure-time curve. The shock impulse, however, has to be obtained by integrating the recorded overpressures over the positive portion of the shock wave. This can also be done with the aid of digital recording instru-

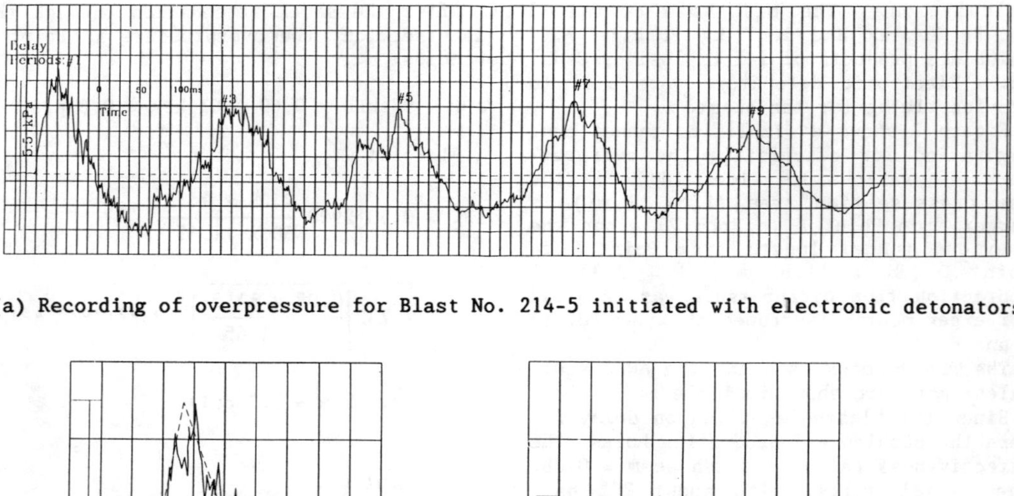

(a) Recording of overpressure for Blast No. 214-5 initiated with electronic detonators

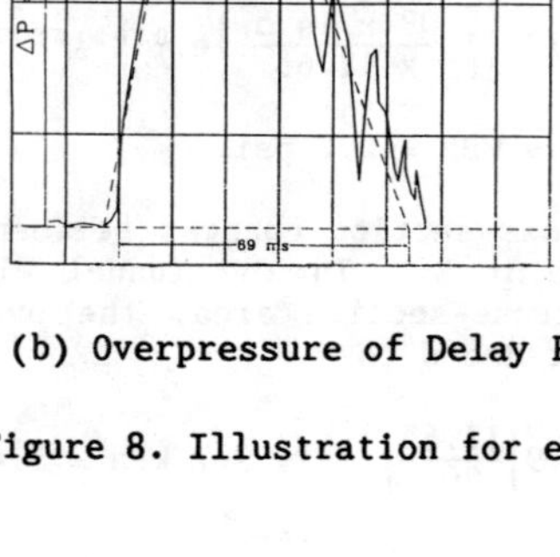

(b) Overpressure of Delay Period #1

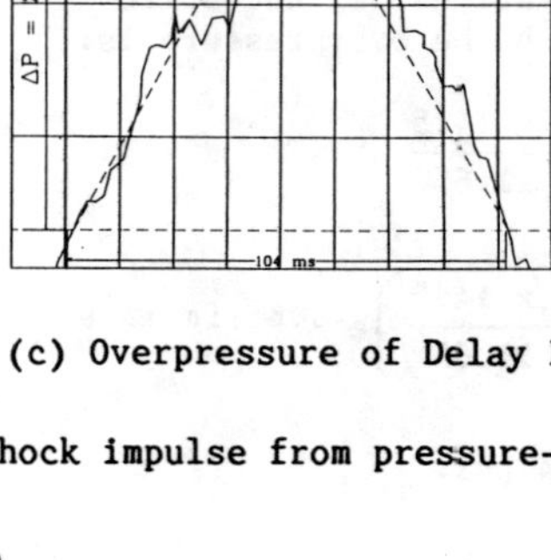

(c) Overpressure of Delay Period #7

Figure 8. Illustration for estimating shock impulse from pressure-time trace ments, such as a digital oscilloscope or data acquisition system and computing software.

If the recording is only available in analog plottings, a manual approximation can be used to obtain the impulse values. Figure 8 illustrates such a method. The estimation of a shock impulse is done by calculating the area framed by approximation lines (e.g., the dashed lines in Figures 8(b) and (c)). The impulse equals the area of the triangle. The height and the width of the triangle can be used to calculate the impulse using the following equation:

$$I_a = \frac{\Delta P_a \bullet t_{da}}{2} \tag{8}$$

where I_a is the approximated shock impulse (kPa•ms),
ΔP_a is the approximated peak overpressure (kPa), and
t_{da} is the approximated positive duration (ms).

For the overpressure recording from Blast 214-5 (Figure 8(a)), the first peak has the highest overpressure and the fourth peak has the second highest overpressure. However, the positive duration of the fourth peak is longer and thus the impulse may be greater. The estimation of the impulses helps to compare the destructive power of the two peaks.

As shown in Figures 8(b) and (c), the impulse values are almost equal for both the first peak and the fourth peak: I_1 = 100 kPa•ms and I_4 = 104 kPa•ms. This estimation method is convenient for processing the important peaks with analog plottings. The overpressure and impulse values can be used to compare the effects of bulkheads and intersections on the attenuation of air shock waves.

To get an impression of the impulse magnitude on damage, the following values give a relative quantity for shock waves with the wave form shown in Figure 1. An impulse of 250 kPa•ms will result in wood-framed buildings being badly damaged, while 2000 kPa•ms will result in human fatality (Kinney and Graham 1985).

6 PREVENTION AND PROTECTION

6.1 *Prevention*

Whenever explosives are used in rock blasting, it is impossible to eliminate air shock waves completely. However, it is possible to design blasts to limit the source of the shock-wave generation so the shock magnitude can be reduced.

First of all, the magnitude of the air shock wave is directly related to the mass of the explosive charge. By reducing the charge mass of explosive per delay, both the overpressure and the shock impulse can be reduced effectively. Thus, when blasting a tunnel round, the more delay periods the better. However, there is a limit of delay periods available. Therefore, it is very important to average the total explosive mass over the delay periods available. Many holes are often fired on the same delay in the design of the firing sequence for cushion and production holes (between the cut and the perimeter holes). In this case, it is critical to limit the number of holes fired by each delay and to average the total number of holes with the available delay periods whenever possible.

Secondly, the delay period can be long enough to separate the peaks of shock overpressure from successive delays. Short delay periods may cause the shock waves to accumulate and form a stronger shock wave. The positive duration of shock waves is found to be in the range of 75-100 ms for most shock peaks on the basis of overpressure recordings made during the blasting in Room 214. Thus, the peaks with delay periods over 100 ms can be separated. However, if the delay interval is shorter than 75 ms, the shock peaks may overlap. This implies that stronger shock waves may be generated because of insufficient delay intervals. Therefore, long period delays should be used whenever possible. In general, a longer delay period is required for blast designs using a higher charge mass of explosives per delay.

Thirdly, stemming can help to confine the detonation products inside the blasthole to do more work on rock fragmentation rather than to be ejected into the air to form shock waves. With the loading practice in tunnel blasting, the detonator is usually placed at the bottom of the blasthole. The detonation wave in this case propagates towards the collar of the blasthole. Thus, the detonation front carries the reaction products in the same direction, with a particle velocity about one-quarter the velocity of detonation (VOD) of the explosive column. If there is no stemming at the collar of the blasthole, the detonation products are ejected into the air at this high velocity (e.g., about 1000 m/s for an explosive with VOD = 4000 m/s). The waste of explosive energy and the generation of air shock waves can therefore be significant without stemming, particularly for blastholes with shorter lengths or larger diameters.

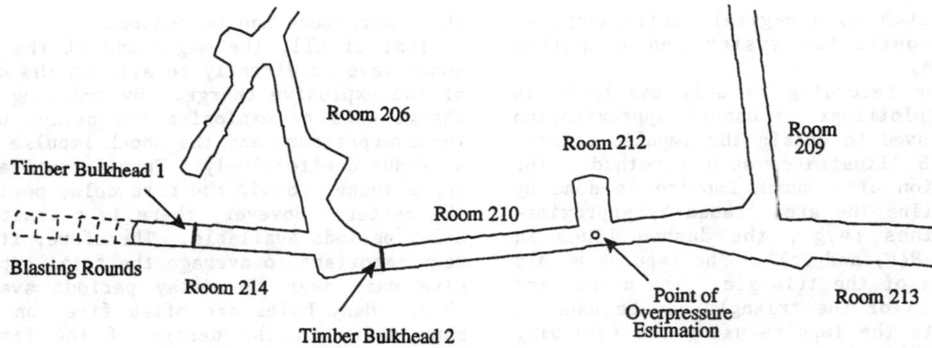

Figure 9. Plan view of shock wave protection for further blasting in Room 214

Finally, the use of low-density explosives will help to reduce the shock energy from detonation and subsequently the overpressure of the air shock waves. The pressure of detonation can be calculated by:

$$P = \frac{1}{4}\,\rho(\mathrm{VOD})^2 \qquad (9)$$

where P is the detonation pressure (MPa),
ρ is the density of explosive (g/cm^3), and
VOD is the velocity of detonation of the explosive (km/s).

For a certain type of explosive, the VOD usually has a linear relationship with density:

$$\mathrm{VOD} = a + b\rho \qquad (10)$$

where VOD and ρ are in m/s and g/cm^3 respectively, while a and b are the constants determined experimentally (e.g., a = 777 and b = 4383 for an emulsion explosive (Liu 1991)).

Equations (9) and (10) show that the detonation pressure is a function of the third power of the explosive density. Therefore, the most effective way to reduce the detonation pressure, or the shock energy of explosives, is to reduce the density of the explosives.

6.2 *Protection Principles and Design*

Protection against the damage caused by air shock waves depends on the force that the shock waves exert on a target, on how long the force is applied, and on the ability of the target to withstand its effects or absorb the air shock energy. Thus, the principle of designing protection against an air shock wave is, first, to reduce the strength of the shock wave before it arrives at the target. If the shock wave cannot be weakened sufficiently before its arrival, the next step is to reinforce the structure of the target to withstand the shock wave. For a structure on the surface, the second step is perhaps the most important. In underground tunnels, it is possible to weaken the shock wave by constructing obstacles such as timber bulkheads, wooden or steel gates and walls. In this case, it becomes much easier and sometimes unnecessary to reinforce the target structure.

The principal consideration in the design of obstacles is to flatten the sharp peak overpressure by a full-face curtain that should be flexible, able to swing and have a relatively large mass. Such an obstacle can reduce the shock pressure by absorbing the shock energy in deformation and movement (e.g., swinging of the curtain). The shock wave will have lower overpressure but longer positive duration after passing the obstacle. Thus, the design of the second obstacle should be to withstand the long impulse of the shock wave, requiring the structure to be more rigid, stronger and fixed. The location of the second obstacle should be either close to the target or at an intersection to enhance its blocking effect by reflecting into alternate tunnels.

Figure 9 shows a plan view of the blasting and bulkhead locations for further excavation of Room 214. Optimum bulkhead placements are given to protect Rooms 212 and 213. The structure of the bulkhead, shown in Figure 10 (Kuzyk et al. 1991), is applicable for shock protection. The first blast round will cause the highest overpressure. For this round, the use of 12 kg of explosive per delay in production holes will generate an overpressure of 14 kPa at Room 212 without any bulkheads, as

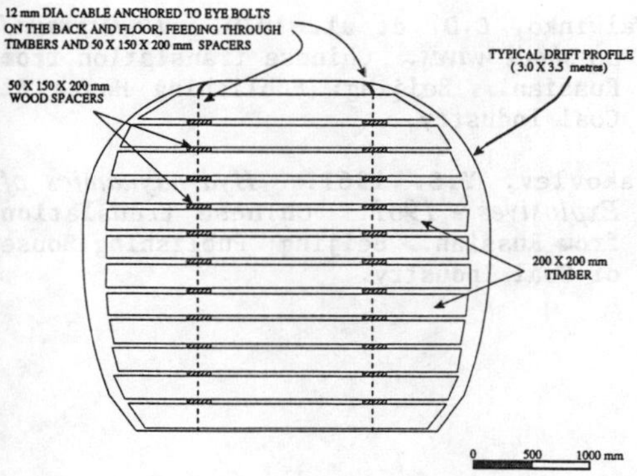

Figure 10. Details of timber curtain

determined using Equation (3). If the bulkheads can reduce the overpressure by 2/3, the overpressure at Room 212 should be reduced under 5 kPa, which is acceptably safe according to previous experience.

7 SUMMARY AND RECOMMENDATIONS

Air shock waves from tunnel blasting are generated by Mach reflections on the tunnel walls. The reflection zone is 4 to 8 times the tunnel width. The plane shock wave beyond the reflection zone propagates in the one-dimensional tunnel and attenuates more slowly than a similar blast in open air, which is three-dimensional. The amplitude of overpressure depends on the mass of explosive charge fired per delay, the type of blasthole, the distance from the blast face, the cross-section area of the tunnel and the roughness of the tunnel walls. Intersections and changes in tunnel direction significantly reduce overpressure. The wave form of the first shock peak usually has a sharp pressure rise, whereas subsequent delays have a slower pressure rise but a longer duration of positive wave form when travelling in air disturbed by earlier firing charges.

The generation of air shock waves from tunnel blasting, although unavoidable, can be minimized through the proper use of explosives. The mass of explosive fired per delay should be averaged among the available delay periods whenever possible. Delay intervals should be long enough to avoid overlapping of overpressure peaks, i.e., the delay intervals should be longer than 50 ms. Blastholes with high loading densities should be stemmed to avoid excessive ejection of detonation gases into the air, since they form strong shock waves. The use of low-density explosives is expected to produce less overpressures from tunnel blasting because of the lower detonation pressures.

Protection against overpressure can be achieved by constructing obstacles in the path to reduce its strength before it reaches the target. The design of the timber curtain used at the URL has the characteristics of being flexible, strong, heavy and easy to repair. The spaces between the timbers provide more surface area for the wood to absorb energy. The curtain also reduces the shock energy by its swinging movement.

The understanding of shock waves generated from tunnel blasting is limited since not much research has been done to detail the shock-wave propagation, damaging criteria and protection principles and designs. Further work should include:

- The study of the contribution of explosives on shock-wave generation from the energy partition (i.e., shock vs. gas energy) point of view as a part of rock-blasting mechanism.

- Obtaining practical rock wall roughness factors (n) and studying the influence of rock type and structure on the n value.

- Establishing reliable data based on the effect of intersections and the change in cross-sectional areas.

- Investigating the effect of delay periods on shock-wave accumulation.

- Investigating the overpressure change when a shock wave is propagating from a large tunnel to a small one. This study can be applied to obtain overpressures from stope blast.

- Testing the effectiveness of different bulkhead designs in terms of material type, structure, stiffness, etc.

- Establishing damage criteria on underground structures and objects.

This research would improve the understanding of the characteristics of the air shock waves from tunnel blasting and protection principles. The findings would have applications in both underground civil construction and mining engineering projects.

ACKNOWLEDGEMENT

The Canadian Nuclear Fuel Waste Management

Program is jointly funded by AECL Research and Ontario Hydro under the auspices of the CANDU Owners Group.

REFERENCES

Dormuth, K.W. & P.A. Gillespie 1990. Nuclear fuel waste disposal in Canada–The generic research program. Atomic Energy of Canada Limited Report, AECL-10183.

Hiismaki, P. & V.J. Kamarainen 1987. A hygrometer based on measuring the speed of sound and the temperature. *J. Phys.* E 20: 1465-1468.

Kinney, G.F. & K.J. Graham 1985. *Explosive shocks in air*, 2nd edition. New York: Springer-Verlag, Inc.

Kjartanson, B.H., M.N. Gary & B.C.M. Pulles 1991. Development for in situ tests on compacted bentonite-based buffer materials. *Materials Research Society Symposium Proceedings (Scientific Basis for Nuclear Waste Management XIV)* 212: 467-474. Pittsburgh: Materials Research Society.

Kuzyk, G.W., P.J. Babulic, P.A. Lang & R.A. Morin 1987. Blast design and quality control procedures at Atomic Energy of Canada Limited's underground research facility. *Proc. 13th Conf. of Explosives and Blasting Techniques:* 214-223. Montville: Society of Explosives Engineers.

Kuzyk, G.W., B. Mohanty & D.P. Onagi 1991. Control of blast overpressure and vibrations at the Underground Research Laboratory. *Presented at the 17th Conference on Explosives and Blasting Techniques*, Las Vegas, Nevada. Montville: Society of Explosives Engineers.

Liu, Q. 1991. *Investigation of the explosibility of sulphide dusts*. Ph.D. Thesis, Department of Mining Engineering, Queen's University, Canada.

Olson, J.J. & L.L. Fletcher 1971. *Airblast overpressure levels from confined underground production blasts*. U.S. Bureau of Mines Report, RI7574.

Potier, J-M., J-M. Hoorelbeke, G.W. Kuzyk, Y. Sifre & B. Mohanty 1992. Engineered blast feasibility study using low-shock energy explosives. *Proc. Int. Symp. on Geological Disposal of Spent Fuel, High Level and Alpha-bearing Wastes*. Vienna: International Atomic Energy Agency.

Valwinko, C.D. et al. 1979. *Underground air shock waves*. Chinese translation from Russian. Beijing: Publishing House of Coal Industry.

Yakovlev, Y.S. 1961. *Hydrodynamics of Explosives - 1961*. Chinese translation from Russian. Beijing: Publishing House of Coal Industry.

Innovative Mine Design for the 21st Century, Bawden & Archibald (eds) © 1993 Balkema, Rotterdam, ISBN 90 5410 325 6

Further developments of the concept of electro magnetic induction initiation

Liqing Liu
Department of Mining Engineering, Queen's University, Kingston, Ont., Canada

ABSTRACT: A series of new detonators characterized by the introduction of a small ferrite core in the firing mechanism have been developed. Based on the principle that an electric detonator with a mini ferrite core will elimininate the hazards of stray currents and static electricity, different models of electro magnetic induction initiating detonators have been designed to meet different application requirements. With the replacement of the primary explosives by secondary explosives from the detonator, the comprehensive safety features of the detonator have been greatly improved and the detonator is not only safe to electricity but also much safer to mechanical shock. To meet the special need for casing perforation blasting in oil well production where high temperatures and high pressures are expected, ferrite cores with higher Curie temperature are adopted. The detonator with a combination of inner and outer ferrite cores provides satisfactory safety features as well as the simplicity in connection. In some situations iron instead of copper legwires are used to offset part of the extra cost of the ferrite core. In underground large diameter borehole mining, a versatile primer has been developed. A pair of detonators with very short legwires are used in the primer and the weight of the primer can be adjusted according to the sensitivity of the explosives being used. To fire the detonators, a powerful high frequency blasting machine has been invented.

1 INTRODUCTION

In a blasting project, the safety of blasting operations is often of great concern. Since the explosives being used today are relatively insensitive to external energy such as mechanical shock, friction and electrical energy, the safety level of a blasting project is to some degree determined by that of the initiation system or the detonators chosen for it. Recently, significant improvements have been made on detonators. These improvements may be summarised in two aspects: electric and non-electric. Although the most striking development in non-electric detonators, represented by the nonel shock tube initiation method, has been widely accepted in the recent years, electric detonators have their position that is not replaceable by other kinds of detonators. However, conventional electric detonators do not have the ability to distinguish firing currents from stray currents and any kind of electric energy of sufficient value running through an electric detonator will cause it to initiate. Conventional electric detonators are not safe because of the presence of electrical energy around a blasting site. Electric detonators immune to electric hazards have been pursued for many years. The emergence of the electro magnetic detonator, typically the Magnadet, can provide protection against these hazards satisfactorily. The obvious safety features and simplicity in handling have made the electro magnetic detonator very suitable to mine drifting, shafting, tunneling as well as underground and open pit mining. However, there are two main reasons to do further research work into the new detonator. One is that there are some intrinsic limitations associated with the early design of the initiation system, such as the vulnerability to stray currents of the products with long legwires, and the use of primary explosives; the other is that the Detonator can extend its application to other technical fields after appropriate modifications, In underground mining with large diameter boreholes, the assembled primer should be rebuilt to provide the flexibility for different initiation requirements of explosives for the different choices to lower an

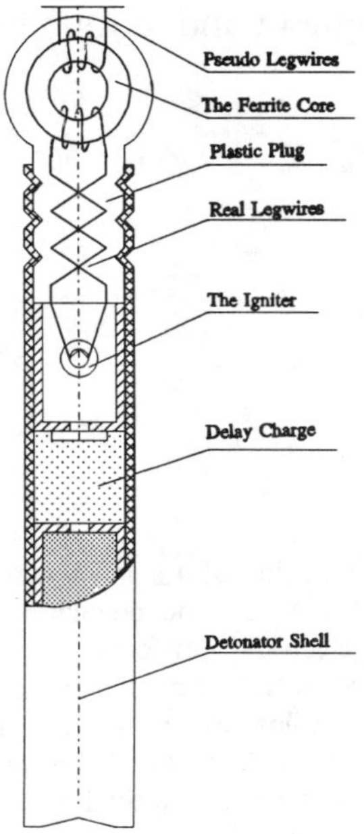

Figure 1 An Electro Magnetic Detonator With Built-in Ferrite Core

assembled primer into the blasthole. Furthermore, a more powerful blasting machine is necessary for the further developments and extented applications. The work described in this paper has been aimed to modify the early designs and to meet the new application requirements.

2 ELECTRO MAGNETIC DETONATOR WITH BUILT-IN FERRITE CORE

The electro magnetic induction detonator, or simply the electro magnetic detonator, adopts a special magnetic material in its initiating energy transmitting mechanism. This magnetic material is a man-made soft magnet often consists of Manganese Oxide, Zinc Oxide and Ferrous Oxide (Mn-Zn Ferrite). The Mn-Zn Ferrite eliminates low frequency but transmits high frequency electric currents when it is used as a transformer core. Since electric energy in blasting operations concentrates on the low frequency range, e.g., an AC power source has a frequency of 50 or 60 Hz, and a static discharge also concentrates its energy in a relatively low frequency range, the Mn-Zn Ferrite is used to distinguish between low frequency extraneous electric energy and high frequency firing currents. This idea was put forward no sooner than the emergence of the Mn-Zn Ferrite in the 1940's. Since a powerful blasting machine is required to provide a high frequency firing current, the electro magnetic detonator was not put into practical use till the late 1970's when power semiconductors were developed. The electro magnetic detonator, typically Magnadet (which is a trademark of ICI), comprises a magnetic core and an electric detonator. The legwires of the electrical detonator circle the ferrite core 3 or 4 times and are then electrically shorted. External electrical energy has no access to enter the detonator body. The detonator is simply a combination of a common electric detonator and a ferrite core. This means that there is still potential hazards associated with it, especially when the legwires are very long. For example, if the legwires of the detonator are unexpectedly cut by the edge of the stemming rock or by charging tools during the charging operation of the blastholes, the legwires are then exposed to external electrical energy around the blasting site. It must be stressed that the igniter of an electro magnetic detonator is often more sensitive to electric currents than that of a conventional electric detonator.This is necessary to offset the high frequency energy loss along the primary wire and the long legwires and to ensure that an adequate number of detonators that can be initiated in each round. When the long legwires of an electrio magnetic detonator are cut and exposed, stray currents, static electricity may enter the detonator and cause it to explode just as with conventional electric detonators. This is even more serious when the blastholes are pneumatically loaded where the accumulated static energy is very high. It is very hard to find and locate one or two such legwires cut in a blasting net by eye assay or using instsruments.

To make the electro magnetic detonator even safer to external electric energy, a new model characterised by a built -in ferrite core has been developed. Figure 1 shows the section view of such a detonator. As is shown in the figure, a much smaller ferrite core is used and integrated in the detonator. The ferrite core is sealed in the plastic plug when the latter is formed in an injection mould. In the new detonator, the traditional concept of legwires is no more valid. Typically, the legwires are the conducting wires connected with the firing bridge and extending from the detonator. In a conventional detonator, the legwires form the only access for external electric energy to enter the detonator. In the electro magnetic detonator with built-in ferrite core, the real legwires are now static-shorted and minimised in length to its possible value and wholly sealed in the plastic plug. If it is viewed as a transformer, the secondary winding along with the firing bridge wire is now totally isolated and protected by the plastic plug and the metallic casing. Therefore, any damage to the wire coming out of the detonator (pseudo legwires) will not affect its ability to protect against extraneous electric energy. It has been tested that the detonator will not fire even when (1). an AC power source of 220 volts, 50 Hz or (2). an AC power source of 380 volts, 50 Hz or (3). 226 joule static energy (from a capacitor) is applied to the pseudo legwires. 220-volt and 380-volt power sources are upper limitations of stray currents. It is interesting that the detonator does not fire even when the ferrite core is cracked by overheat generated by the strong currents running through the pseudo legwires.

In practical applications, it is recommented to fix the mini ferrite core at the collar of the detonator shell instead of within it. When fixed at the collar, the only change to the production line of conventional detonators is to use a new injection mould that can integrate the mini ferrite core and the plastic plug. If it is fixed within the detonator shell, not only should the injection mould be changed, the detonator shell should also be lengthened and a series of other changes in the production line is required because of the change in shell length. The outer diameter of the ferrite core is only 5(mm) when it is totally within the detonator shell, or 7(mm) when it is at the collar of the shell. When a ferrite core of 7*4*4(outer diameter*inner diameter*thickness,in mm) is used, it has a volume only 1/17 of that of a Magnadet which uses a ferrite core sized 18*8*10(outer diameter*inner diameter*thickness, in mm), therefore, much less ferrite material is used for each detonator and the extra cost can be reduced.

The electro magnetic detonator with built-in ferrite core can provide more satisfactory protection against static electricity. Since the casing of the detonator is made from copper or other metallic materials, the isolated firing bridge is in fact electrostatically shielded by it. Theoretically, the firing bridgewire will be electrically neutral even when the shell is charged. Therefore, the firing bridgewire will not be an electrode and the static discharge between the bridgewire and the shell will never happen. A better protection against radio frequency energy and lightning can also be expected for an electro magnetic detonator with built-in ferrite core. However, as with other initiation systems, all personnels still have to be evacuated from the blasting site in case of thunderstorm activity.

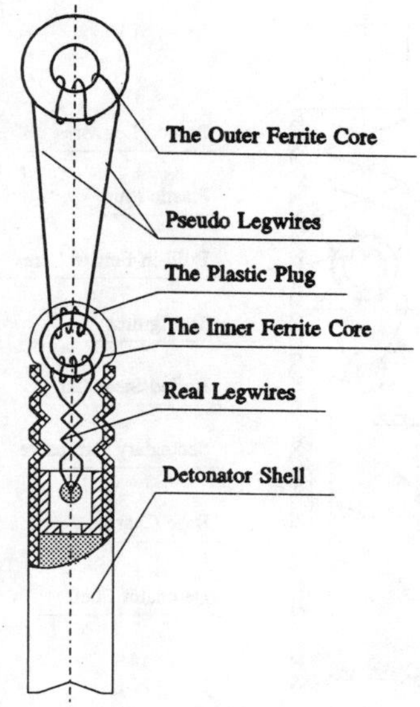

Figure 2 An Electro Magnetic Detonator With Two Ferrite Cores

3 ELECTRO MAGNETIC DETONATOR WITH TWO FERRITE CORES

One of the most striking advantages of electro magnetic detonators over conventional electric detonators is that they are very simple to use. In order to construct a blasting net, all required to do is to put a conducting wire called primary wire through the ferite cores protruding from the blastholes one by one. The procedure is time saving and there are no joints except where primary wires are connected to form a loop and to lead to the blasting machine. However, when the mini ferrite core is integrated with the detonator, as described in the foregoing paragraph, the ferrite core can no more be used to connect with others by simply putting a conducting wire through it. This problem is solved by adding another ferrite core to the detonator. An electro magnetic detonator with two ferrite cores is shown in Figure 2. The detonator looks like a common Magnadet. The outer ferrite core is used to make jointless connection with other detonators, and the inner one is used to isolate the firing bridge wire with extraneous electric energy. An electro magnetic detonator so structured makes it not only very safe to external electric activities but also possesses the simplicity of ordinary Magnadets in use. However, during the initiation process, the electric current running through the primary wire is greatly attenuated since it must first

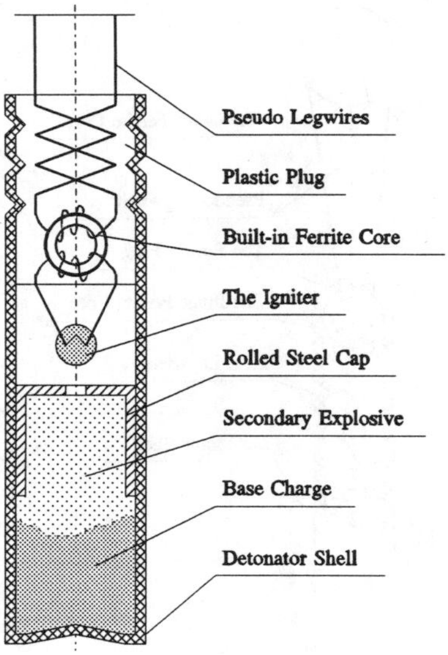

Figure 3 An Electro Magnetic Detonator Containing Only Secondary Explosives

induce a current in the pseudo legwires, then induce a current in the bridge wire to fire the detonator. Therefore, a more powerful blasting machine is required to initiate a certain number of detonators in a round. The use of this detonator can give us the simplicity in handling as well as more confidence in safety especially when the pseudo legwires are relatively long.

4 ELECTRO MAGNETIC DETONATOR CONTAINING ONLY SECONDARY EXPLOSIVES

Common electro magnetic detonators have the same charge configuration as conventional detonators, both primary and secondary explosives are used. Secondary explosives are relatively dull to external electric and mechanical energy, but primary explosives are more sensitive to electrostatic discharge, mechanical shock and friction. Many accidents have been occured because of the unsafeness of primary explosives, both in the manufacturing of detonators and in their field uses. To make the electro magnetic detonators even safer, it is meaningful to eliminate the use of primary explosives in the detonator, that is, to manufacture the detonator with only secondary explosives (and some pyrotechnic materials). Up to present, there are many ways to make a detonator containing only secondary explosives, some of them use a very fast flying impactor to shock initiate the base charge, others use a very high voltage electric discharge to cause the explosives to detonate. These methods are not suitable for an electro magnetic detonator. It has been found that secondary explosives such as RDX,PETN and Tetryl can turn from inflammation to detonation under strong confinement. In our experiments, a cap rolled from steel is used to provide the strong confinement. The cap is similar in shape to that used in a primary explosive detonator, except that it is made from steel and has a greater length instead of being a short copper or aluminum one. The materials charged in the cap is a mixed combination of secondary explosives and pyrotechnic compositions. Since the cap is relatively long, it is charged independently to obtain the required charging density before it is assembled with the base charge. Primary explosives are also removed from the chemical compositions of the igniter, it simply contains pyrotechnic materials such as lead thiocyanate and potassium chlorate or potassium perchlorate($PbCNS+KClO3$, $PbCNS+KClO4$). Upon ignition, the rapid release of the gases from the reaction creates a high temperature and high pressure environment within the cap. With the increase of temperature and pressure within the cap, inflammation of materials turns to low order detonation, the detonating velocity increases along the cap and finally reaches high order detonation and detonates the base charge under the cap. If ignition delay is required for the detonator, a delay element which gives designed delay time can be used between the igniter and the cap. The time interval for the materials contained in the cap to turn from inflammation to high order detonation is within 1 milisecond.

An electro magnetic detonator containing only secondary explosives provides satisfactory protection against extraneous electric energy as well as mechanical shock. Its behaviour against extraneous electricity is the same as the electro magnetic detonator with built-in ferrite core which is already described. The ability of a detonator to protect against mechanical shock is important before as well as after a blast. Experience shows that a misfired detonator along with unreacted explosives remained in a blasthole can be very dangerous for subsequent drillig operations. To observe the behaviour of the detonator to protect against mechanical shock after removal of primary explosives from its compositions, the detonator for impact sentivity was tested with a 10-kilogram hammer falling from different heights. To concentrate all the impacting energy of the falling hammer on the cap which is the most sensitive part of the detonator, a small triangular steel plate was placed on the position of the cap. Detonators with paper and metallic casing are tested. Test results show that the detonator containing only secondary explosive has a critical impact energy of 156.8(J) (paper casing), or under 137.2(J)(metallic casing). As a comparison, detonators containing primary explosives will explode under the impaction of only 58.8(J). The impact work in a stroked hand operated drilling machine is about 60(joules) but other types of drilling machines may have a higher value. The use of detonators containing only secondary explosives will notably decrease the danger associated with drilling a remained blasthole although not finally eliminate it.

5 ELECTRO MAGNETIC DETONATOR FOR OIL WELL CASING PERFORATION BLASTING

Oil (or gas) well production is a new field with attractive potential where the electro magnetic detonator can be applied. In metal mines, one can chose between the electric and non-electric initiation systems, but in oil well casing perforation blasting, electric detonators are always used. However,electric detonators used in casing perforation are quite different from those used in metal mines. In oil well production, holes are drilled where underground oil is found. While being drilled, the hole is cased with cement and steel tube to control the outflow of oil. When jet perforators are used, the electric detonator is used to start a chain reaction which successively detonates the primacord, the high velocity booster in the charge, and finally, the main explosive. The drillhole where the detonator is used may be several thousand meters in depth from the hole mouth. The detonator must stand the temperature as high as 180 degrees(Centigrade) and the high pressure from cement slurry. When an electro magnetic detonator is used, the detonator is fired by the current induced by the ferrite core attached to it. However, if the magnetic behaviour of the ferrite core varies with the ambient temperature. If the ambient temperature equals to or is higher than the critical temperature--the Curie temperature of the ferrite core, it will lose its magnetism. Therefore, ferrite cores used for this purpose must have a Curie temperature higher than the temperature expected.

The detonator is often connected to the blasting machine at the mouth of the well by a coaxial cable. The resistence of the cable depends on its type and length and it is often much higher than the DC resistance of a common blasting net of electro magnetic detonators. The distributed capacitance along the coaxial cable has a value higher than common conducting wires. Much of the initiation energy carried by the high frequency current is lost along the cable and only a small fraction reaches the detonator. If the current frequency is low, the efficiency of the ferrite core will also be low; however the current frequency is too high, the loss of initiation energy along the cable will be very great. The best frequency range must be experimented or calculated to determine the corresponding frequency parameters of the blasting machine. Because of the above factors contributing to the difficulties of initiation, a powerful blasting machine is fundamental for the successful use in this area.

Since electric activities around a well rig are very frequent, the electric hazards associated with oil well casing perforation blasting are more serious than in metal mines. Because electric blasting is the only choice for this purpose, it can be forecast that the electro magnetic detonator will be accepted and developed rapidly in this field.

6 ELECTRO MAGNETIC DETONATOR WITH IRON LEGWIRES

Although electro magnetic detonators generally have copper legwires, many conventional electric detonators use iron legwires to maintain a lower cost. To achieve wider acception, it is important to develop electro magnetic detonators with iron legwires (or pseudo legwires).

Contradiction exists between high frequency initiation and iron conducting wires. Most electro magnetic initiation systems of present design are not suitable for detonators with iron legwires. The DC electric resistance of iron conducting wires is about 5 times as high as that of copper conducting wires with the same diameter and length. However, the most striking differences between copper and iron legwires are the serious skin effect of iron legwires and its inductance under high frequencies. The tendency of high frequency currents to concentrate on the surface of the conducting wire builds up an additional resistance called skin resistance. Conducting wires which are permeable to magnetism will result a high skin resistance under a certain current frequency. Iron conducting wires often have a magnetic permeability coefficient as

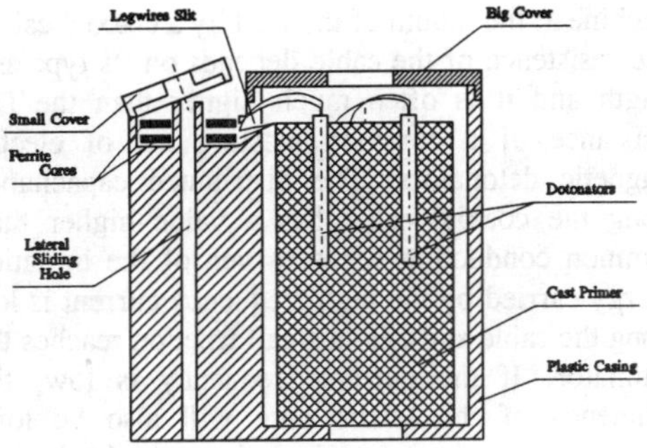

Figure 4 An Assembled Primer With Lateral Sliding Hole

high as 10,000. This is not only much higher than that of copper wires but also higher than the initial magnetic permeability coefficient of the ferrite core used for the detonator. Also because of the high magnetic permeability, the legwires of each detonator are in fact an inductor when connected in a blasting net. When iron legwires are used, the high skin resistance and inductance act together to make the detonator very hard to initiate. Tests have shown that it is almost impossible to fire such detonators with less powerful high frequency blasing machines. With the specially designed model KDC-1 high frequency blasting machine (which is described in a later paragraph of this paper), as many as 30 detonators can be fired in each round with each detonator having a pair of 2 m iron pseudo legwires. The mini ferrite core used in the detonator is sized 5*2.5*3(outer diameter*inner diameter*thickness, in mm). The DC firing current of the igniter used in the detonator is 0.36 (amperes) and the transmission distance 100 meters. The ability to fire 30 detonators each round is just competent for blasting projects of routine small scale blasting in coal mines.

7 APPLICATION IN UNDERGROUND LARGE DIAMETER BOREHOLE MINING

Underground large diameter borehole mining is an application in which the electro magnetic induction initiation system has obvious advantages over traditional electric or non-electric initiation systems. Large diameter borehole mining methods have become very popular since ten years ago. In underground mining, the depth of the borehole is normally 30-40 m, sometimes 80-100 m. In-the-hole multiple delays are commonly used in blasting. The charging process with traditional electric or non-electric initiation systems is troublesome and time consuming. If there are more than one delay charges in a hole, each one must have its own electric conducting wires or plastic shock tubes from the position of the delay charge to the collar of the borehole. If six delay charges are used in the borehole, there are as many as 12 conducting wires or plastic shock tubes at the collar of the borehole if each one uses a pair of detonators. It is certainly troublesome for electric initiation apart from the disadvantage of extra consumption of blasting supplies. With the electro magnetic system, the charging process of the borehole can be greatly simplified. In this application, A pair of detonators with very short legwires (only 10 cm) is often used in a specially designed plastic casing and it is protected along with the cast primer (RDX/TNT) by the plastic casing. The assembled primer (which is a combination of a pair of electro magnetic detonators with delay, a cast primer and the plastic casing) may be lowered to anywhere intented in the blasthole. The Magna primer which uses Magnadets is such an assembled primer. When it is used, only two primary wires are needed (down and up the hole, which are in fact one conducting wire) in the blasthole no matter how many delay charges are used in the blasthole. When the blasthole is being charged, to form the delay charges in the hole all required to do is to lower the primers along the primary wire one after another. However,the original type of the Magna primer is not suitable for extended applications.

In some cases,to lower the assembled primer, one has to choose between independent sliding and simultaneous lowering with explosive cartridges. The Magna primer has its sliding hole in the center of the primer, which makes it possible to be lowered only individually along the primary wire. The primer of such design is convenient for blastholes which are pneumatically or bulk loaded and do not have a very great depth. But when the blastholes are charged with explosive cartridges (such as emulsion explosive) or have a great depth, individual sliding of the primer becomes problematic as to its safety and reliability. With the increase in blasthole depth, the primer lowering in it would acquire a very high speed. It is analysed that the collision of the primer with the blasthole wall or the botttom would damage the primer or even result in unexpected explosion. Such a high speed collision would detach the cover of the plastic casing, then the cast primer along with the detonators would be exposed to outside environment. Since the cast primer is a brittle material and the detonators (especially those containing primary explosives) are sensitive to mechanic shock, further collisions would break the primer or initiate the detonator. Furthermore, if the primer is individually lowered, it is possible that it falls into the cavities along the blasthole ; and when the blastholes are filled with water, independent sliding is very difficult to locate the primer in the place as designed. On account of the above reasons, it is prudent not to slide the primers individually but to lower them simultaneously with explosive cartridges.

Assembled primers with central sliding hole are not suitable for simultaneous lowering. The newly developed assembled primer with lateral sliding hole is versatile for both choices, as shown in Fig.4. The sliding hole along which the primary wire runs through is at one side of the plastic casing, the ferrite cores of two detonators are pressed into the preset hole and protected by the small cover, the legwires are pulled through the slit in the plastic casing and the detonators are inserted in the cast primer. The assembled primer with lateral sliding hole shown in the figure is designed to be used with emulsion explosive cartridges. The explosive cartridge along with the assembled primer is lowered with a piece of lowering hose and the primer moves steadily, simultaneously with the cartridge. Thus lowering speed of the system is controlled and adjusted by the holder of the lowering hose, see Fig.5. In this way the primer is lowered safely and reliably. Experiments have shown that when the blasthole favors individual sliding of primers, the assembled primer with lateral sliding hole can be used just as conveniently as with primers with central sliding hole since the displacement of the sliding hole from the center to one side essentially does not affect the sliding process of the primer. The cast primer in the plastic casing can have different size and weight representing different initiation power. Therefore, an assembled primer with lateral sliding hole is adjustable to explosives with different boosting requirements.

For cap sensitive explosives, a sliding block is developed on the basis of an assembled primer with lateral sliding hole. The sliding block is a plastic bar which can hold and protect a pair of electro magnetic detonators shown in Fig.6. It may be viewed as the left part of the assembled primer as shown in Fig.4 and is used in a way similar to it. Fig.6 shows the top view of the sliding block with its small cover open and turned aside. To achieve uniform detonation and shorten the time interval for detonation growth of the explosive, a short segment

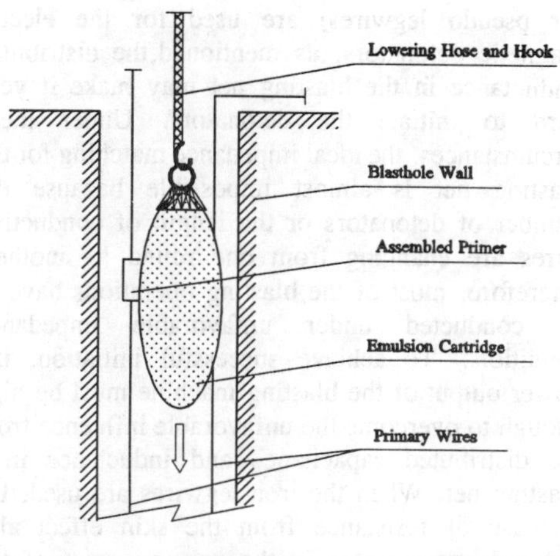

Figure 5 An Assembled Primer Being Lowered With An Emulsion Cartridge

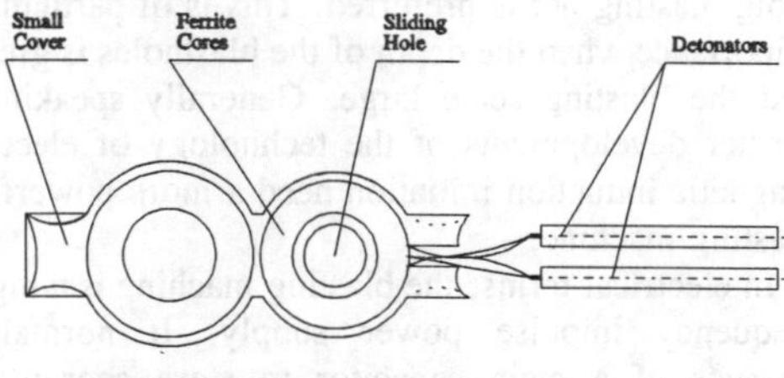

Figure 6 Top View Of A Sliding Block For Cap Sensitive Explosives

of detonating cord (about 1 meter in length) is bound against the pair of detonators, buried in the cartridge and functions along with the detonators.

8 THE MODEL KDC-1 HIGH FREQUENCY BLASTING MACHINE

In the electro magnetic detonator with built-in ferrite core or two ferrite cores, the reduction of the number of winding turns around the mini ferrite cores partly depends on the power output of the blasting machine. Since the winding of the ferrite cores is temporarily manual, less number of winding turns is strongly preferred.Only a blasting machine with high voltage and strong current output can initiate electro magnetic detonators with fewer winding turns when other related parameters (e.g., the size and material of the ferrite core) are determined. In oil-well perforation blasting, the influence of the distributed capacitance along the coaxial cable is very striking; when iron legwires (or pseudo legwires) are used for the electro magnetic detonators, as mentioned,the distributed inductance in the blasting net may make it very hard to initiate the detonators. Under these circumstances, the ideal impedance matching for the blasting net is almost impossible because the number of detonators or the length of conducting wires are changing from one round to another. Therefore, most of the blasting operations have to be conducted under unfavorable impedance conditions. To achieve successful initiation, the power output of the blasting machine must be high enough to overcome the unfavorable influence from the distributed capacitance and inductance in a blasting net. When the iron legwires are used, the increase of resistance from the skin effect also demands an increase in the power output of the blasting machine. Furthermore, in the application of electro magnetic detonators in large diameter borehole mining, a blasting machine which can fire a big blasting net is preferred. This is of particular importance when the depth of the blastholes is great and the blasting scale large. Generally speaking, further developments of the technology of electro magnetic induction initiation need a more powerful blasting machine.

In electrical terms, the blasting machine is a high frequency impulse power supply. It normally consists of a main capacitor to store energy, a charging circuit to supply electric energy to the main capacitor, and an inversion circuit to transfer the DC current discharged from the main capacitor to high frequency alternating current, which is the output electric current of the blasting machine. The key to the success of a blasting machine is the inversion circuit. Present inversion circuits that can be used in a high frequency blasting machine can supply very limited power and are difficult to extend the power output. An inversion circuit may or not have a ferrite transformer as its component. In a common four-armed inversion circuit, as many as 8 power transistors (power transistors occupy the main part of the cost of a blasting machine) are used , with 2 transistors in each arm. The number of transistors may be reduced if a push-pull power amplifier transformer is used in the inversion circuit, but a series of other problems may arise. Since the transformer uses basically the same ferrite material as that used for detonators, its existence in the inversion circuit forms a big electric inductor in the blasting net when it is connected to the blasting machine. This makes the impedance in a blasting net complicated. The adoption of the transformer also limits the maximum power output by its size. Therefore, relatively powerful blasting machines have to use very big transformers that are heavy in weight and high in cost. It may be concluded that blasting machines of present design are not suitable for the extended applications of the technology of electro magnetic induction initiation.

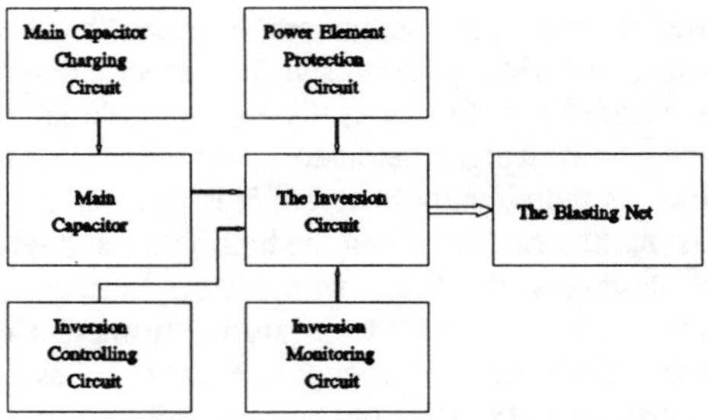

Figure 7 Circuit Configuration Of The Model KDC-1 High Frequency Blasting Machine

The model KDC-1 high frequency blasting machine uses a novel inversion circuit. The principles on which the new inversion circuit works is totally new. The electric elements of the circuit is arranged on a circuit board having the size of a palm and no output transformer is included. The inversion circuit works under a voltage of 500 volts and has a maximum current output of 20 amperes while the inversion frequency is adjustable from 10 KHz--1 MHz. The inversion circuit is used to form a high frequency blasting machine along with some periphery circuits, as shown in Fig.7. The periphery circuits are designed to charge the main capacitor, to control and to monitor the inversion process, and to protect the power semiconductors in the inversion circuit against possible electric breakdown due to the high voltage and strong currents. The main capacitor of the blasting machine may be charged by an external alternative power supply that is available in most blasting sites, or be charged by the dry batteries carried by the blasting machine. Since no output transformer is used and the novel inversion circuit is simple and cheap, as compared to common high frequency blasting machines, the cost, size, and weight of the new one have been greatly decreased and a higher efficiency and great power output have been achieved. The main parameters of the model KDC-1 high frequency blasting machine is listed below:

(1). Output voltage: 500 volts
(2). Maximum output current: 20 amperes
(3). Frequency: 20--60 KHz
(4). Power supply:36 or 220 volts, 50 Hz
(5). Charging time: 10 seconds
(6). Dimension:

length*width*height=180*120*190,mm

(7). Weight:1.5 Kg.

The blasting machine can be used to initiate common electro magnetic detonators, electro magnetic detonators with a built-in ferrite core or two ferrite cores. It is also powerful enough to be used in oil well casing perforation blasting, in underground large diameter borehole mining or to initiate electro magnetic detonators with iron legwires. If the blast operations demand the use of an even more powerful machine, the power output can be increased by increasing the output voltage and current on the basis of the model KDC-1 blasting machine. Such increase in output parameters will not notably increase the size and cost of the blasting machine.

CONCLUSION

It is about 12 years since the emergence of the electro magnetic detonator and field uses have proved its marked safety features and operational advantages.Further developments described in this paper are based on the early design of the electro magnetic initiation system. The work has been aimed to improve the early design and to extend the application of the technology of electro magnetic induction initiation. Since these developments may be viewed as the results of the collected experience in the past years, a great degree of improvement is possible in the future years. Still further research work on the initiation system and its theoretical aspects is needed to advance the deveplopment.

ACKNOWLEDGEMENTS

The author wishes to express sincere thanks to his thesis supervisor Dr. P.D. Katsabanis, whose encouragement and help have made this paper possible. The research work was carried out in Beijing General Research Institute of Mining and Metallurgy, Pepole's Repulic of China. State Owned Weixing Chemical Factory in Heibei Province and Fankou Lead Zinc Mine in Guangdong Province offered cooperation and assistance in the development and application of the new detonators.

REFERENCES

FOX,R., and WATT, R., A Novel Initiating System for Blasting Explosives; 7th Annual Conference of Explosives Engineers, Phoenix, Arizona, U.S.A., Jan. 1981

Morry, W.B., The MAGNADET electric initiation system; CIM Bulletin, vol. 75, No.847, pp61-65, 1982

BALL, M.J., and WATT, R., Field Experience with New Methods of Electric Shotfi. ıg; 8th Annual Conference of the Society of Explosives Engineers, New Orleans, Louisiana, U.S.A., Feb. 1982

NAGAOKA, AKIHIKO, and SAKAGUCHI, MICHIAKI, Theoretical Study of Electromagnetic Inductive Detonator; Kogyo Kayaku, vol.45, No.1, 1984

A Report on the Research of the Model BJ-1 Safe Electric Detonator, Beijing General Research Institute of Mining and Metallurgy, 1987

A Report on the Research of the Model BJ-2 Safe Electric Detonator, Beijing General Research Institute of Mining and Metallurgy, 1991

A Report on the Application of the Electro Magnetic Initiation System in Underground Large Diameter Borehole Mining, Beijing General Research Institute of Mining and Metallurgy and Fankou Lead Zinc Mine, 1990

A Report on the Research of the Model KDC-1 High Frequency Blasting Machine, Beijing General Research Institute of Mining and Metallurgy, 1991

CN87,103,475

CN2,095,396U

Innovative Mine Design for the 21st Century, Bawden & Archibald (eds) © 1993 Balkema, Rotterdam, ISBN 90 5410 325 6

SCB technology for new electronic and fiber optic-based detonators and initiation systems

W.B.Gregg, Jr
Thiokol Corporation, Elkton Division, Md., USA

1 INTRODUCTION

The rock blasting and mining industries use a system of detonators and explosives for mining operations. Many detonators use small metal bridgewires, which are heated by passing a current through the wire, heating and igniting the pyrotechnic powder in contact with the wire. Typical bridgewires require a current of at least 3 amps, applied for a few milliseconds, for an explosive output. The semiconductor bridge (SCB), in comparison, initiates an explosive in 0.01 to 0.1 of a millisecond.

In the 1980s, Sandia National Laboratories developed a new approach for the ignition of explosive material using an SCB, or electronic match, to replace the conventional bridgewire. This new device offers distinct advantages in fast function time (time from ignition signal input to fusion/ignition), precision and repeatability, low-energy consumption, and safety.

Thiokol Corporation began learning about SCB technology in January 1988. The next year was spent with Sandia National Laboratories and a new company, SCB Technologies, Inc., which was formed to be the exclusive licensee of SCB technology for exploring certain commercial applications. SCB Technologies, Inc., in turn, granted an exclusive SCB license to Thiokol in May 1989 for use in vehicle air bags. The agreement was expanded in January 1992 to include the rock blasting and mining industry.

Over the past 4 years, Thiokol has been developing and testing initiator configurations with the SCB chip, or SCB, for military and commercial applications. In the last year, for the mining industry, we expanded our SCB design activities to include an all-electronic delay cap for operations requiring precise timing delays at low cost. Concurrently, we are refining SCB processing for high-volume production. This design is now being tested both at our Elkton, MD, facility and in field operations in the United States and Canada.

2 KEYS TO THE SCB

The SCB is an innovative method of initiating pyrotechnics and detonating explosives that has specific technical advantages compared to a conventional bridgewire. These technical advantages are responsible for the SCB's success in military and commercial markets. The most important advantages are the following:

- Fast function time - The SCB initiator ignites in 10 to 100 microseconds, vs 1 to 4 milliseconds for bridgewire devices.
- Precision - The SCB ignition is highly precise and repeatable, in the 0.1-millisecond range.
- Low-energy operation - The SCB requires only 1 to 3 millijoules of energy for ignition, vs 30 to 35 millijoules for a conventional bridgewire.
- Meets military safety requirements:
  - MIL-STD-1385B hazards of electromagnetic radiation to ordnance (HERO)
  - DoD and DoE electrostatic discharge (ESD)
  - 1-amp/1-watt no-fire

The SCB has other technical advantages that

Figure 1. Semiconductor Bridge (SCB) Chip

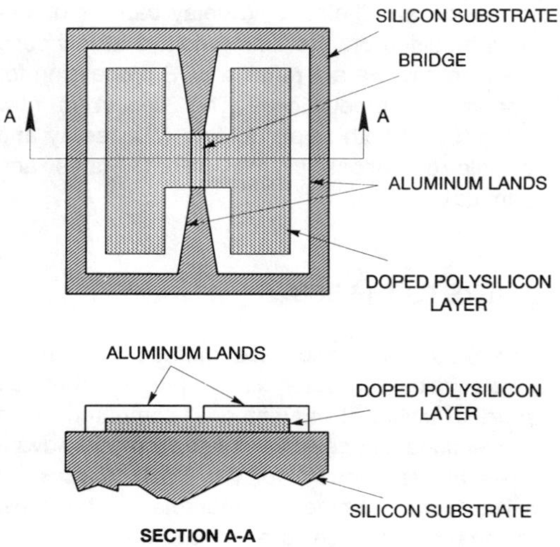

Figure 2. Typical 1-Ohm SCB

contribute to its uniqueness:

- Low energy and compactness make integration with electronic microcircuits easy.
- The SCB is tailorable to meet various no-fire or ESD requirements.
- Hot plasma output (5500 K) initiates insensitive explosive and pyrotechnic materials, whereas sensitive materials are required with bridgewire ignition systems.
- The SCB chip and initiator design lends itself to automated continuous processing for high-volume production.

3 THE SCB CHIP

The SCB (Figures 1 and 2) is composed of an electrical material or doped polysilicone that defines the bridge. The bridge is mounted on a 1.5-mm-square silicon or sapphire substrate. Two aluminum lands cover most of the doped polysilicone area to complete the bridge and to serve as the contact for electrical connection of the SCB to the header pins.

SCBs are mass-produced (Figure 3) on wafers of silicon substrate using microchip processing methods. A doping layer is applied to the top polysilicon layer, and defines the bridge area through a photoresist and etching process. An aluminum layer is then deposited on and around the silicon to form the aluminum lands and provide for electrical contacts.

The SCB, on 2- to 8-inch wafers, is made using state-of-the-art foundry operations also routinely used for the production of semiconductor devices. Production runs of 30 million units per year or higher are common for similar devices.

4 SCB OPERATION

Application of an electrical pulse to an SCB initiates a burn process that results in a plasma discharge between the aluminum lands. When the pulse is applied, the bridge between the aluminum lands begins to melt into molten silicon channels, producing a weakly ionized vapor of silicon above the bridge. As the current continues, wider channels form, melting and vaporizing the entire bridge region between the aluminum lands. When the bridge vaporizes, it bursts into a plasma with an equivalent temperature of 5500 K, causing ignition of the explosive powder pressed against the bridge.

In contrast, metal bridgewires heat the surrounding powder by a thermal conductive mechanism, dependent on the mechanical contact between the bridge and the powder. Consequently, their operation is subject to variations in the powder parameters and the explosive used. The SCB heat transfer is a convective process in which the plasma penetrates into the powder bed and condenses onto the powder particles, heating them to their ignition temperature. This efficient convective process permits device function time in a few tens of microseconds or less.

5 SCB INITIATOR/DETONATOR

For initiator applications, the SCB chip is electrically attached to a conventional glass-to-metal seal header. A cup loaded with pyrotechnic material is laser-welded to the header, producing an SCB initiator (Figure 4). This serves as the core device for both squib and detonator applications. Design modifications are made as necessary to accommodate new requirements or to optimize for high-volume production. The manufacturing process yields a robust finished product that has been demonstrated to operate reliably after exposure to both military and commercial environments.

To form a detonator, the core device is installed into a blasting cap with a detonating explosive and time-delay circuitry. The timer can be preprogrammed or programmed in the field for delay times from 1 to 500 milliseconds, in 1-millisecond increments.

6 SCB ADVANTAGES FOR MINING

The SCB's distinct advantages of precision and repeatability provide the missing link for precise control of blasting operations to the end user. Use of the SCB eliminates the 1- to 4-millisecond variability in function time found in conventional bridgewire devices. Coupling the SCB to an electronic timer provides the ultimate in precision and flexibility by offering millisecond delays with microsecond precision. The flexibility enables the end user to structure his blast pattern according to the latest analytical software available, which produces the following operating benefits:

- Elimination of surface delays by using a programmable down-hole delay
- Elimination of cutoffs - Each hole is independent of the other when fired.
- Simultaneity for mid-column air deck
- Multiple decking with discrete delays
- Versatility with delays that are programmable on site
- Enhanced safety - The device is insensitive tc ESD and electromagnetic (EM) and radio frequency (RF) interference.

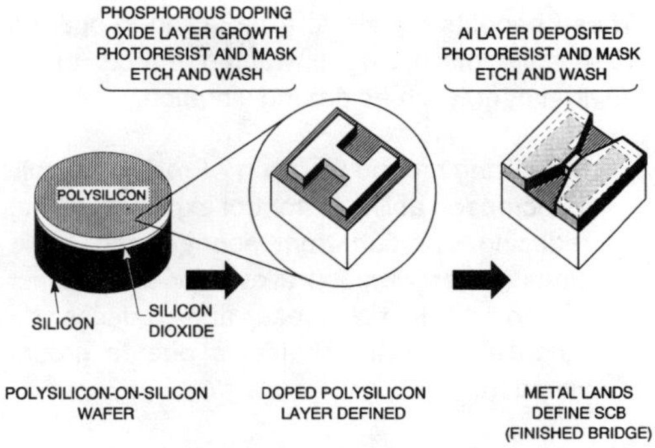

Figure 3. SCB Processing Uses Established Microchip Methods

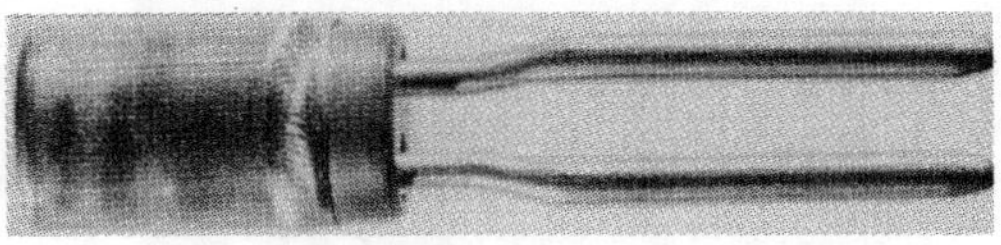

Figure 4. SCB Initiator Baseline or Core Design

7 FIBER OPTICS - THE ULTIMATE IN SAFETY

The next generation of SCBs can offer fiber optics to replace the conventional ignition wire system. In a fiber-optic system, light travels through clear optical cabling from a light source (detonation signal) to the detonator, where the light is converted to energy, igniting the SCB. Even though the SCB has already been proven safe to RF, ESD, and EMI hazards, fiber optics further reduces the possibility of being affected by these environments by eliminating the electrical wires.

The powering and controlling of the SCB via fiber optics has already been proven by Sandia National Laboratories and has been further demonstrated by Thiokol under military contract. The primary reason the SCB is compatible with fiber optics is its low-energy consumption compared to a conventional bridgewire (1 - 2 millijoules vs 30 - 35 millijoules). This key SCB attribute makes the fiber-optic concept a possibility.

8 COST BENEFITS TO THE USER

The SCB's versatility, precision, and safety translate to value-added benefits to the user.

These benefits equate to a reduction in operator cost while minimizing unwanted effects to the environment, such as ground vibration.

- Minimizing ground vibration - Precision results in increased ability to control explosion timing, reducing vibration from poorly timed explosions. Less vibration allows for closer blasting to residential areas and reduces the potential for plant shutdown due to ground vibration.

- Energy cost - Greater timing precision results in closer-timed explosions, better fragmentation, and fewer oversized particles, leading to less energy consumption (and less time) required at the crushers.

- Flexibility - Any combination of delays can be preprogrammed or programmed in the field, providing flexibility on site and reducing the time to make changes because of field conditions.

- Simplicity for field loading - The SCB delay detonator package (with explosive and timer) is ready to be connected to the fireset, eliminating surface delays.

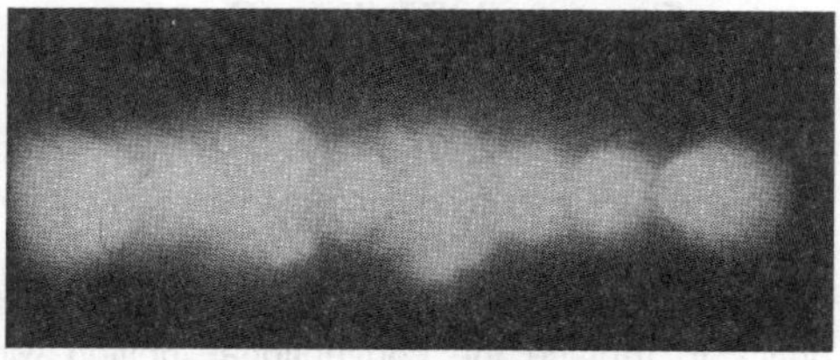

Still Frame of Test

- Initiation deviation ±32 microseconds
- Cable length distance from fireset to initiator: 590 to 990 ft
  - First unit at 590 ft
  - Subsequent units at average 50-ft spacing

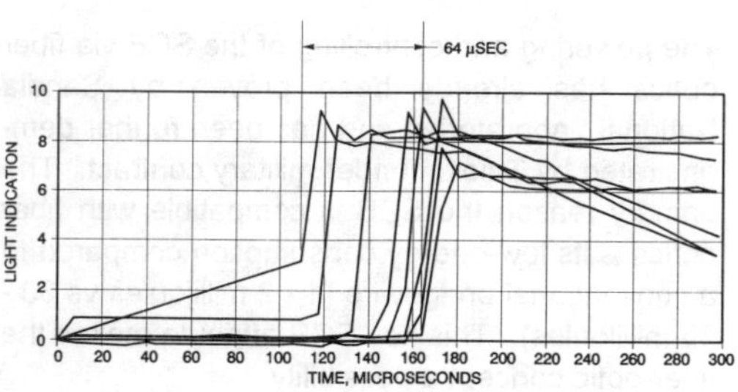

Test Results

Figure 5. Eight SCB Initiators Demonstrate Simultaneous Detonation Within Microseconds

9 DEMONSTRATION TESTING

Over the past year, Thiokol, in conjunction with SCB Technologies, Inc., has been involved in demonstration testing to systematically prove the SCB's precision and repeatability when coupled to a blasting cap. Testing has progressed to include the use of a trigger circuit and also an electronic timer.

In February 1992, simultaneous initiation was achieved using eight SCB initiators, with all peak outputs within a ±32-microsecond range (Figure 5). This demonstrated precision is three times better than the capability of a bridgewire. In additional laboratory testing, the SCB initiator has demonstrated an average function time of 76 microseconds. The initiator has also undergone environmental testing, such as vibration, thermal shock, humidity, and drop tests.

In May 1992, SCB initiators were successfully fired in their first field test (Figure 6). Three types of explosives (ANFO, emulsion, and a combined ANFO/emulsion) were successfully detonated using SCBs. Further field testing is ongoing.

10 SUMMARY

The SCB concept of an all-electronic device has now been taken through laboratory and initial field testing. Test results look very promising regarding the ability to better control fragmentation and unwanted vibration through the precision offered by an SCB electronic delay. These benefits translate to operational cost savings to the end user.

REFERENCES

Benson, D. A., Larsen, M. E., Renlund, A. M., Trott, W. M., and Bickes, R. W., Jr., "Semiconductor bridge (SCB): a plasma gen-

Figure 6. SCB Prototype and Primer Being Loaded for First Field Test

erator for the ignition of explosives," Journ. Appl. Phys. 62, 1622, September 1987.

Benson, D. A., and Rose, B. H., "An electro-optic explosive igniter," Sandia National Laboratories report SAND87-2597, December 1987.

Bickes, R. W., Jr., "Smart semiconductor bridge (SCB) igniter for explosives," Third Canadian symposium on mining automation, Montreal, Quebec, Canada, September 14-16, 1988. Available as Sandia National Laboratories report SAND88-1438C.

Bickes, R. W., Jr., and McCampbell, C. B., "Semiconductor bridge (SCB) research and development," Third high-tech seminar, blasting technology, instrumentation and explosives applications, San Diego, CA, June 2-7, 1991.

Bickes, R. W., Jr., Schlobohm, S. L., and Ewick, D. W., "Semiconductor bridge (SCB) igniter studies: I. Comparison of SCB and hot-wire pyrotechnic actuators," 13th international pyrotechnic seminar, Grand Junction Colorado, July 1988.

Chen, H. Y. and Jungling, K., Chapt. 2 in "Semiconductor bridge (SCB) development -- technology transfer symposium," R. W. Bickes, Jr., editor, Sandia National Laboratories report SAND86-2688, January 1987.

Cooper, J. W., and Atkeson P. L., "Feasibility demonstration of an electro-optical semiconductor bridge safe/arm/fire initiation system," 1990 JANNAF propulsion meeting, Anaheim, CA, October 2-4, 1990.

Headley, P. S. and Bickes, R. W., Jr., "Semiconductor bridge (SCB) primary explosive detonator," SAND86-2045, November 1985.

McCampbell, C. B., Bickes, R. W., Jr., and Schlobohm, S. L., "Very low energy ignition of pyrotechnics using a semiconductor bridge (SCB)," Sandia National Laboratories report SAND89-2259, November 1989.

Reil, J., "Controlling ground vibration and air blast," BAI blasting technology seminar, Orlando, FL, May 17, 1989.

Schwarz, A. C. and Bickes, R. W., Jr., "Feasibility study of a semiconductor bridge (SCB) for initiating pyrotechnics and explosives: reprint of internal memorandum RS2515/83/3," Sandia National Laboratories report SAND86-2617, January 1987.

Gregg, W. Barry Jr., "SCB technology for new electronic and fiber optic-based detonators and initiation system," fourth high-tech seminar on blasting technology, instrumentation and explosives applications, Nashville, TN, USA, June 20-25, 1992.

Innovative Mine Design for the 21st Century, Bawden & Archibald (eds) © 1993 Balkema, Rotterdam, ISBN 90 5410 325 6

Long round development

W.R. Dengler
Dynatec International Limited, Richmond Hill, Ont., Canada

ABSTRACT: The competitive nature of the mine contracting industry in Canada led Dynatec to focus on productivity improvements through the use of long rounds. This paper describes the development of long rounds in shaft sinking and lateral excavation. The benefits in productivity for both shaft sinking and lateral excavation are documented.

INTRODUCTION

In Canada, most of the rock being broken in underground metal mines is being carried out using traditional drilling and blasting methods. Despite advances being made in hard rock boring and other, more exotic techniques, it is Dynatec's belief that the traditional methods will continue to prevail for many years to come. It is also our belief that modest research efforts expended on the traditional methods will result in substantial financial benefits to the mining and rock tunnelling businesses.

Dynatec has pursued productivity improvements through the use of long rounds in both shaft sinking and lateral development. The concept of longer rounds was developed by Dynatec on the Craig Shaft Sinking Project (RETC 1991). This paper reviews the Craig Shaft Long Round success. This technology was subsequently utilized to sink the two shafts at Aur Resources with great success. The development of the Dynatec Long Round Jumbo for tunnelling was a natural progression of the long round technology developed for the Craig Shaft.

RECENT HISTORY OF SHAFT SINKING

Shaft sinking is one of the oldest and most difficult mining methods. Due to the relatively small amount of shaft sinking in proportion to tunnelling and mining, technological improvements have been slow to develop. Historically, European shaft sinking techniques have generally focused on the control of unstable or poor ground conditions. For South African shaft sinking technology, however, the focus has been on sinking extremely deep shafts at high speeds. South African shaft sinking methods have made use of very large labour forces, and technological developments have been restricted to shaft bottom cleaning or mucking. More recently, shaft jumbos have been successfully introduced to South African shaft sinking projects.

Approximately 40 years ago, the first mechanization of North American shaft sinking involved the development of mechanical excavating equipment. During this period, shafts sunk in North America tended to be rather shallow, which was not conducive to the high capital cost of mechanizing the drilling cycles. In addition, Canadian shaft sinking practices generally involved benching with hand-held sinker drills rather than full face excavating making mechanization of drilling even more difficult. These shafts were traditionally sunk with timber support, where the timber is kept reasonably close to the shaft bottom during sinking.

As the depths of North America shafts have increased, so has the use of concrete linings. Except in the case of ventilation shafts, most new shafts are equipped with steel buntons, timber guides for cages, and steel guides for skips. Usually, the use of shaft drilling jumbos has been restricted to drilling a full face bottom with a V cut. The use of a V cut, however, limits the length of round possible in the shaft. For example, normal practice with a shaft drilling jumbo has been to blast full face rounds that are 2.5 to 3 metres (8-10 feet) in length. An additional disadvantage is that the sinking stage must be moved at least 60 metres (200 feet) from the shaft bottom before blasting. Despite these problems, shaft jumbos do contribute to an improved working environment especially in the areas of noise reduction and safety.

DYNATEC'S CRAIG SHAFT SINKING DESIGN

Dynatec's experience with the Ansil Vent Shaft (Dengler and Shaver, SME 1990) indicated that North American shaft sinking could be conducted at high rates of advance over an extended time period. The Ansil vent shaft was sunk to a depth of 1,360 metres (4,460 feet) in 361 calendar days, for an average advance of 3.7 metres (12.1 feet) per calendar day, or 4.1 metres (13.3 feet) per sinking day.

In analyzing the performance data, the criteria for high speed shaft sinking can be classified in terms of three distinct functions:

* mucking
* drilling and blasting
* lining and equipping

Modification of all three cycles was required to allow the Craig Shaft to achieve the same rate of advance as the Ansil Shaft.

While the Ansil vent shaft was only a 4.0 metre (13 foot) diameter, concrete-lined shaft, the Craig Shaft's diameter was more than double that size on a volume basis (6.3 metres, 20.5 feet). The Ansil vent shaft was mucked using a Cryderman Shaft Mucker with a capacity of approximately 50 tonnes (55 T) per hour. To achieve a similar performance for the Craig shaft, a shaft mucking capability exceeding 100 tonnes (110 T) per hour was required. This prompted the use of the stage-mounted Cactus Grab.

For the Craig project, maintaining a drilling rate comparable to that experienced in the Ansil Vent shaft would require a 10-driller crew. However, the added labour costs associated with such a large drill crew would be excessive. Thus, to reduce the shaft bottom crew and the associated labour costs, the shaft sinking system was designed around a drill jumbo.

For the full face rounds, the V cut drilling pattern was considered undesirable since the muck would be thrown up the shaft during blasting which necessitates moving the sinking stage prior to each blast. Consequently, simultaneous shaft equipping is precluded. This desire to equip the shaft simultaneously with sinking led to choosing a burn cut rather than a V cut. Since the burn cut does not throw the muck, the need to move the stage was virtually eliminated. However, in drilling a burn cut with a series of small holes, it is absolutely necessary that these holes be parallel if the cut is to shatter properly and break the full drilled length. This problem was solved by using an ITH drill to create a large diameter drilled cut.

The owner's 4.6 metre (15 foot) hoist allowed the use of 12.5 tonne (14T) buckets that facilitated high production rates for the shaft mucking operations. In addition, the shaft's cage guide spacing permitted large diameter buckets that would remain stable during the shaft mucking operations. While pouring the shaft lining, high production rates were also possible using $4.6m^3$ ($162.4ft^3$) concrete buckets.

The stage was a five-deck, 50 ton (55T) working platform (Figure 3) suspended on three single-drum winches that were electrically synchronized for simultaneous operation. Equipped with 47mm (1 7/8 in) non-rotating rope, each winch had a rope pull of 34,000 kg (75,000 lb). One of the winches was located on the surface, and the other two were in the ventilation adit 15 m (50 ft) below the surface.

A four-boom, electric-hydraulic Continuous Mine Systems (CMS) jumbo was built into the stage. This jumbo design was a joint effort between Dynatec and CMS. The four-boom jumbo actually consisted of two units, each with two booms and an individual power pack. These two units were mounted together and suspended from electric winches located under the top deck.

Each pair of booms would be lowered into the drilling position at the bottom deck. The stage was positioned such that when the booms were lowered to the bottom deck, the jumbo was in the drilling position. Electric-over-hydraulic controls were located at the bottom of each slide to facilitate drilling control by an operator standing on the shaft bottom. The booms were originally equipped with 5 metre (16 ft) slides, but these were changed to 6 metre (20 ft) slides in November of 1990.

Emergency exit from the shaft was provided by a man-hoisting winch located in the hoistroom, and emergency power was available from a diesel generator. The emergency escape winch was capable of hoisting five men in a small cage suspended in the skip compartment.

The concrete forms were suspended on eight Dwidag hanging rods. Pouring was accomplished using concrete buckets with a concrete distribution system mounted on the stage. The concrete was introduced into the system from the concrete buckets at the top deck.

Steel installation was conducted from the top deck. To make a safe work platform, hinged doors were used to close the bucket holes in the stage's top deck.

PROBLEMS ENCOUNTERED AND RESOLVED

The primary problem with the Craig shaft sinking system was the crew's inexperience with electric hydraulic drill jumbos for shaft sinking operations. In addition, Canadian crews were generally not experienced in sinking full faced shafts.

To further complicate matters, the use of a small crew made shaft bottom cleaning difficult. Thus, the Dynavac, a vacuum system, was developed for the final bottom cleaning prior to drilling. While this vacuum cleaner was functional, it was quickly proved too slow to be effectively utilized. Despite the relatively small crew size, it was possible to hand clean the shaft bottom in a few hours by using blow pipes and mucking, with shovels, to the Cactus Grab.

Initially, due to the equipment congestion at the shaft bottom, the jumbo drilling and the ITH hammer drilling were not conducted simultaneously as had been planned. The problem was solved be removing two of the jumbo's booms and replacing the jumbo's hammers with Montabert model HC80. These replacement hammers were from the same Montabert line as the HC40 hammers used to start the shaft project. The increased hammer size allowed the drilling of a larger diameter blasthole, 44mm (1.75 inch) versus 38mm (1.50 inch). Therefore the original drill plan's 126 holes per round were reduced to 90.

The larger diameter blast holes, the increased efficiency of using two jumbo booms rather than four, and the simultaneous use of the jumbo and the in-the-hole hammer drill allowed a drilling cycle that met the original design. A problem arose with the initial ITH cut design that required aligning two 152mm (6 inch) diameter cut holes. When one 203mm (8 inch) diameter cut hole was used instead, the improved efficiency allowed the length of the full face bench to be extended from 3.5 metres (11.5 ft) to over 4.5 metres (15.7 ft).

One of the problems not considered prior to the project's startup was blast hole cleaning. There was always some fine muck left on the shaft bottom following the mucking cycle. In addition, the large volume of flushing water used by the jumbo drills washed this fine muck and the drill cuttings into completed drill holes. It became apparent that collar pipes were required to reduce the amount of blast hole cleaning time. After trying many types, a very thin plastic collar pipe was selected as the most efficient. These plastic collar pipes do not require that the holes be drilled a larger size at the collar.

Various explosive packages were also tested during the early shaft sinking efforts before settling on 1200mm (3.9 ft) X 40mm (1.6 in) power gel cartridges for all but the perimeter holes. The perimeter holes were loaded with 1200mm (3.9 ft) X 25mm (1 in) power gel cartridges. For priming, magnadets were used. This detonating system, while more expensive than traditional electric or non-electric blasting systems, is the safest, most reliable and most efficient system available today. It should be noted that a primer and cartridge must be used to blast water from the cut hole before the first cut helpers or blast holes are fired.

One problem in utilizing an electric hydraulic drill jumbo and transferring high voltage power to the sinking stage is the presence of high moisture levels from water leaking into the shaft. From the project's outset, this problem was anticipated, and plans were made to provide as dry a shaft as possible. To this end, Falconbridge cooperated with Dynatec in installing three water rings between surface and the first level. With these water rings and some grouting conducted during the initial sinking, the shaft was virtually dry. The importance of having a dry shaft in a highly mechanized and electrified system, such as the Craig sinking project, cannot be over emphasized. A dry shaft is also better for the mine's long term operation.

SHAFT SINKING SAFETY

Shaft sinking has traditionally been a high risk environment. Accident statistics demonstrate that a substantial improvement in safety related to shaft sinking projects has been achieved in Canada over the past 34 years. Improvements in shaft sinking equipment involving the use of winch supported stages, mechanized mucking and drilling equipment, has led to improved safety results. Training programs involving common core training, Workplace Hazardous Materials Information Systems training and first aid training contribute to an improved working environment. The use of a joint health and safety committee (workers and management), as well as weekly safety meetings on shaft sinking projects, have improved safety standards. Loss control audits improve the performance of the safety committees.

In addition to all of these programs, a (monetary) safety bonus was in place for the Craig Shaft and has been developed in conjunction with Aur for the Louvicourt project. There is no question that the safety bonus contributes to safety awareness on the job site. The safety bonus is a graduated bonus increasing for each month that the crew works without a lost time accident reaching a maximum after three months of experience with no lost time accidents. All bonus is lost on the occurrence of a single lost time accident.

PERFORMANCE

The target performance of 3.7 metres (12 feet) per day was met within eight months (Figure 1). Three months later, performance improved to 4.6m (15 ft) per day because of longer rounds. Delays in achieving optimum performance were generally associated with the normal gearing-up process associated with any shaft sinking operation. With the Craig project, the startup period was complicated because the system was based on a relatively new and untried electric hydraulic drill jumbo, and an in-the-hole hammer drill used for the cut. In addition, the project was started with a temporary single drum hoist and 7-tonne (7.7st) buckets. The subsequent change to the permanent double drum hoist and 12.5-tonne (14.05 st) buckets required another adjustment for the workers. The learning curve for each hoisting system is apparent from Figure 1.

Traditionally, the most difficult portion of the shaft sinking process has been the hand-held drilling or sinker drilling operations. The replacement of these procedures with a highly efficient, electric hydraulic drilling system was rewarding. For example, it made an extended drilling cycle a relatively easy process. In addition, the drill jumbo reduced the crew's risk of injury during the drilling cycle by eliminating the drill-steel failures and the constant heavy lifting associated with the hand-held sinker-drilling operations.

LATERAL DEVELOPMENT

The impressive performances achieved during the sinking of the Craig Shaft prompted Dynatec personnel to investigate the possibility of adapting the experience gained through shaft sinking to lateral development.

Primary development headings, which consist of ramps, drifts, cross cuts and raises, are difficult openings to excavate. Full face development of these openings is a demanding blasting problem as there is only one free face available. Thus, the first step in blasting a round in one of these faces is to

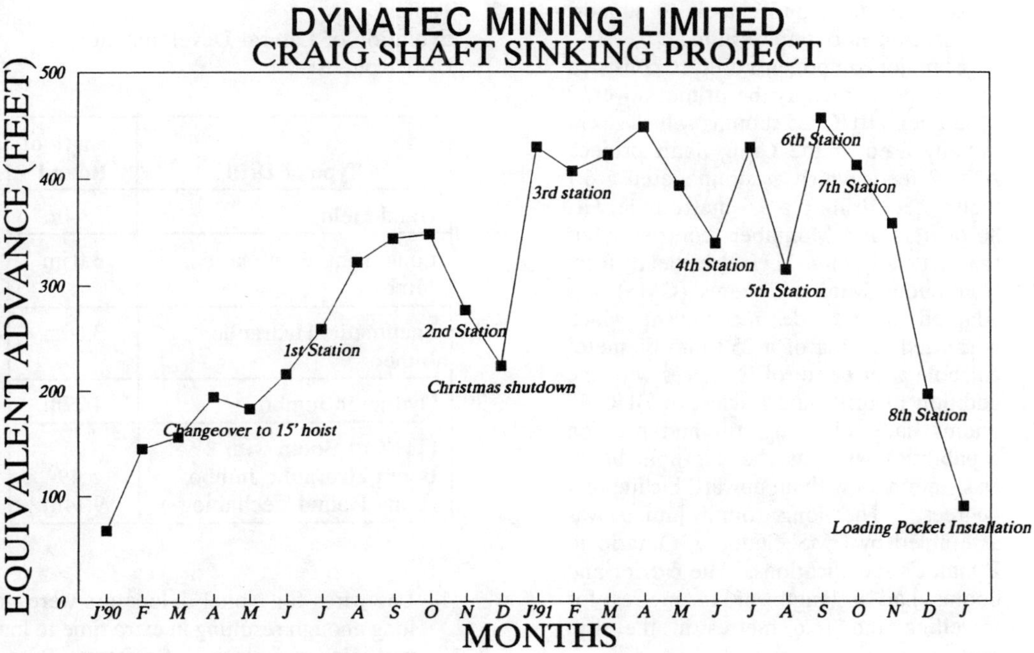

YEARS 1991 & 1992
FIGURE NO.1

create a second free face by producing an opening at least as deep as the designed depth of advance. The angled cut and burn or shatter cut are basically two types of blasting patterns or cuts that can be used to produce this opening. The burn cut is the most commonly used drill pattern and, depending upon the type of drilling equipment used, results in advance rates of 2.5 meters for jackleg drills to 4.5 metres for jumbo drills.

By experimentation, Dynatec has found that it is possible to drill a cut and blast a round that will allow the face to advance by as much as 10 metres with a single blast. Test work was carried out by Dynatec at two mining sites under full scale production conditions. The drilling equipment is standard, off-the-shelf items mounted on a specially designed jumbo carrier. The technology has been named Long Round Technology.

Prior to the development of Long Round Technology, the necessity to drill the cut holes parallel has been the major drawback in preventing the drilling and blasting of long rounds. Hand held drills, such as jacklegs, drill holes that have a small diameter and, in order to provide sufficient void space, must be drilled very close to one another. Thus, even a small error in mis-alignment would prevent the successful blasting of long rounds. When jacklegs are replaced by jumbo carriers and larger drills with booms which enhance parallel drilling of holes, the length of round that can be drilled and blasted increases. The success of Long Round Technology lies in the fact that the cluster of small diameter blast holes which normally constitute the cut portion of the drilling pattern has been replaced by a single large diameter hole. This hole can have a diameter of 20 to 25 cm and is drilled with an in-the-hole drill. The necessity for exact parallelism of cut holes decreases and the probability of breaking the full length of the round increases.

Generally, the expected depth per round, using the burn cut and various types of drilling equipment, has been tabulated below.

In 1991 it was decided to build a drift development jumbo (Figure 2) with the capacity of drilling the cut to a diameter of 250 mm. This required the use of an in-the-hole hammer and in turn a substantial boom and chassis to support this drill for horizontal drilling.

The major components utilized consisted of a Wagner truck as the prime mover, 2 Montabert BUC 35 booms which were initially used at the Craig Shaft project, ATC 8 feeds which accommodated a 6.1 metre steel drilling a 5.5 metre hole, HC 80 drifters and Montabert controls. For the cut hole, an ITH drivehead from Continuous Mining Systems (CMS) was adapted to a slide mechanism which permitted drilling of a 250 mm diameter cut hole to a depth of 9 metres with the addition of drill pipe. The two BUC 35 booms had full swing, tilt and rotation capabilities whereas the cut hole boom was stationary with an upward incline of 4 degrees. The long round jumbo was assembled by CMS, Sudbury, Ontario to Dynatec's specifications. The carrier and booms had a length of 14 metres for travelling and 16.6 metres in the drill ready position.

The construction of the long round jumbo was completed in September 1991 and tested in October and November 1991 at Mines Gaspe, property of Noranda Minerals Inc., Murdochville, Quebec. The test site was a number of drifts which formed part of the development required for the new CI orebody. The rock is metamorphic with a compressive strength of 32,000 psi.

The cross-sectional area of the drifts varied from 4 metres x 4.3 metres to 4.6 metres x 4.6 metres. With the latter dimensions being predominant. Ground support consisted of 2.4 metre long grouted rebar. The slope of the drift varied from 2% to 4%. There were no turns or curves.

During the testing period (October 7 - November 12), the Dynatec Long Round Jumbo was used to drill 20 rounds as tabulated in Table 2.

The diameter of the blast holes was 45 mm for the 18 foot long rounds and 57 mm for the 25 foot and 30 foot long rounds which were drilled with extension steel.

The cut hole had a diameter of 250 mm for 12 rounds and 200 mm for 8 rounds. Separate drilling patterns were developed for the 18 foot and 25 - 30 foot long rounds.

During the course of carrying out the above test, there were deficiencies noted in the following areas:

Table 1
Evolution of Lateral Development Drift Rounds:

Type of Drill	**Length of Round**
Hand Held	2.4m
Long Tom, Pneumatic Jumbo	3.05m
Pneumatic, Hydraulic Jumbo	3.66m
Hydraulic Jumbo	4.27m
ITH Cut Boom with 2 Boom Hydraulic Jumbo (Long Round Technology)	5.49-9.14m

1. Leg wires for Nonel detonators were not long enough resulting in extra time to load and improper placing of primers. Proper length leg wires will improve blasting cycle time and break.
2. The use of extension steel resulted in manual handling of rods. Future jumbos should have automatic rod change mechanism and storage carousel.
3. The jumbo carrier was quite long and difficult to manoeuvre.
4. Feed rail extension cylinders were too short requiring manual relocation of cylinder yokes to complete the feed rail extension from the travel position to the drill position.

Although the above deficiencies resulted in loss of efficiency during the tests, it will be quite easy to overcome the deficiencies in the future.

As a result of the tests, Dynatec has proven the viability of its Long Round Technology and now has the experience to design drift jumbos which have the capability of drilling and breaking rounds having lengths of 8 to 12 metres.

Time studies, taken during the above tests, show that by increasing the length of rounds, from 12 feet to 25 feet, overall savings in the development cycle time of over 20% can be achieved. Results of these studies are given in Table 3.

FIGURE 2
DYNATEC LONG ROUND JUMBO

Table 2
Long Round Jumbo Advance

Date	Length Drilled (Feet)	Broken Length Surveyed (Feet)	Date	Length Drilled (Feet)	Broken Length Surveyed (Feet)
Oct. 8	13	12	Oct. 25	16	17
Oct. 9	18	18	Oct. 26	17	13
Oct. 10	18	13	Oct. 30	18	16
Oct. 15	18	18	Oct. 31	25	24
Oct. 16	18	15	Oct. 31	18	18
Oct. 18	18	18	Nov. 2	18	17.5
Oct. 19	18	17	Nov. 4	18	17
Oct. 21	18	17	Nov. 7	30	27
Oct. 22	18	18	Nov. 11	18	17
Oct. 23	18	17	Nov. 12	18	17
TOTALS				**371**	**346.5**

Table 3
Development Cycle Time (Hours)

Length Of Round	12 Ft. (4 M.)	18 Ft. (6 M.)	25 Ft. (8 M.) (Estimated)	38 Ft. (12 M.) (Estimated)
Muck Removal	3.0	4.1	5.3	7.6
Ground Support	1.6	2.1	2.7	3.7
Drill and Blast	4.7	5.6	7.7	10.6
Total	9.3	11.8	15.7	21.9
Hours Per Foot	.78	.65	.63	.58
Productivity Improvement		20%	24%	34%

APPLICATION OF LONG ROUND TECHNOLOGY

In December of 1991 Aur Resources Inc. requested tenders for the preproduction development of the 30 million tonne copper zinc orebody near Val d'Or, Quebec. The development of the Louvicourt project includes the sinking of two shafts and the lateral development and raising required to bring the new mine to the production stage. Dynatec tendered the project on the basis of using Long Round Technology. In March of 1992 this project was awarded to Dynatec Mining Limited. The project will take approximately 30 months to complete and will have a completed value of approximately $45,000,000. When the mine is in full production, ore will be hoisted at a rate of approximately 900,000 tonnes per year.

SCOPE OF THE PROJECT

Exploration Shaft (E1)

The exploration shaft is a 5.5 metre diameter concrete shaft that will be sunk to a depth of 725 metres. The shaft will have stations at the 475, 565 and 655 levels. A loading pocket will be located below the 655 level. The shaft will be equipped with steel sets and timber guides. The shaft has two hoisting compartments and a manway and ventilation pipe compartment.

Production Shaft (P1)

The production shaft is a 7.0 metre diameter concrete shaft that will have an ultimate depth of 965 metres. The shaft will have stations at the 230, 475, 565, 655, 770 and 860 levels. A loading pocket and shaft bottom clean-out ramp will access the shaft below the 860 level.

This shaft will have 3 hoists servicing it. The production hoist will hoist muck in the two compartments. The service hoist will service the mine via a large cage which will be used to bring personnel and materials to each level. An auxiliary hoist will provide access to the various levels for personnel and small materials. This is the first shaft in northern Quebec that is not equipped with a manway. The auxiliary hoist provides the access normally provided by a manway.

Exploration Shaft Lateral Development (E2)

The exploration shaft sinking was started in June 1992 and the shaft will be completed in April 1993. The development work will start in June 1993 after the shaft sinking equipment is removed and the changeover to development skips has taken place.

The development from the exploration shaft will take place on the 475, 565 and 655 levels. This development will include production development drifts, ramps, ore passes, ventilation and fill raises. Development will be carried out with trackless equipment. Raising will be carried out using raise drills and alimak equipment.

The main part of this development will be completed in early 1994.

Production Shaft Lateral Development (P2)

The production shaft sinking was started in September 1992 and is scheduled to be completed in January 1994. The development work will start in March 1994 after the sinking equipment is removed and the changeover to the production hoist and skips is completed.

The development from the production shaft will take place on the 770 and 860 levels. This development will include production development drifts, ramps, ore passes, ventilation and fill raises. The 860 level development will include a crusher station, ore and waste bins as well as sumps for clean and dirty water.

Development will be completed in late 1994 and the mine will then go into full production.

EXPLORATION SHAFT SINKING

The exploration shaft plant included a 120" hoist equipped with a 1.5 inch non-rotating hoisting rope capable of hoisting a 7 ton payload. A 3000 cfm compressor plant is installed in the hoistroom for the contractor's use.

Dynatec equipped the headframe with conventional dump doors and safety doors. Muck is dumped on the ground.

The hoisting capability during sinking operations (70 tph) indicated that one Cryderman shaft mucker would be insufficient to keep up to the hoist. The Cactus Grab shaft mucking equipment is unnecessarily large (120 tph), so a decision was made to use two Cryderman Shaft Muckers. Each Cryderman is suspended from a winch which controls its up and down requirements. The Cryderman muckers are mounted on the forms.

Although the shaft is smaller than the Craig shaft, the decision was made to use a two-boom jumbo. An ITH drill is used for cut hole drilling (Figure 3).

The stage is a 5-deck 30-tonne working platform suspended on three single drum winches electrically synchronized for simultaneous operation. Winches are equipped with non-rotating rope and have a rope pull of 50,000 lbs. for each winch. The winches are located in buildings adjacent to the headframe.

The two-boom electric hydraulic Continuous Mine systems jumbo is built into the stage. The design of the jumbo was a joint effort between Dynatec and CMS. The two-boom unit with its own power packs is suspended from an electric winch located under the top deck of the stage. During the drilling cycle, the stage is positioned such that when the jumbo is lowered to the bottom deck, the jumbo is in a drilling position. Electric over hydraulic controls are located at the bottom of each slide to facilitate drilling control while standing on the shaft bottom. The booms are equipped with 6-metre slides and Montabert HC-80 drills.

The cut is drilled by an ITH drill that is lowered down the shaft during the drilling cycle and then returned to surface.

Concrete lining equipment involves steel forms suspended on 5 Dwidag hanging rods. The curb ring includes one metre of forms which is poured prior to lowering the balance (4 metres) of the concrete forms. The concrete buckets are equipped with pneumatically operated Garbo gates feeding a funnel and concrete placement hose to the shaft forms.

Steel installation is carried out from the top deck. Hinged doors are used to close the bucket holes in the top deck of the stage to make a safe work platform.

Power is transferred to the stage with a 3/C - 4/0 5 k.v. trailing cable. One hundred feet of trailing cable is stored on a reel on the second deck of the stage. Additional trailing cable is stored in a loop suspended in the shaft and lowered as required to supplement the reel storage on the stage. Permanent power is transferred to each station as the station is equipped and the trailing cable recoiled on the stage and looped in the shaft.

As of March 1, 1993, the shaft has been sunk to the 665 metre level. Shaft sinking has

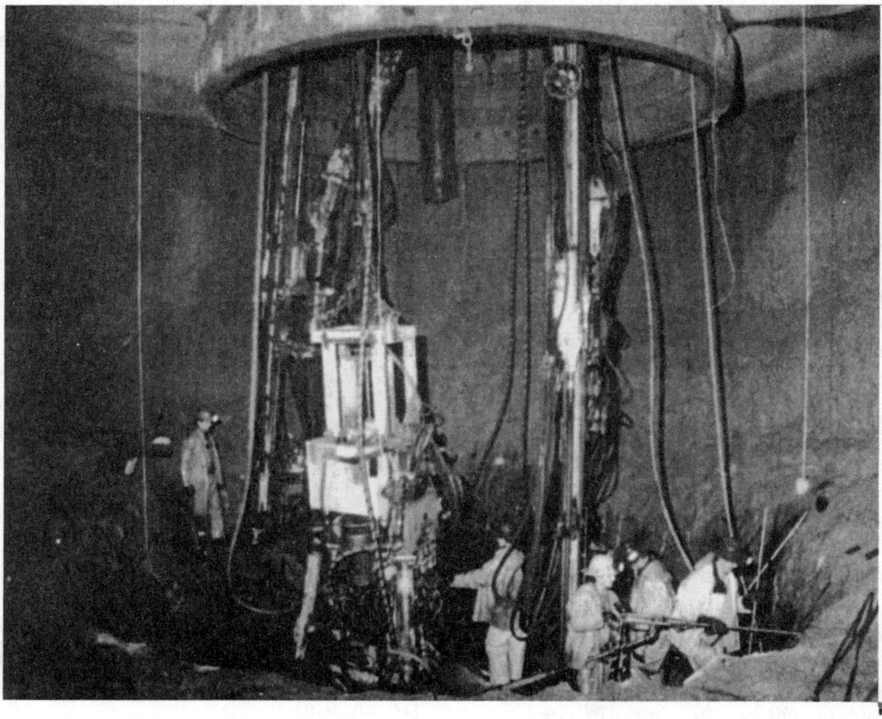

FIGURE 3
AUR VENTILATION SHAFT
2-BOOM DRILL JUMBO AND ITH DRILL

progressed at an average rate of about 4.4 metres per day and station excavation at a rate of in excess of 100 cubic metres per day. The best advance over one month was 4.7 metres per day.

PRODUCTION SHAFT SINKING

The sinking hoist is a 142 inch hoist with a 1.75 inch non-rotating hoist rope capable of hoisting 8.5 ton buckets. This hoist will be the service hoist in the final arrangement.

Dynatec equipped the headframe with conventional safety doors and dump doors. The buckets dump into a rock box and the muck is then conveyed laterally via a chute to the side of the headframe. This is to allow for the permanent bin installation during shaft sinking.

The limited hoisting capability (90 tph) again indicated that two Cryderman shaft muckers would be the best choice for sinking this shaft. The Cryderman shaft muckers are suspended from winches.

The size of the shaft and the estimated number of holes per bench indicated that a two-boom jumbo could drill off the shaft in the estimated time. However, a three-boom jumbo was installed to provide more capability.

The stage is a 5-deck 40-tonne working platform suspended on three single drum winches electrically synchronized for simultaneous operation. Winches are equipped with 47mm (1 7/8") non-rotating rope and have a rope pull of 34,000 Kg (75,000 lbs.) for each winch. A three-boom electric hydraulic Continuous Mine systems jumbo is built into the stage.
The three-boom jumbo with power packs is mounted together and suspended from an electric winch located under the top deck. During the drilling cycle, the stage is positioned such that when the booms are lowered to the bottom deck, the jumbo is in a drilling position. Electric over

hydraulic controls are located at the bottom of each slide to facilitate drilling control while standing on the shaft bottom. The booms are equipped with 6-metre slides and Montabert HC 80 drills.

The cut is drilled by a separate ITH drill that comes down the shaft after the start of blast hole drilling. The cut is drilled and the ITH drill returns to surface. Two 200 mm cut holes are presently being used.

As of March 1, 1993, the shaft has been sunk to the 475 metre level. The production shaft has not as yet achieved its scheduled average performance. In November, the target shaft advance of 4.0 metres per day was achieved. The average shaft advance is just over 3.5 metres per day.

GOOD THINGS HAPPENING AT AUR

The experience of the contractor was essential in getting both shafts started on time and allowing each shaft to remain on or ahead of schedule. The fast schedule was due primarily to the long round technology.

CONCLUSION

The new shaft sinking technology developed by Dynatec at Craig is being used with great success at Aur by Dynatec. Dynatec is achieving rates 50% faster than conventional sinking with crews that are 25% smaller. This has resulted in shafts that are less expensive to sink at rates faster than previously known.

REFERENCES

Dengler, W.R., Shaver, W.M., Letourneau, R. 1991. *New Technology in Shaft Sinking - The Craig Project.* RETC 1991 Seattle, Washington.

Dengler, W.R., Shaver, W.M. *High Speed Shaft Sinking With Conventional Sinking Methods.* SME 1990.

Innovative Mine Design for the 21st Century, Bawden & Archibald (eds) © 1993 Balkema, Rotterdam, ISBN 90 5410 325 6

Blast damage study by measurement and numerical modelling of blast damage and vibration in the area adjacent to blast hole

W.F. Bawden & P. Katsabanis
Department of Mining Engineering, Queen's University, Ont., Canada
R.L. Yang
Technical Centre, ICI Explosives Canada, McMasterville, Que., Canada

ABSTRACT: This paper reports the major achievements of a research project carried out at Queen's University to study blast induced damage in hard rock. New techniques and concepts have been developed in the areas of near field blast vibration monitoring, blast damage assessment and numerical modelling.

1. Introduction

In underground hard rock mining, drilling and blasting with conventional explosives remains the primary excavation technique. This is most likely to remain so for the foreseeable future, especially for bulk mining. Advances in equipment technology over the past decades have led to the introduction of large blast hole drilling equipment and bulk explosive loading in the underground environment. These new equipment technologies have resulted in the development of new, higher productivity bulk mining methods. These methods utilize, in many cases, larger explosive charges per delay than the older, lower productivity methods.

Blast design is a critical factor in determining stope stability in underground hard rock mining . Conventional blast design focuses on the ability to provide adequately broken rock. While less attention is given to the effect of blast induced damage. However, a blast design resulting in satisfactory fragmentation may induce excessive shock loading to the rockmass adjacent to the excavation. This may generate new fractures and/or movement and loosening along existing fractures, resulting in excessive support requirements and/or dilution of the ore grade.

This paper reports the major achievements of a research project carried out at Queen's University on blast induced damage. The objective of the study was to understand the mechanisms controlling blast damage, as an initial step towards developing practical guidelines to control blast damage in mining operations. The study involved are 1) monitoring and interpretation of stress wave propagation in a rock mass near an explosive charge; 2) assessment of blast induced damage in rock mass; 3) development of a damage criterion. In the following, the experiments conducted, the experimental results, the numerical model and the numerical modelling results are presented.

2 Experimental field layout

The experiment was conducted at the Queen's University blast test site, located some 65km north-west of Kingston. The rock mass at the site consists of a moderately jointed granite gneiss. The joint fabric was mapped using conventional geomechanical techniques. A borehole camera was used to log fractures exposed in the boreholes before and after the blast. One major joint set striking north-west to south-east and dipping about 85 degrees N-E was observed in the area before the blast. Joint apertures varied from tight to 5cm with some fillings at the ground surface. Joint spacing averaged 1.5m and the rock density is 2.55g/cc. Figure 1 shows the surface expression of the jointing relative to the borehole locations (Rocque. 1992).

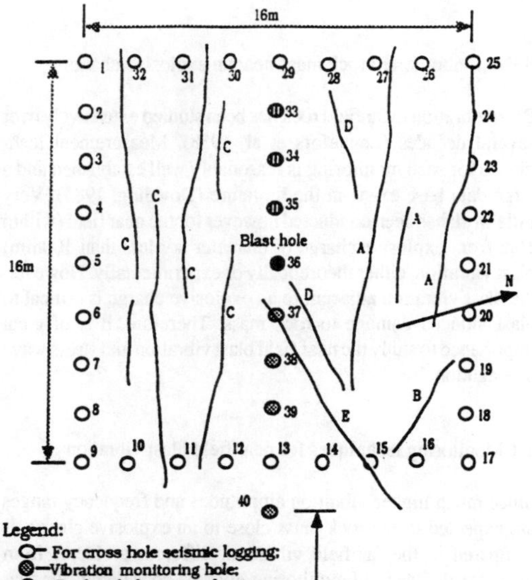

Figure 1 Experiment design and relative location of several joints to boreholes, Joint identifications (A, B, C, D and E refer to Table 1).

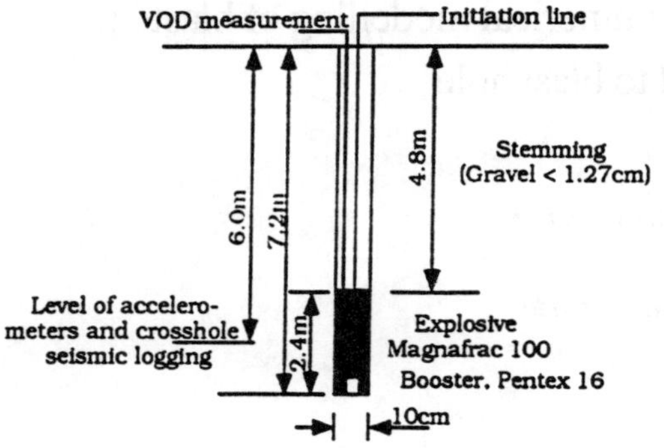

Figure 2. Configuration of blast hole loading and the levels of seismic logging and accelerometer relative to the explosive location.

A total of 41 holes were drilled for blast vibration monitoring, bore hole camera observation and seismic tomographic imaging (Figure 1). Hole #36 in the centre of the pattern was used as the blast hole. The blast was confined with the only free face at the ground surface. All holes were 4 inches in diameter. Hole #36 was drilled 7.2m deep, loaded 2.4m at the bottom with an emulsion explosive (Magnafrac-100) and stemmed 4.8m at the top to minimize fly rock. Holes #37-#41 were monitoring holes for blast vibration. The blast vibration monitoring holes were designed to align approximately with the strike direction of the major joint set (as seen in the Figure 1) in order to keep the rock mass condition similar for the different measurement holes. The cross section of the explosive hole and the depth of the recording instrumentation are given in Figure 2.

3 Blast vibration in rock mass near an explosive charge

Blast vibration in far field rock has been studied extensively over several decades (Langefors et al, 1958). Measurement techniques for such monitoring is reasonably well established and a large data base exists in the literature (Dowding, 1985). Very little work has been conducted however for the near field (within 10m from explosive charge of diameter not less than 100mm) blast vibration, either theoretically or experimentally. However, the blast vibration adjacent to an explosive charge is critical to blast induced damage to rock mass. Therefore, it is of great importance to study the near field blast vibration and stress wave propagation.

3.1 Monitoring techniques for near field blast vibration

Since much higher vibration amplitudes and frequency ranges are expected in the rock mass close to an explosive charge as compared to the far field vibration, different problems from those for the far field monitoring must be considered for near field blast vibration monitoring, including selection and installation of transducers, design of a monitoring system and signal recording (Yang et al 1992).

Sensor selection: Selection of sensors were based on an evaluation of ranges of the maximum frequency and amplitude of signal expected to be generated by the blast at the monitoring locations. The highest frequency was estimated to be 50kHz and maximum amplitude of acceleration to be 500,000m/s^2 (i.e. 50,000g). Such estimates are largely based on intuition since there was no data availabe in literature for the experiments of a similar scale. Accelerometers were chosen as monitoring sensors. Each accelerometer is connected to a charge amplifier, which converts the charge concentration into voltage. The signal is then recorded using a suitable data acquisition unit. Specification of charge amplifiers and connection cables were carefully selected to be suitable to the measurement of high frequency signals (Yang et al 1992).

Coupling sensor to rock: For measurements in an area very near the blast holes, it is believed that coupling between rock mass and cement becomes important since the wave length is small compared to the borehole radius and the amplitude of vibration is much higher. A series of experiments were conducted with different mixing materials. It was found that aluminum oxide powder mixed with Portland #30 cement (ratio of 2.6:1) achieved about 75% of the p-wave impedance of the granite gneiss at the test site. Otherwise, cement without additive materials has only about 48% of the p-wave impedance of the rock. In order to install the accelerometer in the monitoring holes, an aluminum cylinder was used for each accelerometer. Sensors were fixed inside the cylinders using screws and the cylinders were then filled with the selected cement mixture. The aluminum oxide powder mixed with Portland #30 cement was used to fill the aluminum cylinders and fix the accelerometers as well as to cement the entire assembly in the monitoring borehole.

3.2 Monitoring results and interpretation

Blast vibrations at different distances were recorded with a sampling interval of 25 microseconds, i.e, 40kHz sampling frequency. In the data processing, an anti-aliasing filter with low pass frequency of 10 kHz was used. A high pass filter to limit the 60Hz noise from power lines was also used.

Figures 3 a) and 3 b) show the particle acceleration records from hole at 2m and 4m from the blast hole, respectively. The wave forms recorded at other three monitoring holes are not displayed here due to the space limitation in this paper. There is a high frequency signal superimposed on the acceleration wave form at 2m from the blast hole. However, there is no such high frequency signal superimposed on acceleration wave forms at 4m, 6m and 15m from the blast hole. The magnitude of the acceleration at 2m from the blast hole is 88,408m/s^2, near to the estimate made during sensor selection.

From the acceleration records, the corresponding velocity history can be obtained by integration, since the phase shift of the selected accelerometers is insignificant at frequencies below the mounted resonance (>28kHz). Table 1 lists the peak values of particle acceleration and velocity at different monitoring holes.

3.2.1 Comparison with prediction from the charge weight scaling law

Figure 4 is the data curve fit and the resulting equation of the peak value of first velocity arrivals (PPV) as a function of distance

Table 1 peak values of particle acceleration and velocity at different monitoring holes

Distance (m)	2	4	6	15
Acceleration (m/s^2)	88408	22012	19318	3746
Velocity (m/s)	6207	801	1139	360

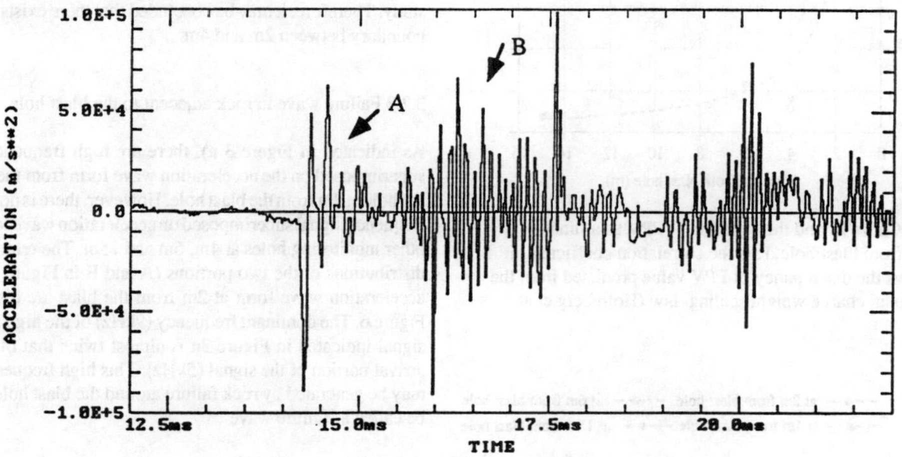

Figure 3a. A - early arrival of low frequency; B - high frequency signal

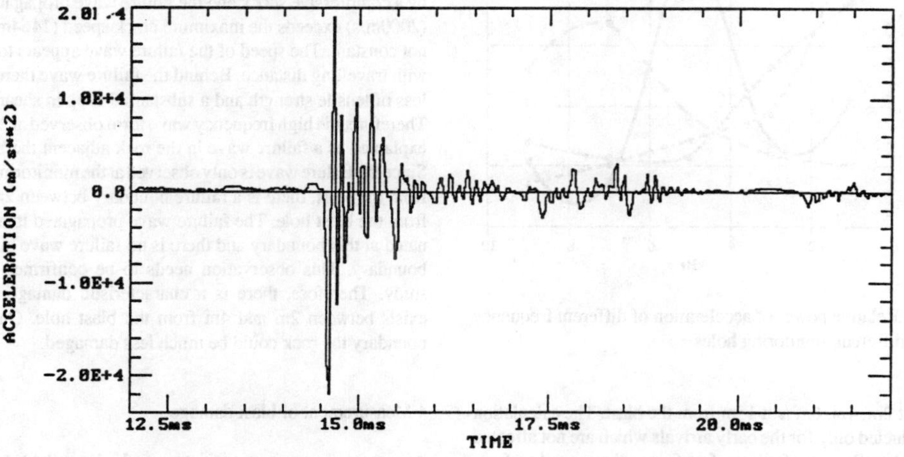

Figure 3b. Particle acceleration at 4m from blast

Figure 3. The particle acceleration records from holes at 2m and 4m.

from the blast hole. It also shows the comparison of the PPV between our measurements and predictions using the conven--tional charge weight scaling law (Holmberg et al, 1979). The value of the site parameter for the charge weight scaling law selected was typical of the value recommended for many far field vibration predictions. The equation $PPV=700\ Q^{0.7}/R^{1.5}$ was used and integration over the charge length was performed, (Q & R are the charge weight of an incremental length of explosive and the distance, respectively). It is evident (Figure 4) that vibration amplitudes predicted by the conventional charge weight scaling law are significantly lower than the measured values. This shows the importance of obtaining near field blast vibration measure-ments to establish valid prediction capabilities.

3.2.2 Power spectrum of acceleration signal

The power spectrum of acceleration signal was calculated for

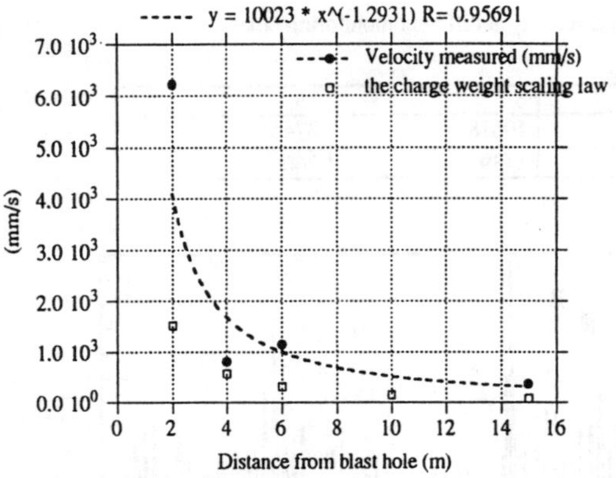

Figure 4 Curve fit and the equation of PPV as a function of distance from blast hole. R is the correlation coefficient. It also shows the discrepancy of PPV value predicted from the conventional charge weight scaling law (Holmberg et al, 1979).

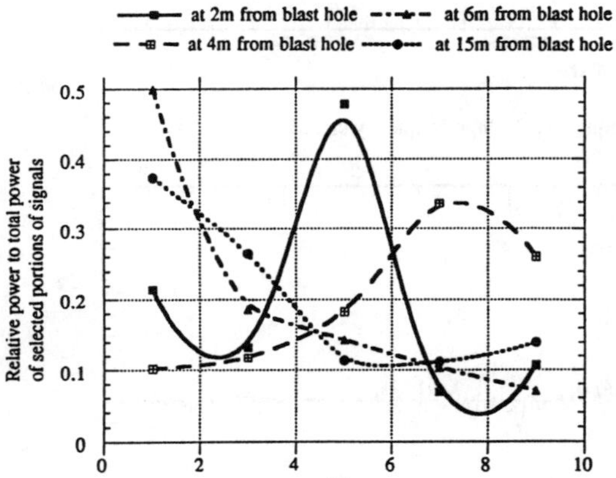

Figure 5. Relative power of acceleration of different frequency bands at different monitoring holes

records at 2m, 4m, 6m and 15m from the blast. The calculation was conducted only for the early arrivals which are not affected by possible reflections from the free face at the ground surface. A time window was used to select the early arrival portion of the signal. Figure 5 shows the power spectrums from the four monitoring holes, from which the dominant frequency at each monitoring hole can be determined. It can seen that the dominant frequency (7kHz) at 4m is higher than that (5kHz) at 2m. This implies that attenuation of frequency components of 5kHz is faster than that of the components of 7kHz from 2m to 4m. This is different from the attenuation of far field vibration which is predominantly dependent on frequency, i.e. higher frequency components in a signal will be attenuated more quickly than lower frequency components over a travelling distance (Blair, 1990). It is therefore clear that the attenuation of the vibration from 2m to 4m is not only dependent on frequency but also on amplitude. It may be concluded that frequency components of about 5kHz produced more damage between 2m and 4m from the blast due to higher amplitudes as indicated in Figure 5. By comparing the dominant frequency at 4m (7kHz) to that at 6m (1kHz) and at 15m (1kHz), it can be seen that the basic feature of attenuation from 4m to 6m or to 15m is similar to the far field vibration attenuation, i.e. the higher frequency components attenuate faster. This observation needs to be confirmed in future study. Therefore, it may be concluded that there exists a damage boundary between 2m and 4m.

3.2.3 Failure wave in rock adjacent to the blast hole

As indicated in Figure 3 a), there are high frequency signals superimposed on the acceleration wave form from the monitoring hole at 2m from the blast hole. However, there is no such high frequency signal superimposed on acceleration wave forms from other monitoring holes at 4m, 6m and 15m. The relative power distributions of the two portions (A and B in Figure 3a) of the acceleration wave form at 2m from the blast are displayed in Figure 6. The dominant frequency (9kHz) of the high frequency signal indicated in Figure 3a is almost twice that of the early arrival portion of the signal (5kHz). This high frequency signal may be generated by rock failure around the blast hole and may be called a 'failure wave'.

Bless et al (1992) observed failure waves in glass plates and rods struck by high-speed projectiles. They concluded that failure waves propagate into material that has been prestressed by a compressive shock and the failure wave propagation speed (2000m/s) exceeds the maximum crack speed (1464m/s) and is not constant. The speed of the failure wave appears to decrease with travelling distance. Behind the failure wave there is a total loss of tensile strength and a substantial drop in shear strength. Therefore, the high frequency wave form observed at 2m may be explained as a failure wave in the rock adjacent the blast hole. Since the failure wave is only observed at the monitoring hole 2m from the blast, there is a failure boundary between 2m and 4m from the blast hole. The failure wave propagated to and terminated at this boundary and there is no failure wave outside this boundary. This observation needs to be confirmed in future study. Therefore, there is a characteristic damage boundary exists between 2m and 4m from the blast hole. Outside this boundary the rock could be much less damaged.

4 Measurement of blast damage

At present, there are no consistent methods available to measure and quantify blast induced damage. It is difficult to measure the damage inside the rock mass with conventional measuring methods. For example, joint mapping is only applicable to measure joints on exposed rock surfaces. It is not easy to measure the extent of damage into a rock mass. Borehole camera surveys may be used to map visible joints intersecting with a borehole before and after a blast. Conventional extensometers are seldom used in the region close to an explosive charge because they may be easily damaged during blasting. Seismic logging seems to provide a method since information can be obtained about the internal structure of the rock mass. It has potential for very wide application in underground mining rock mechanics, but limited experience exists in using this technology to measure blast damage adjacent to blast holes as well as close to a free face. Therefore, it would appear useful to test different methods at the same time and to compare the results.

4.1 Cross hole seismic logging

The change in P-wave travel times across the blast damaged zone were measured from the cross hole seismic logging of the rock. P-waves generated by the detonation of Seismic Mark II caps were monitored using hydrophone sensors, with the caps and hydrophone sensors deployed in seismic monitoring holes (hole #18-25 and #26-#1) at depths of 6m (refer to Figure 2). Crosshole seismic logging was also performed with caps in hole #25 and hydrophone sensors in hole #9 at 1m intervals before and after the blast.

The change in travel times were measured in a plane at a depth of 6m, for raypaths travelling across the plane. Figure 7 shows the change in travel time (in milliseconds) plotted as a function of the relative position of the ray to the blast hole (Location is 0 for the blast hole). The error estimates are approximately 0.08 ms (5 sample points), relatively large in comparison to the observed changes. Nevertheless, the data shows an increase in travel time (or decreased velocity) near the blast, with substantial scatter in the data resulting from limited data accuracy.

4.2 Borehole camera

The Borehole camera recording before and after the blast from hole #35 (2m from the blast hole) shows an apparent increase in crack density after the blast, but recordings in other holes showed no obvious increase in crack density.

4.3 Damage assessment

According to the borehole camera observation and the failure wave observed at only 2m from the blast hole, a more severe damage region about 3m in radius exists around the blast hole. The analysis of dominant frequencies at the different monitoring holes also supports the above assessment. According to the crosshole seismic logging, there is a progressively damaged zone from the blast hole to about 8m in radius. However, outside the region of 3m radius the rock is apparently significantly less damaged.

5 Damage criterion

Peak particle velocity has been used for sometime in practice as a measure of blast damage to structures or the rock mass. In this criterion the shape of the wave form and duration of dynamic loading is not taken into account. In fact, the response of rock material to dynamic loading is strongly affected by the strain rate (Kipp and Grady, 1979). In dynamic loading, the stress can exceed the material strength, but it may not damage the material if the duration of this stress is too short. Blast damage is accumulated as a function of time and the applied stress. A single continuum mechanics field variable, (e.g. stress, strain, particle velocity, etc.), at any time can not be expected to characterize the

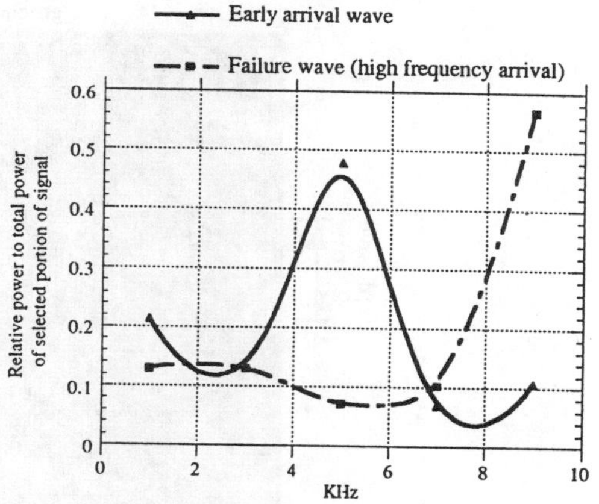

Figure 6. Comparison of the dominant frequency between the early arrival (A in Figure 3a) and the high frequency failure wave (B in Figure 3a)

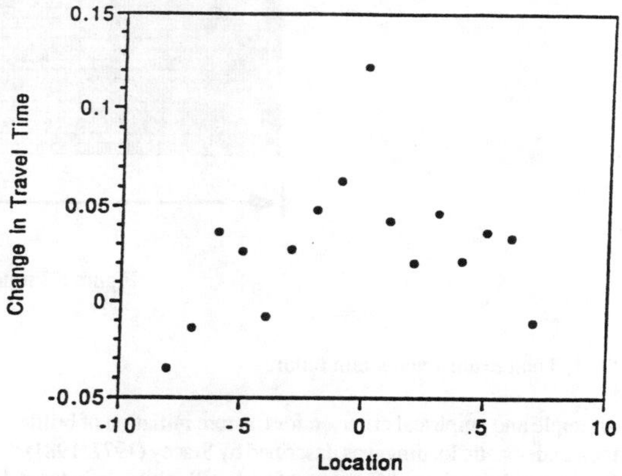

Figure 7. Change in travel time (ms) as a function of the relative position (m) to the blast hole (0m) after the blast

dynamic fracture process. A damage model based on extensional strain initiation, cumulative crack propagation and continuous damage of rock was developed to model the tensile failure of rock with strain rate dependence. The basic concepts and a typical modelling result are reported as follows.

5.1 A new constitutive model for blast damage

A New Constitutive Model for Blast Damage was developed in the present study, which consists of the following three concepts. It models initiation of cracks, crack propagation or generation and material softening. Cumulative damage and the effect of strain-rate are also modeled.

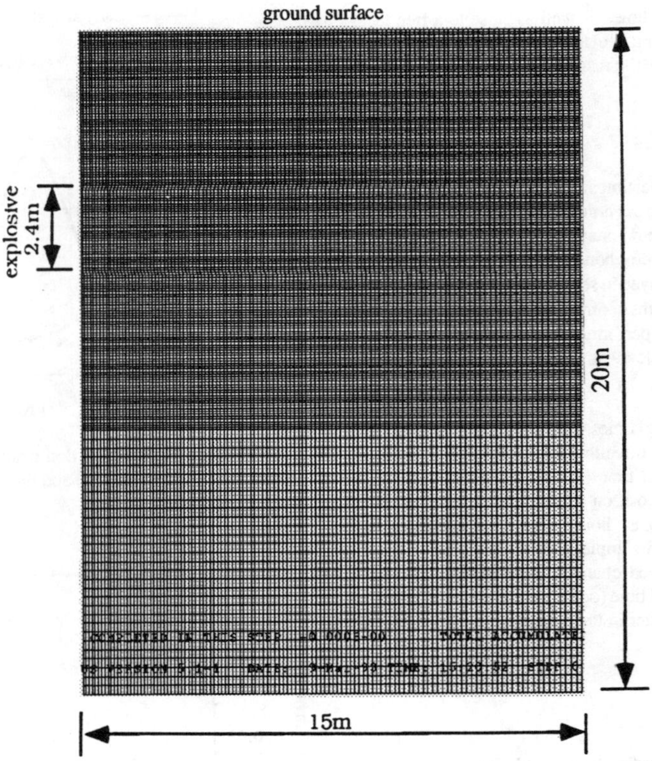

Figure 8 Finite element layout

5.1.1. Total extensional strain failure

A simple and empirical criterion for fracture initiation of brittle rock under static loading was described by Stacey (1977, 1981). It was stated simply as: " Fracture of rock will initiate in indirect tension when the tensile strain exceeds a limiting value which is dependent on the properties of the rock type." The terminology was changed from tensile strain to extensional strain by Stacey (1981) since the former term might imply the presence of a tensile stress. This is not a necessary condition for extensional strain to occur, since, even though all three principal stresses may be compressive extensional strain may occur. The criterion automatically takes account of the three-dimensional state of stress and strain and hence no confusion arises regarding the effect of the intermediate principal stress, since the extensional strain will be affected directly or indirectly by Poisson's effect. It was demonstrated by Waldeck (1979) and Bawden and Young (1991) that potential fracture zones predicted around certain underground excavations using an elastic stress analysis and an extensional strain criterion yielded better agreement with the observed fracture zones than those predicted using conventional elastic-plastic models.

In the present damage model, extensional strain is defined as the sum of all the principal tensile strains (logarithmic strain) at a point. In the damage model, it is assumed that tensile failure will occur, i.e. cracks will develop at a point if the extensional strain exceeds some critical value at that point in the rock mass. The material stiffness at the point will be reduced according to how much the material becomes cracked. On the other hand, if extensional strain does not exceed a critical value, elasto-plastic theory can be used to model strains and stresses in the rock mass. Volumetric stress was used by several investigators to classify tensile failure (Taylor et al, 1986, Kuszmaul 1987 and Thorne et al, 1990). It was assumed that tensile failure occurred only if the volumetric stress was tensile. Although the volumetric stress based criterion takes all principal stresses into consideration, it has a drawback because even if the vulumeteric stress is compressive, there may exist tensile stresses causing rock to fail in tension.

5.1.2. Fictitious crack

Within a rockmass flaws and cracks exist. Under tensile strain, these flaws or cracks will grow and new ones may be generated. These will reduce the stiffness and strength of the rockmass. A fictitious crack concept proposed herein assumes that a fictitious crack can be assumed to exist at each point within the rock, which has the same effect on the mechanical properties of rock as the actual flaws and cracks. The propagation and increase of actual flaws and cracks can be represented by the growth of the fictitious crack. The growth of the fictitious crack depends on the deformation of the rock at that point. The length of the fictitious crack is equivalent to the density of cracks in the rock.

Table 2 Physical parameters of rock used in the modelling

E (GPa)	μ	ρ (t/m^3)	σ_c (MPa)	α $X10^6$	β	θ_c (1/1000)
51.8	0.33	2.55	215	3.15	1.	0.2

E - Young's modulus; μ - possion's ratio; ρ - density of rock; σ_c - compressive yield stress of rock; α, β and θ_c are parameters defined in equation (1).

Table 3 Parameters used in the modelling of explosive detonation by the JWL equation

A GPa	B GPa	R1	R2	ω	E_0 X 10^6 J/kg	VOD m/s	ρ_0 g/mm^3
214.36	0.182	4.2	0.9	0.15	3.20	5500	1.31

5.1.3. Cumulative damage

Rock damage by strain waves from blasting is different from rock failure under static stresses. Rock damage from blast induced strain waves depends not only on the wave amplitude but also its duration since work done by the stress wave requires time to complete, i.e. the damage is strain rate dependent. Realistically blast damage should be considered as a cumulative effect of blast-induced stress waves. The fatigue and cumulative damage of metals under cyclic loading has been studied extensively in the metal fracture and strength studies. Some of the concepts in metal damage under the cyclic loading could then be used to model the tensile strain induced cracking in rock.

The following presents a proposed model for calculation of propagation of the fictitious crack. It is assumed that:

$$\frac{dl}{dt} = \alpha\,(\theta - \theta_c)^{\beta} \qquad (1)$$

Where l is the crack length; t is time; α and β are material constants; and θ and θ_c are the extensional strain and the critical extensional strain, respectively. Therefore, the crack length at a point is increased by

$$dl = \alpha\,(\theta - \theta_c)^{\beta} dt \qquad (2)$$

Formulas (1) has a similar form to that used in modelling of metal damage under the cyclic loading (Kedward and Beaumont, 1992). The fatigue and cumulative damage of metals under cyclic loading has been studied extensively in metal fracture and strength studies.

It is assumed that the Young's modulus and the shear modulus of the damaged rock are:

$$E_d = E\,(1 - D) \qquad (4)$$

$$G_d = G\,(1 - D) \qquad (5)$$

Where E and G are the Young's modulus and the Shear modulus of the undamaged rock, respectively. D is the damage parameter. It is assumed that

$$D = 1 - e^{-l^2} \qquad (6)$$

where, l is the crack length. The effect of strain rate on the damage is taken into account in (2). Cumulative damage is accounted for in this model.

When extensional strains are less than the critical extensional strain, elasto-plastic theory is used to model strain and stress in the rock mass. The Von Mises yield criterion with kinematic hardening is used for the plasticity model. This is used to confine the stress magnitude. Plastic strains are not used to characterize damage. A further development of the model to incorporate the Dracker-Prager yield criterion with strain rate dependent, kinematic hardening and non-associative models will be studied. Combined damage from the extension and the yield will also be studied.

5.1.4 Modelling results

One of the modelling results is described in this section to illustrate the model simulation. In order to model the experimental blast, axisymmetric finite element calculations were performed using the new constitutive model and the ABAQUS/Explicit finite element program. An axis of symmetry was assumed along the centre of the borehole. The 0.09m diameter borehole dictated that the column of elements representing the explosive charge have a radial dimension 0.045m. The radial dimension of elements increases with radius. The radius of the element boundary is set as 15m for a non reflection boundary. A vertical dimension of 0.08m was used from the top boundary of the element mesh to 4m below the explosive charge. Beyond 4m below the explosive charge, elements with a large vertical dimension (0.267m) were used to provide an 8m thick medium to eliminate possible reflections from the lower boundary of the element mesh to the region of interest. Total of 21600 elements were used including 30 elements for the explosive charge. The distance from the top of the explosive charge to the top boundary of the mesh is 4.8m. The length of the explosive charge is 2.4m. The layout of the finite element mesh is shown in Figure 8. Parameters for rock properties used in the modelling are listed in Table 2.

The explosive elements were modelled using the JWL equation of state in ABAQUS/Explicit with parameters shown in Table 3. Detonation of the explosive charge was assumed to start at the bottom of the charge. The volume of rock having a damage value of at least 0.22 is displayed in Figure 9. This value was

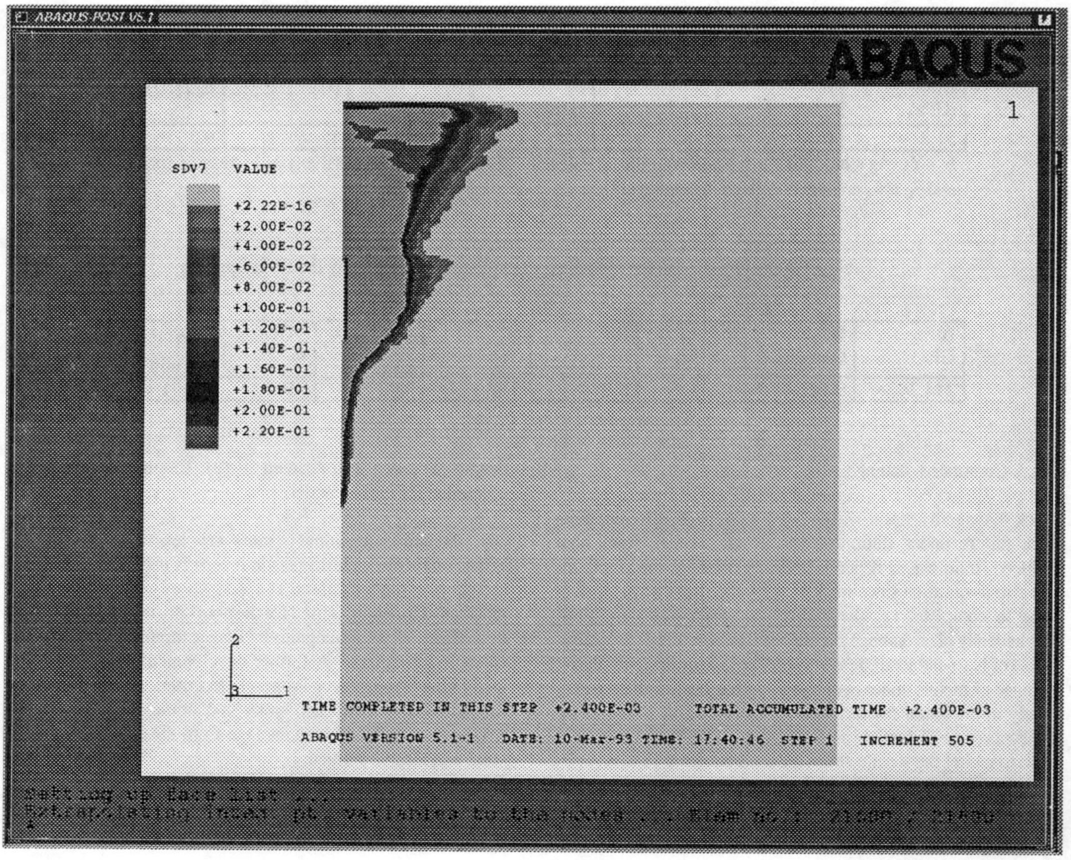

Figure 9 Modelling result of damaged zone

chosen because the 0.22 damage contour provides a reasonable match to the crater which was excavated in the field. As shown in Figure 9, outside the 0.22 damage contour are contours of less severe damage with microcracking up to 8-10 meters from the blast hole.

As shown in Figure 9, the damage radius at the level of the middle of the explosive charge is about 2.3m, which is very close to the estimation (3m) made in the section 4.3. The calculated damage radius at ground surface is about 3.8 meters, compared to the measured radii of the severe damage of 2.5m and the apparent crack opening of 5m from the blast hole (Rocque, 1992). Therefore, the model seems to give reasonable predictions. However more testing is required.

6. Conclusions

A monitoring technique for blast vibration in the area immediately adjacent to a blast hole was developed. Blast vibrations and attenuation within a rock mass near a blast hole (2m up to 15m from the blast hole) were monitored for a blast hole diameter of 100mm and a 2.40m column of emulsion explosive. Attenuation of different frequency bands of the signal during propagation from the 2m to the 4m monitoring locations was not only dependent on the frequency but also on its amplitude. A failure wave was postulated from observations at the monitoring hole 2m from the blast. These observations need to be confirmed in future study. A damage zone in the rock, analyzed according to the acceleration wave forms, has a radius about 3m. A combination of different techniques used to measure blast damage seems to provide a reasonable assessment of the damage zone.

A damage model based on the crack initiation by extensional strain, crack propagation by time cumulative effects of the extensional strain and continuous damage of rock seems to provide reasonable predictions. It may provide a useful tool to simulate effects of different design parameters on blast damage. Success of the research project has been achieved in three critical aspects of blast damage, which are: near field blast vibration monitoring; blast damage assessment; and numerical modelling with an advanced damage criterion. The research results will help form an improved basis for optimazing blast design in mining practice.

Acknowledgement: Research funding was provided by URIF under the research contract "Support of Underground Excavations in Hard Rock".

Reference

Bawden, W.F. and Young, R.P., Geomechanical analysis in mine design: An integrated approach using microseismicity, numerical modelling and structural geology, Rock mechanics as a multidisciplinary science, Roegiers (ed.), 1991 Balkema, Rotterdam

Blair, D.P., Some problems associated with standard charge weight vibration scaling laws, The third international symposium on rock fragmentation by blasting, Brisbane, Australia, 26-31 August, 1990

Bless, S.J., Brar, N.S., Kanel, G., Rosenberg, Z., Failure Wave in Glass, J. Am. Ceram. Soc.,V75, 1992, pp1002-1004.

Dowding C.H., Blast vibration monitoring and control, Prentice-Hall, Englewood Cliffs, NS, 1985.

Holmberg, R. and Persson,P.A., Design of tunnel perimeter blasthole patterns to prevent rock damage,IMM Proc.,Tunnelling 1979,March 12-16, 1979, London, pp280-283.

Kedward K.T. and Beaumont P.W.R., The treatment of fatique and damage accumulation in composite design, Int. J. Fatique 14 N0.5, pp283-294, 1992

Kipp, M.E. and Grady, D.E., The Micromechanics of Impact Fracture of Rock, Int. J. Rock Mech. Min. Sci. & Geomech. Abstr. Vol. 16, pp293-302, 1979

Langefors U. , Westerberg H. and Kihlstrom B., Ground vibrations in blasting, Water Power, Sept, 1958.

Rocque, P., Techniques utilized for the detection and characterization of blast induced damage, M.Sc. Eng. Thesis, Queen's University at Kingston, Ontario, Canada, 1992.

Stacey, T.R. & De Jongh C.L. Stress fracturing around a deep level bored tunnel. Jl S. Afr.Ins.Min.Metall 78, 124-133, 1977

Stacey, T.R., A simple extension strain criterion for fracture of brittle rock, Int. J. Rock Mech. Min. Sci. & Geomech.. Abstr. Vol. 18, pp469-474, 1981

Taylor, L.M., E. P. Chen and J. S. Kuszmaul, Microcrack Induced Damage Accumulation in Brittle Rock Under Dynamic Loading, Computer Methods in Applied Mechanics and Engineering 55, 1986, pp301-320.

Waldeck, H. G., the design and support of large underground chambers at depth in gold mines of the gold fields group of South Africa. Proc.4th Int. Congr. ISRM, Montreux, Vol.1, pp.565-571, 1979

Yang, R.L., Bawden, W.F., Talebi, S., Rocque, An integrated technique for vibration monitoring adjacent to a blast hole, submitted to The Canadian Mining and Metallurgical Bulletin, 1992

Innovative Mine Design for the 21st Century, Bawden & Archibald (eds) © 1993 Balkema, Rotterdam, ISBN 90 5410 325 6

Quantification of physical connections between blastholes by water loss measurements

P.A. Rustan
Luleå University of Technology, Division of Mining and Rock Excavation, Sweden

ABSTRACT: Blastholes located close to each other in development headings, and detonated on separate delays can disturb the detonation in neighbouring blastholes by different mechanisms. The joints, faults, cavities or crushed rock between the blastholes are probably a major reason to this disturbance. In full scale tests at LKAB in Kiruna, Sweden the water loss method was used to quantify the connections between blastholes in two tunnel rounds with parallel holes. The test result showed that the water loss method was useful for quantification of physical connections between the blastholes in hard rock. The probability of physical connection was almost 100% at hole distances less than 25 cm and less than 10% at distances greater than 1 m. The result is important for cut designs.

1 INTRODUCTION

The planning of blasthole pattern, and initiation in tunneling is mainly done according to experience. Very few detailed observations of the real detonation of each blasthole in tunneling rounds are known, because of the difficulty to do observations.

In Sweden, detailed studies of the detonation of the blastholes in parallel hole cuts have been made by the Swedish Detonic Research Foundation in open pit bench faces, and have been reported by Persson (1987) and Niklasson et al (1989). These investigations confirmed that many blastholes in a cut do not deliver its full energy, because the explosive may be pushed out of the hole by blast gases from earlier detonating charges. The explosive may also be dead-pressed by the stress wave from earlier detonating charges according to Wieland (1985), Demissý et al (1985), and Nie et al (1991).

The objective of this investigation was to examine if the water loss measurement technique could be used to quantify the physical connections between blastholes which can be caused by joints, faults, cavities, and crushed zones. Subgoals were to find;

* a suitable water loss equipment and its performance
* a measurement strategy
* how many boreholes are tight?
* how large is the leakage in the holes, and where is the leakage located along the boreholes?
* time for measurement?
* suggestions of improvements of the method

This information was needed for another project, where the pressure in the explosive was going to be measured by transducers in the blasthole close to the detonating hole.

The hypothesis was, that when the physical opening between the holes is large, there will be a negative influence on the neighbouring hole and if a joint is tight there will be no negative influence. In the future it is therefore interesting to find out at what water flow rate the negative consequences start.

2 LITERATURE REVIEW

The literature was reviewed about methods to investigate the physical connections in rock regarding to theories, and measurement technique.

Different medias, gas or water, were discussed to be used in the beginning. The test idea was to pressurize one hole, and observe how much gas or water flows into the surrounding holes. Finkel (1977) reported about some tests to use gas penetration for classifying damage in the remaining rock after controlled contour blasting. In some holes it was difficult to penetrate gas through the rock, so one conclusion was that the method was not good for classification of rock damage.

In this project it was therefore decided to use water loss measurements to quantify the connection between the blastholes.

According to Aven (1984) the rock mass permeability k_i (m/s) to water flow (hydraulic conductivity) is commonly based on the water flow which can be pressed into the rock mass within a known length of a borehole. These measurements are called "water loss measurements". The equipment used consists of one single packer or one double packer which are moved into the borehole, and can be tightend against the borehole wall with a mechanic or hydraulic operation outside the borehole, see Figure 1. The packers are connected with a hose to a water pump with measurement equipment for water pressure, and water flow.

When making rough measurements, or measurements in a tight rock mass, a single packer often gives sufficient accuracy. The water loss between the packer, and the borehole

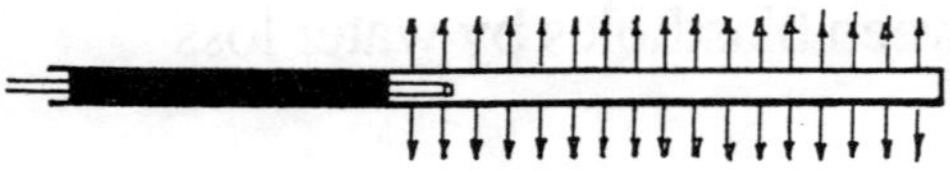

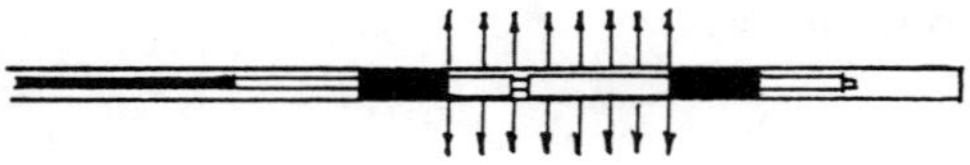

Figure 1. Principle for water loss measurement.

bottom is measured. When more precise measurements are needed of small water losses, or avoidance of large water losses is needed due to the fact that the capacity of the water pump is overrun, the double packer is used. In this case the water is pressed into the borehole between two packers which are connected with a perforated tube with a length of 2-5 m. The length can also be shorter. Measurements are started at the borehole bottom, whereafter the packer unit is moved certain steps, and new measurements are taken.

Measurements with a single packer, placed just below the rock surface, should start the measurement series, because in many cases one could immediately determine if the rock is tight. If the rock is not tight, the packer is moved to half the borehole depth, and new measurements are taken.

The water losses should be measured at different pressures for example 0.3, 0.5, and 0.7 MPa and at both increased, and decreased pressures. This could reveal if the joints or cracks are cleaned from fine particles.

The permeability of the rock mass can be calculated, if the water loss, the measurement section length, the ground water pressure, and the measurement time are known.

The permeability in each measurement section can be calculated according to Ave'n (1984) with the following formula;

$$k_i = \frac{(1 + \ln(l/d))\ V}{2\ l\ p\ t} \qquad (1)$$

where k_i is the permeability or hydraulic conductivity (m/sec), l is the pressurized length of the borehole (m), d is the borehole diameter (m), V is the water loss volume (m^3), p is the induced overpressure in m water gauge, and finally t is measurement time (sec).

The permeability in chrystalline rock, from surface and down to 150-200 m below the surface, varies normally between 10^{-5} to 10^{-8} m/sec. This corresponds to 2.4-0.0024 m/24 hour. Effective porosity in chrystaline rock is about 0.01 %, and the water flow depends fully on the joint intensity. The hydraulic conductivity for hard chrystaline rock without joints or induced fractures is 10^{-12} m/sec. Corresponding effective porosity for cambrium sandstone is about 1%, and for gravel sediments about 15%.

The measured water flow could also be represented by the water volume pressed in per meter of borehole length, time, and pressure unit. The unit for water loss is then expressed in m^3/m sec MPa.

Investigations by Pusch and Stanfors (1992) of the hydraulic conductivity around a test drift, in the nuclear waste deposite laboratory in Stripa, south Sweden, shows the following values of the hydralic conductivity in the disturbed zone from blasting. The axial conductivity (along the drift) was measured to 10^{-8} m/sec, and this should be compared to the virgin rock mass conductivity of 10^{-11}-10^{-10} m/sec.

3 FIELD TEST AREA, EQUIPMENT, TEST- AND EVALUATION PROCEDURE

3.1 Field test area

The field test area was located at the southern end of the Kirunavaara orebody in Kiruna in connection to a smaller orebody, called Sofia, where sublevel stoping was under development between the levels 598, and 650 m. This orebody was used for a big development project where the goal was to develop technique to blast 7.5 m deep rounds with 64 mm diameter holes, and to reduce damage to the remaining rock by testing different blasting technique, including electronic detonators, and water jet cutted slots along the rim of the perimeter blastholes.

Two development rounds were investigated in this project, both located in waste rock, one in the loading crosscut No. 56 on the 650 m level, and the other round in the gallery head on the 598 m level. The depth is counted from the top of the Kirunavaara mountain, and these numbers have to be subtracted by approximatively 260 m to get the depth below surface. The rock consisted of granitic porphyry, and breccia on the 650 m level, and breccia and syenite porphyry on the 598 m level bouth rounds representing hard rock. The variation of the rock properties are shown in Table 1.

The joint pattern consisted of three joint sets representing normal conditions for the Kiruna mine.

3.2 Equipment

The objective was to quantify how high pressure blast gases flows through the joints. As mentioned earlier, the selected water loss measurement method can not reveal fully how high pressure gas will flow, but it will give a quantification of the major possible pathways for the gases. By radial cracking, and shearing of the rock in the real blast situation the number of pathways will be much greater in the actual blast situation.

The water pressure loss equipment, hired from Stabilator AB in Skellefteå, is shown in Figure 2.

The equipment includes the following;
1 injection pump with pressure variation range 0-10 MPa (0-100 bar), weight 380 kg, and voltage 380 V, and a pressure meter, 0-15 MPa, after the pump. 1 handpump used to increase the packer pressure between 0-6 MPa. During the test it was found that 2-3 MPa was enough packer pressure. 1 single packer, and one double packer were used, each with 20 m pressure hose. Finally one water flow meter was used for measurement of water flow. The smallest flow, which could be measured within

Table 1. Strenght of rock

Rock type	Uniaxial compressive strength (MPa)		Tensile strength (MPa)	
Breccia	130 - 280	m.v. 190	8 - 17	m.v. 14
Syenite porphyry	180 - 320	m.v. 222	11 - 21	m.v. 15
Granitic porphyry	200 - 370	m.v. 250	11 - 20	m.v. 16

a resonable time was 1 l/min. The length of the rubber packer, when it was pressurized, depends on the diameter of the borehole. The larger the diameter of the borehole the shorter is the packer length.

3.3 Test procedure

Two development rounds were tested, one crosscut round (width 6 m and height 4.5 m) including 53 blastholes 48 mm in diameter, and one uncharged opening hole 100 mm in diameter, and one gallery head round (width 6 m and height 5.3 m including) 43 blastholes 64 mm in diameter, and one uncharged opening hole 100 mm in diameter. Approximately 20 to 30 holes were pressurized in each round, however not including the bottom and top holes of the rounds, see Figure 3 and 6.

All holes were tested, first with the single packer, as close as possible to the surface, to check if the borehole is tight or not. If the hole is not tight, the pressure and flow of water is noted. After that, the single packer is placed as deep as possible in the borehole to measure if there is any leakage at the bottom of the hole because this part can not be pressurized by the double packer.

All holes which showed a leakage was later on investigated in detail with the double packer to find out where along the borehole the leakage was located. With the double packer it was therefore possible to see if the connection between boreholes was increased in the vicinity of the blast front where a higher damage from the bottom charges could be expected.

When pressurizing one hole, observation of waterflow were made in all other holes. This means that a large amount of observation data are available from only two tested rounds.

3.4 Evaluation procedure

In the evaluation of connections between boreholes, all pressurized holes without leakage is included in the statistical calculation of probabilities of connection by including all distances to the other holes in the round as a measurement value. When a borehole has a leakage to other holes, only the distances to those holes have been included in the statistics, because there may be connections to other holes but because of the leakage the water pressure will not be large enough to indicate this.

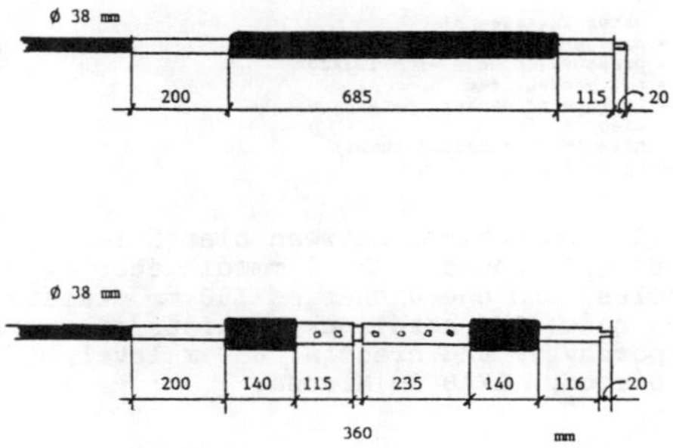

Figure 2. Water loss measurement equipment used in the field tests at LKAB in Kiruna.

The distances between the holes have been measured on the surface, and is the centre to centre distance of holes.

4 TEST RESULT

4.1 Experiences from water loss measurement equipment

First some practical experiences of the handling of the equipment. As shown earlier the pump equipment was heavy, and therefore it was necessary to use a lorry for its transportation. The pressure in the packers had to be increased by a hand pump, and this was time consuming. The increase of pressure in the borehole was done manually by turning a valve, and this was also time consuming. The effective time for water loss measurement of one round was 8 hours for two men.

The packer could not be used very close to the front because of the hydraulic breaking power of the rock caused by the high water pressure. This might have caused rock fall on the operator or the equipment. The packer should therefore normally always be located at larger depth than 0.5 m when pressurized.

The holes have to be cleaned very carefully before entering the packer, because if a small rock piece enters the hole it can be difficult to remove the packer from the hole.

The water pressure used was normally 2-3 MPa (20-30 bar), but in some cases with an almost open connection between the holes this pressure could not be reached, because of the large water flow, which sometimes was greater than the capacity of the pump.

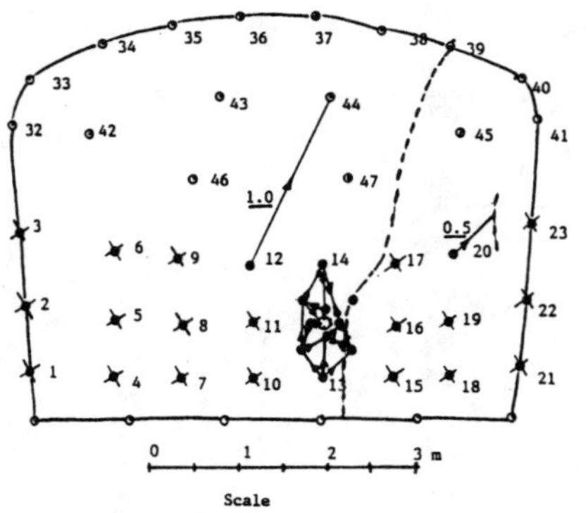

Figure 3. Connections between blastholes in a parallel hole round with 48 mm diameter blastholes, and one uncharged 100 mm diameter opening hole. The hard rock consists of granitic porphyry, and breccia. 650 m level, Sofia orebody, LKAB in Kiruna.

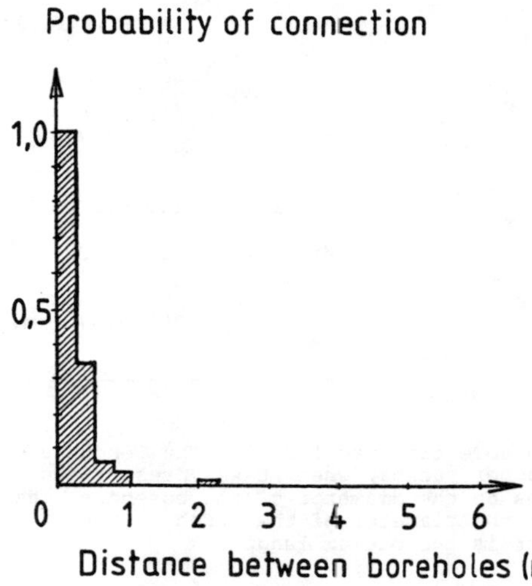

Figure 5. Probability of connection between blastholes. Blasthole diameter 48 mm, and one uncharged 100 mm in diameter opening hole. 650 m level, Sofia orebody, LKAB in Kiruna.

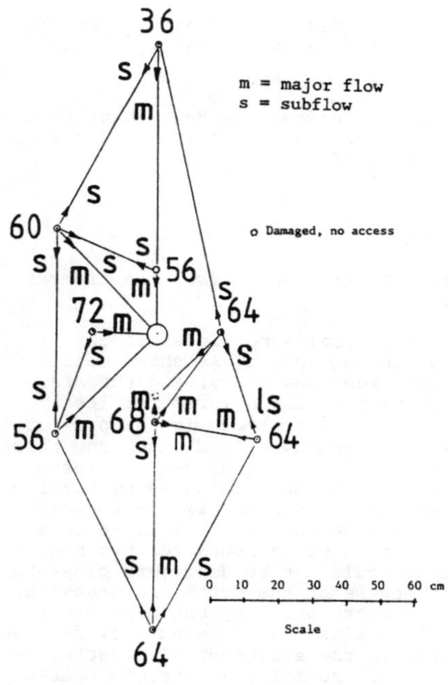

Figure 4. Water losses, l/min in the cut, and their schematic pathways. The hard rock consists of granitic porphyry and breccia. 650 m level, Sofia orebody, LKAB in Kiruna.

4.2 Real borehole positions and borehole deviations in the cut

The borehole alignment measurement of the cut holes showed that the cut holes were not crossing each other.

4.3 48 mm diameter blastholes

The result from conductivity test in the crosscut is shown in Figure 3.
From 30 tested holes, 19 were tight. The holes showing connections were mainly located in the cut, except two holes, one with a leakage of 1 l/min at 5 Mpa pressure to a borehole 2 m away, and another hole with a leakage of 0.5 l/min to a joint 0.75 m away. The details of leakage in the cut is shown in Figure 4.

Each hole leaks to one, two or three neighbouring holes in the cut, and the leakage varies between 56-72 l/min. It is obvious, that if the blastholes are set under different delays, some holes will achieve sympathetic detonations, and the risk of cut offs, and removement of explosive from not yet detonated holes will be large.

The probability of connection between two blastholes at different distances can be calculated. The distances have been grouped into the following classes, 0-25, >25-50, >50-75, >75-100 cm, and so on. The result from the probability calculation is shown in Figure 5.

From Figure 5 it is clear that the probabilty is 1, for connection of blastholes with distances between 0-25 cm and the probability is low at distances larger than 1 m.

The location of leakage along the borehole was done by the double packer, and the result is shown in Table 2.

Table 2. Result from water loss measurements in 48 mm diameter blastholes in a tunnel round with parallel holes. 650 m level, Sofia orebody, LKAB in Kiruna.

Blasthole number	The location of the connection along the blasthole										
	Double packer (l/min)										Single packer (l/min)
	0	0.5	1.0	1.5	2.0	2.5	3.0	3.5	4.0	4.5 m	1.0-4.5 m
12										1.0	1
13						60					64
24								60			56
25						60					60
26								64			64
28									68		72
29							68				68
30											56
31	The hole was damaged, and can not be used										

In Table 2 it is shown that the connections were located at depths between 2.5 and 4.5 m and within the 36 cm distance of the packers.

4.4 64 mm diameter boreholes

In the gallery head round 19, holes were pressurized. 9 of them showed a leakage and 7 of the holes were located in the cut and the rest outside the cut, see Figure 6.

Outside the cut, one hole had a flow of 32 l/min at 0.85 m distance between the holes, and another hole had a flow of 36 l/min at 1.1 m distance between the holes. One hole in the cut had a connection to a bottom hole with a flow of 8 l/min at a distance of 2 m. All holes in the cut had connections, see Figure 7.

The water flow between the cut holes was sometimes devided into several flows, and the quantity varied from 0.4 to 72 l/min. The mean water flow in the cut with 64 mm diameter holes was generally lower, 18 l/min compared to the 48 mm blastholes where the mean flow was 60 l/min. The reason to the difference may be the increase of hole distance when the borehole diameter is increased, but it may also be due to the different rock structure in the two compared rounds. In Figure 8 the probability of connection between blastholes is shown.

According to Figure 8, and compared to Figure 5, the probability for the 64 mm diameter blastholes is almost the same as for the 48 mm blastholes. The probability of connection is 1 for distances up to 25 cm, and it is low for distances larger than 1 m.

The results from the single and double packer measurements are shown in Table 3. Observe that the values to quantify the different flows in one borehole can not be added to calculate the total flow, because some of the connections may have some extension along the hole.

In the development round with 64 mm blastholes, the connections are more evenly distributed along the borehole compared to

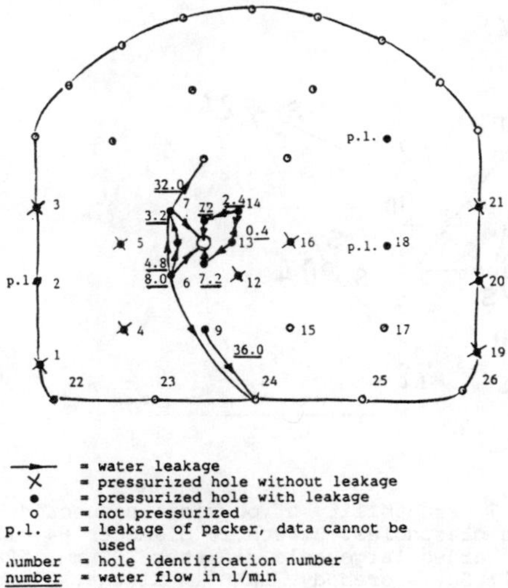

Figure 6. Connections between blastholes in a parallel hole tunnel round with 64 mm diameter blastholes, and a uncharged 100 mm diameter opening hole. The hard rock consists of syenite porphyry. 598 m level, Sofia orebody, LKAB in Kiruna.

Table 3. Result from water loss measurements in 64 mm diameter blastholes in a tunnel round with parallel holes. 598 m level, Sofia orebody, LKAB in Kiruna.

Borehole number	The location of the connection along the blasthole Double packer 0 0.5 1.0 1.5 2.0 2.5 3.0 3.5 4.0 4.5 m (l/min)	Single packer 0.9-4.5 m (l/min)
6	4.8,3.6.0,5.2,0, 3.6,0,10, 12, 6, 8	8.0
7	4.4, 0, 0, 6,4.8, 0.5,6,30,32, 0,1.6	32.0
8	0, 0,3.2,1.6,1.6,4.0,6,0, 0, 0	4.8
9	0, 0, 0, 0,4.8, 0,3.6, 50,0 -	36.0
10	0, 0, 0, 3.6, 2.4,8,40, 0, 0, 0	7.2
11	0,1.0,1.0,0 2.8,76,0, 0, 0, 0	72.0
14	2.0,0, 2.8,3.6, 0, 0, 0, 0, 0, 0	2.4

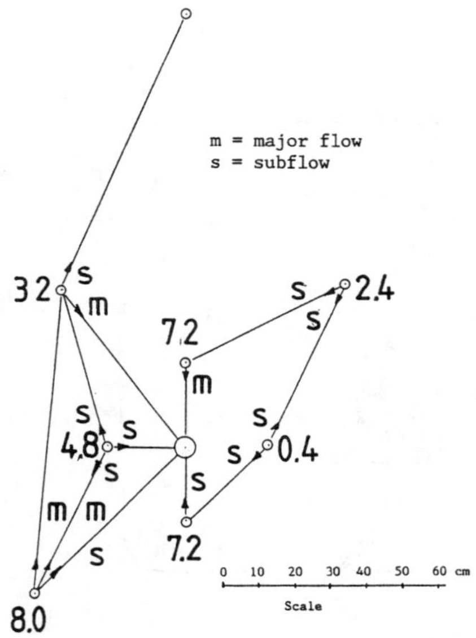

Figure 7. Probability of physical connections between blastholes. Blasthole diameter 64 mm, and uncharged large hole diameter 100 mm. 598 m level, Sofia orebody, LKAB in Kiruna.

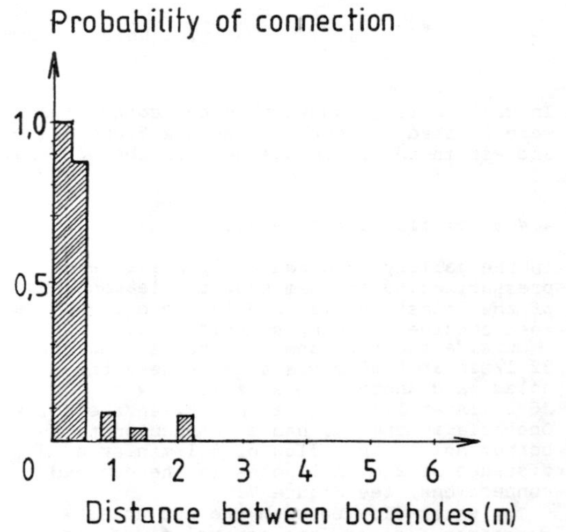

Figure 8. Probability of connection between blastholes. Blasthole diameter 64 mm, and one uncharged 100 mm in diameter opening hole. 598 m level, Sofia orebody, LKAB in Kiruna.

the developent round with 48 mm diameter blastholes. It is important to point out that the measurement method used does not allow readings closer than 35 cm to the blast front, because of direct leakage of water to the front.

4.5 Conductivity

The conductivity calculated according to formula (1) varied from 0.78 10^{-7} to 2.9 m/s. The lower values are in accordance to the values found by Pusch and Stanfors (1992) for the disturbed zone around a blasted tunnel.

5 CONCLUSIONS

Water loss measurements was shown to be a quantitative technique to classify the physical connections between boreholes under static conditions (2-3 MPa) pressure.

The cause for physical connections between boreholes can be joints, faults, cavities, and crushed volumes of rock. For closed spaced holes, < 25 cm it was judged that some of the connections can be caused by the stresses formed during the mechanical drilling, and flushing of the boreholes.

The leakage along the borehole can be localized with the double packer but not closer than 35 cm to the surface.

The mean flow for all measured flows in the cut was 60 l/min for the 48 mm diameter holes, and 18 l/min for the 64 mm diameter holes. The lower flow at larger borehole diameters may be due to the larger hole spacings at larger hole diameters.

Only 10% of the boreholes outside the two first squares in the cut (8 blastholes) showed a leakage.

Because of the 100% probability of connection between blastholes in hard rock at distances less than 25 cm, the conclusion is that it is neccessary to design the cut, so small distances between blastholes in the cut are avoided. The probability for borehole connections at distances larger than 1 m was namely very low for the two rounds tested.

The logical action after the new findings is to use the spiral cut design. This cut is however not used at tunneling work in Sweden. In the literature it was shown by Gustafsson (1979), that the spiral cut can be used for blasting smaller pole holes in rock. The use of the spiral cut in tunneling has also been mentioned by Hagan (1980).

6 FUTURE WORK

In a thesis by Koorem (1992), initiated by the result of this research, it was shown by testing three identical spiral cuts in a 30 m^2 tunnel round, 4.5 m deep and with 45 mm parallel holes, that the advance was very high, 102% after scaling. It is therefore now of interest to study the spiral cut on a production basis in Sweden.

ACKNOWLEDGEMENT

This project was financed by the 50 MSEK Governmental Swedish Development Program for Mines "Mining Technique 2000", running from 1987 to 1992. The following companies, and personel were involved in the project reported here. From LKAB, Mats Keisu acted as coordinator to LKAB, and Bengt Niklasson from the Swedish Detonic Research Foundation was coordinating the project to the Sofia tunneling project. Mats Hansson, Stabilator AB instructed us how to use the water loss equipment, which was hired from Stabilator AB. Jan Öqvist, Luleå University of Technology performed the water loss measurements, and Per Näsström SKANSKA measured the real borehole positions on surface, and the alignment of the boreholes in the cut. All the participants are hereby thanked for a good team work which resulted in the proof that the water loss measurement method is useful for the quantification of physical connections between boreholes.

REFERENCES

Ave'n, S. editor, 1984. *Geoteknik Handboken Bygg. (Geotechnique Handbook, Build)*. In Swedish. Liber förlag, Stockholm.

Demissy, M. and Bigourd, J. 1985. Compression-induced desensitization in explosives. Proceedings of the 21th Int. Conf. of Safety in Mines. Research Institutes. 21-25 Oct., pp. 757-761.

Finkel, M. 1977. Metoder att utvärdera sprängskador i berg. Delrapport nr 1. (Methods to evaluate blast damage in rock. Part report No. 1). In Swedish. Luleå University of Technology, Technical Report No. 1977:72 T, pp. 1-43.

Hagan, T.N. 1980. Understanding the burn cut - a key to greater advance rates. Institution of Mining and Metallurgy, Jan., pp. A30-A36.

Koorem, B. 1992. Spiralkil vid ort och tunneldrivning. (Spiral cut in drifting and tunneling). In Swedish. Luleå University of Technology, Master of Science Thesis No. 1992:088 E, pp. 1-18.

Nie, S.L. 1991. Tryckpåverkan på sprängämne i borrhål. (Influence of pressure on explosive in a borehole). In Swedish. Swedish Detonic Research Foundation, report No. DS 1991:5G, pp. 1-74.

Niklasson, B., Karlsson, L., Nordquist, A. and Persson, G. Öppningsförsök för ort-och tunnelsalvor i Nukutusvaara dagbrott 1989. (Cut tests for drift- and tunnel rounds in the Nukutusvaara open pit 1989). In Swedish. Swedish Detonic Research Foundation, Report No. DS 1989:5, pp. 1-82.

Persson, G. 1987. Studier av parallellhålskilars funktion medelst höghastighetsfilmning. (Studies of the function of parallel hole cuts with high speed photography). In Swedish. Swedish Detonic Research Foundation report No. DS 1987:9, pp. 1-42.

Pusch, R. and Stanfors, R. 1992. The zone of disturbance around blasted tunnels at depth. *Int. J. Rock. Mech. Min. Sci. and Geomech. Abstr. Vol.29, No.5, pp. 447-456.*

Wieland, M.S. 1985. The relative sensitivity of permissible explosives to dynamic pressure desensitization. Proceedings of the 21th Int. Conf. of Safety in Mines, 21-25 Oct, Sydney, pp. 751-755.

10 Mine management into the 21st century

Innovative Mine Design for the 21st Century, Bawden & Archibald (eds) © 1993 Balkema, Rotterdam, ISBN 90 5410 325 6

Long term scheduling at Olympic Dam Operations, South Australia

M.A.Van Leuven
WMC (Olympic Dam Operations) Pty Ltd, Roxby Downs, Australia

ABSTRACT: In the past long term production scheduling has been performed on PC based spreadsheets such as Lotus123 or Quattro Pro. With the commencement of work on a long term review it was decided to use a database scheduling package called XPAC developed by Runge Mining (Australia). Weaknesses in the spreadsheet packages were the main reasons for selecting the XPAC system. The database is capable of storing up to 300 datafields in each record. XPAC has its own programming language (XCM) which can be used to write procedures that can be performed on the database. After the basic production schedule has been performed XCM's are run which assign the type of backfill required and the appropriate amount of capital development to each record in the scheduling path. Once the database has been built it becomes possible to run various "what if ?" schedules quite easily. The output from XPAC for each option can then be used in a cost model to determine the ultimate scale of expansion suitable for the Olympic Dam Operations.

1 INTRODUCTION

The Olympic Dam copper-uranium-gold-siver deposit, see Figure 1, was discovered in 1975 by Western Mining Corporation and was subjected to intensive evaluation during the period from early 1979 until late 1985. In 1979 a joint venture was set up between WMC (51%), who are the managers and operators of the project and British Petroleum Australia (49%). In March 1993 the Olympic Dam Operations became 100% owned by WMC.

Extensive drilling, both surface and underground, mine development and metallurgical testing were undertaken as part of a feasibility study for the project. The decision to proceed with the project was taken in December 1985 and production commenced in mid 1988. Olympic Dam Operations (ODO) has a major world class resource of copper and uranium. The deposit is made up of numerous ore bodies, each of different copper/uranium grade combinations.

Current production consists of the mining of 2.3 Mt/yr from the underground operation, with finished products of 70,000 t of copper cathode, 1,400 t of uranium ore concentrate (yellowcake), 20,000 oz of gold and 460,000

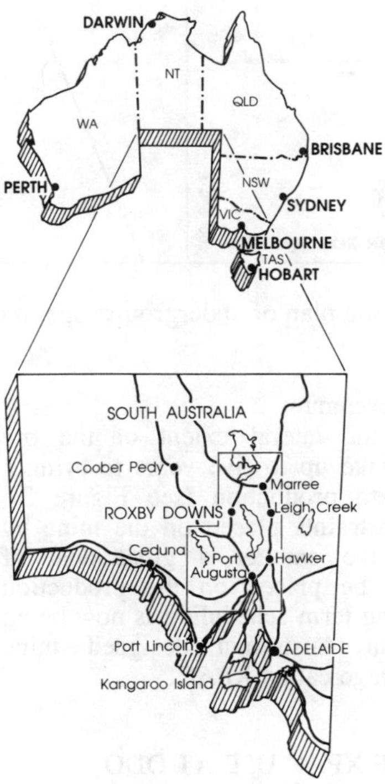

Fig.1 Location of Olympic Dam Operations

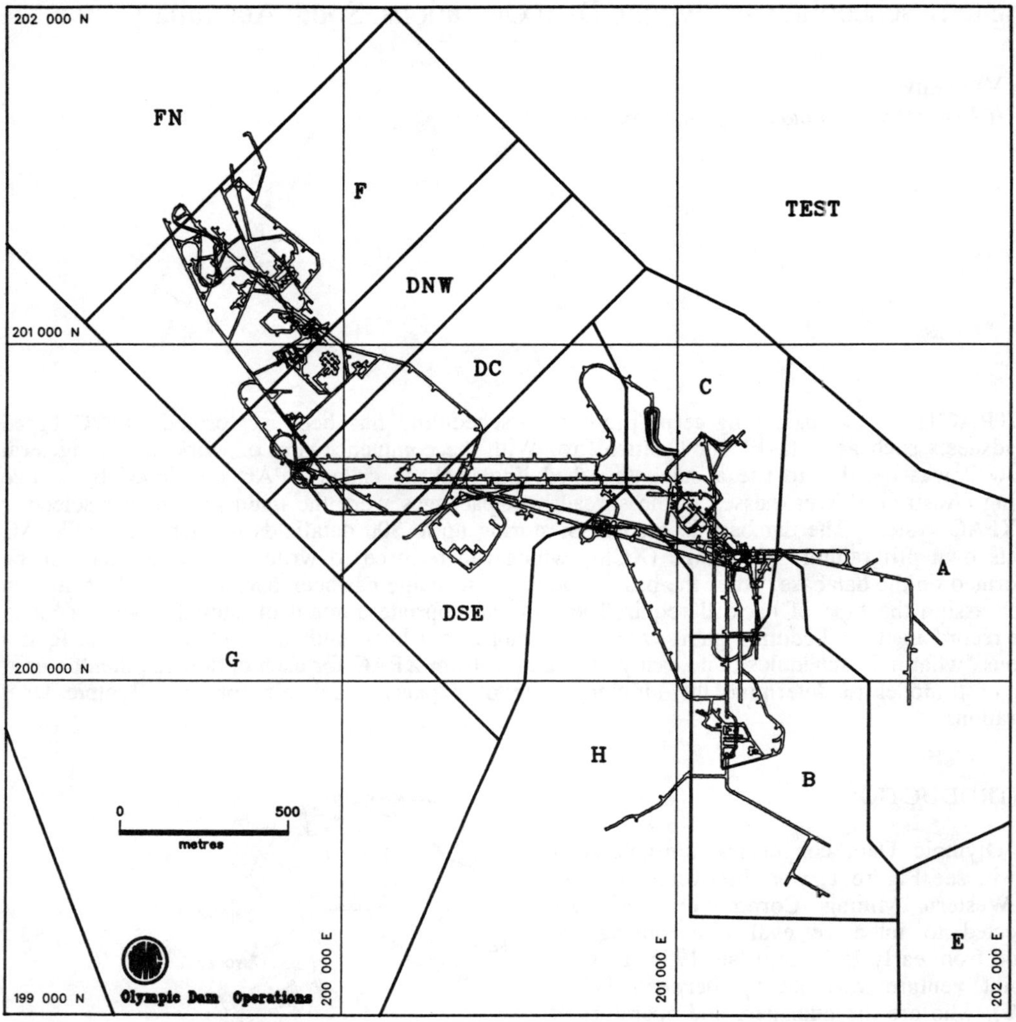

Fig.2 Generalised plan of underground operations

oz of silver, per annum.

Because of the lateral extent of the ore bodies, it can take up to five years to bring a mining area into production (see Figure 2). Due to the constraints placed on the mine by the mill and the smelter a great deal of emphasis must be placed on the production scheduling. Long term scheduling is now being done using an Australian designed mine scheduling package called XPAC.

2 HISTORY OF XPAC USE AT ODO

XPAC was first used on the Olympic Dam Project feasibility study back in the early 80's when it was a VAX based computer package.

After the approval to proceed with the project had been given all the detailed scheduling had been carried out. It was during this period that spreadsheet packages appeared on the market. With the additional flexibility that desktop PC's gave, the bulk of the scheduling began to be performed with PC based packages.

With modifications to the XPAC package it was again used for the Pre-feasibility Expansion Study in 1990. At this stage the program was still a Vax based package.

In 1991 ODO obtained Version 5 which was

a PC based package. This version was much more user friendly than the Vax based versions.

The current version being used is Version 6b Advanced which has a number of scheduling and reporting refinements over the previous versions.

3 DESIGNING OF STOPES AT ODO

Initially the various ore bodies are modelled by the geologists using Datamine. Because of the complex make up of the ore bodies, and the need to make the stope designing procedure as uncomplicated as possible, stopes are generally designed to a specific dollar value cut off.

An algorithm is used to accumulate the in-situ value of the Cu, U_3O_8, Au and Ag. Individual stopes can then be designed on these accumulated values based on a particular dollar value cut off.

As a result, stopes are rarely designed to a specific Cu or U_3O_8 cut off grade ,but to a combined in-situ value of the four metals mentioned above.

Once designed, the stope outlines are then used in Datamine to calculate the in-situ tonnes and grades. These are then reviewed by the planning engineer to see if the result is acceptable. The engineer may then decide to alter the shape in order to improve the tonnage and/or grade of the proposed stope.

After final design of the stopes the Senior Ore Reserves Geologist will issue a memorandum listing the official tonnes and grades of the particular stopes, after applying dilution and recovery factors to the in-situ design.

Once the official figures have been issued the scheduling engineer can place the newly designed stopes in the production scheduling spreadsheet.

4 SCHEDULING CRITERIA

Because of the complex nature of the deposit and the varied end products it is necessary to schedule more than one datafield.

The main product is copper. Due to the association of uranium with copper in the deposit it was decided during the pre-feasibility stage that a smelter and refinery would be needed on site to produce a copper product that would not be disadvantaged in the market place, i.e., a "radioactively clean" product.

The main mineralisation being mined at the moment is a chalcocite/bornite ore which has a high Cu:S ratio. This allows a copper concentrate of approximately 60% Cu to be produced by the mill.

Because of the grade and Cu:S ratio of the concentrate, a flash smelting process can be used where the concentrate is flashed from concentrate directly to blister without the need to produce matte.

With this direct flash furnace on site it is crucial that the mine delivers a product that has a high Cu:S ratio. As a result this ratio is also an important scheduling criterion.

In the long term the mineralogy of the ore mined will be predominantly chalcopyrite which has a Cu:S ratio well below that which is capable of sustaining the flash smelting process. This will involve the addition of a second stage to the smelting process, hence the need to know when the Cu:S ratio drops below a certain level.

The uranium grade must also be considered although not as much emphasis is placed on this because of the current reduced demand for uranium. The average grade of the uranium is generally high enough to cover the operations existing uranium contracts.

Gold and silver currently contribute 6% of total revenue on site. In most mine areas the gold and silver grades are not scheduled. The production of gold and silver from these areas is considered to be an added bonus.

Although gold and silver are not scheduled, the in-situ value of these metals are included in the algorithm to calculate an in-situ dollar value of the mineralisation. The Au and Ag grades are also carried to calculate the production of these metals.

5 PRODUCTION SCHEDULING

Until recently all production scheduling has been carried out using PC based spreadsheet packages such as Lotus 123 and Quattro Pro.

When creating a production schedule on the spreadsheet, it takes into account the Cu head grade and the Cu:S ratio. Considering that the number of stopes that may be in the spreadsheet available for scheduling and the formulae that need to be built in to calculate the weighted averages of Cu, U, Au, Ag and Cu:S, the spreadsheets can soon become unwieldly in size. These spreadsheets are only capable of providing a basic production schedule.

A number of other spreadsheets have to be utilised to carry out the scheduling of other functions such as raise drilling, production drilling, cable bolting, etc..

After a production schedule has been conceptually layed out, the stopes have to be located on plans to determine the inter-relationship with other stopes on the schedule. This is done to see if there are any constraints to mining of the stope such as backfilling of adjacent stopes.

Until recently long term production schedules have also been performed on spreadsheets. However, with the commencement of work on the long term review it was decided that XPAC Version 6b should be trialled as a scheduling tool. The aim of the review was to work out the timing and scale of additional expansions at ODO. This work involved the creation of a number of scheduling options.

5.1 Description of XPAC

The XPAC Mining System and associated software products are copyright of Runge Mining (Australia). Version 6b of XPAC is currently installed at ODO on a 486 33MHz PC with 8 Mbyte of RAM and a 220 Mbyte hard disk.

The XPAC Database and Scheduling System is heirarchically structured and can handle up to seven levels. For long term scheduling the database has been set up at ODO with three levels (see Figure 3):

1. Mine Area.
2. Suite.
3. Stope.

In short term planning five levels would be used with "Level" and "Ring" being added to the three mentioned above.

Each record can contain up to 300 data fields that can be accumulated as sums, averages and weighted averages. Records which represent the smallest seperately defined part of the mine layout (stopes in the long term schedule) are designated as the lowest level records. The upper levels in the heirarchy represent the aggregate of all the lower level records under them.

Each record contains three basic types of data:

1. Record number.
2. Name.
3. Record information.

```
AREA 4           115: B
 |
 |----- SUITE 1              116: B/YELLOW
 |        |
 |        |----- STOPE 1        117: B/YELLOW/1
 |        |----- STOPE 2        118: B/YELLOW/7
 |        |----- STOPE 3        119: B/YELLOW/8
 |        |----- STOPE 4        120: B/YELLOW/12
 |        |----- STOPE 5        121: B/YELLOW/14
 |        |----- STOPE 6        122: B/YELLOW/15
 |        |----- STOPE 7        123: B/YELLOW/16
 |        |----- STOPE 8        124: B/YELLOW/17
 |        |----- STOPE 9        125: B/YELLOW/18
 |        |----- STOPE 10       126: B/YELLOW/19
 |        |----- STOPE 11       127: B/YELLOW/20
 |        |----- STOPE 12       128: B/YELLOW/21
 |
 |----- SUITE 2              129: B/68D-A
 |----- SUITE 3              130: B/70D-B
 |----- SUITE 4              131: B/68D-C
 |----- SUITE 5              132: B/71D-D
 |----- SUITE 6              133: B/71D-E
```

Fig.3 XPAC heirarchical structure

The record information can be either numeric or descriptive. It is the numeric data that are accumulated in the database structure.

5.2 Setting up the main database

A great deal of thought went into the structure of the database before setting it up. Though it is not difficult to modify the structure after it has been set up, once all the data has been input it can be troublesome to add additional data fields within the structure. Provided there are blank datafields at the end of the database structure there is no problem in adding additional fields.

The database for the long term schedules consists of 250 data fields. The data field descriptions are grouped together in the following categories:

1. R1-R14 - Geological data.
2. R15-R21 - Metallurgical data.
3. R22-R51 - Stope development, including grades, met recoveries, etc..
4. R52-R80 - Raise drilling.
5. R81-R84 - Cable bolt drilling.
6. R85-R89 - 89 mm production drilling.
7. R90-R93 - 102 mm production drilling - top drill levels.
8. R94-R97 - 102 mm production drilling - sub drill levels.

9. R98-R123 - Extraction (i.e. production).

10. R124-R131 - Backfill - Cemented Aggregate Fill (CAF).

11. R132-R136 - Backfill - Rock Fill (RF).

12. R137-R154 - Scheduling controls for backfill.

13. R155-R165 - General summary of datafields.

14. R166-R234 - Scheduling controls for capital and stope development.

15. R235-R238 - Total metres and tonnes of development.

16. R239-R250 - Blank.

After the database structure has been set up the geological, mining and metallurgical data can be imported via ascii files or spreadsheet files. Not all the data is imported, most of it is calculated using XCM procedures. XCM is the programming language used by XPAC.

As mentioned in Section 5.1 the structure has three levels, however, not all ore bodies used in the schedule go down to the "Stope" level of detail (see Figure 2). Calculations are only performed on the lowest level records in each branch so calculations will only be performed on the "Suite" level if no records at "Stope" level are present. Likewise, calculations will only be performed at the "Mine Area" level if no records at the "Suite" level are present.

5.3 Calendar database

Before any schedules can be performed an additional database (calendar) has to be established which contains the target information. The calendar has the same heirarchical structure as the main database, although the structure is time based rather than mining block based.

For long term scheduling a calendar database was created with the first year going down to the third level of detail (Months), the second year going down to the second level (Quarters) and from year three onwards only the first level of detail (Years) was required.

The calendar database can be set up with numerous targets such as different scheduling rates. Running a small XCM it is very simple to set up different yearly production rates for various options.

5.4 Extraction scheduling

Production, or extraction, scheduling using XPAC is still time consuming as the current version of the software will not automatically schedule to a particular head grade. This is not seen as a major disadvantage as the engineer still has control of the selection of individual stopes.

When choosing a record to schedule one has the choice to schedule the whole record or a percentage of it. An array schedule can be performed that will schedule a record over several time periods at a fixed percentage until the record has been completely "mined".

A basic production schedule is controlled by the targeted tonnes in the calendar database. The scheduler needs to keep an eye on the Cu grade and the Cu:S ratio as he/she chooses the various steps for mining. When the tonnage target is reached in a particular scheduling period the schedule will move to the next period automatically.

Once the schedule has been completed the scheduling sequence chosen can be saved as a defined path. This path can then be used by various XCM procedures to set backfill and development flags. This is an advantage over scheduling with a spreadsheet package.

5.5 Determining backfill type

There are two main types of backfill used in the underground operations, cemented aggregate fill (CAF) and rockfill (RF).

The backfill scheduling control datafields are displayed in Table 1. R139 and R140 are classified fields which are used to assign text flags to database records. When using these

Table 1. Backfill scheduling control data fields

Data Field	Description	Unit	Field Type
137	SCHEDULING CONTROLS		Title
138	--		Title
139	Fill Type Fixed		Classification
140	Fill Type		Classification
141	% CAF	%	Weight_Avg,3
142	Precedence - Stope 1		Note
143	Precedence - Stope 2		Note
144	Precedence - Stope 3		Note
145	Precedence - Stope 4		Note
146	Precedence - Stope 5		Note
147	Precedence - Stope 6		Note
148	Precedence - Stope 7		Note
149	Precedence - Stope 8		Note

classifications in XCM's a test is done on the datafield to determine what flag each record has been assigned. R139 is used to set the text flag to "CAF", "RF" or "None" if it is known what it will be regardless of the scheduling path. As can be seen in Table 1 there are eight data fields available for adjoining stopes (R142 - R149). In the Mine Areas that do go down to the third level of detail (Stope) it is possible to list the adjoining stopes in these data fields.

An XCM procedure has been written which loops through the scheduling periods and the steps in each period. At each scheduling step it moves to the selected record and first checks to see if the fill type is fixed (R139). If not then it checks to see if any of the stopes in the list of that record are present in the scheduling path prior to that point in time. If the answer is yes then the backfill type (R140) is set to CAF. On the other hand, if all stopes in the list have been scheduled prior to that point in time then backfill type is set to RF. As the fill type classification is generally dependent on the scheduling path, both types of fill data have to be carried in the record (R124-R136).

After the backfill type flag has been set another subroutine calculates the volumes of each fill type required for each record and the cement/flyash requirements for the CAF. For other mine blocks that don't have stopes designed the flag at R139 can be set to "BOTH". The XCM will then set the CAF/RF ratio in accordance with the factor in R141.

Because the XCM procedure modifies the text flag in the individual records of the main database it is necessary to re-run the production schedule using the stored scheduling path after the backfill XCM has been run. This will ensure that the correct backfill requirements are carried into the results database.

5.6 Assigning development metres

The same philosophy has been used for assigning development as for choosing backfill type. Because of the large number of mining areas that are available to select in production schedules, assigning development metres to areas can be difficult.

In some Mine Areas there are a number of unmined blocks that have not been developed yet. Because of their relative positions to each other it will be possible to access several ore blocks with only one local ramp system. The same principle holds true for assigning perimeter development on individual levels to specific stopes.

When setting up the initial database it was necessary to detail the amount of stope development needed for each stope (record). Capital development was also calculated. Because of the complexity of tracking the total development for an area and distributing it amongst the stopes it was necessary to build into the database five sets of scheduling controls for development, similar to the backfill controls.

The complexity of the orebodies mean that even within the suite, stopes can have different top and bottom RL's. For example, a suite may have five levels but individual stopes generally only have three levels. Therefore, when scheduling a particular stope in a suite it will not be necessary to develop all five levels; however, the next stope scheduled from that suite may require a level that was not needed for the previous stope. As a result scheduling of the development can be difficult.

The first of the five sets of controls has been used for assigning ramp systems to suites, where as the other four are required to assign perimeter and stope development to suites and stopes. Table 2 shows the structure of the first of the five scheduling controls for development; all five having identical structures.

R168 is a classified field which sets the text flag to "Yes", "No" or "None". The main database was initially set up with the flag set to "None" in all records. R169 - R178 are note fields that contain the list of stopes or suites

Table 2. Development scheduling control data fields

Data Field	Description	Unit	Field Type
166	DEVT CONTROLS A		Title
167	--		Title
168	Access Development		Classification
169	Precedence - Stope 1		Note
170	Precedence - Stope 2		Note
171	Precedence - Stope 3		Note
172	Precedence - Stope 4		Note
173	Precedence - Stope 5		Note
174	Precedence - Stope 6		Note
175	Precedence - Stope 7		Note
176	Precedence - Stope 8		Note
177	Access Development	m	Additive
178	Access Development	t	Additive

that the development is linked to. A list of suites that are dependent on the ramp development is placed in each record.

In each of these scheduling controls there are only eight data fields available to list dependant stopes. Generally there are more than eight stopes in the region, therefore, it is necessary to list the stopes that are most likely to be mined first in the schedule, ie., the higher grade stopes.

After the basic production schedule has been performed an XCM procedure is run to assign the development to the correct records. The XCM follows the scheduling path, jumping to each record as it meets it in the path. Firstly the XCM checks the access development flag (R168). If it is set to "No" then the XCM does not need to perform any further checks or calculations and moves on to the next record in the path. If set to "None" then it checks the note data fields (R169-R176) to see if any of the stopes in the list have been scheduled previously. The access development flag will be set to "Yes" only when none of the records in the list have been scheduled previously.

After the flag has been set the XCM moves on to the next record in the schedule and continues the same procedure throughout the entire scheduling path. Once the flags have been set another subroutine will calculate the development required for each record in each of the five scheduling controls where the flags have been set to "Yes".

As with the assignment of backfill types, in order to carry the correct development metres into the results database it is neccessary to re-run the basic extraction schedule after the development XCM has been run.

5.7 Output of Results

There are several forms of output available with XPAC. The production schedule can be produced as either a series of Gantt charts or as a table listing scheduling periods, records scheduled and percentages of records scheduled.

A printout of the results database, which has the same format as the main database, was the main output required for the review, combined with graphs of the backfill and development requirements for each particular option. Examples of these are appended at the end of the paper.

6 CHOOSING A LONG TERM SCHEDULE

The work carried out using XPAC involved scheduling a large number of options, the main ones being:

1. Baseline case of 2.4 Mt of ore per year.
2. Baseline case with reduced development requirements.
3. Baseline case with absolute minimum development.
4. Production rate of 4.0 Mt of ore per year.
5. Production rate of 6.5 Mt of ore per year.
6. Production of 100,000 t of Cu metal per year.
7. Production of 200,000 t of Cu metal per year.
8. Production of 300,000 t of Cu metal per year.

With the results from the above schedules it was relatively easy to choose the options that warranted more detailed analyses. The relevant data from the suitable options have been placed into a cost model that has been developed for the site operations incorporating operating cost power curves and replacement capital power curves. Output from this work will produce an option which can then be used in a full feasibility study on future expansion.

7 CONCLUSIONS

The aim of this long term review was to produce a number of options that would be physically possible for the mine. It took several months to get the main database set up with the correct lists of scheduling controls for backfill and development, as well as the writing of numerous XCM procedures. Although the time taken to initially set up the database was quite tedious, the benefits in the long run were that it became relatively quick to run out a number of "what if ?" schedules for various options.

The use of XPAC and the cost model has enabled ODO to move closer to an individual option that can then be used as the basis of a detailed feasibility study for future expansion at Olympic Dam.

8 ACKNOWLEDGMENTS

The author wishes to thank the Joint Venture Management Committee for the approval to publish this paper and the various staff members of ODO who helped in the preparation and editing of the paper; thankyou.

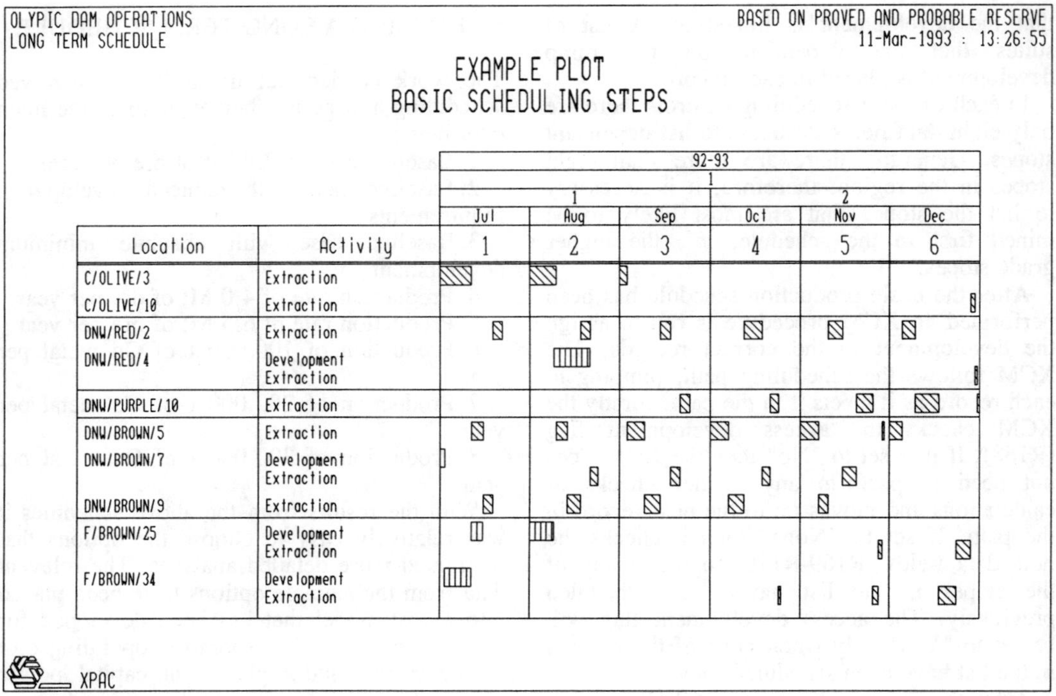
OLYPIC DAM OPERATIONS
LONG TERM SCHEDULE
BASED ON PROVED AND PROBABLE RESERVE
11-Mar-1993 : 13:26:55
EXAMPLE PLOT
BASIC SCHEDULING STEPS
92-93
Jul
Aug
Sep
Oct
Nov
Dec
1
2
3
4
5
6
Location
Activity
C/OLIVE/3
C/OLIVE/10
DNW/RED/2
DNW/RED/4
DNW/PURPLE/10
DNW/BROWN/5
DNW/BROWN/7
DNW/BROWN/9
F/BROWN/25
F/BROWN/34
Extraction
Development
XPAC

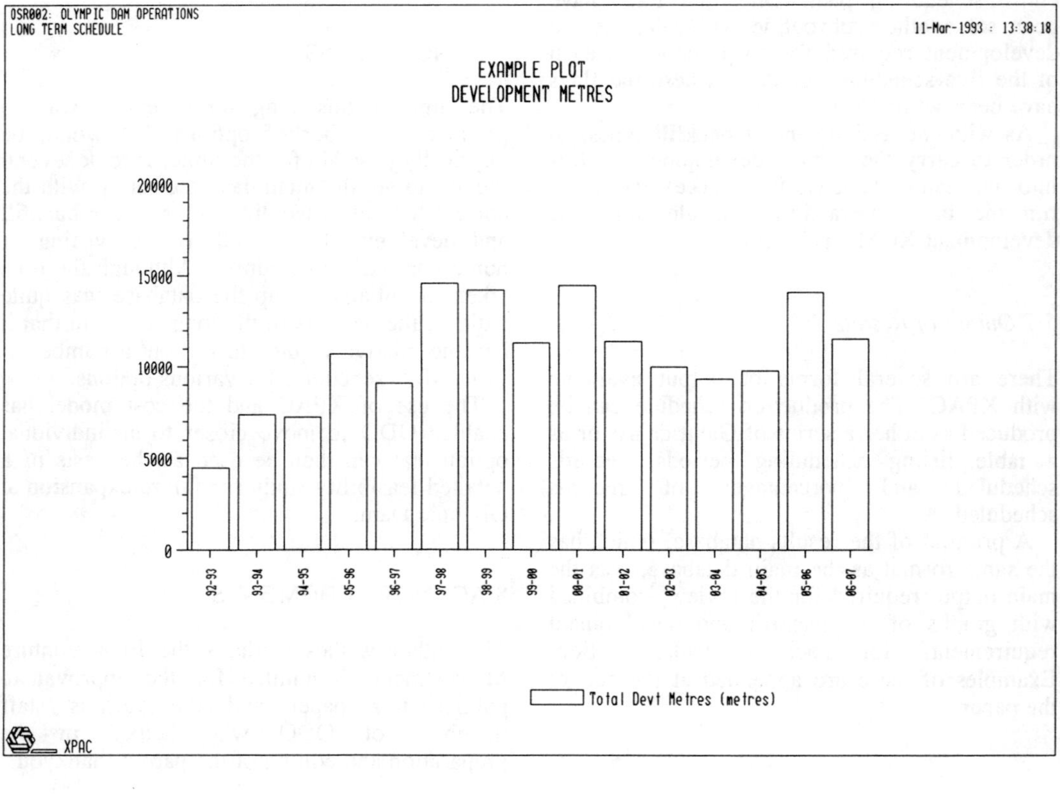
OSR002: OLYMPIC DAM OPERATIONS
LONG TERM SCHEDULE
11-Mar-1993 : 13:38:18
EXAMPLE PLOT
DEVELOPMENT METRES
20000
15000
10000
5000
0
92-93
93-94
94-95
95-96
96-97
97-98
98-99
99-00
00-01
01-02
02-03
03-04
04-05
05-06
06-07
Total Devt Metres (metres)
XPAC

XPAC Mining Systems -- Database: OSR002 --------------------------- Page 0001
OLYPIC DAM OPERATIONS 11-Mar-1993 13:47:01
LONG TERM SCHEDULE BASED ON PROVED AND PROBABLE RESERVE

Data Field Names and Formats
From Data Field 1 To Data Field 178

Data Field			Field Type	Format	Activity	Protect
1	GEOLOGICAL DATA		Title		0 - Not Applicable	No
2	---------------	-----------	Title	\-	0 - Not Applicable	No
3	Stope Volume	cu.metres	Additive	0	0 - Not Applicable	No
4	Ore Quantity	tonnes	Additive	0	0 - Not Applicable	No
5	Cu Grade	%	Weight_Avg,4	2	0 - Not Applicable	No
6	U308 Grade	kg/tonne	Weight_Avg,4	2	0 - Not Applicable	No
7	Au Grade	g/tonne	Weight_Avg,4	2	0 - Not Applicable	No
8	Ag Grade	g/tonne	Weight_Avg,4	2	0 - Not Applicable	No
9	Cu/S Ratio		Weight_Avg,12	2	0 - Not Applicable	No
10	Ore S.G.		Weight_Avg,3	2	0 - Not Applicable	No
11	Sulphur	%	Weight_Avg,4	2	0 - Not Applicable	No
12	Contained Sulphur	tonnes	Additive	0	0 - Not Applicable	No
13	Drilling		Classification		0 - Not Applicable	No
14			Title		0 - Not Applicable	No
15	METALLURGICAL DATA		Title		0 - Not Applicable	No
16	------------------		Title		0 - Not Applicable	No
17	Cu Recovery		Weight_Avg,4	2	0 - Not Applicable	No
18	U308 Recovery		Weight_Avg,4	2	0 - Not Applicable	No
19	Au Recovery		Weight_Avg,4	2	0 - Not Applicable	No
20	Ag Recovery		Weight_Avg,4	2	0 - Not Applicable	No
21			Title		0 - Not Applicable	No
22	DEVELOPMENT		Classification		0 - Not Applicable	No
23	-----------	-----------	Title	\-	0 - Not Applicable	No
24	Stope Development	metres	Additive	,0	1 - Development	No
25	Development Quantity	tonnes	Additive	,0	1 - Development	No
26	- Proportion Ore	%	Additive	2	1 - Development	No
27	Mullock Quantity	tonnes	Additive	,0	1 - Development	No
28	Ore Quantity	tonnes	Additive	,0	1 - Development	No
29	Cu Grade	%	Weight_Avg,28	2	1 - Development	No
30	U308 Grade	kg/tonne	Weight_Avg,28	2	1 - Development	No
31	Au Grade	g/tonne	Weight_Avg,28	2	1 - Development	No
32	Ag Grade	g/tonne	Weight_Avg,28	2	1 - Development	No
33	Cu/S Ratio		Weight_Avg,28	2	1 - Development	No
34	Sulphur		Weight_Avg,28	2	1 - Development	No
35			Title		0 - Not Applicable	No
36	Contained Cu	tonnes	Additive	,0	1 - Development	No
37	Contained U308	tonnes	Additive	,0	1 - Development	No
38	Contained Au	ounces	Additive	,0	1 - Development	No
39	Contained Ag	ounces	Additive	,0	1 - Development	No
40	Contained S	tonnes	Additive	,0	1 - Development	No
41			Title		0 - Not Applicable	No
42	Cu Recovered Metal	tonnes	Additive	,0	1 - Development	No
43	U308 Recovered Metal	tonnes	Additive	,0	1 - Development	No
44	Au Recovered Metal	ounces	Additive	,0	1 - Development	No
45	Ag Recovered Metal	ounces	Additive	,0	1 - Development	No
46			Title		0 - Not Applicable	No
47	Recovered Cu Grade	%	Weight_Avg,28	2	1 - Development	No
48	Recovered U308 Grade	kg/tonne	Weight_Avg,28	2	1 - Development	No
49	Recovered Au Grade	g/tonne	Weight_Avg,28	2	1 - Development	No
50	Recovered Ag Grade	g/tonne	Weight_Avg,28	2	1 - Development	No
51			Title		0 - Not Applicable	No
52	RAISE DRILLING		Title		0 - Not Applicable	No

Data Field Names and Formats
From Data Field 1 To Data Field 178

Data Field			Field Type	Format	Activity	Protect
53	--------------	-----------	Title	\-	0 - Not Applicable	No
54	Drilling	metres	Additive	,0	2 - Raise Drilling	No
55	Raise Diameter	metres	Weight_Avg,54	1	2 - Raise Drilling	No
56	Drilled Volume	cu.metres	Additive	,0	2 - Raise Drilling	No
57	Drilled Quantity	tonnes	Additive	,0	2 - Raise Drilling	No
58	Cu Grade	%	Weight_Avg,57	2	2 - Raise Drilling	No
59	U308 Grade	kg/tonne	Weight_Avg,57	2	2 - Raise Drilling	No
60	Au Grade	g/tonne	Weight_Avg,57	2	2 - Raise Drilling	No
61	Ag Grade	g/tonne	Weight_Avg,57	2	2 - Raise Drilling	No
62	Cu/S Ratio		Weight_Avg,57	2	2 - Raise Drilling	No
63	Sulphur		Weight_Avg,57	2	2 - Raise Drilling	No
64			Title		0 - Not Applicable	No
65	Contained Cu	tonnes	Additive	,0	2 - Raise Drilling	No
66	Contained U308	tonnes	Additive	,0	2 - Raise Drilling	No
67	Contained Au	ounces	Additive	,0	2 - Raise Drilling	No
68	Contained Ag	ounces	Additive	,0	2 - Raise Drilling	No
69	Contained S	tonnes	Additive	,0	2 - Raise Drilling	No
70			Title		0 - Not Applicable	No
71	Cu Recovered Metal	tonnes	Additive	,0	2 - Raise Drilling	No
72	U308 Recovered Metal	tonnes	Additive	,0	2 - Raise Drilling	No
73	Au Recovered Metal	ounces	Additive	,0	2 - Raise Drilling	No
74	Ag Recovered Metal	ounces	Additive	,0	2 - Raise Drilling	No
75			Title		0 - Not Applicable	No
76	Recovered Cu Grade	%	Weight_Avg,57	2	2 - Raise Drilling	No
77	Recovered U308 Grade	kg/tonne	Weight_Avg,57	2	2 - Raise Drilling	No
78	Recovered Au Grade	g/tonne	Weight_Avg,57	2	2 - Raise Drilling	No
79	Recovered Ag Grade	g/tonne	Weight_Avg,57	2	2 - Raise Drilling	No
80			Title		0 - Not Applicable	No
81	CABLE BOLT DRILLING		Title		0 - Not Applicable	No
82	--------------------	----------	Title	\-	0 - Not Applicable	No
83	Drilling	metres	Additive	,0	3 - Cable Bolt Drilling	No
84			Title		0 - Not Applicable	No
85	89mm DRILLING		Title		0 - Not Applicable	No
86	-------------	-----------	Title	\-	0 - Not Applicable	No
87	Drilling	metres	Additive	,0	4 - 89mm Drilling	No
88			Title		0 - Not Applicable	No
89			Title		0 - Not Applicable	No
90	102mm DRILLING - TOP		Title		0 - Not Applicable	No
91	--------------------		Title	\-	0 - Not Applicable	No
92	Drilling	metres	Additive	,0	5 - 102mm Drilling - TOP	No
93			Title		0 - Not Applicable	No
94	102mm DRILLING - SUB		Title		0 - Not Applicable	No
95	--------------------		Title	\-	0 - Not Applicable	No
96	Drilling	metres	Additive	,0	6 - 102mm Drilling - SUB	No
97			Title		0 - Not Applicable	No
98	EXTRACTION		Title		0 - Not Applicable	No
99	----------	------------	Title	\-	0 - Not Applicable	No
100	Ore Quantity	tonnes	Additive	0	7 - Extraction	No
101	Cu Grade	%	Weight_Avg,100	2	7 - Extraction	No
102	U308 Grade	kg/tonne	Weight_Avg,100	2	7 - Extraction	No
103	Au Grade	g/tonne	Weight_Avg,100	2	7 - Extraction	No
104	Ag Grade	g/tonne	Weight_Avg,100	2	7 - Extraction	No

Data Field Names and Formats
From Data Field 1 To Data Field 178

Data Field			Field Type	Format	Activity	Protect
105	Cu/S Ratio		Weight_Avg,100	2	7 - Extraction	No
106	Sulphur	%	Weight_Avg,100	2	7 - Extraction	No
107			Title		0 - Not Applicable	No
108	Contained Cu	tonnes	Additive	,0	7 - Extraction	No
109	Contained U308	tonnes	Additive	,0	7 - Extraction	No
110	Contained Au	ounces	Additive	,0	7 - Extraction	No
111	Contained Ag	ounces	Additive	,0	7 - Extraction	No
112	Contained S	tonnes	Additive	,0	7 - Extraction	No
113			Title		0 - Not Applicable	No
114	Cu Recovered Metal	tonnes	Additive	,0	7 - Extraction	No
115	U308 Recovered Metal	tonnes	Additive	,0	7 - Extraction	No
116	Au Recovered Metal	ounces	Additive	,0	7 - Extraction	No
117	Ag Recovered Metal	ounces	Additive	,0	7 - Extraction	No
118			Title		0 - Not Applicable	No
119	Recovered Cu Grade	%	Weight_Avg,100	2	7 - Extraction	No
120	Recovered U308 Grade	kg/tonne	Weight_Avg,100	2	7 - Extraction	No
121	Recovered Au Grade	g/tonne	Weight_Avg,100	2	7 - Extraction	No
122	Recovered Ag Grade	g/tonne	Weight_Avg,100	2	7 - Extraction	No
123			Title		0 - Not Applicable	No
124	BACKFILL - CAF		Title		0 - Not Applicable	No
125	--------------		Title		0 - Not Applicable	No
126	Volume	cu.metres	Additive	,0	8 - Backfill CAF	No
127	Quantity	tonnes	Additive	,0	8 - Backfill CAF	No
128	Cement	tonnes	Additive	,0	8 - Backfill CAF	No
129	Fly Ash	tonnes	Additive	,0	8 - Backfill CAF	No
130	Mullock	tonnes	Additive	,0	8 - Backfill CAF	No
131			Title		0 - Not Applicable	No
132	BACKFILL ROCK		Title		0 - Not Applicable	No
133	-------------		Title		0 - Not Applicable	No
134	Volume	cu.metres	Additive	,0	9 - Backfill RF	No
135	Quantity	tonnes	Additive	,0	9 - Backfill RF	No
136			Title		0 - Not Applicable	No
137	SCHEDULING CONTROLS		Title		0 - Not Applicable	No
138	-------------------	-----------	Title	\-	0 - Not Applicable	No
139	Fill Type Fixed		Classification		0 - Not Applicable	No
140	Fill Type		Classification		0 - Not Applicable	No
141	- % CAF	%	Weight_Avg,3	0	0 - Not Applicable	No
142	Precedence - Stope 1		Note		0 - Not Applicable	No
143	Precedence - Stope 2		Note		0 - Not Applicable	No
144	Precedence - Stope 3		Note		0 - Not Applicable	No
145	Precedence - Stope 4		Note		0 - Not Applicable	No
146	Precedence - Stope 5		Note		0 - Not Applicable	No
147	Precedence - Stope 6		Note		0 - Not Applicable	No
148	Precedence - Stope 7		Note		0 - Not Applicable	No
149	Precedence - Stope 8		Note		0 - Not Applicable	No
150	Finish Fill		Date	dd-mon-yyyy hh:mi	0 - Not Applicable	No
151	Start Excavation		Date	dd-mon-yyyy hh:mi	0 - Not Applicable	No
152	Available		Date	dd-mon-yyyy hh:mi	0 - Not Applicable	No
153	Lead -/ Lag +	days	Additive	,1	0 - Not Applicable	No
154	SUMMARY		Title		0 - Not Applicable	No
155	Au Recovered Metal	ounces	Additive	,0	0 - Not Applicable	No
156	Ag Recovered Metal	ounces	Additive	,0	0 - Not Applicable	No

Data Field Names and Formats
From Data Field 1 To Data Field 178

Data Field			Field Type	Format	Activity	Protect
157	Ore Quantity	tonnes	Additive	,0	0 - Not Applicable	No
158	Cu Grade	%	Weight_Avg,157	2	0 - Not Applicable	No
159	Cu Recovered Metal	tonnes	Additive	,0	0 - Not Applicable	No
160	Cu/S Ratio		Weight_Avg,164	2	0 - Not Applicable	No
161	U308 Grade		Weight_Avg,157	2	0 - Not Applicable	No
162	U308 Recovered Metal		Additive	,0	0 - Not Applicable	No
163	Au Grade	g/tonne	Weight_Avg,157	2	0 - Not Applicable	No
164	Contained Sulphur	tonnes	Additive	0	0 - Not Applicable	No
165	Ag Grade	g/tonne	Weight_Avg,157	2	0 - Not Applicable	No
166	DEVELOPMENT CONTROLS A		Title		0 - Not Applicable	No
167	----------------------	-----------	Title	\-	0 - Not Applicable	No
168	Access Development		Classification		0 - Not Applicable	No
169	Precedence - Stope 1		Note		0 - Not Applicable	No
170	Precedence - Stope 2		Note		0 - Not Applicable	No
171	Precedence - Stope 3		Note		0 - Not Applicable	No
172	Precedence - Stope 4		Note		0 - Not Applicable	No
173	Precedence - Stope 5		Note		0 - Not Applicable	No
174	Precedence - Stope 6		Note		0 - Not Applicable	No
175	Precedence - Stope 7		Note		0 - Not Applicable	No
176	Precedence - Stope 8		Note		0 - Not Applicable	No
177	Access Development	metres	Additive	,0	1 - Development	No
178	Access Development	tonnes	Additive	,0	1 - Development	No

Innovative Mine Design for the 21st Century, Bawden & Archibald (eds) © 1993 Balkema, Rotterdam, ISBN 90 5410 325 6

Mine profitability and the application of new technology in deep level South African gold mines

R.G.P. Pickering, G.L. Smith & D. Campbell
Division of Mining Technology, CSIR, South Africa

ABSTRACT: The long term viability of deep level gold mining operations in South Africa is dependent on the ability of the industry to maintain acceptable financial returns in a climate of increasingly difficult mining conditions, escalating working costs and fluctuating gold price. Within this context, the combination of new technology with alternative mining systems could provide the means for safer and more profitable extraction of deep level ore reserves vital to the long term survival of the industry.

Owing to the high interdependence of activities that influence profitability, consideration must be given to an integrated approach when evaluating the economic impact of changes to mining systems. Evaluation of the potential profitability of mining systems is thus dependent on the ability to identify areas from which maximum returns can be gained using changes in technology and their interrelationship.

This paper reviews work carried out by COMRO (Chamber of Mines Research Organization), now the Division of Mining Technology, CSIR, to identify and evaluate key system parameters which influence profitability. To this end, a specially developed economic model is used to assess the sensitivity of various strategies that could be employed to increase profitability of deep level gold mining. Parameters of dilution, face advance rate, development ratios and operational costs are identified as having the greatest impact on profitability in the stoping activity area, whilst strategies for improved rockbreaking, rockhandling, materials handling and support techniques are explored. In addition, the impact on profitability of these key parameters is considered, on the basis of current production trials, for the impact ripper mining system - a mechanised mining method suitable for hard rock, deep level gold mines.

1 INTRODUCTION

The gold mining industry in South Africa is justifiably proud of its many technological achievements. Through the need to mine at progressively greater depths, it has been confronted with new challenges of ever-increasing magnitude, with continuing efforts being made to address the challenges. These efforts have met with considerable success in certain aspects, but the improvement of stoping methods in deep narrow reefs has proved less tractable. The conditions prevailing in deep narrow stopes are such that only rudimentary machines have been used to any significant extent. Of the four kinds of machine that are in use, there is one for each of the operations of rock-breaking, rockhandling, materials handling, and hangingwall support[1].

In rockbreaking, pneumatic rockdrills, mounted on thrustlegs, are used to drill small diameter blast holes. Hammer-drills were, however, first used soon after the commencement of mining on the Witwatersrand. These early drills were massive and had to be rig mounted; they drilled unnecessarily large diameter holes very slowly and produced prodigious quantities of dust. The jack-hammer type of rockdrill, using a hollow drill steel and an integral rotation mechanism, was conceived shortly after the discovery of gold on the Witwatersrand, and gold mining stimulated and played a significant role in the development of the drill. As its development proceeded, it was introduced progressively, and by the 1950s it was in use throughout the industry. At that time, thrust for the drills was provided manually, but soon thereafter, air-legs were introduced, and today they are used almost universally.

In rockhandling, broken ore is removed from the stope faces and gullies by means of scraper scoops driven from electrically powered winches with power ranges from 30 kW to 90 kW. When mining on the Witwatersrand commenced, the reefs were relatively wide and steeply inclined, and it was only in the 1920s that the first scrapers were introduced, to move ore from the narrow, less steeply inclined reefs on the East Rand. The early winches were of low power and were driven by compressed air. Over the proceeding decades, scrapers were introduced gradually and did not gain wide acceptance until the push towards concentrated mining in the late 1950s. Today they are used universally in deep narrow-reef mining.

Rope-conveyors were introduced in the late 1960s to transport timber into stopes for hangingwall support. They gained rapid acceptance and now are used quite widely in deep stopes, most of them being driven electrically.

The first test of hydraulic props for hangingwall support in gold mines took place in 1956. These early props failed badly under rockburst conditions, and, not until the development of technology which enabled the props to yield rapidly in response to sudden movements of the hangingwall, did props prove effective. These rapid yielding hydraulic props gained acceptance in the early 1970s and are now used widely in deep stopes. The props are set by means of water-emulsion pumps driven from the compressed air supply.

The introduction of new technologies in alternative mining systems could provide the means for more efficient and profitable extraction of deep ore reserves which is vital to the long term survival of the gold mining industry. In this context, efficient extraction can be defined as maximizing percentage gold recovery and minimizing working costs without compromising safety.

2 WORKING COST DISTRIBUTION

Mine working cost control procedures tend to be structured around management responsibility areas. Using this system, for example, direct mining costs associated with the mining production departments may be categorised broadly into stoping, development and transport costs. Services costs associated with these activities, such as ventilation and refrigeration costs, are not considered as direct mining costs and are considered as environmental services costs whilst charges relating to supervisory labour are regarded as general or management services costs.

Although appropriate for operational control purposes, responsibility-based approaches such as this, because of their generalized nature, are not suitable for determining the true cost of mining activities required for the financial evaluation of new technology. This, rather, requires a cost allocation system which acknowledges the actual mining activities taking place rather than how the operation is being managed.

The generalized nature of the "responsibility-based" cost accounting system is illustrated in Figure 1; the values used in this figure are taken from a representative deep level West Witwatersrand gold mine for the period April 1990 to June 1990. A high percentage (46 %) of these costs, is thus, allocated to services, and, therefore, to ascertain the true distribution of working costs between the primary mining activities of development, stoping, transport, reduction and genuine overheads, a series of linked LOTUS 1-2-3 spread sheets has been created[2]. This model adopts a total cost approach and accounts for all costs, direct and indirect. Results from using this model to redistribute the service costs are presented in Figure 2, which shows that the true cost of stoping and development is 64 % of total mining cost compared with 39 %, when using the 'responsibility-based' approach.

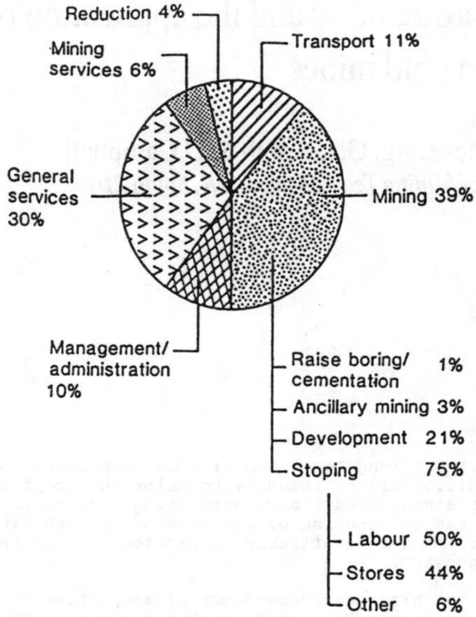

Figure 1. Total mining cost of R167/ton milled distributed according to responsibility costing

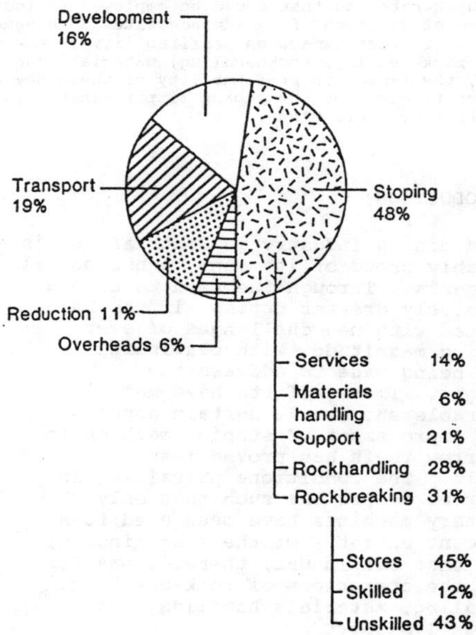

Figure 2. Total mining cost of R167/ton milled distributed according to activity costing

3 ECONOMIC MODELLING

Having developed a method of redistributing both the management cost and the statistics information from responsibility costing to activity based costing, it was then possible to develop an incremental analysis type economic model which would provide a rapid means of estimating changes to working cost and revenue following changes to the mining system. The model compares a proposed change to a system with an existing technique. As changes in mining activities resulting from the introduction of the proposed change to the system are used to modify existing costs, the economic implications of the proposed change can be expressed in terms of a familiar base case. To facilitate comparison, the model is structured around a "Minimum Production Unit" consisting of a single sided raise connection. This unit of comparison can be varied to suit local requirements but must remain small enough to allow comparison of the component parts of the new mining system, layout or technique.

It must be stressed that using this approach results in the evaluation of proposed changes relative to a performance "snapshot" which is relevant only to the mine under consideration. However, it is anticipated that the trends which are established will be applicable to similar operations within the mining industry.

Using the model, the profit sensitivity to the parameters of dilution, face advance rate and development ratios has been evaluated for a marginal West Rand gold mine using conventional stoping techniques.

3.1 Dilution

In narrow reef mining, the main causes of dilution are the mining of a stoping width in excess of the channel width and the digging of large gullies. Improved stope width control can be achieved through the use of controlled rockbreaking techniques, either explosive or non-explosive, and effective stope support systems which control fractured rock under both dynamic and static loading conditions. The size and extent of the gully system is almost entirely dependent on the rock handling method used. Figure 3 shows the impact on profitability of variation in both stoping width and gully cross section. Stoping width shows the highest sensitivity, a 10 % change resulting in a R10/ton change in working profit. For this particular mine, gullies are small and their excavation has a limited effect on profitability.

3.2 Face advance rate

Increased rates of face advance are desirable as they lead to a concentration of mining activities and associated cost savings, which are achieved through minimizing the reticulation of services to the mining activity and through intensified utilization of equipment. Figure 4 shows the improvements in face advance rates, and their effect on profitability, following changes to mining cycles, lost blast rates

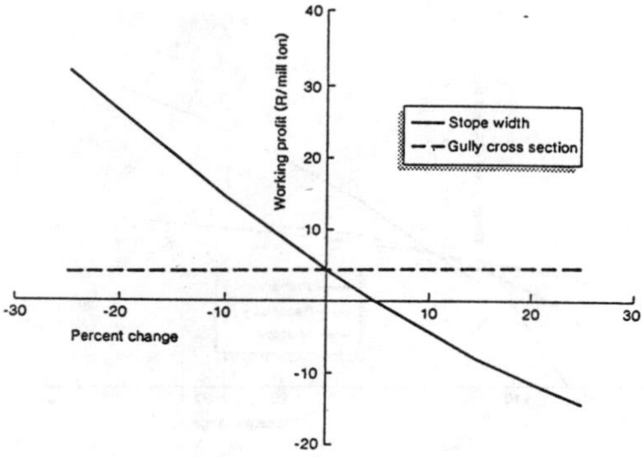

Figure 3. Effect of stope width and gully cross section on profitability

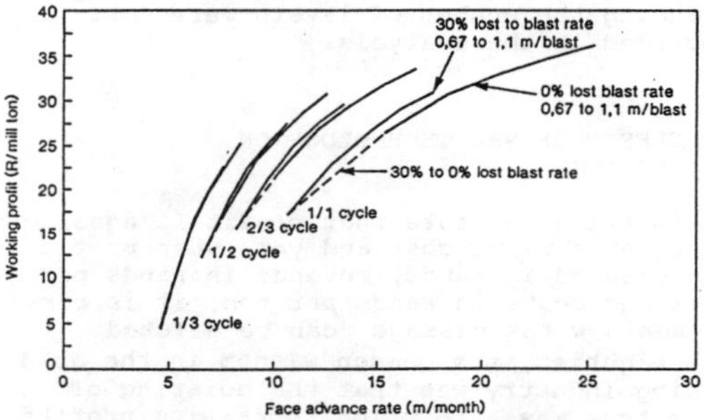

Figure 4. Effect of lost blast rate, blast cycle and advance per blast on profitability

and advance per blast. It is interesting to note that a change from a blast cycle of once every three days to twice every three days would increase profitability by R12/ton milled. By obtaining the maximum advance which could be achieved from 1,2 m long holes drilled at 70° to the face, a cycle of blasting every second day would increase profitability by R14/ton milled.

3.3 Reduced off-reef development

A reduction in waste mining will result in reduced working costs through improved labour utilization and lower use of consumables, coupled with increased ore hoisting capacity. Development ratios, measured in m^2 (of stoping)/m (of development), can be reduced by improving mining layouts through the utilization of new technologies that increase the distances, in both dip and strike, over which rock and material can be transported. Figure 5 shows the effect of changes in primary development (footwall and cross-cut excavations), secondary development (raising and boxholes) and in-stope development (dip and strike gullies)). In this example, the maximum benefit comes from primary development with a 20 % change in development ratio resulting in a profit increase of less than R3/ton milled.

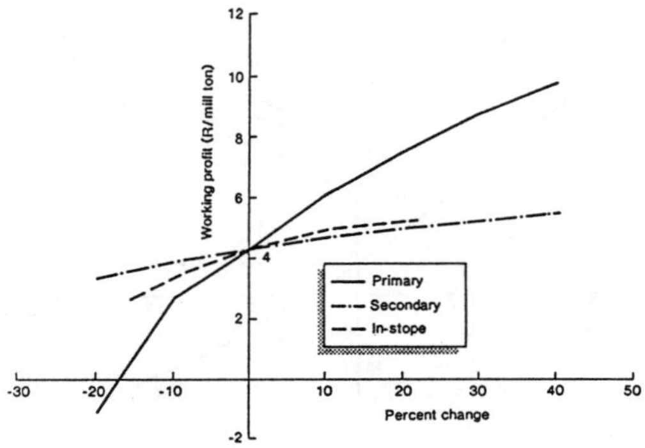

Figure 5. Effect of primary, secondary and in-stope development ratio on profitability

However, the major savings possible from reducing the number of levels were not included in the analysis.

4 EFFECT OF NEW TECHNOLOGY ON PROFITS

It is trite to state that profit is equal to revenue minus cost and yet, when profit is measured in rands, revenue in rands per gram and costs in rands per ton, it is easy to see how the message can be diluted. In the eighties, the common wisdom in the gold mining industry was that the hoisting of more tons was a solution to falling profits as, with higher tonnage, overhead costs would be distributed over a larger base and costs per ton hoisted would be reduced. The argument is correct but the flaw was that the grams of gold in each ton hoisted were also reduced, resulting in revenue decreasing at a faster rate than costs; the nett result of lower working costs was thus lower profits. Currently, substantial effort is being spent convincing the authorities of the need to work a seven day week with the argument being similar: spread the overheads over a greater base. However, a more effective way to increase profits is through the introduction of new technology, and in this context new technology is any technology which is not being used by the mines. The new technology can be introduced as new tools or new mining methods.

4.1 New tools

In the early 1900s one man using a steel jumper and a 4 lb hammer was able to drill one blast hole, 1 m in length and 40 mm diameter, in one shift. Using pneumatic rockdrills mounted on thrustlegs, the drilling rate increased to 10 holes per hour, while the use of the new leg-mounted water hydraulic rockdrills has made it possible to drill 25 to 30 holes per hour. The cost of drilling with both pneumatic and water hydraulic rockdrills is shown in Table 1. From this table it can be seen that, while the new technology in the form of the water hydraulic rockdrill is more expensive to maintain than is the pneumatic

Table 1

	Pneumatic drill 2 man crew	Water rockdrill 2 man crew
Performance		
Drilling Rate (m/min)	0,2	0,49
Drilling Rate (m/hr)	10,0	24,0
Costs (c/m)		
Compressors	21	
Pumping system		62
Pipes hoses and fittings	50	32,5
Rockdrill and thrustleg	40	138
Drill steel	45	45
Electrical power	188	18
Operating labour	329	137
Total	673	432,5

rockdrill, it is far more efficient - drilling at more than twice the rate and consuming one tenth of the electrical power - thus reducing costs. On the assumption of one hole per stoping ton broken, changing to hydraulic rockdrills would increase profits by 241 c/ton milled.

4.2 New mining methods

A more effective means of applying new technology is to reach beyond a single activity and to introduce a number of complementary technologies which, together, will reduce costs and increase revenue, thus having a double impact on profit. An example is the non-explosive impact mining system developed for use in South African gold mines[3].

4.2.1 Description of the present impact mining system

The impact mining system face layout is shown in Figure 6. The face conveyor is installed parallel to the face with a short stand-off to enable the installation of face side props, if required. The conveyor is articulated to follow an uneven footwall and minor faults, and also has horizontal flexibility for advancing continuously behind the mining machine towards the face. The conveyor is a reciprocating flight conveyor, which consists of an articulated drive bar along the rear end of the conveyor to which hinged flights are attached to move the rock on a steel pan at a rate of 30 t/h. From the tail end of the face conveyor the rock is discharged via a chute onto a gully conveyor. The face conveyor also incorporates a guide rail for the mining machine and a line of pegs for the drive mechanism of the machine to

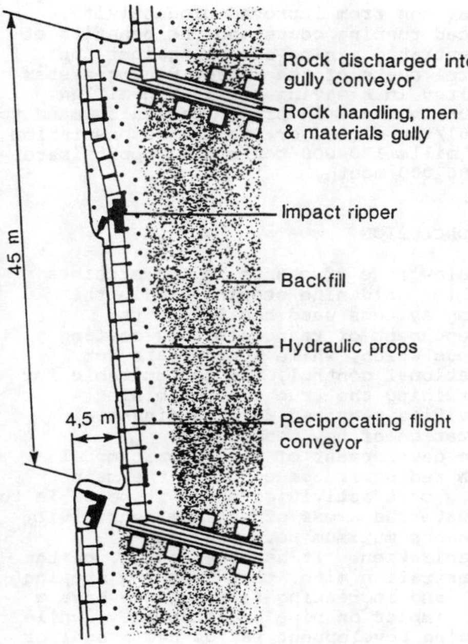

Figure 6. Face layout of the impact mining system

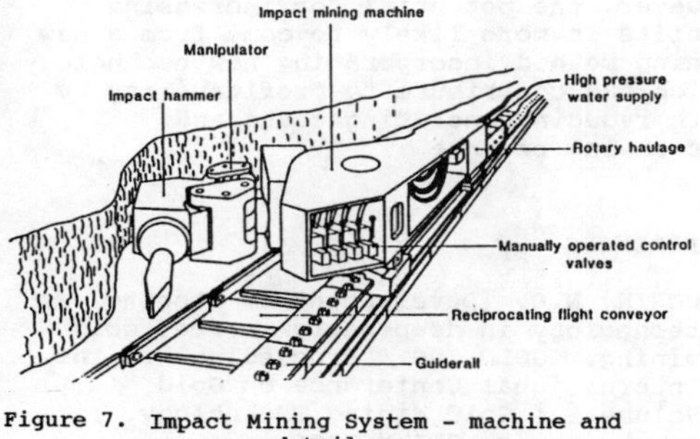

Figure 7. Impact Mining System - machine and conveyor details

engage in; the general arrangement is shown in Figure 7.

The mining machine is located on the conveyor guide rail and supported on the conveyor front edge. The manipulation mechanism which carries the impact hammer is attached to the face side of the machine. Mining takes place with a horizontal circular motion, which results in a 200 mm thick slice of rock approximately 800 mm deep being removed. Typically three cuts are made, one along the centre of the face and one each along the hanging- and the footwall. While the hammer is swung back to make a new pass, a retraction cylinder pulls the tool tip away from the rock face. During the mining of the footwall, the broken rock is pushed onto the conveyor by the hammer. The impact ripper is controlled by the driver via four control valves, three of which are for the manipulation functions and one for

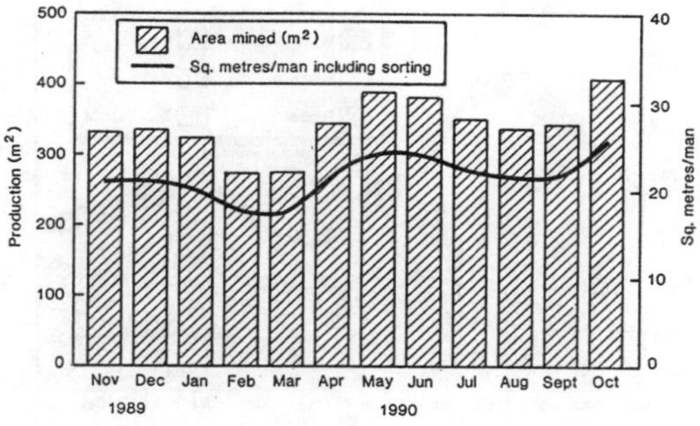

Figure 8. Production and labour productivity achieved at Doornfontein

the haulage, which is a water hydraulic motor. Power supply to the entire mining system is high pressure water, which may either be supplied by a remote pumping system or by hydro-power.

Once the machine has mined to the bottom of the panel, it is driven to the top of the panel onto a special top end section. This dead length mining system allows the machine to mine the top end of the face, despite the fact that, for operational reasons, the hammer has to be mounted on the down dip side of the machine.

Behind the conveyor, two rows of single hydraulic props are installed, which are leap-frogged forward as the face conveyor is pushed forward with a hand-held ram. This ram is anchored against the props as are a number of repositioning cylinders which are attached to the rear end of the conveyor and used to position the conveyor relative to the gullies.

In cases where the reef is narrower than the stoping width and is easily distinguishable from the waste rock (e.g. Carbon Leader Reef), the waste rock can be manually sorted from the slow-moving face conveyor and immediately packed into the back area. This results in good backfill close to the face and the additional advantage that less rock leaves the stope, which has the same effect as a reduced stoping width. Although this operation is labour-intensive, it is considerably less costly than the use of conventional backfill and permits worthwhile savings from reduced dilution.

Better than normal protection is provided for the operators, first because they normally work between the line of props and second because of the closeness of the backfill.

The face is fully accessible while mining takes place, as are the gullies in which rock is removed on a specially developed high reliability belt conveyor. Material handling to the face is also mechanized with a monorail in the dip gully and a transport car, which uses the belt conveyor structure as a rail, for movement of materials in the strike gully.

4.2.2 Impact mining system performance

Figure 8 shows the production and stope labour productivity per month derived from using the impact mining system throughout

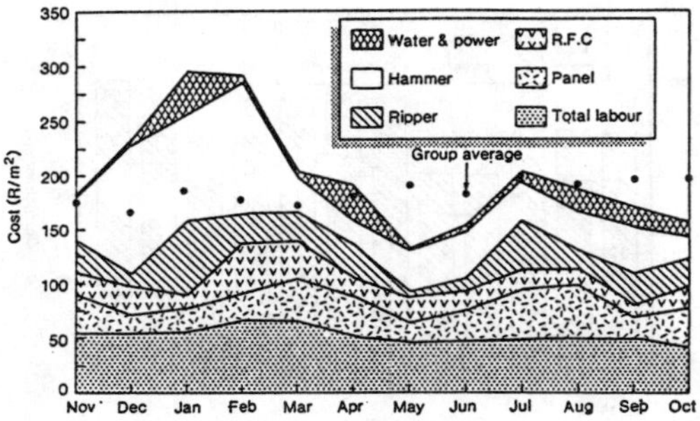

Figure 9. Impact Mining System running costs throughout 1990

1990. The average performance achieved in the six month period May to October was 370 m^2/month at a labour productivity of 23 m^2/man month and an average system utilization of 48 %. The maximum monthly performance was 410 m^2/month with a corresponding labour productivity of 25 m^2/man month. This performance was achieved working one six hour face shift per day and compares well with conventional mining practice where a mining rate of 360 m^2/month (a 30 m panel advanced 12 m) and a productivity of 15 m^2/man month would be considered good when working two shifts per day.

As important as productivity is the operating cost and Figure 9 shows how this has varied over the same period. Running costs were high in the early part of the year owing to some major problems with the impact hammer. These were systematically eliminated resulting in the running costs dropping dramatically. As the year continued, costs continued to show a downward trend while the average costs for the mining group hosting the experiment continued to rise. The peak in July was caused by the failure of a drive gear box which has subsequently been modified to incorporate a shear pin, thus reducing the cost of future failure to a few cents.

The benefits from impact ripping include continuous mining, high rates of face advance, concentrated mining, better utilization of infrastructure, better labour productivity and reduced dilution, and it is the financial implications of these benefits which are of prime importance. Using the model described in the early part of this paper, account was taken of how these factors affect profitability.

The existing machine system operated at a cost of R126/m^2 and broke 355 m^2/month of reef on a single shift basis; 40 % of the stoping rock was sorted on the face and packed in the back area as support. It was assumed that a production system would achieve the same performance but would operate two shifts per day and that the mine would maintain a constant tonnage feed to the mill.

The increase in profit was calculated as follows:

- underground sorting increased the gold value from 5,2 to 6,6 g/ton and revenue by R48/ton milled.
- savings from improved productivity, reduced running costs and the benefits of concentrated mining offset against the capital costs of the impact mining system resulted in a saving of R13/ton milled.

Thus increase in profit was calculated to be R61/ton milled or, as the demonstration mine mills 130 000 ton/month, approximately R8 000 000/month.

5 CONCLUSION

The slow rate of change in the machines used for gold mine stoping and in the mining systems used has led to the entrenchment of responsibility costing methods which, while appropriate for operational control, are not suitable for determining the true cost of mining activities required for the financial evaluation of new technology.

The development of an economic model which redistributes costs to the main mining cost activities makes it possible to evaluate the areas of maximum sensitivity and hence maximum potential for mechanization. It has been shown, on the demonstration mine, that reducing stoping width and increasing face advance have a major impact on mine profitability, while changing development ratios has a smaller impact.

The introduction of a new tool, such as the hydraulic rockdrill instead of the pneumatic rockdrill, will reduce total drilling costs and hence increase profit. However, the potential for increasing profits is more likely to come from a new mining method incorporating new technology which can contribute to profitability by both reducing operating costs and increasing revenue.

REFERENCES

JOUGHIN, N.C. Development of stoping technology in deep narrow - reef gold mining. GOLD 100. Proceedings of the International Conference on Gold, Volume 1 : Gold Mining Technology. Johannesburg, SAIMM 1986.

SMITH, G.L. A model for the economic evaluation of alternative stoping systems for deep level, narrow reef, gold mining operations. MSc Thesis University of the Witwatersrand 1992.

CAMPBELL, D, HAASE, H.H. & OZBAY, M.U. Impact mining for South African Gold Mines. International Deep Mining Conference : Technical Challenges in Deep Level Mining. Johannesburg, SAIMM 1990.

Innovative Mine Design for the 21st Century, Bawden & Archibald (eds) © 1993 Balkema, Rotterdam, ISBN 90 5410 325 6

Selection of media for a high bandwidth utility to serve as the backbone of a mine information system

H.J.Hackwood
Automated Mining Systems Inc., Sudbury, Ont., Canada
R.Poole
Ainsworth Automation Inc., Toronto, Ont., Canada

ABSTRACT: As a result of five years of research and development on its Future Ore Manufacturing (FOREMAN) project, INCO Mines Research and its partners have developed a communications and information system that supports the current and future demands for underground mining automation and control. The system is capable of supporting simultaneously, over a single cable system, mobile and fixed information services for voice, video and data. As the "backbone" for the system, a high bandwidth broadband coaxial cable system has been selected. The following paper discusses the criteria around which this selection was made. Consideration of high bandwidth architecture utilizing fibre optics cable is discussed. Other media types, such as unshielded twisted pair, shielded twisted pair and simple coaxial cable are not discussed because of the relative low bandwidth.

1 INTRODUCTION

1.1 *The challenge to mine operators*

Mining companies are facing serious challenges. Uncontrollable costs such as taxes, power and compensation continue to rise and pose serious challenges to survival. In order to remain competitive, mining companies are searching for ways to trim controllable costs by increasing productivity, improving safety, and creating a higher quality product using fewer resources.

To achieve these goals, traditional manufacturing automation techniques have been applied successfully to the fixed confines of a smelters, mills, and refineries. These techniques are comparable to the traditional manufacturing plant environment. To apply similar techniques to the underground environment, one must overcome the challenges of the ever changing network of drifts, shafts and ramp systems.

In the relatively small confines of a surface manufacturing plant, the transmission of information for control and automation purposes is comparatively easy. Most of the equipment is stationary, and the environment is bounded and well defined; therefore, interconnecting equipment for process control and communications is uncomplicated. The barrier to this kind of information delivery system in the mining environment is the high cost of installing and maintaining the various cable systems for the many different services required. As well, mining often involves mobile equipment operating in an unstructured environment.

1.2 *The foundation of automation*

Without a system capable of supporting the automation functions of voice, video and data, the automation of the underground mine will not be economically or technically feasible.

The Mine Automation Architecture developed overcomes the challenges facing the mine operator by supporting simultaneously, over a single cable system, mobile and fixed information services for control, data acquisition, voice communications, and video monitoring.

A well designed, functional and practical broadband coaxial cable utility forms the backbone of the Mine Automation Architecture. It is designed to support high bandwidth real-time voice, video and data from fixed equipment. The choice of the fixed utility service media is affected by criteria, used for the selection of computer Local Area Networks (LAN) for the manufacturing environment as well as the physical, and data requirement constraints of the underground mine environment.

2 UTILITY SELECTION CHARACTERISTICS

When choosing a transmission utility to serve as the main information carrying network for underground mine automation functions, many of the desirable features used for factory communication schemes apply. These include:

1. Support real-time control.
2. High data integrity.
3. High noise immunity.
4. High reliability in harsh environment.
5. Suitable for large installations.

Additional features are essential for the mining industry. The following section discusses the main characteristics and considerations.

2.1 *High bandwidth*

Automation architectures available to the factory floor have capabilities for telemetry, high-speed data, low-speed data and video. Each capability has different bandwidth requirements. The diverse requirements of the automated mine, for voice, video, high speed data, low-speed data, energy management, and other services, establishes the need for a communication network with high bandwidth (Baiden 1991). The most demanding requirement is the need for multiple video channels; therefore the selected media must support these functions

In addition to supporting the existing needs, extra bandwidth capacity within the network must be available so that future expensive expansion will not be required. If we use an analogy to the electrical wiring in the home, the total requirement of the service in the home should exceed the demand of any current and future appliances. If adequate provision is not made for these devices, extra wiring will be required.

2.2 *Readily expanded with widespread access*

The nature of the ever-changing layout of the mine, requires a system that is easily expanded as the mine grows. As new headings, ramps and mining horizons are developed, expansion of the system should be as easy as running a new power service. A one-time installation that is feasible in other surface applications does not apply to the underground utility.

It is essential to provide widespread access. The provision of system outlets at virtually every location where one may be needed minimizes the costs associated with even the smallest amount of rewiring. Moving devices from one location to the other should be as simple as plugging in a lamp, if we consider the house electrical service analogy once again.

2.3 *Application independence*

Regardless of the device that is "plugged" into the system; camera, monitor, PLC or other device, it is desirable that it performs as designed without changes, adapters or other rework to the device or the network. In the electrical analogy, when a lamp is plugged into the power socket, you would expect it to illuminate when you turn on the switch, regardless of which socket you plugged into.

2.4 *Simplicity and Flexibility*

Ease of maintenance and administration is possible if one utility supports all the applications required. Total system costs are reduced and overall administration is greatly simplified. One set of maintenance procedures, one training plan and one type of hardware and software minimize the costs as compared with multiple network systems.

In order to achieve vendor independence and flexibility, the ability of the network to support existing network schemes, as well as future developments, minimizes the cost of obsolescence.

The cost of flexibility includes moving existing devices, adding new devices, and changing existing applications and functions. The ability to purchase from multiple vendors based on vendor reputation, service, or quality as opposed to system compatibility, is also important.

2.5 *Reliable and robust*

The ideal mining utility is reliable and robust against the harsh mining environment. This includes immunity to Electromagnetic Interference (EMI), operation in high humidity and ability to withstand minor physical damage.

2.6 *Topology*

The arrangement of the components in a mine utility, like a computer LAN, is described in two ways. The physical topology describes where the components are located. The logical topology or architecture describes how the components are connected to each other by the network.

2.6.1 *Physical topology*

The placement of the components and distribution network determines the physical topology of the system. Examples of the patterns of connection of the devices located on the network are star, bus or ring as illustrated in Figures 1, 2 and 3, respectively.

The star topology can be implemented with a simple point-to-point protocol; however, it does not lend itself to distributed processing due to the dependence on a central hub. Other disadvantages are the high costs for wiring in large installations, limited throughput, and the fact that hub failure results in the system crashing.

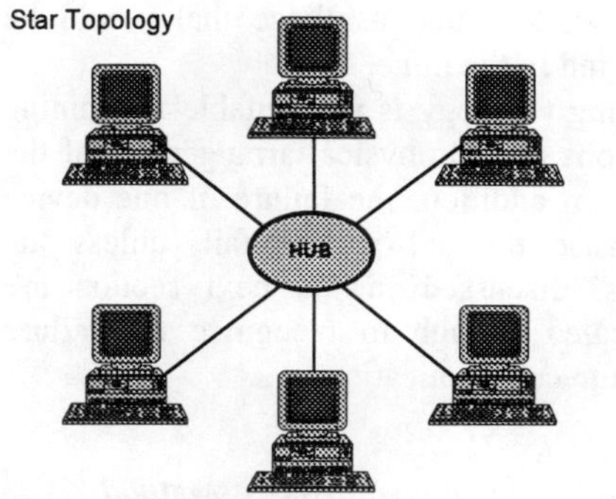

Figure 1 Star Topology

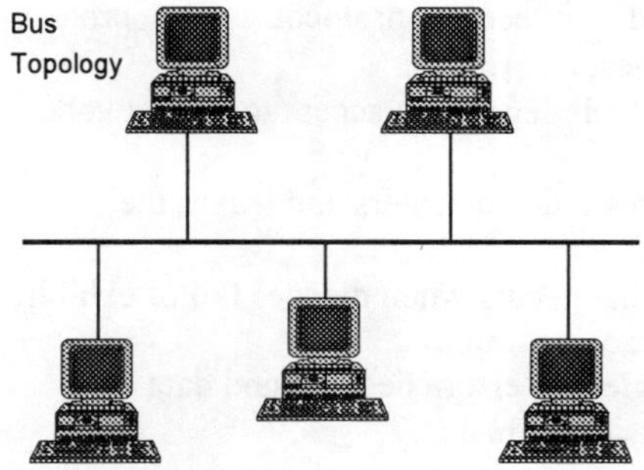

Figure 2 Bus Topology

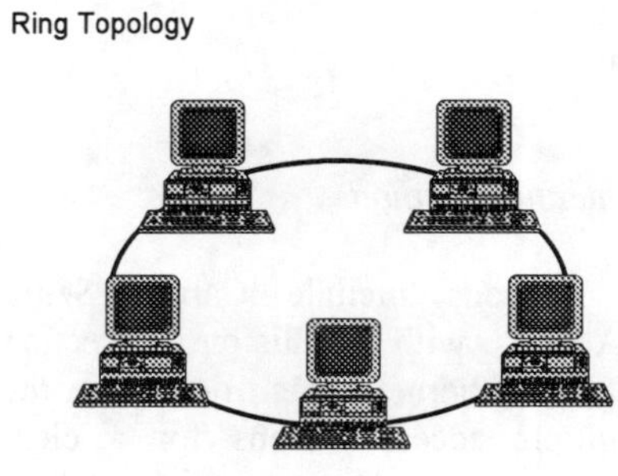

Figure 3 Ring Topology

The bus topology consists of a main trunk line in which individual devices are connected in a multi-drop fashion. Communication can take place between any two nodes without passing information through a central hub. In this way, distributed control applications are easily implemented. The main disadvantage is the dependence of a shared bus to service all the devices. A break in the main trunk line could affect many nodes. On the other hand this is usually the most cable-efficient approach in large networks, such as those that would be encountered in the mine.

The ring topology is not suitable for mining applications due to physical arrangement of the cabling. In addition, the failure of one device could cause the network to fail, unless the protocols, discussed in the next section are sophisticated enough to recognize the failure and reroute communications.

2.6.2 *Logical topology or Access Method*

The rules by which devices access the network is called the access protocol. The protocol determines:

1. Which device has access to the network next.
2. How a device enters and leaves the network.
3. What occurs when devices fail or exhibit incorrect behaviour.
4. Rules for error checking and data transmission formats.

There are many ways to accomplish these rules of behaviour; however, they can generally be divided into two methods:

1. Contention.
2. Polling.

2.6.2.1 *Contention methods*

Contention methods include Carrier Sense Multiple Access with Collision Detection (CSMA/CD), or Ethernet. Each device on the system (multiple access) listens for a clear network (carrier sense) then transmits its data. If two devices detect that they are transmitting at one time (collision detection), they stop and wait a random amount of time before trying again. Collisions occur only if two devices begin transmitting within a time interval equal to the propagation delay between the devices, the collision window. When two distant devices transmit at the same time the propagation delay can be such that the collision is not detected. Therefore, the physical layout of the cable network is restricted to a distance where the collision window is not exceeded. The practical limitation of contention systems at 10Mbps is 1.6km; therefore contention methods may not service larger mining networks.

In addition, each collision and detection takes time which can not be used for transmission; therefore, throughput drops off and access time increases. Where real-time control and priority transmission capabilities are required, the data rate and throughput must not be compromised because of safety concerns.

2.6.2.2 *Polling methods*

Polling techniques include master-slave access and token passing. Master-slave involves a master polling each device in sequence, to determine if it has data to transmit. The device would either send its data or send a message advising that it has no data to send. This technique is inefficient for real-time control because each transmission must first be sent to the master, then to the slave. However this scheme could be used for less critical control.

With token passing, each interface device attached to the network has permission to transmit when it receives a unique pattern of data called a token. When a device has information to send, it captures the token, transmits its data and replaces the token on the network.

Ring and bus topologies dominate token networks. Token passing networks are deterministic and as such have no limitations in distance, other than distortion from noise. For distributed control, where access times are critical, token passing is preferred.

2.7 *Transmission method*

For each transmission media there are one or more basic transmission methods, including:

1. Analog broadband.
2. Analog baseband.
3. Digital baseband.
4. Digital fibre optics.

Each has its advantages and disadvantages.

2.7.1 *Analog broadband*

Broadband is a transmission scheme that uses frequency division multiplexing to divide a "broad" range of frequencies into multiple smaller independent frequency channels. In this way, a broadband system can carry a great deal of data, provided the cable is able to handle the bandwidth.

The cable bandwidth is divided into channels, much like the cable television system. These channels are then used in a variety of ways, for voice, video and data transmissions.

Broadband schemes often use two channel ranges because devices on the network transmit at one frequency and receive at another frequency. The transmit and receive frequencies are divided by a range of frequencies that act as a "buffer" or "guard", so that signals do not interfere with each other.

All transmitted signals travel along the cable to the head-end. At the head-end, the transmitted frequency is converted to the receive frequency and retransmitted over the cable. Devices tuned to the proper receive frequency respond to the retransmission.

2.7.2 *Analog baseband*

Baseband uses only one frequency channel to transmit data over the selected media. Only one signal is transmitted at a time. This has an advantage in that the design of the modem is simplified and there is no retransmission facility, because data is sent and received on the same channel. However, there is a limit to the length of a baseband system and also the number of stations that can coexist on the network. Finally, baseband systems can not be used for any other transmissions, other than those which it is already carrying.

2.7.3 *Digital baseband*

Digital baseband transmits a digital wave form over a wire as fluctuations in voltage. The baseband signaling technique transmits a signal in its original form, without modulation. The entire purpose of the cable is to transmit a single signal. Only one signal is transmitted at a time. Therefore, digital baseband has the same limitations as analog baseband.

2.7.4 *Digital fibre optics*

Digital transmission over fibre optics is accomplished through on/off pulses of light. Each digital electrical signal from any device must be converted into light signals.

A Light Emitting Diode (LED) flashes at extremely high rates yielding a very high data rate over long distances. Transmitted digital information from a device must be converted from an electrical pulse to a light pulse and then back again at the receiving station.

To send information over the fibre, electronic signals from a device are converted to optical signals, light, by transducing devices called optical emitters, transmitters, and optical detectors, receivers. Therefore, a single point-to-point link between two devices, consists of a fibre optic cable pair together with a matched pair of optical transmitters and optical receivers.

2.8 *Media Types*

As discussed in the Introduction, this paper limits its discussion to broadband coaxial cable and fibre optics cable. Due to the relative low data speed and distance restriction associated with twisted pair and lower capacity coaxial cable, they are not topics here.

2.8.1 *Broadband Coaxial Cable*

Broadband coaxial cable is used extensively in the North American cable television industry. This wide use results in a widely available cable which is relatively inexpensive. In addition, installation tools, diagnostic equipment and wide-spread services are available. These broadband coaxial cables are available in different carrying capacities, sizes and special designs. See Figure 4.

The center conductor is solid copper or copper clad aluminum wire. The dielectric is electrical insulation used to maintain position of the center conductor and is normally composed of foamed polyethylene or evenly spaced polyethylene discs. The shield or outer conductor is constructed of an aluminum tube. The cable size 412, 500, 750 and 1000 is

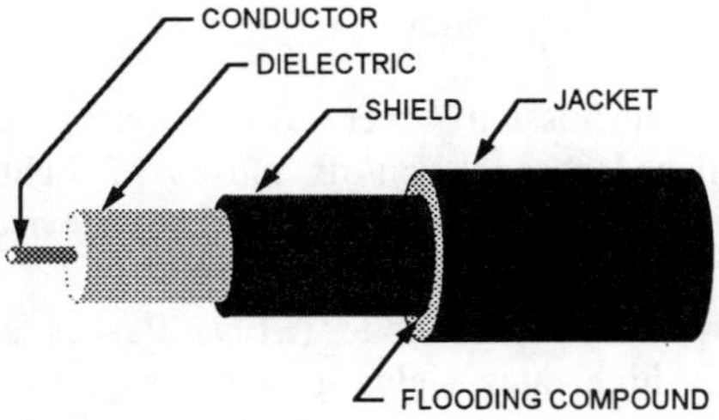

Figure 4 Coaxial cable

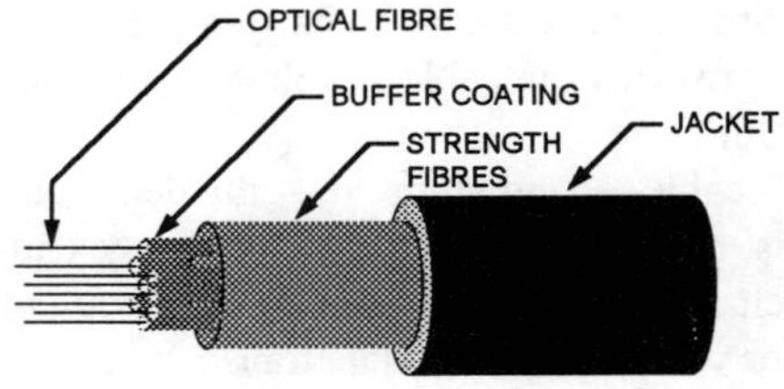

Figure 5 Fibre optic cable

derived from its outside diameter (500 representing 0.500 inch for example). A viscous substance placed between the outer conductor and the jacket maintains a protective seal should the jacket contain or develop any cuts or openings. The jacket is normally a black polyethylene coating over the aluminum outer conductor to provide a environment-tight seal.

2.8.2 *Fibre Optic Cable*

The fibre optic cable is usually a group of glass or plastic fibre cores, each covered by a plastic buffer coating cladding, including kevlar fibre strands for strength and a final protective sheath or jacket of plastic or Teflon. See Figure 5.

The thin glass fibre consists of an inner glass core and an outer glass cladding. The transmitted light travels inside the transparent inner core, reflecting off the outer cladding. How the light travels inside the core can be divided into two categories:

1. Single mode.
2. Multi mode.

These modes are a function of the core diameter; the multi-mode being the larger of the two.

3 BACKBONE UTILITY SELECTION

The following section discusses the selection of the appropriate utility media for the backbone architecture considering the many criteria and background information discussed in the last section.

3.1 *Broadband Coaxial Utility*

A single broadband coaxial cable can provide several hundred MHz of bandwidth supporting tens of independent frequency division multiplexed (FDM) channels. Each FDM channel can in turn be subdivided further by using a variety of channel access techniques, yielding hundreds of independent communications channels. Broadband supports simultaneous low-speed, high-speed, narrow band and wide band communications services.

Using this media, a mine is wired with outlets (taps) installed at predetermined locations, similar to installing electrical outlets in your home. Once a mine is thus wired, devices can be attached anywhere a standard tap is installed. The installation and repair of the coaxial cable is straight forward and can be accomplished by trained tradesmen, such as the mine electrical staff.

The broadband design ensures that each tap has the same signal level across the entire bandwidth. In this way, an attached device performs as specified regardless of where it resides in the network. It is difficult to use a fibre link for more than one service: high bandwidth, multiple service fibre modems are still quite expensive, for now.

A device installed on the broadband network has access to the full system bandwidth. It is the devices responsibility to only transmit/receive in the spectrum allocated to it. The available bandwidth is divided according to the applications' requirements, whether high-speed data, low-speed data or a complete video channel.

A broadband system can support several different types of local networks. The physical topology of the broadband network in the mine is "branched tree" or "bus". This represents the layout of the typical mine. In the logical topology, token passing access techniques are usually used due to the large distances to be covered underground.

This arrangement offers the advantage of minimizing the infrastructure required in the initial development of the operation and allows for future expansion as the operation grows.

The design of the broadband system provides excellent shielding from such primary EMI sources as transformers, fluorescent lighting, radio transmitters, etc., by virtue of the cable shielding techniques.

3.2 *Fibre optic Utility*

The bandwidth of fibre optic cable is typically 10 to 100 times that of coaxial cable ordinarily used in local area networks (Raasch 1990). The fibre has uniformly low attenuation across its bandwidth from the lowest to the highest data rates, even above 200 Mbps.

With fibre optics, the cumulative fibre losses from relatively high loss taps, and the difficulty in tapping fibre creates problems. Technically and economically, is is difficult to create a distributed, multiple outlet bus topology similar to the broadband approach.

In addition, to achieve the extended performance required by the token bus architecture in the mine, the use of passive star segments interconnected by repeaters has been proposed for large buildings (Scoll 1988). The distance of this arrangement is limited to four optical repeaters in any signal path, thus limiting its effectiveness in some mines.

The star topology, a point-to-point configuration, requires a dedicated fibre pair to extend to the passive or active hub, from each network device. To achieve the same functionality of broadband approach, several multiplexed baseband fibre pairs would be required.

An unlimited number of devices may be required to be added to the broadband system. Communications approaches using fibre optic links, are suitable for point-to-point systems, or buses with a limited number of taps. Fibre optic systems do not suit a flexible system where attached devices may have to be added later, often without their locations being pre-determined.

Fibre optic transmission systems will continue to decrease in costs, increase in bandwidth and data rate, driven by technological developments in the competitive telecommunications industry. For example, as demands for a system that supports multiple services grow in this business, the use of wavelength division multiplex (WDM) systems, will mature. This is a technique whereby two optical carriers are transmitted on a single fibre at two different wavelengths. For now, new developments that support the higher data rates and multiple service environment are expensive, and require sophisticated testing equipment.

With respect to the environment, optical fibres are immune to EMI; there are no electrical connections along the cable because the connection is glass. As such, it does not exhibit any interference modes that could be associated with copper conductor based systems. However, in the event of a lightning strike, the electronic hubs, optical repeaters, and other active devices are equally susceptible to failure, as are the electronic components of other systems.

Another point to consider is that optical fibre degrades with time. When exposed to gamma radiation, that is found in uranium mines, the degradation accelerates.

The fibre cable is lighter and more flexible than the broadband cable. However this does not make it any easier to install. Fibre and broadband cable require the same care with respect to bending or tensioning forces. Cable terminations and the installation of connectors is considered by many to be more difficult with fibre optic cable. Although many others claim to be able to install connectors in as little as five minutes, technicians familiar with splicing fibre on a day-to -day basis average 20 to 30 minutes per splice, in a surface environment.

Table 1 summarizes the above discussions.

4 MINE AUTOMATION ARCHITECTURE SUMMARY

The key to enabling practical automation

Table 1 Backbone utility comparison

Criteria	Broadband Coaxial	Fibre Optic Cable
Bandwidth	High	Very High
Distributed access	Yes	Difficult
Access methods	Many simultaneously over one cable	Many over multiple fibres
Installation	Easy coaxial base	More difficult connectors
Maintenance	Standard RF techniques	Sophisticated equipment required
Reliability	High	High
Transmission	Broadband, multi-drop	Point to point
Physical topology	Branching bus	Point to point
EMI immunity	High	Very high

projects in the mine workplace is a communications system flexible enough to adapt to the rapidly changing conditions underground. In the Mine Automation Architecture, this is accomplished in a two-part system that links a robust fixed utility with flexible distributed antenna coverage for mobile users.

A distributed antenna system provides coverage using radiating coaxial cable to the mine working areas. The two sub-systems are linked at multiple locations throughout the mine using Distributed Antenna Translators (DATs), which pass specific services back and forth through a set of electronic support modules, installed within the DATs.

4.1 *Broadband utility summary*

It is for the reasons noted in Table 1 that the selected backbone is the broadband coaxial utility. The basic hardware and cable used in a broadband cable system are similar to that used to provide cable television service to homes in your community. Cable, connectors, amplifiers and other hardware are all designed for outdoor use in extreme environments. The reliability of the system components is typically over twenty years mean time before failure (MTBF) (Nickols 1992); longer than the life of some mines. The equipment for the utility itself, as well as support tools, test, and terminal equipment is readily available from a variety of suppliers to the competitive cable television (CATV) industry. This cabling system is designed for use over large geographic areas.

The components are installed and aligned to become an IEEE 802.7 compliant cable system that has been adapted from the CATV industry. It is this standard that has been adopted by large industrial automation users such as GM, Ford, a variety of processing companies, utility and government agencies.

The primary flexibility of the system lies in the ability to provide a variety of simultaneous services on different "channels" over the same cabling system. It is typically equipped with a built-in monitoring and fail-safe facility to monitor performance of the utility. This can be supported with fail-safe amplifiers that are enabled if primary amplifiers fail.

The installation of a broadband cable utility reduces is more economical than a cable per service, while providing expandability for future services. The total costs for a "cable-per-service" installation easily exceeds the cost of a broadband installation if cable cost, installation, personnel training, documentation, multiple inventories, and other associated costs are considered. Further, the fact that the communications utility monitors itself and reports faults and potential problems can increase the utilization of an automation system and reduce down-time costs, through easier and quicker maintenance.

In addition to providing a single, well-supported and maintainable system, future services can be added to the mine without any additional cable costs by plugging into the installed broadband at any tap. Any service available at a tap on the broadband system is also available at any other tap in the entire mine, facilitating migration and re-organization of users or equipment, quickly and easily.

Future developments in LAN technologies can be mated to the broadband system to allow evolution of services without the need to rewire (Cooper 1986).

5 CONCLUSION

The architecture allows the mine to be viewed as a large manufacturing plant. It permits the application of standard manufacturing automation techniques to the mining process. These techniques will enable mining companies to realize the same benefits of cost reduction and increased productivity and safety that traditional surface manufacturing companies have received in recent years.

The main impact of this technology is the significant improvement in productivity gained, as automation services are implemented in the mine. Similar advancements in safety are achieved as better communications are established. Operators are relocated from potentially hazardous areas underground, performing remote operations from the surface. The key to all of these improvements is the increased information flow from all facets of the mining operation made possible by implementing effective high capacity communications that forms the building block of mining automation.

An IEEE 802.7 compliant broadband coaxial network has been selected to form the backbone of the Mine Automation Architecture. The key factors which relate to the selection include:

1. Physical topology suitable for the large extent of the underground mine.
2. Token passing bus access techniques to allow virtually unlimited expansion.
3. High bandwidth that is easily tapped.
4. Installation and maintenance can be performed by moderately skilled workers.
5. Multiple simultaneous services including voice, video and data.

The broadband CATV technology of the communication utility is well able to withstand the harsh environments of a mine. It is a mature technology that supports a variety of data rates, low connection costs, and ease of implementation for multiple services.

While the cost to implement a single application system is slightly higher than a single serive cable system, the existence of a broadband utility facilitates further automation without the need to re-engineer and support another discrete solution. In an industry where constant change and development of the work site is the norm, the broadband cable approach gives unique flexibility in allowing new systems to be added and attached. An essentially unlimited number of devices can be added to the system, depending on the service provided, without regard to physical distance, cable losses, branching topology, or other considerations.

For today's mine automation backbone, the solution available now, is a broadband coaxial cable utility: the architecture that forms the foundation of mine prosperity.

REFERENCES

Baiden, G.R. 1991. Mine-wide information system development at INCO Limited. *Proceedings International Symposium on Mine Mechanization and Automation.* Golden: 3-1 - 3-14.

Cooper, E. 1986. *Broadband network technology - An overview for the data and telecommunications industries*: 14. Mountain View: Sytek, Inc.

Du Toit, D. 1989. A Passively Tapped Linear Fibre Optic Data Bus. *Proceedings International Congress and Exposition.* Warrendale, PA, SAE Technical Paper Series.

Hawkes, W.C. 1989. A step forward. *Proceedings Ninth Underground Operators' Conference.* Sudbury.

Nickols, D. 1992. Broadband network communications and associated applications for mines. *Proceedings Underground Monitoring and Communications Workshop.* Australian Coal Association.

Raasch, E.M., Hatfield, W.B. 1990. *The fibre optic LAN handbook - Second edition.* 135. Yonkers, Codenoll Technology Corporation.

Innovative Mine Design for the 21st Century, Bawden & Archibald (eds) © 1993 Balkema, Rotterdam, ISBN 90 5410 325 6

Towards a gender-free mining industry

M. Phyllis Bray
Kingston, Ont., Canada

ABSTRACT: Mining has the reputation of being a very "macho" industry. Recently, wider educational opportunities for women in nontraditional occupations and changes in mine technology have combined to make mining a viable career choice for women also. To benefit from this new resource, the mining industry faces the challenge of developing a different work environment from the one now burdened with sexism. Management-supported initiatives are needed to facilitate such a transformation. The same rethinking of gender attitudes might also be extended to mining women in their traditional role as "wife of." The assumption that the male worker's wife is an unpaid adjunct is being questioned. It is suggested that women's separate concerns both as helpmate and as part of the mining workforce should be considered when planning for mining into the 21st century.

Without science and engineering, women lose.
Without women, science and engineering lose.
(CCWE 1992)

INTRODUCTION

To discuss the place of women in the mining industry of the 21st century may not seem appropriate in this Mine Management Section of the Congress. The mining industry has traditionally been very male-oriented and certainly the gender division of the Congress registration would indicate that the gender composition of the industry has changed very little. Why should it change? The industry has apparently managed very well without feeling the need of women's direct participation and without acknowledging any debt to the indirect contribution of mining wives.

This paper will attempt to demonstrate that such a complacent attitude should be reappraised as we approach the 21st century. In reality, gender relations in the mining industry are one of the critical management issues smouldering below the surface. It will be further argued that unless this problem is addressed by senior management with some degree of immediacy, it will escalate causing significant human and economic damage. In the following discussion "Toward a gender-free mining industry," the present discriminatory practices will be described in some detail and an attempt made to inject a sense of urgency into suggestions for change in the coming century. Canada will be used as an example of the industry and the engineering profession as a whole as a model from which to extrapolate problems more specific to mining. Women's direct participation in the mining industry as professional and nonprofessional members of the workforce will be discussed as will proposed innovations in methods of mine development as they affect women in their traditional role as mining wives.

EVIDENCE OF GENDER BIAS

The problem of gender bias in the engineering profession as a whole and mining as a part has been identified in detail in the April 1992 report of the Canadian Committee on Women in Engineering (CCWE). With the aid of statistics, the Committee wastes no time on moral suasion. The Report states bluntly that unless this misogyny is modified and the industry opened to women, all indicators show that Canada will be seriously disadvantaged in world competition in the next century. Already Canada produces fewer engineers proportionately than almost every other industrial country. The Organization for Economic Co-operation and Development (OECD) in a recent survey placed Canada last among seven leading industrialized nations (McKay 1992). The shortfall used to be made up through immigration, but this source is drying up as other countries are anticipating a shortage of engineers and hanging on to the supply they have (McKay 1992). The Canadian Council of Professional Engineers estimates that the number of engineering jobs will increase dramatically before the end of this century and on in to the next, far outdistancing the number of engineers coming in to the labour market. While the shortage will probably not be as acute in the currently depressed mining industry, it does not need a crystal ball or Ms. Popcorn's predictions to conclude from these forecasts that Canada will face an actute shortage of engineers in the coming century (McKay 1992). Making up the projected shortfall by attracting many more bright young women into the field would seem an obvious solution (McKay 1992). Yet in spite of these deeply troubling forecasts, women, who in Canada make up about 50% of the population, are still being excluded from meaningful participation in the engineering profession and in the workforce. The question has to be asked why this profession seems to have such difficulty in accepting women as working partners.

IDENTIFYING THE PROBLEM

Some answers came out of the horrifying events of the Montreal Massacre of December 1989 when 13 young women engineering students at Ecole Polytechnique were murdered. The intense misogyny which fueled this event was so disturbing that the Canadian Committee on Women in Engineering (CCWE) was formed to examine the event and its implications for the engineering profession. This was a joint federal government, professional association and industry project and the gender-balanced Committee included engineers, educators, employers and association representatives. The Committee's report, "More Than Just Numbers," was published in April 1992. The Report's conclusion is that the engineering profession is at best unwelcoming to women, at worst contemptuous and hostile (McKay, 1992). Ihe causes of this attitude are examined in some detail as are its effects on young women as they move through school, university and on in to the workforce. The Report concludes that there must be a major change in the collective thinking of the engineering profession from the overt sexism which sours the atmosphere in engineering classrooms, workplaces and association boardrooms (McKay 1992). Only when this takes place will women feel encouraged to enter engineering as an opportunity to which they are entitled (McKay 1992).

POSSIBILITIES FOR CHANGE

How to bring about this major change in collective thinking on the part of the engineering profession? It seems agreed that some of that change must begin in the home and in the school system starting with the elementary grades. Long before career decisions are made, it is apparent that there are cultural influences at work which give young women the perception that mathematics and physics, the foundation of engineering courses, are too difficult for women, that engineering is a 'male' profession, that nontraditional jobs are not appropriate. There are very few female role models to make nonsense of these myths, which leaves family and the educational establishment to undertake the major responsibility for changing these attitudes. But a committed engineering profession could also do much towards promoting science education for women in public schools, funding college level scholarships, encouraging women engineers to serve as role models for young women.

DISCRIMINATION: PROFESSIONAL JOBS

Although these outreach programs will have some effect in attracting women to the profession, more fundamental attitudinal changes are needed among male engineers before women will feel at home in the profession. Some idea of the problems which have to be overcome may be seen from an interchange of Letters to the Editor in 1991-1992 issues of "Engineering Dimensions," the official journal of the Association of Professional Engineers of Ontario. This interchange was fueled by several articles discussing the women-in-engineering problem. Comments ran the gamut from the suggestion that the Montreal massacre would not have taken place if the women engineering students involved had been encouraged to stay home (a classic example of blaming the victims of violent crime), through statements that women should stop whining about lack of opportunity with the implication that their wish for job opportunities was fueled by greed. Another correspondent suggested that the reason some men were opposed to women working in non-traditional jobs was their own insecurity about their masculinity, and finally, the end-the-discussion comment that when male engineers put their emotions aside and applied their training in rational thinking to the issue, some solutions would be found.

Rational thinking alone would question limiting the choice of employees by any other criterion than ability. Nor would it seem rational to exclude arbitrarily half the population from consideration in choosing a workforce. But the endemic problem of gender bias is a very emotional issue arising from entrenched social and cultural beliefs which support restrictive stereotypes of women. Genuine solutions would require fundamental attitudinal changes about women and their freedom of choice. The engineering profession is lagging far behind other previously male-dominated professions in accepting women as equals even though it has long been acknowledged that there are really no physical or intellectual barriers preventing women from entering this profession. The Committee on Women in Engineering found that women represent just 14% of engineering students at Canadian universities and in mining the enrolment was 11.6% in 1985 and 13% in 1990, one of the lower rankings (CCWE 1992). Yet in law, medicine and business, formerly male-dominated professions, enrolment is now almost equal so that women are no longer in a minority position.

DISCRIMINATION: NONPROFESSIONAL JOBS

The same reluctance to accept women as equals applies even more to women in nonprofessional jobs in the industry. Here again, there is really no reason why women cannot be considered for nontraditional nonprofessional jobs. In mining for instance, push button operations are the norm and it does not take great strength to control the largest pieces of equipment. Indeed women perform very well handling heavy equipment both underground and on the surface. But many women who have taken up the challenge to perform these nontraditional jobs have faced even a more punishing atmosphere than in the professional ranks of female jobs. In this workplace, sexism is not just covert but overt,- where sexual terminology is used for virtually every machine, tool and object, where the language of the workplace is full of sexual innuendo, where pornographic pinups are ever present, and where the presence of women disrupts and destroys the exclusively male-defined sexual nature of the workplace (Braid 1981). With strong union support, or strong management support, women perform up to male standards. Indeed, there is evidence that despite the masculine environment, women derive great satisfaction in performing well in nontraditional jobs because of the variety and challenge, the opportunity to learn, the crew camaraderie and of course, the pay, which is much higher than for traditional female jobs (Martin 1986). However, fundamental changes would be needed before the work environment would be comfortable enough to make nontraditional mining jobs attractive to women.

THE RMC INTEGRATION MODEL

To bring about a gender-free mining industry as part of a move to open up engineering to women is not just a matter of identifying the problem and then hoping that time will take care of the solution. To boost the percentage of women in a workforce requires a plan, the will to carry it out and definite numerical goals. It might be useful to examine the transformation which took place in the early 1980s at the Royal Military College of Canada here in Kingston as a possible example of the way an all-male organization was moved toward gender neutrality. The Royal Military College (RMC) is one of the officer training universities for the Canadian armed forces and until recently had an all-male student enrolment, the majority in engineering programs. The integration of women into the RMC student body took place as the result of a government directive as part of official policy to eliminate gender bias in the Canadian forces.

In the course of a conversation, Major-General Stewart, RMC Commandant at the time of integration, explained that there were already many women in the support branches of the armed forces and it seemed only fair to give them the same opportunity for officer training as had been available to men. There had been some prior consultation with the RMC senior alumni about the proposed integration. Many were very much opposed to the proposal and these mounted a vigorous campaign to convince the government not to proceed. However, the decision was made to go ahead and 1980 saw the first class in which females were accepted as officer cadets. There were 32 in that first class. This number was chosen specifically so that there would be groups of four women in each of the eight squadrons into which the college was divided.

With a firm commitment to equality of treatment, RMC settled the 32 women, four to a squadron, two to a room, sharing dormitories with the men. To monitor the process, a special "Minerva" Committee was set up chaired by the Commandant and including among others the Principal, two women officers, and the Director of Cadets. This high level group met at least once a month during the early stages of the integration. This offered a continuous monitoring of what was happening not only to prevent problems from developing but to deal with those already on hand. In addition, one Squadron Commander, a woman officer, was designated as the person to whom women cadets could go with problems and this was publicized. For serious complaints resulting from sexual harassment, a clear policy was developed on the part of administration: dismisssal prompt and unequivocal for a cadet guilty of the offence and disciplinary action for those in charge who were aware of the offence but took no action. Now after more than a decade, women are an accepted part of the College community and, looking back, many of the participants wonder what all the fuss was about! When asked what features of the RMC integration might apply in the industrial world, General Stewart listed: strong motivation to carry out the process, commitment to equal treatment, clear guidelines, well-defined penalties for noncompliance and a senior monitoring organization empowered to set policy and take appropriate action.

CIVILIAN APPLICATION OF RMC MODEL

It could be argued that this model would be inappropriate outside the military establishment. Certainly RMC had a powerful incentive for change: a government directive reinforced by the instrument of army discipline based on following lawful orders without question. In other words, the will to change was imposed and army discipline made implementation of the change obligatory. This is not such an easy model to transfer to the civilian workplace. But the industrial workforce is also very well motivated to comply with directives from the upper echelons. Jobs are jobs after all,- a powerful incentive to compliance with managerial directives. What industry lacks is the will. At the moment, there is nothing but public opinion to force change, backed by dire predictions of what will happen if the potential of half the population continues to be ignored. Militating against action are firmly held sexist notions and behaviors among those within the industry. These reinforce the idea of traditional jobs for women and discourage any move to change the current ratio of male and female in the workforce.

THE CHALLENGE

Without will to change, is there any chance that women will find a place in the male-dominated engineering enclave? Monique Frize, chairperson of the Canadian Committee on Women in Engineering takes heart from the fact that "when engineers are faced with a problem, their inclination is to solve it on time and within budget" (McKay 1992). This Congress offers an ideal opportunity to challenge the engineering profession in general and the mining industry in particular to recognize the problem of endemic discrimination against women so that the collective mind can be put to finding a solution. And what more appropriate venue than the Mine Management Division in which to plan ways to achieve this visionary workforce based on ability not gender. The RMC experience clearly shows that strong support and commitment from the upper levels of the organization are absolutely vital to the success of such an operation. There is no way that gender bias will solve itself in due course nor will it react to bandaid solutions of hiring a few token females as a gesture toward equity. Rather, to solve this problem management must undertake the much more difficult task of changing people's minds and entrenched attitudes. This is a revolutionary, not an evolutionary process. Etta Wharton (Ontario Hydro's director of employment equity and herself an engineer) who has been through the process, gives this advice: recognize that gender bias is inherent in the system, stop rationalizing and start working on a solution (Eng Dim Nov./Dec. 1990).

MEETING THE CHALLENGE

There are guidelines. The RMC experience of integration has already been outlined. Hydro, in moving toward a gender-free workplace, has developed well-defined and well-tested guidelines to implement, without unnecessary disruption of the workforce, changes which may seem revolutionary. Rather than using a military model, Hydro has used basic business practise to achieve its objectives but with some modifications because of the emotional and controversial nature of the changes being introduced. Basic to the change is a well-defined chain of responsibility in which management has its objectives clear, responsibility is assigned, goals set, people in the organization made responsible for performance and rewarded accordingly. Further Wharton suggests that guidelines for recruitment and promotion be clearly laid out with penalties for nonconformance spelled out. She stresses that the success of this operation depends on the behaviour of supervisors at all levels. To this end, she believes that part of the selection criteria for supervisory positions should be the requirement to support the program and demonstrate results. This in itself would help develop a supportive cadre to fill the gap between what an organization says and what it does and to reward desired behaviour. (Eng Dim May/June 1991).

However, the whole operation falters unless, in all the plans for implementing integration, there is a numerical objective. It is important to state clearly what proportion of women in the work force would be considered equitable, how management plans to reach the objective and within what time frame. (1) This can develop into a very touchy area of quotas, hiring procedures tainted by number crunching, and equal opportunity considerations skewing hiring criteria away from merit. If an organization is serious about developing a gender-free workforce then probably at the beginning quotas will be necessary to "kick start" the process. Great care needs to be taken in this event that ability is still the first criterion so that the women do not suffer a backlash from male colleagues who perceive gender and not qualifications as the reason for being hired. In spite of the high level of performance of women engineers, this reaction has to be expected and should be addressed. The process is difficult but already there is a fund of experience on which to draw and companies need not go in to this process without support.

ROADBLOCKS

The experience of RMC and Hydro gives guidelines for ways an industry seriously concerned with employment equity for women can reach its objectives. There must be awareness, however, that the best guidelines may falter before entrenched attitudes. Directives alone will not necessarily alter mind sets which can create an intimidating, hostile or offensive work environment for those who do not fit the traditional norm. Discrimination can be as subtle as intolerance of work styles different from the male norm or unthinking isolation from informal after-hour activities, sports, even professional organizations so important for careers (CCWE 1992). Discrimination can also be as blatant as statements that engineering is not an appropriate profession for women, or marginally concealed reluctance to take orders from women who are often made to feel bossy rather than decisive. Objectionable conduct can be found as easily in the boardroom as on the shop floor or underground and it is the duty of management to define clearly what conduct is unacceptable.

Firms like Hydro, Northern Telecom, Nova Corp and GE Canada have, in their equity programs, set strict standards on gender relations with a policy to hire only graduates prepared to work in an atmosphere of zero tolerance for sexism. Even so, change has been slow. Wharton states that "Hydro has still a long way to go before it can eradicate a fundamental attitude of discrimination that, in spite of our policies, still exists" (McKay 1992). During the slow process before an equitable gender balance has been achieved, it has been found that hiring a "token woman" is not the way to start. Only extremely self-confident or highly motivated women will attempt to train and work in nontraditional jobs as the only woman in a male workforce. In the face of often blatant sexual discrimination, many simply leave,- an unacceptable loss of a promising resource. Numbers do help. Another safeguard, especially at the beginning of the integration process, is a "Minerva Committee" to monitor integration. This group should include high ranking members of the organization, representatives of middle management and of the minority groups to monitor the integration. Also, some grievance procedure should be set up to receive complaints such as sexual harassment and a high ranking female in the organization should be designated to whom other women could go for advice and counselling in confidence. (2)

BONUS BENEFITS OF INTEGRATION

Thus far we have suggested the need for opening up the profession to women and outlined the challenge of introducing gender equality into the mining and other engineering professions. We have also suggested methods of bringing this revolutionary change about. Along the way, notice has been given that the process is difficult, requires a very firm managerial hand and can have results that are difficult to assess. But from the experience of other professions, the mining industry will be enriched by the addition of women. The new recruits will not just be more of the same. At least in the initial stages of integration, women will tend to be the top achievers representing the brightest and best.

There is also a very interesting suggestion that the infusion of women into the industry might bring a distinctive feminine perspective on technological issues,- a social awareness of the so-called 'female' values concerned with issues of environment, health, occupational safety and family heretofore sadly lacking in the industry. Not that mining would suddenly become a kinder, gentler profession if there were more women in the workforce. But women because of their socialization, could give a much needed human and social dimension to a profession which which one prominent woman engineer sees as having "little time for human or social values" (McKay 1992). Heretofore the engineering profession has prided itself on being rational and pragmatic encouraging in their members tough, aggressive behavior more suited to taming Nature for economic benefit than formulating a socially responsible approach to technical problems. How else can we understand an industry that has left behind pristine northern lakes filled with tailings, clear streams used as conduits for noxious effluent, countrysides reduced to moonscapes by poisonous fumes and whole landscapes destroyed by poorly controlled open pit operations. Here women, traditionally socialized to be nurturers, could make a major contribution by helping to develop a more humane and socially beneficial approach to future mine planning. In this view, women as part of planning teams, would be interested not only in questions of feasibility but in the social and environmental implications of the project in both the immediate and long term. (3)

While opening the mining and other engineering professions to women is numerically a necessity if we are to meet the projected demands of the 21st century, there is an added benefit which in the long run may be even more important. The expected infusion of social and environmental conscience which women will bring must have only positive results for an industry wishing to improve its public image and compete more effectively in the markets of the world. It has even been suggested that the most successful companies in the next century will be those who through enlightened hiring practices have the greatest range and depth of technical skills. It would be prudent to have a strategy to meet this objective in place before the end of this century. Given the logic of the gender solution to the supply crisis and the possibility of unique benefits from this solution there is a growing impatience on the part of women, the government and some male engineers with the profession's apparent difficulty in dealing with the problem.

"WIFE OF"

So far we have been discussing the need to open the mining profession to women if the industry is to prosper in the next century. But a second gender problem just waiting to surface in this period is the role of the mining wife. Until now, she has been an "add on," an indispensable person who came along with each mine worker as an unacknowledged and unpaid support in keeping the workforce operational. No one seemed to question the demands the mining life-style made on women as they attempted to maintain a domestic base, run the home and bring up the children in frequently changing locations often not of their choice. Compliance with the demands of the husband's job was taken for granted in the male-centred industry which operated under the assumption that "we hire workers not families."

What resulted was often a punishing life for women when the demands of the husband's job could require moving to isolated single industry towns where no job opportunities for women were provided, or keeping a family together through many changes of residence, or juggling family routines around a demanding and rigid work schedule, or hiding their feelings of resentment at the loss of their homes in the heartbreaking scenarios of Schefferville, Elliot Lake and others. If many mine wives felt exploited and resentful, there was really no alternative to acceptance of the status quo. This was a fact of mining life. In a one-income family, with few if any job opportunities for women, the wife had little choice but to "shape up." And of course many of them did, magnificently. But for all the wives who were successful in meeting such demands, there were many more who were simply unsuited to the life to which they were expected to adjust. Their failures were less a matter of not "shaping up" than of being faced with a life for which they were physically or psychologically unfitted. Some attention to their problems might have prevented much human damage.

CHANGING FAMILY RELATIONSHIPS

The change in family relationships, which is just now showing up statistically, will require mining management to regard the wives of their employees in a much different light in the next century. Increasingly women are training for careers, have been independent before marriage and plan to continue afterwards. The single income nuclear family is fast becoming the exception rather than the rule. As more women are financially independent there comes a desire for more participation in family/work decisions and some say in the family future. The mining industry in the next century must expect to adjust from its current state of regarding the male workers in isolation. The "wife problem" which is emerging indicates a fundamental tension between organizational life and family obligations and demands. Is the wife a helper, an unpaid worker or an independent person on whom the organization has no claims but whose needs have to be considered? What responsibilities do organizations have when their policies and practices impact on the family? How is a two-career family accommodated within the industry? These are questions which will have to be answered as we enter the next century. Suddenly, management decisions are being questioned which affect mining families as a whole yet arrived at without consultation with or input from those concerned. The impact of employment on family life cannot be avoided and companies who are looking for a competitive edge in the future will take note (Crowe c1985).

LDC - THE HUMAN DIMENSION

A case in point might be the new fly-in fly-out method of mining. On the mine management side there are many pluses to this new arrangement. But on the human side, it has many negative aspects none of which seem to have been considered in the planning stage. The Long Distance Commute (LDC) is an abnormal way of life, fluctuating between an isolated, high pressure, extremely structured macho work environment and an unstructured period away from the mine site. The success of this newest method of resource development depends on the invisible workforce at home supporting the visible workforce at the worksite. Family back-up has simply been taken for granted. But the strains to maintain the illusion of a normal and self-sufficient household through this unrelenting and rapid cycle of transition from one family lifestyle to another seems to fall disproportionately on the wife. There are really no guidelines for wives of LDC workers in coping with the discontinuities of this lifestyle and no recognition in planning these operations of the effect on the wife and family. Mine operators need to recognize the stresses which this new form of mining are likely to place on both the worker and the family and to inaugurate and support the research necessary to establish safeguards and formulate coping mechanisms.

In some way, the LDC does offer a solution to the two-career mining family. If the pick-up point is a city, as it so often is, there is the possibility of the wife carrying on her career concurrent with her husband's work in the far distance mine. This entails a great amount of accommodation on both sides since the wife's work schedule and the husband's are discontinuous. Another solution would be husband and wife openings in the mines themselves. One scenario to avoid is hiring a young worker who leaves a wife and small children behind to cope in a strange city while he is working miles away, unavailable even in times of crisis. Certainly in the case of fly-in fly-out operations some discussion of these problems needs to take place and some planning on the part of the industry to formulate solutions sensitive to the needs of all concerned.

CONCLUSION

The purpose of this discussion is to alert mine management to the gender problems in the industry which are likely to arise in the next century and to suggest ways to use the contribution of women to the benefit of the industry as a whole, whether as trained personnel or in the traditional role. Initiatives will require radical changes in attitude and a sense of urgency. However, the mining profession has always been one to accept challenges and the same initiative which make possible mine development in inhospitable locations, if applied to this human problem, will find acceptable solutions. "All" that is needed is the will.

NOTES

1. One practical approach has been to match the percentage of women engineers hired with the percentage of women graduating in the relevant discipline of engineering faculties and schools (CCWE 1992).

2. It is interesting to note that "denial" of any gender problem is prevalent among women engineering students and young women engineers beginning careers. Many do not see the reporting of discrimatory behaviour as a viable course of action. "Denial" such as this is a well-known coping mechanism used by minorities anxious to be accepted in a strange and often hostile workplace. Unless the reason for this coping behaviour is recognized, assessment of discriminatory practises in any organization is difficult. It is no wonder that management, if not firmly committed to gender equity, finds support for the status quo.

3. Women might also raise the issue of more humane engineering practice in proposing and encouraging alternative work arrangements to make it easier to balance family and professional responsibilities and career development (CCWE 1992). While this might be seen as benefiting women especially, there are many male engineers who would welcome a less rigid work atmosphere. (Eng.Dim Sept./Oct. 1992). In board rooms, women might inject a different form of discussion and decision making which would have less abrasive and damaging effects down the line. The currently fashionable participatory and consultative style of management is one well suited to women and one they have long used in arriving at group decisions. They could well share their expertise. Also women have not been conditioned by old management structures based on hierarchies determined by other criteria than ability. They are in a good position to bring fresh thinking to an organization especially on matters of gender equality.

REFERENCES

Braid, K. 1981. Women in men's territory: the sexuality of the non-traditional work place. Canadian Women's Studies 3 (2): 65-71.

Bray, M.P. 1991. The long distance commute in the mining industry: the human dimension. CIM Bulletin 84 (953): 62-64

Bray, M.P. 1993. The Sancta Barbara Order of Merit: a retrospective. CIM Bulletin 88 (968).

Canadian Committee on Women in Engineering (CCWE). 1992. More than just numbers. Report.

Crowe, D.S., Davis, K. & Pratt, V. c1985. Middle management in the mineral industry - the challenge of the 1980s. Proc. 13th CMMI Congress: 135-142.

Engineering Dimensions, official journal, Association of Professional Engineers of Ontario. 1990-1992. [abbr. Eng Dim]

Martin, L. E. 1986. Women workers in a masculine domain in Lundy, K.L.A. and Warme, B. Work in the Canadian context. 2d ed. Toronto: Butterworth.

McKay, S. 1992. Boys'club. Report on Business Magazine Sept.: 37-44.

Stewart, Major General (ret'd) J. A. 1992. [Conversation]

Innovative Mine Design for the 21st Century, Bawden & Archibald (eds) © 1993 Balkema, Rotterdam, ISBN 90 5410 325 6

Modern concepts in supervisory training

J.Tribe & J.H.Decker
Norwest Mine Services Ltd., Calgary, Alberta, Canada

ABSTRACT: The most critical link between management and production is the Front Line Supervisor. This person is relied upon to enforce company policies, promote the corporate image, and most importantly, achieve production goals; in short, he is the vital element in putting theory into practice. For the above reason, the Front Line Supervisor becomes the central core of any operation. Traditionally, little real effort has been put into supplying Front Line Supervisors with the training needed to do their job. This paper describes how this training gap can be bridged through courses designed to provide insight into techniques and methods of supervision, the structure of organizations, and the critical elements required to accomplish the daily open pit mining routine. This approach combines theory with practice without undue emphasis on either one.

1 INTRODUCTION

The front line supervisor is central and absolutely critical to any mining company achieving success. He (he/she used interchangeably in this paper) is the primary and vital link between paper simulations based on theory and the daily accomplishments expected on the job. Management relies on the supervisor to uphold policies, to communicate company conditions and desired changes, and to promote the company image; all of this is secondary to achieving the company production goals. The supervisor is expected to accomplish this without any formal training or background in the company policies. He is usually given little or no training in how to accomplish the job. In addition, there is almost no definition of success criteria.

Another complication is the fact that in the mining industry it is common for operations to be conducted 24 hours/day, 7 days/week; however, professional managerial support for the supervisor is readily available only on week days from 8 AM to 4 PM. Thus, instead of a learning, supportive situation, the supervisor is constantly exposed to a workforce which may not share his objectives. In his normal work condition he can only look to a peer group that is equally as well prepared for their specific roles as he is. In short, management structures an almost perfect scenario for maximum stress, frustration and a high probability for failure.

In the business environment we encounter terms and titles such as "foreman", "supervisor" and more recently "team leader" or "coordinator". Additionally we are confronted by the titles "coach", "captain" or "mentor". New words are appearing almost daily that seem to put a different slant on the supervisor's job. These new terms are not simply being introduced to confuse and confound management and the workforce, but rather to demonstrate the need to re-think and re-evaluate changes of the job and the work involved. While all these terms are used to present a particular approach or changing viewpoint to the management in an organization, it seems that the essentials of supervisory skills are being lost. We will be talking about the differences and the need to be different while at the same time maintaining these traditional aspects which need to be preserved.

This paper presents concepts in supervisory training derived from the lifelong careers of the authors. These concepts are based on insights of how to improve an industry facing extremely intense international competition. Put bluntly, these concepts reflect what is seen as a major shortfall in todays business environment.

2 ANALYSIS OF EXISTING CONDITIONS

2.1 *What's New?*

"There is nothing so constant as change" is an old adage that applies equally well to the workforce as it does to the society in which we live. There is a very real difference in today's workforce. It is not correct to say that the younger generation is not motivated or that they do not have a work ethic. It must be recognized that motivation and work ethic are being molded by the world in which the workforce lives.

In order to understand that there is a need for change we need to do some additional examinations. One will see that changes are in effect whether we desire them, and in fact, whether we are involved in them or not. Workers and members of society in all walks of life are indicating that beyond the minimum requirements, there needs to be something referred to as "quality of worklife." Worker mobility is a reality and families move not only to seek work but also to improve living conditions, cultural and educational opportunities. Some of the various cultural shifts that supervisors must be aware of involve equal opportunity, sexual harassment, racial and minority discrimination, to name a few. In addition, environmental considerations are forefront in all areas of work.

Today's workforce is more knowledgeable about what is available. The world keeps shrinking as much a result of communications as of rapid and effective travel. The effects of increased knowledge are being experienced both by management and the workforce in general. Our reaction and analysis of information is reflected in our ambitions, moral values, motivation, and work ethic.

2.2 *What's Old?*

The basic premise of mining, and indeed any business venture, has not changed. The application of capital and physical resources under the guidance of a human resource is used to exploit a useful natural resource at a profit. The work to be done in many cases appears the same despite cultural and social changes.

There are the same expectations of return on capital investment and the same expectation of a day's work for a day's pay and that people will be accountable for their work. There still is, on the part of management at least, a sense of urgency based on the knowledge that resources are not only finite, but a resource lost is lost forever and cannot be recovered.

Into this scenario we put the supervisor who is now management's representative. The supervisor is still expected to "get results". He is expected to be committed, dedicated, responsible, motivated, and a self starter. He is expected to be fully accountable with only a minimum of authority. Through his actions and words he must maintain a sense of urgency and economy. The delivery of a quality product at a minimum unit cost while at the same time enhancing the "quality of worklife" is a common expression of a corporate goal.

Such a mandate does not provide much guidance on a cold winter night for the supervisor who is at his wits end about what to do next!

2.3 *What's Changed*

There are changes in both the social and cultural background in which business is operating, but there are also business practice and business expectation changes. As the world develops there is an ever-increasing body of people and countries entering the competition to supply the necessary resources to fuel a maturing civilization. It is becoming increasingly difficult to achieve high rates of return on projects whether they are located in a developed or developing country; simply put, the return on investment has reduced in recent times. New organizations are willing to work on increasingly reduced margins in order to achieve their position in limited markets. Many governments play an intrusive role with respect to new projects. As a result of these factors the market for our products is worldwide, in itself a mixed blessing. Not only are we able to sell our products any place in the world, but also the producers from all over the world are able to compete in those same markets.

3. BODY

3.1 *Part 1*

3.1.1 *Available Resources*

When we reduce our industry to its most basic components we see that there are only three main factors or resources. First, the natural resource which has been provided by nature and, in the case of mining, is finite. Second, the physical resource is applied by man's ingenuity in the form of machines and related infrastructure. Third, there is the human resource which applies ability and expertise to achieve the objectives related to the other two. Within the time frame relative to business, the relationship of these three resources is critical but often overlooked. At any instant in time no company or business truly has a controllable competitive edge in terms of the first two resources with which it has to work.

More and more companies are beginning to focus on the fact that the single largest area for improvements in productivity and cost reduction lies in the area of effective utilization of its human resource, where there is an opportunity to gain a truly competitive edge. It has been shown in numerous cases that even companies considered to be efficient in the traditional sense can achieve up to 40% productivity gains through a willing workforce.

We will return to this key to productivity a little later, but first we need to continue being introspective.

3.1.2 *Supervisor's Role*

The supervisor's role has not changed in one very important aspect and that is that he is expected to make it happen. When all else is considered the supervisor is expected to safely achieve the maximum amount of production at a minimum cost given the conditions, human resources and equipment. In this respect the supervisor is the very foundation of management. His principles, dedication, and sense of urgency are those of the company. To the employees reporting to him, he is "the company".

The unique position of the supervisor in an open pit mine can be summed up as follows:

- On an 8 hour shift basis there are 21 shifts in a week of continuous operation and management is scheduled for only 5 of these. This means that management advice, direction, and help is available less than 25% of the working hours in a week.
- In a continuous operation the supervision is carried out day and night which means that 50% of the time it's pitch black at the work place.
- From the supervisor's perspective this makes the job a lonely one and also means that he must be self reliant.
- From his workers' perspective the supervisor more often, obviously and firmly represents the company and how it operates than anyone else!

For the most part the relationship a supervisor has with his crew is the true relationship between management and union. The workers reporting to him often have divided loyalties. He is subjected to pressures from other supervisors who are in one fashion or another working with constrained resources just the same as he is and in some circumstances are in competition for the same services and resources. Finally, management appears to be constantly wanting more from the employees and the supervisor. The supervisor very much is the "ham in the sandwich", receiving pressure from above and below; at the same time, he is the main substance of the sandwich.

3.1.3 *Standard Practice*

Now that we have looked briefly at the resources available to management and laid out the common objective of nearly all supervisors, we will now look at how they are normally chosen for the position. In order to fill the position of supervisor, management usually follows very traditional approaches. The first attribute will be a hard worker which is probably evident in a person who requires very little direction and has had the least exposure to it. In some cases there will be an opportunity to see the candidate in action. Acting as a relief foreman while another is on vacation will provide at least a limited opportunity to experience some of the job requirements. Finally, supervisors are sometimes chosen as a reward for service or some other recognition. There is nothing wrong with the selection technique except that it often is viewed as the termination of a process rather than the initialization that it really is.

As incredible as it may sound, the majority of supervisors that we work with had the "weekend" training approach. This is a Friday affair which includes the announcement to the employees and the suggestion that the newly appointed supervisor "bone up" on the company procedure manual over the weekend.

The new supervisor is thrust into a management position and management is willing to make "allowance" for the new role; however, it should be clearly understood that there is a loftly expectation for rapid improvement. While certain management flaws may be tolerated, production targets are "real".

Management in many cases does not expend the effort to define what every supervisor needs. Without criteria it is difficult to plan the development. In other words, without a defined goal all directions are equally valid.

Often management can see the problem much more clearly than they can see the solution. A poll of the average operation would show the three predominant features:

1. The frustration for the new supervisor is that he really would like to do a superlative job - if he could only figure out what that might be and how he might do it. This frustration is often expressed as "Why is it that management doesn't understand the problems on the floor or in the field?"
2. The frustration of management is that it knows that it has not adequately prepared the supervisor for his role, and further, if we could just "stop the clock" we could impart a lot of help to the new supervisor.
3. The frustration for the employees is that the person who advanced from within the ranks, and for whom they held such high hopes, is obviously floundering and they are suffering because of it.

The standards of performance in the work place are perhaps the least understood and most abused expectation of any in the workplace. Performance objectives and expectations must relate to aspects that each worker can feel he is in control of and can make a personal contribution toward its achievement.

3.2 *Part 2*

3.2.1 *The Fix*

The most well known fix has always been the same and that is to wait until experience teaches the supervisor how to do the job. This is the "sink or swim" method of supervisory training.

There is the "in a perfect world" dodge which is used to condone performance we would otherwise deem unacceptable. By accepting supervisor shortfalls without seeking remedy is to agree to reduce to the lowest common denominator. The new supervisor is monitored for performance standards that are usually held secret from him. Management tends to confuse objectives with actions. Actions the employee can control; objectives serve a multitude of purposes, few of which are beneficial to either the company or the individual.

3.2.2 *Proactive Training*

The real need in today's situation is to better prepare our supervisors for the job they are expected to do.

It can be difficult to convince management that this is the best tack to take because the payback is not easy to calculate or demonstrate. It is a rare company indeed that will undertake to "educate" prospective supervisors at its leisure during the periods when no vacancy exists, yet this is exactly what needs to be done to achieve and maintain a true competitive edge.

Management has been relatively proficient at recognizing a supervisor's specific problem with the job, analyzing and determining where weaknesses lie and providing special enhancements to alleviate the problem. This is analogous to increasing the voltage of a motor to increase its power - while it certainly does for a short time the underlying problem remains intact, and in no case is it a panacea.

The effect is much the same as ordering parts only when they are clearly needed, as demonstrated by broken equipment. Management development programs have to be more than lip service.

3.2.3 *Teaching Resources*

Through the process of natural selection those in upper management positions have a good understanding of what is required but are generally not available to pass on their knowledge to those who most need it.

Since much of management received its schooling through "hard knocks" there is rarely any unified directed training or educational program in place and being practised at the operation. In many organizations the knowledge is disjointed and may never have been put into logical and understandable order.

There should be a clear distinction between courses designed to provide a skill and the managerial education that provides the basis for an organization to prosper.

3.2.4 *Training Resources*

The necessary training resources are actually close at hand to almost any mining operation. The members of the senior management group are normally well experienced people. Progression through the ranks is still very common and the learning experiences are plentiful. With senior foremen, superintendents, and area managers there really is no shortage of experience; they are an extremely valuable resource that is available to the front line supervisor.

The unfortunate situation for most operations is that this wealth of experience cannot be effectively drawn on to improve the supervisor's effectiveness. Two of the most time consuming activities that senior staff members are involved in are performance analysis or variance analysis and planning exercises. Analysis functions are extremely valuable in terms of understanding what is happening to the business, but in reality do almost nothing to ensure operational improvements for the future. The other activity that occupies senior staff is planning for the future in terms of budgeting or activity planning. This again is a vital function but does little to help the supervisor in his daily activities.

As a result, management is faced with an extreme over demand on its experienced resource. Management really has only three options:

1. Do as much of the training, coaching, mentoring that they can.
2. Do nothing and hope that osmosis achieves the objective.
3. Hire the appropriate resource to do the job for them.

It has been our experience that the first two options are chosen far too often. The third option is being provided by Norwest.

3.3 *Part 3*

3.3.1 *What Has to Happen*

Some key components need to be common throughout all management and supervisor members in the entire organization:

- The purpose of the organization;
- The method that the company will follow to achieve that purpose;
- Long term goals;
- Short term goals; and
- Actions and measures.

This last step is indeed the only controllable item for any supervisor. Actions are absolutely critical if each supervisor is to really be effective since, by definition, they are items that are individually controllable. The measures naturally will reflect those same action items.

3.3.2 *Customized Training*

Sending people to courses on "decision making" or "problem solving" are excellent IF the employee has a need for this skill. In general there is much less problem with the course than there is for the expectations of the results to be achieved after returning from the course. In many instances we are trying to cure the symptom rather than diagnosing the ailment.

3.3.3 *Unique Offer*

The outside resource can dedicate the time and effort to ensure that the proper questions are asked. He will also be able to pass to all management in a very short period of time the same message at essentially the same time to all the employees. The outside resource can examine on a candid level the individual members of the management team to ascertain conflicts and weaknesses.

Jim Decker and John Tribe of The Norwest Group have been concerned with the problem and dedicated to the solution for some time. Both professionals spent the early part of their careers as front line supervisors. The examples cited in this paper are related from actual experience and observation over extensive practical operating careers. In direct response, and in conjunction with Queen's University Mining Department, a course was designed which serves two critical purposes. The first is to provide guidance to the front line supervisor in areas which are universal in their application and acceptance. The second purpose is to provide individual operations with the opportunity to customize the package so that emphasis can be made in areas of special concern.

The course, entitled "The Professional Supervisor," is a dynamic instrument which responds to the needs of the specific group and is modified according to a feedback cycle. The course provides an overall view of organizations and how they work, the position and role of the supervisor and some direct operating tips that can make him more successful in his job. For the new supervisor the course is necessary in developing the basic skills available; for the more experienced foreman it provides a fresh approach to managing people and situations as well as a means of assessing his own effectiveness. The overall effect is that the supervisor becomes more comfortable with his organization, his personal role in that organization, what he can expect from it, what it expects from him, and finally, some real assistance on how to improve his own on-the-job performance.

In our training programs we agree with the concepts of empowerment, high involvement, total quality, and coaching because they all try to tap the latent talents and abilities of employees. We provide a focus and a tempering of theory with the practical aspects that come from having been there.

3.3.4 *Documentation*

Documents such as a foreman's manual, an engineering manual or a policy manual all serve the same purpose of achieving consistency. They also allow the opportunity to study, the ability to review and probably most importantly, an avenue to challenge the system. The value of the documentation lies in its availability for the new person to read and understand what has gone on before and in a sense catch up with his peer group. An equally important aspect is the formalization and legitimization of the company policies, procedures, and practices. In written form they will also provide the opportunity to be challenged and changed as appropriate.

The company will have provided formal information on its structure, its philosophy, its beliefs, its values, and its principles; this may be an exercise that is new to the organization.

4 CONCLUSION

The modern concept in supervisor training is that successful companies will recognize that cultural shifts make it imperative that the supervisor be educated. The individual components that a supervisor should be familiar with are not new nor is the utilization of the information. What is new is taking the time to put it all together and presenting it to the front line supervisor as a complete package. The supervisor must be exposed to, and have made available to him, much more sophisticated knowledge of his organization in order to apply his common sense to the best advantage. This can almost be defined as "sensitivity training". Companies need to prepare larger numbers of qualified people before their talents and services are actually required. Only in this way will we have supervisors that know what the company wants to do, how it wants those things done, and exactly what individual actions are expected to be. In this manner we will prepare our organizations to be truly competitive in a world situation.

Innovative Mine Design for the 21st Century, Bawden & Archibald (eds) © 1993 Balkema, Rotterdam, ISBN 90 5410 325 6

The management approach to the research and commercial development activities of the Mining Industry Research Organization of Canada

Peter N. Calder & Charles F.A. Hews
MIROC, c/o Department of Mining Engineering, Queen's University, Kingston, Ont., Canada

ABSTRACT: The Mining Industry Research Organization of Canada (MIROC), incorporated in 1975, is a non-profit organization sponsored by seven member companies from within the Canadian Mining Industry. MIROC's research activities are funded from member company support and about thirty percent from royalties generated from previous research activities. MIROC's mission is to encourage the formulation of ideas to improve mining technology and methods by any source, and to foster the developing and testing of promising initiatives in progressive steps from the early stages of conceptual and feasibility research through manufacture to delivery of new technology to the Mining Industry. In this paper MIROC's method of operation involving its Board of Directors, Advisory Committee, Research Directors, University Participants, Consultants and Manufacturer's is described as a brief overview of its active research and commercial activities.

MIROC AND THE MINING INDUSTRY

The Mining Industry Research Organization of Canada (MIROC), incorporated in 1975, is a non-profit organization sponsored by member companies from within the Canadian Mining Industry. The Canadian Mining industry is a world leader in technological development and in organizational and human resources development. A prime objective within the industry is to retain this leadership and also improve it competitive position worldwide.

MIROC is dedicated to support the objectives of its member companies and their common interest, and the objectives of the industry. In this content: MIROC's mission is to encourage the formulation of ideas to improve mining technology and methods by any source, and to foster the development and testing of promising initiatives in progressive steps from the early stages of conceptual and feasibility research through manufacture to delivery of new technology to the Mining industry. In so doing MIROC plays an important role in the development of individuals skilled in mining research.

MANAGEMENT APPROACH

MIROC's research program is driven by member companies who have interests throughout Canada. The Board of Directors approve projects that are acceptable to its members. While MIROC's basic research funding is provided by its member companies, about thirty percent of its annual research budget is now derived from royalties generated from previous research activities.

Ron Lord was MIROC's first Research Director. In 1989, the Research Administrative office was moved to the Mining Engineering Department of Queen's University, with Peter Calder as Research Director and Charles Hews as Associate Research Director.

MIROC's management approach include realistic hands-on evaluation of new ideas by its Advisory Committee, its Research Directors and its Board of Directors, all of whom have broad experience and contacts throughout the Mining industry. A multi-disciplinary approach is often utilized bringing together teams from member companies and a number of universities and other research agencies to tackle a particular problem. Once consensus regarding an idea is obtained from the Advisory Committee and approved by the Board of Directors, the conceptual and early stages of feasibility research is carried out under contract, usually to a university, and by a senior investigator.

When sufficient feasibility is determined additional progressive stages of research are then carried out, also under contract, in laboratories or in a member company's mine where the necessary in-kind support such as a test-site and the required infrastructure is made available. This stage of investigation could involve, in addition to the principal investigator, consultants, graduate students, research assistants, member company engineers and management, joint safety committees and the appropriate government agency. Prototype testing and early stages of commercial development could be carried out in a similar fashion.

If and when commercial development is indicated and if the intellectual property developed is MIROC's, MIROC will patent the idea and then licence a manufacturer, through a contract agreement, to commercially develop the resulting product for the mining industry. MIROC will also provide funding under a commercial contract agreement to manufacturers who own and retain the specific intellectual property.

MIROC's Board of Directors, who are at executive or senior managerial or technical level, meet four times a year with the Research Directors and are closely involved in all projects.

The Research Directors and Advisory Committee, which is composed of senior mine personnel, also meet four times a year at mine or research sites where testing and/or application of current and new ideas can be thoroughly discussed. These meetings, which include underground or facility tours, also provide excellent opportunities, not otherwise available to these mine operators, to share experiences and exchange information regarding common interests.

The Board of Directors and the Advisory Committee usually meet together once a year and are updated directly by the principal investigators on the status of projects.

The Advisory Committee is kept informed of the development of commercial agreements which, of course, must be approved by the Board of Directors.

In 1992, MIROC for the first time, entered into joint agreements, one with the Ontario government and one with the Federal government, both under the Federal Provincial Mineral Development Agreement called the Northern Ontario Development Agreement (NODA). Agreement items such as confidentiality, patenting, membership of intellectual property, ownership of reports, commercial development and the use of project fundings for academic purposes were mutually resolved. These particular agreements will also take two projects, previously brought to feasibility stage by MIROC, to commercial development, with a level of funding beyond MIROC's scope to itself provide.

Thus, MIROC plays an important role in enhancing relationships and cooperation between personnel in the industry, university, governmental, consulting, equipment manufacturing and service sectors and in general, provides a national focus for mining research in Canada.

Since its inception in 1975, MIROC has funded forty research projects on a wide range of subjects related to production, health and safety, mine environment and mine communications.

RESEARCH PROJECTS

I would like now to bring you up to date on the status of five example projects.

Northern Light

An earlier MIROC program lead to the development and patenting of an improved cap lamp for miners. LEVITT SAFETY LIMITED was granted an exclusive license by MIROC to manufacture and market the lamp. LEVITT'S NL TECHNOLOGY DIVISION is successfully marketing the Northern Light in Canada and abroad. NL TECHNOLOGY continues to carry out research and development activities regarding the system and have made significant improvements to the lamp, the battery and the maintenance system. This was substantiated by a MIROC sub-committee which, in 1991, and again in 1992, in conjunction with NL TECHNOLOGY, studied the application, usage, maintenance and costs of the system. The Northern Light is recommended as a superior product.

Vibration damping in rock drill handle design

In 1989, an initial study was carried out at Queen's University to determine the effectiveness of a new handle design under laboratory conditions and to set up a protocol to develop standardized vibration measuring techniques that could be used in a scientific manner.

The protocol was established and the test results indicated that for the new handle design, the overall acceleration level experienced by the hand can be reduced by as much as a factor of three, relative to the regular handle. This reduction, which takes place mainly in the frequency range above 500 Hz, could be a significant factor in reducing exposure to vibration induced neurological damage.

Field test confirmed the laboratory results and material durability. A design change, for operator acceptance, was indicated. A further project leading to commercial development by TELEDYNE CANADA MINING PRODUCTS, the handle manufacturer, is currently well underway under the NODA agreement.

MIROC workboot

MIROC owns patents covering a certain successful toe and metatarsal protective device. MIROC has licensed A. LAMBERT INTERNATIONAL INC. (manufacturers of the ACTON workboot) to develop and market the MIROC WORKBOOT, incorporating the device, in Canada, the United States, the United Kingdom and elsewhere.

Testing of the prototype boot, was carried out at INCO's Crean Hill Mine and Falconbridge's Kidd Creek Mine in two stages.

Design modifications based on the feedback from the prototype test and on the recommendations of a podiatric specialist have resulted in an approved, lighter, more comfortable workboot which is in final stages of commercial development.

Dry drilling and dust control

Phase I of this project was carried out, under a contract to Queen's University, on surface at Inco's Crean Hill Mine in 1989, to scientifically determine the effectiveness of using dry drilling in place of wet drilling, in "In-the-hole Drilling". Substantial increases in penetration rate and bit life and thus reduced overall drilling costs for dry drilling were indicated using both regular and high pressure air.

The Phase II test, carried out in 1990, again at Crean Hill, investigated the effectiveness of dry drilling and certain dust control technology under production conditions. Results from 4500 feet of drilling indicated penetration rate increases in excess of 35% with dust concentrations below TLV levels.

MIROC has an agreement with J.R. ENGLAND CONSULTING LTD. under which ENGLAND acquired and applied a DUST CONTROL SYSTEM for the Phase II test and under which ENGLAND will commercially develop the SYSTEM relative to MIROC, ENGLAND has patents pending regarding the SYSTEM.

Spray on Linings

Caisson support of loose rock has been effectively demonstrated within simulated mine settings. Air tight coatings placed about simulated rock walls have resisted local gravity falls of loose rock, where no other means of support existed, when differential air pressure between the membrane sides was induced by natural rock conveyance. It is thought he traditional applications of gunite and shotcrete in underground locations creates a similar support mechanism. Those traditional methods are costly and time consuming.

The initial phase of a multi disciplinary study was carried out at Queen's University and indicated great promise for certain spray on linings as a viable alternative for existing underground support techniques and as radon impermeable barriers. However, significant concern regarding flammability and health and safety suitability had to be resolved.

The coatings that have been the most successful are solvent free, polyurethane two component materials that are sprayed using special purpose high pressure equipment. As the two component streams meet in the head of the spray gun a chemical reaction commences which continues as the material hits the rock face. Speed of setting is very rapid and the optimum thickness is therefore achieved very quickly.

In subsequent project test phases steps were taken to both assess and mitigate flammability and toxicity associated with certain lining materials and to develop techniques to safely apply the systems in the underground environment. Comprehensive environmental tests were carried out by Queen's, Ontario Ministry of Labour (MOL), Health Officials and Consultants in conjunction with the field application and the performance trials. Initial field testing was carried out at Kidd Creek Mine in December 1990. Further performance testing was carried out at Inco's Copper Cliff North Mine in May and July 1991 and again in April 1992.

MOL, INCO and Queen's tests indicated environment readings well under TLV limits. MOL, however, will require the continuation of the installation protocol including the wearing of an in line breathing apparatus at the application site and the wearing of charcoal filter masks downstream of the site.

Flammability ratings, determined from tests conducted at the Underwriter's Test Laboratory at the University of Toronto, indicate flame spread ratings approaching that of gypsum wall board which is normally considered to be non flammable.

Laboratory tests have also indicated that a reduction in radon gas flow in excess of 99% is possible.

The original underground tests at Inco were not satisfactory due to component reaction to high humidity levels. Further laboratory work to better define and improve the adhesion and tensile strength characteristics of the varying formulations under high humidity conditions were successfully carried out. The subsequent underground test on April 15, 1992, at Inco's Copper Cliff North was most encouraging.

MIROC has Patents pending for the range of linings and their commercial applications and has licensed URYLON PLASTICS INC. of Guelph, Ontario, who will develop the markets for the lining formulations under the trademark MIROC MINGUARD™. Urylon has broad industrial experience in applying coatings to a wide range of sub-strata materials. Production testing at Lac Minerals Bousquet Mines has been very encouraging.

Other projects

MIROC also has projects underway on Noise Reduction, Blasting, and a major project, just begun, on High Frequency Microseismic Monitoring.

I would like to thank the members of the Board of Directors of MIROC, the members of the Advisory Committee and the Sub-committee for their great support and contributions. I also especially want to recognize the cooperation and support of all the people at the member company mines where our projects are tested: the management, supervisors, engineers, safety committee members and the workers who carried out the tests, all of whom contributed their ideas, particularly those at Inco's Crean Hill and Copper Cliff North Mines at Falconbridge's Kidd Creek Mine and at Lac's Bousquet Mine where we carried out repetitive testing.

DIRECTORS AND OFFICERS

Joe W. Loring
President
Inco Ltd.

Andre Piche
Director
Noranda Minerals Inc.

Peter M. Rowlandson
Vice-President
Williams Operating Corp.

Lawrence Davies
Treasurer
Falconbridge Limited

Louis Dionne
Director
American Barrick

Tom Leishman
Secretary
Inco Ltd.

Phil Donaldson
Director
Falconbridge Ltd.

Peter N. Calder
Research Director
Queen's University

Peter C. Jones
Director
Hudson Bay Mining and Smelting Co. Ltd.

Charles F.A, Hews
Associate Research Director
Queen's University

Gerald J. Gauthier
Director
Lac Minerals Ltd.

Ronald D. Lord
Honorary Research Director

ADVISORY COMMITTEE

Stewart Brown, Chairman
Williams Operating Corp.

John Haflidson, Member
American Barrick
Resources Corporation

Gilles Brousseau, Member
Lac Minerals Ltd.

John Vary, Memeber
Falconbridge Limited

Lewis Cluett, Member
Hudson Bay Mining and
Smelting Co. Limited

Don Gibson, Secretary
Inco Ltd.

ADMINISTRATION

MIROC's Research Administrative Office was moved to Queen's University, Department of Mining Engineering, in January 1989.

Information concerning membership and participation in MIROC is available from the Research Directors.

Contact:

Peter N. Calder
or Charles F.A. Hews

Queen's University
Department of Mining Engineering
Kingston, Ontario
K7L 3N6

TEL: 613 - 545-6655
FAX: 613 - 545-6597

Innovative Mine Design for the 21st Century, Bawden & Archibald (eds) © 1993 Balkema, Rotterdam, ISBN 90 5410 325 6

Mine management into the 21st century: Perspectives from Ashanti

Sam E. Jonah
Lonrho Plc, London, UK & Ashanti Goldfields Corporation, Accra, Ghana
Kofi O. Nti
Penn State University, University Park, Pa., USA

ABSTRACT: The challenges of the twenty first century require that mine management be responsive to the demands of globalization, information revolution, lead time reduction, and worldwide pressures for more acceptable extraction and environmental programs. Working within a socioeconomic environment similar to conditions elsewhere in Africa, Ashanti Goldfields Corporation has managed to rejuvenate an old mine, embarked on a major expansion program, and is well poised to be among the leaders in mining as we enter the twenty first century. In this paper, we discuss the evolving impact of the following trends on Ashanti's strategy and performance: 1) The settled ideological climate and the new fiscal regime; 2) Globalization and the information revolution; 3) Lead time reduction; and 4) Extraction and environment policies.

1 THE SIGNIFICANCE AND RELEVANCE OF ASHANTI

Ashanti Goldfields Corporation (Ghana) Limited operates one of the mining world's best known gold mines. Based at Obuasi, about 180 kilometers from Accra, the capital of Ghana, Ashanti has been in continuous large scale production since 1897. Since 1972 the company has been operated as a joint venture between the government of Ghana and Lonrho Plc of the United Kingdom. Ghana holds 55 percent ownership and Lonrho holds 45 percent. Lonrho has an agreement for the management of the company and assists with technical advise and procurement.

Ashanti has both underground and open pit operations within a total concession of 256 square kilometers. It employs 9,250 people, including 640 senior staff of which 12 percent are expatriates. Ghanaians are represented at all levels of management, including the managing director, a general manager, and several divisional managers. Ashanti has successfully raised US$289 million from international lenders in the last seven years to fund capital projects valued at US$830 million. Ashanti has funded the balance of US$541 million from internally generated funds. In 1992 Ashanti produced 654,000 ounces of gold and remains on target to produce 1,000,000 ounces by 1995-96, which will make it one of the largest gold mining companies in the world.

To fully appreciate the significance of Ashanti's recent achievements, it is necessary to recall that gold production had been on a downward trend from 1972, dropping to 246,000 ounces in 1987. Thus production has more than doubled within the last five years. What has made all this possible? We believe the successes at Ashanti have come about because it has worked systematically with all its major stakeholders, namely the shareholders, the government, lenders, suppliers, employees, and the local community, to remove some of the barriers to mining development in Ghana. Ashanti's unique experience over the last few years offers an example for other companies who might wish to operate in a similar environment.

Although Africa is noted for its considerable mineral wealth, it is significant that mining opportunities have not been fully explored on the continent. For a long time political risk considerations have diverted exploration and development effort away from Africa, towards Australia, Canada, and other countries. However, existing geological information as well as mining practice suggest that Africa is the minerals continent. Angola, Burkina Faso, Ghana, Namibia, Zaire, Zimbabwe, and other African countries remain unexplored and un-mined yet Africa as a whole appears to have the potential to produce some of the greatest finds in the mining world.

The experiences at Ashanti are quite relevant because, working within a socioeconomic environment similar to conditions elsewhere in Africa, it has managed to rejuvenate an old mine in a difficult part of the world, put it at the cutting edge of mining, and is well poised to be among the leaders in mining as we enter the 21st century.

2 THE 21ST CENTURY BUSINESS ENVIRONMENT

In working out strategies at Ashanti, senior management has had to be acutely conscious of the fundamental changes that are taking place in the business and the economic world during the last decades of the twentieth century. We expect these changes to deepen during the early years of the twenty first century and become fully accepted as part of the normal business environment during most of the 21st century. It is particularly encouraging that Ashanti, situated in a remote area in a small developing country in Africa, has sensed the broad strokes of the 21st century and has gained relevant experience during the transitional years to enable it to become one of the leaders in mining during the early decades of the 21st century.

The most notable characteristic of the new business environment that we wish to focus on in this paper are: 1) The ideological climate and the new fiscal regime; 2) Globalization and the information revolution; 3) Lead time reduction; and 4) Extraction policy and the environment.

In the rest of the paper we discuss each of these developments, and how it has impacted on Ashanti's strategy and performance. More significantly, we show how Ashanti's strategy is being shaped by its understanding and projections of the nature of the mining business in the 21st century

3 IDEOLOGICAL CLIMATE AND THE NEW FISCAL REGIME

3.1 *The settled ideological climate*

Since independence, it has been quite difficult for Ghanaians to fully appreciate the role of private initiative in the processes of economic development. This difficulty was exacerbated by the larger debate in the world at large stimulated by the cold war between the West and the East. Partly due to severe resource constraints and partly because the institutions inherited from the colonial administration restricted the growth of an indigenous private sector, the post-independence governments proceeded to expand the state sector and erect numerous barriers to impede the growth of the private sector, indigenous and foreign. Among other things, there was a complicated system of exchange controls, import licensing, high taxation of export earnings, and the creation of inefficient government bureaucracy and state owned enterprises.

In the existing ideological climate, these policies appeared quite normal. But by the late 70's it had become clear that the Ghanaian economy was grinding to a halt. The infrastructure was badly deteriorated, no new investments were being made, and agricultural output had fallen significantly. In the early 80's the government broke with the ideological past and began implementing a series of measures to put the Ghanaian economy back on track. Working closely with the International Monetary Fund and the World Bank, the government designed and implemented an economic recovery program which helped to remove some of the major barriers which impeded mining development in Ghana. Similar programs have been implemented by other African countries. Incidentally, these programs were implemented in Africa long before we could all absorb and draw valuable lessons from the collapse of socialism in Eastern Europe.

Now there is a clear consensus that private initiative, not government, is the main engine for economic development in Africa. The government's role is to design the proper macroeconomic framework to enable private initiative to flourish and create jobs and incomes to raise living standards. The net effect has been a continuous reduction of the government's role in production, relaxation of controls, minimization of government intervention in economic decisions, and an increasingly favorable climate for private investment on the continent.

These developments have had a tremendous impact on the performance of Ashanti, as will be evident throughout this paper, and are continuing to attract investments into the mining sector in Ghana.

3.2 *The new fiscal regime*

Prior to the implementation of a structural adjustment and economic recovery program in 1983-86, the mining industry in Ghana was subjected to a wide range of taxes and legislative barriers which discouraged expansion and development. Since 1984 the government has enacted a unified minerals law and established a single minerals commission to process applications for mineral rights and negotiate mineral agreements with investors.

The fiscal regime has also changed. Previously, the minimum front-end charges amounted to 12 percent of gross revenue. Corporate tax rate was 50 percent and capital allowance was quite low. Before, companies were constrained by import licensing and foreign exchange quota system. The new Ghana law has reduced the minimum front-end charges to 3 percent of gross revenue. The corporate tax rate for mining companies has been reduced to 45 percent. The capital allowance includes a 75 percent write-off during the first year. The mining industry is exempt from payment of import duties on plant, equipment, spares and operating supplies. The new fiscal regime also allows a mining company to set up trustee operated off-shore foreign currency accounts to facilitate payment of loans, equipment, supplies, expatriate salaries, and dividends from earnings.

The new fiscal regime has helped to attract over US$1 billion of mining investments into the country, leading to the establishment of seven additional gold mines. A significant portion of the new investment involves the development and processing of lower grade ores which would not have been economical under the old fiscal regime.

For Ashanti, the new fiscal regime has made it possible to explore new concepts in international procurement and borrowing, and to take advantage of certain risk management opportunities.

4 GLOBALIZATION AND THE INFORMATION REVOLUTION

The main issue to consider here is the total linking of markets globally, facilitated by the information revolution and satellite technology. This means that with a modest capital investment one can access databases throughout the world and get the latest information on developments in the mining and financial world. While Ashanti has not yet taken full advantage of the information revolution to connect with the rest of the world, we expect this to happen in due course. Already, price information at some of major commodity and futures exchanges are available to Ashanti instantaneously. Thus in due course, Ashanti's remote location will not be such a constraining factor. It will then be possible for managers in Ashanti to interact on demand with suppliers, lenders, and other overseas based constituencies to make timely decisions. This vision should be contrasted with the situation five years ago when a telephone contact with London was considered a major achievement.

4.1 *Price instability and risk management*

An important aspect of globalization and the information revolution is the tremendous volume of decisions being made almost simultaneously by major economic actors. This has created a situation where prices and financial variables have become extremely volatile and unstable.

Ashanti is based in Africa, where critical inputs such as professional labor, technology, capital, and management are in short supply. These critical factors have to be imported and paid for in one or more foreign currencies while the output of the mine may be sold in another currency. Fluctuations in foreign exchange markets directly affects the mine's cash flow and value. Gold mining has the additional problem that the output is also held as a financial asset and so the price of the product is affected by movements in real interest rates. We should also mention that gold prices used to be sensitive to political insecurity, but now this appears not be the case. Thus the cash flow and value of the mining operation is dependent on the volatile path of exchange rates, interest rates, and commodity prices.

To be a world class mine manager in the 21st century, it will no longer be sufficient to invest in the best technology, discover and mine the richest ore body, or use the cheapest labor and energy. Mine managers will need to become more sophisticated about managing the tremendous risks arising from increased volatility of commodity prices, interest rates, and foreign exchange rates. The name of the game is to learn to manage these risks by separating and exchanging undesirable risks with other parties. Mine managers will have to follow closely, understand, and use to their advantage the variety of new financial products that have been designed to separate and repackage risks. We have in mind the use of forward contracts, futures contracts, swaps, options, and other instruments in the financial engineer's toolbox.

In addition to selling its output on the spot market, Ashanti is increasingly making use of forward sales and put options to hedge its gold revenue. Some of the important raw materials are bought on forward contracts. Ashanti is also arranging to set up longer term hedging contracts involving the use of gold swaps and gold lease rate swaps. In a recently arranged loan to fund a major capital project, Ashanti and its lenders structured the loan as a synthetic gold loan, providing Ashanti with further opportunities to hedge additional gold revenue.

We expect that Ashanti's successful use of some of these innovative financial instruments will contribute to an increasing acceptance by lenders that the use of these instruments is legitimate and cost effective for financing African mining projects. When this development is combined with the emergence of stock exchanges in Africa and the new thinking on foreign exchange restrictions, the prospects for the 21st century appear promising. Bankers who wish to participate in these opportunities will have to become more focused and innovative in funding African mining ventures.

5 LEAD TIME REDUCTION

Increasingly lead time reduction is being recognized as necessary for success in the new business environment. Lead time refers to the time it takes from the recognition of a need to the availability of the matching product or service. The trend is to achieve zero lead time when feasible. Reducing lead time to zero means the business can operate in real time in delivering its product to market or in securing its supplies. Already some financial institutions are using information technology to operate practically in real time. They can put together within minutes any type of financial product required by its customers. The Japanese automobile manufacturers are operating in real time with respect to both their customers and suppliers. Using just-in-time suppliers, the kanban control system, and flexible manufacturing technology they can customize production to customers' specification within days. Reduced lead time, or the limiting case of real time production, makes the business more capable of adapting to changing production or market conditions.

We now discuss how Ashanti is being positioned to reduce lead time, focussing on project implementation and the redesign of the supply function.

5.1 *Implementation of capital projects*

The fast pace of change in the new business environment requires a new approach to the development and implementation of capital projects. The time from conception to implementation must be reduced. That is, lead time reduction principles also apply to strategy and capital projects. It is well known that organization is the mechanism for implementing strategy. An opportunity may disappear if you cannot quickly put the right organization in place to exploit it. More often the costs of a sleepy organization is the value lost due to unattained performance targets.

It is impossible to quickly implement capital projects if the organizational principles are not based on lead time reduction assumptions. In the 21st century, we believe there will be a premium for mine managers who can create flexible organizational structures to enable them to quickly implement and realize strategic goals and major capital plans.

In the last seven years Ashanti has completed two major capital projects and is currently in the process of implementing a third one.

The first project was a US$160 million rehabilitation and expansion program initiated in 1986. That project enabled the company to replace plant and equipment and develop the necessary infrastructure to expand underground mining. Specifically, a tailings treatment plant was built, deep-level shafts were sunk, mechanized mining equipment were installed, and the existing treatment plant was redesigned and upgraded to make it more efficient. The successful implementation of the project raised gold production from 242,000 ounces in 1985 to 400,000 ounces in 1990.

The second project was a US$115 million Sansu project begun in 1990 and completed in 1991. The Sansu project enabled Ashanti to develop a surface mine and install a heap leach and an oxide plant to process some of the lower grade ores on the concession. The completion of the Sansu project also helped to introduce some flexibility into the operations, permitting a cross transfer of ore feed between different treatment plants. The Sansu project made it possible to increase gold production by an additional 230,000 ounces by 1992.

The third and most ambitious project is the Ashanti mine expansion project (AMEP). AMEP is a US$555 million project of which US$295 million will be spent during 1993-94. The project was initiated in 1992 and is designed to raise gold production to over 1,000,000 ounces by 1995-96. The center piece of AMEP involves the adoption of mechanized mining and the construction of a sulfide treatment plant to enable Ashanti to develop several large sulfide deposits on the concession. In addition, AMEP funds will be used to support the required increase of working capital and to upgrade and extend the company's infrastructure.

It is evident that Ashanti has gained considerable project implementation experience during the last seven years. A variety of new organizational forms have been used to enable Ashanti to quickly plan, implement, and operate the projects. Previously, projects of this magnitude would have been implemented as pure turnkey projects, without an effective mechanism for technology transfer and commitment.

A project team concept was used for the Sansu project, but the key personnel, who would later operate and manage the project, were included in the project team. This enabled them to assume ownership of the project and develop the necessary commitment to bring the project to completion on budget and in a timely manner. Because AMEP is not a stand alone project like Sansu, the challenge is to implement it without affecting operations adversely. This requires gaining the commitment of operational managers without overburdening them with AMEP responsibilities. To achieve this a system of working groups consisting of planners and operations people was created to advise and collaborate with the AMEP project team.

Ashanti implementation strategy also involves the conditioning and training of personnel involved in these major projects by having them visit or train at other mines operating similar plants and processes.

5.2 *Overhaul and redesign of the supply function*

For a mine like Ashanti, which is far from the centers where major inputs are manufactured, senior management has had to grapple with the problem of reducing lead time in the procurement of raw material, spares, and capital equipment. Typically, it took about 18 months from time of order placement to receipt of goods on the mine. The long lead time made it extremely difficult to plan and properly manage the supply function.

In the last two years, Ashanti has undertaken a systematic study to overhaul and redesign the supply function. The objectives are to reduce lead time; realize potential economies in the supply of consumables and spares; reduce excess stockholding at the mine; and improve service levels and information flow within the supply system. The entire supply system is being reorganized towards continuous improvement and elimination of bottlenecks. In particular, Ashanti is developing and implementing a system to improve planning and coordination within the supply chain, extending the use of supplier support agreements, and reducing lead times on overseas procurement.

A major part of Ashanti's strategy is to encourage overseas suppliers to set up local operations and stores within the concession to service its requirements. Ashanti will no longer spend millions of dollars to buy equipment and supplies from suppliers who do not make a commitment in the form of local back-up services.

We believe that Ashanti's successes in the procurement area will enable it to contribute to the removal of another of the important geographical constraints to mining in Ghana. As more suppliers, attracted by Ashanti's business, set up shops in Ghana, it will be far easier for other mines to operate without the need for the scale of working capital Ashanti has had to tie up. Ashanti will also free up working capital to support other worthwhile ventures, within or outside mining.

6 EXTRACTION POLICY AND THE ENVIRONMENT

We all know the great concern people throughout the world have about diminishing resources and degradation of the environment. It should be expected that increasing pressure will be brought to bear on mining management to implement more acceptable extraction and environmental programs. We see these developments as part of the larger question of defining the new rules of the game. We would like to comment on how the management at Ashanti is adapting to the new environmental challenges, broadly interpreted.

6.1 *Extraction policy*

For most of its life as a large scale mining operation, Ashanti practised a "high grading" policy. That is, it concentrated on mining only the high grade reserves. The cut-off grade was set at 10 grams per metric ton. Eventually, the known high grade reserves were depleted inducing management to process progressively lower grade ores. However, the existing processing facilities were incapable of achieving adequate gold recovery rates with the lower grade ores. Output began to decrease and continued on a downward path. Not much exploration activity occurred either. It was only a matter of time before the mine would have reached the end of its useful life. Fortunately, the situation was reversed with the introduction of the rehabilitation and expansion project discussed earlier.

The main insight from this experience is that high grading is unsustainable and untenable. Mine management should use the most appropriate technology to facilitate the development of the lowest possible ore grade in the most economical manner. By introducing technology that enabled the lowering of its cut-off grade to 3 grams per metric ton, Ashanti's proven and probable reserves is currently 16 million ounces of gold. A vigorous exploration program is also in place to help extend the asset base.

Although the change in Ashanti's extraction policy came about autonomously, we believe it is in the interest of mine managers in Ghana and elsewhere in Africa to support and lobby for extraction and augmentation policies that are compatible with the best technology available. The coalition of stakeholders can endure only when each party is reasonably sure that the mine has a long useful life ahead.

6.2 *Environmental outlook*

Mining is an environmentally stressful activity. In the past, mine management did not take adequate measures to deal with the passing of toxic discharges into the environment, stripping of the landscape, destruction of wildlife sanctuaries, and other disruptions. The environment was considered to be inexhaustible and free. It was also felt that people would always be willing to exchange environmental degradation for economic growth and income.

Now, the times have changed. Pollution control regulations have been passed in most countries. The recycling movement is gaining momentum and new technology is available to control pollution and assist in clean-ups. Although the environmental lobby is not as organized in Africa as it is in the West, there is no doubt that environmental issues are on the agenda throughout Africa. It is now difficult to secure external funding for African mining projects without an environmental impact study. Environmental protection agencies are active, compliance to environmental standards are monitored, and a rising environmental consciousness is evident in the popular press of most African countries.

In response to pressure to control airborne emissions of arsenic trioxide and sulphur dioxide from its oldest treatment plant, Ashanti has recently constructed a US$7 million arsenic recovery plant to safely collect the arsenic in sealed containers for export. Ashanti also has an environmental management plan to address the control of water runoff and dust from surface mining, handling of cyanide from treatment plants, disposal of tailings, rehabilitation of mines areas, and worker safety practices, especially in underground mining.

Environmental considerations also played a role in Ashanti's choice of technology for the new sulphur treatment plant to be constructed as part of the AMEP project. The relatively new bio-oxidation process selected has built in pollution controls whereby arsenic is produced as an inert compound.

Finally, it should be mentioned that Ashanti is also conscious of the social and cultural environment within which it operates. Part of its success is explained by the significant investments and contributions it has made to improve the quality of life on the mine and the surrounding community. Ashanti investments in social infrastructure include housing projects, schools, sports facilities, and a hospital. There is a direct relationship between Ashanti's productivity increases and the quality of life on the mine. There has also been a considerable improvement in industrial relations in recent years.

7 CONCLUSION

In this paper, we identified some of the most important characteristics of the business environment into the twenty first century. For mine management, we focused on the settled ideological climate, improved fiscal regime, demands of globalization and the information revolution, price instability, lead time reduction, and worldwide pressures for more acceptable extraction and environmental programs and related them to the specific situation of Ashanti. As we enter the twenty first century, mine management around the world will have to develop innovative strategies to cope with these demands but each will have to make its strategies relevant to its specific situation.

Innovative Mine Design for the 21st Century, Bawden & Archibald (eds) © 1993 Balkema, Rotterdam, ISBN 90 5410 325 6

Continuous improvement at Kidd Creek

W.W. Holmes
Falconbridge Limited, Kidd Creek Division, Timmins, Ont., Canada

ABSTRACT: The Kidd Creek mine and metallurgical plants are internationally respected for commitment to excellence, technical development and product quality control. Continuous improvement has been a way of life.

Financially, however, unit costs have been eroded due to normal inflationary forces and to the imposition of government mandated costs. The management of Kidd Creek decided late in 1990 that The Continuous Improvement Philosophy needed to be directed to flattening unit costs, i.e. to achieving enough improvement to offset inflation. Having achieved this goal, and with lower inflation rates, it has since been decided to work to continuously lowering unit costs.

In 1991 costs were flat in comparison with 1990. As we looked ahead to 1992 it became obvious that some additional management emphasis and tools were required. The Continuous Improvement and Total Quality Management models were reviewed, and adopted. In 1992 unit costs started to decline, and this trend is expected to continue.

The paper will focus on how Kidd Creek is using Continuous Improvement to achieve the objectives of the business.

CONTINUOUS IMPROVEMENT AT KIDD CREEK

The Kidd Creek orebody, mine, and metallurgical plants have been the subject of many papers by Kidd Creek personnel, and studies by many academics and consultants. An overview of production to date is explanation for this interest.

Exhibit I. Kidd Creek production history

		1992	Operations To Date
Metal in concentrate			
Copper	MT	109,400	2.254,400
Zinc	MT	137,800	5,485,800
Silver (ooo's)	MT	119,400	5,567,400
Zinc plant			
Zinc Metal	MT	138,300	2,733,000
Cadmium	MT	790	10,590
Acid Plants.			
Zinc Plant	MT	191,300	3,694,890
Refinery	MT	483,100	3,518,100
Copper refinery			
Cathodes	MT	94,660	869,860

The history of the operation, is that of a series of major projects to build the mine and concentrator, to expand these, and to build additional major downstream processing capacity i.e. the zinc plant, copper smelter and refinery, and the indium plant.

Exhibit II. Plant start up dates

Discovery	April 16, 1964
Open pit production	October, 1965
Concentrator	November 16, 1966
No.1 Mine production	March, 1972
Zinc plant	April, 1972
Cadmium plant	April, 1972

No.2 Mine production	October, 1974
Copper smelter & refinery	June 25, 1981
Indium plant	April 16, 1991
No.3 Mine production	1992
Liquid SO_2 plant	1993

There have been few technical mineral industry meetings or conferences in North America over the past 25 years since the original plant start-up that have not covered at some point the current progress of plant development, environmental improvement, safety management, employee relations or some technical innovation at the Kidd Operation.

CONTINUOUS IMPROVEMENT - A TRADITION AT KIDD CREEK

The operation has been internationally respected for commitment to excellence, technical development and product quality control. Continuous improvement has been a way of life in the area of Safety, Environment, Process Improvement, Product Quality.

Safety

Different people saw the operation from different perspectives. One of these was the safety perspective. Kidd Creek consistently had among the best safety performances in Canadian mining.

Exhibit III. Kidd Creek safety performance vs Ontario mining

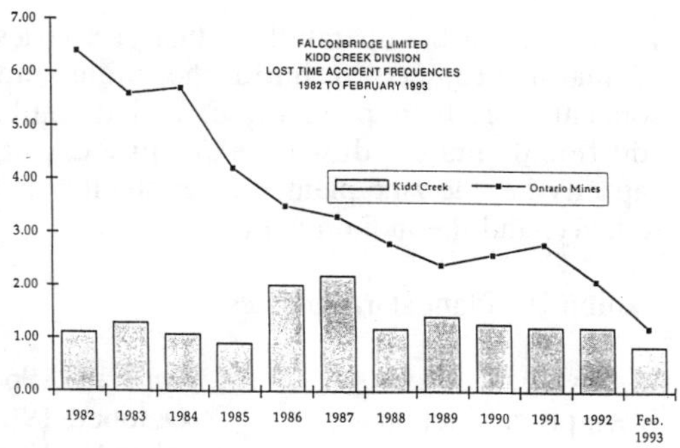

This leadership in safety was rewarded by the winning of 11 National John T. Ryan Safety Trophies as the safest metalliferous operation in Canada. The most recent of these awards was for the year 1991.

The exhibit indicates the excellence and consistency of Kidd Creek performance over the years, and also the dramatic improvements made by the Ontario industry.

Environment

Excellence has also been a trademark of the Environmental control program. The zinc operation and the copper smelter were both designed to capture a very high percentage of the SO_2 emissions. However there were some exceedences above the special limit, .25 ppm for a 30 minute average, set for the Kidd operation. A determined and successful effort has been made to reduce the frequency of these exceedences. In 1992 the results plateaued as expected. But in true Continuous Improvement fashion, objectives, procedures and protocols have now been reviewed and the downward trend will be renewed.

Exhibit IV. Annual total minutes of exceedence

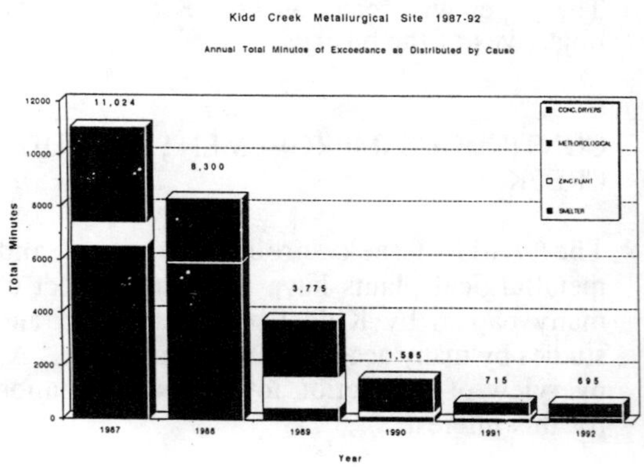

In both the safety and environmental areas, significant improvement must continue, and priority is put on these activities. Safety performance is being improved by introducing loss control programs and the internal responsibility system into each area of the plant. The environmental improvement program is being guided by a five year plan and by increasing the level of employee awareness of environmental law through use of ELAP (environmental law awareness program). In both the safety and environmental area the issues have been addressed technically and with extensive involvement of employees.

Exhibit V. Management of safety and environmental issues

MANAGEMENT OF SAFETY AND ENVIRONMENTAL ISSUES

Technical
- Safety - Loss Control
- Environment - Five year plan to achieve environmental targets

Employee Involvement
- Safety - Occuptational Health & Safety Committees
  - Internal Responsibility System
- Environment - Elap (Environmental Law Awareness Program)

Progress and Product Quality Improvement

Safety and Environment are clearly our number one and two priorities, and have been subject to a long term continuous improvement effort, the technical side of the operation has also benefited from the same discipline and approach. The following exhibits are meant as a small sample of this effort.

Exhibit VI and VII illustrate the results of process improvements projects which have yielded cost savings.

Exhibit VI. Current efficiency - 1985 to date

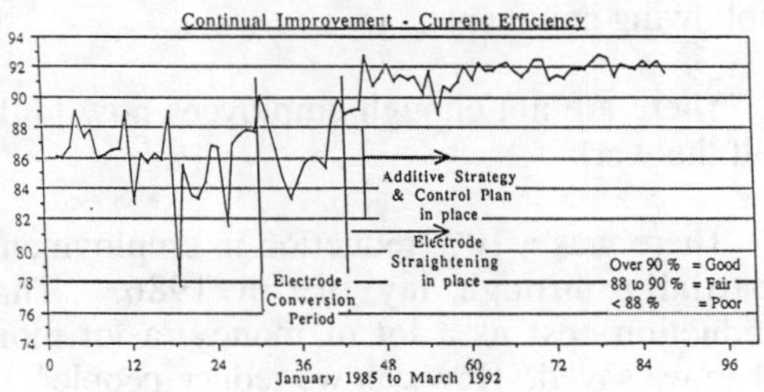

Exhibit VII. Power consumption - Kwhr/T

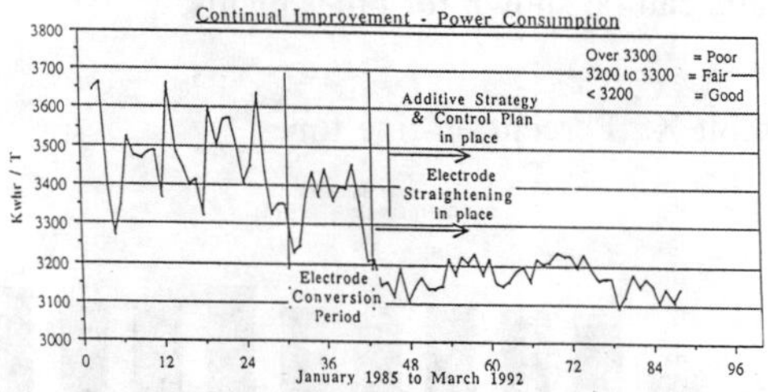

Exhibits VIII and IX illustrate the results of efforts to improve product quality. This has been a major thrust at Kidd Creek. Both these exhibits illustrate efforts in the area of zinc quality where Kidd is about to achieve ISO 9002 standard. Similar attention has been directed to all final products i.e. copper, cadmium, acid, indium.

Exhibit VIII. Continual improvement-cathode lead

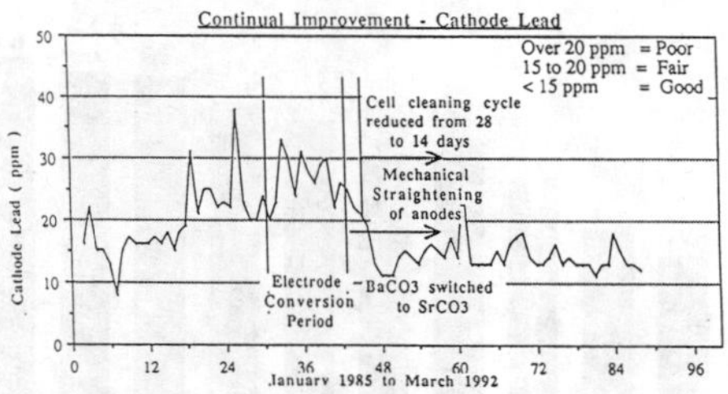

Exhibit IX. #2 Zinc casting line - dross reduction program

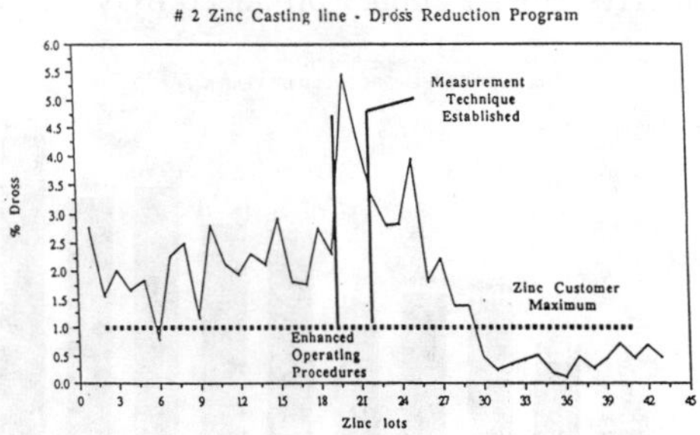

Exhibits X, XI, XII illustrate the impact of a multitude of process improvements on production throughput. While these exhibits are based on copper plant results, similar results can be shown for other plants.

Exhibit X. Percent on-line time

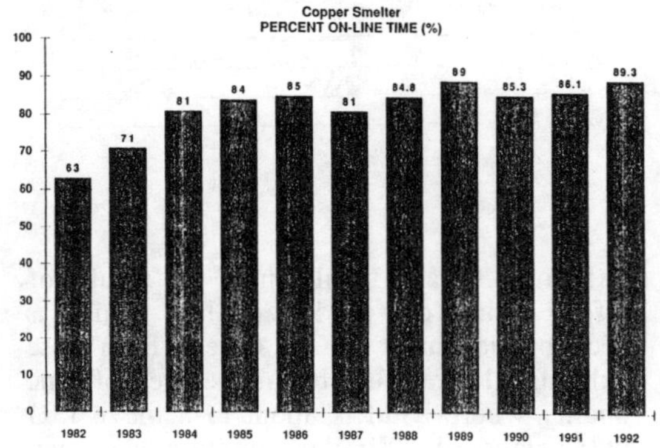

Exhibit XI. Instantaneous concentrate feed rate

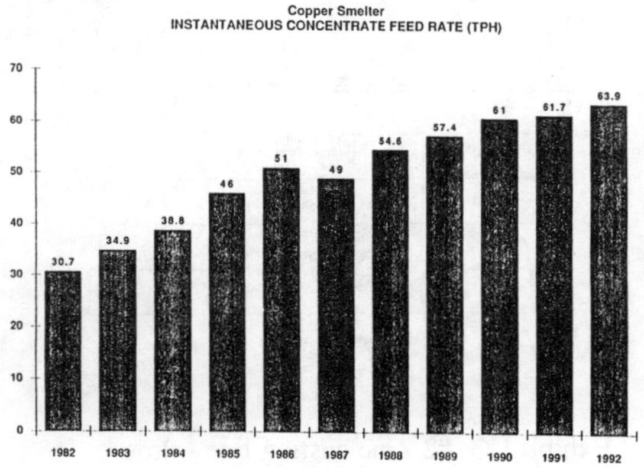

Exhibit XII. Blister copper produced by year

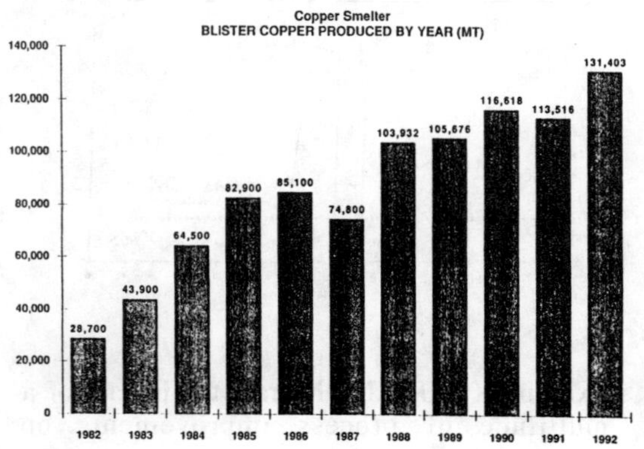

IMPACT OF CONTINUOUS IMPROVEMENT ON UNIT COSTS

However despite the fact that Kidd Creek operations had a history of quality improvement initiatives, production costs per unit of metal of output had continued to rise at an alarming eight per cent per year through the last half of the 80's. The reasons for this increase included

- declining mine production due in part to decreasing tonnes per vertical foot with depth
- more difficult mining conditions with depth
- inflationary impact on the cost of people and supplies
- the increasing level of government mandated costs such as

W.C.B.
Electrical Energy
Environmental Requirements
etc.

CHANGING THE FOCUS

This adverse cost trend could not be allowed to continue. The management group therefore made three very important but simple decisions

1. Unit costs would not be allowed to increase!
2. Since a significant portion of costs were employment costs, that attrition would be allowed to impact significantly on employment. For every five people who left the employ of the division, consideration would be given to replacing one.
3. Quality of products and processes must continue to improve.

These simple decisions were communicated throughout the division along with the reason for them. Employees generally understood that unit costs could no longer be allowed to escalate. The focus of the division changed quickly. However several cultural problems emerged immediately as indicated by the following comments:

"There are not enough employees now to do all the work"

"There was a 10% reduction in employment, partially through lay off in 1986. That reduction cost us a lot of money, a lot more than we saved. How can we reduce people".

"Every decrease in employment is going to require an equal and offsetting increase in the number of contractors employed (normally 100 - 50 contractors worked in operations on the site, plus an equal number on capital projects)"

"Why focus on the number of employees (head count) all the time. Lets concentrate on doing the job right and maximizing revenue."

"The Kidd continuous improvement process is so advanced that there is little room left for cost improvement."

These statements were confronted by

- the realities of the history of increasing costs
- the obvious need to control costs to preserve financial margins to ensure long term viability
- the importance of improving Kidd's world competitive position to offset some natural disadvantages
- the benefits of achieving improvement continuously through a consistent long term effort, versus the alternative trauma of a sudden step change.
- the preference for achieving decreased employment costs and increased productivity through maximum use of attrition.

While it was obvious that the flat unit cost objective would require continued reductions in employment, increases in productivity and continuous improvement over a long period of time, it was recognized as a desirable objective.

Improving Cost and Quality Performance

Many organizations, undertaking cost improvements, particularly if the projects involve decreasing employment, find that many tasks previously done, are left undone. As these items accumulate, costs increase because of breakdowns. At Kidd, this has been avoided through the use of attrition with broad based support, coordinated with a determined effort to use job redesign and productivity improvement techniques to ensure that all necessary tasks were being continued.

An operating superintendent was assigned a role, reporting to the vice president and general manager, to facilitate the change. This was a Key Appointment and Organizational Change. It telegraphed to the total organization that the Change was Real and not just Words. The incumbent was given the responsibility for working with the department managers, superintendents and throughout the organization to assist in the job redesign, job enlargement, etc. that was required to allow attrition to take its course. As opposed to hiring one for every five who left the company, Kidd Creek experienced a need to hire only one for every ten who left.

Revisiting Continuous Improvement

While this process was on-going, the management team started looking at Total Quality Management and Continuous Improvement Initiatives being run by other companies. While the concept was attractive it was recognized that

1. some companies had spent a large amount of time and money on such initiatives and not made much progress for many years
2. some companies had maintained a well planned and coordinated effort and achieved attractive results only after an extended time period.

The Kidd Creek management team decided that the concept of Continuous Improvement existed within the operation, and that it would be easiest to build on this concept rather than facing the pitfalls of embarking on a New Program Mandated From On High. In order to build on the concept of Continuous Improvement it was decided

1. to revisit The Idea Awards Program with the objective of stimulating interest in a program which was still generating significant savings for the Division, although the trend of results was downward

2. to move the Continuous Improvement concepts from the engineering and technical levels in the Division, to the level of employee groups working on problems and projects

3. to hire a consultant to help us educate the management and employee teams on an "As Needed" basis to tackle continuous improvement projects

4. to hire a organizational consultant to assist groups at whatever level, facilitate the organizational change

5. to continue to build on this base, as the organization gained experience.

Idea Awards

The Idea Awards Program was reviewed and objectives updated for the coordinators and the management team. As shown by Exhibit XIII, the results have been very encouraging. The percentage of ideas accepted has not changed but the number of ideas submitted has increased because of the additional interest and emphasis. The change in organizational commitment became particularly evident in 1992.

Exhibit XIII a,b,c. Idea award program results - 1988-1992

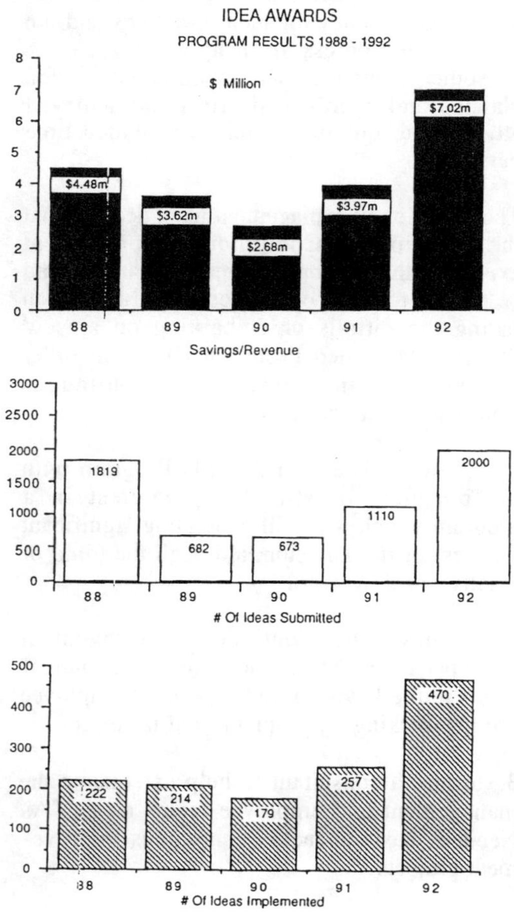

Management Training

Management understanding and commitment are critical to success. Many in the management group had been exposed to Quality Management and Continuous Improvement readings and courses. A consultant was selected by the management group and utilized to provide management group exposure to the Continuous Improvement principles and to provide the on-going employee training as required.

Employee Training and Participation

Employee input and ownership of improvement programs is important to the success of any continuous improvement effort. For this reason project teams will normally have representation from a cross section of the organization. Often such groups require some formal training in continuous improvement concepts and this is provided to facilitators by a consultant, and then through the facilitator to the teams. The emphasis however is providing the training as required.

Training To Facilitate Organization Change

Economies have been achieved throughout the operation by downsizing using attrition, and also by having the regular operating employees replace contractors. This has resulted in organizational change in almost every department. The extent of organizational change has varied between departments but has included

a) elimination of an organizational layer
b) decentralizing the maintenance function
c) combining maintenance and operating jobs
d) reducing the number of supervisory levels and the number of supervisors at a particular level

These organizational changes required a lot of training.

Training has consisted of:

cross trades training - to train tradespeople with one skill area to work to handle the basic functions of another trade. i.e. electricians to handle basic mechanic functions.

cross function training - to train operators to handle basic maintenance functions and vice versa organizational behaviour training - in some areas the change in organization required significantly different communication channels and skills and it has been necessary to have a consultant work with the department organization to ensure that everyone understands their new roles and can function effectively.

- several service departments have changed their culture dramatically to become Customer Oriented.

The objective of a quality product, is first and foremost. The change process necessary to achieve the organizations productivity and cost objectives must be designed to ensure that the people are properly trained to handle the change whether that be new technical skills that are required to handle an enlarged job, or new soft skills that are required as a result of organizational change.

Communicating Goals

Many companies have sophisticated systems for developing company goals which are customer oriented, and ensuring that these goals are communicated throughout the organization, and that employee efforts and the total management system are directed towards achieving these goals. The process is often called policy deployment.

At Kidd Creek since we are at a very early stage in our Continuous Improvement program, we elected for reasons of simplicity, to utilize current management systems to ensure harmony of goals throughout the company. As an example, a management by objectives (MBO) system has been utilized successfully for several years to identify the key objectives and focus of the senior management people in the Division. The focus of the MBO system has been achievement of the current years budget. The MBO tool offered potential to change the focus of the management team from this years operation, to the achievement of longer term goals. While traditional MBO and the policy deployment process used in total quality management companies are different in many ways, the Kidd Creek team chose to ignore these differences. However as time progresses each individuals objectives have started to reflect more team goals.

The priorities of the Division were identified as Continuous Improvement in the areas of

Safety
Environment
Product Quality
Metallurgical Efficiency
Idea Awards
Unit Costs
Business Development

and the divisional goals in each of these areas were identified. Although some are short range in nature, with each iteration of the plan, the goals identified become longer term. This process has served to change the focus of the management team from this years operation and the preparation of next years budget, to the longer term focus. Managements job has become one of identifying the change required to achieve the goals for 2, 3, 4 and 5 years from now, and of implementing the action plans at the appropriate time, that are required to achieve the longer range goals.

The development of the Divisional Goals and the MBO process which focuses on departmental goals, ensure that departmental goals and divisional goals are coordinated.

While approximately forty people are currently involved in the MBO process now, it is obviously necessary that much larger number of employees are also aware of, and working towards the achievement of the Divisional and Departmental Goals. The first step of this communication is fairly easy and is handled in quarterly supervisory and technical meetings, the process of having the whole organization identify with the Divisional Goals is somewhat more difficult.

Currently a series of departmental meetings involving all employees is being held to communicate Divisional and Departmental Goals. While these are quite large and open meetings, Managers also meet quarterly with all employees in small groups, to provide feedback on the progress of the Division and the Department. This is in addition to the regular communications that occur between supervisor and employee.

RESULTS

Exhibits XIV to XX are provided to indicate results to date. The results are very satisfactory. Despite the inflationary impact on employment costs per person and major costs such as Ontario Hydro, and the continued increase in government mandated costs, costs in 1990 and 1991 were essentially flat when compared to 1989 and unit costs in 1992 were lower. While costs are lower, product quality, and environmental and safety results continue to improve. The action plans developed, and being developed by management and by groups throughout the division working on continuous improvement indicate that these results will continue. The more Kidd Creek people work on improving the process, the more Kidd Creek people recognize the potential for further change, to improve the competitive position of the organization.

Exhibit XIV. Kidd mine productivity

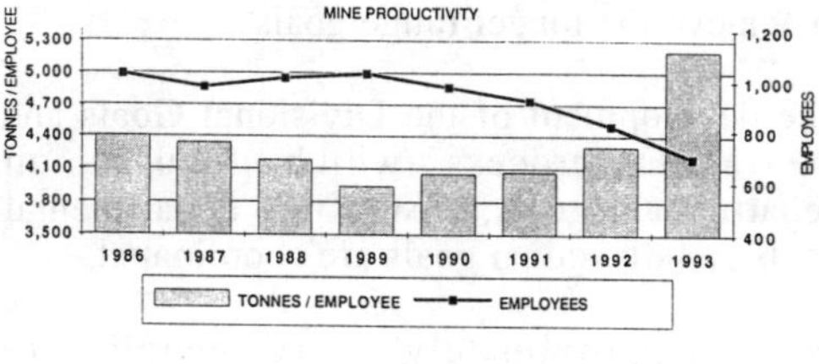

Exhibit XV. Mill unit cost index

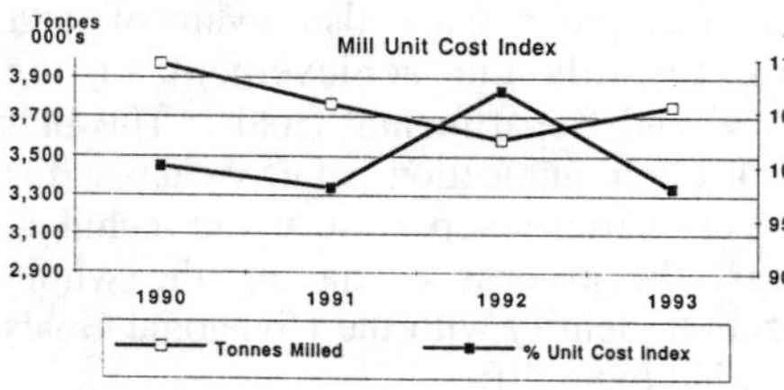

Exhibit XVI. Zinc plant production & unit costs

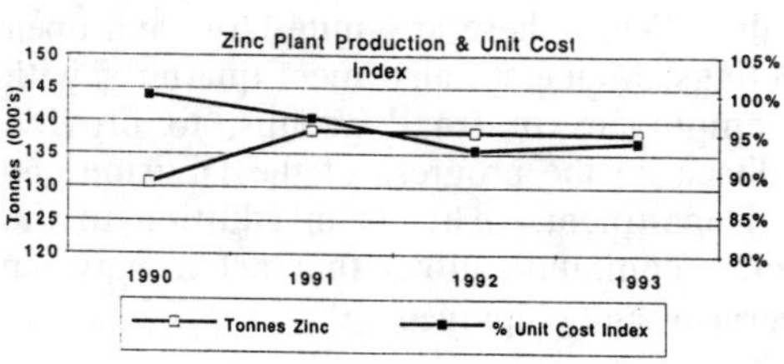

Exhibit XVII. Smelter production & unit cost index

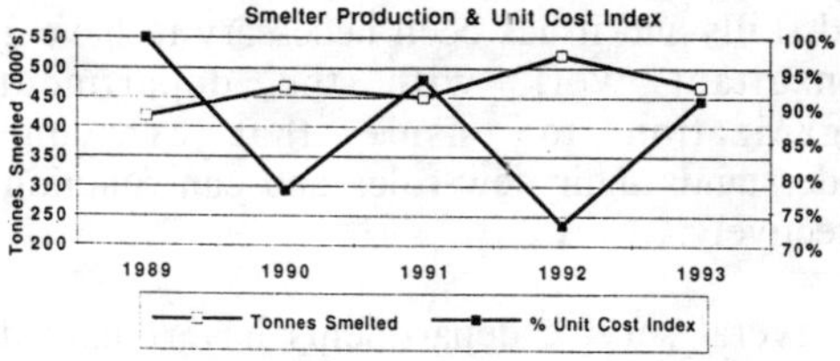

Exhibit XVIII. Refinery Production & unit cost index

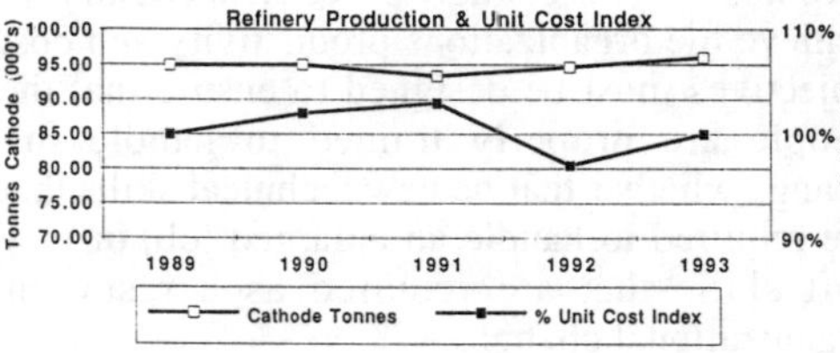

Exhibit XVIX. Metallurgical site weekly (average) maintenance costs

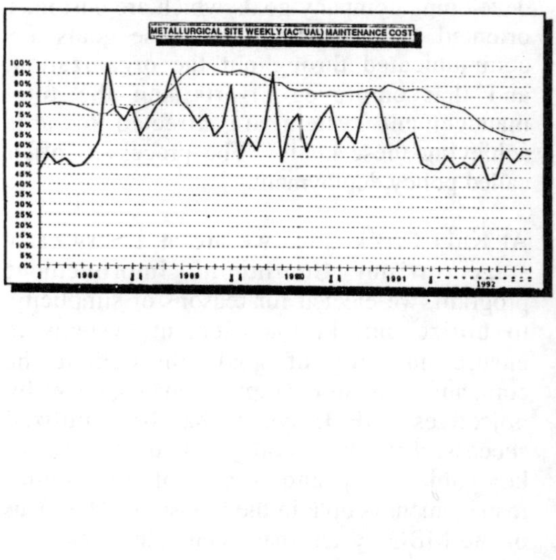

Innovative Mine Design for the 21st Century, Bawden & Archibald (eds) © 1993 Balkema, Rotterdam, ISBN 90 5410 325 6

Emerging strategies in the mineral industries: Global challenges

Peter R. Richardson
School of Business, Queen's University, Kingston, Ont., Canada

ABSTRACT: Financial returns for many mining companies have been low in the last decade. While current efforts aimed at improving quality, cost and capital utilization in the industry are important, more fundamental changes underway in the global economy portend more significant changes to mineral industry strategies yet to come. This paper examines the nature of these changes, and identifies opportunities for future investment, growth and diversification that are open to firms in this important industry.

FORMULATING STRATEGIES IN A TURBULENT ENVIRONMENT

Corporations in the mining industry have to formulate strategy in an environment which can only be charitably described as turbulent unpredictable, uncertain and uncontrollable. Executives in the industry are used to dealing with many of these characteristics - exploration risk, cyclical prices, fluctuating demand and political risk. Other trends have proved more problematic - environmental issues, emerging competition from substitute materials, and changing workforce expectations - all continue to be major concerns. The purpose of this paper is to present some views on how mining companies must respond to the challenges - and opportunities - present in the contemporary global economy.

Many mining companies have failed to cope well with emerging global economic competition. Financial returns for many of the larger firms are low (Table 1). They have a poor public image, are unable to diversify successfully into related material businesses and fail to fully exploit the capabilities of their employees. Mining remains a relatively hazardous business in which to work.

As a reference point in this paper, I will use the strategy of what is probably the world's most consistently profitable mining company: the Minnesota, Mining and Manufacturing Corporation, better known to us all as 3-M. Their average returns over the last decade are better than the profits of any Canadian mining company. Inspite of its name, 3-M has never operated a mine, but it's an important benchmark for mining companies. Nowadays in the fiercely competitive business of modern materials, mining companies don't just compete with each other, they compete with the likes of 3-M and Dupont and ICI, companies which manage their strategies very differently from those of most mining companies. There are a number of facts about 3-M that are particularly relevant to this paper:

* of its U.S. $14 billion revenues in 1991, 25 percent came from products new in the last five years,
* it has 70,000 products produced and sold around the world,
* it spends nearly 7 percent of revenues on research and development,
* it consistently has profits after tax of about 10 percent of revenues,
* it is recognized as an industry leader in environmental responsibility, and
* in many markets its products compete with those of the Canadian mineral industry.

One other thing worth noting is that although its products are highly differentiated and command substantial profit margins, the corporation publishes aggressive cost reduction goals for itself.

This paper will go beyond the immediate financial and competitive pressures facing the mining industry and address the fundamental strategic challenges and opportunities which I believe will confront the mineral industry, and particularly Canadian producers, over the next decade or so. These can be summarized as:

* obtaining adequate returns on capital employed in the industry,
* investing sufficient funds in future-oriented activities (exploration, technology, human resource development) to maintain a satisfactory growth trajectory,
* ensuring that Canadian firms are not by-passed in the emerging global mineral industry structure,
* capitalizing on opportunities for expansion and growth into related material-technology businesses, so

that Canada will have a number of world-scale materials corporations in the twenty-first century, and
* developing a pool of executive talent which is capable of managing these strategies.

PROFITS AND SATISFACTORY RETURNS

In the past few years, earnings for many major mining corporations, not just in Canada, have been less-than-satisfactory. Even in 1988 and 1989, the best years since 1981 for most major metals producers, returns on capital were still not outstanding. For the last decade as a whole, financial returns in the mineral industry have been dismal, as shown in Table 1, particularly given the level of inherent risk.

In the long-run, a capital intensive business like mining has to generate a good rate of return if it is to attract funds for re-investment and growth. The challenge facing mining executives is how to maintain a reasonable margin, even in periods of low prices, to provide for continued investment in exploration, new technology and market development.

In the face of these dismal returns, mining industry executives have 'stuck to their knitting' and focused on strengthening their core businesses. Unfortunately, investment in the future has to be funded either from profit margins or increased debt, and the weak balance sheets of most of Canada's major producers have detracted from their ability to:

* invest in exploration to ensure an adequate reserve position,
* invest in market and product development to offset substitution by substitute materials such as plastics,
* invest in major new process technologies in order to remain cost competitive to maximize margins in a cyclical industry,
* build a high level of employee involvement, motivation and trust to maximize productivity in a business which is often not only capital, but labour intensive, and
* maintain a strong balance sheet to finance new developments and cushion the impact of troughs in the price and demand cycle.

Executives in most mining companies know that these are the key success factors in this business. The major Canadian integrated metals companies, such as Alcan, Cominco, Falconbridge, Inco, and Noranda, all increased investment levels in exploration, mine development and new technology during the price boom of the late 1980s, but found it difficult to maintain these levels as prices declined after 1989. Instead, they have renewed their focus on cost reduction, seeking to become among the industry's lowest cost producers through the traditional approaches of workforce reduction, and reduced investments in exploration and new technology.

In contrast, the major Australian producers - BHP, CRA, Western Mining and Mount Isa Mines all have relatively strong financial positions. In spite of a similar domestic political and economic climate to that facing Canadian firms, these companies panies have maintained, and even expanded their exploration, technology development and human resource programs. Several major discoveries have been resulted, and Australian producers are set to be internationally competitive in the long-term.

In spite of their protestations to the contrary, many Canadian mining executives did not learn a few vital lessons during the lean years of the 1980s in respect of these key success factors:

* investment in exploration and new technology must not be treated as period costs and slashed during down-turns, but maintained at relatively consistent levels throughout the economic cycle,
* cost reduction through being 'lean and mean' can only be accomplished over short periods, and that what I term 'lean and keen' cost management strategies are required to bring long-run costs down,
* the market encroachment of substitute products can be successfully resisted by cooperative market development programs and renewed product development thrusts,
* investment in marginal mines and facilities during boom times brings a premature close to high-price cycles and financial disaster in the subsequent troughs.

In the recent price down-turn, Canadian mining executives have resorted to many of their traditional approaches to cost-cutting which have caused problems in the past. Admittedly, thin profit margins, weak balance sheets and high debt loads have made these actions necessary. However, while these actions arguably improve the efficiency of Canadian mining firms in the short-run, the reality is that they continue to fall further behind the rapid pace of change elsewhere in modern economies.

Unfortunately, mining executives' traditional view of competition and corporate strategy is no longer appropriate. The mining industry now has a number of new competitors:

* wilderness and environmental groups - for alternative land use,
* new industries - such as computers, telecommunications and services - for scarce skilled employees,
* government and unions - for the trust and loyalty of employees, and
* producers of engineered materials - for markets and capital.

Competition is now no longer simply between metal producers, vying with each other to be the lowest cost, but also takes place across a broad spectrum of different dimensions. In future, to generate the margins and earn the returns necessary for reinvestment and growth, Canada's mineral producers will have to adopt competitive strategies other than cost minimization.

As shown in Exhibit 1, the alternative to pure cost competition is some form of product differentiation (or, of course, diversification into completely new business areas). Differentiation can be achieved in a variety of ways, through service, quality, innovation or intangible product characteristics. In the materials business, the only form of differentiation for which customers are likely to pay consistent premiums are superior performance characteristics.

Canada's mining firms must increase their efforts to differentiate themselves and their products from those of the competition.

GROWTH IN CORE BUSINESS AREAS

One way to grow is through developing and opening new mines in Canada, but in base metals it's now twenty years since the last elephant (Kidd Creek) was discovered. In fact, several of Canada's integrated mining corporations such as Noranda, Hudson Bay and Cominco have extremely weak mineral inventory positions in Canada, and might not be able to sustain current levels of production into the next decade without major new discoveries being made.

Of course, growth can also be achieved through acquisition, which can at the same time provide diversification across a variety of metals. In recent years, Falconbridge acquired Kidd Creek, Teck and Metallgesellschaft acquired Cominco, and Placer merged with Dome Mines. Acquisition offers the potential of immediate growth, but many studies have now shown that this course of action often benefits the shareholders of the acquired firm more than the purchaser.

Moreover, post-merger re-structuring and re-positioning can be a lengthy and costly task, especially if less than 100 percent ownership is obtained. During the two or three years that this process often takes, management time and attention is necessarily taken away from real growth opportunities.

GLOBAL OPPORTUNITIES

Growth within the mineral industries outside Canada is an attractive prospect for many firms. There is little doubt that there is a significant trend towards domination of the world mineral industries by a relatively small number of globally positioned corporations. RTZ, BHP, Outokumpu, Trelleborg, Anglo-American Corporation and Mitsui are examples of this type of corporation. They frequently take major equity positions in large, national mining firms and often form strategic alliances and joint ventures to take on the world's large new mining developments.

With very few exceptions, Canadian mining firms have not followed this route. As a result, only Alcan and Noranda of our 'large' mining firms ranked among the world's top ten (according to market capitalization) in a recent study, and Noranda's position on this list is only because of its forest products business. In an industry where scale and access to capital is important, corporate size is significant and there is a danger here of Canadian firms being gradually displaced from the first rank of world mineral producers.

Moreover, when Canadian firms have gone international, the success rates have not been very high. Alcan, Noranda, Inco and Falconbridge have all suffered serious setbacks in overseas operations. Of our major producers, only Alcan implements a truly global strategy. Placer-Dome has had considerable success in recent years in the Pacific region, but for the most-part, the others are primarily North-American based with some international operations, sales and marketing functions.

OPPORTUNITIES FOR VERTICAL INTEGRATION

Given low rates of return, diversification out of the minerals business appears an attractive growth strategy for a number of corporations. As shown in Exhibit 1, there are forward and backward integration opportunities, as well as related diversification options which might be exploited.

Forward integration into fabrication is tempting, in order to capture additional added value. Alcan has integrated forward into fabrication activities. Inco has a significant alloys business, while Sherritt Gordon has exited mining completely and become a value-added metals producer (although it has yet to record a high level of profitability). In contrast Noranda has almost completely withdrawn from the semi-finished copper products industry.

One drawback to this approach is that it puts the corporation in competition with its customers. In addition, the margins and profit returns in these businesses have generally not been spectacular, and in some cases have been disastrous. Nevertheless, when a corporation's mines are gone and there is a desire to stay in business, as happened to Sherritt Gordon, revenues and profits can be generated from a down-stream product driven strategy.

Backward integration opportunities for mining corporations are somewhat limited since they are at the start of the value chain. Advocates of a Canadian industrial strategy based on economic linkages have always argued that mining companies should enter the mining equipment business. However, there is no Canadian company which presently derives a significant part of its revenues from this source. Inco does have a relatively small subsidiary, Continuous Mining Systems, in this field but it is hard to envisage it ever being a major business for the corporation. Moreover, given the cyclical nature of capital investment in mining, and the propensity of mining firms not to buy from competitors, equipment manufacturing is viewed by many people as a high risk venture.

DIVERSIFICATION OPPORTUNITIES

Diversification is now the predominant growth strategy for major corporations. In fact, around the world, among major industrial firms, for every corporation, such as Alcan, that does business in only a single product category, there are approximately three that are diversified, such as 3-M, General Electric and Procter and Gamble.

When considering diversification, it is extremely important to differentiate between corporations like 3-M that have some form of relatedness among their products and markets, and others, such as Canadian Pacific and Bell Canada Enterprises, which have adopted an unrelated diversification strategy.

Related product firms have been responsible for much of the economic growth in the U.S., Japan and Europe during the last few decades. They spend heavily on product and market development to expand into adjacent business areas. In Canada, there are very few corporations of this type. Instead the industrial landscape is dominated by either large single product firms such as Alcan and Inco, or by unrelated conglomerates, such as Edper, Power Corporation, Brascan, and Canadian Pacific.

Unrelated firms, which often adopt a holding company structure, exhibit much lower growth rates. They also appear to have a harder time generating superior financial performance than related product firms, and it is well known that unrelated acquisitions have a high failure rate. In short, the unrelated product strategy is difficult to implement well. In fact many of the present spate of major acquisitions in the U.S. are of this type of sprawling conglomerate, which are subsequently broken up and managed more effectively.

In the mineral industries, there are two major related product diversification options. One is to become a diversified natural resources corporation. Noranda has adopted this approach and its stated mission is to be the world's premier diversified natural resource corporation. It competes in this ambition with corporations such as Broken Hill Proprietary and Mitsui.

The second related diversification option is to become a diversified materials producer, operating across the spectrum of metals, plastics, ceramics and other engineered materials. Alcoa has adopted such a strategy, aggressively diversifying away from its aluminum base. In Canada, Stelco is targeting plastics as an area of future growth potential.

The third diversification option is to use the cash flow generated by mineral activities to fund diversification into totally unrelated business areas. RTZ, the world's largest mineral producer has adopted this strategy and has moved into chemicals and electrical components. Lonhro is another British-based conglomerate with substantial mining interests.

Traditionally, Canadian mining companies have not been great diversifiers, sticking instead to the minerals business. As Bill James, the CEO of Falconbridge was fond of saying "We make enough mistakes in the business we know, so let's not get into something we know little about." Where they have diversified, the results have been mixed, at best. For example Noranda has built a world-scale forest products business, but its returns have not been spectacular. Inco's foray into the battery business was disastrous.

STRATEGIES FOR MAXIMIZING RETURN ON CAPITAL

To generate superior returns in the mineral industry all a company has to do is to find large, high-grade mineral deposits near to surface and adjacent to existing treatment facilities. As noted earlier, this strategy while easy to formulate, has proved hard to implement in recent years. In fact, increasing amounts of feed to Canada's metallurgical plants are being sourced from off-shore.

As noted earlier, even during the period of high metal prices in the late 1980s, with the exception of Inco, Canada's mineral corporations did not earn outstanding returns on capital employed. In part this performance was due to the relatively large amounts of capital tied up in the business and the length of time it takes to get new mines operating at full capacity. Mining has an extremely high level of investment required per dollar of annual revenue generated. One strategic challenge for the industry is to reduce the capital intensity of the business, which will only be achieved through continuing investment in new technologies capable of greater capital productivity.

More importantly perhaps, the industry has to work on margin improvement. Continuing to market and sell minerals as bulk commodity products will not achieve this goal. Instead, Canada's mineral producers must aim to develop strategies which can differentiate their products in a way that generates premium margins or allows them to capture more added value. This strategy will probably require the development: of new product families aimed at small, specialized markets and customers in much the same way that 3-M does with its 60,000 products.

There are a number of ways in which these goals can be achieved. For example, Inco and Falconbridge have traditionally held the major share of the premium-price nickel plating market because they have developed processes and products that other nickel producers cannot replicate. Alcan has gained a substantial share of the beverage can market by installing a full scale can line in its Kingston Research Centre to do product research. Sherritt Gordon has taken a major share of the coinage business with its products developed specifically for that market. These premium-priced market segments, though possibly small, tend to exhibit greater price stability than the tonnage segments.

INVESTING IN THE FUTURE

To maintain growth rates and margins, firms in the mineral industries will have to invest more in future-oriented activities than in the past.

In recent years, the industry sharply curtailed investment in exploration and R & D. Executives cite the maturity of their markets and expenditures on exploration as reasons for spending on R & D that only averages about 0.5 percent of revenues for the industry. However, in the present-decade until 1988, the major mining companies only spent in the range of 1-2 percent of revenues on exploration. Investment in employee development and training, essential to future growth, has been negligible until very recently.

So, in total, Canadian mineral producers have invested somewhere between two and three percent of their revenues on these activities. In fact, the position is worse than it appears, because the majority of these expenditures are not targeted at growth - creating new processes, products and markets - but at sustaining the current business. Industry investment in activities that generate real growth has been very small indeed.

In contrast, as noted earlier, 3-M invests approximately 7 percent of its revenues on R & D alone. Dupont, another materials competitor, spends over 5 percent. These figures do not include significant additional amounts that these corporations spend on market development and employee training. In Canada, DuPont spends an amount equal to 7 percent of its salary and wage bill on training and development programs. In total, these corporations invest closer to 10 percent of revenues in activities intended to create new long-term growth.

In the long-run, at current levels of investment in growth-generating activities, Canadian mineral producers cannot reasonably expect to maintain their market share, let alone increase it. Australian metals producers are, relatively speaking, investing far more. To compete effectively for global markets for materials, the Canadian mineral industry has to become market and technology driven, a change which will require major changes to corporate strategy and spending patterns.

Significant and sustained increases in the level of investment in product and process technology will be a cornerstone of the strategies for successful mineral producers. These are necessary if more sophisticated new exploration technologies are to be developed. They are necessary to develop engineered products based on mineral technology which can compete with and complement plastics, ceramics and other advanced materials. They are required to spearhead diversification into new material technologies and market applications.

Increased levels of investment in exploration are also essential in order to re-build the mineral inventory base of the Canadian mineral industry. Precious metals are in relatively good shape after the spending boom in the mid-1980s, but base metals need renewed investment if production levels are to be maintained beyond the end of the current century.

To maintain their position over the next decade, the major mining corporations in Canada will, as their Australian counterparts are already doing, have to invest in the order of 5 percent of revenues in exploration, between 2 and 3 percent on new technology and a further 2-3 percent in the development and training of employees. These investments are not huge and could easily be financed by relatively modest savings out of the industry's large capital expenditure budgets.

ADOPTING RELATED PRODUCT GROWTH STRATEGIES

Complementing these increased levels of investment in future-oriented activities, will be corporate strategies that emphasize related product growth, either into other resource businesses or into other materials. Dome Mines showed how this could be done when they financed Jack Gallagher's start-up of Dome Petroleum.

As demonstrated by 3-M and other similar corporations, there are outstanding growth and profit opportunities in markets and technologies adjacent to those already served by mineral producers. With appropriate executive direction, the gap between mineral products and other resource industries or other material technologies can be bridged successfully. The technologies are understandable, and the risks are manageable. Mining companies have a number of strengths upon which to build these strategies. They are good at managing risk. They are excellent operators. In Canada, there are excellent technical resources in corporate technology centres and in the broad materials expertise within CANMET. A judicious mix of outside management brought in to manage this diversification, internal growth and educational acquisitions could spark tremendous growth.

DEVELOPING GLOBAL STRATEGIES

The world mineral industry is coming to be dominated by a relatively small group of global corporations. Canadian mineral producers, other than those that are already subsidiaries of these firms, have been left out of this trend. Corporations like Noranda, Inco, Falconbridge and Cominco have overseas operations and sell throughout the world, but they are not global corporations. Of Canadian mineral producers, only Alcan and possibly Inco, can be said to have such a strategy.

Canadian producers ignore this trend at the risk of forfeiting the mantle of world leadership to which they have been accustomed. Our industry has to be willing to enter into strategic alliances, joint ventures and global partnerships. We must establish global production and marketing strategies that enable us to compete with the likes of 3-M, Mitsui, Mitsubishi and Metallgesellschaft which view the entire world as their theatre of operations.

DEVELOPING THE NECESSARY EXECUTIVE TALENT

Successful implementation of these strategies will require a new breed of executive in the mineral industry. Traditionally, executives in mining firms have been promoted from within after having spent a good portion of their careers in the field. This approach certainly had its advantages when mining firms competed with each other. However, now that competition in the materials business is wide open, much broader perspectives are needed in the strategic councils of the industry.

I will validate my point by three salient examples. Noranda's diversification program blossomed under the astute leadership of Alf Powis, an accountant with no mining background. One of the most successful sales organizations in the business, Noranda sales Corporation is run by Bill Deeks, a Procter and Gamble alumnus. The vision for Alcan's dramatic product development drive was originally provided by Hugh Wynn-Edwards, former university professor, federal government executive and a member of the Royal Society.

There are many outstanding individuals in Canadian mineral firms, but too often their career development has not been carefully planned. The industry is notorious for leaving bright young people in the same job for too long. The result is narrow vision and lack of drive. When they go and look at other businesses, most mining executives visit other mining firms. They rarely go and look at firms in other industries which may be leading in critical technologies or practices.

Spending on management development is low. Little is spent on educating managers in strategic management techniques. For an industry which does much of its business outside the country, there is a sad lack of awareness of international business practice. Marketing, in its true sense, is a scarce skill in this business. Education in these fields should, in my opinion, commence at university - as integral parts of mining and metallurgical engineering programs, especially at the undergraduate level.

For the industry to prosper and grow in the next decade, the next generation of executives must have a much broader world view, a much more open attitude to new growth paths, and the skills to take our mineral corporations in these directions. For this to occur we need to significantly boost spending on management development, pay more attention to executive career planning, and hire more executives from outside the industry, perhaps even from other materials businesses.

The present generation of management has steered the industry through tough times, but has been unable to build a strong vision on which to base future growth. Instead, the industry in Canada remains focused on short-run issues and survival. In contrast, sustained success in the future will depend on being able to formulate and implement strategies that make Canadian mineral producers world-class competitors in a complex, global materials industry.

Table 1. Selected financial performance - international mineral producers

in Millions of Canadian dollars

	1991 Earnings (After tax)	5 Year Average Return on Capital	10 Year Average Return on Capital
Australia			
BHP	1,270	6.80	6.57
CRA	(29)	6.08	2.94
MOUNT ISA MINES	N/A	3.82	2.30
Canada			
COMINCO	(41)	6.86	1.29
FALCONBRIDGE	21	7.40	2.80
INCO	83	16.54	7.70
NORANDA INC.	(133)	2.84	0.70
Other			
ASARCO	56	9.05	2.30
CODELCO	3.40	15.62	9.43
OUTOKUMPU OY	(145)	2.20	1.70
RTZ	6.20	6.80	9.50

SOURCE: Annual Reports and Financial Post

Net earnings after income taxes and minority interests but before extraordinary items

EXHIBIT 1.

SOURCES OF COMPETITIVE ADVANTAGE

COST \ DIFFERENTIATION	LOW	HIGH
Low	Commodities - Non-Fabricated Products	Engineered Products
High	Commodities - Not viable in the Long-run	High-Performance Materials

11 Research and development

Innovative Mine Design for the 21st Century, Bawden & Archibald (eds) © 1993 Balkema, Rotterdam, ISBN 90 5410 325 6

The industrial revolution in mining – A view towards the future

J.E.Udd
Mining Research Laboratories, CANMET, Energy, Mines and Resources, Ottawa, Ont., Canada

ABSTRACT: During the past decade the Canadian mining industry has experienced the beginning of an industrial revolution. Pressures from the international market place and the will to survive have forced radical changes to the technology employed. Given the rate of change that is also taking place internationally it is likely that, in order to successfully compete, Canadian industry must be in the vanguard.

In this paper, the author provides an overview of some of the key features of the radical changes which have occurred thus far, and offers some opinions of the future changes that will be experienced as we enter the third millennium.

INTRODUCTION

To say that the past decade has been traumatic for Canadian industry is both a gross understatement and a cliché. Having been faced with two severe recessions - the latest one from which we are now emerging (hopefully) being compared with the great depression - the key word has been "survival". Plant closures, relocations, shut-downs, lay-offs, downsizings, have been all-too-common as our society re-adjusts both its expectations and its living standards in a global economy which is increasingly competitive. Every type of organization, government included, is attempting to "do more with less" and to eliminate every possible waste or mis-use of resources.

Canada's mining industry has taken the full brunt of both recessions. In 1982-1983, for example, the prices of many of our most important mineral commodities were near historic lows (Fig. 1). Simultaneously, productivity in the metal mining sector had reached a plateau at an average of about 3 tonnes mined per person-hour paid (Fig. 2). Clearly, the industry was in trouble.

The only possible solution, both then and now, lay in improving productivity in our mines. Quite evidently, the days of exerting a major influence on minerals prices are long past (except in rare exceptions). There are simply too many nations and producers in the international marketplace for any one to dominate. The sole way of remaining competitive and profitable is to control and reduce costs and to become more productive and competitive than the competition.

Generalizations are always dangerous, but it is probably safe to say that Canadian underground mining, until the 1980's, was characterized by highly-selective high-cost methods of exploitation. The first of the two recessions of the decade changed that by forcing a widespread conversion to lower cost methods of mining. The reasons are very simple: only the mining part of the production cycle offered very rapid paybacks through investments in new technology. Both mineral treatment plants and smelters had been modernized and mechanized to the point at which, while improvements would be (and are being) made, these were much less than could be obtained from changing mining methods -- which were very traditional and which were based as much on "art" as on "science".

The first half of the 1980's in the mining

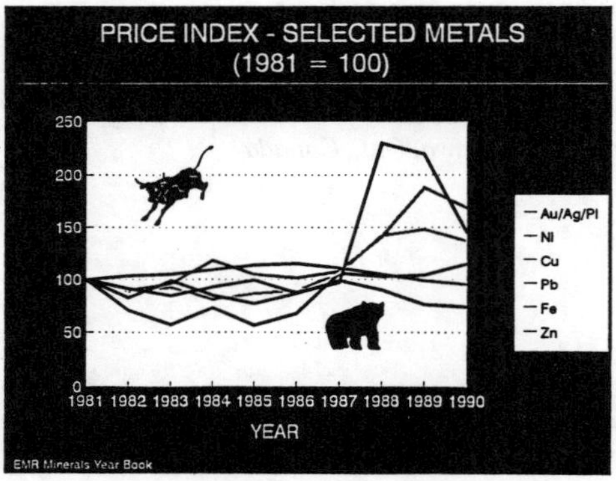

FIGURE 1: PRICE INDEX

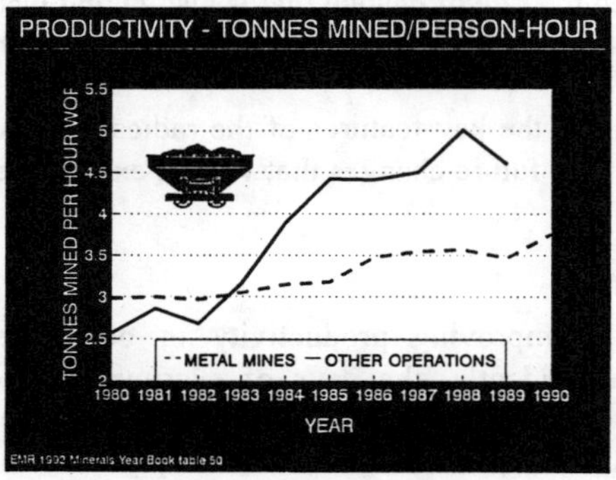

FIGURE 2: PRODUCTIVITY

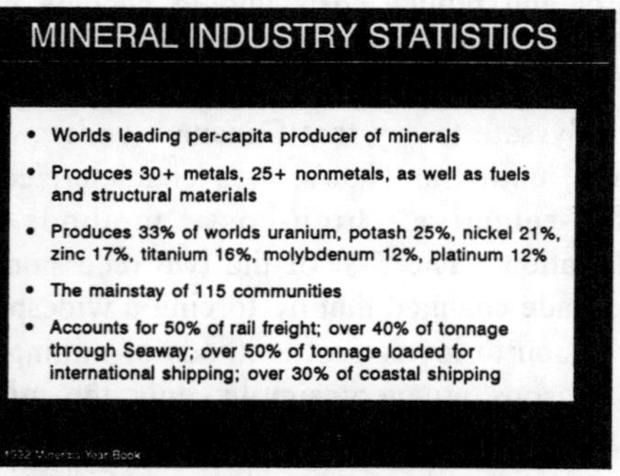

FIGURE 3: MINERAL INDUSTRY STATISTICS

industry, therefore, was characterized by: increased productivity, decreased employment, cost control, mechanization, and the will to survive.

During the second half of the decade a general recovery in metals prices (Fig. 1) enabled many, if not most, operators to recoup the operating losses which had been accrued earlier. Unfortunately, though, the euphoria was short-lived as a second recession - much more severe than the first - immersed the economy. Today, in the early 1990's, the language of the early 1980's is with us again. It is all too evident that cost-reductions and increasing productivity are not the effects of a momentary passing storm but, rather, the features of a permanent way of life. Society is undergoing a major readjustment of its values and expectations.

THE CANADIAN MINING INDUSTRY

Before proceeding, it is to the point that we should be reminded of the importance of our industry.

Canada has been, and continues to be, the world's leading producer of minerals on a per-capita basis (Fig. 3). Mining is the mainstay of no less than 115 communities. Our national minerals production accounts for: 50% of all rail freight; 40% of the tonnage shipped through the St. Lawrence Seaway; and over 50% of the tonnage loaded for international shipping. The resources industries account for almost 7% of our Gross Domestic product - and mining contributes over half of the total for all of these industries combined - more than from fisheries plus forestry plus agriculture (Fig. 4).

At present, it is estimated that there are about 350 underground and open pit mines in Canada - together with several hundred quarries and placers. In 1990, employment in mining was about 74,000 - representing about one-fifth of the total employment in mining, smelting, and refining, and mineral manufacturing (Fig. 5).

It is a backbone of the Canadian economy.

Some Key Results of the 1980's

As mentioned, during the 1980's our industry became much more efficient. This is reflected in both Figs. 2 and 6 which show that, concurrently with an increase in productivity (Fig. 2), there has been a decrease in overall employment (Fig. 6) - most particularly in the metal mining sector of the industry. Meanwhile, the total value of the Canadian production of metallic minerals (Fig. 7) has risen.

A statistic of some concern, however, is the severe decrease in the nation's ore resources which occurred during the decade. The causes include both decreased exploration expenditures (and resulting discoveries) and previously minable deposits being rendered uneconomic due to decreased minerals prices and/or increased costs of production.

During the 1980's, there was also an increased commitment to research and development. This was especially evident during the latter part of the decade (Fig. 8) and might have been driven by the decrease in the value of the production of metallic minerals which took place during the same period (Fig. 7). From a research perspective one would like to think that the decrease in production spurred a greater commitment to R and D.

During the 1980's, as the needs for mining research became more widely accepted, one of the most notable developments is that the industry began to talk amongst itself and to seek increasing collaboration and coordination of efforts. This had happened many years previously between mineral processors and metallurgists but was a very new phenomenon among mining engineers.

It can be said truthfully that while great improvement has been made, there is still a long way to go! We are still located on the early part of a "learning curve".

Some Key Developments in Mining Technology in the 1980's

Since the most effective means of improving economic viability has been through the control of expenditures, the industry has become very adept at the use of cost management. In the larger organizations there may be no aspect of expense that has not been scrutinized with a view towards reduction, replacement, substitution, or elimination. Indeed, "doing more with less" has been a common catch-phrase of the last decade.

To those familiar with the appearance of mining operations as recently as the 1970's, the changes have been revolutionary. The more evident include:

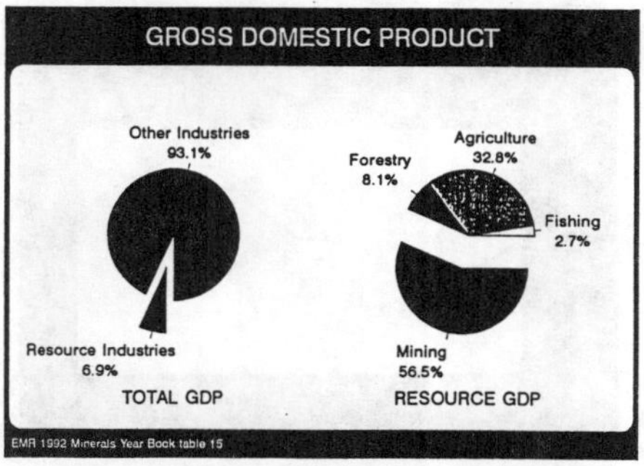

FIGURE 4: GROSS DOMESTIC PRODUCT

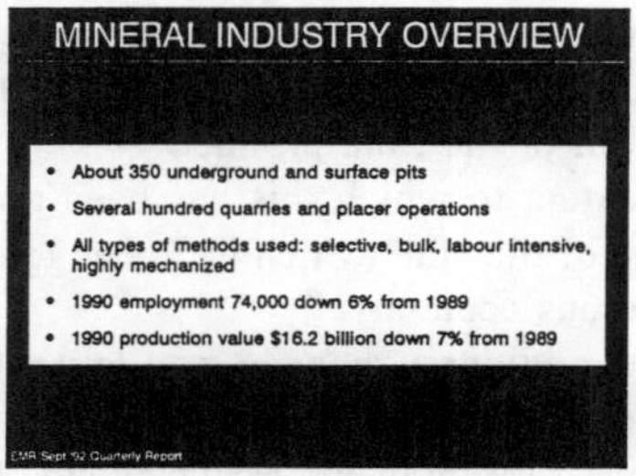

FIGURE 5: MINERAL INDUSTRY OVERVIEW

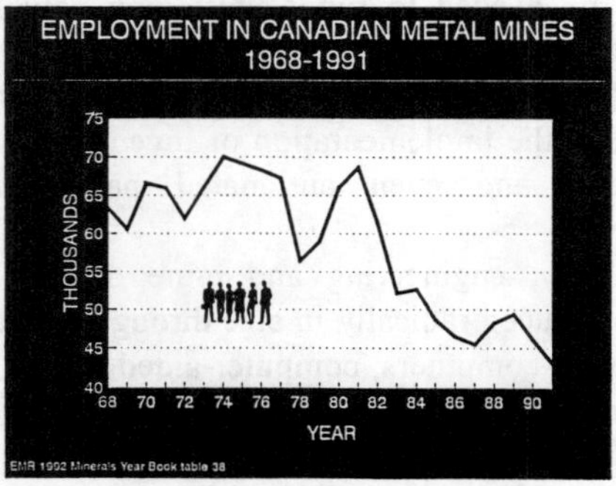

FIGURE 6: EMPLOYMENT IN CANADIAN METAL MINES

♦ organizational structures have been simplified through de-layering of the levels in the "chain of command". In today's flatter and broader organizational pyramid there are many fewer supervisors, with more people reporting to each.

There is no question but that people are, on the average, working far harder than they did a decade ago.

♦ organizational structures have also been thinned out as industry has sought to eliminate

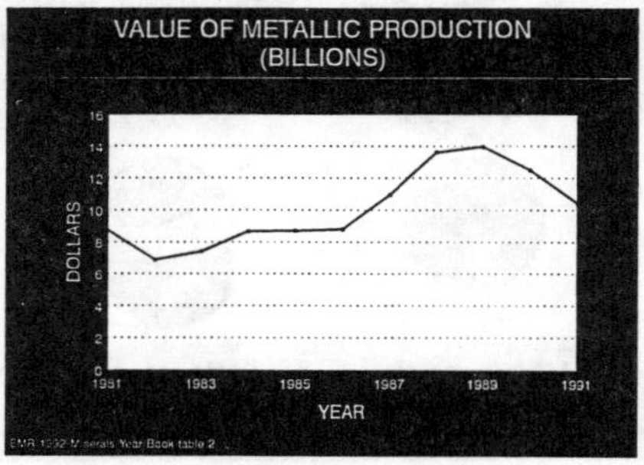

FIGURE 7: VALUE OF METALLIC PRODUCTION

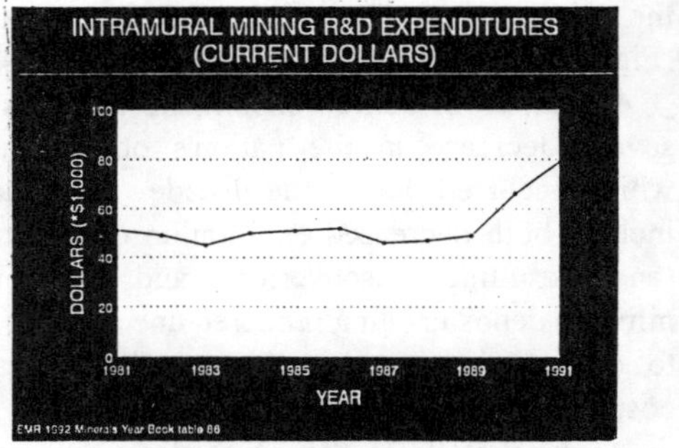

FIGURE 8: INTRAMURAL MINING R&D EXPENDITURES

positions which do not make a direct contribution, in a value-added sense, to either production or the final product.

The extent to which this has been possible is evidence of the "fat" which had been tolerated in a prosperous economy.

♦a revolutionary change has taken place in both the surface plants and offices at mining operations. Mineral processing plants, or mills, have been automated and equipped with computer-controlled process control systems to the point at which only a dozen or so persons may be needed to run a shift in a very large plant.

Warehouse-keeping has also been transformed through the implementation of inventory-control systems and even automated parts-retrieval systems!

Finally, engineering and mine offices have been reduced radically in size through the use of desk-top computers, computer-aided design, and automated data acquisition and logging systems. In the larger operations mine surveying and drafting offices have all but disappeared. With three-dimensional software packages now being readily available for micro-, mini- and even personal computers, the process of mine design has become much more challenging and sophisticated. With computer-interactive graphics, for example, one can now examine both the economic and engineering aspects of many design possibilities relatively quickly. There is much more "engineering" in the design of underground structures than was heretofore imaginable.

♦Likewise, in order to achieve better economics of scale and to permit increased productivity, there has been a trend towards larger underground production openings. Because of the increased dimensions it has become essential that the design of such openings should incorporate the geotechnical as well as the economic aspects. A major part of the "industrial revolution" to which I refer has been the extent to which the many facets of rock mechanics and ground control has entered into routine engineering practice. Neither computers nor rock mechanics remain any longer in the exclusive domains of specialists!

♦The trend towards larger stopes and pillars has also driven a need for greater accuracy in drilling; control of fragmentation and; ease of ore handling. The inevitable result has been the development of highly specialized mining equipment, which is increasingly automated. Mucking by remote control and automated haulage systems are becoming increasingly common. So too, are items of equipment with on-board sensors for performance monitoring.

♦These, in turn, have driven research and development into real-time mine-wide voice and data communications systems. Once a suitable mine-wide communications and control system has been produced there will likely be a quantum leap in the extent to which underground mining systems are automated.

All of the above advances have been accomplished in a climate of open dialogue and increasing collaboration. One of the most remarkable effects of the decade has been that, for the first time, the industry as a whole has begun to discuss and to priorize its mining research needs and to take concrete action to drive technological innovation. The result has been the formation of several research directorates and centres of expertise.

At the national level, for example, the Mining Technology Council of Canada (MITEC) was formed by the Mining Association of Canada (MAC). At the provincial level, the Ontario Mining Association (OMA) formed the Mining Research Directorate (MRD) while, very recently, the Quebec Mining Association (QMA) formed la Société de recherche et développement minier (SOREDEM).

The MRD, located in Sudbury, has concerned itself principally with a geomechanics research - which is the particular expertise being developed in Ontario through a tri-partite linkage between Laurentian and Queen's Universities and the University of Toronto. Likewise, SOREDEM, will develop a specialization in narrow vein mining, or "gisements filonien", which, at present, appears to be the particular expertise developing in Québec. Both Ecole Polytechnique and McGill University, in Montréal, are participating in the Canadian Centre for Automation and Robotization in Mining (CCARM), while, at Laval University, in Québec City, research is taking place in narrow-vein mining.

Industrially-sponsored and co-financed chairs (by either the National Science and Engineering Research Council (NSERC) or the Ontario Government) have been introduced in most of the universities mentioned.

Also at the provincial level, research committees have been formed by: the British Columbia and Yukon Chamber of Mines; the Québec Mining Association; and the Saskatchewan Potash Producers Association (SPPA).

Finally, to complete the picture, two other industrially-driven research organizations are also functioning: HDRK, which was originally co-owned by INCO, Falconbridge, Noranda, and Kidd Creek Mines; and MIROC (the Mining Research Organization of Canada), which is now located at Queen's University.

There has been significant growth not only in technological developments but also in the number of organizations that are involved. Naturally, and inevitably, some confusion in the market place has resulted. There is presently a need for increased collaboration and rationalization and, in the self-correcting way of the market place, it appears that that is now taking place.

TOWARDS THE THIRD MILLENNIUM

The great Canadian philosopher, Marshall McLuhan, once stated that the dangers in forecasting are akin to controlling a vehicle through observations in the rear-view mirror! One simply can not predict the abrupt turns ahead!

Notwithstanding the dangers, then, I believe that there are some generalizations that can be made that are very safe assumptions; namely that change directed towards the increasing automation of mines and the increased sophistication of engineering design will continue at an ever-increasing pace.

From the point of view of an underground mine operator some of the most important roadblocks are: the irregular and unpredictable shapes and sizes of hard-rock ore bodies; the dependency upon the traditional drill-blast-muck-support mining cycle; economic changes which force changes upon ore limits and reserves; the lack of communications and control systems; the limitations imposed by men and equipment; the dependency upon available energy; and the variability of conditions in the mine.

In some respects, the hard rock miner can only envy the degree to which automation has been achieved in stratiform deposits - notably in potash, salt, and coal mining.

Nonetheless, I believe that there are substantial technological advances which could be made very rapidly and that we will be driven, for economic reasons to achieve these. Controlled blasting to minimize overbreak, dilution, and needless damage to the rock mass and automated sensor-based ventilation circuits in order to optimize both mine air quality and energy consumption are but two examples. In both instances the technology is available. All that is lacking is the drive to implement it and to make it work.

Technologically, the breakthrough that will open many paths for innovation will be the development of a successful mine-wide voice and data communications systems that can be used for process control. The dream of underground miners in hard rock is to break free of the traditional mining cycle and, somehow, to be able to mine in an ever-changing physical environment from a remote location. It sounds far-fetched and impossible but, then, in hindsight, so did many of today's common conveniences only a generation ago!

A shopping list of some of the most desirable items would include:

◆sensors which could allow one to differentiate between rock and ore even under the most hostile conditions of dirt, water, vibration, and physical abuse.

◆communication systems which would enable the data to be received at locations remote from the sensors and permit remote-control of mining machinery.

◆control systems which would permit one to monitor, and control, every controllable aspect of the operation - which could include a long list of items, including as only examples: following ore limits; controlling energy consumption; monitoring the pollutants present; regulating ventilating air, etc.

◆machines, designed and constructed according to fatigue criteria. In theory, with continuous production, the entire output of a mine could be derived from very few continuously-operating pieces of equipment. The equipment, however, would have to be designed to withstand continuous, and hard, operation. It is doubtful if the equipment used at present could withstand the abuse.

CONCLUSION

In terms of the change which has taken place, driven by economic factors, the past decade has been dramatic for the Canadian mining industry. One can truly see that we are witnessing no less than an industrial revolution in the way in which we mine.

And yet, because the entire global economy is becoming much more competitive, it is probably safe to say that "the future will mean even more change from the past" and that the revolution has only just begun!

Thus far, because the industry has been very traditional and conservative in outlook, change has been by slow and progressive evolution.

Given the changes which have been forced upon it in only the past decade, future rapid change by a revolution in technology is likely.

Innovative Mine Design for the 21st Century, Bawden & Archibald (eds) © 1993 Balkema, Rotterdam, ISBN 90 5410 325 6

Mining research, development and application in the Saskatchewan potash industry

A. M. Coode
Central Canada Potash, Noranda Minerals Inc., Colonsay, Sask., Canada
G. Strathdee
Potash Corporation of Saskatchewan, Saskatoon, Sask., Canada

ABSTRACT: Events in the mid-1980s drove the need for co-operative research and development in the Saskatchewan potash industry. A body, the SPPA Steering Committee, was set up to govern the direction of the research and development. Three umbrella committees were established as a priority: mining, processing and environment. In mining, the needs were concentrated in a limited number of areas, all defined as non-competitive. Projects were part of either brine inflow prediction, protection or control, rock mechanics or new mining technology. Projects, which fall under more than one umbrella, such as backfilling, are reviewed and discussed by more than one technical committee. Mining methods have either incorporated the results of the present research or are being changed. Future directions for the Saskatchewan potash industry are discussed looking into the 21st Century.

1 BACKGROUND

1.1 Geology

Saskatchewan potash deposits exist in beds in the upper 60 metres of a thick sequence of halite and anhydrite known as the Prairie Evaporite Formation. The formation represents a time nearly 400 million years ago (in the middle Devonian age) when the sea stretched diagonally across what is now the Prairies from the Peace River area of Alberta, through Saskatchewan to the south west corner of Manitoba. The open sea lay to the northwest. The climate was hot and dry.

The Prairie Evaporite is underlain by the limestones of the Winnipegosis Formation and overlain by the limestones of the Dawson Bay Formation.

There are immense mounds or stalagmites resting on the Winnipegosis. These have been surrounded by gypsum, which later turned to anhydrite. These mounds rise to half the height of the existing Prairie Evaporite and are capped by a bed of anhydrite and dolomite, called the Shell Lake Member.

These mounds are important for two reasons. First, they are generally associated with poor ground conditions and, secondly, seismic surveys show the Shell Lake Member sags in a trough over their flanks, indicating that some of the salt below has been solutioned away and the ground above may have collapsed in the distant past.

The four groups of potash-bearing beds are, in ascending order, the Esterhazy, White Bear, Belle Plaine and Patience Lake Members. Each member consists of one or more beds interbedded with more or less barren halite.

The Prairie Evaporite in Saskatchewan varies in thickness between 100 and 200 metres. The strike of the Formation is east-west at the Alberta-Saskatchewan border to northwest - southeast in Manitoba. The shallow dip increases marginally from west to east and increases with depth. It is closest to the surface near the Alberta-Saskatchewan border and deepest at the Saskatchewan-North Dakota border. Over the area of a mine permit, the mining horizon varies in height because of solutioning but little of the regional dip can be seen.

1.2 Salt solutioning

Fresh water dissolves halite to form sodium rich brines. Sodium rich brines dissolve sylvite to form potassium rich brines. Consequently, nearly all the material of the Prairie Evaporite Formation is readily dissolved by waters or brines. The effects of solutioning can be divided into those occurring at the time of creation and those occurring after the Dawson Bay limestone cap had been formed. The size of the anomalies vary from microscopic to massive where miles of the complete formation have been dissolved. When solutioning is large enough to collapse the capping limestone beds, a chimney forms to the surface. When the cavity is smaller, the beds flex or fracture. Solutioning has taken place throughout geologic time to the present.

1.3 Brine inflows

Originally there were nine potash shaft mines and one potash solution mine operating in Saskatchewan. Of the nine shaft mines, all but Potash Corporation of Saskatchewan (Allan Division) have experienced water inflows.

Both Cominco and Central Canada Potash have experienced inflows of less than 100 gallons per minute. Central Canada Potash had two inflows from the same source of brine. The first occurred in 1974 in Panel 109, the second in 1978 in Panel 209. Grouting of the Second Red Beds in Panel 109 proved unsuccessful. Subsequent "blow back protection" drilling and grouting into the Dawson Bay was successful. An extensive rehabilitation, drilling and grouting program was completed in 1986. Cominco experienced a similar sized brine inflow in the late eighties. They are still drilling and grouting the site.

Potash Corporation of Saskatchewan (Rocanville Division) inadvertently cut into a collapse anomaly in November, 1984. They ceased to cut when they realized that they had intercepted blocks of Dawson Bay limestone. Brine started to leak into the mine. It increased and although work was pressed as fast as possible, it proved to be a substantial inflow measuring thousands rather than hundreds of gallons per minute. A bulkhead was installed in collapse anomaly salt and, as the brine was saturated in sodium, it held - proving that the technology exists to install bulkheads in soft rock environments.

In late 1985, International Minerals & Chemical Corporation experienced an inflow of Dawson Bay brine into the mined out workings of its K-2 mine. Although the inflow point was not observed until approximately ten months later, estimates indicate that the inflow rate peaked at approximately 9,000 gallons per minute. Although the inflow rate has been reduced through injection of a chemical grout into the Dawson Bay formation, the inflow continues to this day. The use of isolation bulkheads was tested but because of concerns about the effects of dissolution they were not widely installed.

In 1976, mining at Potash Company of America was following the potash beds when it was realized that the miner had cut into the Second Red Beds. Apparently the Prairie Evaporite Formation had been dissolved down to the top of the potash zone. Containment activity resulted in the minor inflow being controlled. In late December, 1985 the inflow began to increase at the old inflow site. Seismic surveys showed that a fault, passing through the Blairmore Formation aquifer, passed near the site. They rehabilitated the area, drilled the formations above the Prairie Evaporite and tried to grout, without success. They tried to install a bulkhead without success. The size of the inflow increased with time and resulted in flooding of the mine. Potash Company of America decided to convert their dry mine into a solution mine.

1.4 Rock mechanics

Ground conditions at the Saskatchewan potash mines differ considerably from location to location. In general the potash mines in the Saskatoon area have similar conditions and the two in the east of Saskatchewan have comparable conditions.

Whether in the east or around Saskatoon, the evaporites behave in a similar fashion. The rock immediately adjacent to the opening fractures and cracks. The profile around the opening of these fractures is that of a diamond surrounding a rectangle. Surrounding the fracture zone, there is a microfracture zone, where the salt and potash crystals exhibit hair line cracks. Surrounding the microfracture zone the evaporites creep. Typically the carnallite creeps an order of magnitude more than the potash and the potash creeps an order of magnitude more than the salt. The clay seams, the Second Red Beds and the Dawson Bay do not creep. The effect of the clay seams is to allow the salt and potash beams to separate from each other (i.e. the effect of pressure on a stack of salt beams separated by clay seams is the same as a stack of paper sheets with air in between when pushed from the sides). Clay seams introduce a level of instability that would otherwise not be present.

The mines in the east have few clay seams and the salt above the mining horizon is approximately 27 metres (90 ft.). The absence of clay seams and a substantial salt back result in stable ground. Typically the mines can cut rooms up to 18 metres (60 ft.) wide which, although creep slowly closes, stay open for extended periods of time. Rehabilitation with a boom miner and rock bolting is the most frequent ground control technique used.

The mines in the Saskatoon area, with extensive clay seams, all cut to a salt beam two and a half to three feet thick immediately above three clay seams. When this salt beam starts to show signs of deterioration, the mines initially "slot" the back with an undercutter and rock bolt or rebar (with epoxy) the back. If, or when, this further deteriorates the mines are left with little option but to cut a minimum of eight feet from the back of the entry and sometimes they are forced to go as high as thirteen feet. At eight feet, there is a three foot potash bed and, at thirteen feet, there is a ten foot clean salt bed. This form of rehabilitation leads to large quantities of waste material which, if rooms have closed nearby, force the operator to place it on the conveyor which loads more waste salt and clays into the mill circuits than would otherwise be the case.

1.5 Mining methods

The mining methods utilized are constrained by two factors: first, the thickness of the bed and second, the ground conditions. The ground conditions lead to a low overall extraction ratio of between 35% to 45%. Attempts to increase the ratio have been buffered by the uncertainty of ensuring the integrity of the Dawson Bay Formation and have not increased substantially.

The eastern mines have approximately eight feet of ore. Choice of mining equipment has been four rotor eight feet high miners. By choice, these mines have both chosen to cut with a straight ahead mining method. This means that the mining block or panel is first outlined by four or five entry development headings from which it is possible to use the latest geophysical techniques to scan the interior potash for anomalies. Should the area be considered clear, they start cutting entries from one side to the other using the same techniques and equipment used in the development. The entry width in the production rooms differs from that in the development headings but the methodology remains the same.

The Saskatoon area mines cut potash lain down in the centre of the Elk Point Basin as opposed to the margins. Five times as much insoluble materials is present in the form of disseminated clays and clay seams. These make a substantial and significant difference in ground control.

The mines have typically three zones. The upper is the richest and a minimum of eleven feet thick. Approximately twenty feet below that is another eleven foot thick ore zone. Both are part of the Patience Lake Member and are capped by a two to three feet thick salt beam. At Potash Corporation of Saskatchewan (Lanigan Division) the two seams are very close to each other so that the intervening salt is almost non-existent. Twenty to thirty feet below the middle zone, there is a lower zone, which is part of the Belle Plaine Member. It is eight feet thick and contains some carnallite which is rarely, if ever, seen in the upper two zones.

Because of the different clay content, the Saskatoon mines cannot cut as wide as the eastern mines. Typically, entries are cut no wider than thirty to thirty three feet wide without the threat of back collapse. Even in the Saskatoon area thickness of salt beams and locations of clay seams, although generally similar, differ from mine to mine.

It is believed that two conditions give rise to these differences: first, there are small but significant differences in the deposition of the Patience Lake Member from one mine site to another; second, the subsequent leaching and solutioning of the original beds from secondary replacement to massive collapse structures vary from mine to mine and these have an effect on the sequence. It is possible to recognize differences within a single mine.

Initially the mines started to use the same mining methods as coal mines. It was quickly found that these methods were not transferable to the potash mines. Evolution took place and all the mines centred on one of two mining methods: first, straight ahead rooms or, second, variations on chevron rooms. Three mines chose chevron rooms: Central Canada Potash, Potash Corporation of Saskatchewan (Allan Division) and Potash Corporation of Saskatchewan (Cory Division). The methods were often constrained by capital equipment purchased in the first years of development with a reluctance to change the method being governed by financial considerations. Other mines made minimum capital expenditure and mined straight ahead (i.e. Cominco and Potash Corporation of Saskatchewan (Lanigan Division)).

2 POTASH RESEARCH TO THE PRESENT

2.1 Research activity overview

Challenging and expensive brine inflows in the middle eighties forced consideration of brine inflow prediction, protection, containment and control. Activities of potash companies experiencing inflows offered insight and information to all operators on what needed to be done to avoid similar occurrences. These were quickly and efficiently implemented. There remained outstanding questions, however, which had not been answered in the hectic and energy-filled months following the inflows.

The potash operators established a lobby group, the Saskatchewan Potash Producers Association (SPPA). In the early eighties, a steering committee was established to guide research and development needs within the potash industry concentrating on research of a non-competitive nature. The question of security of potash mines was ideally suited for this research program.

A series of sub-committees were established, which eventually evolved to cover three areas of expertise. These committees were responsible for the technical review of projects in their areas:

- Brine prediction, containment & control and new mining technology,
- Environment, and
- Processing and product quality.

Emphasis, in this paper, is placed on the history and development of the Technical Review Committee (TRC) for brine prediction, containment & control and new mining technology, referred to as the Mining TRC. Committee roles were:

- TRCs make recommendations for projects to the Steering Committee and direct research,
- Steering Committee approves projects and establishes the budget. Three companies must participate before project approval is given.

2.2 Financing

One of the Steering Committee original goals was to invite both the federal and provincial government to participate with the potash mines on an equal footing. Each would assume one third of the cost. Although the Saskatchewan government quickly agreed to deduct the cost of research and development from a form of royalty/taxation they impose, called the PRPA, alleviating the financial burden on projects by 40%, it proved difficult to establish a satisfactory working relationship with the federal government. In the early eighties, financing with the federal government was established under the Mineral Development Agreements.

An attempt was made to utilize CANMET facilities and personnel but this was not successful. Finally the federal government, through the offices of the Western Economic Development Canada (WEDC), approved an arrangement with the potash mines and the Province of Saskatchewan which, although complicated and emphasizing the importance of obtaining a target level of expenditures for research, came close to the original idea of "equal-share-of-cost".

The agreement with the federal government started in 1989 and was scheduled to finish five years later with an expenditure of $8,000,000. The demands of research in which the participating mines were willing to contribute were such that annual expenditures were not as high as initially projected. The agreement has, therefore, been renegotiated and extended for an additional two years ending in 1995.

As may be expected, budgeted amounts are never exactly met. In some instances the budgeted amount is exceeded and in some it is below plan. These figures reflect the constraints exercised by the Steering Committee. As time passed, mine operators became more confident and satisfied that mine inflows could be avoided or predicted.

At this time Saskatchewan Environment and Public Safety was requesting that as much emphasis be placed on environmental issues as possible. In 1991 environmental projects absorbed over one million dollars of the budget. Projects matured from the laboratory stage. Several reached pilot plant and field trial stage requiring a large influx of funds.

2.3 Mining technical review committee activity

The goals of research and development in Earth Sciences and Mining were formed as a direct result of the brine inflows occurring in the middle eighties. Subject matter for projects were divided approximately as follows:

- Brine prediction: geology, geochemistry geophysics;
- Brine control: rock mechanics, grout studies, bulkhead materials studies, permeability studies, chemical dissolutioning control, and
- New mining technology: projects of a non-competitive nature.

Table 1. SPPA R & D Program
Summary of Budgeted/Actual Expenditures for the period 1989 to 1993
(Thousands of dollars)

Technical Review Committee		1989	1990	1991	1992	1993	Total
Earth sciences & mining	Budget	393	395	537	584	639	2,548
	Actual	236	266	463	539	N/A	1,504
Environmental research	Budget	311	715	1,096	541	686	3,349
	Actual	253	795	834	545	N/A	2,427
Process, material & offshore product quality	Budget	50	375	105	335	308	1,173
	Actual	38	273	58	245	N/A	614
Project management	Budget	25	75	58	27	27	212
	Actual	15	20	34	25	N/A	94
Total	Budget	779	1,560	1,796	1,487	1,660	7,282
Total	Actual	542	1,354	1,389	1,354	N/A	4,639

Each project is led by a representative of the lead company and has representatives from the participating companies, who serve as a Project Advisory Group and ensure that the original project proposal and objectives are met and original budget is adhered to +/- 10%.

2.3.1 Geology

The simplest question, "Can an operator recognize a potential brine inflow from looking at the rock geology?" is often the most difficult to answer.

After laboratory work, it is possible to relate the presence of certain minerals to the increased risk of a sodium rich brine inflow but it has not been possible to suggest that the operator of a machine can use any visual clue as the warning of a brine inflow.

Work has and is going on in the geology of the Prairie Evaporite Formation, the underlying Winnipegosis Formation and the overlying Dawson Bay Formation. Research has looked at the genesis of the Dawson Bay Formation in both the Saskatoon and Esterhazy areas.

One project integrated the methodologies of geology, seismic surveys, palynological studies, geochemistry of isotopes in hydrogen and oxygen, radar and EM studies and microseismic monitoring work.

Another project reviewed the presence of palynomorphs within the clays seams of the Saskatoon areas mines. The work produced some interesting concepts and is continuing.

2.3.2 Geochemistry

There are two ideas which have dominated geochemistry over the years. First, measurement of hydrogen and oxygen isotopes give an indication of how long water has been in vapour form, how fresh it is and how long it has been in the ground, in other words the water's or brine's age. Secondly, it has been necessary to work on improving the quality of chemical analysis in saturated brines, especially for trace amounts and low concentrations of certain elements.

2.3.3 Geophysics

Geophysics covers a very large range of projects and subjects from seismic work to EM to acoustic emissions.

In seismic work, participating companies have looked at seismic tomography over an area of 1.5 km. by 1.5 km. and concluded that the signal to noise ratio was too small for the technique to be used as an effective tool in anomaly diagnosis.

Participating companies are now investigating the possibility of improving the results of standard seismic work either through use of superior analytical and interpretive methods or three component seismic surveying on either 2D or 3D seismic surveying. One of the more challenging problems is defining the Dawson Bay status from surface. Work is progressing.

EM34 equipment is used throughout the mining industry to establish the presence of sodium rich brine in the Dawson Bay and underground developments. We have revised graphs used to interpret measurements made using this instrumentation.

A successful and useful project has been the application of tomographic techniques to the data from Radio Imaging Methods, or EM Tomography. Frequencies in the order of 100 and 300 kilohertz have allowed operators to penetrate distances up to 2 km through the potash horizon. The presence of massive or disseminated brine within a block is clearly seen. Interpretive work is being done to clarify the analyses.

Other projects cover studies of resistivity, acoustic emissions, pulsed EM and high resolution seismic, which have all contributed to better understanding of the symptomatic information generated in geophysics.

2.3.4 Rock mechanics

The most important project has been to conduct tri-axial creep tests on cores from both potash and salt horizons. These have taken time and been spread out over a number of years. The results have been used to generate possible constitutive power laws for salt and potash from different mines.

Comparisons have been made between different Finite Element Modelling packages and results circulated to participating mines.

A search is underway to find the optimum non-ferrous composite rock bolt (or "rebar"), which can be used in entry backs and will not impede back cutting with rehabilitation machines.

2.3.5 Brine control

Projects under this heading fall into one of three groups: grouting from underground, grouting from surface and construction of a bulkhead.

In 1987, equipment was prepared to test long term behaviour of grouts adjacent to saturated brine under 0 psi, 500 psi and 1,000 psi. This work continues and has given some interesting results. Participating companies have tested superfine grouts with mixed reviews.

Work has developed on producing a solution which when injected from surface into brines in the Dawson Bay results in crystal formation which catches and clogs in cracks. The work looks very promising.

As stated earlier, construction of the bulkhead at Potash Corporation of Saskatchewan (Rocanville Division) showed that bulkhead construction in soft rock does not present any serious physical challenges.

Bulkheads placed in a potash environment can be threatened by sodium rich brines dissolving the adjacent potash. The control of potash leaching around a bulkhead is still being addressed.

2.3.6 New mine technology

Two projects have been approved under the heading of new mine technology. First, non-destructive testing of shaft linings (an essential project to define the health of shaft concrete liners ultra-sonically) and, second, the non-destructive testing of hoist ropes - a MITEC project, for which Potash Corporation of Saskatchewan is lead company.

3. POTASH IN THE FUTURE

Direction of mining industry in the future will be primarily governed by three important issues: product demand, availability of supply, and the constraints of society.

Much has been said, and I am sure that much will continue to be said, about supply and demand of products, especially as it pertains to the mining industry.

The Saskatchewan potash supply and demand picture primarily centres on the United States market and the ability of the American farmer to make a profit. The financial health of the United States farmer is critical to the health of the potash industry. Beyond the United States, the industry sells internationally to the Far East, Australia, South America and beyond where competitors (especially from the C.I.S. countries) abound. At the moment the marketplace can be easily over supplied, the demand is not sufficient to drive an expanding potash industry. Saskatchewan potash producers have voluntarily and unilaterally reduced annual production, however, supply potential continues to outweigh demand.

The United Nations predictions of world population increase over the next few decades, especially in Africa, are both frightening and stunning. If the increase in population is not slowed by either AIDS or some other new world epidemic (like drug resistant TB, for instance), world farming communities are going to be expected to provide these people with food (or the catastrophes we have seen on the television showing peoples starving to death by the thousands in areas of Ethiopia, Somalia, Cambodia and Mozambique will pale by comparison). It is difficult to see how agricultural research can provide increases in food production without full use of fertilizers over larger and larger areas to provide better food production per acre.

The Saskatchewan potash mines currently operate at approximately fifty percent of capacity. With continually increasing demand, provided that vendor financing can be maintained, the future of the local industry looks cautiously optimistic.

With these constraints, it would appear that there might be little pressure to adopt innovative technologies and improve mining methods, however, there is the third constraint, mentioned above: the constraint of society which takes the form of pressure from a number of independent sources. Some tend to increasing with time, others increase cyclically and yet others tend to increase with randomly occurring peaks. These are:

The highest hourly pay rate in the world's mining industry,
Public demand for improved environmental responsiveness from all aspects of industry, including the mining industry.
Steady integration of the Canadian economy into the North American Free Trade Agreement (NAFTA),
Determination of provincial/federal governments to control and reduce the deficit with its related higher taxes,
Determined and ruthless competition from countries hungry for international currencies,
Replacement commodities,
Media programmes and the perception of the industry, which distort reality and fertilize the ground requiring guidelines and regulations can be introduced.

For the Saskatchewan potash industry, these will be the pressures for technological innovation and development into the 21st Century.

We see that pressures exerted on mines by the high cost of Canadian labour can only result in an increasing level of automation, improved communication and enhanced monitoring and control systems. Mines will invest capital into these areas to maximize profitability and minimize operating costs. Manpower levels will be subject to further and further cuts until the demand for manpower from maintenance crews, addressing the needs of electrical and mechanical preventive and emergency repair, ventilation and ground control will match the loss in operating crews and a new equilibrium in manpower levels is attained. Without doubt, integrated automation of the underground environment will be the most important strategic thrust over the next ten to twenty years (in the same way that mine safety and rock mechanics has been for the last twenty years).

NAFTA, which is designed to encourage a tariff free border, also encourages capital movement due to inequities existing in government regulations.

These issues together with higher taxes and increasing environmental concern by government has resulted in increasing pressure on all mines. Neither federal or provincial governments want to close mines down using environmental regulations, however, in spite of this position, all governments continue to introduce more and more restrictive regulations and continue to add financial burdens to Canadian mines, discouraging exploration in Canada and encouraging it outside North America. There is little doubt that provincial governments desperately want new mines to open but environmental regulations and higher taxes place a burden on Canadian mining. To compete with his Mexican or Chilean or Brazilian cousins, the Canadian miner must consider:

Increasing his investment in mines outside Canada,
Finding and bringing into production properties of sufficiently high grade to provide a satisfactory rate of return after absorbing the cost of regulations,
Accepting a lower level of profits,
Using research and development technology to reduce mine operating costs and capital requirements for new mines to maintain an acceptable rate of return for stockholders.

It is in this last area that research groups should concentrate. New project proposals can only be justified by improving the rate of return. The proposed capital and operating costs have to be shown to safely improve productivity by either increasing revenue or reducing operating costs.

Only in high grade mines, where recovered mineral content is in excess of 33%, is it possible to dispose the concentrator waste underground as backfill. Lower grades result in a need for storage elsewhere. In the potash industry, even with grades in excess of 33% sylvite, creep and closure of the back, floor and walls together with poor ground conditions lead to problems in returning all material underground.

Other venues may exist and are presently being investigated through the co-operative research of the Saskatchewan potash industry.

Very often use and disposal of backfill is dependent on the mining method employed. An operator will not get a license to develop a new mine in most jurisdictions unless an approved decommissioning plan has been developed. MAC's MEND programme and some integrated mining companies' backfill and tailings research and development efforts are aimed at addressing these questions but it will be at the individual mines that "the rubber meets the road". Future shock may arise over the use of financial penalties for the use of surface ground to store tailings. Alternate mining methods may have to be adopted to most efficiently dispose of the steady waste stream from the concentrator.

Although two solution mines exist in Saskatchewan, one west of Regina and the other east of Saskatoon, the latest proposal for a new mine does not utilize this technology but continues to use conventional dry mining practice. It is interesting to speculate whether this will continue over the next twenty years or whether more mines will migrate from dry to solution mining methods.

In the potash industry, where the extraction ratio is between 35% and 45%, there has and continues to be great pressure to overcome some of the problems allowing the use of short wall or longwall mining. These involve research on: rock mechanical behaviour of the capping limestone bed (the Dawson Bay Formation) to ensure that it does not fracture but that its flexing is limited by backfill; the best means of introducing backfill (paste, heavy density or hydraulic); and, the strength of backfill *in situ*. The advantages of successfully introducing backfill and cutting either short wall or long wall lead to: disposal of a significant part of the concentrator's waste stream; increased extraction ratio - to a figure of 70%; and, the minimization of development.

4. CONCLUSIONS

The Saskatchewan potash industry has had a research and development programme in place since the early eighties. It has undergone two major changes in direction. The first, resulted from brine inflows in the middle eighties and, the second, from increasing environmental awareness at the end of the eighties.

The research and development of the Mining TRC has concentrated on: geology; geochemistry; geophysics of brine prediction and prevention; brine control; rock mechanics and new mining technology. None of these impinge upon a mine's competitive nature.

Future direction of potash mining and mine design will be governed by high pay rates, higher taxes and government regulations (primarily related to environmental concerns). Ancillary constraints will be concerns arising from NAFTA, international competition for foreign currency by hungry nations and public perception of the industry as a whole. These pressures are likely to result in:

Increased automation, improved mine wide communications, both through the use of paging systems as well as voice, data and video transmissions, and monitoring and control systems.

Resolution of the challenges facing the industry regarding tailings disposal, and

Investigation of methods to increase extraction ratios.

Innovative Mine Design for the 21st Century, Bawden & Archibald (eds) © 1993 Balkema, Rotterdam, ISBN 90 5410 325 6

On the fracture mechanism and creep behaviour of rock under compressive load

S.Yamashita & F.Sugimoto
Mining College, Akita University, Japan

K.Otsuka
Institute for Mining Research and Studies, Akita Prefecture, Japan

ABSTRACT: A reliable method is required for the estimation of mechanical properties of rock and for observation of the stability of rock-structures. In this study, creep tests under uniaxial compressive load are carried out with the aim of estimating the long term stability of 'Ohya-stone' (porous tuff, in Japan). In the experiment, axial and lateral strains are measured and AE activities are monitored, at the same time. From these results, the creep behavior and AE activity behavior are discussed using the failure mechanism hypothesis proposed by Bieniawski and by others. Furthermore, predictions for the critical time to failure are attempted for stress fields of 'Ohya-stone'.

1 INTRODUCTION

Recently, underground rock mass construction has been attempted to construct rock structures such as large scale petroleum reservoirs in order to make use of excellent insulation, wall stability against earthquake, long-term stability of the rock structure, constant temperature or humidity conditions and the like.

On the other hand, unexpected degradation of rock-structures (such as 'Ohya-stone' pit fields ; porous tuff, located in Tochigi prefecture , north Japan), have often been occurred. Therefore, it has been necessary to establish methods for estimation of suitable mechanical properties of rock and rock-mass characteristics, and to devise methods to permit observation of stability conditions of rock-structures, with the goal being to optimize design, construction, maintenance and control operations in rock excavation.

One of the problems for estimating the long-term stability character of rock, concerns the acquisition of data, not only of the stress-strain relations and strength properties, but also of the time-dependent deformation properties.

In this study, drained creep tests were carried out on Ohya-stone to estimate the critical failure intervals under prescribed stresses and to develop methods for predicting such intervals.

In these experiments, relationship between elapsed time and strain behavior, under sustained creep stress, and relationship between creep time and acoustic emission activity (AE) were observed.

From the results of these tests, the creep behavior of the rock was assessed as was a hypothesis for rock fracture. On this basis, critical times to failure were predicted by extrapolating relationship shown to exist between the logarithm of time and the stress ratio of the sustained axial creep load.

The loading times maintained in these creep tests varied between several minuets to one month.

2 FRACTURE MECHANISM AND AE ACTIVITIES OF ROCK

It has been proposed by Bieniawski (1962) that failure of rock does not occur at the critical rock stress of the rock, but that failure takes place over a four step process occurring between beginning of loading and to prior to critical stress loading of the rock.

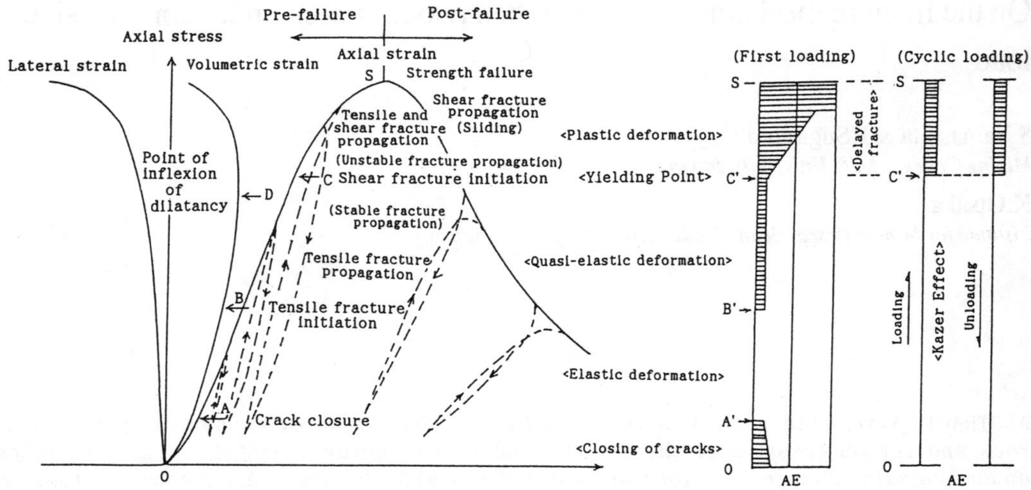

Fig.1 Fracture process and mechanism of brittle rock in compression.

In previous research studies, uniaxial and triaxial compression tests had been carried out on cylindrical specimens of various rock types (Yamashita 1984, 1985, 1986). In these tests, both axial and lateral strain versus stress, as well as AE activity versus stress measurements were simultaneously observed.

The main results obtained are as follows:

a) AE activity counts rates were shown to initiate when the stress level reached to approximately 40 % to 45 % of the rock failure strength, and a level corresponding to the point where tensile micro-fracture growth initiates in the rock.

b) The micro-fracture propagates only if the stress is increased, and AE activity is slightly detected in the duration to next step c).

c) The AE counts rate increase rapidly the stress level reaches about 70 % of the failure strength, and where distinct plastic strain behavior appears in the stress-axial strain curve.

In the same manner, uniaxial cyclic loading tests were carried out on the same rocks. In these tests, we have newly pointed out that the Kaiser Effect could be detected on the AE activity behavior at points below the stress level where the plastic strain is observed on the axial stress-strain curve.

Where the stress condition can be sustained constant below these levels, no AE activity is detected and no plastic strain develops. If the sustained stress exceeds this level, AE activity and plastic strains are observed to occur simultaneously.

Accordingly, we have defined this stress level as the yielding stress of a rock.

And, at the yielding point there exists a possibility of predicting the rock failure by observing the AE activities of the rock-structure.

Besides these experiments, simulations for the stress-strain behaviors associated with the fracture initiation and propagation from an initial crack in the specimen model with the aid of the Finite Element Method were carried out (Yamashita 1985).

Finaly, the fracture process and mechanism of rock failure under compressive load has been proposed by our studies as the hypothesis shown in Fig.1.

In this study, one of the primary project goals was to assess AE activity behavior in creep tests as a means to predict rock failure conditions.

3 ROCK SAMPLES AND TESTING PROCEDURE

3.1 Rock samples and fundamental properties of rocks

The rock material which was used in this study was 'Ohya-stone', which is a porous tuff deposited during the Miocene Epoch of the Tertiary Period. Four samples were collected from two different sites in the Ohya-stone fields.

Samples 0 and 0-1 were collected in the same working pit which was situated at a depth of about 50 meters from ground surface. T series samples were collected by vertical boring from the ground surface of the deposit. Samples T-11 and T-12 were collected at depth of at about 50 meters and 70 meters, respectively, from the ground surface.

The physical properties of samples are given in Table 1. The averaged specific gravity, γa, value for all samples was 1.77 in natural condition,and the average sample porosity, n, was determined as 31.4 %.

Cylindrical test specimens were prepared for uniaxial compression and creep tests by using a boring machine and diamond rock cutter to be 50 mm in diameter and 100 mm in height.

Mechanical properties of the samples are shown in Table 2. The averaged uniaxial compressive strength, Sc, value for all samples was 110 kg/cm^2; for tensile strength, St, by the Brazilian Test, was 13.6 kg/cm^2; for Young's Modulus, Es, was 26,700 kg/cm^2; and for Poisson's Ratio, ν, was 0.29.

These values were found to exist at levels approximately two times higher than the minimum measured values. It can therefore be said that 'Ohya-stone' has essentially a hetero-geneous nature. Also remarkable was the fact that the Poisson's Ratio values of these samples were very large compared with usual sedimentary rock values.

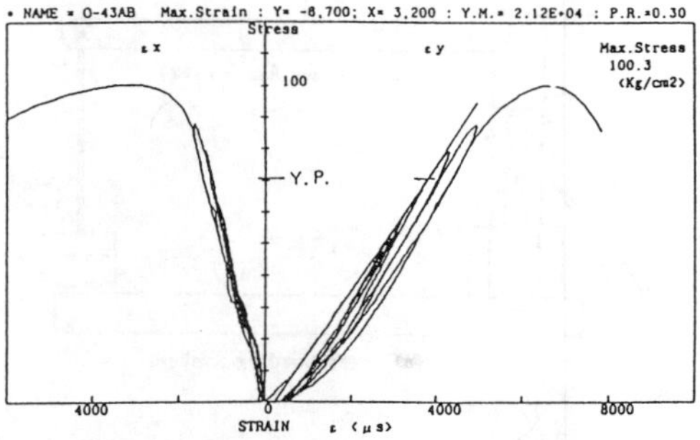

Fig.2 Stress–axial and lateral strains curves in cyclic loading test.

Fig.2 shows typical relations existing between stress and strains (εy ; axial, εx ;lateral) obtained during uniaxial compression tests with cyclic loading. By the figure, the stress at the yield point is estimated to be about 70 % of the uniaxial compressive strength of the sample.

3.2 Testing procedure

Creep tests were carried out under natural water content and drained conditions and at a sustained creep stress, σcr, which was set at a prescribed stress level set as a fraction of uniaxial compressive strength, Sc, of the sample.

The normalized stress ratio, $\sigma cr/Sc$, is defined by the term of 'creep stress ratio' in this study.

An illustration of the testing apparatus is shown in Fig.3 (a) and (b). The creep loading system consists of an axial loading piston and a cantilever beam which provides a 20 times creep load multiplied by applying dead weights at the end of the beam. The maximum load capability of this system is 3000 kgf.

Table 1 Physical properties of tested rock (Ohya-stone)

Sample	Spacific Gravity	Water Content	Water Absorption	Porosity	Wave Velocity	
					P wave	S wave
Name	γa	ab (%)	W (%)	n (%)	VP (m/sec)	VS (m/sec)
0	1.71	22.6	20.8	30.4	1,886	846
0-1	1.77	14.2	9.1	23.6	1,903	792
T-11	1.77	27.2	36.9	38.2	2,550	1,085
T-12	1.83	21.9	30.0	33.5	2,250	1,143
Av.	1.77	21.5	24.2	31.4	2,150	967

Table 2 Mechanical properties of tested rock (Ohya-stone)

Sample Name	Uniaxial Compressive Strength	Brazilian Tensile Strength	Brittleness	Young's Modulus (Statical)	Poisson's Ratio (Statical)
	Sc (kg/cm^2)	St (kg/cm^2)	Br	Es (kg/cm^2)	νs
0	95.8	11.4	8.4	19,200	0.25
0-1	138.7	11.8	11.8	44,300	0.39
T-11	119.7	17.4	6.9	26,500	0.26
T-12	85.1	13.6	6.3	16,800	0.24
Av.	109.8	13.6	8.1	26,700	0.29

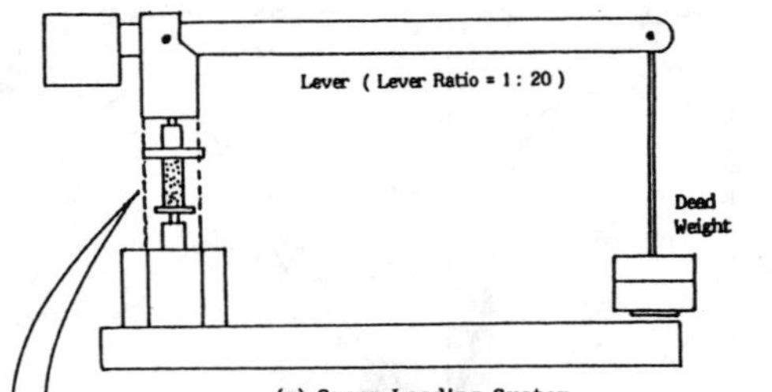

(a) Creep Loading System.

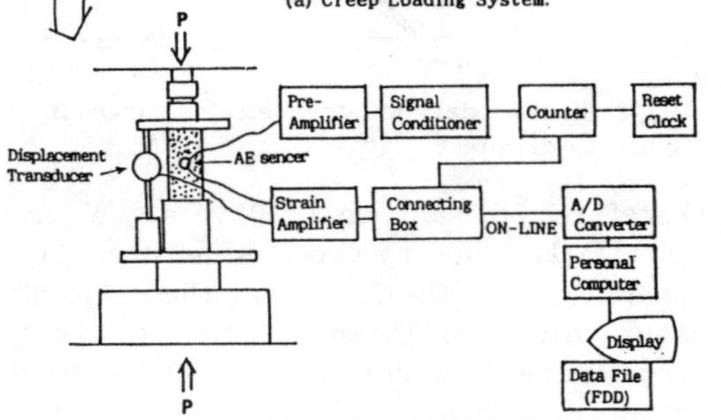

(b) Outline of Measuring System.

Fig.3 Schematic diagram of creep testing apparatus.

This system was set up in a chamber to maintain a constant temperature of 24 ℃ and a constant humidity of about 60 % by using an air conditioner.

Each test specimen was coated with paraffin wax and wrapped in thin plastic film to maintain natural specimen conditions during the period of creep loading.

Test measurements were made on the changes of axial strain, εy, lateral strain, εx, and the rate of AE activity during the entire creep interval time. A micro-computer system was used to acquire and store data into a FDD disk.

4 RESULTS OF CREEP TESTS

In this section, the results of 0 and 0-1 samples will be reported and discussed, since both axial and lateral strains were measured only on these samples.

From the results, next three kinds of figures are drawn.

(a) elapsed time-strain curve (creep curve) (axis of abscissa ; logarithm of elapsed time, axis of ordinate ; strains)

(b) elapsed time-strain rate curve (creep rate curve) (axis of abscissa ; logarithm of elapsed time, axis of ordinate ; logarithm of strain rate)

(c) elapsed time-strain and strain rate curves (normal time creep curve) (axis abscissa; elapsed time, axis of ordinate ; strains, axial strain rate, AE counts rate)

4.1 Typical experimental result

Fig.4 (a), (b), (c) show a set of typical experimental results obtained on sample 0 a creep stress ratio of 79.5 %.

Fig.4 (a) shows the change of strains, εy, $-\varepsilon x$ and volumetric strain, $-\varepsilon v$ ($= \varepsilon y - 2 * \varepsilon x$) with logarithm of the elapsed time, t. In the figure, the change AE count rate activity is shown by the height of the histogram. On these creep curves, several

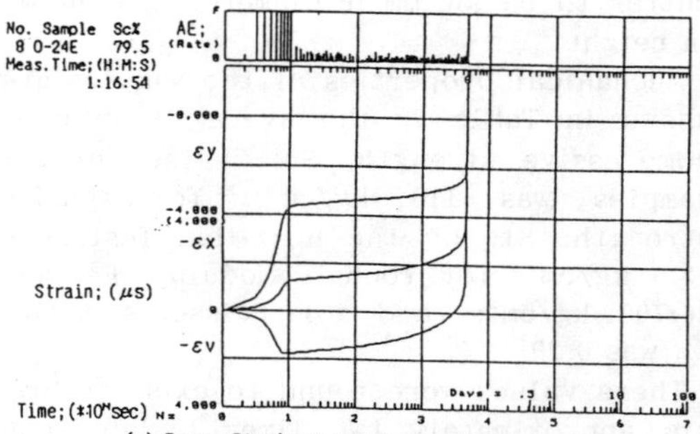

(a) Creep Strain versus Logarithm-Time Curve.

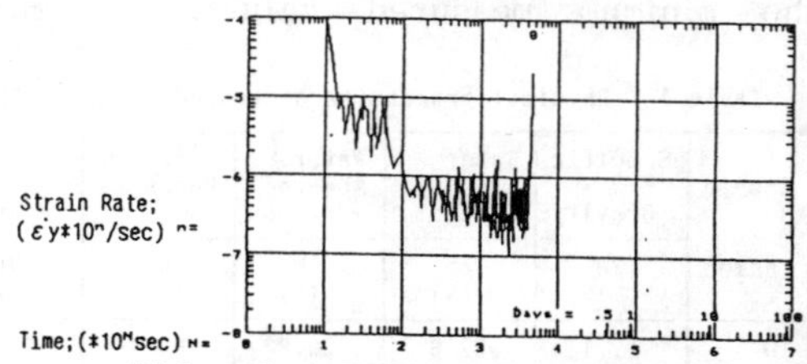

(b) Creep Strain Rate versus Logarithm-Time Curve.

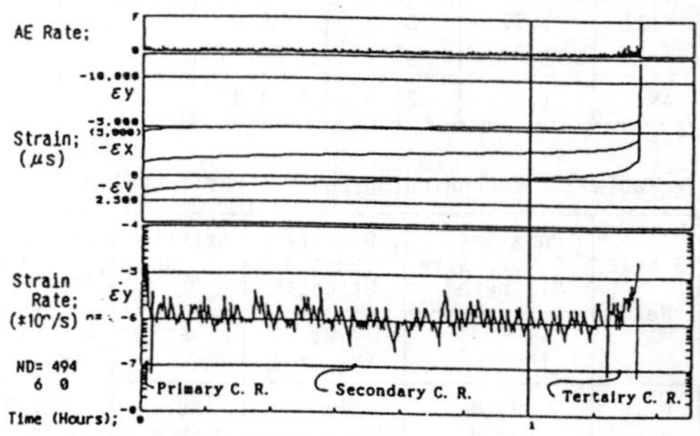

(c) Creep Strain & Strain Rate versus Real-Time Curves.

Fig.4 Curves obtained in creep test. (Sample of O series, 79.5%Sc)

processes could be seen; 1) as soon as the prescribed creep stress is applied, someinstantaneous strains are observed on these strain curves, 2) after that, when aconstant load is sustained, all strains are shown to increase with increasing elapsed time, 3) then, all strains are shown to rapidly increase, followed by specimen failure.

In Fig.4 (b), showing the relationship between the logarithm of elapsed time versus axial strain rate (Ɛy; expressed as 10^n strain/sec), additional processes could be seen; 1) the strain rate is shown to decrease during the time to 300 sec. by a nearly constant rate, 2) after that, the averaged axial strain rate decreases to a minimum value, 3) then, it rapidly increases to failure.

Fig.4 (c) shows the normal time creep curves of the same specimen. In this figure, it is clear that the time period when the strain rate is nearly constant occupies the larger part of the entire test interval.

On these figures, the primary creep region, the secondary creep region and the tertiary creep region can be easily recognized.

4.2 Behavior of strain

Fig.5 (a), (b) and Fig.6 (a), (b) show all of the creep curves obtained for samples 0 and 0-1. In these figures, the point of curve interruption indicates the point where the creep stress is unloaded, either by breaking of the specimen or by stopping of the test.

In this study, the elapsed time to failure is defined as the 'critical creep time'.

By these curves, the test specimens were shown to fail when the creep stress ratio exceeded 70 % of the uniaxial compressive strength, but the critical creep time was not found to be proportional to the creep stress ratio.

With regard to axial strains, σ y, the instantaneous strain is nearly propor-

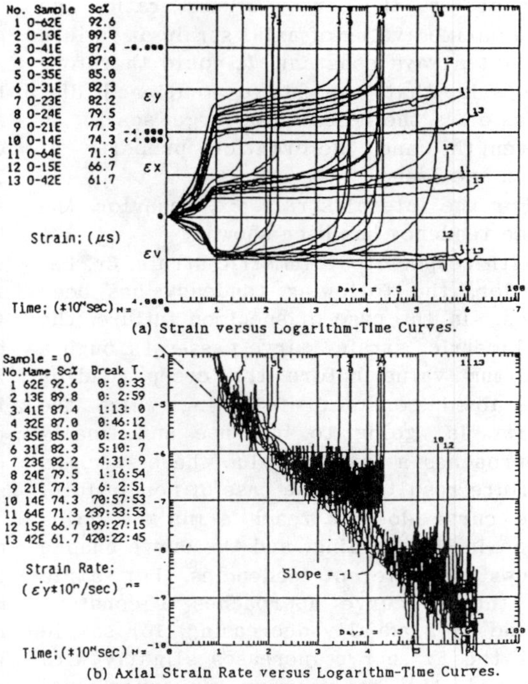

(a) Strain versus Logarithm-Time Curves.

(b) Axial Strain Rate versus Logarithm-Time Curves.

Fig.5 All curves obtained for sample of O series.

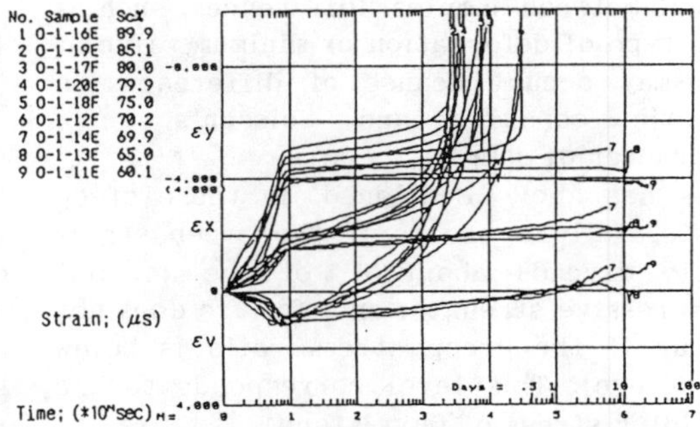

(a) Strain versus Logarithm-Time Curves.

(b) Axial Strain Rate versus Logarithm-Time Curves.

Fig.6 All curves obtained for sample of O-1 series.

tional to the creep stress ratio, and the maximum value of axial strain at failure is not always constant. It could therefore be said that the above mentioned dispersion is the result of a large scatter in strength and deformation property of each specimen.

For the lateral strain, εx, behavior, the same tendency has been shown.

With respect to volumetric strain, εv, behavior, the following tendency has been seen. In the case of specimen failure, the volumetric strain curve passes through a minimum value before the creep load is sustained constant, following which the curve is going to increase and finally approaches a maximum value where specimen failure results. In the case of non-failure, the curve doesn't reach a minimum value upon initial loading, and the curve change shows two different tendencies. For sample 0, the εv curve approaches a constant value by gradually decreasing; for sample 0-1 the εv curve increases slightly with elapsed time to stopping the experiment. The minimum value of εv is almost constant for each sample, whereas this value is not equal between samples. Differences, such as the type of deformation or minimum value of εv, may occur because of differences in elastic constant and Poisson's ratio character of samples.

As has above mentioned, in these creep tests, failure occurs if the creep stress ratio exceeds about 70 % of the uniaxial compressive strength, and failure does not occur if the creep stress ratio is below this level. This level corresponds to the yielding stress of Ohya-stone.

It is recognized that, when a creep stress exceeds the yielding stress of the rock, the load induced fracture mechanism approaches the 'unstable fracture propagating region', suggested by Bieniawski (1967), or the 'shear fracture propagating region', have suggested by the authors (1985).

With regard to failure in creep, it could be considered that the yielding stress in uniaxial compression corresponds to the boundary stress condition which separates the failure state of specimens from failure or non-failure in short term creep test.

4.3 Behaviour of strain rate

Figs.5 (b) and Fig.6 (b) show the axial creep strain rate curves for of specimens 0 and 0-1, respectively.

These curves evidence the same conditions shown in Fig.4 (b), but the relationship between the minimum strain rate value and the elapsed time are different from the applied creep stress ratio. In these cases, time intervals where the strain rate changes in narrow range of the value are fairly long in entire creep time. Also, it is clear that relationship between the points of minimum strain rate and that elapsed time becomes a straight line which slop is -1.

These linear relations have been recognized in soil studies by Saito (1961) and in rock studies by Adachi (1981). Also, Kodama (1991) has reported a similar relationship between minimum inelastic volumetric strain rate and elapsed time.

Those relations will be discussed in section 5.

4.4 Discussion on creep curve by normal time range

A creep curve, which expresses the relationship between strain or strain rate and logarithm of time, is useful to permit comparison between creep curves developed using different creep times and stresses. From these curves, it is difficult to investigate the details of time in those relations. Therefore, normal time creep curves, as shown in Fig.4 (c), were drawn for each of the test results obtained.

Fig.7 (a) - (d) show some of the creep curves for sample 0, in which the time to failure or elapsed time varied between 46 minutes and up to 240 hours. In the last of these figures, failure did not occur during the test period shown.

According to these curves, it is clearthat the change in strain rate is very rapid over the short term. But, in terms of averaged values, it could be said that the period where creep strain rate is nearly constant, that is, the secondary creep region, occupies a large part of whole creep time in any case.

Where failure doesn't occur, the creep strain rate remains constant.

The failure mechanism within the secondary creep region, could be explained by the stick-slip oscillating phenomena. Namely, the strain increases by a repeating process of slow but stable sliding and locking over a short time intervals, as have reported by Paterson (1978).

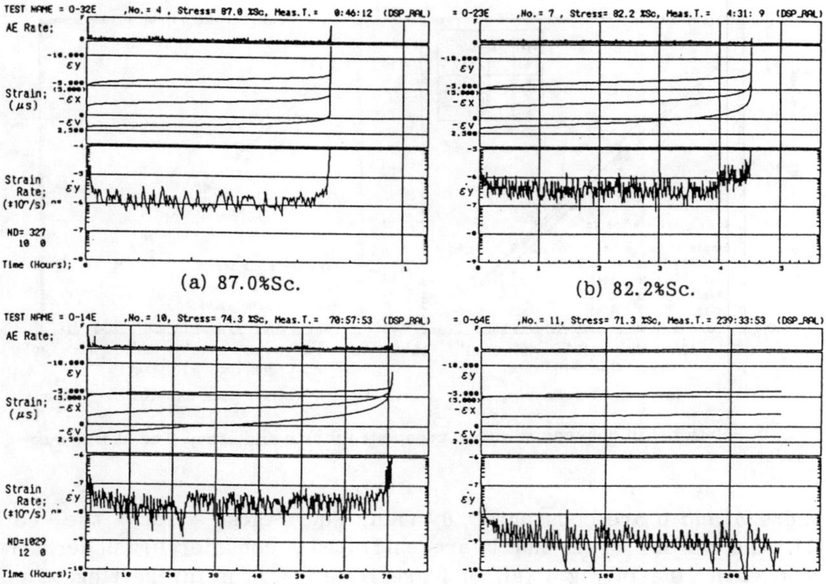

(a) 87.0%Sc. (b) 82.2%Sc. (c) 74.3%Sc. (d) 71.3%Sc.

Fig.7 Creep strain & strain rate versus real-time curves for sample of O series.

4.5 Behavior of AE activity

From each of the tests performed, it was clear that a fair amount of AE activity appeared in the primary creep region in all cases. In the secondary creep region in the case of failure, the AE signal is often not able to be separated from noise.

AE activity was found to be high in the tertiary creep region. In the case of non-failure, other than some back-ground noise, no AE activity was observed after the onset of the secondary creep region. These behavior make sure the facts that the stress state in the failure hypothesis in the case of failure is in the 'unstable' region and in the case of non-failure is in the 'stable' region.

Possibly due to the large porosity of Ohya-stone, it is very difficult to monitor AE activity in normal fashion using AE sensors attached to the surface of test specimens.

Therefore, should an AE monitoring program of underground Ohya-stone gallary fields be planed, it would be feasible to apply the Steel-Rod Method suggested by Nakajima(1985) to collect AE information.

However, since AE activates only at the onset of the tertiary creep region, and this time interval is very short duration in the creep process to failure, then a more accurate investigation of the time-dependent behavior of rock is required to permit failure prediction, even if the method of AE monitoring is improved and the accuracy of measurement improved.

5. RELATIONSHIP BETWEEN AXIAL STRAIN RATE AND TIME TO FAILURE

It is shown in Fig.5 (b) and 6 (b) that the relationship between minimum axial strain rate and elapsed time becomes linear with a slope of -1.

Fig.8 (a) and (b) show the relation between minimum strain rate, $\dot{\varepsilon}c$, and time to failure, Tc, on a log-log graph, and straight lines are drawn by using the least squares method in these figures.

The equation of this straight line is expressed as;

$$\log \dot{\varepsilon}c = a + b \log Tc \qquad (1)$$

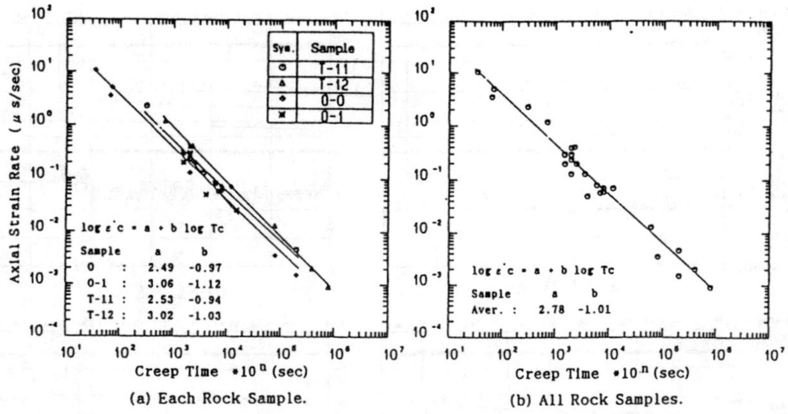

Fig.8 Relationship between minimum creep strain rate and creep time at minimum value.

where a and b are constants. By using Eq. (1), the values of a and b are estimated from each line in Fig.8 (a), and are listed in Table 3. Constant b is the slope of each line and its value varies between -1.12 to -0.94, with an averaged value of -1.01 (Fig.8 (b)). Constant a represents the axis intercept of each straight line and its value ranges between 2.49 to 3.06.

In these creep curves, the time duration of the secondary creep region occupies a large part of whole creep time, and the time ratios of the secondary creep region to the entire time of test are 94.1 % and 93.5 %, on sample 0 and sample 0-1, respectively, except where sample failure occurred rapidly within several minutes after load initiation.

Consequently, it could be considered that the minimum strain rate of the creep test corresponds to the average strain rate over the whole creep interval.

For this consideration, strain, εa, at failure is expressed as;

$$\varepsilon a \fallingdotseq \varepsilon c * Tc . \qquad (2)$$

Eq. (1) is transformed to;

$$\varepsilon a = 10^{a} * Tc^{b} . \qquad (3)$$

Eq. (3) is substituted in Eq. (2);

$$\varepsilon a \fallingdotseq 10^{a} * Tc^{(b+1)} . \qquad (4)$$

From the experimental results, the averaged value of b is about -1.

Accordingly, Eq. (4) becomes;

$$\varepsilon a \fallingdotseq 10^{a} . \qquad (5)$$

From Eq. (5), it is found that creep strain in the creep interval can be related using with only the constant a.

The value of the creep strain, which is obtained from Eq. (4), is given in Table 3. It is thought that the values of creep strain for samples 0-1 and T-1 are reasonable ones, but those for the other samples are considerably underestimated.

It is interesting to note that if the value of b in Eq. (4) is -1, rock failure will occur when the creep strain reaches a constant value in the creep test. Then, the values of creep strain of un-failed rocks, are not so large as to make a failure in creep time.

6 ESTIMATION OF TIME TO FAILURE

As has been mentioned above, there is a high possibility that the failure process and the mechanism of rock failure in sustained load will differ on each side of the yielding stress boundary. For non-failure, it is required to complete creep tests at stress level below 70 % creep stress ratio.

But, it will be impossible, practically, to carry out such an experiment, because the expected time to failure by this test will take several years or more, and possibly decades, to complete. Therefore, attempts to predict the critical time to failure have been made by applying the experimental

Table 3 Creep strain calculated by the tested results.

Sample Name	Constants a	Constants b	Creep Strain εa (μs)	Note
0	2.49	-0.97	310	
0-1	3.06	-1.12	1,150	by Fig.8(a)
T-11	2.55	-0.94	340	
T-12	3.02	-1.03	1,050	
Average	2.78	-1.01	595	by Fig.8(b)

equations derived from sample tests performed above the 70 % creep stress ratio.

Two functional equations have shown high correlation for this relationship; one is a power function equation suggested by Wilkins (1973) and the other is an exponential function equation proposed by the authors. The correlation coefficients are 0.96 and 0.98, respectively, for these two functional equations.

The times to failure for some of the testedrocks in the case of a 50 % creep stress ratio are estimated by the extrapolated line of the exponential equation in Fig.9 (a). It has been estimated that the time to failure will be several hundred years for sample T-12 and about 10 years (= $3*10^8$sec) for sample 0. For all other test samples, time to failure will be shorter than for sample 0.

Considering that the extraction ratio of the 'Ohya' district pit is about 75 %, that the uniaxial compressive strength is about 110kg/cm^2 and that the specific gravity of the rock is nearly 1.8, an underground excavation where the thickness of overburden exceeds the value of 70 meters from ground surface has a possibility of failure within 20 years to 30 years. Additionally, it must be taken into consideration that weakening of the rock mass will take place by the influence of variables such as humidity in the ventilation air, by mine water, etc..

This prediction of the time to failure would be severe because it is based on results obtained from rock tests conducted of stress levels well above the yielding stress limit of the rock.

To confirm the method for predicting the long-term stability interval of rock, an additional experimental program of creep testing is now in progress in our laboratory.

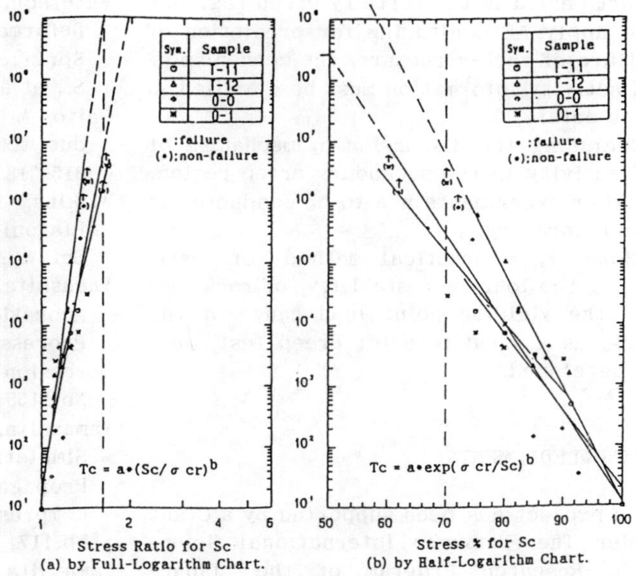

Fig.9 Predicted time of creep failure by the experimental results.

7. CONCLUSION

To evaluate the long-term stability of Ohya-stone rock, creep behaviour of the rock, such as the elapsed time versus strain, strain ratio and AE activity of samples under sustained creep stress, were investigated in this study.

Some general observations made are:

1) The creep stress in any case of failure must exceed 70 % of the uniaxial compressive strength, which corresponds to the yielding stress of the rock.

2) The minimum value of volumetric strain for this rock type in the creep process is nearly constant in each sample.

3) The relationship between the minimum strain rate and the time to failure in log-log plotting is linear.

4) The strain rate in the secondary creep region is almost constant and the creep time in this region occupies over 90 % of entire test intervals.

5) Consequently, it has high possibility that failure of rock will occur when the amount of strain in creep region reaches a constant value in each sample.

6) The AE activity noted to develop during the creep process is more active only for a

short period in the tertiary creep region.

To apply AE monitoring for prediction of failure of rock-structure, the mechanism of AE activity information must be clarified in more detail.

Including the discussion of mechanism of AE activity in the secondary creep region, further experiments are to be conducted in our laboratory.

However, a practical method for estimating the long-term stability of rock below the yielding point load range of the rock, as a substitute for creep test, must be developed.

ACKNOWLEDGEMENTS

This project has been supported by a Grant under The Monbusho International Scientific Research Program of the 'Japan-Canada Co-operative Research Project Between Akita University and Queen's University' and by The Tokyo Electric Power Services Co.,Ltd.

The authors would like to express their sincere thanks to all students and graduated students who assisted in the experimental works.

And, the authors gratefully acknowledge to Dr. J.F. Archibald, chairman of the Technical Committee of ICMD, who have kindly advised on the expressions of this report in details.

REFERENCE

Adachi, T. & A. Takase 1981. Prediction of long term strength of sedimentary rock, Proc. Int. Sym. Weak Rock, Tokyo: 99-104.

Bieniawski, Z. T. 1967. Mechanism of Brittle Fracture of Rock, PART-I, PART-II, PART-III, Int. J. Rock Mech. & Mining Sci., Vol.4, 395-430.

Kodama, J. & Y. Ishijima,et al.1992. Fatigue, Creep Properties and Long-term Strength of Granite under Uniaxial Compression, J. Min. Metal. Ins. Japan, No.3, Vol.108, 182- 186.

Nakajima, I. et al. 1983. The Monitoring of The Acoustic Emission, Accompanied with Advance Boring, J. Min. & Metal. Ins. Japan, No.1150, Vol.99, 1041-1048.

Paterson, M.S. 1978. Experimental Rock Deformation -The Brittle Field, Berlin, Springer-Verlag. (Trans.Japanese, Ogino, S., et al.,1986)

Saito, M & H.Uezawa, 1961. Failure of soil due to creep, Proc. 5th ICSMFE, Paris: 315-318.

Wilkins, 1987, The Long-Term Strength of Platonic Rock, Int. J. Rock Mech. & Min. Sci. & Geomch. Abs., 24, 379-380.

Yamashita, S. & K. Otsuka et al. 1984. Acoustic Emission Activities and Fracture Process of Brittle Rock in Uniaxial Compression Tests, J. Min. & Metal. Ins. Japan. No.1155, Vol.100, 385-390.

Yamashita, S. & K. Otsuka et al. 1985. Simulation for Fracture Initiation and Propagation of Rock under Uniaxial Compression, j. Min. & Metal. Ins. Japan. No.1171, Vol.101, 503-509.

Yamashita, S. & K. Otsuka et al. 1985. Discussion on a Fracture Process of Rock by Uniaxial Cyclic-loading Tests, J. Min. & Metal. Ins. Japan. No.1174, Vol.101, 753-759.

Yamashita, S. & K. Otsuka et al. 1986. Studies on a Mechanism of Brittle Fracture of Rock in Compression, Int. Sym. on Modern Coal Mining Technology, Fuxin, China. Vol.2, 84-89.

Innovative Mine Design for the 21st Century, Bawden & Archibald (eds) © 1993 Balkema, Rotterdam, ISBN 90 5410 325 6

Risk-Cost-Benefit Analysis applied to the design of support in burst-prone mines

R.K.Brummer, J-P.Nosé, L.Hay & P.K.Kaiser
Geomechanics Research Centre, Laurentian University, Sudbury, Ont., Canada

ABSTRACT: The paper describes a decision-making procedure (under development at the Geomechanics Research Centre) for selecting rockburst-resistant tunnel support in burst-prone mines. This procedure is based on Risk-Cost-Benefit Analysis and makes use of support cost data, costs associated with the failure of the support or loss of the excavation, and benefits achieved by installing rockburst-resistant support. This enables the rock mechanics engineer to select an upgraded or rockburst-resistant support system which minimises the overall cost of the support while maximising the benefit. The method accounts for the risk of damage to the support system due to rockbursts, potential loss of production and the need to rehabilitate in the event of a damaging rockburst.

INTRODUCTION

In addition to environmental considerations, Canadian Mines are currently facing two very severe problems - the weak price of most minerals on world markets, together with increasing public concern about the safety of the mine as a work place. While mines attempt to reduce their costs in order to survive, they cannot afford to do so at the expense of safety considerations. This means that it is imperative for mines to optimize the investment they make in the form of underground support in order to maximise the safety and economic benefits which result. This not easy to do under conditions of uncertainty, especially where mines are exposed to the hazards of largely unpredictable events such as rockbursts.

Canada currently produces 7.6 billion dollars worth of gold, copper, nickel and zinc (in order, the country's four most important metallic minerals) every year. Most of the nickel and zinc and a large proportion of the gold and copper is produced by mines which experience rockbursts. These mines require special rock support, capable of withstanding the damaging effects of these sudden energy releases. However, rockburst-resistant support is generally more expensive than conventional support, and most mines cannot afford to install rockburst-resistant support in every underground excavation. In practice this support is only installed in areas where, based on actual rockburst experience, it is considered to be necessary to undertake remedial preventative action. Unfortunately, some of these rules are based on limited experience so that decisions are often made on an ad-hoc basis.

SUPPORT OPTIONS

The problem of rockbursts in mining is usually attacked on two fronts: a strategic front, where mining layouts which minimise

Table 1 - Support Levels and Costs

Support Level	Description	Cost $/m
Level 1	Spot bolts only - density 1 bolt per 3 m^2	$107
Level 2	Pattern bolts - 1.2 m by 1.2 m	$214
Level 3	Pattern bolts - 1.2 m by 1.2 m and weldmesh screen	$383
Level 4	Pattern bolts - 1.2 m by 1.2 m, weldmesh screen, shotcrete	$1299
Level 5	Pattern bolts, 1.2 m by 1.2 m, shepherd's crooks, 1.2 m by 1.2 m, chainlink screen & cable lacing	$1397

the incidence of rockbursts are chosen, and a tactical front, where support which is resistant to rockbursts is selected. The tactical decisions which must be made are concerned with choosing the appropriate level of support. Several different types of rockburst-resistant support are available in the arsenal of the rock mechanics engineer. These range from the simple addition of mesh or screen to more elaborate schemes involving yielding bolts, cable lacing, shotcrete, or a combination of these and other elements. Naturally, each increasing level of support comes at a price, in terms of both labour and material costs. However, the increased burst-resistance of the support brings with it benefits, in terms of reduced risk of damage, elimination of rehabilitation costs, fewer production losses during rehabilitation and reduced risk of loss of the orebody or stope.

For this paper it is assumed that the decisions which must be made regard the selection of support for a drift. For illustration, five different categories or levels of support are considered (Table 1). The different support levels range from Level 1 (a one-tier system of spot bolts) to Level 5 (a three-tier system of cable lacing over rockbolts, shepherd's crooks, and chain-link mesh). The categories range from the inexpensive (Level 1) to the most expensive (Level 5) and it is presumed that each increasing level of support has increased effectiveness as a support system under rockburst conditions.

In order to facilitate the decision methodology, several groups of information are required, namely the costs of the various support options, the risk of damage due to rockbursts, the degree to which each support option reduces the risk of damage due to rockbursts, and some measure of the consequences of rockburst damage.

SUPPORT COSTS

Our previous work (Kaiser et al, 1991) and an examination of past studies of the cost of different ground control methods revealed great differences in the approach taken to arrive at a meaningful cost figure for each type of support. Some studies considered all costs, direct and indirect, including the capital cost of major equipment such as drilling jumbos and scissor trucks. However, they did not deal with handling and transport costs underground, which can be considerable, depending on the location of the drift. Other studies (internal studies by mining firms) included only the direct labour, drilling, and material costs.

Mining companies do not consider large capital equipment charges as part of the cost of ground control; such charges are amortized on a "per ton of ore" basis as a cost of production. In addition, since the equipment usually consists of multi-purpose machines, it is difficult to estimate accurately the proportion of their available time that is used for rockbolting. For example, a scissor lift truck may be only used 30-60% of its working time for bolting.

Accounting procedures also vary widely; some firms depreciate capital equipment on a five year straight line basis, while others have a policy of expecting a two-year payback on large capital outlays. The cost of borrowing capital is also highly variable, as well as the internal rate of return companies use in their feasibility studies. Any attempt at including indirect costs must also include transport and handling costs of the ground support elements, which can also be extremely variable and thus very difficult to determine as an average cost.

Costs also differ widely depending on whether support or rehabilitation work is executed by a contractor or done by the mine itself. In order to arrive at a common base for comparison, the costs considered in this study are the direct costs of labour, drilling and materials. This is in line with the practice in the mining industry and mining contracting firms. The exception was for shotcrete, in which case it was necessary to include equipment rental charges. Mining companies do not, as a rule, calculate the cost of first-pass systematic ground control separately during a development round cycle; all costs of drilling, blasting, mucking and rockbolting of a drift advance are usually lumped together and charged to a crew or heading. However, detailed time studies are carried out on the installation of ground control elements for bonus calculations during extensive rehabilitation work. These studies, and other estimates, are the primary source of information for the determination of a standard installation time for rockbolts, screen or shotcrete. Since labour hourly rates (including benefits) also vary widely, the cost of labour was calculated on a man-hour per unit basis. All required materials and equipment are assumed to be on site.

Direct Cost of Labour

Some mining companies that were contacted supplied the results of detailed time studies for the installation of various types and various lengths of rockbolts, based on a two-man crew using a stoper and a jackleg. The standards included standard delays, set-up and scaling times. Other companies quoted the number of a certain type and length of bolt that would be installed per shift at a sustainable rate. An analysis of the data showed that a linear relationship existed between bolt length and installation time, and a standard was derived for man-hours per foot of bolt. Based on the information we obtained, the standard installation time for mechanical rockbolts was calculated to be 0.0414 man-hours per foot of bolt (0.136 man-hours/m - mines still largely use costs per foot; we have converted where appropriate and all costs are given in 1992 values).

It was then a straightforward exercise to arrive at a total bolt length per round of advance, according to the desired unit bolt length and density, and hence the total expected man-hours per round. The applicable labour rate was then used to calculate the total cost of ground-reinforcing labour time per length of advance.

Direct Costs of Drilling

These include the cost of drilling bits and steel, lubrication oil, and maintenance on drilling equipment. The figures were supplied by companies on a cost per foot drilled and were averaged. The drilling cost for a 1.25" borehole was found to be $0.65/ft ($2.13/m).

Cost of Mechanical Rockbolts

The costs of various lengths of mechanical rockbolts delivered to the mine, as supplied by the firms, was found to be reasonably linear (including plates and shells), and were used as a cost/ft. The bolt cost was approximately $1.00/ft ($3.30/m).

SUPPORT COSTS FOR A STANDARD 16 ft by 16 ft DRIFT

The following average costs were used in the initial RCB analysis:

Level 1 Spot Bolting

This support level is defined as having an intensity of 50% of a standard 4' x 4' pattern of mechanical rockbolts. The assumption is that support requirements are 8' bolts in the back and 6' bolts in the walls, down to floor level, and that spot bolting a 16' x 16' x 8' round requires 3 8' bolts in the back and 10 6' bolts in the sides.

Cost of labour:	
((3x8+10x6)x0.0414)@$35.00	= $121.80
Cost of drilling:	
(84 x 0.65)	= $54.60
Cost of support elements:	
(84 x 1.00)	= $84.00
Total cost/8' round	= $260.40
Cost/foot of advance	= $32.55
Cost/m of advance	= $107.00

Level 2 Pattern Bolting (4' x 4') without mesh

This support level is assumed to consist of a systematic 4' x 4' pattern of 8' mechanical bolts in the back and 6' bolts in the walls, down to the floor (no screen).

Cost of labour:	= $243.60
Cost of drilling:	= $109.20
Cost of support elements:	= $168.00
Total cost/8' round:	= $520.80
Cost/foot of advance:	= $65.10
Cost/m of advance:	= $214.00

Level 3 Pattern bolting (4' x 4') with #6 screen

This support level consists of level 2 with complete coverage of the bolted area with #6 (medium) welded wire screen down to the drift floor. The industry standard for the installation of #6 screen was found to be 0.01097 man-hours/ft^2. The material cost of screen is $13.40 per 5' x 11' section (including extra plates for installation). The assumed installation procedure for screen is a 6" overlap on the sides and 18" overlap on the ends.

Bolting costs/round	= $520.80
Screening costs:	
Labour = (0.01097x660)x$35.00	= $253.40
Material = (12 sections)x$13.40	= $160.80
Total screen cost	= $414.20
Total cost/8' round	= $935.00
Cost/foot of advance	= $116.88
Cost/m of advance	= $383.00

Level 4 Screened pattern bolting with a 4" layer of dry mix shotcrete.

The cost of shotcreting is highly dependent on the competence of the crew, as the rebound rate can vary from 10-40%, depending on proper moisture control and skill of application. For costing purposes, it is reasonable to assume a 20% rebound rate and a wall roughness factor (which accounts for additional material required to cover rough walls) of 1.30. The average standard for three mining firms and a shotcreting contractor was 1.5 man-hrs/bag of dry mix shotcrete installed. This is based on a three-man crew using a 020 shotcreting machine and 1000 kg dry mix bags. For a 16' x 16' x 8' round and a 4" layer, approximately 13 bags are required, at $100.00 per bag. Two scenarios are considered:

1) Assuming the crew has more than one heading available for shotcreting, the costs involved are:

(a) Cost of shotcreting:

Labour = (13 x 1.5) x $35.00	= $682.50
Material = (13 x 100.00)	= $1300.00
Equipment rental (including 2 wear plates/day)	= $250.00
Total shotcreting cost per 8' round	= $2232.50
(b) Cost of bolting and screening per 8' round	= $935.00
Total cost per 8' round	= $3167.50
Total cost/foot of advance	= $396.00
Total cost/m of advance	= $1299.00

2) If, due to mining constraints, only one heading is available per day for shotcreting, the labour costs increase to 24 x $35.00 = $840.00/round, and the total shotcreting costs increase to $2390.00/round. Thus the total costs of level 4 support (if one available heading/day) are $3325.00/round or:

Total cost /foot of advance	= $415.60
Total cost/m of advance	= $1363.00

Level 5 Cable Lacing (diagonal 4.25' bolt pattern, including chain link screen)

This support level consists of installing 2" woven (chain-link) mesh screen on the back and on the sides down to the floor, and 0.5" diameter cable laced diagonally and horizontally through each 8' bolt in the back and each 6' bolt in the walls. The rockbolts (called eyebolts or shepherd's crooks) are installed on a staggered 6' pattern such that the diagonal distance between bolts is 4.25'. Each eyebolt has three cables laced through it. The costs of cable lacing were obtained from the results of two separate detailed studies carried out in 1985 and 1989. The first quoted a cost of $6.10/ft^2 (1985 dollars) and the second quoted a cost of $6.73/ft^2 (1989 dollars). Applying the average Canadian long term inflation factor of 1.055/year resulted in a present cost of $9.36/ft^2 for the 1985 study and $8.24/ft^2 for the 1989 results. On a per foot of advance basis, the costs were on average, for a 16' x 16' x 8' round:

Cost/foot of advance	= $426.00
Cost/m of advance	= $1397.00

This excludes the cost of any initial support installed on development.

The costs of the various support alternatives are also included in Table 1.

UNCERTAINTY AND RISK

For many problems, it is not possible to state with certainty what consequences a particular action will have. Ground control decisions match this description well since one typically tries to answer questions such as: "If we support this highly stressed drift with Level 3 support, will we have ground control problems during the life of the excavation?". Since one cannot answer such a question with certainty, the treatment of risk in the decision process usually involves two major actions:

- identifying and quantifying the possible hazards (i.e. "What ground control problems could we have at this location, and what is the probability or risk that they will occur?") and
- deciding how much risk is acceptable (i.e. "Are we satisfied with this level of risk, or should we invest in

improved support in order to reduce the chance of damage?)

The use of probabilistic design for problems of this nature is well advanced in some fields (Harr, 1987).

We have followed two alternative paths in order to estimate the risk of damage to a mining excavation, namely an evaluation of the seismic hazard (the probability of occurrence of a seismic event larger than a given magnitude during a given time period) as indicated by past seismic records, and a direct evaluation of the hazardous events due to rockbursts at mines (in terms of severity of damage per unit length of drift per unit time period) based on an analysis of damage records. In the present context, the hazardous events are arbitrary divisions of rockburst severity such as the Damage Levels (DL1 to DL5) as described by Kaiser et.al., 1992, although other hazardous events can also be accommodated within the decision-making procedure. How decisions are made on how much risk is acceptable will be described later.

The support decisions must be made under conditions of uncertainty, since each action has several possible outcomes for which the probabilities may or may not be known.

Probabilistic decision theory enables one to make decisions like this in a rational way.

For completeness, we distinguish three types of probability:

- **a-priori** probability - which is established by the definition of the system (e.g. a "fair coin", where the chance of a head is defined to be 0.5)

- **a-posteriori** probability - in which the probability of an event happening is the proportion of times the event would occur in a long run of repeated experiments (e.g. a real coin tossed many times to establish how many heads do fall in a large number of tosses)

- **subjective** probability - where the probabilities are based on a person's experience or specialized knowledge. When the person is an expert or has some special experience, he is able to provide an opinion or estimate of the chance of an event's occurrence. Many engineering decisions which rely on judgement fall into this category.

The latter two categories are the most useful in the context of ground support. A ground control specialist should have a good idea of the chance of damage to an excavation with a given level of installed support, based on his personal experience under the given conditions at the particular mine. This experience may even take the form of a quantitative database of mine damage reports which enable him to make statements such as: "During the past year, we had to rehabilitate 10% of drifts supported with bolts only, but we rehabilitated none supported with cable lacing." Therefore an educated estimate can be made on how much rehabilitation is likely to be required under equivalent ground conditions. In this way, local knowledge and experience regarding support effectiveness can be quantified for use in support design decisions. The drawback of this type of approach is that mines actually install upgraded support in areas where they anticipate damage - hence some form of extrapolation is necessary in order to estimate how much damage would have occurred had conventional support been installed. One way of performing such an extrapolation is discussed later.

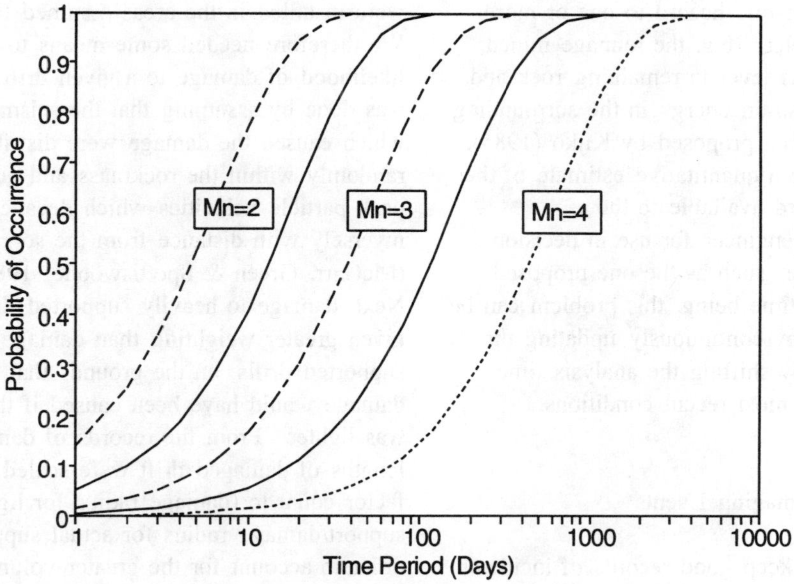

Figure 1 Magnitude/Hazard/Time Plot for a large Canadian mine. The figure shows the probability of occurrence of a seismic event larger than the indicated magnitude during a given time period.

Seismic Hazard Analysis

Mine seismic systems can provide a meaningful estimate of the seismic hazard (the probability of occurrence of an event) if the data which they provide is interpreted appropriately. From the hazard estimate, the risk of damage (which depends on the hazard and the degree of vulnerability or potential loss) can be assessed. Several methods for determining seismic hazard have been developed - the unconditional probability of occurrence of Benjamin (1968) and the conditional probability of occurrence as described by Fernandez and du Plessis (1988) are good examples. Figure 1 is a "Magnitude/Hazard/Time Plot", based on a double-truncated Gutenberg & Richter frequency-magnitude relationship and the unconditional probability model of Benjamin (1968), prepared for a typical burst-prone Canadian mine. The figure shows the probability of occurrence of a seismic event of magnitude equal to or greater than a chosen magnitude during a given future time period, based on the past seismic record. Naturally, this involves the analysis of past events, and projecting this past history into the future. The acceptability of this logic depends on the assumption that past history is a good indicator of future events. While this is often acceptable in most engineering fields concerning equipment reliability and is routinely used in the field of earthquake hazard analysis, this approach has severe limitations in the analysis of a mine. The major reason for this shortcoming is that as mining progresses, the stress levels change, and the probability of occurrence of seismic events can increase, or even decrease, for example as a highly stressed pillar is finally removed. As our understanding of the mechanisms which cause mining-induced seismicity improves, it must be expected that these methods for hazard assessment will also improve. In future then, we expect to be able to relate the seismic hazard to one or more mining parameters (e.g. the tonnage mined, change in stress level in remaining rock and ore, or stored strain energy in the surrounding rock) as has been proposed by Kijko (1985). This will make a quantitative estimate of the rockburst hazard available to the geomechanics engineer for use in decision making schemes such as the one proposed here. For the time being, this problem can be circumvented by continuously updating the database and by shifting the analysis time window to the most recent conditions.

Records of Damaging Events

Mines usually keep good records of incidents of damage. These can be used to estimate the probability of occurrence of damaging events. Preliminary results from an exercise in estimating the required probability numbers are discussed here. In this case a large number of incidents of damage due to rockbursts were recorded at a Canadian mine over a two year period. Damage was divided into three categories using the Damage Level (DL) scale of Kaiser et al (1992): Minor Damage (DL1/DL2), Moderate Damage (DL3/DL4) and Major Damage (DL5). Five different levels of support (Level 1 to Level 5) were used at various locations in the mine. Since upgraded support had been installed in areas where problems were anticipated the data appear anomalous in that the least damage occurred to the lowest support levels. This is of course not surprising since the lightest levels of support were installed in the areas assumed to be safe. We therefore needed some means to assess the likelihood of damage to a given drift. This was done by assuming that the seismic events which caused the damage were distributed randomly within the rockmass and generated peak particle velocities which decayed inversely with distance from the source (McGarr, Green & Spottiswoode, 1981). Next, damage to heavily supported drifts was given greater weighting than damage to lightly supported drifts, on the grounds that more damage would have been caused if the support was lighter. From the records of damage, the lengths of damaged drift were scaled up by a factor equal to (damage radius for lightest support/damage radius for actual support)^3 in order to account for the greater volume of rock which would have been affected by the event. The damage radii, or distances from a given seismic source at which damage would be caused to the given support system, were taken from Kaiser (1993) and provided a convenient and systematic way to account for the differences (or burst resistance) between the different support systems. The total equivalent damage (which would have occurred had no upgraded support been specified) was then determined by dividing the total length of equivalent damage by the total length of drift exposed to damage during the time period in question. The risk of damage to drifts with upgraded support was then determined by scaling back using the same factors which were used to scale up from upgraded to standard support. The results of this exercise, for a large Ontario mine, are presented in Table 2. These figures are

Table 2 - Percentages of drift lengths damaged per year

Damage	Level 1	Level 2	Level 3	Level 4	Level 5
Minor	6.410	1.899	0.563	0.237	0.070
Moderate	2.095	0.621	0.184	0.078	0.023
Major	0.433	0.128	0.038	0.016	0.005

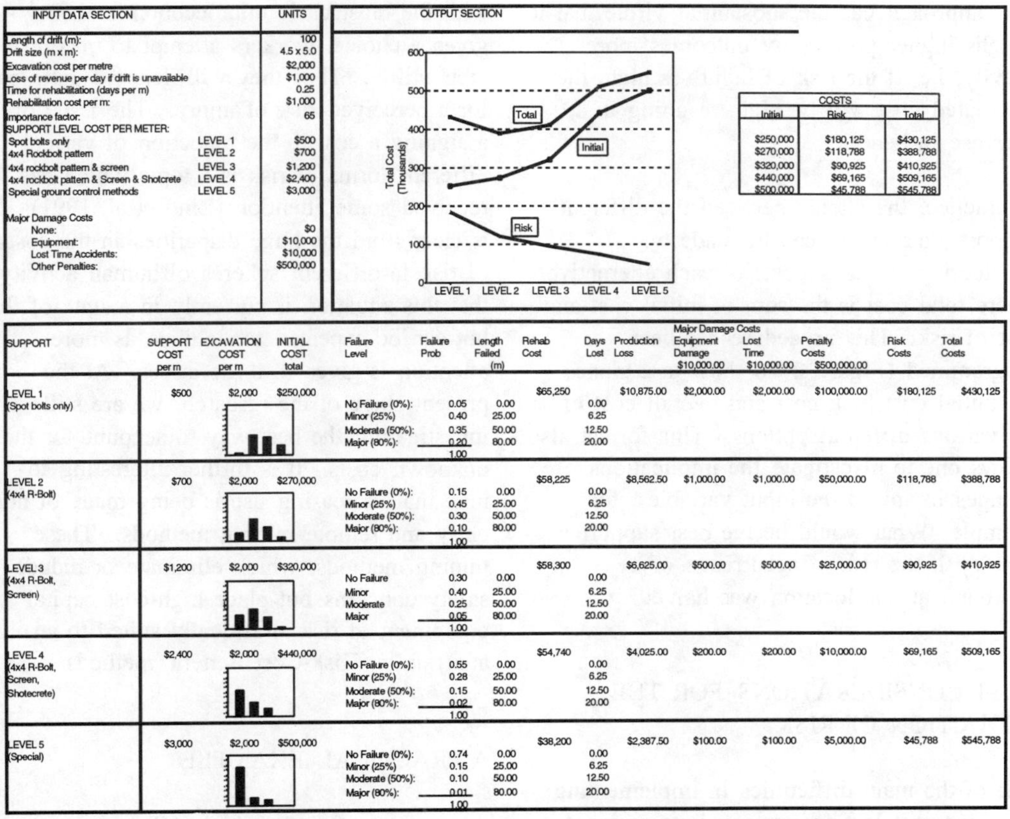

INPUT DATA SECTION		UNITS
Length of drift (m):		100
Drift size (m x m):		4.5 x 5.0
Excavation cost per metre		$2,000
Loss of revenue per day if drift is unavailable		$1,000
Time for rehabilitation (days per m)		0.25
Rehabilitation cost per m:		$1,000
Importance factor:		65
SUPPORT LEVEL COST PER METER:		
Spot bolts only	LEVEL 1	$500
4x4 Rockbolt pattern	LEVEL 2	$700
4x4 rockbolt pattern & screen	LEVEL 3	$1,200
4x4 rockbolt pattern & Screen & Shotcrete	LEVEL 4	$2,400
Special ground control methods	LEVEL 5	$3,000
Major Damage Costs		
None:		$0
Equipment:		$10,000
Lost Time Accidents:		$10,000
Other Penalties:		$500,000

COSTS		
Initial	Risk	Total
$250,000	$180,125	$430,125
$270,000	$118,788	$388,788
$320,000	$90,925	$410,925
$440,000	$69,165	$509,165
$500,000	$45,788	$545,788

SUPPORT	SUPPORT COST per m	EXCAVATION COST per m	INITIAL COST total	Failure Level	Failure Prob	Length Failed (m)	Rehab Cost	Days Lost	Production Loss	Major Damage Costs Equipment Damage $10,000.00	Lost Time $10,000.00	Penalty Costs $500,000.00	Risk Costs	Total Costs
LEVEL 1 (Spot bolts only)	$500	$2,000	$250,000				$65,250		$10,875.00	$2,000.00	$2,000.00	$100,000.00	$180,125	$430,125
				No Failure (0%):	0.05	0.00		0.00						
				Minor (25%)	0.40	25.00		6.25						
				Moderate (50%):	0.35	50.00		12.50						
				Major (80%):	0.20	80.00		20.00						
					1.00									
LEVEL 2 (4x4 R-Bolt)	$700	$2,000	$270,000				$58,225		$8,562.50	$1,000.00	$1,000.00	$50,000.00	$118,788	$388,788
				No Failure (0%):	0.15	0.00		0.00						
				Minor (25%)	0.45	25.00		6.25						
				Moderate (50%):	0.30	50.00		12.50						
				Major (80%):	0.10	80.00		20.00						
					1.00									
LEVEL 3 (4x4 R-Bolt, Screen)	$1,200	$2,000	$320,000				$58,300		$6,625.00	$500.00	$500.00	$25,000.00	$90,925	$410,925
				No Failure	0.30	0.00		0.00						
				Minor	0.40	25.00		6.25						
				Moderate	0.25	50.00		12.50						
				Major	0.05	80.00		20.00						
					1.00									
LEVEL 4 (4x4 R-Bolt, Screen, Shotecrete)	$2,400	$2,000	$440,000				$54,740		$4,025.00	$200.00	$200.00	$10,000.00	$69,165	$509,165
				No Failure (0%):	0.55	0.00		0.00						
				Minor (25%)	0.28	25.00		6.25						
				Moderate (50%):	0.15	50.00		12.50						
				Major (80%):	0.02	80.00		20.00						
					1.00									
LEVEL 5 (Special)	$3,000	$2,000	$500,000				$38,200		$2,387.50	$100.00	$100.00	$5,000.00	$45,788	$545,788
				No Failure (0%):	0.74	0.00		0.00						
				Minor (25%)	0.15	25.00		6.25						
				Moderate (50%):	0.10	50.00		12.50						
				Major (80%):	0.01	80.00		20.00						
					1.00									

Figure 2 Example of a detailed spreadsheet for Risk-Cost-Benefit Analysis of support options

intended as preliminary estimates; research is still in progress to refine the estimates further and verify the results.

THE DECISION-MAKING PROCEDURE

The treatment which follows is based on the decision matrix (Siddall, 1972). For each of the chosen range of support options (Support Levels 1 to 5) there exists a finite probability of some outcome (in this case Damage Levels 1 to 5 or No Damage). For example, if we choose Support Level 1 for a particular drift, there will be a certain chance of suffering Damage Level 1, a chance of suffering Damage Level 2, etc. These different combinations can be arranged in the form of a matrix or table, where each position in the table represents a particular combination of support strategies and damage potentials. In practice, we do not know which level of damage will occur, but we should have some understanding (based on experience at the particular mine) of the probability of each level of damage, given a particular support strategy, as described earlier. Armed with this information we can estimate the value (benefits) of a particular policy by summing the cost of each outcome where each outcome is weighted by its probability of occurrence.

The final step is to select the policy with the highest expected present value of benefits. In the ground support application the highest benefits are achieved where the sum of the risk costs and support costs are a minimum.

This approach has the substantial virtue that it weighs higher probability outcomes more heavily, i.e. if the risk of failure is high, the associated cost will be high, resulting in a decrease in benefits.

In practice, the choice between the different support alternatives can be made by considering the total costs of each alternative, where total cost is the sum of initial cost and cost of risk. This procedure has been programmed (Figure 2) to show at a glance the initial cost, risk cost and overall cost of the various different options. This format also allows one to investigate the implications of changes in any of the input variables, for example "What would be the best support strategy if the risk of occurrence of a rockburst at this location was halved?"

COST CONSIDERATIONS FOR THE REDUCTION OF RISK

One of the main difficulties in implementing this support selection strategy is to assign a cost for accidents, particularly accidents involving injury or death to workers. Companies have serious reservations about assigning a cost to a major accident, since this could be construed as equating the loss of human life to a dollar loss or the "selling" of a human life. This reluctance to come face-to-face with the figures is to some extent understandable. However, to say that there is simply no way to value a human life is to be unrealistic. Anyone who enters a motor vehicle in order to drive to work or to a corner store knows that this action involves the assumption of some risk, usually referred to as a voluntary risk. That some people continue to smoke (while acknowledging that the action is bad for their health) is a stark reminder of how much additional risk some people are willing to assume. Risky occupations nominally pay higher wages than otherwise equivalent alternatives. This is in accordance with the most basic microeconomic theory - given a choice, workers attempt to maximise their utility. Thus they will demand extra pay for a perceived risk of injury. The issue of assigning a cost to the reduction of various different forms of risk is a topic which has received some attention (Lind et al, 1991). It is clear from the huge disparities in the costs of risk in different spheres of human activity that this situation is currently in a state of flux but can be expected to stabilize as more attention is given to these issues. At the present stage of the research, we are still investigating the best way to account for these unknown costs. It is further interesting to note that increasing use is being made of non-entry and remote mining methods. These mining methods, which eliminate or reduce safety concerns but place high-cost capital equipment at risk, are ideally suited to an analysis by Risk-Cost-Benefit methods.

A PRACTICAL EXAMPLE

A practical example which will serve to illustrate the principles involved in the decision process is as follows:

A mine considers the crosscut access to a new stope. The crosscut has already been developed and is supported with pattern bolts and screen. The question is raised: "Is it necessary to upgrade the support (using shotcrete at a total cost of $48 000.00) in order to ensure the access to the stope?

From past experience at the mine (during the previous 7.5 years) 12 incidents occurred where equivalent crosscuts were damaged. At any one time, a total of ten crosscuts were kept open and were thus exposed to the same risk. The stope has a design life of one year.

In order to make the decision, it is assumed (conservatively) that if an incident of rock damage occurs in the crosscut, it will be necessary to close the crosscut and rehabilitate it completely (with shotcrete) in order to ensure its further stability. The crosscut is 55 metres long and it is estimated that it will take 11 days to rehabilitate it. During this period, a revenue loss of $9000 per day will be incurred. The upgrading option will cost the same as the cost of shotcreting, and it is assumed that the crosscut will then be stable.

The risk was estimated as follows:

Chance of a damaging event during period exposed =

$$\frac{(12_{incidents})x(12_{months})x(1_{drifts})}{(80_{months})x(10_{drift})}$$

$$= 0.18 \text{ or } 18\%.$$

Thus there is an 18% chance of suffering some damage to the stope entrance during the life of the crosscut (based solely on previous experience at the mine). The cost of this risk is the product of the probability of occurrence and the cost, i.e.
0.18 x (11days x $9,000 + $48,00) = $26,460.

It then remains to estimate how much safer the entrance will be if it is shotcreted. In this case, even if the risk of damage is entirely eliminated by shotcreting (which it clearly is not) then the shotcreting option is still the most expensive, since the choices are:

Option 1 - Support crosscut with shotcrete:

Cost of shotcrete:	$ 48,000.00
Cost of risk (estimate):	$ 0.00

Total cost:	$ 48,000.00

Option 2 - Don't support crosscut with shotcrete:

Cost of shotcrete:	$ 0.00
Cost of risk (estimate):	$ 26,480.00

Total cost:	$ 26,480.00

In this simple example the lowest cost solution would be to take the chance of damage, since the risks are not high. In this particular example, it was assumed that safety concerns were automatically satisfied, since all previous seismic events had occurred during blast times, and the mining method employed was a no-entry method.

CONCLUSIONS

The decision-making procedures described are intended to improve mine safety and efficiency. Even if the method is not implemented in its entirety, great benefit is usually derived from the exercise of evaluating the risks and costs involved in the various support options. Most importantly, the approach provides a method for evaluating the need for upgraded support in the light of the possible risks due to rockbursts which could exist in mines, since by implementing the method it is necessary to investigate:

- The hazard or likelihood of damage due to rockbursts
- The costs of the various support alternatives
- The cost of the risk in terms related to rehabilitation costs or loss-of-production costs.

Clearly, it is not necessary to provide expensive upgraded support if the risk of rockburst damage is small. This method enables one to make the necessary decisions in a quantitative and optimal fashion. Risk-Cost-Benefit Analysis is expected, once fully developed, to greatly enhance the value of several allied research projects in progress at GRC as part of the Canadian Rockburst Research Programme.

ACKNOWLEDGEMENTS

This work was partially funded by means of grants from the Institute for Risk Research, University of Waterloo, and a Laurentian University Seed funding grant.

REFERENCES

Benjamin, J. R. (1968) Probabilistic Models for Seismic Force Design. Journal of the Structural Division of the ASCE, May, 1968, pp. 1175-1196.

Fernandez, L. M. & Du Plessis, A. (1988) The Conditional Probability of Occurrence of Rockbursts in the Gold Fields of South Africa. SANGORM Symposium, Rock Mechanics in Africa, November 1988.

Harr, M. E. (1987) Reliability-Based Design in Civil Engineering. McGraw-Hill, New York.

Kaiser, P. K., Maloney, S. M. & Singh, S. P. (1990) Feasibility of long-term ground control with stainless steel friction rock bolts. CIM Bulletin, August 1990, pp. 55-60.

Kaiser, P. K., Tannant, D. D., McCreath, D. R. & Jesenak, P. (1992) Rockburst Damage Assessment Procedure. Rock Support in Mining and Underground Construction, pp. 639-647, Kaiser & McCreath (Eds.), Balkema, Rotterdam.

Kaiser, P. K. (1993) Support of Tunnels in Burst-Prone Ground - Toward a Rational Design Methodology. Keynote Lecture at this Conference.

Kijko, A. (1985) Theoretical Model for a Relationship between Mining Seismicity and Excavation Area. Acta Geophysica Polonica, Vol. XXXIII, No. 3, pp. 231-241.

Lind, N. C., Nathwani, J. S. & Siddall, E. (1991) Managing Risks in the Public Interest. Institute for Risk Research, Waterloo.

McGarr, A., Green, R. W. E & Spottiswoode, S. M. (1981) Strong ground motion of mine tremors: Some implications for near-source ground motion parameters. Bulletin of the Seismological Society of America, vol. 71, 1981, pp. 597-612.

Siddall, J. N. (1972) Analytical Decision-Making in Engineering Design, Prentice-Hall, Inc.

Innovative Mine Design for the 21st Century, Bawden & Archibald (eds) © 1993 Balkema, Rotterdam, ISBN 90 5410 325 6

Vibration control of a pipe string for mining manganese nodules in the deep sea

K.Aso, K.Kan, H.Doki & T.Ohkoshi
Mining College, Akita University, Japan

ABSTRACT: In order to control the longitudinal vibration of the pipe string for mining manganese nodules from deep sea bottoms by a pump-lift system, first, the shape effects of the pump modules and buffer equipped with the pipe string on its longitudinal vibration were examined. The modules and buffer whose shapes cause higher drag forces proved to be more useful for reducing the resonance amplitudes of the vibration. Then, a vibration absorber, which was composed of a mass, springs and dampers, was attached to the pump modules as well as the buffer, and the effect of the absorber was analyzed theoretically. The result indicated that the absorber attached to the buffer could almost reduce its resonance amplitude by half as compared with that pertaining to the buffer without any vibration-absorbers.

1 INTRODUCTION

For mining manganese nodules from deep sea bottoms, three types of the mining systems, as shown in Fig.1, have been developed so far. The continuous line bucket system (CLB system Fig.1-a) is the one which mines the nodules by circulating a loop of wire rope equipped with buckets. On the other hand, the hydraulic dredge systems (Fig.1-b,c) use a pipe string for lifting the nodules from deep sea bottoms to the mining ship on the sea surface. The dredge system is classified into the towed miner system (Fig.1-b) and the self-propelled miner system (Fig.1-c). From the viewpoint

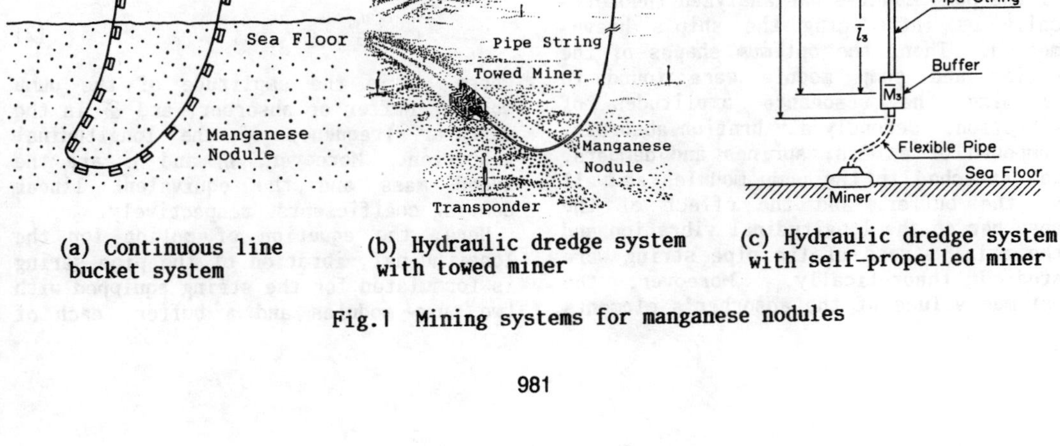

(a) Continuous line bucket system (b) Hydraulic dredge system with towed miner (c) Hydraulic dredge system with self-propelled miner

Fig.1 Mining systems for manganese nodules

of hydro-transportation of ore, the system is also divided into the air-lift system and the pump-lift system. At present, the hydraulic dredge systems are considered preferable, because of their higher productivity and better controllability, although the CLB system is less expensive and simpler in its mechanism. Among the hydraulic dredge systems, the pump-lift system with a self-propelled miner (mining robot) will be the most promissing system in future, because of its highest controllability.

Up to the present, the air-lift system with a self-propelled miner has been studied by Chung et al.(1981a,1981b,1983), Felippa (1981), Whitney et al.(1981), Sparks et al.(1982),Tikhonov et al.(1984), Aso et al.(1987,1988) and Yuan et al. (1988,1989a,1989b,1990a,1990b). However, studies on the pump-lift system with a self-propelled miner have not been reoprted except by the authors (1991a, 1992a,1992b).

In this pump-lift system,the pipe string must be equipped with the pump modules for pumping the nodules up to the ship and with the buffer which regulates the density of nodule-fluid mixture in the pipe string and plays the role of the weight to stabilize the mechanical behavior of the string. The pipe string in this system is apt to vibrate longitudinally and laterally due to the ship motions. Hence, these vibrations must be analyzed for the design of the above-mentioned mining system. From the previous reports (Aso et al. 1987,1988,1991a), the longitudinal vibration is considered more destructive as far as the strength of the pipe string is concerned, and so must be controlled as much as possible.

In this study, firstly, various buffers and pump modules designed to cause various drag forces were selected, and the forced longitudinal vibration of the pipe string equipped with one of the buffers and two of the pump modules was analyzed theoretically by introducing the ship's heave-motion. Then, the optimum shapes of the buffer and pump module were found to minimize the resonance amplitudes of vibration. Secondly,a vibration absorber, composed of a mass, springs and dampers, was attached to the pump modules as well as the buffer, and the effect of the absorber on the longitudinal vibration and the axial stress of the pipe string were studied theoretically. Moreover, the optimum values of the absorber's elements were determined for minimizing the vibration and the axial stress induced in the pipe string. Finally, the mutual effect of the shape of the buffer and its vibration absorber on the longitudinal vibration of the pipe string was examined.

2 ANALYSIS

The analytical model of this problems is shown in Fig.1-c, in which the vertical pipe string of length, L, equipped with a buffer (mass; $\overline{m}_3$) at its lower end (at $x=l_3$) and two pump modules (mass; $\overline{m}_1$, $\overline{m}_2$) in between (at $x=l_1, l_2$), is suspended from the mining ship on the sea surface and connected to the miner on the sea floor through a flexible pipe. Here, it is assumed that the top of the pipe string is forced to vibrate vertically in the harmonic function of $u_0(t)=a\sin\omega t$ by the heaving motion of the ship.

At the beginning of this analysis, the fluid forces acting on the pump module and buffer must be determined. These forces can be calculated using Morison's equation (Morison et al.1950):

$$F(t) = C_m m_a \frac{d^2 u_m}{dt^2} + 0.5\rho C_d \bar{S} \frac{du_m}{dt}\left|\frac{du_m}{dt}\right|, \quad (1)$$

where C_m, C_d, m_a, ρ, $\overline{S}$, u_m are an added mass coefficient, drag coefficient, the mass of water displaced by the pump module, buffer or absorber, the density of the ambient water, the cross-sectional area, the longitudinal displacement of the pump module, buffer or absorber, respectively. When the second term of Eq.1 is linearized by the energy method (Timoshenko 1955), the fluid force $F(t)$ can be expressed as follows:

$$F(t) \cong C_m m_a \frac{d^2 u_m}{dt^2} + (4\rho\bar{S}C_d\bar{A}\tilde{\omega}/3\pi)\frac{du_m}{dt}$$
$$= \tilde{m}\frac{d^2 u_m}{dt^2} + c\frac{du_m}{dt}, \quad (2)$$

where $\overline{A}$ is the amplitude of the pump module, buffer or absorber, and $\tilde{\omega}$ is the circular frequency of the longitudinal vibration. Moreover, $\tilde{m}$ and c are the added mass and the equivalent linear damping coefficients, respectively.

Here, the equation of motion for the longitudinal vibration of the pipe string is formulated for the string equipped with two pump modules and a buffer, each of

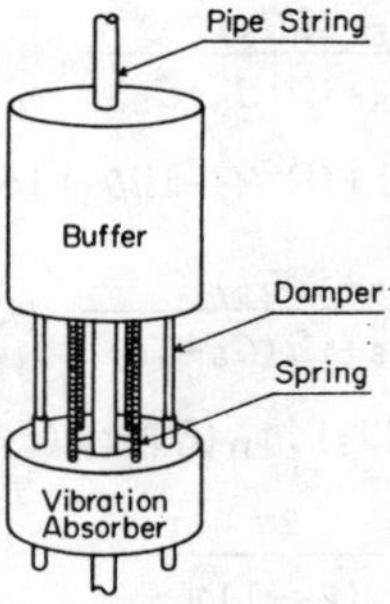

Fig.2 Schematic diagram of the vibration absorber

which has a vibration absorber, as shown in Fig.2, since this equation includes the case for those pump modules and buffers with no vibration absorbers. The absorber is composed of a mass ($\overline{m}_4, \overline{m}_5$ for pump modules or $\overline{m}_6$ for the buffer), springs with the resultant spring constant k_4, k_5 or k_6 (k_4, k_5 for pump modules or k_6 for the buffer) and dampers with the resultant viscous-damping coefficient g_4, g_5 or g_6 (g_4, g_5 for pump modules or g_6 for the buffer).

Furthermore, it has been assumed that both the pump modules and buffer attached to the pipe string can be considered as concentrated masses, the fluid inside the pipe string is uniform in density and vibrates together with the string, and that the flexible pipe does not affect the vibration of the pipe string because of its much smaller mass and rigidity compared to the pipe string.

Here, $\widetilde{m}_i$ and c_i (i=1~6) are defined as the added masses and the equivalent linear damping coefficients, respectively; suffixes 1,2 stand for the quantities pertaining to the pump modules 1,2, respectively; suffixes 4,5 for those pertaining to their absorbers, respectively; suffix 3 for the buffer; and suffix 6 pertaining to its absorber.

Thus, if the sum of $\overline{m}_i$ and $\widetilde{m}_i$ is defined as m_i, the equations of motion for the longitudinal vibration of the pipe string and for the absorber masses m_i (i=4~6) in this problem can be represented as follows:

$$EA\frac{\partial^2 u}{\partial x^2} + g_4(\dot{x}_4 - \dot{u})\delta(x - l_1) + g_5(\dot{x}_5 - \dot{u})\delta(x - l_2) + g_6(\dot{x}_6 - \dot{u})\delta(x - l_3) + k_4(x_4 - u)\delta(x - l_1) + k_5(x_5 - u)\delta(x - l_2) + k_6(x_6 - u)\delta(x - l_3)$$

$$= \{m_0 + m_1\delta(x - l_1) + m_2\delta(x - l_2) + m_3\delta(x - l_3)\}\ddot{u} + \{c_0 + c_1\delta(x - l_1) + c_2\delta(x - l_2) + c_3\delta(x - l_3)\}\dot{u} \qquad (3)$$

$$m_i\ddot{x}_i + g_i\{\dot{x}_i - \dot{u}]_{x=l_{i-3}}\} + c_i\dot{x}_i + k_i\{x_i - u]_{x=l_{i-3}}\} = 0, \quad (i=4\sim6) \qquad (4)$$

where m_0, c_0, E and A are the mass per unit length of the pipe string, the coefficient of viscous damping of the string, Young's modulus of the pipe material and the cross-sectional area of the pipe string, respectively. In addition, the dot attached on the variables represents the differentiation with respect to time, $\delta(\cdot)$ is the Dirac delta function, u is the displacement of the pipe string in the x-direction, and t is the time.

Thus, the equation for the longitudinal vibration of the pipe string equipped with two pump modules and a buffer without any vibration absorbers can be obtained as the limiting equation when m_i, g_i and k_i (i=4~6) go to zero.

The boundary conditions for this problem are expressed in the following equations:

$$u(0, t) = u_0(t), \quad EA\frac{\partial u(L, t)}{\partial x} = 0 \qquad (5)$$

where $u_0(t)$ is the forced displacement applied at the top of the pipe string, which is equal to $a\sin\omega t$, as mentioned previously.

Now, the solution to Eq.3 can be assumed to take the following form:

$$u(x, t) = u_V(x, t) + u_0(t). \qquad (6)$$

Then, substituting Eq.6 into Eqs.3–5 and defining the following nondimensional quantities,

$$\xi = \frac{x}{L},\ \xi_1 = \frac{l_1}{L},\ \xi_2 = \frac{l_2}{L},\ \xi_3 = \frac{l_3}{L},\ X_4 = \frac{x_4}{L},$$

$$X_5 = \frac{x_5}{L},\ X_6 = \frac{x_6}{L},\ U_V = \frac{u_V}{L},\ U_0 = \frac{u_0}{L},$$

$$\tau = \frac{t\sqrt{EA}}{L\sqrt{m_0}},\ \Omega = \omega L\sqrt{\frac{m_0}{EA}},\ \gamma = \frac{a}{L},\ M_1 = \frac{m_1}{m_0 L},$$

$$M_2 = \frac{m_2}{m_0 L},\ M_3 = \frac{m_3}{m_0 L},\ M_4 = \frac{m_4}{m_0 L},\ M_5 = \frac{m_5}{m_0 L},$$

$$M_6 = \frac{m_6}{m_0 L},\ K_4 = \frac{k_4 L}{EA},\ K_5 = \frac{k_5 L}{EA},\ K_6 = \frac{k_6 L}{EA},$$

$$G_4=\frac{g_4}{\sqrt{EAm_0}},\ G_5=\frac{g_5}{\sqrt{EAm_0}},\ G_6=\frac{g_6}{\sqrt{EAm_0}},$$

$$C_4=\frac{c_4}{\sqrt{EAm_0}},\ C_5=\frac{c_5}{\sqrt{EAm_0}},\ C_6=\frac{c_6}{\sqrt{EAm_0}},$$

$$D_0=\frac{c_0 L}{\sqrt{EAm_0}},\ D_1=\frac{c_1}{\sqrt{EAm_0}},\ D_2=\frac{c_2}{\sqrt{EAm_0}},$$

$$D_3=\frac{c_3}{\sqrt{EAm_0}}, \tag{7}$$

the nondimensional equations of motion for U_V, X_i ($i=4\sim6$) and the nondimensional boundary consitions are obtained. Furthermore the approximate solutions for the equations can be assumed to take the following form if $U_0(\tau)$ is represented as $\gamma\cdot\exp(i\Omega\tau)$.

$$U_0(\tau)=\gamma e^{i\Omega\tau},$$

$$U_V(\xi,\tau)=\sum_{n=1}^{N} b_n\phi_n(\xi)e^{i\Omega\tau},$$

$$X_i(\tau)=S_i e^{i\Omega\tau}, \tag{8}$$

where b_n, S_i ($i=4\sim6$) are complex constants to be determined, and $\phi_n(\xi)$ could be assumed as the following function which satisfies the nondimensional boundary conditions:

$$\phi_n(\xi)=\sin\{(2n-1)\pi\xi/2\},\quad (n=1,2,\cdots,N) \tag{9}$$

Substituting the approximate solutions of Eq.8 into the nondimensional equations of motion, and applying Galerkin's method to the resulting equations, the following relations between S_i and b_n as well as the simultaneous equations with respect to the complex constants b_n are obtained:

$$S_i=\frac{[K_i+i\Omega G_i][\sum_{n=1}^{N} b_n\phi_n(\xi_{i-3})+\gamma]}{K_i-\Omega^2 M_i+i\Omega(G_i+C_i)},\quad (i=4\sim6), \tag{10}$$

$$\begin{aligned}
&[\Omega^2-\frac{(2k-1)^2\pi^2}{4}-i\Omega D_0]\frac{b_k}{2}\\
&+\sum_{n=1}^{N} b_n[\{\frac{(K_4+i\Omega G_4)^2}{K_4-\Omega^2 M_4+i\Omega(G_4+C_4)}\\
&-(K_4+i\Omega G_4)+(\Omega^2 M_1-i\Omega D_1)\}\phi_n(\xi_1)\phi_k(\xi_1)\\
&+\{\frac{(K_5+i\Omega G_5)^2}{K_5-\Omega^2 M_5+i\Omega(G_5+C_5)}\\
&-(K_5+i\Omega G_5)+(\Omega^2 M_2-i\Omega D_2)\}\phi_n(\xi_2)\phi_k(\xi_2)]\\
&+\{\frac{(K_6+i\Omega G_6)^2}{K_6-\Omega^2 M_6+i\Omega(G_6+C_6)}-(K_6+i\Omega G_6)\\
&+(\Omega^2 M_3-i\Omega D_3)\}\phi_n(\xi_3)\phi_k(\xi_3)]\\
&=(-\Omega^2+i\Omega D_0)\frac{2\gamma}{(2k-1)\pi}\\
&-[\frac{(K_4+i\Omega G_4)^2}{K_4-\Omega^2 M_4+i\Omega(G_4+C_4)}-(K_4+i\Omega G_4)\\
&+(\Omega^2 M_1-i\Omega D_1)]\gamma\phi_k(\xi_1)\\
&-[\frac{(K_5+i\Omega G_5)^2}{K_5-\Omega^2 M_5+i\Omega(G_5+C_5)}-(K_5+i\Omega G_5)\\
&+(\Omega^2 M_2-i\Omega D_2)]\gamma\phi_k(\xi_2)\\
&-[\frac{(K_6+i\Omega G_6)^2}{K_6-\Omega^2 M_6+i\Omega(G_6+C_6)}-(K_6+i\Omega G_6)\\
&+(\Omega^2 M_3-i\Omega D_3)]\gamma\phi_k(\xi_3),
\end{aligned}$$

$$(k=1,2,\cdots,N) \tag{11}$$

Solving equation 11, b_n ($n=1\sim N$) are determined and, after substituting these constants into Eqs.10 and 8, $U_V(\xi,\tau)$ and $X_i(\tau)$ ($i=4\sim6$) are obtained. Moreover, if these values are substituted into Eqs.6 and 7, and if the imaginary parts of the resulting quantities are taken, the displacement of the pipe string, $u(x,t)$, the amplitudes of the vibration absorbers attached to the pump modules, $x_4(t)$ and $x_5(t)$, and the amplitude of the absorber attached to the buffer, $x_6(t)$, can be obtained.

Furthermore, the axial stress, $\sigma(x,t)$, produced in the pipe string can be obtained by the following equation:

$$\sigma(x,t)=E\frac{\partial u}{\partial x}+\frac{W_b+w_0(L-x)}{A} \tag{12}$$

where W_b stands for the sum of the submerged weights of the pump modules, buffer and vibration absorbers equipped below the position considered along the pipe string, and w_0 represents the submerged weight per unit length of the pipe string. It must be noted that the above-mentioned calculations of the displacement and stress of the pipe string as well as the amplitudes of the absorbers need the values of the added mass $\widetilde{m}_i$ and the equivalent damping coefficient c_i, each of which is a function of the amplitude of the pump module, buffer or absorber. The amplitude, however, is not known in advance. Therefore, an iterative method started with the assumption of an initial amplitude must be used in these calculations.

3 CONTROL BY THE SHAPE OF BUFFER

At the beginning of this analysis, the shapes of the pump module and buffer were selected from the models shown in Fig.3. Then the added-mass and drag coefficients for the above-mentioned shapes of pump module and buffer were found by the authors' method (Aso et al. 1991b) to determine the fluid forces acting on them by Eq.2. The coefficients obtained are

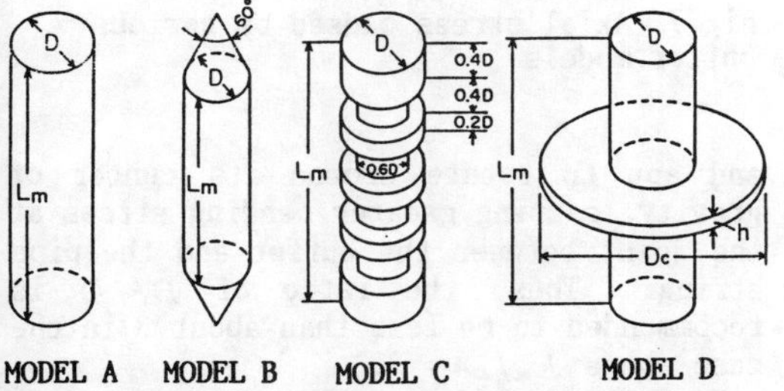

Fig.3 Shapes of pump module and buffer

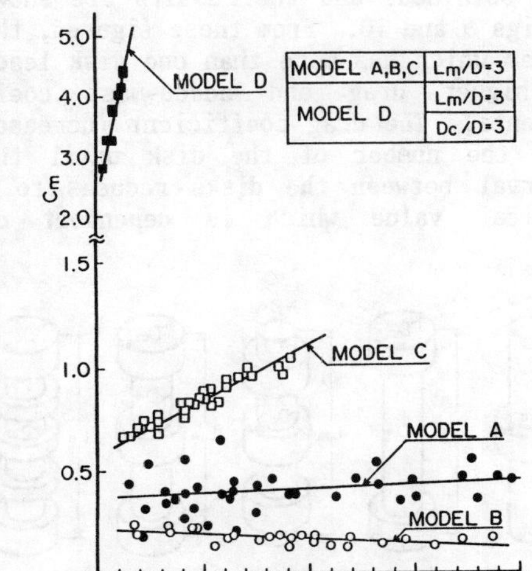

Fig.4 The relationship between C_m and K_C

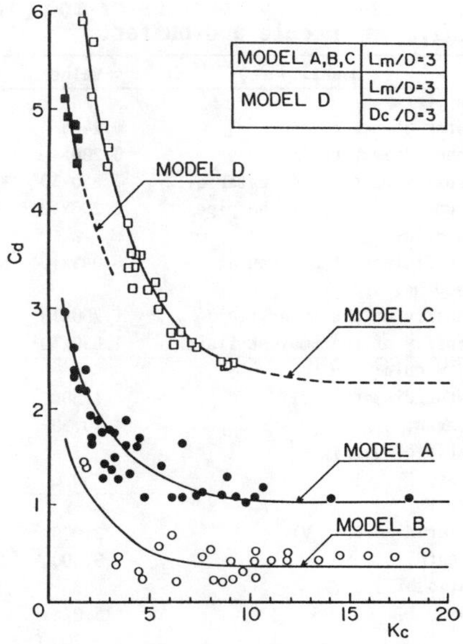

Fig.5 The relationship between C_d and K_C

shown in Figs.4 and 5, where the ordinate represents the added-mass coefficient C_m or the drag coefficient C_d and the abscissa represents the Keulegan-Carpenter number K_C (Keulegan et al. 1958). This number is expressed as $\dot{U}_m T/D$ (D_C instead of D for the case of MODEL D), if $\dot{U}_m (=A\omega)$ is denoted as the maximum velocity of the vibrating body and T as its period.

Now, based upon the aforementioned analytical method, the longitudinal displacement, $u(x,t)$, and the axial stress, $\sigma(x,t)$, produced in the pipe string were analyzed for the various shapes of pump modules and buffer without any vibration absorbers. In these cases, the pipe string, pump module and buffer were assumed to have the physical properties shown in Table 1. Various layouts of the pump modules and buffer along the pipe string (Table 2) were considered. Furthermore, the heaving amplitude of the ship, a, has been fixed as 1m throughout this study.

The results obtained are shown in Figs.6 and 7. Fig.6 indicates the shape-effect of buffer on the amplitude of vibration of pipe string, and Fig.7 shows the same effect on the axial stress produced at the top of the pipe string when only the shape of buffer is changed and the shapes of pump modules are fixed as MODEL A. From

Table 1. Physical properties of the pipe string, pump module and buffer.

Quantity	Value	Unit
Pipe string:		
Outer diameter, d_o	0.2445	m
Inner diameter, d_i	0.2005	m
Density of the pipe material, ρ_p	7.850x10³	Kg/m³
Young's modulus of the pipe material, E	0.206x10¹²	N/m²
Coefficient of structural damping, c_o	0.049x10³	Ns/m²
Density of the inner fluid, ρ_f	1.200x10³	Kg/m³
Density of the ambient fluid, ρ	1.030x10³	Kg/m³
Pump module:		
Outer diameter, D	1.500	m
Length, L_m	10.50	m
Ratio of L_m/D	7	
Mass, $\overline{m}_1$	50.0	t
Buffer:		
Outer diameter, D	3.000	m
Length, L_m	9.000	m
Ratio of L_m/D	3	
Mass, $\overline{m}_2$	75.0 ~300.0	t

Table 2. Layout of the pump modules and buffer.

	$\bar{l}_1$	$\bar{l}_2$	$\bar{l}_3$	l_1	l_2	l_3
Case 1	1000 m	1000 m	3000 m	1000 m	2000 m	5000 m
Case 2	1000 m	2000 m	2000 m	1000 m	3000 m	5000 m
Case 3	3000 m	1000 m	1000 m	3000 m	4000 m	5000 m
Total length of the pipe-string, L :					5000 m	

Fig.6, it can be seen that the buffer whose shape causes a higher drag force makes the resonance amplitude smaller, and that the amplitude produced by the highest-drag buffer model (MODEL D) among the models used in this study is about one third of that produced by the lowest-drag model (MODEL B). The maximum axial stress produced in the pipe string (Fig.7) is also smaller in case of the buffer causing a higher drag force, and the stress produced by the highest-drag model (MODEL D) is about 67 % of that produced by the lowest-drag model (MODEL B).

Hence, in practical cases, it is recommended to use a buffer whose shape causes a higher drag force, considering the smaller contribution of the added mass to the increase of the longitudinal vibration of the pipe string. The buffer in MODEL D is very effective for reducing the longitudinal vibration of the pipe string, and the buffer with a disk of larger diameter further reduces the vibration. However, the author's experiment suggested that the buffer with larger disk was more unstable

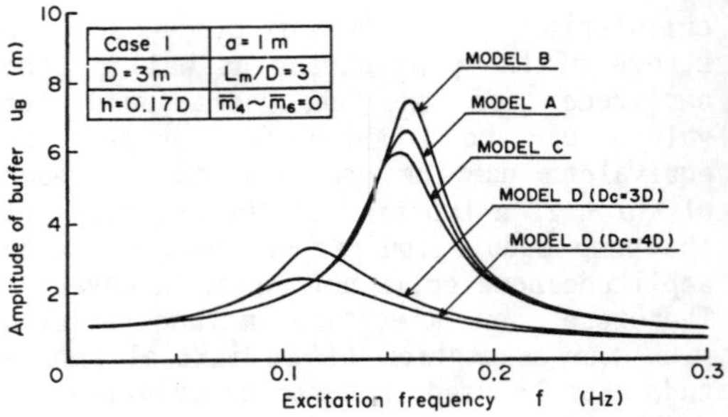

Fig.6 Amplitudes of various buffer models

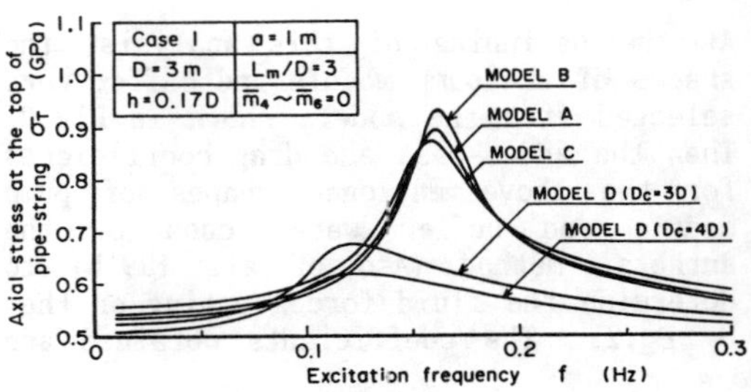

Fig.7 Axial stress caused by various buffer models

and apt to rotate around its center of gravity, causing greater bending stress at the joint between the buffer and the pipe string. Thus, the ratio of D_C/D is recommended to be less than about 3 in the case where $L_m/D = 3$.

Next, the added-mass and drag coefficients for the models shown in Fig.8 were obtained, and the results are shown in Figs 9 and 10. From these figures, the buffer which has more than one disk leads to higher drag and added-mass coefficients. The drag coefficient increases with the number of the disk until the interval between the disks reduces to a critical value which is dependent on

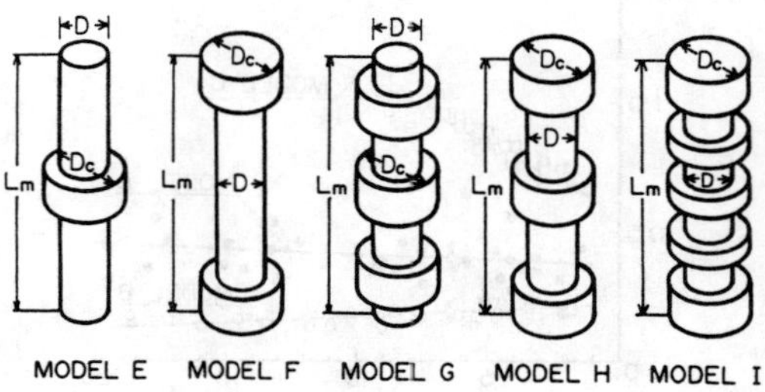

Fig.8 Various buffer models with disks

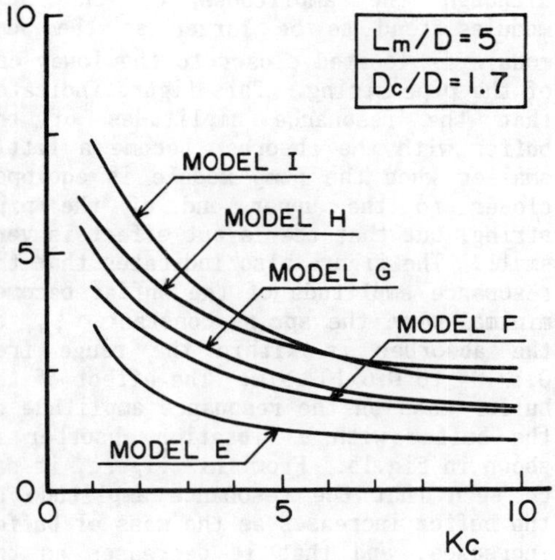

Fig.9 The relationship between C_d and K_C for buffer models with disks

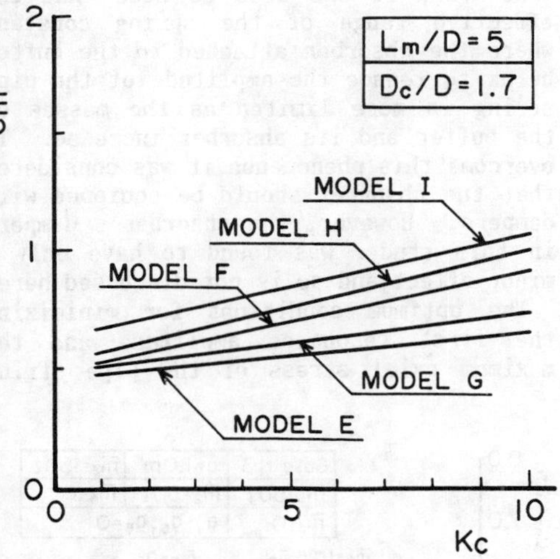

Fig.10 The relationship between C_m and K_C for buffer models with disks

D_C/D, L_m/D and K_C. Hence, it can be shown that, if these quantities are given, an optimum number of the disks attached to the buffer exists to minimize the longitudinal vibration of the pipe string.

The shape-effect of the pump module on the longitudinal vibration of the pipe string was also examined, and the effect was found to be rather small, compared with that of the buffer, because of the smaller mass and diameter of the pump module.

Table 3. Physical properties of the vibration absorbers.

Quantity	Value	Unit
Vibration Absorber of Pump Module:		
Shape; Hollow cylinder		
Outer diameter	1.500	m
Inner diameter	0.500	m
Mass	$\overline{m}_1/5 \sim \overline{m}_1/10$	t
Added-mass coefficient	1.00	
Drag coefficient	1.16	
Vibration absorber of Buffer:		
Shape; Hollow cylinder		
Outer diameter	3.000	m
Inner diameter	0.750	m
Mass	$\overline{m}_3/5 \sim \overline{m}_3/20$	t
Added-mass coefficient	1.00	
Drag coefficient	1.16	

4 CONTROL BY THE VIBRATION ABSORBER

A vibration absorber, shown in Fig.2, which had the physical properties given in Table 3, was attached to the buffer and pump module of the pipe string for the aforementioned practical cases. Then, the longitudinal displacement, $u(x,t)$, and the axial stress, $\sigma(x,t)$, produced in the pipe string as well as the amplitudes of the vibration absorbers were analyzed according to the analytical method mentioned previously. Here, the added mass and drag coefficients of the absorbers were fixed since only coefficients pertaining to hollow cylinders in steady flow had been reported. On the other hand coefficients for the cylindrical pump modules and buffer, reported by Aso et al. (1991b), were used in the study.

The results obtained are shown in Figs. 11-13. Fig.11 shows the frequency response of the pump modules and buffer without any vibration absorbers in case 2 of the aforementioned layout and Fig.12 shows the corresponding amplitudes of the pump modules and buffer in the case of an absorber attached only to the buffer. This absorber has a mass of 60 t, which is one fifth of $\overline{m}_3$, a spring constant of 0.70×10^5 N/m and has no dampers. From these figures, it can be seen that both the resonance amplitudes of the pump modules and buffer are reduced by almost 50% by the absorber and that the amplitude of the buffer is much greater than the amplitudes of the pump modules. On the other hand, Fig.13 shows the peak axial stress produced at the top of the pipe string in this case and indicates that

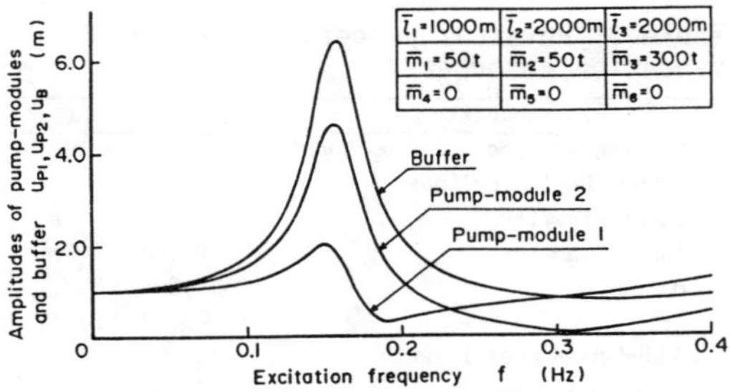

Fig.11 Frequency-response curve for pump modules and buffer without any vibration absorbers

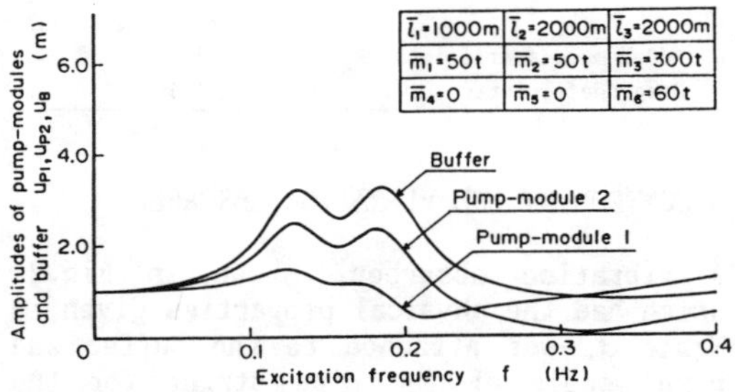

Fig.12 Frequency-response curve for pump modules and buffer with a vibration absorber

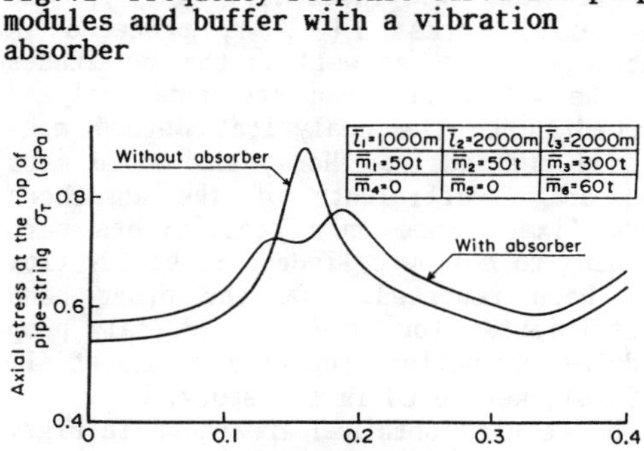

Fig.13 Frequency vs. axial stress at the top of pipe string in case of the buffer with or without a vibration absorber

this stress could be reduced by about 13% by the absorber.

Next, the effect of the layout of the pump modules and the buffer with a vibration absorber on the resonance amplitudes was examined. Fig.14 shows the resonance amplitudes, u_{BR}, of the 150 t buffer with a vibration absorber for the two cases of the layouts 1 and 3. In this figure, only the resonance amplitudes of buffer are shown because these amplitudes are always greater than those of the pump modules, although the amplitudes of the pump modules tend to be larger as the pump module is located closer to the lower end of the pipe string. This figure indicates that the resonance amplitudes of the buffer with the absorber become a little smaller when the pump module is equipped closer to the upper end of the pipe string, but that the layout effect is very small. The figure also indicates that the resonance amplitude of the buffer becomes minimal when the spring constant, k_6, of the absorber is within the range from 0.3×10^5 to 0.6×10^5 N/m. The effect of the buffer mass on the resonance amplitude of the buffer with a vibration absorber is shown in Fig.15. From this figure, it can be seen that the resonance amplitude of the buffer increases as the mass of buffer increases, and that it decreases as the mass of the absorber increases within the mass range considered. Moreover, the rate of decrease of the amplitude is greater in case of the buffer with a larger mass.

However, it can also be seen that the effective range of the spring constant where the absorber attached to the buffer helps to reduce the amplitude of the pipe string is more limited as the masses of the buffer and its absorber increase. To overcome this phenomenon,it was considered that the absorber should be equipped with dampers. However, the absorber's damper, in this study, was found to have only a minor effect and so is not discussed here.

The optimum conditions for minimizing the first resonance amplitude and the maximum axial stress of the pipe string

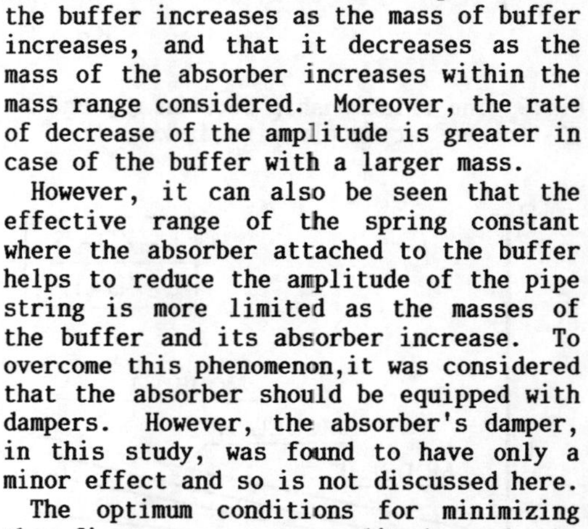

Fig.14 The layout-effect on the resonance amplitude of buffer (Cases 1 and 3)

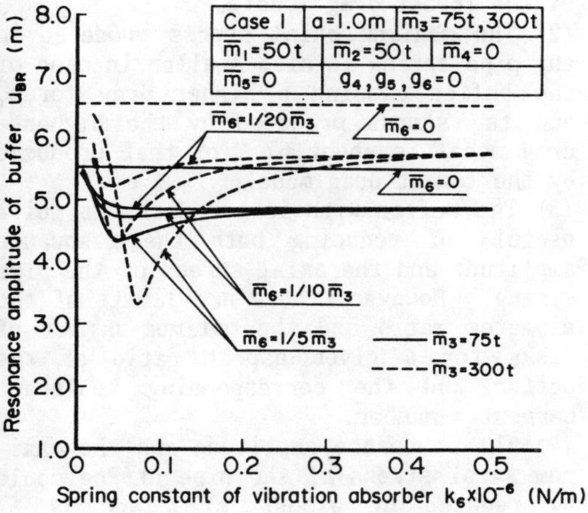

Fig.15 The mass-effect of buffer on its resonance amplitude ($\overline{m}_3$=75t,300t)

were then determined based on the mass and spring constant of the vibration absorber attached only to the buffer. The condition for the amplitude in case 1 of the layout is shown in Fig.16. In this figure, the broken line indicates the condition on the spring constant of the absorber with a given mass in order to minimize the first resonance amplitude of the buffer with the prescribed mass. The solid line indicates the condition for the smallest amplitude corresponding to that broken line, representing the optimum condition for the combination of the mass and the spring constant of the absorber. Under the solid line, there is the reduction ratio which is the percentage of the first-resonance amplitude of buffer with the vibration absorber under the optimum condition to that of the buffer with no vibration absorbers. Furthermore, the dot-dash line indicates the condition for which the first-resonance amplitude of buffer becomes 105% of that in the optimum condition. Fig.17 shows the corresponding lines for the maximum axial stress produced at the top of the pipe string to those for the resonance amplitude of buffer in Fig.16, if the reduction ratio for the stress is defined in the same way as that for the amplitude. From these figures, it can be seen that a greater mass of buffer, under optimum conditions, leads to a smaller reduction ratio in the amplitude of the buffer but a larger reduction ratio of the axial stress produced in the pipe string. However, these reduction ratios for the amplitude and stress do not vary remarkably in the practical mass-range of the buffer. When a larger mass of buffer is used a combination of larger mass and smaller spring constant of the vibration absorber is needed to obtain optimum values for the amplitude and axial stress. Here, it must be noted that the optimum condition for the amplitude does not exactly coincide with that for the stress, and that the former condition needs a larger mass for the absorber than the latter in the range where the buffer's mass is less than 300t, although the spring constant remains almost the same. In addition, it can be seen that the upper range, between the upper dot-dash line and the optimum line,

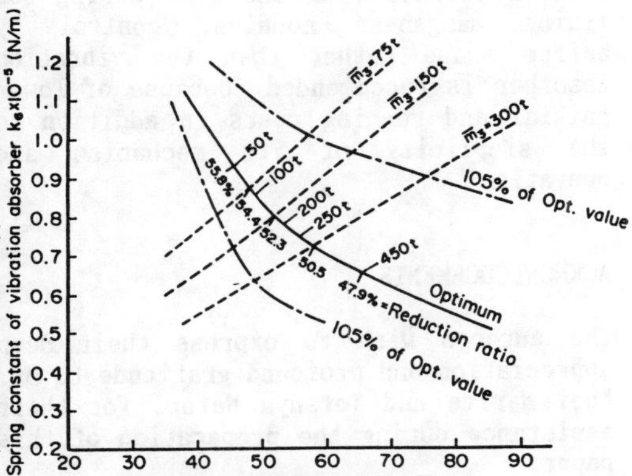

Fig.16 The optimum condition for the amplitude of buffer (Case 1)

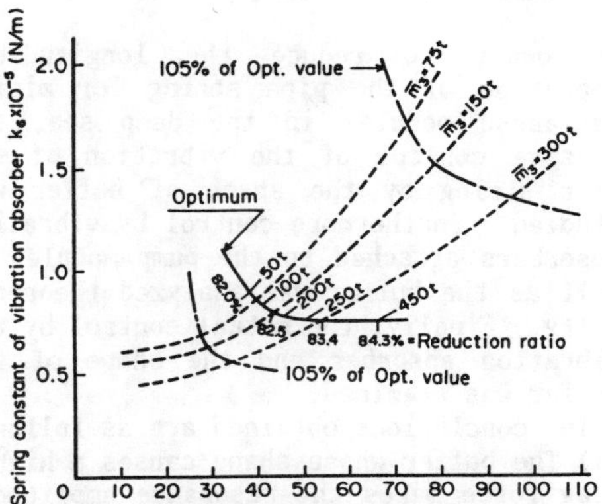

Fig.17 The optimum condition for the maximum axial stress (Case 1)

is wider than the lower one, between the optimum line and the lower dot-dash line, and that changes to the absorber's elements in this upper range are less sensitive to increases of the amplitude and axial stress of the pipe string. Hence, the combination of the mass and spring constant of the absorber in the upper range is safer for use when optimum conditions can not be realized.

Finally, the effect of the vibration absorber attached to the pump module was studied, and the result indicated that the resonance amplitudes of the pump modules and buffer were reduced very little by the absorber in comparison with those obtained previously. Therefore, it can be said that the vibration absorber attached to the pump module causes a much smaller effect on the vibration of the pipe string than that attached to the buffer.

5 THE MUTUAL EFFECT OF THE SHAPE OF BUFFER AND ITS VIBRATION ABSORBER

The mutual effect of both the shape of the buffer and its vibration absorber on the longitudinal vibration of the pipe string was analyzed and proved to be very small. In the previous case, shown in Figs.6 and 7 for example, the buffer (MODEL D) with the disk (D_C/D= 3) and with the vibration absorber, decreased the resonance amplitude of buffer at most by about 0.5 m as compared with that of the same buffer without any vibration absorbers.

6 CONCLUSIONS

In order to reduce the longitudinal vibration of the pipe string for mining manganese nodules in the deep sea, the passive control of the vibration of the pipe string by the shape of buffer was studied. Furthermore control by vibration absorbers attached to the pump modules as well as the buffer was analyzed theoretically. Finally, the mutual control by the vibration absorber and the shape of the buffer was examined.

The conclusions obtained are as follows:

(1) The buffer whose shape causes a higher drag force makes the resonance amplitudes of the buffer and pump modules smaller, and the first-resonance amplitude produced by the highest-drag model used in this study is about one third of that produced by the lowest-drag model.

(2) The maximum axial stress produced in the pipe string is also smaller in case of the buffer causing a higher drag force, and the stress produced by the highest-drag model is about 67 % of that produced by the lowest-drag model.

(3) The buffer with larger disks is quite useful for reducing both the resonance amplitude and the axial stress of the pipe string. However, there are limits of the diameter ratio and the optimum number of disks for a given aspect ratio of the buffer and the corresponding Keulegan-Carpenter number.

(4) The resonance amplitude and the maximum axial stress of the pipe string could be reduced by almost 50 % and 17 %, respectively by the absorber attached to the buffer.

(5) The layout effect of the pump modules and buffer on the longitudinal vibration of the pipe string is quite small as far as the effect of the absorber is concerned.

(6) There is an optimum combination of mass and spring constant for the absorber for minimizing the resonance amplitude of the buffer and the maximum axial stress of the pipe string. However, the optimum combination for the amplitude does not exactly coincide with that for the stress.

(7) The effect of the vibration absorber attached to the buffer is much greater than that of the absorber attached to the pump module.

(8) The mutual effect by the shape of buffer and its vibration absorber is very small.

(9) For the passive control of the longitudinal vibration of the pipe string for mining manganese nodules, control by buffer shape rather than the vibration absorber is recommended because of lower initial and running costs in addition to the simplicity of its mechanism and operation.

ACKNOWLEDGEMENTS

The authors wish to express their deep appreciation and profound gratitude to Mr. Koji Narita and Tetsuya Natori for their assistance during the preparation of this paper.

REFERENCES

Aso,K. & Kan,K. 1987. Longitudinal vibration of a pipe string with a buffer at its lower end: J. Min. Metal. Inst. Japan, Vol.103, No.1193: 443-448.

Aso,K., Kan,K. & Yuan,S. 1988. Dynamic behavior of a stepped pipe string with a buffer at its lower end: J. Min. Metal. Inst. Japan, Vol.104, No.1205: 443-448.

Aso,K., Kan,K., Doki H. & Mori,M. 1991a. The shape-effect of buffer on the longitudinal vibration of a pipe-string in the deep sea: Int. J. Offshore & Polar Eng., Vol.1, No.2: 154-160.

Aso,K., Kan,K., Doki H. & Mori,M. 1991b. The drag and added-mass coefficients of various buffers vibrating axially in water: Trans. ASME, J. Energy Resour. Tech., Vol.113: 80-86

Aso,K., Kan,K., Doki,H. & Iwato,K. 1992a. The effects of vibration absorbers on the longitudinal vibration of a pipe string in the deep sea: Int. J. Offshore & Polar Eng., Vol.2, No.4: 214-249.

Aso,K.,Kan,K.,Doki,H. & Ohkoshi,T. 1992b. The effects of vibration absorbers on the longitudinal vibration of a pipe string for mining manganese nodules in the deep sea. Proc. 2nd Int.Offshore & Polar Eng. Conf., San Francisco, Vol.I: 512-519.

Chung,J.S. & Felippa,C.A. 1981a. Nonlinear static analysis of deep ocean mining pipe --Part II: Trans. ASME, J. Energy Resour. Tech., Vol.103: 16-25.

Chung,J.S.,Whitney,A.K. & Loden,W.A.1981b. Nonlinear transient motion of deep ocean mining pipe: Trans. ASME, J. Energy Resou. Tech., Vol.103: 2-10.

Chung,J.S. & Whitney,A.K. 1983. Axial stretching oscillation of an 18,000-ft vertical pipe in the ocean: Trans. ASME, J. Energy Resour.Tech.,Vol.105: 195-200.

Felippa,C.A. & Chung,J.S. 1981. Nonlinear static analysis of deep ocean mining pipe --Part I: Trans. ASME, J. Energy Resour. Tech., Vol.103: 11-15.

Morison,J.R., O'brien,M.P., Johnson,J.W. & Schaaf,S.A. 1950. The force exerted by surface waves on piles: Petro. Trans. AIME, Vol.189: 149-154.

Sparks,C.P., Cabilic,J.P. & Schawann,J.C. 1982. Longitudinal resonant behavior of very deep water risers. Proc. 14th Annual OTC, Houston, Texas : 201-211.

Tikhonov,V.S., Fisher,A.S. & Zubarev,V.K. 1984. An efficient procedure of static analysis of long ocean mining pipe: Trans. ASME, J. Energy Resour.Tech., Vol.106: 329-335.

Timoshenko,S.P. 1955. Vibration problems in engineering, 3rd ed., p.92-97. Van Nostrand.

Whitney,A.K., Chung,J.S. & Yu,B.K. 1981. Vibrations of long marine pipes due to vortex shedding: Trans. ASME, J. Energy Resour. Tech., Vol.103: 231-236.

Yuan,S., Aso,K. & Tani,J. 1988. Static behavior of a pipe string designed for mining manganese nodules: Trans. JSME, Ser.A, Vol.54, No.507: 1943-1948.

Yuan,S.,Aso,K. & Tani,J. 1989a. A solution of free lateral vibrations of a pipe string designed for mining manganese nodules: Trans. JSME, Ser.C, Vol.55, No. 517: 2313-2320.

Yuan,S., Aso,K. & Tani,J. 1989b. Dynamic instability and postflutter vibrations of a pipe string designed for mining manganese nodules: Trans. JSME, Ser.C, Vol.55, No.520: 2958-2966.

Yuan,S., Tani,J. & Aso,K. 1990a. Longitudinal vibration of a pipe string designed for mining manganese nodules in the deep ocean: Trans. JSME, Ser.C, Vol.56, No.530: 13-17.

Yuan,S., Tani,J. & Aso,K. 1990b. The responses of a long marine pipe string under parametric excitation: Trans.JSME, Ser.C, Vol.56, No.530: 18-24.

Innovative Mine Design for the 21st Century, Bawden & Archibald (eds) © 1993 Balkema, Rotterdam, ISBN 90 5410 325 6

Dual-spin rock cutting research

Bohuslav Šolc
Havířov, Czech Republic

ABSTRACT: Dual-spin rock cutting technology opens up the prospect of cutting very hard and abrasive rock by means of relatively light equipment and at an acceptable cost of tools. This effect is achieved by reducing the duration of contact between the rock and the bit through a simultaneous second spin of the cutting tool, in order to prevent the thermal degradation of the hard metal. The dual spin and a particular position of the axes of both simultaneous rotations result in multiplying the actual rock cutting force to the level that would be hard to attain with any of the existing rock cutting technologies. The paper reports of bench tests where the above features were observed and recorded.

INTRODUCTION

Despite the fact that the twentieth century has brought about a successful development of the mechanical rock cutting techniques, there remains a quite vast area of rock excavation, where this new technology could not yet establish itself. It is the mining and the medium to short distance tunnelling in abrasive rock with a strength in excess of 100 MPa, where the TBM technology is not applicable in the above activities for a number of reasons and where the roadheaders using fixed picks wear out the tools at an unacceptable rate.

Over the past decades there have been many attempts to use roadheaders for tunnelling in the above mentioned abrasive rock. However, most of them proved this method of mechanisation to be too costly, due to the excessive consumption of picks . There are still some cases where efforts are being made to use special slowly rotating cutting heads with picks for excavation in the hard rock at subcritical cutting speeds (i.e. speeds at which the bit temperature does not exceed the critical temperature described later). In these cases, the output in terms of volume of rock excavated is considerably reduced when compared with conventional techniques, which restricts the application of this technology.

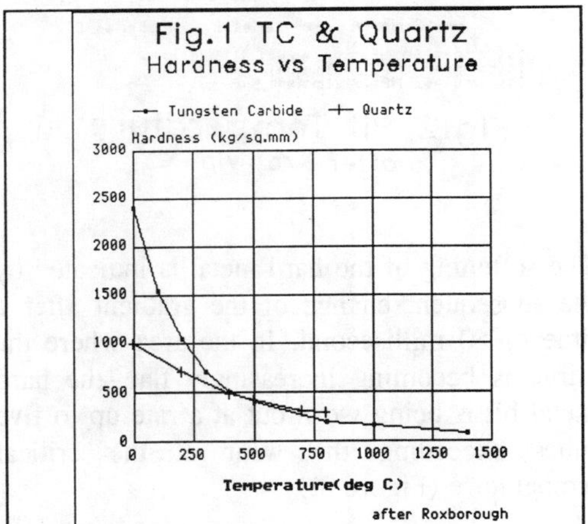

The principal reason for the failure of attempts to excavate the hard and abrasive rock by using fixed picks, lies in the heat generated by friction of the pick against the grains of rock. The temperature of the hard-metal bit tip rises in a fraction of second to the values, where its components become softer than the quartz grains, as described by Roxborough [1]quite a time ago (Figure 1).

The temperature curve of the TC tips during the rock cutting was measured by Krapivin et al.[2],[3], in the nineteen-seventies. Figure 2 shows the resulting curves for various rock strengths. Our interest is in the curve for the hardest rock with the unconfined compressive strength of about 160 MPa. The curve shows that the temperature exceeds the critical value of 500°C in about 30 milliseconds.

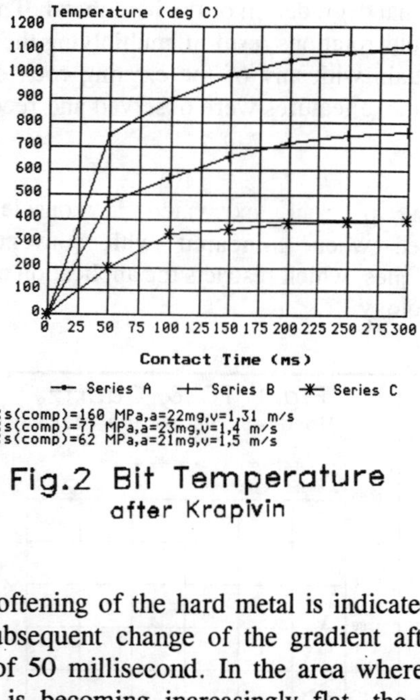

Fig.2 Bit Temperature
after Krapivin

The softening of the hard metal is indicated by the subsequent change of the gradient after a time of 50 millisecond. In the area where the curve is becoming increasingly flat, the hard metal bit is being worn out at a rate up to five times exceeding the wear at the critical temperature (Figure 3).

The impact of the frictional heat can be eliminated or reduced by regularly interrupting the contact between the pick and the rock at approximately the critical temperature of 500°C, and thus keeping the TC bit in a good condition for the repeated attacks on the hard rock.

The theoretical approach to solving the problem was outlined in an earlier paper[4] . The present paper being aimed at reporting on some practical R & D steps carried out in order to achieve the above goal. However, two basic notions are to

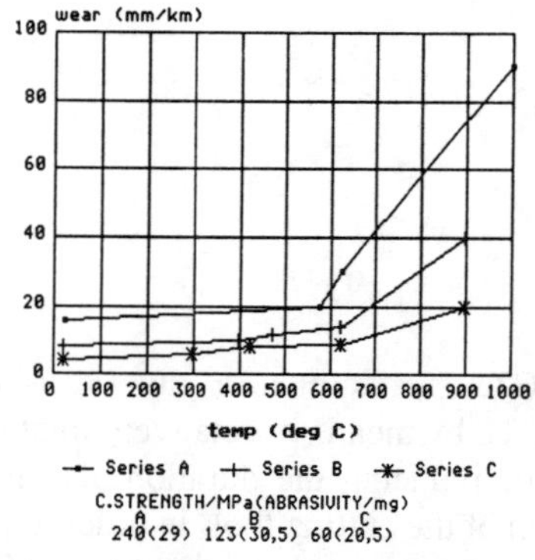

Fig.3 WEAR / TEMPERATURE
of TC bits in rock cutting
(after Krapivin)

be clarified here also in order to facilitate the understanding of the practical experiments. They are: dual spin and autorotation.

DUAL SPIN AND AUTOROTATION

The cutting heads of the present roadheaders, keep usually the picks in contact with the rock for several hundred milliseconds at cutting speeds reaching up to 4 mps, thus exceeding ten and more times the time limit for attaining the critical temperature. The picks are being moved on a circular path (R1 in Figure 4) around the axis of the cutting head. If they could, at the same time, be rotated in a plane normal to that of R1, their contact with the rock would be interrupted and reinstated in regular intervals, much shorter than the conventional contact interval of a roadheader pick.

There are various ways to introduce a second spin to the pick, but for practical reasons the autorotation method was considered to be the most suitable.

Figure 5 shows the principle of autorotation applied throughout the experiments reported in this paper.

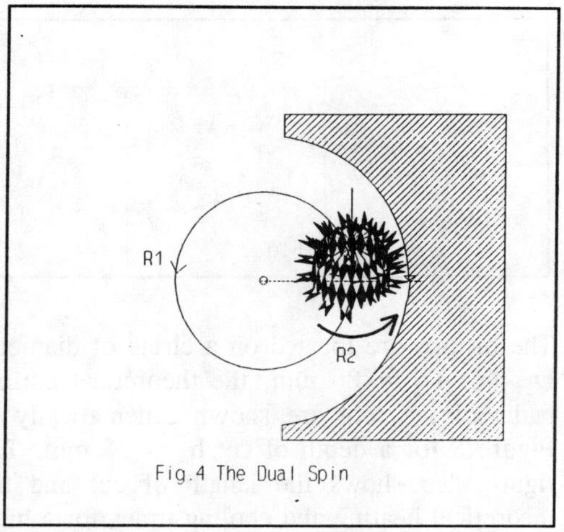

Fig.4 The Dual Spin

The left portion of Figure 5 shows a disk capable of rotating about its axis and fitted with a rigidly attached pick. The pick is in contact with exerts a resistance R to its movement in any direction and a horizontal force F is applied to the centre of the disk at an acute angle.

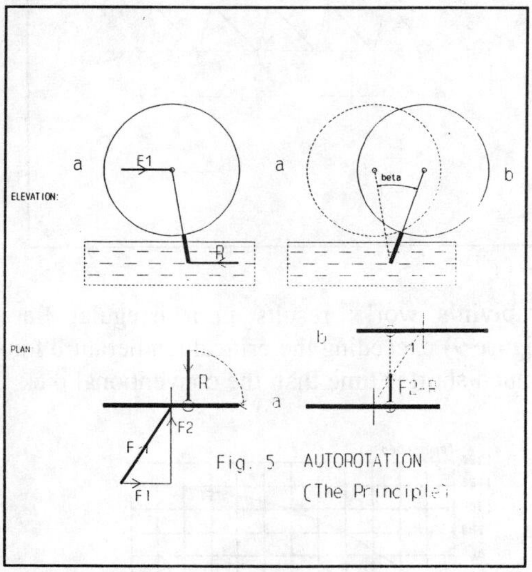

Fig. 5 AUTOROTATION (The Principle)

The application of the force F tends to move the disk in the direction shown. However, due to resistance R exerted by the contact with the surface, the pick will only partly follow the movement of the disk. It will move in the vector direction of the resulting difference of forces F2 - R, because of the rigidity of the pick fixed on the disk. The other component of the force F (F1) works in couple with the resistance R and turns the disk around its axis by angle beta while it moves from position "a" to position"b" (right portion of Figure 5).

This induced rotation can continue as long as the force F is active, provided another pick is in contact with the surface before the one shown in Figure 5 is withdrawn.

Figure 6 shows this arrangement with a disk having multiple picks rotating around its centre C2 due to a force F. The disk is travelling along a circular path around a centre point C1. Assuming that the material being cut has the rheological characteristics of a soft rock, the picks would leave visible traces on its surface as shown in Figure 6 (right bottom portion). This serial pattern of cuts has a width corresponding to the depth of the cut and is a characteristic of the Dual Spin rock cutting, as we will be seen on later slides.

THE DUAL SPIN TOOL

The principle of autorotation described was tested on an industrial test bench. A special 6-bit tool was developed, the stem of which fits into a conventional tool holder of a point-attack pick. This tool holder was fixed on a positioning table rotated on a diameter identical to that of a

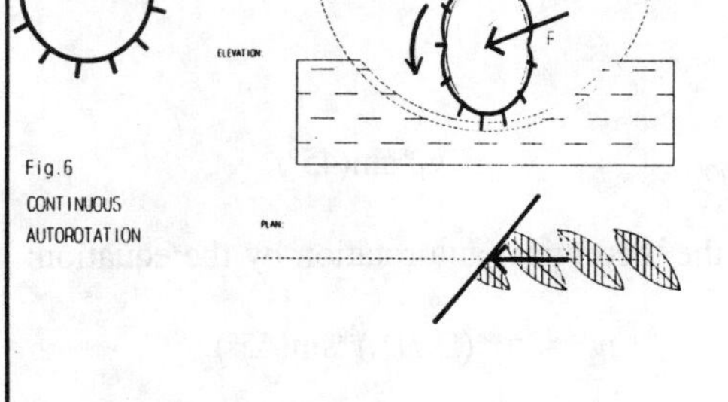

Fig.6 CONTINUOUS AUTOROTATION

cutting head. A block of sandstone of 130 MPa compressive strength was mechanically moved to interfere with the path of the pick in a position, where it was cut at a desired depth. During these tests, the DS tool was followed visually and the paths of the bits on the rock surface were photographed.

The observations confirmed that:

(a) the autorotation of the pick takes place in the manner as theoretically predicted.

(b) during the autorotation, the hard metal temperature is lower than that of the conventional bits (with the rotating experimental tool there was little generation of sparks although sparks occurred constantly with the conventional picks in the same rock)

During the tests it was evident that the shape of the upper part of the body of the experimental tool would not allow deeper penetration of the picks into the rock. It is planned to remedy this with the use of a re-designed shape of the DS tool as proposed in Figure 7.

A DS tool fixed on a conventional cutting head of a diameter D_1 = 1300mm and revolving at 1 rev/s has a tangential velocity v_1 = 4.084 m/s. If a DS tool is fixed into a holder, deviating 45° from the vector direction of the speed of primary rotation, the speed component, producing the autorotation of the DS tool is given by the equation

$$v_2 = v_1 * \sin(45^o)$$

and the theoretical autorotation by the equation:

$$n_2 = n_1 * (D_1/D_2) * \sin(45^o).$$

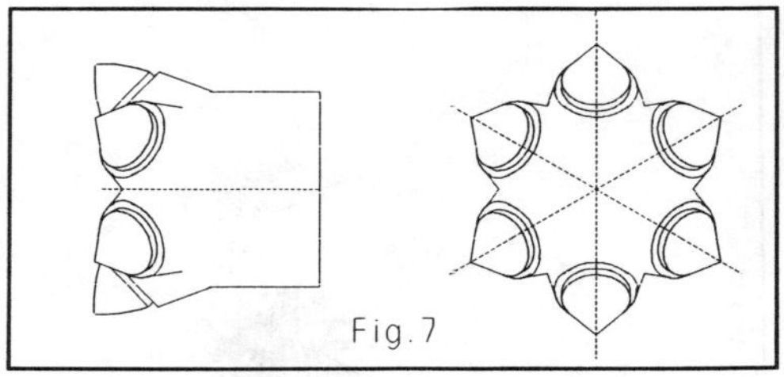

Fig.7

The bit tips are located on a circle of diameter D_2. For D_2 = 86 mm, the theoretical cutting paths of the tool are shown schematically in Figure 8 for a depth of cut h_c = 25 mm. The figure also shows the length of cut and the theoretical heating and cooling times for a bit.

The projection of the cutting and cooling time onto the heating and cooling curves from

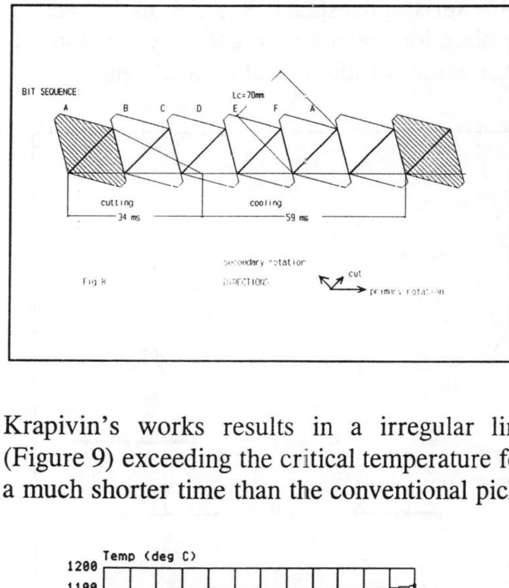

Fig.8

Krapivin's works results in a irregular line (Figure 9) exceeding the critical temperature for a much shorter time than the conventional pick,

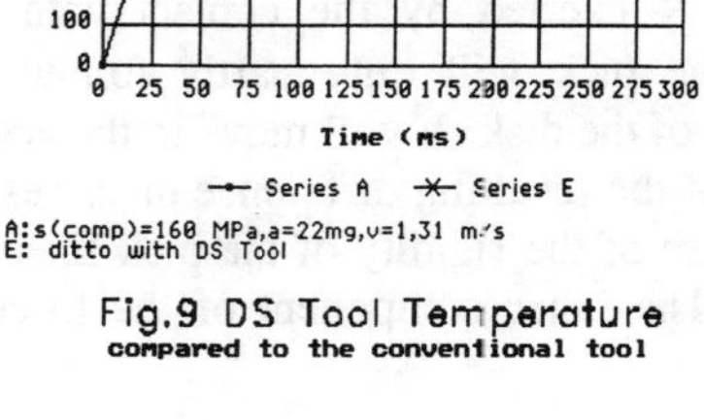

Fig.9 DS Tool Temperature
compared to the conventional tool

thus reducing the wear of the TC tips due to the friction. The width of the serial cut path exceeds the single bit cut one and so the total number of DST on a cutting head can be reduced compared with the conventional tools.

There is a limitation to the use of DS tool however, the desired reduction of the wear can only be achieved on a transversal head, where there is never a 360° exposure of the tool to the friction of the rock, such as occurs on the on-line cutting heads during the initial operation of sumping. In this case, high temperatures would quickly be attained due to the prolonged exposure of the pick to the frictional forces. Therefore, even when sumping, the DS tool requires a pause in its contacts with the rock for a part of the rotation of the cutting head (which occurs regularly on all roadheaders during the normal cutting period). Only the transversal heads comply with this requirement.

THE DUAL-SPIN SATELLITE HEAD

The DS Satellite Head can operate safely even

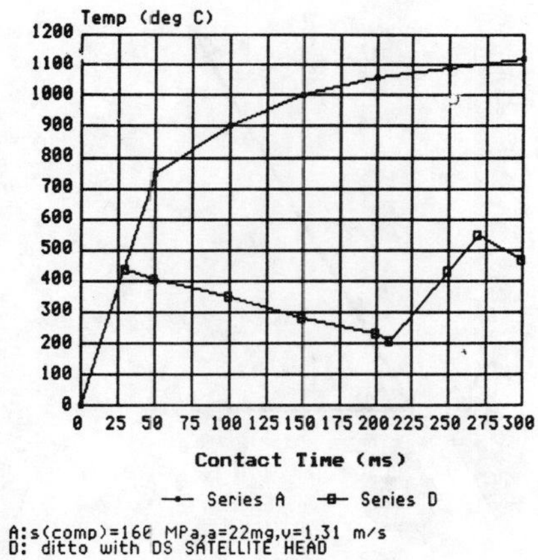

Fig.10 DS Satellite Head
Bit Temperature Comparison vs. the Conventional Head

during the sumping period of an in-line cutting head, as shown during the bench tests. In this case, the diameter of the bit tips circle is much greater than with the DS tool. As a result the cooling time is extended (Figure 10) to more than three times that of the DS tool and so the temperature of the bit can be maintained below the critical area almost continually during a normal operation. In addition, the satellite head could cope with the sumping without showing a substantial increase in temperature.

Tests have confirmed not only the fact of rotation and the restricted temperature growth, but they also provided data concerning the rotational speed of the satellite head, the torque, etc. (Figure 10a). Cutting the rock at various inclinations of the satellite head axis has helped to establish the optimum angle for this type of operation. The actual contact time between the pick and the rock could be established from the recorded data.

FIG. 10A

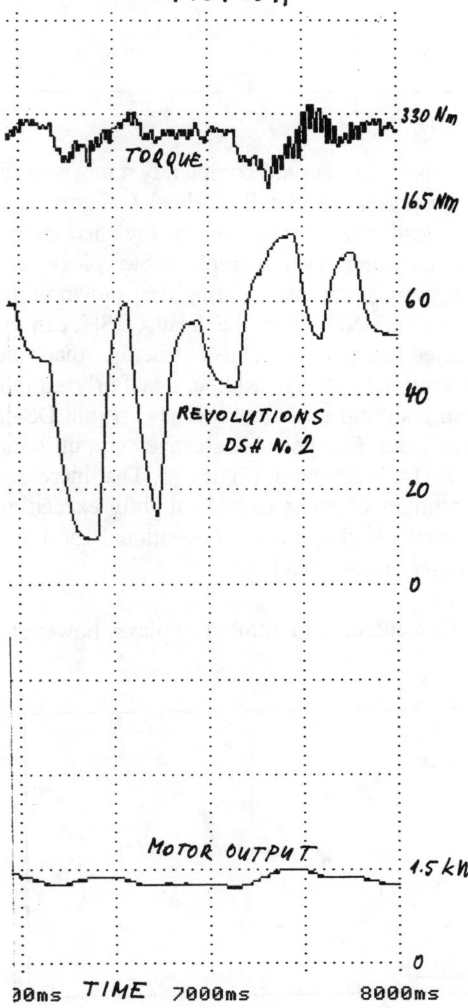

The tests also indicated that the shape of the experimental head rendered the picks mounted at its the top part completely inactive. Another feature that was meant for the test purposes only, was the absence of a convenient method to replace the picks.

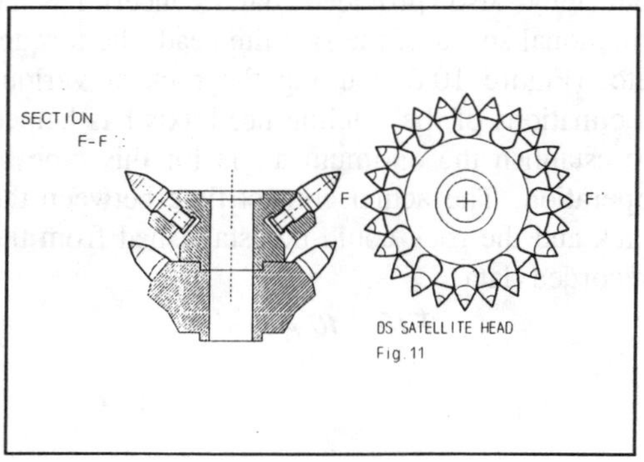

DS SATELLITE HEAD
Fig. 11

Both disadvantages are completely removed with the proposed DS Satellite Head (Figure 11). This head consists of two sandwiched disks, each carrying twelve replaceable picks and rotating on a common bearing (not shown in the sketch) housed inside the head. DSH can be mounted on a less robust structure than the conventional cutting head, e.g. a fairly simple rotating structure of steel beams carrying DSHs on each end. One possible arrangement, utilizing six DSHs, is shown in Figure 12. That increases the number of picks to 144, slightly exceeding the usual 130 found on a conventional 1300 mm diameter cutting head.

The 10% increase in number of picks, however, is more than outweighed by a substantial advantage of the DSH that the autorotation brings about a steep increase of the rock cutting force with DSH axis inclination of more than 50 degrees from horizontal (Figure 11a).

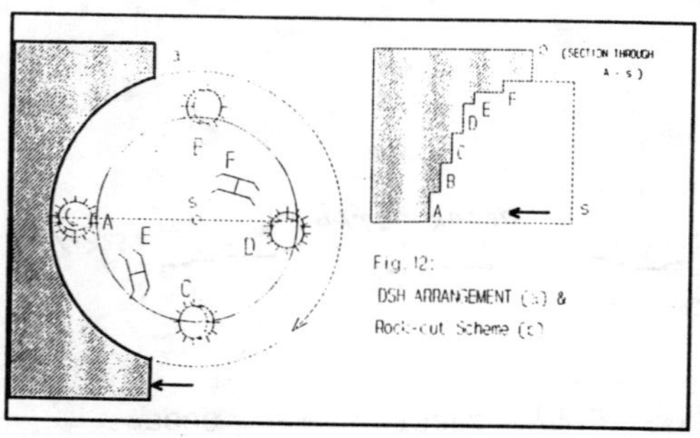

Fig. 12:
DSH ARRANGEMENT (a) &
Rock-cut Scheme (c)

In this way, the actual cutting force exceeds the conventional one e.g. at 60 deg. inclination by a factor of two. Since according to the layout in Figure 12 the number of picks actually engaged with the rock is only 18 (as against more than 40 with the conventional cutting head), the actual multiplication factor for the DS cutting force will be, at the inclination angle shown above more than 4. The conclusion: DS rock cutting can not only prevent the excessive thermal wear in hard and abrasive rock in the strength area immediately above 100 MPa: it can also successfully cut the rock up to now completely out of reach for the tunnelling by roadheaders.

THE OUTLOOK

With the increasing demand for introducing a

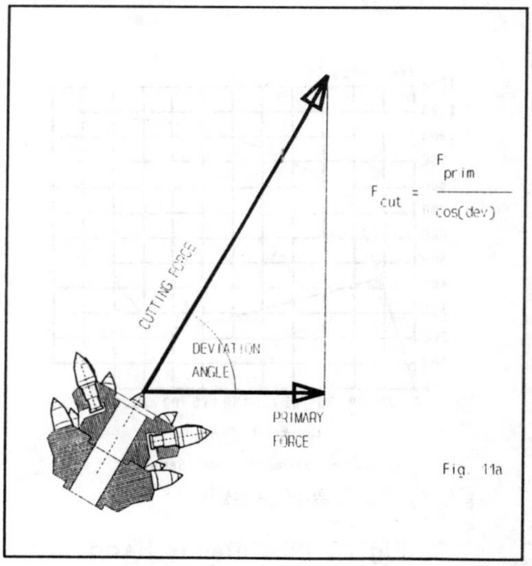

Fig. 11a

high degree of mechanization - and even automation, which is hardly imaginable while we have to drill, blast and muck - into the hard rock mining and short-distance tunnelling, the DS rock cutting can find a good and fast growing market. It can be applied to the remotely controlled excavation in operations automation, which is hardly imaginable while we have to drill, blast and muck - into the hard rock mining and short-distance tunnelling, the DS rock cutting can find a good and fast growing market. It can be applied to the remotely controlled excavation in operations where the presence of men is unacceptable for reasons of safety and health or for reasons of inaccessibility etc. (e.g. in very deep and hot mines). DS rock cutting can also be of interest in urban areas, where rock blasting creates many problems. Many mines loose thousands of operation hours each year due to the "blasting and ventilation time", which is usually performed twice a day. Looking even further into the future, the application of rock cutting could make these operations much more profitable.

ACKNOWLEDGEMENTS

The author wishes to express his gratitude to the management of the Voest-Alpine Bergtechnik in Zeltweg, Austria, for their active support rendering the tests possible. Special thanks belong to Mr A. Zitz, for his leading role in the matter of the DS tests and to the late Mr Franz Schoeffmann, who managed the preparation of testing. The substantial contribution of Mr Wiesenbauer, who built the DSH cutting head model, and that of the test bench team with Mr Stanek and the absolutely indispensable Mr Pichlmair, should not be forgotten. The list of those who helped at several occasions would be incomplete without Mr Gerer, Mr Siebenhofer, Mr Helmuth Wlach who helped to establish the first contacts, and - last, but by far not least - without Frau Peinhopf, the secretary of the Technical Department.

REFERENCES

1 Roxborough, F.F. Rock cutting research Tunnels and Tunnelling 1969, Sept-Oct.

2 Krapivin, M.G. 1976. Gornye instrumenty. (The Mining Tools) Nedra Publishers, Moscow

3 Krapivin, M.G, Manakov, V.M., Mikhaylov, V.G. Research into the effects of interruptions on the temperature and the wear of the rock cutting tool. IVUZ Gornyj zhurnal 1967, 12, 93-97.

4 Šolc, B. 1991. Control of frictional heat in rock cutting. Transactions, Tunnelling '91 Conference, Elsevier, Amsterdam

Innovative Mine Design for the 21st Century, Bawden & Archibald (eds) © 1993 Balkema, Rotterdam, ISBN 90 5410 325 6

Application of cable bolts in underground coal mines

J.M.Goris
Spokane Research Center, US Bureau of Mines, Wash., USA

S.C.Tadolini
Denver Research Center, US Bureau of Mines, Colo., USA

ABSTRACT: The use of cable bolt supports in underground coal mines is gaining greater attention in the United States, especially with the development of techniques to use rock bolters and resin cartridges for drilling and anchoring cable supports. This paper outlines research and development being conducted by the U.S. Bureau of Mines in cooperation with several mines to advance cable bolt support technology, especially in underground coal mines. Research efforts include the study of support and strata monitoring techniques, evaluation of strata behavior, study of shear properties of cables in horizontal beds, development of safe and efficient cable handling and insertion techniques, and development of mine roof design theories.

1 INTRODUCTION

The use of cable bolts has been widely accepted as an effective ground control technique in metal and nonmetal mines in many countries. They can be installed as either passive or active supports; they can behave as either stiff or yieldable supports, depending on their configuration; and they can be placed at any angle. They can contribute to safe, productive mining without the need to make changes in existing mining methods.

In recent years, ground control engineers in Australia have been successful at utilizing cable bolt supports in underground coal mines to reinforce roof strata (Gale 1986; Gale and Fabjanczyk 1987), and engineers in the United Kingdom, India, and Canada have installed cable bolts in underground coal mines on a trial basis (Daws 1991; Singh et al. 1986).

In 1992, the U.S. Bureau of Mines began studying the use of cable bolts in coal mines in the United States with an emphasis on using resin cartridges for grouting the cables. The primary objective of these studies has been to design and evaluate cable support systems that will be effective in redistributing stresses in longwall gateroads.

The roof in gateroads is usually controlled by placing crib supports throughout the openings. Placement of cribs is expensive, time consuming, and hazardous, and the supports themselves restrict the flow of air, which is vital for ventilating the mine. A recent analysis of the number of reported accidents related to placing cribs, timbers, and blocks reveals that there were 457 accidents associated with these activities in 1989 and 458 in 1990. This amounts to approximately 4.25 pct of all underground accidents. Additional analyses revealed that the number of accidents in underground longwall operations in western U.S. coal fields was higher than in eastern fields, accounting for approximately 8 pct of all underground accidents (Mine Safety and Health Administration 1990, 1991). This dramatic difference in accidents between western and eastern coal mines is directly related to thicker coal seams in western mines, which require the use of ladders to place heavy and cumbersome cribbing materials in high entries.

Removing cribs from gateroads would help reduce the fire hazard associated with wood and improve ventilation in a mine. In addition, fewer cribs would reduce the harvest of trees in our national forests. One square kilometer of trees is required to support a single 2,530-m-long gateroad with 0.2- by 0.2- by 1.8-m crib blocks in a 2.9-m-high entry. Shortages of reliable supplies of quality mine timbers are common in the United States.

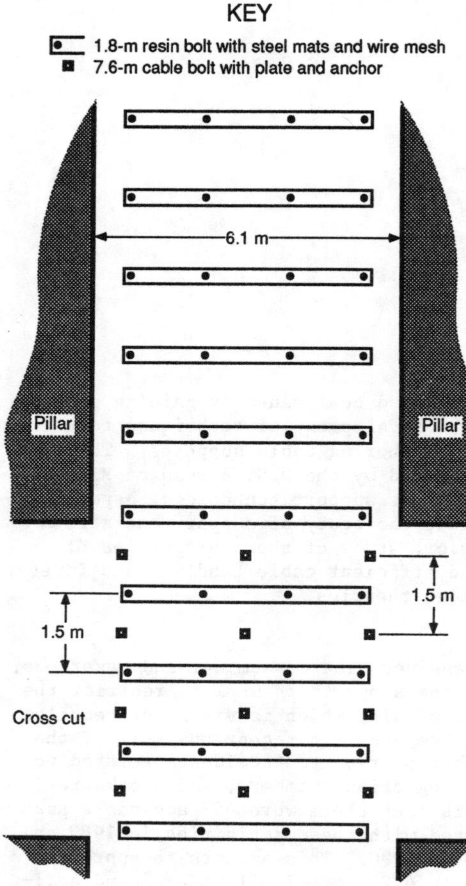

Figure 1. Plan view of intersection showing spacing of cables and primary supports.

2 RESEARCH APPROACH

The investigation of the behavior of coal mine strata when supported by cable bolts consists of four major efforts: (1) Field evaluation of cable support behavior in gateroad, (2) determination of the shear properties of cables in horizontal beds, (3) development of safe and efficient cable handling and insertion techniques, and (4) development of mine roof design theories.

2.1 Field evaluation of cable support behavior

A field evaluation of cable bolt supports in a gateroad is underway at an underground coal mine in Colorado. The primary support system is 2.4-m-long, full-column, resin-grouted rock bolts on a 1.5-m spacing pattern installed with steel mats and wire meshing. Secondary support consists of wooden cribs. Under a cooperative agreement between the U.S. Bureau of Mines and the coal mining company, a limited number of cable bolt supports were installed in a small area of a longwall gateroad. An initial set of twelve 7.6-m-long cables was installed by Bureau personnel to serve as secondary support (fig. 1). Two purposes were served. The first was to demonstrate proper techniques and procedures involved in installing cable bolts, and the second was to monitor loads on the cables as the longwall panel was mined.

The mine has a 4.3-m-high minable coal seam, but the gateroad entries are driven 2.9 m high. A generalized geologic summary of the immediate roof in the test area included about 1.1 m of top coal overlain by about 0.5 m of silty shale. The roof above the silty shale graded vertically into an interbedded unit of siltstone, shale, and sandstone (fig. 2). The thickness of the sandstone in the test area was approximately 4.9 m thick.

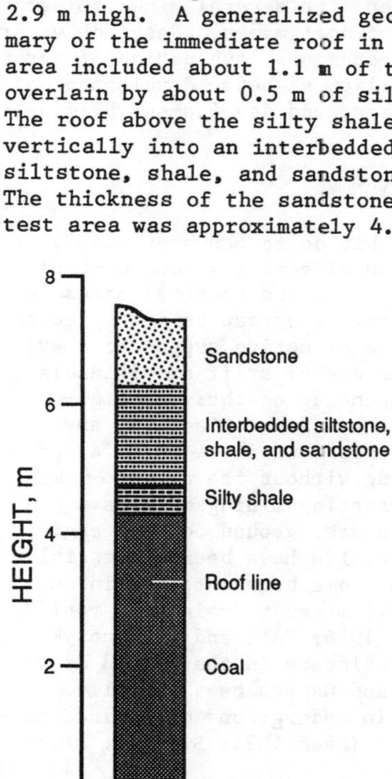

Figure 2. Geologic column of mine strata.

The 7.6-m-long cables were installed by hand along with a 12.7-mm-diam plastic breather tube that allowed displaced air in the holes to escape. The grout consisted of cement and water at a ratio of 0.35 kg of water per 1 kg of cement. A

Figure 3. Goodyear pressure pad (left) and pressure load cell (right).

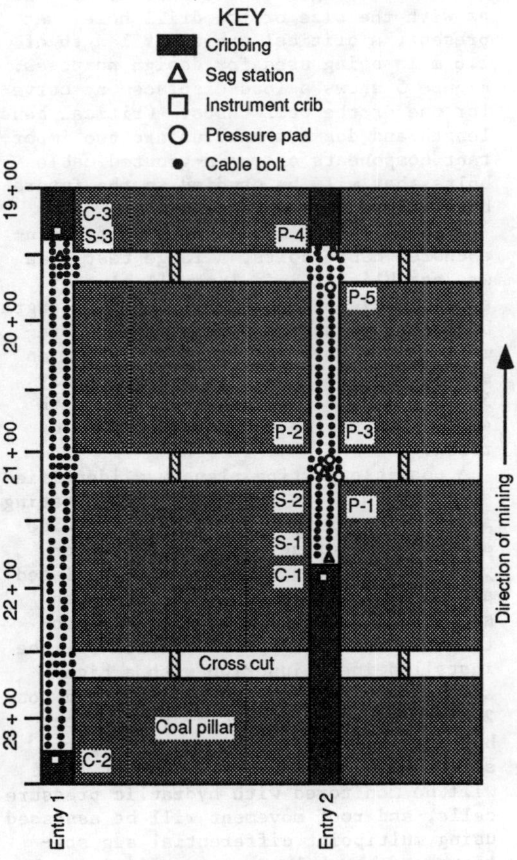

Figure 4. Plan view of test area and location of monitoring instruments.

water-reducing agent (3.9 mL per 1 kg of cement) was used to improve the pumpability of the grout. The collars of the holes were plugged with cotton waste and expanding foam, and the grout was pumped into the holes through a 19-mm-diam grout tube.

Pressure pads and pressure load cells were used to monitor cable loads (fig. 3). These instruments are easy to install, do not have electrical components, and are easy to read. The pressure on the gauge is directly related to the load on the cable. While the roof in a tailgate bleeder system can be subjected to high loads depending on the number of mined panels, pillar configuration, and the localized geology, loads on the individual cables installed in the test area never exceeded 13.3 kN. This was attributed to the large pillars that had been developed and the competent sandstone roof above the coal and shale.

2.2 Full-scale test site

Based on the success of the initial 12 cable bolts installed, ground control engineers at the mine began a program to replace wood cribs with cable bolts as secondary supports in a full-scale test. This would provide an opportunity to assess support performance of the cable bolts over a long period of time. Two entries closest to the longwall panel were selected (fig. 4). In entry 1, 110 m of roadway was supported with

7.6-m-long cable bolt support on 2.1-m spacings. In entry 2, 76.2 m of entry was supported with 5.5-m-long cables installed on a 2.1-m spacings. The cables were installed using cement-based grout in the same manner as the original 12 cables.

Individual cable loads were monitored with Goodyear pressure pads and pressure load cells (fig. 3). The roof behavior was simultaneously monitored with differential magnetic sag stations and closure meters on 7.6-m spacings during the development of the longwall panel to evaluate the response of the immediate and main roofs.

The field evaluation of cable bolts was deemed very successful by managers at the mine and has prompted the use of cable supports in other areas where crib-type supports had been used. Details of this investigation are available in Tadolini (1993).

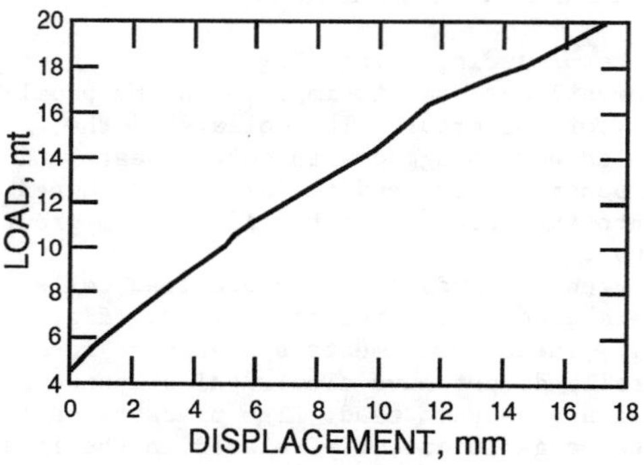

Figure 5. Load-displacement curve for resin-grouted cable bolt.

2.3 Use of resin grout with cable bolts

Installing cable bolt supports using cement-based grouts is time consuming because the grout must be mixed, pumped, and allowed to harden for 24 hr. In underground coal mines, the delay could be critical because plates and grips are placed on the lower ends of the cable, thereby requiring that the grips be set at least 24 hr after the grout has been placed. A new approach to installing cable bolts in coal mines is to use existing roof bolting machines and resin cartridges. The resin cartridges are easy to handle and the resin can set in less than 30 sec. The need to install cable bolts rapidly has prompted rock bolt manufacturers to devise methods of securing nuts on the end of a 15.2-mm-diam cable that will fit into the chuck of a roof bolter. Cables can now be installed by using a roof bolter to drill a hole, inserting a resin cartridge into the hole along with a cable, placing the cable into the chuck of the bolter, and then pushing the cable into the hole while spinning it to mix the resin. Depending on the setting time of the resin, anchorage can be achieved in as little as 30 sec.

The major advantage of using resin rather than cement-based grout is the quick installation time and the instantaneous support achieved. Depending on the hardness of the strata and the setting time of the resin, a 3- to 5-m-long cable can be installed and ready to carry a load in less than 5 min.

A number of pull tests have been conducted on resin-grouted cable bolts. Preliminary test results indicate that the critical embedment length is approximately 1.2 m; however, this will vary with the strength of the strata, as well as with the size of the drill hole. At present, a critical embedment length of 1.8 m is being used for design purposes. Figure 5 shows a load-displacement curve for one of the pull tests. Critical bond length and desired annulus are two important components of resin-grouted cable bolts that will be studied in the future by the Bureau.

To compliment investigations of resin-anchored cable bolts, a large test area was established in a longwall bleeder entry. This entry runs the entire length behind the longwall panel. It must remain open to provide ventilation and an escapeway. The entry is approximately 5.9 m wide and is being supported with 4.9-m-long cables embedded in 3.7 m of resin.

A possible parting plane was identified about 2.1 m above the back, necessitating an embedment length of 3.7 m to provide adequate anchorage. The installation pattern consisted of three cables placed across the width of the opening; each set of cables was spaced on 1.5-m centers down the entry. The cables will be installed in conjunction with a high-strength, 9.5-mm-thick steel mat and four 2.4-m, full-column, resin-grouted rock bolts to form a high-strength support system (fig. 6). Individual bolt loads will be monitored with hydraulic pressure cells, and roof movement will be assessed using multipoint differential sag stations and borehole camera equipment. The test area will be subjected to maximum pressures when the longwall panel is extracted to about 60 m. However, the accessibility of the test area will permit long-term evaluation of the performance of resin-grouted cable supports as the longwall face retreats.

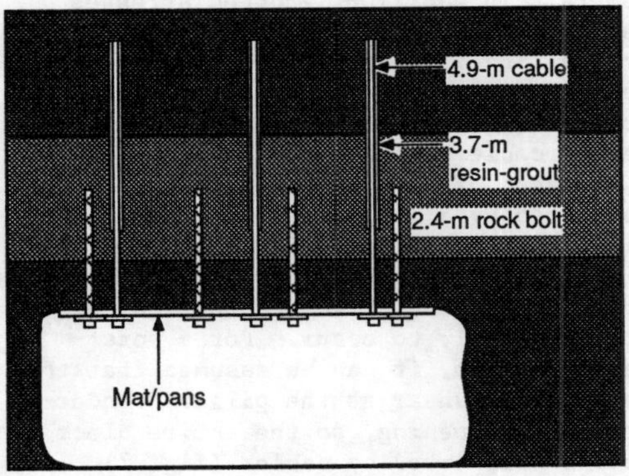

Figure 6. Profile of support system using resin-grouted cable bolts in bleeder entry.

3 DETERMINING THE SHEAR PROPERTIES OF CABLE BOLTS

In addition to studying the axial stability of cable bolts in horizontal strata, Bureau personnel are also investigating the shear properties of cable supports in cooperation with a second underground coal mine in Colorado. Ground-control engineers at this mine report that as much as 32 mm of horizontal displacement have been measured about 0.6 m above the mine roof. It is believed that although cable bolts will not stop this displacement, the cable bolts will still be intact when horizontal movement stops and will therefore still offer axial restraint. This is because of the cable's great flexibility.

The strengthening effect of both tensioned and untensioned rock bolts across joints has been studied using numerical models and by conducting direct shear tests under laboratory conditions. Haas (1981) determined that both conventional and resin-grouted roof bolts significantly increased shear resistance along a joint when the average normal compressive stress on that plane was on the order of 0.17 MPa and bolt orientation was between zero and +45° normal to the joint surface. However, the shear properties of cable bolts in coal strata have not been studied. The data collected to date on rock bolts does have some application to cable bolts, but there are major differences in physical properties and configurations of cables versus rock bolts and differences in how cables and rock bolts perform under load. Therefore, a major effort by Bureau personnel will be to study shear properties of cable bolts embedded with resin grouts.

Another Bureau research project is presently investigating the shear properties of cable bolts with respect to wedge failures and displacement of large rock blocks in shallow ore deposits. Information from this work may also provide important information to use in coal mines because the tests simulate depths of less than 300 m and involve various joint roughnesses.

4 DEVELOPMENT OF SAFE AND EFFICIENT CABLE INSTALLATION TECHNIQUES

The installation of cable bolt supports using resin grouts is different in many respects from those used when cement-based grouts are involved and will require an accurate assessment of safe and efficient installation techniques. As previously mentioned, each cable bolt can be installed in about 5 min when resin is used, depending on the length of the cable and the strength of the strata. Consequently, cable bolt crews will require large inventories of cables on the bolter machine. Handling such large volumes will require some ingenuity, as well as safe and coordinated practices. Shipping and handling techniques will depend on cable lengths. Cable bolts 4.6 m or so long do not necessarily have to be coiled because at many underground coal mines, these cables can be shipped underground uncoiled on the beds of rubber-tired rigs. Such conditions will greatly simplify cable handling.

There will be cases, however, where the cables must be coiled. Cutting the bands around coiled cables can be a hazardous job. Each coil is secured by banding the cable with either steel or plastic straps and then placing the coiled cables on shipping pallets. Once the cables arrive at the mine site, they are usually shipped underground on the same pallet and uncoiled as needed. The straps used to band each cable are placed on the coil in sequence in the factory and must be cut in the reverse order prior to use to limit the amount of cable released with each strap. Uncoiling cables is a dangerous process because the cable tends to whip as the straps are cut. Bureau personnel are investigating methods to allow cable bolt crews to uncoil these cables while the cables are still confined within the shipping pallet.

Installation of some cable bolts will involve the use of cement-based grouts. Depending on the orientation of the holes, these cables may require anchoring while the hole is being grouted. Bureau personnel are studying various anchoring techniques and documenting their effectiveness. Presently, tests are being conducted to determine how drill hole size and cable length influence anchorage capacity for several anchoring techniques.

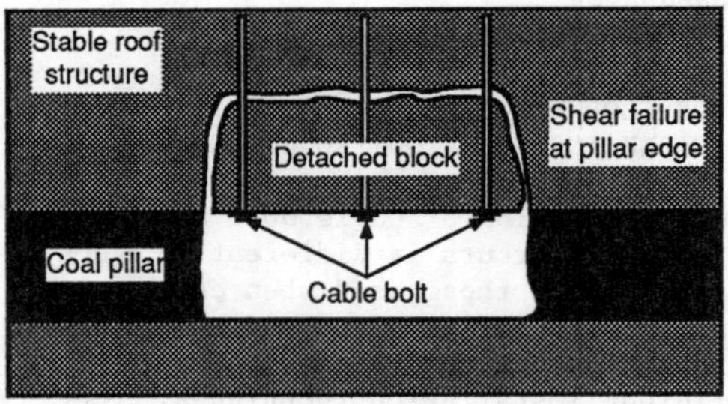

Figure 7. Cross section showing block failure in coal mine.

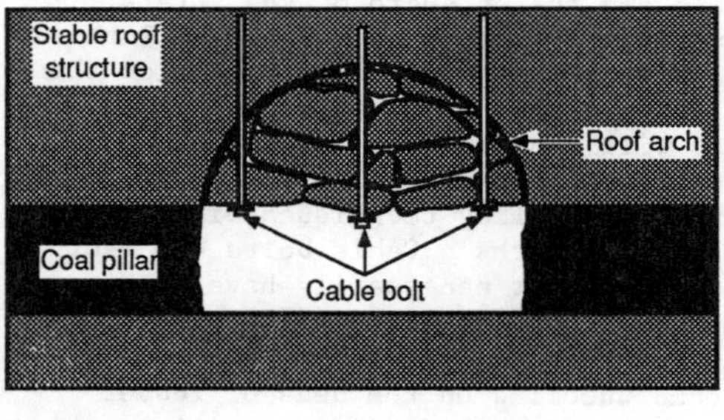

Figure 8. Cross section showing zone of failure according to arch failure.

5 DEVELOPMENT OF MINE ROOF DESIGN THEORIES

A simple approach to designing cable bolt density and length for coal mines is to apply beam theory, because the roof strata are generally parallel to the surface to be supported. A major effort under the Bureau's Research and Development program is to update design procedures.

At present, the first step of the design procedure is to gather detailed information on ground conditions in the mine. This includes a general estimation of the rock mass quality, the geological composition, and strengths of the immediate and main roof members. These data can be obtained from core samples and supplemented with a borescope or borehole camera. In addition, induced stresses can be estimated using empirical methods or modeling. The weight of the rock mass can be expressed per linear meter of roof for mine entries to determine the required cable spacing.

The zone of rock material that must be supported by a cable system can be determined in several ways. The simplest approach is to identify a parting plane where separation above the roof-bolted zone is likely to occur. For a worst-case scenario, it can be assumed that the strata will shear at the pillar boundaries of the opening, so the entire block must be supported by cables (fig. 7). The weight of the material can be determined by using the equation

$$F_w = W_e \times H_p \times \delta \quad (1)$$

where F_w = weight of the rock per linear meter, kg/m,
W_e = effective opening width, m,
H_p = distance from the coal mine roof to the parting plane, m,
and δ = average rock density, kg/m^3.

This method is conservative because it is more likely that the coal pillars will provide additional support to the roof structure, forcing the formation of a pressure arch of the failed material. As shown in figure 8, the cables must be capable of supporting only the material within the boundaries of the pressure arch. The height of the arch can be correlated to the vertical and horizontal stresses acting in the immediate roof and is believed to increase as the in situ horizontal stress increases. An acceptable criterion for determining the height of the pressure arch is that the failure height is 0.5 to 1.5 times the height of the seam (Adler and Sun 1976). The weight of the material within the arch can be estimated with respect to opening width and height of the arch by the equation

$$F_a = 1/2 \times \pi \times W_e/2 \times H_a \times \delta \quad (2)$$

where F_a = weight of the rock within the pressure arch per linear meter,
and H_a = height of the pressure arch, m.

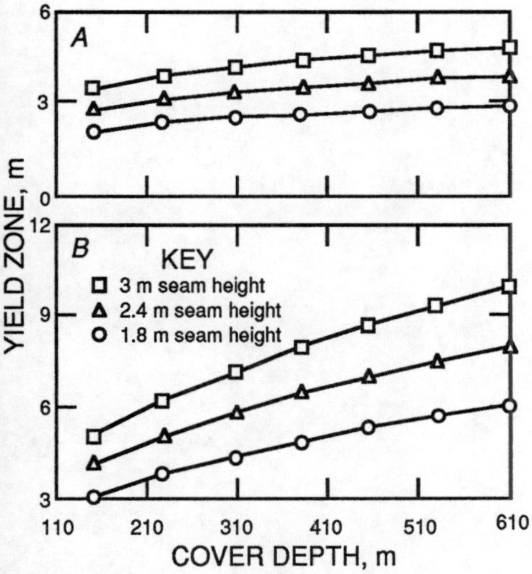

Figure 9. Yield zone estimates using Wilson's equations.

The behavior of the pillar under loaded conditions determines the effective opening width W_e. As the pillars yield, the effective opening span increases proportionately. Wilson (1972) defines the depth of the yield zone as the depth at which the coal strength in the entire pillar is exceeded by the loads imposed. Wilson's equations can be used to estimate the depth of this yield or approximated using charts A and B in figure 9. Chart A is used for rigid roof conditions and chart B for yielding roof conditions. Both charts were created using a value of 35° for the angle of internal friction. The effective opening width can then be calculated by using

$$W_e = W + Y_{p1} + Y_{p2} \tag{3}$$

where W = mined width of the opening, m,

Y_{p1} = yield zone for pillar 1, m,

and Y_{p2} = yield zone for pillar 2, m.

Once the volume and weight of the material to be supported with cables is determined, then the number and spacings of the cables required to support the gateroad entry can also be determined. Using a cable capacity of 26,300 kg per cable and varying the number of cables across the opening, the best design for a specific application and operation can be supplied. Figure 10 shows a design chart constructed for two, three, and four cables installed for an effective opening width of 7.6 m, a cable strength of

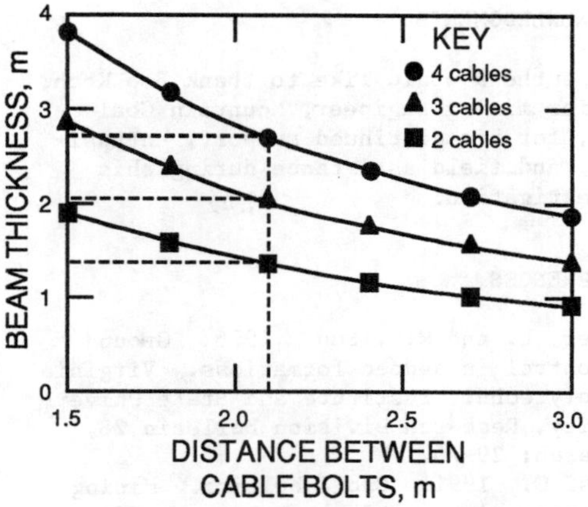

Figure 10. Design chart using two, three, and four cables across entry.

26,300 kg, and a material weighing 2,400 kg/m^3. The intersection of X and Y values at a given point (fig. 10) indicates the thickness of a failed beam member that can be entirely supported with installed cables. In the example, two, three, and four cables at 2.1-m spacings across the entry would have the capacity to support 1.37, 2.04, and 2.71 m of roof, respectively.

6 CONCLUSIONS

It is anticipated that the successful completion of this research will provide the following benefits to the coal mining industry:

1. A safe and cost-effective alternative to secondary roof support that will reduce accidents and injuries associated with the installation of crib supports.
2. A technique for minimizing restrictions to airflow, which will permit lowering of air velocities while still maintaining adequate ventilation.
3. A ground-control technique for redistributing stress concentrations into gob areas by the design of a system that will fail after mining passes.
4. Identification and correction of major hazards associated with cable bolt support, and recommendations for standard practices for handling and installing such supports.

Bureau personnel, with the assistance of industry, are continuing to examine the effectiveness of cable bolt supports under a wide range of geological and mining conditions. The ultimate goal of this investigation is to provide a safe, economic support system.

ACKNOWLEDGMENTS

The authors would like to thank Bob Koch, senior mining engineer, Mountain Coal Co., for his continued support, enthusiasm, and field assistance during this investigation.

REFERENCES

Adler, L. and M.S. Sun. 1976. Ground control in bedded formations. Virginia Polytechnic Institute and State University, Research Division Bulletin 28, March: 29-76.

Daws, G. 1991. Cable bolting. Mining Engineering (London), February: 261-266.

Gale, W.J. 1986. Design considerations for reinforcement of coal mine roadways in the Illawarra Coal Measures. Symposium, Ground Movement and Control Related to Coal Mining, Aus. IMM.

Gale, W.J. and M.W. Fabjanczyk. 1987. Application of field measurement techniques to the design of roof reinforcement systems in underground coal mines. CSIRO.

Haas, C.J. 1981. Analysis of rock bolting to prevent shear movement in fractured ground. Mining Engineering (Littleton), v. 33, No 6: 698-704.

Mine Safety and Health Administration. 1990. IR 1196.

Mine Safety and Health Administration. 1991. IR 1205.

Singh, T.N., R.N. Gupta, B.K. Dubey, B.P. Verma, and B. Singh. 1986. Cable bolted slicing of thick seam standing on pillars. Journal of Mines, Metals & Fuels, July: 350-358.

Tadolini, S.C. 1993. Field evaluation of resin grouted cable bolts, 12th International Ground Control Conference. In press.

Wilson, A.H. 1972. A hypothesis concerning pillar stability. Mining Engineering (London), v. 131, No. 141: 409-417.

Innovative Mine Design for the 21st Century, Bawden & Archibald (eds) © 1993 Balkema, Rotterdam, ISBN 90 5410 325 6

Seismic refraction and tomography studies in the El Teniente copper mine

José Blondel & Emilio Aguirre
Corporación Nacional del Cobre, División El Teniente, Chile

Gerardo Behn & Carlos Toledo
Geoexploraciones Limitada, Santiago, Chile

ABSTRACT: Refraction and tomography surveys were used with good results for rock mass damage classification in pillars in the El Teniente underground mine. The mine uses the Panel Caving System and the damage is caused by the advancing caving front in hard primary ore prone to rockbursting phenomena.

1 OBJECTIVES OF THE SEISMIC WORK UNDERTAKEN

The research into underground pillar damage was undertaken in the North 1 FW sector of level Teniente Sub-6 between November 1990 and January 1991.

The results of the seismic surveys form part of a medium term research program that is continuing with the application of ground probing radar and direct observations of damage using borehole camera techniques.

The objective of this research is to contribute to the improvement of geomechanical analysis of the modifications which may have to be made in the mining sequence and in the selection of pillars to be repaired using cement or polymer grout injection, all with the objective of stabilizing the mine design to enable massive mining in the zones with a propensity to rockbursts.

2 GEOTECHNICAL CONSIDERATIONS

2.1 Structural Geology

The El Teniente porphyry copper orebody is formed by andesites diorites and hydrothermal breccias of the miocene era. Polydirectional veinlets and stockworks of the primary ore zone prone to rockburst are cemented principally with anhydrite, quartz and sulphides.

The chimney of subvolcanic breccias known as the "Braden Pipe" postdates the principal copper/molybdenum mineralization. It has the form of an inverted cone with outliers distributed over a 1 km radius.

The main zones of shear slip faulting correspond to regional features and have strikes of 065 Degrees and 305 Degrees.

The area studied falls within an intersection of these fault zones as shown in Fig. 1.

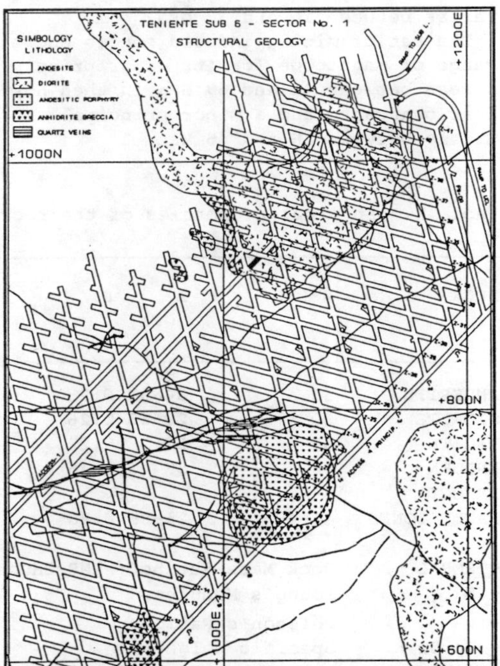

Figure 1 Structural Geology of the Area Studied

2.2 Physical properties

The physical properties of the rocks comprising the El Teniente orebody are detailed in Table 1

2.3 Stress Measurements

The moduli of elasticity measured in the biaxial chamber have a high variability due to the effect of the hydrothermal alteration and the frequency of veinlets and microstructures along the drill holes where the measurements were taken. For this reason the principal stress magnitudes are indicated as ranges, as detailed in table 2 which were obtained by the USBM and CSIR doorstopper methods.

The tectonic field stresses recognized in the area studied, located between approximately 800 and 1000 m below the surface, has a major principal stress magnitude between 34 and 60 MPa, dipping subhorizontally in then north west quadrant. The minor principal stress is in the range of 16 to 33 MPa.

The set of stresses categorized as being the transition between the tectonic stress and the mining induced stress has its major principal stress between 41 and 61 MPa inthe north west quadrant dipping subvertically. The minor principal stress varies between 16 and 33 MPa.

The set of mining induced forces shows a range of magnitude for its major principal stress between 44 and 69 MPa in then north west quadrant, and a minor principal stress between 18 and 46 MPa.

The last stress measurement done, in August 1990 was carried out under caved area, obtaining a maximum magnitude of 17 MPa for the major principal stress an minimum magnitude of 12 MPa.

3 MICROSEISMIC ACTIVITY

Currently 60% of the production at the El Teniente Mine comes from the primary ore zone or rockburst prone zone using the methods of panel caving and sublevel caving. This type of mass mining induces an important amount of microseismic activity around it. The statistics detailed in Table 3 represent the seismic behavior of the rock mass in the 6 month period between August 1992 and January 1993.

This data is represented in Figure 2 following the Guttenberg-Richter convention. The greatest occurrence of microseismic activity manifested itself during the periods of production, however the existence of significant background activity is clearly shown.

Table 1. Physical properties of the rocks comprising the El Teniente Ore Body.

ROCK	RMR L	E (GPa)	ν	Sp Gr (t/m3)	UCS (Mpa)	HOEK-BROWN m	s
ANDESITE	55	34-37	0.28	2.807	124-160	25.57	1
DIORITE	60	29-56	0.26	2.668	140-200	24.96	1
DACITE	65	50	0.22	2.640	154	51.25	1
BRADEN BRECCIA	70	38	0.26	2.500	167	15.12	1
FAULT ZONES	44-48	25-59	0.28	2.700	47-82		

- RMR L : Rock Mass Rating (=RMR Laubscher.(1991))
- E : Young's Modulus.
- ν : Poisson's Ratio.
- Sp Gr : Specific Weight.
- UCS : Uniaxial Compressive Strength (MPa)

Table 2 Stress Measurements (U.S.B.M. & C.S.I.R Methods) Sector N1 FW TTE SUB-6

STATION Date,Loc	STATION DESCRIPTION Technique :Overcoring		Youngs Mod.	Poisson Ratio		Mag[MPa]	Stress Field Az.	Inc.
ST-6	Rock Type: ANDESITE	N : 1262			S1	41 - 60	315°	-24°
Jul 1983	Soil height : 0m	east : 640			S2	22 - 32	212°	-25°
T-6	Rock height :460m	elvn.: 2162	32-47	0.28	S3	16 - 23	082°	-54°
Transit	Crater Structural control							
Stress	Struct.Domain:limit F.Z. T.W-W N=17				Sz	21 - 31	LP: 13[MPa]	
ST-8	Rock Type: ANDESITE	north: 1045			S1	51 - 61	321°	- 8°
Nov 1983	Soil height : 0m	east : 938			S2	45 - 55	230°	- 3°
T-6	Rock height :630m	elvn.: 2162	45-54	0.28	S3	27 - 32	120°	-81°
Transit	Crater Structural control							
Stress	Struct.Domain:Fault Z. T. W-W N=8				Sz	28 - 33	LP: 18[MPa]	
ST-9	Rock : ANDESITE	N : 1200			S1	44 - 58	275°	-16°
Apr 1984	Soil height : 0m	E : 995			S2	31 - 41	178°	-23°
T-6	Rock height :680m	elvn.: 2162	34-45	0.28	S3	22 - 30	036°	-62°
Transit	Crater Structural control							
Stress	Struct.Domain:30m F.Z. T. W-W N=15				Sz	25 - 33	LP: 19[MPa]	
ST-16	Rock : ANDESITE	N : 870			S1	44 - 56	172°	- 7°
Mar 1985	Soil height :205m	E : 892			S2	42 - 54	269°	-43°
T-7	Rock height :490m	elvn.: 2043	31-40	0.28	S3	18 - 23	074°	-46°
Induced	Crater Geometry							
Stress	Struct.Domain:limit F.Z. T.W-W N=22				Sz	30 - 38	LP: 17[MPa]	
ST-18	Rock : ANDESITE	N : 870			S1	53 - 68	293°	-39°
May 1985	Soil height :205m	E : 892			S2	49 - 63	181°	-25°
T-7	Rock height :490m	elvn.: 2043	31-40	0.28	S3	35 - 46	067°	-41°
Induced	Crater geometry							
Stress	Struct.Domain:limit F.Z. T.W-W				Sz	45 - 57	LP: 17[MPa]	
ST-23	Rock : ANDESITE	N : 877			S1	36 - 46	313°	-50°
Ago 1985	Soil height : 0m	E : 1273			S2	18 - 24	060°	-14°
T-7	Rock height :970m	elvn.: 2043	31-40	0.28	S3	14 - 18	160°	-37°
Tectonic								
Stress	Struct.Domain:limit F.Z. T.W-W				Sz	27 - 35	LP: 27[MPa]	
ST-39	Rock : ANDESITE	N : 931			S1	34 - 50	334°	-36°
Jul 1989	Soil height : 0m	E : 1201			S2	29 - 43	217°	-32°
T-S6	Rock height :870m	elvn.: 2108	32-47	0.28	S3	19 - 28	098°	-38°
Tectonic	Mined Area :608m²							
Stress	Struct.Domain:limit F.Z. T.W-W				Sz	27 - 40	LP: 24[MPa]	
ST-41	Rock : ANDESITE	N : 931			S1	41 - 60	295°	-40°
Dec 1989	Soil height : 0m	E : 1201			S2	25 - 36	170°	-34°
T-S6	Rock height :870m	elvn.: 2108	32-47	0.28	S3	13 - 20	056°	-32°
Tectonic	Mined Area :12.950m²							
Stress	Struct.Domain:limit F.Z. T.W-W				Sz	28 - 42	LP: 24[MPa]	
ST-42	Rock : ANDESITE	N : 877			S1	14 - 17	105°	-04°
Apr 1990	Extracted height:44m	E : 1181			S2	10 - 13	198°	-42°
T-S6	Rock height : 18m	elvn.: 2107	45-54	0.28	S3	10 - 12	010°	-48°
Unconfin	Underneath Mined Area							
Stress	Struct.Domain:out Fault Zones				Sz	10 - 12	LP: 11[MPa]	

LP=Litostatic Pressure; Soil Specific Weight:1.8 [gr/cc]; F.Z.=Fault Zone; N: Nº of samples Inc: positive over horizontal plane; elvn.= altitude above mean sea level

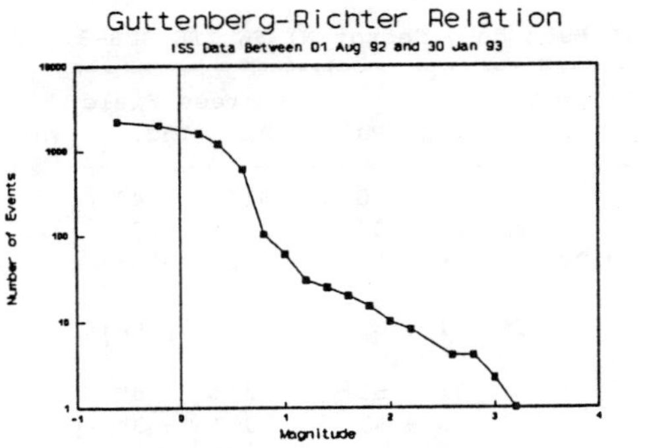

Figure 2 Graph of Guttenberg-Richter relation

Table 3 Important seismic events

ISS SEISMIC NETWORK
01-08-92 TO 31-01-93

Range of Magnitude	Number of Events
M < 1.0	2078
1.0 <= M < 2.0	48
M >= 2.0	8

Date	Time	Magn.	East	North	Vert.
04 08 92	16:24	3.04	1015	548	2423
04 08 92	16:24	2.15	1658	-1678	2357
18 09 92	08:53	2.23	1358	-692	1668
18 09 92	14:16	2.39	1163	-577	1463
24 10 92	07:58	2.57	1350	-608	2360
07 11 92	15:19	2.49	1769	15	2264
08 12 92	08:31	2.38	1710	159	2415
15 01 93	23:50	3.71	313	-888	2433

4 MINING DESIGN

The El Teniente Mine is currently producing 89000 MTPD of ore with an average grade of 1.1% Cu coming from productive sectors using the methods of block caving in the secondary ore zone and panel caving and sub level caving in the primary ore zone prone to rockburst phenomena.

The sector studied is of the panel caving type with its corresponding undercut level, production level, ventilation sublevel, pick hammer level and haulage level, detailed in Fig. 3. The geometry of the production level is detailed in Fig. 4 and is comprised of a mesh of LHD drives (calles) and extraction crosscuts (zanjas) containing the drawpoints.

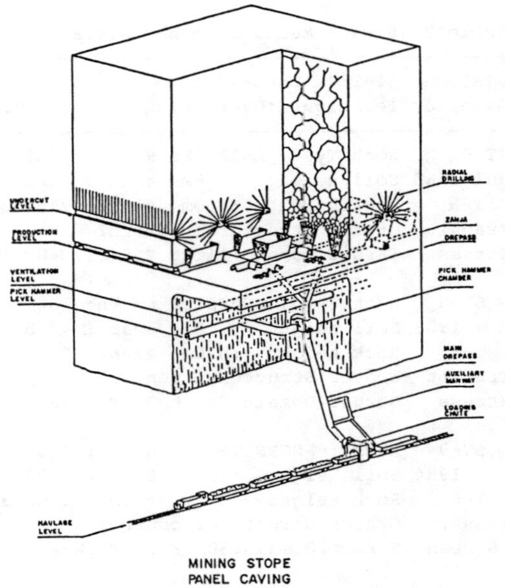

Figure 3 The panel caving method

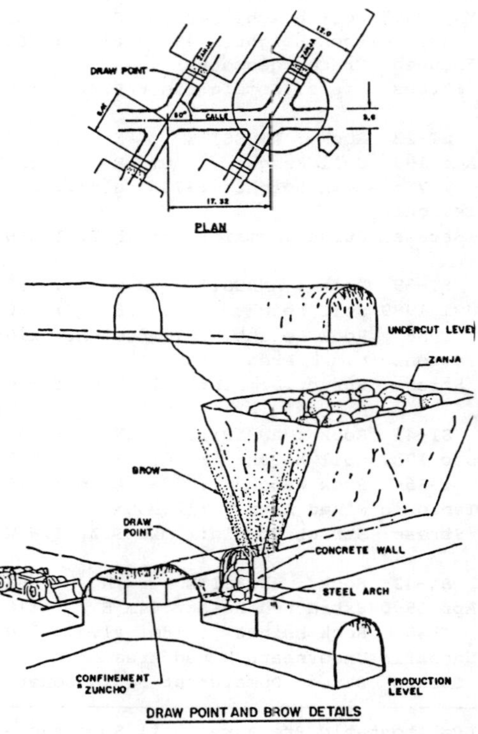

Figure 4 Drawpoint details

5 SEISMIC SURVEY

5.1 Instrumentation

During the refraction and tomography studies the seismic signals were collected using an EG+G Geometrics model ES 1225 12 channel portable digital seismograph, with a time resolution of 0.025 milliseconds.

The equipment pre-amplifies the analog signals coming from each geophone, performs an analog-to-digital conversion, displays the results on a CRT screen and optionally outputs through a standard RS-232 interface to a portable (laptop P.C.) Computer.

The geophones used were vertically sensitive and with an internal frequency of 14 Hz. To ensure a good coupling to the rock specially adapted G-clamps were used clamping onto the ends of pre-existing fully grouted rockbolts.

To generate the necessary seismic energy good results were obtained using a 32lb sledgehammer striking the rock directly at selected points. This method has the desired characteristics of repeatability, simplicity and effectiveness.

Finally, the technique of stacking or summing the signals proved to be a key factor in arriving at an adequate signal/noise ratio in the noisy operating mine environment.

5.2 Seismic Refraction

The processing of the seismic refraction information included the stages of seismograms, construction of path/time diagrams and adjustment of a radial seismic model. For this the first arrival times were used, corresponding to the compressional waves.

The radial seismic model corresponds to a radial volume of rock surrounding the tunnel within this volume it is postulated that the seismic velocity increases linearly away from the tunnel, in a relation with the competence of the rock from the relatively fractured zone immediately surrounding the tunnel to the less damaged and more highly stressed rock further away from the excavation.

$$V\ (d) = V_i + K * D \qquad \text{(Eq 1)}$$

with;

V_i = Compressional velocity at the interface.

K = Linear velocity increase factor.

D = Radial distance from the profile.

Using this function it can be addressed shown that the travel time for a seismic ray between the striking point (source) and a geophone at distance x to given by the equation.

$$T(x) = 2\ /\ K * \sinh^{-1}\ (K * X)\ /\ 2 * V_i \qquad \text{(Eq 2)}$$

This function is then adjusted by a least-squares method to the points of the experimental path/time curve.

By knowing the values of K and V_i it is possible to calculate the theoretical compressional velocity at a given distance or, conversely, given a compressional velocity it is possible to calculate the theoretical distance from the profile to the rock mass that has this velocity, using the formula

$$RAD = (V_r - V_i)/K \qquad \text{(Eq 3)}$$

Where;

RAD = The radial distance at which the compressional velocity takes the value V_r.

The parameter RAD is included in the results and plans and was calculated for a velocity (V_r) of 5800 m/s, being the mean velocity estimated by linear regression in the test conducted in the undercut level (UCL) in Teniente Sub-6, in which there is the largest relative quantity of samples (24).

In this manner RAD (5800) signifies the thickness of rock with a velocity lower than the mean, and represents to some extent a volumetric measure of the rock that has geotechnical and seismic/elastic properties inferior to this value.

The extent of the validity of the hypothesis of the linear increase in the velocity with the radial distance from the tunnel wall has practical and theoretical limit. It is obvious that the velocity does not increase infinitely, nor that the penetration of the profile is unlimited.In reality the penetration depends on the actual profile length and the values of the parameters V_i and K.

As an approximation, in general terms the penetration is of the order of 1/3 of the profile length. For a survey of 60 m this means that the extent of the area explored is around 20 m.

For each seismic test taken a compressional velocity was calculated by linear regression (VLR) adjusted by least-squares to the points of the path/time diagram.

This parameter represents an elementary measure of the average compressional wave velocity determined in a test, and with this it is possible to empirically estimate the in-situ Young's modulus using the formula.

$$E = 1.16 * V * \exp [2.34] \qquad \text{(Eq 4)}$$

With;
E = Young's Modulus (in GPa)
V = Velocity (in km/s)

In this calculation the rock density is unimportant this is only an empirical estimation and is used only as a point of reference.

To arrive at empirical relations that are specifically valid for the mine it is recommended that the correlation between the compressional velocity and the Young's modulus calculated analytically by the P wave/S wave relation be studied.

Another possibility is that tests be conducted using samples or drill cores and extrapolate this to the actual mean. The relationship between the V_p/V_s radio and the measured compressional velocity allows the Youngs modulus to be calculated.

This system was used with good results in a similar study done in an open-pit mine where around 700 samples were available.

In Table 4 the processed information is summarized for each of the tests performed. The coordinates correspond to the point situated at 1/3 of the array, measured from the respective source. In this manner a representative point is given for the values both measured and calculated for each hammer blow.

The chance is arbitrary, but has the advantage of allowing the objective comparison of the values obtained between different sources and arrays.

5.3 Seismic Tomography

Tomography is a technique for reconstructing internal images of a medium by the determination of the distribution of a physical property. The most well known example is the so called CAT Scanner (Computerized Axial Tomography) in which the body may be examined by rays generated by an X ray source and the information generated is used to construct images that

Table 4 Calculated Parameters Teniente Sub 6-Undercut Level

Point Number	North (Mine)	East	Level masl	V1 [m/s]	K [1/s]	VRL [m/s]	RAD [m]	E Gpa
HC0201N	1001	1116	2123	5720	46	5620	1.7	66
HC0201S	978	1122	2123	4990	96	5750	8.4	70
HC0202N	97	1125	2123	5010	94	5460	8.4	62
HC0202S	952	1133	2123	4060	215	5970	8.1	76
HC0301N	1065	1068	2123	5780	0.5	5820	40.0	72
HC0301S	1043	1074	2123	5640	45	5510	3.6	63
HC0302N	1126	1052	2123	5990	26	6310	0	87
HC0302S	1102	1058	2123	6030	0.5	6090	0	80
HC0303N	995	1086	2123	5050	153	5730	4.9	69
HC0303S	976	1092	2123	4950	173	5930	4.9	75
HC0304N	948	1099	2123	5180	145	5120	4.3	53
HC0304S	926	1105	2123	3980	196	5020	9.3	51
HC0501N	1050	1009	2123	5590	356	8510	0.6	174
HC0501S	1028	1015	2123	6460	176	7000	0	110
HC0502N	1092	998	2123	5060	239	6030	3.1	78
HC0502S	1072	1003	2123	5960	206	8300	0	164
HC0503N	975	1030	2123	3050	95	3260	28.9	18
HC0503S	952	1036	2123	5010	91	5470	8.7	62
HC0504N	910	1047	2123	4980	146	5560	5.6	64
HC0504S	886	1053	2123	2840	6.5	2900	45.5	14
HXC301E	1023	1099	2123	5430	45	5380	8.2	60
HXC301W	1017	1078	2123	5050	165	5840	4.5	72
HXC302E	1002	1023	2123	5960	116	6330	0	87
HXC302W	995	995	2123	6120	95	6290	0	86

by an X ray source and the information generated is used to construct images that represent changes in the properties of the tissues.

In seismic tomography the spatial distribution of the compressional velocity is associated with the elastic/mechanical properties of the medium, permitting the delineation of anomalous zones, whether this caused by fractures, changes over time or stress concentration.

To construct the images using wave arrival times the ART or Algebraic Reconstruction Technique was used.

The travel time t_v of the compressional wave is related to the velocity (V) by the equation

$$t_k = \int_L n \, ds \qquad \text{(Eq 5)}$$

Where;
L = The wave path
ds = The differential path length
n = The slowness = $1/_V$

Discretizing the medium into rectangular cells, the integral may be transformed into the discrete sum

$$T_k = \sum_{i=1}^{N} \sum_{j=1}^{N} \Delta S_{ijk} * n_{ij} \qquad k = 1,N \qquad \text{(Eq 6)}$$

The system of linear equations is non-linear, owing to the effect that the trajectory of the rays depends on the distribution of η. That is the ray trajectories are curved. To simplify the problem and, make possible the numeric solution, however, it may be assumed that the rays follow a straight line. Using this assumption the problem becomes linear.

The validity of this assumption was proven by a synthetic example using velocities between 4000 and 6000 m/s.

N_{11} ... ... N_{1j}
... ...
N_{i1} ... ... N_{ij}

i = Row number
j = Column number

The inversion problem consists of estimating the values of the η (slowness) values, given the travel times T_k.

The solution of the equation is accomplished using the ART method which consists of the following steps.

1) An initial slowness model (a) is defined (eg using the average value calculated by all the times and ray paths).
2) The theoretical travel time (t_c) for each ray is calculated using the slowness model (η) and the trajectories.
3) The calculated times are compared with the observed and the error vector is determined $e_k = (t_k - t^c_k)$
4) The slowness (η) is corrected by the equation

$$n_{i,j}^{q+1} = n_{i,j}^{q} + \frac{\Delta S_{ijk}}{e_k} \cdot * \Sigma(\Delta S_{ijk}) \qquad \text{(Eq 7)}$$

Where;

q = The number of iterations.

5) Each slowness value (η) is averaged with the neighboring cells to smooth out noise.
6) The variance of slowness (η) is compared between the iterations q + 1 and q. if the percentage difference is greater than the the tolerance allocated (1%) steps 2-6 are repeated.
7) If the difference in the variation is less than the tolerance the final model has been arrived at and from it the velocity distribution.

The resolution of the spacial definition of the slowness is controlled by the wavelength. Theoretically the maximum resolution is $\lambda/4$.

Where;

λ = wavelength (m)

During the tests it was determined that the dominant wavelength was of the order for 3.5 m, giving a theoretical resolution of 0.9 m.

The tomographic studies were carried out in the undercut level (UCL) and production level of Teniente Sub-6 sector N1 FW. Fig. 5 shows an example of the distribution of geophone, sources and raypaths in one of the production level pillars (PT12). It may be observed that the coverage of the centre of the pillar is better than at the periphery.

Figure 6 Shows the result of the tomography inversion with the velocity

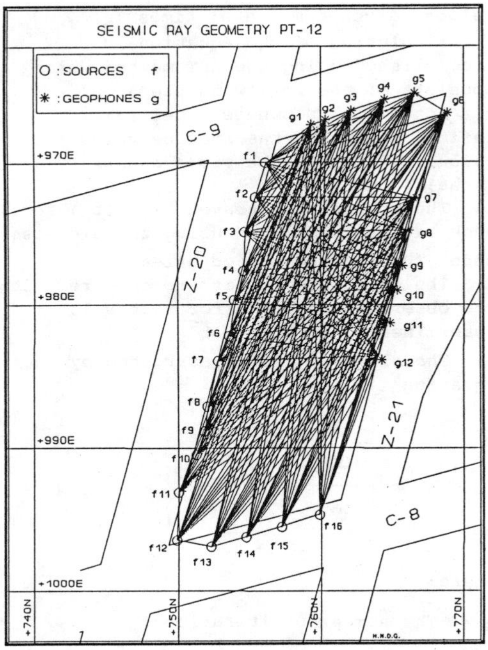

Figure 5 Ray paths in pillar PT12

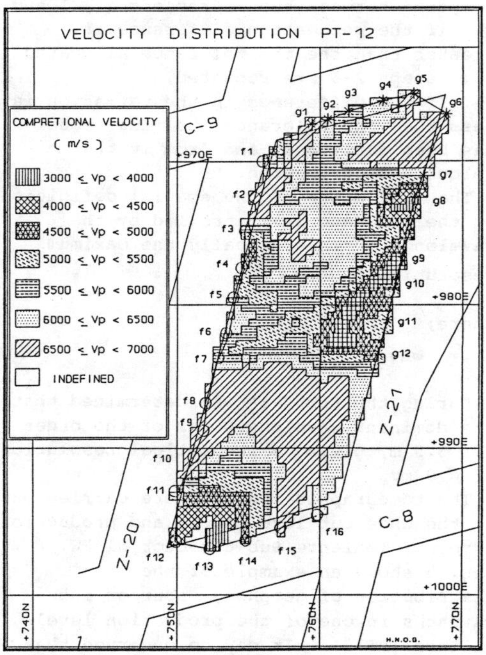

Figure 6 Tomography results (velocities)

distribution coded in shades of grey. The reliability of the results is better in the zone where the concentration of rays is higher. Fig. 7 shows statistics of the velocity distribution within this pillar.

Table 5 is a statistical summary of all the studies considering only the cells intersected by 3 or more rays.

Table 5 Pillar tomography results

Tomog. (Pillar)	V.MIN (m/s)	V.max (m/s)	V.Avg (m/s)	S.Dv. (m/s)	N
HT01	3005	7472	5901	926	UCL
HT02	3368	7268	5615	629	UCL
HT03	3111	7310	5513	618	UCL
HT04	3007	7495	5842	934	UCL
HT05	3637	7480	6029	787	UCL
HT06	3037	7496	5513	897	UCL
PT01	3152	7465	5065	680	PROD
PT02	3224	7484	6228	996	PROD
PT03	3000	7055	4245	1049	PROD
PT04	3472	7117	4499	792	PROD
PT05	4014	6935	5296	544	PROD
PT06	3004	7068	3609	751	PROD
PT07	3055	7500	5754	1092	PROD
PT08	3004	7500	4594	1209	PROD
PT09	3000	7500	4493	1084	PROD
PT10	3141	7500	4844	1060	PROD
PT11	3713	7500	5844	776	PROD
PT12	3567	7353	5972	817	PROD
PT13	3640	7466	6247	612	PROD
PT14	3801	7499	6373	711	PROD
PT15	3450	7366	6409	797	PROD
PT16	3090	7493	5803	1338	PROD
PT17	3000	7368	4231	973	PROD
PT18	3080	7461	4831	973	PROD
PT19	3084	7471	4874	1055	PROD
PT20	3211	7323	4849	939	PROD
PT21	3423	7430	5648	920	PROD
PT22	3483	7233	5354	1010	PROD
PT23	3005	7374	5292	1441	PROD
PT24	3375	7302	5857	930	PROD
HPT01	3075	7497	6002	1233	UCL-PROD

Mean Velocity UCL ..... : 5943 m/s
Mean Velocity Production : 5456 m/s

6 SEISMIC CATEGORIZATION OF ROCK MASS DAMAGE

6.1 Compressional Velocity in Contact Rock

The compressional velocities determined in intact diorite and andesite cores were found to be in the range of 4300 to 5700 m/s this velocity range was used to provide a reference for the bounds of the velocities corresponding to damaged rock.

6.2 Seismic Refraction in the Undercut Level Drives

Using geometric information for 1,579.25 m of refraction survey and arrival times and path-time diagrams were constructed allowing the definition of a radial seismic behavior model of the rock mass. It is postulated that the seismic velocity increases linearly with respect to distance from the excavation.

The distribution of the ranges of the V1 parameter measured in the field was compared with the information of a survey of the frequency of open fractures (FF/m^3) existing in the abutment zone and with the distance to the cave front. The comparison obtained is detailed in Table 5 and Fig. 8 respectively.

Table 5. Relation Between FF/m^3 and V1

FF/m^3	V1 (m/s)
>3	< 4500
1 to 3	4500 - 5000
Closed up veinlet	> 5000

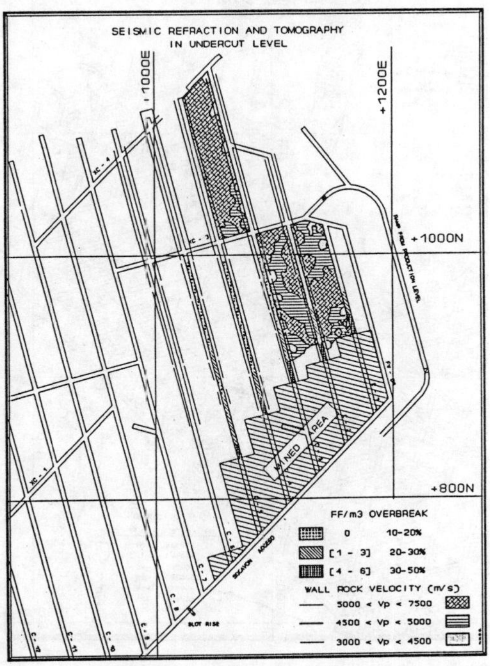

Figure 7 Refraction survey area UCL

The combination of the compressional velocity (V1) at the perimeter of the excavation and the thickness of the rock with a velocity lower than the defined average of 5800 m/s (the so-called RAD (5800)) give an acceptable definition of the damaged abutment zone ahead of the caving front.

In the abutment zone the extension of the loosening of the rock mass (V1 < 4500 m/s) was determined to be in the region of 50 m.

The parameter of RAD (5800) represents to a certain extent a volumetric estimation of the rock with seismoelastic and geomechanical properties below the average velocity of 5800 m/s.

The extent of the RAD (5800) parameter shows a correlation with the influence of the abutment zone in the area studied. This is strongly controlled by the zones of shear-slip faulting in the North East and North-West directions.

The extent of the abutment zone of a distance of the order of 200 m ahead of the cave front is in agreement within the distribution of open fractures, perimeter velocities of less than 5000 m/s and the distribution of microseismic events. With this zone exists a higher degree of loosening ahead of the cave front as shown in Table 6.

Table 6.

RAD (5800)(m)	Description
< 6	Outside the abutment zone
7-8	Abutment zone with rock mass loosening

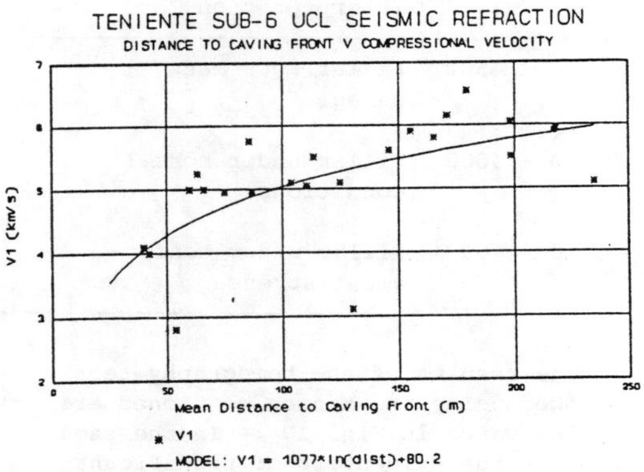

Figure 8 Graph of V1 v cave front distance

The refraction seismic method allows the collection of objective geotechnical information by measuring the relationship of parameters that reflect the volumetric properties "in situ" surrounding the excavations.

The limitations of the method must be considered, however, in the sense that there is no clear directional resolution and it cannot be determined from which specific area the refracted signals are being received. In addition it must be taken into account that the signals received correspond to first arrivals, that is, the wave propagation normally follows the zones of highest relative velocity. Cavities or very damaged zones are not identified by this method.

6.3 Pillar Tomography in the Production Level

Tomography may be understood as being like a photograph of the stress and damage conditions that exist at the time the measurements are taken.

The seismic tomography survey was performed on 23 pillars, identified as PT-02 to PT-24 in Fig. 9 From these measurements the results of pillar PT-02 to PT-11 were discarded because of insufficient seismic ray coverage. The distribution of the compressional velocity values detailed at Table 7, were grouped into intervals of 500 m/s, with values in the range of 3,000 to 7,500 m/s. Compressional velocities of greater than 5,000 m/s are interpreted as being characteristic of pillars under high confinement stress.

Table 7 Compressional velocities and their interpretation

VP (m/s)	Interpretation
< 4500	Pillar Rock Mass Damage
4500 - 5000	Pillar under normal Conditions
5000 - 7500	Pillar under confinement stress

The results of the tomography test showed using the ranges mentioned are illustrated in Fig. 10 . In the same figure the epicenters of significant rockburst events previous to the study are shown.

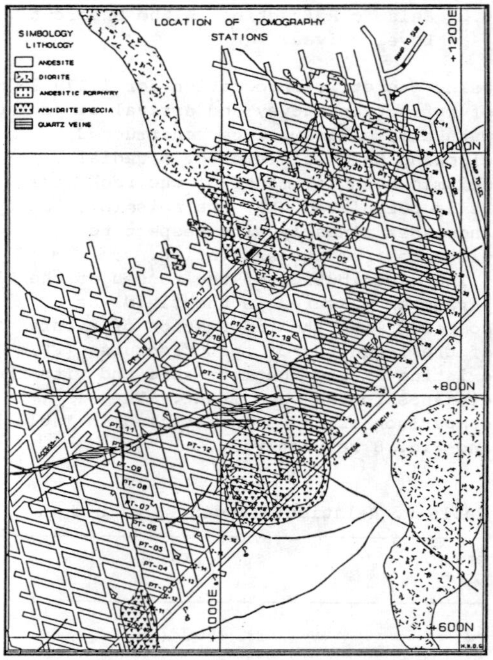

Figure 9 Locations of pillars studied

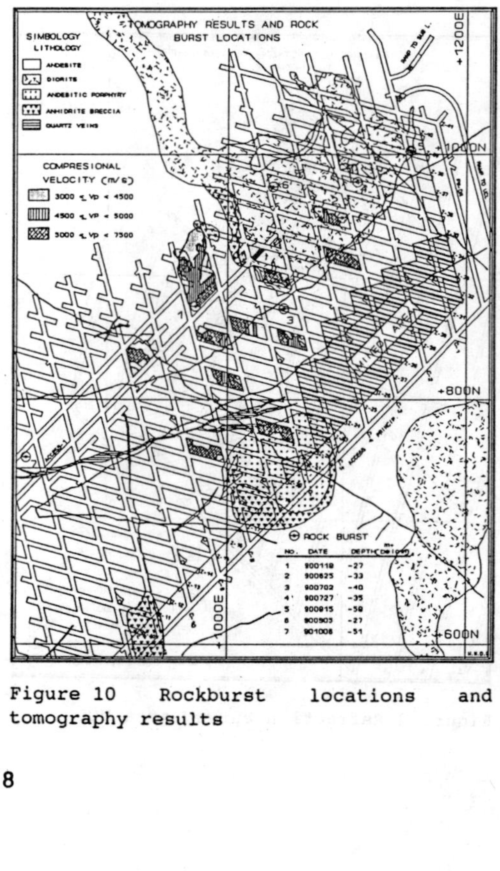

Figure 10 Rockburst locations and tomography results

The compressional velocity has, on one hand a monotonically increasing relationship with applied stress and on the other hand a decreasing relationship with the frequency of open fractures.

7 CONCLUSIONS.

The application of the seismic refraction method to the approximate classification of rock mass using profiles performed in underground excavations permits objective quantitative information to be obtained which will allow the comparison of the zones with respect to tune and formulate a measure of the state of the rock mass.

The average compressional velocity (5800 m/s) determined in the mine is noticably high when compared with similar rock exposed at the surface, whose values area normally below 5000 m/s. This situation certainly reflects to a large degree the effect of confinement to which the rock is subjected.

The V1 and RAD parameters are consistent in showing the variation of the geotechnical characteristics of the rock sumounding the excavations in the undercut level and at differing distances ahead of the cave front.

The subsidence crater in the area studied has not yet broken through to the surface. The seismic characteristics of the abutment zone are the following:

In the zone 50m ahead of the cave front the lowest (3000 to 4500 m/s) velocities were observed, indicating rock damage due to the abutment stress. This same area shows a higher incidence of open fractures observed. A little further to the North, further away from the cave front, velocities of between 4500 and 5000 m/s were measured for V1, indicating a more intact rock mass. Damage mapping in this same area shows that the rock does not present open fractures. At a greater distance from the cave front the highest velocities were measured (5000 to 7000 m/s), showing that this area is under compression. This may be explained by the location underneath the rim of the subsidence crater of older production areas, typically an area of very high stress concentrations.

Seismic refraction showed itself to be an efficient method for sampling large distances of tunnel in a short period of time and did not suffer due to noise during the mine development phase. This is important if this technique is to be incorporated as a normal part of operations. It should be noted however that the process of extraction is more noisy and future surveys may have to be scheduled during shift changes, equipment maintenance or other times when the noise level is lower.

Tomography proved to be an effective diagnostic tool for showing the differences in rock quality in individual pillars. Zones of low velocity indicate damage, while high velocities indicate good rock quality and applied stress. It must be taken into account that tomography indicates the conditions when the tests were done and these will vary over time.

The seismic methods may be applied to the determination of deterioration of the rock mass and pillars over time and the identification of pillars that have effectively failed.

It may also be possible to design fortification criteria based in part on seismic information, and to optimize the geometry and extracion sequence of the zone studied.

8 ACKNOWLEDGEMENTS.

The authors wish to thank CODELCO-CHILE for the authorization to publish this article, which formed part of the initial work for the purpose of characterizing the rock mass damage and to be able to make the corresponding revisions.

We extend our sincere thanks to Mrs. Leticia Florio for her work in typing this article, Mr. John Bonisch for the translation from Spanish to English, the official language of the event, and to Messrs. Sergio Saguas and Jaime Pinto for the preparation of the illustrations.

9 REFERENCES

Howell F.M. & Molloy J.S 1960. Geology of the Braden Orebody, Chile, South America. Economic Geology Vol 55 August 1960 pp. 863-905

Geoexploraciones Ltda 1991. Caracterizacion Geosismica de Rocas Mina El Teniente Tomo 1 Refraccion Sismica

Geoexploraciones Ltda 1991. Caracterizacion Geosismica de Rocas Mina El Teniente Tomo 2 Tomografia Sismica

12 Student category

Innovative Mine Design for the 21st Century, Bawden & Archibald (eds) © 1993 Balkema, Rotterdam, ISBN 90 5410 325 6

A back analysis of a fall-of-ground addressing the effect of stress change on cable bolt capacity

E. Pieterse
Department of Mining Engineering, Laurentian University, Sudbury, Ont., Canada

ABSTRACT: This paper examines a fall-of-ground which occurred in a cable bolted area at a Sudbury mine, to determine if this event was consistent with predictions made by a computer model of cable bolt capacity. The back analysis consists of three major components. The first component is the determination of the anticipated demand placed on the cable bolts, using the computer programs DIPS and UNWEDGE. For this case, which demand is defined as the dead load exerted on the cables due to a structurally defined wedge which decouples from the rock mass. The second component evaluates the capacity of the cables, which is determined by calculating the ultimate bond strength which the cables can develop, using EXAMINE^{2d}, CSTRESS and CABLEBND, then comparing this to the ultimate tensile strength of the cables. The third component of the analysis is a discussion of how a factor of safety relates the cable demand to the cable capacity. Also presented is a conceptual idea explaining the stress re-distribution which occurs inside a wedge as it de-couples from a rock mass.

1 INTRODUCTION

Cable bolts were first introduced in Australia where they were installed for support of cut-and-fill stopes. This usage proved to be ideal since mining induced stresses, in the area where the cables were installed, increased after each cut. The redistribution of the stresses had a clamping effect on the cables. These trials proved to be very successful. Cable bolts are easy to install, are relatively cheap, and have the added benefit that they act as rock mass pre-reinforcement. They are now widely used.

However, there are many reported cases in which mining failures occur in situations where support designs seem to be conservative. Initially, failures resulted from poor quality control of the installation and grouting of the cables. These problems have been eliminated, in most cases, by the development of special grout pumps, fail-proof installation procedures, and improved crew training. Instances still occur, however, where cables lose their original capacity during the course of surrounding mining activities. Ground falls occur as a result.

One probable cause of cable bolt capacity loss is the effect of stress changes in the rock mass containing the cables. These changes may result in a reduction in bond strength, and hence, a loss of cable bolt capacity.

The mechanical aspects of cable bolt failure have been considered from the point of view of the properties of grout and cable, as well as the rock in which it is anchored. Results from investigations indicate that most cable failures occur at the cable-grout interface (Stillborg, 1984) . Bond strength of a fully grouted cable bolt is primarily frictional and depends on the pressure across the cable-grout interface.

1.1 GRC bond strength model (BSM)

The Bond Strength Model (BSM) introduced by Yazici and Kaiser (1991) focuses on the performance of the cable-grout interface, and seeks to explain the development of bond strength at the cable-grout interface. Bond strength is created as a result of friction between steel and grout. Any radial pressure change at this interface causes a change in the frictional forces that retain the rock on the cable. Interface pressure depends on the elastic properties of both grout and rock, the geometry of the borehole and the surface of the cable. Bond strength develops when the pressure at the cable-grout interface increases due to the forced dilation of the cable surface against the grout and confining rock. Dilation is generated as the flutes of the seven stranded cable interfere with the corresponding spiral asperities in the surrounding grout. Bond strength increases until the shear strength of interlocked asperities of grout is reached (Kaiser, Diederichs and Yazici, 1992). Dilation stops as the cable shears through the grout.

The bond strength model relates axial and lateral displacements, pressure at the cable-grout interface and bond strength. The relationship between bond strength (τ) and radial pressure p_1 at the cable-grout interface is given by:

$$\tau = p_1 \tan(i + \phi)$$

Where:

τ = ultimate bond strength

p_1 = radial interface pressure at grout-cable contact

i = dilation angle which is functionally related to p_1

ϕ = cable-grout friction angle

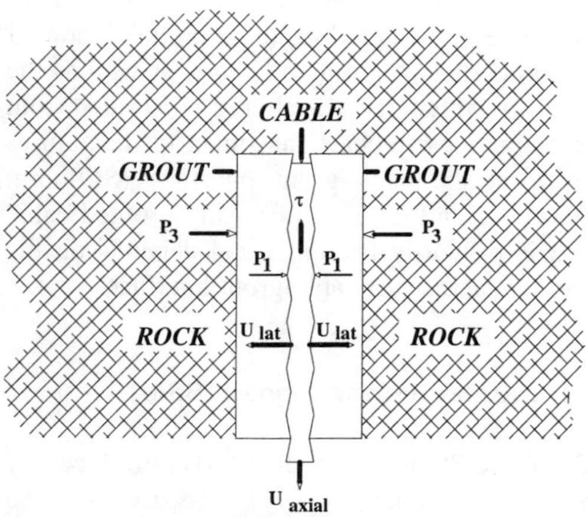

Figure 1 Relationship between radial pressure and lateral displacement.

The relationship of the radial pressure p_1 to the dilation or lateral displacement U_{lat} is a function of grout/rock elastic properties, cable/hole geometries and compressive and tensile strengths of the grout (Yazici and Kaiser, 1992) .

It has become apparent that the capacity of a cable bolt system depends on radial pressure. Generally, this radial pressure is developed by dilation and from any external force on the cable bolt. However, stresses and stress changes that occur in the parent rock mass also play an important role. In fact, the stress changes which occur may be a governing factor. Radial stress on the column of grout immediately after installation is essentially zero. This would indicate that initial field stresses do not play a part in the capacity of the cable. However, stress shadows or concentrations can result from subsequent mining activities, and any changes in the rock stresses near the cable bolt after installation will effect the interface pressure.

A change in the interface pressure, p_1, will have a corresponding influence on the frictional bond strength (Kaiser, Yazici and Nose, 1992). If stress increases, interface pressure increases, hence ultimate bond strength increases. Conversely, if the stress decreases the bore hole may enlarge and the grout may move outward, separating from the cable. As a result, the interface pressure becomes smaller, or remains at zero, counteracting the effects of radial displacement due to axial displacement of the cable. The ultimate bond strength, therefore, decreases.

2 DESIGN METHODOLOGY OF CABLE BOLTING

Broadly, there are two types of applications for cable bolts.

These are:

1. Pre-reinforcement of the rock mass.
2. Holding and/or suspending structurally defined blocks or wedges in place.

Improved understanding of rock mass behaviour and support systems has lead to a more efficient technique of supporting ground, called pre-reinforcement. Which consists of installing support prior to excavating the rock adjacent to the excavation. The principal effect is to limit the displacement of the rock mass to small values. This minimizes the shear and dilation along existing geological structures and preserves their in-situ cohesion and friction angle. The rock mass becomes "self-supporting". The mechanics of pre-reinforcement that are presently being studied at the Geomechanics Research Centre (GRC) at Laurentian University are beyond the scope of this paper. This discussion will focus on the second application of cable bolts as discussed above, i.e., the support of structurally defined blocks.

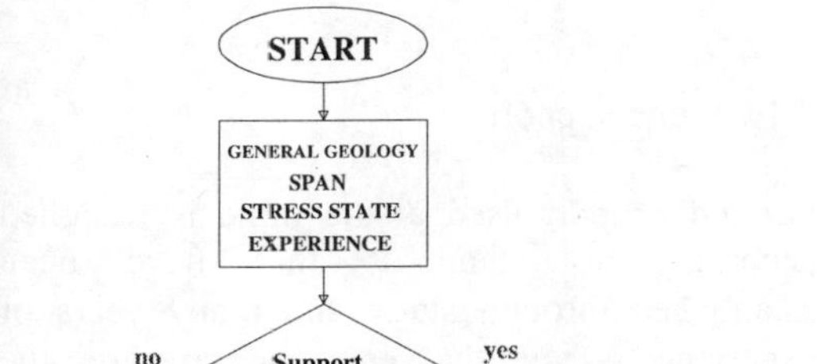

Figure 2. Flow chart for planning cable bolt requirements.

The basic mechanics of holding and/or suspending structurally defined blocks or wedges are well understood. In terms of cable bolt design, the weight of a block or wedge is the "demand," placed on the cables where as the "capacity" is the strength of either the cable bond or the cable steel, whichever is less. Therefore, one can determine if the cable will support its load or will fail, by calculating the cable capacity and comparing it to the demand on the cable.

In order to calculate demand it is necessary to have knowledge of the structural geology. This information is required to identify potential wedges or blocks with respect to opening geometry. The weight of a block can be determined by multiplying its volume by its density. The weight of the block divided by the number of cables in the block, then gives the demand per cable.

Calculating cable capacity requires calculation of bond strength (i.e. cable-grout interface strength) since this usually is the weakest link in cable bolt support. As previously noted bond strength depends on the stress state at the grout-cable interface. Ultimate bond strength depends on mining induced stress changes.

Modeling the stress field twice, once for the time of cable installation and again after mining has occurred, will determine if there is any significant stress change. Then, using the GRC's Bond Strength Model, one can determine if the stress change has had a significant effect on the cable bond strength. With the bond strength known capacity can be determined by multiplying the embedment length by the cable radius by the bond strength.

The steps taken in the planning of a cable bolt installation are summarized in Figure 2. The first part of the cable bolt planning flow chart deals with general structural geology, span, stress state and experience. This refers to the information that a planner would have available, to determine if cable bolt support is required. Obviously if no support is needed then there is no need to follow the cable bolt planning flow chart any further. If cable bolting is required then there is a series of steps to be followed. First, there is the demand determination, followed by the capacity determination.

Starting the demand rationale requires a preliminary design to be done, identifying the number of cables and the pattern used. These patterns are usually a standard for any opening larger than some specific span. In conjunction with the preliminary design, structural geology studies must be done. After the preliminary design is completed computer programs are used to relate the structural geology and the geometry of the preliminary design. The computer programs DIPS and UNWEDGE were used to determine demand. Demand is the weight of a wedge or block that forms from the governing structural geology and the geometry of the designed opening. If the demand is greater than the breaking strength of the cable steel then the number of cables used or the pattern should be changed. If the bond strength is inadequate, based on experience or on the GRC's Bond Strength Model, mix design (w:c ratio) should be reviewed. This brings the design path back to the preliminary design. If demand is not greater than the breaking strength of the steel or the ultimate bond capacity then the design path leads to the next steps, which are concerned with cable bolt capacity.

It is important to note that this paper deals with ultimate capacity and deadload demand. Accommodation of initial stress driven rock displacement is beyond the scope of this work.

The weakest part of a cable bolt is the cable-to-grout bond, and this bond strength depends on the stress change which occur after installation. Hence, computer stress modelling is the first of the capacity calculating steps. Calculating pre- and post- mining stress fields will enable one to predict any cable capacity decrease. If the decrease in cable capacity is insignificant, so that the capacity as mining proceeds continues to exceed then no further design steps are needed. If, however, the new capacity is less than demand, orientation and/or timing of the cable bolting must be altered or additional cables must be installed. This brings the design path back to the computer stress modelling step.

Cable bolt planning procedures allow for designs that consider both demand and capacity of cable bolts. The preceding methodology will be used to study a case history of a fall of ground (FOG) in a Sudbury area mine.

3 BACKGROUND OF MINE STUDIED

The mine where the FOG occurred is located of an off-set from the main elliptical nickel irruptive of the Sudbury basin. The mine is currently seismically active, and has a maximum depth of 1330 m. The mine was originally a shrinkage and blast hole operation which closed in 1978. In 1983 it was re-opened as a vertical retreat mining (VRM) operation, mining several distinct ore bodies. The principal ore zone is a steeply dipping, tabular ore body which varies in width from 8-25m.

The primary VRM stopes were originally designed to be 14 m along strike, 130 m high and separated by 25 m long pillars which were to be mined on a second pass. The stopes were to be filled with cemented back fill, and when the pillars were mined they would be filled with un-cemented rock fill.

3.1 Ground support

Ground support used at the mine is designed according to standards that have been established through study and many years of experience. In the area of concern the following support was installed.

Screen (#6 gauge) was placed with 2 m mechanical rock bolts placed in a 1.25 m x 0.75 m pattern. The whole area was also supported with 2.5 m resin grouted rebar on a 1 m x 0.75 m pattern. This area was also cable bolted. Cables 7.5 m and 10 m in length were installed in a 2 m x 2 m pattern.

3.2 Fall of Ground (FOG)

In early 1991, two rock bursts having Nutli (local Richter) magnitudes of 3.4 and 2.4 respectively occurred at the mine. These bursts were located approximately 110 m below the FOG. They occurred in a stiff pillar with dimensions of 40 m x 7.5 m x 110 m. Accompanying the rock bursts there was a 1500 ton FOG in a top sill on 1000 m level. Figure 3 shows a schematic section through the FOG.

The ore type can be classified as massive to disseminated sulfides. The host wall rock was quartz diorite and was strongly jointed.

A geological controlled wedge seemed to have been present in the area, and the cable supports of this wedge were approximately 30% stripped.

4. ANALYSIS OF THE FALL-OF-GROUND

The methodology outlined in Section 3 was used to study the FOG. The preliminary design stated the cable spacing and pattern used, and this information was then used for the determination of demand and capacity according to the methodology.

4.1 Cable bolt demand.

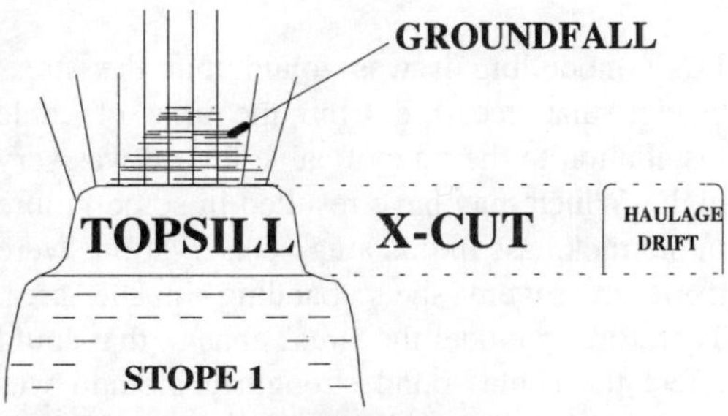

Figure 3 Section through FOG.

Careful geological mapping of the area of concern was undertaken. Eight traverses were established in the X-cuts near the FOG and mapping measurements were taken. This mapping included such features as predominant joint sets, faults and shears. Figure 4 is an equal angle, lower hemisphere stereonet plot of the geological structures mapped in the x-cut adjacent to the area of the FOG. The data collected showed that there are 5 dominate joint sets.

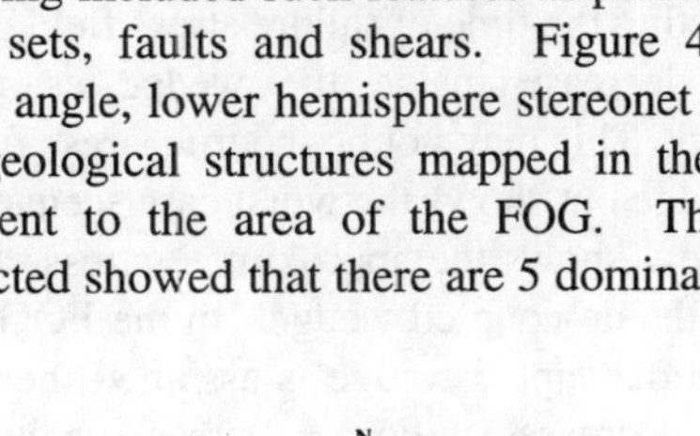

Figure 4 Stereonet of geological structure.

All of these joint sets have an average spacing of 0.5 to 3 m, with slightly rough, hard surfaces, and are normally tight. In addition to these major joint sets there are occasional random joints.

A critical part of cable bolt planning is determining the potential for wedge failure. The computer program UNWEDGE developed by the Rock Engineering Group, Department of Civil Engineering, University of Toronto, was used to identify possible wedges. A total of ten analyses of all combinations of the five joint sets was run for each of the orientations of the x-cuts. This analysis determined that many wedges can form, but only few lend themselves to further study. Many of these wedges form in the walls or were extremely long and thin and were not considered.

Another wedge was formed by the intersection of a shear (referred to as plane 6) and planes 2 and 5 in Figure 5. The wedge formed by these planes was found using UNWEDGE. A tunnel width of 10 m, and orientation of 340° was used which formed a 481 ton wedge. The volume of this wedge was relatively small compared to the amount of ground that actually fell. However, since UNWEDGE uses the width of the tunnel to determine the size of the wedge, and as the FOG actually occurred at an intersection, further analysis was necessary.

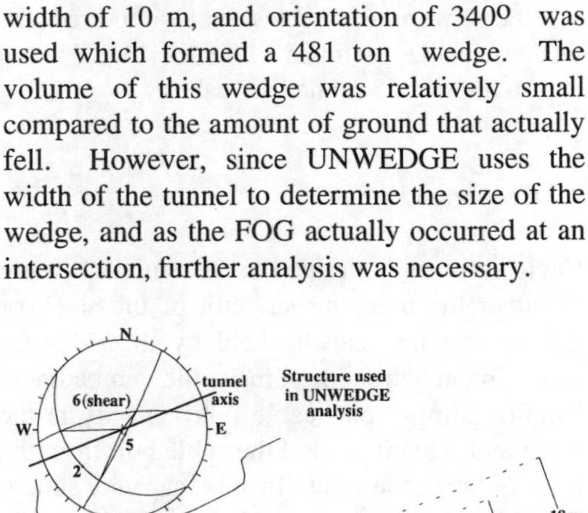

Figure 5 Wedge formation in intersection.

A stereonet of the shear and intersecting planes, with the tunnel direction, was plotted on a plan of the area of the ground fall. This is shown in figure 5. The wedge that formed with a tunnel width of 10 m was then plotted in the intersection. Then this wedge was expanded to fill the intersection. An equivalent tunnel width, which would form the larger wedge, was found to be 18 m. Finally UNWEDGE was run with the new tunnel width. This resulted in a wedge with a weight of 2512 tonnes and an approximate height od 17 m forming in the intersection.

Since the actual FOG was observed to be truncated at 5 m height, the upper part of the model wedge was subtracted. This small top wedge had a weight of 744 t, therefore the weight of the bottom part is 1768 t. This is similar to the actual amount of ground that fell in the area of concern.

Once the geometry and mass of the wedge were determined it was possible to determine demand for a cable. The wedge that formed at the intersection of the x-cut superimposed on the cable bolt locations showed the number of cable bolts that intersected the 1768 t wedge. As there were 63 cables that supported the wedge, cable demand for the FOG was approximately 28 tonnes per cable.

4.2 Cable bond capacity

Cable capacity is defined as the smallest figure of either the breaking strength of the steel, or the force which can be held by the cable-to-grout bond strength times the embedment length. Simply put, cable bolt capacity is the force that will either pull the cable bolt from the hole, or break the steel. Ideally, capacity should be greater than the breaking strength of the steel. When this is the case, the governing factor in the support design is the strength of the cable steel (in a gravity loading situation) such that optimum use is made of each cable.

4.3 Modelling in 2D

Field observations suggested that there had been
a bond strength decrease which lead to the stripping of the cables. Using EXAMINE 2D, a model was run to determine the stress change in a wedge de-coupling from a rock mass. The 2D stress fields do not truly represent the actual stress in the rock mass, but can be used to give an approximation of the internal stress change of the de-coupling wedge. Cable parameters that were used in the models were:

grout modulus	- 10500 MPa
grout comp. strength	- 61 MPa
grout tensile strength	- 3.94 MPa
grout-steel friction	- 28 deg.
cable radius	- 0.016 m
borehole radius	- 0.0254 m

From modelling it was found that the stress change that occurred from the time of cable installation to the time of cable failure was very high. Which may have resulted in some failure in the rockmass in this area. Observations were made of severe shear banding in the area. Therefore to model the stress change that could effect the cables bond strength a method was used that subtracted the stress field at the time of cable failure from a zero stress field. This datum field with no stress is shown in Figure 6. The stress field with the de-coupled wedge was needed so that it could be used as a datum when subtracting the stress fields. This gave an area of zero stress initially inside the wedge then subtracting the time of failure stress field gave a stress decrease inside the wedge as it de-coupled. This may not be the true stress field in the wedge but shows the worst case scenario.

Figure 7 shows the approximate stress change inside the de-coupled wedge. In the FOG area, when a decoupled wedge is assumed there is a stress decrease. For a stress change of approximately -15 MPa inside the wedge, there is a 50 to 100% decrease in cable bond strength.

Since the total de-coupling of the wedge is the worst case an average loss of bond strength for the wedge of 75% will be used in determining average bond capacity. The original cable bond strength was 7.5 MPa. Reduced to 25%, the cable bond strength inside the de-coupled wedge, is 1.9 MPa. Since there was a dominating wedge in this area the bond strength capacity was determined on an individual cable basis. The embedment length of each cable was determined so that the cable bond capacity (t/cable) could be found.

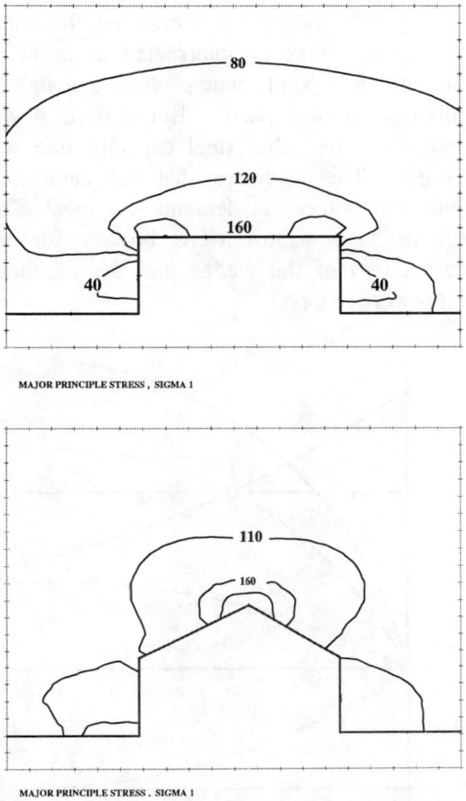

Figure 6 Stress fields at time of cable failures.

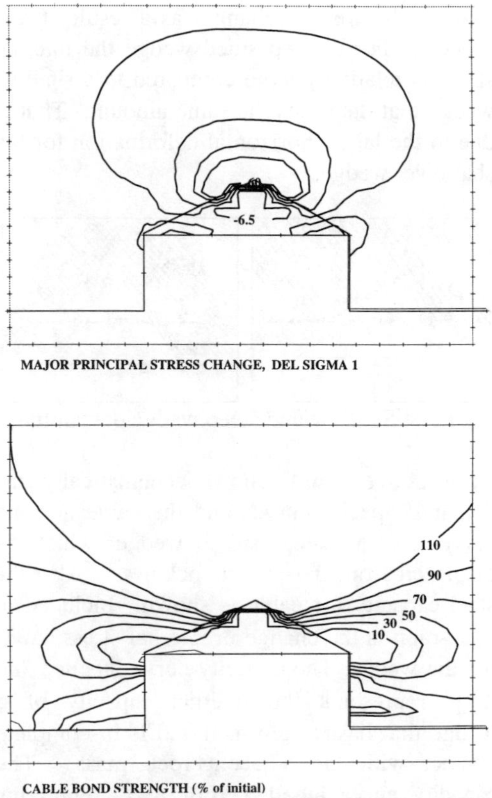

Figure 7 Stress change inside the decoupled wedge and the cable bond strength as a %.

The cable bond capacity was determined by multiplying the double cable circumference by the embedment length to find the surface area, and multiplying by the bond strength. Three bond capacities was calculated Initial cable bond capacity was calculated using a bond strength of 7.5 MPa. A final cable bond capacity was calculated using a bond strength of 1.9 MPa. As the wedge de-couples from the rock mass the cables with a cable bond strength less than the breaking point of steel start to strip. This movement allows for dilation to occur, increasing the frictional resistance, and finally the cable is at its maximum bond strength. As this occurs the other cables that have a cable bond strength above the breaking strength of steel take the weight of the de-coupling wedge. At this time these cables break and the wedge's weight is transferred to the perimeter cables, which have lower cable bond strength. These cables then strip from the wedge and the whole thing comes down.

The value for the probable cable bond capacity is 2350 t for all cables passing through the 1768 t wedge. Since there are 63 cables the average cable bond capacity, for the FOG area is 37 t.

5. WEDGE DE-COUPLING THEORY

The wedge de-coupling theory provides an explanation of why cables either strip or break in the FOG area. To help visualise this concept Figure 8 shows a simplified representation of the mechanism. This wedge is shown decoupled from the surrounding rock mass. Clearly, for any downward deformation (δ) there is also a horizontal deformation.

For the case considered in this paper, the wedge does not de-couple from the rock mass immediately but deforms with the rockmass. This deformation continues until the wedge decouples from the rock mass. Stress that flows through the wedges changes as a result of this process. For a steep sided wedge the internal stress is relatively large compared to a shallow wedge that displaces the same amount. That is due to the larger horizontal deformation for the shallower wedge.

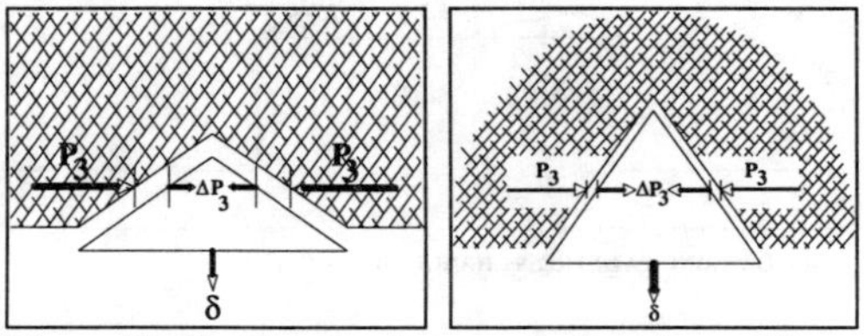

Figure 8 Shallow and steep wedge decoupling

The above figure shows schematically the demand/capacity curves and the cable demand curve for a steep sided wedge which is displaced from the parent rockmass. Also the steel capacity for cable is shown. Included in this graph is the change in internal stress (ΔP_3) for the wedge. The capacity curve labelled "no ΔP_3" represents the internal capacity of a wedge that hasn't deformed and is in complete contact with the adjacent rock mass. The capacity curve labelled "full ΔP_3" represents the capacity of a wedge in which all internal stresses have been relieved. This would occur if the wedge was totally de-coupled from the rock mass. These two curves define a region bounded by the upper and lower ranges in which capacity may fall as internal stress changes occur. The demand curve represents the amount of force required to hold the deforming wedge in place. As the wedge becomes completely de-coupled from the rock mass the demand equals the dead load.

In the lower quadrant of the graph this curve represents the change in internal stress (ΔP_3) of the wedge as it deforms. This change is gradual because the change in horizontal distance for some vertical deformation is small for a steep sided wedge. Hence the stress propagating from the rock mass to the wedge slowly becomes smaller as the wedge deforms.

To determine the actual capacity curve the ΔP_3 curve is subtracted from the "no ΔP_3" curve resulting in the capacity curve. This curve lies between the "no ΔP_3" curve and the "ΔP_3" curve as expected. Following this curve the demand curve is intersected at point A. This is the point where the demand is equivalent to the capacity. But as point A was approached the cable steel capacity line was crossed. This indicates that the cable bolt would fail before the demand was met. This explains why cables were broken for the internal part of the wedge and the perimeter cables were stripped.

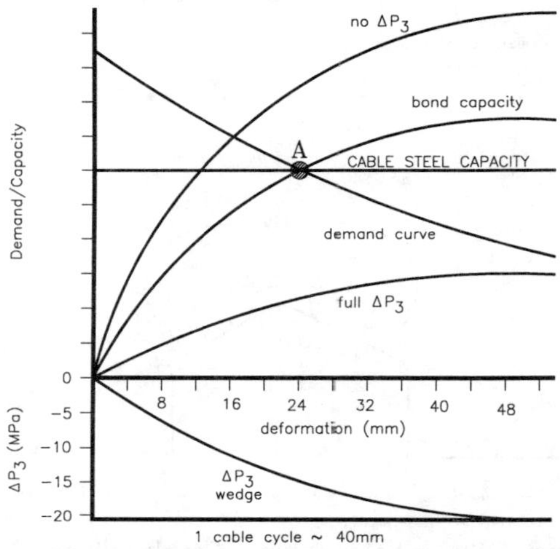

Figure 9 Demand/capacity curves and the cable demand curve for the steep sided wedge

A number of mechanisms which occur simultaneously, some of which increase demand, some decrease bond capacity are briefly summarised:

- Stress field increases causing rockmass deformation. The demand on the cables at this point includes the driving forces due to rock mass divergence.
- Wedge deformation occurs to a point where it decouples reducing the demand to the dead load of the wedge.
- During the above displacement reduction in stress inside the wedge causes a loss in the bond strength.
- A shallow sided wedge stripes all cables because of the large stress reduction and shorter embedment lengths resulting in a lower bond strength.
- A steep sided wedge requires more deformation before decoupling and

before any significant stress reduction occurs.

- This required deformation is prevented by the inner cables due to the long embedment lengths. These cables break and the wedge's weight is transferred to the outside cables which strip.

Ideally the capacity curve, at its maximum, should intersect the demand curve under the steel capacity line. This would result in the wedge and rock mass being supported.

6. CONCLUSION

There have been indications that stress changes in the rock mass induced by mining may significantly reduce cable capacity. Using the computer model, based on the GRC's Bond Strength Model (BSM) the effect of induced stress changes was evaluated. The model assumes that the bond strength between a cable and the surrounding annulus of grout depends on frictional resistance, and this assumption was supported by observations and by recent testing. Since frictional resistance is directly proportional to normal stress, increases in radial compressive stress will generally improve cable capacity. Conversely a decrease in radial stress will reduce cable capacity.

Using this model, an example of an actual fall of ground (FOG) in a cable bolted area at a Sudbury mine was studied to determine if the effect of stress change on cable bolt capacity could be incorporated into design.

The demand on the cable bolts was determined based on an extensive structural analysis using the computer programs DIPS and UNWEDGE. From this, it was seen that a significant wedge formed in the FOG area. With this information the demand on each cable was determined.

To determine the predicted cable bolt capacity stress redistribution due to mining was analysed and utilised within the CSTRESS and CABLEBOND computer programs. Capacity decrease could be explained based on the mechanics of a rock wedge which de-couples from the parent rock mass. A model was run in 2D to determine the approximate stress change in a wedge de-coupling from the rock mass. This analysis demonstrated that there could be a 50 to 100% decrease in cable capacity due to this mechanism of wedge de-coupling.

It is clear that the actual performance of a cable bolt will depend on a number of complex and inter-related factors which can change the radial stresses acting on the cable, and hence, can change the cable bolt capacity. As these radial stress changes are not fully incorporated into current design analyses of cable bolts, a careful and prudent approach must be taken. This analysis, which attempted to incorporate the effects of stress change, indicated that a factor of safety between 1.3 and 1.5 would be calculated for the areas in which falls of ground actually occurred. These are relatively low values, given the complexity of the mechanisms involved, and it is suggested that a factor of safety of at least 2.0 should be used for design purposes.

REFERENCES

Kaiser, P.K., Maloney, S. and Yazici, S., (1991), A New Perspective on Bolt Design, Proc. of the 93rd CIM AGM, Van., 16 p.

Kaiser, P.K., Diederichs, M.S. and Yazici, S., (1992), Cable Bolt Preformance During Mining Induced Stress Change - Three Case Examples. International Symposium on Rock Support, Sudbury.

Kaiser, P.K., Yazici, S., nad Nose, J., (1992), Effect of Stress Change on Fully Grouted Bolts and Cables International Journal for Rock Mechanics and Mining Sciences & Geomechanics Abstracts, 29p.

Stillborg, B., (1984), Experimental Investigation of Steel Cables for Rock Reinforcement in Hard Rock, Ph. D., Lulea University, Sweden, 127 p.

Innovative Mine Design for the 21st Century, Bawden & Archibald (eds) © 1993 Balkema, Rotterdam, ISBN 90 5410 325 6

Analysis of new flexible and rigid composite tendons for mining

U.H. Khan & F.P. Hassani
Department of Mining and Metallurgical Engineering, McGill University, Montreal, Que., Canada

ABSTRACT: Significant potential applications for flexible and rigid composite tendons exist in the mining industry. This paper presents an analysis of the characteristics of fourteen commercially available composite tendons along with their pull out test response. It also considers the effect of different surface profiles on their support behaviour. In addition, various comparable applications of composite tendons in civil engineering are mentioned. This new generation of support system not only offers the potential to ensure the stability of underground excavations but also eliminates some existing concern of conventional steel cable-bolt support systems such as corrosion, transportation and storage, non-cuttablity.

1 INTRODUCTION

The potential application of composite materials has been realized since World War II. The aerospace industry pioneered the use of such material, for example, to replace steel and aluminum by glass fiber reinforced composites. A prime reason was that steel and aluminum alloy performed poorly against corrosion and fatigue.
High market demand, increased production and advanced technology has opened other areas of composites in the sports, recreational and electronics industries. Generally, interest in the development of specific use for composites has been initiated by industry. This paper aims to outline the characteristics of some high performance composites and points to potential applications to the mining industry.
Composites can be defined as [Rosato, 1982] " the combination of two or more materials, present as separate phases and combined to form structures that take advantage of certain desirable properties of each components. The constituent materials can be organic, inorganic or metallic in the shape of particles, rods, fibers, plates, foam, etc." More specifically, suitable fibers are combined together and held in any desirable shape by a binder material, known as the matrix. Fibers are the load bearing component and the matrix materials are responsible for the dimensional stability of the product and the protection of the fibers. Fiber-reinforced composites are of two types - discontinuous or short-fiber composites, i.e. the properties of composites change with fiber length. Continuous fiber composites are those in which there is no change in the elastic modulus of the composite caused by any increase in fiber length. The application of short fibers is in plates, shells, beams, column, etc., where structural elements should sustain both compression and tension forces. The cross-sectional areas of short-fiber composites are more than their counter part. Continuous fiber composites are primarily manufactured to replace heavy, tensile bearing structural elements. Such geometry makes composites weaker under compression load. The other

advantage of combing continuous fiber is that it makes the end-product a redundant material i.e. failure starts with the breaking of a few fibers and simultaneously redistributes the load onto other fibers; this thus prevents a catastrophic failure.

In the beginning, composites were not very attractive due to their very high cost and poor mechanical properties. Since 1970, however, improved raw materials as well as low cost, and efficient manufacturing has resulted in increased production of composites with much improved mechanical properties. Figures 1 and 2 give some idea of the drastic changes in composite production and cost over the last two decades. The development of aramid fiber, a man-made fiber, played a very vital role in the increased production.

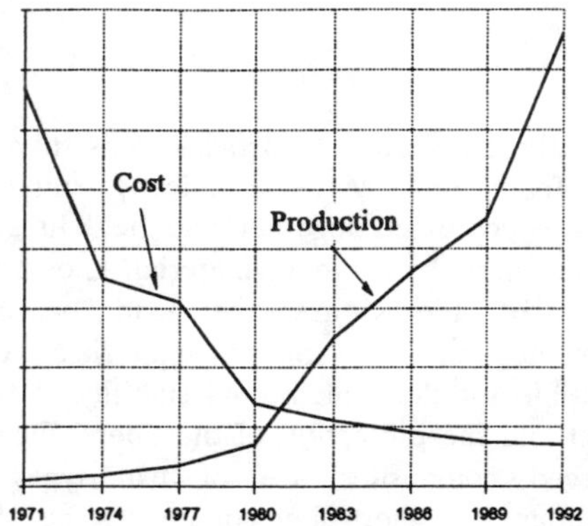

Figure 1: Cost versus production of composite (Technical brochure, BFGoodrich, 1986)

2 RAW MATERIALS

High strength glass fiber, developed in the 1960's, was a key technological development. The uses of fiber-glass reinforced composite range from non-structural, low-performance applications, such as panels in aircraft and appliances, to high-performance applications, such as rocket motor cases and pressure vessels. Competitive prices, availability, ease of processing, and high strength are the reasons

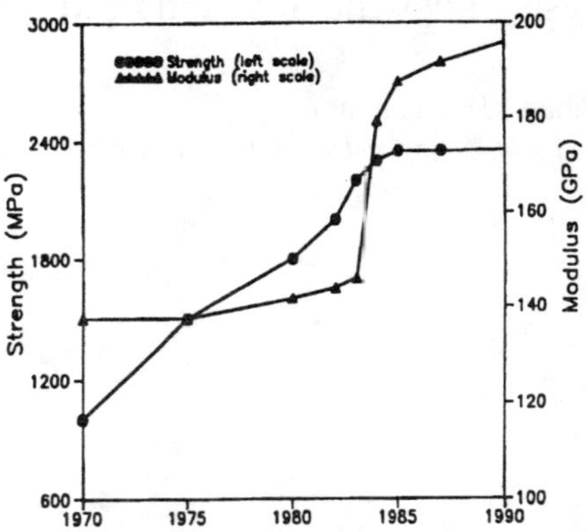

Figure 2: Improvement in the strength and modulus of composites, since 1970 (Technical brochure, BFGoodrich, 1986)

for widespread glass-fiber use. Its performance, however, was still poor in aggressive environments and against long-term loading at stress levels exceeding threshold failure stresses. There are many types of glass-fibers available for specific applications. For example, the products of high-alkali glass (A-glass) have high chemical resistance; low-alkali glass (E-glass) gives excellent electrical insulation properties; whilst magnesium alumuno silicate (S-glass) gives high tensile strength. Research has progressed considerably in combining the qualities of different materials in order to produce a universal and optimum composite material. Thermoset resin matrices such as polyester, vinyl ester and epoxy are mostly used as matrix material in glass fiber products. A recent innovation in the manufacturing process of glass fiber made it possible to produce any shape and size of end-products, partially or fully reinforced. Aramid, man-made fibers are five times stronger than steel of the same weight and ten times stronger than aluminum. High strength, high modulus, and heat and corrosion resistance are fundamental properties of aramid fibers. These were commercially introduced as Kevlar by DuPont in 1960. Unlike glass-fibers, aramid fibers are used only as continuous fibers, so their applications are limited to tires, hoses, conveyors, cables, and other applications where tensile load bearing is the main function. Due to its high cohesiveness, aramid can absorb greater energy than brittle fibers, making them suitable for lightweight armour. The recent development of Twaron, another form of aramid fiber, and its end-products, Arapree, are finding wide application in prestressing in place of steel tendons. The end-products of aramid are manufactured through a pultrusion process. The potential applications of aramid cable bolts and rockbolts in underground mining are considered to be significant.

Carbon or graphite composites though superior in strength and E-modulus, did not gain popularity outside the military, aerospace and recreational industries due to their excessive cost. The product of carbon/graphite fibers have not been commercially available in civil or mining engineering. However, the Japanese are working presently to produce carbon prestressing tendons (Minosaku, 1992).

There are other fibers such as boron, silicon carbide, polyethylene, etc. but they do not meet the requirements of the mining industry. There are two categories of bonding matrices; thermoplastic and thermoset. Thermoplastic resins are capable of being modified by variation of temperature, and are more environmentally friendly. Thermoplastic resins, due to their insufficient strength, have not been widely used. Their use was limited to applications where ductility was required. However, since 1980, thermoset resin was combined with thermoplastic, to optimize its strength, ductility and environment friendliness.

Thermoset resins are very widely used because of their high performance. Epoxy resin, polyester and vinyl ester are some examples. These resins are brittle and non-recyclable because, once cured, they cannot be reused.

3 MANUFACTURING PROCESSES

There are three methods of producing composite end-products: lay-up, pultrusion and filament winding.

Hand lay-up or spray-up is the easiest method where chopped or continuous fibers are sprayed in different layers on the mold surface. These layers are bonded together by matrix materials. Numerous reinforcement additives and fillers can be combined with the resin to customize the mixture to meet specific requirements.

Pultrusion or a Continuous Manufacturing Process is a continuous feed of fibers, in selected directions, which is then impregnated with resin and pulled through a heated die to give the shape of the final section (e.g. tubes or I-beams). Partial or complete curing occurs during passage through the die. Braiding is further developed in the pultrusion process. In braiding, continuous fiber in the shape of threads are mounted (as spindles) on the "carrier" (carrier refers to the piece of machine that holds the spool of raw material to be braided). Continuous threads are pulled through the pre-sized, pre-arranged orifice and carrier spines at the same time. In this process flexible tubular tendons are impregnated into epoxy, which, on curing, gives a solid tendon. Stiffness of the tubular structure depends on the matrix, while mechanical properties depend on the orientation of fibers.

Filament Winding Continuous reinforcement in the form of roving or mono-filaments are wound over a rotating mandrel. Specially designed machines, traversing at speeds synchronized with the mandrel rotation, control the winding angles and the placement of the reinforcement.

4 APPLICATIONS OF COMPOSITE MATERIALS IN MINING

Composite materials have several potentially significant applications in mining. They have already been used in milling processes and in conveyer belts. The following are some of the potentially significant applications in mining:

1. As underground support structures, composite materials have the capacity to replace heavy steel structures like H or I section beams and columns. Moreover, composites can also be fabricated into rigid arches (curved H-sections), yieldable arches, posts and linear plates for shafts. Rock bolts and cable bolts (flexible tendons) of composites and their applications are discussed in the following sections. In addition, composites qualify to be used as ore pass liners

to replace timber, steel, concrete or the various combinations of these materials currently used.

2. Material transport is one of the higher cost operations at the mine site. Haulage equipment such as trucks, scrapers, mine cars and shuttle cars are very heavy and expensive to operate. Composite materials could significantly reduce the weight and increase the performance of haulage equipment. For example, the ratio of payload weight to truck tare weight could be increased from 1.5 to 2.0, with the result that a full 170-tonne truck would carry 30 tonnes less dead weight and require 10 percent less horsepower. Similarly, filament wound drive shafts and suspension systems of reinforced plastic might save as much as 50 percent over conventional steel component weight.

3. The recent generation of robotics systems have heavy structural members fabricated from commercial metals. These members provide rigidity and structural integrity but generate large internal loads that inhibit their performance. For example, current robots are characterized by payload weight-to-arm weight ratios of only about 1:20. However, if a superior range of robotics products are to be developed for the mining industry, then high strength, light weight articulating structural members must be fabricated to increase payload, reduce energy consumption, improve the dynamic response of the system and increase the operating speed, at the same time ensuring the structural stiffness is not compromised.

4. A flywheel of composite materials should be able to store many times more energy per unit weight than metal flywheels.

5. The conveyor belt industry has already used high performance composite materials. Load bearing steel components have been replaced by thermoplastic and filament-wound components. In longwall tunnels, a chain conveyer is used in transportation. Productivity depends on undistributed long testing performance of the chain conveyer. The existing metal chain conveyer cannot be guaranteed against corrosion, blockage and economy (due to heavy weight). All the existing metal components like rails, wheels, transportation equipment etc. degrade due to corrosion in salt mines. Composite components work successfully at low cost in such salt mines in place of existing metals. A durable, light weight and economical chain conveyer can be fabricated from polyethylene thermoplastic composite material.

6. Underground storage rooms to store chemicals can be manufactured by fire and corrosion proof composites.

5 CIVIL ENGINEERING APPLICATIONS

The focus for more immediate application of composites in mining is in underground support. Therefore, all the examples reviewed here are from civil fields of relevance to support systems.

Fiber-reinforced prestressing tendons were used in the construction of Ulengergstrane Highway bridge in Dusseldorf, Germany, in 1986. The two-span slab bridge (max. dead load of 60 tonne per vehicle) has 21.3 and 25.6 metre (69.9 and 84.0 ft) spans. The bridge has been monitored since its opening for traffic and it has been working as a normal structure (Wolff and Miesseler, 1987).

Beams with a span of 9.0 m length and 60.0 cm height were prestressed by a 19-bar tendon with a permissible prestressing force of 600.0 kN (134.9 lbs). Test results showed that the beam was more ductile than the conventional prestressed version and it returned to its initial straight position after a full relaxation of the test load (Wolff and Miesseler, 1987).

In Japan, a pretensioned prestressed concrete slab highway bridge with a 5.76 m (18.9 ft) span and 7.0 m (22.9 ft) width was constructed with carbon tendons (Minosaku, 1992). There are several other bridges constructed in Japan and Germany using composite tendons. The capacity and performance of these bridges are equivalent to concrete bridges.

Nani et al. (1992) investigated and determined the transfer length of composite tendons, made of braided epoxy-impregnated aramid fiber, embedded in concrete members. Ballinger (1990) documented applications of composites as structural components in four 35 ft high by 35 ft square rooftop turrets that house radio antennae erected on glass fiber-reinforced I-shape. Bedard (1992) discussed highway bridges damaged by corrosion and composite reinforcing bars as a remedial measure.

6 COMMERCIALLY AVAILABLE COMPOSITE TENDONS

An extensive market survey was conducted to evaluate composite tendons compatible with mining constraints. After scrutinizing this data, fourteen tendons were selected for further research and testing. The merit of selection was based on cost, as well as the mechanical and physical properties of the tendons. Although tendons can be made of identical material, their properties frequently vary widely. Therefore, each patented composite product is briefly discussed.

6.1 Arapree

Arapree is a braided epoxy impregnated aramid fiber. It is manufactured exclusively to be used in place of steel tendons in the prestressing concrete industry. The combination of low density (1250 kg/m3), high strength (2800 MPa) and high elastic modulus (125 GPa) makes Arapree an ideal choice for reinforcing potentially unstable rock masses. Arapree is available both in sanded and helical wound round shape and in rectangular strip shape with a regular pattern of small nobs. The rectangular shape is more suitable for mining. Arapree can be used both in place of (flexible) steel cable bolts and (rigid) rockbolts. The cross-section dimension of Arapree is 20 x 5 mm and its breaking load is more than 13 tonnes.

6.2 Weldgrip Fiberglass Rockbolts

These are manufactured specifically as rockbolts and their potential imminent use is in UK coal mines. The load required to break a 25 mm diameter solid Weldgrip rockbolt is more than 30 tonnes. The surface profile is very rough with engraved nobs. These properties make this product also potentially ideal for bolts in continuous mining applications in hardrock mines. In addition to high strength, Weldgrip bolts are of high E-modulus, anti-static, non-corrosive and dimensionally stable, and come with all accessories such as end-plates, nuts, bolts, etc. Weldgrip bolts are manufactured by pultrusion processes with glass fibers and polyester resin being the main constituents.

6.3 Weidmann

Weidmann products are fibreglass reinforced thermoplastic and thermosetting plastics. Weidmann rockbolts are developed exclusively to be used as cuttable rockbolts in mining support systems. Rod breaking load ranges from 160 kN to 350 kN. Design parameters and an installation method are recommended by the Weidmann company, however, there seems to be little remedy for the problems relating to the smooth surface profile. Weidmann rockbolts have only been used in soft rock mines, e.g. coal mines. Parameters relating to hardrock applications need to be developed. The other disadvantage is the very high cost of Weidmann rockbolts.

6.4 Polystal

This is another member of the polyester resin and (E-) glass fibre family. Polystal tendons can be used as rockbolts due to their high tensile strength and chemical, thermal and corrosion resistance properties. The disadvantage associated with Polystal also relates to its smooth surface.

6.5 Polyglas

Polyglas is a combination of polyester resin and glass fibres. Structural shapes such as angles, channels, I-beams, and round and square solid and hollow tendons are easily pultruded with Polyglas. The Polyglas ultimate tensile strength is 210 MPa with a 50 MPa tensile modulus. Polyglas tendons have a smooth surface profile which leads to weak bonding between the grout and rockbolt interface. A modification would therefore be required, either in the surface profile of tendons or in the installation techniques.

6.6 Celtite Rockbolts

These comprise a compound of fibreglass reinforced thermoplastic and polyester resin. Celtite rockbolts have high strength and stiffness, creep and fatigue resistance, with a low coefficient of expansion and dimensional stability. In field tests, a maximum of 100 kN pull-out load was achieved with resin grout as well as cement grout, whereas the steel rockbolt pull-out load was 125 kN (Lapointe, 1990). The main disadvantage of Celtite rockbolts appears to derive from the associated loose fibres which cause skin irritation and which can be inhaled during handling and installation.

6.7 Kodiak Fiberglass-Reinforced Plastic Rebars

These bars are another product of fiberglass and resin. They were developed to be used in the concrete industry to replace existing deformed steel bars. In the manufacturing process, a band of glass fibers is wound around Kodiak FRP in a spiral shape to create a "deformed" surface. It is expected that this deformed shape would provide the same rebar-to-grout bonding as obtained in the existing steel deformed rebar-to-concrete bond strength.

6.8 Isorod Composite Rockbolts

Isorod tendons are glass fiber-reinforced composite rockbolts, manufactured in Canada. These products have been specifically developed as reinforcing bars for concrete construction. Isorod bars have a spiral shaped surface profile.

6.9 PSI Fiberbar Composite Rockbolts

PSI Fiberbar are glass fiber-reinforced composite rockbolts, manufactured in the USA. These products have been specifically developed as reinforcing bars for concrete construction. PSI Fiberbar bolts have a deformed (like steel rebars) surface profile.

6.10 Thiessen Fiberglass Rockbolts

Thiessen rockbolts are also made up of glass fibers and have a deformed surface profile i.e. a band of glass fibers spirally wound around the bolts. These bolts are presently used for rock reinforcement in ore passes by Inco Ltd. in their, Manitoba Division.

6.11 Kevlar

Kevlar is a new type of fibre, introduced by DuPont. Kevlar fibres are high strength (more than 3500 MPa) and high modulus fibres, being chemically and mechanically stable over a wide range of temperature. Kevlar ropes have been successfully used in marine applications because of its high strength and light weight. Kevlar is more expensive than other fibers and its surface is also very smooth.

6.12 Extern fibreglass

Extern fiberglass is a combination of fibreglass and thermoset polyester or vinyl ester resin. Extern fibreglass is high strength and lightweight. It is a dimensionally stable, corrosion resistant and anti-skid surface profile product. Extern is produced in numerous structural shapes like I-beam, L-beam, flat-plate, threaded nut and bolts, etc. Extern can be used as rockbolts only. It is much more expensive than steel products.

6.13 Newport Composites

Newport composites (NC) are manufactured through a braiding process and end products are in the shape of rigid or flexible tubular tendons. High strength, low or high elasticity modulus (can be modified), chemical and corrosion resistance, and light weight are the properties of NC tubular tendons. Cement or chemical grout can be injected from inside and outside of the hollow NC tendon, which can give a high bonding strength. NC is more expensive than several other composite materials and because of its unique flexible nature, the existing parameters of support systems cannot be applied to it.

6.14 Ryton-PPS

Ryton-PPS is a pultruded glass or carbon fibre and polyphenylene sulfide-based composite. Ryton-PPS has numerous structural shape products like solid and hollow pipes, flat and curved plates, molded shapes, etc. Ryton-PPS tendons are very expensive compared to other composites.

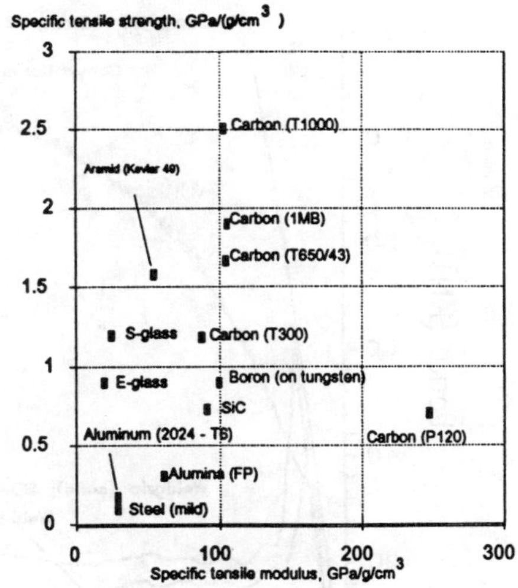

Figure 3: Comparison of different fibres (Reinhard & Clement, 1987)

7 LABORATORY EXPERIMENTS

The basic mechanical properties of all the above mentioned patented composite tendons are available in the commercial literature. These values, however, depend on the net cross-section of the fibers rather then the tendon itself. Figure 3 shows graphically specific tensile strength versus specific tensile modulus of various types of fibers. In order to evaluate the performance of tendons, pull out tests have been conducted in the laboratory of the McGill Rock Mechanics Research Group. Since the chucks (or grips) were not available for all types of tendons, a technique as shown in Figure 4, has been developed. A relatively small embedded length is used so that shear strength developed in at tendon-grout interface should be uniform throughout the embedded length. An embedment length of approximately 110 mm was selected. Results of pull out testing of thirteen different types of tendons is shown graphically in Figures 5, 6 and 7. In order to have an over all comparison, pull out behaviour of conventional steel is shown in Figure 5, and test results of all the conventional bolts is shown in Figure 6 (Stillborg, 1986). A direct comparison of these bolts is not possible because tendon variables such as diameter and surface roughness are not constant. A direct compression is most likely possible by comparing the shear strength of the tendon-grout interface. The following equation (Littlejohn, 1992) is used to find the shear strength:

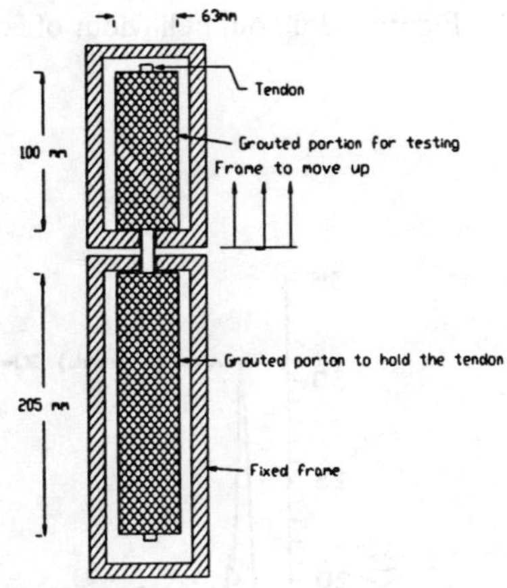

Figure 4: Pull out technique used in the experiments

$$\tau_{peak} = \text{Maximum pull force} / \pi DL$$

Where, D and L are the diameter and length of tendon.

Table 1 shows various parameters as well as the interface shear strength of composite tendons.

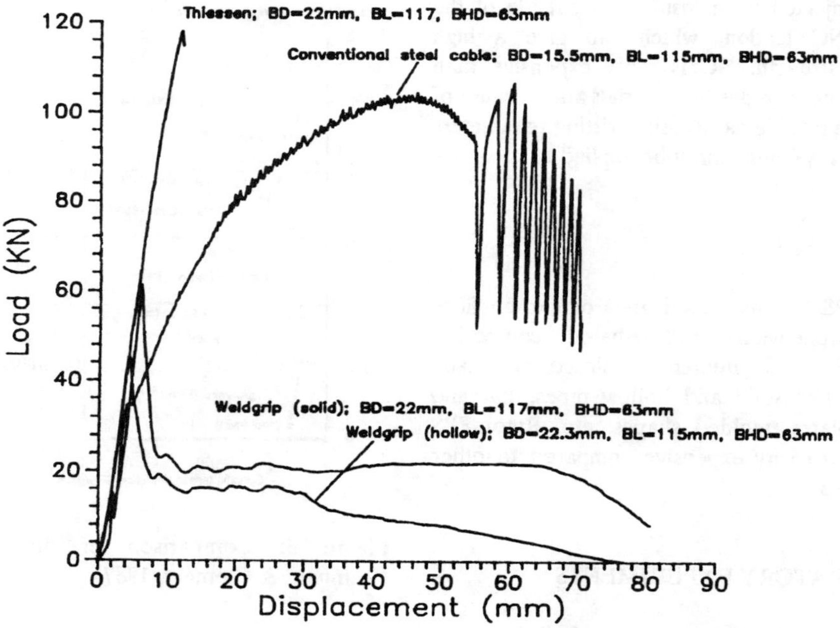

Figure 5: Pull out behaviour of composite tendons (when pull out capacity > 40 kN)

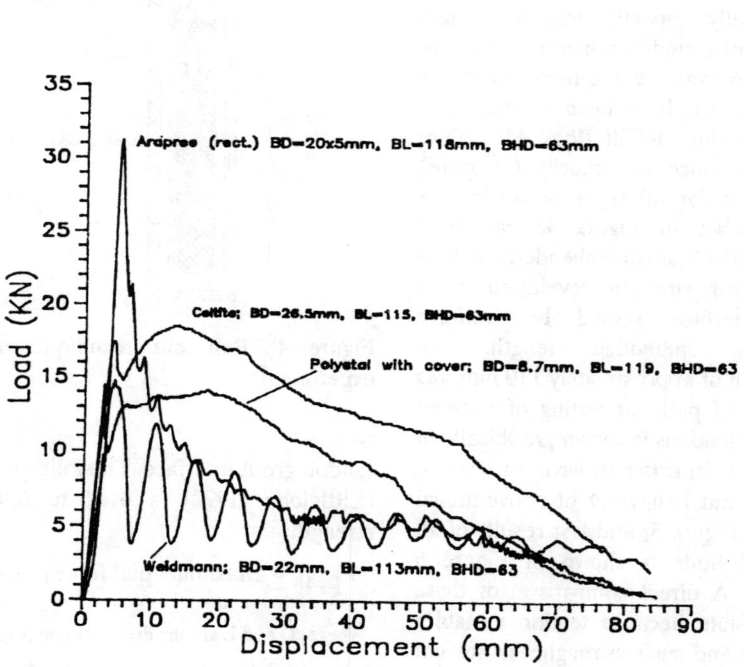

Figure 6: Pull out behaviour of composite tendons (when pull out capacity > 15 kN)

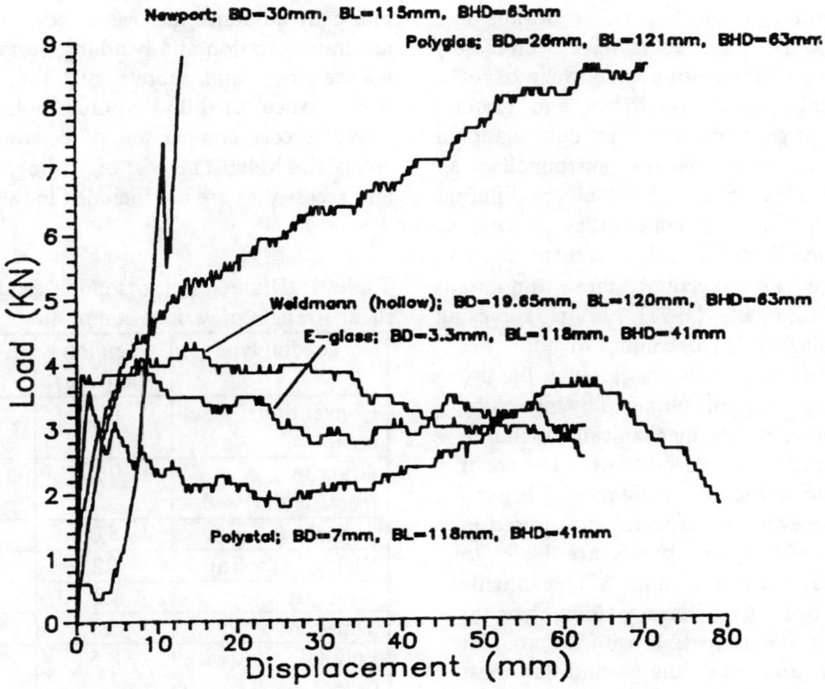

Figure 7: Pull out behaviour of composite tendons (when pull out capacity < 10 kN)

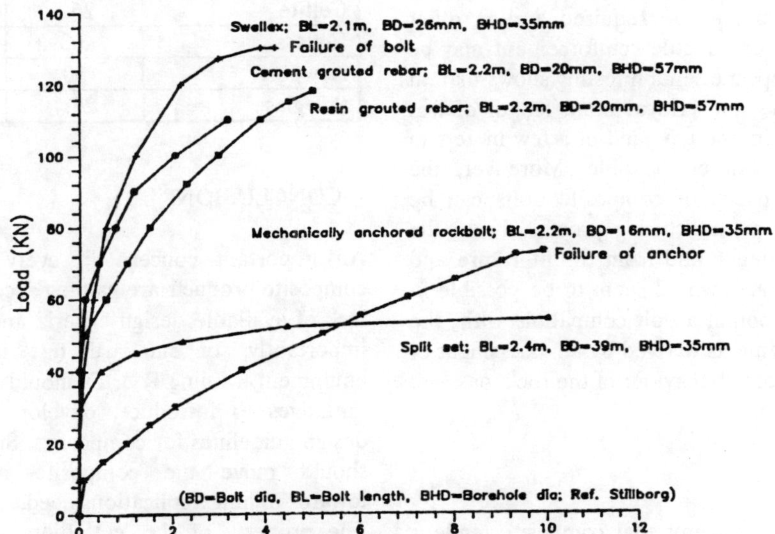

Figure 8: Pull out behaviour of conventional tendons (Stillborg, 1986)

The application of tendons, either flexible like cable or rigid like steel rebars, have been mainly as reinforcing elements in a wide range of rocks such as coal, potash, sandstone and granite. There are large variations in deformational behaviour of rock masses surrounding an excavation. This is mainly due to different degrees of inherent discontinuities as well as stress regime. Therefore reinforcement of rocks should take into account these important variables. Choquet [1991], after surveying support in thirteen Quebec mines wrote:

> "For a given mine, it is possible that the use of more rigid support obtained by increasing the diameter of the mechanically anchored bolts or resin-grouted bolts would make it possible to reduce the incidence of broken bolts. However, the experience acquired in locations where rock bursts are likely to occur shows that the mine will eventually have to opt for a more yielding type of support, more compatible with the amount of energy released by the ground that must be absorbed by the support"

For a given mine, it is obvious that in future, in rock mass classification systems, in addition to existing parameters, the ductility of both bolt material and grout-bolt interface will be mentioned. In some cases, a more brittle type of support system may be required and in other situations a more ductile reinforcement may be suitable. Composite tendon testing shows that, at the same time, the critical bond length of less than 100 mm in one type and of a few meters in another type can be possible. Moreover, the elongation capacity of composite bolts can be varied at the manufacturing process. From the authors' experience and from the literature and laboratory data, it would seem to be possible in the future to obtain a bolt compatible with the design requirements derived by an assessment of the deformational behaviour of the rock mass.

8 COST

At first glance, it seems that composite tendons are more expensive than conventional tendons. But, there are other factors which directly or indirectly influence the overall cost. Such factors are transportation and handling, storage life and maintenance and, above all, the long term performance of fully grouted bolts. Table 2 shows a cost comparison of conventional and composite bolts. The cost of drilling, installation and accessories are not included in this table.

Table 1: Diameter and tendon-grout interface shear strength of various composites

Tendon type	Tendon dia (mm)	τ_{max} (MPa)
Conventional steel cable	15.5	18.21
Arapree (rectangular)	20 x 5	10.8
Arapree (round)	8.0	6.35
Weldgrip (solid)	22	7.67
Weldgrip	22	6.33
Thiessen	22	14.34
Polystal (with cover)	8.5	4.3
Polystal (without cover)	6.0	2.0
Weidmann (solid)	22	1.92
Weidmann (hollow)	20	0.58
Celtite	26.5	2.58
E-glass	3.3	3.51
Newport	30	0.692
Polyglas	25.4	0.85

9 CONCLUSIONS

An important concern in every field where composite products are gaining acceptance is the lack of available design criteria and, even more importantly, of standard test methods and equipment. Mining R & D should also focus on initiatives to introduce, develop and produce design guidelines for composites. Such initiatives should move the composite manufacturers towards mining applications needs.

The property of the 'cuttability' of composite tendons should encourage further the development of continuous mining based on a continuous support system.

The 'tailorable' properties of composites, such as stiffness and surface roughness, facilitates the concept of varying the stiffness of the interface of grout and tendon. A support system compatible with the mechanical deformational behaviour of a rock mass should be achievable in the future. Test results show that critical lengths of less than 100 mm and more than a few meters will be achievable in the future with composite tendons. In short, composite materials potentially will display much superior performance than existing metal structures. At present, composites are expensive. It is foreseeable, however, that as volume production and competition increases, then so will costs decrease to acceptable levels. A timely initiative is required in Mining R&D to open a new horizon of application to the industry.

Table 2: Cost of composite tendons along with conventional tendons

Bolt type	Bolt dia (mm)	Bolt length (m)	Cost(c$)
Mechanical anchor	25	2.1	10.0
Resin grouted bolt	25	2.1	11.0
Cement grouted bolt	25	2.1	9.0
Split set	39	1.8	11.0
Swellex	26	2.1	17.0
Arapree (round)	7.9	2.0	5.2
Arapree (rectangular)	20 x 5	2.0	10.8
Weldgrip	22	1.8	16.0
Polystal	7	2.0	4.5
Celtite	26	2.0	18.0
Weidmann	22	2.0	40.0
Extern	25.4	2.0	32.0
Fiberloc	25.4	2.0	120.0

REFERENCES

Ballinger, C. 1990. Structural FRP composites, Civil Engineering, July , pp 63 - 65.

Bedard, C. 1992. Composite reinforcing bars: assessing their use in construction, Concrete International, January, pp 55 - 59.

Choquet, P. 1991. Rock bolting practical guide, CANMET, Energy, Mines and Resources Canada.

Handbook of Composites; Edited by George Lubin 1982. Van Nostrand Reinhold Company.

Hassani, F.P., and Khan, U.H. 1991. "Continuous Hardrock Support Systems", Rock Mechanics Research Group, Department of Mining and Metallurgical Engineering, McGill University.

Lapointe, B. 1990. Essais de treaction comparatifs de tiges d'ancrage en figre de verre et de tiges d'acier crenelees d'armature, Project STM-89TT36, Le service de la technologie miniere, Quebec.

Littlejohn, G.S. 1992. Keynote lecture: Rock anchorage practice in civil engineering, Proc: Rock support in mining and underground construction, Sudbury, Canada. pp 257 - 268.

Minosaku, K. 1992. Using FRP materials in prestressed concrete structures, Concrete International, August. pp 41 - 44.

Nani, A., and Tanigaki, M. 1992. Pretensioned prestressed concrete members with bonded fiber reinforced tendons: development and flexural lengths (static), ACI Structural Journal, V. 89, No. July-August. pp 433 - 441.

Reinhard, T.J., and Clement, L.L. 1987. Introduction to composites. Engineering Materials Handbook, vol. 1. ASM International, Metals Park, OH.

Rosato, D.V. 1982. An overview of composites, Handbook of Composites; Edited by George Lubin, Van Nostrand Reinhold Company,

Stillborg, B. 1986. Professional users handbook for rock bolting, Trans Tech Publications.

Wolff, R., and Miesseler, H. 1989. New material for prestressing and monitoring heavy structures. Concrete International, Sept. pp 86 - 89.

"Fiber for polymer composite", 1986. Technical brochure, BFGoodrich, Ohio, USA.

Innovative Mine Design for the 21st Century, Bawden & Archibald (eds) © 1993 Balkema, Rotterdam, ISBN 90 5410 325 6

Author index